Trigonometric Formulas

$\sin\theta = \dfrac{y}{r}$ $\qquad \cos\theta = \dfrac{x}{r}$ $\qquad \tan\theta = \dfrac{y}{x}$

$\csc\theta = \dfrac{r}{y}$ $\qquad \sec\theta = \dfrac{r}{x}$ $\qquad \cot\theta = \dfrac{x}{y}$

$\sin^2\theta + \cos^2\theta = 1$
$1 + \tan^2\theta = \sec^2\theta$
$1 + \cot^2\theta = \csc^2\theta$

y (x, y) r θ x

Cosine law- $c^2 = a^2 + b^2 - 2ab\cos\theta$

Sine law- $\dfrac{\sin A}{a} = \dfrac{\sin B}{b} = \dfrac{\sin C}{c}$

A c b B C a

$\sin(A + B) = \sin A\cos B + \cos A\sin B$
$\sin(A - B) = \sin A\cos B - \cos A\sin B$
$\cos(A + B) = \cos A\cos B - \sin A\sin B$
$\cos(A - B) = \cos A\cos B + \sin A\sin B$

$\tan(A + B) = \dfrac{\tan A + \tan B}{1 - \tan A\tan B}$ $\qquad \tan(A - B) = \dfrac{\tan A - \tan B}{1 + \tan A\tan B}$

$\sin 2A = 2\sin A\cos A$
$\cos 2A = \cos^2 A - \sin^2 A = 1 - 2\sin^2 A = 2\cos^2 A - 1$
$\tan 2A = \dfrac{2\tan A}{1 - \tan^2 A}$

$\sin A\sin B = \dfrac{1}{2}[-\cos(A + B) + \cos(A - B)]$

$\sin A\cos B = \dfrac{1}{2}[\sin(A + B) + \sin(A - B)]$

$\cos A\cos B = \dfrac{1}{2}[\cos(A + B) + \cos(A - B)]$

$\sin A + \sin B = 2\sin\left(\dfrac{A+B}{2}\right)\cos\left(\dfrac{A-B}{2}\right)$

$\sin A - \sin B = 2\cos\left(\dfrac{A+B}{2}\right)\sin\left(\dfrac{A-B}{2}\right)$

$\cos A + \cos B = 2\cos\left(\dfrac{A+B}{2}\right)\cos\left(\dfrac{A-B}{2}\right)$

$\cos A - \cos B = -2\sin\left(\dfrac{A+B}{2}\right)\sin\left(\dfrac{A-B}{2}\right)$

Inverse Trigonometric Function	**Principal Values**
$\text{Sin}^{-1}x$	$-\dfrac{\pi}{2} \le y \le \dfrac{\pi}{2}$
$\text{Tan}^{-1}x$	$-\dfrac{\pi}{2} < y < \dfrac{\pi}{2}$
$\text{Cos}^{-1}x$	$0 \le y \le \pi$

Inverse Trigonometric Function	**Principal Values**
$\text{Cot}^{-1}x$	$0 < y < \pi$
$\text{Csc}^{-1}x$	$-\pi < y \le -\dfrac{\pi}{2}, \quad 0 < y \le \dfrac{\pi}{2}$
$\text{Sec}^{-1}x$	$-\pi \le y < -\dfrac{\pi}{2}, \quad 0 \le y < \dfrac{\pi}{2}$

DAN BOAKES
(519) 747-0966

Calculus

Calculus

Donald Trim
University of Manitoba

Prentice Hall Canada Inc.
Scarborough, Ontario

Canadian Cataloguing in Publication Data

Trim, Donald W.
Calculus

Includes Index.

ISBN 0–13-117276–X

1. Calculus. I. Title.

QA303.T75 1993 515 C93–093112-2

Prentice Hall, Inc., Englewood Cliffs, New Jersey
Prentice-Hall International, Inc., London
Prentice-Hall of Australia, Pty., Ltd., Sydney
Prentice-Hall of India Pvt., Ltd., New Delhi
Prentice-Hall of Japan, Inc., Tokyo
Prentice-Hall of Southeast Asia (Pte.) Ltd., Singapore
Editora Prentice-Hall do Brasil Ltda., Rio de Janeiro
Prentice-Hall Hispanoamericana, S.A., Mexico

ISBN 0–13-117276–X

Acquisitions Editor: Jacqueline Wood
Developmental Editor: David Jolliffe
Copy Editor: Santo D'Agostino
Production Editor: Kelly Dickson
Production Coordinator: Florence Rousseau
Interior and Cover Design: Full Spectrum Art
Cover Image Credit: Mark Tomalty/Masterfile
Page Layout: Pronk&Associates

1 2 3 4 5 AGC 97 96 95 94 93

Printed and bound in the U.S.A. by Arcata Graphics

Contents

Donald W. Trim received his honours degree in mathematics and physics in 1965. Masters and doctorate degrees followed with a dissertation in general relativity. As a graduate student at the University of Waterloo, he discovered that teaching was his life's ambition. He quickly became known for his teaching, and in 1971 he was invited to become the first member of the Department of Applied Mathematics in the Faculty of Science at the University of Manitoba. In 1976, after only five years at the university, he received one of the university's highest awards for teaching excellence, based on submissions by faculty and graduating students. At the University of Manitoba no individual can receive the other award within ten years of receiving the first. In 1988, Professor Trim received the second award. In addition, the Faculty of Engineering presented him with a gold replica of the engineer's iron ring in appreciation for his service.

Professor Trim's skills in teaching are reflected in the notes and books that he has written. These include an Introduction to Applied Mathematics, an Introduction to the Theory and Applications of Complex Functions, and Applied Partial Differential Equations, and this calculus text. All have received most favourable reviews from students.

Professor Trim is a member of the Canadian Applied Mathematics Society and the Mathematical Association of America. His present interests are in partial differential equations and mathematical education.

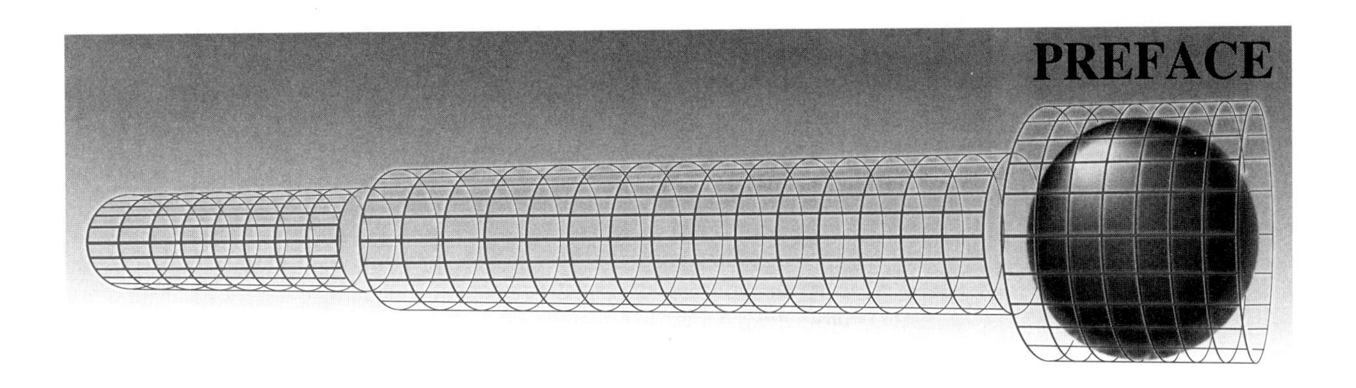

To the Instructor

Approach

This calculus book evolved from a set of notes developed by the author over a number of years of teaching freshman and sophomore calculus, and an earlier version of the text. For the most part our approach is intuitive, making free use of geometry and familiar physical settings to motivate and illustrate concepts. For instance, limits are introduced intuitively, but an optional section is included to introduce more advanced students to the formal definition of a limit. Derivatives are motivated and defined algebraically as instantaneous rates of change, and interpreted geometrically as slopes of tangent lines. The definite integral is motivated through two geometric and two physical problems, but its definition is algebraic.

The order of topics in the book is fairly standard with a few exceptions:

1. Conic sections are introduced in Chapter 1 as illustrations of how the form of the equation of a curve dictates its shape and conversely how the shape of a curve influences its equation. Detailed discussions of properties of conic sections are delayed until Chapter 10.
2. Limits at infinity and infinite limits are presented in Chapter 2 along with general discussions on limits. Students have little difficulty mastering the three situations (at least on an intuitive level), and completion of the topic allows for an early introduction to the use of limits in curve sketching.
3. It is assumed that students are familiar with trigonometric, exponential, and logarithm functions, but for those students whose background is weak, Appendices A and B review this material. Derivatives of these functions are developed in Chapter 3 making them available for the applications of differentiation in Chapter 4.
4. In-depth discussions on inverse trigonometric functions and hyperbolic functions, and their derivatives, are delayed until Chapter 8. To discuss these functions in Chapter 3 would certainly bring the topic of differentiation to an early conclusion, but this argument is overshadowed by what we call the "one-at-a-time" principle. *Do not do too many new things at one time; two or more new ideas presented simultaneously leads to confusion.* By restricting the domain of functions that can be differentiated to relatively simple ones in Chapter 3, students can concentrate on differentiation itself. With difficult functions, such as inverse trigonometric and hyperbolic, differentiation principles could be lost in the complexity of the functions.

5. Newton's iterative procedure for solving equations is the first application of differentiation. Numbers in subsequent applications need not then be contrived in order to lead to equations with elementary solutions. We use the intermediate value theorem to discuss accuracy of approximations from two viewpoints — number of decimal places and maximum possible error — and demand that the accuracy of every approximation be clearly identified.
6. Velocity and acceleration are treated as applications of differentiation in Chapter 4, and from an integration point of view in Chapter 5. We do not encourage the use of standard physics formulas; instead we expect students to state time conventions, choose coordinate systems, and develop whatever equations are necessary to solve a problem.
7. Differential equations are introduced early in the book in order that students become familiar with the ideas surrounding this important branch of calculus. Only the simplest of equations are solved in the early chapters, however; we use them more as an illustration of mathematical modelling. Differential equations are discussed in detail in Chapter 16.
8. We place applications of the definite integral before techniques of integration. In this way the student sees the diversity of applications of integration on simple functions, but can appreciate why it is necessary to antidifferentiate more difficult functions. At the same time, we can review the applications in the examples and exercises of the techniques chapter.

We have, as does every instructor, certain views on the best pedagogical approach to calculus. Some of these are listed below:

1. "Curve sketching" is developed one step at a time as calculus unfolds. We begin with basic ideas of symmetries, translations, and addition and multiplication of ordinates in Chapter 1. Limits in Chapter 2 lead to discontinuities, and vertical and horizontal asymptotes. The geometric interpretation of the derivative as the slope of the tangent line in Chapter 3 hints at its value as a sketching tool. With critical points, relative extrema, and points of inflection, we bring curve sketching techniques together for a final discussion in Section 4.5. It is not the purpose of these techniques to develop an extremely accurate graph of a function; a computer can do this more easily. We use calculus to obtain as detailed a graph of a function as is required for a particular application. We see how the graph becomes more and more refined as additional techniques are brought into play, and we discover how the techniques complement one another. In addition, we see how calculus can verify what a calculator or computer may suggest — the existence of vertical, horizontal, or slanted asymptotes, or vertical or horizontal points of inflection.
2. Relative extrema and absolute extrema are discussed in separate sections, the former in Section 4.3 and the latter in Section 4.6. Most applied extrema problems involve absolute extrema; curve sketching uses relative extrema. They are different and they are used differently. Let the student see them "one-at-a-time." We continue this separation in multivariable work. Relative extrema are discussed in Section 13.10 and absolute extrema in Section 13.11. Lagrange multipliers provide an important alternative for constrained extrema problems.
3. We have included a section with hints on how to manage word problems. We stress how to think about a problem, how to key on certain words, and how a series of questions and answers leads from problem to solution.
4. Antiderivatives or indefinite integrals are treated in a separate chapter. Relegating them to a section in the chapter on differentiation does not do them justice. Conceptually integration is simple — backwards differentiation — but this operation is so important, and it is so much more difficult to perform, only a separate chapter

will bring students to this realization. In Chapter 5 we introduce the three most basic integration techniques: straightforward recognition, "guess-and-fix-up", and simple substitutions. Other techniques (integration by parts, trigonometric substitutions, partial fractions, etc.) are discussed in Chapter 9.

5. We treat antidifferentiation from an organizational viewpoint, rather than from the viewpoint of formulas. The student is taught to organize an integrand into a form in which integration is obvious; he is encouraged not to memorize formulas unnecessarily.

6. Definite integrals are motivated with four problems—area, volume, work, and fluid flow—but they are defined algebraically as limits of Riemann sums. This permits us to use the definition of a definite integral in a wide variety of applications without fear that it has been associated with any one in particular. Definitions of double, triple, line, and surface integrals in multivariable calculus as limit-summations follow quite naturally.

7. Antiderivatives are used to evaluate definite integrals; this is the fundamental theorem of integral calculus. The fact that a definite integral with a variable upper limit can be regarded as an antiderivative is given secondary importance. Students do not, at the earliest stage, appreciate or understand why we would want to make the upper limit of a definite integral a variable quantity. Thus, we do not detract from the importance of the antiderivative as a calculational tool for definite integrals (one-thing-at-a-time).

8. Applications of definite integrals (area, volume, lengths of curves, etc.) are approached from a "representative element" viewpoint. Instead of providing formulas for such problems, we stress the need for understanding both the application and the definite integral. We establish representative elements to approximate the required quantity and use definite integrals to add over all elements. Students will then be able to introduce definite integrals into other applications, and the transition to double, triple, line, and surface integrals will be reasonably smooth.

9. We choose to keep the polar coordinate r nonnegative. Nothing is lost by this convention; curves may have slightly different equations depending on which convention is used, but by choosing $r > 0$, considerable simplification is achieved when equations of curves in polar coordinates are rewritten in Cartesian coordinates.

10. The limit ratio test for convergence of infinite series is applied only to series with positive terms. It is not used directly as a test for convergence of series with positive and negative terms. Many students have difficulty with the concept of convergence and divergence of series. Let us take it one step at a time. First we discuss convergence of series of positive terms using the comparison, limit comparison, integral, limit ratio, and limit root tests. Then we introduce series with positive and negative terms, and use the above tests as tests for absolute convergence.

11. The basis for power series expansions is Taylor's remainder formula; when remainders approach zero, the Taylor series converges to the function. This is discussed first. It then becomes a matter of illustrating that various techniques exist to find power series (such as integrating and differentiating known series), and these techniques conveniently avoid remainders.

12. Many students take entire courses in differential equations, a topic in single-variable calculus. For those who need only an introduction to the topic and knowledge of some of the more important techniques for solving differential equations we have included Chapter 16. There are sections on separable, linear first order, and second order equations easily reduced to first order equations; the exercises introduce (first order) homogeneous and Bernoulli equations. Considerable emphasis is placed on the important topic of linear differential equations (four sections). Applications to Newtonian mechanics, population dynamics, and vibrating mass-spring systems

and electric circuits are discussed in detail. Many other applications are introduced through examples and exercises.

13. Three-dimensional analytic geometry and vectors in Chapter 12 provide the tools for multivariable calculus. We stress the value of drawing curves and surfaces in space, using the curve sketching tools learned in Chapters 1–4. Such diagrams are essential to the evaluation of double, triple, line, and surface integrals, and to an appreciation of many of the ideas of differential calculus. Vectors are handled algebraically and geometrically, since neither approach by itself is satisfactory. Every algebraic definition is interpreted geometrically, and every geometric definition is followed by an algebraic equivalent. Differentiation and integration of vectors dependent on a single parameter lead to discussions of tangent and normal vectors to curves, curvature and arc length, and three-dimensional kinematics.

14. Gradients are useful in many areas of applied mathematics. Introducing them in the context of an application leads students to associate them with that particular application. For this reason we introduce gradients in Section 13.4, and apply them to directional derivatives and normal vectors to curves and surfaces in Sections 13.8 and 13.9.

15. Chain rules for composite, multivariable functions are endless in variations and applications. We show students how to appreciate each term in a chain rule as a contribution of particular variables to the overall rate of change of a function, and then provide them with a schematic diagram to handle the most complicated functional situations.

16. Chain rules are used to find systems of equations for partial derivatives of implicitly defined functions. Cramer's rule and Jacobians facilitate solving the equations.

17. Our definitions of double and triple integrals as "limit-summations" are completely analogous to the definition of the definite integral; evaluation by double and triple iterated integrals is geometric. Through representative boxes, rectangles, strips, and columns, the student visualizes the summation process and affixes appropriate limits to integrals. There is no algebraic manipulation of inequalities; a thoughtfully prepared diagram does it all. The geometric approach also helps students to visualize integrations in polar, cylindrical, and spherical coordinates. We demonstrate that many of the applications of the definite integral in Chapter 7 are handled much more easily with double integrals.

18. One of the biggest difficulties for students is to decide whether a definite, double, triple, line, or surface integral should be used to solve a given problem. We suggest that they ask what it is that they are integrating over, not what they are finding. To integrate over an area in a coordinate plane, use a double integral; over a volume, use a triple integral; along a curve, use a line integral; and over a surface, use a surface integral. It seems natural, then, to find volumes in space with triple integrals rather than double integrals.

19. Students believe that topics with very few definitions are inherently simpler than those with many definitions. With this in mind, we tell students that there is one kind of line integral, $\int_C f(x,y,z)\,ds$ and that integrals of the form $\int_C P\,dx + Q\,dy + R\,dz$ are a special case when $f(x,y,z)$ is the tangential component of some vector function defined along C. They can all be evaluated by substituting from parametric equations of the curve. Alternatively, it may be expedient to use Green's or Stokes's thereoms or determine whether the integral is independent of path.

20. Likewise, there is only one surface integral $\iint_S f(x,y,z)\,dS$, but in many applications, $f(x,y,z)$ is the normal component of some vector function defined on S.

21. Appendix E contains 585 review problems. They are divided into 370 problems on Chapters 1–11 and 215 problems on Chapters 12–16. Students should use them

in preparation for their final exams in single and multivariable calculus. For the single-variable problems, we have provided answers to the first 150 even-numbered problems. The last 70 or so become more and more challenging. We usually throw some of them out to the class, one at a time, for a bonus of 3, 4, or 5 marks. Perhaps the first correct solution, or the best solution, earns the bonus. This often sparks the interest of the more advanced student. For the multivariable problems we have given answers to the first 101 even-numbered problems.

Two supplements have been prepared to aid instructors and students. Available to students is a manual containing detailed solutions to all even-numbered exercises. For instructors there is a manual with answers to odd-numbered exercises. Any errors in the text or supplements are the responsibility of the author, and we would appreciate having them brought to our attention.

I wish to thank the following individuals who reviewed the manuscript and offered many helpful suggestions: R. Biggs, The University of Western Ontario; K. Dunn, Dalhousie University; P. Ponzo, University of Waterloo; R. Ross, University of Toronto; C. Roth, McGill University; and G.P. Wright, University of Ottawa.

Thanks also to D. Greenwood, and K. Zhan, P. Siregar and K. Hand, University of Guelph, who proofed the calculations.

I also wish to express my appreciation to the staff at Prentice Hall Canada Inc. for a superb effort in editing, design, and illustration. Special thanks to J. Wood, D. Jolliffe, C. Chan, S. D'Agostino, K. Dickson, and A. Lui-Ma for their suggestions and encouragement.

To the Student

This book was written for you, the student of calculus. It begins with an introduction to the fundamentals of plane analytic geometry in order to give you a way to visualize problems geometrically. You will be surprised at the number and variety of problems that you will be able to solve using calculus, but as it so often happens in mathematics, the most difficult part of the solution to a problem is the initial step. Once started, the solution frequently unfolds smoothly and easily, but that first step sometimes seems impossible. One of the best ways to start a problem is with a diagram. A picture, no matter how rough, is invaluable in giving you a "feeling" for what is going on. It displays the known facts surrounding the problem; it permits you to see what the problem really is, and how it relates to the known facts; and it often suggests that all-important first step. Don't underestimate the value of a diagram; get into the habit of making one at every opportunity — not just to solve problems, but to understand what calculus is all about.

Our study of calculus really begins in Chapter 2 with a discussion of limits. The idea of limits is fundamental to all of calculus, and it is surprising that the founders of calculus, Gottfried Leibniz (1646–1716) and Isaac Newton (1642–1727), had not formulated the idea of a limit. Their development of the subject was based on the concept of "infinitesimals" — quantities that are extremely close to zero, but not equal to zero. It was the work of such mathematicians as L. Euler (1707–1783), J. D'Alembert (1717–1783), and later, A. Cauchy (1789–1857) that led to the rigorous formulation of limits and a firm foundation on which to develop the rest of calculus.

All calculus originates with two basic ideas — the derivative and the definite integral. If one quantity depends on another, and the first quantity is in a state of change, so also will be the second. For example, the speed of a car depends on how far the accelerator is depressed — more depression, more speed. As soon as the accelerator depression is changed, so also is the speed of the car, and derivatives describe how these changes are related. When a furnace is turned on, the amount of heat it produces affects room termperature — more heat, higher temperature. Derivatives relate the rate at which temperature rises to the rate at which heat is produced.

The definite integral is a completely different kind of entity. Suppose, for instance, that we use a pair of scissors to cut some strange shape from a piece of paper and ask for its area. This could be very difficult since there are very few shapes for which we have an area formula. We could, however, estimate its area by drawing rectangles on the piece of paper and adding together their areas. By drawing smaller and smaller rectangles near the curved edge, we could get a better and better estimate of the true area. In essence, this is the idea of the definite integral; it is a sum of terms where the number of terms is increased indefinitely.

Surprising as it may seem, these two concepts which appear so different are intimately related, and it is this connection that makes calculus so fascinating. In fact, calculus is a subject wherein every topic depends heavily on those that went before, and at the same time leads naturally into the next. Once we identify these interdependences, we begin to understand the inner beauty of calculus.

You will find that we place considerable emphasis on problem solving, and we do this because you will never learn calculus by only going to class and/or reading the textbook. You must also involve yourself in the subject, and the best way to do this is with problems. Only when you solve problems do you begin to think about what your instructor has said and what you have read, and then you begin to understand. The exercises in each section begin with routine, drill-type problems designed to reinforce fundamentals. It follows, therefore, that if you have difficulties with early exercises, then you have not understood the fundamentals, and you should therefore review material in that section. In later exercises we make you think more. Some of the exercises are of the same type as were found earlier, but with more involved calculations; some require you to think about a concept from a different point of view; others may draw on ideas from previous sections; still others may do all of these. It is in these exercises that you find most of the applications, for only when you have mastered fundamentals can you begin to apply calculus to problems from other fields. Certain exercises are marked with a calculator icon to indicate the need for an electronic calculator. There are many other exercises for which a calculator could eliminate laborious hand calculations and use of trigonometry and logarithm tables, but exercises marked with the icon definitely require a calculator. In addition, a calculator is essential for those sections that deal with approximations and iterations (Sections 4.1, 9.9, 11.2, 11.7, and 11.12).

At the end of each chapter there is a brief summary of the results in that chapter and a set of review exercises. Don't underestimate the value of these exercises; they test your understanding of the material in the entire chapter. Section exercises develop expertise in small areas; chapter exercises test knowledge of major blocks of subject matter. In addition, Appendix E contains exercises on the entire book. Use these exercises as a review when you have completed your calculus studies, or in preparation for your examinations.

Answers to even-numbered problems can be found in Appendix D. In addition, a student supplement containing detailed solutions to all even-numbered problems is available.

Throughout the book you will encounter an icon to warn you about pitfalls which frequently entrap students. These pitfalls may be inappropriate calculations or misinterpretations of ideas. Pay close attention to each one.

The key word in our approach to calculus is "think". Above all else, we hope that you will train your mind to think and to think logically. We want you to learn how to organize facts and interpret them mathematically. If there is a problem, we want you to be able to decide exactly what that problem is, and finally, to produce a step-by-step procedure by which to solve it. Do not, therefore, read this book expecting formulas that can be used to solve all problems; we won't give them to you. We will give you a few formulas and, we hope, a great deal of insight into the ideas surrounding these formulas. Then you begin to think for yourself. You use what you know to solve the first problem. Having gained from that experience, you then move on to the next problem. In this way, your calculus studies will be a true learning experience — not an exercise in memorization and regurgitation of facts and formulas, but a developmental period during which you both learn calculus and increase your mental capacity. When you reach the end of the road and successfully complete your calculus studies, think back and compare how much you knew at the beginning to how much you know at the end. Congratulate yourself on accomplishing a great deal, and mostly through your own efforts.

Winnipeg, Canada
1993

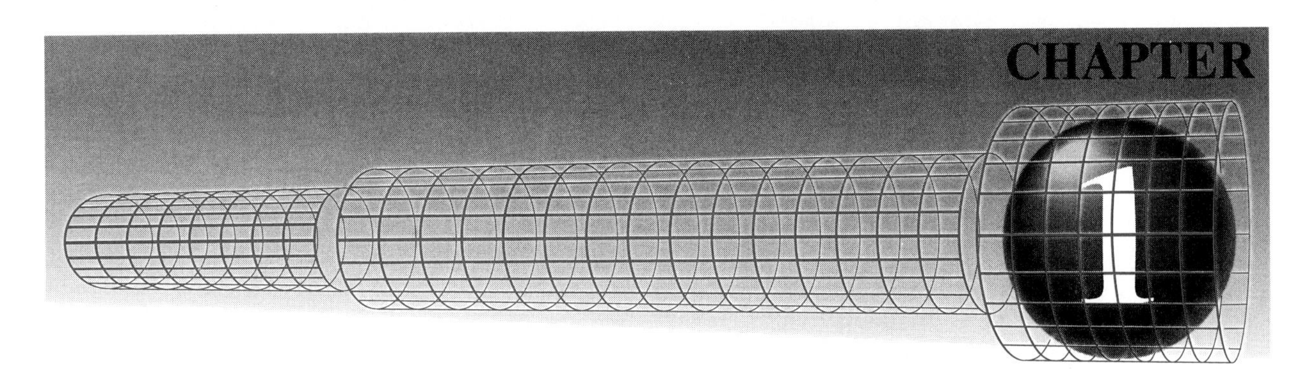

Plane Analytic Geometry and Functions

Analytic geometry is a union of algebra and geometry. On the one hand, it provides a way to describe geometric objects algebraically; on the other hand, it permits a geometric visualization of algebraic statements. In the first four sections of this chapter we introduce the principles of plane analytic geometry, proceeding from coordinates along a line in Section 1.1 to coordinates in a plane in Section 1.2. In Sections 1.3 and 1.4 we discuss curves and their representation by use of equations, paying particular attention to straight lines, circles, parabolas, ellipses, and hyperbolas. Students who have previously studied analytic geometry could skip these sections and begin with functions in Section 1.5, but we strongly recommend against this. Even the best-prepared student can benefit by a cursory reading of these sections. Not only is it helpful to refresh one's memory on concepts learned some time ago, it is also wise to become familiar with the terminology, notation, and conventions set forth in these sections.

SECTION 1.1 Real Numbers and Coordinates

At an early stage in our mathematical education we learned how the real number line serves as a geometric representation of the real-number system. In this section we reverse these roles and use the real-number system to *coordinatize* or identify points on a line. Essentially the two approaches are one and the same; they set up a correspondence between real numbers and points on a line. To coordinatize a line, which we call the x-axis, we begin by choosing some point of reference O on the line. This point, called the **origin**, is identified or coordinatized by the real number zero. We now start at O and mark off equal distances along the line in both directions (Figure 1.1). These marks could, for example, be one centimetre apart or one metre apart. The first point to the right of O is one unit of distance from O and is coordinatized by the real number 1; the second point is two units from O and is identified by the number 2, and so on. Unit distances to the left of O are identified by the negative integers -1, -2, -3, and so on. Fractional numbers such as $3/4$ and $-5/3$ identify points that are $3/4$ of our unit distance to the right of O, and $5/3$ of our unit distance to the left of O, respectively. Continuation of this process leads to a coordinatization of the entire x-axis. Every point on the axis is identified by a real number that we denote by x, and this number is the **directed distance** from the origin O to the point. For example, point Q in Figure 1.1 is two and one-half units of distance to the right of O, and therefore its x-coordinate is $5/2$. Point R is three and one-quarter units to the left of O; its x-coordinate is -3.25.

FIGURE 1.1

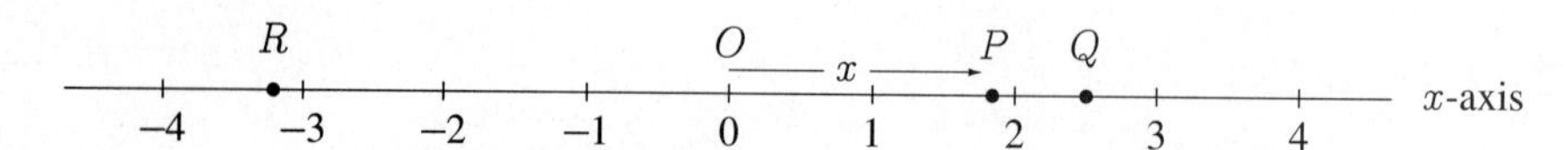

To describe Q we should say "the point whose x-coordinate is $5/2$," but this is somewhat cumbersome. We usually say "the point $x = 5/2$." To refer to a general point P on the x-axis, we say "the point x" meaning, of course, the point with coordinate x. In this book we use the convention that directed distances from one point to another are represented by line segments which have an arrowhead on the "to" end but not on the "from" end. In Figure 1.1 we have the directed distance from O to P with an arrowhead at P. Were we interested in the length of the line segment joining O and P, rather than the directed distance from O to P, we would place arrowheads on both ends of the line segment (Figure 1.2). We denote this length by $||OP||$. When P is to the right of O, $||OP||$ and x are equal, but when P is to left of O, $||OP||$ is the negative of x. Coordinates are always directed distances and should not be confused with lengths.

FIGURE 1.2

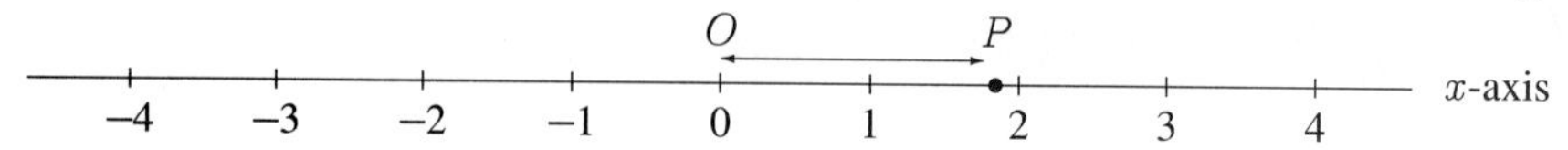

Figure 1.3 shows a second coordinatization of the x-axis using a unit of distance smaller than that in Figure 1.1. Since the unit of length can be chosen arbitrarily, as can the point of reference O, the number of ways to coordinatize a line is infinite.

FIGURE 1.3

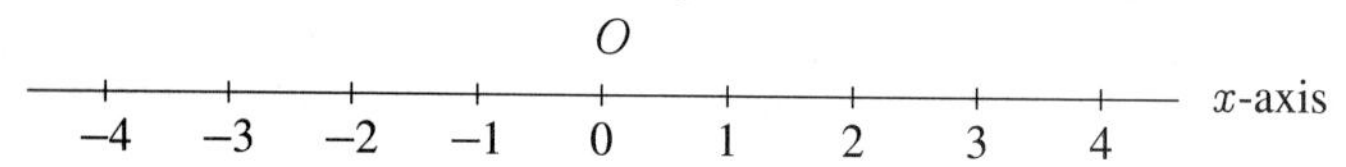

To represent sets of points on the x-axis we use open and closed intervals as defined below.

Definition 1.1

The set of points on the x-axis between two given points $x = a$ and $x = b$, where $a < b$, but not including these points, is called an **open interval**. It is denoted by (a, b).

The open interval (a, b) is described algebraically as all points on the x-axis whose x-coordinates satisfy the inequality $a < x < b$. For example, $1 < x < 2$ describes the open interval $(1, 2)$. We have shown this interval in Figure 1.4, where the open circles at $x = 1$ and $x = 2$ indicate that these points are not included in the set.

FIGURE 1.4

The interval (1, 2)

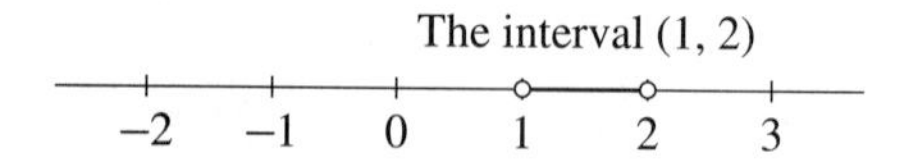

Definition 1.2

The set of points on the x-axis between two given points $x = a$ and $x = b$, where $a < b$, including these points, is called a **closed interval**. It is denoted by $[a, b]$.

Algebraically, $[a, b]$ is the set of points whose x-coordinates satisfy $a \leq x \leq b$. The closed interval $[1, 2]$ in Figure 1.5 is therefore described by $1 \leq x \leq 2$. The solid dots at $x = 1$ and $x = 2$ indicate that these points are included in the interval. A closed interval contains its end points, whereas an open interval does not.

FIGURE 1.5

The interval (1, 2]

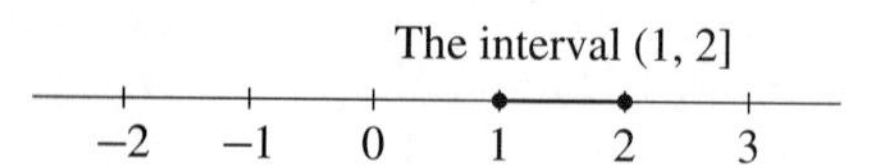

Intervals need not be open or closed. For example, the set of points whose coordinates satisfy $1 < x \leq 2$ is denoted by $(1,2]$. It is a half-open and half-closed interval; it contains one end but not the other (Figure 1.6). Intervals need not be of finite length. The set of points whose coordinates satisfy $x > 2$ (or $2 < x < \infty$) has only one end, namely, $x = 2$, which is not contained in the set; the set is also denoted by $(2,\infty)$ (Figure 1.7). The notations $2 < x < \infty$ and $(2,\infty)$ have introduced the symbol ∞, called *infinity*. Infinity is *not* a number; it is a symbol that is notationally convenient in many parts of calculus. In $2 < x < \infty$, it means that there is no largest value for x; x can be chosen arbitrarily large. The interval described by $x \leq 2$ is also denoted by $(-\infty, 2]$; it contains its only end point $x = 2$ (Figure 1.8). A square bracket is never placed on the end of an interval corresponding to ∞ or $-\infty$; that is, we never write $(2,\infty]$ or $[-\infty, 2)$. A square bracket indicates that the point next to it is included in the interval. Since neither $-\infty$ or ∞ is a point, they cannot be included.

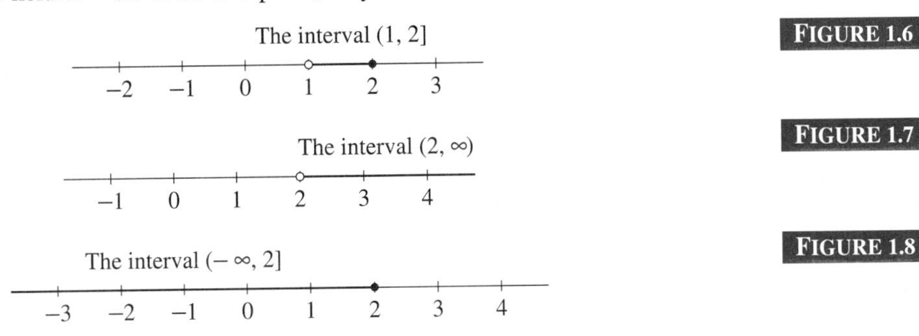

The interval (1, 2]

FIGURE 1.6

The interval (2, ∞)

FIGURE 1.7

The interval (− ∞, 2]

FIGURE 1.8

Intervals are used to describe more complicated sets of points on the x-axis. For example, the set of points between $x = 1$ and $x = 3$ but not including $x = 1$, $x = 2$, or $x = 3$ consists of all points in the open intervals $1 < x < 2$ and $2 < x < 3$ (Figure 1.9). The set of points satisfying the inequality $x^2 \geq 9$ consists of all points in the intervals $x \geq 3$ and $x \leq -3$ (Figure 1.10). Inequalities that are even more complicated can often be simplified to one or more of the four basic inequalities $a < x < b$, $a \leq x < b$, $a < x \leq b$, $a \leq x \leq b$ using the following rules:

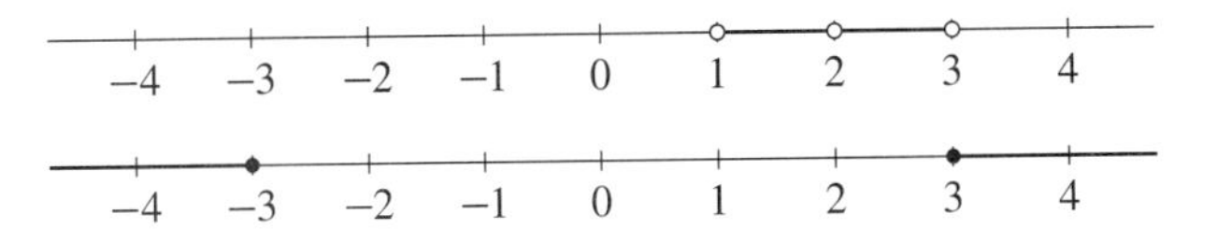

FIGURE 1.9

FIGURE 1.10

If $a < b$, then

$$a + c < b + c \quad \text{for any real } c. \tag{1.1}$$

If $a < b$, then

$$ac < bc \quad \text{when } c > 0, \tag{1.2 a}$$

$$ac > bc \quad \text{when } c < 0. \tag{1.2 b}$$

If $0 < a < b$, then

$$\frac{1}{a} > \frac{1}{b}. \tag{1.3}$$

Each inequality $<$ in 1.1 and 1.2 may be replaced by $\leq$ provided that $>$ is at the same time replaced by $\geq$.

EXAMPLE 1.1

Simplify the following inequalities:

$$\text{(a) } 4 - 2x \le 5 \qquad \text{(b) } 1 + 2x^2 < 5 \qquad \text{(c) } \frac{2x - 1}{x + 2} > 1$$

SOLUTION

(a) If we add -4 to both sides of the inequality (using 1.1), we obtain

$$-2x \le 1.$$

Multiplication by $-1/2$ then gives the interval

$$x \ge -1/2.$$

Thus, equivalent to the inequality $4 - 2x \le 5$ is the simpler statement $x \ge -1/2$.

(b) If we subtract 1 from both sides of the inequality, which is equivalent to adding -1,

$$2x^2 < 4,$$

and divide by 2 (this is equivalent to multiplying by $1/2$),

$$x^2 < 2.$$

Points must therefore lie in the open interval $-\sqrt{2} < x < \sqrt{2}$.

(c) For fractional inequalities it is often useful to bring all terms to one side of the inequality. To do this we add -1 to both sides,

$$0 < \frac{2x - 1}{x + 2} - 1 = \frac{(2x - 1) - (x + 2)}{x + 2} = \frac{x - 3}{x + 2}.$$

Figure 1.11 contains what we call a "sign diagram". It is a convenient way to analyze the sign of an expression which is the product and/or quotient of a number of factors. The first line indicates that $x - 3$ is negative for $x < 3$ and positive for $x > 3$. The second line shows that $x + 2$ is negative for $x < -2$ and positive for $x > -2$. The last line combines these two. The vertical bars at $x = 3$ and $x = -2$ in the first two lines are brought down, thus dividing the x-axis into three intervals $x < -2$, $-2 < x < 3$, and $x > 3$. The sign of $(x - 3)/(x + 2)$ in each of these intervals is determined by counting the corresponding number of negative signs in lines one and two. It follows that $(x - 3)/(x + 2) > 0$ [or $(2x - 1)/(x + 2) > 1$] when $x < -2$ or $x > 3$.

FIGURE 1.11

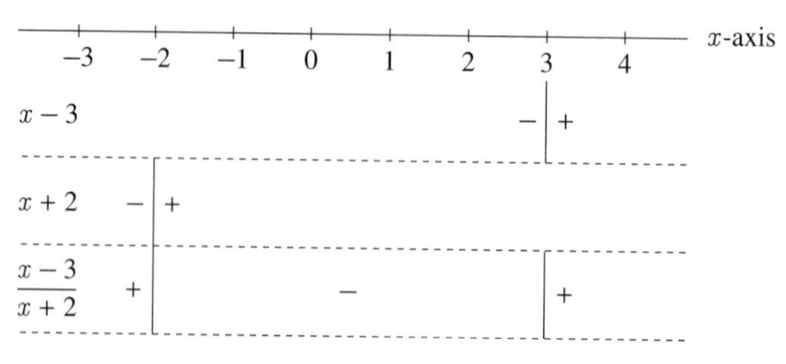

■

In many applications of mathematics we find it necessary to discuss the "sign" of a quantity, and sign diagrams like that in Figure 1.11 are very helpful. We shall use them in many places in succeeding chapters.

When we say that the coordinate of a point P on the x-axis is x, we mean that the directed distance from the origin O to P is x. The quantity $x - a$ for any real number a has a similar geometric interpretation. It is the directed distance from the point $x = a$ to P. For instance, $x - 5$ is the directed distance from the point Q at $x = 5$ in Figure 1.12 to any point P. If P is at $x = -4$, the directed distance from Q to P is $-4 - 5 = -9$. Similarly, in Figure 1.13, $x + 2$ is the directed distance from the point Q at $x = -2$ to any point P.

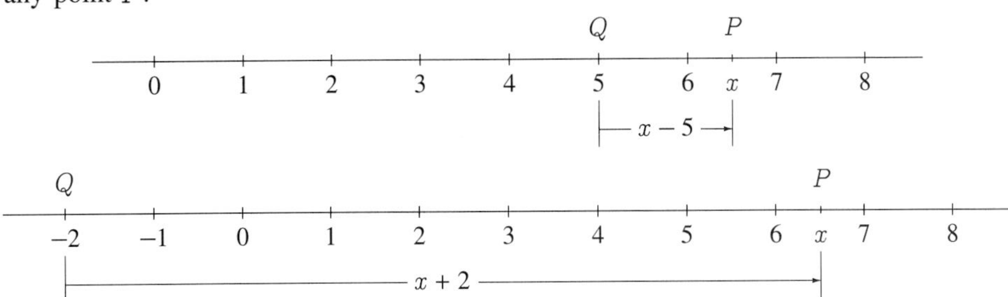

FIGURE 1.12

FIGURE 1.13

One of the most useful notations for specifying sets of points on the x-axis is provided by the next definition.

Definition 1.3

The **absolute value** of a real number x is

$$|x| = \begin{cases} x & x \geq 0 \\ -x & x < 0. \end{cases} \tag{1.4}$$

For example, $|10| = 10$, $|-9| = 9$, and $|2(4)^2 - 9(5)| = |32 - 45| = |-13| = 13$. The absolute value then is an operation that defines the size or "magnitude" of a number without regard for sign.

To see how the absolute value can be used to specify sets of points on the x-axis, consider first the set described by the inequality $|x| < 2$. This inequality states that the magnitude of x must be less than 2. It means that x must lie between -2 and 2; hence $|x| < 2$ describes all points in the open interval

$$-2 < x < 2.$$

Similarly, the inequality $|x - 5| < 2$ states that the magnitude of $x - 5$ must be less than 2, and therefore

$$-2 < x - 5 < 2.$$

Addition of 5 to each of these terms gives the interval

$$3 < x < 7.$$

Alternatively, we could have reasoned that $x - 5$ is the directed distance from 5 to x on the x-axis (Figure 1.12). The inequality $|x - 5| < 2$ requires this distance to be less in magnitude than 2; in other words, the distance from 5 to x must be less than 2. Hence x must lie between 3 and 7.

The inequality

$$|x^2 + 1| \leq 2$$

requires that $x^2 + 1$ be between -2 and 2, including ± 2;

$$-2 \leq x^2 + 1 \leq 2.$$

When 1 is subtracted from each of these terms,

$$-3 \leq x^2 \leq 1.$$

Now the left-hand inequality $-3 \leq x^2$ is really not a restriction on x at all — every real x satisfies this inequality. Consequently, x must satisfy only $x^2 \leq 1$, or $-1 \leq x \leq 1$. The original inequality $|x^2 + 1| \leq 2$ is therefore equivalent to $-1 \leq x \leq 1$ or $|x| \leq 1$.

EXAMPLE 1.2

Find all points satisfying each of the following inequalities:

(a) $|3 - 5x| \leq 2$ (b) $|4x^2 - 7| > 2$

SOLUTION (a) The absolute value may be replaced by the double inequality

$$-2 \leq 3 - 5x \leq 2,$$

and when we subtract 3, we find

$$-5 \leq -5x \leq -1.$$

Multiplication of each term in these inequalities by $-1/5$ gives

$$1 \geq x \geq \frac{1}{5} \quad \text{or} \quad \frac{1}{5} \leq x \leq 1.$$

(b) The inequality $|4x^2 - 7| > 2$ is satisfied if x satisfies either $4x^2 - 7 > 2$ or $4x^2 - 7 < -2$. From the first, $4x^2 > 9$ or $x^2 > 9/4$. This implies that either $x > 3/2$ or $x < -3/2$. From the inequality $4x^2 - 7 < -2$, we find that $x^2 < 5/4$ or $-\sqrt{5}/2 < x < \sqrt{5}/2$. Consequently, in order to satisfy $|4x^2 - 7| > 2$, x must lie in one of the three intervals $x < -3/2$, $x > 3/2$, or $-\sqrt{5}/2 < x < \sqrt{5}/2$ (Figure 1.14). These may be stated more compactly as $|x| < \sqrt{5}/2$ and $|x| > 3/2$.

FIGURE 1.14

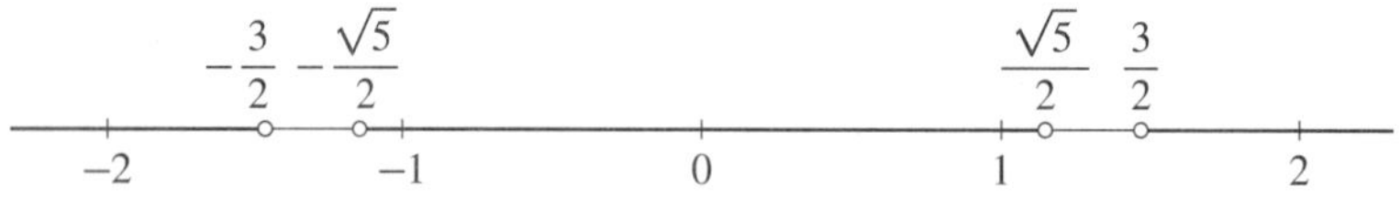

■

This example uses a convention adopted by all mathematicians. The square root symbol "$\sqrt{\ }$" means positive square root. For instance $\sqrt{4} = 2$; it is not equal to -2. This convention is often forgotten in situations like $\sqrt{x^2}$. This is equal to x, if x is positive. But if x is negative, it is equal to $-x$. For example, when $x = -3$, $\sqrt{x^2} = \sqrt{(-3)^2} = 3 = -x$. Thus, we can write that

$$\sqrt{x^2} = \begin{cases} x & x \geq 0 \\ -x & x < 0. \end{cases}$$

Absolute values combine these conveniently into

$$\sqrt{x^2} = |x|, \tag{1.5}$$

valid for all x.

EXAMPLE 1.3

Describe the set of points between 1 and 3, but not including 1, 2 or 3, in terms of absolute values.

SOLUTION To specify that x must be within a distance of 1 on either side of 2, we write

$$|x-2|<1.$$

The condition that $x \neq 2$ is now easily incorporated

$$0<|x-2|<1.$$

The absolute value of the product of two real numbers is always equal to the product of their absolute values:

$$|ab| = |a||b|. \tag{1.6a}$$

The absolute value of a quotient is the quotient of the absolute values:

$$\left|\frac{a}{b}\right| = \frac{|a|}{|b|}. \tag{1.6b}$$

These results are not valid for sums and differences. The absolute value of the sum of a and b is equal to the sum of their absolute values when a and b have the same sign; that is,

$$|a+b| = |a| + |b|$$

when a and b are either both positive or both negative. When a and b have opposite signs, the correct statement is

$$|a+b| < |a| + |b|.$$

In general, then, we may state that for any two real numbers a and b,

$$|a+b| \le |a| + |b|. \tag{1.6c}$$

This inequality finds its way into many applications of mathematics. It can also be used to prove another inequality that is quite useful,

$$|a+b| \ge \big||a| - |b|\big|, \tag{1.6d}$$

(see Exercise 45).

EXERCISES 1.1

1. Coordinatize the x-axis with unit distance equal to (a) 2 cm; (b) 3/4 cm. Mark on the axis points with coordinates $-3, -2.5, -2, -1.5, -1, -0.5, 0, 0.5, 1, 1.5, 2, 2.5$, and 3.

In Exercises 2–9 express the interval as an inequality.

2. $(-3,4)$ **3.** $[5,8]$

4. $(-1,6]$ **5.** $[-12,4)$

6. $[3,\infty)$ **7.** $(-\infty,-4)$

8. $(-\infty,\infty)$ **9.** $(-10,\infty)$

10. Is $(3,2)$ an acceptable interval?

11. Is $(2,\infty]$ an acceptable interval?

In Exercises 12–21 express the inequality in interval notation.

12. $3 \le x \le 10$ **13.** $-2 \le x < 5$

14. $4 < x \le 22$ **15.** $x > 1$

16. $x \le -2$ **17.** $1 < x < \infty$

18. $-\infty < x \le -2$

19. $x \ge -3$

20. $-\infty < x < \infty$

21. $-0.2 < x \le -0.05$

In Exercises 22–39 find all points that satisfy the inequality.

22. $3x - 2 < 7$

23. $2x^2 + 12 \ge 19$

24. $3/x > 5$

25. $3/x > -5$

26. $x^2 - 4 \le 5$

27. $15 + x^2 < 14$

28. $|3x - 2| < 7$

29. $|2 - 3x| \ge 7$

30. $|2x^2 + 5| \le 6$

31. $|2x^2 + 6| \le 5$

32. $|2x + 6| \le 5$

33. $|x^2 - 5| > 4$

34. $|x^2 - 4| \ge 5$

35. $|x^2 + 5| > 4$

36. $|x^2 + 4| \ge 5$

37. $\dfrac{2x - 6}{x - 2} > 3$

38. $\dfrac{3x + 5}{4x - 2} < 3$

39. $\dfrac{9x + 5}{3x - 2} < 2$

40. Verify that the inequality

$$x^2 + 4x < 6$$

can be written in the form

$$(x + 2)^2 < 10,$$

and thereby find all points satisfying the inequality.

In Exercises 41–44 use the technique of Exercise 40 to find all points that satisfy the inequality.

41. $x^2 + 6x + 1 \le 0$

42. $2x^2 + 4x - 5 > 0$

43. $-x^2 - 2x \ge 1$

44. $|x^2 + 2x + 5| \ge 3$

45. Use inequality 1.6c to verify 1.6d.

46. Is

$$|a - b| \ge \big||a| - |b|\big|$$

a correct variation of 1.6d?

In Exercises 47–60 find all points that satisfy the inequality.

47. $x + \dfrac{1}{x} \ge 2$

48. $\dfrac{1}{x + 2} - \dfrac{1}{x - 2} < 5$

49. $11 < x^2 + 6x + 4 < 20$

50. $|x^2 + 2x - 3| \le 1$

51. $\dfrac{3x^2 + 5}{x^2 - 4} > 1$

52. $\dfrac{4x^2 - 64}{x^2 - 9} > 3$

53. $\left|\dfrac{3x + 4}{x - 2}\right| < 4$

54. $\dfrac{x + 5}{x - 3} < \dfrac{x + 2}{x + 1}$

55. $|x^2 + 2x + 3| \ge 5$

56. $|3x + 5| \le 4x$

57. $|3x + 5| \le |4x|$

58. $|x^2 + 5| + |x - 2| \le 8$

59. $\dfrac{x - 3}{x + 1} > \dfrac{2x + 4}{x}$

60. $\dfrac{2x + 3}{x - 5} > x + 4$

In Exercises 61–64 express the condition algebraically and find all points that satisfy the condition.

61. The sum of the distances from a point x to $x = 1$ and $x = 5$ must be less than 7.

62. The sum of the distances from a point x to $x = -1$ and $x = 2$ must be greater than 4.

63. The absolute value of the difference of the distances from a point x to $x = 5$ and $x = 12$ must be less than 14.

64. The absolute value of the difference of the distances from a point x to $x = 2$ and $x = -6$ must be less than 3.

65. The **arithmetic mean** (average) of any two real numbers a and b is defined as $(a + b)/2$. The **geometric mean** of any two positive numbers a and b is $\sqrt{ab}$. Show that the arithmetic mean of any two distinct positive numbers is always greater than their geometric mean.

66. Find all points that satisfy the following inequalities:

(a) $x^2 + 12x + 11 < 0$

(b) $\dfrac{2x^2 + 11x + 5}{x + 2} < x - 3$

(c) $2x + 1 < \dfrac{x^2 - x - 6}{x + 5}$

(d) $\dfrac{2x + 1}{x + 2} < \dfrac{x - 3}{x + 5}$

67. Find all points that satisfy

$$\frac{|3x + 1|}{x + 5} < x.$$

SECTION 1.2

Rectangular Coordinates in the Plane

The technique of coordinatizing a line by means of real numbers can be extended to define rectangular coordinates in a plane and rectangular coordinates in space. This turns out to be indispensable to our approach to calculus. At every stage in our development of the subject we shall have pictures — pictures to introduce ideas, pictures to illustrate concepts, and pictures to reinforce principles. These pictures can only be described adequately with coordinate systems.

To coordinatize a plane, we begin by choosing any point O in the plane, called the origin, and any line through O, called the x-axis. This line, which we coordinatize as in Section 1.1 is drawn horizontally with positive direction to the right (denoted by an arrowhead in Figure 1.15). Next we draw through O a line perpendicular to the x-axis called the y-axis. This line is coordinatized with positive direction usually upward. In Figure 1.15, we coordinatized the x- and y-axes with exactly the same unit of distance, but we shall see in Section 1.3 that it is sometimes more convenient to have different scales along the two axes. Unless we indicate otherwise, unit distances along the axes are chosen equal.

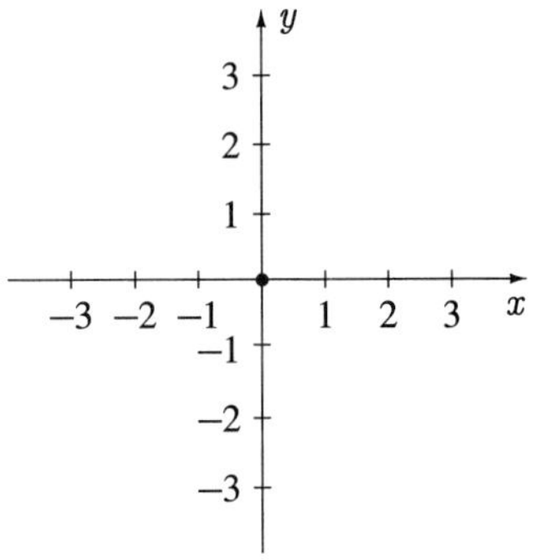

FIGURE 1.15

The two coordinatizations of the axes can be used to identify every point in the plane. Through the point P in Figure 1.16 we draw lines parallel to the x- and y-axes to intersect the axes at points A and B. The x-coordinate of A is $5/2$, and the y-coordinate of B is $3/2$. We call these the x- and y-coordinates of P, and write them together in the form $(5/2, 3/2)$. The x-coordinate of P, also called the **abscissa** of P, is therefore the distance cut off on the x-axis by the line through P parallel to the y-axis (Figure 1.17). Since the x- and y-axes are perpendicular, we can state alternatively that the x-coordinate of P is the directed (perpendicular) distance from the y-axis to P. Similarly, the y-coordinate of P, also called the **ordinate** of P, is the directed distance from the x-axis to P, or the distance cut off on the y-axis by the line through P parallel to the x-axis.

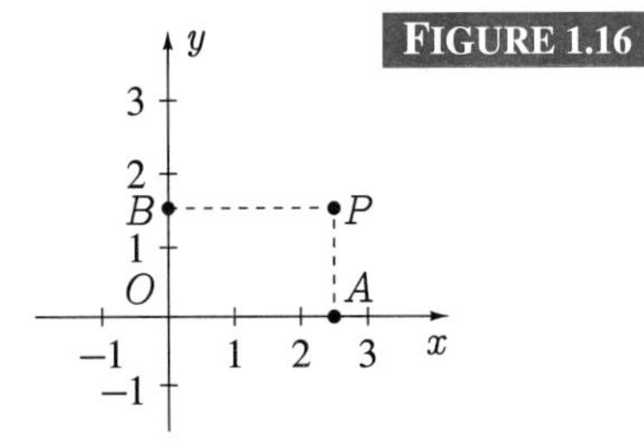

FIGURE 1.16

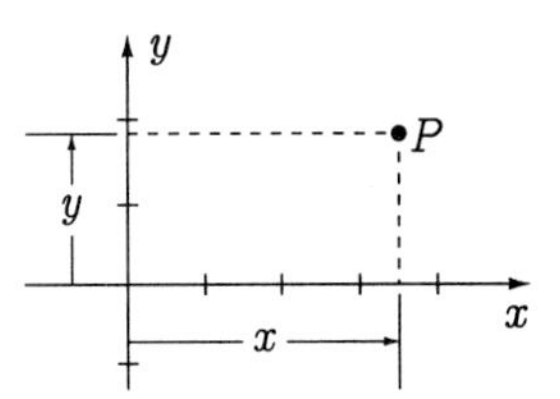

FIGURE 1.17

By this construction, every point in a plane can be identified or coordinatized by an ordered pair of real numbers (x, y) called its x- and y-coordinates. Although the ordered pair notation for the coordinates of a point is identical to that for an open interval, context always makes it clear which interpretation is intended. Conversely, given any pair of real numbers x and y, there is one and only one point in the plane with these as x- and y-coordinates. Since coordinates (x, y) are defined in terms of a pair of perpendicular axes, we call them **rectangular** or **Cartesian coordinates**. Figure 1.18 shows four points together with their Cartesian coordinates. They are called Cartesian coordinates after the French mathematician René Descartes (1596-1650), one of the founders of analytic geometry.

The axes divide the plane into four parts which, beginning at the upper right and proceeding counterclockwise, are called respectively the first, second, third, and fourth **quadrants**. The axes themselves are not considered part of any quadrant. Points in the first quadrant have x- and y-coordinates that are both positive, points in the second quadrant have a negative x-coordinate and a positive y-coordinate, and so on (Figure 1.19).

FIGURE 1.18

FIGURE 1.19

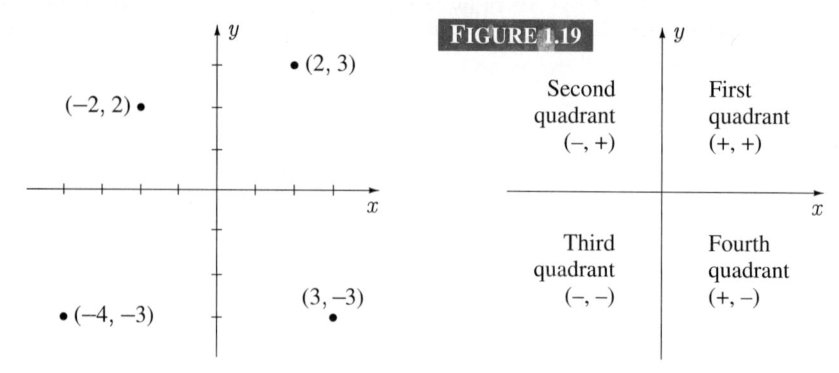

If P and Q are any two points with coordinates (x_1, y_1) and (x_2, y_2) respectively (Figure 1.20), then by the Pythagorean relation for the right-angled triangle PQS,

$$||PQ||^2 = ||PS||^2 + ||QS||^2,$$

where $||PQ||$ denotes the length of the line segment joining P and Q, and similarly for $||PS||$ and $||QS||$. But

$$||PS|| = |x_2 - x_1| \quad \text{and} \quad ||QS|| = |y_2 - y_1|,$$

and therefore

$$||PQ||^2 = |x_2 - x_1|^2 + |y_2 - y_1|^2 \quad \text{or}$$

$$||PQ|| = \sqrt{(x_2 - x_1)^2 + (y_2 - y_1)^2}. \tag{1.7}$$

This very important result expresses the length of a line segment joining two points P and Q in terms of their coordinates (x_1, y_1) and (x_2, y_2). For example, if the coordinates of P and Q are $(-3, 4)$ and $(2, -1)$, then

$$||PQ|| = \sqrt{(2+3)^2 + (-1-4)^2} = 5\sqrt{2}$$

(Figure 1.21).

FIGURE 1.20

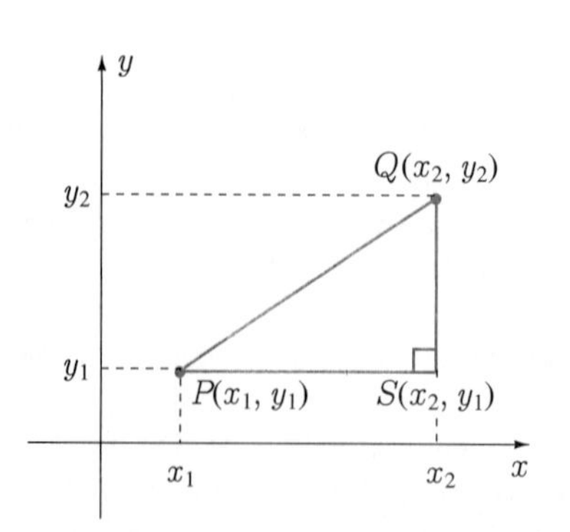

FIGURE 1.21

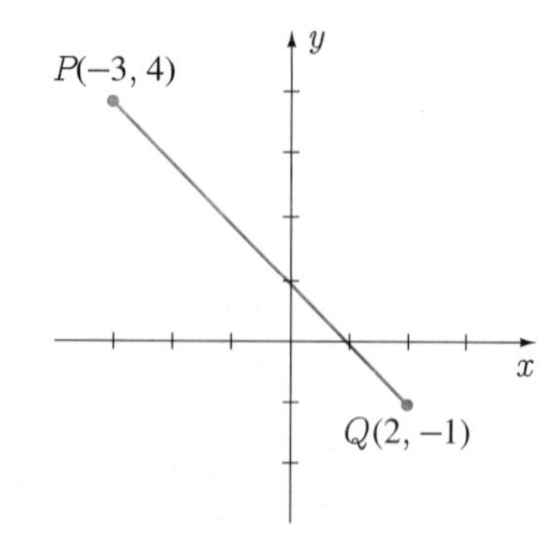

EXAMPLE 1.4

Points with coordinates $(0,0)$, $(7,0)$, and $(3,5)$ form three vertices of a parallelogram. What are possible coordinates for the fourth vertex?

SOLUTION Figures 1.22(a,b,c) show that there are three possibilities. The coordinates of R in Figures (a) and (b) should be fairly clear. Since line QR in (a) is horizontal and of length 7, the coordinates of R are $(10,5)$. Similarly, in Figure (b), R is the point $(-4,5)$. To find the coordinates (x,y) of R in Figure (c), we express the facts that $\|OR\| = \|QP\|$ and $\|OQ\| = \|PR\|$ algebraically,

$$\sqrt{(x-0)^2 + (y-0)^2} = \sqrt{(7-3)^2 + (0-5)^2},$$
$$\sqrt{(3-0)^2 + (5-0)^2} = \sqrt{(x-7)^2 + (y-0)^2}.$$

When these equations are squared, and simplified, the result is

$$x^2 + y^2 = 41,$$
$$x^2 - 14x + y^2 = -15.$$

If we subtract the second of these from the first, we obtain $14x = 56$, or $x = 4$. Substitution of this into either of the above equations gives $y = \pm 5$. But only $x = 4$ and $y = -5$ yield a parallelogram. The coordinates of R are therefore $(4,-5)$.

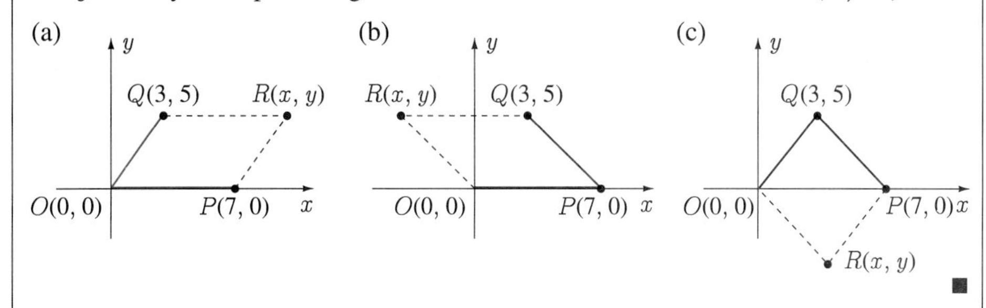

FIGURE 1.22

■

Suppose point $R(x,y)$ in Figure 1.23 is the midpoint of the line segment joining $P(x_1,y_1)$ and $Q(x_2,y_2)$.

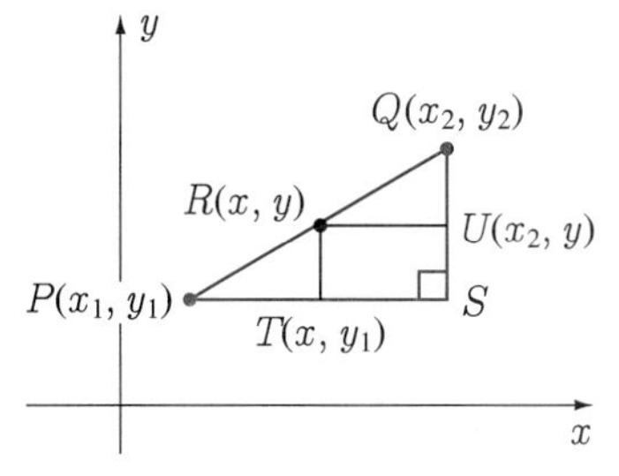

FIGURE 1.23

Since triangles PTR and RUQ are congruent, it follows that $\|RU\| = \|PT\|$, and for P and Q as shown in Figure 1.23, this condition becomes $x_2 - x = x - x_1$. Consequently,

$$x = \frac{x_1 + x_2}{2}. \tag{1.8a}$$

Similarly, equality of $\|RT\|$ and $\|QU\|$ gives

$$y = \frac{y_1 + y_2}{2}. \tag{1.8b}$$

Derivation of these results assumed that $x_2 > x_1$ and $y_2 > y_1$, but the same formulas are valid for any x_1, x_2, y_1, and y_2 whatsoever. Coordinates of the midpoint of a line segment are therefore averages of coordinates of the ends of the line segment.

EXERCISES 1.2

1. Show a plane coordinatized with Cartesian coordinates in each of the following ways:

(a) equal unit distances of 1 cm along the x- and y-axes;

(b) unit distances along the x- and y-axes of $1/2$ and 1 cm respectively;

(c) unit distances along the x- and y-axes of 1 cm and 1 mm respectively.

2. On a plane with rectangular coordinates show the points $(2,5)$, $(-3,6)$, $(7,-2)$, $(6,-3)$, $(-4,1)$, $(1,3)$, $(-1/3,1)$, $(2,3/2)$, $(-1/2,-3/4)$.

In Exercises 3–6 find the distance between the points.

3. $(1,3)$, $(3,4)$ **4.** $(-2,1)$, $(4,-2)$

5. $(-1,-2)$, $(-3,-8)$ **6.** $(3,2)$, $(-4,-1)$

7.–10. Find the midpoint of the line segment joining the points in Exercises 3–6.

11. A square has sides of length 4 units. Find the coordinates of its vertices if:

(a) one vertex is at the origin, two of its sides lie along the coordinate axes, and one vertex lies in the third quadrant;

(b) its diagonals lie along the axes.

12. Show all points in a plane with Cartesian coordinates x and y that have each of the following:

(a) an x-coordinate equal to 4;

(b) a y-coordinate equal to -1;

(c) equal x- and y-coordinates;

(d) a y-coordinate equal to the negative of the x-coordinate.

13. Determine whether each of the triangles with the following vertices is right-angled:

(a) $A(2,9)$, $B(-1,3)$, $C(6,7)$

(b) $A(-1,4)$, $B(3,6)$, $C(1,1)$

14. A ship sails N30°E from a pier at 5 km/h. If coordinates are set up with the pier as origin and the positive x- and y-axes in the east and north directions, respectively, what are the coordinates of the ship after 3 hours?

15. A ladder 10 m long rests against a wall with its feet 4 m from the base of the wall (Figure 1.24). If a man stands on the middle rung of the ladder, what are the coordinates of his feet?

FIGURE 1.24

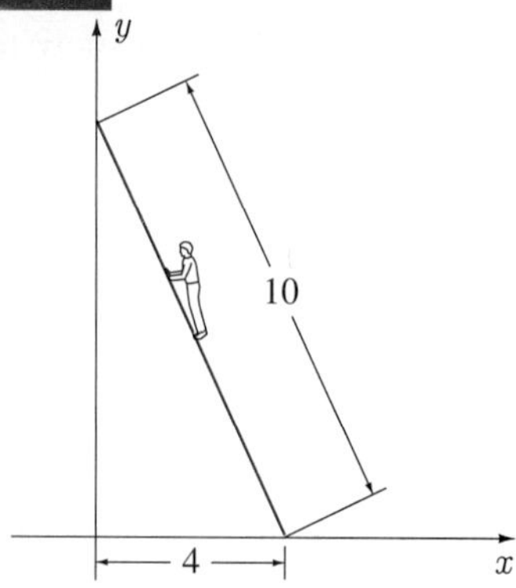

16. What are the coordinates of points A, B, and C in Figure 1.25 if BC and BA form a right angle and the side lengths are as shown?

FIGURE 1.25

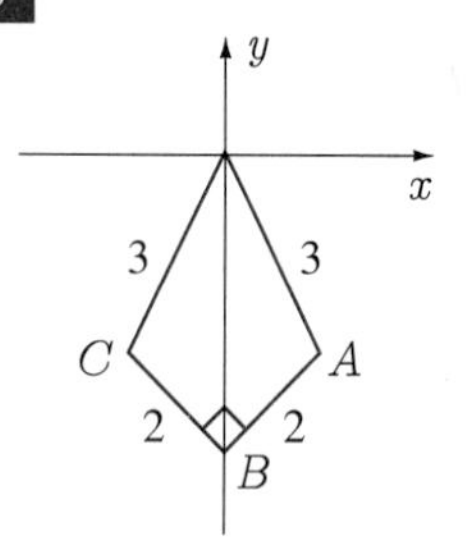

17. Lines that intersect at angles other than right angles (Figure 1.26) can also be used to coordinatize the plane. If l_1 and l_2 are coordinatized as shown, and P is any point in the plane, we draw through P lines parallel to l_2 and l_1. The lengths x and y cut off on l_1 and l_2 are then called the coordinates (x,y) of P.

FIGURE 1.26

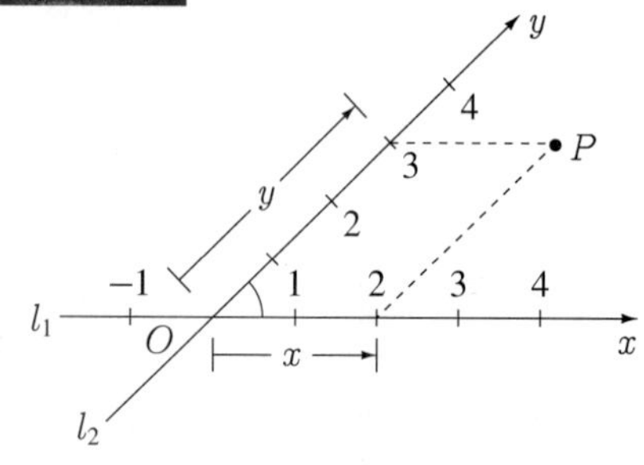

(a) Show the plane coordinatized with equal unit distances of 1 cm along two lines that intersect at $\pi/6$ radians.

(b) On your diagram mark the points $(1,3)$, $(-2,3)$, $(4,-5)$, $(-1,-2)$, $(1/2,-1/2)$, $(-2,1/2)$.

18. If the radius of the circle in Figure 1.27 is 10 cm, find the length of line segment AB.

FIGURE 1.27

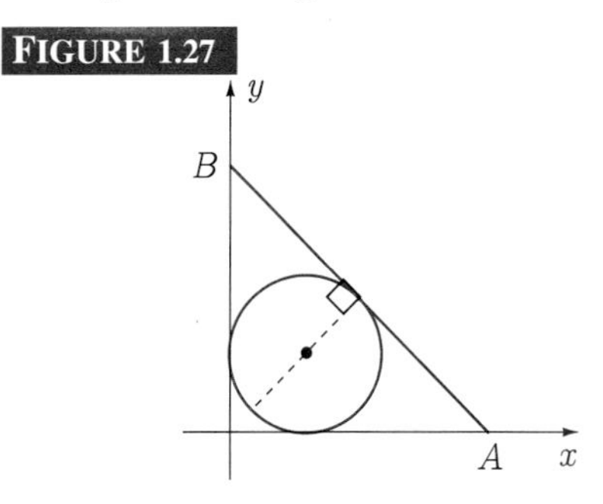

19. If the minute and hour hands of a clock are, respectively, 10 cm and 7.5 cm long (Figure 1.28), find coordinates for the tips of the hands when the time is:

(a) 9:00 (b) 4:00 (c) 7:30 (d) 1:40

FIGURE 1.28

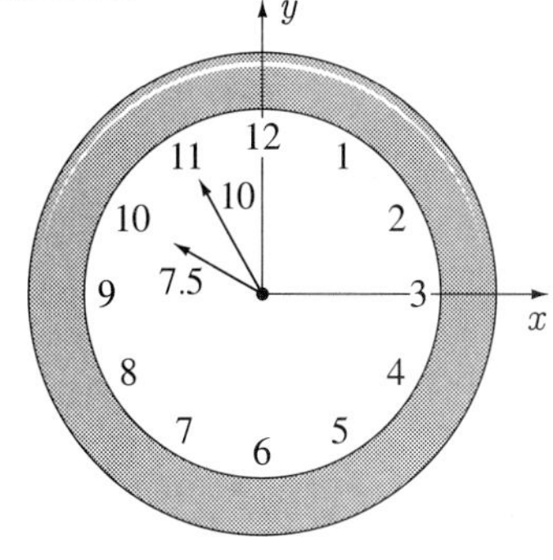

20. If a circle passes through three points $(0,0)$, $(4,0)$, and $(0,3)$, what are the coordinates of its centre?

21. Plot points $P(2,1)$ and $Q(5,5)$ on a Cartesian plane coordinatized with unit distances of 1 cm and 0.5 cm along the x- and y-axes, respectively. According to distance formula 1.7, $\|PQ\| = \sqrt{(5-2)^2 + (5-1)^2} = 5$. Measure $\|PQ\|$ with a ruler. Is it 5 cm? Explain.

22. A man 2 m tall walks along the edge of a straight road 10 m wide (Figure 1.29). On the other edge of the road stands a streetlight 8 m high. If rectangular coordinates (with unit distances equal to 1 m) are set up as shown, find coordinates for the tip of the man's shadow when he is 10 m from the point directly opposite the light.

FIGURE 1.29

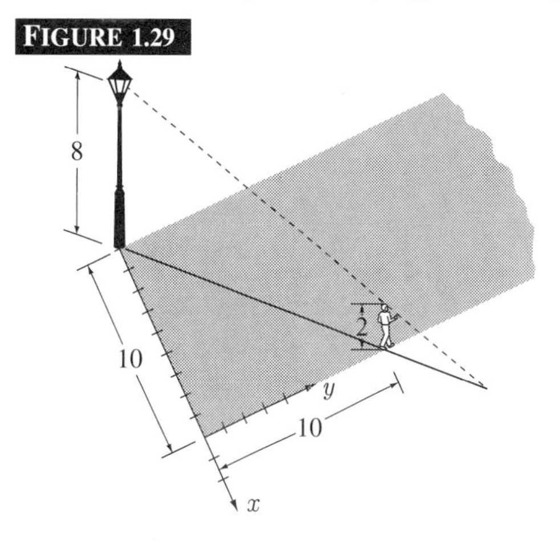

23. Generalize the result of equations 1.8 to prove that if a point R divides the length PQ so that

$$\frac{\text{length } PR}{\text{length } RQ} = \frac{r_1}{r_2},$$

where r_1 and r_2 are positive integers, then the coordinates of R are

$$x = \frac{r_1 x_2 + r_2 x_1}{r_1 + r_2} \quad \text{and} \quad y = \frac{r_1 y_2 + r_2 y_1}{r_1 + r_2}.$$

24. Prove that in any triangle the sum of the squares of the lengths of the medians is equal to three-quarters of the sum of the squares of the lengths of the sides. (A median of a triangle is a line segment drawn from one vertex to the midpoint of the opposite side.)

SECTION 1.3

Equations and Curves

In Section 1.2 we began the study of plane analytic geometry by using ordered pairs of real numbers (x, y) to describe the positions of points in a plane, an algebraic description of a geometric object. In this section and Section 1.4 we bring out the other aspect of analytic geometry: that algebraic statements can be visualized geometrically.

An equation such as $y = x^2$ specifies a relationship between numbers represented by the letter x and those represented by the letter y — y must always be the square of x. Algebraically, we speak of pairs of values x and y that satisfy this equation. To illustrate this we have constructed Table 1.1, called a **table of values**.

TABLE 1.1

x	−6	−5	−4	−3	−2	−1	0	1	2	3	4	5	6
y	36	25	16	9	4	1	0	1	4	9	16	25	36

Each pair of entries in this table represents a pair of values of x and y that satisfies the equation $y = x^2$. If we write these pairs in the form (x, y), and interpret each as coordinates of a point, we obtain the thirteen points in Figure 1.30(a). Note that since the values of y range from 0 to 36 and those for x range only from −6 to 6, we have used different scales along the x- and y-axes. We could add more points to this illustration by choosing noninteger values of x between −6 and 6, as well as values of x beyond −6 and 6. We would find that all such points lie on the curve in Figure 1.30(b). This curve then is a geometric visualization of solution pairs of the equation $y = x^2$. Every pair of values x and y that satisfies $y = x^2$ is represented by a point on the curve. Conversely, the coordinates (x, y) of every point on the curve provide a pair of numbers that satisfies $y = x^2$. Algebraic solutions then are represented geometrically as points on a curve; points on the curve provide algebraic solutions.

FIGURE 1.30

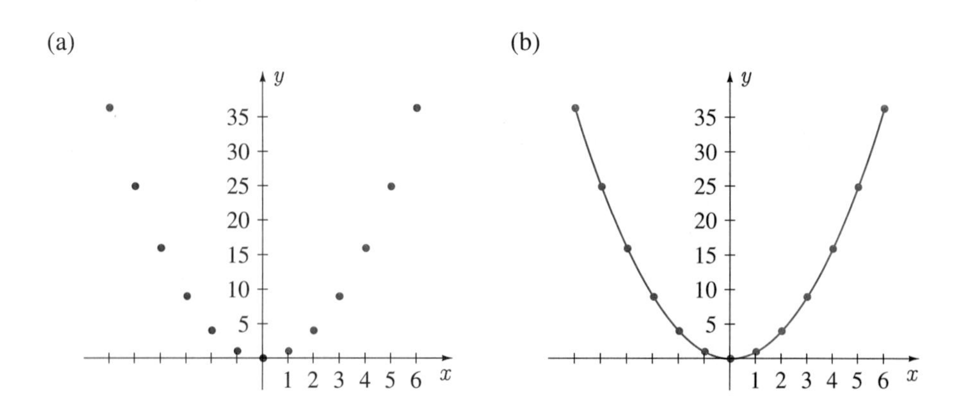

If (x, y) are the coordinates of a point in the plane, then the point is on the curve in Figure 1.30(b) if and only if x and y satisfy the equation $y = x^2$. In other words, this equation completely characterizes all points on the curve. We therefore call $y = x^2$ the **equation of the curve**. The equation of a curve then is an equation that the coordinates of every point on the curve satisfy and at the same time is an equation that no point off the curve satisfies. In the remainder of this section and Section 4 we discuss five classes of curves — straight lines, parabolas, circles, ellipses, and hyperbolas — and show that each can be described by a particular type of equation. Once again note the algebraic-geometric interplay; the form of an equation dictates the shape of the curve, and conversely, the shape of a curve determines the form of its equation.

The Straight Line

Every point on the vertical line in Figure 1.31 has an x-coordinate equal to 2; the equation of the line is therefore $x = 2$. The equation of the horizontal line in the same diagram is $y = -5/2$.

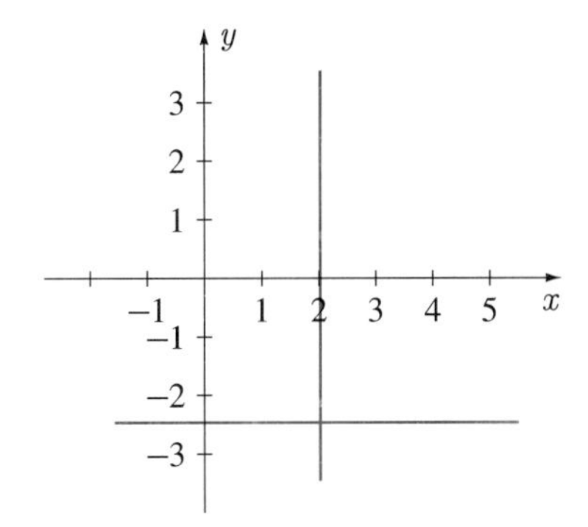

FIGURE 1.31

To find the equation of a straight line that is not parallel to either axis, we introduce the idea of *slope* in Figure 1.32(a).

Definition 1.4

The **slope** of a line through points (x_1, y_1) and (x_2, y_2) is

$$m = \frac{y_2 - y_1}{x_2 - x_1}. \tag{1.9}$$

It is customary to call the difference $y_2 - y_1$ the **rise** because it represents the vertical distance between the points, and $x_2 - x_1$ the **run**, the horizontal distance between the points. It is easy to show using similar triangles that m is independent of the two points chosen on the line; that is, no matter what two points we choose on the line to evaluate m, the result is always the same. The four numbers x_1, x_2, y_1, and y_2 vary, but the ratio $(y_2 - y_1)/(x_2 - x_1)$ remains unchanged. Note in Figure 1.32(b) that a horizontal line has slope zero (since $y_2 - y_1 = 0$), whereas the slope of a vertical line is undefined (since $x_2 - x_1 = 0$). In Figure 1.32(c), line l_1, which leans to the right, has a positive slope, and line l_2, which leans to the left, has a negative slope.

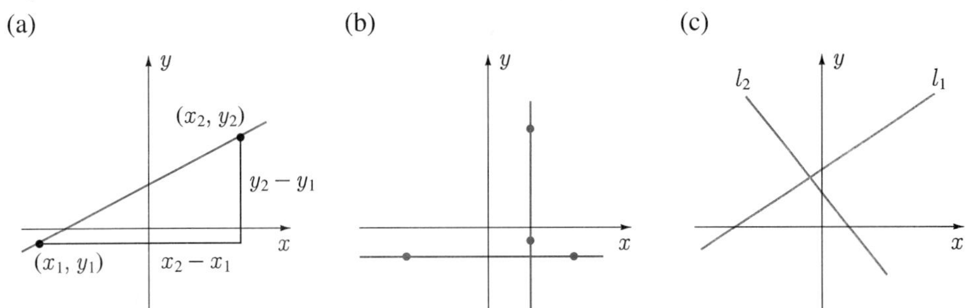

FIGURE 1.32

In Figure 1.33 we have shown the line through the points $(1, 2)$ and $(3, 5)$. Its slope is $(5 - 2)/(3 - 1) = 3/2$.

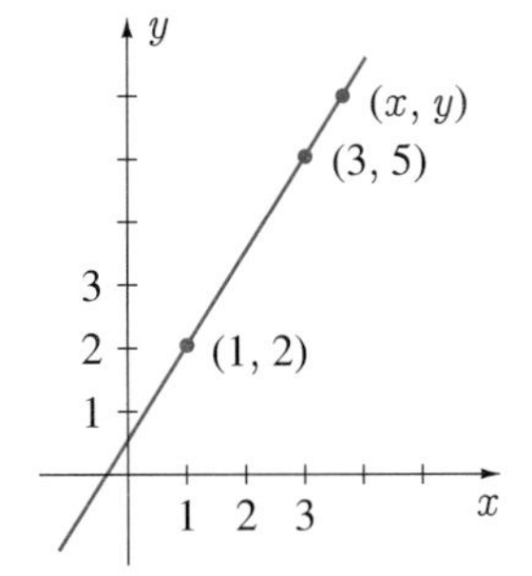

FIGURE 1.33

To find the equation for this line, we let (x, y) be the coordinates of any other point on the line. Since the slope of the line must also be given by $(y-2)/(x-1)$, it follows that

$$\frac{y-2}{x-1} = \frac{3}{2},$$

and when this equation is simplified

$$2y = 3x + 1.$$

If (x, y) are the coordinates of any point not on this line, then they do not satisfy this equation because the slope $(y-2)/(x-1)$ joining (x, y) to $(1, 2)$ is not equal to $3/2$. Thus, $2y = 3x + 1$ is the equation for the straight line through $(1, 2)$ and $(3, 5)$.

We can use this procedure to find the equation for the straight line through any point (x_1, y_1) with any slope m (Figure 1.34). If (x, y) are the coordinates of any other point on the line, then the slope of the line is $(y - y_1)/(x - x_1)$; therefore

$$\frac{y - y_1}{x - x_1} = m \quad \text{or}$$

$$y - y_1 = m(x - x_1). \tag{1.10}$$

This is called the **point-slope** formula for the equation of a straight line; it uses the slope m of the line and a point (x_1, y_1) on the line to determine the equation of the line.

FIGURE 1.34

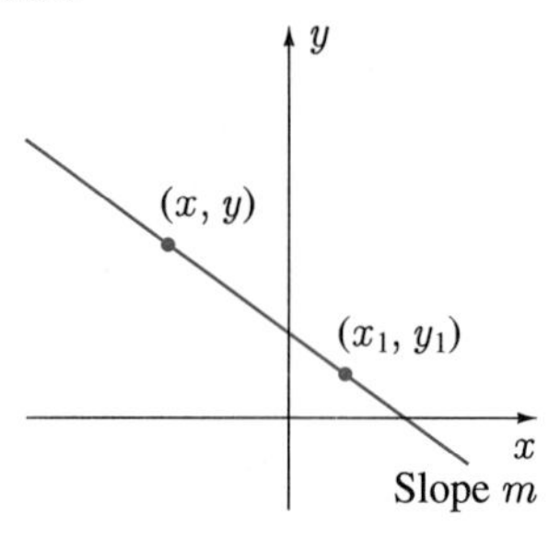

Other formulas for the equation of a line are used when different characteristics of the line are given. They are listed below:

Form of Equation	Name of Equation	Description of Characteristics Determining the Line
$\frac{y - y_1}{y_2 - y_1} = \frac{x - x_1}{x_2 - x_1}$	Two-point formula	Two points (x_1, y_1) and (x_2, y_2) on the line
$y = mx + b$	Slope y-intercept formula	m is the slope; b is the y-intercept (point where line crosses y-axis)
$y = m(x - a)$	Slope x-intercept formula	m is the slope; a is the x-intercept (point where line crosses x-axis)
$\frac{x}{a} + \frac{y}{b} = 1$	Two-intercept formula	a and b are x- and y-intercepts

Each of these formulas immediately determines the equation of the line when the specified characteristics are known. Alternatively, with minimal calculations, point-slope formula 1.10 can be used in each situation. For example, if we know that the y-intercept of a line is b, then a point on the line is $(0, b)$. Hence, 1.10 gives $y - b = m(x - 0)$, or, $y = mx + b$, the slope y-intercept formula.

There is one other equation that is worth mentioning. Vertical lines, which do not have slopes, cannot be represented in form 1.10; they have the form $x = k$, for some constant k. On the other hand, all lines can, for various values of the constants A, B, and C, be represented in the form

$$Ax + By + C = 0. \tag{1.11}$$

This is often called the **general equation** of a line.

EXAMPLE 1.5

Find the equation of the line through the points $(-1, 1)$ and $(2, 3)$.

SOLUTION Since the slope of the line is $(3 - 1)/(2 + 1) = 2/3$, we can use the point $(-1, 1)$ and the slope $2/3$ in point-slope formula 1.10 to give

$$y - 1 = \frac{2}{3}(x + 1) \qquad \text{or} \qquad 3y = 2x + 5.$$

The same result is obtained if the point $(2, 3)$ is used in place of $(-1, 1)$. Furthermore, the two-point formula could have been used. ■

EXAMPLE 1.6

Find equations for the lines in Figure 1.35.

FIGURE 1.35

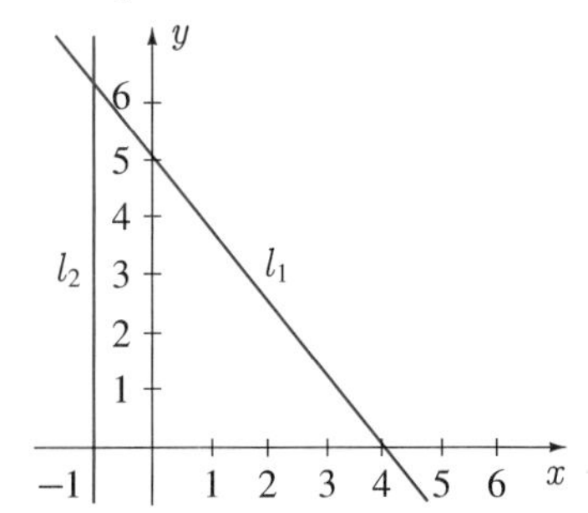

SOLUTION Since the slope of l_1 is $-5/4$ and its y-intercept is 5, we use the slope y-intercept formula,

$$y = -\frac{5}{4}x + 5 \qquad \text{or} \qquad 4y = 20 - 5x.$$

The equation of l_2 is clearly $x = -1$. ■

EXAMPLE 1.7

What is the slope of the line $3x + 4y = 2$?

SOLUTION By solving this equation for y in terms of x,

$$y = -\frac{3}{4}x + \frac{1}{2},$$

we express the equation in slope y-intercept form. Thus the slope of the line must be $-3/4$. ■

To find the point of intersection of two straight lines — say, $3y = 2x + 5$ and $4y + x = 14$ (Figure 1.36) — we must find the point whose coordinates (x, y) satisfy both equations. If we solve each equation for x and equate the expressions, we obtain $(3y - 5)/2 = 14 - 4y$, which immediately yields $y = 3$. Either of the original equations then gives $x = 2$, and the point of intersection of the lines is $(2, 3)$. Note once again the algebraic-geometric interplay. Geometrically, $(2, 3)$ is the point of intersection of two straight lines. Algebraically, the two numbers constitute the solution of the equations $3y = 2x + 5$ and $4y + x = 14$.

FIGURE 1.36

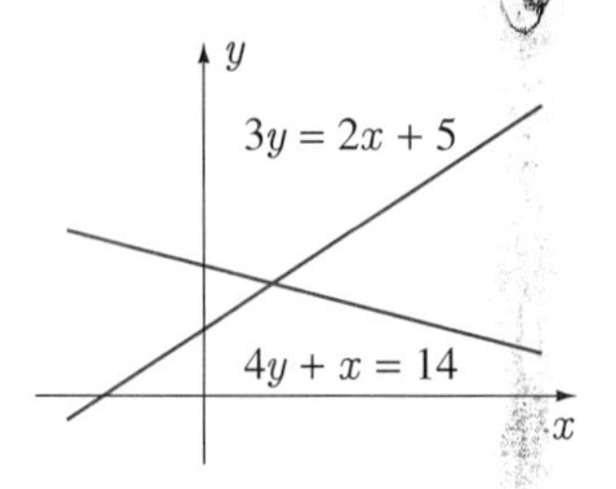

Of particular importance to the study of lines are the concepts of parallelism and perpendicularity.

Definition 1.5 Two lines are said to be **parallel** if they have no point of intersection.

EXAMPLE 1.8 Verify that the lines with equations $2x - y = 4$ and $4x - 2y = 7$ are parallel.

SOLUTION If we solve each of these equations for y and equate the resulting expressions, we obtain

$$2x - 4 = 2x - \frac{7}{2},$$

an obvious impossibility. Consequently, the lines do not intersect. ■

Geometrically, the following is clear.

Theorem 1.1 *Two lines are parallel if and only if they have the same slope.*

For example, if we write the equations of the lines in Example 1.8 in slope y-intercept form $y = 2x - 4$ and $y = 2x - 7/2$, coefficients of the x-terms identify slopes of the lines. Since each has slope 2, the lines are parallel.

Definition 1.6 Two lines are said to be **perpendicular** if they intersect at right angles.

The following theorem gives a test for perpendicularity of straight lines in terms of slopes.

Theorem 1.2

Two lines with nonzero slopes m_1 and m_2 are perpendicular if and only if

$$m_1 = -\frac{1}{m_2} \qquad \text{or} \qquad m_1 m_2 = -1. \tag{1.12}$$

Proof Suppose nonparallel lines l_1 and l_2 with nonzero slopes m_1 and m_2 intersect at a point Q (Figure 1.37).

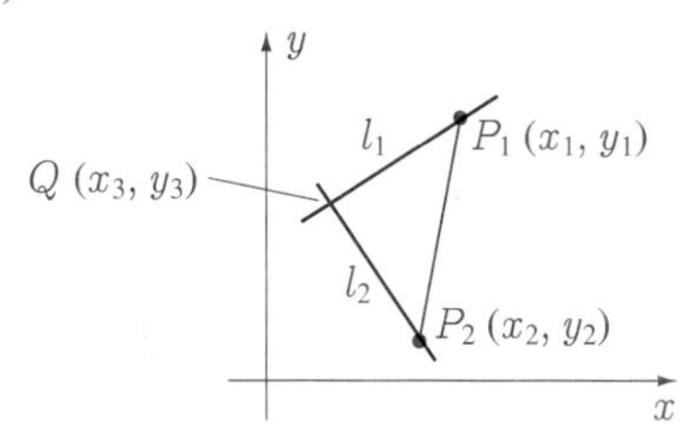

FIGURE 1.37

Suppose the coordinates of Q are (x_3, y_3), and let $P_1(x_1, y_1)$ and $P_2(x_2, y_2)$ be any two points on l_1 and l_2 respectively. Lines l_1 and l_2 are perpendicular if and only if triangle QP_1P_2 is right-angled at Q; that is, if and only if,

$$\|P_1P_2\|^2 = \|QP_1\|^2 + \|QP_2\|^2.$$

Using distance formula 1.7, this condition can be expressed as

$$(x_1 - x_2)^2 + (y_1 - y_2)^2 = (x_1 - x_3)^2 + (y_1 - y_3)^2 + (x_2 - x_3)^2 + (y_2 - y_3)^2.$$

When terms are expanded, and the equation is simplified, the result is

$$x_3^2 - x_1x_3 + x_1x_2 - x_2x_3 = -y_3^2 + y_1y_3 - y_1y_2 + y_2y_3.$$

After factoring each side, this equation can be rearranged into the form

$$\frac{y_3 - y_2}{x_3 - x_2} = -\frac{x_3 - x_1}{y_3 - y_1} = -\frac{1}{\dfrac{y_3 - y_1}{x_3 - x_1}}.$$

Since $m_1 = (y_3 - y_1)/(x_3 - x_1)$ and $m_2 = (y_3 - y_2)/(x_3 - x_2)$, we have shown that lines l_1 and l_2 are perpendicular if and only if

$$m_2 = -\frac{1}{m_1}.$$

EXAMPLE 1.9

Find the equation of the straight line that passes through the point $(2, 4)$ and is perpendicular to the line $3x + y = 5$.

SOLUTION Since the slope of the given line $y = -3x + 5$ is -3, the required line has slope $1/3$. Using point-slope formula 1.10, we find that the equation of the required line is

$$y - 4 = \frac{1}{3}(x - 2) \qquad \text{or} \qquad x - 3y = -10.$$

■

Perpendicularity and parallelism deal with lines that make a right angle at their point of intersection, or make no angle, as parallel lines do not intersect. Lines that intersect usually do so at angles other that $\pi/2$ radians. In this book all angles are measured in radians. We shall see why in Chapter 3. In order to determine the angle at which two lines intersect, we first define the inclination of a line.

Definition 1.7

The **inclination** of a line l is the angle of rotation ϕ $(0 \leq \phi < \pi)$ from the positive x-direction to the line.

The line $y = x - 3$ in Figure 1.38(a) has an inclination of $\pi/4$ radians, and the line $y = -x - 4$ in Figure 1.38(b) has an inclination of $3\pi/4$ radians.

FIGURE 1.38

(a)

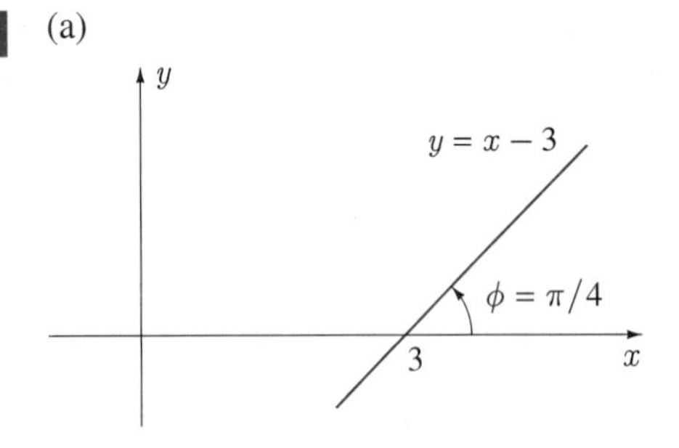

(b)

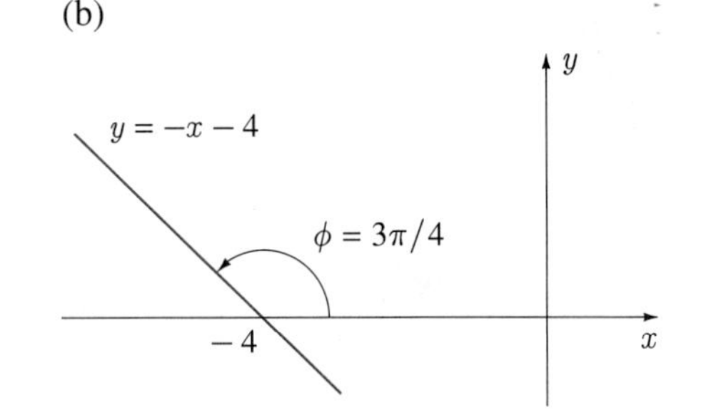

When the slope m of a line is positive, as shown in Figure 1.39(a), it is clear that ϕ must be in the interval $0 < \phi < \pi/2$, and m and ϕ are related by $\tan\phi = m$. When the slope of l is negative as in Figure 1.39(b), we use identity A.12b in Appendix A to write

$$m = -\tan(\pi - \phi) = -\frac{\tan\pi - \tan\phi}{1 + \tan\pi\tan\phi} = \tan\phi.$$

FIGURE 1.39

(a)

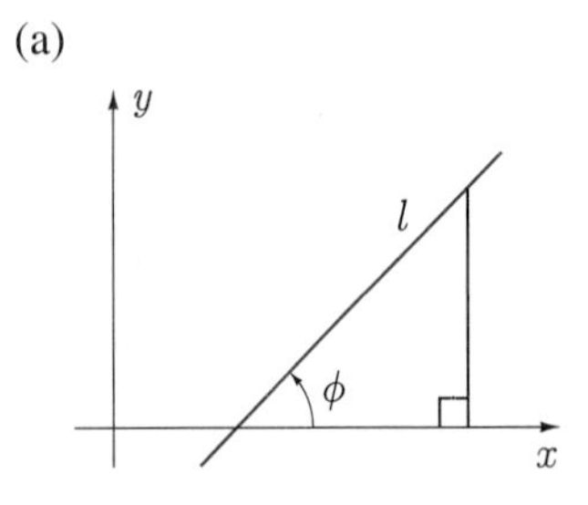

(b)

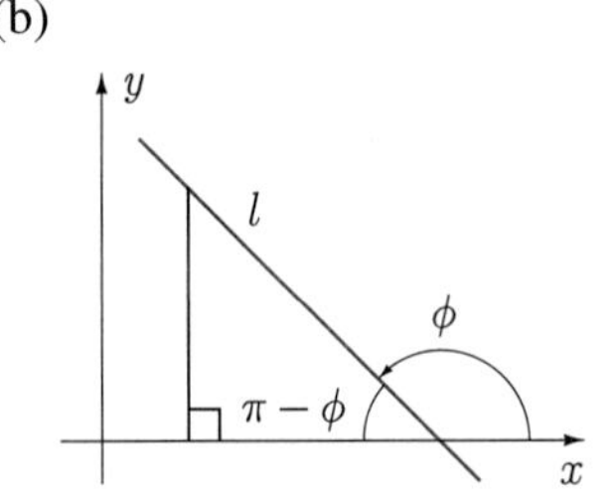

Thus, whenever the slope m of a line is defined, the angle of inclination is related to m by the equation

$$\tan\phi = m. \tag{1.13}$$

In some sense this equation is true even when m is not defined. This occurs for vertical lines, which have no slope. For a vertical line, $\phi = \pi/2$, and $\tan\phi$ is undefined. Thus, equation 1.13 is also valid for vertical lines from the point of view that neither side of the equation is defined.

EXAMPLE 1.10

What is the angle of inclination of the line $2x - 3y = 4$?

SOLUTION From $y = 2x/3 - 4/3$, the slope of the line is $2/3$. The angle whose tangent is equal to $2/3$ is $\phi = 0.588$ radians. ■

When two lines l_1 and l_2 with nonzero slopes m_1 and m_2 intersect (Figure 1.40), the angle θ $(0 < \theta < \pi)$ between the lines is given by the equation

$$\phi_1 = \theta + \phi_2 \qquad \text{or} \qquad \theta = \phi_1 - \phi_2 .$$

By applying the tangent function to both sides of this equation and using identity A.12b from Appendix A once again, we can express θ in terms of m_1 and m_2,

$$\tan\theta = \tan(\phi_1 - \phi_2) = \frac{\tan\phi_1 - \tan\phi_2}{1 + \tan\phi_1 \tan\phi_2} = \frac{m_1 - m_2}{1 + m_1 m_2}.$$

This equation determines θ when $\phi_1 > \phi_2$. When $\phi_1 < \phi_2$, the equation is replaced by

$$\tan\theta = \frac{m_2 - m_1}{1 + m_1 m_2}.$$

In both cases we may write

$$\tan\theta = \left| \frac{m_1 - m_2}{1 + m_1 m_2} \right|, \tag{1.14}$$

for the acute angle between the lines.

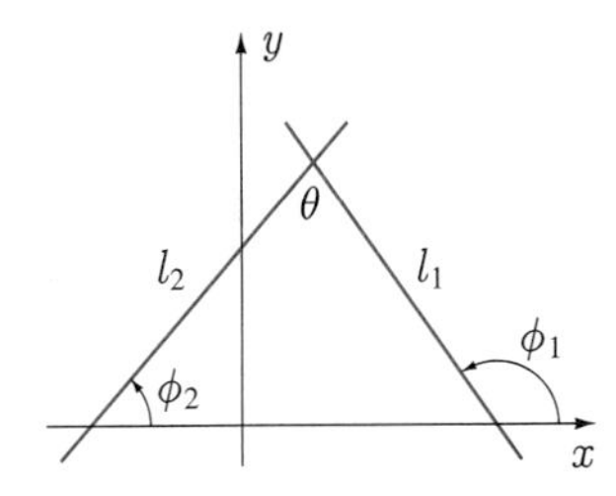

FIGURE 1.40

EXAMPLE 1.11

Find the angle between the lines $2x - 3y = 4$ and $x + 4y = 6$.

SOLUTION Since slopes of these lines are $2/3$ and $-1/4$, it follows that

$$\tan\theta = \left| \frac{2/3 - (-1/4)}{1 + (2/3)(-1/4)} \right| = \frac{11}{10}.$$

The acute angle whose tangent is equal to $11/10$ is $\theta = 0.833$ radians. ■

EXERCISES 1.3

In Exercises 1–10 draw the line described and find its equation.

1. Through the points $(1, 2)$ and $(-3, 4)$

2. Through the points $(3, -6)$ and $(5, -6)$

3. Through the point $(-2, -3)$ with slope 3

4. Through the point $(1, -3/2)$ with slope $-1/2$

5. The y-axis

6. The x-axis

7. Through the point $(4, 3)$ and crossing the x-axis at -2

8. Through the point $(-1, -2)$ and crossing the y-axis at 4

9. Crossing the x- and y-axes at 1 and -3, respectively

10. Through the origin and the midpoint of the line segment joining $(3,4)$ and $(-7,8)$

In Exercises 11–16 find the angle of inclination of the line.

11. $x - y + 1 = 0$

12. $x + 2y = 3$

13. $3x - 2y = 1$

14. $y - 3x = 4$

15. $x = 4$

16. $y = 2$

In Exercises 17–24 determine whether the lines are perpendicular, parallel, or neither perpendicular nor parallel. In the last case, find the angle between the lines.

17. $y = -x + 4,\ y = x + 6$

18. $x + 3y = 4,\ 2x + 6y = 7$

19. $x = 3y + 4,\ y = x/3 - 2$

20. $2x + 3y = 1,\ 3x - 2y = 5$

21. $y = 3x + 2,\ y = -x/2 + 1$

22. $x - y = 5,\ 2x + 3y = 4$

23. $x = 0,\ y = 5$

24. $x + y + 2 = 0,\ 3x - y = 4$

In Exercises 25–29 find the point of intersection of the lines.

25. $x + y = 0,\ x - 2y = -3$

26. $x = 1,\ y = 2$

27. $3x + 4y = 6,\ x - 6y = 3$

28. $y = 2x + 6,\ x = y + 4$

29. $x/2 + y/3 = 1,\ 2x - y/4 = 15$

30. $14x - 2y = 5,\ 3x + 10y = 12$

In Exercises 31–38 find the equation of the line described.

31. Parallel to $x + 2y = 15$, and through the point of intersection of $2x - y = 5$ and $x + y = 4$

32. Perpendicular to $x - y = 4$, and through the point of intersection of $2x + 3y = 3$ and $x - y = 4$

33. Parallel to the line through $(1,2)$ and $(-3,0)$, and through the point $(5,6)$

34. Perpendicular to the line through $(-3,4)$ and $(1,-2)$, and through the point $(-3,-2)$

35. Crossing through the first quadrant to form an isosceles triangle with area 8 square units

36. Through $(3,5)$ and crossing through the first quadrant to form a triangle with area 30 square units

37. Has slope 2, and that part in the second quadrant has length 3

38. Parallel to the x-axis, below the point of intersection of the lines $x = y$ and $x + y = 4$, and forms with these lines a triangle with area 9 square units

39. A median of a triangle is a line segment drawn from a vertex to the midpoint of the opposite side. Find equations for the three medians of the triangle with vertices $(1,1)$, $(3,5)$, and $(0,4)$. Show that all three medians intersect in a point called the centroid of the triangle.

40. If (x_1, y_1), (x_2, y_2), (x_3, y_3), and (x_4, y_4) are vertices of any quadrilateral, show that the line segments joining the midpoints of adjacent sides form a parallelogram.

41. Find the equation of the perpendicular bisector of the line segment joining $(-1,2)$ and $(3,-4)$. The perpendicular bisector is the line that cuts the line segment in half and is perpendicular to it.

42. Find coordinates of the point that is equidistant from the three points $(1,2)$, $(-1,4)$, and $(-3,1)$.

43. Prove Theorem 1.1.

44. Let P be any point inside an equilateral triangle (Figure 1.41). Show that the sum of the distances of P from the three sides is always equal to the height h of the triangle.

FIGURE 1.41

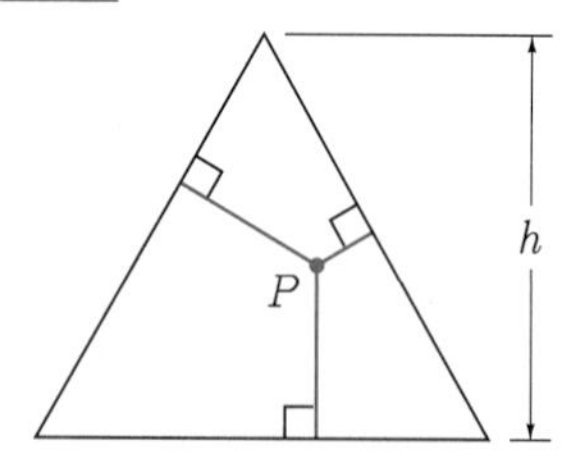

45. Prove that the perpendicular distance from a point (x_1, y_1) to a line $Ax + By + C = 0$ is given by

$$\frac{|Ax_1 + By_1 + C|}{\sqrt{A^2 + B^2}}.$$

Conic Sections

In this section we discuss parabolas, circles, ellipses, and hyperbolas, the so-called **conic sections**. As we are only interested in illustrating the algebraic-geometric interplay of analytic geometry, we do not give a complete development of these curves together with their many properties. We show only how the form of the equation for each conic section dictates its shape.

The Parabola

When the y-coordinate of a point (x, y) on a curve is related to its x-coordinate by an equation of the form

$$y = ax^2 + bx + c, \qquad (1.15)$$

where a, b, and c are constants (with $a \neq 0$), the curve is called a **parabola**. To determine the shape of a parabola we consider a number of examples. In Section 1.3 we used Table 1.1 to draw the simplest of all parabolas $y = x^2$ (Figure 1.30(b)). For every point (x, y) to the right of the y-axis on this curve, there is a point equidistant to the left of the y-axis which has the same y-coordinate; that is, the point $(-x, y)$ is also on the curve. Putting it another way, that part of the parabola to the left of the y-axis is the image in the y-axis (thought of as a mirror) of that part to the right of the y-axis. Such a curve is said to be symmetric about the y-axis. It happens whenever the equation of a curve is unchanged when each x therein is replaced by $-x$. In other words, we have the test:

A curve is **symmetric** *about the y-axis if its equation remains unchanged when x is replaced by* $-x$.

The values in Table 1.2 for the parabola $y = 9 - x^2$ lead to the curve in Figure 1.42. This parabola is also symmetric about the y-axis. It is said to "open downward", whereas the parabola in Figure 1.30(b) "opens upward". The sign in front of the x^2-term dictates which way a parabola opens.

TABLE 1.2

x	−4	−3	−2	−1	0	1	2	3	4
y	−7	0	5	8	9	8	5	0	−7

A table of values for the parabola $y = 2x^2 + 4x - 6$ leads to the curve in Figure 1.43. It is not symmetric about the y-axis; notice that its equation does change if x is replaced by $-x$. The parabola appears to be symmetric about the line $x = -1$, and its lowest point is $(-1, -8)$. To prove that this is indeed the case, we rewrite the equation as

$$y = 2(x^2 + 2x) - 6 = 2(x + 1)^2 - 8.$$

This form clearly indicates that the smallest value for y is -8, and it occurs when $x = -1$. In addition, whenever the values $x = -1 \pm k$, where k is some fixed number, are substituted into the equation of the parabola, the same value for y is obtained, namely $y = 2k^2 - 8$. This means that the points $(-1 + k, 2k^2 - 8)$ and $(-1 - k, 2k^2 - 8)$ are both on the parabola, and these points are equidistant from the line $x = -1$; that is, the parabola is symmetric about $x = -1$.

The technique used above to rewrite $y = 2x^2 + 4x - 6$ as $2(x + 1)^2 - 8$ is called *completing the square*. If we apply the same technique to the general parabola $y = ax^2 + bx + c$, we obtain

$$y = a\left(x^2 + \frac{b}{a}x\right) + c = a\left(x + \frac{b}{2a}\right)^2 + \left(c - \frac{b^2}{4a}\right). \qquad (1.16)$$

This form for the equation of the parabola shows the following:

1. When $a > 0$, the parabola opens upward and has a minimum at the point

$$\left(-\frac{b}{2a}, c - \frac{b^2}{4a}\right).$$

2. When $a < 0$, the parabola opens downward and has a maximum at the point

$$\left(-\frac{b}{2a}, c - \frac{b^2}{4a}\right).$$

3. The parabola is symmetric about the line $x = -b/(2a)$ (Figure 1.44); that is, for every point P on one side of the line, there is a point Q on the other side that is the mirror image of P in the line.

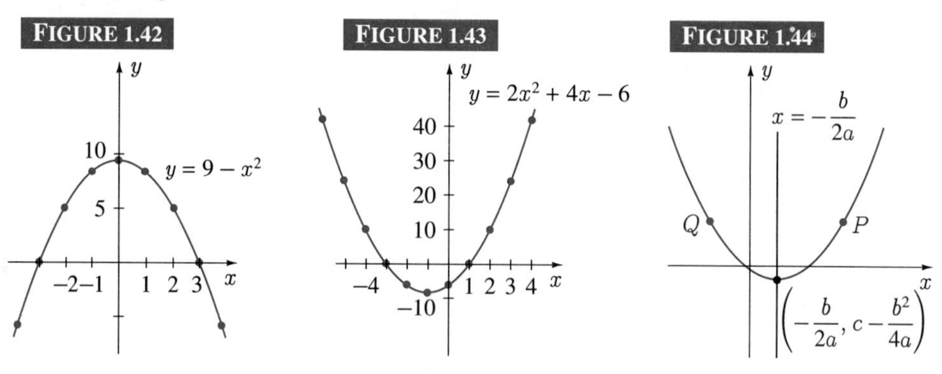

4. The parabola crosses the x-axis when

$$0 = y = a\left(x + \frac{b}{2a}\right)^2 + \left(c - \frac{b^2}{4a}\right).$$

To solve this equation for x, we first write

$$\left(x + \frac{b}{2a}\right)^2 = \frac{b^2}{4a^2} - \frac{c}{a},$$

and then take square roots of each side,

$$x + \frac{b}{2a} = \pm\sqrt{\frac{b^2}{4a^2} - \frac{c}{a}}.$$

Finally then,

$$x = -\frac{b}{2a} \pm \sqrt{\frac{b^2}{4a^2} - \frac{c}{a}} = \frac{-b \pm \sqrt{b^2 - 4ac}}{2a}. \tag{1.17}$$

This is called the **quadratic formula**. It determines points where the parabola $y = ax^2 + bx + c$ crosses the x-axis. Equivalently, it defines roots of the quadratic equation $ax^2 + bx + c = 0$. Note that when $b^2 - 4ac > 0$, there are two solutions (Figure 1.45(a)); when $b^2 - 4ac = 0$, there is only one solution (Figure 1.45(b)); and when $b^2 - 4ac < 0$, there is no solution (Figure 1.45(c)); that is, no real solution.

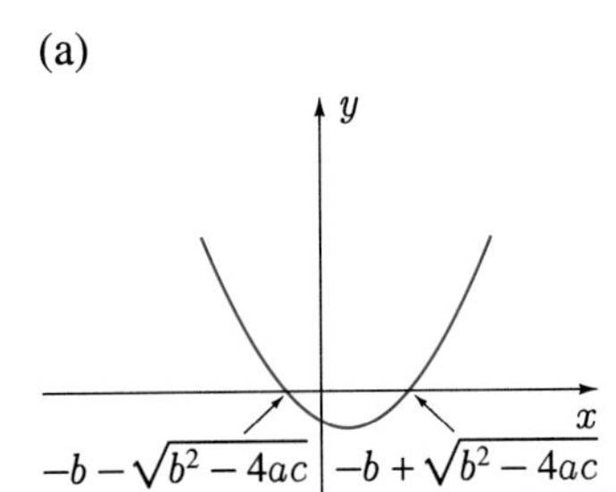

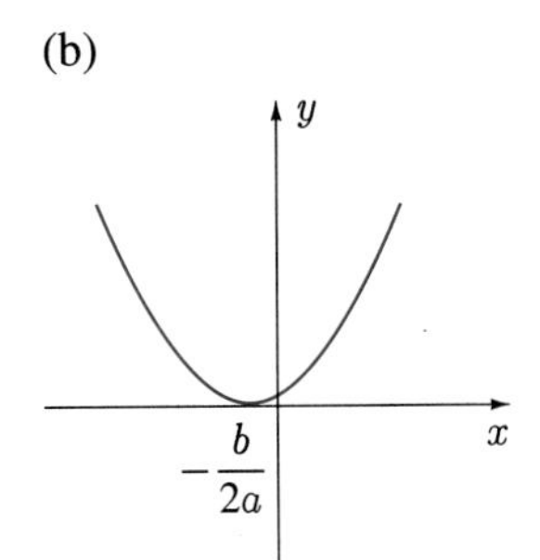

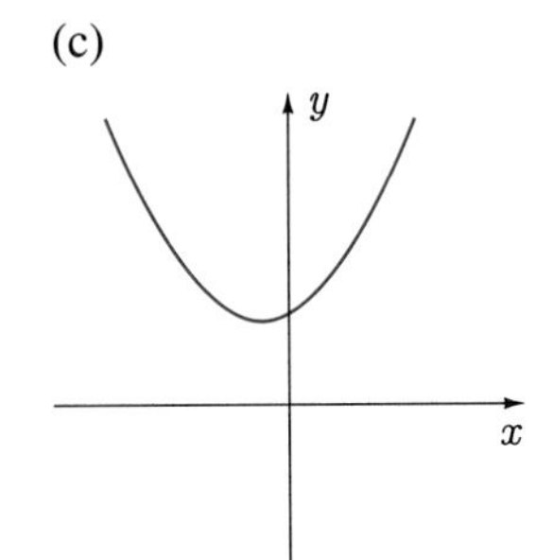

FIGURE 1.45

When x and y in equation 1.15 are interchanged, the resulting equation

$$x = ay^2 + by + c \tag{1.18}$$

still defines a parabola. In this case we construct a table of values by listing various values of y and calculating corresponding values for x. For example, Tables 1.3 and 1.4 lead to the curves in Figures 1.46 and 1.47 for the parabolas $x = y^2 + 1$ and $x = -y^2 + 4y - 4$ respectively. The parabola $x = y^2 + 1$ is symmetric about the x-axis; for any point (x, y) on the parabola, so also is $(x, -y)$. In general, *a curve is symmetric about the x-axis if its equation remains unchanged when y is replaced by* $-y$.

TABLE 1.3

y	−4	−3	−2	−1	0	1	2	3	4
x	17	10	5	2	1	2	5	10	17

TABLE 1.4

y	−1	0	1	2	3	4	5
x	−9	−4	−1	0	−1	−4	−9

FIGURE 1.46

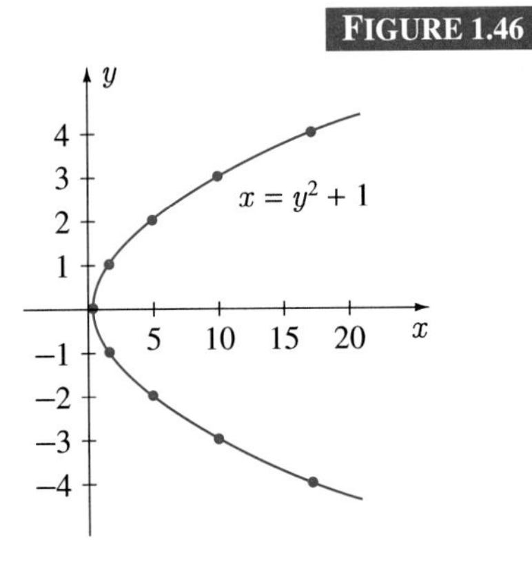

FIGURE 1.47

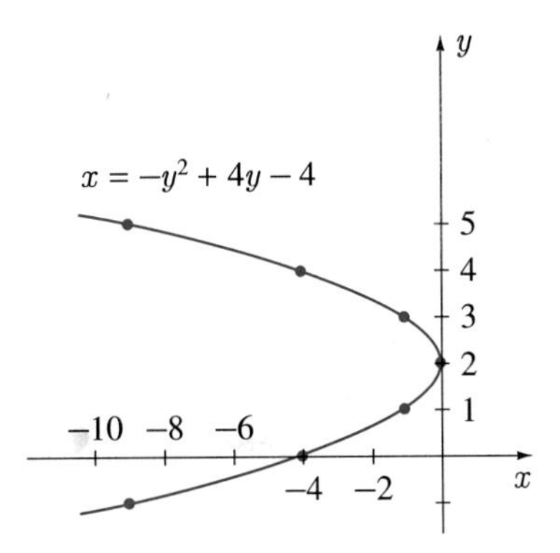

EXAMPLE 1.12

Find equations for the parabolas in Figures 1.48 and 1.49.

FIGURE 1.48

SOLUTION The fact that the parabola in Figure 1.48 is symmetric about the x-axis means that b in equation 1.18 must vanish; that is, its equation must be of the form $x = ay^2 + c$. Since the points $(2, 0)$ and $(0, 3)$ are on the parabola, their coordinates must satisfy the equation of the parabola,

$$2 = a(0)^2 + c,$$
$$0 = a(3)^2 + c.$$

These imply that $c = 2$ and $a = -c/9 = -2/9$. Thus, the equation of the parabola is $x = -2y^2/9 + 2$.

The parabola in Figure 1.49 has no special attributes that we can utilize (such as the position of the line of symmetry in Figure 1.48). We therefore use the facts that its equation must be of the form $y = ax^2 + bx + c$, and the three points $(-1, 3)$, $(0, 2)$, and $(4, 6)$ are on the parabola. Substitution of the coordinates of these points into the equation gives

FIGURE 1.49

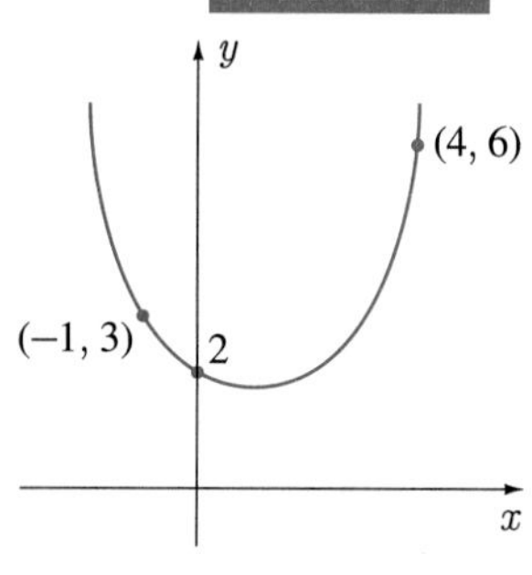

$$3 = a(-1)^2 + b(-1) + c,$$
$$2 = a(0)^2 + b(0) + c,$$
$$6 = a(4)^2 + b(4) + c.$$

The second equation yields $c = 2$, and when this is substituted into the other two equations,

$$a - b = 1,$$
$$16a + 4b = 4.$$

These can be solved to obtain $a = 2/5$ and $b = -3/5$, and therefore the required equation is $y = 2x^2/5 - 3x/5 + 2$. ■

EXERCISES 1.4A

In Exercises 1–12 draw the parabola.

1. $y = 2x^2 - 1$
2. $y = -x^2 + 4x - 3$
3. $y = x^2 - 2x + 1$
4. $3x = 4y^2 - 1$
5. $x = y^2 + 2y$
6. $2y = -x^2 + 3x + 4$
7. $x + y^2 = 1$
8. $2y^2 + x = 3y + 5$
9. $y = 4x^2 + 5x + 10$
10. $x = 10y^2$
11. $y = -x^2 + 6x - 9$
12. $x = -(4 + y)^2$

13. Find x- and y-intercepts for the following parabolas:
(a) $y = x^2 - 2x - 5$
(b) $x = 4y^2 - 8y + 4$

In Exercises 14–17 find the equation for the parabola shown.

14.

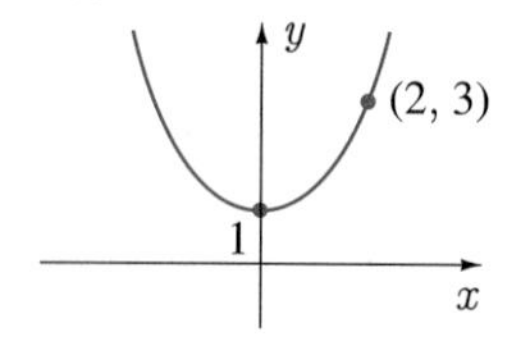

15.

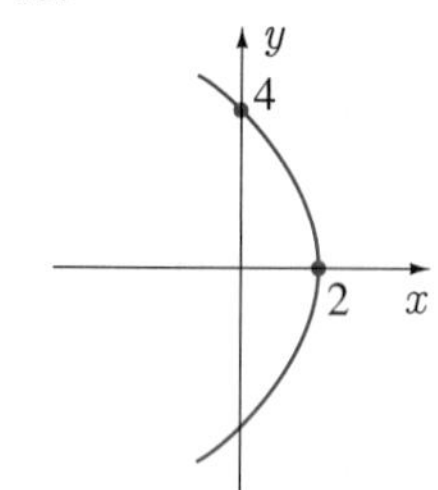

16.

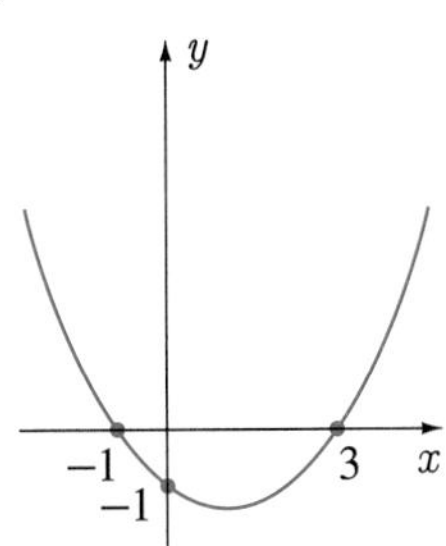

17.

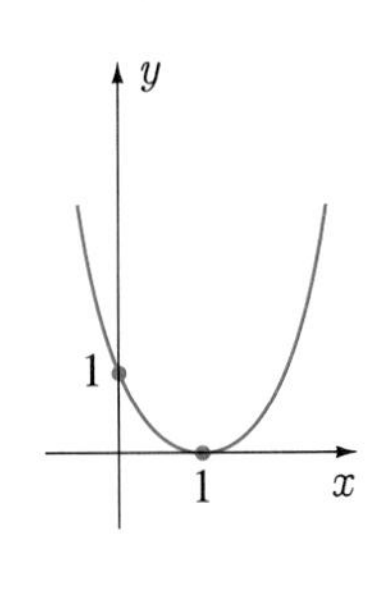

In Exercises 18–23 find all points of intersection for the curves. In each case draw the curves.

18. $y = 1 - x^2$, $y = x + 1$ **19.** $y + 2x = 0$, $y = 1 + x^2$

20. $y = 2x - x^2 - 6$, $25 + x = 5y$

21. $x = y(y - 1)$, $2y = 2x + 1$

22. $x = -y^2 + 1$, $x = y^2 + 2y - 3$

23. $y = 6x^2 - 2$, $y = x^2 + x + 1$

24. When a shell is fired from an artillery gun (Figure 1.50), it follows a parabolic path

$$y = -\frac{4.905}{v^2 \cos^2 \theta} x^2 + x \tan \theta,$$

where v is the muzzle velocity of the shell.

FIGURE 1.50

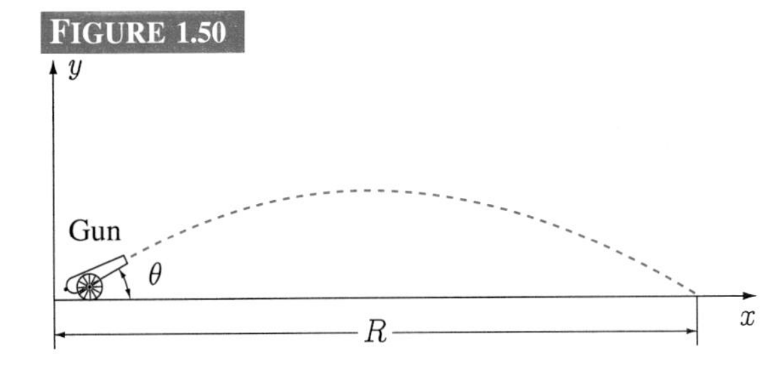

(a) Find the range R of the shell.

(b) What is the maximum height attained by the shell?

25. The cable of the suspension bridge in Figure 1.51 hangs in the shape of a parabola. The towers are 200 m apart and extend 50 m above the roadway. If the cable is 10 m above the roadway at its lowest point, find the length of the supporting rods 30 m from the towers.

FIGURE 1.51

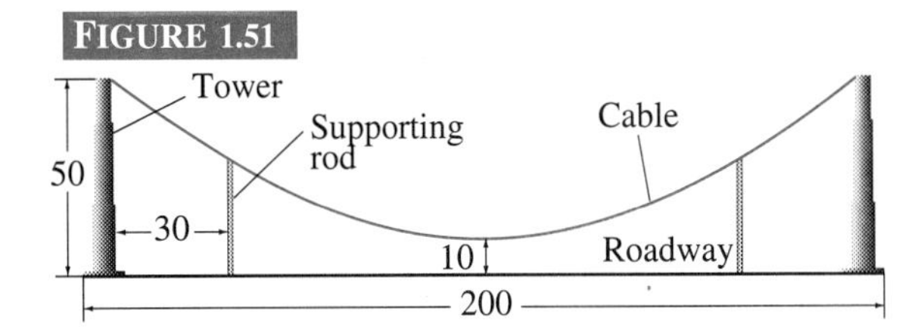

26. Find points of intersection for the parabolas

$$y = (x - 2)^2, \qquad 5x = y^2 + 4.$$

27. Find the height of the parabolic arch in Figure 1.52.

FIGURE 1.52

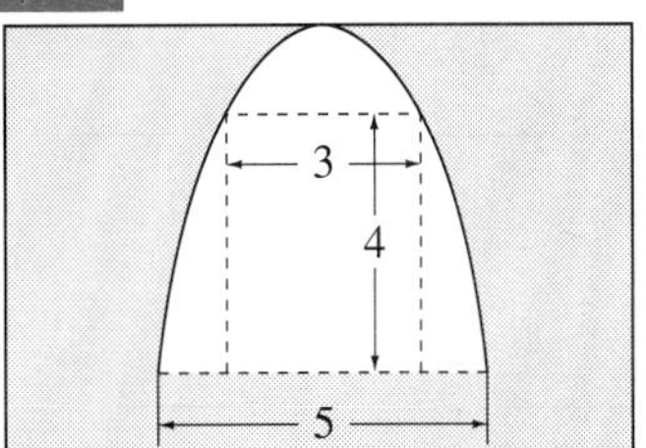

28. Find a parabola of type 1.15 passing through the points $(1, 2)$, $(-3, 10)$, and $(3, 4)$.

29. Find lines through the point $(3, 4)$ that intersect the parabola $y = x^2 - 1$ at only one point.

The Circle

When the x- and y-coordinates of points on a curve are related by an equation of the form

$$(x - h)^2 + (y - k)^2 = r^2, \tag{1.19}$$

where h, k, and $r > 0$ are constant, the curve is called a **circle**. It takes but a quick recollection of distance formula 1.7 to convince ourselves that this definition of a circle coincides with our intuitive idea of a circle. If we write equation 1.19 in the form

$$\sqrt{(x - h)^2 + (y - k)^2} = r,$$

the left side is the distance from the point (x, y) to the point (h, k). Equation 1.19 therefore describes all points (x, y) at a fixed distance r from (h, k), a circle centred at (h, k) with radius r (Figure 1.53). For example, the radius of the circle in Figure 1.54 is equal to 2, and its equation is therefore

$$(x+1)^2 + (y-2)^2 = 4.$$

FIGURE 1.53

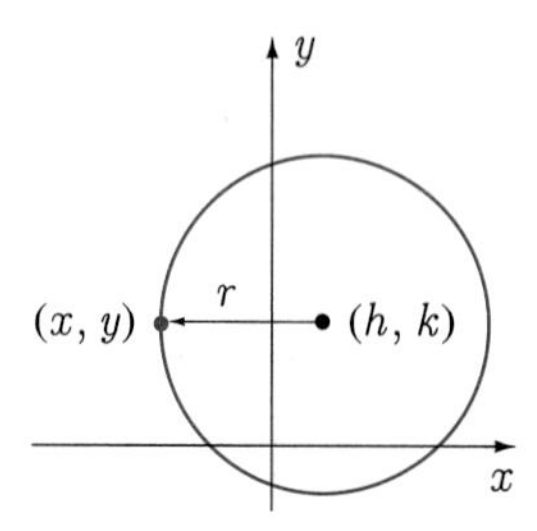

FIGURE 1.54

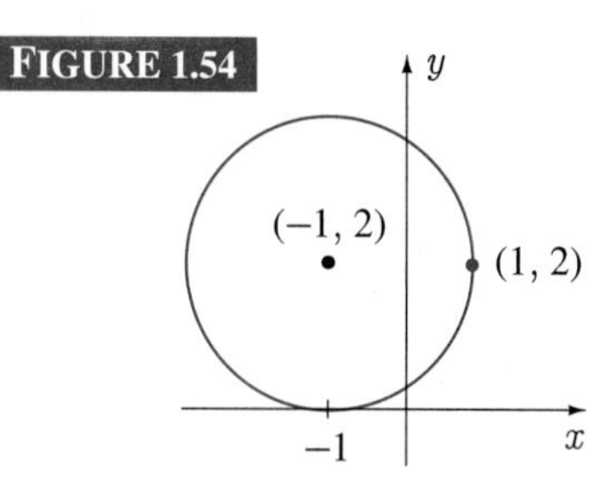

When equation 1.19 is expanded, we have

$$x^2 - 2hx + h^2 + y^2 - 2ky + k^2 = r^2 \qquad \text{or}$$

$$x^2 + y^2 - 2hx - 2ky + h^2 + k^2 - r^2 = 0.$$

This shows that the equation of a circle may be given in another form, namely,

$$x^2 + y^2 + fx + gy + e = 0. \tag{1.20}$$

Given this equation, the centre and radius can be identified by reversing the expansion and completing the squares of $x^2 + fx$ and $y^2 + gy$. For instance, if $x^2 + y^2 + 2x - 3y - 5 = 0$, then

$$0 = (x+1)^2 + (y-3/2)^2 - 5 - 1 - 9/4 = (x+1)^2 + (y-3/2)^2 - 33/4.$$

The centre of the circle is therefore $(-1, 3/2)$ and its radius is $\sqrt{33}/2$.

When the centre of a circle is the origin $(0, 0)$, equation 1.19 simplifies to

$$x^2 + y^2 = r^2. \tag{1.21}$$

Be careful to use equation 1.19, not 1.21, when the centre of the circle is not the origin. It is a common error to use (1.21).

EXAMPLE 1.13

Figure 1.55 shows an arc of a circle. Find the equation for the circle.

FIGURE 1.55

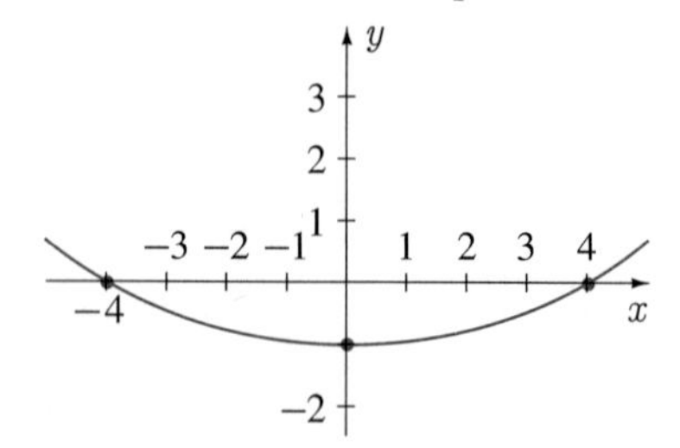

SOLUTION From the symmetry of the figure, we see that the centre of the circle is on the y-axis. Its equation must be of the form

$$x^2 + (y - k)^2 = r^2 .$$

Because $(4, 0)$ is a point on the circle, these coordinates must satisfy its equation; that is,

$$16 + (0 - k)^2 = r^2 \qquad \text{or} \qquad 16 + k^2 = r^2 .$$

Similarly, since $(0, -1)$ is on the circle,

$$(-1 - k)^2 = r^2 \qquad \text{or} \qquad 1 + 2k + k^2 = r^2 .$$

If we subtract these two equations, we obtain $2k - 15 = 0$, from which we see that $k = 15/2$. Consequently, $r^2 = 16 + k^2 = 16 + 225/4 = 289/4$, and the equation of the circle is

$$x^2 + (y - 15/2)^2 = 289/4 .$$

■

EXERCISES 1.4B

In Exercises 1–10 draw the circle.

1. $x^2 + y^2 = 50$

2. $(x + 5)^2 + (y - 2)^2 = 6$

3. $x^2 + 2x + y^2 = 15$

4. $x^2 + y^2 - 4y + 1 = 0$

5. $x^2 - 2x + y^2 - 2y + 1 = 0$

6. $2x^2 + 2y^2 + 6x = 25$

7. $3x^2 + 3y^2 + 4x - 2y = 6$

8. $x^2 + 4x + y^2 - 2y = 5$

9. $x^2 + y^2 - 2x - 4y + 5 = 0$

10. $x^2 + y^2 + 6x + 3y + 20 = 0$

In Exercises 11–16 find an equation for the circle.

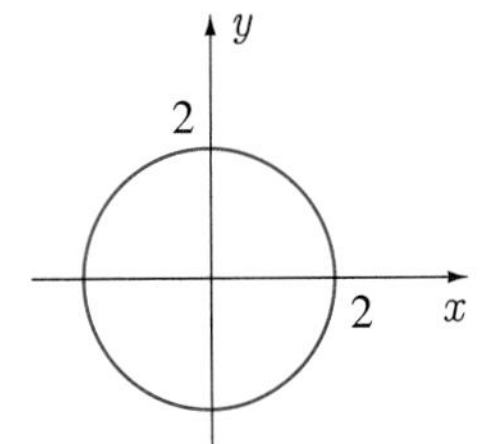

12.

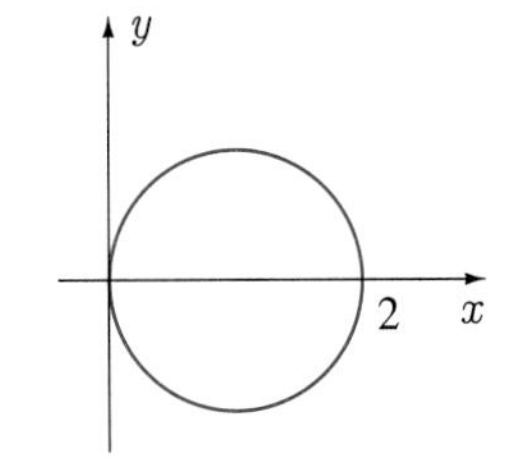

13.

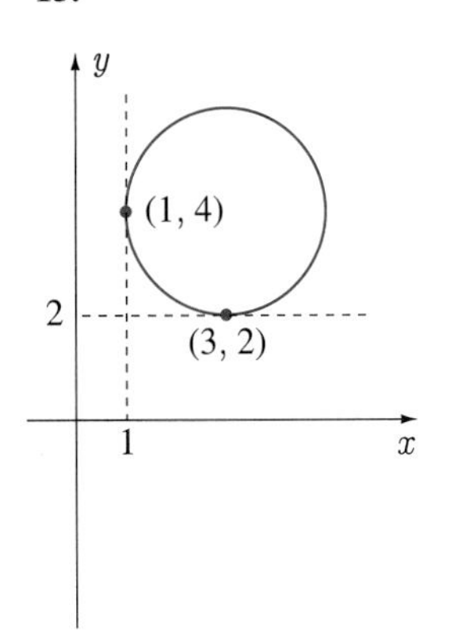

14.

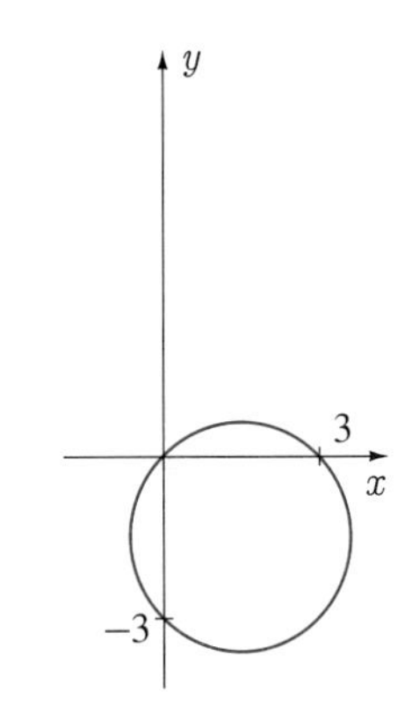

15.

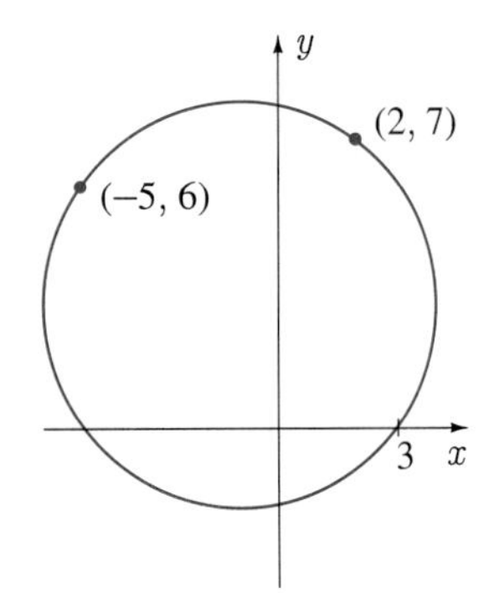

16.

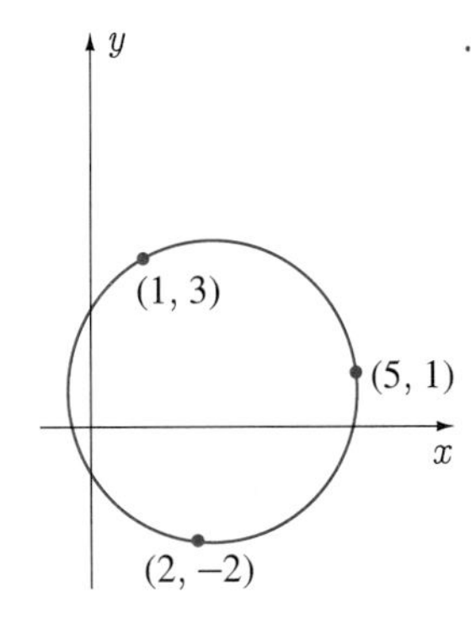

17. Find the equation of a circle that passes through the points $(3,4)$ and $(1,-10)$, and has its centre:

(a) on the line $2x+3y+16=0$;

(b) on the line $x+7y+19=0$.

In Exercises 18–21 find points of intersection for the curves.

18. $x^2+2x+y^2=4$, $y=3x+2$

19. $x^2+y^2-4y+1=0$, $2x+y=1$

20. $x^2+y^2=9$, $y=3x^2+4$

21. $(x+3)^2+y^2=25$, $y^2=16(x+1)$

22. Show that every equation of the form 1.20 represents a circle, a point, or nothing at all.

23. Prove that the perpendicular bisector of a chord of a circle always passes through the centre of the circle.

24. The *circumcircle* for a triangle is that circle which passes through all three of its vertices. Find the circumcircle for the triangle with vertices $A(1,1)$, $B(-3,3)$, and $C(2,4)$ by:

(a) finding its centre and radius (see Exercise 23);

(b) taking the equation of the circle in form 1.19 and requiring A, B, and C to be on the circle.

25. Prove that the three altitudes of the triangle in Exercise 24 intersect in a point called the *orthocentre* of the triangle.

26. Show that if a line $Ax+By+C=0$ and a circle $(x-h)^2+(y-k)^2=r^2$ do not intersect, then the shortest distance between them is the smaller of the two numbers

$$\frac{|(Ah+Bk+C)\pm r\sqrt{A^2+B^2}|}{\sqrt{A^2+B^2}}.$$

Hint: Use the result of Exercise 45 in Section 1.3.

27. The *incircle* of a triangle is that circle which lies interior to the triangle but touches all three sides. The centre of the incircle is called the *incentre*. Use the result of Exercise 45 in Section 1.3 to show that the incentre (x,y) of the triangle with vertices $(0,0)$, $(2,0)$, and $(0,1)$ must satisfy the equations

$$|x|=|y|=\frac{|x+2y-2|}{\sqrt{5}}.$$

Solve these equations for the incentre, and explain why there are four points that satisfy these equations.

The Ellipse

The set of points whose coordinates (x,y) satisfy an equation of the form

$$\frac{x^2}{a^2}+\frac{y^2}{b^2}=1, \qquad (1.22)$$

where a and b are positive constants, is said to constitute an **ellipse**. Since this equation is so similar to equation 1.21, and is exactly the same when $a=b=r$, it is not unreasonable to expect that the shape of this curve might be similar to a circle, especially when values of a and b are close together. To see that this is the case, we first note that x- and y-intercepts of the ellipse are $\pm a$ and $\pm b$, respectively. Next we write equation 1.22 in the form

$$\frac{x^2}{a^2}=1-\frac{y^2}{b^2}=\frac{b^2-y^2}{b^2}.$$

Since the left side is nonnegative, so also must be the right side. But this implies that values of y must be confined to the interval $-b\le y\le b$. Similarly, by writing

$$\frac{y^2}{b^2}=\frac{a^2-x^2}{a^2},$$

we conclude that x-coordinates of points on the ellipse must satisfy $-a\le x\le a$. What we have shown is that the ellipse must lie in the rectangle of Figure 1.56, and that the midpoints of the four sides of the rectangle are on the ellipse. Now the ellipse must be symmetric about both the x-axis and the y-axis because replacing x by $-x$ and y by $-y$ in 1.22 changes nothing. As a result, if we determine the shape of the ellipse in the first quadrant portion of the rectangle in Figure 1.56, mirror images lead to the remaining three parts of the curve.

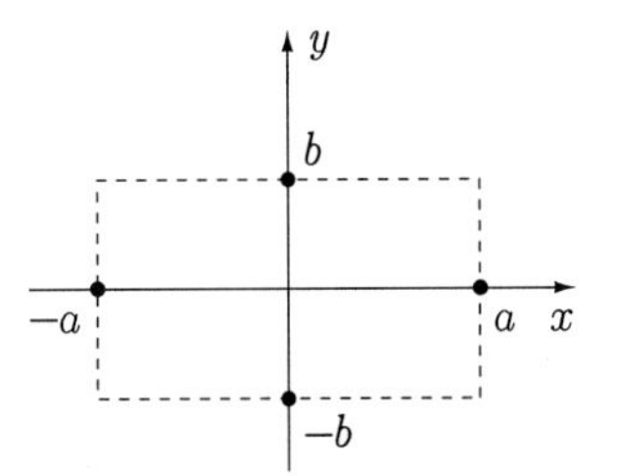

FIGURE 1.56

When we take positive square roots of the last equation, the result is

$$y = \frac{b}{a}\sqrt{a^2 - x^2}.$$

What does this equation tell us? As x increases from 0 to a, y steadily decreases from b to 0; y values do not go up and down, they drop steadily from b to 0. Similarly as y decreases from b to 0, x steadily increases from 0 to a. We have drawn this in Figure 1.57(a) giving the curve an outward "bulge" to resemble a circle. Reflections in the axes give the full ellipse in Figure 1.57(b).

FIGURE 1.57

(a)

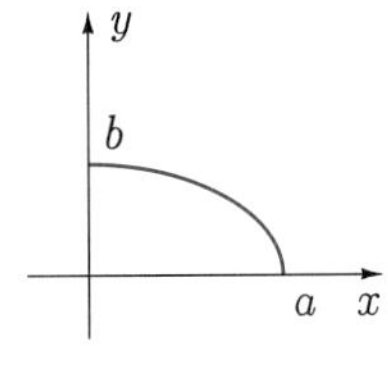

(b)

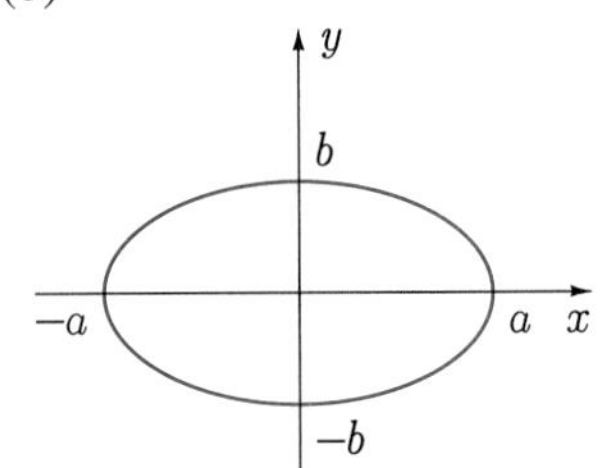

FIGURE 1.58

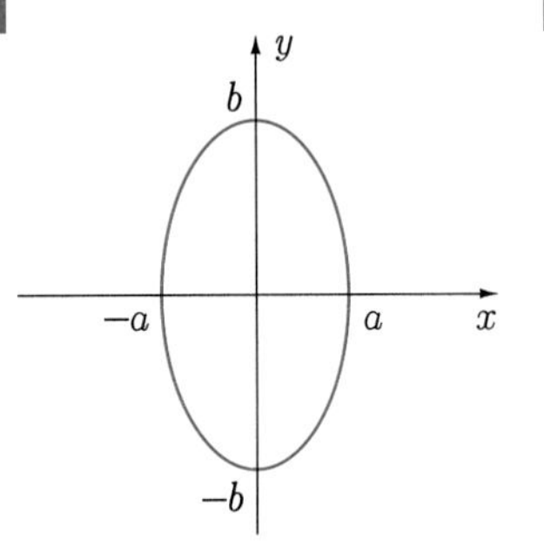

When $b > a$, the ellipse is elongated in the y-direction rather than the x-direction (Figure 1.58). The point of intersection of the lines of symmetry of an ellipse is called the **centre** of the ellipse. For equation 1.22 the centre is the origin since the x- and y-axes are the lines of symmetry.

EXAMPLE 1.14

Find the equation of the ellipse that has its centre at the origin, the x- and y-axes as axes of symmetry, and passes through the points $(4, 21/5)$ and $(-15\sqrt{5}/7, 2)$.

SOLUTION If we substitute the coordinates of the two given points into equation 1.22 (since they are both on the ellipse),

$$\frac{4^2}{a^2} + \frac{(21/5)^2}{b^2} = 1,$$
$$\frac{(-15\sqrt{5}/7)^2}{a^2} + \frac{2^2}{b^2} = 1.$$

When the first equation is divided by 16 and solved for $1/a^2$, the result is

$$\frac{1}{a^2} = \frac{1}{16} - \frac{441}{400\,b^2}.$$

This is now substituted into the second equation,

$$\frac{1125}{49}\left(\frac{1}{16}-\frac{441}{400b^2}\right)+\frac{4}{b^2}=1 \qquad \text{or}$$

$$\frac{1}{b^2}\left(4-\frac{1125\cdot 441}{49\cdot 400}\right)=1-\frac{1125}{49\cdot 16}.$$

Thus,

$$b^2=\left(\frac{4\cdot 49\cdot 400-1125\cdot 441}{49\cdot 400}\right)\left(\frac{49\cdot 16}{49\cdot 16-1125}\right)=49 \qquad \text{and}$$

$$\frac{1}{a^2}=\frac{1}{16}-\frac{441}{49\cdot 400}=\frac{1}{25}.$$

The required equation is therefore $x^2/25+y^2/49=1$. ■

When equation 1.22 is changed to

$$\frac{(x-h)^2}{a^2}+\frac{(y-k)^2}{b^2}=1, \tag{1.23}$$

where h and k are constants, the curve is still an ellipse; its shape remains the same. Just as a change from equation 1.19 to 1.21 for a circle moves the centre of the circle from $(0,0)$ to (h,k), equation 1.23 moves the centre of the ellipse to (h,k). Lines $x=h$ and $y=k$ are the new lines of symmetry (Figure 1.59), and a and b are the distances between the centre and where the ellipse crosses the lines of symmetry.

FIGURE 1.59

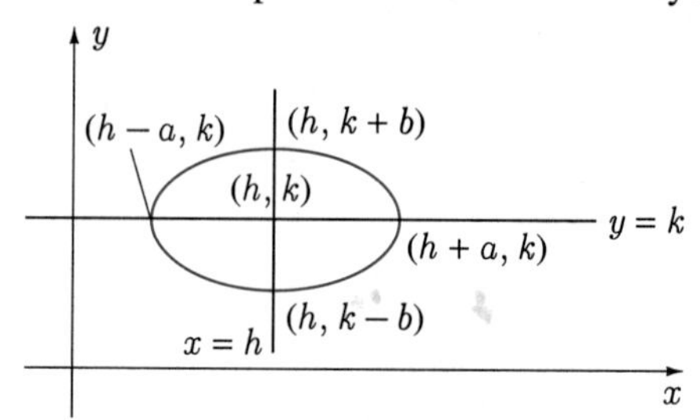

EXAMPLE 1.15

Sketch the ellipse $16x^2+25y^2-160x+50y=1175$.

SOLUTION When we complete squares on the x- and y-terms,

$$16(x-5)^2+25(y+1)^2=1600 \qquad \text{or} \qquad \frac{(x-5)^2}{100}+\frac{(y+1)^2}{64}=1.$$

The centre of the ellipse is $(5,-1)$; it cuts the lines $y=-1$ and $x=5$ at distances of 10 and 8 units from the centre respectively (Figure 1.60).

FIGURE 1.60

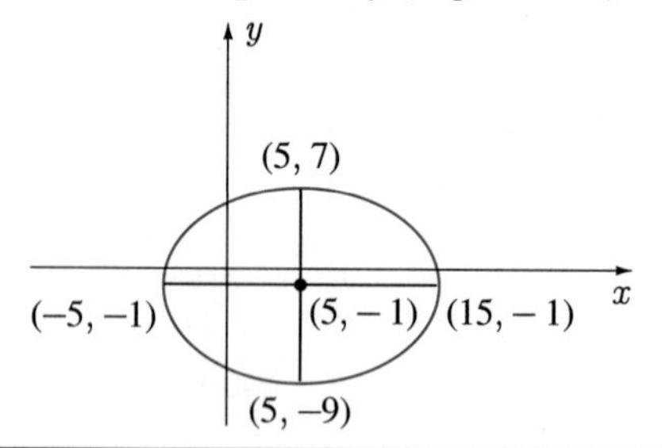

■

EXERCISES 1.4C

In Exercises 1–8 draw the ellipse.

1. $x^2/25 + y^2/36 = 1$

2. $7x^2 + 3y^2 = 16$

3. $9x^2 + 289y^2 = 2601$

4. $3x^2 + 6y^2 = 21$

5. $x^2 + 16y^2 = 2$

6. $x^2 + 2x + 4y^2 - 16y + 13 = 0$

7. $9x^2 + y^2 - 18x - 6y = 26$

8. $x^2 + 4x + 2y^2 + 16y + 32 = 0$

9. Find the equation of an ellipse through the points $(-2, 4)$ and $(3, 1)$.

10. Find the width of the elliptic arch in Figure 1.61.

FIGURE 1.61

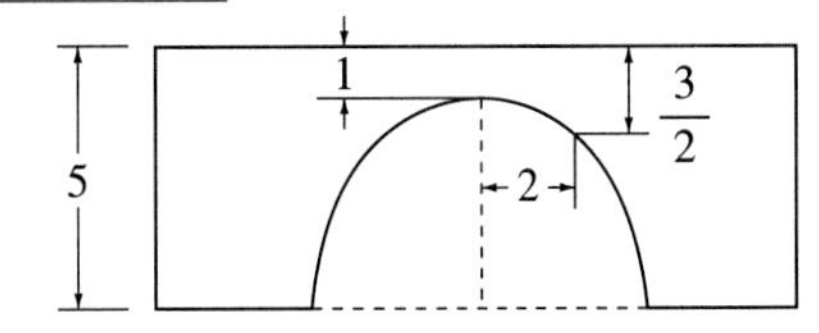

In Exercises 11–16 find all points of intersection for the curves. In each case draw the curves.

11. $x^2 + 4y^2 = 4,\ y = x$

12. $16x^2 + 9y^2 = 144,\ y = x + 3$

13. $9x^2 - 18x + 4y^2 = 27,\ 2y = \sqrt{3}x + \sqrt{3}$

14. $9x^2 - 18x + 4y^2 = 27,\ 2y = -\sqrt{3}x + 5\sqrt{3}$

15. $x^2 + 4y^2 - 8y = 0,\ y = x^2$

16. $x^2 + 4y^2 = 4,\ y = x^2 - 4$

The Hyperbola

Changing one sign in the equation of an ellipse leads to a curve with totally different characteristics. The set of points whose coordinates (x, y) satisfy an equation of the form

$$\frac{x^2}{a^2} - \frac{y^2}{b^2} = 1 \quad \text{or} \tag{1.24a}$$

$$\frac{y^2}{b^2} - \frac{x^2}{a^2} = 1, \tag{1.24b}$$

where a and b are positive constants, is called a **hyperbola**. All hyperbolas have similar properties. Let us discover them with the specific example

$$\frac{x^2}{9} - \frac{y^2}{36} = 1.$$

This hyperbola crosses the x-axis at $x = \pm 3$, but does not cross the y-axis. Since the equation remains unchanged when x is replaced by $-x$ and y is replaced by $-y$, the hyperbola is symmetric about both the x-axis and the y-axis. This means that we can concentrate on drawing that part of the hyperbola in the first quadrant, and obtain the parts in the remaining three quadrants by reflections in the axes. To discover its shape in the first quadrant we begin by solving the equation for y in terms of x. By taking positive square roots of

$$\frac{y^2}{36} = \frac{x^2}{9} - 1,$$

we obtain

$$\frac{y}{6} = \sqrt{\frac{x^2}{9} - 1},$$

from which

$$y = 2\sqrt{x^2 - 9}.$$

What does this equation, for that part of the hyperbola in the first quadrant, tell us? First, it indicates that there can be no point on the curve with x-coordinate between 0 and 3. Secondly, as values of x get bigger and bigger, so also do corresponding values of y. In other words, the curve must proceed upwards and to the right from the point $(3, 0)$. Finally, when large values of x are considered, say, 1000, 10 000, or 100 000, values of y are approximately equal to $2\sqrt{x^2} = 2x$ — the 9 becoming less and less significant the larger the value of x. This means that for large values of x, the hyperbola is very close to the line $y = 2x$. We have shown these facts in Figure 1.62(a). The complete hyperbola is shown in Figure 1.62(b). The lines $y = \pm 2x$ which the hyperbola approaches for large positive and negative values of x are called **asymptotes** of the hyperbola.

FIGURE 1.62

(a)

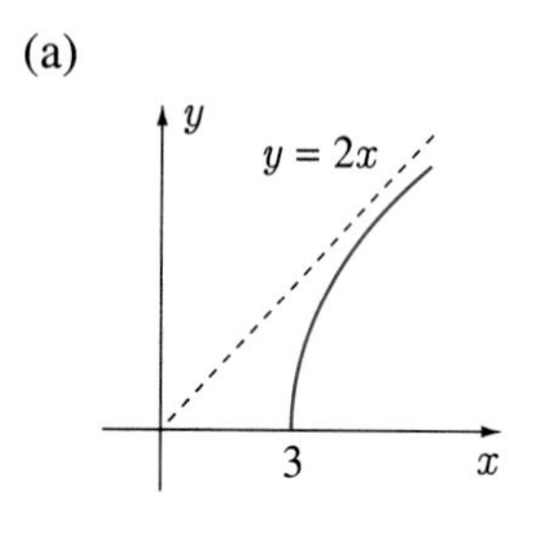

(b)

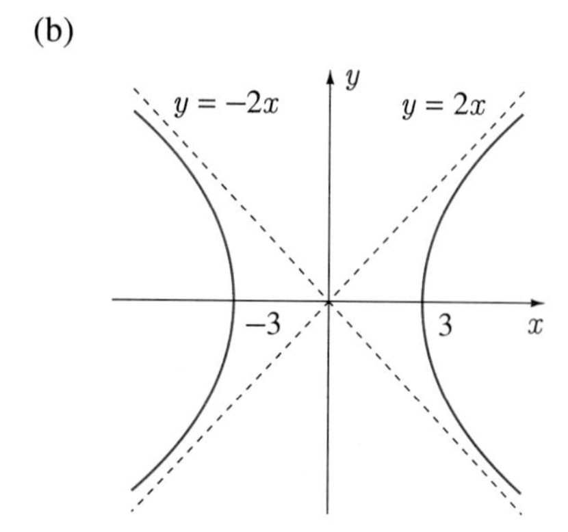

Let us now return to the more general equation 1.24a of a hyperbola. The x-intercepts of this hyperbola are $\pm a$, and it is symmetric about both axes. By writing the equation in the form

$$y = \pm \frac{b}{a}\sqrt{x^2 - a^2},$$

we see that for very large positive or negative x, y can be approximated by $\pm bx/a$; that is, the asymptotes of the hyperbola are the lines $y = \pm bx/a$. The hyperbola is shown in Figure 1.63. A similar analysis of equation 1.24b leads to the hyperbola in Figure 1.64.

FIGURE 1.63

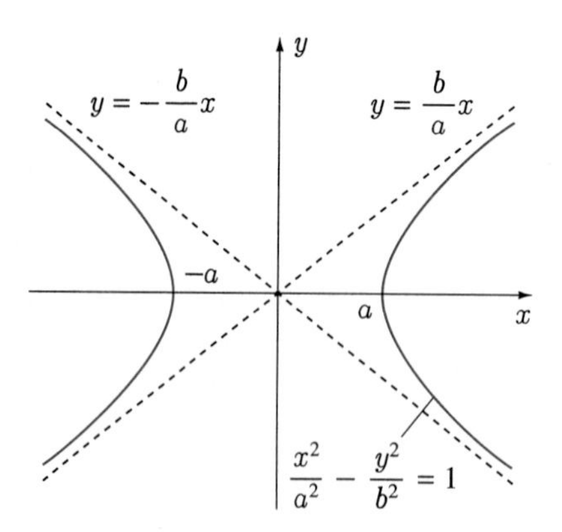

FIGURE 1.64

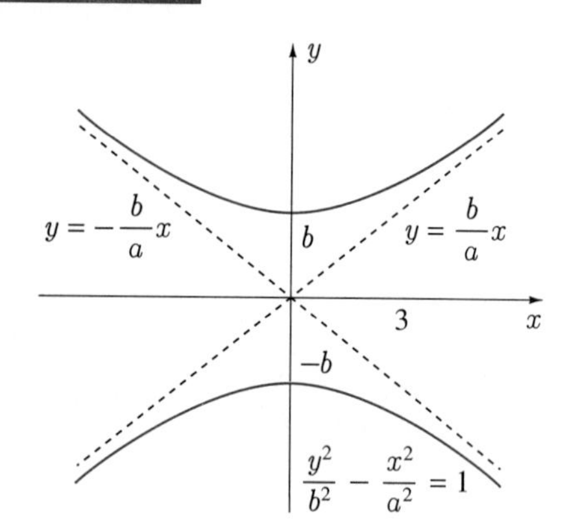

EXAMPLE 1.16

Find the equation of a hyperbola that cuts the y-axis at $y = \pm 5$ and has the lines $y = \pm x/\sqrt{3}$ as asymptotes.

SOLUTION Since the hyperbola crosses the y-axis at $y = \pm 5$, we take its equation in form 1.24b with $b = 5$,

$$\frac{y^2}{25} - \frac{x^2}{a^2} = 1.$$

When we solve this equation for y in terms of x, the result is

$$y = \pm \frac{5}{a}\sqrt{x^2 + a^2}.$$

Since asymptotes of this hyperbola are $\pm 5x/a$, it follows that $5/a = 1/\sqrt{3}$, or, $a = 5\sqrt{3}$. The equation of the hyperbola is therefore

$$\frac{y^2}{25} - \frac{x^2}{75} = 1.$$

■

When x and y in equations 1.24a,b are replaced by $x - h$ and $y - k$, the resulting equations

$$\frac{(x-h)^2}{a^2} - \frac{(y-k)^2}{b^2} = 1 \quad \text{and} \tag{1.25a}$$

$$\frac{(y-k)^2}{b^2} - \frac{(x-h)^2}{a^2} = 1, \tag{1.25b}$$

still describe hyperbolas. They are shown in Figures 1.65 and 1.66. These are the hyperbolas of Figures 1.63 and 1.64 shifted so that the asymptotes intersect at the point (h, k). Equations of the asymptotes are $y = k \pm b(x - h)/a$, and the lines $x = h$ and $y = k$ are now axes of symmetry.

FIGURE 1.65

FIGURE 1.66

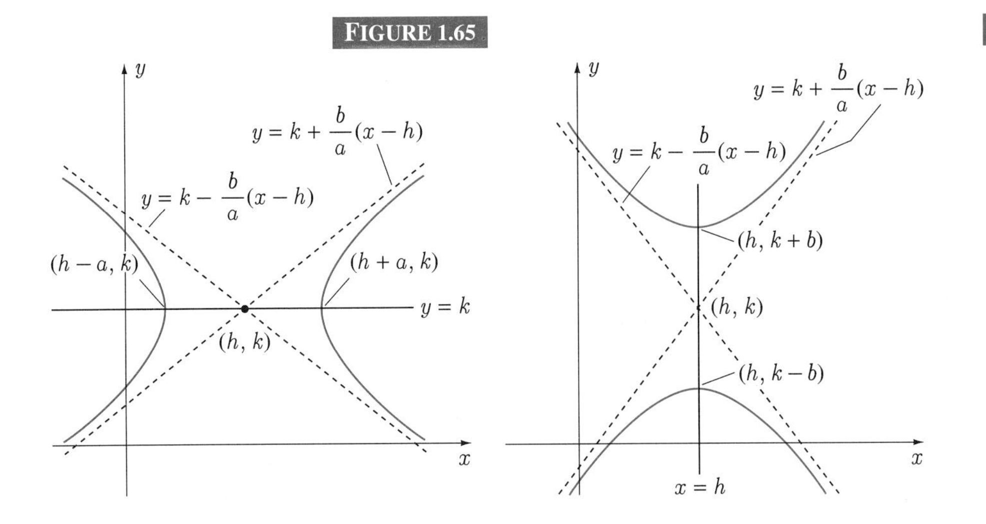

EXAMPLE 1.17 Sketch the hyperbola $x^2 - y^2 + 4x + 10y = 5$.

SOLUTION If we complete squares on the x- and y-terms, we obtain

$$(x+2)^2 - (y-5)^2 = -16 \qquad \text{or} \qquad \frac{(y-5)^2}{16} - \frac{(x+2)^2}{16} = 1.$$

The axes of symmetry of the hyperbola are $x = -2$ and $y = 5$, intersecting at the point $(-2, 5)$. When we solve for y in terms of x, the result is

$$y = 5 \pm \sqrt{(x+2)^2 + 16}.$$

Asymptotes are therefore $y = 5 \pm (x+2)$, and the hyperbola can be drawn as in Figure 1.67.

FIGURE 1.67

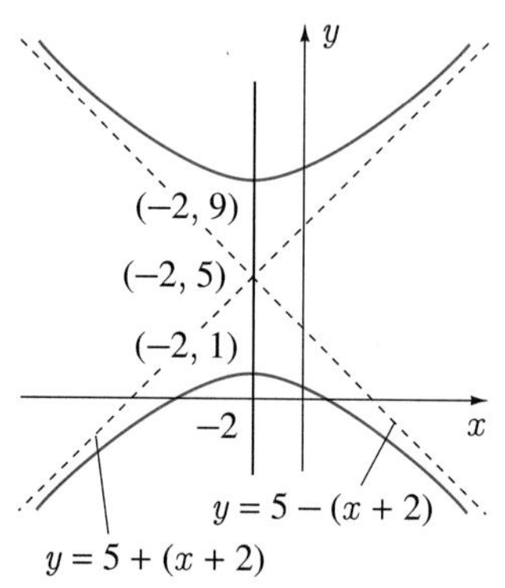

■

EXERCISES 1.4D

In Exercises 1–10 draw the hyperbola.

1. $y^2 - x^2 = 1$
2. $x^2 - y^2 = 1$
3. $x^2 - y^2/16 = 1$
4. $25y^2 - 4x^2 = 100$
5. $y^2 = 10(2 + x^2)$
6. $3x^2 - 4y^2 = 25$
7. $x^2 - 6x - 4y^2 - 24y = 11$
8. $9x^2 - 16y^2 - 18x - 64y = 91$
9. $4y^2 - 5x^2 + 8y - 10x = 21$
10. $x^2 + 2x - 16y^2 + 64y = 79$
11. Find the equation of a hyperbola that passes through the point $(1, 2)$ and has asymptotes $y = \pm 4x$.

In Exercises 12–18 find all points of intersection for the curves. In each case draw the curves.

12. $x^2 - 2y^2 = 1$, $x = 2y$
13. $9y^2 - 4x^2 = 36$, $y = x$
14. $9y^2 - 4x^2 = 36$, $x = 3y$
15. $3x^2 - y^2 = 3$, $2x + y = 1$
16. $x^2 - 2x - y^2 = 0$, $x = y^2$
17. $x^2 - 2x - y^2 = 0$, $x = -y^2$
18. $9(x-1)^2 - 4(y-1)^2 = 36$, $27x = 5(y-1)^2$

SECTION 1.5

Functions and Their Graphs

Most quantities that we encounter in everyday life are dependent on many, many other quantities. For example, think of what might be affecting room temperature as you read this sentence — thermostat setting, outside temperature, wind conditions, insulation of the walls, ceiling, and floors, and perhaps other factors that you can think of. Functional notation allows interdependences of such quantities to be represented in a very simple way.

When one quantity depends on a second quantity, we say that the first quantity is a function of the second. For example, the volume V of a sphere depends on its radius r; in particular, $V = 4\pi r^3/3$. We say that V is a function of r. When an object is dropped, the distance d (metres) that it falls in time t (seconds) is given by the formula $d = 4.905t^2$. We say that d is a function of t. Mathematically, we have the following definition.

Definition 1.8

A quantity y is said to be a function of a quantity x if there exists a rule by which we can associate exactly one value of y with each value of x. The rule that associates the value of y with each value of x is called the **function**.

If we denote the rule or function in this definition by the letter f, then the value that f assigns to x is denoted by $f(x)$, and we write

$$y = f(x). \tag{1.26}$$

We should read this equation as "y is the value that f assigns to x", but usually we simply say "y equals f of x". In our first example above, we write $V = f(r) = 4\pi r^3/3$, and the function f is the operation of cubing a number and then multiplying the result by $4\pi/3$. For $d = f(t) = 4.905t^2$, the function f is the operation of squaring a number and multiplying the result by 4.905.

We call x in equation 1.26 the **independent variable** because values of x are substituted into the function, and y the **dependent variable** because its values depend on the assigned values of x. The **domain** of a function is the set of all specified (real) values for the independent variable. It is an essential part of a function and should always be specified. Whenever the domain of a function is not mentioned, we assume that it consists of all possible values for which $f(x)$ is a real number.

As the independent variable x takes on values in the domain, a set of values of the dependent variable is obtained. This set is called the **range** of the function. For the function $V = f(r) = 4\pi r^3/3$, which represents the volume of a sphere, the largest possible domain is $r > 0$, and the corresponding range is $V > 0$. Note that mathematically the function $f(r)$ is defined for negative as well as positive values of r, and $r = 0$; it is because of our interpretation of r as the radius of a sphere that we restrict $r > 0$. The function $d = f(t) = 4.905t^2$ represents the distance fallen in time t by an object that is dropped at time $t = 0$. If it is dropped from a height of 20 m, then it is clear that the range of this function is $0 \le d \le 20$. The domain that gives rise to this range is $0 \le t \le \sqrt{20/4.905}$.

EXAMPLE 1.18

If $f(x) = \sqrt{x^2 - 3}$, calculate each of the following:

(a) $f(2)$ (b) $f(-4)$ (c) $f(t)$ (d) $f(h+4)$

(e) $\dfrac{f(x)}{f(1-x)}$ (f) $[f(x)]^2 + 3f(x)f(-x)$

SOLUTION (a) Setting $x = 2$ gives

$$f(2) = \sqrt{2^2 - 3} = 1.$$

(b) Setting $x = -4$ gives

$$f(-4) = \sqrt{(-4)^2 - 3} = \sqrt{13}.$$

(c) To find $f(t)$ we substitute t for each x in $f(x)$:

$$f(t) = \sqrt{t^2 - 3}.$$

(d) For $f(h+4)$, we replace each x in $f(x)$ with $h+4$,

$$f(h+4) = \sqrt{(h+4)^2 - 3} = \sqrt{h^2 + 8h + 13}.$$

(e)

$$\frac{f(x)}{f(1-x)} = \frac{\sqrt{x^2-3}}{\sqrt{(1-x)^2-3}} = \frac{\sqrt{x^2-3}}{\sqrt{x^2-2x-2}}$$

(f)

$$\begin{aligned}[f(x)]^2 + 3f(x)f(-x) &= \left[\sqrt{x^2-3}\right]^2 + 3\sqrt{x^2-3}\sqrt{(-x)^2-3} \\ &= x^2 - 3 + 3(x^2-3) = 4(x^2-3)\end{aligned}$$

■

EXAMPLE 1.19

Find the largest possible domain for the function

$$f(x) = \sqrt{\frac{x-4}{x+1}}.$$

SOLUTION The sign diagram in Figure 1.68 shows that the quotient $(x-4)/(x+1)$ is positive when $x > 4$ and when $x < -1$. Since the quotient is equal to zero when $x = 4$ and is undefined when $x = -1$, the largest possible domain for $f(x)$ consists of all values of x in the intervals $x < -1$ and $x \geq 4$.

FIGURE 1.68

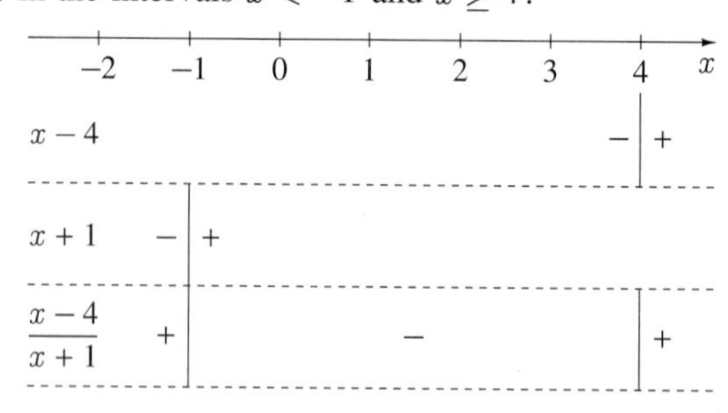

■

EXAMPLE 1.20

Does the equation $x - 4 - y^2 = 0$ define y as a function of x for $x \geq 4$?

SOLUTION For any $x > 4$, the equation has two solutions for y:

$$y = \pm\sqrt{x - 4}.$$

Since the equation does not define exactly one value of y for each value of x, it does not define y as a function of x. ■

If we add to the equation in Example 1.20 an additional restriction such as $y \geq 0$, then y is defined as a function of x, namely,

$$y = \sqrt{x - 4}.$$

Note that $x - 4 - y^2 = 0$ does, however, define x as a function of y:

$$x = y^2 + 4.$$

In this case, y is the independent variable and x is the dependent variable. In other words, whenever an equation (such as $x - 4 - y^2 = 0$) is to be regarded as defining a function, it must be made clear which variable is to be considered as independent and which as dependent.

In the study of calculus and its applications we are interested in the behaviour of functions; that is, for certain values of the independent variable, what can we say about the dependent variable? The simplest and most revealing method for displaying characteristics of a function is a graph. To obtain the graph of a function $f(x)$ we use a plane coordinatized with Cartesian coordinates x and y (in short, the Cartesian xy-plane). The **graph** of the function $f(x)$ is defined to be the curve with equation $y = f(x)$, where x is limited to the domain of the function. We therefore take each value of x in the domain of the function, calculate $y = f(x)$, and plot the point (x, y) in the Cartesian plane. For example, if $f(x) = x^3 - 27x + 1$, the values in Table 1.5 lead to the curve in Figure 1.69.

TABLE 1.5

x	−6	−5	−4	−3	−2	−1	0	1	2	3	4	5	6
y	−53	11	45	55	47	27	1	−25	−45	−53	−43	−9	55

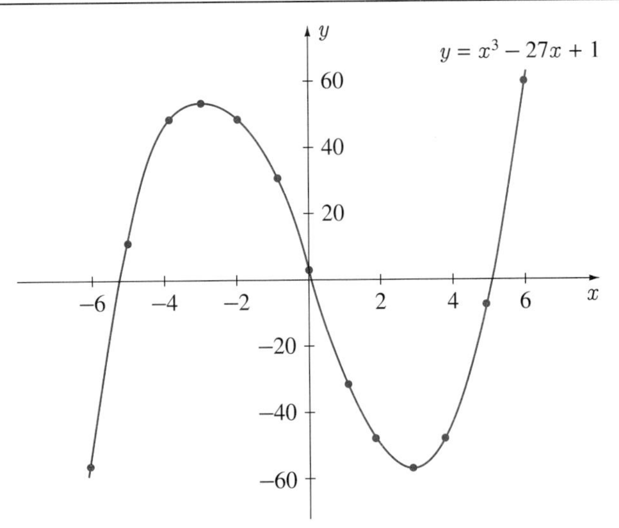

FIGURE 1.69

This curve is a pictorial or geometric representation of the function $f(x) = x^3 - 27x + 1$. The value of the function for given x is visually displayed as the y-coordinate of that point on the curve with x-coordinate equal to the given x. This graph clearly illustrates properties of the function that may not be so readily obvious from the algebraic definition $f(x) = x^3 - 27x + 1$. For instance,

1. For negative values of x, the largest value of $f(x)$ is $f(-3) = 55$; for positive values of x, the smallest value of $f(x)$ is $f(3) = -53$.
2. As x increases, values of $f(x)$ increase for $x < -3$ and $x > 3$, and values of $f(x)$ decrease for $-3 < x < 3$.
3. $f(x)$ is equal to zero for three values of x — one a little less than -5, one a little larger than 0, and one a little larger than 5.

Graphs of functions can be drawn very accurately or very roughly. An accurate graph of a function is usually obtained through an extensive table of values; this is more or less the method emphasized in Sections 1.3 and 1.4, and the above example. Although an accurate graph certainly has its advantages, and is indeed necessary in some instances, for many problems its construction is too time-consuming. Quite often we need only a rough sketch of the graph of a function. A sketch gives a general idea of the behaviour of the function; that is, a general indication of the shape of the graph. In the following examples we illustrate some useful techniques for obtaining sketches. Others will be introduced in Chapters 2, 3, and 4.

EXAMPLE 1.21

Sketch a graph of the function $f(x) = |x^3 + 5|$.

SOLUTION To obtain the curve $y = |x^3 + 5|$, we first sketch $y = x^3$, as in Figure 1.70(a). Next we sketch the curve $y = x^3 + 5$ by adding 5 to every ordinate in Figure 1.70(a). The result in Figure 1.70(b) is the curve in Figure 1.70(a) shifted upward 5 units. It crosses the x-axis at $-5^{1/3}$. The last step is to take the absolute value of every ordinate on the curve $y = x^3 + 5$. This changes no ordinate that is already positive, but changes the sign of any ordinate that is negative. The result and final sketch is shown in Figure 1.70(c).

FIGURE 1.70

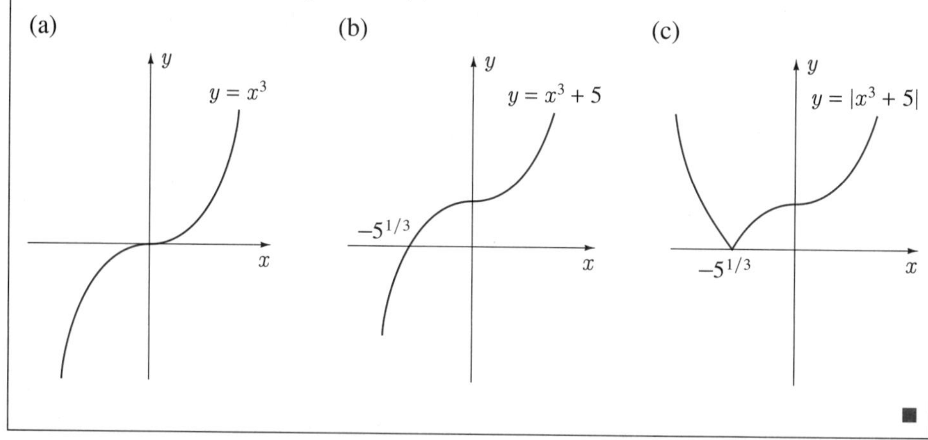

■

Note that in our "building" procedure we did not attempt to sketch $y = |x^3 + 5|$ in one step. By a series of steps we began with the curve $y = x^3$, proceeded to $y = x^3 + 5$, and finally obtained $y = |x^3 + 5|$. An essential feature of this graph is the sharp point where $x = -5^{1/3}$. For those readers with plotting calculators or computer access, ask the calculator or computer to draw this function. Does the curve touch the x-axis, and does there appear to be a sharp point on the curve at $x = -5^{1/3}$? The answer is probably no to both questions. This is a serious shortcoming of plotting routines in calculators and computers. They sometimes fail to divulge important features of functions. We shall see that one of the greatest accomplishments of calculus is its ability to recognize every characteristic of a function.

EXAMPLE 1.22

Sketch a graph of the function $f(x) = 2\sqrt{4 - x^2}$.

SOLUTION If we square $y = 2\sqrt{4 - x^2}$, we find that $y^2 = 4(4 - x^2)$, or, $4x^2 + y^2 = 16$. We recognize this equation as that for the ellipse in Figure 1.71(a). This is not the required graph, however, since it is not even the graph of a function, there being two y-values for each x in $-2 < x < 2$. The required graph is the upper half of the ellipse in Figure 1.71(b). We know this because the square root in $y = 2\sqrt{4 - x^2}$ demands that y be nonnegative. By squaring this equation we introduced the lower half of the ellipse.

(a)

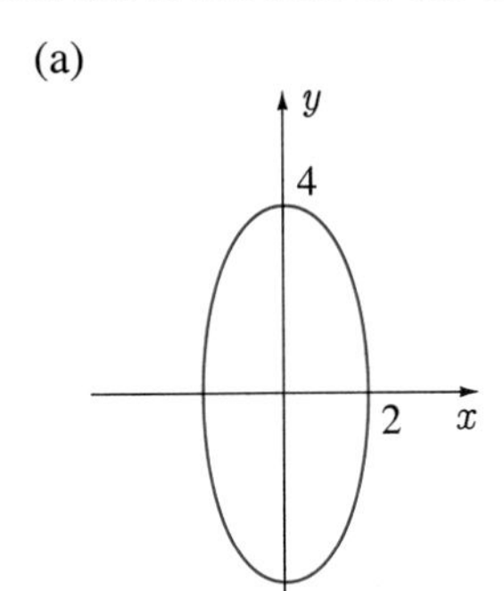

(b)

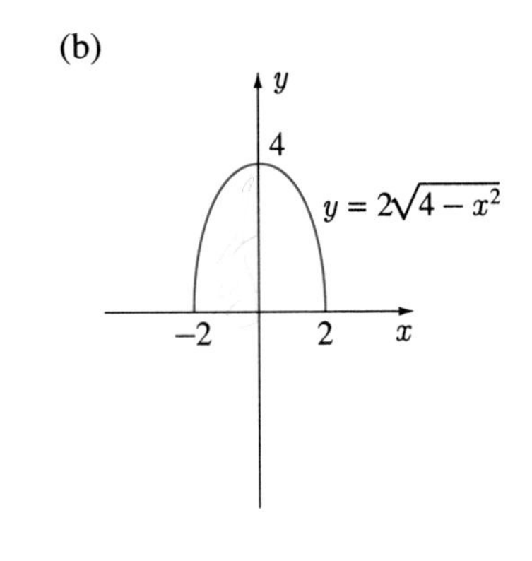

FIGURE 1.71

■

EXAMPLE 1.23

Sketch a graph of the function $f(x) = x + |x|$.

SOLUTION To sketch this function we first draw the straight line $y = x$ in Figure 1.72(a). The curve $y = |x|$ is obtained therefrom (shown dotted) by taking absolute values of all ordinates. The required curve $y = x + |x|$ has as its ordinates the sums of the ordinates x and $|x|$ just sketched. Consequently, to obtain $y = x + |x|$, we add corresponding ordinates of these two curves in Figure 1.72(b). The final result is shown in Figure 1.72(c).

(a)

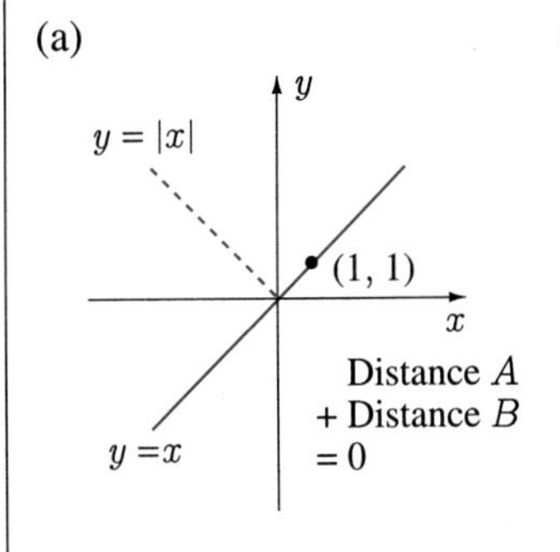

(b)

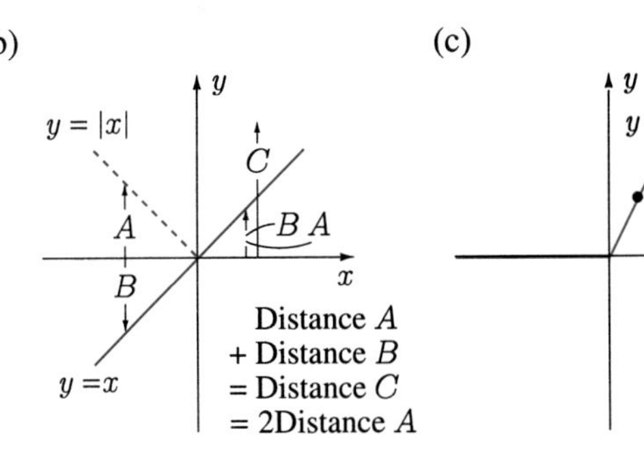

(c)

FIGURE 1.72

■

This building procedure, called **addition of ordinates**, is used again in the following example.

EXAMPLE 1.24

Sketch a graph of the function $f(x) = x + \sqrt{x^2 - 4}$.

SOLUTION If we square $y = \sqrt{x^2 - 4}$, then $y^2 = x^2 - 4$ or $x^2 - y^2 = 4$. This equation describes a hyperbola, the top of half of which is shown in Figure 1.73(a). On the same axes we draw the straight line $y = x$. Addition of ordinates now gives the required graph in Figure 1.73(b). Since the hyperbola is asymptotic to the line $y = x$ for $x > 0$, the graph of $y = f(x)$ is asymptotic to the line $y = 2x$ for $x > 0$. It is also asymptotic to the negative real axis.

FIGURE 1.73

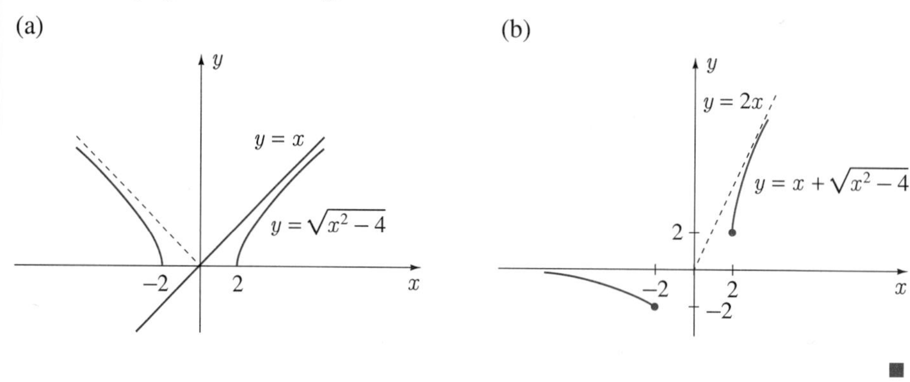

■

EXAMPLE 1.25

Sketch a graph of the function $f(x) = x^2\sqrt{1 - x}$.

SOLUTION We first sketch the parabola $y = x^2$ in Figure 1.74(a). Next we consider a sketch of $y = \sqrt{1 - x}$, which implies that $y^2 = 1 - x$ or $x = 1 - y^2$. The top half of this parabola is shown on the same axes. The required curve $y = x^2\sqrt{1 - x}$ has ordinates which are products of the ordinates x^2 and $\sqrt{1 - x}$ just sketched. Thus to obtain $y = x^2\sqrt{1 - x}$ we multiply corresponding ordinates of the two curves in Figure 1.74(a). The result is shown in Figure 1.74(b).

FIGURE 1.74

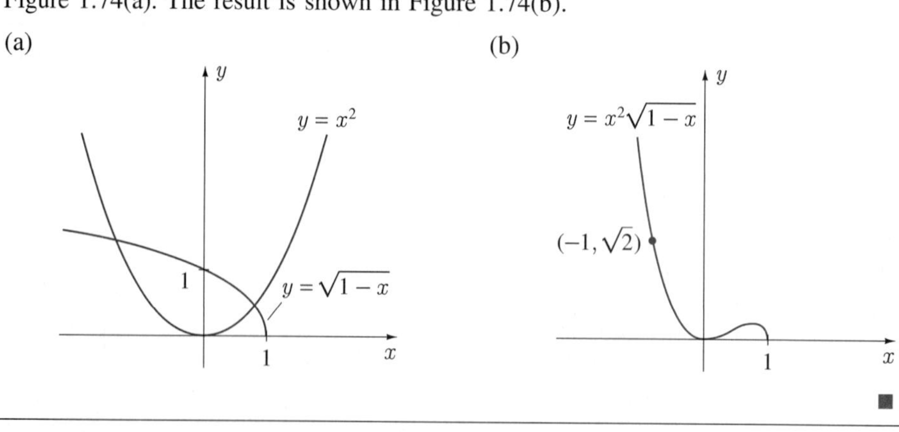

■

This building procedure is called **multiplication of ordinates**.

EXAMPLE 1.26

When a ball is dropped from the top of a building 20 m high at time $t = 0$, the distance d (in metres) that it falls in time t is given by the function $d = f(t) = 4.905t^2$. Sketch its graph.

SOLUTION In this example independent and dependent variables are denoted by letters t and d which suggest their physical meaning — time and distance — rather than the generic labels x and y. With the axes labeled correspondingly as the t-axis and d-axis, we sketch that part of the parabola $d = 4.905t^2$ shown in Figure 1.75. The remainder of the parabola has no physical significance in the context of this problem.

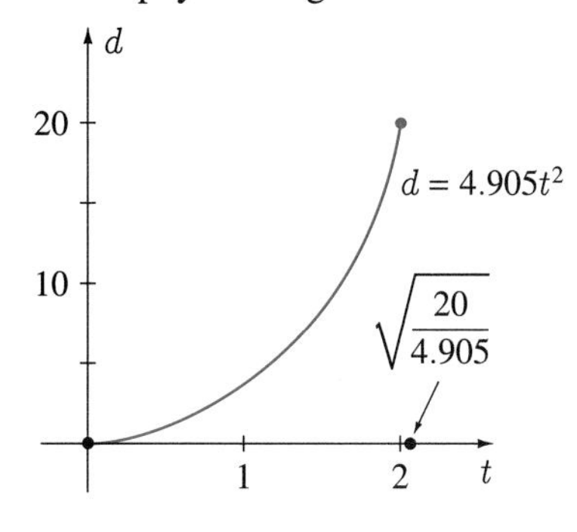

FIGURE 1.75

Polynomials and Rational Functions

Two very important classes of functions are polynomials and rational functions. A **polynomial of degree n** is a function of the form

$$f(x) = a_n x^n + a_{n-1} x^{n-1} + \cdots + a_2 x^2 + a_1 x + a_0 \qquad (1.27)$$

where n is a nonnegative integer, and $a_0, a_1, \ldots, a_n$ are real numbers ($a_n \neq 0$). Special names have become associated with some of the lower-degree polynomials because of the frequency with which they arise:

$f(x) = a_1 x + a_0$	polynomial of degree 1, linear polynomial, linear function
$f(x) = a_2 x^2 + a_1 x + a_0$	polynomial of degree 2, quadratic polynomial, quadratic function
$f(x) = a_3 x^3 + a_2 x^2 + a_1 x + a_0$	polynomial of degree 3, cubic polynomial, cubic function

Polynomials are defined for all values of x. Graphs of linear and quadratic polynomials are straight lines and parabolas. Cubic and higher degree polynomials are easy to graph when they can be factored into linear polynomials. For example, the cubic polynomial $f(x) = x^3 - 2x^2 - x + 2$ can be factored into $f(x) = (x-1)(x+1)(x-2)$. In this form we see that the x-intercepts of the graph of $f(x)$ are ± 1 and 2, and the sign diagram of Figure 1.76 leads to the curve in Figure 1.77. The factored form of the "quartic" polynomial $f(x) = x^4 - 13x^2 + 36 = (x-2)(x+2)(x-3)(x+3)$ leads to the graph in Figure 1.78.

FIGURE 1.76

	$x<-1$	$-1<x<1$	$1<x<2$	$x>2$
$x-1$		$-$	$+$	
$x+1$	$-$	$+$		
$x-2$			$-$	$+$
$f(x)$	$-$	$+$	$-$	$+$

FIGURE 1.77

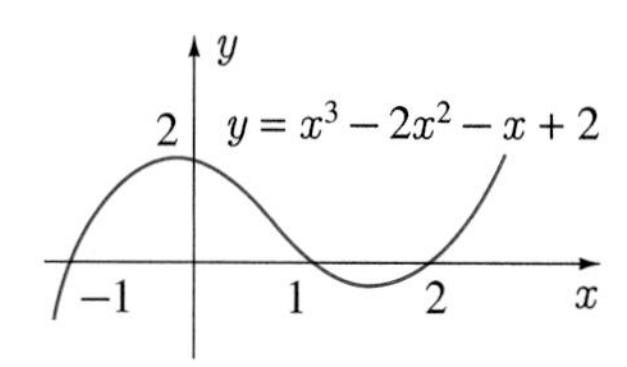

FIGURE 1.78

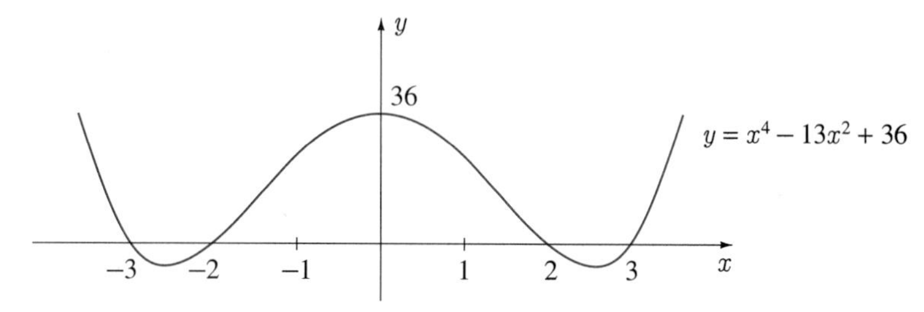

Graphing polynomials that cannot be factored completely into linear terms is more difficult. In Chapter 4 we shall see that calculus provides a complete set of tools for doing this. One might ask, however, whether addition of ordinates could be used to sketch polynomials. For example, could $f(x) = x^3 - 2x^2 - x + 1$ be sketched by adding ordinates of x^3 and $-2x^2 - x + 1$, or of x^3, $-2x^2$, and $1 - x$. In general the answer is no. (Try it.) Usually, so many questions arise in the process, that one is not at all confident in the result.

A **rational function** $R(x)$ is defined as the quotient of two polynomials $P(x)$ and $Q(x)$:

$$R(x) = \frac{P(x)}{Q(x)}. \tag{1.28}$$

Rational functions are undefined only at points where $Q(x) = 0$. To graph many rational functions we require "limits" from Chapter 2 and "derivatives" from Chapter 3.

Every function $f(x)$ can be represented pictorially by its graph, the curve with equation $y = f(x)$. But what about the reverse situation? Does every curve in the xy-plane represent a function $f(x)$? The curves in Figure 1.79, which both extend between $x = a$ and $x = b$, illustrate that the answer is no. The curve in Figure 1.79(a) represents a function, whereas the curve in Figure 1.79(b) does not, because for values of x between a and c there are two possible values of y. In other words, *a curve represents a function f(x) if every vertical line that intersects the curve does so at exactly one point.*

FIGURE 1.79

(a)

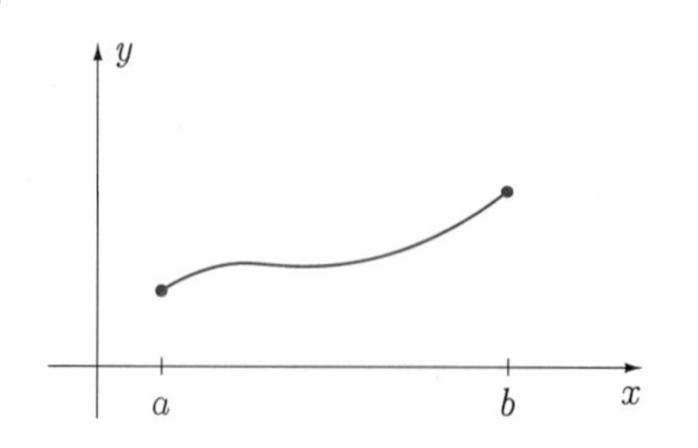

(b)

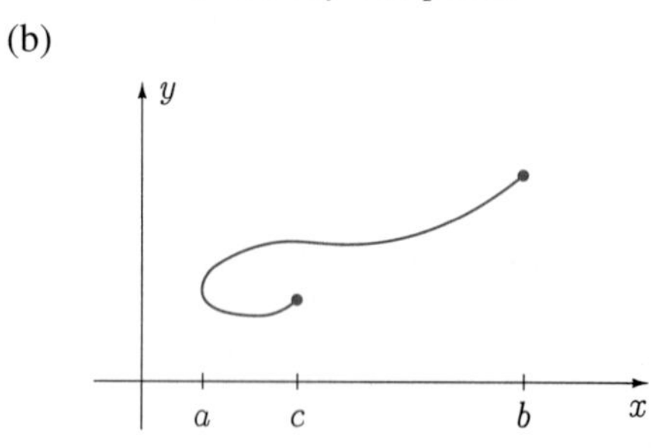

EXERCISES 1.5

In Exercises 1–6 find the largest possible domain of the function.

1. $f(x) = \sqrt{9 - x^2}$ **2.** $f(x) = 1/(x - 2)$

3. $f(x) = \dfrac{1}{x\sqrt{x^2 + 4}}$ **4.** $f(x) = \dfrac{1}{x\sqrt{x^2 - 4}}$

5. $f(x) = \dfrac{1}{x\sqrt{4 - x^2}}$ **6.** $f(x) = \sqrt{\dfrac{x^2 - 4}{9 - x^2}}$

In Exercises 7–34 sketch a graph of the function.

7. $f(x) = x^4 - 3$ **8.** $f(x) = 3x^3 + 2$

9. $f(x) = x^2 - x$ **10.** $f(x) = -x^3 + 3x - 2$

11. $f(x) = \sqrt{5 - x^2}$ **12.** $f(x) = -\sqrt{5 - x^2}$

13. $f(x) = -\sqrt{x^2 - 5}$ **14.** $f(x) = \sqrt{x^2 - 5}$

15. $f(x) = \sqrt{5 - 4x^2}$ **16.** $f(x) = -\sqrt{5 - 4x^2}$

17. $f(x) = |x| + 2x$ **18.** $f(x) = |x| - 2x$

19. $f(x) = 5x + 2\sqrt{x}$ **20.** $f(x) = x^{1/3}$

21. $f(x) = x^{2/3}$ **22.** $f(x) = 3x^{3/2}$

23. $f(x) = 3|x|^{3/2}$ **24.** $f(x) = x\sqrt{x + 1}$

25. $f(x) = x\sqrt{x^2 - 1}$ **26.** $f(x) = -x\sqrt{4 - 9x^2}$

27. $f(x) = x^2\sqrt{4 - x}$ **28.** $f(x) = x^2\sqrt{x^2 - 4}$

29. $f(x) = x^2\sqrt{4 - 9x^2}$ **30.** $f(x) = x^3 + 3x^2 - 4x - 12$

31. $f(x) = x^3 - x$ **32.** $f(x) = 10 - 3x - 6x^2 - x^3$

33. $f(x) = x^3(x - 1)^4$ **34.** $f(x) = |x^2 - x - 12|$

35. What is the condition that a curve in the xy-plane represent a function $x = f(y)$?

36. The *greatest integer function* is denoted by $f(x) = [[x]]$ = greatest integer that does not exceed x.

(a) Sketch a graph of the greatest integer function.

(b) If first-class postage is 40¢ for each 50 gm, or fraction thereof, up to and including 500 gm, sketch a graph of this cost function.

(c) Express the cost function in (b) in terms of the greatest integer function.

In Exercises 37–47 sketch a graph of the function.

37. $f(x) = \sqrt{2x - x^2}$ **38.** $f(x) = \sqrt{2x - 4x^2}$

39. $f(x) = \sqrt{4x^2 - 2x}$ **40.** $f(x) = \sqrt{2x - x^2 - 4}$

41. $f(x) = x + 2 + \sqrt{x}$

42. $f(x) = \sqrt{9 - 4x^2} + \sqrt{4x^2 - 9}$

43. $f(x) = x^2 + |x| - 2$

44. $f(x) = \sqrt{(x^2 - 4)^2}$

45. $f(x) = (x^2 + x - 2)^2$

46. $f(x) = \sqrt{(x^2 - 1)^2} - (x^2 - 1)$

47. $f(x) = \sqrt{2 - \sqrt{1 + x}}$

48. Use graphs of the functions $f(x) = x^3$ and $g(x) = 2 - x$ to show that the equation

$$x^3 + x - 2 = 0$$

has exactly one real solution.

49. Use graphs to illustrate the number of solutions of the equation

$$\sqrt{4 - x} + x^2 - x^6 = 0.$$

50. A cherry orchard has 255 trees, each of which produces on the average 25 baskets of cherries. For each additional tree planted, the yield per tree decreases by one-twelfth of a basket. If x represents the number of extra trees (beyond 255) and Y the total yield, find Y as a function of x, and sketch its graph. How many more trees should be planted for maximum yield?

51. A rectangle with sides parallel to the axes is inscribed inside the ellipse $b^2x^2 + a^2y^2 = a^2b^2$ (Figure 1.80). Find a formula for the area A of the rectangle in terms of x. Sketch a graph of this function.

FIGURE 1.80

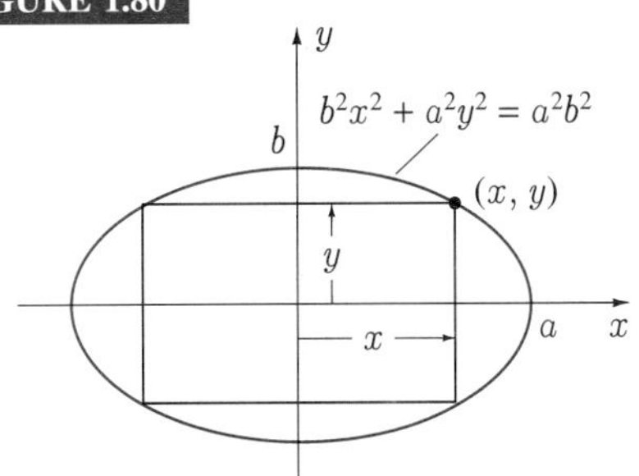

52. A man 2 m tall walks along the edge of a straight road that is 5 m wide. On the other edge of the road stands a street light 10 m high. Find a functional relationship for the length of the man's shadow in terms of his distance from the point on his side of the road directly across from the light. Sketch a graph of this function.

53. When two substances A and B are brought together, a chemical reaction takes place to form a new substance C. It requires 2 L of A for each litre of B to produce 3 L of C. The rate R at which A and B react to form C is proportional to the product of the amounts of A and B present at that instant. If the original amounts of A and B are 20 L and 40 L, respectively, and if x represents the amount of C present in the reaction at any given time, find a formula for R as a function of x. Sketch a graph of this function, and determine when the reaction rate is highest.

54. In ordinary physics and engineering, the mass m of an object is constant, independent of how fast it is moving. In special relativity, however, m is given by the formula

$$m = \frac{m_0}{\sqrt{1 - (v^2/c^2)}},$$

where m_0 is the mass of the object when it is not moving, v is the speed of the mass, and c is a constant (the speed of light). Sketch a graph of this function, and draw any conclusions that you feel are suggested.

55. A square plate 4 m on each side is slowly submerged in a large tank of water. One diagonal is kept vertical and lowered at a rate of 0.5 m/s, entering the water at time $t = 0$. If A is the area of the submerged portion of the surface (one side only) at time t until complete submersion occurs, find A as a function of t and sketch its graph.

56. Due to construction, no passing is permitted on a 10 km stretch of highway. If cars travel at v km/h along this stretch, a safe distance between them must be maintained, and this distance increases as v increases. In particular, the highway traffic commission has determined that for speeds over 50 km/h, the distance in metres between cars should be at least

$$d = \frac{3v^2}{500}.$$

If it is supposed that everyone maintains the safe distance, and the same constant speed v through the stretch, find the number q of cars leaving the "bottleneck" per hour as a function of speed v. Sketch a graph of this function for $50 \leq v \leq 100$, and determine the speed that maximizes q.

57. A box measuring 1 m on each side is attached to a rope. The rope passes over a pulley 10 m from the ground and a truck pulls on the other end in a horizontal direction along a line 1 m above the ground (Figure 1.81). We denote positions of the truck and the bottom of the box by x and y, respectively.

FIGURE 1.81

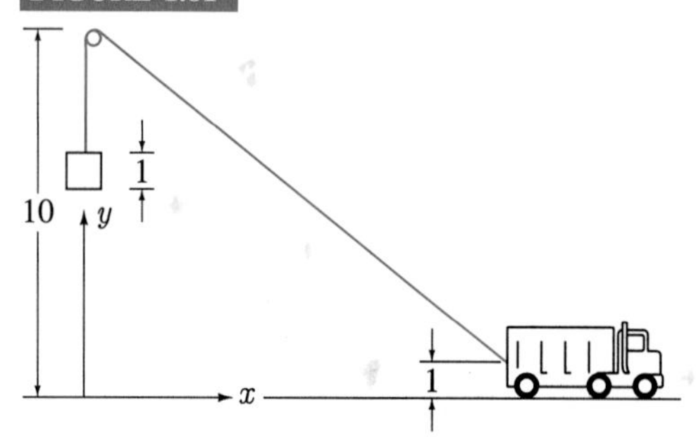

Find y as a function of x if the truck starts at position $x = 5$ m and stops when the top of the box touches the pulley. Assume that the length of rope between truck and box is 25 m. Sketch a graph of the function.

In Exercises 58–61 sketch a graph of the function where $[[x]]$ is the greatest integer function of Exercise 36.

58. $f(x) = [[2x]]$

59. $f(x) = x + [[x]]$

60. $f(x) = x[[x]]$

61. $f(x) = [[x + [[x]]]]$

62. Is $[[f(x) + g(x)]] = [[f(x)]] + [[g(x)]]$?

SECTION 1.6

Further Graphing Techniques

In this section we introduce additional graphing techniques. They are simple in concept but powerful in practice.

Even and Odd Functions

The parabola $y = x^2$ in Figure 1.30(b) is the graph of the function $f(x) = x^2$; it is symmetric about the y-axis. The curve in Figure 1.78 is the graph of the polynomial $y = x^4 - 13x^2 + 36$; it is also symmetric about the y-axis. These are examples of a special class of functions identified in the following definition.

Definition 1.9

A function $f(x)$ is said to be **even** if for each x in its domain

$$f(-x) = f(x); \tag{1.29a}$$

it is said to be **odd** if

$$f(-x) = -f(x). \tag{1.29b}$$

The first of these implies that the equation $y = f(x)$ for the graph of an even function is unchanged when x is replaced by $-x$; therefore the graph of an even function is symmetric about the y-axis. As a result, $f(x) = x^2$ and $f(x) = x^4 - 13x^2 + 36$ are even functions.

Equation 1.29b implies that if (x, y) is any point on the graph of an odd function, so too is the point $(-x, -y)$. This is illustrated by the graph of the odd function $f(x) = x^3 - 4x$ in Figure 1.82. Another way to describe the graph of an odd function is to note that either half ($x < 0$ or $x > 0$) is the result of two reflections of the other half, first in the y-axis and then in the x-axis. Alternatively, either half of the graph is a result of rotating the other half by π radians (one-half a revolution) around the origin.

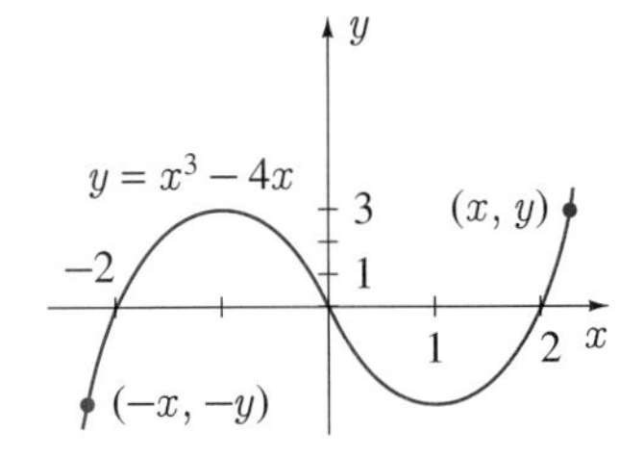

FIGURE 1.82

The six trigonometric functions are prominent examples of even and odd functions. The sine, tangent, cosecant, and cotangent are odd functions, while cosine and secant are even. Their graphs can be found in the trigonometry review of Appendix A.

EXAMPLE 1.27

Which of the following functions are even, odd, or neither even nor odd?

(a) $f(x) = \sqrt{|x|}$ (b) $f(x) = x^5 - x$ (c) $f(x) = x^2 + x$

Sketch a graph of each function.

SOLUTION (a) Since

$$f(-x) = \sqrt{|-x|} = \sqrt{|x|} = f(x),$$

this function is even. Its graph, the curve $y = \sqrt{|x|}$, is symmetric about the y-axis. When $x > 0$, this equation becomes $y = \sqrt{x}$, the half-parabola in Figure 1.83(a). The complete graph of the function is in Figure 1.83(b).

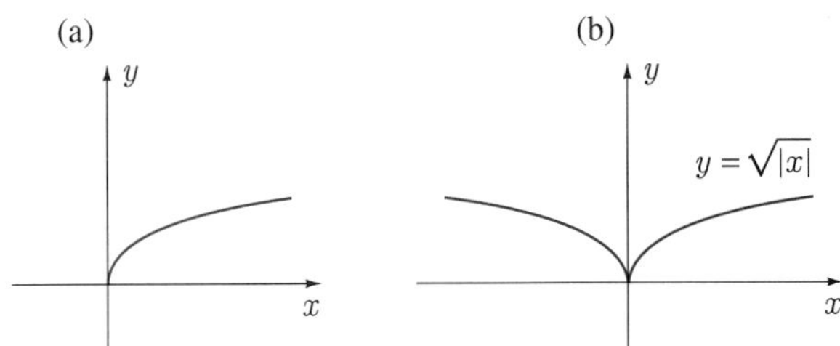

FIGURE 1.83

(b) Since

$$f(-x) = (-x)^5 - (-x) = -x^5 + x = -(x^5 - x) = -f(x),$$

this function is odd. To sketch its graph we write

$$f(x) = x(x^4 - 1) = x(x^2 - 1)(x^2 + 1) = x(x - 1)(x + 1)(x^2 + 1).$$

For $x \geq 0$, the sign diagram in Figure 1.84 suggests the graph in Figure 1.85(a). Two reflections of this graph lead to the complete picture of $y = x^5 - x$ in Figure 1.85(b).

FIGURE 1.84

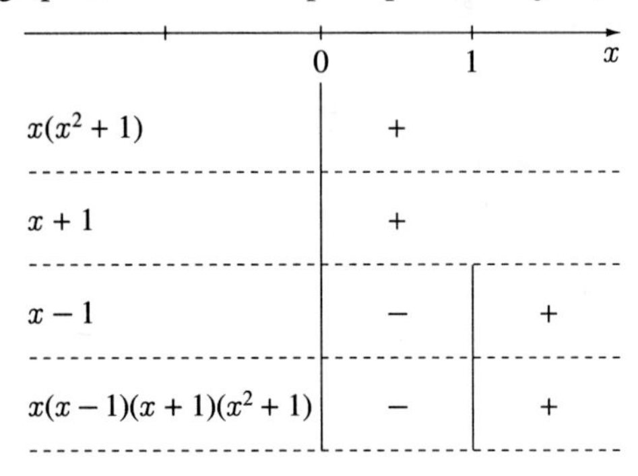

FIGURE 1.85

(a)

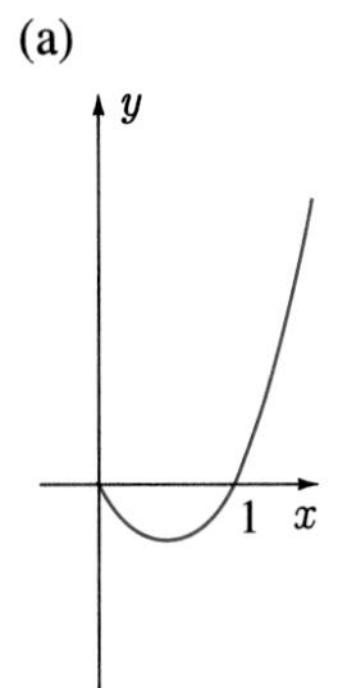

(b)

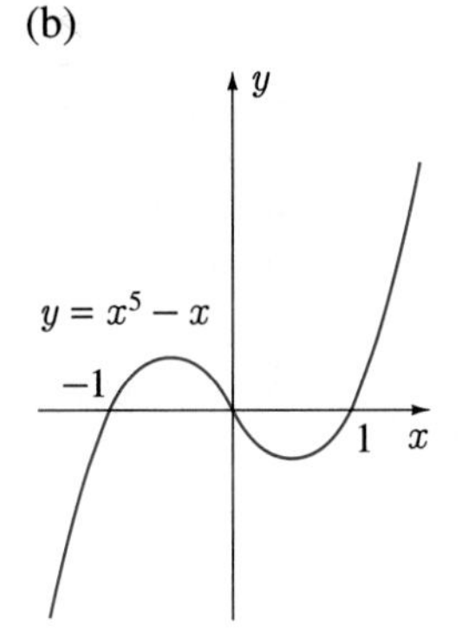

(c) The function $f(x) = x^2 + x$ is neither even nor odd. Its graph is a parabola which opens upward, crossing the x-axis at $x = 0$ and $x = -1$ (Figure 1.86).

FIGURE 1.86

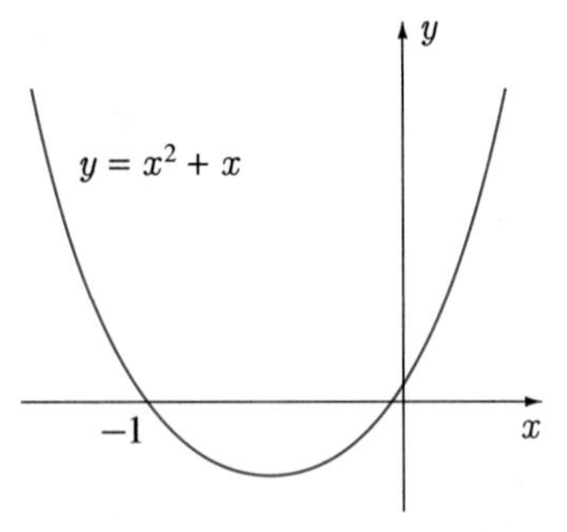

■

Exponential and logarithm functions are reviewed in Appendix B. Neither of the functions a^x nor $\log_a x$ is even or odd. but functions derived from them may be even or odd. This is illustrated in the following example.

EXAMPLE 1.28

Is the function $f(x) = e^{-x^2}$ even or odd? Sketch its graph.

SOLUTION Since

$$f(-x) = e^{-(-x)^2} = e^{-x^2} = f(x),$$

the function e^{-x^2} is even; its graph is symmetric about the y-axis. To draw the graph we begin with the fact that $f(0) = 1$. If we increase values of x beginning at 0, values of e^{x^2} get larger and larger, and therefore values of $e^{-x^2} = 1/e^{x^2}$ get smaller and smaller. Since e^{-x^2} can never be negative, the graph must approach the positive x-axis for large values of x (Figure 1.87(a)). Symmetry yields the final graph in Figure 1.87(b). This curve is very important in statistics. It is often called the *bell curve* or *normal distribution*.

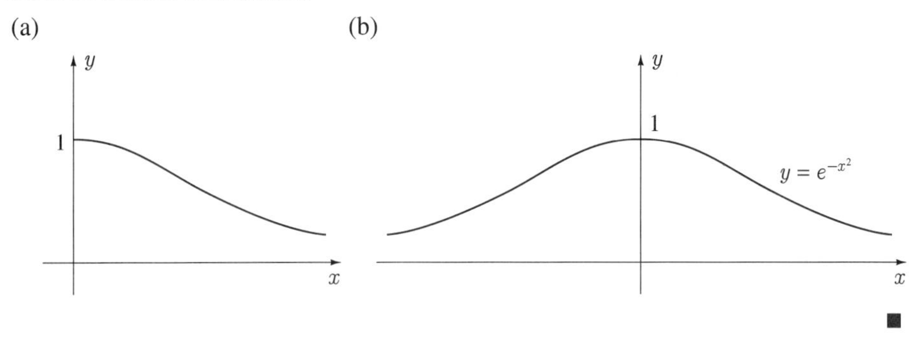

FIGURE 1.87

■

It is worthwhile pointing out that for the graph of $f(x) = e^{-x^2}$ in Example 1.28, we rounded the graph at $x = 0$ as opposed to giving it a sharp point as say in Figure 1.88. This adheres to the general principle that we avoid points on the graph of a function unless there are good reasons to do otherwise. The three graphs in which we have seen points are Figures 1.70(c), 1.72(c), and 1.83(b). Discussions leading to these figures made it abundantly clear that points must be present.

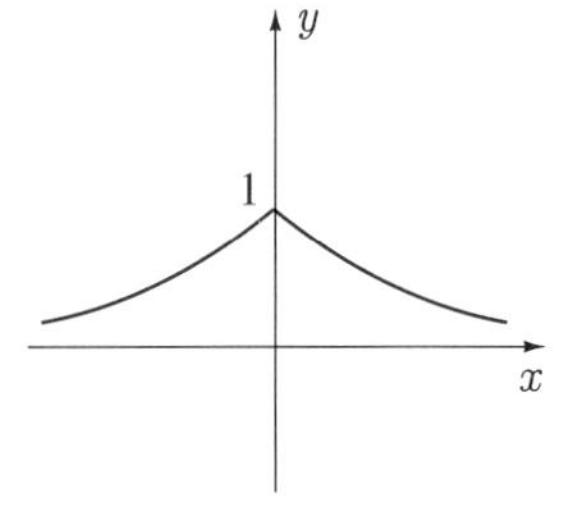

FIGURE 1.88

Can a function be both even and odd? Such a function would have to satisfy both 1.29a and 1.29b, and hence

$$f(x) = -f(x).$$

But this implies that $f(x) = 0$, and this therefore is the only even and odd function.

Translation of Curves

In Section 1.4 we saw how graphs of conic sections can be shifted in the xy-plane, and how these shifts, or **translations** as they are called, are reflected in the equations of the curves. This principle applies to all curves, not just conic sections. When every x in the equation of a curve is replaced by $x - c$, where c is a constant, the curve is shifted c units to the right. When every x is replaced by $x + c$, the curve is shifted c units to the left. For example, when x in the equation $x^2 + y^2 = r^2$ is replaced by $x - c$, the centre of the resulting circle is shifted from the origin to the point $(c, 0)$ (Figure 1.89).

FIGURE 1.89

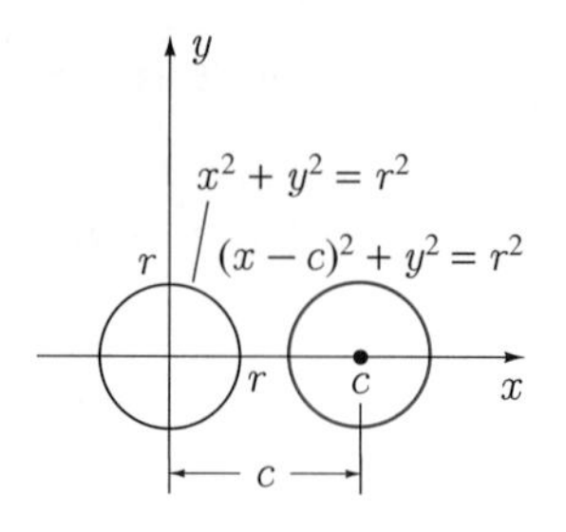

When each x in the parabola $y = x^2 + x$ is replaced by $x + c$,

$$y = (x + c)^2 + (x + c) = (x + c)(x + c + 1),$$

the parabola is shifted c units to the left (Figure 1.90).

FIGURE 1.90

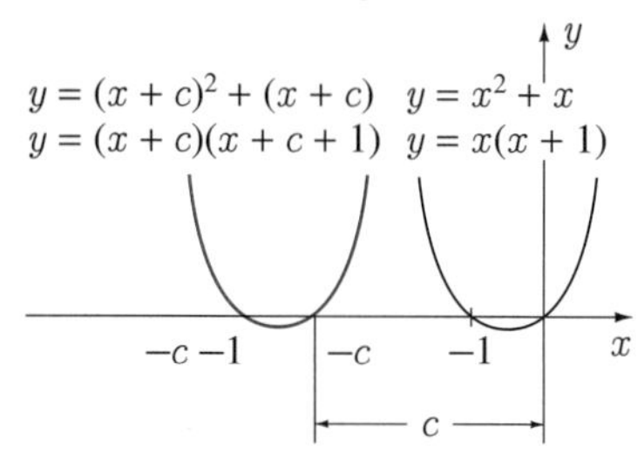

Vertical shifts result when y is replaced by $y \pm c$ in the equation of a curve. For instance, the curve $x^2 - (y + 1)^2 = 1$ is the hyperbola $x^2 - y^2 = 1$ in Figure 1.91(a) shifted downward 1 unit (Figure 1.91(b)).

FIGURE 1.91

(a)

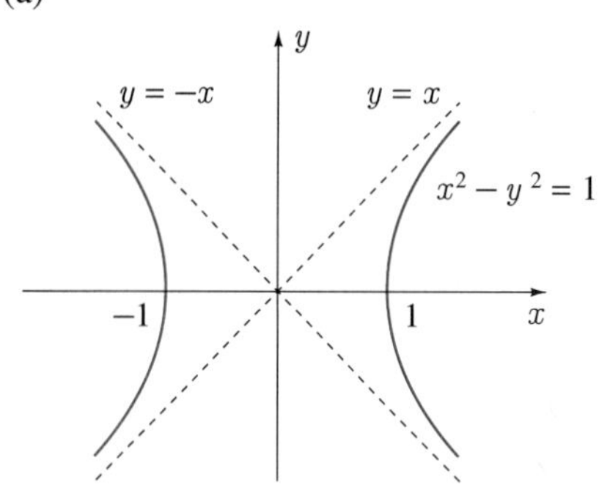

(b)

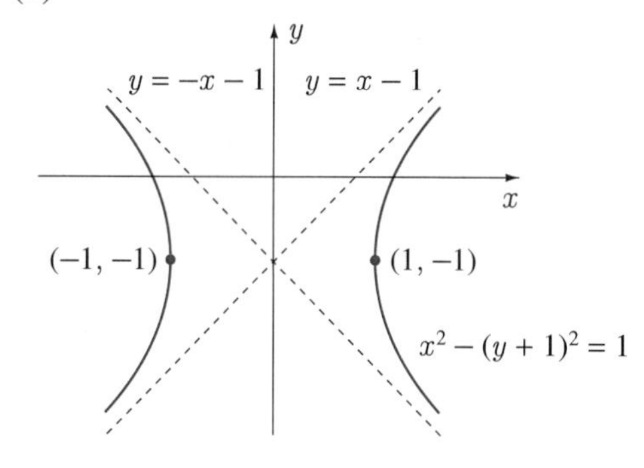

EXAMPLE 1.29

Sketch the curve $|x| + |y - 2| = 1$.

SOLUTION First we sketch $|x| + |y| = 1$. Since this equation remains unchanged when x is replaced by $-x$ and y is replaced by $-y$, the curve $|x| + |y| = 1$ is symmetric about both the x-axis and the y-axis. This means that we can concentrate our efforts on sketching the graph in the first quadrant, where the equation reduces to $x + y = 1$. The segment of this straight line in quadrant one is shown in Figure 1.92(a). To obtain $|x| + |y| = 1$ (Figure 1.92(b)), we reflect this curve in the axes. Finally, $|x| + |y - 2| = 1$ may be obtained by shifting $|x| + |y| = 1$ upward 2 units (Figure 1.92(c)).

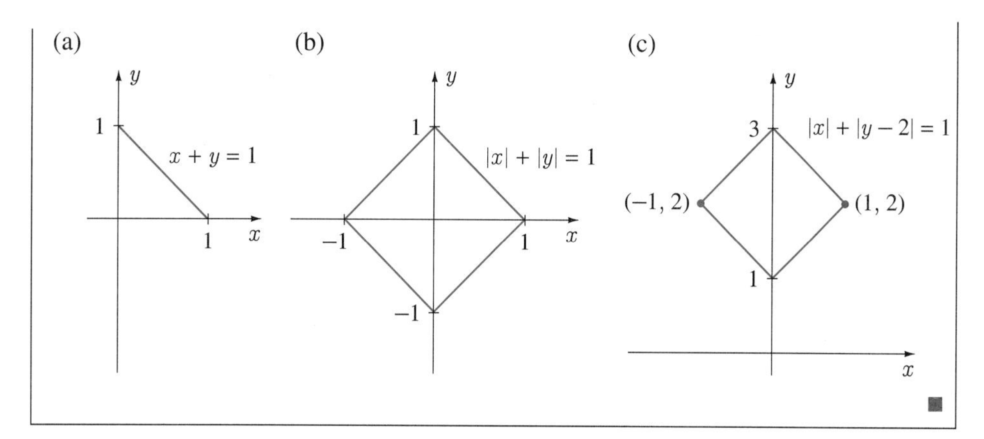

FIGURE 1.92

In calculus we make the agreement that arguments of trigonometric functions are always expressed in radians, never degrees. The reason for this will become apparent in Chapter 3 when we *differentiate* trigonometric functions.

EXAMPLE 1.30

Sketch the graph of the function $f(x) = \cos(x - \pi/3)$.

SOLUTION The graph of this function is that of $\cos x$ shifted $\pi/3$ units to the right (Figure 1.93).

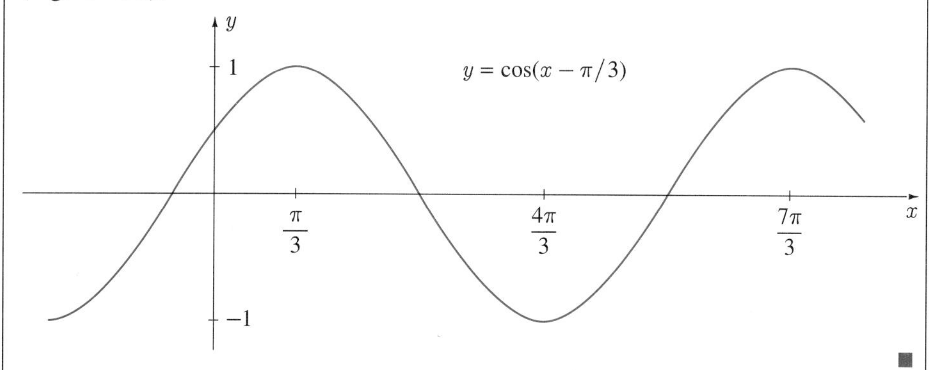

FIGURE 1.93

EXAMPLE 1.31

Sketch the graph of $f(x) = 3\sin 2x + 3\sqrt{3}\cos 2x$.

SOLUTION We could sketch graphs of $3\sin 2x$ and $3\sqrt{3}\cos 2x$, and add ordinates, but a superior technique is to write $f(x)$ in the form $f(x) = R\sin(2x + \phi)$, where R and ϕ are constants. When we equate this to the given expression for $f(x)$, and expand $\sin(2x + \phi)$ with a compound angle formula, we obtain

$$3\sin 2x + 3\sqrt{3}\cos 2x = R\sin(2x + \phi) = R[\sin 2x\cos\phi + \cos 2x\sin\phi].$$

For this equation to be valid for all x, it must certainly be true for $x = 0$ and $x = \pi/4$. Substitution of these gives the two equations

$$3\sqrt{3} = R\sin\phi,$$
$$3 = R\cos\phi.$$

To solve these for R and ϕ, we square each equation and add the results,

$$27 + 9 = R^2 \sin^2 \phi + R^2 \cos^2 \phi = R^2.$$

This implies that $R = \pm 6$. If we choose $R = 6$ ($R = -6$ works equally well), then

$$3\sqrt{3} = 6 \sin \phi, \qquad 3 = 6 \cos \phi.$$

These equations are satisfied by $\phi = \pi/3$ (there are other angles also), and therefore $f(x)$ can be expressed in the form

$$f(x) = 6 \sin(2x + \pi/3) = 6 \sin[2(x + \pi/6)].$$

The function is most easily graphed by shifting the graph of $g(x) = \sin 2x$ in Figure 1.94(a) to the left by $\pi/6$ units and modifying the scale on the y-axis to change the amplitude from 1 to 6. The result is shown in Figure 1.94(b).

FIGURE 1.94

(a)

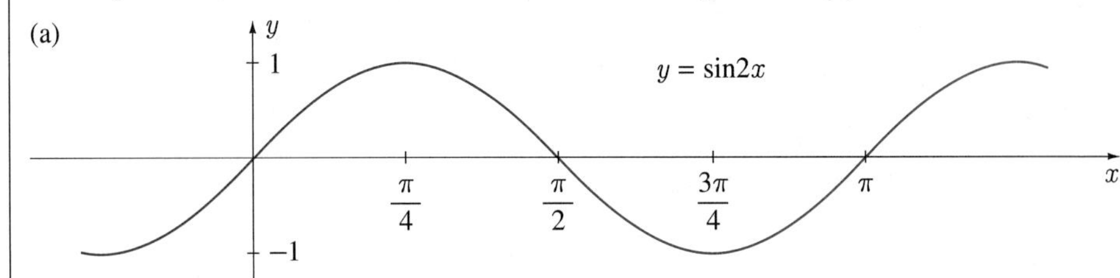

(b)

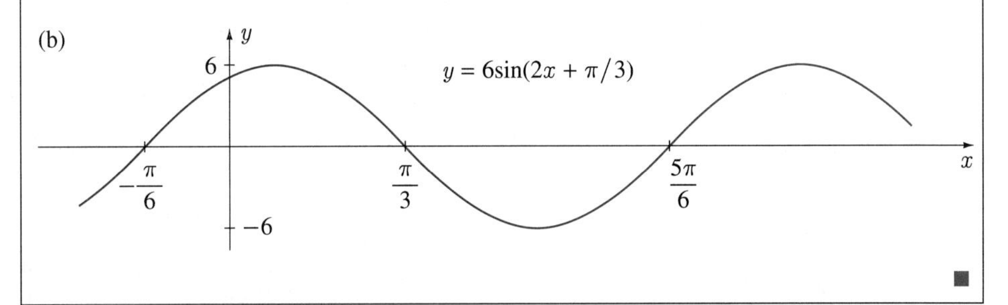

■

EXAMPLE 1.32 Sketch a graph of the function $f(x) = \ln(x + 3)$.

SOLUTION By translating the graph of $f(x) = \ln x$ in Figure 1.95(a) 3 units to the left, we obtain the graph of $f(x) = \ln(x + 3)$ in Figure 1.95(b).

FIGURE 1.95

(a) (b)

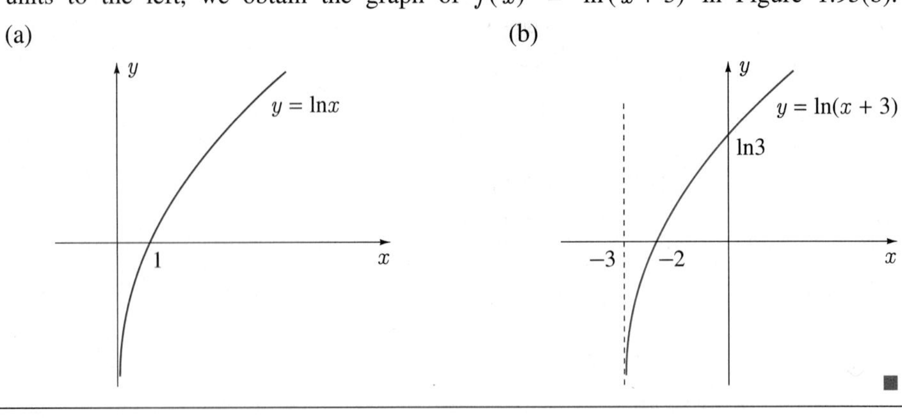

■

EXERCISES 1.6

In Exercises 1–10 determine whether the function is even, odd, or neither even nor odd. Sketch each function.

1. $f(x) = 1 + x^2 + 2x^4$ **2.** $f(x) = x^5 - x$

3. $f(x) = 12x^2 + 2x$ **4.** $f(x) = x^{1/5}$

5. $f(x) = (x - 1)/(x + 1)$ **6.** $f(x) = x(x^2 + x)$

7. $f(x) = x \sin x$ **8.** $f(x) = x^2 \sin x$

9. $f(x) = \log_{10}(x^2 + 5)$

10. $f(x) = xe^{-x^2}, \quad -2 \le x \le 2$

11. Verify that when an odd function $f(x)$ is defined at $x = 0$, its value must be $f(0) = 0$.

12. Prove each of the following:

(a) The product of two even functions or two odd functions is an even function.

(b) The product of an even and an odd function is an odd function.

In Exercises 13–28 sketch the curve. Indicate whether the curve defines y as a function of x.

13. $y = x^2 - x^6$ **14.** $y = (x + 1)^2 - 2(x + 1)$

15. $(x - 2)^2 + y^2 = 4$ **16.** $y = \sqrt{4 - (x - 2)^2}$

17. $x^2 - (y + 5)^2 = 1$ **18.** $x = -\sqrt{64 + 9(y + 5)^2}$

19. $y = (x + 1)^4 - (x + 1)^2$

20. $|x| + |y + 2| = 1$

21. $y = \big|3 - |x + 2|\big| - 1$

22. $|x| - |y| = 1$

23. $x = 4 - |y|$

24. $(y - 1)^2 = (x - 1)^2[1 - (x - 1)^2]$

25. $y = \cos^2 x - \cos x$ **26.** $x = \cos^2 y - \sin^2 y$

27. $y = \sin|x|$ **28.** $y^2 = \cos x$

In Exercises 29–50 sketch a graph of the function.

29. $f(x) = 3 \sin x$ **30.** $f(x) = \sin 2x$

31. $f(x) = 3 \sin 2x$ **32.** $f(x) = \sin(x + \pi/4)$

33. $f(x) = 3 \sin(x + \pi/4)$

34. $f(x) = \sin(2x + \pi/4)$

35. $f(x) = 3 \sin(2x + \pi/4)$

36. $f(x) = 4 \cos(x/3)$

37. $f(x) = 2 \sin(x/2 - \pi)$

38. $f(x) = 5 \cos(\pi/2 - 3x)$

39. $f(x) = \sec 2x$ **40.** $f(x) = \tan 3x$

41. $f(x) = \csc(x - \pi/3)$ **42.** $f(x) = \cot(x + \pi/4)$

43. $f(x) = \tan^2 x$ **44.** $f(x) = \sqrt{1 - \cos^2 x}$

45. $f(x) = \sqrt{1 + \tan^2 x}$ **46.** $f(x) = 5 - 2 \sec x$

47. $f(x) = 4 + 2 \tan x$ **48.** $f(x) = \tan|x|$

49. $f(x) = -|\cot 2x|$ **50.** $f(x) = 3 \csc(x/2)$

In Exercises 51-57 use the technique of Example 1.31 to sketch a graph of the function.

51. $f(x) = \sin 2x + \cos 2x$

52. $f(x) = -\sin 3x + \cos 3x$

53. $f(x) = -\sqrt{3} \sin 2x + \cos 2x$

54. $f(x) = 4 \sin 3x + 4\sqrt{3} \cos 3x$

55. $f(x) = 4 \sin 2x + \cos 2x$

56. $f(x) = -2 \sin 2x + 4 \cos 2x$

57. $f(x) = 4 \sin x + 3 \cos x$

In Exercises 58–67 sketch a graph of the function.

58. $f(x) = \log_{10}(4x)$ **59.** $f(x) = \ln(1 - x)$

60. $f(x) = \ln(1 - x^2)$ **61.** $f(x) = \ln(x^2 - 1)$

62. $f(x) = 10^{x+2}$ **63.** $f(x) = e^{2-x}$

64. $f(x) = e^{x^2}$ **65.** $f(x) = e^{-x^2} \sin x$

66. $f(x) = e^{-x^2} \cos x$ **67.** $f(x) = x^2 e^x, \quad -1 \le x \le 1$

68. The Heaviside unit step function is defined by

$$H(x) = \begin{cases} 0 & x < 0 \\ 1 & x > 0. \end{cases}$$

(a) Sketch graphs of the functions $H(x)$ and $H(x - a)$, where a is some constant.

(b) Sketch graphs of the functions $f(x) = x^{1/3}$ and

$$g(x) = \begin{cases} 0 & x < a \\ f(x - a) & x > a. \end{cases}$$

(c) Write $g(x)$ in terms of $f(x)$ and $H(x - a)$.

(d) What in essence does multiplication of $f(x - a)$ by $H(x - a)$ do?

69. If a thermal nuclear reactor is built in the shape of a right circular cylinder of radius r and height h, then neutron diffusion theory requires r and h to satisfy an equation of the form

$$\frac{a^2}{r^2} + \frac{b^2}{h^2} = 1,$$

where a and b are positive constants. Sketch a graph of the function $r = f(h)$ defined by this equation for $h \ge 2b$.

70. Is there a difference between the graphs of the functions $f(x) = \ln(x^2)$ and $g(x) = 2\ln x$?

In Exercises 71–73 sketch the curve.

71. $x + y + \sqrt{2xy} = 1$

72. $(x^2 + 2y - 1)^2 = y^2(1 - x^2)$

73. $x^2 + 4x^2y = x^4 + y^2$

74. Prove that every function can be expressed as the sum of an even function and an odd function, and that there is only one way to do this.

SUMMARY

In the first four sections of this chapter we developed the principles of plane analytic geometry. Analytic geometry is a combination of geometry and algebra. Algebraic equations are used to describe geometric curves and curves are the geometric representation of equations. The form of an equation dictates the shape of the curve, and conversely, the shape of a curve influences its equation. To illustrate this fact we discussed straight lines, circles, parabolas, ellipses, and hyperbolas. The most common forms for equations of these curves are as follows:

$$\text{Straight line}\begin{cases} y - y_1 = m(x - x_1) & \text{Point-slope} \\ \dfrac{x - x_1}{x_2 - x_1} = \dfrac{y - y_1}{y_2 - y_1} & \text{Two-point} \\ y = mx + b & \text{Slope } y\text{-intercept} \\ y = m(x - a) & \text{Slope } x\text{-intercept} \\ \dfrac{x}{a} + \dfrac{y}{b} = 1 & \text{Two-intercept} \\ Ax + By + C = 0 & \text{General} \end{cases}$$

$$\text{Parabola}\begin{cases} y = ax^2 + bx + c & \text{Vertical axis of symmetry} \\ x = ay^2 + by + c & \text{Horizontal axis of symmetry} \end{cases}$$

$$\text{Circle}\begin{cases} x^2 + y^2 + fx + gy + e = 0 \\ (x - h)^2 + (y - k)^2 = r^2 \end{cases}$$

$$\text{Ellipse}\begin{cases} \dfrac{x^2}{a^2} + \dfrac{y^2}{b^2} = 1 \\ \dfrac{(x - h)^2}{a^2} + \dfrac{(y - k)^2}{b^2} = 1 \end{cases}$$

$$\text{Hyperbola}\begin{cases} \dfrac{x^2}{a^2} - \dfrac{y^2}{b^2} = 1 \\ \dfrac{y^2}{b^2} - \dfrac{x^2}{a^2} = 1 \\ \dfrac{(x - h)^2}{a^2} - \dfrac{(y - k)^2}{b^2} = 1 \\ \dfrac{(y - k)^2}{b^2} - \dfrac{(x - h)^2}{a^2} = 1 \end{cases}$$

Basic to all mathematics is the concept of a function, a rule that assigns to each number x in a domain, a unique number y in the range. A function is simply another way of saying "a quantity y depends on x". The notation $y = f(x)$ for a function immediately suggests that a function can be represented geometrically by a curve — the curve with equation $y = f(x)$ — and we call this curve the graph of the function. Although graphs can always be drawn by plotting points from tables of values and joining these points smoothly, other methods such as addition and multiplication of ordinates, symmetry properties, and translations are frequently more expedient. In Chapters 2 and 4, we will discover further curve-sketching aids, and these will all be combined for a complete discussion in Section 4.5.

Key Terms and Formulas

In reviewing this chapter, you should be able to define or discuss the following key terms:

Open interval
Closed interval
Infinity
Sign diagram
Absolute value
Origin
Abscissa
Ordinate
Coordinates
Quadrant
Length of line segment
Midpoint of line segment
Equation of a curve
Slope of a line
Rise
Run
Point-slope formula
Parallel lines
Perpendicular lines
Inclination of a line
Parabola
Symmetry about x-axis
Symmetry about y-axis
Quadratic formula
Circle
Ellipse
Hyperbola
Asymptotes
Function
Domain
Range
Graph
Addition of ordinates
Multiplication of ordinates
Polynomial
Rational function
Even function
Odd function
Translation of a curve

REVIEW EXERCISES

In Exercises 1–6 find all values of x satisfying the inequality.

1. $2x - 6 < 3x + 5$ **2.** $|x^2 + 1| \leq 5$

3. $\dfrac{x - 4}{2 - x} \geq 5$ **4.** $\dfrac{x^2 - 14}{x - 3} < 2$

5. $|x^2 - 4| \geq 5$ **6.** $|x^2 - 5| > 4$

In Exercises 7 and 8 find the distance between the points and the midpoint of the line segment joining the points.

7. $(-1, 3)$, $(4, 2)$ **8.** $(2, 1)$, $(-3, -4)$

In Exercises 9–12 find the equation for the line described.

9. Parallel to the line $x - 2y = 4$ and through the point $(2, 3)$

10. Perpendicular to the line joining $(-2, 1)$ to the origin and through the midpoint of the line segment joining $(1, 3)$ and $(-1, 5)$

11. Perpendicular to the line $x = 4y - 11$ and through the point of intersection of this line and $x = \sqrt{y^2 + 9}$

12. Joining the points of intersection of the curves $y = x^2$ and $5x = 6 - y^2$

In Exercises 13–22 find the largest possible domain for the function.

13. $f(x) = \sqrt{x^2 + 5}$ **14.** $f(x) = \sqrt{x^2 - 5}$

15. $f(x) = \dfrac{1}{x^2 + 3x + 2}$ **16.** $f(x) = \dfrac{x + 4}{x^3 + 2x^2 + x}$

17. $f(x) = (x^3 - 8)^{1/3}$ **18.** $f(x) = x^{3/2}$

19. $f(x) = \sqrt{x^2 + 4x - 6}$ **20.** $f(x) = \dfrac{1}{\sqrt{2x^2 + 4x - 5}}$

21. $f(x) = \sqrt{\dfrac{2x+1}{x-3} + 2}$ 22. $f(x) = \sqrt{x - \dfrac{1}{x}}$

In Exercises 23–34 identify the curve as a straight line, parabola, circle, ellipse, hyperbola, or none of these.

23. $x + 2y = 4$
24. $x = y^2 - 2y + 3$
25. $y = x^3 + 3$
26. $x^2 + 2y^2 = 4$
27. $y^2 - x^2 = x$
28. $x^2 + y^2 + 5 = 0$
29. $x^2 - 2x + y^2 = 16$
30. $x + y^2 = 3$
31. $x^2 + 2y^2 + y = 2x$
32. $x^2 - x + y^2 + y = 0$
33. $2x^2 + 20x + 38 = 3y^2 + 12y$
34. $2y^2 - x = 3x^2 - y$

In Exercises 35–60 sketch the curve.

35. $y = 2x^2 + 3$
36. $x^2 = 4 - y^2$
37. $y = x^3 - 1$
38. $|y| = |x|$
39. $y = x^3\sqrt{1 - x^2}$
40. $y = \sqrt{-x^2 + 4x + 4}$
41. $4x^2 + y^2 = 0$
42. $x^2 + 3y^2 = 6$
43. $2y^2 - x^2 = 3$
44. $x^2 - 2x - y^2 + 4y = 1$
45. $x = y\sqrt{y + 3}$
46. $x^2 - 4y + 2 = 4x - 2y^2$
47. $y = \sin 3x$
48. $y = \cos(2x + \pi/2)$
49. $y = \cos(2x - \pi/4)$
50. $y = 2\sin(3x + \pi/2)$
51. $y = |x| + |x - 1|$
52. $y = \sqrt{|x - 1| - 1}$
53. $x = \tan y$
54. $y = 2\ln(3x + 4)$
55. $x = e^{-y}$
56. $y = \sec^2(x + \pi/4)$
57. $y = xe^{-x}, \quad x \geq 0$
58. $y = \sin|x|$
59. $|y| = |\sin x|$
60. $x = \sqrt{\sin y}$

In Exercises 61-64 give an example of a function $y = f(x)$ with the indicated properties.

61. The range of the function consists of one number only.

62. The largest possible domain of the function is $-1 \leq x \leq 2$.

63. The domain of the function consists of all reals except $x = \pm 1$.

64. The domain of the function is $x \leq 0$ and the range is $y \geq 1$.

65. Is the triangle with vertices $(1,3)$, $(2,-1)$, and $(-2,-2)$ right-angled?

66. Show that the lines that successively join the points $(0,1)$, $(3,4)$, $(3 - \sqrt{2}, 0)$, and $(-\sqrt{2}, -3)$ form the sides of a rhombus, which is a parallelogram with all sides of equal length.

67. Find the distance from the point $(2,1)$ to the line $2x + 3y = 4$.

68. A lighthouse is 6 km off shore and a cabin on the straight shoreline is 9 km from the point on the shore nearest the lighthouse (Figure 1.96).

FIGURE 1.96

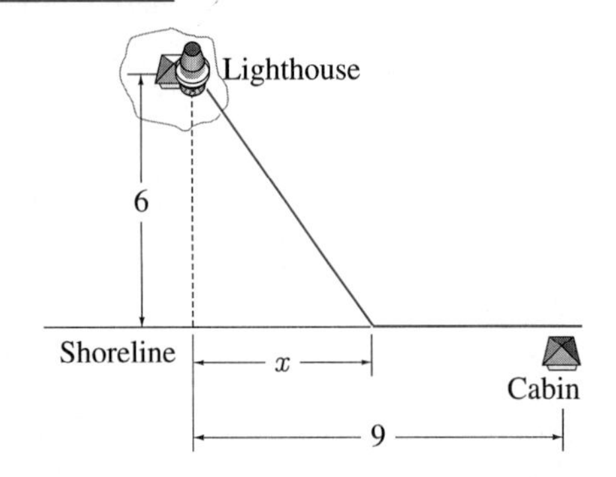

Show that if a man rows at 3 km/h and walks at 5 km/h, and he beaches the boat at x, then the total travel time from lighthouse to cabin is

$$t = f(x) = \frac{\sqrt{x^2 + 36}}{3} + \frac{9 - x}{5}, \quad 0 \leq x \leq 9.$$

Sketch a graph of this function.

69. Prove that the diagonals of a rhombus intersect at right angles.

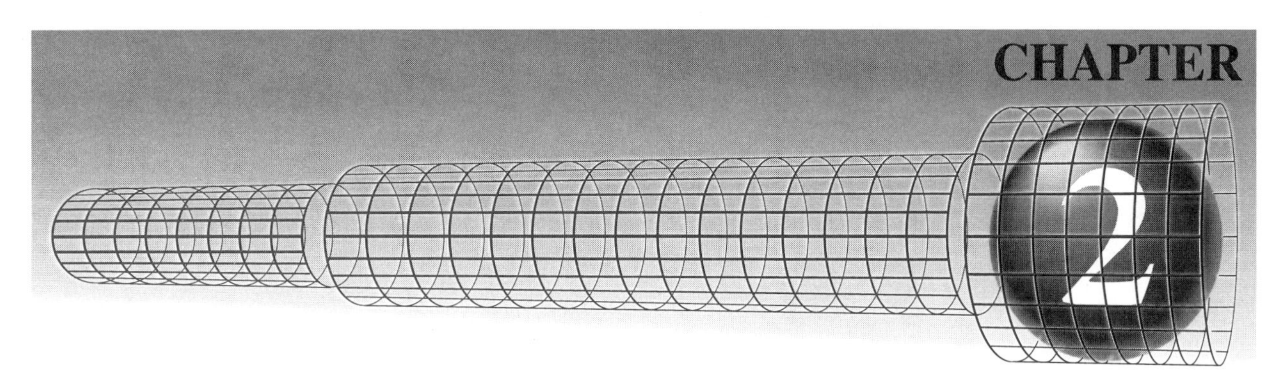

Limits and Continuity

The concept of a "limit" is crucial to calculus, for the two basic operations in calculus are differentiation and integration, each of which is defined in terms of a limit. For this reason we must have a clear understanding of limits from the beginning. In Sections 2.1–2.4 we give an intuitive discussion of limits of functions; in Section 2.5 we show how these ideas can be formalized mathematically.

SECTION 2.1 Limits

The value of the function $f(x) = x^2 - 4x + 5$ at $x = 2$ is $f(2) = 1$. A completely different consideration is contained in the question,"What number does $f(x) = x^2 - 4x + 5$ get closer and closer to as x gets closer and closer to 2?" Table 2.1 shows that *as x gets closer and closer to* 2, $x^2 - 4x + 5$ *gets close to* 1 *and stays close to* 1.

TABLE 2.1

x	$f(x) = x^2 - 4x + 5$	x
2.1	1.01	1.9
2.01	1.0001	1.99
2.001	1.000001	1.999
2.0001	1.00000001	1.9999

Likewise, the graph of the function (Figure 2.1) clearly shows that $f(x)$ approaches 1 as x approaches 2. This statement is not precise enough for our purposes. For instance, the graph also indicates that as x gets closer and closer to 2, $f(x)$ gets closer and closer to 0. It does not get very close to 0, but nonetheless, $f(x)$ does get closer and closer to 0 as x gets closer and closer to 2. In fact, we can make this statement for any number less than 1.

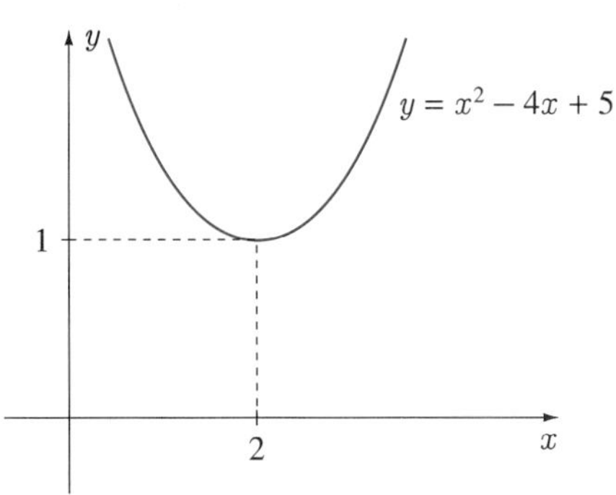

FIGURE 2.1

To distinguish 1 from all numbers less than 1, we say that $x^2 - 4x + 5$ *can be made arbitrarily close to* 1 *by choosing* x *sufficiently close to* 2. Numbers less than 1 do not satisfy this statement. For instance, it is not true that $x^2 - 4x + 5$ can be made arbitrarily close to 0 by choosing x sufficiently close to 2. The closest the function gets to 0 is one unit when $x = 2$. In calculus we say that **the limit of** $x^2 - 4x + 5$ **as** x **approaches** 2 **is** 1 to represent the more lengthy statement "$x^2 - 4x + 5$ can be made arbitrarily close to 1 by choosing x sufficiently close to 2". In addition, we have a notation to represent both statements:

$$\lim_{x \to 2} (x^2 - 4x + 5) = 1. \tag{2.1}$$

This notation is read "the limit of (the function) $x^2 - 4x + 5$ as x approaches 2 is equal to 1", and this stands for the statement "$x^2 - 4x + 5$ can be made arbitrarily close to 1 by choosing x sufficiently close to 2".

We emphasize that the limit in 2.1 is not concerned with the value of $x^2 - 4x + 5$ at $x = 2$. It is concerned with the number that $x^2 - 4x + 5$ approaches as x approaches 2. These numbers are not always the same.

Generally we say that a function $f(x)$ has limit L as x approaches a, and write

$$\lim_{x \to a} f(x) = L, \tag{2.2}$$

if $f(x)$ can be made arbitrarily close to L by choosing x sufficiently close to a. Sometimes it is more convenient to write $f(x) \to L$ as $x \to a$ to mean that $f(x)$ approaches L as x approaches a. This is especially so in the middle of a paragraph, as opposed to a displayed equation such as 2.2.

EXAMPLE 2.1

Evaluate $\lim_{x \to 1} (x^2 + 2x + 5)$.

SOLUTION As x gets closer and closer to 1, $x^2 + 2x + 5$ gets closer and closer to 8, and we therefore write

$$\lim_{x \to 1} (x^2 + 2x + 5) = 8.$$

Once again this is corroborated by the graph of the function in Figure 2.2.

FIGURE 2.2

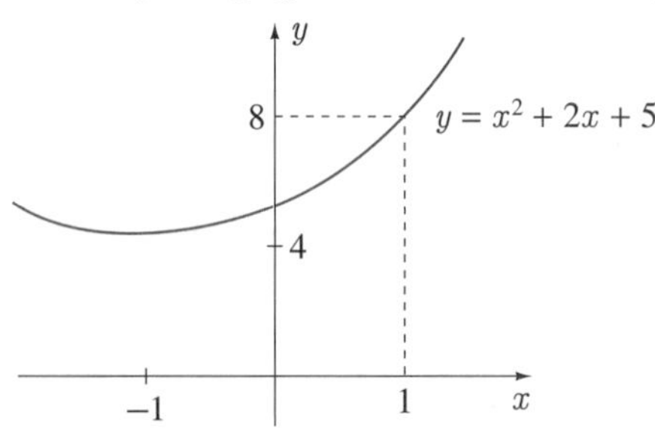

■

To calculate the limit of a function $f(x)$ as x approaches a, we evaluate $f(x)$ at values of x that get closer and closer to a. For limits of complicated functions such as

$$\lim_{x \to 5} \frac{x^2(3 - x)}{x^3 + x},$$

it would be tedious to evaluate $x^2(3 - x)/(x^3 + x)$ at many values of x approaching 5. The following theorem provides a much easier method.

Theorem 2.1

If $\lim_{x\to a} f(x) = F$ *and* $\lim_{x\to a} g(x) = G$, *then*

$$\text{(i)} \quad \lim_{x\to a} [f(x) + g(x)] = F + G; \tag{2.3}$$

$$\text{(ii)} \quad \lim_{x\to a} [f(x) - g(x)] = F - G; \tag{2.4}$$

$$\text{(iii)} \quad \lim_{x\to a} [cf(x)] = cF, \quad \textit{when } c \textit{ is a constant}; \tag{2.5}$$

$$\text{(iv)} \quad \lim_{x\to a} [f(x)g(x)] = FG; \tag{2.6}$$

$$\text{(v)} \quad \lim_{x\to a} \frac{f(x)}{g(x)} = \frac{F}{G}, \quad \textit{provided } G \neq 0. \tag{2.7}$$

What this theorem says is that a limit such as $\lim_{x\to 5} [x^2(3-x)/(x^3+x)]$ can be broken down into smaller problems and reassembled later. For instance, since

$$\lim_{x\to 5} x^2 = 25, \qquad \lim_{x\to 5} (3-x) = -2, \qquad \lim_{x\to 5} x^3 = 125, \qquad \lim_{x\to 5} x = 5,$$

we may write

$$\lim_{x\to 5} \frac{x^2(3-x)}{x^3+x} = \frac{25(-2)}{125+5} = \frac{-50}{130} = -\frac{5}{13}.$$

Although the results of Theorem 2.1 may seem evident, to prove them mathematically is not a simple task. In fact, because we have not yet given a precise definition for limits, a proof at this time is impossible. When we give definitions for limits in Section 2.5, it will then be possible to verify the theorem (see Exercises 31–35 in Section 2.5).

EXAMPLE 2.2

Evaluate $\lim_{x\to -2} [(x+2)/(x^2+9)]$.

SOLUTION Since $\lim_{x\to -2} (x+2) = 0$ and $\lim_{x\to -2} (x^2+9) = 13$, part (v) of Theorem 2.1 gives

$$\lim_{x\to -2} \frac{x+2}{x^2+9} = \frac{0}{13} = 0.$$

■

EXAMPLE 2.3

Evaluate $\lim_{x\to -1} [x^2(1-x^3)/(2x^2+x+1)]$.

SOLUTION Using Theorem 2.1, we can write

$$\lim_{x\to -1} \frac{x^2(1-x^3)}{2x^2+x+1} = \frac{(1)(2)}{2+(-1)+1} = 1.$$

Be sure that you understand how we obtained the expression

$$\frac{(1)(2)}{2+(-1)+1}.$$

In particular, we *did not* set $x = -1$ in $x^2(1-x^3)/(2x^2+x+1)$. Indeed, this is not permitted because to evaluate a limit as x approaches -1, we are not to set $x = -1$; we are to let x get closer and closer to -1. What we did do is take limits of x^2, $1-x^3$, $2x^2$, and x as x approaches -1, and then use Theorem 2.1.

■

The following example illustrates what can happen if we substitute $x = a$ into $f(x)$ in the evaluation of $\lim_{x \to a} f(x)$.

EXAMPLE 2.4

Evaluate $\lim_{x \to 3} [(x^2 - 9)/(x - 3)]$.

SOLUTION Because $\lim_{x \to 3} (x - 3) = 0$, we cannot use Theorem 2.1. Nor can we set $x = 3$ in $(x^2 - 9)/(x - 3)$ because it is inherent in the limiting procedure that we do not put $x = 3$. Besides, if we did, we would obtain the meaningless expression $0/0$. The correct procedure is to factor $x^2 - 9$ into $(x - 3)(x + 3)$ and cancel a factor of $x - 3$ from numerator and denominator:

$$\lim_{x \to 3} \frac{x^2 - 9}{x - 3} = \lim_{x \to 3} \frac{(x - 3)(x + 3)}{x - 3} = \lim_{x \to 3} (x + 3) = 6.$$

Cancellation of the factor $x - 3$ would not be permissible if $x - 3$ were equal to 0; that is, if x were equal to 3. But once again this cannot happen, because in the limiting operation we let x get closer and closer to 3, but do not set $x = 3$. ■

Note in this example that although the limit is 6, there is no value of x for which the function $(x^2 - 9)/(x - 3)$ is ever equal to 6. The graph of the function in Figure 2.3 is a straight line with the point at $x = 3$ removed. The function is undefined at $x = 3$, but as x approaches 3, $f(x)$ clearly gets arbitrarily close to 6.

FIGURE 2.3

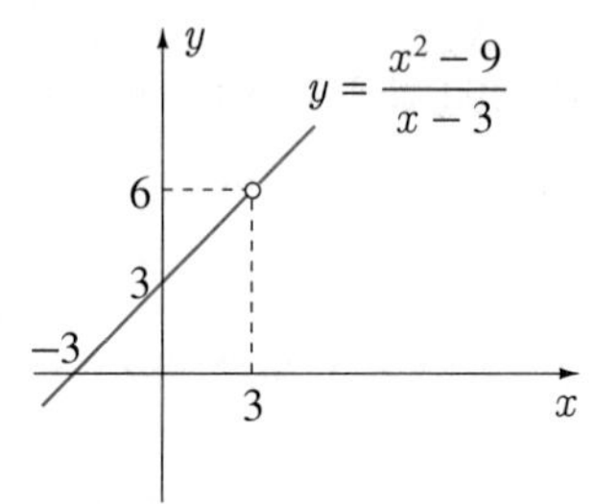

EXAMPLE 2.5

Evaluate $\lim_{x \to 0} [(-2x + 3x^2)/(4x - 5x^2)]$.

SOLUTION We have

$$\lim_{x \to 0} \frac{-2x + 3x^2}{4x - 5x^2} = \lim_{x \to 0} \frac{x(-2 + 3x)}{x(4 - 5x)} = \lim_{x \to 0} \frac{3x - 2}{4 - 5x} = \frac{-2}{4} = -\frac{1}{2}.$$

■

EXAMPLE 2.6

Evaluate

$$\lim_{x \to 0} \frac{\sqrt{1 + x} - 1}{\sqrt{x}}.$$

SOLUTION Since the limit of the denominator as x approaches 0 is 0, we cannot immediately use Theorem 2.1. The best procedure is to rationalize the numerator, that is, rid the numerator of the square root by multiplying numerator and denominator by $\sqrt{1 + x} + 1$:

$$\lim_{x\to 0} \frac{\sqrt{1+x}-1}{\sqrt{x}} = \lim_{x\to 0}\left(\frac{\sqrt{1+x}-1}{\sqrt{x}}\,\frac{\sqrt{1+x}+1}{\sqrt{1+x}+1}\right)$$
$$= \lim_{x\to 0} \frac{x}{\sqrt{x}\left(\sqrt{1+x}+1\right)}$$
$$= \lim_{x\to 0} \frac{\sqrt{x}}{\sqrt{1+x}+1}$$
$$= 0.$$

■

EXAMPLE 2.7

Evaluate

$$\lim_{x\to 0} \frac{\sin 2x}{\sin x}.$$

SOLUTION Once again we cannot immediately use Theorem 2.1 since the limit of the denominator is zero. But using the double-angle formula $\sin 2x = 2 \sin x \cos x$, we find that

$$\lim_{x\to 0} \frac{\sin 2x}{\sin x} = \lim_{x\to 0} \frac{2\sin x \cos x}{\sin x} = \lim_{x\to 0} (2\cos x) = 2.$$

■

EXAMPLE 2.8

Do the functions

$$f(x) = \sin\left(\frac{1}{x}\right) \qquad \text{and} \qquad g(x) = x \sin\left(\frac{1}{x}\right)$$

have limits as x approaches 0?

SOLUTION The best way to find out whether these particular functions have limits as x approaches 0 is to sketch their graphs. Suppose we divide the positive x-axis into two parts $0 < x \leq 1$ and $1 < x < \infty$. Now the function $\sin x$, with period 2π, has part of a cycle in the interval $0 < x \leq 1$, and an infinite number of oscillations in $1 < x < \infty$ (Figure 2.4).

FIGURE 2.4

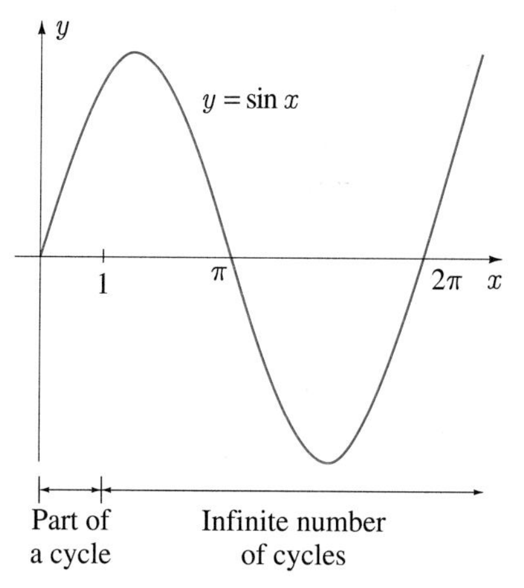

The function $\sin(1/x)$ performs two operations; it first inverts a value of x to give $1/x$ and then takes the sine of this number. Given a value of x in the interval $0 < x < 1$, $\sin(1/x)$ first evaluates $1/x$ which is in the interval $1 < 1/x < \infty$, and then takes the sine of $1/x$. Conversely, given an x in $1 < x < \infty$, $\sin(1/x)$ first finds $1/x$, which is in the interval $0 < 1/x < 1$, and then calculates $\sin(1/x)$. The infinite number of oscillations of $\sin x$ in the interval $1 < x < \infty$ take place in the interval $0 < x < 1$ for $\sin(1/x)$. Furthermore, the part cycle of $\sin x$ in $0 < x < 1$ is spread out over $1 < x < \infty$ by $\sin(1/x)$. This is shown in Figure 2.5(a). Notice in particular that $\sin(1/x)$ is equal to 0 when $x = 1/(n\pi)$ for every positive integer n. Since $\sin(1/x)$ is an odd function, its full graph is shown in Figure 2.5(b). The function oscillates more and more rapidly as x approaches 0, and clearly $f(x) = \sin(1/x)$ cannot have a limit there.

FIGURE 2.5

FIGURE 2.6

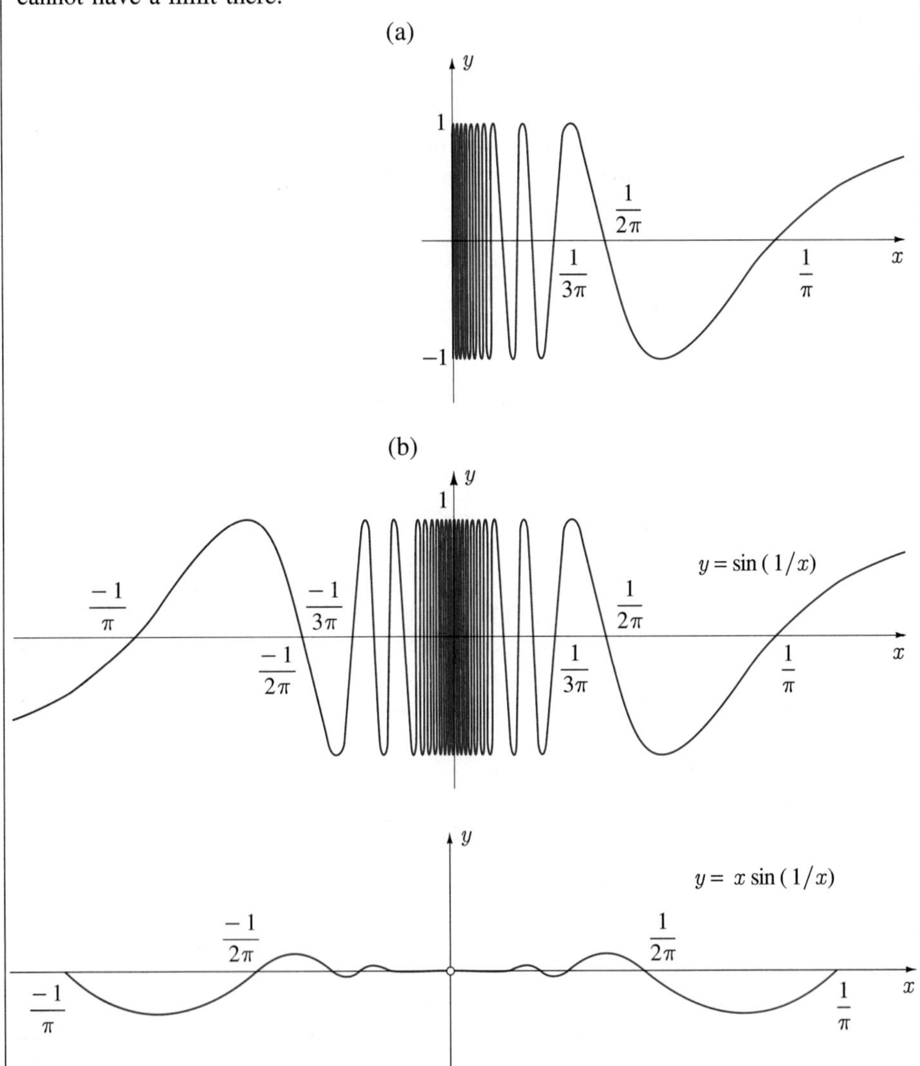

The graph of $g(x) = x\sin(1/x)$ is easily obtained by multiplication of ordinates of $y = x$ and $y = \sin(1/x)$. The result is shown in Figure 2.6 between $-1/\pi$ and $1/\pi$. Once again there is an infinite number of oscillations, but the factor x before $\sin(1/x)$ in $g(x)$ causes the "amplitude" to approach 0 as x approaches 0. In other words, $\lim_{x \to 0} g(x) = 0$. ■

One-sided Limits

When we write $L = \lim_{x \to a} f(x)$, we mean that $f(x)$ gets arbitrarily close to L as x gets closer and closer to a. But how is x to approach a? Does x approach a through numbers larger than a, or does it approach a through numbers smaller than a? Or does x jump back and forth between numbers larger than a and numbers smaller than a, gradually getting closer and closer to a? We have not previously mentioned "mode" of approach simply because it would have made no difference to our discussion. In each of the preceding examples, all possible modes of approach lead to the same limit. In particular, Table 2.1 and Figure 2.1 illustrate that 1 is the limit of $f(x) = x^2 - 4x + 5$ as x approaches 2 whether x approaches 2 through numbers larger than 2 or through numbers smaller than 2.

Approaching a number a either through numbers larger than a or through numbers smaller than a are two modes of approach that will be very important, and we therefore give them special notations:

$\lim_{x \to a^-} f(x)$ indicates that x approaches a through numbers smaller than a (often called a **left-hand limit** since x approaches a along the x-axis from the left of a)

$\lim_{x \to a^+} f(x)$ indicates that x approaches a through numbers larger than a (often called a **right-hand limit** since x approaches a along the x-axis from the right of a)

Example 2.6 should, in fact, be designated a right-hand limit,

$$\lim_{x \to 0^+} \frac{\sqrt{1+x} - 1}{\sqrt{x}} = 0,$$

since the presence of $\sqrt{x}$ in the denominator demands that x be positive.

A natural question to ask is "What should we conclude if for a function $f(x)$

$$\lim_{x \to a^+} f(x) \neq \lim_{x \to a^-} f(x)?\text{"}$$

Our entire discussion has suggested (and indeed it can be proved; see Exercise 20 in Section 2.5) that if a function has a limit as x approaches a, then it has only one such limit; that is, the limit must be the same for every possible method of approach. Consequently, if we arrive at two different results depending on the mode of approach, then we conclude that the function does not have a limit. This situation is illustrated in the following example.

EXAMPLE 2.9

Evaluate, if possible, $\lim_{x \to 0} (|x|/x)$.

SOLUTION If $x < 0$, then $|x| = -x$, and

$$\lim_{x \to 0^-} \frac{|x|}{x} = \lim_{x \to 0^-} \frac{-x}{x} = -1.$$

If $x > 0$ then $|x| = x$, and

$$\lim_{x \to 0^+} \frac{|x|}{x} = \lim_{x \to 0^+} \frac{x}{x} = 1.$$

The function has a right-hand limit and a left-hand limit at $x = 0$, but because they are not the same, $\lim_{x\to 0}(|x|/x)$ does not exist. The graph of $f(x) = |x|/x$ in Figure 2.7 clearly illustrates the situation.

FIGURE 2.7

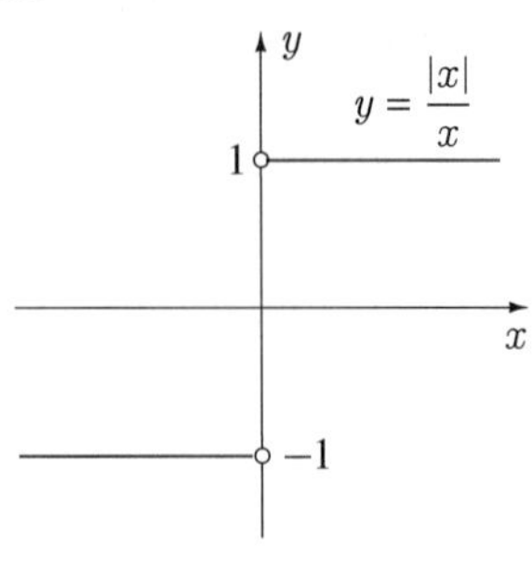

EXERCISES 2.1

In Exercises 1–40 find the indicated limit, if it exists.

1. $\lim_{x\to 7} \frac{x^2-5}{x+2}$

2. $\lim_{x\to -2} \frac{x^3+8}{x+5}$

3. $\lim_{x\to -5} \frac{x^2+3x+2}{x^2+25}$

4. $\lim_{x\to 0} \frac{x^2+3x}{3x^2-2x}$

5. $\lim_{x\to 3^+} \frac{2x-3}{x^2-5}$

6. $\lim_{x\to 2^-} \frac{2x-4}{3x+2}$

7. $\lim_{x\to 0^-} \frac{x^4+5x^3}{3x^4-x^3}$

8. $\lim_{x\to 2^+} \frac{x^2+2x+4}{x-3}$

9. $\lim_{x\to 2} \frac{x^2-4}{x-2}$

10. $\lim_{x\to 3^+} \frac{x^2-9}{x-3}$

11. $\lim_{x\to 5^-} \frac{x^2-25}{x-5}$

12. $\lim_{x\to 3} \frac{x^2-2x-3}{3-x}$

13. $\lim_{x\to 2} \frac{x^2-4x+4}{x-2}$

14. $\lim_{x\to 2} \frac{x^3-6x^2+12x-8}{x^2-4x+4}$

15. $\lim_{x\to 1} \frac{x^3-6x^2+11x-6}{x^2-3x+2}$

16. $\lim_{x\to 2} \frac{x^3-6x^2+11x-6}{x^2-3x+2}$

17. $\lim_{x\to 3^+} \frac{x^3-6x^2+11x-6}{x^2-3x+2}$

18. $\lim_{x\to 3^-} \frac{x^3-6x^2+11x-6}{x^2-3x+2}$

19. $\lim_{x\to 0} \frac{x^3-6x^2+11x-6}{x^2-3x+2}$

20. $\lim_{x\to -1} \frac{12x+5}{x^2-2x+1}$

21. $\lim_{x\to 1} \sqrt{\frac{2-x}{2+x}}$

22. $\lim_{x\to 5} \frac{\sqrt{1-x^2}}{3x+2}$

23. $\lim_{x\to 0} \frac{\sin 4x}{\sin 2x}$

24. $\lim_{x\to 0^+} \frac{\sin 6x}{\sin 3x}$

25. $\lim_{x\to 0} \frac{\tan x}{\sin x}$

26. $\lim_{x\to \pi/4} \frac{\sin x}{\tan x}$

27. $\lim_{x\to 0^+} \frac{\sin 2x}{\tan x}$

28. $\lim_{x\to 2} \frac{x-2}{\sqrt{x}-\sqrt{2}}$

29. $\lim_{x\to 0} \frac{\sqrt{1-x}-\sqrt{1+x}}{x}$

30. $\lim_{x\to 5^+} \frac{|x^2-25|}{x^2-25}$

31. $\lim_{x\to 5^-} \frac{|x^2-25|}{x^2-25}$

32. $\lim_{x\to 5} \frac{|x^2-25|}{x^2-25}$

33. $\lim_{x\to 0^+} \frac{\sqrt{x+2}-2}{\sqrt{x}}$

34. $\lim_{x\to 0} \frac{1-\sqrt{x^2+1}}{2x^2}$

35. $\lim_{x\to -2} \frac{\sqrt{-x}+\sqrt{2}}{x+2}$

36. $\lim_{x\to 0} \frac{\sqrt{1+x}-\sqrt{1-x}}{x}$

37. $\lim_{x\to -2^+} \frac{\sqrt{x+3}-\sqrt{-x-1}}{\sqrt{x+2}}$

38. $\lim_{x\to 0} \frac{x}{\sqrt{x+4}-2}$

39. $\lim_{x\to 0} \frac{\sqrt{1+x}-\sqrt{1-x}}{\sqrt{2+x}-\sqrt{2-x}}$

40. $\lim_{x\to 0} \frac{\sqrt{x+1}-\sqrt{2x+1}}{\sqrt{3x+4}-\sqrt{2x+4}}$

41. Does the greatest integer function $[[x]]$ of Exercise 36 in Section 1.5 have a limit, a right-hand limit, or a left-hand limit as x approaches integer values?

42. Prove or disprove the following statement: If

$f(x) < g(x)$ for all $x \neq a$, then

$$\lim_{x\to a} f(x) < \lim_{x\to a} g(x).$$

43. If n is a positive integer, evaluate

$$\lim_{h\to 0} \frac{(x+h)^n - x^n}{h}.$$

Hint: Use either the binomial theorem or the result that

$$a^n - b^n = (a-b)(a^{n-1} + a^{n-2}b + \cdots + ab^{n-2} + b^{n-1}).$$

44. Evaluate

$$\lim_{h\to 0} \frac{\sqrt{x+h} - \sqrt{x}}{h}.$$

45. The weight W of an object depends on its distance d from the centre of the earth. If d is less than the radius R of the earth, then W is directly proportional to d; and if d is greater than or equal to R, then W is inversely proportional to d^2. If the weight of the object on the earth's surface is W_0, find a definition for W as a function of d, and sketch its graph.

46. At the present time it is impossible for us to calculate algebraically

$$\lim_{x\to 0} \frac{\sin x - x}{x^3}.$$

What can we do? One suggestion might be to use a calculator to evaluate the function $(\sin x - x)/x^3$ for various values of x that approach 0. Try this with $x = 0.1, 0.001, 0.00001, 0.0000001, 0.000000001$, and make any conclusion you feel is justified. How would you feel about the use of the calculator if you knew that the value of the limit were $-1/6$?

47. If

$$\lim_{x\to a} \frac{f(x) - g(x)}{x - a}$$

exists, and $\lim_{x\to a} f(x) = L$, find $\lim_{x\to a} g(x)$.

48. If $f(x)$ is an even function and $\lim_{x\to a} f(x) = L$, find, if possible, $\lim_{x\to -a} f(x)$.

49. If $f(x)$ is an even function and $\lim_{x\to a^+} f(x) = L$, find, if possible, $\lim_{x\to -a^-} f(x)$.

50. If $f(x)$ is an even function and $\lim_{x\to a^+} f(x) = L$, find, if possible, $\lim_{x\to -a^+} f(x)$.

51.–53. Repeat Exercises 48.–50. for an odd function $f(x)$.

SECTION 2.2

Infinite Limits

Functions do not always have limits. Sometimes this is due to the fact that right-hand and left-hand limits are not identical; sometimes it is a result of erratic oscillations. Examples of both of these situations were discussed in Section 2.1 (see Examples 2.9 and 2.8). Nonexistence of a limit may also be due to excessively large values of the function. For instance, consider the function $f(x) = 1/(x-2)^2$ which is not defined at $x = 2$. Does it have a limit as x approaches 2? Table 2.2 indicates that as x approaches 2, values of $1/(x-2)^2$ become very large; in fact, values of the function can be made arbitrarily large by choosing x sufficiently close to 2. Thus, the function does not have a limit as x approaches 2. To express this fact we write

$$\lim_{x\to 2} \frac{1}{(x-2)^2} = \infty,$$

where ∞ once again represents infinity. This statement does not say that the limit exists, since ∞ is not a number. It states that the *limit does not exist*, and indicates that the reason it does not exist is that values of the function become arbitrarily large as x approaches 2. The graph of $f(x)$ in Figure 2.8 further illustrates this point. We say that the line $x = 2$ is a **vertical asymptote** for the graph.

TABLE 2.2

x	$f(x)$	x
1.9	10^2	2.1
1.99	10^4	2.01
1.999	10^6	2.001
1.9999	10^8	2.0001
1.99999	10^{10}	2.00001

FIGURE 2.8

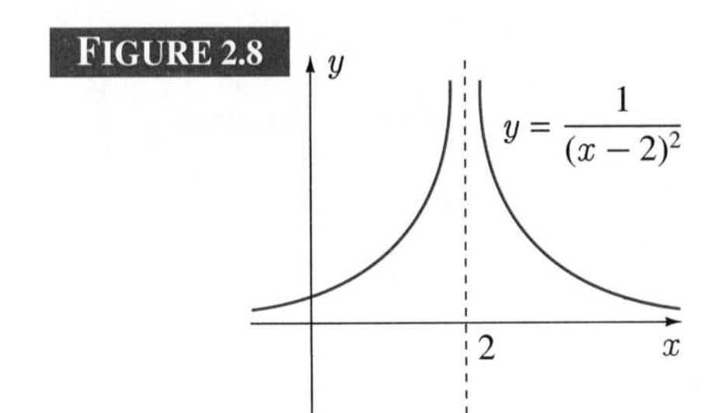

EXAMPLE 2.10

Evaluate $\lim_{x\to 1}[1/(x-1)]$, if it exists.

SOLUTION In this example we consider right-hand and left-hand limits as x approaches 1. We find that

$$\lim_{x\to 1^+} \frac{1}{x-1} = \infty \qquad \text{and} \qquad \lim_{x\to 1^-} \frac{1}{x-1} = -\infty,$$

the latter meaning that as x approaches 1 from the left, the function takes on arbitrarily "large" negative values. Either one of these expressions is sufficient to conclude that the function $1/(x-1)$ does not have a limit as x approaches 1. The function is shown in Figure 2.9; the line $x = 1$ is a vertical asymptote.

FIGURE 2.9

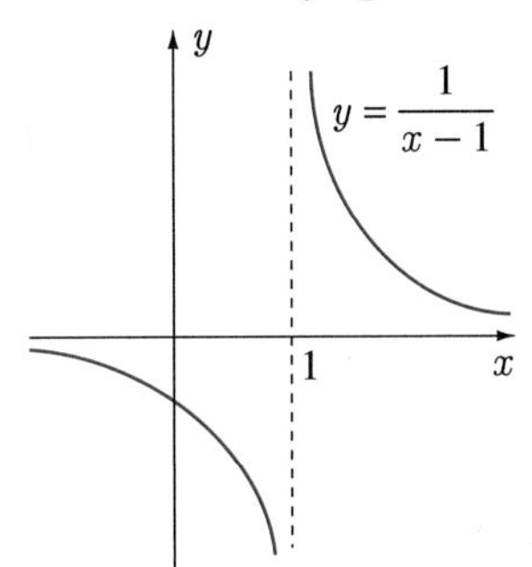

■

EXAMPLE 2.11

What are left-hand and right-hand limits of the function

$$f(x) = \frac{x^2 - 9}{x^2 + x - 2}$$

at $x = 1$ and $x = -2$?

SOLUTION The function is undefined at $x = 1$ and $x = -2$ (why?). To take limits as $x \to 1$ and $x \to -2$, it is advantageous to factor numerator and denominator of $f(x)$ as much as possible,

$$f(x) = \frac{(x+3)(x-3)}{(x+2)(x-1)}.$$

Consider now the (right-hand) limit as $x \to 1^+$. Each of the four factors has a limit as $x \to 1^+$:

$$x + 3 \to 4, \qquad x - 3 \to -2, \qquad x + 2 \to 3, \qquad x - 1 \to 0.$$

Were we to combine them according to Theorem 2.1, we would write

$$\frac{4(-2)}{3(0)}.$$

In spite of the fact that this is not correct — part (v) of the theorem does not allow a 0 in the denominator, this expression does tell us what is happening to $f(x)$ as $x \to 1^+$. It states that as $x \to 1^+$, the numerator of $f(x)$ approaches -8 and the denominator approaches 0. But this means that $f(x)$ must be taking on larger and larger values as $x \to 1^+$. Are these values positive or negative? The numerator is clearly negative, but the sign of the denominator depends on whether the 0 is approached through positive or negative numbers. Recalling that 0 arose from the fact that $x - 1 \to 0$ as $x \to 1^+$, we can be more specific; $x-1$ must approach 0 through positive numbers since $x > 1$ for $x \to 1^+$. We indicate this by writing $x - 1 \to 0^+$as $x \to 1^+$. The above fraction is therefore replaced by

$$\frac{4(-2)}{3(0^+)},$$

and it is clearly negative. In other words,

$$\lim_{x \to 1^+} \frac{x^2 - 9}{x^2 + x - 2} = \lim_{x \to 1^+} \frac{(x+3)(x-3)}{(x+2)(x-1)} = -\infty.$$

Similarly, as $x \to 1^-$, the fraction

$$\frac{4(-2)}{3(0^-)},$$

which shows limits of the four factors of $f(x)$, indicates that

$$\lim_{x \to 1^-} \frac{(x+3)(x-3)}{(x+2)(x-1)} = \infty.$$

In the following, the expressions at the right of the page yield limits as $x \to -2^+$ and $x \to -2^-$:

$$\lim_{x \to -2^+} \frac{(x+3)(x-3)}{(x+2)(x-1)} = \infty, \qquad \frac{(1)(-5)}{(0^+)(-3)}$$

$$\lim_{x \to -2^-} \frac{(x+3)(x-3)}{(x+2)(x-1)} = -\infty. \qquad \frac{(1)(-5)}{(0^-)(-3)}$$

These expressions will be useful when limits are used to sketch a graph of this function in Section 2.4. ■

EXERCISES 2.2

In Exercises 1–24 evaluate the limit, if it exists.

1. $\lim_{x\to 2^+} \dfrac{1}{x-2}$

2. $\lim_{x\to 2^-} \dfrac{1}{x-2}$

3. $\lim_{x\to 2} \dfrac{1}{x-2}$

4. $\lim_{x\to 2^+} \dfrac{1}{(x-2)^2}$

5. $\lim_{x\to 2^-} \dfrac{1}{(x-2)^2}$

6. $\lim_{x\to 2} \dfrac{1}{(x-2)^2}$

7. $\lim_{x\to 1} \dfrac{5x}{(x-1)^3}$

8. $\lim_{x\to 1/2} \dfrac{6x^2+7x-5}{2x-1}$

9. $\lim_{x\to 1} \dfrac{2x+3}{x^2-2x+1}$

10. $\lim_{x\to 2} \dfrac{x-2}{x^2-4x+4}$

11. $\lim_{x\to \pi/2^+} \tan x$

12. $\lim_{x\to \pi/2^-} \tan x$

13. $\lim_{x\to 0} \csc x$

14. $\lim_{x\to \pi/4} \sec(x-\pi/4)$

15. $\lim_{x\to 3\pi/4} \sec(x-\pi/4)$

16. $\lim_{x\to 0^+} \cot x$

17. $\lim_{x\to 1} \dfrac{x^2-2x+1}{x^3-3x^2+3x-1}$

18. $\lim_{x\to 0} \dfrac{\sqrt{1+x}-1}{x^2}$

19. $\lim_{x\to 0} \dfrac{2x}{1-\sqrt{x^2+1}}$

20. $\lim_{x\to 4} \dfrac{|4-x|}{x^2-8x+16}$

21. $\lim_{x\to 0^+} \ln(4x)$

22. $\lim_{x\to 1} \dfrac{1}{\ln|x-1|}$

23. $\lim_{x\to 0} e^{1/x}$

24. $\lim_{x\to 0} e^{1/|x|}$

25. It is not clear whether the limit

$$\lim_{x\to 0^+} x^2 \ln x$$

exists due to the fact that $\lim_{x\to 0^+} x^2 = 0$ and $\lim_{x\to 0^+} \ln x = -\infty$. It depends on which term is more dominant in the product, x^2 or $\ln x$. Calculate x^2, $\ln x$, and $x^2 \ln x$ for $x = 10^{-2}$, 10^{-4}, 10^{-6}, and 10^{-8}, and use this information to decide on a value for the limit.

26. Repeat Exercise 25 for the limit

$$\lim_{x\to 0^+} x^{10} e^{1/x},$$

but pick your own values of x at which to evaluate x^{10} and $e^{1/x}$.

SECTION 2.3

Limits at Infinity

In many applications we are concerned with the behaviour of a function as its independent variable takes on very large values, positively or negatively. For instance, consider finding, if possible, a number that the function $f(x) = (2x^2+3)/(x^2+1)$, with domain $x \geq 0$, gets closer and closer to as x becomes larger and larger and larger. When x is very large, it is certainly not equal to zero, and we may therefore divide numerator and denominator by x^2:

$$f(x) = \frac{2x^2+3}{x^2+1} = \frac{2+\dfrac{3}{x^2}}{1+\dfrac{1}{x^2}}.$$

For very large x, the terms $3/x^2$ and $1/x^2$ are very close to zero, and therefore $f(x)$ is approximately equal to 2. Indeed $f(x)$ can be made arbitrarily close to 2 by choosing x sufficiently large. In calculus we express this fact by saying **the limit of** $(2x^2+3)/(x^2+1)$ **as x approaches infinity is** 2, and write

$$\lim_{x\to\infty} \frac{2x^2+3}{x^2+1} = 2.$$

Once again we stress that ∞ is not a number. The notation $x \to \infty$ simply means "as x gets larger and larger and larger". The graph of this function is shown in Figure 2.10. We say that the line $y = 2$ is a **horizontal asymptote**.

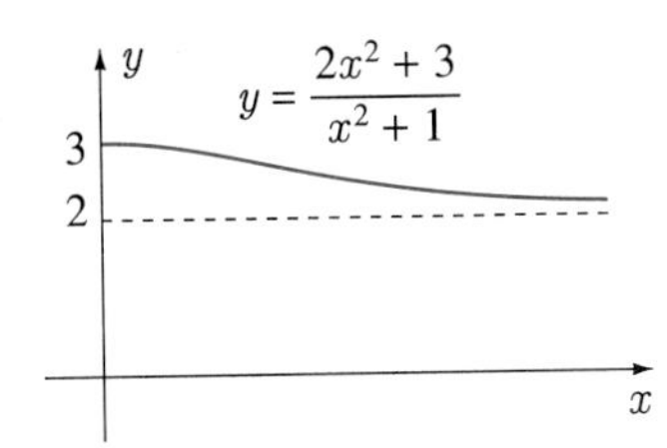

FIGURE 2.10

We may also find limits of functions as x takes on arbitrarily large negative numbers, denoted $x \to -\infty$.

EXAMPLE 2.12

Evaluate

$$\lim_{x \to -\infty} \frac{5x^4 - 3x + 5}{x^4 - 2x^2 + 5},$$

if it exists.

SOLUTION Division of numerator and denominator by x^4 leads to

$$\lim_{x \to -\infty} \frac{5 - \dfrac{3}{x^3} + \dfrac{5}{x^4}}{1 - \dfrac{2}{x^2} + \dfrac{5}{x^4}} = 5.$$

■

If the limit in the above example is taken as $x \to \infty$, the same result is obtained. The graph of this function is shown in Figure 2.11. The line $y = 5$ is a horizontal asymptote.

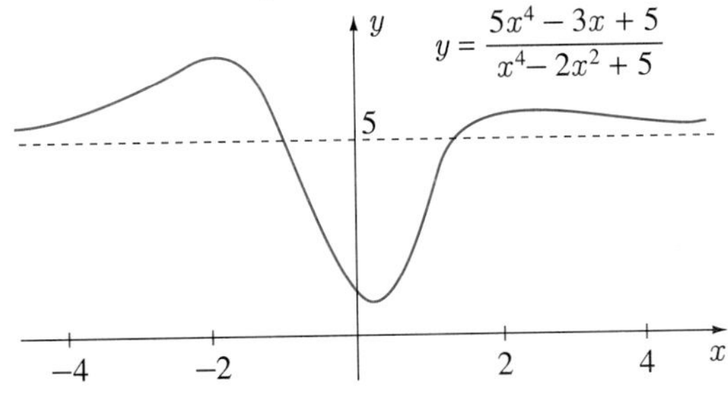

FIGURE 2.11

In general we say that a line $y = L$ is a horizontal asymptote for the graph of a function $f(x)$ if either of the following situations exists:

$$\lim_{x \to -\infty} f(x) = L \qquad \text{or} \qquad \lim_{x \to \infty} f(x) = L.$$

In Example 2.12, both of these conditions are satisfied for $L = 5$.

EXAMPLE 2.13

Evaluate the following limits:

(a) $\displaystyle\lim_{x \to \infty} \frac{2x^3 - 4}{x^3 + x^2 + 2}$ (b) $\displaystyle\lim_{x \to -\infty} \frac{2x^2 - 14}{3x^3 + 5x}$ (c) $\displaystyle\lim_{x \to \infty} \frac{2x^4 - 14}{3x^3 + 5x}$

SOLUTION (a) To obtain this limit we divide numerator and denominator of the fraction by x^3:

$$\lim_{x\to\infty} \frac{2x^3 - 4}{x^3 + x^2 + 2} = \lim_{x\to\infty} \frac{2 - \dfrac{4}{x^3}}{1 + \dfrac{1}{x} + \dfrac{2}{x^3}} = 2.$$

(b) Once again we divide numerator and denominator by x^3:

$$\lim_{x\to-\infty} \frac{2x^2 - 14}{3x^3 + 5x} = \lim_{x\to-\infty} \frac{\dfrac{2}{x} - \dfrac{14}{x^3}}{3 + \dfrac{5}{x^2}}.$$

Since the numerator approaches 0 and the denominator approaches 3, we conclude that

$$\lim_{x\to-\infty} \frac{2x^2 - 14}{3x^3 + 5x} = 0.$$

We could also have obtained this limit by dividing numerator and denominator by x^2 instead of x^3,

$$\lim_{x\to-\infty} \frac{2x^2 - 14}{3x^3 + 5x} = \lim_{x\to-\infty} \frac{2 - \dfrac{14}{x^2}}{3x + \dfrac{5}{x}}.$$

Now the numerator approaches 2, but since the denominator becomes very large, the fraction once again approaches zero.

(c) Division by x^3 in this case gives

$$\lim_{x\to\infty} \frac{2x^4 - 14}{3x^3 + 5x} = \lim_{x\to\infty} \frac{2x - \dfrac{14}{x^3}}{3 + \dfrac{5}{x^2}}.$$

Since the numerator becomes arbitrarily large as $x \to \infty$ and the denominator approaches 3, it follows that

$$\lim_{x\to\infty} \frac{2x^4 - 14}{3x^3 + 5x} = \infty.$$

■

EXAMPLE 2.14

Sketch a graph of the function

$$f(x) = \frac{1}{x}\sin x, \quad x \geq \pi.$$

Does it have a horizontal asymptote?

SOLUTION Multiplication of ordinates of the graphs of $y = 1/x$ and $y = \sin x$ in Figure 2.12(a) gives the graph in Figure 2.12(b). The positive x-axis is a horizontal asymptote. The graph actually crosses the asymptote an infinite number of times.

FIGURE 2.12

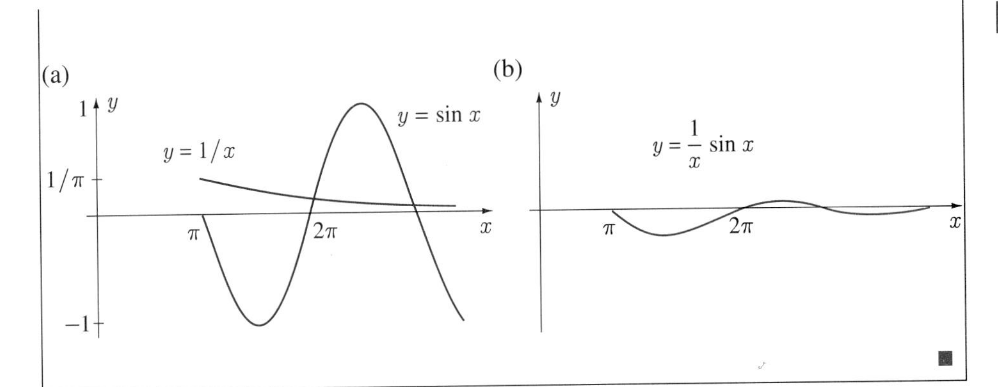

EXAMPLE 2.15

Evaluate

$$\lim_{x\to\infty} \frac{\sqrt{2x^2+4}}{x+5},$$

if it exists. What is the limit as $x \to -\infty$?

SOLUTION When we divide numerator and denominator by x, and take the x inside the square root as x^2,

$$\lim_{x\to\infty} \frac{\sqrt{2x^2+4}}{x+5} = \lim_{x\to\infty} \frac{\frac{1}{x}\sqrt{2x^2+4}}{\frac{1}{x}(x+5)} = \lim_{x\to\infty} \frac{\sqrt{\frac{2x^2+4}{x^2}}}{1+\frac{5}{x}}$$

$$= \lim_{x\to\infty} \frac{\sqrt{2+\frac{4}{x^2}}}{1+\frac{5}{x}} = \sqrt{2}.$$

In evaluating the limit as $x \to -\infty$, we must be more careful,

$$\lim_{x\to-\infty} \frac{\sqrt{2x^2+4}}{x+5} = \lim_{x\to-\infty} \frac{\frac{\sqrt{2x^2+4}}{x}}{1+\frac{5}{x}}.$$

It is not correct in this case to take the x inside the square root as

$$\frac{\sqrt{2x^2+4}}{x} = \sqrt{\frac{2x^2+4}{x^2}}$$

since for negative x, the expression on the left is negative and that on the right is positive. In this case, we should replace x by $-\sqrt{x^2}$, and write

$$\frac{\sqrt{2x^2+4}}{x} = \frac{\sqrt{2x^2+4}}{-\sqrt{x^2}} = -\sqrt{\frac{2x^2+4}{x^2}}.$$

Hence,

$$\lim_{x\to-\infty}\frac{\sqrt{2x^2+4}}{x+5}=\lim_{x\to-\infty}\frac{-\sqrt{\dfrac{2x^2+4}{x^2}}}{1+\dfrac{5}{x}}=\lim_{x\to-\infty}\frac{-\sqrt{2+\dfrac{4}{x^2}}}{1+\dfrac{5}{x}}=-\sqrt{2}.$$

The graph of this function (Figure 2.13) has two horizontal asymptotes, $y=\sqrt{2}$ and $y=-\sqrt{2}$.

FIGURE 2.13

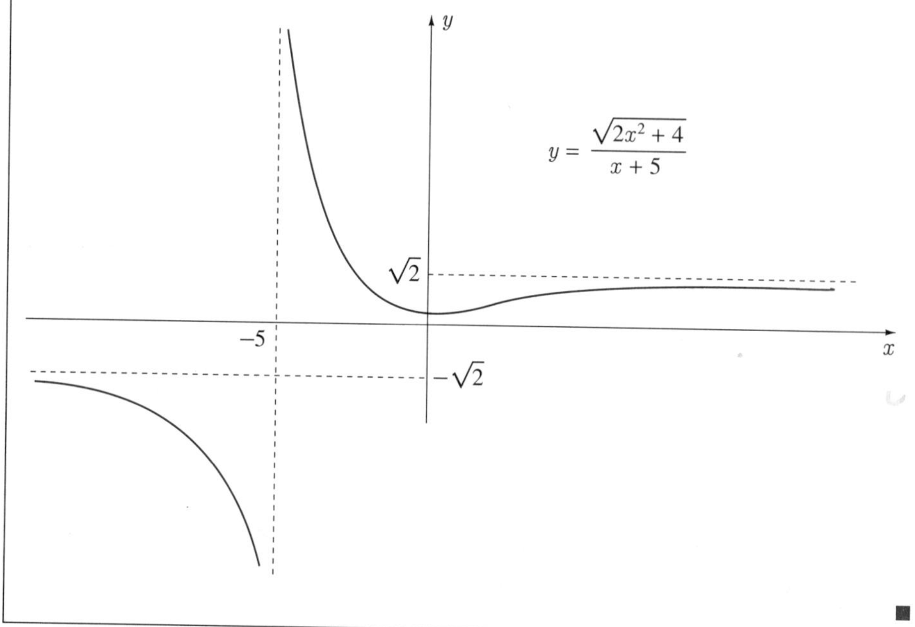

■

EXERCISES 2.3

In Exercises 1–38 evaluate the limit, if it exists.

1. $\lim_{x\to\infty}\frac{x+1}{2x-1}$
2. $\lim_{x\to\infty}\frac{1-x}{3+2x}$
3. $\lim_{x\to\infty}\frac{x^2+1}{2x^3+5}$
4. $\lim_{x\to\infty}\frac{1-4x^3}{3+2x-x^2}$
5. $\lim_{x\to-\infty}\frac{2+x-x^2}{3+4x^2}$
6. $\lim_{x\to-\infty}\frac{x^3-2x^2}{3x^3+4x^2}$
7. $\lim_{x\to-\infty}\frac{x^3-2x^2+x+1}{x^4+3x}$
8. $\lim_{x\to-\infty}\frac{x^3-2x^2+x+1}{x^2-x+1}$
9. $\lim_{x\to\infty}\frac{\sqrt{x^2+1}}{2x+1}$
10. $\lim_{x\to\infty}\frac{3x-1}{\sqrt{5+4x^2}}$
11. $\lim_{x\to-\infty}\frac{\sqrt{1-2x^2}}{x+2}$
12. $\lim_{x\to-\infty}\frac{\sqrt{1-2x}}{x+2}$
13. $\lim_{x\to\infty}\sqrt{\frac{2+x}{x-2}}$
14. $\lim_{x\to\infty}\frac{\sqrt{3+x}}{\sqrt{x}}$
15. $\lim_{x\to\infty}(x^2-x^3)$
16. $\lim_{x\to\infty}\left(x+\frac{1}{x}\right)$
17. $\lim_{x\to\infty}\frac{x}{\sqrt{x+5}}$
18. $\lim_{x\to-\infty}\frac{x^2}{\sqrt{3-x}}$
19. $\lim_{x\to-\infty}\frac{x}{\sqrt[3]{4+x^3}}$
20. $\lim_{x\to\infty}\frac{3x}{\sqrt[3]{2+4x^3}}$
21. $\lim_{x\to\infty}\frac{1}{2x}\cos x$
22. $\lim_{x\to-\infty}\frac{1}{2x}\cos x$
23. $\lim_{x\to\infty}\frac{\sin 4x}{x^2}$
24. $\lim_{x\to\infty}\frac{\sin^2 x}{x}$
25. $\lim_{x\to-\infty}\tan x$
26. $\lim_{x\to\infty}\frac{1}{x}\tan x$
27. $\lim_{x\to\infty}\left(\sqrt{x^2+1}-x\right)$
28. $\lim_{x\to\infty}\left(\sqrt{x^2+4}-x\right)$
29. $\lim_{x\to\infty}\left(\sqrt{2x^2+1}-x\right)$
30. $\lim_{x\to-\infty}\left(\sqrt{2x^2+1}-x\right)$

31. $\lim_{x\to\infty} \dfrac{\sqrt{3x^2+2}}{x+4}$ **32.** $\lim_{x\to\infty} \dfrac{\sqrt{4x^2+7}}{2x+3}$

33. $\lim_{x\to-\infty} \dfrac{\sqrt{3x^2+2}}{x+4}$ **34.** $\lim_{x\to-\infty} \dfrac{\sqrt{4x^2+7}}{2x+3}$

35. $\lim_{x\to\infty} \left(\sqrt{x^2+4} - \sqrt{x^2-1}\right)$

36. $\lim_{x\to\infty} \left(\sqrt[3]{1+x} - \sqrt[3]{x}\right)$

37. $\lim_{x\to\infty} \left(\sqrt{x^2+x} - x\right)$ **38.** $\lim_{x\to-\infty} \left(\sqrt{x^2+x} - x\right)$

39. What is the value of

$$\lim_{x\to-\infty} \frac{\sqrt{ax^2+bx+c}}{dx+e}$$

where $a > 0$, b, c, d, and e are constants?

40. What conditions on the constants a, b, c, d, e, and f will ensure that the

$$\lim_{x\to\infty} \left(\sqrt{ax^2+bx+c} - \sqrt{dx^2+ex+f}\right)$$

exists? What is the value of the limit in this case?

SECTION 2.4

Continuity

We have noticed that sometimes the limit of a function $f(x)$ as $x \to a$ is the same as the value $f(a)$ of the function at $x = a$. This property is described in the following definition.

Definition 2.1

A function $f(x)$ is said to be **continuous** at $x = a$ if

$$\lim_{x\to a} f(x) = f(a). \tag{2.8}$$

There are three conditions imposed by this definition:

(1) $f(x)$ must be defined at $x = a$;

(2) $\lim_{x\to a} f(x)$ must exist;

(3) the numbers in (1) and (2) must be the same.

The graph in Figure 2.14 illustrates a function that is discontinuous (that is, not continuous) at $x = a$, b, c, d, and e. For instance, at $x = a$ conditions (1) and (2) are satisfied but (3) is violated; at $x = b$, (2) is satisfied, but (1) and (3) are not. Figure 2.14 suggests that discontinuities of a function are characterized geometrically by separations in its graph. This is indeed true, and it is often a very informative way to illustrate the nature of a discontinuity.

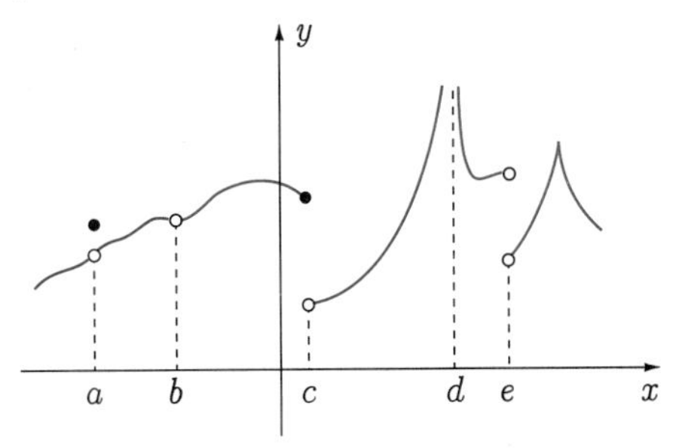

FIGURE 2.14

When a function is defined on a closed interval $a \le x \le b$, Definition 2.1 must be rephrased in terms of right- and left-hand limits for continuity at $x = a$ and $x = b$. Specifically, we say that $f(x)$ is **continuous from the right** at $x = a$ if $\lim_{x\to a^+} f(x) = f(a)$; and that $f(x)$ is **continuous from the left** at $x = b$ if $\lim_{x\to b^-} f(x) = f(b)$.

A function is said to be **continuous on an interval** if it is continuous at each point of that interval. In the event that the interval is closed, $a \le x \le b$, continuity at $x = a$ and $x = b$ is interpreted as continuity from the right and left, respectively. Geometrically, a function is continuous on an interval if a pencil can completely trace its graph without being lifted from the page.

EXAMPLE 2.16

Sketch graphs of the following functions indicating any discontinuities:

(a) $f(x) = \dfrac{x^2 - 16}{x - 4}$ (b) $f(x) = \dfrac{1}{(x-4)^2}$ (c) $f(x) = \dfrac{|x^2 - 25|}{x^2 - 25}$

SOLUTION (a) Since $f(x)$ is undefined at $x = 4$, the function is discontinuous there. For $x \ne 4$,

$$y = f(x) = \frac{(x+4)(x-4)}{x-4} = x + 4.$$

Consequently, the graph of $f(x)$ is a straight line with a hole at $x = 4$ (Figure 2.15). Note that $\lim_{x\to 4} [(x^2 - 16)/(x-4)]$ exists and is equal to 8.

FIGURE 2.15

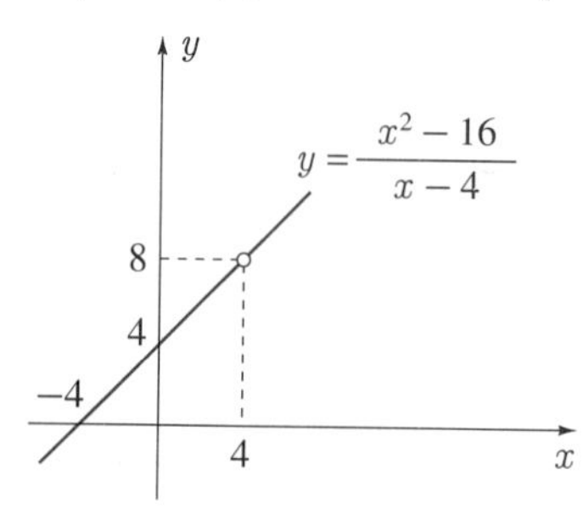

(b) Since $1/(x-4)^2$ is undefined at $x = 4$, the function is discontinuous there. The limits

$$\lim_{x\to 4^+} \frac{1}{(x-4)^2} = \infty, \qquad \lim_{x\to 4^-} \frac{1}{(x-4)^2} = \infty,$$

$$\lim_{x\to \infty} \frac{1}{(x-4)^2} = 0, \qquad \lim_{x\to -\infty} \frac{1}{(x-4)^2} = 0$$

and the fact that $f(x)$ is always positive lead to the graph in Figure 2.16.

FIGURE 2.16

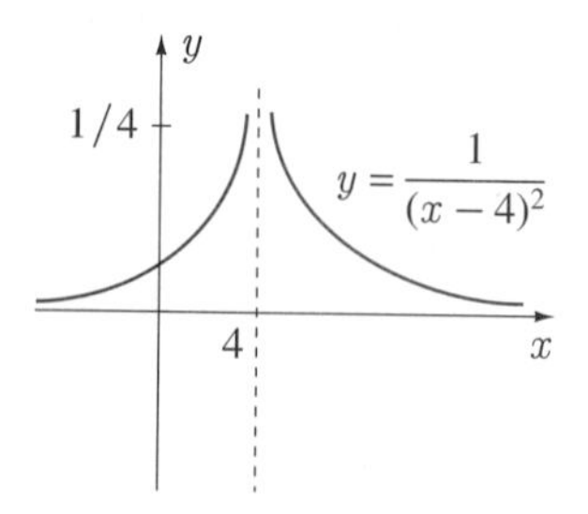

(c) The function $f(x) = |x^2 - 25|/(x^2 - 25)$ is undefined at $x = \pm 5$, and is therefore discontinuous at these values. When $-5 < x < 5$,

$$f(x) = \frac{25 - x^2}{x^2 - 25} = -1,$$

and when $|x| > 5$,

$$f(x) = \frac{x^2 - 25}{x^2 - 25} = 1.$$

The graph is shown in Figure 2.17. In this example, neither of the following limits exists:

$$\lim_{x \to 5} \frac{|x^2 - 25|}{x^2 - 25} \quad \text{or} \quad \lim_{x \to -5} \frac{|x^2 - 25|}{x^2 - 25}.$$

FIGURE 2.17

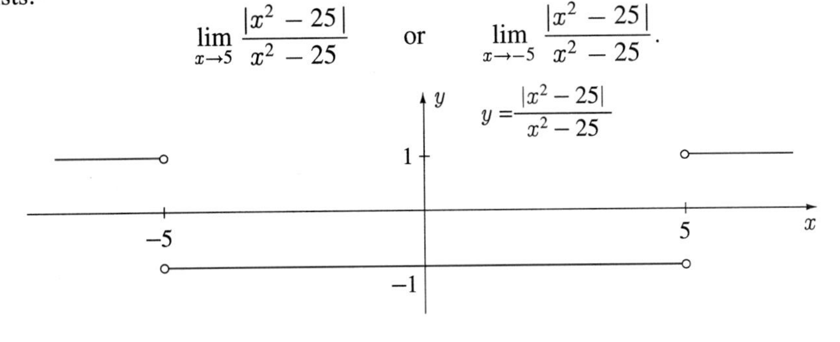

■

In Chapter 1 we introduced various techniques for sketching curves — addition and multiplication of ordinates, symmetries, and translations to name a few. Right-hand limits, left-hand limits, limits at infinity, and infinite limits can all be useful in sketching graphs of functions or curves. Part (b) of Example 2.16 illustrates this, and the following two examples contain further illustrations.

EXAMPLE 2.17

Sketch a graph of the function $f(x) = (3x - 6)/(x^2 + 5)$.

SOLUTION We first note that the graph has a horizontal asymptote given by

$$y = \lim_{x \to \infty} \frac{3x - 6}{x^2 + 5} = \lim_{x \to \infty} \frac{3 - \frac{6}{x}}{x + \frac{5}{x}} = 0.$$

Figure 2.18 shows two ways that $y = 0$ could be a horizontal asymptote as $x \to \infty$.

FIGURE 2.18

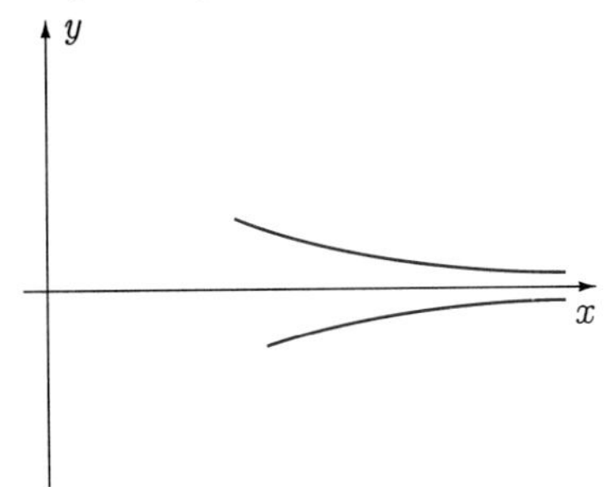

We need to decide whether $f(x) \to 0$ through positive or negative values. But this is easy to decide. For large positive x, $3 - 6/x$ is clearly positive, as is $x + 5/x$, and therefore $f(x)$ must approach 0 through positive numbers. We indicate this by writing

$$\lim_{x\to\infty} \frac{3x-6}{x^2+5} = 0^+.$$

Similarly, for large negative values of x,

$$\lim_{x\to-\infty} \frac{3x-6}{x^2+5} = \lim_{x\to-\infty} \frac{3 - \dfrac{6}{x}}{x + \dfrac{5}{x}} = 0^-$$

as the numerator is positive and the denominator is negative for large negative x. This information is shown in Figure 2.19(a), where we have also added the points $(-1, -3/2)$ and $(4, 2/7)$ and x- and y-intercepts of 2 and $-6/5$, respectively. There are many ways we could finish the graph; the simplest, and correct, way is shown in Figure 2.19(b). In Chapter 3, when we deal with differentiation, verification that this is indeed the correct graph will be straightforward. We will be able to show that the lowest point on the graph is at $x = -1$ and the highest point is at $x = 5$.

FIGURE 2.19

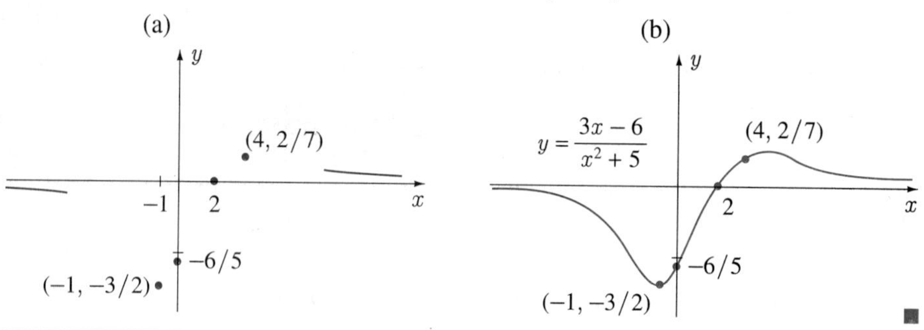

EXAMPLE 2.18

Sketch a graph of the function $f(x) = (x^2 - 9)/(x^2 + x - 2)$.

SOLUTION This function is discontinuous at $x = 1$ and $x = -2$. We should therefore calculate limits of $f(x)$ as $x \to -2^-$, $x \to -2^+$, $x \to 1^-$, and $x \to 1^+$. This was done in Example 2.11, the results being

$$\lim_{x\to1^+} f(x) = -\infty, \quad \lim_{x\to1^-} f(x) = \infty, \quad \lim_{x\to-2^+} f(x) = \infty, \quad \lim_{x\to-2^-} f(x) = -\infty.$$

We have shown these limits in Figure 2.20(a) together with a y-intercept of $9/2$, and x-intercepts of ±3. We now calculate limits as $x \to \pm\infty$. First we note that

$$\lim_{x\to\infty} \frac{x^2-9}{x^2+x-2} = \lim_{x\to\infty} \frac{1 - \dfrac{9}{x^2}}{1 + \dfrac{1}{x} - \dfrac{2}{x^2}} = 1.$$

To commit this to a graph we must know whether 1 is approached through numbers larger than 1 or smaller than 1. This is most easily determined by noting that for very large x, the 9 and 2 in $f(x)$ are insignificant relative to x^2 and x, and therefore for such x, $f(x)$ is very closely approximated by

$$f(x) \approx \frac{x^2}{x^2 + x}.$$

But clearly the numerator is less than the denominator, and we conclude that

$$\lim_{x \to \infty} \frac{x^2 - 9}{x^2 + x - 2} = 1^-.$$

On the other hand, for large negative values of x, the numerator of $x^2/(x^2 + x)$ is greater than its denominator, and therefore

$$\lim_{x \to -\infty} \frac{x^2 - 9}{x^2 + x - 2} = 1^+.$$

These facts are all shown in Figure 2.20(b). The graph is completed in Figure 2.20(c).

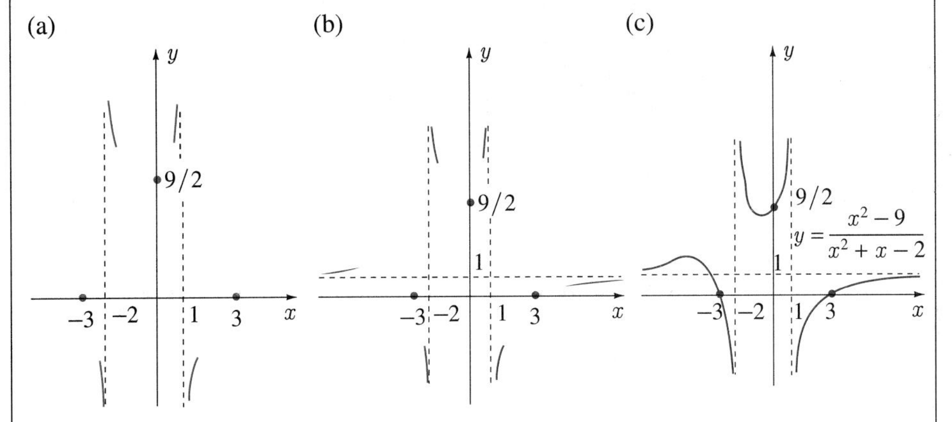

FIGURE 2.20

Once again determination of the exact positions of the high point to the left of $x = -2$ and the low point between $x = -2$ and $x = 1$ requires differentiation from Chapter 3. ■

According to the following theorem, when continuous functions are added, subtracted, multiplied, and divided, the result is a continuous function.

Theorem 2.2

If functions $f(x)$ and $g(x)$ are continuous at $x = a$, then so also are the functions $f(x) \pm g(x)$, $f(x)g(x)$, and $f(x)/g(x)$ (provided $g(a) \neq 0$ in the case of division).

This is easily established with Theorem 2.1. For instance, to verify that $f(x) + g(x)$ is continuous, we note that since $\lim_{x \to a} f(x) = f(a)$ and $\lim_{x \to a} g(x) = g(a)$,

$$\lim_{x \to a} [f(x) + g(x)] = \lim_{x \to a} f(x) + \lim_{x \to a} g(x) \qquad \text{(by Theorem 2.1)}$$

$$= f(a) + g(a) \qquad \Big(\text{by continuity of } f(x) \text{ and } g(x)\Big).$$

Corollary *If functions $f(x)$ and $g(x)$ are continuous on an interval, then so also are the functions $f(x) \pm g(x)$, $f(x)g(x)$, and $f(x)/g(x)$ (provided $g(x)$ never vanishes in the interval in the case of division).*

It is an immediate consequence of this corollary that polynomials are continuous for all real numbers. Rational functions $P(x)/Q(x)$, where P and Q are polynomials, are continuous on intervals in which $Q(x) \neq 0$.

Besides adding, subtracting, multiplying, and dividing functions, functions can also be "composed", or substituted one into another. For instance, when $f(x) = \sqrt{x + 1/x}$ and $g(x) = x^2 + 3$, the function obtained by replacing each x in $f(x)$ by $g(x)$ is called the **composition** of f and g. It is often denoted by $(f \circ g)$, so that

$$(f \circ g)(x) = f\big(g(x)\big) = \sqrt{x^2 + 3 + \frac{1}{x^2 + 3}}.$$

Were we required to find the limit of this function as $x \to 1$, say, we would probably write nonchalantly that

$$\lim_{x\to 1} \sqrt{x^2 + 3 + \frac{1}{x^2 + 3}} = \sqrt{4 + \frac{1}{4}} = \frac{\sqrt{17}}{2}.$$

Effectively, we have interchanged the operation of taking square roots and that of taking limits; that is, without thinking we have written

$$\lim_{x\to 1} \sqrt{x^2 + 3 + \frac{1}{x^2 + 3}} = \sqrt{\lim_{x\to 1} \left(x^2 + 3 + \frac{1}{x^2 + 3}\right)} = \sqrt{4 + \frac{1}{4}} = \frac{\sqrt{17}}{2}.$$

This is correct according to the following theorem because the square root function is continuous.

Theorem 2.3

If $\lim_{x\to a} g(x) = L$, and $f(y)$ is a function that is continuous at $y = L$, then

$$\lim_{x\to a} f(g(x)) = f\left[\lim_{x\to a} g(x)\right] = f(L). \tag{2.9}$$

An immediate consequence of this result is that the composition of a continuous function with a continuous function yields a continuous function. We state this as a corollary.

Corollary *If $g(x)$ is continuous at $x = a$, and $f(y)$ is continuous at $g(a)$, then $f(g(x))$ is continuous at $x = a$.*

EXAMPLE 2.19

Evaluate

$$\lim_{x\to 2} \sin\left(\frac{x^2 - 2}{2x - 3}\right).$$

SOLUTION Since the sine function is continuous for all values of its argument, we may take the limit operation inside the function and write

$$\lim_{x\to 2} \sin\left(\frac{x^2 - 2}{2x - 3}\right) = \sin\left[\lim_{x\to 2} \frac{x^2 - 2}{2x - 3}\right] = \sin 2.$$

■

Discontinuous functions are encountered in many applications of mathematics. To illustrate this, consider the phenomenon of radioactive disintegration, during which radioactive elements such as uranium-238 emit high-energy particles and in the process become new elements. If we analyze an ore sample that contains, along with various impurities, an amount A_0 of uranium, then disintegrations of uranium nuclei into thorium-234 nuclei gradually reduce the amount of uranium in the sample. Suppose we let $A = f(t)$ represent the amount of uranium in the sample at any given time, and choose time $t = 0$ when $A = A_0$. If we were to carefully measure the amount of uranium in the sample over a very short time interval (perhaps 10^{-10} second), a graph of $A = f(t)$ might appear as shown in Figure 2.21.

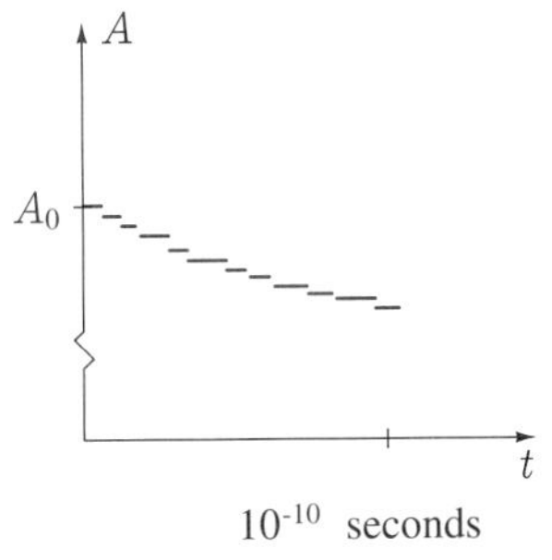

FIGURE 2.21

Each jump (discontinuity) in this graph signals the disintegration of a uranium nucleus into a thorium nucleus. Were we to complete this graph, and assume that no two disintegrations would occur at exactly the same instant, then the total number of discontinuities would be equal to the original number of uranium nuclei in the sample. In practice, we cannot hope to make such infinitesimal measurements. Instead, we must take measurements of the amount of uranium in the sample at much larger time intervals, and plot the data as a sequence of points, as in Figure 2.22(a). If these points are joined with a smooth curve as in Figure 2.22(b), we obtain a reasonable picture of A as a function of time. This macroscopic view (view in the large) of $A = f(t)$ suppresses the true nature of the graph, but it would be entirely adequate if we were interested in knowing, perhaps, when disintegrations would reduce the uranium to half its original amount. Questions such as these will be answered when we continue our discussion of radioactive disintegration in Section 9.2.

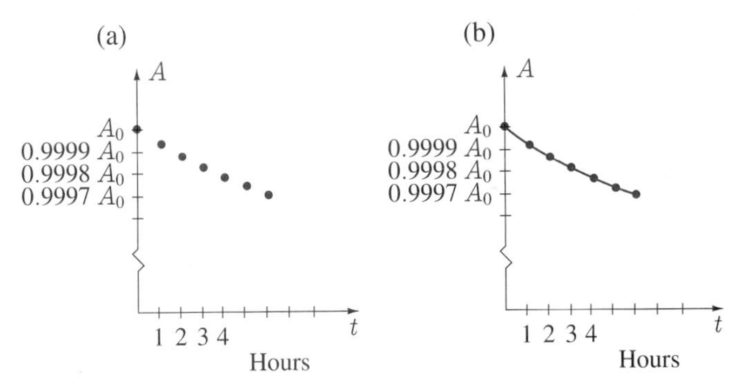

FIGURE 2.22

Our second example is taken from economics. A sugar manufacturer sells sugar in one-kilogram bags at the following prices:

40¢ per bag for the first 100 bags;

35¢ per bag for each bag after 100 up to 1000;

30¢ per bag for each bag after 1000.

If we denote the selling price of the xth bag by $f(x)$, then

$$f(x) = \begin{cases} 40 & 1 \leq x \leq 100 \\ 35 & 101 \leq x \leq 1000 \\ 30 & x \geq 1001. \end{cases}$$

This function is unlike any we have previously seen; its domain consists of only the positive integers. Its "graph", shown in Figure 2.23, therefore consists of dots above these integers. In economics, the domain of the function is extended to all positive values of x according to

$$f(x) = \begin{cases} 40 & 0 < x \leq 100 \\ 35 & 100 < x \leq 1000 \\ 30 & x > 1000, \end{cases}$$

with the resulting graph shown in Figure 2.24.

FIGURE 2.23

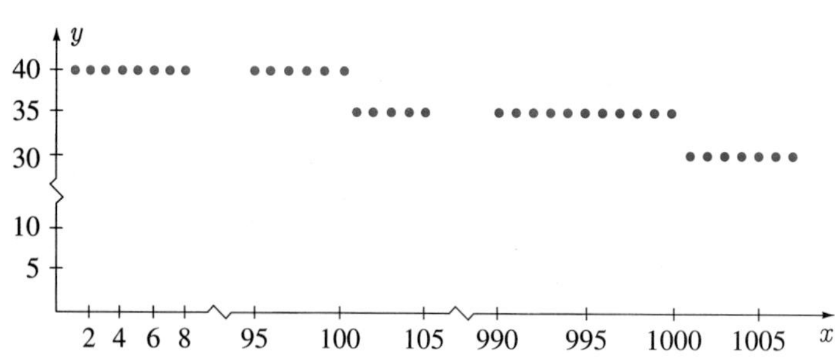

FIGURE 2.24

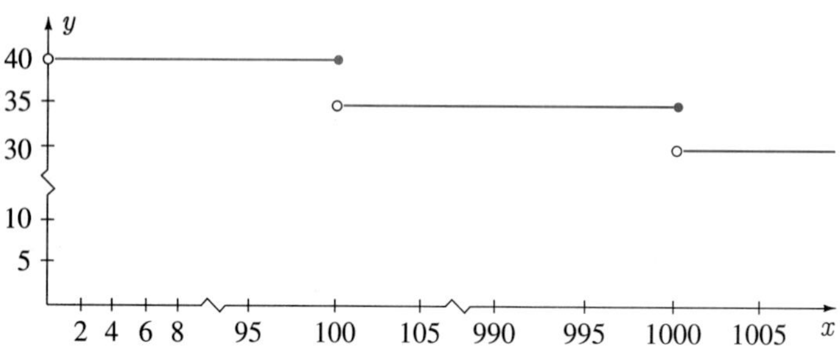

In spite of the fact that $f(x)$ is now defined for all $x > 0$, it still has significance only for x a positive integer. For instance, $f(2.1)$ would indicate the price of a tenth of a bag of sugar after two bags have already been purchased. To say that this fraction of a bag would cost 40¢ is pointless, since the manufacturer doesn't sell sugar in one-tenth kilogram bags. The advantage of extending the domain of $f(x)$ to noninteger values is therefore not one of physical significance. The real reason is that calculus can be applied to the extended function in order to analyze the economics of sugar sales. For the present, it is clear that the function $f(x)$ in Figure 2.24 is discontinuous at $x = 100$ and $x = 1000$.

We can derive from $f(x)$ a function that is far more useful in economic theory; the average price $r(x)$. It is the average price of a one-kilogram bag of sugar when x bags are purchased. Using the fact that average price is total selling price divided by total number of bags, we calculate that

$$r(x) = \begin{cases} 40 & 0 < x \leq 100 \\ \dfrac{40(100) + 35(x - 100)}{x} & 100 < x \leq 1000 \\ \dfrac{40(100) + 35(900) + 30(x - 1000)}{x} & x > 1000 \end{cases}$$

$$= \begin{cases} 40 & 0 < x \le 100 \\ 35 + \dfrac{500}{x} & 100 < x \le 1000 \\ 30 + \dfrac{5500}{x} & x > 1000\,. \end{cases}$$

Again, $r(x)$ is meaningful only for x an integer, but as shown in Figure 2.25 we extend its domain to all positive values of x. In addition, note that unlike $f(x)$, $r(x)$ has no discontinuities for $x > 0$.

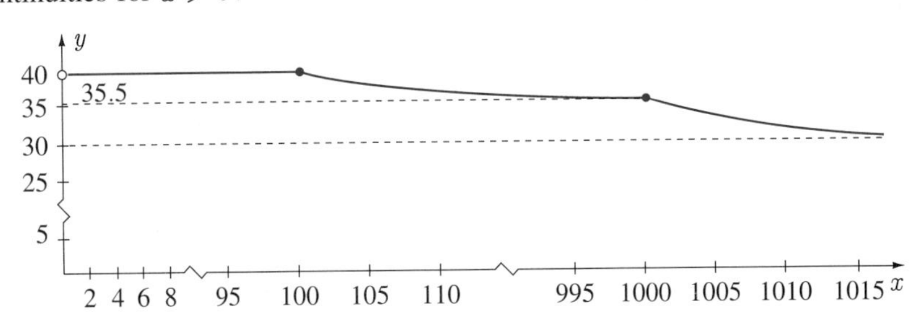

FIGURE 2.25

EXERCISES 2.4

In Exercises 1–30 sketch a graph of the function indicating any discontinuities.

1. $f(x) = \dfrac{1}{x+2}$ **2.** $f(x) = \dfrac{16-x^2}{x+4}$

3. $f(x) = |x^2 - 5|$ **4.** $f(x) = \dfrac{12}{x^2+2}$

5. $f(x) = \dfrac{12}{x^2+2x}$ **6.** $f(x) = \dfrac{12}{x^2+2x+2}$

7. $f(x) = \dfrac{3+2x-x^2}{x+1}$ **8.** $f(x) = \dfrac{x^3+x^2-2x}{x^2-x}$

9. $f(x) = \dfrac{x^3-2x^2+5x-10}{x-2}$

10. $f(x) = \tan x$

11. $f(x) = \sec 2x$ **12.** $f(x) = \sin(1/x)$

13. $f(x) = x^2 \sin(1/x)$ **14.** $f(x) = \dfrac{x+12}{x^2-9}, \quad -5 \le x \le 4$

15. $f(x) = \dfrac{x^2+2x}{x^2-9}, \quad -3 \le x \le 6$

16. $f(x) = \dfrac{x^3-27}{|x-3|}$

17. $f(x) = \dfrac{x}{x^2-1}$ **18.** $f(x) = \dfrac{3x+2}{x^2-x-2}$

19. $f(x) = \dfrac{x^2-3x+2}{x^2+4x-5}$ **20.** $f(x) = \dfrac{3x^2-6x}{x^2-6x-7}$

21. $f(x) = \dfrac{x^2+3x+2}{x+4}$ **22.** $f(x) = \dfrac{x^2-2x+4}{x-1}$

23. $f(x) = \dfrac{1}{\sqrt{x-1}}$ **24.** $f(x) = \dfrac{1}{\sqrt{5+x}}$

25. $f(x) = \dfrac{1}{\sqrt{5-x}}$ **26.** $f(x) = \sqrt{\dfrac{x-3}{x+2}}$

27. $f(x) = \dfrac{1}{x^3-4x}$ **28.** $f(x) = \dfrac{2x^3-2}{x^3+5x^2}$

29. $f(x) = \dfrac{1}{x^4+3x^2}$ **30.** $f(x) = \dfrac{|3x+1|}{x+5}$

31. Is the function

$$f(x) = \begin{cases} x\sin(1/x) & x \ne 0 \\ 0 & x = 0 \end{cases}$$

continuous at $x = 0$?

32. A function $f(x)$ is said to have a *removable discontinuity* at $x = a$ if $\lim_{x\to a} f(x)$ exists, but either $f(a)$ is not defined or $\lim_{x\to a} f(x) \ne f(a)$. Which of the discontinuities in Exercises 1–30 are removable?

33. Can rational functions have removable discontinuities?

34. Verify that the condition

$$\lim_{h\to 0} f(a+h) = f(a)$$

is equivalent to 2.8.

35. The number of bacteria in a culture is a function of time t, say, $N = f(t)$. Make two graphs of this function, one "in the small" to show birth and death situations, and a second "in the large" to get an overview of the rate of growth of the culture.

36. The *signum function* (or *sign function*), denoted by sgn, is defined by

$$\operatorname{sgn} x = \begin{cases} -1 & x < 0 \\ 0 & x = 0 \\ 1 & x > 0. \end{cases}$$

Sketch its graph, indicating any discontinuities.

37. Sketch a graph of the function $f(x) = \operatorname{sgn}(x + 1) - \operatorname{sgn}(x - 1)$, indicating any discontinuities.

38. Three of the most important functions in physics and engineering are defined below. Sketch a graph of each, indicating any discontinuities.

(a) The Heaviside unit step function $H(x - a)$:

$$H(x - a) = \begin{cases} 0 & x < a \\ 1 & x > a. \end{cases}$$

(b) The unit pulse function $P_\epsilon(x - a)$:

$$P_\epsilon(x - a) = \frac{1}{\epsilon}[H(x - a) - H(x - a - \epsilon)].$$

(c) The Dirac delta function $\delta(x - a)$:

$$\delta(x - a) = \lim_{\epsilon \to 0} P_\epsilon(x - a).$$

39. Where is the greatest integer function $[[x]]$ of Exercise 36 in Section 1.5 discontinuous?

40. (a) Sketch a graph of the function $f(x) = [[10x]]/10$. Where is the function discontinuous?

(b) Prove that $f(x)$ truncates positive numbers after the first decimal.

41. What function truncates negative numbers after two decimals?

42. (a) Sketch a graph of the function $f(x) = [[x + 1/2]]$. Where is the function discontinuous?

(b) Prove that $f(x)$ rounds positive numbers to the nearest integer.

43. What function rounds positive numbers to:

(a) the nearest tenth;

(b) the nearest hundredth;

(c) 10^{-n} where n is a positive integer?

44. A toy manufacturer has two pricing policies for one of his dolls:

Policy I: $15.00 each for the first 10 dolls,
$12.50 for each doll after 10 up to 100,
$10.00 for each doll after 100;

Policy II: $15.00 each for the first 10 dolls,
10% off the total price on orders of 11 or more,
20% off the total price on orders of 101 or more.

Sketch average price functions for each of these policies on the same graph. Which policy is better for a retailer?

45. A retailer of building supplies sells sand by the bag and in bulk. He quotes prices as follows:

$3.00 per 50-kilogram bag up to 20 bags,

$50.00 per tonne delivered only in integer tonnes.

If, for example, a customer wants 2.2 tonness of sand, he receives 2 tonnes in bulk and 4 bags. Find an algebraic expression for the average price function and sketch its graph. Hint: The greatest integer function may be useful.

46. Determine points of continuity, if any, for the function

$$f(x) = \begin{cases} 1 & x \text{ is a rational number} \\ 0 & x \text{ is an irrational number.} \end{cases}$$

47. Determine points of continuity, if any, for the function

$$f(x) = \begin{cases} x & x \text{ is a rational number} \\ 0 & x \text{ is an irrational number.} \end{cases}$$

SECTION 2.5

A Mathematical Definition of Limits

Our work on limits has been intuitive, but some later topics in this book require a precise definition of limits. To obtain such a definition, we begin with our intuitive statement of a limit and make a succession of paraphrases, each of which is one step closer to the definition. We do this because the definition of a limit is at first sight quite overwhelming, and we wish to show that it can be obtained by a fairly straightforward sequence of steps. We hope that this will be a convincing argument that the definition of a limit does indeed describe in mathematical terms the concept of a limit. Our intuitive statement is:

- A function $f(x)$ has limit L as x approaches a if $f(x)$ can be made arbitrarily close to L by choosing x sufficiently close to a.

  Next we paraphrase "$f(x)$ can be made arbitrarily close to L".

- A function $f(x)$ has limit L as x approaches a if the difference $|f(x) - L|$ can be made arbitrarily close to zero by choosing x sufficiently close to a.

Now we take the important step — make "$|f(x) - L|$ can be made arbitrarily close to zero" mathematical.

- A function $f(x)$ has limit L as x approaches a if given any real number $\epsilon > 0$, no matter how small, we can make the difference $|f(x) - L|$ less than ϵ by choosing x sufficiently close to a.

  Penultimately, we paraphrase "by choosing x sufficiently close to a".

- A function $f(x)$ has limit L as x approaches a if given any $\epsilon > 0$, we can make $|f(x) - L| < \epsilon$ by choosing $|x - a|$ sufficiently close to zero.

  Finally by making "choosing $|x - a|$ sufficiently close to zero" precise, we arrive at the definition of a limit.

Definition 2.2

A function $f(x)$ has limit L as x approaches a if given any $\epsilon > 0$, we can find a $\delta > 0$ such that

$$|f(x) - L| < \epsilon$$

whenever $0 < |x - a| < \delta$.

Notice that by requiring $0 < |x - a|$, this definition states explicitly that as far as limits are concerned, the value of $f(x)$ at $x = a$ is irrelevant. In taking limits we consider values of x closer and closer to a, but we do not consider the value of $f(x)$ at $x = a$. This definition states in precise terms our intuitive idea of a limit; that $f(x)$ can be made arbitrarily close to L by choosing x sufficiently close to a. A graphical interpretation may help to consolidate the definition. Figure 2.26(a) indicates a function that has limit L as x approaches a. In practice, we will not have a graph of the function, and verification that $\lim_{x \to a} f(x) = L$ will have to be done algebraically using Definition 2.2. Let us illustrate what must be done. We are given a value $\epsilon > 0$, which we should envisage as being very small, but we are never told exactly what it is. We are to restrict x so that $|f(x) - L| < \epsilon$. This inequality is equivalent to $-\epsilon < f(x) - L < \epsilon$ or $L - \epsilon < f(x) < L + \epsilon$, and the latter describes a horizontal band of width 2ϵ around the line $y = L$ [shaded in Figure 2.26(b)]. What Definition 2.2 requires is that we find an interval of width 2δ around $x = a$, as $|x - a| < \delta$ is equivalent to $a - \delta < x < a + \delta$, such that whenever x is in this interval, the values of $f(x)$ are all within the shaded horizontal band around $y = L$. Such an interval is shown in Figure 2.26(c) for the given ϵ. Now Definition 2.2 requires us to verify that the δ-interval can always be found no matter how small ϵ is chosen. This is always possible for the function illustrated in Figure 2.26, and it is clear that the smaller the given value of ϵ, the smaller δ will have to be chosen. For instance, for the value of ϵ in Figure 2.26(d), δ is smaller than that in Figure 2.26(c). In other words, the value of δ depends on the value of ϵ. Herein lies the difficulty in using Definition 2.2. We must determine precisely how δ depends on ϵ in order to ensure that δ can always be found. We illustrate with two examples.

FIGURE 2.26

(a)

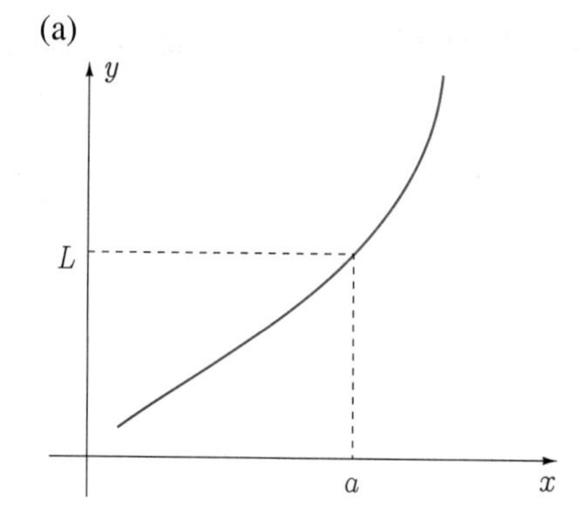

(b)

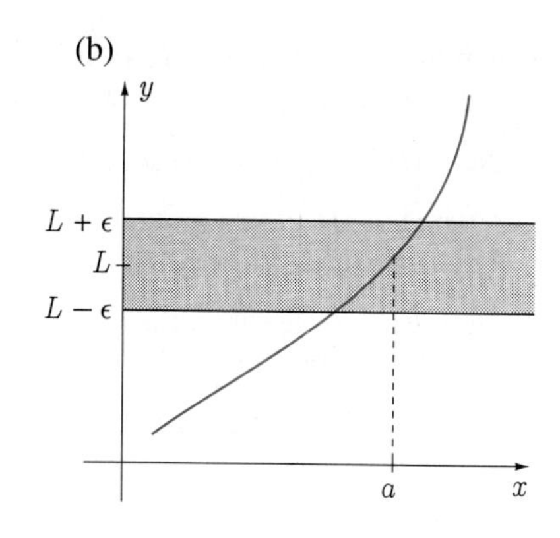

(c)

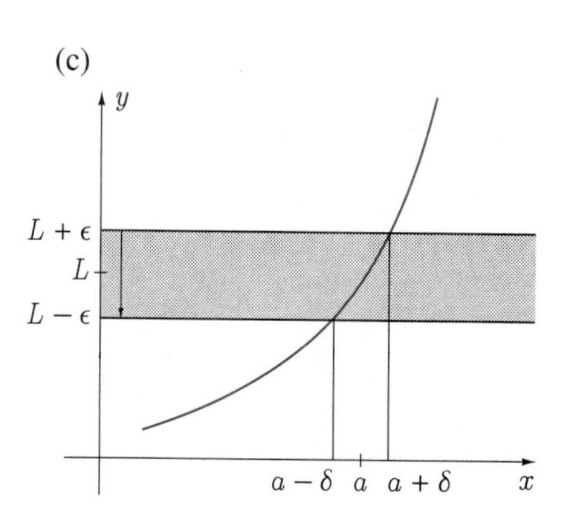

(d)

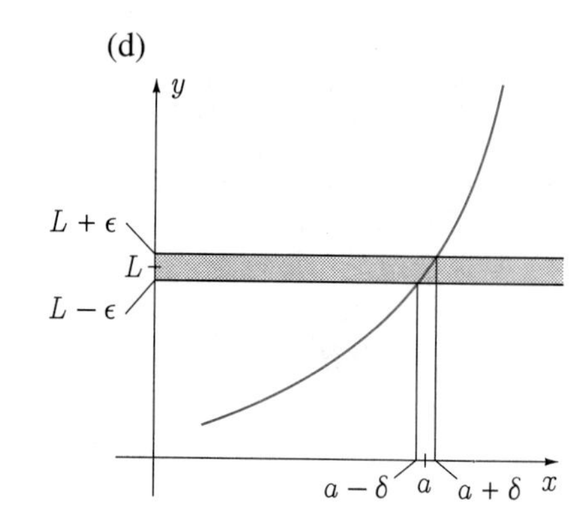

Consider proving first that

$$\lim_{x \to 3} (2x + 4) = 10.$$

Now, it is true that based on Theorem 2.1, this is quite obvious, but remember, this theorem has not yet been verified. For the next two examples we want to verify the limits, assuming that Theorem 2.1 is not at our disposal; verification must be through Definition 2.2. We must show that given any $\epsilon > 0$, there is a $\delta > 0$ such that

$$|(2x + 4) - 10| < \epsilon$$

whenever $0 < |x - 3| < \delta$; i.e., we must find a δ. Since

$$|2x + 4 - 10| = |2x - 6| = 2|x - 3|,$$

we must find a δ such that

$$2|x - 3| < \epsilon$$

whenever $0 < |x - 3| < \delta$. But this will be true if $\delta = \epsilon/2$; in other words, if we choose x to satisfy $0 < |x - 3| < \delta = \epsilon/2$, then

$$|(2x + 4) - 10| = 2|x - 3| < 2(\epsilon/2) = \epsilon.$$

The verification is now complete. We have shown that we can make $2x + 4$ as close to 10 as we want (within ϵ) by choosing x sufficiently close to 2 (within $\delta = \epsilon/2$).

Verifying that

$$\lim_{x \to 2} (x^2 + 5) = 9$$

is somewhat more complicated, but the idea is the same. We must show that given any $\epsilon > 0$, there is a $\delta > 0$ such that

$$|(x^2 + 5) - 9| < \epsilon \qquad \text{or} \qquad |x^2 - 4| < \epsilon$$

whenever $0 < |x - 2| < \delta$; i.e., we must find a δ. By working backward from the ϵ-inequality we can see how to choose δ. In addition, because the δ-inequality involves x in the combination $x - 2$, we rework the ϵ-inequality into the same combination. Since

$$|x^2 - 4| = |(x - 2)^2 + 4x - 8| = |(x - 2)^2 + 4(x - 2)|,$$

we must choose δ such that

$$|(x - 2)^2 + 4(x - 2)| < \epsilon$$

whenever $0 < |x - 2| < \delta$. Recalling inequality 1.6c from Section 1.1, we can state that

$$|(x - 2)^2 + 4(x - 2)| \leq |x - 2|^2 + 4|x - 2|.$$

As a result, if we can find a $\delta > 0$ such that

$$|x - 2|^2 + 4|x - 2| < \epsilon$$

whenever $0 < |x - 2| < \delta$, then we will have a satisfactory δ. But now we have the key. Suppose we set $z = |x - 2|$ and consider the parabola $Q(z) = z^2 + 4z - \epsilon$ shown in Figure 2.27.

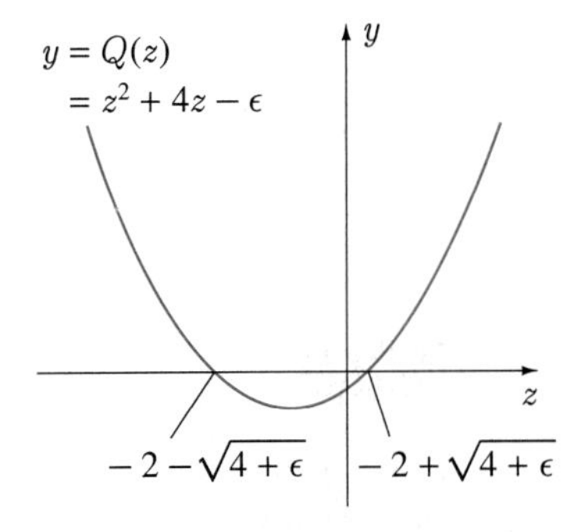

FIGURE 2.27

It crosses the z-axis when

$$z^2 + 4z - \epsilon = 0,$$

a quadratic with solutions

$$z = \frac{-4 \pm \sqrt{16 + 4\epsilon}}{2} = -2 \pm \sqrt{4 + \epsilon}.$$

The graph shows that whenever $0 < z < -2 + \sqrt{4 + \epsilon}$, then

$$z^2 + 4z - \epsilon < 0.$$

In other words, if $0 < z < \delta = \sqrt{4 + \epsilon} - 2$, then

$$z^2 + 4z < \epsilon.$$

Since $z = |x - 2|$, we can say that if $0 < |x - 2| < \delta = \sqrt{4 + \epsilon} - 2$, then

$$|x - 2|^2 + 4|x - 2| < \epsilon,$$

and therefore

$$|(x - 2)^2 + 4(x - 2)| < \epsilon.$$

The verification is now complete, We have shown that we can make $x^2 + 5$ as close to 9 as we want (within ϵ) by choosing x sufficiently close to 2 (within $\delta = \sqrt{4 + \epsilon} - 2$).

These examples have indicated the difference between "evaluation of" and "verification of" a limit. If you have learned only an appreciation for this difference, then our discussion has been worthwhile. If your comprehension goes beyond this and includes a working understanding of the definition of a limit, we could hope for no more. Results in succeeding chapters rely heavily on our intuitive understanding of limits and the ability to calculate limits, but use of Definition 2.2 is kept to a minimum.

EXERCISES 2.5

In Exercises 1–9 use Definition 2.2 to verify the limit.

1. $\lim_{x \to 1} (x + 5) = 6$

2. $\lim_{x \to 2} (2x - 3) = 1$

3. $\lim_{x \to 0} (x^2 + 3) = 3$

4. $\lim_{x \to 1} (x^2 + 4) = 5$

5. $\lim_{x \to -2} (3 - x^2) = 1$

6. $\lim_{x \to 3} (x^2 - 7x) = -12$

7. $\lim_{x \to -1} (x^2 - 3x + 4) = 7$

8. $\lim_{x \to 1} (x^2 + 3x + 5) = 9$

9. $\lim_{x \to 2} \dfrac{x + 2}{x - 1} = 4$

In Exercises 10–19 give a mathematical definition for each statement.

10. $\lim_{x \to a^+} f(x) = L$

11. $\lim_{x \to a^-} f(x) = L$

12. $\lim_{x \to \infty} f(x) = L$

13. $\lim_{x \to -\infty} f(x) = L$

14. $\lim_{x \to a} f(x) = \infty$

15. $\lim_{x \to a} f(x) = -\infty$

16. $\lim_{x \to \infty} f(x) = \infty$

17. $\lim_{x \to \infty} f(x) = -\infty$

18. $\lim_{x \to -\infty} f(x) = \infty$

19. $\lim_{x \to -\infty} f(x) = -\infty$

20. Use Definition 2.2 to prove that a function $f(x)$ cannot have two limits as x approaches a.

In Exercises 21–28 use the appropriate definition from Exercises 10–19 to verify the limit.

21. $\lim_{x \to 1} \dfrac{1}{(x - 1)^2} = \infty$

22. $\lim_{x \to -2} \dfrac{-1}{(x + 2)^2} = -\infty$

23. $\lim_{x \to \infty} (x + 5) = \infty$

24. $\lim_{x \to \infty} (5 - x^2) = -\infty$

25. $\lim_{x \to \infty} \dfrac{x + 2}{x - 1} = 1$

26. $\lim_{x \to -\infty} \dfrac{x + 2}{x - 1} = 1$

27. $\lim_{x \to -\infty} (5 - x) = \infty$

28. $\lim_{x \to -\infty} (3 + x - x^2) = -\infty$

29. Use the definitions in Exercises 10 and 11 to verify that the Heaviside function of Exercise 38 in Section 2.4 has a right- and left-hand limit at $x = a$.

30. Prove that if $\lim_{x \to a} f(x) = L > 0$, then there exists an open interval I containing a in which $f(x) > 0$ except possibly at $x = a$.

In Exercises 31–35 we use Definition 2.2 to prove Theorem 2.1. In each exercise assume that $\lim_{x \to a} f(x) = F$ and that $\lim_{x \to a} g(x) = G$.

31. Show that given any $\epsilon > 0$, there exists numbers $\delta_1 > 0$ and $\delta_2 > 0$ such that

$$|f(x) - F| < \epsilon/2 \quad \text{whenever } 0 < |x - a| < \delta_1 \quad \text{and}$$
$$|g(x) - G| < \epsilon/2 \quad \text{whenever } 0 < |x - a| < \delta_2.$$

Use these results along with identity 1.6c to prove part (i) of Theorem 2.1.

32. Use a proof similar to that in Exercise 31 to verify part (ii) of Theorem 2.1.

33. Verify part (iii) of Theorem 2.1.

34. (a) Verify that

$$|f(x)g(x) - FG| \le |f(x)||g(x) - G| + |G||f(x) - F|.$$

(b) Show that given any $\epsilon > 0$, there exists numbers $\delta_1 > 0$, $\delta_2 > 0$, and $\delta_3 > 0$ such that

$$|f(x)| < |F| + 1 \quad \text{whenever } 0 < |x - a| < \delta_1,$$
$$|g(x) - G| < \frac{\epsilon}{2(|F| + 1)} \quad \text{whenever } 0 < |x - a| < \delta_2,$$
$$|f(x) - F| < \frac{\epsilon}{2|G| + 1} \quad \text{whenever } 0 < |x - a| < \delta_3.$$

(c) Use these results to prove part (iv) of Theorem 2.1.

35. (a) Verify that when $G \neq 0$,

$$\left|\frac{f(x)}{g(x)} - \frac{F}{G}\right| \leq \frac{|f(x) - F|}{|g(x)|} + \frac{|F||G - g(x)|}{|G||g(x)|}.$$

(b) Show that given any $\epsilon > 0$, there exist numbers $\delta_1 > 0$, $\delta_2 > 0$, and $\delta_3 > 0$ such that

$$|g(x)| > \frac{|G|}{2} \quad \text{whenever } 0 < |x - a| < \delta_1,$$

$$|f(x) - F| < \frac{\epsilon|G|}{4} \quad \text{whenever } 0 < |x - a| < \delta_2,$$

$$|g(x) - G| < \frac{\epsilon|G|^2}{4|F| + 1} \quad \text{whenever } 0 < |x - a| < \delta_3.$$

(c) Now prove part (v) of Theorem 2.1.

36. Does the function

$$f(x) = \begin{cases} x\sin(1/x) & x \neq 1/(n\pi) \\ 1 & x = 1/(n\pi), \end{cases}$$

where n is an integer, have a limit as $x \to 0$?

SUMMARY

In Section 2.1 we introduced limits of functions. For the most part our discussion was intuitive, beginning with the statement "$\lim_{x\to a} f(x) = L$ if $f(x)$ can be made arbitrarily close to L by choosing x sufficiently close to a." This idea was then extended to include the following:

Right-hand limits: $\lim_{x\to a^+} f(x)$;

Left-hand limits: $\lim_{x\to a^-} f(x)$;

Limits at infinity: $\lim_{x\to\infty} f(x)$, $\lim_{x\to-\infty} f(x)$;

Infinite limits: $\lim_{x\to a} f(x) = \infty$, $\lim_{x\to a} f(x) = -\infty$.

Keep in mind that the term "infinite limits" is somewhat of a misnomer since in both situations the limit does not exist.

A function $f(x)$ is continuous at a point $x = a$ if

$$\lim_{x\to a} f(x) = f(a).$$

Note that the conditions that $f(x)$ is defined at $x = a$ and $\lim_{x\to a} f(x)$ exists are inherent in this definition. The function is continuous on an interval if it is continuous at each point of that interval. Geometrically, this means that one must be able to completely trace its graph without lifting pencil from page.

We illustrated that limits can be used to sketch the graph of a function near a point of discontinuity, as well as to indicate its behaviour for very large positive and negative values of x. To our list of techniques for sketching curves which already contains such methods as addition and multiplication of ordinates, symmetries, and translations, we therefore add the use of limits.

In Section 2.5 we developed the mathematical definition of limits.

Key Terms and Formulas

In reviewing this chapter, you should be able to define or discuss the following key terms:

Limit
Right-hand limit
Left-hand limit
Infinite limit
Vertical asymptote
Limit at infinity
Horizontal asymptote
Continuity at a point
Continuity on an interval
Composite function

REVIEW EXERCISES

In Exercises 1–20 evaluate the limit, if it exists.

1. $\lim_{x\to 1} \dfrac{x^2 - 2x}{x+5}$

2. $\lim_{x\to -1} \dfrac{x^2 - 1}{x+1}$

3. $\lim_{x\to -2} \dfrac{x^2 + 4x + 4}{x+3}$

4. $\lim_{x\to \infty} \dfrac{x+5}{x-3}$

5. $\lim_{x\to -\infty} \dfrac{x^2 + 3x + 2}{2x^2 - 5}$

6. $\lim_{x\to -\infty} \dfrac{5 - x^3}{3 + 4x^3}$

7. $\lim_{x\to \infty} \dfrac{3x^3 + 2x - 5}{x^2 + 5x}$

8. $\lim_{x\to \infty} \dfrac{4 - 3x + x^2}{3 + 5x^3}$

9. $\lim_{x\to 2^+} \dfrac{x^2 - 2x}{x^2 + 2x}$

10. $\lim_{x\to 2^-} \dfrac{x^2 - 4x + 4}{x - 2}$

11. $\lim_{x\to 0} \dfrac{x^2 + 2x}{3x - 2x^2}$

12. $\lim_{x\to 1} \dfrac{x^2 + 5x}{(x-1)^2}$

13. $\lim_{x\to 1} \dfrac{\sqrt{x} - 1}{x}$

14. $\lim_{x\to 1} \dfrac{\sqrt{x} - 1}{x - 1}$

15. $\lim_{x\to 1/2} \dfrac{(2 - 4x)^3}{x(2x-1)^2}$

16. $\lim_{x\to \infty} \dfrac{\cos 5x}{x}$

17. $\lim_{x\to -\infty} x \sin x$

18. $\lim_{x\to -\infty} \dfrac{\sqrt{3x^2 + 4}}{2x + 5}$

19. $\lim_{x\to \infty} \dfrac{\sqrt{3x^2 + 4}}{2x + 5}$

20. $\lim_{x\to \infty} \left(\sqrt{2x+1} - \sqrt{3x-1}\right)$

In Exercises 21–32 sketch a graph of the function, indicating any discontinuities.

21. $f(x) = \dfrac{1}{x-2}$

22. $f(x) = \dfrac{x}{x-2}$

23. $f(x) = \dfrac{x^2}{x-2}$

24. $f(x) = \dfrac{x^2 - 36}{x - 6}$

25. $f(x) = \dfrac{x+1}{x-1}$

26. $f(x) = \left|\dfrac{x+1}{x-1}\right|$

27. $f(x) = \dfrac{|x+1|}{x-1}$

28. $f(x) = \dfrac{x+1}{|x-1|}$

29. $f(x) = \dfrac{2x}{x^2 - 3x - 4}$

30. $f(x) = \dfrac{2x^2}{x^2 - 3x - 4}$

31. $f(x) = \dfrac{x^3 - 3x^2 + 3x - 1}{x - 1}$

32. $f(x) = \dfrac{x^3 - 3x^2 + 3x - 1}{x^2 - 2x + 1}$

33. For what values of x is the function $f(x) = [[x^2]]$ discontinuous? Sketch its graph.

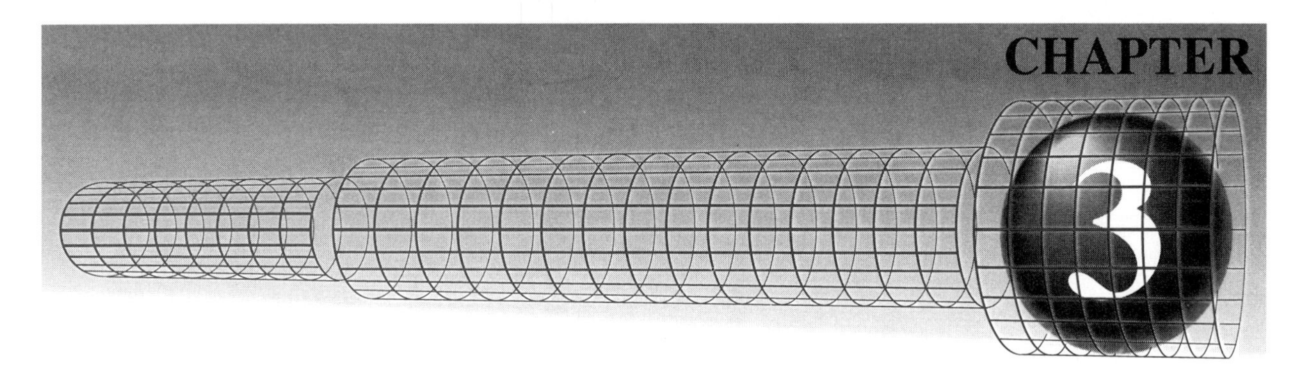

Differentiation

Chapters 1 and 2 have prepared the way for calculus. The functions and curves in these chapters yield a wealth of examples for our discussions, and limits from Chapter 2 provide the tool by which calculus is developed. Calculus has two major components, *differentiation* and *integration*. In this chapter we study differentiation. We learn what a *derivative* is algebraically and geometrically and develop some of its properties; we learn how to differentiate polynomials, rational functions, trigonometric functions, and exponential and logarithm functions; and we see glimpses of the applications that are to follow in Chapter 4.

SECTION 3.1 The Derivative

Very few quantities in real life remain constant; most are in a state of change. For example, room temperature, the speed of a car, and the angle of elevation of the sun are three commonplace quantities which are constantly changing. A few more technical ones are the current in a transmission line, barometric pressure, moisture content of the soil, and stress in vertical members of tall buildings during hurricanes. Rates at which these quantities change are called derivatives. To study the concept of a rate of change more thoroughly, we consider two commonplace situations. First, suppose a car moves along a straight street between the two stop lights in Figure 3.1.

FIGURE 3.1

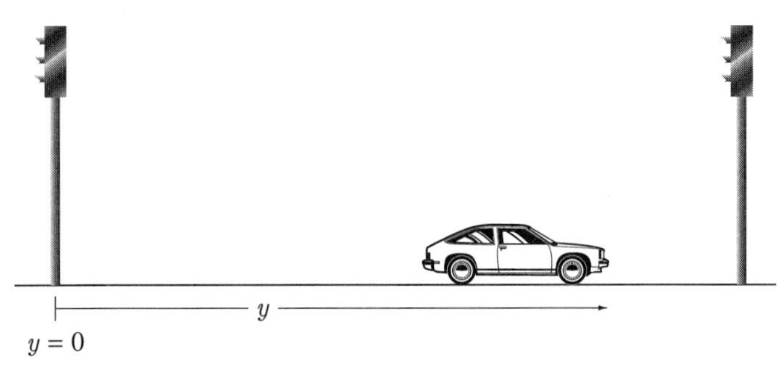

If its distance from the first light is denoted by y, then y is a function of time t, say, $y = f(t)$. When $y_1 = f(t_1)$ and $y_2 = f(t_2)$ are positions of the car at times t_1 and t_2, where $t_2 > t_1$, the quotient

$$\frac{y_2 - y_1}{t_2 - t_1} = \frac{f(t_2) - f(t_1)}{t_2 - t_1}$$

is called the *average velocity* of the car between y_1 and y_2. It is the average rate of change of y with respect to t between t_1 and t_2. Suppose time t_1 is kept fixed, and we take a number of times t_2 that are closer and closer to t_1 so that the time interval $t_2 - t_1$ gets smaller and smaller. If we demand that t_2 get arbitrarily close to t_1, then we are really asking for the limit of the average velocity as t_2 approaches t_1, written

$$\lim_{t_2 \to t_1} \frac{y_2 - y_1}{t_2 - t_1} = \lim_{t_2 \to t_1} \frac{f(t_2) - f(t_1)}{t_2 - t_1}.$$

This quantity is called the *instantaneous velocity* of the car at time t_1. It is the instantaneous rate of change of y with respect to t at time t_1.

As a second example, consider an open cylindrical container placed outside a house. The depth D of water in the container collected during a rainstorm (Figure 3.2) is a function of time t, say, $D = f(t)$.

FIGURE 3.2

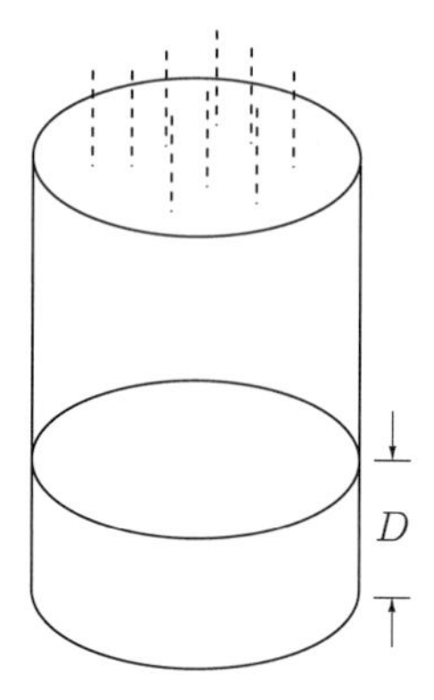

The quotient

$$\frac{D_2 - D_1}{t_2 - t_1} = \frac{f(t_2) - f(t_1)}{t_2 - t_1}$$

represents the *average rate of rainfall* during the time interval from t_1 to t_2. If we take the limit of this quotient as t_2 approaches t_1,

$$\lim_{t_2 \to t_1} \frac{D_2 - D_1}{t_2 - t_1} = \lim_{t_2 \to t_1} \frac{f(t_2) - f(t_1)}{t_2 - t_1},$$

we obtain the *instantaneous rate of rainfall* at time t_1. It is the instantaneous rate of change of D with respect to t at time t_1.

These rates of change are examples of rates of change that occur in a multitude of applications — applications from such diverse fields as physics, economics, psychology, and medicine, to name a few. Let us establish a mathematical framework within which to examine rates of change more closely, at the same time introducing notation which eliminates subscripts. Suppose the function $f(x)$ is defined for all x, and $x = a$ and $x = a + h$ are two values of x. The quotient

$$\frac{f(a+h) - f(a)}{h}$$

is the **average rate of change** of $f(x)$ with respect to x in the interval $a \leq x \leq a + h$. The limit of the quotient as h approaches zero,

$$\lim_{h \to 0} \frac{f(a+h) - f(a)}{h}$$

is called the **instantaneous rate of change** of $f(x)$ with respect to x at $x = a$. It is also called the derivative of $f(x)$ at $x = a$.

Definition 3.1

The **derivative** of a function $f(x)$ with respect to x at a point $x = a$, denoted by $f'(a)$, is defined as

$$f'(a) = \lim_{h \to 0} \frac{f(a+h) - f(a)}{h}, \qquad (3.1)$$

provided the limit exists.

The operation of taking the derivative of a function is called **differentiation**. We say that we differentiate the function when we find its derivative.

Tangent Lines and the Geometric Interpretation of the Derivative

We have defined the derivative of a function algebraically as its instantaneous rate of change. Its geometric interpretation is also important. Derivatives are intimately connected to tangent lines to curves, and most students of calculus have an intuitive idea of what it means for a line to be tangent to a curve. Often it is the idea of "touching". A line is tangent to a curve if it touches the curve at exactly one point. For curves such as circles, ellipses and parabolas, this notion is adequate, but in general it is unsatisfactory. For instance, in Figure 3.3 we have drawn what would look like the tangent line to the curve $y = x^3$ at the point $P(1, 1)$. But this line intersects the curve again at the point $Q(-2, -8)$. The tangent line at P does not touch the curve at precisely one point; it intersects the curve at a second point.

The idea that a tangent line touches a curve is good, but it needs to be phrased properly. Consider defining what is meant by the tangent line at the point P on the curve $y = f(x)$ in Figure 3.4. When P is joined to another point Q_1 on the curve by a straight line l_1, certainly l_1 is not the tangent line to $y = f(x)$ at P. If we join P to a point Q_2 on $y = f(x)$ closer to P than Q_1, then l_2 is not the tangent line at P either, but it is closer to it than l_1. A point Q_3 even closer to P yields a line l_3 that is even closer to the tangent line than l_2. Repeating this process over and over again, leads to a set of lines $l_1, l_2, l_3, \ldots$ which get closer and closer to what we feel is the tangent line to $y = f(x)$ at P. We therefore define the **tangent line** to $y = f(x)$ at P as the limiting position of these lines as points $Q_1, Q_2, Q_3, \ldots$ get arbitrarily close to P.

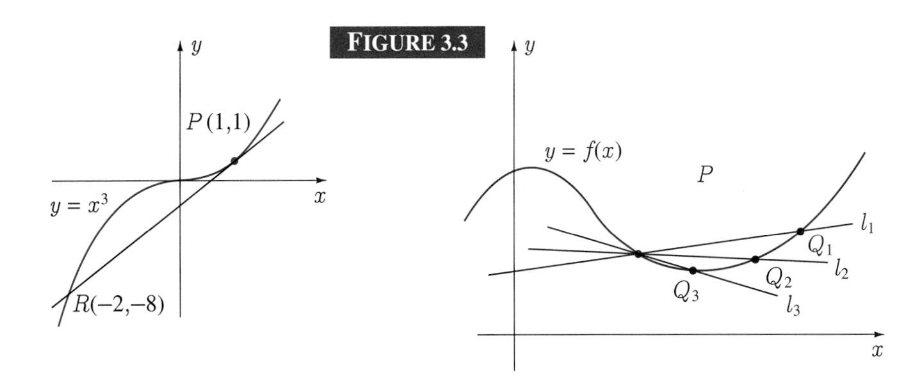

FIGURE 3.3

FIGURE 3.4

The line in Figure 3.3 satisfies this definition; it is the limiting position of lines joining P to other points on the curve which approach P. It is irrelevant whether this line intersects the curve again at some distance from P. What is the tangent line to the curve $y = x^3$ in Figure 3.3 at the origin $(0, 0)$? According to the above definition the limiting position of lines joining $(0, 0)$ to other points on the curve is the x-axis; that is, the x-axis is tangent to $y = x^3$ at $(0, 0)$ (Figure 3.5). Notice that the tangent line actually crosses from one side of the curve to the other at $(0, 0)$. To the left of $x = 0$, the tangent line is above the curve whereas to the right of $x = 0$, it is below the curve.

FIGURE 3.5

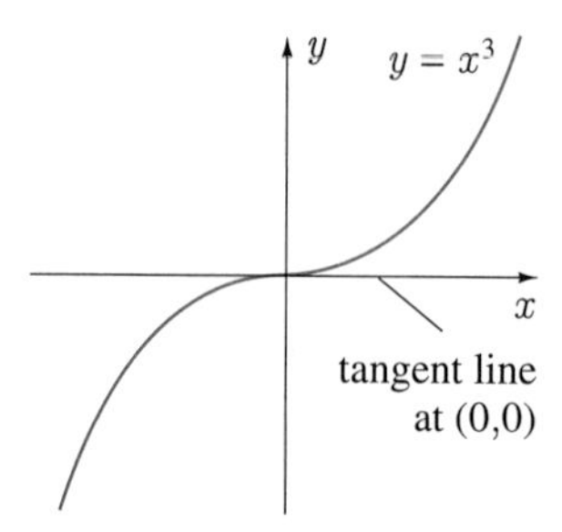

Now that we understand what it means for a line to be tangent to a curve, let us make the connection with derivatives. Suppose $P(a, f(a))$ is a point on the curve $y = f(x)$ in Figure 3.6.

If $Q(a + h, f(a + h))$ is another point on the curve, then the quotient

$$\frac{f(a + h) - f(a)}{h}$$

in Definition 3.1 is the slope of the line joining P and Q. As $h \to 0$, point Q moves along the curve toward P, and the line joining P and Q moves toward the tangent line at P. It follows that limit 3.1, the derivative $f'(a)$, is the slope of the tangent line to the curve $y = f(x)$ at $x = a$ (Figure 3.7).

FIGURE 3.6 **FIGURE 3.7**

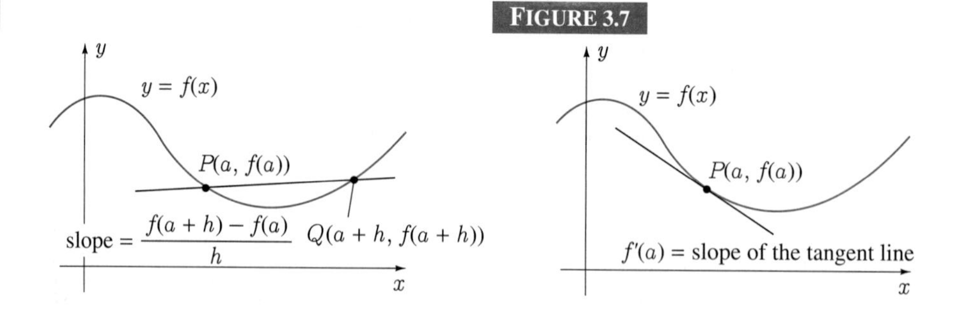

To summarize, algebraically the derivative $f'(a)$ of a function $f(x)$ at $x = a$ is its instantaneous rate of change; geometrically it is the slope of the tangent line to the graph of $f(x)$ at the point $(a, f(a))$.

Let us use Definition 3.1 to calculate derivatives of some simple functions.

EXAMPLE 3.1

Find $f'(1)$ if $f(x) = x^2$.

SOLUTION According to equation 3.1,

$$f'(1) = \lim_{h\to 0} \frac{f(1+h) - f(1)}{h} = \lim_{h\to 0} \frac{(1+h)^2 - 1}{h}$$
$$= \lim_{h\to 0} \frac{(1+2h+h^2) - 1}{h} = \lim_{h\to 0} \frac{2h+h^2}{h} = \lim_{h\to 0} (2+h) = 2.$$

Algebraically, the instantaneous rate of change of $f(x) = x^2$ when $x = 1$ is equal to 2. Geometrically, the slope of the tangent line to the curve $y = x^2$ at the point $(1, 1)$ is 2 (see Figure 3.8).

FIGURE 3.8

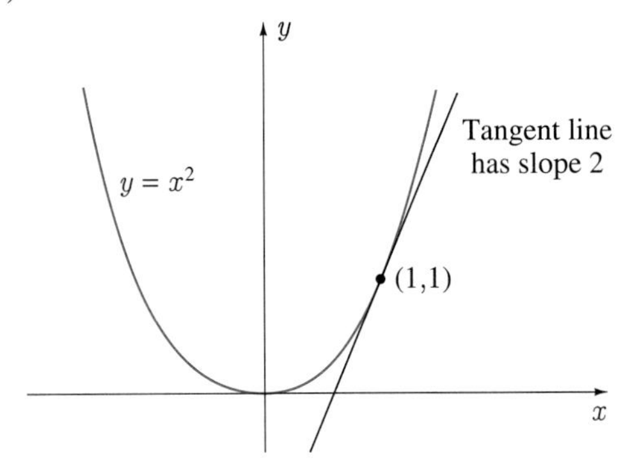

■

EXAMPLE 3.2

Find $f'(2)$ if $f(x) = x^3 - 12x$.

SOLUTION Using equation 3.1,

$$f'(2) = \lim_{h\to 0} \frac{f(2+h) - f(2)}{h} = \lim_{h\to 0} \frac{[(2+h)^3 - 12(2+h)] - (8-24)}{h}$$
$$= \lim_{h\to 0} \frac{6h^2 + h^3}{h} = \lim_{h\to 0} (6h + h^2) = 0.$$

This result is substantiated in Figure 3.9 where we see that the tangent line to $y = x^3 - 12x$ is horizontal (has zero slope) at $x = 2$.

FIGURE 3.9

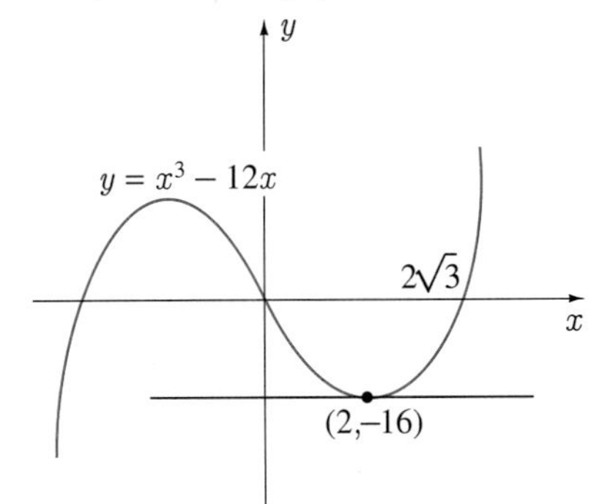

■

In Examples 3.1 and 3.2 we required the derivative for the function at only one value of a, and therefore set a in equation 3.1 equal to this value. An alternative, which would be far more advantageous, especially in an example where the derivative were required at a number of points, would be to evaluate $f'(a)$ and then set a to its desired value later. For instance, in Example 1 we could calculate that

$$f'(a) = \lim_{h\to 0} \frac{f(a+h) - f(a)}{h} = \lim_{h\to 0} \frac{(a+h)^2 - a^2}{h}$$
$$= \lim_{h\to 0} \frac{2ah + h^2}{h} = \lim_{h\to 0} (2a + h) = 2a.$$

This is the derivative of $f(x) = x^2$ at any value $x = a$. For $a = 1$, we obtain $f'(1) = 2(1) = 2$. But it is also easy to calculate $f'(a)$ at other values of a. For example, $f'(0) = 0$, $f'(-1) = -2$, and $f'(4) = 8$.

We now carry this idea to its logical conclusion. The derivative of $f(x)$ at a is denoted by $f'(a)$. But what is a? It is a specific value of x at which to calculate the derivative, and it can be any value of x. Why not simply drop references to a, and talk about the derivative of $f(x)$ at values of x? Following this suggestion, we denote by $f'(x)$ the derivative of the function $f(x)$ at any value of x. With this notation, equation 3.1 is replaced by

$$f'(x) = \lim_{h\to 0} \frac{f(x+h) - f(x)}{h}, \qquad (3.2)$$

for the derivative of $f(x)$ at x. We call $f'(x)$ the **derivative function**.

When a function is represented by the letter y, as in $y = f(x)$, another common notation for the derivative is $\frac{dy}{dx}$. Be careful when using this notation. Do not interpret it as a quotient; it is one symbol representing an accumulation of all the operations in equation 3.2. It is not therefore to be read as "dy divided by dx". Typographically, it is easier to print dy/dx rather than $\frac{dy}{dx}$, and we will take this liberty whenever it is convenient to do so. But remember, dy/dx is not a quotient, it is one symbol representing the limit operation in 3.2.

Sometimes it is more convenient to use parts of each of these notations and write

$$\frac{d}{dx} f(x).$$

In this form we understand that d/dx means to differentiate with respect to x whatever follows it, in this case $f(x)$. Let us use these new notations in calculating two more derivatives.

EXAMPLE 3.3

Find dy/dx if $y = f(x) = (x-1)/(x+2)$.

SOLUTION Using 3.2,

$$\frac{dy}{dx} = \lim_{h\to 0} \frac{f(x+h) - f(x)}{h}$$
$$= \lim_{h\to 0} \frac{1}{h}\left(\frac{x+h-1}{x+h+2} - \frac{x-1}{x+2}\right).$$

If we bring the terms in parentheses to a common denominator, the result is

$$\frac{dy}{dx} = \lim_{h \to 0} \frac{1}{h} \left(\frac{(x+h-1)(x+2) - (x+h+2)(x-1)}{(x+h+2)(x+2)} \right).$$

When we simplify the numerator, we find

$$\begin{aligned}
\frac{dy}{dx} &= \lim_{h \to 0} \frac{1}{h} \left(\frac{3h}{(x+h+2)(x+2)} \right) \\
&= \lim_{h \to 0} \frac{3}{(x+h+2)(x+2)} \\
&= \frac{3}{(x+2)^2}.
\end{aligned}$$

■

EXAMPLE 3.4

Find dv/dt if $v = f(t) = 1/t$.

SOLUTION In terms of variables v and t, equation 3.2 takes the form

$$\begin{aligned}
\frac{dv}{dt} &= \lim_{h \to 0} \frac{f(t+h) - f(t)}{h} \\
&= \lim_{h \to 0} \frac{1}{h} \left(\frac{1}{t+h} - \frac{1}{t} \right) \\
&= \lim_{h \to 0} \frac{t - (t+h)}{t(t+h)h} \\
&= \lim_{h \to 0} \frac{-1}{t(t+h)} \\
&= -\frac{1}{t^2}.
\end{aligned}$$

■

In Section 1.3 we proved that the inclination ϕ of a line l is related to its slope m by the equation $\tan \phi = m$ (equation 1.13). When l is the tangent line to a curve $y = f(x)$ at point (x_0, y_0), as shown in Figure 3.10, its slope is $f'(x_0)$.

FIGURE 3.10

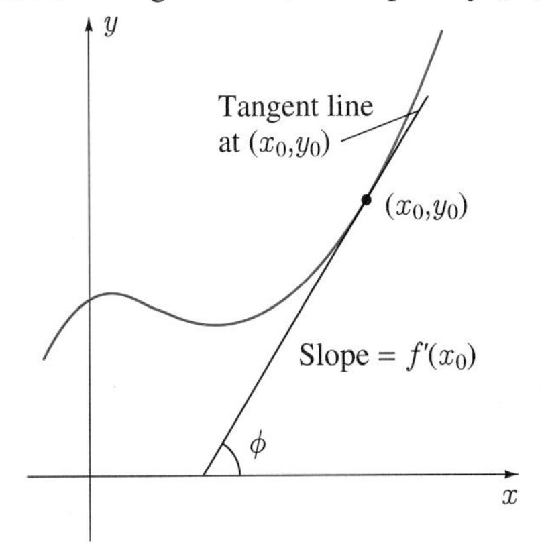

Hence, the inclination ϕ of the tangent line to a curve at a point (x_0, y_0) is given by the equation

$$\tan \phi = f'(x_0). \tag{3.3}$$

EXAMPLE 3.5

Find the inclination of the tangent line to the curve $y = (x-1)/(x+2)$ at the point $(4, 1/2)$.

SOLUTION In Example 3.3, we calculated the slope of the tangent line to this curve at any point as

$$\frac{dy}{dx} = \frac{3}{(x+2)^2}.$$

According to equation 3.3, the inclination ϕ of the tangent line at $(4, 1/2)$ is given by

$$\tan\phi = \frac{3}{(4+2)^2} = \frac{1}{12}.$$

The angle with tangent equal $1/12$ is $\phi = 0.083$ radians. ■

EXERCISES 3.1

In Exercises 1–10 use equation 3.2 to find $f'(x)$.

1. $f(x) = x + 2$

2. $f(x) = 3x^2 + 5$

3. $f(x) = 1 + 2x - x^2$

4. $f(x) = x^3 + 2x^2$

5. $f(x) = x^4 + 4x - 12$

6. $f(x) = \dfrac{x+4}{x-5}$

7. $f(x) = \dfrac{x^2+2}{x+3}$

8. $f(x) = x^2(x+2)$

9. $f(x) = \dfrac{3x-2}{4-x}$

10. $f(x) = \dfrac{x^2-x+1}{x^2+x+1}$

In Exercises 11–14 find the specified rate of change.

11. The rate of change of the circumference C of a circle with respect to its radius r.

12. The rate of change of the area A of a circle with respect to its radius r.

13. The rate of change of the area A of a sphere with respect to its radius r.

14. The rate of change of the volume V of a sphere with respect to its radius r.

Answer Exercises 15–19 by drawing a graph of the function. Do not calculate the required derivative.

15. What is $f'(x)$ if $f(x) = 2x - 4$?

16. What is $f'(x)$ if $f(x) = mx + b$, where m and b are constants?

17. What is $f'(0)$ if $f(x) = x^2$?

18. What is $f'(1)$ if $f(x) = (x-1)^2$?

19. What is $f'(0)$ if $f(x) = x^{1/3}$?

In Exercises 20–23 find the equation of the tangent line to the curve at the given point.

20. $y = x^2 + 3$ at $(1, 4)$

21. $y = 3 - 2x - x^2$ at $(4, -21)$

22. $y = 1/x^2$ at $(2, 1/4)$

23. $y = (x+1)/(x+2)$ at $(0, 1/2)$

In Exercises 24–27 find the inclination of the tangent line to the curve at the given point.

24. $y = x^2$ at $(1, 1)$

25. $y = x^3 - 6x$ at $(2, -4)$

26. $y = 1/x^2$ at $(2, 1/4)$

27. $y = 1/(x+1)$ at $(0, 1)$

In Exercises 28–32 find $f'(x)$.

28. $f(x) = x^8$

29. $f(x) = \sqrt{x+1}$

30. $f(x) = \dfrac{1}{(x-2)^4}$

31. $f(x) = \dfrac{1}{\sqrt{x-3}}$

32. $f(x) = x\sqrt{x+1}$

In Exercises 33 and 34 find the specified rate of change.

33. The rate of change of the radius r of a circle with respect to its area A.

34. The rate of change of the volume V of a sphere with respect to its area A.

35. Find the angle between the tangent lines to the curves $y = x^2$ and $x = y^2$ at their point of intersection.

36. Find $f'(x)$ if $f(x) = |x|$.

37. A sphere of radius R has a uniform charge distribution of ρ coulombs per cubic metre. The electrostatic potential V at a distance r from the centre of the sphere is defined by

$$V = f(r) = \begin{cases} \dfrac{\rho}{6\epsilon_0}(3R^2 - r^2) & 0 \le r \le R \\ \dfrac{R^3\rho}{3\epsilon_0 r} & r > R, \end{cases}$$

where ϵ_0 is a constant. Sketch a graph of this function. Does $f(r)$ appear to have a derivative at the surface of the sphere? That is, does $f'(R)$ exist? Prove your conjecture using Definition 3.1.

38. Repeat Exercise 37 for the magnitude E of the electrostatic field:

$$E = f(r) = \begin{cases} \dfrac{\rho r}{3\epsilon_0} & 0 \le r \le R \\ \dfrac{\rho R^3}{3\epsilon_0 r^2} & r > R. \end{cases}$$

39. Find $f'(x)$ if $f(x) = x^{1/3}$.

40. Let $f(x)$ be a function with the property that $f(x+z) = f(x)f(z)$ for all x and z, and be such that $f(0) = f'(0) = 1$. Prove that $f'(x) = f(x)$ for all x.

SECTION 3.2

Rules for Differentiation

Since calculus plays a key role in many branches of applied science, we need to differentiate many types of functions: polynomials, rational functions, trigonometric functions, exponential and logarithmic functions, etc. To use equation 3.2 each time would be extremely laborious, and we might soon doubt the usefulness of differentiation. Fortunately, however, we can develop a number of rules for taking derivatives that eliminate the necessity of using the definition each time. We state each of these formulas in the form of a theorem.

Theorem 3.1

If $f(x) = c$, where c is a constant, then $f'(x) = 0$.

Proof By equation 3.2,

$$f'(x) = \lim_{h\to 0} \frac{f(x+h) - f(x)}{h} = \lim_{h\to 0} \frac{c - c}{h} = 0.$$

In short, the derivative of a constant function is zero;

$$\frac{d}{dx}(c) = 0. \tag{3.4}$$

Theorem 3.2

If $f(x) = x$, then $f'(x) = 1$.

Proof With equation 3.2,

$$f'(x) = \lim_{h\to 0} \frac{f(x+h) - f(x)}{h} = \lim_{h\to 0} \frac{(x+h) - x}{h} = \lim_{h\to 0} \frac{h}{h} = 1.$$

In short,

$$\frac{d}{dx}(x) = 1. \tag{3.5}$$

Note that graphs of the functions $f(x) = c$ and $f(x) = x$ in Figure 3.11 confirm the results of equations 3.4 and 3.5. The tangent line to $y = c$ always has slope zero, whereas the tangent line to $y = x$ always has slope equal to one.

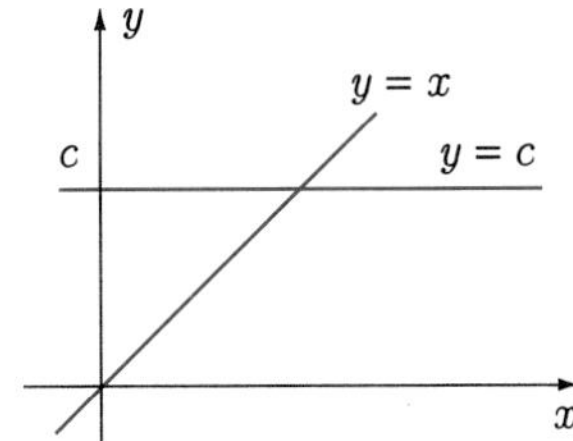

FIGURE 3.11

Theorem 3.3

If $f(x) = x^n$, where n is a positive integer, then $f'(x) = nx^{n-1}$.

Proof Equation 3.2 gives

$$\frac{dy}{dx} = f'(x) = \lim_{h\to 0} \frac{f(x+h) - f(x)}{h} = \lim_{h\to 0} \frac{(x+h)^n - x^n}{h}.$$

If we expand $(x+h)^n$ by means of the binomial theorem, we have

$$\frac{dy}{dx} = \lim_{h\to 0} \frac{1}{h}\left[x^n + nx^{n-1}h + \frac{n(n-1)}{2}x^{n-2}h^2 + \cdots + nxh^{n-1} + h^n - x^n\right].$$

The first and last terms in brackets cancel, and dividing h into the remaining terms gives

$$\frac{dy}{dx} = \lim_{h\to 0}\left[nx^{n-1} + \frac{n(n-1)}{2}x^{n-2}h + \cdots + h^{n-1}\right] = nx^{n-1}.$$

In short,

$$\frac{d}{dx}(x^n) = nx^{n-1}. \tag{3.6}$$

This is called the **power rule** for differentiation. Although we have proved the power rule only for n a positive integer, it is in fact true for every real number n. We will assume that equation (3.6) can be used for any real number n, and will prove this more general result in Section 3.9. For example,

$$\frac{d}{dx}(x^3) = 3x^2 \qquad \text{and} \qquad \frac{d}{dx}\left(x^{1/3}\right) = \frac{1}{3}x^{-2/3}.$$

Theorem 3.4

If $g(x) = cf(x)$, where c is a constant, then

$$g'(x) = cf'(x). \tag{3.7a}$$

Proof By equation 3.2 and Theorem 2.1,

$$g'(x) = \lim_{h\to 0} \frac{g(x+h) - g(x)}{h} = \lim_{h\to 0} \frac{cf(x+h) - cf(x)}{h}$$
$$= c\lim_{h\to 0} \frac{f(x+h) - f(x)}{h} = cf'(x).$$

Thus, for $y = f(x)$, we may write that

$$\frac{d}{dx}(cy) = c\frac{dy}{dx}. \tag{3.7b}$$

Theorem 3.5

If $p(x) = f(x) + g(x)$, then

$$p'(x) = f'(x) + g'(x). \tag{3.8a}$$

Proof Equation 3.2 gives

$$\begin{aligned} p'(x) &= \lim_{h\to 0} \frac{p(x+h) - p(x)}{h} \\ &= \lim_{h\to 0} \frac{[f(x+h) + g(x+h)] - [f(x) + g(x)]}{h} \\ &= \lim_{h\to 0} \left[\frac{f(x+h) - f(x)}{h} + \frac{g(x+h) - g(x)}{h}\right] \\ &= f'(x) + g'(x). \end{aligned}$$

In short, if we set $u = f(x)$ and $v = g(x)$,

$$\frac{d}{dx}(u+v) = \frac{du}{dx} + \frac{dv}{dx}, \tag{3.8b}$$

or, in words, the derivative of a sum is the sum of the derivatives.

We now use these formulas to calculate derivatives in the following examples.

EXAMPLE 3.6

Find dy/dx if

(a) $y = x^4$ (b) $y = 3x^6 - x^{-2}$ (c) $y = \dfrac{x^4 - 6x^2}{3x^3}$

SOLUTION (a) By equation 3.6,

$$\frac{dy}{dx} = 4x^3.$$

(b) Formula 3.8 allows us to differentiate each term separately; and by formulas 3.7 and 3.6 it follows that

$$\frac{dy}{dx} = 3(6x^5) - (-2x^{-3}) = 18x^5 + 2x^{-3}.$$

(c) If we write y in the form $y = (x/3) - 2x^{-1}$, then we can proceed as in (b):

$$\frac{dy}{dx} = \frac{1}{3}(1) - 2(-x^{-2}) = \frac{1}{3} + \frac{2}{x^2}.$$

■

EXAMPLE 3.7

If $f(x) = 3x^4 - 2$, evaluate $f'(1)$.

SOLUTION Since $f'(x) = 12x^3$, it follows that $f'(1) = 12$. Geometrically, 12 is the slope of the tangent line to the curve $y = 3x^4 - 2$ at the point $(1, 1)$ in Figure 3.12. Algebraically, the result implies that at $x = 1$, y changes 12 times as fast as x.

FIGURE 3.12

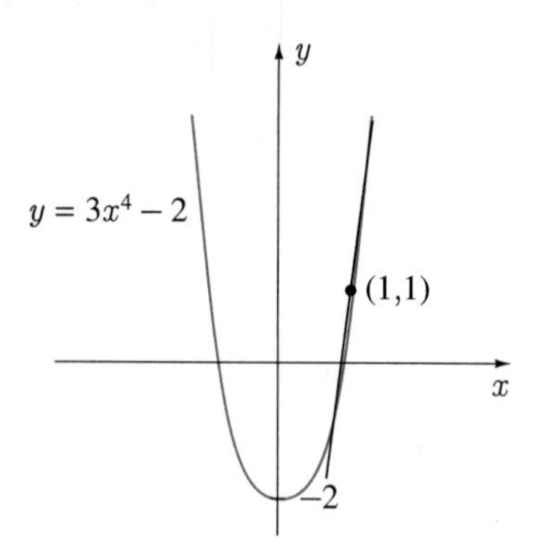

EXAMPLE 3.8

Find the equation of the tangent line to the curve $y = f(x) = x^3 + 5x$ at the point $(1,6)$.

SOLUTION Since $f'(x) = 3x^2 + 5$, the slope of the tangent line to the curve at $(1,6)$ is

$$f'(1) = 3(1)^2 + 5 = 8.$$

Using point-slope formula 1.10 for a straight line, we obtain for the equation of the tangent line at $(1,6)$

$$y - 6 = 8(x - 1) \qquad \text{or} \qquad 8x - y = 2.$$

The line through $(1,6)$ perpendicular to the tangent line in Figure 3.13 is called the **normal line** to the curve at $(1,6)$.

FIGURE 3.13

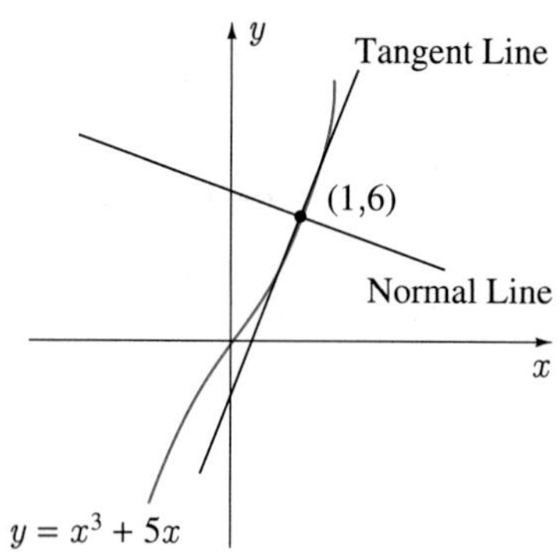

Since two lines are perpendicular only if their slopes are negative reciprocals (see equation 1.12), the normal line at $(1,6)$ must have slope $-1/8$. The equation of the normal line to $y = x^3 + 5x$ at $(1,6)$ is therefore

$$y - 6 = -\frac{1}{8}(x - 1) \qquad \text{or}$$

$$x + 8y = 49.$$

EXAMPLE 3.9

Find points on the curve $y = x^4/4 - 2x^3/3 - x^2/2 + x$ at which the slope of the tangent line is -1.

SOLUTION The slope of the tangent line to the curve is -1 whenever

$$-1 = \frac{dy}{dx} = x^3 - 2x^2 - x + 1.$$

When the -1 is transposed to the right side, and the cubic polynomial is factored, the result is

$$0 = x^3 - 2x^2 - x + 2 = (x-1)(x+1)(x-2).$$

This equation has solutions $x = 1, -1$, and 2. Thus the points at which the slope of the tangent line is -1 are $(1, 1/12)$, $(-1, -7/12)$, and $(2, -4/3)$. ■

In applications, independent and dependent variables of functions $y = f(x)$ represent physical quantities and have units associated with them. Units for the derivative $f'(x)$ are units of y divided by units of x. We see this from equation 3.2 where units of the quotient $[f(x+h) - f(x)]/h$ are clearly units of y divided by units of x, and the limit of this quotient as $h \to 0$ does not alter these units. For example, if x measures length in metres and y measures mass in kilograms, the units of dy/dx are kilograms per metre (kilograms divided by metres).

Increment Notation

We have used the letter h to represent a small change in x when calculating derivatives. An alternative notation is sometimes more suggestive. A small change in x, called an **increment** in x, is often denoted by Δx. It is pronounced "delta x" since Δ is the capital letter delta in the Greek alphabet. When x is given an increment Δx in the function $y = f(x)$, the corresponding change or increment in y is denoted by Δy. It is equal to

$$\Delta y = f(x + \Delta x) - f(x). \tag{3.9}$$

With this notation, definition 3.2 takes the form

$$f'(x) = \lim_{\Delta x \to 0} \frac{\Delta y}{\Delta x}. \tag{3.10a}$$

The notation dy/dx for the derivative fits very nicely with increment notation,

$$\frac{dy}{dx} = \lim_{\Delta x \to 0} \frac{\Delta y}{\Delta x}; \tag{3.10b}$$

the derivative of y with respect to x is the limit of the change in y divided by the change in x as the change in x approaches zero. We use this notation in the following example.

EXAMPLE 3.10

Use equation 3.10 to calculate the derivative of $y = f(x) = 3x^2 - 2x$ and check your answer using the differentiation rules discussed at the beginning of this section.

SOLUTION Since

$$\begin{aligned}\Delta y &= f(x + \Delta x) - f(x)\\ &= [3(x + \Delta x)^2 - 2(x + \Delta x)] - [3x^2 - 2x]\\ &= 3[x^2 + 2x\Delta x + (\Delta x)^2] - 2x - 2\Delta x - 3x^2 + 2x\\ &= \Delta x(6x - 2 + 3\Delta x),\end{aligned}$$

equation 3.10 gives

$$\frac{dy}{dx} = \lim_{\Delta x \to 0} \frac{\Delta y}{\Delta x} = \lim_{\Delta x \to 0} \frac{\Delta x(6x - 2 + 3\Delta x)}{\Delta x} = \lim_{\Delta x \to 0} (6x - 2 + 3\Delta x) = 6x - 2.$$

Rules 3.8, 3.7, and 3.6 for differentiation of $3x^2 - 2x$ also yield $6x - 2$. ■

EXERCISES 3.2

In Exercises 1–20 find $f'(x)$.

1. $f(x) = 2x^2 - 3$

2. $f(x) = 3x^3 + 4x + 5$

3. $f(x) = 10x^2 - 3x$

4. $f(x) = 4x^5 - 10x^3 + 3x$

5. $f(x) = 1/x^2$

6. $f(x) = 2/x^3$

7. $f(x) = 5x^4 - 3x^3 + 1/x$

8. $f(x) = -\frac{1}{2x^2} + \frac{3}{x^4}$

9. $f(x) = x^{10} - \frac{1}{x^{10}}$

10. $f(x) = 5x^4 + \frac{1}{4x^5}$

11. $f(x) = 5x^{-4} + \frac{1}{4x^{-5}}$

12. $f(x) = \sqrt{x}$

13. $f(x) = \frac{3}{x^2} + \frac{2}{\sqrt{x}}$

14. $f(x) = \frac{1}{x^{3/2}} + x^{3/2}$

15. $f(x) = 2x^{1/3} - 3x^{2/3}$

16. $f(x) = \pi x^{\pi}$

17. $f(x) = (x^2 + 2)^2$

18. $f(x) = (4x^6 - x^2)/x^5$

19. $f(x) = x^{5/3} - x^{2/3} + 3$

20. $f(x) = (2x + 5)^3$

In Exercises 21–24 find equations for the tangent and normal lines to the curve at the point indicated. In each case, sketch the curve and lines.

21. $y = x^2 - 2x + 5$ at $(2,5)$

22. $y = \sqrt{x} + 5$ at $(4,7)$

23. $y = 2x^3 - 3x^2 - 12x$ at $(2,-20)$

24. $x = \sqrt{y+1}$ at $(3,8)$

25. Find the points on the curve $y = x^4/4 - 2x^3/3 - 19x^2/2 + 22x$ at which the slope of the tangent line is 2.

26. An electronics firm manufactures a device called a "gadget". The money that it receives from the sale of x gadgets is the total revenue and is denoted by $R(x)$. The total cost $C(x)$ of producing these x gadgets also depends on x. The total profit realized on the sale is

$$P(x) = R(x) - C(x).$$

The derivatives of these functions, $R'(x)$, $C'(x)$, and $P'(x)$ are called respectively, marginal revenue, marginal cost, and marginal profit. Calculate these marginal functions if

$$R(x) = 75x - \frac{7x^2}{100} \quad \text{and}$$

$$C(x) = 50x - \frac{x^2}{30} - \frac{x^3}{60\,000},$$

both functions being defined for $1 \le x \le 800$.

27. Sketch a graph of the function $x = f(t) = t^3 - 8t^2$. Find the value(s) of t at which the tangent line to this curve is parallel to the line $x = 6t - 3$.

28. At what point(s) on the curve $y = x^3 + x^2 - 22x + 20$ does the tangent line pass through the origin?

29. At what point(s) on the parabola $y = x^2$ does the normal line pass through the point $(2, 5)$? Can you suggest an application of this result?

30. Show that the sum of the x- and y-intercepts of the tangent line to the curve $\sqrt{x} + \sqrt{y} = \sqrt{a}$ is always equal to a.

31. Prove that

$$\frac{d}{dx}(ax + b)^n = an(ax + b)^{n-1}$$

when a and b are constants.

32. Find a formula for

$$\frac{d}{dx}|x|^n,$$

when $n > 1$ is an integer.

33. Find the two points on the curve $y = x(1 + 2x - x^3)$ that share a common tangent line.

SECTION 3.3

Differentiability and Continuity

Many functions fail to have a derivative at isolated points. We examine such behaviour in this section. For example, consider the function $f(x) = |x|$ shown in Figure 3.14.

It is clear that for $x > 0$, $f'(x) = 1$, and for $x < 0$, $f'(x) = -1$. At $x = 0$, however, there is a problem. If $f(x)$ is to have a derivative at $x = 0$, it must be given by

$$\lim_{h \to 0} \frac{f(0 + h) - f(0)}{h} = \lim_{h \to 0} \frac{|h|}{h}.$$

But this limit does not exist since

$$\lim_{h \to 0^-} \frac{|h|}{h} = -1 \qquad \text{and} \qquad \lim_{h \to 0^+} \frac{|h|}{h} = 1.$$

Consequently, $f(x) = |x|$ does not have a derivative at $x = 0$.

The same conclusion can be drawn at any point at which the graph of a function takes an abrupt change in direction. Such a point is often called a *corner*. As a result, the function in Figure 3.15 does not have a derivative at $x = a, b, c$, or d. It is also true that a function cannot have a derivative at a point where the function is discontinuous (see, for example, the discontinuities in Figure 2.14). This result is an immediate consequence of the following theorem.

FIGURE 3.14

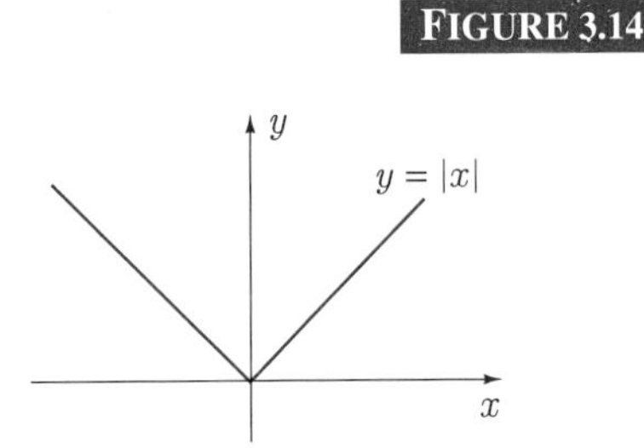

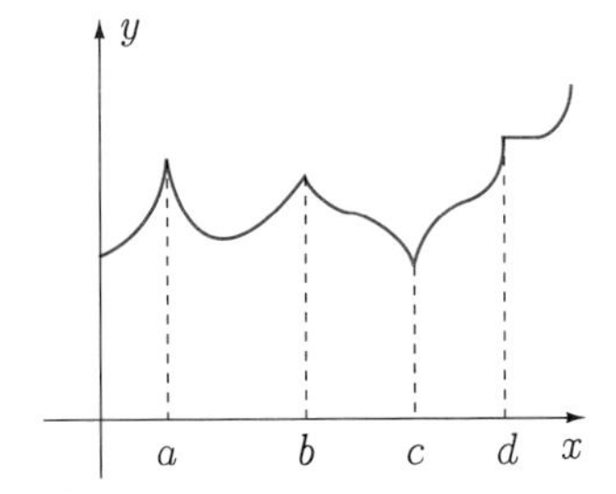

FIGURE 3.15

Theorem 3.6

If a function $f(x)$ has a derivative at $x = a$, then $f(x)$ is continuous at $x = a$.

Proof To prove this theorem we show that $\lim_{x \to a} f(x) = f(a)$, the condition that defines continuity of $f(x)$ at $x = a$ (Definition 2.1 in Section 2.4). The derivative $f'(a)$ is defined by

$$f'(a) = \lim_{h \to 0} \frac{f(a + h) - f(a)}{h}.$$

No one will argue with the fact that

$$0 = \lim_{h \to 0} h.$$

But note that if we multiply the left-hand and right-hand sides of these two equations, we obtain

$$0 \cdot f'(a) = \left[\lim_{h \to 0} h\right]\left[\lim_{h \to 0} \frac{f(a+h) - f(a)}{h}\right].$$

Using part (iv) of Theorem 2.1, we may bring the limits on the right together,

$$0 = \lim_{h \to 0} [f(a+h) - f(a)], \qquad \text{or}$$

$$\lim_{h \to 0} f(a+h) = f(a).$$

But this is just another way of saying that

$$\lim_{x \to a} f(x) = f(a),$$

(see Exercises 34 in Section 2.4).

As a corollary to this theorem we have the following.

Corollary *If $f(x)$ is discontinuous at $x = a$, then $f'(a)$ does not exist.*

We introduced this section by showing that the derivative of $|x|$ is 1 when $x > 0$; it is -1 when $x < 0$; and it does not exist at $x = 0$. These can be combined into the simple formula

$$\frac{d}{dx}|x| = \frac{|x|}{x}. \tag{3.11}$$

It is straightforward to generalize 3.11 and obtain the derivative of $|f(x)|$ at any point at which $f(x) \neq 0$ and $f'(x)$ exists. When $f(x) > 0$, we may write

$$\frac{d}{dx}|f(x)| = \frac{d}{dx}f(x) = f'(x).$$

On the other hand, when $f(x) < 0$, we have

$$\frac{d}{dx}|f(x)| = \frac{d}{dx}[-f(x)] = -f'(x).$$

Both of these results are contained in the one equation

$$\frac{d}{dx}|f(x)| = \frac{|f(x)|}{f(x)} f'(x). \tag{3.12}$$

This then is the derivative of $|f(x)|$ at any point at which $f(x) \neq 0$. The exceptional case when $f(x) = 0$ is discussed in Exercise 20.

Right-hand and Left-hand Derivatives

When a function $f(x)$ is defined on a closed interval $b \leq x \leq c$, equation 3.2 can be used to calculate $f'(x)$ only at points in the open interval $b < x < c$. For instance, it is impossible to evaluate

$$f'(b) = \lim_{h \to 0} \frac{f(b+h) - f(b)}{h}$$

since $f(x)$ is not defined for $x < b$ and therefore $f(b+h)$ is not defined for $h < 0$. When a function $f(x)$ is defined only to the right of a point, we define a right-hand derivative at the point; and when $f(x)$ is defined only to the left of a point, we define its left-hand derivative.

Definition 3.2

The **right-hand derivative** of $f(x)$ with respect to x is defined as

$$f'_+(x) = \lim_{h \to 0^+} \frac{f(x+h) - f(x)}{h}, \tag{3.13a}$$

provided the limit exists. The **left-hand derivative** of $f(x)$ is

$$f'_-(x) = \lim_{h \to 0^-} \frac{f(x+h) - f(x)}{h}, \tag{3.13b}$$

if the limit exists.

When a function has a derivative at a point x, its right- and left-hand derivatives both exist at x and are equal to $f'(x)$. It is possible, however, for a function to have both a left-hand and a right-hand derivative at a point but not a derivative. An example of this is the absolute value function $f(x) = |x|$ at $x = 0$ (see Figure 3.14). Its right-hand derivative at $x = 0$ is equal to 1 and its left-hand derivative there is -1.

When a function has a derivative at $x = a$, we say that it is **differentiable** at $x = a$. When it has a derivative at every point in some interval, we say that it is differentiable on that interval. In the event that the interval is closed, $b \le x \le c$, we understand that derivatives at $x = b$ and $x = c$ mean right-hand and left-hand derivatives respectively.

EXAMPLE 3.11

Is the function

$$g(x) = \begin{cases} x \sin(1/x) & x \neq 0 \\ 0 & x = 0 \end{cases}$$

differentiable at $x = 0$?

SOLUTION We encountered the function $x \sin(1/x)$ in Example 2.8 of Section 2.1 and drew its graph (except for the point at $x = 0$) in Figure 2.6. According to equation 3.2, the derivative of $g(x)$ at $x = 0$ is

$$g'(0) = \lim_{h \to 0} \frac{g(0+h) - g(0)}{h},$$

provided the limit exists. When we substitute from the definition of $g(x)$ for $g(h)$ and $g(0)$, this limit takes the form

$$g'(0) = \lim_{h \to 0} \frac{h \sin(1/h) - 0}{h} = \lim_{h \to 0} \sin\left(\frac{1}{h}\right).$$

Since this limit does not exist (see Example 2.8), $g(x)$ is not differentiable at $x = 0$. ■

EXERCISES 3.3

In Exercises 1–6 determine whether the function has a right-hand derivative, a left-hand derivative, and a derivative at the given value of x.

1. $f(x) = |x-5|$ at $x = 5$

2. $f(x) = x^{3/2}$ at $x = 0$

3. $f(x) = |x-5|^3$ at $x = 5$

4. $f(x) = \operatorname{sgn} x$ at $x = 0$ (see Exercise 36 in Section 2.4)

5. $f(x) = (x^2 - 1)/(x-1)$ at $x = 1$

6. $f(x) = H(x-a)$ at $x = a$ (see Exercise 38 in Section 2.4)

7. Does it make any difference in Exercise 6 if we define $H(a) = 0$?

8. Does it make any difference in Exercise 6 if we define $H(a) = 1$?

9. Does it make any difference in Exercise 4 if $\operatorname{sgn} x$ does not have a value at $x = 0$?

In Exercises 10–13 determine whether the statement is true or false.

10. If a function $f(x)$ has a derivative at $x = a$, then its graph has a tangent line at the point $(a, f(a))$.

11. If a function $f(x)$ has a tangent line at a point $(a, f(a))$, then it has a derivative at $x = a$.

12. If a function does not have a tangent line at $(a, f(a))$, then it does not have a derivative at $x = a$.

13. If a function $f(x)$ does not have a derivative at $x = a$, then it does not have a tangent line at $(a, f(a))$.

In Exercises 14–16 show algebraically that $f'(0)$ does not exist. Sketch a graph of the function.

14. $f(x) = x^{1/3}$

15. $f(x) = x^{2/3}$

16. $f(x) = x^{1/4}$

17. Sketch a graph of the function $f'(x)$ when $f(x) = [[x]]$ is the greatest integer function (Exercise 36 in Section 1.5).

18. Is the function $f(x) = x|x|$ differentiable at $x = 0$?

19. Find $f'(x)$ if $f(x) = |x| + |x-1|$. Sketch graphs of $f(x)$ and $f'(x)$.

20. If $f(x)$ is a differentiable function, does $|f(x)|$ have a derivative at points where $f(x) = 0$? Hint: Draw some pictures.

21. If $\lim_{x\to\infty} f(x) = L$ (so that $y = L$ is a horizontal asymptote for the graph of the function), is it necessary that $\lim_{x\to\infty} f'(x) = 0$?

22. Is the function

$$f(x) = \begin{cases} x^2 \sin(1/x) & x \neq 0 \\ 0 & x = 0 \end{cases}$$

differentiable at $x = 0$?

23. For what values of the real number n is the function

$$f(x) = \begin{cases} x^n \sin(1/x) & x \neq 0 \\ 0 & x = 0 \end{cases}$$

differentiable at $x = 0$?

24. Is the function

$$f(x) = \begin{cases} x^2 & x \text{ a rational number} \\ 0 & x \text{ an irrational number} \end{cases}$$

differentiable at $x = 0$?

25. Show that the derivative of the function in Exercise 22 is not continuous at $x = 0$.

SECTION 3.4

Product and Quotient Rules

In this section we add two more formulas to those of Section 2 for calculating derivatives. The first is a rule for differentiating a function that is the product of two other functions.

Theorem 3.7

If $p(x) = f(x)g(x)$, *where* $f(x)$ *and* $g(x)$ *are differentiable, then*

$$p'(x) = f(x)g'(x) + f'(x)g(x). \tag{3.14a}$$

Proof By equation 3.2,

$$\begin{aligned} p'(x) &= \lim_{h\to 0} \frac{p(x+h) - p(x)}{h} \\ &= \lim_{h\to 0} \frac{f(x+h)g(x+h) - f(x)g(x)}{h}. \end{aligned}$$

To organize this quotient further, we add and subtract the quantity $f(x+h)g(x)$ in the numerator:

$$\begin{aligned} p'(x) &= \lim_{h\to 0} \left\{ \frac{[f(x+h)g(x+h) - f(x+h)g(x)] + [f(x+h)g(x) - f(x)g(x)]}{h} \right\} \\ &= \lim_{h\to 0} \left[f(x+h)\frac{g(x+h) - g(x)}{h} + g(x)\frac{f(x+h) - f(x)}{h} \right] \\ &= f(x)g'(x) + g(x)f'(x). \end{aligned}$$

In taking the limit of the first term, we have used the fact that $\lim_{h\to 0} f(x+h) = f(x)$, which follows from continuity of $f(x)$ (see Theorem 3.6 and Exercise 34 in Section 2.4).

This result is called the **product rule** for differentiation. If we set $u = f(x)$ and $v = g(x)$, then the product rule may also be expressed in the form

$$\frac{d}{dx}(uv) = u\frac{dv}{dx} + v\frac{du}{dx}. \tag{3.14b}$$

We use increment notation to prove the **quotient rule**. Use of h in place of Δx works equally well.

Theorem 3.8

If $p(x) = f(x)/g(x)$, *where* $f(x)$ *and* $g(x)$ *are differentiable, then*

$$p'(x) = \frac{g(x)f'(x) - f(x)g'(x)}{[g(x)]^2}. \tag{3.15a}$$

Proof Using equation 3.10,

$$\begin{aligned} p'(x) &= \lim_{\Delta x\to 0} \frac{p(x+\Delta x) - p(x)}{\Delta x} \\ &= \lim_{\Delta x\to 0} \frac{1}{\Delta x}\left[\frac{f(x+\Delta x)}{g(x+\Delta x)} - \frac{f(x)}{g(x)}\right] \\ &= \lim_{\Delta x\to 0} \frac{f(x+\Delta x)g(x) - g(x+\Delta x)f(x)}{\Delta x g(x)g(x+\Delta x)}. \end{aligned}$$

To simplify this limit we add and subtract $f(x)g(x)$ in the numerator:

$$p'(x) = \lim_{\Delta x \to 0} \frac{[f(x+\Delta x)g(x) - f(x)g(x)] - [g(x+\Delta x)f(x) - f(x)g(x)]}{\Delta x g(x)g(x+\Delta x)}$$

$$= \lim_{\Delta x \to 0} \frac{1}{g(x)g(x+\Delta x)} \left\{ g(x)\left[\frac{f(x+\Delta x) - f(x)}{\Delta x}\right] - f(x)\left[\frac{g(x+\Delta x) - g(x)}{\Delta x}\right]\right\}$$

$$= \frac{1}{[g(x)]^2} \left\{ g(x) \lim_{\Delta x \to 0} \frac{f(x+\Delta x) - f(x)}{\Delta x} - f(x) \lim_{\Delta x \to 0} \frac{g(x+\Delta x) - g(x)}{\Delta x} \right\}$$

$$= \frac{g(x)f'(x) - f(x)g'(x)}{[g(x)]^2}.$$

If we set $u = f(x)$ and $v = g(x)$, then quotient rule 3.15a can be expressed in the form

$$\frac{d}{dx}\left(\frac{u}{v}\right) = \frac{v\dfrac{du}{dx} - u\dfrac{dv}{dx}}{v^2}. \tag{3.15b}$$

It makes no difference which term in product rule 3.14 is written first; it does make a difference in the quotient rule. Do not interchange the terms in 3.15b.

EXAMPLE 3.12

For the following three functions, find $f'(x)$ in simplified form:

(a) $f(x) = (x^2+2)(x^4+5x^2+1)$ (b) $f(x) = \dfrac{\sqrt{x}}{3x^2-2}$ (c) $f(x) = \dfrac{1}{(5x^3-2)^2}$

SOLUTION (a) With product rule 3.14a,

$$f'(x) = (x^2+2)\frac{d}{dx}(x^4+5x^2+1) + (x^4+5x^2+1)\frac{d}{dx}(x^2+2)$$
$$= (x^2+2)(4x^3+10x) + (x^4+5x^2+1)(2x)$$
$$= 6x^5 + 28x^3 + 22x.$$

(b) With quotient rule 3.15,

$$f'(x) = \frac{(3x^2-2)\left(\dfrac{1}{2\sqrt{x}}\right) - \sqrt{x}(6x)}{(3x^2-2)^2}$$
$$= \frac{\dfrac{3x^2 - 2 - 12x^2}{2\sqrt{x}}}{(3x^2-2)^2}$$
$$= -\frac{9x^2+2}{2\sqrt{x}(3x^2-2)^2}.$$

(c) If we express $f(x)$ in the form

$$f(x) = \left(\frac{1}{5x^3-2}\right)\left(\frac{1}{5x^3-2}\right),$$

and use the product rule,

$$f'(x) = \frac{1}{5x^3 - 2}\frac{d}{dx}\left(\frac{1}{5x^3 - 2}\right) + \frac{1}{5x^3 - 2}\frac{d}{dx}\left(\frac{1}{5x^3 - 2}\right)$$
$$= \frac{2}{5x^3 - 2}\frac{d}{dx}\left(\frac{1}{5x^3 - 2}\right).$$

The quotient rule now gives

$$f'(x) = \frac{2}{5x^3 - 2}\left[\frac{(5x^3 - 2)(0) - 1(15x^2)}{(5x^3 - 2)^2}\right]$$
$$= \frac{-30x^2}{(5x^3 - 2)^3}.$$

■

Differentiation in part (c) of this example was somewhat cumbersome considering the simplicity of the function. Imagine the difficulty had the power been 12 instead of 2. Fortunately, in Section 3.6 we develop a very simple rule for differentiation of functions like this.

EXAMPLE 3.13

Sketch a graph of the function $y = f(x) = (x - 1)/(x + 2)$. Calculate dy/dx and show qualitatively that the graph agrees with your calculation.

SOLUTION The x- and y-intercepts of the graph are 1 and $-1/2$ respectively. Since $f(x)$ is discontinuous at $x = -2$, we calculate

$$\lim_{x \to -2^-} f(x) = \infty \qquad \text{and} \qquad \lim_{x \to -2^+} f(x) = -\infty$$

This information, together with the following limits leads to the graph in Figure 3.16,

$$\lim_{x \to -\infty} f(x) = 1^+ \qquad \text{and} \qquad \lim_{x \to \infty} f(x) = 1^-.$$

Using quotient rule 3.15b, we find

$$\frac{dy}{dx} = \frac{(x + 2)(1) - (x - 1)(1)}{(x + 2)^2} = \frac{3}{(x + 2)^2}.$$

The sketch in Figure 3.16 and dy/dx agree that:

(a) The slope of the curve is always positive;

(b) The slope becomes larger and larger as x approaches -2; i.e.,

$$\lim_{x \to -2} f'(x) = \infty;$$

(c) The slope approaches zero as x approaches $\pm\infty$; i.e.,

$$\lim_{x \to \pm\infty} f'(x) = 0.$$

FIGURE 3.16

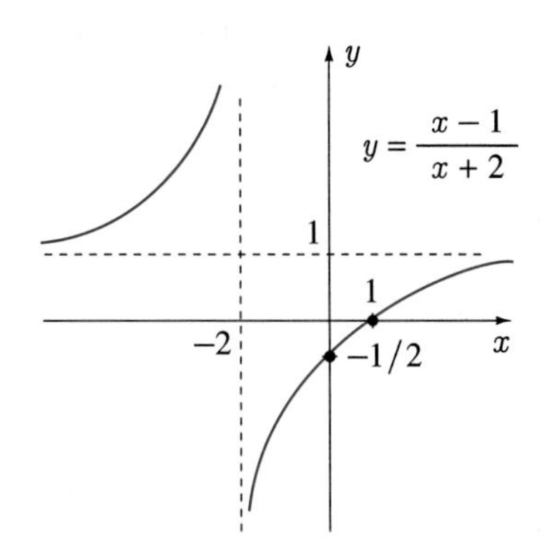

EXAMPLE 3.14

Sketch a graph of the function $f(x) = (x - 4)/x^2$.

SOLUTION The x-intercept of the graph is $x = 4$. Since the function is discontinuous at $x = 0$, we calculate

$$\lim_{x\to 0^-} f(x) = -\infty \qquad \text{and} \qquad \lim_{x\to 0^+} f(x) = -\infty.$$

Limits as $x \to \pm\infty$ are

$$\lim_{x\to -\infty} f(x) = 0^- \qquad \text{and} \qquad \lim_{x\to \infty} f(x) = 0^+.$$

This information, shown in Figure 3.17(a), would lead us to suspect that the graph should be completed as the dashed lines indicate. To show that this is indeed the case, we calculate

$$f'(x) = \frac{x^2(1) - (x-4)(2x)}{x^4} = \frac{8-x}{x^3}.$$

Since

$$\begin{aligned} f'(x) &< 0 \quad \text{for } x < 0, \\ f'(x) &> 0 \quad \text{for } 0 < x < 8, \\ f'(x) &= 0 \quad \text{for } x = 8, \\ f'(x) &< 0 \quad \text{for } x > 8, \end{aligned}$$

our suspicions were correct, and the final graph is shown in Figure 3.17(b).

FIGURE 3.17

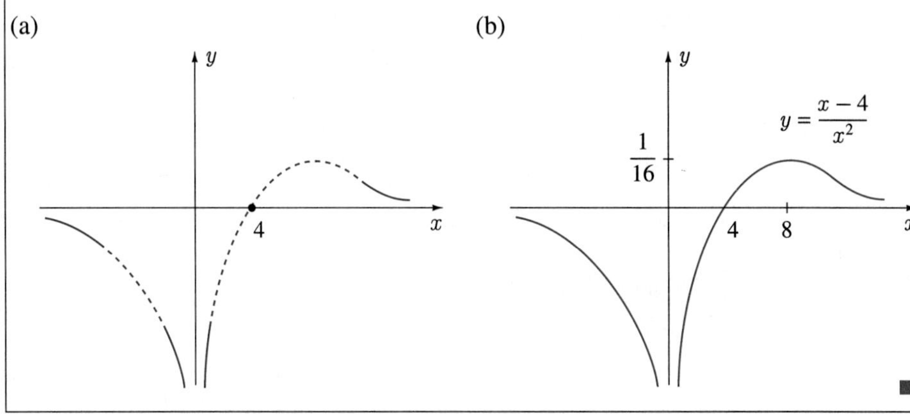

Examples 3.13 and 3.14 illustrate that derivatives can be helpful in sketching graphs. A complete discussion of the application of derivatives to curve sketching will be given in Section 4.5.

Angle Between Intersecting Curves

The angle θ between two curves which intersect at a point (x_0, y_0) (Figure 3.18) is defined as the angle between the tangent lines to the curves at (x_0, y_0). This can be calculated using formula 1.14 once slopes of the tangent lines are known. In the event that $\theta = \pi/2$ (Figure 3.19), the curves are said to be **orthogonal** or perpendicular at (x_0, y_0). Should $\theta = 0$, the curves are said to be tangent at (x_0, y_0) (Figure 3.20).

FIGURE 3.18

FIGURE 3.19

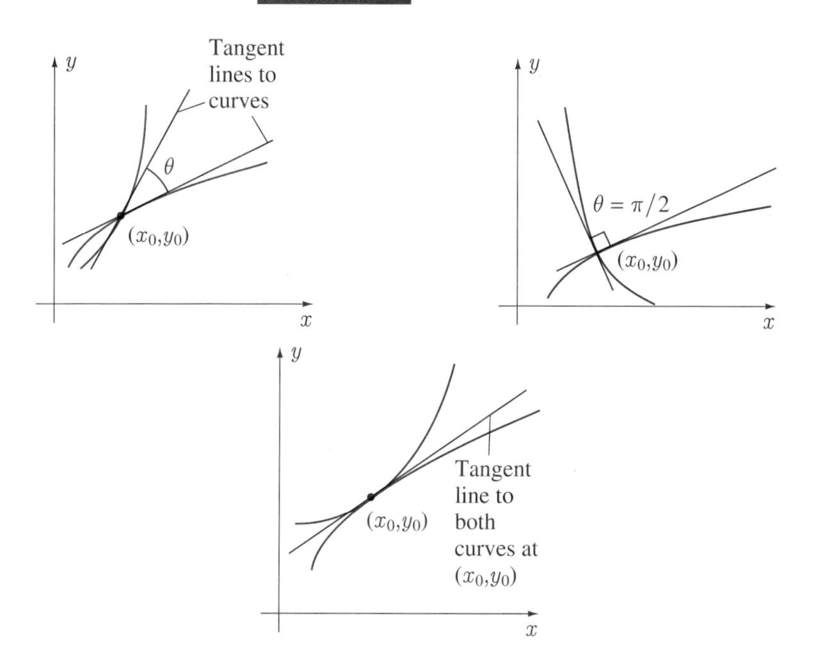

FIGURE 3.20

EXAMPLE 3.15

Find the angle between the curves $y = 6 - x^2$ and $y = 3x/(x + 1)$ at their point of intersection.

SOLUTION The curves intersect at the point, or points, with x-coordinate satisfying

$$6 - x^2 = \frac{3x}{x + 1}.$$

When we multiply both sides by $x + 1$, and simplify, the result is

$$0 = x^3 + x^2 - 3x - 6 = (x - 2)(x^2 + 3x + 3).$$

Since the only (real) solution of this equation is $x = 2$, the point of intersection of the curves is $(2, 2)$ (Figure 3.21).

FIGURE 3.21

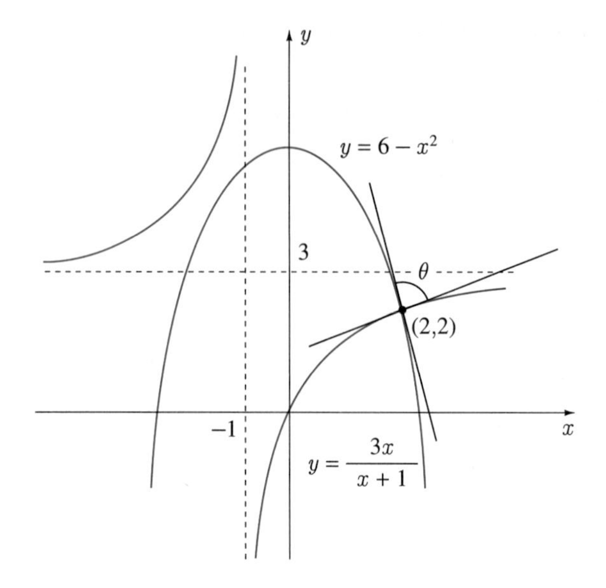

The slope of $y = f(x) = 6 - x^2$ at $(2,2)$ is

$$f'(2) = -2x_{|x=2} = -4.$$

When a quantity is suffixed at the subscript level with an expression like $|x = 2$, it means that the quantity is to be evaluated at $x = 2$. The slope of $y = g(x) = 3x/(x+1)$ at this point is

$$g'(2) = \frac{(x+1)(3) - 3x(1)}{(x+1)^2}_{|x=2} = \frac{1}{3}.$$

Using formula 1.14 with $m_1 = -4$ and $m_2 = 1/3$, the angle θ between the curves at $(2,2)$ is given by

$$\tan\theta = \left|\frac{-4 - 1/3}{1 + (-4)(1/3)}\right| = 13.$$

Hence, $\theta = 1.494$ radians. ■

EXERCISES 3.4

In Exercises 1–16 find $f'(x)$ in simplified form.

1. $f(x) = (x^2 + 2)(x + 3)$

2. $f(x) = (2 - x^2)(x^2 + 4x + 2)$

3. $f(x) = \dfrac{x}{3x + 2}$

4. $f(x) = \dfrac{x^2}{4x^2 - 5}$

5. $f(x) = \dfrac{x^2}{2x - 1}$

6. $f(x) = \dfrac{x^3}{4x^2 + 1}$

7. $f(x) = \sqrt{x}(x + 1)$

8. $f(x) = x\sqrt{x + 1}$

9. $f(x) = \dfrac{2x^2 - 5}{3x + 4}$

10. $f(x) = \dfrac{x + 5}{2x^2 - 1}$

11. $f(x) = \dfrac{x(x + 1)}{1 - 3x}$

12. $f(x) = \dfrac{x^2 + 2x + 3}{x^2 - 5x + 1}$

13. $f(x) = \dfrac{x^{1/3}}{1 - \sqrt{x}}$

14. $f(x) = \dfrac{\sqrt{x} + 2x}{\sqrt{x} - 4}$

15. $f(x) = \dfrac{1}{x^3 - 3x^2 + 2x + 5}$

16. $f(x) = \dfrac{x^3 + 3x^2 + 3x + 10}{(x + 1)^3}$

17. Find equations for the tangent and normal lines to the curve $y = (x + 3)/(x - 4)$ at the point $(1, -4/3)$. Sketch the curve and lines.

18. (a) Sketch a graph of the function $f(x) = x^2/(x^2 + x - 2)$.

(b) Find $f'(x)$ and show that it agrees with your sketch.

19. If the total cost of producing x items of a commodity is given by the equation

$$C(x) = ax\left(\frac{x+b}{x+c}\right),$$

where a, b, and c are constants, show that the marginal cost $C'(x)$ is

$$a\left[1 + \frac{c(b-c)}{(x+c)^2}\right].$$

20. Find a rule for the derivative of the product of three functions $f(x)$, $g(x)$, and $h(x)$.

In Exercises 21–24 find the angle (or angles) between the curves at their point (or points) of intersection.

21. $y = x^2, \quad x + y = 2$ **22.** $y = x^2, \quad y = 1 - x^2$

23. $y = x(x-5), \quad y = x\sqrt{4-x}$

24. $y = 2x + 2, \quad y = x^2/(x-1)$

In Exercises 25 and 26 show that the curves are orthogonal.

25. $x - 2y + 1 = 0, \quad y = 2 - x^2$

26. $y = 3 - x^2, \quad 4y - 7 = x^2$

In Exercises 27 and 28 show that the curves are tangent at the indicated point.

27. $y = x - 2x^2, \quad y = x^3 + 2x$ at $(-1, -3)$

28. $y = x^3, \quad y = x^2 + x - 1$ at $(1, 1)$

29. (a) The equation $x + 2y = C_1$, where C_1 is a constant, represents a *one-parameter family of curves*. For each value of C_1, the parameter, a different curve in the family is obtained. Draw curves corresponding to $C_1 = -2, -1, 0, 1, 2$.

(b) Draw a few curves from the one-parameter family $2x - y = C_2$ on the graph in (a).

(c) Two families of curves are said to be *orthogonal trajectories* if every curve from one family intersects every curve from the other family orthogonally. Are the families in (a) and (b) orthogonal trajectories?

30. For what value of k are the one-parameter families $2x - 3y = C_1$ and $x + ky = C_2$ orthogonal trajectories?

31. (a) A manufacturer's profit from the sale of x kilograms of a commodity per week is given by

$$P(x) = \frac{3x - 200}{x + 400}.$$

Sketch a graph of this function.

(b) The average profit per kilogram when x kilograms are sold is given by

$$p(x) = \frac{P(x)}{x}.$$

If a point (x, P) on the total profit curve is joined to the origin, the slope of this line is the average profit $p(x)$ for that x. Use this idea to find the sales level for highest average profit.

32. Find all points on the curve $y = (5 - x)/(6 + x)$ at which the tangent line passes through the origin.

33. Show that the ellipse $b^2x^2 + a^2y^2 = a^2b^2$ and the hyperbola $d^2x^2 - c^2y^2 = c^2d^2$ intersect orthogonally if $a^2 - b^2 = c^2 + d^2$.

SECTION 3.5

Higher-Order Derivatives

When $y = f(x)$ is a function of x, its derivative $f'(x)$ is also a function of x. We can therefore take the derivative of the derivative to get what is called the **second derivative** of the function. This can be repeated over and over again. For instance, if $y = f(x) = x^3 + 1/x$, then

$$\frac{dy}{dx} = 3x^2 - \frac{1}{x^2}.$$

We denote the second derivative of y with respect to x by d^2y/dx^2 or $f''(x)$:

$$f''(x) = \frac{d^2y}{dx^2} = \frac{d}{dx}\left(\frac{dy}{dx}\right) = 6x + \frac{2}{x^3}.$$

Similarly,

$$f'''(x) = \frac{d^3y}{dx^3} = 6 - \frac{6}{x^4}.$$

This is called the **third derivative** or the derivative of order three. Clearly we can continue the differentiation process indefinitely to produce derivatives of any positive integer order whatsoever.

EXAMPLE 3.16

Find a formula for the nth derivative of $y = x/(x+1)$.

SOLUTION When $n = 1$,

$$\frac{dy}{dx} = \frac{(x+1)(1) - x(1)}{(x+1)^2} = \frac{1}{(x+1)^2}.$$

When $n = 2$,

$$\frac{d^2y}{dx^2} = \frac{d}{dx}\left[\frac{1}{(x+1)^2}\right] = \frac{(x+1)^2(0) - \frac{d}{dx}(x+1)^2}{(x+1)^4}.$$

According to Exercise 31 in Section 3.2,

$$\frac{d}{dx}(x+1)^n = n(x+1)^{n-1}.$$

Thus,

$$\frac{d^2y}{dx^2} = \frac{-2(x+1)}{(x+1)^4} = \frac{-2}{(x+1)^3}.$$

When $n = 3$,

$$\frac{d^3y}{dx^3} = \frac{(x+1)^3(0) + 2(3)(x+1)^2}{(x+1)^6} = \frac{2 \cdot 3}{(x+1)^4}.$$

When $n = 4$,

$$\frac{d^4y}{dx^4} = \frac{(x+1)^4(0) - 2(3)(4)(x+1)^3}{(x+1)^8} = \frac{-2 \cdot 3 \cdot 4}{(x+1)^5}.$$

It is fairly clear that the pattern emerging is

$$\frac{d^ny}{dx^n} = \frac{(-1)^{n+1}n!}{(x+1)^{n+1}},$$

where the notation $n!$, called "n factorial," represents the product of the first n positive integers: $n! = 1(2)(3)\cdots(n-1)(n)$.

■

EXAMPLE 3.17

Find a formula for $d^2(uv)/dx^2$ if $u = f(x)$ and $v = g(x)$.

SOLUTION Since

$$\frac{d}{dx}(uv) = v\frac{du}{dx} + u\frac{dv}{dx},$$

then

$$\begin{aligned}\frac{d^2}{dx^2}(uv) &= \frac{d}{dx}\left(v\frac{du}{dx} + u\frac{dv}{dx}\right)\\ &= v\frac{d^2u}{dx^2} + \frac{dv}{dx}\frac{du}{dx} + u\frac{d^2v}{dx^2} + \frac{du}{dx}\frac{dv}{dx}\\ &= v\frac{d^2u}{dx^2} + 2\frac{du}{dx}\frac{dv}{dx} + u\frac{d^2v}{dx^2}.\end{aligned}$$

■

EXAMPLE 3.18

How many derivatives does $f(x) = x^{8/3}$ have at $x = 0$?

SOLUTION Since

$$f'(x) = \frac{8}{3}x^{5/3}, \qquad f''(x) = \left(\frac{8}{3}\right)\left(\frac{5}{3}\right)x^{2/3}, \qquad f'''(x) = \left(\frac{8}{3}\right)\left(\frac{5}{3}\right)\left(\frac{2}{3}\right)x^{-1/3},$$

and $f'''(0)$ is not defined, $f(x)$ has only a first and second derivative at $x = 0$. ■

EXERCISES 3.5

In Exercises 1–10 find the indicated derivative.

1. $f''(x)$ if $f(x) = x^3 + 5x^4$
2. $f'''(x)$ if $f(x) = x^3 - 3x^2 + 2x + 1$
3. $f''(2)$ if $f(x) = (x + 1)(x^3 + 3x + 2)$
4. $f'''(1)$ if $f(x) = x^4 - 3x^2 + 1/x$
5. $f''(x)$ if $f(x) = (x + 1)/\sqrt{x}$
6. $f'''(t)$ if $f(t) = t^3 - 1/t^3$
7. d^9y/dx^9 if $y = x^{10}$
8. $f''(u)$ if $f(u) = \sqrt{u}/(u + 1)$
9. d^6t/dx^6 if $t = x/(2x - 6)$
10. $f''(x)$ if $f(x) = x/(\sqrt{x} + 1)$
11. When a beam of length L and uniform mass m kilograms per metre is supported as shown in Figure 3.22, it bends under its own weight.

FIGURE 3.22

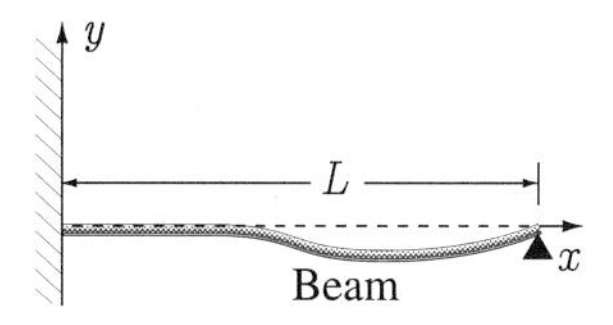

It can be shown that the deflection $y = f(x)$ from the horizontal must satisfy the equation

$$(EI)\frac{d^2y}{dx^2} = -\frac{mg}{2}(L - x)^2 + \frac{3}{8}mgL(L - x),$$

where E, I, and g are constants. Show that

$$y = f(x) = -\frac{mg}{48EI}(2x^4 - 5Lx^3 + 3L^2x^2)$$

satisfies this equation.

12. Steady-state temperature T in a region bounded by two concentric spheres of radii a and b (where $a < b$) must satisfy the equation

$$\frac{d^2T}{dr^2} + \frac{2}{r}\frac{dT}{dr} = 0,$$

where r is the radial distance from the common centre of the spheres.

(a) Show that for any constants c and d, the function

$$T = f(r) = c + \frac{d}{r}$$

satisfies the equation.

(b) If temperatures on the two spheres are maintained at constant values T_a and T_b, calculate c and d in terms of a, b, T_a, and T_b.

13. Find a and b in order that the function $y = f(x) = ax^3 + bx^2$ has its second derivative equal to zero when $x = 1$ and $y = 2$.

14. Find constants a, b, c, and d in order that the function $y = f(x) = ax^3+bx^2+cx+d$ has its first derivative equal to 4 when $x = 1$ and $y = 0$, and its second derivative equal to 5 when $x = 2$ and $y = 4$.

15. (a) Find a formula for $d^3(uv)/dx^3$ if $u = f(x)$ and $v = g(x)$.

(b) On the basis of part (a) and Example 3.17, could you hazard a guess at a formula for $d^4(uv)/dx^4$; i.e., do you see the pattern emerging? Exercise 18 asks you to verify the correct formula.

16. If n is a positive integer, find a formula for

$$\frac{d^n}{dx^n}\left(\frac{2x}{4-x^2}\right).$$

Hint: Decompose $2x/(4 - x^2)$ into two simpler fractions and proceed as in Example 3.16.

17. Evaluate

$$\frac{d^{2n}}{dx^{2n}}(x^2 - 1)^n$$

for n a positive integer.

18. If n is a positive integer, and u and v are functions of x, show by mathematical induction that

$$\frac{d^n}{dx^n}(uv) = \sum_{r=0}^{n}\binom{n}{r}\left(\frac{d^r u}{dx^r}\right)\left(\frac{d^{n-r}v}{dx^{n-r}}\right),$$

where $d^0u/dx^0 = u$, and $\binom{n}{r}$ are the binomial coefficients

$$\binom{n}{r} = \frac{n!}{r!\,(n-r)!}.$$

(If you are not familiar with sigma notation, see Section 6.1. Mathematical induction is discussed in Appendix C.)

SECTION 3.6

The Chain Rule and the Extended Power Rule

With the differentiation rules in Sections 3.2 and 3.4, we can differentiate any polynomial whatsoever. For instance, if $f(x) = x^3 - 3x^2 + 2x + 1$, then $f'(x) = 3x^2 - 6x + 2$. Consider the polynomial $f(x) = (2x^2 - 3)^8$, which has been conveniently factored for us. To find its derivative, we could expand $(2x^2 - 3)^8$, by the binomial theorem say, then differentiate, and finally simplify. We could also consider using the product rule over and over and over again. Thus, in spite of the fact that the rules of Sections 3.2 and 3.4 permit differentiation of $(2x^2 - 3)^8$, they are not convenient to use. Even more disturbing would be differentiation of, say, $g(x) = 1/(3x^2 + 8)^{12}$. In this section, we obtain results which enable us to differentiate much wider classes of functions, which include $(2x^2 - 3)^8$ and $1/(3x^2 + 8)^{12}$.

Suppose that y is defined as a function of u by

$$y = f(u) = \frac{u}{u+1},$$

and u, in turn, is defined as a function of x by

$$u = g(x) = \frac{\sqrt{x}}{x+2}.$$

These equations imply that y is a function of x; indeed, y is the composition of f and g,

$$y = (f \circ g)(x) = f(g(x)) = \frac{\dfrac{\sqrt{x}}{x+2}}{\dfrac{\sqrt{x}}{x+2} + 1}.$$

After some algebraic simplification we find that

$$y = \frac{\sqrt{x}}{\sqrt{x} + x + 2},$$

and we can therefore calculate

$$\frac{dy}{dx} = \frac{(\sqrt{x} + x + 2)\left(\dfrac{1}{2\sqrt{x}}\right) - \sqrt{x}\left(\dfrac{1}{2\sqrt{x}} + 1\right)}{(\sqrt{x} + x + 2)^2}.$$

This can be reduced to

$$\frac{dy}{dx} = \frac{2 - x}{2\sqrt{x}(\sqrt{x} + x + 2)^2}.$$

But notice that if we differentiate the original functions,

$$\frac{dy}{du} = \frac{(u+1)(1) - u(1)}{(u+1)^2} = \frac{1}{(u+1)^2} \quad \text{and}$$

$$\frac{du}{dx} = \frac{(x+2)\left(\dfrac{1}{2\sqrt{x}}\right) - \sqrt{x}(1)}{(x+2)^2} = \frac{2 - x}{2\sqrt{x}(x+2)^2}.$$

The product of these derivatives is

$$\frac{dy}{du}\frac{du}{dx} = \frac{2 - x}{2\sqrt{x}(x+2)^2(u+1)^2} = \frac{2 - x}{2\sqrt{x}(x+2)^2\left(\dfrac{\sqrt{x}}{x+2} + 1\right)^2} = \frac{2 - x}{2\sqrt{x}(\sqrt{x} + x + 2)^2}.$$

Consequently, for this example, we can write

$$\frac{dy}{dx} = \frac{dy}{du}\frac{du}{dx}.$$

The derivative of a composite function $y = f(g(x))$ can always be calculated by the above formula, according to the following theorem.

Theorem 3.9

(The chain rule) *If $y = f(u)$ and $u = g(x)$ are differentiable functions, then the derivative of the composite function $y = f(g(x))$ is*

$$\frac{dy}{dx} = \frac{dy}{du}\frac{du}{dx}. \tag{3.16}$$

Increment notation is particularly useful in proving the chain rule.

Proof By equation 3.10, the derivative of the composite function is

$$\frac{dy}{dx} = \lim_{\Delta x \to 0} \frac{\Delta y}{\Delta x} = \lim_{\Delta x \to 0} \frac{f(g(x + \Delta x)) - f(g(x))}{\Delta x}.$$

Now a change Δx in x produces a change $g(x + \Delta x) - g(x)$ in u. If we denote this change by Δu, then it in turn produces the change $f(u + \Delta u) - f(u)$ in y. We may write

$$\frac{dy}{dx} = \lim_{\Delta x \to 0} \frac{f(u + \Delta u) - f(u)}{\Delta x}.$$

Since $f(u)$ is differentiable, its derivative exists and is defined by

$$f'(u) = \lim_{\Delta u \to 0} \frac{f(u + \Delta u) - f(u)}{\Delta u}.$$

An equivalent way to express the fact that this limit is $f'(u)$ is to say that

$$\frac{f(u + \Delta u) - f(u)}{\Delta u} = f'(u) + \epsilon,$$

where ϵ must satisfy the condition $\lim_{\Delta u \to 0} \epsilon = 0$. We may write

$$f(u + \Delta u) - f(u) = [f'(u) + \epsilon]\Delta u,$$

and if we substitute this into the second expression for dy/dx above, we obtain

$$\frac{dy}{dx} = \lim_{\Delta x \to 0} \frac{[f'(u) + \epsilon]\Delta u}{\Delta x} = \lim_{\Delta x \to 0} \left\{[f'(u) + \epsilon]\frac{\Delta u}{\Delta x}\right\}.$$

But

$$\lim_{\Delta x \to 0} \frac{\Delta u}{\Delta x} = \lim_{\Delta x \to 0} \frac{g(x + \Delta x) - g(x)}{\Delta x} = \frac{du}{dx}.$$

Furthermore, since $g(x)$ is differentiable, it is continuous (Theorem 3.6), and this implies that $\Delta u \to 0$ as $\Delta x \to 0$. Consequently, $\lim_{\Delta x \to 0} \epsilon = \lim_{\Delta u \to 0} \epsilon = 0$, and these results give

$$\frac{dy}{dx} = f'(u)\frac{du}{dx} = \frac{dy}{du}\frac{du}{dx}.$$

This result is called the **chain rule** for the derivative of a composite function. It expresses the derivative of a composite function as the product of the derivatives of the functions in the composition. From the point of view of rates of change, the chain rule seems quite reasonable. It states that if a variable is defined in terms of a second variable, which is in turn defined in terms of a third variable, then the rate of change of the first variable with respect to the third is the rate of change of the first with respect to the second multiplied by the rate of change of the second with respect to the third. For example, if car A travels twice as fast as car B, and car B travels three times as fast as car C, then car A travels six times as fast as car C.

It is essential to understand the difference between the derivatives dy/dx and dy/du in 3.16. The second, dy/du, is the derivative of y regarded as a function of u, the given function $y = f(u)$. On the other hand, dy/dx is the derivative of y with respect to x, the derivative of the composite function $f(g(x))$.

EXAMPLE 3.19

Find dy/dt at $t = 4$ when $y = x^2 - x$ and $x = \sqrt{t}/(t+1)$.

SOLUTION By the chain rule

$$\frac{dy}{dt} = \frac{dy}{dx}\frac{dx}{dt} = (2x-1)\left[\frac{(t+1)(1/2)t^{-1/2} - \sqrt{t}(1)}{(t+1)^2}\right].$$

When $t = 4$, we find $x = \sqrt{4}/(4+1) = 2/5$. Once again we use the notation $\frac{dy}{dt}_{|t=4}$ to represent dy/dt evaluated at $t = 4$,

$$\frac{dy}{dt}_{|t=4} = [2(2/5) - 1]\left[\frac{(4+1)(1/2)(4)^{-1/2} - \sqrt{4}}{(4+1)^2}\right] = \frac{3}{500}.$$

■

The Extended Power Rule

When the function $f(u)$ in Theorem 3.9 is a power function, the chain rule gives what we call the **extended power rule**, often considered the most important differentiation formula of calculus.

Corollary *When* $u = g(x)$ *is differentiable,*

$$\frac{d}{dx}u^n = nu^{n-1}\frac{du}{dx}. \qquad (3.17)$$

It is essentially power rule 3.6 in Section 3.2 with an extra factor du/dx to account for the fact that what is under the power (u) is not just x, it is a function of x. In the special case that u is equal to x, 3.17 reduces to 3.6. It is important not to confuse these rules. Rule 3.6 can be used only for x^n; if anything other than x is raised to a power, formula 3.17 should be used. The most common error in using 3.17 is to forget the du/dx. With 3.17 it is a simple matter to differentiate the functions in the first paragraph of this section:

$$\frac{d}{dx}(2x^2-3)^8 = 8(2x^2-3)^7\frac{d}{dx}(2x^2-3) = 8(2x^2-3)^7(4x) = 32x(2x^2-3)^7 \quad \text{and}$$

$$\frac{d}{dx}\left[\frac{1}{(3x^2+8)^{12}}\right] = \frac{d}{dx}(3x^2+8)^{-12} = -12(3x^2+8)^{-13}\frac{d}{dx}(3x^2+8)$$
$$= \frac{-12}{(3x^2+8)^{13}}(6x) = \frac{-72x}{(3x^2+8)^{13}}.$$

Notice that in neither of these examples do you see the letter u, although rule 3.17 is stated in terms of u's. We could have introduced u, in the first example, say, by setting $u = 2x^2 - 3$, and proceeded as follows: With $u = 2x^2 - 3$,

$$\frac{d}{dx}(2x^2-3)^8 = \frac{d}{dx}u^8 = \frac{d}{du}(u^8)\frac{du}{dx} = 8u^7(4x) = 32xu^7 = 32x(2x^2-3)^7.$$

But this is unnecessary; with an understanding of 3.17, we should proceed directly to the derivatives without defining an intermediate variable. With a little practice, the writing should be shortened even more. For example, calculation of the derivative of $(2x^2-3)^8$ should appear as

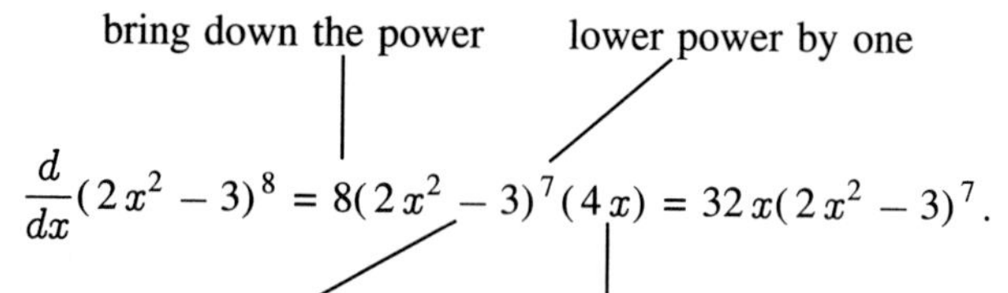

Here is an example to illustrate power rules within power rules within power rules.

EXAMPLE 3.20

Find the derivative of the function $f(x) = \sqrt{1+\left(2+\sqrt{1-3x}\right)^2}$.

SOLUTION If we write the function in the form

$$f(x) = \left[1+\left(2+\sqrt{1-3x}\right)^2\right]^{1/2},$$

then

$$\begin{aligned} f'(x) &= \frac{1}{2}\left[1+\left(2+\sqrt{1-3x}\right)^2\right]^{-1/2} \frac{d}{dx}\left[1+\left(2+\sqrt{1-3x}\right)^2\right] \\ &= \frac{1}{2\sqrt{1+\left(2+\sqrt{1-3x}\right)^2}}\left[2\left(2+\sqrt{1-3x}\right)\frac{d}{dx}\left(2+\sqrt{1-3x}\right)\right] \\ &= \frac{2+\sqrt{1-3x}}{\sqrt{1+\left(2+\sqrt{1-3x}\right)^2}}\left[\frac{1}{2}(1-3x)^{-1/2}\frac{d}{dx}(1-3x)\right] \\ &= \frac{2+\sqrt{1-3x}}{2\sqrt{1-3x}\sqrt{1+\left(2+\sqrt{1-3x}\right)^2}}(-3) \\ &= \frac{-3\left(2+\sqrt{1-3x}\right)}{2\sqrt{1-3x}\sqrt{1+\left(2+\sqrt{1-3x}\right)^2}}. \end{aligned}$$

■

EXERCISES 3.6

In Exercises 1–8 use the chain rule to find dy/dx.

1. $y = t^2 + \frac{1}{t}, \quad t = x^2 + 1$

2. $y = \frac{u}{u+1}, \quad u = \sqrt{x} + 1$

3. $y = (u^2+1)(u+1), \quad u = \frac{1}{\sqrt{x}-4}$

4. $y = \frac{s}{s^2-2}, \quad s = x^2 - 2x + 3$

5. $y = (v^2+v)(\sqrt{v}+1), \quad v = \frac{x}{x^2-1}$

6. $y = \frac{t+3}{t-4}, \quad t = \frac{x-2}{x+1}$

7. $y = \frac{t^2+3}{t-4}, \quad t = (3x+2)(x^2+4x)$

8. $y = u^2\left(1+\sqrt{u}\right), \quad u = \dfrac{x+1}{x-x^2}$

In Exercises 9–36 find dy/dx, where $y = f(x)$.

9. $f(x) = x(x^3+3)^4$

10. $f(x) = x\sqrt{x+1}$

11. $f(x) = x^2(2x+1)^2$

12. $f(x) = \dfrac{x}{\sqrt{2x+1}}$

13. $f(x) = (x+2)^2(x^2+3)$

14. $f(x) = \dfrac{(2x-1)^2}{3x+5}$

15. $f(x) = \dfrac{3x+5}{(2x-1)^2}$

16. $f(x) = x^3(2-5x^2)^{1/3}$

17. $f(x) = \dfrac{x^3}{(2-5x^2)^{1/3}}$

18. $f(x) = (x+1)^2(3x+1)^3$

19. $f(x) = \dfrac{x^{1/3}}{1-\sqrt{x}}$

20. $f(x) = \dfrac{\sqrt{2-3x}}{x^2}$

21. $f(x) = \left(\dfrac{x^3-1}{2x^3+1}\right)^4$

22. $f(x) = \sqrt[4]{\dfrac{2-x}{2+x}}$

23. $f(x) = (x^3-2x^2)^3(x^4-2x)^5$

24. $f(x) = (x+5)^4\sqrt{1+x^3}$

25. $f(x) = \dfrac{x\sqrt{1-x^2}}{(3+x)^{1/3}}$

26. $f(x) = x(x+5)^4\sqrt{1+x^3}$

27. $f(x) = \dfrac{x^2(x^3+3)^2}{(x-2)(x+5)^2}$

28. $f(x) = x\sqrt{1+x\sqrt{1+x}}$

29. $y = t^2 + \dfrac{1}{t^3}, \quad t = \sqrt{4-x^2}$

30. $y = (2s-s^2)^{1/3}, \quad s = \dfrac{1}{x^2+5}$

31. $y = \dfrac{v^2}{v^3-1}, \quad v = x\sqrt{x^2-1}$

32. $y = \dfrac{u}{u+5}, \quad u = \dfrac{\sqrt{x-1}}{x}$

33. $y = u^4(u^3-2u)^2, \quad u = \sqrt{x-2x^2}$

34. $y = t + \sqrt{t+\sqrt{t}}, \quad t = \dfrac{x^2+1}{x^2-1}$

35. $y = \left(\dfrac{v^2+1}{1-v^3}\right)^3, \quad v = \dfrac{1}{x^3+3x^2+2}$

36. $y = \dfrac{\sqrt{k}}{1+k+k^2}, \quad k = x(x^2+5)^5$

37. If $y = f(u)$, $u = g(s)$, and $s = h(x)$, show that

$$\frac{dy}{dx} = \frac{dy}{du}\frac{du}{ds}\frac{ds}{dx}.$$

38. When an electrostatic charge $q = 5 \times 10^{-6}$ coulombs is at a distance r metres from a stationary charge $Q = 3 \times 10^{-6}$ coulombs, the magnitude of the force of repulsion of Q on q is

$$F = \frac{Qq}{4\pi\epsilon_0 r^2} \text{ newtons,}$$

where $\epsilon_0 = 8.85 \times 10^{-12}$. If q is moved directly away from Q at 2 m/s, how fast is F changing when $r = 2$ m?

39. When a mass m of 5 kg is r metres from the centre of the earth, the magnitude of the force of attraction of the earth on m is

$$F = \frac{GmM}{r^2} \text{ newtons,}$$

where M is the mass of the earth and $G = 6.67 \times 10^{-11}$. If m is falling at 100 km/h when it is 5 km above the surface of the earth, how fast is F changing? The mean density of the earth is 5.52×10^3 kg/m^3, and its mean radius is 6370 km.

40. Show that if an equation can be solved for both y in terms of x, $y = f(x)$, and x in terms of y, $x = g(y)$, then

$$\frac{dy}{dx} = \frac{1}{\dfrac{dx}{dy}}.$$

41. Prove that the derivative of an odd function is an even function, and that the derivative of an even function is an odd function. Recall Definition 1.9 in Section 1.6.

42. Determine whether the following very simple proof of the chain rule has a flaw? If Δu denotes the change in

$u = g(x)$ resulting from a change Δx in x, and Δy is the change in $y = f(u)$ resulting from Δu, then

$$\begin{aligned}\frac{dy}{dx} &= \lim_{\Delta x \to 0} \frac{\Delta y}{\Delta x} = \lim_{\Delta x \to 0} \frac{\Delta y}{\Delta u}\frac{\Delta u}{\Delta x} \\ &= \left(\lim_{\Delta x \to 0} \frac{\Delta y}{\Delta u}\right)\left(\lim_{\Delta x \to 0} \frac{\Delta u}{\Delta x}\right) \\ &= \left(\lim_{\Delta u \to 0} \frac{\Delta y}{\Delta u}\right)\left(\lim_{\Delta x \to 0} \frac{\Delta u}{\Delta x}\right) \\ &= \frac{dy}{du}\frac{du}{dx}.\end{aligned}$$

In Exercises 43–46 find d^2y/dx^2.

43. $y = v^2 + v, \quad v = \dfrac{x}{x+1}$

44. $y = (u+1)^3 - \dfrac{1}{u}, \quad u = x + \sqrt{x+1}$

45. $y = \sqrt{t-1}, \quad t = (x + x^2)^2$

46. $y = \dfrac{s}{s+6}, \quad s = \dfrac{\sqrt{x}}{1+\sqrt{x}}$

In Exercises 47–54 assume that $f(u)$ is a differentiable function of u. Find the derivative of the given function with respect to x in as simplified a form as possible. Then set $f(u) = u^3 - 2u$ and simplify further.

47. $f(2x+3)$

48. $[f(3-4x)]^2$

49. $f(1-x^2)$

50. $f(x + 1/x)$

51. $f(f(x))$

52. $\sqrt{3 - 4[f(1-3x)]^2}$

53. $\dfrac{f(-x)}{3 + 2f(x^2)}$

54. $f(x - f(x))$

In Exercises 55–58 show that the families of curves are orthogonal trajectories (see question 29 in Exercises 3.4). Sketch both families of curves.

55. $y = mx, \quad x^2 + y^2 = r^2$

56. $y = ax^2, \quad x^2 + 2y^2 = c^2$

57. $x^2 - y^2 = C_1, \quad xy = C_2$

58. $2x^2 + 3y^2 = C^2, \quad y^2 = ax^3$

59. If $y = f(u)$ and $u = g(x)$, use the chain rule to show that

$$\frac{d^2y}{dx^2} = \frac{d^2u}{dx^2}\frac{dy}{du} + \frac{d^2y}{du^2}\left(\frac{du}{dx}\right)^2.$$

60. Use the result of Exercise 59 to find d^2y/dx^2 at $x = 1$ when

$$y = (u+1)^3, \quad u = 3x - \frac{2}{x^2}.$$

61. Find the rate of change of $y = f(x) = x^9 + x^6$ with respect to x^3.

62. Find the rate of change of $y = f(x) = \sqrt{1-x^2}$ with respect to $x/(x+1)$.

63. If $y = f(u)$ and $u = g(x)$, show that

$$\frac{d^3y}{dx^3} = \frac{d^3u}{dx^3}\frac{dy}{du} + 3\frac{d^2y}{du^2}\frac{d^2u}{dx^2}\frac{du}{dx} + \frac{d^3y}{du^3}\left(\frac{du}{dx}\right)^3.$$

64. At what point(s) on the hyperbola $x^2 - 16y^2 = 16$ do tangent lines pass through the point $(2,3)$?

65. Generalize the result of equation 3.12 to find a formula for

$$\frac{d}{dx}|f(x)|^n, \qquad \text{for } n > 1 \text{ an integer}.$$

66. The curve $x^2y^2 = (x+1)^2(4-x^2)$ is called a *conchoid of Nicomedes*. Sketch the curve and find points where its tangent line is horizontal.

SECTION 3.7

Implicit Differentiation

We say that y is defined **explicitly** as a function of x if the dependence of y on x is given in the form

$$y = f(x). \tag{3.18}$$

Examples are $y = x^2$, $y = 3x + \sin x$, and $y = 1/(x+1)$. In each case, the dependent variable stands alone on the left side of the equation. The differentiation rules in Theorems 3.1 to 3.5, 3.7, and 3.8 are applicable to explicitly defined functions.

An equation in x and y may define y as a function of x even when it is not in form (3.18). Equations for which y has not been separated out are often written in the generic form

$$F(x,y) = 0. \tag{3.19}$$

The notation $F(x, y)$ is used to denote an expression that depends on two variables x and y. For example, the volume of a right circular cylinder depends on its radius r and its height h; in particular, $V = \pi r^2 h$. In such a case we write that $V = F(r, h) = \pi r^2 h$. Examples of equations of form 3.19 are $y - x^3 = 0$ and $y^3 - 3y - 2x = 0$. Equation 3.19 is said to define y **implicitly** as a function of x in some domain D, if for each x in D, there is one, and only one, value of y for which (x, y) satisfies 3.19. The equation $y - x^3 = 0$ defines y implicitly as a function of x for all (real) x. The explicit definition of the function is $y = x^3$. The equation $y^3 - 3y - 2x = 0$ does not define y as a function of x. For each x in the interval $-1 < x < 1$, there are three solutions of the equation for y. For $x < -1$ and $x > 1$, the equation has only one value of y corresponding to each x. This is most easily seen from the graph of the curve in Figure 3.23. We shall have more to say about the equation $y^3 - 3y - 2x = 0$ later in this section.

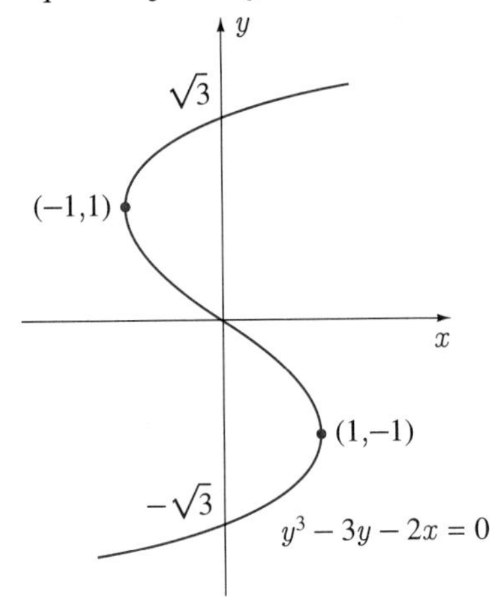

FIGURE 3.23

Each of the following equations defines y as a function of x, and does so implicitly:

$$y + x^2 - 2x + 5 = 0;$$
$$2y + x^2 + y^2 - 25 = 0, \quad y \geq -1;$$
$$x + y^5 + x^2 + y = 0.$$

It is easy to solve the first equation for the explicit definition of the function,

$$y = -x^2 + 2x - 5,$$

and find its derivative,

$$\frac{dy}{dx} = -2x + 2.$$

We can also obtain an explicit definition of the function defined by the second equation, but not so simply. If we write

$$y^2 + 2y + (x^2 - 25) = 0,$$

and use quadratic formula 1.17, we obtain

$$y = \frac{-2 \pm \sqrt{4 - 4(x^2 - 25)}}{2} = -1 \pm \sqrt{26 - x^2}.$$

Since y is required to be greater than or equal to -1, we must choose

$$y = -1 + \sqrt{26 - x^2}.$$

This is the explicit definition of the function, and it is now easy to find the derivative of y with respect to x:

$$\frac{dy}{dx} = \frac{1}{2\sqrt{26 - x^2}}(-2x) = \frac{-x}{\sqrt{26 - x^2}}.$$

The third equation is quite different, for it is impossible to solve this equation for an explicit definition of the function. Does this mean that it is also impossible to obtain dy/dx? The answer is no! To see this, we differentiate both sides of the equation with respect to x, keeping in mind that y is a function of x. Only the term in y^5 presents any difficulty, but its derivative can be calculated using extended power rule 3.17,

$$1 + 5y^4\frac{dy}{dx} + 2x + \frac{dy}{dx} = 0.$$

We can now solve this equation for dy/dx by grouping the two terms in dy/dx on one side of the equation, and transposing the remaining two terms:

$$(5y^4 + 1)\frac{dy}{dx} = -1 - 2x.$$

Division by $5y^4 + 1$ gives the required derivative,

$$\frac{dy}{dx} = -\frac{2x + 1}{5y^4 + 1}.$$

This process of differentiating an equation that implicitly defines a function is called **implicit differentiation**. It could also have been used in each of the first two examples. If $y + x^2 - 2x + 5 = 0$ is differentiated with respect to x, we have

$$\frac{dy}{dx} + 2x - 2 = 0,$$

from which, as before

$$\frac{dy}{dx} = -2x + 2.$$

When $2y + x^2 + y^2 - 25 = 0$ is differentiated with respect to x, we find

$$2\frac{dy}{dx} + 2x + 2y\frac{dy}{dx} = 0,$$

from which,

$$(2 + 2y)\frac{dy}{dx} = -2x,$$

and therefore

$$\frac{dy}{dx} = \frac{-x}{y + 1}.$$

Although this result appears different from the previous expression for dy/dx, when we recall that $y = -1 + \sqrt{26 - x^2}$, we find

$$\frac{dy}{dx} = \frac{-x}{-1 + \sqrt{26 - x^2} + 1} = \frac{-x}{\sqrt{26 - x^2}}.$$

These examples illustrate that if implicit differentiation is used to obtain a derivative, then the result may depend on y as well as x. Naturally if we require the derivative at a certain value of x, then the y-value used is determined by the original equation defining y implicitly as a function of x.

EXAMPLE 3.21

Assuming that y is defined implicitly as a function of x by the equation

$$x^3y^3 + x^2y + 2x = 12,$$

find dy/dx when $x = 1$.

SOLUTION When we differentiate both sides of the equation with respect to x, using the product rule on the first two terms, we find

$$3x^2y^3 + 3x^3y^2\frac{dy}{dx} + 2xy + x^2\frac{dy}{dx} + 2 = 0.$$

Thus,

$$(3x^3y^2 + x^2)\frac{dy}{dx} = -(2 + 2xy + 3x^2y^3) \qquad \text{and}$$

$$\frac{dy}{dx} = -\frac{2 + 2xy + 3x^2y^3}{3x^3y^2 + x^2}.$$

When $x = 1$ is substituted into the given equation defining y as a function of x, the result is

$$0 = y^3 + y - 10 = (y-2)(y^2 + 2y + 5),$$

and the only solution of this equation is $y = 2$. We now substitute $x = 1$ and $y = 2$ into the formula for dy/dx to calculate the derivative at $x = 1$,

$$\frac{dy}{dx}_{|x=1} = -\frac{2 + 2(1)(2) + 3(1)^2(2)^3}{3(1)^3(2)^2 + (1)^2} = -\frac{30}{13}.$$

■

Implicit differentiation can also be used to find second and higher order derivatives of functions that are defined implicitly. Calculations can be messy, but principles are the same. For example, when the equation $x^5 + y^3 + y^2 = 1$ defines y implicitly as a function of x, it is straightforward to calculate that

$$\frac{dy}{dx} = \frac{-5x^4}{3y^2 + 2y}.$$

To find the second derivative d^2y/dx^2, we differentiate both sides of this equation with respect to x. We use the quotient rule on the right, and when differentiating the denominator we once again keep in mind that y is a function of x. The result is

$$\frac{d^2y}{dx^2} = \frac{(3y^2 + 2y)(-20x^3) + 5x^4\left(6y\frac{dy}{dx} + 2\frac{dy}{dx}\right)}{(3y^2 + 2y)^2}.$$

We now replace dy/dx by its expression in terms of x and y,

$$\frac{d^2y}{dx^2} = \frac{-20x^3(3y^2 + 2y) + 5x^4(6y + 2)\left(\frac{-5x^4}{3y^2 + 2y}\right)}{(3y^2 + 2y)^2},$$

and bring the two terms in the numerator to a common denominator,

$$\frac{d^2y}{dx^2} = \frac{\dfrac{-20x^3(3y^2 + 2y)^2 - 25x^8(6y + 2)}{(3y^2 + 2y)}}{(3y^2 + 2y)^2}$$

$$= -\frac{20x^3(3y^2+2y)^2 + 25x^8(6y+2)}{(3y^2+2y)^3}.$$

It would not be a pleasant task to proceed to the third derivative of y with respect to x.

In the above examples we seem to have adopted the principle that both sides of an equation can be differentiated with respect to the same variable. But can all equations be differentiated? The answer is an emphatic no! For example, if we differentiate both sides of the equation $4x = 2x$, we obtain the ludicrous result that $4 = 2$. Obviously, then, this equation cannot be differentiated with respect to x.

Possibly the reason differentiation fails in the above example is that the equation contains only one variable, namely, x. Perhaps a more reasonable question might be: "Can every equation containing two variables be differentiated?" To answer this question we consider the situation of a ship heading northeast from a pier. Let us choose east as the positive x-direction and north as the positive y-direction, both originating from the pier (Figure 3.24). We will find the rate of change of the y-coordinate of the ship with respect to its x-coordinate when the ship is 2 km from the pier. Obviously the answer is 1, since the x- and y-coordinates change at the same rate. But consider the following argument.

FIGURE 3.24

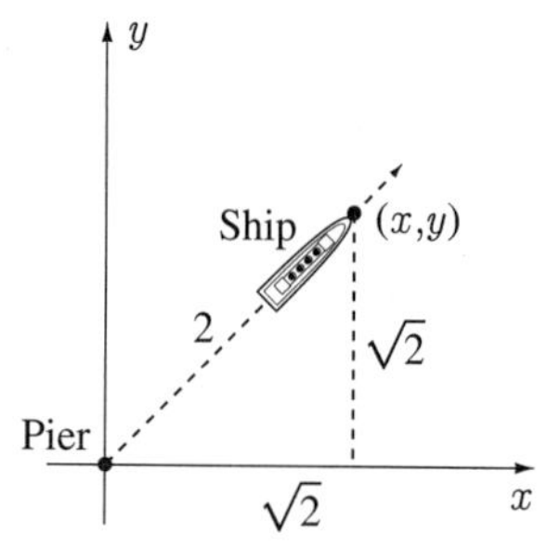

When the ship is 2 km from the pier, $x = y = \sqrt{2}$, and therefore

$$x^2 + y^2 = 4.$$

If we differentiate this equation with respect to x, we find that

$$2x + 2y\frac{dy}{dx} = 0, \qquad \text{or} \qquad \frac{dy}{dx} = -\frac{x}{y},$$

and with $x = y = \sqrt{2}$,

$$\frac{dy}{dx} = -1.$$

Differentiation of the equation $x^2 + y^2 = 4$, which contains two variables, has again led to an erroneous result.

These two examples have certainly illustrated that not all equations can be differentiated. What then distinguishes an equation that can be differentiated from one that cannot? The answer is that an equation can be differentiated with respect to a variable only if the equation is valid for a continuous range of values of that variable. The equation $4x = 2x$ cannot be differentiated because it is true only for $x = 0$. The equation $x^2 + y^2 = 4$ in the above example cannot be differentiated because *in that example* it is valid only when the ship is 2 km from the pier; i.e., only when $x = y = \sqrt{2}$. In other situations it might be acceptable to differentiate $x^2 + y^2 = 4$. For example, if at position $(\sqrt{2}, \sqrt{2})$, the ship is following the circular path defined by the equation $x^2 + y^2 = 4$, then it is indeed correct to differentiate $x^2 + y^2 = 4$ to find $dy/dx = -1$. This is an extremely important principle, and we will return to it many times. To emphasize it once again, *we may differentiate an equation with respect to a variable only if the equation is valid for a continuous range of values of that variable.*

Each of the three equations at the beginning of this section was given as defining y as a function of x, and as such defines the function for some domain of values for x. Differentiation of these equations was therefore acceptable according to the above principle.

When we are told that equation 3.19 defines y as a function of x, implicit differentiation leads to the derivative dy/dx. But how can we tell whether an equation $F(x,y) = 0$ defines y implicitly as a function of x? Additionally, given that $F(x,y) = 0$ does define y as a function of x, how do we know that the expression for dy/dx obtained by implicit differentiation is a valid representation of the derivative of the function? After all, functions do not always have derivatives at all points in their domains.

Answers to these questions are intimately related. To see how, suppose an equation $F(x,y) = 0$ in x and y is satisfied by a point (x_0, y_0), and when the equation is differentiated implicitly it leads to a quotient for dy/dx,

$$\frac{dy}{dx} = \frac{P(x,y)}{Q(x,y)}.$$

Mathematicians have shown that if $Q(x_0, y_0) \neq 0$, then the equation $F(x,y) = 0$ defines y implicitly as a function of x for some open interval containing x_0, and the derivative of this function at x_0 is $P(x_0, y_0)/Q(x_0, y_0)$. Proofs are usually given in advanced books on mathematical analysis. When $Q(x_0, y_0) = 0$, two possibilities exist. First, the equation $F(x,y) = 0$ might not define y as a function of x in an interval around x_0. Second, the equation might define a function, but the function does not have a derivative at x_0. To illustrate consider first the equation $y^3 - 3y - 2x = 0$, which we introduced earlier in this section. The curve defined by this equation is shown in Figure 3.23, and it clearly illustrates that the equation does not define y as a function of x. It does define the three functions of x in Figure 3.25. Suppose we differentiate the equation with respect to x,

$$3y^2\frac{dy}{dx} - 3\frac{dy}{dx} - 2 = 0,$$

and solve for dy/dx,

$$\frac{dy}{dx} = \frac{2}{3(y^2 - 1)}.$$

This derivative is obviously undefined at the points $(-1, 1)$ and $(1, -1)$, and these are precisely the points which separate the original curve into three parts. It is impossible to find a portion of the curve around either of these points which defines y as a function of x. At any other point on the curve, the formula $dy/dx = (2/3)(y^2 - 1)^{-1}$ is a valid representation for the derivative for whichever function of Figure 3.25 contains the point.

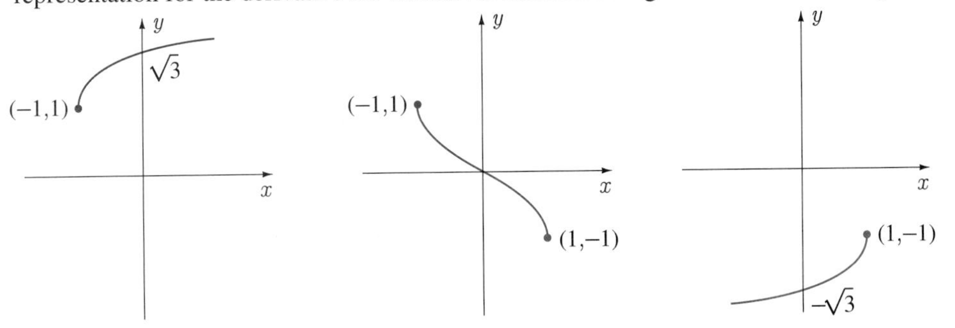

FIGURE 3.25

Consider now the equation $x = y^3$, shown graphically in Figure 3.26. Implicit differentiation leads to

$$\frac{dy}{dx} = \frac{1}{3y^2}.$$

Clearly, y is a function of x for all x, but dy/dx is undefined at $(0,0)$ since the tangent line is vertical at this point.

FIGURE 3.26

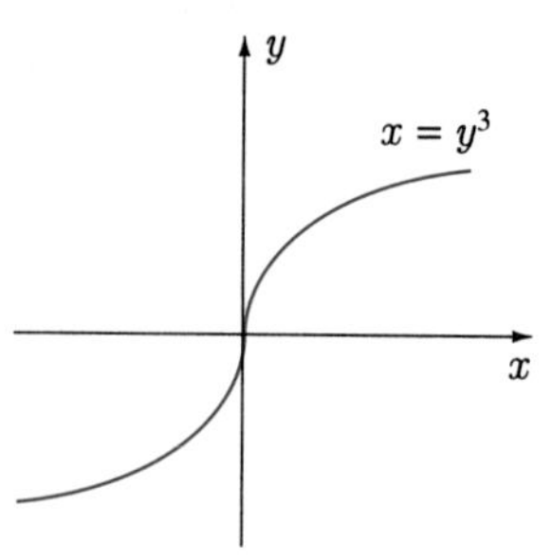

EXAMPLE 3.22

Sketch the curve defined by the equation $y^2 = x^2 - x^4$. Use implicit differentiation to find dy/dx and discuss what happens when the point $(0,0)$ is substituted into the result.

SOLUTION The curve is symmetric about both the x- and y-axes. To draw that portion in the first quadrant, we write $y = x\sqrt{1-x^2}$ and multiply ordinates of $y = x$ and $y = \sqrt{1-x^2}$ (Figure 3.27(a)). The entire curve is shown in Figure 3.27(b).

Implicit differentiation of $y^2 = x^2 - x^4$ gives

$$2y\frac{dy}{dx} = 2x - 4x^3,$$

from which

$$\frac{dy}{dx} = \frac{x(1-2x^2)}{y}.$$

This result is undefined at $(0,0)$, and Figure 3.27(b) indicates why. The equation does not define y as a function of x around $x = 0$.

FIGURE 3.27

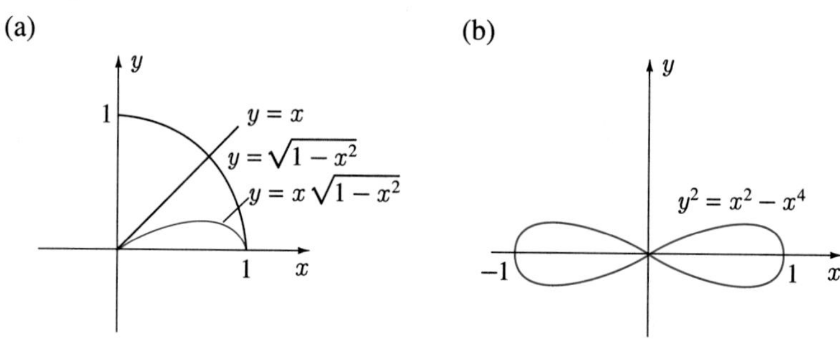

■

EXERCISES 3.7

In Exercises 1–10 find dy/dx wherever y is defined as a function of x.

1. $y^4 + y = 4x^3$
2. $x^4 + y^2 + y^3 = 1$
3. $xy + 2x = 4y^2 + 2$
4. $2x^3 - 3xy^4 + 5xy - 10 = 0$
5. $x + xy^5 + x^2y^3 = 3$
6. $(x+y)^2 = 2x$
7. $x(x-y) - 4y^3 = 2x + 5$
8. $\sqrt{x+y} + y^2 = 12x^2 + y$
9. $\sqrt{1+xy} - xy = 15$
10. $\dfrac{x}{x+y} - \dfrac{y}{x} = 4$

11. Find equations for the tangent and normal lines to the curve $xy^2 + y^3 = 2$ at the point $(1, 1)$.

In Exercises 12–15 find d^2y/dx^2 wherever y is defined as a function of x.

12. $x^2 + y^3 + y = 1$ **13.** $2x^2 - y^3 = 4 - xy$

14. $y^2 + 2y = 5x$ **15.** $(x + y)^2 = x$

16. Find dy/dx at $x = 1$ when $x^3y + xy^3 = 2$.

In Exercises 17–20 find dy/dx and d^2y/dx^2.

17. $(x + y)^2 = x^2 + y^2$ **18.** $x^2y^3 + 2x + 4y = 5$

19. $xy^2 - 3x^2y = x + 1$ **20.** $x = y\sqrt{1 - y^2}$

21. Find dy/dx when $x = 0$ if $\sqrt{1 - xy} + 3y = 4$.

22. Find dy/dx when $x = 1$ if $x^2y^3 + xy = 2$.

23. Find dy/dx and d^2y/dx^2 when $x = 2$ if $y^5 + (x-2)y = 1$.

24. Find dy/dx and d^2y/dx^2 when $y = 1$ if $x^2 + 2xy + 3y^2 = 2$.

25. Find point(s) on the curve $xy^2 + x^2y = 16$ at which the slope of the tangent line is equal to zero.

26. Find point(s) on the curve $x^2 + y^{2/3} = 2$ at which the second derivative is equal to zero.

27. If a thermal nuclear reactor is built in the shape of a right circular cylinder of radius r and height h, then according to neutron diffusion theory, r and h must satisfy the equation

$$\left(\frac{2.4048}{r}\right)^2 + \left(\frac{\pi}{h}\right)^2 = k = \text{constant.}$$

Find dr/dh.

28. The elasticity of a function $y = f(x)$ is defined as

$$\frac{Ey}{Ex} = \frac{x}{y}\frac{dy}{dx}.$$

Calculate elasticity for the function defined by each of the following equations:

(a) $f(x) = x\left(\dfrac{x + 1}{x + 2}\right)$ (b) $x = \dfrac{400y + 200}{3 - y}$

29. Show that the elasticity of a function is equal to one if and only if the tangent line to its graph passes through the origin.

30. Find that point $P(a, b)$ on the first-quadrant part of the ellipse $2x^2 + 3y^2 = 14$ at which the tangent line at P is perpendicular to the line joining P and $(2, 5)$.

31. Show that the equation of the tangent line to the hyperbola $b^2x^2 - a^2y^2 = a^2b^2$ at the point (x_0, y_0) is $b^2xx_0 - a^2yy_0 = a^2b^2$.

32. Prove that for any circle $(x - h)^2 + (y - k)^2 = r^2$,

$$\left|\frac{d^2y/dx^2}{[1 + (dy/dx)^2]^{3/2}}\right| = \frac{1}{r}.$$

33. A solution passes through a conical filter 24 cm deep and 16 cm across the top into a cylindrical vessel of diameter 12 cm. Find an equation relating the depth h of solution in the filter and depth H of solution in the cylinder. What is the rate of change of h with respect to H?

34. If x objects are sold at a price of $r(x)$ per object, the total revenue is $R(x) = xr(x)$. Find the marginal revenue $R'(x)$ if price is defined implicitly by the equation

$$x = 4a^3 - 3ar^2 + r^3,$$

where $a > 0$ is a constant, and $0 < x < 4a^3$.

35. The general polynomial of degree n is

$$a_0 + a_1x + a_2x^2 + \cdots + a_nx^n,$$

where $a_0, a_1, \ldots, a_n$ are constants. Show that two polynomials of degree n can be equal for all x,

$$a_0 + a_1x + a_2x^2 + \cdots + a_nx^n = b_0 + b_1x + b_2x^2 + \cdots + b_nx^n,$$

if and only if $a_0 = b_0, a_1 = b_1, \ldots, a_n = b_n$.

36. (a) Find $f'(0)$ if $y = f(x)$ is defined implicitly as a function of x by

$$x\sqrt{1 + 2y} = x^2 - y.$$

(b) Show that by squaring this equation

$$x^2 + 4x^2y = x^4 + y^2.$$

Differentiate this equation with respect to x to find $f'(0)$. Do you have any difficulties? Explain.

37.–40. Use implicit differentiation to redo questions 55–58 in Exercises 3.6.

41. (a) Use implicit differentiation to find dy/dx if $y^2 = x^2 - 4x^4$.

(b) Can you calculate dy/dx at $x = 0$ using the result of part (a)? Sketch the curve $y^2 = x^2 - 4x^4$ in order to explain this difficulty.

42. (a) Find dy/dx if $\sqrt{x} + \sqrt{y} = 1$, where $0 \leq x \leq 1$, defines y implicitly as a function of x.

(b) Sketch the curve $\sqrt{|x|} + \sqrt{|y|} = 1$.

43. What is dy/dx if $x^2 + 4x + y^2 + 6y + 15 = 0$?

44. Find dy/dx if $y = u/\sqrt{u^2 - 1}$ and u is defined implicitly as a function of x by $x^2u^2 + \sqrt{u^2 - 1} = 4$.

45. Given that the equations

$$y^4 + yv^3 = 3,$$
$$x^2v + 3xv^2 = 2x^3y + 1$$

define y as a function of v and v as a function of x, find dy/dx in terms of x,y, and v.

46. Show that any function defined implicitly by the equation

$$\frac{x^2}{y^3} - x = C,$$

where C is a constant, satisfies the equation

$$3x^2\frac{dy}{dx} + y^4 = 2xy.$$

47. Show that any function defined implicitly by the equation

$$2x^2 - 3y = Cx^2y^3,$$

where C is a constant, satisfies the equation

$$(xy - x^3)\frac{dy}{dx} + y^2 = 0.$$

48. Verify power rule 3.6 in the case that n is a rational number.

49. Find points on the curve $xy^2 + x^2y = 2$ at which the slope of the tangent line is equal to one.

50. Show that the families of curves $y^2 = x^3/(a - x)$ and $(x^2 + y^2)^2 = b(2x^2 + y^2)$ are orthogonal trajectories.

51. The curve described by the equation $x^3 + y^3 = 3axy$, where $a > 0$ is a constant, is called the *folium of Descartes* (Figure 3.28). Find points where the slope of the tangent line is equal to -1.

FIGURE 3.28

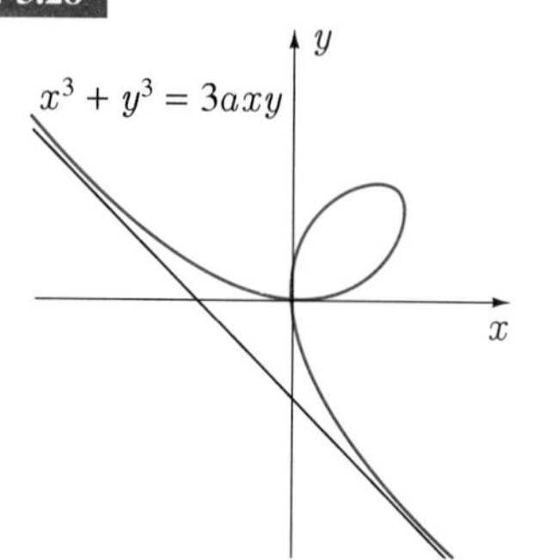

52. (a) When a point (x, y) moves in the xy-plane so that the product of its distances from the points $(a, 0)$ and $(-a, 0)$ is always equal to a constant which we denote by c^2 $(c > a)$, the curve that it follows is called the *ovals of Cassini*. Verify that the equation for these ovals is $(x^2 + y^2 + a^2)^2 = c^4 + 4a^2x^2$.

(b) Show that when $c < \sqrt{2}a$, there are six points on the ovals at which the tangent line is horizontal, but when $c \geq \sqrt{2}a$, there are only two such points.

SECTION 3.8

Derivatives of the Trigonometric Functions

The trigonometric functions, their properties, and the identities that they satisfy are discussed in Appendix A. We assume in this section and the remainder of the book that the reader is familiar with this material. For those readers who feel comfortable with their grasp of trigonometry and do not intend reading the appendix, we emphasize one extremely important convention in calculus. Arguments of trigonometric functions are always measured in radians, never degrees. With this in mind we prove the following theorem.

Theorem 3.10

$$\lim_{\theta \to 0} \frac{\sin \theta}{\theta} = 1. \tag{3.20}$$

Proof If θ is a positive acute angle as shown in Figure 3.29, then

area of triangle $BOP <$ area of sector $BOA <$ area of triangle OBT.

FIGURE 3.29

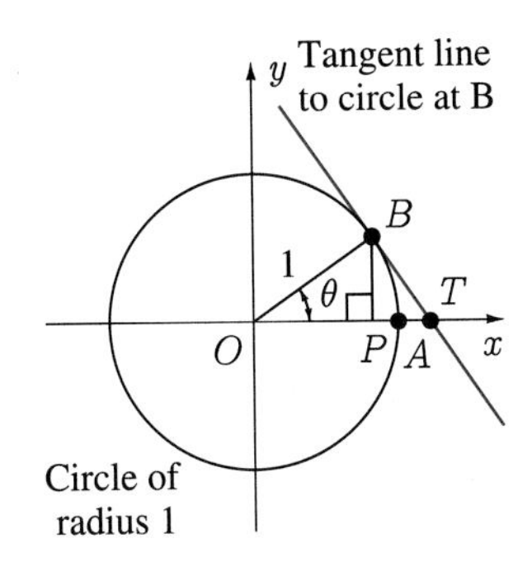

With θ expressed in radians, the area of a sector of a circle is $(r^2\theta)/2$. Consequently,

$$\frac{1}{2}||BP||||OP|| < \frac{1}{2}(1)^2\theta < \frac{1}{2}||OB||||BT||.$$

When we multiply each term in these inequalitites by 2 and express lengths of the line segments in terms of θ,

$$(\sin\theta)(\cos\theta) < \theta < (1)(\tan\theta).$$

Division by $\sin\theta$ gives

$$\cos\theta < \frac{\theta}{\sin\theta} < \frac{1}{\cos\theta},$$

and when each term is inverted, the inequality signs are reversed, so

$$\frac{1}{\cos\theta} > \frac{\sin\theta}{\theta} > \cos\theta.$$

We now take limits as $\theta \to 0^+$. Since $\cos\theta \to 1$ and $1/\cos\theta \to 1$, it follows that $(\sin\theta)/\theta$ which is between $1/\cos\theta$ and $\cos\theta$ must also approach 1,

$$\lim_{\theta \to 0^+} \frac{\sin\theta}{\theta} = 1.$$

When $\theta < 0$, we set $\phi = -\theta$, in which case

$$\lim_{\theta \to 0^-} \frac{\sin\theta}{\theta} = \lim_{\phi \to 0^+} \frac{\sin(-\phi)}{-\phi} = \lim_{\phi \to 0^+} \frac{\sin\phi}{\phi} = 1.$$

Since left-hand and right-hand limits of $(\sin\theta)/\theta$ are both equal to 1, it follows that

$$\lim_{\theta \to 0} \frac{\sin\theta}{\theta} = 1.$$

With this result we can now find the derivative of the sine function. Derivatives of the other five trigonometric functions follow easily.

Theorem 3.11

$$\frac{d}{dx}\sin x = \cos x \tag{3.21}$$

Proof If we set $f(x) = \sin x$ in equation 3.2 and use the trigonometric identity

$$\sin A - \sin B = 2\cos\left(\frac{A+B}{2}\right)\sin\left(\frac{A-B}{2}\right),$$

we obtain

$$\begin{aligned}
f'(x) &= \lim_{h\to 0}\frac{f(x+h)-f(x)}{h} \\
&= \lim_{h\to 0}\frac{\sin(x+h)-\sin x}{h} \\
&= \lim_{h\to 0}\frac{1}{h}\left[2\cos\left(\frac{2x+h}{2}\right)\sin\left(\frac{h}{2}\right)\right] \\
&= \lim_{h\to 0}\left[\cos\left(x+\frac{h}{2}\right)\frac{\sin(h/2)}{h/2}\right] \\
&= \lim_{h\to 0}\cos\left(x+\frac{h}{2}\right)\lim_{h/2\to 0}\frac{\sin(h/2)}{h/2} \\
&= \cos x.
\end{aligned}$$

It is a simple application of the chain rule to prove the following corollary.

Corollary *If* $u = f(x)$ *is a differentiable function,*

$$\frac{d}{dx}\sin u = \cos u\frac{du}{dx}. \tag{3.22}$$

EXAMPLE 3.23

Find derivatives for the following functions:

$$\text{(a) } f(x) = \sin 3x \qquad \text{(b) } f(x) = \sin^3 4x$$

SOLUTION (a) According to formula 3.22,

$$f'(x) = \cos 3x\frac{d}{dx}(3x) = 3\cos 3x.$$

(b) For this function we must use extended power rule 3.17 before 3.22,

$$f'(x) = 3\sin^2 4x\frac{d}{dx}\sin 4x = 3\sin^2 4x\cos 4x\frac{d}{dx}(4x) = 12\sin^2 4x\cos 4x.$$

■

Since the cosine function can be expressed in terms of the sine function, it is straightforward to find its derivative.

Theorem 3.12

$$\frac{d}{dx}\cos x = -\sin x \tag{3.23}$$

Proof Since $\cos x$ can always be expressed in the form $\cos x = \sin(\pi/2 - x)$, it follows that

$$\frac{d}{dx}\cos x = \frac{d}{dx}\sin\left(\frac{\pi}{2} - x\right) = -\cos\left(\frac{\pi}{2} - x\right) = -\sin x.$$

Corollary *If* $u = f(x)$ *is a differentiable function,*

$$\frac{d}{dx}\cos u = -\sin u\frac{du}{dx}. \tag{3.24}$$

Derivatives of the other trigonometric functions are obtained by expressing them in terms of the sine and cosine functions. For the tangent function, we use the chain rule and the quotient rule to calculate

$$\frac{d}{dx}\tan u = \frac{d}{du}\left(\frac{\sin u}{\cos u}\right)\frac{du}{dx} = \frac{\cos u(\cos u) - \sin u(-\sin u)}{\cos^2 u}\frac{du}{dx} = \frac{1}{\cos^2 u}\frac{du}{dx}.$$

Consequently,

$$\frac{d}{dx}\tan u = \sec^2 u\frac{du}{dx}. \tag{3.25}$$

A similar calculation gives

$$\frac{d}{dx}\cot u = -\csc^2 u\frac{du}{dx}. \tag{3.26}$$

For the secant function, we obtain

$$\frac{d}{dx}\sec u = \frac{d}{du}\left(\frac{1}{\cos u}\right)\frac{du}{dx} = (-1)(\cos u)^{-2}(-\sin u)\frac{du}{dx} = \frac{1}{\cos u}\frac{\sin u}{\cos u}\frac{du}{dx} \quad \text{or}$$

$$\frac{d}{dx}\sec u = \sec u\tan u\frac{du}{dx}. \tag{3.27}$$

Similarly,

$$\frac{d}{dx}\csc u = -\csc u\cot u\frac{du}{dx}. \tag{3.28}$$

Notice the relationship between derivatives of the *cofunctions* — cosine is the cofunction of sine, cotangent of tangent, and cosecant of secant — and corresponding derivatives for the functions. Each function is replaced by its cofunction, and a negative sign is added.

EXAMPLE 3.24

Find dy/dx if y is defined as a function of x in each of the following:

(a) $y = \sin 2x$ (b) $y = 4\sec(2x^3 + 5)$ (c) $y = \tan^2 4x$
(d) $y = 2\csc 3x^2 + 5x\sin x$ (e) $x^2 \tan y + y \sin x = 5$

SOLUTION

(a) $$\frac{dy}{dx} = (\cos 2x)\frac{d}{dx}(2x) = 2\cos 2x$$

(b) $$\frac{dy}{dx} = 4\sec(2x^3 + 5)\tan(2x^3 + 5)\frac{d}{dx}(2x^3 + 5) = 24x^2 \sec(2x^3 + 5)\tan(2x^3 + 5)$$

(c) $$\frac{dy}{dx} = 2\tan 4x\frac{d}{dx}\tan 4x = 2\tan 4x(4\sec^2 4x) = 8\tan 4x\sec^2 4x$$

(d) $$\frac{dy}{dx} = -2\csc 3x^2 \cot 3x^2 \frac{d}{dx}(3x^2) + 5\sin x + 5x\cos x = -12x\csc 3x^2\cot 3x^2 + 5\sin x + 5x\cos x$$

(e) If we differentiate the equation (implicitly) with respect to x, we obtain

$$2x\tan y + x^2\sec^2 y\frac{dy}{dx} + \frac{dy}{dx}\sin x + y\cos x = 0.$$

When we solve this equation for dy/dx, the result is

$$\frac{dy}{dx} = -\frac{2x\tan y + y\cos x}{x^2\sec^2 y + \sin x}.$$

■

EXAMPLE 3.25

Find values of x for which the derivative of the function $f(x) = \cos(x + 1/x)$ is equal to zero.

SOLUTION For

$$0 = f'(x) = -\left(1 - \frac{1}{x^2}\right)\sin\left(x + \frac{1}{x}\right),$$

there are two possibilities:

$$1 - \frac{1}{x^2} = 0 \qquad \text{or} \qquad \sin\left(x + \frac{1}{x}\right) = 0.$$

The first gives $x = \pm 1$. The second implies that

$$x + \frac{1}{x} = n\pi,$$

where n is an integer. Multiplication by x leads to the quadratic equation

$$x^2 - n\pi x + 1 = 0,$$

with solutions

$$x = \frac{n\pi \pm \sqrt{n^2\pi^2 - 4}}{2}.$$

Clearly, we must choose $n \neq 0$ (else $n^2\pi^2 - 4 < 0$). ■

EXERCISES 3.8

In Exercises 1–30 find dy/dx.

1. $y = 2\sin 3x$
2. $y = \cos x - 4\sin 5x$
3. $y = \sin^2 x$
4. $y = \tan^{-3} 3x$
5. $y = \sec^4 10x$
6. $y = \csc(4 - 2x)$
7. $y = \sin^2(3 - 2x^2)$
8. $y = x\cot x^2$
9. $y = \dfrac{\sin 2x}{\cos 5x}$
10. $y = \dfrac{x\sin x}{x + 1}$
11. $y = \sqrt{\sin 3x}$
12. $y = \left(1 + \tan^3 x\right)^{1/4}$
13. $y = \sin^3 x + \cos x$
14. $y = \sin 2x \cos 2x$
15. $2\sin y + 3\cos x = 1$
16. $x\cos y - y\cos x = 3$
17. $4\sin^2 x - 3\cos^3 y = 1$
18. $\tan(x + y) = y$
19. $x + \sec xy = 5$
20. $x^3 y + \tan^2 y = 3x$
21. $x = y^3 \csc^3 y$
22. $y = \cos(\tan x)$
23. $y = x^3 - x^2\cos x + 2x\sin x + 2\cos x$
24. $y = \sin^4 x^2 - \cos^4 x^2$
25. $y = u^3 \sec u, \quad u = x\tan(x + 1)$
26. $y = \sqrt{3 - \sec v}, \quad v = \tan\sqrt{x}$
27. $y = \sqrt{t^2 + 1}, \quad t = \sin(\sin x)$
28. $y = (1 + \sec^3 u)^{1/3}, \quad u = \sqrt{1 + \cos x^2}$
29. $y = \dfrac{\sin^2 x}{1 + \cos^3 x}$
30. $y = \dfrac{1 + \tan^3(3x^2 - 4)}{x^2 \sin x}$

In Exercises 31 and 32 find d^2y/dx^2.

31. $\sin y = x^2 + y$
32. $\tan y = x + xy$

In Exercises 33–38 evaluate the limit if it exists.

33. $\displaystyle\lim_{x\to 0} \frac{\tan x}{x}$
34. $\displaystyle\lim_{x\to 0} \frac{1 - \cos x}{x}$
35. $\displaystyle\lim_{x\to 0} \frac{\sin 2x}{x}$
36. $\displaystyle\lim_{x\to\infty} \frac{\sin(2/x)}{\sin(1/x)}$
37. $\displaystyle\lim_{x\to 0} \frac{(x + 1)^2 \sin x}{3x^3}$
38. $\displaystyle\lim_{x\to\pi/2} \frac{\cos x}{(x - \pi/2)^2}$

39. Sketch graphs of the functions $f(x) = |\sin x|$ and $g(x) = \sin|x|$. Where do these functions fail to be differentiable?

40. Does $\lim_{x\to 0} g'(x)$ exist for the function $g(x)$ in Example 3.11 of Section 3.3?

41. The angular displacement of the pendulum in Figure 3.30 at time t is given by

$$\theta = f(t) = A\cos(\omega t + \phi), \quad t \geq 0,$$

where A, ω, and ϕ are constants. Show that $\theta = f(t)$ satisfies the equation

$$\frac{d^2\theta}{dt^2} + \omega^2\theta = 0.$$

FIGURE 3.30

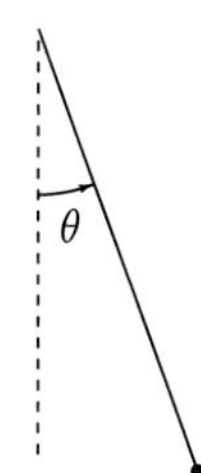

42. When the mass m in Figure 3.31 moves vertically on the end of the spring, its displacement y must satisfy the equation

$$m\frac{d^2y}{dt^2} + ky = 0,$$

where $k > 0$ is the constant of elasticity for the spring. Show that the function

$$y = f(t) = A\sin\left(\sqrt{\frac{k}{m}}\,t\right) + B\cos\left(\sqrt{\frac{k}{m}}\,t\right)$$

satisfies this equation for any constants A and B.

FIGURE 3.31

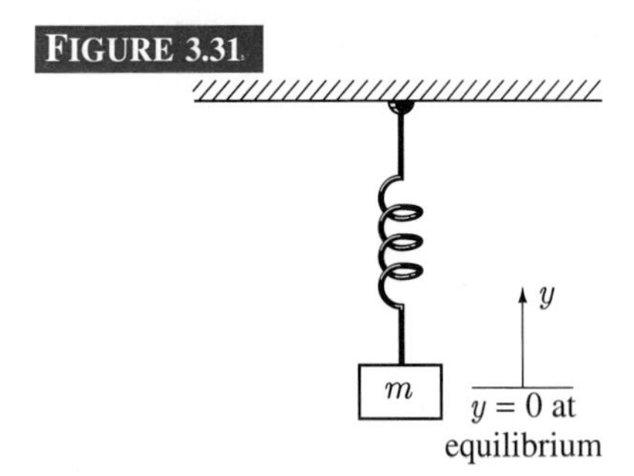

43. If the voltage source labeled E in the circuit of Figure 3.32 is suddenly short-circuited by switching the switch labeled S, the current I in the circuit thereafter must satisfy the equation

$$L\frac{d^2 I}{dt^2} + \frac{1}{C}I = 0,$$

where L and C are the sizes of the inductor and capacitor, respectively, and t is time. Show that

$$I = f(t) = A\cos\left(\frac{t}{\sqrt{LC}}\right) + B\sin\left(\frac{t}{\sqrt{LC}}\right)$$

satisfies this equation for any constants A and B whatsoever.

FIGURE 3.32

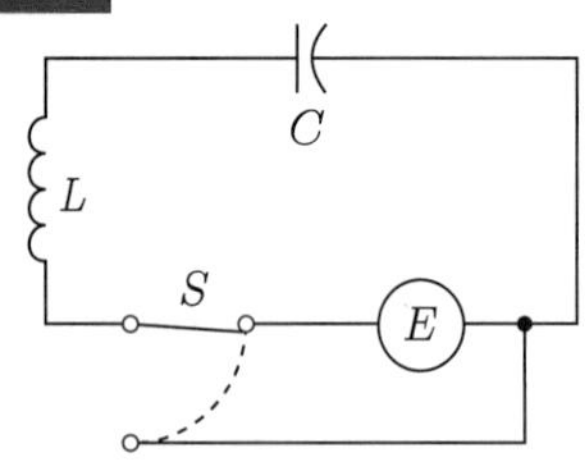

44. Find

$$\frac{d}{dx}|\sin x|.$$

45. If n is a positive integer, find formulas for

$$\frac{d^n}{dx^n}(\sin x) \quad \text{and} \quad \frac{d^n}{dx^n}(\cos x).$$

46. Show that the function

$$f(x) = \begin{cases} x^2 \sin(1/x) & x \neq 0 \\ 0 & x = 0 \end{cases}$$

has a derivative at $x = 0$, but $f'(x)$ is not continuous at $x = 0$.

47. Any cross-section of the reflector in a car headlight is in the form of a parabola

$$y^2 = 4x, \quad 0 \leq x \leq 2,$$

with the bulb at the point $(1, 0)$ called the *focus* of the parabola. It is a principle of optics that all light rays from the bulb are reflected by the mirror so that the angle between the incident ray and the normal to the mirror is equal to the angle between the reflected ray and the normal. Show that all rays are reflected parallel to the x-axis.

SECTION 3.9

Derivatives of the Exponential and Logarithm Functions

Exponential functions abound in applied mathematics, and much of this is due to the fact that the derivative of an exponential function is itself, varied by a constant in some cases. We obtain derivatives of exponential and logarithm functions in this section, but many of the applications which rely on these derivatives are delayed until Section 9.2. Exponential and logarithm functions and their properties are discussed in Appendix B. We assume that the reader is familiar with this material. To differentiate logarithm and exponential functions, we must return to the definition of the derivative (see equation 3.2 in Section 3.1). We can apply the definition to differentiate either $\log_a x$ or a^x, and once one of these derivatives is obtained, implicit differentiation can be used to find the other. We choose to differentiate $\log_a x$.

Theorem 3.13

The derivative of the logarithm function $\log_a x$ *is*

$$\frac{d}{dx}\log_a x = \frac{1}{x}\log_a e, \tag{3.29a}$$

where

$$e = \lim_{z\to\infty}\left(1+\frac{1}{z}\right)^z = \lim_{v\to 0}(1+v)^{1/v}. \tag{3.29b}$$

Proof By equation 3.2,

$$\begin{aligned}
\frac{d}{dx}\log_a x &= \lim_{h\to 0}\frac{\log_a(x+h)-\log_a x}{h} \\
&= \lim_{h\to 0}\frac{1}{h}\left[\log_a\left(\frac{x+h}{x}\right)\right] && [\text{since } \log_a b - \log_a c = \log_a(b/c)] \\
&= \lim_{h\to 0}\frac{1}{h}\left[\log_a\left(1+\frac{h}{x}\right)\right] \\
&= \frac{1}{x}\lim_{h\to 0}\frac{x}{h}\left[\log_a\left(1+\frac{h}{x}\right)\right] \\
&= \frac{1}{x}\lim_{h\to 0}\log_a\left(1+\frac{h}{x}\right)^{x/h} && [\text{since } c\log_a b = \log_a b^c].
\end{aligned}$$

Since the logarithm function is continuous, we may interchange the limit and logarithm operations (see Theorem 2.3 in Section 2.4),

$$\frac{d}{dx}\log_a x = \frac{1}{x}\log_a\left[\lim_{h\to 0}\left(1+\frac{h}{x}\right)^{x/h}\right].$$

But

$$\lim_{h\to 0}\left(1+\frac{h}{x}\right)^{x/h} = \lim_{h/x\to 0}\left(1+\frac{h}{x}\right)^{x/h} = e,$$

and therefore

$$\frac{d}{dx}\log_a x = \frac{1}{x}\log_a e.$$

If we choose e as the base of logarithms, that is, set $a = e$, derivative 3.29 simplifes to

$$\frac{d}{dx}\log_e x = \frac{1}{x}.$$

Because of the simplicity of this result, logarithms to base e are used more extensively than those to any other base. With the notation $\ln x = \log_e x$ for logarithms to base e, we may state the following corollary.

Corollary 1

$$\frac{d}{dx}\ln x = \frac{1}{x} \tag{3.30}$$

The chain rule gives the next two results.

Corollary 2 *If* $u(x)$ *is a differentiable function of* x, *then*

$$\frac{d}{dx}\log_a u = \frac{1}{u}\frac{du}{dx}\log_a e. \tag{3.31}$$

Corollary 3 *If* $u(x)$ *is differentiable, then*

$$\frac{d}{dx}\ln u = \frac{1}{u}\frac{du}{dx}. \tag{3.32}$$

We now use implicit differentiation to find the derivative of the exponential function.

Theorem 3.14

$$\frac{d}{dx}a^x = a^x \ln a \tag{3.33}$$

Proof If we set $y = a^x$, then $x = \log_a y$, and we can differentiate (implicitly) with respect to x using 3.31,

$$1 = \frac{1}{y}\frac{dy}{dx}\log_a e.$$

Thus,

$$\frac{dy}{dx} = \frac{y}{\log_a e} = \frac{a^x}{\log_a e}.$$

When it is necessary to change bases of logarithms, the rule is

$$\log_b x = (\log_a x)(\log_b a),$$

(see equation B.8 in Appendix B). If we set $x = b = e$ in this equation, the result is

$$\log_e e = (\log_a e)(\log_e a) \qquad \text{or} \qquad 1 = (\log_a e)(\ln a).$$

Consequently, $1/\log_a e = \ln a$, and our differentiation formula becomes

$$\frac{dy}{dx} = a^x \ln a.$$

Certainly it is again clear that should we use exponentials to base e, the differentiation formula in Theorem 3.14 simplifies.

Corollary 1

$$\frac{d}{dx}e^x = e^x \tag{3.34}$$

The exponential function e^x is therefore its own derivative. In fact, for any constant C whatsoever, the function Ce^x differentiates to give itself. In Chapter 5 we shall see that Ce^x is the only function that is its own derivative.

Corollary 2 *For a differentiable function* $u(x)$,

$$\frac{d}{dx}a^u = a^u\frac{du}{dx}\ln a. \tag{3.35}$$

Corollary 3 *For a differentiable function* $u(x)$,

$$\frac{d}{dx}e^u = e^u\frac{du}{dx}. \tag{3.36}$$

EXAMPLE 3.26

Find dy/dx if y is defined as a function of x in each of the following:

(a) $y = 2^{3x}$ (b) $y = \log_{10}(3x^2 + 4)$ (c) $y = x^2 e^{-2x}$
(d) $y = (\ln x)/x$ (e) $y = \sqrt{1 + e^{2x}}$

SOLUTION (a) Using 3.35,

$$\frac{dy}{dx} = 2^{3x}\frac{d}{dx}(3x)\ln 2 = (3\ln 2)2^{3x}.$$

(b) With 3.31,

$$\frac{dy}{dx} = \frac{1}{3x^2 + 4}\frac{d}{dx}(3x^2 + 4)\log_{10} e = \frac{6x}{3x^2 + 4}\log_{10} e.$$

(c) The product rule and 3.36 give

$$\frac{dy}{dx} = 2xe^{-2x} + x^2 e^{-2x}(-2) = 2x(1 - x)e^{-2x}.$$

(d) The quotient rule and 3.30 yield

$$\frac{dy}{dx} = \frac{x(1/x) - \ln x}{x^2} = \frac{1 - \ln x}{x^2}.$$

(e) With extended power rule 3.17 and 3.36,

$$\frac{dy}{dx} = \frac{1}{2\sqrt{1 + e^{2x}}}\frac{d}{dx}(1 + e^{2x}) = \frac{2e^{2x}}{2\sqrt{1 + e^{2x}}} = \frac{e^{2x}}{\sqrt{1 + e^{2x}}}.$$

■

EXAMPLE 3.27

Find values of x for which the first derivative of the function $f(x) = x^2 \ln x$ is equal to zero, and values of x for which the second derivative is equal to zero.

SOLUTION The first derivative of $f(x)$ vanishes when

$$0 = f'(x) = 2x\ln x + x^2\left(\frac{1}{x}\right) = x(2\ln x + 1).$$

Since the function is undefined at $x = 0$, we set

$$2\ln x + 1 = 0, \quad \text{or} \quad \ln x = -\frac{1}{2}.$$

Thus, $x = e^{-1/2} = 1/\sqrt{e}$.

The second derivative is equal to zero when

$$0 = f''(x) = (2\ln x + 1) + x\left(\frac{2}{x}\right) = 2\ln x + 3.$$

The only solution of this equation is $x = e^{-3/2}$.

■

In Section 3.2 we proved power rule 3.6 only in the case that n was a nonnegative integer. It is now easy to prove it for all real n, at least when $x > 0$. To do this we write x^n in the form $x^n = e^{n \ln x}$ (see equation B.13a). Formulas 3.36 and 3.30 give

$$\frac{d}{dx}x^n = \frac{d}{dx}e^{n \ln x} = e^{n \ln x}\frac{d}{dx}(n \ln x) = x^n\left(\frac{n}{x}\right) = nx^{n-1}.$$

A discussion of the power rule in the case that $x \leq 0$ depends on the value of n (see Exercise 39).

EXERCISES 3.9

In Exercises 1–30, y is defined as a function of x. Find dy/dx in as simplified a form as possible.

1. $y = 3^{2x}$

2. $y = \ln(3x^2 + 1)$

3. $y = \log_{10}(2x + 1)$

4. $y = e^{1-2x}$

5. $y = xe^{2x}$

6. $y = x \ln x$

7. $y = e^{2 \ln x}$

8. $y = \log_{10}(3 - 4x)$

9. $y = \ln(\sin x)$

10. $y = \ln(3 \cos x)$

11. $y = x \ln(x + 1)$

12. $y = x^2 + x^3 e^{4x}$

13. $y = \dfrac{e^{1-x}}{1 - x}$

14. $y = \sin(e^{2x})$

15. $y = \ln(\ln x)$

16. $y = e^{-2x} \sin 3x$

17. $y = \ln(x^2 e^{4x})$

18. $y = \dfrac{e^x - e^{-x}}{e^x + e^{-x}}$

19. $y = e^{-x} \ln x$

20. $y = x[\sin(\ln x) - \cos(\ln x)]$

21. $y = e^{\sin u}, \quad u = e^{1/x}$

22. $y = \ln(\cos v), \quad v = \sin^2 x$

23. $\ln(x + y) = x^2 y$

24. $xe^y + x^2 \ln y + y \sin x = 0$

25. $y = \ln(\sec x + \tan x)$

26. $y = x\sqrt{x^2 + 1} - \ln\left(x + \sqrt{x^2 + 1}\right)$

27. $y = \ln\left(x + 4 + \sqrt{8x + x^2}\right)$

28. $y = x - \dfrac{1}{4}\ln(1 + 5e^{4x})$

29. $e^{xy} = (x + y)^2$

30. $e^{1/x} + e^{1/y} = \dfrac{1}{x} + \dfrac{1}{y}$

31. A spring with constant 5 N/m is attached to a fixed wall on one end and a 1 kg mass m on the other (Figure 3.33).

FIGURE 3.33

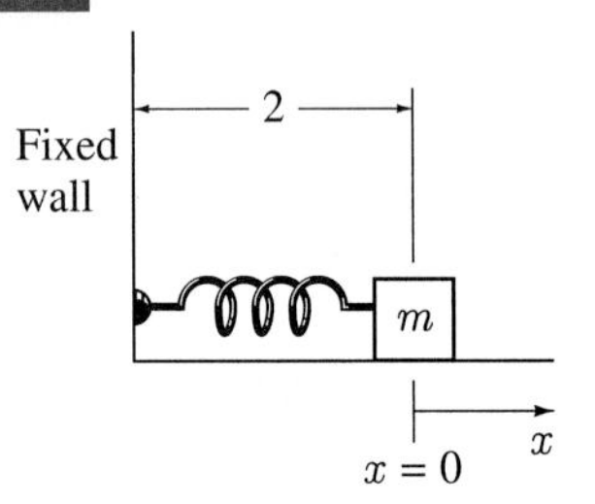

We choose a coordinate system with x positive to the right and with $x = 0$ at the centre of mass of m when the spring is in the unstretched position. At time $t = 0$, m is pulled 1 m to the left of $x = 0$ and given a speed of 3 m/s to the right. During the subsequent motion, a frictional force equal in newtons to twice the velocity of the mass acts on m. It can be shown that if $x(t)$ is the position of m, then $x(t)$ must satisfy the equation

$$\frac{d^2 x}{dt^2} + 2\frac{dx}{dt} + 5x = 0$$

and the initial conditions $x(0) = -1$ and $x'(0) = 3$. Verify that $x(t) = e^{-t}(\sin 2t - \cos 2t)$ satisfies the equation and initial conditions.

32. Two long parallel rectangular loops lying in the same plane have lengths l and L and widths w and W, respectively (Figure 3.34).

FIGURE 3.34

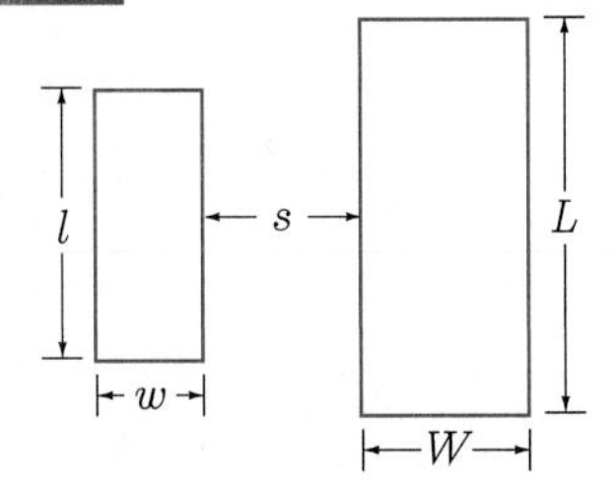

If the loops do not overlap, and the distance between the near sides is s, the mutual inductance between the loops is

$$M = \frac{\mu_0 l}{2\pi} \ln \left[\frac{s + W}{s\left(1 + \dfrac{W}{s + w}\right)} \right]$$

where $\mu_0 > 0$ is a constant. Show that the derivative of M as a function of s is negative for $s > 0$.

33. Two parallel wires carrying currents I (Figure 3.35) create a magnetic field.

FIGURE 3.35

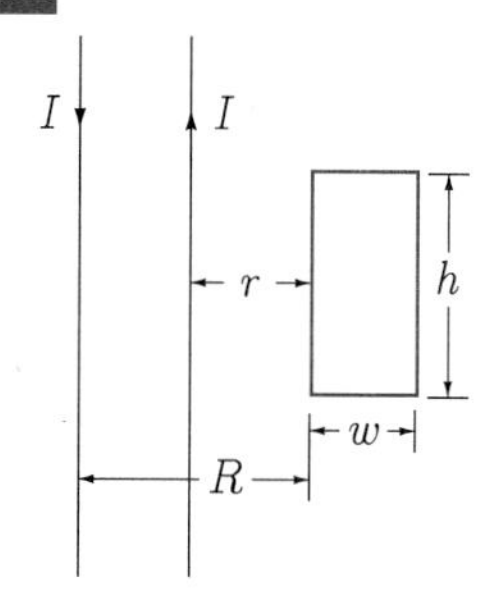

The flux Φ of this field through the loop of dimensions h and w is

$$\Phi = \frac{\mu_0 h I}{2\pi} \ln \left[\frac{R(r + w)}{r(R + w)} \right],$$

where $\mu_0 > 0$ is a constant. Sketch a graph of Φ first as a function of h and then as a function of w.

34. Show that the function

$$y = f(x) = Ax + B\left[\frac{x}{2} \ln\left(\frac{x - 1}{x + 1}\right) + 1\right]$$

satisfies the equation

$$(1 - x^2)\frac{d^2 y}{dx^2} - 2x\frac{dy}{dx} + 2y = 0$$

for any constants A and B.

35. The equation $x = e^y - e^{-y}$ defines y implicitly as a function of x. Find dy/dx in terms of x in two ways:

(a) Solve first for an explicit definition for $y = f(x)$.

(b) Differentiate implicitly with respect to x.

36. The current I in the RC-circuit of Figure 3.36 must satisfy the equation

$$R\frac{dI}{dt} + \frac{I}{C} = \frac{dE}{dt}.$$

FIGURE 3.36

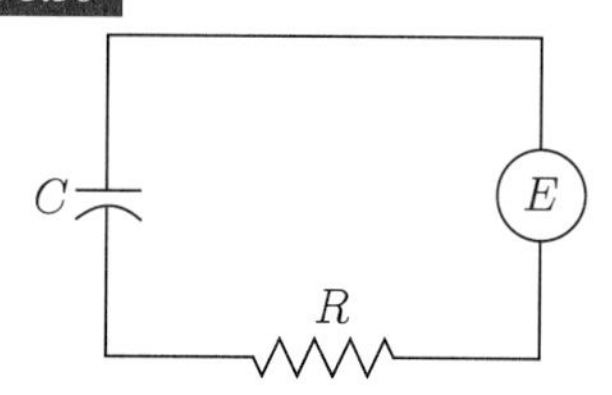

If $E = E_0 \sin \omega t$, where E_0 and ω are constants, show that a solution is

$$I = f(t) = Ae^{-t/(RC)} + \frac{E_0}{Z} \sin(\omega t - \phi),$$

where A is any constant whatsoever, and

$$Z = \sqrt{R^2 + \frac{1}{\omega^2 C^2}}, \qquad \tan \phi = -\frac{1}{\omega C R}.$$

37. The current I in the LR-circuit of Figure 3.37 must satisfy the equation

$$L\frac{dI}{dt} + RI = E.$$

FIGURE 3.37

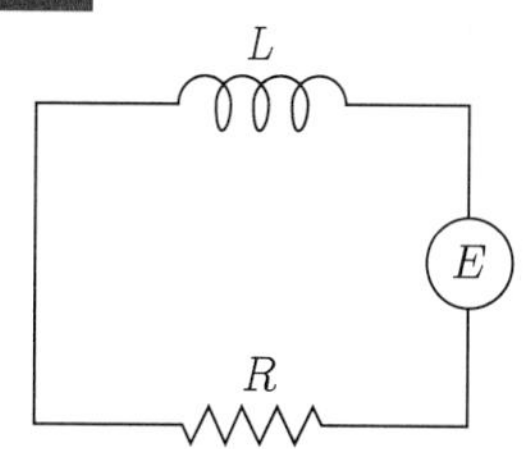

If E is as in Exercise 36, show that a solution is

$$I = f(t) = Ae^{-Rt/L} + \frac{E_0}{Z} \sin(\omega t - \phi),$$

where A is any constant, and

$$Z = \sqrt{R^2 + \omega^2 L^2}, \qquad \tan \phi = \frac{\omega L}{R}.$$

38. The current in the RCL-circuit of Figure 3.38 must satisfy the equation

$$L\frac{d^2 I}{dt^2} + R\frac{dI}{dt} + \frac{I}{C} = \frac{dE}{dt}.$$

FIGURE 3.38

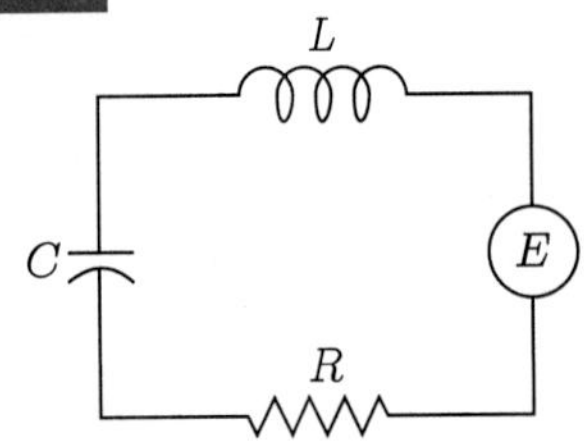

If E is as in Exercise 36, show that a solution is

$$I = f(t) = e^{-Rt/(2L)}\,(A\cos\nu t + B\sin\nu t) + \frac{E_0}{Z}\sin(\omega t - \phi),$$

where A and B are any constants whatsoever, and

$$\nu = \sqrt{\frac{1}{LC} - \frac{R^2}{4L^2}}, \quad Z = \sqrt{R^2 + \left(\omega L - \frac{1}{\omega C}\right)^2},$$

$$\tan\phi = \frac{\omega L - 1/(\omega C)}{R}.$$

39. (a) The function $f(x) = x^n$ is not defined for $x < 0$ if n is irrational, that is, is not rational. Verify power rule 3.6 for $x < 0$ if n is rational and x^n is defined.

(b) Discuss the derivative of $f(x) = x^n$ at $x = 0$.

SECTION 3.10

Logarithmic Differentiation

Compare the three functions

$$x^n, \qquad a^x, \qquad x^x.$$

The first, in which the exponent is constant and the base is variable, is called a **power function**. The second, in which the base is constant and the exponent is variable, is called an **exponential function**. We have differentiation rules for power functions and exponential functions, namely

$$\frac{d}{dx}x^n = nx^{n-1} \qquad \text{and} \qquad \frac{d}{dx}a^x = a^x \ln a,$$

respectively. The third function x^x, which we consider only for $x > 0$, is neither a power nor an exponential function; therefore, we cannot use either of the above formulas to find its derivative. Instead, we set $y = f(x) = x^x$, and take natural logarithms,

$$\ln y = x \ln x.$$

Implicit differentiation with respect to x now gives

$$\frac{1}{y}\frac{dy}{dx} = \ln x + \frac{x}{x} = \ln x + 1.$$

Consequently,

$$\frac{dy}{dx} = y(\ln x + 1) \qquad \text{or} \qquad \frac{d}{dx}x^x = x^x(\ln x + 1).$$

This process of taking logarithms and then differentiating is called **logarithmic differentiation**.

EXAMPLE 3.28

Find the derivative of the function $f(x) = x^{\sin x}$ when $x > 0$.

SOLUTION If we set $y = x^{\sin x}$, then

$$\ln y = \sin x \ln x.$$

Differentiation with respect to x gives

$$\frac{1}{y}\frac{dy}{dx} = \cos x \ln x + \frac{1}{x}\sin x,$$

and this can be solved for dy/dx,

$$\frac{dy}{dx} = x^{\sin x}\left(\cos x \ln x + \frac{1}{x}\sin x\right).$$

■

Logarithmic differentiation can also be used to differentiate complicated products or quotients, as in the following example.

EXAMPLE 3.29

Find the derivative of the function

$$y = f(x) = \frac{x^3(x^2+1)^{2/3}}{\sin^3 x}$$

on the interval $0 < x < \pi$. Extend the result to other values of x.

SOLUTION When $0 < x < \pi$, we take natural logarithms of both sides of the definition for y:

$$\ln y = 3\ln x + \frac{2}{3}\ln(x^2+1) - 3\ln(\sin x).$$

Differentiation with respect to x gives

$$\frac{1}{y}\frac{dy}{dx} = \frac{3}{x} + \frac{2}{3}\frac{2x}{x^2+1} - \frac{3}{\sin x}\cos x,$$

and therefore

$$\frac{dy}{dx} = y\left[\frac{3}{x} + \frac{4x}{3(x^2+1)} - 3\cot x\right] = \frac{x^3(x^2+1)^{2/3}}{\sin^3 x}\left[\frac{3}{x} + \frac{4x}{3(x^2+1)} - 3\cot x\right].$$

When x is not in the interval $0 < x < \pi$, this derivation may not be valid. For instance, when $x < 0$, it is not acceptable to write $\ln x$, and when x is in the interval $\pi < x < 2\pi$, the term $\ln(\sin x)$ is not defined. These difficulties are easily overcome by first taking absolute values,

$$|y| = \frac{|x|^3(x^2+1)^{2/3}}{|\sin x|^3}.$$

Logarithms now give

$$\ln|y| = 3\ln|x| + \frac{2}{3}\ln(x^2+1) - 3\ln|\sin x|.$$

To differentiate this equation, we use equation 3.12 which states that when $f(x)$ is differentiable, and $f(x) \neq 0$,

$$\frac{d}{dx}|f(x)| = \frac{|f(x)|}{f(x)}f'(x).$$

When this is combined with formula 3.32, we obtain

$$\frac{d}{dx}\ln|f(x)| = \frac{1}{|f(x)|}\frac{d}{dx}|f(x)| = \frac{1}{|f(x)|}\frac{|f(x)|}{f(x)}f'(x) = \frac{f'(x)}{f(x)}. \tag{3.37}$$

Application of this result to the equation in $\ln|y|$ gives

$$\frac{1}{y}\frac{dy}{dx} = \frac{3}{x} + \frac{2}{3}\frac{2x}{x^2+1} - 3\frac{\cos x}{\sin x}.$$

This equation is identical to that obtained for dy/dx when $0 < x < \pi$, but its derivation here shows that it is valid even when x is not in the interval $0 < x < \pi$. ■

EXERCISES 3.10

In Exercises 1–24 use logarithmic differentiation to find $f'(x)$.

1. $f(x) = x^{-x}, \quad x > 0$
2. $f(x) = x^{4\cos x}, \quad x > 0$
3. $f(x) = x^{4x}, \quad x > 0$
4. $f(x) = (\sin x)^x, \quad 0 < x < \pi$
5. $f(x) = \left(1 + \frac{1}{x}\right)^x, \quad x > 0$
6. $f(x) = \left(1 + \frac{1}{x}\right)^{x^2}, \quad x > 0$
7. $f(x) = \left(\frac{1}{x}\right)^{1/x}, \quad x > 0$
8. $f(x) = \left(\frac{2}{x}\right)^{3/x}, \quad x > 0$
9. $f(x) = (\sin x)^{\sin x}, \quad 0 < x < \pi$
10. $f(x) = (\ln x)^{\ln x}, \quad x > 1$
11. $f(x) = (x^2 + 3x^4)^3(x^2+5)^4$
12. $f(x) = \dfrac{\sqrt{x}(1+2x^2)}{\sqrt{1+x^2}}$
13. $f(x) = x\sqrt[3]{1-\sin x}$
14. $f(x) = (x^2+3x)^3(x^2+5)^4$
15. $f(x) = x^2 e^{4x}$
16. $f(x) = x^{3/2}e^{-2x}$
17. $f(x) = x^2 \ln x$
18. $f(x) = \dfrac{e^x}{\ln(x-1)}$
19. $f(x) = (x^3+3)^3(x^2-2x)$
20. $f(x) = \dfrac{\sqrt{x}(1-x^2)}{\sqrt{1+x^2}}$
21. $f(x) = \dfrac{x^2-1}{x\sqrt{1-4\tan^2 x}}$
22. $f(x) = x^3(x^2-4x)\sqrt{1+x^3}$
23. $f(x) = \dfrac{\sin^3 3x}{\tan^5 2x}$

24. $f(x) = \dfrac{\sin 2x \sec 5x}{(1 - 2\cot x)^3}$

25. If $u(x)$ is positive for all x, find a formula for the derivative of u^u with respect to x.

26. If a company sells a certain commodity at price r, the market demands

$$x = r^a e^{-b(r+c)},$$

items per week, where $r > a/b$, and a, b, and c are positive constants.

(a) Show that the demand increases as the price decreases.

(b) Calculate the elasticity of demand defined by

$$\frac{Er}{Ex} = \frac{x}{r}\frac{dr}{dx}.$$

SECTION 3.11

Rolle's Theorem and the Mean Value Theorems

Certain results in calculus are immediately seen to be important. For example, the power, product, and quotient rules, which eliminate the necessity of using equation 3.2 to calculate derivatives are clearly indispensable. Even the algebraic and geometric interpretations of the derivative itself are recognized as useful. Through various examples and exercises of this chapter, we have hinted at the variety and quantity of applications of the derivative. These will be dealt with at length in Chapter 4. Other results in calculus, especially those of a theoretical nature, are regarded as less important, or even unimportant, often because it is not obvious how they will be used. In this section we consider three very important theorems, without which we would encounter serious difficulty in treating many of the topics in the remainder of this book. The first theorem is needed to prove the second, and the second leads immediately to the third.

Theorem 3.15

(Rolle's Theorem) *Suppose $f(x)$ is continuous on the closed interval $a \le x \le b$ and has a first derivative at each point in the open interval $a < x < b$. If $f(a) = f(b)$, then there exists at least one point c in the open interval at which $f'(c) = 0$.*

For the function in Figure 3.39 there are two possible choices for c.

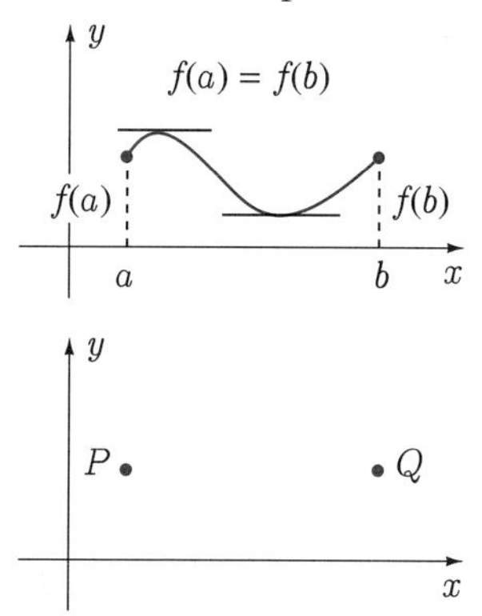

FIGURE 3.39

FIGURE 3.40

Geometrically, Rolle's theorem seems quite evident. Begin with two points, say P and Q in Figure 3.40, which have the same y-coordinate. Now try to join these points by a curve which never has a horizontal tangent line while satisfying the following two conditions:

(a) do not lift pencil from page — continuity of $f(x)$;

(b) the curve must have a tangent line at all points, and this tangent line must not be vertical — $f'(x)$ exists at all points.

It is impossible; therefore a point c where $f'(c) = 0$ must exist.

To verify this theorem directly requires Theorem 4.3 in Section 4.6. Since the latter result is quoted without proof, it seems as reasonable to accept Rolle's theorem on obvious geometric grounds as to base a proof on a theorem which itself is stated without proof. But for those who would like to see a proof based on Theorem 4.3 in Section 4.6, see Exercise 10 in that section.

Rolle's theorem can be used to prove the following result.

Theorem 3.16

(Cauchy's generalized mean value theorem) *Suppose functions $f(x)$ and $g(x)$ are continous on the closed interval $a \le x \le b$ and differentiable on the open interval $a < x < b$. If $g'(x) \neq 0$ on $a < x < b$, then there exists at least one point c in the open interval for which*

$$\frac{f(b) - f(a)}{g(b) - g(a)} = \frac{f'(c)}{g'(c)}. \tag{3.38}$$

Proof First note that $g(b) - g(a)$ cannot equal zero. If it did, then $g(a)$ would equal $g(b)$, and Rolle's theorem applied to $g(x)$ on the interval $a \le x \le b$ would imply the existence of a point c at which $g'(c) = 0$, contrary to the given assumption. To prove the theorem, we construct a function $h(x)$ to satisfy the conditions of Rolle's theorem. Specifically, we consider

$$h(x) = f(x) - f(a) - \frac{f(b) - f(a)}{g(b) - g(a)}[g(x) - g(a)].$$

Since $f(x)$ and $g(x)$ are continuous for $a \le x \le b$, so too is $h(x)$. In addition,

$$h'(x) = f'(x) - \frac{f(b) - f(a)}{g(b) - g(a)}g'(x);$$

therefore, $h'(x)$ exists for $a < x < b$. Finally, since $h(a) = h(b) = 0$, we may conclude from Rolle's theorem that there exists a number c such that $a < c < b$, and

$$0 = h'(c) = f'(c) - \frac{f(b) - f(a)}{g(b) - g(a)}g'(c);$$

that is,

$$\frac{f'(c)}{g'(c)} = \frac{f(b) - f(a)}{g(b) - g(a)}.$$

The next theorem states an important special case of this result that occurs when $g(x) = x$.

Theorem 3.17

(The mean value theorem) *Suppose $f(x)$ is continuous on the closed interval $a \le x \le b$ and has a first derivative at each point of the open interval $a < x < b$. Then there exists at least one point c in the open interval for which*

$$f'(c) = \frac{f(b) - f(a)}{b - a}. \tag{3.39}$$

From a geometric point of view 3.39 seems as obvious as Rolle's theorem. Figure 3.41 illustrates that the quotient $[f(b) - f(a)]/(b - a)$ is the slope of the line l joining the points $(a, f(a))$ and $(b, f(b))$ on the graph $y = f(x)$. The mean value theorem states that there is at least one point c between a and b at which the tangent line is parallel to l. In Figure 3.41 there are clearly two such points. Algebraically, 3.39 states that at some point between a and b, the instantaneous rate of change of the function $f(x)$ is equal to its average rate of change over the interval $a \leq x \leq b$. Similar interpretations of Theorem 3.16 are given in Section 10.6.

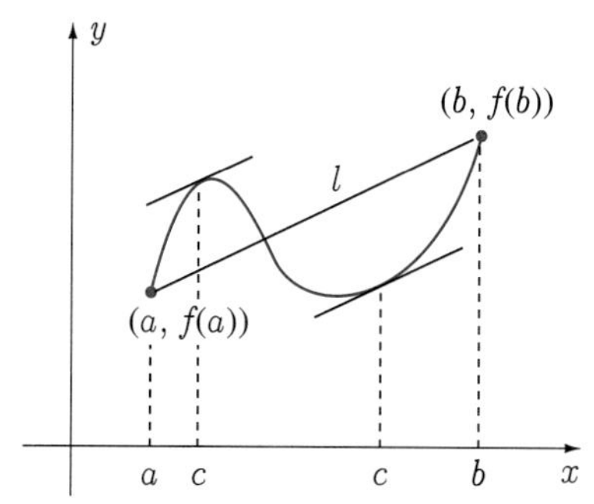

FIGURE 3.41

EXAMPLE 3.30

Find all values of c satisfying the mean value theorem for the function $f(x) = x^3 - 4x$ on the interval $-1 \leq x \leq 3$.

SOLUTION Since $f(x)$ is differentiable, and therefore continuous, at each point in $-1 \leq x \leq 3$, we can indeed apply the mean value theorem and claim the existence of at least one number c in $-1 < x < 3$ such that

$$f'(c) = \frac{f(3) - f(-1)}{3 - (-1)};$$

that is,

$$3c^2 - 4 = \frac{15 - 3}{4} = 3.$$

Consequently, $c = \pm\sqrt{7/3}$. Since $-\sqrt{7/3} < -1$, the only value of c in the interval $-1 < x < 3$ is $c = \sqrt{7/3}$ (Figure 3.42).

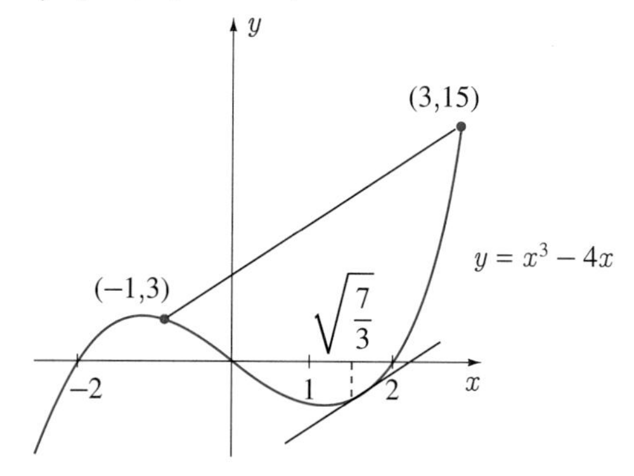

FIGURE 3.42

■

EXAMPLE 3.31

A traffic plane (Figure 3.43) measures the time that it takes a car to travel between points A and B as 15 seconds, and radios this information to a patrol car. What is the maximum speed at which the police officer can claim the car was traveling between A and B?

FIGURE 3.43

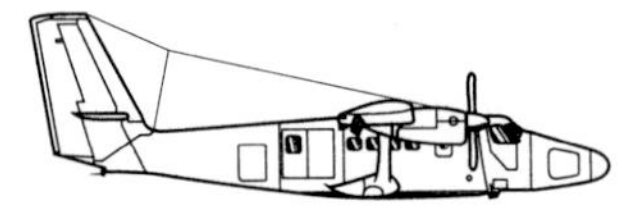

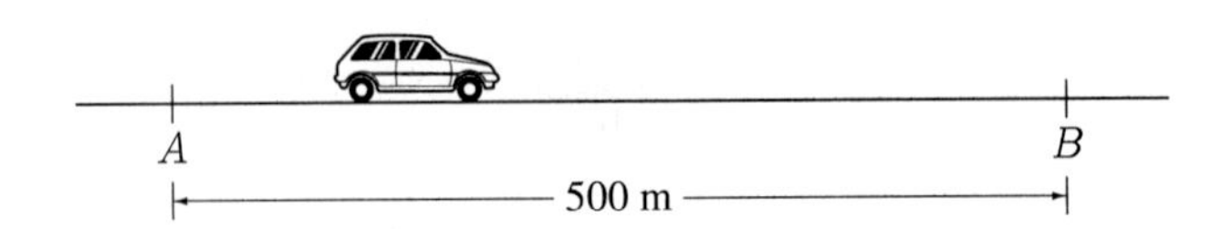

SOLUTION The average speed of the car between A and B is $500/15 = 100/3$ m/s. According to the mean value theorem, the instantaneous speed of the car must also have been $100/3$ m/s at least once. This is the maximum speed attributable to the car between A and B. It may have been traveling much faster at some points, but from the information given, no speed greater than $100/3$ m/s can be claimed by the officer. ■

EXERCISES 3.11

In Exercises 1–14 decide whether the mean value theorem can be applied to the function on the interval. If it cannot, explain why not. If it can, find all values of c in the interval that satisfy equation 3.39.

1. $f(x) = x^2 + 2x, \quad -3 \le x \le 2$

2. $f(x) = 4 + 3x - 2x^2, \quad 1 \le x \le 3$

3. $f(x) = x + 5, \quad 2 \le x \le 3$

4. $f(x) = |x|, \quad -1 \le x \le 1$

5. $f(x) = |x|, \quad 0 \le x \le 1$

6. $f(x) = x^3 + 2x^2 - x - 2, \quad -3 \le x \le 2$

7. $f(x) = x^3 + 2x^2 - x - 2, \quad -1 \le x \le 2$

8. $f(x) = (x+2)/(x-1), \quad 2 \le x \le 4$

9. $f(x) = (x+1)/(x+2), \quad -3 \le x \le 2$

10. $f(x) = x^2/(x+3), \quad -2 \le x \le 3$

11. $f(x) = \sin x, \quad 0 \le x \le 2\pi$

12. $f(x) = \ln(2x+1), \quad 0 \le x \le 2$

13. $f(x) = e^{-x}, \quad -1 \le x \le 1$

14. $f(x) = \sec x, \quad 0 \le x \le \pi$

In Exercises 15–18 decide whether Cauchy's generalized mean value theorem can be applied to the function on the interval. If it cannot, explain why not. If it can, find all values of c in the interval that satisfy equation 3.38.

15. $f(x) = x^2, \quad g(x) = x, \quad 1 \le x \le 2$

16. $f(x) = x + 1, \quad g(x) = |x|^{3/2}, \quad -1 \le x \le 1$

17. $f(x) = x^2 + 3x - 1, \quad g(x) = x^3 + 5x + 4, \quad 0 \le x \le 2$

18. $f(x) = x/(x+1), \quad g(x) = x/(x-1), \quad -3 \le x \le -2$

19. Show that if $|f'(x)| \le M$ on $a \le x \le b$, then

$$|f(b) - f(a)| \le M(b-a).$$

20. Show that the value c that satisfies the mean value theorem for any quadratic function $f(x) = dx^2 + ex + g$ on any interval $a \le x \le b$ whatsoever is $c = (a+b)/2$.

21. Use the mean value theorem to show that

$$|\sin a - \sin b| \le |a - b|$$

for all real a and b. Is the same inequality valid for the cosine function?

22. Let $f(x)$ and $g(x)$ be two functions that are differentiable at each point of the interval $a \le x \le b$. Prove that if $f(a) = g(a)$ and $f(b) = g(b)$, then there exists c in the open interval $a < x < b$ for which $f'(c) = g'(c)$.

23. Verify that, for a cubic polynomial $f(x) = dx^3 + ex^2 + gx + h$ defined on any interval $a \le x \le b$, the values of c that satisfy equation 3.39 are equidistant from $x = -e/(3d)$.

SUMMARY

In this chapter we defined the derivative of a function $y = f(x)$ as

$$\frac{dy}{dx} = f'(x) = \lim_{h \to 0} \frac{f(x+h) - f(x)}{h}.$$

Algebraically it is the instantaneous rate of change of y with respect to x; geometrically it is the slope of the tangent line to the graph of $f(x)$. To eliminate the necessity of using this definition over and over again, we derived the sum, product, quotient, and power rules:

$$\frac{d}{dx}(u+v) = \frac{du}{dx} + \frac{dv}{dx},$$

$$\frac{d}{dx}(uv) = u\frac{dv}{dx} + v\frac{du}{dx},$$

$$\frac{d}{dx}\left(\frac{u}{v}\right) = \frac{v\dfrac{du}{dx} - u\dfrac{dv}{dx}}{v^2},$$

$$\frac{d}{dx}(u^n) = nu^{n-1}\frac{du}{dx}.$$

These four simple rules are fundamental to all calculus.

When a function $y = f(x)$ is defined implicitly by some equation $F(x,y) = 0$, we use implicit differentiation to find its derivative. We differentiate each term in the equation with respect to x, and then solve the resulting equation for dy/dx. We pointed out that care must be taken in differentiating equations. An equation can be differentiated with respect to a variable only if it is valid for a continuous range of values of that variable.

The chain rule defines the derivative of a composite function $y = f(g(x))$ as the product of the derivatives of $y = f(u)$ and $u = g(x)$:

$$\frac{dy}{dx} = \frac{dy}{du}\frac{du}{dx}.$$

These rules and techniques form the basis for the rest of differential calculus. When they are combined with the derivatives of the trigonometric functions, the exponential and logarithm functions, and the functions of Chapter 8, we will be well prepared to handle those applications of calculus that involve differentiation. Derivative formulas for the trigonometric, exponential and logarithm functions are listed below:

$$\frac{d}{dx}\sin u = \cos u\,\frac{du}{dx}, \qquad \frac{d}{dx}\cos u = -\sin u\,\frac{du}{dx},$$

$$\frac{d}{dx}\tan u = \sec^2 u\,\frac{du}{dx}, \qquad \frac{d}{dx}\cot u = -\csc^2 u\,\frac{du}{dx},$$

$$\frac{d}{dx}\sec u = \sec u \tan u \frac{du}{dx}, \qquad \frac{d}{dx}\csc u = -\csc u \cot u \frac{du}{dx},$$

$$\frac{d}{dx}\log_a u = \frac{1}{u}\frac{du}{dx}\log_a e, \qquad \frac{d}{dx}\ln u = \frac{1}{u}\frac{du}{dx},$$

$$\frac{d}{dx}a^u = a^u \frac{du}{dx}\ln a, \qquad \frac{d}{dx}e^u = e^u \frac{du}{dx}.$$

We completed the chapter by using Rolle's theorem to prove two mean value theorems. When $f(x)$ and $g(x)$ are continuous for $a \le x \le b$ and differentiable for $a < x < b$, Cauchy's generalized mean value theorem guarantees the existence of at least one point c between a and b such that

$$\frac{f(b) - f(a)}{g(b) - g(a)} = \frac{f'(c)}{g'(c)},$$

provided also that $g'(x) \neq 0$ for $a < x < b$. When $g(x) = x$, we obtain as a special case, the mean value theorem

$$f'(c) = \frac{f(b) - f(a)}{b - a}.$$

Key Terms and Formulas

In reviewing this chapter, you should be able to define or discuss the following key terms:

Average rate of change
Instantaneous rate of change
Derivative
Tangent line
Derivative function
Power rule
Normal line
Increment
Differentiable
Right-hand derivative
Left-hand derivative
Product rule
Quotient rule
Angle between curves
Orthogonal
Chain rule
Extended power rule
Explicit definition of a function
Implicit definition of a function
Implicit differentiation
Logarithmic differentiation
Rolle's theorem
Generalized mean value theorem
Mean value theorem

REVIEW EXERCISES

In Exercises 1–48 assume that y is defined as a function of x and find dy/dx in as simplified a form as possible.

1. $y = x^3 + \dfrac{1}{x^2}$

2. $y = 3x^2 + 2x + \dfrac{1}{x}$

3. $y = 2x - \dfrac{1}{3x^2} + \dfrac{1}{\sqrt{x}}$

4. $y = x^{1/3} - \dfrac{2}{3}x^{5/3}$

5. $y = x(x^2 + 5)^4$

6. $y = (x^2 + 2)^2(x^3 - 3)^3$

7. $y = \dfrac{3x^2}{x^3 - 5}$

8. $y = \dfrac{3x - 2}{x + 5}$

9. $y = \dfrac{x^2 + 2x + 2}{x^2 + 2x - 1}$

10. $y = \dfrac{4x}{x^2 + 5x - 2}$

11. $xy + 3y^3 = x + 1$

12. $\dfrac{x}{y} + \dfrac{y}{x} = x$

13. $x^2y^2 - 3y\sin x = 14$

14. $x^2y + y\sqrt{1 + x} = 3$

15. $y = \tan^3(3x + 2)$

16. $y = \sec^2(1 - 4x)$

17. $y = \dfrac{\sin 2x}{\cos 3x}$

18. $y = \sec(\tan 2x)$

19. $y = x^2 \cos x^2$

20. $y = \sin^2 x \cos^2 x$

21. $y = u^2 - 2u, \quad u = (1 + 2x)^{5/3}$

22. $y = t + \cos 2t, \quad t = x - \cos 2x$

23. $y = \sqrt{1 - t^3}, \quad t = \sqrt{1 + x^2}$

24. $y = v\cos^2 v, \quad v = \sqrt{1 - x^2}$

25. $y = \sqrt{1 + \sqrt{1 + x}}$ 26. $x = e^{2y}$

27. $\dfrac{2xy}{3x + 4} = x^2 + 2$ 28. $y = (x^2 + 1)\ln(x^2 - 1)$

29. $x\sin y + 2xy = 4$ 30. $5\cos(x - y) = 1$

31. $x = \sqrt{1 + x\cot y^2}$ 32. $x = \sqrt{\dfrac{4 + y}{4 - y}}$

33. $y = \sqrt{\dfrac{4 + x^2}{4 - x^2}}$ 34. $y = \dfrac{x^2\sqrt{1 - x}}{x + 5}$

35. $y = \sqrt{7 - \sqrt{7 - \sqrt{x}}}$ 36. $\dfrac{x}{x + y} = \dfrac{y}{x - y}$

37. $y = \sqrt{\dfrac{4 + t}{4 - t}}, \quad t = \tan x$

38. $x = \dfrac{y^2 - 2y}{y^3 + 4y + 6}$

39. $y = \dfrac{x^3 - 6x^2 + 12x - 8}{x^2 - 4x + 4}$

40. $y = \dfrac{x - 2}{\sqrt{x} - \sqrt{2}}$

41. $y = x^{2x}$ 42. $y = (\cos x)^x, \ 0 < x < \pi/2$

43. $y = \dfrac{e^x}{e^x + 1}$ 44. $y = \log_{10}(\log_{10} x)$

45. $y = e^x \ln x$ 46. $x = e^y + e^{-y}$

47. $xye^{xy} = 1$ 48. $x^2 y + \ln(x + y) = x + 2$

In Exercises 49–52 find equations for the tangent and normal lines to the curve at the point indicated.

49. $y = x^3 + 3x - 2$ at $(1, 2)$

50. $y = \dfrac{1}{x + 5}$ at $(0, 1/5)$

51. $y = \cos 2x$ at $(\pi/2, -1)$

52. $y = \dfrac{x^2 + 3x}{2x - 5}$ at $(1, -4/3)$

In Exercises 53–55 find d^2y/dx^2 assuming that y is defined implicitly as a function of x.

53. $x^2 - y^2 + 2(x - y) = y^3$

54. $(x - y)^2 = 3xy$

55. $\sin(x + y) = x$

56. Sketch a graph of $f(x) = \sin x^2$. Is it periodic?

57. Find all points on the curve $y = x^3 + x^2$ at which the tangent line passes through the origin.

58. Find that point on the curve $x = y^2 - 4$ at which the normal line passes through the point $(-6, 7)$. What application could be made of this result?

59. (a)How many functions with domain $-1 \leq x \leq 1$ are defined implicitly by the equation $x^2 + y^2 = 1$?

(b) How many continuous functions with domain $-1 \leq x \leq 1$ are defined implicitly by the equation?

60. What is the rate of change of the area A of an equilateral triangle with respect to its side length l?

61. In a heated house, the temperature varies as the thermostat continually engages and disengages the furnace. Suppose that at the thermostat, the temperature T in degress Celsius over a four-hour time interval $0 \leq t \leq 4$ is given by

$$T = f(t) = 20 + 3\sin(4\pi t - \pi/2).$$

(a) Sketch a graph of $f(t)$.

(b) How many times is the furnace on during the four-hour period?

(c) What is the maximum time rate of change of temperature?

62. The curve defined by the equation $(x^2 + y^2)^2 = x^2 - y^2$ and shown in Figure 3.44 is called a *lemniscate*. Find the four points at which the tangent line is horizontal.

FIGURE 3.44

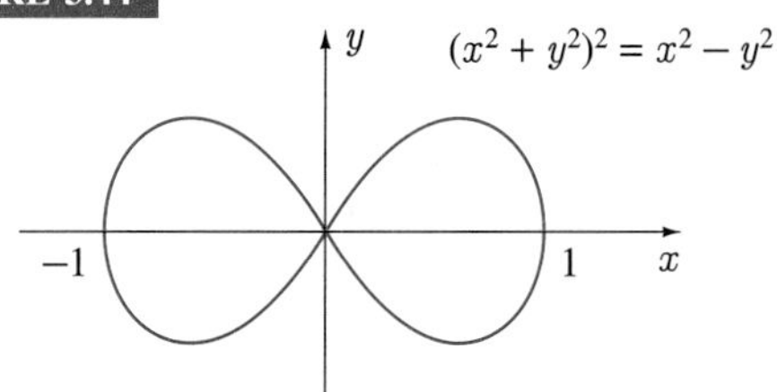

63. Find all points c in the interval $3 \leq x \leq 6$ that satisfy equation (3.39) when $f(x) = x^3 + 3x - 2$.

64. Find all points c in the interval $-1 \leq x \leq 1$ that satisfy equation (3.38) when $f(x) = 3x^2 - 2x + 4$ and $g(x) = x^3 + 2x$.

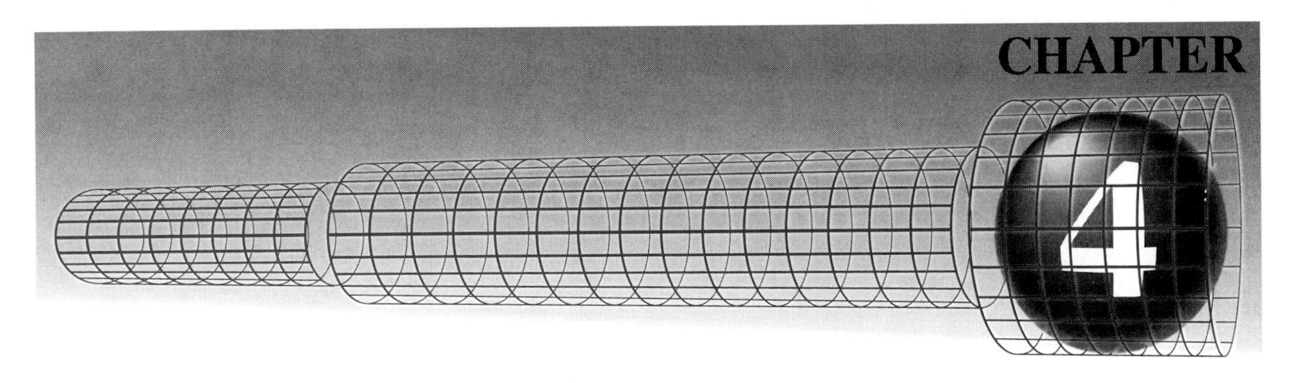

Applications of Differentiation

In chapter 3 we hinted at some of the applications of the derivative; in this chapter we discuss them in detail.

Equations must be solved in almost every area of applied mathematics, and most equations do not have analytic solutions; solutions must be approximated to some specified accuracy. In Section 4.1 we show that derivatives provide one of the most potent methods for doing this. Much of this chapter (Sections 4.2–4.6) is devoted to the topic of optimization and its theory and applications. The first derivative of a function is used to determine where the values of a function are increasing or decreasing, and where the graph of a function attains a maximum or minimum value relative to nearby points. The second derivative determines intervals on which the first derivative is increasing or decreasing, and the transition points between these intervals. Such information can be invaluable in drawing the graph of a function. In Section 4.5 we combine these techniques with the tools from Chapters 1 and 2 for a final discussion on curve sketching. Sections 4.6–4.8, with their wealth of applied problems, show the real power of calculus. In Section 4.6 we illustrate the simplicity that calculus brings to solving applied maxima-minima problems; in Section 4.7 we use the results from Sections 4.2– 4.6 to develop a deeper understanding for the already familiar notions of velocity and acceleration; and in Section 4.8 we use the interpretation of the derivative as a rate of change to investigate interdependences of related quantities in a wide variety of applications. The calculation of many otherwise intractable limits becomes relatively straightforward with L'Hôpital's rule in Section 4.9. In Section 4.10 we discuss differentials, quantities essential to the topic of integration which begins in Chapter 5. To close the chapter in Section 4.11, we offer suggestions on how to organize and solve complicated multistage problems.

SECTION 4.1

Newton's Iterative Procedure for Solving Equations

In almost every area of application of mathematics, it is necessary to solve equations. When the problem is to solve one equation in one unknown, the equation can be expressed in the form

$$f(x) = 0, \tag{4.1}$$

where $f(x)$ is usually a differentiable function of x. It might be a polynomial, a trigonometric function, an exponential or a logarithm function, or a complicated combination of these. The equation may have one solution, or many solutions, and these solutions are also called roots of the equation, or zeros of the function. Few equations can be solved by formula. Even when $f(x)$ is a polynomial, the only simple solution formula is the quadratic formula which solves the equation for second degree polynomials. Functions other than polynomials rarely have formulas for their zeros.

Since we can seldom find a formula for solving 4.1, a different approach must be taken. In this section we discuss one of many techniques for approximating solutions to equations. To illustrate, consider finding the depth of water that half fills (by volume) the hemispherical tank in Figure 4.1. It can be shown (see Exercise 15 in Section 7.4) that when the water is x metres deep, the volume of water in the tank is given by

FIGURE 4.1

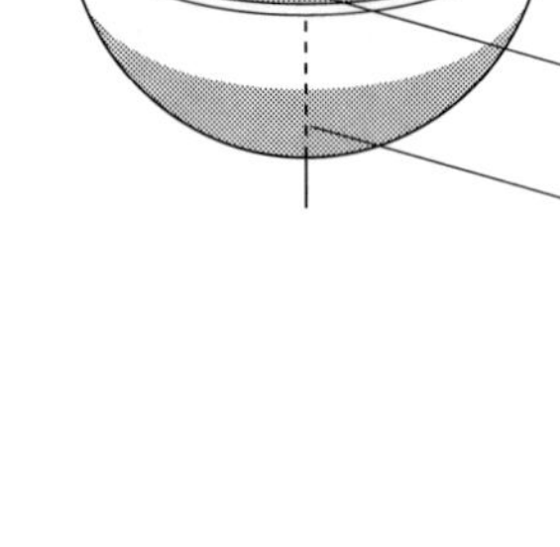

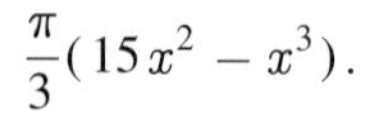

$$\frac{\pi}{3}(15x^2 - x^3).$$

Since the volume of one-half the tank is one-quarter that of a sphere of radius 5, it follows that the tank is half full when x satisfies the equation

$$\frac{1}{4}\left[\frac{4}{3}\pi(5)^3\right] = \frac{\pi}{3}(15x^2 - x^3),$$

and this equation simplifies to

$$f(x) = x^3 - 15x^2 + 125 = 0.$$

How do we solve this equation when we know that the solution we want is somewhere around $x = 3$? One of many possible methods is called **Newton's iterative procedure**.

A sketch of $f(x)$ between $x = 3$ and $x = 3.5$ is shown in Figure 4.2. The curvature has been exagerated in order to more clearly depict the following geometric construction. If we set $x_1 = 3$, then x_1 is an approximation to the solution of the equation — not a good approximation, but an approximation nonetheless. Suppose we draw the tangent line to $y = f(x)$ at $(x_1, f(x_1))$. If x_2 is the point of intersection of this tangent line with the x-axis, it is clear that x_2 is a better approximation to the solution of the equation than x_1. If we draw the tangent line to $y = f(x)$ at $(x_2, f(x_2))$, its intersection point x_3 with the x-axis is an even better approximation. Continuation of this process leads to a succession of numbers $x_1, x_2, x_3, \ldots$, each of which is closer to the solution of the equation $f(x) = 0$ than the preceding numbers. This procedure for finding a better approximation to the solution of an equation is called Newton's iterative procedure (or the Newton-Raphson iterative procedure). We say that the numbers $x_1, x_2, x_3, \ldots$ converge to the root of the equation.

FIGURE 4.2

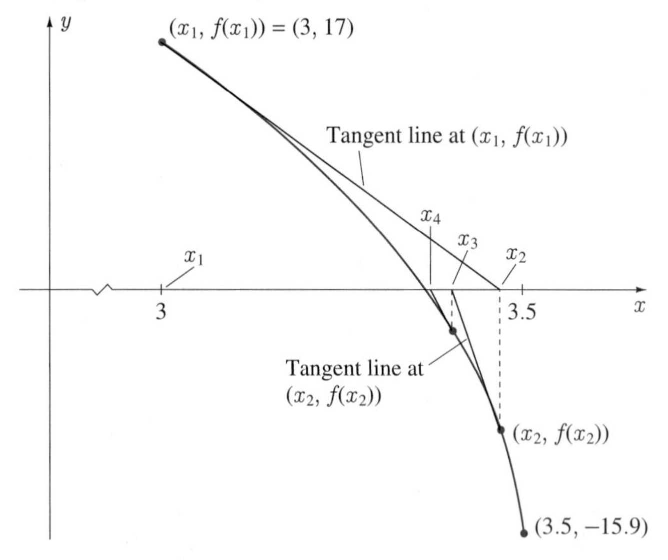

What we need now is an algebraic formula by which to calculate the numbers $x_1, x_2, x_3, \ldots$. The equation of the tangent line to $y = f(x)$ at $(x_1, f(x_1))$ is

$$y - f(x_1) = f'(x_1)(x - x_1).$$

To find the point of intersection of this tangent line with the x-axis, we set $y = 0$,

$$-f(x_1) = f'(x_1)(x - x_1),$$

and solve for x,

$$x = x_1 - \frac{f(x_1)}{f'(x_1)}.$$

But the point of intersection of the tangent line at $(x_1, f(x_1))$ with the x-axis is the second approximation x_2; that is,

$$x_2 = x_1 - \frac{f(x_1)}{f'(x_1)}.$$

To find x_3 we repeat this procedure with x_2 replacing x_1; the result is

$$x_3 = x_2 - \frac{f(x_2)}{f'(x_2)}.$$

As we repeat this process over and over again, the following formula for the $(n+1)^{\text{th}}$ approximation x_{n+1} in terms of the nth approximation x_n emerges

$$x_{n+1} = x_n - \frac{f(x_n)}{f'(x_n)}. \tag{4.2}$$

This formula defines each approximation in Newton's iterative procedure in terms of its predecessor.

The derivative of $f(x) = x^3 - 15x^2 + 125$ in the above example is $f'(x) = 3x^2 - 30x$, and formula 4.2 becomes

$$x_{n+1} = x_n - \frac{x_n^3 - 15x_n^2 + 125}{3x_n^2 - 30x_n}.$$

Calculation of the first five approximations gives

$$x_2 = x_1 - \frac{x_1^3 - 15x_1^2 + 125}{3x_1^2 - 30x_1} = 3 - \frac{3^3 - 15(3)^2 + 125}{3(3)^2 - 30(3)} = 3.26984;$$

$$x_3 = x_2 - \frac{x_2^3 - 15x_2^2 + 125}{3x_2^2 - 30x_2} = 3.26352;$$

$$x_4 = x_3 - \frac{x_3^3 - 15x_3^2 + 125}{3x_3^2 - 30x_3} = 3.2635182;$$

$$x_5 = x_4 - \frac{x_4^3 - 15x_4^2 + 125}{3x_4^2 - 30x_4} = 3.2635182.$$

Newton's iterative procedure has therefore produced 3.263518 as an approximate solution to the equation $x^3 - 15x^2 + 125 = 0$. In spite of the fact that we have written six decimals in this final answer, our analysis in no way guarantees this degree of accuracy; we have simply judged on the basis that $x_4 = x_5$ that 3.263518 might be accurate to six decimals. The following theorem enables us to confirm this accuracy.

Theorem 4.1 (Intermediate Value Theorem) *If* $f(a)f(b) < 0$ *for a function* $f(x)$ *continuous on* $a \le x \le b$*, then there exists at least one number* c *between* a *and* b *for which* $f(c) = 0$.

Figure 4.3 illustrates the situation. The condition $f(a)f(b) < 0$ requires that one of $f(a)$ and $f(b)$ be positive and the other be negative. We have shown $f(a) > 0$ and $f(b) < 0$. As $f(x)$ is continuous for $a \le x \le b$, its graph can be traced between a and b without lifting pencil from paper. But then it must cross the x-axis at least once, and such a point is c in the theorem.

FIGURE 4.3

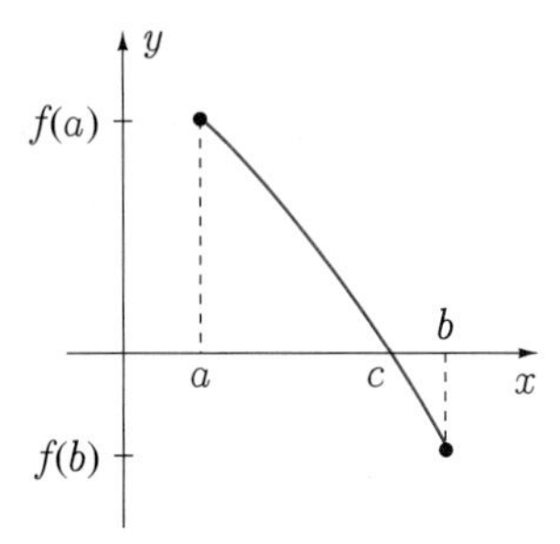

To show that 3.263518 is a root of the equation $f(x) = x^3 - 15x^2 + 125 = 0$ accurate to six decimals, we calculate $f(3.2635175) = 4.77 \times 10^{-5}$ and $f(3.2635185) = -1.83 \times 10^{-5}$. Since these function values are opposite in sign, the intermediate value theorem guarantees a root between 3.2635175 and 3.2635185. Consequently, when rounded to six decimals, the root must be 3.263518.

EXAMPLE 4.1

Use Newton's iterative procedure to find the only positive root of the equation

$$3x^4 + 15x^3 - 125x - 1500 = 0$$

accurate to 5 decimals. (This equation will be encountered in Exercise 44 of Section 4.6.)

SOLUTION

If we set $f(x) = 3x^4 + 15x^3 - 125x - 1500$, then a quick calculation shows that $f(4) = -272$ and $f(5) = 1625$. Thus the required solution is between $x = 4$ and $x = 5$. The nth approximation predicted by Newton's method is defined by

$$x_{n+1} = x_n - \frac{f(x_n)}{f'(x_n)} = x_n - \frac{3x_n^4 + 15x_n^3 - 125x_n - 1500}{12x_n^3 + 45x_n^2 - 125}.$$

When we choose $x_1 = 4$ ($f(4)$ is closer to 0 than $f(5)$), we find

$$x_2 = 4.19956, \qquad x_3 = 4.187268, \qquad x_4 = 4.1872187, \qquad x_5 = 4.1872187.$$

Since

$$f(4.187215) = -5.73 \times 10^{-3} \qquad \text{and} \qquad f(4.187225) = 9.72 \times 10^{-3},$$

it follows that the root is $x = 4.18722$, accurate to five decimals. ■

In the following example we again use Newton's iterative procedure to approximate the solution of an equation but specify the required accuracy in a different way.

EXAMPLE 4.2

Use Newton's method to find the smallest root of the cubic equation $x^3 - 3x + 1 = 0$ with an error no greater than 10^{-5}.

SOLUTION

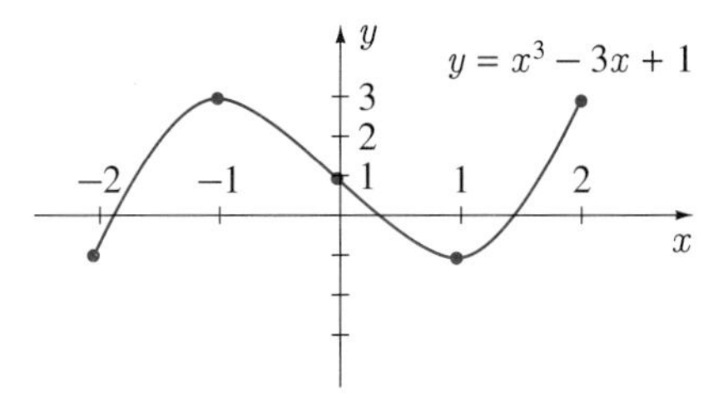

FIGURE 4.4

To find an estimate of the smallest root of the equation, we sketch a quick graph of the function $f(x) = x^3 - 3x + 1$ using the five points in Figure 4.4. The required root is clearly between -2 and -1. We take $x_1 = -2$ and use Newton's iterative procedure to define the nth approximation

$$x_{n+1} = x_n - \frac{f(x_n)}{f'(x_n)} = x_n - \frac{x_n^3 - 3x_n + 1}{3x_n^2 - 3}.$$

The next four approximations are

$$x_2 = -1.88889, \qquad x_3 = -1.87945, \qquad x_4 = -1.879385, \qquad x_5 = -1.8793852.$$

To determine the root to the required accuracy, we must find an interval of length 2×10^{-5} inside of which the root is known to lie. Since

$$f(-1.87939) = -3.6 \times 10^{-5} \qquad \text{and} \qquad f(-1.87937) = 1.16 \times 10^{-4},$$

such an interval is $-1.87939 < x < -1.87937$. Because the required solution must lie in this interval, it follows that -1.87938 is a solution with error no greater than 10^{-5}. ■

Of the many techniques for approximating solutions to equations, Newton's method is the most popular. It converges to a solution a of $f(x) = 0$ provided $f'(a) \neq 0$ and x_1 is chosen sufficiently close to a (see Exercise 33 in Section 11.2). The condition $f'(a) \neq 0$ is included so that the denominator in 4.2 does not approach 0 as $x_n \to a$. However, $f(x_n)$ ususally approaches 0 faster than $f'(x_n)$ so that even when $f'(a) = 0$, Newton's method is successful. It is impossible to indicate how close x_1 must be to a in order to guarantee convergence to a. In some examples x_1 can be any number whatsoever, but we are also aware of examples where $|x_1 - a|$ must be less than 0.01 (see Exercise 31).

When x_1 is not sufficiently close to a, Newton's iterative procedure may not converge, or may converge to a solution other than expected. For instance, were we to attempt to approximate the largest root in Example 4.2, and inadvertently choose $x_1 = 1$, we could not find the second approximation: algebraically because $3x_1^2 - 3 = 0$, and geometrically because at $x_1 = 1$, the tangent line is horizontal, and does not intersect the x-axis. The function $f(x)$ in Figure 4.5 has zeros near $x = 1/2$ and $x = 1$. Were we to attempt to find the smaller zero using an initial approximation $x_1 = 0$, we would find the second approximation x_2 to be larger than 1. Further iterations would then converge to the zero near $x = 1$, not the zero near $x = 1/2$. We conclude, therefore, that the initial approximation in Newton's iterative procedure is most important. A poor choice for x_1 may lead to numbers that either converge to the wrong root or do not converge at all.

FIGURE 4.5

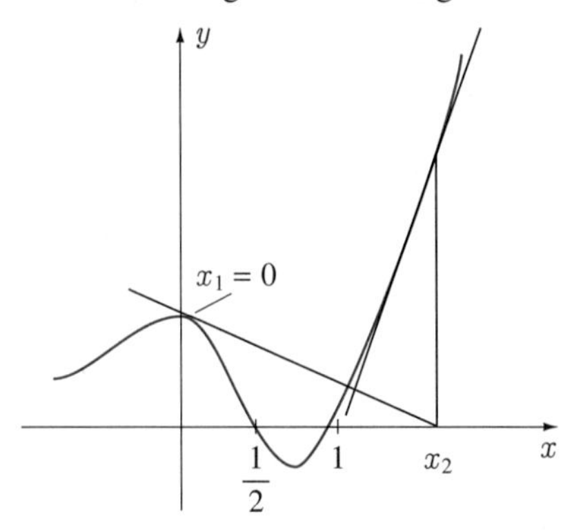

It is sometimes difficult to determine the number of solutions to an equation $f(x) = 0$ and initial approximations to these solutions. Even the roughest sketch of the function $f(x)$ can be instrumental in making such decisions. This was done in our introductory example concerning the hemispherical tank and again in Example 4.2. For some equations, it proves advantageous to arbitrarily separate the function $f(x)$ into two parts and write equation 4.1 in the form $g(x) = h(x)$. Solutions of this equation can be interpreted as x-coordinates of points of intersection of the curves $y = g(x)$ and $y = h(x)$. Sketches of these curves may quickly yield the number of solutions of the equation and their approximate values. For example, given the equation $2 \sin x - (x-1)^2 = 0$, we write $2 \sin x = (x-1)^2$ and sketch the curves $y = 2 \sin x$ and $y = (x-1)^2$ (Figure 4.6). Clearly there are two points of intersection, one between $x = 0$ and $x = 1$, and the other between $x = 2$ and $x = 3$.

FIGURE 4.6

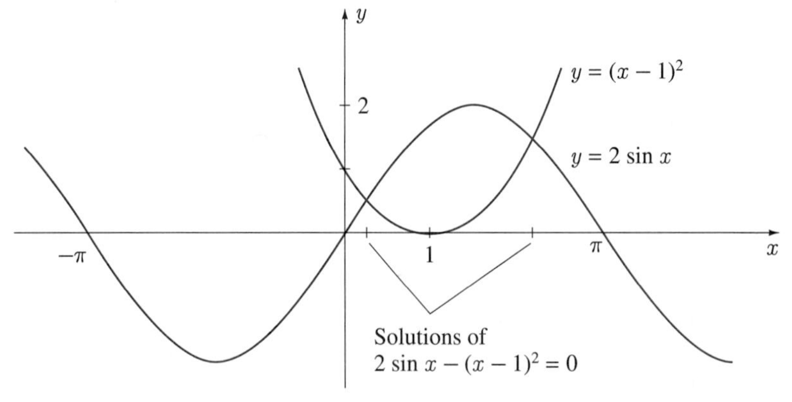

EXERCISES 4.1

In Exercises 1–16 use Newton's iterative procedure to find approximations to all roots of the equation accurate to 6 decimal places. In each case, we suggest making a sketch in order to obtain an initial approximation to each root.

1. $x^2 + 3x + 1 = 0$

2. $x^2 - x - 4 = 0$

3. $x^3 + x - 3 = 0$

4. $x^3 - x^2 + x - 22 = 0$

5. $x^3 - 5x^2 - x + 4 = 0$

6. $x^5 + x - 1 = 0$

7. $x^4 + 3x^2 - 7 = 0$

8. $\dfrac{x+1}{x-2} = x^2 + 1$

9. $x - 10 \sin x = 0$

10. $\sec x = \dfrac{2}{1 + x^4}$

11. $(x+1)^2 = \sin 4x$

12. $(x+1)^2 = 5 \sin 4x$

13. $x + 4 \ln x = 0$

14. $x \ln x = 6$

15. $e^x + e^{-x} = 10x$

16. $x^2 - 4e^{-2x} = 0$

In Exercises 17–24 use Newton's iterative procedure to find approximations to all roots of the equation with an error no greater than that specified.

17. $x^3 - 5x - 1 = 0, \quad 10^{-3}$

18. $x^4 - x^3 + 2x^2 + 6x = 0, \quad 10^{-4}$

19. $\dfrac{x}{x+1} = x^2 + 2, \quad 10^{-5}$

20. $(x+1)^2 = x^3 - 4x, \quad 10^{-3}$

21. $(x+1)^2 = 5\sin 4x, \quad 10^{-3}$

22. $\cos^2 x = x^2 - 1, \quad 10^{-4}$

23. $x + (\ln x)^2 = 0, \quad 10^{-3}$

24. $e^{3x} + e^x = 4, \quad 10^{-4}$

In Exercises 25–28 find all points of intersection for the curves accurate to 4 decimals.

25. $y = x^3, \quad y = x + 5$

26. $y = (x+1)^2, \quad y = x^3 - 4x$

27. $y = x^4 - 20, \quad y = x^3 - 2x^2$

28. $y = \dfrac{x}{x+1}, \quad y = x^2 + 2$

29. When the beam in Figure 4.7 vibrates vertically, there are certain frequencies of vibration, called natural frequencies.

FIGURE 4.7

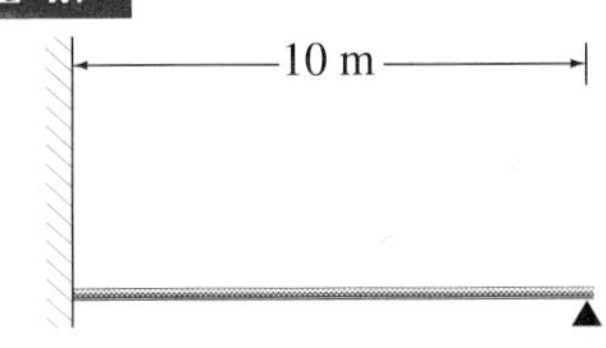

They are solutions of the equation

$$\tan x = \frac{e^x - e^{-x}}{e^x + e^{-x}}$$

divided by 20π. Find the two smallest frequencies correct to 4 decimals.

30. A stone of mass 100 gm is thrown vertically upward with speed 20 m/s. Air exerts a resistive force on the stone proportional to its speed, and has magnitude 0.1 N when the speed of the stone is 10 m/s. It can be shown that the height y above the projection point attained by the stone is given by

$$y = -98.1t + 1181\left(1 - e^{-t/10}\right) \text{ m},$$

where t is time (measured in seconds with $t = 0$ at the instant of projection).

(a) The time taken for the stone to return to its projection point can be obtained by setting $y = 0$ and solving the equation for t. Do so (correct to 2 decimals).

(b) When air resistance is neglected, the formula for y is

$$y = 20t - 4.905t^2 \text{ m}.$$

What is elapsed time in this case from the instant the stone is projected until it returns to the projection point?

31. A uniform hydro cable $P = 80$ m long with mass per unit length $\rho = 0.5$ kg/m is hung from two supports at the same level $L = 70$ m apart (Figure 4.8).

FIGURE 4.8

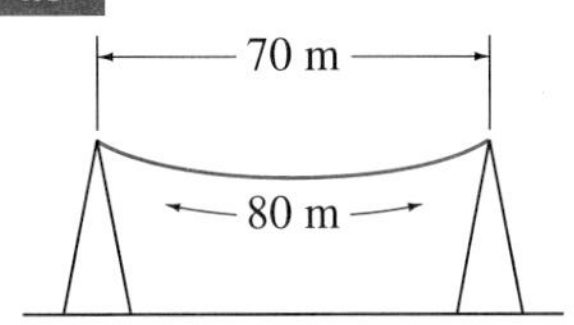

The tension T in the cable at its lowest point must satisfy the equation

$$\frac{\rho g P}{T} = e^{\rho g L/(2T)} - e^{-\rho g L/(2T)},$$

where $g = 9.81$. If we set $z = \rho g/(2T)$, then z must satisfy

$$2Pz = e^{Lz} - e^{-Lz}.$$

Solve this equation for z and hence find T correct to 1 decimal.

32. Planck's law for the energy density E of blackbody radiation at 1000°K states that

$$E = E(\lambda) = \frac{k\lambda^{-5}}{e^{c/\lambda} - 1}$$

where $k > 0$ is a constant and $c = 0.0014386$. This function is shown in Figure 4.9.

FIGURE 4.9

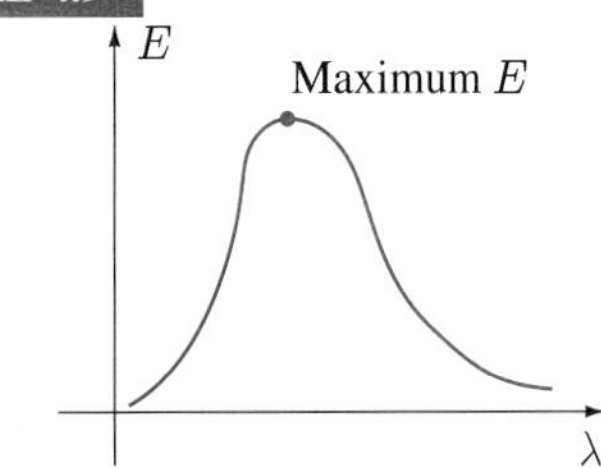

The value of λ at which E is a maximum must satisfy the equation

$$(5\lambda - c)e^{c/\lambda} - 5\lambda = 0.$$

Find this value of λ correct to 7 decimals.

33. Let $f(x)$ be a continuous function with domain and range both equal to the interval $[a, b]$. Show that there is at least one value of x in (a, b) for which $f(x) = x$.

34. (a) Use the intermediate value theorem to prove that when the domain of a continuous function is an interval, so also is its range. Hint: Use the idea that a set S of points on the y-axis constitutes an interval if for any two points c and d in S, the points $c < y < d$ are all in S.

(b) If the domain is an open interval, is the range an open interval?

35. Use the Intermediate Value Theorem to prove that at any given time there is a pair of points directly opposite each other on the equator of the earth that have exactly the same temperature. Hint: Take the equator to be the circle $x^2 + y^2 = r^2$. Let $f(x)$ be the temperature on the upper semicircle and $g(x)$ be the temperature on the lower semicircle. Consider the function $F(x) = f(x) - g(-x)$.

36. A marathoner runs the 26 odd miles from point A to point B starting at 7:00 a.m. Saturday morning. Starting at 7:00 a.m. Sunday morning he runs the course again, but this time from point B to point A. Prove that there is a point on the course that he passed at exactly the same time on both days.

SECTION 4.2

Increasing and Decreasing Functions

Many mathematical concepts have their origin in very intuitive ideas. In this section we analyze the intuitive idea of one quantity "getting larger" and another "getting smaller". To describe what it means for a quantity to be increasing (getting larger) or decreasing (getting smaller), we first suppose that the quantity is represented by some function $f(x)$ of a variable x. The mathematical definition for $f(x)$ to be increasing or decreasing is as follows.

Definition 4.1

A function $f(x)$ is said to be **increasing** on an interval I if for all $x_1 > x_2$ in I

$$f(x_1) > f(x_2). \tag{4.3}$$

A function $f(x)$ is said to be **decreasing** on I if for all $x_1 > x_2$ in I

$$f(x_1) < f(x_2). \tag{4.4}$$

The continuous function in Figure 4.10 is increasing on the intervals

$$a \le x \le b, \qquad c \le x \le d, \qquad e \le x \le f,$$

and decreasing on the intervals

$$b \le x \le c, \qquad d \le x \le e.$$

When the function has points of discontinuity, such as in Figure 4.11, the situation is somewhat more complicated. This function is increasing on the intervals

$$a \le x < b, \qquad d \le x < e, \qquad e < x \le f,$$

and decreasing for

$$b < x < d.$$

Pay special attention to whether end points of each interval are included.

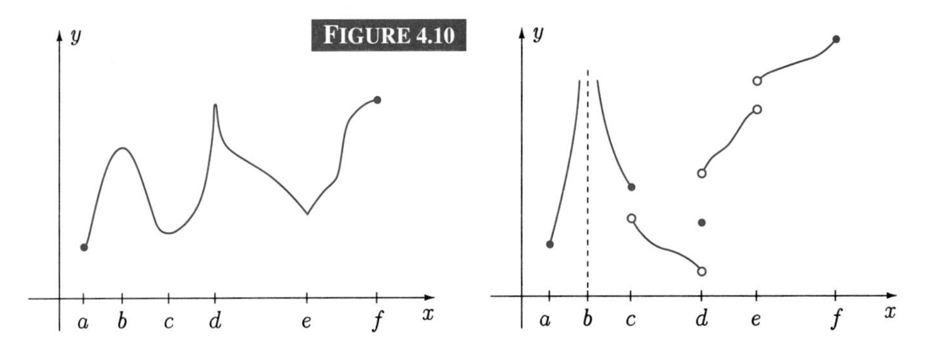

FIGURE 4.10

FIGURE 4.11

For continuous functions the graph in Figure 4.10 indicates that the sign of $f'(x)$ determines whether a function is increasing or decreasing on an interval. A precise statement of the situation for continuous functions is as follows.

Increasing and Decreasing Test for Continuous Functions

(i) A continuous function $f(x)$ is increasing on an interval I if on I

$$f'(x) \geq 0 \tag{4.5}$$

and is equal to zero at only a finite number of points.

(ii) A continuous function $f(x)$ is decreasing on an interval I if on I

$$f'(x) \leq 0 \tag{4.6}$$

and is equal to zero at only a finite number of points.

In conditions 4.5 and 4.6 we permit $f'(x)$ to vanish at only a finite number of points and do not, therefore, allow it to vanish on an interval. The reason for this is illustrated in Figures 4.12 and 4.13. In Figure 4.12, $f'(x)$ is equal to zero on the interval $a \leq x \leq b$, and certainly $f(x)$ is not increasing on any interval that contains these points. The function $f(x)$ in Figure 4.13 has $f'(0) = 0 = f'(1)$, and yet $f(x)$ is increasing on the interval $a \leq x \leq b$.

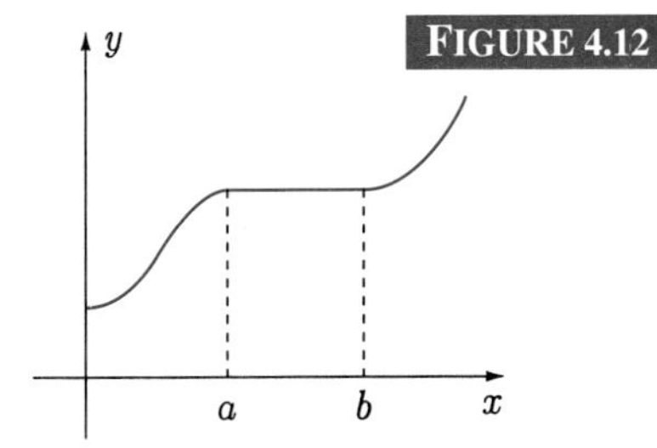

FIGURE 4.12

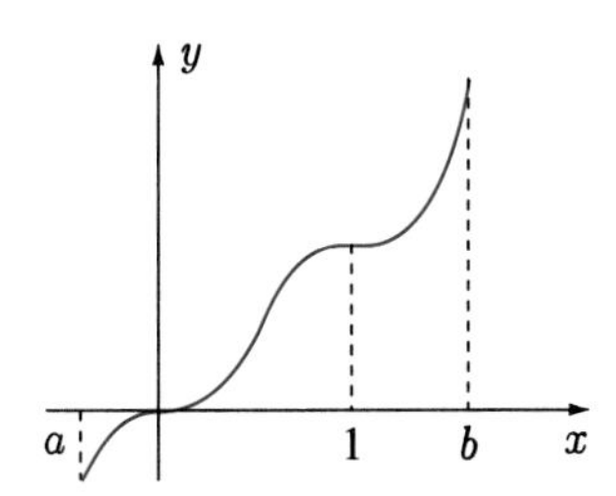

FIGURE 4.13

With the mean value theorem from Section 3.11, it is easy to verify this test. We prove (i); the proof of (ii) is similar. Suppose $f'(x) \geq 0$ on an interval $a \leq x \leq b$ and $d_1, d_2, \ldots, d_n$ are the points at which $f'(x) = 0$. Take any two points x_1 and x_2 in the interval $d_1 \leq x \leq d_2$, where $x_2 > x_1$. By the mean value theorem there exists a point c between x_1 and x_2 such that

$$f'(c) = \frac{f(x_2) - f(x_1)}{x_2 - x_1} \qquad \text{or} \qquad f(x_2) - f(x_1) = f'(c)(x_2 - x_1).$$

Since $x_2 > x_1$ and $f'(c) > 0$, it follows that $f(x_2)$ must be greater than $f(x_1)$. In other words, $f(x)$ is increasing on the interval $d_1 \leq x \leq d_2$. But by this argument $f(x)$

is increasing on each of the intervals $a \le x \le d_1$, $d_1 \le x \le d_2$, ..., $d_{n-1} \le x \le d_n$, $d_n \le x \le b$, and therefore must be increasing for $a \le x \le b$.

According to the above test, the sign of $f'(x)$ dictates where a function $f(x)$ is increasing and where it is decreasing. Sign diagrams introduced in Section 1.1 are useful in analyzing positivity and negativity of $f'(x)$.

EXAMPLE 4.3

Find intervals on which the following functions are increasing and decreasing:

$$\text{(a)}\quad f(x) = 2x^3 + 3x^2 - 12x + 4 \qquad \text{(b)}\quad f(x) = \frac{x}{x+1}$$

SOLUTION (a) Since $f'(x) = 6x^2 + 6x - 12 = 6(x-1)(x+2)$, the sign diagram in Figure 4.14 indicates that $f'(x) \ge 0$ when $x \le -2$ and when $x \ge 1$, and that $f'(x) \le 0$ when $-2 \le x \le 1$. Consequently, $f(x)$ is increasing on the intervals $x \le -2$ and $x \ge 1$, and decreasing on $-2 \le x \le 1$.

FIGURE 4.14

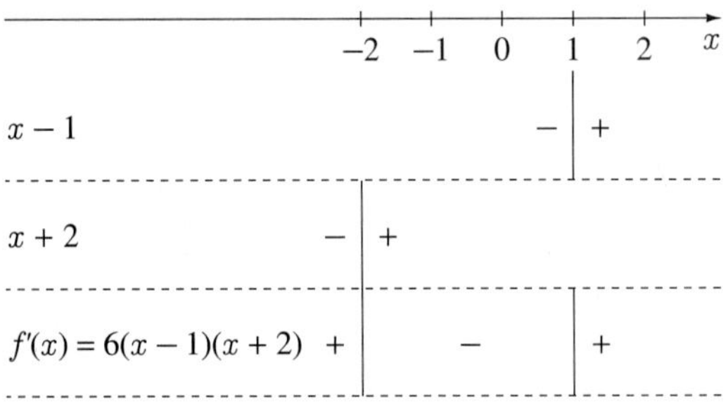

	$x < -2$	$-2 < x < 1$	$x > 1$
$x - 1$		$-$	$+$
$x + 2$	$-$	$+$	
$f'(x) = 6(x-1)(x+2)$	$+$	$-$	$+$

(b) Because

$$f'(x) = \frac{(x+1)(1) - x(1)}{(x+1)^2} = \frac{1}{(x+1)^2},$$

$f'(x)$ is always positive, except at $x = -1$ where it does not exist. We conclude that $f(x)$ is increasing on the intervals $x < -1$ and $x > -1$. The function is not, however, increasing for all x [since, for example, $f(0) < f(-2)$]. The graph of the function in Figure 4.15 clarifies the situation.

FIGURE 4.15

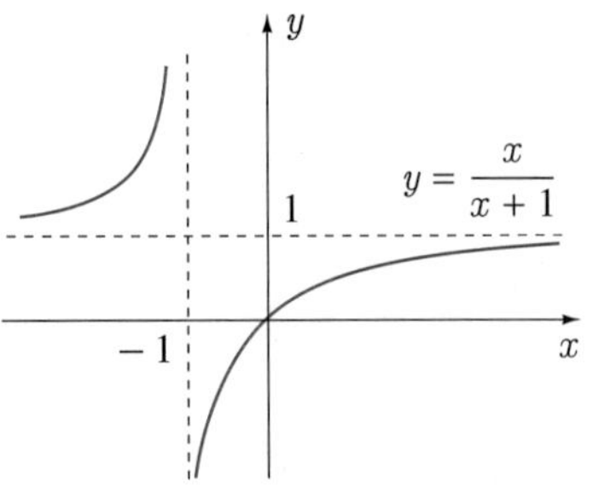

■

EXERCISES 4.2

In Exercises 1–24 determine intervals on which the function is increasing and decreasing.

1. $f(x) = 2x - 3$

2. $f(x) = 4 - 5x$

3. $f(x) = x^2 - 3x + 4$

4. $f(x) = -2x^2 + 5x$

5. $f(x) = 3x^2 + 6x - 2$

6. $f(x) = 5 + 2x - 4x^2$

7. $f(x) = 2x^3 - 18x^2 + 48x + 1$

8. $f(x) = x^3 + 6x^2 + 12x + 5$

9. $f(x) = 4x^3 - 18x^2 + 1$

10. $f(x) = 4 - 18x - 9x^2 - 2x^3$

11. $f(x) = 3x^4 + 4x^3 - 24x + 2$

12. $f(x) = 3x^4 - 4x^3 + 24x^2 - 48x$

13. $f(x) = x^4 - 4x^3 - 8x^2 + 48x + 24$

14. $f(x) = x^5 - 5x + 2$

15. $f(x) = x + \dfrac{1}{x}$ 16. $f(x) = x^2 + \dfrac{1}{x^2}$

17. $f(x) = \dfrac{x}{2 - x}$ 18. $f(x) = \dfrac{x^2 + 4}{x^2 - 1}$

19. $f(x) = \dfrac{x^3}{x + 1}$ 20. $f(x) = |x^2 - 1| + 1$

21. $f(x) = xe^{-x}$ 22. $f(x) = x^2 e^{-x}$

23. $f(x) = \ln(x^2 + 5)$ 24. $f(x) = x \ln x$

25. If the price of a certain gadget is set at r, then the market demands x gadgets per week, where

$$x = 4a^3 - 3ar^2 + r^3,$$

where $a > 0$ is a constant, and $0 < r < 2a$. Show that the price function $r = f(x)$ defined implicitly by this equation is a decreasing function.

In Figure 4.16(a) we have sketched the graph of a function $f(x)$ and in Figure 4.16(b), the graph of $f'(x)$. Note that $f'(x) \geq 0$ when $f(x)$ is increasing and $f'(x) \leq 0$ when $f(x)$ is decreasing. In addition, the corner in $f(x)$ at $x = 2$ is reflected in the discontinuity in $f'(x)$. In Exercises 26–33 draw similar graphs for the function and its derivative.

FIGURE 4.16

(a)

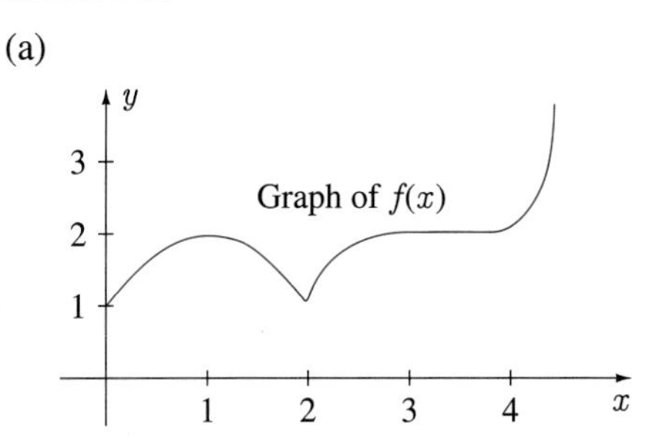

(b)

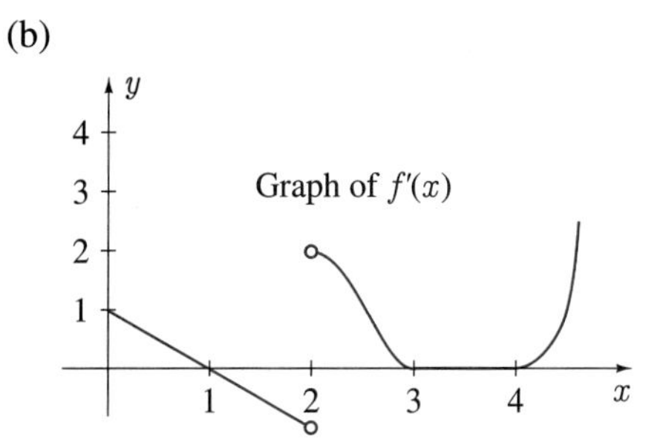

26. $f(x) = x^2 + 2x$ 27. $f(x) = x^4 - x^2$

28. $f(x) = \dfrac{x - 1}{x + 2}$ 29. $f(x) = \dfrac{x - 4}{x^2}$

30. $f(x) = \dfrac{|x|}{x}$ 31. $f(x) = |x^2 - 4|$

32. $f(x) = \dfrac{|x^2 - 4|}{x - 2}$ 33. $f(x) = \sin 3x$

In Exercises 34–37 find (accurate to 4 decimals) intervals on which the function is increasing and decreasing.

34. $f(x) = x^4 + 2x^2 - 6x + 5$

35. $f(x) = 3x^4 - 20x^3 - 24x^2 + 48x$

36. $f(x) = x^2 \sin x, \quad -\pi \leq x \leq \pi$

37. $f(x) = \tan x - x(x + 2), -\pi/2 < x < \pi/2$

38. Show that $\sin x < x$ for all $x > 0$. Hint: Calculate $f(0)$ and $f'(x)$ for $f(x) = x - \sin x$.

39. Show that $\cos x > 1 - x^2/2$ for all $x > 0$. Hint: See the technique of Exercise 38.

40. Use the result of Exercise 39 to prove that for $x > 0$

$$\sin x > x - \frac{x^3}{6}.$$

41. Use the result of Exercise 40 to prove that for $x > 0$

$$\cos x < 1 - \frac{x^2}{2} + \frac{x^4}{24}.$$

42. Use the technique of Exercise 38 to verify that for $x > 0$

$$\frac{1}{\sqrt{1 + 3x}} > 1 - \frac{3x}{2}.$$

43. If $f(x)$ and $g(x)$ are differentiable and increasing on an interval I, is $f(x)g(x)$ increasing on I?

44. If positive functions $f(x)$ and $g(x)$ are differentiable and increasing on an interval I, is $f(x)g(x)$ increasing on I?

SECTION 4.3

Relative Maxima and Minima

One of the most important applications of calculus is in the field of optimization, the study of maxima and minima. In this section we begin discussions of this topic, which continue through to Section 4.6. Fundamental to discussions on optimization are critical points.

Definition 4.2

A **critical point** of a function is a point in the domain of the function at which the first derivative either is equal to zero or does not exist.

Specifically, $x = c$ is a critical point for $f(x)$ if $f'(c) = 0$ or $f'(c)$ does not exist, but in the latter $f(c)$ must exist. Interpreting derivatives as slopes of tangent lines, we can state that a critical point of a function is a point at which the graph of the function has a horizontal tangent line, a vertical tangent line, or no tangent line at all. For example, the eight points a through h on the x-axis in Figure 4.17 are all critical. At a, b, c, and d, the tangent line is horizontal, at e and f, the tangent line is vertical, and at g and h, there is no tangent line.

FIGURE 4.17

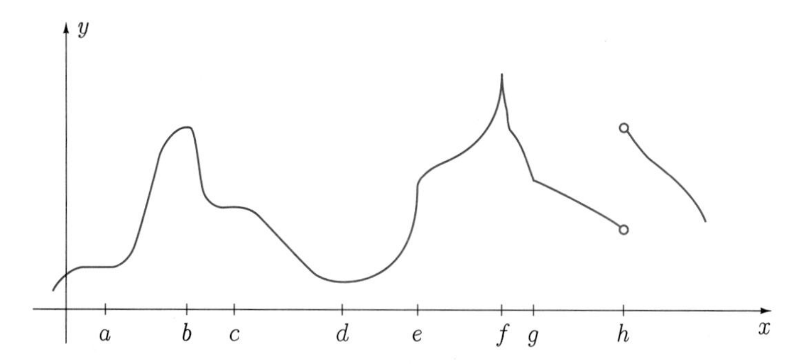

When the domain of a function $f(x)$ is a closed interval $a \le x \le b$, the end point $x = a$ is critical if the right-hand derivative of $f(x)$ at $x = a$ is equal to zero or does not exist. Likewise, $x = b$ is critical if the left-hand derivative vanishes or does not exist there. This is consistent with the definition of differentiability in Section 3.3. The function in Figure 4.18 has domain $1 \le x \le 3$. Both end points are critical — $x = 1$ because $f'_+(1)$ does not exist, and $x = 3$ because $f'_-(3) = 0$.

FIGURE 4.18

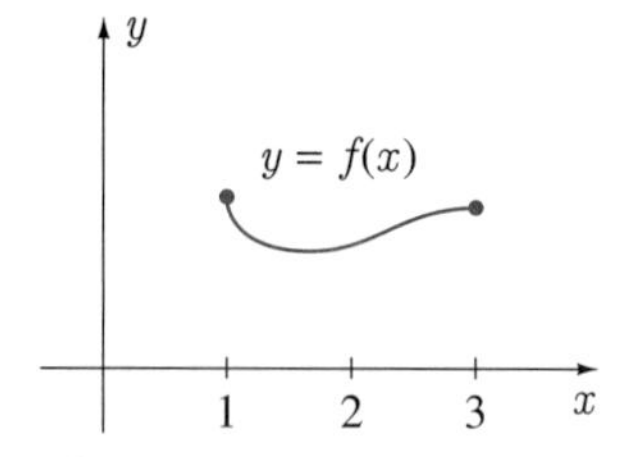

The function in Figure 4.17 is discontinuous at $x = h$, and we know that a function cannot have a derivative at a point of discontinuity (Theorem 3.6). Does this mean that every point of discontinuity of a function is critical? Not necessarily. The function must be defined at a point for that point to be critical. Thus, points of discontinuity are critical only if the function is defined at the point. In the remainder of this section we consider only critical points at which a function is continuous and which are not end points of its domain of definition.

EXAMPLE 4.4

Find critical points for the following functions:

(a) $f(x) = x^3 - 7x^2 + 11x + 6$ (b) $f(x) = \dfrac{x^2}{x^3 - 1}$ (c) $f(x) = x \ln x$

SOLUTION (a) For critical points we first solve

$$0 = f'(x) = 3x^2 - 14x + 11 = (3x - 11)(x - 1),$$

and obtain $x = 11/3$ and $x = 1$. These are the only critical points since there are no points where $f'(x)$ does not exist.

(b) For critical points we calculate

$$f'(x) = \frac{(x^3 - 1)(2x) - x^2(3x^2)}{(x^3 - 1)^2} = \frac{-x(x^3 + 2)}{(x^3 - 1)^2}.$$

Clearly $f'(x) = 0$ when $x = 0$, and when $x^3 + 2 = 0$, which implies that $x = -2^{1/3}$. The derivative does not exist when $x^3 - 1 = 0$; i.e., when $x = 1$. But $f(x)$ is not defined at $x = 1$ either, and therefore $x = 1$ is not a critical point. There are only two critical points: $x = 0$ and $x = -2^{1/3}$.

(c) For critical points we first solve

$$0 = f'(x) = \ln x + \frac{x}{x} = \ln x + 1.$$

The only solution of this equation is $x = 1/e$. Since $f'(x)$ exists for all x in the domain of $f(x)$, namely, $x > 0$, the function has no other critical points. ■

At the critical points b and f in Figure 4.19, the graph of the function has "high" points.

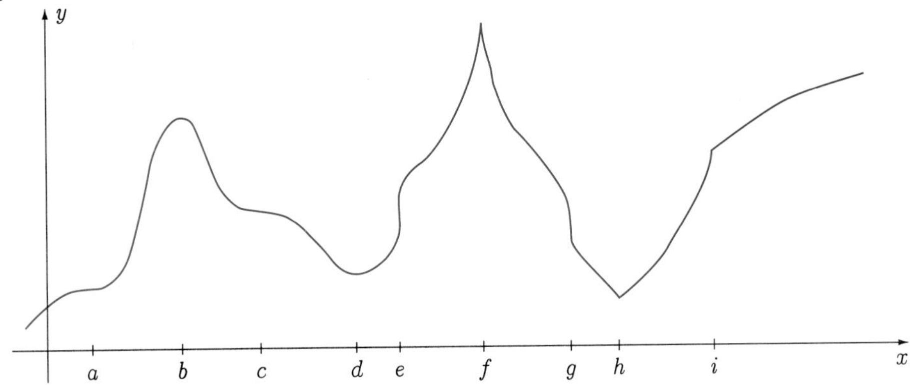

FIGURE 4.19

Definition 4.3

A function $f(x)$ is said to have a **relative maximum** $f(x_0)$ at $x = x_0$ if there exists an open interval I containing x_0 such that for all x in I

$$f(x) \le f(x_0). \tag{4.7}$$

Since such intervals can be drawn around $x = b$ and $x = f$, relative or local maxima occur at these points. At a relative or local maximum, the graph of the function is highest relative to nearby points.

Critical points d and h in Figure 4.19 , where the graph has "low" points, are described in a similar definition.

Definition 4.4

A function $f(x)$ is said to have a **relative minimum** $f(x_0)$ at $x = x_0$ if there exists an open interval I containing x_0 such that for all x in I

$$f(x) \geq f(x_0). \tag{4.8}$$

Relative (or local) minima therefore occur at $x = d$ and $x = h$.

The critical points $x = a$ and $x = c$, where $f'(x) = 0$ and $x = e$ and $x = g$, where $f'(x)$ does not exist will be discussed in Section 4.4.

At $x = i$, the graph takes an abrupt change in direction. The function has a left-hand derivative and a right-hand derivative, but $f'(i)$ does not exist. In Section 3.3 we called the corresponding point on the curve a corner. Corners can sometimes be relative extrema ($x = h$ yields a corner and a relative minimum).

Relative maxima and minima represent high and low points on the graph of a function relative to points near them. It is not coincidence in Figure 4.19 that the two relative maxima and the two relative minima occur at critical points. According to the following theorem, this is always the case.

Theorem 4.2

Relative maxima and relative minima of a function must occur at critical points of the function.

Proof To verify this we prove that at any point at which the derivative $f'(x)$ of a function $f(x)$ exists and is not zero, it is impossible for $f(x)$ to have a relative extremum, that is, a relative maximum or a relative minimum. Suppose that at some point $x = a$, the derivative $f'(a)$ exists and is positive (Figure 4.20),

$$0 < f'(a) = \lim_{h \to 0} \frac{f(a+h) - f(a)}{h}.$$

According to Exercise 30 in Section 2.5, there exists an open interval $I : b < x < c$ around $x = a$ in which

$$\frac{f(a+h) - f(a)}{h} > 0.$$

This implies that when $h > 0$ (and $a + h$ is in I), $f(a+h) - f(a)$ must also be positive, and therefore $f(a+h) > f(a)$. But when $h < 0$ (and $a + h$ is in I), $f(a+h) - f(a)$ must be negative, and therefore $f(a+h) < f(a)$. There is an interval $b < x < a$ in which $f(x) < f(a)$, and an interval $a < x < c$ in which $f(x) > f(a)$. Thus, $x = a$ cannot yield a relative extremum.

A similar proof holds when $f'(a) < 0$.

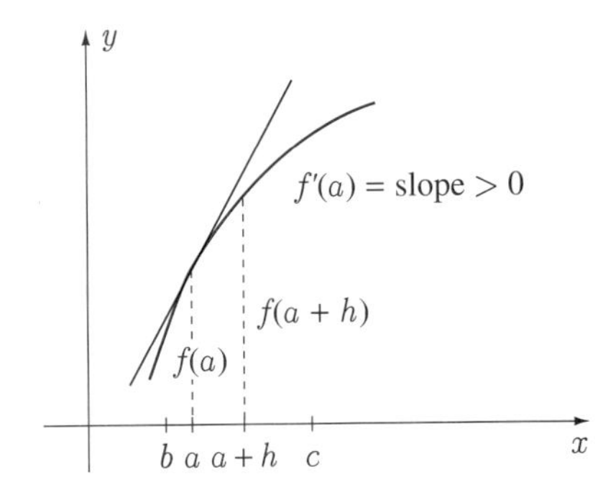

FIGURE 4.20

Although every relative extremum of a function must occur at a critical point, not all critical points give relative extrema. For continuous functions, there is a simple test to determine whether a critical point gives a relative maximum or a relative minimum. To understand this test, consider the critical points in Figure 4.19 as you read the statements.

First-derivative Test for Relative Extrema of Continuous Functions

(i) If $f'(x)$ (slope of a graph) changes from a positive quantity to a negative quantity as x increases through a critical point, then that critical point yields a relative maximum of $f(x)$.

(ii) If $f'(x)$ changes from a negative quantity to a positive quantity as x increases through a critical point, then that critical point yields a relative minimum of $f(x)$.

We prove part (i) of this test and leave verification of (ii) to the reader. Let $x = c$ be a critical point of $f(x)$ and let $f'(x)$ change from a positive quantity to a negative quantity as x increases through c. The fact that $f'(x)$ is positive to the left of c indicates that $f(x)$ must be increasing on some interval $a < x \leq c$ (Figure 4.21). Likewise, for $f'(x)$ to be negative to the right of c, $f(x)$ must be decreasing on some interval $c \leq x < b$. It follows that the value of $f(x)$ at $x = c$ must be larger than its value at any other point in the interval $a < x < b$.

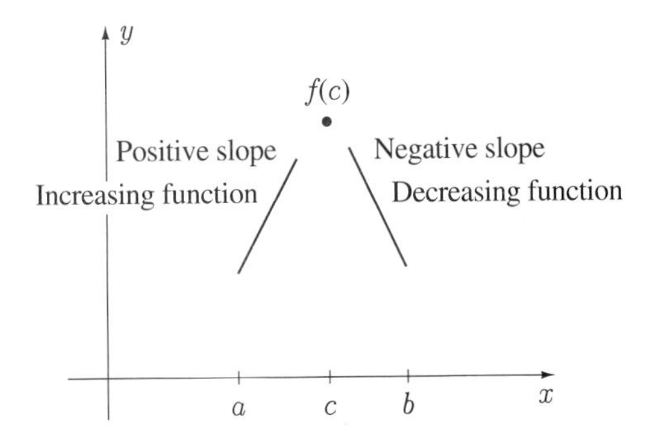

FIGURE 4.21

According to the first derivative test, a critical point at which $f(x)$ is continuous yields a relative maximum or a relative minimum when $f'(x)$ changes sign as x passes through the critical point. Sign diagrams are once again useful in making these decisions.

EXAMPLE 4.5

Find critical points for the following functions and determine whether they yield relative maxima or relative minima for the function.

(a) $f(x) = 2x^3 - 9x^2 - 24x + 6$ (b) $f(x) = \dfrac{x^2}{x^3 - 1}$ (c) $f(x) = x^{5/3} - x^{2/3}$

SOLUTION (a) For critical points we first solve

$$0 = f'(x) = 6x^2 - 18x - 24 = 6(x - 4)(x + 1),$$

and obtain $x = -1$ and $x = 4$. These are the only critical points as there are no points at which $f'(x)$ is undefined. The sign diagram in Figure 4.22 illustrates that as x increases through -1, $f'(x)$ changes from positive to negative; therefore $x = -1$ yields a relative maximum. Since $f'(x)$ changes from negative to positive as x increases through 4, $x = 4$ yields a relative minimum.

FIGURE 4.22

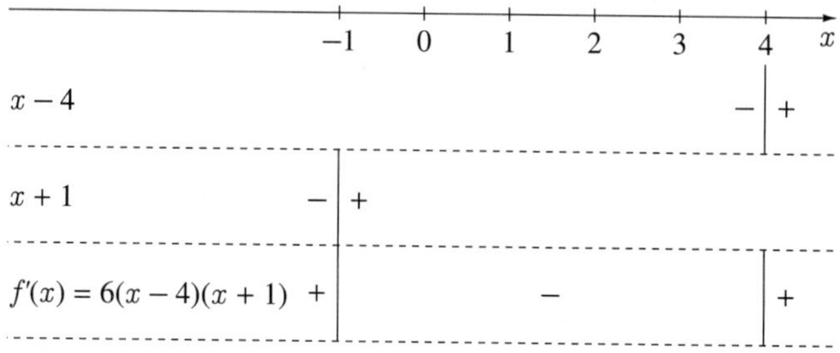

x	$x<-1$	$-1<x<4$	$x>4$
$x - 4$	$-$	$-$	$+$
$x + 1$	$-$	$+$	$+$
$f'(x) = 6(x-4)(x+1)$	$+$	$-$	$+$

(b) In Example 4.4 we found that

$$f'(x) = \frac{-x(x^3 + 2)}{(x^3 - 1)^2},$$

and that the only critical points are $x = 0$ and $x = -2^{1/3}$. Because the denominator is always positive, the sign of $f'(x)$ is completely determined by the sign of $-x(x^3+2)$. This is analyzed in Figure 4.23. Since $f'(x)$ changes from negative to positive as x increases through $-2^{1/3}$, $x = -2^{1/3}$ gives a relative minimum. A relative maximum exists at $x = 0$ since $f'(x)$ changes from positive to negative as x increases through 0.

FIGURE 4.23

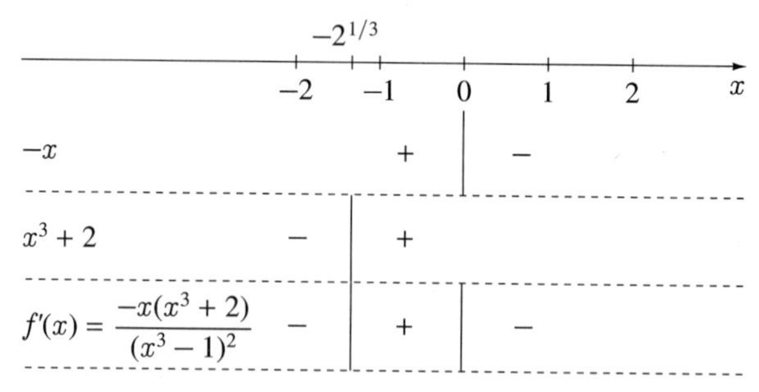

(c) For critical points we first solve

$$0 = f'(x) = \frac{5}{3}x^{2/3} - \frac{2}{3}x^{-1/3} = \frac{5x - 2}{3x^{1/3}}.$$

Clearly, $x = 2/5$ is a critical point since $f'(2/5) = 0$. In addition, because $f'(0)$ does not exist, but $f(0)$ does $(= 0)$, it follows that $x = 0$ is also a critical point. From the sign diagram in Figure 4.24, we see that $f'(x)$ changes from negative to positive as x increases through $x = 2/5$, which implies a relative minumum at this value of x. Since $f(x)$ is continuous at $x = 0$, and $f'(x)$ changes from positive to negative as x increases through 0, a relative maximum occurs at $x = 0$.

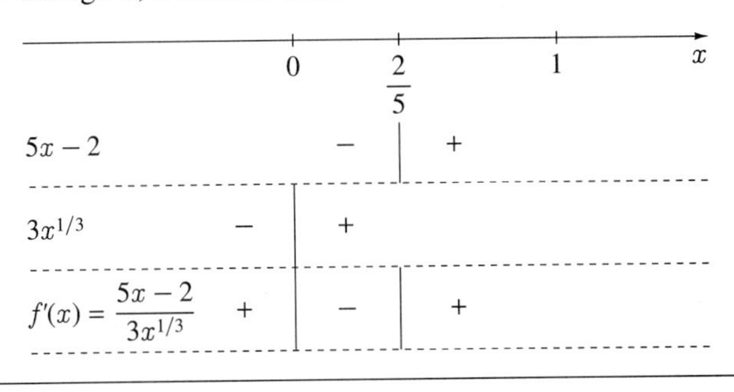

FIGURE 4.24

It is perhaps evident that were we required to sketch the graph of a function, it would be very helpful to know the function's critical points and whether these points give relative maxima or relative minima. This is indeed true, but before we consider the entire topic of curve sketching in Section 4.5, we introduce one further topic in Section 4.4.

EXERCISES 4.3

In Exercises 1–40 find all critical points of the function and determine whether they yield relative maxima or relative minima.

1. $f(x) = x^2 - 2x + 6$

2. $f(x) = 2x^3 + 15x^2 + 24x + 1$

3. $f(x) = x^4 - 4x^3/3 + 2x^2 - 24x$

4. $f(x) = (x-1)^5$

5. $f(x) = \dfrac{x+1}{x^2+8}$

6. $f(x) = \dfrac{x^2+1}{x-1}$

7. $f(x) = \dfrac{x}{\sqrt{1-x}}$

8. $f(x) = \dfrac{x^3}{x^4+1}$

9. $f(x) = \sin^2 x$

10. $f(x) = x^{1/3}$

11. $f(x) = x^3/3 - x^2/2 - 2x$

12. $f(x) = x^3 - 6x^2 + 12x + 9$

13. $f(x) = 3x^4 - 16x^3 + 18x^2 + 2$

14. $f(x) = x + \dfrac{1}{x}$

15. $f(x) = 2x^3 - 15x^2 + 6x + 4$

16. $f(x) = |x| + x$

17. $f(x) = (x-1)^{2/3}$

18. $f(x) = \dfrac{(x-1)^3}{(x+1)^4}$

19. $f(x) = (x+2)^3(x-4)^3$

20. $f(x) = x + 2\sin x$

21. $f(x) = x^{1/5} + x$

22. $f(x) = x^2 + \dfrac{25x^2}{(x-2)^2}$

23. $f(x) = \left(\dfrac{x+8}{x}\right)\sqrt{x^2+100}$

24. $f(x) = \dfrac{1+x+x^2+x^3}{1+x^3}$

25. $f(x) = x^{5/4} - x^{1/4}$

26. $f(x) = \dfrac{x^2}{x^2-4}$

27. $f(x) = \dfrac{x^2}{x^3-1}$

28. $f(x) = \dfrac{(2x-1)(x-8)}{(x-1)(x-4)}$

29. $f(x) = \sin^2 x \cos x, \quad 0 \le x \le 2\pi$

30. $f(x) = \sin x + \cos x$

31. $f(x) = 2\csc x - \cot x, \quad 0 < x < \pi/2$

32. $f(x) = \csc x + 8\sec x, \quad 0 < x < \pi/2$

33. $f(x) = \dfrac{\tan x}{x}$

34. $f(x) = x + \sin^2 x, \quad 0 < x < 2\pi$

35. $f(x) = e^{1/x}$

36. $f(x) = x \ln x$

37. $f(x) = x^2 \ln x$

38. $f(x) = xe^{-2x}$

39. $f(x) = xe^{-x^2}$

40. $f(x) = x^2 e^{3x}$

41. Does the function $f(x) = x^2$ defined only for $0 \le x \le 1$ have a relative minimum of $f(0) = 0$ and a relative maximum of $f(1) = 1$?

42. If $f(x) = x^2$ is defined only for $0 \le x \le 1$, are $x = 0$ and $x = 1$ critical points?

In Exercises 43–50 determine whether the statement is true or false.

43. Points of discontinuity of a function are critical if, and only if, the function is defined at the point of discontinuity.

44. When a function is defined only on the interval $a \le x \le b$, the ends $x = a$ and $x = b$ must yield relative maxima or minima for the function.

45. A function is discontinuous at a point if, and only if, it has no derivative at the point.

46. A function can have a relative maximum and a relative minimum at the same point.

47. If the derivative of a function changes sign when passing through a point, the point must yield a relative extrema for the function.

48. If a function has two relative maxima, it must have a relative minimum between them.

49. It is possible for every point in the domain of a function to be critical.

50. On an interval of finite length, a nonconstant function can have only a finite number of critical points at which its derivative is equal to zero.

51. Verify that the function $f(x) = x^3 + \cos x$ has two critical points, one of which is $x = 0$. Find the other critical point correct to three decimal places.

52. (a) Use multiplication of ordinates to sketch a graph of the function $f(x) = e^{-x}\sin x$, $x \ge 0$.

(b) For what values of x does $f(x)$ have relative extrema? Does your graph in (a) agree?

In Exercises 53–56 find all critical points of the function correct to four decimals. Determine whether each critical point yields a relative maximum or a relative minimum.

53. $f(x) = x^4 + 6x^2 + 4x + 1$

54. $f(x) = x^4 - 10x^2 - 4x + 5$

55. $f(x) = x^3 - 2\cos x$

56. $f(x) = \dfrac{x^2 - 4}{(x^2 - 5x + 4)^2}$

57. The distance from the origin to any point (x, y) on the curve $C : y = f(x)$ is given by $D = \sqrt{x^2 + y^2}$. Show that if $P(x_0, y_0)$ is a point on C for which D has a relative extremum, then the line OP is perpendicular to the tangent line to C at P. Assume that $f(x)$ is differentiable, and that the curve does not pass through the origin.

58. An experiment is performed n times, resulting in n measurements $x_1, x_2, \ldots, x_n$ of the variable x. To obtain the best estimate of the value of x, we choose that value $\overline{x}$ which makes the sum of the squares of the differences between the x_i and $\overline{x}$ as small as possible; i.e., we choose $\overline{x}$ to minimize

$$S(x) = (x - x_1)^2 + (x - x_2)^2 + \cdots + (x - x_n)^2.$$

Show that $\overline{x}$ is the ordinary arithmetic mean of the x_i:

$$\overline{x} = \frac{1}{n}(x_1 + x_2 + \cdots + x_n).$$

59. A 200-gram mass is attached to a spring with elasticity constant 20 N/m (newtons per metre) in Figure 4.25.

FIGURE 4.25

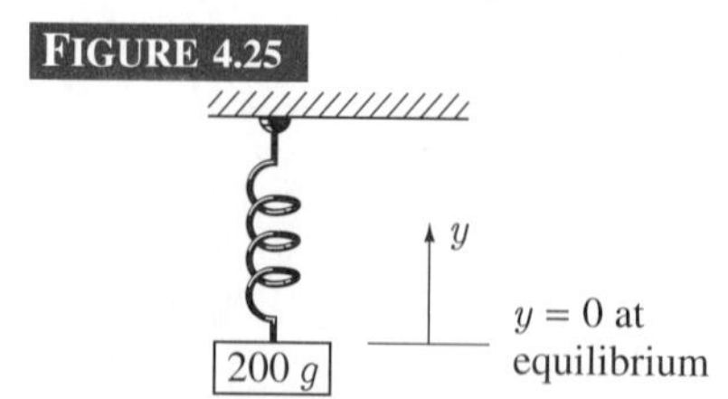

If the mass is pulled 10 cm below the position where it would hang motionless ($y = 0$), and given an upward velocity of 2 m/s, then its position as a function of time t (in seconds) is given by

$$y(t) = -\frac{1}{10}\cos(10t) + \frac{1}{5}\sin(10t) \text{ m}.$$

Find the maximum distance from $y = 0$ attained by the mass.

60. An inductor L, a resistor R, and a capacitor C are connected in series with a generator, producing an oscillatory voltage $E = E_0 \cos \omega t$, for $t \geq 0$ (Figure 4.26).

FIGURE 4.26

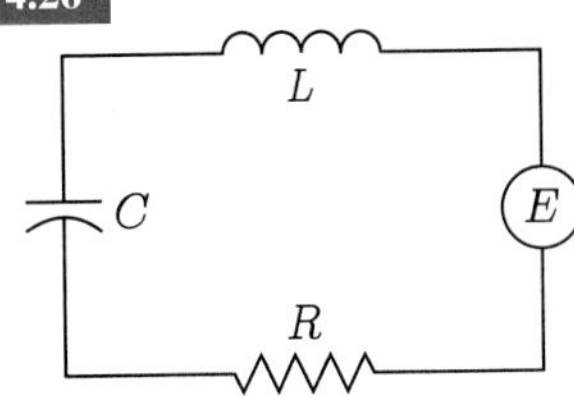

If L, C, R, E_0, and ω are all constants, the steady-state current I in the circuit is given by

$$I = \frac{E_0}{Z}\cos(\omega t - \phi),$$

where

$$Z = \sqrt{R^2 + \left(\omega L - \frac{1}{\omega C}\right)^2} \quad \text{and} \quad \tan\phi = \frac{\omega L - 1/(\omega C)}{R}.$$

Find the value of ω that makes the amplitude E_0/Z of the current a maximum. What is the maximum amplitude?

61. A landing approach is to be shaped generally as shown in Figure 4.27.

FIGURE 4.27

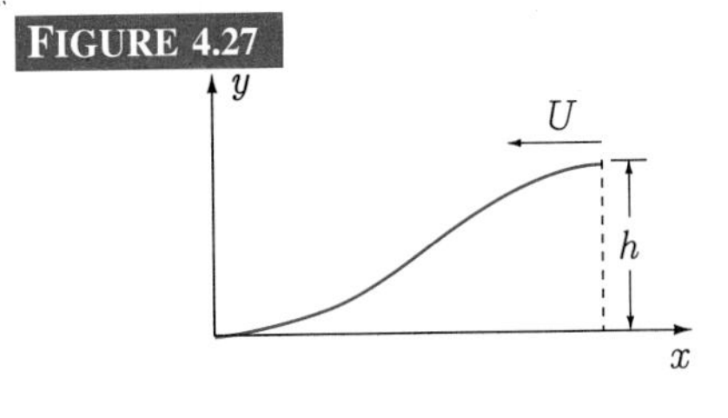

The following conditions are imposed on the approach pattern:

(a) The altitude must be h when descent commences;

(b) Touchdown must occur at $x = 0$;

(c) A constant horizontal speed U must be maintained throughout;

(d) At no time must vertical acceleration in absolute value exceed some fixed positive constant k.

Find a cubic polynomial satisfying these requirements.

62. The equation $(x^2 + y^2 + x)^2 = x^2 + y^2$ describes a *cardioid* (Figure 4.28). Find the maximum y-coordinate for points on the curve.

FIGURE 4.28

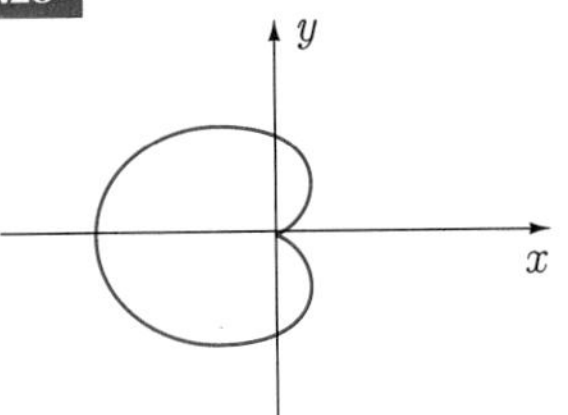

63. The equation $(x^2 + y^2)^2 = x^2 y$ describes a *bifolium* (Figure 4.29). Find the points on the curve farthest from the origin.

FIGURE 4.29

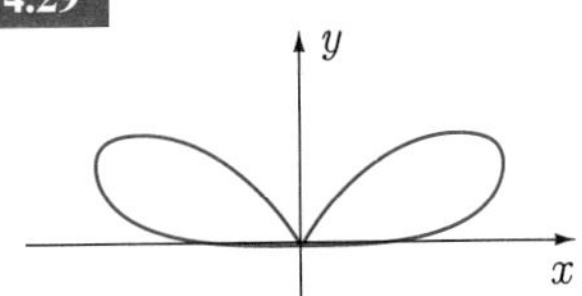

SECTION 4.4 Concavity and Points of Inflection

In the first three sections of this chapter we concentrated on the first derivative of a function. In this section we turn our attention to the second derivative.

Consider the function in Figure 4.30. If we draw tangent lines to the graph at the five points c_1, c_2, c_3, c_4, and c_5, it is clear that the slope is greater at c_5 than it is at c_4, greater at c_4 than at c_3, and so on. In fact, given any two points x_1 and x_2 in the interval $a < x < b$, where $x_2 > x_1$, the slope at x_2 is greater than at x_1,

$$f'(x_2) > f'(x_1).$$

What we are saying is that the function $f'(x)$ is increasing on the interval $a < x < b$.

FIGURE 4.30

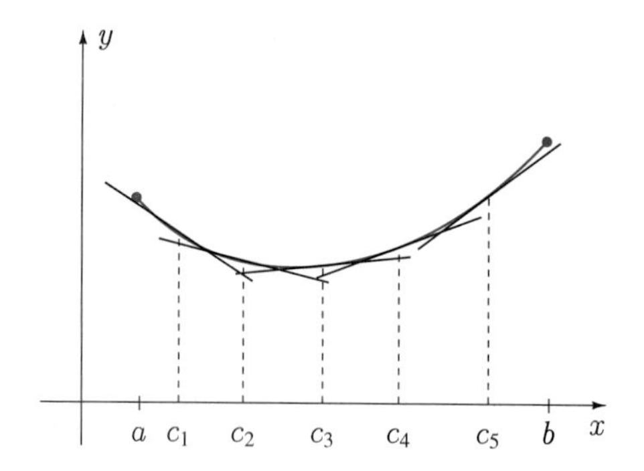

The first derivative of the function $f(x)$ in Figure 4.31 is decreasing on the interval $b < x < c$. In Figure 4.32, we have pieced together the functions in Figures 4.30 and 4.31 to form a function which has $f'(x)$ increasing on $a < x \leq b$ and decreasing on $b \leq x < c$. Since the left-hand derivative $f'_-(b)$ in Figure 4.30 and the right-hand derivative $f'_+(b)$ in Figure 4.31 are the same, it follows that $f'(b)$ exists in Figure 4.32, and that is why b is included in the intervals in which $f'(x)$ is increasing and decreasing. We give names to intervals on which the first derivative of a function is increasing and decreasing, and points which separate such intervals in the following definition.

FIGURE 4.31

FIGURE 4.32

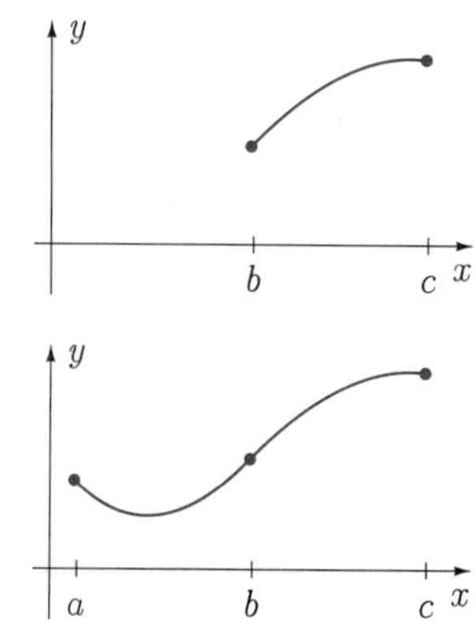

Definition 4.5

The graph of a function $f(x)$ is said to be **concave upward** on an interval I if $f'(x)$ is increasing on I, and **concave downward** on I if $f'(x)$ is decreasing on I. Points on a graph that separate intervals of opposite concavity are called **points of inflection**.

The graph in Figure 4.33 is concave upward on the intervals

$$b \leq x \leq c, \qquad d \leq x < e, \qquad e < x \leq f,$$

and concave downward for

$$a < x \leq b, \qquad c \leq x \leq d, \qquad f \leq x < g, \qquad g < x < h.$$

The points $x = e$ and $x = g$ are not included in these intervals because $f'(x)$ is not defined at these points. Points on the curve corresponding to $x = b$, c, d, and f are points of inflection; those at $x = e$ and g are not. They do not separate intervals of opposite concavity.

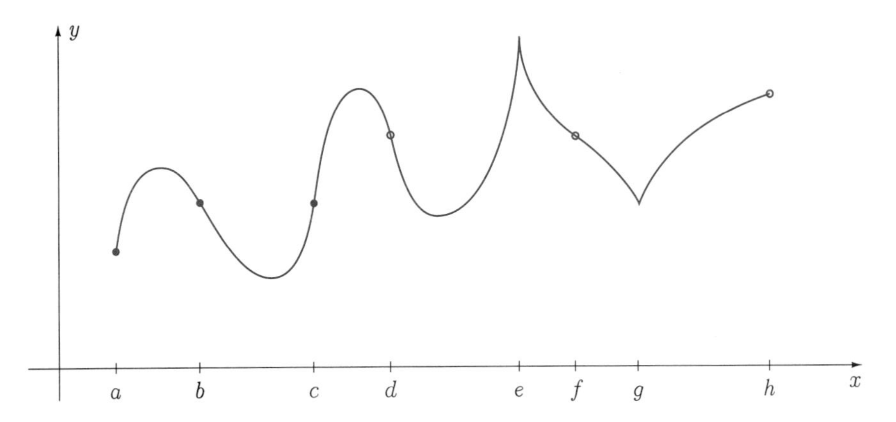

FIGURE 4.33

The tangent line has a peculiar property at a point of inflection. To see it, draw the tangent line at the four points of inflection in Figure 4.33. The tangent line actually crosses from one side of the curve to the other (something many readers think a tangent line should not do).

The critical points $x = a$ and $x = c$ in Figure 4.19, where $f'(x) = 0$, and the critical points $x = e$ and $x = g$ where $f'(x)$ does not exist, do not yield relative extrema for $f(x)$. Critical points of these types are illustrated again in Figures 4.34 and 4.35. The tangent line at $x = a$ is horizontal in Figure 4.34 ($f'(a) = 0$). Since the graph is concave downward to the left of $x = a$ and concave upward to the right, the point $(a, f(a))$ is a point of inflection. We call it a **horizontal point of inflection**. The point $(a, f(a))$ in Figure 4.35 is also a point of inflection. As the tangent line at this point is vertical, we call it a **vertical point of inflection**.

FIGURE 4.34

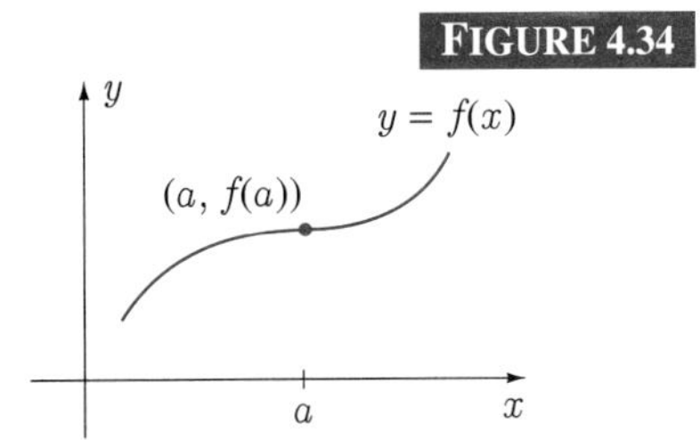

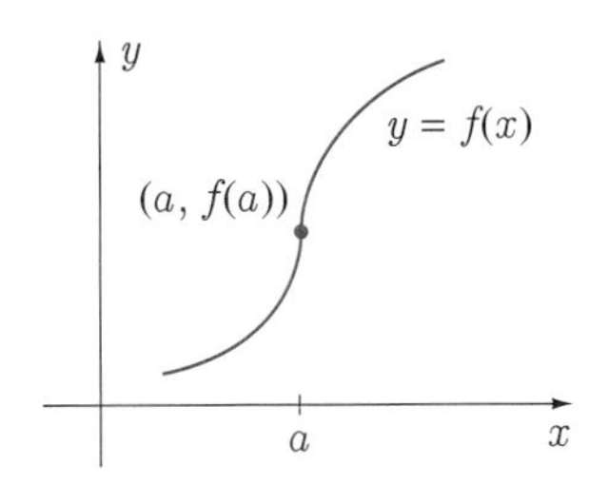

FIGURE 4.35

In Section 4.2 we stated that a function is increasing on an interval if its first derivative is greater than or equal to zero on that interval, and equal to zero at only a finite number of points; it is decreasing if its first derivative is less than or equal to zero. Furthermore, points that separate intervals on which the function is increasing and decreasing are points at which the first derivative is either equal to zero (and changes sign) or does not exist. We can use these ideas to locate analogously points of inflection and intervals on which the graph of a function is concave upward and concave downward.

Test for Concavity

(i) The graph of a function $f(x)$ is concave upward on an interval I if on I

$$f''(x) \geq 0 \tag{4.9}$$

and is equal to zero at only a finite number of points.

(ii) The graph of a function $f(x)$ is concave downward on an interval I if on I

$$f''(x) \leq 0 \tag{4.10}$$

and is equal to zero at only a finite number of points.

(iii) Points of inflection occur where

$$f''(x) = 0 \tag{4.11}$$

and $f''(x)$ changes sign, or perhaps also where $f''(x)$ does not exist.

Consequently, just as the first derivative of a function is used to determine its relative extrema and intervals on which it is increasing and decreasing, the second derivative is used to find points of inflection and intervals on which the function is concave upward and concave downward. This parallelism between the first and second derivative is reiterated in the following table.

First Derivative	**Second Derivative**
1. A relative extremum occurs at a point where $f'(x) = 0$ and $f'(x)$ changes sign as x increases through the point.	**1.** A point of inflection occurs at a point where $f''(x) = 0$ and $f''(x)$ changes sign as x increases through the point.
2. A relative extremum may also occur at a point where $f'(x)$ does not exist.	**2.** A point of inflection may also occur at a point where $f''(x)$ does not exist.
3. $f(x)$ is increasing on an interval if $f'(x) \geq 0$; $f(x)$ is decreasing on an interval if $f'(x) \leq 0$. $f'(x)$ may be equal to zero at only a finite number of points in either case.	**3.** $f(x)$ is concave upward on an interval if $f''(x) \geq 0$; $f(x)$ is concave downward on an interval if $f''(x) \leq 0$. $f''(x)$ may be equal to zero at only a finite number of points in either case.

These discussions lead to what is often called the *second-derivative test* for determining whether a critical point at which $f'(x) = 0$ yields a relative maximum or a relative minimum.

Second-derivative Test for Relative Extrema

(i) If $f'(x_0) = 0$ and $f''(x_0) > 0$, then $x = x_0$ yields a relative minimum for $f(x)$.

(ii) If $f'(x_0) = 0$ and $f''(x_0) < 0$, then $x = x_0$ yields a relative maximum for $f(x)$.

(iii) If $f'(x_0) = 0$ and $f''(x_0) = 0$, then no conclusion can be made.

In (i), $f'(x_0) = 0$ implies that $x = x_0$ is a critical point with a horizontal tangent line; $f''(x_0) > 0$ implies that around $x = x_0$ the curve is concave upward, and $x = x_0$ therefore yields a relative minimum. Similarly, $f''(x_0) < 0$ at a critical point in (ii) implies that the curve is concave downward and therefore has a relative maximum.

To show that the test fails in case (iii), consider the three functions $f(x) = x^4$, $f(x) = -x^4$, and $f(x) = x^3$ shown in Figures 4.36, 4.37, and 4.38 respectively. For each function $f'(0) = f''(0) = 0$; yet the first has a relative minimum, the second a relative maximum, and the third a horizontal point of inflection at $x = 0$.

FIGURE 4.36

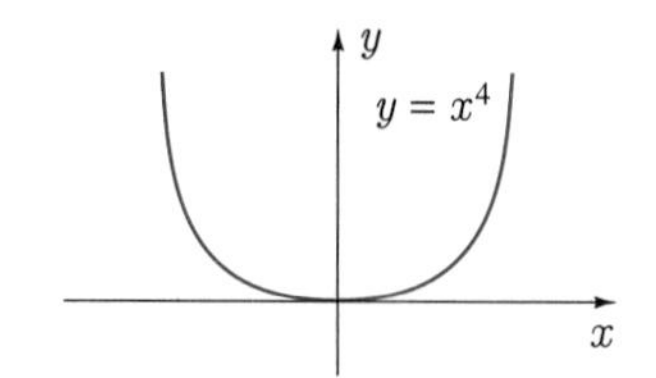

FIGURE 4.37

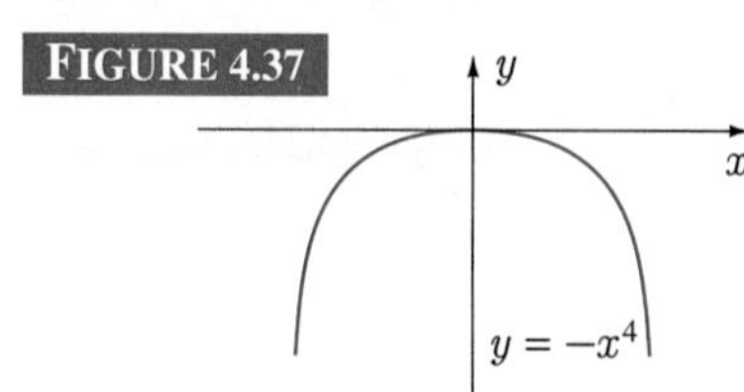

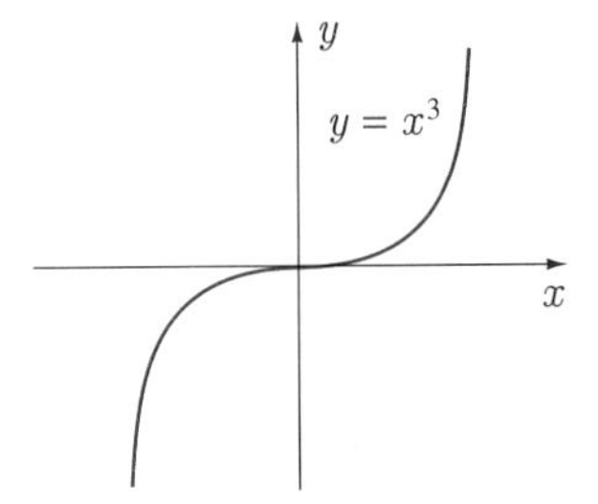

FIGURE 4.38

The second-derivative test very quickly determines the nature of a critical point x_0 when $f'(x_0) = 0$ and $f''(x_0) \neq 0$. For instance, the second derivative of the function $f(x) = 2x^3 - 9x^2 - 24x + 6$ in Example 4.5(a) of Section 4.3 is $f''(x) = 12x - 18$. Since $f''(-1) = -30$, the critical point $x = -1$ yields a relative maximum, and because $f''(4) = 30$, $x = 4$ gives a relative minimum.

A more refined test to determine the nature of a critical point x_0 when $f'(x_0) = f''(x_0) = 0$ is discussed in Exercise 34.

EXAMPLE 4.6

For each of the following functions, find all points of inflection and intervals on which the graph of the function is concave upward and concave downward.

(a) $f(x) = \dfrac{2}{3}x^3 - 6x^2 + 16x + 1$ (b) $f(x) = x^{6/5} + x^{1/5}$

SOLUTION (a) For points of inflection we first solve

$$0 = f''(x) = \frac{d}{dx}(2x^2 - 12x + 16) = 4x - 12,$$

and obtain $x = 3$. Since $f''(x)$ changes sign as x passes through 3, $x = 3$ gives the point of inflection $(3, 13)$. The graph $y = f(x)$ is concave downward for $x \leq 3$ (since $f''(x) \leq 0$) and concave upward for $x \geq 3$ ($f''(x) \geq 0$). Since $f''(x)$ is defined for all x, there can be no other points of inflection.

(b) For points of inflection we first solve

$$0 = f''(x) = \frac{d}{dx}\left(\frac{6}{5}x^{1/5} + \frac{1}{5}x^{-4/5}\right) = \frac{6}{25}x^{-4/5} - \frac{4}{25}x^{-9/5} = \frac{6x - 4}{25x^{9/5}},$$

and find $x = 2/3$. Since $f''(x)$ changes sign as x passes through $2/3$ (Figure 4.39), $x = 2/3$ gives a point of inflection with $f(2/3) \approx 1.5$.

We also note that $f'(x)$ and $f''(x)$ do not exist at $x = 0$, although $f(0)$ does $(= 0)$. Since $f''(x)$ changes sign as x passes through 0, $(0, 0)$ must also be a point of inflection. Finally, the sign diagram in Figure 4.39 indicates that the graph $y = f(x)$ is concave upward for $x < 0$ and $x \geq 2/3$ and concave downward for $0 < x \leq 2/3$.

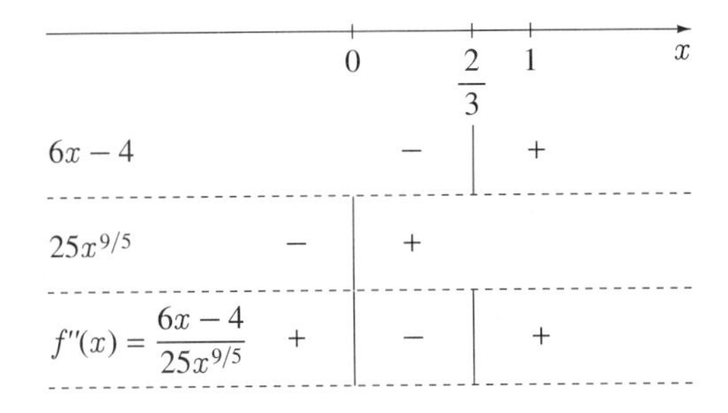

FIGURE 4.39

■

In Section 4.6 we will see the importance of critical points in the solution of applied maxima-minima problems. In some problems the significance of inflection points is equally important. As an example, consider how an infectious disease spreads through a population of susceptible individuals. To simplify the problem, we assume that all individuals are equally susceptible and that the disease is introduced to the population by one infective. Contact between individuals spreads the disease slowly at first, then more rapidly as more individuals become infected, and finally, when most of the population has become infected, the spread tapers off. To describe this process in more mathematical terms, we denote by N the total number of individuals in the population, and by $x(t)$ the number of infectives at any given time t. Since the disease is spread through contacts between infectives and noninfectives, the rate at which individuals become infected depends on both the number of infectives x and the number of noninfectives $N - x$ at any given time. In practice, the rate of infection, represented by dx/dt, is often assumed proportional to the product of x and $N - x$:

$$\frac{dx}{dt} = kx(N - x),$$

where k is a constant. Given that this equation describes the rate at which the disease spreads, it is possible to determine the number of infectives $x(t)$ present at any given time. We cannot solve this problem completely at this point, but we can obtain some information on its graph.

First, we note that dx/dt is always nonnegative so that the graph of $x(t)$ is increasing. Second, since points of inflection on the graph are defined by a vanishing second derivative, we set

$$0 = x''(t) = \frac{d}{dt}[kNx - kx^2] = kN\frac{dx}{dt} - 2kx\frac{dx}{dt} = k(N - 2x)\frac{dx}{dt}.$$

But $k \neq 0$, and $dx/dt = 0$ only when the population is either totally infected or not at all infected; therefore, a point of inflection occurs only when $N - 2x = 0$ or $x = N/2$, that is, when the population is half infected. Since $x''(t) > 0$ when $x < N/2$ and $x''(t) < 0$ when $x > N/2$, the graph of $x(t)$ is concave upward for $x < N/2$ and concave downward for $x > N/2$. These are shown in Figure 4.40; we have chosen a time convention of $t = 0$ when the first infective is introduced. In other words, we have completely determined the shape of the graph of $x(t)$. What remains is the insertion of a time scale and this we shall do in Section 9.7.

FIGURE 4.40

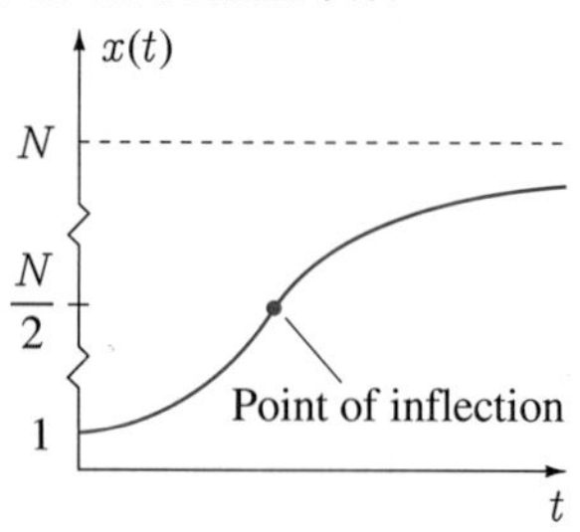

At this point the significance of the point of inflection at $x = N/2$ is clear. Since the slope of the curve (dx/dt) is increasing for $x < N/2$ and decreasing for $x > N/2$, it follows that when $x = N/2$, the rate of spread of the disease is at its peak.

EXERCISES 4.4

In Exercises 1–14 determine where the graph of the function is concave upward, is concave downward, and has points of inflection.

1. $f(x) = x^3 - 3x^2 - 3x + 5$

2. $f(x) = 3x^4 + 4x^3 - 24x + 2$

3. $f(x) = x^2 + \dfrac{1}{x^2}$

4. $f(x) = \dfrac{x^2 + 4}{x^2 - 1}$

5. $f(x) = x + \cos x, \quad |x| < 2\pi$

6. $f(x) = x^2 - \sin x$

7. $f(x) = x^2 - 2\sin x$

8. $f(x) = x^2 - 4\sin x, \quad |x| < 2\pi$

9. $f(x) = x \ln x$

10. $f(x) = x^2 \ln x$

11. $f(x) = e^{1/x}$

12. $f(x) = xe^{-2x}$

13. $f(x) = x^2 e^{3x}$

14. $f(x) = x^2 - e^{-x}$

In Exercises 15–22 use the second-derivative test to determine whether critical points where $f'(x) = 0$ yield relative maxima or relative minima.

15. $f(x) = x^3 - 3x^2 - 3x + 5$

16. $f(x) = x + \dfrac{1}{x}$

17. $f(x) = 3x^4 - 16x^3 + 18x^2 + 2$

18. $f(x) = x^{5/4} - x^{1/4}$

19. $f(x) = x \ln x$

20. $f(x) = x^2 \ln x$

21. $f(x) = xe^{2x}$

22. $f(x) = x^2 e^{-2x}$

23. Is the graph of a function concave upward, concave downward, both, or neither on an interval I if on I its second derivative is always equal to zero?

24. Prove that the curve $y = 2\cos x$ passes through all points of inflection of the curve $y = x \sin x$.

25. Show that every cubic polynomial has exactly one point of inflection on its graph.

Sketch the graph of a function $f(x)$ that is defined everywhere on the interval $I : a \le x \le b$, and that possesses the properties in Exercises 26–31. In each case assume that $a < x_0 < b$.

26. $f(x)$ is increasing on I and discontinuous at x_0.

27. $f''(x) \ge 0$ on I and $f(x)$ is not concave upward on I.

28. $f(x)$ is decreasing on I, $f(x)$ is continuous at x_0, but $f'(x)$ is discontinuous at x_0.

29. $f(x)$ is increasing on $a \le x \le x_0$, increasing on $x_0 < x \le b$, but not increasing on I.

30. $f(x)$ is concave downward on $a \le x < x_0$, concave downward on $x_0 < x \le b$, but not concave downward on I.

31. $f(x)$ is increasing on $a \le x \le x_0$, concave downward on $a \le x < x_0$, decreasing on $x_0 \le x \le b$, and concave upward on $x_0 < x \le b$.

32. If $f(x)$ and $g(x)$ are twice differentiable and concave upward on an interval I, is $f(x)g(x)$ concave upward on I?

33. If the functions in Exercise 32 are also increasing on I, is $f(x)g(x)$ concave upward on I?

34. The second-derivative test fails to classify a critical point x_0 of a function $f(x)$ as yielding a relative maximum, a relative minimum, or a horizontal point of inflection if $f'(x_0) = f''(x_0) = 0$. The following test can be used in such cases. Suppose $f(x)$ has derivatives of all orders in an open interval around x_0, and the first n derivatives all vanish at x_0, but the $(n+1)^{\text{th}}$ derivative at x_0 is not zero. If we denote the nth derivative of $f(x)$ at x_0 by $f^{(n)}(x_0)$, these conditions are

$$0 = f'(x_0) = f''(x_0) = \cdots = f^{(n)}(x_0), \qquad f^{(n+1)}(x_0) \neq 0.$$

Then:

(i) if n is even, $f(x)$ has a horizontal point of inflection at x_0;

(ii) if n is odd and $f^{(n+1)}(x_0) > 0$, $f(x)$ has a relative minimum at x_0;

(ii) if n is odd and $f^{(n+1)}(x_0) < 0$, $f(x)$ has a relative maximum at x_0.

A proof of this result requires the use of material from Chapter 11 and is therefore delayed until that time (see Exercise 10 in Section 11.9). Note that the second-derivative test is the special case when $n = 1$. Use this test to determine whether critical points of the following functions yield relative maxima, relative minima, or horizontal points of inflection.

(a) $f(x) = (x^2 - 1)^3$ (b) $f(x) = x^2\sqrt{1 - x}$

35. Show that the points of inflection of $f(x) = (k - x)/(x^2 + k^2)$, where k is a constant, all lie on a straight line.

36. Prove that if a cubic polynomial has both a relative maximum and a relative minimum, then the point of inflection between these extrema is the midpoint of the line segment joining them.

37. (a) In Exercise 22 of Section 3.5, the derivative of the function

$$f(x) = \begin{cases} x^2 \sin\left(\dfrac{1}{x}\right), & x \neq 0 \\ 0 & x = 0 \end{cases}$$

is zero at $x = 0$; that is, $f'(0) = 0$. It follows that the graph of the function has a horizontal tangent line at $(0,0)$. Sketch the graph and the tangent line.

(b) Show that $f''(0)$ does not exist. Hint: See Exercise 25 in Section 3.3.

(c) Is the point $(0,0)$ a relative maximum, a relative minimum, a horizontal point of inflection, or none of these?

SECTION 4.5

Curve Sketching

Some years ago, one of the major applications of differentiation was to draw graphs of functions. With computers and advanced electronic calculators that provide visual displays of these graphs, many users of mathematics feel that this application is now obsolete. Such is not the case, and for a number of reasons. First, plotting programs in calculators and computers have built in smoothing routines which eliminate anomolies on curves. This may not be appropriate. For example, a corner on the graph of a function represents a point where the function does not have a derivative. A plotting program which "rounds" the corner is misleading; it removes the evidence that the function has a point where its derivative does not exist. Second, calculators may suggest, but do not offer conclusive evidence, that a graph has vertical, horizontal, or "slanted" asymptotes, and discontinuities are not handled well by calculators. Finally, the electronic display may suggest that a point on a curve is a horizontal or vertical point of inflection, but cannot be a conclusive argument.

What we suggest is that although the use of calculus in curve sketching may not require the stress that it once enjoyed, there are definitely situations in which it is essential.

Curve sketching, as a topic, began in Sections 1.3 and 1.4 when we used the basic idea of a table of values. Following this, we illustrated in Section 1.5 that addition and multiplication of ordinates often yield rough but very quick sketches. We also saw that symmetries of a curve about the x- and y-axes can be helpful in many cases. We used limits in Section 2.4 to discover the behaviour of the graph of a function near a point of discontinuity or for very large positive and negative values of the independent variable. Each of these techniques has its value and none should be neglected.

In this chapter we have defined critical points, relative extrema, increasing and decreasing functions, concavity, and points of inflection, and each of these concepts can add information to a sketch. The natural question to ask at this point is: "Given a curve to sketch, which of the many tools that are now available should we use?" There is no simple answer to this question, except to say that we should use those tools that give us a suitable sketch with the least amount of effort. But a sketch suitable for one problem may not be suitable for another. For some problems, a very rough sketch might be sufficient, in which case we might need to use only addition or multiplication of ordinates; in other problems, a fairly detailed graph might be appropriate, and we would need further information such as relative extrema and points of inflection. In this section we illustrate how further detail can be added to a sketch by using more and more tools on a specific example. At the same time, we see how the various techniques complement each other.

EXAMPLE 4.7

Sketch a graph of the function $f(x) = (x-4)/x^2$.

SOLUTION With an x-intercept equal to 4, and the limits

$$\lim_{x\to 0^-} f(x) = -\infty, \qquad \lim_{x\to 0^+} f(x) = -\infty, \qquad \lim_{x\to -\infty} f(x) = 0^-, \qquad \lim_{x\to \infty} f(x) = 0^+,$$

we begin our sketch as shown in Figure 4.41(a). This information would lead us to suspect that the graph should be completed as shown in Figure 4.41(b). To verify this we find critical points for $f(x)$:

$$0 = f'(x) = \frac{x^2(1) - (x-4)(2x)}{x^4} = \frac{x(8-x)}{x^4} = \frac{8-x}{x^3}.$$

Clearly, $x = 8$ is the only critical point of the function, and the first derivative test indicates that it gives a relative maximum. Our sketch is as shown in Figure 4.41(c).

Figure 4.41(c) makes it clear that there is a point of inflection to the right of $x = 8$, which we could pinpoint with $f''(x)$:

$$f''(x) = \frac{x^3(-1) - (8-x)(3x^2)}{x^6} = \frac{2(x-12)}{x^4}.$$

Since $f''(12) = 0$ and $f''(x)$ changes sign as x passes through 12, $(12, 1/18)$ is the point of inflection. Figure 4.41(d) contains the final graph.

FIGURE 4.41

(a)

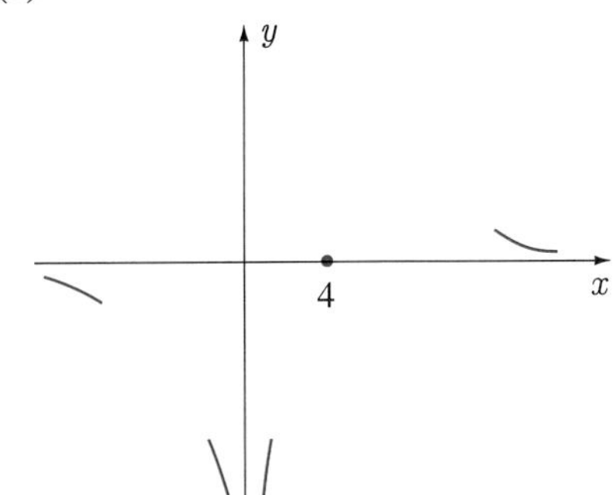

(b)

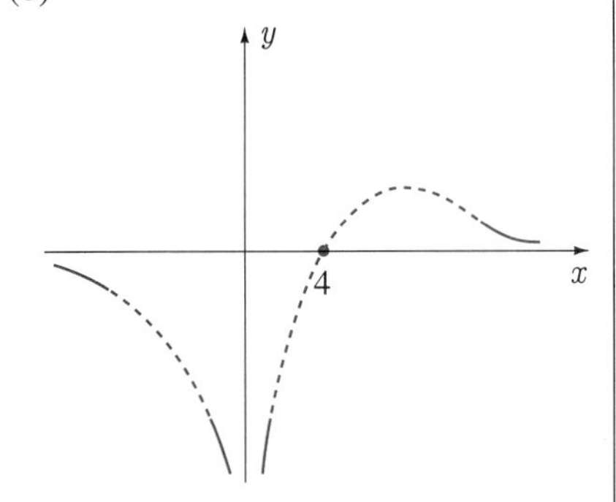

(c)

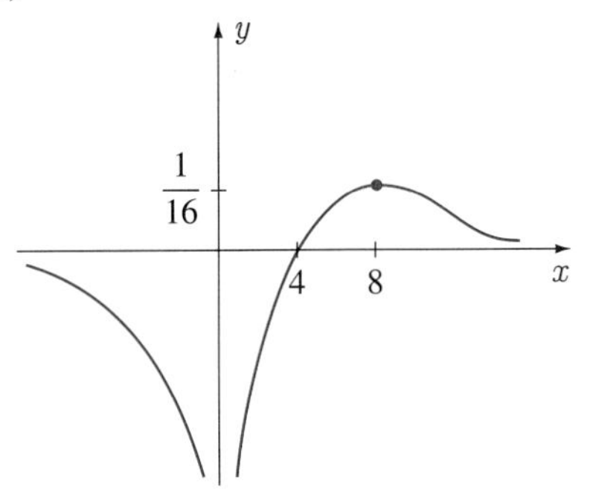

(d)

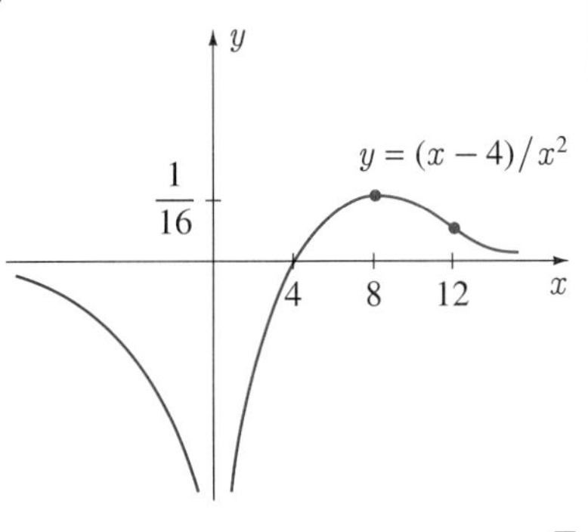

■

EXAMPLE 4.8

Sketch a graph of the function $f(x) = x + \sin x$.

SOLUTION We might try addition of ordinates for the curves $y = x$ and $y = \sin x$ in Figure 4.42(a). This could result in the sketches in Figures 4.42(b),(c), or (d). To decide which of these curves is correct, we consider the critical points of $f(x)$, defined by

$$0 = f'(x) = 1 + \cos x.$$

The solutions of this equation are $x = (2n+1)\pi$, where n is an integer. Since $f'(x)$ does not change sign as x passes through these points, they cannot yield relative maxima or minima. They must give horizontal points of inflection, and the graph in Figure 4.42(c) is correct.

FIGURE 4.42

(a)

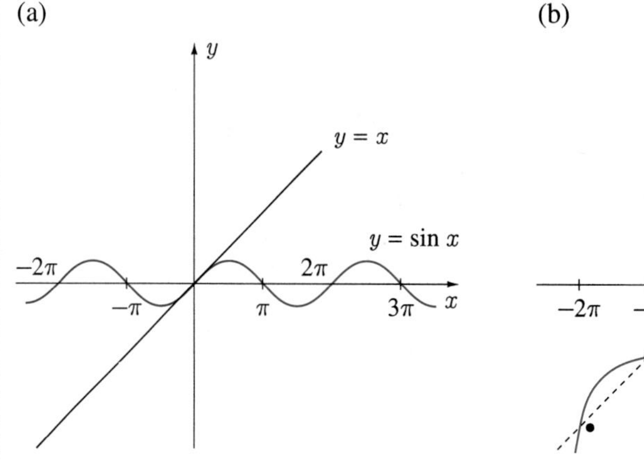

(b)

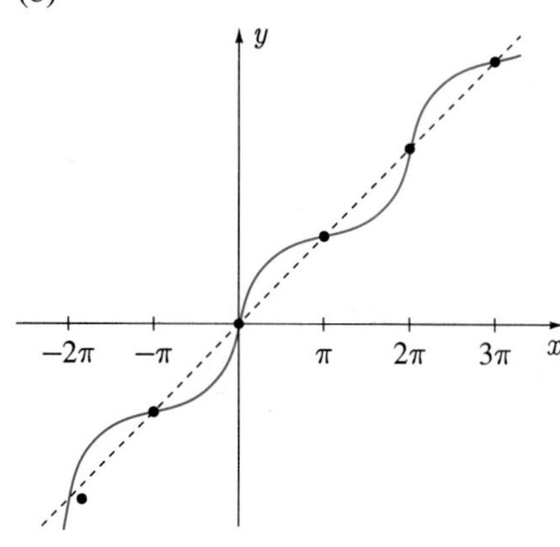

(c)

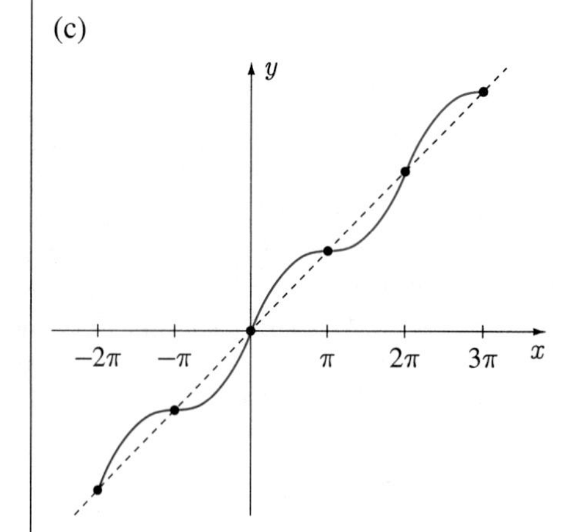

(d)

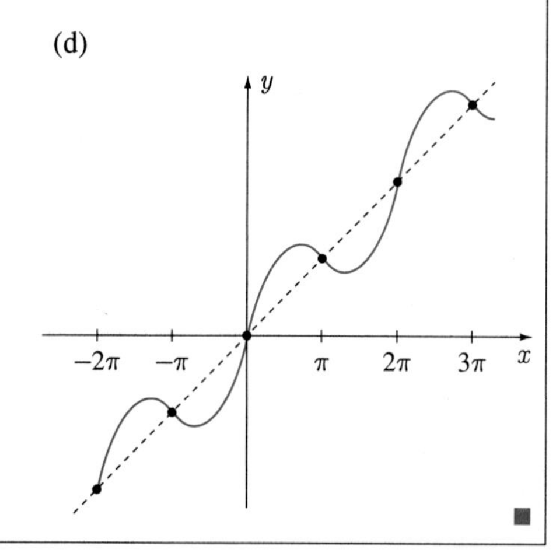

■

EXAMPLE 4.9

Sketch a graph of the function $f(x) = x^{5/3} - x^{2/3}$.

SOLUTION We first note that the three points $(0,0)$, $(1,0)$, and $(-1,-2)$ are on the graph. Next we add the facts that

$$\lim_{x\to-\infty} f(x) = -\infty \qquad \text{and} \qquad \lim_{x\to\infty} f(x) = \infty,$$

as shown in Figure 4.43(a). We cannot be sure that the concavity is as indicated, but this will be verified shortly. To find critical points of $f(x)$, we first solve

$$0 = f'(x) = \frac{5}{3}x^{2/3} - \frac{2}{3}x^{-1/3} = \frac{5x - 2}{3x^{1/3}}.$$

The only solution is $x = 2/5$, and because $f'(x)$ changes from negative to positive as x increases through $2/5$, we have a relative minimum of $f(2/5) = (2/5)^{5/3} - (2/5)^{2/3} = -0.33$. Since $f'(x)$ does not exist at $x = 0$, this is also a critical point, and we calculate that

$$\lim_{x \to 0^-} f'(x) = \infty \qquad \text{and} \qquad \lim_{x \to 0^+} f'(x) = -\infty.$$

This information, along with the relative minimum, is shown in Figure 4.43(b). We now join these parts smoothly to produce the sketch in Figure 4.43(c). To verify that the concavity is as indicated we calculate

$$f''(x) = \left(\frac{5}{3}\right)\left(\frac{2}{3}\right)x^{-1/3} - \left(\frac{2}{3}\right)\left(-\frac{1}{3}\right)x^{-4/3} = \frac{2(5x + 1)}{9x^{4/3}}.$$

Clearly there is only one point of inflection, namely when $x = -1/5$. The sign of $f''(x)$ on each side of this point indicates that the concavity is indeed as shown in Figure 4.43(c), and the final sketch is in Figure 4.43(d).

(a)

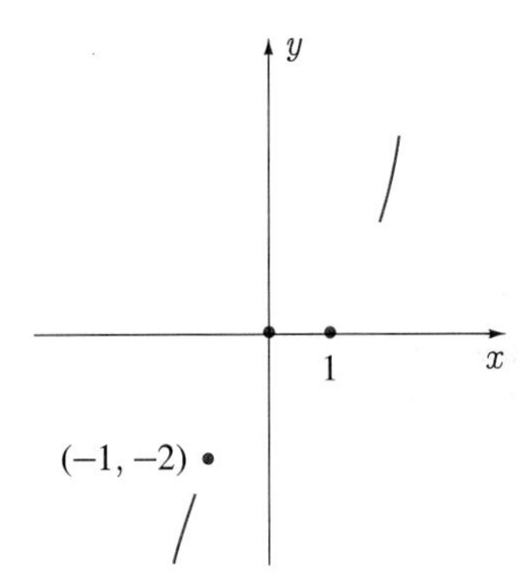

(b)

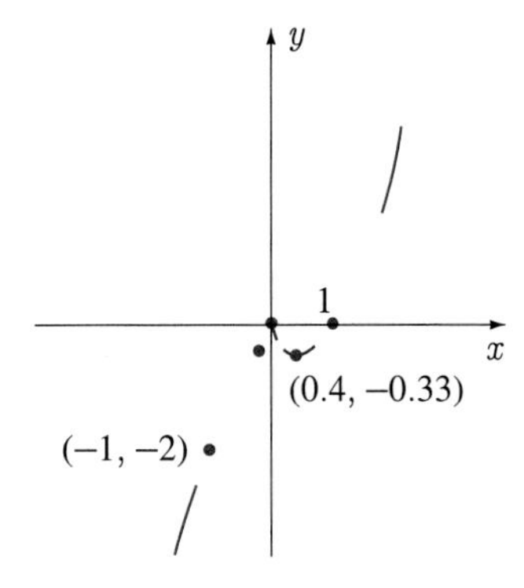

(c)

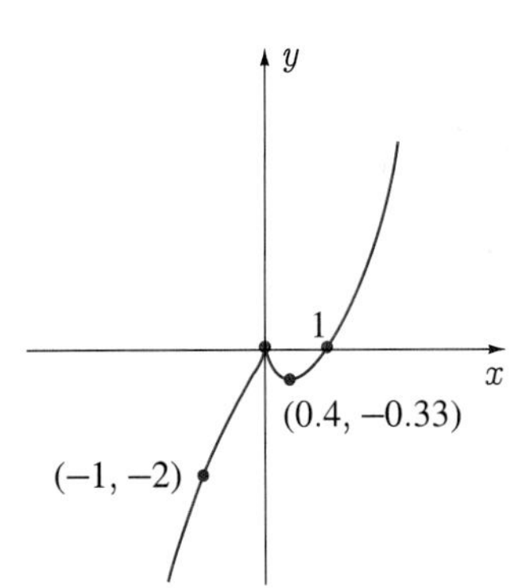

(d)

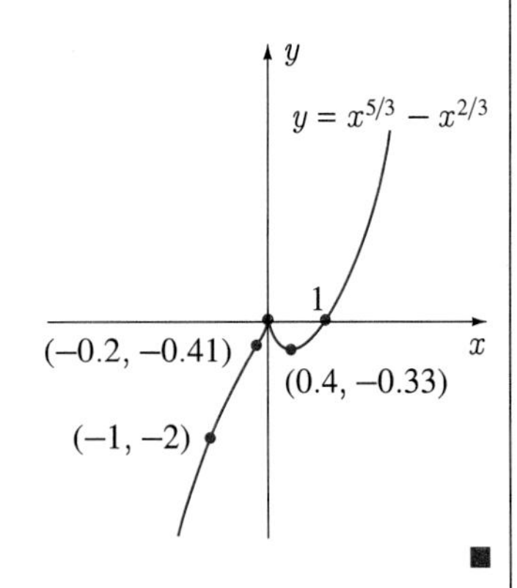

FIGURE 4.43

■

EXAMPLE 4.10

Sketch a graph of the function $f(x) = (x+2)^2/x^2$.

SOLUTION We write

$$f(x) = \frac{(x+2)^2}{x^2} = \left(\frac{x+2}{x}\right)^2 = \left(1+\frac{2}{x}\right)^2.$$

A sketch of $y = 2/x$ is shown in Figure 4.44(a), and $y = 1+2/x$ requires only a shift of 1 unit in the y-direction. If we now square all ordinates of the curve $y = 1+2/x$, we obtain the sketch of $y = (1+2/x)^2$ in Figure 4.44(b). The only information that could be added is the exact position of the point of inflection to the left of $x = -2$.

FIGURE 4.44

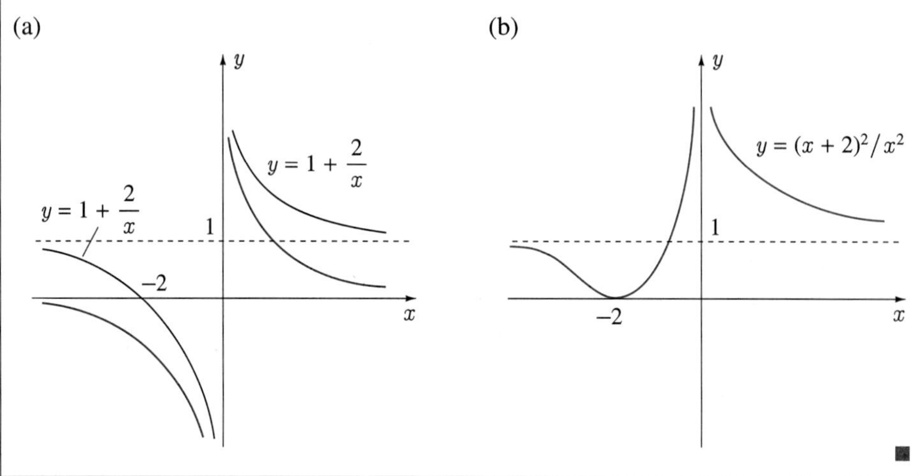

■

EXAMPLE 4.11

Sketch a graph of the function $f(x) = (x^2+1)/(x-1)$.

SOLUTION If we divide x^2+1 by $x-1$, we find

$$f(x) = (x+1) + \frac{2}{x-1}.$$

Sketches of $y = x+1$ and $y = 2/(x-1)$ are shown in Figure 4.45(a), and by addition of ordinates, we obtain the graph in Figure 4.45(b).

To add the positions of the relative extrema we find critical points. Since

$$f'(x) = \frac{(x-1)(2x) - (x^2+1)(1)}{(x-1)^2} = \frac{x^2-2x-1}{(x-1)^2},$$

we set

$$x^2 - 2x - 1 = 0.$$

Critical points are therefore

$$x = \frac{2 \pm \sqrt{4+4}}{2} = 1 \pm \sqrt{2}.$$

Our sketch in Figure 4.45(b) clearly indicates that $x = 1 + \sqrt{2}$ gives a relative minimum and $x = 1 - \sqrt{2}$ a relative maximum. This could also be verified by the first-derivative test. Would you consider the second-derivative test? The final graph is shown in Figure 4.45(c). As $x \to \pm\infty$, the graph gets arbitrarily close to the line $y = x + 1$. We call this line a **slanted asymptote**.

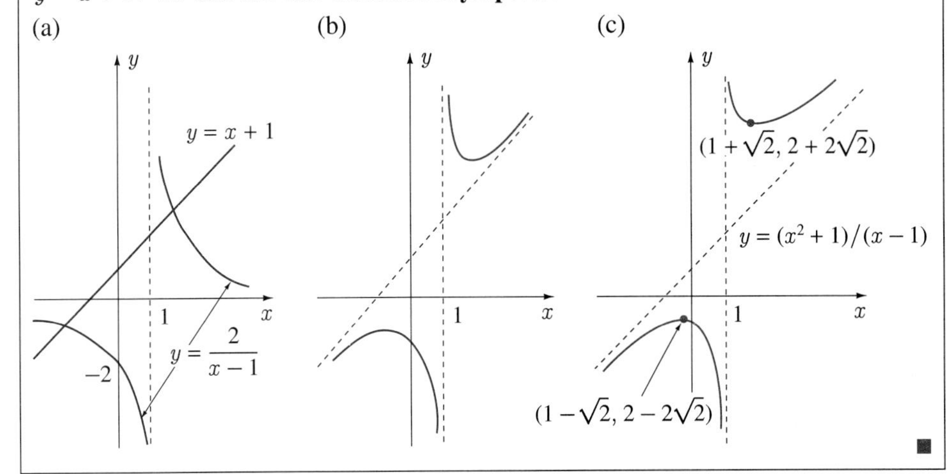

FIGURE 4.45

For rational functions, such as that in Example 4.11, slanted asymptotes occur when the numerator has degree exactly one higher than the denominator.

EXERCISES 4.5

In Exercises 1–10 find all relative maxima and minima for the function and points of inflection for its graph. Use this information to sketch a graph of the function.

1. $f(x) = \dfrac{x^3}{3} - \dfrac{x^2}{2} - 2x$

2. $f(x) = x^3 - 6x^2 + 12x + 9$

3. $f(x) = 3x^4 - 16x^3 + 18x^2 + 2$

4. $f(x) = x - 3x^{1/3}$

5. $f(x) = 2x^3 - 15x^2 + 6x + 4$

6. $f(x) = \dfrac{x}{x^2 + 4}$

7. $f(x) = (x-2)^3(x+2)$

8. $f(x) = x^{2/3}(8 - x)$

9. $f(x) = \dfrac{(x+2)^2}{x^3}$

10. $f(x) = 2x^{3/2} - 9x + 12x^{1/2}$

In Exercises 11–20 sketch a graph of the function.

11. $f(x) = \dfrac{x^3 + 16}{x}$ **12.** $f(x) = \dfrac{x^2 + x + 1}{x}$

13. $f(x) = x + \dfrac{1}{x}$ **14.** $f(x) = \dfrac{x^3}{x^2 - 4}$

15. $f(x) = \dfrac{2x^2}{x^2 - 8x + 12}$

16. $f(x) = \dfrac{x^2 + 1}{x^2 - 1}$

17. $f(x) = \dfrac{(x-1)^3}{(x+1)^4}$

18. $f(x) = (x-1)^{2/3}$

19. $f(x) = (x+2)^3(x-4)^3$

20. $f(x) = x + 2\sin x$

21. A voltage source with constant electromotive force E and constant internal resistance r maintains a current I through a circuit with resistance R (Figure 4.46), where

$$E = I(r + R).$$

The power (work per unit time) necessary to maintain this current in r and R is given by

$$P = I^2(r + R) = I^2 r + I^2 R.$$

(a) If we define $P_R = I^2 R$ and $P_r = I^2 r$ as the power dissipated in R and r respectively, sketch P_R and P_r as functions of R.

(b) Sketch a graph of $P(R) = P_R(R) + P_r(R)$.

FIGURE 4.46

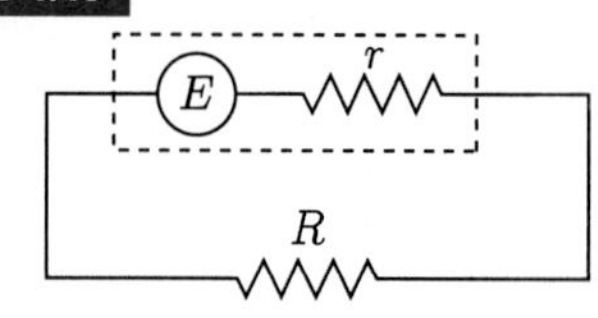

22. The rate of photosynthesis P in a leaf depends on the intensity of light I on the leaf according to

$$P = \frac{MI}{I+K} - R,$$

where $M > R$ and K are all positive constants. Sketch a graph of this function.

In Exercises 23–34 sketch a graph of the function.

23. $f(x) = x/(x^2+3)$

24. $f(x) = x^4 + 10x^3 + 6x^2 - 64x + 5$

25. $f(x) = x^4 - 2x^3 + 2x$

26. $f(x) = \dfrac{x^2-8}{x-5}$

27. $f(x) = \dfrac{x^2}{x^2-4}$

28. $f(x) = x^{5/4} - x^{1/4}$

29. $f(x) = |x^2 - 9| + 2$

30. $f(x) = \dfrac{(2x-1)(x-8)}{(x-1)(x-4)}$

31. $f(x) = \sin^2 x \cos x, \quad 0 \le x \le 2\pi$

32. $f(x) = \sin x + \cos x$

33. $f(x) = \dfrac{x}{\sqrt{1-x}}$

34. $f(x) = \left(\dfrac{x+8}{x}\right)\sqrt{x^2+100}$

35. (a) A company produces x kilograms of a commodity per day at a total cost of

$$C(x) = \frac{x^2}{300}\left(\frac{x+100}{x+300}\right) + 60, \quad 1 \le x \le 200.$$

Show that the cost graph is always concave upward. Sketch the graph.

(b) The average production cost per kilogram when x kilograms are produced is given by $c(x) = C(x)/x$. Sketch a graph of this function.

(c) Show that the output at which the average cost is least satisfies the equation

$$(x+300)^2(x^2 - 18\,000) - 60\,000x^2 = 0.$$

Do this in two ways:

(i) by finding critical points for $c(x)$;

(ii) by noting that the average cost is the slope of the line joining a point $(x, C(x))$ to the origin; hence minimum average cost occurs when the tangent line to the $C(x)$ graph passes through the origin.

36. The function

$$f(x) = \frac{1}{\sqrt{2\pi}\sigma} e^{-(x-\mu)^2/(2\sigma^2)}$$

where $\sigma > 0$ and μ are constants is called the *normal probability density function*. Sketch its graph.

37. When a drug is injected into the blood at time $t = 0$, it is sometimes assumed that the concentration of the drug thereafter is given by the function

$$f(t) = k\left(e^{-at} - e^{-bt}\right)$$

where k, a, and b are positive constants ($b > a$). Sketch a graph of this function indicating any relative extrema and points of inflection.

38. Sketch a graph of the function

$$f(x) = \frac{1 + x + x^2 + x^3}{1 + x^3}.$$

SECTION 4.6

Absolute Maxima and Minima

Seldom in applications of maxima and minima theory do we hear questions such as: What are the relative maxima of this quantity, or what are the relative minima of that quantity? More likely it is: What is the biggest of these, the smallest of those, the best way to do this, the cheapest way to do that? In other words, when a function which represents a measureable quantity is to be maximized (or minimized), it is not relative maxima (or minima) that are required; a different kind of extrema is involved—considering all points in the domain of the function, what is the largest (or smallest) value of the function?

Definition 4.6

The **absolute maximum** of a function $f(x)$ on an interval I is $f(x_0)$ if x_0 is in I and if for all x in I

$$f(x) \leq f(x_0); \tag{4.12}$$

$f(x_0)$ is said to be the **absolute minimum** of $f(x)$ on I if for all x in I

$$f(x) \geq f(x_0). \tag{4.13}$$

For the function in Figure 4.47, the absolute maximum of $f(x)$ on the interval $a \leq x \leq b$ is $f(c)$, and the absolute minimum is $f(d)$. For the function in Figure 4.48, the absolute maximum on $a \leq x \leq b$ is $f(b)$, and the absolute minimum is $f(a)$. For the function in Figure 4.49, the absolute maximum on $a \leq x \leq b$ is $f(b)$, and the absolute minimum is $f(c)$. Note that we speak of absolute maxima and minima (absolute extrema) of a function only on some specified interval; that is, we do not ask for the absolute maximum or minimum of a function $f(x)$ without specifying the interval I.

FIGURE 4.47

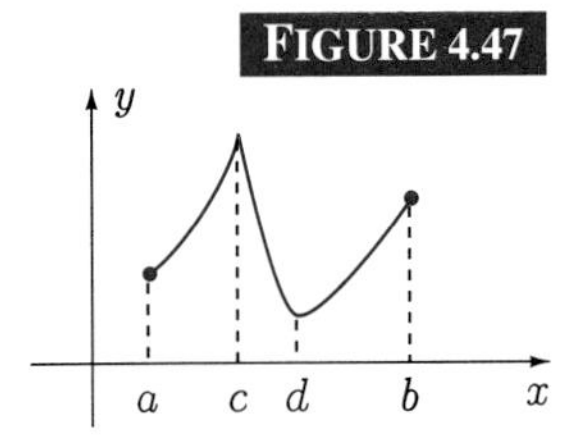

FIGURE 4.48

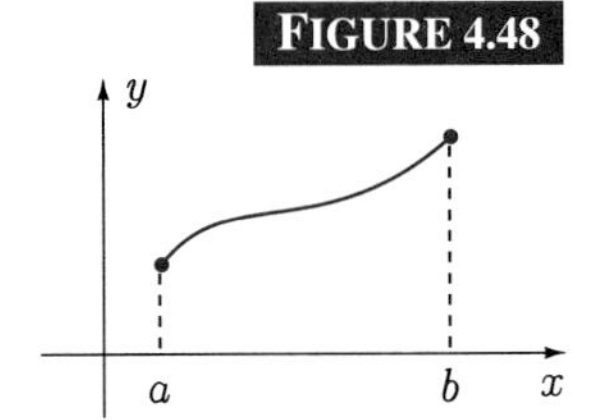

FIGURE 4.49

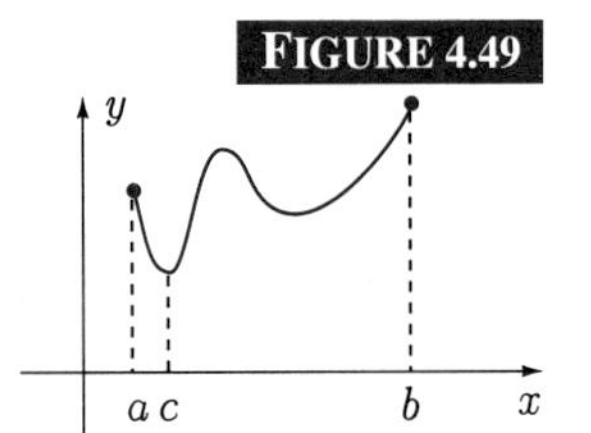

Each of the functions in Figures 4.47–4.49 is continuous on a closed interval $a \leq x \leq b$. The following theorem asserts that every such function has absolute extrema. For a proof of this result, the interested reader should consult books on advanced analysis.

Theorem 4.3

A function that is continuous on a closed interval must attain an absolute maximum and an absolute minimum on that interval.

The conditions of this theorem are sufficient to guarantee existence of absolute extrema; that is, if a function is continuous on a closed interval, then it *must* have absolute extrema on that interval. However, they are not necessary. If these conditions are not met, the function may or may not have absolute extrema. For instance, if the function in Figure 4.47 is confined to the open interval $a < x < b$, it still attains its absolute extrema at $x = c$ and $x = d$. On the other hand, the function in Figure 4.48 does not have absolute extrema on the open interval $a < x < b$. The function in Figure 4.50 is not continuous on the closed interval $a \leq x \leq d$; it has no absolute maximum on this interval, but it does have absolute minimum $f(a)$. This function is not continuous on $d \leq x \leq b$, but it has absolute maximum $f(d)$ and absolute minimum $f(b)$.

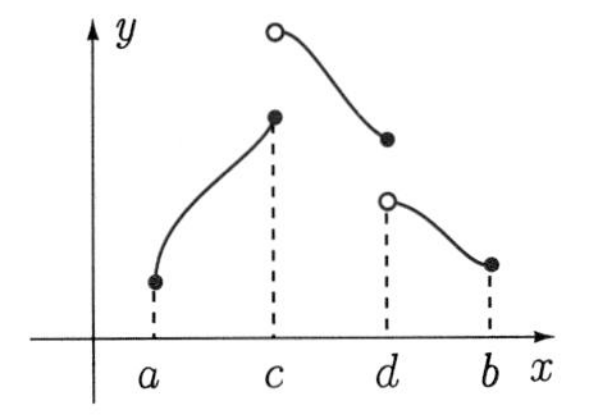

FIGURE 4.50

Absolute extrema of the functions in Figures 4.47–4.49 always occur either at a critical point of the function or at an end of the interval $a \leq x \leq b$. This result is always true for continuous functions.

Theorem 4.4

A function that is continuous on a closed interval always attains its absolute extrema either at critical points or at the ends of the interval.

Proof Let $f(x)$ be a continuous function on a closed interval $a \le x \le b$. To verify the theorem, we show that $f(x)$ cannot have an absolute extremum at any point x_0 in $a < x < b$ that is not critical. If $a < x_0 < b$ and x_0 is not critical, then x_0 cannot yield a relative extremum for $f(x)$ (see Theorem 4.2). Definitions 4.3 and 4.4 now imply that in every open interval around x_0, $f(x)$ takes on values that are larger than $f(x_0)$ and values that are smaller than $f(x_0)$. It follows that $f(x_0)$ cannot possibly be an absolute extremum for $f(x)$ on $a \le x \le b$.

Because of this theorem, there is a simple procedure to determine the absolute extrema of a *continuous* function $f(x)$ on a *closed* interval $a \le x \le b$.

Finding Absolute Extrema for a Continuous Function $f(x)$ on a Closed Interval $a \le x \le b$

(i) Find all critical points $x_1, x_2, \ldots, x_n$ of $f(x)$ in $a < x < b$.

(ii) Evaluate

$$f(a), \quad f(x_1), \quad f(x_2), \quad \ldots, \quad f(x_n), \quad f(b).$$

(iii) The absolute maximum of $f(x)$ on $a \le x \le b$ is the largest of the numbers in (ii); the absolute minimum is the smallest of these numbers.

Note that it is not necessary to classify the critical points of $f(x)$ as yielding relative maxima, relative minima, horizontal points of inflection, vertical points of inflection, corners, or none of these. We need only evaluate $f(x)$ at $x = a$, $x = b$, and at its critical points. The largest and smallest of these numbers must be the absolute extrema.

EXAMPLE 4.12

Find the absolute extrema for the function $f(x) = (x^2 + 2x + 21)/(x + 5)$ on the interval $0 \le x \le 6$.

SOLUTION For critical points of $f(x)$, we first solve

$$\begin{aligned} 0 = f'(x) &= \frac{(x+5)(2x+2) - (x^2 + 2x + 21)(1)}{(x+5)^2} \\ &= \frac{x^2 + 10x - 11}{(x+5)^2} = \frac{(x+11)(x-1)}{(x+5)^2}. \end{aligned}$$

The only critical point between $x = 0$ and $x = 6$ is $x = 1$. Since

$$f(0) = \frac{21}{5}, \qquad f(1) = 4, \qquad f(6) = \frac{69}{11},$$

the absolute maximum is $69/11$ and the absolute minimum is 4. ∎

In many applied problems the function is continuous, but its domain is not a finite interval. In this case we replace evaluation of the function at its end points with limits as the independent variable becomes very large negatively, positively, or both.

EXAMPLE 4.13

What are the absolute extrema of the function

$$f(x) = \frac{(2-x)^2}{x^2}, \quad x \geq \frac{1}{2}.$$

SOLUTION For critical points of $f(x)$, we first solve

$$0 = f'(x) = \frac{x^2(-2)(2-x) - (2-x)^2(2x)}{x^4} = \frac{4(x-2)}{x^3}.$$

The only solution is $x = 2$. Furthermore, $f'(x)$ exists for all $x > 1/2$. We now evaluate

$$f(1/2) = 9, \qquad f(2) = 0, \qquad \lim_{x\to\infty} f(x) = 1.$$

Maximum and minimum values of $f(x)$ are therefore 9 and 0 respectively. ■

We now show how absolute extrema can be used to solve applied problems.

EXAMPLE 4.14

A rectangular field is to be fenced on three sides with 1000 m of fencing (the fourth side being a straight river's edge). Find the dimensions of the field in order that the area be as large as possible.

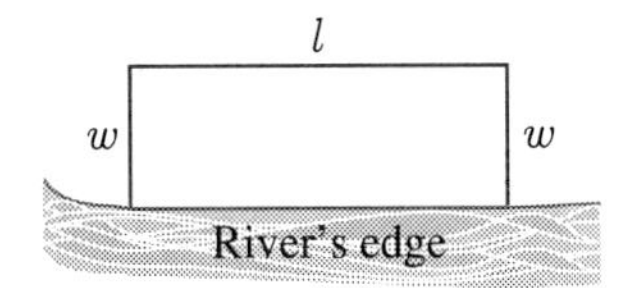

FIGURE 4.51

SOLUTION Since the area of the field is to be maximized, we first define a function representing this area. The area of a field of width w and length l (Figure 4.51) is

$$A = lw.$$

This function represents the area of a field with arbitrary length l and arbitrary width w. Our problem states that only 1000 m of fencing are available for the three sides; therefore, l and w must satisfy the equation

$$2w + l = 1000.$$

With this equation we can express A completely in terms of w (or l):

$$A(w) = w(1000 - 2w) = 1000w - 2w^2.$$

To maximize the area of the field, we must therefore maximize the function $A(w)$. But what are the values of w under consideration? Clearly, w cannot be negative, and in order to satisfy the restriction $2w + l = 1000$, w cannot exceed 500. The physical problem has now been modeled mathematically. Find the absolute maximum of the (continuous) function $A(w)$ on the (closed) interval $0 \leq w \leq 500$.

We first find the critical points of $A(w)$ by solving

$$0 = A'(w) = 1000 - 4w;$$

the only solution is $w = 250$. Since $A'(w)$ exists for all $0 \le w \le 500$, we evaluate $A(w)$ at $w = 250$ and the end points:

$$A(0) = 0, \qquad A(500) = 0, \qquad A(250) = 1000(250) - 2(250)^2 > 0.$$

We conclude that the largest possible area is obtained when the width of the field is 250 m and its length is 500 m. ■

EXAMPLE 4.15

Pop cans to hold 300 mL are made in the shape of right-circular cylinders. Find dimensions of the can that minimize its surface area.

SOLUTION The surface area of the can (Figure 4.52(a)) consists of a circular top, an identical bottom, and a rectangular piece formed into the side of the can. The total area of these three pieces is

$$A = 2\pi r^2 + 2\pi r h,$$

where we measure r and h in centimetres (see Figure 4.52(b)). Since the can must hold 300 mL of pop, it follows that

$$\pi r^2 h = 300 \qquad \text{or} \qquad h = \frac{300}{\pi r^2}.$$

This equation can be used to express A completely in terms of r,

$$A(r) = 2\pi r^2 + 2\pi r\left(\frac{300}{\pi r^2}\right).$$

For what values of r is $A(r)$ defined? The radius of the can must be positive and therefore $r > 0$. How large can r be? In effect, it can be as large as desired. The height of the cylinder can always be chosen sufficiently small when r is large to satisfy the volume condition $\pi r^2 h = 300$.

FIGURE 4.52

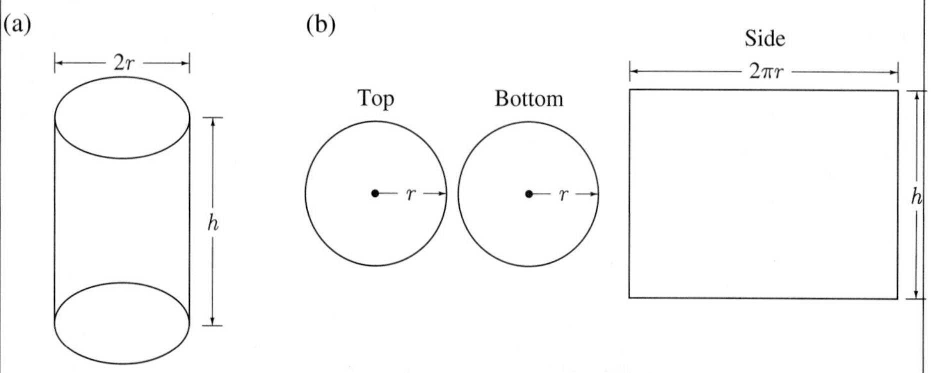

The can may not be esthetically pleasing for large r and small h, but no mathematical difficulties occur in this situation. To solve the problem we must therefore maximize

$$A(r) = 2\pi r^2 + \frac{600}{r}, \qquad r > 0.$$

For critical points of $A(r)$, we solve

$$0 = A'(r) = 4\pi r - \frac{600}{r^2} \qquad \text{or} \qquad 4\pi r^3 - 600 = 0.$$

The only solution of this equation is $r = (150/\pi)^{1/3}$. Were $A(r)$ defined on a closed interval we would evaluate it at the critical point and the ends of the interval. With a continuous function on the interval $r > 0$, we replace evaluation at the ends of the interval with limits as r approaches the ends:

$$\lim_{r \to 0^+} A(r) = \infty, \qquad A\left((150/\pi)^{1/3}\right) > 0, \qquad \lim_{r \to \infty} A(r) = \infty.$$

Thus, minimum surface area occurs when the radius of the can is $(150/\pi)^{1/3}$ cm and its height, obtained from the fact that $h = 300/(\pi r^2)$, is $2(150/\pi)^{1/3}$ cm. ■

On the basis of these two examples, we suggest the following steps in solving applied maxima-minima problems:

1. Sketch a diagram illustrating the situation.
2. Identify the quantity that is to be maximized or minimized, choose a letter to represent it, and find an expression for this quantity.
3. If necessary, use information in the problem to rewrite the expression in 2 as a function of only one variable.
4. Determine the domain of the function in 3.
5. Maximize or minimize the function in 3 on the domain in 4.
6. Interpret the maximum or minimum values in terms of the original problem.

Step 2 is crucial. Do not consider subsidiary information in the problem until the quantity to be maximized or minimized is clearly identified, labelled, and an expression found for it. Only then should other information be considered. For instance, in Example 4.14, the restriction $2w + l = 1000$ for the length of fencing available was not introduced until the expression $A = lw$ had been identified. Likewise in Example 4.15, volume condition $\pi r^2 h = 300$ was introduced after surface area $A = 2\pi r^2 + 2\pi r h$.

We now take a look at three additional examples.

A lighthouse is 6 km off shore and a cabin on the straight shoreline is 9 km from the point on the shore nearest the lighthouse. If a man rows at a rate of 3 km/h, and walks at a rate of 5 km/h, where should he beach his boat in order to get from the lighthouse to the cabin as quickly as possible? Repeat the problem in the case that the cabin is 4 km along the shoreline.

SOLUTION Figure 4.53 illustrates the path followed by the man when he beaches the boat x kilometres from the point on land closest to the lighthouse. His travel time t for this path is the sum of his time t_1 in the water and his time t_2 on land. Since velocities are constant, each of these times may be calculated by dividing distance by velocity; that is, $t_1 = \sqrt{x^2 + 36}/3$ and $t_2 = (9 - x)/5$. To minimize t for some value of x between 0 and 9, we minimize the function

$$t(x) = t_1 + t_2 = \frac{\sqrt{x^2 + 36}}{3} + \frac{9 - x}{5}, \qquad 0 \le x \le 9.$$

For critical points of $t(x)$ we solve

$$0 = t'(x) = \frac{x}{3\sqrt{x^2 + 36}} - \frac{1}{5} \qquad \text{or} \qquad 5x = 3\sqrt{x^2 + 36}.$$

Thus,

$$25x^2 = 9(x^2 + 36),$$

from which we accept only the positive solution $x = 9/2$. Since $t'(x)$ exists for all x, and

$$t(0) = \frac{\sqrt{36}}{3} + \frac{9}{5} = \frac{19}{5},$$

$$t(9) = \frac{\sqrt{81 + 36}}{3} = \sqrt{13},$$

$$t\left(\frac{9}{2}\right) = \frac{\sqrt{81/4 + 36}}{3} + \frac{9 - 9/2}{5} = \frac{17}{5} < \sqrt{13},$$

the boat should therefore be beached 4.5 km from the cabin.

When the cabin is 4 km along the shoreline, then

$$t(x) = \frac{\sqrt{x^2 + 36}}{3} + \frac{4 - x}{5}, \qquad 0 \le x \le 4.$$

Critical points are again given by the equation

$$\frac{x}{3\sqrt{x^2 + 36}} - \frac{1}{5} = 0,$$

but the solution $x = 9/2$ must be rejected since it does not fall in the interval $0 \le x \le 4$. Travel time must therefore be minimized by heading directly to shore or directly to the cabin. Since

$$t(0) = \frac{\sqrt{36}}{3} + \frac{4}{5} = \frac{14}{5},$$

$$t(4) = \frac{\sqrt{16 + 36}}{3} = \frac{2\sqrt{13}}{3} < \frac{14}{5},$$

the boat should head directly toward the cabin.

FIGURE 4.53

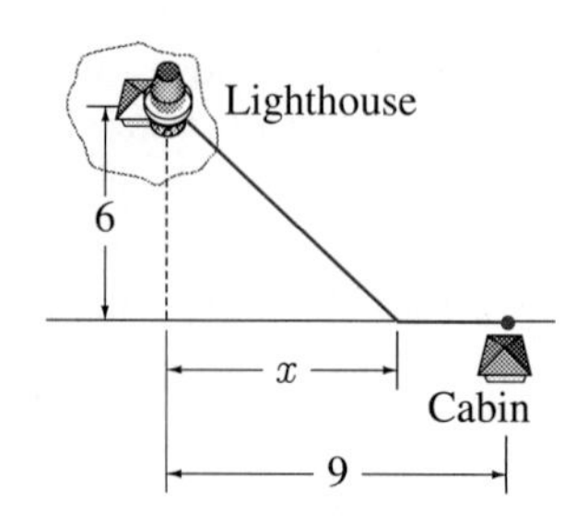

EXAMPLE 4.17

A wall of a building is to be braced by a beam that must pass over a parallel wall 5 metres high and 2 metres from the building (Figure 4.54). Find the length of the shortest beam that can be used.

FIGURE 4.54

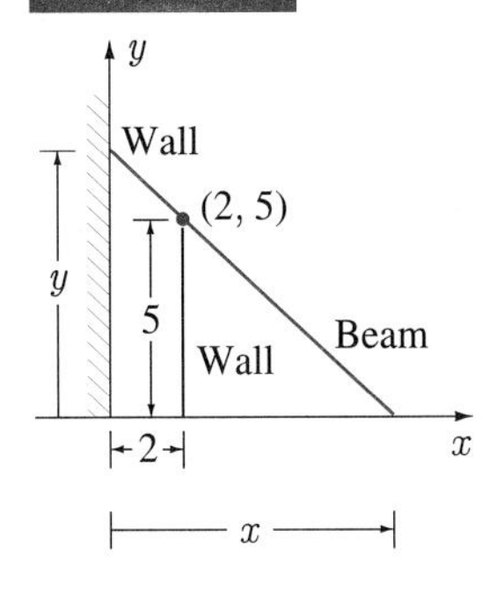

SOLUTION If we introduce coordinates as shown in Figure 4.54, we see that the shortest beam is the shortest of all line segments that stretch from points on the positive x-axis to points on the positive y-axis, and pass through the point $(2, 5)$. The length of the line segment that has intercepts x and y is given by

$$D = \sqrt{x^2 + y^2}.$$

To express D in terms of only one variable, we must find an equation relating x and y. Since slopes of the line segments joining $(0, y)$ to $(x, 0)$ and $(2, 5)$ to $(x, 0)$ must be identical, we have

$$\frac{y-0}{0-x} = \frac{0-5}{x-2} \qquad \text{or} \qquad y = \frac{5x}{x-2}.$$

Thus,

$$D = \sqrt{x^2 + \frac{25x^2}{(x-2)^2}}.$$

Since D is a minimum when D^2 is a minimum, we consider minimizing

$$D^2 = f(x) = x^2 + \frac{25x^2}{(x-2)^2}, \qquad x > 2.$$

For critical points of $f(x)$, we solve

$$\begin{aligned} 0 = f'(x) &= 2x + \frac{50x}{(x-2)^2} - \frac{50x^2}{(x-2)^3} \\ &= \frac{2x[(x-2)^3 + 25(x-2) - 25x]}{(x-2)^3} \\ &= \frac{2x}{(x-2)^3}[(x-2)^3 - 50]. \end{aligned}$$

Since x cannot be equal to zero, we set

$$(x-2)^3 = 50,$$

and this implies that $x = 2 + 50^{1/3}$. Once again since $f(x)$ is defined on the infinite interval $x > 2$, we evaluate it at the critical point and take limits as $x \to 2^+$ and $x \to \infty$:

$$\lim_{x \to 2^+} f(x) = \infty, \qquad f(2 + 50^{1/3}) > 0, \qquad \lim_{x \to \infty} f(x) = \infty.$$

We conclude that D^2, and therefore D, must be minimized for $x = 2 + 50^{1/3}$, and the shortest beam has length

$$\sqrt{f\left(2+50^{1/3}\right)} = \left(2+50^{1/3}\right)\sqrt{1+\frac{25}{50^{2/3}}} = 9.6 \text{ m}.$$

FIGURE 4.55

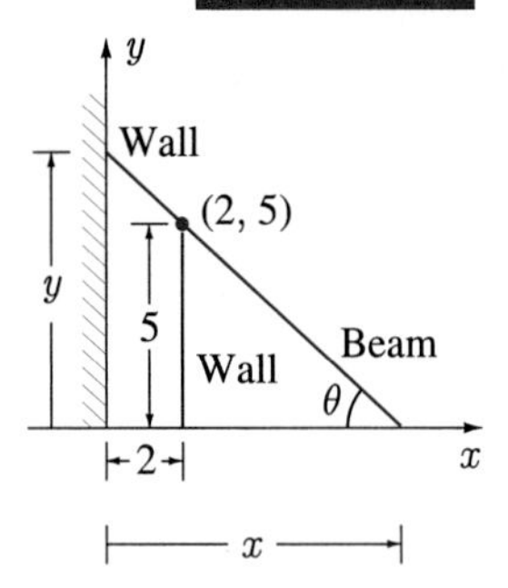

We could also have solved this problem by introducing the angle θ in Figure 4.55. The length D of the beam can be expressed in terms of θ by

$$D = x \sec\theta.$$

Since $x - 2 = 5\cot\theta$, it follows that

$$D = g(\theta) = \sec\theta(2 + 5\cot\theta), \quad 0 < \theta < \pi/2.$$

For critical points of this function, we set

$$0 = g'(\theta) = \sec\theta\tan\theta(2 + 5\cot\theta) + \sec\theta(-5\csc^2\theta).$$

When this equation is expressed in the form

$$\frac{\cos\theta}{\sin^2\theta}(2\tan^3\theta - 5) = 0,$$

we see that the only critical point is given by

$$\tan\theta = \left(\frac{5}{2}\right)^{1/3}.$$

The angle between 0 and $\pi/2$ satisfying this equation is 0.9358 radians. Since

$$\lim_{\theta\to 0^+} g(\theta) = \infty, \qquad g(0.9358) = 9.6, \qquad \lim_{\theta\to\pi/2^-} g(\theta) = \infty,$$

it follows once again that the shortest beam has length 9.6 m. ■

EXAMPLE 4.18

A printer contracts to print 200 000 copies of a membership card. It costs \$10 per hour to run the press, and the press produces 1000 impressions per hour. The printer is at liberty to choose the number of set types per impression to a limit of 40, each set type costing \$2. If x set types are chosen, then each impression yields x cards. How many set types should be used?

SOLUTION The printer will choose that number of set types which will minimize production costs. If x set types are used, then the cost for the set types themselves is $2x$ dollars. With this number of set types, the press prints $1000x$ cards per hour. In order to produce 200 000 cards, the press must therefore run for $200\,000/(1000x)$ hours at a cost of $[10][200\,000/(1000x)]$ dollars. The total cost of producing the cards when x set types are used is therefore given by

$$C(x) = 2x + \frac{2000}{x}.$$

Now, $C(x)$ is different than most functions that we have encountered; it is defined only for integer values between 1 and 40 inclusively since the printer must use at least one set type, cannot use more than 40, and cannot use a fraction of a set type. To minimize $C(x)$ we could therefore evaluate it at the first 40 positive integers, and pick the smallest such number. Then there would be no need for calculus at all.

On the other hand, if this were a different problem in which $C(x)$ were again only defined for integer values, but, say, a billion of them, we might not be so complacent about avoiding calculus. But it is impossible to differentiate a function that is defined only for integers. To remedy this, we extend the domain of definition of $C(x)$ from integer values to all real numbers between 1 and 40, and minimize $C(x)$ on this interval, $1 \le x \le 40$.

Our customary procedure is to first find critical points of $C(x)$ by setting

$$0 = C'(x) = 2 - \frac{2000}{x^2}.$$

The only solution of this equation in the interval $1 \le x \le 40$ is $x = 10\sqrt{10} = 31.6$. As a function of a continuous variable x, $C(x)$ has one critical point, namely, 31.6, at which its derivative is zero. But this is an unacceptable number to the printing problem since x must be an integer. We therefore do not evaluate $C(x)$ at 31.6, but rather at the integer values 31 and 32 on either side; that is, we calculate

$$C(1) = 2002, \qquad C(31) = 126.52, \qquad C(32) = 126.50, \qquad C(40) = 130.$$

The printer should therefore use 32 set types. ■

EXERCISES 4.6

In Exercises 1–9 find absolute extrema for the function on the interval, if they exist.

1. $f(x) = x^3 - x^2 - 5x + 4, \quad -2 \le x \le 3$

2. $f(x) = \dfrac{x-4}{x+1}, \quad 0 \le x \le 10$

3. $f(x) = x + \dfrac{1}{x}, \quad \dfrac{1}{2} \le x \le 5$

4. $f(x) = x - 2\sin x, \quad 0 \le x \le 4\pi$

5. $f(x) = x\sqrt{x+1}, -1 \le x \le 1$

6. $f(x) = \dfrac{12}{x^2 + 2x + 2}, \quad x < 0$

7. $f(x) = \dfrac{x+1}{x-1}, \quad x > 1$

8. $f(x) = \dfrac{x}{x^2+3}, \quad x > 0$

9. $f(x) = \dfrac{(2x-1)(x-8)}{(x-1)(x-4)}, \quad x \le -2$

10. Use Theorem 4.3 to prove Theorem 3.15 in Section 3.11.

11. One end of a uniform beam of length L is built into a wall, and the other end is simply supported (Figure 4.56). If the beam has constant mass per unit length m, its deflection y from the horizontal at a distance x from the built-in end is given by

$$(48EI)y = -mg(2x^4 - 5Lx^3 + 3L^2x^2),$$

where E and I are constants depending on the material and cross-section of the beam, and $g > 0$ is a constant. How far from the built-in end does maximum deflection occur?

FIGURE 4.56

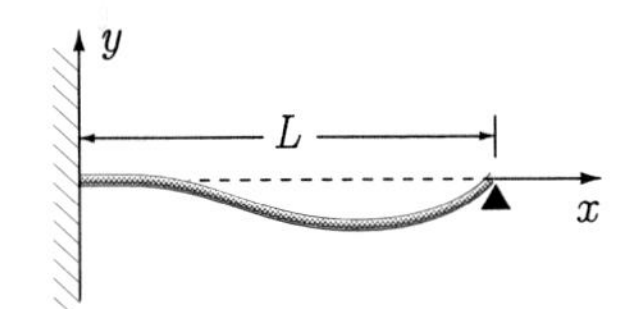

12. An agronomist wishes to fence eight rectangular plots for experimentation as shown in Figure 4.57. If each plot must contain 9000 m^2, find the minimum amount of fencing that can be used.

FIGURE 4.57

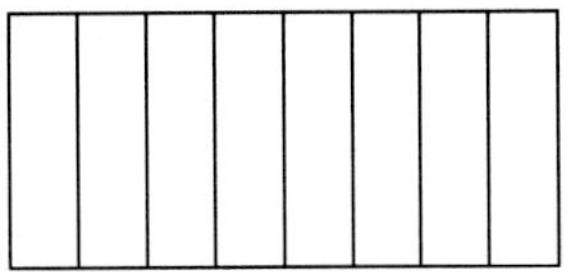

13. A closed box is to have length equal to three times its width and total surface area of $30\ \mathrm{m}^2$. Find the dimensions that produce maximum volume.

14. A square box with open top is to have a volume of 6000 L. Find the dimensions of the box that minimize the amount of material used.

15. A cherry orchard has 255 trees, each of which produces on the average 25 baskets of cherries. For each additional tree planted, the yield per tree decreases by one-twelfth of a basket. How many more trees will produce a maximum crop?

16. A high school football field is being designed to accommodate a 400 m track. The track runs along the length of the field and has semicircles beyond the end zones with diameters equal to the width of the field. If the end zones must be 10 m deep, find the dimensions of the field that maximize playing area (not including the end zones).

17. When a manufacturing company sells x objects per month, it sets the price r of each object at

$$r(x) = 100 - \frac{x^2}{10\,000}.$$

The total cost C of producing the x objects per month is

$$C(x) = \frac{x^2}{10} + 2x + 20.$$

Find the number of objects the company should sell per month in order to realize maximum profits.

18. The base of an isosceles triangle, which is not one of the equal sides, has length b, and its altitude has length a. Find the area of the largest rectangle that can be placed inside the triangle if one of the sides of the rectangle must lie on the base of the triangle.

19. A manufacturer builds cylindrical metal cans that hold $1000\ \mathrm{cm}^3$. There is no waste involved in cutting material for the curved surface of the can. However, each circular end piece is cut from a square piece of metal, leaving four waste pieces. Find the height and radius of the can that uses the least amount of metal, including all waste materials.

20. Sides AB and AC of an isosceles triangle have equal length. The base BC has length $2a$, as does the altitude AD from A to BC. Find the height of a point P on AD at which the sum of the distances AP, BP, and CP is a minimum.

21. Two poles are driven into the ground 3 m apart. One pole protrudes 2 m above the ground and the other pole 1 m above the ground. A single piece of rope is attached to the top of one pole, passed through a loop on the ground, pulled taut, and attached to the top of the other pole. Where should the loop be placed in order that the rope be as short as possible?

22. In designing pages for a book, a publisher decides that the rectangular printed region on each page must have area $150\ \mathrm{cm}^2$. If the page must have 2.5 cm margins on each side and 3.75 cm margins at top and bottom, find the dimensions of the page of smallest possible area.

23. Find the points on the hyperbola $y^2 - x^2 = 9$ closest to $(4,0)$.

24. Find the point on the parabola $y = x^2$ closest to $(-2,5)$.

25. A light source is to be placed directly above the centre of a circular area of radius r (Figure 4.58). The illumination at any point on the edge of the circle is directly proportional to the cosine of the angle θ and inversely proportional to the square of the distance d from the source. Find the height h above the circle at which illumination on the edge of the table is maximized.

FIGURE 4.58

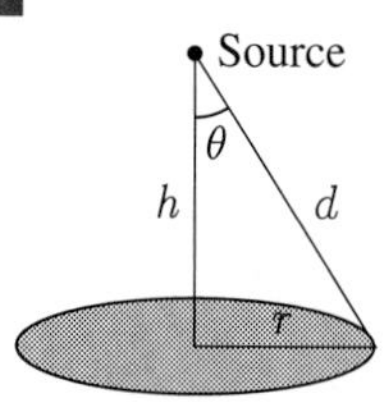

26. Among all line segments that stretch from points on the positive x-axis to points on the positive y-axis and pass through the point $(2,5)$, find that one which makes with the positive x- and y-axes the triangle with least possible area.

27. A long piece of metal one metre wide is to be bent in two places to form a spillway so that its cross-section is an isosceles trapezoid (Figure 4.59). Find the angle θ at which the bends should be formed in order to obtain maximum possible flow along the spillway if the lengths of AB, BC, and CD are all one-third metre.

FIGURE 4.59

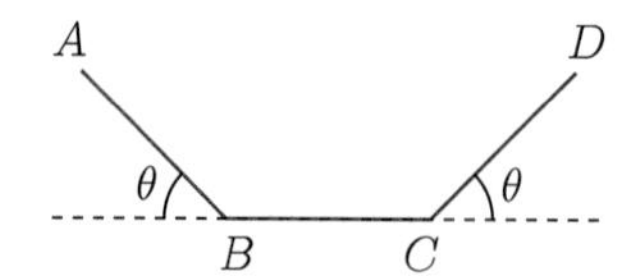

28. The strength of a rectangular beam varies as the product of its width and the square of its depth. Find the dimensions of the strongest rectangular beam that can be cut from a circular log of radius 25 cm.

29. At noon a ship S_1 is 20 km north of ship S_2. If S_1 sails south at 6 km/h, and S_2 east at 8 km/h, find when the two ships are closest together.

30. A military courier is located on a desert 6 km from a point P, which is the point on a long straight road nearest him. He is ordered to report to a point Q on the road. If we assume that he can travel at a rate of 14 km/h on the desert and a rate of 50 km/h on the road, find the point where he should reach the road in order to get to Q in the least possible time when:

(a) Q is 3 km from P;

(b) Q is 1 km from P.

31. Find the area of the largest rectangle that can be inscribed inside a circle of radius r.

32. Among all rectangles that can be inscribed inside the ellipse $b^2x^2 + a^2y^2 = a^2b^2$ and have sides parallel to the axes, find the one with largest area.

33. Among all rectangles that can be inscribed inside the ellipse $b^2x^2 + a^2y^2 = a^2b^2$ and have sides parallel to the axes, find the one with largest perimeter.

34. Find the area of the largest rectangle that has one side on the x-axis and two vertices on the curve $y = e^{-x^2}$.

35. Find the volume of the largest right circular cylinder that can be inscribed in a sphere of radius r.

36. Find the volume of the largest right circular cylinder that can be inscribed inside a right circular cone of radius r and height h.

37. Two corridors, one 3 m wide and the other 6 m wide, meet at right angles. Find the length of the longest beam that can be transported horizontally around the corner. Ignore the dimensions of the beam.

38. Repeat Exercise 37 taking into account that the beam has square cross-section $1/3$ m on each side.

39. A bee's cell is always constructed in the shape of a regular hexagonal cylinder open on one end and a trihedral apex at the other (Figure 4.60).

FIGURE 4.60

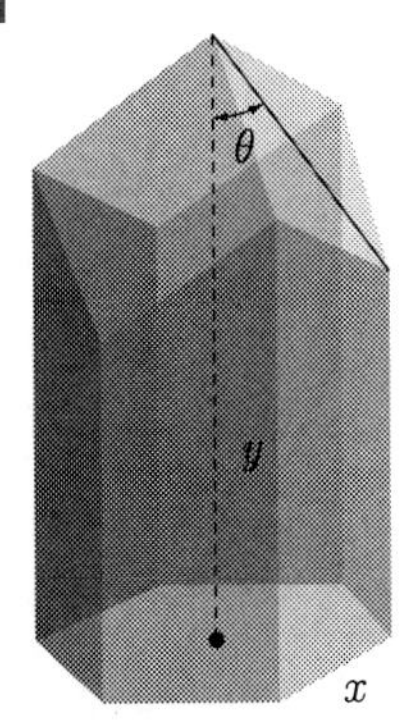

It can be shown that the total area of the nine faces is given by

$$A = 6xy + \frac{3}{2}x^2(\sqrt{3}\csc\theta - \cot\theta).$$

Find the angle that minimizes A.

40. A rental company buys a new machine for p dollars, which it then rents to customers. If the company keeps the machine for t years (before replacing it), the average replacement cost per year for the t years is p/t. During these t years, the company must make repairs on the machine, the number n depending on t as given by $n(t) = t^\alpha/\beta$, where $\alpha > 1$ and $\beta > 0$ are constants. If r is the average cost per repair, then the average maintenance cost per year over the life of the machine is nr/t. The total yearly expense associated with the machine if it is kept for t years is therefore

$$C(t) = \frac{p}{t} + \frac{nr}{t}.$$

Find the optimum time at which to replace the machine.

41. Discuss Exercise 40 in the cases that $\alpha = 1$ and $0 < \alpha < 1$.

42. The illumination at a point is inversely proportional to the square of the distance from the light source and directly proportional to the intensity of the source. If two light sources of intensities I_1 and I_2 are a distance d apart, at what point on the line segment joining the sources is the sum of their illuminations a minimum — relative to all other points on the line segment?

43. A window is in the form of a rectangle surmounted by a semicircle with diameter equal to the width of the window. If the rectangle is of clear glass while the semicircle is of coloured glass that transmits only half as much light per unit area as the clear glass, and if the total perimeter is fixed, find the proportions of the rectangular and semicircular part of the window that admit the most light.

44. A company wishes to construct a storage tank in the form of a rectangular parallelepiped with a square horizontal cross-section. The volume of the tank must be 100 m^3.

(a) If material for the sides and top costs \$1.25 per square metre, and material for the bottom costs \$4.75 per square metre, find the dimensions that minimize material costs.

(b) Repeat (a) if the 12 edges must be welded at a cost of \$7.50 per metre of weld.

45. When an unloaded die is thrown, there is a probability of $1/6$ that it will come up "two". If the die is loaded, on the other hand, the probability that a "two" will appear is not $1/6$, but is some number p between zero and one. To find p we could roll the die a large number of times, say n, and count the number of times "two" appears, say, m. It seems reasonable that an estimate for p is m/n. Mathematicians define a likelihood function, which for the present situation turns out to be

$$f(x) = \frac{n!}{m!(n-m)!} x^m (1-x)^{n-m}.$$

The value of x that maximizes $f(x)$ on the interval $0 \le x \le 1$ is called the maximum likelihood estimate of p. Show that this estimate is m/n.

46. An underground pipeline is to be constructed between two cities A and B (Figure 4.61).

FIGURE 4.61

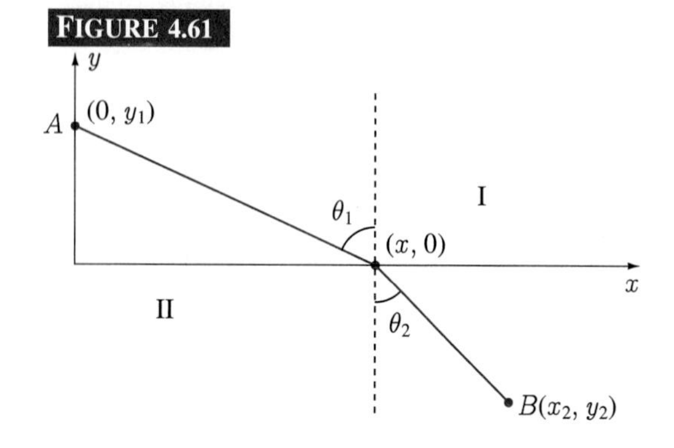

An analysis of the substructure indicates that construction costs per kilometre in regions I($y > 0$) and II($y < 0$) are c_1 and c_2 respectively. Show that the total construction cost is minimized when x is chosen so that

$$c_1 \sin\theta_1 = c_2 \sin\theta_2 .$$

47. A submarine is sailing on the surface due east at a rate of s km/h. It is to pass 1 km north of a point of land on an island at midnight. Soldiers on the island wishing to escape the enemy plan to intercept the submarine by rowing a rubber raft in a straight-line course at a rate of v km/h ($v < s$). What is the last instant that they can leave the island and expect to make contact with the submarine?

48. A packing company wishes to form the 1 m by 2 m piece of cardboard in Figure 4.62(a) into a box as shown in Figure 4.62(b). Cuts are to be made along solid lines and folds along dotted lines, and two sides are to be taped together as shown. If the outer flaps on top and bottom must meet in the centre but the inner flaps need not, find the dimensions of the box holding the most volume. How far apart will the inner flaps be?

FIGURE 4.62

(a)

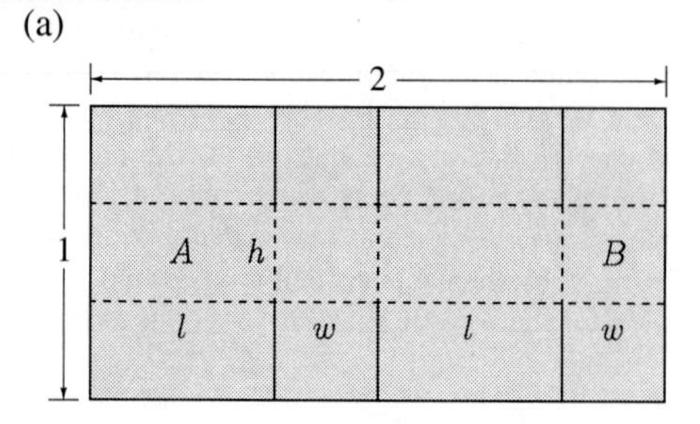

(b)

Bottom is same as top

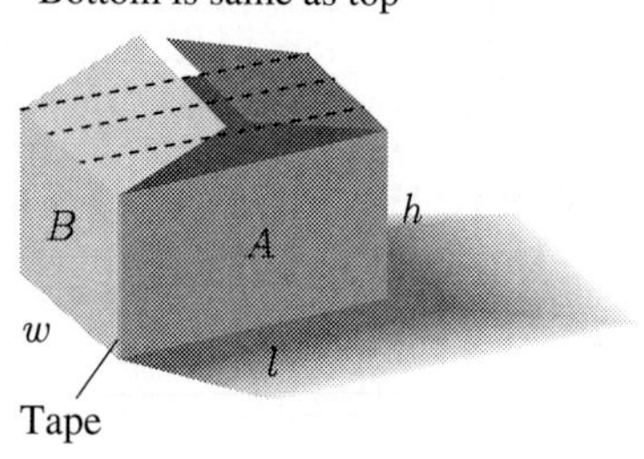

49. The cost of fuel per hour for running a ship varies directly as the cube of the speed, and is \$$B$ per hour when the speed is b kilometres per hour. There are also fixed costs of \$$A$ per hour. Find the most economical speed at which to make a trip.

50. A paper drinking cup in the form of a right-circular cone can be made from a circular piece of paper by removing a sector and joining edges OA and OB, as shown in Figure 4.63. If the radius of the circle is R, what choice of θ yields a cup of maximum volume?

FIGURE 4.63

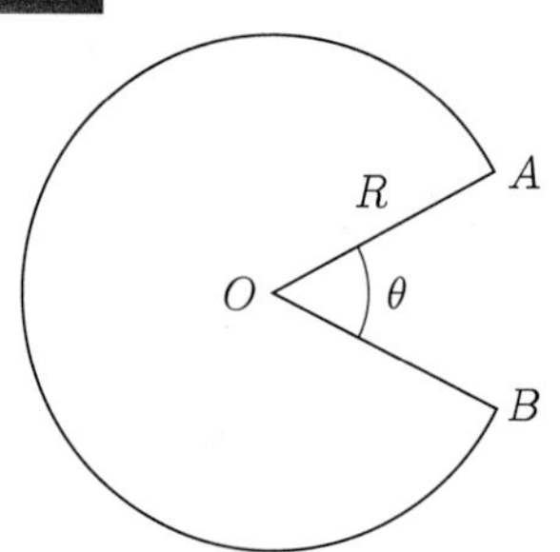

51. A truck driver estimates that the cost per kilometre of operating his rig is $40 + (50 - v)^2/10$ cents, where v is his average speed in kilometres per hour. In order to cover this cost he charges \$20 per hour for rental of his truck. For his driving he charges \$10 per hour. For a 1000 km nonstop haul, what average speed would you recommend to the driver?

52. Prove that the minimum distance from a point (x_1, y_1) to a line $Ax + By + C = 0$ is

$$\frac{|Ax_1 + By_1 + C|}{\sqrt{A^2 + B^2}}.$$

53. A window is in the form of a rectangle surmounted by a semicircle with diameter equal to the width of the rectangle. The rectangle is of clear glass costing \$$a$ per unit area, while the semicircle is of coloured glass costing \$$b$ per unit area. The coloured glass transmits only a fraction p ($0 < p < 1$) as much light per unit area as the clear glass. In addition, the curved portion of the window is surmounted by a special frame at a cost of \$$c$ per unit length. If the total cost of the window must not exceed \$$A$, find the dimensions of the window that admit the most light.

54. Find the point on the ellipse $4x^2 + 9y^2 = 36$ closest to $(4, 13\sqrt{5}/6)$.

55. The frame for a kite is to be made from six pieces of wood as shown in Figure 4.64. The four outside pieces have predetermined lengths a and b ($b > a$). Yet to be cut are the two diagonal pieces. How long should they be in order to make the area of the kite as large as possible?

FIGURE 4.64

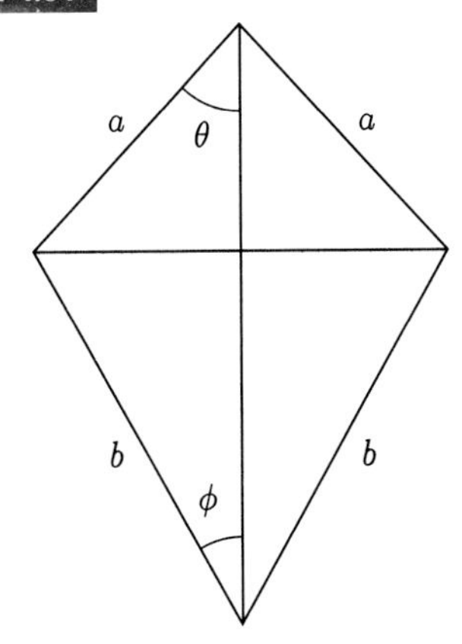

56. Suppose that the line $Ax + By + C = 0$ intersects the parabola $y = ax^2 + bx + c$ at points P and Q (Figure 4.65). Find the point R on the parabola between P and Q that maximizes the area of triangle PQR.

FIGURE 4.65

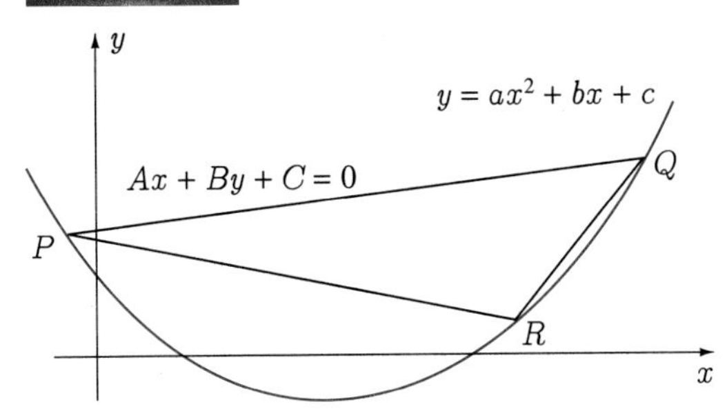

57. There are n red stakes securely driven into the ground in a straight line. A blue stake is to be added to the line, and each red stake is to be joined to the blue one by a string. Where should the blue stake be placed in order that the total length of all strings be as small as possible?

58. When blood flows through a vein or artery, it encounters resistance due to friction with the walls of the blood vessel and the viscosity of the blood itself. *Poiseuille's law* for laminar blood flow states that for a circular vessel, resistance R to blood flow is proportional to the length L of the blood vessel and inversely proportional to the fourth power of its radius r:

$$R = k\frac{L}{r^4},$$

where k is a constant. Figure 4.66 shows a blood vessel of radius r_1 from A to B and a branching vessel from D to C of radius $r_2 < r_1$.

FIGURE 4.66

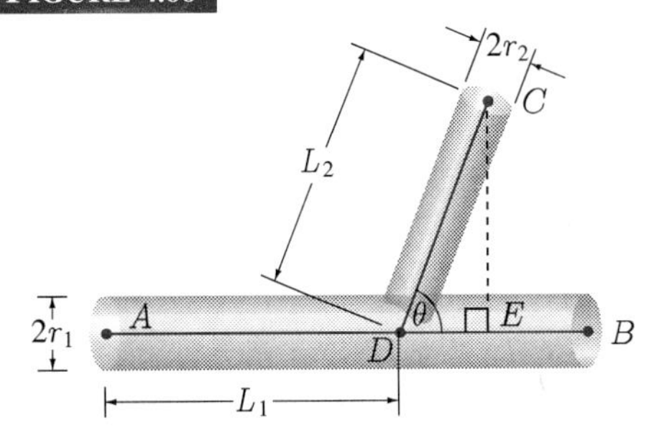

The resistance encountered by the blood in flowing from A to C is given by

$$R = k\frac{L_1}{r_1^4} + k\frac{L_2}{r_2^4},$$

where L_1 and L_2 are the lengths of AD and DC respectively.

(a) If B is assumed to the right of E, show that R can be expressed in terms of θ as

$$R = f(\theta) = \frac{k}{r_1^4}(X - Y\cot\theta) + \frac{k}{r_2^4}Y\csc\theta,$$

where X and Y are the lengths of AE and CE respectively.

(b) Show that $f(\theta)$ has only one critical point $\overline{\theta}$ in the range $0 < \theta < \pi/2$, and $\overline{\theta}$ is defined by

$$\cos\overline{\theta} = \frac{r_2^4}{r_1^4}.$$

(c) Verify that $\overline{\theta}$ yields a relative minimum for $f(\theta)$.

(d) Show that

$$f(\overline{\theta}) = \frac{kX}{r_1^4} + \frac{kY}{r_2^4}\sqrt{1 - \left(\frac{r_2}{r_1}\right)^8} < f(\pi/2).$$

(e) Does $\overline{\theta}$ provide an absolute minimum for $f(\theta)$?

59. If the fencing in Example 4.14 is to form the arc of a circle, what is the maximum possible area?

60. The upper left corner of a piece of paper a units wide and b units long ($b > a$) is folded to the right edge as shown in Figure 4.67. Calculate length x in order that the length y of the fold be a minimum.

FIGURE 4.67

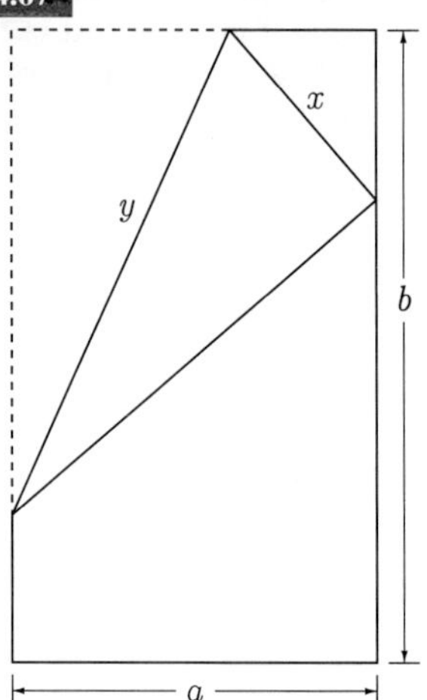

SECTION 4.7

Velocity and Acceleration

One of the most important applications of calculus and one that can be traced back to its founder, Sir Isaac Newton, is the field of kinematics — the study of the relationships among displacement, velocity, and acceleration. In this section we treat kinematics of one-dimensional motion, but do so only from a differentiation point of view; given the position of a particle moving along a straight line, find its velocity and acceleration and use these quantities to describe the motion of the particle. In Section 5.2 we reverse these operations; beginning with the acceleration, we find velocity and position. This process is essential to physics and engineering, where acceleration of a particle is determined by the forces acting on it.

In Section 3.1 we indicated that when the position of a particle moving along the x-axis is known as a function of time t, say, $x(t)$, its **instantaneous velocity** is the derivative of $x(t)$ with respect to t,

$$v(t) = \frac{dx}{dt}. \tag{4.14}$$

The **instantaneous acceleration** of the particle is defined as the rate of change of velocity with respect to time:

$$a(t) = \frac{dv}{dt} = \frac{d^2x}{dt^2}. \tag{4.15}$$

In actual fact, $x(t)$, $v(t)$, and $a(t)$ are the x-components of the displacement, velocity, and acceleration vectors, respectively. In the absence of a complete discussion of vectors, we omit the terms "vector" and "component", and simply call $x(t)$, $v(t)$, and $a(t)$ displacement, velocity, and acceleration.

When distance is measured in metres and time in seconds, velocity is measured in metres per second (m/s). Since acceleration is the time derivative of velocity, its units must be units of velocity divided by units of time, i.e., metres per second per second (m/s/s). Usually we shorten this by saying "metres per second squared," and write m/s^2.

Suppose that the position of a particle moving along the x-axis is

$$x(t) = t^3 - 27t^2 + 168t + 46, \quad t \geq 0,$$

where x is measured in metres and t in seconds, and we wish to describe the motion of the particle. The best way to do this is to draw a graph of the displacement function. With calculus we can then show how geometric properties of the graph reflect important features about the velocity and acceleration of the particle. The velocity and acceleration of the particle are quickly calculated to be

$$v(t) = \frac{dx}{dt} = 3t^2 - 54t + 168 = 3(t-4)(t-14) \quad \text{m/s},$$

$$a(t) = \frac{d^2x}{dt^2} = 6t - 54 = 6(t-9) \quad \text{m/s}^2.$$

Ignoring the physical interpretations of dx/dt and d^2x/dt^2 as velocity and acceleration for the moment, and concentrating only on the fact that they are the first and second derivatives of the function $x(t)$, we immediately find that $x(t)$ has a relative maximum of $x(4) = 350$ and a relative minimum of $x(14) = -150$ at the critical points $t = 4$ and $t = 14$. Add to this a point of inflection at $(9, 100)$ and we obtain the graph of $x(t)$ in Figure 4.68.

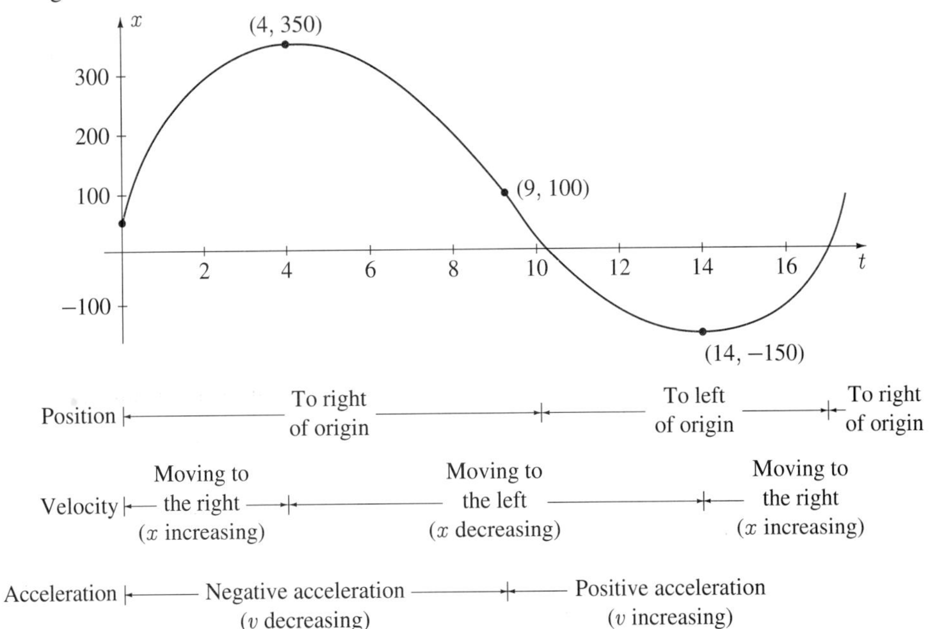

FIGURE 4.68

Let us now discuss what this graph tells us about the motion of the particle. Ordinates of this graph represent horizontal distances of the particle from the origin $x = 0$. When an ordinate is positive, the particle is that distance to the right of the origin; when an ordinate is negative, the particle is that distance to the left of the origin. For instance, at time $t = 0$, we calculate $x = 46$ and therefore the particle begins 46 m to the right of $x = 0$. At time $t = 4$, it is 350 m to the right of $x = 0$, and at $t = 14$, it is 150 m to the left of the origin.

The slope of the graph represents velocity of the particle. When slope is positive, namely, in the intervals $0 < t < 4$ and $t > 14$, the particle is moving to the right along the x-axis; for $4 < t < 14$, velocity is negative, indicating that the particle is moving to the left. At times $t = 4$ and $t = 14$, the particle is instantaneously at rest.

The concavity of the graph reflects the sign of the acceleration. For $0 < t < 9$, the graph is concave downward, its slope is decreasing. Physically this means that it has a negative acceleration; that is, its velocity is decreasing. For $t > 9$, the graph is concave upward, its slope is increasing. Physically, acceleration is positive; that is, velocity is increasing. At the point of inflection $(9, 100)$, the acceleration changes sign. Notice that the acceleration is not zero at $t = 4$ and $t = 14$. In spite of the fact that the velocity is zero at these times, the acceleration does not vanish. You might ask yourself what feature of the graph would reflect coincident zeros for velocity and acceleration (see Exercise 21).

Speed is defined to be the magnitude of velocity. It represents how fast the particle is moving without regard for direction. For instance, at time $t = 0$, the velocity and speed are both 168 m/s, whereas at time $t = 10$, velocity is -72 m/s and speed is 72 m/s. Geometrically, speed is represented by the slope of the graph without regard for sign. Do not confuse velocity and speed; they are closely related, but they are not interchangeable.

With the above ideas in mind, let us detail the history of the particle's motion. At time $t = 0$, it begins 46 m to the right of the origin, moving to the right with velocity (or speed) 168 m/s. Since the acceleration is negative (to the left), the particle is slowing down (both velocity and speed are decreasing), until at time $t = 4$ s, it comes to an instantaneous stop 350 m to the right of the origin. Because the acceleration continues to be negative, the particle moves to the left, its velocity decreasing, but its speed increasing. At time $t = 9$ s, when the particle is 100 m to the right of the origin, the acceleration changes sign. At this instant, the velocity has attained a (relative) minimum value, but speed is a (relative) maximum. With acceleration to the right for $t > 9$ s, the particle continues to move left, but slows down, its velocity increasing until at time $t = 14$ s, when it once again comes to a stop 150 m to the left of the origin. For time $t > 14$ s, it moves to the right, picking up speed, and passes through the origin just before $t = 17$ s.

Further analysis of the interdependences of displacement, velocity, and acceleration is contained in the following example.

EXAMPLE 4.19

The position of a particle moving along the x-axis is given by the function

$$x(t) = 3t^4 - 32t^3 + 114t^2 - 144t + 40, \quad 0 \le t \le 5,$$

where x is measured in metres and t in seconds. Answer the following questions concerning its motion:

(a) What is its velocity and speed at $t = 1/2$ s?

(b) When is its acceleration increasing?

(c) Is the velocity increasing or decreasing at $t = 2$ s?

(d) Is the particle speeding up or slowing down at $t = 2$ s?

(e) What is the maximum velocity of the particle in the time interval $0 \le t \le 2$?

(f) What is the maximum distance the particle ever attains from the origin?

SOLUTION Taking into consideration that there are six parts to this example, which examine much of the particle's motion, a graph of the displacement function could be very helpful. However, had only one or two parts of the example been required, a graphical approach might not have been particularly efficient. Let us take the approach here that each question had been asked in isolation, and that we shall find answers using the displacement, velocity and acceleration functions. The velocity and acceleration are

$$v(t) = 12t^3 - 96t^2 + 228t - 144 = 12(t-1)(t-3)(t-4) \quad \text{m/s},$$
$$a(t) = 12(3t^2 - 16t + 19) \quad \text{m/s}^2.$$

(a) The velocity at $t = 1/2$ is $v(1/2) = -105/2$ m/s, and the speed is $105/2$ m/s.

(b) The acceleration is increasing when its derivative is nonnegative. Since $da/dt = 12(6t - 16) = 24(3t - 8)$, acceleration is increasing for $8/3 \le t \le 5$.

(c) Since $a(2) = -12$ m/s^2, the velocity is decreasing at $t = 2$ s.

(d) Since $v(2) = 24$ m/s, the particle is moving to the right. But according to part (c), its acceleration is to the left at this time. As a result, it must be slowing down.

(e) To maximize $v(t)$ on $0 \le t \le 2$, we first find critical points of $v(t)$; that is, solve for zeros of the acceleration

$$0 = 12(3t^2 - 16t + 19).$$

Of the two solutions $(8 \pm \sqrt{7})/3$, only $(8 - \sqrt{7})/3$ is in the given interval $0 \le t \le 2$. Since

$$v(0) = -144, \qquad v\left((8 - \sqrt{7})/3\right) = 25.35, \qquad v(2) = 24,$$

maximum velocity is 25.35 m/s occuring at $t = (8 - \sqrt{7})/3$ s.

(f) Critical points for $x(t)$ occur at $t = 1$, 3, and 4, where velocity is zero. Since

$$x(0) = 40, \quad x(1) = -19, \quad x(3) = 13, \quad x(4) = 8, \quad x(5) = 45,$$

maximum distance from the origin is 45 m. ■

EXERCISES 4.7

1. Figure 4.69 shows the graph of the displacement function $x(t)$ of a particle moving along the x-axis during the time interval $0 \le t \le 6$. Answer the following questions, estimating values from the graph when necessary.

FIGURE 4.69

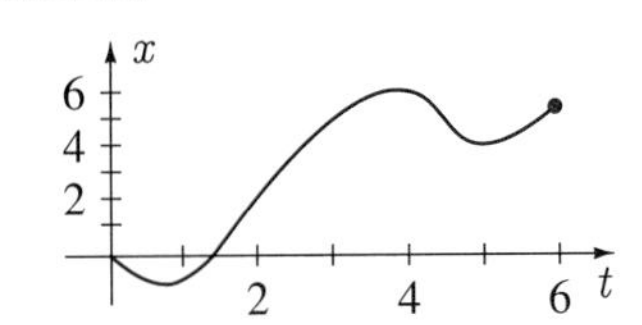

(a) Is the particle to the left or right of the origin at times $t = 1$ and $t = 4$?

(b) By drawing a horizontal line at $x = 2$ determine whether the particle is more often to the right or to the left of the point $x = 2$.

(c) What is the maximum distance from the origin achieved by the particle?

(d) What is the maximum distance from the point $x = 3$ achieved by the particle?

(e) Is the particle moving to the right or to the left at times $t = 1/2$ and $t = 3$?

(f) How many times does the particle change direction?

(g) Is the velocity greater at $t = 7/2$ or at $t = 9/2$?

(h) Is the speed greater at $t = 7/2$ or at $t = 9/2$?

(i) Is the acceleration positive or negative at $t = 1$ and $t = 4$?

(j) How many times does the acceleration change sign?

2. Repeat Exercise 1 for the graph in Figure 4.70.

FIGURE 4.70

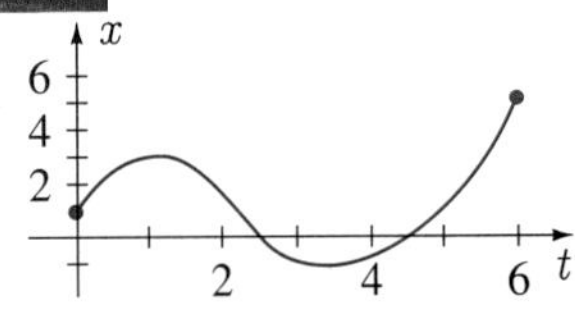

In Exercises 3–12 find the velocity and acceleration of an object that moves along the x-axis with the given position function. In each exercise, discuss the motion, including in your discussion a graph of the function $x(t)$. Assume that x is measured in metres and t in seconds.

3. $x(t) = 2t + 5, \quad t \ge 5$

4. $x(t) = t^2 - 7t + 6, \quad t \ge 0$

5. $x(t) = t^2 + 5t + 10, \quad t \ge 1$

6. $x(t) = -2t^3 + 2t^2 + 16t - 1, \quad t \ge 0$

7. $x(t) = t^3 - 9t^2 + 15t + 3, \quad t \geq 2$

8. $x(t) = 3\cos 4t, \quad t \geq 0$

9. $x(t) = 1/t, \quad t \geq 1$

10. $x(t) = t + 4/t, \quad t \geq 1$

11. $x(t) = (t-4)/t^2, \quad t \geq 2$

12. $x(t) = t^2\sqrt{1-t}, \quad 0 \leq t \leq 1$

13. An object moving along the x-axis has position given by

$$x(t) = t^3 - 9t^2 + 24t + 1,$$

where x is measured in metres and $t \geq 0$ is time in seconds. Determine at $t = 1$ s:

(a) the position of the object;

(b) whether the object is moving to the right or to the left;

(c) the acceleration of the object;

(d) whether the object's speed is increasing or decreasing.

14. The position of a particle moving along the x-axis is measured in metres to be

$$x(t) = t^3 - 9t^2 + 15t - 2,$$

where $t \geq 0$ is time in seconds.

(a) What is the velocity and acceleration of the particle?

(b) Is the particle ever instantaneously at rest?

(c) Is the particle speeding up or slowing down at $t = 4$ s?

(d) When is the velocity equal to -10 m/s?

(e) Is the particle's acceleration increasing or decreasing at $t = 3$ s?

15. A particle moves along the x-axis in such a way that its position as a function of time t is given by

$$x(t) = t\sin t, \quad 0 \leq t \leq 9.$$

Determine graphically each of the following.

(a) The number of times the particle reverses direction.

(b) The number of times the particle has zero acceleration.

(c) The direction of motion at time $t = 6$.

16. Can the position curve $x = x(t)$ of a realistic particle moving along the x-axis be represented by a function $x(t)$ that has a discontinuity? Explain.

In Exercises 17–20 we have defined the position function of an object moving along the x-axis. Sketch graphs of the position, velocity, and acceleration functions. Also draw a graph of speed as a function of time.

17. $x(t) = t^2 + 2t - 15, \quad t \geq 0$

18. $x(t) = t^3 - 3t, \quad t \geq 0$

19. $x(t) = t + 4/t, \quad t \geq 1$

20. $x(t) = 5\sin t, \quad t \geq 0$

21. What feature on the displacement graph would indicate a time when the velocity and acceleration are simultaneously zero?

22. Are critical points for the velocity function the same as those for the speed function? Explain using the graphs from Exercise 20.

23. What is the maximum speed of the particle in Example 4.19 over the time interval $0 \leq t \leq 2$?

24. In many velocity and acceleration problems it is more convenient to express acceleration in terms of a derivative with respect to position, as opposed to a derivative with respect to time. Show that acceleration can be written in the form

$$a = v\frac{dv}{dx}.$$

25. When an object moves with constant acceleration a along the x-axis, its position as a function of time t must be of the form

$$x = x(t) = \frac{1}{2}at^2 + bt + c,$$

where b and c are constants.

(a) If the object is at positions x_1 and x_2 at times t_1 and t_2, what is its average velocity over the time interval $t_1 \leq t \leq t_2$?

(b) At what time in this time interval is the instantaneous velocity equal to the average velocity? Where is the object at this time? Is it at the midpoint of the interval between x_1 and x_2, is it closer to x_1, or is it closer to x_2? Assume in this part of the problem that $a > 0$ and that the velocity of the object is positive at time t_1.

In Exercises 26–29 assume that a particle moves along the x-axis in such a way that its position, velocity, and acceleration are continuous functions on the interval $a \leq t \leq b$. Discuss the validity of each statement.

26. When position has a relative maximum so also does the absolute value of the distance from the origin to the particle.

27. When position has a relative minimum so also does the absolute value of the distance from the origin to the particle.

28. When velocity has a relative minimum so does speed.

29. When velocity has a relative maximum so does speed.

SECTION 4.8

Related Rates

Many interesting and practical problems involving rates of change are commonly referred to as *related rate problems*. In these problems, two or more quantities are related to each other and rates at which some of them change are known. It is required to find rates at which the others change. Related rate problems deal almost exclusively with rates of change of quantities with respect to time. To solve these problems we first consider three examples. These will suggest the general procedures by which all related rate problems can be analyzed. We shall then discuss two somewhat more complicated problems.

EXAMPLE 4.20

A man 2 m tall walks directly away from a streetlight that is 8 m high at the rate of 3/2 m/s. How fast is the length of his shadow changing?

FIGURE 4.71

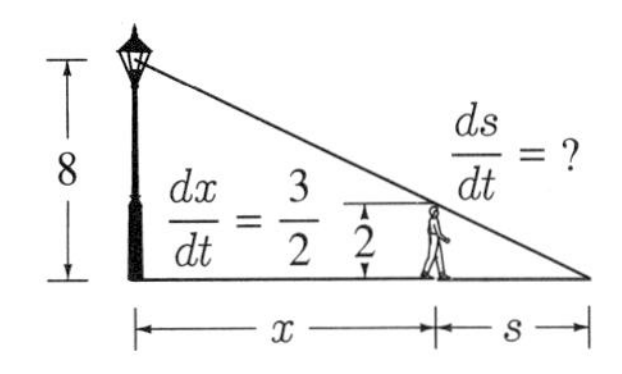

SOLUTION When x denotes the distance between the man and the lightpost (Figure 4.71), the fact that he walks directly away from the light at 3/2 m/s means that x is changing at a rate of 3/2 m/s; that is, $dx/dt = 3/2$ m/s. If s represents the length of the man's shadow, then we are searching for ds/dt. Similar triangles in Figure 4.71 enable us to relate s and x,

$$\frac{x+s}{s} = \frac{8}{2},$$

and this equation can be solved for s in terms of x:

$$s = \frac{x}{3}.$$

Now, s and x are each functions of time t,

$$s = f(t) \qquad \text{and} \qquad x = g(t),$$

although we have not calculated the exact form of these functions. Indeed, the essence of the related rate problem is to find ds/dt without ever knowing $f(t)$ explicitly. To do this we note that since the equation $s = x/3$ is valid at any time t when the man is walking away from the light, we may differentiate with respect to t to obtain

$$\frac{ds}{dt} = \frac{1}{3}\frac{dx}{dt}.$$

This equation relates the known rate $dx/dt = 3/2$ with the unknown rate ds/dt. It follows that

$$\frac{ds}{dt} = \left(\frac{1}{3}\right)\left(\frac{3}{2}\right) = \frac{1}{2},$$

and the man's shadow is therefore getting longer at the rate of 1/2 m/s. ■

Knowing dx/dt in this example, we have calculated ds/dt, and have done so without finding s explicitly as a function of time t. This is the essence of a related rate problem. Since dx/dt is a constant value, it is quite easy to find s as a function of t, and hence ds/dt. Indeed, if we choose time $t = 0$ when the man starts to walk away from the streetlight, then his distance from the light at any given time is $x = 3t/2$ metres. Combine this with the fact that $s = x/3$ and we may write

$$s = \frac{1}{3}\left(\frac{3t}{2}\right) = \frac{t}{2} \text{ m.}$$

With this explicit formula for s, it is clear that $ds/dt = 1/2$. What is important to realize is that the solution in this paragraph is possible only because the man walks at a constant rate. Were his speed not constant, it might be impossible to find s explicitly in terms of t. The next example illustrates this point in that the given rate is known only at one instant in time.

EXAMPLE 4.21

A ladder leaning against a house (Figure 4.72) is prevented from moving by a young child. Suddenly, something distracts the child and he releases the ladder. The ladder begins slipping down the wall of the house, picking up speed as it falls. If the top end of the ladder is moving at 1 m/s when the lower end is 15 m from the house, how fast is the foot of the ladder moving away from the house at this instant?

FIGURE 4.72

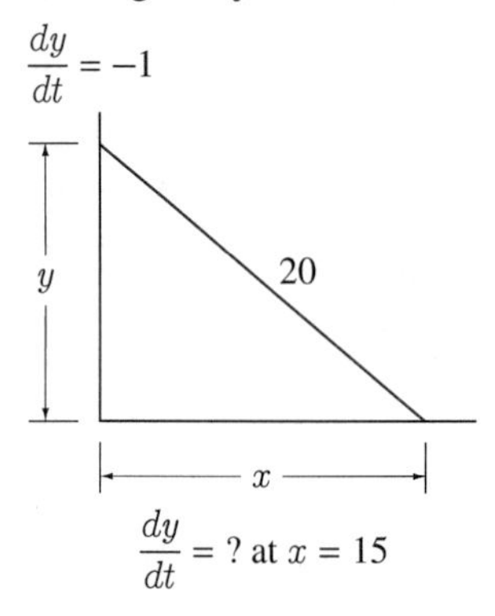

SOLUTION Figure 4.72 indicates that when y denotes the height of the top of the ladder above the ground, then $dy/dt = -1$ m/s when $x = 15$ m (the negative sign because y is decreasing). We emphasize here that $dy/dt = -1$ *only* when $x = 15$. What is required is dx/dt when $x = 15$ m. Because the triangle in the figure is right-angled, we may write

$$x^2 + y^2 = 20^2,$$

and this equation is valid at any time during which the ladder is slipping. If we differentiate with respect to time t, using extended power rule 3.17,

$$2x\frac{dx}{dt} + 2y\frac{dy}{dt} = 0.$$

When $x = 15$, we calculate that $y = \sqrt{400 - 225} = 5\sqrt{7}$, and therefore at this instant

$$15\frac{dx}{dt} + 5\sqrt{7}(-1) = 0.$$

This yields $dx/dt = \sqrt{7}/3$, and we can say that when the foot of the ladder is 15 m from the wall, it is moving away from the wall at $\sqrt{7}/3$ m/s. ■

EXAMPLE 4.22

A tank in the form of a right circular cone with altitude 6 m and base radius 3 m (Figure 4.73) is being filled with water at a rate of 4000 L/min. How fast is the surface of the water rising when the depth is 3 m?

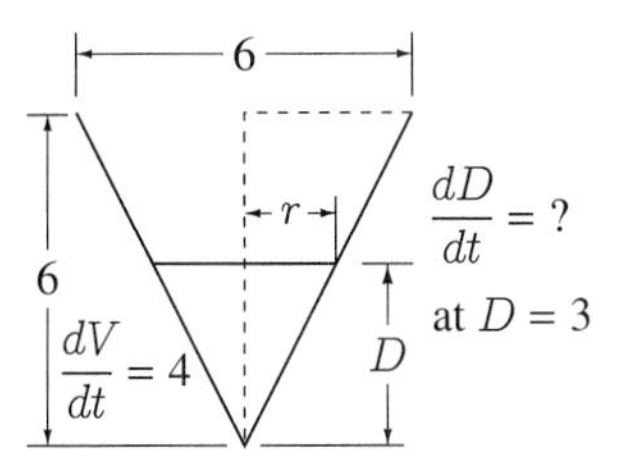

FIGURE 4.73

SOLUTION Figure 4.73, which illustrates a cross-section of the tank, indicates that when the depth of water in the tank is D, the volume V of water is

$$V = \frac{1}{3}\pi r^2 D.$$

Of the three variables V, r, and D in this equation, we are concerned only with V and D, since dV/dt is given and dD/dt is what we want. This suggests that we eliminate r using similar triangles. Since $r/D = 3/6$, we have $r = D/2$; therefore,

$$V = \frac{1}{3}\pi\left(\frac{D}{2}\right)^2 D = \frac{1}{12}\pi D^3.$$

Because this result is valid for all time t during the filling process, we can differentiate with respect to t, once again using extended power rule 3.17:

$$\frac{dV}{dt} = \frac{1}{4}\pi D^2 \frac{dD}{dt}.$$

Since $dV/dt = 4\ \mathrm{m}^3/\mathrm{min}$ which is converted from the rate of 4000 L/min since the litre is not an acceptable unit of measure for volume, we find that when $D = 3$,

$$4 = \frac{1}{4}\pi(3)^2\frac{dD}{dt},$$

from which $dD/dt = 16/(9\pi)$. The surface is therefore rising at a rate of $16/(9\pi)$ m/min. ■

Examples 4.20–4.22 illustrate the following general procedure for solving related-rate problems:

1. **Sketch** a diagram illustrating all given information, especially given rates of change and desired rates of change.
2. **Find** an equation valid for all time (in some interval about the instant in question) that involves only variables whose rates of change are given or required.
3. **Differentiate** the equation in 2 and solve for the required rate.

Steps 1 and 3 are usually quite straightforward; step 2, on the other hand, may tax your ingenuity. To find the equation in the appropriate variables, it may be necessary to introduce and substitute for additional variables. Finding these substitutions requires you to analyze the problem very closely.

Be careful not to substitute numerical data that represent the instant at which the derivative is required before differentiation has taken place. Numerical data must be substituted after differentiation. For instance, in Example 4.22, the radius r of the surface of the water when $D = 3$ is $3/2$. If we substitute this into $V = \pi r^2 D/3$, we obtain a function $V = \pi(3/2)^2 D/3 = 3\pi D/4$ which is valid only when $D = 3$. It cannot therefore be differentiated; only equations that are valid for a range of values of t can be differentiated with respect to t.

We now apply this procedure to two further examples. The first is an extension of Example 4.20.

EXAMPLE 4.23

A man 2 m tall walks along the edge of a straight road 10 m wide. On the other edge of the road stands a streetlight 8 m high. If the man walks at $3/2$ m/s, how fast is his shadow lengthening when he is 10 m from the point directly opposite the light?

FIGURE 4.74

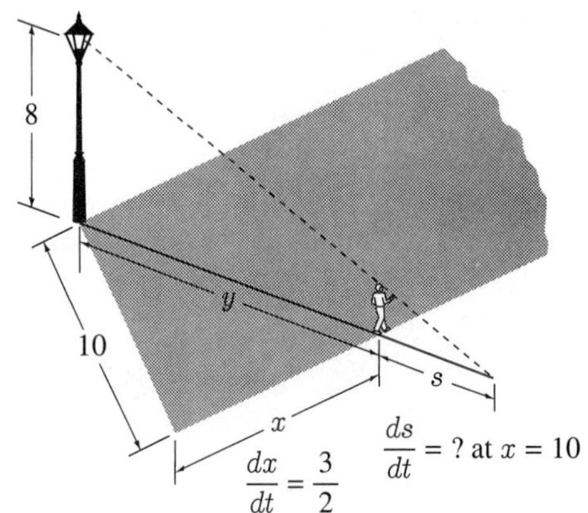

SOLUTION First we draw Figure 4.74 wherein the man's speed is represented as the time rate of change of his distance x from the point on his side of the road directly opposite the light. What is required is the rate of change ds/dt of the length of his shadow when $x = 10$. To find an equation relating x and s, we first use similar (vertical) triangles to write

$$\frac{y+s}{s} = \frac{8}{2} = 4,$$

from which

$$y = 3s,$$

an equation that relates s to y, rather than s to x. However, since

$$y^2 = x^2 + 100,$$

we substitute to obtain

$$9s^2 = x^2 + 100.$$

The derivative of this equation with respect to time t gives

$$18s\frac{ds}{dt} = 2x\frac{dx}{dt}.$$

When $x = 10$, we obtain $s = \sqrt{100+100}/3 = 10\sqrt{2}/3$, and, at this instant,

$$18\left(\frac{10\sqrt{2}}{3}\right)\frac{ds}{dt} = 2(10)\left(\frac{3}{2}\right).$$

Thus,

$$\frac{ds}{dt} = \frac{30}{60\sqrt{2}} = \frac{\sqrt{2}}{4},$$

and the man's shadow is therefore lengthening at the rate of $\sqrt{2}/4$ m/s. ■

It is worthwhile noting in this example that were we to solve for ds/dt before substituting $x = 10$ and $s = 10\sqrt{2}/3$, then we would have

$$\frac{ds}{dt} = \frac{x}{9s}\frac{dx}{dt}.$$

Since dx/dt is always equal to $3/2$, and $s = \sqrt{x^2 + 100}/3$, we can write

$$\frac{ds}{dt} = \frac{x}{3\sqrt{x^2 + 100}}\left(\frac{3}{2}\right) = \frac{x}{2\sqrt{x^2 + 100}},$$

a general formula for ds/dt. The limit of this rate as x becomes very large is

$$\lim_{x \to \infty} \frac{ds}{dt} = \lim_{x \to \infty} \frac{x}{2\sqrt{x^2 + 100}} = \frac{1}{2}.$$

But for very large x, the man essentially walks directly away from the light, and this answer, as we might expect, is identical to that in Example 4.20.

EXAMPLE 4.24

One end of a rope is tied to a box. The other end is passed over a pulley 5 m above the floor and tied at a level 1 m above the floor to the back of a truck. If the rope is taut and the truck moves at $1/2$ m/s, how fast is the box rising when the truck is 3 m from the plumb line through the pulley?

FIGURE 4.75

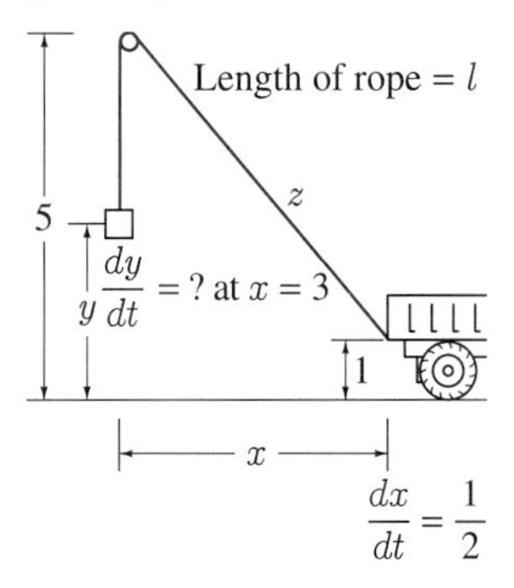

SOLUTION In Figure 4.75, we have represented the speed of the truck as the rate of change of length x, $dx/dt = 1/2$ m/s. What is required is dy/dt when $x = 3$. To find an equation relating x and y, we first use the fact that the length z of rope between pulley and truck is the hypotenuse of a right-angled triangle with sides of lengths x and 4,

$$z^2 = x^2 + 16.$$

For an equation relating y and z, we note that the length of the rope, call it L, remains constant, and is equal to the sum of z and $5 - y$,

$$L = z + (5 - y).$$

These two equations can be combined into

$$(L - 5 + y)^2 = x^2 + 16,$$

and differentiation with respect to time t now gives

$$2(L - 5 + y)\frac{dy}{dt} = 2x\frac{dx}{dt}.$$

When $x = 3$, we may write that

$$(L - 5 + y)^2 = 9 + 16 = 25.$$

We could solve this equation for y (in terms of L), but it is really not y that is needed to obtain dy/dt from the previous equation. It is $L - 5 + y$, and this is clearly equal to 5. Thus, when $x = 3$, we have

$$2(5)\frac{dy}{dt} = 2(3)\left(\frac{1}{2}\right);$$

that is, $dy/dt = 3/10$, and the box is rising at a rate of $3/10$ m/s. ■

The reader should compare this example with the problem in Exercise 2. They may appear similar, but are really quite different.

EXERCISES 4.8

1. A convertible is travelling along a straight highway at 100 km/h. A child in the car accidentally releases a helium-filled balloon, which then rises vertically at 10 m/s. How fast are the child and balloon separating 4 seconds after the balloon is released?

2. A rope passes over a pulley and one end is attached to a cart as shown in Figure 4.76. If the rope is pulled vertically downward at 2 m/s, how fast is the cart moving when $s = 6$ m?

FIGURE 4.76

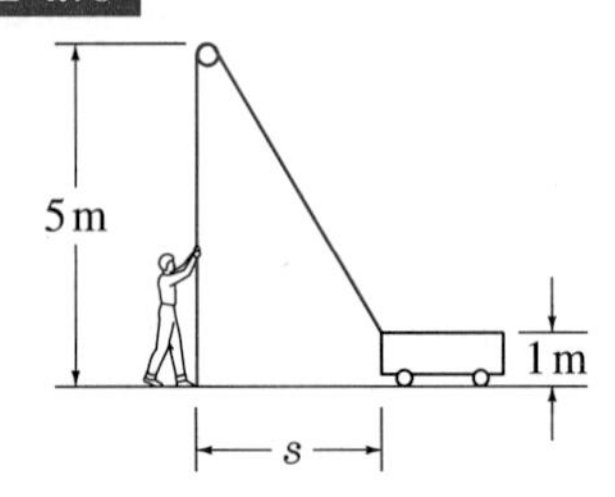

3. A light is on the ground 20 m from a building. A man 2 m tall walks from the light directly toward the building at 3 m/s. How fast is the length of his shadow on the building changing when he is 8 m from the building?

4. A funnel in the shape of a right circular cone is 15 cm across the top and 30 cm deep. A liquid is flowing in at the rate of 80 mL/s and flowing out at 15 mL/s. At what rate is the surface of the liquid rising when the liquid fills the funnel to a depth of 20 cm?

5. A water tank is in the form of a right circular cylinder of diameter 3 m and height 3 m on top of a right circular cone of diameter 3 m and height 1 m. If water is being drawn from the bottom at the rate of 1 L/min, how fast is the water level falling when

(a) it is 1 m from the top of the tank?

(b) it is 3.5 m from the top of the tank?

6. A point P moves along the curve $y = x^2 + x + 4$ where x and y are measured in metres. Its x-coordinate decreases at 2 m/s. If the perpendicular from P to the x-axis intersects this axis at point Q, how fast is the area of the triangle with vertices P, Q, and the origin changing when the x-coordinate of P is 2 m?

7. Water is being pumped into a swimming pool which is 10 m wide, 20 m long, 1 m deep at the shallow end, and 3 m deep at the deep end. If the water level is rising at 1 cm/min when the depth is 1 m at the deep end, at what rate is water being pumped into the pool?

8. Boyle's law for a perfect gas states that the pressure exerted by the gas on its containing vessel is inversely proportional to the volume occupied by the gas. If when the volume is 10 L and the pressure is 50 N/m^2, the volume is increasing at $1/2$ L/s, find the rate of change of the pressure of the gas.

9. A woman driving 100 km/h along a straight highway notes that the shadow of a cloud is keeping pace with her. What can she conclude about the speed of the cloud?

10. A fisherman is trolling at a rate of 2 m/s with his lure 100 m behind the boat and on the surface. Suddenly a fish strikes and dives vertically at a rate of 3 m/s. If the fisherman permits the line to run freely and it always remains straight, how fast is the line being played out when the reel is 50 m from its position at the time of the strike?

11. Air expands adiabatically in accordance with the law $PV^{7/5}$ = constant. If, at a given time, the volume V is 100 L and the pressure P is 40 N/cm^2, at what rate is the pressure changing when the volume is decreasing at 1 L/s?

12. Sand is poured into a right-circular cylinder of radius $1/2$ m along its axis (Figure 4.77). Once sand completely covers the bottom, a right circular cone is formed on the top.

FIGURE 4.77

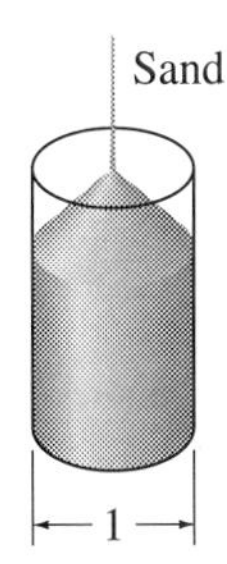

(a) If 0.02 m^3 of sand enters the container every minute, how fast is the top of the sand pile rising?

(b) How fast is the sand rising along the side of the cylinder?

13. A balloon has the shape of a right-circular cylinder of radius r and length l with a hemisphere at each end of radius r. The balloon is being filled at a rate of 10 mL/s in such a way that l increases twice as fast as r. Find the rate of change of r when $r = 8$ cm and $l = 20$ cm.

14. An oval racetrack has a straight stretch 100 m long and two semicircles, each of radius 50 m (Figure 4.78). Car 1, on the infield, moves along the x-axis from O to B. It accelerates from rest at O, attains a speed of 10 m/s at C, and maintains this speed along CB. Car 2 travels along the quarter oval $ADEB$. It is at D when Car 1 is at C, and between D and B, Car 2 maintains the same rate of change of its x-coordinate as does Car 1.

FIGURE 4.78

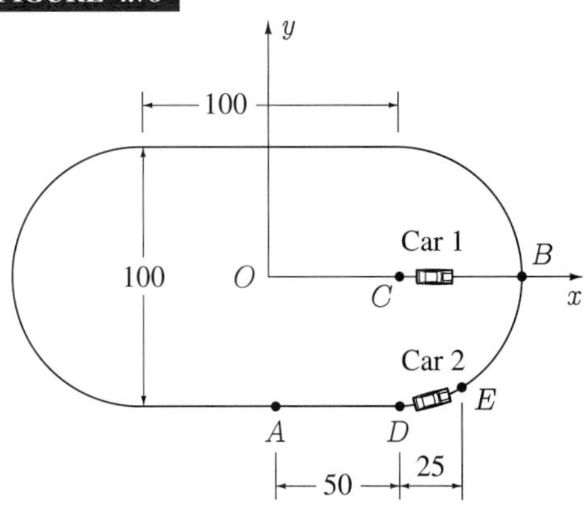

(a) Find a formula for the rate of change of the y-coordinate of Car 2 between D and B.

(b) How fast is the y-coordinate of Car 2 changing when it is at point E?

(c) If the cars collide at B, which car suffers the most damage?

15. A ship is 1 km north of a pier and is travelling N30°E at 3 km/h. A second ship is $3/4$ km east of the pier and is travelling east at 7 km/h. How fast are the ships separating?

16. The circle in Figure 4.79 represents a long-playing record which is rotating clockwise at $100/3$ revolutions per minute. A bug is walking away from the centre of the record directly toward point P on the rim of the record at 1 cm/s. When the bug is at position R, 10 cm from O, angle θ is $\pi/4$ radians. Find the rate at which the distance from the bug to the fixed point Q is changing when the bug is at R.

FIGURE 4.79

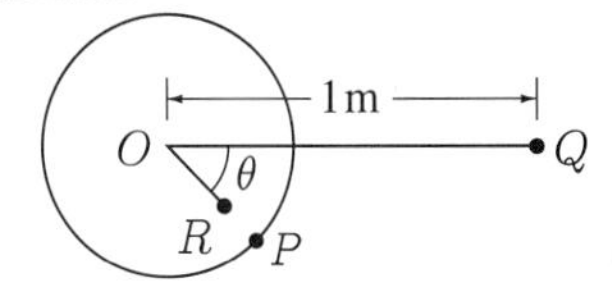

17. Eight skaters form a "whip". Show that the seventh person on the whip travels twice as fast as the fourth person.

18. A particle moves counterclockwise around a circle of radius 5 cm centred at the origin making 4 revolutions each second. How fast is the particle moving away from the point with coordinates $(5, 6)$ when it is at position $(-3, 4)$?

19. Two people (A and B in Figure 4.80) walk along opposite sides of a road 10 m wide. A walks to the right at 1 m/s, and B walks to the left at 2 m/s. A third person C walks along a sidewalk 5 m from the road in such a way that B is always on the line joining A and C. Find the speed of C.

FIGURE 4.80

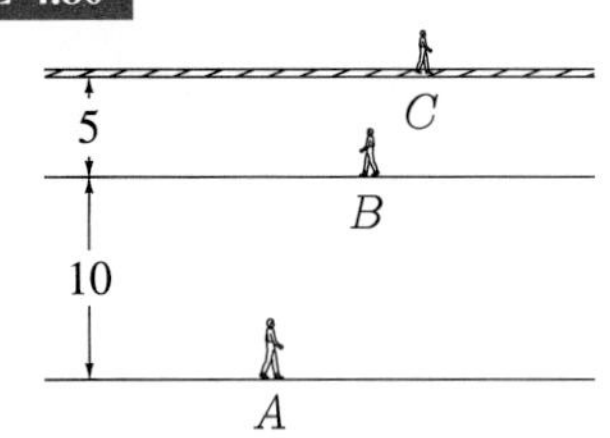

20. Let $P(x, y)$ be a point on the first quadrant portion of the hyperbola $x^2 - y^2 = 1$. Let R be the foot of the perpendicular from P to the x-axis, and $Q(x^*, 0)$ be the x-intercept of the normal line to the hyperbola at P.

(a) Show that $x^* = 2x$.

(b) If P moves along the hyperbola so that its x-coordinate is decreasing at 3 units per unit time, how fast is the area of triangle QPR changing when $x = 4$?

21. A solution passes from a conical filter 24 cm deep and 16 cm across the top into a cylindrical container of diameter 12 cm. When the depth of solution in the filter is 12 cm, its level is falling at the rate of 1 cm/min. How fast is the level of solution rising in the cylinder at this instant?

22. A light is at the top of a pole 25 m high and a ball is dropped at the same height from a point 10 m from the light. How fast is the shadow of the ball moving along the ground 1 second later? The distance fallen by the ball t seconds after it has been dropped is $d = 4.905t^2$ metres.

23. A point moves along the parabola $y = x^2 - 3x$ (x and y measured in metres) in such a way that its x-coordinate changes at the rate of 2 m/s. How fast is its distance from the point $(1, 2)$ changing when it is at $(4, 4)$?

24. Repeat Exercise 23 given that the parabola is replaced by the curve $(x + y)^2 = 16x$.

25. The volume of wood in the trunk of a tree is sometimes calculated by considering it as a frustrum of a right-circular cone (Figure 4.81).

(a) Verify that the volume of the trunk is

$$V = \frac{1}{3}\pi h(R^2 + rR + r^2).$$

(b) Suppose that at the present time the radii of the top and bottom are $r = 10$ and $R = 50$ cm, and the height is $h = 30$ m. If the tree continues to grow so that ratios r/R and r/h always remain the same as they are now, and R increases at a rate of $1/2$ cm/year, how fast will the volume be changing in 2 years?

FIGURE 4.81

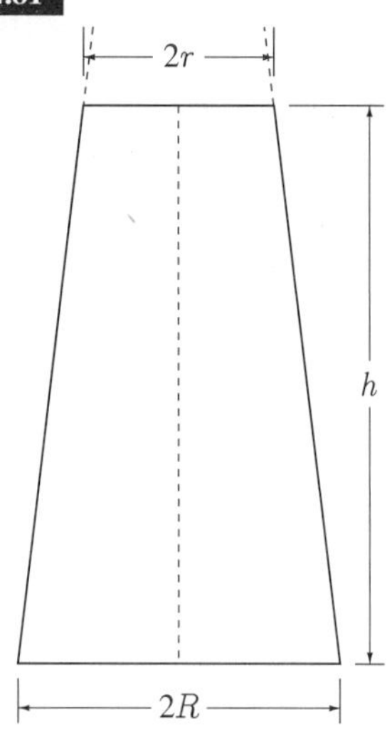

26. Sand is poured into a right-circular cone of radius 2 m and height 3 m along its axis (Figure 4.82).

FIGURE 4.82

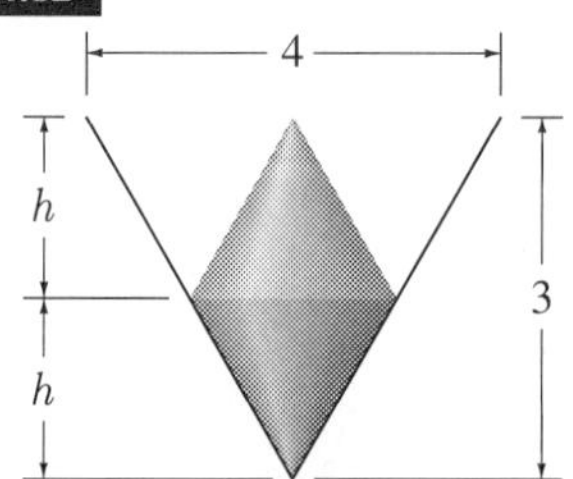

The sand forms two cones of equal height h, one inverted on top of the other.

(a) If 0.02 m^3 of sand enters the container every minute, how fast is the top of the pile rising when it is just level with the top of the container?

(b) How fast is the sand rising along the side of the container at this instant?

27. Figure 4.83 shows a rod OB rotating counterclockwise in the xy-plane about O at a rate of 2 revolutions per second. Attached to OB is a rod AB, and A is confined to sliding horizontally along the x-axis. We are given that OB has length l metres, and AB has length L metres ($L > 2l$).

FIGURE 4.83

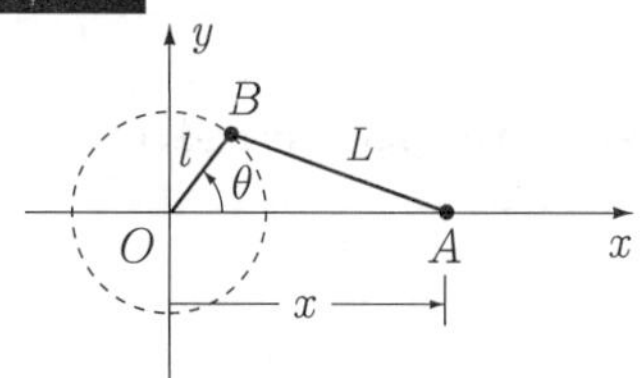

(a) Find the velocity of A as a function of x and θ.

(b) Find the acceleration of A.

28. If to the mechanism in Figure 4.83 we add a rod AC (Figure 4.84), where C is confined to sliding vertically, find the velocity of C in terms of x, θ, and y.

FIGURE 4.84

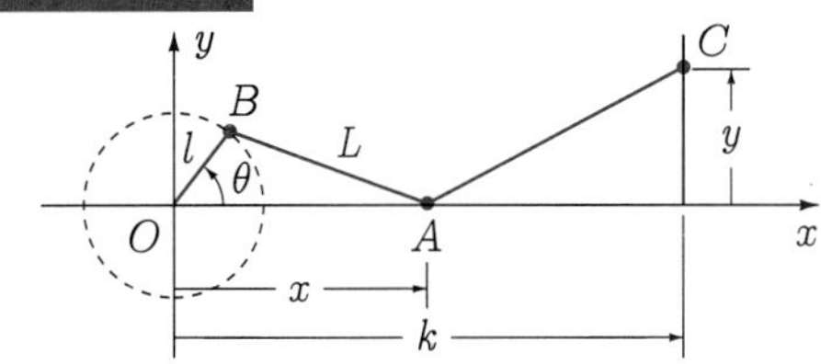

29. In Figure 4.85 the boy's feet make 1 revolution per second around a sprocket of radius R metres. The chain travels around a sprocket of radius r metres on the back wheel, which itself has radius $\overline{R}$ metres. If a stone imbedded in the tire becomes dislodged, how fast is it travelling when it leaves the tire?

FIGURE 4.85

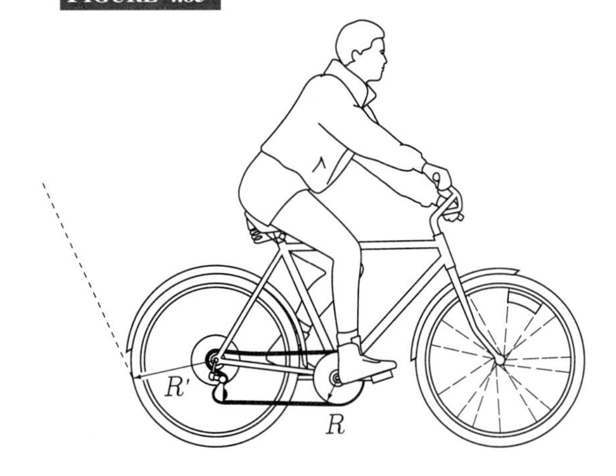

30. (a) If θ is the angle formed by the minute and hour hands of a clock, what is the time rate of change of θ (in radians per minute)?

(b) If the lengths of the hands on the clock are 10 cm and 7.5 cm, find the rate at which their tips approach each other at 3:00 (i.e., find dz/dt in Figure 4.86).

(c) Repeat (b) but replace time 3:00 with 8:05.

FIGURE 4.86

31. If the sides of a triangle have lengths a, b, and c, its area is given by

$$A = \sqrt{s(s-a)(s-b)(s-c)},$$

where $s = (a+b+c)/2$ is one-half its perimeter. If the length of each side increases at a rate of 1 cm/min, how fast is A changing when $a = 3$ cm, $b = 4$ cm, and $c = 5$ cm?

SECTION 4.9

Indeterminate Forms and L'Hôpital's Rule

Derivatives are instantaneous rates of change, defined as limits of average rates of change. As a result, it was necessary to discuss limits in Chapter 2 prior to the introduction of derivatives in Section 3.1. Now that we have derivatives, it may seem quite surprising that they can be used to evaluate certain limits.

The Indeterminate Form $0/0$

If we let x approach zero in numerator and denominator of the limit

$$\lim_{x \to 0^+} \frac{\sqrt{1+x}-1}{\sqrt{x}}, \qquad (4.16)$$

we find that

$$\lim_{x\to 0^+}\left(\sqrt{1+x}-1\right)=0 \qquad \text{and} \qquad \lim_{x\to 0^+}\sqrt{x}=0.$$

We say that limit 4.16 is of the **indeterminate form** $0/0$. Likewise, the limit

$$\lim_{x\to 2}\frac{x^3-x^2-8x+12}{x^2-4x+4} \tag{4.17}$$

is of the indeterminate form $0/0$ since both numerator and denominator approach zero as x approaches 2.

In Section 2.1 we evaluated limit 4.16 by rationalizing the numerator (see Example 2.6). Factoring numerator and denominator in limit 4.17 gives

$$\lim_{x\to 2}\frac{x^3-x^2-8x+12}{x^2-4x+4}=\lim_{x\to 2}\frac{(x-2)^2(x+3)}{(x-2)^2}=\lim_{x\to 2}(x+3)=5.$$

These two examples illustrate the "trickery" to which we resorted in Chapter 2 in order to evaluate limits. With Cauchy's generalized mean value theorem from Section 3.11, however, we can prove a result called L'Hôpital's rule, which makes evaluation of many limits of the indeterminate form $0/0$ quite simple.

Theorem 4.5

(L'Hôpital's rule) *Let $f(x)$ and $g(x)$ be functions that are differentiable on an open interval I except possibly at the point $x=a$ in I. Suppose further that $g'(x)\neq 0$ in I, except possibly at a. If $\lim_{x\to a} f(x)=0$ and $\lim_{x\to a} g(x)=0$, and if*

$$\lim_{x\to a}\frac{f'(x)}{g'(x)}=L,$$

then

$$\lim_{x\to a}\frac{f(x)}{g(x)}=L.$$

Proof We define two functions $F(x)$ and $G(x)$ that are identical to $f(x)$ and $g(x)$ on I but have value zero at $x=a$:

$$F(x)=\begin{cases} f(x) & x\neq a\\ 0 & x=a,\end{cases} \qquad G(x)=\begin{cases} g(x) & x\neq a\\ 0 & x=a.\end{cases}$$

Since $F'(x)=f'(x)$ and $G'(x)=g'(x)$ for x in I except possibly at $x=a$, $F(x)$ and $G(x)$ are therefore differentiable on I except possibly at $x=a$. In addition, differentiability of a function implies continuity of the function (Theorem 3.6), so that $F(x)$ and $G(x)$ must certainly be continuous on I except possibly at $x=a$. But

$$\lim_{x\to a}F(x)=\lim_{x\to a}f(x)=0=F(a),$$

and the same is true for $G(x)$; hence, $F(x)$ and $G(x)$ are continuous for all x in I. Consequently, $F(x)$ and $G(x)$ are identical to $f(x)$ and $g(x)$ in every respect, except that they have been assigned a value at $x=a$ to guarantee their continuity there. This extra condition permits us to apply Cauchy's generalized mean value theorem to $F(x)$ and $G(x)$ on the interval between a and x, so long as x is in I. We can conclude that there exists a number c between a and x such that

$$\frac{F(x)-F(a)}{G(x)-G(a)}=\frac{F'(c)}{G'(c)},$$

or, since $F(a) = G(a) = 0$,

$$\frac{F(x)}{G(x)} = \frac{F'(c)}{G'(c)}.$$

Since x and c are points in I, we can also write that $F(x) = f(x)$, $F'(c) = f'(c)$, $G(x) = g(x)$, $G'(c) = g'(c)$, and therefore

$$\frac{f(x)}{g(x)} = \frac{F(x)}{G(x)} = \frac{F'(c)}{G'(c)} = \frac{f'(c)}{g'(c)}.$$

If we now let x approach a, then c must also approach a since it is always between a and x. Consequently,

$$\lim_{x\to a}\frac{f(x)}{g(x)} = \lim_{c\to a}\frac{f'(c)}{g'(c)} = \lim_{x\to a}\frac{f'(x)}{g'(x)},$$

and if

$$L = \lim_{x\to a}\frac{f'(x)}{g'(x)},$$

it follows that

$$\lim_{x\to a}\frac{f(x)}{g(x)} = L.$$

This theorem is also valid if L is replaced by ∞ or $-\infty$. The only difference in the proof is to make the same change in the last sentence.

Theorem 4.5 is also valid if $x \to a$ is replaced by either a right-hand limit, $x \to a^+$, or a left-hand limit, $x \to a^-$. The only difference in these cases is that interval I is replaced by open intervals $a < x < b$ and $b < x < a$ respectively, and the proofs are almost identical. In addition, the following theorem indicates that $x \to a$ can be replaced by $x \to \infty$ (or $x \to -\infty$).

Theorem 4.6

(L'Hôpital's rule) *Let $f(x)$ and $g(x)$ be functions that are differentiable for some interval $x > b > 0$, and suppose that $g'(x) \neq 0$ on this interval. If $\lim_{x\to\infty} f(x) = 0$ and $\lim_{x\to\infty} g(x) = 0$, and if*

$$\lim_{x\to\infty}\frac{f'(x)}{g'(x)} = L \quad (\text{or } \pm\infty),$$

then

$$\lim_{x\to\infty}\frac{f(x)}{g(x)} = L \quad (\text{or } \pm\infty).$$

In other words, L'Hôpital's rule applies to any type of limit that yields the indeterminate form $0/0$ (be it $x \to a$, $x \to a^+$, $x \to a^-$, $x \to \infty$, or $x \to -\infty$). A common error when using L'Hôpital's rule is to differentiate $f(x)/g(x)$ with the quotient rule and then take the limit of the resulting derivative. L'Hôpital's rule calls for the limit of $f'(x)/g'(x)$; $f(x)$ and $g(x)$ are differentiated separately.

If we use L'Hôpital's rule on limit 4.16, we find

$$\lim_{x\to 0^+}\frac{\sqrt{1+x}-1}{\sqrt{x}} = \lim_{x\to 0^+}\frac{\dfrac{1}{2\sqrt{1+x}}}{\dfrac{1}{2\sqrt{x}}} = \lim_{x\to 0^+}\frac{\sqrt{x}}{\sqrt{1+x}} = 0.$$

For limit 4.17, we have

$$\lim_{x\to 2} \frac{x^3 - x^2 - 8x + 12}{x^2 - 4x + 4} = \lim_{x\to 2} \frac{3x^2 - 2x - 8}{2x - 4},$$

which is still a limit of the indeterminate form $0/0$. Note that this is a conditional equation; that is, it says that the limit on the left is equal to the limit on the right, provided the limit on the right exists. If we apply L'Hôpital's rule a second time, to the limit on the right, we obtain

$$\lim_{x\to 2} \frac{x^3 - x^2 - 8x + 12}{x^2 - 4x + 4} = \lim_{x\to 2} \frac{6x - 2}{2} = 5.$$

EXAMPLE 4.25

Evaluate the following limits:

(a) $\displaystyle\lim_{x\to -3} \frac{3 + x}{\sqrt{3} - \sqrt{-x}}$ (b) $\displaystyle\lim_{x\to 0} \frac{\tan x}{x}$

(c) $\displaystyle\lim_{x\to 4} \frac{x - 4}{x^2 - 8x + 16}$ (d) $\displaystyle\lim_{x\to 4} \frac{x^3 - 4x^2 + 9x - 36}{x^2 + 5}$

SOLUTION (a) Since we have the indeterminate form $0/0$, we use L'Hôpital's rule to write

$$\lim_{x\to -3} \frac{3 + x}{\sqrt{3} - \sqrt{-x}} = \lim_{x\to -3} \frac{1}{\dfrac{1}{2\sqrt{-x}}} = \lim_{x\to -3} 2\sqrt{-x} = 2\sqrt{3}.$$

(b) If we use L'Hôpital's rule, we have

$$\lim_{x\to 0} \frac{\tan x}{x} = \lim_{x\to 0} \frac{\sec^2 x}{1} = 1.$$

(c) By L'Hôpital's rule,

$$\lim_{x\to 4} \frac{x - 4}{x^2 - 8x + 16} = \lim_{x\to 4} \frac{1}{2x - 8}.$$

Since

$$\lim_{x\to 4^+} \frac{1}{2x - 8} = \infty \quad \text{and} \quad \lim_{x\to 4^-} \frac{1}{2x - 8} = -\infty,$$

we conclude that

$$\lim_{x\to 4^+} \frac{x - 4}{x^2 - 8x + 16} = \infty \quad \text{and} \quad \lim_{x\to 4^-} \frac{x - 4}{x^2 - 8x + 16} = -\infty.$$

(d) This limit is not of the indeterminate form $0/0$ since $\lim_{x\to 4} (x^2 + 5) = 21$; thus we cannot use L'Hôpital's rule. Since $\lim_{x\to 4} (x^3 - 4x^2 + 9x - 36) = 0$,

$$\lim_{x\to 4} \frac{x^3 - 4x^2 + 9x - 36}{x^2 + 5} = 0.$$

■

Had we used L'Hôpital's rule in part (d) of Example 4.25, we would have obtained an incorrect answer:

$$\lim_{x\to 4}\frac{x^3-4x^2+9x-36}{x^2+5}=\lim_{x\to 4}\frac{3x^2-8x+9}{2x}=\frac{25}{8}.$$

In other words, L'Hôpital's rule is not to be used indiscriminately; *it must be used only on the indeterminate forms for which it is designed.*

The Indeterminate Form ∞/∞

The limit

$$\lim_{x\to\infty}\frac{1+\sqrt{x-1}}{2x+5}$$

is said to be of the **indeterminate form** ∞/∞ since numerator and denominator become increasingly large as $x\to\infty$. Theorems 4.5 and 4.6 for L'Hôpital's rule can be adapted to this indeterminate form also; hence we calculate that

$$\lim_{x\to\infty}\frac{1+\sqrt{x-1}}{2x+5}=\lim_{x\to\infty}\frac{\dfrac{1}{2\sqrt{x-1}}}{2}=\lim_{x\to\infty}\frac{1}{4\sqrt{x-1}}=0.$$

EXAMPLE 4.26

Evaluate the following limits, if they exist:

$$\text{(a)}\quad \lim_{x\to\infty}\frac{x^2}{e^x}\qquad \text{(b)}\quad \lim_{x\to-\infty}\frac{\sqrt{2x^2+3x+2}}{1-x}$$

SOLUTION (a) Since this limit exhibits the indeterminate form ∞/∞, we use L'Hôpital's rule to write

$$\lim_{x\to\infty}\frac{x^2}{e^x}=\lim_{x\to\infty}\frac{2x}{e^x}.$$

Since this limit is still of the form ∞/∞, we use L'Hôpital's rule again,

$$\lim_{x\to\infty}\frac{x^2}{e^x}=\lim_{x\to\infty}\frac{2}{e^x}=0.$$

(b) By L'Hôpital's rule,

$$\lim_{x\to-\infty}\frac{\sqrt{2x^2+3x+2}}{1-x}=\lim_{x\to-\infty}\frac{\dfrac{4x+3}{2\sqrt{2x^2+3x+2}}}{-1}=\lim_{x\to-\infty}\frac{4x+3}{2\sqrt{2x^2+3x+2}}.$$

This limit is also of the indeterminate form ∞/∞. Further applications of L'Hôpital's rule do not lead to a simpler form for the limit. Thus, L'Hôpital's rule does not prove advantageous on this limit. It is better to divide numerator and denominator by x,

$$\lim_{x\to-\infty}\frac{\sqrt{2x^2+3x+2}}{1-x}=\lim_{x\to-\infty}\frac{\dfrac{\sqrt{2x^2+3x+2}}{x}}{\dfrac{1}{x}-1}=\lim_{x\to-\infty}\frac{-\sqrt{2+\dfrac{3}{x}+\dfrac{2}{x^2}}}{\dfrac{1}{x}-1}=\sqrt{2}.$$

■

The Indeterminate Form $0 \cdot \infty$

The limits

$$\lim_{x\to\infty} xe^{-2x} \qquad \text{and} \qquad \lim_{x\to 0^+} x^2 \ln x$$

are said to be of the **indeterminate form** $0 \cdot \infty$. L'Hôpital's rule can again be used if we first rearrange the limits into one of the forms $0/0$ or ∞/∞:

$$\lim_{x\to\infty} xe^{-2x} = \lim_{x\to\infty} \frac{x}{e^{2x}} = \lim_{x\to\infty} \frac{1}{2e^{2x}} = 0;$$

$$\lim_{x\to 0^+} x^2 \ln x = \lim_{x\to 0^+} \frac{\ln x}{\dfrac{1}{x^2}} = \lim_{x\to 0^+} \frac{\dfrac{1}{x}}{\dfrac{-2}{x^3}} = \lim_{x\to 0^+} \left(-\frac{x^2}{2}\right) = 0.$$

Note that had we converted the second limit into the $0/0$ form, we would have had

$$\lim_{x\to 0^+} x^2 \ln x = \lim_{x\to 0^+} \frac{x^2}{\dfrac{1}{\ln x}} = \lim_{x\to 0^+} \frac{2x}{\dfrac{-1}{x(\ln x)^2}} = \lim_{x\to 0^+} -2x^2(\ln x)^2.$$

Although this is correct, the limit on the right is more difficult to evaluate than the original. In other words, we must be judicious in converting a limit from the $0 \cdot \infty$ indeterminate form to either $0/0$ or ∞/∞.

EXAMPLE 4.27

Evaluate the following limits if they exist:

$$\text{(a)} \quad \lim_{x\to\pi/2} (x - \pi/2)\sec x \qquad \text{(b)} \quad \lim_{x\to 0^+} xe^{1/x}$$

SOLUTION (a) $\displaystyle \lim_{x\to\pi/2} (x - \pi/2)\sec x = \lim_{x\to\pi/2} \frac{x - \pi/2}{\cos x} = \lim_{x\to\pi/2} -\frac{1}{-\sin x} = -1$

(b) $\displaystyle \lim_{x\to 0^+} xe^{1/x} = \lim_{x\to 0^+} \frac{e^{1/x}}{1/x} = \lim_{x\to 0^+} \frac{e^{1/x}(-1/x^2)}{-1/x^2} = \lim_{x\to 0^+} e^{1/x} = \infty.$ ■

The Indeterminate Forms 0^0, 1^∞, ∞^0, and $\infty - \infty$

Various other indeterminate forms arise in the evaluation of limits, and many of these can be reduced to the $0/0$ and ∞/∞ forms by introducing logarithms. In particular, the limits

$$\lim_{x\to 0^+} x^x, \quad \lim_{x\to\infty} \left(1 + \frac{1}{x}\right)^{x^2}, \quad \lim_{x\to\pi/2^-} (\sec x)^{\cos x} \quad \text{and} \quad \lim_{x\to\pi/2} (\sec x - \tan x) \qquad (4.18)$$

are said to display the indeterminate forms 0^0, 1^∞, ∞^0, and $\infty - \infty$, respectively. To evaluate $\lim_{x\to 0^+} x^x$, we set

$$L = \lim_{x\to 0^+} x^x$$

and take natural logarithms of both sides,

$$\ln L = \ln\left(\lim_{x\to 0^+} x^x\right).$$

As the logarithm function is continuous, we may interchange the limit and logarithm operations (see Theorem 2.3),

$$\ln L = \lim_{x\to 0^+} (\ln x^x) = \lim_{x\to 0^+} x \ln x = \lim_{x\to 0^+} \frac{\ln x}{\frac{1}{x}}.$$

We are now in a position to use L'Hôpital's rule:

$$\ln L = \lim_{x\to 0^+} \frac{\frac{1}{x}}{\frac{-1}{x^2}} = \lim_{x\to 0^+} (-x) = 0.$$

Exponentiation of both sides of $\ln L = 0$ now gives $L = e^0 = 1$; that is,

$$\lim_{x\to 0^+} x^x = 1.$$

For the second limit in 4.18 we again set

$$L = \lim_{x\to\infty} \left(1 + \frac{1}{x}\right)^{x^2}$$

and take natural logarithms:

$$\ln L = \ln \left[\lim_{x\to\infty} \left(1 + \frac{1}{x}\right)^{x^2}\right] = \lim_{x\to\infty} \left[\ln \left(1 + \frac{1}{x}\right)^{x^2}\right]$$

$$= \lim_{x\to\infty} \left[x^2 \ln \left(1 + \frac{1}{x}\right)\right] = \lim_{x\to\infty} \left[\frac{\ln \left(\frac{x+1}{x}\right)}{\frac{1}{x^2}}\right].$$

By L'Hôpital's rule, we have

$$\ln L = \lim_{x\to\infty} \left[\frac{\frac{x}{x+1}\left(\frac{-1}{x^2}\right)}{\frac{-2}{x^3}}\right] = \lim_{x\to\infty} \frac{x^2}{2(x+1)} = \infty.$$

Consequently,

$$L = \lim_{x\to\infty} \left(1 + \frac{1}{x}\right)^{x^2} = \infty.$$

In the third limit of 4.18 we set

$$L = \lim_{x\to\pi/2^-} (\sec x)^{\cos x},$$

in which case

$$\ln L = \ln \left[\lim_{x\to\pi/2^-} (\sec x)^{\cos x}\right] = \lim_{x\to\pi/2^-} [\cos x \ln(\sec x)]$$

$$= \lim_{x\to\pi/2^-} \left[\frac{\ln(\sec x)}{\sec x}\right] = \lim_{x\to\pi/2^-} \left[\frac{\frac{1}{\sec x} \sec x \tan x}{\sec x \tan x}\right]$$

$$= \lim_{x\to\pi/2^-} \cos x = 0.$$

Thus,

$$L = \lim_{x \to \pi/2^-} (\sec x)^{\cos x} = e^0 = 1.$$

Finally, the last limit in 4.18 is evaluated by rewriting it in the $0/0$ form,

$$\lim_{x \to \pi/2} (\sec x - \tan x) = \lim_{x \to \pi/2} \left(\frac{1 - \sin x}{\cos x} \right) = \lim_{x \to \pi/2} \frac{-\cos x}{-\sin x} = 0.$$

EXAMPLE 4.28

Sketch a graph of the function $f(x) = x \ln x$.

SOLUTION For critical points of $f(x)$, we solve

$$0 = f'(x) = \frac{x}{x} + \ln x.$$

The only critical point is $x = 1/e$. Since $f''(x) = 1/x$, it follows that $f''(1/e) = e > 0$, and $x = 1/e$ yields a relative minimum of $f(1/e) = -1/e$. Furthermore, since $f''(x)$ is never equal to zero, there are no points of inflection on the graph, which is always concave upward. This information, along with the point $(1, 0)$, is shown in Figure 4.87(a). To complete the sketch, we use l'Hôpital's rule to calculate

$$\lim_{x \to 0^+} x \ln x = \lim_{x \to 0^+} \frac{\ln x}{\frac{1}{x}} = \lim_{x \to 0^+} \frac{\frac{1}{x}}{-\frac{1}{x^2}} = \lim_{x \to 0^+} (-x) = 0^-,$$

and add the extra information that

$$\lim_{x \to 0^+} f'(x) = -\infty.$$

The final graph is shown in Figure 4.87(b).

FIGURE 4.87

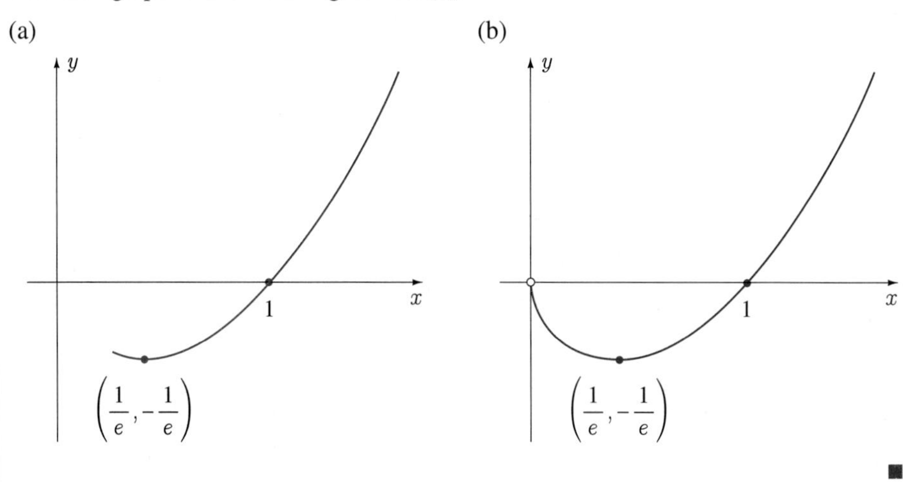

■

EXERCISES 4.9

In Exercises 1–42 evaluate the limit if it exists.

1. $\lim_{x\to 0} \dfrac{x^2+3x}{x^3+5x^2}$

2. $\lim_{x\to 3} \dfrac{x^2-9}{x-3}$

3. $\lim_{x\to -\infty} \dfrac{x^3+3x-2}{x^2+5x+1}$

4. $\lim_{x\to \infty} \dfrac{2x^2+3x}{5x^3+4}$

5. $\lim_{x\to 5} \dfrac{x^2-10x+25}{x^3-125}$

6. $\lim_{x\to 1} \dfrac{1}{(x-1)^2}$

7. $\lim_{x\to \infty} \dfrac{\sqrt{x^2+1}}{2x+5}$

8. $\lim_{x\to -\infty} \dfrac{\sin x}{x}$

9. $\lim_{x\to \infty} \dfrac{\sin(2/x)}{\sin(1/x)}$

10. $\lim_{x\to \pi/2} \dfrac{\cos x}{(x-\pi/2)^2}$

11. $\lim_{x\to 1^+} \dfrac{(1-1/x)^3}{\sqrt{x}-1}$

12. $\lim_{x\to \infty} \dfrac{\sin(1/x)}{1/x^2}$

13. $\lim_{x\to 9^-} \dfrac{\sqrt{x}-3}{\sqrt{9-x}}$

14. $\lim_{x\to 0} \dfrac{\sqrt{5+x}-\sqrt{5-x}}{x}$

15. $\lim_{x\to 0} \dfrac{x-\sin x}{x^3}$

16. $\lim_{x\to a} \dfrac{x^n-a^n}{x-a}$

17. $\lim_{x\to 0} \dfrac{(1-\cos x)^2}{3x^2}$

18. $\lim_{x\to 0} \dfrac{\tan x}{x}$

19. $\lim_{x\to 0} \dfrac{\sin 3x}{\tan 2x}$

20. $\lim_{x\to 1} \dfrac{\left(1-\sqrt{2-x}\right)^{3/2}}{x-1}$

21. $\lim_{x\to 0} \dfrac{\sqrt{x+1}-\sqrt{2x+1}}{\sqrt{3x+4}-\sqrt{2x+4}}$

22. $\lim_{x\to 0} \dfrac{(1-\cos x)^2}{3x^4}$

23. $\lim_{x\to \infty} x\sin\left(\dfrac{1}{x}\right)$

24. $\lim_{x\to 2} \dfrac{(x-2)^{10}}{\left(\sqrt{x}-\sqrt{2}\right)^{10}}$

25. $\lim_{x\to 0} \left(\dfrac{4}{x^2}-\dfrac{2}{1-\cos x}\right)$

26. $\lim_{x\to -\infty} xe^x$

27. $\lim_{x\to \infty} x^2e^{-4x}$

28. $\lim_{x\to -\infty} x\sin\left(\dfrac{4}{x}\right)$

29. $\lim_{x\to 0} x\cot x$

30. $\lim_{x\to 0} \csc x(1-\cos x)$

31. $\lim_{x\to 0^+} (\sin x)^x$

32. $\lim_{x\to 0^+} x^{\sin x}$

33. $\lim_{x\to \infty} \left(\dfrac{x+5}{x+3}\right)^x$

34. $\lim_{x\to 0} (1+x)^{\cot x}$

35. $\lim_{x\to \infty} x^{1/x}$

36. $\lim_{x\to 0^+} |\ln x|^{\sin x}$

37. $\lim_{x\to 0^+} xe^{1/x}$

38. $\lim_{x\to 0} (\tan x-\csc x)$

39. $\lim_{x\to 0} (\csc x-\cot x)$

40. $\lim_{x\to 1} \left(\dfrac{x}{\ln x}-\dfrac{1}{x\ln x}\right)$

41. $\lim_{x\to 1} \left(\dfrac{x}{x-1}-\dfrac{1}{\ln x}\right)$

42. $\lim_{x\to 0} \left(\dfrac{1}{x^2}-\dfrac{1}{\sin^2 x}\right)$

In Exercises 43–54 sketch a graph of the function.

43. $f(x)=xe^{-2x}$

44. $f(x)=x^2e^{3x}$

45. $f(x)=xe^{-x^2}$

46. $f(x)=e^{1/x}$

47. $f(x)=\dfrac{\ln x}{x}$

48. $f(x)=x^2\ln x$

49. $f(x)=xe^{1/x}$

50. $f(x)=\dfrac{x^2}{\ln x}$

51. $f(x)=x^x,\quad x>0$

52. $f(x)=x^{10}e^{-x}$

53. $f(x)=e^{-x}\ln x,\quad x>0$

54. $f(x)=2\csc x-\cot x,\quad 0<x<\pi/2$

55. Evaluate

$$\lim_{x\to\infty}\left(\frac{x+a}{x+b}\right)^{cx}$$

for any constants a, b, and c.

56. The indeterminate forms 0^0, 1^∞, and ∞^0 are often evaluated by introducing logarithms. Show that the limit

$$\lim_{x\to\infty}(x-\ln x)$$

can be evaluated by introducing exponentials.

57. When an electrostatic field E is applied to a gaseous or liquid polar dielectric, a net dipole moment P per unit volume is set up, where

$$P(E)=\frac{e^E+e^{-E}}{e^E-e^{-E}}-\frac{1}{E}.$$

Show that $\lim_{E\to 0^+} P(E)=0$.

58. Planck's law for the energy density ψ of blackbody radiation states that

$$\psi=\psi(\lambda)=\frac{k\lambda^{-5}}{e^{c/\lambda}-1},$$

where k and c are positive constants and λ is the wavelength of the radiation.

(a) Show that

$$\lim_{\lambda\to 0^+}\psi(\lambda)=0 \qquad\text{and}\qquad \lim_{\lambda\to\infty}\psi(\lambda)=0.$$

(b) Show that $\psi(\lambda)$ has one critical point which must satisfy the equation

$$(5\lambda - c)e^{c/\lambda} = 5\lambda.$$

Find the critical point accurate to seven decimals when $c = 0.001\,438\,6$.

(c) Sketch a graph of the function $\psi(\lambda)$.

59. The following limit arises in the calculation of the electric field intensity for a half-wave antenna:

$$\lim_{\theta \to 0} f(\theta),$$

where

$$f(\theta) = \sin\theta\left\{\frac{\sin[\pi/2(\cos\theta - 1)]}{\cos\theta - 1} + \frac{\sin[\pi/2(\cos\theta + 1)]}{\cos\theta + 1}\right\}.$$

Evaluate this limit by first showing that $f(\theta)$ can be written in the form

$$f(\theta) = \frac{2\cos\left(\dfrac{\pi}{2}\cos\theta\right)}{\sin\theta},$$

and then using L'Hôpital's rule.

60. Find all values of a, b, and c for which

$$\lim_{x \to 0} \frac{e^{ax} - bx - \cos(x + cx^2)}{2x^3 + 5x^2} = 5.$$

61. (a) Sketch a graph of the function

$$f(x) = \begin{cases} e^{-1/x^2} & x \neq 0 \\ 0 & x = 0. \end{cases}$$

(b) Show that for every positive integer n,

$$\lim_{x \to 0} \frac{e^{-1/x^2}}{x^n} = 0.$$

(c) Prove by mathematical induction that $f^{(n)}(0) = 0$, where $f^{(n)}(0)$ is the nth derivative of $f(x)$ evaluated at $x = 0$.

SECTION 4.10

Differentials

In Section 3.1 we pointed out that the notation dy/dx for the derivative of a function $y = f(x)$ should not be considered a quotient. In this section, we change this. It is not that we lied to you in Chapter 3; it's just that we were not prepared to tell you the whole story. Besides, nothing in Chapters 3 and 4 would have been any easier had we regarded dy/dx as "dy divided by dx". Beginning in Chapter 5, however, it is essential that we be able to do this, and therefore in this section we define "differentials" dx and dy so that dy/dx can be regarded as a quotient.

When we use the notation

$$\frac{dy}{dx} = \lim_{\Delta x \to 0} \frac{f(x + \Delta x) - f(x)}{\Delta x}$$

for the derivative of a function $y = f(x)$, we call Δx an increment in x. It represents a change in the value of the independent variable from some value x to another value $x + \Delta x$. This change can be positive or negative depending on whether we want $x + \Delta x$ to be larger or smaller than x. When the independent variable changes from x to $x + \Delta x$, the dependent variable changes by an amount Δy where

$$\Delta y = f(x + \Delta x) - f(x). \tag{4.19}$$

In other words, Δy is the change in y resulting from the change Δx in x. For the function in Figure 4.88, Δy is positive when Δx is positive, and Δy is negative when Δx is negative.

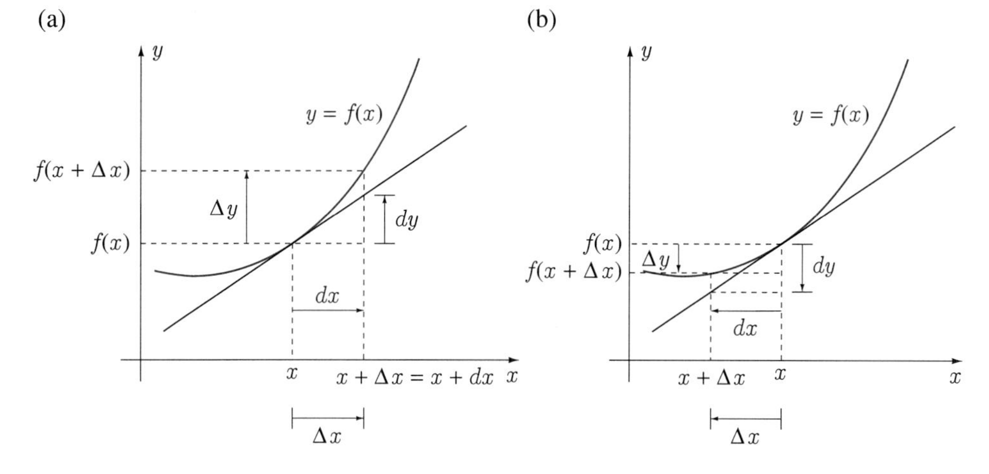

FIGURE 4.88

For example, when $y = x^2 - 2x$, the change Δy in y when x is changed from 3 to 3.2 is

$$\Delta y = [(3.2)^2 - 2(3.2)] - [3^2 - 2(3)] = 0.84.$$

The function increases by 0.84 when x increases from 3 to 3.2.

For purposes of integration, a topic which begins in Chapter 5 and continues in every chapter thereafter, an alternative notation for an increment in the independent variable x is more suggestive.

Definition 4.7

An increment Δx in the independent variable x is denoted by

$$dx = \Delta x, \tag{4.20}$$

and when written as dx, it is called the **differential** of x.

The differential dx is synonymous with the increment Δx; it represents a change in x (in most applications a very small change). The differential of the dependent variable is *not* synonymous with Δy.

Definition 4.8

The **differential** of $y = f(x)$, corresponding to the differential dx in x, is denoted by dy and is defined by

$$dy = f'(x)\, dx. \tag{4.21}$$

The difference between Δy and dy is most easily seen in Figure 4.88. We know that Δy is the exact change in the function $y = f(x)$ when x is changed by an amount Δx or dx. It is the difference in the height of the curve at x and $x + \Delta x = x + dx$. Now the slope of the tangent line to the graph at the point (x, y) is $f'(x)$. Definition 4.8 indicates that dy can be interpreted as the difference in the height of this tangent line at x and at $x + dx$. In other words, dy is the change in y corresponding to the change dx in x if we follow the tangent line to $y = f(x)$ at (x, y) rather than the curve itself.

Figure 4.88 also suggests that when dx is very small (close to zero), dy is approximately equal to Δy; that is,

$$dy \approx \Delta y \qquad \text{when } dx \approx 0.$$

This is also illustrated in the following numerical example.

EXAMPLE 4.29 Find Δy and dy for the function $y = f(x) = \sqrt{x^2+1}$ when $x = 2$ and $dx = 0.1$.

SOLUTION According to equation 4.19, the change in y as x increases from 2 to 2.1 is

$$\Delta y = f(2.1) - f(2) = \sqrt{(2.1)^2+1} - \sqrt{2^2+1} = 0.08987.$$

Since $f'(x) = x/\sqrt{x^2+1}$, the differential of y for $x = 2$ and $dx = 0.1$ is

$$dy = f'(2)(0.1) = \frac{2}{\sqrt{2^2+1}}(0.1) = 0.08944.$$

The difference between dy and Δy is therefore 0.00043, a difference of 43 parts in 8987. ■

Before the invention of electronic calculators, differentials were used to approximate a function near points at which it was easily evaluated. For example, imagine trying to evaluate the function $f(x) = x^{1/3}$ at $x = 126$ without a calculator. We could use differentials to approximate $126^{1/3}$ as follows:

$$\begin{aligned}126^{1/3} &= f(126) = f(125) + \Delta y \approx f(125) + dy \\ &= 5 + f'(5)(1) = 5 + \frac{1}{3(125)^{2/3}} = 5 + \frac{1}{75} = \frac{376}{75}.\end{aligned}$$

With the advent of the electronic calculator, problems of this type exist, but it is ridiculous to attack them with differentials. Differentials can be useful when we examine changes in a function without specifying values for the independent variable. Very prominent in this context are the "relative" and "percentage" changes.

When a quantity y undergoes a change Δy, then its *relative change* is defined as

$$\frac{\Delta y}{y}, \tag{4.22}$$

and its *percentage change* is given by

$$100\frac{\Delta y}{y}. \tag{4.23}$$

Sometimes relative and percentage changes are more important than actual changes. To illustrate this, consider the function $V = 4\pi r^3/3$ which represents the volume of a sphere. If the radius of the sphere is increased from 0.10 m to 0.11 m, then the change in the volume of the sphere is

$$\Delta V = \frac{4}{3}\pi(0.11)^3 - \frac{4}{3}\pi(0.10)^3 = 4.4\pi \times 10^{-4}\ \text{m}^3.$$

This is not a very large quantity, but in relation to the original size of the sphere, we have a relative change of

$$\frac{\Delta V}{V} = \frac{4.4\pi \times 10^{-4}}{4\pi(0.10)^3/3} = 0.33,$$

and a percentage change of

$$100\frac{\Delta V}{V} = 33\%.$$

Suppose the same increase of 0.01 m is applied to a sphere with radius 100 m. The change in the volume is

$$\Delta V = \frac{4}{3}\pi(100.01)^3 - \frac{4}{3}\pi(100)^3 = 4.0\pi \times 10^2 \text{ m}^3.$$

This is quite a large change in volume, but the relative change is

$$\frac{\Delta V}{V} = \frac{4.0\pi \times 10^2}{4\pi(100)^3/3} = 3.0 \times 10^{-4},$$

and the percentage change is

$$100\frac{\Delta V}{V} = 0.03\%.$$

Although the change 400π in V when $r = 100$ is much larger than the change 0.00044π when $r = 0.1$, the relative and percentage changes are much smaller when $r = 100$. We see, then, that in certain cases, it may be relative and percentage changes that are significant rather than actual changes.

In the above example, when $r = 100$ m, the change $dr = 0.01$ is certainly small compared to r. We should therefore be able to use the differential $dV = V'(r)\,dr = 4\pi r^2\,dr$ to approximate ΔV. With $r = 100$ and $dr = 0.01$, we have

$$dV = 4\pi(100)^2(0.01) = 4.0\pi \times 10^2 \text{ m}^3.$$

(To two significant figures, dV is equal to ΔV.) Thus relative change in V is approximately equal to dV/V and percentage change $100\,dV/V$. So for small changes in an independent variable, the differential of the dependent variable may be used in place of its increment in the calculation of relative and percentage changes; that is, equations (4.22) and (4.23) can be replaced by

$$\frac{dy}{y} \quad \text{and} \quad 100\frac{dy}{y}.$$

We do this in the following example.

EXAMPLE 4.30

When a pendulum swings, the frequency (number of cycles per second) of its oscillations is given by

$$f = h(l) = 2\pi\sqrt{\frac{g}{l}},$$

where l is the length of the pendulum and $g > 0$ is a constant. If the length of the pendulum is increased by $\frac{1}{4}\%$, calculate the approximate percentage change in f.

SOLUTION The approximate change in f is given by

$$df = h'(l)\,dl = 2\pi\sqrt{g}\left(-\frac{1}{2}\right)l^{-3/2}\,dl = -\pi\sqrt{g}\frac{dl}{l^{3/2}};$$

hence the approximate percentage change in f is

$$100\frac{df}{f} = 100\left(\frac{-\pi\sqrt{g}\,dl}{l^{3/2}}\right)\left(\frac{l^{1/2}}{2\pi\sqrt{g}}\right) = -\frac{1}{2}\left(100\frac{dl}{l}\right).$$

But because l increases by $\frac{1}{4}\%$, it follows that $100(dl/l) = 1/4$, and

$$100\,\frac{df}{f} = -\frac{1}{8}.$$

The frequency therefore changes by $-\frac{1}{8}\%$, the negative sign indicating that because l increases, f decreases. ■

The differential dy cannot always be used as an approximation for the actual change Δy in a function $y = f(x)$. Sometimes it cannot be used even when dx is very close to zero. For example, if $f(x) = 2x^3 + 9x^2 - 24x + 6$, then

$$dy = f'(x)\,dx = (6x^2 + 18x - 24)\,dx.$$

If x is changed from 1 to 1.01, then the approximate change in y as predicted by the differential is

$$dy = (6 + 18 - 24)(0.01) = 0.$$

In fact, for any dx whatsoever, we find that $dy = 0$. Geometrically speaking, we can see why. Since $f'(1) = 0$, $x = 1$ is a critical point of $f(x)$ (Figure 4.89); therefore dy, which is the tangent line approximation to Δy, will always be zero. We cannot use differentials to approximate function changes at critical points.

FIGURE 4.89

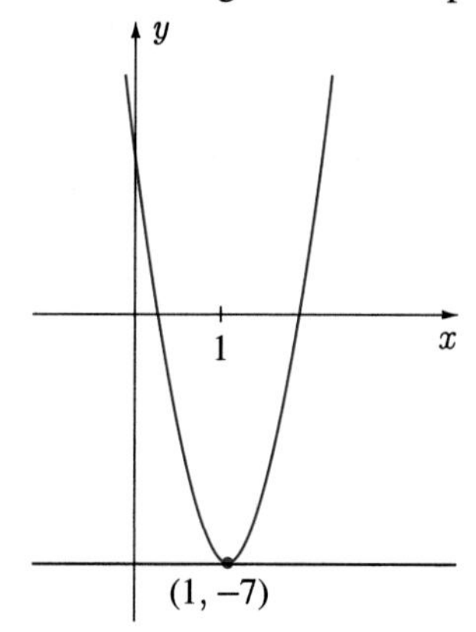

The differential for the function $y = f(x) = x^{100}$ is

$$dy = 100\,x^{99}\,dx.$$

If x is changed by 1% from $x = 1$ to $x = 1.01$, then $dx = 0.01$ and

$$dy = 100(1)^{99}(0.01) = 1.0.$$

The actual change in y is

$$\Delta y = (1.01)^{100} - 1^{100} = 1.7.$$

We would hardly regard this dy as a very good approximation for Δy even though the change in x is only one percent.

This latter example begs the question."How small in general must dx be in order that dy be a reasonable approximation for Δy?" This is not a simple question to answer. We will discuss approximations in more detail in Chapter 11, and then be able to answer the more important question, "How good an approximation to Δy is dy?" After all, it is not much use to say that $dy \approx \Delta y$ if we cannot say to how many decimal places the approximation is accurate. Suffice it to say now that use of the differential dy to approximate Δy is to be regarded with some reservation. This is not to say that differentials are useless. We will see in Chapters 5–7 that differentials are indispensable to the topic of integration.

We make one last comment before leaving this section. If equation 4.21 is divided by differential dx, then

$$\frac{dy}{dx} = f'(x).$$

Now, the left side of this equation is the differential of y divided by the differential of x. The quotient of differentials dy and dx is equal to the derivative $f'(x)$. The entity dy/dx can henceforth be regarded either as "the derivative of y with respect to x", or "dy divided by dx", whichever is appropriate for the discussion at hand.

EXERCISES 4.10

In Exercises 1–10 find dy in terms of x and dx.

1. $y = x^2 + 3x - 2$

2. $y = \dfrac{x+1}{x-1}$

3. $y = \sqrt{x^2 - 2x}$

4. $y = \sin(x^2 + 2) - \cos x$

5. $y = x^{1/3} - x^{5/3}$

6. $y = x^3\sqrt{3 - 4x^2}$

7. $y = x^2 \sin x$

8. $y = \dfrac{x^3 - 3x^2 + 3x + 5}{x^2 - 2x + 1}$

9. $y = \sqrt{1 + \sqrt{1 - x}}$

10. $y = \dfrac{x^2(x-2)}{x^3 + 5x}$

11. The momentum M and kinetic energy K of a mass m moving with speed v are given by

$$M = mv \quad \text{and} \quad K = \frac{1}{2}mv^2.$$

If v is changed by 1%, what are approximate percentage changes in M and K?

12. The magnitude of the gravitational force of attraction between two point masses m and M is given by

$$F = \frac{GmM}{r^2},$$

where $G > 0$ is a constant, and r is the distance between the masses. If r changes by 2%, by how much does F change approximately?

13. Under adiabatic expansion, a gas obeys the law

$$PV^{7/5} = \text{a constant},$$

where P is pressure and V is volume. If the pressure is increased by 2%, find the approximate percentage change in the volume.

14. The magnitude of the gravitational force of attraction between two point masses m and M is defined in Exercise 12. If the earth is considered a perfect sphere (radius 6.37×10^6 m), then this law predicts a gravitational attraction of $9.81m$ N on a mass m on its surface. Use differentials to determine the height above the surface of the earth at which the gravitational attraction decreases to $9.80m$ N.

SECTION 4.11

Problem Solving

Differentiation is a powerful tool for solving problems in many fields — physics, chemistry, engineering, psychology, medecine, and economics, to name a few. Many of these applications involve related rate and applied maxima-minima problems, and are examples of *word* problems. In this section we give some suggestions on how to handle word problems, which may be helpful in other mathematical problems as well. As we follow these suggestions through the examples, we illustrate the logical thought processes that lead from problem to solution.

Three suggestions important to many mathematical problems, and in particular to those that are calculus-based, are:

Suggestion 1: Make a diagram whenever possible to illustrate all given information.

Suggestion 2: Identify the nature of the problem. Is it a related rate problem, a maxima-minima problem, or some other kind of problem. There are always key words in the statement of the problem to help make this decision. Find them.

This suggestion is extremely important because further steps in the solution are intimately connected to the nature of the problem. However, most problems will also follow the next suggestion.

Suggestion 3: Choose letters to represent key variables and find appropriate formulas or equations involving these variables.

Problem 1 The slope and y-intercept of the line $y = mx + b$ are such that the line intersects the parabola $y = x^2$ in two points P and Q. A point R on the parabola between P and Q is joined to P and Q to create a triangle PQR. Where should R be chosen in order that the area of the triangle be as large as possible?

Certainly we can follow Suggestion 1 and begin with Figure 4.90(a). To identify the nature of the problem (Suggestion 2), we key on the last four words "as large as possible"; they identify that we have a maximum problem. Find the position of R in order to maximize the area of triangle PQR. According to Suggestion 3 we should choose a letter, say A, to represent the area of triangle PQR, and now find a formula for A. What should A be expressed in terms of? Since we want to find the position of R that maximizes A, A should be expressed in terms of R. How do we do this? By giving coordinates to R. If its x-coordinate is denoted by x, then its y-coordinate is x^2 (since it is on the parabola). Notice that we have not assigned coordinates to P and Q; we will do this only if it becomes necessary. We now express A in terms of x since x identifies the position of R. Because the area of a triangle is one-half the product of its base and height, we drop a perpendicular from R to PQ to intersect PQ at S (Figure 4.90(b)). Then

$$A = \frac{1}{2}||PQ||||RS||.$$

A key point to realize here is that P and Q are fixed, and therefore so also is length $||PQ||$ of the base of the triangle. It is not necessary to find a formula for $||PQ||$; simply regard it as a constant.

FIGURE 4.90 (a) (b)

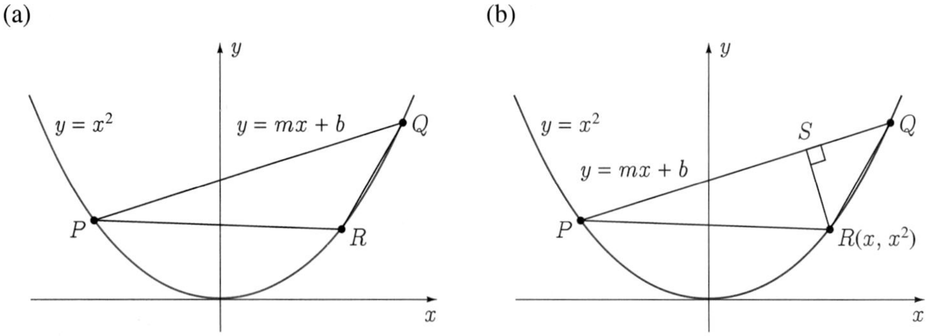

We must find $||RS||$ in terms of x. The simplest way is to use the distance formula developed in Exercise 45 of Section 1.3 and Exercise 52 of Section 4.6. With the equation of the line expressed in the form $mx - y + b = 0$, the distance from $R(x, x^2)$ to S is

$$||RS|| = \frac{|mx - y + b|}{\sqrt{m^2 + 1}} = \frac{|mx - x^2 + b|}{\sqrt{m^2 + 1}}.$$

Thus,

$$A = \frac{1}{2}||PQ||\frac{|mx - x^2 + b|}{\sqrt{m^2 + 1}}.$$

Are we ready to differentiate? No. This is an absolute maximum problem — among all points on the parabola between P and Q, find the one that maximizes A. It follows that x must be restricted to the interval $x_P \le x \le x_Q$ where x_P and x_Q are x-coordinates of P and Q. Our mathematical problem therefore is to find the absolute maximum of the function

$$A(x) = \frac{\|PQ\|}{2\sqrt{m^2+1}}|mx - x^2 + b|, \quad x_P \le x \le x_Q.$$

To find critical points of $A(x)$, we first set $A'(x) = 0$, using differentiation formula 3.12,

$$0 = A'(x) = \frac{\|PQ\|}{2\sqrt{m^2+1}} \frac{|mx - x^2 + b|}{mx - x^2 + b}(m - 2x).$$

The only solution of this equation is $x = m/2$. We should now evaluate A at x_P, at the critical point $m/2$, and at x_Q. Since the triangle degenerates to a straight line when R is at either P or Q (that is, when $x = x_P$ or $x = x_Q$), it follows that $x = m/2$ must yield a maximum for A. Thus, R should be chosen as the point $(m/2, m^2/4)$.

Problem 2 A hemispherical tank of radius 3 m has a light on its upper edge as in Figure 4.91(a). A stone falls vertically along the axis of symmetry of the tank, and when it is 1 m from the bottom of the tank it is falling at 2 m/s. How fast is its shadow moving along the surface of the tank at this instant?

Although Figure 4.91(a) would seem to fullfil Suggestion 1, there is really no need to draw a 3-dimensional diagram. The shadow follows a circular path down the side of the tank which we have shown in Figure 4.91(b).

FIGURE 4.91

(a)

Light

Stone

Shadow of stone

(b)

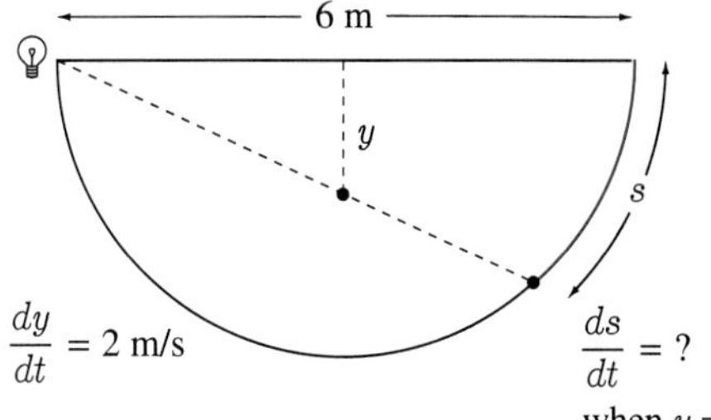

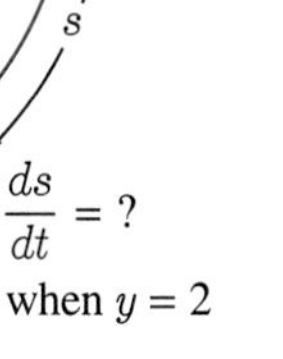

(c)

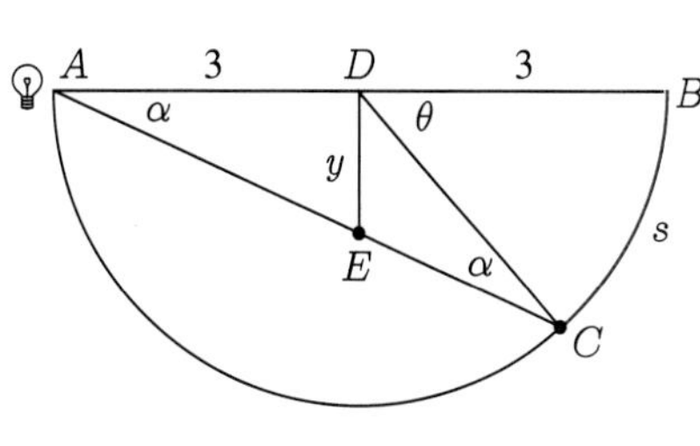

The nature of the problem is identified in the last sentence "How fast is the shadow moving". It is a related rate problem; we want the rate at which the shadow is moving, given the rate at which the stone is falling.

To follow Suggestion 3, we introduce the distance s along the circle as the distance that the shadow has travelled since the stone passed the diameter of the circle. Its rate of change with respect to time t, namely ds/dt, is what is required. To express the rate at which the stone is falling as a derivative, we let y be the distance from the diameter of the circle to the stone. Then $dy/dt = 2$ m/s. These facts have been added to Figure 4.91(b).

For a related rate problem, we must find an equation relating y and s. First, arc length along a circle is the product of the radius of the circle and the angle θ that it subtends at the centre of the circle. We therefore name some key points in Figure 4.91(c), join C and D, and write

$$s = 3\theta.$$

We wanted an equation relating s and y, not s and θ. We must therefore relate y and θ. Since triangle ADC is isosceles, angles DAC and DCA are equal — called α in Figure 4.91(c), and $\theta = 2\alpha$. Hence

$$s = 3(2\alpha) = 6\alpha.$$

From triangle ADE, we may write $y = 3\tan\alpha$. Combine this with $s = 6\alpha$, and we finally relate s and y,

$$y = 3\tan\left(\frac{s}{6}\right).$$

Differentiation with respect to t gives

$$\frac{dy}{dt} = 3\sec^2\left(\frac{s}{6}\right)\left(\frac{1}{6}\frac{ds}{dt}\right).$$

When the stone is 1 m from the bottom of the tank, $y = 2$ m, and therefore

$$2 = 3\tan\left(\frac{s}{6}\right).$$

It follows that at this instant

$$\sec^2\left(\frac{s}{6}\right) = 1 + \tan^2\left(\frac{s}{6}\right) = 1 + \left(\frac{2}{3}\right)^2 = \frac{13}{9}.$$

Substitution of this into the equation that relates dy/dt and ds/dt, along with $dy/dt = 2$, yields

$$2 = 3\left(\frac{13}{9}\right)\left(\frac{1}{6}\frac{ds}{dt}\right).$$

Hence, $ds/dt = 36/13$, and the shadow is moving at $36/13$ m/s.

Problem 3 Prove that when $0 < a < b < \pi/2$,

$$\frac{\tan b}{\tan a} > \frac{b}{a}. \tag{4.18}$$

This is not a word problem; it is included to once again illustrate the logic by which we proceed from problem to solution. We do not begin with $0 < a < b < \pi/2$; the important requirement is to verify the tangent inequality. As there is no obvious connection between the two sides of the inequality, perhaps it would be beneficial to rewrite it in the form

$$\frac{1}{b}\tan b > \frac{1}{a}\tan a. \tag{4.19}$$

This looks promising; both sides now have exactly the same form. Effectively, we have multiplied 4.18 by $b^{-1}\tan a$ to obtain 4.19. Inequality 4.19 remains in the same direction only if $b^{-1}\tan a$ is positive, and this is indeed the case because we are considering only values of a and b between 0 and $\pi/2$.

We have noted that both sides of 4.19 have the same form. In fact, they can be obtained by setting $x = b$ and $x = a$ in the function

$$f(x) = \frac{1}{x} \tan x.$$

We can now paraphrase the problem: Show that for any two points a and b in the interval $0 < x < \pi/2$, with $b > a$,

$$f(b) > f(a).$$

We must show that the function $f(x)$ is increasing on the interval $I : 0 < x < \pi/2$. But according to 4.5, this is established if $f'(x) \geq 0$ on I. Using the product rule we calculate

$$\begin{aligned} f'(x) &= -\frac{1}{x^2} \tan x + \frac{1}{x} \sec^2 x = \frac{1}{x^2}(x \sec^2 x - \tan x) \\ &= \frac{1}{x^2}\left(\frac{x}{\cos^2 x} - \frac{\sin x}{\cos x}\right) = \frac{1}{x^2 \cos^2 x}(x - \sin x \cos x). \end{aligned}$$

Since $x^2 \cos^2 x$ is positive on I, the sign of $f'(x)$ is determined by that of $x - \sin x \cos x$. We must therefore show that on I,

$$0 \leq x - \sin x \cos x.$$

This inequality can be simplified if it is multiplied by 2,

$$0 \leq 2x - 2 \sin x \cos x = 2x - \sin 2x \qquad \text{or} \qquad 2x \geq \sin 2x. \qquad (4.20)$$

Inequality 4.18 is verified once 4.20 is established. Graphs of the functions $2x$ and $\sin 2x$ on I are shown in Figure 4.92. Important to this picture are the facts that the line $y = 2x$ and the curve $y = \sin 2x$ both have slope 2 at $x = 0$, but the slope of $y = \sin 2x$ is less than 2 for $x > 0$. Clearly, then, $2x \geq \sin 2x$, and our proof is complete.

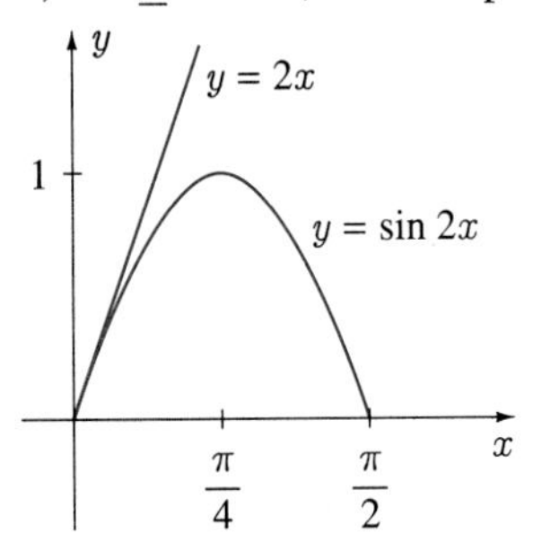

FIGURE 4.92

We have solved Problem 3 backwards. We have replaced inequality 4.18 with 4.19, which is tantamount to showing that $f(x) = x^{-1} \tan x$ is increasing on $I : 0 < x < \pi/2$. But this is established once 4.20 is verified, and Figure 4.92 verifies 4.20.

We could now rewrite the solution to this problem in a very elegant way. We could begin by saying:

"Consider the function $f(x) = x^{-1} \tan x$ on the interval $0 < x < \pi/2$." We could show that $f(x)$ is increasing and 4.19 and 4.18 would immediately follow. This is exactly how the solution would appear in most calculus books and/or solutions manuals. The author would appear brilliant; out of nowhere he would produce the function $f(x) = x^{-1} \tan x$. He actually solved the problem backwards, as above, and then rewrote it.

Problems 1, 2, and 3 have been chosen as involved, multistage problems to emphasize the importance of Suggestions 1, 2, and 3, and to illustrate how to organize our thinking into a logical sequence.

SUMMARY

In this chapter we discussed a number of applications of differentiation, the first of which was Newton's iterative procedure for approximating the roots of equations. It is perhaps the most popular of all approximation methods, because of its speed, simplicity, and accuracy.

In Section 4.3 we defined a critical point of a function as a point in its domain where its first derivative either vanishes or does not exist. Geometrically, this corresponds to a point where the graph of the function has a horizontal tangent line, a vertical tangent line, or no tangent line at all. The first derivative test indicates if critical points yield relative maxima or relative minima. A function is increasing (or decreasing) on an interval if its graph slopes upward to the right (respectively left), and this is characterized by a nonnegative (respectively nonpositive) derivative. It is concave upward (or downward) if its slope is increasing (respectively decreasing), and consequently if its second derivative is nonnegative (respectively nonpositive). Points that separate intervals of opposite concavity are called points of inflection. In Section 4.5 we showed how our curve sketching techniques, which now include addition and multiplication of ordinates, symmetries, limits, relative extrema, and points of inflection, complement one another in drawing graphs.

In Section 4.6 we illustrated that many applied extrema problems require absolute extrema rather than relative extrema. Absolute extrema of a continuous function on a closed interval must occur at either critical points or the ends of the interval. This fact implies that to find the absolute extrema of a continuous function $f(x)$ on a closed interval $a \leq x \leq b$, we evaluate $f(x)$ at its critical points between a and b and at a and b. The largest and smallest of these values are the absolute extrema of $f(x)$ on $a \leq x \leq b$.

When an object moves along a straight line, its velocity and acceleration are the first and second derivatives respectively of its displacement with respect to time. In other words, if we observe straightline motion of an object, and record its position as a function of time, then we can calculate the velocity and acceleration of that object at any instant.

Changes in a number of interrelated quantities usually produce changes in the others—sometimes small, sometimes large. How the rates of change of these variables relate to each other was the subject of Section 4.8. Related-rate problems made us acutely aware of the importance of differentiating an equation with respect to a variable only if the equation is valid for a continuous range of values of that variable.

Cauchy's generalized mean value theorem enabled us to develop L'Hôpital's rule in Section 4.9 for evaluation of various indeterminate forms such as $0/0$, ∞/∞, $0 \cdot \infty$, 0^0, 1^∞, ∞^0, and $\infty - \infty$.

Key Terms and Formulas

In reviewing this chapter, you should be able to define or discuss the following key terms:

Newton's iterative procedure
Increasing function
Decreasing function
Critical point
Relative maximum
Relative minimum
First derivative test
Concave upward
Concave downward
Point of inflection
Vertical point of inflection
Absolute maximum
Absolute minimum
Instantaneous velocity
Instantaneous acceleration
Instantaneous speed
Related rates
Indeterminate form
L'Hôpital's rule
Differential

Second derivative test
Horizontal point of inflection

Relative change
Percentage change

REVIEW EXERCISES

1. (a) Prove that the area of the isosceles triangle in Figure 4.93 is

$$A = \frac{l^2 \sin\theta}{2}.$$

FIGURE 4.93

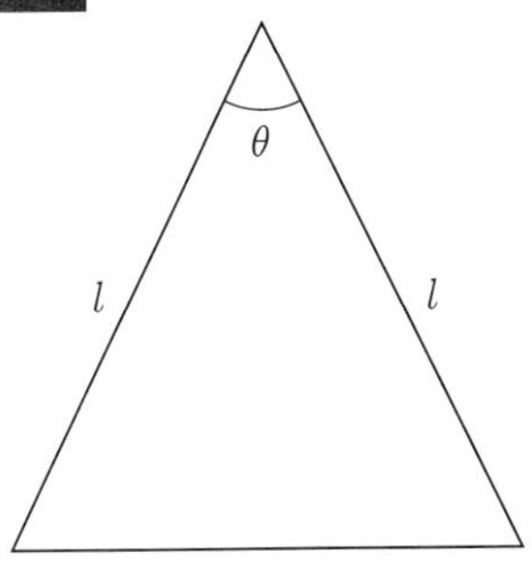

(b) If the angle θ is increasing at $1/2$ radian per minute, but l remains constant, how fast is the area of the triangle changing? What does your answer predict when $\theta = 0,\ \pi/2$, and π?

(c) When does A change most rapidly and most slowly; that is, when is $|dA/dt|$ largest and smallest?

2. Sketch a graph of each of the following functions, indicating all relative maxima and minima and points of inflection.

(a) $f(x) = 4x^3 + x^2 - 2x + 1$

(b) $f(x) = \dfrac{x^2 - 2x + 4}{x^2 - 2x + 1}$

3. Of all pairs of positive numbers that add to some given constant $c > 0$, find that pair which has the largest product.

4. Solve Exercise 3 for the smallest product.

5. Of all pairs of positive numbers that multiply to some given constant $c > 0$, find that pair which has the smallest sum.

6. Solve Exercise 5 for the largest sum.

7. Two sides of the triangle in Figure 4.94 maintain constant lengths of 3 cm and 4 cm but the length l of the third side decreases at the rate of 1 cm/min. How fast is angle θ changing when l is 4 cm?

FIGURE 4.94

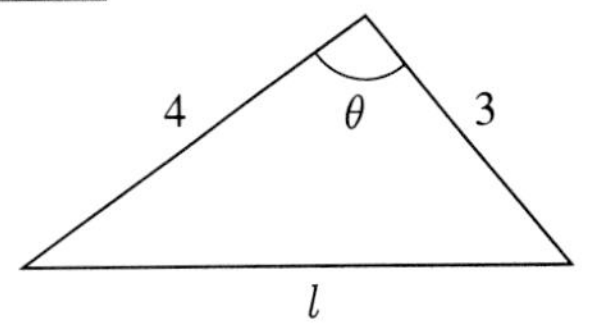

In Exercises 8–19, evaluate the limit if it exists.

8. $\displaystyle\lim_{x\to 0} \frac{3x^2 + 2x^3}{3x^3 - 2x^2}$

9. $\displaystyle\lim_{x\to\infty} \frac{\sin 3x}{2x}$

10. $\displaystyle\lim_{x\to 4} \frac{x^2 - 16}{x - 4}$

11. $\displaystyle\lim_{x\to 0} \frac{\sin 3x}{2x}$

12. $\displaystyle\lim_{x\to -\infty} \frac{\sin x^2}{2x}$

13. $\displaystyle\lim_{x\to 2^+} \frac{\sqrt{x-2}}{\sqrt{x} - \sqrt{2}}$

14. $\displaystyle\lim_{x\to\infty} x^2 e^{-3x}$

15. $\displaystyle\lim_{x\to 0^+} x^{2x}$

16. $\displaystyle\lim_{x\to 0^+} x^4 \ln x$

17. $\displaystyle\lim_{x\to 0} \frac{\sin 2x}{\tan 3x}$

18. $\displaystyle\lim_{x\to\infty} \left(\frac{x+1}{x-1}\right)^x$

19. $\displaystyle\lim_{x\to -\infty} xe^x$

20. Use Newton's method to find all critical points for the following functions accurate to six decimals.

(a) $f(x) = x^4 + 3x^2 - 2x + 5$

(b) $f(x) = \dfrac{x^3 + 1}{3x^3 + 5x + 1}$

21. An object moves along the x-axis with its position defined as a function of time t by

$$x = x(t) = t^4 - \frac{44}{3}t^3 + 62t^2 - 84t, \quad t \geq 0.$$

Sketch a graph of this function, indicating times when the velocity and acceleration of the object are equal to zero.

22. If the graph in Figure 4.95 represents the position $x(t)$ of an object moving along the x-axis, what could physically cause the corner at time t_0?

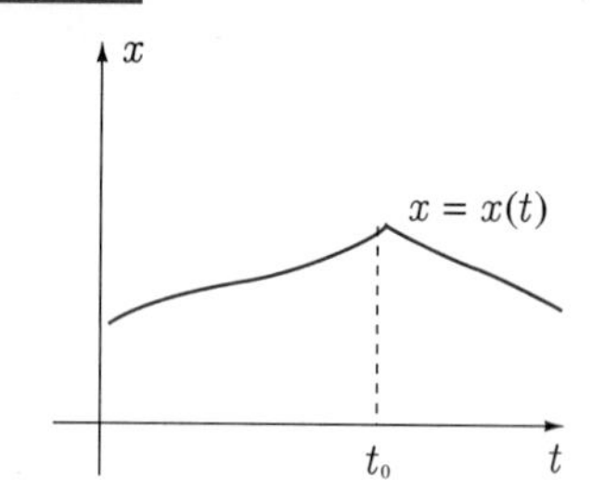

FIGURE 4.95

23. If an object moves along the x-axis with constant acceleration, is it possible for a graph of its position function $x(t)$ to have a point of inflection?

24. An open box is formed from a square piece of cardboard (l units long on each side) by cutting out a square at each corner and folding up the sides. What is the maximum possible volume for the box?

25. (a) If the average speed of a car for a trip is 80 km/h, must it at some time have had an instantaneous speed of 80 km/h? Explain.

(b) Must the car at some instant have had a speed of 83 km/h?

26. Sketch graphs of the following functions, indicating all relative maxima and minima and points of inflection:

(a) $f(x) = \dfrac{x^3}{x^2 - 1}$ (b) $f(x) = x^2 + \sin^2 x$

27. If at some instant of time sides a and b of the triangle in Figure 4.96 form a right angle, and if these sides are increasing at equal rates, does it remain a right-angled triangle?

FIGURE 4.96

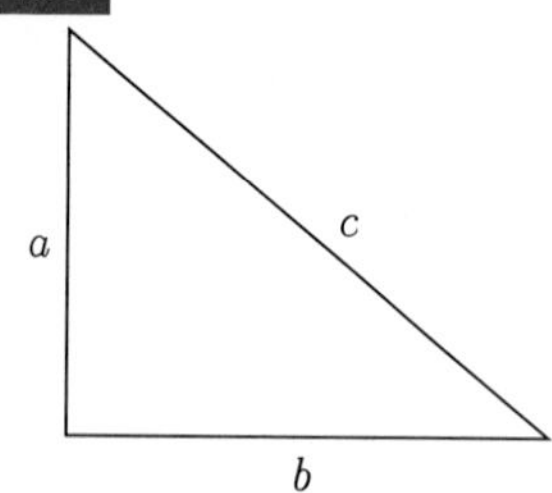

28. A footbal team presently sells tickets at prices of \$8, \$9, and \$10 per seat, depending on the position of the seat. At these prices it averages sales of 10 000 at \$10, 20 000 at \$9, and 30 000 at \$8. The team wishes to raise the price of each ticket by the same amount, but feels that for every dollar the price is raised, 10% fewer tickets of each type will be sold. What price increase per ticket will maximize revenue?

29. Repeat Exercise 28 given that the team takes into account the fact that profit from concession sales for each person at the game is fifty cents.

30. If a particle moves away from the origin along the positive x-axis with a constant speed of 10 m/s, how fast is its distance from the curve $y = x^2$ changing when it is at $x = 3$ m?

31. Each evening a cow in a pasture returns to its barn at point B (Figure 4.97). But it always does so by first walking to the river for a drink. If the cow walks at 2 km/h and stops to drink for 2 minutes, what is the minimum time it takes for the cow to get from the pasture to the barn? What is the minimum time if the cow walks twice as fast?

FIGURE 4.97

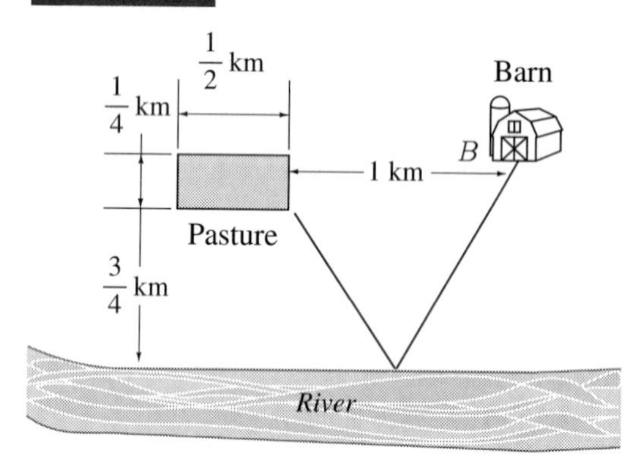

32. A farmer has 100 hectares to plant in corn and potatoes. Undamaged corn yields \$$p$ per hectare and potatoes \$$q$ per hectare. For each crop, the loss due to disease and pests per unit hectare is directly proportional to the area planted. If the farmer plants x hectares of corn, the loss due to disease and pests is equal to ax per hectare. The total loss of corn is therefore ax^2 hectares. Similarly, the total loss of potatoes is by^2, if the area planted in potatoes is y hectares. Find the areas that should be planted in corn and potatoes in order to minimize monetary loss. Substitute sample values for a, b, p, and q to see whether your results look reasonable.

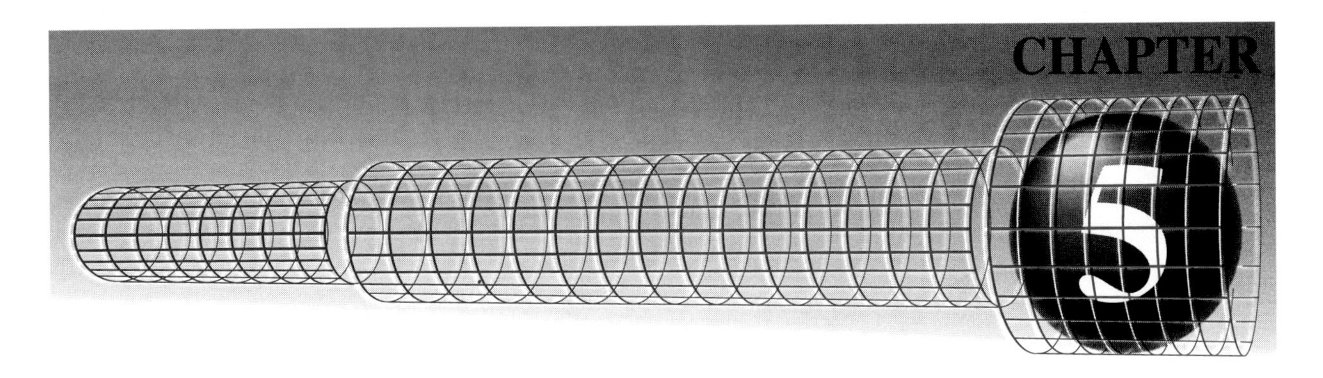

The Indefinite Integral or Antiderivative

In Chapter 3 we introduced the derivative and ways to differentiate functions defined explicitly and implicitly. We then discussed various applications of calculus, including velocity and acceleration, related rates, maxima and minima, Newton's iterative procedure, and L'Hôpital's rule. In this chapter, we reverse the differentiation process. Instead of giving you a function and asking for its derivative, we give you the derivative and ask you to find the function. This process of "backwards differentiation" or "antidifferentiation" has such diverse applications that antidifferentiation is as important to calculus as differentiation. In many problems we find ourselves differentiating at one stage and antidifferentiating at another.

Antidifferentiation is a much more difficult process than differentiation. All but one of the rules for differentiation were developed in Chapter 3; we have a few more functions to differentiate in Chapter 8, and one more rule in Section 10.1. In contrast to this, the list of formulas and techniques for finding antiderivatives is endless. In this chapter we introduce the three simplest, but most important, techniques; in Chapter 9 we discuss many others.

SECTION 5.1

The Reverse Operation of Differentiation

In our discussions on velocity and acceleration in Section 4.7, we showed that when an object moves along the x-axis with its position described by the function $x(t)$, its velocity and acceleration are the first and second derivatives of $x(t)$ with respect to time t:

$$v(t) = \frac{dx}{dt} \qquad \text{and} \qquad a(t) = \frac{d^2x}{dt^2}.$$

For example, if $x(t) = t^3 + 3t^2$, then

$$v(t) = 3t^2 + 6t \qquad \text{and} \qquad a(t) = 6t + 6.$$

When engineers and physicists study the motions of objects, a more common type of problem is to determine the position of an object; that is, the position is not given. What they might know, however, is the acceleration of the object (perhaps through Newton's second law, which states that acceleration is proportional to the resultant force on the object). So the question we must now ask is: If we know the acceleration $a(t)$ of an object as a function of time, can we obtain its velocity and position by reversing the differentiations? In the above example, if we know that

$$a(t) = \frac{dv}{dt} = 6t + 6,$$

can we find the function $v(t)$ that differentiates to give $6t+6$? We know that to arrive at the terms $6t$ and 6 after differentiation, $v(t)$ might have contained the terms $3t^2$ and $6t$. In other words, one possible velocity function that differentiates to give $a(t) = 6t + 6$ is $v(t) = 3t^2 + 6t$. It is not, however, the only one; $v(t) = 3t^2 + 6t + 10$ and $v(t) = 3t^2 + 6t - 22$ also have derivative $6t+6$. In fact, for any constant C whatsoever, the derivative of

$$v(t) = 3t^2 + 6t + C$$

is $a(t) = 6t + 6$. In Theorem 5.1 we shall show that this velocity function represents all functions that have $6t+6$ as their derivative. We shall also demonstrate how to evaluate the constant C.

For the present, let us set $C = 0$ so that

$$v(t) = \frac{dx}{dt} = 3t^2 + 6t.$$

We now ask what position function $x(t)$ differentiates to give $3t^2 + 6t$. One possibility is $x(t) = t^3 + 3t^2$, and for the same reason as above, so is

$$x(t) = t^3 + 3t^2 + D$$

for any constant D whatsoever.

Thus, by reversing the differentiation operation in this example we have proceeded from the acceleration $a(t) = 6t + 6$ to possible velocity functions $v(t)$, and then to possible position functions $x(t)$. This process of antidifferentiation has applications far beyond velocity and acceleration problems, and we shall see many of them as we progress through this book. We begin our formal study of antidifferentiation with the following definition.

Definition 5.1

A function $F(x)$ is called an **antiderivative** or **indefinite integral** of $f(x)$ on an interval I if on I

$$F'(x) = f(x). \tag{5.1}$$

For example, since

$$\frac{d}{dx}x^4 = 4x^3,$$

we say that x^4 is an antiderivative or an indefinite integral of $4x^3$ for all x. But for any constant C, the function $x^4 + C$ is also an antiderivative of $4x^3$. The following theorem indicates that these are the only antiderivatives of $4x^3$.

Theorem 5.1

If $F(x)$ is an indefinite integral of $f(x)$ on an interval I, then every indefinite integral of $f(x)$ on I is of the form

$$F(x) + C, \qquad \text{where } C \text{ is a constant.}$$

Proof Suppose that $F(x)$ and $G(x)$ are two antiderivatives of $f(x)$ on I. If we define a function $D(x) = G(x) - F(x)$, then on I

$$D'(x) = G'(x) - F'(x) = f(x) - f(x) = 0.$$

If x_1 and x_2 are any two points in the interval I, then certainly $D'(x) = 0$ on the interval $x_1 \leq x \leq x_2$. But differentiability of $D(x)$ on $x_1 \leq x \leq x_2$ implies continuity of $D(x)$ thereon also (Theorem 3.6). We may therefore apply the mean value theorem (Theorem 3.15) to $D(x)$ on the interval $x_1 \leq x \leq x_2$, and conclude that there exists a number c between x_1 and x_2 such that

$$D'(c) = \frac{D(x_2) - D(x_1)}{x_2 - x_1}.$$

But because $D'(c) = 0$, it follows that $D(x_1) = D(x_2)$. Since x_1 and x_2 are any two points in I, we conclude that $D(x)$ must have the same value at every point in I; that is, on I,

$$D(x) = G(x) - F(x) = C,$$

where C is a constant. Consequently,

$$G(x) = F(x) + C,$$

and our proof is complete.

Because of this theorem, if we find one antiderivative $F(x)$ of $f(x)$ by any means whatsoever, then we have found every antiderivative of $f(x)$, since every antiderivative can be written as $F(x)$ plus a constant C. As every indefinite integral of $f(x)$ is of the form $F(x) + C$, we call $F(x) + C$ *the* indefinite integral of $f(x)$. The operation of taking the indefinite integral is denoted by

$$\int f(x)\,dx = F(x) + C, \tag{5.2}$$

where the differential dx indicates that antidifferentiation is with respect to x. We call $f(x)$ the **integrand** of the indefinite integral. For example, we write

$$\int x^2\,dx = \frac{x^3}{3} + C \qquad \text{and} \qquad \int \frac{1}{x^3}\,dx = -\frac{1}{2x^2} + C.$$

The following theorem is fundamental to the calculation of antiderivatives. Its proof is a straightforward exercise in differentiation and the use of Definition 5.1.

Theorem 5.2

If $f(x)$ and $g(x)$ have antiderivatives on an interval I, then on I:

(i) $$\int [f(x) + g(x)]\,dx = \int f(x)\,dx + \int g(x)\,dx; \tag{5.3}$$

(ii) $$\int kf(x)\,dx = k\int f(x)\,dx, \qquad k \text{ a constant.} \tag{5.4}$$

For example, to evaluate the antiderivative of $2x^3 - 4x$, we write

$$\int (2x^3 - 4x)\,dx = \int 2x^3\,dx + \int -4x\,dx \qquad \text{(by part (i) of Theorem 5.2)}$$

$$= 2\int x^3\,dx - 4\int x\,dx \qquad \text{(by part (ii) of Theorem 5.2)}$$

$$= 2\left(\frac{x^4}{4}\right) - 4\left(\frac{x^2}{2}\right) + C$$

$$= \frac{x^4}{2} - 2x^2 + C.$$

Every differentiation formula developed in Chapter 3 can be expressed as an integration formula. In fact, equations 5.3 and 5.4 are integral counterparts of equations 3.8 and 3.7. Some of the differentiation formulas for trigonometric, exponential, and logarithm functions are restated below as integration formulas.

$$\int \sin x\,dx = -\cos x + C, \tag{5.5a}$$

$$\int \cos x\,dx = \sin x + C, \tag{5.5b}$$

$$\int \sec^2 x\,dx = \tan x + C, \tag{5.5c}$$

$$\int \sec x \tan x\,dx = \sec x + C, \tag{5.5d}$$

$$\int \csc^2 x\,dx = -\cot x + C, \tag{5.5e}$$

$$\int \csc x \cot x\,dx = -\csc x + C, \tag{5.5f}$$

$$\int e^x\,dx = e^x + C, \tag{5.5g}$$

$$\int a^x\,dx = a^x \log_a e + C, \tag{5.5h}$$

$$\int \frac{1}{x}\,dx = \ln|x| + C. \tag{5.5i}$$

The last equation follows immediately from 3.37 with $f(x) = x$.

Perhaps the most important integration formula is the counterpart of power rule 3.6. Since

$$\frac{d}{dx}x^{n+1} = (n+1)x^n,$$

it follows that

$$\int (n+1)x^n\,dx = x^{n+1} + \overline{C}.$$

With property 5.4, the $n+1$ can be removed from the integral and taken to the other side of the equation,

$$\int x^n\,dx = \frac{1}{n+1}x^{n+1} + C, \qquad n \neq -1, \tag{5.6}$$

where $C = \overline{C}/(n+1)$. As indicated, this result is valid provided $n \neq -1$, but formula 5.5i takes care of this exceptional case.

EXAMPLE 5.1

Evaluate

$$\int \left(2x + \frac{1}{x^2}\right) dx.$$

SOLUTION Using Theorem 5.2 and formula (5.6), we find

$$\int \left(2x + \frac{1}{x^2}\right) dx = 2\int x\,dx + \int \frac{1}{x^2}\,dx = 2\left(\frac{x^2}{2}\right) - \frac{1}{x} + C = x^2 - \frac{1}{x} + C.$$

Where is $x^2 - 1/x + C$, the indefinite integral of $2x + 1/x^2$? It is valid on the intervals $x < 0$ and $x > 0$, but not for all x because the function is not defined at $x = 0$. To make this clear, we should write

$$\int \left(2x + \frac{1}{x^2}\right) dx = \begin{cases} x^2 - \dfrac{1}{x} + C_1 & x < 0 \\ x^2 - \dfrac{1}{x} + C_2 & x > 0 \end{cases}$$

where the constants C_1 and C_2 need not be the same. For brevity we often write

$$\int \left(2x + \frac{1}{x^2}\right) dx = x^2 - \frac{1}{x} + C,$$

thereby suppressing the complete description of the indefinite integral. When the context demands that we distinguish various intervals on which the indefinite integral is defined, we shall be careful to give the extended version. ■

EXAMPLE 5.2

Find a curve that passes through the point $(1, 5)$ and whose tangent line at each point (x, y) has slope $5x^4 - 3x^2 + 2$.

SOLUTION If $y = f(x)$ is the equation of the curve, then

$$\frac{dy}{dx} = 5x^4 - 3x^2 + 2.$$

If we take antiderivatives of both sides of this equation with respect to x, we obtain

$$\int \frac{dy}{dx}\,dx = \int (5x^4 - 3x^2 + 2)\,dx \qquad \text{or}$$

$$y = x^5 - x^3 + 2x + C.$$

Since $(1, 5)$ is a point on the curve, its coordinates must satisfy the equation of the curve:

$$5 = 1^5 - 1^3 + 2(1) + C.$$

Thus $C = 3$, and the required curve is

$$y = x^5 - x^3 + 2x + 3.$$

■

In taking the antiderivative of each side of the equation

$$\frac{dy}{dx} = 5x^4 - 3x^2 + 2$$

in the above example, we added an arbitrary constant C to the right-hand side. You might question why we did not add a constant to the left-hand side. Had we done so, the result would have been

$$y + D = x^5 - x^3 + 2x + E.$$

If we then wrote

$$y = x^5 - x^3 + 2x + (E - D)$$

and defined $C = E - D$, we would have obtained exactly the same result. Hence, nothing is gained by adding an arbitrary constant to both sides; a constant on one side is sufficient.

The problem in Example 5.2 was geometric: find the equation of a curve satisfying certain properties. We quickly recast it as the problem of finding the function $y = f(x)$ that satisfies the equation

$$\frac{dy}{dx} = 5x^4 - 3x^2 + 2,$$

subject to the additional condition that $f(1) = 5$. Many applications of calculus lead to problems of this form. Find a function that satisfies an equation which involves the derivative of the function. Such an equation is called a **differential equation**. Differential equations are discussed again in Section 5.4, and before studyng them in detail in Chapter 16, we shall encounter them repeatedly in our discussions. When we solve a differential equation with no subsidiary conditions, say,

$$\frac{dy}{dx} = 4x^2 + 7x,$$

we do not get a function, but rather a one-parameter family of functions. For this differential equation, we obtain

$$y = \frac{4}{3}x^3 + \frac{7}{2}x^2 + C,$$

FIGURE 5.1

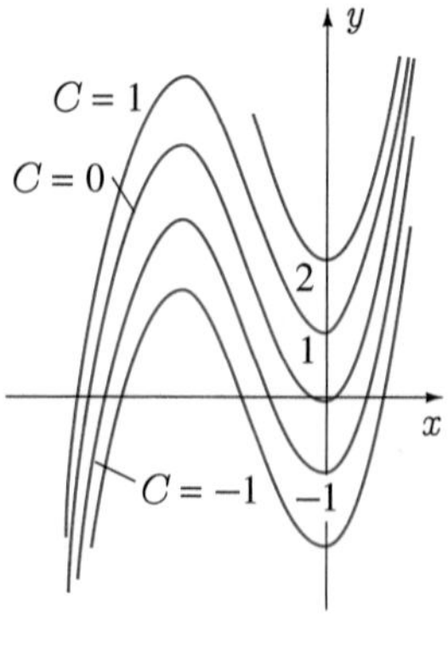

y $C = 1$ $C = 0$ 2 1 x $C = -1$ -1

a one-parameter family of cubic polynomials, C being the parameter. Geometrically, we have the one-parameter family of curves in Figure 5.1. Parameter C represents a vertical shift of one curve relative to another. Note that if a vertical line is drawn at any position x to intersect these curves, then at the points of intersection, every curve has exactly the same slope, namely, $4x^2 + 7x$. The slope of each cubic at $x = 0$ is zero. If an extra condition is added to the differential equation, such as to demand that y be equal to 4 when $x = 2$, then

$$4 = \frac{4}{3}(2)^3 + \frac{7}{2}(2)^2 + C,$$

or, $C = -62/3$. This condition singles out one particular function from the family, namely, $y = 4x^3/3 + 7x^2/2 - 62/3$. Geometrically, it determines that curve in the family which passes through the point $(2, 4)$.

In this chapter, we discuss three basic ways to find antiderivatives. First, some antiderivatives are obvious, and the better you are at differentiation, the more likely this will be. For example, you should have no trouble recognizing that

$$\int 7x^5\,dx = \frac{7x^6}{6} + C \qquad \text{and} \qquad \int \frac{1}{x^3}\,dx = \frac{-1}{2x^2} + C.$$

Our second method results from the answer to the following question: How do we check that a function $F(x)$ is an antiderivative of $f(x)$? We differentiate $F(x)$, of course. This simple fact suggests an approach to slightly more complex problems, say,

$$\int (2x+3)^5\,dx.$$

We might reason that in order to have $2x + 3$ raised to power 5 after differentiation, we had $2x + 3$ to power 6 before differentiation; that is, a logical "guess" for the antiderivative is $(2x+3)^6$. Differentiation of this function gives

$$\frac{d}{dx}(2x+3)^6 = 6(2x+3)^5(2) = 12(2x+3)^5,$$

and we see that $(2x+3)^6$ is not the correct antiderivative. It has produced $(2x+3)^5$, as required, but it has also given an undesireable factor of 12. We therefore "fix up" our original guess by multiplying it by $1/12$; that is, the correct antiderivative is

$$\int (2x+3)^5\,dx = \frac{1}{12}(2x+3)^6 + C.$$

This is what we call "guess and fix up": we guess the antiderivative closely enough to have it fixed up. Our guess must be correct within a *multiplicative constant.* We cannot be out by x's. Let us illustrate this with two very similar problems,

$$\int \frac{1}{(5x+2)^5}\,dx \qquad \text{and} \qquad \int \frac{1}{(5x^2+2)^5}\,dx.$$

For the first problem we might guess at $(5x+2)^{-4}$. Differentiation gives

$$\frac{d}{dx}\left[\frac{1}{(5x+2)^4}\right] = \frac{-4}{(5x+2)^5}(5) = \frac{-20}{(5x+2)^5}.$$

Since we are out by a factor of -20, we fix up the original guess,

$$\int \frac{1}{(5x+2)^5}\,dx = \frac{-1}{20(5x+2)^4} + C.$$

It might seem just as logical to guess at $(5x^2+2)^{-4}$ for the second problem, but differentiation yields

$$\frac{d}{dx}\left[\frac{1}{(5x^2+2)^4}\right] = \frac{-4}{(5x^2+2)^5}(10x) = \frac{-40x}{(5x^2+2)^5}.$$

This time the discrepancy is $-40x$, which involves x. We must not try to fix up the original guess; it must be abandoned totally. This indefinite integral is quite difficult, and we will have to wait for the more powerful techniques of Chapter 9. To emphasize once again, do not try to fix up x's, only constants.

EXAMPLE 5.3

Evaluate the following indefinite integrals:

$$\text{(a)} \int \cos 3x\, dx \qquad \text{(b)} \int e^{-2x}\, dx \qquad \text{(c)} \int \frac{x}{3x^2 - 4}\, dx$$

Solutions (a) To obtain $\cos 3x$ after differentiation, we guess $\sin 3x$ as the antiderivative. Since

$$\frac{d}{dx} \sin 3x = 3 \cos 3x,$$

it is necessary to fix up a 3,

$$\int \cos 3x\, dx = \frac{1}{3} \sin 3x + C.$$

(b) With an initial guess of e^{-2x} for the antiderivative based on the fact that the derivative of an exponential function always returns the same exponential, we calculate

$$\frac{d}{dx} e^{-2x} = -2e^{-2x}.$$

Consequently, we fix up a -2,

$$\int e^{-2x}\, dx = -\frac{1}{2} e^{-2x} + C.$$

(c) Since

$$\frac{d}{dx} \ln(3x^2 - 4) = \frac{1}{3x^2 - 4}(6x),$$

the required antiderivative is

$$\int \frac{x}{3x^2 - 4}\, dx = \frac{1}{6} \ln|3x^2 - 4| + C.$$

We have inserted absolute values as suggested by formula 5.5i. ■

The third technique for finding antiderivatives is discussed in Section 5.3.

EXERCISES 5.1

In Exercises 1–20 evaluate the indefinite integral.

1. $\int (x^3 - 2x)\, dx$

2. $\int (x^4 + 3x^2 + 5x)\, dx$

3. $\int (2x^3 - 3x^2 + 6x + 6)\, dx$

4. $\int \sin x\, dx$

5. $\int 3 \cos x\, dx$

6. $\int \sqrt{x}\, dx$

7. $\int \left(x^{10} - \frac{1}{x^3}\right) dx$

8. $\int \left(\frac{1}{x^2} - \frac{2}{x^4}\right) dx$

9. $\int \left(x^{3/2} - x^{2/7}\right) dx$

10. $\int \left(\frac{1}{x^2} + \frac{1}{2\sqrt{x}}\right) dx$

11. $\int \left(\frac{4}{x^{3/2}} + 2x^{1/3}\right) dx$

12. $\int \left(-\frac{1}{2x^2} + 3x^3\right) dx$

13. $\int \frac{1}{x^\pi}\, dx$

14. $\int \left(2\sqrt{x} + 3x^{3/2} - 5x^{5/2}\right) dx$

15. $\int x^2(x^2 - 3)\, dx$

16. $\int \sqrt{x}(x + 1)\, dx$

17. $\int \left(\frac{x-2}{x^3} \right) dx$

18. $\int x^2 (1 + x^2)^2 \, dx$

19. $\int (x^2 + 1)^3 \, dx$

20. $\int \frac{(x-1)^2}{\sqrt{x}} \, dx$

In Exercises 21–24 find the curve $y = f(x)$ that passes through the given point and whose slope at each point (x, y) is defined by the indicated derivative.

21. $dy/dx = x^2 - 3x + 2, \quad (2, 1)$

22. $dy/dx = 2x^3 + 4x, \quad (0, 5)$

23. $dy/dx = -2x^4 + 3x^2 + 6, \quad (1, 0)$

24. $dy/dx = 2 - 4x + 8x^7, \quad (1, 1)$

25. Find the equation of the curve that has a second derivative equal to $6x^2$ and passes through the points $(0, 2)$ and $(-1, 3)$.

26. Find a function $f(x)$ that has a relative maximum $f(2) = 3$ and has a second derivative equal to $-5x$.

27. Is it possible to find a function $f(x)$ that has a relative minimum $f(2) = 3$ and has a second derivative equal to $-5x$?

In Exercises 28–59 use "guess and fix up" to evaluate the antiderivative.

28. $\int \sqrt{x+2} \, dx$

29. $\int (x+5)^{3/2} \, dx$

30. $\int \sqrt{2-x} \, dx$

31. $\int \frac{1}{\sqrt{4x+3}} \, dx$

32. $\int (2x-3)^{3/2} \, dx$

33. $\int (3x+1)^5 \, dx$

34. $\int (1-2x)^7 \, dx$

35. $\int \frac{1}{(x+4)^2} \, dx$

36. $\int \frac{1}{(1+3x)^6} \, dx$

37. $\int x(x^2+1)^3 \, dx$

38. $\int x^2 (2 + 3x^3)^7 \, dx$

39. $\int \frac{x}{(2+x^2)^2} \, dx$

40. $\int \cos 2x \, dx$

41. $\int \cos^2 x \sin x \, dx$

42. $\int 3 \sin 2x \cos 2x \, dx$

43. $\int \sec 12x \tan 12x \, dx$

44. $\int \csc^2 4x \, dx$

45. $\int e^{4x} \, dx$

46. $\int x e^{-x^2} \, dx$

47. $\int \frac{e^{3/x}}{x^2} \, dx$

48. $\int e^{4x-3} \, dx$

49. $\int \frac{1}{3x+2} \, dx$

50. $\int \frac{2}{7-5x} \, dx$

51. $\int \frac{x}{1-x^2} \, dx$

52. $\int \frac{3x^2}{1-4x^3} \, dx$

53. $\int 2^x \, dx$

54. $\int 3^{2x} \, dx$

55. $\int \frac{e^x}{e^x+1} \, dx$

56. $\int \sin x (1 + \cos x)^4 \, dx$

57. $\int \frac{\cos x}{\sin^3 x} \, dx$

58. $\int e^{2x} (1 + e^{2x})^3 \, dx$

59. $\int \frac{\sec^2 x}{\tan^2 x} \, dx$

In Exercises 60–65 find a one-parameter family of functions satisfying the differential equation.

60. $\frac{dy}{dx} = x^3 - \frac{1}{x^2}$

61. $\frac{dy}{dx} = \sqrt{3-4x}$

62. $\frac{dy}{dx} = \frac{1}{(3x+5)^{3/2}}$

63. $\frac{dy}{dx} = x^2 (2x^3 + 4)^4$

64. $\frac{dy}{dx} = \frac{x^3}{(2+3x^4)^2}$

65. $\frac{dy}{dx} = \sin x (1 + \cos^2 x)$

66. Find a function $y = f(x)$ that satisfies the differential equation

$$\frac{dy}{dx} = \frac{1}{x^2}$$

and passes through the two points $(1, 1)$ and $(-1, -2)$.

67. (a) Find the indefinite integral of the signum function of Exercise 36 in Section 2.4.

(b) Prove that $\operatorname{sgn} x$ cannot have an antiderivative on any interval containing $x = 0$.

68. Find the indefinite integral of the Heaviside unit step function of Exercise 38(a) in Section 2.4. Sketch a graph of the indefinite integral function.

SECTION 5.2

Velocity and Acceleration Revisited

In Section 4.7 we discussed relationships among position, velocity, and acceleration from the viewpoint of derivatives. We now consider these relationships through antiderivatives, an approach providing a far more practical viewpoint when it comes to applications.

For motion along the x-axis, velocity is the derivative of position with respect to time:

$v(t) = dx/dt$. We can say, therefore, that position is the antiderivative of velocity:

$$x(t) = \int v(t)\, dt. \tag{5.7}$$

Similarly, as acceleration is the derivative of velocity,

$$v(t) = \int a(t)\, dt. \tag{5.8}$$

Thus, given the acceleration of an object moving along a straight line, we can antidifferentiate to find its velocity and antidifferentiate again for its position. Since each antidifferentiation introduces an arbitrary constant, additional information must be specified in order to evaluate these constants.

EXAMPLE 5.4

The car in Figure 5.2 accelerates from rest when the light turns green. Initially the acceleration is 10 m/s^2, but it decreases linearly to zero after 10 s. Find the velocity and position of the car during this time interval.

SOLUTION Let us set up the coordinate system in Figure 5.2 and choose time $t = 0$ when the car pulls away from the light. Given this time convention, the acceleration of the car as a function of time t (see Figure 5.3) is

$$a(t) = 10 - t, \quad 0 \le t \le 10.$$

If $v(t)$ denotes the velocity of the car during the time interval, then

$$\frac{dv}{dt} = 10 - t,$$

FIGURE 5.2

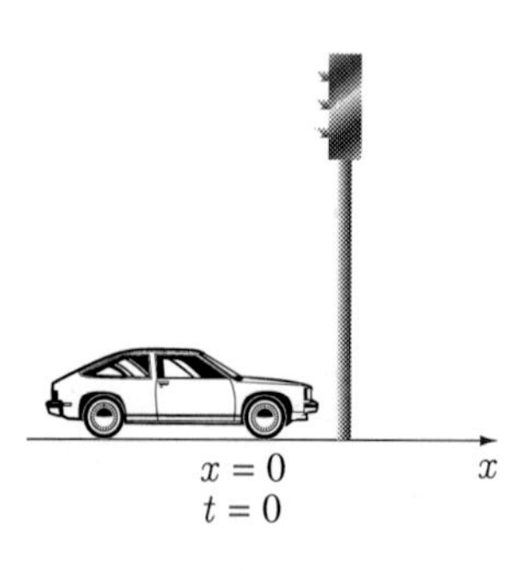

FIGURE 5.3

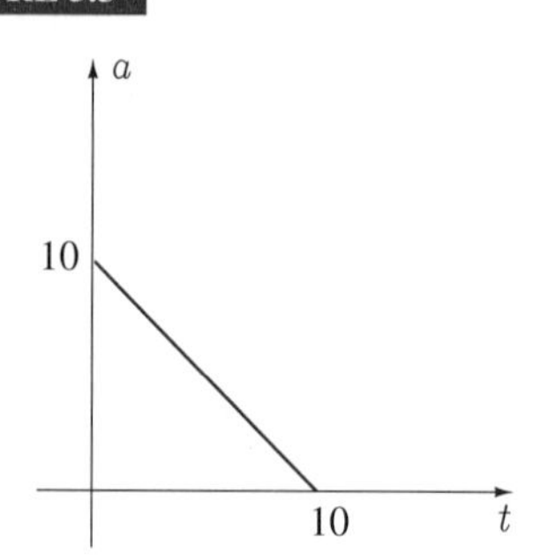

and antidifferentiation gives

$$v(t) = 10t - \frac{t^2}{2} + C.$$

By our time convention, the velocity of the car is zero at time $t = 0$; that is, $v(0) = 0$, and from this we obtain

$$0 = 10(0) - \frac{(0)^2}{2} + C.$$

Consequently, $C = 0$, and

$$v(t) = 10t - \frac{t^2}{2}.$$

This is the velocity of the car during the time interval $0 \leq t \leq 10$, measured in metres per second.

Since the position of the car with respect to its original position at time $t = 0$ is denoted by x,

$$v = \frac{dx}{dt} = 10t - \frac{t^2}{2}.$$

Thus,

$$x(t) = 5t^2 - \frac{t^3}{6} + D.$$

Since $x(0) = 0$, the constant D must also be zero, and the position of the car indicated by its distance in metres from the stoplight, for $0 \leq t \leq 10$, is

$$x(t) = 5t^2 - \frac{t^3}{6}.$$

■

EXAMPLE 5.5

A stone is thrown vertically upward over the edge of a cliff at 25 m/s. When does it hit the base of the cliff if the cliff is 100 m high?

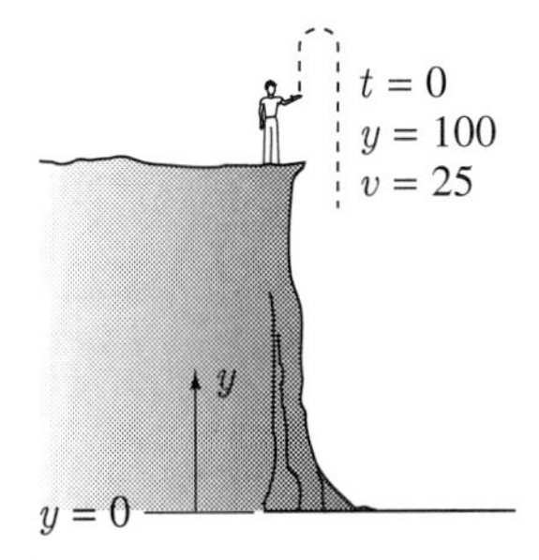

FIGURE 5.4

SOLUTION Let us measure y as positive upward, taking $y = 100$ and $t = 0$ at the point and instant of projection (Figure 5.4). A law of physics states that when an object near the earth's surface is acted on by gravity alone, it experiences an acceleration whose magnitude is 9.81 m/s^2. If a denotes the acceleration of the stone and v its velocity, then

$$a = \frac{dv}{dt} = -9.81$$

(a is negative since it is in the negative y-direction). To obtain v we take antiderivatives with respect to t:

$$v(t) = -9.81t + C.$$

By our time convention, $v(0) = 25$ so that

$$25 = -9.81(0) + C.$$

Thus, $C = 25$, and

$$v(t) = -9.81t + 25.$$

We now have the velocity of the stone at any given instant. To find the position of the stone we set $v = dy/dt$,

$$\frac{dy}{dt} = -9.81t + 25,$$

and take antiderivatives once again:

$$y(t) = -4.905t^2 + 25t + D.$$

Since $y = 100$ when $t = 0$,

$$100 = -4.905(0)^2 + 25(0) + D.$$

Hence, $D = 100$, and

$$y(t) = -4.905t^2 + 25t + 100.$$

We have found the equation that tells us exactly where the stone is at any given time. To determine when the stone strikes the base of the cliff, we set $y(t) = 0$; that is,

$$0 = -4.905t^2 + 25t + 100,$$

a quadratic equation with solutions

$$t = \frac{-25 \pm \sqrt{25^2 - 4(-4.905)(100)}}{-9.81} = \frac{25 \pm \sqrt{2587}}{9.81} = 7.7 \quad \text{or} \quad -2.6.$$

Since the negative root must be rejected, we find that the stone strikes the base of the cliff after 7.7 seconds. ■

EXAMPLE 5.6

Two stones are thrown vertically upward one second apart over the edge of the cliff in Example 5.5. The first is thrown at 25 m/s, the second at 20 m/s. Determine if and when they ever pass each other.

FIGURE 5.5

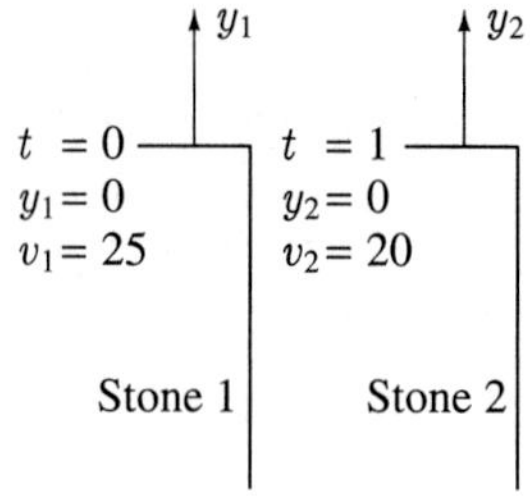

SOLUTION Let us choose a different coordinate system to that of Example 5.5 by taking y as positive upward with $y = 0$ and $t = 0$ at the point and instant of projection of the first stone (Figure 5.5). The acceleration of stone 1 is

$$a_1 = -9.81 = \frac{dv_1}{dt},$$

from which

$$v_1(t) = -9.81t + C.$$

Since $v_1(0) = 25$, it follows that $25 = C$, and

$$v_1(t) = -9.81t + 25, \quad t \geq 0.$$

Thus,

$$\frac{dy_1}{dt} = -9.81t + 25,$$

from which we have

$$y_1(t) = -4.905t^2 + 25t + D.$$

Since $y_1(0) = 0$, we find that $0 = D$, and

$$y_1(t) = -4.905t^2 + 25t, \quad t \geq 0.$$

The acceleration of stone 2 is also -9.81; hence we have

$$a_2 = -9.81 = \frac{dv_2}{dt},$$

from which

$$v_2(t) = -9.81t + E.$$

Because $v_2(1) = 20$, we must have $20 = -9.81(1) + E$, or $E = 29.81$. Thus,

$$v_2(t) = -9.81t + 29.81, \quad t \geq 1.$$

Consequently,

$$\frac{dy_2}{dt} = -9.81t + 29.81,$$

from which

$$y_2(t) = -4.905t^2 + 29.81t + F.$$

Since $y_2(1) = 0$, it follows that $0 = -4.905(1)^2 + 29.81(1) + F$, or $F = -24.905$. Thus,

$$y_2(t) = -4.905t^2 + 29.81t - 24.905, \quad t \geq 1.$$

The stones will pass each other if y_1 and y_2 are ever equal for the same time t; that is, if

$$-4.905t^2 + 25t = -4.905t^2 + 29.81t - 24.905.$$

The solution of this equation is

$$t = \frac{24.905}{4.81} = 5.2.$$

Because the first stone does not strike the base of the cliff for 7.7 seconds (Example 5.5), it follows that the stones do indeed pass each other 5.2 seconds after the first stone is projected. ■

An important point to note in Examples 5.4, 5.5, and 5.6 is that the coordinate system and time convention were specified immediately; that is, we decided, and did so at the start of the problem, where to place the origin of our coordinate system, which direction to choose as positive, and when to choose $t = 0$. Only then were we able to specify the correct sign for acceleration. Furthermore, we determined constants from antidifferentiations using initial conditions expressed in terms of our coordinate system and time convention. Throughout the solutions, we were careful to refer everything to our choice of coordinates and time. Remember, then, to specify clearly the coordinate system and time convention at the beginning of a problem.

We should also note that in each of the examples we antidifferentiated with respect to time, *only those equations that were valid for a range of values of time.* It is a common error to antidifferentiate equations that are only valid at one instant. For instance, students would reason in Example 5.5 that the initial velocity is 25, $v = dy/dt = 25$, and hence $y(t) = 25t + C$. This is incorrect, because the equation $dy/dt = 25$ is valid only at time $t = 0$, and cannot be antidifferentiated.

EXERCISES 5.2

In Exercises 1–8 we have defined acceleration $a(t)$ of an object moving along the x-axis during some time interval and specified the initial conditions $x(0)$ and $v(0)$. Find the velocity $v(t)$ and position $x(t)$ of the object as functions of time.

1. $a(t) = t + 2, \quad 0 \le t \le 3; \quad v(0) = 0, x(0) = 0$

2. $a(t) = 6 - 2t, \quad 0 \le t \le 3; \quad v(0) = 5, x(0) = 0$

3. $a(t) = 6 - 2t \quad 0 \le t \le 4; \quad v(0) = 5, x(0) = 0$

4. $a(t) = 120t - 12t^2, \quad 0 \le t \le 10; \quad v(0) = 0, x(0) = 4$

5. $a(t) = t^2 + 1, \quad 0 \le t \le 5; \quad v(0) = -1, x(0) = 1$

6. $a(t) = t^2 + 5t + 4, \quad 0 \le t \le 15; \quad v(0) = -2, x(0) = -3$

7. $a(t) = \cos t, \quad t \ge 0; \quad v(0) = 0, x(0) = 0$

8. $a(t) = 3 \sin t, \quad t \ge 0; \quad v(0) = 1, x(0) = 4$

9. The velocity of an object moving along the x-axis is given in metres per second by

$$v(t) = 3t^2 - 9t + 6$$

where t is time in seconds. If the object starts from position $x = 1$ m at time $t = 0$, answer each of the following questions:

(a) What is the acceleration of the object at $t = 5$ s?

(b) What is the position of the object at $t = 2$ s?

(c) Is the object speeding up or slowing down at $t = 5/4$ s?

(d) What is the closest the object ever comes to the origin?

10. A particle moving along the x-axis has acceleration

$$a(t) = 6t - 2$$

in metres per second per second for time $t \ge 0$.

(a) If the particle starts at the point $x = 1$ moving to the left with speed 3 m/s, find its position as a function of time t.

(b) At what time does the particle have zero velocity (if any)?

11. The acceleration of a particle moving along the x-axis is given in metres per second per second by

$$a(t) = 6t - 15,$$

where $t \ge 0$ is time in seconds.

(a) If the velocity of the particle at $t = 2$ s is 6 m/s, what is its velocity at $t = 1$ s?

(b) If the particle is 10 m to the right of the origin at time $t = 0$, what is its position as a function of time t?

(c) What is the closest the particle ever comes to the origin?

12. A car is sitting at rest at a stoplight. When the light turns green at time $t = 0$, the driver immediately presses the accelerator, imparting an acceleration of $a(t) = 3 - t/5$ m/s^2 to the car for 10 s.

(a) Where is the car after the 10 s?

(b) If the driver applies the brakes at $t = 10$ s, and the car experiences a constant deceleration of 2 m/s^2, where and when does the car come to a stop?

13. Find how far a plane will move when landing if, in t seconds after touching the ground, its speed in metres per second is given by $180 - 18t$.

14. A stone is thrown directly upward with an initial speed of 10 m/s. How high will it rise?

15. You are standing on a bridge 25 m above a river. If you wish to drop a stone onto a piece of floating wood, how soon before the wood reaches the appropriate spot should you drop the stone?

16. You are standing at the base of a building and wish to throw a ball to a friend on the roof 20 m above you. With what minimum speed must you throw the ball?

17. A car is travelling at 20 m/s when the brakes are applied. What constant deceleration must the car experience if it is to stop before striking a tree that is 50 m from the car at the instant the brakes are applied? Assume that the car travels in a straight line.

18. The position of an object moving along the x-axis is given by

$$x(t) = t^3 - 6t^2 + 9t - 20, \quad t \geq 0.$$

Sketch graphs of the position, velocity, and acceleration functions for this motion. Pay special attention to the fact that $v(t)$ represents the slope of $x = x(t)$ and $a(t)$ is the slope of $v = v(t)$.

19. Repeat Exercise 18 given that the acceleration of the object is

$$a(t) = 6t - 30, \quad t \geq 0,$$

and $v(0) = -33$ and $x(0) = 400$.

20. You are called on as an expert to testify at a traffic hearing. The question concerns the speed of a car that made an emergency stop with brakes locked and wheels sliding. The skid mark on the road measured 9 m. Assuming that the deceleration of the car was constant and could not exceed the acceleration due to gravity of a freely falling body (and this is indeed a reasonable assumption), what can you say about the speed of the car before the brakes were applied? Are you testifying for the prosecution or the defence?

21. When the brakes of an automobile are applied, they produce a constant deceleration of 5 m/s^2. What is the distance, from the point of application of brakes, required to stop a car travelling at:

(a) 100 km/hr?

(b) 50 km/hr?

(c) What is the ratio of these distances?

(d) Repeat (a), (b), and (c) given that the reaction time of the driver to get his foot from accelerator to brake is $3/4$ s, and distances are calculated taking this reaction time into account.

22. Two trains, one travelling at 100 km/hr and the other at 60 km/hr, are headed toward each other along a straight level track. When they are 2 km apart, each engineer sees the other's train and locks his wheels.

(a) If the deceleration of each train has magnitude $1/4$ m/s^2, determine whether a collision occurs.

(b) Repeat (a) given that the deceleration is caused by the wheels being reversed rather than locked.

(c) Illustrate graphically the difference between the situations in (a) and (b).

23. A steel bearing is dropped from the roof of a building. An observer standing in front of a window 1 m high notes that the bearing takes $1/8$ s to fall from the top to the bottom of the window. The bearing continues to fall, makes a completely elastic collision with a horizontal sidewalk, and reappears at the bottom of the window 2 s after passing it on the way down. After a completely elastic collision, the bearing will have the same speed at a point going up as it had going down. How tall is the building?

24. In the theory of special relativity, Newton's second law ($F = ma$) is replaced by

$$F = m_0 \frac{d}{dt}\left[\frac{v}{\sqrt{1 - (v^2/c^2)}}\right],$$

where F is the applied force, m_0 the mass of the particle measured at rest, v its speed, and c the speed of light — a constant. Show that if we set $a = dv/dt$, then

$$F = \frac{m_0 a}{[1 - (v^2/c^2)]^{3/2}}.$$

Explain the difference between this law and Newton's second law.

25. Two stones are thrown vertically upward over the edge of a bottomless abyss, the second stone t_0 units of time after the first. The first stone has an initial speed of v_0', and the second an initial speed of v_0''.

(a) Show that if the stones are ever to pass each other during their motions, two conditions must be satisfied:

$$gt_0 > v_0' - v_0'' \qquad \text{and} \qquad v_0' > gt_0/2,$$

where $g > 0$ is the acceleration due to gravity.

(b) Show that the first condition is equivalent to the requirement that stone 1 must begin its downward trajectory before stone 2.

(c) Show that the second condition is equivalent to the requirement that stone 1 must not pass its original projection point before the projection of stone 2.

SECTION 5.3

Change of Variable in the Indefinite Integral

In Section 5.1 we suggested two methods for evaluating indefinite integrals—"recognition" and "guess and fix up". In this section we show how a "change of variable" can often replace a complex integration problem with a simpler one.

Consider the indefinite integral

$$\int x\sqrt{2x+1}\,dx.$$

What is annoying about this integrand is the sum of two terms $2x+1$ under a square root. This can be changed by setting $u = 2x + 1$. As a result, $\sqrt{2x+1} = \sqrt{u}$, and the x in front of the square root is equal to $(u-1)/2$. Now the differential dx is used to indicate antidifferentiation in the problem with respect to x. But surely there must be another reason why we have chosen the differential to denote this. If we regard $x\sqrt{2x+1}\,dx$ as a product of $x\sqrt{2x+1}$ and dx, then perhaps we should obtain an expression for dx in terms of du. To do this we note that since $u = 2x + 1$, then

$$\frac{du}{dx} = 2.$$

As derivatives can be regarded as quotients of differentials, we can rewrite this equation in the form

$$dx = \frac{du}{2}.$$

If we make all these substitutions into the indefinite integral

$$\int x\sqrt{2x+1}\,dx,$$

the result is an integration problem in the variable u, which is easy to evaluate,

$$\int \left(\frac{u-1}{2}\right)\sqrt{u}\,\frac{du}{2} = \frac{1}{4}\int (u^{3/2} - u^{1/2})\,du = \frac{1}{4}\left(\frac{2}{5}u^{5/2} - \frac{2}{3}u^{3/2}\right) + C.$$

If u is now replaced by $2x + 1$, the result is

$$\frac{1}{10}(2x+1)^{5/2} - \frac{1}{6}(2x+1)^{3/2} + C.$$

Differentiation of this function quickly indicates that its derivative is indeed $x\sqrt{2x+1}$, and therefore,

$$\int x\sqrt{2x+1}\,dx = \frac{1}{10}(2x+1)^{5/2} - \frac{1}{6}(2x+1)^{3/2} + C.$$

In this example, the substitution $u = 2x + 1$ replaces a complex integration in x with a simple one in u. Once the problem in u is solved, replacement of u's with x's gives the solution to the original indefinite integral. This method is generally applicable and is justified in the following theorem.

Theorem 5.3

If by the substitution $u = g(x)$, the product $f(x)\,dx$ becomes $h(u)\,du$, and if

$$\int h(u)\,du = H(u) + C,$$

then

$$\int f(x)\,dx = H(g(x)) + C.$$

Proof To prove this theorem we must show that

$$\frac{d}{dx}[H(g(x))] = f(x).$$

If we set $y = H(u)$, then the chain rule gives

$$\frac{dy}{dx} = \frac{dy}{du}\frac{du}{dx} = H'(u)\frac{du}{dx} = h(u)\frac{du}{dx}.$$

But $h(u)\,du = f(x)\,dx$, so that

$$\frac{dy}{dx} = \frac{f(x)\,dx}{dx} = f(x),$$

and the proof is complete.

EXAMPLE 5.7

Evaluate the following indefinite integrals:

$$\text{(a)}\quad \int \sqrt[5]{2x+4}\,dx \qquad \text{(b)}\quad \int \frac{x}{\sqrt{x+1}}\,dx \qquad \text{(c)}\quad \int \sin^3 x \cos^2 x\,dx$$

SOLUTION (a) We could use "guess and fix up" in this case. With an initial guess of $(2x+4)^{6/5}$, we calculate

$$\frac{d}{dx}(2x+4)^{6/5} = \frac{6}{5}(2x+4)^{1/5}(2).$$

Thus,

$$\int (2x+4)^{1/5}\,dx = \frac{5}{12}(2x+4)^{6/5} + C.$$

Alternatively, if we set $u = 2x+4$, then $du = 2\,dx$, and

$$\int (2x+4)^{1/5}\,dx = \int u^{1/5}\frac{du}{2} = \frac{1}{2}\left\{\frac{5}{6}u^{6/5}\right\} + C = \frac{5}{12}(2x+4)^{6/5} + C.$$

(b) If we set $u = x+1$, then $du = dx$, and

$$\int \frac{x}{\sqrt{x+1}}\,dx = \int \frac{(u-1)}{\sqrt{u}}\,du = \int (u^{1/2} - u^{-1/2})\,du = \frac{2}{3}u^{3/2} - 2u^{1/2} + C$$
$$= \frac{2}{3}(x+1)^{3/2} - 2(x+1)^{1/2} + C.$$

A different substitution is also possible. If we set $u = \sqrt{x+1}$, then

$$du = \frac{1}{2\sqrt{x+1}}\,dx \qquad \text{and}$$

$$\begin{aligned}\int \frac{x}{\sqrt{x+1}}\,dx &= \int (u^2-1)(2\,du) = 2\left(\frac{u^3}{3} - u\right) + C \\ &= \frac{2}{3}(x+1)^{3/2} - 2(x+1)^{1/2} + C.\end{aligned}$$

(c) If we set $u = \cos x$, then $du = -\sin x\,dx$ and

$$\begin{aligned}\int \sin^3 x \cos^2 x\,dx &= \int \sin^2 x \cos^2 x \sin x\,dx \\ &= \int (1-\cos^2 x)\cos^2 x \sin x\,dx \\ &= \int (1-u^2)u^2(-du) = \int (u^4 - u^2)\,du \\ &= \frac{u^5}{5} - \frac{u^3}{3} + C = \frac{1}{5}\cos^5 x - \frac{1}{3}\cos^3 x + C.\end{aligned}$$

■

EXERCISES 5.3

In Exercises 1–30 evaluate the indefinite integral.

1. $\int (5x+14)^9\,dx$
2. $\int \sqrt{1-2x}\,dx$
3. $\int \frac{1}{(3y-12)^{1/4}}\,dy$
4. $\int \frac{5}{(5-42x)^{1/4}}\,dx$
5. $\int x^2(3x^3+10)^4\,dx$
6. $\int \frac{x}{(x^2+4)^2}\,dx$
7. $\int \sin^4 x \cos x\,dx$
8. $\int \frac{x^2}{(x-2)^4}\,dx$
9. $\int z\sqrt{1-3z}\,dz$
10. $\int \frac{x}{\sqrt{2x+3}}\,dx$
11. $\int \frac{1+\sqrt{x}}{\sqrt{x}}\,dx$
12. $\int s^3\sqrt{s^2+5}\,ds$
13. $\int \sin^2 x \cos^3 x\,dx$
14. $\int \sqrt{1-\cos x}\,\sin x\,dx$
15. $\int \frac{x^3}{(3-x^2)^3}\,dx$
16. $\int y^2\sqrt{y-4}\,dy$
17. $\int \frac{(1+\sqrt{u})^{1/2}}{\sqrt{u}}\,du$
18. $\int x^8(3x^3-5)^6\,dx$
19. $\int \frac{1+z^{1/4}}{\sqrt{z}}\,dz$
20. $\int \frac{x+1}{(x^2+2x+2)^{1/3}}\,dx$
21. $\int \frac{(x-1)(x+2)}{\sqrt{x}}\,dx$
22. $\int \frac{\cos^3 x}{(3-4\sin x)^4}\,dx$
23. $\int \sqrt{1+\sin 4t}\,\cos^3 4t\,dt$
24. $\int \sqrt{1+\sqrt{x}}\,dx$
25. $\int \tan^2 x \sec^2 x\,dx$
26. $\int \tan x \sec^2 x\,dx$
27. $\int \frac{e^{2x}}{e^{2x}+1}\,dx$
28. $\int \frac{\ln x}{x}\,dx$
29. $\int \frac{1}{x\ln x}\,dx$
30. $\int \frac{x}{(x^2+1)[\ln(x^2+1)]^2}\,dx$
31. Evaluate

$$\int \sin^3 x \cos^3 x\,dx$$

by:

(a) making the substitution $u = \sin x$;
(b) making the substitution $u = \cos x$.

(c) Verify that these answers are the same.

32. Evaluate

$$\int \sqrt{\frac{x^2}{1+x}}\,dx.$$

In Exercises 33–36 use the suggested change of variable to evaluate the indefinite integral.

33. $\displaystyle\int \frac{\sqrt{4x-x^2}}{x^3}\,dx;$ set $u = 2/x$

34. $\displaystyle\int \frac{\sqrt{x-x^2}}{x^4}\,dx;$ set $u = 1/x$

35. $\displaystyle\int \frac{x}{(5-4x-x^2)^{3/2}}\,dx;$ set $u^2 = (5+x)/(1-x)$

36. $\displaystyle\int \frac{1}{3(1-x^2)-(5+4x)\sqrt{1-x^2}}\,dx;$ set $u^2 = (1-x)/(1+x)$

37. Show that the substitution $u - x = \sqrt{x^2+x+4}$ replaces the integral

$$\int \sqrt{x^2+x+4}\,dx$$

with the integral of a rational function of u.

38. Show that the substitution $(x+1)u = \sqrt{4+3x-x^2}$ replaces the integral

$$\int \frac{1}{\sqrt{4+3x-x^2}}\,dx$$

with the integral of a rational function of u.

SECTION 5.4

An Introduction to Differential Equations

A differential equation contains derivatives of some unknown function; the equation must be solved for this function. In this section we use calculus to derive mathematical models to describe four applied problems. These models turn out to be differential equations that are reasonably easy to solve.

Before doing so, however, let us express extended power rule 3.17 in an integral form that is particularly useful for differential equations. Equation 3.17 states that when $u(x)$ is a differentiable function of x,

$$\frac{d}{dx}u^n = nu^{n-1}\frac{du}{dx}.$$

It follows, therefore, that the antiderivative of $nu^{n-1}\,du/dx$ is u^n,

$$\int nu^{n-1}\frac{du}{dx}\,dx = u^n + C.$$

Property 5.4 permits us to take the constant outside the integral, and the equation can then be divided by this quantity:

$$\int u^{n-1}\frac{du}{dx}\,dx = \frac{1}{n}u^n + C, \quad n \neq 0.$$

Now n is an arbitrary number in this formula, and therefore there is no loss in generality if we replace each n with $n+1$. The result is

$$\int u^n\frac{du}{dx}\,dx = \frac{1}{n+1}u^{n+1} + C, \quad n \neq -1. \qquad (5.9\text{a})$$

The $n = -1$ case is the integral counterpart of formula 3.37,

$$\int \frac{1}{u}\frac{du}{dx}\,dx = \ln|u| + C. \qquad (5.9\text{b})$$

EXAMPLE 5.8

Find all curves that satisfy the condition that at every point on the curve the normal line passes through the origin.

FIGURE 5.6

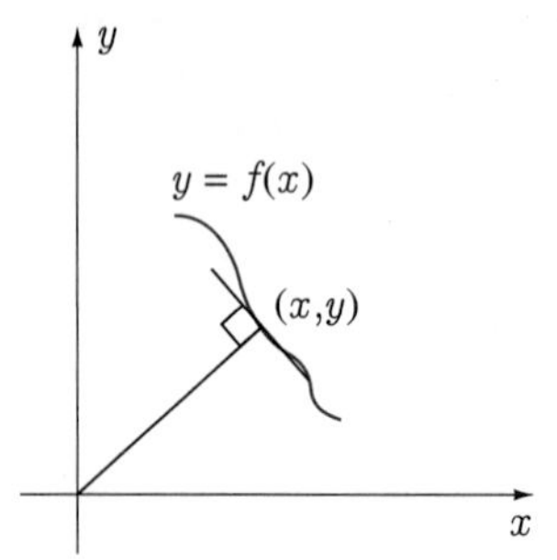

SOLUTION If $y = f(x)$ is the equation of any curve, then the slope of the tangent line at a point (x, y) on this curve is dy/dx, and the slope of the normal line is the negative reciprocal thereof. As derivatives can now be regarded as quotients of differentials, the negative reciprocal of dy/dx is $-dx/dy$. But when the normal line passes through the origin, its slope must also be given by the quotient $(y-0)/(x-0)$ (Figure 5.6). It follows that the required curve $y = f(x)$ must satisfy

$$\frac{y}{x} = -\frac{dx}{dy} \quad \text{or} \quad \frac{dy}{dx} = -\frac{x}{y}.$$

To solve this differential equation, we write

$$y\frac{dy}{dx} = -x$$

and antidifferentiate both sides with respect to x:

$$\int \left(y\frac{dy}{dx} \right) dx = \int -x\, dx.$$

The right integral is obviously $-x^2/2$, and by formula (5.9a), the antiderivative on the left is $y^2/2$. When a constant of integration is added, we have

$$\frac{y^2}{2} = -\frac{x^2}{2} + C.$$

If we set $D^2 = 2C$, then

$$x^2 + y^2 = D^2;$$

thus, the only curves satisfying the given condition are circles centred at the origin. ■

EXAMPLE 5.9

An upright cylindrical container (Figure 5.7) has height 2 m and radius 1 m. Water flows out through a hole in the bottom with speed in metres per second given by

$$v = \frac{3\sqrt{2gD}}{5},$$

where D (in metres) is the depth of water in the container and $g = 9.81$ m/s^2 is the acceleration due to gravity. This formula indicates that water comes out much more quickly when the container is full than when it is almost empty. Find a formula for D as a function of time if the tank is originally full and if the area of the hole is 3 cm^2.

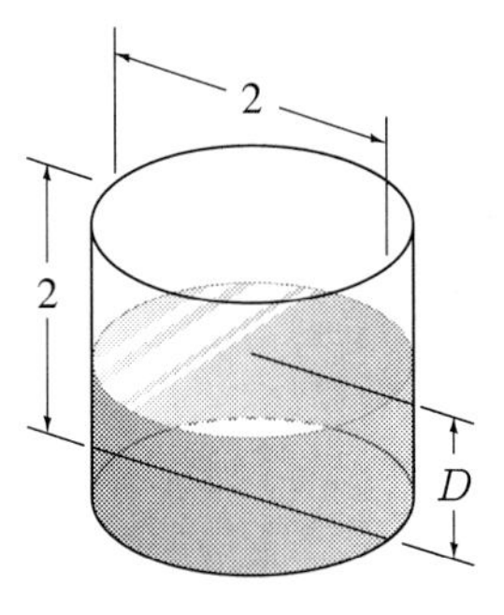

FIGURE 5.7

SOLUTION If V is the volume of water in the container at any time t, then

$$V = \pi(1)^2 D = \pi D.$$

The derivative

$$\frac{dV}{dt} = \pi \frac{dD}{dt}$$

represents the rate of change of the volume of water in the container, where it is clear that $dV/dt < 0$ since the volume is decreasing. But this rate of change must equal the rate at which the water exits through the hole. Since the area of the hole is 0.0003 m^2, and water exits with speed v m/s, water leaves the container at a rate of $0.0003v$ m^3/s. Consequently,

$$-\pi \frac{dD}{dt} = 0.0003v = 0.00018\sqrt{2gD},$$

and the unknown function $D(t)$ must satisfy the differential equation

$$\frac{1}{\sqrt{D}} \frac{dD}{dt} = -\frac{0.00018\sqrt{2g}}{\pi}.$$

Antiderivatives with respect to t yield

$$\int \frac{1}{\sqrt{D}} \frac{dD}{dt}\, dt = \int \frac{-0.00018\sqrt{2g}}{\pi}\, dt.$$

Since formula (5.9a) can be used on the left integral, and the integrand on the right is a constant, we obtain

$$2\sqrt{D} = -\frac{0.00018\sqrt{2g}}{\pi}t + C.$$

If we choose time $t = 0$ when the container is full ($D = 2$), substitution of these values into the above equation gives

$$2\sqrt{2} = C.$$

The solution of the differential equation for $D(t)$ is defined implicitly by the equation

$$2\sqrt{D} = -\frac{0.00018\sqrt{2g}}{\pi}t + 2\sqrt{2}.$$

By solving this equation for D in terms of t, we obtain the explicit function

$$D(t) = \left(\sqrt{2} - \frac{0.00009\sqrt{2g}}{\pi}t\right)^2.$$

This solution tells us that we can determine how long it takes for the container to empty. If we set $D = 0$, we obtain

$$0 = \left(\sqrt{2} - \frac{0.00009\sqrt{2g}}{\pi}t\right)^2.$$

Thus,

$$t = \frac{\pi}{0.00009\sqrt{g}} = \frac{\pi}{0.00009\sqrt{9.81}} = 11\,145,$$

and the container therefore empties in 186 minutes. ■

EXAMPLE 5.10

When a mothball is exposed to the air, it slowly evaporates. If the mothball remains spherical at all times, and its radius after one year is only half of its original radius, when will it completely disappear?

SOLUTION The size of the mothball can be measured in various ways—by its volume, its surface area, or its radius. Since we were given information about its radius, let us denote by $r(t)$ the radius of the mothball as a function of time t, and choose time $t = 0$ when the mothball is originally exposed to the air. If we let R represent the original radius of the mothball, then we can say that $r(0) = R$ and $r(1) = R/2$. This is the only information about $r(t)$ that we were given, and it certainly is not enough to solve the problem.

We need a statement on the instantaneous rate of evaporation of the mothball. Without such a statement, we must make our own reasonable assumption about how fast the naphthalene evaporates from the surface of the mothball. Since evaporation takes place only at the surface, let us assume that the amount of naphthalene evaporating per unit time is proportional to the surface area of the mothball at that time. If V and A represent volume and area of the mothball at any time, then this assumption becomes

$$\frac{dV}{dt} = kA,$$

where k is a constant. We have four variables in the problem: t, r, A, and V. By substituting $V = 4\pi r^3/3$ and $A = 4\pi r^2$, we can eliminate V and A:

$$\frac{d}{dt}\left(\frac{4}{3}\pi r^3\right) = 4\pi r^2 k,$$

and this equation reduces to

$$\frac{dr}{dt} = k.$$

Now we have a differential equation for $r(t)$, with solution

$$r(t) = kt + C.$$

Using the conditions $r(0) = R$ and $r(1) = R/2$, we find $C = R$ and $k = -R/2$. Consequently,

$$r(t) = R - \frac{Rt}{2} = R\left(1 - \frac{t}{2}\right).$$

The mothball completely disappears when $r = 0$, and this occurs in $t = 2$ years. ■

EXAMPLE 5.11

A tank originally contains 1000 L of water, in which 5 kg of salt has been dissolved.

(a) If a brine mixture containing 2 kg of salt for each 100 L of solution is poured into the tank at 10 mL/s, find the amount of salt in the tank as a function of time.

(b) If at the same time brine is being added, the mixture in the tank is being drawn off at 5 mL/s, find the amount of salt in the tank as a function of time. Assume that the mixture is stirred constantly.

SOLUTION (a) Suppose we let $S(t)$ represent the number of grams of salt in the tank at time t. If we choose time $t = 0$ at the instant the brine mixture begins entering the original solution, then $S(0) = 5000$. Since 10 mL of mixture enter the tank each second, and each millilitre contains 0.02 gm of salt, it follows that 0.2 gm of salt enter the tank each second. Consequently, after t seconds, $0.2t$ grams of salt have been added to the tank, and the total amount of salt in the solution is, in grams,

$$S(t) = 5000 + 0.2t.$$

(b) Once again we let $S(t)$ represent the number of grams of salt in the tank at time t. Its derivative dS/dt, the rate of change of $S(t)$, is the difference between how fast salt is being added to the tank in the brine, and how fast salt is being removed as the mixture is removed. As in part (a), salt is being added at the constant rate of 0.2 gm/s. The rate at which salt leaves the tank, on the other hand, is not constant; it depends on the concentration of salt in the tank. Since the tank originally contained 10^6 mL of solution, and a net 5 mL enters each second, the amount of solution in the tank at time t is $10^6 + 5t$ mL. It follows that the concentration of salt in the solution at time t is, in grams per millilitre,

$$\frac{S}{10^6 + 5t},$$

where $S = S(t)$ is the amount of salt in the tank at that time. As solution is being drawn off at the rate of 5 mL/s, the rate at which salt leaves the tank is, in grams per second,

$$\frac{S}{10^6 + 5t}(5).$$

The net rate at which salt enters the tank is therefore

$$0.2 - \frac{5S}{10^6 + 5t}.$$

Since dS/dt also represents the rate of change of the amount of salt in the tank, we set

$$\frac{dS}{dt} = \frac{1}{5} - \frac{5S}{10^6 + 5t}.$$

To find $S(t)$ we must solve this differential equation subject to the condition that $S(0) = 5000$. We can clear the term $10^6 + 5t$ from the denominator if we multiply the equation by this factor:

$$(10^6 + 5t)\frac{dS}{dt} + 5S = \frac{1}{5}(10^6 + 5t).$$

But note that the left-hand side of this differential equation is the derivative of the product of S and $10^6 + 5t$; that is,

$$\frac{d}{dt}[S(10^6 + 5t)] = (10^6 + 5t)\frac{dS}{dt} + 5S.$$

Consequently, we may express the differential equation in the form

$$\frac{d}{dt}[S(10^6 + 5t)] = \frac{1}{5}(10^6 + 5t),$$

and take antiderivatives with respect to t on both sides:

$$S(10^6 + 5t) = \frac{1}{5}\int(10^6 + 5t)\,dt = \frac{1}{50}(10^6 + 5t)^2 + C.$$

Since $S(0) = 5000$, we find

$$5000(10^6) = \frac{1}{50}(10^6)^2 + C,$$

and hence, $C = -15 \times 10^9$. Thus, $S(t)$ is defined implicitly by

$$S(10^6 + 5t) = \frac{1}{50}(10^6 + 5t)^2 - 15 \times 10^9,$$

or explicitly by

$$S(t) = \frac{1}{50}(10^6 + 5t) - \frac{15 \times 10^9}{10^6 + 5t}.$$

A sketch of this function is shown in Figure 5.8. It is asymptotic to the line $S = (10^6 + 5t)/50$. ■

FIGURE 5.8

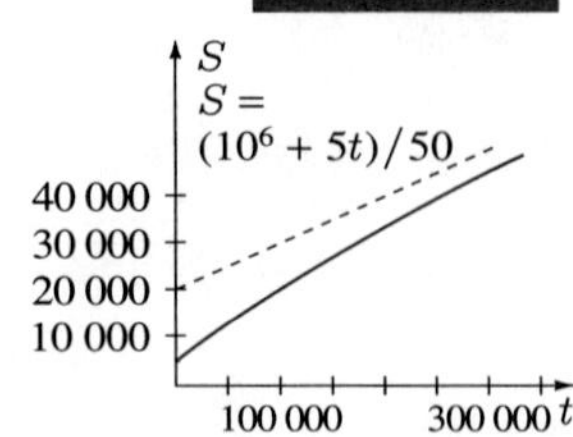

EXERCISES 5.4

1. A boy lives 6 km from school. He decides to walk to school at a speed that is always proportional to the square of his distance from the school. If he is half-way to school after one hour, find his distance from school at any time. How long does it take him to reach school?

2. A diving board of length L and uniform mass per unit length m is fixed horizontally (Figure 5.9).

FIGURE 5.9

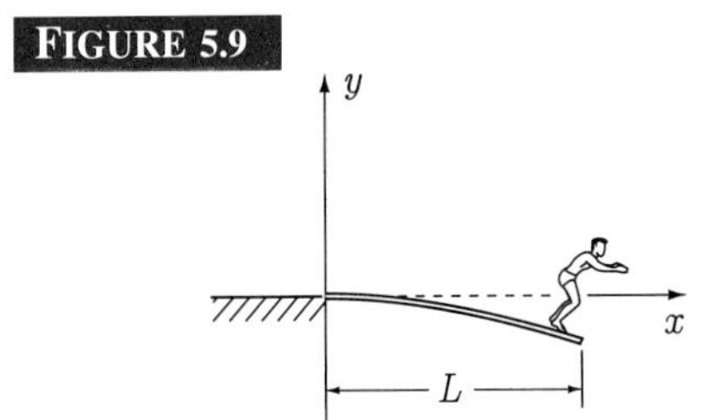

If a diver of mass M stands on the end of the board, the deflection $y = f(x)$ of the board must satisfy the differential equation

$$(EI)\frac{d^2 y}{dx^2} = Mg(L-x) + \frac{mg}{2}(L-x)^2,$$

where E and I are constants depending on the shape and material of the board, and $g < 0$ is the acceleration due to gravity.

(a) Find $y = f(x)$.

(b) What is the maximum deflection of the board from the horizontal?

3. The water trough in Figure 5.10 is 4 m long. Its cross-section is an isosceles triangle with a half-metre base and a half-metre altitude.

FIGURE 5.10

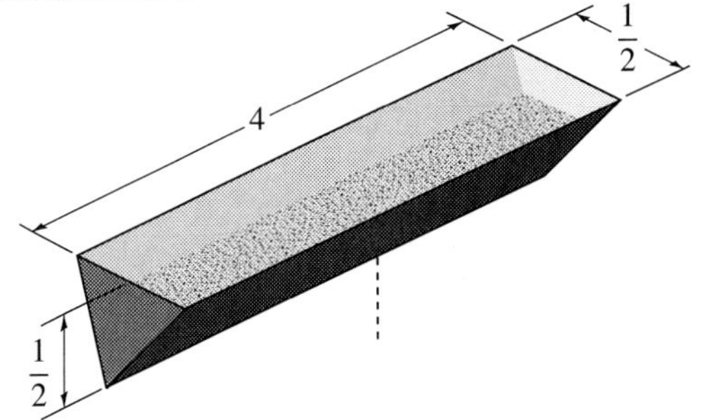

Water leaks out through a hole of area 1 cm^2 in the bottom with speed in metres per second given by $v = \sqrt{gD/2}$, where D (in metres) is the depth of water in the trough, and $g > 0$ is the acceleration due to gravity. Find how long a full trough takes to empty.

4. The y-intercept of the normal line to a curve at any point is always equal to 2. If the curve passes through the point $(3,4)$, find its equation.

5. A tank has 100 L of solution containing 4 kg of sugar. A mixture with 10 gm of sugar per litre of solution is added at 200 mL/min. At the same time, 100 mL of well-stirred mixture is removed each minute. Find the amount of sugar in the tank as a function of time.

6. A tank in the form of an inverted right-circular cone of height H and radius R is filled with water. Water escapes through a hole of cross-sectional area A at the vertex with speed $v = c\sqrt{2gD}$, where $0 < c < 1$ is a constant called the discharge coefficient, $g > 0$ is the acceleration due to gravity, and D is the depth of water in the tank. Find a formula for the time the tank takes to empty.

7. A spring of negligible mass and elasticity constant $k > 0$ is attached to a wall at one end and a mass M at the other (Figure 5.11).

FIGURE 5.11

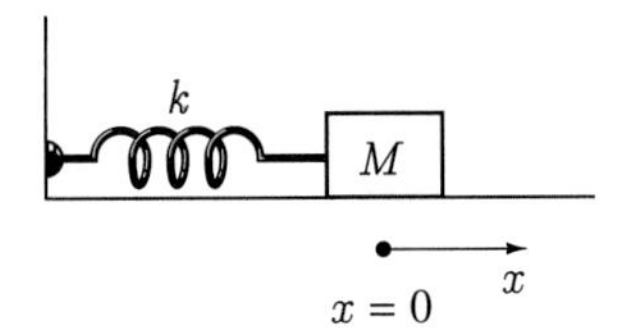

The mass is free to slide horizontally along a frictionless surface. If $x = 0$ is taken as the position of M when the spring is unstretched and M is set into motion, the differential equation describing the position $x(t)$ of M is

$$\frac{d^2 x}{dt^2} = -\frac{k}{M}x.$$

If motion is initiated by imparting a speed v_0 in the positive x-direction to M at position $x = 0$, find the velocity of M as a function of position. Note that

$$\frac{d^2 x}{dt^2} = \frac{dv}{dt} = \frac{dv}{dx}\frac{dx}{dt} = v\frac{dv}{dx}.$$

8. When a mass m falls under the influence of gravity alone, it experiences an acceleration $d^2 r/dt^2$ described by

$$m\frac{d^2 r}{dt^2} = -\frac{GmM}{r^2},$$

where M is the mass of the earth, G is a positive constant, and r is the distance from m to the centre of the earth (Figure 5.12). If m is dropped from a height h above the surface of the earth, find its velocity when it strikes the earth. What is the maximum attainable speed of m?

FIGURE 5.12

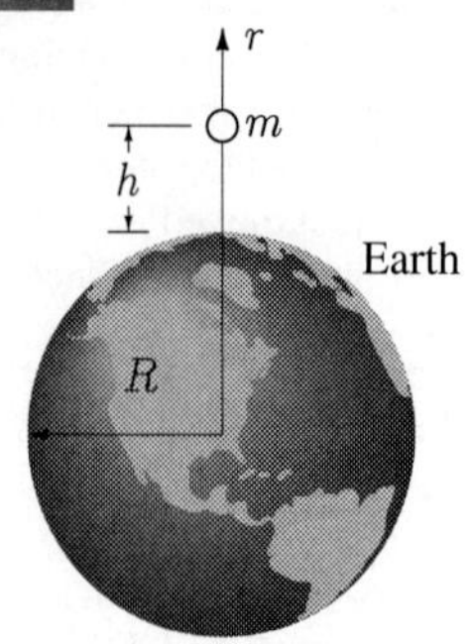

FIGURE 5.13

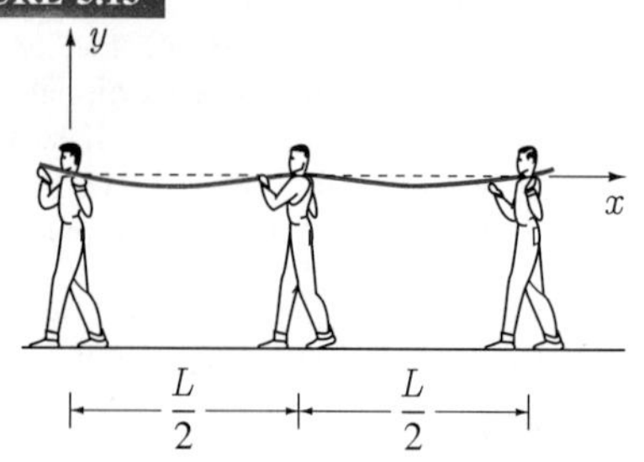

9. A container in the form of an inverted right-circular cone of radius 4 cm and height 10 cm is full of water. Water evaporates from the surface at a rate proportional to the area of the surface. If the water level drops 1 cm in the first 5 days, how long will it take for the water to evaporate completely?

10. Allometry is the quantitative study of the size of one part of a biological object in relation to the size of another part. For example, a researcher might be interested in the rate at which a child's eyes grow in relation to the rate of growth of his or her head. It is clear that the head grows more rapidly than the eyes, but just how much more rapidly? If $x(t)$ and $y(t)$ represent volumes of the eye and head respectively, at any given time t, then $(1/x)dx/dt$ and $(1/y)dy/dt$ are called *relative growth rates*. Numerous experimental results have suggested that these relative growth rates are proportional

$$\frac{1}{y}\frac{dy}{dt} = k\frac{1}{x}\frac{dx}{dt}.$$

This is called the *allometric law*. If we eliminate t from this equation, we obtain a differential equation for y as a function of x:

$$\frac{dy}{dx} = k\frac{y}{x}.$$

Solve this differential equation. (Hint: Multiply by x^k).

11. In psychology, it is sometimes assumed that if R represents the reaction of a subject to an amount S of stimulus, then relative increases of R and S are proportional. In other words, if S is changed at rate dS/dt, and the resulting rate of change of R is dR/dt, then

$$\frac{1}{R}\frac{dR}{dt} = k\frac{1}{S}\frac{dS}{dt}.$$

Although this is often called the Brentano-Stevens law, it is essentially the allometric law of Exercise 10 in a different setting. Find R as a function of S.

12. Three boys of the same height carry a board of length L and uniform mass per unit length m horizontally as shown in Figure 5.13.

The weight of the board causes it to bend, and its shape is the same on either side of the middle boy. Between the first two boys, for $0 \le x \le L/2$, the displacement of the board from the horizontal $y = f(x)$ must satisfy the differential equation

$$(EI)\frac{d^2y}{dx^2} = Ax + \frac{mg}{2}x^2,$$

where E and I are constants depending on the cross section and material of the board, A is the weight supported by the boy at $x = 0$, and $g < 0$ is the acceleration due to gravity.

(a) Solve this equation along with the conditions $f(0) = f(L/2) = 0$ to find $f(x)$ as a function of x, E, I, m, g, and A.

(b) Find a condition that permits determination of A.

13. The steady-state temperature T in a region bounded by two concentric spheres of radii 1 m and 2 m must satisfy the differential equation

$$\frac{d^2T}{dr^2} + \frac{2}{r}\frac{dT}{dr} = 0,$$

where r is the radial distance from the common centre of the spheres. If temperatures on the inner and outer spheres are maintained at 10°C and 20°C respectively, find the temperature distribution $T(r)$ between the spheres.

14. A room with volume 100 m^3 initially contains 0.1% carbon dioxide. Beginning at time $t = 0$, fresher air containing 0.05% carbon dioxide flows into the room at 5 m^3/min. The well-mixed air in the room flows out at the same rate. Find the amount of carbon dioxide in the room as a function of time. What is the limit of the function as $t \to \infty$?

15. Find all curves such that if the normal is drawn at any point (x, y) on the curve, then that part of the normal between (x, y) and the x-axis is bisected by the y-axis.

16. It is sometimes assumed that the density ρ of the atmosphere is related to height h above sea level according to the differential equation

$$\frac{d\rho}{dh} = -\frac{\rho^{2-\delta}}{k\delta},$$

where $\delta > 1$ and $k > 0$ are constants.

(a) Show that if ρ_0 is density at sea level, then

$$\rho^{\delta-1} = -\frac{h}{k}\left(\frac{\delta - 1}{\delta}\right) + \rho_0^{\delta-1}.$$

(b) If air pressure P and density are related by the equation $P = k\rho^{\delta}$, prove that

$$P^{1-1/\delta} = P_0^{1-1/\delta} - [(1 - 1/\delta)\rho_0 P_0^{-1/\delta}]h,$$

where P_0 is air pressure at sea level.

(c) Show that the effective height of the atmosphere is

$$\frac{\delta P_0}{(\delta - 1)\rho_0}.$$

SUMMARY

The indefinite integral or antiderivative of a function $f(x)$ is a family of functions, each of which has $f(x)$ as its first derivative. If we can find one antiderivative of $f(x)$, then every other antiderivative in the family differs from this one by at most an additive constant. According to Theorem 5.2, the antiderivative of a sum of two functions is the sum of their antiderivatives, and multiplicative constants may be bypassed when finding antiderivatives

$$\int [f(x) + g(x)]\,dx = \int f(x)\,dx + \int g(x)\,dx; \qquad \int cf(x)\,dx = c\int f(x)\,dx.$$

The most important integration formula is for powers of x,

$$\int x^n\,dx = \begin{cases} \dfrac{1}{n+1}x^{n+1} + C & n \neq -1 \\ \ln|x| + C & n = -1. \end{cases}$$

Other integration formulas that arise from differentiations of trigonometric and exponential functions are listed in equations 5.5.

In this chapter we have studied three ways to evaluate indefinite integrals. (Others will follow in Chapter 9.) First, due to our expertise in differentiation, some antiderivatives are immediately recognizable. Second, sometimes an antiderivative can be guessed to within a multiplicative constant, and then fixed up. Third, a change of variable can often replace a complex integration problem with a simpler one.

Integration plays a fundamental role in kinematics. Since velocity is the derivative of position, and acceleration is the derivative of velocity, it follows that position is the antiderivative of velocity, and velocity is the antiderivative of acceleration,

$$v(t) = \int a(t)\,dt \qquad \text{and} \qquad x(t) = \int v(t)\,dt.$$

Many physical systems are modelled by differential equations, and the solution of a differential equation usually involves one or more integrations.

Key Terms and Formulas

In reviewing this chapter, you should be able to define or discuss the following key terms:

Indefinite integral
Antiderivative
Integrand
Differential equation
Change of variable

REVIEW EXERCISES

In Exercises 1–24 evaluate the indefinite integral.

1. $\int (3x^3 - 4x^2 + 5)\, dx$ 2. $\int \left(\frac{1}{x^5} + 2x - \frac{1}{x^3}\right) dx$

3. $\int (2x^2 - 3x + 7x^6)\, dx$ 4. $\int \left(\frac{1}{x^2} - 2\sqrt{x}\right) dx$

5. $\int \sqrt{x-2}\, dx$ 6. $\int x(1+3x^2)^4\, dx$

7. $\int \left(\sqrt{x} - \frac{1}{\sqrt{x}}\right) dx$ 8. $\int \left(\frac{x^2+5}{\sqrt{x}}\right) dx$

9. $\int \frac{1}{(x+5)^4}\, dx$ 10. $\int \left(\frac{\sqrt{x}}{x^2} - \frac{15}{\sqrt{x}}\right) dx$

11. $\int \sin 3x\, dx$ 12. $\int x\sqrt{1-x^2}\, dx$

13. $\int x \cos x^2\, dx$ 14. $\int x^2(1-2x^2)^2\, dx$

15. $\int x\sqrt{1+x}\, dx$ 16. $\int \frac{x}{\sqrt{2-x}}\, dx$

17. $\int \frac{1}{(1+x)^2}\, dx$ 18. $\int (2+\sqrt{x})^2\, dx$

19. $\int \frac{1}{\sqrt{x}(2+\sqrt{x})^2}\, dx$ 20. $\int \sin^4 x \cos x\, dx$

21. $\int e^{3-5x}\, dx$ 22. $\int x e^{-4x^2}\, dx$

23. $\int \frac{e^x - 1}{e^{2x}}\, dx$ 24. $\int \frac{1}{5x \ln x}\, dx$

25. Find the curve $y = f(x)$ for which $f''(x) = x^2 + 1$, and that passes through the point $(1, 1)$ with slope 4.

26. Find the curve $y = f(x)$ for which $f''(x) = 12x^2$, and that passes through the two points $(1, 4)$ and $(-1, -3)$.

27. Find the curve $y = f(x)$ for which $f''(x) = 24x^2 + 6x$, and that is tangent to the line $y = 4x + 4$ at $(1, 8)$.

28. A boy lives 6 km from school. He decides to walk to school at a speed that is always proportional to the square root of his distance from the school. If he is halfway to school after one hour, find his distance from school at any time. How long does it take him to reach school?

29. If a ball is thrown vertically upward with a speed of 30 m/s, how high will it rise?

30. A stone is thrown vertically downward over the edge of a bridge 50 m above a river. If the stone strikes the water in 2.2 s, what was its initial speed?

In Exercises 31–38 evaluate the indefinite integral.

31. $\int \frac{1}{\sqrt{1+\sqrt{x}}}\, dx$ 32. $\int \frac{x}{\sqrt{1+x}+1}\, dx$

33. $\int \frac{\sin x}{\sqrt{4+3\cos x}}\, dx$ 34. $\int x^8(3-2x^3)^6\, dx$

35. $\int \frac{(2+x)^4}{x^6}\, dx$ 36. $\int \sin^3 x \cos^3 x\, dx$

37. $\int \frac{1}{x\sqrt{1+3\ln x}}\, dx$ 38. $\int \tan x\, dx$

39. A graph of the acceleration $a(t)$ of an object is shown in Figure 5.14. Find its velocity $v(t)$ and position $x(t)$ in the time interval $0 \le t \le 15$ if $v(0) = 0 = x(0)$, and sketch graphs of each.

FIGURE 5.14

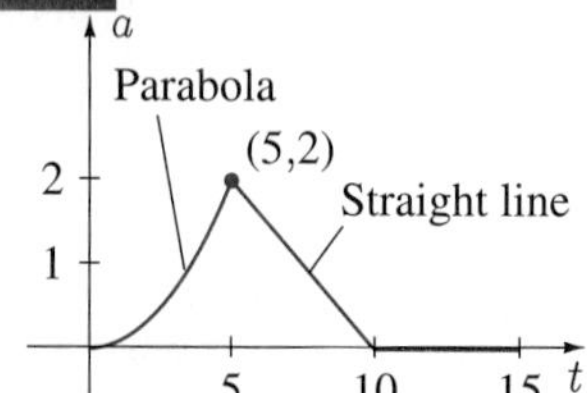

40. Find the equation of the curve that passes through the point $(1, 1)$ such that the slope of the tangent line at any point (x, y) is half the square of the slope of the line from the origin to (x, y).

41. The equation $y = x^3 + C$ describes a family of cubics where C represents the distance of its horizontal point of inflection above the x-axis. Find the equation of the curve that passes through the point $(1, 1)$ and intersects each of these curves at right angles.

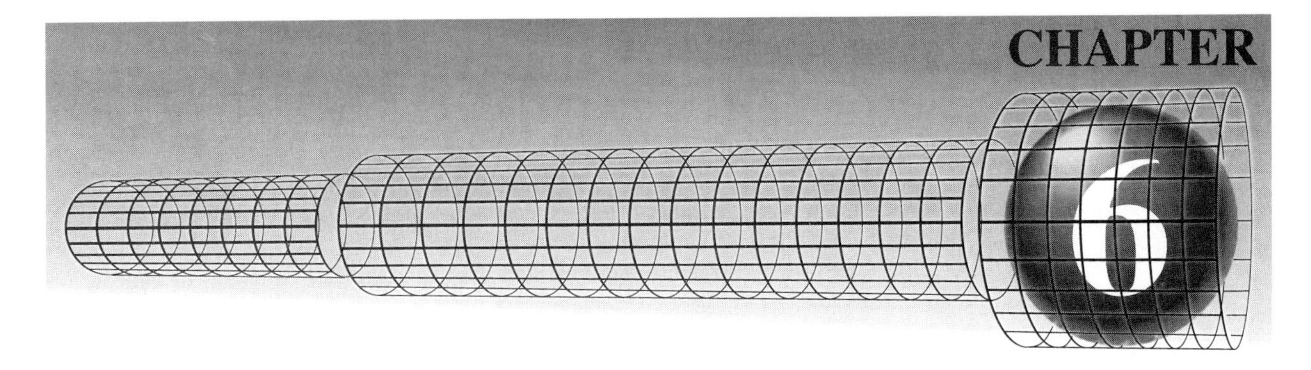

The Definite Integral

There are two aspects to calculus: differentiation and integration. We dealt with differentiation and its applications in Chapters 3 and 4. Integration in Chapter 5 was synonymous with antidifferentiation. In this chapter we investigate a new type of integral called the "definite integral". Before doing that, we introduce sigma notation, a compact notation for sums of terms, particularly useful for definite integrals. In Section 6.2 we discuss four different types of problems that motivate the concept of the definite integral. In subsequent sections we develop the definition for the integral and discuss various ways to evaluate it. Section 6.6 presents an application of the definite integral — finding the average value of a function.

The definite integral is, by definition, very different from the indefinite integral, yet the two are intimately related through the Fundamental Theorem of Integral Calculus (Sections 6.4 and 6.5).

SECTION 6.1 Sigma Notation

One of the most important notations in calculus, *sigma notation*, is used to represent a sum of terms, all of which are similar in form. For example, the six terms in the sum

$$\frac{1}{1+2^2}+\frac{2}{1+3^2}+\frac{3}{1+4^2}+\frac{4}{1+5^2}+\frac{5}{1+6^2}+\frac{6}{1+7^2}$$

are all formed in the same way: each is an integer divided by 1 plus the square of the next integer. If k represents an integer, we can say that every term has the form $k/[1+(k+1)^2]$; the first term is obtained by setting $k=1$, the second by setting $k=2$, and so on. As $k/[1+(k+1)^2]$ represents each and every term in the sum, we can describe the sum in words by saying, "Assign k in $k/[1+(k+1)^2]$ the integer values between 1 and 6, inclusively, and add the resulting numbers together." The notation used to represent this statement is

$$\sum_{k=1}^{6}\frac{k}{1+(k+1)^2}.$$

The Σ symbol is the Greek letter sigma which, in this case, means "sum". Summed are expressions of the form $k/[1+(k+1)^2]$, and the "$k=1$" and "6" indicate that every integer from 1 to 6 is substituted into $k/[1+(k+1)^2]$. We call $k/[1+(k+1)^2]$ the **general term** of the sum, since it represents each and every term therein. The letter k is called the **index of summation** or **variable of summation**, and 1 and 6 are called the **limits of summation**. Any letter may be used to represent the index of summation; most commonly used are i, j, k, l, m, and n.

Here is another example:

$$\sum_{n=5}^{15} n^3 = 5^3 + 6^3 + 7^3 + 8^3 + 9^3 + 10^3 + 11^3 + 12^3 + 13^3 + 14^3 + 15^3.$$

In summing a large number of terms, it is quite cumbersome to write them all down. One way around this difficulty is to write the first few terms to indicate the pattern by which the terms are formed, three dots, and the last term. For example, to indicate the sum of the cubes of the positive integers less than or equal to 100, we write

$$1^3 + 2^3 + 3^3 + 4^3 + 5^3 + \cdots + 100^3,$$

where the dots indicate that all numbers between 5^3 and 100^3 are to be filled in according to the same pattern suggested. Obviously, it would be preferable to express the sum in sigma notation

$$\sum_{k=1}^{100} k^3,$$

which is both precise and compact.

EXAMPLE 6.1

Write each of the following sums in sigma notation:

$$(a)\quad \frac{1}{2\cdot 3} + \frac{4}{3\cdot 4} + \frac{9}{4\cdot 5} + \frac{16}{5\cdot 6} + \cdots + \frac{169}{14\cdot 15};$$

$$(b)\quad \frac{16}{\sqrt{2}} + \frac{32}{\sqrt{3}} + \frac{64}{\sqrt{4}} + \frac{128}{\sqrt{5}} + \cdots + \frac{4096}{\sqrt{10}}.$$

SOLUTION (a) We examine each term carefully to determine the pattern. If we use i as the index of summation, the general term is $i^2/[(i+1)(i+2)]$. Consequently,

$$\frac{1}{2\cdot 3} + \frac{4}{3\cdot 4} + \frac{9}{4\cdot 5} + \frac{16}{5\cdot 6} + \cdots + \frac{169}{14\cdot 15} = \sum_{i=1}^{13} \frac{i^2}{(i+1)(i+2)}.$$

(b) If n is chosen as the index of summation, then

$$\frac{16}{\sqrt{2}} + \frac{32}{\sqrt{3}} + \frac{64}{\sqrt{4}} + \frac{128}{\sqrt{5}} + \cdots + \frac{4096}{\sqrt{10}} = \sum_{n=4}^{12} \frac{2^n}{\sqrt{n-2}}.$$

■

The representations for the sums in Example 6.1 in terms of sigma notation are not unique. In fact, there is an infinite number of representations for each sum. Consider, for example, the sum represented by

$$\sum_{i=1}^{9} \frac{2^{i+3}}{\sqrt{i+1}}.$$

If we write out some of the terms in the summation, we find

$$\sum_{i=1}^{9} \frac{2^{i+3}}{\sqrt{i+1}} = \frac{16}{\sqrt{2}} + \frac{32}{\sqrt{3}} + \frac{64}{\sqrt{4}} + \frac{128}{\sqrt{5}} + \cdots + \frac{4096}{\sqrt{10}},$$

the same sum as that in Example 6.1(b). This sum can also be represented by

$$\sum_{j=-2}^{6} \frac{2^{j+6}}{\sqrt{j+4}} \quad \text{and} \quad \sum_{m=15}^{23} \frac{2^{m-11}}{\sqrt{m-13}}.$$

We can transform any one of these representations into any other by making a change of variable of summation. For example, if in the summation of Example 6.1(b) we set $i = n - 3$, then $n = i + 3$, and

$$\frac{2^n}{\sqrt{n-2}} = \frac{2^{i+3}}{\sqrt{i+1}}.$$

For the limits, we find that $i = 1$ when $n = 4$, and $i = 9$ when $n = 12$. It follows that

$$\sum_{n=4}^{12} \frac{2^n}{\sqrt{n-2}} = \sum_{i=1}^{9} \frac{2^{i+3}}{\sqrt{i+1}}.$$

Similarly, the changes $j = n - 6$ and $m = n + 11$ transform

$$\sum_{n=4}^{12} \frac{2^n}{\sqrt{n-2}}$$

into

$$\sum_{j=-2}^{6} \frac{2^{j+6}}{\sqrt{j+4}} \quad \text{and} \quad \sum_{m=15}^{23} \frac{2^{m-11}}{\sqrt{m-13}}.$$

EXAMPLE 6.2

Change each of the following summations into representations that are initiated with the integer 1:

(a) $\displaystyle\sum_{i=4}^{26} \frac{i^{2/3}}{i^2 + i + 1}$ (b) $\displaystyle\sum_{j=-3}^{102} \frac{j^2 + 2j + 5}{\sin(j+5)}$

SOLUTION (a) To initiate the summation at 1, we want $n = 1$ when $i = 4$, so we set $n = i - 3$. Then $i = n + 3$, and by substitution, we have

$$\sum_{i=4}^{26} \frac{i^{2/3}}{i^2 + i + 1} = \sum_{n=1}^{23} \frac{(n+3)^{2/3}}{(n+3)^2 + (n+3) + 1} = \sum_{n=1}^{23} \frac{(n+3)^{2/3}}{n^2 + 7n + 13}.$$

(b) In this case, we set $n = j + 4$. Then, $j = n - 4$, and

$$\sum_{j=-3}^{102} \frac{j^2 + 2j + 5}{\sin(j+5)} = \sum_{n=1}^{106} \frac{(n-4)^2 + 2(n-4) + 5}{\sin(n+1)} = \sum_{n=1}^{106} \frac{n^2 - 6n + 13}{\sin(n+1)}.$$

■

If we examine the results of Example 6.2 and the summations immediately preceding this example, we soon come to realize that there is a very simple way to change variables. To illustrate, consider once again

$$\sum_{n=4}^{12} \frac{2^n}{\sqrt{n-2}}.$$

Should we wish to initiate the summation with 1, rather than 4, we lower both limits by 3. To compensate, we replace each n in the general term by $n+3$; the result is

$$\sum_{n=1}^{9} \frac{2^{n+3}}{\sqrt{(n+3)-2}} = \sum_{n=1}^{9} \frac{2^{n+3}}{\sqrt{n+1}}.$$

Likewise, for simplicity in the summation

$$\sum_{n=1}^{10} \frac{(n+4)^2}{e^{n+4}},$$

it would be advisable to lower each n in the general term by 4. This can be done provided we raise each limit by 4,

$$\sum_{n=1}^{10} \frac{(n+4)^2}{e^{n+4}} = \sum_{n=5}^{14} \frac{n^2}{e^n}.$$

Every summation represented in sigma notation is of the form

$$\sum_{i=m}^{n} f(i), \tag{6.1}$$

where m and n are integers ($n > m$), and $f(i)$ is some function of the index of summation i. In Example 6.2(a), $f(i) = i^{2/3}/(i^2 + i + 1)$, $m = 4$, and $n = 26$; in Example 6.1(a), $f(i) = i^2/[(i+1)(i+2)]$, $m = 1$, and $n = 13$. The following properties of sigma notation are easily proved by writing out each summation.

Theorem 6.1

If $f(i)$ and $g(i)$ are functions of i, and m and n are positive integers such that $n > m$, then

$$\sum_{i=m}^{n} [f(i) + g(i)] = \sum_{i=m}^{n} f(i) + \sum_{i=m}^{n} g(i); \tag{6.2}$$

$$\sum_{i=m}^{n} cf(i) = c\sum_{i=m}^{n} f(i), \tag{6.3}$$

if c is a constant independent of i.

Compare this theorem with Theorem 5.2 in Section 5.1; notice the similarities between properties of summations in sigma notation and those of indefinite integrals.

We emphasize that sigma notation is simply a concise symbolism used to represent a sum of terms; it does not evaluate the sum. In the following, we develop formulas for certain sums of terms.

Summation $\sum_{i=1}^{n} i$ represents the sum of the first n integers:

$$\sum_{i=1}^{n} i = 1 + 2 + 3 + 4 + \cdots + (n-1) + n.$$

If we write the terms on the right in reverse order, we have

$$\sum_{i=1}^{n} i = n + (n-1) + (n-2) + \cdots + 4 + 3 + 2 + 1.$$

Addition of these two equations gives us

$$2\sum_{i=1}^{n} i = (n+1) + (n+1) + (n+1) + \cdots + (n+1) + (n+1) = n(n+1).$$

Consequently,

$$\sum_{i=1}^{n} i = \frac{n(n+1)}{2}. \tag{6.4}$$

This result can be used to develop formulas for the sums of the squares, cubes, etc., of the positive integers. To find the sum of the squares of the first n integers, we note that by expansion and simplification

$$i^3 - (i-1)^3 = 3i^2 - 3i + 1$$

for any integer i whatsoever. It follows that

$$\sum_{i=1}^{n} [i^3 - (i-1)^3] = \sum_{i=1}^{n} (3i^2 - 3i + 1).$$

But if we write the left-hand side in full, we find

$$\sum_{i=1}^{n} [i^3 - (i-1)^3] = [1^3 - 0^3] + [2^3 - 1^3] + [3^3 - 2^3] + \cdots [n^3 - (n-1)^3].$$

Most of these terms cancel one another, leaving only n^3; that is,

$$\sum_{i=1}^{n} [i^3 - (i-1)^3] = n^3.$$

Thus,

$$\begin{aligned}
n^3 &= \sum_{i=1}^{n} (3i^2 - 3i + 1) \\
&= 3\sum_{i=1}^{n} i^2 - 3\sum_{i=1}^{n} i + \sum_{i=1}^{n} 1 && \text{(using Theorem 6.1)} \\
&= 3\sum_{i=1}^{n} i^2 - 3\frac{n(n+1)}{2} + n && \text{(using formula 6.4).}
\end{aligned}$$

We can solve this equation for $\sum_{i=1}^{n} i^2$:

$$\sum_{i=1}^{n} i^2 = \frac{1}{3}\left[n^3 + \frac{3n(n+1)}{2} - n\right] = \frac{n(n+1)(2n+1)}{6}. \tag{6.5}$$

A similar procedure beginning with the identity $i^4 - (i-1)^4 = 4i^3 - 6i^2 + 4i - 1$ yields

$$\sum_{i=1}^{n} i^3 = \frac{n^2(n+1)^2}{4}. \qquad (6.6)$$

These results can also be established independently of one another by mathematical induction.

EXERCISES 6.1

In Exercises 1–10 express the sum in sigma notation. Initiate the summation with the integer 1.

1. $2 \cdot 3 + 3 \cdot 4 + 4 \cdot 5 + 5 \cdot 6 + \cdots + 99 \cdot 100$

2. $\frac{1}{2} + \frac{2}{4} + \frac{3}{8} + \frac{4}{16} + \frac{5}{32} + \cdots + \frac{10}{1024}$

3. $\frac{16}{14+15} + \frac{17}{15+16} + \frac{18}{16+17} + \cdots + \frac{199}{197+198}$

4. $1 + \sqrt{2} + \sqrt{3} + 2 + \sqrt{5} + \sqrt{6} + \sqrt{7} + \sqrt{8} + 3 + \cdots + 121$

5. $1 + \frac{1}{2} + \frac{1}{2 \cdot 3} + \frac{1}{2 \cdot 3 \cdot 4} + \cdots + \frac{1}{2 \cdot 3 \cdot 4 \cdot 5 \cdot \cdots \cdot 16}$

6. $-2 + 3 - 4 + 5 - 6 + 7 - 8 + \cdots - 1020$

7. $\frac{2 \cdot 3}{1 \cdot 4} + \frac{6 \cdot 7}{5 \cdot 8} + \frac{10 \cdot 11}{9 \cdot 12} + \frac{14 \cdot 15}{13 \cdot 16} + \cdots \frac{414 \cdot 415}{413 \cdot 416}$

8. $\frac{\tan 1}{2} + \frac{\tan 2}{1+2^2} + \frac{\tan 3}{1+3^2} + \frac{\tan 4}{1+4^2} + \cdots + \frac{\tan 225}{1+225^2}$

9. $4^3 + 5^2 + 6 + 1 + \frac{1}{8} + \frac{1}{9^2} + \cdots + \frac{1}{25^{18}}$

10. $0.9 + 0.99 + 0.999 + \cdots + 0.999999999$

In Exercises 11–15 verify by a change of variable of summation that the two summations are identical.

11. $\sum_{n=1}^{24} \frac{n^2}{2n+1} \qquad \sum_{i=4}^{27} \frac{i^2 - 6i + 9}{2i - 5}$

12. $\sum_{k=2}^{101} \frac{3k - k^2}{\sqrt{k+5}} \qquad \sum_{m=0}^{99} \frac{2 - m - m^2}{\sqrt{7+m}}$

13. $\sum_{n=5}^{20} (-1)^n \frac{2^n}{n^2+1} \qquad \sum_{j=1}^{16} 16(-1)^j \frac{2^j}{j^2 + 8j + 17}$

14. $\sum_{i=0}^{37} \frac{3^{3i}}{i!} \qquad \sum_{m=2}^{39} \frac{3^{3m}}{729(m-2)!}$

15. $\sum_{r=15}^{225} \frac{1}{r^2 - 10r} \qquad \sum_{n=10}^{220} \frac{1}{n^2 - 25}$

In Exercises 16–21 use Theorem 6.1 and the formulas 6.4–6.6 to evaluate the sum.

16. $\sum_{n=1}^{12} (3n+2)$ **17.** $\sum_{j=1}^{21} (2j^2 + 3j)$

18. $\sum_{m=1}^{n} (4m-2)^2$ **19.** $\sum_{k=2}^{29} (k^3 - 3k^2)$

20. $\sum_{n=1}^{25} (n+5)(n-4)$ **21.** $\sum_{i=1}^{n} i(i-3)^2$

22. Verify Theorem 6.1.

23. Find a formula for

$$\sum_{k=1}^{n} \frac{1}{k(k+1)}.$$

Hint:

$$\frac{1}{k(k+1)} = \frac{1}{k} - \frac{1}{k+1}.$$

24. If $f(x)$ is a function of x, defined for all x, simplify the sum

$$\sum_{i=1}^{n} [f(i) - f(i-1)].$$

25. Prove formula 6.6.

26. Is

$$\sum_{i=1}^{n} [f(i)g(i)] \quad \text{equal to} \quad \left[\sum_{i=1}^{n} f(i)\right]\left[\sum_{i=1}^{n} g(i)\right]?$$

27. A geometric series is a sum of terms of the form

$$a + ar + ar^2 + ar^3 + \cdots + ar^{n-1}.$$

There is a first term a, and every term thereafter is obtained by multiplying the preceding term by r (called the *common ratio*).

(a) If S_n represents this sum, express S_n in sigma notation.

(b) Prove that

$$S_n = \frac{a(1-r^n)}{1-r}.$$

Use the formula in Exercise 27 to sum the geometric series in Exercises 28–31.

28. $\frac{1}{8} + \frac{1}{16} + \frac{1}{32} + \frac{1}{64} + \cdots + \frac{1}{1\,048\,576}$

29. $1 - \frac{1}{3} + \frac{1}{9} - \frac{1}{27} + \frac{1}{81} - \cdots - \frac{1}{19\,683}$

30. $40(0.99) + 40(0.99)^2 + 40(0.99)^3 + \cdots 40(0.99)^{15}$

31. $\sqrt{0.99} + 0.99 + (0.99)^{3/2} + (0.99)^2 + \cdots + (0.99)^{10}$

32. Prove that

$$\left|\sum_{i=1}^{n} f(i)\right| \le \sum_{i=1}^{n} |f(i)|.$$

33. Express the following summation in sigma notation:

$$1 + \frac{1}{2} - \frac{1}{4} - \frac{1}{8} + \frac{1}{16} + \frac{1}{32} - \frac{1}{64} - \frac{1}{128} + \cdots + \frac{1}{4096}.$$

SECTION 6.2

The Need for the Definite Integral

In this section we consider four inherently different problems: on area, on volume, on blood flow, and on work; but we shall see that a common method of solution exists for all four of them. This common theme leads to the definition of the definite integral in Section 6.3.

Problem 1

We first consider the problem of finding the area A in Figure 6.1(a). At present, we have formulas for areas of very few geometric shapes — rectangles, triangles, polygons, and circles, and the shape in Figure 6.1(a) is not one of them. We can, however, find an approximation to the area by constructing rectangles in the following way. Between a and b we pick $n-1$ points $x_1, x_2, x_3, \ldots, x_{n-1}$ on the x-axis such that

$$a = x_0 < x_1 < x_2 < \cdots < x_{n-1} < x_n = b.$$

We draw vertical lines through each of these $n-1$ points to intersect the curve $y = f(x)$ and form rectangles, as shown in Figure 6.1(b).

FIGURE 6.1

(a)

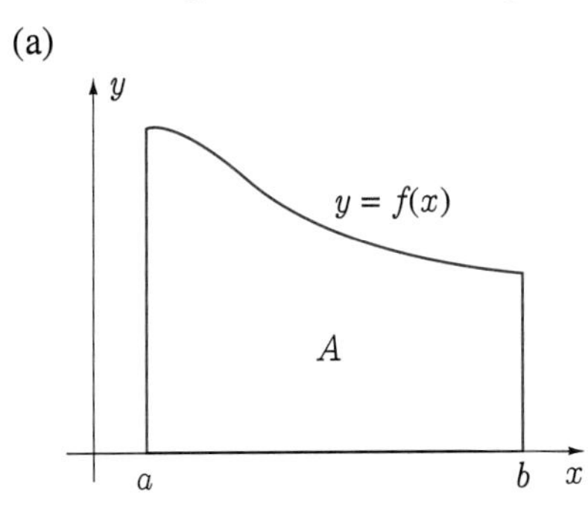

(b)

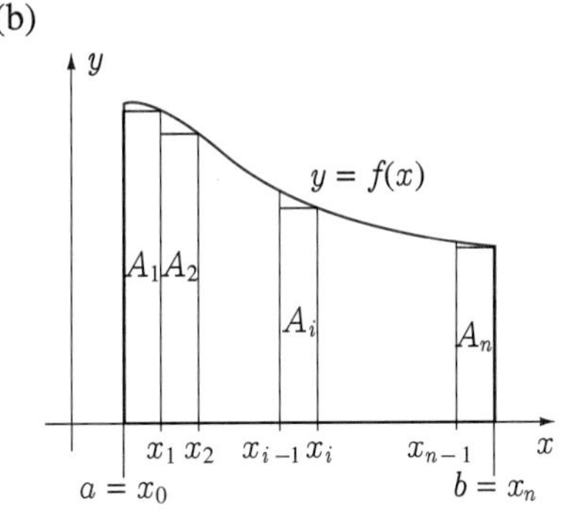

If we denote by A_i the area of the ith such rectangle ($i = 1, \ldots, n$), then

$$\begin{aligned} A_i &= (\text{height of rectangle})(\text{width of rectangle}) \\ &= f(x_i)(x_i - x_{i-1}). \end{aligned}$$

The sum of these n rectangular areas is an approximation to the required area; that is, A is approximately equal to

$$\sum_{i=1}^{n} A_i = \sum_{i=1}^{n} f(x_i)(x_i - x_{i-1}).$$

If we let the number of these rectangles get larger and larger (and at the same time require each to have a smaller and smaller width that eventually approaches zero), the approximation appears to get better and better. In fact, we expect that

$$A = \lim_{n\to\infty} \sum_{i=1}^{n} A_i = \lim_{n\to\infty} \sum_{i=1}^{n} f(x_i)(x_i - x_{i-1}). \tag{6.7}$$

Problem 2

If the area in Figure 6.1(a) is rotated about the x-axis, it traces out a volume V. This is certainly not a standard shape for which we have a volume formula; therefore, to calculate V we first find an approximation. We take the rectangles in Figure 6.1(b) (which approximate the area) and rotate them about the x-axis (Figure 6.2).

FIGURE 6.2

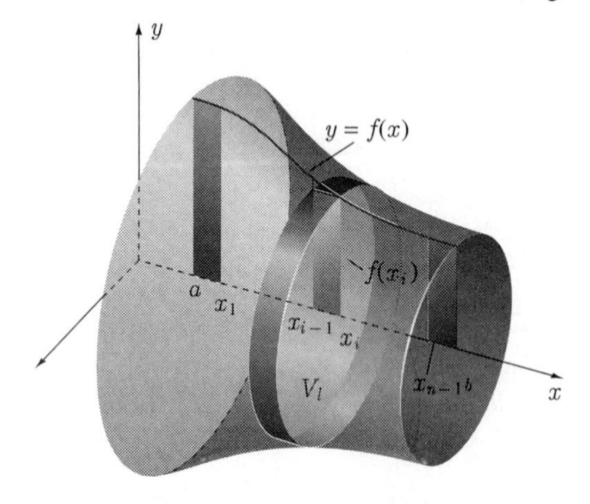

Each A_i traces out a disc of volume V_i ($i = 1, \dots, n$), where

$$\begin{aligned} V_i &= \text{(surface area of disc)(thickness of disc)} \\ &= \pi\text{(radius of disc)}^2\text{(thickness of disc)} \\ &= \pi[f(x_i)]^2(x_i - x_{i-1}). \end{aligned}$$

An approximation to the required volume is then

$$\sum_{i=1}^{n} V_i = \sum_{i=1}^{n} \pi[f(x_i)]^2(x_i - x_{i-1}).$$

Again we feel intuitively that as the number of discs becomes larger and larger (and the width of each approaches zero), the approximation becomes better and better, and

$$V = \lim_{n\to\infty} \sum_{i=1}^{n} V_i = \lim_{n\to\infty} \sum_{i=1}^{n} \pi[f(x_i)]^2(x_i - x_{i-1}). \tag{6.8}$$

Problem 3

When blood flows through a vein or artery, it encounters resistance due to friction with the walls of the blood vessel and due to the viscosity of the blood itself. As a result, the velocity of the blood is not constant across a cross-section of the vessel; blood flows more quickly near the centre of the vessel than near its walls. It has been shown that for laminar blood flow in a vessel of circular cross-section (Figure 6.3), the velocity of blood is given by

$$v = v(r) = c(R^2 - r^2), \quad 0 \leq r \leq R,$$

where $c > 0$ is a constant, R is the radius of the blood vessel, and r is radial distance measured from the centre of the vessel. We wish to find the rate of blood flow through the vessel, that is, the volume of blood flowing through the cross-section per unit time.

FIGURE 6.3

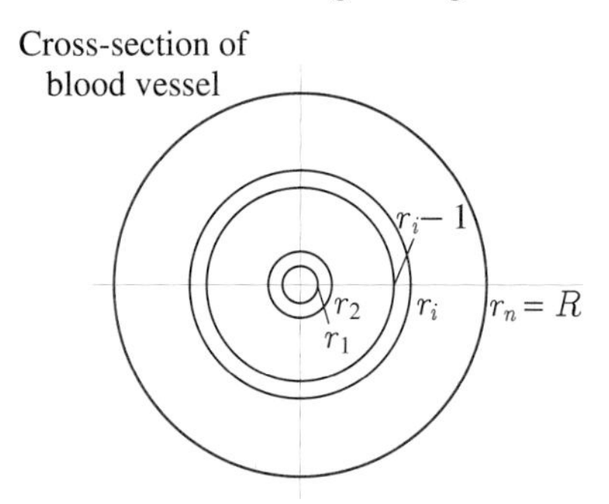

If v were constant over the cross-section, then flow per unit time would be the product of v and the cross-sectional area. Unfortunately this is not the case, but we can still use the idea that flow is velocity multiplied by area. We divide the cross-section into rings with radii

$$0 = r_0 < r_1 < r_2 < \cdots < r_{n-1} < r_n = R.$$

Over the ith ring, the variation in v is small and v can be approximated by $v(r_i)$. The flow through the ith ring can therefore be approximated by

$$\begin{aligned} F_i &= \text{(area of ring)(velocity at outer radius of ring)} \\ &= (\pi r_i^2 - \pi r_{i-1}^2)v(r_i). \end{aligned}$$

An approximation to the required flow F is the sum of these F_i:

$$\sum_{i=1}^{n} F_i = \sum_{i=1}^{n} (\pi r_i^2 - \pi r_{i-1}^2)v(r_i),$$

and it seems reasonable that as the number of rings increases, so too does the accuracy of the approximation; that is, we anticipate that

$$F = \lim_{n \to \infty} \sum_{i=1}^{n} F_i = \lim_{n \to \infty} \sum_{i=1}^{n} (\pi r_i^2 - \pi r_{i-1}^2)v(r_i). \tag{6.9}$$

Problem 4

A spring is fixed horizontally into a wall at one end, and the other end is free. Consider finding the work done in stretching the spring three centimetres by pulling on its free end (Figure 6.4(a)).

FIGURE 6.4

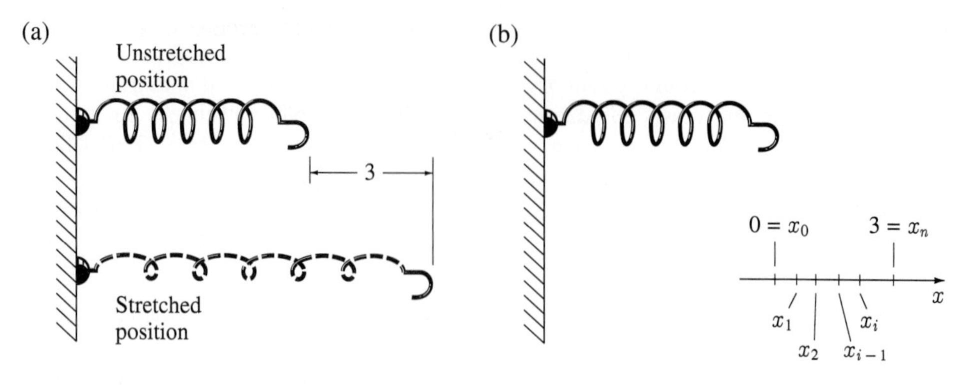

Let us choose an x-axis positive to the right with $x = 0$ at the position of the free end of the spring when it is in the unstretched position (Figure 6.4(b)). In order to calculate the work to stretch the spring, we must know something about the forces involved. It has been shown experimentally that the force F that must be exerted on the free end of the spring in order to maintain a stretch x in the spring is proportional to x:

$$F = F(x) = kx,$$

where $k > 0$ is a constant. This then is the force that will perform the work. The basic definition of work W done by a constant force F acting along a straight line segment of length d is $W = Fd$. Unfortunately, our force $F(x)$ is not constant; it depends on x, and we cannot therefore simply multiply force by distance. What we can find, however, is an approximation to the required work by dividing the length between $x = 0$ and $x = 3$ into n subintervals by $n - 1$ points $x_1, x_2, \ldots, x_{n-1}$ such that

$$0 = x_0 < x_1 < x_2 < \ldots < x_{n-1} < x_n = 3.$$

When the spring is stretched between x_{i-1} and x_i, the force necessary to maintain this stretch does not vary greatly and can be approximated by $F(x_i)$. It follows then that the work necessary to stretch the spring from x_{i-1} to x_i is approximately equal to

$$W_i = F(x_i)(x_i - x_{i-1}).$$

As a result, an approximation to the total work required to pull the free end of the spring from $x = 0$ to $x = 3$ is

$$\sum_{i=1}^{n} W_i = \sum_{i=1}^{n} F(x_i)(x_i - x_{i-1}).$$

Once again we expect that as n becomes indefinitely large, this approximation approaches the required work W, and

$$W = \lim_{n\to\infty} \sum_{i=1}^{n} W_i = \lim_{n\to\infty} \sum_{i=1}^{n} F(x_i)(x_i - x_{i-1}). \tag{6.10}$$

Each of these four problems on area, volume, blood flow, and work has been tackled in the same way, and the method can be described qualitatively.

The quantity to be calculated, say W, cannot be obtained for the object G given because no formula exists. As a result, n smaller objects, say G_i, are constructed. The G_i are chosen in such a way that the quantity W can be calculated, exactly or approximately, for each G_i, say W_i. Then an approximation for W is

$$\sum_{i=1}^{n} W_i.$$

If the number of G_i is increased indefinitely, this approximation becomes more and more accurate and

$$W = \lim_{n\to\infty} \sum_{i=1}^{n} W_i.$$

It is this "limit-summation" process that we shall discuss throughout the remainder of the chapter. We begin in Section 6.3 with a mathematical description of the process, and by doing so, we obtain a unified approach to the whole idea. We then discover that there is a very simple way to calculate these limits. At that point we will be ready to use the technique in a multitude of applications, including the four problems in this section.

SECTION 6.3

The Definite Integral

The four problems of Section 6.2 have a common theme: the limit of a summation. By means of a summation we approximated some quantity (area, volume, blood flow, work), and the limit led, at least intuitively, to an exact value for the quantity. In this section, we investigate the mathematics of the limit-summation — but only its mathematics. We concentrate here on what a definite integral is, and how to evaluate it, while interpretation of the definite integral as area, volume, work, etc. is made in Chapter 7.

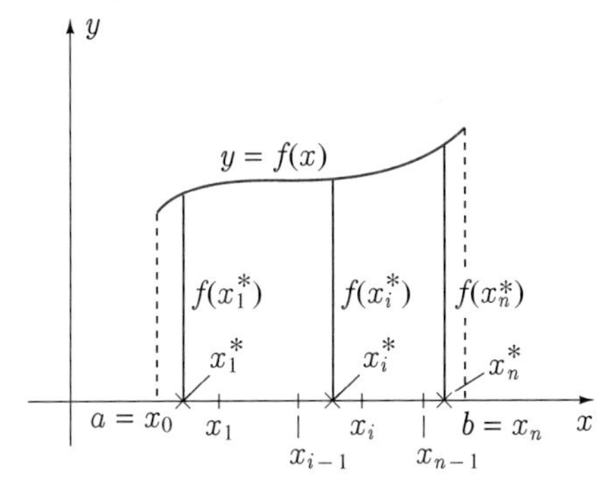

FIGURE 6.5

To define the definite integral of a function $f(x)$ on an interval $a \le x \le b$ (Figure 6.5), we divide the interval into n subintervals by any $n-1$ points:

$$a = x_0 < x_1 < x_2 < \cdots < x_i < \cdots < x_{n-1} < x_n = b.$$

Next we choose in each subinterval $x_{i-1} \le x \le x_i$ any point x_i^* whatsoever, and evaluate $f(x_i^*)$. We now form the sum

$$f(x_1^*)(x_1 - x_0) + f(x_2^*)(x_2 - x_1) + \cdots + f(x_n^*)(x_n - x_{n-1}) = \sum_{i=1}^{n} f(x_i^*)(x_i - x_{i-1})$$

$$= \sum_{i=1}^{n} f(x_i^*)\Delta x_i,$$

where we have set $\Delta x_i = x_i - x_{i-1}$. We denote by $\|\Delta x_i\|$ the length of the longest of the n subintervals,

$$\|\Delta x_i\| = \max_{i=1,\ldots,n} |\Delta x_i|.$$

It is often called the **norm** of the particular partition of $a \le x \le b$ into the subintervals Δx_i. With this notation, we are ready to define the definite integral of $f(x)$. It is the limit of the above summation as the number of subintervals becomes increasingly large and every subinterval shrinks to a point. An easier way to say this is to take the limit as the norm of the partition approaches zero. In other words, we define the **definite integral** of $f(x)$ with respect to x from $x = a$ to $x = b$ as

$$\int_a^b f(x)\,dx = \lim_{\|\Delta x_i\| \to 0} \sum_{i=1}^{n} f(x_i^*)\Delta x_i, \tag{6.11}$$

provided the limit exists. If the limit exists, but is dependent on the choice of subdivision Δx_i or star-points x_i^*, then the definite integral is of little use. We stipulate, therefore, that in order for the definite integral to exist, the limit of the sum in equation 6.11 must be independent of the manner of subdivision of the interval $a \leq x \leq b$ and choice of star-points in the subintervals. At first sight this requirement might seem rather severe, since we must now check that all subdivisions and all choices of star-points lead to the same limit before concluding that the definite integral exists. Fortunately, however, the following theorem indicates that, for continuous functions, this is unnecessary. A proof of this theorem can be found in advanced books on mathematical analysis.

Theorem 6.2

If a function $f(x)$ is continuous on a finite interval $a \leq x \leq b$, then the definite integral of $f(x)$ with respect to x from $x = a$ to $x = b$ exists.

For a continuous function, the definite integral exists, and any choice of subdivision and star-points leads to its correct value through the limiting process. We call $f(x)$ on the left-hand side of equation 6.11 the integrand, and a and b the **lower** and **upper limits of integration**, respectively. The sum $\sum_{i=1}^{n} f(x_i^*)\Delta x_i$ is called a **Riemann sum**, and because of this, definite integral (6.11) is also called the **Riemann integral**. The integral was named after German mathematician G.F.B. Riemann (1826-1866), who introduced the notion of the definite integral as a sum.

EXAMPLE 6.3

Evaluate the definite integral

$$\int_0^1 x^2\,dx.$$

SOLUTION Since $f(x) = x^2$ is continuous on the interval $0 \leq x \leq 1$, the definite integral exists, and we may choose any subdivision and star-points in its evaluation. The simplest partition is into n equal subintervals of length $1/n$ by the points (Figure 6.6)

$$x_i = \frac{i}{n}, \qquad i = 0, \ldots, n.$$

We choose for star-points the right end of each subinterval; that is, in $x_{i-1} \leq x \leq x_i$, we choose $x_i^* = x_i = i/n$. Then, by equation 6.11, we have

$$\int_0^1 x^2\,dx = \lim_{\|\Delta x_i\| \to 0} \sum_{i=1}^{n} f(x_i^*)\Delta x_i = \lim_{\|\Delta x_i\| \to 0} \sum_{i=1}^{n} (x_i^*)^2 \Delta x_i.$$

Since all subintervals have equal length $\Delta x_i = 1/n$, the norm of the partition is $\|\Delta x_i\| = 1/n$, and taking the limit as $\|\Delta x_i\| \to 0$ is tantamount to letting $n \to \infty$. Thus,

$$\int_0^1 x^2\,dx = \lim_{n\to\infty}\sum_{i=1}^{n}\left(\frac{i}{n}\right)^2\left(\frac{1}{n}\right)$$
$$= \lim_{n\to\infty}\frac{1}{n^3}\sum_{i=1}^{n} i^2.$$

If we now use equation 6.5 for the sum of the squares of the first n positive integers, we obtain

$$\int_0^1 x^2\,dx = \lim_{n\to\infty}\frac{1}{n^3}\frac{n(n+1)(2n+1)}{6} = \frac{1}{3}.$$

FIGURE 6.6

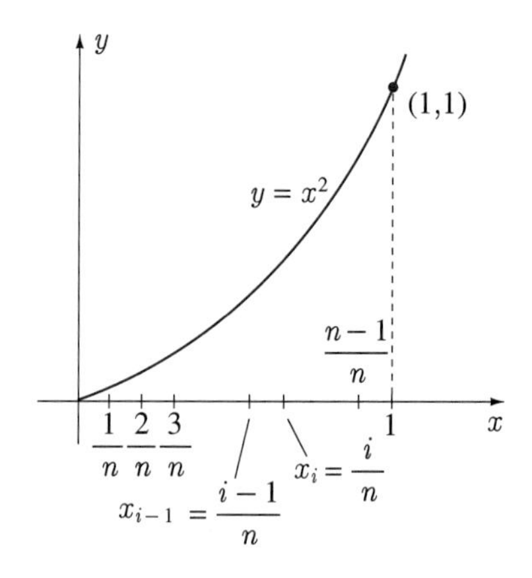

■

This example illustrates that even for a simple function such as $f(x) = x^2$, evaluation of the definite integral by equation 6.11 is not really simple. In fact, had we not known formula 6.5 for the sum of the squares of the integers, we would not have been able to complete the calculation. Imagine the magnitude of the problem were the integrand equal to $f(x) = x(x+1)^{-2/3}$. In other words, if definite integrals are to be at all useful, we must find a simpler way to evaluate them. This we do in Section 6.4, but in order to stress the definite integral as a limit-summation, we consider one more example.

EXAMPLE 6.4

Evaluate the definite integral

$$\int_{-1}^{1}(5x-2)\,dx.$$

SOLUTION Since $f(x) = 5x - 2$ is continuous on the interval $-1 \le x \le 1$, the definite integral exists, and we may choose any partition and star-points in its evaluation. For n equal subdivisions of length $2/n$, we use points (Figure 6.7)

$$x_i = -1 + \frac{2i}{n}, \qquad i = 0, \ldots, n.$$

If we choose the right end of each subinterval as a star-point, that is, $x_i^* = x_i = -1 + 2i/n$, then equation 6.11 gives

$$\int_{-1}^{1}(5x-2)\,dx7 = \lim_{||\Delta x_i||\to 0}\sum_{i=1}^{n} f(x_i^*)\Delta x_i$$

$$= \lim_{||\Delta x_i||\to 0}\sum_{i=1}^{n}(5x_i^*-2)\Delta x_i .$$

Once again all subintervals have equal length $\Delta x_i = 2/n$, and therefore we may replace $||\Delta x_i|| \to 0$ with $n \to \infty$,

$$\int_{-1}^{1}(5x-2)\,dx = \lim_{n\to\infty}\sum_{i=1}^{n}\left[5\left(-1+\frac{2i}{n}\right)-2\right]\left(\frac{2}{n}\right) = \lim_{n\to\infty}\sum_{i=1}^{n}\left[\frac{20i-14n}{n^2}\right].$$

Applying Theorem 6.1, we can break the summation into two parts, and take constants outside each summation to obtain

$$\int_{-1}^{1}(5x-2)\,dx = \lim_{n\to\infty}\left[\sum_{i=1}^{n}\frac{20i}{n^2}-\sum_{i=1}^{n}\frac{14}{n}\right]$$

$$= \lim_{n\to\infty}\left[\frac{20}{n^2}\sum_{i=1}^{n} i-\frac{14}{n}\sum_{i=1}^{n}1\right]$$

$$= \lim_{n\to\infty}\left[\frac{20}{n^2}\frac{n(n+1)}{2}-\frac{14}{n}n\right] \qquad \text{(using formula 6.4)}$$

$$= \lim_{n\to\infty}\left[\frac{10-4n}{n}\right]$$

$$= -4 .$$

FIGURE 6.7

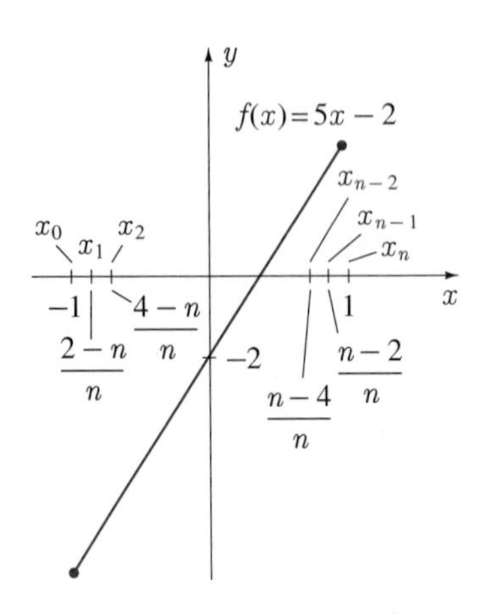

■

Note that in Example 6.3 the value of the definite integral is positive, and in Example 6.4 it is negative. The value of the definite integral in Exercise 8 is zero. In other words, the value of a definite integral can be positive, negative, or zero, depending on the limits and the integrand.

EXERCISES 6.3

In Exercises 1–8 use equation 6.11 to evaluate the definite integral.

1. $\int_0^1 x\,dx$ **2.** $\int_0^2 3x\,dx$

3. $\int_0^1 (3x+2)\,dx$ **4.** $\int_0^2 x^3\,dx$

5. $\int_1^2 (x^2+2x)\,dx$ **6.** $\int_{-1}^0 (-x+1)\,dx$

7. $\int_{-1}^1 x^2\,dx$ **8.** $\int_{-1}^1 x^3\,dx$

9. Evaluate the definite integral

$$\int_{-1}^1 x^{15}\,dx.$$

10. (a) Consider the definite integral

$$\int_0^1 2^x\,dx.$$

Show that when the interval $0 \le x \le 1$ is subdivided into n equal subintervals, and star-points are chosen as right-hand end points in each subinterval, equation 6.11 leads to

$$\int_0^1 2^x\,dx = \lim_{n\to\infty} \frac{1}{n}\sum_{i=1}^{n} 2^{i/n}.$$

(b) Use the formula in question 27(b) from Exercises 6.1 to express the summation in closed form,

$$\int_0^1 2^x\,dx = \lim_{n\to\infty} \frac{2^{1/n}}{n(2^{1/n}-1)}.$$

(c) Use L'Hôpital's rule to evaluate this limit, and hence find the value of the definite integral.

11. Use the technique of Exercise 10 to evaluate

$$\int_1^3 e^x\,dx.$$

12. Use the formula

$$\sum_{i=1}^{n} \sin i\theta = \frac{\sin\frac{(n+1)\theta}{2}\sin\frac{n\theta}{2}}{\sin\frac{\theta}{2}}$$

to evaluate

$$\int_0^{\pi} \sin x\,dx.$$

13. Use the formula

$$\sum_{i=1}^{n} \cos i\theta = \frac{\cos\frac{(n+1)\theta}{2}\sin\frac{n\theta}{2}}{\sin\frac{\theta}{2}}$$

to evaluate

$$\int_0^{\pi/2} \cos x\,dx.$$

14. In this exercise we evaluate

$$\int_a^b x^k\,dx$$

for $b > a > 0$ and any $k \neq -1$.

(a) Let $h = (b/a)^{1/n}$ and subdivide the interval $a \le x \le b$ into n subintervals by the points

$$x_0 = a, \quad x_1 = ah, \quad x_2 = ah^2, \ldots, x_i = ah^i, \ldots,$$
$$x_n = ah^n = b.$$

Show that with the choice of $x_i^* = x_i$ in the ith subinterval $x_{i-1} \le x \le x_i$, equation 6.11 gives

$$\int_a^b x^k\,dx = a^{k+1} \lim_{n\to\infty}\left[\left(\frac{h-1}{h}\right)\sum_{i=1}^{n} (h^{k+1})^i\right].$$

(b) Use the result of Exercise 27 in Section 6.1 to write the summation in closed form, and hence show that

$$\int_a^b x^k\,dx = (b^{k+1} - a^{k+1}) \lim_{n\to\infty} \frac{\left(\frac{b}{a}\right)^{k/n}\left[\left(\frac{b}{a}\right)^{1/n} - 1\right]}{\left(\frac{b}{a}\right)^{(k+1)/n} - 1}.$$

(c) Use L'Hôpital's rule to evaluate this limit, and hence obtain

$$\int_a^b x^k\,dx = \frac{b^{k+1} - a^{k+1}}{k+1}.$$

SECTION 6.4

The Fundamental Theorem of Integral Calculus

In Section 6.3 we demonstrated how to evaluate definite integrals by using their definition in equation 6.11. Integrands x^2 and $5x - 2$ in Examples 6.3 and 6.4 are very simple polynomials, as are the integrands in questions 1–9 of Exercise set 6.3, but in spite of this, calculations were frequently laborious, and invariably required formulas 6.4–6.6. We promised a very simple technique that would replace these calculations, and this is the substance of the *fundamental theorem of integral calculus.*

Theorem 6.3

(The fundamental theorem of integral calculus) *If* $f(x)$ *is continuous on the interval* $a \le x \le b$, *and* $F(x)$ *is an antiderivative of* $f(x)$ *thereon, then*

$$\int_a^b f(x)\,dx = F(b) - F(a). \tag{6.12}$$

Proof Since $f(x)$ is continuous on $a \le x \le b$, the definite integral of $f(x)$ from $x = a$ to $x = b$ exists and is defined by equation 6.11, where we are at liberty to choose the Δx_i and x_i^* in any way whatsoever. For any choice of Δx_i, a convenient choice for the x_i^* can be found by applying the mean value theorem n times to $F(x)$, once on each subinterval $x_{i-1} \le x \le x_i$. This is possible since $F'(x) = f(x)$ is continuous for $a \le x \le b$. The mean value theorem states that for each subinterval, there exists at least one point c_i between x_{i-1} and x_i such that

$$\frac{F(x_i) - F(x_{i-1})}{x_i - x_{i-1}} = F'(c_i), \qquad i = 1, \ldots, n.$$

But $F'(c_i) = f(c_i)$, so that

$$F(x_i) - F(x_{i-1}) = f(c_i)\Delta x_i, \qquad i = 1, \ldots, n.$$

If we now choose $x_i^* = c_i$, then

$$f(x_i^*)\Delta x_i = F(x_i) - F(x_{i-1}),$$

and equation 6.11 gives

$$\int_a^b f(x)\,dx = \lim_{\|\Delta x_i\| \to 0} \sum_{i=1}^n f(x_i^*)\Delta x_i = \lim_{\|\Delta x_i\| \to 0} \sum_{i=1}^n [F(x_i) - F(x_{i-1})].$$

When we write out all terms in the summation, we find that many cancellations take place,

$$\begin{aligned}\int_a^b f(x)\,dx &= \lim_{\|\Delta x_i\| \to 0} \{[F(x_1) - F(x_0)] + [F(x_2) - F(x_1)] + \cdots + [F(x_n) - F(x_{n-1})]\} \\ &= \lim_{\|\Delta x_i\| \to 0} \{F(x_n) - F(x_0)\} \\ &= \lim_{\|\Delta x_i\| \to 0} \{F(b) - F(a)\} \\ &= F(b) - F(a).\end{aligned}$$

If we introduce the notation

$$\{F(x)\}_a^b$$

to represent the difference $F(b) - F(a)$, then Theorem 6.3 can be expressed in the form

$$\int_a^b f(x)\,dx = \left\{\int f(x)\,dx\right\}_a^b. \tag{6.13}$$

This is a fantastic result. No longer is it necessary to consider limits of summations in order to evaluate definite integrals. We simply find an antiderivative of the integrand, substitute $x = b$ and $x = a$, and subtract. For instance, to evaluate the definite integral in Example 6.3, we easily write

$$\int_0^1 x^2\,dx = \left\{\frac{x^3}{3}\right\}_0^1 = \frac{1}{3} - 0 = \frac{1}{3}.$$

Note that had we used *the* indefinite integral $x^3/3 + C$ for x^2, we would have had

$$\int_0^1 x^2\,dx = \left\{\frac{x^3}{3} + C\right\}_0^1 = \left\{\frac{1}{3} + C\right\} - C = \frac{1}{3}.$$

Because the arbitrary constant always vanishes in the evaluation of definite integrals, we omit the constant in this context.

Similarly, for Example 6.4, we obtain

$$\int_{-1}^1 (5x - 2)\,dx = \left\{\frac{5x^2}{2} - 2x\right\}_{-1}^1 = \left\{\frac{5}{2} - 2\right\} - \left\{\frac{5}{2} + 2\right\} = -4.$$

EXAMPLE 6.5

Evaluate the following definite integrals:

$$\text{(a)} \int_1^2 (3x^2 - x + 4)\,dx \qquad \text{(b)} \int_{-2}^4 \sqrt{x+4}\,dx$$

SOLUTION

$$\text{(a)} \int_1^2 (3x^2 - x + 4)\,dx = \left\{x^3 - \frac{x^2}{2} + 4x\right\}_1^2 = \{8 - 2 + 8\} - \left\{1 - \frac{1}{2} + 4\right\} = \frac{19}{2}$$

$$\text{(b)} \int_{-2}^4 \sqrt{x+4}\,dx = \left\{\frac{2}{3}(x+4)^{3/2}\right\}_{-2}^4 = \frac{2}{3}(8)^{3/2} - \frac{2}{3}(2)^{3/2} = \frac{28\sqrt{2}}{3}$$

■

EXAMPLE 6.6

Evaluate

$$\int_1^3 \left(\frac{1}{x^2} + 3x^3\right) dx.$$

SOLUTION Since $-1/x + 3x^4/4$ is an antiderivative for $1/x^2 + 3x^3$,

$$\int_1^3 \left(\frac{1}{x^2} + 3x^3\right) dx = \left\{-\frac{1}{x} + \frac{3x^4}{4}\right\}_1^3 = \left\{-\frac{1}{3} + \frac{243}{4}\right\} - \left\{-1 + \frac{3}{4}\right\} = \frac{182}{3}.$$

■

EXAMPLE 6.7

Can the following definite integral be evaluated with Theorem 6.3,

$$\int_{-1}^{1} \frac{1}{x^2}\, dx\ ?$$

SOLUTION No! Theorem 6.3 requires the integrand $1/x^2$ to be continuous on the interval $-1 \le x \le 1$, and this is not the case. The function is discontinuous at $x = 0$. ■

There is a difficulty with Theorem 6.3. It is subtle, but important. The theorem states that, to evaluate definite integrals of continuous functions, we use antiderivatives. But how do we know that continuous functions have antiderivatives? We don't yet. This fact will be established in Section 6.5 when we verify an alternative version of the fundamental theorem. You might ask why this alternative version is not proved first. Would it not be more logical to first establish existence of antiderivatives, and then use this fact to prove Theorem 6.3? From a logic point of view, the answer is yes. However, from a practical point of view, Theorem 6.3 is infinitely more useful than its alternative version, and we therefore want to give it every possible emphasis. To prove the alternative version first would detract from the importance and simplicity of Theorem 6.3.

Before we move on to the alternative version of the fundamental theorem, we present the following theorems which describe properties of the definite integral.

Theorem 6.4

If $f(x)$ is continuous on $a \le x \le b$, then:

$$\text{(i)} \quad \int_{b}^{a} f(x)\, dx = -\int_{a}^{b} f(x)\, dx; \tag{6.14}$$

$$\text{(ii)} \quad \int_{a}^{b} f(x)\, dx = \int_{a}^{c} f(x)\, dx + \int_{c}^{b} f(x)\, dx. \tag{6.15}$$

Theorem 6.5

If $f(x)$ and $g(x)$ are continuous on $a \le x \le b$, then:

$$\text{(i)} \quad \int_{a}^{b} [f(x) + g(x)]\, dx = \int_{a}^{b} f(x)\, dx + \int_{a}^{b} g(x)\, dx; \tag{6.16}$$

$$\text{(ii)} \quad \int_{a}^{b} kf(x)\, dx = k\int_{a}^{b} f(x)\, dx, \tag{6.17}$$

when k is a constant.

Properties 6.16 and 6.17 are analogous to 5.3 and 5.4 for indefinite integrals. Theorems 6.4 and 6.5 can be proved using either Theorem 6.3 or equation 6.11.

We can also establish the following property.

Theorem 6.6

When $f(x)$ is continuous on $a \le x \le b$ and $m \le f(x) \le M$ on this interval,

$$m(b-a) \le \int_a^b f(x)\,dx \le M(b-a). \qquad (6.18)$$

Proof By equation 6.11, we can write

$$\begin{aligned}\int_a^b f(x)\,dx &= \lim_{\|\Delta x_i\|\to 0} \sum_{i=1}^{n} f(x_i^*)\Delta x_i \\ &\le \lim_{\|\Delta x_i\|\to 0} \sum_{i=1}^{n} M\Delta x_i \\ &= M \lim_{\|\Delta x_i\|\to 0} \sum_{i=1}^{n} \Delta x_i \\ &= M \lim_{\|\Delta x_i\|\to 0} [(x_1-a)+(x_2-x_1)+(x_3-x_2)+\cdots+(b-x_{n-1})] \\ &= M \lim_{\|\Delta x_i\|\to 0} (b-a) \\ &= M(b-a).\end{aligned}$$

A similar proof establishes the inequality involving m.

EXAMPLE 6.8

Use Theorem 6.6 to find a maximum possible value for

$$\int_1^4 \frac{\sin x^2}{1+x^2}\,dx.$$

SOLUTION Clearly $\sin x^2 \le 1$ for all x, and on the interval $1 \le x \le 4$, the largest value of $1/(1+x^2)$ is $1/2$. Consequently, $(\sin x^2)/(1+x^2) \le 1/2$ for $1 \le x \le 4$, and by Theorem 6.6,

$$\int_1^4 \frac{\sin x^2}{1+x^2}\,dx \le \frac{1}{2}(4-1) = \frac{3}{2}.$$

■

EXERCISES 6.4

In Exercises 1–34 evaluate the definite integral.

1. $\displaystyle\int_3^4 (x^3+3)\,dx$

2. $\displaystyle\int_1^3 (x^2-2x+3)\,dx$

3. $\displaystyle\int_{-1}^1 (4x^3+2x)\,dx$

4. $\displaystyle\int_{-3}^{-1} \frac{1}{x^2}\,dx$

5. $\displaystyle\int_4^2 \left(x^2+\frac{3}{x^3}\right)dx$

6. $\displaystyle\int_0^{\pi/2} \sin x\,dx$

7. $\displaystyle\int_{-1}^1 (x^2-1-x^4)\,dx$

8. $\displaystyle\int_{-1}^{-2} \left(\frac{1}{x^2}-2x\right)dx$

9. $\displaystyle\int_1^2 (x^4+3x^2+2)\,dx$

10. $\displaystyle\int_0^1 x(x^2+1)\,dx$

11. $\displaystyle\int_0^1 x^2(x^2+1)^2\,dx$

12. $\displaystyle\int_0^{2\pi} \cos 2x\, dx$

13. $\displaystyle\int_1^3 \frac{x^2+3}{x^2}\, dx$

14. $\displaystyle\int_0^1 (x^{2.2} - x^{\pi})\, dx$

15. $\displaystyle\int_{-1}^1 x^2(x^3 - x)\, dx$

16. $\displaystyle\int_3^4 \frac{(x^2-1)^2}{x^2}\, dx$

17. $\displaystyle\int_1^2 \left(\sqrt{x} - \frac{1}{\sqrt{x}}\right) dx$

18. $\displaystyle\int_{-2}^3 (x-1)^3\, dx$

19. $\displaystyle\int_2^4 \frac{(x^2-1)(x^2+1)}{x^2}\, dx$

20. $\displaystyle\int_0^{\pi/4} 3\cos x\, dx$

21. $\displaystyle\int_0^{\pi/4} \sec^2 x\, dx$

22. $\displaystyle\int_{\pi/2}^{\pi} \sin x \cos x\, dx$

23. $\displaystyle\int_{\pi/6}^{\pi/3} \csc^2 3x\, dx$

24. $\displaystyle\int_{-\pi/4}^{\pi/4} \sec x \tan x\, dx$

25. $\displaystyle\int_0^2 2^x\, dx$

26. $\displaystyle\int_{-1}^2 e^x\, dx$

27. $\displaystyle\int_0^1 e^{3x}\, dx$

28. $\displaystyle\int_{-3}^{-2} \frac{1}{x}\, dx$

29. $\displaystyle\int_1^3 \frac{(x+1)^2}{x}\, dx$

30. $\displaystyle\int_0^1 3^{4x}\, dx$

31. $\displaystyle\int_0^5 |x|\, dx$

32. $\displaystyle\int_0^4 x|x+1|\, dx$

33. $\displaystyle\int_{-5}^5 |x|\, dx$

34. $\displaystyle\int_{-2}^1 x|x+1|\, dx$

35. If $v(t) = 3t^2 - 6t - 105$ represents the velocity (in metres per second) of a particle moving along the x-axis, evaluate and interpret physically:

(a) $\displaystyle\int_0^{12} v\, dt$ (b) $\displaystyle\int_0^{12} |v|\, dt$

In Exercises 36–41 use Theorem 6.6 to find maximum and minimum values for the integral.

36. $\displaystyle\int_0^{\pi/4} \frac{\sin x}{1+x^2}\, dx$

37. $\displaystyle\int_0^{\pi/2} \frac{\sin x}{1+x}\, dx$

38. $\displaystyle\int_0^{\pi} \frac{\sin x}{2+x^2}\, dx$

39. $\displaystyle\int_{\pi/4}^{\pi/2} \frac{\sin 2x}{10+x^2}\, dx$

40. $\displaystyle\int_0^1 (1+4x^4)\cos(x^2)\, dx$

41. $\displaystyle\int_1^3 \sqrt{4+x^3}\, dx$

SECTION 6.5

An Alternative Version of the Fundamental Theorem

The fundamental theorem of integral calculus in Section 6.4 allows us to use antiderivatives to evaluate definite integrals. To complete the picture, we now show that in some sense, definite integrals can be regarded as antiderivatives.

When $f(t)$ is continuous for $a \le t \le b$, its definite integral

$$\int_a^b f(t)\, dt$$

is a number. If b is changed but a is kept fixed, the value of the definite integral changes; for each value of b, a new value for the definite integral. In other words, the value of the definite integral is a function of its upper limit. Suppose we replace b by x, and denote the resulting function by $F(x)$,

$$F(x) = \int_a^x f(t)\, dt.$$

We now show that the derivative of $F(x)$ is $f(x)$; that is, $F(x)$ is an antiderivative of $f(x)$.

Theorem 6.7

(Alternative Version of the Fundamental Theorem) *When $f(x)$ is continuous for $a \le x \le b$, the function*

$$F(x) = \int_a^x f(t)\,dt \tag{6.19}$$

is differentiable for $a \le x \le b$, and $F'(x) = f(x)$.

Proof If x is any point in the open interval $a < x < b$, then h can always be chosen sufficiently small that $x + h$ is also in the interval $a < x < b$. By definition 3.1, the derivative of $F(x)$ at this x is defined as

$$\begin{aligned} F'(x) &= \lim_{h \to 0} \frac{F(x+h) - F(x)}{h} \\ &= \lim_{h \to 0} \frac{1}{h} \left[\int_a^{x+h} f(t)\,dt - \int_a^x f(t)\,dt \right] \\ &= \lim_{h \to 0} \frac{1}{h} \left[\int_a^{x+h} f(t)\,dt + \int_x^a f(t)\,dt \right] \qquad \text{(by property 6.14)} \\ &= \lim_{h \to 0} \frac{1}{h} \int_x^{x+h} f(t)\,dt \qquad \text{(by property 6.15).} \end{aligned}$$

According to property 6.18,

$$mh \le \int_x^{x+h} f(t)\,dt \le Mh$$

where m and M are the absolute minimum and absolute maximum of $f(x)$ on the interval between x and $x + h$. Division by h gives

$$m \le \frac{1}{h} \int_x^{x+h} f(t)\,dt \le M.$$

Consider what happens as we let $h \to 0$. The limit of the middle term is $F'(x)$. Furthermore, the numbers m and M must approach one another, and in the limit must both be equal to $f(x)$. They are minimum and maximum values of $f(x)$ on the interval between x and $x + h$, and h is approaching zero. We conclude therefore that

$$F'(x) = \lim_{h \to 0} \int_x^{x+h} f(t)\,dt = f(x).$$

This argument can also be used to establish that $F(x)$ has a right-hand derivative $f(a)$ at $x = a$, and a left-hand derivative $f(b)$ at $x = b$.

Recall that Theorem 6.3 assumed the existence of antiderivatives of continuous functions. Theorem 6.7 establishes this fact, and states that definite integral 6.19 with a variable upper limit is one such antiderivative. In practice, we do not use 6.19 to find antiderivatives, but the theorem does establish existence of antiderivatives of continuous functions.

Symbolically, we may write the result of Theorem 6.7 in the form

$$\frac{d}{dx}\int_a^x f(t)\,dt = f(x). \tag{6.20}$$

The lower limit in this result need not be a; it can also be b,

$$\frac{d}{dx}\int_b^x f(t)\,dt = f(x).$$

The lower limit can also be any number c between a and b,

$$\frac{d}{dx}\int_c^x f(t)\,dt = f(x).$$

EXAMPLE 6.9

Evaluate

$$\frac{d}{dx}\int_0^x \sqrt{1-t^2}\,dt.$$

SOLUTION According to 6.20,

$$\frac{d}{dx}\int_0^x \sqrt{1-t^2}\,dt = \sqrt{1-x^2}.$$

This is valid for $-1 \le x \le 1$. ■

EXAMPLE 6.10

Evaluate

$$\frac{d}{dx}\int_1^{2x^2} \frac{\sin t}{1+t^2}\,dt.$$

SOLUTION For this problem we set $u = 2x^2$, and invoke the chain rule,

$$\frac{d}{dx}\int_1^{2x^2} \frac{\sin t}{1+t^2}\,dt = \frac{d}{dx}\int_1^{u} \frac{\sin t}{1+t^2}\,dt = \left[\frac{d}{du}\int_1^{u} \frac{\sin t}{1+t^2}\,dt\right]\frac{du}{dx}$$

$$= \frac{\sin u}{1+u^2}(4x) = \frac{4x\sin(2x^2)}{1+4x^4}.$$

■

EXAMPLE 6.11

Evaluate

$$\frac{d}{dx}\int_x^5 (1+t^3)^{2/3}\,dt.$$

SOLUTION We can solve this problem by reversing the limits on the integral,

$$\frac{d}{dx}\int_x^5 (1+t^3)^{2/3}\,dt = -\frac{d}{dx}\int_5^x (1+t^3)^{2/3}\,dt = -(1+x^3)^{2/3}.$$

■

In the following example, the variable x appears in both limits.

EXAMPLE 6.12

Evaluate

$$\frac{d}{dx}\int_{x^2}^{2x} \cos(2t^3+1)\,dt.$$

SOLUTION Since the integrand is continuous for all real numbers, property 6.15 permits us to write

$$\int_{x^2}^{2x} \cos(2t^3+1)\,dt = \int_{x^2}^{a} \cos(2t^3+1)\,dt + \int_{a}^{2x} \cos(2t^3+1)\,dt,$$

for any real number a whatsoever. To find the derivative of the second integral on the right, we set $v = 2x$ and use the chain rule (as in Example 6.10), and for the derivative of the first integral on the right, we reverse the limits (as in Example 6.11) and then use the chain rule with $u = x^2$:

$$\begin{aligned}
\frac{d}{dx}\int_{x^2}^{2x} \cos(2t^3+1)\,dt &= -\frac{d}{dx}\int_{a}^{x^2} \cos(2t^3+1)\,dt + \frac{d}{dx}\int_{a}^{2x} \cos(2t^3+1)\,dt \\
&= \left[-\frac{d}{du}\int_{a}^{u} \cos(2t^3+1)\,dt\right]\frac{du}{dx} + \left[\frac{d}{dv}\int_{a}^{v} \cos(2t^3+1)\,dt\right]\frac{dv}{dx} \\
&= -2x\cos(2u^3+1) + 2\cos(2v^3+1) \\
&= -2x\cos(2x^6+1) + 2\cos(16x^3+1).
\end{aligned}$$

■

EXERCISES 6.5

In Exercises 1–20 differentiate the definite integral with respect to x.

1. $\int_0^x (3t^2+t)\,dt$

2. $\int_1^x \frac{1}{\sqrt{t^2+1}}\,dt$

3. $\int_x^2 \sin(t^2)\,dt$

4. $\int_x^{-1} t^3\cos t\,dt$

5. $\int_0^{3x} (2t-t^4)^2\,dt$

6. $\int_1^{2x} \sqrt{t+1}\,dt$

7. $\int_4^{3x^2} \sin(3t+4)\,dt$

8. $\int_{-2}^{5x+4} \sqrt{t^3+1}\,dt$

9. $\int_x^{2x} (3\sqrt{t}-2t)\,dt$

10. $\int_{4x}^{4x+4} \left(t^3-\frac{1}{\sqrt{t}}\right)dt$

11. $\int_{-2x}^{x} \tan(3t+1)\,dt$

12. $\int_{-x^2}^{-2x^2} \sec(1-t)\,dt$

13. $\int_0^{\sin x} \cos(t^2)\,dt$

14. $\int_{\cos x}^{\sin x} \frac{1}{\sqrt{t+1}}\,dt$

15. $\int_0^{2\sqrt{x}} \sqrt{t}\,dt$

16. $\int_{\sqrt{x}}^{2\sqrt{x}} \sqrt{t}\,dt$

17. $\int_1^{x^2} t^2 e^{4t}\,dt$

18. $\int_x^2 \ln(t^2+1)\,dt$

19. $\int_x^{2x} t\ln t\,dt$

20. $\int_{-2x}^{3x} e^{-4t^2}\,dt$

21. Verify that when $a(x)$ and $b(x)$ are differentiable functions of x,

$$\frac{d}{dx}\int_{a(x)}^{b(x)} f(t)\,dt = f[b(x)]\frac{db}{dx} - f[a(x)]\frac{da}{dx}.$$

Is equation 6.19 a special case of this result?

22.–28. Use the result of Exercise 21 to redo Exercises 8, 10, 12, 14, 16, 18, and 20.

SECTION 6.6

Average Values

The average value of two numbers c and d is defined as $(c+d)/2$. The average value of a set of n numbers $y_1, y_2, \ldots, y_n$ is $(y_1 + y_2 + \cdots + y_n)/n$. We would like to extend this idea to define the average value of a function $f(x)$ over an interval $a \le x \le b$. By beginning with some simple functions we can see eventually how average values should be defined.

The function $f(x)$ in Figure 6.8(a) is equal to 1 for the first third of the interval shown, 2 for the second third, and 3 for the last third. We suggest that its average value over the interval $0 \le x \le 6$ is 2. The function in Figure 6.8(b) takes on the same function values, namely, 1, 2, and 3, but not on the same subintervals. The fact that it has value 2 for $2 < x \le 5$ and value 3 for $5 < x \le 6$ would indicate that its average value should be somewhat less than the average value of 2 for the function in Figure 6.8(a). The average value of the function in Figure 6.8(c) should be even less than that in Figure 6.8(b). These three functions suggest that two factors are important when considering average values of functions: values that the function takes on, and lengths of the intervals on which they take these values. Perhaps what should be done to calculate average values for these functions is to add together the products obtained by multiplying each of the function values by the length of the interval in which it has this value, and then divide this sum by the length of the overall interval.

FIGURE 6.8

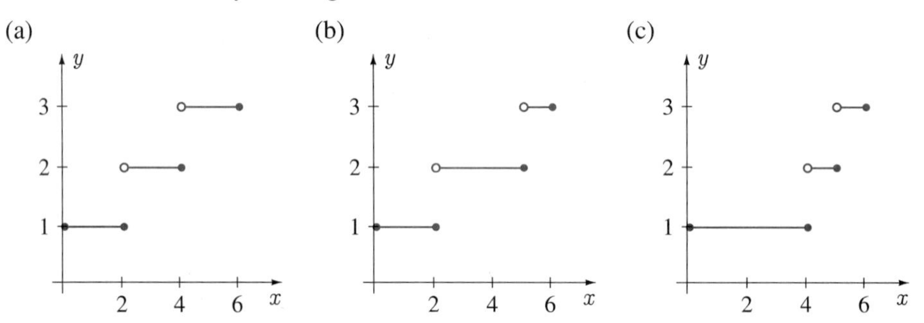

For the function in Figure 6.8(a) this would yield an average value of

$$\frac{1}{6-0}[1(2-0) + 2(4-2) + 3(6-4)] = 2;$$

for the function in Figure 6.8(b),

$$\frac{1}{6-0}[1(2-0) + 2(5-2) + 3(6-5)] = \frac{11}{6};$$

and for the function in Figure 6.8(c),

$$\frac{1}{6-0}[1(4-0) + 2(5-4) + 3(6-5)] = \frac{3}{2}.$$

This procedure is applicable only to a function whose domain can be subdivided into a finite number of subintervals inside each of which the function has a constant value. Such a function is said to be *piecewise constant*. What shall we do for functions which are not piecewise constant? By rephrasing the above procedure, it will become obvious. The same three average values are obtained if we adopt the following approach [illustrated for the function $f(x)$ in Figure 6.8(c)]. Divide the interval $0 \le x \le 6$ into three subintervals $0 < x \le 4$, $4 < x \le 5$, and $5 < x \le 6$. Pick a point in each subinterval; call the points x_1^*, x_2^*, and x_3^*. Evaluate $f(x)$ at each point and multiply by the length of the subinterval in which the point is found. Add these results, and divide by the length of the interval to obtain the average value,

$$\frac{1}{6}[f(x_1^*)(4-0) + f(x_2^*)(5-4) + f(x_3^*)(6-5)] = \frac{1}{6}[1(4) + 2(1) + 3(1)] = \frac{3}{2}.$$

But this is the procedure used to define the definite integral of a function; it lacks the limit because the function is piecewise constant. In other words, for a function $f(x)$ which has a value at every point in the interval $a \le x \le b$ (but is not necessarily piecewise constant), we define its average value as

$$\text{Average Value} = \frac{1}{b-a}\int_a^b f(x)\,dx = \frac{1}{b-a}\lim_{\|\Delta x_i\|\to 0}\sum_{i=1}^{n} f(x_i^*)\Delta x_i. \quad (6.21)$$

EXAMPLE 6.13

What is the average value of the function $f(x) = x^2$ on the interval $0 \le x \le 2$?

SOLUTION By equation 6.21,

$$\text{Average Value} = \frac{1}{2}\int_0^2 x^2\,dx = \frac{1}{2}\left\{\frac{x^3}{3}\right\}_0^2 = \frac{1}{2}\left(\frac{8}{3}\right) = \frac{4}{3}.$$

■

EXAMPLE 6.14

Find the average value of $f(x) = \sin x$ on the intervals: (a) $0 \le x \le \pi/2$ (b) $0 \le x \le \pi$ (c) $0 \le x \le 2\pi$.

SOLUTION We calculate average values on these intervals as:

$$\text{(a)}\quad \frac{1}{\pi/2}\int_0^{\pi/2} \sin x\,dx = \frac{2}{\pi}\{-\cos x\}_0^{\pi/2} = \frac{2}{\pi}[0+1] = \frac{2}{\pi};$$

$$\text{(b)}\quad \frac{1}{\pi}\int_0^{\pi} \sin x\,dx = \frac{1}{\pi}\{-\cos x\}_0^{\pi} = \frac{1}{\pi}[1+1] = \frac{2}{\pi};$$

$$\text{(c)}\quad \frac{1}{2\pi}\int_0^{2\pi} \sin x\,dx = \frac{1}{2\pi}\{-\cos x\}_0^{2\pi} = \frac{1}{2\pi}[-1+1] = 0.$$

The graph of $\sin x$ in Figure 6.9 also suggests that the average values in (a) and (b) should be the same, and that the average value in (c) should be zero.

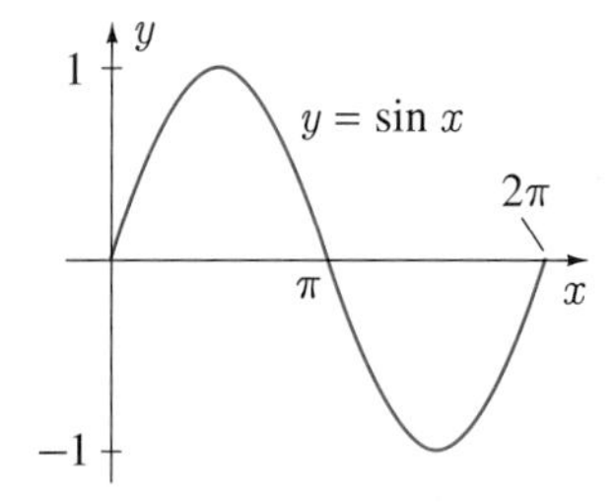

FIGURE 6.9

■

EXAMPLE 6.15 What is the average value of a linear function?

SOLUTION The average value of the linear function $f(x) = mx+c$ over an interval $a \le x \le b$ is

$$\begin{aligned}\frac{1}{b-a}\int_a^b (mx+c)\,dx &= \frac{1}{b-a}\left\{\frac{mx^2}{2}+cx\right\}_a^b \\ &= \frac{1}{b-a}\left[\left(\frac{mb^2}{2}+cb\right)-\left(\frac{ma^2}{2}+ca\right)\right] \\ &= \frac{1}{b-a}\left[\frac{m}{2}(b^2-a^2)+c(b-a)\right] \\ &= \frac{m}{2}(b+a)+c \\ &= m\left(\frac{b+a}{2}\right)+c.\end{aligned}$$

This is the value of the function midway between $x = a$ and $x = b$. ■

Theorem 6.7 can be used to establish the next theorem.

Theorem 6.8 (Mean Value Theorem for Definite Integrals) *If $f(x)$ is continuous for $a \le x \le b$, then there exists at least one number c between a and b such that*

$$\int_a^b f(x)\,dx = (b-a)f(c). \tag{6.22}$$

Proof By Theorem 6.7, the function

$$F(x) = \int_a^x f(t)\,dt$$

is continuous for $a \le x \le b$, and has derivative $f(x)$ for $a < x < b$. Mean Value Theorem 3.15 applied to $F(x)$ guarantees at least one number c between a and b such that

$$F(b) - F(a) = (b-a)F'(c).$$

By substitution,

$$\int_a^b f(t)\,dt - \int_a^a f(t)\,dt = (b-a)f(c).$$

Since the second integral vanishes, we have

$$\int_a^b f(x)\,dx = (b-a)f(c).$$

By writing equation 6.22 in the form

$$f(c) = \frac{1}{b-a}\int_a^b f(x)\,dx,$$

Theorem 6.8 states that the function must take on its average value at least once in the interval.

EXAMPLE 6.16

Find all values of c satisfying Theorem 6.8 for the function $f(x) = x^2$ on the interval $0 \le x \le 1$.

SOLUTION Substituting $a = 0$ and $b = 1$ in equation 6.22 gives

$$(1 - 0)c^2 = \int_0^1 x^2\, dx = \left\{\frac{x^3}{3}\right\}_0^1 = \frac{1}{3} - 0 = \frac{1}{3}.$$

Of the two solutions $c = \pm 1/\sqrt{3}$ for this equation, only the positive one is between 0 and 1. Thus, only $c = 1/\sqrt{3}$ satisfies Theorem 6.8 for the function $f(x) = x^2$ on the interval $0 \le x \le 1$. ■

EXAMPLE 6.17

Find all values of c satisfying Theorem 6.8 for the function $f(x) = \sin x$ on the interval $\pi/4 \le x \le 3\pi/4$.

SOLUTION Substituting $a = \pi/4$ and $b = 3\pi/4$ in equation 6.22 gives

$$\left(\frac{3\pi}{4} - \frac{\pi}{4}\right)\sin c = \int_{\pi/4}^{3\pi/4} \sin x\, dx = \{-\cos x\}_{\pi/4}^{3\pi/4} = \frac{1}{\sqrt{2}} + \frac{1}{\sqrt{2}} = \sqrt{2}.$$

Thus,

$$\sin c = \frac{2\sqrt{2}}{\pi}.$$

There are two angles between $\pi/4$ and $3\pi/4$ with a sine equal to $2\sqrt{2}/\pi$, namely $c = 1.12$ and $c = 2.02$. ■

EXERCISES 6.6

In Exercises 1–20 find the average value of the function over the interval.

1. $f(x) = x^2 - 2x, \quad 0 \le x \le 2$
2. $f(x) = x^3 - x, \quad -1 \le x \le 1$
3. $f(x) = x^3 - x, \quad 0 \le x \le 1$
4. $f(x) = x^4, \quad 1 \le x \le 2$
5. $f(x) = \sqrt{x + 1}, \quad 0 \le x \le 1$
6. $f(x) = \sqrt{x + 1}, \quad -1 \le x \le 1$
7. $f(x) = x^4 - 1, \quad 0 \le x \le 1$
8. $f(x) = x^4 - 1, \quad 0 \le x \le 2$
9. $f(x) = \cos x, \quad -\pi/2 \le x \le \pi/2$
10. $f(x) = \cos x, \quad 0 \le x \le \pi/2$
11. $f(x) = |x|, \quad -2 \le x \le 2$
12. $f(x) = |x|, \quad 0 \le x \le 2$
13. $f(x) = |x^2 - 4|, \quad 0 \le x \le 3$
14. $f(x) = |x^2 - 4|, \quad -3 \le x \le 3$
15. $f(x) = \operatorname{sgn} x, \quad -1 \le x \le 1$ (see question 36 in Exercises 2.4)
16. $f(x) = \operatorname{sgn} x, \quad -1 \le x \le 3$
17. $f(x) = H(x - 1), \quad 0 \le x \le 2$ (see question 38 in Exercises 2.4)
18. $f(x) = H(x - 4), \quad 0 \le x \le 2$
19. $f(x) = [[x]], \quad 0 \le x \le 3$ (see question 36 in Exercises 1.5)
20. $f(x) = [[x]], \quad 0 \le x \le 3.5$

21. If a particle moving along the x-axis is at position x_1 at time t_1 and at position x_2 at time t_2, its average velocity over this time interval is

$$\frac{x_2 - x_1}{t_2 - t_1}.$$

Verify that this is the same as the average of the velocity function over the time interval $t_1 \le t \le t_2$.

22. The velocity v of blood flowing through a circular vein or artery of radius R at a distance r from the centre of the blood vessel is

$$v(r) = c(R^2 - r^2).$$

(See Problem 3 in Section 6.2.) What is the average value of $v(r)$ with respect to r?

In Exercises 23–30 find all values of c satisfying equation 6.22 for the function $f(x)$ on the specified interval.

23. $f(x) = 2x - x^2, \quad 0 \le x \le 2$

24. $f(x) = x^3 - 8x, \quad -2 \le x \le 2$

25. $f(x) = \cos x, \quad 0 \le x \le \pi/2$

26. $f(x) = \cos x, \quad 0 \le x \le \pi$

27. $f(x) = \sqrt{x+1}, \quad 1 \le x \le 3$

28. $f(x) = x^2(x+1), \quad 0 \le x \le 1$

29. $f(x) = x\sqrt{x^2+1}, \quad 0 \le x \le 2$

30. $f(x) = 1/x^2 + 1/x^3, \quad 1 \le x \le 2$

SECTION 6.7

Change of Variable in the Definite Integral

The fundamental theorem of integral calculus indicates that to evaluate the definite integral

$$\int_{-4}^{-1} \frac{x}{\sqrt{x+5}}\,dx$$

we should first find an indefinite integral for $x/\sqrt{x+5}$. To do this we set $u = x + 5$, in which case $du = dx$, and

$$\begin{aligned}\int \frac{x}{\sqrt{x+5}}\,dx &= \int \frac{u-5}{\sqrt{u}}\,du = \int (u^{1/2} - 5u^{-1/2})\,du \\ &= \frac{2}{3}u^{3/2} - 10u^{1/2} + C = \frac{2}{3}(x+5)^{3/2} - 10(x+5)^{1/2} + C.\end{aligned}$$

Consequently,

$$\begin{aligned}\int_{-4}^{-1} \frac{x}{\sqrt{x+5}}\,dx &= \left\{\frac{2}{3}(x+5)^{3/2} - 10(x+5)^{1/2}\right\}_{-4}^{-1} \\ &= \left[\frac{2}{3}(4)^{3/2} - 10(4)^{1/2}\right] - \left[\frac{2}{3}(1)^{3/2} - 10(1)^{1/2}\right] \\ &= -\frac{16}{3}.\end{aligned}$$

An alternative approach, that often turns out to be less work, is to make the change of variable $u = x + 5$ directly in the definite integral. In this case we again replace

$$\frac{x}{\sqrt{x+5}}\,dx \quad \text{by} \quad \frac{u-5}{\sqrt{u}}\,du.$$

In addition, we replace the limits $x = -4$ and $x = -1$ by those values of u that correspond to these values of x, namely, $u = 1$ and $u = 4$ respectively. We then obtain

$$\int_{-4}^{-1} \frac{x}{\sqrt{x+5}}\,dx = \int_1^4 \frac{u-5}{\sqrt{u}}\,du = \int_1^4 (u^{1/2} - 5u^{-1/2})\,du$$

$$= \left\{\frac{2}{3}u^{3/2} - 10\,u^{1/2}\right\}_1^4$$
$$= \left[\frac{2}{3}(4)^{3/2} - 10(4)^{1/2}\right] - \left[\frac{2}{3}(1)^{3/2} - 10(1)^{1/2}\right]$$
$$= -\frac{16}{3}.$$

That this method is generally acceptable is proved in the following theorem.

Theorem 6.9

Suppose $f(x)$ is continuous on $a \le x \le b$, and we set $x = g(u)$, where $a = g(\alpha)$ and $b = g(\beta)$. Then

$$\int_a^b f(x)\,dx = \int_\alpha^\beta f(g(u))\,g'(u)\,du, \tag{6.23}$$

if $g'(u)$ is continuous on $\alpha \le u \le \beta$, and if when u is between α and β, $g(u)$ is between a and b.

Proof If we define

$$F(x) = \int f(x)\,dx \qquad \text{and} \qquad G(u) = \int f(g(u))\,g'(u)\,du,$$

then by Theorem 6.3,

$$\int_a^b f(x)\,dx = F(b) - F(a) \qquad \text{and} \qquad \int_\alpha^\beta f(g(u))\,g'(u)\,du = G(\beta) - G(\alpha).$$

So we must show that

$$F(b) - F(a) = G(\beta) - G(\alpha).$$

Now the chain rule applied to $y = F(x)$, $x = g(u)$ yields

$$\frac{dy}{du} = \frac{dy}{dx}\frac{dx}{du} = f(x)g'(u);$$

that is,

$$\frac{d}{du}F(g(u)) = f(g(u))\,g'(u).$$

Since this is an identity in u, we may integrate with respect to u from $u = \alpha$ to $u = \beta$:

$$\int_\alpha^\beta \frac{d}{du}F(g(u))\,du = \int_\alpha^\beta f(g(u))\,g'(u)\,du.$$

Thus,

$$\{F(g(u))\}_\alpha^\beta = \{G(u)\}_\alpha^\beta \qquad \text{or}$$

$$G(\beta) - G(\alpha) = F(g(\beta)) - F(g(\alpha)) = F(b) - F(a).$$

EXAMPLE 6.18 Evaluate

$$\int_2^4 \frac{\sqrt{x}}{(\sqrt{x}-1)^4}\,dx.$$

SOLUTION If we set $u = \sqrt{x} - 1$, then $du = [1/(2\sqrt{x})]\,dx$, and

$$\begin{aligned}
\int_2^4 \frac{\sqrt{x}}{(\sqrt{x}-1)^4}\,dx &= \int_{\sqrt{2}-1}^1 \frac{u+1}{u^4} 2(u+1)\,du \\
&= 2\int_{\sqrt{2}-1}^1 \left(\frac{1}{u^2} + \frac{2}{u^3} + \frac{1}{u^4}\right) du \\
&= 2\left\{-\frac{1}{u} - \frac{1}{u^2} - \frac{1}{3u^3}\right\}_{\sqrt{2}-1}^1 \\
&= -2\left[1 + 1 + \frac{1}{3}\right] + 2\left[\frac{1}{\sqrt{2}-1} + \frac{1}{(\sqrt{2}-1)^2} + \frac{1}{3(\sqrt{2}-1)^3}\right] \\
&= 21.2.
\end{aligned}$$

EXERCISES 6.7

In Exercises 1–22 evaluate the definite integral.

1. $\int_1^2 x(3x^2-2)^4\,dx$ **2.** $\int_0^1 z\sqrt{1-z}\,dz$

3. $\int_{-1}^0 \frac{x}{\sqrt{x+3}}\,dx$ **4.** $\int_{\pi/4}^{\pi/3} \cos^5 x \sin x\,dx$

5. $\int_1^3 x^3\sqrt{9-x^2}\,dx$ **6.** $\int_{-5}^6 \frac{x}{\sqrt{x^2-12}}\,dx$

7. $\int_4^5 y^2\sqrt{y-4}\,dy$ **8.** $\int_{1/2}^1 \sqrt{\frac{x^2}{1+x}}\,dx$

9. $\int_1^4 \frac{\sqrt{1+\sqrt{u}}}{\sqrt{u}}\,du$ **10.** $\int_{-2}^1 \frac{x+1}{(x^2+2x+2)^{1/3}}\,dx$

11. $\int_3^4 \frac{x^2}{(x-2)^4}\,dx$ **12.** $\int_0^{\pi/6} \sqrt{2+3\sin x}\cos x\,dx$

13. $\int_{\pi/4}^{\pi/2} \frac{\sin^3 x}{(1+\cos x)^4}\,dx$

14. $\int_1^4 \frac{(x+1)(x-1)}{\sqrt{x}}\,dx$

15. $\int_4^9 \sqrt{1+\sqrt{x}}\,dx$ **16.** $\int_{-1/2}^1 \sqrt{\frac{x^2}{1+x}}\,dx$

17. $\int_{-1}^1 \frac{|x|}{(x+2)^3}\,dx$ **18.** $\int_{-1}^1 \left|\frac{x}{(x+2)^3}\right|\,dx$

19. $\int_0^1 x^2 e^{x^3}\,dx$ **20.** $\int_1^2 \frac{(\ln x)^2}{x}\,dx$

21. $\int_2^4 \frac{1}{x\ln x}\,dx$ **22.** $\int_{-\pi/4}^{\pi/4} \frac{\sec^2 x}{\sqrt{4+3\tan x}}\,dx$

23. Show that if $f(x)$ is an odd function, then

$$\int_{-a}^a f(x)\,dx = 0;$$

and that if $f(x)$ is an even function,

$$\int_{-a}^a f(x)\,dx = 2\int_0^a f(x)\,dx.$$

In Exercises 24–26 use the suggested substitution to evaluate the definite integral.

24. $\int_1^3 \frac{1}{x^{3/2}\sqrt{4-x}}\,dx \qquad u = \frac{1}{\sqrt{x}}$

25. $\int_{-6}^{-1} \frac{\sqrt{x^2-6x}}{x^4}\,dx \qquad u = \frac{1}{x}$

26. $\int_{-4}^0 \frac{x}{(5-4x-x^2)^{3/2}}\,dx \qquad u^2 = \frac{1-x}{5+x}$

27. Show that

$$\int_0^\pi \frac{\cos^2[(\pi/2)\cos\theta]}{\sin\theta}\,d\theta = \frac{\pi}{2}\int_0^{2\pi} \frac{1-\cos\phi}{\phi(2\pi-\phi)}\,d\phi.$$

This integral is used in calculating radiated power from a half-wave antenna.

SUMMARY

The definite integral of a continuous function $f(x)$ from $x = a$ to $x = b$ is a number, one that depends on the function $f(x)$ and the limits a and b. We have defined the definite integral by subdividing the interval $a \leq x \leq b$ into n parts by $n + 1$ points $a = x_0 < x_1 < \cdots < x_{n-1} < x_n = b$, and choosing a point x_i^* in each subinterval $x_{i-1} \leq x \leq x_i$. The definite integral is then the limit of the summation

$$\int_a^b f(x)\,dx = \lim_{\|\Delta x_i\| \to 0} \sum_{i=1}^{n} f(x_i^*)\Delta x_i,$$

where $\Delta x_i = x_i - x_{i-1}$. Since all calculations in this limit take place on the x-axis, we regard the definite integral as an integration along the x-axis from $x = a$ to $x = b$.

The fundamental theorem of integral calculus allows us to calculate definite integrals using antiderivatives,

$$\int_a^b f(x)\,dx = \left\{ \int f(x)\,dx \right\}_a^b .$$

This presupposes that a continuous function has an antiderivative, as verified in Section 6.5. It was shown that when the definite integral is given a variable upper limit x, the resulting function

$$F(x) = \int_a^x f(t)\,dt$$

is an antiderivative of $f(x)$, that is, $F'(x) = f(x)$.

The average value of a function $f(x)$ over an interval $a \leq x \leq b$ is defined as

$$\frac{1}{b-a} \int_a^b f(x)\,dx.$$

The mean value theorem for definite integrals guarantees the existence of at least one number c between a and b, at which $f(x)$ takes on its average value,

$$f(c) = \frac{1}{b-a} \int_a^b f(x)\,dx.$$

Evaluation of a definite integral with a complex integrand can sometimes be simplified with an appropriate change of variable.

Key Terms and Formulas

In reviewing this chapter, you should be able to define or discuss the following key terms:

Sigma notation
Definite integral
Norm of a partition
Fundamental theorem
Average value of a function
Mean value theorem for integrals
Change of variable

REVIEW EXERCISES

In Exercises 1–20 evaluate the definite integral.

1. $\int_0^3 (x^2 + 3x - 2)\, dx$

2. $\int_{-1}^1 (x^2 - x^4)\, dx$

3. $\int_{-1}^1 (x^3 - 3x)\, dx$

4. $\int_0^2 (x^2 - 2x)\, dx$

5. $\int_1^2 (x + 1)^2\, dx$

6. $\int_{-3}^{-2} \frac{1}{x^2}\, dx$

7. $\int_4^9 \left(\frac{1}{\sqrt{x}} - \sqrt{x}\right) dx$

8. $\int_0^{\pi} \cos x\, dx$

9. $\int_{-1}^1 x(x + 1)^2\, dx$

10. $\int_1^2 x^2(x^2 + 3)\, dx$

11. $\int_0^3 \sqrt{x + 1}\, dx$

12. $\int_1^5 x\sqrt{x^2 - 1}\, dx$

13. $\int_1^4 \left(\frac{\sqrt{x} + 1}{\sqrt{x}}\right) dx$

14. $\int_{-1}^0 x\sqrt{x + 1}\, dx$

15. $\int_1^2 \frac{x^2 + 1}{(x + 1)^4}\, dx$

16. $\int_{-4}^{-2} x^2\sqrt{2 - x}\, dx$

17. $\int_0^{\pi/4} \frac{\cos x}{(1 + \sin x)^2}\, dx$

18. $\int_2^3 x(1 + 2x^2)^4\, dx$

19. $\int_1^8 \frac{(1 + x^{1/3})^2}{x^{2/3}}\, dx$

20. $\int_{-4}^4 |x + 2|\, dx$

21. Use equation 6.11 to evaluate the following definite integrals:

(a) $\int_0^2 (x - 5)\, dx$ (b) $\int_0^3 (x^2 + 3)\, dx$

22. Prove that each of the following answers is incorrect, but do so without evaluating the definite integral. Think about what the definite integral represents.

(a) $\int_0^4 (x^2 + 3x)\, dx = -3$ (b) $\int_{-3}^{-2} \frac{1}{x}\, dx = 5$

In Exercises 23–26 find the average value of the function on the interval.

23. $f(x) = \sqrt{x + 4}, \quad 0 \le x \le 1$

24. $f(x) = 1/x^2 - x, \quad -2 \le x \le -1$

25. $f(x) = x\sqrt{x + 1}, \quad 0 \le x \le 1$

26. $f(x) = \cos^3 x \sin^2 x, \quad 0 \le x \le \pi/2$

In Exercises 27–32 differentiate the integral with respect to x.

27. $\int_1^x t\sqrt{t^3 + 1}\, dt$

28. $\int_x^{-3} t^2(t + 1)^3\, dt$

29. $\int_1^{x^2} \sqrt{t^2 + 1}\, dt$

30. $\int_{2x}^4 t\cos t\, dt$

31. $\int_{2x+3}^{1-x} \frac{1}{t^2 + 1}\, dt$

32. $\int_{-x^2}^{x^2} \sin^2 t\, dt$

In Exercises 33–40 evaluate the definite integral.

33. $\int_0^1 \frac{x^3}{(x^2 + 1)^{3/2}}\, dx$

34. $\int_0^{\pi/6} \frac{\cos^3 x}{\sqrt{1 + \sin x}}\, dx$

35. $\int_{-1}^1 x^3\sqrt{1 - x^2}\, dx$

36. $\int_{-1}^2 \left|\frac{x}{\sqrt{3 + x}}\right| dx$

37. $\int_{-1}^2 x^2(4 - x^3)^5\, dx$

38. $\int_1^5 \frac{6x^2 + 8x + 2}{\sqrt{x^3 + 2x^2 + x}}\, dx$

39. $\int_1^2 \frac{x - 25}{\sqrt{x} - 5}\, dx$

40. $\int_0^1 \frac{1}{\sqrt{2 + x} + \sqrt{x}}\, dx$

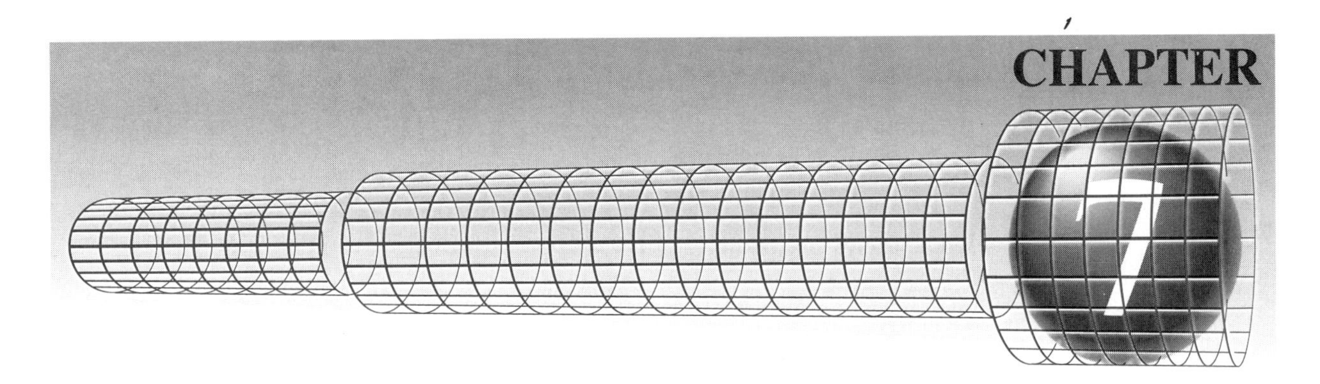

Applications of the Definite Integral

In Chapter 6 we defined the definite integral of a function $f(x)$ as the limit of a summation

$$\int_a^b f(x)\,dx = \lim_{\|\Delta x_i\| \to 0} \sum_{i=1}^{n} f(x_i^*)\,\Delta x_i.$$

In this chapter we use this equation to evaluate certain geometric and physical quantities. Each of these quantities is expressed as a limit-summation of the form in this equation. The limit-summation can immediately be interpreted as a definite integral. But then the definite integral can be evaluated by means of the corresponding indefinite integral.

SECTION 7.1

Area

We have formulas for areas of very few shapes, such as squares, rectangles, triangles, and polygons of any shape, since they can be divided into rectangles and triangles. We also know that the area of a circle is π times the square of its radius (but do you remember where this formula came from?). Now consider finding the area in Figure 7.1(a). Each of us has intuitive ideas about this area. In this section we take these intuitive ideas of what the area *ought to be*, and make a precise mathematical definition of what the area *is*.

FIGURE 7.1

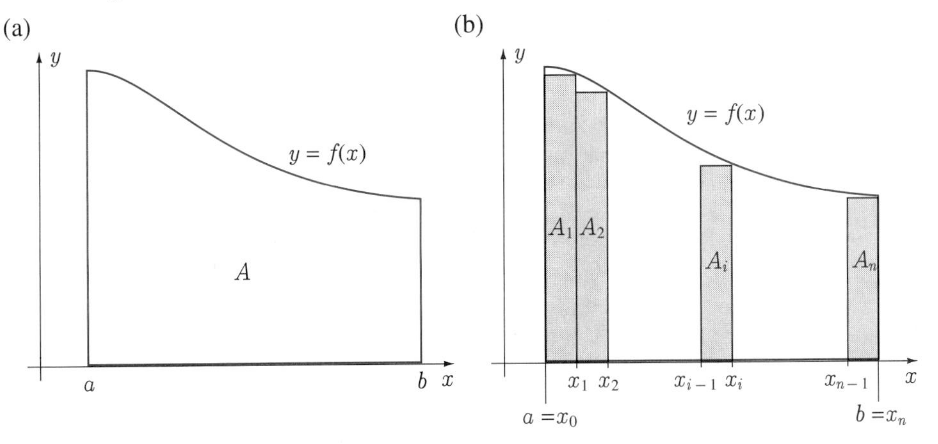

Recall from Problem 1 of Section 6.2 that the area in Figure 7.1(a) can be approximated by rectangles. Specifically, we partition the interval $a \le x \le b$ into

n parts by points

$$a = x_0 < x_1 < x_2 < \cdots < x_{n-1} < x_n = b,$$

and construct rectangles as shown in Figure 7.1(b). The area A_i of the ith rectangle is

$$A_i = f(x_i)(x_i - x_{i-1}) = f(x_i)\Delta x_i,$$

and an approximation to the required area A is therefore

$$\sum_{i=1}^{n} A_i = \sum_{i=1}^{n} f(x_i)\Delta x_i.$$

If we were to let the number of these rectangles get larger and larger, and at the same time require each to have smaller and smaller width that eventually approaches zero, we feel that a better and better approximation would be obtained. In fact, were we to take the limit as the norm $||\Delta x_i||$ of the partition approaches zero, we would get what we think is A. Since we have as yet no formal definition for the area of odd-shaped figures, we take this opportunity to make our own. We define the area A in Figure 7.1(a) as

$$A = \lim_{||\Delta x_i|| \to 0} \sum_{i=1}^{n} f(x_i)\Delta x_i. \tag{7.1}$$

Area A then has been defined as the limit of a sum of rectangular areas.

The right-hand side of equation 7.1 is strikingly similar to the definition of the definite integral of $f(x)$ from $x = a$ to $x = b$ (equation 6.11),

$$\int_a^b f(x)\, dx = \lim_{||\Delta x_i|| \to 0} \sum_{i=1}^{n} f(x_i^*)\Delta x_i. \tag{6.11}$$

The only difference is the absence of *'s in 7.1. But when we recall that in 6.11, x_i^* may be chosen as any point in the subinterval $x_{i-1} \leq x \leq x_i$, we see that by choosing $x_i^* = x_i$,

$$\int_a^b f(x)\, dx = \lim_{||\Delta x_i|| \to 0} \sum_{i=1}^{n} f(x_i)\Delta x_i.$$

It follows that area A of Figure 7.1(a) may be calculated by means of the definite integral

$$A = \int_a^b f(x)\, dx. \tag{7.2}$$

It is important to realize that equation 7.2 does not imply that a definite integral should always be thought of as an area; on the contrary, we will find that definite integrals can represent many other quantities. What we have said is that the area in Figure 7.1(a) is defined by limit 7.1. But this limit may also be interpreted as the definite integral of $f(x)$ with respect to x from $x = a$ to $x = b$; hence the area may be calculated by definite integral 7.2. Since definite integrals can be evaluated using antiderivatives, it seems that we have a very simple way to find areas.

It is simple to extend this result to the problem of finding the area in Figure 7.2. With our interpretation of equation 7.2, we can state that the area under the curve $y = f(x)$, above the x-axis, and between the vertical lines $x = a$ and $x = b$ is given by

$$A_2 = \int_a^b f(x)\, dx.$$

Since the area under the curve $y = g(x)$ is given similarly by

$$A_1 = \int_a^b g(x)\,dx,$$

it follows that the required area is

$$A = A_2 - A_1 = \int_a^b f(x)\,dx - \int_a^b g(x)\,dx = \int_a^b [f(x) - g(x)]\,dx. \quad (7.3)$$

We now have two formulas for finding areas: equation 7.2 for the area under a curve, above the x-axis, and between two vertical lines; and equation 7.3 for the area between two curves and two vertical lines. Note that 7.2 is a special case of 7.3. At this point we could solve a number of area problems using these two results, but to do so would strongly suggest that integration should be approached from a "formula" point of view; and if there is any point of view that we wish to adopt, it is completely the opposite. By the end of this chapter we hope to have developed a sufficiently clear understanding of the limit-summation process that use of integration in situations other than those discussed here will be straightforward.

FIGURE 7.2

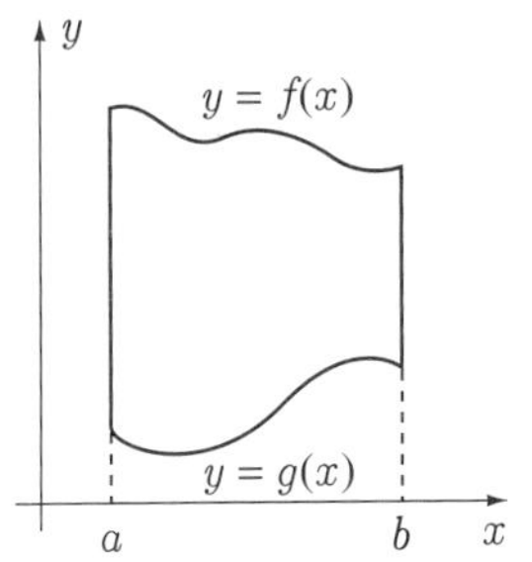

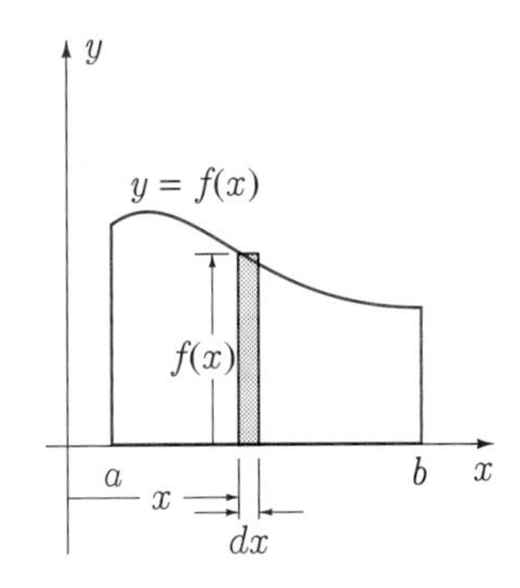

FIGURE 7.3

To illustrate how we arrive at the correct definite integral for area problems without memorizing either 7.2 or 7.3, consider again finding the area in Figure 7.1(a). We draw a rectangle of width dx at position x as shown in Figure 7.3. The area of this rectangle is

$$f(x)\,dx.$$

We visualize this rectangle as a representative for a large number of such rectangles between a and b. To find the required area we add together all such rectangular areas and take the limit as their widths approach zero. But this is precisely the concept of the definite integral, so that we write for the limit-summation process

$$A = \int_a^b f(x)\,dx.$$

The limits a and b identify the x-positions of the first and last rectangles, respectively.

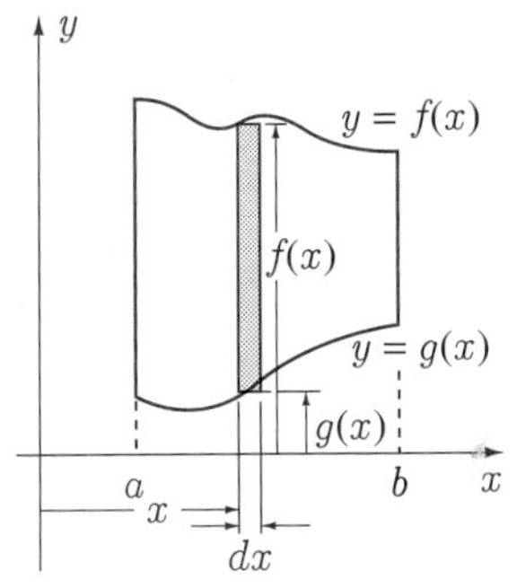

FIGURE 7.4

Similarly, for the area in Figure 7.4 we draw a rectangle of width dx and length $f(x) - g(x)$ and, therefore, of area

$$[f(x) - g(x)]\,dx.$$

To add the areas of all such rectangles between a and b, and at the same time to take the limit as their widths approach zero, we once again use the definite integral,

$$A = \int_a^b [f(x) - g(x)]\,dx.$$

For area problems then, we start with the area of a representative rectangle and proceed to the required area by summation with the definite integral. We might write symbolically that

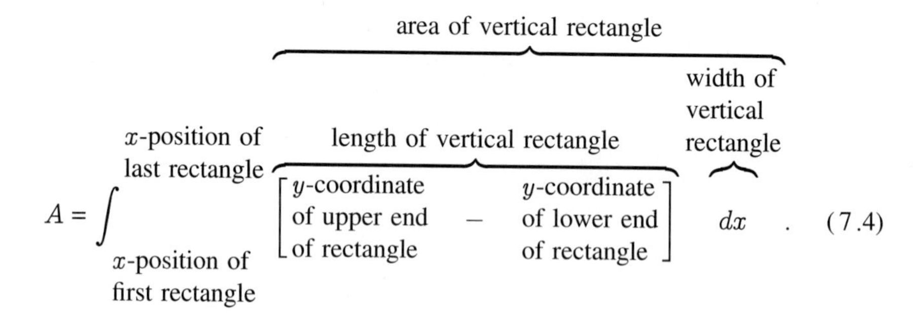

EXAMPLE 7.1

Find the area enclosed by the curves $y = x^2$ and $y = x^3$.

SOLUTION The area of the representative rectangle in Figure 7.5 is

$$(x^2 - x^3)\,dx;$$

hence,

$$A = \int_0^1 (x^2 - x^3)\,dx = \left\{\frac{x^3}{3} - \frac{x^4}{4}\right\}_0^1 = \frac{1}{12}.$$

FIGURE 7.5

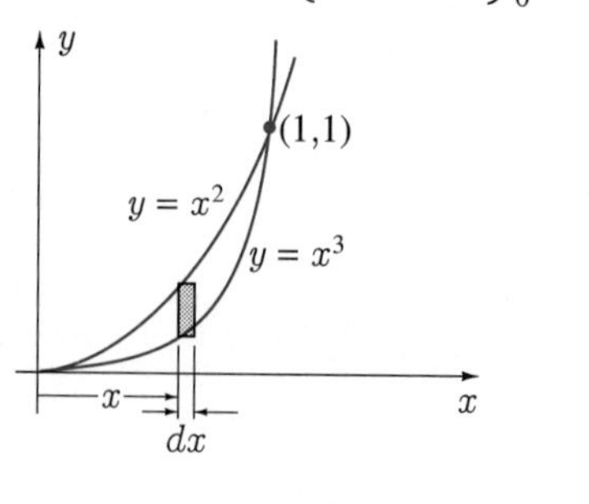

■

In the next two examples, the expression for the area of the representative rectangle varies within the required area. In such cases, we must set up different integrals corresponding to the different representative rectangles.

EXAMPLE 7.2

Find the area bounded by the x-axis and the curve $y = x^3 - x$.

SOLUTION

FIGURE 7.6

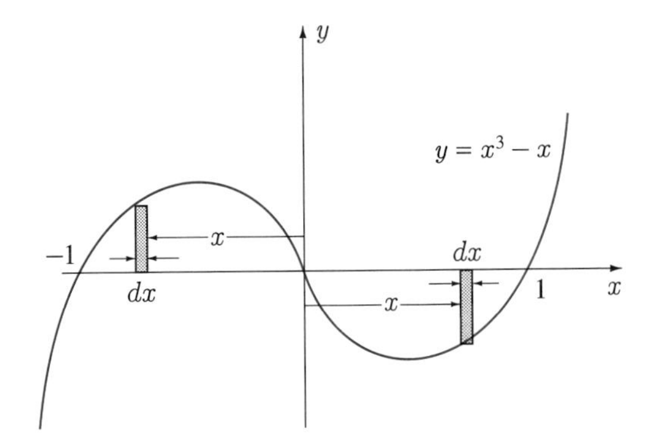

The area of any rectangle between $x = -1$ and $x = 0$ (Figure 7.6) is

$$[(x^3 - x) - 0]\,dx,$$

whereas the area of a rectangle between $x = 0$ and $x = 1$ is

$$[0 - (x^3 - x)]\,dx.$$

Consequently,

$$\begin{aligned} A &= \int_{-1}^{0} (x^3 - x)\,dx + \int_{0}^{1} (-x^3 + x)\,dx \\ &= \left\{\frac{x^4}{4} - \frac{x^2}{2}\right\}_{-1}^{0} + \left\{-\frac{x^4}{4} + \frac{x^2}{2}\right\}_{0}^{1} \\ &= -\left(\frac{1}{4} - \frac{1}{2}\right) + \left(-\frac{1}{4} + \frac{1}{2}\right) \\ &= \frac{1}{2}. \end{aligned}$$

We could have saved ourselves some calculations in this example by noting that because of the symmetry of the diagram ($x^3 - x$ is an odd function), the two areas are identical. Hence, we could find the left area and double it,

$$A = 2\int_{-1}^{0} (x^3 - x)\,dx = 2\left\{\frac{x^4}{4} - \frac{x^2}{2}\right\}_{-1}^{0} = -2\left(\frac{1}{4} - \frac{1}{2}\right) = \frac{1}{2}.$$

■

EXAMPLE 7.3

Find the area of the triangle with edges $y = x$, $y = -x/2$, and $y = 5x - 44$.

SOLUTION

FIGURE 7.7

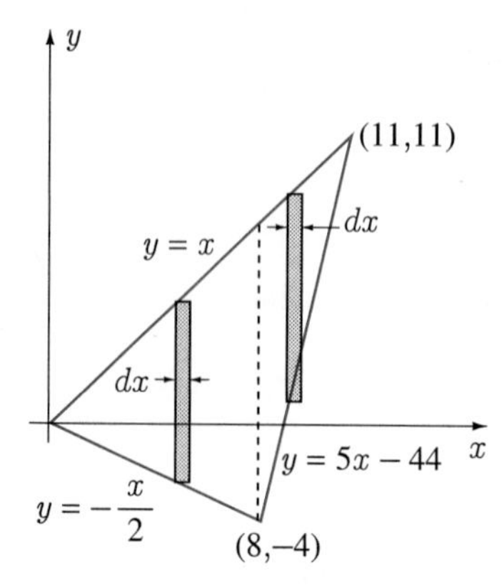

Areas of representative rectangles to the left and right of $x = 8$ (Figure 7.7) are, respectively,

$$[x - (-x/2)]\,dx \qquad \text{and} \qquad [x - (5x - 44)]\,dx;$$

therefore

$$\begin{aligned} A &= \int_0^8 [x - (-x/2)]\,dx + \int_8^{11} [x - (5x - 44)]\,dx \\ &= \frac{3}{2}\int_0^8 x\,dx + \int_8^{11} (44 - 4x)\,dx \\ &= \frac{3}{2}\left\{\frac{x^2}{2}\right\}_0^8 + \left\{44x - 2x^2\right\}_8^{11} \\ &= 48 + (484 - 242) - (352 - 128) \\ &= 66. \end{aligned}$$

■

We see from the above examples that the length of a representative rectangle in equation 7.4 as "upper y minus lower y" is valid whether the rectangle is in the first quadrant (Figure 7.5), the second and fourth quadrants (Figure 7.6), or partially in the first and partially in the fourth (Figure 7.7). In fact it is valid for rectangles in all quadrants.

FIGURE 7.8

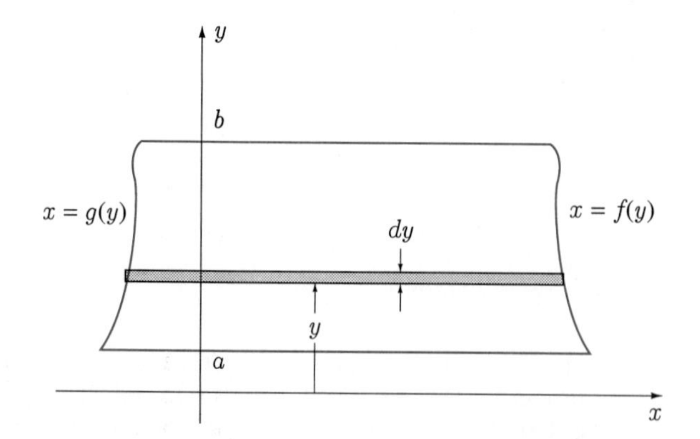

There is nothing special about vertical rectangles. Sometimes it is more convenient to subdivide an area into horizontal rectangles. For example, to find the area in Figure 7.8, we draw a representative rectangle at position y of width dy. Its length is $[f(y)-g(y)]$, and therefore its area is

$$[f(y) - g(y)]\, dy.$$

Adding over all rectangles gives

$$A = \int_a^b [f(y) - g(y)]\, dy. \tag{7.5}$$

Corresponding to 7.4, we could write that for a subdivision into horizontal rectangles

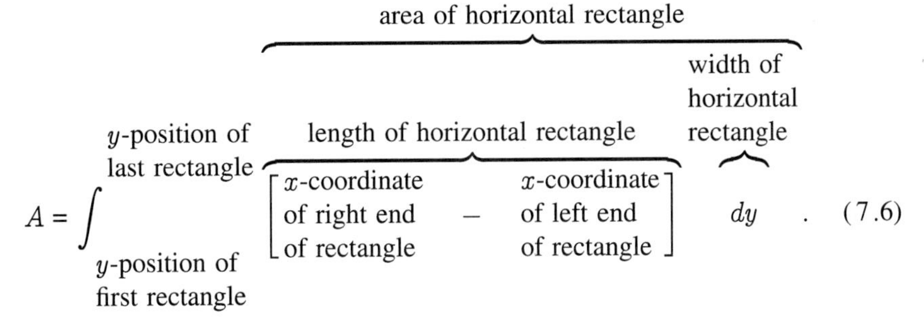

$$A = \overbrace{\int_{\substack{y\text{-position of}\\ \text{first rectangle}}}^{\substack{y\text{-position of}\\ \text{last rectangle}}} \overbrace{\left[\begin{matrix}x\text{-coordinate}\\ \text{of right end}\\ \text{of rectangle}\end{matrix} - \begin{matrix}x\text{-coordinate}\\ \text{of left end}\\ \text{of rectangle}\end{matrix}\right]}^{\text{length of horizontal rectangle}} \overbrace{dy}^{\substack{\text{width of}\\ \text{horizontal}\\ \text{rectangle}}}}^{\text{area of horizontal rectangle}}. \tag{7.6}$$

EXAMPLE 7.4

Find the area bounded by the curves $y = \sqrt{x + 14}$, $x = \sqrt{y}$, and $y = 0$.

SOLUTION Subdivision of the area (Figure 7.9) into vertical rectangles results in two integrations: one to the left and the other to the right of the y-axis. On the other hand, throughout the required area, the area of a horizontal rectangle is

$$[\sqrt{y} - (y^2 - 14)]\, dy,$$

and therefore

$$A = \int_0^4 \left(\sqrt{y} - y^2 + 14\right) dy = \left\{\frac{2}{3}y^{3/2} - \frac{1}{3}y^3 + 14y\right\}_0^4 = \frac{16}{3} - \frac{64}{3} + 56 = 40.$$

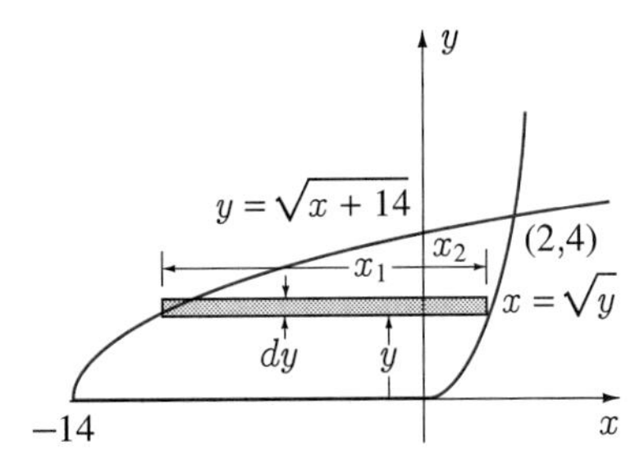

FIGURE 7.9

In choosing between horizontal and vertical rectangles, two objectives must be considered:

1. to minimize the number of integrations; and

2. to obtain simple definite integrals.

For instance, by choosing one type of rectangle we may obtain only one definite integral, but that may be very difficult to evaluate. If the other type of rectangle leads to two simple definite integrals, then it would be wise to choose the two simple integrals.

EXAMPLE 7.5 Find the area enclosed by the curves

$$y = \frac{x}{\sqrt{x^2 - 16}}, \qquad y = \frac{x^2}{15}, \qquad y = -x^2, \qquad x = 6, \qquad x \geq 0.$$

SOLUTION Examination of Figure 7.10 indicates that a choice of horizontal rectangles would necessitate three definite integrals. In addition, they would require us to solve the equations $y = x^2/15$ and $y = x/\sqrt{x^2 - 16}$ for x in terms of y. We therefore opt for vertical rectangles.

$$\begin{aligned} A &= \int_0^5 \left[\frac{x^2}{15} - (-x^2)\right] dx + \int_5^6 \left[\frac{x}{\sqrt{x^2 - 16}} - (-x^2)\right] dx \\ &= \frac{16}{15} \int_0^5 x^2\, dx + \int_5^6 \left[\frac{x}{\sqrt{x^2 - 16}} + x^2\right] dx \\ &= \frac{16}{15} \left\{\frac{x^3}{3}\right\}_0^5 + \left\{\sqrt{x^2 - 16} + \frac{x^3}{3}\right\}_5^6 \\ &= \frac{16}{15}\left(\frac{125}{3}\right) + \left(\sqrt{20} + 72\right) - \left(3 + \frac{125}{3}\right) \\ &= 76.2. \end{aligned}$$

FIGURE 7.10

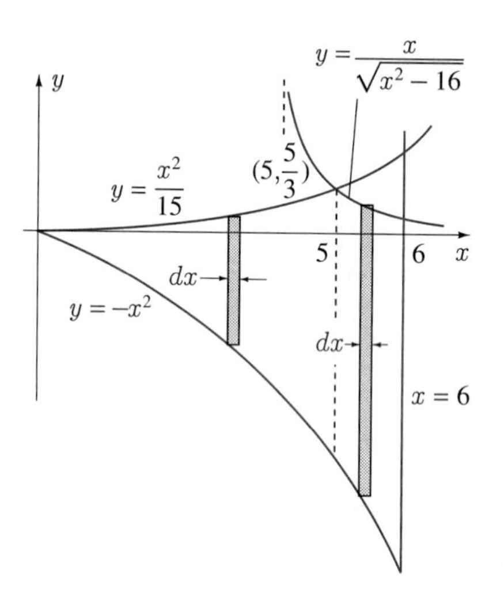

Recall the mean value theorem for definite integrals (Theorem 6.8),

$$\int_a^b f(x)\, dx = (b - a) f(c).$$

It has a very simple interpretation in terms of area — at least when $f(x) \geq 0$ for $a \leq x \leq b$. Since $f(x) \geq 0$ for $a \leq x \leq b$, the definite integral may be interpreted as the area under the curve $y = f(x)$, above the x-axis, and between vertical lines at $x = a$ and $x = b$ (Figure 7.11). The right side is the area of the rectangle of width $b - a$ and height $f(c)$ shaded in Figure 7.11. The mean value theorem guarantees at least one point c between a and b for which the rectangular area is equal to the area under the curve. For the curve in Figure 7.11, there is exactly one such point c; for the curve in Figure 7.12, there are three choices for c.

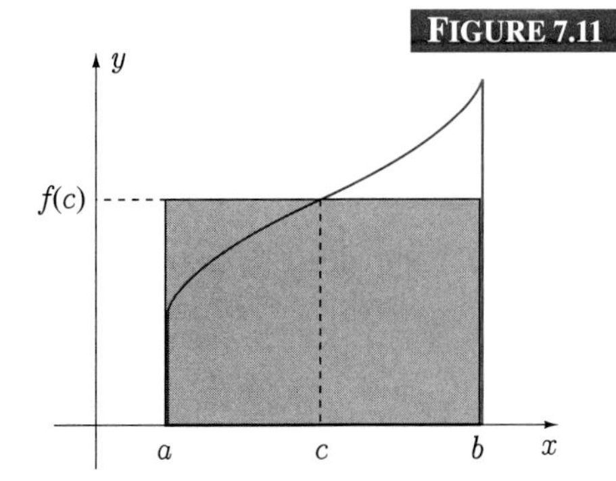

FIGURE 7.11

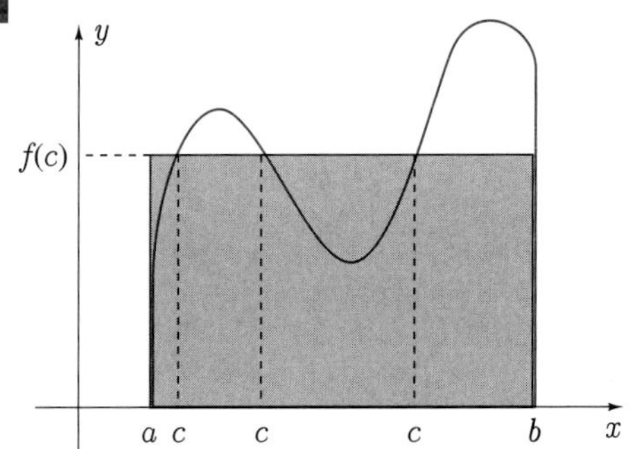

FIGURE 7.12

EXERCISES 7.1

In Exercises 1–16 find the area bounded by the curves.

1. $y^2 = 4x, \quad x^2 = 4y$

2. $y = x^3 + 8, \quad y = 4x + 8$

3. $yx^2 = 4, \quad y = 5 - x^2$

4. $x = y(y - 2), \quad x + y = 12$

5. $x = 4y - 4y^2, \quad y = x - 3, \quad y = 1, \quad y = 0$

6. $y = e^{3x}, \quad x = 1, \quad x = 2, \quad y = -x$

7. $12y = 7 - x^2, \quad y = 1/(2x)$

8. $y = \sqrt{x + 4}, \quad y = (x + 4)^2/8$

9. $y = (x - 1)^5, \quad x = 0, \quad y = 0$

10. $y = \sin x \ (0 \leq x \leq \pi), \quad y = 0$

11. $y = x^5 - x, \quad y = 0$

12. $x + y = 1, \quad x + y = 5, \quad y = 2x + 1, \quad y = 2x + 6$

13. $x = \sec^2 y, \quad y = 0, \quad y = \pi/4, \quad x = 0$

14. $xy = e, \quad y = x^2, \quad y = 2$ (smaller area)

15. $y = e^2(2 - x^2), \quad y = e^{2x}, \quad y = e^{-2x}, \quad y \geq 1$

16. $x = |y| + 1, \quad x + (y - 1)^2 = 4, \quad y = 0$ (above the x-axis)

17. (a) Find the area A bounded by the curves

$$y = 16 - x^2, \qquad y = 0, \qquad x = 1, \qquad x = 3.$$

In the remainder of this problem we approximate A by rectangles and show that the accuracy of the approximation increases as the number of rectangles increases. To do this you will need a large graph of the function $f(x) = 16 - x^2$ on the interval $1 \leq x \leq 3$.

(b) On the graph draw two rectangles of equal width (one unit) and with heights determined in the same way as in Figure 7.1(b). If A_2 denotes the sum of the areas of these two rectangles, what is A_2? What is the error in the approximation of A by A_2? Illustrate this error on the graph.

(c) Repeat part (b) with four rectangles all of width one-half, denoting the sum of the areas of the four rectangles by A_4. Show on the graph the extra precision of A_4 over A_2.

(d) Repeat part (b) with eight rectangles of equal width, denoting the sum of the areas of the eight rectangles by A_8. Show on the graph the extra precision of A_8 over A_4.

This discussion is continued in Exercise 46.

18. Repeat Exercise 17 with the area bounded by the curves

$$y = x^3 + 1, \qquad x = 1, \qquad x = 3, \qquad y = 0.$$

In Exercises 19–22 set up (but do not evaluate) definite integral(s) for the area bounded by the curves.

19. $x = 1/\sqrt{4 - y^2}, \quad 4x = -y^2, \quad y = -1, \quad y = 1$

20. $x^2 + y^2 = 4, \quad x^2 + y^2 = 4x$ (interior to both)

21. $x^2 + y^2 = 4, \quad x^2 + y^2 = 6x$ (interior to both)

22. $x^2 + y^2 = 16, \quad x = y^2$ (smaller area)

23. Find the area bounded by $y^2 = 4ax$ and $x^2 = 4ay$, where $a > 0$ is a constant.

In Exercises 24–35 find the area bounded by the curves.

24. $y = x/\sqrt{x + 3}, \quad x = 1, \quad x = 6, \quad y = -x^2$

25. $x = y^2 + 2, \quad x = -(y - 4)^2, \quad y = -x + 4, \quad y = 0$

26. $y = x^3 - x, \quad x + y + 1 = 0, \quad x = \sqrt{y + 1}$

27. $y = \left|\dfrac{x}{(x-2)^3}\right|, \quad y = 0, \quad x = -1, \quad x = 1$

28. $x = 2ye^{-y^2}, \quad y = x$

29. $y = \sin^3 x, \quad y = 1/8, \quad 0 \le x \le 2\pi$

30. $y = \ln x^2, \quad y = 1 - x^2, \quad y = 1$

31. $|x|^{1/2} + |y|^{1/2} = 1$

32. $y^2 = x^2(4 - x^2)$

33. $y^2 = x^4(9 + x)$

34. $y^2 = x^2(x^2 - 4), \quad x = 5$

35. $(2x - y)^2 = x^3, \quad x = 4$

In Exercises 36–43, it is necessary to use a calculator or computer to find points of intersection of the curves. Find the area bounded by the curves (to 3 decimal places).

36. $y = x^3 + 3x^2 + 2x + 1, \quad x = 0, \quad y = 0$

37. $y = x^3 - 4x, \quad y = 2 - x - x^2$

38. $y = x^4 - 5x^2 + 5, \quad y = 0$

39. $y = e^x, \quad y = 2 - x^2$

40. $y = \cos x, \quad 4y = x + 2$

41. $y = \dfrac{2}{x+2}, \quad y = x^3 + 3x - 1, \quad x = 0$

42. $y = x^3 - 3x^2 + 4x - 2, \quad x = 4 - y^2$

43. $x = y^3 - y^2 - 2y, \quad x = \sqrt{2y + 1}$

44. For what values of m do the curves

$$y = \frac{x}{3x^2 + 1} \quad \text{and} \quad y = mx$$

bound an area? Find the area.

45. Find a point (a, b) on the curve $y = x/\sqrt{x^2 + 1}$ such that the area bounded by this curve, the x-axis, and the line $x = a$ is equal to twice the area bounded by the curve, the y-axis, and the line $y = b$.

46. If 2^n $(n = 1, 2, \ldots)$ rectangles (all of equal width) are drawn in Exercise 17 to approximate A, and A_{2^n} denotes the sum of the areas of these rectangles, show that

$$A_{2^n} = A - \left[\frac{1}{2^{n-2}} + \frac{1}{6}\left(\frac{3}{2^{n-3}} + \frac{1}{2^{2n-3}}\right)\right],$$

and hence that

$$\lim_{n \to \infty} A_{2^n} = A.$$

47. Show that curves

$$y = \frac{x^3}{x^4 + 16} \quad \text{and} \quad 204y = 13x^2 - 1$$

bound three areas. Find the largest area.

48. Let P be a point on the cubic curve $y = f(x) = ax^3$. Let the tangent line at P intersect $y = f(x)$ again at Q, and let A be the area of the region bounded by $y = f(x)$ and the line PQ. Let B be the area of the region defined in the same way by starting with Q instead of P. Show that B is 16 times as large as A.

49. Prove that the result in Exercise 48 is valid for any cubic $y = f(x) = ax^3 + bx^2 + cx + d$.

SECTION 7.2

Volumes of Solids of Revolution

In Section 6.2 we discussed the idea of rotating flat areas about coplanar lines to produce volumes of solids of revolution. To find the volume generated when the area in Figure 7.1(a) is revolved about the x-axis, we again approximate the area by n rectangles as in Figure 7.1(b). If each of these rectangles is rotated about the x-axis, then n discs are formed. Since the radius of the ith disc is $f(x_i)$ (Figure 7.13), its volume is given by

$$V_i = \pi[f(x_i)]^2(x_i - x_{i-1}) = \pi[f(x_i)]^2 \Delta x_i.$$

FIGURE 7.13

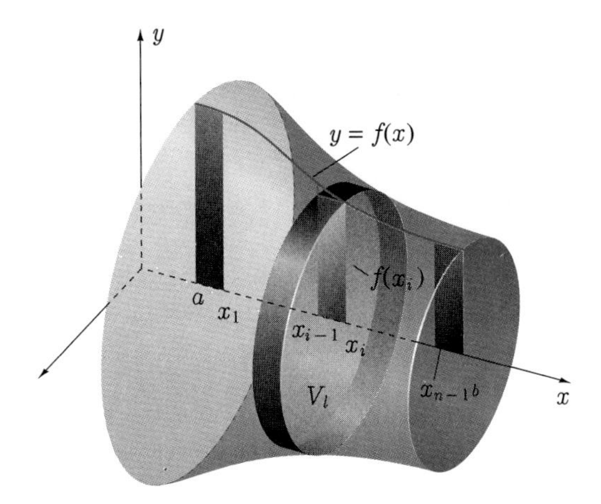

An approximation to the required volume V is therefore

$$\sum_{i=1}^{n} V_i = \sum_{i=1}^{n} \pi[f(x_i)]^2 \Delta x_i.$$

If we let the number of rectangles get larger and larger, and at the same time require the width of each to approach zero, it seems reasonable that we will obtain a better approximation. Furthermore, were we to take the limit as the norm of the partition approaches zero, we should get what we think is V. Since we have as yet no formal definition for such volumes, we make our own. We define

$$V = \lim_{\|\Delta x_i\| \to 0} \sum_{i=1}^{n} \pi[f(x_i)]^2 \Delta x_i. \tag{7.7}$$

But we can interpret the right-hand side of this definition as the definite integral of $\pi[f(x)]^2$ with respect to x from $x = a$ to $x = b$; that is,

$$\int_a^b \pi[f(x)]^2 \, dx = \lim_{\|\Delta x_i\| \to 0} \sum_{i=1}^{n} \pi[f(x_i)]^2 \Delta x_i.$$

Consequently, we can calculate the volume of the solid of revolution generated by rotating the area in Figure 7.1(a) about the x-axis, by means of the definite integral

$$V = \int_a^b \pi[f(x)]^2 \, dx. \tag{7.8}$$

The volume of the solid of revolution has been defined by limit 7.7, but for evaluation of this limit we use definite integral 7.8.

To avoid memorizing 7.8 as a formula, we use the technique introduced in Section 7.1. For the volume obtained by rotating the area in Figure 7.1(a) about the x-axis, we construct a rectangle of width dx at position x, as shown in either Figure 7.3 or Figure 7.14. When this rectangle is rotated around the x-axis, the volume of the disc generated is

$$\pi[f(x)]^2 \, dx,$$

where dx is the thickness of the disc and $\pi[f(x)]^2$ is the area of its flat surface. This disc is pictured as representing a large number of such discs between a and b. We find the required volume by adding the volumes of all such discs and taking the limit as their widths approach zero. But this is precisely the concept of the definite integral, so that we write for the limit-summation process

$$V = \int_a^b \pi[f(x)]^2\, dx,$$

where the limits $x = a$ and $x = b$ identify the x-positions of first and last discs respectively. This is called the **disc method** for finding the volume of a solid of revolution.

FIGURE 7.14

FIGURE 7.15

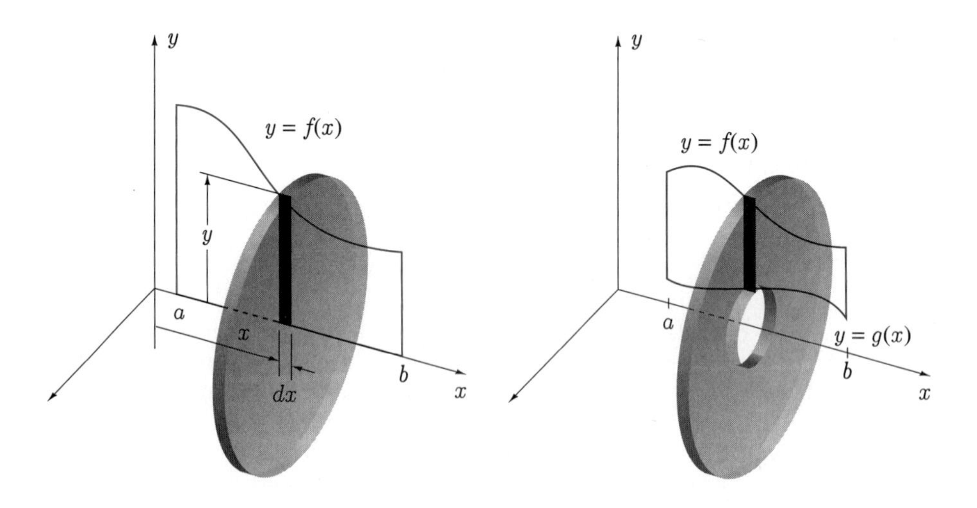

A slightly more general problem is that of finding the volume of the solid of revolution generated by rotating an area bounded by two curves and two vertical lines as shown in Figure 7.15 about the x-axis. If a representative rectangle of width dx at position x is rotated around the x-axis, the volume formed is a **washer**. Since outer and inner radii of this washer are $f(x)$ and $g(x)$ respectively, its volume is

$$\{\pi[f(x)]^2 - \pi[g(x)]^2\}\, dx.$$

To add the volumes of all such washers between a and b, and at the same time take limits as their widths approach zero, we use the definite integral

$$V = \int_a^b \{\pi[f(x)]^2 - \pi[g(x)]^2\}\, dx. \tag{7.9}$$

EXAMPLE 7.6

Prove that the volume of a sphere of radius r is $4\pi r^3/3$.

SOLUTION A sphere of radius r is formed when the semicircle $x^2 + y^2 \le r^2$ ($y \ge 0$) is rotated around the x-axis. The volume of the representative disc generated by rotating the rectangle in Figure 7.16 about the x-axis is

$$\pi y^2\, dx = \pi(r^2 - x^2)\, dx.$$

Since the volume formed by the left quarter circle is the same as for the right quarter circle, we calculate the volume generated by the right quarter and double the result:

$$V = 2\int_0^r \pi(r^2 - x^2)\,dx = 2\pi\left\{r^2x - \frac{x^3}{3}\right\}_0^r = \frac{4}{3}\pi r^3.$$

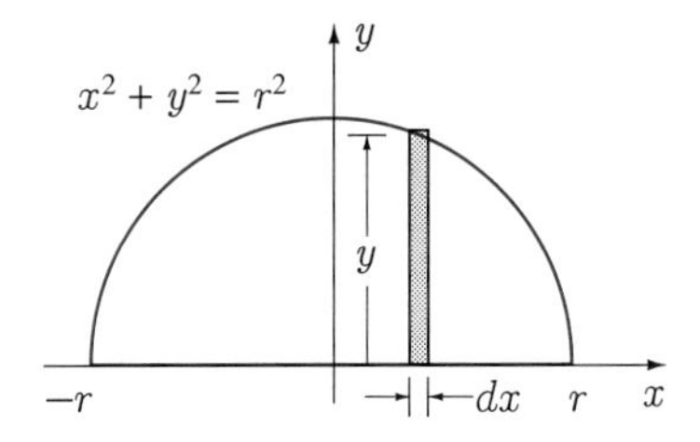

FIGURE 7.16

■

EXAMPLE 7.7

Prove that the volume of a right-circular cone of base radius r and height h is $\pi r^2 h/3$.

SOLUTION The cone can be generated by rotating the triangle in Figure 7.17 around the x-axis. The volume of the representative disc formed by rotating the rectangle shown is given by

$$\pi y^2\,dx = \pi\left(\frac{rx}{h}\right)^2 dx;$$

hence,

$$V = \int_0^h \frac{\pi r^2 x^2}{h^2}\,dx = \frac{\pi r^2}{h^2}\left\{\frac{x^3}{3}\right\}_0^h = \frac{1}{3}\pi r^2 h.$$

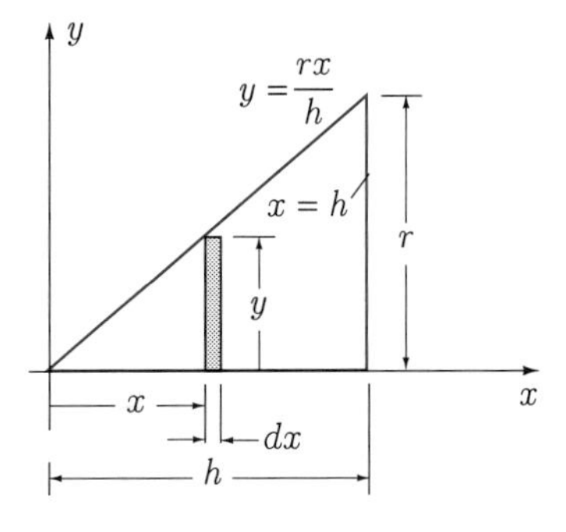

FIGURE 7.17

■

EXAMPLE 7.8

Find the volume of the solid of revolution obtained by rotating the area bounded by the curves $y = 1 - x^2$ and $y = 4 - 4x^2$ about (a) the x-axis and (b) the line $y = -1$.

SOLUTION (a) When the rectangle in Figure 7.18 is rotated around the x-axis, the volume of the washer formed is

$$(\pi y_2^2 - \pi y_1^2)\, dx = \pi[(4 - 4x^2)^2 - (1 - x^2)^2]\, dx = 15\pi(1 - 2x^2 + x^4)\, dx.$$

FIGURE 7.18

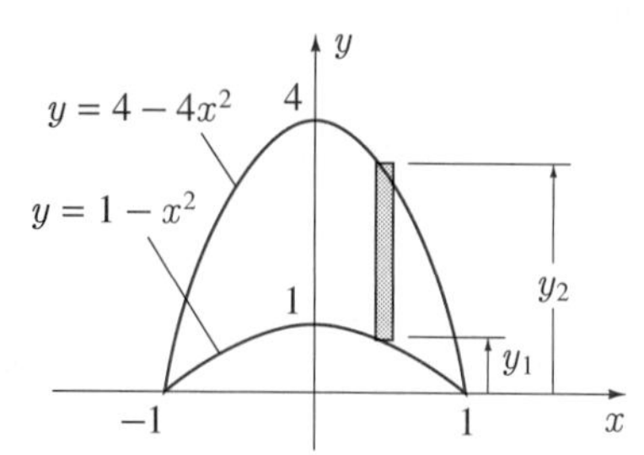

Because of the symmetry of the area, we rotate only the right half and double the result:

$$V = 2\int_0^1 15\pi(1 - 2x^2 + x^4)\, dx = 30\pi\left\{x - \frac{2x^3}{3} + \frac{x^5}{5}\right\}_0^1 = 16\pi.$$

(b) When the rectangle in Figure 7.19 is rotated about the line $y = -1$, inner and outer radii of the washer are r_1 and r_2 respectively. Now r_1 is a length in the y-direction, and in Section 7.1 we learned that to calculate lengths in the y-direction, we take upper y minus lower y. Hence, $r_1 = (1 - x^2) - (-1) = 2 - x^2$. Similarly, $r_2 = (4 - 4x^2) - (-1) = 5 - 4x^2$. The volume of the washer is therefore

$$(\pi r_2^2 - \pi r_1^2)\, dx = [\pi(5 - 4x^2)^2 - \pi(2 - x^2)^2]\, dx = 3\pi(7 - 12x^2 + 5x^4)\, dx.$$

Consequently,

$$V = 2\int_0^1 3\pi(7 - 12x^2 + 5x^4)\, dx = 6\pi\left\{7x - 4x^3 + x^5\right\}_0^1 = 24\pi.$$

FIGURE 7.19

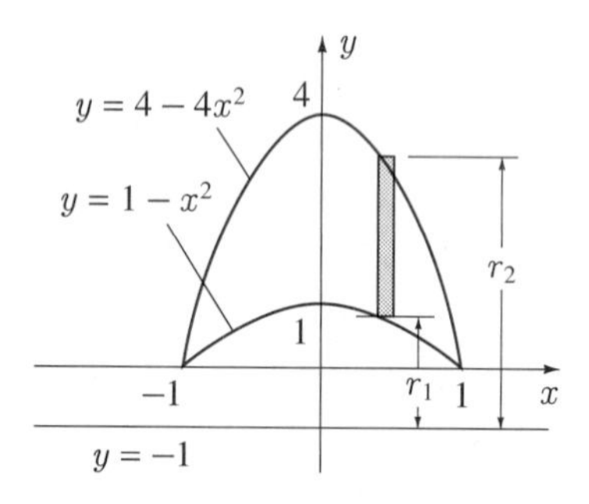

■

We can also rotate horizontal rectangles about vertical lines to produce discs and washers as illustrated in the next two examples.

EXAMPLE 7.9

Find the volume of the solid of revolution when the area enclosed by the curves $y = \ln x$, $y = 0$, $y = 1$, and $x = 0$ is revolved around the y-axis.

SOLUTION When the rectangle in Figure 7.20 is rotated around the y-axis, the volume of the disc formed is

$$\pi x^2\, dy = \pi(e^y)^2\, dy = \pi e^{2y}\, dy.$$

The required volume is therefore

$$V = \int_0^1 \pi e^{2y}\, dy = \pi \left\{ \frac{1}{2} e^{2y} \right\}_0^1 = \frac{\pi}{2}(e^2 - 1).$$

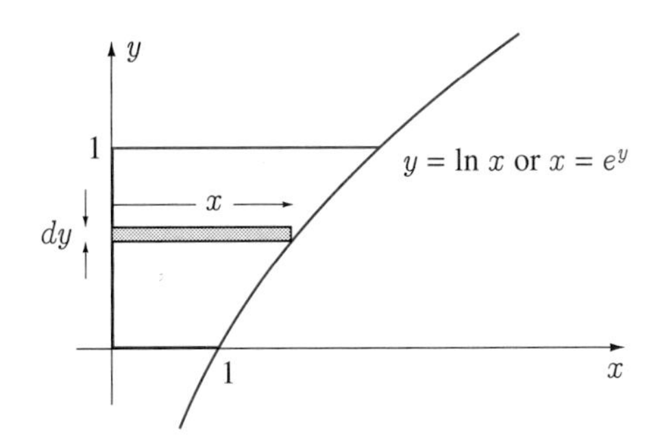

FIGURE 7.20

EXAMPLE 7.10

Find the volume of the solid of revolution obtained by rotating the area enclosed by the curves $y = x^2 - 1$ and $y = 0$ about the line $x = 5$.

SOLUTION When the horizontal rectangle in Figure 7.21 is rotated around the line $x = 5$, the volume of the washer formed is

$$(\pi r_2^2 - \pi r_1^2)\, dy = \pi[(5 - x_2)^2 - (5 - x_1)^2]\, dy.$$

Since x_1 and x_2 are x-coordinates of points on the curve $y = x^2 - 1$, we solve this equation for $x = \pm\sqrt{y + 1}$. Since $x_1 > 0$ and $x_2 < 0$, we set $x_1 = \sqrt{y + 1}$ and $x_2 = -\sqrt{y + 1}$. Thus, the volume of the washer can be expressed as

$$\pi \left[\left(5 + \sqrt{y + 1}\right)^2 - \left(5 - \sqrt{y + 1}\right)^2 \right] dy = 20\pi\sqrt{y + 1}\, dy.$$

The required volume is therefore

$$V = \int_{-1}^0 20\pi\sqrt{y + 1}\, dy = 20\pi \left\{ \frac{2}{3}(y + 1)^{3/2} \right\}_{-1}^0 = \frac{40\pi}{3}.$$

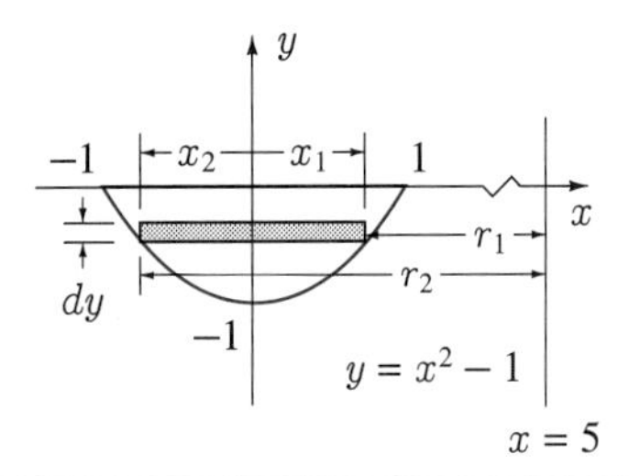

FIGURE 7.21

The solution of Example 7.10 by washers is not as straightforward as in previous examples. For some problems, washers are totally inappropriate. Consider, for example, rotating the area in Figure 7.22 about the x-axis. Each of the rectangles shown yields a washer with a volume formula different from the others. As a result, use of washers requires six definite integrals and only one of these is easy to set up. Determination of this volume seems to lend itself to the use of horizontal rather than vertical rectangles.

To see that this is indeed true, we divide the interval $a \leq y \leq b$ into n parts by the points

$$a = y_0 < y_1 < y_2 < \cdots < y_{n-1} < y_n = b.$$

In each subinterval $y_{i-1} \leq y \leq y_i$, we find the midpoint

$$y_i^* = \frac{y_{i-1} + y_i}{2}$$

and construct a rectangle of length $f(y_i^*) - g(y_i^*)$ and width $y_i - y_{i-1}$ as shown in Figure 7.23. When this i^{th} rectangle is rotated around the x-axis, a cylindrical shell is formed. Since the length of the shell is $f(y_i^*) - g(y_i^*)$, and its inner and outer radii are y_{i-1} and y_i respectively, its volume is

$$\begin{aligned}(\pi y_i^2 - \pi y_{i-1}^2)[f(y_i^*) - g(y_i^*)] &= \pi(y_i + y_{i-1})(y_i - y_{i-1})[f(y_i^*) - g(y_i^*)] \\ &= 2\pi y_i^*[f(y_i^*) - g(y_i^*)]\Delta y_i,\end{aligned}$$

where $\Delta y_i = y_i - y_{i-1}$. When we add the volumes of all such shells, we obtain an approximation to the required volume V:

$$\sum_{i=1}^{n} 2\pi y_i^*[f(y_i^*) - g(y_i^*)]\Delta y_i.$$

If we let the number of rectangles get larger and larger, and at the same time require the width of each to approach zero, it seems reasonable to expect that we will get a better approximation. As the norm $||\Delta y_i||$ approaches zero, the limit should yield what we think is V. We therefore define the required volume as

$$V = \lim_{||\Delta y_i|| \to 0} \sum_{i=1}^{n} 2\pi y_i^*[f(y_i^*) - g(y_i^*)]\Delta y_i. \tag{7.10}$$

FIGURE 7.22

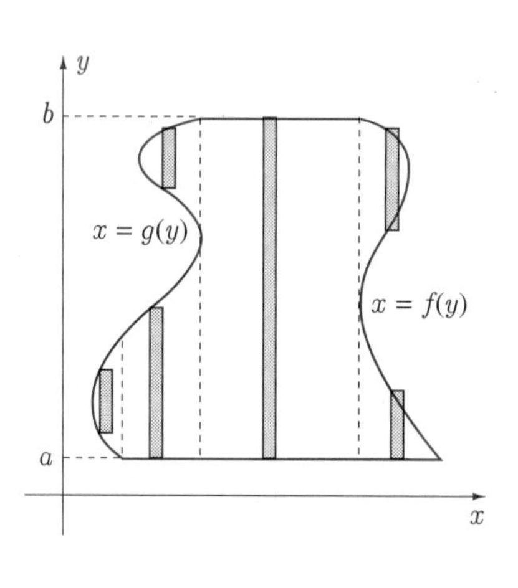

FIGURE 7.23

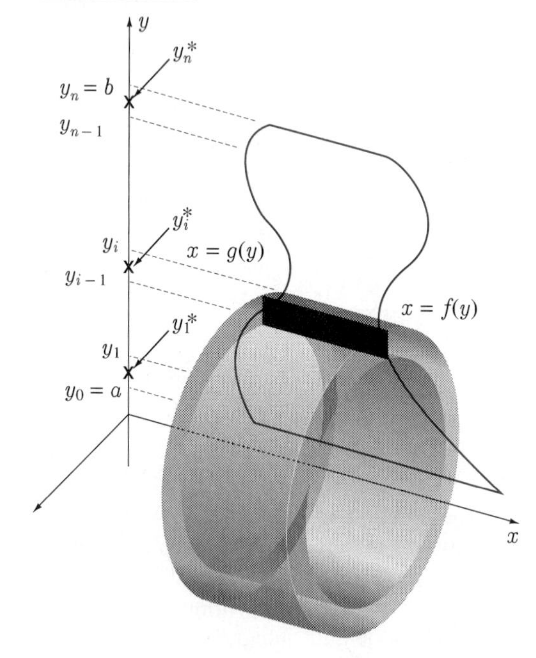

But the right-hand side of this definition may be interpreted as the definite integral of $2\pi y[f(y) - g(y)]$ with respect to y from $y = a$ to $y = b$; that is,

$$\int_a^b 2\pi y[f(y) - g(y)]\,dy = \lim_{\|\Delta y_i\| \to 0} \sum_{i=1}^{n} 2\pi y_i^*[f(y_i^*) - g(y_i^*)]\Delta y_i.$$

Consequently, the volume of the solid of revolution generated by rotating the area in Figure 7.22 around the x-axis is given by the definite integral

$$V = \int_a^b 2\pi y[f(y) - g(y)]\,dy. \tag{7.11}$$

In practice we develop 7.11 by drawing a rectangle of width dy at position y as shown in Figure 7.24(a). When this rectangle is rotated about the x-axis, the volume of the cylindrical shell generated is approximately

$$2\pi y[f(y) - g(y)]\,dy.$$

We obtain this by picturing the shell as being cut along the rectangle and opened up into a slab with dimensions dy, $f(y) - g(y)$,and $2\pi y$ as in Figure 7.24(b). The thickness dy of the shell corresponds to the thickness of the slab; the length $f(y) - g(y)$ of the shell corresponds to that of the slab; and the inner circumference $2\pi y$ of the shell corresponds to the width of the slab. If we now add the volumes of all such cylindrical shells, and at the same time take the limit as their widths approach zero, we obtain the required volume. But this limit-summation defines the definite integral

$$\int_a^b 2\pi y[f(y) - g(y)]\,dy.$$

It follows that

$$V = \int_a^b 2\pi y[f(y) - g(y)]\,dy.$$

The method described is called the **cylindrical-shell method** for finding the volume of a solid of revolution.

(a)

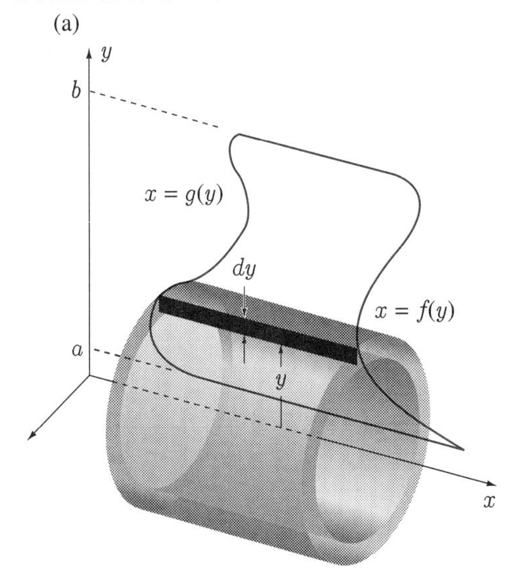

(b)

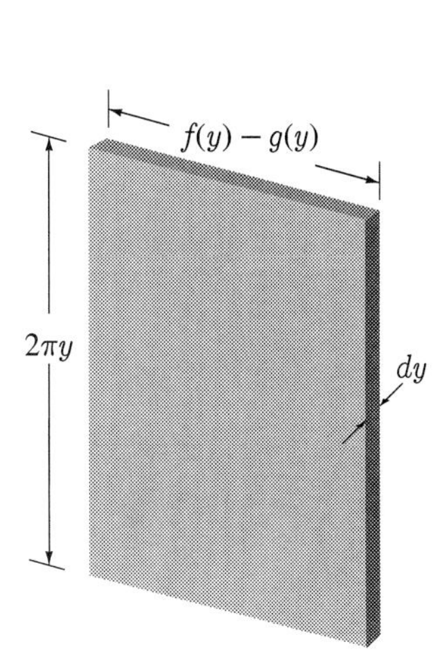

FIGURE 7.24

EXAMPLE 7.11

Find the volume of the solid of revolution when the area enclosed by $x = 2y - y^2$ and the y-axis is revolved around the x-axis.

SOLUTION The volume of the cylindrical shell formed by rotating the rectangle in Figure 7.25 around the x-axis is approximately

$$(2\pi y)(x)\,dy = 2\pi y(2y - y^2)\,dy = 2\pi(2y^2 - y^3)\,dy.$$

Hence,

$$V = \int_0^2 2\pi(2y^2 - y^3)\,dy = 2\pi\left\{\frac{2y^3}{3} - \frac{y^4}{4}\right\}_0^2 = 2\pi\left(\frac{16}{3} - 4\right) = \frac{8\pi}{3}.$$

FIGURE 7.25

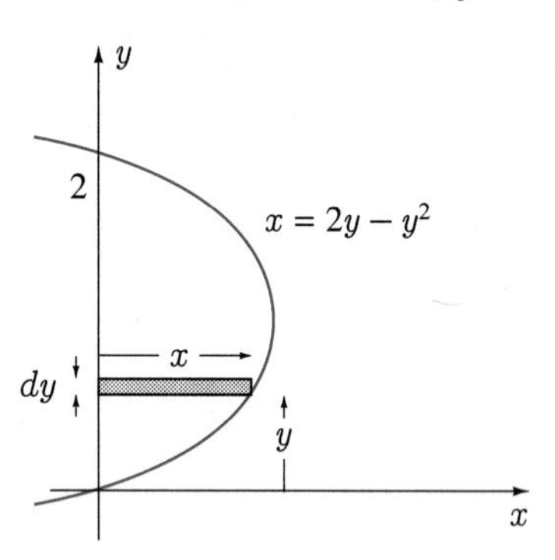

■

EXAMPLE 7.12

Use cylindrical shells to calculate the volume in Example 7.10.

SOLUTION When we rotate the rectangle in Figure 7.26 around the line $x = 5$, the volume of the cylindrical shell is approximately

$$2\pi(5 - x)(0 - y)\,dx = 2\pi(5 - x)(1 - x^2)\,dx = 2\pi(x^3 - 5x^2 - x + 5)\,dx.$$

The total volume is therefore

$$V = \int_{-1}^1 2\pi(x^3 - 5x^2 - x + 5)\,dx = 2\pi\left\{\frac{x^4}{4} - \frac{5x^3}{3} - \frac{x^2}{2} + 5x\right\}_{-1}^1$$
$$= 2\pi\left(\frac{1}{4} - \frac{5}{3} - \frac{1}{2} + 5\right) - 2\pi\left(\frac{1}{4} + \frac{5}{3} - \frac{1}{2} - 5\right) = \frac{40\pi}{3}.$$

This is the result obtained by the washer method in Example 7.10.

FIGURE 7.26

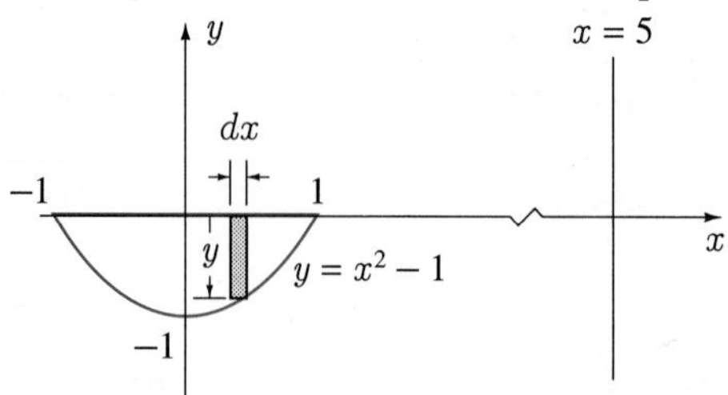

■

Even though the area in this example is symmetric about the y-axis, the volume generated by the right half is less than that generated by the left. For this reason we cannot integrate from $x = 0$ to $x = 1$ and double the result. If rotation were performed about the x-axis, we could indeed integrate over either half of the interval and double. Finally, for rotation about the y-axis, we would integrate over only one half of the interval and neglect the other half in order to eliminate duplications.

EXAMPLE 7.13

Find the volume of the solid of revolution obtained by rotating the area enclosed by the curves $y = x^3$, $y = -x$, and $y = 1$ about the line $y = -2$.

SOLUTION When we rotate the rectangle in Figure 7.27 around the line $y = -2$, the volume of the representative cylindrical shell is approximately

$$2\pi(y + 2)(x_2 - x_1)\,dy = 2\pi(y + 2)(y^{1/3} + y)\,dy.$$

The total volume is therefore

$$V = \int_0^1 2\pi(y + 2)(y^{1/3} + y)\,dy = 2\pi \int_0^1 (y^{4/3} + 2y^{1/3} + y^2 + 2y)\,dy$$

$$= 2\pi \left\{ \frac{3}{7}y^{7/3} + \frac{3}{2}y^{4/3} + \frac{y^3}{3} + y^2 \right\}_0^1 = 2\pi\left(\frac{3}{7} + \frac{3}{2} + \frac{1}{3} + 1\right) = \frac{137\pi}{21}.$$

■

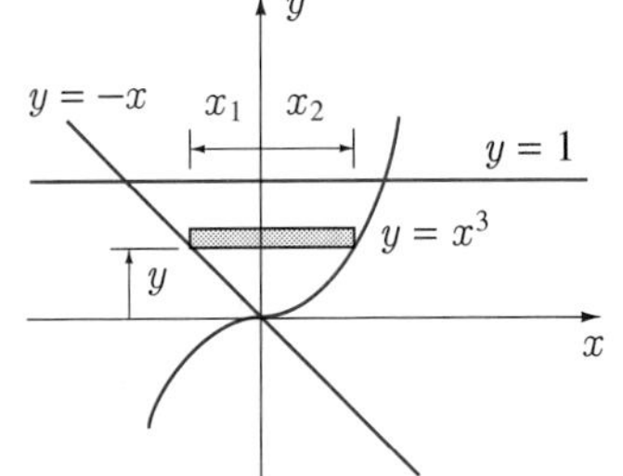

FIGURE 7.27

FIGURE 7.28

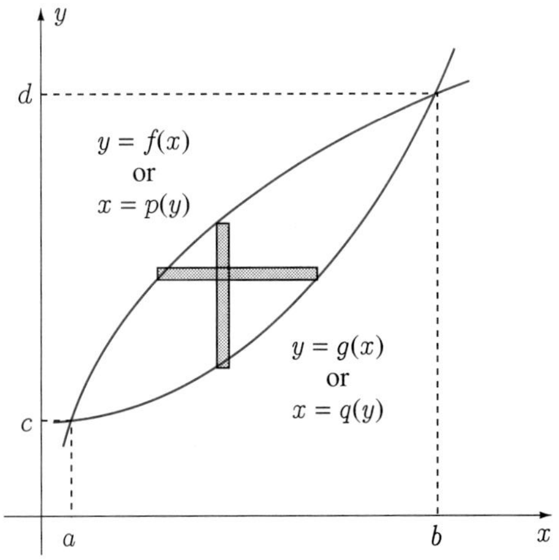

Equations 7.9 and 7.11 provide two methods for calculating volumes of solids of revolution: washers and shells. Where applicable both methods give the same results. We illustrate this for the area bounded by the curves in Figure 7.28. As indicated in the figure, each curve defines y as a function of x, and x as a function of y. When this area is rotated around the x-axis, cylindrical shells can be used to calculate the volume of the solid of revolution

$$V = \int_c^d 2\pi y[q(y) - p(y)]\,dy = 2\pi \int_c^d yq(y)\,dy - 2\pi \int_c^d yp(y)\,dy. \quad (7.12)$$

Suppose we make the change of variable $x = q(y)$ in the first integral. When this equation is solved for y in terms of x, the result is $y = g(x)$, and therefore $dy = g'(x)\,dx$. Because $x = a$ when $y = c$ and $x = b$ when $y = d$, we obtain

$$\int_c^d yq(y)\,dy = \int_a^b g(x)xg'(x)\,dx.$$

Now, the product rule for differentiation gives

$$\frac{d}{dx}\{x[g(x)]^2\} = 2xg(x)g'(x) + [g(x)]^2,$$

and therefore

$$g(x)xg'(x) = \frac{1}{2}\frac{d}{dx}\{x[g(x)]^2\} - \frac{1}{2}[g(x)]^2.$$

So,

$$\begin{aligned}
\int_c^d yq(y)\,dy &= \int_a^b \left\{\frac{1}{2}\frac{d}{dx}\{x[g(x)]^2\} - \frac{1}{2}[g(x)]^2\right\} dx \\
&= \left\{\frac{1}{2}x[g(x)]^2\right\}_a^b - \frac{1}{2}\int_a^b [g(x)]^2\,dx \\
&= \frac{1}{2}b[g(b)]^2 - \frac{1}{2}a[g(a)]^2 - \frac{1}{2}\int_a^b [g(x)]^2\,dx \\
&= \frac{1}{2}bd^2 - \frac{1}{2}ac^2 - \frac{1}{2}\int_a^b [g(x)]^2\,dx.
\end{aligned}$$

Similarly, the change of variable $x = p(y)$ on the second integral in 7.12 leads to

$$\int_c^d yp(y)\,dy = \frac{1}{2}bd^2 - \frac{1}{2}ac^2 - \frac{1}{2}\int_a^b [f(x)]^2\,dx.$$

Substitution of these into 7.12 gives

$$\begin{aligned}
V &= 2\pi\left\{\frac{1}{2}bd^2 - \frac{1}{2}ac^2 - \frac{1}{2}\int_a^b [g(x)]^2\,dx\right\} - 2\pi\left\{\frac{1}{2}bd^2 - \frac{1}{2}ac^2 - \frac{1}{2}\int_a^b [f(x)]^2\,dx\right\} \\
&= -\pi\int_a^b [g(x)]^2\,dx + \pi\int_a^b [f(x)]^2\,dx \\
&= \int_a^b \{\pi[f(x)]^2 - \pi[g(x)]^2\}\,dx.
\end{aligned}$$

But this is precisely the integral obtained when the washer method is used to find the volume.

EXERCISES 7.2

In Exercises 1–12 use the disc or washer method to find the volume of the solid of revolution obtained by rotating the area bounded by the curves about the line.

1. $x^2 + y^2 = 36$, about $y = 0$

2. $y^2 = 5 - x$, $x = 0$, about $x = 0$

3. $y = x^2 + 4$, $y = 2x^2$, about $y = 0$

4. $x - y^2 = 16$, $x = 20$, about $x = 0$

5. $x - 1 = y^2$, $x = 5$, about $x = 1$

6. $x + y + 1 = 0$, $2y = x - 2$, $y = 0$, about $y = 0$

7. $y = 4x^2 - 4x$, $y = x^3$, about $y = -2$

8. $x = 2y - y^2 - 2$, $x = -5$, about $x = 0$

9. $y^2 = 5 - x$, $x = 0$, about $x = 6$

10. $y = x^2 - 2x$, $y = 2x - x^2$, about $y = 2$

11. $y = \csc x$, $y = 0$, $x = \pi/4$, $x = 3\pi/4$, about $y = 0$

12. $y = \ln(x + 1)$, $y = 1$, $x = 0$, about $x = 0$

In Exercises 13–24 use the cylindrical shell method to find the volume of the solid of revolution obtained by rotating the area bounded by the curves about the line.

13. $y = 1 - x^3$, $x = 0$, $y = 0$, about $x = 0$

14. $y = -\sqrt{4 - x}$, $x = 0$, $y = 0$, about $y = 0$

15. $y = (x - 1)^2$, $y = 1$, about $x = 0$

16. $x + y = 4$, $y = 2\sqrt{x - 1}$, $y = 0$, about $y = 0$

17. $y = 3x - x^2$, $y = x^2 - 3x$, about $x = 4$

18. $y = 2 - |x|$, $y = 0$, about $y = -1$

19. $x = y^3$, $x = 2 - y^2$, $y = 0$, about $y = 1$

20. $y = x^2$, $y = -x^2$, $x = -1$, about $x = -1$

21. $y = -\sqrt{9 - x}$, $x = 0$, $y = 0$, about $y = 0$

22. $y = x$, $xy = 9$, $x + y = 10$, $(x \geq y)$, about $x = 0$

23. $y = 0$, $(x + 1)y = \sin x$, $0 \leq x \leq 2\pi$, about $x = -1$

24. $y = 10 - x^2$, $x^2 y = 9$, about $y = 0$

In Exercises 25–32 use the most appropriate method to find the volume of the solid of revolution obtained by rotating the area bounded by the curves about the line.

25. $(x^2 + 1)^2 y = 4$, $y = 1$, about $x = 0$

26. $y = (x - 1)^2 - 4$, $5y = 12x$, $x = 0$, $(x \geq 0)$, about $x = 0$

27. $x^2 - y^2 = 5$, $9y = x^2 + 9$, $9y + x^2 + 9 = 0$, $(-3 \leq x \leq 3)$, about $x = 0$

28. $y = |x^2 - 1|$, $x = -2$, $x = 2$, $y = -1$, about $y = -1$

29. $y = x^2 - 2$, $y = 0$, about $y = -1$

30. $x = \sqrt{4 + 12y^2}$, $x - 20y = 24$, $y = 0$, about $y = 0$

31. $y = (x + 1)^{1/4}$, $y = -(x + 1)^2$, $x = 0$, about $x = 0$

32. $y = x^4 - 3$, $y = 0$, about $y = -1$

In Exercises 33–36 a calculator or computer is needed to find the points of intersection of the curves. Find the volume of the solid of revolution when the area bounded by the curves is rotated about the line correct to three decimal places.

33. $y = x^3 - x$, $y = \sqrt{x}$, about $x = 0$

34. $y = e^{-2x}$, $y = 4 - x^2$, about $y = -1$

35. $y = \dfrac{1}{\sqrt{x - 1}}$, $y = 16 - x^2$, about $y = -1$

36. $y = \sqrt{4 - x}$, $y = x^3 + 1$, $y = 0$, about $y = 0$

37. A tapered rod of length L has circular cross sections. If the radii of its ends are a and b, what is the volume of the rod?

38. During one revolution an airplane propeller displaces an amount of air that can be calculated as a volume of a solid of revolution. If the area yielding the volume is that bounded by the curves $64x = y(y - 4)$ and $64x = y(4 - y)$, and is rotated about the x-axis, calculate the volume of air displaced.

39. If a sphere of radius r is sliced a distance h from its centre, show that the volume of the smaller piece is

$$V = \frac{\pi}{3}(r - h)^2(2r + h).$$

40. An embankment is to be built around the circular wading pool in Figure 7.29(a). Figure 7.29(b) shows a cross-section of the embankment.

(a) Find a cubic polynomial $y = ax^3 + bx^2 + cx + d$ to fit the three points $(0,0)$, $(4,2)$, and $(6,0)$.

(b) Determine the amount of fill required to build the embankment.

FIGURE 7.29

(a)

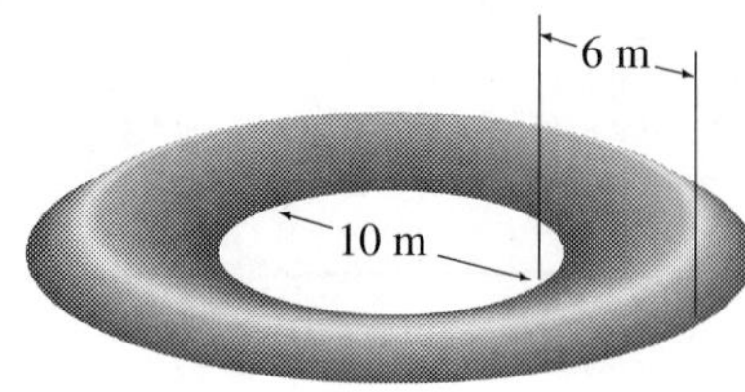

(b)

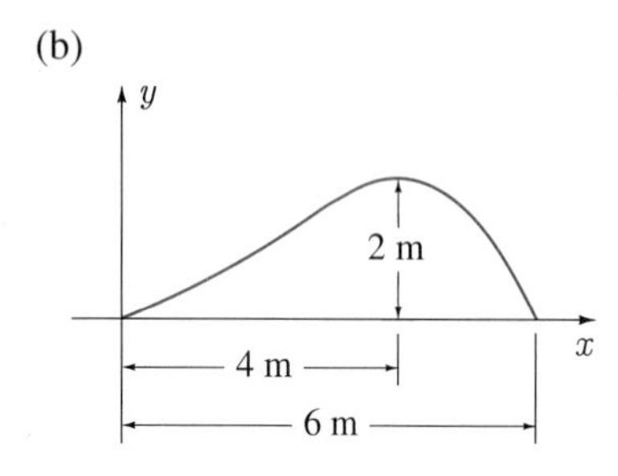

41. Find the volume of the donut obtained by rotating the circle in Figure 7.30 about the y-axis.

FIGURE 7.30

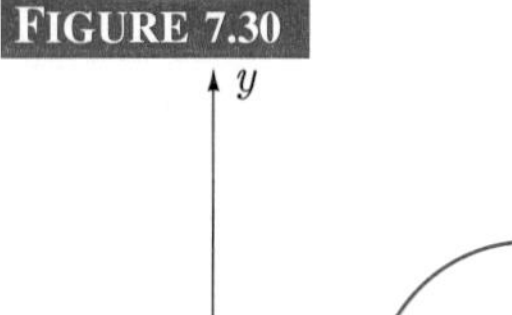

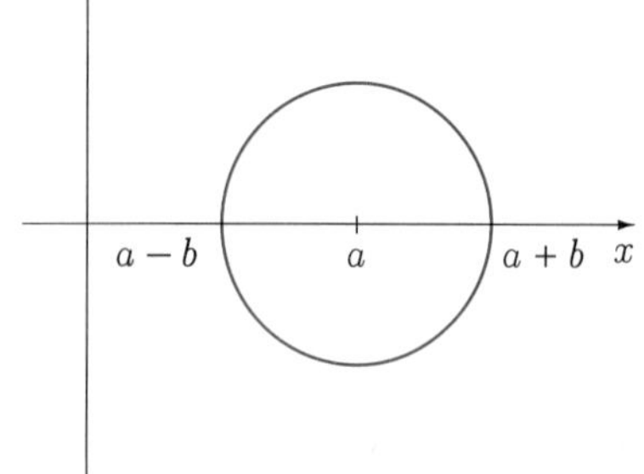

42. A cylindrical hole is bored through the centre of a sphere, the length of the hole being L. Show that no matter what the radius of the sphere, the volume of the sphere that remains is always the same and equal to the volume of a sphere of diameter L.

43. Water half fills a cylindrical pail of radius a and height L. When the pail is rotated about its axis of symmetry with angular speed ω (Figure 7.31), the surface of the water assumes a parabolic shape, the cross-section of which is given by

$$y = H + \frac{\omega^2 x^2}{2g},$$

where $g > 0$ is the acceleration due to gravity and H is a constant. Find the speed ω in terms of L, a, and g, at which water spills over the top.

FIGURE 7.31

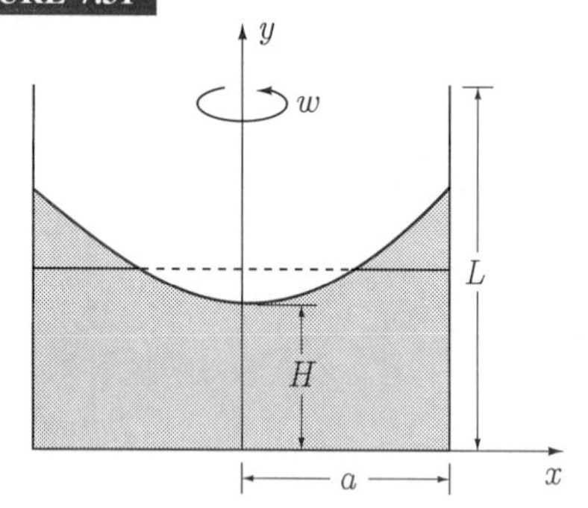

44. A right circular cone of height H and base radius r has its vertex at the centre of a sphere of radius R ($R < H$). Find that part of the volume of the sphere inside the cone.

45. Devise a way to calculate the volume of the solid of revolution when the area in the first quadrant bounded by

$$y = 1 - x^2, \quad x = 0, \quad y = 0$$

is rotated about the line $y = x + 1$. Hint: You may want to use the distance formula in question 45 of Exercises 1.3.

SECTION 7.3

Lengths of Curves

Formula 1.7 defines the length of the straight-line segment joining two points $P(x_1, y_1)$ and $Q(x_2, y_2)$. The formula $s = r\theta$ gives the length of the arc of a circle of radius r subtended by an angle θ at the centre of the circle. In this section we derive a result that will theoretically enable us to find the length of any curve.

Consider finding the length of the curve $C : y = f(x)$ in Figure 7.32(a) joining points $A\big(a, f(a)\big)$ and $B\big(b, f(b)\big)$. To find its length L we begin by approximating C with a series of straight-line segments. Specifically, we choose $n - 1$ consecutive points on C

between A and B,

$$A = P_0, P_1, P_2, \ldots, P_{n-1}, P_n = B,$$

and join each P_{i-1} to P_i ($i = 1, \ldots, n$) by means of a straight-line segment, as in Figure 7.32(b). If the coordinates of P_i are denoted by (x_i, y_i), then the length of the line segment joining P_{i-1} and P_i is

$$\|P_{i-1}P_i\| = \sqrt{(x_i - x_{i-1})^2 + (y_i - y_{i-1})^2}.$$

For a large number of these line segments, it is reasonable to approximate L by the sum of the lengths of the segments:

$$\sum_{i=1}^{n} \|P_{i-1}P_i\| = \sum_{i=1}^{n} \sqrt{(x_i - x_{i-1})^2 + (y_i - y_{i-1})^2}.$$

In fact, as we increase n and at the same time decrease the length of each segment, we expect the approximation to become more and more accurate. We therefore define

$$L = \lim_{\|\Delta x_i\| \to 0} \sum_{i=1}^{n} \|P_{i-1}P_i\| = \lim_{\|\Delta x_i\| \to 0} \sum_{i=1}^{n} \sqrt{(x_i - x_{i-1})^2 + (y_i - y_{i-1})^2}. \quad (7.13)$$

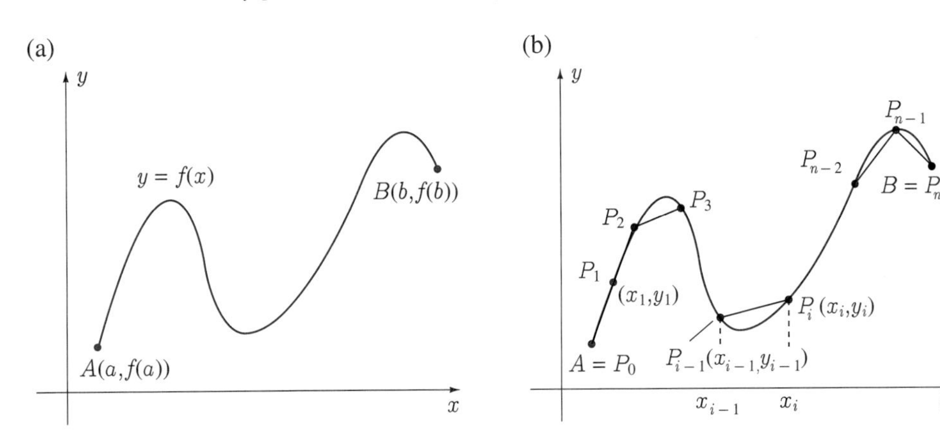

FIGURE 7.32

If we set $\Delta x_i = x_i - x_{i-1}$, and note that $y_{i-1} = f(x_{i-1})$ and $y_i = f(x_i)$, we have

$$\begin{aligned} L &= \lim_{\|\Delta x_i\| \to 0} \sum_{i=1}^{n} \sqrt{(\Delta x_i)^2 + [f(x_i) - f(x_{i-1})]^2} \\ &= \lim_{\|\Delta x_i\| \to 0} \sum_{i=1}^{n} \sqrt{1 + \left[\frac{f(x_i) - f(x_{i-1})}{\Delta x_i}\right]^2}\, \Delta x_i. \end{aligned}$$

In order that 7.13 be a useful definition from a calculational point of view, we must find a convenient way to evaluate this limit-summation. To do this we assume that $f'(x)$ exists at every point in the interval $a \le x \le b$, and apply the mean value theorem to $f(x)$ on each subinterval $x_{i-1} \le x \le x_i$ (Figure 7.33). The theorem guarantees the existence of at least one point x_i^* between x_{i-1} and x_i such that

$$f'(x_i^*) = \frac{f(x_i) - f(x_{i-1})}{x_i - x_{i-1}} = \frac{f(x_i) - f(x_{i-1})}{\Delta x_i}.$$

Consequently, using these points x_i^*, we can express the length of C in the form

$$L = \lim_{\|\Delta x_i\| \to 0} \sum_{i=1}^{n} \sqrt{1 + [f'(x_i^*)]^2}\, \Delta x_i. \quad (7.14)$$

But the right-hand side of this equation is the definition of the definite integral of the function $\sqrt{1+[f'(x)]^2}$ with respect to x from $x = a$ to $x = b$:

$$\int_a^b \sqrt{1+[f'(x)]^2}\,dx = \lim_{\|\Delta x_i\| \to 0} \sum_{i=1}^{n} \sqrt{1+[f'(x_i^*)]^2}\,\Delta x_i.$$

Thus we can evaluate the length of C by the definite integral

$$L = \int_a^b \sqrt{1+\left(\frac{dy}{dx}\right)^2}\,dx. \tag{7.15}$$

FIGURE 7.33

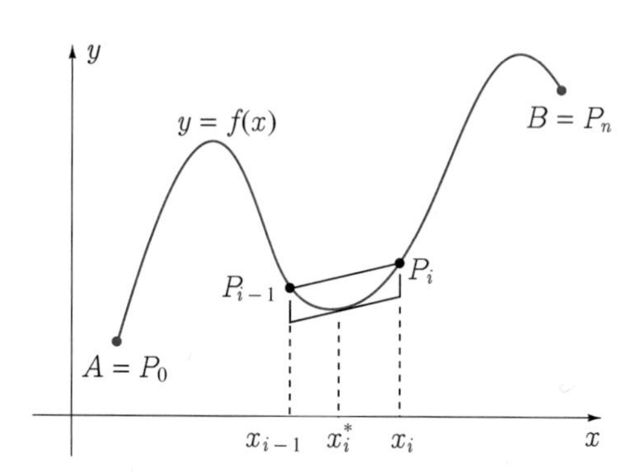

In this derivation we assumed that $f'(x)$ was defined at each point in the interval $a \le x \le b$, necessary for the mean value theorem to apply. But in order to guarantee existence of the definite integral in 7.15, Theorem 6.2 requires continuity of the integrand $\sqrt{1+(dy/dx)^2}$. Consequently, to ensure that the length of a curve can be calculated by means of 7.15, we assume that $f(x)$ has a continuous first derivative on the interval $a \le x \le b$.

EXAMPLE 7.14

Find the length of the curve $y = x^{3/2}$ from $(1, 1)$ to $(2, 2\sqrt{2})$ (Figure 7.34).

FIGURE 7.34

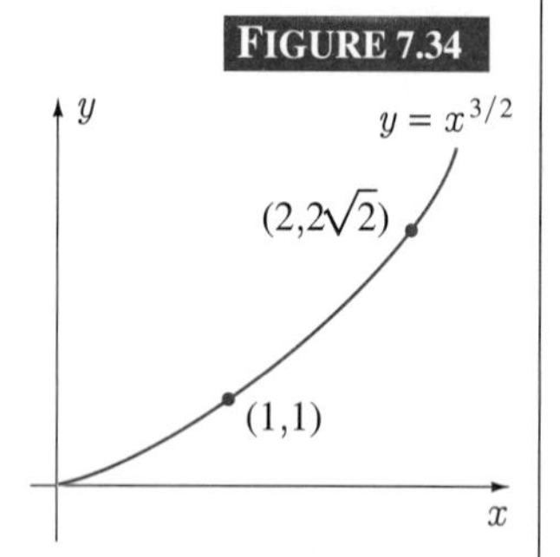

SOLUTION Since

$$\frac{dy}{dx} = \frac{3}{2}x^{1/2},$$

equation 7.15 gives

$$L = \int_1^2 \sqrt{1+\left(\frac{3}{2}x^{1/2}\right)^2}\,dx = \int_1^2 \sqrt{1+\frac{9}{4}x}\,dx$$

$$= \left\{\frac{8}{27}\left(1+\frac{9}{4}x\right)^{3/2}\right\}_1^2 = \frac{8}{27}\left[\left(\frac{11}{2}\right)^{3/2} - \left(\frac{13}{4}\right)^{3/2}\right] = \frac{1}{27}(22^{3/2} - 13^{3/2}).$$

■

FIGURE 7.35

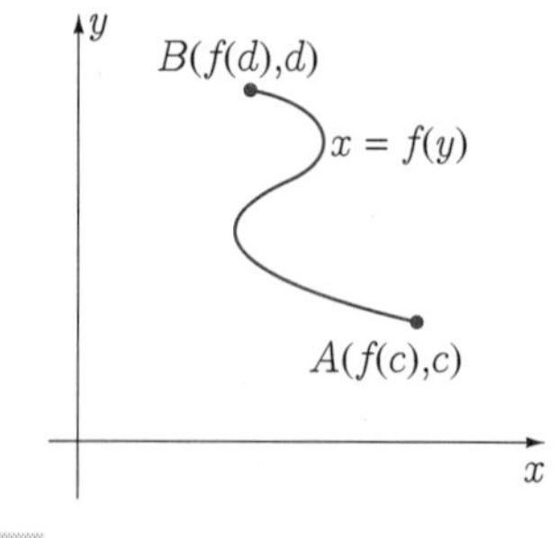

The result in 7.15 is clearly useful when the equation of the curve is expressed in the form $y = f(x)$. If the curve is defined in the form $x = g(y)$ (Figure 7.35), a similar analysis gives

$$L = \int_c^d \sqrt{1+\left(\frac{dx}{dy}\right)^2}\,dy. \tag{7.16}$$

For example, we may write the equation of the curve in Example 7.14 in the form $x = y^{2/3}$, in which case $dx/dy = (2/3)y^{-1/3}$, and

$$L = \int_1^{2\sqrt{2}} \sqrt{1 + \left(\frac{2}{3}y^{-1/3}\right)^2}\, dy = \int_1^{2\sqrt{2}} \sqrt{1 + \frac{4}{9}y^{-2/3}}\, dy$$

$$= \int_1^{2\sqrt{2}} \sqrt{\frac{4 + 9y^{2/3}}{9y^{2/3}}}\, dy = \int_1^{2\sqrt{2}} \frac{\sqrt{4 + 9y^{2/3}}}{3y^{1/3}}\, dy.$$

If we set $u = 4 + 9y^{2/3}$, then $du = 6y^{-1/3}\, dy$, and

$$L = \frac{1}{3}\int_{13}^{22} \sqrt{u}\,\frac{du}{6} = \frac{1}{18}\left\{\frac{2}{3}u^{3/2}\right\}_{13}^{22} = \frac{1}{27}(22^{3/2} - 13^{3/2}).$$

Integrals 7.15 and 7.16 can both be interpreted geometrically. This interpretation will be very useful when we study parametric equations of curves in Section 10.1, and line integrals in Chapter 15. Figure 7.36 shows the curve C of Figure 7.32(a). At a point (x, y) on C, we draw the tangent line to the curve. If dx is a short length along the x-axis at position x, the length of that part of the tangent line between vertical lines at x and $x + dx$ is given by

$$\sqrt{(dx)^2 + (dy)^2}.$$

This length along the tangent line closely approximates the length of the curve between the same two vertical lines, the approximation being more accurate the shorter the length dx. If we picture a large number of these tangential line segments between a and b, we can find the total length of C by adding together all such lengths and taking the limit as each approaches zero. But this process is represented by the definite integral, and we therefore write

$$L = \int_{x=a}^{x=b} \sqrt{(dx)^2 + (dy)^2} \qquad \text{or} \qquad L = \int_{y=c}^{y=d} \sqrt{(dx)^2 + (dy)^2}. \qquad (7.17)$$

The first integral is chosen when it is more convenient to integrate with respect to x, in which case dx is taken outside the square root. We set

$$L = \int_a^b \sqrt{1 + \left(\frac{dy}{dx}\right)^2}\, dx,$$

which is equation 7.15. The second corresponds to a more convenient integral with respect to y, in which case dy is taken outside the square root. We obtain

$$L = \int_c^d \sqrt{1 + \left(\frac{dx}{dy}\right)^2}\, dy,$$

which is equation 7.16.

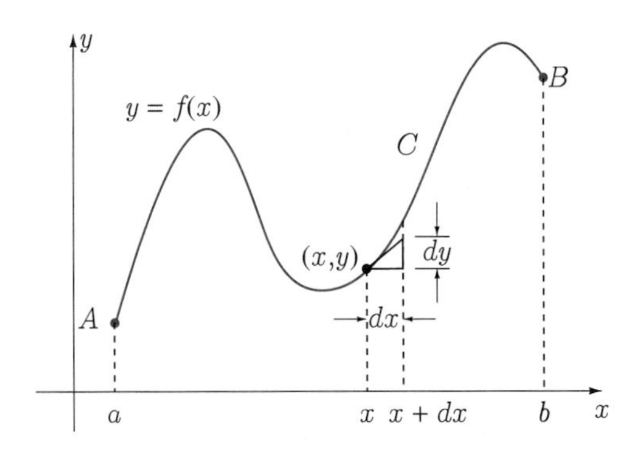

FIGURE 7.36

EXAMPLE 7.15

Find the length of the curve $24xy = x^4 + 48$ between $(2, 4/3)$ and $(3, 43/24)$.

SOLUTION It is easier to express y in terms of x (rather than x in terms of y),

$$y = \frac{x^4 + 48}{24x} = \frac{x^3}{24} + \frac{2}{x}.$$

We now use the fact that small lengths along the curve are approximated by

$$\sqrt{(dx)^2 + (dy)^2} = \sqrt{1 + \left(\frac{dy}{dx}\right)^2}\, dx = \sqrt{1 + \left(\frac{x^2}{8} - \frac{2}{x^2}\right)^2}\, dx$$
$$= \sqrt{1 + \frac{x^4}{64} - \frac{1}{2} + \frac{4}{x^4}}\, dx = \sqrt{\frac{1}{64x^4}(x^8 + 32x^4 + 256)}\, dx$$
$$= \frac{1}{8x^2}\sqrt{(x^4 + 16)^2}\, dx = \frac{x^4 + 16}{8x^2}\, dx.$$

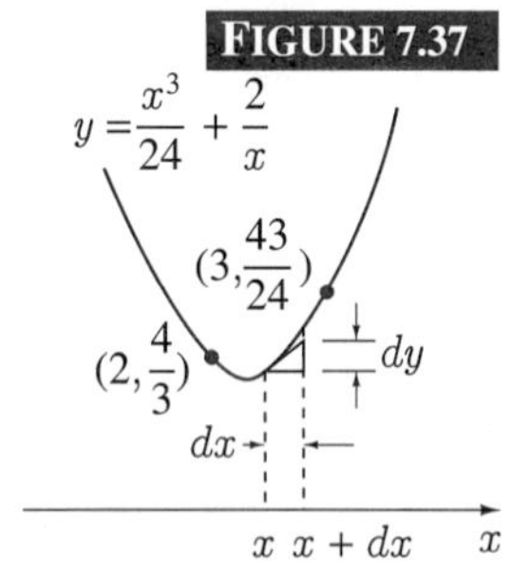

FIGURE 7.37

The total length along the curve is therefore (Figure 7.37)

$$L = \int_2^3 \frac{x^4 + 16}{8x^2}\, dx = \frac{1}{8}\int_2^3 \left(x^2 + \frac{16}{x^2}\right) dx$$
$$= \frac{1}{8}\left\{\frac{x^3}{3} - \frac{16}{x}\right\}_2^3 = \frac{9}{8}.$$

■

EXERCISES 7.3

In Exercises 1–10 find the length of the curve.

1. $8x^2 = y^3$ from $(1, 2)$ to $(2\sqrt{2}, 4)$
2. $3y = 2(x^2 + 1)^{3/2}$ from $(-2, 10\sqrt{5}/3)$ to $(-1, 4\sqrt{2}/3)$
3. $x = 2(y - 2)^{3/2}$ from $(0, 2)$ to $(2, 3)$
4. $y = (x - 1)^{3/2}$ from $(2, 1)$ to $(10, 27)$
5. $y = \dfrac{x^3}{4} + \dfrac{1}{3x}$ from $(1, 7/12)$ to $(2, 13/16)$
6. $y = (e^x + e^{-x})/2$ between the lines $x = 0$ and $x = 1$
7. $y = \dfrac{x^4}{4} + \dfrac{1}{8x^2}$ from $(2, 257/64)$ to $(1/2, 65/64)$
8. $36xy = x^4 + 108$ from $(3, 7/4)$ to $(4, 91/36)$
9. $x = \dfrac{y^7}{20} + \dfrac{1}{7y^5}$ between the lines $y = -1$ and $y = -2$
10. $y = x^5/5 + 1/(12x^3)$ from $(1, 17/60)$ to $(2, 3077/480)$

In Exercises 11–20 set up (but do not evaluate) a definite integral for the length of the curve.

11. $y = x^2$ from $(0, 0)$ to $(1, 1)$
12. $y = 3x^2 - 4x$ from $(1, -1)$ to $(2, 4)$
13. $x^2 - y^2 = 1$ from $(1, 0)$ to $(2, \sqrt{3})$
14. $x^2 - y^2 = 1$ from $(2, -\sqrt{3})$ to $(3, 2\sqrt{2})$
15. $y = \sin x$ from $(0, 0)$ to $(\pi, 0)$
16. $y = \ln(\cos x)$ between the lines $x = 0$ and $x = \pi/4$
17. $y = \ln x$ from $(1, 0)$ to any other point on the curve
18. $8y^2 = x^2(1 - x^2)$ (complete length)
19. $x = y^2 - 2y$ from $(0, 0)$ to $(0, 2)$
20. $x^2/4 + y^2/9 = 1$ (complete length)
21. Find the length of the curve $3x = \sqrt{y}(y - 3)$ from $(-2/3, 1)$ to $(2/3, 4)$.
22. Find the length of the curve $x^{2/3} + y^{2/3} = 1$.

23. If n is any number other than 1 or -1, and $0 < a < b$, find the length of the curve

$$y = \frac{x^{n+1}}{n+1} + \frac{1}{4(n-1)x^{n-1}}$$

between $x = a$ and $x = b$.

24. If n is any number greater than $1/2$, and $0 < a < b$, find the length of the curve

$$y = \frac{x^{2n+1}}{4(2n-1)} + \frac{1}{(2n+1)x^{2n-1}}$$

between $x = a$ and $x = b$.

SECTION 7.4

Work

When a body moves a distance d along a straight line l under the action of a constant force F, which acts in the same direction as the motion (Figure 7.38), the work done on the body by F is defined as

$$W = Fd. \tag{7.18}$$

FIGURE 7.38

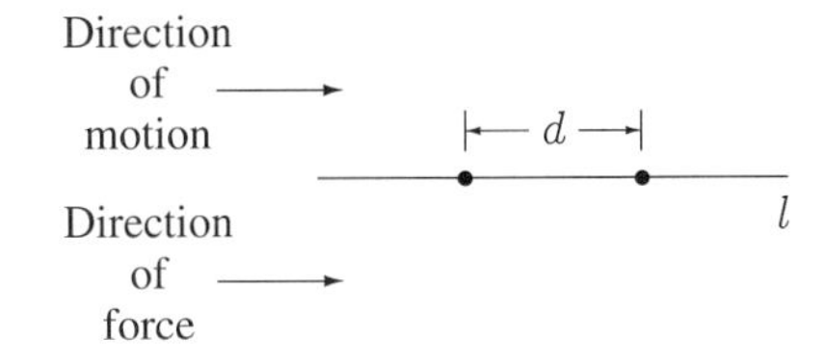

Forces often vary in either magnitude or direction, or both, and when such forces act on the body, calculation of the work done is not as simple as using equation (7.18). In this section we show how we can use definite integrals to calculate work done by forces that always act along the direction of motion, but do not have constant magnitude. Specifically, let us consider a particle that moves along the x-axis from $x = a$ to $x = b$, where $b > a$, under the action of some force. Suppose the force always acts in the positive x-direction, but its size, which we denote by $F(x)$, is not constant along the x-axis. To find the work done by the force as the particle moves from $x = a$ to $x = b$, we cannot simply multiply force by distance. (Where would we evaluate $F(x)$?) Instead, we divide the interval $a \le x \le b$ into n subintervals by points (Figure 7.39)

$$a = x_0 < x_1 < x_2 < \cdots < x_{n-1} < x_n = b.$$

FIGURE 7.39

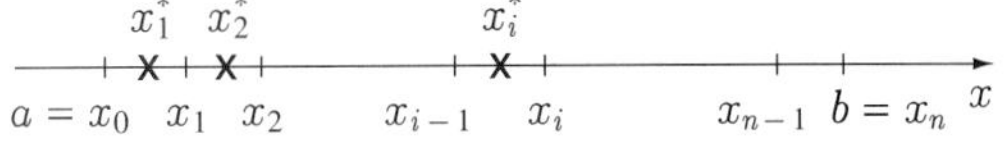

In each subinterval $x_{i-1} \le x \le x_i$ we choose a point x_i^*. If lengths of the subintervals are small and $F(x)$ is continuous, then $F(x)$ does not vary greatly over any given subinterval. In such a situation, we may approximate $F(x)$ by a constant force $F(x_i^*)$ on each subinterval $x_{i-1} \le x \le x_i$. It follows that the work done by the force over the ith subinterval is approximately

$$F(x_i^*)(x_i - x_{i-1}) = F(x_i^*)\Delta x_i.$$

Further, an approximation to the total work done by the force as the particle moves from $x = a$ to $x = b$ is

$$\sum_{i=1}^{n} F(x_i^*)\Delta x_i.$$

As n becomes larger and each Δx_i approaches zero, this approximation becomes better, and we therefore define

$$W = \lim_{\|\Delta x_i\| \to 0} \sum_{i=1}^{n} F(x_i^*)\Delta x_i. \tag{7.19}$$

But this limit may also be interpreted as the definite integral of $F(x)$ with respect to x from $x = a$ to $x = b$. Consequently, W may be calculated with the definite integral

$$W = \int_a^b F(x)\, dx. \tag{7.20}$$

It is simple enough to interpret the various parts of 7.20, and in so doing we begin to feel the definite integral at work. The integrand $F(x)$ is the force at position x, and dx is a small distance along the x-axis. The product $F(x)\, dx$ is therefore interpreted as the (approximate) work done by the force along dx. The definite integral then adds over all dx's, beginning at $x = a$ and ending at $x = b$, to give the total work.

What is important about equation 7.20 is not the particular form of the definite integral, but the fact that work done by a force can be evaluated by means of a definite integral. As we solve work problems in this section, we find that the form of the definite integral varies considerably from one problem to another, but the underlying fact remains that each problem is solved with a definite integral.

With the exception of work problems that involve emptying tanks (Example 7.18), we recommend that you always make a diagram illustrating the physical setup at some intermediate stage between start and finish. Determine forces at this position in order to set up the work integral.

In the derivation of 7.20 we assumed continuity of $F(x)$, and this guarantees existence of the definite integral.

EXAMPLE 7.16

Find the work necessary to expand a spring from a stretch of 5 cm to a stretch of 15 cm if a force of 200 N stretches it 10 cm.

SOLUTION Let the spring be stretched in the positive x-direction, and let $x = 0$ correspond to the free end of the spring in the unstretched position in the top half of Figure 7.40. In the bottom half of this figure we have shown the spring stretched to an intermediate position.

FIGURE 7.40

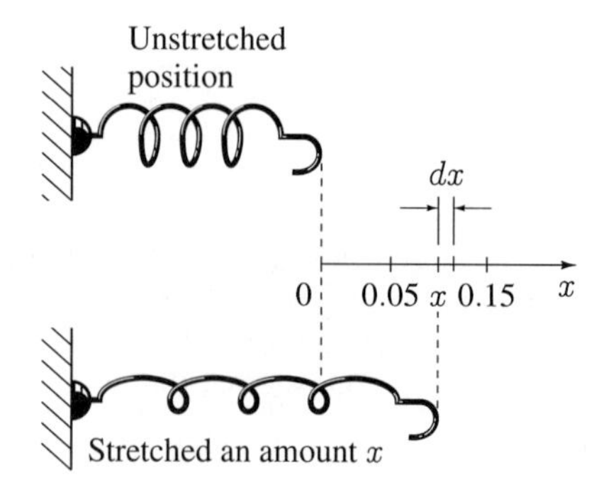

According to Hooke's law, when the spring is stretched an amount x, the restoring force in the spring is proportional to x:

$$F_s = -kx,$$

where $k > 0$ is a constant. The negative sign indicates that the force is in the negative x-direction. Since $F_s = -200$ N when $x = 0.10$ m,

$$-200 = -k(0.10),$$

it follows that $k = 2000$ N/m. The force required to counteract the restoring force of the spring when it is stretched an amount x is therefore

$$F(x) = 2000x.$$

The work done by this force in stretching the spring from position x a further amount dx is approximately

$$2000x\,dx,$$

and hence the total work to increase the stretch from 5 cm to 15 cm is

$$W = \int_{0.05}^{0.15} 2000x\,dx = 2000\left\{\frac{x^2}{2}\right\}_{0.05}^{0.15} = 20.$$

The total work done is therefore 20 N-m (or joules). ■

EXAMPLE 7.17

A cable of length 100 m and mass 300 kg hangs vertically from the top of a building. What work is required to lift the entire cable to the top of the building?

SOLUTION

FIGURE 7.41

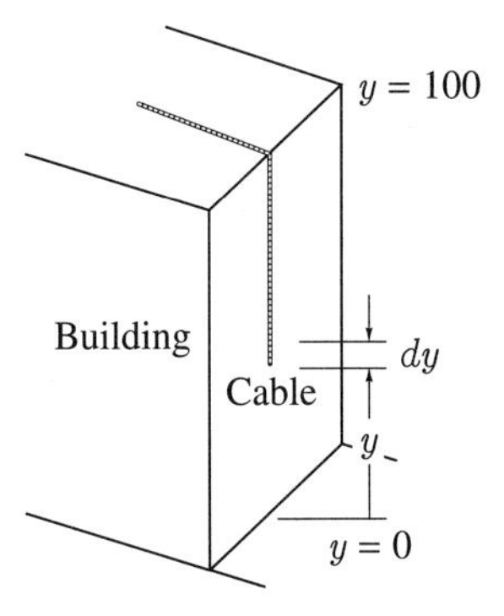

When the cable has been lifted to an intermediate point where its lower end is y metres above its original position (Figure 7.41), the length of cable still hanging is $100 - y$ m. Since each metre of cable has mass 3 kg, the mass of $100 - y$ m of cable is $3(100 - y)$ kg. It follows that the force that must be exerted to overcome gravity on this much cable is $9.81(3)(100 - y)$ N. The work that this force does in raising the end of the cable an additional amount dy is approximately

$$9.81(3)(100 - y)\,dy,$$

and the total work to raise the entire cable is therefore

$$W = \int_0^{100} 9.81(3)(100 - y)\,dy = 29.43\left\{100y - \frac{y^2}{2}\right\}_0^{100} = 147\,150 \ J.$$

■

EXAMPLE 7.18

A tank in the form of an inverted right-circular cone of depth 10 m and radius 4 m is full of water. Find the work required to pump the water to a level 1 m above the top of the tank.

SOLUTION

FIGURE 7.42

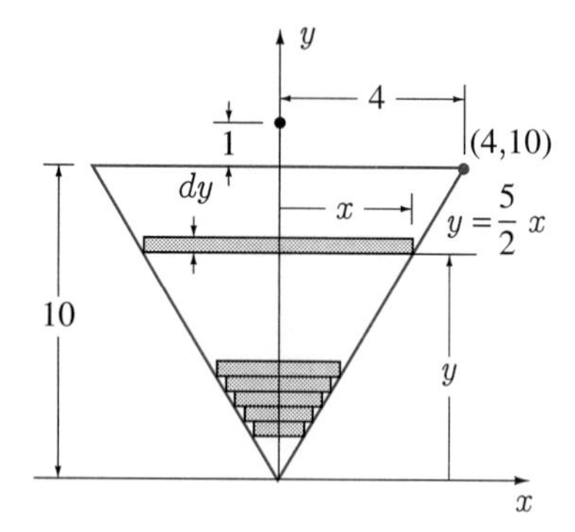

A cross-section of the tank is shown in Figure 7.42. Suppose we approximate the water in the tank with circular discs formed by rotating the rectangle shown around the y-axis. The force of gravity on the representative disc at position y is its volume $\pi x^2\,dy$, multiplied by the density of water (1000 kg/m^3), multiplied by the acceleration due to gravity (−9.81),

$$-9.81(1000)\pi x^2\,dy.$$

It is negative because gravity acts in the negative y-direction. The work done by an equal and opposite force in lifting the disc to a level 1 m above the top of the tank (a distance $11 - y$) is approximately

$$(11 - y)9.81(1000)\pi x^2\,dy.$$

The total work to empty the tank is therefore

$$W = \int_0^{10} (11 - y)9810\pi x^2\,dy.$$

To express x in terms of y, we note that x and y are the coordinates of points on the straight line through the origin and the point $(4, 10)$. Since the equation of this line is $y = 5x/2$, we obtain

$$W = 9810\pi \int_0^{10} (11 - y)\left(\frac{2y}{5}\right)^2 dy = 1569.6\pi \int_0^{10} (11y^2 - y^3)\, dy$$

$$= 1569.6\pi \left\{\frac{11y^3}{3} - \frac{y^4}{4}\right\}_0^{10} = 5.75 \times 10^6 \text{ J.}$$

■

Two observations are noteworthy:

1. In Examples 7.16 and 7.17 the differential represents distance moved and the integrand represents force, and we applied equation 7.20 directly. In Example 7.18 the differential is part of the force, and the distance moved is part of the integrand. This is why we stated earlier that it is not advisable to use equation 7.20 as a formula. The integrand is not always force and the differential is not always distance moved.
2. In each of these examples we set up the coordinate system; it was not given. We can use any coordinate system whatsoever; but once we have chosen our coordinates, we must refer everything to that system. For instance, were we to use the coordinate system in Figure 7.43 for Example 7.18, the definite integral would be

$$W = \int_{-10}^{0} (1 - y)9810\pi \left[\frac{2}{5}(10 + y)\right]^2 dy.$$

Evaluation of this definite integral would again lead to the result $W = 5.75 \times 10^6$ J.

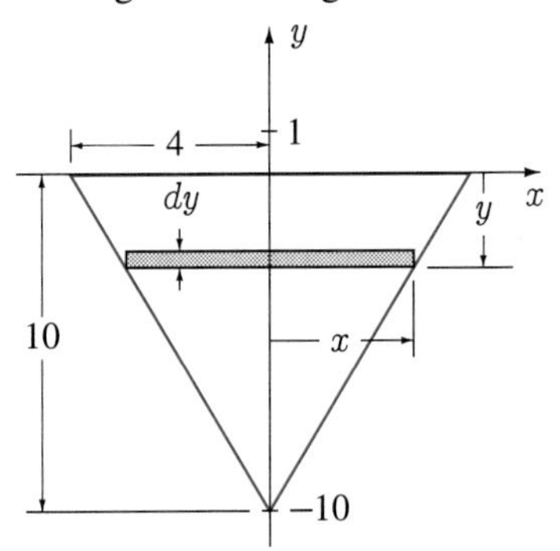

FIGURE 7.43

EXERCISES 7.4

1. A spring requires a 10 N force to stretch it 3 cm. Find the work to increase the stretch of the spring from 5 cm to 7 cm.
2. Find the work to increase the stretch of the spring in Exercise 1 from 7 cm to 9 cm.
3. A cage of mass M kilograms is to be lifted from the bottom of a mine shaft h metres deep. If the mass of the cable used to hoist the cage is m kilograms per metre, find the work done.
4. A uniform cable of length 50 m and mass 100 kg hangs vertically from the top of a building 100 m high. How much work is required to get 10 m of the cable on top of the building?
5. A 2 m chain of mass 20 kg lies on the floor. If friction between floor and chain is ignored, how much work is required to lift one end of the chain 2 m straight up?
6. How much work is required to lift the end of the chain in Exercise 5 only 1 m?
7. How much work is required to lift the end of the chain in Exercise 5 a distance of 4 m?

8. A 5 m chain of mass 15 kg hangs vertically. It is required to lift the lower end of the chain 5 m so that it is level with the upper end. Calculate the work done using each of the coordinate systems in Figure 7.44.

FIGURE 7.44

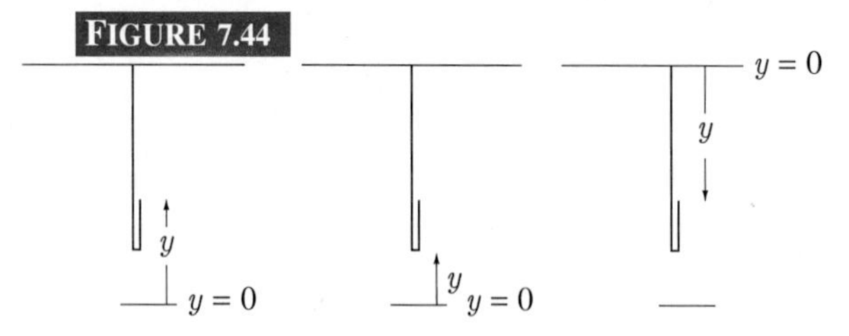

9. A tank filled with water has the form of a paraboloid of revolution with vertical axis (Figure 7.45). If the depth of the tank is 12 m and the diameter of the top is 8 m, find the work in pumping the water to the top of the tank.

FIGURE 7.45

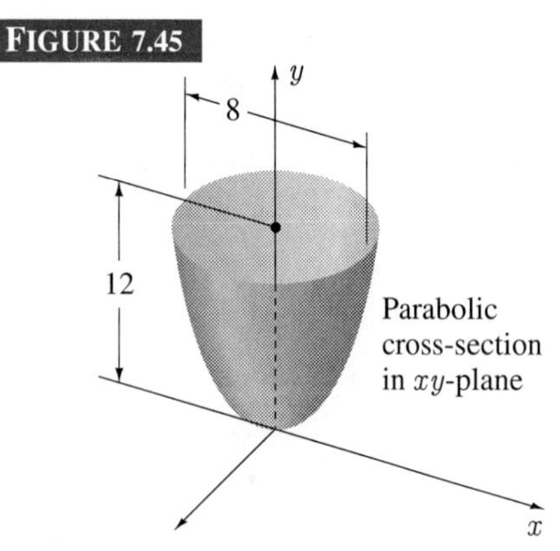

10. In Exercise 9, find the work to empty the tank through an outlet 2 m above the top of the tank.

11. The ends of a trough are isosceles triangles with width 2 m and depth 1 m. The trough is 5 m long and it is full of water. How much work is required to lift all of the water to the top of the trough?

12. How much work is required to lift the water in the trough of Exercise 11 to a height of 2 m above the top of the trough?

13. A rectangular swimming pool full of water is 25 m long and 10 m wide. The depth is 3 m for the first 10 m of length, decreasing linearly to 1 m at the shallow end. How much work is required to lower the level of the water in the pool by 1/2 m?

14. How much work is required to empty the pool of Exercise 13 over its edge?

15. The force of repulsion between two point charges of like sign, one of size q and the other of size Q, has magnitude

$$F = \frac{qQ}{4\pi\epsilon_0 r^2},$$

where ϵ_0 is a constant and r is the distance between the charges. If Q is placed at the origin, and q is moved along the x-axis from $x = 2$ to $x = 5$, find the work done by the electrostatic force.

16. Two positive charges q_1 and q_2 are placed at positions $x = 5$ and $x = -2$ on the x-axis. A third positive charge q_3 is moved along the x-axis from $x = 1$ to $x = -1$. Find the work done by the electrostatic forces of q_1 and q_2 on q_3 (see Exercise 15).

17. If the chain in Exercise 5 is stretched out straight on the floor, and if the coefficient of friction between floor and chain is 0.01, what work is required to raise one end of the chain 2 m? The force of friction on that part of the chain on the floor is 0.01 times the weight of the chain on the floor.

18. Two similar springs, each 1 m long in an unstretched position, have spring constant k N/m. The springs are joined together at P (Figure 7.46), and their free ends are fastened to two posts 4 m apart. What work is done in moving the midpoint P a distance b m to the right?

FIGURE 7.46

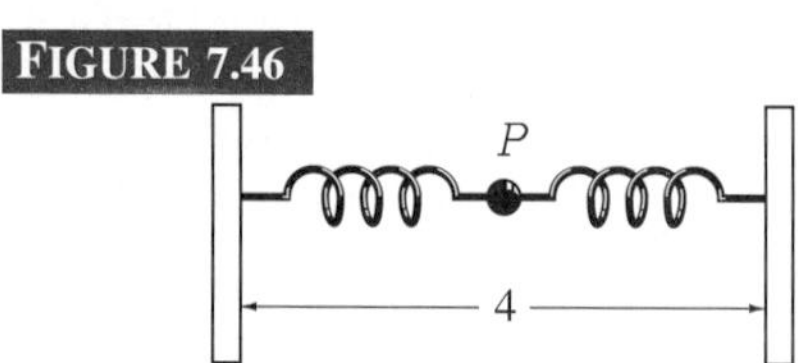

19. If the force of a crossbow (in newtons) is proportional to the draw (in metres), and it is 200 N at a full draw of 50 cm, what is the speed of a 20-gm arrow from a fully drawn crossbow? Hint: The work necessary to draw the bow is converted into kinetic energy K, where $K = mv^2/2$ (m is the mass of the arrow and v is the speed of the arrow).

20. A bucket of water with mass 100 kg is on the ground attached to one end of a cable with mass per unit length 5 kg/m. The other end of the cable is attached to a windlass 100 m above the bucket. If the bucket is raised at a constant speed, water runs out through a hole in the bottom at a constant rate to the extent that the bucket has mass 80 kg when it reaches the top. To further complicate matters, a pigeon of mass 2 kg lands on the bucket when it is 50 m above the ground. He immediately begins taking a bath, splashing water over the side of the bucket at the rate of 1 kg/m. Find the work done by the windlass in raising the bucket 100 m.

21. A hemispherical tank with diametral plane at the top has radius 5 m.

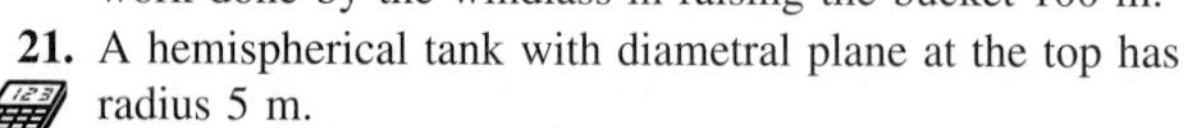

(a) Show that if the depth of oil in the tank is h metres when the tank is one-half full by volume, then h must satisfy the equation

$$h^3 - 15h^2 + 125 = 0.$$

Find h to four decimals.

(b) Find the work to empty the half-full tank if the oil has density 750 kg/m^3, and the outlet is at the top of the tank.

22. Newton's universal law of gravitation states that the force of attraction between two point masses m and M has magnitude

$$F = \frac{GmM}{r^2},$$

where r is the distance between the masses and $G = 6.67 \times 10^{-11}$ Nm^2/kg^2 is a constant.

(a) If M represents the mass of the earth and we regard it as a point mass concentrated at its centre, show that Newton's universal law of gravitation at the earth's surface reduces to $F = mg$, where $g = 9.82$ m/s^2. Assume for the calculation of M that the earth is a sphere with radius 6370 km and mean density 5.52×10^3 kg/m^3.

(b) Use the original law $F = GmM/r^2$ with the earth regarded as a point mass to calculate the work required to lift a mass of 10 kg from the earth's surface to a height of 10 km.

(c) Calculate the work in (b) using the constant gravitational force $F = mg$ in (a). Is there a significant difference?

23. A right circular cylinder with horizontal axis has radius r metres and length h metres. If it is full of oil with density ρ kilograms per cubic metre, how much work is required to empty the tank through an outlet at its top?

24. A gas is confined in a cylinder, closed on one end with a piston on the other. If the temperature is held constant, then $PV = C$, where P is the pressure of the gas in the cylinder, V is its volume, and C is a constant. The force needed to move the piston is $F = PA$, where A is the area of the face of the piston. If the radius of the cylinder is 3 units, find the work done in moving the piston from a point 10 units to a point 5 units from the closed end.

25. A gas is confined in a cylinder, closed on one end with a piston on the other. If the gas expands adiabatically, then the pressure P and volume V of the gas obey the law

$$PV^{7/5} = C,$$

where C is a constant. The force on the piston face due to the pressure of the gas is given by $F = PA$, where A is the area of the face of the piston. Show that if the piston is moved so as to reduce the volume occupied by the gas from V_0 to $V_0/2$, then the work done by the piston is

$$\frac{5}{2}\left(2^{2/5} - 1\right) CV_0^{-2/5}.$$

26. A particle is moved along the x-axis by a force $F(x)$ in the x-direction. Use Newton's second law to show that if $F(x)$ is the total resultant force on the particle, then the work done by $F(x)$ is equal to the change in kinetic energy of the particle.

27. A diving bell of mass 10 000 kilograms is attached to a chain of mass 5 kg/m, and the bell sits on the bottom of the ocean 100 m below the surface (Figure 7.47). How much work is required to lift the bell to deck level 5 m above the surface? Take into account the fact that when the bell and chain are below the surface, they weigh less than when above. The apparent loss in weight is equal to the weight of water displaced by the bell and chain. Assume that the bell is a perfect cube, 2 m on each side; therefore, when completely submerged, it displaces 8 m^3 of water. Assume also that the chain displaces 1 litre of water per metre of length.

FIGURE 7.47

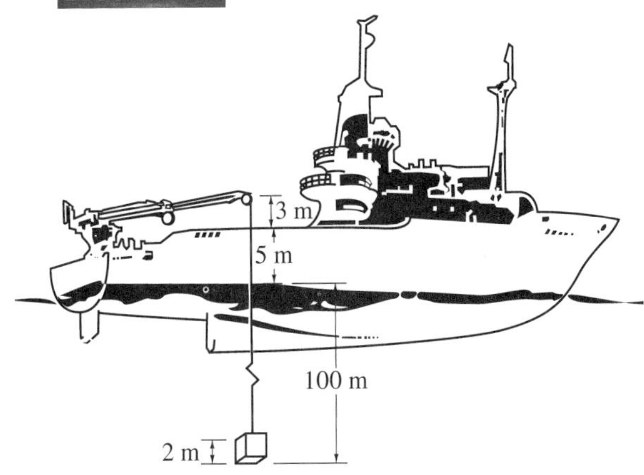

28. Two springs with constants k_1 and k_2 are joined together, and then one end is fastened to a wall (Figure 7.48). It is shown in physics that when a horizontal force is applied to the free end, the ratio of the stretches s_1 and s_2 in the two springs is inversely proportional to the ratio of their spring constants:

$$\frac{s_1}{s_2} = \frac{k_2}{k_1}.$$

Show that if the free end is moved so as to produce a total stretch in the springs of L, the amount of work done is

$$\frac{k_1 k_2 L^2}{k_1 + k_2}.$$

FIGURE 7.48

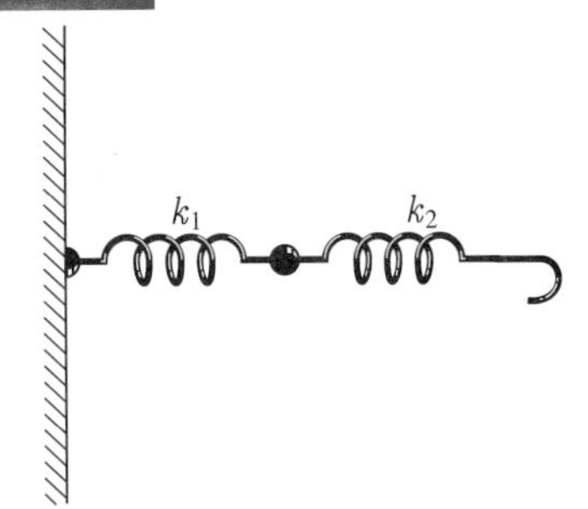

SECTION 7.5

Fluid Pressure

When an object is immersed in a fluid, it is acted on by fluid forces. These forces are independent of the object, and are therefore a property of the fluid itself. They always act perpendicular to the surface of the submerged object. We use the concept of pressure to describe these fluid forces. We define **pressure** at a point in a fluid as the *magnitude of the force per unit area* that would act on a surface at that point in the fluid. Because pressure is the magnitude of the fluid force at a point, it is therefore a positive quantity.

FIGURE 7.49

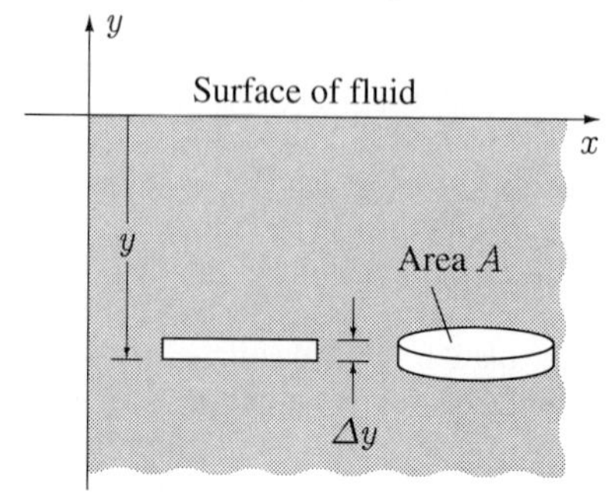

Experience suggests that pressure depends on two factors: depth below the surface of the fluid and the type of fluid itself. In order to discover the precise dependence, we consider a small horizontal disc of the fluid (Figure 7.49). Suppose we denote by $P(y)$ the functional dependence of pressure P on depth y. Then the pressure at the bottom of the disc is $P(y)$ and the pressure at the top of the disc is $P(y + \Delta y)$. If the fluid is stationary, then the sum of all vertical forces on the disc must be zero. There are three vertical forces acting on the disc: fluid forces on its top and bottom faces, and gravity. Since pressure $P(y)$, which is force per unit area, is the same at all points on the bottom of the disc, it follows that the force on the bottom of the disc must be $AP(y)$; that is, the product of the area A of the bottom of the disc and pressure at points on the bottom of the disc. Similarly, the fluid force on the top of the disc is $-AP(y + \Delta y)$; it is negative because it is in the negative y-direction. Finally, if ρ is the density of the fluid (mass per unit volume), the force of gravity on the disc is $-9.81\rho(A\Delta y)$. Since the sum of these three forces must be zero, we set

$$AP(y) - AP(y + \Delta y) - 9.81\rho(A\Delta y) = 0.$$

Rearrangement of this equation yields

$$\frac{P(y + \Delta y) - P(y)}{\Delta y} = -9.81\rho,$$

and if we take limits of both sides as $\Delta y \to 0$, we obtain

$$\frac{dP}{dy} = -9.81\rho.$$

This differential equation for $P(y)$ is immediately integrable:

$$P(y) = -9.81\rho y + C.$$

Since fluid pressure at the surface of the fluid is equal to zero ($P(0) = 0$), C must be equal to zero; hence,

$$P = -9.81\rho y. \tag{7.21}$$

Since $-y$ is a measure of depth d below the surface of the fluid, we have shown that

$$P = 9.81\rho d, \tag{7.22}$$

where d is always taken as positive.

An illuminating interpretation of formula 7.22 is suggested by Figure 7.50. Above a point at depth d below the surface of a fluid, we consider a column of fluid of unit cross-sectional area. The weight of this column of fluid is its volume multiplied by 9.81ρ:

$$W = 9.81\rho V.$$

But V is the product of the length of the column, d, and the (unit) cross-sectional area; that is, $V = (1)d = d$, and hence

$$W = 9.81\rho d. \tag{7.23}$$

A comparison of equations 7.22 and 7.23 suggests that pressure at any point is precisely the weight of a column of fluid of unit cross-sectional area above that point.

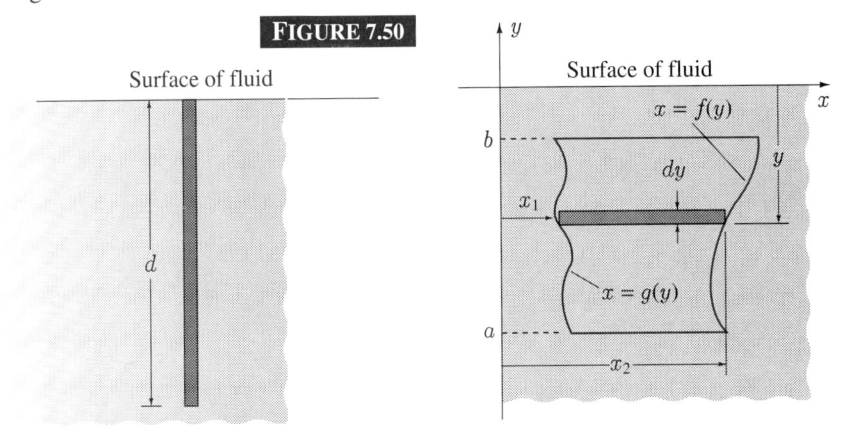

FIGURE 7.50

FIGURE 7.51

We now consider the problem of determining total force on one side of a flat vertical plate (Figure 7.51), when it is immersed in a fluid of density ρ. If the area of the plate is subdivided into horizontal rectangles of width dy, then the pressure at each point of this rectangle is approximately $P = -9.81\rho y$. This is an approximation because slight variations in pressure do occur over vertical displacements within the rectangle. It follows that the force on the representative rectangle is approximately equal to its area multiplied by $-9.81\rho y$,

$$-9.81\rho y(x_2 - x_1)\,dy = -9.81\rho y[f(y) - g(y)]\,dy.$$

Total force on the plate is found by adding the forces on all such rectangles and taking the limit as their widths approach zero. We obtain the required force, therefore, as

$$F = \int_a^b -9.81\rho y[f(y) - g(y)]\,dy. \tag{7.24}$$

Once again we do not suggest that equation 7.24 be memorized because it is closely connected with the choice of coordinates in Figure 7.51. For a different coordinate system, the definite integral would be correspondingly different (see Example 7.19). What is important is the procedure: subdivide the area of the plate into horizontal rectangles, find the force on a representative rectangle, and finally add over all rectangles with the definite integral.

Note that vertical rectangles cannot be used without further discussion since it is not evident how to calculate the force on such a rectangle. Consideration of vertical rectangles is given in Exercise 8.

EXAMPLE 7.19

The vertical face of a dam is parabolic with breadth 100 m and height 50 m. Find the total force due to fluid pressure on the face.

SOLUTION If we set up the coordinate system in Figure 7.52, we see that the edge of the dam has an equation of the form $y = kx^2$. Since $(50, 50)$ is a point on this curve, it follows that $k = 1/50$, and $y = x^2/50$. The area of the representative rectangle shown is $2x\,dy$, and it is at a depth $50 - y$ below the surface of the water. Since the density of water is 1000 kg/m^3, the force on the representative rectangle is approximately equal to

$$(9.81)(1000)(50 - y)2x\,dy = 19\,620(50 - y)\sqrt{50y}\,dy.$$

The total force on the dam must therefore be

$$F = \int_0^{50} 19\,620(50 - y)5\sqrt{2}y^{1/2}\,dy = 98\,100\sqrt{2}\int_0^{50}(50y^{1/2} - y^{3/2})\,dy$$

$$= 98\,100\sqrt{2}\left\{\frac{100y^{3/2}}{3} - \frac{2y^{5/2}}{5}\right\}_0^{50} = 6.54 \times 10^8 \text{ N}.$$

FIGURE 7.52

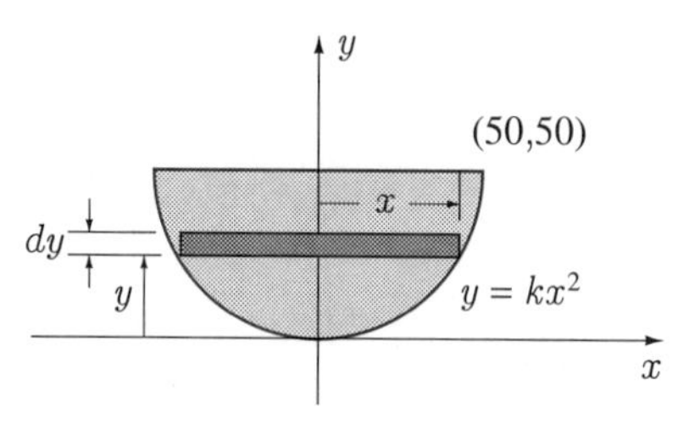

■

EXAMPLE 7.20

A tank in the form of a right-circular cylinder of radius 2 m and length 10 m lies on its side. If it is half filled with oil of density ρ kilograms per cubic metre, find the force on each end of the tank.

SOLUTION The force on the representative rectangle in Figure 7.53 is approximately equal to the pressure $9.81\rho(-y)$ multiplied by the area $2x\,dy$ of the rectangle,

$$9.81\rho(-y)2x\,dy = -19.62\rho y\sqrt{4 - y^2}\,dy.$$

The total force on each end of the tank is therefore

$$F = \int_{-2}^{0} -19.62\rho y\sqrt{4 - y^2}\,dy = 19.62\rho\left\{\frac{1}{3}(4 - y^2)^{3/2}\right\}_{-2}^{0} = 52.32\rho \text{ N}.$$

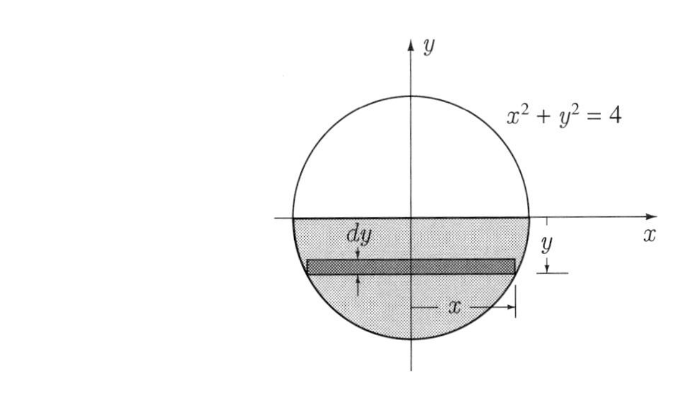

FIGURE 7.53

EXERCISES 7.5

1. A tropical fish tank has length 1 m, width 0.5 m, and depth 0.5 m. Find the force due to water pressure on each of the sides and bottom when the tank is full.

2. The vertical surface of a dam exposed to the water of a lake has the shape shown in Figure 7.54. Find the force of the water on the face of the dam.

FIGURE 7.54

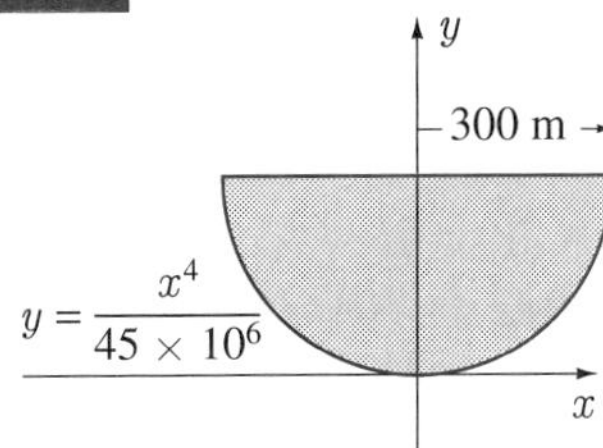

3. The vertical end of a water trough is an isosceles triangle with width 2 m and depth 1 m. Find the force of the water on each end when the trough is half full by volume.

4. A square plate, 2 m on each side, has one diagonal vertical. If it is one-half submerged in water, what is the force due to water pressure on each side of the plate?

5. A cylindrical oil tank of radius r and height h has its axis vertical. If the density of the oil is ρ, find the force on the bottom of the tank when it is full.

6. A rectangular swimming pool full of water is 25 m long and 10 m wide. The depth is 3 m for the first 10 m at the deep end, decreasing linearly to 1 m at the shallow end. Find the force due to the weight of the water on each of the sides and ends of the pool.

7. The vertical face of a dam across a river has the shape of a parabola 36 m across the top and 9 m deep at the centre. What is the force that the river exerts on the dam if the water is 0.5 m from the top?

8. Show that the force due to fluid pressure on the vertical rectangle in Figure 7.55 is

$$F = \frac{9.81\rho}{2} h(y_1^2 - y_2^2),$$

where ρ is the density of the fluid.

FIGURE 7.55

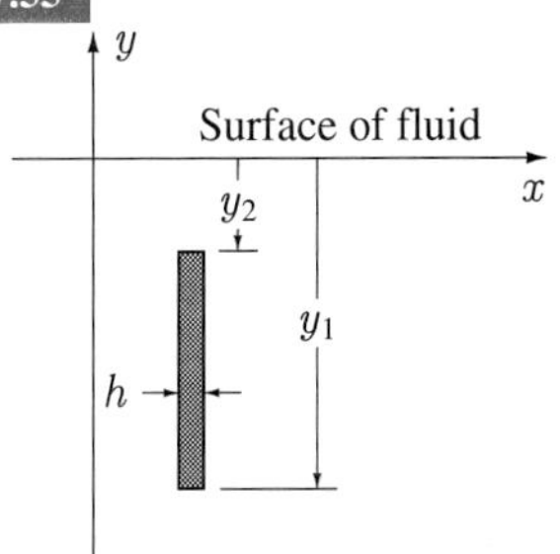

9. A flat plate in the shape of a trapezoid is submerged vertically in a fluid with density ρ. The plate has two parallel vertical sides of lengths 6 and 8 and a third side of length 5 that is perpendicular to the parallel sides and at a depth of 1 below the surface of the fluid (Figure 7.56). Find the force due to fluid pressure on each side of the plate, using both horizontal and vertical rectangles (see Exercise 8 for vertical rectangles).

FIGURE 7.56

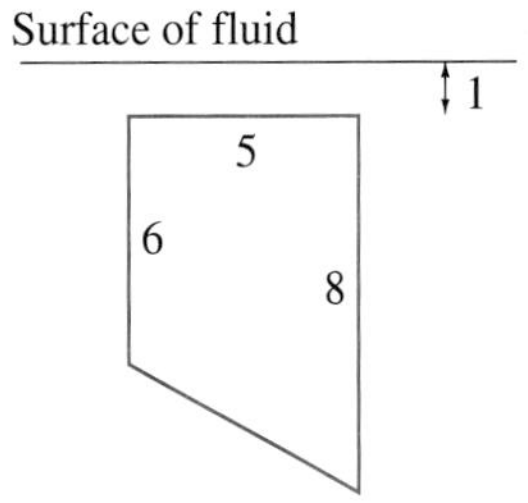

Surface of fluid
1
5
6
8

10. The base of a triangular plate, of length a, lies in the surface of a fluid of density ρ. The third vertex of the triangle is at depth b below the surface (Figure 7.57). Show that the force due to fluid pressure on each side of the plate is $9.81\rho ab^2/6$, no matter what the shape of the triangle.

FIGURE 7.57

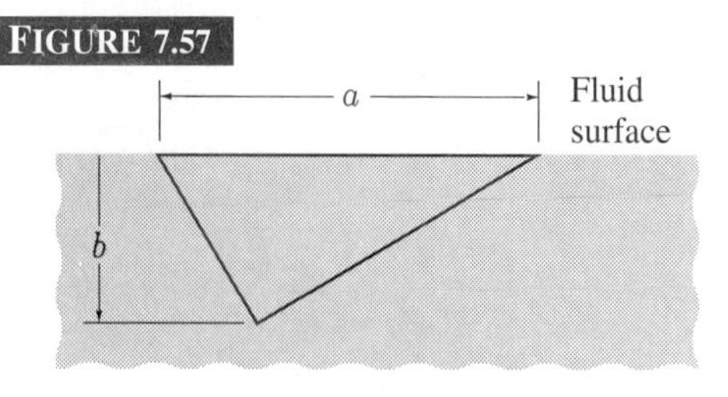

11. Set up (but do not evaluate) a definite integral(s) to find the force due to water pressure on each side of the flat vertical plate in Figure 7.58.

FIGURE 7.58

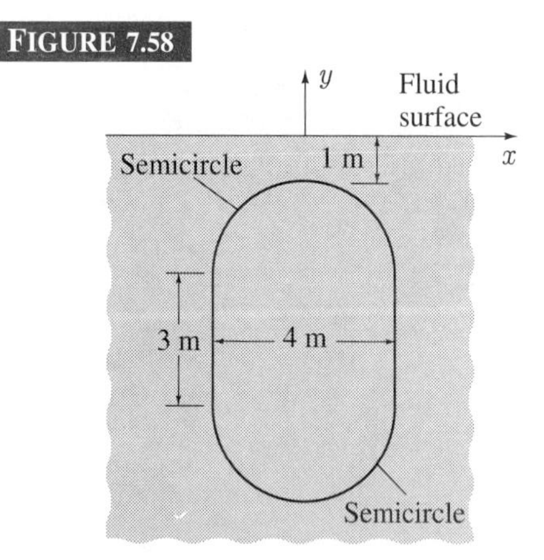

12. In Exercise 6, find the force due to water pressure on each part of the bottom of the pool.

13. The bow of a landing barge (Figure 7.59) consists of a rectangular flat plate A metres wide and B metres long. When the barge is stationary, this plate makes an angle of $\pi/6$ radians with the surface of the water. Find the maximum force of the water on the bow.

FIGURE 7.59

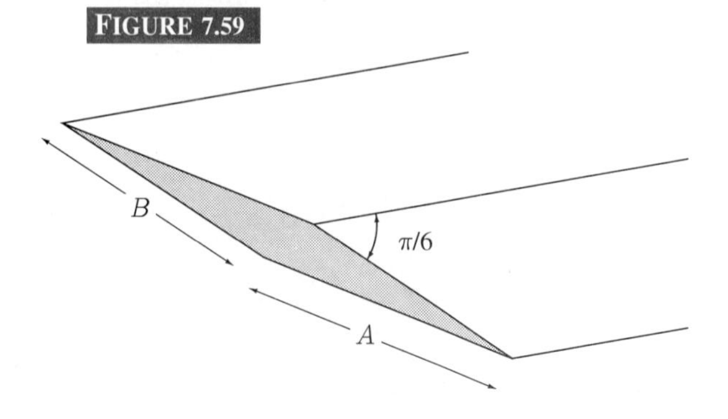

14. A water tank is to be built in the form of a rectangular box with all six sides welded along their joins (Figure 7.60). Sides $ABCD$ and $EFGH$ are to be square, and all sides except $ABCD$ are supported from the outside. If the maximum force that $ABCD$ can withstand is 20 000 newtons, what is the largest cross-section that can be built, assuming that at some stage the tank will be full?

FIGURE 7.60

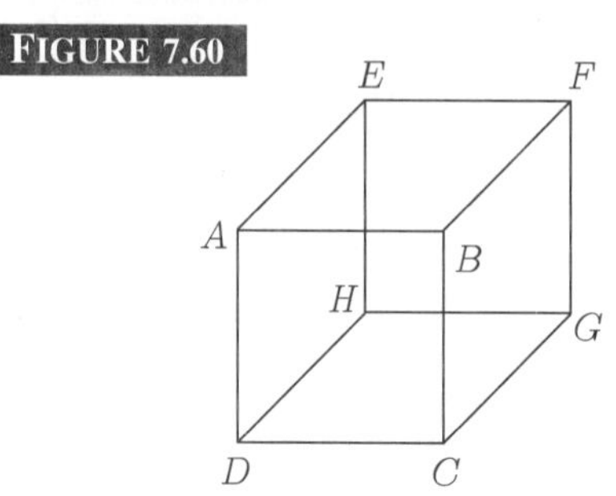

15. A cylindrical oil tank of radius r and length h has its axis horizontal. If the density of the oil is ρ, find the force on each end of the tank when it is half full.

16. A right-circular cylinder of radius r and height h is immersed in a fluid of density ρ with its axis vertical. Show that the buoyant force on the cylinder due to the pressure of the fluid is equal to the weight of the fluid displaced. This is known as *Archimedes' principle*, and is valid for an object of any shape.

17. The lower half of a cubical tank 2 m on each side is occupied by water, and the upper half by oil (density $0.90\ \text{gm/cm}^3$).

(a) What is the force on each side of the tank due to the pressure of the water and the oil?

(b) If the oil and water are stirred to create a uniform mixture, does the force on each side change? If not explain why not. If so, by how much does it increase or decrease?

18. Archimedes' principle states that the buoyant force on an object when immersed or partially immersed in a fluid is equal to the weight of the fluid displaced by the object.

(a) Show that if an object floats partially submerged in water, the percentage of the volume of the object above water is

$$100\,\frac{\rho_w - \rho_o}{\rho_w},$$

where ρ_o and ρ_w are the densities of the object and water respectively.

(b) If the densities of ice and water are 915 and 1000 kg/m^3 respectively, show that only 8.5% of the volume of an iceberg is above water.

19. If a full tube of mercury is inverted in a large container of mercury, the level of mercury in the tube will fall, but it will stabilize at a point higher than that in the container (Figure 7.61). This is due to the fact that air pressure acts on the surface of the mercury in the container but not on the surface of the mercury in the tube. The extra column of mercury, of height h, creates a force at A that counteracts the atmospheric pressure transmitted through the mercury in the container to the tube so that the total pressure at A is equal to the total pressure at B.

FIGURE 7.61

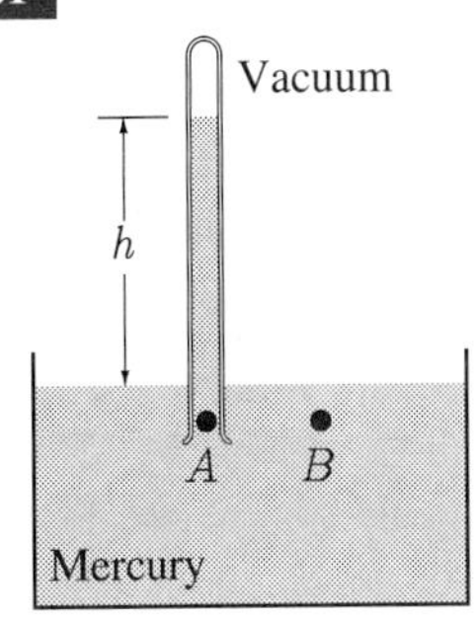

(a) Show that if the density of mercury is 13.6 gm/mL, then the atmospheric pressure at the surface of the mercury in the container is $1.33 \times 10^5 h$ N/m^2, provided h is measured in metres.

(b) If h is measured as 761 millimetres, find the atmospheric pressure.

20. (a) A block of wood (density 0.40 gm/cm^3) is cubical (0.25 m on each side). If it floats in water, how deep is its lowest point below the surface? Refer to Archimedes' principle in Exercise 18.

(b) Repeat part (a) given that the block is a sphere of radius 0.25 m.

21. A tank is to be built in the form of a right-circular cylinder with horizontal axis. The ends are to be joined to the cylindrical side by a continuous weld. One end of the tank and the cylindrical side are supported from the outside. The remaining unsupported end can withstand a total force of 40 000 N less 1000 N for each metre of weld on that end. What is the maximum radius for the tank if it is to hold a fluid weighing 1.019×10^3 N/m^3?

22. Find the ratio L/R such that the forces due to fluid pressure on the rectangular and semicircular parts of the plate in Figure 7.62 are equal.

FIGURE 7.62

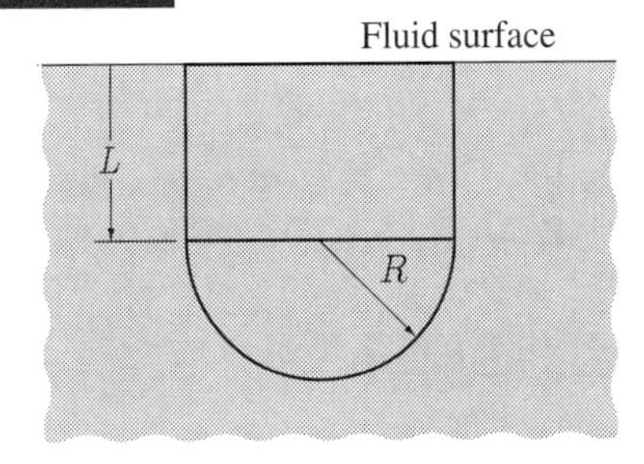

SECTION 7.6

Centres of Mass and Centroids

Everyone is acquainted with the action of a teeter-totter or seesaw. Two children of unequal masses can pass many hours rocking, provided the child with greater mass sits closer to the fulcrum. In this section we discuss the mathematics of the seesaw. This requires a definition of "moments of masses", and moments lead to the idea of the "centre of mass" of distributions of masses, lumped or continuous.

To discuss the mathematics of a seesaw, we consider in Figure 7.63 a uniform seesaw of length $2L$ balanced at its centre, with a child of mass m at one end. If a second child of equal mass is placed at the other end, the ideal seesaw situation is created. If, however, the mass of the second child is $M > m$, then this child must be placed somewhat closer to the fulcrum. To find the exact position, we must determine what might be called *rocking power* of a mass. A little experimentation shows that when $M = 2m$, M must be placed halfway between the end and the fulcrum; when $M = 3m$, M must be placed a distance $L/3$ from the fulcrum; and in general, when $M = am$ $(a > 1)$, M must be placed L/a from the fulcrum. Now the rocking power of the child of mass m is constant, and for each mass $M = am$ we have found an equal and opposite rocking power if M is placed at L/a.

FIGURE 7.63

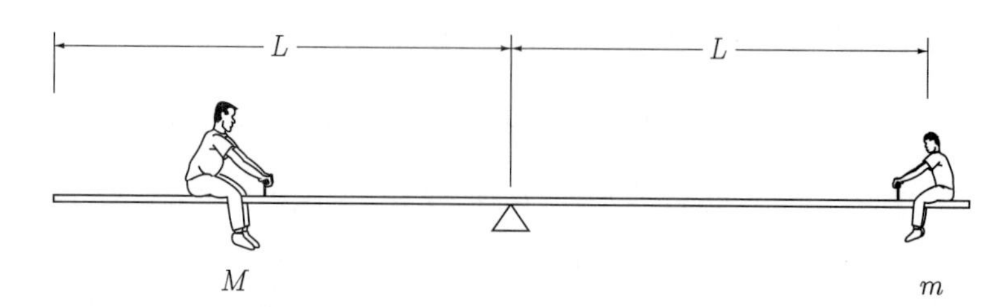

Clearly, then, rocking power depends on both mass and distance from the fulcrum. A little thought shows that the mathematical quantity that remains constant for the various masses $M = am$ is the product of M and distance to the fulcrum; in each case, this product is $(am)(L/a) = mL$, the same product as for the child of mass m. It would appear then, that rocking power should be defined as the product of mass and distance. We do this in the following definition, and at the same time give rocking power a new name.

Definition 7.1

The **first moment** of a point mass m about a point P is the product md where d is the directed distance from P to m.

If directed distances to the right of point P in Figure 7.64 are chosen as positive and distances to the left are negative, then d_1 is positive and d_2 is negative. Mass m_1 has a positive first moment $m_1 d_1$ about P, and m_2 has a negative first moment $m_2 d_2$.

FIGURE 7.64

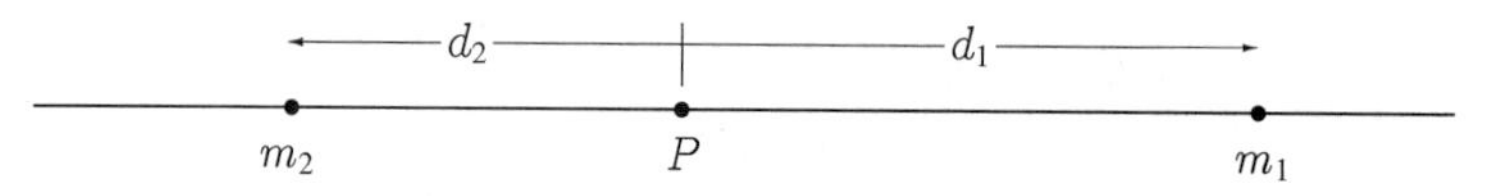

In Figure 7.65, we have placed five children of masses m_1, m_2, m_3, m_4, and m_5 on the same seesaw. A sixth child of mass m_6 is to be placed somewhere on the seesaw so that all six children form the ideal seesaw.

FIGURE 7.65

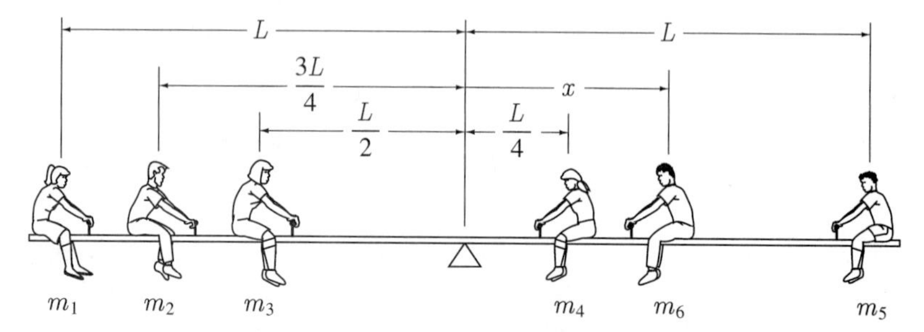

To find the appropriate position for the sixth child, we let x be the directed distance from the fulcrum to the point where this child should be placed. The total first moment of all six children about the fulcrum, choosing distances to the right as positive and to the left as negative, is

$$m_1(-L) + m_2(-3L/4) + m_3(-L/2) + m_4(L/4) + m_5(L) + m_6(x).$$

We regard this as the resultant first moment of all six children attempting to turn the seesaw — clockwise if the moment is positive, counterclockwise if the moment is negative. Balance occurs if this resultant first moment is zero:

$$0 = -m_1 L - \frac{3}{4} m_2 L - \frac{1}{2} m_3 L + \frac{1}{4} m_4 L + m_5 L + m_6 x.$$

We may solve this equation for the position of m_6:

$$x = \frac{L}{4m_6}(4m_1 + 3m_2 + 2m_3 - m_4 - 4m_5).$$

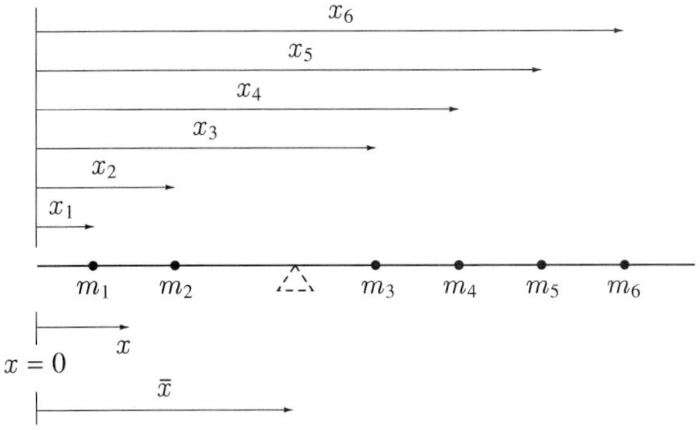

FIGURE 7.66

We now turn this problem around and place the six children at distances from the left end as shown in Figure 7.66. If the mass of the seesaw itself is neglected, where should the fulcrum be placed in order to create the ideal seesaw? To solve this problem, we let the distance that the fulcrum should be placed from the left end be represented by $\overline{x}$. In order for balance to occur, the total first moment of all six children about the fulcrum must vanish; hence,

$$0 = m_1(x_1 - \overline{x}) + m_2(x_2 - \overline{x}) + m_3(x_3 - \overline{x}) + m_4(x_4 - \overline{x}) + m_5(x_5 - \overline{x}) + m_6(x_6 - \overline{x}).$$

The solution of this equation is

$$\overline{x} = \frac{m_1x_1 + m_2x_2 + m_3x_3 + m_4x_4 + m_5x_5 + m_6x_6}{m_1 + m_2 + m_3 + m_4 + m_5 + m_6} = \frac{1}{M}\sum_{i=1}^{6} m_ix_i, \quad (7.25)$$

where $M = \sum_{i=1}^{6} m_i$ is the total mass of all six children. This point $\overline{x}$ at which the fulcrum creates a balancing position is called the *centre of mass* for the six children. It is a point where masses to the right are balanced by masses to the left.

In the remainder of this section we extend the idea of a centre of mass of point masses along a line (the seesaw) to the centre of mass of a distribution of point masses in a plane, and then to the centre of mass of a continuous distribution of mass. Our first step is to define the first moment of a point mass in a plane about a line in the plane (Figure 7.67).

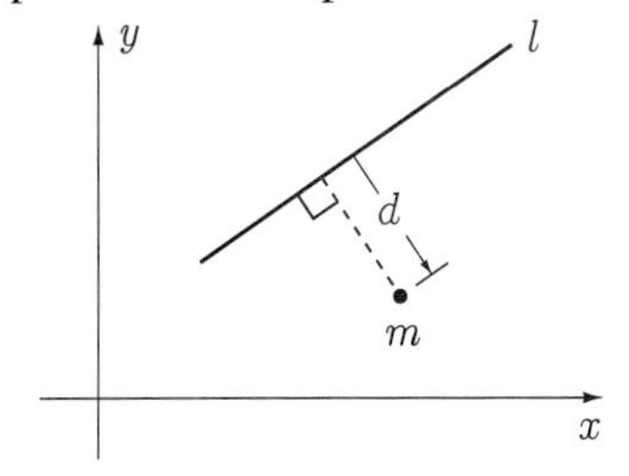

FIGURE 7.67

Definition 7.2

The first moment of a mass m about a line l is md where d is the directed distance from l to m.

Once again directed distances are used in calculating first moments, and therefore distances on one side of the line must be chosen as positive and distances on the other side as negative. For vertical and horizontal lines, there is a natural convention for doing this. Distances to the right of a vertical line are chosen as positive, and distances to the left are negative. Distance upward from a horizontal line are positive, and distances downward are negative. In particular, when a mass m is located at position (x, y) in the xy-plane, its first moments about the x- and y-axes are my and mx, respectively.

The first moments of a system of n point masses $m_1, m_2, \ldots, m_n$ located at points $(x_1, y_1), (x_2, y_2), \ldots, (x_n, y_n)$ respectively (Figure 7.68(a)) about the x- and y-axes are defined as the sums of the first moments of the individual masses about these lines:

$$\text{First moment of system about } x\text{-axis} = \sum_{i=1}^{n} m_i y_i \quad \text{and} \tag{7.26a}$$

$$\text{First moment of system about } y\text{-axis} = \sum_{i=1}^{n} m_i x_i. \tag{7.26b}$$

FIGURE 7.68

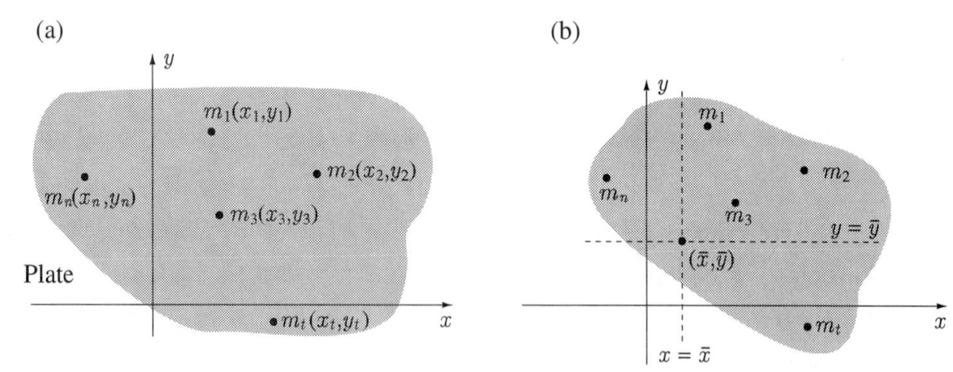

What is the physical meaning of these first moments? Do they, for instance, play the same role that first moments did for the seesaw? To see this we imagine that each point mass is embedded in a thin plastic plate in the xy-plane. The plate itself is massless and extends to include all n masses m_i. Picture now that the plate is horizontal, and a sharp edge is placed along the y-axis. Does the plate rotate about this edge, or does it balance? It is clear that the plate balances if the first moment of the system about the y-axis is equal to zero, and rotates otherwise, the direction depending on whether the first moment is positive or negative. Similarly, the plate balances on a sharp edge placed along the x-axis only if the first moment of the system about the x-axis vanishes. In general, the plate balances along any straight edge only if the first moment of the system about that edge vanishes.

We defined the centre of mass for a given distribution of children on the seesaw as the point at which to place the fulcrum in order to obtain balance. Analogously, we define the *centre of mass* $(\overline{x}, \overline{y})$ of the distribution of point masses in Figure 7.68(a) as the position to place a sharp point in order to obtain balance. Remember that the plastic itself is massless and only the point masses m_i can create moments. Balance occurs at a point $(\overline{x}, \overline{y})$ only if balance occurs about every straight line through $(\overline{x}, \overline{y})$. In particular, balance must occur about the lines $x = \overline{x}$ and $y = \overline{y}$ parallel to the y- and x-axes as in Figure 7.68(b). Since balance occurs about $x = \overline{x}$ only if the total first moment of the system about this line vanishes, we obtain the condition that

$$0 = m_1(x_1 - \overline{x}) + m_2(x_2 - \overline{x}) + \cdots + m_n(x_n - \overline{x}),$$

which can be solved for $\overline{x}$,

$$\overline{x} = \frac{1}{M}\sum_{i=1}^{n} m_i x_i, \tag{7.27}$$

where $M = \sum_{i=1}^{n} m_i$ is the total mass of the system. Note that this equation is identical to 7.25. Similarly, for balance about $y = \overline{y}$, we find that $\overline{y}$ must be

$$\overline{y} = \frac{1}{M} \sum_{i=1}^{n} m_i y_i. \tag{7.28}$$

We have obtained a unique point $(\overline{x}, \overline{y})$ based on conditions of balance about the lines $x = \overline{x}$ and $y = \overline{y}$. Does this necessarily imply that balance occurs about every straight line through $(\overline{x}, \overline{y})$? The answer is yes, and the proof is found in Exercise 34.

We have now proved that every planar point mass distribution has a centre of mass $(\overline{x}, \overline{y})$, where $\overline{x}$ and $\overline{y}$ are defined by equations 7.27 and 7.28. Our derivation has shown that the first moment of the system about any line through $(\overline{x}, \overline{y})$ must be equal to zero. Are there any other significant features of this point? To answer this question we note that if a fictitious particle of mass M (the total mass of the system) is located at the centre of mass $(\overline{x}, \overline{y})$, then its first moment about the y-axis is $M\overline{x}$. But from equation 7.27, we have

$$M\overline{x} = \sum_{i=1}^{n} m_i x_i, \tag{7.29}$$

and we conclude that the first moment of this fictitious particle M about the y-axis is exactly the same as the first moment of the system about the y-axis. Similarly, the first moment of M about the x-axis is $M\overline{y}$, and from 7.28,

$$M\overline{y} = \sum_{i=1}^{n} m_i y_i. \tag{7.30}$$

Thus, the centre of mass of a system of point masses m_i is a point at which a single particle of mass $M = \sum_{i=1}^{n} m_i$ has the same first moments about the x- and y-axes as the system. It can be shown further (Exercise 34) that the first moment of M about any line is the same as the first moment of the system about that line.

In summary, we defined the centre of mass of a system of point masses as a balance point. We found as a result that the centre of mass is a point at which a single particle of mass equal to the total mass of the system has the same first moment about any line as the system itself. This argument is reversible. Were we to define the centre of mass as a point to place the mass of the system for equivalent first moments, it would be a balance point. In other words, we have two equivalent definitions of the centre of mass of a system of point masses—a balance point or an equivalent point for first moments.

We now make the transition from a discrete system of particles to a continuous distribution of mass in the form of a thin plate of constant mass per unit area ρ (Figure 7.69). In order to find the mass of the plate, we proceed in exactly the same way that we did for areas. We divide the plate into vertical rectangles, the mass in a representative rectangles of width dx at position x being

$$\rho[f(x) - g(x)]\,dx.$$

To find the total mass of the plate, we add over all such rectangles of ever-diminishing widths to obtain

$$M = \int_a^b \rho[f(x) - g(x)]\,dx. \tag{7.31}$$

Based on our discussion for systems of point masses, we define the centre of mass of a continuous distribution of mass as that point $(\overline{x}, \overline{y})$ where a particle of mass M

has the same first moments about the x- and y-axes as the distribution. In algebraic terms, we note that $M\overline{x}$ is the first moment about the y-axis of a particle of mass M at $(\overline{x}, \overline{y})$. To this we must equate the first moment of the original distribution about the y-axis. Now each point in the representative rectangle in Figure 7.69 is approximately the same distance x from the y-axis — approximately, because the rectangle does have finite, though very small, width. The first moment, then, of this rectangle about the y-axis is approximately

$$x\rho[f(x) - g(x)]\,dx.$$

FIGURE 7.69

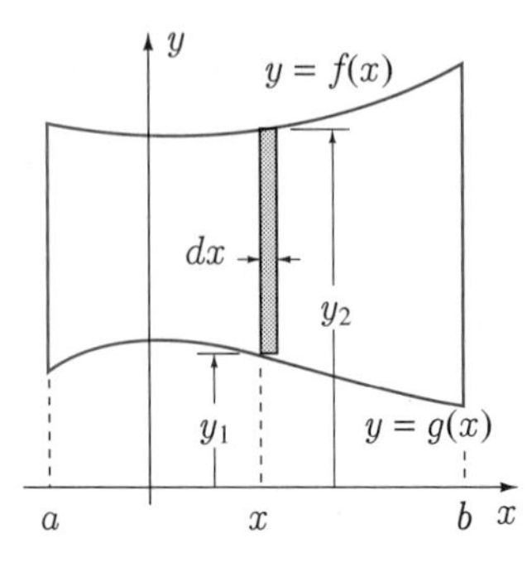

We find the first moment of the entire plate about the y-axis by adding first moments of all such rectangles and taking the limit as their widths approach zero. But once again this is the process defined by the definite integral, and we obtain, therefore, for the first moment of the plate about the y-axis

$$\int_a^b x\rho[f(x) - g(x)]\,dx.$$

Consequently,

$$M\overline{x} = \int_a^b x\rho[f(x) - g(x)]\,dx, \qquad (7.32)$$

and this equation can be solved for $\overline{x}$ once M and the integral on the right have been evaluated. Equation 7.32 represents for continuous distributions what equation 7.27 does for discrete distributions.

To find $\overline{y}$ we must equate the product $M\overline{y}$ to the first moment of the plate about the x-axis. If we consider the representative rectangle in Figure 7.69 we see that not all points therein are the same distance from the x-axis. To circumvent this problem, we consider all of the mass of the rectangle to be concentrated at its centre of mass. Since the centre of mass is the midpoint of the rectangle — a point distant $[f(x) + g(x)]/2$ from the x-axis, it follows that the first moment of this rectangle about the x-axis is

$$\frac{1}{2}[f(x) + g(x)]\rho[f(x) - g(x)]\,dx.$$

The total first moment of the plate about the x-axis is the definite integral of this expression, and we set

$$M\overline{y} = \int_a^b \frac{1}{2}[f(x) + g(x)]\rho[f(x) - g(x)]\,dx. \qquad (7.33)$$

This equation is used to evaluate $\overline{y}$.

Equations 7.32 and 7.33 can be memorized as formulas for $\overline{x}$ and $\overline{y}$, but it is far easier to perform the above operations mentally and arrive at these equations. Besides, for various shapes of plates, we might use horizontal rectangles or combinations of horizontal and vertical rectangles, and in such cases 7.32 and 7.33 would have to be modified.

On the basis of definitions 7.32 and 7.33 for $(\overline{x}, \overline{y})$, we can show that the first moment of M at $(\overline{x}, \overline{y})$ about any line is the same as the first moment of the plate about that same line. This implies that the plate balances along any line through $(\overline{x}, \overline{y})$ and therefore at $(\overline{x}, \overline{y})$.

EXAMPLE 7.21

Find the centre of mass of a thin plate of constant mass per unit area ρ if its edges are defined by the curves

$$y = 2 - x^2, \qquad y = 0, \qquad x = 0, \qquad x \geq 0.$$

SOLUTION

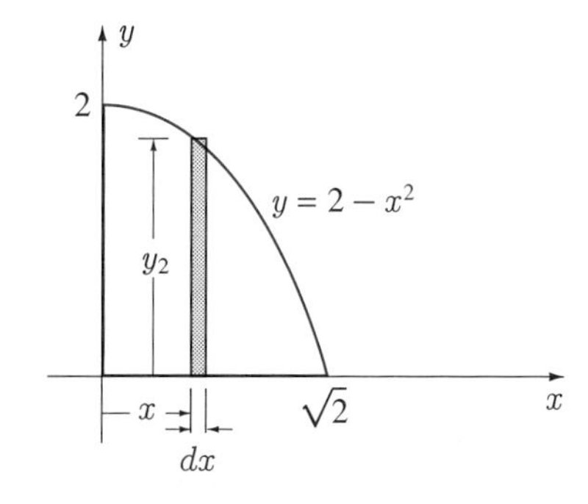

FIGURE 7.70

Using vertical rectangles (Figure 7.70), we find

$$M = \int_0^{\sqrt{2}} \rho y \, dx = \rho \int_0^{\sqrt{2}} (2 - x^2) \, dx = \rho \left\{ 2x - \frac{x^3}{3} \right\}_0^{\sqrt{2}} = \frac{4\sqrt{2}\rho}{3}.$$

If $(\overline{x}, \overline{y})$ is the centre of mass of the plate, then $M\overline{x}$ is the first moment of the single particle of mass M about the y-axis. This must be equated to the first moment of the plate about the y-axis. Since $x\rho(2 - x^2) \, dx$ is approximately the first moment of the rectangle in Figure 7.70 about the y-axis, the following integral gives the first moment of the plate about the y-axis,

$$\int_0^{\sqrt{2}} x\rho(2 - x^2) \, dx = \rho \int_0^{\sqrt{2}} (2x - x^3) \, dx = \rho \left\{ x^2 - \frac{x^4}{4} \right\}_0^{\sqrt{2}} = \rho.$$

Hence we set $M\overline{x} = \rho$, and solve this equation for $\overline{x}$,

$$\overline{x} = \frac{\rho}{M} = \rho \frac{3}{4\sqrt{2}\rho} = \frac{3}{4\sqrt{2}}.$$

To find $\overline{y}$, we calculate the first moment of the plate about the x-axis. Since the centre of mass of the rectangle in Figure 7.70 is $y/2$ units above the x-axis, it follows that the first moment of this rectangle about the x-axis is $(y/2)\rho y \, dx$. When we integrate this to find the first moment of the plate about the x-axis, and equate it to $M\overline{y}$, the result is

$$M\overline{y} = \int_0^{\sqrt{2}} \frac{y}{2}\rho y\, dx = \frac{\rho}{2}\int_0^{\sqrt{2}} (2 - x^2)^2\, dx = \frac{\rho}{2}\int_0^{\sqrt{2}} (4 - 4x^2 + x^4)\, dx$$

$$= \frac{\rho}{2}\left\{4x - \frac{4x^3}{3} + \frac{x^5}{5}\right\}_0^{\sqrt{2}} = \frac{16\sqrt{2}\rho}{15}.$$

Thus,

$$\overline{y} = \frac{16\sqrt{2}\rho}{15}\frac{3}{4\sqrt{2}\rho} = \frac{4}{5}.$$

■

EXAMPLE 7.22

Find the centre of mass of a thin plate of constant mass per unit area ρ if its edges are defined by the curves

$$y = 2x - x^2, \qquad y = x^2 - 4.$$

SOLUTION

FIGURE 7.71

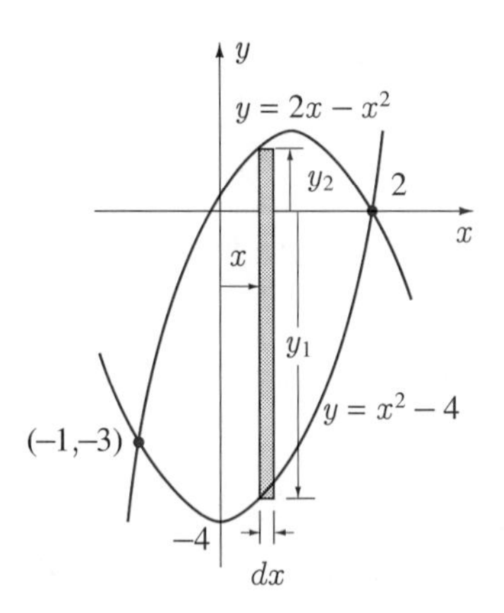

Using vertical rectangles (Figure 7.71),

$$M = \int_{-1}^{2} \rho(y_2 - y_1)\, dx = \rho\int_{-1}^{2} [(2x - x^2) - (x^2 - 4)]\, dx$$

$$= \rho\int_{-1}^{2} (4 + 2x - 2x^2)\, dx = \rho\left\{4x + x^2 - \frac{2x^3}{3}\right\}_{-1}^{2} = 9\rho.$$

If $(\overline{x}, \overline{y})$ is the centre of mass of the plate, then first moments of the plate and M about the y-axis give

$$M\overline{x} = \int_{-1}^{2} x\rho(y_2 - y_1)\, dx = \rho\int_{-1}^{2} (4x + 2x^2 - 2x^3)\, dx$$

$$= \rho\left\{2x^2 + \frac{2x^3}{3} - \frac{x^4}{2}\right\}_{-1}^{2} = \frac{9\rho}{2}.$$

Thus,

$$\overline{x} = \frac{9\rho}{2}\frac{1}{9\rho} = \frac{1}{2}.$$

To find $\overline{y}$, we use moments about the x-axis to write

$$\begin{aligned}
M\overline{y} &= \int_{-1}^{2} \frac{1}{2}(y_1 + y_2)\rho(y_2 - y_1)\,dx = \frac{\rho}{2}\int_{-1}^{2}(y_2^2 - y_1^2)\,dx \\
&= \frac{\rho}{2}\int_{-1}^{2}[(2x - x^2)^2 - (x^2 - 4)^2]\,dx = 2\rho\int_{-1}^{2}(-x^3 + 3x^2 - 4)\,dx \\
&= 2\rho\left\{-\frac{x^4}{4} + x^3 - 4x\right\}_{-1}^{2} = -\frac{27\rho}{2}.
\end{aligned}$$

Consequently,

$$\overline{y} = -\frac{27\rho}{2}\frac{1}{9\rho} = -\frac{3}{2}.$$

■

EXAMPLE 7.23

Find the first moments of a thin plate of constant mass per unit area ρ about the lines (a) $y = 0$, (b) $y = -2$, and (c) $x = -2$, if its edges are defined by the curves

$$x = |y|^3, \qquad x = 2 - y^2.$$

SOLUTION (a) Since the mass is distributed symmetrically about the x-axis (Figure 7.72), the first moment about $y = 0$ is zero.

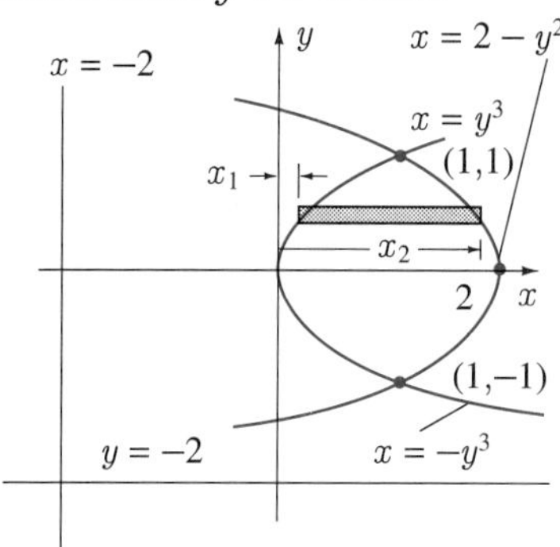

FIGURE 7.72

(b) Since the centre of mass of the plate is on the x-axis, its first moment about the line $y = -2$ is

$$\begin{aligned}
(2)(\text{mass of plate}) &= 2(2)(\text{mass of plate above the } x\text{-axis}) \\
&= 4\int_0^1 \rho(x_2 - x_1)\,dy = 4\rho\int_0^1 (2 - y^2 - y^3)\,dy \\
&= 4\rho\left\{2y - \frac{y^3}{3} - \frac{y^4}{4}\right\}_0^1 \\
&= \frac{17\rho}{3}.
\end{aligned}$$

(c) The first moment of the plate about the line $x = -2$ is twice the first moment of its upper half. Since the x-coordinate of the centre of mass of the representative rectangle in Figure 7.72 is $(x_1 + x_2)/2$, the distance from the line $x = -2$ to the centre of mass of the rectangle is $2 + (x_1 + x_2)/2$. The first moment of the horizontal rectangle about $x = -2$ is therefore

$$\left(2 + \frac{x_1 + x_2}{2}\right) \rho(x_2 - x_1)\, dy,$$

and the first moment of the plate is

$$\begin{aligned} 2\int_0^1 \left(2 + \frac{x_1 + x_2}{2}\right) \rho(x_2 - x_1)\, dy &= \rho \int_0^1 [4(x_2 - x_1) + (x_2^2 - x_1^2)]\, dy \\ &= 4\rho \int_0^1 (x_2 - x_1)\, dy \\ &\quad + \rho \int_0^1 [(2 - y^2)^2 - (y^3)^2]\, dy \\ &= \frac{17\rho}{3} + \rho \int_0^1 (4 - 4y^2 + y^4 - y^6)\, dy \\ &= \frac{17\rho}{3} + \rho \left\{4y - \frac{4y^3}{3} + \frac{y^5}{5} - \frac{y^7}{7}\right\}_0^1 \\ &= \frac{881\rho}{105}. \end{aligned}$$

■

It has become apparent through our discussions and examples that in calculating the centre of mass of a thin plate with constant mass per unit area ρ, ρ is really unnecessary. As a constant it is taken out of each integration and cancels in the final division. The location of the centre of mass depends only on the geometric shape of the plate, and for this reason we could replace all references to mass by area. In particular, the mass of the plate M can be replaced by its area A, first moments (of mass) $M\overline{x}$ and $M\overline{y}$ can be replaced by first moments (of area) $A\overline{x}$ and $A\overline{y}$, and equations 7.32 and 7.33 then take the form

$$A\overline{x} = \int_a^b x[f(x) - g(x)]\, dx, \qquad (7.34)$$

$$A\overline{y} = \int_a^b \frac{1}{2}[f(x) + g(x)][f(x) - g(x)]\, dx. \qquad (7.35)$$

It is customary when using first moments of area to call $(\overline{x}, \overline{y})$ the centroid of the area rather than the centre of mass of the plate, simply because all references to mass have been deleted. We emphasize, however, that the statements in this paragraph apply only when mass per unit area is constant.

EXERCISES 7.6

In Exercise 1–5 find the centre of mass of the thin plate with constant mass per unit area.

1.

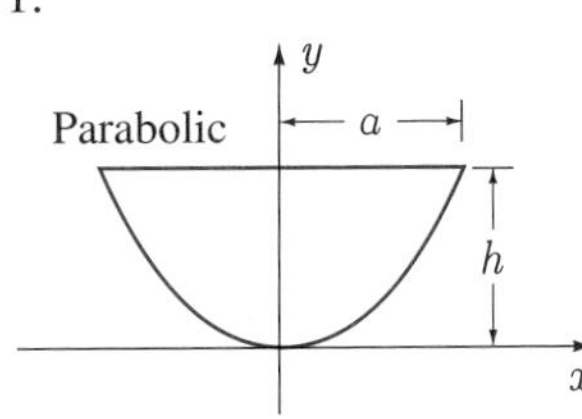

2.

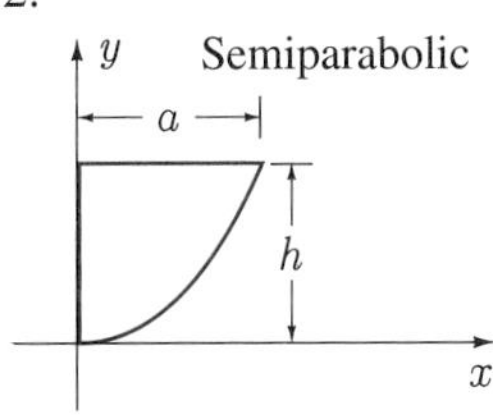

3.

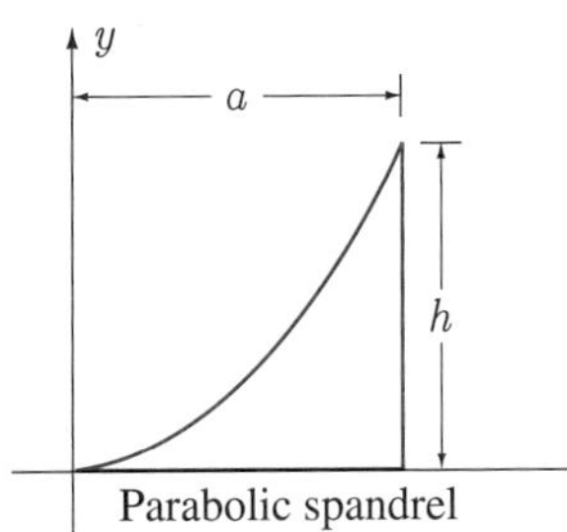

4.

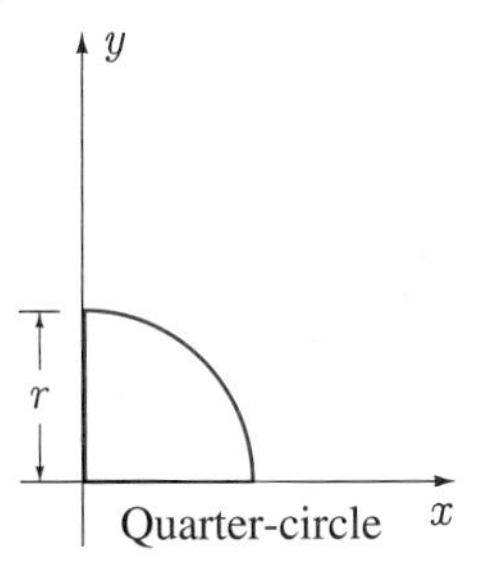

5.

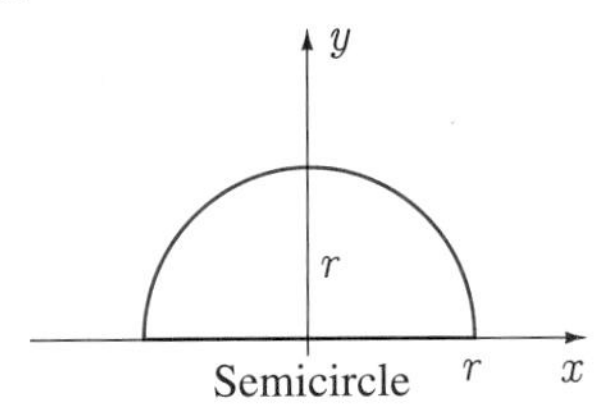

In Exercise 6–13 find the centroid of the area bounded by the curves.

6. $y = x^2 - 1, \quad y = -x^2 - 2x - 1$

7. $y = \sqrt{|x|}, \quad y = 2 - x^2$

8. $y = x^3, \quad x = y^3$

9. $y = x, \quad y = 2x, \quad 2y = x + 3$

10. $x = y^2 - 2y, \quad x + y = 12$

11. $y = \sqrt{2 - x}, \quad x + y = 2$

12. $x = 4y - 4y^2, \quad y = x - 3, \quad y = 1, \quad y = 0$

13. $x^3 y = 8, \quad y = 9 - x^3$

14. The edges of a thin plate with constant mass per unit area ρ are defined by the curves

$$y = |x|^{1/2}, \qquad y = x + 2, \qquad y = 2 - x.$$

Find its first moment about the line $x = -5$.

15. Show that the centroids of areas A and B in Figure 7.73 have coordinates:

$$\bar{x}_A = \left(\frac{n+1}{n+2}\right)a, \quad \bar{y}_A = \left(\frac{n+1}{4n+2}\right)b,$$

$$\bar{x}_B = \left(\frac{n+1}{2n+4}\right)a, \quad \bar{y}_B = \left(\frac{n+1}{2n+1}\right)b.$$

FIGURE 7.73

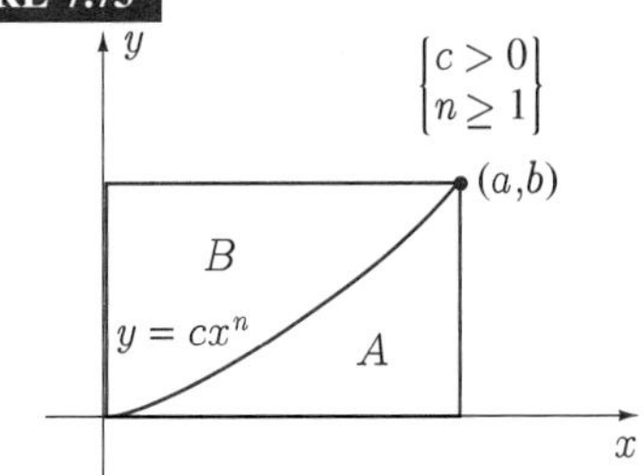

16. Find the centre of mass of the seesaw in Figure 7.66 if the mass of the seesaw is not neglected. Assume that it has uniform mass per unit length ρ and length $2L$.

In Exercises 17–19 find the centroid of the area bounded by the curves.

17. $x = \sqrt{y + 2}, \quad y = x, \quad y = 0$

18. $y + x^2 = 0, \quad x = y + 2, \quad x + y + 2 = 0, \quad y = 2$ (above $y + x^2 = 0$)

19. $y = \sqrt{2 - x}, \quad 15y = x^2 - 4$

20. Find the centre of mass of the thin plate in Figure 7.74 if it has constant mass per unit area.

FIGURE 7.74

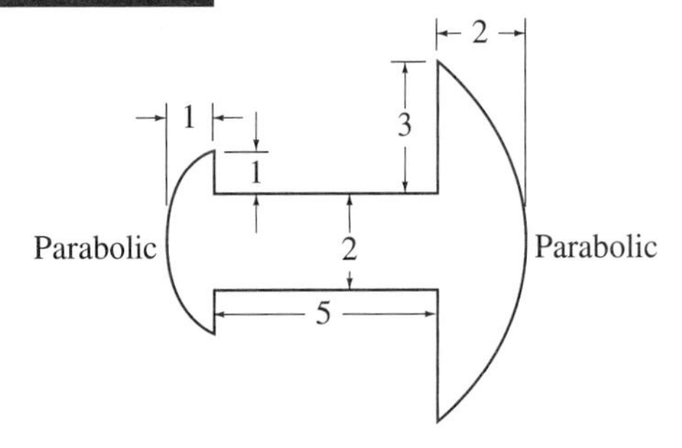

21. Find the first moment about the line $x + y = 1$ for the area defined by the curves

$$x + 2 = y^2, \qquad y = x.$$

22. If an area A can be subdivided into n subareas A_i $(i = 1, \ldots, n)$ such that the centroid of each A_i is $(\overline{x}_i, \overline{y}_i)$, show that the centroid $(\overline{x}, \overline{y})$ of A is given by

$$\overline{x} = \frac{1}{A}\sum_{i=1}^{n} A_i\overline{x}_i, \qquad \overline{y} = \frac{1}{A}\sum_{i=1}^{n} A_i\overline{y}_i.$$

In Exercises 23–26 use the technique suggested in Exercise 22 to find the centroid of the area.

23.

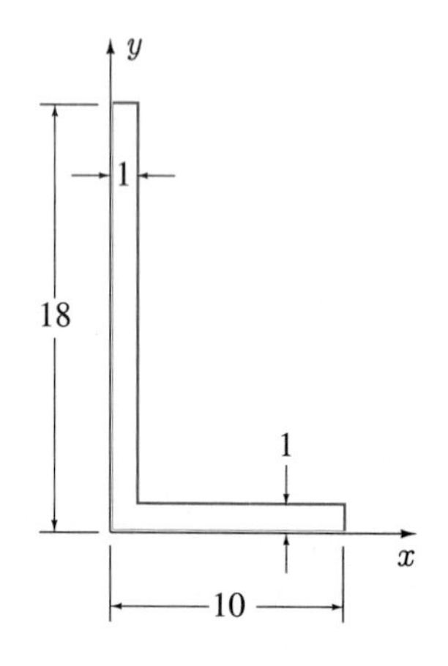

24.

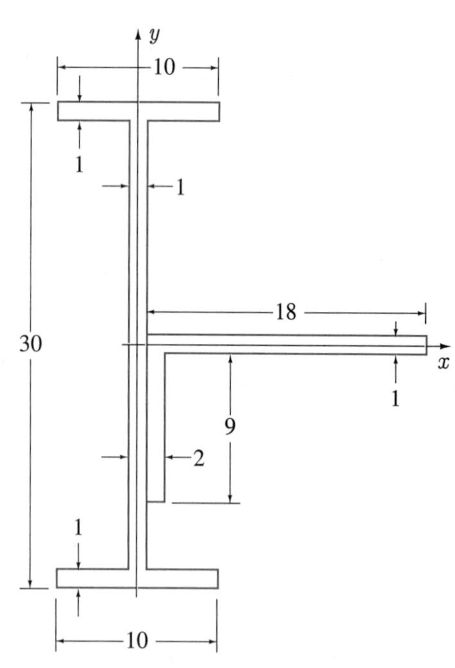

25.

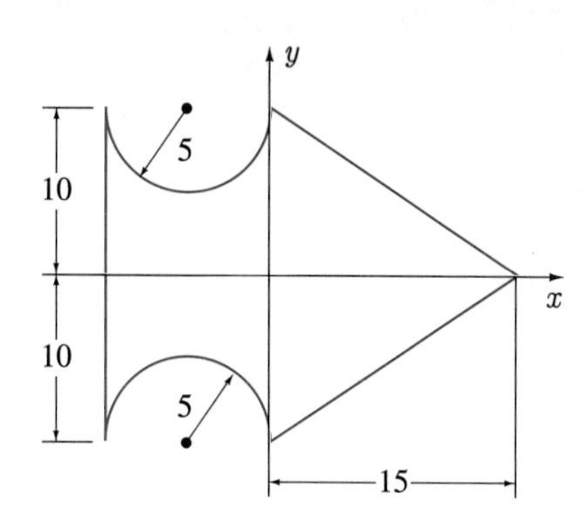

26.

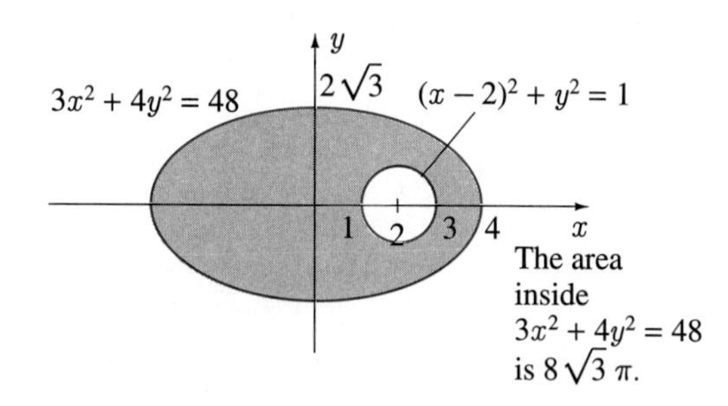

In Exercises 27–30 find coordinates of the centroid of the area accurate to three decimal places.

27. $y = x^3 + 3x^2 + 2x + 1, \quad x = 0, \quad y = 0$

28. $y = x^4 - 5x^2 + 5, \quad y = 0$ (above the x-axis)

29. $x = y^3 - y^2 - 2y, \quad x = \sqrt{2y + 1}$

30. $y = x^3 - x, \quad y = \sqrt{x}$

31. Prove the following *theorem of Pappus*: If a plane area is revolved about a coplanar axis not crossing the area, the volume generated is equal to the product of the area and the circumference of the circle described by the centroid of the area.

32. Use the result of Exercise 31 to find the volume of the donut in Exercise 41 of Section 7.2.

33. Use the result of Exercise 31 to find the volume in Exercise 45 of Section 7.2.

34. (a) Show that the first moment of the system of point masses in Figure 7.68(b) about any line is the same as the first moment of a point mass $M = \sum_{i=1}^{n} m_i$ at $(\overline{x}, \overline{y})$ about that line. Hint: See Exercise 45 in Section 1.3 for a formula for the distance from a point to a line.

(b) Does it follow that the first moment of the system about any line through $(\overline{x}, \overline{y})$ is zero?

35. A thin flat plate of area A is immersed vertically in a fluid of density ρ. Show that the total force due to fluid pressure on each side of the plate is equal to the product of 9.81ρ, A, and the depth of the centroid of the plate below the surface of the fluid.

36. Show that centroid of a triangle with vertices (x_1, y_1), (x_2, y_2), and (x_3, y_3) is

$$\overline{x} = \frac{x_1 + x_2 + x_3}{3}, \qquad \overline{y} = \frac{y_1 + y_2 + y_3}{3}.$$

SECTION 7.7

Moments of Inertia

Newton's second law $F = ma$ is fundamental to the study of translational motion of bodies. For rotational motion of bodies, its counterpart is $\tau = I\alpha$ where τ is torque, α is angular acceleration, and I is the "moment of inertia" of the body. The kinetic energy of a body of mass m moving with velocity v is $mv^2/2$. The kinetic energy of a body rotating with angular velocity ω is $I\omega^2/2$. Thus, for rotational motion, there is a quantity called the "moment of inertia" of a body which is analogous to mass in translational equations. In this section we define and calculate moments of inertia.

To define moments of inertia of bodies, we begin with the moment of inertia of a point mass.

Definition 7.3

The **moment of inertia** or **second moment** of a point mass m about a line l (Figure 7.67) is the product md^2 where d is the directed distance from l to m.

In particular, if m is at position (x, y) in the xy-plane, its moments of inertia about the x- and y-axes are my^2 and mx^2 respectively. For a system of n particles of masses m_1, m_2, ..., m_n located at points (x_1, y_1), (x_2, y_2), ..., (x_n, y_n) as in Figure 7.68(a), the moments of inertia of the system about the x- and y-axes are the sums of the moments of inertia of the particles about the x- and y-axes:

$$\text{Moment of inertia about } x\text{-axis} = \sum_{i=1}^{n} m_i y_i^2, \tag{7.36a}$$

$$\text{Moment of inertia about } y\text{-axis} = \sum_{i=1}^{n} m_i x_i^2. \tag{7.36b}$$

The transition from the discrete case to a continuous distribution in the form of a thin plate with constant mass per unit area ρ is not always so simple as for first moments. First consider the moment of inertia of the plate in Figure 7.69 in Section 7.6 about the y-axis. The mass of the representative rectangle shown is

$$\rho[f(x) - g(x)]\,dx,$$

and each point of the rectangle is approximately the same distance x from the y-axis. The moment of inertia, then, of this rectangle about the y-axis is approximately

$$x^2\rho[f(x) - g(x)]\,dx.$$

The moment of inertia of the plate about the y-axis is found by adding moments of inertia of all such rectangles and taking the limit as their widths approach zero. But again this process defines a definite integral, and therefore the moment of inertia of the plate in Figure 7.69 about the y-axis is

$$I = \int_a^b x^2\rho[f(x) - g(x)]\,dx. \tag{7.37}$$

EXAMPLE 7.24

Find the moment of inertia about the y-axis of a thin plate with constant mass per unit area ρ if its edges are defined by the curves

$$y = x^3, \qquad y = \sqrt{2 - x}, \qquad x = 0.$$

SOLUTION

FIGURE 7.75

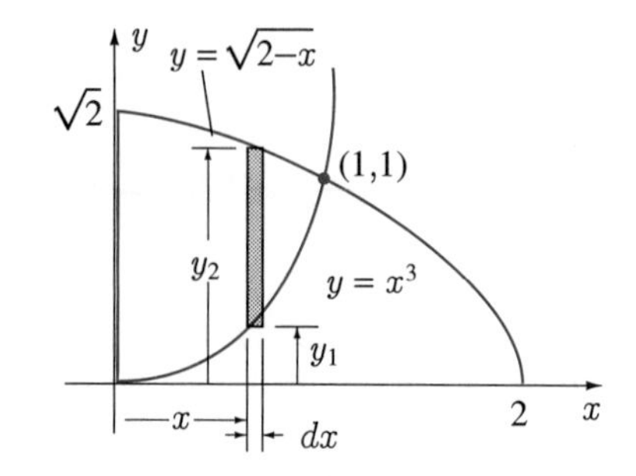

Since the moment of inertia of the vertical rectangle in Figure 7.75 is

$$x^2 \rho(y_2 - y_1)\,dx = \rho x^2 \left(\sqrt{2 - x} - x^3\right) dx,$$

the moment of inertia of the plate is

$$I = \int_0^1 \rho x^2 \left(\sqrt{2 - x} - x^3\right) dx = \rho \int_0^1 x^2 \sqrt{2 - x}\,dx - \rho \int_0^1 x^5\,dx.$$

In the first integral we set $u = 2 - x$, in which case $du = -dx$, and

$$I = \rho \int_2^1 (2 - u)^2 u^{1/2}(-du) - \rho \left\{ \frac{x^6}{6} \right\}_0^1 = \rho \int_1^2 (4u^{1/2} - 4u^{3/2} + u^{5/2})\,du - \frac{\rho}{6}$$

$$= \rho \left\{ \frac{8u^{3/2}}{3} - \frac{8u^{5/2}}{5} + \frac{2u^{7/2}}{7} \right\}_1^2 - \frac{\rho}{6} = \frac{256\sqrt{2} - 319}{210}\rho.$$

■

EXAMPLE 7.25

Find the moment of inertia about the line $y = -1$ of a thin plate of constant mass per unit area ρ if its edges are defined by the curves

$$x = y^2, \qquad x = 2y.$$

SOLUTION Since the directed distance from the line $y = -1$ to all points in the horizontal rectangle in Figure 7.76 is approximately $y + 1$, the moment of inertia of the rectangle about the line $y = -1$ is approximately $(y+1)^2 \rho(x_2 - x_1)\,dy$. It follows that the moment of inertia of the plate is

$$I = \int_0^2 (y + 1)^2 \rho(x_2 - x_1)\,dy = \rho \int_0^2 (y + 1)^2 (2y - y^2)\,dy$$

$$= \rho \int_0^2 (-y^4 + 3y^2 + 2y)\,dy = \rho \left\{ -\frac{y^5}{5} + y^3 + y^2 \right\}_0^2 = \frac{28\rho}{5}.$$

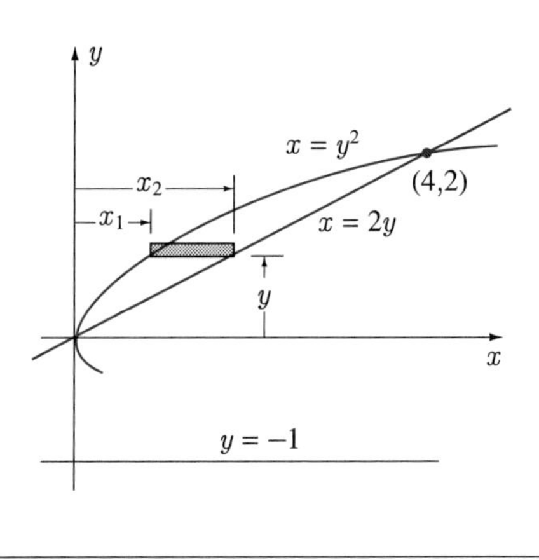

FIGURE 7.76

In Examples 7.24 and 7.25, and in the discussion leading to equation (7.37), we chose rectangles that had lengths parallel to the line about which we required the moment of inertia. This is not just coincidence; the use of perpendicular rectangles is more complicated. Consider, for instance, finding the moment of inertia about the x-axis of the plate in Example 7.24. To use the vertical rectangles in Figure 7.75, we first require the moment of inertia of such a rectangle about the x-axis. The mass of the rectangle, as in Example 7.24, must now be multiplied by the square of the distance from the x-axis to the rectangle. Unfortunately, different points in the rectangle are at different distances. One suggestion might be to concentrate all of the mass of the rectangle at its centre of mass and use the distance from the x-axis to the centre of mass. This is *incorrect*. The centre of mass is a point at which mass can be concentrated if we are discussing first moments. We are discussing second moments. (In Exercise 12 we show that the moment of inertia of the rectangle in Figure 7.77 cannot be obtained by concentrating its mass at its centre of mass.) What are we to do then? To use this type of rectangle, we must first develop a formula for its moment of inertia. To obtain this formula, let us consider the moment of inertia about the x-axis of the rectangle of width h and length $y_2 - y_1$ in Figure 7.77. If we subdivide this rectangle into smaller rectangles of width dy, the moment of inertia of the tiny rectangle about the x-axis is approximately

$$y^2 \rho h \, dy.$$

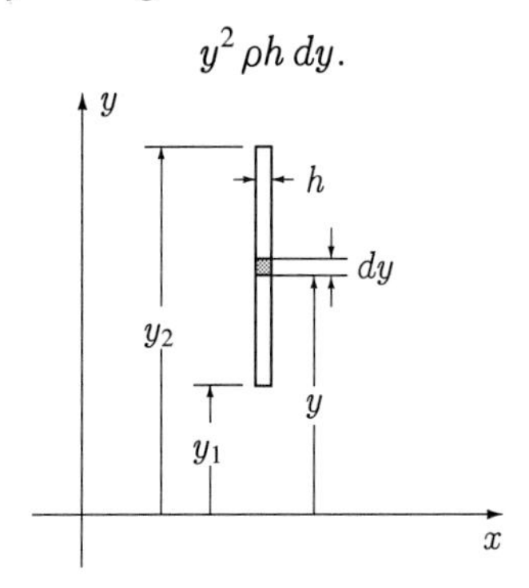

FIGURE 7.77

The moment of inertia of the long, vertical rectangle can be obtained by adding over all the tiny rectangles as their widths dy approach zero,

$$\int_{y_1}^{y_2} y^2 \rho h \, dy = \rho h \int_{y_1}^{y_2} y^2 \, dy = \frac{\rho h}{3}(y_2^3 - y_1^3). \tag{7.38}$$

We can use this formula to state that the moment of inertia about the x-axis of the vertical rectangle in Figure 7.75 is

$$\frac{\rho}{3}(y_2^3 - y_1^3) \, dx = \frac{\rho}{3}[(2 - x)^{3/2} - x^9] \, dx.$$

The moment of inertia of the plate about the x-axis is therefore

$$I = \int_0^1 \frac{\rho}{3}[(2-x)^{3/2} - x^9]\,dx = \frac{\rho}{3}\left\{-\frac{2}{5}(2-x)^{5/2} - \frac{x^{10}}{10}\right\}_0^1 = \frac{(16\sqrt{2}-5)\rho}{30}.$$

The alternative procedure for this problem is to use horizontal rectangles which are parallel to the x-axis and obtain two definite integrals:

$$I = \int_0^1 y^2 \rho y^{1/3}\,dy + \int_1^{\sqrt{2}} y^2 \rho(2-y^2)\,dy.$$

In summary, we have two methods for determining moments of inertia of thin plates:

1. Choose rectangles parallel to the line about which the moment of inertia is required, in which case only the basic idea of a moment of inertia (mass times distance squared) is needed.
2. Choose rectangles perpendicular to the line about which the moment of inertia is required, in which case formula 7.38, or a similar formula, is needed.

As for finding centres of mass, we could, in the special case of uniform mass distributions, drop all references to mass and talk about second moments of area about a line.

EXAMPLE 7.26

Find the moment of inertia about the line $x = 2$ of a plate with mass per unit area ρ if its edges are defined by the curves

$$x = 2y - y^2, \qquad x = y^2 - 2y.$$

SOLUTION

FIGURE 7.78

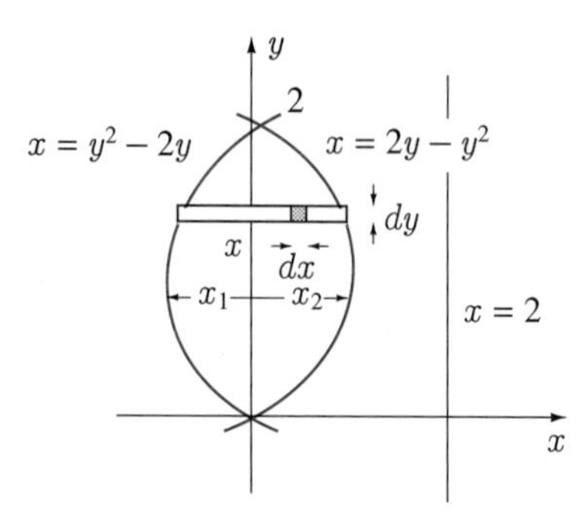

We first divide the plate into horizontal rectangles of width dy, and then subdivide this rectangle into smaller rectangles of width dx (Figure 7.78). Since the directed distance from the line $x = 2$ to the tiny rectangle is $x - 2$, the moment of inertia of the tiny rectangle about $x = 2$ is

$$(x-2)^2 \rho\,dy\,dx.$$

It follows that the moment of inertia of the long horizontal rectangle about $x = 2$ is

$$\int_{x_1}^{x_2} (x-2)^2 \rho\,dy\,dx = \rho dy\left\{\frac{1}{3}(x-2)^3\right\}_{x_1}^{x_2} = \frac{\rho}{3}[(x_2-2)^3 - (x_1-2)^3]\,dy.$$

The moment of inertia of the entire plate is now

$$\begin{aligned}
I &= \int_0^2 \frac{\rho}{3}[(x_2-2)^3-(x_1-2)^3]\,dy \\
&= \frac{\rho}{3}\int_0^2 [(2y-y^2-2)^3-(y^2-2y-2)^3]\,dy \\
&= \frac{2\rho}{3}\int_0^2 (24y-12y^2+8y^3-12y^4+6y^5-y^6)\,dy \\
&= \frac{2\rho}{3}\left\{12y^2-4y^3+2y^4-\frac{12y^5}{5}+y^6-\frac{y^7}{7}\right\}_0^2 \\
&= \frac{1184\rho}{105}.
\end{aligned}$$

■

EXERCISES 7.7

In Exercises 1–10 the curves define a thin plate with constant mass per unit area ρ. Find its moment of inertia about the line.

1. $y = x^2, \quad y = x^3, \quad$ about the y-axis

2. $y = x, \quad y = 2x + 4, \quad y = 0, \quad$ about the x-axis

3. $y = x^2, \quad 2y = x^2 + 4, \quad$ about $y = 0$

4. $y = x^2 - 4, \quad y = 2x - x^2, \quad$ about $x = -2$

5. $xy\sqrt{y^2+12} = 1, \quad x = 0, \quad y = 1, \quad y = 1/2, \quad$ about $y = 0$

6. $y = |x|^{1/3}, \quad y = 2 - |x|^{1/3}, \quad$ about $x = 0$

7. $x + y = 2, \quad y - x = 2, \quad y = 0, \quad$ about $x = -2$

8. $x = 1 - y^2, \quad x = y^2 - 1, \quad$ about $x = -1$

9. $x = 1 - y^2, \quad x = y^2 - 1, \quad$ about $y = 1$

10. $x = y^2, \quad x + y = 2, \quad$ about $y = 3$

11. (a) If I_{Ax} and I_{Bx} represent the second moments of area about the x-axis of the areas in Figure 7.73, show that

$$I_{Bx} = 3nI_{Ax}.$$

(b) Show that if I_{Ay} and I_{By} represent the second moments of area about the y-axis, then

$$nI_{Ay} = 3I_{By}.$$

12. What is the product of the mass of the rectangle in Figure 7.77 and the square of the distance from the x-axis to the centre of mass of the rectangle? Is it equal to the expression in equation 7.38?

13. The radius of gyration r of a thin plate with constant mass per unit area about a line is defined by $I = Mr^2$, where M is the mass of the plate and I is its moment of inertia about that line. Find radii of gyration about the x- and y-axes for the plate with edges

$$y = 2x^3, \qquad y + x^3 = 0, \qquad 2y = x + 3.$$

Explain the physical significance of r.

14. Show that the kinetic energy of a long-playing record is equal to one-half the product of the moment of inertia of the record about a line through its centre and perpendicular to its face, and the square of its angular speed.

15. Prove the *parallel axis theorem*: The moment of inertia of a thin plate (with constant mass per unit area) with respect to any coplanar line is equal to the moment of inertia with respect to the parallel line through the centre of mass plus the mass multiplied by the square of the distance between the lines.

16. If a line $x = \tilde{x}$ is drawn through the area in Figure 7.69, what integral represents the second moment of area about this line? What should be the value of $\tilde{x}$ for the smallest possible second moment?

17. The polar moment of inertia of a point mass m at (x, y) is defined as the product of m and the square of its distance from the origin

$$J_0 = m(x^2 + y^2).$$

For the thin plate (with constant mass per unit area) in Figure 7.79, let I_x and I_y be its moments of inertia about the x- and y-axes. Show that

$$J_0 = I_x + I_y.$$

FIGURE 7.79

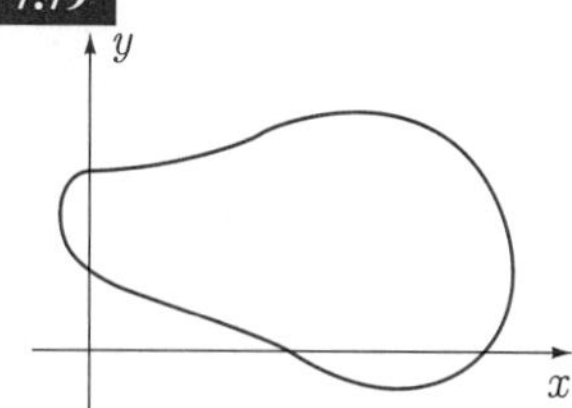

In Exercises 18–20 the curves define a plate with mass per unit area equal to 2. Find its moment of inertia about the line accurate to three decimals.

18. $y = 1 - x^2, \quad x = y^2 \quad$ about $y = 0$

19. $y = x^3 - x, \quad y = \sqrt{x} \quad$ about $x = 0$

20. $y = x^3 - x, \quad y = \sqrt{x} \quad$ about $y = 0$

21. Find the second moment of area of a rectangle about its diagonal.

SECTION 7.8

Additional Applications

Volumes by Slicing

If we can represent the area of parallel cross-sections of a volume as a function of one variable, we can use a definite integral to calculate the volume. In particular, when we use the disc or washer method to determine the volume of a solid of revolution, parallel cross-sections are circles. In the following example, parallel cross-sections are squares.

EXAMPLE 7.27

A uniformly tapered rod of length 2 m has square cross-sections. If the areas of its ends are 4 and 16 cm^2 as in Figure 7.80(a), what is the volume of the rod?

FIGURE 7.80

(a)

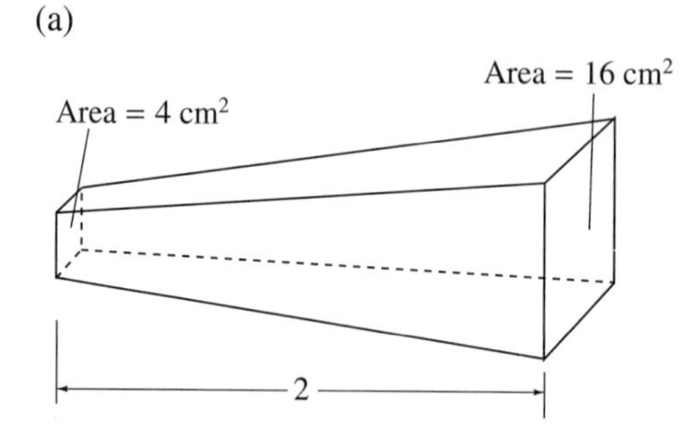

(b)

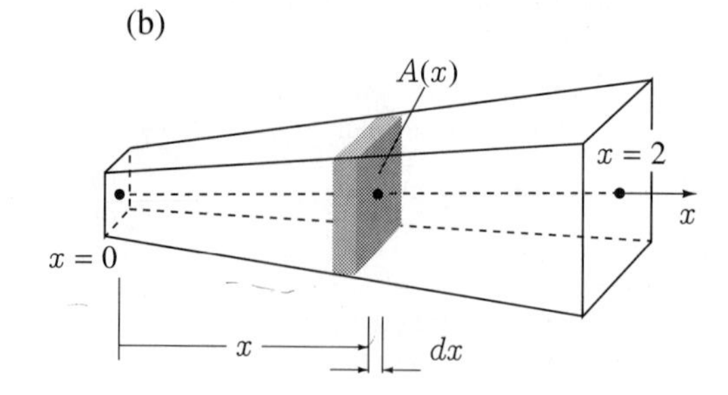

SOLUTION If we define an x-coordinate perpendicular to the square cross-sections as in Figure 7.80(b), then the side lengths of the cross-sections at $x = 0$ and $x = 2$ are 0.02 and 0.04 m respectively. Since the rod is uniformly tapered, the side length of the cross-section at x is

$$0.02 + x\left(\frac{0.04 - 0.02}{2}\right) = 0.02 + 0.01x = \frac{2 + x}{100}.$$

The area of the cross-section at x is therefore

$$A(x) = 10^{-4}(2 + x)^2.$$

If we construct at x a "slab" of cross-sectional area $A(x)$ and width dx, the volume of the slab is

$$A(x)\,dx = 10^{-4}(2 + x)^2\,dx.$$

To obtain the volume of the rod, we add the volumes of all such slabs between $x = 0$ and $x = 2$, and take the limit as their widths approach zero:

$$\begin{aligned} V &= \int_0^2 A(x)\,dx = \int_0^2 10^{-4}(2 + x)^2\,dx \\ &= 10^{-4}\left\{\frac{(2 + x)^3}{3}\right\}_0^2 = \frac{10^{-4}}{3}(64 - 8) = \frac{56}{3} \times 10^{-4}\ \text{m}^3. \end{aligned}$$

■

Area of a Surface of Revolution

If a curve in the xy-plane is rotated about the x- or y-axis (or a line parallel to the x- or y-axis) in order to produce a surface, we can calculate the area of this surface using the ideas of lengths along curves discussed in Section 7.3.

EXAMPLE 7.28

If that part of the parabola $y = x^2$ between $x = 0$ and $x = 1$ is rotated around the y-axis, find the area of the surface of revolution traced out by the curve (Figure 7.81).

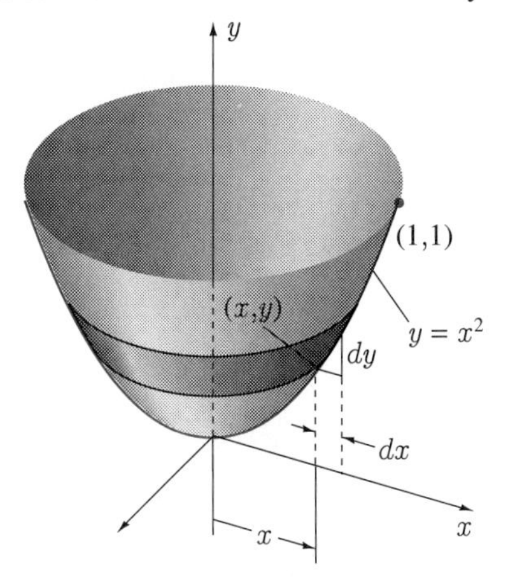

FIGURE 7.81

SOLUTION We can approximate length along the parabola corresponding to a change dx in x by the tangential straight-line length:

$$\sqrt{(dx)^2 + (dy)^2} = \sqrt{1 + \left(\frac{dy}{dx}\right)^2}\, dx = \sqrt{1 + (2x)^2}\, dx = \sqrt{1 + 4x^2}\, dx.$$

If this straight-line segment is rotated around the y-axis, each point follows a circular path of radius approximately equal to x; therefore, the area traced out by the line segment is approximately equal to

$$(2\pi x)\left(\sqrt{1 + 4x^2}\, dx\right).$$

We can find the total surface area by adding all such areas, and taking the limit as the widths dx approach zero:

$$A = \int_0^1 2\pi x\sqrt{1 + 4x^2}\, dx = 2\pi\left\{\frac{(1 + 4x^2)^{3/2}}{12}\right\}_0^1 = \frac{(5\sqrt{5} - 1)\pi}{6}.$$

■

EXAMPLE 7.29

Find the area of the surface of revolution traced out by rotating that part of the curve $y = x^3$ between $x = 1$ and $x = 2$ about the x-axis.

SOLUTION We can approximate length along the cubic corresponding to a change dx in x by the tangential straight-line length:

$$\sqrt{(dx)^2 + (dy)^2} = \sqrt{1 + \left(\frac{dy}{dx}\right)^2}\, dx = \sqrt{1 + (3x^2)^2}\, dx = \sqrt{1 + 9x^4}\, dx.$$

FIGURE 7.82

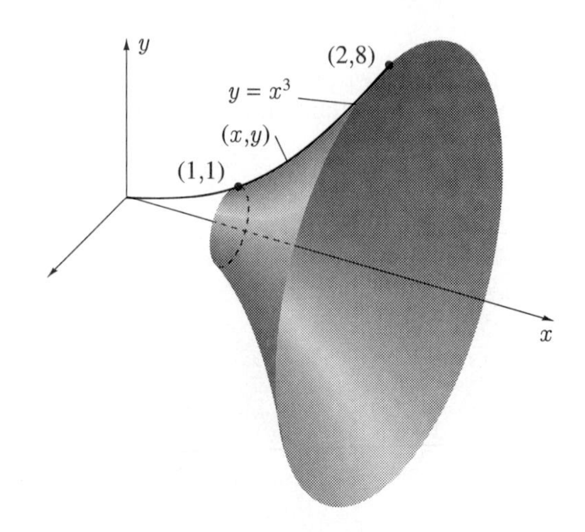

If this straight-line segment is rotated about the x-axis (Figure 7.82), each point follows a circular path of radius approximately equal to y; therefore, the area traced out by the line segment is approximately equal to

$$(2\pi y)\left(\sqrt{1+9x^4}\,dx\right).$$

By adding over all such areas,

$$A = \int_1^2 2\pi y\sqrt{1+9x^4}\,dx = 2\pi\int_1^2 x^3\sqrt{1+9x^4}\,dx$$
$$= 2\pi\left\{\frac{(1+9x^4)^{3/2}}{54}\right\}_1^2 = \frac{(145^{3/2}-10^{3/2})\pi}{27}.$$

■

Rates of Flow

In Problem 3 of Section 6.2 we considered laminar blood flow in a circular vessel. Specifically, the velocity of blood through the cross-section in Figure 7.83(a) is a function of radial distance r from the centre of the vessel:

$$v(r) = c(R^2 - r^2), \quad 0 \le r \le R,$$

where $c > 0$ is a constant and R is the radius of the vessel.

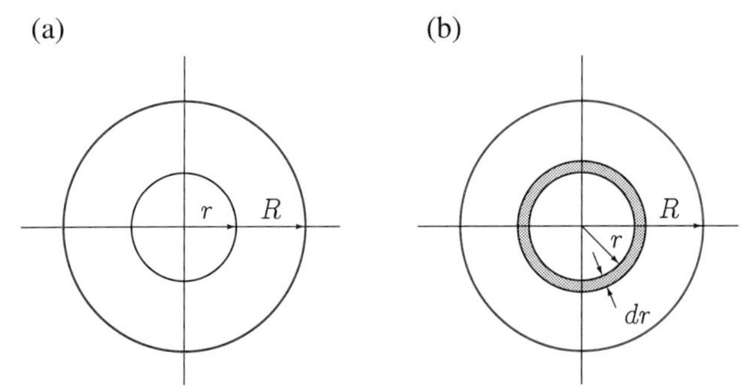

FIGURE 7.83

If we construct, at radius r, a thin ring of width dr as in Figure 7.83(b), then the area of this ring is approximately $(2\pi r)\,dr$. Since v does not vary greatly over this ring, the amount of blood flowing through the ring per unit time is approximately $v(r)$ multiplied by the area of the ring:

$$v(r)(2\pi r\,dr).$$

We can find the total flow through the blood vessel by adding the flows through all rings and taking the limit as the widths dr of the rings approach zero:

$$F = \int_0^R v(r)2\pi r\,dr = 2\pi\int_0^R rc(R^2 - r^2)\,dr = 2\pi c\left\{\frac{R^2r^2}{2} - \frac{r^4}{4}\right\}_0^R = \frac{\pi cR^4}{2}.$$

Repairs, Depreciation, and Replacement

In most businesses, depreciation and eventual replacement of equipment must be considered. Take, for example, a salesman who must decide when to replace his car. When he buys a new car, he knows that its value depreciates quickly at first and less rapidly after a few years. On the other hand, the longer he keeps the car, the more costly repairs become. The problem then is to decide on the optimum time to replace the car. Let us consider a situation in which the rate of depreciation of the car is $f(t)$, and the rate of accumulation of maintenance costs is $g(t)$. We have shown a possible $f(t)$ in Figure 7.84, where the ordinate at time t represents the number of dollars per year the car depreciates at that time. If we draw a rectangle of width dt at time t, then the area of this rectangle, $f(t)\,dt$, represents the amount the car depreciates during the time interval dt. Consequently, the total depreciation of the car from the time of purchase ($t = 0$) to time T is

$$\int_0^T f(t)\,dt.$$

In a similar way, the total maintenance cost from time $t = 0$ to $t = T$ (Figure 7.85) is

$$\int_0^T g(t)\,dt.$$

The cost associated with owning and maintaining the car over a period of T years is

$$\int_0^T f(t)\,dt + \int_0^T g(t)\,dt = \int_0^T [f(t) + g(t)]\,dt,$$

and the average yearly cost is

$$C(T) = \frac{1}{T}\int_0^T [f(t) + g(t)]\,dt.$$

FIGURE 7.84

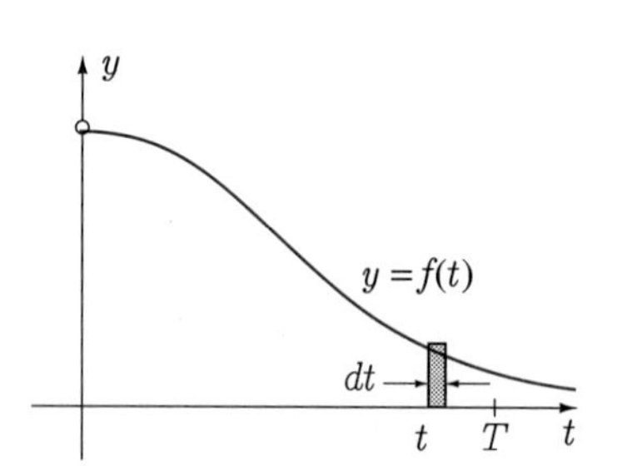

FIGURE 7.85

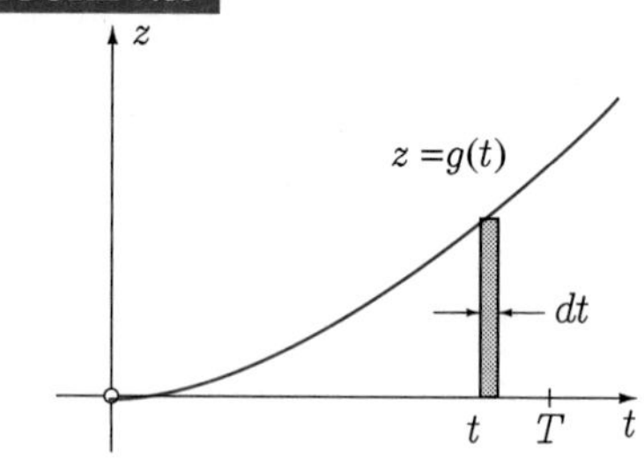

The best time to replace the car is that value T which minimizes this average yearly cost. Critical points of $C(T)$ are defined by

$$0 = C'(T) = -\frac{1}{T^2}\int_0^T [f(t) + g(t)]\,dt + \frac{1}{T}\frac{d}{dT}\int_0^T [f(t) + g(t)]\,dt.$$

Equation 6.20 in Section 6.5 can be used to evaluate the derivative of the integral; the result is

$$0 = -\frac{1}{T^2}\int_0^T [f(t) + g(t)]\,dt + \frac{1}{T}[f(T) + g(T)], \qquad \text{or}$$

$$\frac{1}{T}\int_0^T [f(t) + g(t)]\,dt = f(T) + g(T).$$

Since the left side of this equation is $C(T)$, we can say that critical points of $C(T)$ occur when

$$C(T) = f(T) + g(T); \tag{7.39}$$

that is, when average yearly costs are equal to the rate at which costs are changing. At one of these points — if there is more than one, $C(T)$ will take on its minimum value.

EXAMPLE 7.30

Find the optimum time to replace a car if rates of depreciation and accumulation of maintenance costs are, respectively,

$$f(t) = -\frac{V}{50}(t - 10) \qquad \text{and} \qquad g(t) = \frac{V}{400}t^2,$$

where V is the value of the new car. Assume that the car must be replaced by the time it has depreciated to no value.

SOLUTION Since the total depreciation of the car in T years is given by

$$\int_0^T f(t)\,dt = \int_0^T -\frac{V}{50}(t - 10)\,dt = -\frac{V}{50}\left\{\frac{t^2}{2} - 10t\right\}_0^T = -\frac{V}{100}(T^2 - 20T),$$

it follows that the car completely depreciates when

$$V = -\frac{V}{100}(T^2 - 20T).$$

The only solution of this equation is $T = 10$ years. Do not make the mistake of assuming that complete depreciation occurs when $f(t) = 0$; this is not always true. Complete depreciation occurs when the area under $f(t)$ is equal to the initial value of the car.

The average yearly cost associated with the car if it is kept for T years is

$$\begin{aligned} C(T) &= \frac{1}{T}\int_0^T [f(t) + g(t)]\,dt \\ &= \frac{1}{T}\int_0^T \left(-\frac{Vt}{50} + \frac{V}{5} + \frac{Vt^2}{400}\right) dt \\ &= \frac{1}{T}\left(-\frac{VT^2}{100} + \frac{VT}{5} + \frac{VT^3}{1200}\right) \\ &= -\frac{VT}{100} + \frac{V}{5} + \frac{VT^2}{1200}, \qquad 0 < T \le 10. \end{aligned}$$

For critical point of $C(T)$, we solve

$$0 = C'(T) = -\frac{V}{100} + \frac{VT}{600},$$

the solution of which is $T = 6$. We could also have obtained $T = 6$ by solving equation 7.39. Since

$$\lim_{T \to 0^+} C(T) = \frac{V}{5}, \qquad C(6) = \frac{17V}{100}, \qquad C(10) = \frac{11V}{60},$$

it follows that the car should be replaced in 6 years. ■

EXERCISES 7.8

1. Verify that the surface area of a sphere of radius r is $4\pi r^2$.

2. Find the area of the curved surface of a right-circular cone of radius r and height h.

3. Calculate the rate of flow of blood through a circular vessel of radius R if the velocity profile is
(a) $v = f(r) = cR\sqrt{R^2 - r^2}$ and
(b) $v = f(r) = (c/R^2)(R^2 - r^2)^2$.

4. Find the volume of the pyramid in Figure 7.86.

FIGURE 7.86

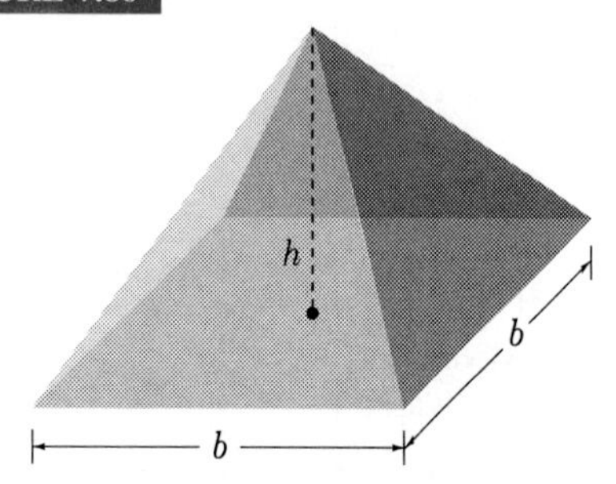

5. The amount of water consumed by a community varies throughout the day, peaking, naturally, around meal hours. During the six-hour period between 12:00 noon ($t = 0$) and 6:00 p.m. ($t = 6$), we find that the number of cubic metres of water consumed per hour at time t is given by the function

$$f(t) = 5000 + 21.65t^2 - 249.7t^3 + 97.52t^4 - 9.680t^5.$$

Find the total consumption during this 6-hour period.

6. The number of bees per unit area at a distance x from a hive is given by

$$\rho(x) = \frac{600\,000}{31\pi R^5}(R^3 + 2R^2x - Rx^2 - 2x^3), \quad 0 \le x \le R,$$

where R is the maximum distance travelled by the bees.
(a) What is the number of bees in the colony?
(b) How many bees are within a distance $R/2$ of the hive?

7. If the radius of the blood vessel in Figure 7.83 is reduced to $R/2$ because of arteriosclerosis, the velocity profile is

$$v(r) = c(R^2 - 4r^2), \quad 0 \le r \le R/2.$$

What percentage of the normal flow $(\pi cR^4/2)$ gets through the hardened vessel?

8. A tree trunk of diameter 50 cm (Figure 7.87) has a wedge cut from it by two planes. The lower plane is perpendicular to the axis of the trunk, and together the planes make an angle $\pi/3$ radians, meeting along a diameter of the circular cross-section of the trunk. Find the volume of the wedge.

FIGURE 7.87

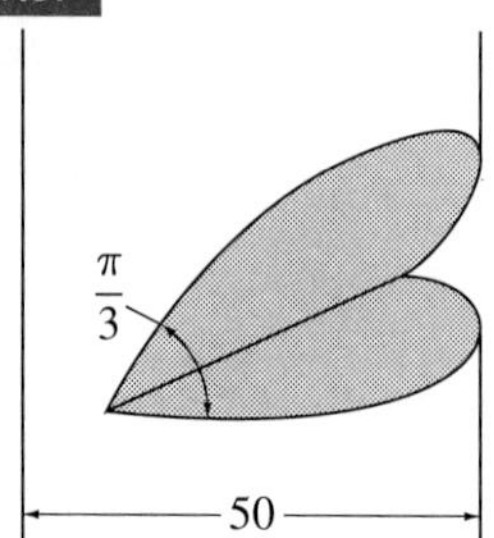

9. Find the area of the surface of revolution formed by rotating that part of the curve $24xy = x^4 + 48$ between $x = 1$ and $x = 2$ about the x-axis.

10. Find the area of the surface of revolution generated by rotating the curve $8y^2 = x^2(1 - x^2)$ about the x-axis.

11. Find the area of the surface of revolution generated by rotating the loop of the curve $9y^2 = x(3 - x)^2$ about (a) the y-axis (b) the x-axis.

12. The base of a solid is the circle $x^2 + y^2 = r^2$, and every plane section perpendicular to the x-axis is an isosceles triangle (Figure 7.88). Find the volume of the solid.

FIGURE 7.88

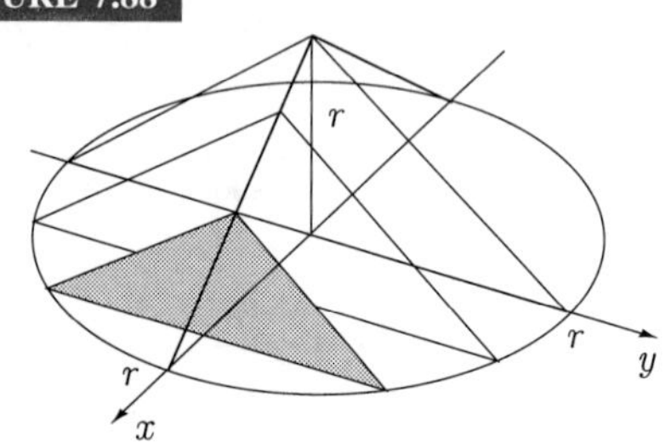

13. Electrons are fired from an electron gun at a target (Figure 7.89). The probability that an electron strikes the target in a ring of unit area at a distance x from the centre of the target is given by

$$p = f(x) = \frac{5}{3\pi R^5}(R^3 - x^3), \quad 0 \le x \le R.$$

What percentage of a cascade of electrons hits within a distance r from the centre of the target?

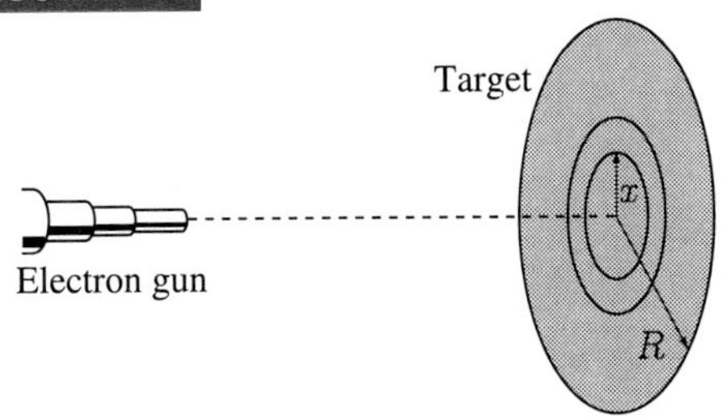

FIGURE 7.89

In Exercises 14-16, $f(t)$ and $g(t)$ represent rate of depreciation and rate of accumulation of maintenance costs for a piece of machinery. Sketch graphs of these functions, and show that the total area under $y = f(t)$ is equal to V (the original value of the machine). Determine the optimum time at which to replace the machine. Sketch a graph of $C(T)$ (the average yearly expense) to illustrate your answer.

14. $f(t) = \begin{cases} \dfrac{V}{10} - \dfrac{Vt}{200} & 0 < t \le 20 \\ 0 & t > 20 \end{cases} \qquad g(t) = \dfrac{Vt^2}{4000}$

15. $f(t) = \begin{cases} -\dfrac{V}{50}(t - 10) & 0 < t \le 10 \\ 0 & t > 10 \end{cases} \qquad g(t) = \dfrac{Vt}{400}$

16. $f(t) = \begin{cases} \dfrac{V}{10} - \dfrac{6Vt^2}{15\,625} & 0 < t \le 12.5 \\ 0 & t > 12.5 \end{cases} \qquad g(t) = \dfrac{Vt^2}{3000}$

17. Two right-circular cylinders, each of radius r, have axes that intersect at right angles. Find the volume common to the two cylinders.

18. What is the volume of air trapped in the attic of a house if the roof has shape shown in Figure 7.90? The peak of the roof is 2 m above the base.

FIGURE 7.90

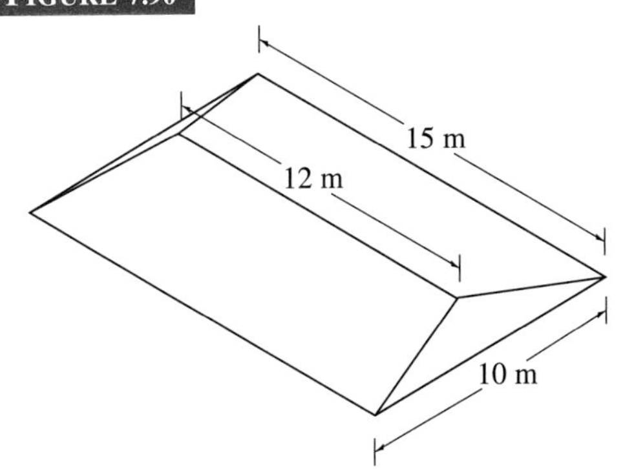

19. Suppose that at time $t = 0$ the number of people in a certain population is N_0, and the probability that each individual lives to age t is $p(t)$. For time $t > 0$, the birth rate is $r(t)$ individuals per unit time. Show that at time T the number of people in the population is

$$N(T) = N_0 p(T) + \int_0^T p(T - t) r(t)\, dt.$$

20. The depth of water in the container in Figure 7.91 is originally 10 cm. Water evaporates from the surface at a rate proportional to the area of the surface. If the water level drops 1 cm in 5 days, how long does it take for the water to evaporate completely? Note that Exercise 9 in Section 5.4 is this same problem given a container that is a right-circular cone. Now we ask you to repeat the problem with no knowledge of the shape of the container.

FIGURE 7.91

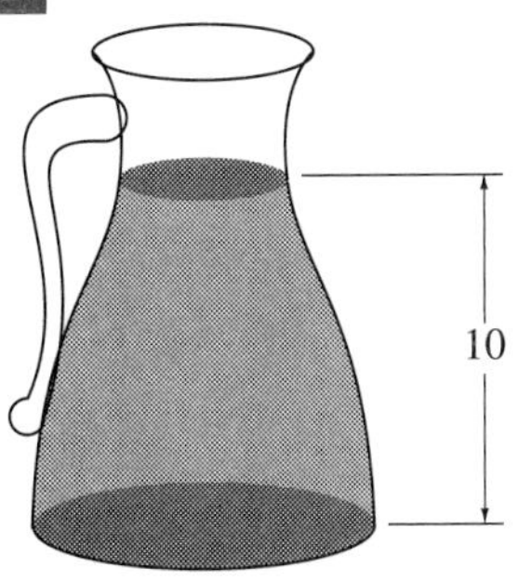

21. An appliance retailer must be concerned with her inventory costs. For example, given that she sells N refrigerators per year, she must decide whether to order the year's supply at one time, in which case a large storage area would be necessary, or to make periodic orders throughout the year, perhaps of $N/12$ at the first of each month. In the latter case she would incur costs due to paper work, delivery charges, etc. Let us suppose that the retailer decides to order in equal-lot sizes, x, at equally spaced intervals, N/x times per year.

(a) If each order has fixed costs of F dollars, plus f dollars for each refrigerator, what are the total yearly ordering costs?

(b) Between successive deliveries, the retailer's stock dwindles from x to 0. If she assumes that the number of refrigerators in stock decreases linearly in time, and the yearly stocking cost per refrigerator is p dollars, what are the yearly stocking costs?

(c) If the retailer's total yearly inventory costs are her ordering costs in part (a) plus her stocking costs in part (b), what value of x minimzes inventory costs?

SECTION 7.9

Improper Integrals

According to the fundamental theorem of integral calculus (Theorem 6.3), definite integrals are evaluated with indefinite integrals,

$$\int_a^b f(x)\,dx = \left\{\int f(x)\,dx\right\}_a^b,$$

provided $f(x)$ is continuous on $a \le x \le b$. In this section we investigate what to do when $f(x)$ is not continuous on $a \le x \le b$, or when either a or b is infinite.

Consider whether a reasonable meaning can be given to the integrals

$$\int_1^\infty \frac{1}{x^2}\,dx \qquad \text{and} \qquad \int_1^\infty \frac{1}{\sqrt{x}}\,dx.$$

Both integrals are "improper" in the sense that their upper limits are not finite, and we therefore call them **improper integrals**. If $b > 1$, there is no difficulty with the evaluation and possible interpretation of either

$$\int_1^b \frac{1}{x^2}\,dx \qquad \text{or} \qquad \int_1^b \frac{1}{\sqrt{x}}\,dx.$$

Clearly,

$$\int_1^b \frac{1}{x^2}\,dx = \left\{-\frac{1}{x}\right\}_1^b = 1 - \frac{1}{b} \qquad \text{and}$$

$$\int_1^b \frac{1}{\sqrt{x}}\,dx = \{2\sqrt{x}\}_1^b = 2\sqrt{b} - 2.$$

We can interpret the first integral as the area under the curve $y = 1/x^2$, above the x-axis, and between the vertical lines $x = 1$ and $x = b$ (Figure 7.92). The second integral has exactly the same interpretation but uses the curve $y = 1/\sqrt{x}$ in place of $y = 1/x^2$ (Figure 7.93).

FIGURE 7.92

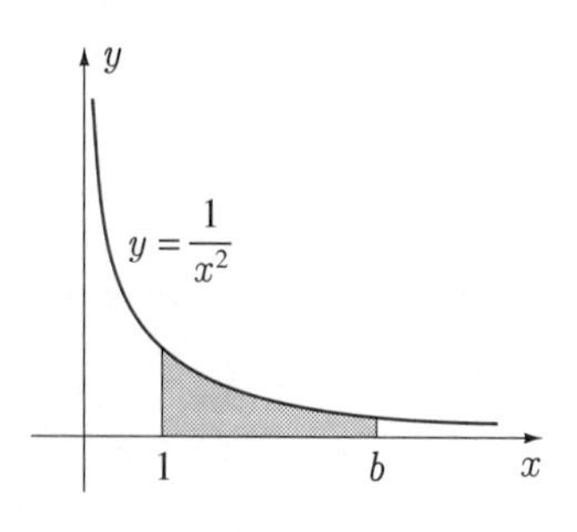

FIGURE 7.93

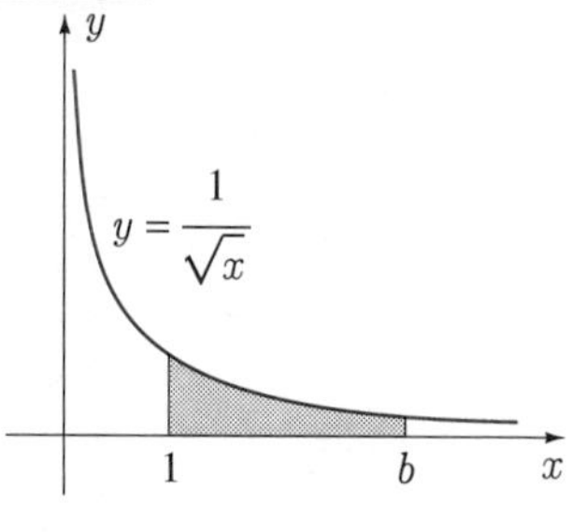

In Table 7.1 we have listed values of these definite integrals corresponding to various values of b. It is clear that the two integrals display completely different characteristics. As b is made very large, the integral of $1/x^2$ is always less than 1, but gets closer to 1 as b increases. In other words,

$$\lim_{b\to\infty} \int_1^b \frac{1}{x^2}\,dx = 1.$$

Geometrically, the area in Figure 7.92 is always less than 1 but approaches 1 as b approaches infinity.

Contrast this with the integral of $1/\sqrt{x}$, which becomes indefinitely large as b increases:

$$\lim_{b\to\infty}\int_1^b \frac{1}{\sqrt{x}}\,dx = \infty;$$

that is, the area in Figure 7.93 can be made as large as desired by choosing b sufficiently large.

TABLE 7.1

b	$\int_1^b \frac{1}{x^2}\,dx$	$\int_1^b \frac{1}{\sqrt{x}}\,dx$
100	0.99	18
10 000	0.9999	198
1 000 000	0.999999	1 998
100 000 000	0.99999999	19 998

On the basis of these calculations, we would like to say that the improper integral of $1/x^2$ has value 1, whereas the improper integral of $1/\sqrt{x}$ has no value. In other words, if these improper integrals are to have values, they should be defined by the limits

$$\int_1^\infty \frac{1}{x^2}\,dx = \lim_{b\to\infty}\int_1^b \frac{1}{x^2}\,dx \qquad \text{and} \qquad \int_1^\infty \frac{1}{\sqrt{x}}\,dx = \lim_{b\to\infty}\int_1^b \frac{1}{\sqrt{x}}\,dx.$$

In the second case the limit does not exist, and we interpret this to mean that the improper integral does not exist or has no value.

In general, then, improper integrals are defined in terms of limits as follows.

Definition 7.4

If $f(x)$ is continuous for $x \geq a$, we define

$$\int_a^\infty f(x)\,dx = \lim_{b\to\infty}\int_a^b f(x)\,dx, \tag{7.40}$$

provided the limit exists.

If the limit exists, we say that the improper integral is equal to the limit, or that the improper integral *converges* to the limit. If the limit does not exist, we say that the improper integral has no value, or that it *diverges*.

EXAMPLE 7.31

Determine whether the following improper integrals converge or diverge:

(a) $\displaystyle\int_3^\infty \frac{1}{(x-2)^3}\,dx$ (b) $\displaystyle\int_0^\infty \frac{x^2}{\sqrt{1+x^3}}\,dx$ (c) $\displaystyle\int_1^\infty \frac{2x^2+6}{3x^2+5}\,dx$

SOLUTION (a) Using equation 7.40,

$$\int_3^\infty \frac{1}{(x-2)^3}\,dx = \lim_{b\to\infty}\int_3^b \frac{1}{(x-2)^3}\,dx = \lim_{b\to\infty}\left\{\frac{-1}{2(x-2)^2}\right\}_3^b$$
$$= \lim_{b\to\infty}\left[\frac{1}{2} - \frac{1}{2(b-2)^2}\right] = \frac{1}{2},$$

and the improper integral therefore converges to $\frac{1}{2}$.

(b) Once again 7.40 gives

$$\int_0^\infty \frac{x^2}{\sqrt{1+x^3}}\,dx = \lim_{b\to\infty}\int_0^b \frac{x^2}{\sqrt{1+x^3}}\,dx = \lim_{b\to\infty}\left\{\frac{2}{3}\sqrt{1+x^3}\right\}_0^b$$
$$= \lim_{b\to\infty}\left[\frac{2}{3}\sqrt{1+b^3} - \frac{2}{3}\right] = \infty,$$

and the improper integral therefore diverges.

(c) Although we cannot at this time find an antiderivative for $f(x) = (2x^2 + 6)/(3x^2 + 5)$, we can solve this problem by interpreting the improper integral as an area. The graph of $f(x)$ in Figure 7.94 indicates that the curve is asymptotic to the line $y = \frac{2}{3}$. Clearly, then, the improper integral

$$\int_1^\infty \frac{2x^2+6}{3x^2+5}\,dx$$

can have no value since the area under the curve is larger than the area of a rectangle of width $\frac{2}{3}$ and infinite length.

FIGURE 7.94

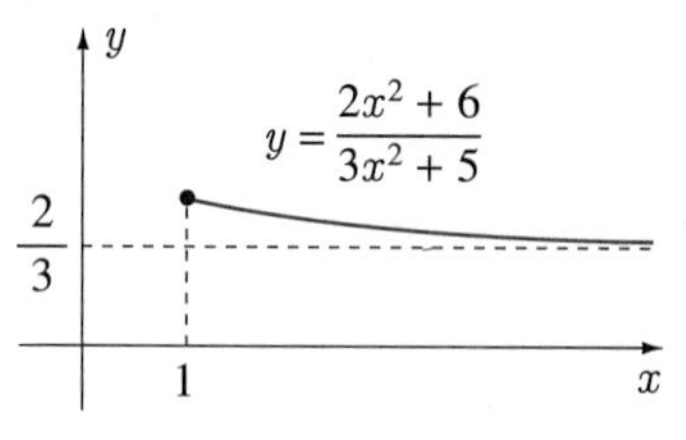

■

We can define improper integrals with infinite lower limits in exactly the same way as for improper integrals with infinite upper limits. Specifically, if $f(x)$ is continuous for $x \le b$, we define

$$\int_{-\infty}^b f(x)\,dx = \lim_{a\to-\infty}\int_a^b f(x)\,dx, \tag{7.41}$$

provided the limit exists. For example,

$$\int_{-\infty}^5 \frac{1}{(x-6)^5}\,dx = \lim_{a\to-\infty}\int_a^5 \frac{1}{(x-6)^5}\,dx = \lim_{a\to-\infty}\left\{\frac{-1}{4(x-6)^4}\right\}_a^5$$
$$= \lim_{a\to-\infty}\left[-\frac{1}{4} + \frac{1}{4(a-6)^4}\right] = -\frac{1}{4}.$$

When $f(x)$ is continuous for all x, we define

$$\int_{-\infty}^\infty f(x)\,dx = \lim_{a\to-\infty}\int_a^c f(x)\,dx + \lim_{b\to\infty}\int_c^b f(x)\,dx, \tag{7.42}$$

provided both limits exist. The number c is arbitrary; existence of the limits is independent of c (see Exercise 33). For example, if we choose $c = 0$, then

$$\begin{aligned}
\int_{-\infty}^{\infty} \frac{x}{(1+x^2)^2}\,dx &= \lim_{a\to-\infty}\int_a^0 \frac{x}{(1+x^2)^2}\,dx + \lim_{b\to\infty}\int_0^b \frac{x}{(1+x^2)^2}\,dx \\
&= \lim_{a\to-\infty}\left\{\frac{-1}{2(1+x^2)}\right\}_a^0 + \lim_{b\to\infty}\left\{\frac{-1}{2(1+x^2)}\right\}_0^b \\
&= \lim_{a\to-\infty}\left[\frac{1}{2(1+a^2)} - \frac{1}{2}\right] + \lim_{b\to\infty}\left[\frac{1}{2} - \frac{1}{2(1+b^2)}\right] \\
&= -\frac{1}{2} + \frac{1}{2} = 0.
\end{aligned}$$

EXAMPLE 7.32

(a) Is it possible to assign a number to represent the area bounded by the curves $y = 1/x$, $x = 1$, and $y = 0$?
(b) If the region in (a) is rotated around the x-axis, is it possible to assign a number to represent the volume of the solid so generated?

SOLUTION (a) If a number can be assigned to the area of the region bounded by the curves (Figure 7.95), it must be defined by the improper integral

$$A = \int_1^{\infty} \frac{1}{x}\,dx = \lim_{b\to\infty}\int_1^b \frac{1}{x}\,dx = \lim_{b\to\infty}\{\ln|x|\}_1^b = \lim_{b\to\infty}\ln b = \infty.$$

Consequently, we cannot assign an area to the region.

(b) If we use circular discs to calculate the volume in the solid formed by rotating the region in Figure 7.95 around the x-axis, then

$$V = \int_1^{\infty} \pi y^2\,dx = \lim_{b\to\infty}\int_1^b \frac{\pi}{x^2}\,dx = \lim_{b\to\infty}\left\{\frac{-\pi}{x}\right\}_1^b = \lim_{b\to\infty}\left(\pi - \frac{\pi}{b}\right) = \pi.$$

Thus we should assign to the solid a volume of π cubic units. Does this bother you? The area in Figure 7.95 bounded by the curves $y = 1/x$ the x-axis, and vertical lines at $x = 1$ and $x = b$ can be made larger than any number whatsoever by choosing b sufficiently large. On the other hand, when this area is revolved around the x-axis, no matter how large b is, the volume of the solid of revolution is never greater than π. This may seem to defy your intuition at first, but think about it very carefully.

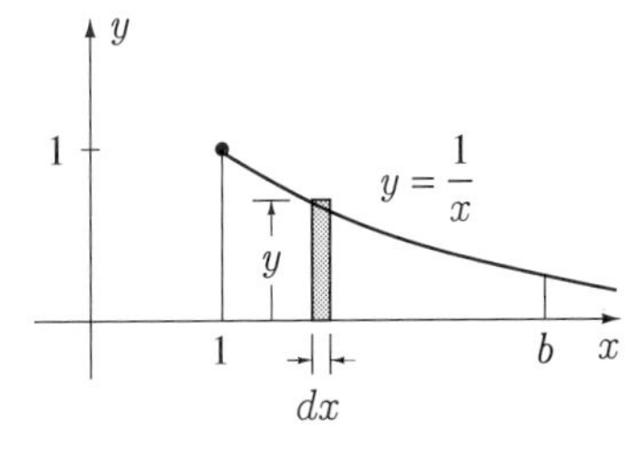

FIGURE 7.95

■

EXAMPLE 7.33

Find the escape velocity of a projectile from the earth's surface.

FIGURE 7.96

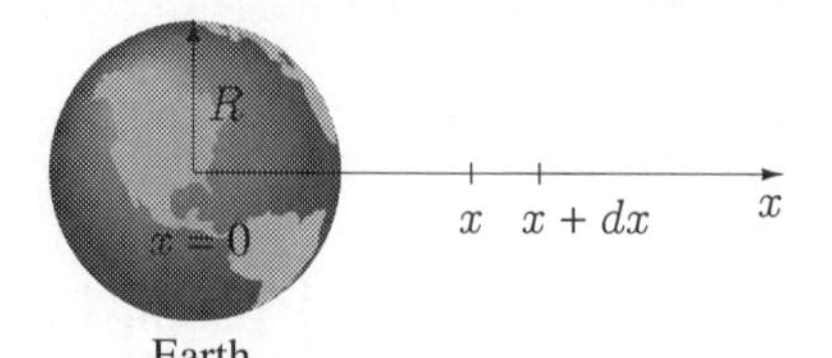

SOLUTION Suppose the projectile is fired from the earth's surface with speed v in the x-direction, as shown in Figure 7.96. If the projectile is to escape the earth's gravitational pull, its initial kinetic energy must be greater than or equal to the work done against gravity as the projectile travels from the earth's surface ($x = R$) to a point where it is free of gravity ($x = \infty$). When the projectile is a distance x from the centre of the earth, the force of attraction on it is

$$F(x) = -\frac{GMm}{x^2},$$

where $G > 0$ is a constant, m is the mass of the projectile, and M is the mass of the earth. This is known as *Newton's universal law of gravitation*. The work done against gravity in travelling from $x = R$ to $x = \infty$ is therefore

$$\begin{aligned} W &= \int_R^\infty \frac{GmM}{x^2}\,dx = \lim_{b\to\infty}\int_R^b \frac{GmM}{x^2}\,dx \\ &= \lim_{b\to\infty}\left\{-\frac{GmM}{x}\right\}_R^b = \lim_{b\to\infty}\left[\frac{GmM}{R} - \frac{GmM}{b}\right] = \frac{GmM}{R}. \end{aligned}$$

Since the initial kinetic energy of the projectile is $mv^2/2$, it escapes the gravitational pull of the earth if

$$\frac{1}{2}mv^2 \geq \frac{GmM}{R};$$

that is, if

$$v \geq \sqrt{\frac{2GM}{R}}.$$

Hence $\sqrt{2GM/R}$ is the escape velocity (and notice that it is independent of the mass of the projectile). If we take the mean radius of the earth as 6370 km, its mean density as $\rho = 5.52\times 10^3$ kg/m^3, and $G = 6.67\times 10^{-11}$, we obtain $v = 11.2$ km/s. (Compare this with a 308 Winchester that fires a 150-grain bullet with a muzzle velocity of only 0.81 km/s.)

■

A second type of improper integral occurs when the integrand is discontinuous at a point or points in the interval of integration. For example, integrands of the improper integrals

$$\int_1^5 \frac{1}{\sqrt{x-1}}\,dx \qquad \text{and} \qquad \int_1^5 \frac{1}{(x-1)^4}\,dx$$

each have infinite discontinuities at $x = 1$. If c is a number between 1 and 5, it is easy to calculate

$$\int_c^5 \frac{1}{\sqrt{x-1}}\,dx = \left\{2\sqrt{x-1}\right\}_c^5 = 4 - 2\sqrt{c-1} \qquad \text{and}$$

$$\int_c^5 \frac{1}{(x-1)^4}\,dx = \left\{\frac{-1}{3(x-1)^3}\right\}_c^5 = \frac{1}{3(c-1)^3} - \frac{1}{192},$$

and one possible interpretation of these definite integrals is the areas in Figures 7.97 and 7.98.

FIGURE 7.97

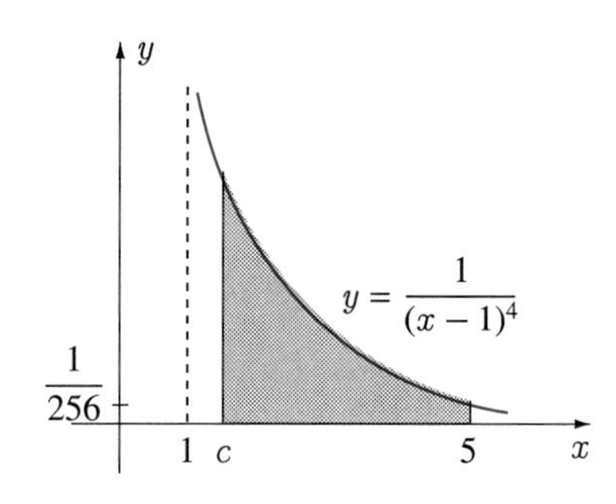

FIGURE 7.98

If we let c approach 1 from the right in the first integral, we find that

$$\lim_{c\to 1^+}\int_c^5 \frac{1}{\sqrt{x-1}}\,dx = \lim_{c\to 1^+}\left(4 - 2\sqrt{c-1}\right) = 4;$$

and in the second integral,

$$\lim_{c\to 1^+}\int_c^5 \frac{1}{(x-1)^4}\,dx = \lim_{c\to 1^+}\left[\frac{1}{3(c-1)^3} - \frac{1}{192}\right] = \infty.$$

It would seem reasonable then to define the area under the curve $y = 1/\sqrt{x-1}$ above the x-axis, and between the vertical lines $x = 1$ and $x = 5$ as 4 square units. On the other hand, the area bounded by $y = 1/(x-1)^4$, $y = 0$, $x = 5$, and $x = c$ becomes increasingly large as $x = c$ moves closer to $x = 1$, and we cannot therefore assign a value to this area when $c = 1$. In other words, if we define

$$\int_1^5 \frac{1}{\sqrt{x-1}}\,dx = \lim_{c\to 1^+}\int_c^5 \frac{1}{\sqrt{x-1}}\,dx \qquad \text{and}$$

$$\int_1^5 \frac{1}{(x-1)^4}\,dx = \lim_{c\to 1^+}\int_c^5 \frac{1}{(x-1)^4}\,dx,$$

then the first improper integral has a value of 4, whereas the second has no value.

These improper integrals are examples of the general situation described in the following definition and illustrated in Figure 7.99.

(a)

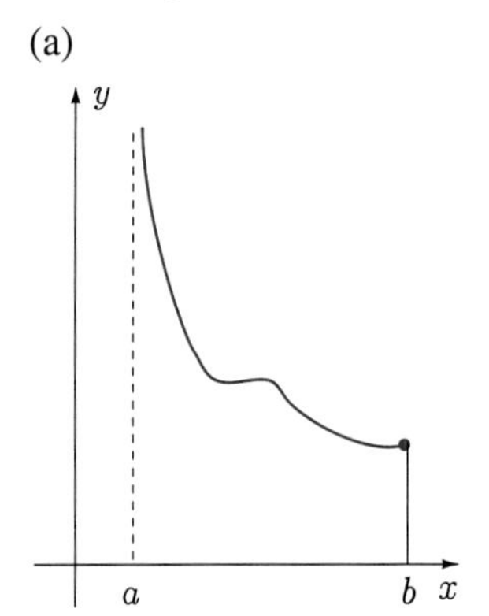

(b)

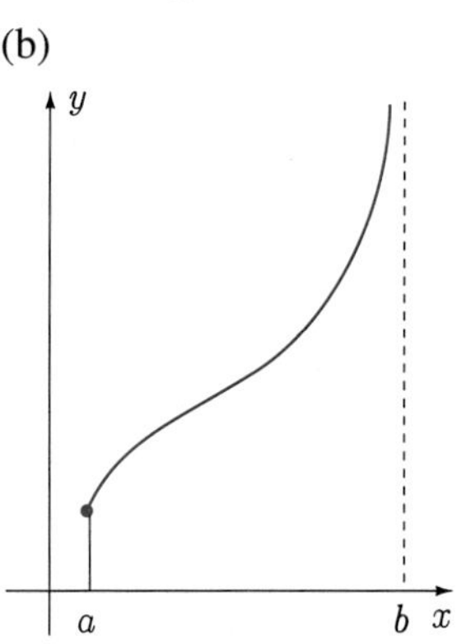

(c)

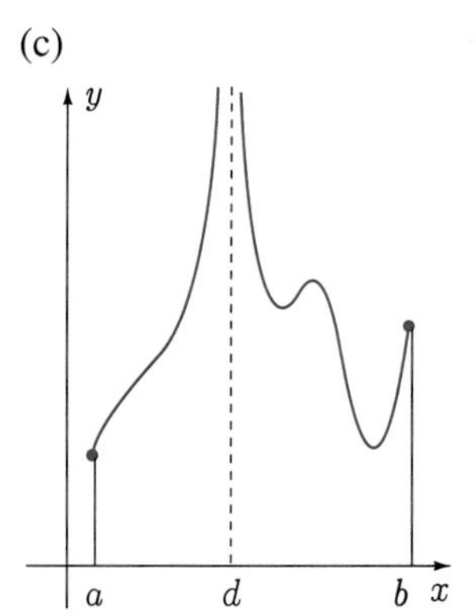

FIGURE 7.99

Definition 7.5

If $f(x)$ is continuous at every point in the interval $a \le x \le b$ except at $x = a$, then

$$\int_a^b f(x)\,dx = \lim_{c\to a^+} \int_c^b f(x)\,dx, \tag{7.43}$$

provided the limit exists (Figure 7.99(a)). If $f(x)$ is continuous at every point in the interval $a \le x \le b$ except $x = b$, then

$$\int_a^b f(x)\,dx = \lim_{c\to b^-} \int_a^c f(x)\,dx, \tag{7.44}$$

provided the limit exists (Figure 7.99(b)). If $f(x)$ is continuous at every point in the interval $a \le x \le b$ except $x = d$ where $a < d < b$, then

$$\int_a^b f(x)\,dx = \lim_{c\to d^-} \int_a^c f(x)\,dx + \lim_{c\to d^+} \int_c^b f(x)\,dx, \tag{7.45}$$

provided both limits exist (Figure 7.99(c)).

EXAMPLE 7.34

Determine whether the following improper integrals converge or diverge:

(a) $\displaystyle\int_{-2}^{2} \frac{1}{(x-2)^3}\,dx$ (b) $\displaystyle\int_{-1}^{1} \frac{1}{x^4}\,dx$ (c) $\displaystyle\int_{1}^{5} \frac{x}{\sqrt{x-1}}\,dx$

FIGURE 7.100

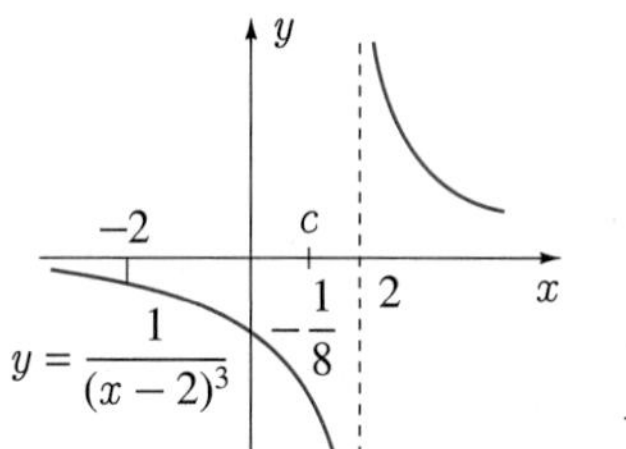

FIGURE 7.101

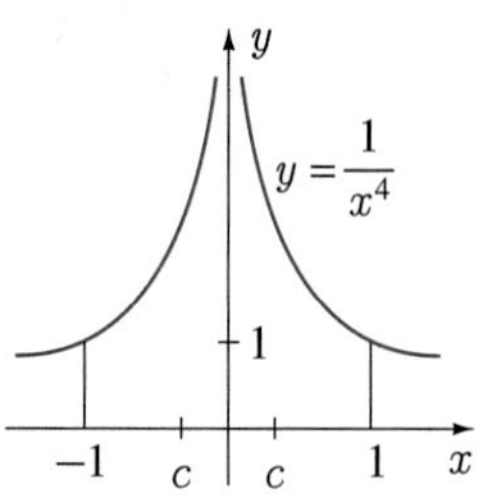

FIGURE 7.102

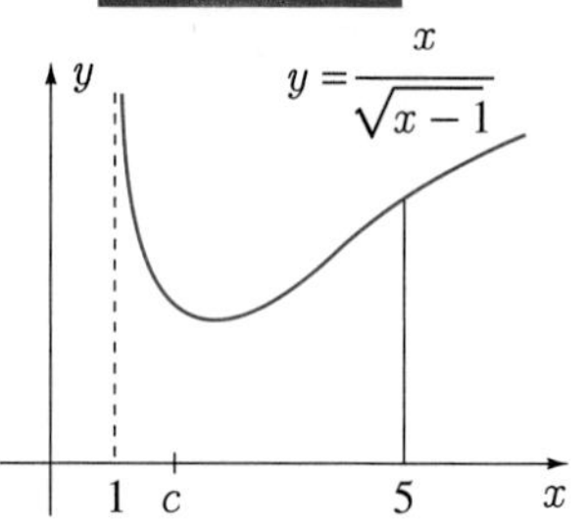

SOLUTION

(a) The graph of the integrand (Figure 7.100) shows a discontinuity at $x = 2$. Hence,

$$\int_{-2}^{2} \frac{1}{(x-2)^3}\,dx = \lim_{c\to 2^-} \int_{-2}^{c} \frac{1}{(x-2)^3}\,dx = \lim_{c\to 2^-} \left\{ \frac{-1}{2(x-2)^2} \right\}_{-2}^{c}$$

$$= \lim_{c\to 2^-} \left[\frac{-1}{2(c-2)^2} + \frac{1}{32} \right] = -\infty.$$

The improper integral therefore diverges.

(b) Since the discontinuity is interior to the interval of integration (Figure 7.101), we use equation 7.45,

$$\int_{-1}^{1} \frac{1}{x^4}\,dx = \lim_{c\to 0^-} \int_{-1}^{c} \frac{1}{x^4}\,dx + \lim_{c\to 0^+} \int_{c}^{1} \frac{1}{x^4}\,dx$$

$$= \lim_{c\to 0^-} \left\{ \frac{-1}{3x^3} \right\}_{-1}^{c} + \lim_{c\to 0^+} \left\{ \frac{-1}{3x^3} \right\}_{c}^{1}$$

$$= \lim_{c\to 0^-} \left[-\frac{1}{3} - \frac{1}{3c^3} \right] + \lim_{c\to 0^+} \left[\frac{1}{3c^3} - \frac{1}{3} \right].$$

Since neither of these limits exists, the improper integral diverges.

(c) Since the integrand is discontinuous only at $x = 1$ (Figure 7.102),

$$\int_{1}^{5} \frac{x}{\sqrt{x-1}}\,dx = \lim_{c\to 1^+} \int_{c}^{5} \frac{x}{\sqrt{x-1}}\,dx.$$

If we set $u = x - 1$, then $du = dx$, and

$$\int \frac{x}{\sqrt{x-1}}\,dx = \int \frac{u+1}{\sqrt{u}}\,du = \int (u^{1/2} + u^{-1/2})\,du$$

$$= \frac{2}{3}u^{3/2} + 2u^{1/2} + C = \frac{2}{3}(x-1)^{3/2} + 2\sqrt{x-1} + C.$$

Thus,

$$\int_{1}^{5} \frac{x}{\sqrt{x-1}}\,dx = \lim_{c\to 1^+} \left\{ \frac{2}{3}(x-1)^{3/2} + 2\sqrt{x-1} \right\}_{c}^{5}$$

$$= \lim_{c\to 1^+} \left[\frac{28}{3} - \frac{2}{3}(c-1)^{3/2} - 2\sqrt{c-1} \right] = \frac{28}{3},$$

and the improper integral converges. ■

Various types of improper integrals may occur in the same problem. For example, Figure 7.103 shows that the integral of $f(x) = 1/(x^2 - 1)$ from $x = -3$ to $x = \infty$ involves the use of five limits:

$$\int_{-3}^{\infty} \frac{1}{x^2-1}\,dx = \lim_{c\to -1^-} \int_{-3}^{c} \frac{1}{x^2-1}\,dx + \lim_{c\to -1^+} \int_{c}^{0} \frac{1}{x^2-1}\,dx + \lim_{c\to 1^-} \int_{0}^{c} \frac{1}{x^2-1}\,dx$$

$$+ \lim_{c\to 1^+} \int_{c}^{10} \frac{1}{x^2-1}\,dx + \lim_{b\to\infty} \int_{10}^{b} \frac{1}{x^2-1}\,dx.$$

If just one of these limits fails to exist, then the improper integral diverges.

If the integrand has a finite discontinuity, such as the function $f(x)$ in Figure 7.104, then the improper integral of $f(x)$ from $x = a$ to $x = b$ is defined in terms of two limits,

$$\int_{a}^{b} f(x)\,dx = \lim_{c\to d^-} \int_{a}^{c} f(x)\,dx + \lim_{c\to d^+} \int_{c}^{b} f(x)\,dx,$$

but there is no question in this case that the improper integral converges.

FIGURE 7.103

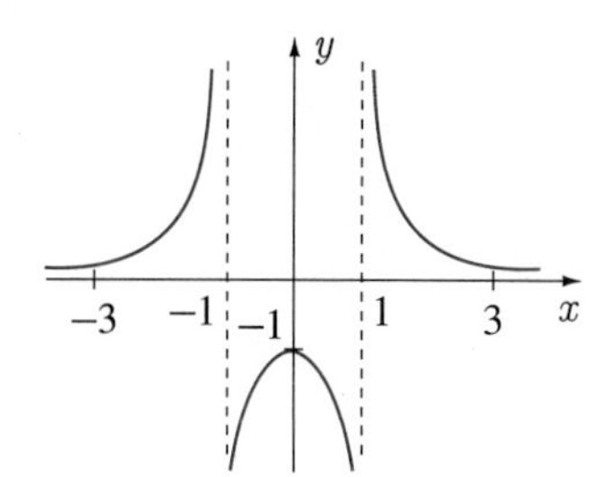

FIGURE 7.104

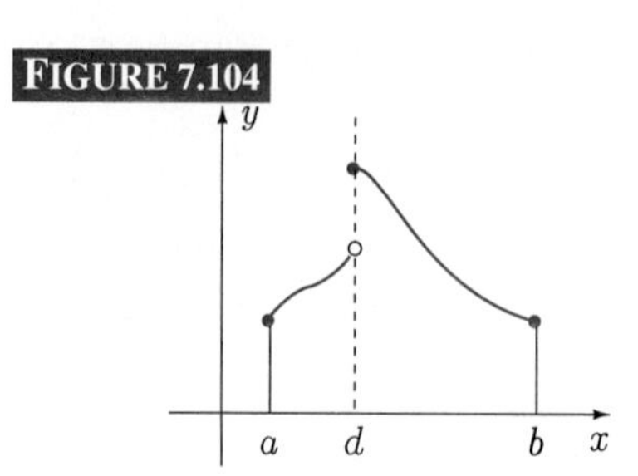

EXERCISES 7.9

In Exercises 1–18 determine whether the improper integral converges or diverges. Find the value for each convergent integral.

1. $\displaystyle\int_3^{\infty} \frac{1}{(x+4)^2}\,dx$ **2.** $\displaystyle\int_3^{\infty} \frac{1}{(x+4)^{1/3}}\,dx$

3. $\displaystyle\int_{-\infty}^{-4} \frac{x}{\sqrt{x^2-2}}\,dx$ **4.** $\displaystyle\int_{-\infty}^{-4} \frac{x}{(x^2-2)^4}\,dx$

5. $\displaystyle\int_{-\infty}^{\infty} \frac{10^{10}x^3}{(x^4+5)^2}\,dx$ **6.** $\displaystyle\int_{-\infty}^{\infty} \frac{x^3}{(x^4+5)^{1/4}}\,dx$

7. $\displaystyle\int_0^{1} \frac{1}{(1-x)^{5/3}}\,dx$ **8.** $\displaystyle\int_0^{1} \frac{1}{\sqrt{1-x}}\,dx$

9. $\displaystyle\int_1^{\infty} x\sqrt{x^2-1}\,dx$ **10.** $\displaystyle\int_2^{5} \frac{x}{\sqrt{x^2-4}}\,dx$

11. $\displaystyle\int_{-1}^{1} \frac{x}{(1-x^2)^2}\,dx$ **12.** $\displaystyle\int_{-\infty}^{\infty} \frac{1}{x^2}\,dx$

13. $\displaystyle\int_0^{\infty} \frac{1}{\sqrt{x}}\,dx$ **14.** $\displaystyle\int_{-\infty}^{\pi/2} \frac{x}{(x^2-4)^2}\,dx$

15. $\displaystyle\int_4^{\infty} \cos x\,dx$ **16.** $\displaystyle\int_{-\infty}^{\infty} \sin x\,dx$

17. $\displaystyle\int_0^{\infty} \frac{x}{\sqrt{x+3}}\,dx$ **18.** $\displaystyle\int_2^{3} \frac{x^3}{\sqrt{x^2-4}}\,dx$

19. If $f(x)$ is a continuous odd function, is it necessarily true that

$$\int_{-\infty}^{\infty} f(x)\,dx = 0?$$

20. (a) Is it possible to assign a number to represent the area bounded by the curves

$$y = 1 - x^{-1/4}, \qquad y = 1, \qquad x = 1\ ?$$

(b) If the region in (a) is rotated about the line $y = 1$, is it possible to assign a number to represent its volume?

21. Repeat Exercise 20 if $y = 1 - x^{-1/4}$ is replaced by $y = 1 - x^{-2/3}$.

22. Repeat Exercise 20 if $y = 1 - x^{-1/4}$ is replaced by $y = 1 - x^{-3}$.

23. A function $f(x)$ qualifies as a *probability density function* (pdf) on the interval $x \geq 0$ if it satisfies two conditions: $f(x) \geq 0$ for $x \geq 0$, and

$$\int_0^{\infty} f(x)\,dx = 1.$$

(a) Show that each of the following functions qualifies as a pdf:

$$f(x) = \frac{6x}{(1+3x^2)^2}; \qquad f(x) = \frac{2x}{(1+x)^3}$$

(b) If a variable x has pdf $f(x)$ defined for $x \geq 0$, then the probability that x lies in an interval I is the definite integral of $f(x)$ over I. In particular, the probability that x is greater than or equal to a is

$$P(x \geq a) = \int_a^{\infty} f(x)\,dx.$$

Calculate $P(x \geq 3)$ for each pdf in part (a).

24. Verify that

$$\int_1^{\infty} \frac{1}{x^p}\,dx$$

converges if $p > 1$ and diverges if $p < 1$.

25. The force of repulsion between two point charges of like sign, one of size q and the other of size unity, has magnitude

$$F = \frac{q}{4\pi\epsilon_0 r^2},$$

where ϵ_0 is a constant and r is the distance between the charges. The potential V at any point P due to charge q is defined as the work required to bring the unit charge to P from infinity along the straight line joining q and P. Find a formula for V.

26. Verify that if $f(x)$ is continuous on $a \leq x \leq b$ except for a finite discontinuity at d (Figure 7.104), then the improper integral of $f(x)$ from $x = a$ to $x = b$ must converge.

27. Find the length of the loop of the curve $9y^2 = x(3 - x)^2$.

28. Is equation 7.42 equivalent to the following equation?

$$\int_{-\infty}^{\infty} f(x)\, dx = \lim_{a \to \infty} \int_{-a}^{a} f(x)\, dx$$

In Exercises 29–32 determine whether the improper integral converges or diverges. Hint: Compare each integral to a known convergent or divergent integral.

29. $\displaystyle\int_{2}^{\infty} \frac{x^2}{\sqrt{x^2 - 1}}\, dx$ **30.** $\displaystyle\int_{1}^{3} \frac{x^3}{(27 - x^3)^2}\, dx$

31. $\displaystyle\int_{0}^{1} \frac{x^2}{\sqrt{1 - x^2}}\, dx$ **32.** $\displaystyle\int_{-\infty}^{-2} \frac{\sqrt{-x}}{(x^2 + 5)^2}\, dx$

33. Show that existence of the limits in equation 7.42 is independent of the choice of c.

34. One of the most important functions in physics and engineering is the *Dirac delta function* $\delta(x - a)$ introduced in Exercise 24 of Section 2.5 as the limit of the unit pulse function (Figure 7.105):

$$\delta(x - a) = \lim_{\epsilon \to 0} P_\epsilon(x - a).$$

For a continuous function $f(x)$,

$$\int_{-\infty}^{\infty} f(x)\delta(x - a)\, dx = \int_{-\infty}^{\infty} f(x)\left[\lim_{\epsilon \to 0} P_\epsilon(x - a)\right] dx.$$

Assuming that the order of integration and the process of taking the limit as ϵ approaches zero can be interchanged, that is,

$$\int_{-\infty}^{\infty} f(x)\left[\lim_{\epsilon \to 0} P_\epsilon(x - a)\right] dx = \lim_{\epsilon \to 0}\int_{-\infty}^{\infty} f(x) P_\epsilon(x - a)\, dx,$$

show that

$$\int_{-\infty}^{\infty} f(x)\delta(x - a)\, dx = f(a).$$

FIGURE 7.105

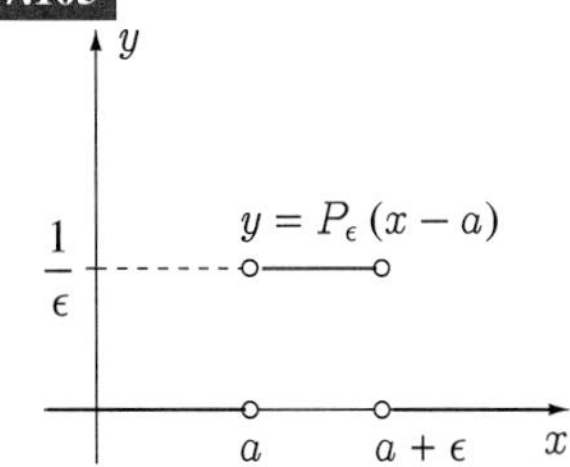

SUMMARY

In this chapter we used definite integrals in a wide variety of applications. With differentials and representative elements we were able to avoid memorization of formulas. To summarize the use of representative elements in various applications, consider the region R in Figure 7.106.

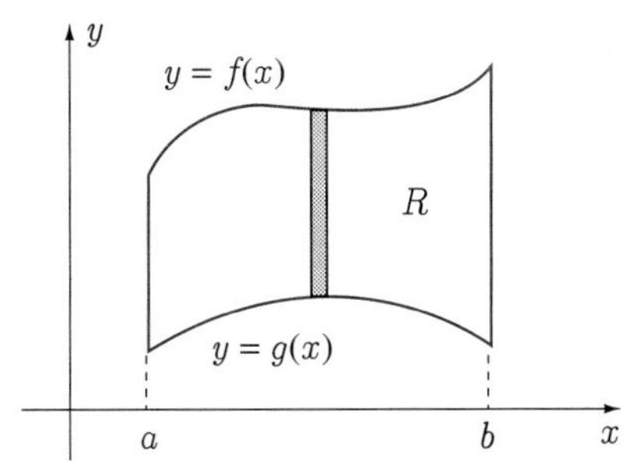

FIGURE 7.106

To find the area of R, we draw at position x a rectangle of length $f(x) - g(x)$ and width dx. The areas $[f(x) - g(x)]\, dx$ of all such rectangles are then added to find the total area:

$$\int_{a}^{b} [f(x) - g(x)]\, dx.$$

If R is rotated around the x-axis to form a solid of revolution, the rectangle generates a washer with volume $\{\pi[f(x)]^2 - \pi[g(x)]^2\}\,dx$, and therefore the total volume is

$$\int_a^b \left\{\pi[f(x)]^2 - \pi[g(x)]^2\right\}\,dx.$$

If R is rotated about the y-axis, the rectangle traces out a cylindrical shell of volume $2\pi x[f(x) - g(x)]\,dx$, and the total volume is

$$\int_a^b 2\pi x[f(x) - g(x)]\,dx.$$

First moments of this rectangle about the x-and y-axes are

$$\frac{1}{2}[f(x) + g(x)][f(x) - g(x)]\,dx \qquad \text{and} \qquad x[f(x) - g(x)]\,dx,$$

and these lead to the definite integrals

$$\int_a^b \frac{1}{2}\left\{[f(x)]^2 - [g(x)]^2\right\}\,dx \qquad \text{and} \qquad \int_a^b x[f(x) - g(x)]\,dx$$

for first moments of area of R about the x- and y-axes. Second moments of area of region R about the x- and y-axes are

$$\int_a^b \frac{1}{3}\left\{[f(x)]^3 - [g(x)]^3\right\}\,dx \qquad \text{and} \qquad \int_a^b x^2[f(x) - g(x)]\,dx,$$

where in the first integral it was necessary to use formula 7.38 for the second moment of the rectangle.

The key, then, to the use of the definite integral is to divide the region into smaller elements, calculate the required quantity for a representative element, and then use a definite integral to add over all elements.

An important point to keep in mind is that for lengths in the y-direction, we always take "upper y minus lower y", and for lengths in the x-direction, we use "larger x minus smaller x". This way many sign errors can be avoided when calculating *lengths*. For instance, note that this rule is used to find the length of a rectangle (in areas, volumes, fluid pressure), the radius of a circle (in volumes, surface area), depth (fluid pressure), and some distances (work).

An improper integral is a definite integral with an infinite limit and/or a point of discontinuity in the interval of integration. The value of an improper integral is defined by first calculating it over a finite interval in which the integrand is continuous, and then letting the limit on the integral approach either infinity or the point of discontinuity.

Key Terms and Formulas

In reviewing this chapter, you should be able to define or discuss the following key terms:

Area
Representative rectangle
Volume of solid of revolution
Disc method
Washer method
Cylindrical shell method
Length of a curve
Work
Fluid pressure
Fluid force
First moment of mass
First moment of area
Centre of mass
Centroid
Moment of inertia
Second moment
Volume by slicing
Area of surface of revolution
Improper integral

REVIEW EXERCISES

In Exercises 1–5, calculate for the region bounded by the curves: (a) its area; (b) the volumes of the solids of revolution obtained by rotating the region about the x- and y-axes; (c) its centroid; and (d) its second moments of area about the x- and y-axes.

1. $y = 9 - x^2, \quad y = 0$

2. $y = x^3, \quad y = 2 - x, \quad x = 0$

3. $x = y^2 - 1, \quad x = 4 - 4y^2$

4. $2y = x, \quad y + 1 = x, \quad y = 0$

5. $2x + y = 2, \quad y - 2x = 2, \quad x = y + 1, \quad y + x + 1 = 0$

In Exercises 6–10 find the volume of the solid of revolution obtained by rotating the area bounded by the curves about the line.

6. $y = |x| + 2, \quad y = 3, \quad$ about $y = 1$

7. $x = y^4, \quad x = 4, \quad$ about $x = 4$

8. $y = x^3 - x, \quad y = 0, \quad$ about $y = 0$

9. $yx^2 = 1, \quad x = 1, \quad x = 2, \quad y = 0 \quad$ about $y = -1$

10. $y = x + 2, \quad y = \sqrt{4 - 2x}, \quad y = 0, \quad$ about $y = 1$

In Exercises 11–15 find the first and second moments of area of the region bounded by the curves about the line.

11. $y = 4x^2, \quad y = 2 + 2x^2, \quad$ about $y = -1$

12. $2x = \sqrt{y}, \quad \sqrt{2}x = \sqrt{y - 2}, \quad x = 0, \quad$ about $x = 1$

13. $3y = 2x + 6, \quad 2y + x = 4, \quad y = 0, \quad$ about $y = -2$

14. $3y = 2x + 6, \quad 2y + x = 4, \quad y = 0, \quad$ about $y = 2$

15. $y = x^2 - 2x \ (y \leq 0), \quad y = 2x, \quad y = 4 - 2x, \quad$ about $x = 1$

16. A water tank in the form of a right-circular cylinder with radius 1 m and height 3 m has its axis vertical. If it is half full, how much work is required to empty the tank through an outlet at the top of the tank?

17. A car runs off a bridge into a river and submerges. If the tops of its front and rear side windows are 2 m below the surface of the water (Figure 7.107), what is the force due to water pressure on each window?

FIGURE 7.107

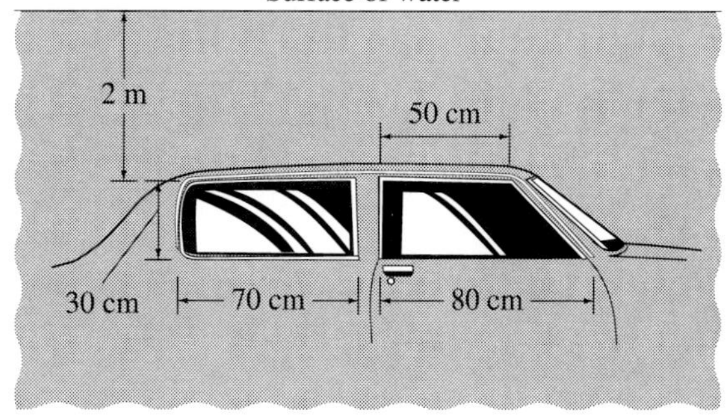

18. Find the surface area of the volume of the solid of revolution obtained by rotating the area bounded by the curves

$$2x + y = 2, \qquad y - 2x = 2, \qquad y = 0$$

about the x-axis.

19. A tank in the form of an inverted right-circular cone of depth 10 m and radius 4 m is full of water. How much work is required to lower the water level in the tank by 2 m by pumping water out through an outlet 1 m above the top of the tank?

20. Show that if it takes W units of work to stretch a spring from equilibrium to a certain stretch, then it requires $3W$ more units of work to double that stretch.

In Exercises 21–28 determine whether the improper integral converges or diverges. Evaluate each convergent integral.

21. $\displaystyle\int_1^{\infty} \frac{1}{x^{3/2}}\, dx$ **22.** $\displaystyle\int_0^{3} \frac{1}{\sqrt{3 - x}}\, dx$

23. $\displaystyle\int_{-1}^{0} \frac{1}{(x + 1)^2}\, dx$ **24.** $\displaystyle\int_{-2}^{2} \frac{x}{\sqrt{4 - x^2}}\, dx$

25. $\displaystyle\int_{-\infty}^{\infty} \frac{1}{(x + 3)^3}\, dx$ **26.** $\displaystyle\int_{-\infty}^{-3} \frac{1}{\sqrt{-x}}\, dx$

27. $\displaystyle\int_{-6}^{\infty} x\sqrt{x^2 + 4}\, dx$ **28.** $\displaystyle\int_{-\infty}^{\infty} \frac{x}{(x^2 - 1)^2}\, dx$

29. A uniformly tapered rod of length 1 m has circular cross-sections, the radii of its ends being 1 cm and 2 cm. Find the volume of the rod by:

(a) using the fact that every cross-section is circular;

(b) considering the rod as a volume of a solid of revolution.

30. In Exercise 16, how much work is required to empty the tank if its axis is horizontal and the outlet is at the top of the tank?

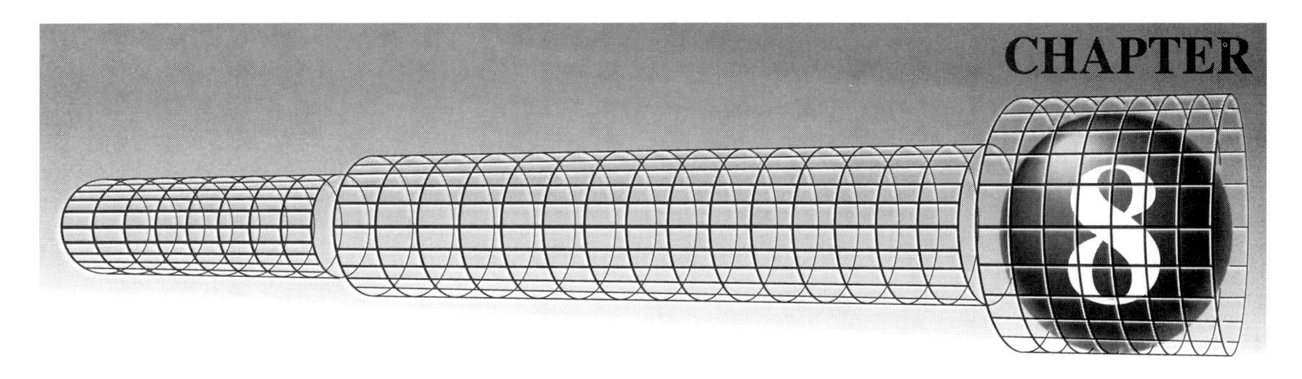

Further Transcendental Functions and their Derivatives

In Chapter 3 we learned to differentiate functions that involve a finite number of additions, subtractions, multiplications, divisions, and roots. These functions fall into a wider class called *algebraic functions*. A function $y = f(x)$ is said to be **algebraic** if, for all x in its domain, it satisfies an equation of the form

$$P_0(x)y^n + P_1(x)y^{n-1} + \cdots + P_{n-1}(x)y + P_n(x) = 0, \qquad (8.1)$$

where $P_0(x),\ldots,P_n(x)$ are polynomials in x, and n is a positive integer. A polynomial $P(x)$ is itself algebraic since it satisfies $y - P(x) = 0$; that is, it satisfies 8.1 with $n = 1$, $P_0(x) = 1$, and $P_1(x) = -P(x)$. Rational functions $P(x)/Q(x)$ are also algebraic ($n = 1$, $P_0(x) = Q(x)$, and $P_1(x) = -P(x)$). The function $f(x) = x^{1/3}$ is algebraic since it satisfies $y^3 - x = 0$. The equation $y^3 + y = x$ defines y implicitly as a function of x. We cannot find an explicit form for the function, but according to 8.1, the function so defined is algebraic.

A function that is not algebraic is called **transcendental**. The trigonometric, exponential, and logarithm functions which we differentiated in Chapter 3 are transcendental. In this chapter we consider two other classes of transcendental functions: inverse trigonometric and hyperbolic functions.

SECTION 8.1 Inverse Functions

When we speak of a function $y = g(x)$, it is inherent that there is a unique y associated with each x; that is, given a value of x in the domain of $g(x)$, the function associates one — and only one — value y in the range. However, it may happen, and quite often does, that a value of y in the range of the function may be associated with more than one value of x. For example, each $y > 0$ in the range of the function $y = g(x) = x^2$ (Figure 8.1) is associated with two values of x, namely, $x = \pm\sqrt{y}$.

Some functions have the property that each value of y in the range arises from only one x in the domain. For instance, given any value y in the range of the function $y = f(x) = x^3$ in Figure 8.2, there is a unique x such that $y = x^3$, namely, $x = y^{1/3}$. Such a function is said to be "one-to-one". Formally, we say that a function $f(x)$ is **one-to-one** if, for any two values x_1 and x_2 in the domain of $f(x)$, it follows that $f(x_1) \neq f(x_2)$. To repeat, a one-to-one function $f(x)$ has the property that given any y in its range, there is one — and only one — x in its domain for which $y = f(x)$. We can therefore define a function which maps values in the range of $f(x)$ onto values in the domain, a function that maps y onto x if x is mapped by $f(x)$ onto y. We call this function the inverse function of $f(x)$. It reverses the action of $f(x)$. For the function $f(x) = x^3$ in Figure 8.2, the inverse function is the function that takes cube roots of real numbers.

FIGURE 8.1

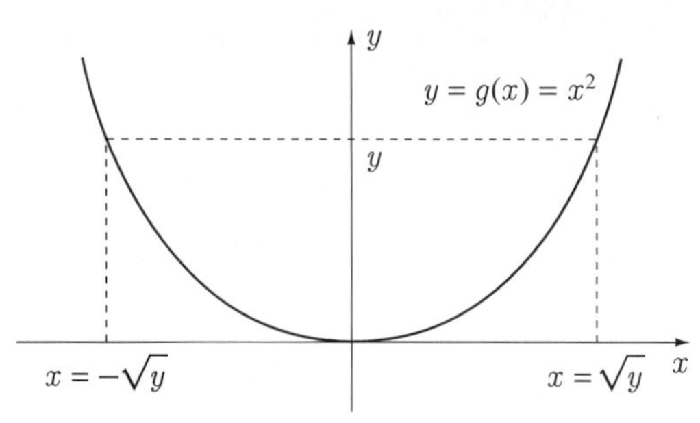

FIGURE 8.2

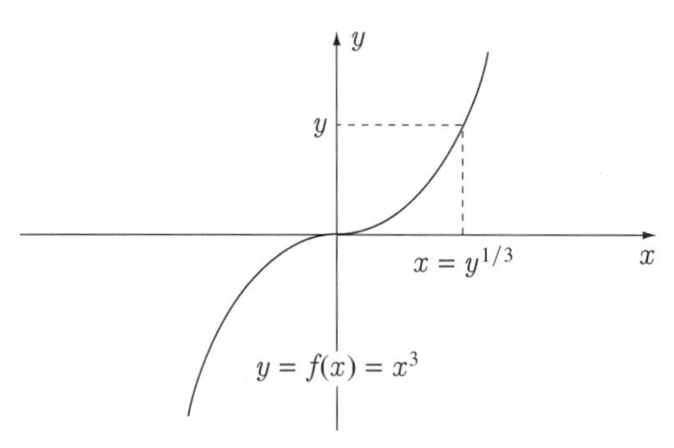

Unless there is a good reason to do otherwise, it is our custom to denote the independent variable of a function by the letter x and the dependent variable by y. For the inverse function of a function $y = f(x)$, there is a natural tendency to denote the independent variable by y and the dependent variable by x since the inverse function maps the range of $f(x)$ onto its domain. Not for obvious reasons, it is better to maintain our usual practice of using x as independent variable and y as dependent variable even for inverse functions. When a function $f(x)$ has an inverse function, we adopt the notation $f^{-1}(x)$ to represent the inverse function. For example, when $f(x) = x^3$, the inverse function is $f^{-1}(x) = x^{1/3}$. Be careful with this notation. Do not interpret the "-1" as a power and write $f^{-1}(x)$ as $1/f(x)$. This is not correct. The notation f^{-1} represents a function, just as tan represents the tangent function and $\sqrt{\ }$ represents the positive square root function. It represents the inverse function of $f(x)$.

The inverse function $f^{-1}(x)$ of a function $f(x)$ "undoes" what $f(x)$ "does"; it reverses the effect of $f(x)$. For example, the function $f(x) = x^2,\ x \geq 0$ in Figure 8.3 is one-to-one; it squares nonnegative numbers. Its inverse is $f^{-1}(x) = \sqrt{x}$, the positive square root function (Figure 8.4). Squaring an x and then taking the positive square root of the result returns the original x. Likewise, the inverse of $g(x) = x^2,\ x \leq 0$ (Figure 8.5) is the negative square root function $g^{-1}(x) = -\sqrt{x}$ (Figure 8.6).

FIGURE 8.3

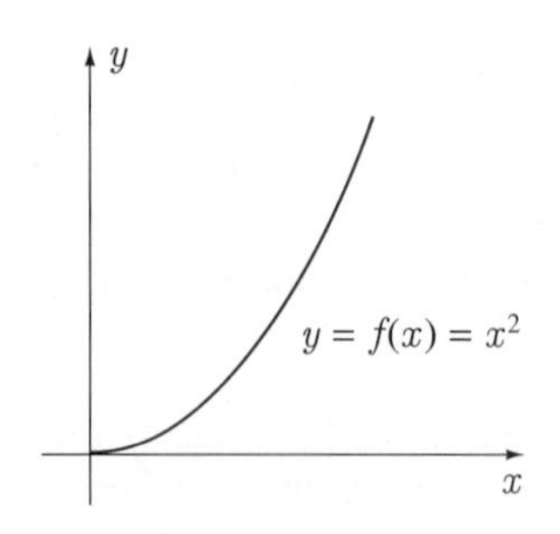

FIGURE 8.4

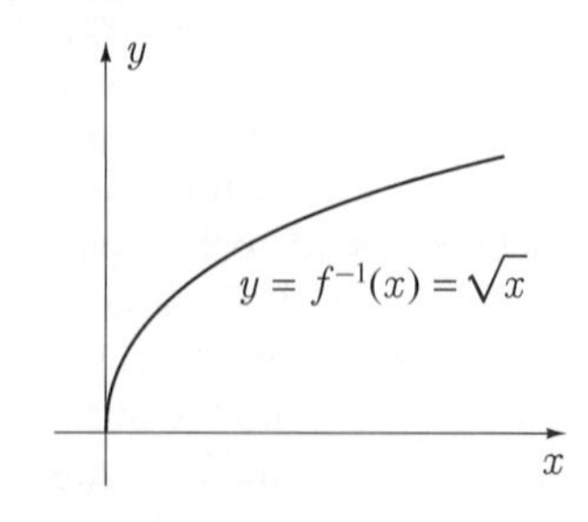

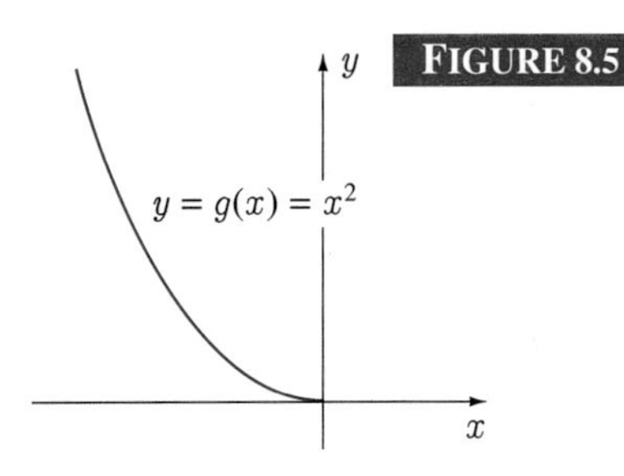

FIGURE 8.5

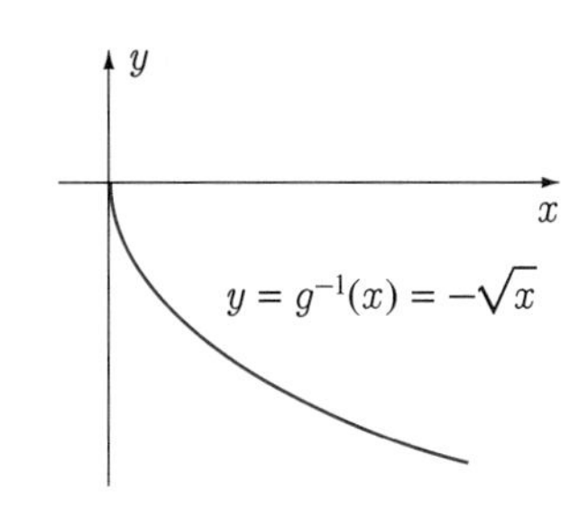

FIGURE 8.6

The fact that $f^{-1}(x)$ undoes $f(x)$ can be stated algebraically as follows. For each x in the domain of $f(x)$,

$$f^{-1}(f(x)) = x. \tag{8.2}$$

This is the defining relation for an inverse function. We understand that the domain of $f^{-1}(x)$ is the same as the range of $f(x)$ so that $f^{-1}(x)$ operates on all outputs of $f(x)$.

EXAMPLE 8.1

What is the inverse function of $f(x) = (x + 1)/(x - 2)$.

SOLUTION To find the algebraic definition of $f^{-1}(x)$, we solve the equation $y = (x + 1)/(x - 2)$ for x in terms of y. First we cross-multiply,

$$x + 1 = y(x - 2);$$

then group terms in x,

$$x(y - 1) = 2y + 1;$$

and finally divide by $y - 1$,

$$x = \frac{2y + 1}{y - 1}.$$

What have we accomplished by solving for x in terms of y? If we take an x in the domain of $f(x)$, then $f(x)$ produces $y = (x + 1)/(x - 2)$. Now this same x and y satisfy the equation $x = (2y + 1)/(y - 1)$ because this equation is simply a rearrangement of $y = (x + 1)/(x - 2)$. Consequently, if we substitute y into the right-hand side of the equation,

$$x = \frac{2y + 1}{y - 1},$$

we obtain the original x. This equation must therefore define the inverse function of $f(x)$. In other words, if we denote the independent variable by x, the inverse function is

$$f^{-1}(x) = \frac{2x + 1}{x - 1}.$$

■

Example 8.1 has illustrated that to find the inverse of a function $y = f(x)$, the equation should be solved for x in terms of y, and then variables should be renamed. If the equation does not have a unique solution for x in terms of y, then $f(x)$ does not have an inverse function. Such is the case for the function $g(x) = x^2$ in Figure 8.1. Solving $y = g(x) = x^2$ for x gives two solutions $x = \pm\sqrt{y}$.

So far our discussion of inverse functions has been algebraic. The geometry of inverse functions is most revealing. You may have noticed a relationship between the graphs of

$f(x)$ and $f^{-1}(x)$ in Figures 8.3 and 8.4, and of $g(x)$ and $g^{-1}(x)$ in Figures 8.5 and 8.6. The inverse function is the mirror image of the function in the line $y = x$; that is, graphs of inverse pairs are symmetric about the line $y = x$. These two examples are not mere coincidence, but as we now show, graphs of inverse functions are always mirror images of each other in the line $y = x$.

Suppose $f^{-1}(x)$ is the inverse of $f(x)$, and that (x, y) is any point on the graph of $y = f(x)$. Since $f^{-1}(x)$ is the inverse of $f(x)$, the result of substituting y into $f^{-1}(x)$ is x; that is, the point (y, x) is on the graph of $f^{-1}(x)$. But the point (y, x) is the mirror image of (x, y) in the line $y = x$. When this is combined with the fact that the domain of $f^{-1}(x)$ is the range of $f(x)$, our conjecture is verified.

This suggests a very simple way to graphically determine the inverse of a function $f(x)$: take its mirror image in the line $y = x$. Note, too, that if the mirror image does not represent a function, then no inverse function for $f(x)$ exists. For example, the mirror image of $y = x^2$, $-\infty < x < \infty$, in the line $y = x$ is shown in Figure 8.7 and does not represent a function. We conclude as before that this function does not have an inverse.

FIGURE 8.7

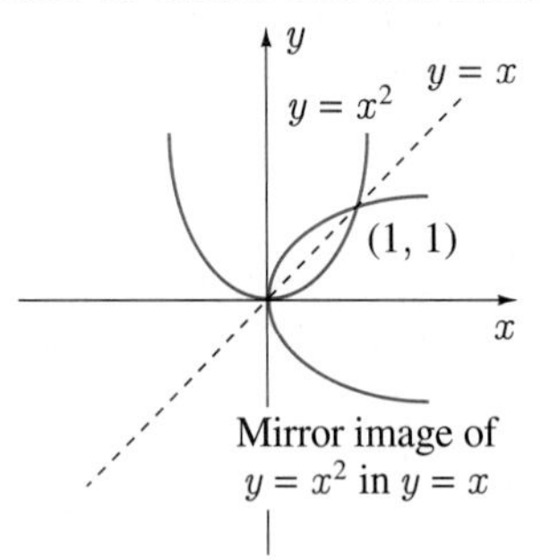

If $y = f^{-1}(x)$ is the mirror image of $y = f(x)$ in the line $y = x$, then $y = f(x)$ is the mirror image of $y = f^{-1}(x)$. This means that $f(x)$ is the inverse of $f^{-1}(x)$, and that $f(x)$ undoes what $f^{-1}(x)$ does:

$$f\left(f^{-1}(x)\right) = x. \tag{8.3}$$

Geometrically, a function $f(x)$ has an inverse if the reflection of its graph $y = f(x)$ in the line $y = x$ represents a function. But we know that a curve represents a function if every vertical line that intersects it does so at exactly one point. Furthermore, a vertical line intersects the reflected curve at exactly one point only if its horizontal reflection intersects $y = f(x)$ at exactly one point. These two facts enable us to state that a function $f(x)$ has an inverse function if — and only if — every horizontal line that intersects it does so at exactly one point. This is a geometric interpretation of a function being one-to-one. See, for example, the functions in Figures 8.3 and 8.5. Horizontal lines that intersect the curves do so at exactly one point. The function in Figure 8.1, which has no inverse, is intersected in two points by every horizontal line $y = c > 0$. When a function is increasing on an interval I (Figure 8.8), it passes the horizontal line test (and is one-to-one). Likewise, a function which is decreasing on I (Figure 8.9) is one-to-one. A function which is either increasing on an interval I or decreasing on I is said to be **monotonic** on I. What we have shown is the following result.

FIGURE 8.8

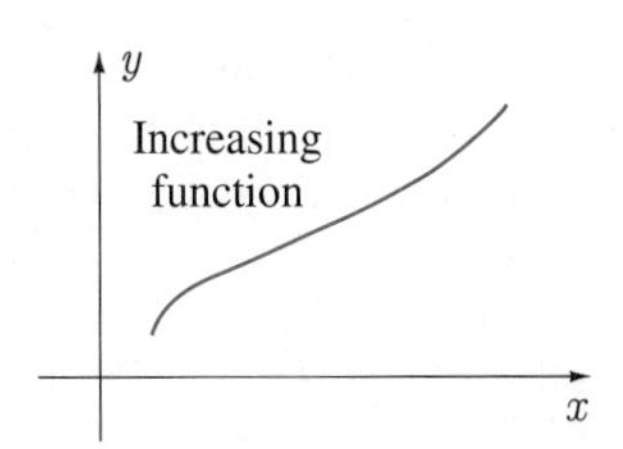

FIGURE 8.9

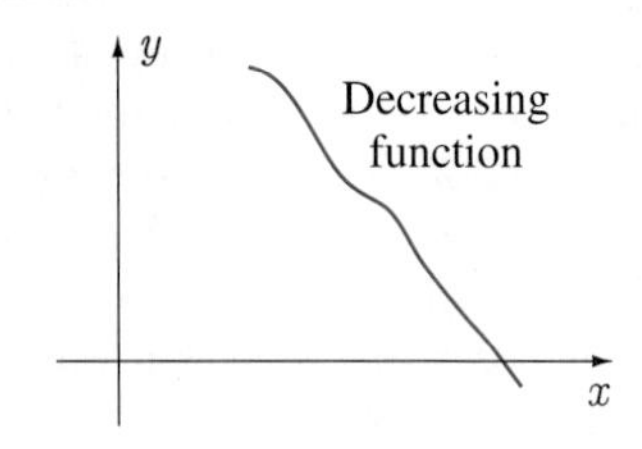

Theorem 8.1

> *A function which is monotonic on an interval has an inverse function on that interval.*

It is important to realize that monotonicity is a sufficient condition for existence of an inverse function; that is, if a function is monotonic, then it has an inverse. It is not, however, a necessary condition. The function in Figure 8.10(a) is not monotonic on the interval $-1 \le x \le 1$, but it does have the inverse function in Figure 8.10(b).

FIGURE 8.10

(a)

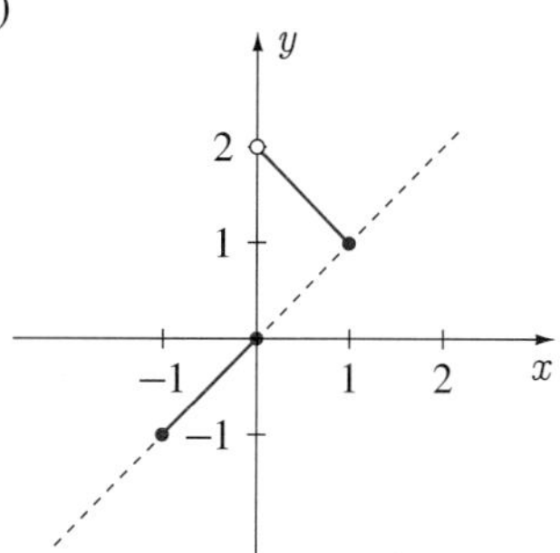

(b)

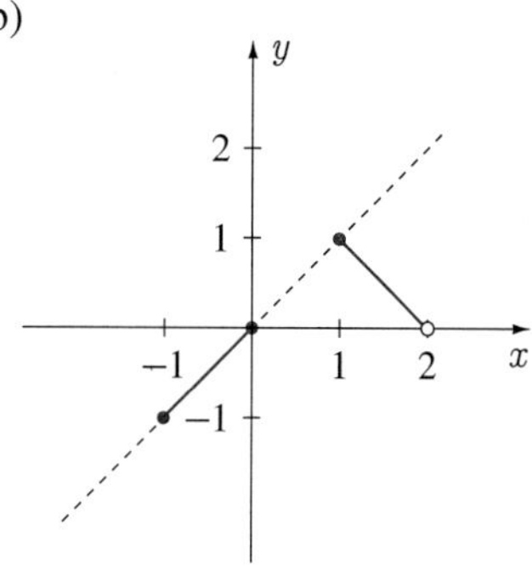

EXAMPLE 8.2

Does the function $f(x) = x^2 - 2x$ have an inverse? Does it have an inverse on the interval $x \ge 1$, and on the interval $x \le 1$?

SOLUTION The graph of $f(x)$ in Figure 8.11 indicates that $f(x)$ does not have an inverse. When restricted to the interval $x \ge 1$, the function is one-to-one, and does have an inverse. To find it, we solve $y = x^2 - 2x$ for x in terms of y. The quadratic formula applied to

$$x^2 - 2x - y = 0$$

gives

$$x = \frac{2 \pm \sqrt{4 + 4y}}{2} = 1 \pm \sqrt{y + 1}.$$

Since x must be greater than or equal to one, we must choose $x = 1 + \sqrt{y + 1}$. Hence, the inverse of $f(x) = x^2 - 2x,\ x \ge 1$ is

$$f^{-1}(x) = 1 + \sqrt{x + 1}.$$

Similarly, when restricted to the interval $x \le 1$, $f(x)$ has inverse

$$f^{-1}(x) = 1 - \sqrt{x + 1}.$$

FIGURE 8.11

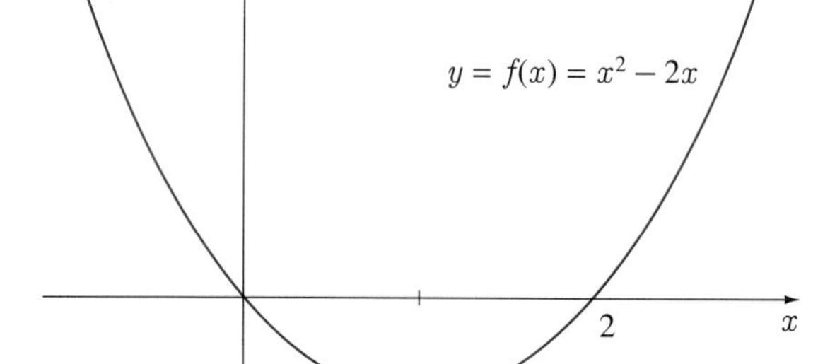

■

This example illustrates that when a function is not monotonic on an interval I, the interval can usually be subdivided into subintervals on which the function is monotonic, and on each such subinterval the function has an inverse.

EXERCISES 8.1

In Exercises 1–14 determine graphically whether the function has an inverse. Find each inverse function.

1. $f(x) = 2x + 3$

2. $f(x) = \sqrt{x+1}$

3. $f(x) = x^2 + x$

4. $f(x) = \dfrac{x+5}{2x+4}$

5. $f(x) = 1/x$

6. $f(x) = 3x^3 + 2$

7. $f(x) = \sqrt{4 - x^2}, \quad 0 \le x \le 2$

8. $f(x) = 2x + |x|$

9. $f(x) = x + |x|$

10. $f(x) = \sqrt{1 - x^2}$

11. $f(x) = x^4 + 2x^2 + 2, \quad x \le 0$

12. $f(x) = x^2 - 2x + 4, \quad x \ge 1$

13. $f(x) = \left(\dfrac{x+2}{x-2}\right)^3$

14. $f(x) = \dfrac{x}{3 + x^2}$

In Exercises 15–20 show that the function does not have an inverse. Subdivide its domain of definition into subintervals on which the function has an inverse, and find the inverse function on each subinterval.

15. $f(x) = x^4$

16. $f(x) = 1/x^4$

17. $f(x) = x^2 + 2x + 3$

18. $f(x) = x^4 + 4x^2 + 2$

19. $f(x) = \dfrac{x^2}{x^2 + 4}$

20. $f(x) = \dfrac{x^4}{x^2 + 4}$

21. If a manufacturing firm sells x objects of a certain commodity per week, it sells them at a price of r per object, and r depends on x, $r = f(x)$. In economic theory this function is usually considered decreasing, as shown in Figure 8.12(a); hence it has an inverse function $x = f^{-1}(r)$. In this function x depends on r, indicating that if the price of the object is set at r, then the market will demand x of them per week. This function is therefore called the *demand function* (Figure 8.12(b)). Find demand functions if

(a) $r = \dfrac{a}{x+b} + c$, $\quad a$, b, and c positive constants; and

(b) $r = \dfrac{a}{x^2+b} + c$, $\quad a$, b, and c positive constants.

Sketch the demand function $x = f^{-1}(r)$ and given function $r = f(x)$ in each case.

FIGURE 8.12

(a)

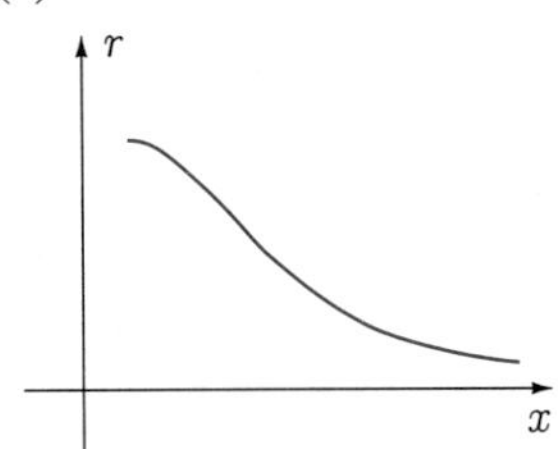

(b)

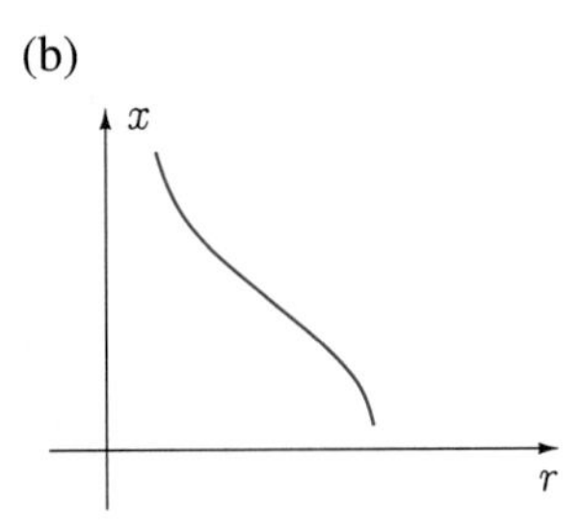

22. Show that the demand function

$$x = f(r) = 4a^3 - 3ar^2 + r^3, \quad 0 < r < 2a$$

has an inverse function $r = f^{-1}(x)$. What is the domain of $f^{-1}(x)$? Sketch its graph.

23. Find the inverse function for

$$f(x) = \frac{x^2}{(1+x)^2}, \quad -1 < x \le 0.$$

SECTION 8.2

The Inverse Trigonometric Functions

The function $f(x) = \sin x$, defined for all real x, does not have an inverse. This is evident from the fact that the function is not one-to-one, or that it fails the horizontal line test. By restricting the domain of $\sin x$, however, the function can be made one-to-one, and this can be done in many ways. In particular, when restricted to the domain

$-\pi/2 \le x \le \pi/2$, $\sin x$ is one-to-one, and therefore has an inverse function. This function, denoted by

$$y = \mathrm{Sin}^{-1}\, x, \tag{8.4}$$

and called the **inverse sine function**, is shown in Figure 8.13. The range of the function is

$$-\frac{\pi}{2} \le \mathrm{Sin}^{-1}\, x \le \frac{\pi}{2}. \tag{8.5}$$

The values that fall in this range are called the **principal values** of the inverse sine function. They have resulted from our restriction of the domain of $\sin x$ to this same interval.

An equivalent way to derive the inverse sine function is as follows. The reflection of the graph of $y = \sin x$ in the line $y = x$ does not represent a function, but by restricting the range of values of the reflected curve, we can produce a single-valued function (Figure 8.14). Once again this can be done in many ways, and when we do so by restricting the y-values to the interval $[-\pi/2, \pi/2]$, the resulting function is called the inverse sine function $\mathrm{Sin}^{-1}\, x$. Note very carefully that $\mathrm{Sin}^{-1}\, x$ is not the inverse function of $f(x) = \sin x$, because the sine function has no inverse. It is, however, the inverse of the sine function restricted to the domain $-\pi/2 \le x \le \pi/2$. It is perhaps then a misnomer to call the function $\mathrm{Sin}^{-1}\, x$ the inverse sine function, but this has become the accepted terminology.

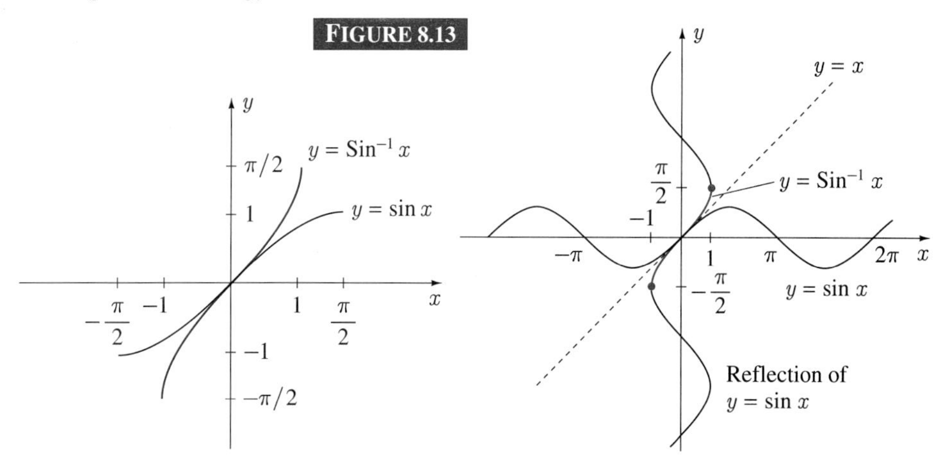

FIGURE 8.13

FIGURE 8.14

We now know what the inverse sine function looks like graphically, but what does it mean to say that $y = \mathrm{Sin}^{-1}\, x$? Certainly given any value of x, we can push a few buttons on an electronic calculator and find $\mathrm{Sin}^{-1}\, x$ for that x. To use inverse trigonometric functions in practice, we must have a feeling for what they do. To obtain this insight, we note that if (x, y) is a point on the curve $y = \mathrm{Sin}^{-1}\, x$, then (y, x) is a point on the sine curve; that is,

$$y = \mathrm{Sin}^{-1}\, x \quad \text{only if} \quad x = \sin y. \tag{8.6}$$

In the latter equation y is an angle and x is the sine of that angle. Thus, when we see $y = \mathrm{Sin}^{-1}\, x$, we should not read "$y$ equals inverse sine x", but rather "y is an angle whose sine is x." For instance, if $x = \frac{1}{2}$, then $y = \mathrm{Sin}^{-1}\left(\frac{1}{2}\right)$ is read "y is an angle whose sine is $\frac{1}{2}$." Clearly the angle whose sine is $\frac{1}{2}$ is $\pi/6$, and we write $y = \mathrm{Sin}^{-1}\left(\frac{1}{2}\right) = \pi/6$. Remember that we must choose the principal value $\pi/6$; there are many angles that have a sine equal to $\frac{1}{2}$, but the inverse sine function demands that we choose that angle in the range $-\pi/2 \le y \le \pi/2$, and state it with a number in this interval. For example, $2\pi + \pi/6$ may represent the same angle with the positive x-axis as $\pi/6$, but the number

$2\pi + \pi/6$ does not lie between $-\pi/2$ and $\pi/2$. In summary, the function $\sin x$ regards x as an angle and assigns to x the sine of the angle; the inverse sine function $\text{Sin}^{-1} x$ regards x as the sine of an angle and assigns to x the angle with that sine

Another notation that is commonly used for the inverse sine function is $\arcsin x$. One reason the notation $\arcsin x$ is preferable to $\text{Sin}^{-1} x$ is the possible misinterpretation of $\text{Sin}^{-1} x$. Sometimes students regard the "-1" as a power and write $\text{Sin}^{-1} x$ as $1/\sin x$. This is *not* correct, a fact that you were warned about in Section 8.1. Sin^{-1} is the name of a function, just as sin is the name of the sine function, and $\sqrt{\ }$ is the notation for the positive square root function. The capital S in Sin^{-1} should also warn you that this is the inverse sine function.

EXAMPLE 8.3

Simplify each of the following expressions:

(a) $\text{Sin}^{-1}(-\sqrt{3}/2)$ (b) $\text{Sin}^{-1}(1)$ (c) $\text{Sin}^{-1}(3)$
(d) $\text{Sin}^{-1}(3/5) + \text{Sin}^{-1}(4/5)$ (e) $\sin\left[\text{Sin}^{-1}(\sqrt{3}/2)\right]$

SOLUTION (a) $\text{Sin}^{-1}(-\sqrt{3}/2)$ asks for the angle whose sine is equal to $-\sqrt{3}/2$. Clearly,

$$\text{Sin}^{-1}(-\sqrt{3}/2) = -\pi/3.$$

(b) $\text{Sin}^{-1}(1) = \pi/2$.

(c) $\text{Sin}^{-1}(3)$ is not defined since the domain of $\text{Sin}^{-1} x$ is $-1 \leq x \leq 1$.

(d) If $\phi = \text{Sin}^{-1}(3/5)$, then ϕ is illustrated in the triangle in Figure 8.15. Since the third side must have length 4, it follows that $\text{Sin}^{-1}(4/5) = \pi/2 - \phi$, and

$$\text{Sin}^{-1}\left(\frac{3}{5}\right) + \text{Sin}^{-1}\left(\frac{4}{5}\right) = \phi + \left(\frac{\pi}{2} - \phi\right) = \frac{\pi}{2}.$$

(e) $\sin\left[\text{Sin}^{-1}(\sqrt{3}/2)\right] = \sin(\pi/3) = \sqrt{3}/2$.

FIGURE 8.15

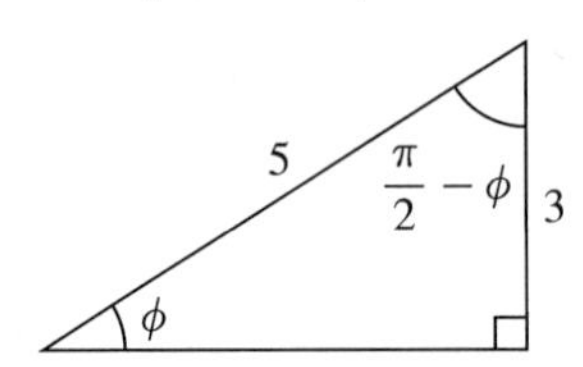

■

EXAMPLE 8.4

For what values of x are the following valid:

(a) $\sin\left(\text{Sin}^{-1} x\right) = x$ (b) $\text{Sin}^{-1}(\sin x) = x$

SOLUTION These two equations express the fact that the sine function and the inverse sine function are inverses, provided we are careful about domains:

(a) The equation $\sin\left(\text{Sin}^{-1} x\right) = x$ is valid for $-1 \leq x \leq 1$. Given an x in this interval, $\text{Sin}^{-1} x$ finds that angle (in the principal value range) which has x as its sine. Then $\sin\left(\text{Sin}^{-1} x\right)$ takes the sign of this angle. Naturally it returns the original number x.

(b) The composite function $\text{Sin}^{-1}(\sin x)$ is defined for all x, but only on the domain $-\pi/2 \leq x \leq \pi/2$ is it equal to x. ■

Our analysis of the inverse sine function is now complete. We could repeat a similar discussion five times, once for each of the inverse functions associated with the other five trigonometric functions. However, we shall give an abbreviated version for the inverse cosine function and tabulate results for the remaining four functions.

The reflection of the graph of the function $f(x) = \cos x$ is shown in Figure 8.16, and it does not represent a function. If we restrict the y-values on the reflected curve to

$$0 \leq y \leq \pi, \tag{8.7}$$

then we do obtain a function called the inverse cosine function, denoted by

$$y = \text{Cos}^{-1}\, x. \tag{8.8}$$

The values in 8.7 are called the principal values of the inverse cosine function. Note again that $\text{Cos}^{-1}\, x$ is not the inverse function of $\cos x$, but is the inverse function of $f(x) = \cos x,\ 0 \leq x \leq \pi$.

When we write $y = \text{Cos}^{-1}\, x$, we do not read this as "y equals inverse cosine x," but rather as "y is an angle whose cosine is x,", for if $y = \text{Cos}^{-1}\, x$, then $x = \cos y$.

FIGURE 8.16

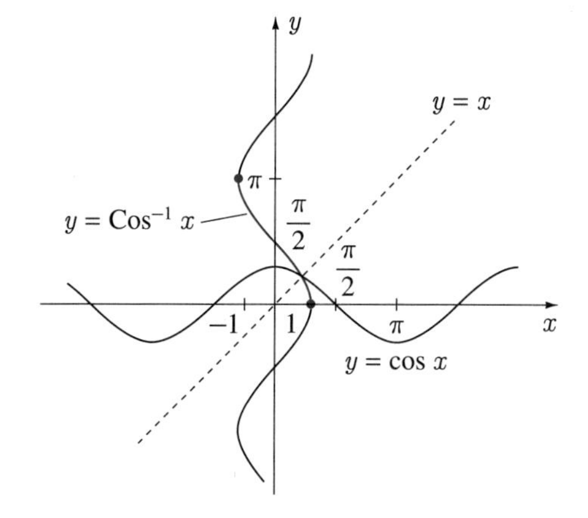

The remaining four inverse trigonometric functions are shown in Figures 8.17(c)–(f). Lighter curves represent reflections in the line $y = x$ of the trigonometric functions $\tan x$, $\cot x$, $\csc x$, and $\sec x$. Heavier curves represent principal values of the inverse trigonometric functions $\text{Tan}^{-1}\, x$, $\text{Cot}^{-1}\, x$, $\text{Csc}^{-1}\, x$, and $\text{Sec}^{-1}\, x$. Principal values of the six inverse trigonometric functions are listed in Table 8.1.

TABLE 8.1

Inverse Trigonometric Function	Principal Values
$\text{Sin}^{-1}\, x$	$-\frac{\pi}{2} \leq y \leq \frac{\pi}{2}$
$\text{Tan}^{-1}\, x$	$-\frac{\pi}{2} < y < \frac{\pi}{2}$
$\text{Cos}^{-1}\, x$	$0 \leq y \leq \pi$
$\text{Cot}^{-1}\, x$	$0 < y < \pi$
$\text{Csc}^{-1}\, x$	$-\pi < y \leq -\frac{\pi}{2},\ 0 < y \leq \frac{\pi}{2}$
$\text{Sec}^{-1}\, x$	$-\pi \leq y < -\frac{\pi}{2},\ 0 \leq y < \frac{\pi}{2}$

FIGURE 8.17

(a)

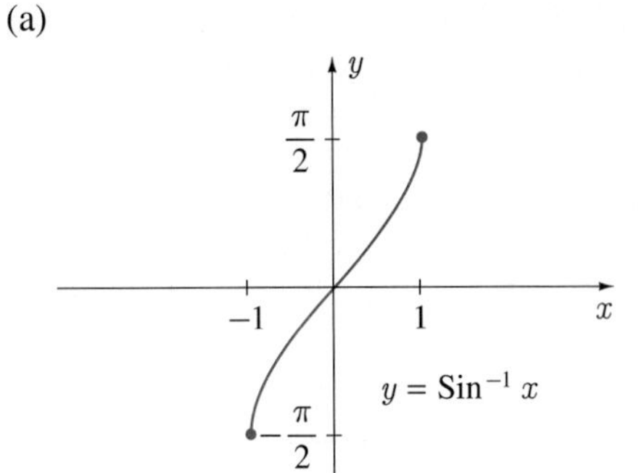

(b)

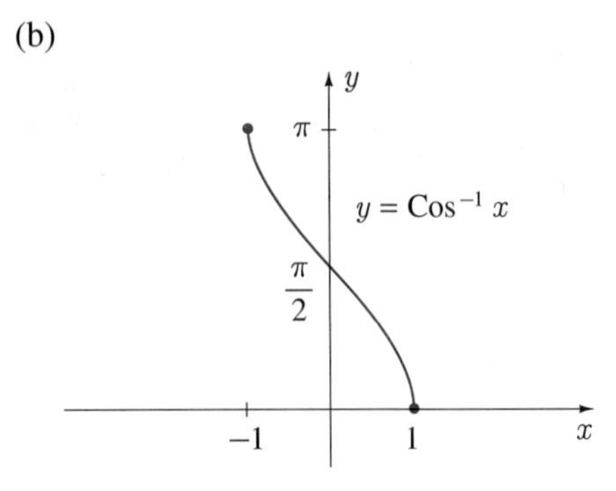

(c)

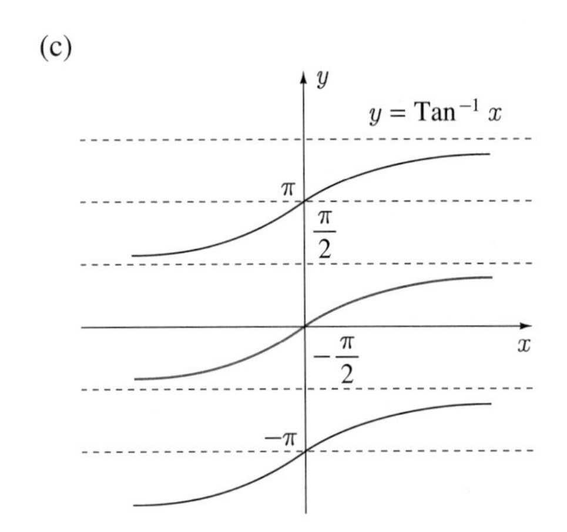

(d)

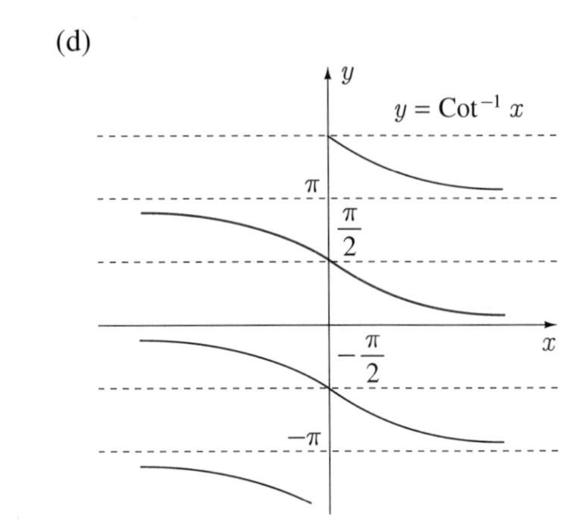

(e)

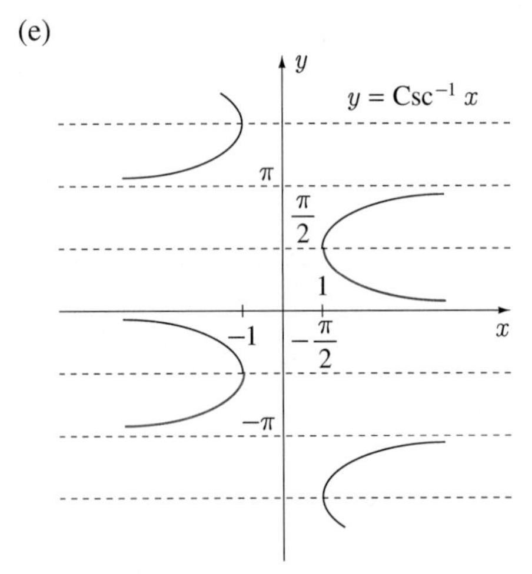

(f)

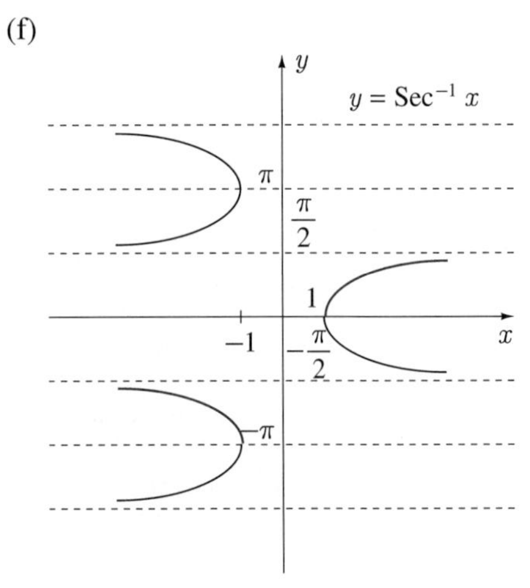

It would be reasonable to ask why principal values of $\text{Sec}^{-1}\, x$ were not chosen as $0 \le y < \pi/2,\ \pi/2 < y \le \pi$. Had they been chosen in this way, they would have been very similar to those of $\text{Cos}^{-1}\, x$ and $\text{Cot}^{-1}\, x$. Likewise, why are the principal values of $\text{Csc}^{-1}\, x$ not $-\pi/2 \le y < 0,\ 0 < y \le \pi/2$? The answer is that they could have been selected in this way, and some authors do indeed make this choice; it is simply a matter of preference. Each choice does, however, create corresponding changes in later work. Specifically, in the exercises of Section 8.3, if principal values of $\text{Sec}^{-1}\, x$ and $\text{Csc}^{-1}\, x$ are selected in this alternative way, then derivatives of these functions are modified correspondingly.

In the remainder of this section we solve a number of problems which illustrate some of the properties of the inverse trigonometric functions and their applications.

EXAMPLE 8.5

Show that

$$\operatorname{Csc}^{-1} x = \begin{cases} \operatorname{Sin}^{-1}\left(\dfrac{1}{x}\right) & x \geq 1 \\ -\pi - \operatorname{Sin}^{-1}\left(\dfrac{1}{x}\right) & x \leq -1. \end{cases}$$

SOLUTION When $x \geq 1$, we set $y = \operatorname{Csc}^{-1} x$, in which case $0 < y \leq \pi/2$. It follows that $x = \csc y$, and

$$\frac{1}{x} = \frac{1}{\csc y} = \sin y.$$

If we apply the inverse sine function to both sides of this equation, the result is

$$\operatorname{Sin}^{-1}\left(\frac{1}{x}\right) = \operatorname{Sin}^{-1}(\sin y) = y,$$

because y is in the principal value range of the inverse sine function. Hence, when $x \geq 1$,

$$\operatorname{Csc}^{-1} x = \operatorname{Sin}^{-1}\left(\frac{1}{x}\right).$$

When $x \leq -1$, we again set $y = \operatorname{Csc}^{-1} x$, and obtain

$$\operatorname{Sin}^{-1}\left(\frac{1}{x}\right) = \operatorname{Sin}^{-1}(\sin y).$$

But in this case the right side is not equal to y, because $-\pi < y \leq -\pi/2$. To remedy this, we note that when $-\pi < y \leq -\pi/2$, we may write $\sin y = \sin(-\pi - y)$. Since $-\pi - y$ is in the principal range for the inverse sine function ($-\pi/2 \leq -\pi - y < 0$), it follows that

$$\operatorname{Sin}^{-1}\left(\frac{1}{x}\right) = \operatorname{Sin}^{-1}(\sin y) = \operatorname{Sin}^{-1}[\sin(-\pi - y)] = -\pi - y = -\pi - \operatorname{Csc}^{-1} x.$$

■

Similar relations involving the inverse secant and inverse cotangent functions are presented in Exercises 45 and 46.

EXAMPLE 8.6

Show that

$$\operatorname{Sin}^{-1} x = \begin{cases} -\operatorname{Cos}^{-1}\sqrt{1 - x^2} & -1 \leq x < 0 \\ \operatorname{Cos}^{-1}\sqrt{1 - x^2} & 0 \leq x \leq 1. \end{cases}$$

SOLUTION When $0 \leq x \leq 1$, we set $y = \operatorname{Sin}^{-1} x$, in which case $0 \leq y \leq \pi/2$. It follows that $x = \sin y$, and because $\sin^2 y + \cos^2 y = 1$, we have

$$\cos y = \pm\sqrt{1 - \sin^2 y} = \pm\sqrt{1 - x^2}.$$

Since y is an angle in the first quadrant, its cosine must be nonnegative, and therefore

$$\cos y = \sqrt{1 - x^2}.$$

When we apply the inverse cosine function to both sides of this equation, we obtain

$$\mathrm{Cos}^{-1}(\cos y) = y = \mathrm{Cos}^{-1}\sqrt{1-x^2}.$$

When $-1 \le x < 0$, we continue to set $y = \mathrm{Sin}^{-1} x$, and once again obtain

$$\cos y = \sqrt{1-x^2},$$

because $-\pi/2 \le y < 0$. Application of the inverse cosine function gives

$$\mathrm{Cos}^{-1}(\cos y) = \mathrm{Cos}^{-1}\sqrt{1-x^2},$$

but the left side is not equal to y because y is not in the principal value range of the inverse cosine function. This is easily adjusted by noting that with $-\pi/2 \le y < 0$, we have $\cos y = \cos(-y)$. Hence,

$$\mathrm{Cos}^{-1}(\cos y) = \mathrm{Cos}^{-1}[\cos(-y)] = -y = \mathrm{Cos}^{-1}\sqrt{1-x^2};$$

that is,

$$y = -\mathrm{Cos}^{-1}\sqrt{1-x^2}.$$

■

Similar relations among other inverse trigonometric functions are presented in Exercises 47 and 48. We now consider a few examples wherein inverse trigonometric functions are used to solve equations involving trigonometric functions.

EXAMPLE 8.7

Find all solutions of the equation

$$\sin^2 x + 3\sin x = 2.$$

SOLUTION The quadratic formula applied to the equation

$$(\sin x)^2 + 3(\sin x) - 2 = 0$$

yields

$$\sin x = \frac{-3 \pm \sqrt{9+8}}{2} = \frac{-3 \pm \sqrt{17}}{2}.$$

Since $\sin x$ can only take on values in the interval $-1 \le \sin x \le 1$, the possibility $\sin x = (-3 - \sqrt{17})/2$ must be rejected, and

$$\sin x = \frac{-3 + \sqrt{17}}{2}.$$

One solution of this equation is

$$x = \mathrm{Sin}^{-1}\left(\frac{\sqrt{17} - 3}{2}\right) = 0.60 \text{ radians}.$$

This is not the only solution, however; there are many other angles with a sine equal to $(\sqrt{17}-3)/2$. Since $\sin x$ is 2π-periodic, the angles $2n\pi + 0.60$, for n any integer, all have sine equal to $(\sqrt{17}-3)/2$. Because $\sin(\pi - x) = \sin x$, it follows that $\sin(\pi - 0.60) = (\sqrt{17}-3)/2$, and therefore $\pi - 0.60$ is another solution. When multiples of 2π are added to this angle, $2n\pi + (\pi - 0.60)$ are also solutions. Thus, the complete set of solutions is

$$2n\pi + 0.60, \quad 2n\pi + (\pi - 0.60),$$

where n is an integer. Figure 8.18 suggests that this set of numbers can be represented more compactly as an initial rotation of $\pi/2$, plus or minus $\pi/2 - 0.60 = 0.97$, and possible multiples of 2π; that is,

$$\frac{\pi}{2} \pm 0.97 + 2n\pi = \left(\frac{4n+1}{2}\right)\pi \pm 0.97.$$

FIGURE 8.18

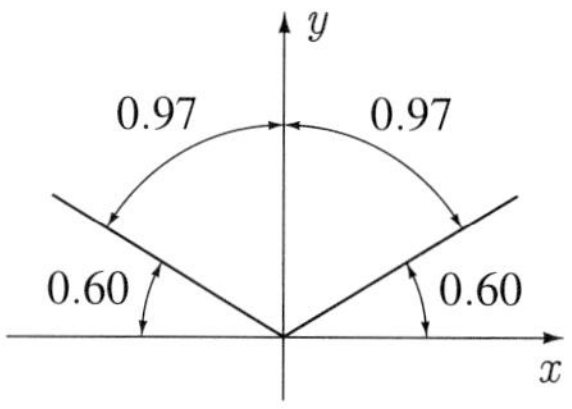

■

Often in problems like these, $x = \text{Sin}^{-1}[(\sqrt{17}-3)/2] = 0.60$ is the only solution given. We can see the reasoning behind this conclusion. There is only one principal value of the inverse sine function, and therefore one solution to the equation. But the equation $\sin x = (\sqrt{17}-3)/2$ says nothing about the inverse sine function. We have introduced it simply as a convenience. When we obtain the solution 0.60 we must ask whether there are other solutions to the original equation. For the above example, there are other solutions; in some problems, the principal value is the only acceptable solution. Remember, then, if we introduce an inverse trigonometric function into a problem along with its corresponding principal values, we must ask whether there are other possibilities besides the principal values. This point is emphasized again in the following example.

EXAMPLE 8.8

Find all solutions of the equation

$$\tan(\cos x) = \frac{1}{\sqrt{3}}.$$

SOLUTION If we set $y = \cos x$, then $\tan y = 1/\sqrt{3}$, and one solution of this equation is $y = \text{Tan}^{-1}(1/\sqrt{3}) = \pi/6$. But there are many other solutions for y, namely,

$$y = \frac{\pi}{6} + n\pi,$$

where n is an integer. But $y = \cos x$, and $\cos x$ must take on values in the interval $-1 \le \cos x \le 1$. There is only one possibility for n, namely, $n = 0$; hence,

$$\cos x = \frac{\pi}{6}.$$

From the solution $x = \text{Cos}^{-1}(\pi/6) = 1.02$ and Figure 8.19, we obtain

$$x = \pm 1.02 + 2n\pi,$$

where n is an integer, as the complete set of solutions.

FIGURE 8.19

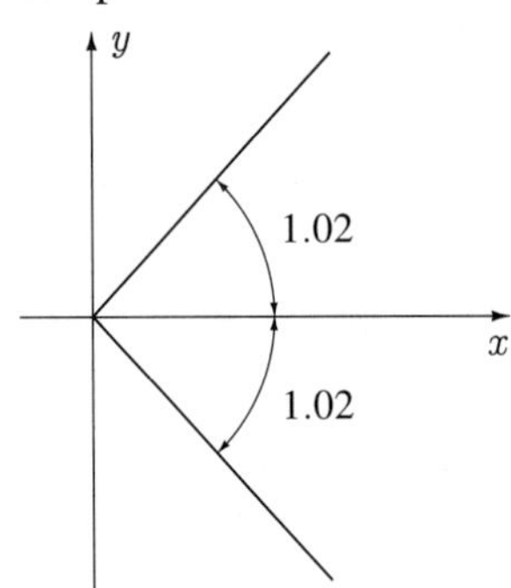

■

EXAMPLE 8.9

Find constants $R > 0$ and $0 < \phi < \pi$ so that the function $f(x) = 3\cos\omega x - 4\sin\omega x$, where $\omega > 0$ is a constant, can be expressed in the form $R\sin(\omega x + \phi)$ for all real x.

SOLUTION If we expand $R\sin(\omega x + \phi)$ by means of a compound angle formula, and equate it to $f(x)$, we have

$$R[\sin\omega x\cos\phi + \cos\omega x\sin\phi] = 3\cos\omega x - 4\sin\omega x.$$

This equation will be valid for all x if we can find values of R and ϕ so that

$$R\cos\phi = -4 \qquad \text{and} \qquad R\sin\phi = 3.$$

When we square and add these equations, the result is

$$R^2\cos^2\phi + R^2\sin^2\phi = R^2 = (-4)^2 + (3)^2 = 25.$$

Consequently, $R = 5$, and

$$\cos\phi = -\frac{4}{5} \qquad \text{and} \qquad \sin\phi = \frac{3}{5}.$$

The only angle in the range $0 < \phi < \pi$ satisfying these equations is $\phi = \text{Cos}^{-1}(-4/5) = 2.50$ radians. Notice that $\text{Sin}^{-1}(3/5)$ does not give this angle. Thus, $f(x)$ can be expressed in the form

$$f(x) = 5\sin(\omega x + 2.50).$$

■

EXERCISES 8.2

In Exercises 1–16 evaluate the expression (if it has a value).

1. $\operatorname{Tan}^{-1}(-1/3)$ **2.** $\operatorname{Sin}^{-1}(1/4)$

3. $\operatorname{Sec}^{-1}(\sqrt{3})$ **4.** $\operatorname{Csc}^{-1}(-2/\sqrt{3})$

5. $\operatorname{Cot}^{-1}(1)$ **6.** $\operatorname{Cos}^{-1}(3/2)$

7. $\operatorname{Sin}^{-1}(\pi/2)$ **8.** $\operatorname{Tan}^{-1}(-1)$

9. $\sin\left(\operatorname{Tan}^{-1}\sqrt{3}\right)$ **10.** $\tan\left(\operatorname{Sin}^{-1} 3\right)$

11. $\operatorname{Sin}^{-1}[\tan(1/6)]$ **12.** $\operatorname{Tan}^{-1}[\sin(1/6)]$

13. $\sec\left[\operatorname{Cos}^{-1}(1/2)\right]$ **14.** $\operatorname{Sin}^{-1}[\sin(3\pi/4)]$

15. $\sin\left[\operatorname{Sin}^{-1}(1/\sqrt{2})\right]$ **16.** $\operatorname{Sin}^{-1}\left[\cos\left(\operatorname{Sec}^{-1}(-\sqrt{2})\right)\right]$

In Exercises 17–22 find all solutions of the equation.

17. $\sin x = 1/3$

18. $\tan x = -1.2$

19. $4\sin^2 x - 2\cos^2 x = 1$

20. $4\sin^2 x + 2\cos^2 x = 1$

21. $\cos^2 x - 3\cos x + 1 = 0$

22. $\sin^2 x - 3\sin x - 5 = 0$

In Exercises 23–26 sketch a graph of the function.

23. $f(x) = 2 + \left(\operatorname{Csc}^{-1} x\right)^2$

24. $f(x) = \sqrt{\operatorname{Tan}^{-1} x} + \sqrt{\operatorname{Sec}^{-1} x}$

25. $f(x) = \operatorname{Sin}^{-1}(x - 3)$

26. $f(x) = \operatorname{Sin}^{-1} x + \operatorname{Csc}^{-1} x$

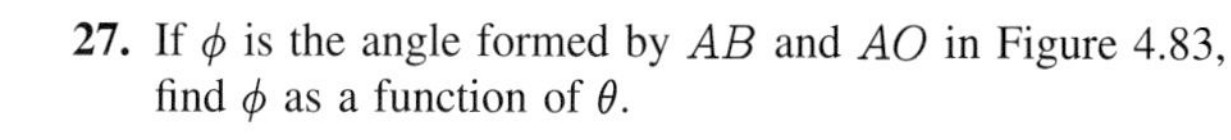

27. If ϕ is the angle formed by AB and AO in Figure 4.83, find ϕ as a function of θ.

In Exercises 28–33 find all solutions of the equation.

28. $\sin x \tan^2 x - 3 + \tan^2 x - 3\sin x = 0$

29. $\sin x + \cos x = 1$

30. $\sec(\sin x) = -\sqrt{2}$

31. $\cos\left(\operatorname{Sin}^{-1} x\right) = 1/2$

32. $\sec(\tan x) = -\sqrt{2}$

33. $\operatorname{Cos}^{-1}[\tan(x^2 + 4)] = 2\pi - 5$

34. Sketch graphs of the following functions:

(a) $f(x) = \sin\left(\operatorname{Sin}^{-1} x\right)$ (b) $f(x) = \operatorname{Sin}^{-1}(\sin x)$

35. Sketch graphs of the following functions:

(a) $f(x) = \cos\left(\operatorname{Cos}^{-1} x\right)$ (b) $f(x) = \operatorname{Cos}^{-1}(\cos x)$

36. Express the function $f(x) = 4\sin 2x + \cos 2x$ in the form $R\sin(2x + \phi)$ where $R > 0$ and $0 < \phi < \pi$.

37. Express the function $f(x) = -2\sin 3x + 4\cos 3x$ in the form $R\cos(3x + \phi)$ where $R > 0$ and $0 < \phi < \pi$.

38. Repeat Exercise 36 for $f(x) = -2\sin 2x + 4\cos 2x$.

39. Repeat Exercise 37 for $f(x) = -4\sin 3x + 5\cos 3x$.

40. A pendulum consists of a mass m suspended from a string of length L (Figure 8.20). At time $t = 0$, the mass is pulled through a small angle θ_0 — to the right when $\theta_0 > 0$, and to the left when $\theta_0 < 0$, and given an initial speed v_0 to the right. Its subsequent angular displacements are given by

$$\theta = \theta(t) = \theta_0 \cos\omega t + \frac{v_0}{\omega L}\sin\omega t, \quad t \geq 0,$$

where $\omega = \sqrt{9.81/L}$, provided any resistance due to the air is neglected. Show that $\theta(t)$ can be expressed in the form

$$\theta(t) = \sqrt{\theta_0^2 + \frac{v_0^2}{\omega^2 L^2}}\,\sin(\omega t + \phi),$$

where $\phi = \operatorname{Tan}^{-1}(\omega L\theta_0/v_0)$.

FIGURE 8.20

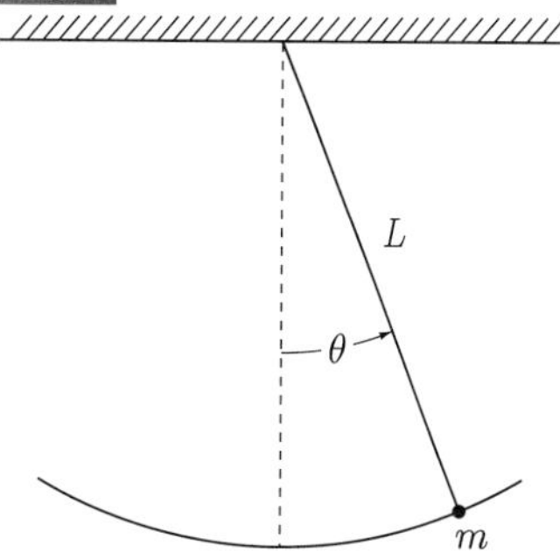

41. What changes, if any, must be made in Exercise 40 if the initial speed v_0 is to the left rather than the right?

42. A mass m is suspended from a spring with constant k (Figure 8.21). If, at time $t = 0$, the mass is given an initial displacement y_0 and an upward speed v_0, its subsequent displacements are given by

$$y = y(t) = y_0 \cos \omega t + \frac{v_0}{\omega} \sin \omega t, \quad t \geq 0,$$

where $\omega = \sqrt{k/m}$, provided any resistance due to the air is neglected. Show that $y(t)$ can be expressed in the form

$$y(t) = \sqrt{y_0^2 + \frac{v_0^2}{\omega^2}} \sin(\omega t + \phi),$$

where $\phi = \text{Tan}^{-1}(\omega y_0 / v_0)$.

FIGURE 8.21

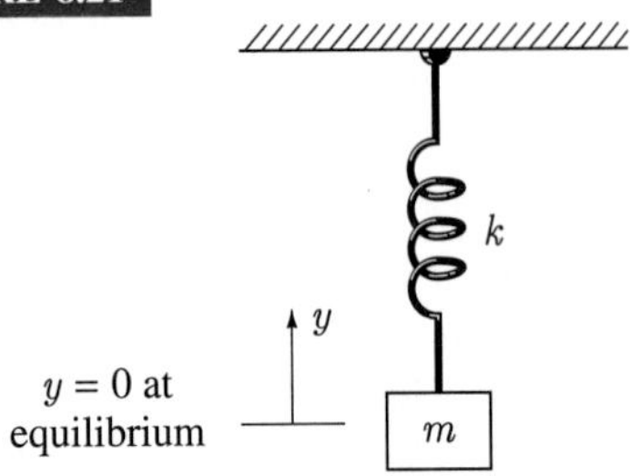

43. What changes, if any, must be made in Exercise 42 if the initial speed v_0 is downward rather than upward?

44. An inductance L, resistance R, and capacitance C are connected in series with a generator producing an oscillatory voltage $E = E_0 \cos \omega t$, $t \geq 0$ (Figure 8.22). If $L, C, R, E_0 > 0$, and $\omega > 0$ are all constants, the steady-state current I in the circuit is

$$I = \frac{E_0}{\sqrt{R^2 + \left(\omega L - \dfrac{1}{\omega C}\right)^2}} \left[R \cos \omega t + \left(\omega L - \frac{1}{\omega C}\right) \sin \omega t\right].$$

Express I in the form

$$I = A \cos(\omega t - \phi),$$

where $A > 0$ and $-\pi/2 \leq \phi \leq \pi/2$.

FIGURE 8.22

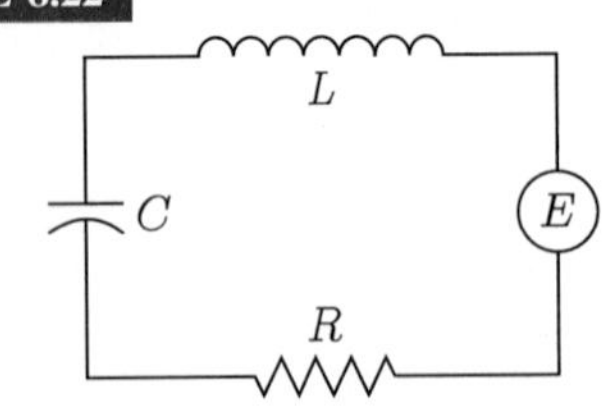

45. Prove that

$$\text{Sec}^{-1} x = \begin{cases} \text{Cos}^{-1}\left(\dfrac{1}{x}\right) & x \geq 1 \\ -\text{Cos}^{-1}\left(\dfrac{1}{x}\right) & x \leq -1. \end{cases}$$

46. Prove that

$$\text{Cot}^{-1} x = \begin{cases} \text{Tan}^{-1}\left(\dfrac{1}{x}\right) & x > 0 \\ \pi + \text{Tan}^{-1}\left(\dfrac{1}{x}\right) & x < 0. \end{cases}$$

47. Verify that

$$\text{Sec}^{-1} x = \begin{cases} \text{Tan}^{-1}\sqrt{x^2 - 1} & x \geq 1 \\ -\pi + \text{Tan}^{-1}\sqrt{x^2 - 1} & x \leq -1. \end{cases}$$

48. Verify that

$$\text{Csc}^{-1} x = \begin{cases} \text{Cot}^{-1}\sqrt{x^2 - 1} & x \geq 1 \\ -\pi + \text{Cot}^{-1}\sqrt{x^2 - 1} & x \leq -1. \end{cases}$$

49. Verify that

$$\text{Cos}^{-1} x = \begin{cases} \text{Sin}^{-1}\sqrt{1 - x^2} & 0 \leq x \leq 1 \\ \pi - \text{Sin}^{-1}\sqrt{1 - x^2} & -1 \leq x < 0. \end{cases}$$

50. In Figure 8.23, a ray of light is incident on a prism at angle i. The light is refracted and leaves the prism at an angle of deviation ψ relative to the incident direction.

FIGURE 8.23

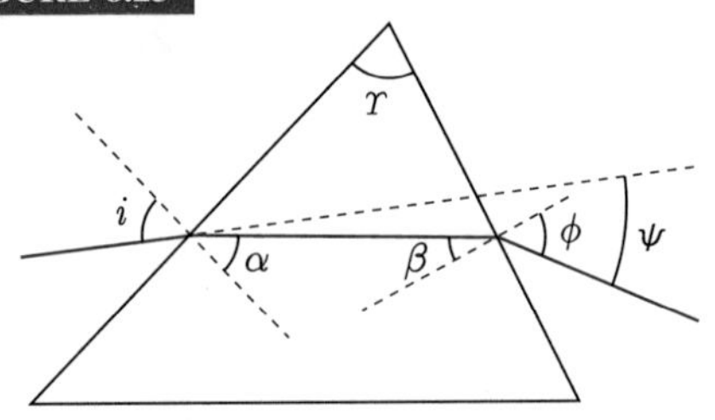

(a) Show that $\gamma = \alpha + \beta$, and use this to derive the expression

$$\psi = i + \phi - \gamma.$$

(b) To express ψ in terms of i and properties of the prism, we use Snell's law, which states that if the prism is in air, then the relationships between the incident and refracted light at the two faces are

$$n \sin \alpha = \sin i \qquad \text{and} \qquad n \sin \beta = \sin \phi,$$

where n is the refractive index of the prism. Use these equations to show that ψ can be expressed in the form

$$\psi = i - \gamma + \text{Sin}^{-1}\left(\sqrt{n^2 - \sin^2 i}\, \sin \gamma - \sin i \cos \gamma\right).$$

51. Show that for the prism in Exercise 50 the angle of deviation is a minimum when $i = \phi$. If ψ_m is the minimum angle of deviation, prove that

$$n = \frac{\sin\left[(\psi_m + \gamma)/2\right]}{\sin(\gamma/2)}.$$

In practice, this equation is used to determine the index of refraction of the prism.

52. Verify that if $0 \le x < 1$, then

$$2\,\mathrm{Tan}^{-1}\sqrt{\frac{1+x}{1-x}} = \pi - \mathrm{Cos}^{-1}\,x.$$

53. Find intervals on which the function

$$f(x) = x + \sec\left(\frac{x}{2} - 1\right), \quad 0 \le x \le 5,$$

is increasing and decreasing.

SECTION 8.3

Derivatives of the Inverse Trigonometric Functions

Derivatives of inverse trigonometric functions are most easily obtained with implicit differentiation. We begin with the inverse sine function.

Theorem 8.2

$$\frac{d}{dx}\mathrm{Sin}^{-1}\,x = \frac{1}{\sqrt{1-x^2}} \tag{8.9}$$

Proof If we set $y = \mathrm{Sin}^{-1}\,x$, then $x = \sin y$, and we can differentiate implicitly with respect to x:

$$1 = \cos y \frac{dy}{dx}.$$

Thus,

$$\frac{dy}{dx} = \frac{1}{\cos y}.$$

To express $\cos y$ in terms of x we can proceed in two ways:

(i) From the trigonometric identity $\sin^2 y + \cos^2 y = 1$, we obtain

$$\cos y = \pm\sqrt{1 - \sin^2 y} = \pm\sqrt{1-x^2}.$$

We know that $y = \mathrm{Sin}^{-1}\,x$ and the principal values of $\mathrm{Sin}^{-1}\,x$ are $-\pi/2 \le y \le \pi/2$. Therefore y is an angle in either the first or fourth quadrant. It follows that $\cos y \ge 0$, and we must choose

$$\cos y = \sqrt{1-x^2}.$$

(ii) We use the triangle in Figure 8.24 to replace the trigonometric identity. The triangle is designed to fit the equation $x = \sin y$. The third side is then $\sqrt{1-x^2}$. It follows that

$$\cos y = \sqrt{1-x^2}.$$

But $\cos y$ is not always positive as indicated here. For example, if y is between $\pi/2$ and π, then $\cos y < 0$. To decide on the correct sign for $\cos y$, we use the discussion of principal values in (i), and obtain again $\cos y = \sqrt{1-x^2}$.

FIGURE 8.24

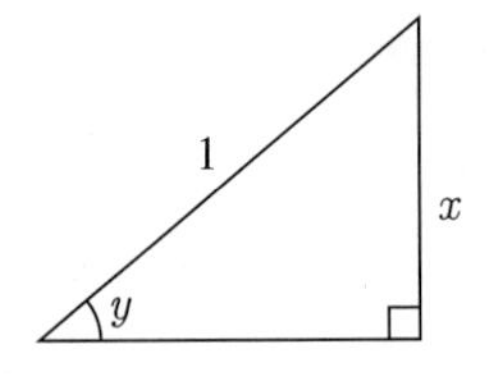

Finally, then, we have

$$\frac{dy}{dx} = \frac{1}{\sqrt{1-x^2}}.$$

It is a straightforward application of the chain rule to obtain the following corollary.

Corollary *If* $u(x)$ *is a differentiable function, then*

$$\frac{d}{dx}\text{Sin}^{-1} u = \frac{1}{\sqrt{1-u^2}}\frac{du}{dx}. \tag{8.10}$$

Next we obtain the derivative of the inverse cosine function.

Theorem 8.3

$$\frac{d}{dx}\text{Cos}^{-1} x = \frac{-1}{\sqrt{1-x^2}} \tag{8.11}$$

Proof If we set $y = \text{Cos}^{-1} x$, then $x = \cos y$, and we can differentiate with respect to x,

$$1 = -\sin y \frac{dy}{dx},$$

and solve for

$$\frac{dy}{dx} = \frac{-1}{\sin y}.$$

The triangle in Figure 8.25, obtained from $x = \cos y$, yields $\sin y = \sqrt{1-x^2}$. Since the principal values of $y = \text{Cos}^{-1} x$ are $0 \le y \le \pi$, it follows that $\sin y$ is indeed nonnegative, and therefore

$$\frac{dy}{dx} = \frac{-1}{\sqrt{1-x^2}}.$$

FIGURE 8.25

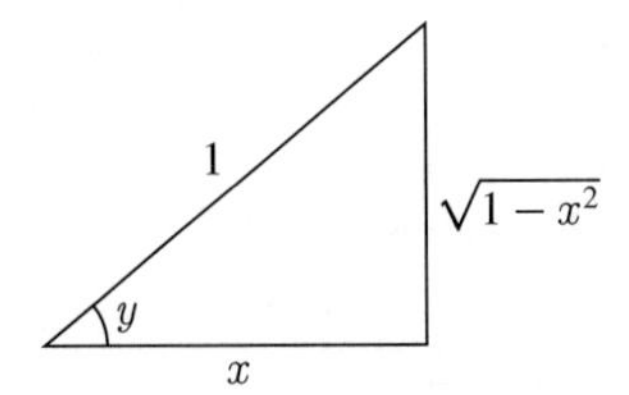

Corollary *If* $u(x)$ *is differentiable, then*

$$\frac{d}{dx}\text{Cos}^{-1} u = \frac{-1}{\sqrt{1-u^2}}\frac{du}{dx}. \tag{8.12}$$

Derivatives for the other inverse trigonometric functions can be derived in a similar way. We list them below and include derivatives of $\mathrm{Sin}^{-1} u$ and $\mathrm{Cos}^{-1} u$ for completeness:

$$\frac{d}{dx}\mathrm{Sin}^{-1} u = \frac{1}{\sqrt{1-u^2}}\frac{du}{dx}; \tag{8.10}$$

$$\frac{d}{dx}\mathrm{Cos}^{-1} u = \frac{-1}{\sqrt{1-u^2}}\frac{du}{dx}; \tag{8.12}$$

$$\frac{d}{dx}\mathrm{Tan}^{-1} u = \frac{1}{1+u^2}\frac{du}{dx}; \tag{8.13}$$

$$\frac{d}{dx}\mathrm{Cot}^{-1} u = \frac{-1}{1+u^2}\frac{du}{dx}; \tag{8.14}$$

$$\frac{d}{dx}\mathrm{Sec}^{-1} u = \frac{1}{u\sqrt{u^2-1}}\frac{du}{dx}; \tag{8.15}$$

$$\frac{d}{dx}\mathrm{Csc}^{-1} u = \frac{-1}{u\sqrt{u^2-1}}\frac{du}{dx}. \tag{8.16}$$

Note that the derivative of an inverse cofunction is the negative of the derivative of the corresponding inverse function; that is, derivatives of $\mathrm{Cos}^{-1} u$, $\mathrm{Cot}^{-1} u$, and $\mathrm{Csc}^{-1} u$ are the negatives of the derivatives of $\mathrm{Sin}^{-1} u$, $\mathrm{Tan}^{-1} u$, and $\mathrm{Sec}^{-1} u$.

EXAMPLE 8.10

Find dy/dx if y is defined as a function of x in each of the following:

(a) $y = \mathrm{Sec}^{-1}(x^2)$ (b) $y = 2x\mathrm{Cot}^{-1}(3x)$ (c) $y = \left[\mathrm{Sin}^{-1}(x^3)\right]^2$

(d) $y = \dfrac{\mathrm{Sin}^{-1} x}{\mathrm{Cos}^{-1} x}$ (e) $\mathrm{Cos}^{-1}(xy) + x^2 y + 5 = 0$

SOLUTION (a) $$\frac{dy}{dx} = \frac{1}{x^2\sqrt{(x^2)^2-1}}\frac{d}{dx}(x^2) = \frac{2x}{x^2\sqrt{x^4-1}} = \frac{2}{x\sqrt{x^4-1}}$$

(b) $$\frac{dy}{dx} = 2\,\mathrm{Cot}^{-1}(3x) + 2x\left(\frac{-1}{1+9x^2}\right)(3) = 2\,\mathrm{Cot}^{-1}(3x) - \frac{6x}{1+9x^2}$$

(c) $$\frac{dy}{dx} = 2\,\mathrm{Sin}^{-1}(x^3)\frac{d}{dx}\mathrm{Sin}^{-1}(x^3) = 2\,\mathrm{Sin}^{-1}(x^3)\frac{1}{\sqrt{1-x^6}}(3x^2) = \frac{6x^2\,\mathrm{Sin}^{-1}(x^3)}{\sqrt{1-x^6}}$$

(d) $$\frac{dy}{dx} = \frac{\mathrm{Cos}^{-1} x\left(\dfrac{1}{\sqrt{1-x^2}}\right) - \mathrm{Sin}^{-1} x\left(\dfrac{-1}{\sqrt{1-x^2}}\right)}{(\mathrm{Cos}^{-1} x)^2} = \frac{\mathrm{Cos}^{-1} x + \mathrm{Sin}^{-1} x}{\sqrt{1-x^2}\,(\mathrm{Cos}^{-1} x)^2}$$

(e) Differentiation with respect to x gives

$$\frac{-1}{\sqrt{1-x^2y^2}}\left(y + x\frac{dy}{dx}\right) + 2xy + x^2\frac{dy}{dx} = 0.$$

Thus,

$$\frac{dy}{dx}\left(\frac{-x}{\sqrt{1-x^2y^2}} + x^2\right) = \frac{y}{\sqrt{1-x^2y^2}} - 2xy,$$

from which

$$\frac{dy}{dx} = \frac{y - 2xy\sqrt{1-x^2y^2}}{\sqrt{1-x^2y^2}}\;\frac{\sqrt{1-x^2y^2}}{-x + x^2\sqrt{1-x^2y^2}} = \frac{y - 2xy\sqrt{1-x^2y^2}}{-x + x^2\sqrt{1-x^2y^2}}.$$

■

EXAMPLE 8.11

Find the derivative of the function $f(x) = \text{Sin}^{-1} x + \text{Cos}^{-1} x$. What is your conclusion?

SOLUTION The derivative of the function is

$$f'(x) = \frac{1}{\sqrt{1-x^2}} + \frac{-1}{\sqrt{1-x^2}} = 0.$$

Since this result is valid for the entire domain $-1 \le x \le 1$ of $f(x)$, we can antidifferentiate both sides of the equation,

$$f(x) = C = \text{ a constant.}$$

Since $f(0) = \text{Sin}^{-1}(0) + \text{Cos}^{-1}(0) = 0 + \pi/2 = \pi/2$, it follows that $C = \pi/2$. Consequently,

$$\text{Sin}^{-1} x + \text{Cos}^{-1} x = \frac{\pi}{2}.$$

■

EXAMPLE 8.12

A road information sign specifying the distance to the next city is 1 m high and sits 5 m above road level (Figure 8.26). If the average motorist's eye level is 1.5 m above ground level, at what distance from the sign do the letters appear tallest?

FIGURE 8.26

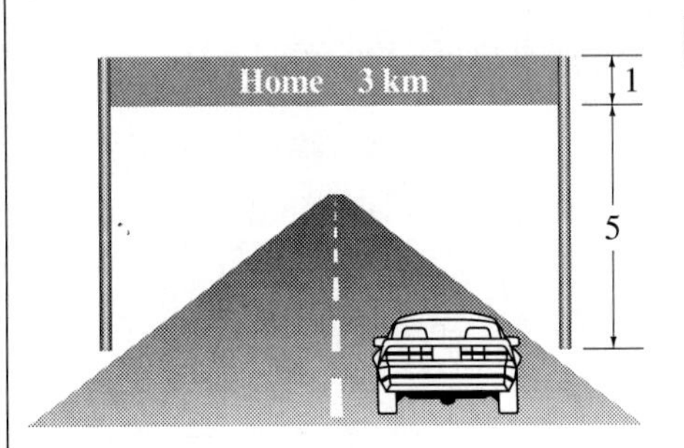

FIGURE 8.27

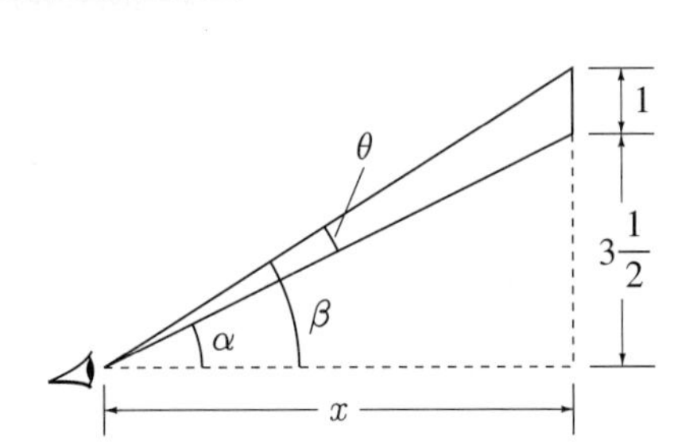

SOLUTION The letters appear tallest when the angle θ that they subtend at the eye (Figure 8.27) is greatest. We assume that the letters are 1 m high. Since

$$\theta = \beta - \alpha = \text{Tan}^{-1}\left(\frac{9/2}{x}\right) - \text{Tan}^{-1}\left(\frac{7/2}{x}\right), \quad 0 < x < \infty,$$

critical points of $\theta(x)$ are given by

$$0 = \frac{d\theta}{dx} = \frac{1}{1+\left(\frac{9}{2x}\right)^2}\left(-\frac{9}{2x^2}\right) - \frac{1}{1+\left(\frac{7}{2x}\right)^2}\left(-\frac{7}{2x^2}\right)$$

$$= \frac{-18}{4x^2+81} + \frac{14}{4x^2+49}.$$

Equivalently,

$$18(4x^2+49) = 14(4x^2+81),$$

which has only one positive solution, $x = 3\sqrt{7}/2$. Since $\theta(x)$ approaches zero as x approaches zero and as x becomes very large, it follows that θ is maximized when the motorist is $3\sqrt{7}/2$ m from the sign.

■

EXERCISES 8.3

In Exercises 1–30 y is defined as a function of x. Find dy/dx in as simplified a form as possible.

1. $y = \mathrm{Cos}^{-1}(2x+3)$ **2.** $y = \mathrm{Cot}^{-1}(x^2+2)$

3. $y = \mathrm{Csc}^{-1}(3-4x)$ **4.** $y = \mathrm{Tan}^{-1}(2-x^2)$

5. $y = \mathrm{Sec}^{-1}(3-2x^2)$ **6.** $y = x\mathrm{Csc}^{-1}(x^2+5)$

7. $y = (x^2+2)\mathrm{Sin}^{-1}(2x)$

8. $y = \mathrm{Tan}^{-1}\sqrt{x+2}$

9. $y = \mathrm{Sin}^{-1}\sqrt{1-x^2}$ **10.** $y = \mathrm{Cot}^{-1}\sqrt{x^2-1}$

11. $y = \left[\mathrm{Tan}^{-1}(x^2)\right]^2$ **12.** $y = x^2\mathrm{Sec}^{-1}x$

13. $y = \tan\left(3\mathrm{Sin}^{-1}x\right)$ **14.** $y = \mathrm{Cot}^{-1}[(1+x)/(1-x)]$

15. $y = \mathrm{Csc}^{-1}(1/x)$ **16.** $y = \mathrm{Sin}^{-1}[(1-x)/(1+x)]$

17. $y = \mathrm{Tan}^{-1}(u^2+1/u)$, $u = \tan(x^2+4)$

18. $y = t\mathrm{Cos}^{-1}t$, $t = \sqrt{1-x^2}$

19. $y^2\sin x + y = \mathrm{Tan}^{-1}x$

20. $\mathrm{Sin}^{-1}(xy) = 5x + 2y$

21. $y = \sqrt{x^2-1} - \mathrm{Sec}^{-1}x$

22. $y = \dfrac{1}{3}\mathrm{Csc}^{-1}\left(\dfrac{x}{3}\right) - \dfrac{\sqrt{x^2-9}}{x^2}$

23. $y = x\mathrm{Cos}^{-1}\left(\dfrac{x}{2}\right) - \sqrt{4-x^2}$

24. $y = \dfrac{\mathrm{Csc}^{-1}(3x)}{x} - \dfrac{\sqrt{9x^2-1}}{x}$

25. $y = x^2\mathrm{Sec}^{-1}x - \sqrt{x^2-1}$

26. $y = x\left(\mathrm{Cos}^{-1}x\right)^2 - 2x - 2\sqrt{1-x^2}$

27. $y = (1+9x^2)\mathrm{Cot}^{-1}(3x) + 3x$

28. $y = (x-2)\sqrt{4x-x^2} + 4\mathrm{Sin}^{-1}\left(\dfrac{x-2}{2}\right)$

29. $y = (2x^2-1)\mathrm{Sin}^{-1}x + x\sqrt{1-x^2}$

30. $y = \mathrm{Tan}^{-1}\left(\dfrac{\sqrt{2}x}{\sqrt{1+x^4}}\right)$

31. Evaluate the derivative of

$$f(x) = \mathrm{Sec}^{-1}x + \mathrm{Cot}^{-1}\sqrt{x^2-1}.$$

What is your conclusion?

32. Find all relative extrema and points of inflection for the function

$$f(x) = x\mathrm{Sin}^{-1}x + \sqrt{1-x^2}.$$

Sketch its graph.

33. (a) Show that if the principal values of $f(x) = \mathrm{Sec}^{-1}x$ are chosen as $0 \le y \le \pi$, $y \ne \pi/2$, then the derivative of the function is

$$\frac{d}{dx}\mathrm{Sec}^{-1}x = \frac{1}{|x|\sqrt{x^2-1}}.$$

(b) What is the derivative of $f(x) = \mathrm{Csc}^{-1}x$ if its principal values are chosen as $-\pi/2 \le y \le \pi/2$, $y \ne 0$?

34. Find the angle between the curves $y = \mathrm{Sin}^{-1}x$ and $y = \mathrm{Cos}^{-1}x$ at their point of intersection.

35. Verify the results in equations 8.13–8.16.

36. When a force F is applied to the object of mass m in Figure 8.28, three other forces act on m: the force of gravity mg ($g > 0$) directly downward, a reactional force of the supporting surface, and a horizontal, frictional force opposing F. The least force F that will overcome friction and produce motion is given by

$$F = \frac{\mu mg}{\cos\theta + \mu\sin\theta},$$

where μ is a constant called the *coefficient of static friction*. Find the angle θ for which F is minimal.

FIGURE 8.28

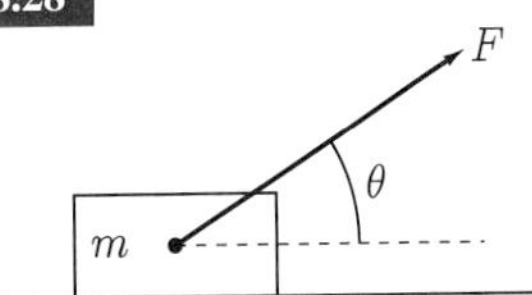

37. Find d^2y/dx^2 if y is defined implicitly as a function of x by

$$\ln(x^2+y^2) = 2\mathrm{Tan}^{-1}\left(\frac{y}{x}\right).$$

38. A runner moves counterclockwise around the track in Figure 8.29 at a rate of 4 m/s. A camera at the centre of the track is placed on a swivel so that it can follow the runner. Find the rate at which the camera turns when (a) the runner is at A and (b) the runner is at B.

FIGURE 8.29

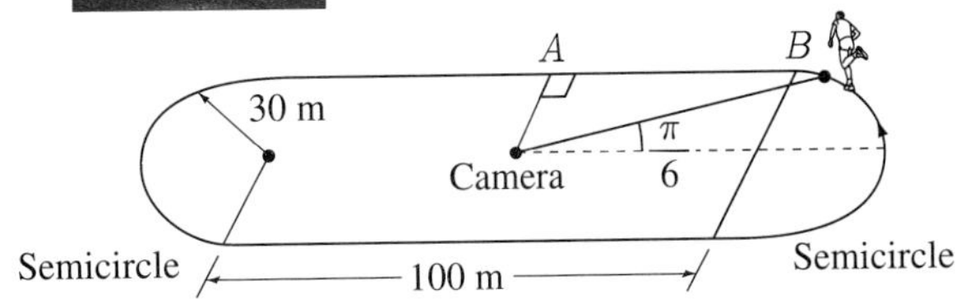

39. A rope with a ring at one end is passed through two fixed rings at the same level (Figure 8.30). The end of the rope without the ring is then passed through the ring at the other end, and a mass m is attached to it. If the rope moves so as to maximize the distance from m to the line through the fixed rings, find angle θ.

FIGURE 8.30

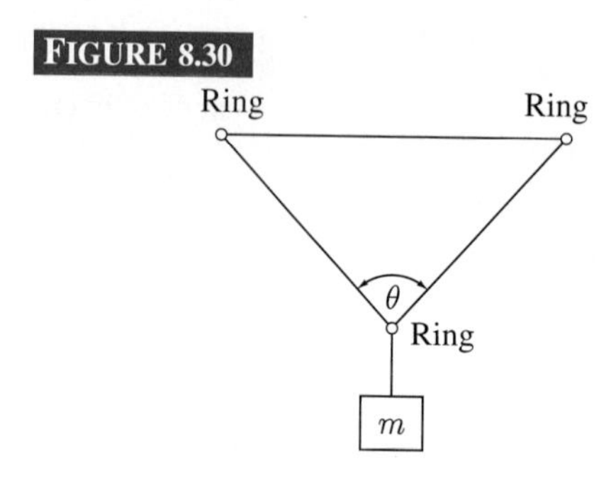

SECTION 8.4

Hyperbolic Functions and Their Derivatives

Certain combinations of the exponential function occur so often in physical applications that they are given special names. Specifically, half the difference of e^x and e^{-x} is defined as the hyperbolic sine function and half their sum is the hyperbolic cosine function. These functions are denoted as follows:

$$\sinh x = \frac{e^x - e^{-x}}{2} \quad \text{and} \quad \cosh x = \frac{e^x + e^{-x}}{2}. \tag{8.17}$$

The names of these functions and their notations bear a striking resemblance to those for the trigonometric functions, and there are reasons for this. First, the hyperbolic functions $\sinh x$ and $\cosh x$ are related to the curve $x^2 - y^2 = 1$, called the *unit hyperbola*, in much the same way as the trigonometric functions $\sin x$ and $\cos x$ are related to the unit circle $x^2 + y^2 = 1$. We will point out two of these similarities in Examples 8.16 and 8.17. Second, for each identity satisfied by the trigonometric functions, there is a corresponding identity satisfied by the hyperbolic functions — not the same identity, but one very similar. For example, using equations 8.17, we have

$$\begin{aligned}(\cosh x)^2 - (\sinh x)^2 &= \left(\frac{e^x + e^{-x}}{2}\right)^2 - \left(\frac{e^x - e^{-x}}{2}\right)^2 \\ &= \frac{1}{4}\left[\left(e^{2x} + 2 + e^{-2x}\right) - \left(e^{2x} - 2 + e^{-2x}\right)\right] \\ &= 1.\end{aligned}$$

Thus the hyperbolic sine and cosine functions satisfy the identity

$$\cosh^2 x - \sinh^2 x = 1, \tag{8.18}$$

which is reminiscent of the identity $\cos^2 x + \sin^2 x = 1$ for the trigonometric functions.

Just as four other trigonometric functions are defined in terms of $\sin x$ and $\cos x$, four corresponding hyperbolic functions are defined as follows:

$$\begin{aligned}\tanh x &= \frac{\sinh x}{\cosh x} = \frac{e^x - e^{-x}}{e^x + e^{-x}}, \quad & \coth x &= \frac{\cosh x}{\sinh x} = \frac{e^x + e^{-x}}{e^x - e^{-x}}, \\ \operatorname{sech} x &= \frac{1}{\cosh x} = \frac{2}{e^x + e^{-x}}, \quad & \operatorname{csch} x &= \frac{1}{\sinh x} = \frac{2}{e^x - e^{-x}}.\end{aligned} \tag{8.19}$$

These definitions and 8.18 immediately imply that

$$1 - \tanh^2 x = \operatorname{sech}^2 x, \tag{8.20a}$$

$$\coth^2 x - 1 = \operatorname{csch}^2 x, \tag{8.20b}$$

analogous to $1 + \tan^2 x = \sec^2 x$ and $1 + \cot^2 x = \csc^2 x$ respectively.

Graphs of $\cosh x$ and $\sinh x$ in Figures 8.31(a) and (b) are easily obtained by addition and subtraction of ordinates of e^x and e^{-x}. Note that $\cosh x$ is an even function, whereas $\sinh x$ is odd, and that neither is periodic. To sketch $\tanh x$, we use division of ordinates, together with the facts that the function is odd and that $\lim_{x\to\infty} \tanh x = 1$ in Figure 8.31(c). These three graphs then lead to the graphs for $\coth x$, $\operatorname{sech} x$, and $\operatorname{csch} x$ in Figures 8.31(d)–(f).

FIGURE 8.31

(a)

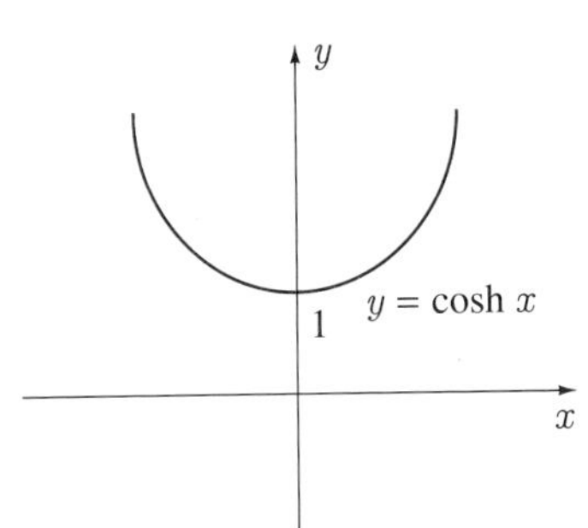

(b)

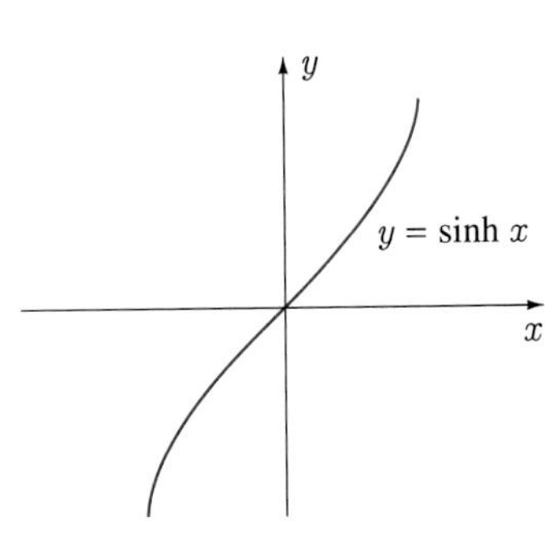

(c)

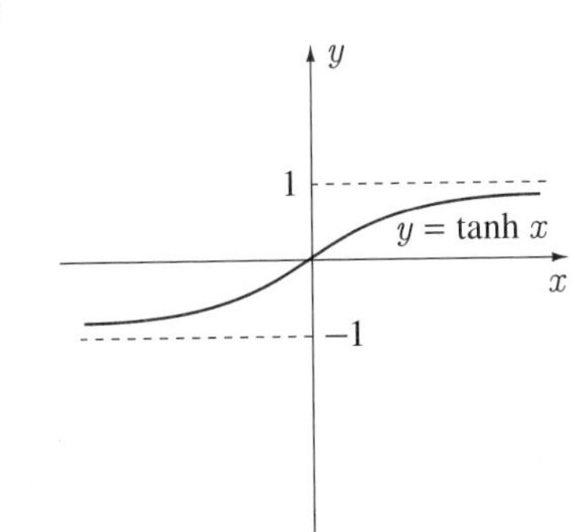

(d)

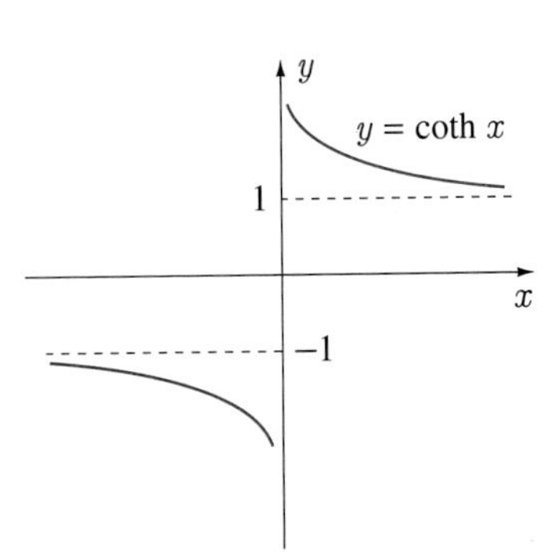

(e)

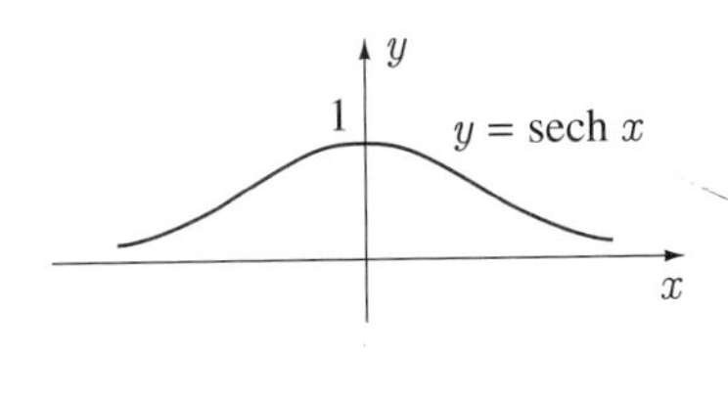

(f)

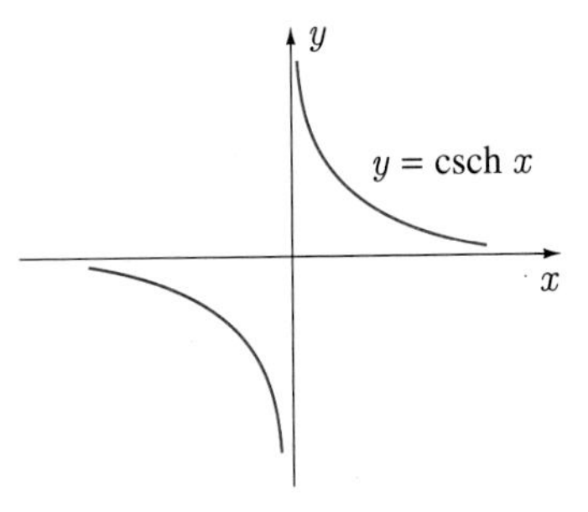

In Appendix A we point out that most trigonometric identities can be derived from the compound angle formulas for $\sin(A \pm B)$ and $\cos(A \pm B)$. It is easy to verify similar formulas for the hyperbolic functions:

$$\sinh(A \pm B) = \sinh A \cosh B \pm \cosh A \sinh B \quad \text{and} \tag{8.21a}$$

$$\cosh(A \pm B) = \cosh A \cosh B \pm \sinh A \sinh B. \tag{8.21b}$$

For example, equations 8.17 give

$$\begin{aligned}
\cosh A \cosh B - \sinh A \sinh B &= \left(\frac{e^{A} + e^{-A}}{2}\right)\left(\frac{e^{B} + e^{-B}}{2}\right) \\
&\quad - \left(\frac{e^{A} - e^{-A}}{2}\right)\left(\frac{e^{B} - e^{-B}}{2}\right) \\
&= \frac{1}{4}\Big[\left(e^{A+B} + e^{A-B} + e^{B-A} + e^{-A-B}\right) - \\
&\quad \left(e^{A+B} - e^{A-B} - e^{B-A} + e^{-A-B}\right)\Big] \\
&= \frac{1}{2}\left[e^{A-B} + e^{-(A-B)}\right] \\
&= \cosh(A - B).
\end{aligned}$$

With these formulas, we can derive hyperbolic identities analogous to trigonometric identities (A.12)–(A.17):

$$\tanh(A \pm B) = \frac{\tanh A \pm \tanh B}{1 \pm \tanh A \tanh B}, \tag{8.21c}$$

$$\sinh 2A = 2 \sinh A \cosh A, \tag{8.21d}$$

$$\cosh 2A = \cosh^2 A + \sinh^2 A \tag{8.21e}$$

$$= 2\cosh^2 A - 1 \tag{8.21f}$$

$$= 1 + 2\sinh^2 A, \tag{8.21g}$$

$$\tanh 2A = \frac{2 \tanh A}{1 + \tanh^2 A}, \tag{8.21h}$$

$$\sinh A \sinh B = \frac{1}{2}\cosh(A + B) - \frac{1}{2}\cosh(A - B), \tag{8.21i}$$

$$\sinh A \cosh B = \frac{1}{2}\sinh(A + B) + \frac{1}{2}\sinh(A - B), \tag{8.21j}$$

$$\cosh A \cosh B = \frac{1}{2}\cosh(A + B) + \frac{1}{2}\cosh(A - B), \tag{8.21k}$$

$$\sinh A + \sinh B = 2 \sinh\left(\frac{A + B}{2}\right)\cosh\left(\frac{A - B}{2}\right), \tag{8.21l}$$

$$\sinh A - \sinh B = 2 \cosh\left(\frac{A + B}{2}\right)\sinh\left(\frac{A - B}{2}\right), \tag{8.21m}$$

$$\cosh A + \cosh B = 2 \cosh\left(\frac{A + B}{2}\right)\cosh\left(\frac{A - B}{2}\right), \tag{8.21n}$$

$$\cosh A - \cosh B = 2 \sinh\left(\frac{A + B}{2}\right)\sinh\left(\frac{A - B}{2}\right). \tag{8.21o}$$

Since the hyperboic sine and cosine are defined in terms of the exponential function, for which we know the derivative, and the remaining hyperbolic functions are defined in terms of the hyperbolic sine and cosine, it follows that calculation of the derivatives of the hyperbolic functions should be straightforward. Indeed, if $u(x)$ is a differentiable function of x, then

$$\frac{d}{dx}\sinh u = \cosh u \frac{du}{dx}, \tag{8.22a}$$

$$\frac{d}{dx}\cosh u = \sinh u \frac{du}{dx}, \tag{8.22b}$$

$$\frac{d}{dx}\tanh u = \operatorname{sech}^2 u\,\frac{du}{dx}, \tag{8.22c}$$

$$\frac{d}{dx}\coth u = -\operatorname{csch}^2 u\,\frac{du}{dx}, \tag{8.22d}$$

$$\frac{d}{dx}\operatorname{sech} u = -\operatorname{sech} u \tanh u\,\frac{du}{dx}, \tag{8.22e}$$

$$\frac{d}{dx}\operatorname{csch} u = -\operatorname{csch} u \coth u\,\frac{du}{dx}. \tag{8.22f}$$

EXAMPLE 8.13

Find dy/dx if y is defined as a function of x by:

(a) $y = \operatorname{sech}(3x^2)$ (b) $y = \tanh(1 - 4x)$

(c) $y = \cos 2x \sinh 2x$ (d) $y = \cosh\left(\operatorname{Tan}^{-1} x^2\right)$

SOLUTION

(a) $$\frac{dy}{dx} = -\operatorname{sech}(3x^2)\tanh(3x^2)\frac{d}{dx}(3x^2) = -6x\operatorname{sech}(3x^2)\tanh(3x^2)$$

(b) $$\frac{dy}{dx} = \operatorname{sech}^2(1-4x)\frac{d}{dx}(1-4x) = -4\operatorname{sech}^2(1-4x)$$

(c) $$\frac{dy}{dx} = \cos 2x(2\cosh 2x) + (-2\sin 2x)\sinh 2x = 2(\cos 2x\cosh 2x - \sin 2x \sinh 2x)$$

(d) $$\frac{dy}{dx} = \sinh\left(\operatorname{Tan}^{-1} x^2\right)\frac{d}{dx}\operatorname{Tan}^{-1} x^2 = \sinh\left(\operatorname{Tan}^{-1} x^2\right)\frac{2x}{1+x^4}$$

■

EXAMPLE 8.14

If a uniform cable is hung between two fixed supports (Figure 8.32), the shape of the curve $y = f(x)$ must satisfy the differential equation

$$\frac{d^2y}{dx^2} = \frac{\rho g}{H}\sqrt{1 + \left(\frac{dy}{dx}\right)^2},$$

where ρ is the mass per unit length of the cable, $g > 0$ is the acceleration due to gravity, and $H > 0$ is the tension in the cable at its lowest point. Show that a solution of the equation is

$$y = f(x) = \frac{H}{\rho g}\cosh\left(\frac{\rho g x}{H}\right) + C,$$

where C is a constant.

FIGURE 8.32

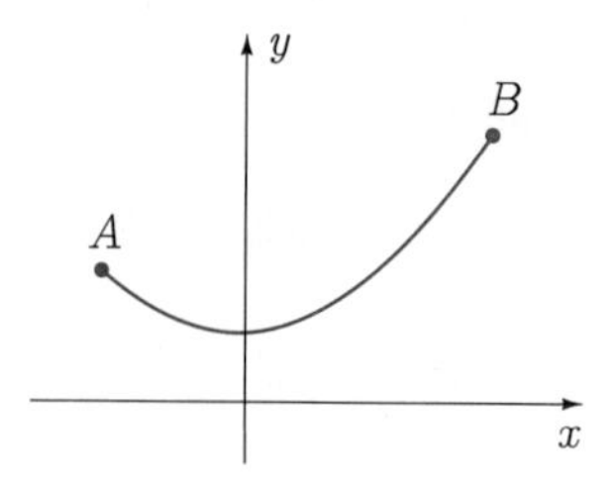

SOLUTION In Exercises 41 of Section 9.5 we derive this solution. For now we simply wish to verify that the hyperbolic cosine is indeed a solution. The first derivative of the function is

$$\frac{dy}{dx} = \sinh\left(\frac{\rho g x}{H}\right),$$

and therefore its second derivative is

$$\frac{d^2 y}{dx^2} = \frac{\rho g}{H}\cosh\left(\frac{\rho g x}{H}\right).$$

On the other hand,

$$\frac{\rho g}{H}\sqrt{1 + \left(\frac{dy}{dx}\right)^2} = \frac{\rho g}{H}\sqrt{1 + \sinh^2\left(\frac{\rho g x}{H}\right)} = \frac{\rho g}{H}\sqrt{\cosh^2\left(\frac{\rho g x}{H}\right)} = \frac{\rho g}{H}\cosh\left(\frac{\rho g x}{H}\right).$$

Thus, the function $y = \big(H/(\rho g)\big)\cosh(\rho g x/H) + C$ does indeed satisfy the given differential equation. ■

Example 8.14 shows that the many telephone and hydro wires crisscrossing the country hang in the form of hyperbolic cosines. Engineers often call this curve a *catenary*.

EXAMPLE 8.15

In studying wave guides, the electrical engineer often encounters the differential equation

$$\frac{d^2 y}{dx^2} - ky = 0,$$

where $k > 0$ is a constant. Show that $y = f(x) = A\cosh\sqrt{k}x + B\sinh\sqrt{k}x$ is a solution for any constants A and B.

SOLUTION The first derivative of $y = f(x)$ is

$$\frac{dy}{dx} = \sqrt{k}A\sinh\sqrt{k}x + \sqrt{k}B\cosh\sqrt{k}x,$$

and therefore

$$\frac{d^2 y}{dx^2} = kA\cosh\sqrt{k}x + kB\sinh\sqrt{k}x = k(A\cosh\sqrt{k}x + B\sinh\sqrt{k}x).$$

Thus,

$$\frac{d^2 y}{dx^2} = ky.$$

■

In Examples 8.16 and 8.17 we illustrate geometric parallels between the trigonometric sine and cosine functions and the hyperbolic sine and cosine functions.

EXAMPLE 8.16

Show that:

(a) every point (x, y) on the unit circle $x^2 + y^2 = 1$ can be expressed in the form $x = \cos t$, $y = \sin t$ for some real number t in the interval $0 \le t \le 2\pi$;

(b) every point (x, y) on the right half of the unit hyperbola $x^2 - y^2 = 1$ can be expressed in the form $x = \cosh t$, $y = \sinh t$ for some real number t.

SOLUTION

(a) If t is the angle shown in Figure 8.33, then clearly the coordinates of P are $x = \cos t$ and $y = \sin t$. As angle t ranges for 0 to 2π, P traces the circle exactly once.

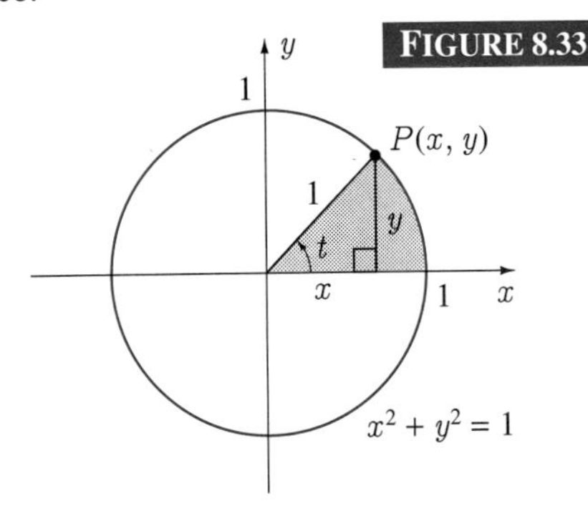

FIGURE 8.33

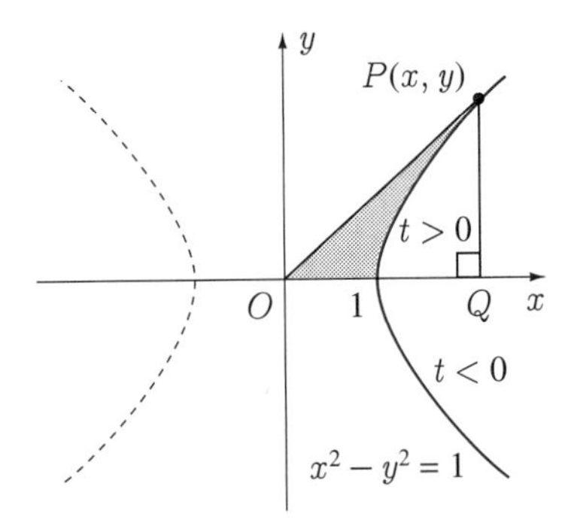

FIGURE 8.34

(b) A sketch of the unit hyperbola $x^2 - y^2 = 1$ is shown in Figure 8.34. If $x = \cosh t$ and $y = \sinh t$ are coordinates of a point P in the plane, where t is some real number, then identity 8.18 implies that $x^2 - y^2 = \cosh^2 t - \sinh^2 t = 1$. In other words, P is on the unit hyperbola. Furthermore, since $x = \cosh t$ is always positive, P must be on the right half of the hyperbola. Finally, because the range of $x = \cosh t$ is $x \ge 1$ in Figure 8.31(a), and the range of $y = \sinh t$ is $-\infty < y < \infty$ in Figure 8.31(b), it follows that every point on the right half of the hyperbola can be obtained from some value of t.

Note that t is *not* the angle formed by the positive x-axis and the line joining the origin to (x, y). ■

EXAMPLE 8.17

Show that the area bounded by the positive x-axis, the line joining the origin to the point $P(x, y)$ on the curves in Figures 8.33 and 8.34, and the curves themselves is $t/2$.

SOLUTION (a) Since the area of a sector of a circle is one-half the product of the square of the radius and the angle at the centre of the circle, the shaded area in Figure 8.33 is $(1/2)(1)^2(t) = t/2$.

(b) The shaded area in Figure 8.34 is the area of triangle OQP less the area beneath the hyperbola between $x = 1$ and Q:

$$\text{Area} = \frac{1}{2}(x)(y) - \int_1^x y\,dx.$$

Since the coordinates of P can be expressed in the form $x = \cosh t$ and $y = \sinh t$ (Example 8.16), we may set

$$\text{Area} = \frac{1}{2}\cosh t \sinh t - \int_0^t \sinh t(\sinh t\, dt).$$

Now, by identity 8.21d, $\cosh t \sinh t = (1/2)\sinh 2t$, and by identity 8.21f, $\sinh^2 t = (1/2)(\cosh 2t - 1)$. Therefore

$$\text{Area} = \frac{1}{4}\sinh 2t - \frac{1}{2}\int_0^t (\cosh 2t - 1)\, dt.$$

Since the derivative of $\sinh 2t$ is $2\cosh 2t$, it follows that an antiderivative of $\cosh 2t$ is $(1/2)\sinh 2t$, and

$$\text{Area} = \frac{1}{4}\sinh 2t - \frac{1}{2}\left\{\frac{1}{2}\sinh 2t - t\right\}_0^t = \frac{t}{2}.$$

■

EXERCISES 8.4

In Exercises 1–10 evaluate the expression (if it has a value).

1. $3\cosh 1$

2. $\sinh(\pi/2)$

3. $\tanh\sqrt{1 - \sin 3}$

4. $\mathrm{Sin}^{-1}(\operatorname{sech} 10)$

5. $\mathrm{Cos}^{-1}(2\operatorname{csch} 1)$

6. $\coth(\sinh 5)$

7. $\sqrt{\ln|\sinh(-3)|}$

8. $\operatorname{sech}[\sec(\pi/3)]$

9. $e^{-2\cosh e}$

10. $\sinh[\mathrm{Cot}^{-1}(-3\pi/10)]$

In Exercises 11–20 y is defined as a function of x. Find dy/dx in as simplified a form as possible.

11. $y = \operatorname{csch}(2x + 3)$

12. $y = x\sinh(x/2)$

13. $y = \sqrt{1 - \operatorname{sech} x}$

14. $y = \tanh(\ln x)$

15. $\cosh(x + y) = 2x$

16. $y + \coth x = \sqrt{1 + y}$

17. $y = u\cosh u,\ u = e^x + e^{-x}$

18. $y = \tan(\cosh t),\ t = \cos(\tanh x)$

19. $y = \mathrm{Tan}^{-1}(\sinh x)$

20. $y = \ln\sqrt{\tanh 2x}$

21. Verify the results in identities 8.21c–8.21o.

22. Verify differentiation formulas 8.22.

23. Show that $y = \sinh x$ and $y = \tanh x$ have points of inflection at $x = 0$.

24. To analyze vertical vibrations of the beam in Figure 8.35, we must solve the differential equation

$$\frac{d^4 y}{dx^4} - k^4 y = 0,$$

where $k > 0$ is a constant.

FIGURE 8.35

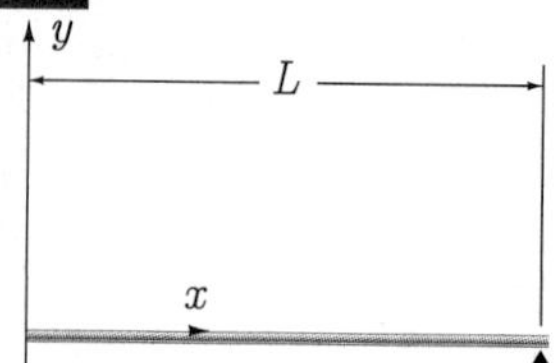

(a) Show that a solution is

$$y = f(x) = A\cos kx + B\sin kx + C\cosh kx + D\sinh kx$$

for any constants A, B, C, and D.

(b) If the left end ($x = 0$) is fastened horizontally and the right end ($x = L$) is pinned, then $f(x)$ must satisfy the conditions

$$f(0) = f'(0) = f(L) = f''(L) = 0.$$

Show that these restrictions imply that $C = -A$, $D = -B$, and A and B must satisfy the equations

$$A(\cos kL - \cosh kL) + B(\sin kL - \sinh kL) = 0,$$
$$A(\cos kL + \cosh kL) + B(\sin kL + \sinh kL) = 0.$$

(c) Eliminate A and B between these equations to show that k must satisfy the condition

$$\tan kL = \tanh kL.$$

25. Show that $y = \operatorname{sech} x$ has points of inflection for $x = \pm \ln(\sqrt{2} + 1)$.

26. A uniform hydro cable P metres long with mass ρ kilograms per metre of length is hung from two supports at the same level L metres apart (Figure 8.36). If the supports are b metres higher than the lowest point in the cable and H is the tension in the cable at the lowest point, it can be shown that

$$\rho g P = 2H \sinh \frac{\rho g L}{2H}.$$

FIGURE 8.36

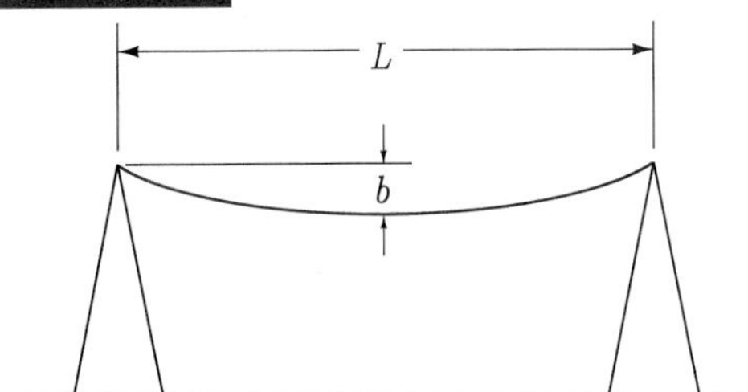

(a) If $L = 70$ m, $\rho = 0.5$ kg/m, and $P = 80$ m, find H. Hint: Set $z = \rho g/(2H)$, and use Newton's iterative procedue with $z_1 = 0.01$ to solve $Pz = \sinh(Lz)$ for z.

(b) Use the result in (a) to find b. See Example 8.14 for the equation of the curve describing the cable.

27. Each hyperbolic function has associated with it an inverse hyperbolic function.

(a) Draw the inverse functions $\operatorname{Sinh}^{-1} x$, $\operatorname{Tanh}^{-1} x$, $\operatorname{Coth}^{-1} x$, and $\operatorname{Csch}^{-1} x$ for $\sinh x$, $\tanh x$, $\coth x$, and $\operatorname{csch} x$.

(b) Why do $\cosh x$ and $\operatorname{sech} x$ not have inverse functions? It is customary to associate functions $\operatorname{Cosh}^{-1} x$ and $\operatorname{Sech}^{-1} x$ with $\cosh x$ and $\operatorname{sech} x$ by restricting their domains to nonnegative numbers. Draw graphs of $\operatorname{Cosh}^{-1} x$ and $\operatorname{Sech}^{-1} x$.

(c) Obtain the following derivatives of the inverse hyperbolic functions:

$$\frac{d}{dx}\operatorname{Sinh}^{-1} x = \frac{1}{\sqrt{x^2+1}};$$

$$\frac{d}{dx}\operatorname{Cosh}^{-1} x = \frac{1}{\sqrt{x^2-1}};$$

$$\frac{d}{dx}\operatorname{Tanh}^{-1} x = \frac{1}{1-x^2}, \quad |x| < 1;$$

$$\frac{d}{dx}\operatorname{Coth}^{-1} x = \frac{1}{1-x^2}, \quad |x| > 1;$$

$$\frac{d}{dx}\operatorname{Sech}^{-1} x = \frac{-1}{x\sqrt{1-x^2}};$$

$$\frac{d}{dx}\operatorname{Csch}^{-1} x = \frac{-1}{|x|\sqrt{1+x^2}}.$$

(d) Show that

$$\operatorname{Sinh}^{-1} x = \ln\left(x + \sqrt{x^2+1}\right);$$

$$\operatorname{Cosh}^{-1} x = \ln\left(x + \sqrt{x^2-1}\right);$$

$$\operatorname{Tanh}^{-1} x = \frac{1}{2}\ln\left(\frac{1+x}{1-x}\right), \quad |x| < 1.$$

28. When an object of mass m fall from rest at time $t = 0$ under the influence of gravity and an air resistance proportional to the square of velocity, its velocity v as a function of time is defined by the equation

$$\frac{1}{2}\sqrt{\frac{\beta}{mg}}\ln\left(\frac{\sqrt{mg/\beta} - v}{\sqrt{mg/\beta} + v}\right) = -\frac{\beta t}{m},$$

where $\beta > 0$ is a constant and $g > 0$ is the acceleration due to gravity.

(a) Show that when this equation is solved for v in terms of t, the result is

$$v(t) = \sqrt{\frac{mg}{\beta}}\tanh\left(\sqrt{\frac{\beta g}{m}}t\right).$$

(b) What is the limit of v for large t, called the *limiting velocity*?

SUMMARY

With derivatives of trigonometric, exponential, and logarithm functions in Chapter 3, and derivatives of inverse trigonometric and hyperbolic functions in this chapter, we have completed the list of differentiation formulas.

Inverse trigonometric functions reverse the roles of trigonometric functions. A trigonometric function such as the sine function associates a value called $\sin x$ with a real number (angle) x. The corresponding inverse function, $\mathrm{Sin}^{-1}\, x$, regards x as the sine of an angle, and yields the angle.

Hyperbolic functions are special combinations of exponential functions that arise sufficiently often in applications to warrant special consideration. They satisfy identities and have derivatives very similar to the trigonometric functions.

The transcendental functions that we have discussed fall into three pairs: trigonometric and inverse trigonometric, exponential and logarithm (the inverse of the exponential), and hyperbolic and inverse hyperbolic (Exercise 27 in Section 8.4). The relationship between a function and its inverse is very important. For instance, evaluation of the indeterminate forms 0^0, 1^∞, and ∞^0 in Section 4.9 was based on the fact that taking logarithms and exponentials are inverse operations. By taking logarithms, we transformed each indeterminate form into a limit to which we could apply L'Hôpital's rule. We then used exponentials to recapture the original limit. In Section 9.2, the inverse character of exponential and logarithm functions is indispensable to the solution of many differential equations. The integration technique of Section 9.5 is based on the relation of an inverse trigonometric function to the corresponding trigonometric function.

Key Terms and Formulas

In reviewing this chapter, you should be able to define or discuss the following key terms:

Algebraic function
Transcendental function
One-to-one function
Inverse function
Horizontal line test
Monotonic function
Inverse trigonometric function
Principal values
Hyperbolic function

REVIEW EXERCISES

In Exercises 1–10 find all solutions of the equation.

1. $\cos^2 x + 5\cos x - 1 = 0$

2. $4\sin 2x = 1$

3. $\csc^2(x+1) = 3$

4. $\mathrm{Tan}^{-1}(3x+2) = 5 - 2\pi$

5. $\cos 2x = \sin x$

6. $\ln(\sin x) + \ln(1 + \sin x) = \ln 3 - \ln 2$

7. $3\,\mathrm{Sin}^{-1}\left(e^{x+2}\right) = 2$

8. $3\sin\left(e^{x+2}\right) = 2$

9. $\tan(x\cosh 2) = 1/4$

10. $\sinh x = 4$

In Exercises 11–20 y is defined as a function of x. Find dy/dx in as simplified a form as possible.

11. $y = \mathrm{Sin}^{-1}(2 - 3x)$

12. $y = 3\sinh(x^2)$

13. $y = \dfrac{\mathrm{Cos}^{-1}\, x}{\mathrm{Sin}^{-1}\, x}$

14. $y = \mathrm{Tan}^{-1}\left(\dfrac{1}{x} + x\right)$

15. $y = e^{\cosh x}$

16. $\sinh y = \sin x$

17. $\mathrm{Sec}^{-1}(x+y) = xy$

18. $y = x\,\mathrm{Csc}^{-1}\left(\dfrac{1}{x^2}\right)$

19. $y = e^{2x}\cosh 2x$

20. $\ln\left[\mathrm{Tan}^{-1}(x+y)\right] = 1$

21. Sketch graphs of the following functions:

(a) $f(x) = \tan\left(\mathrm{Tan}^{-1}\, x\right)$ (b) $f(x) = \mathrm{Tan}^{-1}(\tan x)$

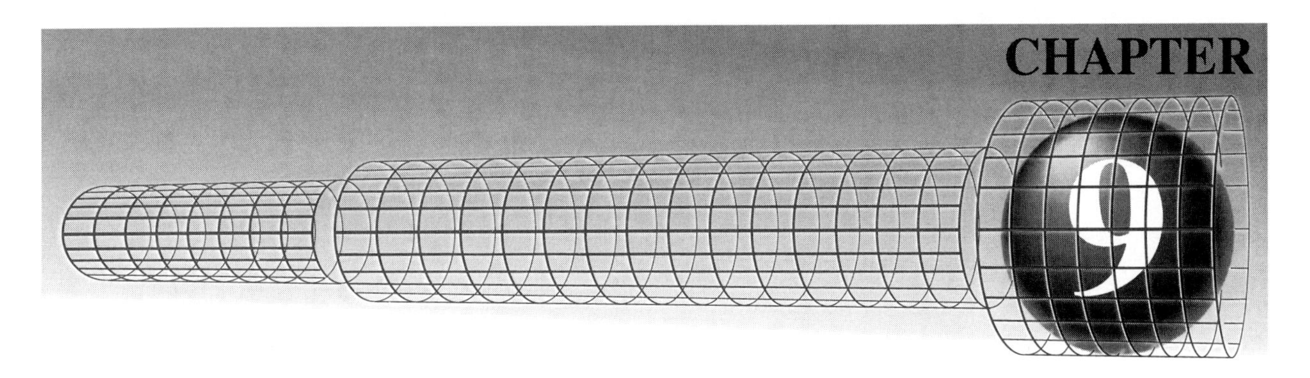

Techniques for Integration

In Chapter 5, when we defined indefinite integrals as antiderivatives, we discussed three techniques for evaluating antiderivatives. First, since every differentiation formula can be restated as an antidifferentiation formula, it follows that the more competent we are at differentiation, the more likely we are to recognize an antiderivative. Our second technique was called "guess and fix up." It is applicable to sufficiently simple integrands that we can immediately determine the antiderivative to within a multiplicative constant. It is then a matter of adjusting the constant. The third technique was to change the variable of integration. By a suitable transformation, replace a complicated antiderivative with a simpler one.

In this chapter we develop additional techniques applicable to more complex integration problems. At the same time we take the opportunity to review the applications of definite integrals in Chapter 7.

SECTION 9.1

Integration Formulas and Substitutions

Every differentiation formula can be restated as an antidifferentiation formula. A partial list of important formulas is as follows:

$$\int u^n\,du = \frac{u^{n+1}}{n+1} + C, \quad n \neq -1; \tag{9.1}$$

$$\int \frac{1}{u}\,du = \ln|u| + C; \tag{9.2}$$

$$\int e^u\,du = e^u + C; \tag{9.3}$$

$$\int a^u\,du = a^u \log_a e + C; \tag{9.4}$$

$$\int \cos u\,du = \sin u + C; \tag{9.5}$$

$$\int \sin u\,du = -\cos u + C; \tag{9.6}$$

$$\int \sec^2 u\,du = \tan u + C; \tag{9.7}$$

$$\int \csc^2 u\,du = -\cot u + C; \tag{9.8}$$

$$\int \sec u \tan u \, du = \sec u + C; \tag{9.9}$$

$$\int \csc u \cot u \, du = -\csc u + C; \tag{9.10}$$

$$\int \tan u \, du = \ln|\sec u| + C = -\ln|\cos u| + C; \tag{9.11}$$

$$\int \cot u \, du = -\ln|\csc u| + C = \ln|\sin u| + C; \tag{9.12}$$

$$\int \sec u \, du = \ln|\sec u + \tan u| + C; \tag{9.13}$$

$$\int \csc u \, du = \ln|\csc u - \cot u| + C. \tag{9.14}$$

Formulas 9.1–9.10 are restatements of differentiation formulas 5.5 and 5.6 in Chapter 5. Formulas 9.11 and 9.13 are verified in Example 9.1 below. Formulas 9.12 and 9.14 are similar.

EXAMPLE 9.1

Verify formulas 9.11 and 9.13.

SOLUTION To verify 9.11 we express $\tan u$ in terms of $\sin u$ and $\cos u$,

$$\int \tan u \, du = \int \frac{\sin u}{\cos u} \, du = -\ln|\cos u| + C = \ln|\sec u| + C.$$

If integration to $\ln|\cos u|$ is not obvious, try the substitution $v = \cos u$.

To obtain the antiderivative of $\sec u$ requires a subtle trick. We first multiply numerator and denominator by $\sec u + \tan u$,

$$\int \sec u \, du = \int \sec u \frac{\sec u + \tan u}{\sec u + \tan u} \, du = \int \frac{\sec^2 u + \sec u \tan u}{\sec u + \tan u} \, du.$$

If we now set $v = \sec u + \tan u$, then $dv = (\sec u \tan u + \sec^2 u) \, du$, and

$$\int \sec u \, du = \int \frac{1}{v} \, dv = \ln|v| + C = \ln|\sec u + \tan u| + C.$$

■

We could list many other integration formulas, but one of the purposes of this chapter is to develop integration techniques that will eliminate the need for excessive memorization of formulas. Our emphasis will be on the development of antiderivatives rather than the memorization of formulas, and we prefer to keep the list of formulas short.

In Chapter 5 we demonstrated how a change of variable could sometimes replace a complex integration problem with a simpler one. When the integrand contains $x^{1/n}$, where n is a positive integer, it is often useful to set $u = x^{1/n}$, or, equivalently, $x = u^n$. We illustrate in the following example.

EXAMPLE 9.2

Evaluate the following indefinite integrals:

$$\text{(a)} \quad \int \frac{2-\sqrt{x}}{2+\sqrt{x}}\,dx \qquad \text{(b)} \quad \int \frac{x^{1/3}}{1+x^{2/3}}\,dx \qquad \text{(c)} \quad \int \frac{\sqrt{x}}{1+x^{1/3}}\,dx$$

SOLUTION (a) If we set $u = \sqrt{x}$ or $x = u^2$, then $dx = 2\,u\,du$, and

$$\int \frac{2-\sqrt{x}}{2+\sqrt{x}}\,dx = \int \frac{2-u}{2+u}(2\,u\,du) = 2\int \frac{2\,u-u^2}{2+u}\,du.$$

Long division of $-u^2 + 2\,u$ by $u + 2$ immediately gives

$$\begin{aligned}\int \frac{2-\sqrt{x}}{2+\sqrt{x}}\,dx &= 2\int \left(-u+4-\frac{8}{u+2}\right)du \\ &= 2\left(-\frac{u^2}{2}+4\,u-8\ln|u+2|\right)+C \\ &= -x+8\sqrt{x}-16\ln\left(2+\sqrt{x}\right)+C.\end{aligned}$$

(b) If we set $u = x^{1/3}$ or $x = u^3$, then $dx = 3\,u^2\,du$, and

$$\int \frac{x^{1/3}}{1+x^{2/3}}\,dx = \int \frac{u}{1+u^2}(3\,u^2\,du) = 3\int \frac{u^3}{1+u^2}\,du.$$

Division of u^3 by $u^2 + 1$ leads to

$$\begin{aligned}\int \frac{x^{1/3}}{1+x^{2/3}}\,dx &= 3\int \left(u-\frac{u}{u^2+1}\right)du \\ &= 3\left(\frac{u^2}{2}-\frac{1}{2}\ln|u^2+1|\right)+C \\ &= \frac{3}{2}x^{2/3}-\frac{3}{2}\ln(x^{2/3}+1)+C.\end{aligned}$$

(c) If we set $u = x^{1/6}$ or $x = u^6$, then $dx = 6\,u^5\,du$, and

$$\begin{aligned}\int \frac{\sqrt{x}}{1+x^{1/3}}\,dx &= \int \frac{u^3}{1+u^2}(6\,u^5\,du) \\ &= 6\int \left(u^6-u^4+u^2-1+\frac{1}{1+u^2}\right)du \\ &= 6\left(\frac{u^7}{7}-\frac{u^5}{5}+\frac{u^3}{3}-u+\operatorname{Tan}^{-1}u\right)+C \\ &= \frac{6}{7}x^{7/6}-\frac{6}{5}x^{5/6}+2\,x^{1/2}-6\,x^{1/6}+6\operatorname{Tan}^{-1}(x^{1/6})+C.\end{aligned}$$

■

EXERCISES 9.1

In Exercises 1–24 evaluate the indefinite integral.

1. $\displaystyle\int \frac{x^2}{5 - 3x^3}\,dx$

2. $\displaystyle\int x e^{-2x^2}\,dx$

3. $\displaystyle\int \frac{x}{(x^2 + 2)^{1/3}}\,dx$

4. $\displaystyle\int \frac{e^x}{1 + e^x}\,dx$

5. $\displaystyle\int \frac{4t + 8}{t^2 + 4t + 5}\,dt$

6. $\displaystyle\int x^2\sqrt{1 - 3x^3}\,dx$

7. $\displaystyle\int (x + 1)(x^2 + 2x)^{1/3}\,dx$

8. $\displaystyle\int \frac{x^2}{(1 + x^3)^3}\,dx$

9. $\displaystyle\int \frac{\sqrt{x}}{1 - \sqrt{x}}\,dx$

10. $\displaystyle\int \frac{1 - \sqrt{x}}{\sqrt{x}}\,dx$

11. $\displaystyle\int \frac{x + 2}{x + 1}\,dx$

12. $\displaystyle\int \frac{x^2 + 2}{x^2 + 1}\,dx$

13. $\displaystyle\int \frac{\sin\theta}{\cos\theta - 1}\,d\theta$

14. $\displaystyle\int \frac{x + 3}{\sqrt{2x + 4}}\,dx$

15. $\displaystyle\int \frac{e^x}{1 + e^{2x}}\,dx$

16. $\displaystyle\int \sin^3 2x \cos 2x\,dx$

17. $\displaystyle\int x^5(2x^2 - 5)^4\,dx$

18. $\displaystyle\int \frac{x^3}{(x + 5)^2}\,dx$

19. $\displaystyle\int z^2\sqrt{3 - z}\,dz$

20. $\displaystyle\int \tan 3x\,dx$

21. $\displaystyle\int \frac{(x - 3)^{2/3}}{(x - 3)^{2/3} + 1}\,dx$

22. $\displaystyle\int \frac{\sqrt{x}}{1 + x^{1/4}}\,dx$

23. $\displaystyle\int \frac{1}{x^2 - 4}\,dx$

24. $\displaystyle\int \frac{e^x - 1}{e^x + 1}\,dx$

25. Find the length of the curve $y = \ln(\cos x)$ from $(0, 0)$ to $(\pi/4, -(\ln 2)/2)$.

26. Find the area bounded by the curves

$$y = \frac{x^2 + 1}{x + 1}, \qquad x + 3y = 7.$$

27. (a) Find the centroid of the area bounded by the curves

$$y = (3 + x)^{3/2}, \qquad x = 1, \qquad y = 0.$$

(b) What is its second moment of area about the line $x = 1$?

28. If $f(x)$ is a continuous even function, prove that

$$\int_{-a}^{a} f(x)\,dx = 2\int_{0}^{a} f(x)\,dx;$$

and if $f(x)$ is a continuous odd function,

$$\int_{-a}^{a} f(x)\,dx = 0.$$

29. (a) Show that the function $f(x) = \lambda e^{-\lambda x}$, where $\lambda > 0$ is a constant, qualifies as a probability density function on the interval $x \geq 0$. See Exercise 23 in Section 7.9 for the definition of a pdf.

(b) If the probability that $x \geq 3$ is 0.5, what is λ?

30. A paratrooper and his parachute fall from rest with a combined mass of 100 kg. At any instant during descent, the parachute has an air resistance force acting on it that in newtons is equal to one-half its speed (in metres per second). Assuming that the paratrooper falls vertically downward and that the parachute is already open when the jump takes place, the differential equation describing his motion is

$$200\,\frac{dv}{dt} = 1962 - v,$$

where v is velocity and t is time.

(a) By writing this equation in the form

$$\frac{1}{1962 - v}\,\frac{dv}{dt} = \frac{1}{200},$$

and antidifferentiating with respect to t, find v as a function of t.

(b) If x measures vertical distance fallen, then $v = dx/dt$. Use this to find the position of the paratrooper as a function of time.

31. Evaluate

$$\int_0^{4\pi} \sqrt{1 + \cos x}\,dx.$$

Hint: $\cos x = 2\cos^2(x/2) - 1$.

32. Evaluate

$$\int \frac{1}{x(3 + 2x^n)}\,dx$$

for any $n > 0$. Hint: Find A and B in order that

$$\frac{1}{x(3 + 2x^n)} = \frac{A}{x} + \frac{Bx^{n-1}}{3 + 2x^n}.$$

33. Show that the substitution $u - x = \sqrt{x^2 + bx + c}$ replaces the integral

$$\int \sqrt{x^2 + bx + c}\,dx$$

with the integral of a rational function of u.

34. Use the substitution $u - x = \sqrt{x^2 + 3x + 4}$ to evaluate

$$\int \frac{1}{(x^2 + 3x + 4)^{3/2}}\, dx.$$

35. Show that when the quadratic $c + bx - x^2$ factors as $c + bx - x^2 = (p + x)(q - x)$, then either of the substitutions $(p + x)u = \sqrt{c + bx - x^2}$ or $(q - x)u = \sqrt{c + bx - x^2}$ replaces the integral

$$\int \frac{1}{\sqrt{c + bx - x^2}}\, dx$$

with the integral of a rational function of u.

36. If a, b, and n are constants, evaluate

$$\int x(a + bx)^n\, dx.$$

37. The function

$$f(x) = \frac{1}{\sqrt{2\pi}\sigma} e^{-(x-\mu)^2/(2\sigma^2)}, \quad -\infty < x < \infty,$$

where μ and σ are constants, is called the *normal probability density function*. Consequently,

$$\int_{-\infty}^{\infty} f(x)\, dx = 1.$$

If the expected value of x is defined as

$$\int_{-\infty}^{\infty} x f(x)\, dx,$$

show that μ is the expected value.

38. Evaluate, if possible,

$$\int_{-\infty}^{\infty} e^{-|a-x|} e^{-|x|}\, dx,$$

where a is a constant.

SECTION 9.2

Exponential Growth and Decay

Exponential functions play a prominent role in many areas of applied mathematics. This is often due to the fact that the derivative of the exponential function e^{kx} is ke^{kx}; that is, the derivative of e^{kx} is proportional to itself. In this section we exploit this property in a wide variety of applications.

Many quantities have functional dependence on time t which takes the form Ce^{kt} where $C > 0$ and k are constants. When $k > 0$, the quantities are said to experience exponential growth; and when $k < 0$, they experience exponential decay. We consider an example of each.

Suppose that at some initial time t, which we take as $t = 0$, the number of bacteria in a culture is 2500. Given a proper environment, the culture will grow; and by studying this growth over a limited time span, we may be able to predict the number $N(t)$ of bacteria present at any given time. If we carefully observe the culture over a small time interval, perhaps 40 seconds, starting at time $t = 0$, we could draw a graph of $N = N(t)$ as shown in Figure 9.1. Each jump (discontinuity) in this graph indicates the birth of a new bacterium. Such a "microscopic" analysis of $N(t)$, in which we attempt to predict the birth of each bacterium, is unrealistic. We are interested in a more "macroscopic" analysis, with which we might be able to predict when the number of bacteria increases by, say, 25%, or doubles. In order to do so, we take measurements of the number of bacteria at much larger time intervals than those in Figure 9.1. The result would be a sequence of points such as that in Table 9.1 which we have plotted in Figure 9.2(a).

TABLE 9.1

t (minutes)	0	5	10	15	20	25	30	35	40	45	50
N (number)	2500	2650	2810	2980	3150	3340	3540	3750	3980	4210	4470

FIGURE 9.1

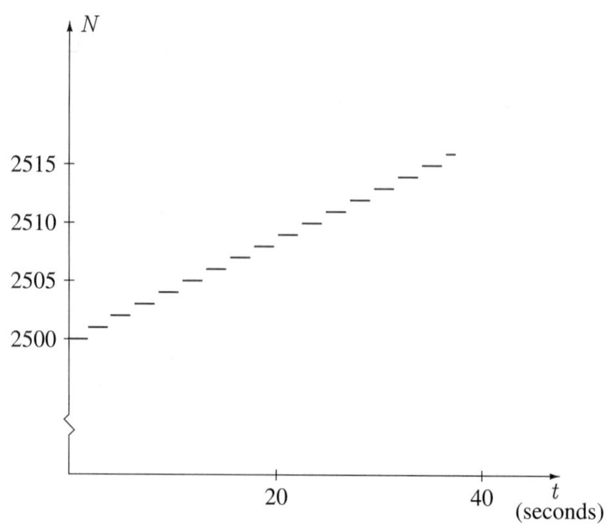

FIGURE 9.2

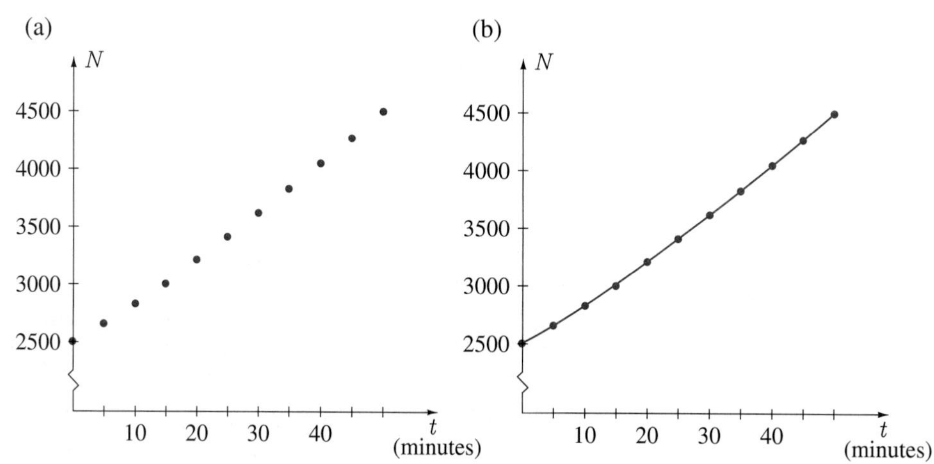

If we join these points with a smooth curve as in Figure 9.2(b), we obtain a reasonable overview of N as a function of time. Figure 9.2(b) is, however, only a graphical representation of physical measurements. It is useful to fit a mathematical formula to this graph; that is, to find an algebraic expression for $N(t)$. To do this often requires a great deal of experimentation with different types of curves. For this particular problem, the key is in the following measurements. At any time t we measure, for a very small interval Δt, the corresponding change $N(t + \Delta t) - N(t)$ in N (Figure 9.3). It we take a number of such measurements, we find that $N(t + \Delta t) - N(t)$ is always proportional to the product of N and Δt; that is

$$N(t + \Delta t) - N(t) = kN\,\Delta t, \qquad (9.15)$$

where $k > 0$ is some constant. In Table 9.2 we have used $\Delta t = 5$ to calculate $N(t + \Delta t) - N(t)$ and $N\Delta t = 5N$. In each case, $N(t + 5) - N(t)$ is approximately 0.012 times $5N$.

TABLE 9.2

t (minutes)	0	5	10	15	20	25	30	35	40	45
$N(t+5) - N(t)$	150	160	170	170	190	200	210	230	230	260
$5N$	12500	13250	14050	14900	15750	16700	17700	18750	19900	21050
$\frac{N(t+5) - N(t)}{5N}$	0.012	0.012	0.012	0.011	0.012	0.012	0.012	0.012	0.012	0.012

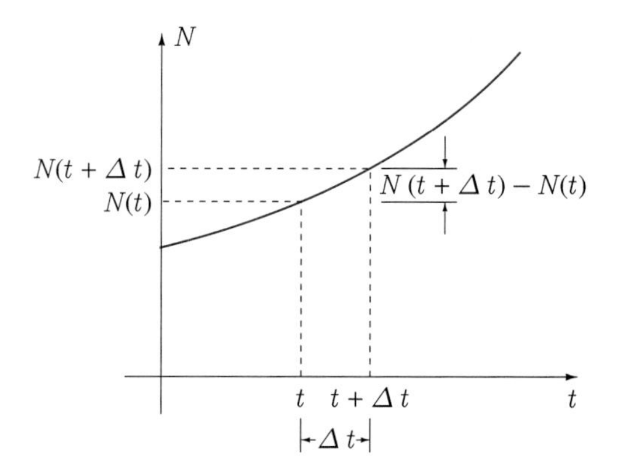

FIGURE 9.3

Experiments have shown that this is not a peculiarity of just one type of bacterium, but is characteristic of the growth pattern for many bacteria, the value of k varying from bacterium to bacterium. If we divide both sides of equation 9.15 by Δt,

$$\frac{N(t+\Delta t) - N(t)}{\Delta t} = kN,$$

and take limits as Δt approaches zero, we obtain

$$\frac{dN}{dt} = kN. \tag{9.16}$$

Experimental measurements have therefore led to mathematical law 9.16 for the rate of growth of bacteria in a culture. The time rate of change of the number of bacteria is proportional to the number of bacteria present at that time. The more bacteria present, the faster their number increases.

To find $N(t)$, we must solve this differential equation subject to the condition that $N(0) = N_0$. If we rewrite equation 9.16 in the form

$$k = \frac{1}{N}\frac{dN}{dt}$$

and take antiderivatives with respect to t, formula 9.2 gives

$$kt + C = \ln|N| = \ln N.$$

We may dispense with the absolute values since N is always positive in this case. Exponentiation of both sides to base e yields

$$e^{kt+C} = e^{\ln N} \qquad \text{or} \qquad N = e^C e^{kt} = De^{kt},$$

where we have set $D = e^C$. The initial condition $N(0) = N_0$ implies that $D = N_0$, and hence,

$$N = N_0 e^{kt}.$$

On the basis of experimental evidence, we have shown that the number of bacteria in a culture increases exponentially. To obtain the constant k, we must specify one additional piece of information. For example, if measurements indicate that the number of bacteria increases by 10% in 50 minutes, then $N(50) = (11/10)N_0$; that is

$$\frac{11}{10}N_0 = N_0 e^{50k}.$$

When we cancel the N_0 on each side and take natural logarithms, we have

$$\ln\left(\frac{11}{10}\right) = 50k, \qquad \text{or} \qquad k = \frac{1}{50}\ln\left(\frac{11}{10}\right) = 0.0019.$$

Finally, the number of bacteria in this particular culture is given by

$$N(t) = N_0 e^{(t/50)\ln(1.1)} = N_0 e^{0.0019t}.$$

An important result of this example is that *all* solutions of the differential equation

$$\frac{dN}{dt} = kN \tag{9.17a}$$

take the form

$$N = Ce^{kt}, \tag{9.17b}$$

where C is a constant. This is simply another statement of the property of the exponential function e^{kt} that its derivative is k times itself. But our derivation shows that the *only* functions that satisfy this property are constants multiplied by e^{kt}; no other function has a derivative which is proportional to itself.

As an example of a decay problem, we continue the discussion of radioactive disintegration begun in Section 2.4. Consider an ore sample that contains, along with various impurities, an amount A_0 of radioactive material, say, uranium. Disintegrations gradually reduce this amount of uranium. To describe the rate at which these disintegrations occur, we let $A(t)$ represent the amount of uranium in the sample at any given time, and choose time $t = 0$ when $A = A_0$. If we were to take measurements and perform an analysis similar to that for bacteria in a culture (see also Section 2.4), we would obtain the following *law of radioactive disintegration*: The time rate of change of the amount of radioactive material is proportional at any instant to the amount of radioactive material present. Algebraically, therefore,

$$\frac{dA}{dt} = kA, \tag{9.18}$$

where k is a constant. Since A is decreasing, dA/dt and hence k must be negative. Except for the sign of k, this is differential equation 9.17a, and its solution is therefore given by 9.17b,

$$A = Ce^{kt}.$$

The initial condition $A(0) = A_0$ requires $C = A_0$, and hence,

$$A = A_0 e^{kt}. \tag{9.19}$$

The amount of radioactive material therefore decreases exponentially, as shown in Figure 9.4. To find k we need to know A at one additional time. For example, if we know that one ten-millionth of one percent of the original amount of uranium decays in 6.5 years, then we have

$$0.999\,999\,999\,A_0 = A_0 e^{13k/2}.$$

FIGURE 9.4

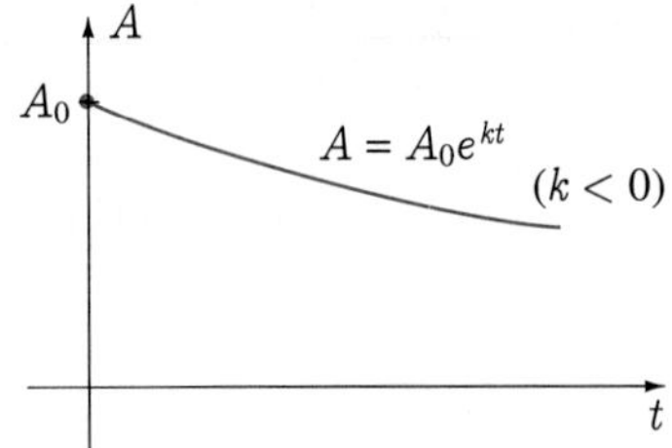

If we solve this equation for k, we obtain

$$k = \frac{2}{13}\ln(0.999\,999\,999) = -1.54 \times 10^{-10},$$

and therefore,

$$A = A_0 e^{-1.54\times 10^{-10}t}.$$

The law of radioactive disintegration has an important application in the dating of once-living plants and animals. All living tissue contains two isotopes of carbon: C^{14} (carbon-14), which is radioactive, and C^{12} (carbon-12), which is stable. In living tissue, the ratio of the amount of C^{14} to that of C^{12} is $1/10\,000$ for all fragments of the tissue. When the tissue dies, however, the ratio changes due to the fact that no more carbon is produced, and the original C^{14} present decays radioactively into an element other than C^{12}. Thus, as the dead tissue ages, the ratio of C^{14} to C^{12} decreases, and by measuring this ratio, it is possible to predict how long ago the tissue was alive. Suppose, for example, the present ratio of C^{14} to C^{12} in a specimen is $1/100\,000$; that is, one-tenth that for a living tissue. Then 90% of the original amount of C^{14} in the specimen has disintegrated.

If we let $A(t)$ be the amount of C^{14} in the specimen at time t, taking $A = A_0$ to be the amount present in the living specimen at its death ($t = 0$), then $A = A_0 e^{kt}$. To determine k, we use the fact that the half-life of C^{14} is approximately 5550 years. The *half-life* of a radioactive element is the time required for one-half an original sample of the material to disintegrate. For carbon-14, this means that A is equal to $A_0/2$ when $t = 5550$,

$$\frac{A_0}{2} = A_0 e^{5550k}.$$

When we divide by A_0 and take natural logarithms,

$$k = -\frac{1}{5550}\ln 2 = -0.000125.$$

Consequently, the amount of C^{14} in the specimen of dead tissue at any time t is given by

$$A = A_0 e^{-0.000125t}.$$

If T is the present time, when the amount of C^{14} is known to be 10% of its original amount, then at this time

$$0.1A_0 = A_0 e^{-0.000125T}.$$

The solution of this equation is

$$T = \frac{\ln 10}{0.000125} = 18\,400,$$

and we conclude that the tissue died about 18 400 years ago.

EXERCISES 9.2

1. Bacteria in a culture increase at a rate proportional to the number present. If the original number increases by 25% in 2 hours, when will it double?

2. If the number of bacteria in a culture doubles in 3 hours, when will it triple?

3. If one-half of a sample of radioactive substance decays in 15 days, how long does it take for 90% of the sample to decay?

4. If 10% of a sample of radioactive material decays in 3 seconds, what is its half-life?

5. After 4 half-lives of a radioactive substance, what percentage of the original amount remains?

6. In psychology, it is sometimes assumed that if R represents the reaction of a subject to an amount S of stimulus, then dR/dS is inversely proportional to S; that is,

$$\frac{dR}{dS} = \frac{k}{S},$$

where $k > 0$ is a constant. This is called the *Weber-Fechner law*. Find R as a function of S if S_0 is the smallest stimulus detectable by the subject.

7. If the Weber-Fechner law in Exercise 6 is replaced by the Brentano-Stevens law,

$$\frac{dR}{dS} = k\frac{R}{S},$$

find $R(S)$. Can the detection threshold S_0 of Exercise 6 be used to determine the constant of integration in this case?

8. Suppose the amount of a drug injected into the body decreases at a rate proportional to the amount still present. If a dose decreases by 5% in the first hour, when will it decrease to one-half its original amount?

9. In the arctic, the snowy owl preys on the lemming, and the number of owls and lemmings in any given area fluctuate in time. If the number of lemmings increases, so does the number of owls, but with some delay. Eventually an overabundance of owls results, and a scarcity of food (lemmings) to support such a population causes a decrease in the number of owls. The lemming population consequently begins to rise and the cycle repeats. One model to describe the relationship between L, the number of lemmings, and s, the number of snowy owls, states that $s(L)$ must satisfy the differential equation

$$\frac{ds}{dL} = \frac{s(AL - B)}{L(C - Ds)},$$

where A, B, C, and D are positive constants. Solve this differential equation for an implicit definition of $s(L)$.

10. A sugar cube 1 cm on each side is dropped into a cup of coffee. If the sugar dissolves in such a way that the cube always remains a cube, compare the times for the cube to completely dissolve under the following conditions:

(a) Dissolving occurs at a rate proportional to the surface area of the remaining cube; and

(b) Dissolving occurs at a rate proportional to the amount of sugar remaining.

11. Solve Exercise 10 if the sugar is not in the form of a cube, but rather in free form from the sugar bowl. Assume that the sugar consists of n spherical particles each of radius r_0 centimetres.

12. An analysis of a sample of fossil remains shows that it contains only 1.51% of the original C^{14} in the living creature. When did the creature die?

13. If a fossilized creature died 100 000 years ago, what percentage of the original C^{14} remains?

14. Glucose is administered intravenously to the bloodstream at a constant rate of R units per unit time. As the glucose is added, it is converted by the body into other substances at a rate proportional to the amount of glucose in the blood at that time. Show that the amount of glucose in the blood as a function of time t is given by

$$C(t) = \frac{R}{k}(1 - e^{-kt}) + C_0 e^{-kt},$$

where k is a constant and C_0 is the amount at time $t = 0$ when the intravenous feeding is initiated. Sketch a graph of this function for $C_0 < R/k$ and $C_0 > R/k$.

15. Prove that if a quantity decreases at a rate proportional to its present amount, and if its percentage decrease in some interval of time is i%, then its percentage decrease in any interval of time of the same length is i%.

16. Water at temperature 90 °C is placed in a room at constant temperature 20 °C. Newton's law of cooling states that the time rate of change of the temperature T of the water is proportional to the difference between T and the temperature of the environment:

$$\frac{dT}{dt} = k(T - 20),$$

where k is a constant. If the water cools to 60 °C in 40 minutes, find T as a function of t.

The exponential growth of bacteria as defined by the equation $N = N_0 e^{kt}$ is unrealistic for large t in that there is no maximum for $N(t)$. Sketch the growth curve in Exercises 17–20 and show that each has a maximum.

17. The *logistic* growth curve

$$N(t) = \frac{N_0(1 + b)}{1 + be^{-kt}}, \quad t \geq 0,$$

where b, k, and N_0 are positive constants.

18. The *Gompertz* growth curve

$$N(t) = Ae^{-be^{-kt}}, \quad t \geq 0,$$

where A, b, and k are positive constants.

19. The *monomolecular* growth curve

$$N(t) = \frac{N_0}{1 - b}(1 - be^{-kt}), \quad t \geq 0,$$

where $b < 1$, k, and N_0 are positive constants.

20. The *Von Bertalanffy* growth curve

$$N(t) = N_0(1 - be^{-kt})^3, \quad t \geq 0,$$

where $b < 1$, k, and N_0 are positive constants.

21. The differential equation

$$\frac{dN}{dt} = kN(t)\cos t, \quad t \geq 0,$$

where $k > 0$ is a constant, describes a population that undergoes periodic growths and declines. If $N(0) = N_0$, find $N(t)$ and sketch its graph.

22. Sketch a graph of the sigmoid survival curve

$$s(x) = 1 - \left(1 - e^{-kx}\right)^n, \quad x \geq 0,$$

where $k > 0$ is a constant and n is a positive integer.

23. When a deep-sea diver inhales air, his body tissues absorb extra amounts of nitrogen. Suppose the diver enters the water at time $t = 0$, drops very quickly to depth d, and remains at this depth for a very long time. The amount N of nitrogen in his body tissues increases as he remains at this depth until a maximum amount $\overline{N}$ is reached. The time rate of change of N is proportional to the difference $\overline{N} - N$. Show that if N_0 is the amount of nitrogen in his body tissues when he enters the water, then

$$N = N_0 e^{-kt} + \overline{N}(1 - e^{-kt}),$$

where $k > 0$ is a constant.

24. A dead body is found after the final calculus examination. The coroner arrives on the scene at 9:00 A.M. and immediately records the temperature of the body as 30.1°C. One hour later she notes that the body's temperature has dropped to 29.2°C, and that during the hour, room temperature has remained at 20°C. Assuming that the body's temperature obeys Newton's law of cooling (Exercise 16), determine when death occurred. Normal body temperature is 37°C.

25. (a) In Section 7.5 we proved that the rate of change of pressure P with respect to vertical distance y within a fluid with density ρ is $dP/dy = -\rho g$, where $g = 9.81$. Since air may be considered a fluid, air pressure above sea level ($y = 0$) must satisfy the equation

$$\frac{dP}{dy} = -\rho g,$$

where $\rho = \rho(y)$ is air density at altitude y. If air is assumed to be a perfect gas, the pressure of a given volume V of gas must satisfy the equation

$$P = \rho RT,$$

where $T = T(y)$ is the temperature of the air and $R > 0$ is a constant. To a reasonable approximation, temperature varies linearly with respect to y:

$$T = T_0 - \lambda y$$

where T_0 is sea level temperature and λ is a constant (about 6.10×10^{-4}°C per 100 m). Show that $P(y)$ must satisfy the differential equation

$$\frac{dP}{dy} = -\frac{Pg}{R(T_0 - \lambda y)},$$

and solve the equation for

$$P(y) = P_0\left(1 - \frac{\lambda y}{T_0}\right)^{g/(\lambda R)},$$

where P_0 is sea level pressure.

(b) Find the altitude at which air pressure is one one-millionth that of sea level pressure, if $T_0 = 293$°K, $g = 9.81$ m/s^2, and $R = 287$ m·N/kg·°K.

26. When light passes through a material, such as water or glass, its intensity decreases with depth of penetration. The Bouguer-Lambert law for absorption of light states that the amount of light absorbed by a small thickness of the material is proportional to the product of this thickness and the intensity of light incident on it.

(a) Show that if I is the intensity of light, then

$$I = I(x) = I_0 e^{-kx},$$

where $k > 0$ is a constant, I_0 is the intensity at the surface of the material, and x is the penetration distance from the surface of the material (Figure 9.5).

FIGURE 9.5

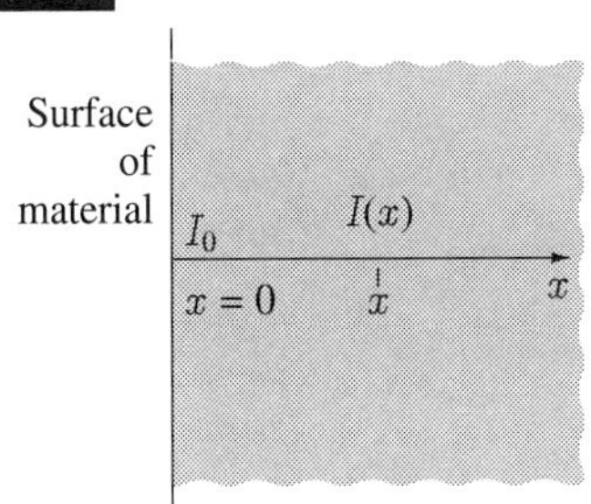

(b) Give an alternative statement of the Bouguer-Lambert law in terms of rates of change.

27. (a) To perform a one-hour operation on a dog a veterinarian anesthetizes the dog with sodium pentobarbital. During the operation the dog's body breaks down the drug at a rate proportional to the amount present, and only half the original dose remains after 5 hours. If the dog has mass 20 kg, and 0.02 gm of sodium pentobarbital per kilogram of body mass are required to maintain surgical anesthesia, what original dose is required?

(b) If after the hour, the operation is unfinished and the dog begins to stir, a further dose of anesthetic must be administered. What extra dose will guarantee surgical anesthesia for an additional 20 minutes?

SECTION 9.3

Integration by Parts

One of the most important integration techniques is that of **integration by parts**. It results from the product rule for differentiation:

$$\frac{d}{dx}(uv) = u\frac{dv}{dx} + v\frac{du}{dx}.$$

If we take antiderivatives of both sides of this equation with respect to x, we obtain

$$\int \frac{d}{dx}(uv)\,dx = \int u\frac{dv}{dx}\,dx + \int v\frac{du}{dx}\,dx.$$

Since the antiderivative on the left is uv, this equation may be rewritten in the form

$$\int u\,dv = uv - \int v\,du. \tag{9.20}$$

Equation 9.20 is called the **integration-by-parts formula**. It says that if we are given an unknown integral (the left-hand side of 9.20), and we arbitrarily divide the function after the integral sign into two parts — one called u and the other called dv, then we can set the unknown integral equal to the right-hand side of 9.20. What this does is replace the integral

$$\int u\,dv \qquad \text{with} \qquad \int v\,du.$$

This will be useful if the new integral is easier to evaluate than the original. The problem is to choose u and dv so that the new integral is simpler than the original. For many examples, it is advantageous to let dv be the *most complicated part of the integrand (plus differential) that we can integrate mentally*. The example

$$\int xe^x\,dx$$

will clarify these ideas. To use integration by parts on this problem we must define u and dv in such a way that $u\,dv = xe^x\,dx$. There are four possibilities:

$$\begin{aligned} u &= 1, & dv &= xe^x\,dx;\\ u &= x, & dv &= e^x\,dx;\\ u &= e^x, & dv &= x\,dx;\\ u &= xe^x, & dv &= dx. \end{aligned}$$

Our rule suggests the choice

$$u = x, \qquad dv = e^x\,dx,$$

in which case

$$du = dx, \qquad v = e^x.$$

Formula 9.20 then gives

$$\int xe^x\,dx = xe^x - \int e^x\,dx.$$

We have therefore replaced the integration of xe^x with the integration of e^x, a definite simplification. The final solution is therefore

$$\int xe^x\,dx = xe^x - e^x + C.$$

This example illustrates that integration by parts replaces an integral with two others: one mental integration (from dv to v) and a second integration that is hopefully simpler than the original. We should also note that in the evaluation of the mental integration we did not include a constant of integration. For example, in the integration from $dv = e^x\,dx$ to $v = e^x$, we did not write $v = e^x + D$. Had we done so, the solution would have proceeded as follows:

$$\begin{aligned}\int xe^x\,dx &= x(e^x + D) - \int (e^x + D)\,dx \\ &= x(e^x + D) - e^x - Dx + C \\ &= xe^x - e^x + C.\end{aligned}$$

The constant D has therefore disappeared from the eventual solution. This always occurs, and consequently, we do not include an arbitrary constant in the mental integration for v.

EXAMPLE 9.3

Evaluate the following integrals:

$$\text{(a)} \int \ln x\,dx \qquad \text{(b)} \int x^2 \cos x\,dx \qquad \text{(c)} \int e^x \sin x\,dx$$

SOLUTION (a) Integration by parts leaves no choice for u and dv in this integral; they must be

$$u = \ln x, \qquad dv = dx,$$

in which case

$$du = \frac{1}{x}\,dx, \qquad v = x.$$

Then we have

$$\int \ln x\,dx = x \ln x - \int x\frac{1}{x}\,dx = x\ln x - x + C.$$

(b) If we set

$$u = x^2, \qquad dv = \cos x\,dx,$$

then

$$du = 2\,x\,dx, \qquad v = \sin x, \qquad \text{and}$$

$$\int x^2 \cos x\,dx = x^2 \sin x - \int 2\,x \sin x\,dx.$$

Integration by parts has therefore reduced the power on x from 2 to 1. To eliminate the x in front of the trigonometric function completely, we perform integration by parts once again, this time with

$$u = x, \qquad dv = \sin x\,dx.$$

Then,

$$du = dx, \qquad v = -\cos x, \qquad \text{and}$$

$$\begin{aligned}\int x^2 \cos x\,dx &= x^2 \sin x - 2\left(-x\cos x - \int -\cos x\,dx\right) \\ &= x^2 \sin x + 2\,x\cos x - 2\,\sin x + C.\end{aligned}$$

(c) If we set

$$u = e^x, \qquad dv = \sin x\,dx,$$

then

$$du = e^x\,dx, \qquad v = -\cos x, \qquad \text{and}$$

$$\int e^x \sin x\,dx = -e^x \cos x - \int -e^x \cos x\,dx.$$

Integration by parts appears to have led nowhere; the integration of $e^x \cos x$ is essentially as difficult as that of $e^x \sin x$. If, however, we persevere and integrate by parts once again with

$$u = e^x, \qquad dv = \cos x\,dx,$$

then

$$du = e^x\,dx, \qquad v = \sin x, \qquad \text{and}$$

$$\int e^x \sin x\,dx = -e^x \cos x + e^x \sin x - \int e^x \sin x\,dx.$$

This equation can now be solved for the unknown integral. By transposing the last term on the right to the left, we find that

$$2\int e^x \sin x\,dx = e^x(\sin x - \cos x),$$

and therefore,

$$\int e^x \sin x\,dx = \frac{1}{2}e^x(\sin x - \cos x) + C,$$

to which we have added the arbitrary constant C. ■

The technique in part (c) of this example is often called **integration by reproduction**. Two applications of integration by parts enabled us to reproduce the unknown integral and then solve the equation for it.

EXAMPLE 9.4

Evaluate

$$\int \frac{x^3}{\sqrt{9 + x^2}}\,dx.$$

Solution Methods

(i) If we set $u = 9 + x^2$, then $du = 2\,x\,dx$, and

$$\int \frac{x^3}{\sqrt{9 + x^2}}\,dx = \int \frac{u - 9}{\sqrt{u}}\frac{du}{2} = \frac{1}{2}\int (u^{1/2} - 9\,u^{-1/2})\,du$$

$$= \frac{1}{2}\left(\frac{2}{3}u^{3/2} - 18\,u^{1/2}\right) + C = \frac{1}{3}(9 + x^2)^{3/2} - 9\sqrt{9 + x^2} + C.$$

(ii) If we set $u = \sqrt{9 + x^2}$, then $du = \left(x/\sqrt{9 + x^2}\right) dx$, and

$$\int \frac{x^3}{\sqrt{9 + x^2}}\, dx = \int (u^2 - 9)\, du = \frac{u^3}{3} - 9u + C$$
$$= \frac{1}{3}(9 + x^2)^{3/2} - 9\sqrt{9 + x^2} + C.$$

(iii) If we set

$$u = x^2, \qquad dv = \frac{x}{\sqrt{9 + x^2}}\, dx,$$

then

$$du = 2x\, dx, \qquad v = \sqrt{9 + x^2}, \qquad \text{and}$$

$$\int \frac{x^3}{\sqrt{9 + x^2}}\, dx = x^2\sqrt{9 + x^2} - \int 2x\sqrt{9 + x^2}\, dx$$
$$= x^2\sqrt{9 + x^2} - 2\left[\frac{1}{3}(9 + x^2)^{3/2}\right] + C$$
$$= x^2\sqrt{9 + x^2} - \frac{2}{3}(9 + x^2)^{3/2} + C.$$

■

Example 9.4 illustrates that there may be more than one way to evaluate an indefinite integral, and the answer may appear different depending on the technique used. According to Theorem 5.1, these solutions must be identical, up to an additive constant. To illustrate the case, note that the solution obtained by method (iii) can be written as

$$\sqrt{9 + x^2}\left[x^2 - \frac{2}{3}(9 + x^2)\right] + C = \frac{1}{3}\sqrt{9 + x^2}\left[-18 + x^2\right] + C$$
$$= \frac{1}{3}\sqrt{9 + x^2}\left[-27 + (9 + x^2)\right] + C$$
$$= -9\sqrt{9 + x^2} + \frac{1}{3}(9 + x^2)^{3/2} + C,$$

which is the solution obtained by the other two methods. It is also worth noting that the three methods work very well provided the power on x in the numerator of the integrand is odd. When it is even, they fail miserably. In such cases, trigonometric substitutions (Section 9.5) come to the rescue (see Example 9.12).

Integration-by-parts formula 9.20 applies to indefinite integrals. For definite integrals, it is replaced by

$$\int_a^b u\, dv = \left\{uv\right\}_a^b - \int_a^b v\, du. \tag{9.21}$$

EXAMPLE 9.5

Evaluate

$$\int_0^4 \frac{x}{\sqrt{x+8}}\,dx.$$

SOLUTION If we set

$$u = x, \qquad dv = \frac{1}{\sqrt{x+8}}\,dx,$$

then

$$du = dx, \qquad v = 2\sqrt{x+8}, \qquad \text{and}$$

$$\int_0^4 \frac{x}{\sqrt{x+8}}\,dx = \left\{2x\sqrt{x+8}\right\}_0^4 - \int_0^4 2\sqrt{x+8}\,dx$$

$$= 8\sqrt{12} - 2\left\{\frac{2}{3}(x+8)^{3/2}\right\}_0^4$$

$$= \frac{16}{3}(4\sqrt{2} - 3\sqrt{3}).$$

■

EXERCISES 9.3

In Exercises 1–18 evaluate the indefinite integral.

1. $\int x \sin x\,dx$
2. $\int x^2 e^{2x}\,dx$
3. $\int x^4 \ln x\,dx$
4. $\int \sqrt{x}\ln(2x)\,dx$
5. $\int z\sec^2(z/3)\,dz$
6. $\int x\sqrt{3-x}\,dx$
7. $\int \mathrm{Sin}^{-1}\,x\,dx$
8. $\int x^2\sqrt{x+5}\,dx$
9. $\int \frac{x}{\sqrt{2+x}}\,dx$
10. $\int \frac{x^2}{\sqrt{2+x}}\,dx$
11. $\int \frac{x}{\sqrt{2+x^2}}\,dx$
12. $\int (x-1)^2 \ln x\,dx$
13. $\int e^x \cos x\,dx$
14. $\int \mathrm{Tan}^{-1}\,x\,dx$
15. $\int \cos(\ln x)\,dx$
16. $\int e^{2x}\cos 3x\,dx$
17. $\int \frac{x^3}{\sqrt{5+3x^2}}\,dx$
18. $\int \ln(x^2+4)\,dx$
19. Consider the evaluation of $\int x^5 e^x\,dx$. Were we to use integration by parts the answer would eventually be of the form

$$\int x^5 e^x dx = Ax^5e^x + Bx^4e^x + Cx^3e^x + Dx^2e^x + Exe^x + Fe^x + G.$$

Differentiate this equation, and thereby obtain A, B, C, D, E, and F (with no integration).

20. The face of a dam is shown in Figure 9.6, where the curve has equation

$$y = e^{k|x|} - 1, \qquad k = \frac{1}{100}\ln 201.$$

Find the total force on the dam due to water pressure when the water level on the dam is 100 m.

FIGURE 9.6

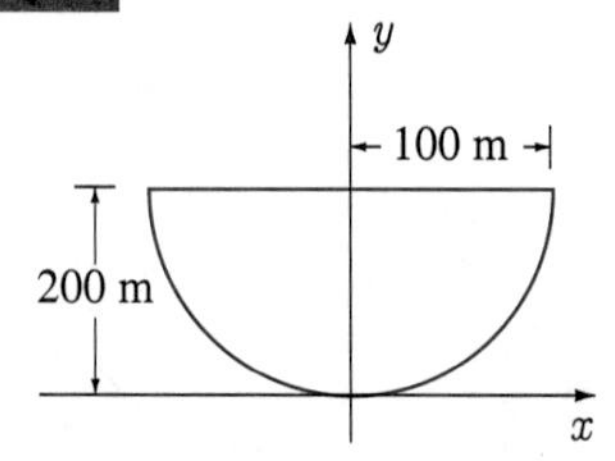

21. If we set $u = x^{-1}$ and $dv = dx$, then $du = -x^{-2}\,dx$ and $v = x$, and

$$\int \frac{1}{x}\,dx = \left(\frac{1}{x}\right)x - \int x\left(-\frac{1}{x^2}\,dx\right) = 1 + \int \frac{1}{x}\,dx.$$

Subtraction of $\int x^{-1}\,dx$ from each side of this equation now gives $0 = 1$. What is wrong with the argument?

22. The *gamma function* $\Gamma(n)$ for $n > 0$ is defined by the improper integral

$$\Gamma(n) = \int_0^{\infty} x^{n-1} e^{-x}\, dx.$$

Show that when n is a positive integer, $\Gamma(n) = (n-1)!$.

23. The *Laplace transform* of a function $F(t)$ is the function $f(s)$ defined by the improper integral

$$f(s) = \int_0^{\infty} e^{-st} F(t)\, dt.$$

Find Laplace transforms of the following functions:

(a) e^{3t} (b) t^2 (c) $\sin t$ (d) te^{-t}

24. Show that the function

$$f(x) = \frac{x^{\alpha-1} e^{-x/\beta}}{\Gamma(\alpha)\beta^{\alpha}}$$

for positive constants α and β qualifies as a probability density function on the interval $x \geq 0$. See Exercise 23 in Section 7.9 for the definition of a pdf and Exercise 22 for the definition of $\Gamma(\alpha)$.

In Exercises 25–27 evaluate the indefinite integral.

25. $\displaystyle\int \mathrm{Tan}^{-1} \sqrt{x}\, dx$ **26.** $\displaystyle\int x^2 \cos^2 x\, dx$

27. $\displaystyle\int x e^x \sin x\, dx$

SECTION 9.4

Trigonometric Integrals

Differentiation formulas for trigonometric functions suggest three pairings of these functions: sine and cosine, tangent and secant, and cotangent and cosecant. The derivative of either function in a given pair involves at most functions of that pair. Note that integration formulas 9.5–9.14 also substantiate this pairing. Each antiderivative gives functions in the same pairing. Because of this, a first step in every integral involving trigonometric functions is to rewrite the integrand so that each term involves only one pair. For example, to evaluate

$$\int \frac{\tan x + \sec x \cot^2 x}{\sin^3 x}\, dx,$$

we would first rewrite it in the form

$$\int \left(\frac{\csc^3 x}{\cot x} + \frac{\cos x}{\sin^5 x} \right) dx.$$

The first term contains cosecants and cotangents, and the second has sines and cosines. We now consider integrals involving each of these three pairings individually.

Integrals Involving the Sine and Cosine

We frequently encounter integrals of the form

$$\int \cos^n x \sin^m x\, dx,$$

where m and n may or may not be integers. The key to evaluation of this type of integral is recognition of the fact that when either $n = 1$ or $m = 1$, the integrations become straightforward. To illustrate this, consider differentiation of $\sin^{n+1} x$ by power rule 3.16,

$$\frac{d}{dx} \sin^{n+1} x = (n+1) \sin^n x \cos x.$$

This means that differentiation of a power of $\sin x$ leads to a power of $\sin x$ one lower than the original, multiplied by $\cos x$. In order to reverse the procedure and antidifferentiate a power of $\sin x$, we need a factor of $\cos x$. It then becomes a matter of raising the power on $\sin x$ by one and dividing by the new power; that is, we can say

$$\int \sin^n x \cos x \, dx = \frac{1}{n+1} \sin^{n+1} x + C, \quad n \neq -1. \tag{9.22a}$$

Similarly,

$$\int \cos^n x \sin x \, dx = -\frac{1}{n+1} \cos^{n+1} x + C, \quad n \neq -1. \tag{9.22b}$$

Formulas 9.12 and 9.11 contain the $n = -1$ cases for these integrals. Of course, a simple substitution can always be made to evaluate integrals 9.22a,b. For example, if in 9.22a we set $u = \sin x$, then $du = \cos x \, dx$, and

$$\int \sin^n x \cos x \, dx = \int u^n \, du = \frac{u^{n+1}}{n+1} + C = \frac{\sin^{n+1} x}{n+1} + C.$$

Also useful are the formulas

$$\int \sin nx \, dx = -\frac{1}{n} \cos nx + C, \qquad \text{and} \tag{9.22c}$$

$$\int \cos nx \, dx = \frac{1}{n} \sin nx + C. \tag{9.22d}$$

Trigonometric identities are used to write integrands in the forms contained in equations 9.22. Particularly helpful are

$$\sin^2 x + \cos^2 x = 1, \tag{9.23a}$$

$$\sin 2x = 2 \sin x \cos x, \tag{9.23b}$$

$$\cos 2x = 2 \cos^2 x - 1, \tag{9.23c}$$

$$\cos 2x = 1 - 2 \sin^2 x. \tag{9.23d}$$

EXAMPLE 9.6

Evaluate the following integrals:

(a) $\int \sin^3 x \, dx$ (b) $\int \sin^3 x \cos^5 x \, dx$ (c) $\int \sin 2x \cos^6 2x \, dx$

(d) $\int \sqrt{\sin x} \cos^3 x \, dx$ (e) $\int \frac{\cos^3 3x}{\sin^2 3x} \, dx$ (f) $\int \sin^2 x \, dx$

(g) $\int \cos^4 x \, dx$ (h) $\int \sin^2 x \cos^2 x \, dx$

SOLUTION (a) With identity 9.23a, we may write

$$\int \sin^3 x \, dx = \int \sin x (1 - \cos^2 x) \, dx = \int (\sin x - \cos^2 x \sin x) \, dx,$$

and the second term is in form 9.22b. Thus,

$$\int \sin^3 x\, dx = -\cos x + \frac{1}{3}\cos^3 x + C.$$

(b) Once again we use 9.23a to rewrite the integrand as two easily integrated terms,

$$\begin{aligned}\int \sin^3 x \cos^5 x\, dx &= \int \sin x(1-\cos^2 x)\cos^5 x\, dx\\ &= \int (\cos^5 x \sin x - \cos^7 x \sin x)\, dx\\ &= -\frac{1}{6}\cos^6 x + \frac{1}{8}\cos^8 x + C.\end{aligned}$$

(c) This integrand is already in a convenient form for integration. Guess and fix up leads to

$$\int \sin 2x \cos^6 2x\, dx = -\frac{1}{14}\cos^7 2x + C.$$

(d) Using 9.23a, we obtain

$$\begin{aligned}\int \sqrt{\sin x}\cos^3 x\, dx &= \int \sin^{1/2} x \cos x\,(1-\sin^2 x)\, dx\\ &= \int (\sin^{1/2} x \cos x - \sin^{5/2} x \cos x)\, dx\\ &= \frac{2}{3}\sin^{3/2} x - \frac{2}{7}\sin^{7/2} x + C.\end{aligned}$$

(e) Again we use 9.23a,

$$\begin{aligned}\int \frac{\cos^3 3x}{\sin^2 3x}\, dx &= \int \frac{\cos 3x\,(1-\sin^2 3x)}{\sin^2 3x}\, dx = \int \left(\frac{\cos 3x}{\sin^2 3x} - \cos 3x\right) dx\\ &= \frac{-1}{3\sin 3x} - \frac{1}{3}\sin 3x + C.\end{aligned}$$

In the five examples above, at least one of the powers on the sine and cosine functions was odd, and in each case we used identity 9.23a. In the next three examples, where all powers are even, identities 9.23b,c,d are useful.

(f) Double-angle formula 9.23d can be rearranged as $\sin^2 x = (1-\cos 2x)/2$, and therefore

$$\int \sin^2 x\, dx = \frac{1}{2}\int (1-\cos 2x)\, dx.$$

Formula 9.22d can now be used on the second term

$$\int \sin^2 x\, dx = \frac{1}{2}\left(x - \frac{1}{2}\sin 2x\right) + C.$$

(g) With double-angle formula 9.23c rewritten in the form $\cos^2 x = (1+\cos 2x)/2$, we obtain

$$\int \cos^4 x\, dx = \frac{1}{4}\int (1+\cos 2x)^2\, dx = \frac{1}{4}\int (1 + 2\cos 2x + \cos^2 2x)\, dx.$$

When x is replaced by $2x$ in 9.23c, the result is $\cos 4x = 2\cos^2 2x - 1$. This can be rearranged as $\cos^2 2x = (1 + \cos 4x)/2$, and therefore

$$\int \cos^4 x\, dx = \frac{1}{4}\int \left(1 + 2\cos 2x + \frac{1 + \cos 4x}{2}\right) dx.$$

We now use 9.22d on the two cosine terms,

$$\int \cos^4 x\, dx = \frac{1}{4}\left(\frac{3x}{2} + \sin 2x + \frac{1}{8}\sin 4x\right) + C.$$

(h) In this integral we use identities 9.23b,c,

$$\int \sin^2 x \cos^2 x\, dx = \int (\sin x \cos x)^2\, dx = \frac{1}{4}\int \sin^2 2x\, dx$$
$$= \frac{1}{4}\int \left(\frac{1 - \cos 4x}{2}\right) dx = \frac{1}{8}\left(x - \frac{1}{4}\sin 4x\right) + C.$$

■

EXAMPLE 9.7

Evaluate

$$\int \sin 5x \cos 2x\, dx.$$

SOLUTION Unlike Example 9.6, in which all trigonometric functions had the same argument, this integrand is the product of trigonometric functions with different arguments. Our first step is to rewrite the integrand so that we do not have such a product. Note that a sum is obviously integrable, but not a product. The product formula

$$\sin A \cos B = \frac{1}{2}[\sin(A + B) + \sin(A - B)],$$

can be used to advantage here. We can rewrite the integrand in the form

$$\int \sin 5x \cos 2x\, dx = \int \frac{1}{2}(\sin 7x + \sin 3x)\, dx,$$

and now integration is easily handled by 9.22c,

$$\int \sin 5x \cos 2x\, dx = \frac{1}{2}\left(-\frac{1}{7}\cos 7x - \frac{1}{3}\cos 3x\right) + C = -\frac{1}{6}\cos 3x - \frac{1}{14}\cos 7x + C.$$

Two integration by parts (and reproduction) can also be used to evaluate this integral, but the above method is much simpler.

■

Integrals Involving Tangent and Secant

For integrals involving tangent and secant, we have two alternatives: rewrite the integrand in terms of sines and cosines, or express the integrand in terms of easily integrated combinations of tangent and secant. It will usually be more fruitful to investigate the second possibility before expressing the integrand completely in terms of sines and cosines. In this regard we note that combinations of tangent and secant that can be integrated immediately are

$$\int \tan^n x \sec^2 x\, dx = \frac{\tan^{n+1} x}{n+1} + C, \quad n \neq -1; \qquad (9.24\text{a})$$

$$\int \sec^n x \tan x\, dx = \frac{\sec^n x}{n} + C, \quad n \neq 0. \qquad (9.24\text{b})$$

Once again these results are suggested by differentiations of powers of $\tan x$ and $\sec x$. Alternatively, these integrals can be evaluated with substitutions: $u = \tan x$ for 9.24a and $u = \sec x$ for 9.24b.

To rearrange integrands into these integrable combinations, we use the identity

$$1 + \tan^2 x = \sec^2 x. \qquad (9.25)$$

EXAMPLE 9.8

Evaluate the following integrals:

(a) $\int \tan^2 x\, dx$ (b) $\int \tan^3 x\, dx$ (c) $\int \sec^4 x\, dx$

(d) $\int \tan^4 3x \sec^2 3x\, dx$ (e) $\int \tan^3 x \sec^3 x\, dx$ (f) $\int \frac{\sec^4 x}{\tan^2 x}\, dx$

SOLUTION

(a) $$\int \tan^2 x\, dx = \int (\sec^2 x - 1)\, dx = \tan x - x + C$$

(b) $$\int \tan^3 x\, dx = \int \tan x (\sec^2 x - 1)\, dx$$
$$= \int (\sec^2 x \tan x - \tan x)\, dx$$
$$= \frac{1}{2}\sec^2 x + \ln|\cos x| + C$$

(c) $$\int \sec^4 x\, dx = \int \sec^2 x (1 + \tan^2 x)\, dx$$
$$= \int (\sec^2 x + \tan^2 x \sec^2 x)\, dx$$
$$= \tan x + \frac{1}{3}\tan^3 x + C$$

(d) $$\int \tan^4 3x \sec^2 3x\, dx = \frac{1}{15}\tan^5 3x + C$$

(e) $$\int \tan^3 x \sec^3 x\,dx = \int \tan x\,(\sec^2 x - 1)\sec^3 x\,dx$$
$$= \int (\sec^5 x \tan x - \sec^3 x \tan x)\,dx$$
$$= \frac{1}{5}\sec^5 x - \frac{1}{3}\sec^3 x + C$$

(f) $$\int \frac{\sec^4 x}{\tan^2 x}\,dx = \int \frac{\sec^2 x\,(1 + \tan^2 x)}{\tan^2 x}\,dx$$
$$= \int \left(\frac{\sec^2 x}{\tan^2 x} + \sec^2 x\right) dx$$
$$= -\frac{1}{\tan x} + \tan x + C$$

■

EXAMPLE 9.9 Evaluate

$$\int \sec^3 x\,dx.$$

SOLUTION We use integration by parts and reproduction. By setting

$$u = \sec x, \qquad dv = \sec^2\,dx,$$

then

$$du = \sec x \tan x\,dx, \qquad v = \tan x,$$

and formula 9.20 gives

$$\int \sec^3 x\,dx = \sec x \tan x - \int \sec x \tan^2 x\,dx.$$

Since $\tan^2 x = \sec^2 x - 1$, we can write

$$\int \sec^3 x\,dx = \sec x \tan x - \int \sec x\,(\sec^2 x - 1)\,dx$$
$$= \sec x \tan x - \int \sec^3 x\,dx + \ln|\sec x + \tan x|.$$

We now solve for

$$\int \sec^3 x\,dx = \frac{1}{2}(\sec x \tan x + \ln|\sec x + \tan x|) + C.$$

■

Integrands in the following example cannot be transformed by means of identity 9.25 into forms 9.24, and we therefore express them in terms of sines and cosines.

EXAMPLE 9.10

Evaluate the following integrals:

$$\text{(a)} \int \frac{\tan^2 2x}{\sec^3 2x}\,dx \qquad \text{(b)} \int \frac{1}{\sec x \tan^2 x}\,dx$$

SOLUTION

(a)
$$\int \frac{\tan^2 2x}{\sec^3 2x}\,dx = \int \sin^2 2x \cos 2x\,dx = \frac{1}{6}\sin^3 2x + C$$

(b)
$$\int \frac{1}{\sec x \tan^2 x}\,dx = \int \frac{\cos^3 x}{\sin^2 x}\,dx = \int \frac{\cos x\,(1 - \sin^2 x)}{\sin^2 x}\,dx$$
$$= \int \left(\frac{\cos x}{\sin^2 x} - \cos x\right) dx = -\frac{1}{\sin x} - \sin x + C$$

■

Integrals Involving Cotangent and Cosecant

Since derivatives of the cotangent and cosecant pair

$$\frac{d}{dx}\cot x = -\csc^2 x, \qquad \frac{d}{dx}\csc x = -\csc x \cot x$$

are analogous to those of the tangent and secant pair

$$\frac{d}{dx}\tan x = \sec^2 x, \qquad \frac{d}{dx}\sec x = \sec x \tan x,$$

a discussion parallel to that concerning the tangent and secant could be made here. Aside from the fact that the cotangent replaces the tangent and the cosecant replaces the secant, there are also sign changes. We leave examples of integrations of the cotangent and cosecant pair to the exercises.

EXERCISES 9.4

In Exercises 1–20 evaluate the indefinite integral.

1. $\int \cos^3 x \sin x\,dx$

2. $\int \frac{\cos x}{\sin^3 x}\,dx$

3. $\int \tan^5 x \sec^2 x\,dx$

4. $\int \csc^3 x \cot x\,dx$

5. $\int \cos^3 (x+2)\,dx$

6. $\int \sqrt{\tan x}\sec^4 x\,dx$

7. $\int \frac{1}{\sin^4 t}\,dt$

8. $\int \sec^6 3x \tan 3x\,dx$

9. $\int \cos^2 x\,dx$

10. $\int \frac{\tan^3 x \sec^2 x}{\sin^2 x}\,dx$

11. $\int \sin^3 y \cos^2 y\,dy$

12. $\int \frac{\csc^2 \theta}{\cot^2 \theta}\,d\theta$

13. $\int \frac{\sin\theta}{1 + \cos\theta}\,d\theta$

14. $\int \frac{\sec^2 x}{\sqrt{1 + \tan x}}\,dx$

15. $\int \cos\theta \sin 2\theta\,d\theta$

16. $\int \frac{3 + 4\csc^2 x}{\cot^2 x}\,dx$

17. $\int \sin^5 x \cos^5 x\,dx$

18. $\int \sin^4 x\,dx$

19. $\int \frac{\tan^3 x}{\sec^4 x}\,dx$

20. $\int \frac{\csc^4 x}{\cot^3 x}\,dx$

21. Find the area between the x-axis and $y = \sin x$ for $0 \le x \le \pi$.

22. Find the volume of the solid of revolution obtained by rotating the area bounded by $y = \tan x$, $y = 0$, and $x = \pi/4$ about the line $y = -1$.

23. Evaluate

$$\int_0^{\pi} \sqrt{1 - \sin^2 x}\, dx.$$

In Exercises 24–31 evaluate the indefinite integral.

24. $\int \cot^4 z\, dz$

25. $\int \frac{\cos^3 \theta}{3 + \sin \theta}\, d\theta$

26. $\int \frac{\cos^4 \theta}{1 + \sin \theta}\, d\theta$

27. $\int \sin^4 x \cos^2 x\, dx$

28. $\int \cos 6x \cos 2x\, dx$

29. $\int \cos^2 2x \sin 3x\, dx$

30. $\int \frac{1}{\sin x \cos^2 x}\, dx$

31. $\int \sec^5 x\, dx$

32. The alternating current in a power line is given by

$$I = f(t) = A \cos \omega t + B \sin \omega t,$$

where A, B, and ω are constants, and t is time. The root-mean-square (rms) current I_{rms} is defined by

$$(I_{\text{rms}})^2 = \frac{1}{T} \int_0^T I^2\, dt,$$

where the interval $0 \le t \le T$ represents any number of complete oscillations of the current. Show that I_{rms} is $1/\sqrt{2}$ times the amplitude of the current.

33. Evaluate

$$\int (1 + \sin \theta + \sin^2 \theta + \sin^3 \theta + \cdots)\, d\theta.$$

34. If n is a positive even integer, show that

$$\int \sec^n x\, dx = \sum_{r=0}^{n/2-1} \binom{n/2 - 1}{r} \frac{\tan^{2r+1} x}{2r + 1} + C.$$

35. A set of functions $f_1(x)$, $f_2(x)$, ... is said to be orthonormal over an interval $a \le x \le b$ if

$$\int_a^b f_n(x) f_m(x)\, dx = \begin{cases} 0 & m \neq n \\ 1 & m = n. \end{cases}$$

Show that the set of functions

$$\frac{1}{\sqrt{2\pi}},\quad \frac{1}{\sqrt{\pi}} \sin x,\quad \frac{1}{\sqrt{\pi}} \cos x,\quad \frac{1}{\sqrt{\pi}} \sin 2x,$$

$$\frac{1}{\sqrt{\pi}} \cos 2x,\quad \frac{1}{\sqrt{\pi}} \sin 3x,\quad \ldots$$

is orthonormal over the interval $0 \le x \le 2\pi$.

SECTION 9.5

Trigonometric Substitutions

Physical and geometric problems arising from circles, ellipses and hyperbolas often give rise to integrals involving the square roots $\sqrt{a^2 - b^2 x^2}$, $\sqrt{a^2 + b^2 x^2}$, and $\sqrt{b^2 x^2 - a^2}$. *Trigonometric substitutions* replace them with integrals involving trigonometric functions.

Let us first consider two integrals that involve square roots of the form $\sqrt{a^2 - b^2 x^2}$:

$$\text{(i)} \quad \int \frac{1}{\sqrt{1 - x^2}}\, dx \qquad \text{and} \qquad \text{(ii)} \quad \int \frac{1}{\sqrt{4 - 9x^2}}\, dx.$$

Integral (i) is evident from differentiation of the inverse trigonometric functions in Section 8.3,

$$\int \frac{1}{\sqrt{1 - x^2}}\, dx = \operatorname{Sin}^{-1} x + C.$$

We now show that a substitution can be made on this integral which is applicable to much more difficult problems involving the square root $\sqrt{a^2 - b^2 x^2}$, including (ii). If instead of the square root $\sqrt{1 - x^2}$ in (i), we had the square root $\sqrt{1 - \sin^2 \theta}$, we could immediately simplify the latter root to $\sqrt{\cos^2 \theta} = |\cos \theta|$. This prompts us to make the substitution $x = \sin \theta$, from which we have $dx = \cos \theta\, d\theta$. Integral (i) then becomes

$$\int \frac{1}{\sqrt{1 - x^2}}\, dx = \int \frac{1}{|\cos \theta|} \cos \theta\, d\theta.$$

To eliminate the absolute values, we need to know whether $\cos\theta$ is positive or negative. But this means that we must know the possible values of θ. There is a problem. The equation $x = \sin\theta$ does not really define θ; for any given x, we do not have a unique θ, but an infinite number of possibilities. We must therefore restrict the values of θ, and we do this by specifying that $\theta = \operatorname{Sin}^{-1} x$. In other words, although we have used the equation $x = \sin\theta$ to change variables, and will continue to do so, it is really the equation $\theta = \operatorname{Sin}^{-1} x$ that properly defines the change.

Since θ so defined must lie in the interval $-\pi/2 \le \theta \le \pi/2$ (the principal values of the inverse sine function), it follows that $\cos\theta \ge 0$, and the absolute values may be dropped. Consequently,

$$\int \frac{1}{\sqrt{1-x^2}}\,dx = \int \frac{\cos\theta}{|\cos\theta|}\,d\theta = \int d\theta = \theta + C = \operatorname{Sin}^{-1} x + C.$$

For integral (ii), the substitution $x = \sin\theta$ is of little use since the square root $\sqrt{4 - 9\sin^2\theta}$ does not simplify. If, however, the square root were $\sqrt{4 - 4\sin^2\theta}$, it would immediately reduce to $\sqrt{4 - 4\sin^2\theta} = \sqrt{4\cos^2\theta} = 2|\cos\theta|$. We can obtain this result if the substitution is modified to $x = (2/3)\sin\theta$ (or, more properly, $\theta = \operatorname{Sin}^{-1}(3x/2)$). Since $dx = (2/3)\cos\theta\,d\theta$, we have

$$\begin{aligned}\int \frac{1}{\sqrt{4-9x^2}}\,dx &= \int \frac{(2/3)\cos\theta}{\sqrt{4-4\sin^2\theta}}\,d\theta = \frac{2}{3}\int \frac{\cos\theta}{2\cos\theta}\,d\theta \\ &= \frac{1}{3}\int d\theta = \frac{\theta}{3} + C = \frac{1}{3}\operatorname{Sin}^{-1}\left(\frac{3x}{2}\right) + C.\end{aligned}$$

For integrals containing square roots of the form $\sqrt{a^2 + b^2x^2}$, consider the pair of integrals

$$\text{(iii)} \quad \int \frac{1}{1+x^2}\,dx \qquad \text{(iv)} \quad \int \frac{1}{(3+5x^2)^{3/2}}\,dx.$$

Integral (iii) obviously has the answer $\operatorname{Tan}^{-1} x + C$. Alternatively, if we had $1 + \tan^2\theta$ instead of $1+x^2$, the two terms in the denominator would simplify to the one term $\sec^2\theta$. We therefore substitute $x = \tan\theta$ or, more properly, $\theta = \operatorname{Tan}^{-1} x$, from which we have $dx = \sec^2\theta\,d\theta$. With this substitution, we obtain

$$\int \frac{1}{1+x^2}\,dx = \int \frac{\sec^2\theta}{\sec^2\theta}\,d\theta = \int d\theta = \theta + C = \operatorname{Tan}^{-1} x + C.$$

The substitution $x = \tan\theta$ in (iv) yields $(3+5x^2)^{3/2} = (3 + 5\tan^2\theta)^{3/2}$. For simplification, we need a 3 in front of the $\tan^2\theta$ and not a 5. This can be accomplished by modification of our substitution to

$$x = \sqrt{\frac{3}{5}}\tan\theta,$$

from which

$$dx = \sqrt{\frac{3}{5}}\sec^2\theta\,d\theta.$$

Then,

$$\begin{aligned}\int \frac{1}{(3+5x^2)^{3/2}}\,dx &= \int \frac{1}{(3+3\tan^2\theta)^{3/2}}\sqrt{\frac{3}{5}}\sec^2\theta\,d\theta = \frac{1}{3\sqrt{5}}\int \frac{\sec^2\theta}{(\sec^2\theta)^{3/2}}\,d\theta \\ &= \frac{1}{3\sqrt{5}}\int \frac{1}{\sec\theta}\,d\theta = \frac{1}{3\sqrt{5}}\int \cos\theta\,d\theta \\ &= \frac{1}{3\sqrt{5}}\sin\theta + C.\end{aligned}$$

FIGURE 9.7

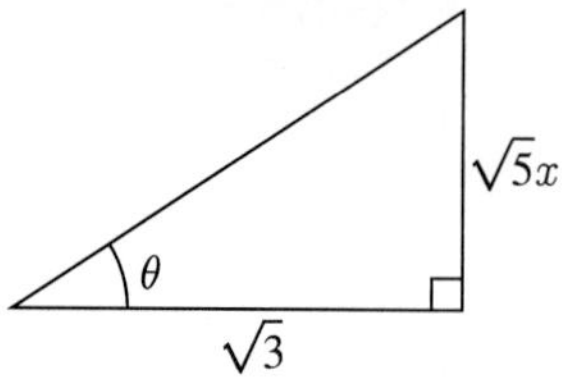

To express $\sin\theta$ in terms of x, we draw the triangle in Figure 9.7 to fit the change of variable $\sqrt{5}\,x/\sqrt{3} = \tan\theta$. Since the hypotenuse of the triangle is $\sqrt{3+5x^2}$, we obtain

$$\int \frac{1}{(3+5x^2)^{3/2}}\,dx = \frac{1}{3\sqrt{5}}\frac{\sqrt{5}\,x}{\sqrt{3+5x^2}} + C = \frac{x}{3\sqrt{3+5x^2}} + C.$$

For integrals containing square roots of the form $\sqrt{b^2x^2 - a^2}$, consider

$$\int \frac{1}{\sqrt{x^2-4}}\,dx.$$

If we had $\sqrt{4\sec^2\theta - 4}$ instead of $\sqrt{x^2-4}$, an immediate simplification would occur. We therefore substitute $x = 2\sec\theta$, from which $dx = 2\sec\theta\tan\theta\,d\theta$. Once again the real substitution is $\theta = \mathrm{Sec}^{-1}(x/2)$, in which case θ is an angle in either the first or third quadrant. With this change,

$$\sqrt{x^2-4} = \sqrt{4\sec^2\theta - 4} = \sqrt{4\tan^2\theta} = 2\,|\tan\theta| = 2\tan\theta \qquad \text{and}$$

$$\begin{aligned}\int \frac{1}{\sqrt{x^2-4}}\,dx &= \int \frac{2\sec\theta\tan\theta}{2\tan\theta}\,d\theta \\ &= \int \sec\theta\,d\theta \\ &= \ln|\sec\theta + \tan\theta| + C.\end{aligned}$$

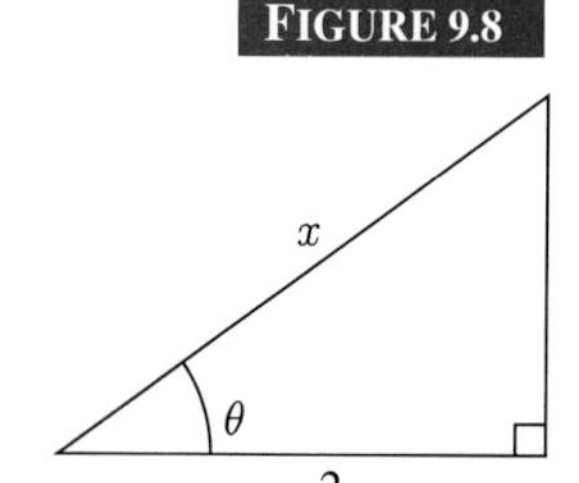

FIGURE 9.8

To express $\tan\theta$ in terms of x, we draw the triangle in Figure 9.8 to fit the change of variable $x/2 = \sec\theta$. Since the third side is $\sqrt{x^2-4}$, we have

$$\int \frac{1}{\sqrt{x^2-4}}\,dx = \ln\left|\frac{x}{2} + \frac{\sqrt{x^2-4}}{2}\right| + C = \ln\left|x + \sqrt{x^2-4}\right| + D,$$

where $D = C - \ln 2$.

We have illustrated by simple examples that trigonometric substitutions can be useful in the evaluation of integrals involving square roots of the forms $\sqrt{a^2 - b^2x^2}$, $\sqrt{a^2 + b^2x^2}$, and $\sqrt{b^2x^2 - a^2}$. Essentially, the method replaces the terms under the square root by a perfect square and thereby rids the integrand of the square root. To obtain the trigonometric substitution appropriate to each square root, we suggest the following procedure. We have been working with the trigonometric identities:

$$\begin{aligned}1 - \sin^2\theta &= \cos^2\theta,\\ 1 + \tan^2\theta &= \sec^2\theta,\\ \sec^2\theta - 1 &= \tan^2\theta.\end{aligned}$$

To determine which trigonometric substitution is appropriate to, say $\sqrt{a^2 - b^2x^2}$, we mentally set $a = b = 1$ and obtain $\sqrt{1-x^2}$. We note that $1 - x^2$ resembles $1 - \sin^2\theta$, and immediately set $x = \sin\theta$. To simplify the original square root $\sqrt{a^2 - b^2x^2}$, we then modify this substitution to

$$x = \frac{a}{b}\sin\theta \qquad \left[\text{or } \theta = \mathrm{Sin}^{-1}\left(\frac{bx}{a}\right)\right].$$

Similarly, for the square root $\sqrt{a^2 + b^2 x^2}$, we initially set $x = \tan\theta$, having mentally set $a = b = 1$, and noted that $1 + x^2$ resembles $1 + \tan^2\theta$. We then modify the substitution to

$$x = \frac{a}{b}\tan\theta \qquad \left[\text{or } \theta = \text{Tan}^{-1}\left(\frac{bx}{a}\right)\right].$$

Finally, for the square root $\sqrt{b^2 x^2 - a^2}$, we initially set $x = \sec\theta$ and then modify to

$$x = \frac{a}{b}\sec\theta \qquad \left[\text{or } \theta = \text{Sec}^{-1}\left(\frac{bx}{a}\right)\right].$$

The square roots $\sqrt{a^2 - b^2 x^2}$, $\sqrt{a^2 + b^2 x^2}$, and $\sqrt{b^2 x^2 - a^2}$ are replaced respectively by

$$\begin{aligned} &\sqrt{a^2 - a^2\sin^2\theta} \\ &= \sqrt{a^2\cos^2\theta} \\ &= a|\cos\theta| \\ &= a\cos\theta; \end{aligned} \qquad \begin{aligned} &\sqrt{a^2 + a^2\tan^2\theta} \\ &= \sqrt{a^2\sec^2\theta} \\ &= a|\sec\theta| \\ &= a\sec\theta; \end{aligned} \qquad \begin{aligned} &\sqrt{a^2\sec^2\theta - a^2} \\ &= \sqrt{a^2\tan^2\theta} \\ &= a|\tan\theta| \\ &= a\tan\theta. \end{aligned}$$

In each case it is our choice of principal values for the inverse trigonometric functions that enables us to neglect the absolute value.

We now consider a number of more complex examples.

EXAMPLE 9.11

Evaluate the following integrals:

$$\text{(a)} \int \frac{1}{x^2(9x^2+2)}\,dx \qquad \text{(b)} \int x^2\sqrt{3 - x^2}\,dx$$

SOLUTION (a) If we set $x = \left(\sqrt{2}/3\right)\tan\theta$, then $dx = \left(\sqrt{2}/3\right)\sec^2\theta\,d\theta$, and

$$\begin{aligned}
\int \frac{1}{x^2(9x^2+2)}\,dx &= \int \frac{1}{\left(\frac{2}{9}\tan^2\theta\right)(2\tan^2\theta + 2)}\left(\frac{\sqrt{2}}{3}\right)\sec^2\theta\,d\theta \\
&= \frac{3}{2\sqrt{2}}\int \frac{\sec^2\theta}{\tan^2\theta\sec^2\theta}\,d\theta \\
&= \frac{3}{2\sqrt{2}}\int \cot^2\theta\,d\theta = \frac{3}{2\sqrt{2}}\int(\csc^2\theta - 1)\,d\theta \\
&= \frac{3}{2\sqrt{2}}(-\cot\theta - \theta) + C \\
&= \frac{-3}{2\sqrt{2}}\left[\frac{\sqrt{2}}{3x} + \text{Tan}^{-1}\left(\frac{3x}{\sqrt{2}}\right)\right] + C \qquad \text{(Figure 9.9)} \\
&= \frac{-1}{2x} - \frac{3}{2\sqrt{2}}\text{Tan}^{-1}\left(\frac{3x}{\sqrt{2}}\right) + C.
\end{aligned}$$

FIGURE 9.9

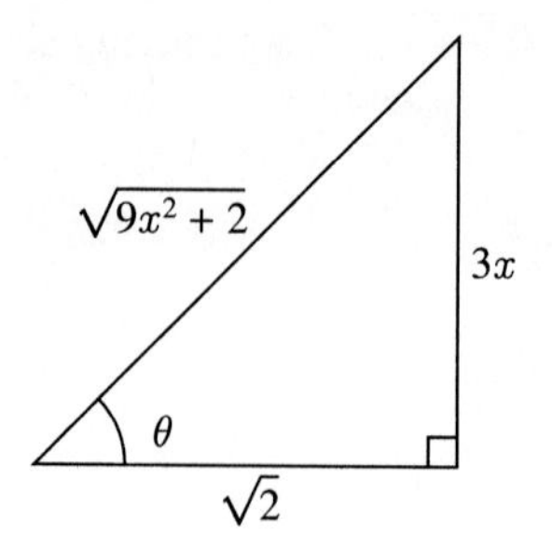

(b) If we set $x = \sqrt{3}\sin\theta$, then $dx = \sqrt{3}\cos\theta\, d\theta$, and

$$\int x^2\sqrt{3-x^2}\,dx = \int (3\sin^2\theta)\sqrt{3-3\sin^2\theta}(\sqrt{3}\cos\theta\, d\theta)$$
$$= 9\int \sin^2\theta\sqrt{1-\sin^2\theta}\cos\theta\, d\theta$$
$$= 9\int \sin^2\theta\cos^2\theta\, d\theta$$
$$= \frac{9}{4}\int \sin^2 2\theta\, d\theta \quad \text{(by 9.23b)}$$
$$= \frac{9}{4}\int\left(\frac{1-\cos 4\theta}{2}\right)d\theta \quad \text{(by 9.23d)}$$
$$= \frac{9}{8}\left(\theta - \frac{1}{4}\sin 4\theta\right) + C \quad \text{(by 9.22d)}$$
$$= \frac{9}{8}\text{Sin}^{-1}\left(\frac{x}{\sqrt{3}}\right) - \frac{9}{32}(2\sin 2\theta\cos 2\theta) + C \quad \text{(by 9.23b)}$$
$$= \frac{9}{8}\text{Sin}^{-1}\left(\frac{x}{\sqrt{3}}\right) - \frac{9}{16}(2\sin\theta\cos\theta)(1-2\sin^2\theta) + C \quad \text{(by 9.23b,d)}$$
$$= \frac{9}{8}\text{Sin}^{-1}\left(\frac{x}{\sqrt{3}}\right) - \frac{9}{8}\left(\frac{x}{\sqrt{3}}\right)\frac{\sqrt{3-x^2}}{\sqrt{3}}\left(1-\frac{2x^2}{3}\right) + C \quad \text{(Figure 9.10)}$$
$$= \frac{9}{8}\text{Sin}^{-1}\left(\frac{x}{\sqrt{3}}\right) - \frac{x}{8}\sqrt{3-x^2}(3-2x^2) + C.$$

FIGURE 9.10

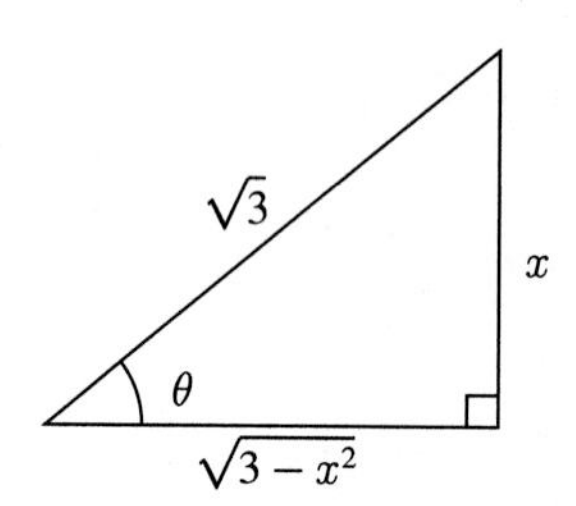

■

EXAMPLE 9.12

Use a trigonometric substitution to evaluate the integral in Example 4.

SOLUTION If we set $x = 3\tan\theta$, then $dx = 3\sec^2\theta\, d\theta$, and

$$\begin{aligned}
\int \frac{x^3}{\sqrt{9+x^2}}\,dx &= \int \frac{27\tan^3\theta}{\sqrt{9+9\tan^2\theta}} 3\sec^2\theta\,d\theta \\
&= 27\int \frac{\tan^3\theta\sec^2\theta}{\sec\theta}\,d\theta \\
&= 27\int \tan\theta\,(\sec^2\theta - 1)\sec\theta\,d\theta && \text{(by 9.25)} \\
&= 27\int (\sec^3\theta\tan\theta - \sec\theta\tan\theta)\,d\theta \\
&= 27\left(\frac{1}{3}\sec^3\theta - \sec\theta\right) + C && \text{(by 9.24b)} \\
&= 9\left(\frac{\sqrt{9+x^2}}{3}\right)^3 - 27\frac{\sqrt{9+x^2}}{3} + C && \text{(Figure 9.11)} \\
&= \frac{1}{3}(9+x^2)^{3/2} - 9\sqrt{9+x^2} + C.
\end{aligned}$$

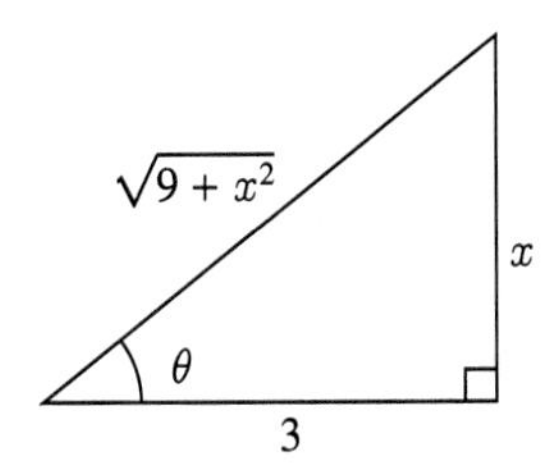

FIGURE 9.11

■

EXAMPLE 9.13

Verify the formula πr^2 for the area of a circle of radius r.

SOLUTION If we consider the circle $x^2 + y^2 = r^2$ in Figure 9.12, we see that its area is four times that in the first quadrant:

$$A = 4\int_0^r \sqrt{r^2 - x^2}\,dx.$$

If we set $x = r\sin\theta$, then $dx = r\cos\theta\,d\theta$, and

$$\begin{aligned}
A &= 4\int_0^{\pi/2} \sqrt{r^2 - r^2\sin^2\theta}\, r\cos\theta\,d\theta \\
&= 4r^2\int_0^{\pi/2} \cos^2\theta\,d\theta \\
&= 4r^2\int_0^{\pi/2} \left(\frac{1+\cos 2\theta}{2}\right) d\theta && \text{(by 9.23c)} \\
&= 2r^2\left\{\theta + \frac{1}{2}\sin 2\theta\right\}_0^{\pi/2} && \text{(by 9.22d)} \\
&= 2r^2(\pi/2) = \pi r^2.
\end{aligned}$$

FIGURE 9.12

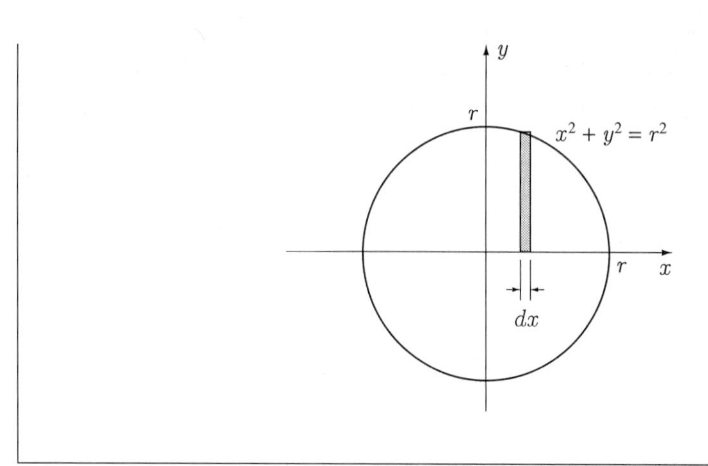

EXERCISES 9.5

In Exercises 1–20 evaluate the indefinite integral.

1. $\displaystyle\int \frac{1}{x\sqrt{2x^2-4}}\,dx$ **2.** $\displaystyle\int \frac{1}{\sqrt{9-5x^2}}\,dx$

3. $\displaystyle\int \frac{1}{10+x^2}\,dx$ **4.** $\displaystyle\int \frac{1}{x^2\sqrt{4-x^2}}\,dx$

5. $\displaystyle\int \sqrt{7-x^2}\,dx$ **6.** $\displaystyle\int x\sqrt{5x^2+3}\,dx$

7. $\displaystyle\int x^3\sqrt{4+x^2}\,dx$ **8.** $\displaystyle\int \frac{1}{1-x^2}\,dx$

9. $\displaystyle\int \frac{1}{\sqrt{x^2-5}}\,dx$ **10.** $\displaystyle\int \frac{x+5}{10x^2+2}\,dx$

11. $\displaystyle\int \frac{1}{x\sqrt{x^2+3}}\,dx$ **12.** $\displaystyle\int \frac{\sqrt{4-x^2}}{x}\,dx$

13. $\displaystyle\int \frac{x^2}{(2-9x^2)^{3/2}}\,dx$ **14.** $\displaystyle\int \frac{\sqrt{x^2-16}}{x^2}\,dx$

15. $\displaystyle\int \frac{1}{x^2\sqrt{2x^2+7}}\,dx$ **16.** $\displaystyle\int \frac{1}{x^3\sqrt{x^2-4}}\,dx$

17. $\displaystyle\int \frac{\sqrt{9-z^2}}{z^4}\,dz$ **18.** $\displaystyle\int \frac{y^3}{\sqrt{y^2+4}}\,dy$

19. $\displaystyle\int \frac{1}{(4x^2-9)^{3/2}}\,dx$ **20.** $\displaystyle\int \frac{x^2+2}{x^3+x}\,dx$

21. Show that if $a > 0$, then

$$\int \frac{1}{a^2-x^2}\,dx = \frac{1}{2a}\ln\left|\frac{a+x}{a-x}\right| + C.$$

22. Verify that the area inside the ellipse $b^2x^2+a^2y^2 = a^2b^2$ is πab.

23. Find the area common to the circles $x^2+y^2 = 4$ and $x^2+y^2 = 4x$.

24. A boy initially at O (Figure 9.13) walks along the edge of a pier (the y-axis) towing his sailboat by a string of length L.

(a) If the boat starts at Q and the string always remains straight, show that the equation of the curved path $y = f(x)$ followed by the boat must satisfy the differential equation

$$\frac{dy}{dx} = -\frac{\sqrt{L^2-x^2}}{x}.$$

(b) Find $y = f(x)$.

FIGURE 9.13

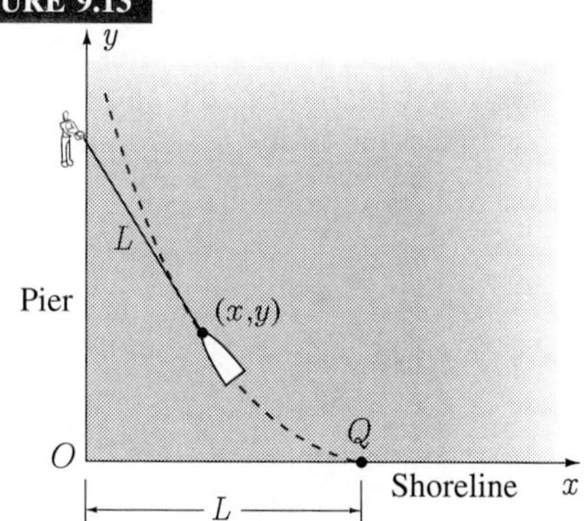

25. A dog at position $(L, 0)$ in the xy-plane spots a rabbit at position $(0, 0)$ running in the positive y-direction. If the dog runs at the same speed as the rabbit and always moves directly toward the rabbit, find the equation of the path followed by the dog.

26. The parabola $x = y^2$ divides the circle $x^2+y^2 = 4$ into two parts. Find the second moment of area of the smaller part about the x-axis.

27. Find the centroid of the area in Exercise 26.

28. Find the horizontal line that divides the ellipse $b^2x^2 + a^2y^2 = a^2b^2$ into two parts so that the area of the lower part is twice the area of the upper part.

29. If a thin circular plate has radius 2 units and constant mass per unit area ρ, find its moment of inertia about any tangent line to its edge.

30. A cylindrical oil can with horizontal axis has radius r and length h. If the density of the oil is ρ, find the force on each end of the can when it is full.

In Exercises 31–37 evaluate the indefinite integral.

31. $\displaystyle\int \frac{2x^4 - x^2}{2x^2 + 1}\,dx$

32. $\displaystyle\int (7 - x^2)^{3/2}\,dx$

33. $\displaystyle\int \frac{1}{x - x^3}\,dx$

34. $\displaystyle\int \frac{1}{x^3(4x^2 - 1)^{3/2}}\,dx$

35. $\displaystyle\int \sqrt{x^2 - 4}\,dx$

36. $\displaystyle\int \sqrt{1 + 3x^2}\,dx$

37. $\displaystyle\int \frac{x^2}{\sqrt{x^2 - 5}}\,dx$

38. Find the length of the curve $8y^2 = x^2(1 - x^2)$.

39. Find the length of the parabola $y = x^2$ from $(0,0)$ to $(1,1)$.

40. At what distance from the centre of a circle should a line be drawn in order that the second moment of area of the circle about that line be equal to twice the second moment of the circle about a line through its centre?

41. When a flexible cable of constant mass per unit length ρ hangs between two fixed points A and B (Figure 9.14), the shape $y = f(x)$ of the cable must satisfy the differential equation

$$\frac{d^2y}{dx^2} = \frac{\rho g}{H}\sqrt{1 + \left(\frac{dy}{dx}\right)^2},$$

where $g > 0$ and $H > 0$ are constants. If we set $k = \rho g/H$ and $p = dy/dx$, then

$$\frac{dp}{dx} = k\sqrt{1 + p^2} \qquad \text{or} \qquad \frac{1}{\sqrt{1 + p^2}}\frac{dp}{dx} = k.$$

It follows that

$$\int \frac{1}{\sqrt{1 + p^2}}\,dp = kx + C.$$

(a) Evaluate the integral shown to find $p = dy/dx$.

(b) Integrate once more to find the shape of the cable (see also Example 8.14 in Section 8.4).

FIGURE 9.14

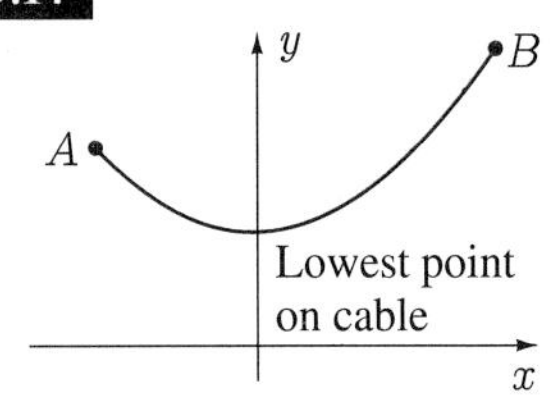

42. Find the area inside the loop of the strophoid $y^2(a + x) = x^2(a - x)$.

43. (a) By rationalizing the denominator, show that

$$\frac{1}{x + 1 + \sqrt{x^2 + 1}} = \frac{x + 1 - \sqrt{x^2 + 1}}{2x}.$$

(b) Use (a) to prove that

$$\int_0^1 \frac{1}{x + 1 + \sqrt{x^2 + 1}}\,dx = 1 - \frac{\sqrt{2}}{2} + \frac{1}{2}\ln\left(\frac{1 + \sqrt{2}}{2}\right).$$

44. Show that

$$\int_{-1}^{1} \sqrt{\frac{1 + x}{1 - x}}\,dx = \pi.$$

SECTION 9.6

Completing the Square and Trigonometric Substitutions

Trigonometric substitutions reduce integrals containing square roots of the form $\sqrt{a^2 - b^2x^2}$, $\sqrt{a^2 + b^2x^2}$, and $\sqrt{b^2x^2 - a^2}$ to trigonometric integrals. For integrals containing square roots of the form $\sqrt{ax^2 + bx + c}$ we can again reduce the integral to a trigonometric integral by a trigonometric substitution, if we first complete the square.

Consider the integral

$$\int \frac{1}{(x^2 + 2x + 5)^{3/2}}\,dx.$$

If we complete the square in the denominator, we obtain

$$\int \frac{1}{(x^2 + 2x + 5)^{3/2}}\, dx = \int \frac{1}{[(x+1)^2 + 4]^{3/2}}\, dx.$$

Had the denominator been $(x^2 + 4)^{3/2}$, we would have set $x = 2\tan\theta$. It is natural, then, for the denominator $[(x+1)^2 + 4]^{3/2}$ to set

$$x + 1 = 2\tan\theta,$$

in which case $dx = 2\sec^2\theta\, d\theta$, and

FIGURE 9.15

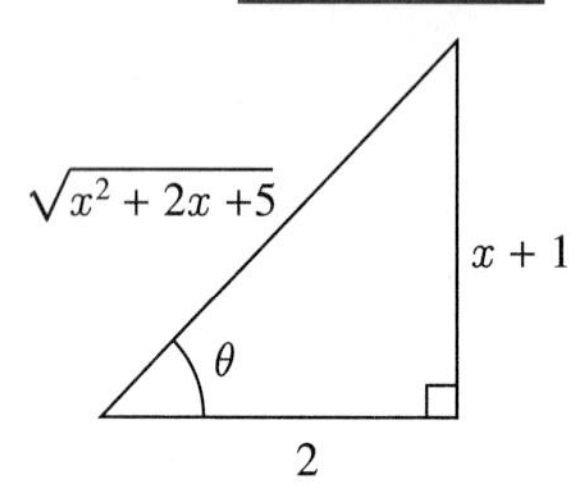

$$\begin{aligned}\int \frac{1}{(x^2 + 2x + 5)^{3/2}}\, dx &= \int \frac{1}{(4\tan^2\theta + 4)^{3/2}} 2\sec^2\theta\, d\theta = \int \frac{2\sec^2\theta}{8\sec^3\theta}\, d\theta \\ &= \frac{1}{4}\int \cos\theta\, d\theta = \frac{1}{4}\sin\theta + C \\ &= \frac{x+1}{4\sqrt{x^2 + 2x + 5}} + C. \qquad \text{(Figure 9.15)}\end{aligned}$$

Consider another example,

$$\int \sqrt{4 + x - x^2}\, dx.$$

Completion of the square leads to

$$\int \sqrt{4 + x - x^2}\, dx = \int \sqrt{-(x - 1/2)^2 + 17/4}\, dx.$$

If we now set

$$x - \frac{1}{2} = \frac{\sqrt{17}}{2}\sin\theta,$$

then

$$dx = \frac{\sqrt{17}}{2}\cos\theta\, d\theta, \qquad \text{and}$$

FIGURE 9.16

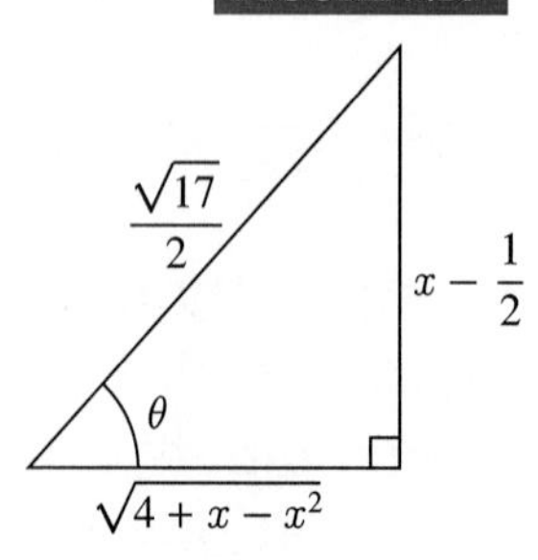

$$\begin{aligned}\int \sqrt{4 + x - x^2}\, dx &= \int \sqrt{-\frac{17}{4}\sin^2\theta + \frac{17}{4}}\, \frac{\sqrt{17}}{2}\cos\theta\, d\theta \\ &= \int \frac{\sqrt{17}}{2}\cos\theta \frac{\sqrt{17}}{2}\cos\theta\, d\theta \\ &= \frac{17}{4}\int \cos^2\theta\, d\theta = \frac{17}{4}\int \left(\frac{1 + \cos 2\theta}{2}\right) d\theta \\ &= \frac{17}{8}\left(\theta + \frac{1}{2}\sin 2\theta\right) + C = \frac{17}{8}(\theta + \sin\theta\cos\theta) + C \\ &= \frac{17}{8}\mathrm{Sin}^{-1}\left(\frac{2x-1}{\sqrt{17}}\right) + \frac{17}{8}\left[\frac{2}{\sqrt{17}}\left(x - \frac{1}{2}\right)\right]\frac{\sqrt{4 + x - x^2}}{\sqrt{17}/2} + C \quad \text{(Figure 9.16)} \\ &= \frac{17}{8}\mathrm{Sin}^{-1}\left(\frac{2x-1}{\sqrt{17}}\right) + \frac{1}{4}(2x - 1)\sqrt{4 + x - x^2} + C.\end{aligned}$$

As a final illustrative example, we consider

$$\int \frac{x}{\sqrt{2x^2+3x-6}}\,dx = \frac{1}{\sqrt{2}}\int \frac{x}{\sqrt{x^2+\dfrac{3x}{2}-3}}\,dx = \frac{1}{\sqrt{2}}\int \frac{x}{\sqrt{\left(x+\dfrac{3}{4}\right)^2-\dfrac{57}{16}}}\,dx.$$

If we set

$$x+\frac{3}{4}=\frac{\sqrt{57}}{4}\sec\theta,$$

then

$$dx=\frac{\sqrt{57}}{4}\sec\theta\tan\theta\,d\theta, \qquad \text{and}$$

$$\begin{aligned}
\int \frac{x}{\sqrt{2x^2+3x-6}}\,dx &= \frac{1}{\sqrt{2}}\int \frac{(\sqrt{57}/4)\sec\theta-3/4}{\sqrt{(57/16)\sec^2\theta-57/16}}\,\frac{\sqrt{57}}{4}\sec\theta\tan\theta\,d\theta\\
&= \frac{1}{\sqrt{2}}\int \frac{(\sqrt{57}/4)\sec\theta-3/4}{(\sqrt{57}/4)\tan\theta}\,\frac{\sqrt{57}}{4}\sec\theta\tan\theta\,d\theta\\
&= \frac{1}{\sqrt{2}}\int\left(\frac{\sqrt{57}}{4}\sec^2\theta-\frac{3}{4}\sec\theta\right)d\theta\\
&= \frac{\sqrt{57}}{4\sqrt{2}}\tan\theta-\frac{3}{4\sqrt{2}}\ln|\sec\theta+\tan\theta|+C\\
&= \frac{\sqrt{57}}{4\sqrt{2}}\left[\frac{\sqrt{x^2+(3/2)x-3}}{\sqrt{57}/4}\right]\\
&\quad -\frac{3}{4\sqrt{2}}\ln\left|\frac{x+3/4}{\sqrt{57}/4}+\frac{\sqrt{x^2+(3/2)x-3}}{\sqrt{57}/4}\right|+C \quad \text{(Figure 9.17)}\\
&= \frac{1}{2}\sqrt{2x^2+3x-6}\\
&\quad -\frac{3}{4\sqrt{2}}\ln\left|4x+3+2\sqrt{2}\sqrt{2x^2+3x-6}\right|+D,
\end{aligned}$$

FIGURE 9.17

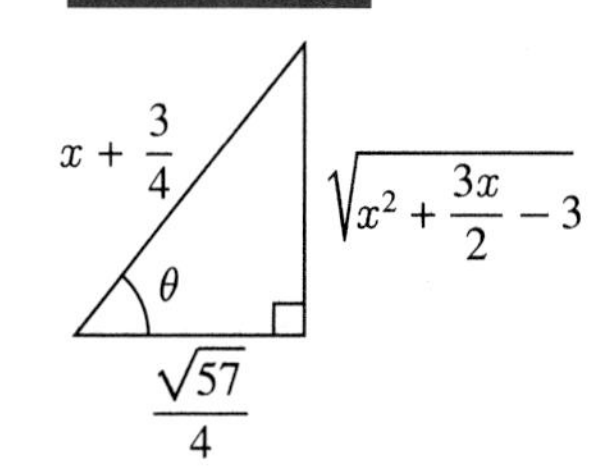

where

$$D=C+\frac{3}{8\sqrt{2}}\ln 57.$$

We have shown that the technique of completing the square is the natural generalization of the technique of trigonometric substitutions. It reduces integrals containing square roots of the form $\sqrt{ax^2+bx+c}$ to trigonometric integrals.

EXAMPLE 9.14

Find the volume of a donut.

SOLUTION A donut is generated (as the volume of the solid of revolution) when the circle

$$(x-a)^2 + y^2 = b^2, \qquad (a > b),$$

is rotated around the y-axis (Figure 9.18). Cylindrical shells for the vertical rectangle shown yield the volume generated by the upper semicircle, which can be doubled to give

$$V = 2\int_{a-b}^{a+b} 2\pi xy\, dx.$$

If we solve the equation of the circle for $y = \pm\sqrt{b^2 - (x-a)^2}$, then

$$V = 2\int_{a-b}^{a+b} 2\pi x\sqrt{b^2 - (x-a)^2}\, dx.$$

We now set $x - a = b\sin\theta$, in which case $dx = b\cos\theta\, d\theta$, and

$$\begin{aligned} V &= 4\pi\int_{-\pi/2}^{\pi/2} (a + b\sin\theta)b\cos\theta\, b\cos\theta\, d\theta \\ &= 4\pi b^2\int_{-\pi/2}^{\pi/2} (a\cos^2\theta + b\cos^2\theta\sin\theta)\, d\theta \\ &= 4\pi b^2\int_{-\pi/2}^{\pi/2} \left[a\left(\frac{1+\cos 2\theta}{2}\right) + b\cos^2\theta\sin\theta\right] d\theta \\ &= 4\pi b^2\left\{\frac{a}{2}\left(\theta + \frac{1}{2}\sin 2\theta\right) - \frac{b}{3}\cos^3\theta\right\}_{-\pi/2}^{\pi/2} \\ &= 4\pi b^2\left[\frac{a}{2}\left(\frac{\pi}{2} + \frac{\pi}{2}\right)\right] = 2\pi^2 ab^2. \end{aligned}$$

FIGURE 9.18

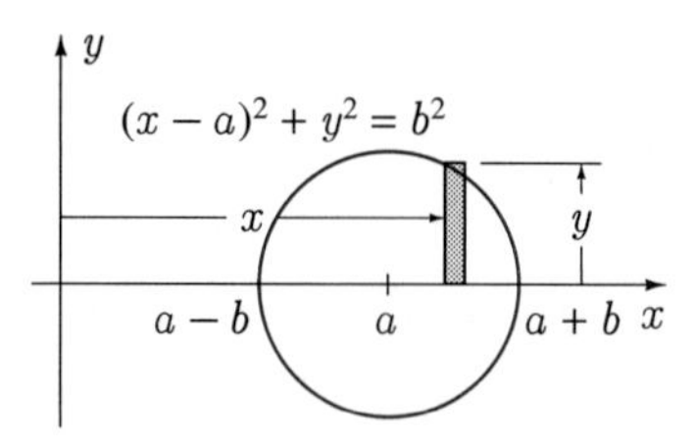

■

Should we use horizontal rectangles, as in Figure 9.19, the washer method yields

$$V = 2\int_0^b (\pi x_2^2 - \pi x_1^2)\, dy.$$

We can solve the equation of the circle for $x = a \pm \sqrt{b^2 - y^2}$; hence we have

$$x_1 = a - \sqrt{b^2 - y^2}, \qquad x_2 = a + \sqrt{b^2 - y^2}.$$

Thus

$$V = 2\pi\int_0^b \left[\left(a + \sqrt{b^2 - y^2}\right)^2 - \left(a - \sqrt{b^2 - y^2}\right)^2\right] dy,$$

and this simplifies to

$$V = 2\pi \int_0^b 4a\sqrt{b^2 - y^2}\, dy.$$

If we set $y = b\sin\theta$, then $dy = b\cos\theta\, d\theta$, and

$$\begin{aligned} V &= 8\pi a \int_0^{\pi/2} b\cos\theta\, b\cos\theta\, d\theta \\ &= 8\pi ab^2 \int_0^{\pi/2} \left(\frac{1 + \cos 2\theta}{2}\right) d\theta \\ &= 4\pi ab^2 \left\{\theta + \frac{1}{2}\sin 2\theta\right\}_0^{\pi/2} \\ &= 4\pi ab^2(\pi/2) = 2\pi^2 ab^2. \end{aligned}$$

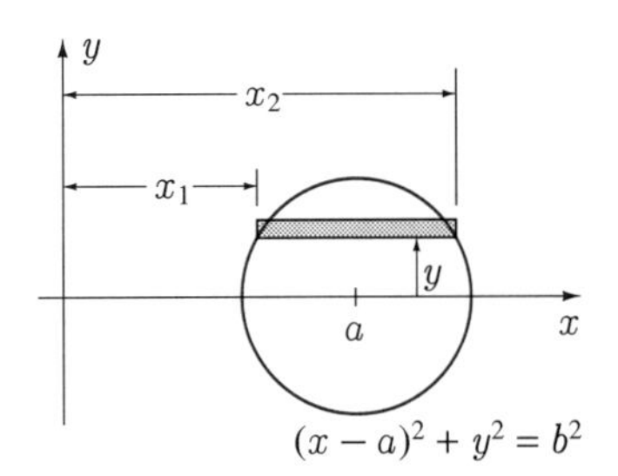

FIGURE 9.19

EXERCISES 9.6

In Exercises 1–10 evaluate the indefinite integral.

1. $\displaystyle\int \frac{x}{\sqrt{27 + 6x - x^2}}\, dx$ **2.** $\displaystyle\int \frac{1}{\sqrt{x^2 + 2x + 2}}\, dx$

3. $\displaystyle\int \frac{1}{(y^2 + 4y)^{3/2}}\, dy$ **4.** $\displaystyle\int \frac{1}{3x - x^2 - 4}\, dx$

5. $\displaystyle\int \frac{\sqrt{x^2 + 2x - 3}}{x + 1}\, dx$ **6.** $\displaystyle\int \frac{x}{(4x - x^2)^{3/2}}\, dx$

7. $\displaystyle\int \sqrt{-y^2 + 6y}\, dy$ **8.** $\displaystyle\int \frac{2x - 3}{x^2 + 6x + 13}\, dx$

9. $\displaystyle\int \frac{5 - 4x}{\sqrt{12x - 4x^2 - 8}}\, dx$

10. $\displaystyle\int \frac{1}{x\sqrt{6 + 4\ln x + (\ln x)^2}}\, dx$

11. Use the substitution $z = 1/x$ to evaluate

$$\int \frac{1}{x\sqrt{x^2 + 6x + 3}}\, dx.$$

12. One of the gates in a dam is circular with radius 1 m. If the gate is closed and the surface of the water is 3 m above the top of the gate, find the force due to water pressure on the gate.

13. Evaluate

$$\int \frac{1}{3x - x^2}\, dx$$

(a) by completing the square, and setting $x - 3/2 = (3/2)\sin\theta$;

(b) by multiplying numerator and denominator by -1, completing the square, and setting $x - 3/2 = (3/2)\sec\theta$.

(c) Explain the difference in the two answers.

In Exercises 14–16 evaluate the integral.

14. $\displaystyle\int \sqrt{x^2 - 2x - 3}\, dx$ **15.** $\displaystyle\int \frac{1}{x\sqrt{2x - x^2}}\, dx$

16. $\displaystyle\int \frac{1}{(2x + 5)\sqrt{2x - 3} + 8x - 12}\, dx$

SECTION 9.7

Partial Fractions

Partial fractions is a method that we apply to integrals of the form

$$\int \frac{N(x)}{D(x)}\,dx, \tag{9.26}$$

where $N(x)$ and $D(x)$ are polynomials in x, and the degree of $N(x)$ is less than the degree of $D(x)$. When the degree of $N(x)$ is greater than or equal to that of $D(x)$, we divide $D(x)$ into $N(x)$. For example, in the integral

$$\int \frac{x^4 + 4x^3 + 2x + 4}{x^3 + 1}\,dx,$$

the numerator has degree 4 and the denominator degree 3. By long division, we obtain

$$\int \frac{x^4 + 4x^3 + 2x + 4}{x^3 + 1}\,dx = \int \left(x + 4 + \frac{x}{x^3 + 1}\right) dx.$$

The first two terms on the right, namely, $x+4$, can be integrated immediately, and partial fractions can be applied to the remaining term $x/(x^3 + 1)$.

A general rule is that if the degree of $N(x)$ is greater than or equal to the degree of $D(x)$, divide $D(x)$ into $N(x)$ to produce a quotient polynomial $Q(x)$ and a remainder polynomial $R(x)$ of degree less than that of $D(x)$; that is,

$$\frac{N(x)}{D(x)} = Q(x) + \frac{R(x)}{D(x)},$$

where $\deg R < \deg D$.

Integral 9.26 can then be written in the form

$$\int \frac{N(x)}{D(x)}\,dx = \int Q(x)\,dx + \int \frac{R(x)}{D(x)}\,dx.$$

The first integral on the right is trivial. This leaves integration of the rational function

$$\int \frac{R(x)}{D(x)}\,dx, \tag{9.27}$$

where $\deg R < \deg D$.

To use partial fractions on integral 9.27 we must factor the denominator $D(x)$. It is an established fact of algebra that every real polynomial, and specifically $D(x)$, can be factored into real linear factors $ax + b$ and irreducible quadratic factors $ax^2 + bx + c$. Note that a quadratic factor is irreducible if $b^2 - 4ac < 0$. In the complete factorization of $D(x)$, these factors may or may not be repeated; that is, the factorization of $D(x)$ contains terms of the form

$$(ax + b)^n \qquad \text{and} \qquad (ax^2 + bx + c)^n,$$

where $n \geq 1$ is an integer. When $n = 1$ the factor is nonrepeated, and when $n > 1$, it is repeated. For example, in the factorization

$$D(x) = (x-1)(2x+1)^3(3x^2+4x+5)(x^2+1)^2,$$

$x-1$ and $3x^2+4x+5$ are nonrepeated, and $2x+1$ and x^2+1 are repeated.

Having factored $D(x)$, we can break down the rational function in 9.27 into frational components. We call this the **partial fraction decomposition** of the integrand. We illustrate it with the rational function $(x^2+1)/[(x-1)(2x+1)^3(3x^2+4x+5)(x^2+1)^2]$, and then state general rules for all decompositions. The partial fraction decomposition of this particular rational function is

$$\frac{x^2+1}{(x-1)(2x+1)^3(3x^2+4x+5)(x^2+1)^2} = \frac{A}{x-1} + \frac{B}{2x+1} + \frac{C}{(2x+1)^2} + \frac{D}{(2x+1)^3} + \frac{Ex+F}{3x^2+4x+5} + \frac{Gx+H}{x^2+1} + \frac{Ix+J}{(x^2+1)^2},$$

where $A, B, \ldots, J$ are constants. The first term corresponds to the nonrepeated linear factor $x-1$, the next three corresponds to the repeated linear factor $2x+1$, the fifth term corresponds to the nonrepeated quadratic factor $3x^2+4x+5$, and the last two terms result from the repeated quadratic x^2+1.

Let us now state general rules for the partial fraction decomposition of rational functions $R(x)/D(x)$. There are three rules:

1. For each repeated or nonrepeated linear factor $(ax+b)^n$ in $D(x)$, include the following terms in the decomposition

$$\frac{A_1}{ax+b} + \frac{A_2}{(ax+b)^2} + \cdots + \frac{A_n}{(ax+b)^n}. \qquad (9.28\text{a})$$

   The number of terms corresponds to the power n on $(ax+b)^n$.

2. For each repeated or nonrepeated irreducible quadratic factor $(ax^2+bx+c)^n$ in $D(x)$, include the following terms in the decomposition

$$\frac{B_1x+C_1}{ax^2+bx+c} + \frac{B_2x+C_2}{(ax^2+bx+c)^2} + \cdots + \frac{B_nx+C_n}{(ax^2+bx+c)^n}. \qquad (9.28\text{b})$$

   Again the number of terms corresponds to the power n on $(ax^2+bx+c)^n$.

3. The complete decomposition is the sum of all terms in 9.28a and all terms in 9.28b.

The following examples will clarify these rules.

EXAMPLE 9.15

What form do partial fraction decompositions for the following rational functions take:

(a) $\dfrac{x^2+2x+3}{3x^3-x^2-3x+1}$ (b) $\dfrac{x^2+3x-1}{x^4+x^3+x^2+x}$

(c) $\dfrac{x^3+3x^2-x}{x^5+x^4+2x^3+2x^2+x+1}$ (d) $\dfrac{3x^6-1}{(3x^2+5)(x^2+2x+3)(2x-1)^2}$

SOLUTION (a) Since $3x^3-x^2-3x+1 = (3x-1)(x-1)(x+1)$, we have nonrepeated linear factors. Rules 1. and 3. give the partial fraction decomposition

$$\frac{x^2+2x+3}{3x^3-x^2-3x+1} = \frac{A}{3x-1} + \frac{B}{x-1} + \frac{C}{x+1}.$$

(b) The factorization $x^4+x^3+x^2+x = x(x+1)(x^2+1)$ has two nonrepeated linear factors, and a nonrepeated quadratic factor. The three rules lead to

$$\frac{x^2+3x-1}{x^4+x^3+x^2+x} = \frac{A}{x} + \frac{B}{x+1} + \frac{Cx+D}{x^2+1}.$$

(c) Since $x^5+x^4+2x^3+2x^2+x+1 = (x+1)(x^2+1)^2$, the partial fraction decomposition is

$$\frac{x^3+3x^2-x}{x^5+x^4+2x^3+2x^2+x+1} = \frac{A}{x+1} + \frac{Cx+D}{x^2+1} + \frac{Ex+F}{(x^2+1)^2}.$$

(d) Since x^2+2x+3 can be factored no further, the decomposition is

$$\frac{3x^6-1}{(3x^2+5)(x^2+2x+3)(2x-1)^2} = \frac{Ax+B}{3x^2+5} + \frac{Cx+D}{x^2+2x+3} + \frac{E}{2x-1} + \frac{F}{(2x-1)^2}.$$

■

To illustrate how to calculate coefficients in partial fraction decompositions, we use part (b) of Example 9.15. We bring the right side of the decomposition to a common denominator,

$$\begin{aligned}\frac{x^2+3x-1}{x^4+x^3+x^2+x} &= \frac{A}{x} + \frac{B}{x+1} + \frac{Cx+D}{x^2+1} \\ &= \frac{A(x+1)(x^2+1) + Bx(x^2+1) + x(x+1)(Cx+D)}{x^4+x^3+x^2+x},\end{aligned}$$

and equate numerators

$$x^2+3x-1 = A(x+1)(x^2+1) + Bx(x^2+1) + x(x+1)(Cx+D).$$

There are two methods for finding the constants A, B, C, and D which satisfy this equation.

Method 1 – First, we gather together terms in the various powers of x on the right side of the equation,

$$x^2+3x-1 = (A+B+C)x^3 + (A+C+D)x^2 + (A+B+D)x + A.$$

Now, Exercise 35 in Section 3.7 states that two polynomials of the same degree can be equal for all values of x if and only if coefficients of corresponding powers of x are identical. Since we have equal cubic polynomials in the above equation, we equate coefficients of x^3, x^2, x, and x^0 (meaning terms with no x's):

$$\begin{aligned} x^3 &: & 0 &= A+B+C, \\ x^2 &: & 1 &= A+C+D, \\ x &: & 3 &= A+B+D, \\ x^0 &: & -1 &= A. \end{aligned}$$

Solutions of these four linear equations in four unknowns are $A = -1$, $B = 3/2$, $C = -1/2$, and $D = 5/2$. Hence, the partial fraction decomposition of $(x^2 + 3x - 1)/(x^4 + x^3 + x^2 + x)$ is

$$\frac{x^2 + 3x - 1}{x^4 + x^3 + x^2 + x} = -\frac{1}{x} + \frac{3/2}{x + 1} + \frac{-x/2 + 5/2}{x^2 + 1}.$$

Method 2 – In this method we substitute convenient values of x into the equation

$$x^2 + 3x - 1 = A(x + 1)(x^2 + 1) + Bx(x^2 + 1) + x(x + 1)(Cx + D).$$

Clearly $x = 0$ is most convenient, since it yields the value of A,

$$-1 = A(1)(1),$$

that is, $A = -1$. Convenient also is $x = -1$,

$$(-1)^2 + 3(-1) - 1 = B(-1)(2);$$

it gives $B = 3/2$. The values $x = 1$ and $x = 2$ yield the equations

$$\begin{aligned} 1 + 3(1) - 1 &= A(2)(2) + B(1)(2) + 1(2)(C + D), \\ 4 + 3(2) - 1 &= A(3)(5) + B(2)(5) + 2(3)(2C + D). \end{aligned}$$

When $A = -1$ and $B = 3/2$ are substituted into these, the resulting equations are

$$\begin{aligned} C + D &= 2, \\ 4C + 2D &= 3. \end{aligned}$$

Solutions of these are $C = -1/2$ and $D = 5/2$.

It is also possible to use a combination of Methods 1 and 2 for finding coefficients in a partial fraction decomposition; that is, substitute some values of x and equate some coefficients. The total number must be equal to the number of unknown coefficients in the decomposition.

Once we have completed the partial fraction decomposition of a rational function, we can integrate the function by finding antiderivatives of the component fractions. For example, with the partial fraction decomposition of $(x^2 + 3x - 1)/(x^4 + x^3 + x^2 + x)$, it is very simple to find its antiderivative,

$$\begin{aligned} \frac{x^2+3x-1}{x^4+x^3+x^2+x}\,dx &= \left(-\frac{1}{x} + \frac{3/2}{x+1} + \frac{-x/2+5/2}{x^2+1}\right) dx \\ &= -\frac{1}{x}\,dx + \frac{3}{2}\frac{1}{x+1}\,dx - \frac{1}{2}\frac{x}{x^2+1}\,dx + \frac{5}{2}\frac{1}{x^2+1}\,dx \\ &= -\ln|x| + \frac{3}{2}\ln|x+1| - \frac{1}{4}\ln(x^2+1) + \frac{5}{2}\mathrm{Tan}^{-1}\,x + C. \end{aligned}$$

Other integrations by partial fraction decompositions are illustrated in the following examples.

EXAMPLE 9.16

Evaluate the following indefinite integrals:

$$\text{(a)} \int \frac{x}{x^4 + 6x^2 + 5}\,dx \qquad \text{(b)} \int \frac{x^2 + 1}{2x^3 - 5x^2 + 4x - 1}\,dx$$

$$\text{(c)} \int \frac{1}{x^3 + 8}\,dx \qquad \text{(d)} \int \frac{x^2}{x^3 + 8}\,dx$$

SOLUTION (a) Since $x^4 + 6x^2 + 5 = (x^2 + 1)(x^2 + 5)$, the partial fraction decomposition of $x/(x^4 + 6x^2 + 5)$ has the form

$$\frac{x}{x^4 + 6x^2 + 5} = \frac{Ax + B}{x^2 + 1} + \frac{Cx + D}{x^2 + 5}.$$

When we bring the right side to a common denominator and equate numerators, the result is

$$x = (Ax + B)(x^2 + 5) + (Cx + D)(x^2 + 1).$$

We now multiply out the right side and equate coefficients of like powers of x,

$$\begin{aligned} x^3 &: & 0 &= A + C, \\ x^2 &: & 0 &= B + D, \\ x &: & 1 &= 5A + C, \\ x^0 &: & 0 &= 5B + D. \end{aligned}$$

Solutions of these equations are $A = 1/4$, $B = 0$, $C = -1/4$, and $D = 0$, and therefore

$$\begin{aligned} \int \frac{x}{x^4 + 6x^2 + 5}\,dx &= \int \left(\frac{x/4}{x^2 + 1} + \frac{-x/4}{x^2 + 5} \right) dx \\ &= \frac{1}{8} \ln(x^2 + 1) - \frac{1}{8} \ln(x^2 + 5) + C. \end{aligned}$$

(b) Since $2x^3 - 5x^2 + 4x - 1 = (2x - 1)(x - 1)^2$, the partial fraction decomposition takes the form

$$\begin{aligned} \frac{x^2 + 1}{2x^3 - 5x^2 + 4x - 1} &= \frac{A}{2x - 1} + \frac{B}{x - 1} + \frac{C}{(x - 1)^2} \\ &= \frac{A(x - 1)^2 + B(x - 1)(2x - 1) + C(2x - 1)}{(2x - 1)(x - 1)^2}. \end{aligned}$$

When we equate numerators,

$$x^2 + 1 = A(x - 1)^2 + B(x - 1)(2x - 1) + C(2x - 1).$$

We now set $x = 0$, $x = 1$, and $x = 1/2$:

$$\begin{aligned} x = 0 &: & 1 &= A + B - C, \\ x = 1 &: & 2 &= C(1), \\ x = 1/2 &: & 5/4 &= A(1/4). \end{aligned}$$

These give $A = 5$, $B = -2$, and $C = 2$, and therefore

$$\int \frac{x^2 + 1}{2x^3 - 5x^2 + 4x - 1}\,dx = \int \left(\frac{5}{2x - 1} - \frac{2}{x - 1} + \frac{2}{(x-1)^2}\right) dx$$
$$= \frac{5}{2}\ln|2x - 1| - 2\ln|x - 1| - \frac{2}{x - 1} + C.$$

(c) Since $x^3 + 8 = (x + 2)(x^2 - 2x + 4)$, we set

$$\frac{1}{x^3 + 8} = \frac{1}{(x + 2)(x^2 - 2x + 4)} = \frac{A}{x + 2} + \frac{Bx + C}{x^2 - 2x + 4}$$
$$= \frac{A(x^2 - 2x + 4) + (Bx + C)(x + 2)}{(x + 2)(x^2 - 2x + 4)},$$

and now equate numerators,

$$1 = A(x^2 - 2x + 4) + (Bx + C)(x + 2).$$

We set $x = -2$ and equate coefficients of x^2 and 1:

$$\begin{aligned} x = -2: &\qquad 1 = 12A, \\ x^2: &\qquad 0 = A + B, \\ 1: &\qquad 1 = 4A + 2C. \end{aligned}$$

These give $A = 1/12$, $B = -1/12$, and $C = 1/3$, and therefore

$$\int \frac{1}{x^3 + 8}\,dx = \int \left(\frac{1/12}{x + 2} + \frac{-x/12 + 1/3}{x^2 - 2x + 4}\right) dx$$
$$= \frac{1}{12}\ln|x + 2| + \frac{1}{12}\int \frac{4 - x}{(x - 1)^2 + 3}\,dx.$$

For the integral on the right, we set $x - 1 = \sqrt{3}\tan\theta$, in which case $dx = \sqrt{3}\sec^2\theta\,d\theta$, and

$$\int \frac{1}{x^3 + 8}\,dx = \frac{1}{12}\ln|x + 2| + \frac{1}{12}\int \frac{4 - (1 + \sqrt{3}\tan\theta)}{3\sec^2\theta}\sqrt{3}\sec^2\theta\,d\theta$$
$$= \frac{1}{12}\ln|x + 2| + \frac{1}{12\sqrt{3}}\int \left(3 - \sqrt{3}\tan\theta\right) d\theta$$
$$= \frac{1}{12}\ln|x + 2| + \frac{1}{12\sqrt{3}}\left(3\theta + \sqrt{3}\ln|\cos\theta|\right) + C$$
$$= \frac{1}{12}\ln|x + 2| + \frac{\sqrt{3}}{12}\mathrm{Tan}^{-1}\left(\frac{x - 1}{\sqrt{3}}\right) + \frac{1}{12}\ln\left|\frac{\sqrt{3}}{\sqrt{x^2 - 2x + 4}}\right| + C$$

(Figure 9.20)

$$= \frac{1}{12}\ln|x + 2| + \frac{\sqrt{3}}{12}\mathrm{Tan}^{-1}\left(\frac{x - 1}{\sqrt{3}}\right) - \frac{1}{24}\ln(x^2 - 2x + 4) + D,$$

where $D = C + (1/24)\ln 3$.

FIGURE 9.20

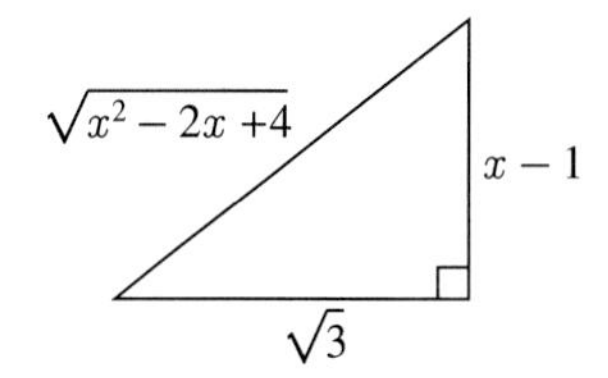

(d) Do not be misled into partial fractions in this example; the given rational function is immediately integrable,

$$\int \frac{x^2}{x^3+8}\,dx = \frac{1}{3}\ln|x^3+8| + C.$$

■

In Section 4.4 we discussed the rate at which an infectious disease might spread through a population of susceptible individuals. If N represents the number of individuals in the population and $x(t)$ the number infected at any given time t, it is often assumed that the rate of infection is proportional to the numbers both infected and not infected:

$$\frac{dx}{dt} = kx(N-x),$$

where k is a constant. Thus, $x(t)$ must satisfy this differential equation subject to an initial condition such as, perhaps, $x(0) = 1$, where therefore one infected individual is introduced into the population at time $t = 0$. To solve this differential equation, we write

$$\frac{1}{x(N-x)}\frac{dx}{dt} = k,$$

and determine the partial fraction decomposition for the left side, namely,

$$\left(\frac{1/N}{x} + \frac{1/N}{N-x}\right)\frac{dx}{dt} = k.$$

Integration of both sides of this equation with respect to time gives

$$\frac{1}{N}\left(\ln|x| - \ln|N-x|\right) = kt + C.$$

Since x and $N-x$ are both positive, we may drop the absolute value signs and write

$$\ln\left(\frac{x}{N-x}\right) = N(kt + C).$$

To solve this equation for an explicit definition of $x(t)$, we first exponentiate both sides of the equation,

$$\frac{x}{N-x} = De^{Ft},$$

where we have substituted $D = e^{NC}$ and $F = kN$. Multiplication by $N-x$ gives

$$x = De^{Ft}(N-x) \qquad \text{or}$$

$$x\left(1 + De^{Ft}\right) = NDe^{Ft}.$$

Thus,

$$x(t) = \frac{NDe^{Ft}}{1+De^{Ft}} = \frac{ND}{D+e^{-Ft}}.$$

Since $x(0) = 1$,

$$1 = \frac{ND}{1+D},$$

which implies that $D = 1/(N-1)$, and

$$x(t) = \frac{\dfrac{N}{N-1}}{\dfrac{1}{N-1} + e^{-Ft}} = \frac{N}{1+(N-1)e^{-Ft}}.$$

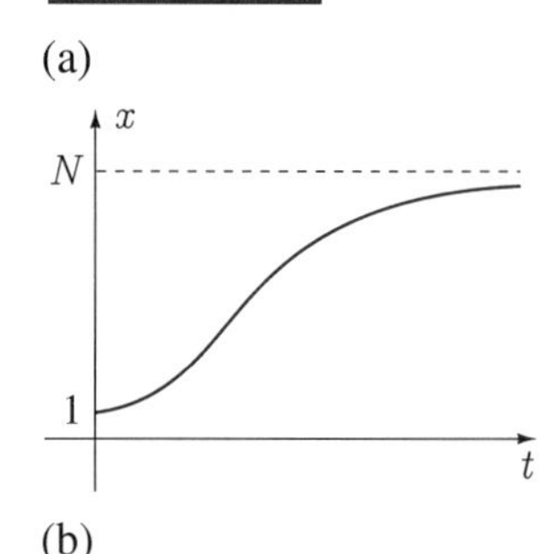

The shape of this curve is illustrated in Figure 9.21(a) (see also the discussion in Section 4.4). The time scale can be affixed to the diagram once one further piece of information is given. For instance, if it is known that one-half the population will be infected in 20 days, then $x(20) = N/2$, and the time scale is as shown in Figure 9.21(b). This added information determines F, since

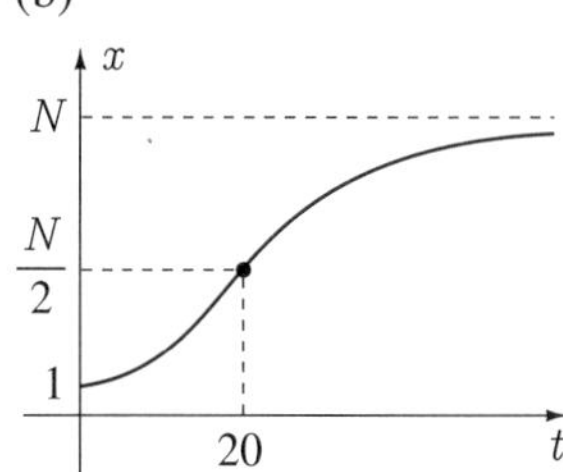

$$\frac{N}{2} = \frac{N}{1+(N-1)e^{-20F}},$$

and this equation gives

$$F = \frac{1}{20}\ln(N-1).$$

EXERCISES 9.7

In Exercises 1–16 evaluate the indefinite integral.

1. $\displaystyle\int \frac{x+2}{x^2-2x+1}\,dx$

2. $\displaystyle\int \frac{1}{y^3+3y^2+3y+1}\,dy$

3. $\displaystyle\int \frac{1}{z^3+z}\,dz$

4. $\displaystyle\int \frac{x^2+2x-4}{x^2-2x-8}\,dx$

5. $\displaystyle\int \frac{x}{(x-4)^2}\,dx$

6. $\displaystyle\int \frac{y+1}{y^3+y^2-6y}\,dy$

7. $\displaystyle\int \frac{3x+5}{x^3-x^2-x+1}\,dx$

8. $\displaystyle\int \frac{x^3}{(x^2+2)^2}\,dx$

9. $\displaystyle\int \frac{1}{x^2-3}\,dx$

10. $\displaystyle\int \frac{y^2}{y^2+3y+2}\,dy$

11. $\displaystyle\int \frac{z^2+3z-2}{z^3+5z}\,dz$

12. $\displaystyle\int \frac{y^2+6y+4}{y^4+5y^2+4}\,dy$

13. $\displaystyle\int \frac{x}{x^4+7x^2+6}\,dx$

14. $\displaystyle\int \frac{x^2+3}{x^4+x^2-2}\,dx$

15. $\displaystyle\int \frac{3t+4}{t^4-3t^3+3t^2-t}\,dt$

16. $\displaystyle\int \frac{x^3+6}{x^4+2x^3-3x^2-4x+4}\,dx$

17. Find the length of the curve $y = \ln(1-x^2)$ from $x=0$ to $x=1/2$.

18. Natural factors such as space and available food supply inhibit the growth of any population. Let N represent the population of a species and M the maximum size of a stable population for which the death rate is just equal to the birth rate. If the time rate of change of N is assumed proportional to both N and $M-N$, then N must satisfy the differential equation

$$\frac{dN}{dt} = kN(M-N).$$

If $N(0) = N_0 < M$, solve this equation for $N(t)$, and sketch its graph.

19. Find the centroid of the area bounded by the curves

$$(x+2)^2 y = 4-x, \qquad x=0, \qquad y=0, \qquad (x, y \geq 0).$$

20. Two substances A and B react to form a third substance C in such a way that 1 gm of A reacts with 1 gm of B to produce 2 gm of C. The rate at which C is formed is proportional to the product of the amounts of A and B present in the mixture.

(a) Show that if 10 gm of A and 15 gm of B are brought together at time $t=0$, the amount $x(t)$ of C present in the mixture must satisfy the differential equation

$$\frac{dx}{dt} = k(20-x)(30-x), \qquad x(0) = 0,$$

where k is a constant.

(b) Solve this equation for $x(t)$.

In Exercises 21–26 evaluate the indefinite integral.

21. $\displaystyle\int \frac{x^3+x+2}{x^5+2x^3+x}\,dx$

22. $\displaystyle\int \frac{1}{x^5 + x^4 + 2x^3 + 2x^2 + x + 1}\,dx$

23. $\displaystyle\int \frac{1}{(x^2 + 5)(x^2 + 2x + 3)}\,dx$

24. $\displaystyle\int \frac{1}{1 + x^3}\,dx$

25. $\displaystyle\int \frac{\sin x}{\cos x(1 + \cos^2 x)}\,dx$

26. $\displaystyle\int \frac{x^4 + 8x^3 - x^2 + 2x + 1}{x^5 + x^4 + x^2 + x}\,dx$

27. If an integrand is a rational function of $\sin x$ and $\cos x$, it can be reduced to a rational function of t by the substitution

$$t = \tan\left(\frac{x}{2}\right).$$

Show that with this substitution

$$dx = \frac{2}{1 + t^2}\,dt, \qquad \sin x = \frac{2t}{1 + t^2}, \qquad \cos x = \frac{1 - t^2}{1 + t^2}.$$

In Exercises 28–31 use the substitution of Exercise 27 to evaluate the integral.

28. $\displaystyle\int \sec x\,dx$

29. $\displaystyle\int \frac{1}{3 + 5\sin x}\,dx$

30. $\displaystyle\int \frac{1}{1 - 2\cos x}\,dx$

31. $\displaystyle\int \frac{1}{\sin x + \cos x}\,dx$

32. Show that the answer to Exercise 28 can be expressed in the usual form $\ln|\sec x + \tan x| + C$.

33. Evaluate

$$\int \frac{x^2 + x + 3}{x^4 + x^3 + 2x^2 + 11x - 5}\,dx.$$

SECTION 9.8

Using Tables and Reduction Formulas

For antiderivatives more difficult than we have considered in this chapter, it may be necessary to consult a set of integration tables; and for integrals which are amenable to our techniques, it may be more expeditious to use tables. Sets of integration tables can be found in *Standard Mathematical Tables* published by the Chemical Rubber Publishing Company, and *Handbook of Mathematical, Scientific, and Engineering Formulas, Tables, Functions, Graphs, Transforms* published by the Research and Education Association.

To illustrate the use of tables, consider the integral

$$\int \frac{1}{x^2(3 - 4x)^2}\,dx.$$

Certainly we could evaluate this integral by finding the partial fraction decomposition of the integrand. Alternatively, in tables we find the integration formula

$$\int \frac{1}{x^2(a + bx)^2}\,dx = -\frac{a + 2bx}{a^2x(a + bx)} + \frac{2b}{a^3}\ln\left(\frac{a + bx}{x}\right).$$

Hence, if we set $a = 3$ and $b = -4$ in this formula, we obtain

$$\int \frac{1}{x^2(3 - 4x)^2}\,dx = -\frac{3 - 8x}{9x(3 - 4x)} - \frac{8}{27}\ln\left(\frac{3 - 4x}{x}\right) + C.$$

Tables do not normally include a constant of integration. Remember to add it. There is an additional difficulty with tables; they do not always place absolute value signs around arguments of logarithm functions. The above answer should read

$$\int \frac{1}{x^2(3 - 4x)^2}\,dx = \frac{8x - 3}{9x(3 - 4x)} - \frac{8}{27}\ln\left|\frac{3 - 4x}{x}\right| + C$$

in order to accommodate definite integrals wherein $(3 - 4x)/x$ is negative.

EXAMPLE 9.17

Use tables to evaluate the integral

$$\int \frac{1}{x^2 + 2x + 3}\, dx.$$

SOLUTION Tables indicate that for an integral of the form

$$\int \frac{1}{X}\, dx$$

where $X = ax^2 + bx + c$ and $q = 4ac - b^2$, there are three possibilities:

$$\int \frac{1}{X}\, dx = \frac{2}{\sqrt{q}} \text{Tan}^{-1}\left(\frac{2ax + b}{\sqrt{q}}\right), \tag{9.29a}$$

$$\int \frac{1}{X}\, dx = \frac{-2}{\sqrt{-q}} \text{Tanh}^{-1}\left(\frac{2ax + b}{\sqrt{-q}}\right), \tag{9.29b}$$

$$\int \frac{1}{X}\, dx = \frac{1}{\sqrt{-q}} \ln\left(\frac{2ax + b - \sqrt{-q}}{2ax + b + \sqrt{-q}}\right). \tag{9.29c}$$

No explanation is given as to when each should be used; it is up to the reader to make this decision. Certainly, we would use 9.29a when $q > 0$, and 9.29b,c when $q < 0$ in order to keep $\sqrt{q}$ and $\sqrt{-q}$ real. In our example, $q = 4(1)(3) - (2)^2 = 8$, and we therefore use 9.29a,

$$\int \frac{1}{x^2 + 2x + 3}\, dx = \frac{2}{\sqrt{8}} \text{Tan}^{-1}\left(\frac{2(1)x + 2}{\sqrt{8}}\right) + C = \frac{1}{\sqrt{2}} \text{Tan}^{-1}\left(\frac{x + 1}{\sqrt{2}}\right) + C.$$

■

EXAMPLE 9.18

Use tables to evaluate the integral

$$\int \frac{1}{x\sqrt{3 - 2x - 2x^2}}\, dx.$$

SOLUTION It should be noted that none of our integration techniques seem particularly suited to this integral. Tables list the formulas

$$\int \frac{1}{x\sqrt{X}}\, dx = -\frac{1}{\sqrt{c}} \ln\left(\frac{\sqrt{X} + \sqrt{c}}{x} + \frac{b}{2\sqrt{c}}\right), \quad \text{and} \tag{9.30a}$$

$$\int \frac{1}{x\sqrt{X}}\, dx = \frac{1}{\sqrt{-c}} \text{Sin}^{-1}\left(\frac{bx + 2c}{x\sqrt{b^2 - 4ac}}\right), \tag{9.30b}$$

for $X = ax^2 + bx + c$. Since $c = 3$ in our example, we use 9.30a,

$$\int \frac{1}{x\sqrt{3 - 2x - 2x^2}}\, dx = -\frac{1}{\sqrt{3}} \ln\left(\frac{\sqrt{3 - 2x - 2x^2} + \sqrt{3}}{x} + \frac{-2}{2\sqrt{3}}\right) + C.$$

To check that 9.30a is the correct formula, and that we have used it correctly, we should calculate the derivative of this function:

$$\frac{d}{dx}\left[\frac{-1}{\sqrt{3}}\ln\left(\frac{\sqrt{3-2x-2x^2}+\sqrt{3}}{x}-\frac{1}{\sqrt{3}}\right)\right] = \frac{-1/\sqrt{3}}{\dfrac{\sqrt{3-2x-2x^2}+\sqrt{3}}{x}-\dfrac{1}{\sqrt{3}}}$$

$$\left[\frac{x(1/2)(3-2x-2x^2)^{-1/2}(-2-4x)-\left(\sqrt{3-2x-2x^2}+\sqrt{3}\right)(1)}{x^2}\right]$$

$$= \frac{-1}{\sqrt{3}}\frac{\sqrt{3}x}{\sqrt{3}\sqrt{3-2x-2x^2}+3-x}$$
$$\left[\frac{-x(1+2x)-(3-2x-2x^2)-\sqrt{3}\sqrt{3-2x-2x^2}}{x^2\sqrt{3-2x-2x^2}}\right]$$

$$= \frac{-x}{\sqrt{3}\sqrt{3-2x-2x^2}+3-x}\left[\frac{-3+x-\sqrt{3}\sqrt{3-2x-2x^2}}{x^2\sqrt{3-2x-2x^2}}\right]$$

$$= \frac{1}{x\sqrt{3-2x-2x^2}}.$$

Once again, until we can be sure that the argument of the logarithm function is positive, it would be wise to insert absolute value signs,

$$\int \frac{1}{x\sqrt{3-2x-2x^2}}\,dx = -\frac{1}{\sqrt{3}}\ln\left|\frac{\sqrt{3-2x-2x^2}+\sqrt{3}}{x}-\frac{1}{\sqrt{3}}\right| + C.$$

■

Integration tables contain a large number of entries called **reduction formulas**. To introduce these, consider the integral

$$\int x^3 e^{2x}\,dx.$$

We could use integration by parts three times to eliminate the x^3 before e^{2x}. A better procedure is to use integration by parts on

$$\int x^n e^{2x}\,dx$$

where n represents a positive integer. With

$$u = x^n, \qquad dv = e^{2x}\,dx, \qquad du = nx^{n-1}\,dx, \qquad v = \frac{1}{2}e^{2x},$$

integration by parts formula 9.20 gives

$$\begin{aligned}\int x^n e^{2x}\,dx &= \frac{1}{2}x^n e^{2x} - \int \frac{n}{2}x^{n-1}e^{2x}\,dx \\ &= \frac{1}{2}x^n e^{2x} - \frac{n}{2}\int x^{n-1}e^{2x}\,dx. \qquad (9.31)\end{aligned}$$

We now use this formula with $n = 3$ on the given integral,

$$\int x^3 e^{2x}\,dx = \frac{1}{2}x^3 e^{2x} - \frac{3}{2}\int x^2 e^{2x}\,dx.$$

But formula 9.31 can be used again with $n = 2$ to simplify the integral on the right,

$$\begin{aligned}\int x^3 e^{2x}\,dx &= \frac{1}{2}x^3 e^{2x} - \frac{3}{2}\left(\frac{1}{2}x^2 e^{2x} - \frac{2}{2}\int xe^{2x}\,dx\right)\\ &= \frac{1}{2}x^3 e^{2x} - \frac{3}{4}x^2 e^{2x} - \frac{3}{2}\int xe^{2x}\,dx.\end{aligned}$$

One more application of 9.31 with $n = 1$ rids the integrand of the pesky factor in front of the exponential,

$$\begin{aligned}\int x^3 e^{2x}\,dx &= \frac{1}{2}x^3 e^{2x} - \frac{3}{4}x^2 e^{2x} - \frac{3}{2}\left(\frac{1}{2}xe^{2x} - \frac{1}{2}\int e^{2x}\,dx\right)\\ &= \frac{1}{2}x^3 e^{2x} - \frac{3}{4}x^2 e^{2x} - \frac{3}{4}xe^{2x} + \frac{3}{8}e^{2x} + C.\end{aligned}$$

Equation 9.31 is an example of a reduction formula. By using it on the antiderivative of $x^n e^{2x}$, we reduce the power on x from n to $n - 1$. Successive reductions eventually eliminate this term.

Many reduction formulas can be found in tables. They have the same purpose as 9.31, to reduce, and perhaps eliminate, powers of certain factors in an integrand.

EXAMPLE 9.19

Use a reduction formula to evaluate

$$\int \cos^6 x\,dx.$$

SOLUTION The following reduction formula can be found in tables,

$$\int \cos^n x\,dx = \frac{1}{n}\cos^{n-1} x \sin x + \frac{n-1}{n}\int \cos^{n-2} x\,dx. \qquad (9.32)$$

It reduces the power on $\cos x$ by two each time it is used. We begin with $n = 6$,

$$\int \cos^6 x\,dx = \frac{1}{6}\cos^5 x \sin x + \frac{5}{6}\int \cos^4 x\,dx.$$

We now set $n = 4$ in 9.32,

$$\int \cos^6 x\,dx = \frac{1}{6}\cos^5 x \sin x + \frac{5}{6}\left(\frac{1}{4}\cos^3 x \sin x + \frac{3}{4}\int \cos^2 x\,dx\right).$$

Finally, with $n = 2$ we obtain

$$\begin{aligned}\int \cos^6 x\,dx &= \frac{1}{6}\cos^5 x \sin x + \frac{5}{24}\cos^3 x \sin x + \frac{5}{8}\left(\frac{1}{2}\cos x \sin x + \frac{1}{2}\int dx\right)\\ &= \frac{1}{6}\cos^5 x \sin x + \frac{5}{24}\cos^3 x \sin x + \frac{5}{16}\cos x \sin x + \frac{5x}{16} + C.\end{aligned}$$

■

Sometimes it is necessary to use more than one entry from tables on an integral, as the following example illustrates.

EXAMPLE 9.20 Evaluate

$$\int \frac{\sqrt{6x - x^2}}{x}\, dx.$$

SOLUTION We use the reduction formula

$$\int \frac{\sqrt{2ax - x^2}}{x^n}\, dx = \frac{(2ax - x^2)^{3/2}}{(3 - 2n)ax^n} + \frac{n-3}{(2n-3)a} \int \frac{\sqrt{2ax - x^2}}{x^{n-1}}\, dx.$$

If we set $n = 1$ and $a = 3$ in this identity,

$$\begin{aligned}\int \frac{\sqrt{6x - x^2}}{x}\, dx &= \frac{(6x - x^2)^{3/2}}{(3-2)(3)x} + \frac{1-3}{(2-3)(3)} \int \sqrt{6x - x^2}\, dx \\ &= \frac{(6x - x^2)^{3/2}}{3x} + \frac{2}{3} \int \sqrt{6x - x^2}\, dx.\end{aligned}$$

We can now use the formula

$$\int \sqrt{2ax - x^2}\, dx = \frac{1}{2}\left[(x - a)\sqrt{2ax - x^2} + a^2 \operatorname{Sin}^{-1}\left(\frac{x - a}{|a|}\right)\right]$$

to finish the integration,

$$\begin{aligned}\int \frac{\sqrt{6x-x^2}}{x}\, dx &= \frac{(6x - x^2)^{3/2}}{3x} + \frac{2}{3}\frac{1}{2}\left[(x-3)\sqrt{6x - x^2} + 9 \operatorname{Sin}^{-1}\left(\frac{x-3}{3}\right)\right] + C \\ &= \frac{(6x - x^2)^{3/2}}{3x} + \frac{1}{3}(x-3)\sqrt{6x - x^2} + 3\operatorname{Sin}^{-1}\left(\frac{x-3}{3}\right) + C \\ &= \sqrt{6x - x^2} + 3\operatorname{Sin}^{-1}\left(\frac{x-3}{3}\right) + C.\end{aligned}$$

■

Tables are an excellent companion to partial fractions. Partial fractions decompose complicated rational functions into simpler ones which can often be found in tables. This is illustrated in the following example.

EXAMPLE 9.21 Evaluate

$$\int \frac{3x^4 - x^3 + 18x^2 - 5x + 17}{x^5 - x^4 + 6x^3 - 6x^2 + 9x - 9}\, dx.$$

SOLUTION Since $x^5 - x^4 + 6x^3 - 6x^2 + 9x - 9 = (x - 1)(x^2 + 3)^2$, the partial fraction decomposition of the integrand is

$$\frac{3x^4 - x^3 + 18x^2 - 5x + 17}{(x-1)(x^2+3)^2} = \frac{A}{x-1} + \frac{Bx + C}{x^2 + 3} + \frac{Dx + E}{(x^2+3)^2}.$$

We find $A = 2$, $B = 1$, $C = 0$, $D = 3$, and $E = 1$, so that

$$\int \frac{3x^4 - x^3 + 18x^2 - 5x + 17}{x^5 - x^4 + 6x^3 - 6x^2 + 9x - 9}\,dx = \int \left(\frac{2}{x-1} + \frac{x}{x^2+3} + \frac{3x+1}{(x^2+3)^2} \right) dx$$
$$= 2\ln|x-1| + \frac{1}{2}\ln(x^2+3) - \frac{3}{2(x^2+3)} + \int \frac{1}{(x^2+3)^2}\,dx.$$

Using the reduction formula

$$\int \frac{1}{(a^2+x^2)^n}\,dx = \frac{x}{2a^2(n-1)(a^2+x^2)^{n-1}} + \frac{2n-3}{2a^2(n-1)} \int \frac{1}{(a^2+x^2)^{n-1}}\,dx, \quad n \geq 2,$$

(with n set equal to 2), we obtain

$$\int \frac{1}{(x^2+3)^2}\,dx = \frac{x}{2(3)(1)(3+x^2)} + \frac{1}{2(3)(1)} \int \frac{1}{3+x^2}\,dx.$$

With the integration formula

$$\int \frac{1}{a^2+x^2}\,dx = \frac{1}{a}\mathrm{Tan}^{-1}\left(\frac{x}{a}\right),$$

we can finish the problem,

$$\int \frac{3x^4 - x^3 + 18x^2 - 5x + 17}{x^5 - x^4 + 6x^3 - 6x^2 + 9x - 9}\,dx = 2\ln|x-1| + \frac{1}{2}\ln(x^2+3) - \frac{3}{2(x^2+3)}$$
$$+ \frac{x}{6(x^2+3)} + \frac{1}{6}\left[\frac{1}{\sqrt{3}}\mathrm{Tan}^{-1}\left(\frac{x}{\sqrt{3}}\right)\right] + C$$
$$= 2\ln|x-1| + \frac{1}{2}\ln(x^2+3) + \frac{x-9}{6(x^2+3)} + \frac{1}{6\sqrt{3}}\mathrm{Tan}^{-1}\left(\frac{x}{\sqrt{3}}\right) + C.$$

■

EXERCISES 9.8

In Exercises 1–20 use tables to evaluate the indefinite integral.

1. $\displaystyle\int \frac{1}{x^2+9}\,dx$

2. $\displaystyle\int \frac{1}{x^2-16}\,dx$

3. $\displaystyle\int \frac{1}{4-3x^2}\,dx$

4. $\displaystyle\int \frac{1}{x\sqrt{x^2+5}}\,dx$

5. $\displaystyle\int \frac{x}{\sqrt{3x-2}}\,dx$

6. $\displaystyle\int \frac{\sqrt{2x+3}}{x}\,dx$

7. $\displaystyle\int \frac{1}{\sqrt{x^2+4x}}\,dx$

8. $\displaystyle\int \frac{1}{\sqrt{4x-9x^2}}\,dx$

9. $\displaystyle\int \sqrt{3+2x+4x^2}\,dx$

10. $\displaystyle\int \sqrt{3+2x-4x^2}\,dx$

11. $\displaystyle\int \frac{1}{3+\sin x}\,dx$

12. $\displaystyle\int \frac{1}{1+3\sin x}\,dx$

13. $\displaystyle\int x\mathrm{Sin}^{-1}x\,dx$

14. $\displaystyle\int \frac{1}{x^2}\mathrm{Sin}^{-1}x\,dx$

15. $\displaystyle\int e^{4x}\cos 2x\,dx$

16. $\displaystyle\int e^{-2x}\sin 3x\,dx$

17. $\displaystyle\int x^7 \ln x\,dx$

18. $\displaystyle\int (\ln x)^2\,dx$

19. $\displaystyle\int \frac{1}{e^x + 2e^{-2x}}\,dx$ 20. $\displaystyle\int \frac{1}{2 + x^3}\,dx$

In Exercises 21–30 use a reduction formula to aid in the evaluation of the integral.

21. $\displaystyle\int x^4 e^{3x}\,dx$ 22. $\displaystyle\int x^3 \cos 2x\,dx$

23. $\displaystyle\int \frac{x}{\sqrt{9 + x + x^2}}\,dx$ 24. $\displaystyle\int \frac{x}{\sqrt{9 + x - x^2}}\,dx$

25. $\displaystyle\int (4 + x^2)^{3/2}\,dx$ 26. $\displaystyle\int \frac{x^2}{\sqrt{2x - x^2}}\,dx$

27. $\displaystyle\int \sin^6 x\,dx$ 28. $\displaystyle\int \frac{1}{\cos^6 x}\,dx$

29. $\displaystyle\int \cot^5 x\,dx$ 30. $\displaystyle\int x^3\,\mathrm{Sin}^{-1}\,x\,dx$

31. Evaluate

$$\int \sec^9 x\,dx$$

by first developing the reduction formula

$$\int \sec^n x\,dx = \frac{\tan x \sec^{n-2} x}{n-1} + \frac{n-2}{n-1}\int \sec^{n-2} x\,dx.$$

32. (a) Use a trigonometric substitution to evaluate

$$\int \frac{5x + 2}{(3 + x^2)^4}\,dx.$$

(b) Verify the reduction formula in Example 9.21 and use it to evaluate the integral in (a).

33. If n is a positive integer, define

$$I_n = \int_{-1}^{1} (1 - x^2)^n\,dx.$$

(a) Show that

$$(2n + 1) I_n = 2nI_{n-1}, \quad n \geq 2.$$

(b) Prove that

$$I_n = \frac{2^{2n+1}(n!)^2}{(2n + 1)!}, \quad n \geq 1.$$

(c) Evaluate

$$\int_0^{\pi/2} \cos^{2n+1} x\,dx.$$

SECTION 9.9

Numerical Integration

To evaluate the definite integral of a continuous function $f(x)$ with respect to x from $x = a$ to $x = b$, we have used the fundamental theorem of integral calculus: Find an indefinite integral for $f(x)$, substitute the limits $x = b$ and $x = a$, and subtract. The evaluation procedure depends on our ability to produce an indefinite integral for $f(x)$. When a function $f(x)$ is complicated, it may be difficult, or even impossible to find its antiderivative. In such a case, it may be necessary to approximate the definite integral of $f(x)$ on some interval $a \leq x \leq b$, rather than evaluate it analytically. We discuss three methods for doing this, the rectangular, trapezoidal, and Simpson's rules. Each method divides the interval $a \leq x \leq b$ into subintervals and approximates $f(x)$ with an easily integrated function on each subinterval.

FIGURE 9.22

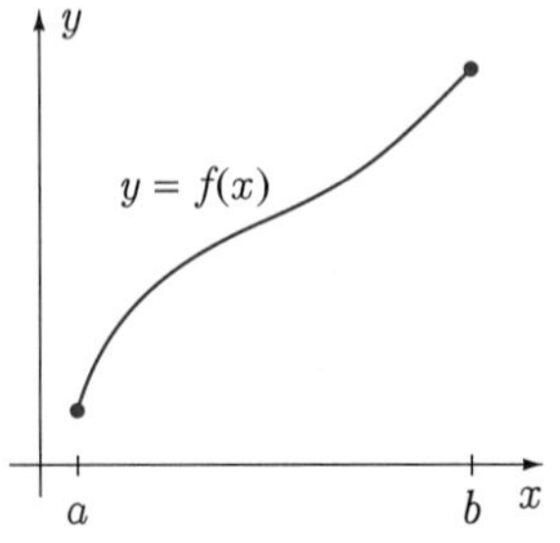

If $f(x) \geq 0$ on $a \leq x \leq b$ (Figure 9.22), the definite integral of $f(x)$ with respect to x can always be interpreted as the area bounded by $y = f(x)$, $y = 0$, $x = a$, and $x = b$. We have carefully pointed out that it is not always wise to think of a definite integral as area, but for our discussion here it is convenient to do so. Our problem is to approximate the area in Figure 9.22 when it is difficult or impossible to find an antiderivative for $f(x)$ because of the complexity of $f(x)$.

Rectangular Rule

The first method is essentially to return to equation 6.11 which defines the definite integral. This means that we subdivide the interval $a \leq x \leq b$ into n subintervals by points $a = x_0 < x_1 < \cdots < x_{n-1} < x_n = b$, and pick a point x_i^* in each subinterval. For simplicity, we choose n equal subdivisions, in which case

$$x_i = a + i\left(\frac{b-a}{n}\right),$$

and select $x_i^* = x_i$. If we denote by h the length of each subdivision, then $h = \Delta x_i = (b-a)/n$, and equation 6.11 reduces to

$$\int_a^b f(x)\,dx = \lim_{\|\Delta x_i\| \to 0} \sum_{i=1}^{n} f(x_i^*)\Delta x_i = \lim_{n \to \infty} h \sum_{i=1}^{n} f(x_i).$$

If we delete the limit from this equation, the summation becomes an approximation to the definite integral, and we write

$$\int_a^b f(x)\,dx \approx h \sum_{i=1}^{n} f(x_i). \tag{9.33}$$

This is the **rectangular rule** for approximating the definite integral; we have replaced the area under $y = f(x)$ with n rectangles (Figure 9.23). In other words, we have replaced the original function $f(x)$ by a function that is constant on each subinterval (Figure 9.24), but the constant value may vary from subinterval to subinterval. Such a function is called a *step function*.

Graphically, it is reasonable to expect that an increase in the number of subdivisions results in an increase in the accuracy of the approximation.

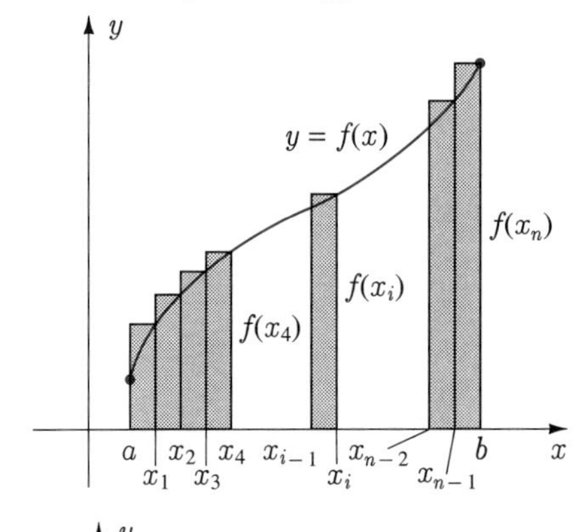

FIGURE 9.23

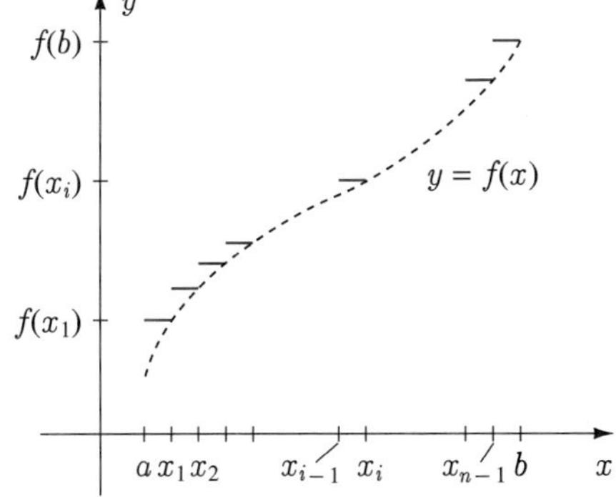

FIGURE 9.24

EXAMPLE 9.22 Use a subdivision of the interval $1 \le x \le 3$ into 5, 10, and 20 equal parts to approximate the definite integral

$$\int_1^3 \sin x\,dx$$

with the rectangular rule.

SOLUTION With 5 equal parts, $h = 2/5$ and $x_i = 1 + 2i/5$. Consequently,

$$\begin{aligned}\int_1^3 \sin x\,dx &\approx \frac{2}{5}\sum_{i=1}^{5}\sin\left(1+\frac{2i}{5}\right)\\ &= \frac{2}{5}\left(\sin\frac{7}{5}+\sin\frac{9}{5}+\sin\frac{11}{5}+\sin\frac{13}{5}+\sin 3\right)\\ &= 1.370.\end{aligned}$$

With 10 equal parts, $h = 2/10 = 1/5$ and $x_i = 1 + i/5$ so that

$$\begin{aligned}\int_1^3 \sin x\,dx &\approx \frac{1}{5}\sum_{i=1}^{10}\sin\left(1+\frac{i}{5}\right)\\ &= \frac{1}{5}\left(\sin\frac{6}{5}+\sin\frac{7}{5}+\sin\frac{8}{5}+\sin\frac{9}{5}+\sin 2+\sin\frac{11}{5}\right.\\ &\qquad\left.+\sin\frac{12}{5}+\sin\frac{13}{5}+\sin\frac{14}{5}+\sin 3\right)\\ &= 1.455.\end{aligned}$$

With 20 equal parts we have

$$\begin{aligned}\int_1^3 \sin x\,dx &\approx \frac{1}{10}\sum_{i=1}^{20}\sin\left(1+\frac{i}{10}\right)\\ &= \frac{1}{10}\left(\sin\frac{11}{10}+\sin\frac{6}{5}+\sin\frac{13}{10}+\cdots+\sin\frac{29}{10}+\sin 3\right)\\ &= 1.494.\end{aligned}$$

■

When we compare these results with the correct answer

$$\int_1^3 \sin x\,dx = \Big\{-\cos x\Big\}_1^3 = -\cos 3 + \cos 1 = 1.530\,294\,8,$$

we see that as n increases from 5 to 10 to 20, the approximation improves, but it must be increased even further to give a reasonable approximation to this integral.

Trapezoidal Rule

Regarding area, it is clear in Figure 9.23 that were we to join successive points $(x_i, f(x_i))$ on the curve $y = f(x)$ with straight line segments as in Figure 9.25(a), the area under this broken straight line would be a much better approximation to the area under $y = f(x)$ than that provided by the rectangular rule. Effectively, we now approximate the area by n trapezoids.

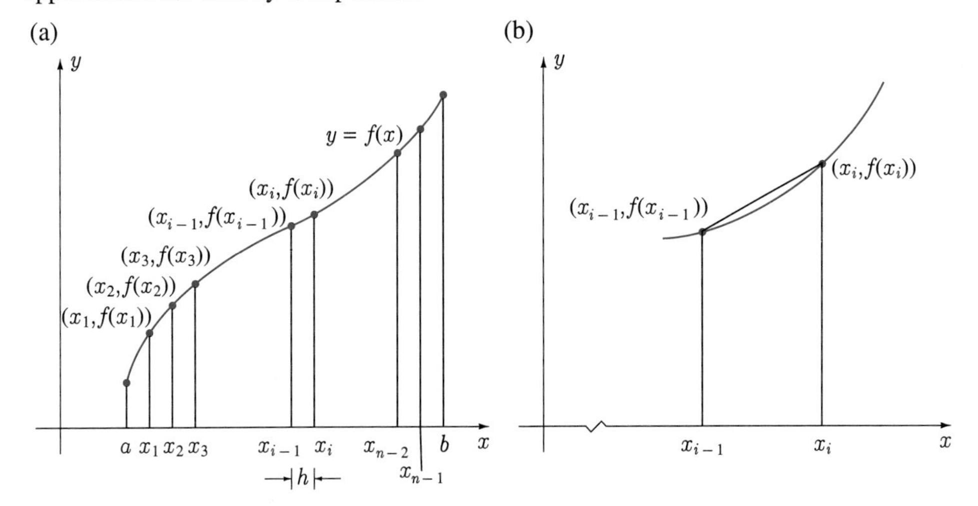

FIGURE 9.25

Since the area of a trapezoid is its width multiplied by the average of its parallel lengths, it follows that the area of the ith trapezoid in Figure 9.25(b) is given by

$$h\left(\frac{f(x_i) + f(x_{i-1})}{2}\right),$$

where again $h = x_i - x_{i-1}$. As a result, the area under $y = f(x)$ can be approximated by the sum

$$\begin{aligned} &h\left[\frac{f(x_1) + f(x_0)}{2} + \frac{f(x_2) + f(x_1)}{2} + \cdots + \frac{f(x_n) + f(x_{n-1})}{2}\right] \\ &= \frac{h}{2}\left[f(a) + 2f(x_1) + 2f(x_2) + \cdots + 2f(x_i) + \cdots + 2f(x_{n-1}) + f(b)\right] \\ &= \frac{h}{2}\left[f(a) + 2\sum_{i=1}^{n-1} f(x_i) + f(b)\right]. \end{aligned}$$

We write therefore that

$$\int_a^b f(x)\,dx \approx \frac{h}{2}\left[f(a) + 2\sum_{i=1}^{n-1} f(x_i) + f(b)\right], \qquad (9.34\text{a})$$

where $x_i = a + ih = a + i(b - a)/n$, and call this the **trapezoidal rule** for approximating a definite integral. Note that if 9.34a is written in the form

$$\int_a^b f(x)\,dx \approx h\left[\frac{f(a) - f(b)}{2}\right] + h\sum_{i=1}^{n} f(x_i), \qquad (9.34\text{b})$$

the summation on the right, except for the first two terms, is the rectangular rule. In other words, the extra numerical calculation involved in using the trapezoidal rule rather than the rectangular rule is minimal, but it would appear that the accuracy is significantly increased. For this reason, the trapezoidal rule supplants the rectangular rule in most applications.

EXAMPLE 9.23

Use the trapezoidal rule to approximate the definite integral in Example 9.22.

SOLUTION With 5 equal partitions,

$$\int_1^3 \sin x \, dx \approx \frac{(2/5)}{2} \left[\sin 1 + 2 \sum_{i=1}^{4} \sin\left(1 + \frac{2i}{5}\right) + \sin 3 \right] = 1.5098 .$$

With 10 equal parts,

$$\int_1^3 \sin x \, dx \approx \frac{(1/5)}{2} \left[\sin 1 + 2 \sum_{i=1}^{9} \sin\left(1 + \frac{i}{5}\right) + \sin 3 \right] = 1.5252 .$$

Finally, with $n = 20$, we have

$$\int_1^3 \sin x \, dx \approx \frac{(1/10)}{2} \left[\sin 1 + 2 \sum_{i=1}^{19} \sin\left(1 + \frac{i}{10}\right) + \sin 3 \right] = 1.5290 .$$

As expected, these approximations are significantly better than corresponding results using the rectangular rule. ■

Simpson's Rule

The rectangular rule replaces a function $f(x)$ with a step function; the trapezoidal rule replaces $f(x)$ with a succession of linear functions — geometrically, a broken straight line. So far as ease of integration is concerned, the next simplest function is a quadratic function. Consider, then, replacing the curve $y = f(x)$ by a succession of parabolas on the subintervals $x_{i-1} \leq x \leq x_i$. Now the equation of a parabola with a vertical axis of symmetry is of the form $y = ax^2 + bx + c$ with three constants a, b, and c to be determined. If this parabola is to approximate $y = f(x)$ on $x_{i-1} \leq x \leq x_i$, we should have the parabola pass through the end points $(x_{i-1}, f(x_{i-1}))$ and$(x_i, f(x_i))$. But this imposes only two conditions on a, b, and c, not three. To take advantage of this flexibility, we demand that the parabola also pass through the point $(x_{i+1}, f(x_{i+1}))$ (Figure 9.26).

FIGURE 9.26

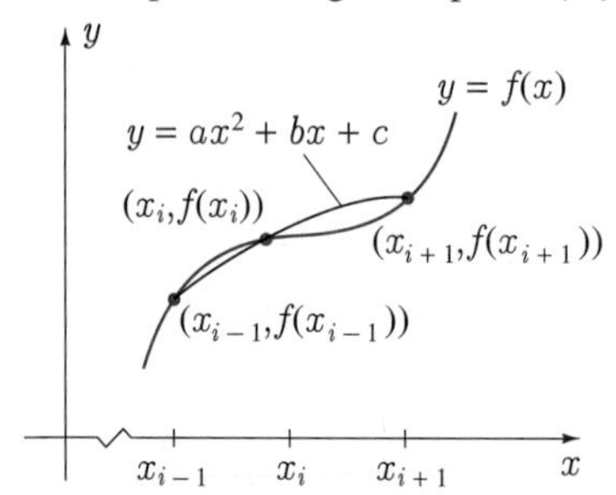

In other words, instead of replacing $y = f(x)$ with n parabolas, one on each subinterval $x_{i-1} \leq x \leq x_i$, we replace it with $n/2$ parabolas, one on each pair of subintervals $x_{i-1} \leq x \leq x_{i+1}$. Note that this requires n to be an even integer. These three conditions imply that a, b, and c must satisfy the equations

$$ax_{i-1}^2 + bx_{i-1} + c = f(x_{i-1}), \tag{9.35a}$$

$$ax_i^2 + bx_i + c = f(x_i), \tag{9.35b}$$

$$ax_{i+1}^2 + bx_{i+1} + c = f(x_{i+1}). \tag{9.35c}$$

These equations determine the values for a, b, and c, but it will not be necessary to actually solve them. Suppose for the moment that we have solved equations 9.35 for a, b, and c, and we continue our main discussion. With the parabola $y = ax^2 + bx + c$ replacing the curve $y = f(x)$ on the interval $x_{i-1} \le x \le x_{i+1}$, we approximate the area under $y = f(x)$ with that under the parabola, namely,

$$\begin{aligned}\int_{x_{i-1}}^{x_{i+1}} (ax^2 + bx + c)\,dx &= \frac{a}{3}(x_{i+1}^3 - x_{i-1}^3) + \frac{b}{2}(x_{i+1}^2 - x_{i-1}^2) + c(x_{i+1} - x_{i-1}) \\ &= \frac{x_{i+1} - x_{i-1}}{6}[2a(x_{i+1}^2 + x_{i+1}x_{i-1} + x_{i-1}^2) \\ &\quad + 3b(x_{i+1} + x_{i-1}) + 6c].\end{aligned}$$

Now if we use $x_{i+1} - x_{i-1} = 2h$, $x_{i+1} = x_i + h$, and $x_{i-1} = x_i - h$ to write everything in terms of h and x_i, we obtain

$$\begin{aligned}\int_{x_{i-1}}^{x_{i+1}} (ax^2 + bx + c)\,dx &= \frac{2h}{6}\Big\{2a\left[(x_i + h)^2 + (x_i + h)(x_i - h) + (x_i - h)^2\right] \\ &\quad + 3b[(x_i + h) + (x_i - h)] + 6c\Big\} \\ &= \frac{h}{3}\left\{a(6x_i^2 + 2h^2) + 6bx_i + 6c\right\}.\end{aligned}$$

But if we add equation 9.35a, equation 9.35c, and four times equation 9.35b, we find that

$$\begin{aligned}f(x_{i-1}) + 4f(x_i) + f(x_{i+1}) &= a(x_{i-1}^2 + 4x_i^2 + x_{i+1}^2) + b(x_{i-1} + 4x_i + x_{i+1}) + 6c \\ &= a\left[(x_i - h)^2 + 4x_i^2 + (x_i + h)^2\right] \\ &\quad + b[(x_i - h) + 4x_i + (x_i + h)] + 6c \\ &= a(6x_i^2 + 2h^2) + 6bx_i + 6c.\end{aligned}$$

Thus the area under the parabola may be written in the form

$$\int_{x_{i-1}}^{x_{i+1}} (ax^2 + bx + c)\,dx = \frac{h}{3}[f(x_{i-1}) + 4f(x_i) + f(x_{i+1})],$$

and the right-hand side is free of a, b, and c. The same expression would have resulted had we solved equations 9.35 for a, b, and c and then evaluated the integral of ax^2+bx+c from x_{i-1} to x_{i+1}. The above derivation, however, is much simpler.

When we add all such integrals over the $n/2$ subintervals $x_{i-1} \le x \le x_{i+1}$ between $x = a$ and $x = b$, we obtain

$$\begin{aligned}&\frac{h}{3}[f(x_0) + 4f(x_1) + f(x_2)] + \frac{h}{3}[f(x_2) + 4f(x_3) + f(x_4)] + \cdots \\ &\qquad + \frac{h}{3}[f(x_{n-2}) + 4f(x_{n-1}) + f(x_n)] \\ &= \frac{h}{3}[f(x_0) + 4f(x_1) + 2f(x_2) + 4f(x_3) + \cdots + 2f(x_{n-2}) + 4f(x_{n-1}) + f(x_n)].\end{aligned}$$

In other words, the definite integral of $f(x)$ from $x = a$ to $x = b$ can be approximated by

$$\begin{aligned}\int_a^b f(x)\,dx \approx \frac{h}{3}\Big[f(a) + 4f(x_1) + 2f(x_2) + 4f(x_3) + \cdots \\ + 2f(x_{n-2}) + 4f(x_{n-1}) + f(b)\Big],\end{aligned} \tag{9.36}$$

where $x_i = a + ih = a + i(b-a)/n$. This result is called **Simpson's rule** for approximating a definite integral. Although the formula does not explicitly display it (except by counting terms), *do not forget that n must be an even integer.*

EXAMPLE 9.24

Approximate the definite integral in Example 9.22 using Simpson's rule with $n = 10$ and $n = 20$.

SOLUTION With $n = 10$,

$$\int_1^3 \sin x\, dx \approx \frac{1/5}{3}\left[\sin 1 + 4\sin\frac{6}{5} + 2\sin\frac{7}{5} + 4\sin\frac{8}{5} + \cdots + 2\sin\frac{13}{5} + 4\sin\frac{14}{5} + \sin 3\right] = 1.530\,308\,5.$$

With $n = 20$,

$$\int_1^3 \sin x\, dx \approx \frac{1/10}{3}\left[\sin 1 + 4\sin\frac{11}{10} + 2\sin\frac{6}{5} + 4\sin\frac{13}{10} + \cdots + 2\sin\frac{14}{5} + 4\sin\frac{29}{10} + \sin 3\right] = 1.530\,295\,7.$$

■

Table 9.3 lists the approximations in Examples 9.22–24. The correct answer for the integral is 1.530 294 8 (to 7 decimals). It is clear that each method gives a better approximation as the value of n increases, and that Simpson's rule is by far the most accurate.

TABLE 9.3

	Rectangular rule	Trapezoidal rule	Simpson's rule
$n = 5$	1.370	1.5098	
$n = 10$	1.455	1.5252	1.530 308 5
$n = 20$	1.494	1.5290	1.530 295 7

In practice, we use the rectangular, trapezoidal, and Simpson's rules to approximate definite integrals that cannot be handled analytically, and we will not therefore have the correct answer with which to compare the approximation. We would still like to make some statement about the accuracy of the approximation, however, since what good is the approximation otherwise? The following two theorems give error estimates for the trapezoidal rule and for Simpson's rule.

Theorem 9.1

If $f''(x)$ exists on $a \le x \le b$, and T_n is the error in approximating the definite integral of $f(x)$ from $x = a$ to $x = b$ using the trapezoidal rule with n equal subdivisions,

$$T_n = \int_a^b f(x)\,dx - \frac{h}{2}\left[f(a) + 2\sum_{i=1}^{n-1} f(x_i) + f(b)\right],$$

then

$$|T_n| \le \frac{M(b-a)^3}{12n^2}, \tag{9.37}$$

where M is the maximum value of $|f''(x)|$ on $a \le x \le b$.

Theorem 9.2

If $f''''(x)$ exists on $a \leq x \leq b$, and S_n is the error in approximating the definite integral of $f(x)$ from $x = a$ to $x = b$ using Simpson's rule with n equal subdivisions,

$$S_n = \int_a^b f(x)\,dx - \frac{h}{3}\left[f(a) + 4\sum_{i=1}^{n/2} f(x_{2i-1}) + 2\sum_{i=1}^{n/2-1} f(x_{2i}) + f(b)\right],$$

then

$$|S_n| \leq \frac{M(b-a)^5}{180\,n^4}, \tag{9.38}$$

where M is the maximum value of $|f''''(x)|$ on $a \leq x \leq b$.

Proofs of these theorems can be found in books on numerical analysis. Note that because of the n^4 factor in the denominator of 9.38, the accuracy of Simpson's rule increases much more rapidly than does that of the trapezoidal rule.

For the function $f(x) = \sin x$ in Examples 9.22–24,

$$f''(x) = -\sin x \qquad \text{and} \qquad f''''(x) = \sin x.$$

Consequently, in both cases we can state that $M = 1$, and therefore

$$|T_n| \leq \frac{(3-1)^3}{12\,n^2} = \frac{2}{3\,n^2} \qquad \text{and} \qquad |S_n| \leq \frac{(3-1)^5}{180\,n^4} = \frac{8}{45\,n^4}.$$

For $n = 10$ and $n = 20$, we find

$$|T_{10}| \leq \frac{2}{3(10)^2} < 0.0067 \qquad \text{and} \qquad |S_{10}| \leq \frac{8}{45(10)^4} < 0.000018;$$

$$|T_{20}| \leq \frac{2}{3(20)^2} < 0.0017 \qquad \text{and} \qquad |S_{20}| \leq \frac{8}{45(20)^4} < 0.000012.$$

Differences between the correct value for the integral and the approximations listed in Table 9.3 corroborate these predictions.

EXAMPLE 9.25

What is the maximum possible error in using the trapezoidal rule with 100 equal subdivisions to approximate

$$\int_1^3 \frac{1}{\sqrt{1+x^3}}\,dx.$$

SOLUTION According to formula 9.37, if T_{100} is the maximum possible error, then

$$|T_{100}| \leq \frac{M(3-1)^3}{12(100)^2} = \frac{2\,M}{3 \times 10^4},$$

where M is the maximum of (the absolute value of) the second derivative of $1/\sqrt{1+x^3}$ on $1 \leq x \leq 3$. Now,

$$\frac{d^2}{dx^2}\frac{1}{\sqrt{1+x^3}} = \frac{d}{dx}\left(\frac{-3x^2}{2(1+x^3)^{3/2}}\right) = \frac{3x(5x^3-4)}{4(1+x^3)^{5/2}}.$$

Instead of maximizing the absolute value of this function on the interval $1 \le x \le 3$, which would require another derivative, we note that the maximum value of the numerator is obtained for $x = 3$, and the minimum value of the denominator occurs at $x = 1$. It follows therefore that the second derivative cannot possibly be larger than

$$\frac{3(3)[5(3)^3-4]}{4(1+1)^{5/2}} = \frac{1179}{16\sqrt{2}}.$$

Thus, M must be less than or equal to $1179/(16\sqrt{2})$, and we can state that

$$|T_{100}| \le \frac{2}{3\times 10^4}\left(\frac{1179}{16\sqrt{2}}\right) \le 0.0035;$$

that is, the error in using the trapezoidal rule with 100 equal subdivisions to approximate the definite integral cannot be any larger than 0.0035. ■

EXAMPLE 9.26

How many equal subdivisions of the interval $0 \le x \le 2$ guarantee an error of less than 10^{-5} in the approximation of the definite integral

$$\int_0^2 e^{-x^2}\,dx$$

using Simpson's rule?

SOLUTION According to Theorem 9.2, the error in using Simpson's rule with n equal subdivisions to approximate this definite integral is

$$|S_n| \le \frac{M(2-0)^5}{180n^4},$$

where M is the maximum of the (absolute value of the) fourth derivative of e^{-x^2} on the interval $0 \le x \le 2$. It is a short calculation to find

$$\frac{d^4}{dx^4}\left(e^{-x^2}\right) = 4(3-12x^2+4x^4)e^{-x^2}.$$

Instead of maximizing the absolute value of this function, we note that e^{-x^2} has a maximum value of 1 (when $x = 0$). Furthermore, because $|3 - 12x^2 + 4x^4| \le 3 + 12x^2 + 4x^4$, which has a maximum at $x = 2$, it follows that

$$M \le 4[3 + 12(2)^2 + 4(2)^4](1) = 460.$$

Consequently,

$$|S_n| \leq \frac{460(2)^5}{180n^4} = \frac{3680}{45n^4}.$$

The error is less than 10^{-5} if n is chosen sufficiently large that

$$\frac{3680}{45n^4} < 10^{-5};$$

that is, if

$$n > \left(\frac{3680}{45 \times 10^{-5}}\right)^{1/4} = 53.5.$$

Since n must be an even integer, the required accuracy is guaranteed if n is chosen greater than or equal to 54. ■

Numerical techniques are indispensable in situations where we know that y is a function of x, say $y = f(x)$, but we do not know the exact form of the function $f(x)$. What might be available is a set of experimental data listing values of y corresponding to various values of x (Table 9.4). We have plotted these points in Figure 9.27. The trapezoidal rule and Simpson's rule can be used to approximate the definite integral of $f(x)$ from $x = 1$ to $x = 3$ even though we do not have a formula for $f(x)$. Indeed, formulas 9.34 and 9.36 do not use the form of the function $f(x)$ to be integrated, only its values at the subdivision points x_i. We have these points in Table 9.4 or Figure 9.27. Since there are 11 points, and therefore 10 subdivisions of the interval $1 \leq x \leq 3$, the trapezoidal rule gives

$$\begin{aligned}\int_1^3 f(x)\,dx &\approx \frac{1/5}{2}[f(1.0)+2f(1.2)+2f(1.4)+\cdots+2f(2.6)+2f(2.8)+f(3.0)]\\ &= \frac{1}{10}[0.84+2(1.12)+2(1.40)+\cdots+2(1.34)+2(0.94)+0.42]\\ &= 2.80.\end{aligned}$$

Simpson's rule gives

$$\begin{aligned}\int_1^3 f(x)\,dx &\approx \frac{1/5}{3}[f(1.0)+4f(1.2)+2f(1.4)+\cdots+2f(2.6)+4f(2.8)+f(3.0)]\\ &= \frac{1}{15}[0.84+4(1.12)+2(1.40)+\cdots+2(1.34)+4(0.94)+0.42]\\ &= 2.81.\end{aligned}$$

What we lose in this type of application is the ability to predict a maximum possible error in the approximation since there is no way to estimate M in Theorems 9.1 and 9.2.

TABLE 9.4

x	1.0	1.2	1.4	1.6	1.8	2.0	2.2	2.4	2.6	2.8	3.0
y	0.84	1.12	1.40	1.60	1.75	1.82	1.79	1.62	1.34	0.94	0.42

FIGURE 9.27

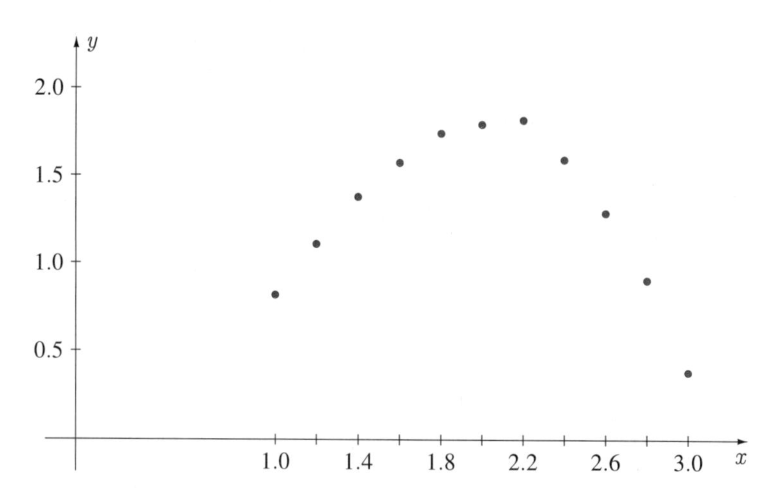

The error bounds in Theorems 9.1 and 9.2 are somewhat idealistic in the sense that they are error predictions based on the use of exact numbers. For instance, they predict that by increasing n indefinitely, any degree of accuracy is attainable. Theoretically this is true, but practically it is not. No matter how we choose to evaluate the summations in 9.34 and 9.36, be it by hand, by an electronic hand calculator, or by a high-speed computer, each calculation is rounded off to a certain number of decimals. The final sum takes into account many, many of these "approximate numbers", and must therefore be inherently inaccurate. We call this *round-off error* and it is very difficult to predict how extensive it is. It depends on both the number and nature of the operations involved in 9.34 and 9.36. In the approximation of a definite integral by the trapezoidal rule or Simpson's rule, there are two sources of error. Theorems 9.1 and 9.2 predict errors due to the methods themselves; round-off errors may also be appreciable for large n.

We should emphasize once again that although we have used area as a convenient vehicle by which to explain the approximation of definite integrals by the rectangular, trapezoidal, and Simpson's rules, it is not necessary for $f(x)$ to be nonnegative. All three methods can be used to approximate the definite integral of a function $f(x)$, be it positive, negative, or sometimes positive and sometimes negative on the interval of integration. The only condition that we have imposed is that $f(x)$ be continuous. In view of our discussion of improper integrals in Section 7.9, even this is not always necessary.

EXERCISES 9.9

In Exercises 1–10 use the trapezoidal rule and Simpson's rule with 10 equal subdivisions to approximate the definite integral. In each case, evaluate the integral analytically to get an idea of the accuracy of the approximation.

1. $\displaystyle\int_1^2 \frac{1}{x}\,dx$

2. $\displaystyle\int_2^3 \frac{1}{\sqrt{x+2}}\,dx$

3. $\displaystyle\int_0^1 \tan x\,dx$

4. $\displaystyle\int_0^{1/2} e^x\,dx$

5. $\displaystyle\int_{-1}^1 \sqrt{x+1}\,dx$

6. $\displaystyle\int_{-3}^{-2} \frac{1}{x^3}\,dx$

7. $\displaystyle\int_{1/2}^1 \cos x\,dx$

8. $\displaystyle\int_0^1 \frac{1}{3+x^2}\,dx$

9. $\displaystyle\int_1^3 \frac{1}{x^2+x}\,dx$

10. $\displaystyle\int_0^{1/2} xe^{x^2}\,dx$

In Exercises 11–14 use the trapezoidal rule and Simpson's rule with 10 equal subdivisions to approximate the definite integral.

11. $\displaystyle\int_0^2 \frac{1}{1+x^3}\,dx$

12. $\displaystyle\int_0^1 e^{x^2}\,dx$

13. $\displaystyle\int_1^2 \sqrt{1+x^4}\,dx$

14. $\displaystyle\int_{-1}^0 \sin(x^2)\,dx$

15. Show graphically that if $y = f(x)$ is concave downward on the interval $a \le x \le b$, then the trapezoidal rule underestimates the definite integral of $f(x)$ on this interval.

16. What happens to the errors in 9.37 and 9.38 when the number of partitions is doubled?

17. The definite integral

$$\int_a^b e^{-x^2}\,dx$$

is very important in mathematical statistics. Use Simpson's rule with $n = 16$ to evaluate the integral for $a = 0$ and $b = 1$.

18. Use Simpson's rule with 10 equal intervals to approximate the definite integral for the length of the parabola $y = x^2$ between $x = 0$ and $x = 1$. Compare the answer to that of Exercise 39 in Section 9.5.

19. Use the trapezoidal rule and Simpson's rule with 10 equal subdivisions to approximate the definite integral for the length of the curve $y = \sin x$ from $x = 0$ to $x = \pi/2$.

20. The numerical techniques of this section can also be used to approximate many improper integrals. Consider

$$\int_0^4 \frac{e^x}{\sqrt{x}}\,dx.$$

(a) Why can the trapezoidal rule and Simpson's rule not be used directly to approximate this integral?

(b) Show that the change of variable $u = \sqrt{x}$ replaces this improper integral with an integral that is not improper and use Simpson's rule with 20 equal subdivisions to approximate its value.

(c) Could you use the rectangular rule on the improper integral?

21. (a) Show that the definite integral

$$L = \frac{4}{3}\int_0^3 \sqrt{\frac{81 - 5x^2}{9 - x^2}}\,dx$$

represents the length L of the ellipse $4x^2 + 9y^2 = 36$.

(b) Show that when we set $x = 3\sin\theta$, the θ-integral is no longer improper.

(c) Use the trapezoidal rule and Simpson's rule with 8 equal subdivisions to approximate the θ-integral in (b).

22. To approximate the improper integral

$$\int_1^\infty \frac{1}{1 + x^4}\,dx$$

set $x = 1/t$, and use the trapezoidal rule and Simpson's rule with 10 equal subdivisions on the resulting integral.

23. Use the technique of Exercise 22 to approximate

$$\int_1^\infty \frac{x^2}{x^4 + x^2 + 1}\,dx.$$

24. Use the trapezoidal rule and Simpson's rule to approximate the definite integral from $x = -1$ to $x = 4$ for the function tabulated in Table 9.5.

25. When the truss in Figure 9.28 is subjected to a force F at its centre, point A deflects an amount y from its equilibrium position. Forces (in kilonewtons) required to produce deflections from 0 to 5 cm at intervals of 0.5 cm are listed in Table 9.6. Plot these points on a set of axes with y horizontal and F vertical. Use Simpson's rule to approximate the work done by F in deflecting A by a full 5 cm.

FIGURE 9.28

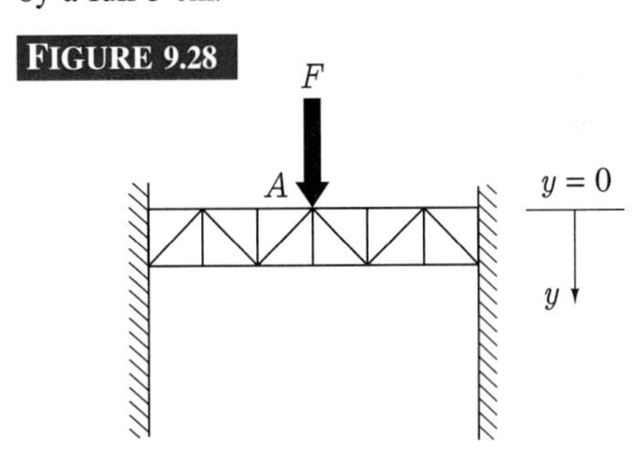

26. Show that when $f(x)$ is a cubic polynomial, evaluation of the definite integral of $f(x)$ from $x = a$ to $x = b$ by Simpson's rule always gives the *exact* answer. Illustrate with an example.

In Exercises 27–30 how many equal subdivisions of the interval of integration guarantee an error of less than 10^{-4} in the approximation of the definite integral using (a) the trapezoidal rule and (b) Simpson's rule?

27. $\displaystyle\int_1^4 \frac{1}{x}\,dx$

28. $\displaystyle\int_0^{\pi/4} \cos x\,dx$

29. $\displaystyle\int_0^{1/3} e^{2x}\,dx$

30. $\displaystyle\int_4^5 \frac{1}{\sqrt{x+2}}\,dx$

TABLE 9.5

x	−1.0	−0.5	0	0.5	1.0	1.5	2.0	2.5	3.0	3.5	4.0
y	2.287	0.395	0	0.145	0.310	0.334	1.819	0.123	0.021	−0.037	−0.055

TABLE 9.6

y	0	0.5	1.0	1.5	2.0	2.5	3.0	3.5	4.0	4.5	5.0
F	0	1.45	2.90	4.40	5.90	7.43	9.05	10.7	13.2	15.3	18.0

SUMMARY

Antidifferentiation is a far more complicated process than differentiation; it is not possible to state a set of rules and formulas that will suffice for most functions. Certainly there are integration formulas that we must learn, but most of those given in this chapter have been differentiation formulas listed in Chapters 3 and 8 written in terms of integrals rather than derivatives. We have stressed the importance of knowing these obvious integration formulas, but beyond this it becomes an organizational problem — organizing a difficult integral into a form that utilizes these simple formulas. The three most important techniques for doing this are substitutions, integration by parts, and partial fractions.

There is a variety of substitutions for the evaluation of indefinite integrals, many suggested by the form of the integrand. For example, a term in $x^{1/n}$ (n an integer) suggests the substitution $u = x^{1/n}$. The trigonometric substitution is most important. It eliminates square roots of the form $\sqrt{a^2 \pm b^2 x^2}$ and $\sqrt{b^2 x^2 - a^2}$ and the general root $\sqrt{ax^2 + bx + c}$.

Integration by parts is a powerful integration technique. It is used to evaluate antiderivatives of transcendental functions that are multiplied by powers of x, it leads to the method of integration by reproduction, and it is used to develop reduction formulas.

The method of partial fractions decomposes complicated rational functions into simple fractions that are either immediately integrable or amenable to other methods such as trigonometric substitutions.

Tables of integrals provide formulas for simple antiderivatives as well as very complex ones. They contain reduction formulas, and are an excellent companion to partial fractions.

Even with the techniques that we have studied and tables, there are many functions that either cannot be antidifferentiated at all or can be antidifferentiated only with extreme difficulty. To approximate definite integrals of such functions, numerical techniques such as the rectangular rule, the trapezoidal rule, and Simpson's rule are essential.

Key Terms and Formulas

In reviewing this chapter, you should be able to define or discuss the following key terms:

Exponential growth
Exponential decay
Integration by parts
Integration by reproduction
Trigonometric substitution
Partial fraction decomposition
Rectangular rule
Trapezoidal rule
Simpson's rule

REVIEW EXERCISES

In Exercises 1–50 evaluate the indefinite integral.

1. $\int \sqrt{2 - x}\, dx$

2. $\int \frac{1}{(x + 3)^2}\, dx$

3. $\int \frac{x^2 + 3}{x}\, dx$

4. $\int \frac{x^2 + 3}{x + 1}\, dx$

5. $\int \frac{x^2 + 3}{x^2 + 1}\, dx$

6. $\int \frac{x}{\sqrt{x + 3}}\, dx$

7. $\int \sin^2 x \cos^3 x\, dx$

8. $\int x \sin x\, dx$

9. $\int \tan^2 (2x)\, dx$

10. $\int \frac{x}{x^2 + 2x - 3}\, dx$

11. $\int \frac{1}{\sqrt{4 - 3x^2}}\, dx$

12. $\int \frac{2 - \sqrt{x}}{\sqrt{x} + 5}\, dx$

13. $\int \frac{x}{3x^2 + 4}\, dx$

14. $\int \frac{e^x}{\sqrt{1 - e^{2x}}}\, dx$

15. $\int x^2 \ln x\, dx$

16. $\int \frac{x}{(x^2 + 1)^2}\, dx$

17. $\displaystyle\int \frac{x^2}{(x^2+1)^2}\,dx$

18. $\displaystyle\int \frac{x^3}{(x^2+1)^2}\,dx$

19. $\displaystyle\int \frac{x+1}{x^3-4x}\,dx$

20. $\displaystyle\int \left(\frac{x+1}{x-1}\right)^2 dx$

21. $\displaystyle\int \frac{x^2}{(1+3x^3)^4}\,dx$

22. $\displaystyle\int \text{Cos}^{-1}\, x\,dx$

23. $\displaystyle\int \sin x \cos 2x\,dx$

24. $\displaystyle\int \sin x \cos 5x\,dx$

25. $\displaystyle\int e^{3x}\cos 2x\,dx$

26. $\displaystyle\int \frac{1}{\sqrt{x^2+4x-5}}\,dx$

27. $\displaystyle\int \frac{1}{x^2+4x-5}\,dx$

28. $\displaystyle\int x^3\sqrt{4-x^2}\,dx$

29. $\displaystyle\int \frac{\cos 2x}{1-\sin 2x}\,dx$

30. $\displaystyle\int \frac{6x}{4-x^2}\,dx$

31. $\displaystyle\int \frac{1}{x\sqrt{\ln x}}\,dx$

32. $\displaystyle\int \frac{1}{x^2+4x-4}\,dx$

33. $\displaystyle\int \frac{\sin x}{1+\cos^2 x}\,dx$

34. $\displaystyle\int \frac{1}{x^4+x^3}\,dx$

35. $\displaystyle\int x\sec^2(3x)\,dx$

36. $\displaystyle\int \frac{1}{\sqrt{16-3x+x^2}}\,dx$

37. $\displaystyle\int \frac{\sqrt{x^2-4}}{x^2}\,dx$

38. $\displaystyle\int x^2\,\text{Tan}^{-1}\, x\,dx$

39. $\displaystyle\int \frac{x^2}{x^3+3x^2+3x+1}\,dx$

40. $\displaystyle\int \frac{\ln x}{x}\,dx$

41. $\displaystyle\int \frac{x^2}{1+4x^6}\,dx$

42. $\displaystyle\int \frac{1}{x(9+x^2)^2}\,dx$

43. $\displaystyle\int \frac{x^2+2}{x^3+5x^2+4x}\,dx$

44. $\displaystyle\int \frac{x^2+2}{x^3+4x^2+4x}\,dx$

45. $\displaystyle\int \frac{x^2+2}{x^3+x^2+4x}\,dx$

46. $\displaystyle\int \frac{3x^2+2x+4}{x^3+x^2+4x}\,dx$

47. $\displaystyle\int x\text{Sin}^{-1}\, x\,dx$

48. $\displaystyle\int \sqrt{\cot x}\csc^4 x\,dx$

49. $\displaystyle\int \ln(\sqrt{x}+1)\,dx$

50. $\displaystyle\int \frac{1}{(4x-x^2)^{3/2}}\,dx$

In Exercises 51–54 use the trapezoidal rule and Simpson's rule with ten equal partitions to approximate the definite integral.

51. $\displaystyle\int_1^2 \frac{\sin x}{x}\,dx$

52. $\displaystyle\int_0^1 \sqrt{\sin x}\,dx$

53. $\displaystyle\int_2^4 \frac{1}{\ln x}\,dx$

54. $\displaystyle\int_{-1}^3 \frac{1}{1+e^x}\,dx$

55. If 10% of a sample of radioactive material decays in 10 days, what is its half-life?

In Exercises 56–71 evaluate the indefinite integral.

56. $\displaystyle\int \frac{1}{x^{1/3}-\sqrt{x}}\,dx$

57. $\displaystyle\int \ln(1+x^2)\,dx$

58. $\displaystyle\int \frac{x}{x^4+16}\,dx$

59. $\displaystyle\int \csc^3 x\,dx$

60. $\displaystyle\int \frac{1}{(3x-x^2)^{3/2}}\,dx$

61. $\displaystyle\int \frac{1}{x^3\sqrt{x^2-9}}\,dx$

62. $\displaystyle\int \sin\sqrt{x}\,dx$

63. $\displaystyle\int \sin(\ln x)\,dx$

64. $\displaystyle\int x\cos x\sin 3x\,dx$

65. $\displaystyle\int \frac{x^4+3x^2+1}{x(x^2+1)^2}\,dx$

66. $\displaystyle\int \frac{1}{1+\cos 2x}\,dx$

67. $\displaystyle\int \frac{x^4+3x^2-2x+5}{x^2-3x+7}\,dx$

68. $\displaystyle\int \sin^2 x\cos 3x\,dx$

69. $\displaystyle\int \frac{1}{x^3(4-x^2)^{3/2}}\,dx$

70. $\displaystyle\int \sqrt{1-x^2}\,\text{Sin}^{-1}\, x\,dx$

71. $\displaystyle\int \frac{1}{x+\sqrt{x^2+4}}\,dx$

72. (a) Use the substitution $z^2 = (1+x)/(1-x)$ to show that

$$\int \sqrt{\frac{1+x}{1-x}}\,dx = 2\,\text{Tan}^{-1}\sqrt{\frac{1+x}{1-x}} - \sqrt{1-x^2} + C.$$

(b) Evaluate the integral in (a) by multiplying numerator and denominator of the integrand by $\sqrt{1+x}$. Verify that this answer is the same as that in (a).

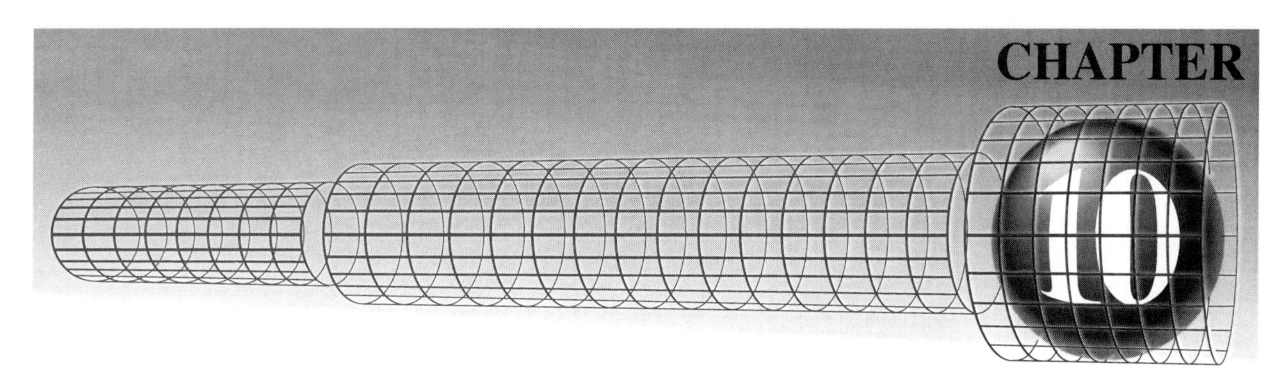

Parametric Equations and Polar Coordinates

In this chapter we introduce parametric representations for curves, and polar coordinates. Parametric representations of curves offer an alternative to the explicit and implicit forms used in Chapters 1–9. Polar coordinates provide an alternative way to identify points in a plane. They are more efficient than Cartesian coordinates in many applications. In particular, we shall see how they provide a unified approach to conic sections.

SECTION 10.1 Parametric Equations

A curve is defined explicitly by equations of the form $y = f(x)$ or $x = g(y)$, and implicitly by an equation $F(x, y) = 0$. A third method is described in the following definition.

Definition 10.1

A curve is said to be defined **parametrically** if it is given in the form

$$x = x(t), \qquad y = y(t), \qquad \alpha \le t \le \beta. \tag{10.1}$$

Variable t is called a *parameter*; it is a connecting link between x and y. Each value of t in the interval $\alpha \le t \le \beta$ is substituted into equations 10.1, and the pair $(x, y) = (x(t), y(t))$ represents a point on the curve. When the interval for t is unspecified, we assume that it consists of all values for which both $x(t)$ and $y(t)$ are defined.

EXAMPLE 10.1

Sketch the curve defined parametrically by $x = 3 - t$, $y = t^2 - 2$.

SOLUTION We can construct a table of values (Table 10.1) by substituting various values of t and calculating corresponding values of x and y. This leads to the points in Figure 10.1(a), which can be joined as in Figure 10.1(b). Alternatively, if we solve the first equation for t, we obtain $t = 3 - x$, and substituting this expresssion into the second equation, we have

TABLE 10.1

t	−3	−2	−1	0	1	2	3
x	6	5	4	3	2	1	0
y	7	2	−1	−2	−1	2	7

$$y = (3 - x)^2 - 2 = x^2 - 6x + 7,$$

an explicit definition of the curve. The given equations are therefore parametric equations for this parabola.

FIGURE 10.1

(a)

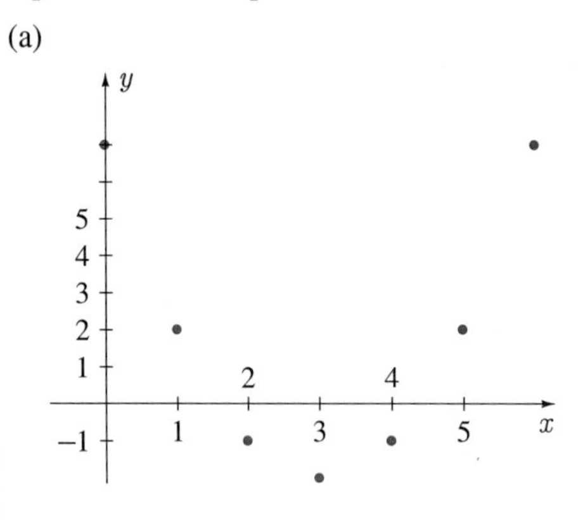

(b)

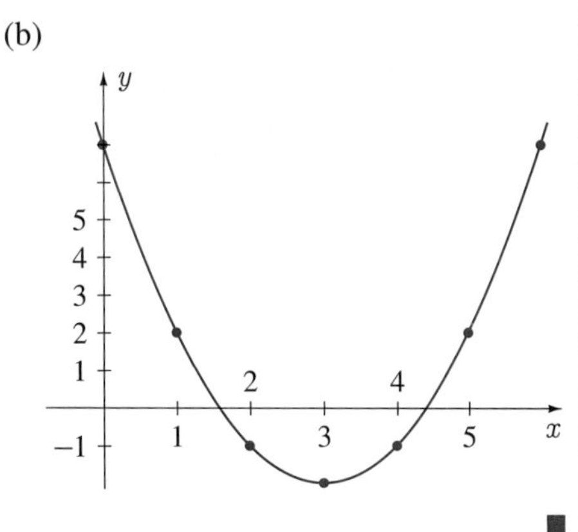

■

EXAMPLE 10.2

Discuss the curve defined parametrically by

$$x = 2\cos\theta, \quad y = 2\sin\theta, \quad 0 \le \theta < 2\pi.$$

SOLUTION

If the given equations are squared and added, the result is

$$x^2 + y^2 = 4\cos^2\theta + 4\sin^2\theta = 4,$$

an implicit definition of the curve. The given equations therefore define this circle parametrically. In this case we can interpret the parameter θ as the angle in Figure 10.2. The values $0 \le \theta < 2\pi$ guarantee that the parametric equations describe the complete circle. Additional values of θ would only duplicate points on this circle.

FIGURE 10.2

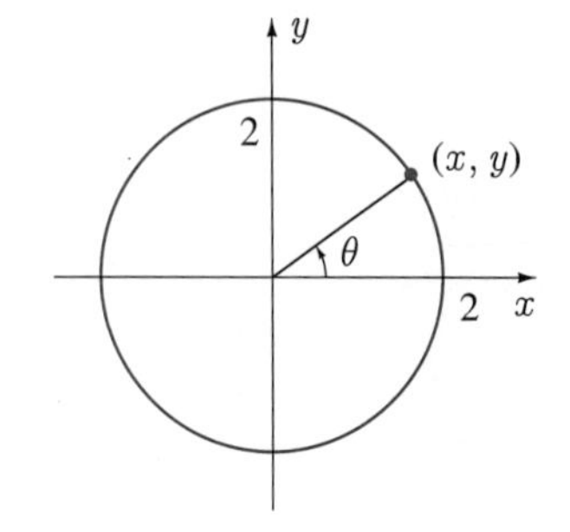

■

EXAMPLE 10.3

Sketch the curve defined parametrically by

$$x = t - t^3, \qquad y = t + t^3.$$

SOLUTION It is possible once again to eliminate the parameter. By adding and subtracting the given equations, we have

$$x + y = 2t, \qquad \text{and} \qquad y - x = 2t^3.$$

It follows from the first of these that $t = (x + y)/2$, and when this is substituted into the second,

$$y - x = \frac{(x + y)^3}{4}.$$

This is an implicit definition of the curve, but not a particularly simple one. An explicit definition is very cumbersome. To sketch the curve using the parametric representation, we could make a table of values as in Example 10.1. A more efficient method — one that is very helpful for complex curves — is to draw separate graphs of the functions $x = t - t^3$ and $y = t + t^3$ using the same scale on the t-axes (Figure 10.3).

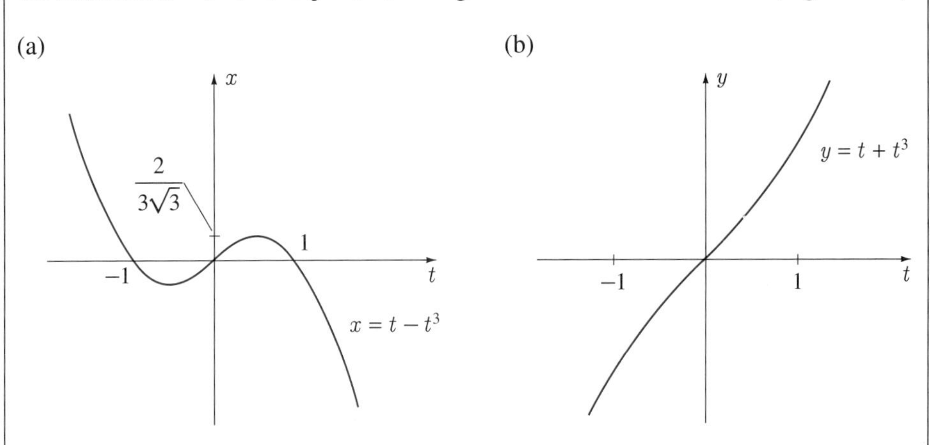

FIGURE 10.3

From these graphs we read pairs of values of x and y for the same t, and interpret them as coordinates (x, y) of a point. Beginning with $t = 0$, we obtain $x = 0$ and $y = 0$, and therefore the point $(0, 0)$. As t increases from $t = 0$ to $t = 1$, values of x increase from 0 to a maximum value $2/(3\sqrt{3})$ at $t = 1/\sqrt{3}$, and then decrease to 0. Simultaneously, values of y increase steadily from 0 to 2. This is shown in Figure 10.4(a). As t increases beyond 1, values of x decrease through negative numbers while y continues to increase. This is reflected in that part of the graph in Figure 10.4(b) to the left of the y-axis. A similar procedure using negative values of t leads to the final graph in Figure 10.4(c). Alternatively, for negative t, we note that because $x = t - t^3$ and $y = t + t^3$ are odd functions, replacing t by $-t$ reverses the signs of x and y. This means that corresponding to each point (x, y) on the graph for which $t > 0$, we must have the point$(-x, -y)$.

FIGURE 10.4

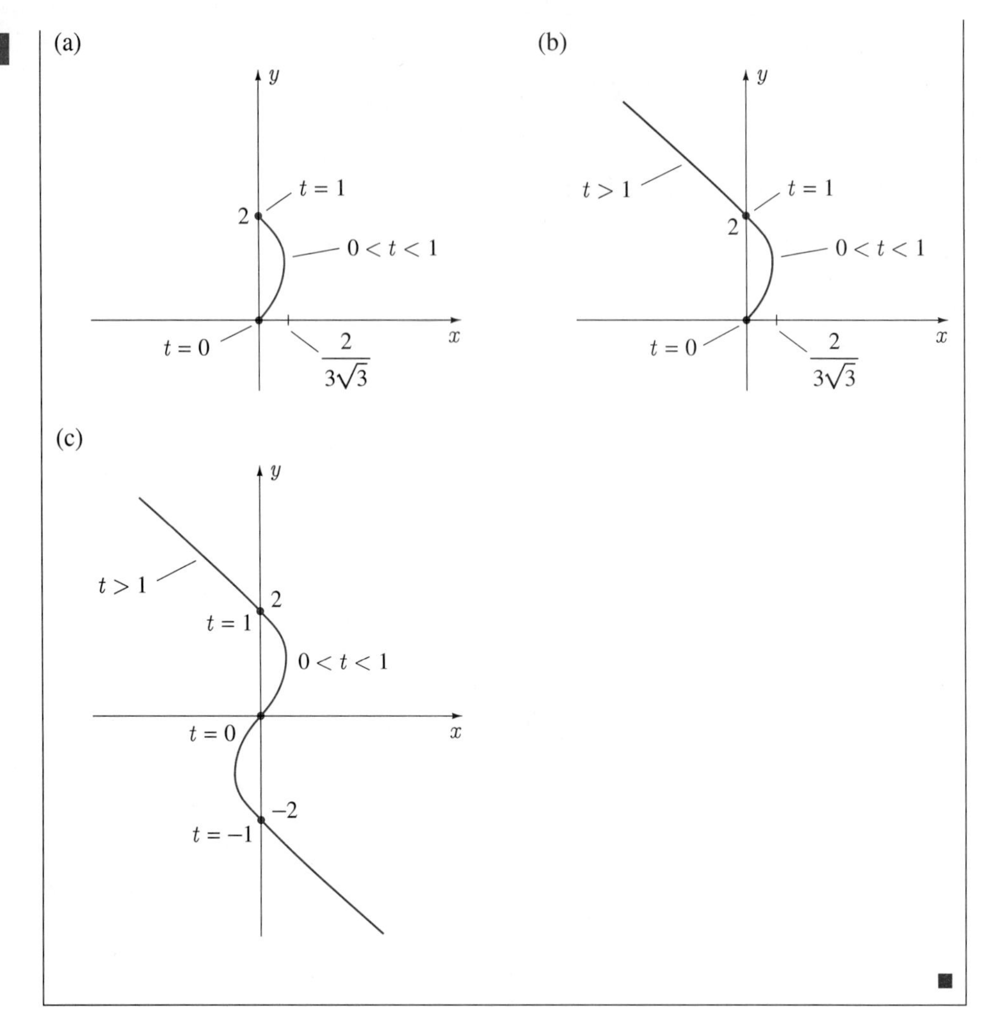

Parametric equations are frequently used in problems concerning the motion of objects. For example, suppose a stone is thrown horizontally over the edge of a cliff 100 m above a river (Figure 10.5). If the initial speed of the stone is 30 m/s, the path followed by the stone is described parametrically by

$$x = 30t, \qquad y = 100 - 4.905t^2, \quad 0 \le t \le \sqrt{\frac{100}{4.905}},$$

where t is time in seconds and x and y are in metres. For these equations, t has been chosen equal to zero at the instant of projection, and $t = \sqrt{100/4.905}$ is the time at which the stone strikes the river.

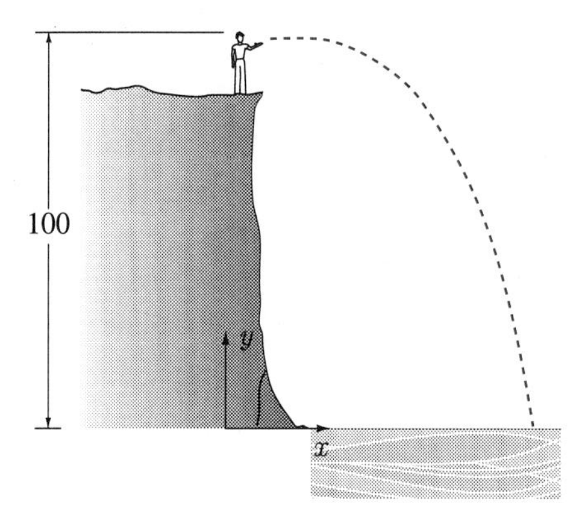

FIGURE 10.5

When parameter t is eliminated from this parametric representation of the trajectory, the resulting equation is

$$y = 100 - 4.905\left(\frac{x}{30}\right)^2 = 100 - 0.00545x^2.$$

It clearly indicates that the stone follows a parabolic path. Were we to discuss the motion of the stone as regards, say, velocity and acceleration, however, we would find it necessary to return to the parametric representation which describes the motion in horizontal and vertical directions.

We say that equations 10.1 define y **parametrically as a function of** x if the curve so defined represents a function; that is, if every vertical line that intersects the curve does so exactly once. The parametric equations in Example 10.1 therefore define a function parametrically, but in Examples 10.2 and 10.3 they do not. However, each of the curves in these latter examples can be divided into subcurves that do represent functions. For instance, in Example 10.3, each of the subcurves

$$\begin{array}{lll} x = t - t^3, & y = t + t^3, & t \le -1/\sqrt{3}; \\ x = t - t^3, & y = t + t^3, & -1/\sqrt{3} \le t \le 1/\sqrt{3}; \\ x = t - t^3, & y = t + t^3, & t \ge 1/\sqrt{3}; \end{array}$$

defines a function, and does so parametrically; together all three curves make up the original curve.

When equations 10.1 do define y parametrically as a function of x, it is straightforward to find the derivative of the function. If we denote this function by $y = f(x)$, then the chain rule applied to $y = f(x)$, $x = x(t)$ gives

$$\frac{dy}{dt} = \frac{dy}{dx}\frac{dx}{dt}.$$

Provided then that $dx/dt \neq 0$, we may solve for

$$\frac{dy}{dx} = \frac{\frac{dy}{dt}}{\frac{dx}{dt}}. \tag{10.2}$$

This formula is called the **parametric rule** for differentiation of a function defined parametrically by equations 10.1; it defines the derivative of y with respect to x in terms of derivatives of the given functions $x(t)$ and $y(t)$ with respect to t.

For example, if y is defined parametrically as a function of x by

$$x(t) = \frac{1}{t^2+1}, \qquad y(t) = \frac{2}{t(t^2+1)}, \qquad t \ge 0,$$

then

$$\frac{dy}{dx} = \frac{\dfrac{dy}{dt}}{\dfrac{dx}{dt}} = \frac{\dfrac{-2(3t^2+1)}{(t^3+t)^2}}{\dfrac{-2t}{(t^2+1)^2}} = \frac{3t^2+1}{t^3}.$$

We can also calculate d^2y/dx^2 in spite of the fact that dy/dx is in terms of t rather than x. Once again it is the chain rule that comes to the rescue:

$$\frac{d^2y}{dx^2} = \frac{d}{dx}\left(\frac{dy}{dx}\right) = \frac{d}{dt}\left(\frac{dy}{dx}\right)\frac{dt}{dx} = \frac{\dfrac{d}{dt}\left(\dfrac{dy}{dx}\right)}{\dfrac{dx}{dt}}$$

$$= \frac{\dfrac{t^3(6t)-(3t^2+1)(3t^2)}{t^6}}{\dfrac{-2t}{(t^2+1)^2}} = \frac{-3t^4-3t^2}{t^6}\cdot\frac{(t^2+1)^2}{-2t} = \frac{3(t^2+1)^3}{2t^5}.$$

EXAMPLE 10.4

Sketch the strophoid

$$x = \frac{1-t^2}{1+t^2}, \qquad y = \frac{t(1-t^2)}{1+t^2}$$

and find the points at which its tangent line is horizontal.

SOLUTION

To sketch the strophoid we first sketch graphs of the functions $x(t)$ and $y(t)$. For $x(t)$, we write $x = -1 + 2/(1+t^2)$, and first draw $2/(1+t^2)$ in Figure 10.6(a). The graph of $x(t)$ is this curve shifted downward one unit. Since $y(t) = tx(t)$, we sketch a graph of $y(t)$ by multiplication of ordinates. The result is shown in Figure 10.6(b). To sketch the strophoid, we proceed as in Example 10.3 by reading pairs of values from the graphs in Figure 10.6. The curve is shown in Figure 10.7.

FIGURE 10.6

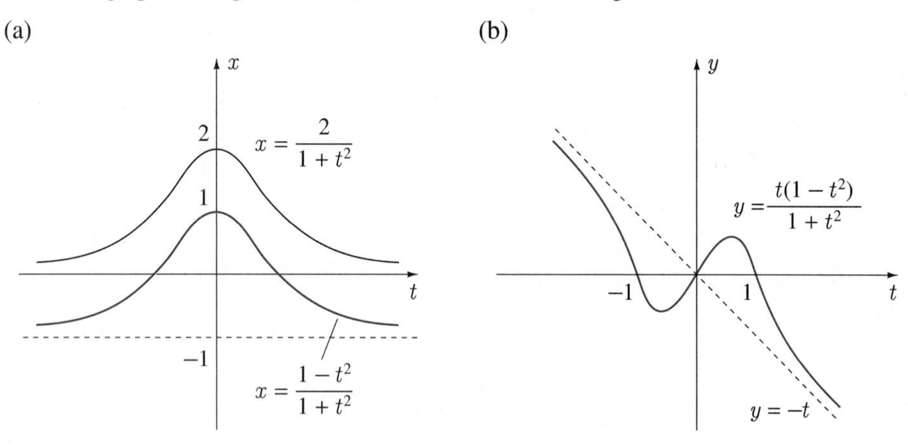

The points at which the tangent line is horizontal can be found by setting

$$0=\frac{dy}{dx}=\frac{\dfrac{dy}{dt}}{\dfrac{dx}{dt}}=\frac{\dfrac{(1+t^2)(1-3t^2)-(t-t^3)(2t)}{(1+t^2)^2}}{\dfrac{(1+t^2)(-2t)-(1-t^2)(2t)}{(1+t^2)^2}}=\frac{1-4t^2-t^4}{-4t}.$$

But this equation implies that

$$t^4+4t^2-1=0.$$

This is a quadratic equation in t^2, so that

$$t^2=\frac{-4\pm\sqrt{16+4}}{2}=-2\pm\sqrt{5}.$$

Since t^2 must be nonnegative, it follows that $t^2=\sqrt{5}-2$, and therefore

$$t=\pm\sqrt{\sqrt{5}-2}.$$

For these values of t,

$$x=\frac{1-(\sqrt{5}-2)}{1+(\sqrt{5}-2)}=\frac{3-\sqrt{5}}{\sqrt{5}-1}\cdot\frac{\sqrt{5}+1}{\sqrt{5}+1}=\frac{\sqrt{5}-1}{2}=0.62, \qquad \text{and}$$

$$y=tx=\left(\pm\sqrt{\sqrt{5}-2}\right)\left(\frac{\sqrt{5}-1}{2}\right)=\pm 0.30.$$

The points with a horizontal tangent line are therefore $(0.62, \pm 0.30)$. ■

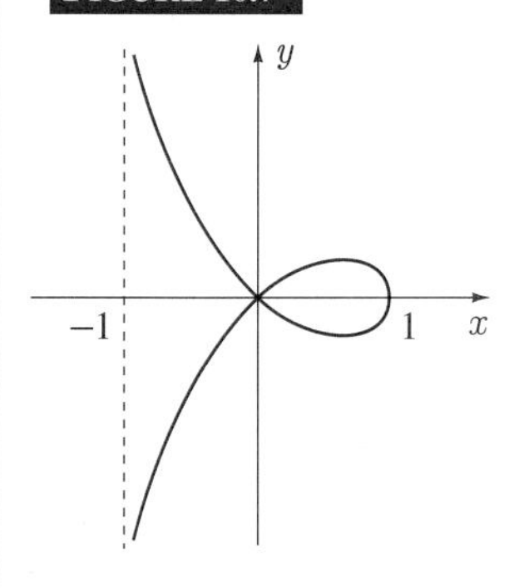

FIGURE 10.7

Recall formula 7.17 from Section 7.3 for the length of a curve. We express it in a slightly different form,

$$L=\int_A^B\sqrt{(dx)^2+(dy)^2},$$

where A and B are the initial and final points on the curve (Figure 10.8). When the equation of the curve is $y=f(x)$, we rewrite this formula as

$$L=\int_a^b\sqrt{1+\left(\frac{dy}{dx}\right)^2}\,dx$$

where a and b are x-coordinates of A and B, provided $a<b$. When the curve is given in the form $x=g(y)$, we write

$$L=\int_c^d\sqrt{1+\left(\frac{dx}{dy}\right)^2}\,dy,$$

provided $c<d$.

When the curve is defined parametrically by 10.1, the above formulas are replaced by

$$L=\int_\alpha^\beta\sqrt{\left(\frac{dx}{dt}\right)^2+\left(\frac{dy}{dt}\right)^2}\,dt. \qquad (10.3)$$

Verification of this formula is left to Exercise 42.

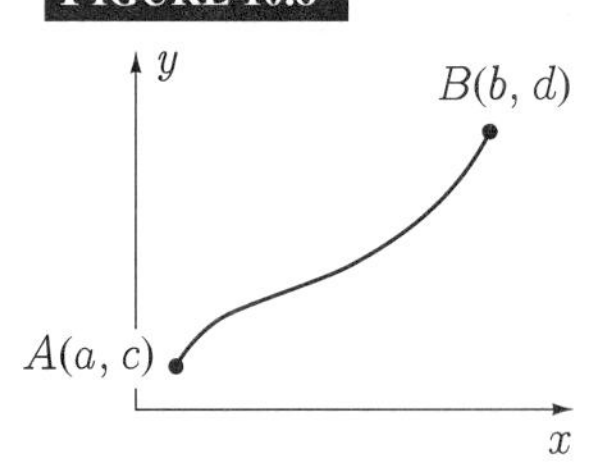

FIGURE 10.8

EXAMPLE 10.5

Find the length of the curve

$$x = e^t \cos t, \qquad y = e^t \sin t, \qquad 0 \le t \le \pi/2.$$

SOLUTION According to 10.3,

$$\begin{aligned} L &= \int_0^{\pi/2} \sqrt{(e^t \cos t - e^t \sin t)^2 + (e^t \sin t + e^t \cos t)^2}\, dt \\ &= \int_0^{\pi/2} \sqrt{e^{2t}(2\cos^2 t + 2\sin^2 t)}\, dt \\ &= \int_0^{\pi/2} \sqrt{2}\, e^t\, dt = \sqrt{2}\left\{e^t\right\}_0^{\pi/2} = \sqrt{2}(e^{\pi/2} - 1). \end{aligned}$$

■

EXERCISES 10.1

In Exercises 1–12 sketch the curve.

1. $x = 2 + t, \quad y = 3t - 1$

2. $x = t^2 + 3t + 4, \quad y = 1 - t$

3. $x = 1 + 2\cos t, \quad y = 2 + 2\sin t, \quad 0 \le t \le 2\pi$

4. $x = -2 + 4\cos t, \quad y = 3 + 4\sin t, \quad 0 \le t \le \pi$

5. $x = 1 + \cos t, \quad y = -1 - \sin t, \quad 0 \le t \le \pi/4$

6. $x = t + 1/t, \quad y = t - 1/t$

7. $x = 2\cos t, \quad y = 4\sin t, \quad 0 \le t < 2\pi$

8. $x = 1 + 3\cos t, \quad y = -2 + 2\sin t, \quad 0 \le t \le \pi$

9. $x = -1 + \sin t, \quad y = -1 - 3\cos t, \quad 0 \le t \le \pi/2$

10. $x = t - t^2, \quad y = t + t^2$

11. $x = t^2 + 1, \quad y = t^3 + 3$

12. $x = 2\cot\theta, \quad y = 2\sin^2\theta, \quad -\pi/2 \le \theta \le \pi/2, \quad \theta \ne 0$

In Exercises 13–20 assume that y is defined parametrically as a function of x, and find dy/dx.

13. $x = t^3 + 3t - 2, \quad y = t^2 - 1$

14. $x = \dfrac{u}{u-1}, \quad y = \dfrac{u^2}{u^2 - 1}$

15. $x = (v^3 + 2)\sqrt{v - 1}, \quad y = 2v^3 + 3$

16. $x = \sqrt{\dfrac{2+t}{2-t}}, \quad y = \sqrt{\dfrac{2-t}{2+t}}$

17. $x = s^{3/2} - s^{2/3}, \quad y = s^2 + 2s$

18. $x = (2t + 3)^4, \quad y = \dfrac{t}{t + 6}$

19. $x = \left(\dfrac{1+u}{1-u}\right)^{1/3}, \quad y = \left(\dfrac{1-u}{1+u}\right)^4$

20. $x = \sqrt{-t^2 + 3t + 5}, \quad y = \dfrac{1}{t^2 + 2t - 5}$

In Exercises 21–24 assume that y is defined parametrically as a function of x, and find dy/dx and d^2y/dx^2.

21. $x = t^2 + \dfrac{1}{t}, \quad y = t^2 - \dfrac{1}{t}$

22. $x = \sqrt{t - 1}, \quad y = \sqrt{t + 1}$

23. $x = 2u + 5, \quad y = 7 - 14u$

24. $x = v^2 + 2v + 3, \quad y = 2v - 4$

25. Find equations for the tangent and normal lines to the curve

$$x = t + \frac{1}{t}, \quad y = t - \frac{1}{t}$$

at the point corresponding to $t = 4$.

26. Find point(s) on the curve

$$x = \frac{t^3}{3} - 3t, \quad y = \frac{3t^2}{2} + t$$

where the slope of the tangent line to the curve is equal to 1.

27. Is there a difference between the two curves

$$y = 2x^2 - 1 \qquad \text{and} \qquad x = \cos t, \quad y = 2\cos^2 t - 1?$$

28. What curve is described by the parametric equations

$$x = h + a\cos\theta, \quad y = k + b\sin\theta, \quad 0 \le \theta < 2\pi?$$

29. Find parametric equations for a circle with centre (h, k) and radius r.

30. Show that the straight line through two points $P_1(x_1, y_1)$ and $P_2(x_2, y_2)$ has parametric equations

$$x = x_1 + (x_2 - x_1)t, \quad y = y_1 + (y_2 - y_1)t.$$

31. Sketch the following curves and determine whether they are related:

(a) $x = \sec\theta, \quad y = \tan\theta, \quad -\pi/2 < \theta < \pi/2$

(b) $x = \cosh\phi, \quad y = \sinh\phi$

(c) $x = \dfrac{1}{2}\left(t + \dfrac{1}{t}\right), \quad y = \dfrac{1}{2}\left(t - \dfrac{1}{t}\right), \quad t \ge 0$

In Exercises 32–35 find parametric equations for the curve.

32. $y = \dfrac{x+1}{x-2}$

33. $x + y^3 + xy = 5y^2$

34. $x^2 + y^2 + 2x - 4y = 0$

35. $4 - x^2 + 2y^2 = 0$

36. Two particles move along straight lines l_1 and l_2 defined parametrically by

$$l_1: \quad x = 1 - t, \quad y = t, \quad t \ge 0;$$
$$l_2: \quad x = 4t - 5, \quad y = 2t - 1, \quad t \ge 0;$$

where t is time. When are they closest together?

37. If $x = x(t)$ and $y = y(t)$ define y as a function of x, show that

$$\frac{d^2y}{dx^2} = \frac{\dfrac{dx}{dt}\dfrac{d^2y}{dt^2} - \dfrac{dy}{dt}\dfrac{d^2x}{dt^2}}{\left(\dfrac{dx}{dt}\right)^3}.$$

In Exercises 38–40 find the length of the curve.

38. $x = 3 + 4\cos t, \quad y = -2 + 4\sin t, \quad 0 \le t < 2\pi$

39. $x = e^{-t}\sin t, \quad y = e^{-t}\cos t, \quad 0 \le t \le 1$

40. $x = t + \ln t, \quad y = t - \ln t, \quad 1 \le t \le 2$

41. Set up, but do not evaluate, a definite integral representing the length of the ellipse

$$x = a\cos\theta, \quad y = b\sin\theta, \quad 0 \le \theta < 2\pi.$$

42. Verify formula 10.3 for the length of a curve.

43. The equations $x = t^2 + 2t - 1$, $y = t + 5$, $1 \le t \le 4$ define a curve parametrically. Find parametric equations that describe this curve, but have values of the parameter in the intervals (a) $0 \le t \le 3$; (b) $0 \le t \le 1$.

44. Suppose $x(t)$ and $y(t)$ in equations 10.1 are continuous on $\alpha \le t \le \beta$ and have derivatives on $\alpha < t < \beta$, and that $x'(t) \ne 0$ on $\alpha < t < \beta$. Show that Cauchy's generalized mean value theorem (Theorem 3.16) implies that there exists a number c between α and β such that

$$\frac{y(\beta) - y(\alpha)}{x(\beta) - x(\alpha)} = \frac{y'(c)}{x'(c)}.$$

Interpret this result geometrically.

45. A particle travels around the circle $x^2 + y^2 = 4$ counterclockwise at constant speed, making 2 revolutions each second. If the particle starts at point $(2, 0)$ at time $t = 0$, find parametric equations for its position in terms of t.

46. (a) A stone is embedded in the tread of a tire and the tire rolls, without slipping, along the x-axis (Figure 10.9). If the stone starts at the origin, the path that it follows is called a *cycloid*. Show that if θ is the angle through which the stone has turned, and R is the radius of the tire, parametric equations for the cycloid are

$$x = R(\theta - \sin\theta), \quad y = R(1 - \cos\theta).$$

(b) Find the area under one arch of the cycloid.

(c) Find the length of one arch of the cycloid.

FIGURE 10.9

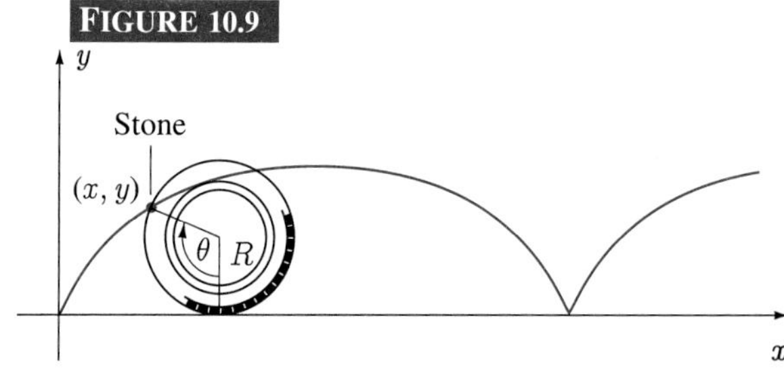

47. If the stone in Exercise 46 is embedded in the side of the tire rather than the tread, its path is called a *trochoid* (Figure 10.10). Show that if the distance from the centre of the tire to the stone is b, parametric equations for the trochoid are

$$x = R\theta - b\sin\theta, \quad y = R - b\cos\theta.$$

FIGURE 10.10

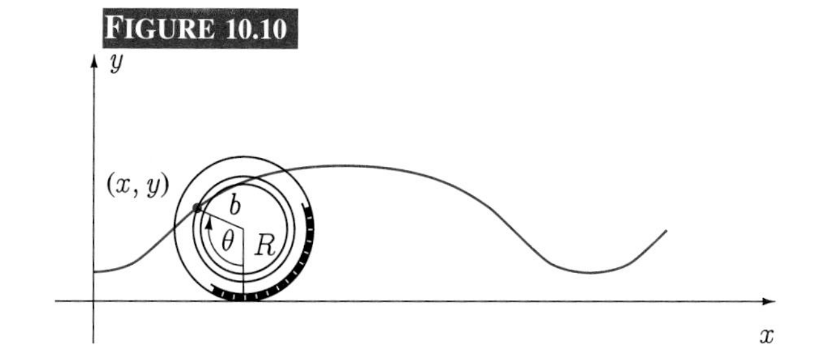

In Exercises 48–50 sketch the curve.

48. $x = \cos\theta, \quad y = \sin 3\theta$ (Curve of Lissajous)

49. $x = \dfrac{3t}{1+t^3}, \quad y = \dfrac{3t^2}{1+t^3}$ (Folium of Descartes)

50. $x = \cos^3\theta, \quad y = \sin^3\theta$ (Astroid or hypocycloid of four cusps)

SECTION 10.2

Polar Coordinates

In this section we introduce polar coordinates, an alternative coordinate system for the plane. Many problems that have complex solutions using Cartesian coordinates become much simpler to solve using polar coordinates.

Polar coordinates are defined by choosing a point O in the plane called the **pole** and a half-line originating at O called the **polar axis** (Figure 10.11). If P is a point in the plane, we join O and P. The first polar coordinate of P, denoted by r, is the length of the line segment OP. The other polar coordinate is the angle θ through which the polar axis must be rotated to coincide with line segment OP. Counterclockwise rotations are regarded as positive, and clockwise rotations as negative. In Figure 10.12, position OQ is reached through a positive rotation of $\pi/6$ radians; therefore, for point Q, $\theta = \pi/6$. But clearly we could arrive at this position in many other ways. We could, for instance, rotate the polar axis counterclockwise through any number of complete revolutions, bringing it back to its original position, and then rotate a further $\pi/6$ radians. Alternatively, we could rotate in a clockwise direction any number of complete revolutions, and then a further $-11\pi/6$ radians. In other words, polar coordinate θ for Q could be any of the values $\pi/6 + 2n\pi$, where n is an integer. Possible values of θ for point R in Figure 10.12 are $3\pi/4 + 2n\pi$. For point Q, $r = 2$ and for R, $r = 1$. Polar coordinates r and θ for a point are written in the form (r, θ) so that possible polar coordinates for Q and R are $(2, \pi/6 + 2n\pi)$ and $(1, 3\pi/4 + 2n\pi)$.

FIGURE 10.11

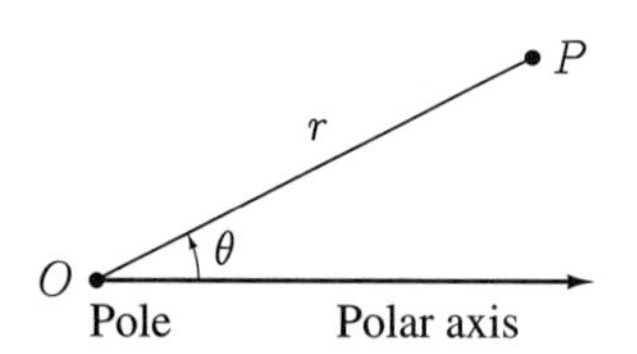

FIGURE 10.12

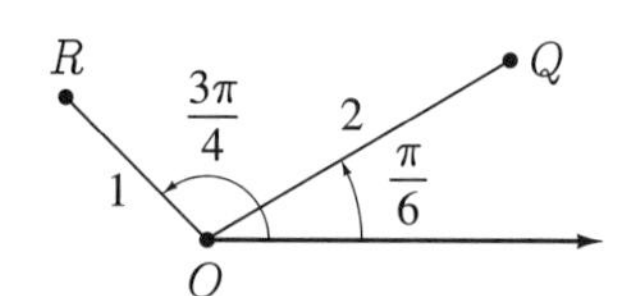

This situation is not like that for Cartesian coordinates where each point has only one set of coordinates (x, y), and every ordered pair of real numbers specifies one point. With polar coordinates, every ordered pair of real numbers (r, θ), where r must be nonnegative, represents one and only one point; but every point has an infinity of possible representations. We should point out that in some applications this is not a desirable situation. For instance, in the branch of mathematics called *tensor analysis*, it is necessary that polar coordinates assign exactly one pair of coordinates to each point. This can be accomplished in any region that does not contain the pole by demanding, for instance, that $-\pi < \theta \le \pi$. When we use polar coordinates to find areas in Section 10.4, we must also be particular about our choice of θ. For now, however, no advantage is gained by imposing restrictions on θ, and we therefore accept the fact that if (r, θ) are polar coordinates of a point, so are $(r, \theta + 2n\pi)$ for any integer n. Note also that polar coordinates for the pole are $(0, \theta)$ for any θ whatsoever.

If we introduce into a plane both a system of Cartesian coordinates (x, y) and a system of polar coordinates (r, θ), then relations will exist between the two. Suppose, for example, the pole of the polar coordinates and the origin of the Cartesian coordinates are chosen as the same point, and that the polar axis is chosen as the positive x-axis (Figure 10.13). In this case, Cartesian and polar coordinates of any point P are related by the equations

$$x = r\cos\theta, \quad y = r\sin\theta. \tag{10.4}$$

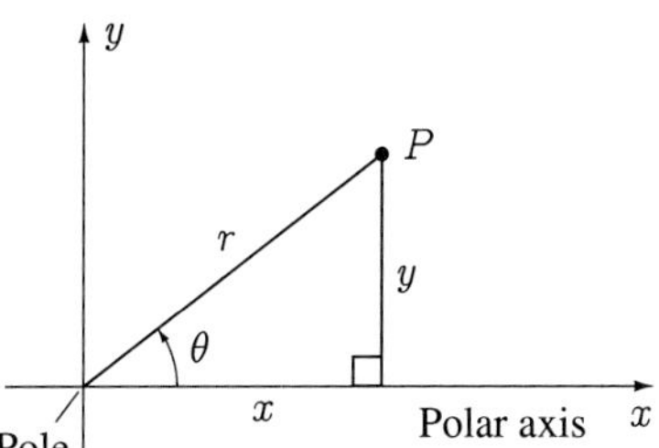

FIGURE 10.13

These equations define the Cartesian coordinates of a point in terms of its polar coordinates; that is, given its polar coordinates (r, θ), we can calculate its Cartesian coordinates (x, y) by means of 10.4. For example, if the polar coordinates of a point are $(3, 2)$, then its Cartesian coordinates are

$$x = 3\cos 2 = -1.25, \quad y = 3\sin 2 = 2.73.$$

Equations 10.4 implicitly define the polar coordinates of a point in terms of its Cartesian coordinates. For instance, if the Cartesian coordinates of a point are $(1, 1)$ (Figure 10.14), its polar coordinates must satisfy

$$1 = r\cos\theta, \quad 1 = r\sin\theta.$$

If we square these equations and add, we have

$$1 + 1 = r^2\cos^2\theta + r^2\sin^2\theta = r^2,$$

and therefore $r = \sqrt{2}$. It follows that $\cos\theta = \sin\theta = 1/\sqrt{2}$, from which we get $\theta = \pi/4 + 2n\pi$. Thus, polar coordinates of the point are $(\sqrt{2}, \pi/4 + 2n\pi)$.

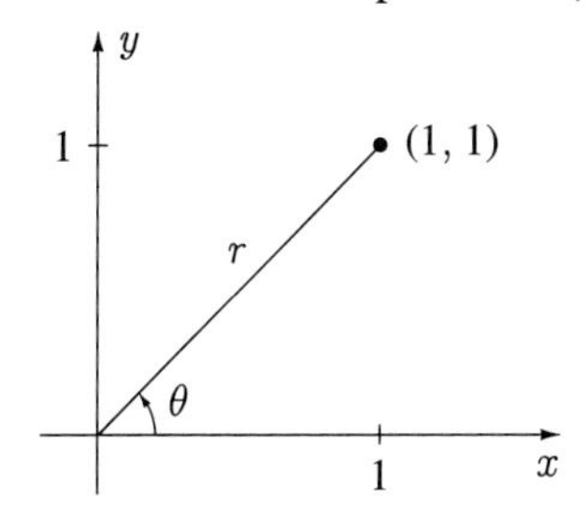

FIGURE 10.14

Equations 10.4 define r and θ implicitly in terms of x and y, but obviously it would be preferable to have explicit definitions. There is no problem expressing r explicitly in terms of x and y,

$$r = \sqrt{x^2 + y^2}, \tag{10.5}$$

but the case for θ is not so simple. If we substitute expression 10.5 for r into equations 10.4, we obtain

$$\cos\theta = \frac{x}{\sqrt{x^2 + y^2}}, \quad \sin\theta = \frac{y}{\sqrt{x^2 + y^2}}. \tag{10.6}$$

Except for the pole, these two equations determine all possible values of θ for given x and y. What we would like to do is obtain one equation, if possible, that defines θ. If we divide the second of these equations by the first, we have

$$\tan\theta = \frac{y}{x}, \tag{10.7}$$

and this equation suggests that we set

$$\theta = \text{Tan}^{-1}\left(\frac{y}{x}\right). \tag{10.8}$$

Unfortunately, neither equation 10.7 nor 10.8 is satisfactory. For instance, given the point P with Cartesian coordinates $(-1, 1)$ (Figure 10.15), equation 10.7 yields $\tan\theta = -1$, the solutions of which are $\theta = -\pi/4 + n\pi$. These are angles in the second and fourth quadrants, so only half of them are acceptable polar angles for P. Equation 10.8 gives $\theta = -\pi/4$, which is not a possible polar angle for P.

FIGURE 10.15

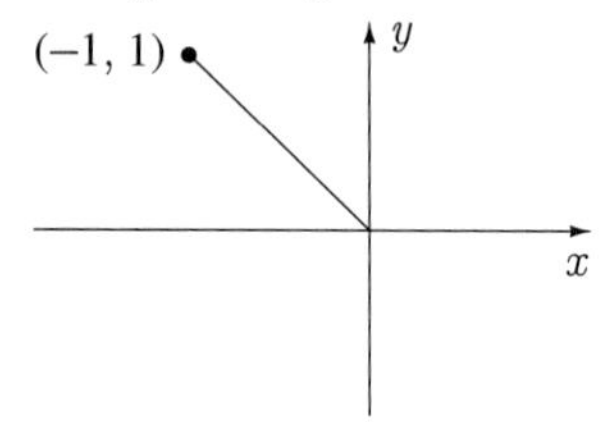

We suggest all angles satisfying 10.7 be found, and then a diagram be used to determine those angles that are acceptable values for θ. We illustrate in the following example.

EXAMPLE 10.6

Find all polar coordinates for points with the following Cartesian coordinates:

(a) $(1,2)$ (b) $(-2,3)$ (c) $(3,-1)$ (d) $(-2,-4)$

SOLUTION (a) $r = \sqrt{1^2 + 2^2} = \sqrt{5}$. Angles which satisfy $\tan\theta = 2/1$ are $1.11 + n\pi$. Since the point is in the first quadrant (Figure 10.16), acceptable values for θ are $1.11 + 2n\pi$. Polar coordinates of the point are therefore $(\sqrt{5}, 1.11 + 2n\pi)$.

(b) $r = \sqrt{(-2)^2 + 3^2} = \sqrt{13}$. Angles satisfying $\tan\theta = -3/2$ are $-0.98 + n\pi$. Since the point is in the second quadrant, possible values for θ are $(\pi - 0.98) + 2n\pi = 2.16 + 2n\pi$. Polar coordinates are therefore $(\sqrt{13}, 2.16 + 2n\pi)$.

(c) $r = \sqrt{3^2 + (-1)^2} = \sqrt{10}$. Since values of θ satisfying $\tan\theta = -1/3$ are $-0.32 + n\pi$, and the point is in the fourth quadrant, it follows that polar coordinates are $(\sqrt{10}, -0.32 + 2n\pi)$.

(d) $r = \sqrt{(-2)^2 + (-4)^2} = 2\sqrt{5}$. Since angles satisfying $\tan\theta = 2$ are $1.11 + n\pi$, and the point is in the third quadrant, polar coordinates are $(2\sqrt{5}, 4.25 + 2n\pi)$.

FIGURE 10.16

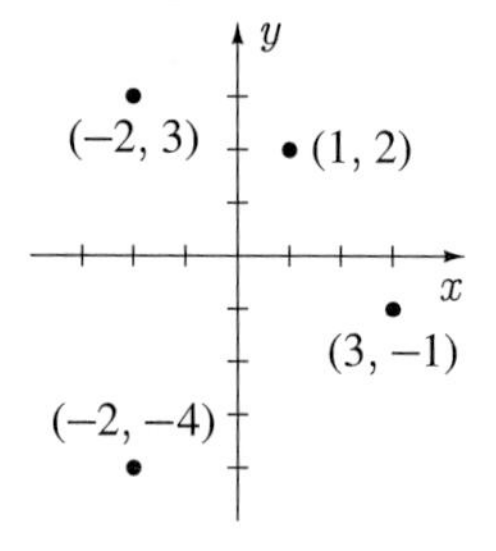

■

The results in equations 10.4–10.8 are valid only when the pole and origin coincide, and the polar axis and positive x-axis are identical. For a different arrangement, these relations would be changed accordingly. For example, if the pole is at the point with Cartesian coordinates (h, k) and the polar axis is as shown in Figure 10.17, equations 10.4 are replaced by

$$x = h + r\cos(\theta + \alpha), \quad y = k + r\sin(\theta + \alpha). \tag{10.9}$$

You will verify these equations in Exercise 13.

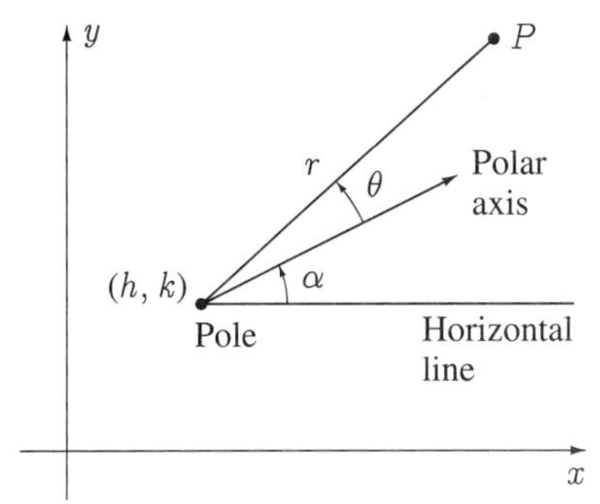

FIGURE 10.17

The usual choice of pole and polar axis is that in Figure 10.13, and unless otherwise stipulated, we assume this to be the case.

EXERCISES 10.2

In Exercises 1–8 plot the point having the given set of Cartesian coordinates, and find all possible polar coordinates.

1. $(1, -1)$ **2.** $(-1, \sqrt{3})$

3. $(4, 3)$ **4.** $(-2\sqrt{3}, 2)$

5. $(2, 6)$ **6.** $(-1, -4)$

7. $(7, -5)$ **8.** $(-5, 2)$

In Exercises 9–12 plot the point having the given set of polar coordinates, and find its Cartesian coordinates.

9. $(2, \pi/4)$ **10.** $(6, -\pi/6)$

11. $(7, 1)$ **12.** $(3, -2.4)$

13. Verify that polar and Cartesian coordinates as shown in Figure 10.17 are related by equations 10.9.

SECTION 10.3

Curves in Polar Coordinates

A curve is defined explicitly in Cartesian coordinates by equations $y = f(x)$ or $x = g(y)$, and implicitly by $F(x, y) = 0$. A point is on a curve if and only if its Cartesian coordinates (x, y) satisfy the equation of the curve.

Analogously, a curve is defined explicitly in polar coordinates if its equation is expressed in either of the forms

$$r = f(\theta) \qquad \text{or} \qquad \theta = g(r), \tag{10.10}$$

and is defined implicitly when its equation is given in the form

$$F(r, \theta) = 0. \tag{10.11}$$

As discussed in the preceding section, a point may be represented by an infinite number of sets of polar coordinates. A point is on a curve if at least one of these sets of polar coordinates (r, θ) satisfies the equation of the curve. It follows then, that all sets of polar coordinates for a point do not necessarily satisfy the equation. For example, the origin or pole has polar coordinates $(0, \theta)$ for any θ whatsoever. But only those coordinates of the form $(0, (2n+1)\pi)$ satisfy the equation $r = 1 + \cos\theta$. Likewise, the polar coordinates $(1, \pi)$ satisfy the equation $r = \sin(\theta/2)$, but the coordinates $(1, 3\pi)$ of the same point do not satisfy this equation.

Often we are required to transform the equation of a curve from Cartesian coordinates to polar coordinates, and vice versa. To transform from Cartesian to polar is straightforward: replace each x with $r\cos\theta$ and each y with $r\sin\theta$. For example, the equation $x^2 + y^2 = 9$ describes a circle centred at the origin with radius 3. In polar coordinates, its equation is

$$9 = (r\cos\theta)^2 + (r\sin\theta)^2 = r^2\cos^2\theta + r^2\sin^2\theta = r^2.$$

Consequently, $r = 3$ is the equation of this circle in polar coordinates.

EXAMPLE 10.7

Find equations in polar coordinates for the following curves:

(a) $2x + 3y = 3$ (b) $x^2 - 2x + y^2 = 0$ (c) $x^2 + y^2 = \sqrt{x^2 + y^2} - 4x$

SOLUTION (a) For $2x + 3y = 3$, we obtain

$$3 = 2r\cos\theta + 3r\sin\theta, \qquad \text{or} \qquad r = \frac{3}{2\cos\theta + 3\sin\theta}.$$

(b) For the equation $x^2 - 2x + y^2 = 0$, we have

$$0 = -2x + (x^2 + y^2) = -2r\cos\theta + r^2 = r(r - 2\cos\theta).$$

Thus,

$$r = 0 \qquad \text{or} \qquad r = 2\cos\theta.$$

Since $r = 0$ defines the pole, and this point also satisfies $r = 2\cos\theta$ (for $\theta = \pi/2$), it follows that we need only write $r = 2\cos\theta$.

(c) For the curve with equation $x^2 + y^2 = \sqrt{x^2 + y^2} - 4x$, we obtain

$$r^2 = r - 4r\cos\theta = r(1 - 4\cos\theta).$$

Thus,

$$0 = r^2 - r(1 - 4\cos\theta) = r(r - 1 + 4\cos\theta),$$

from which we have

$$r = 0 \qquad \text{or} \qquad r = 1 - 4\cos\theta.$$

Again the pole satisfies the second of these equations, and therefore the equation of the curve in polar coordinates is $r = 1 - 4\cos\theta$. ■

To transform equations of curves from polar to Cartesian coordinates can sometimes be more difficult, principally because we have no substitution for θ. If, however, the equation involves $\cos\theta$ and/or $\sin\theta$, we use equations 10.6.

EXAMPLE 10.8

Find equations in Cartesian coordinates for the following curves:

(a) $r = 1 + \cos\theta$ (b) $r^2 \cos 2\theta = 1$ (c) $r^2 = 9 \sin 2\theta$

SOLUTION (a) We use equations 10.5 and 10.6 to obtain

$$\sqrt{x^2 + y^2} = 1 + \frac{x}{\sqrt{x^2 + y^2}},$$

and this equation simplifies to

$$(x^2 + y^2 - x)^2 = x^2 + y^2.$$

(b) For $r^2 \cos 2\theta = 1$, we use double-angle formula 9.23c to write the equation in terms of $\cos\theta$ rather than $\cos 2\theta$, and then use equations 10.5 and 10.6,

$$1 = r^2(2\cos^2\theta - 1) = (x^2 + y^2)\left(2\frac{x^2}{x^2 + y^2} - 1\right) = (x^2 + y^2)\frac{x^2 - y^2}{x^2 + y^2} = x^2 - y^2.$$

(c) This time we use double-angle formula 9.23a on $\sin 2\theta$,

$$x^2 + y^2 = 18 \sin\theta\cos\theta = 18\frac{y}{\sqrt{x^2 + y^2}}\frac{x}{\sqrt{x^2 + y^2}} \quad \text{or}$$

$$(x^2 + y^2)^2 = 18xy.$$

■

Examples 10.7 and 10.8 illustrate that equations for some curves become much simpler when expressed in polar coordinates. It is also true that some curves are more easily sketched using their polar representation. For instance, consider the curve $r = 1 + \cos\theta$ of Example 10.8. To sketch this curve, we could construct a table of values (Table 10.2) and obtain the set of points in Figure 10.18(a). These points could then be joined to produce the correct curve in Figure 10.18(b). It is called a *cardioid*. The task would be much more complicated if we were to use the corresponding equation in Cartesian coordinates.

TABLE 10.2

θ	$\frac{-3\pi}{4}$	$-\frac{\pi}{2}$	$-\frac{\pi}{4}$	0	$\frac{\pi}{4}$	$\frac{\pi}{2}$	$\frac{3\pi}{4}$	π
r	0.29	1	1.71	2	1.71	1	0.29	0

FIGURE 10.18

(a)

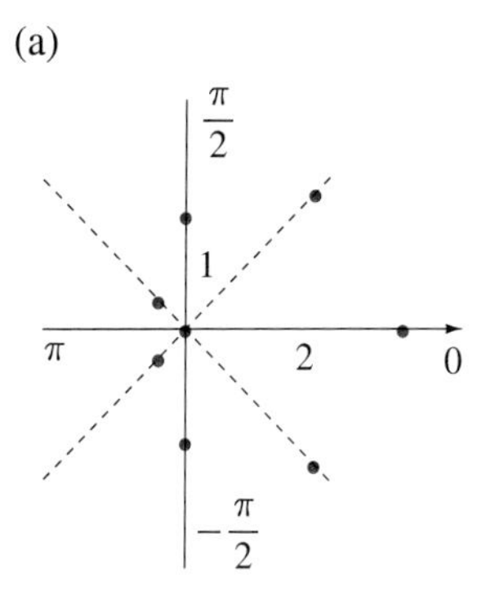

(b)

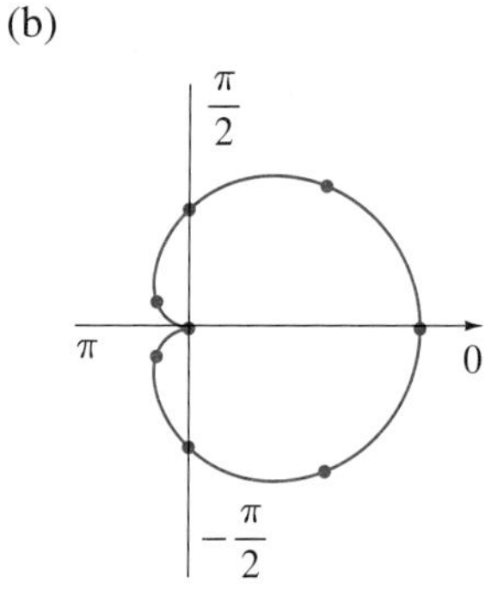

(c)

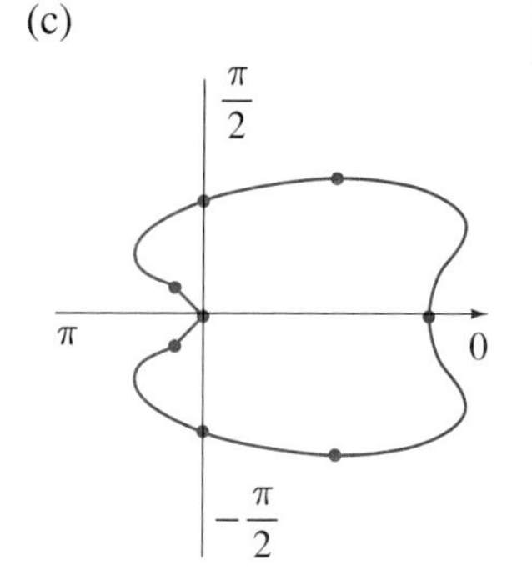

The curve shown in Figure 10.18(b) is not the only way to join the eight points in Figure 10.18(a). For instance, Figure 10.18(c) shows a curve which also passes through these eight points, but which bears little resemblance to the cardioid. This could be rectified by constructing an increasingly larger table of values, but there is an alternative method that is much more efficient.

We draw a Cartesian coordinate system consisting of a horizontal θ-axis and a vertical r-axis. On this set of axes we sketch a graph of the *function* $r = f(\theta) = 1 + \cos\theta$ (Figure 10.19). This is *not* the required curve; it is a sketch of the function $f(\theta)$, illustrating values of r for various values of θ. We might say that this sketch represents an "infinite table of values."

FIGURE 10.19

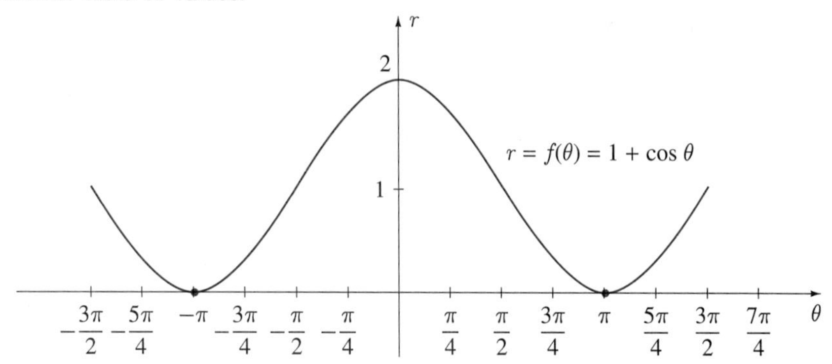

To draw the cardioid $r = 1 + \cos\theta$, we now read pairs of polar coordinates (r, θ) from this sketch, and interpret r as radial distance and θ as rotation. Suppose we begin with the two points $(2, 0)$ and $(1, \pi/2)$. Figure 10.19 indicates that as θ increases from 0 to $\pi/2$, values of r decrease from 2 to 1. This means that as we rotate from the $\theta = 0$ line to the $\theta = \pi/2$ line in the first quadrant, radial distances from the origin become smaller. This is shown in Figure 10.20(a). As rotation is increased from $\pi/2$ to π, Figure 10.19 shows that radial distances continue to decrease, eventually reaching 0 at an angle of π radians as in Figure 10.20(b). Notice that the line $\theta = \pi$ is tangent to the curve at the pole, reflecting the fact that the pole is attained for an angle of π radians. Consideration of the graph in Figure 10.19 to the left of $\theta = 0$ leads to that part of the cardioid below the $\theta = 0$ and $\theta = \pi$ lines in Figure 10.20(c). The symmetry of Figure 10.19 about the r-axis is reflected in the symmetry of Figure 10.20(c) about the $\theta = 0$ and $\theta = \pi$ lines. Since the function $r = f(\theta) = 1 + \cos\theta$ is 2π-periodic, only values of θ in the interval $-\pi < \theta \leq \pi$ need be considered. Values outside this interval retrace previous points.

FIGURE 10.20

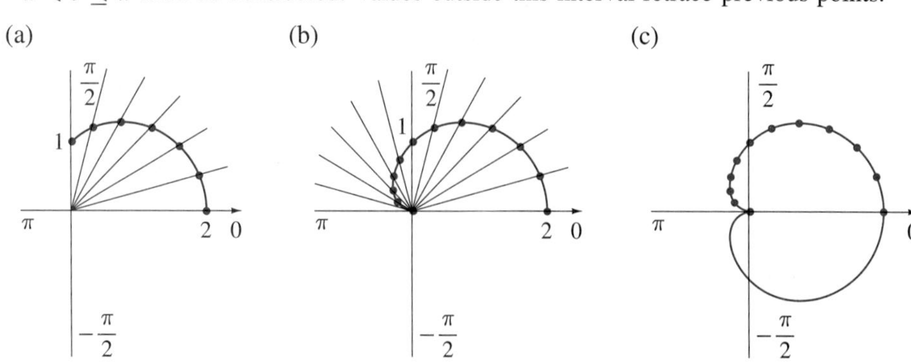

EXAMPLE 10.9

Sketch the following curves:

(a) $r^2 = 9\sin 2\theta$ (b) $r = |1 - 4\cos\theta|$ (c) $r = \theta, \quad \theta \geq 0$

SOLUTION (a) If $r^2 = 9\sin 2\theta$, then $r = 3\sqrt{\sin 2\theta}$. A sketch of the function $\sin 2\theta$ in Cartesian coordinates is shown in Figure 10.21(a), and from this we obtain the sketch of $f(\theta) = 3\sqrt{\sin 2\theta}$ in Figure 10.21(b). The curve $r = 3\sqrt{\sin 2\theta}$ can now be obtained by reading polar coordinate pairs (r, θ) from this graph, the result being shown in Figure 10.22. It is called a *lemniscate*.

FIGURE 10.21

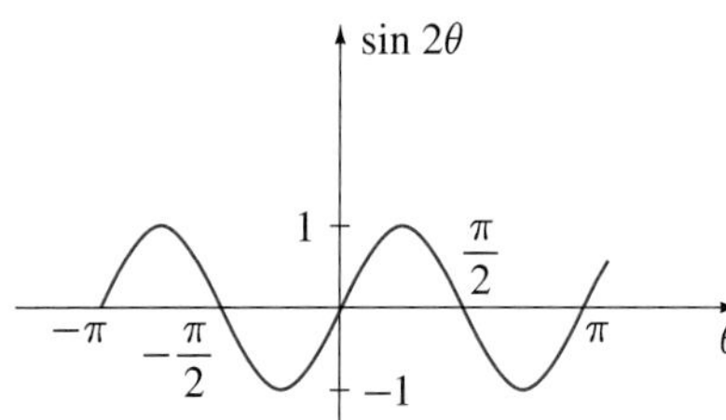

(b)

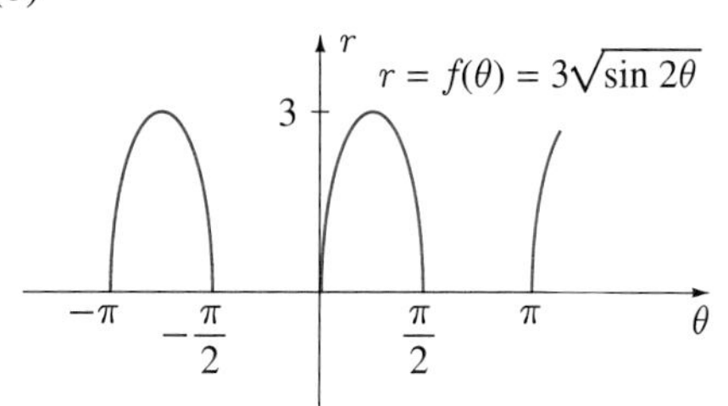

FIGURE 10.22

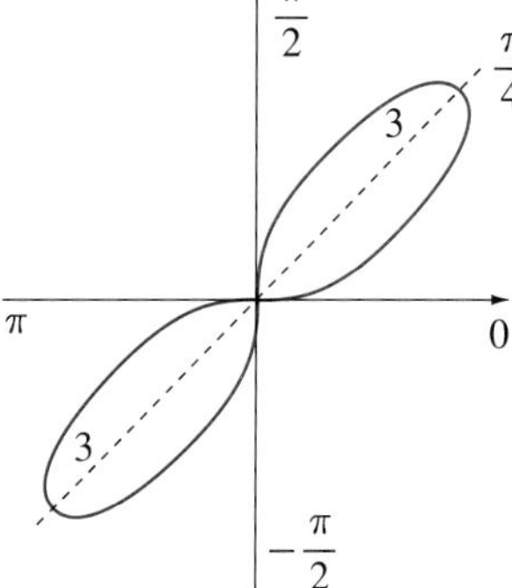

(b) A sketch of the function $-4\cos\theta$ is shown in Figure 10.23(a), and a vertical shift of one unit yields the sketch of $f(\theta) = 1 - 4\cos\theta$ in Figure 10.23(b). When that portion of the curve below the θ-axis is turned upside down, a sketch of the function $r = f(\theta) = |1 - 4\cos\theta|$ is obtained as in Figure 10.23(c). The values of θ at which $r = 0$ are important in sketching the curve $r = |1 - 4\cos\theta|$; they have been identified in this diagram. From Figure 10.23(c) we obtain the curve in Figure 10.24.

FIGURE 10.23

(a)

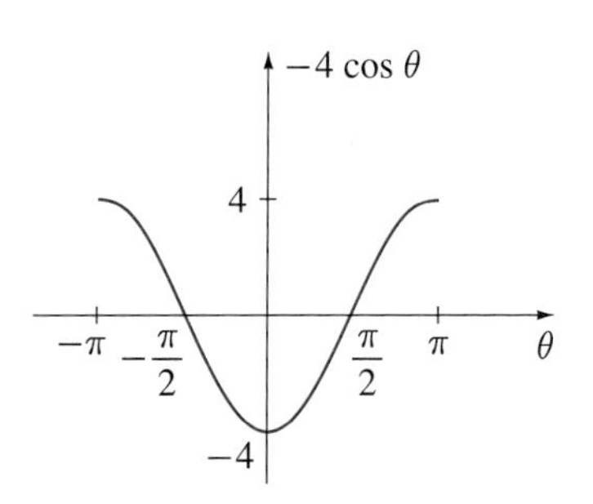

(b)

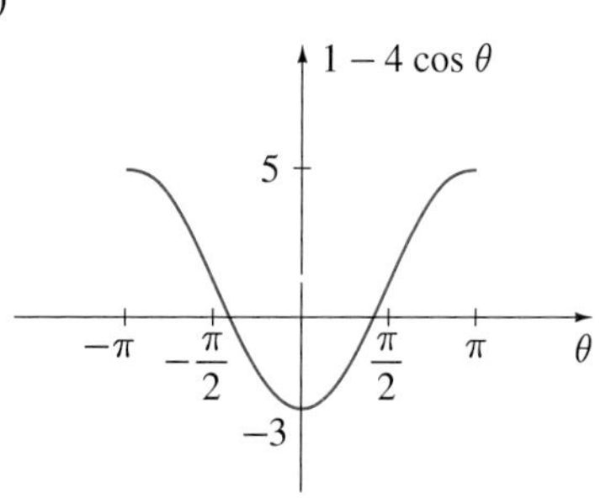

(c)

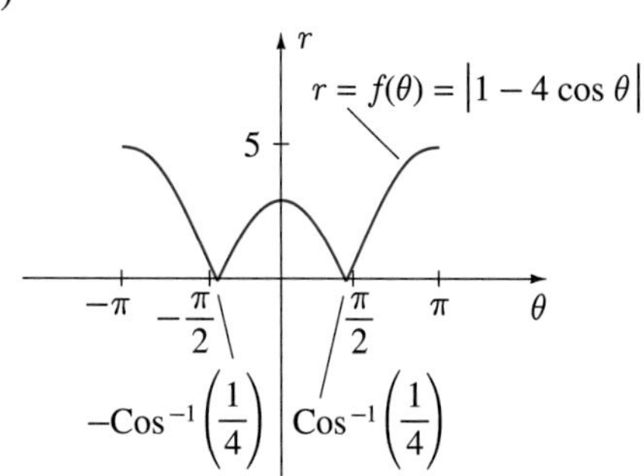

FIGURE 10.24

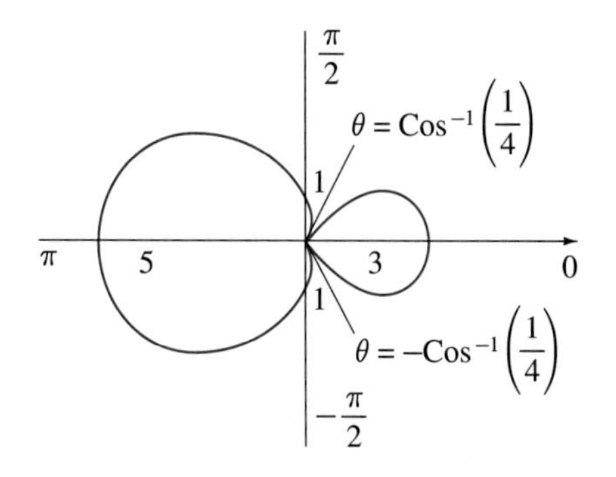

(c) The straight line $r = \theta$ in Figure 10.25 leads to the spiral in Figure 10.26.

FIGURE 10.25

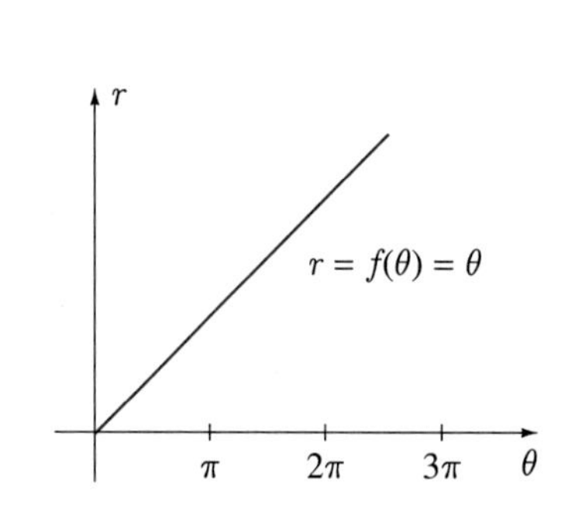

FIGURE 10.26

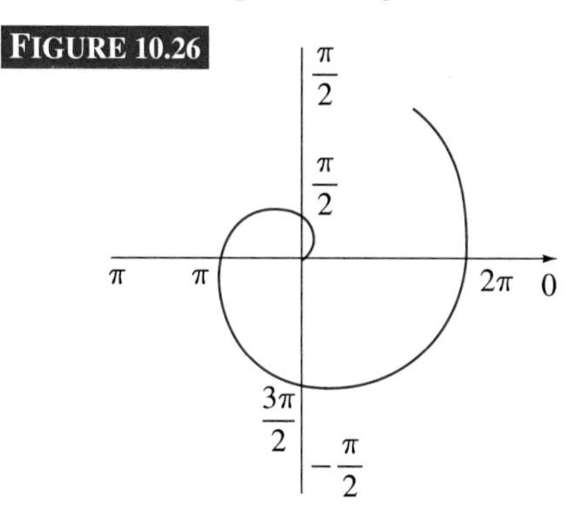

■

To find points of intersection of two curves whose equations are given in Cartesian coordinates, we solve the two equations simultaneously for all (real) solutions. Each solution represents a distinct point of intersection. For curves whose equations are given in polar coordinates, the situation is somewhat more complex because we have multiple names for points. To find the points of intersection, we again solve the equations simultaneously for all solutions. Each solution represents a point of intersection; but as points have many sets of polar coordinates, some of these solutions may represent the same point. In addition, it may also happen that one set of polar coordinates for a point of intersection satisfies one equation, whereas a different set satisfies the other equation. Particularly troublesome in this respect is the pole, which has so many sets of polar coordinates. The best way to handle these difficulties is to sketch the two curves.

EXAMPLE 10.10

Find points of intersection for the curves $r = \sin\theta$ and $r = 1 - \sin\theta$.

SOLUTION If we set $\sin\theta = 1 - \sin\theta$, then $\sin\theta = 1/2$. All solutions of this equation are defined by

$$\theta = \begin{cases} \dfrac{\pi}{6} + 2n\pi \\ \dfrac{5\pi}{6} + 2n\pi \end{cases}$$

where n is an integer. Sketches of the curves in Figure 10.27 indicate that these give the points of intersection $(1/2, \pi/6)$ and $(1/2, 5\pi/6)$. The figure also indicates that the origin is a point of intersection of the curves. We did not obtain this point by solving $r = \sin\theta$ and $r = 1 - \sin\theta$ because different values of θ yield $r = 0$ in $r = \sin\theta$ and $r = 1 - \sin\theta$. To obtain $r = 0$ from $r = \sin\theta$, θ must be one of the values $n\pi$, whereas to obtain $r = 0$ from $r = 1 - \sin\theta$, θ must be one of the values $\pi/2 + 2n\pi$. Thus, both curves pass through the pole, but the pole cannot be obtained by simultaneously solving the equations of the curves.

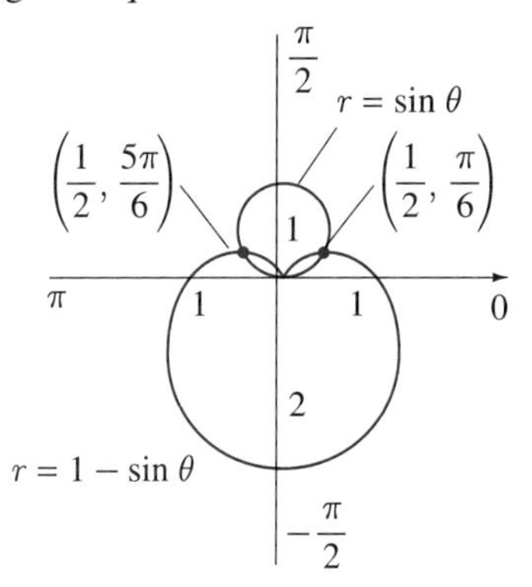

FIGURE 10.27

■

Slopes of Curves in Polar Coordinates

When a curve has polar equation $r = f(\theta)$, $\alpha \le \theta \le \beta$, substitution into equations 10.4 gives

$$x = f(\theta)\cos\theta, \quad y = f(\theta)\sin\theta, \quad \alpha \le \theta \le \beta. \tag{10.12}$$

These are parametric equations for the curve, where the parameter is the polar angle θ. Equation 10.2 gives

$$\frac{dy}{dx} = \frac{\dfrac{dy}{d\theta}}{\dfrac{dx}{d\theta}} = \frac{f'(\theta)\sin\theta + f(\theta)\cos\theta}{f'(\theta)\cos\theta - f(\theta)\sin\theta}. \tag{10.13}$$

This formula defines the slope of the tangent line to a curve which has polar equation $r = f(\theta)$.

EXAMPLE 10.11

Find the points on the cardioid $r = 1 + \sin\theta$ at which the tangent line is horizontal.

SOLUTION The cardioid is shown in Figure 10.28, and there are clearly three points at which the tangent line is horizontal. To find them we use equation 10.13 to write

$$0 = \frac{dy}{dx} = \frac{(\cos\theta)\sin\theta + (1 + \sin\theta)\cos\theta}{(\cos\theta)\cos\theta - (1 + \sin\theta)\sin\theta}.$$

Since the numerator must vanish, we set

$$0 = \cos\theta\sin\theta + \cos\theta + \sin\theta\cos\theta = \cos\theta(1 + 2\sin\theta).$$

From $\cos\theta = 0$, we obtain $\theta = \pi/2$ and from $1 + 2\sin\theta = 0$, we take $\theta = -\pi/6$, $-5\pi/6$. Thus, the points at which the cardioid has a horizontal tangent line have Cartesian coordinates $(0, 2)$ and $(\pm\sqrt{3}/4, -1/4)$.

FIGURE 10.28

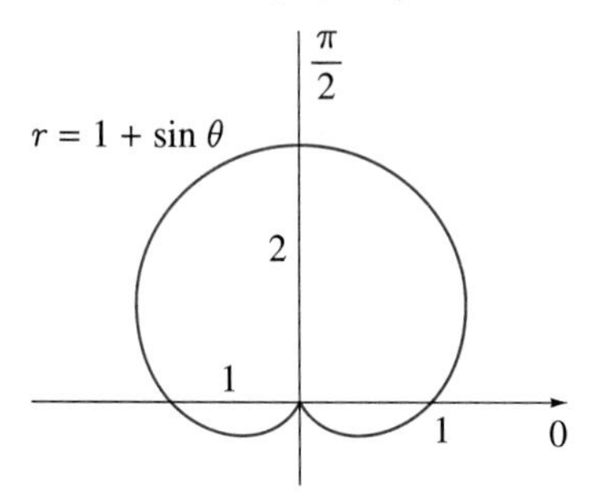

■

Lengths of Curves in Polar Coordinates

Recall that the length of a curve defined parametrically by $x = x(t)$ and $y = y(t)$ is given by formula 10.3:

$$L = \int_\alpha^\beta \sqrt{\left(\frac{dx}{dt}\right)^2 + \left(\frac{dy}{dt}\right)^2}\, dt.$$

If we substitute from equations 10.12 into this formula with t replaced by θ, we obtain

$$\begin{aligned} L &= \int_\alpha^\beta \sqrt{[f'(\theta)\cos\theta - f(\theta)\sin\theta]^2 + [f'(\theta)\sin\theta + f(\theta)\cos\theta]^2}\, d\theta \\ &= \int_\alpha^\beta \sqrt{[f'(\theta)]^2 + [f(\theta)]^2}\, d\theta. \end{aligned}$$

Thus, we may write for the length of a curve $r = f(\theta)$, $\alpha \le \theta \le \beta$,

$$L = \int_\alpha^\beta \sqrt{r^2 + \left(\frac{dr}{d\theta}\right)^2}\, d\theta. \qquad (10.14)$$

EXAMPLE 10.12

Find the length of the cardioid $r = 1 - \cos\theta$.

SOLUTION According to equation 10.14 (see Figure 10.29),

$$\begin{aligned}
L &= \int_0^{2\pi} \sqrt{(1-\cos\theta)^2 + (\sin\theta)^2}\, d\theta \\
&= \int_0^{2\pi} \sqrt{2}\sqrt{1-\cos\theta}\, d\theta \\
&= \sqrt{2}\int_0^{2\pi} \sqrt{1 - \left[1 - 2\sin^2\left(\frac{\theta}{2}\right)\right]}\, d\theta \\
&= 2\int_0^{2\pi} \sin\left(\frac{\theta}{2}\right) d\theta \\
&= 2\left\{-2\cos\left(\frac{\theta}{2}\right)\right\}_0^{2\pi} \\
&= 8.
\end{aligned}$$

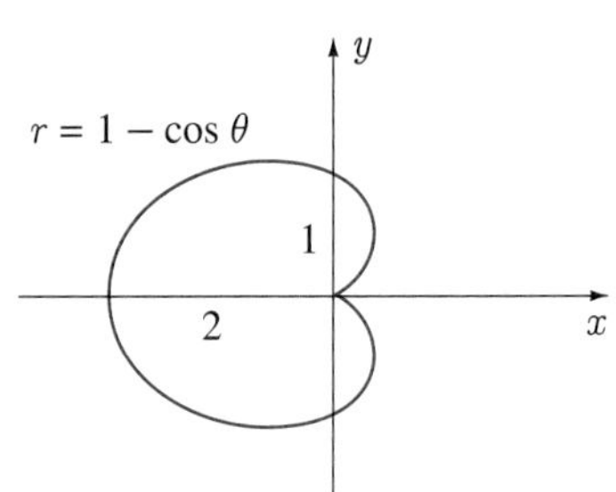

FIGURE 10.29

EXERCISES 10.3

In Exercises 1–10 find an equation for the curve in polar coordinates, and sketch the curve.

1. $x + 2y = 5$ **2.** $y = -x$

3. $x^2 + y^2 = 3$ **4.** $x^2 - 2x + y^2 - 2y + 1 = 0$

5. $y = 4x^2$ **6.** $x^2 + 2y^2 = 3$

7. $x^2 + y^2 = x$ **8.** $x^2 + y^2 = \sqrt{x^2 + y^2} - x$

9. $(x^2 + y^2)^2 = x$ **10.** $y = 1/x^2$

In Exercises 11–20 find the equation of the curve in Cartesian coordinates, and sketch the curve.

11. $r = 5$ **12.** $\theta = 1$

13. $r = 3\sin\theta$ **14.** $r^2 = 4\sin 2\theta$

15. $r = 3 + 3\sin\theta$ **16.** $r = 2\sin 2\theta$

17. $r^2 = -4\cos 2\theta$ **18.** $r = 3 - 4\cos\theta$

19. $r = 5\csc\theta$ **20.** $r = \cot^2\theta\csc\theta$

21. Sketch the curves:

(a) $r = 2 + 2\sin\theta$ (b) $r = 2 + 4\sin\theta$

(c) $r = 4 + 2\sin\theta$

In Exercises 22–25 find all points of intersection for the curves.

22. $r = 2,\quad r^2 = 8\cos 2\theta$

23. $r = \cos\theta,\quad r = 1 + \cos\theta$

24. $r = 1 + \cos\theta,\quad r = 2 - 2\cos\theta$

25. $r = 1,\quad r = 2\cos 2\theta$

26. Show that if $f(\theta)$ is an even function, then the curve $r = f(\theta)$ is symmetric about the lines $\theta = 0$ and $\theta = \pi$ (or x-axis). Illustrate with two examples.

In Exercises 27–34 sketch the curve.

27. $r = \sin 3\theta$ **28.** $r = \cos 2\theta$

29. $r = \sin 4\theta$ **30.** $r^2 = \theta$

31. $r = e^\theta$ **32.** $r = 2\sin(\theta/2)$

33. $r = -2\cos(\theta/2)$ **34.** $r = 1 + \cos(\theta + \pi/6)$

In Exercises 35–38 find the slope of the curve at the given value of θ.

35. $r = 9\cos 2\theta$ at $\theta = \pi/6$

36. $r^2 = 9\sin 2\theta$ at $\theta = -5\pi/6$

37. $r = 3 - 5\cos\theta$ at $\theta = 3\pi/4$

38. $r = 2\cos(\theta/2)$ at $\theta = \pi/2$

39. Find the slope of the tangent line to the curve $r = 3/(1 - \sin\theta)$ at the point with polar coordinates $(6, \pi/6)$ in two ways:
    (a) by using 10.13;
    (b) by finding the equation of the curve in Cartesian coordinates, and calculating dy/dx.

40. Find the length of the cardioid $r = a(1 + \sin\theta)$. (a is a constant.)

41. (a) The electrostatic charge distribution consisting of a charge $q > 0$ at the point with polar coordinates $(s, 0)$ and a charge $-q$ at (s, π) is called a *dipole*. When s is very small, the lines of force for the dipole are defined by the equation $r = A\sin^2\theta$, where each value of the constant $A > 0$ defines a particular line of force. Sketch lines of force for $A = 1$, 2, and 3.

    (b) The equipotential lines for the dipole are defined by $r^2 = B\cos\theta$, where B is a constant. Sketch equipotential lines for $B = \pm 1, \pm 2$, and ± 3.

42. Sketch the *bifolium* $(x^2 + y^2)^2 = x^2 y$.

43. Curves with equations of the form $r = a(1 \pm \cos\theta)$ or $r = a(1 \pm \sin\theta)$ ($a > 0$ a constant) are called *cardioids*.
    (a) Sketch all such curves.
    (b) Find equations for the cardioids in Cartesian coordinates.

44. Curves with equations of the form $r^2 = a^2\cos 2\theta$ or $r^2 = a^2\sin 2\theta$ ($a > 0$ a constant) are called *lemniscates*.
    (a) Sketch all such curves.
    (b) Find equations for the lemniscates in Cartesian coordinates.

45. (a) Sketch the curves $r = b \pm a\cos\theta$ and $r = b \pm a\sin\theta$ where a and b are positive constants in the three cases $a < b$, $a = b$, and $a > b$.

    (b) Find equations for the curves in Cartesian coordinates.

    (c) Compare these curves with the cardioids of Exercise 43 when $a = b$.

46. A curve with equation of the form $r = a\sin n\theta$ or $r = a\cos n\theta$ where $a > 0$ is a constant and $n > 0$ is an integer is called a *rose*. Show that the rose has n petals.

47. Show that the roses $r = |a\sin n\theta|$ and $r = |a\cos n\theta|$, where $a > 0$ is a constant and $n > 0$ is an integer, have $2n$ petals.

SECTION 10.4

Areas in Polar Coordinates

In Section 7.1 we used definite integrals to find areas bounded by curves whose equations are conveniently expressed in Cartesian coordinates. In this section we indicate how to find areas bounded by curves whose equations are expressed in polar coordinates. We require the formula

$$\frac{1}{2}r^2(\theta_2 - \theta_1) \tag{10.15}$$

for the area of the shaded sector of the circle in Figure 10.30. This formula results from the fact that the area of the sector is the fractional part $(\theta_2 - \theta_1)/(2\pi)$ of the area πr^2 of the circle.

FIGURE 10.30

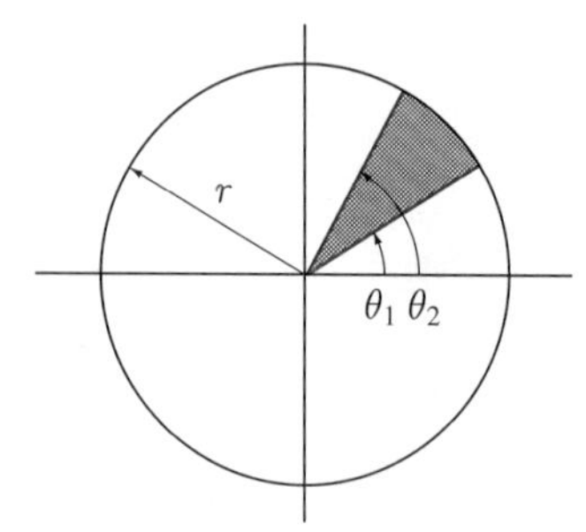

Consider finding the area in Figure 10.31(a) bounded by the radial lines $\theta = \alpha$ and $\theta = \beta$ and the curve $r = f(\theta)$. We divide the area into subareas by means of $n+1$ radial lines $\theta = \theta_i$ where

$$\alpha = \theta_0 < \theta_1 < \theta_2 < \cdots < \theta_{n-1} < \theta_n = \beta.$$

On that part of the curve $r = f(\theta)$ between $\theta = \theta_{i-1}$ and $\theta = \theta_i$, we pick any point with polar coordinates $\left(f(\theta_i^*), \theta_i^*\right)$ as in Figure 10.31(b). If between the lines $\theta = \theta_{i-1}$ and $\theta = \theta_i$ we draw the arc of a circle with centre at the pole and radius $f(\theta_i^*)$, a sector is formed with area

$$\Delta A_i = \frac{1}{2}[f(\theta_i^*)]^2 \Delta\theta_i,$$

where $\Delta\theta_i = \theta_i - \theta_{i-1}$. Since this sector approximates that part of the required area between the radial lines $\theta = \theta_{i-1}$ and $\theta = \theta_i$, we can say that an approximation to the required area is

$$\sum_{i=1}^{n} \Delta A_i = \sum_{i=1}^{n} \frac{1}{2}[f(\theta_i^*)]^2 \Delta\theta_i.$$

FIGURE 10.31

(a)

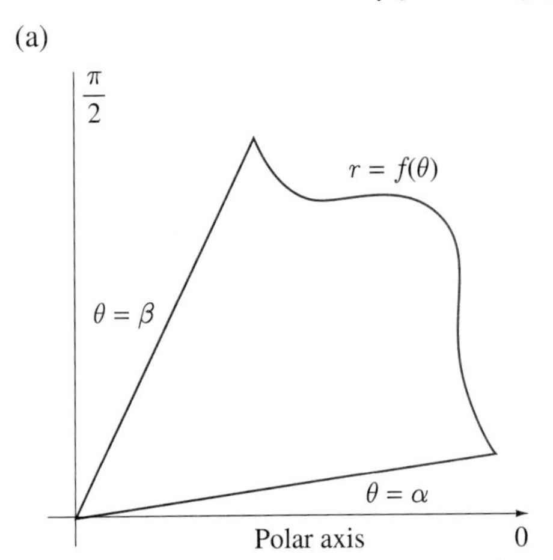

(b)

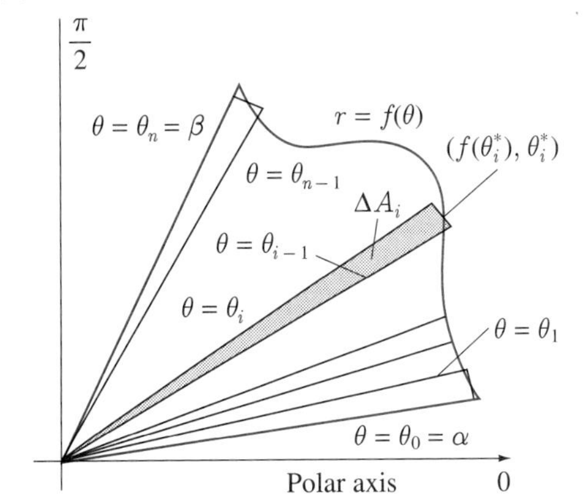

By increasing the number of sectors indefinitely, and at the same time requiring each of the $\Delta\theta_i$ to approach zero, we obtain a better and better approximation, and in the limit

$$\text{Area} = \lim_{\|\Delta\theta_i\| \to 0} \sum_{i=1}^{n} \frac{1}{2}[f(\theta_i^*)]^2 \Delta\theta_i.$$

But this limit is the definition of the definite integral of the function $(1/2)[f(\theta)]^2$ with respect to θ from $\theta = \alpha$ to $\theta = \beta$, and we therefore write

$$\text{Area} = \int_{\alpha}^{\beta} \frac{1}{2}[f(\theta)]^2 \, d\theta. \tag{10.16}$$

In order to arrive at this integral in any given problem, without memorizing it, we use the procedure adopted for definite integrals in Cartesian coordinates discussed in Chapter 7. We draw at angle θ a representative sector of angular width $d\theta$ and radius r (Figure 10.32). The area of this sector is

$$\frac{1}{2}r^2 \, d\theta = \frac{1}{2}[f(\theta)]^2 \, d\theta.$$

If areas of all such sectors from angle α to angle β are added together, and the limit is taken as their widths approach zero, the required area is obtained. But this is the process defined by the definite integral, and we therefore write equation 10.16 for the area. Definite integral 10.16 exists when $f(\theta)$ is continuous for $\alpha \le \theta \le \beta$.

FIGURE 10.32

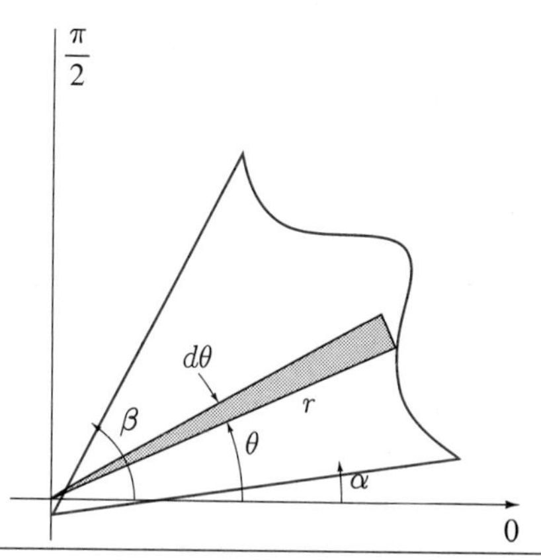

EXAMPLE 10.13

Find the area inside the cardioid $r = 1 + \sin\theta$.

SOLUTION The area of the representative sector in Figure 10.33 is

$$\frac{1}{2}r^2\,d\theta = \frac{1}{2}(1 + \sin\theta)^2\,d\theta,$$

and we must add over all sectors interior to the cardioid. Since areas on either side of the $\theta = \pi/2$ line are identical, we calculate the area to the right and double the result. To find the area to the right of the line $\theta = \pi/2$, we must identify angular positions of the first and last sectors. The first sector is at the pole, and the equation of the cardioid indicates that $r = 0$ when $\sin\theta = -1$; that is, when $\theta = -\pi/2 + 2n\pi$. But which of these values of θ shall we choose? Similarly, the last sector occurs when $r = 2$, in which case $\sin\theta = 1$, and θ could be any of the values $\pi/2 + 2n\pi$. Again, which shall we choose? If we choose $\alpha = -\pi/2$ and $\beta = \pi/2$, then all values of θ in the interval $-\pi/2 \le \theta \le \pi/2$ yield points on the right half of the cardioid with no duplications. Consequently,

$$\begin{aligned}\text{Area} &= 2\int_{-\pi/2}^{\pi/2}\frac{1}{2}(1 + \sin\theta)^2\,d\theta = \int_{-\pi/2}^{\pi/2}(1 + 2\sin\theta + \sin^2\theta)\,d\theta\\ &= \int_{-\pi/2}^{\pi/2}\left(1 + 2\sin\theta + \frac{1 - \cos 2\theta}{2}\right)d\theta\\ &= \left\{\frac{3\theta}{2} - 2\cos\theta - \frac{\sin 2\theta}{4}\right\}_{-\pi/2}^{\pi/2}\\ &= \frac{3\pi}{2}.\end{aligned}$$

FIGURE 10.33

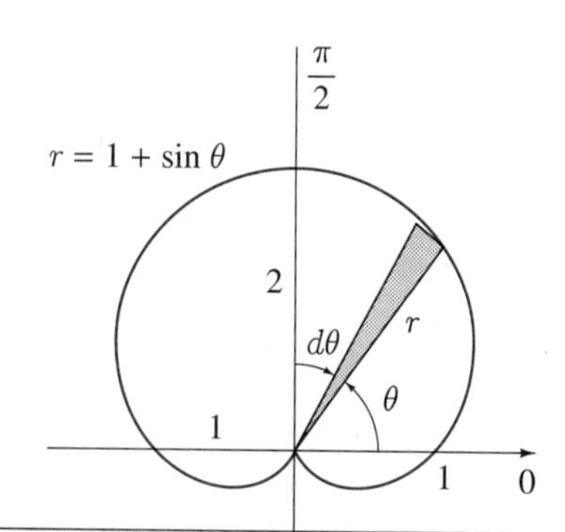

■

EXAMPLE 10.14

Find the area common to the circles $x^2 + y^2 = 4$ and $x^2 + y^2 = 4x$.

SOLUTION Equations for the circles in polar coordinates are $r = 2$ and $r = 4\cos\theta$, and they intersect in the points with polar coordinates $(2, \pm\pi/3)$ (Figure 10.34). If A_s is the area above the x-axis, outside $x^2 + y^2 = 4$ and inside $x^2 + y^2 = 4x$, then the area common to the circles is the area of either circle less twice A_s:

$$\text{Area} = \pi(2)^2 - 2A_s.$$

The area of the representative element is the difference in the areas of two sectors:

$$\frac{1}{2}(4\cos\theta)^2\,d\theta - \frac{1}{2}(2)^2\,d\theta = 2(4\cos^2\theta - 1)\,d\theta.$$

Since all sectors in A_s can be identified by values of θ between 0 and $\pi/3$, the required area is

$$\text{Area} = 4\pi - 2\int_0^{\pi/3} 2(4\cos^2\theta - 1)\,d\theta = 4\pi - 4\int_0^{\pi/3}[2(1+\cos 2\theta) - 1]\,d\theta$$

$$= 4\pi - 4\{\theta + \sin 2\theta\}_0^{\pi/3} = 4\pi - 4\left(\frac{\pi}{3} + \frac{\sqrt{3}}{2}\right) = \frac{8\pi}{3} - 2\sqrt{3}.$$

■

FIGURE 10.34

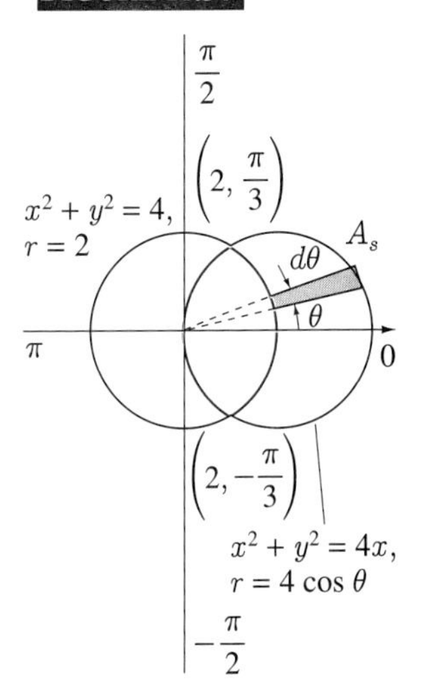

EXERCISES 10.4

In Exercises 1–10 find the area enclosed by the curve.

1. $r = 3\sin\theta$

2. $r = -6\cos\theta$

3. $r = 2\sin 2\theta$

4. $r^2 = 2\sin 2\theta$

5. $r^2 = -\cos\theta$

6. $r = 2 - 2\cos\theta$

7. $r = 4 - 4\cos\theta$

8. $r = 4 - 2\cos\theta$

9. $r = \sin 3\theta$

10. $r = 2(\cos\theta + \sin\theta)$

In Exercises 11–21 find the indicated area.

11. Outside $r = 3$ but inside $r = 6\sin\theta$

12. Inside both $r = 1$ and $r = 1 - \sin\theta$

13. Inside $r = 2\sin 2\theta$ but outside $r = 1$

14. Inside both $r = 2 + 2\cos\theta$ and $r = 2 - 2\cos\theta$

15. Inside both $r = \sin\theta$ and $r = \cos\theta$

16. Inside both $r = \cos\theta$ and $r = 1 - \cos\theta$

17. Inside $r = 1 - 4\cos\theta$

18. Inside $r = 4 + 3\sin\theta$ but outside $r = 2$

19. Inside $r = |1 - 4\cos\theta|$

20. Inside the bifolium $r = \sin\theta\cos^2\theta$

21. Bounded by $\theta = \pi$ and $r = \theta$, $0 \le \theta \le \pi$

22. (a) Show that in polar coordinates the strophoid

$$y^2 = x^2\frac{a - x}{a + x},$$

where $a > 0$ is a constant, takes the form $r = a\cos 2\theta\sec\theta$.

(b) Sketch the curve and find the area inside its loop.

SECTION 10.5

Conic Sections

In Section 1.4 we introduced conic sections to illustrate the algebraic-geometric interplay of plane analytic geometry. Each of the conic sections can be visualized as the curve resulting from the intersection of a plane with a pair of right-circular cones (Figure 10.35). Certainly this suggests why the conic sections are so-named, but because of the three-dimensional nature of the cone, an analysis of conic sections from this point of view is not yet possible. In this section we use plane analytic geometry to develop definitions for parabolas, ellipses, and hyperbolas.

FIGURE 10.35

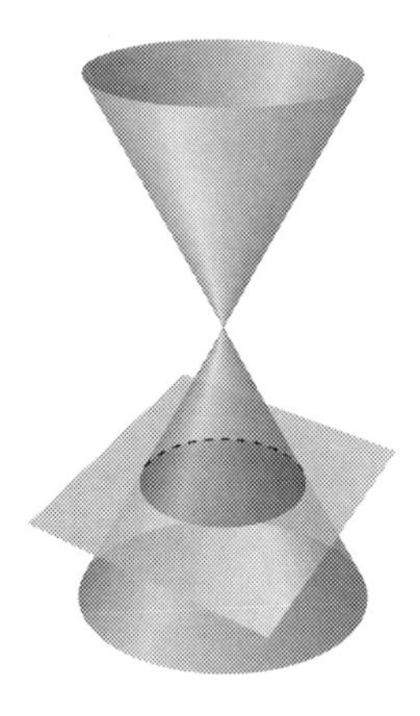

(a) Circle: Plane perpendicular to axis of cones

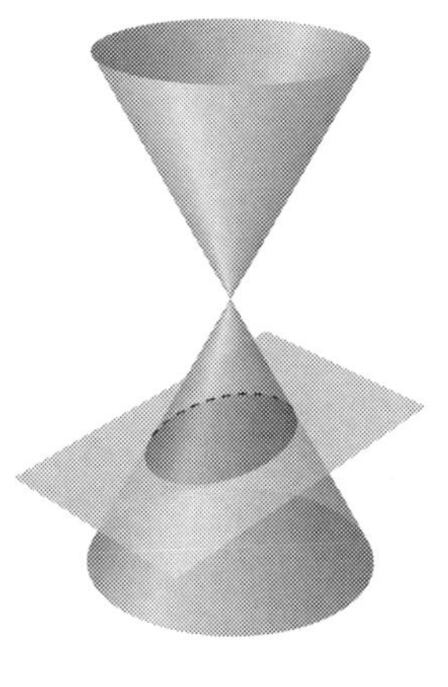

(b) Ellipse: Plane cuts completely across one cone but not perpendicular to axis

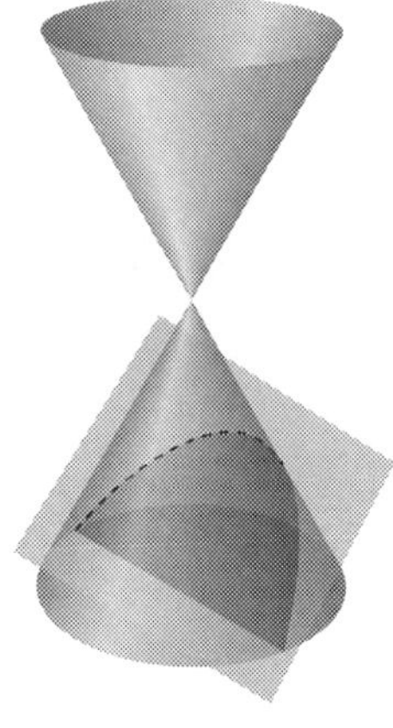

(c) Parabola: Plane cuts only one cone but not completely across

(d) Hyperbola: Plane cuts both cones

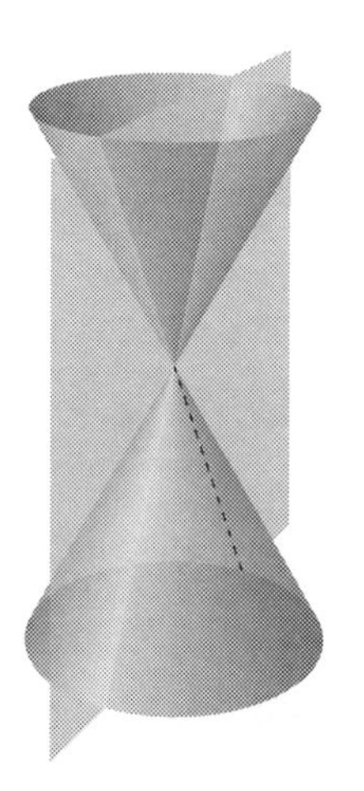

(e) Pair of straight lines: Plane passes through vertex and cuts through both cones

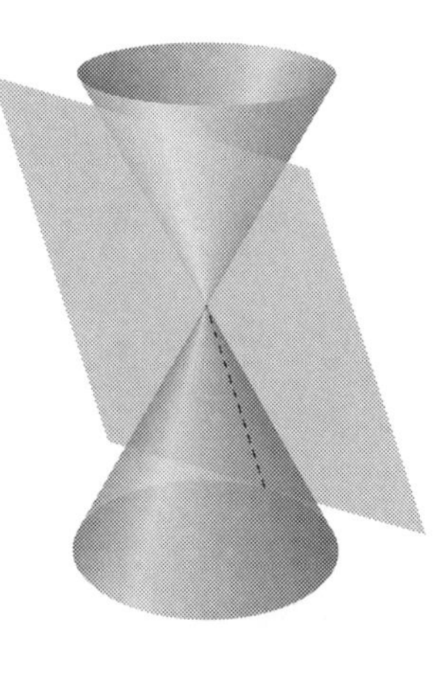

(f) One straight line: Plane passes through vertex and touches both cones

The Parabola

Definition 10.2

A **parabola** is the curve traced out by a point that moves in a plane so that its distances from a fixed point called the *focus* and a fixed line called the *directrix* are always the same.

Suppose the focus of a parabola is the point (p, q) and the directrix is a line $y = r$ parallel to the x-axis as in Figure 10.36(a). If $P(x, y)$ is any point on the parabola, then the fact that its distance from (p, q) must be equal to its distance from $y = r$ is expressed as

$$\sqrt{(x-p)^2 + (y-q)^2} = |y - r|. \tag{10.17}$$

With the absolute values, this equation includes the case of a directrix above the focus as in Figure 10.36(b). If we square both sides of the equation and rearrange terms, we obtain

$$(x-p)^2 = (y-r)^2 - (y-q)^2 = 2y(q-r) + r^2 - q^2.$$

FIGURE 10.36

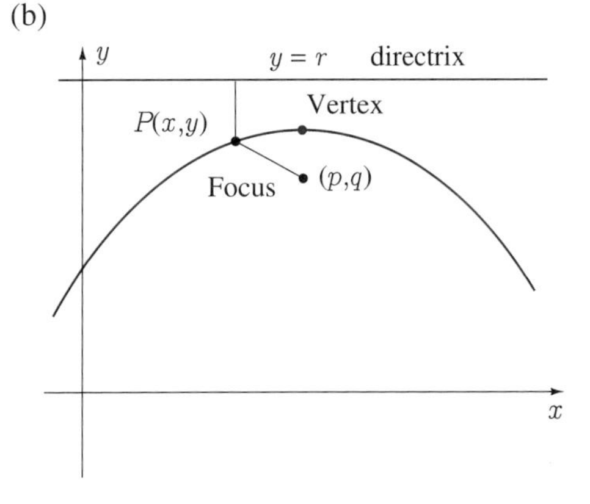

We can solve this equation for y in terms of x; the result is

$$y = \frac{1}{2(q-r)}[(x-p)^2 + (q^2 - r^2)]. \tag{10.18}$$

We could rewrite this equation in our accustomed form $y = ax^2 + bx + c$ for a parabola, but the present form is more informative. Firstly, the line $x = p$ through the focus and perpendicular to the directrix is the line of symmetry for the parabola. Secondly, the parabola opens upward if $q > r$, in which case the focus is above the directrix, and opens downward if $q < r$. Finally, the vertex of the parabola is found by setting $x = p$, in which case $y = (q + r)/2$, halfway between the focus and directrix.

A similar analysis shows that when the directrix is parallel to the y-axis (Figure 10.37), the equation of the parabola is of the form

$$x = \frac{1}{2(p-r)}[(y-q)^2 + (p^2 - r^2)]. \tag{10.19}$$

FIGURE 10.37

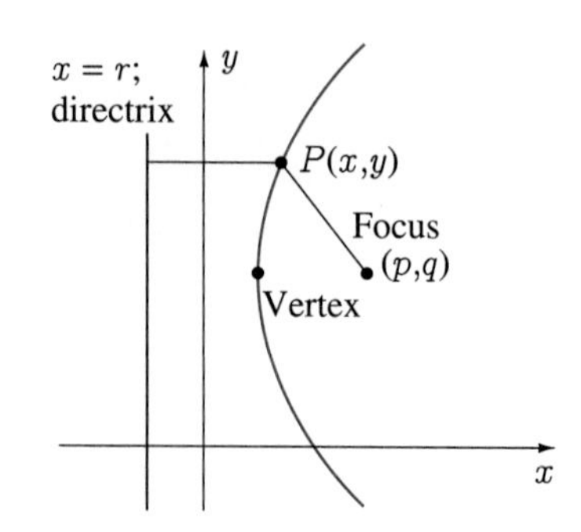

Given the focus and directrix (parallel to a coordinate axis) of a parabola, we can easily find its equation: use formulas 10.18 or 10.19, or follow the algebraic steps leading from 10.17 to 10.18. Conversely, given the equation of a parabola in the form $y = ax^2 + bx + c$ or $x = ay^2 + by + c$, we can identify its focus and directrix (see Exercises 52 and 53).

EXAMPLE 10.15

Find the equation of the parabola that has focus $(2, 4)$ and directrix $x = 6$.

SOLUTION If (x, y) is any point on the parabola (Figure 10.38), the fact that its distance from $(2, 4)$ is equal to its distance from $x = 6$ is expressed as

$$\sqrt{(x-2)^2 + (y-4)^2} = 6 - x.$$

If we square both sides and simplify, the result is $x = (16 + 8y - y^2)/8$.

FIGURE 10.38

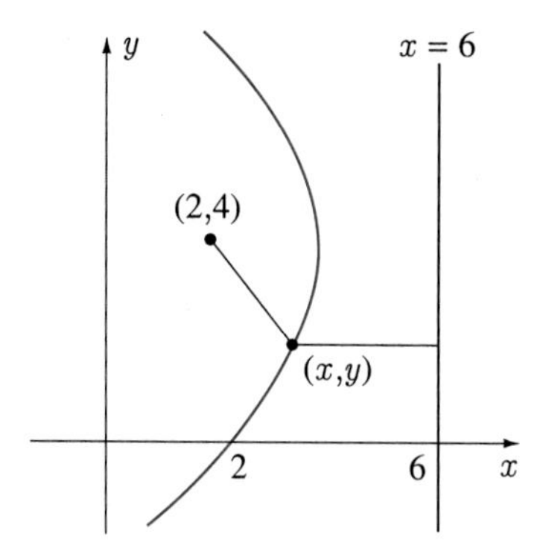

■

The Ellipse

Definition 10.3

An **ellipse** is the curve traced out by a point that moves in a plane so that the sum of its distances from two fixed points called *foci* remains constant.

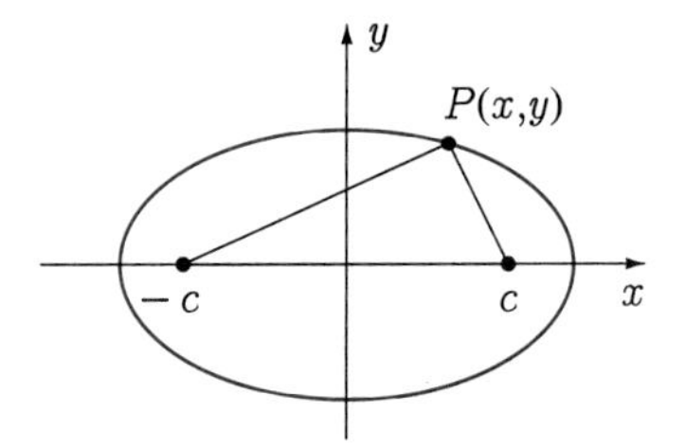

FIGURE 10.39

The equation of an ellipse is simplest when the foci lie on either the x-axis or the y-axis and are equidistant from the origin. Suppose the foci are $(c, 0)$ and $(-c, 0)$ (Figure 10.39), and the sum of the distances from these foci to a point on the ellipse is $2a$, where $a > c \geq 0$. If $P(x, y)$ is any point on the ellipse, Definition 10.3 implies that

$$\sqrt{(x+c)^2 + y^2} + \sqrt{(x-c)^2 + y^2} = 2a. \tag{10.20}$$

If we transpose the second term to the right-hand side and square both sides, we obtain

$$(x+c)^2 + y^2 = 4a^2 - 4a\sqrt{(x-c)^2 + y^2} + (x-c)^2 + y^2,$$

and this equation simplifies to

$$a^2 - cx = a\sqrt{(x-c)^2 + y^2}.$$

Squaring once again leads to

$$a^4 - 2a^2cx + c^2x^2 = a^2(x^2 - 2cx + c^2 + y^2) \qquad \text{or}$$

$$x^2(a^2 - c^2) + a^2y^2 = a^4 - a^2c^2.$$

Division by $a^2(a^2 - c^2)$ gives

$$\frac{x^2}{a^2} + \frac{y^2}{a^2 - c^2} = 1. \tag{10.21}$$

It is customary to denote the y-intercepts of an ellipse by $\pm b$ $(b > 0)$ (Figure 10.40), in which case $b^2 = a^2 - c^2$, and the equation of the ellipse becomes

$$\frac{x^2}{a^2} + \frac{y^2}{b^2} = 1, \qquad \text{or} \qquad b^2x^2 + a^2y^2 = a^2b^2. \tag{10.22}$$

FIGURE 10.40

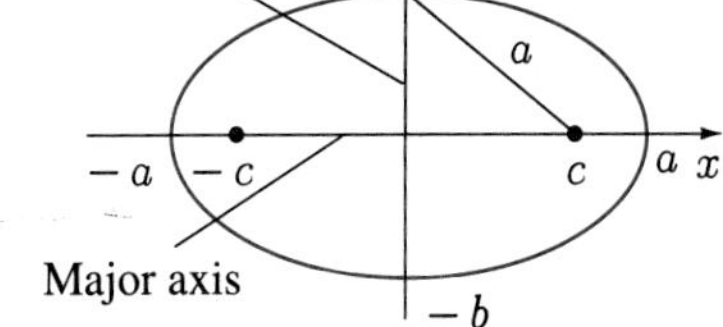

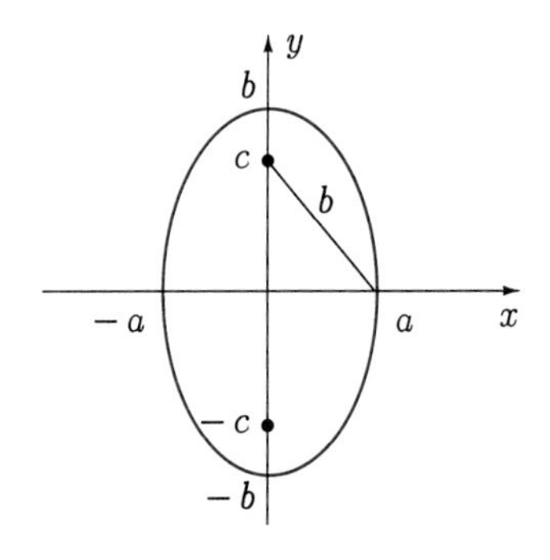

FIGURE 10.41

The line segment across the ellipse and through the foci is called the **major axis** of the ellipse; it has length $2a$ (see Figure 10.40). The midpoint of the major axis is called the **centre** of the ellipse. The line segment across the ellipse, through its centre, and perpendicular to the major axis is called the **minor axis**; it has length $2b$. Note that the line segment joining either end of the minor axis to a focus (Figure 10.40) has length a, and the triangle formed specifies the relationship among a, b, and c, namely, $a^2 = b^2 + c^2$.

A similar analysis shows that when the foci of the ellipse are on the y-axis, equidistant from the origin, the equation of the ellipse is again in form 10.22. In this case, $2b$ is the length of the major axis, $2a$ is the length of the minor axis, and $b^2 = a^2 + c^2$ (Figure 10.41).

What we should remember is that an equation of form 10.22 always specifies an ellipse. The foci are on the longer axis and can be located using $c^2 = |a^2 - b^2|$. The length of the major axis represents the sum of the distances from any point on the ellipse to the foci.

EXAMPLE 10.16

Sketch the ellipse $16x^2 + 9y^2 = 144$, indicating its foci.

SOLUTION If we write the ellipse in the form $x^2/9 + y^2/16 = 1$, its x- and y-intercepts are ± 3 and ± 4. A sketch of the ellipse is therefore as shown in Figure 10.42. The foci must lie on the y-axis at distances $\pm c = \pm\sqrt{4^2 - 3^2} = \pm\sqrt{7}$ from the origin. ■

FIGURE 10.42

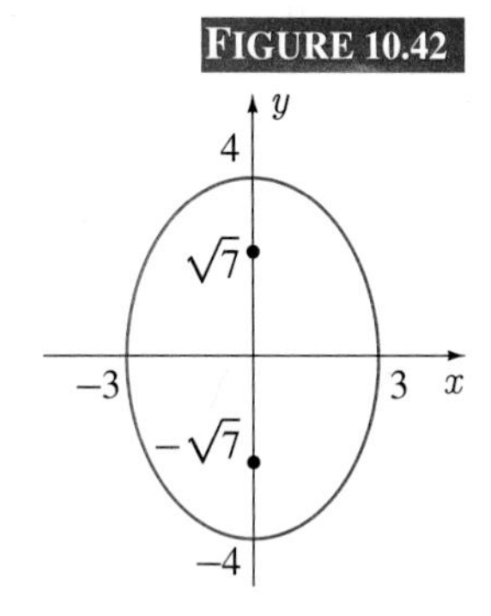

If $a = b$ in equation 10.22, then $x^2 + y^2 = a^2$, and this is the equation for a circle with radius a and centre at the origin. But if $a = b$, the distance from the origin to each focus of the ellipse must be $c = 0$. In other words, a circle may be regarded as a degenerate ellipse whose foci are at one and the same point. It is also true that when c is very small compared to half the length of the major axis (Figure 10.43), the ellipse is shaped very much like a circle. On the other hand, when these lengths are almost equal (Figure 10.44), the ellipse is long and narrow.

FIGURE 10.43

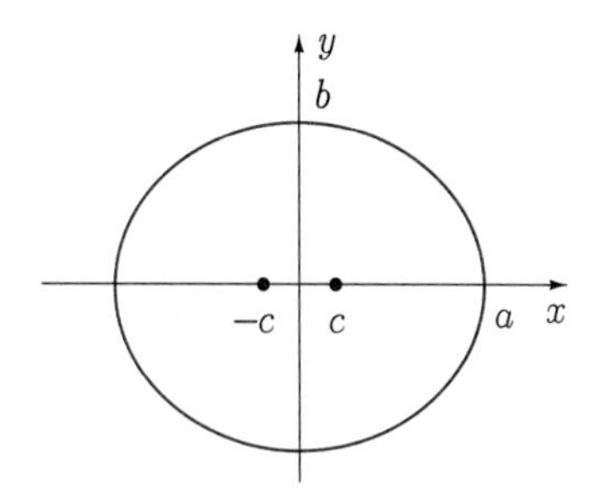

FIGURE 10.44

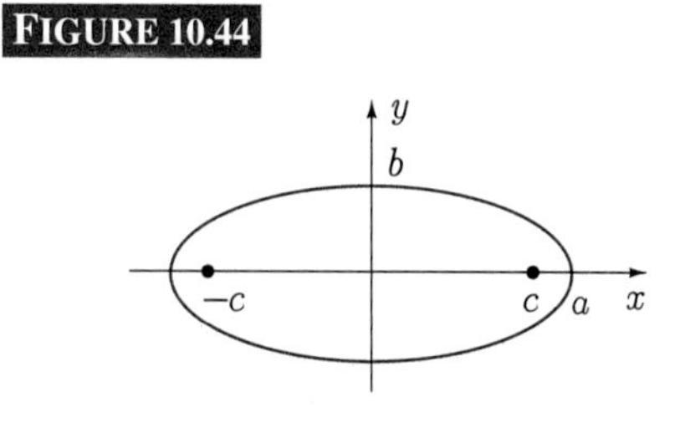

When the centre of an ellipse is at point (h, k) and the foci lie on either the line $x = h$ or $y = k$ (Figure 10.45), the equation for the ellipse is somewhat more complex that 10.22. If $2a$ and $2b$ are again the lengths of the axes of the ellipse, a calculation similar to that leading from 10.20 to 10.22 gives (see Exercise 55)

$$\frac{(x-h)^2}{a^2} + \frac{(y-k)^2}{b^2} = 1. \tag{10.23}$$

Alternatively, the curves in Figure 10.45 are those in Figures 10.39 and 10.41 translated h units in the x-direction and k units in the y-direction. According to Section 1.5, equations for the translated ellipses can be obtained by replacing x and y in $x^2/a^2 + y^2/b^2 = 1$ by $x - h$ and $y - k$ respectively.

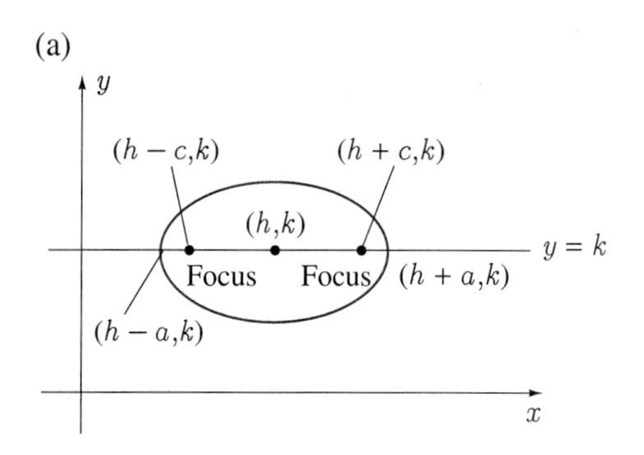

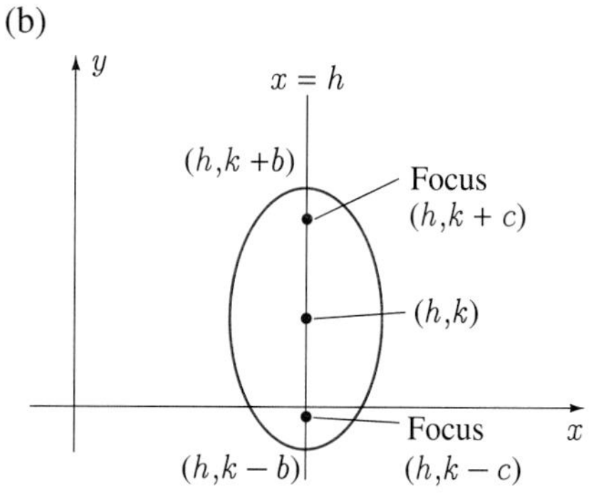

FIGURE 10.45

EXAMPLE 10.17

Sketch the ellipse $16x^2 + 25y^2 - 160x + 50y = 1175$.

SOLUTION

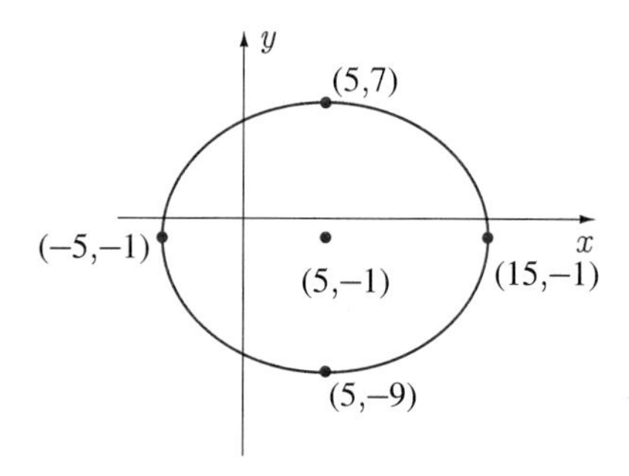

FIGURE 10.46

If we complete the squares on the x- and y-terms, we obtain

$$16(x-5)^2 + 25(y+1)^2 = 1600 \qquad \text{or}$$

$$\frac{(x-5)^2}{100} + \frac{(y+1)^2}{64} = 1.$$

The centre of the ellipse is $(5, -1)$, and lengths of its major and minor axes are 20 and 16 respectively (Figure 10.46). ■

The Hyperbola

Definition 10.4

A **hyperbola** is the path traced out by a point that moves in a plane so that the difference of its distances from two fixed points called *foci* remains constant.

Like the ellipse, the simplest hyperbolas have foci on either the x- or y-axis, equidistant from the origin. Suppose the foci are $(\pm c, 0)$ (Figure 10.47) and the difference in the distances from $P(x, y)$ to these foci is $2a$. Then Definition 10.4 implies that

$$\left|\sqrt{(x+c)^2 + y^2} - \sqrt{(x-c)^2 + y^2}\right| = 2a. \qquad (10.24)$$

This equation can be simplified by a calculation similar to that leading to 10.22; the result is

$$\frac{x^2}{a^2} - \frac{y^2}{b^2} = 1, \tag{10.25}$$

where $b^2 = c^2 - a^2$. The hyperbola has x-intercepts equal to $\pm a$.

When the foci are on the y-axis (Figure 10.48), the equation of the hyperbola becomes

$$\frac{y^2}{b^2} - \frac{x^2}{a^2} = 1, \tag{10.26}$$

where $2b$ is the constant difference in the distances from a point (x, y) to the foci, and $a^2 = c^2 - b^2$. This hyperbola intersects the y-axis at $\pm b$.

FIGURE 10.47

FIGURE 10.48

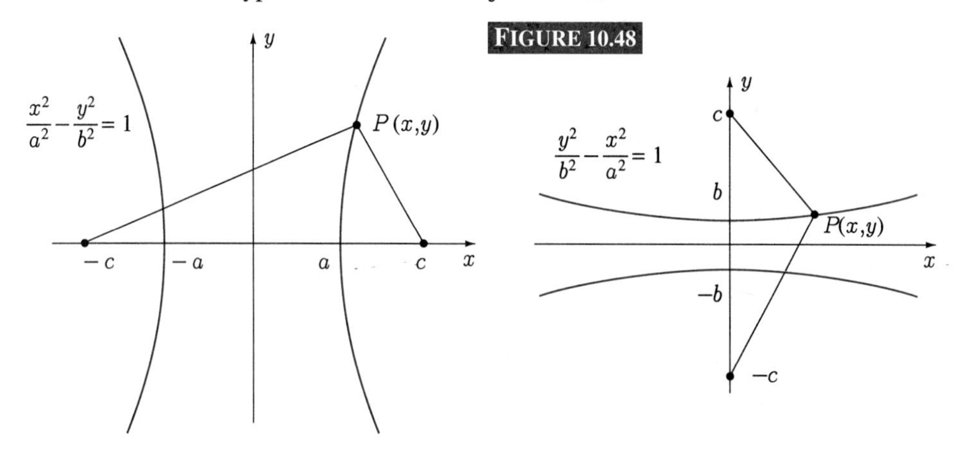

That part of the line segment joining the foci of a hyperbola which is between the two branches of the curve is called the **transverse axis** of the hyperbola; it has length $2a$ as in Figure 10.49(a) or $2b$ as in Figure 10.49(b). The midpoint of the transverse axis is called the **centre** of the hyperbola. The line segment perpendicular to the transverse axis, through its centre, and of length $2b$ as in Figure 10.49(a) or $2a$ as in Figure 10.49(b) is called the **conjugate axis**. Asymptotes of both hyperbolas are the lines $y = \pm bx/a$.

What we should remember is that an equation of form 10.25 or 10.26 specifies a hyperbola. The foci lie on the extension of the transverse axis and can be located using $c^2 = a^2 + b^2$. The length of the transverse axis represents the difference of the distances from any point on the hyperbola to the foci.

FIGURE 10.49

(a)

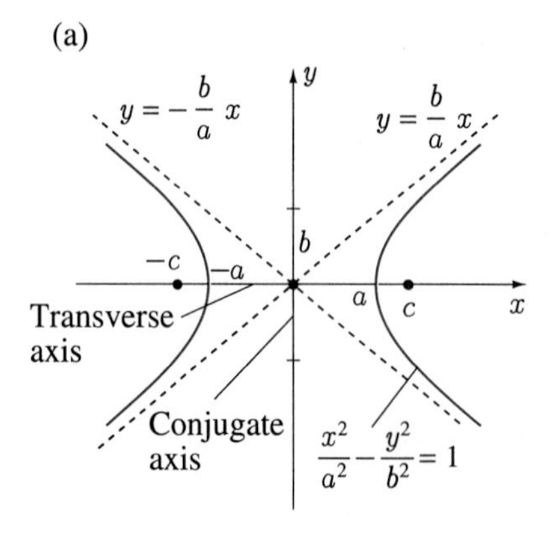

(b)

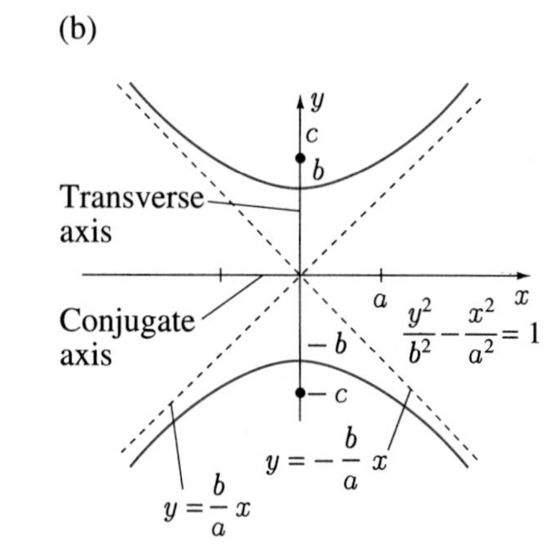

EXAMPLE 10.18

Sketch the hyperbola $16x^2 - 9y^2 = 144$, indicating its foci.

SOLUTION If we express the hyperbola in the form $x^2/9 - y^2/16 = 1$, its x-intercepts are ± 3. With asymptotes $y = \pm 4x/3$, we obtain the sketch in Figure 10.50. The foci must lie on the x-axis at distances $\pm c = \pm\sqrt{4^2 + 3^2} = \pm 5$ from the origin.

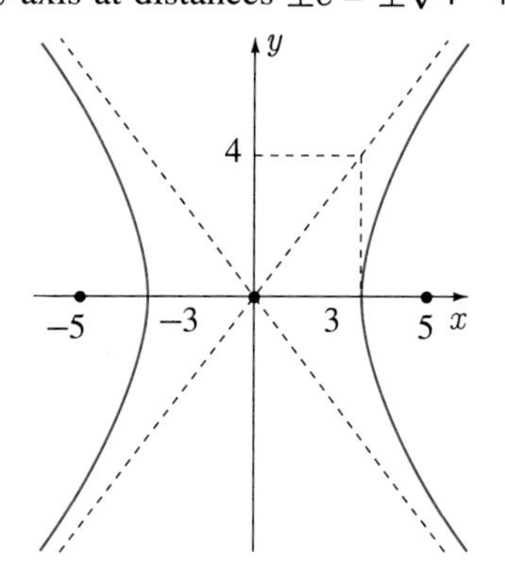

FIGURE 10.50

When the centre of a hyperbola is at point (h, k) and its foci are on the lines $x = h$ or $y = k$, equations 10.25 and 10.26 are modified in exactly the same way as equation 10.22 was modified for an ellipse. We replace each x by $x - h$ and each y by $y - k$ (see also Exercise 56). Consequently, equations for the hyperbolas in Figures 10.51 are

$$\frac{(x-h)^2}{a^2} - \frac{(y-k)^2}{b^2} = 1 \quad \text{and} \quad \frac{(y-k)^2}{b^2} - \frac{(x-h)^2}{a^2} = 1. \quad (10.27)$$

(a)

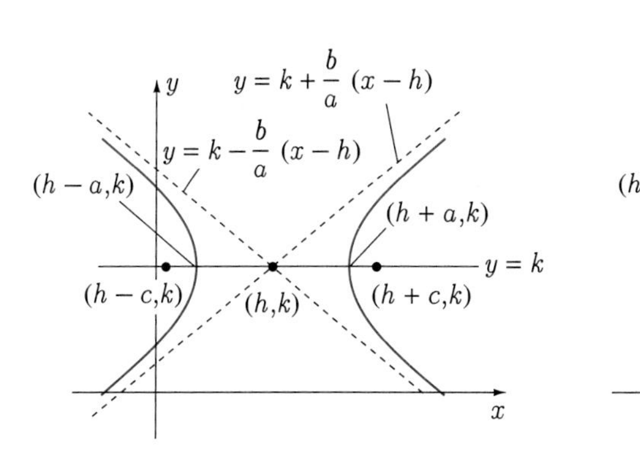

(b)

y
$y = k + \frac{b}{a}(x - h)$
$y = k - \frac{b}{a}(x - h)$
(h,k + c)
(h,k + b)
(h,k)
(h,k − b)
(h,k − c)
x = h
x

FIGURE 10.51

EXAMPLE 10.19

Sketch the hyperbola $x^2 - y^2 + 4x + 10y = 5$.

SOLUTION If we complete squares on the x- and y-terms, we find

$$(x+2)^2 - (y-5)^2 = -16 \quad \text{or}$$

$$\frac{(y-5)^2}{16} - \frac{(x+2)^2}{16} = 1.$$

The centre of the hyperbola is $(-2, 5)$, and the length of its transverse axis (along $x = -2$) is 8. If we solve the equation for y, we obtain

$$y = 5 \pm \sqrt{(x + 2)^2 + 16};$$

the asymptotes of the hyperbola are then $y = 5 \pm (x + 2)$. The hyperbola can now be sketched as in Figure 10.52. Its foci are at the points $(-2, 5 + 4\sqrt{2})$ and $(-2, 5 - 4\sqrt{2})$.

FIGURE 10.52

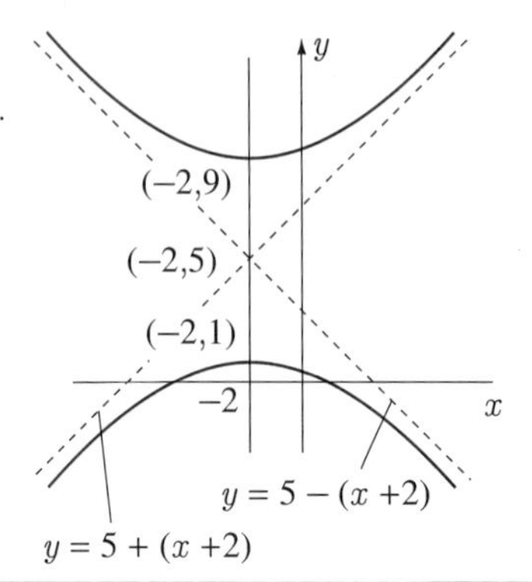

If P is a point on a conic section, the *focal radii* at P are the lines joining P to the foci (Figure 10.53). As a result, a parabola has one focal radius at each point, and an ellipse and hyperbola each have two. One of the properties of conics that makes them so useful is the fact that the normal line to the conic at any point bisects the angle between the focal radii. For the parabola, the normal bisects the angle between the focal radius and the line through P parallel to the axis of symmetry of the parabola. We will verify these facts in Exercises 59 and 60. To obtain one physical significance of these results, suppose each conic in Figure 10.53 is rotated about the x-axis to form a surface of revolution, which we regard as a mirror. It is a law of optics that when a ray of light strikes a reflecting surface, the angle between incident light and the normal to the surface is always equal to the angle between reflected light and the normal. Consequently, if a beam of light travels in the negative x-direction and strikes the parabolic mirror in Figure 10.53(a), all light is reflected toward the focus. Conversely, if F is a source of light, all light striking the mirror is reflected parallel to the x-axis. If either focus of the ellipse in Figure 10.53(b) is a light source, all light striking the elliptic mirror is reflected toward the other focus. Similarly, if either focus of the hyperbola in Figure 10.53(c) is a source, all light striking the mirror is reflected in a direction that would make it seem to originate at the other focus. Conversely, if light that is directed at one focus first strikes the mirror, it is reflected toward the other focus. This is precisely why we have parabolic reflectors in automobile headlights and searchlights, parabolic and hyperbolic reflectors in telescopes, and elliptic ceilings in whispering rooms.

FIGURE 10.53

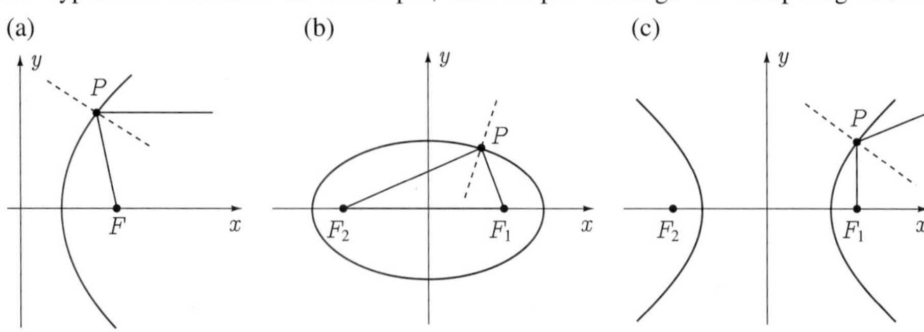

Thus far, we have defined parabolas, ellipse, and hyperbolas in terms of distances; for the parabola we use a focus and a directrix, and for the ellipse and hyperbola two foci. In Section 10.6 we show that ellipses and hyperbolas can also be defined in terms of a focus and directrix.

EXERCISES 10.5

In Exercises 1–14 identify the equation as representing a straight line, circle, parabola, ellipse, hyperbola, or none of these.

1. $2x + 3y = y^2$ **2.** $x^2 + y^2 - 3x + 2y = 25$

3. $2x - y = 3$ **4.** $x^2 + y^3 = 3x + 2$

5. $5x^2 = 11 - 2y^2$ **6.** $2x^2 - 3y^2 + 5 = 0$

7. $y^2 - x + 3y = 14 - x^2$

8. $x^2 + 2x = 3y + 4$

9. $y^2 + x^2 - 2x + 6y + 15 = 0$

10. $x^2 + 2y^2 + 24 = 0$

11. $5 + y^2 = 3x^2$ **12.** $x^2 + 2y^2 = 24$

13. $y^3 = 3x + 4$ **14.** $3 - x = 4y$

In Exercises 15–36 sketch the curve. Identify foci for each ellipse and hyperbola.

15. $y = 2x^2 - 1$ **16.** $\dfrac{x^2}{25} + \dfrac{y^2}{36} = 1$

17. $x^2 - \dfrac{y^2}{16} = 1$ **18.** $3x = 4y^2 - 1$

19. $\dfrac{y^2}{4} - \dfrac{x^2}{25} = 1$ **20.** $7x^2 + 3y^2 = 16$

21. $x + y^2 = 1$ **22.** $2y^2 + x = 3y + 5$

23. $9x^2 + 289y^2 = 2601$

24. $y^2 = 10(2 - x^2)$

25. $3x^2 - 4y^2 = 25$ **26.** $y^2 - x^2 = 5$

27. $y = -x^2 + 6x - 9$ **28.** $2x^2 - 3y^2 = 5$

29. $3x^2 + 6y^2 = 21$ **30.** $x^2 + 16y^2 = 2$

31. $y^2 - 3x^2 = 1$ **32.** $x = -(4 + y)^2$

33. $x^2 + 2x + 4y^2 - 16y + 13 = 0$

34. $x^2 - 6x - 4y^2 - 24y = 11$

35. $9x^2 + y^2 - 18x - 6y = 0$

36. $9x^2 - 16y^2 - 18x - 64y = 91$

37. Find the equation of a hyperbola that passes through the point $(1, 2)$ and has asymptotes $y = \pm 4x$.

38. Find the equation of an ellipse through the points $(-2, 4)$ and $(3, 1)$.

39. Find the width of the elliptic arch in Figure10.54.

FIGURE 10.54

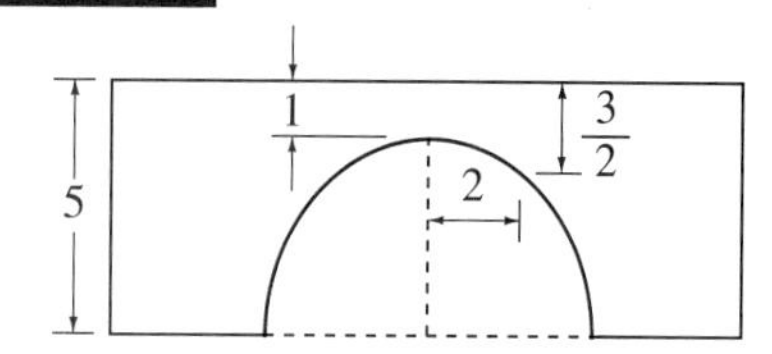

40. Find the height of the parabolic arch in Figure 10.55.

FIGURE 10.55

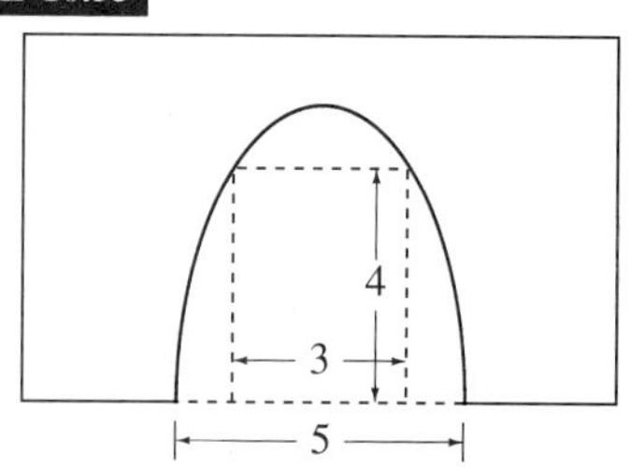

41. Explain how an ellipse can be drawn with a piece of string, two tacks, and a pencil.

42. Find the equation of the ellipse traced out by a point that moves so that the sum of its distances from $(\pm 4, 0)$ is always equal to 10:

(a) by using equation 10.22 with suitable values for a and b;

(b) by establishing and simplifying an equation similar to 10.20.

43. Find the equation of the hyperbola traced out by a point that moves so that the difference of its distances from $(0, \pm 3)$ is always equal to 1:

(a) by using equation 10.26 with suitable values for a and b;

(b) by establishing and simplifying an equation similar to 10.24.

44. Show that the equation of every straight line, every circle, and every conic section discussed in this section can be obtained by appropriate choices of constants A, C, D, E, and F in the equation

$$Ax^2 + Cy^2 + Dx + Ey + F = 0.$$

45. Show that the equation of the tangent line to the ellipse $b^2x^2 + a^2y^2 = a^2b^2$ at a point (x_0, y_0) is $b^2xx_0 + a^2yy_0 = a^2b^2$.

46. Show that the equation of the tangent line to the hyperbola $b^2x^2 - a^2y^2 = a^2b^2$ at a point (x_0, y_0) is $b^2xx_0 - a^2yy_0 = a^2b^2$.

47. Find the point P on that part of the ellipse $2x^2 + 3y^2 = 14$ in the first quadrant where the tangent line at P is perpendicular to the line joining P and $(2, 5)$.

48. Find the area inside the ellipse $b^2x^2 + a^2y^2 = a^2b^2$.

49. Among all rectangles that can be inscribed inside the ellipse $b^2x^2 + a^2y^2 = a^2b^2$ and have sides parallel to the axes, find the one with largest possible area.

50. A *prolate spheroid* is the solid of revolution obtained by rotating an ellipse about its major axis. An *oblate spheroid* is obtained by rotating the ellipse about its minor axis. Find volumes for the prolate and oblate spheroids generated by the ellipse $b^2x^2 + a^2y^2 = a^2b^2$ if $a > b$.

51. A sharp noise originating at one focus F_1 of an ellipse is reflected by the ellipse toward the other focus F_2. Explain why all reflected noise arrives at F_2 at exactly the same time.

52. Use equation 10.18 to show that when a parabola is written in the form $y = ax^2 + bx + c$, the following formulas identify its focus (p, q) and directrix $y = r$:

$$p = -\frac{b}{2a}, \qquad q = \frac{1}{4a}(1 + 4ac - b^2),$$
$$r = \frac{1}{4a}(-1 + 4ac - b^2).$$

53. What are formulas for the focus (p, q) and directrix $x = r$ for a parabola of the type $x = ay^2 + by + c$?

54. Use the formulas in Exercises 52 and 53 to identify the focus and directrix for any parabolas in Exercises 15–36.

55. Show that when the centre of an ellipse is at point (h, k) and its foci are on the line $x = h$ or $y = k$, Definition 10.3 leads to equation 10.23.

56. Show that when the centre of a hyperbola is at point (h, k) and its foci are on the line $x = h$ or $y = k$, Definition 10.4 leads to equation 10.27.

57. Prove that the normal line to the parabola $x = ay^2$ in Figure 10.56 bisects the angle between the focal radius FP and the line PQ parallel to the x-axis. Hint: Draw the tangent line PR at P and show that $||PF|| = ||RF||$.

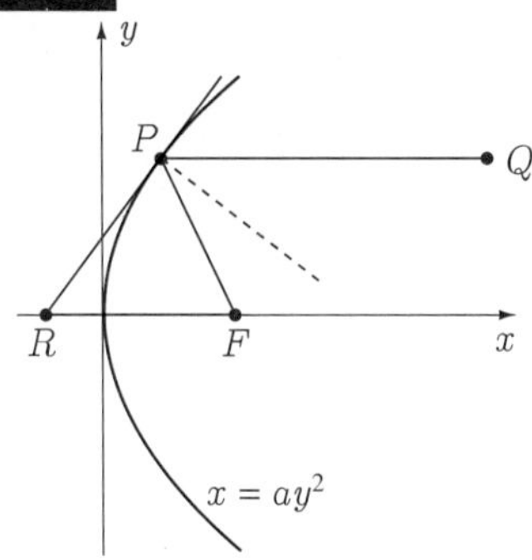

58. When a beam of light travelling in the negative x-direction (Figure 10.57) strikes a parabolic mirror with cross-section represented by $x = ay^2 + c$, all light rays are reflected to the focus F of the mirror. Show that all photons that pass simultaneously through $x = d$ arrive at F at the same time.

FIGURE 10.57

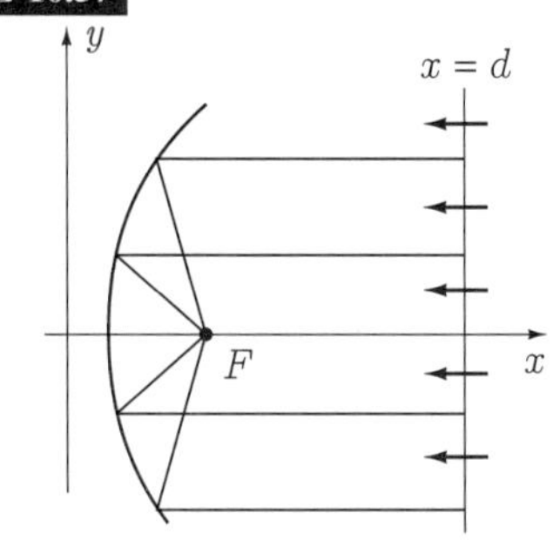

59. Prove that the normal line to an ellipse or hyperbola bisects the angle between the focal radii.

60. Prove that the normal line to the parabola $y = ax^2 + bx + c$ at any point P bisects the angle between the focal radius and the line through P parallel to the y-axis.

61. A line segment through the focus of a parabola with ends on the parabola is called a *focal chord*. It is an established result that tangents to a parabola at the ends of a focal chord are perpendicular to each other and intersect on the directrix. Prove this for the parabola $y = ax^2$.

Conic Sections in Polar Coordinates

In Section 10.5 we defined parabolas using a focus and directrix and ellipses and hyperbolas using two foci. In this section we show that all three conics can be defined using a focus and directrix, and that in polar coordinates one equation represents all three conics.

Let F be a fixed point (the *focus*), and l be a fixed line (the *directrix*) that does not pass through F as in Figure 10.58(a). We propose to find the equation of the curve traced out by a point that moves so that its undirected distances from F and l always remain in a constant ratio ϵ called the *eccentricity*. To do this we set up polar coordinates with F as pole and polar axis directed away from l and perpendicular to l as in Figure 10.58(b).

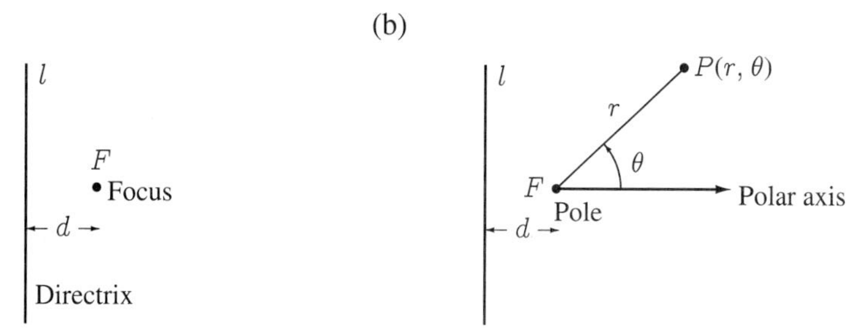

FIGURE 10.58

If (r, θ) are polar coordinates for any point P on the required curve, the fact that the ratio of the distances from F and l to P is equal to ϵ is expressed as

$$\frac{r}{d + r\cos\theta} = \epsilon. \qquad (10.28)$$

When this equation is solved for r, we have

$$r = \frac{\epsilon d}{1 - \epsilon\cos\theta}, \qquad (10.29)$$

and this is the polar equation of the curve traced out by the point. Certainly this curve should be a parabola when $\epsilon = 1$. We now verify this, and show that the curve is an ellipse when $\epsilon < 1$ and a hyperbola when $\epsilon > 1$. To do this we transform the equation into the usual Cartesian coordinates $x = r\cos\theta$ and $y = r\sin\theta$. From equations 10.5 and 10.6, we obtain

$$\sqrt{x^2 + y^2} = \frac{\epsilon d}{1 - \dfrac{\epsilon x}{\sqrt{x^2 + y^2}}},$$

and this equation simplifies to

$$\sqrt{x^2 + y^2} = \epsilon(d + x), \qquad (10.30\text{a})$$

or when squared,

$$x^2 + y^2 = \epsilon^2(d + x)^2. \qquad (10.30\text{b})$$

When $\epsilon = 1$, the x^2-terms in 10.30b cancel, and the equation reduces to that for a parabola:

$$x = \frac{1}{2d}(y^2 - d^2), \qquad (10.31)$$

as in Figure 10.59(a). When $\epsilon \neq 1$, we write

$$x^2 + y^2 = \epsilon^2(d^2 + 2dx + x^2) \qquad \text{or}$$

$$(1-\epsilon^2)x^2 - 2d\epsilon^2 x + y^2 = \epsilon^2 d^2.$$

When $\epsilon < 1$, we divide by $1-\epsilon^2$,

$$x^2 - \frac{2d\epsilon^2}{1-\epsilon^2}x + \frac{y^2}{1-\epsilon^2} = \frac{\epsilon^2 d^2}{1-\epsilon^2},$$

and complete the square on the x-terms,

$$\left(x - \frac{d\epsilon^2}{1-\epsilon^2}\right)^2 + \frac{y^2}{1-\epsilon^2} = \frac{\epsilon^2 d^2}{1-\epsilon^2} + \frac{d^2\epsilon^4}{(1-\epsilon^2)^2} = \left(\frac{\epsilon d}{1-\epsilon^2}\right)^2. \qquad (10.32)$$

Comparing equation 10.32 with 10.23, we conclude that 10.32 is the equation for an ellipse with centre at position $x = d\epsilon^2/(1-\epsilon^2)$ on the x-axis. Since one focus is at the origin, it follows that the other must also be on the x-axis at $x = 2d\epsilon^2/(1-\epsilon^2)$ as in Figure 10.59(b).

FIGURE 10.59

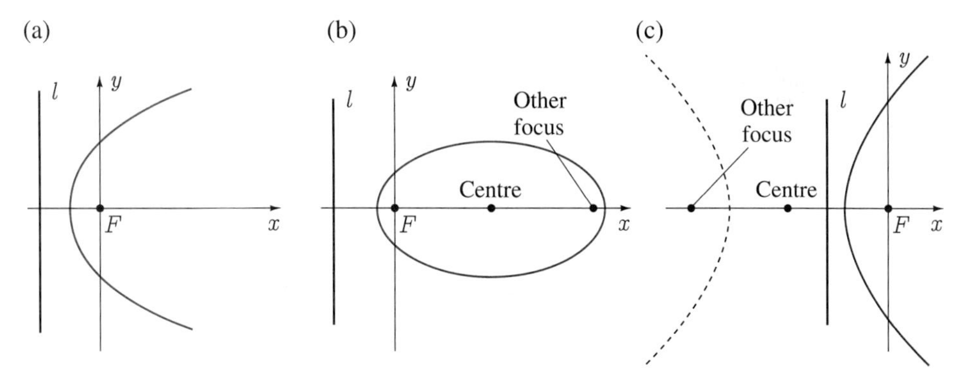

A similar calculation shows that when $\epsilon > 1$, points on the right half of the hyperbola

$$\left(x + \frac{d\epsilon^2}{\epsilon^2-1}\right)^2 - \frac{y^2}{\epsilon^2-1} = \left(\frac{\epsilon d}{\epsilon^2-1}\right)^2 \qquad (10.33)$$

are obtained as in Figure 10.59(c). Equation 10.30a is not satisfied by points with x-coordinates less than $-d$, and therefore points on the left half of the hyperbola, shown dotted, do not satisfy 10.29.

We have shown that equation 10.29 defines an ellipse when $0 < \epsilon < 1$, a parabola when $\epsilon = 1$, and a hyperbola when $\epsilon > 1$, and this provides a unifying approach to conic sections. All three conics can be studied using a focus and a directrix.

It is clear that equation 10.29 can only yield a parabola that opens to the right and has its focus on the x-axis, and an ellipse and hyperbola with foci on the x-axis, one at the origin. To obtain parabolas that open to the left, or up, or down, and ellipses and hyperbolas with foci on the y-axis, we must change the position of the directrix. The conic sections in Figures 10.60a-c have a directrix to the right of the focus. They have equations

$$r = \frac{\epsilon d}{1 + \epsilon\cos\theta}. \qquad (10.34\,\text{a})$$

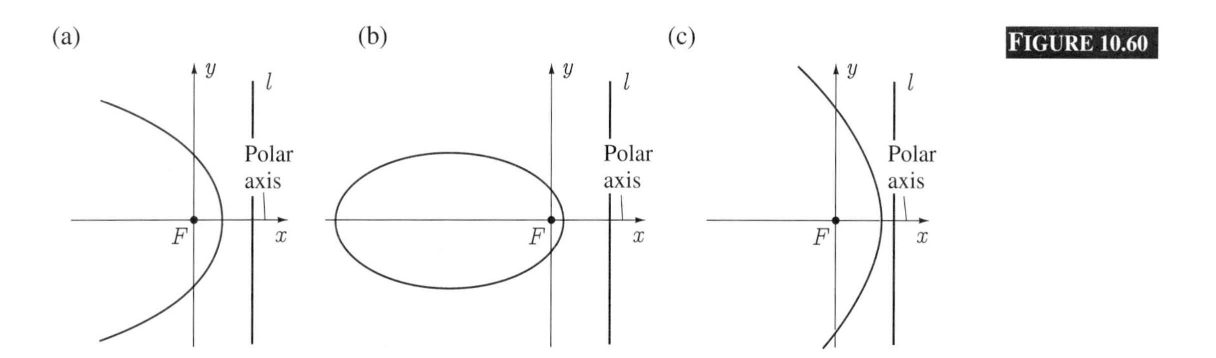

FIGURE 10.60

The conic sections in Figures 10.61 and 10.62 have equations of the form respectively,

$$r = \frac{\epsilon d}{1 - \epsilon \sin\theta} \quad \text{and} \quad r = \frac{\epsilon d}{1 + \epsilon \sin\theta}. \qquad (10.34\text{b})$$

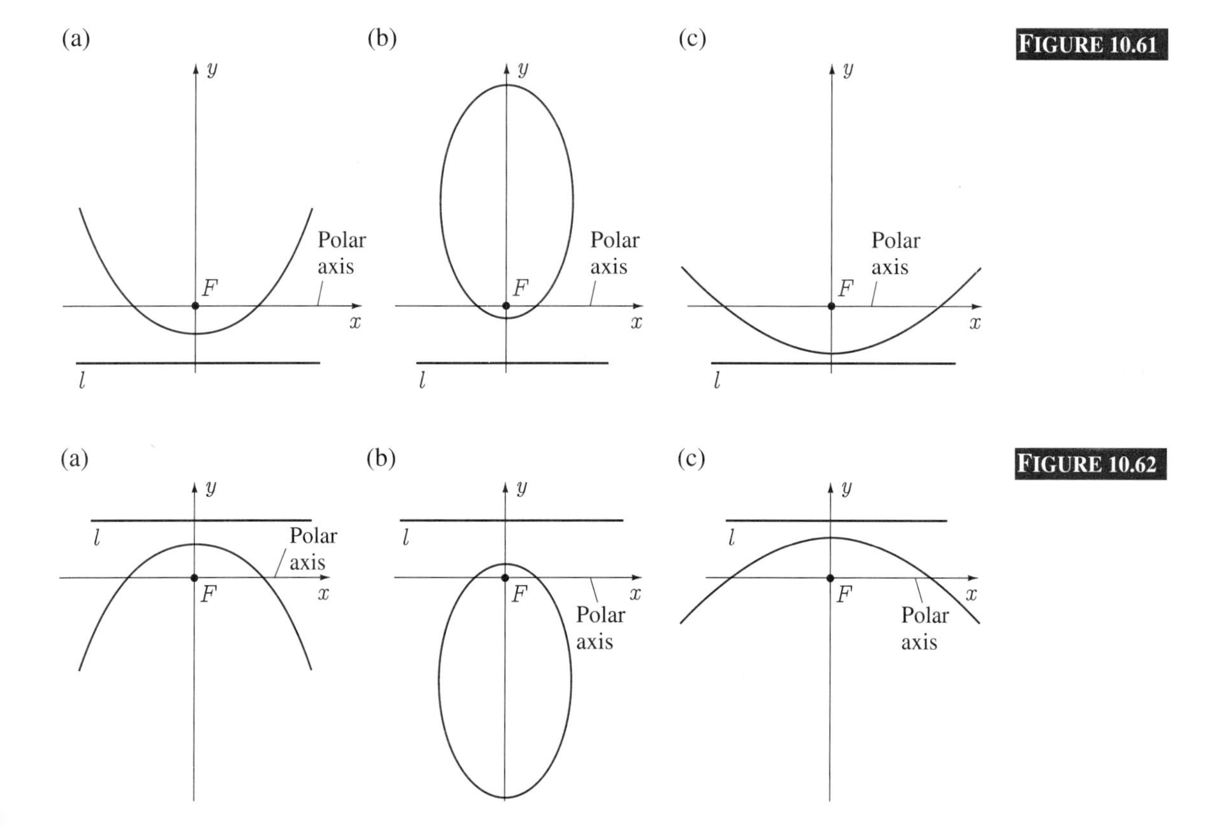

FIGURE 10.61

FIGURE 10.62

In Figures 10.59-62, one focus of the conic is chosen as the pole. In other words, the simplicity of equations 10.29 and 10.34 to describe conic sections is a direct consequence of the fact that the pole is at a focus, and the directrix is either parallel or perpendicular to the polar axis.

EXAMPLE 10.20 Sketch the curve $r = 15/(3 + 2\cos\theta)$.

SOLUTION If we write the equation in form 10.34a,

$$r = \frac{5}{1 + \left(\frac{2}{3}\right)\cos\theta},$$

the eccentricity $\epsilon = 2/3$ indicates that the curve is an ellipse. Both foci lie on the x-axis, and one is at the origin. The ends of the major axis occur when $\theta = 0$ and $\theta = \pi$, and for these values $r = 3$ and $r = 15$ (Figure 10.63). It now follows that the centre of the ellipse is at $x = -6$, and its other focus is at $x = -12$. If $b > 0$ denotes half the length of the minor axis (it is also the maximum y-value on the ellipse, occuring when $x = -6$), then $b^2 = a^2 - c^2 = 9^2 - 6^2 = 45$. Consequently, $b = 3\sqrt{5}$, and the ellipse is as shown in Figure 10.63. This information now permits us to write the equation of the ellipse in Cartesian coordinates:

$$\frac{(x+6)^2}{81} + \frac{y^2}{45} = 1.$$

FIGURE 10.63

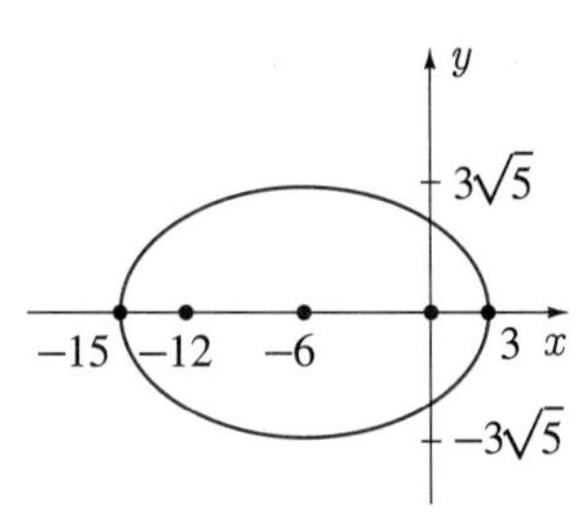

■

An alternative approach in Example 10.20 would be to substitute $r = \sqrt{x^2 + y^2}$ and $\cos\theta = x/\sqrt{x^2 + y^2}$ into the polar equation for the conic, simplify the equation to $(x+6)^2/81 + y^2/45 = 1$, and then sketch the ellipse from this equation. We illustrate this method in the following example.

EXAMPLE 10.21 Sketch the curve $r = 2/(3 - 4\sin\theta)$.

SOLUTION If we set $r = \sqrt{x^2 + y^2}$ and $\sin\theta = y/\sqrt{x^2 + y^2}$, then

$$\sqrt{x^2 + y^2} = \frac{2}{3 - \dfrac{4y}{\sqrt{x^2 + y^2}}} = \frac{2\sqrt{x^2 + y^2}}{3\sqrt{x^2 + y^2} - 4y}.$$

Division by $\sqrt{x^2 + y^2}$ leads to

$$3\sqrt{x^2 + y^2} - 4y = 2 \qquad \text{or}$$
$$3\sqrt{x^2 + y^2} = 4y + 2.$$

If we now square both sides, we obtain

$$9(x^2 + y^2) = 4 + 16y + 16y^2 \qquad \text{or}$$

$$9x^2 - 7y^2 - 16y = 4.$$

If we complete the square on the y-terms, we obtain

$$9x^2 - 7\left(y + \frac{8}{7}\right)^2 = 4 - \frac{64}{7} = -\frac{36}{7}.$$

Division by $-36/7$ yields the equation

$$\frac{\left(y + \frac{8}{7}\right)^2}{\frac{36}{49}} - \frac{x^2}{\frac{4}{7}} = 1.$$

This equation describes a hyperbola with centre $(0, -8/7)$ and y-intercepts equal to $-8/7 \pm 6/7 = -2/7, -2$. Its asymptotes are

$$y = -\frac{8}{7} \pm \frac{6}{7}\sqrt{\frac{7x^2}{4}} = -\frac{8}{7} \pm \frac{3x}{\sqrt{7}}.$$

The hyperbola is shown in Figure 10.64, but only the top half is described by $r = 2/(3 - 4\sin\theta)$. The equation $3\sqrt{x^2 + y^2} = 4y + 2$ does not permit $y \leq -2$.

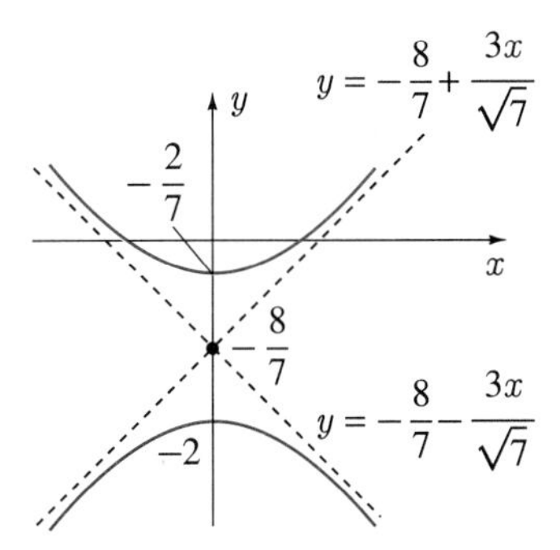

FIGURE 10.64

■

EXAMPLE 10.22

Find a polar representation for the ellipse

$$\frac{(x-1)^2}{4} + \frac{y^2}{9} = 1.$$

FIGURE 10.65

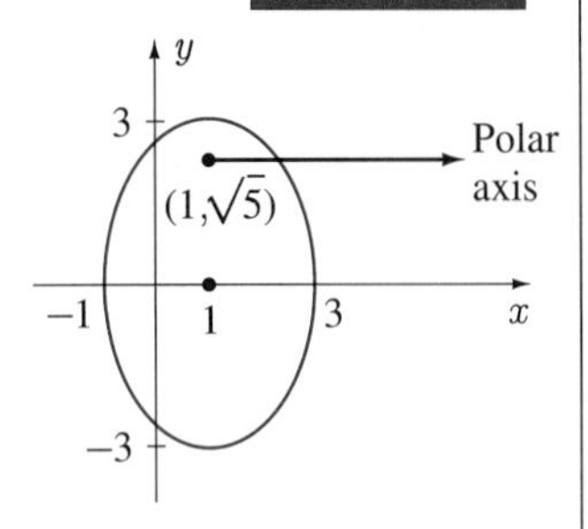

SOLUTION To find a polar representation for the ellipse (Figure 10.65), we could use the usual polar coordinates defined by $x = r\cos\theta$ and $y = r\sin\theta$, but the resulting equation would not be simple. Try it. We know that a simple polar representation must result if the pole is chosen at a focus of the ellipse and polar axis either parallel or perpendicular to the directrix. For this ellipse, foci are on the line $x = 1$ at distances $\pm c = \pm\sqrt{9-4} = \pm\sqrt{5}$ from the x-axis. Let us choose the pole at position $(1, \sqrt{5})$ and the polar axis parallel to the x-axis, and therefore parallel to the directrix. According to equation 10.9, polar and Cartesian coordinates are related by $x = 1 + r\cos\theta$ and $y = \sqrt{5} + r\sin\theta$. If we substitute these into the equation for the ellipse, we obtain

$$\frac{r^2\cos^2\theta}{4} + \frac{(\sqrt{5} + r\sin\theta)^2}{9} = 1,$$

or

$$9r^2\cos^2\theta + 4(5 + 2\sqrt{5}r\sin\theta + r^2\sin^2\theta) = 36.$$

This equation can be expressed as a quadratic equation in r:

$$r^2(9\cos^2\theta + 4\sin^2\theta) + r(8\sqrt{5}\sin\theta) - 16 = 0.$$

Solutions for r are

$$\begin{aligned} r &= \frac{-8\sqrt{5}\sin\theta \pm \sqrt{320\sin^2\theta + 64(9\cos^2\theta + 4\sin^2\theta)}}{2(9\cos^2\theta + 4\sin^2\theta)} \\ &= \frac{-8\sqrt{5}\sin\theta \pm \sqrt{576(\cos^2\theta + \sin^2\theta)}}{2(9\cos^2\theta + 4\sin^2\theta)} \\ &= \frac{4(\pm 3 - \sqrt{5}\sin\theta)}{9\cos^2\theta + 4\sin^2\theta}. \end{aligned}$$

Since r must be nonnegative, we must choose $+3$, and not -3, and therefore

$$\begin{aligned} r &= \frac{4(3 - \sqrt{5}\sin\theta)}{9(1 - \sin^2\theta) + 4\sin^2\theta} = \frac{4(3 - \sqrt{5}\sin\theta)}{9 - 5\sin^2\theta} \\ &= \frac{4(3 - \sqrt{5}\sin\theta)}{(3 - \sqrt{5}\sin\theta)(3 + \sqrt{5}\sin\theta)} = \frac{4}{3 + \sqrt{5}\sin\theta}. \end{aligned}$$

■

EXERCISES 10.6

In Exercises 1–10 sketch the conic section.

1. $r = \dfrac{3}{1 + \cos\theta}$

2. $r = \dfrac{16}{3 + 5\cos\theta}$

3. $r = \dfrac{4}{3 - 3\sin\theta}$

4. $r = \dfrac{16}{5 + 3\cos\theta}$

5. $r = \dfrac{4}{3 - 4\sin\theta}$

6. $r = \dfrac{4}{4 - 3\sin\theta}$

7. $r = \dfrac{1}{2 - 2\cos\theta}$

8. $r = \dfrac{1}{2 + \sin\theta}$

9. $r = \dfrac{\sec\theta}{3 + 6\sec\theta}$

10. $r = \dfrac{4\csc\theta}{7\csc\theta - 2}$

In Exercises 11–16 find the Cartesian equation for the curve.

11. $r = \dfrac{3}{1 - \sin\theta}$

12. $r = \dfrac{1}{3 + \cos\theta}$

13. $r = \dfrac{1}{1 + 2\cos\theta}$ **14.** $r = \dfrac{2}{1 - 3\cos\theta}$

15. $r = \dfrac{4}{6 - 3\sin\theta}$ **16.** $r = \dfrac{4}{5 + 5\cos\theta}$

In Exercises 17–20 find a polar equation in one of the four forms of 10.29 or 10.34 for the conic.

17. $x^2 - 16y^2 = 16$ **18.** $4x^2 + 9y^2 = 36$

19. $(x - 1)^2 + \dfrac{(y + 1)^2}{4} = 1$

20. $x^2 - 9(y - 2)^2 = 9$

21. (a) Show that the eccentricity ϵ of an ellipse or hyperbola is always equal to the distance from its centre to either focus, divided by half the length of the major or transverse axis.

(b) Discuss the eccentricities of the ellipses in Figures 10.43 and 10.44.

22. A circle has been described as a degenerate ellipse in the sense that its foci coincide. What happens to the eccentricity of an ellipse if the distance between its foci approaches zero? Hint: See Exercise 21.

SUMMARY

In Chapter 1 we defined curves explicitly and implicitly. In this chapter we added a third description — the parametric definition

$$x = x(t), \quad y = y(t), \quad \alpha \leq t \leq \beta.$$

When such a curve defines a function, the derivative of the function can be calculated using the parametric rule:

$$\frac{dy}{dx} = \frac{\dfrac{dy}{dt}}{\dfrac{dx}{dt}};$$

second and higher-order derivatives can be calculated using the chain rule. The length of such a curve is defined by the definite integral

$$\int_\alpha^\beta \sqrt{\left(\frac{dx}{dt}\right)^2 + \left(\frac{dy}{dt}\right)^2}\, dt.$$

Polar coordinates provide an alternative way to identify the positions of points in a plane. They use distance from a point, called the pole, and rotation of a half line, called the polar axis. Many curves that have complex equations in Cartesian coordinates can be represented very simply using polar coordinates. Particularly simple are multileaved roses, cardioids, lemniscates, and some circles. Polar coordinates also provide a unified approach to parabolas, ellipses, and hyperbolas. Each curve can be described as the path traced out by a point that moves so that the ratio ϵ of its distances from a fixed point and a fixed line remains constant. The curve is an ellipse, a parabola, or a hyperbola depending on whether $0 < \epsilon < 1$, $\epsilon = 1$, or $\epsilon > 1$ respectively. Ellipses and hyperbolas can also be defined as curves traced out by a point that moves so that the sum and difference of its distances from two fixed points remains constant.

When a curve $r = f(\theta)$ in polar coordinates encloses a region R, the definite integral

$$\int_\alpha^\beta \frac{1}{2}[f(\theta)]^2\, d\theta$$

with appropriate choices of α and β can be used to find the area of R.

Parametric equations for a polar curve $r = f(\theta)$, $\alpha \leq \theta \leq \beta$ are

$$x(\theta) = f(\theta)\cos\theta, \quad y(\theta) = f(\theta)\sin\theta, \quad \alpha \leq \theta \leq \beta.$$

Its slope is given by

$$\frac{dy}{dx} = \frac{f'(\theta)\sin\theta + f(\theta)\cos\theta}{f'(\theta)\cos\theta - f(\theta)\sin\theta},$$

and its length can be calculated with the definite integral

$$\int_{\alpha}^{\beta} \sqrt{r^2 + \left(\frac{dr}{d\theta}\right)^2}\, d\theta.$$

Key Terms and Formulas

In reviewing this chapter, you should be able to define or discuss the following key terms:

Parametric equations
Parametric rule
Length of parametric curve
Pole
Polar axis
Polar coordinates
Slope of polar curve
Length of polar curve
Area in polar coordinates
Parabola
Focus
Directrix
Ellipse
Major axis
Minor axis
Hyperbola
Transverse axis
Conjugate axis
Eccentricity

REVIEW EXERCISES

In Exercises 1–30 sketch the curve.

1. $r = \cos\theta$

2. $r = -\sin\theta$

3. $r = \dfrac{3}{1 + 2\sin\theta}$

4. $r = \dfrac{3}{2 + \sin\theta}$

5. $r = \dfrac{1}{1 + \sin\theta}$

6. $r = \dfrac{3}{2 - 2\cos\theta}$

7. $r + 1 = 2\sin\theta$

8. $r^2 = 4\cos\theta$

9. $x = 2\cos t,\ y = 3\sin t,\ 0 \le t < 2\pi$

10. $x = 4 + t,\ y = 5 - 3t^2$

11. $x = \sin^2 t,\ y = \cos^2 t,\ 0 \le t \le 2\pi$

12. $x = 1 + 4\sin t,\ y = -2 + 4\cos t,\ 0 \le t \le \pi$

13. $(x^2 + y^2)^3 = x$

14. $x^2 + y^2 = 2\sqrt{x^2 + y^2}$

15. $r^2 = 4\cos^2\theta$

16. $r = 3\cos 2\theta$

17. $x + y = \sqrt{x^2 + y^2}$

18. $x^2 + y^2 = x + y$

19. $r = 4\cos 3\theta$

20. $r = \sin^2\theta - \cos^2\theta$

21. $x^2 + y^2 - 3x + 2y = 1$

22. $x^2 - 3y^2 + 4 = 0$

23. $2x^2 + 3y^2 - 6y = 0$

24. $y^2 + x + 2y = 3$

25. $r(2\cos\theta - \sin\theta) = 3$

26. $r = \cos(\theta/2)$

27. $r = \sin^2\theta$

28. $x^2 = y^2 + y$

29. $x = e^{-t},\ y = \ln t,\ t > 0$

30. $x = \sin\theta,\ y = \sin 2\theta$

In Exercises 31–35 find the indicated area.

31. Inside $r = 2 + 2\cos\theta$

32. Inside $r = 4$ but outside $r = 4\sin 2\theta$

33. Common to $r = 2$ and $r^2 = 9\cos 2\theta$

34. Inside $r = \sin^2\theta$

35. Common to $r = 1 + \sin\theta$ and $r = 2 - 2\sin\theta$

In Exercises 36 and 37 assume that y is defined as a function of x, and find dy/dx and d^2y/dx^2.

36. $x = t^3 + 2t, \quad y = 3t - t^3$

37. $x = 2\sin u, \quad y = 3\cos u$

In Exercises 38–40 find the length of the curve.

38. $r = 2 + 2\cos\theta$

39. $x = t^2, \quad y = t^3, \quad 0 \le t \le 1$

40. $x = e^t\cos t, \quad y = e^t\sin t, \quad 0 \le t \le \pi/2$

41. Find the equation of the tangent line to the curve $r = 2 - 2\sin\theta$ at the point with polar coordinates $(1, \pi/6)$.

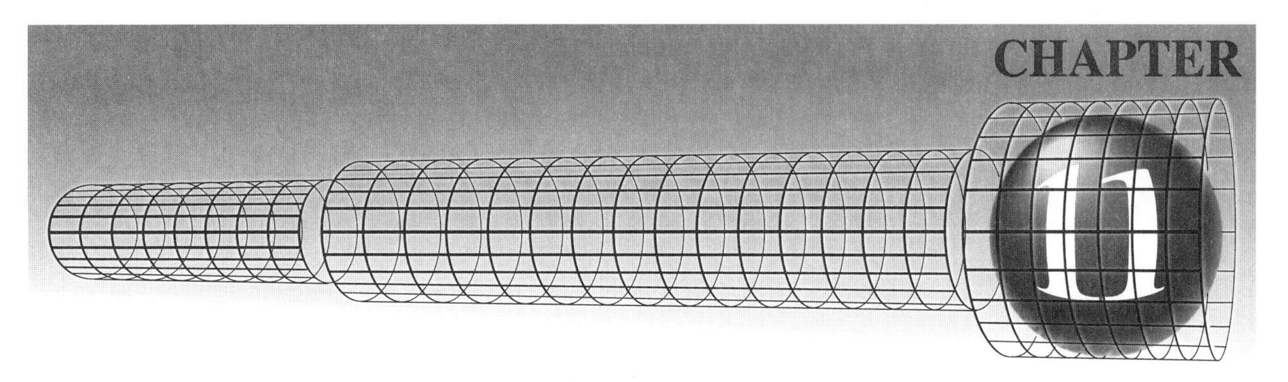

Infinite Sequences and Series

Sequences and series play an important role in many areas of applied mathematics. Sequences were first encountered in Chapter 4, although we did not use the term "sequences" at the time. In Section 4.1, Newton's iterative procedure was used to develop a set of numbers $x_1, x_2, x_3, \ldots$ to approximate a root of an equation $f(x) = 0$. The first number is chosen as some initial approximation to the solution of the equation, and subsequent numbers, defined by the formula

$$x_{n+1} = x_n - \frac{f(x_n)}{f'(x_n)},$$

are better and better approximations. This ordered set of numbers is called a sequence. Each number in the set corresponds to a positive integer, and each is calculated according to a stated formula.

A series is the sum of the numbers in a sequence. If the numbers are $x_1, x_2, x_3, \ldots$, the corresponding series is denoted symbolically by

$$\sum_{n=1}^{\infty} x_n = x_1 + x_2 + x_3 + \cdots,$$

where the three dots indicate that the addition is neverending. It is very well to write an expression like this, but it does not seem to have any meaning. No matter how fast we add, or how fast a calculator adds, or even how fast a supercomputer adds, an infinity of numbers can never be added together in a finite amount of time. We shall give meaning to such expressions, and show that they are really the only sensible way to define many of the more common transcendental functions such as trigonometric, exponential, and hyperbolic.

The first two sections of this chapter are devoted to sequences and the remaining ten to series. This is not to say that series are more important than sequences; they are not. Discussions on series invariably become discussions on sequences associated with series. We have found that difficulties with this chapter can almost always be traced back to a failure to distinguish between the two concepts. Special attention to the material in Sections 11.1 and 11.2 will be rewarded, while a cursory treatment of these sections leads to confusion in later sections.

SECTION 11.1

Infinite Sequences of Constants

Sequences are defined as follows.

Definition 11.1 An **infinite sequence of constants** is a function $f(n)$ whose domain is the set of positive integers.

For example, when $f(n) = 1/n$, the following numbers are associated with the positive integers,

$$1, \quad \frac{1}{2}, \quad \frac{1}{3}, \quad \frac{1}{4}, \ldots .$$

The word "infinite" simply indicates that an infinity of numbers is defined by the sequence as there is an infinity of positive integers, but it indicates nothing about the nature of the numbers. Often we write the numbers $f(n)$ in a line separated by commas,

$$f(1), f(2), \ldots, f(n), \ldots \tag{11.1a}$$

and refer to this array as the sequence rather than the rule by which it is formed. Since this notation is somewhat cumbersome, we adopt a notation similar to that used for the sequence defined by Newton's iterative procedure. We set $c_1 = f(1)$, $c_2 = f(2)$, $\ldots$, $c_n = f(n)$, $\ldots$, and write for 11.1a

$$c_1, c_2, c_3, \ldots, c_n, \ldots . \tag{11.1b}$$

The first number c_1 is called the first **term** of the sequence, c_2 the second term, and for general n, c_n is called the nth term (or general term) of the sequence. For the above example, we have

$$c_1 = 1, \quad c_2 = \frac{1}{2}, \quad c_3 = \frac{1}{3}, \quad \text{etc.}$$

In some applications, it is more convenient to define a sequence as a function whose domain is the set of all integers larger than or equal to some fixed integer N, and N can be positive, negative, or zero. Indeed, in Section 11.8, we find it convenient to initiate the assignment with $N = 0$. For now we prefer to use Definition 11.1 where $N = 1$, in which case we have the natural situation where the first term of the sequence corresponds to $n = 1$, the second term to $n = 2$, etc.

EXAMPLE 11.1 The general terms of four sequences are:

$$\text{(a) } \frac{1}{2^{n-1}} \qquad \text{(b) } \frac{n}{n+1} \qquad \text{(c) } (-1)^n|n-3| \qquad \text{(d) } (-1)^{n+1}$$

Write out the first six terms of each sequence.

SOLUTION The first six terms of these sequences are:

$$\text{(a)} \quad 1, \frac{1}{2}, \frac{1}{4}, \frac{1}{8}, \frac{1}{16}, \frac{1}{32};$$

$$\text{(b)} \quad \frac{1}{2}, \frac{2}{3}, \frac{3}{4}, \frac{4}{5}, \frac{5}{6}, \frac{6}{7};$$

$$\text{(c)} \quad -2, 1, 0, 1, -2, 3;$$

$$\text{(d)} \quad 1, -1, 1, -1, 1, -1.$$

■

The sequences in Example 11.1 are said to be defined **explicitly**; we have an explicit formula for the nth term of the sequence in terms of n. This allows easy determination of any term in the sequence. For instance, to find the one-hundredth term, we simply replace n by 100 and perform the resulting arithmetic. Contrast this with the sequence in the following example.

EXAMPLE 11.2

The first term of a sequence is $c_1 = 1$ and every other term is to be obtained from the formula

$$c_{n+1} = 5 + \sqrt{2 + c_n}, \quad n \geq 1.$$

Calculate c_2, c_3, c_4, and c_5.

SOLUTION To obtain c_2 we set $n = 1$ in the formula,

$$c_{1+1} = c_2 = 5 + \sqrt{2 + c_1} = 5 + \sqrt{2 + 1} = 5 + \sqrt{3} = 6.732.$$

To find c_3, we set $n = 2$,

$$c_3 = 5 + \sqrt{2 + c_2} = 5 + \sqrt{2 + \left(5 + \sqrt{3}\right)} = 5 + \sqrt{7 + \sqrt{3}} = 7.955.$$

Similarly,

$$c_4 = 5 + \sqrt{2 + c_3} = 5 + \sqrt{7 + \sqrt{7 + \sqrt{3}}} = 8.155, \qquad \text{and}$$

$$c_5 = 5 + \sqrt{2 + c_4} = 5 + \sqrt{7 + \sqrt{7 + \sqrt{7 + \sqrt{3}}}} = 8.187.$$

■

When the terms of a sequence are defined by a formula such as the one in Example 11.2, the sequence is said to be defined **recursively**. Note that the terms for a sequence obtained from Newton's iterative procedure are so defined. To find the 100th term of a recursively defined sequence, we must know the 99th; to find the 99th, we must know the 98th; to find the 98th, we need the 97th; and so on down the line. In other words, to find a term in the sequence, we must first find every term that precedes it. Obviously it is much more convenient to have an explicit definition for c_n in terms of n, but this is not always possible. It can be very difficult to find an explicit formula for the nth term of a sequence which is defined recursively.

Sometimes it is impossible to give any algebraic formula for the terms of a sequence. This is illustrated in the following example.

EXAMPLE 11.3

The nth term of a sequence is the nth prime integer (greater than 1) when all such primes are listed in ascending order. List its first ten terms.

SOLUTION The first ten terms of the sequence are

$$2,\ 3,\ 5,\ 7,\ 11,\ 13,\ 17,\ 19,\ 23,\ 29.$$

No one has yet discovered a formula for the prime integers, and it is therefore impossible to express terms of the sequence explicitly or recursively. In spite of this, the sequence is well-defined; and with enough perseverence, and perhaps a high-powered computer, we could find any number of terms in this sequence. ■

When the general term of a sequence is known explicitly, any term in the sequence is obtained by substituting the appropriate value of n. In other words, the general term specifies every term in the sequence. We therefore use the general term to abbreviate the notation for a sequence by writing the general term in braces and using this to represent the sequence. Specifically, for the sequence in Example 11.1(a), we write

$$\left\{\frac{1}{2^{n-1}}\right\}_1^\infty = 1, \frac{1}{2}, \frac{1}{4}, \frac{1}{8}, \ldots, \frac{1}{2^{n-1}}, \ldots,$$

where 1 and ∞ indicate that the first term corresponds to the integer $n = 1$, and that there is an infinite number of terms in the sequence. In general, we write

$$\{c_n\}_1^\infty = c_1, c_2, c_3, \ldots, c_n, \ldots . \tag{11.2}$$

If, as is the case in this section, the first term of a sequence corresponds to the integer $n = 1$, we abbreviate the notation further and simply write $\{c_n\}$ in place of $\{c_n\}_1^\infty$.

Since a sequence $\{c_n\}$ is a function whose domain is the set of positive integers, we can represent $\{c_n\}$ graphically. The sequences of Example 11.1 are shown in Figure 11.1.

FIGURE 11.1

(a)

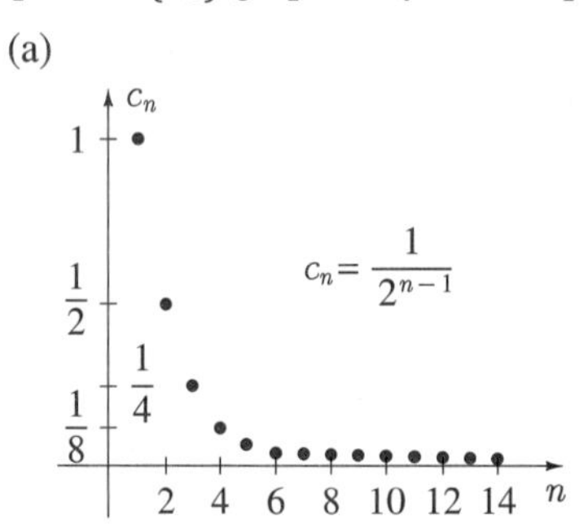

(b)

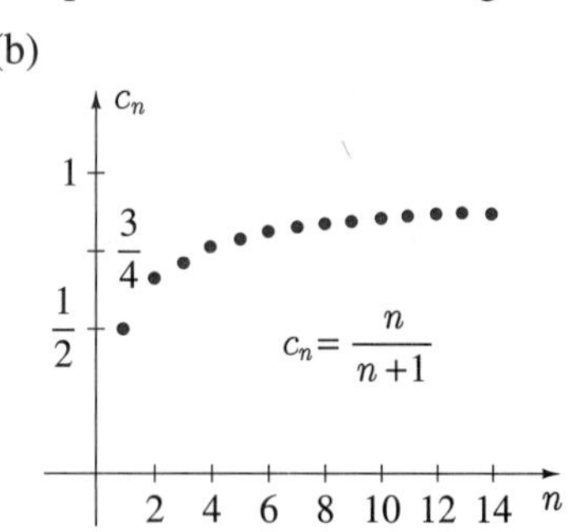

(c)

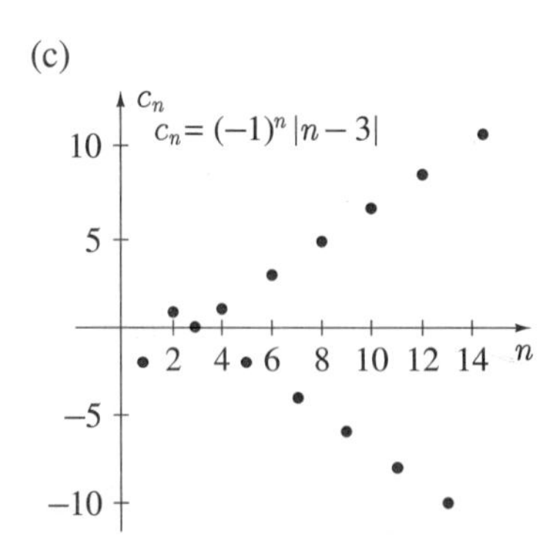

(d)

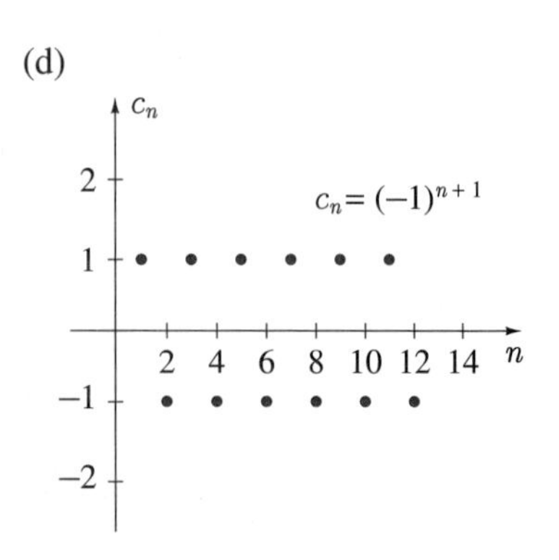

In most applications of sequences we are interested in a number called the **limit** of the sequence. Intuitively, a number L is called the limit of a sequence $\{c_n\}$ if, as we go farther and farther out in the sequence, the terms get arbitrarily close to L, and stay close to L. If such a number L exists, we write

$$L = \lim_{n\to\infty} c_n, \tag{11.3}$$

and say that the sequence $\{c_n\}$ **converges** to L. If no such number exists, we say that the sequence does not have a limit, or that the sequence **diverges**.

For the sequences of Example 11.1, it is evident that:

$$\text{(a)} \quad \lim_{n\to\infty} \frac{1}{2^{n-1}} = 0;$$

$$\text{(b)} \quad \lim_{n\to\infty} \frac{n}{n+1} = 1;$$

$$\text{(c)} \quad \lim_{n\to\infty} (-1)^n |n-3| \text{ does not exist;}$$

$$\text{(d)} \quad \lim_{n\to\infty} (-1)^{n+1} \text{ does not exist;}$$

and in Example 11.3,

$$\lim_{n\to\infty} c_n \text{ does not exist.}$$

Note how the points on the graphs in Figures 11.1(a) and (b) cluster around the limits 0 and 1 as n gets larger and larger. No such clustering occurs in the remaining two figures.

It is usually, but not always, easy to determine whether an explicitly defined sequence has a limit, and what that limit is. It is like finding the limit of a function $f(x)$ as $x \to \infty$. Seldom is this the case, however, for recursive sequences. For instance, it is not at all clear whether the recursive sequence of Example 11.2 has a limit. In spite of the fact that differences between successive terms are approaching zero, and terms of the sequence are therefore getting closer together, the sequence might not have a limit. This is explored further in Exercise 70 at the end of this section. Likewise for recursive sequences defined by Newton's iterative procedure, we pointed out in Section 4.1 that sometimes the sequence does not converge, but gave no conditions that would ensure that it does. What we need, then, are criteria by which to determine whether a given sequence has a limit. The following two definitions lead to Theorem 11.1, which can be very useful in determining whether a recursive sequence has a limit.

Definition 11.2

A sequence $\{c_n\}$ is said to be

(i) **increasing** if $\quad c_{n+1} > c_n \quad$ for all $n \geq 1$; $\quad$ (11.4a)

(ii) **nondecreasing** if $\quad c_{n+1} \geq c_n \quad$ for all $n \geq 1$; $\quad$ (11.4b)

(iii) **decreasing** if $\quad c_{n+1} < c_n \quad$ for all $n \geq 1$; $\quad$ (11.4c)

(iv) **nonincreasing** if $\quad c_{n+1} \leq c_n \quad$ for all $n \geq 1$. $\quad$ (11.4d)

If a sequence satisfies any one of these four properties, it is said to be **monotonic**.

Definition 11.3

A sequence $\{c_n\}$ is said to have an **upper bound** U (be bounded above by U) if

$$c_n \leq U \tag{11.5}$$

for all $n \geq 1$. It has a **lower bound** V (is bounded below by V) if

$$c_n \geq V \tag{11.6}$$

for all $n \geq 1$.

Note that if U is an upper bound for a sequence, then any number greater than U is also an upper bound. If V is a lower bound, so too is any number smaller than V.

For the sequences of Example 11.1, we find that $\{1/2^{n-1}\}$ is decreasing and has an upper bound $U = 1$ and a lower bound $V = 0$; $\{n/(n+1)\}$ is increasing and has an upper bound $U = 5$ and a lower bound $V = -2$; $\{(-1)^n|n-3|\}$ is not monotonic and has neither an upper nor a lower bound; and $\{(-1)^{n+1}\}$ is not monotonic and has an upper bound $U = 1$ and a lower bound $V = -3$.

The sequence of Example 11.3 is increasing and has a lower bound $V = 2$, but has no upper bound (see Exercise 81).

The recursive sequence of Example 11.2 will be discussed in detail in Example 11.4. For a complete discussion of this sequence, however, we require the following theorem.

Theorem 11.1 *A bounded, monotonic sequence has a limit.*

To expand on this statement somewhat, consider a sequence $\{c_n\}$ whose terms are illustrated graphically in Figure 11.2(a). Suppose that the sequence is increasing and therefore monotonic, and that U is an upper bound for the sequence. We have shown the upper bound as a horizontal line in the figure; c_1 is a lower bound. Our intuition suggests that because the terms in the sequence always increase, and they never exceed U, the sequence must have a limit. Theorem 11.1 confirms this. The theorem does not suggest the value of the limit, but obviously it must be less than or equal to U.

Similarly, when a sequence is decreasing or nonincreasing and has a lower bound V (Figure 11.2(b)), it must approach a limit that is greater than or equal to V.

FIGURE 11.2

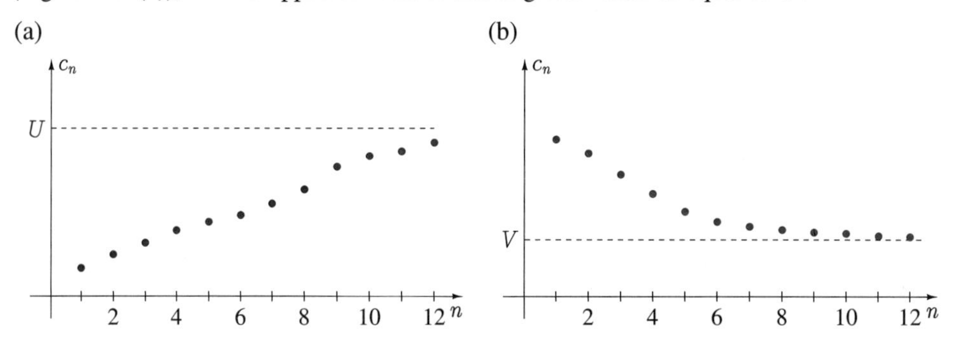

Another way of stating Theorem 11.1 is as follows.

Corollary *A monotonic sequence has a limit if and only if it is bounded.*

We are now prepared to give a complete and typical discussion for the recursive sequence of Example 11.2.

EXAMPLE 11.4

Find the limit of the sequence of Example 11.2, if it has one.

SOLUTION The first five terms of the sequence are

$$c_1 = 1, \quad c_2 = 6.732, \quad c_3 = 7.955, \quad c_4 = 8.155, \quad c_5 = 8.187.$$

They suggest that the sequence is increasing; that is, $c_{n+1} > c_n$. To prove this we use mathematical induction (see Appendix C). Certainly the inequaltiy is valid for $n = 1$ since $c_2 > c_1$. Suppose k is an integer for which $c_{k+1} > c_k$. Then

$$2 + c_{k+1} > 2 + c_k,$$

from which

$$\sqrt{2 + c_{k+1}} > \sqrt{2 + c_k}.$$

It follows that

$$5 + \sqrt{2 + c_{k+1}} > 5 + \sqrt{2 + c_k}.$$

The left side is c_{k+2} and the right side is c_{k+1}. Therefore we have proved that $c_{k+2} > c_{k+1}$. Hence, by mathematical induction $c_{n+1} > c_n$ for all $n \geq 1$. Since the sequence is increasing, its first term $c_1 = 1$ must be a lower bound. Certainly any upper bound, if one exists, must be at least 8.187 (c_5). We can take any number greater than 8.187 and use mathematical induction to test whether it is indeed an upper bound. It appears that $U = 10$ might be a reasonable guess for an upper bound for this sequence, and we verify this by induction as follows. Clearly $c_1 < 10$. We suppose that k is some integer for which $c_k < 10$. Then

$$2 + c_k < 12,$$

from which

$$\sqrt{2 + c_k} < \sqrt{12}.$$

Thus,

$$c_{k+1} = 5 + \sqrt{2 + c_k} < 5 + \sqrt{12} < 10.$$

By mathematical induction, then $c_n < 10$ for $n \geq 1$.

Since the sequence is monotonic and bounded, Theorem 11.1 guarantees that it has a limit, call it L. To evaluate L, we take limits on each side of the equation defining the sequence recursively:

$$\lim_{n\to\infty} c_{n+1} = \lim_{n\to\infty} \left(5 + \sqrt{2 + c_n}\right).$$

It is important to note that this cannot be done until the conditions of Theorem 11.1 have been checked. Since the terms c_n of the sequence approach L as $n \to \infty$, it follows that $5 + \sqrt{2 + c_n}$ approaches $5 + \sqrt{2 + L}$. Furthermore, as $n \to \infty$, c_{n+1} must also approach L. Do not make the mistake of saying that c_{n+1} approaches $L + 1$ as $n \to \infty$. Think about what $\lim_{n\to\infty} c_{n+1}$ means. We conclude therefore that

$$L = 5 + \sqrt{2 + L}.$$

If we transpose the 5 and square both sides of the equation, we obtain the quadratic equation

$$L^2 - 11L + 23 = 0,$$

with solutions

$$L = \frac{11 \pm \sqrt{29}}{2}.$$

Only the positive square root satisfies the original equation $L = 5 + \sqrt{2 + L}$ defining L, so that $L = (11 + \sqrt{29})/2$. The other root, $(11 - \sqrt{29})/2 \approx 2.8$ can also be eliminated on the grounds that all terms beyond the first are greater than 6. ■

Now that we know what it means for a sequence to have a limit, and how to find limits, we can be more precise. To give a mathematical definition for the limit of a sequence, we start with our intuitive description and make a succession of paraphrases, each of which is one step closer to a precise definition:

A sequence $\{c_n\}$ has limit L if its terms get arbitrarily close to L, and stay close to L, as n gets larger and larger.

A sequence $\{c_n\}$ has limit L if its terms can be made arbitrarily close to L by choosing n sufficiently large.

A sequence $\{c_n\}$ has limit L if the difference $|c_n - L|$ can be made arbitrarily close to 0 by choosing n sufficiently large.

A sequence $\{c_n\}$ has limit L if given any real number $\epsilon > 0$, no matter how small, we can make the difference $|c_n - L|$ less than ϵ by choosing n sufficiently large.

Finally, we arrive at the following definition.

Definition 11.4

A sequence $\{c_n\}$ has **limit** L if for any given $\epsilon > 0$, there exists an integer N such that for all $n > N$

$$|c_n - L| < \epsilon.$$

This definition puts in precise terms, and in the simplest possible way, our intuitive idea of a limit. For those who have studied Section 2.5, note the similarity between Definition 11.4 and Definition 2.2. For a better understanding of Definition 11.4, it is helpful to consider its geometric interpretation. The inequality $|c_n - L| < \epsilon$, when written in the form $L - \epsilon < c_n < L + \epsilon$ is interpreted as a horizontal band of width 2ϵ around L (Figure 11.3). Definition 11.4 requires that no matter how small ϵ, we can find a stage, denoted by N, beyond which all terms in the sequence are contained in the horizontal band. For the sequence and ϵ in Figure 11.3, N must be chosen as shown. For the same sequence, but a smaller ϵ, N must be chosen correspondingly larger (Figure 11.4). Proofs of many theorems in this chapter require a working knowledge of this definition. As an example, we use it to verify that a sequence cannot have two limits, a fact that we have implicitly assumed throughout our discussions.

FIGURE 11.3

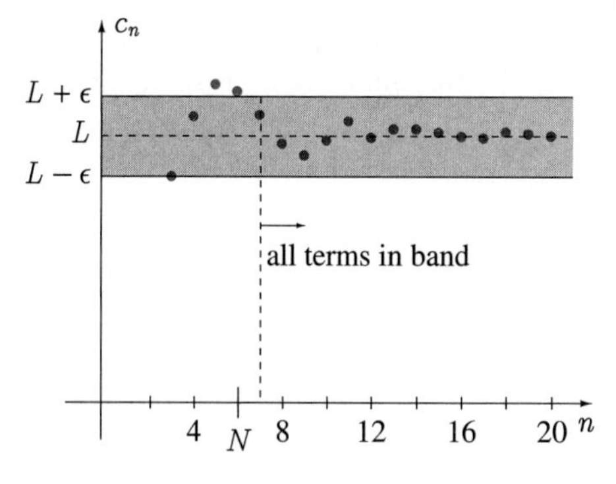

FIGURE 11.4

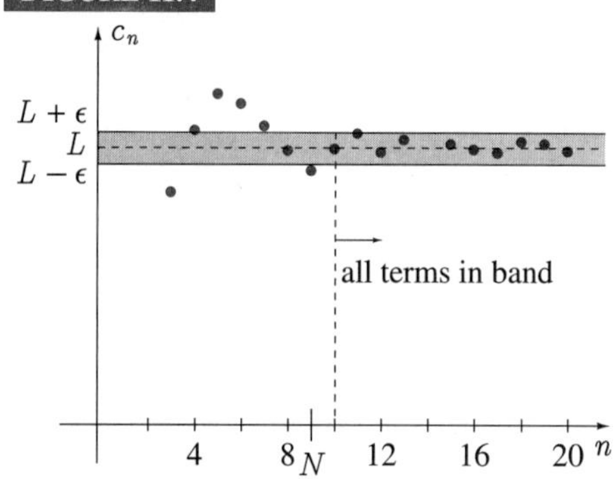

Theorem 11.2

A sequence can have at most one limit.

Proof We prove this theorem by showing that a sequence cannot have two distinct limits. Suppose to the contrary that a sequence $\{c_n\}$ has two distinct limits L_1 and L_2, where $L_2 > L_1$, and let $L_2 - L_1 = \delta$. If we set $\epsilon = \delta/3$, then according to Definition 11.4, there exists an integer N_1 such that for all $n > N_1$

$$|c_n - L_1| < \epsilon = \delta/3;$$

that is, for $n > N_1$, all terms in the sequence are within a distance $\delta/3$ of L_1.

But since L_2 is also supposed to be a limit, there exists an N_2 such that for $n > N_2$, all terms in the sequence are within a distance $\epsilon = \delta/3$ of L_2:

$$|c_n - L_2| < \epsilon = \delta/3.$$

But this is impossible (Figure 11.5), if L_1 and L_2 are a distance δ apart. This contradiction therefore implies that L_1 and L_2 are the same; that is, $\{c_n\}$ cannot have two limits.

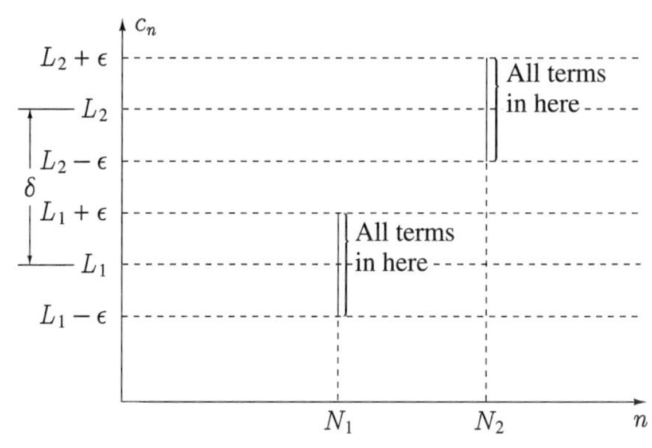

FIGURE 11.5

The following theorem, which states some of the properties of convergent sequences, can also be proved using Definition 11.4. Only the first two parts, however, are straightforward (see Exercises 71, 78, and 79).

Theorem 11.3

If sequences $\{c_n\}$ and $\{d_n\}$ have limits C and D, then:

(i) $\{kc_n\}$ *has limit* kC *if* k *is a constant;*
(ii) $\{c_n \pm d_n\}$ *has limit* $C \pm D$;
(iii) $\{c_n d_n\}$ *has limit* CD;
(iv) $\{c_n/d_n\}$ *has limit* C/D *provided that* $D \neq 0$, *and none of the* $d_n = 0$.

EXERCISES 11.1

In Exercises 1–14 state (without proof) whether the sequence is monotonic, and whether it has an upper bound, a lower bound, and a limit.

1. $\left\{\frac{1}{n}\right\}$

2. $\{3^n + 1\}$

3. $\{3\}$

4. $\left\{\left(\frac{3}{4}\right)^{n+1}\right\}$

5. $\left\{\left(\frac{4}{3}\right)^{n+1}\right\}$

6. $\left\{\left(-\frac{15}{16}\right)^{n+5}\right\}$

7. $\left\{\sin\left(\frac{n\pi}{2}\right)\right\}$

8. $\left\{\frac{n}{n^2 + n + 2}\right\}$

9. $\left\{\frac{(-1)^n}{n}\right\}$

10. $\{\mathrm{Tan}^{-1}\, n\}$

11. $\left\{\frac{\ln n}{n^2 + 1}\right\}$

12. $\left\{(-1)^n\sqrt{n^2 + 1}\right\}$

13. $c_1 = 2, \quad c_{n+1} = \frac{2}{c_n - 1}, \quad n \geq 1$

14. $c_1 = 4, \quad c_{n+1} = -\frac{c_n}{n^2}, \quad n \geq 1$

In Exercises 15–34 determine whether the statement is true or false. Verify any statement that is true, and give a counter example to any statement that is false.

15. A sequence can be increasing and nondecreasing.

16. A monotonic sequence must be bounded.

17. An increasing sequence must have a lower bound.

18. A decreasing sequence must have a lower bound.

19. An increasing sequence with a lower bound must have a limit.

20. An increasing sequence with an upper bound must have a limit.

21. If a sequence diverges, then either $\lim_{n\to\infty} c_n = \infty$, or, $\lim_{n\to\infty} c_n = -\infty$.

22. A sequence cannot be both increasing and decreasing.

23. A sequence can be both nonincreasing and nondecreasing.

24. If a sequence is monotonic and has a limit, it must be bounded.

25. If a sequence is bounded and has a limit, it must be monotonic.

26. If a sequence is bounded, but not monotonic, it cannot have a limit.

27. If a sequence has a limit, it must be bounded.

28. If a sequence is not monotonic, it cannot have a limit.

29. If all terms of a sequence $\{c_n\}$ are less than U, and $L = \lim_{n\to\infty} c_n$ exists, the L must be less than U.

30. A sequence $\{c_n\}$ has a limit if and only if $\{c_n^2\}$ has a limit.

31. A sequence $\{c_n\}$ of positive numbers converges if $c_{n+1} < c_n/2$.

32. A sequence $\{c_n\}$ of numbers converges if $c_{n+1} < c_n/2$.

33. If an increasing sequence has a limit L, then L must be equal to the smallest upper bound for the sequence.

34. If an infinite number of terms of a sequence all have the same value a, then either a is the limit of the sequence or the sequence has no limit.

In Exercises 35–48 discuss, with proofs, whether the sequence is monotonic, and whether it has an upper bound, a lower bound, and a limit.

35. $\left\{\dfrac{n+1}{2n+3}\right\}$ **36.** $\left\{\dfrac{2n+3}{n^2-5}\right\}$

37. $\left\{\dfrac{n^2+5n-4}{n^2+2n-2}\right\}$ **38.** $\left\{ne^{-n}\right\}$

39. $\left\{\dfrac{\ln n}{n}\right\}$ **40.** $\left\{\dfrac{n}{n+1}\operatorname{Tan}^{-1} n\right\}$

41. $c_1 = 1, \quad c_{n+1} = \dfrac{1}{10}\left(c_n^3 + 12\right), \quad n \geq 1$

42. $c_1 = 0, \quad c_{n+1} = \dfrac{1}{12}\left(c_n^4 + 5\right), \quad n \geq 1$

43. $c_1 = 3, \quad c_{n+1} = \sqrt{5 + c_n}, \quad n \geq 1$

44. $c_1 = 1, \quad c_{n+1} = \sqrt{5 + c_n}, \quad n \geq 1$

45. $c_1 = 5, \quad c_{n+1} = 1 + \sqrt{6 + c_n}, \quad n \geq 1$

46. $c_1 = 3, \quad c_{n+1} = 1 + \sqrt{6 + c_n}, \quad n \geq 1$

47. $c_1 = 1, \quad c_{n+1} = 4 - \sqrt{5 - c_n}, \quad n \geq 1$

48. $c_1 = 4, \quad c_{n+1} = 4 - \sqrt{5 - c_n}, \quad n \geq 1$

In Exercises 49–53 find an explicit formula for the general term of the sequence. In each case assume that the remaining terms follow the pattern suggested by the given terms.

49. $\dfrac{1}{2}, \dfrac{3}{4}, \dfrac{7}{8}, \dfrac{15}{16}, \dfrac{31}{32}, \ldots$

50. $4, \dfrac{7}{4}, \dfrac{10}{9}, \dfrac{13}{16}, \dfrac{16}{25}, \dfrac{19}{36}, \ldots$

51. $\dfrac{\ln 2}{\sqrt{2}}, -\dfrac{\ln 3}{\sqrt{3}}, \dfrac{\ln 4}{\sqrt{4}}, \dfrac{-\ln 5}{\sqrt{5}}, \ldots$

52. $1, 0, 1, 0, 1, 0, \ldots$

53. $1, 1, -1, -1, 1, 1, -1, -1, \ldots$

54. Verify that the sequence $\{n^{1/n}\}$ is decreasing for $n \geq 3$ by showing that the function $f(x) = x^{1/x}$ has a negative derivative for $x \geq 3$. Illustrate this result graphically. What is the limit of this sequence?

55. A sequence $\{c_n\}$ is defined by $c_n = 3n/(n^2 - 5n + 1)$.

(a) Consider the following proof that the sequence is decreasing: $c_{n+1} < c_n$ if and only if

$$\frac{3(n+1)}{(n+1)^2 - 5(n+1) + 1} < \frac{3n}{n^2 - 5n + 1}.$$

But this is true if and only if

$$3(n+1)(n^2 - 5n + 1) < 3n(n^2 - 3n - 3) \quad \text{or}$$
$$n^3 - 4n^2 - 4n + 1 < n^3 - 3n^2 - 3n;$$

that is, if and only if

$$1 < n^2 + n.$$

Since this inequality is true for all $n \geq 1$, the sequence is decreasing. Do you see any error in this proof?

(b) Calculate the first five terms in the sequence. What do you now conclude about the argument in (a)?

In Exercises 56–59 show how L'Hôpital's rule (Theorem 4.6, Section 4.9) can be used to evaluate the limit for the sequence.

56. $\left\{\frac{\ln n}{\sqrt{n}}\right\}$

57. $\left\{\frac{n^3+1}{e^n}\right\}$

58. $\left\{n \sin\left(\frac{4}{n}\right)\right\}$

59. $\left\{\left(\frac{n+5}{n+3}\right)^n\right\}$

In Exercises 60–68 discuss, with all necessary proofs, whether the sequence is monotonic, is bounded, and has a limit.

60. $c_1 = 2, \quad c_{n+1} = \frac{1}{3 - c_n}, \quad n \geq 1$

61. $c_1 = 1, \quad c_{n+1} = \frac{1}{4 - 2c_n}, \quad n \geq 1$

62. $c_1 = 1, \quad c_{n+1} = \frac{7}{16 - 8c_n^2}, \quad n \geq 1$

63. $c_1 = 0, \quad c_{n+1} = \frac{7}{16 - 8c_n^2}, \quad n \geq 1$

64. $c_1 = 1, \quad c_{n+1} = \frac{4c_n}{4 + c_n}, \quad n \geq 1$

65. $c_1 = 4, \quad c_{n+1} = \frac{3c_n}{2 + c_n}, \quad n \geq 1$

66. $c_1 = 2, \quad c_{n+1} = \frac{2c_n^2}{3 + c_n}, \quad n \geq 1$

67. $c_1 = \frac{3}{2}, \quad c_{n+1} = \frac{c_n + 2}{4 - c_n}, \quad n \geq 1$

68. $c_1 = 0, \quad c_{n+1} = \frac{3 - c_n}{5 - 2c_n}, \quad n \geq 1$

69. Show that the sequence

$$c_1 = 1, \quad c_{n+1} = \frac{1}{4 - c_n - c_n^2}, \quad n \geq 1$$

is monotonic and bounded. Find an approximation to its limit accurate to 5 decimals.

70. (a) Prove that if a sequence converges, then differences between successive terms in the sequence must approach zero.

(b) The following example illustrates that the converse is not true. Show that differences between successive terms of the sequence $\{\ln n\}$ approach zero, but the sequence itself diverges.

71. Prove parts (i) and (ii) of Theorem 11.3.

72. Find bounds for the sequence

$$c_1 = 1, \quad c_{n+1} = \frac{1 + c_n}{1 + 2c_n}, \quad n \geq 1.$$

73. Determine whether the following sequence is monotonic, has bounds, and has a limit,

$$c_1 = -30, \quad c_2 = -20, \quad c_{n+1} = 5 + \frac{c_n}{2} + \frac{c_{n-1}}{3}, \quad n \geq 2.$$

74. A sequence $\{c_n\}$ has only positive terms. Prove that:

(a) if $\lim_{n\to\infty} c_n = L < 1$, there exists an integer N such that for all $n \geq N$

$$c_n < \frac{L+1}{2}.$$

(b) if $\lim_{n\to\infty} c_n = L > 1$, there exists an integer N such that for all $n \geq N$,

$$c_n > \frac{L+1}{2}.$$

These results are used in Theorem 11.9, Section 11.5.

75. Prove that if $\lim_{n\to\infty} c_n = L$, there exists an integer N such that for all $n \geq N$,

$$c_n < L + 1.$$

This result is used in Theorem 11.8, Section 11.4.

76. The *Fibonacci sequence* found in many areas of applied mathematics is defined by

$$c_1 = 1, \quad c_2 = 1, \quad c_{n+1} = c_n + c_{n-1}, \quad n \geq 2.$$

(a) Evaluate the first ten terms of this sequence.

(b) Is the sequence monotonic, is it bounded, and does it have a limit?

(c) Prove that

$$c_n^2 - c_{n-1}c_{n+1} = (-1)^{n+1}, \quad n \geq 2.$$

(d) Verify that an explicit formula for c_n is

$$c_n = \frac{1}{\sqrt{5}}\left[\left(\frac{1+\sqrt{5}}{2}\right)^n - \left(\frac{1-\sqrt{5}}{2}\right)^n\right].$$

(e) Define a sequence $\{b_n\}$ as the ratio of terms in the Fibonacci sequence

$$b_n = \frac{c_{n+1}}{c_n}.$$

Is this sequence monotonic? Does it have a limit?

77. Prove that the sequence

$$c_1 = d, \quad c_{n+1} = \sqrt{a + 2bc_n}, \quad n \geq 1$$

(a, b, and d all positive constants) is increasing if and only if $d < b + \sqrt{a + b^2}$. What happens when $d = b + \sqrt{a + b^2}$?

78. In this exercise we outline a proof of part (iii) of Theorem 11.3.

(a) Verify that

$$|c_n d_n - CD| \leq |c_n||d_n - D| + |D||c_n - C|.$$

(b) Show that given any $\epsilon > 0$, there exist positive integers N_1, N_2, and N_3 such that

$$|c_n| < |C| + 1 \qquad \text{whenever } n > N_1,$$

$$|d_n - D| < \frac{\epsilon}{2\left(|C| + 1\right)} \qquad \text{whenever } n > N_2,$$

$$|c_n - C| < \frac{\epsilon}{2|D| + 1} \qquad \text{whenever } n > N_3.$$

(c) Use these results to prove part (iii) of Theorem 11.3.

79. In this exercise we outline a proof for part (iv) of Theorem 11.3.

(a) Verify that when $d_n \neq 0$ and $D \neq 0$,

$$\left|\frac{c_n}{d_n} - \frac{C}{D}\right| \leq \frac{|c_n - C|}{|d_n|} + \frac{|C||d_n - D|}{|D||d_n|}.$$

(b) Show that given any $\epsilon > 0$, there exist positive integers N_1, N_2, and N_3 such that

$$|d_n| > \frac{|D|}{2} \qquad \text{whenever } n > N_1,$$

$$|c_n - C| < \frac{\epsilon|D|}{4} \qquad \text{whenever } n > N_2,$$

$$|d_n - D| < \frac{\epsilon|D|^2}{4|C| + 1} \qquad \text{whenever } n > N_3.$$

(c) Now prove part (iv) of Theorem 11.3.

80. Find an explicit formula for the recursive sequence

$$c_1 = 1, \quad c_2 = 2, \quad c_{n+1} = \frac{c_n + c_{n-1}}{2}, \quad n \geq 2.$$

81. Justify the statement that the sequence of Example 11.3 has no upper bound.

SECTION 11.2

Applications of Sequences

When a sequence is monotonic and bounded, it has a limit. Verification that a recursive sequence is monotonic and bounded usually requires mathematical induction. However, when the sequence arises in an application, the derivation of the sequence may make it clear that the sequence has a limit, and it may therefore be possible to proceed directly to its evaluation without verifying monotony and bounds. Such is often the case in the field of numerical analysis, wherein recursive sequences commonly arise in the form of iterative procedures. We have already encountered Newton's iterative procedure as one example. When this method is applied to the equation

$$f(x) = x^3 - 3x + 1 = 0$$

with an initial approximation $x_1 = 0$ to find the root between 0 and 1, the sequence obtained is

$$x_1 = 0, \qquad x_{n+1} = x_n - \frac{f(x_n)}{f'(x_n)} = x_n - \frac{x_n^3 - 3x_n + 1}{3x_n^2 - 3}, \quad n \geq 1.$$

The first three terms of this sequence are illustrated in Figure 11.6, and the tangent line construction by which they are obtained makes it clear that the sequence must converge to the solution of the equation between 0 and 1. In other words, it is not necessary for us to analyze monotony and bounds for the recursive sequence algebraically; they are obvious geometrically. The sequence is increasing and has bounds $U = 1$ and $V = 0$.

To find the limit of a recursive sequence in Section 11.1, we adopted the procedure of letting n become very large on both sides of the recursive definition for the sequence (see Example 11.4). This does not work here as we are simply led back to the equation

$x^3 - 3x + 1 = 0$. Try it. Instead we evaluate terms of the sequence algebraically until they repeat

$$x_1 = 0, \ x_2 = 1/3, \ x_3 = 0.347222, \ x_4 = 0.34729635, \ x_5 = 0.347296355,$$
$$x_6 = 0.347296355.$$

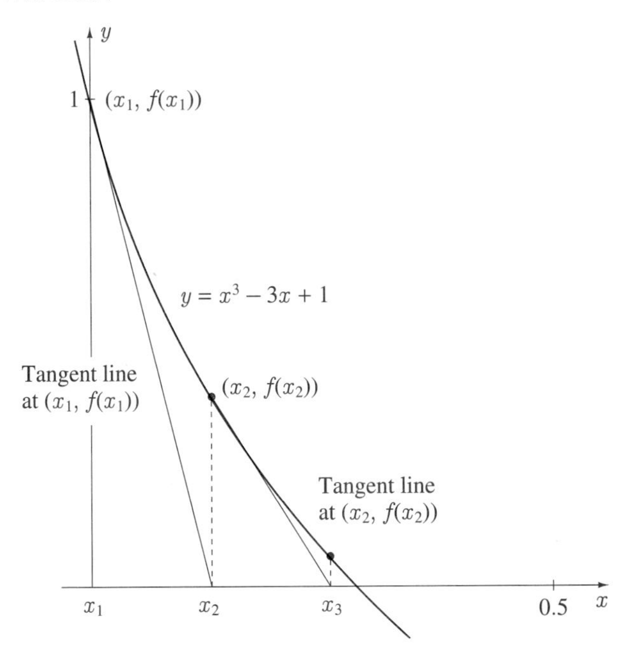

FIGURE 11.6

At this point we might be led to conclude that the required limit and solution of $x^3 - 3x + 1 = 0$ is $x = 0.347296355$. But realizing that our calculations are calculator dependent, we should at least be skeptical of the last digit on the display of our calculator. Suppose the real question were to find the solution accurate to six decimals. This could be accomplished by calculating

$$f(0.3472955) = 2.3 \times 10^{-6} \qquad \text{and} \qquad f(0.3472965) = -3.8 \times 10^{-7}.$$

Since one function value is positive and the other is negative, the intermediate value theorem (see Theorem 4.1) guarantees that the root is between 0.3472955 and 0.3472965. Hence, to six decimals, the solution is 0.347296.

Figure 11.6 illustrated that the sequence defined by Newton's iterative procedure must converge to a root of $f(x) = x^3 - 3x + 1 = 0$, and therefore it was unnecessary to verify existence of a limit of the sequence by proving that the sequence was monotonic and bounded. Are we proposing that a graph of the function $f(x)$ should always be drawn when using Newton's method to solve equations? Not really, although we are of the philosophy that pictures should be drawn whenever possible. Numerical analysts have proved that under very mild restrictions, the sequence defined by Newton's method always converges to a root of the equation provided the initial approximation is sufficiently close to that root. See Exercise 33 for further discussion of this point.

Another way to solve for the same root of $x^3 - 3x + 1 = 0$ is to rewrite the equation in the form

$$x = \frac{1}{3}(x^3 + 1),$$

and use this to define the following recursive sequence

$$x_1 = 0, \qquad x_{n+1} = \frac{1}{3}(x_n^3 + 1), \quad n \geq 1.$$

It is straighforward to show that this sequence is bounded above by 1 and below by 0 and that it is increasing. According to Theorem 11.1, it therefore has a limit L, which must lie in the interval $0 \le L \le 1$. To find L, we set

$$\lim_{n\to\infty} x_{n+1} = \frac{1}{3}\lim_{n\to\infty}(x_n^3+1) \qquad \text{or}$$

$$L = \frac{1}{3}(L^3+1).$$

But this implies that L is the required root of the original equation. In other words, we have defined a recursive sequence $\{x_n\}$ in such a way that its limit is the required solution of $x^3 - 3x + 1 = 0$. The first ten terms of the sequence are:

$$\begin{aligned} x_1 &= 0, & x_6 &= 0.34729353,\\ x_2 &= 1/3, & x_7 &= 0.34729601,\\ x_3 &= 0.345679, & x_8 &= 0.34729631,\\ x_4 &= 0.34710219, & x_9 &= 0.34729635,\\ x_5 &= 0.34727295, \end{aligned}$$

which leads us again to the conclusion that an approximation to the root of $x^3 - 3x + 1 = 0$ between 0 and 1 is $x = 0.347296$.

This method of finding the root of an equation is often called the **method of successive substitutions**. Exercises 21–32 provide more opportunities to practise this method.

In Section 3.9 and Appendix B, we essentially defined the number e as the limit of the sequence

$$e_n = \left(1+\frac{1}{n}\right)^n, \tag{11.7}$$

but gave no proof that the sequence does indeed converge. In spite of the fact that the sequence is defined explicitly, it is far from trivial to prove that it has a limit. To do so, we first find an expanded expression for e_n by means of the binomial theorem,

$$\begin{aligned} e_n = \left(1+\frac{1}{n}\right)^n &= \sum_{r=0}^{n}\binom{n}{r}\left(\frac{1}{n}\right)^r\\ &= 1 + \sum_{r=1}^{n}\frac{n(n-1)\cdots(n-r+1)}{r!\,n^r}, \end{aligned}$$

where we have used the expression $\binom{n}{r} = n(n-1)\cdots(n-r+1)/r!$ for binomial coefficients. Now in the product $n(n-1)\cdots(n-r+1)$, there are r factors. If each is divided by one of the n's in the denominator, e_n can be expressed in the form

$$\begin{aligned} e_n &= 1 + \sum_{r=1}^{n}\frac{1}{r!}\left(\frac{n}{n}\right)\left(\frac{n-1}{n}\right)\left(\frac{n-2}{n}\right)\cdots\left(\frac{n-r+1}{n}\right)\\ &= 1 + \sum_{r=1}^{n}\frac{1}{r!}(1)\left(1-\frac{1}{n}\right)\left(1-\frac{2}{n}\right)\cdots\left(1-\frac{r-1}{n}\right). \end{aligned} \tag{11.8}$$

To show that sequence 11.7 has a limit, we verify that it is increasing ($e_{n+1} > e_n$) and bounded ($2 < e_n < 3$). The expanded form for e_{n+1} is obtained by replacing n by $n+1$ in 11.8,

$$e_{n+1} = 1 + \sum_{r=1}^{n+1}\frac{1}{r!}(1)\left(1-\frac{1}{n+1}\right)\left(1-\frac{2}{n+1}\right)\cdots\left(1-\frac{r-1}{n+1}\right).$$

If we drop the last term from this summation (when $r = n+1$), the right side becomes smaller. In addition, if the $(n+1)$'s in the denominator are all replaced by n's, each factor decreases. Thus we may write

$$e_{n+1} > 1 + \sum_{r=1}^{n} \frac{1}{r!}(1)\left(1 - \frac{1}{n}\right)\left(1 - \frac{2}{n}\right)\cdots\left(1 - \frac{r-1}{n}\right)$$

But this is e_n (see 11.8). In other words, we have verified that $e_{n+1} > e_n$, and the sequence is increasing.

Since the sequence is increasing, it is bounded below by its first term $e_1 = 2$. To verify that 3 is an upper bounded, we use expression 11.8 to write

$$\begin{aligned}
e_n &= 1 + \sum_{r=1}^{n} \frac{1}{r!}(1)\left(1 - \frac{1}{n}\right)\left(1 - \frac{2}{n}\right)\cdots\left(1 - \frac{r-1}{n}\right) \\
&< 1 + \sum_{r=1}^{n} \frac{1}{r!} \qquad \text{(since all omitted factors are less than one)} \\
&= 1 + \left(1 + \frac{1}{2!} + \frac{1}{3!} + \cdots + \frac{1}{n!}\right) \\
&< 1 + \left(1 + \frac{1}{2} + \frac{1}{2^2} + \frac{1}{2^3} + \cdots + \frac{1}{2^{n-1}}\right).
\end{aligned}$$

A little experimentation shows that the sum of the terms in the parentheses is $2 - 1/2^{n-1}$, and therefore

$$e_n < 1 + 2 - \frac{1}{2^{n-1}} = 3 - \frac{1}{2^{n-1}} < 3.$$

With an upper bound established, Theorem 11.1 now guarantees that the sequence $\{e_n\}$ has a limit. The limit can be approximated by setting n in $(1 + 1/n)^n$ equal to a very large number. For example, with $n = 10^6$, we obtain $e_{1\,000\,000} = 2.71828047$. Unfortunately, we have no idea how accurate this approximation is for e. A superior method for approximating e using infinite series is discussed in Section 11.7, and with this method we can make definitive statements about the accuracy of the approximation.

EXERCISES 11.2

In Exercises 1–12 use Newton's iterative procedure with the given initial approximation x_1 to define a sequence of approximations to a solution of the equation. Determine graphically whether the sequence is monotonic, is bounded, and has a limit. Approximate any limit that exists to seven decimals.

1. $x_1 = 1, \quad x^2 + 3x + 1 = 0$

2. $x_1 = -1, \quad x^2 + 3x + 1 = 0$

3. $x_1 = -1.5, \quad x^2 + 3x + 1 = 0$

4. $x_1 = -3, \quad x^2 + 3x + 1 = 0$

5. $x_1 = 4, \quad x^3 - x^2 + x - 22 = 0$

6. $x_1 = 2, \quad x^3 - x^2 + x - 22 = 0$

7. $x_1 = 2, \quad x^5 - 3x + 1 = 0$

8. $x_1 = 1, \quad x^5 - 3x + 1 = 0$

9. $x_1 = 0, \quad x^5 - 3x + 1 = 0$

10. $x_1 = 4/5, \quad x^5 - 3x + 1 = 0$

11. $x_1 = 0.85, \quad x^5 - 3x + 1 = 0$

12. $x_1 = -2, \quad x^5 - 3x + 1 = 0$

13. (a) Use Newton's iterative procedure with $x_1 = 1$ to approximate the root of $x^4 - 15x + 2 = 0$ between 0 and 1.

(b) Use the method of successive substitutions with $x_1 = 1$ and $x_{n+1} = (x_n^4 + 2)/15$ to approximate the root in (a).

(c) Use Newton's method to approximate the root between 2 and 3.

(d) What happens if the sequence in (b) is used to approximate the root between 2 and 3 with $x_1 = 2$ and $x_1 = 3$?

14. A superball is dropped from the top of a building 20 m high. Each time it strikes the ground, it rebounds to 99% of the height from which it fell.

(a) If d_n denotes the distance traveled by the ball between the nth and $(n+1)^{\text{th}}$ bounces, find a formula for d_n.

(b) If t_n denotes the time between the nth and $(n+1)^{\text{th}}$ bounces, find a formula for t_n.

15. A dog sits at a farmhouse patiently watching for his master to return from the fields. When the farmer is 1 km from the farmhouse, the dog immediately takes off for the farmer. When he reaches the farmer, he turns and runs back to the farmhouse, whereupon he again turns and runs to the farmer. The dog continues this frantic action until the farmer reaches the farmhouse. If the dog runs twice as fast as the farmer, find the distance d_n run by the dog from the point when he reaches the farmer for the nth time to the point when he reaches the farmer for the $(n+1)^{\text{th}}$ time. Ignore any accelerations of the dog in the turns.

16. The equilateral triangle in Figure 11.7(a) has perimeter P. If each side of the triangle is divided into three equal parts, an equilateral triangle is drawn on the middle segment of each side, and the figure transformed into Figure 11.7(b), what is the perimeter P_1 of this figure? If each side of Figure 11.7(b) is now subdivided into three equal portions and equilateral triangles are similarly constructed to result in Figure 11.7(c), what is the perimeter P_2 of this figure? If this subdivision process is continued indefinitely, what is the perimeter P_n after the nth subdivision? What is $\lim_{n\to\infty} P_n$?

FIGURE 11.7

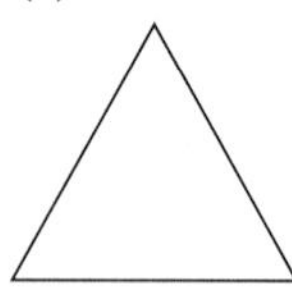

(b)

(c)

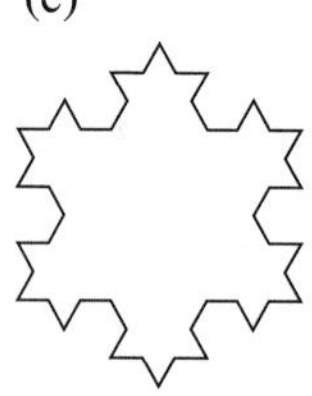

17. A stone of mass 100 gm is thrown vertically upward with speed 20 m/s. Air exerts a resistive force on the stone proportional to its speed, and has magnitude 1/10 N when the speed of the stone is 10 m/s. It can be shown that the height y (in metres) above the projection point attained by the stone is given by

$$y = -98.1t + 1181\left(1 - e^{-t/10}\right),$$

where t is time (measured in seconds with $t = 0$ at the instant of projection).

(a) The time taken for the stone to return to its projection point can be obtained by setting $y = 0$ and solving the equation for t. Do so (correct to 2 decimals).

(b) Find the time for the stone to return if air resistance is ignored.

18. When the beam in Figure 11.8 vibrates vertically, there are certain frequencies of vibration, called *natural frequencies of the system.* They are solutions of the equation

$$\tan x = \frac{e^x - e^{-x}}{e^x + e^{-x}}$$

divided by 20π. Find the two smallest natural frequencies.

FIGURE 11.8

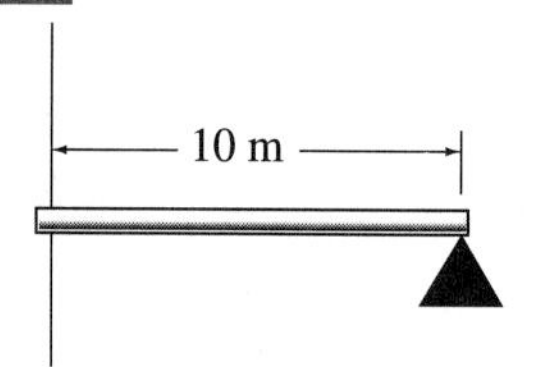

19. Consider the sequence

$$c_1 = 1, \quad c_{n+1} = \frac{1}{2}\left(c_n + \frac{a}{c_n}\right), \quad n \geq 1,$$

where a is any number greater than 1.

(a) Show that $\sqrt{a} < c_n < a$ for $n \geq 2$.

(b) Prove that the sequence is decreasing for $n \geq 2$.

(c) Verify that the sequence converges to $\sqrt{a}$. This result establishes an iterative procedure for finding the square root of real numbers. The sequence converges to $\sqrt{a}$ for $0 < a < 1$ also.

(d) Show that this sequence can be obtained from Newton's iterative procedure for the solution of the equation $x^2 = a$.

20. If A_n is the area of the figure with perimeter P_n in Exercise 16, find a formula for A_n.

To solve an equation $f(x) = 0$ by the method of successive substitutions, we first rearrange the equation into the form $x = g(x)$. We then define a recursive sequence by choosing some initial approximation x_1 and setting

$$x_{n+1} = g(x_n), \quad n \geq 1.$$

Depending on the choice of $g(x)$, this sequence may or may not converge to a root of the equation. In Exercises 21–25 illustrate that the suggested rearrangement of the equation $f(x) = 0$, along with the initial approximation x_1, leads to a sequence which converges to a root of the equation. Find the root accurate to 4 decimals.

21. $f(x) = x^2 - 2x - 1$; $x_1 = 2$; use $x = 2 + 1/x$

22. $f(x) = x^3 + 6x + 3$; $x_1 = -1$; use $x = -(1/6)(x^3 + 3)$

23. $f(x) = x^4 - 120x + 20$; $x_1 = 0$; use $x = (x^4 + 20)/120$

24. $f(x) = x^3 - 2x^2 - 3x + 1$; $x_1 = 3$; use $x = (2x^2 + 3x - 1)/x^2$

25. $f(x) = 8x^3 - x^2 - 1$; $x_1 = 0$; use $x = (1/2)(1 + x^2)^{1/3}$

In Exercises 26–31 find a rearrangement of the equation which leads, through the method of successive substitutions, to a 4 decimal approximation to the root of the equation.

26. $x^3 - 6x^2 + 11x - 7 = 0$ between $x = 3$ and $x = 4$

27. $x^4 - 3x^2 - 3x + 1 = 0$ between $x = 0$ and $x = 1$

28. $x^4 + 4x^3 - 50x^2 + 100x - 50 = 0$ between $x = 0$ and $x = 1$

29. $\sin^2 x = 1 - x^2$ between $x = 0$ and $x = 1$

30. $\sec x = 2/(1 + x^4)$ between $x = 0$ and $x = 1$

31. $e^x + e^{-x} - 10x = 0$ between $x = 0$ and $x = 1$

32. (a) Show that if α is the only root of the equation $x = g(x)$ and Figure 11.9(a) is a graph of $g(x)$, then Figure 11.9(b) exhibits geometrically the sequence of approximations of α determined by the method of successive substitutions.

FIGURE 11.9

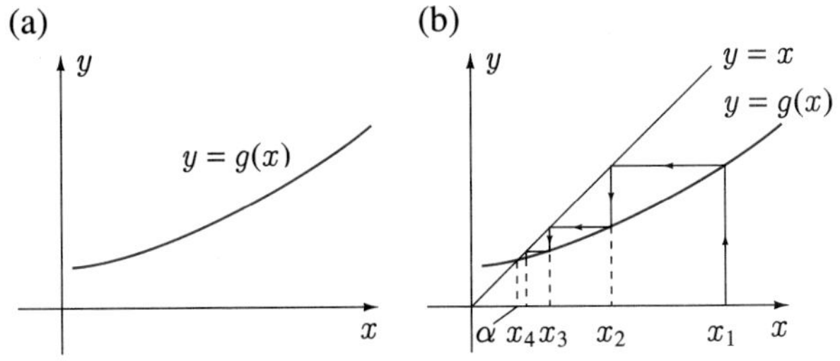

(a)

(b)

(b) Illustrate graphically how the sequence defined by successive substitutions converges to the root $x = \alpha$ for the equation $x = g(x)$ if $g(x)$ is as shown in Figure 11.10.

FIGURE 11.10

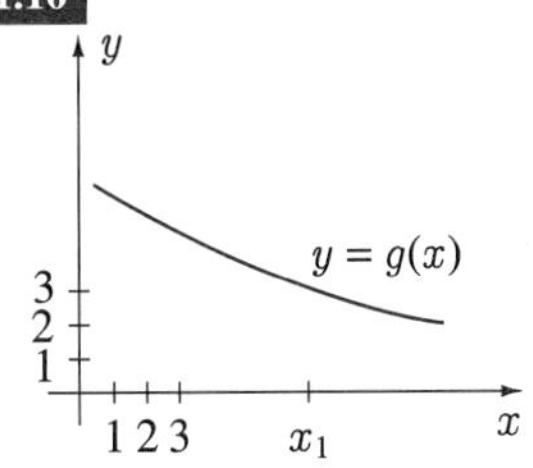

(c) Illustrate graphically that the method fails for the function $g(x)$ in Figure 11.11.

FIGURE 11.11

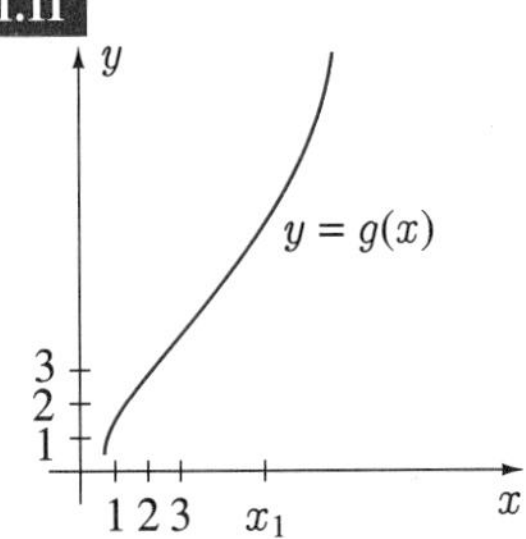

(d) Based on the results of parts (a)–(c), what determines success or failure of the method of successive substitutions? Part (e) provides a proof of the correct answer.

(e) Prove that when $x = g(x)$ has a root $x = \alpha$, the method of successive substitutions with an initial approximation of x_1 always converges to α if $|g'(x)| \leq a < 1$ on the interval $|x - \alpha| \leq |x_1 - \alpha|$. In other words, when an equation $f(x) = 0$ is rearranged into the form $x = g(x)$ for successive subsitutions, success or failure depends on whether the derivative of $g(x)$ is between -1 and 1 near the required root.

33. Suppose that $f''(x)$ exists on an open interval containing a root $x = \alpha$ of the equation $f(x) = 0$. Use the result of Exercises 32 to prove that if $f'(\alpha) \neq 0$, Newton's iterative sequence always converges to α provided the initial approximation x_1 is chosen sufficiently close to α. (Hint: First use Exercise 32(e) to show that Newton's sequence converges to α if on the interval $|x - \alpha| \leq |x_1 - \alpha|$, $|ff''/(f')^2| \leq a < 1$.) In actual fact, Newton's method usually works even when $f'(\alpha) = 0$.

SECTION 11.3

Infinite Series of Constants

If $\{c_n\}$ is an infinite sequence of constants, an expression of the form

$$c_1 + c_2 + c_3 + \cdots + c_n + \cdots \tag{11.9}$$

is called an **infinite series**, or simply a series. We often use sigma notation to represent a series:

$$\sum_{n=1}^{\infty} c_n = c_1 + c_2 + c_3 + \cdots + c_n + \cdots. \tag{11.10}$$

By initiating the sigma notation with $n = 1$, we once again have the convenient situation in which $n = 1$ yields the first term of the series, $n = 2$ the second, etc. We usually abbreviate the notation further by writing $\sum c_n$ in place of $\sum_{n=1}^{\infty} c_n$ whenever the lower limit of summation is chosen as unity.

From the sequences of Example 11.1 we may form the following series.

EXAMPLE 11.5

Write out the first six terms of the following series:

(a) $\sum_{n=1}^{\infty} \frac{1}{2^{n-1}}$ (b) $\sum_{n=1}^{\infty} \frac{n}{n+1}$ (c) $\sum_{n=1}^{\infty} (-1)^n |n-3|$ (d) $\sum_{n=1}^{\infty} (-1)^{n+1}$

SOLUTION The first six terms of these series are shown below:

(a) $\sum_{n=1}^{\infty} \frac{1}{2^{n-1}} = 1 + \frac{1}{2} + \frac{1}{4} + \frac{1}{8} + \frac{1}{16} + \frac{1}{32} + \cdots$

(b) $\sum_{n=1}^{\infty} \frac{n}{n+1} = \frac{1}{2} + \frac{2}{3} + \frac{3}{4} + \frac{4}{5} + \frac{5}{6} + \frac{6}{7} + \cdots$

(c) $\sum_{n=1}^{\infty} (-1)^n |n-3| = -2 + 1 - 0 + 1 - 2 + 3 - \cdots$

(d) $\sum_{n=1}^{\infty} (-1)^{n+1} = 1 - 1 + 1 - 1 + 1 - 1 + \cdots$

■

Perhaps the most frequently encountered series in applications is the **geometric series**; a series of the form

$$a + ar + ar^2 + ar^3 + \cdots.$$

Each term in a geometric series is obtained by multiplying the term that came before it by the same constant r. The number $a \neq 0$ is called the first term of the series and r is called the **common ratio**. In sigma notation,

$$\sum_{n=1}^{\infty} ar^{n-1} = a + ar + ar^2 + ar^3 + \cdots. \tag{11.11}$$

The series in parts (a) and (d) of Example 11.5 are geometric. The first has common ratio $1/2$ the second has common ratio -1.

Another important series is

$$\sum_{n=1}^{\infty} \frac{1}{n} = 1 + \frac{1}{2} + \frac{1}{3} + \frac{1}{4} + \cdots. \tag{11.12}$$

It is called the **harmonic series**. We shall see it many times in our discussions.

It may be quite convenient to write expressions like 11.9 and say that they are infinite series, but since we can add only finitely many numbers, 11.9 is as yet meaningless. To illustrate how we may define a meaning for such a sum, consider the geometric series of Example 11.5(a),

$$\sum_{n=1}^{\infty} \frac{1}{2^{n-1}} = 1 + \frac{1}{2} + \frac{1}{4} + \frac{1}{8} + \cdots.$$

If we nonchalantly start adding terms of this series together, a pattern soon emerges. Indeed, if we denote by S_n the sum of the first n terms of this series, we find that

$$\begin{aligned}
S_1 &= 1, \\
S_2 &= 1 + \frac{1}{2} = \frac{3}{2}, && \text{(sum of first two terms)} \\
S_3 &= 1 + \frac{1}{2} + \frac{1}{4} = \frac{7}{4}, && \text{(sum of first three terms)} \\
S_4 &= 1 + \frac{1}{2} + \frac{1}{4} + \frac{1}{8} = \frac{15}{8}, && \text{(sum of first four terms)} \\
S_5 &= 1 + \frac{1}{2} + \frac{1}{4} + \frac{1}{8} + \frac{1}{16} = \frac{31}{16}, && \text{(sum of first five terms)}
\end{aligned}$$

and so on. By careful examination of the pattern, we can see that the sum of the first n terms of the series is given by the formula

$$S_n = 2 - \frac{1}{2^{n-1}}.$$

As we add more and more terms of this series together, the sum S_n gets closer and closer to 2. It is always less than 2, but S_n can be made arbitrarily close to 2 by choosing n sufficiently large. If this series is to have a sum, the only reasonable sum is 2. In practice, this is precisely what we do; we define the sum of the series $\sum 1/2^{n-1}$ to be 2.

Let us now take this idea and define sums for all infinite series. We begin by defining a sequence $\{S_n\}$ as follows:

$$\begin{aligned}
S_1 &= c_1, \\
S_2 &= c_1 + c_2, \\
S_3 &= c_1 + c_2 + c_3, \\
&\ \vdots \\
S_n &= c_1 + c_2 + c_3 + \cdots + c_n, \\
&\ \vdots
\end{aligned}$$

It is called the **sequence of partial sums** of the series $\sum c_n$. The nth term of the sequence $\{S_n\}$ represents the sum of the first n terms of the series. If this sequence has a limit, say, S, then the more terms of the series that we add together, the closer the sum gets to S. It seems reasonable, then, to call S the sum of the series. We therefore make the following definition.

Definition 11.5

> Let $S_n = \sum_{k=1}^{n} c_k$ be the nth partial sum of a series $\sum_{n=1}^{\infty} c_n$. If the sequence of partial sums $\{S_n\}$ has limit S,
> $$S = \lim_{n\to\infty} S_n,$$
> we call S the sum of the series and write
> $$\sum_{n=1}^{\infty} c_n = S;$$
> if $\{S_n\}$ does not have a limit, we say that the series does not have a sum.

If a series has sum S, we say that the series **converges** to S, which means that its sequence of partial sums converges to S. If a series does not have a sum, we say that the series **diverges**, which means that its sequence of partial sums diverges.

According to this definition, the series $\sum 1/2^{n-1}$ of Example 11.5(a) has sum 2. Since every term of the series $\sum n/(n+1)$ in Example 11.5(b) is greater than or equal to $1/2$, it follows that the sum of the first n terms is $S_n \geq n(1/2) = n/2$. As the sequence of partial sums is therefore (increasing and) unbounded, it cannot possibly have a limit (see the corollary to Theorem 11.1). The series does not therefore have a sum; it diverges. Examination of the first few partial sums of the series $\sum(-1)^n|n-3|$ in Example 11.5(c) leads to the result that for $n > 1$,

$$S_n = \begin{cases} \dfrac{n-4}{2} & \text{if } n \text{ is even} \\[2mm] \dfrac{1-n}{2} & \text{if } n \text{ is odd.} \end{cases}$$

Since the sequence $\{S_n\}$ does not have a limit, the series diverges. The partial sums of the series $\sum(-1)^{n+1}$ are

$$S_1 = 1, \quad S_2 = 0, \quad S_3 = 1, \quad S_4 = 0, \quad \ldots.$$

Since this sequence does not have a limit, the series does not have a sum.

Consider now geometric series 11.11. If $\{S_n\}$ is the sequence of partial sums for this series, then

$$S_n = a + ar + ar^2 + \cdots + ar^{n-1}.$$

If we multiply this equation by r,

$$rS_n = ar + ar^2 + ar^3 + \cdots + ar^n.$$

When we subtract these equations, the result is

$$S_n - rS_n = a - ar^n.$$

Hence, for $r \neq 1$, we obtain

$$S_n = \frac{a(1-r^n)}{1-r}.$$

Furthermore, when $r = 1$,

$$S_n = na,$$

and therefore the sum of the first n terms of a geometric series is

$$S_n = \begin{cases} \dfrac{a(1-r^n)}{1-r} & r \neq 1 \\ na & r = 1. \end{cases} \tag{11.13a}$$

To determine whether a geometric series has a sum, we consider the limit of this sequence. Certainly, $\lim_{n\to\infty} na$ does not exist, unless trivially $a = 0$. In addition, $\lim_{n\to\infty} r^n = 0$ when $|r| < 1$, and does not exist when $|r| > 1$. Thus, we may state that

$$\lim_{n\to\infty} S_n = \begin{cases} \dfrac{a}{1-r} & |r| < 1 \\ \text{does not exist} & |r| \geq 1. \end{cases}$$

The geometric series therefore has sum

$$\sum_{n=1}^{\infty} ar^{n-1} = \frac{a}{1-r} \tag{11.13b}$$

if $|r| < 1$, but otherwise diverges.

Next we consider harmonic series 11.12. This series does not have a sum, and we can show this by considering the following partial sums of the series:

$$\begin{aligned}
S_1 &= 1, \\
S_2 &= 1 + \frac{1}{2} = \frac{3}{2}, \\
S_4 &= S_2 + \frac{1}{3} + \frac{1}{4} > \frac{3}{2} + \frac{1}{4} + \frac{1}{4} = \frac{4}{2}, \\
S_8 &= S_4 + \frac{1}{5} + \frac{1}{6} + \frac{1}{7} + \frac{1}{8} > \frac{4}{2} + \frac{1}{8} + \frac{1}{8} + \frac{1}{8} + \frac{1}{8} = \frac{5}{2}, \\
S_{16} &= S_8 + \frac{1}{9} + \frac{1}{10} + \cdots + \frac{1}{16} > \frac{5}{2} + \frac{1}{16} + \frac{1}{16} + \cdots + \frac{1}{16} = \frac{6}{2}.
\end{aligned}$$

This procedure can be continued indefinitely, and shows that the sequence of partial sums is unbounded. The harmonic series therefore diverges (see once again the corollary to Theorem 11.1).

In each of the above examples, we used Definition 11.5 for the sum of a series to determine whether the series converges or diverges; that is, we formed the sequence of partial sums $\{S_n\}$ in order to consider its limit. For most examples, it is either too difficult or impossible to evaluate S_n in a simple form, and in such cases consideration of the limit of the sequence $\{S_n\}$ is impractical. Consequently, we must develop alternative ways to decide on the convergence of a series. We do this in Sections 11.4–11.6.

We now discuss some fairly simple but important results on convergence of series. If a series $\sum c_n$ has a finite number of its terms altered in any fashion whatsoever, the new series converges if and only if the original series converges. The new series may converge to a different sum, but it converges if the original series converges. For example, if we double the first three terms of the geometric series in Example 11.5, but do not change the remaining terms, the new series is

$$2 + 1 + \frac{1}{2} + \frac{1}{8} + \frac{1}{16} + \cdots.$$

It is not geometric. But its nth partial sum, call it S_n, is very closely related to that of the geometric series, call it T_n. In fact for $n \geq 4$, we can say that $S_n = T_n + \frac{7}{4}$ ($\frac{7}{4}$ is the total change in the first three terms). Since $\lim_{n\to\infty} T_n = 2$, it follows that

$\lim_{n\to\infty} S_n = 2 + \frac{7}{4} = \frac{15}{4}$; that is, the new series converges, but it has a sum which is $\frac{7}{4}$ greater than that of the geometric series.

On the other hand, suppose we change the first 100 terms of the harmonic series, which diverges, to 0, but leave the remaining terms unaltered. The new series is

$$\overbrace{0 + 0 + \cdots + 0}^{100 \text{ terms}} + \frac{1}{101} + \frac{1}{102} + \cdots .$$

We know that the sequence of partial sums $\{T_n\}$ of the harmonic series is increasing and unbounded. The nth partial sum S_n of the new series, for $n \geq 100$, is equal to $T_n - k$, where k is the sum of the first 100 terms of the harmonic series. But because $\lim_{n\to\infty} T_n = \infty$, so also must $\lim_{n\to\infty} S_n = \infty$; that is, the new series diverges.

The following theorem indicates that convergent series can be added and subtracted, and multiplied by constants. These properties can be verified using Definition 11.5.

Theorem 11.4

If series $\sum_{n=1}^{\infty} c_n$ and $\sum_{n=1}^{\infty} d_n$ have sums C and D, then:

$$\text{(i)} \quad \sum_{n=1}^{\infty} kc_n = kC \qquad (\textit{when } k \textit{ is a constant}); \tag{11.14a}$$

$$\text{(ii)} \quad \sum_{n=1}^{\infty} (c_n \pm d_n) = C \pm D. \tag{11.14b}$$

Our first convergence test is a corollary to the following theorem.

Theorem 11.5

If a series $\sum_{n=1}^{\infty} c_n$ converges, then $\lim_{n\to\infty} c_n = 0$.

Proof If series $\sum c_n$ has a sum S, its sequence of partial sums

$$S_1, S_2, \cdots, S_n, \cdots$$

has limit S. The sequence

$$0, S_1, S_2, \cdots, S_{n-1}, \cdots$$

must also have limit S; it is the sequence of partial sums with an additional term equal to 0 at the beginning. According to Theorem 11.3, if we subtract these two sequences, the resulting sequence must have limit $S - S = 0$; that is,

$$S_1 - 0, S_2 - S_1, \cdots, S_n - S_{n-1}, \cdots \to 0.$$

But $S_1 - 0 = S_1 = c_1$, $S_2 - S_1 = c_2$, etc; that is, we have shown that $\lim_{n\to\infty} c_n = 0$.

Theorem 11.5 states that a necessary condition for a series $\sum c_n$ to converge is that the sequence $\{c_n\}$ of its terms must approach zero. What we really want are sufficient conditions to guarantee convergence or divergence of a series. We can take the contrapositive of the theorem and obtain the following.

Corollary (n^{th} term test) *If $\lim_{n\to\infty} c_n \neq 0$, or does not exist, then the series $\sum_{n=1}^{\infty} c_n$ diverges.*

This is our first convergence test, the nth term test. It is in fact a test for divergence rather than convergence, stating that if $\lim_{n\to\infty} c_n$ exists and is equal to anything but zero, or the limit does not exist, then the series $\sum c_n$ diverges. Note well that the nth *term test never indicates that a series converges.* Even if $\lim_{n\to\infty} c_n = 0$, we can conclude nothing about the convergence or divergence of $\sum c_n$; it may converge or it may diverge. For example, the harmonic series $\sum 1/n$ and the geometric series $\sum 1/2^{n-1}$ both satisfy the condition $\lim_{n\to\infty} c_n = 0$, yet one series diverges and the other converges. The nth term test therefore may indicate that a series diverges, but it never indicates that a series converges.

In particular, both series $\sum(-1)^n|n-3|$ and $\sum(-1)^{n+1}$ of Example 11.5 diverge by the nth term test.

To understand series, it is crucial to distinguish clearly among three entities: the series itself, its sequence of partial sums, and its sequence of terms. For any series $\sum c_n$, we can form its sequence of terms $\{c_n\}$ and its sequence of partial sums $\{S_n\}$. Each of these sequences may give information about the sum of the series $\sum c_n$, but in very different ways.

The sum of the series is defined by its sequence of partial sums in that $\sum c_n$ has a sum only if $\{S_n\}$ has a limit. The sequence of partial sums therefore tells us definitely whether the series converges or diverges, provided we can evaluate S_n in a simple form.

The sequence of terms $\{c_n\}$, on the other hand, may or may not tell us whether the series diverges. If $\{c_n\}$ has no limit or has a limit other than zero, we know that the series does not have a sum. If $\{c_n\}$ has limit zero, we obtain no information about convergence of the series, and must continue our investigation.

EXERCISES 11.3

In Exercises 1–10 determine whether the series converges or diverges. Find the sum of each convergent series. To get a feeling for a series, it is helpful to write out its first few terms. Try it.

1. $\displaystyle\sum_{n=1}^{\infty} \frac{n+1}{2n}$ **2.** $\displaystyle\sum_{n=1}^{\infty} \frac{2^n}{5^{n+1}}$

3. $\displaystyle\sum_{n=1}^{\infty} \cos\left(\frac{n\pi}{2}\right)$ **4.** $\displaystyle\sum_{n=1}^{\infty} \left(\frac{n}{n+1}\right)^n$

5. $\displaystyle\sum_{n=1}^{\infty} \frac{7^{2n+3}}{3^{2n-2}}$ **6.** $\displaystyle\sum_{n=1}^{\infty} \frac{7^{n+3}}{3^{2n-2}}$

7. $\displaystyle\sum_{n=1}^{\infty} \sqrt{\frac{n^2-1}{n^2+1}}$ **8.** $\displaystyle\sum_{n=1}^{\infty} \frac{\cos(n\pi)}{2^n}$

9. $\displaystyle\sum_{n=1}^{\infty} \frac{4^n+3^n}{3^n}$ **10.** $\displaystyle\sum_{n=1}^{\infty} \text{Tan}^{-1}\, n$

In Exercises 11–14 we have given a repeating decimal. Express the decimal as a geometric series and use formula 11.13b to express it as a rational number.

11. $0.666666\ldots$ **12.** $0.1313131313\ldots$

13. $1.347346346346\ldots$ **14.** $43.020502050205\ldots$

In Exercises 15–17 complete the statement and give a short proof to substantiate your claim.

15. If $\sum c_n$ and $\sum d_n$ converge, then $\sum(c_n + d_n)\ldots$

16. If $\sum c_n$ converges and $\sum d_n$ diverges, then $\sum(c_n + d_n)\ldots$

17. If $\sum c_n$ and $\sum d_n$ diverge, then $\sum(c_n + d_n)\ldots$

In Exercises 18–21 determine whether the series converges or diverges. Find the sum of each convergent series.

18. $\displaystyle\sum_{n=1}^{\infty} \frac{2^n+3^n}{4^n}$ **19.** $\displaystyle\sum_{n=1}^{\infty} \frac{3^n-1}{2^n}$

20. $\displaystyle\sum_{n=1}^{\infty} \frac{n^2+2^{2n}}{4^n}$ **21.** $\displaystyle\sum_{n=1}^{\infty} \frac{2^n+4^n-8^n}{2^{3n}}$

22. Find the sum of the series

$$\sum_{n=1}^{\infty} \frac{1}{n(n+1)}.$$

Hint: Use partial fractions on the nth term and find the sequence of partial sums.

23. Find the total distance travelled by the superball in Exercise 14 of Section 11.2 before it comes to rest.

24. Find the time taken for the superball in Exercise 14 of Section 11.2 to come to rest.

25. What distance does the dog run from the time when it sees the farmer until the farmer reaches the farmhouse in Exercise 15 of Section 11.2?

26. Find a simplified formula for the area A_n in Exercise 20 of Section 11.2. What is $\lim_{n\to\infty} A_n$?

27. According to equation 11.13a, the nth partial sum of the geometric series $1 + r + r^2 + r^3 + \cdots$ is $S_n = (1 - r^n)/(1 - r)$. If T_n denotes the nth partial sum of the series

$$1 + 2r + 3r^2 + 4r^3 + \cdots,$$

show that

$$T_n - S_n = r(T_n - nr^{n-1}).$$

Solve this equation for T_n and take the limit as $n \to \infty$ to show that

$$\sum_{n=1}^{\infty} nr^{n-1} = \frac{1}{(1-r)^2}, \quad |r| < 1.$$

In Exercises 28–31 use the result of Exercise 27 to find the sum of the series.

28. $\dfrac{1}{2} + \dfrac{2}{2^2} + \dfrac{3}{2^3} + \dfrac{4}{2^4} + \cdots$

29. $\dfrac{2}{5} + \dfrac{4}{25} + \dfrac{6}{125} + \dfrac{8}{625} + \cdots$

30. $\dfrac{2}{3} + \dfrac{3}{27} + \dfrac{4}{243} + \dfrac{5}{2187} + \cdots$

31. $\dfrac{12}{5} + \dfrac{48}{25} + \dfrac{192}{125} + \dfrac{768}{625} + \cdots$

32. Two people flip a single coin to see who can first flip a head. Show that the probability that the first person to flip wins the game is represented by the series

$$\frac{1}{2} + \frac{1}{8} + \frac{1}{32} + \cdots + \frac{1}{2^{2n-1}} + \cdots.$$

What is the sum of this series?

33. Two people throw a die to see who can first throw a six. Find the probability that the person who first throws wins the game.

34. One of *Zeno's paradoxes* describes a race between Achilles and a tortoise. Zeno claims that if the tortoise is given a head start, then no matter how fast Achilles runs, he can never catch the tortoise. He reasons as follows: In order to catch the tortoise, Achilles must first make up the length of the head start. But while he is running this distance, the tortoise "runs" a further distance. While Achilles makes up this distance, the tortoise covers a further distance, and so on. It follows that Achilles is always making up distance covered by the tortoise, and therefore can never catch the tortoise. If the tortoise is given a head start of length L and Achilles runs $c > 1$ times as fast as the tortoise, use infinite series to show that Achilles does in fact catch the tortoise and that the distance he covers in doing so is $cL(c-1)^{-1}$.

35. Find the time between 1:05 and 1:10 when the minute and hour hands of a clock point in the same direction

(a) by reasoning with infinite series as in Exercise 34; and

(b) by finding expressions for the angular displacements of the hands as functions of time.

36. Repeat Exercise 35 for the instant between 10:50 and 10:55 when the hands coincide.

37. A child has a large number of identical cubical blocks, which she stacks as shown in Figure 11.12. The top block protrudes $\frac{1}{2}$ its length over the second block, which protrudes $\frac{1}{4}$ its length over the third block, which protrudes $\frac{1}{6}$ its length over the fourth block, etc. Assuming the centre of mass of each block is at its geometric centre, show that the centre of mass of the top n blocks lies directly over the edge of the $(n+1)^{\text{th}}$ block. Now deduce that if a sufficient number of blocks is piled, the top block can be made to protrude as far over the bottom block as desired without the stack falling.

FIGURE 11.12

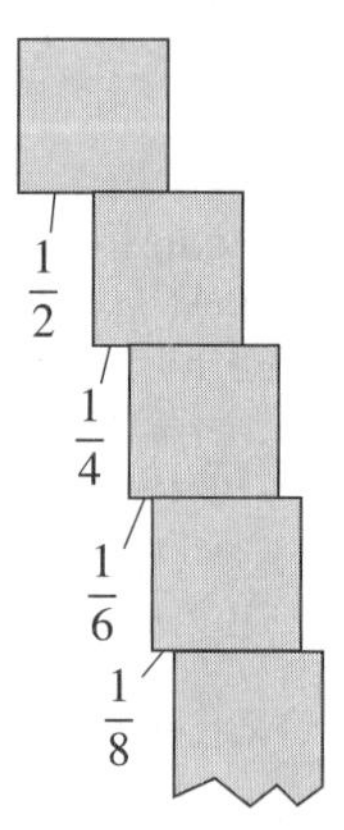

38. It is customary to assume that when a drug is administered to the human body it will be eliminated exponentially; that is, if A represents the amount of drug in the body, then

$$A = A_0 e^{-kt},$$

where $k > 0$ is a constant and A_0 is the amount injected at time $t = 0$. Suppose n successive injections of amount A_0 are administered at equally spaced time intervals T, the first injection at time $t = 0$.

(a) Show that the amount of drug in the body at time t between the nth and $(n+1)^{\text{th}}$ injection is given by

$$A_n(t) = A_0 e^{-kt}\left[\frac{1 - e^{knT}}{1 - e^{kT}}\right], \quad (n-1)T < t < nT.$$

(b) Sketch graphs of these functions on one set of axes.

(c) What is the amount of drug in the body immediately after the nth injection for very large n; that is, what is

$$\lim_{n\to\infty} A_n((n-1)T)\,?$$

39. Prove the following result: If $\sum_{n=1}^{\infty} c_n$ converges, then its terms can be grouped in any manner, and the resulting series is convergent with the same sum as the original series.

SECTION 11.4

Nonnegative Series

In Section 11.3 we derived the nth term test, a test for divergence of a series. In order to develop further convergence and divergence tests, we consider two classes of series:

1. Series with terms that are all nonnegative;
2. Series with both positive and negative terms.

A series with terms that are all nonpositive is the negative of a series of type 1, and therefore any test applicable to series of type 1 is easily adapted to a series with nonpositive terms.

Definition 11.6

A series $\sum_{n=1}^{\infty} c_n$ is said to be **nonnegative** if each term is nonnegative: $c_n \geq 0$.

For example, the harmonic series $\sum 1/n$ and the geometric series $\sum 1/2^{n-1}$ are both nonnegative series. We have already seen that both series have the property that $\lim_{n\to\infty} c_n = 0$, yet the harmonic series diverges and the geometric series converges. Examination of the terms of these series reveals that $1/2^{n-1}$ approaches 0 much more quickly than does $1/n$. In general, whether a nonnegative series does or does not have a sum depends on how quickly its sequence of terms $\{c_n\}$ approaches zero. The study of convergence of nonnegative series is an investigation into the question "How fast must the sequence of terms of a nonnegative series approach zero in order that the series have a sum?" This is not a simple problem; only a partial answer is provided through the tests in this section and the next. We begin with the integral test.

Theorem 11.6

(Integral test) *Suppose the terms in a series $\sum_{n=1}^{\infty} c_n$ are denoted by $c_n = f(n)$ and $f(x)$ is a continuous, positive, decreasing function for $x \geq 1$. Then the series converges if and only if the improper integral $\int_1^{\infty} f(x)\,dx$ converges.*

Proof Suppose first that the improper integral converges to value K. We know that this value can be interpreted as the area under the curve $y = f(x)$, above the x-axis, and to the right of the line $x = 1$ in Figure 11.13. In Figure 11.14 the area of the nth rectangle is $f(n) = c_n$, and clearly it is less than the area under the curve from $x = n-1$ to $x = n$.

FIGURE 11.13

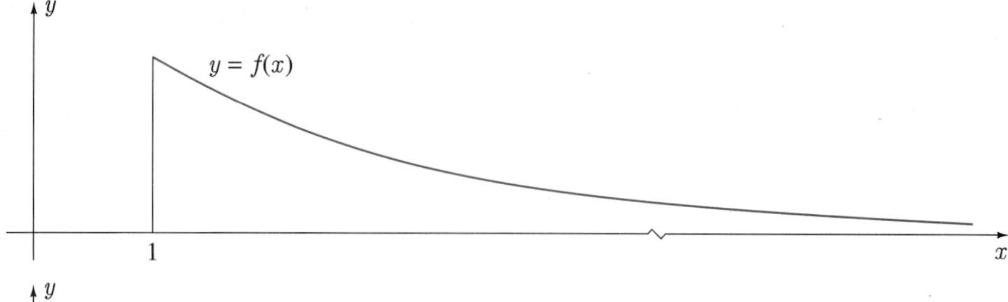

FIGURE 11.14

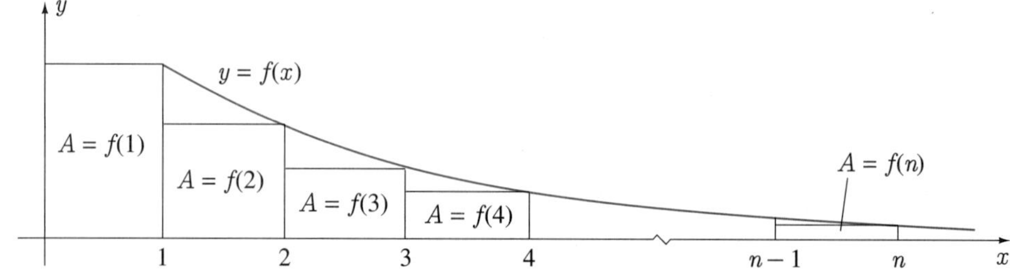

Now, the nth partial sum of the series $\sum c_n$ is

$$\begin{aligned} S_n &= c_1 + c_2 + \cdots + c_n \\ &= f(1) + f(2) + \cdots + f(n) \\ &< f(1) + \int_1^2 f(x)\,dx + \int_2^3 f(x)\,dx + \cdots + \int_{n-1}^n f(x)\,dx \\ &= f(1) + \int_1^n f(x)\,dx \\ &< f(1) + \int_1^\infty f(x)\,dx \\ &= f(1) + K. \end{aligned}$$

What this shows is that the sequence $\{S_n\}$ is bounded (since $f(1) + K$ is independent of n). Because the sequence is also increasing, it follows by Theorem 11.1 that it must have a limit; that is, the series $\sum c_n$ converges.

Conversely, suppose now that the improper integral diverges. This time we draw rectangles to the right of the vertical lines at $n = 1, 2, \ldots$ (Figure 11.15). Then

$$\begin{aligned} S_n &= c_1 + c_2 + \cdots + c_n \\ &= f(1) + f(2) + \cdots + f(n) \\ &> \int_1^2 f(x)\,dx + \int_2^3 f(x)\,dx + \cdots + \int_{n-1}^n f(x)\,dx \\ &= \int_1^n f(x)\,dx. \end{aligned}$$

Since $\lim_{n\to\infty} \int_1^n f(x)\,dx = \infty$, it follows that $\lim_{n\to\infty} S_n = \infty$, and the series therefore diverges.

FIGURE 11.15

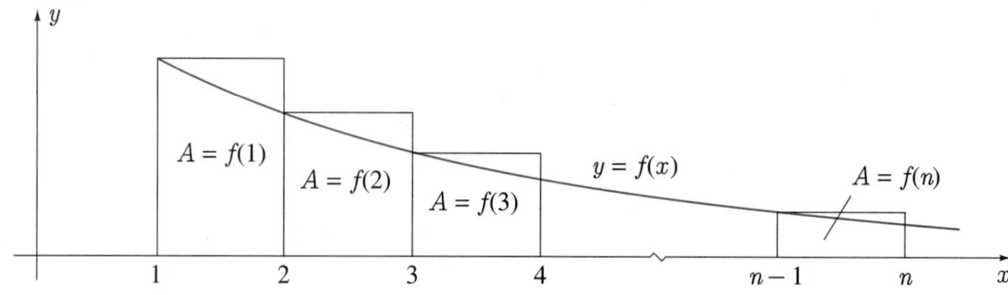

To use the integral test it is necessary to antidifferentiate the function $f(x)$, obtained by replacing n's in c_n with x's. When the antiderivative appears obvious, and other conditions of Theorem 11.6 are met, the integral test may be the easiest way to decide on convergence of the series; when the antiderivative is not obvious, it may be better to try another test.

EXAMPLE 11.6

Determine whether the following series converge or diverge:

$$\text{(a)} \sum_{n=1}^{\infty} \frac{1}{n^2+1} \qquad \text{(b)} \sum_{n=2}^{\infty} \frac{1}{n \ln n} \qquad \text{(c)} \sum_{n=1}^{\infty} ne^{-n}$$

SOLUTION (a) Since $f(x) = 1/(x^2+1)$ is continuous, positive, and decreasing for $x \geq 1$, and

$$\int_1^{\infty} \frac{1}{x^2+1} dx = \left\{\text{Tan}^{-1}\, x\right\}_1^{\infty} = \frac{\pi}{2} - \frac{\pi}{4} = \frac{\pi}{4},$$

it follows that the series $\sum 1/(n^2+1)$ converges.

(b) Since this series begins with $n = 2$, we modify the integral test by considering the improper integral of $f(x) = 1/(x \ln x)$ over the interval $x \geq 2$. Since $f(x)$ is positive, continuous, and decreasing for $x \geq 2$, and

$$\int_2^{\infty} \frac{1}{x \ln x} dx = \{\ln(\ln x)\}_2^{\infty} = \infty,$$

it follows that $\sum_{n=2}^{\infty} 1/(n \ln n)$ diverges.

(c) Since xe^{-x} is continuous, positive, and decreasing for $x \geq 1$, and

$$\int_1^{\infty} xe^{-x} dx = \left\{-xe^{-x} - e^{-x}\right\}_1^{\infty} = \frac{2}{e},$$

the series converges. ■

We mentioned before that two very important series are geometric series and the harmonic series. Convergence of geometric series was discussed in Section 11.3. The harmonic series belongs to a type of series called **p-series** defined as follows

$$\sum_{n=1}^{\infty} \frac{1}{n^p} = 1 + \frac{1}{2^p} + \frac{1}{3^p} + \cdots. \tag{11.15}$$

When $p = 1$, the p-series becomes the harmonic series which we know diverges. The series clearly diverges for $p < 0$ also. Consider the case when $p > 0$, but $p \neq 1$. The function $1/x^p$ is continuous, positive, and decreasing for $x \geq 1$, and

$$\int_1^{\infty} \frac{1}{x^p} dx = \left\{\frac{1}{-(p-1)x^{p-1}}\right\}_1^{\infty} = \begin{cases} \infty & p < 1 \\ \dfrac{1}{p-1} & p > 1. \end{cases}$$

According to the integral test, the p-series converges when $p > 1$ and diverges when $p < 1$. Let us summarize results for geometric and p-series:

Geometric Series $\qquad\qquad$ p-series

$$\sum_{n=1}^{\infty} ar^{n-1} = \begin{cases} \dfrac{a}{1-r} & |r| < 1 \\ \text{diverges} & |r| \geq 1 \end{cases} \qquad \sum_{n=1}^{\infty} \frac{1}{n^p} = \begin{cases} \text{converges} & p > 1 \\ \text{diverges} & p \leq 1 \end{cases} \tag{11.16}$$

It is unfortunate that no general formula can be given for the sum of the p-series when $p > 1$. Some interesting cases which arise frequently are cited below:

$$\sum_{n=1}^{\infty} \frac{1}{n^2} = \frac{\pi^2}{6}, \qquad \sum_{n=1}^{\infty} \frac{1}{n^3} = 1.202\,056\,903\,1, \qquad \sum_{n=1}^{\infty} \frac{1}{n^4} = \frac{\pi^4}{90}. \tag{11.17}$$

Many series can be shown to converge or diverge by comparing them to known convergent and divergent series. This is the essence of the following two tests.

Theorem 11.7

(Comparison Test) *If* $0 \leq c_n \leq a_n$ *for all* n *and* $\sum_{n=1}^{\infty} a_n$ *converges, then* $\sum_{n=1}^{\infty} c_n$ *converges. If* $c_n \geq d_n \geq 0$ *for all* n *and* $\sum_{n=1}^{\infty} d_n$ *diverges, then* $\sum_{n=1}^{\infty} c_n$ *diverges.*

Proof Suppose first that $\sum a_n$ converges and $0 \leq c_n \leq a_n$ for all n. Since $a_n \geq 0$ the sequence of partial sums for the series $\sum a_n$ must be nondecreasing. Since the series converges, the sequence of partial sums must also be bounded (corollary to Theorem 11.1). Because $0 \leq c_n \leq a_n$ for all n, it follows that the sequence of partial sums of $\sum c_n$ is nondecreasing and bounded also. This sequence therefore has a limit and series $\sum c_n$ converges. A similar argument can be made for the divergent case.

Theorem 11.7 states that if the terms of a nonnegative series $\sum c_n$ are smaller than those of a known convergent series, then $\sum c_n$ must converge; if they are larger than a known nonnegative divergent series, then $\sum c_n$ must diverge.

In order for the comparison test to be useful, we require a catalogue of known convergent and divergent series with which we may compare other series. The geometric and p-series in equation 11.16 are extremely useful in this respect.

EXAMPLE 11.7

Determine whether the following series converge or diverge:

$$\text{(a)} \quad \sum_{n=2}^{\infty} \frac{\ln n}{n} \qquad \text{(b)} \quad \sum_{n=1}^{\infty} \frac{2n^2 - 1}{15n^4 + 14} \qquad \text{(c)} \quad \sum_{n=1}^{\infty} \frac{2n^2 + 1}{15n^4 - 14}$$

SOLUTION (a) For this series, we note that when $n \geq 3$,

$$\frac{\ln n}{n} > \frac{1}{n}.$$

Since $\sum_{n=3}^{\infty} 1/n$ diverges (harmonic series with first two terms changed to zero), so does $\sum_{n=3}^{\infty} (\ln n)/n$ by the comparison test. Thus $\sum_{n=2}^{\infty} (\ln n)/n$ diverges. This can also be verified with the integral test.

(b) For the series $\sum (2n^2 - 1)/(15n^4 + 14)$, we note that

$$\frac{2n^2 - 1}{15n^4 + 14} < \frac{2n^2}{15n^4} = \frac{2}{15n^2}.$$

Since $\sum 2/(15n^2) = (2/15)\sum 1/n^2$ converges (p-series with $p = 2$), so does the given series by the comparison test.

(c) For this series we note that the inequality

$$\frac{2n^2 + 1}{15n^4 - 14} \leq \frac{3}{n^2}$$

is valid if and only if

$$45n^4 - 42 \geq 2n^4 + n^2.$$

But this inequality is valid if and only if

$$n^2(43n^2 - 1) \geq 42,$$

which is obviously true for $n \geq 1$. Consequently,

$$\frac{2n^2 + 1}{15n^4 - 14} \leq \frac{3}{n^2},$$

and since $\sum 3/n^2 = 3\sum 1/n^2$ converges, so too does the given series. ■

Two observations about Example 11.7 are worthwhile:

1. To use the comparison test we must first have a suspicion as to whether the given series converges or diverges in order to discuss the correct inequality; that is, if we suspect that the given series converges, we search for a convergent series for the right-hand side of the inequality $\leq$, and if we suspect that the given series diverges, we search for a divergent series for the right-hand side of the opposite inequality $\geq$.
2. Recall that when the terms of a nonnegative series approach zero, whether the series has a sum depends on how fast these terms approach zero. The only difference in the nth terms of Examples 11.7(b) and (c) is the position of the negative sign. For very large n this difference becomes negligible since each nth term can, for large n, be closely approximated by $2/(15n^2)$. We might then expect similar analyses for these examples, and yet they are quite different. In addition, it is natural to ask where we obtained the factor 3 in part (c). The answer is, "by trial and error".

Each of these observations points out weaknesses in the comparison test, but these problems can be eliminated in many examples with the following test.

Theorem 11.8

(Limit Comparison Test) *If $0 \leq c_n$ and $0 < b_n$, and*

$$\lim_{n\to\infty} \frac{c_n}{b_n} = l, \quad 0 < l < \infty, \tag{11.18}$$

then series $\sum_{n=1}^{\infty} c_n$ converges if $\sum_{n=1}^{\infty} b_n$ converges, and diverges if $\sum_{n=1}^{\infty} b_n$ diverges.

Proof Suppose that series $\sum b_n$ converges. Since sequence $\{c_n/b_n\}$ converges to l, we can use the result of Exercise 75 in Section 11.1 and say that for all n greater than or equal to some integer N,

$$\frac{c_n}{b_n} < l + 1, \qquad \text{or} \qquad c_n < (l+1)\,b_n.$$

Since the series $\sum_{n=N}^{\infty} (l+1)\,b_n = (l+1)\sum_{n=N}^{\infty} b_n$ converges, so also must $\sum_{n=N}^{\infty} c_n$ (by the comparison test). Hence, $\sum_{n=1}^{\infty} c_n$ converges. A similar argument can be made for the divergent case.

If $\lim_{n\to\infty} c_n/b_n = l$, then for very large n we can say that $c_n \approx l b_n$. It follows that if $\{b_n\}$ approaches zero, $\{c_n\}$ approaches zero $1/l$ times as fast as $\{b_n\}$. Theorem 11.8 implies then that if the sequence of nth terms of two nonnegative series approach zero at proportional rates, the series converge or diverge together.

To use the limit comparison test, we must find a series $\sum b_n$ so that the limit of the ratio c_n/b_n is finite and greater than zero. To obtain this series, it is sufficient in many examples to simply answer the question, "What does the given series really look like for very large n?". In both Examples 11.7(b) and (c) we see that for large n

$$c_n \approx \frac{2n^2}{15n^4} = \frac{2}{15n^2}.$$

Consequently, we calculate in Example 11.7(c) that

$$l = \lim_{n\to\infty} \frac{\dfrac{2n^2+1}{15n^4-14}}{\dfrac{2}{15n^2}} = \lim_{n\to\infty} \frac{n^2\left(2+\dfrac{1}{n^2}\right)}{n^4\left(15-\dfrac{14}{n^4}\right)} \cdot \frac{15n^2}{2} = 1.$$

Since $\sum 2/(15n^2)$ converges, so too does $\sum (2n^2+1)/(15n^4-14)$.

EXAMPLE 11.8

Determine whether the following series converge or diverge:

$$\text{(a)}\ \sum_{n=1}^{\infty} \frac{\sqrt{n^2+2n-1}}{n^{5/2}+15n-3} \qquad \text{(b)}\ \sum_{n=1}^{\infty} \frac{2^n+1}{3^n+5}$$

SOLUTION (a) For very large n, the nth term of the series can be approximated by

$$\frac{\sqrt{n^2+2n-1}}{n^{5/2}+15n-3} \approx \frac{n}{n^{5/2}} = \frac{1}{n^{3/2}}.$$

We calculate therefore that

$$l = \lim_{n\to\infty} \frac{\dfrac{\sqrt{n^2+2n-1}}{n^{5/2}+15n-3}}{\dfrac{1}{n^{3/2}}} = \lim_{n\to\infty} \frac{n\sqrt{1+\dfrac{2}{n}-\dfrac{1}{n^2}}}{n^{5/2}\left(1+\dfrac{15}{n^{3/2}}-\dfrac{3}{n^{5/2}}\right)} \cdot n^{3/2} = 1.$$

Since $\sum 1/n^{3/2}$ converges (p-series with $p = 3/2$), so too does the given series by the limit comparison test.

(b) For large n,

$$\frac{2^n+1}{3^n+5} \approx \left(\frac{2}{3}\right)^n.$$

We calculate therefore that

$$l = \lim_{n\to\infty} \frac{\dfrac{2^n+1}{3^n+5}}{\left(\dfrac{2}{3}\right)^n} = \lim_{n\to\infty} \frac{2^n\left(1+\dfrac{1}{2^n}\right)}{3^n\left(1+\dfrac{5}{3^n}\right)} \cdot \frac{3^n}{2^n} = 1.$$

Since $\sum (2/3)^n$ converges (a geometric series with $r = 2/3$), the given series converges by the limit comparison test. ■

EXERCISES 11.4

In Exercises 1–22 determine whether the series converges or diverges.

1. $\displaystyle\sum_{n=1}^{\infty} \frac{1}{2n+1}$

2. $\displaystyle\sum_{n=1}^{\infty} \frac{1}{4n-3}$

3. $\displaystyle\sum_{n=1}^{\infty} \frac{1}{2n^2+4}$

4. $\displaystyle\sum_{n=1}^{\infty} \frac{1}{5n^2-3n-1}$

5. $\displaystyle\sum_{n=2}^{\infty} \frac{1}{n^3-1}$

6. $\displaystyle\sum_{n=4}^{\infty} \frac{n^2}{n^4-6n^2+5}$

7. $\displaystyle\sum_{n=1}^{\infty} \frac{1}{(2n-1)(2n+1)}$

8. $\displaystyle\sum_{n=1}^{\infty} \frac{n-5}{n^2+3n-2}$

9. $\displaystyle\sum_{n=2}^{\infty} \frac{1}{\ln n}$

10. $\displaystyle\sum_{n=1}^{\infty} n^2 e^{-2n}$

11. $\displaystyle\sum_{n=2}^{\infty} \frac{\sqrt{n^2+2n-3}}{n^2+5}$

12. $\displaystyle\sum_{n=1}^{\infty} \frac{\sqrt{n+5}}{n^3+3}$

13. $\displaystyle\sum_{n=1}^{\infty} \sqrt{\frac{n^2+2n+3}{2n^4-n}}$

14. $\displaystyle\sum_{n=2}^{\infty} \frac{1}{n^2 \ln n}$

15. $\displaystyle\sum_{n=1}^{\infty} \frac{1}{2^n} \sin\left(\frac{\pi}{n}\right)$

16. $\displaystyle\sum_{n=1}^{\infty} \frac{\sqrt{n^2+1}}{n^3} \operatorname{Tan}^{-1} n$

17. $\displaystyle\sum_{n=1}^{\infty} \frac{2^n+n}{3^n+1}$

18. $\displaystyle\sum_{n=2}^{\infty} \frac{1+\ln^2 n}{n \ln^2 n}$

19. $\displaystyle\sum_{n=1}^{\infty} \frac{1+1/n}{e^n}$

20. $\displaystyle\sum_{n=1}^{\infty} \frac{\ln(n+1)}{n+1}$

21. $\displaystyle\sum_{n=1}^{\infty} ne^{-n^2}$

22. $\displaystyle\sum_{n=2}^{\infty} \frac{1}{n\sqrt[3]{\ln n}}$

In Exercises 23–25 find values of p for which the series converges.

23. $\displaystyle\sum_{n=2}^{\infty} \frac{1}{n^p \ln n}$

24. $\displaystyle\sum_{n=2}^{\infty} \frac{1}{n(\ln n)^p}$

25. $\displaystyle\sum_{n=2}^{\infty} \frac{1}{(\ln n)^p}$

SECTION 11.5

Further Convergence Tests for Nonnegative Series

In this section we consider two additional tests to determine whether nonnegative series converge or diverge. The first test, called the limit ratio test, indicates whether a series $\sum c_n$ resembles a geometric series for large n.

Theorem 11.9 (Limit Ratio Test) *Suppose $c_n > 0$ and*

$$\lim_{n\to\infty} \frac{c_{n+1}}{c_n} = L. \tag{11.19}$$

Then,

(i) *$\sum_{n=1}^{\infty} c_n$ converges if $L < 1$;*

(ii) *$\sum_{n=1}^{\infty} c_n$ diverges if $L > 1$ (or if $\lim_{n\to\infty} c_{n+1}/c_n = \infty$);*

(iii) *$\sum_{n=1}^{\infty} c_n$ may converge or diverge if $L = 1$.*

FIGURE 11.16

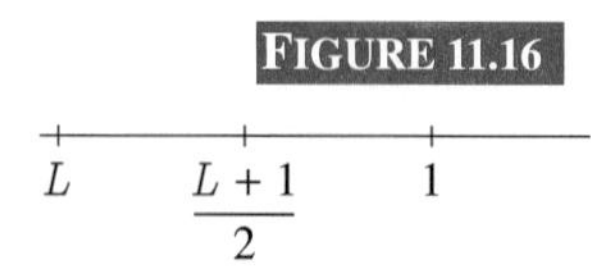

Proof (i) By the result of part (a) in Exercise 74 of Section 11.1 (see also Figure 11.16), we can say that if $\lim_{n\to\infty} c_{n+1}/c_n = L < 1$, there exists an integer N such that for all $n \geq N$,

$$\frac{c_{n+1}}{c_n} < \frac{L+1}{2}.$$

Consequently,

$$c_{N+1} < \left(\frac{L+1}{2}\right) c_N;$$

$$c_{N+2} < \left(\frac{L+1}{2}\right) c_{N+1} < \left(\frac{L+1}{2}\right)^2 c_N;$$

$$c_{N+3} < \left(\frac{L+1}{2}\right) c_{N+2} < \left(\frac{L+1}{2}\right)^3 c_N;$$

and so on. Hence,

$$c_N + c_{N+1} + c_{N+2} + \cdots < c_N + \left(\frac{L+1}{2}\right) c_N + \left(\frac{L+1}{2}\right)^2 c_N + \cdots.$$

Since the right-hand side of this inequality is a geometric series with common ratio $(L+1)/2 < 1$, it follows by the comparison test that $\sum_{n=N}^{\infty} c_n$ converges. Therefore, $\sum_{n=1}^{\infty} c_n$ converges also.

FIGURE 11.17

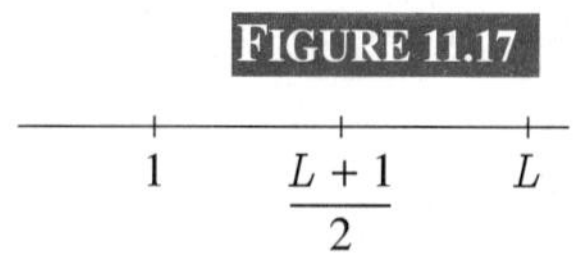

(ii) By part (b) in Exercise 74 of Section 11.1 (see also Figure 11.17), if $\lim_{n\to\infty} c_{n+1}/c_n = L > 1$, there exists an integer N such that for all $n \geq N$,

$$\frac{c_{n+1}}{c_n} > \frac{L+1}{2} > 1.$$

When $\lim_{n\to\infty} c_{n+1}/c_n = \infty$, it is also true that for n greater than or equal to some N, c_{n+1}/c_n must be greater than 1. This implies that for all $n > N$, $c_n > c_N$, and therefore

$$\lim_{n\to\infty} c_n \neq 0.$$

Hence $\sum c_n$ diverges by the nth term test.

(iii) To show that the limit ratio test fails when $L = 1$, consider the two p-series $\sum 1/n$ and $\sum 1/n^2$. For each series, $L = 1$, yet the first series diverges and the second converges.

In a nonrigorous way we can justify the limit ratio test from the following standpoint. If $\lim_{n\to\infty} c_{n+1}/c_n = L$, then for large n, each term of series $\sum c_n$ is essentially L times the term before it; that is, the series resembles a geometric series with common ratio L. We would expect convergence of the series if $L < 1$ and divergence if $L > 1$. We might also anticipate some indecision about the $L = 1$ case, depending on how this limit is reached, since this case corresponds to the common ratio which separates convergent and divergent geometric series.

EXAMPLE 11.9

Determine whether the following series converge or diverge:

(a) $\displaystyle\sum_{n=1}^{\infty} \frac{2^n}{n^4}$ (b) $\displaystyle\sum_{n=1}^{\infty} \frac{n^{100}}{1 \cdot 3 \cdot 5 \cdot \cdots \cdot (2n-1)}$ (c) $\displaystyle\sum_{n=1}^{\infty} \frac{n^n}{n!}$

SOLUTION (a) For the series $\sum 2^n/n^4$,

$$L = \lim_{n\to\infty} \frac{\dfrac{2^{n+1}}{(n+1)^4}}{\dfrac{2^n}{n^4}} = \lim_{n\to\infty} 2\left(\frac{n}{n+1}\right)^4 = 2,$$

and the series therefore diverges by the limit ratio test.

(b) For this series,

$$L = \lim_{n\to\infty} \frac{\dfrac{(n+1)^{100}}{1 \cdot 3 \cdot 5 \cdot \cdots \cdot (2n+1)}}{\dfrac{n^{100}}{1 \cdot 3 \cdot 5 \cdot \cdots \cdot (2n-1)}} = \lim_{n\to\infty} \left(\frac{n+1}{n}\right)^{100} \frac{1}{2n+1} = 0,$$

and the series therefore converges by the limit ratio test.

(c) Since

$$L = \lim_{n\to\infty} \frac{\dfrac{(n+1)^{n+1}}{(n+1)!}}{\dfrac{n^n}{n!}} = \lim_{n\to\infty} \left(\frac{n+1}{n}\right)^n = e > 1$$

(see equation 3.29b), the series $\sum n^n/n!$ diverges. ■

We have one last test for nonnegative series.

Theorem 11.10

(Limit Root Test) *Suppose* $c_n \geq 0$ *and*

$$\lim_{n\to\infty} \sqrt[n]{c_n} = R. \tag{11.20}$$

Then,

(i) $\sum_{n=1}^{\infty} c_n$ *converges if* $R < 1$;

(ii) $\sum_{n=1}^{\infty} c_n$ *diverges if* $R > 1$ *(or if* $\lim_{n\to\infty} \sqrt[n]{c_n} = \infty$*)*;

(iii) $\sum_{n=1}^{\infty} c_n$ *may converge or diverge if* $R = 1$.

Proof (i) By the result of part (a) in Exercise 74 of Section 11.1, if $\lim_{n\to\infty} \sqrt[n]{c_n} = R < 1$, there exists an integer N such that for all $n \geq N$,

$$\sqrt[n]{c_n} < \frac{R+1}{2}.$$

Consequently,

$$\sqrt[N]{c_N} < \frac{R+1}{2}, \qquad \text{or} \qquad c_N < \left(\frac{R+1}{2}\right)^N;$$

$$\sqrt[N+1]{c_{N+1}} < \frac{R+1}{2}, \qquad \text{or} \qquad c_{N+1} < \left(\frac{R+1}{2}\right)^{N+1};$$

$$\sqrt[N+2]{c_{N+2}} < \frac{R+1}{2}, \qquad \text{or} \qquad c_{N+2} < \left(\frac{R+1}{2}\right)^{N+2};$$

and so on. Hence,

$$c_N + c_{N+1} + c_{N+2} + \cdots < \left(\frac{R+1}{2}\right)^{N} + \left(\frac{R+1}{2}\right)^{N+1} + \left(\frac{R+1}{2}\right)^{N+2} + \cdots.$$

Since the right-hand side of this inequality is a convergent geometric series, the left-hand side must be a convergent series also by the comparison test. Thus, $\sum_{n=N}^{\infty} c_n$ converges, and so also must $\sum_{n=1}^{\infty} c_n$.

(ii) When $\lim_{n\to\infty} \sqrt[n]{c_n} = R > 1$, there exists an integer N such that for all $n \geq N$,

$$\sqrt[n]{c_n} > \frac{R+1}{2} > 1 \qquad \text{or} \qquad c_n > 1,$$

as in part (b) of Exercise 74 in Section 11.1. When $\lim_{n\to\infty} \sqrt[n]{c_n} = \infty$, it is also true that for n greater than or equal to some N, $\sqrt[n]{c_n}$ must be greater than 1. But it now follows that $\lim_{n\to\infty} c_n \neq 0$, and the series diverges by the nth term test.

(iii) To show that the test fails when $R = 1$, we show that $R = 1$ for the two p-series $\sum 1/n$ and $\sum 1/n^2$, which diverges and converges respectively. For the harmonic series, let $R = \lim_{n\to\infty} (1/n)^{1/n}$. If we take logarithms, then

$$\begin{aligned}\ln R = \ln\left[\lim_{n\to\infty}\left(\frac{1}{n}\right)^{1/n}\right] &= \lim_{n\to\infty} \ln\left(\frac{1}{n}\right)^{1/n} = \lim_{n\to\infty} \frac{1}{n}\ln\left(\frac{1}{n}\right) = -\lim_{n\to\infty}\frac{1}{n}\ln n \\ &= -\lim_{n\to\infty}\frac{\frac{1}{n}}{1} \qquad \text{(by L'Hôpital's rule)} \\ &= 0.\end{aligned}$$

Hence, $R = 1$ for this divergent series. A similar analysis gives $R = 1$ for the convergent series $\sum 1/n^2$.

EXAMPLE 11.10

Determine whether the following series converge or diverge:

$$\text{(a)} \quad \sum_{n=1}^{\infty} \left(\frac{n+1}{n}\right)^{n^2} \qquad \text{(b)} \quad \sum_{n=1}^{\infty} \frac{n}{(\ln n)^n}$$

SOLUTION (a) Since

$$R = \lim_{n\to\infty}\left[\left(\frac{n+1}{n}\right)^{n^2}\right]^{1/n} = \lim_{n\to\infty}\left(\frac{n+1}{n}\right)^{n} = e > 1,$$

the series diverges by the limit root test.

(b) For this series,

$$R = \lim_{n\to\infty} \left[\frac{n}{(\ln n)^n}\right]^{1/n} = \lim_{n\to\infty} \frac{n^{1/n}}{\ln n}.$$

If we set $L = \lim_{n\to\infty} n^{1/n}$, then

$$\ln L = \ln\left(\lim_{n\to\infty} n^{1/n}\right) = \lim_{n\to\infty} \frac{1}{n} \ln n$$

$$= \lim_{n\to\infty} \frac{\frac{1}{n}}{1} \qquad \text{(by L'Hôpital's rule)}$$

$$= 0.$$

Thus, $L = 1$, and it follows that

$$R = \lim_{n\to\infty} \frac{n^{1/n}}{\ln n} = 0.$$

The series therefore converges. ■

We have developed six tests to determine whether series converge or diverge:

n^{th} term test

integral test

comparison test

limit comparison test

limit ratio test

limit root test

The form of the nth term of a series often suggests which test should be used. Keep the following ideas in mind when choosing a test:

1. The limit ratio test can be effective on factorials, products of the form $1 \cdot 3 \cdot 5 \cdot \cdots \cdot (2n-1)$, and constants raised to powers involving n (2^n, 3^{-n}, etc.)
2. The limit root test thrives on functions of n raised to powers involving n (see Example 11.10).
3. The limit comparison test is successful on rational functions of n, and fractional powers as well ($\sqrt{n}$, $\sqrt[3]{n/(n+1)}$, etc.)
4. The integral test can be effective when the nth term is easily integrated. Logarithms often require the integral test.

By definition, a series $\sum c_n$ converges if and only if its sequence of partial sums $\{S_n\}$ converges. The difficulty with using this definition to discuss convergence of a series is that S_n can seldom be evaluated in a simple form, and therefore consideration of $\lim_{n\to\infty} S_n$ is impossible. The above tests have the advantage of avoiding partial sums. On the other hand, they have one disadvantage. Although they may indicate that a series does indeed have a sum, the tests in no way suggest the value of the sum. The problem of calculating the sum often proves more difficult than showing that it exists in the first place. In Sections 11.7 and 11.11 we discuss various ways to calculate sums for known convergent series.

EXERCISES 11.5

In Exercises 1–20 determine whether the series converges or diverges.

1. $\sum_{n=1}^{\infty} \frac{e^n}{n^4}$

2. $\sum_{n=1}^{\infty} \frac{1}{n!}$

3. $\sum_{n=1}^{\infty} \frac{n^3}{2^n}$

4. $\sum_{n=1}^{\infty} \frac{1}{n^n}$

5. $\sum_{n=1}^{\infty} \frac{(n-1)(n-2)}{n^2 2^n}$

6. $\sum_{n=1}^{\infty} \frac{(2n)!}{(n!)^2}$

7. $\sum_{n=1}^{\infty} \frac{\sqrt{n+1}}{n^{n+1/2}}$

8. $\sum_{n=1}^{\infty} \frac{3^{-n}+2^{-n}}{4^{-n}+5^{-n}}$

9. $\sum_{n=1}^{\infty} \frac{e^{-n}}{\sqrt{n+\pi}}$

10. $\sum_{n=1}^{\infty} \frac{2\cdot 4\cdot \cdots \cdot (2n)}{4\cdot 7\cdot \cdots \cdot (3n+1)}$

11. $\sum_{n=1}^{\infty} \frac{n^{n-1}}{3^{n-1}(n-1)!}$

12. $\sum_{n=1}^{\infty} n\left(\frac{3}{4}\right)^n$

13. $\sum_{n=1}^{\infty} \frac{1+1/n}{e^n}$

14. $\sum_{n=1}^{\infty} \frac{2\cdot 4\cdot \cdots \cdot (2n)}{3\cdot 5\cdot \cdots \cdot (2n+1)}\left(\frac{1}{n^2}\right)$

15. $\sum_{n=1}^{\infty} \frac{n^n}{(n+1)^{n+1}}$

16. $\sum_{n=1}^{\infty} \frac{(n+1)^n}{n^{n+1}}$

17. $\sum_{n=1}^{\infty} \frac{n^4+3}{5^{n/2}}$

18. $\sum_{n=1}^{\infty} \frac{2^n+n^2 3^n}{4^n}$

19. $\sum_{n=1}^{\infty} \frac{n^2 2^n - n}{n^3+1}$

20. $\sum_{n=1}^{\infty} \frac{(2n)!}{(3n)!} 5^{2n}$

21. For what integer values of a is the series

$$\sum_{n=1}^{\infty} \frac{(n!)^2}{(an)!}$$

convergent?

SECTION 11.6

Series with Positive and Negative Terms

The convergence tests of Sections 11.4 and 11.5 are applicable to nonnegative series, series whose terms are all nonnegative. Series with infinitely many positive and negative terms are more complicated. It is fortunate, however, that all our test are still useful in discussing convergence of series with positive and negative terms. What makes this possible is the following definition and Theorem 11.11.

Definition 11.7 A series $\sum_{n=1}^{\infty} c_n$ is said to be **absolutely convergent** if the series of absolute values $\sum_{n=1}^{\infty} |c_n|$ converges.

At first glance it might seem that absolute convergence is a strange concept indeed. What possible good could it do to consider the series of absolute values, which is quite different from the original series? The fact is that when the series of absolute values converges, it automatically follows from the next theorem that the original series converges also. And since the series of absolute values has all nonnegative terms, we can use the comparison, limit comparison, limit ratio, limit root, or integral test to consider its convergence.

Theorem 11.11 *If a series is absolutely convergent, then it is convergent.*

Proof Let $\{S_n\}$ be the sequence of partial sums of the absolutely convergent series $\sum c_n$. Define sequences $\{P_n\}$ and $\{N_n\}$, where P_n is the sum of all positive terms in

S_n, and N_n is the sum of the absolute values of all negative terms in S_n. Then

$$S_n = P_n - N_n.$$

The sequence of partial sums for the series of absolute values $\sum |c_n|$ is

$$\{P_n + N_n\},$$

and this sequence must be nondecreasing and bounded. Since each of the sequences $\{P_n\}$ and $\{N_n\}$ is nondecreasing and a part of $\{P_n + N_n\}$, it follows that each is bounded, and therefore has a limit, say, P and N, respectively. As a result, sequence $\{P_n - N_n\} = \{S_n\}$ has limit $P - N$, and series $\sum c_n$ converges to $P - N$.

EXAMPLE 11.11

Show that the following series are absolutely convergent:

(a) $\displaystyle\sum_{n=1}^{\infty} \frac{(-1)^n n}{(n+1)2^n}$

(b)

$$1 - \frac{1}{2^2} - \frac{1}{3^3} + \frac{1}{4^4} + \frac{1}{5^5} + \frac{1}{6^6} - \frac{1}{7^7} - \frac{1}{8^8} - \frac{1}{9^9} - \frac{1}{10^{10}} + \frac{1}{11^{11}} + \cdots + \frac{1}{15^{15}} - \cdots$$

SOLUTION (a) The series of absolute values is $\sum n/[(n+1)2^n]$. We use the limit comparison test to show that it converges. Since

$$l = \lim_{n\to\infty} \frac{\dfrac{n}{(n+1)2^n}}{\dfrac{1}{2^n}} = 1,$$

and $\sum (1/2)^n$ is convergent (a geometric series with $r = 1/2$), it follows that the series of absolute values converges. The given series therefore converges absolutely.

(b) The series of absolute values is $\sum 1/n^n$. We use the comparison test to verify that this series converges. Since

$$\frac{1}{n^n} \leq \frac{1}{n^2}$$

and $\sum 1/n^2$ converges, so does $\sum 1/n^n$. The given series therefore converges absolutely. ■

In Example 11.11, absolute convergence of the given series implies convergence of the series, but it is customary to omit such a statement. It is important to realize, however, that it is the given series that is being analyzed, and its convergence is guaranteed by Theorem 11.11.

We now ask whether series can converge without converging absolutely. If there are such series, and indeed there are, we must devise new convergence tests. We describe these series as follows.

Definition 11.8

A series that converges but does not converge absolutely is said to **converge conditionally**.

The most important type of series with both positive and negative terms is an alternating series. As the name suggests, an **alternating series** has terms that are alternately positive and negative. For example,

$$\sum_{n=1}^{\infty} \frac{(-1)^{n+1}}{n} = 1 - \frac{1}{2} + \frac{1}{3} - \frac{1}{4} + \frac{1}{5} - \cdots$$

is an alternating series called the **alternating harmonic series.**

Given an alternating series to examine for convergence, we first test for absolute convergence as in Example 11.11(a). Should this fail, we check for conditional convergence with the following test.

Theorem 11.12

(Alternating Series Test) *An alternating series $\sum_{n=1}^{\infty} c_n$ converges if the sequence of absolute values of the terms $\{|c_n|\}$ is nonincreasing and has limit zero.*

Proof Suppose $c_1 > 0$, in which case all odd terms are positive and all even terms are negative. If $\{S_n\}$ is the sequence of partial sums of $\sum c_n$, then the even partial sums S_{2n} can be expressed in two forms:

$$S_{2n} = (c_1 + c_2) + (c_3 + c_4) + \cdots + (c_{2n-1} + c_{2n}),$$
$$S_{2n} = c_1 + (c_2 + c_3) + (c_4 + c_5) + \cdots + (c_{2n-2} + c_{2n-1}) + c_{2n}.$$

Since $\{|c_n|\}$ is nonincreasing ($|c_n| \geq |c_{n+1}|$), each term in the parentheses of the first expression is nonnegative. Consequently, the subsequence $\{S_{2n}\}$ of even partial sums of $\{S_n\}$ is nondecreasing. Since each term in the parentheses in the second expression is nonpositive, as is c_{2n}, it follows that $S_{2n} \leq c_1$ for all n. By Theorem 11.1 then, the sequence $\{S_{2n}\}$ has a limit, say, S.

The subsequence of odd partial sums $\{S_{2n-1}\}$ is such that $S_{2n-1} = S_{2n} - c_{2n}$. Since $\{S_{2n}\}$ has limit S and $\{c_{2n}\}$ has limit zero, it follows that $\{S_{2n-1}\}$ has limit S also. Consequently, $\{S_n\}$ has limit S, and series $\sum c_n$ converges.

A similar proof holds when $c_1 < 0$.

EXAMPLE 11.12

Determine whether the following series converge absolutely, converge conditionally, or diverge:

(a) $\displaystyle\sum_{n=1}^{\infty} \frac{(-1)^{n+1}}{n}$ (b) $\displaystyle\sum_{n=1}^{\infty} (-1)^n \frac{\sqrt{n^2+5n}}{n^{3/2}}$ (c) $\displaystyle\sum_{n=1}^{\infty} (-1)^{n+1} \frac{4^n}{n^5 3^n}$

SOLUTION (a) The alternating harmonic series $\sum (-1)^{n+1}/n$ is not absolutely convergent because the series of absolute values $\sum 1/n$ diverges. Since the sequence of absolute values of the terms $\{1/n\}$ is decreasing with limit zero, series $\sum (-1)^{n+1}/n$ converges conditionally.

(b) For this alternating series, we first consider the series of absolute values

$$\sum_{n=1}^{\infty} \frac{\sqrt{n^2+5n}}{n^{3/2}}.$$

We use the limit comparison test on this series. Since

$$l = \lim_{n\to\infty} \frac{\dfrac{\sqrt{n^2+5n}}{n^{3/2}}}{\dfrac{1}{n^{1/2}}} = \lim_{n\to\infty} \frac{n\sqrt{1+\dfrac{5}{n}}}{n^{3/2}} \cdot n^{1/2} = 1,$$

and $\sum 1/n^{1/2}$ diverges, so does $\sum \sqrt{n^2+5n}/n^{3/2}$. The original series does not therefore converge absolutely. We now resort to the alternating series test. The sequence $\{\sqrt{n^2+5n}/n^{3/2}\}$ of absolute values of the terms of the series is nonincreasing if

$$\frac{\sqrt{(n+1)^2+5(n+1)}}{(n+1)^{3/2}} \le \frac{\sqrt{n^2+5n}}{n^{3/2}}.$$

When we square and cross multiply, the inequality becomes

$$\begin{aligned} n^3(n^2+7n+6) &\le (n^2+5n)(n+1)^3 \\ &= n^5+8n^4+18n^3+16n^2+5n; \end{aligned}$$

that is,

$$n^4+12n^3+16n^2+5n \ge 0,$$

which is obviously valid because $n \ge 1$. Since $\lim_{n\to\infty} \sqrt{n^2+5n}/n^{3/2} = 0$, we conclude that the alternating series $\sum(-1)^n(\sqrt{n^2+5n}/n^{3/2})$ converges conditionally.

(c) If we apply the limit ratio test to the series of absolute values $\sum 4^n/(n^5 3^n)$, we have

$$L = \lim_{n\to\infty} \frac{\dfrac{4^{n+1}}{(n+1)^5 3^{n+1}}}{\dfrac{4^n}{n^5 3^n}} = \lim_{n\to\infty} \frac{4}{3}\left(\frac{n}{n+1}\right)^5 = \frac{4}{3}.$$

Since $L > 1$, the series $\sum 4^n/(n^5 3^n)$ diverges. The original alternating series does not therefore converge absolutely. But $L = \frac{4}{3}$ implies that for large n, each term in the series of absolute values is approximately $\frac{4}{3}$ times the term that precedes it, and therefore

$$\lim_{n\to\infty} \frac{4^n}{n^5 3^n} = \infty.$$

Consequently,

$$\lim_{n\to\infty} (-1)^{n+1} \frac{4^n}{n^5 3^n}$$

cannot possibly exist, and the given series diverges by the nth term test. ■

We have noted several times that the essential question for convergence of a nonnegative series is, "Do the terms approach zero quickly enough to guarantee convergence of the series?" With a series that has infinitely many positive and negative terms, this question is inappropriate. Such a series may converge because of a partial cancelling effect; for example, a negative term may offset the effect of a large positive term. This kind of process may produce a convergent series even though the series would be divergent if all terms were replaced by their absolute values. A specific example is the alternating harmonic series which converges (conditionally) because of this cancelling effect, whereas the harmonic series itself, which has no cancellations, diverges.

In Sections 11.4–6, we have obtained a number of tests for determining whether series converge or diverge. To test a series for convergence, we suggest the following procedure:

1. Try the nth term test for divergence.
2. If $\{c_n\}$ has limit zero and the series is nonnegative , try the comparison, limit comparison, limit ratio, limit root, or integral test.
3. If $\{c_n\}$ has limit zero and the series contains both positive and negative terms, test for absolute convergence using the tests in 2. If this fails and the series is alternating, test for conditional convergence with the alternating series test.

Each of the comparison, limit comparison, limit ratio, limit root, integral, and alternating series tests requires conditions to be satisfied for all terms of the series. Specifically, the comparison, limit comparison, and limit root tests require $c_n \geq 0$ for all n; the limit ratio test requires $c_n > 0$; the integral test requires $f(n)$ to be positive, continuous, and decreasing; and the alternating series test requires $\{|c_n|\}$ to be nonincreasing and $\{c_n\}$ to be alternately positive and negative. None of these requirements is essential for all n; in fact, so long as they are satisfied for all terms in the series beyond some point, say for n greater than or equal to some integer N, the particular test may be used on the series $\sum_{n=N}^{\infty} c_n$. The original series $\sum_{n=1}^{\infty} c_n$ then converges if and only if $\sum_{n=N}^{\infty} c_n$ converges.

EXERCISES 11.6

In Exercises 1–14 determine whether the series converges absolutely, converges conditionally, or diverges.

1. $\displaystyle\sum_{n=1}^{\infty}(-1)^n\frac{n}{n^3+1}$

2. $\displaystyle\sum_{n=1}^{\infty}(-1)^n\frac{n}{n^2+1}$

3. $\displaystyle\sum_{n=1}^{\infty}\frac{\cos(n\pi/2)}{2n^2}$

4. $\displaystyle\sum_{n=1}^{\infty}(-1)^n\frac{n^3}{3^n}$

5. $\displaystyle\sum_{n=1}^{\infty}\frac{(-1)^{n+1}}{\sqrt{n}}$

6. $\displaystyle\sum_{n=1}^{\infty}(-1)^n\frac{3^n}{n^3}$

7. $\displaystyle\sum_{n=1}^{\infty}(-1)^n\frac{n}{n^2+n+1}$

8. $\displaystyle\sum_{n=1}^{\infty}\frac{n\sin(n\pi/4)}{2^n}$

9. $\displaystyle\sum_{n=1}^{\infty}(-1)^{n+1}\left(\frac{n}{n+1}\right)$

10. $\displaystyle\sum_{n=1}^{\infty}(-1)^{n+1}\frac{\sqrt{3n-2}}{n}$

11. $\displaystyle\sum_{n=1}^{\infty}(-1)^n\left(\frac{n}{n+1}\right)^n$

12. $\displaystyle\sum_{n=1}^{\infty}(-1)^n\frac{\sqrt{n^2+3}}{n^2+5}$

13. $\displaystyle\sum_{n=2}^{\infty}(-1)^{n-1}\frac{\ln n}{n}$

14. $\displaystyle\sum_{n=1}^{\infty}\frac{\cos(n\pi/10)\,\mathrm{Cot}^{-1}n}{n^3+5n}$

15. Discuss convergence of the series

$$\sum_{n=1}^{\infty}\frac{\sin(nx)}{n^2}.$$

16. Prove that if $\sum c_n$ converges absolutely, then $\sum c_n^p$ converges absolutely for all integers $p > 1$.

17. Discuss convergence of the series

$$\sum_{n=1}^{\infty}(-1)^n\frac{n^n}{(n+1)^{n+1}}.$$

Approximating the Sum of a Series

We have considered only half of the convergence problem for infinite series of constants. The comparison, limit comparison, limit ratio, limit root, integral, and alternating series tests may determine whether a series converges or diverges, but they do not determine the sum of the series in the case of a convergent series. This part of the problem, as suggested before, can sometimes be more complicated.

If the convergent series is a geometric series, no problem exists; we can use formula 11.13b to find its sum. It may also happen that the nth partial sum S_n of the series can be calculated in a simple form, in which case the sum of the series is $\lim_{n\to\infty} S_n$. Cases of this latter type are very rare. These two methods exhaust our present capabilities for obtaining the sum of a convergent series, but we will find additional methods in Section 11.11.

In many applications of infinite series we only need a reasonable approximation to the sum of a series, and therefore turn our attention to the problem of estimating the sum of a convergent series. The easiest method for estimating the sum S of a series $\sum c_n$ is simply to choose the partial sum S_N for some N as an approximation; that is, truncate the series after N terms and choose

$$S \approx S_N = c_1 + c_2 + \cdots + c_N.$$

But an approximation is of value only if we can make some definitive statement about its accuracy. In truncating the series, we have neglected the infinity of terms $\sum_{n=N+1}^{\infty} c_n$, and the accuracy of the approximation is therefore determined by the size of $\sum_{n=N+1}^{\infty} c_n$; the smaller it is, the better the approximation. The problem is that we do not know the exact value of $\sum_{n=N+1}^{\infty} c_n$; if we did, there would be no need to approximate the sum of the original series in the first place. What we must do is estimate the sum $\sum_{n=N+1}^{\infty} c_n$.

When the integral test or the alternating series test are used to prove that a series converges, simple formulas give accuracy estimates on the truncated series. Let us illustrate these first.

Truncating an Alternating Series

It is very simple to obtain an estimate of the accuracy of a truncated alternating series $\sum c_n$ provided the sequence $\{|c_n|\}$ is nonincreasing with limit zero. In the proof of Theorem 11.12, we showed that when $c_1 > 0$, the subsequence $\{S_{2n}\}$ of even partial sums is nondecreasing and therefore approaches the sum of the series $\sum c_n$ from below (Figure 11.18). In a similar way, we can show that the subsequence $\{S_{2n-1}\}$ of odd partial sums is nonincreasing and approaches the sum of the series from above. It follows that the sum $\sum c_n$ must be between any two terms of the subsequences $\{S_{2n}\}$ and $\{S_{2n-1}\}$. In particular, it must be between any two successive partial sums. Thus, if $\sum c_n$ is approximated by S_N, the maximum possible error is $S_{N+1} - S_N = c_{N+1}$, the next term. *When an alternating series is truncated, the maximum possible error is the next term.*

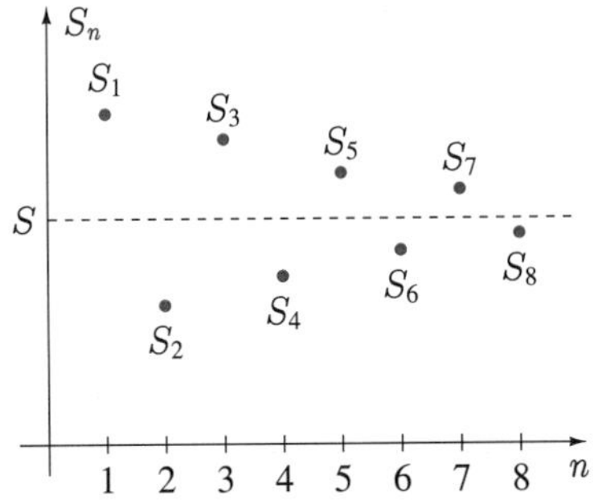

FIGURE 11.18

EXAMPLE 11.13

Use the first twenty terms of the series $\sum_{n=1}^{\infty} (-1)^{n+1}/n^3$ to estimate its sum. Obtain an error estimate.

SOLUTION The sum of the first twenty terms of the series is $0.901\,485$. Since the series is alternating, the maximum possible error in this estimate is the twenty-first term, $1/21^3 < 0.000\,108$. Thus,

$$0.901\,485 < \sum_{n=1}^{\infty} \frac{(-1)^{n+1}}{n^3} < 0.901\,593.$$

■

In practical situations, we often have to decide how many terms of a series to take in order to guarantee a certain degree of accuracy. Once again this is easy for alternating series.

EXAMPLE 11.14

How many terms in the series $\sum_{n=2}^{\infty} (-1)^{n+1}/(n^3 \ln n)$ ensure a truncation error of less than 10^{-5}?

SOLUTION The maximum error in truncating this alternating series when $n = N$ is

$$\frac{(-1)^{N+2}}{(N+1)^3 \ln(N+1)}.$$

The absolute value of this error is less than 10^{-5} when

$$\frac{1}{(N+1)^3 \ln(N+1)} < 10^{-5} \qquad \text{or}$$

$$(N+1)^3 \ln(N+1) > 10^5.$$

A calculator quickly reveals that the smallest integer for which this is valid is $N = 30$. Thus, the truncated series has the required accuracy after the 29th term. Note that the first term corresponds to $n = 2$ not $n = 1$.

■

Truncating a Series Whose Convergence was Established with the Integral Test

Suppose now that a series $\sum_{n=1}^{\infty} c_n$ has been shown to converge with the integral test; that the integral

$$\int_1^{\infty} f(x)\,dx$$

converges where $f(n) = c_n$. If the series is truncated after the Nth term, the error $c_{N+1} + c_{N+2} \cdots$ is shown as the sum of the areas of the rectangles in Figure 11.19. Clearly the sum of these areas is less than the area under $y = f(x)$ to the right of $x = N$. In other words, the error in truncating the series with the Nth term must be less than

$$\int_N^{\infty} f(x)\,dx. \tag{11.21}$$

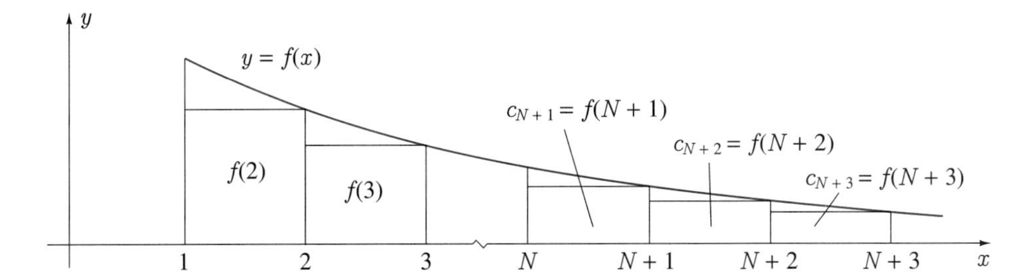

FIGURE 11.19

EXAMPLE 11.15

What is a maximum possible error when the series

$$\sum_{n=2}^{\infty} \frac{1}{n(\ln n)^4}$$

is truncated after 100 terms?

SOLUTION The error cannot be larger than

$$\int_{100}^{\infty} \frac{1}{x(\ln x)^4}\, dx = \left\{ \frac{-1}{3(\ln x)^3} \right\}_{100}^{\infty} = \frac{1}{3(\ln 100)^3} < 0.0035.$$

■

When convergence of a series is established by the comparison, limit comparison, limit ratio, or limit root tests, we often estimate the truncation error $\sum_{n=N+1}^{\infty} c_n$ by comparing it to something that is summable. We illustrate this in the following two examples.

EXAMPLE 11.16

Estimate the error in truncating the series $\sum_{n=0}^{\infty} 1/n!$ after 8 terms. The first term of this series contains 0!. This is defined as 1.

SOLUTION The sum of the first 8 terms of this series is $S_8 = 2.718\,254\,0$. The truncation error in using this approximation is

$$\begin{aligned}
\sum_{n=8}^{\infty} \frac{1}{n!} &= \frac{1}{8!} + \frac{1}{9!} + \frac{1}{10!} + \frac{1}{11!} + \cdots \\
&= \frac{1}{8!}\left(1 + \frac{1}{9} + \frac{1}{9 \cdot 10} + \frac{1}{9 \cdot 10 \cdot 11} + \cdots\right) \\
&< \frac{1}{8!}\left(1 + \frac{1}{9} + \frac{1}{9^2} + \frac{1}{9^3} + \cdots\right) \\
&= \frac{1}{8!}\frac{1}{1 - \frac{1}{9}} \qquad \text{(using equation 11.13b)} \\
&= \frac{9}{8 \cdot 8!} \\
&< 0.000\,028\,0.
\end{aligned}$$

We may write, therefore, that

$$2.718\,254\,0 < \sum_{n=0}^{\infty} \frac{1}{n!} < 2.718\,282\,0 .$$

■

You might recognize that these same digits appeared in the approximation of the number e as the limit of the sequence $\{(1 + 1/n)^n\}$ in Section 11.2. In Section 11.11, we will show that $\sum_{n=0}^{\infty} 1/n!$ converges to e,

$$e = \sum_{n=0}^{\infty} \frac{1}{n!}. \tag{11.22}$$

The advantage of the series definition is obvious; it is very easy to obtain an accuracy statement wherever the series is truncated.

EXAMPLE 11.17

How many terms in the convergent series $\sum_{n=1}^{\infty} n/[(n+1)3^n]$ ensure a truncation error of less than 10^{-5}?

SOLUTION If this series is truncated after the Nth term, the error is

$$\begin{aligned}\sum_{n=N+1}^{\infty} \frac{n}{(n+1)3^n} &= \frac{N+1}{(N+2)3^{N+1}} + \frac{N+2}{(N+3)3^{N+2}} + \cdots \\ &< \frac{1}{3^{N+1}} + \frac{1}{3^{N+2}} + \cdots \\ &= \frac{\frac{1}{3^{N+1}}}{1 - \frac{1}{3}} \qquad \text{(using equation 11.13b)} \\ &= \frac{1}{2 \cdot 3^N}.\end{aligned}$$

Consequently, the error is guaranteed to be less than 10^{-5} if N satisfies the inequality

$$\frac{1}{2 \cdot 3^N} < 10^{-5} \qquad \text{or}$$

$$2 \cdot 3^N > 10^5.$$

With the help of a calculator, we find that this is true if $N \geq 10$, and therefore 10 or more terms yield the required accuracy.

■

EXERCISES 11.7

In Exercises 1–8 use the number of terms indicated to find an approximation to the sum of the series. In each case, obtain an estimate of the truncation error.

1. $\displaystyle\sum_{n=1}^{\infty} \frac{1}{n2^n}$ (10 terms)

2. $\displaystyle\sum_{n=1}^{\infty} \frac{(-1)^n}{n^4}$ (20 terms)

3. $\displaystyle\sum_{n=1}^{\infty} \frac{1}{n^n}$ (5 terms)

4. $\displaystyle\sum_{n=2}^{\infty} \frac{(-1)^{n+1}}{n^3 3^n}$ (3 terms)

5. $\displaystyle\sum_{n=1}^{\infty} \frac{1}{2^n} \sin\left(\frac{\pi}{n}\right)$ (15 terms)

6. $\displaystyle\sum_{n=2}^{\infty} \frac{2^n - 1}{3^n + n}$ (20 terms)

7. $\displaystyle\sum_{n=2}^{\infty} \frac{2^n + 1}{3^n + n}$ (20 terms)

8. $\displaystyle\sum_{n=1}^{\infty} \frac{(-1)^n}{n}$ (100 terms)

In Exercises 9–11 how many terms in the series guarantee an approximation to the sum with a truncation error of less than 10^{-4}?

9. $\displaystyle\sum_{n=1}^{\infty} \frac{(-1)^n}{n^2}$ **10.** $\displaystyle\sum_{n=1}^{\infty} \frac{1}{n^2 4^n}$ **11.** $\displaystyle\sum_{n=1}^{\infty} \frac{2^n}{n!}$

12. This exercise shows that we must be very careful in predicting the accuracy of a result. Consider the series

$$S = 3.125\,100\,1 - 0.000\,090\,18\left(1 + \frac{1}{10} + \frac{1}{10^2} + \frac{1}{10^3} + \cdots\right).$$

(a) Show that the sum of this series is exactly $S = 3.124\,999\,9$. To two decimals, then, the value of S is 3.12.

(b) Verify that the first four partial sums of the series are

$$S_1 = 3.125\,100\,1,$$
$$S_2 = 3.125\,009\,92,$$
$$S_3 = 3.125\,000\,902,$$
$$S_4 = 3.125\,000\,000\,2.$$

(c) If $E_n = S_n - S$ are the differences between the sum of the series and its first four partial sums, show that

$$E_1 = 0.000\,100\,2,$$
$$E_2 = 0.000\,010\,02,$$
$$E_3 = 0.000\,001\,002,$$
$$E_4 = 0.000\,000\,100\,2.$$

What can you say about the accuracy of S_1, S_2, S_3, and S_4 as approximations to S?

(d) If S is approximated by any of S_1, S_2, S_3, or S_4 to two decimals, the result is 3.13, not 3.12 as in (a). Thus, in spite of the accuracy predicted in (c), S_1, S_2, S_3, and S_4 do not predict S correctly to two decimals. Do they predict S correctly to three or four decimals?

SECTION 11.8

Power Series and Intervals of Convergence

Our work on sequences and series of constants in the first seven sections of this chapter has prepared the way for discussions on sequences and series of functions. An infinite sequence of functions is the assignment of functions to integers,

$$\{f_n(x)\}_0^\infty = f_0(x),\ f_1(x),\ \ldots, \tag{11.23}$$

where we have initiated our assignment with the integer $n = 0$ rather than $n = 1$. The reason for this will become apparent shortly. From such a sequence we may form an infinite series of functions:

$$\sum_{n=0}^{\infty} f_n(x) = f_0(x) + f_1(x) + f_2(x) + \cdots. \tag{11.24}$$

When $f_n(x)$ is a constant multiplied by x^n,

$$f_n(x) = a_n x^n, \qquad a_n = \text{ a constant},$$

series 11.24 takes the form

$$\sum_{n=0}^{\infty} a_n x^n = a_0 + a_1 x + a_2 x^2 + \cdots + a_n x^n + \cdots, \tag{11.25}$$

and is called a **power series** in x.

Unlike series of constants, the variable of summation n in 11.25 is one step out of phase with the counting of terms: $n = 0$ identifies the first term, $n = 1$ the second term, etc. Instead, the value of n now identifies the power of x. In the remainder of the chapter we often write $\sum a_n x^n$ in place of $\sum_{n=0}^{\infty} a_n x^n$ whenever the lower limit is $n = 0$.

For each value of x that we substitute into a power series, we obtain a series of constants. For example, if into the power series

$$\sum_{n=0}^{\infty} x^n = 1 + x + x^2 + x^3 + \cdots$$

we substitute $x = 1/2$, we obtain the series of constants

$$\sum_{n=0}^{\infty} \left(\frac{1}{2}\right)^n = 1 + \frac{1}{2} + \frac{1}{2^2} + \frac{1}{2^3} + \cdots,$$

a geometric series that converges to 2. When we substitute $x = -3$, we obtain the divergent geometric series

$$\sum_{n=0}^{\infty} (-3)^n = 1 - 3 + 9 - 27 + \cdots.$$

Definition 11.9

The totality of values of x for which a power series converges is called its **interval of convergence**.

Obviously the interval of convergence for a power series $\sum a_n x^n$ always includes the value $x = 0$, but what other possibilities are there? "Interval of convergence" suggests that the values of x for which a power series converges form some kind of interval. This is indeed true, as we shall soon see.

EXAMPLE 11.18

Find the interval of convergence for the power series $\sum_{n=0}^{\infty} x^n$.

SOLUTION Since the series is a geometric series with common ratio x, its interval of convergence is $-1 < x < 1$. ■

EXAMPLE 11.19

Find intervals of convergence for the following power series:

$$\text{(a)} \quad \sum_{n=1}^{\infty} \frac{x^n}{n^n} \qquad \text{(b)} \quad \sum_{n=1}^{\infty} n!\, x^n \qquad \text{(c)} \quad \sum_{n=1}^{\infty} \frac{(-1)^{n+1}}{n} x^n$$

SOLUTION (a) Consider the series of absolute values $\sum |x|^n/n^n$. We use the limit root test on this series, and calculate

$$R = \lim_{n \to \infty} \left(\frac{|x|^n}{n^n} \right)^{1/n} = \lim_{n \to \infty} \frac{|x|}{n} = 0,$$

and this is true for any x whatsoever. The power series therefore converges absolutely for all x.

(b) Consider again the series of absolute values $\sum n!\, |x|^n$. We use the limit ratio test and obtain

$$L = \lim_{n \to \infty} \frac{(n+1)!\, |x|^{n+1}}{n!\, |x|^n} = |x| \lim_{n \to \infty} (n+1) = \infty.$$

Since $L = \infty$ for all $x \neq 0$, the original power series does not converge absolutely for any $x \neq 0$. But this result implies much more. It shows that the terms in the sequence $\{n!\, |x|^n\}$ become indefinitely large as $n \to \infty$. But then $\lim_{n \to \infty} n!\, x^n$ cannot possibly exist, and the series diverges by the nth term test. The power series therefore converges only for $x = 0$.

(c) Once again for the series of absolute values, we obtain

$$L = \lim_{n \to \infty} \frac{\dfrac{|x|^{n+1}}{n+1}}{\dfrac{|x|^n}{n}} = \lim_{n \to \infty} |x| \frac{n}{n+1} = |x|.$$

Since $L = |x|$, it follows that the series of absolute values converges if $|x| < 1$ and diverges if $|x| > 1$. The original power series therefore converges absolutely for $-1 < x < 1$. Furthermore, when $|x| > 1$, so too is L, and it follows that the terms in the sequence $\{|x|^n/n\}$ must become indefinitely large. But then

$$\lim_{n \to \infty} \frac{(-1)^{n+1}}{n} x^n$$

cannot possibly exist, and the series diverges by the nth term test. If $x < -1$, terms in the sequence $\{(-1)^{n+1} x^n/n\}$ are all negative and "approach negative infinity"; if $x > 1$, the terms are alternately positive and negative with increasing absolute values. When $x = 1$, the power series reduces to the alternating harmonic series

$$\sum_{n=1}^{\infty} \frac{(-1)^{n+1}}{n} = 1 - \frac{1}{2} + \frac{1}{3} - \frac{1}{4} + \cdots,$$

which converges conditionally; and when $x = -1$, it reduces to

$$\sum_{n=1}^{\infty} \frac{(-1)^{n+1}}{n} (-1)^n = -1 - \frac{1}{2} - \frac{1}{3} - \frac{1}{4} - \cdots,$$

which diverges. The interval of convergence is therefore $-1 < x \leq 1$. ■

Example 11.19 has illustrated that there are at least three possible types of intervals of convergence for power series $\sum a_n x^n$:

1. The power series converges only for $x = 0$;
2. The power series converges absolutely for all x;
3. There exists a number $R > 0$ such that the power series converges absolutely for $|x| < R$, diverges for $|x| > R$, and may or may not converge for $x = \pm R$

These are in fact the only possibilities for an interval of convergence, and this is proved in Exercises 32 and 33. In (3) we call R the **radius of convergence** of the power series. It is half the length of the interval of convergence, or, the distance we may proceed in either direction along the x-axis from $x = 0$ and expect absolute convergence of the power series, with the possible exceptions of $x = \pm R$. In order to have a radius of convergence associated with every power series, we say in (1) and (2) that $R = 0$ and $R = \infty$ respectively.

Every power series $\sum a_n x^n$ now has a radius of convergence R. If $R = 0$, the power series converges only for $x = 0$; if $R = \infty$, the power series converges absolutely for all x; and if $0 < R < \infty$, the power series converges absolutely for $|x| < R$, diverges for $|x| > R$, and may or may not converge for $x = \pm R$. For many power series the radius of convergence can be calculated according to the following theorem.

Theorem 11.13 *The radius of convergence of a power series $\sum_{n=0}^{\infty} a_n x^n$ is given by*

$$R = \lim_{n\to\infty} \left| \frac{a_n}{a_{n+1}} \right| \qquad \textit{or,} \qquad (11.26\text{a})$$

$$R = \lim_{n\to\infty} \frac{1}{\sqrt[n]{|a_n|}}, \qquad (11.26\text{b})$$

provided either limit exists or is equal to infinity.

Proof We verify 11.26a using the limit ratio test; verification of 11.26b is similar, using the limit root test.

If the limit ratio test is applied to the series of absolute values $\sum |a_n x^n|$,

$$L = \lim_{n\to\infty} \frac{|a_{n+1} x^{n+1}|}{|a_n x^n|} = |x| \lim_{n\to\infty} \left| \frac{a_{n+1}}{a_n} \right|.$$

Assuming that limit 11.26a exists or is equal to infinity, there are three possibilities:

(i) If $\lim_{n\to\infty} |a_n/a_{n+1}| = 0$, then $\lim_{n\to\infty} |a_{n+1}/a_n| = \infty$. Therefore $L = \infty$, and the power series diverges for all $x \neq 0$. In other words,

$$R = 0 = \lim_{n\to\infty} \left| \frac{a_n}{a_{n+1}} \right|.$$

(ii) If $\lim_{n\to\infty} |a_n/a_{n+1}| = \infty$, then $\lim_{n\to\infty} |a_{n+1}/a_n| = 0$. Therefore $L = 0$, and the power series converges absolutely for all x. Consequently,

$$R = \infty = \lim_{n\to\infty} \left| \frac{a_n}{a_{n+1}} \right|.$$

(iii) If $\lim_{n\to\infty} |a_n/a_{n+1}| = R$, then $\lim_{n\to\infty} |a_{n+1}/a_n| = 1/R$. In this case, $L = |x|/R$. Since the power series converges absolutely for $L < 1$ and diverges for $L > 1$, it follows that absolute convergence occurs for $|x| < R$ and divergence for $|x| > R$. This implies that R is the radius of convergence of the power series.

With this theorem the limit ratio test and limit root test can be avoided in determining intervals of convergence for many power series. For the series in Example 11.19 we proceed as follows:

(a) For the power series $\sum_{n=1}^{\infty} x^n/n^n$, we use 11.26b to obtain

$$R = \lim_{n\to\infty} \frac{1}{\sqrt[n]{|a_n|}} = \lim_{n\to\infty} \frac{1}{(1/n^n)^{1/n}} = \lim_{n\to\infty} n = \infty.$$

The power series therefore converges absolutely for all x.

(b) For the series $\sum_{n=1}^{\infty} n!\, x^n$, we use 11.26a,

$$R = \lim_{n\to\infty} \left|\frac{a_n}{a_{n+1}}\right| = \lim_{n\to\infty} \frac{n!}{(n+1)!} = \lim_{n\to\infty} \frac{1}{n+1} = 0,$$

and the power series converges only for $x = 0$.

(c) For the power series $\sum_{n=1}^{\infty} [(-1)^{n+1}/n]x^n$, 11.26a gives

$$R = \lim_{n\to\infty} \left|\frac{a_n}{a_{n+1}}\right| = \lim_{n\to\infty} \left|\frac{\dfrac{(-1)^{n+1}}{n}}{\dfrac{(-1)^{n+2}}{n+1}}\right| = \lim_{n\to\infty} \left(\frac{n+1}{n}\right) = 1.$$

When $x = 1$, the power series reduces to the alternating harmonic series

$$\sum_{n=1}^{\infty} \frac{(-1)^{n+1}}{n} = 1 - \frac{1}{2} + \frac{1}{3} - \frac{1}{4} + \cdots,$$

which converges conditionally. When $x = -1$, it reduces to

$$\sum_{n=1}^{\infty} \frac{(-1)^{n+1}}{n}(-1)^n = -1 - \frac{1}{2} - \frac{1}{3} - \frac{1}{4} - \cdots,$$

which diverges. The interval of convergence is therefore $-1 < x \le 1$.

EXAMPLE 11.20

Find the interval of convergence for the power series $\sum_{n=1}^{\infty} nx^n/5^{2n}$.

SOLUTION Since

$$R = \lim_{n\to\infty} \left|\frac{a_n}{a_{n+1}}\right| = \lim_{n\to\infty} \frac{\dfrac{n}{5^{2n}}}{\dfrac{n+1}{5^{2n+2}}} = 25,$$

the power series converges absolutely for $-25 < x < 25$. At $x = 25$, the series reduces to $\sum_{n=1}^{\infty} n$, which diverges; and at $x = -25$, it reduces to $\sum_{n=1}^{\infty} n(-1)^n$, which also diverges. The interval of convergence is therefore $-25 < x < 25$. ■

EXAMPLE 11.21 Find the interval of convergence for the power series $\sum_{n=1}^{\infty} x^{2n+1}/(n^2 2^n)$.

SOLUTION Since coefficients of even powers of x are 0 in this power series, the sequence $\{a_n/a_{n+1}\}$ is not defined. We cannot therefore find its radius of convergence directly using Theorem 11.13. Instead we write

$$\sum_{n=1}^{\infty} \frac{1}{n^2 2^n} x^{2n+1} = x \sum_{n=1}^{\infty} \frac{1}{n^2 2^n}(x^2)^n$$

and set $y = x^2$ in the series:

$$\sum_{n=1}^{\infty} \frac{1}{n^2 2^n}(x^2)^n = \sum_{n=1}^{\infty} \frac{1}{n^2 2^n} y^n.$$

According to equation 11.26a the radius of convergence of this series in y is

$$\lim_{n\to\infty} \frac{\dfrac{1}{n^2 2^n}}{\dfrac{1}{(n+1)^2 2^{n+1}}} = \lim_{n\to\infty} 2\left(\frac{n+1}{n}\right)^2 = 2.$$

Since $x = \pm\sqrt{y}$, it follows that the radius of convergence of the power series in x is $R = \sqrt{2}$. When $x = \sqrt{2}$, the series reduces to $\sum_{n=1}^{\infty} \sqrt{2}/n^2 = \sqrt{2}\sum_{n=1}^{\infty} 1/n^2$, which converges; and when $x = -\sqrt{2}$, it reduces to the negative of this series. The interval of convergence is therefore $-\sqrt{2} \leq x \leq \sqrt{2}$. ■

We describe what it means for a function to be the sum of a power series in the following definition.

Definition 11.10 A function $f(x)$ is said to be the **sum** of a power series $\sum_{n=0}^{\infty} a_n x^n$ if at each point x in the interval of convergence of the power series, the sum of the series is the same as the value of $f(x)$ at that x.

We write

$$f(x) = \sum_{n=0}^{\infty} a_n x^n \tag{11.27}$$

to signify that $f(x)$ is the sum of the power series. For example, the sum of the geometric series

$$\sum_{n=0}^{\infty} x^n = 1 + x + x^2 + x^3 + \cdots$$

is $1/(1-x)$ for $|x| < 1$, and we write

$$\frac{1}{1-x} = 1 + x + x^2 + x^3 + \cdots, \qquad |x| < 1.$$

We affix to the equation the interval of convergence of the power series, thus identifying for which values of x the function $1/(1-x)$ is the sum of the series. The function

$1/(1-x)$ is defined for all x, except $x = 1$, but it represents the sum of the series only in the interval $-1 < x < 1$.

In Section 11.3 we defined the sum of an infinite series of constants as the limit of its sequence of partial sums. In the case of power series 11.27, the sequence of partial sums is a sequence of polynomials:

$$\begin{aligned}
S_1(x) &= a_0, \\
S_2(x) &= a_0 + a_1 x, \\
S_3(x) &= a_0 + a_1 x + a_2 x^2, \\
&\ \vdots \\
S_n(x) &= a_0 + a_1 x + a_2 x^2 + \cdots + a_{n-1} x^{n-1}, \\
&\ \vdots
\end{aligned}$$

By equation 11.27 we are saying that for each x in the interval of convergence of the power series,

$$f(x) = \lim_{n\to\infty} S_n(x).$$

For example, the nth partial sum of the geometric series $\sum_{n=0}^{\infty} x^n$ is

$$S_n(x) = 1 + x + x^2 + \cdots + x^{n-1} = \begin{cases} n & x = 1 \\ \dfrac{1-x^n}{1-x} & x \neq 1 \end{cases}$$

(see equation 11.13a). Algebraically, it is clear that for large n, $S_n(x)$ approaches $1/(1-x)$, provided $|x| < 1$, and we write therefore

$$\frac{1}{1-x} = 1 + x + x^2 + x^3 + \cdots, \quad |x| < 1.$$

In Figure 11.20 we show the first five partial sums of this geometric series together with the sum, just to illustrate how the partial sums of the series approximate the sum more closely for increasing n. Note how these figures illustrate the following facts about the geometric series and its partial sums:

1. The partial sums $S_n(x)$ are defined for all x and the sum $1/(1-x)$ is defined for all $x \neq 1$, but the curves $y = S_n(x)$ approach $y = 1/(1-x)$ only for $|x| < 1$.
2. For $x > 0$, the partial sums $S_n(x)$ approach $1/(1-x)$ from below; for $x < 0$, the $S_n(x)$ oscillate about $1/(1-x)$, but gradually approach $1/(1-x)$.

FIGURE 11.20

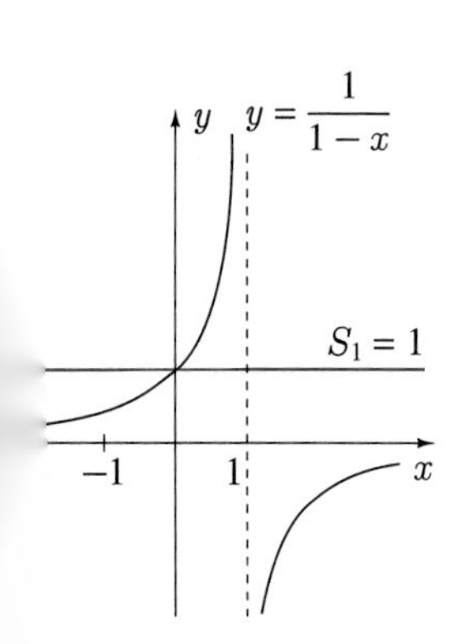

(b)

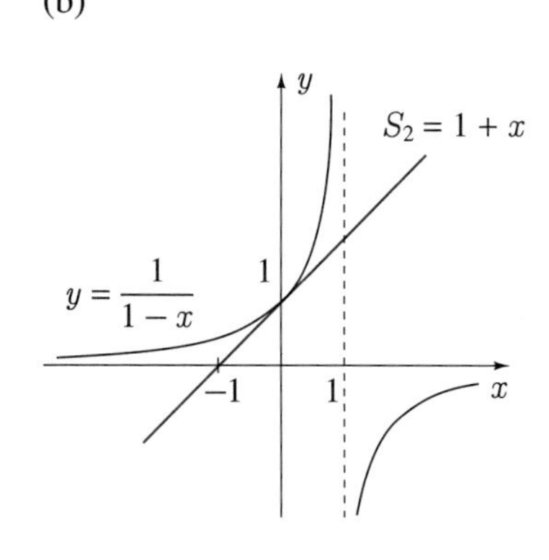

(c)

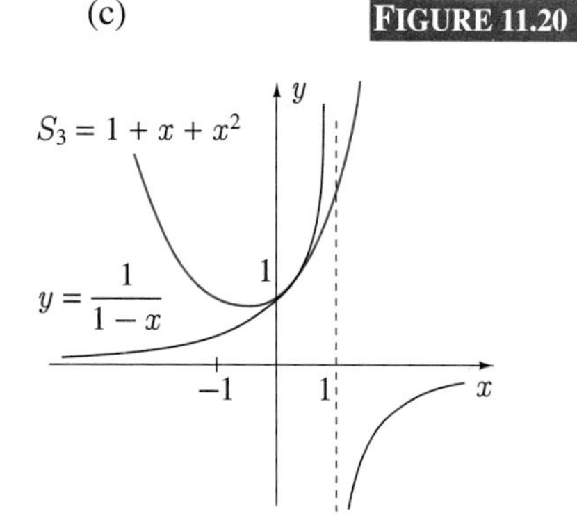

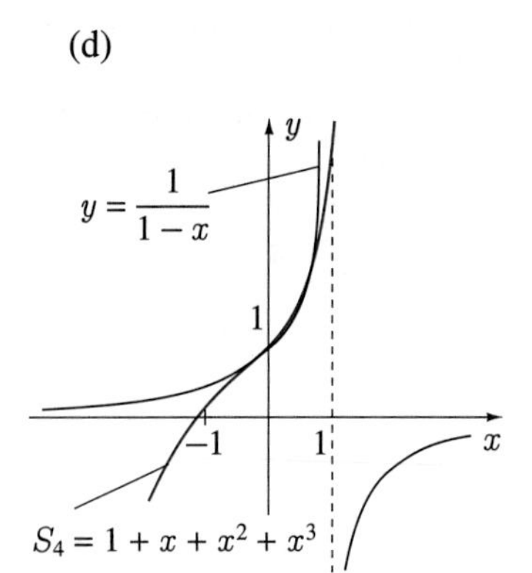

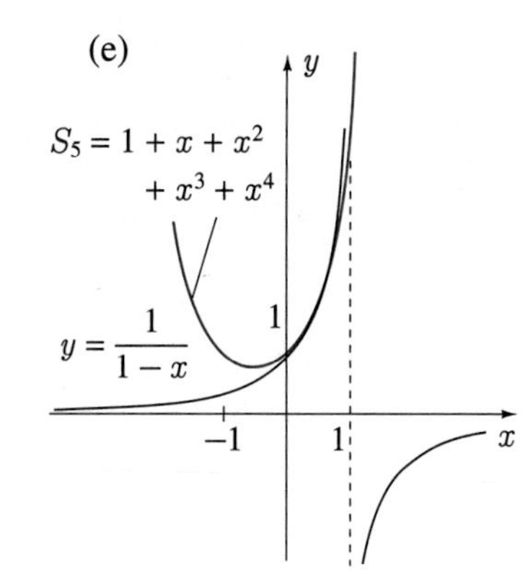

EXAMPLE 11.22 Find the sum of the series

$$\sum_{n=0}^{\infty} \frac{(-1)^n}{2^n} x^{(2n+1)/2}.$$

SOLUTION We write

$$\sum_{n=0}^{\infty} \frac{(-1)^n}{2^n} x^{(2n+1)/2} = x^{1/2} \sum_{n=0}^{\infty} \frac{(-1)^n}{2^n} x^n = x^{1/2} \sum_{n=0}^{\infty} \left(-\frac{x}{2}\right)^n$$

and note that the power series on the right is a geometric series with sum

$$\sum_{n=0}^{\infty} \left(-\frac{x}{2}\right)^n = \frac{1}{1+\frac{x}{2}}, \quad \left|-\frac{x}{2}\right| < 1.$$

Consequently,

$$\sum_{n=0}^{\infty} \frac{(-1)^n}{2^n} x^{(2n+1)/2} = \sqrt{x}\frac{1}{1+\frac{x}{2}} = \frac{2\sqrt{x}}{x+2}, \quad 0 \le x < 2.$$

■

For many power series there is no known function that is equal to the sum of the power series over its interval of convergence. In such cases we continue to write equation 11.27, but say that the power series defines the value of $f(x)$ at each x in the interval of convergence. For instance, a very important series in engineering and physics is

$$\sum_{n=0}^{\infty} \frac{(-1)^n}{2^{2n}(n!)^2} x^{2n},$$

which converges absolutely for all x. It arises so often in applications that it is given a special name, the *Bessel function of the first kind of order zero*, and is denoted by $J_0(x)$. In other words, we write

$$J_0(x) = \sum_{n=0}^{\infty} \frac{(-1)^n}{2^{2n}(n!)^2} x^{2n},$$

and say that the power series defines the value of $J_0(x)$ for each x.

We have called 11.25 a power series in x. It is also said to be a *power series about* 0 (meaning the point 0 on the x-axis), where we note that for any interval of convergence whatsoever, 0 is always at its centre. This suggests that power series about other points might be considered, and this is indeed the case. The general power series about a point c on the x-axis is

$$\sum_{n=0}^{\infty} a_n(x-c)^n = a_0 + a_1(x-c) + a_2(x-c)^2 + \cdots \tag{11.28}$$

and is said to be a power series in $x - c$.

A power series in $x - c$ has an interval of convergence and a radius of convergence analogous to a power series in x. In particular, every power series in $x-c$ has a radius of convergence R such that if $R = 0$, the power series converges only for $x = c$; if $R = \infty$, the power series converges absolutely for all x; and if $0 < R < \infty$, the series converges absolutely for $|x - c| < R$, diverges for $|x - c| > R$, and may or may not converge for $x = c \pm R$. The radius of convergence is again given by equations 11.26, provided the limits exist or are equal to infinity. For power series in $x - c$, then, the point c is the centre of the interval of convergence.

EXAMPLE 11.23

Find the interval of convergence of the power series

$$2^2(x+2) + 2^3(x+2)^2 + 2^4(x+2)^3 + \cdots + 2^{n+1}(x+2)^n + \cdots .$$

SOLUTION By writing the power series in sigma notation,

$$\sum_{n=1}^{\infty} 2^{n+1}(x+2)^n,$$

we can use equation 11.26a (or 11.26b) to calculate its radius of convergence,

$$R = \lim_{n\to\infty}\left|\frac{a_n}{a_{n+1}}\right| = \lim_{n\to\infty}\frac{2^{n+1}}{2^{n+2}} = \frac{1}{2}.$$

Since this is a power series about -2, the ends of the interval of convergence are $x = -5/2$ and $x = -3/2$. At $x = -5/2$, the power series reduces to $\sum_{n=1}^{\infty} 2(-1)^n$, which diverges; and at $x = -3/2$, it reduces to $\sum_{n=1}^{\infty} 2$, which also diverges. The interval of convergence is therefore $-5/2 < x < -3/2$. ■

EXERCISES 11.8

In Exercises 1–22 find the interval of convergence for the power series.

1. $\displaystyle\sum_{n=1}^{\infty} \frac{1}{n}x^n$

2. $\displaystyle\sum_{n=1}^{\infty} n^2 x^n$

3. $\displaystyle\sum_{n=0}^{\infty} \frac{1}{(n+1)^3}x^n$

4. $\displaystyle\sum_{n=0}^{\infty} n^2 3^n x^n$

5. $\displaystyle\sum_{n=0}^{\infty} \frac{1}{2^n}(x-1)^n$

6. $\displaystyle\sum_{n=4}^{\infty} (-1)^n n^3 (x+3)^n$

7. $\displaystyle\sum_{n=1}^{\infty} \frac{1}{\sqrt{n}}(x+2)^n$

8. $\displaystyle\sum_{n=2}^{\infty} 2^n \left(\frac{n-1}{n+2}\right)^2 (x-4)^n$

9. $\displaystyle\sum_{n=1}^{\infty} \frac{1}{n^2}x^{2n}$

10. $\displaystyle\sum_{n=0}^{\infty} (-1)^n x^{3n}$

11. $\displaystyle\sum_{n=0}^{\infty} \frac{n-1}{n+1}(2x)^n$

12. $\displaystyle\sum_{n=0}^{\infty} \frac{1}{\sqrt{n+1}}x^{3n+1}$

13. $\displaystyle\sum_{n=0}^{\infty} \frac{(-1)^n}{3^n} x^{2n+1}$ 14. $\displaystyle\sum_{n=1}^{\infty} \frac{(-e)^n}{n^2} x^n$

15. $\displaystyle\frac{x}{9} + \frac{4}{3^4}x^2 + \frac{9}{3^6}x^3 + \cdots + \frac{n^2}{3^{2n}}x^n + \cdots$

16. $x + 2^2x^2 + 3^3x^3 + \cdots + n^n x^n + \cdots$

17. $\displaystyle\frac{1}{36}(x+10)^6 + \frac{1}{49}(x+10)^7 + \frac{1}{64}(x+10)^8 + \cdots + \frac{1}{n^2}(x+10)^n + \cdots$

18. $3x + 8(3x)^2 + 27(3x)^3 + \cdots + n^3(3x)^n + \cdots$

19. $\displaystyle\frac{3}{4}x^2 + x^4 + \frac{27}{16}x^6 + \cdots + \frac{3^n}{(n+1)^2}x^{2n} + \cdots$

20. $\displaystyle 1 + \frac{1}{5}x^3 + \frac{1}{25}x^6 + \cdots + \frac{1}{5^n}x^{3n} + \cdots$

21. $\displaystyle\sum_{n=2}^{\infty} \frac{1}{\ln n} x^n$ 22. $\displaystyle\sum_{n=2}^{\infty} \frac{1}{n^2 \ln n} x^n$

In Exercises 23–25 find the radius of convergence of the power series.

23. $\displaystyle\sum_{n=1}^{\infty} \frac{(n!)^3}{(3n)!} x^n$

24. $\displaystyle\sum_{n=1}^{\infty} \frac{2 \cdot 4 \cdot 6 \cdot \cdots \cdot (2n)}{3 \cdot 5 \cdot 7 \cdot \cdots \cdot (2n+1)} x^n$

25. $\displaystyle\sum_{n=1}^{\infty} \frac{[1 \cdot 3 \cdot 5 \cdot \cdots \cdot (2n+1)]^2}{2^{2n}(2n)!} x^n$

In Exercises 26–29 find the sum of the power series.

26. $\displaystyle\sum_{n=0}^{\infty} \frac{1}{4^n} x^{3n}$ 27. $\displaystyle\sum_{n=1}^{\infty} (-e)^n x^n$

28. $\displaystyle\sum_{n=1}^{\infty} \frac{1}{3^{2n}} (x-1)^n$ 29. $\displaystyle\sum_{n=2}^{\infty} (x+5)^{2n}$

30. If m is a nonnegative integer, the Bessel function of order m of the first kind is defined by the power series

$$J_m(x) = \sum_{n=0}^{\infty} \frac{(-1)^n}{2^{2n+m} n!\,(n+m)!} x^{2n+m}.$$

(a) Write out the first five terms of $J_0(x)$, $J_1(x)$, and $J_m(x)$.
(b) Find the interval of convergence for each $J_m(x)$.

31. The hypergeometric series is

$$1 + \frac{\alpha\beta}{\gamma} + \frac{\alpha(\alpha+1)\beta(\beta+1)}{2!\,\gamma(\gamma+1)}x^2 + \frac{\alpha(\alpha+1)(\alpha+2)\beta(\beta+1)(\beta+2)}{3!\,\gamma(\gamma+1)(\gamma+2)}x^3 + \cdots,$$

where α, β, and γ are all constants.
(a) Write this series in sigma notation.
(b) What is the radius of convergence of the hypergeometric series if γ is not zero or a negative integer?

32. Show that if a power series $\sum a_n x^n$ converges for $x_1 \neq 0$, then it converges absolutely for all x in the interval $|x| < |x_1|$.

33. Prove that the interval of convergence of a power series $\sum a_n x^n$ must be one of the three possibilities listed on page 554. Use Exercise 32 and the *completeness property* for real numbers, which states: If a nonempty set S of real numbers has an upper bound, then it has a least upper bound; that is, there exists a number l such that all numbers in S are less than or equal to l and there is no upper bound smaller than l.

SECTION 11.9

Taylor's Remainder Formula and Taylor Series

If $f(x)$ is the sum of a power series $\sum_{n=0}^{\infty} a_n(x-c)^n$, we write

$$f(x) = \sum_{n=0}^{\infty} a_n(x-c)^n. \qquad (11.29)$$

In Section 11.8 we considered this equation in situations where the power series is given and the sum is to be determined. An alternative view is to consider the function $f(x)$ as given and the power series as an expansion of the function about the point c. For example, given the function $f(x) = 1/(2x+3)$, it is quite simple to use properties of geometric series to find a power series expansion in x for $f(x)$. If we write

$$f(x) = \frac{1}{2x+3} = \frac{1}{3\left(1+\dfrac{2x}{3}\right)}$$

and interpret the right-hand side as one-third the sum of a geometric series with first term equal to 1 and common ratio $-2x/3$, then

$$\begin{aligned} f(x) &= \frac{1}{3}\sum_{n=0}^{\infty}\left(-\frac{2x}{3}\right)^n, \qquad \left|-\frac{2x}{3}\right| < 1 \\ &= \sum_{n=0}^{\infty}\frac{(-1)^n 2^n}{3^{n+1}}x^n, \qquad |x| < \frac{3}{2}. \end{aligned}$$

Naturally it is not clear at this time why we should want to expand a function in the form of an infinite series, but in Section 11.12, when we consider applications of series, we will find such expansions extremely useful.

In this section we establish two results. First we give conditions that gurarantee that a function $f(x)$ has a power series expansion about a point c and that this power series converges to $f(x)$. Second, we show that if a function has a power series expansion about a point, then it has only one such series. The first result is an immediate consequence of Theorem 11.14.

Theorem 11.14

(Taylor's Remainder Formula) *If $f(x)$ and its first $n-1$ derivatives are continuous on the closed interval between c and x, and if $f(x)$ has an nth derivative on the open interval between c and x, then there exists a point z_n between c and x such that*

$$\begin{aligned} f(x) = f(c) + f'(c)(x-c) + \frac{f''(c)}{2!}(x-c)^2 + \frac{f'''(c)}{3!}(x-c)^3 + \cdots \\ + \frac{f^{(n-1)}(c)}{(n-1)!}(x-c)^{n-1} + \frac{f^{(n)}(z_n)}{n!}(x-c)^n. \end{aligned} \tag{11.30}$$

The notation $f^{(n)}(z_n)$ represents the nth derivative of $f(x)$ evaluated at $x = z_n$. This result can be proved by applying Rolle's theorem (Theorem 3.13) to the function

$$\begin{aligned} F(y) = f(x) - f(y) - f'(y)(x-y) - \frac{f''(y)}{2!}(x-y)^2 - \cdots \\ - \frac{f^{(n-1)}(y)}{(n-1)!}(x-y)^{n-1} - \left(\frac{x-y}{x-c}\right)^n \Big[f(x) - f(c) - f'(c)(x-c) \\ - \frac{f''(c)}{2!}(x-c)^2 - \cdots - \frac{f^{(n-1)}(c)}{(n-1)!}(x-c)^{n-1}\Big] \end{aligned}$$

defined on the interval $c \le y \le x$, or $x \le y \le c$, depending on whether x is larger or smaller than c.

Note that when $n = 1$, Taylor's remainder formula reduces to the mean value theorem of Section 3.9. In other words, the mean value theorem is a special case of Taylor's remainder formula.

Consider now a function $f(x)$ that has derivatives of all orders on the closed interval between c and x. Taylor's remainder formula can therefore be written down for all values of n. Let us write it out for $n = 1, 2, 3$, and 4:

$$f(x) = f(c) + f'(z_1)(x - c), \tag{11.31a}$$

$$f(x) = f(c) + f'(c)(x - c) + \frac{f''(z_2)}{2!}(x - c)^2, \tag{11.31b}$$

$$f(x) = f(c) + f'(c)(x - c) + \frac{f''(c)}{2!}(x - c)^2 + \frac{f'''(z_3)}{3!}(x - c)^3, \tag{11.31c}$$

$$f(x) = f(c) + f'(c)(x - c) + \frac{f''(c)}{2!}(x - c)^2 + \frac{f'''(c)}{3!}(x - c)^3 + \frac{f''''(z_4)}{4!}(x - c)^4. \tag{11.31d}$$

The last terms in these equations are called **remainders**, and when they are denoted by R_1, R_2, R_3, and R_4 respectively,

$$f(x) = f(c) + R_1, \tag{11.32a}$$

$$f(x) = f(c) + f'(c)(x - c) + R_2, \tag{11.32b}$$

$$f(x) = f(c) + f'(c)(x - c) + \frac{f''(c)}{2!}(x - c)^2 + R_3, \tag{11.32c}$$

$$f(x) = f(c) + f'(c)(x - c) + \frac{f''(c)}{2!}(x - c)^2 + \frac{f'''(c)}{3!}(x - c)^3 + R_4, \tag{11.32d}$$

and in general,

$$f(x) = f(c) + f'(c)(x - c) + \frac{f''(c)}{2!}(x - c)^2 + \cdots + \frac{f^{(n-1)}(c)}{(n-1)!}(x - c)^{n-1} + R_n. \tag{11.32e}$$

We call R_n the remainder for the simple reason that if we were to drop the remainders from equations (11.32), then what remains is a sequence of polynomial approximations to $f(x)$:

$$f(x) \approx f(c), \tag{11.33a}$$

$$f(x) \approx f(c) + f'(c)(x - c), \tag{11.33b}$$

$$f(x) \approx f(c) + f'(c)(x - c) + \frac{f''(c)}{2!}(x - c)^2, \tag{11.33c}$$

$$f(x) \approx f(c) + f'(c)(x - c) + \frac{f''(c)}{2!}(x - c)^2 + \frac{f'''(c)}{3!}(x - c)^3. \tag{11.33d}$$

We give a detailed discussion on the accuracy of these approximations in Section 11.12, but for now we can at least appreciate why Theorem 11.14 is called Taylor's remainder formula.

The nth remainder is

$$R_n = R_n(c, x) = \frac{f^{(n)}(z_n)}{n!}(x - c)^n, \tag{11.34}$$

where z_n (which, as the notation suggests, varies with n) is always between c and x. We have written $R_n(c, x)$ to emphasize the fact that the remainders depend on the point of expansion c and the value of x being considered.

Suppose the sequence of remainders has limit zero,

$$\lim_{n \to \infty} R_n = 0. \tag{11.35}$$

What this means is that the sequence of polynomials in 11.33 approximates $f(x)$ more and more closely as n gets larger and larger. In fact, these polynomials are the partial sums for the following infinite series,

$$f(c) + f'(c)(x - c) + \frac{f''(c)}{2!}(x - c)^2 + \frac{f'''(c)}{3!}(x - c)^3 + \cdots. \tag{11.36}$$

Thus, if the sequence of remainders $\{R_n\}$ approaches zero, then power series 11.36 converges to $f(x)$. We may write therefore that

$$f(x) = f(c) + f'(c)(x - c) + \frac{f''(c)}{2!}(x - c)^2 + \frac{f'''(c)}{3!}(x - c)^3 + \cdots, \tag{11.37}$$

and this equation is valid for all those x's for which $\lim_{n \to \infty} R_n(c, x) = 0$. This power series in $x - c$ is called the **Taylor series** of $f(x)$ about the point c. We have shown that *if the sequence of remainders $\{R_n\}$ for a function $f(x)$ exists and has limit zero, the Taylor series converges to $f(x)$ for that x.*

When $c = 0$, the Taylor series takes the form

$$f(x) = f(0) + f'(0)x + \frac{f''(0)}{2!}x^2 + \frac{f'''(0)}{3!}x^3 + \cdots, \tag{11.38}$$

a power series in x called the **Maclaurin series** for $f(x)$.

To obtain the Taylor or Maclaurin series for a function $f(x)$ is quite simple; evaluate all derivatives of $f(x)$ at the point of expansion and use formula 11.37 or 11.38. To show that the series converges to $f(x)$ is a different matter. We must show that the sequence $\{R_n\}$ has limit zero, and this can be very difficult because R_n is defined in terms of an unknown point z_n. This problem is usually circumvented by finding a maximum value for R_n and showing that this maximum value approaches zero. We illustrate this procedure in the next three examples. To do so, however, we require the result of the following theorem. Apart from the fact that Theorem 11.15 is needed for Examples 11.24 and 11.25, its proof is interesting in that it interchanges the usual role of sequences and series. Heretofore, we have used sequences to determine convergence of series. This theorem uses series to verify convergence of a certain sequence.

Theorem 11.15

The sequence $\{|x|^n/n!\}$ has limit 0 for any x whatsoever,

$$\lim_{n \to \infty} \frac{|x|^n}{n!} = 0. \tag{11.39}$$

Proof Consider the series $\sum_{n=1}^{\infty} |x|^n/n!$. Since the ratio test gives

$$L = \lim_{n\to\infty} \frac{\dfrac{|x|^{n+1}}{(n+1)!}}{\dfrac{|x|^n}{n!}} = \lim_{n\to\infty} \frac{|x|}{n+1} = 0,$$

the series converges for any x. But then the nth term test implies that its nth term must have limit 0.

EXAMPLE 11.24 Find the Maclaurin series for $\sin x$ and show that it converges to $\sin x$ for all x.

SOLUTION If $f(x) = \sin x$, then:

$$\begin{aligned} f(0) &= \sin x_{|x=0} = 0; \\ f'(0) &= \cos x_{|x=0} = 1; \\ f''(0) &= -\sin x_{|x=0} = 0; \\ f'''(0) &= -\cos x_{|x=0} = -1; \\ f''''(0) &= \sin x_{|x=0} = 0; \quad \text{etc.} \end{aligned}$$

Taylor's remainder formula for $\sin x$ and $c = 0$ yields

$$\sin x = x - \frac{x^3}{3!} + \frac{x^5}{5!} - \frac{x^7}{7!} + \cdots + (n^{\text{th}}\text{ term}) + R_n,$$

where

$$R_n = \frac{d^n}{dx^n}(\sin x)_{|x=z_n} \frac{x^n}{n!}.$$

But the nth derivative of $\sin x$ is $\pm \sin x$ or $\pm \cos x$, so that

$$\left| \frac{d^n}{dx^n}(\sin x)_{|x=z_n} \right| \leq 1.$$

Hence,

$$|R_n| \leq \frac{|x|^n}{n!}.$$

But according to equation 11.39, $\lim_{n\to\infty} |x|^n/n! = 0$ for any x whatsoever, and therefore

$$\lim_{n\to\infty} R_n = 0.$$

The Maclaurin series for $\sin x$ therefore converges to $\sin x$ for all x, and we may write

$$\begin{aligned} \sin x &= x - \frac{x^3}{3!} + \frac{x^5}{5!} - \frac{x^7}{7!} + \cdots \\ &= \sum_{n=0}^{\infty} \frac{(-1)^n}{(2n+1)!} x^{2n+1}, \qquad -\infty < x < \infty. \end{aligned}$$

■

Often it is necessary to consider separately points on either side of the point of expansion. This is illustrated in the next example.

EXAMPLE 11.25

Find the Maclaurin series for e^x and show that it converges to e^x for all x.

SOLUTION Since

$$\frac{d^n}{dx^n}(e^x)_{|x=0} = e^x{}_{|x=0} = 1,$$

Taylor's remainder formula for e^x and $c = 0$ gives

$$e^x = 1 + x + \frac{x^2}{2!} + \frac{x^3}{3!} + \cdots + \frac{x^{n-1}}{(n-1)!} + R_n,$$

where

$$R_n = \frac{d^n}{dx^n}(e^x)_{|x=z_n}\frac{x^n}{n!} = e^{z_n}\frac{x^n}{n!}.$$

Now, if $x < 0$, then $x < z_n < 0$, and

$$|R_n| < e^0\frac{|x|^n}{n!},$$

which approaches zero as n becomes infinite (see equation 11.39). If $x > 0$, then $0 < z_n < x$, and

$$|R_n| < e^x\frac{|x|^n}{n!},$$

which again has limit zero as n approaches infinity. Thus, for any x whatsoever, $\lim_{n\to\infty} R_n = 0$, and the Maclaurin series for e^x converges to e^x:

$$\begin{aligned} e^x &= 1 + x + \frac{x^2}{2!} + \frac{x^3}{3!} + \cdots \\ &= \sum_{n=0}^{\infty}\frac{1}{n!}x^n, \qquad -\infty < x < \infty. \end{aligned}$$

■

The following example is more difficult. The remainders do not apporach zero for all x so that the Taylor series does not converge to the function for all x. The problem would have been even more difficult had we not suggested the values of x to consider.

EXAMPLE 11.26

Find the Taylor series for $\ln x$ about the point 1 and show that it converges to $\ln x$ for $1/2 \le x \le 2$.

SOLUTION If $f(x) = \ln x$, then:

$$\begin{aligned} f(1) &= \ln x_{|x=1} = 0; \\ f'(1) &= \frac{1}{x}_{|x=1} = 1; \\ f''(1) &= \frac{-1}{x^2}_{|x=1} = -1; \end{aligned}$$

$$f'''(1) = \frac{2}{x^3}\bigg|_{x=1} = 2;$$

$$f''''(1) = \frac{-3!}{x^4}\bigg|_{x=1} = -3!; \quad \text{etc.}$$

Taylor's remainder formula for $\ln x$ and $c = 1$ yields

$$\ln x = (x-1) - \frac{1}{2!}(x-1)^2 + \frac{2!}{3!}(x-1)^3 + \cdots + \frac{(-1)^n(n-2)!}{(n-1)!}(x-1)^{n-1} + R_n,$$

where

$$R_n = \frac{d^n}{dx^n}(\ln x)_{|x=z_n}\frac{(x-1)^n}{n!} = \frac{(-1)^{n+1}(n-1)!}{(z_n)^n}\frac{(x-1)^n}{n!} = \frac{(-1)^{n+1}}{n}\left(\frac{x-1}{z_n}\right)^n,$$

and z_n is between 1 and x.

If $1 < x \le 2$, then the largest value of $x - 1$ is 1. Furthermore, z_n must be larger than 1. It follows that

$$|R_n| < \frac{1}{n}\left(\frac{1}{1}\right)^n = \frac{1}{n}$$

and therefore

$$\lim_{n\to\infty} R_n = 0.$$

If $1/2 \le x < 1$, then $-1/2 \le x - 1 < 0$. Combine this with $x < z_n < 1$, and we can state that $-1 < (x-1)/z_n < 0$. Then,

$$|R_n| < \frac{1}{n} \quad \text{and} \quad \lim_{n\to\infty} R_n = 0.$$

Thus, for $1/2 \le x \le 2$, the sequence of remainders $\{R_n\}$ approaches zero, and the Taylor series converges to $\ln x$ for those values of x:

$$\ln x = (x-1) - \frac{1}{2}(x-1)^2 + \frac{1}{3}(x-1)^3 + \cdots$$
$$= \sum_{n=1}^{\infty} \frac{(-1)^{n+1}}{n}(x-1)^n, \qquad \frac{1}{2} \le x \le 2.$$

This series actually converges to $\ln x$ on the larger interval $0 < x \le 2$. We will prove this in Example 11.30. ■

We have shown that if the sequence of Taylor's remainders $\{R_n\}$ for a function $f(x)$ and a point c has limit zero, then $f(x)$ has at least one power series expansion about c, namely its Taylor series,

$$f(x) = \sum_{n=0}^{\infty} \frac{f^{(n)}(c)}{n!}(x-c)^n, \qquad f^{(0)}(c) = f(c), \tag{11.40}$$

and this series converges to $f(x)$. We now show that this is the *only* power series expansion of $f(x)$ about c. To do this we require the following theorem.

Theorem 11.16

If a function $f(x)$ has a power series expansion $f(x) = \sum_{n=0}^{\infty} a_n(x-c)^n$ with positive radius of convergence R, then each of the following series has radius of convergence R:

$$f'(x) = \sum_{n=0}^{\infty} na_n(x-c)^{n-1}, \tag{11.41}$$

$$\int f(x)\,dx = \sum_{n=0}^{\infty} \frac{a_n}{n+1}(x-c)^{n+1} + C. \tag{11.42}$$

In other words, power series expansions for $f'(x)$ and *$\int f(x)\,dx$ can be obtained by term-by-term differentiation and integration of the power series of $f(x)$, and all three series have the same radius of convergence.*

Due to the difficulty in proving this theorem, and in order to preserve the continuity of our discussion, we omit a proof. Note that the theorem is stated in terms of radii of convergence rather than intervals of convergence. This is due to the fact that in differentiating a power series we may lose the end points of the original interval of convergence, and in integrating we may pick them up.

The next theorem implies that there is at most one power series expansion of a function about a given point.

Theorem 11.17

If a power series expansion of a function has a positive radius of convergence, then it is the Taylor series.

Proof Suppose the power series expansion

$$f(x) = \sum_{n=0}^{\infty} a_n(x-c)^n = a_0 + a_1(x-c) + a_2(x-c)^2 + \cdots$$

has a positive radius of convergence. Substitution of $x = c$ gives

$$f(c) = a_0.$$

If we differentiate the power series according to Theorem 11.16, we obtain

$$f'(x) = a_1 + 2a_2(x-c) + 3a_3(x-c)^2 + \cdots,$$

and if we substitute $x = c$, the result is

$$f'(c) = a_1.$$

If we differentiate the power series for $f'(x)$, we obtain

$$f''(x) = 2a_2 + 3\cdot 2a_3(x-c) + 4\cdot 3a_4(x-c)^2 + \cdots,$$

and substitute $x = c$,

$$f''(c) = 2a_2 \quad \text{or} \quad a_2 = \frac{f''(c)}{2!}.$$

Continued differentiation and substitution leads to the result that for all n,

$$a_n = \frac{f^{(n)}(c)}{n!}.$$

The power series $f(x) = \sum_{n=0}^{\infty} a_n(x-c)^n$ is therefore the Taylor series.

The following corollary is an immediate consequence of this theorem.

Corollary *If two power series* $\sum_{n=0}^{\infty} a_n(x-c)^n$ *and* $\sum_{n=0}^{\infty} b_n(x-c)^n$ *with positive radii of convergence have identical sums,*

$$\sum_{n=0}^{\infty} a_n(x-c)^n = \sum_{n=0}^{\infty} b_n(x-c)^n,$$

then $a_n = b_n$ *for all* n.

Our theory of power series expansions of functions is now complete. Given a function $f(x)$ and a point c, there can only be one power series expansion for $f(x)$ about c, its Taylor series. To show that the Taylor series does indeed have sum $f(x)$, it is sufficient to show that the sequence of Taylor's remainders $\{R_n\}$ has limit zero. When this is the case we write

$$f(x) = \sum_{n=0}^{\infty} \frac{f^{(n)}(c)}{n!}(x-c)^n. \tag{11.43}$$

Equation 11.43 says, then, that the right-hand side is the Taylor series of $f(x)$ and that the Taylor series converges to $f(x)$. This point is most important because there do exist functions $f(x)$ whose Taylor series exist but do not converge to $f(x)$ (see Exercise 12). We therefore write 11.43 only when the Taylor series of $f(x)$ exists and that it converges to $f(x)$.

Only one point of consideration remains. Although we have a formula for calculation of the Taylor series of a function $f(x)$, determination of $f^{(n)}(c)$ could be very complicated. In addition, if a reasonable formula for $f^{(n)}(c)$ cannot be found, how will we ever prove that the sequence of Taylor's remainders has limit zero? Fortunately, in more common situations, special devices exist that enable us to find Taylor series without actually calculating $f^{(n)}(c)$. Indeed, if by any method, we can obtain a power series representation of $f(x)$, then we can be assured by Theorem 11.17 that it is the Taylor series. In Section 11.10 we illustrate some of these techniques.

EXERCISES 11.9

In Exercises 1–6 use Taylor's remainder formula to find Taylor series for the function $f(x)$ about the point indicated. In each case, find the interval on which the Taylor series converges to $f(x)$.

1. $f(x) = \cos x$ about $x = 0$

2. $f(x) = e^{5x}$ about $x = 0$

3. $f(x) = \sin(10x)$ about $x = 0$

4. $f(x) = \sin x$ about $x = \pi/4$

5. $f(x) = e^{2x}$ about $x = 1$

6. $f(x) = 1/(3x+2)$ about $x = 2$

7. Prove the corollary to Theorem 11.17.

8. Prove that if a power series with positive radius of convergence has sum zero, $\sum_{n=0}^{\infty} a_n(x-c)^n = 0$, then $a_n = 0$ for all n.

9. In Section 4.4 we stated the second-derivative test for determining whether a critical point x_0 at which $f'(x_0) = 0$ yields a relative maximum or a relative minimum. Use Taylor's remainder formula to verify this result when $f'(x)$ and $f''(x)$ are continuous on an open interval containing x_0.

10. Extend the result of Exercise 9 to verify the extrema test of Exercise 34 in Section 4.4.

11. There is an integral form for the remainder $R_n(c, x)$ in Taylor's remainder formula that is sometimes more useful than the derivative form in Theorem 11.15: If $f(x)$ and its first n derivatives are continuous on the closed interval between c and x, then

$$f(x) = f(c) + f'(c)(x-c) + \frac{f''(c)}{2!}(x-c)^2 + \cdots + \frac{f^{(n-1)}(c)}{(n-1)!}(x-c)^{n-1} + R_n(c, x),$$

where

$$R_n(c, x) = \frac{1}{(n-1)!}\int_c^x (x-t)^{n-1} f^{(n)}(t)\, dt.$$

Use the following outline of steps to prove this result.

(a) Show that

$$f(x) = f(c) + \int_c^x f'(t)\, dt.$$

(b) Use integration by parts with $u = f'(t)$, $du = f''(t)\, dt$, $dv = dt$, and $v = t - x$ on the integral in (a) to obtain

$$f(x) = f(c) + f'(c)(x-c) + \int_c^x (x-t) f''(t)\, dt.$$

(c) Use integration by parts with $u = f''(t)$, $du = f'''(t)\, dt$, $dv = (x-t)\, dt$, and $v = -(1/2)(x-t)^2$ to obtain

$$f(x) = f(c) + f'(c)(x-c) + \frac{f''(c)}{2!}(x-c)^2 + \frac{1}{2!}\int_c^x (x-t)^2 f'''(t)\, dt.$$

(d) Continue this process to obtain the integral form for Taylor's remainder formula.

12. (a) Sketch a graph of the function

$$f(x) = \begin{cases} e^{-1/x^2} & x \neq 0 \\ 0 & x = 0. \end{cases}$$

(b) Use L'Hôpital's rule to show that for every positive integer n,

$$\lim_{x \to 0} \frac{e^{-1/x^2}}{x^n} = 0.$$

(c) Prove, by mathematical induction, that $f^{(n)}(0) = 0$ for $n \geq 1$.

(d) What is the Maclaurin series for $f(x)$?

(e) For what values of x does the Maclaurin series of $f(x)$ converge to $f(x)$?

SECTION 11.10

Taylor Series Expansions of Functions

In Section 11.9 we pointed out that if by any means we can find a power series representation for a function $f(x)$, then we can be assured by Theorem 11.17 that it is the Taylor series for $f(x)$. In this section we illustrate various techniques for producing power series. This has the advantage of avoiding Taylor remainders in establishing convergence of the series to $f(x)$. Our first example uses geometric series.

EXAMPLE 11.27

Find (a) the Maclaurin series for $1/(4+5x)$; (b) the Taylor series about 5 for $1/(13-2x)$.

SOLUTION (a) If we write

$$\frac{1}{4+5x} = \frac{1}{4\left(1 + \frac{5x}{4}\right)}$$

and interpret the right-hand side as one-quarter of the sum of a geometric series with first term equal to 1 and common ratio $-5x/4$, then

$$\frac{1}{4+5x} = \frac{1}{4}\left[1 + \left(-\frac{5x}{4}\right) + \left(-\frac{5x}{4}\right)^2 + \cdots\right], \qquad \left|-\frac{5x}{4}\right| < 1,$$

$$= \sum_{n=0}^{\infty} \frac{(-1)^n 5^n}{4^{n+1}} x^n, \qquad |x| < \frac{4}{5}.$$

(b) By a similar procedure, we have

$$\frac{1}{13-2x} = \frac{1}{3-2(x-5)}$$

$$= \frac{1}{3\left[1 - \frac{2}{3}(x-5)\right]}$$

$$= \frac{1}{3}\left[1 + \frac{2}{3}(x-5) + \left(\frac{2}{3}\right)^2 (x-5)^2 + \cdots\right], \qquad \left|\frac{2}{3}(x-5)\right| < 1,$$

$$= \sum_{n=0}^{\infty} \frac{2^n}{3^{n+1}} (x-5)^n, \qquad |x-5| < \frac{3}{2}.$$

■

In both examples, properties of geometric series gave not only the required series, but also their intervals of convergence. To appreciate the simplicity of these solutions, we suggest using Taylor remainders in an attempt to obtain the series with the same intervals of convergence. You will quickly abort.

Addition and Subtraction of Power Series

In Section 11.3 we noted that convergent series of constants can be added or subtracted to give convergent series. It follows, therefore, that if $f(x) = \sum a_n(x-c)^n$ and $g(x) = \sum b_n(x-c)^n$, then

$$f(x) \pm g(x) = \sum_{n=0}^{\infty} (a_n \pm b_n)(x-c)^n, \qquad (11.44)$$

and these results are valid for every x that is common to the intervals of convergence of the two series. We use this result in the following example.

EXAMPLE 11.28 Find the Maclaurin series for $f(x) = 5x/(x^2 - 3x - 4)$.

SOLUTION We decompose $f(x)$ into its partial fractions,

$$f(x) = \frac{5x}{x^2 - 3x - 4} = \frac{4}{x-4} + \frac{1}{x+1},$$

and expand each of these terms in a Maclaurin series,

$$\frac{4}{x-4} = \frac{-1}{1-\dfrac{x}{4}} = -\left(1 + \frac{x}{4} + \frac{x^2}{4^2} + \cdots\right), \qquad |x| < 4, \qquad \text{and}$$

$$\frac{1}{1+x} = 1 - x + x^2 - x^3 + \cdots, \qquad |x| < 1.$$

Addition of these series within their common interval of convergence gives the Maclaurin series for $f(x)$:

$$\begin{aligned}
\frac{5x}{x^2 - 3x - 4} &= \left(-1 - \frac{x}{4} - \frac{x^2}{4^2} - \frac{x^3}{4^3} - \cdots\right) + \left(1 - x + x^2 - x^3 + \cdots\right) \\
&= \left(-1 - \frac{1}{4}\right)x + \left(1 - \frac{1}{4^2}\right)x^2 + \left(-1 - \frac{1}{4^3}\right)x^3 + \cdots \\
&= \sum_{n=1}^{\infty}\left[(-1)^n - \frac{1}{4^n}\right]x^n, \qquad |x| < 1.
\end{aligned}$$

■

Differentiation and Integration of Power Series

Perhaps the most useful technique for generating power series expansions is to differentiate or integrate known expansions according to Theorem 11.16.

EXAMPLE 11.29

Find Maclaurin series for the following functions:

(a) $\cos x$ (b) $\dfrac{1}{(1-x)^3}$

SOLUTION (a) Recall from Example 11.24 that

$$\sin x = x - \frac{x^3}{3!} + \frac{x^5}{5!} - \frac{x^7}{7!} + \cdots, \qquad -\infty < x < \infty.$$

Term-by-term differentiation gives

$$\begin{aligned}
\cos x &= 1 - \frac{x^2}{2!} + \frac{x^4}{4!} - \frac{x^6}{6!} + \cdots \\
&= \sum_{n=0}^{\infty}\frac{(-1)^n}{(2n)!}x^{2n}, \qquad -\infty < x < \infty.
\end{aligned}$$

(b) Term-by-term differentiation of the series

$$\frac{1}{1-x} = 1 + x + x^2 + x^3 + \cdots + x^n + \cdots, \qquad |x| < 1$$

gives

$$\frac{1}{(1-x)^2} = 1 + 2x + 3x^2 + \cdots + nx^{n-1} + \cdots, \qquad |x| < 1.$$

Another differentiation yields

$$\frac{2}{(1-x)^3} = 2 + 3 \cdot 2x + 4 \cdot 3x^2 + \cdots + n(n-1)x^{n-2} + \cdots, \qquad |x| < 1.$$

Division by 2 now gives

$$\begin{aligned}\frac{1}{(1-x)^3} &= 1 + 3x + 6x^2 + \cdots + \frac{n(n-1)}{2}x^{n-2} + \cdots \\ &= \sum_{n=2}^{\infty} \frac{n(n-1)}{2}x^{n-2} \\ &= \frac{1}{2}\sum_{n=0}^{\infty}(n+2)(n+1)x^n, \qquad |x| < 1.\end{aligned}$$

■

EXAMPLE 11.30

Find the Taylor series about 1 for $\ln x$.

SOLUTION Noting that $\ln x$ is an antiderivative of $1/x$, we first expand $1/x$ in a Taylor series about 1:

$$\frac{1}{x} = \frac{1}{(x-1)+1} = 1 - (x-1) + (x-1)^2 - (x-1)^3 + \cdots, \qquad |x-1| < 1.$$

If we integrate this series term-by-term, we have

$$\ln|x| = \left[x - \frac{1}{2}(x-1)^2 + \frac{1}{3}(x-1)^3 - \frac{1}{4}(x-1)^4 + \cdots\right] + C.$$

Substitution of $x = 1$ implies that $0 = 1 + C$; that is, $C = -1$, and hence,

$$\begin{aligned}\ln|x| &= (x-1) - \frac{1}{2}(x-1)^2 + \frac{1}{3}(x-1)^3 - \cdots \\ &= \sum_{n=1}^{\infty} \frac{(-1)^{n+1}}{n}(x-1)^n.\end{aligned}$$

According to Theorem 11.16, the radius of convergence of this series is also $R = 1$. So the series certainly converges to $\ln|x|$ for $0 < x < 2$. But when $x = 2$, the series reduces to the alternating harmonic series, which converges conditionally. Consequently, we can extend the interval of convergence to include $x = 2$ and write

$$\ln x = \sum_{n=1}^{\infty} \frac{(-1)^{n+1}}{n}(x-1)^n, \qquad 0 < x \leq 2.$$

Convergence of this series at $x = 2$ does not, by itself, imply convergence to $\ln 2$, as this equation suggests. It is, however, true, and this is a direct application of the following theorem.

■

Theorem 11.18

> *If the Taylor series $\sum_{n=0}^{\infty} a_n(x-c)^n$ of a function $f(x)$ converges at the endpoint $x = c + R$ of its interval of convergence, and if $f(x)$ is continuous at $x = c + R$, then the Taylor series evaluated at $c + R$ converges to $f(c + R)$. The same result is valid at the other endpoint $x = c - R$.*

Comparison of the solutions in Examples 11.26 and 11.30 indicates once again the advantage of avoiding the use of Taylor's remainder formula.

Multiplication of Power Series

In Example 11.28 we added the Maclaurin series for $4/(x-4)$ and $1/(1+x)$ to obtain the Maclaurin series for $5x/(x^2-3x-4)$. An alternative procedure might be to multiply the two series since

$$\begin{aligned}\frac{5x}{x^2-3x-4} &= 5x\left(\frac{1}{x-4}\right)\left(\frac{1}{x+1}\right)\\ &= \frac{5x}{-4}\left(1+\frac{x}{4}+\frac{x^2}{4^2}+\frac{x^3}{4^3}+\cdots\right)\left(1-x+x^2-x^3+\cdots\right).\end{aligned}$$

The rules of algebra would demand that we multiply every term of the first series by every term of the second. If we do this and group all products with like powers of x, we obtain

$$\begin{aligned}\frac{5x}{x^2-3x-4} = \frac{5x}{-4}\bigg[1 &+ \left(-1+\frac{1}{4}\right)x + \left(1-\frac{1}{4}+\frac{1}{4^2}\right)x^2\\ &+ \left(-1+\frac{1}{4}-\frac{1}{4^2}+\frac{1}{4^3}\right)x^3+\cdots\bigg].\end{aligned}$$

It is clear that the coefficient of x^n is a finite geometric series to which we can apply formula 11.13a:

$$\begin{aligned}(-1)^n\left[1-\frac{1}{4}+\frac{1}{4^2}-\cdots+\frac{(-1)^n}{4^n}\right] &= (-1)^n\left[\frac{1-\left(-\frac{1}{4}\right)^{n+1}}{1+\frac{1}{4}}\right]\\ &= (-1)^n\frac{4}{5}\left[1-\left(-\frac{1}{4}\right)^{n+1}\right].\end{aligned}$$

Consequently,

$$\begin{aligned}\frac{5x}{x^2-3x-4} &= \frac{5x}{-4}\sum_{n=0}^{\infty}(-1)^n\frac{4}{5}\left[1-\left(-\frac{1}{4}\right)^{n+1}\right]x^n\\ &= \sum_{n=0}^{\infty}(-1)^{n+1}\left[1-\frac{(-1)^{n+1}}{4^{n+1}}\right]x^{n+1}\\ &= \sum_{n=0}^{\infty}\left[(-1)^{n+1}-\frac{1}{4^{n+1}}\right]x^{n+1}\\ &= \sum_{n=1}^{\infty}\left[(-1)^n-\frac{1}{4^n}\right]x^n.\end{aligned}$$

For this example, then, multiplication as well as addition of power series leads to the Maclaurin series. Clearly, addition of power series is much simpler for this example, but we have at least demonstrated that power series can be multiplied together. That this is generally possible is stated in the following theorem.

Theorem 11.19

If $f(x) = \sum_{n=0}^{\infty} a_n(x-c)^n$ *and* $g(x) = \sum_{n=0}^{\infty} b_n(x-c)^n$, *then*

$$f(x)g(x) = \sum_{n=0}^{\infty} d_n(x-c)^n, \tag{11.45a}$$

where

$$d_n = \sum_{i=0}^{n} a_i b_{n-i} = a_0 b_n + a_1 b_{n-1} + \cdots + a_{n-1} b_1 + a_n b_0, \tag{11.45b}$$

and this series converges absolutely at every point at which the series for $f(x)$ *and* $g(x)$ *converge absolutely.*

EXAMPLE 11.31

Find the Maclaurin series for $f(x) = [1/(x-1)]\ln(1-x)$.

SOLUTION If we integrate the Maclaurin series

$$\frac{1}{1-x} = 1 + x + x^2 + x^3 + \cdots, \qquad |x| < 1,$$

we find

$$-\ln|1-x| = \left(x + \frac{x^2}{2} + \frac{x^3}{3} + \frac{x^4}{4} + \cdots\right) + C.$$

By setting $x = 0$, we obtain $C = 0$, and

$$\ln(1-x) = -x - \frac{x^2}{2} - \frac{x^3}{3} - \frac{x^4}{4} - \cdots.$$

Note that we have dropped the absolute value signs since the radius of convergence of the series is 1. We now form the Maclaurin series for $f(x)$:

$$\begin{aligned}\frac{1}{x-1}\ln(1-x) &= \frac{-1}{1-x}\ln(1-x)\\ &= \left(1 + x + x^2 + x^3 + \cdots\right)\left(x + \frac{x^2}{2} + \frac{x^3}{3} + \cdots\right)\\ &= x + \left(1 + \frac{1}{2}\right)x^2 + \left(1 + \frac{1}{2} + \frac{1}{3}\right)x^3 + \cdots\\ &= \sum_{n=1}^{\infty}\left(1 + \frac{1}{2} + \frac{1}{3} + \cdots + \frac{1}{n}\right)x^n.\end{aligned}$$

Since both of the multiplied series converge absolutely on the interval $|x| < 1$, the Maclaurin series for $(x-1)^{-1}\ln(1-x)$ is also valid on this interval. ■

EXAMPLE 11.32

Find the first three nonzero terms in the Maclaurin series for $\tan x$.

SOLUTION If $\tan x = \sum a_n x^n$, then by setting $\tan x = \sin x / \cos x$, we have

$$\sin x = \cos x \sum_{n=0}^{\infty} a_n x^n.$$

We now substitute the Maclaurin series for $\sin x$ and $\cos x$:

$$x - \frac{x^3}{3!} + \frac{x^5}{5!} - \frac{x^7}{7!} + \cdots = \left(1 - \frac{x^2}{2!} + \frac{x^4}{4!} - \frac{x^6}{6!} + \cdots\right)\left(a_0 + a_1 x + a_2 x^2 + \cdots\right).$$

According to the corollary to Theorem 11.17, two power series can be identical only if their corresponding coefficients are equal. We therefore multiply the right-hand side and equate coefficients of like powers of x:

$$\begin{aligned}
x^0 &: \quad 0 = a_0; \\
x &: \quad 1 = a_1; \\
x^2 &: \quad 0 = a_2 - \frac{a_0}{2!}, \quad \text{which implies } a_2 = 0; \\
x^3 &: \quad -\frac{1}{3!} = a_3 - \frac{a_1}{2!}, \quad \text{which implies } a_3 = \frac{1}{2!} - \frac{1}{3!} = \frac{1}{3}; \\
x^4 &: \quad 0 = a_4 - \frac{a_2}{2!} + \frac{a_0}{4!}, \quad \text{from which } a_4 = 0; \\
x^5 &: \quad \frac{1}{5!} = a_5 - \frac{a_3}{2!} + \frac{a_1}{4!}, \quad \text{from which } a_5 = \frac{1}{5!} + \frac{1}{6} - \frac{1}{4!} = \frac{2}{15}.
\end{aligned}$$

The first three nonzero terms in the Maclaurin series for $\tan x$ are therefore

$$\tan x = x + \frac{1}{3}x^3 + \frac{2}{15}x^5 + \cdots.$$

■

Binomial Expansion

One of the most widely used power series is the binomial expansion. We are well acquainted with the binomial theorem, which predicts the product $(a + b)^m$ for any positive integer m:

$$(a + b)^m = \sum_{n=0}^{m} \binom{m}{n} a^n b^{m-n}. \tag{11.46}$$

With the usual definition of the binomial coefficients,

$$\binom{m}{n} = \frac{m!}{(m-n)!\,n!} = \frac{m(m-1)(m-2)\cdots(m-n+1)}{n!},$$

the binomial theorem becomes

$$(a + b)^m = a^m + ma^{m-1}b + \frac{m(m-1)}{2!}a^{m-2}b^2 + \cdots + mab^{m-1} + b^m. \tag{11.47}$$

Even when m is not a positive integer, this form for the binomial theorem remains almost intact. To show this, we consider the power series

$$1+\sum_{n=1}^{\infty}\frac{m(m-1)(m-2)\cdots(m-n+1)}{n!}x^n$$

for any real number m except a nonnegative integer. The radius of convergence of this power series is

$$\begin{aligned}R&=\lim_{n\to\infty}\left|\frac{m(m-1)(m-2)\cdots(m-n+1)}{n!}\frac{(n+1)!}{m(m-1)(m-2)\cdots(m-n)}\right|\\&=\lim_{n\to\infty}\left|\frac{n+1}{m-n}\right|=1.\end{aligned}$$

The power series therefore converges absolutely for $|x|<1$. Whether the series converges at the end points $x=\pm1$ depends on the value of m. For the time being, we will work on the interval $|x|<1$, and at the end of the discussion, we will state the complete result. Let us denote the sum of the series by

$$f(x)=1+\sum_{n=1}^{\infty}\frac{m(m-1)(m-2)\cdots(m-n+1)}{n!}x^n,\qquad |x|<1.$$

If we differentiate this series term-by-term according to Theorem 11.16,

$$f'(x)=\sum_{n=1}^{\infty}\frac{m(m-1)\cdots(m-n+1)}{(n-1)!}x^{n-1},\qquad |x|<1,$$

and then multiply both sides by x, we have

$$xf'(x)=\sum_{n=1}^{\infty}\frac{m(m-1)\cdots(m-n+1)}{(n-1)!}x^{n},\qquad |x|<1.$$

If we add these results, we obtain

$$f'(x)+xf'(x)=\sum_{n=1}^{\infty}\frac{m(m-1)\cdots(m-n+1)}{(n-1)!}x^{n-1}+\sum_{n=1}^{\infty}\frac{m(m-1)\cdots(m-n+1)}{(n-1)!}x^{n}.$$

We now change the variable of summation in the first sum,

$$(1+x)f'(x)=\sum_{n=0}^{\infty}\frac{m(m-1)\cdots(m-n)}{n!}x^{n}+\sum_{n=1}^{\infty}\frac{m(m-1)\cdots(m-n+1)}{(n-1)!}x^{n}.$$

When these summations are added over their common range, beginning at $n=1$, and the $n=0$ term in the first summation is written out separately, the result is

$$\begin{aligned}(1+x)f'(x)&=m+\sum_{n=1}^{\infty}\frac{m(m-1)\cdots(m-n+1)}{(n-1)!}\left(\frac{m-n}{n}+1\right)x^n\\&=m\left[1+\sum_{n=1}^{\infty}\frac{m(m-1)\cdots(m-n+1)}{n!}x^n\right]\\&=mf(x).\end{aligned}$$

Consequently, the function $f(x)$ must satisfy the differential equation

$$\frac{f'(x)}{f(x)} = \frac{m}{1+x}.$$

Integration immediately gives

$$\ln|f(x)| = m\ln|1+x| + C, \qquad \text{or}$$

$$f(x) = D(1+x)^m.$$

To evaluate D, we note that from the original definition of $f(x)$ as the sum of the power series, $f(0) = 1$, and this implies that $D = 1$. Thus,

$$f(x) = (1+x)^m,$$

and we may write finally that

$$(1+x)^m = 1 + \sum_{n=1}^{\infty} \frac{m(m-1)(m-2)\cdots(m-n+1)}{n!} x^n, \qquad |x| < 1 \quad (11.48\text{a})$$

$$= 1 + mx + \frac{m(m-1)}{2!}x^2 + \frac{m(m-1)(m-2)}{3!}x^3 + \cdots, \quad |x| < 1. \quad (11.48\text{b})$$

This is called the binomial expansion of $(1+x)^m$. We have verified the binomial expansion for m any real number except a nonnegative integer, but in the case of a nonnegative integer, the series terminates after $m+1$ terms and is therefore valid for these values of m also. We mentioned earlier that the binomial expansion may also converge at the end points $x = \pm 1$ depending on the value of m. The complete result states that 11.48 is valid for

$$\begin{array}{rl} -\infty < x < \infty & \text{if } m \text{ is a nonnegative integer,} \\ -1 < x < 1 & \text{if } m \le -1, \\ -1 < x \le 1 & \text{if } -1 < m < 0, \\ -1 \le x \le 1 & \text{if } m > 0 \text{ but not an integer.} \end{array}$$

It is not difficult to generalize this result to expand $(a+b)^m$ for real m. If $|b| < |a|$, we write

$$(a+b)^m = a^m\left(1 + \frac{b}{a}\right)^m$$

and now expand the bracketed term by means of 11.48:

$$(a+b)^m = a^m\left[1 + m\left(\frac{b}{a}\right) + \frac{m(m-1)}{2!}\left(\frac{b}{a}\right)^2 + \cdots\right] \qquad |b| < |a|,$$

$$= a^m + ma^{m-1}b + \frac{m(m-1)}{2!}a^{m-2}b^2 + \cdots \qquad |b| < |a|, \qquad (11.49)$$

which, as we predicted, is equation 11.47 except that the series does not terminate.

EXAMPLE 11.33

Use the binomial expansion to find the Maclaurin series for $1/(1-x)^3$.

SOLUTION By 11.48b, we have

$$\begin{aligned}\frac{1}{(1-x)^3} &= (1-x)^{-3} \\ &= 1 + (-3)(-x) + \frac{(-3)(-4)}{2!}(-x)^2 + \frac{(-3)(-4)(-5)}{3!}(-x)^3 + \cdots \\ &= 1 + 3x + \frac{3\cdot 4}{2}x^2 + \frac{4\cdot 5}{2}x^3 + \frac{5\cdot 6}{2}x^4 + \cdots \\ &= \frac{1}{2}\sum_{n=0}^{\infty}(n+1)(n+2)x^n, \qquad |x| < 1.\end{aligned}$$

■

This result was also obtained in Example 11.29 by differentiation of the Maclaurin series for $1/(1-x)$.

EXAMPLE 11.34

Find the Maclaurin series for $\mathrm{Sin}^{-1}\, x$.

SOLUTION By the binomial expansion, we have

$$\begin{aligned}\frac{1}{\sqrt{1-x^2}} &= 1 + \left(-\frac{1}{2}\right)(-x^2) + \frac{\left(-\frac{1}{2}\right)\left(-\frac{3}{2}\right)}{2!}(-x^2)^2 + \frac{\left(-\frac{1}{2}\right)\left(-\frac{3}{2}\right)\left(-\frac{5}{2}\right)}{3!}(-x^2)^3 + \cdots \\ &= 1 + \frac{1}{2}x^2 + \frac{3}{2^2 2!}x^4 + \frac{3\cdot 5}{2^3 3!}x^6 + \frac{3\cdot 5\cdot 7}{2^4 4!}x^8 + \cdots, \qquad |x| < 1.\end{aligned}$$

Integration of this series gives

$$\mathrm{Sin}^{-1}\, x = \left(x + \frac{1}{2\cdot 3}x^3 + \frac{3}{2^2 2!\, 5}x^5 + \frac{3\cdot 5}{2^3 3!\, 7}x^7 + \frac{3\cdot 5\cdot 7}{2^4 4!\, 9}x^9 + \cdots\right) + C.$$

Evaluation of both sides of this equation at $x = 0$ gives us $C = 0$. According to Theorem 11.16, the radius of convergence of this series must be 1, and we can write

$$\begin{aligned}\mathrm{Sin}^{-1}\, x &= x + \sum_{n=1}^{\infty}\frac{1\cdot 3\cdot 5\cdot \cdots \cdot (2n-1)}{2^n n!\,(2n+1)}x^{2n+1} \\ &= x + \sum_{n=1}^{\infty}\frac{1\cdot 2\cdot 3\cdot 4\cdot 5\cdot \cdots \cdot (2n-2)(2n-1)(2n)}{2\cdot 4\cdot \cdots \cdot (2n)2^n n!\,(2n+1)}x^{2n+1} \\ &= x + \sum_{n=1}^{\infty}\frac{(2n)!}{(2n+1)2^{2n}(n!)^2}x^{2n+1}, \qquad |x| < 1.\end{aligned}$$

In this case, integration of the series for $(1-x^2)^{-1/2}$ does not pick up the end points $x = \pm 1$, although this is difficult to prove.

■

EXERCISES 11.10

In Exercises 1–20 find the power series expansion of the function about the indicated point.

1. $f(x) = \dfrac{1}{3x+2}$ about $x = 0$

2. $f(x) = \dfrac{1}{4+x^2}$ about $x = 0$

3. $f(x) = \dfrac{1}{x+3}$ about $x = 2$

4. $f(x) = \cos(x^2)$ about $x = 0$

5. $f(x) = \dfrac{1}{\sqrt{1+x}}$ about $x = 0$

6. $f(x) = e^{5x}$ about $x = 0$

7. $f(x) = \cosh x$ about $x = 0$

8. $f(x) = \sinh x$ about $x = 0$

9. $f(x) = \ln(1+2x)$ about $x = 0$

10. $f(x) = (1+3x)^{3/2}$ about $x = 0$

11. $f(x) = 1/x$ about $x = 4$

12. $f(x) = x^4 + 3x^2 - 2x + 1$ about $x = 0$

13. $f(x) = \dfrac{1}{(x+2)^3}$ about $x = 0$

14. $f(x) = x^4 + 3x^2 - 2x + 1$ about $x = -2$

15. $f(x) = \dfrac{1}{x^2 + 8x + 15}$ about $x = 0$

16. $f(x) = e^x$ about $x = 3$

17. $f(x) = \operatorname{Tan}^{-1} x$ about $x = 0$

18. $f(x) = \sqrt{x+3}$ about $x = 0$

19. $f(x) = \dfrac{x^2}{(1+x^2)^2}$ about $x = 0$

20. $f(x) = x(1-x)^{1/3}$ about $x = 0$

In Exercises 21–23 find the first four nonzero terms in the Maclaurin series for the function.

21. $f(x) = \tan 2x$

22. $f(x) = \sec x$

23. $f(x) = e^x \sin x$

24. Find the Maclaurin series for $\cos^2 x$.

In Exercises 25–28 find the Maclaurin series for the function.

25. $f(x) = \dfrac{1}{x^6 - 3x^3 - 4}$

26. $f(x) = \operatorname{Sin}^{-1}(x^2)$

27. $f(x) = \dfrac{2x^2 + 4}{x^2 + 4x + 3}$

28. $f(x) = \ln\left[\dfrac{1 + x/\sqrt{2}}{1 - x/\sqrt{2}}\right]$

29. If during a working day, one person drinks from a fountain every 30 seconds (on the average), then the probability that exactly n people drink in a time interval of length t seconds is given by the *Poisson distribution*:

$$P_n(t) = \frac{1}{n!}\left(\frac{t}{30}\right)^n e^{-t/30}.$$

Calculate $\sum_{n=0}^{\infty} P_n(t)$ and interpret the result.

30. A certain experiment is to be performed until it is successful. The probability that it will be successful in any given attempt is p ($0 < p < 1$), and therefore the probability that it will fail is $q = 1 - p$. The expected number of times that the experiment must be performed in order to be successful can be shown to be represented by the infinite series

$$\sum_{n=1}^{\infty} npq^{n-1} = \sum_{n=1}^{\infty} np(1-p)^{n-1}.$$

(a) What is the sum of this series?

(b) If p is the probability that a single die will come up 6, is the answer in (a) what you would expect?

31. Find the Maclaurin series for the *error function* $\operatorname{erf}(x)$ defined by

$$\operatorname{erf}(x) = \frac{2}{\sqrt{\pi}}\int_0^x e^{-t^2}\,dt.$$

32. Find Maclaurin series for the *Fresnel integrals* $C(x)$ and $S(x)$ defined by

$$C(x) = \int_0^x \cos(\pi t^2/2)\,dt, \quad S(x) = \int_0^x \sin(\pi t^2/2)\,dt.$$

Show that Bessel functions of the first kind (defined in Exercise 30 of Section 11.8) satisfy the properties in Exercises 33 and 34.

33. $2mJ_m(x) - xJ_{m-1}(x) = xJ_{m+1}(x)$

34. $J_{m-1}(x) - J_{m+1}(x) = 2J'_m(x)$

35. If the function $(1 - 2\mu x + x^2)^{-1/2}$ is expanded in a Maclaurin series in x,

$$\frac{1}{\sqrt{1-2\mu x+x^2}} = \sum_{n=0}^{\infty} P_n(\mu)x^n,$$

the coefficients $P_n(\mu)$ are called the *Legendre polynomials*. Find $P_0(\mu)$, $P_1(\mu)$, $P_2(\mu)$, and $P_3(\mu)$.

36. (a) If we define $f(x) = x/(e^x - 1)$ at $x = 0$ as $f(0) = 1$, it turns out that $f(x)$ has a Maclaurin series expansion with positive radius of convergence. When this expansion is expressed in the form

$$\frac{x}{e^x - 1} = 1 + B_1 x + \frac{B_2}{2!}x^2 + \frac{B_3}{3!}x^3 + \cdots,$$

the coefficients $B_1, B_2, B_3, \ldots$ are called the *Bernoulli numbers*. Write this equation in the form

$$x = (e^x - 1)\left(1 + B_1 x + \frac{B_2}{2!}x^2 + \cdots\right),$$

and substitute the Maclaurin series for e^x to find the first five Bernoulli numbers.

(b) Show that the odd Bernoulli numbers all vanish for $n \geq 3$.

37. By substituting power series expansions for $e^{xt/2}$ and $e^{-x/(2t)}$ in terms of powers of t and $1/t$ respectively, show that

$$e^{x(t-1/t)/2} = \sum_{n=0}^{\infty} J_n(x)t^n.$$

For a definition of $J_n(x)$, see Exercises 30 in Section 11.8.

SECTION 11.11

Sums of Power Series and Series of Constants

Theorem 11.16 provides an important technique for finding sums of power series. If an unknown series can be reduced to a known series by differentiations or integrations, then its sum can be obtained when these operations are reversed.

EXAMPLE 11.35

Find the sum of the series $\sum_{n=0}^{\infty}(n+1)x^n$.

SOLUTION The series converges absolutely for $|x| < 1$. If we denote the sum of the series by $S(x)$,

$$S(x) = \sum_{n=0}^{\infty}(n+1)x^n,$$

and use Theorem 11.16 to integrate the series term-by-term,

$$\int S(x)\,dx = \sum_{n=0}^{\infty} x^{n+1} + C.$$

But the series on the right is a geometric series with sum $x/(1-x)$, provided $|x| < 1$, and we may write therefore

$$\int S(x)\,dx = \frac{x}{1-x} + C, \qquad |x| < 1.$$

If we now differentiate this equation, we obtain

$$S(x) = \frac{(1-x)(1) - x(-1)}{(1-x)^2} = \frac{1}{(1-x)^2},$$

and therefore

$$\sum_{n=0}^{\infty}(n+1)x^n = \frac{1}{(1-x)^2}, \qquad |x| < 1.$$

■

EXAMPLE 11.36

Find the sum of the series $\sum_{n=1}^{\infty} x^n/n$.

SOLUTION The series converges for $-1 \leq x < 1$. If we set

$$S(x) = \sum_{n=1}^{\infty} \frac{1}{n} x^n,$$

and differentiate the series term-by-term,

$$S'(x) = \sum_{n=1}^{\infty} x^{n-1}.$$

This is a geometric series with sum $1/(1-x)$, so that

$$S'(x) = \frac{1}{1-x}.$$

Antidifferentiation now gives

$$S(x) = -\ln(1-x) + C.$$

Since $S(0) = 0$, it follows that

$$0 = -\ln(1) + C.$$

Hence, $C = 0$, and $S(x) = -\ln(1-x)$. We have shown therefore that

$$\sum_{n=1}^{\infty} \frac{1}{n} x^n = -\ln(1-x), \qquad -1 \leq x < 1.$$

■

By substituting values of x into known power series we obtain formulas for sums of series of constants. For instance, in Example 11.25, we verified that the Maclaurin series for e^x is

$$e^x = \sum_{n=0}^{\infty} \frac{1}{n!} x^n, \qquad -\infty < x < \infty.$$

By substituting $x = 1$, we obtain a series which converges to e,

$$e = \sum_{n=0}^{\infty} \frac{1}{n!} = 1 + \frac{1}{1!} + \frac{1}{2!} + \frac{1}{3!} + \cdots .$$

This was the series that we truncated in Example 11.16.

The following example is another illustration of this idea.

EXAMPLE 11.37

Find the sum of the series

$$\sum_{n=0}^{\infty} \frac{(-1)^n}{(2n+1)2^n}.$$

SOLUTION There are many power series which reduce to the given series upon substitution of a specific value of x. For instance, substitution of $-1/2$, 1, and $1/\sqrt{2}$ into the following power series, respectively, lead to the given series:

$$\sum_{n=0}^{\infty} \frac{1}{2n+1} x^n, \qquad \sum_{n=0}^{\infty} \frac{(-1)^n}{(2n+1)2^n} x^n, \qquad \sum_{n=0}^{\infty} \frac{\sqrt{2}(-1)^n}{2n+1} x^{2n+1}$$

Which should we consider? Although it is not the simplest, the third series looks most promising; the fact that the power on x corresponds to the coefficient in the denominator suggests that we can find the sum of this series. We therefore set

$$S(x) = \sum_{n=0}^{\infty} \frac{\sqrt{2}(-1)^n}{2n+1} x^{2n+1},$$

and this series converges for $-1 \le x \le 1$. If we differentiate the series with respect to x, then

$$S'(x) = \sum_{n=0}^{\infty} \sqrt{2}(-1)^n x^{2n} = \sum_{n=0}^{\infty} \sqrt{2}(-x^2)^n = \frac{\sqrt{2}}{1-(-x^2)} = \frac{\sqrt{2}}{1+x^2}.$$

Antidifferentiation now gives

$$S(x) = \int \frac{\sqrt{2}}{1+x^2}\, dx = \sqrt{2}\,\mathrm{Tan}^{-1}\, x + C.$$

Since $S(0) = 0$, it follows that $C = 0$, and

$$\sum_{n=0}^{\infty} \frac{\sqrt{2}(-1)^n}{2n+1} x^{2n+1} = \sqrt{2}\,\mathrm{Tan}^{-1}\, x, \qquad |x| \le 1.$$

If we now set $x = 1/\sqrt{2}$,

$$\sum_{n=0}^{\infty} \frac{\sqrt{2}(-1)^n}{2n+1} \left(\frac{1}{\sqrt{2}}\right)^{2n+1} = \sqrt{2}\,\mathrm{Tan}^{-1}\left(\frac{1}{\sqrt{2}}\right).$$

Consequently,

$$\sum_{n=0}^{\infty} \frac{(-1)^n}{(2n+1)2^n} = \sqrt{2}\,\mathrm{Tan}^{-1}\left(\frac{1}{\sqrt{2}}\right).$$

■

EXERCISES 11.11

In Exercises 1–10 find the sum of the power series.

1. $\sum_{n=1}^{\infty} nx^{n-1}$

2. $\sum_{n=2}^{\infty} n(n-1)x^{n-2}$

3. $\sum_{n=1}^{\infty} (n+1)x^{n-1}$

4. $\sum_{n=1}^{\infty} n^2 x^{n-1}$

5. $\sum_{n=1}^{\infty} (n^2+2n)x^n$

6. $\sum_{n=0}^{\infty} \frac{1}{n+1} x^n$

7. $\sum_{n=0}^{\infty} \frac{(-1)^n}{2n+1} x^{2n+1}$

8. $\sum_{n=1}^{\infty} \frac{(-1)^n}{n} x^{2n}$

9. $\sum_{n=2}^{\infty} n3^n x^{2n}$

10. $\sum_{n=0}^{\infty} \left(\frac{n+1}{n+2}\right) x^n$

In Exercises 11–20 verify that the sum of the series is as indicated.

11. $\sum_{n=0}^{\infty} \frac{2^n}{n!} = e^2$

12. $\sum_{n=0}^{\infty} \frac{(-1)^n}{(2n+1)!} = \sin 1$

13. $\sum_{n=0}^{\infty} \frac{(-1)^n 3^{2n}}{(2n)!} = \cos 3$

14. $\sum_{n=1}^{\infty} \frac{(-1)^n}{n!} = \frac{1}{e} - 1$

15. $\displaystyle\sum_{n=1}^{\infty} \frac{(-1)^n}{3^{2n}(2n+1)!} = 3\sin\left(\tfrac{1}{3}\right) - 1$

16. $\displaystyle\sum_{n=2}^{\infty} \frac{(-1)^{n+1}2^{2n+3}}{(2n)!} = -8(1+\cos 2)$

17. $\displaystyle\sum_{n=1}^{\infty} \frac{2^n}{n3^n} = \ln 3$

18. $\displaystyle\sum_{n=1}^{\infty} \frac{1}{n2^n} = \ln 2$

19. $\displaystyle\sum_{n=1}^{\infty} \frac{(-1)^n}{2^{2n}} = -\frac{1}{5}$

20. $\displaystyle\sum_{n=1}^{\infty} \frac{n}{2^n} = 2$

21. Find the Maclaurin series for $\mathrm{Tan}^{-1}\,x$ and use it to evaluate

$$\sum_{n=1}^{\infty} \frac{(-1)^n}{2n+1}.$$

22. Find the Maclaurin series for $x/(1+x^2)^2$ and use it to evaluate

$$\sum_{n=1}^{\infty} \frac{n(-1)^n}{3^{2n}}.$$

SECTION 11.12

Applications of Taylor Series and Taylor's Remainder Formula

If Taylor's remainder $R_n(c,x)$ is truncated from remainder formula 11.32e, a polynomial approximation to $f(x)$ is obtained

$$f(x) \approx f(c) + f'(c)(x-c) + \frac{f''(c)}{2!}(x-c)^2 + \cdots + \frac{f^{(n-1)}(c)}{(n-1)!}(x-c)^{n-1}.$$

The remainder

$$R_n(c,x) = \frac{f^{(n)}(z_n)}{n!}(x-c)^n,$$

where z_n is between c and x represents the error of the approximation; the smaller R_n, the better the approximation. In this section we use Taylor series and Taylor's remainder formula in a number of situations that require approximations.

Approximations of Functions

Consider using the first three terms of the Maclaurin series for e^x to approximate e^x on the interval $0 \le x \le 1/2$. Taylor's remainder formula, with $c = 0$, states that

$$e^x = 1 + x + \frac{x^2}{2} + R_3$$

where

$$R_3 = \frac{d^3}{dx^3}(e^x)_{|x=z}\frac{x^3}{3!} = \frac{e^z x^3}{6},$$

and z is between 0 and x. Although z is unknown — except that it is between 0 and x, we can say that because only the values $0 \le x \le \frac{1}{2}$ are under consideration, z must be less than $\frac{1}{2}$. It follows that

$$R_3 < \frac{e^{1/2}x^3}{6} \le \frac{\sqrt{e}(1/2)^3}{6} < 0.035.$$

Thus the quadratic function $1 + x + x^2/2$ approximates e^x on the interval $0 \le x \le \frac{1}{2}$ with an error no greater than 0.035.

In the following example, we determine the number of terms of a Maclaurin series required to guarantee a certain accuracy.

EXAMPLE 11.38

How many terms in the Maclaurin series of $\ln(1+x)$ guarantee a truncation error of less than 10^{-6} for any x in the interval $0 \le x \le 1/2$?

SOLUTION The nth derivative of $\ln(1+x)$ is

$$\frac{d^n}{dx^n}\ln(1+x) = \frac{(-1)^{n+1}(n-1)!}{(x+1)^n}, \qquad n \ge 1,$$

and therefore Taylor's remainder formula with $c = 0$ states that

$$\ln(1+x) = x - \frac{x^2}{2} + \frac{x^3}{3} - \frac{x^4}{4} + \cdots + \frac{(-1)^n}{n-1}x^{n-1} + R_n.$$

Now that we see terms in the Maclaurin series, for $0 \le x \le \frac{1}{2}$, the series is alternating. We could therefore discuss accuracy from an alternating series point of view and avoid Taylor remainders totally. Recall from Section 11.7 that when an alternating series is truncated, the maximum possible error is the next term. Consequently, if the Maclaurin series for $\ln(1+x)$ is truncated after the term $(-1)^n x^{n-1}/(n-1)$, the maximum error is

$$\frac{(-1)^{n+1}}{n}x^n.$$

Considering values of x in the interval $0 \le x \le \frac{1}{2}$, this error is a maximum at $x = \frac{1}{2}$. Thus, the error in truncating the series after $(-1)^n x^{n-1}/(n-1)$ is no greater in absolute value than

$$\frac{(1/2)^n}{n}.$$

This is less than 10^{-6} if

$$\frac{(1/2)^n}{n} < 10^{-6} \qquad \text{or}$$

$$n2^n > 10^6.$$

A calculator quickly indicates that the smallest value of n for which this is true is $n = 16$. Consequently, if $\ln(1+x)$ is approximated by the 15th degree polynomial

$$\ln(1+x) \approx x - \frac{x^2}{2} + \frac{x^3}{3} - \frac{x^4}{4} + \cdots + \frac{x^{15}}{15}$$

on the interval $0 \le x \le 1/2$, the truncation error is less than 10^{-6}.

It is interesting to note that Taylor remainders do no better. To see this we calculate that

$$R_n = \frac{d^n}{dx^n}\ln(1+x)_{|x=z}\frac{x^n}{n!} = \frac{(-1)^{n+1}(n-1)!}{(z+1)^n}\frac{x^n}{n!} = \frac{(-1)^{n+1}}{n(z+1)^n}x^n.$$

Since z is between 0 and x and $0 \le x \le \frac{1}{2}$, we can state that x must be less than or equal to $\frac{1}{2}$, and z must be greater than 0. Hence,

$$|R_n| < \frac{1}{n(1)^n}(1/2)^n = \frac{1}{n2^n},$$

and this is the same error expression previously obtained when analysis was performed from an alternating series point of view. ■

Evaluation of Definite Integrals

In Section 9.9 we developed three numerical techniques for approximating definite integrals of functions $f(x)$ that have no obvious antiderivatives: the rectangular rule, the trapezoidal rule, and Simpson's rule. Each method divides the interval of integration into a number of subintervals and approximates $f(x)$ by a more elementary function on each subinterval. The rectangular rule replaces $f(x)$ by a step function, the trapezoidal rule replaces $f(x)$ by a succession of linear functions, and Simpson's rule uses a sequence of quadratic functions.

Another possibility is to replace $f(x)$ by a truncated power series (a polynomial) over the entire interval of integration. For instance, consider the definite integral

$$\int_0^{1/2} \frac{\sin x}{x}\, dx,$$

where $(\sin x)/x$ is defined as 1 at $x = 0$. Suppose we expand the integrand in a Maclaurin series:

$$\frac{1}{x}\sin x = \frac{1}{x}\left(x - \frac{x^3}{3!} + \frac{x^5}{5!} - \cdots\right) = 1 - \frac{x^2}{3!} + \frac{x^4}{5!} - \cdots, \qquad -\infty < x < \infty.$$

According to Theorem 11.16,

$$\int \frac{\sin x}{x}\, dx = \left(x - \frac{x^3}{3 \cdot 3!} + \frac{x^5}{5 \cdot 5!} - \cdots\right) + C, \qquad -\infty < x < \infty,$$

and therefore

$$\int_0^{1/2} \frac{\sin x}{x}\, dx = \frac{1}{2} - \frac{\left(\frac{1}{2}\right)^3}{3 \cdot 3!} + \frac{\left(\frac{1}{2}\right)^5}{5 \cdot 5!} - \cdots .$$

If we approximate this alternating series with its first three terms, the truncation error is the fourth term. Since

$$\frac{1}{2} - \frac{\left(\frac{1}{2}\right)^3}{3 \cdot 3!} + \frac{\left(\frac{1}{2}\right)^5}{5 \cdot 5!} = 0.493\,107\,6 \qquad \text{and}$$

$$\frac{\left(\frac{1}{2}\right)^7}{7 \cdot 7!} < 0.000\,000\,222,$$

it follows that

$$0.493\,107\,4 < \int_0^{1/2} \frac{\sin x}{x}\, dx < 0.493\,107\,6.$$

Consequently, using only three terms of the Maclaurin series for $(\sin x)/x$, we can say that to five decimals

$$\int_0^{1/2} \frac{\sin x}{x}\, dx = 0.49311.$$

EXAMPLE 11.39 A very important function in engineering and physics is the error function $\text{erf}(x)$ defined by

$$\text{erf}(x) = \frac{2}{\sqrt{\pi}} \int_0^x e^{-t^2}\, dt.$$

Calculate erf(1) correct to three decimals.

SOLUTION If we replace the integrand by its Maclaurin series, we have

$$\begin{aligned}
\frac{\sqrt{\pi}}{2}\text{erf}(1) &= \int_0^1 \left[1 - t^2 + \frac{(-t^2)^2}{2!} + \frac{(-t^2)^3}{3!} + \cdots\right] dt \\
&= \left\{t - \frac{t^3}{3} + \frac{t^5}{5 \cdot 2!} - \frac{t^7}{7 \cdot 3!} + \cdots\right\}_0^1 \\
&= 1 - \frac{1}{3} + \frac{1}{5 \cdot 2!} - \frac{1}{7 \cdot 3!} + \frac{1}{9 \cdot 4!} - \cdots .
\end{aligned}$$

Since

$$\frac{2}{\sqrt{\pi}}\left(1 - \frac{1}{3} + \frac{1}{5 \cdot 2!} - \frac{1}{7 \cdot 3!} + \frac{1}{9 \cdot 4!} - \frac{1}{11 \cdot 5!}\right) = 0.842\,593\,67$$

and the seventh nonzero term is

$$\frac{2}{\sqrt{\pi}} \frac{1}{13 \cdot 6!} < 0.000\,120\,56,$$

it follows that

$$0.842\,593\,67 < \text{erf}(1) < 0.842\,714\,23,$$

and to three decimals

$$\text{erf}(1) = 0.843.$$

■

Limits

We have customarily used L'Hôpital's rule to evaluate limits of the indeterminate form $0/0$. Maclaurin and Taylor series can sometimes be used to advantage. Consider

$$\lim_{x \to 0} \frac{x - \sin x}{x^3}.$$

Three applications of L'Hôpital's rule give a limit of $\frac{1}{6}$. Alternatively, if we substitute the Maclaurin series for $\sin x$,

$$\begin{aligned}
\lim_{x \to 0} \frac{x - \sin x}{x^3} &= \lim_{x \to 0} \frac{1}{x^3}\left[x - \left(x - \frac{x^3}{3!} + \frac{x^5}{5!} - \cdots\right)\right] \\
&= \lim_{x \to 0}\left[\frac{1}{6} - \frac{x^2}{5!} + \cdots\right] \\
&= \frac{1}{6}.
\end{aligned}$$

Here is another example.

EXAMPLE 11.40

Evaluate

$$\lim_{\lambda\to 0^+} \frac{\lambda^{-5}}{e^{c/\lambda} - 1}$$

where $c > 0$ is a constant (see also Exercise 58 in Section 4.9).

SOLUTION We begin by making the change of variable $v = 1/\lambda$ in the limit,

$$\lim_{\lambda\to 0^+} \frac{\lambda^{-5}}{e^{c/\lambda} - 1} = \lim_{v\to\infty} \frac{v^5}{e^{cv} - 1}.$$

We now expand e^{cv} into its Maclaurin series,

$$\begin{aligned}\lim_{\lambda\to 0^+} \frac{\lambda^{-5}}{e^{c/\lambda} - 1} &= \lim_{v\to\infty} \frac{v^5}{\left(1 + cv + \dfrac{c^2 v^2}{2!} + \cdots\right) - 1} \\ &= \lim_{v\to\infty} \frac{v^5}{cv + \dfrac{c^2 v^2}{2!} + \cdots}.\end{aligned}$$

If we now divide numerator and denominator by v^5,

$$\lim_{\lambda\to 0^+} \frac{\lambda^{-5}}{e^{c/\lambda} - 1} = \lim_{v\to\infty} \frac{1}{\dfrac{c}{v^4} + \dfrac{c^2}{2\,v^3} + \dfrac{c^3}{3!\,v^2} + \dfrac{c^4}{4!\,v} + \dfrac{c^5}{5!} + \dfrac{c^6\,v}{6!} + \cdots} = 0.$$

■

Differential Equations

Many differential equations arising in physics and engineering have solutions that can be expressed only in terms of infinite series. One such equation is Bessel's differential equation of order zero for a function $y = f(x)$:

$$xy'' + y' + xy = 0.$$

Before considering this rather difficult differential equation, we introduce the ideas through an easier example.

EXAMPLE 11.41

Determine whether the differential equation

$$\frac{dy}{dx} - 2y = x$$

has a solution which can be expressed as a power series $y = \sum_{n=0}^{\infty} a_n x^n$ with positive radius of convergence.

SOLUTION If

$$y = f(x) = \sum_{n=0}^{\infty} a_n x^n = a_0 + a_1 x + a_2 x^2 + \cdots$$

is to be a solution of the differential equation, we may substitute the power series into the differential equation,

$$\left[a_1 + 2a_2x + 3a_3x^2 + 4a_4x^3 + \cdots\right] - 2\left[a_0 + a_1x + a_2x^2 + \cdots\right] = x.$$

We now gather together like terms in the various powers of x,

$$0 = (a_1 - 2a_0) + (2a_2 - 2a_1 - 1)x + (3a_3 - 2a_2)x^2 + (4a_4 - 2a_3)x^3 + \cdots .$$

Since the power series on the right has sum zero, its coefficients must all vanish (see Exercise 8 in Section 11.9), and therefore we must set

$$\begin{aligned} a_1 - 2a_0 &= 0, \\ 2a_2 - 2a_1 - 1 &= 0, \\ 3a_3 - 2a_2 &= 0, \\ 4a_4 - 2a_3 &= 0, \end{aligned}$$

and so on. These equations imply that

$$\begin{aligned} a_1 &= 2a_0; \\ a_2 &= \frac{1}{2}(1 + 2a_1) = \frac{1}{2}(1 + 4a_0); \\ a_3 &= \frac{2}{3}a_2 = \frac{2}{3!}(1 + 4a_0); \\ a_4 &= \frac{2}{4}a_3 = \frac{2^2}{4!}(1 + 4a_0). \end{aligned}$$

The pattern emerging is

$$a_n = \frac{2^{n-2}}{n!}(1 + 4a_0), \qquad n \geq 2.$$

Thus,

$$\begin{aligned} f(x) &= a_0 + 2a_0x + \frac{1}{2}(1 + 4a_0)x^2 + \cdots + \frac{2^{n-2}}{n!}(1 + 4a_0)x^n + \cdots \\ &= a_0 + 2a_0x + \frac{1}{4}(1 + 4a_0)\left(\frac{2^2}{2!}x^2 + \frac{2^3}{3!}x^3 + \cdots + \frac{2^n}{n!}x^n + \cdots\right). \end{aligned}$$

We can find the sum of the series in parentheses by noting that the Maclaurin series for e^{2x} is

$$e^{2x} = 1 + (2x) + \frac{(2x)^2}{2!} + \frac{(2x)^3}{3!} + \cdots .$$

Therefore the solution of the differential equation is

$$\begin{aligned} y = f(x) &= a_0 + 2a_0x + \frac{1}{4}(1 + 4a_0)\left[e^{2x} - 1 - 2x\right] \\ &= -\frac{1}{4} - \frac{x}{2} + \frac{1}{4}(1 + 4a_0)e^{2x} \\ &= Ce^{2x} - \frac{1}{4} - \frac{x}{2}. \end{aligned}$$

■

Using power series to solve the differential equation in Example 11.41 is certainly not the most expedient method. You will learn better methods if and when you undertake an indepth study of differential equations. But the example clearly illustrated the procedure by which power series are used to solve differential equations. We now apply the procedure to Bessel's differential equation of order zero.

EXAMPLE 11.42

Find a power series solution $y = \sum_{n=0}^{\infty} a_n x^n$, with positive radius of convergence, for Bessel's differential equation of order zero,

$$xy'' + y' + xy = 0\,.$$

SOLUTION In this example we abandon the $\cdots$ notation used in Example 11.41, and maintain sigma notation throughout. When we substitute $y = \sum_{n=0}^{\infty} a_n x^n$ into the differential equation, we obtain

$$\begin{aligned} 0 &= x\sum_{n=2}^{\infty} n(n-1)a_n x^{n-2} + \sum_{n=1}^{\infty} na_n x^{n-1} + x\sum_{n=0}^{\infty} a_n x^n \\ &= \sum_{n=2}^{\infty} n(n-1)a_n x^{n-1} + \sum_{n=1}^{\infty} na_n x^{n-1} + \sum_{n=0}^{\infty} a_n x^{n+1}\,. \end{aligned}$$

In order to bring these three summations together as one, and combine terms in like powers of x, we lower the index of summation in the last term by 2,

$$0 = \sum_{n=2}^{\infty} n(n-1)a_n x^{n-1} + \sum_{n=1}^{\infty} na_n x^{n-1} + \sum_{n=2}^{\infty} a_{n-2}\, x^{n-1}\,.$$

We now combine the three summations over their common interval, beginning at $n = 2$, and write separately the $n = 1$ term in the second summation,

$$0 = a_1 + \sum_{n=2}^{\infty} [n(n-1)a_n + na_n + a_{n-2}]x^{n-1}\,.$$

But the only way a power series can be equal to zero is for all of its coefficients to be equal to zero; that is,

$$a_1 = 0; \qquad n(n-1)a_n + na_n + a_{n-2} = 0\,, \quad n \geq 2\,.$$

Thus,

$$a_n = -\frac{a_{n-2}}{n^2}\,, \qquad n \geq 2\,,$$

a recursive relation defining the unknown coefficient a_n of x^n in terms of the coefficient a_{n-2} of x^{n-2}. Since $a_1 = 0$, it follows that

$$0 = a_1 = a_3 = a_5 = \cdots\,.$$

For $n = 2$,

$$a_2 = -\frac{a_0}{2^2}\,.$$

For $n = 4$,

$$a_4 = -\frac{a_2}{4^2} = \frac{a_0}{2^2 4^2} = \frac{a_0}{2^4(2!)^2}.$$

For $n = 6$,

$$a_6 = -\frac{a_4}{6^2} = \frac{-a_0}{2^4(2!)^2 6^2} = -\frac{a_0}{2^6(3!)^2}.$$

The solution is therefore

$$y = a_0 - \frac{a_0}{2^2}x^2 + \frac{a_0}{2^4(2!)^2}x^4 - \frac{a_0}{2^6(3!)^2}x^6 + \cdots$$

$$= a_0 \sum_{n=0}^{\infty} \frac{(-1)^n}{2^{2n}(n!)^2} x^{2n}.$$

The function defined by the infinite series

$$J_0(x) = \sum_{n=0}^{\infty} \frac{(-1)^n}{2^{2n}(n!)^2} x^{2n}, \qquad -\infty < x < \infty$$

is called the zero-order Bessel function of the first kind. ■

EXERCISES 11.12

In Exercises 1–10 find a maximum possible error in using the given terms of the Taylor series to approximate the function on the interval specified.

1. $e^x \approx 1 + x + \frac{x^2}{2} + \frac{x^3}{6}$ for $0 \le x \le 0.01$

2. $e^x \approx 1 + x + \frac{x^2}{2} + \frac{x^3}{6}$ for $0 \le x < 0.01$

3. $e^x \approx 1 + x + \frac{x^2}{2} + \frac{x^3}{6}$ for $-0.01 \le x \le 0$

4. $e^x \approx 1 + x + \frac{x^2}{2} + \frac{x^3}{6}$ for $|x| \le 0.01$

5. $\sin x \approx x - \frac{x^3}{3!}$ for $0 \le x \le 1$

6. $\cos x \approx 1 - \frac{x^2}{2!} + \frac{x^4}{4!}$ for $|x| \le 0.1$

7. $\ln(1 - x) \approx -x - \frac{x^2}{2} - \frac{x^3}{3}$ $0 \le x \le 0.01$

8. $\frac{1}{(1 - x)^3} \approx 1 + 3x + 6x^2 + 10x^3$ for $|x| < 0.2$

9. $\mathrm{Tan}^{-1}\, x \approx x - \frac{x^3}{3} + \frac{x^5}{5} - \cdots - \frac{x^{99}}{99}$ for $|x| < 1$

10. $\ln x \approx (x - 1) - \frac{1}{2}(x - 1)^2 + \frac{1}{3}(x - 1)^3 - \frac{1}{4}(x - 1)^4$ for $|x - 1| \le 1/2$

In Exercises 11–18 evaluate the integral correct to three decimals.

11. $\int_0^1 \frac{\sin x}{x}\, dx$

12. $\int_0^{1/2} \cos(x^2)\, dx$

13. $\int_0^{2/3} \frac{1}{x^4 + 1}\, dx$

14. $\int_{-1}^{1} x^{11} \sin x\, dx$

15. $\int_0^{1/2} \frac{1}{\sqrt{1 + x^3}}\, dx$

16. $\int_0^{0.3} e^{-x^2}\, dx$

17. $\int_{-0.1}^{0} \frac{1}{x - 1} \ln(1 - x)\, dx$

18. $\int_0^{1/2} \frac{1}{x^6 - 3x^3 - 4}\, dx$

In Exercises 19–24 use series to evaluate the limit.

19. $\lim_{x \to 0} \frac{\tan x}{x}$

20. $\lim_{x \to 0} \frac{1 - \cos x}{x^2}$

21. $\lim_{x \to 0} \frac{(1 - \cos x)^2}{3x^4}$

22. $\lim_{x \to 0} \frac{\sqrt{1 + x} - 1}{x}$

23. $\lim_{x \to \infty} x \sin\left(\frac{1}{x}\right)$

24. $\lim_{x \to 0} \left(\frac{e^x + e^{-x}}{e^x - e^{-x}} - \frac{1}{x}\right)$

In Exercises 25–28 determine where the Maclaurin series for the function may be truncated in order to guarantee the accuracy indicated.

25. $\sin(x/3)$ on $|x| \le 4$ with error less than 10^{-3}

26. $1/\sqrt{1+x^3}$ on $0 < x < 1/2$ with error less than 10^{-4}

27. $\ln(1-x)$ on $|x| < 1/3$ with error less than 10^{-2}

28. $\cos^2 x$ on $|x| < 0.1$ with error less than 10^{-3}

In Exercises 29–34 find a series solution in powers of x for the differential equation.

29. $y' + 3y = 4$ **30.** $y'' + y' = 0$

31. $xy' - 4y = 3x$ **32.** $4xy'' + 2y' + y = 0$

33. $y'' + y = 0$ **34.** $xy'' + y = 0$

35. Find the natural logarithm of 0.999 999 999 9 accurate to 15 decimal places.

36. In special relativity theory, the kinetic energy K of an object moving with speed v is defined by

$$K = c^2(m - m_0),$$

where c is a constant (the speed of light), m_0 is the rest mass of the object, and m is its mass when moving with speed v. The masses m and m_0 are related by

$$m = \frac{m_0}{\sqrt{1 - v^2/c^2}}.$$

Use the binomial expansion to show that

$$K = \frac{1}{2}m_0 v^2 + m_0 c^2\left(\frac{3}{8}\frac{v^4}{c^4} + \frac{5}{16}\frac{v^6}{c^6} + \cdots\right),$$

and hence, to a first approximation, kinetic energy is defined by the classical expression $m_0 v^2/2$.

37. The ellipse $b^2x^2 + a^2y^2 = a^2b^2$ can be represented parametrically by

$$x = a\cos t, \quad y = b\sin t, \quad 0 \le t < 2\pi.$$

(a) Show that the length of the circumference of the ellipse is defined by the definite integral

$$L = 4b\int_0^{\pi/2}\sqrt{1 - k^2\sin^2 t}\,dt, \qquad k^2 = 1 - \frac{a^2}{b^2}.$$

(b) Use the binomial expansion to show that

$$L = 2\pi b\left(1 - \frac{k^2}{4} - \frac{3k^4}{64} - \cdots\right)$$

so that to a first approximation, L is the circumference of a circle of radius b.

38. In determining the radiated power from a half-wave antenna, it is necessary to evaluate

$$\int_0^{2\pi}\frac{1 - \cos\theta}{\theta}\,d\theta.$$

Find a two-decimal approximation for this integral.

39. Planck's law for the energy density Ψ of blackbody radiation of wavelength λ states that

$$\Psi(\lambda) = \frac{8\pi ch\lambda^{-5}}{e^{ch/(\lambda kT)} - 1},$$

where $h > 0$ is Planck's constant, c is the (constant) speed of light, and T is temperature, also assumed constant. Show that for long wavelengths, Planck's law reduces to the Rayleigh-Jeans law:

$$\Psi(\lambda) = \frac{8\pi kT}{\lambda^4}.$$

SUMMARY

An infinite sequence of constants is the assignment of numbers to positive integers. In most applications of sequences, the prime consideration is whether the sequence has a limit. If the sequence has its terms defined explicitly, then our ability to take limits of continuous functions ("limits at infinity" in Chapter 2 and L'Hôpital's rule in Chapter 4) can be very helpful. If the sequence is defined recursively, existence of the limit can sometimes be established by showing that the sequence is monotonic and bounded.

An expression of the form

$$\sum_{n=1}^{\infty} c_n = c_1 + c_2 + \cdots + c_n + \cdots$$

is called an infinite series. We define the sum of this series as the limit of its sequence of partial sums $\{S_n\}$, provided the sequence has a limit. Unfortunately, for most series we cannot find a simple formula for S_n, and therefore analysis of the limit of the sequence $\{S_n\}$ is impossible. To remedy this, we developed various convergence tests that avoided the sequence $\{S_n\}$: n^{th} term, comparison, limit comparison, limit ratio, limit root, integral, and alternating series tests. Note the sequences that are associated with a series $\sum c_n$:

$\{S_n\}$	sequence of partial sums for the definition of a sum;
$\{c_n\}$	sequence of terms for the nth term test;
$\{c_n/b_n\}$	sequence for limit comparison test;
$\{c_{n+1}/c_n\}$	sequence for the limit ratio test;
$\{\sqrt[n]{c_n}\}$	sequence for the limit root test;
$\{\lvert c_n\rvert\}$	sequence for the alternating series test.

Depending on the limits of these sequences — if they exist, we may be able to infer something about convergence of the series.

From infinite sequences and series of constants we proceeded to infinite sequences and series of functions — in particular, power series. We first considered situations where a power series was given, and the sum was to be determined. We saw that every power series $\sum a_n(x-c)^n$ has a radius of convergence R and an associated interval of convergence. If $R = 0$, the interval of convergence consists of only one point $x = c$; if $R = \infty$, the power series converges absolutely for all x; and if $0 < R < \infty$, the interval of convergence must be one of four possibilities: $c-R < x < c+R$, $c-R \leq x < c+R$, $c-R < x \leq c+R$, or $c-R \leq x \leq c+R$. The radius of convergence is given by $\lim_{n\to\infty} \lvert a_n/a_{n+1}\rvert$ or $\lim_{n\to\infty} \lvert a_n\rvert^{-1/n}$ provided the limits exist or are equal to infinity. If at each point in the interval of convergence of the power series the value of a function $f(x)$ is the same as the sum of the series, we write $f(x) = \sum a_n(x-c)^n$ and call $f(x)$ the sum of the series.

We also considered situations where a function $f(x)$ and a point c are given, and ask whether $f(x)$ has a power series expansion about c. We saw that there can be at most one power series expansion of $f(x)$ about c with a positive radius of convergence, and this series must be its Taylor series. One way to verify that $f(x)$ does indeed have a Taylor series about c and that this series converges to $f(x)$ is to show that the sequence of Taylor's remainders $\{R_n(c,x)\}$ exists and has limit zero. Often, however, it is much easier to find Taylor series by adding, multiplying, differentiating, and integrating known series.

When a Taylor series is truncated, Taylor's remainder $R_n(c,x)$ represents the truncation error and, in spite of the fact that R_n is expressed in terms of some unknown point z_n, it is often possible to calculate a maximum value for the error. Sometimes $R_n(c,x)$ can be avoided altogether. For instance, if the Taylor series is an alternating series, then the maximum possible truncation error is the value of the next term.

Power series are often used in situations that require approximations. Taylor series provide polynomial approximations to complicated functions, and they offer an alternative to the numerical techniques of Section 9.9 in the evaluation of definite integrals. Power series are also useful in situations that do not require approximations. They are sometimes helpful in evaluating limits, and they are the only way to solve many differential equations.

Key Terms and Formulas

In reviewing this chapter, you should be able to define or discuss the following key terms:

- Sequence of constants
- Explicit sequence
- Recursive sequence
- Limit of a sequence
- Convergent sequence
- Divergent sequence
- Increasing sequence
- Decreasing sequence
- Nonincreasing sequence
- Nondecreasing sequence
- Monotonic sequence
- Upper bound
- Lower bound
- Successive substitutions
- Series of constants
- Geometric series
- Harmonic series
- Sequence of partial sums
- Convergent series
- Divergent series
- n^{th} term test
- Nonnegative series
- Integral test
- p-series
- Comparison test
- Limit comparison test
- Limit ratio test
- Limit root test
- Absolutely convergent series
- Conditionally convergent series
- Alternating harmonic series
- Truncation error
- Power series
- Interval of convergence
- Radius of convergence
- Sum of a power series
- Taylor remainder formula
- Taylor series
- Maclaurin series
- Binomial expansion

REVIEW EXERCISES

In Exercises 1–4 discuss, with all necessary proofs, whether the sequence is monotonic and has an upper bound, a lower bound, and a limit.

1. $\left\{\dfrac{n^2-5n+3}{n^2+5n+4}\right\}$

2. $c_1 = 1, \quad c_{n+1} = (1/2)\sqrt{c_n^2+1}, \quad n \geq 1$

3. $\left\{\dfrac{\operatorname{Tan}^{-1}(1/n)}{n^2+1}\right\}$

4. $c_1 = 7, \quad c_{n+1} = 15 + \sqrt{c_n - 2}, \quad n \geq 1$

5. Use Newton's iterative procedure and the method of successive substitutions to approximate the root of the equation

$$x = \left(\frac{x+5}{x+4}\right)^2$$

between $x = 1$ and $x = 2$.

6. For what values of k does the sequence

$$c_1 = k, \qquad c_{n+1} = c_n^2, \quad n \geq 1$$

converge?

7. Find an explicit definition for the sequence

$$c_1 = 1, \qquad c_{n+1} = \sqrt{1+c_n^2}, \quad n \geq 1.$$

8. Use the derivative of the function $f(x) = (\ln x)/x$ to prove that the sequence $\{\ln n/n\}$ is decreasing for $n \geq 3$.

In Exercises 9–28 determine whether the series converges or diverges. In the case of a convergent series that has both positive and negative terms, indicate whether it converges absolutely or conditionally.

9. $\displaystyle\sum_{n=1}^{\infty} \frac{n^2-3n+2}{n^3+4n}$

10. $\displaystyle\sum_{n=1}^{\infty} \frac{n^2+5n+3}{n^4-2n+5}$

11. $\displaystyle\sum_{n=1}^{\infty} \frac{5^{2n}}{n!}$

12. $\displaystyle\sum_{n=1}^{\infty} \frac{n^2+3}{n3^n}$

13. $\displaystyle\sum_{n=1}^{\infty} \frac{(\ln n)^2}{\sqrt{n}}$

14. $\displaystyle\sum_{n=1}^{\infty} (-1)^n \left(\frac{n+1}{n^2}\right)$

15. $\displaystyle\sum_{n=1}^{\infty} (-1)^n \left(\frac{n+1}{n^3}\right)$

16. $\displaystyle\sum_{n=1}^{\infty} \operatorname{Cos}^{-1}\left(\frac{1}{n}\right)$

17. $\displaystyle\sum_{n=1}^{\infty} \frac{1}{n}\operatorname{Cos}^{-1}\left(\frac{1}{n}\right)$

18. $\displaystyle\sum_{n=1}^{\infty} \frac{1}{n^2}\operatorname{Cos}^{-1}\left(\frac{1}{n}\right)$

19. $\displaystyle\sum_{n=1}^{\infty} \frac{2\cdot 4\cdot 6\cdot \cdots \cdot (2n)}{n!}$

20. $\displaystyle\sum_{n=1}^{\infty} \frac{3\cdot 6\cdot 9\cdot \cdots \cdot (3n)}{(2n)!}$

21. $\displaystyle\sum_{n=1}^{\infty} \sqrt{\frac{n^2+1}{n^2+5}}$

22. $\displaystyle\sum_{n=1}^{\infty} (-1)^{n+1} \left(1+\frac{1}{n}\right)^3$

23. $\displaystyle\sum_{n=1}^{\infty} \frac{1}{n^2} \sin n$

24. $\displaystyle\sum_{n=1}^{\infty} \frac{10^n}{5^{3n+2}}$

25. $\displaystyle\sum_{n=1}^{\infty} (-1)^n \frac{\ln n}{n}$

26. $\displaystyle\sum_{n=1}^{\infty} \frac{1}{e^{n\pi}}$

27. $\displaystyle\sum_{n=1}^{\infty} \frac{2^n + 2^{-n}}{3^n}$

28. $\displaystyle\sum_{n=1}^{\infty} \frac{1}{\sqrt{n}} \cos(n\pi)$

In Exercises 29–36 find the interval of convergence for the power series.

29. $\displaystyle\sum_{n=0}^{\infty} \frac{n+1}{n^2+1} x^n$

30. $\displaystyle\sum_{n=1}^{\infty} \frac{1}{n^2 2^n} x^n$

31. $\displaystyle\sum_{n=0}^{\infty} (n+1)^3 x^n$

32. $\displaystyle\sum_{n=1}^{\infty} \frac{1}{n^n} x^n$

33. $\displaystyle\sum_{n=0}^{\infty} \frac{1}{4^n} (x-2)^n$

34. $\displaystyle\sum_{n=2}^{\infty} \sqrt{\frac{n+1}{n-1}} (x+3)^n$

35. $\displaystyle\sum_{n=1}^{\infty} n 3^n x^{2n}$

36. $\displaystyle\sum_{n=1}^{\infty} \frac{2^n}{n} x^{3n}$

In Exercises 37–45 find the power series expansion of the function about the indicated point.

37. $f(x) = \sqrt{1+x^2}$, about $x = 0$

38. $f(x) = e^{x+5}$, about $x = 0$

39. $f(x) = \cos(x + \pi/4)$, about $x = 0$

40. $f(x) = x \ln(2x + 1)$, about $x = 0$

41. $f(x) = \sin x$, about $x = \pi/4$

42. $f(x) = x/(x^2 + 4x + 3)$, about $x = 0$

43. $f(x) = e^x$, about $x = 3$

44. $f(x) = (x + 1) \ln(x + 1)$, about $x = 0$

45. $f(x) = x^3 e^{x^2}$, about $x = 0$

46. How many terms in the Maclaurin series for $f(x) = e^{-x^2}$ guarantee a truncation error of less than 10^{-5} for all x in the interval $0 \le x \le 2$?

47. Find a power series solution in powers of x for the differential equation

$$y'' - 4y = 0.$$

48. Find the Maclaurin series for $f(x) = \sqrt{1 + \sin x}$ valid for $-\pi/2 \le x \le \pi/2$ by first showing that $f(x)$ can be written in the form

$$f(x) = \sin(x/2) + \cos(x/2).$$

Why is the restriction $-\pi/2 \le x \le \pi/2$ necessary?

49. On a calculator take the cosine of 1 (radian). Take the cosine of this result, and again, and again, and again, ... What happens? Interpret what is going on.

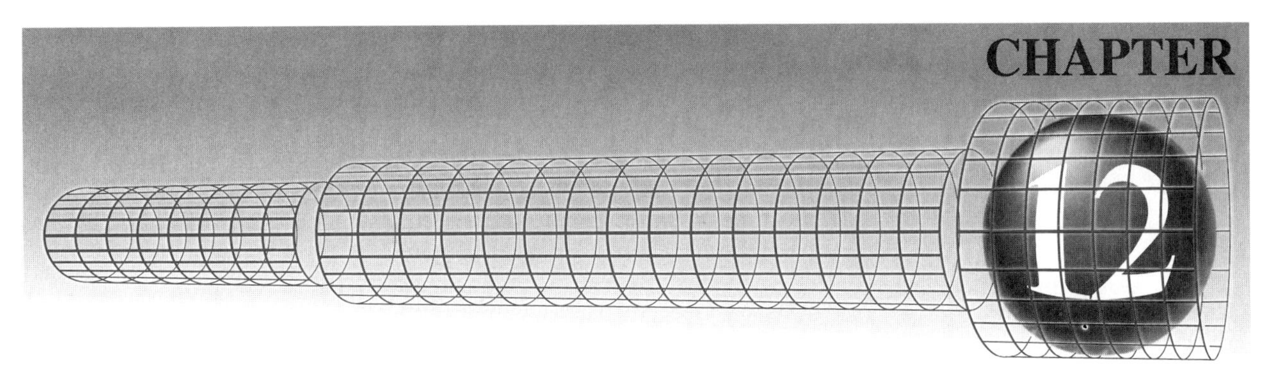

Vectors and Three-dimensional Analytic Geometry

Chapters 1–11 dealt with single-variable calculus–differentiation and integration of functions *f(x)* of one variable. In Chapters 12–15 we study multivariable calculus. Discussions of three-dimensional analytic geometry and vectors in Sections 1–5 of this chapter prepare the way. In Sections 6–10, we differentiate and integrate vector functions, and apply the results to the geometry of curves in space and the motion of objects.

SECTION 12.1

Rectangular Coordinates In Space

The coordinate of a point on the real line is defined as its directed distance from a fixed point called the origin. The Cartesian coordinates of a point in a plane are its directed distances from two fixed lines called the coordinate axes. In space, Cartesian coordinates are directed distances from three fixed planes called the coordinate planes. In particular, we draw through a point O, called the **origin**, three mutually perpendicular lines called the x-, y-, and z-axes (Figure 12.1). Each of the axes is coordinatized with some unit distance (which need not be the same for all three axes). These three coordinate axes determine three planes called **coordinate planes**: The xy-coordinate plane is that plane containing the x- and y-axes, the yz-coordinate plane contains the y- and z-axes, and the xz-coordinate plane contains the x- and z-axes.

FIGURE 12.1

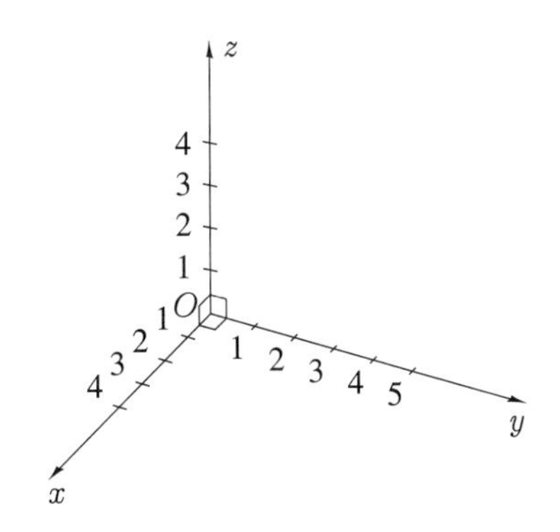

If P is any point in space, we draw lines from P perpendicular to the three coordinate planes (Figure 12.2). The directed distance from the yz-coordinate plane to P is parallel to the x-axis, and is called the x-coordinate of P. Similarly, y- and z-coordinates are defined as directed distances from the xz- and xy-coordinate planes to P. These three coordinates of P, written (x, y, z), are called the **Cartesian** or **rectangular coordinates** of P. Note that if we draw lines through P that are perpendicular to the axes, then the directed distances from O to the points of intersection of these perpendiculars with the axes are also the Cartesian coordinates of P. (Figure 12.3).

FIGURE 12.2

FIGURE 12.3

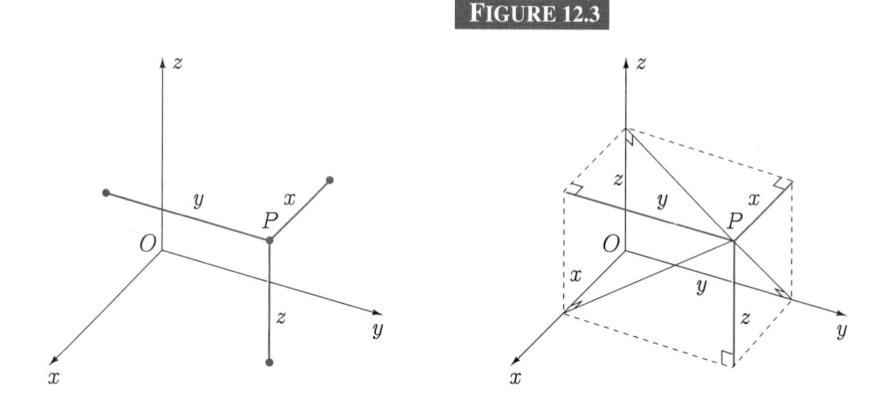

By either definition, each point in space has a unique ordered set of Cartesian coordinates (x, y, z); conversely, every ordered triple of real numbers (x, y, z) is the set of coordinates for one and only one point in space. For example, points with coordinates $(1,1,1)$, $(2,-3,4)$, $(3,4,-1)$, and $(-2,5,3)$ are shown in Figure 12.4.

FIGURE 12.4

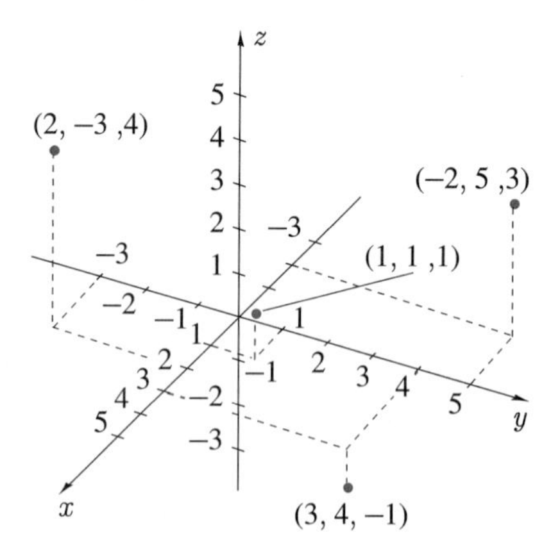

The coordinate systems in Figures 12.1–12.4 are called **right-handed coordinate systems**, because if we curl the fingers on our right hand from the positive x-direction toward the positive y-direction, then the thumb points in the positive z-direction (Figure 12.5). The coordinate system in Figure 12.6, on the other hand, is a **left-handed coordinate system**, since the thumb of the left hand points in the positive z-direction when the fingers of this hand are curled from the positive x-direction to the positive y-direction. We always use right-handed systems in this book.

FIGURE 12.5

FIGURE 12.6

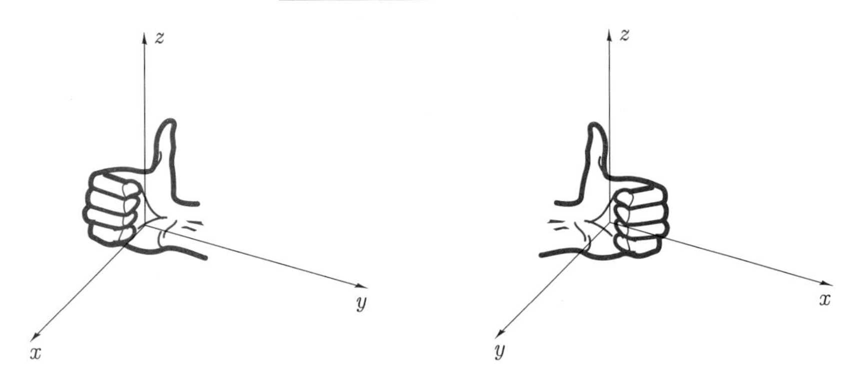

Suppose we construct for any two points P_1 and P_2 with coordinates (x_1, y_1, z_1) and (x_2, y_2, z_2), respectively, a box with sides parallel to the coordinate planes and with line segment P_1P_2 as diagonal (Figure 12.7). Because triangles P_1AB and P_1BP_2 are right-angled, we can write

$$\begin{aligned} ||P_1P_2||^2 &= ||P_1B||^2 + ||BP_2||^2 \\ &= ||P_1A||^2 + ||AB||^2 + ||BP_2||^2 \\ &= (x_2 - x_1)^2 + (y_2 - y_1)^2 + (z_2 - z_1)^2. \end{aligned}$$

In other words, the length of the line segment joining two points $P_1(x_1, y_1, z_1)$ and $P_2(x_2, y_2, z_2)$ is

$$||P_1P_2|| = \sqrt{(x_2 - x_1)^2 + (y_2 - y_1)^2 + (z_2 - z_1)^2}. \qquad (12.1)$$

This is the analogue of formula 1.7 for the length of a line segment joining two points in the xy–plane.

FIGURE 12.7

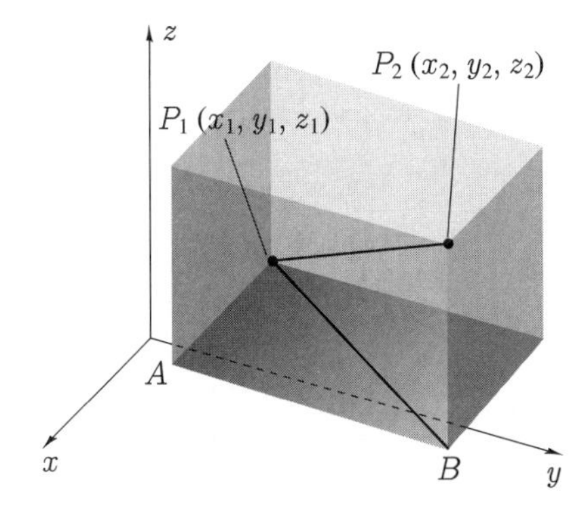

Just as the x- and y-axes divide the xy-plane into four regions called quadrants, the xy-, yz- and xz-coordinate planes divide xyz-space into eight regions called **octants**. The region where x-, y-, and z-coordinates are all positive is called the first octant. There is no commonly accepted way to number the remaining seven octants.

EXERCISES 12.1

1. Draw a Cartesian coordinate system and show the points $(1,2,1)$, $(-1,3,2)$, $(1,-2,4)$, $(3,4,-5)$, $(-1,-2,-3)$, $(-2,-5,4)$, $(8,-3,-6)$, and $(-4,3,-5)$.

2. Find the length of the line segment joining the points $(1,-2,5)$ and $(-3,2,4)$.

3. Prove that the triangle with vertices $(2,0,4\sqrt{2})$, $(3,-1,5\sqrt{2})$, and $(4,-2,4\sqrt{2})$ is right-angled and isosceles.

4. A cube has sides of length 2 units. What are coordinates of its corners if one corner is at the origin, three of its faces lie in the coordinate planes, and one corner has all three coordinates positive?

5. Show that the (undirected, perpendicular) distances from a point (x,y,z) to the x-, y-, and z-axes are, respectively, $\sqrt{y^2+z^2}$, $\sqrt{x^2+z^2}$, and $\sqrt{x^2+y^2}$.

In Exercises 6–9, find the (undirected) distances from the point to **(a)** the origin; **(b)** the x-axis; **(c)** the y-axis; **(d)** the z-axis.

6. $(2,3,-4)$

7. $(1,-5,-6)$

8. $(4,3,0)$

9. $(-2,1,-3)$

10. Prove that the three points $(1,3,5)$, $(-2,0,3)$ and $(7,9,9)$ are collinear.

11. Find that point in the xy-plane that is equidistant from the points $(1,3,2)$ and $(2,4,5)$ and has a y-coordinate equal to three times its x-coordinate.

12. Find an equation describing all points that are equidistant from the points $(-3,0,4)$ and $(2,1,5)$. What does this equation describe geometrically?

13. **(a)** If $(\sqrt{3}-3, 2+2\sqrt{3}, 2\sqrt{3}-1)$ and $(2\sqrt{3}, 4, \sqrt{3}-2)$ are two vertices of an equilateral triangle, and if the third vertex lies on the z-axis, find the third vertex.
    (b) Can you find a third vertex on the x-axis?

14. A birdhouse is built from a box $1/2$ m on each side with a roof as shown in Figure 12.8. If the distance from each corner of the roof to the peak is $3/4$ m, find the coordinates of the nine corners of the house. (The sides of the box are parallel to the coordinate planes.)

FIGURE 12.8

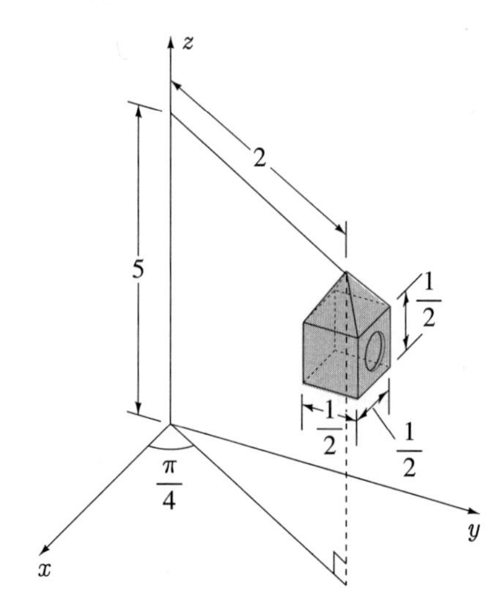

15. If P and Q (Figure 12.9) have coordinates (x_1,y_1,z_1) and (x_2,y_2,z_2), show that coordinates of the point R midway between P and Q are $\left(\frac{x_1+x_2}{2}, \frac{y_1+y_2}{2}, \frac{z_1+z_2}{2}\right)$.

FIGURE 12.9

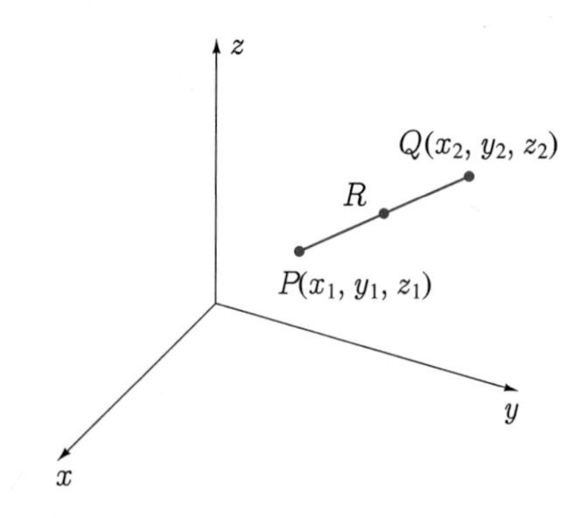

16. **(a)** Find the midpoint of the line segment joining the points $P(1,-1,-3)$ and $Q(3,2,-4)$.
    (b) If the line segment joining P and Q is extended its own length beyond Q to a point R, find the coordinates of R.

17. The four sided object in Figure 12.10 is a *tetrahedron*. If the four vertices of the tetrahedron are as shown, prove that the three lines joining the midpoints of opposite edges (one of which is PQ) meet at a point that bisects each of them.

FIGURE 12.10

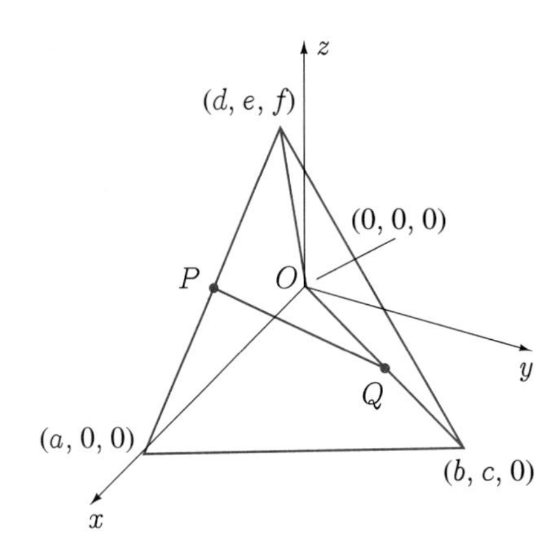

18. Let A, B, C, and D be the vertices of a quadrilateral in space (not necessarily planar). Show that the line segments joining midpoints of opposite sides of the quadrilateral intersect in a point which bisects each.

19. Generalize the result of Exercise 15 to prove that if a point R divides the length PQ so that $\dfrac{\|PR\|}{\|RQ\|} = \dfrac{r_1}{r_2}$, where r_1 and r_2 are positive integers, then the coordinates of R are

$$x = \frac{r_1 x_2 + r_2 x_1}{r_1 + r_2}, \quad y = \frac{r_1 y_2 + r_2 y_1}{r_1 + r_2}, \quad z = \frac{r_1 z_2 + r_2 z_1}{r_1 + r_2}.$$

20. A man 2 m tall walks along the edge of a straight road 10 m wide (Figure 12.11). On the other edge of the road stands a streetlight 8 m high. A building runs parallel to the road and 1 m from it. If Cartesian coordinates are set up as shown (with the x- and y-axes in the plane of the road), find the coordinates of the tip of the man's shadow when he is at the position shown.

FIGURE 12.11

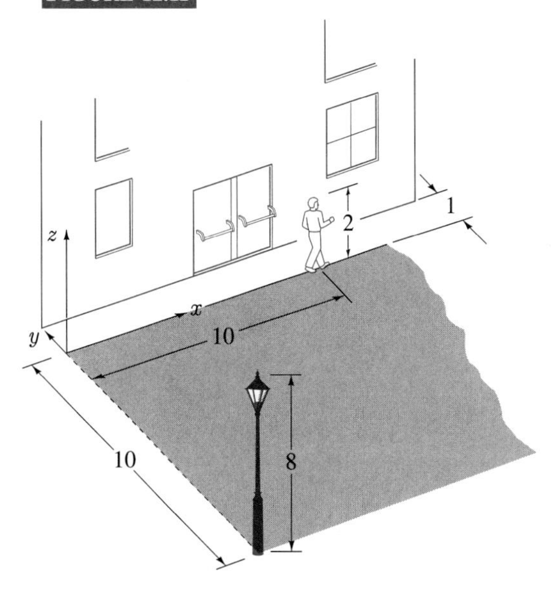

SECTION 12.2

Curves and Surfaces

An equation involving the x- and y-coordinates of points in the xy-plane usually specifies a curve. For example, the equation $x^2+y^2 = 4$ describes a circle of radius 2 centred at the origin (Figure 12.12). We now ask what is defined by an equation involving the Cartesian coordinates (x, y, z) of points in space. For example, the equation $z = 0$ describes all points in the xy-plane since all such points have a z-coordinate equal to zero. Similarly, $y = 2$ describes all points in the plane parallel to and two units to the right of the xz-plane. What does the equation $x^2 + y^2 = 4$ describe? In other words, regarded as a restriction on the x-, y-, and z-coordinates of points in space, rather than a restriction on the x- and y-coordinates of points in the xy-plane, what does it represent? Because the equation says nothing about z, there is no restriction whatsoever on z. In other words, the z-coordinate can take on all possible values, but the x- and y-coordinates must be restricted by $x^2 + y^2 = 4$. If we consider those points in the xy-plane $(z = 0)$ that satisfy $x^2 + y^2 = 4$, we obtain the circle in Figure 12.12. In space, each of these points has coordinates $(x, y, 0)$, where x and y still satisfy $x^2 + y^2 = 4$ (Figure 12.13). If we now take any point Q that is either directly above or directly below a point $P(x, y, 0)$ on this circle, it has exactly the same x- and y-coordinates as P; only its z-coordinate differs. Thus the x- and y-coordinates of Q also satisfy $x^2 + y^2 = 4$. Since we can do this for any point P on the circle, it follows that $x^2 + y^2 = 4$ describes the right-circular cylinder of radius 2 and infinite extent in Figure 12.13.

FIGURE 12.12

FIGURE 12.13

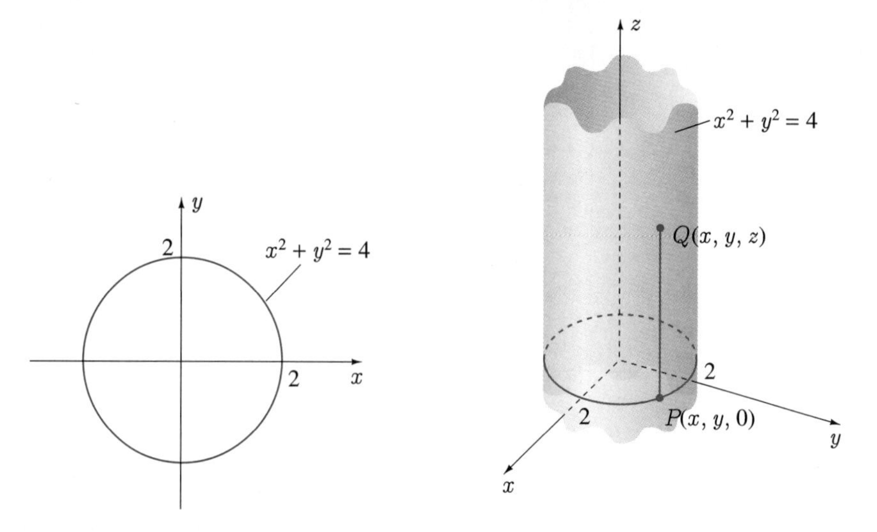

By reasoning similar to that used above, we can show that the equation $2x + y = 2$ describes the plane in Figure 12.14 parallel to the z-axis and standing on the straight line $2x + y = 2$, $z = 0$ in the xy-plane.

FIGURE 12.14

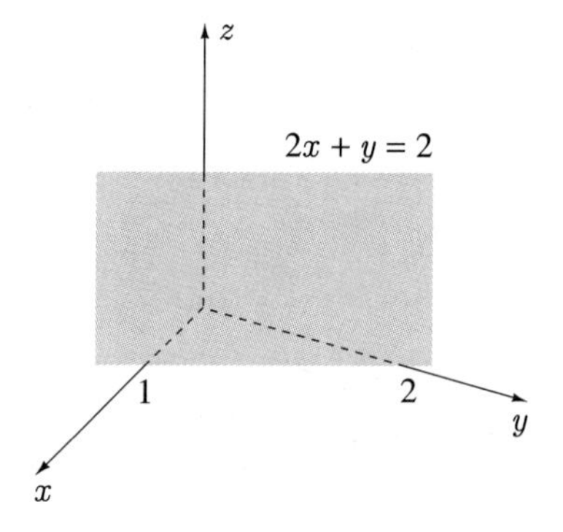

Finally, consider the equation $x^2 + y^2 + z^2 = 9$. Since $\sqrt{x^2 + y^2 + z^2}$ is the distance from the origin to a point with coordinates (x, y, z), this equation describes all points that are three units away from the origin. In other words, $x^2 + y^2 + z^2 = 9$ describes the points on a sphere of radius 3 centred at the origin.

It appears that one equation in the coordinates (x, y, z) of points in space specifies a surface. The shape of the surface is determined by the form of the equation. If one equation in the coordinates (x, y, z) specifies a surface, it is easy to see what two simultaneous equations specify. For instance, suppose we ask for all points in space whose coordinates satisfy both of the equations

$$x^2 + y^2 = 4, \quad z = 1.$$

By itself, $x^2 + y^2 = 4$ describes the cylinder in Figure 12.13. The equation $z = 1$ describes all points in a plane parallel to the xy-plane and one unit above it. To ask for all points that satisfy $x^2 + y^2 = 4$ and $z = 1$ simultaneously is to ask for all points that lie on both surfaces. Consequently, the equations $x^2 + y^2 = 4$, $z = 1$ describe the curve of intersection of the two surfaces—the circle in Figure 12.15.

FIGURE 12.15

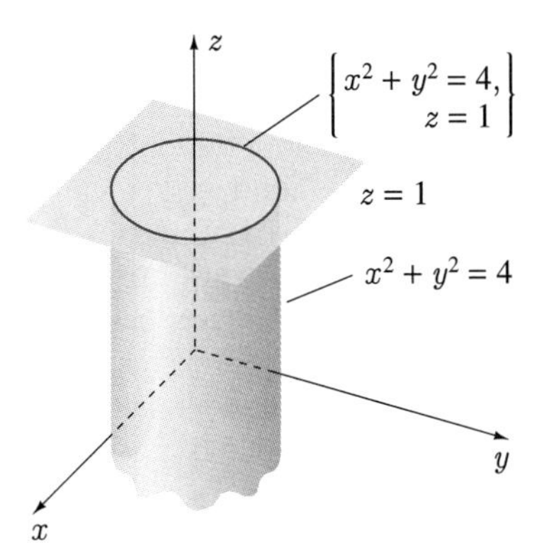

The equation $x = 0$ describes the yz-plane; the equation $y = 0$ describes the xz-plane. If we put the two equations together, $x = 0$ and $y = 0$, we obtain all points that lie on both the yz-plane and the xz-plane; i.e., the z-axis. In other words, equations for the z-axis are $x = 0$, $y = 0$.

Finally, $x^2 + y^2 + z^2 = 9$ is the equation of a sphere of radius 3 centred at the origin, and $y = 2$ is the equation of a plane parallel to the xz-plane and two units to the right. Together, the equations $x^2 + y^2 + z^2 = 9$, $y = 2$ describe the curve of intersection of the two surfaces—the circle in Figure 12.16. Note that by substituting $y = 2$ into the equation of the sphere, we can write alternatively that $x^2 + z^2 = 5$, $y = 2$. This pair of equations is equivalent to the original pair because all points that satisfy $x^2 + y^2 + z^2 = 9$, $y = 2$ also satisfy $x^2 + z^2 = 5$, $y = 2$, and vice versa. This new pair of equations provides an alternative way of visualizing the curve. Again $y = 2$ is the plane of Figure 12.16, but $x^2 + z^2 = 5$ describes a right-circular cylinder of radius $\sqrt{5}$ and infinite extent around the y-axis (Figure 12.17). Our discussion has shown that the cylinder and plane intersect in the same curve as the sphere and plane.

FIGURE 12.16

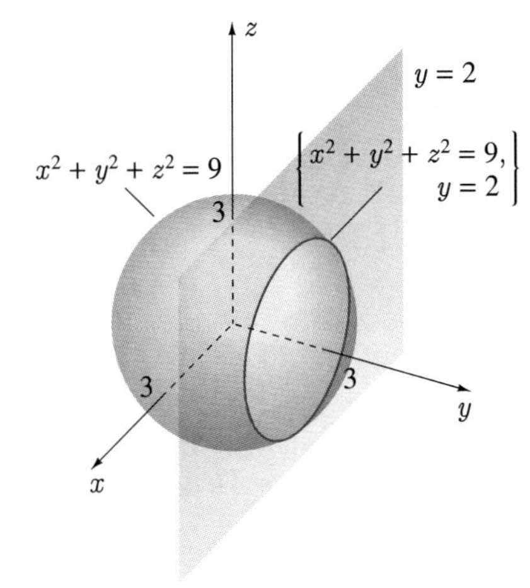

FIGURE 12.17

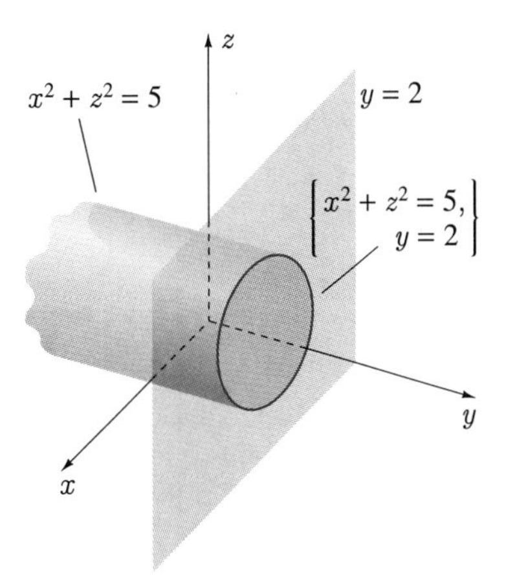

In summary, we have illustrated that one equation in the coordinates (x, y, z) of a point specifies a surface; two simultaneous equations specify a curve, the curve of intersection of the two surfaces (provided, of course, that the surfaces do intersect).

In Chapters 4–10 we learned to appreciate the value of curve sketching in the xy-plane. Sometimes a sketch serves as a device by which we can interpret algebraic statements geometrically (such as the mean value theorem or the interpretation of a critical point of a function as a point where the tangent line to the graph of the function is horizontal, vertical, or does not exist). Sometimes it plays an integral part in the solution of a problem (such as when the definite integral is used to find areas, volumes, etc.). Sometimes a sketch is a complete solution to a problem (such as to determine whether a given function has an inverse). We will find that sketching surfaces can be just as useful for multivariable calculus in Chapters 13–15. Unfortunately, surface sketching is more difficult than curve sketching, principally because we are drawing a space diagram on a page. Our ability to sketch curves, as we will see, can be of immense help.

One of the most helpful techniques for sketching a surface is to imagine the intersection of the surface with various planes—in particular, the coordinate planes. From these cross-sections of the surface, it is sometimes possible to visualize the entire surface. For example, if we intersect the surface $z = x^2 + y^2$ with the yz-plane, we obtain the parabola $z = y^2$, $x = 0$. Similarly, the parabola $z = x^2$, $y = 0$ is the intersection curve with the xz-plane. These curves, shown in Figure 12.18(a), would lead us to suspect that the surface $z = x^2 + y^2$ might be shaped as shown in Figure 12.18(b). To verify this, we intersect the surface with a plane $z = k$ (k a constant), giving the curve

$$z = x^2 + y^2, \quad z = k, \qquad \text{or} \qquad x^2 + y^2 = k, \quad z = k.$$

These latter equations indicate that cross-sections of $z = x^2 + y^2$ with planes $z = k$ are circles centred on the z-axis with radii $\sqrt{k}$ that increase as k increases. This certainly confirms the sketch in Figure 12.18(b).

FIGURE 12.18

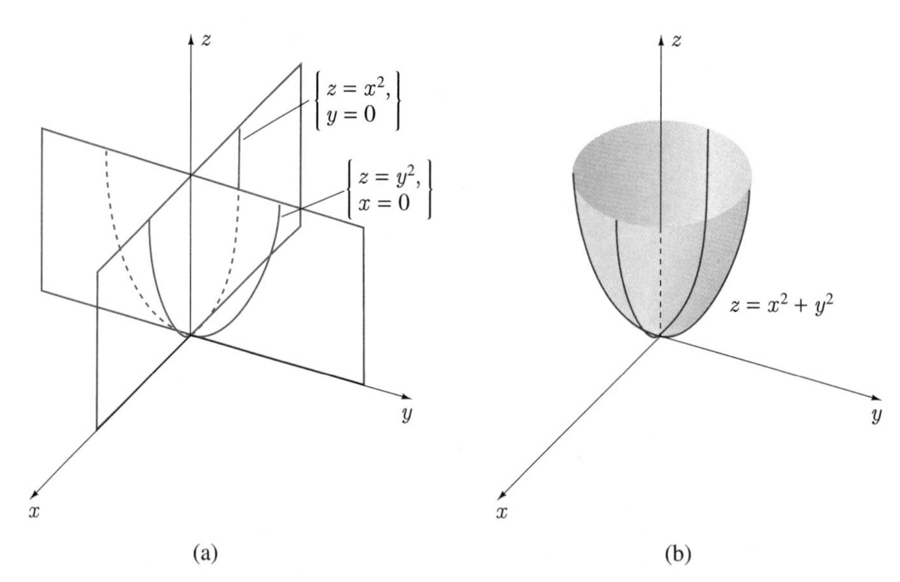

Intersections of the surface $y = z + x^2$ with the xy-, xz-, and yz-coordinate planes give two parabolas and a straight line, shown in Figure 12.19(a). These really do not seem to help us visualize the surface. If, however, we intersect the surface with planes $z = k$, we obtain the parabolas

$$y = z + x^2, \quad z = k, \qquad \text{or} \qquad y = x^2 + k, \quad z = k.$$

These parabolas, shown in Figure 12.19(b), indicate that the surface $y = z + x^2$ should be drawn as in Figure 12.19(c).

FIGURE 12.19

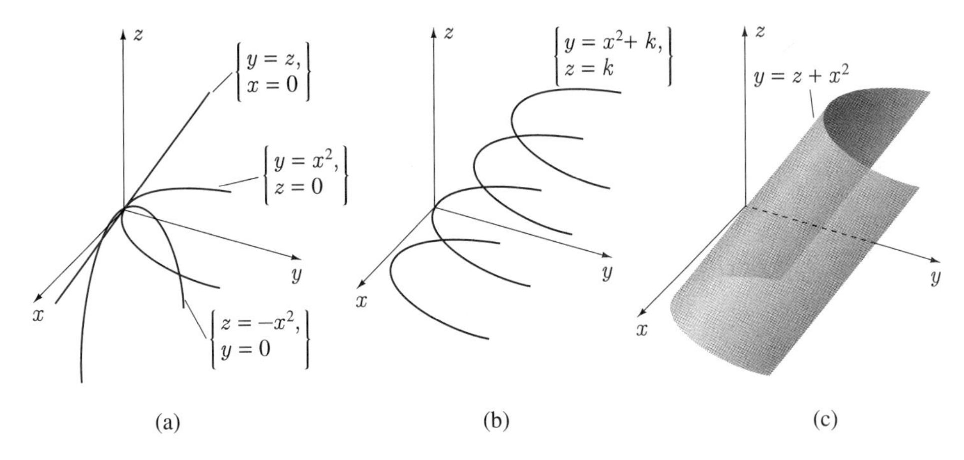

(a) (b) (c)

We can sometimes "build" surfaces in much the same way that we "built" curves in Chapter 1. For the surface $z = 1 - x^2 - y^2$, we first draw the surface $z = x^2 + y^2$ in Figure 12.18(b). To sketch $z = -(x^2 + y^2)$, we turn $z = x^2 + y^2$ upside down (Figure 12.20a), and finally we see that $z = 1 - x^2 - y^2$ is $z = -(x^2 + y^2)$ shifted upward one unit (Figure 12.20b).

FIGURE 12.20

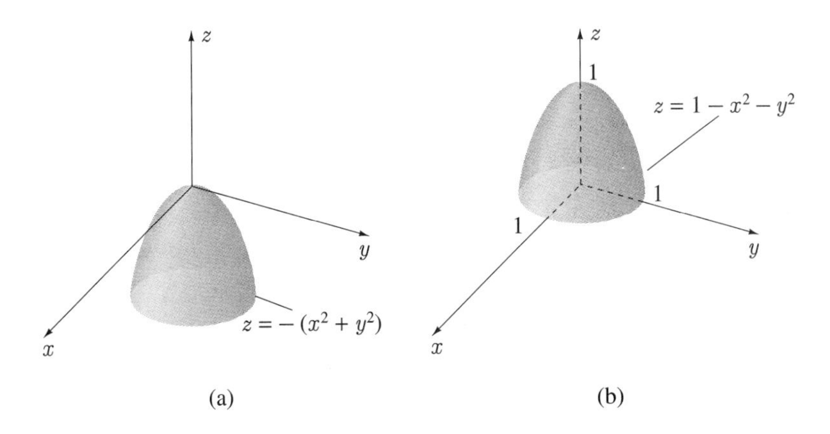

(a) (b)

EXAMPLE 12.1

Sketch the surface defined by each of the following equations:

(a) $z = \sqrt{4x + 2y - x^2 - y^2 - 4}$

(b) $y = 1 + \sqrt{x^2 + z^2}$

SOLUTION

(a) If we square the equation, and at the same time complete the squares on $-x^2 + 4x$ and $-y^2 + 2y$, we have

$$z^2 = -(x-2)^2 - (y-1)^2 + 1,$$

or

$$(x-2)^2 + (y-1)^2 + z^2 = 1.$$

Because $\sqrt{(x-2)^2+(y-1)^2+z^2}$ is the distance from a point (x,y,z) to $(2,1,0)$, this equation states that (x,y,z) must always be a unit distance from $(2,1,0)$; i.e., the equation $(x-2)^2+(y-1)^2+z^2=1$ defines a sphere of radius 1 centred at $(2,1,0)$ (Figure 12.21a). Because the original equation requires z to be nonnegative, the required surface is the upper half of this sphere—the hemisphere in Figure 12.21(b).

FIGURE 12.21

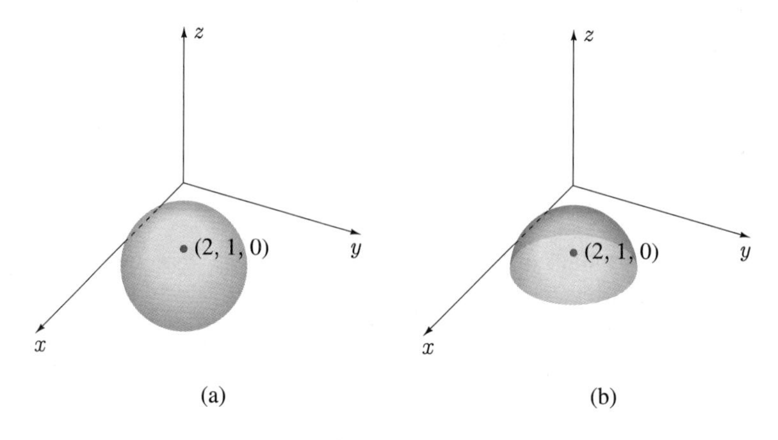

(b) If we intersect the surface $y=\sqrt{x^2+z^2}$ with the xy-plane, we obtain the broken straight line $y=|x|$, $z=0$ in Figure 12.22(a). Intersections of the surface with planes $y=k$ (k a constant) give

$$\begin{cases} y=\sqrt{x^2+z^2}, \\ y=k, \end{cases} \quad \text{or} \quad \begin{cases} x^2+z^2=k^2, \\ y=k. \end{cases}$$

These define circles of radii k in the planes $y=k$ (Figure 12.22b). Consequently, $y=\sqrt{x^2+z^2}$ defines the right-circular cone in Figure 12.22(c). The surface $y=1+\sqrt{x^2+z^2}$ can now be obtained by shifting the cone one unit in the y-direction (Figure 12.22d).

FIGURE 12.22

(a)

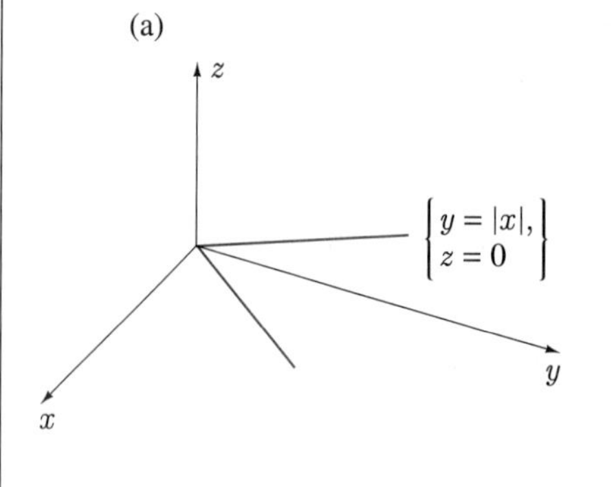

(b)

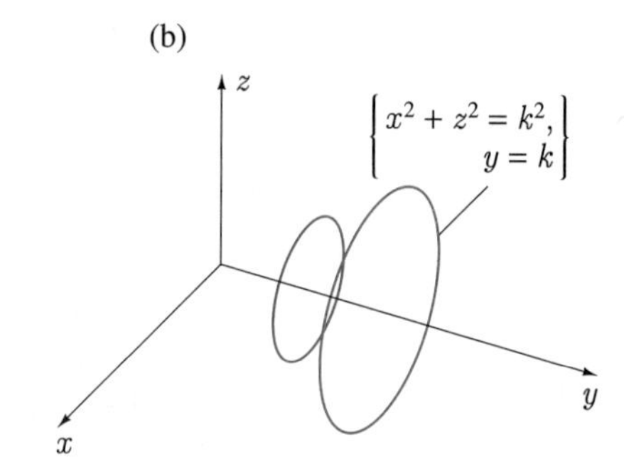

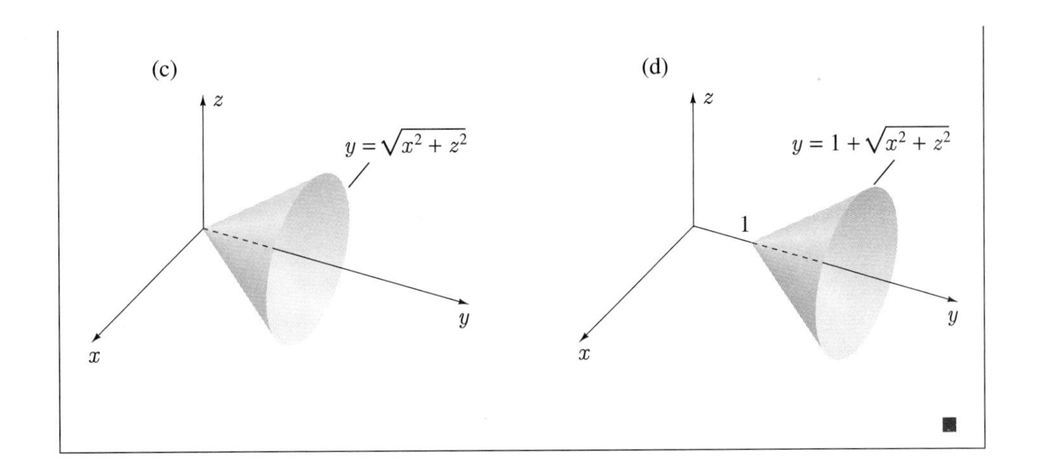

Cylinders

When one of the coordinates is missing from the equation of a surface, cross-sections of the surface with planes perpendicular to the axis of the missing variable are all identical. Such a surface is said to be a **cylinder**. For example, z is missing from the equation $x^2 + y^2 = 4$, and we saw in Figure 12.13 that this is the equation of a right-circular cylinder around the z-axis. Every cross-section of this surface with a plane $z = k$ is a circle of radius 2 centred on the z-axis. The equation $z = x^2$ is free of y. Each cross-section of this surface with a plane $y = k$ is the parabola $z = x^2$ in the plane $y = k$. Consequently, $z = x^2$ is the equation for the parabolic cylinder in Figure 12.23.

FIGURE 12.23

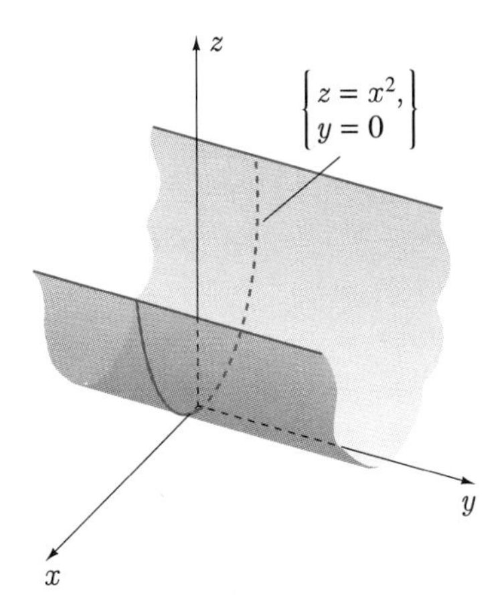

Quadric Surfaces

A **quadric surface** is a surface whose equation is quadratic in x, y, and z, the most general such equation being

$$Ax^2 + By^2 + Cz^2 + Dxy + Eyz + Fxz + Gx + Hy + Iz + J = 0. \qquad (12.2)$$

For the most part, we encounter quadric surfaces whose equations are of the forms

$$Ax^2 + By^2 + Cz^2 = J \quad \text{or} \quad Ax^2 + By^2 = Iz,$$

or these equations with x, y, and z interchanged. Surfaces with these equations fall into nine major classes depending on whether the constants are positive, negative, or zero. They are illustrated in Figures 12.24–12.32.

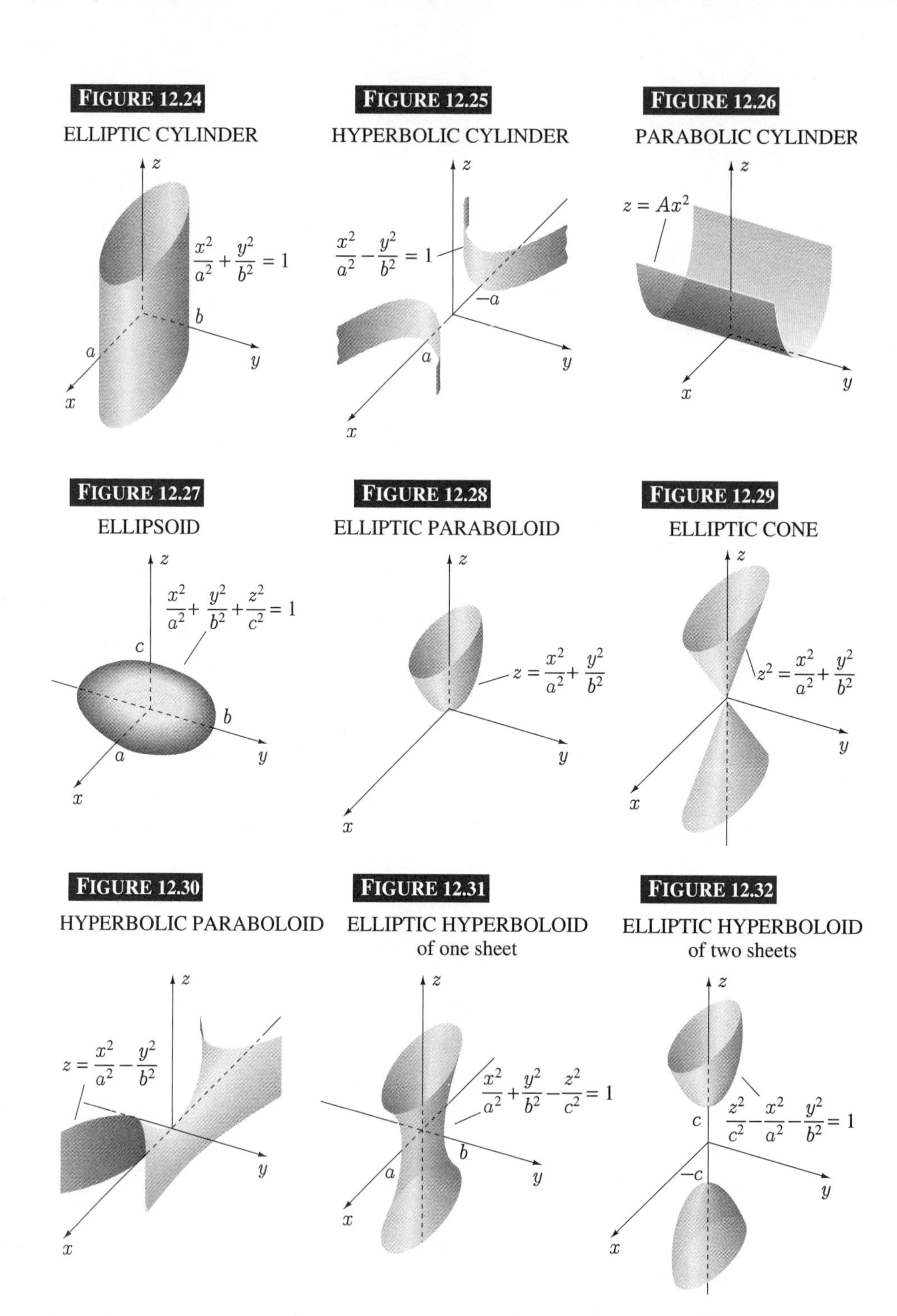

The names of these surfaces are derived from the fact that their cross-sections are ellipses, hyperbolas, or parabolas. For example, cross-sections of the hyperbolic paraboloid with planes $z = k$ are hyperbolas $x^2/a^2 - y^2/b^2 = k$. Cross-sections with planes $x = k$ are parabolas $z = k^2/a^2 - y^2/b^2$, as are cross-sections with planes $y = k$.

In applications of multiple integrals in Chapter 14, it is often necessary to project a space curve into one of the coordinate planes and find the equations of the projection. To illustrate, consider the curve of intersection of the cylinder $x^2 + z^2 = 4$ and the plane $2y + z = 4$ (the first octant part of which is shown in Figure 12.33). Since the curve of intersection lies on the cylinder $x^2 + z^2 = 4$, its projection in the xz-plane is the circle $x^2 + z^2 = 4$, $y = 0$. To find its projection in the xy-plane, we eliminate z between the equations $2y + z = 4$ and $x^2 + z^2 = 4$. The result is $x^2 + (4 - 2y)^2 = 4$, or, $x^2 + 4(y - 2)^2 = 4$. This shows that the curve of intersection lies on the elliptic cylinder $x^2 + 4(y - 2)^2 = 4$, and therefore it projects onto the ellipse $x^2 + 4(y - 2)^2 = 4$, $z = 0$ in the xy-plane. The projection of the curve in the yz-plane is that part of the line $2y + z = 4$, $x = 0$ between the points $(0, 1, 2)$ and $(0, 3, -2)$.

FIGURE 12.33

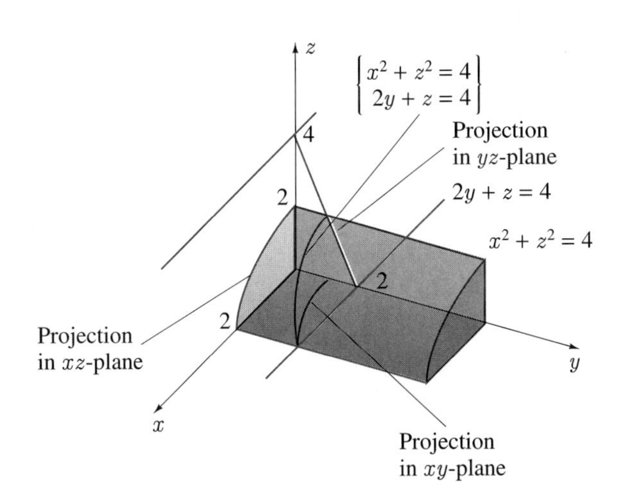

EXERCISES 12.2

In Exercises 1–35 sketch the surface defined by the equation.

1. $2y + 3z = 6$

2. $2x - 3y = 0$

3. $y = x^2 + 2$

4. $z = x^3$

5. $y^2 + z^2 = 1$

6. $x^2 + y^2 + z^2 = 4$

7. $x^2 + 4y^2 = 1$

8. $y^2 - z^2 = 4$

9. $z = 2(x^2 + y^2)$

10. $x = \sqrt{y^2 + z^2}$

11. $x = \sqrt{1 - y^2}$

12. $z = 2 - x$

13. $x^2 = y^2$

14. $x = z^2 + 2$

15. $z = y + 3$

16. $4z = 3\sqrt{x^2 + y^2}$

17. $x^2 - 2x + z^2 = 0$

18. $yz = 1$

19. $x^2 + y^2 + (z - 1)^2 = 3$

20. $z + 5 = 4(x^2 + y^2)$

21. $x^2 + z^2 = y^2$

22. $x^2 + z^2 = y^2 + 1$

23. $y^2 + z^2 = x^6$

24. $x^2 + y^2 + 4z^2 = 1$

25. $9z^2 = x^2 + y^2 + 1$

26. $(y^2 + z^2)^2 = x + 1$

27. $z^2 + 4y^2 = 1$

28. $y - z^2 = 0$

29. $x^2 - z^2 = 4$

30. $x^2 + y^2/4 + z^2/9 = 1$

31. $z = x^2/4 + y^2/25$

32. $x^2 = z^2 + 9y^2$

33. $z = y^2/16 - x^2/4$

34. $x^2 + y^2/4 - z^2/25 = 1$

35. $z^2 - 9x^2 - 16y^2 = 1$

In Exercises 36–45 sketch the curve defined by the equations.

36. $x^2 + y^2 = 2$, $z = 4$

37. $x + 2y = 6$, $y - 2z = 3$

38. $z = x^2 + y^2$, $x^2 + y^2 = 5$

39. $x^2 + y^2 = 2$, $x + z = 1$

40. $z = \sqrt{x^2 + y^2}$, $y = x$

41. $z + 2x^2 = 1$, $y = z$

42. $z = \sqrt{4 - x^2 - y^2}$, $x^2 + y^2 - 2y = 0$

43. $z = y$, $y = x^2$

44. $x^2 + z^2 = 1$, $y^2 + z^2 = 1$

45. $z = x^2$, $z = y^2$

In Exercises 46–55 find equations for the projections of the curve in the xy-, yz-, and xz-coordinate planes. In each case sketch the curve.

46. $x + y = 3$, $2y + 3z = 4$

47. $x + y + z = 4$, $2x - y + z = 6$

48. $x^2 + y^2 = 4$, $z = 4$

49. $x^2 + y^2 = 4$, $y = x$

50. $x^2 + y^2 = 4$, $x = z$

51. $x^2 + y^2 = 4$, $x + y + z = 2$

52. $y^2 + z^2 = 3$, $x^2 + z^2 = 3$

53. $z = x^2 + y^2$, $x + z = 1$

54. $z = \sqrt{x^2 + y^2}$, $z = 6 - x^2 - y^2$

55. $x^2 + y^2 + z^2 = 1$, $y = x$

In Exercises 56–61 find equations for the projection of the curve in the specified plane. Sketch each curve.

56. $z = x^2 - y^2$, $z = 2x + 4y$ in the xy-plane

57. $x^2 + y^2 - 4z^2 = 1$, $x + y = 2$ in the xz-plane

58. $y = z + x^2$, $y + z = 1$ in the xy-plane

59. $x = \sqrt{1 + 2y^2 + 4z^2}$, $x^2 + 9y^2 + 4z^2 = 36$ in the yz-plane

60. $z = x^2 + y^2$, $z = 4(x - 1)^2 + 4(y - 1)^2$ in the xy-plane

61. $x^2 + y^2 - 2y = 0$, $z^2 = x^2 + y^2$ in the xz-plane

In Exercises 62–71 sketch whatever is defined by the equation or equations.

62. $(x - 2)^2 + y^2 + z^2 = 0$

63. $x = 0$, $y = 5$

64. $\sqrt{x} + \sqrt{y} = 1$, $z = x$

65. $x + y = 15$, $y - x = 4$

66. $z = 1 - (x^2 + y^2)^{1/3}$

67. $z = |x|$

68. $z = x^2$, $y = z^2$

69. $x = \ln(y^2 + z^2)$

70. $z = |x - y|$

71. $x = 2$, $y = 4$, $z^2 - 1 = 0$

SECTION 12.3

Vectors

Physical quantities that have associated with them only a magnitude can be represented by real numbers. Some examples are temperature, density, area, moment of inertia, speed, and pressure. They are called **scalars**. There are many quantities, however, that have associated with them both magnitude and direction, and these quantities cannot be described by a single real number. Velocity, acceleration, and force are perhaps the most notable concepts in this category. To represent such quantities mathematically, we introduce **vectors**.

Definition 12.1 A **vector** is defined as a directed line segment.

To denote a vector we use a letter in boldface type, such as $\mathbf{v}$. In Figures 12.34(a), (b), and (c) we show two vectors $\mathbf{u}$ and $\mathbf{v}$ along a line, three vectors $\mathbf{u}$, $\mathbf{v}$, and $\mathbf{w}$ in a plane, and three vectors $\mathbf{u}$, $\mathbf{v}$, and $\mathbf{w}$ in space, respectively. It is customary to place an arrowhead on a vector and call this end the **tip** of the vector. The other end is called the **tail** of the vector, and the direction of the vector is from tail to tip. A vector then has both *direction* and *length*.

FIGURE 12.34

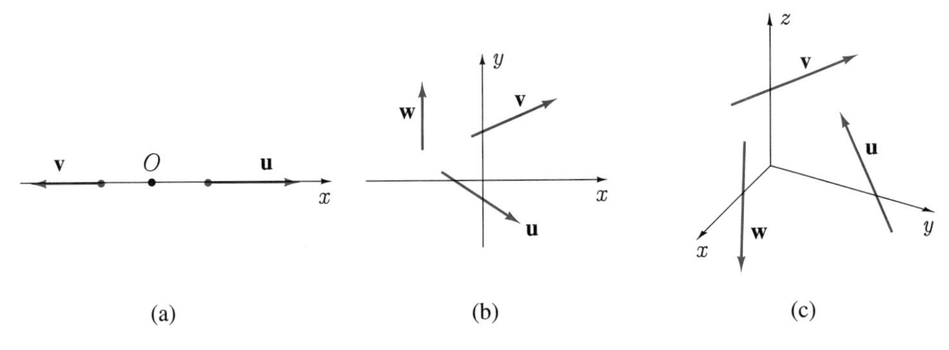

Definition 12.1 for a vector says nothing about its point of application; i.e., where its tail should be placed. This means that we may place the tail anywhere we wish. This suggests the following definition for equality of vectors.

Definition 12.2

Two vectors are equal if and only if they have the same length and direction. Their points of application are irrelevant.

For example, vectors **u** and **v** in Figure12.35 have exactly the same length and direction, and are therefore one and the same. Although the vector **w** in the same figure is parallel to **u** and **v** and has the same length, it points in the opposite direction and is not, therefore, the same as **u** and **v**.

FIGURE 12.35

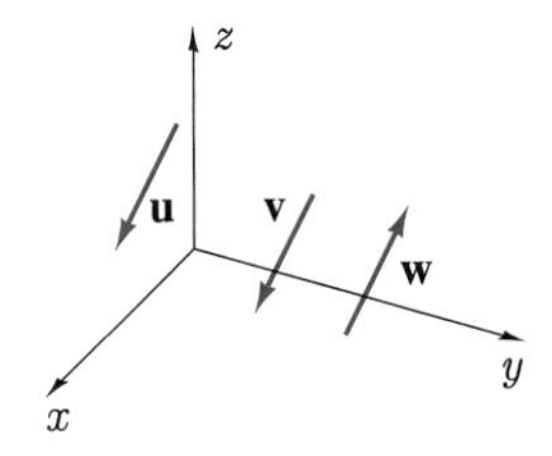

Components of Vectors

We realized in Chapter 1 that to solve geometric problems, it is often helpful to represent them algebraically. In fact, our entire development of single-variable calculus has hinged on our ability to represent a curve by an algebraic equation and also to draw the curve described by an equation. We now show that vectors can be represented algebraically.

Suppose we denote by $\overrightarrow{\mathbf{PQ}}$ the vector from P to Q in Figure 12.36. If P and Q have coordinates (x_1, y_1, z_1) and (x_2, y_2, z_2) in the coordinate system shown, then the length of $\overrightarrow{\mathbf{PQ}}$ is

$$\text{length of } \overrightarrow{\mathbf{PQ}} = \sqrt{(x_2 - x_1)^2 + (y_2 - y_1)^2 + (z_2 - z_1)^2}. \qquad (12.3)$$

FIGURE 12.36

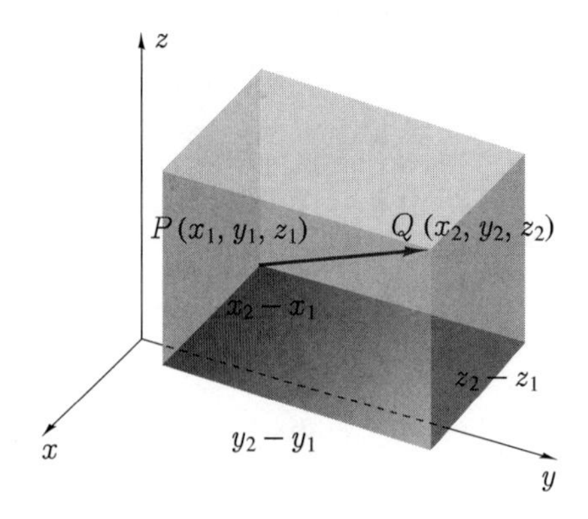

Note also that if we start at point P, proceed $x_2 - x_1$ units in the x-direction, then $y_2 - y_1$ units in the y-direction, and finally $z_2 - z_1$ units in the z-direction, then we arrive at Q. In other words, the three numbers $x_2 - x_1$, $y_2 - y_1$, and $z_2 - z_1$ characterize both the direction and the length of the vector joining P to Q. Because of this we make the following agreement.

Definition 12.3 If the tail of a vector $\mathbf{v}$ is at $P(x_1, y_1, z_1)$ and its tip is at $Q(x_2, y_2, z_2)$, then $\mathbf{v}$ shall be represented by the triple of numbers $x_2 - x_1$, $y_2 - y_1$, $z_2 - z_1$. In such a case we enclose the numbers in parentheses and write

$$\mathbf{v} = (x_2 - x_1, y_2 - y_1, z_2 - z_1). \tag{12.4}$$

The equal sign in 12.4 means "is represented by." The number $x_2 - x_1$ is called the x-**component** of $\mathbf{v}$, $y_2 - y_1$ the y-**component**, and $z_2 - z_1$ the z-**component**. Vectors in the xy-plane have only an x- and a y-component:

$$\mathbf{v} = (x_2 - x_1, y_2 - y_1),$$

where (x_1, y_1) and (x_2, y_2) are the coordinates of the tail and tip of $\mathbf{v}$. Vectors along the x-axis have only an x-component $x_2 - x_1$, where x_1 and x_2 are the coordinates of the tail and tip of $\mathbf{v}$.

We now have an algebraic representation for vectors. Each vector has associated with it a set of components that can be found by subtracting the coordinates of its tail from the coordinates of its tip. Conversely, given a set of real numbers (a, b, c), there is one and only one vector with these numbers as components. We can visualize this vector by placing its tail at the origin and its tip at the point with coordinates (a, b, c) (Figure 12.37). Alternatively, we can place the tail of the vector at any point (x_1, y_1, z_1) and its tip at the point $(x_1 + a, y_1 + b, z_1 + c)$.

FIGURE 12.37

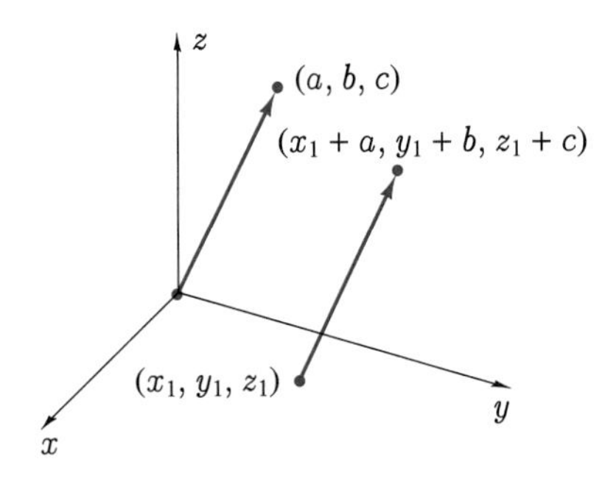

FIGURE 12.38

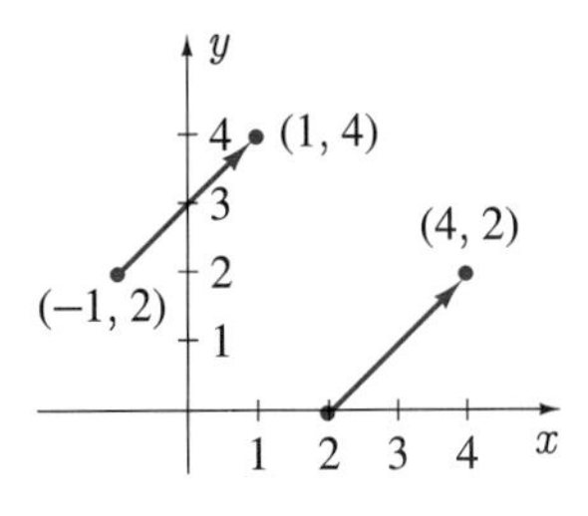

It is worth emphasizing once again that the same components of a vector are obtained for any point of application whatsoever. For example, the two vectors in Figure 12.38 are identical, and in both cases the components $(2, 2)$ are obtained by subtracting the coordinates of the tail from the tip. What we are saying is that Definition 12.2 for equality of vectors can be stated algebraically as follows.

Theorem 12.1

Two vectors are equal if and only if they have the same components.

EXAMPLE 12.2

Find the components of a vector in the xy-plane that has length 5, its tail at the origin, and makes an angle of $\pi/6$ radians with the positive x-axis.

SOLUTION Figure 12.39 illustrates that there are two such vectors, **u** and **v**. From the triangles shown, it is clear that

$$||OP|| = 5\cos(\pi/6) = 5\sqrt{3}/2 \quad \text{and} \quad ||PQ|| = ||PR|| = 5\sin(\pi/6) = \frac{5}{2}.$$

Consequently, Q and R have coordinates $Q = (5\sqrt{3}/2, \frac{5}{2})$ and $R = (5\sqrt{3}/2, -\frac{5}{2})$, and

$$\mathbf{u} = \left(\frac{5\sqrt{3}}{2}, \frac{5}{2}\right), \quad \mathbf{v} = \left(\frac{5\sqrt{3}}{2}, -\frac{5}{2}\right).$$

FIGURE 12.39

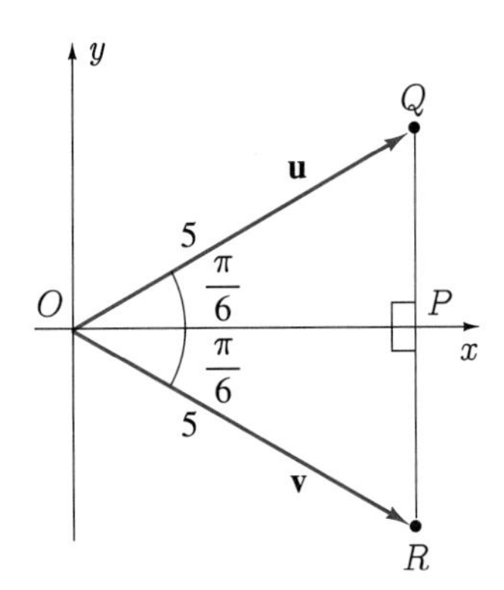

■

EXAMPLE 12.3

Find the components of the vector in the xy-plane that has its tail at the point $(4,5)$, has length 3, and points directly toward the point $(2,-3)$.

SOLUTION In Figure 12.40, $||PQ|| = \sqrt{2^2+8^2} = 2\sqrt{17}$. Because of similar triangles, we can write that

$$\frac{||ST||}{||PS||} = \frac{||QR||}{||PQ||} \quad \text{or} \quad ||ST|| = \frac{3(2)}{2\sqrt{17}} = \frac{3}{\sqrt{17}}.$$

Similarly,

$$||PT|| = ||PS||\frac{||PR||}{||PQ||} = \frac{3(8)}{2\sqrt{17}} = \frac{12}{\sqrt{17}}.$$

Since $||ST||$ and $||PT||$ represent differences in the x- and y-coordinates of P and S (except for signs), the components of $\overrightarrow{\mathbf{PS}}$ are $(-3/\sqrt{17}, -12/\sqrt{17})$.

FIGURE 12.40

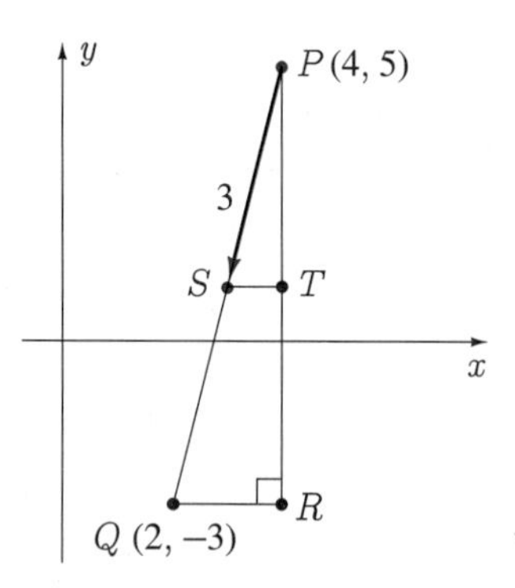

■

Unit Vectors and Scalar Multiplication

If the x-, y-, and z-components of a vector $\mathbf{v}$ are (v_x, v_y, v_z), often called the **Cartesian components** of $\mathbf{v}$, then these components represent the differences in the coordinates of its tip and tail. But then according to equation 12.3, the length of the vector, which we denote by $|\mathbf{v}|$, is

$$|\mathbf{v}| = \sqrt{v_x^2 + v_y^2 + v_z^2}. \tag{12.5}$$

In words, the length of a vector is the square root of the sum of the squares of its components.

Definition 12.4

A vector $\mathbf{v}$ is said to be a **unit vector** if it has length equal to one unit; i.e., $\mathbf{v}$ is a unit vector if

$$v_x^2 + v_y^2 + v_z^2 = 1. \tag{12.6}$$

To indicate that a vector is a unit vector, we place a circumflex ^ above it: $\hat{\mathbf{v}}$.

EXAMPLE 12.4

What is the length of the vector from $(1, -1, 0)$ to $(2, -3, -5)$?

SOLUTION Since the components of the vector are $(1, -2\ -5)$, its length is

$$\sqrt{(1)^2 + (-2)^2 + (-5)^2} = \sqrt{30}.$$

■

We now have vectors, which are directed line segments, and real numbers, which are scalars. We know that scalars can be added, subtracted, multiplied, and divided, but can we do the same with vectors, and can we combine vectors and scalars? In the remainder of this section we show how to add and subtract vectors and multiply vectors by scalars; in Sections 12.4 and 12.5 we define two ways to multiply vectors. Each of these operations can be approached either algebraically or geometrically. The geometric approach uses the geometric properties of vectors, namely, length and direction; the algebraic approach uses components of vectors. Neither method is suitable for all situations. Sometimes an idea is more easily introduced with a geometric approach; sometimes an algebraic approach is more suitable. We will choose whichever we feel expresses the idea more clearly. But, whenever we take a geometric approach, we will be careful to follow it up with the algebraic equivalent; conversely, when an algebraic approach is taken, we will always illustrate the geometric significance of the results.

FIGURE 12.41

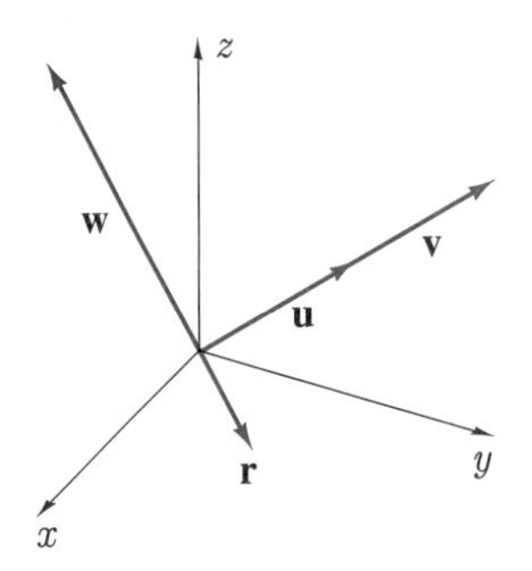

To introduce multiplication of a vector by a scalar, consider the vectors **u** and **v** in Figure 12.41, both of which have their tails at the origin; **v** is in the same direction as **u** but is twice as long as **u**. In such a situation we would like to say that **v** is equal to $2\mathbf{u}$ and write $\mathbf{v} = 2\mathbf{u}$. Vector **w** is in the opposite direction to **r** and is three times as long as **r**, and we would like to denote this vector by $\mathbf{w} = -3\mathbf{r}$. Both of these situations are realized if we adopt the following definition for multiplication of a vector by a scalar.

Definition 12.5

If $\lambda > 0$ is a scalar and **v** is a vector, then $\lambda\mathbf{v}$ is the vector that is in the same direction as **v** and λ times as long as **v**; if $\lambda < 0$, then $\lambda\mathbf{v}$ is the vector that is in the opposite direction to **v** and $|\lambda|$ times as long as **v**.

This is a geometric definition of scalar multiplication; it describes the length and direction of $\lambda\mathbf{v}$. We now show that the components of $\lambda\mathbf{v}$ are λ times the components of **v**. In Figure 12.42 we show a box with faces parallel to the coordinate planes and $\lambda\mathbf{v}$ as diagonal, and have given vector **v** components (v_x, v_y, v_z). From the pairs of similar triangles OAB and OCD, and OBE and ODF, we can write that

$$\frac{||OC||}{v_x} = \frac{||CD||}{v_y} = \frac{||OD||}{||OB||} = \frac{||DF||}{v_x} = \frac{|\lambda\mathbf{v}|}{|\mathbf{v}|} = \lambda.$$

Hence

$$||OC|| = \lambda v_x, \quad ||CD|| = \lambda v_y, \quad ||DF|| = \lambda v_z,$$

where $||OC||$, $||CD||$, and $||DF||$ are the components of $\lambda\mathbf{v}$. In other words, the components of $\lambda\mathbf{v}$ are λ times the components of **v**,

$$\lambda\mathbf{v} = \lambda(v_x, v_y, v_z) = (\lambda v_x, \lambda v_y, \lambda v_z). \qquad (12.7)$$

FIGURE 12.42

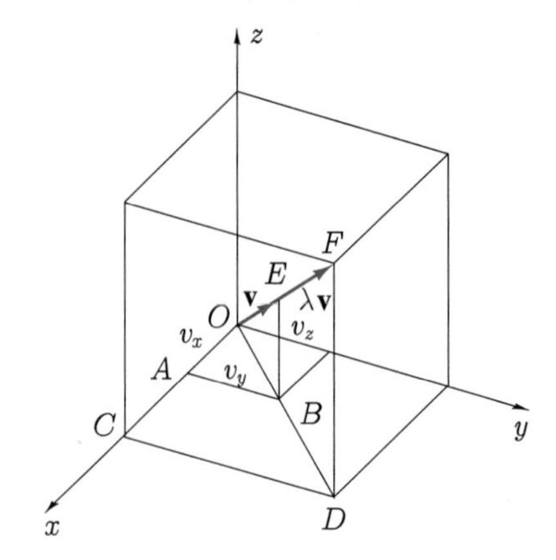

To multiply a vector by a scalar, then, we multiply each component by the scalar.

EXAMPLE 12.5

Find components for the unit vector in the same direction as $\mathbf{v} = (2, -2, 1)$.

SOLUTION The length of $\mathbf{v}$ is $|\mathbf{v}| = \sqrt{(2)^2 + (-2)^2 + 1^2} = 3$. According to our definition of multiplication of a vector by a scalar, the vector $\frac{1}{3}\mathbf{v}$ must have length 1 ($\frac{1}{3}$ that of $\mathbf{v}$) and the same direction as $\mathbf{v}$. Consequently, a unit vector in the same direction as $\mathbf{v}$ is

$$\hat{\mathbf{v}} = \frac{1}{3}\mathbf{v} = \frac{1}{3}(2, -2, 1) = \left(\frac{2}{3}, -\frac{2}{3}, \frac{1}{3}\right).$$

■

This example illustrates that a unit vector in the same direction as a given vector $\mathbf{v}$ is

$$\hat{\mathbf{v}} = \frac{\mathbf{v}}{|\mathbf{v}|}. \tag{12.8}$$

EXAMPLE 12.6

Find components for the vector of length 4 in the direction opposite that of $\mathbf{v} = (1, 2, -3)$.

SOLUTION Since $|\mathbf{v}| = \sqrt{1 + 4 + 9} = \sqrt{14}$, a unit vector in the same direction as $\mathbf{v}$ is

$$\hat{\mathbf{v}} = \frac{1}{\sqrt{14}}\mathbf{v}.$$

The vector of length 4 in the opposite direction to $\mathbf{v}$ must therefore be

$$(-4)\hat{\mathbf{v}} = \left(\frac{-4}{\sqrt{14}}\right)\mathbf{v} = \left(\frac{-4}{\sqrt{14}}\right)(1, 2, -3) = \left(-\frac{4}{\sqrt{14}}, -\frac{8}{\sqrt{14}}, \frac{12}{\sqrt{14}}\right).$$

■

Note that with the operation of scalar multiplication, we can simplify the solution of Example 12.3. The vector that points from $(4, 5)$ to $(2, -3)$ is $\mathbf{v} = (-2, -8)$, and therefore the unit vector in this direction is

$$\hat{\mathbf{v}} = \frac{1}{\sqrt{4+64}}(-2, -8) = \frac{1}{2\sqrt{17}}(-2, -8) = \frac{1}{\sqrt{17}}(-1, -4).$$

The required vector of length 3 is

$$3\hat{\mathbf{v}} = \frac{3}{\sqrt{17}}(-1, -4) = \left(-\frac{3}{\sqrt{17}}, -\frac{12}{\sqrt{17}}\right).$$

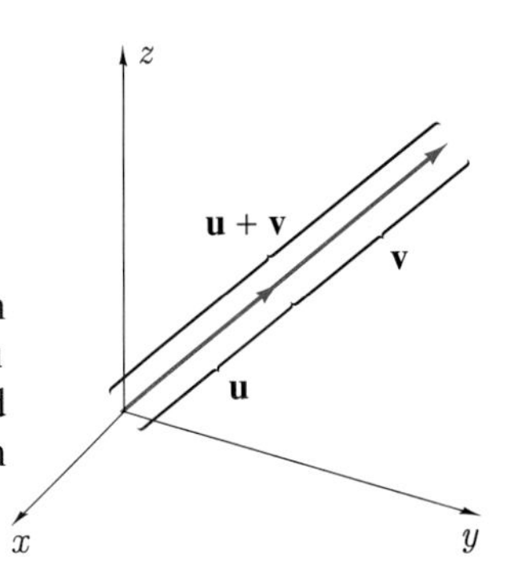

Addition and Subtraction of Vectors

In Figure12.43 we show two parallel vectors **u** and **v** and have placed the tail of **v** on the tip of **u**. It would seem natural to denote the vector that has its tail at the tail of **u** and its tip at the tip of **v** by **u** + **v**. For instance, if **u** and **v** were equal, then we would simply be saying that **u** + **u** = 2**u**. We use this idea to define addition of vectors even when the vectors are not parallel.

Definition 12.6

The sum of two vectors **u** and **v**, denoted by **u** + **v**, is the vector from the tail of **u** to the tip of **v** when the tail of **v** is placed on the tip of **u**.

Because the three vectors **u**, **v** and **u** + **v** then form a triangle (Figure 12.44), we call this **triangular addition** of vectors.

Note that were we to place tails of **u** and **v** both at the same point (Figure 12.45), and complete the parallelogram with **u** and **v** as sides, then the diagonal of this parallelogram would also represent the vector **u** + **v**. This is an equivalent method for geometrically finding **u** + **v**, and it is called **parallelogram addition** of vectors.

FIGURE 12.44

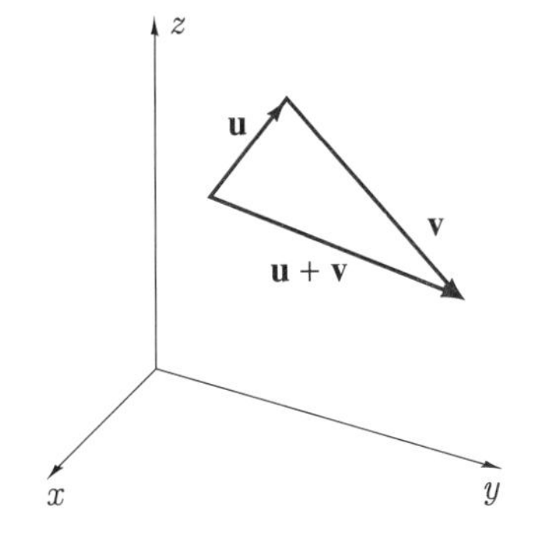

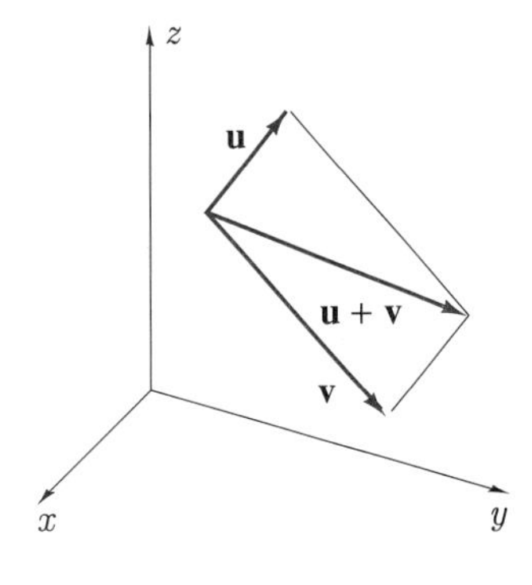

FIGURE 12.45

FIGURE 12.46

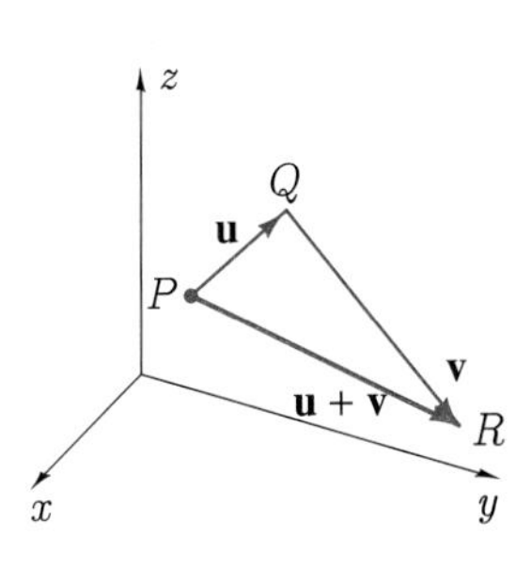

Algebraically, vectors are added component by component; i.e., if $\mathbf{u} = (u_x, u_y, u_z)$ and $\mathbf{v} = (v_x, v_y, v_z)$, then

$$\mathbf{u} + \mathbf{v} = (u_x + v_x, u_y + v_y, u_z + v_z). \tag{12.9}$$

To verify this we simply note that differences in the coordinates of P and Q in Figure 12.46 are (u_x, u_y, u_z), and differences in those of Q and R are (v_x, v_y, v_z). Consequently, differences in the coordinates of P and R must be $(u_x+v_x, u_y+v_y, u_z+v_z)$.

It is not difficult to show (see Exercise 26) that vector addition and scalar multiplication obey the following rules:

$$\mathbf{u} + \mathbf{v} = \mathbf{v} + \mathbf{u}; \tag{12.10a}$$

$$(\mathbf{u} + \mathbf{v}) + \mathbf{w} = \mathbf{u} + (\mathbf{v} + \mathbf{w}); \tag{12.10b}$$

$$\lambda(\mathbf{u} + \mathbf{v}) = \lambda\mathbf{u} + \lambda\mathbf{v}; \tag{12.10c}$$

$$(\lambda + \mu)\mathbf{v} = \lambda\mathbf{v} + \mu\mathbf{v}. \tag{12.10d}$$

If we denote the vector $(-1)\mathbf{v}$ by $-\mathbf{v}$, then the components of $-\mathbf{v}$ are the negatives of those of $\mathbf{v}$:

$$-\mathbf{v} = (-1)\mathbf{v} = (-1)(v_x, v_y, v_z) = (-v_x, -v_y, -v_z). \tag{12.11}$$

This vector has the same length as $\mathbf{v}$, but is opposite in direction to $\mathbf{v}$ (Figure 12.47).

FIGURE 12.47

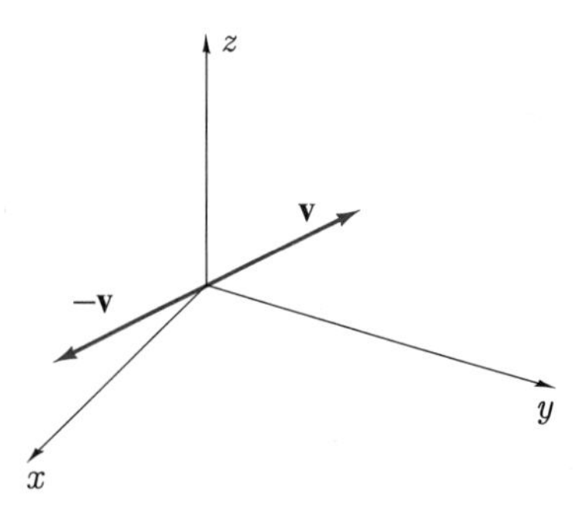

When $\mathbf{v}$ is added to $-\mathbf{v}$, the resultant vector has components that are all zero:

$$\mathbf{v} + (-\mathbf{v}) = (v_x, v_y, v_z) + (-v_x, -v_y, -v_z) = (0, 0, 0).$$

This vector, called the **zero vector**, is denoted by $\mathbf{0}$, and has the property that

$$\mathbf{v} + \mathbf{0} = \mathbf{0} + \mathbf{v} = \mathbf{v} \tag{12.12}$$

for any vector $\mathbf{v}$ whatsoever.

To subtract a vector $\mathbf{v}$ from $\mathbf{u}$, we add $-\mathbf{v}$ to $\mathbf{u}$.

Definition 12.7

The difference $\mathbf{u} - \mathbf{v}$ between two vectors $\mathbf{u}$ and $\mathbf{v}$ is the vector

$$\mathbf{u} - \mathbf{v} = \mathbf{u} + (-\mathbf{v}). \tag{12.13}$$

In Figure 12.48 $\mathbf{u} - \mathbf{v}$ is determined by a triangle, and in Figure 12.49 by a parallelogram. Alternatively, if we denote by $\mathbf{r}$ the vector joining the tip of $\mathbf{v}$ to the tip of $\mathbf{u}$ as in Figure 12.50, then by triangle addition, we have $\mathbf{v} + \mathbf{r} = \mathbf{u}$. Addition of $-\mathbf{v}$ to each side of this equation gives

$$-\mathbf{v} + \mathbf{v} + \mathbf{r} = -\mathbf{v} + \mathbf{u} \quad \text{or} \quad \mathbf{0} + \mathbf{r} = -\mathbf{v} + \mathbf{u}.$$

Thus $\mathbf{r} = \mathbf{u} - \mathbf{v}$, and $\mathbf{u} - \mathbf{v}$ is the vector joining the tip of $\mathbf{v}$ to the tip of $\mathbf{u}$. Definition 12.7 implies that vectors are subtracted component by component:

$$\mathbf{u} - \mathbf{v} = (u_x, u_y, u_z) - (v_x, v_y, v_z) = (u_x - v_x, u_y - v_y, u_z - v_z). \tag{12.14}$$

FIGURE 12.48

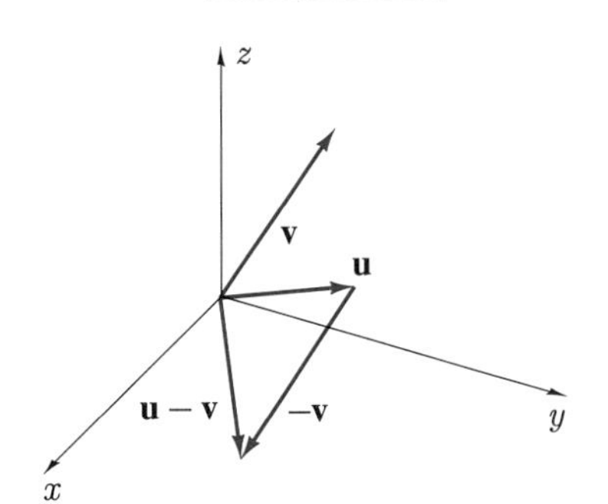

FIGURE 12.49

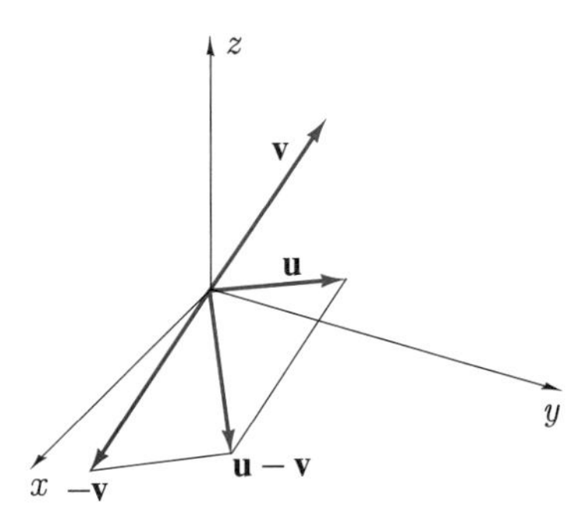

FIGURE 12.50

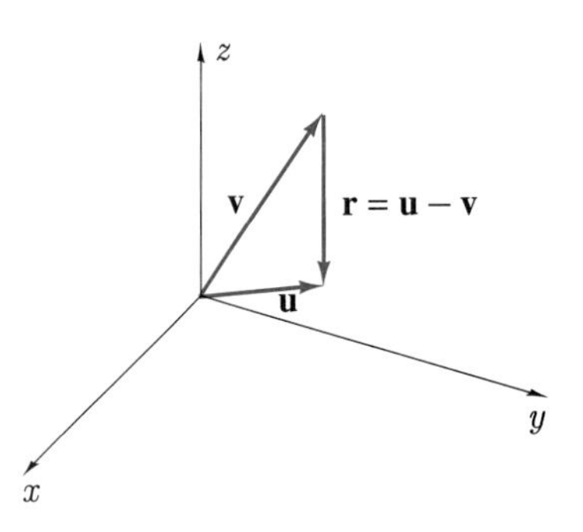

EXAMPLE 12.7

If $\mathbf{u} = (1, 1, 1)$, $\mathbf{v} = (-2, 3, 0)$, and $\mathbf{w} = (-10, 10, -2)$, find:

(a) $3\mathbf{u} + 2\mathbf{v} - \mathbf{w}$; **(b)** $2\mathbf{u} - 4\mathbf{v} + \mathbf{w}$; **(c)** $|\mathbf{u}|\mathbf{v} + \dfrac{4}{|\mathbf{v}|}\mathbf{w}$.

SOLUTION

(a)

$$\begin{aligned} 3\mathbf{u} + 2\mathbf{v} - \mathbf{w} &= 3(1,1,1) + 2(-2,3,0) - (-10,10,-2) \\ &= (3,3,3) + (-4,6,0) + (10,-10,2) \\ &= (9,-1,5) \end{aligned}$$

(b)

$$\begin{aligned} 2\mathbf{u} - 4\mathbf{v} + \mathbf{w} &= 2(1,1,1) - 4(-2,3,0) + (-10,10,-2) \\ &= (0,0,0) = \mathbf{0} \end{aligned}$$

(c) Since $|\mathbf{u}| = \sqrt{1^2 + 1^2 + 1^2} = \sqrt{3}$ and $|\mathbf{v}| = \sqrt{(-2)^2 + 3^2} = \sqrt{13}$,

$$\begin{aligned} |\mathbf{u}|\mathbf{v} + \frac{4}{|\mathbf{v}|}\mathbf{w} &= \sqrt{3}(-2,3,0) + \frac{4}{\sqrt{13}}(-10,10,-2) \\ &= \left(-2\sqrt{3} - \frac{40}{\sqrt{13}}, 3\sqrt{3} + \frac{40}{\sqrt{13}}, -\frac{8}{\sqrt{13}}\right). \end{aligned}$$

■

Forces

We have already mentioned that quantities such as temperature, area, and density have associated with them only a magnitude and are therefore represented by scalars. There are many quantities, however, that have associated with them both magnitude and direction, and these are described by vectors. The most notable of this group are forces. When we speak of a force, we mean a push or pull of some size in some specific direction. For example, when the boy in Figure 12.51(a) pulls his wagon, he exerts a force in the direction indicated by the handle. Suppose that he pulls with a force of ten newtons (10 N) and that the angle between the handle and the horizontal is $\pi/4$ radians. To represent this force as a vector $\mathbf{F}_1$, we choose the coordinate system in Figure 12.51(b), and make the agreement that the *length of* $\mathbf{F}_1$ *be equal to the magnitude of the force.* Since $\mathbf{F}_1$ represents a force of 10 N, it follows that the length of $\mathbf{F}_1$ is 10 units. Furthermore, because $\mathbf{F}_1$ makes an angle of $\pi/4$ radians with the positive x- and y-

axes, the difference in the x-coordinates (and the y-coordinates) of its tip and tail must be $10\cos(\pi/4) = 5\sqrt{2}$. The components of $\mathbf{F}_1$ are therefore $\mathbf{F}_1 = (5\sqrt{2}, 5\sqrt{2})$. If the boy's young sister drags her feet on the ground, then she effectively exerts a force $\mathbf{F}_2$ in the negative x-direction. If the magnitude of this force is 3 N, then its vector representation is $\mathbf{F}_2 = (-3, 0)$. Finally, if the combined weight of the wagon and the girl is 200 N, then the force $\mathbf{F}_3$ of gravity on the wagon and its load is $\mathbf{F}_3 = (0, -200)$. In mechanics we replace the individual forces $\mathbf{F}_1$, $\mathbf{F}_2$ and $\mathbf{F}_3$ by a single force that has the same effect on the wagon as all three forces combined. This force, called the **resultant** of $\mathbf{F}_1$, $\mathbf{F}_2$, and $\mathbf{F}_3$, is represented by the vector $\mathbf{F}$, which is the sum of the vectors $\mathbf{F}_1$, $\mathbf{F}_2$, and $\mathbf{F}_3$:

$$\begin{aligned}\mathbf{F} &= \mathbf{F}_1 + \mathbf{F}_2 + \mathbf{F}_3\\ &= (5\sqrt{2}, 5\sqrt{2}) + (-3, 0) + (0, -200)\\ &= (5\sqrt{2} - 3, 5\sqrt{2} - 200).\end{aligned}$$

The magnitude of this force corresponds to the length of $\mathbf{F}$,

$$|\mathbf{F}| = \sqrt{(5\sqrt{2} - 3)^2 + (5\sqrt{2} - 200)^2} = 193.0,$$

and must therefore be 193.0 N. Its direction is shown in Figure 12.51(c), where

$$\theta = \mathrm{Tan}^{-1}\left(\frac{200 - 5\sqrt{2}}{5\sqrt{2} - 3}\right) = 1.55 \text{ radians}.$$

FIGURE 12.51

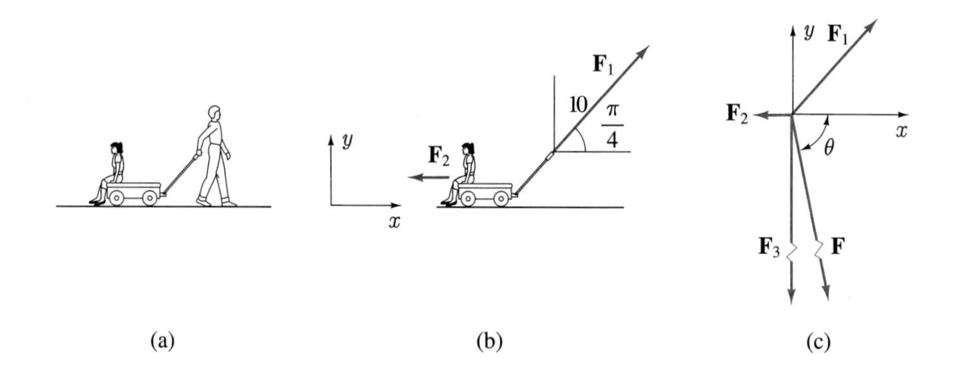

EXAMPLE 12.8

If two water skiers S_1 and S_2 exert forces of 500 N and 600 N, respectively, on the boat shown in Figure 12.52(a), what is the resultant force on the boat?

FIGURE 12.52

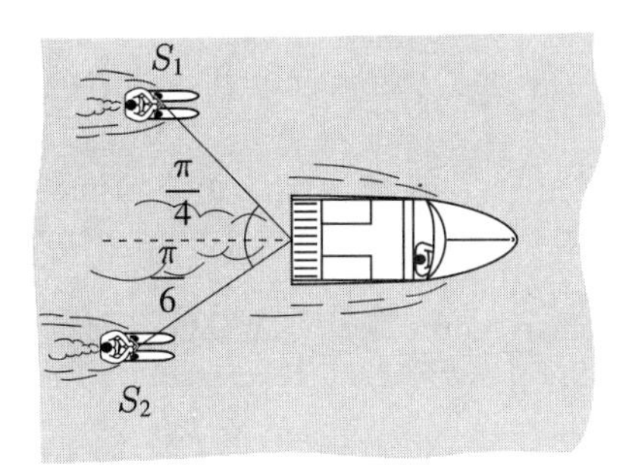

(a)

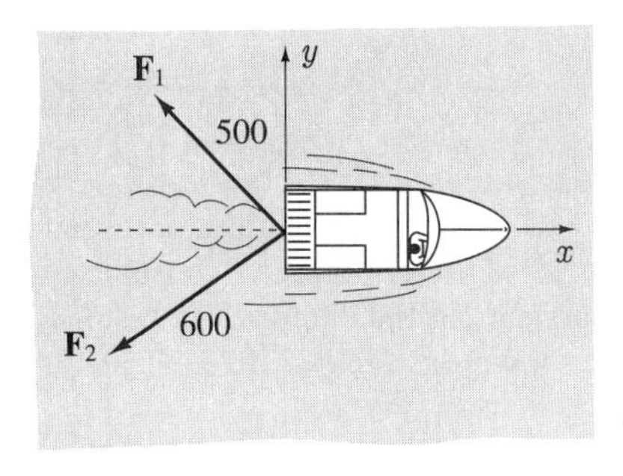

(b)

SOLUTION If we introduce the coordinate system in Figure 12.52(b), then lengths of the force vectors $\mathbf{F}_1$ and $\mathbf{F}_2$ exerted by S_1 and S_2 are 500 and 600 units, respectively. The components of $\mathbf{F}_1$ and $\mathbf{F}_2$ are

$$\begin{aligned}\mathbf{F}_1 &= \left(-\frac{500}{\sqrt{2}}, \frac{500}{\sqrt{2}}\right) \\ &= (-250\sqrt{2}, 250\sqrt{2});\end{aligned} \qquad \begin{aligned}\mathbf{F}_2 &= \left(-\frac{600\sqrt{3}}{2}, -\frac{600}{2}\right) \\ &= (-300\sqrt{3}, -300).\end{aligned}$$

The resultant force $\mathbf{F}$ on the boat is then the sum of $\mathbf{F}_1$ and $\mathbf{F}_2$:

$$\mathbf{F} = \mathbf{F}_1 + \mathbf{F}_2 = (-250\sqrt{2} - 300\sqrt{3}, 250\sqrt{2} - 300)\,\mathrm{N}.$$

■

By the x-, y-, and z-components (v_x, v_y, v_z) of a vector $\mathbf{v}$, we mean that if we start at a point P (Figure 12.53) and proceed v_x units in the x-direction, v_y units in the y-direction, and v_z units in the z-direction to a point Q, then $\mathbf{v}$ is the directed line segment joining P and Q. To phrase this another way, we introduce three special vectors parallel to the coordinate axes. We define $\hat{\mathbf{i}}$ as a unit vector in the positive x-direction, $\hat{\mathbf{j}}$ as a unit vector in the positive y-direction, and $\hat{\mathbf{k}}$ as a unit vector in the positive z-direction. We have shown these vectors with their tails at the origin in Figure 12.54, and it is clear that their components are

$$\hat{\mathbf{i}} = (1,0,0), \quad \hat{\mathbf{j}} = (0,1,0), \quad \hat{\mathbf{k}} = (0,0,1). \tag{12.15}$$

FIGURE 12.53

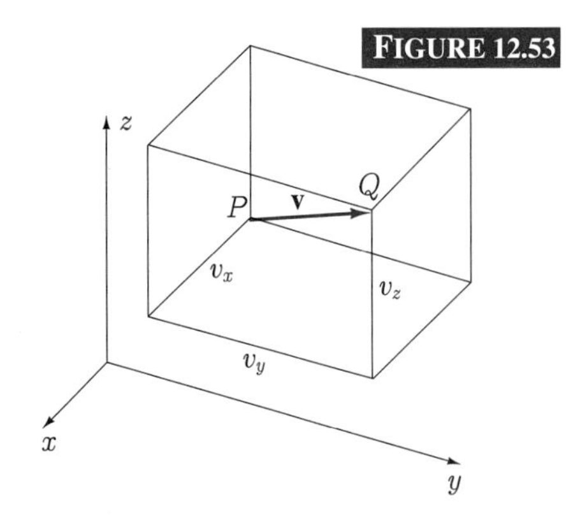

FIGURE 12.54

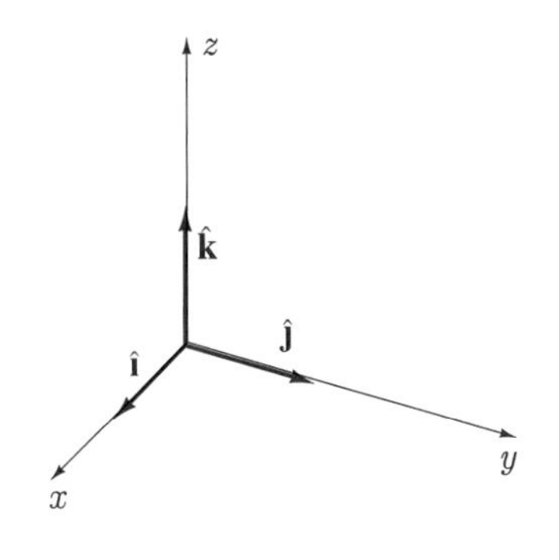

But note, then, that we can write the vector $\mathbf{v} = (v_x, v_y, v_z)$ in the form

$$\begin{aligned}\mathbf{v} &= (v_x, 0, 0) + (0, v_y, 0) + (0, 0, v_z) \\ &= v_x(1, 0, 0) + v_y(0, 1, 0) + v_z(0, 0, 1) \\ &= v_x\hat{\mathbf{i}} + v_y\hat{\mathbf{j}} + v_z\hat{\mathbf{k}}.\end{aligned}$$

In other words, every vector in space can be written as a linear combination of the three vectors $\hat{\mathbf{i}}$, $\hat{\mathbf{j}}$, and $\hat{\mathbf{k}}$ (i.e., as a constant times $\hat{\mathbf{i}}$ plus a constant times $\hat{\mathbf{j}}$ plus a constant times $\hat{\mathbf{k}}$). Furthermore, the constants multiplying $\hat{\mathbf{i}}$, $\hat{\mathbf{j}}$, and $\hat{\mathbf{k}}$ are precisely the Cartesian components of the vector. This result is equally clear geometrically. In Figure 12.53, we have shown the vector $\mathbf{v}$ from $\mathbf{P}$ to $\mathbf{Q}$. If we define points $\mathbf{A}$ and $\mathbf{B}$ as shown in Figure 12.55, then

$$\mathbf{v} = \overrightarrow{\mathbf{PQ}} = \overrightarrow{\mathbf{PB}} + \overrightarrow{\mathbf{BQ}} = \overrightarrow{\mathbf{PA}} + \overrightarrow{\mathbf{AB}} + \overrightarrow{\mathbf{BQ}}.$$

But because $\overrightarrow{\mathbf{PA}}$ is a vector in the positive x-direction and has length v_x, it follows that $\overrightarrow{\mathbf{PA}} = v_x\hat{\mathbf{i}}$. Similarly, $\overrightarrow{\mathbf{AB}} = v_y\hat{\mathbf{j}}$ and $\overrightarrow{\mathbf{BQ}} = v_z\hat{\mathbf{k}}$, and therefore

$$\mathbf{v} = v_x\hat{\mathbf{i}} + v_y\hat{\mathbf{j}} + v_z\hat{\mathbf{k}}. \tag{12.16}$$

To say then that v_x, v_y and v_z are the x-, y-, and z-components of a vector $\mathbf{v}$ is equivalent to saying that $\mathbf{v}$ can be written in the form 12.16.

Some authors refer to v_x, v_y, and v_z as the **scalar components** of the vector $\mathbf{v}$, and the vectors $v_x\hat{\mathbf{i}}$, $v_y\hat{\mathbf{j}}$, and $v_z\hat{\mathbf{k}}$ as the **vector components** of $\mathbf{v}$. By "component," we always mean "scalar component."

Vectors in the xy-plane have only an x- and a y-component, and can therefore be written in terms of $\hat{\mathbf{i}}$ and $\hat{\mathbf{j}}$. If $\mathbf{v} = (v_x, v_y)$, then we write equivalently that $\mathbf{v} = v_x\hat{\mathbf{i}} + v_y\hat{\mathbf{j}}$ (Figure 12.56.). Vectors along the x-axis have only an x-component and can therefore be written in the form $\mathbf{v} = v_x\hat{\mathbf{i}}$.

FIGURE 12.55

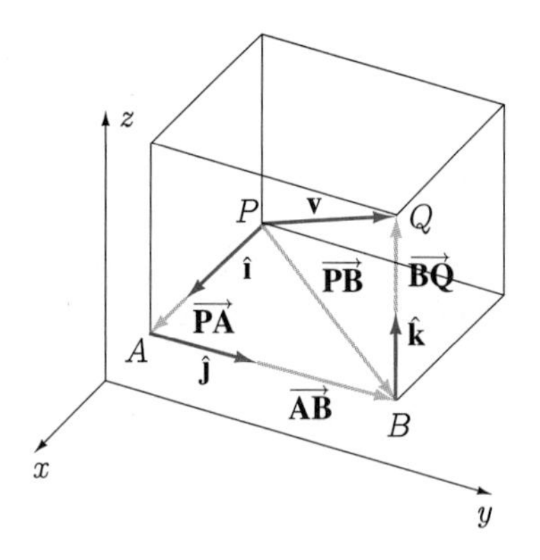

FIGURE 12.56

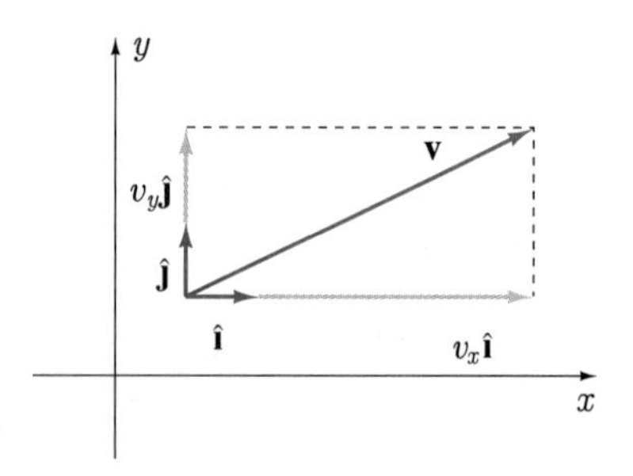

We use this new notation in the following example.

EXAMPLE 12.9

The force $\mathbf{F}$ exerted on a point charge q_1 coulombs by a charge q_2 coulombs is defined by Coulomb's law as

$$\mathbf{F} = \frac{q_1 q_2}{4\pi\epsilon_0 r^2}\hat{\mathbf{r}} \text{ N},$$

where ϵ_0 is a positive constant, r is the distance in metres between the charges, and $\hat{\mathbf{r}}$ is a unit vector in the direction from q_2 to q_1 (Figure 12.57). When q_1 and q_2 are both positive charges or both negative charges, then $\mathbf{F}$ is repulsive, and when one is positive and the other is negative, $\mathbf{F}$ is attractive. In particular, suppose that charges of 2 C and -2 C are placed at $(0,0,0)$ and $(3,0,0)$, respectively, and a third charge of 1 C is placed at $(1,1,1)$. According to Coulomb's law, the 2-C charge will exert a repulsive force on the 1-C charge, and the -2-C charge will exert an attractive force on the 1-C charge. Find the resultant of these two forces on the 1-C charge.

FIGURE 12.57

SOLUTION If $\mathbf{F}_1$ is the force exerted on the 1-C charge by the -2-C charge (Figure12.58), then

$$\mathbf{F}_1 = \frac{(1)(-2)}{4\pi\epsilon_0 r^2}\hat{\mathbf{r}},$$

where the distance between the charges is $r = \sqrt{(-2)^2 + 1^2 + 1^2} = \sqrt{6}$. The vector from $(3,0,0)$ to $(1,1,1)$ is $(-2,1,1)$, and therefore

$$\hat{\mathbf{r}} = \frac{1}{\sqrt{6}}(-2\hat{\mathbf{i}} + \hat{\mathbf{j}} + \hat{\mathbf{k}}).$$

Consequently,

$$\mathbf{F}_1 = \frac{-2}{4\pi\epsilon_0(6)} \cdot \frac{1}{\sqrt{6}}(-2\hat{\mathbf{i}} + \hat{\mathbf{j}} + \hat{\mathbf{k}}) = \frac{1}{12\sqrt{6}\pi\epsilon_0}(2\hat{\mathbf{i}} - \hat{\mathbf{j}} - \hat{\mathbf{k}}).$$

Similarly, the force $\mathbf{F}_2$ exerted on the 1-C charge by the charge at the origin is

$$\mathbf{F}_2 = \frac{(1)(2)}{4\pi\epsilon_0(3)} \cdot \frac{1}{\sqrt{3}}(\hat{\mathbf{i}} + \hat{\mathbf{j}} + \hat{\mathbf{k}}) = \frac{1}{6\sqrt{3}\pi\epsilon_0}(\hat{\mathbf{i}} + \hat{\mathbf{j}} + \hat{\mathbf{k}}).$$

The resultant of these forces is

$$\mathbf{F} = \mathbf{F}_1 + \mathbf{F}_2 = \frac{1}{12\sqrt{6}\pi\epsilon_0}\{(2 + 2\sqrt{2})\hat{\mathbf{i}} + (2\sqrt{2} - 1)\hat{\mathbf{j}} + (2\sqrt{2} - 1)\hat{\mathbf{k}}\} \text{ N}.$$

FIGURE 12.58

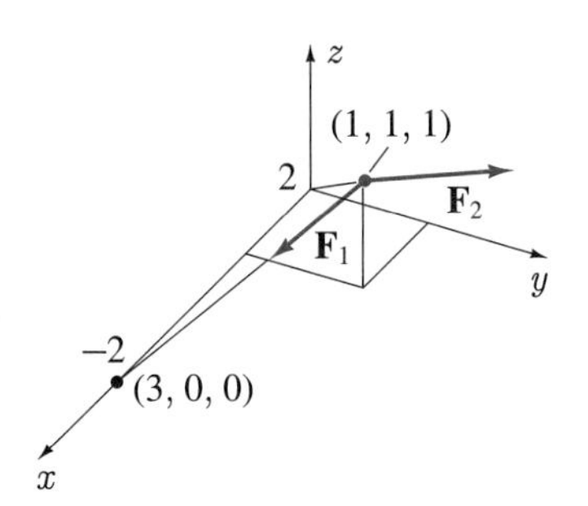

■

EXERCISES 12.3

If $\mathbf{u} = (1,3,6)$, $\mathbf{v} = (-2,0,4)$, and $\mathbf{w} = (4,3,-2)$, find components for the vector in Exercises 1–10.

1. $3\mathbf{u} - 2\mathbf{v}$

2. $2\mathbf{w} + 3\mathbf{v}$

3. $\mathbf{w} - 3\mathbf{u} - 3\mathbf{v}$

4. $\hat{\mathbf{v}}$

5. $2\hat{\mathbf{w}} - 3\mathbf{v}$

6. $|\mathbf{v}|\mathbf{v} - 2|\hat{\mathbf{v}}|\mathbf{w}$

7. $(15 - 2|\mathbf{w}|)(\mathbf{u} + \mathbf{v})$

8. $|3\mathbf{u}|\mathbf{v} - |-2\mathbf{v}|\mathbf{u}$

9. $|2\mathbf{u} + 3\mathbf{v} - \mathbf{w}|\hat{\mathbf{w}}$

10. $\dfrac{\mathbf{v} - \mathbf{w}}{|\mathbf{v} + \mathbf{w}|}$

If $\mathbf{u} = 2\hat{\mathbf{i}} + \hat{\mathbf{j}}$ and $\mathbf{v} = -\hat{\mathbf{i}} + 3\hat{\mathbf{j}}$, find components of the vector in Exercises 11–14 and illustrate the vector geometrically.

11. $\mathbf{u} + \mathbf{v}$

12. $\mathbf{u} - \mathbf{v}$

13. $2\hat{\mathbf{u}}$

14. $\hat{\mathbf{v}} + \hat{\mathbf{u}}$

In Exercises 15–24 find the Cartesian components for the spatial vector described. In each case, draw the vector.

15. From $(1,3,2)$ to $(-1,4,5)$

16. With length 5 in the positive x-direction

17. With length 2 in the negative z-direction

18. With tail at $(1,1,1)$, length 3, and pointing toward the point $(1,3,5)$

19. With positive y-component, length 1, and parallel to the line through $(1,3,6)$ and $(-2,1,4)$

20. In the same direction as the vector from $(1,0,-1)$ to $(3,2,-4)$ but only half as long

21. With positive and equal x- and y-components, length 10, and z-component equal to 4

22. Has its tail at the origin, makes angles of $\pi/3$ and $\pi/4$ radians with the positive x- and y-axes respectively, and has length $5/2$

23. From $(1,3,-2)$ to the midpoint of the line segment joining $(2,4,-3)$ and $(1,5,6)$

24. Has its tail at the origin, makes equal angles with the positive coordinate axes, has all positive components, and has length 2

25. If P, Q, and R are the points with coordinates $(3,2,-1)$, $(0,1,4)$, and $(6,5,-2)$ respectively, find the coordinates of a point S in order that $\overrightarrow{\mathbf{PQ}} = \overrightarrow{\mathbf{RS}}$.

26. Prove that vector addition and scalar multiplication have the properties in equations 12.10.

27. Draw all spatial vectors of length 1 that have equal x- and y-components and tails at the origin.

28. Draw all spatial vectors of length 2 that have their tails at the origin and make an angle of $\pi/4$ with the positive z-axis.

29. If $\mathbf{u} = 3\hat{\mathbf{i}} + 2\hat{\mathbf{j}} - 4\hat{\mathbf{k}}$, and $\mathbf{v} = \hat{\mathbf{i}} + 6\hat{\mathbf{j}} + 5\hat{\mathbf{k}}$, find scalars λ and ρ so that the vector $\mathbf{w} = 5\hat{\mathbf{i}} - 18\hat{\mathbf{j}} - 32\hat{\mathbf{k}}$ can be written in the form $\mathbf{w} = \lambda\mathbf{u} + \rho\mathbf{v}$.

30. Find a vector $\mathbf{T}$ of length 3 along the tangent line to the curve $y = x^2$ at the point $(2,4)$ (Figure 12.59).

FIGURE 12.59

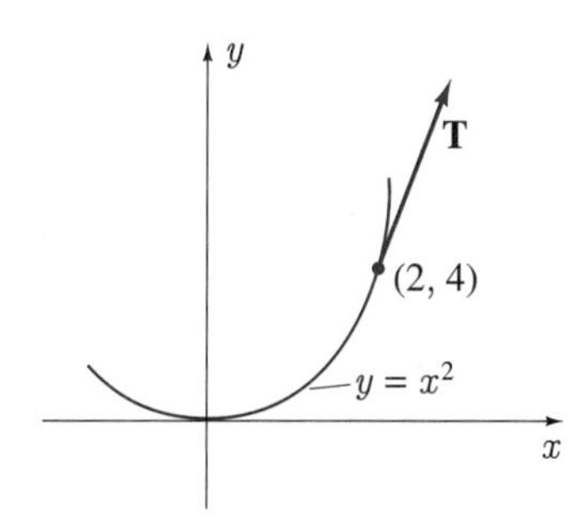

31. Use Coulomb's law (see Example 12.9) to find the force on a charge of 2 coulombs at the origin due to equal charges of 3 coulombs at the points $(1,1,2)$ and $(2,-1,-2)$.

32. Newton's universal law of gravitation states that the force of attraction $\mathbf{F}$, in newtons, exerted on a mass of m kilograms by a mass of M kilograms is

$$\mathbf{F} = \frac{GmM}{r^2}\hat{\mathbf{r}},$$

where $G = 6.67 \times 10^{-11}$ is a constant, r is the distance in metres between the masses, and $\hat{\mathbf{r}}$ is a unit vector in the direction from m to M. If point masses, each of 5 kilograms, are situated at $(5,1,3)$ and $(-1,2,1)$, what is the resultant force on a mass of 10 kilograms at $(2,2,2)$?

33. Illustrate geometrically the triangle inequality for vectors $|\mathbf{u}+\mathbf{v}| \le |\mathbf{u}|+|\mathbf{v}|$. Prove the result algebraically.

Vectors **u**, **v**, and **w** are said to be linearly dependent if there exist three scalars a, b, and c, not all zero, such that $a\mathbf{u} + b\mathbf{v} + c\mathbf{w} = \mathbf{0}$. If this equation can only be satisfied with $a = b = c = 0$, the vectors are said to be linearly independent. In Exercises 34–37 determine whether the vectors are linearly dependent or linearly independent.

34. $\mathbf{u} = (1,1,1)$, $\mathbf{v} = (2,1,3)$, $\mathbf{w} = (4,2,6)$

35. $\mathbf{u} = (1,1,1)$, $\mathbf{v} = (2,1,3)$, $\mathbf{w} = (1,6,4)$

36. $\mathbf{u} = (-1,3,-5)$, $\mathbf{v} = (2,4,-1)$, $\mathbf{w} = (3,11,-7)$

37. $\mathbf{u} = (4,2,6)$, $\mathbf{v} = (1,3,-2)$, $\mathbf{w} = (7,1,4)$

38. Use vectors to show that the line segment joining the midpoints of two sides of a triangle is parallel to the third side and its length is one-half the length of the third side (Figure 12.60).

FIGURE 12.60

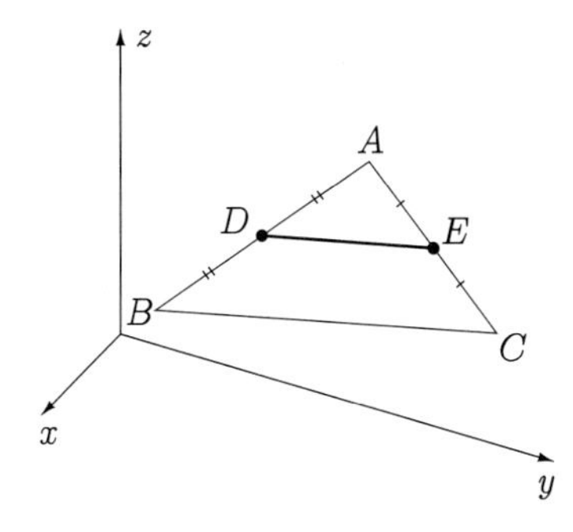

39. Use vectors to show that the medians of a triangle (Figure 12.61) all meet in a point with coordinates

$$\left(\frac{x_1 + x_2 + x_3}{3}, \frac{y_1 + y_2 + y_3}{3}, \frac{z_1 + z_2 + z_3}{3}\right).$$

FIGURE 12.61

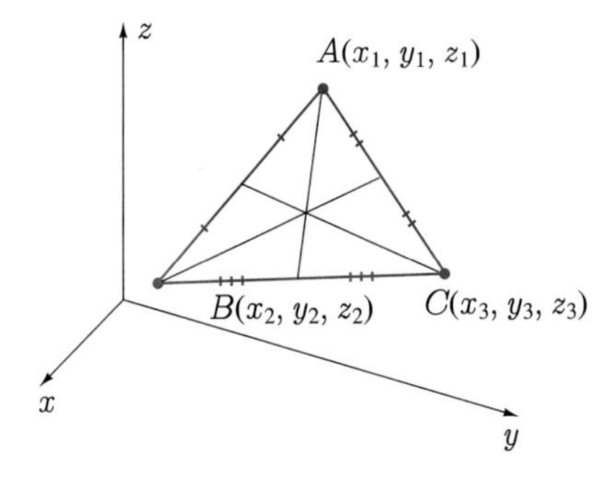

40. If n point masses m_i are located at points (x_i, y_i) in the xy-plane, equations 7.29 and 7.30 define the centre of mass $(\overline{x}, \overline{y})$ of the system. Show that these two scalar equations are represented by the one vector equation

$$M(\overline{x}, \overline{y}) = \sum_{i=1}^{n} m_i \mathbf{r}_i$$

where $\mathbf{r}_i$ is the vector joining the origin to the point (x_i, y_i).

41. If $\mathbf{u} = u_x\hat{\mathbf{i}} + u_y\hat{\mathbf{j}}$ and $\mathbf{v} = v_x\hat{\mathbf{i}} + v_y\hat{\mathbf{j}}$, show that every vector **w** in the xy-plane can be written in the form $\mathbf{w} = \lambda\mathbf{u} + \rho\mathbf{v}$ provided $u_xv_y - u_yv_x \neq 0$.

42. Two springs (with constants k_1 and k_2 and unstretched lengths l) are fixed at points A and B (Figure 12.62). They are joined to a sleeve that slides along the x-axis. Find the resultant force of the springs on the sleeve at any point between O and C.

FIGURE 12.62

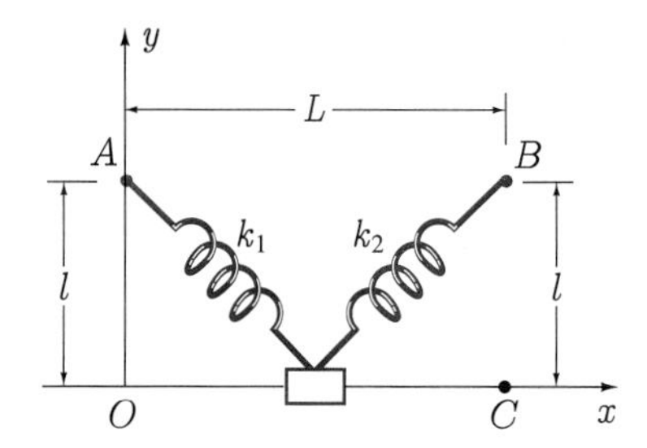

43. In Figure 12.63(a), the xy-plane is the interface between two materials that both transmit light. If a ray of light strikes the surface in a direction defined by the unit vector $\hat{\mathbf{u}} = (u_x, u_y, u_z)$, then some of the light is reflected along a vector $\hat{\mathbf{v}}$, and some is refracted along $\hat{\mathbf{w}}$. The three vectors $\hat{\mathbf{u}}$, $\hat{\mathbf{v}}$, and $\hat{\mathbf{w}}$ all lie in a plane that is perpendicular to the xy-plane.

(a) If the angle of incidence i (Figure 12.63b) is equal to the angle of reflection ϕ, find the components for $\hat{\mathbf{v}}$ in terms of those of $\hat{\mathbf{u}}$.

(b) If the angle of refraction θ is related to the angle of incidence by $n_1 \sin i = n_2 \sin\theta$, where n_1 and n_2 are the indices of refraction of the two materials, find the components of $\hat{\mathbf{w}}$.

FIGURE 12.63

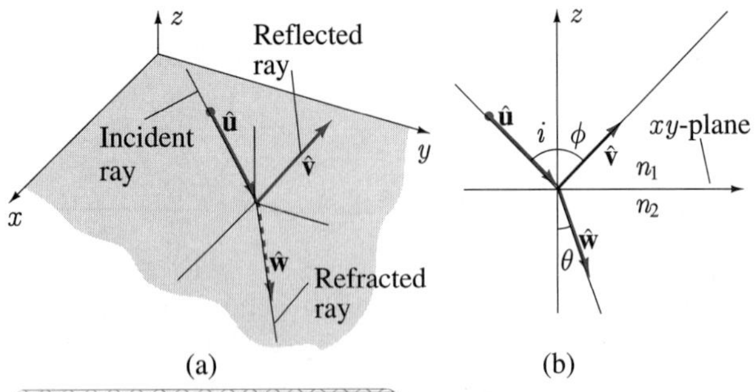

44. Vectors $\mathbf{v}_1, \mathbf{v}_2, \ldots, \mathbf{v}_n$ are drawn from the centre of a regular n-sided polygon in the plane to each of its vertices. Show that the sum of these vectors is the zero vector.

SECTION 12.4

The Scalar Product

In Section 12.3 we learned how to multiply a vector by a scalar and how to add and subtract vectors. The next question naturally is, "Can vectors be multiplied?" The answer is yes, and in fact we define two products for vectors, one of which yields a scalar and the other a vector. The first is defined algebraically as follows.

Definition 12.8

The **scalar product (dot product** or **inner product)** of two vectors **u** and **v** with Cartesian components (u_x, u_y, u_z) and (v_x, v_y, v_z) is defined as

$$\mathbf{u} \cdot \mathbf{v} = u_x v_x + u_y v_y + u_z v_z. \tag{12.17}$$

If **u** and **v** have only x- and y-components, Definition 12.8 reduces to

$$\mathbf{u} \cdot \mathbf{v} = u_x v_x + u_y v_y, \tag{12.18}$$

and if they have only x-components, it becomes

$$\mathbf{u} \cdot \mathbf{v} = u_x v_x. \tag{12.19}$$

It is straightforward to check that

$$\hat{\mathbf{i}} \cdot \hat{\mathbf{i}} = \hat{\mathbf{j}} \cdot \hat{\mathbf{j}} = \hat{\mathbf{k}} \cdot \hat{\mathbf{k}} = 1, \tag{12.20a}$$
$$\hat{\mathbf{i}} \cdot \hat{\mathbf{j}} = \hat{\mathbf{j}} \cdot \hat{\mathbf{k}} = \hat{\mathbf{i}} \cdot \hat{\mathbf{k}} = 0, \tag{12.20b}$$

and that the scalar product is commutative and distributive:

$$\mathbf{u} \cdot \mathbf{v} = \mathbf{v} \cdot \mathbf{u}, \tag{12.21a}$$
$$\mathbf{u} \cdot (\lambda\mathbf{v} + \rho\mathbf{w}) = \lambda(\mathbf{u} \cdot \mathbf{v}) + \rho(\mathbf{u} \cdot \mathbf{w}). \tag{12.21b}$$

EXAMPLE 12.10

If $\mathbf{u} = (-2, 1, 3)$ and $\mathbf{v} = (3, -2, -1)$, evaluate each of the following:

$$\text{(a) } \mathbf{u} \cdot \mathbf{v}; \quad \text{(b) } 3\mathbf{u} \cdot (2\mathbf{u} - 4\mathbf{v}).$$

SOLUTION

(a)
$$\mathbf{u} \cdot \mathbf{v} = (-2)(3) + (1)(-2) + (3)(-1) = -11$$

(b)
$$\begin{aligned} 3\mathbf{u} \cdot (2\mathbf{u} - 4\mathbf{v}) &= (-6, 3, 9) \cdot (-16, 10, 10) \\ &= (-6)(-16) + (3)(10) + (9)(10) \\ &= 216. \end{aligned}$$

■

By taking the scalar product of a vector $\mathbf{v} = (v_x, v_y, v_z)$ with itself, we obtain

$$\mathbf{v} \cdot \mathbf{v} = v_x^2 + v_y^2 + v_z^2 = |\mathbf{v}|^2. \tag{12.22}$$

Because Definition 12.8 for the scalar product of two vectors $\mathbf{u}$ and $\mathbf{v}$ is phrased in terms of the components of $\mathbf{u}$ and $\mathbf{v}$, and these components depend on the coordinate system used, it follows that this definition also depends on the fact that we have used Cartesian coordinates. Were we to use a different set of coordinates (such as polar coordinates), then the definition of $\mathbf{u} \cdot \mathbf{v}$ in terms of components in that coordinate system would be different. For this reason we now find a geometric definition for the scalar product (which is therefore independent of coordinate systems).

Theorem 12.2

If two nonzero vectors $\mathbf{u}$ *and* $\mathbf{v}$ *are placed tail to tail, and* θ *is the angle between these vectors* $(0 \leq \theta \leq \pi)$*, then*

$$\mathbf{u} \cdot \mathbf{v} = |\mathbf{u}||\mathbf{v}|\cos\theta. \tag{12.23}$$

FIGURE 12.64

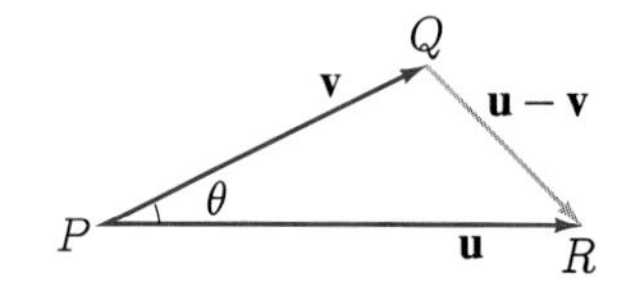

Proof The cosine law applied to the triangle in Figure 12.64 gives

$$|\overrightarrow{\mathbf{QR}}|^2 = |\overrightarrow{\mathbf{PQ}}|^2 + |\overrightarrow{\mathbf{PR}}|^2 - 2|\overrightarrow{\mathbf{PQ}}||\overrightarrow{\mathbf{PR}}|\cos\theta,$$

or

$$|\mathbf{u} - \mathbf{v}|^2 = |\mathbf{v}|^2 + |\mathbf{u}|^2 - 2|\mathbf{v}||\mathbf{u}|\cos\theta.$$

Consequently,

$$|\mathbf{u}||\mathbf{v}|\cos\theta = \frac{1}{2}\{|\mathbf{u}|^2 + |\mathbf{v}|^2 - |\mathbf{u} - \mathbf{v}|^2\},$$

and if (u_x, u_y, u_z) and (v_x, v_y, v_z) are the Cartesian components of $\mathbf{u}$ and $\mathbf{v}$, then

$$\begin{aligned} |\mathbf{u}||\mathbf{v}|\cos\theta &= \frac{1}{2}\{(u_x^2 + u_y^2 + u_z^2) + (v_x^2 + v_y^2 + v_z^2) \\ &\quad - [(u_x - v_x)^2 + (u_y - v_y)^2 + (u_z - v_z)^2]\} \\ &= u_x v_x + u_y v_y + u_z v_z \\ &= \mathbf{u} \cdot \mathbf{v}. \end{aligned}$$

An immediate consequence of this result is the following.

Corollary *Two nonzero vectors* **u** *and* **v** *are perpendicular if and only if*

$$\mathbf{u} \cdot \mathbf{v} = 0. \tag{12.24}$$

For example, the vectors $\mathbf{u} = (1, 2, -1)$ and $\mathbf{v} = (4, 2, 8)$ are perpendicular since $\mathbf{u} \cdot \mathbf{v} = (1)(4) + (2)(2) + (-1)(8) = 0$.

Expression 12.23 doesn't just tell us whether or not the angle between two vectors is $\pi/2$ radians; it can also be used to determine the exact angle between any two nonzero vectors **u** and **v**, simply by solving for θ:

$$\cos\theta = \frac{\mathbf{u} \cdot \mathbf{v}}{|\mathbf{u}||\mathbf{v}|}. \tag{12.25}$$

Since principal values of the inverse cosine function lie between 0 and π, precisely the range for θ, we can write that

$$\theta = \mathrm{Cos}^{-1}\left(\frac{\mathbf{u} \cdot \mathbf{v}}{|\mathbf{u}||\mathbf{v}|}\right). \tag{12.26}$$

EXAMPLE 12.11

Find the angle between the vectors $\mathbf{u} = (2, -3, 1)$ and $\mathbf{v} = (5, 2, 4)$.

SOLUTION According to formula 12.26,

$$\theta = \mathrm{Cos}^{-1}\left(\frac{\mathbf{u} \cdot \mathbf{v}}{|\mathbf{u}||\mathbf{v}|}\right) = \mathrm{Cos}^{-1}\left(\frac{10 - 6 + 4}{\sqrt{4 + 9 + 1}\sqrt{25 + 4 + 16}}\right) = \mathrm{Cos}^{-1}\left(\frac{8}{3\sqrt{70}}\right) = 1.25.$$

■

EXAMPLE 12.12

Find the angle between the lines $x + 2y = 3$ and $4x - 3y = 5$ in the xy-plane.

SOLUTION Since the slope of $x + 2y = 3$ is $-\frac{1}{2}$, a vector along this line is $\mathbf{u} = (-2, 1)$. Similarly, a vector along $4x - 3y = 5$ is $\mathbf{v} = (3, 4)$. If θ is the angle between these vectors, and therefore between the lines (Figure 12.65), then

$$\theta = \mathrm{Cos}^{-1}\left(\frac{\mathbf{u} \cdot \mathbf{v}}{|\mathbf{u}||\mathbf{v}|}\right) = \mathrm{Cos}^{-1}\left(\frac{-6 + 4}{\sqrt{5}\sqrt{25}}\right) = \mathrm{Cos}^{-1}\left(\frac{-2}{5\sqrt{5}}\right) = 1.75$$

radians. The acute angle between the lines is $\pi - 1.75 = 1.39$ radians. Formula 1.14 gives the same result.

FIGURE 12.65

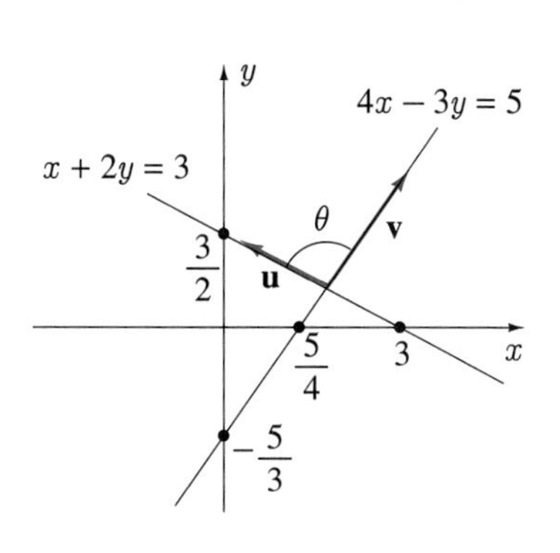

Planes

Scalar products can be used to find equations for planes in space. A plane can be characterized in various ways: by two intersecting lines, by a line and a point not on the line, or by three noncollinear points. For our present purposes, we use the fact that given a point $P(x_1, y_1, z_1)$ and a vector (A, B, C) (Figure 12.66), there is one and only one plane through P that is perpendicular to (A, B, C).

FIGURE 12.66

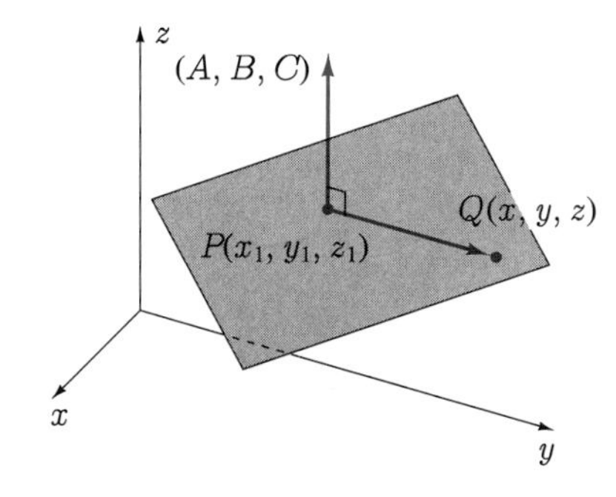

To find the equation of this plane we note that if $Q(x, y, z)$ is any other point in the plane, then the vector $\overrightarrow{\mathbf{PQ}} = (x - x_1, y - y_1, z - z_1)$ lies in the plane. But $\overrightarrow{\mathbf{PQ}}$ must then be perpendicular to (A, B, C); hence by the corollary to Theorem 12.2,

$$(A, B, C) \cdot (x - x_1, y - y_1, z - z_1) = 0.$$

Because this equation must be satisfied by every point (x, y, z) in the plane (and at the same time is not satisfied by any point not in the plane), it must be the equation of the plane through P and perpendicular to (A, B, C). If we expand the scalar product, we obtain the equation of the plane in the form

$$A(x - x_1) + B(y - y_1) + C(z - z_1) = 0.$$

This result is worth stating as a theorem.

Theorem 12.3

An equation for the plane through (x_1, y_1, z_1) perpendicular to the vector (A, B, C) is

$$A(x - x_1) + B(y - y_1) + C(z - z_1) = 0. \qquad (12.27)$$

Equation 12.27 can also be written in the form

$$Ax + By + Cz + D = 0, \tag{12.28}$$

where $D = -(Ax_1 + By_1 + Cz_1)$, and this equation is said to be **linear** in x, y, and z. We have shown then that every plane has a linear equation, and the coefficients A, B, and C of x, y, and z in the equation are the components of a vector (A, B, C) that is perpendicular to the plane. Conversely, every linear equation in the form of 12.28 describes a plane. Instead of saying that (A, B, C) is perpendicular to the plane $Ax + By + Cz + D = 0$, we often say that (A, B, C) is **normal** to the plane or that (A, B, C) is a **normal vector** to the plane.

EXAMPLE 12.13

Find an equation for the plane through the point $(4, -3, 5)$ and normal to the vector $(4, -8, 3)$.

SOLUTION According to 12.27, the equation of the plane is

$$4(x - 4) - 8(y + 3) + 3(z - 5) = 0, \quad \text{or,} \quad 4x - 8y + 3z = 55.$$

■

EXAMPLE 12.14

Determine whether the planes $x + 2y - 4z = 10$ and $2x + 4y - 8z = 11$ are parallel.

SOLUTION Normal vectors to these planes are $\mathbf{N}_1 = (1, 2, -4)$ and $\mathbf{N}_2 = (2, 4, -8)$. Since $\mathbf{N}_2 = 2\mathbf{N}_1$, the normal vectors are in the same direction, and therefore, the planes are parallel. ■

Resolution of Vectors into Perpendicular Directions

In many applications of mathematics, we are called on to find distances between geometric objects. For example, we already have formulas 1.7 and 12.1 for the distances between points in a plane and points in space. In a plane we might want to find the distance from a point to a line or between two parallel lines. In space, there are many other distances that might prove useful: distance from a point to a plane or a line, between parallel planes, between nonintersecting lines, etc. Each of these distances can be calculated in a number of ways, but the easiest method is to use components of vectors.

We have defined what is meant by Cartesian components of a vector. In particular, Cartesian components (v_x, v_y, v_z) of a vector $\mathbf{v}$ are those scalars that multiply $\hat{\mathbf{i}}$, $\hat{\mathbf{j}}$ and $\hat{\mathbf{k}}$ so that $\mathbf{v} = v_x\hat{\mathbf{i}} + v_y\hat{\mathbf{j}} + v_z\hat{\mathbf{k}}$. They can be represented in terms of scalar products as

$$v_x = \mathbf{v} \cdot \hat{\mathbf{i}}, \quad v_y = \mathbf{v} \cdot \hat{\mathbf{j}}, \quad v_z = \mathbf{v} \cdot \hat{\mathbf{k}}. \tag{12.29}$$

Geometrically they can be found by drawing $\mathbf{v}$, $\hat{\mathbf{i}}$, $\hat{\mathbf{j}}$, and $\hat{\mathbf{k}}$ all at the origin and dropping perpendiculars from the tip of $\mathbf{v}$ to the x-, y-, and z-axes (Figure 12.67). We now generalize our definition of a component along an axis to define the component of a vector in any direction whatsoever.

FIGURE 12.67

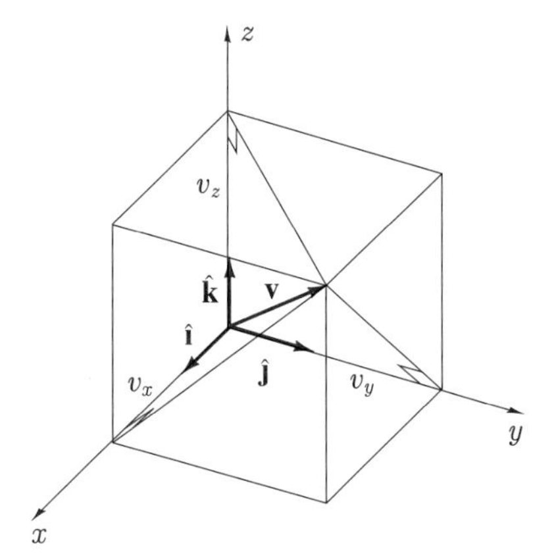

Definition 12.9

To define the component of a vector $\mathbf{v}$ in a direction $\mathbf{u}$, we place $\mathbf{v}$ and $\mathbf{u}$ tail to tail at a point P and draw the perpendicular from the tip of $\mathbf{v}$ to the line containing $\mathbf{u}$ (Figure 12.68). The directed distance PR is called the component of $\mathbf{v}$ in the direction $\mathbf{u}$. If R is on the same side of P as the tip of $\mathbf{u}$, PR is taken as positive; and if R is on that side of P opposite to the tip of $\mathbf{u}$, PR is negative.

FIGURE 12.68

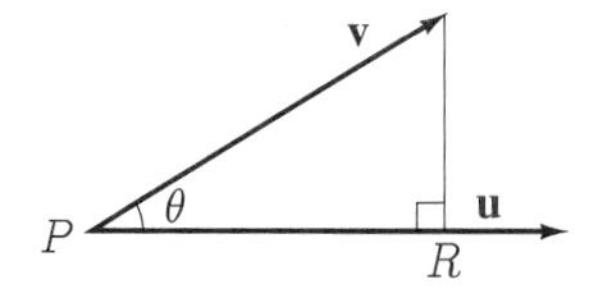

In Figure 12.68 the component of $\mathbf{v}$ in direction $\mathbf{u}$ is positive, and in Figure 12.69 the component of $\mathbf{v}$ in direction $\mathbf{u}$ is negative. Note that the length of $\mathbf{u}$ is irrelevant; it is only the direction of $\mathbf{u}$ that determines the component of $\mathbf{v}$ in the direction $\mathbf{u}$. If θ is the angle between $\mathbf{u}$ and $\mathbf{v}$, then

$$PR = |\mathbf{v}| \cos\theta.$$

FIGURE 12.69

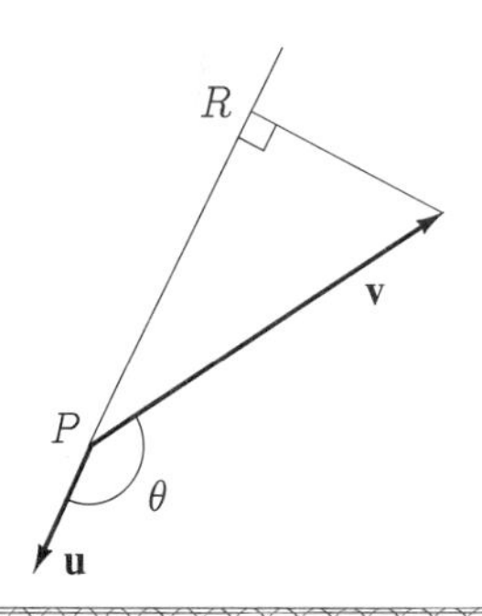

The right-hand side of this equation looks very much like the scalar product of $\mathbf{v}$ and a vector that makes an angle θ with $\mathbf{v}$. It lacks only the length of this second vector. Clearly, $\mathbf{u}$ is a vector that makes an angle θ with $\mathbf{v}$, but we *cannot* write $PR = |\mathbf{v}||\mathbf{u}|\cos\theta$, since the length of $\mathbf{u}$ need not be 1. If, however, $\hat{\mathbf{u}}$ is the unit vector in the same direction as $\mathbf{u}$, then we can write

$$PR = |\mathbf{v}||\hat{\mathbf{u}}|\cos\theta = \mathbf{v} \cdot \hat{\mathbf{u}}. \qquad (12.30)$$

In other words, we have the following theorem.

Theorem 12.4

The component of a vector $\mathbf{v}$ *in a direction* $\mathbf{u}$ *is* $\mathbf{v} \cdot \hat{\mathbf{u}}$, *where* $\hat{\mathbf{u}}$ *is the unit vector in direction* $\mathbf{u}$.

This result agrees with equation 12.29 for the x-, y-, and z- components of $\mathbf{v}$.

But what good does it do to have a component of a vector in every direction? For one thing, we can represent a vector in ways other than equation 12.16. For instance, consider the two perpendicular unit vectors $\hat{\mathbf{e}}_1$ and $\hat{\mathbf{e}}_2$ in Figure 12.70. The components of vector $\mathbf{v} = 3\hat{\mathbf{i}} + 4\hat{\mathbf{j}}$ in the directions of $\hat{\mathbf{e}}_1$ and $\hat{\mathbf{e}}_2$ are

$$\mathbf{v} \cdot \hat{\mathbf{e}}_1 = \frac{3}{\sqrt{2}} + \frac{4}{\sqrt{2}} = \frac{7}{\sqrt{2}}$$

and

$$\mathbf{v} \cdot \hat{\mathbf{e}}_2 = -\frac{3}{\sqrt{2}} + \frac{4}{\sqrt{2}} = \frac{1}{\sqrt{2}}.$$

Geometrically it is clear that we can express $\mathbf{v}$ as the sum

$$\mathbf{v} = \frac{7}{\sqrt{2}}\hat{\mathbf{e}}_1 + \frac{1}{\sqrt{2}}\hat{\mathbf{e}}_2.$$

In other words, $\mathbf{v}$, and every other vector in the plane, can be expressed as a linear combination of the perpendicular vectors $\hat{\mathbf{e}}_1$ and $\hat{\mathbf{e}}_2$ (instead of $\hat{\mathbf{i}}$ and $\hat{\mathbf{j}}$). We say that $\mathbf{v}$ has been **resolved** into components along $\hat{\mathbf{e}}_1$ and $\hat{\mathbf{e}}_2$. We will find this idea of resolving vectors in terms of perpendicular vectors other than $\hat{\mathbf{i}}$, $\hat{\mathbf{j}}$, and $\hat{\mathbf{k}}$ very important in Section 12.10.

FIGURE 12.70

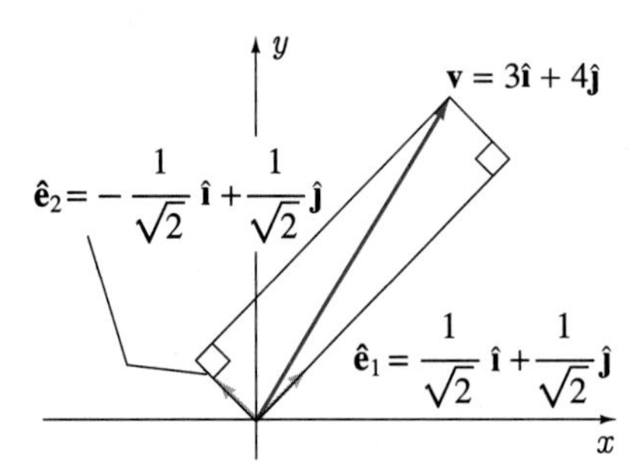

Distances Between Points, Lines, and Planes

We stated earlier that components of vectors in arbitrary directions can be used to calculate distances between various geometric objects. The first problem that we consider is the distance from a point $P(x, y)$ to a line l: $Ax + By + C = 0$ in the xy-plane (Figure 12.71). There are many ways to calculate this distance, and by this distance we mean the perpendicular (or, equivalently, the shortest) distance d from P to l. One way is to use the formula $|Ax+By+C|/\sqrt{A^2+B^2}$ developed in Exercise 45 in Section 1.3. A second way is to find the function that describes the distance from P to any point on l, and then determine the absolute minimum for the function. Another method is to use components of vectors. If Q is any point on l, then d is the component of $\overrightarrow{\mathbf{PQ}}$ in direction $\overrightarrow{\mathbf{PR}}$; i.e.,

$$d = \overrightarrow{\mathbf{PQ}} \cdot \widehat{\mathbf{PR}},$$

where $\widehat{\mathbf{PR}}$ is the unit vector in the direction of $\overrightarrow{\mathbf{PR}}$. To illustrate, suppose that P has coordinates $(3, 5)$ and l is the line $2y = x + 2$ (Figure 12.72). Since $Q(0, 1)$ is a point on l, the vector $\overrightarrow{\mathbf{PQ}}$ has Cartesian components $(-3, -4)$. The distance from P to l is the component of $\overrightarrow{\mathbf{PQ}}$ in the direction of $\overrightarrow{\mathbf{PR}}$. Now we do not know the coordinates of R (otherwise the problem would be finished), but we really do not need them. What we want is the unit vector along $\overrightarrow{\mathbf{PR}}$. Since the slope of l is $\frac{1}{2}$, a vector along l is $(2, 1)$. A vector perpendicular to l is therefore $(1, -2)$ (and note that the scalar product of $(2, 1)$ and $(1, -2)$ vanishes). Consequently, the unit vector along $\overrightarrow{\mathbf{PR}}$ is

$$\widehat{\mathbf{PR}} = \frac{1}{\sqrt{5}}(1, -2) \quad \text{and} \quad d = (-3, -4) \cdot \frac{1}{\sqrt{5}}(1, -2) = \frac{5}{\sqrt{5}} = \sqrt{5}.$$

FIGURE 12.71

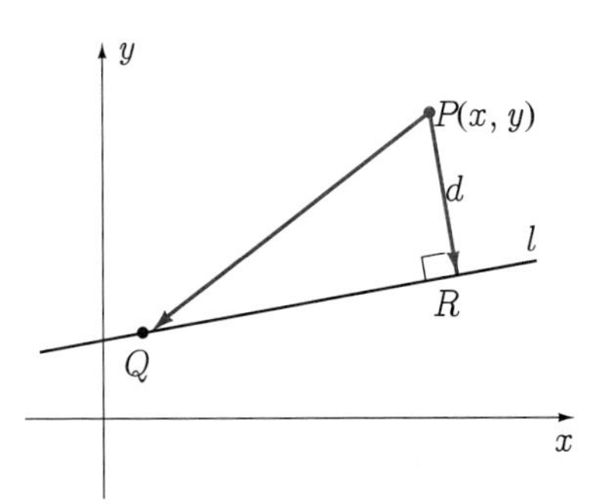

FIGURE 12.72

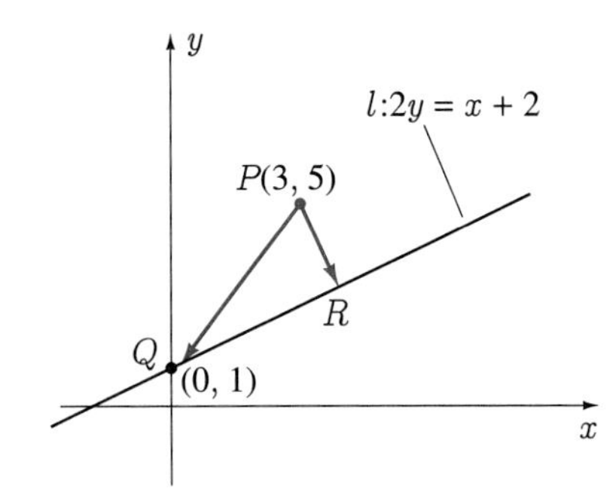

The following examples illustrate that the same technique may be applied to other distances.

EXAMPLE 12.15

Find the distance between the lines $x - 3y = 4$ and $x - 3y = -5$.

SOLUTION The required distance can be calculated as the distance from $P(4,0)$ (a point on $x - 3y = 4$) to $x - 3y = -5$ (Figure 12.73). For $Q(-5,0)$ on $x - 3y = -5$, $\overrightarrow{\mathbf{PQ}} = (-9,0)$. The required distance is the component of $\overrightarrow{\mathbf{PQ}}$ in the direction $\overrightarrow{\mathbf{PR}}$. Since $(3,1)$ is a vector along either of the lines (each has slope $\frac{1}{3}$), a vector perpendicular to the lines is $(-1,3)$. Consequently,

$$d = \overrightarrow{\mathbf{PQ}} \cdot \widehat{\mathbf{PR}} = (-9,0) \cdot \frac{1}{\sqrt{10}}(-1,3) = \frac{9}{\sqrt{10}}.$$

FIGURE 12.73

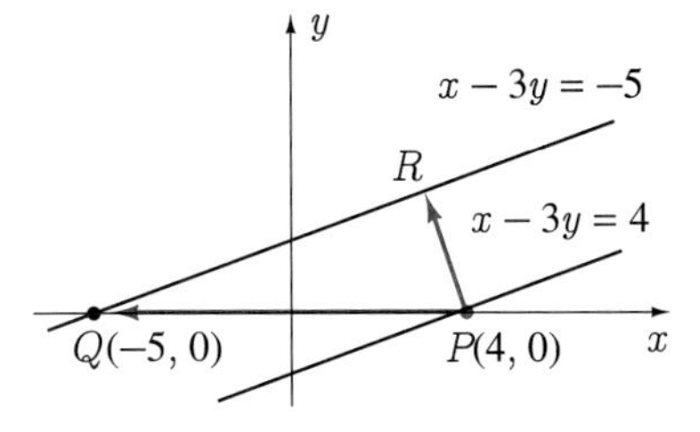

■

EXAMPLE 12.16

Find the distance from the point $(1,2,5)$ to the plane $x + y + 2z = 4$.

SOLUTION If Q is any point in the plane, say $(4,0,0)$, then the distance d from $P(1,2,5)$ to the plane is the component of $\overrightarrow{\mathbf{PQ}}$ in the direction $\overrightarrow{\mathbf{PR}}$ normal to the plane; i.e., $d = \overrightarrow{\mathbf{PQ}} \cdot \widehat{\mathbf{PR}}$ (Figure 12.74). Since $(1,1,2)$ is a vector normal to the plane,

$$\widehat{\mathbf{PR}} = \frac{1}{\sqrt{6}}(-1,-1,-2)$$

and

$$d = (3, -2, -5) \cdot \frac{1}{\sqrt{6}}(-1, -1, -2) = \frac{9}{\sqrt{6}}.$$

FIGURE 12.74

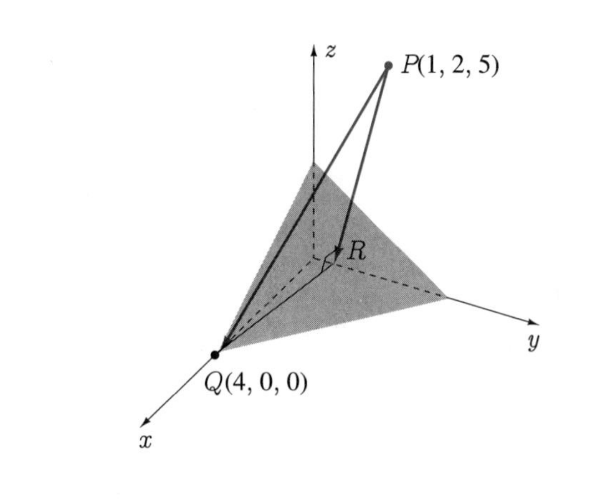

We will find distances between other geometric objects in Section 12.5 when we have defined a "vector product" of vectors. But the approach continues to be through components of vectors in arbitrary directions.

Work

FIGURE 12.75

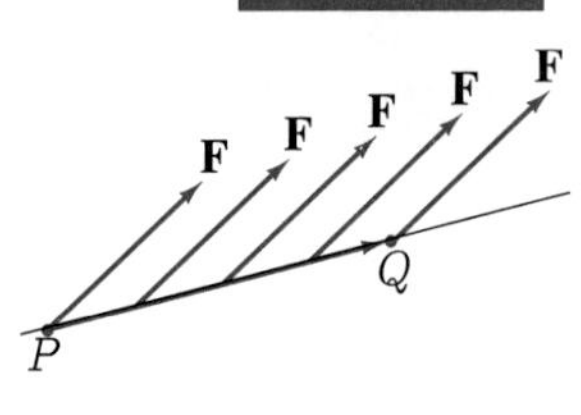

In Section 7.4 we described work as the product of force and distance. Now that we have represented forces by vectors, we can be more precise with our definition. In particular, if a particle moves along the line in Figure 12.75 from P to Q, then vector $\overrightarrow{\mathbf{PQ}}$ represents its displacement. If a constant force $\mathbf{F}$ acts on the particle during this motion, then the work done by $\mathbf{F}$ is defined as

$$W = \mathbf{F} \cdot \overrightarrow{\mathbf{PQ}}. \tag{12.31}$$

It is important to keep in mind exactly when this definition of work can be used: for constant forces acting along straight lines, and by constant $\mathbf{F}$, we mean that $\mathbf{F}$ is constant in both magnitude and direction. Note that when $\mathbf{F}$ and $\overrightarrow{\mathbf{PQ}}$ are both in the same direction, then the angle between the vectors is zero. In this case,

$$W = |\mathbf{F}||\overrightarrow{\mathbf{PQ}}|\cos(0) = |\mathbf{F}||\overrightarrow{\mathbf{PQ}}|,$$

and this is essentially the equation dealt with in Section 7.4. When $\mathbf{F}$ and $\overrightarrow{\mathbf{PQ}}$ are not in the same direction, then

$$W = |\mathbf{F}||\overrightarrow{\mathbf{PQ}}|\cos\theta.$$

Since $|\mathbf{F}|\cos\theta$ is the component of $\mathbf{F}$ along $\overrightarrow{\mathbf{PQ}}$, this equation simply states that when $\mathbf{F}$ and $\overrightarrow{\mathbf{PQ}}$ are not in the same direction, $|\overrightarrow{\mathbf{PQ}}|$ should be multiplied by the component of $\mathbf{F}$ in the direction of $\overrightarrow{\mathbf{PQ}}$.

EXAMPLE 12.17

If the boy in Figure 12.76 pulls the wagon handle with a force of 10 N at an angle of $\pi/4$ radians with the horizontal, how much work does he do in walking a distance of 20 m in a straight line?

SOLUTION The force $\mathbf{F}$ exerted by the boy has magnitude $|\mathbf{F}| = 10$, and points in a direction that makes an angle of $\pi/4$ radians with the displacement vector. If $\mathbf{d}$ is the displacement vector, then

$$W = \mathbf{F} \cdot \mathbf{d} = |\mathbf{F}||\mathbf{d}| \cos\theta = 10(20)\cos\left(\frac{\pi}{4}\right) = \frac{200}{\sqrt{2}} = 100\sqrt{2}\text{ J}.$$

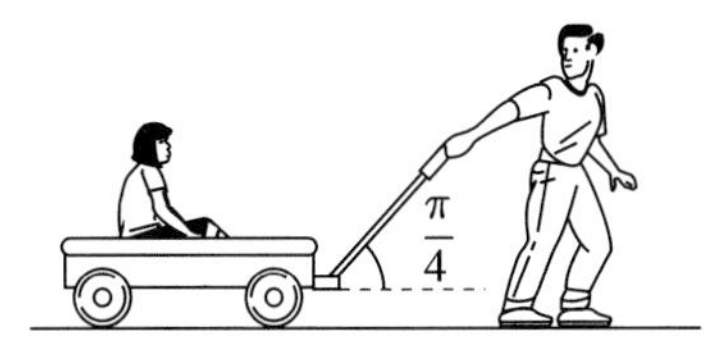

FIGURE 12.76

When motion is along a straight line, but $\mathbf{F}$ is *not* constant in direction or magnitude or both, we must use integration. The following example illustrates such a situation.

EXAMPLE 12.18

The spring in Figure 12.77 is fixed at A and moves the sleeve frictionlessly along the rod from B to C. If the spring is unstretched when the sleeve is at C, find the work done by the spring.

FIGURE 12.77

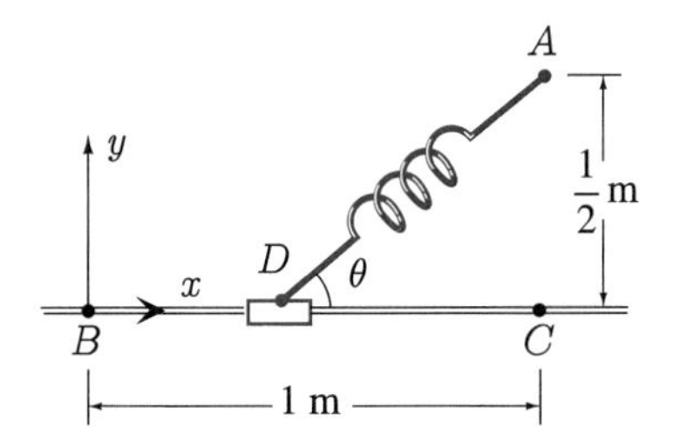

SOLUTION If we set up a coordinate system as shown, then at position D, the force $\mathbf{F}$ exerted by the spring on the sleeve has magnitude

$$|\mathbf{F}| = k\left[\sqrt{(1-x)^2 + \frac{1}{4}} - \frac{1}{2}\right],$$

where k is the spring constant. Since the spring is always stretched during the motion, the direction of $\mathbf{F}$ is along the vector $\overrightarrow{\mathbf{DA}}$. Clearly, then, $\mathbf{F}$ changes in both magnitude and direction as the sleeve moves from B to C. For a small displacement dx at position D, the amount of work done by $\mathbf{F}$ is (approximately)

$$\mathbf{F}\cdot(dx\hat{\mathbf{i}}) = |\mathbf{F}|dx(\cos\theta) = k\left[\sqrt{(1-x)^2+\frac{1}{4}}-\frac{1}{2}\right]dx\frac{1-x}{\sqrt{(1-x)^2+\frac{1}{4}}}$$

$$= k(1-x)\left\{1-\frac{\frac{1}{2}}{\sqrt{(1-x)^2+\frac{1}{4}}}\right\}dx.$$

The total work done by the spring as the sleeve moves from B to C must therefore be

$$W = \int_0^1 k(1-x)\left\{1-\frac{1}{2\sqrt{(1-x)^2+\frac{1}{4}}}\right\}dx$$

$$= k\int_0^1\left\{1-x-\frac{1-x}{2\sqrt{(1-x)^2+\frac{1}{4}}}\right\}dx$$

$$= k\left\{x-\frac{x^2}{2}+\frac{1}{2}\sqrt{(1-x)^2+\frac{1}{4}}\right\}_0^1$$

$$= \frac{k}{4}(3-\sqrt{5})\text{ J.}$$

■

EXERCISES 12.4

If $\mathbf{u} = 2\hat{\mathbf{i}}-3\hat{\mathbf{j}}+\hat{\mathbf{k}}$, $\mathbf{v} = \hat{\mathbf{j}}-\hat{\mathbf{k}}$, and $\mathbf{w} = 6\hat{\mathbf{i}}-2\hat{\mathbf{j}}+3\hat{\mathbf{k}}$, evaluate the scalar or find the components of the vector in Exercises 1–10.

1. $\mathbf{u}\cdot\mathbf{v}$

2. $(\mathbf{v}\cdot\mathbf{w})\mathbf{u}$

3. $(2\mathbf{u}-3\mathbf{v})\cdot\mathbf{w}$

4. $2\hat{\mathbf{i}}\cdot\hat{\mathbf{u}}$

5. $|2\mathbf{u}|\mathbf{v}\cdot\mathbf{w}$

6. $(3\mathbf{u}-4\mathbf{w})\cdot(2\hat{\mathbf{i}}+3\mathbf{u}-2\mathbf{v})$

7. $\mathbf{w}\cdot\hat{\mathbf{w}}$

8. $\dfrac{(105\mathbf{u}+240\mathbf{v})\cdot(105\mathbf{u}+240\mathbf{v})}{|105\mathbf{u}+240\mathbf{v}|^2}$

9. $|\mathbf{u}-\mathbf{v}+\hat{\mathbf{k}}|(\hat{\mathbf{j}}+\mathbf{w})\cdot\hat{\mathbf{k}}$

10. $\mathbf{u}\cdot\mathbf{v}+\mathbf{v}\cdot\mathbf{w}-(\mathbf{u}+\mathbf{w})\cdot\mathbf{v}$

In Exercises 11–14 determine whether the vectors are perpendicular.

11. $(1,2),(3,5)$

12. $(2,4),(-8,4)$

13. $(1,3,6),(-2,1,-4)$

14. $(2,3,-6),(-6,6,1)$

In Exercises 15–20 find the angle between the vectors.

15. $(3,4),(2,-5)$

16. $(1,6),(-4,7)$

17. $(4,2,3),(1,5,6)$

18. $(3,1,-1),(-2,1,4)$

19. $(2,0,5),(0,3,0)$

20. $(1,3,-2),(-2,-6,4)$

Find the equation for the plane in Exercises 21 and 22.

21. Through the point $(1,-1,3)$ and normal to the vector $(4,3,-2)$

22. Through the point $(2,1,5)$ and normal to the vector joining $(2,1,5)$ and $(4,2,3)$

In Exercises 23–28 find the indicated distance.

23. From the point $(1,2)$ to the line $x - y = 1$

24. From the point $(3,2)$ to the line $2x + 3y = 18$

25. From the point $(-1,4)$ to the line $4x + 3y = 8$

26. From the point $(1,3,4)$ to the plane $x + y - 2z = 0$

27. Between the lines $x + 4 = y$ and $3x + 7 = 3y$

28. Between the planes $2x+3y-z = 15$ and $4x+6y-2z = 7$

29. Verify the results in equations 12.21.

30. Show that a vector perpendicular to the line $Ax + By + C = 0$ in the xy-plane is (A,B).

31. The acute angle between two intersecting planes is defined as the acute angle between their normals. Find the acute angle between the planes $x - 2y + 4z = 6$ and $2x + y = z + 4$.

32. Use equation 12.22 to prove the parallelogram law:

$$|\mathbf{u} + \mathbf{v}|^2 + |\mathbf{u} - \mathbf{v}|^2 = 2|\mathbf{u}|^2 + 2|\mathbf{v}|^2.$$

Why is this called the parallelogram law?

33. The angles between a vector $\mathbf{v} = (v_x, v_y, v_z)$ and the vectors $\hat{\mathbf{i}}$, $\hat{\mathbf{j}}$, and $\hat{\mathbf{k}}$ are called the direction angles α, β, and γ of $\mathbf{v}$. Show that

$$\alpha = \text{Cos}^{-1}\left(\frac{\mathbf{v}\cdot\hat{\mathbf{i}}}{|\mathbf{v}|}\right) = \text{Cos}^{-1}\left(\frac{v_x}{|\mathbf{v}|}\right);$$

$$\beta = \text{Cos}^{-1}\left(\frac{\mathbf{v}\cdot\hat{\mathbf{j}}}{|\mathbf{v}|}\right) = \text{Cos}^{-1}\left(\frac{v_y}{|\mathbf{v}|}\right);$$

$$\gamma = \text{Cos}^{-1}\left(\frac{\mathbf{v}\cdot\hat{\mathbf{k}}}{|\mathbf{v}|}\right) = \text{Cos}^{-1}\left(\frac{v_z}{|\mathbf{v}|}\right).$$

34. Find direction angles for the vectors (a) $(1,2,-3)$; (b) $(0,1,-3)$; (c) $(-1,-2,6)$.

35. If $\mathbf{F}$ is a constant force, show that the work done by $\mathbf{F}$ on an object moving around any closed polygon is zero.

In Exercises 36 and 37 verify that $\hat{\mathbf{v}}$ and $\hat{\mathbf{w}}$ are perpendicular, and then resolve $\mathbf{u}$ into components along $\hat{\mathbf{v}}$ and $\hat{\mathbf{w}}$; that is, find λ and ρ so that $\mathbf{u} = \lambda\hat{\mathbf{v}} + \rho\hat{\mathbf{w}}$.

36. $\mathbf{u} = (2,1)$; $\hat{\mathbf{v}} = (1/\sqrt{2}, 1/\sqrt{2})$, $\hat{\mathbf{w}} = (1/\sqrt{2}, -1/\sqrt{2})$

37. $\mathbf{u} = 3\hat{\mathbf{i}} - 2\hat{\mathbf{j}}$; $\hat{\mathbf{v}} = (\hat{\mathbf{i}} - 2\hat{\mathbf{j}})/\sqrt{5}$, $\hat{\mathbf{w}} = (2\hat{\mathbf{i}} + \hat{\mathbf{j}})/\sqrt{5}$

In Exercises 38 and 39 verify that $\hat{\mathbf{u}}$, $\hat{\mathbf{v}}$, and $\hat{\mathbf{w}}$ are mutually perpendicular, and then resolve $\mathbf{r}$ into components along $\hat{\mathbf{u}}$, $\hat{\mathbf{v}}$, and $\hat{\mathbf{w}}$.

38. $\mathbf{r} = (1,3,-4)$; $\hat{\mathbf{u}} = (2,1,0)/\sqrt{5}$, $\hat{\mathbf{v}} = (-1,2,3)/\sqrt{14}$, $\hat{\mathbf{w}} = (3,-6,5)/\sqrt{70}$

39. $\mathbf{r} = 2\hat{\mathbf{i}} - \hat{\mathbf{k}}$; $\hat{\mathbf{u}} = (\hat{\mathbf{i}} + \hat{\mathbf{j}} + \hat{\mathbf{k}})/\sqrt{3}$, $\hat{\mathbf{v}} = (\hat{\mathbf{i}} + \hat{\mathbf{j}} - 2\hat{\mathbf{k}})/\sqrt{6}$, $\hat{\mathbf{w}} = (\hat{\mathbf{i}} - \hat{\mathbf{j}})/\sqrt{2}$

40. If $\mathbf{u} = (3,2)$, $\mathbf{v} = (1,-3)$, and $\mathbf{w} = (6,2)$, verify that $\mathbf{v}$ and $\mathbf{w}$ are perpendicular, and find scalars λ and ρ so that $\mathbf{u} = \lambda\mathbf{v} + \rho\mathbf{w}$.

41. If $\mathbf{u} = (1,0,1)$, $\mathbf{v} = (1,1,-1)$, $\mathbf{w} = (-1,2,1)$, and $\mathbf{r} = (-2,-3,4)$, verify that $\mathbf{u}$, $\mathbf{v}$, and $\mathbf{w}$ are mutually perpendicular, and find scalars λ, ρ, and μ so that $\mathbf{r} = \lambda\mathbf{u} + \rho\mathbf{v} + \mu\mathbf{w}$.

42. Find the equation of a plane normal to $\hat{\mathbf{i}} - 2\hat{\mathbf{j}} + 3\hat{\mathbf{k}}$ and two units from the point $(1,2,3)$.

43. If a, b, and c (all positive constants) are the intercepts of a plane with the x-, y-, and z-axes, and p is the length of the perpendicular from the origin to the plane, show that

$$\frac{1}{p^2} = \frac{1}{a^2} + \frac{1}{b^2} + \frac{1}{c^2}.$$

44. Repeat Example 12.18 if the spring has an unstretched length l that is less than the length of AC.

45. Two positive charges q_1 and q_2 are placed at positions $(5,5)$ and $(-2,3)$ in the xy-plane respectively. A third positive charge q_3 is moved along the x-axis from $x = 1$ to $x = -1$. Find the total work done by the electrostatic forces of q_1 and q_2 on q_3.

46. When the rocket in Figure 12.78 passes close to the spherical asteroid, it is attracted to the asteroid by a gravitational force with magnitude GmM/r^2 where m and M are the masses of the rocket and asteroid, r is the distance from the rocket to the centre of the asteroid, and G is Newton's gravitational constant. Determine the work the rocket must do against this force in order to follow the straight-line path from A to B.

FIGURE 12.78

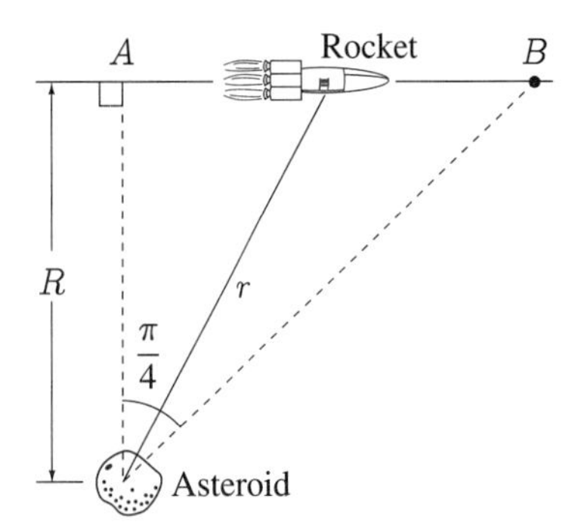

47. Prove that the (undirected) distance from a point $P(x_1, y_1, z_1)$ to the plane $Ax + By + Cz + D = 0$

is

$$\frac{|Ax_1 + By_1 + Cz_1 + D|}{\sqrt{A^2 + B^2 + C^2}}.$$

48. Prove that the (undirected) distance between two parallel planes $Ax + By + Cz + D_1 = 0$ and $Ax + By + Cz + D_2 = 0$ is

$$\frac{|D_1 - D_2|}{\sqrt{A^2 + B^2 + C^2}}.$$

49. Show that the vector $(|\mathbf{v}|\mathbf{u} + |\mathbf{u}|\mathbf{v}) / ||\mathbf{u}|\mathbf{v} + |\mathbf{v}|\mathbf{u}|$ is a unit vector which bisects the angle between $\mathbf{u}$ and $\mathbf{v}$.

SECTION 12.5

The Vector Product

In many applications of vectors we need a vector perpendicular to two given vectors. For instance, suppose we must find the equation for the plane passing through the three given points P_1, P_2 and P_3 in Figure12.79. Clearly, $\overrightarrow{\mathbf{P_1P_2}}$ and $\overrightarrow{\mathbf{P_1P_3}}$ are vectors that lie in the plane, and to find a normal to the plane we therefore require a vector perpendicular to both $\overrightarrow{\mathbf{P_1P_2}}$ and $\overrightarrow{\mathbf{P_1P_3}}$.

FIGURE 12.79

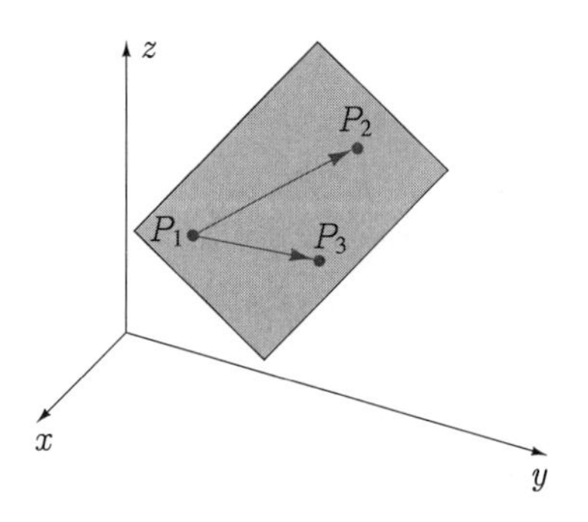

Consider, then, finding a vector $\mathbf{r} = (a, b, c)$ perpendicular to two given vectors $\mathbf{u} = (u_x, u_y, u_z)$ and $\mathbf{v} = (v_x, v_y, v_z)$. The corollary to Theorem 12.2 requires a, b, and c to satisfy

$$0 = \mathbf{r} \cdot \mathbf{u} = au_x + bu_y + cu_z,$$
$$0 = \mathbf{r} \cdot \mathbf{v} = av_x + bv_y + cv_z.$$

When we solve these equations, we find that there is an infinite number of solutions all represented by

$$a = s(u_y v_z - u_z v_y), \qquad b = s(u_z v_x - u_x v_z), \qquad c = s(u_x v_y - u_y v_x),$$

where s is any real number. In other words, any vector of the form

$$\mathbf{r} = s(u_y v_z - u_z v_y, u_z v_x - u_x v_z, u_x v_y - u_y v_x)$$

is perpendicular to the vectors $\mathbf{u} = (u_x, u_y, u_z)$ and $\mathbf{v} = (v_x, v_y, v_z)$. When we choose $s = 1$, the resulting vector is called the vector product of $\mathbf{u}$ and $\mathbf{v}$.

Definition 12.10

The **vector product (cross product** or **outer product)** of two vectors **u** and **v** with Cartesian components (u_x, u_y, u_z) and (v_x, v_y, v_z) is defined as

$$\mathbf{u} \times \mathbf{v} = (u_y v_z - u_z v_y)\hat{\mathbf{i}} + (u_z v_x - u_x v_z)\hat{\mathbf{j}} + (u_x v_y - u_y v_x)\hat{\mathbf{k}}. \tag{12.32}$$

To eliminate the need for memorizing the exact placing of the six components of **u** and **v** in this definition, we borrow the notation for determinants from linear algebra. A brief discussion of determinants and their properties is given in Appendix F. We set up a 3×3 determinant with $\hat{\mathbf{i}}$, $\hat{\mathbf{j}}$, and $\hat{\mathbf{k}}$ across the top row and the components of **u** and **v** across the second and third rows:

$$\begin{vmatrix} \hat{\mathbf{i}} & \hat{\mathbf{j}} & \hat{\mathbf{k}} \\ u_x & u_y & u_z \\ v_x & v_y & v_z \end{vmatrix}.$$

In actual fact, this is not a determinant, since three of the entries are vectors and six are scalars. If we ignore this fact, and apply the rules for the expansion of a 3×3 determinant along its first row (namely, $\hat{\mathbf{i}}$ times the 2×2 determinant obtained by deleting the row and column containing $\hat{\mathbf{i}}$, minus $\hat{\mathbf{j}}$ times the 2×2 determinant obtained by deleting the row and column containing $\hat{\mathbf{j}}$, plus $\hat{\mathbf{k}}$ times the 2×2 determinant obtained by deleting the row and column containing $\hat{\mathbf{k}}$), we obtain

$$\begin{vmatrix} \hat{\mathbf{i}} & \hat{\mathbf{j}} & \hat{\mathbf{k}} \\ u_x & u_y & u_z \\ v_x & v_y & v_z \end{vmatrix} = \begin{vmatrix} u_y & u_z \\ v_y & v_z \end{vmatrix} \hat{\mathbf{i}} - \begin{vmatrix} u_x & u_z \\ v_x & v_z \end{vmatrix} \hat{\mathbf{j}} + \begin{vmatrix} u_x & u_y \\ v_x & v_y \end{vmatrix} \hat{\mathbf{k}}.$$

But the definition for the value of a 2×2 determinant is

$$\begin{vmatrix} a & b \\ c & d \end{vmatrix} = ad - bc.$$

Consequently,

$$\begin{vmatrix} \hat{\mathbf{i}} & \hat{\mathbf{j}} & \hat{\mathbf{k}} \\ u_x & u_y & u_z \\ v_x & v_y & v_z \end{vmatrix} = (u_y v_z - u_z v_y)\hat{\mathbf{i}} - (u_x v_z - u_z v_x)\hat{\mathbf{j}} + (u_x v_y - u_y v_x)\hat{\mathbf{k}},$$

and this is the same as the right-hand side of 12.32. We may therefore write, as a memory-saving device, that

$$\mathbf{u} \times \mathbf{v} = \begin{vmatrix} \hat{\mathbf{i}} & \hat{\mathbf{j}} & \hat{\mathbf{k}} \\ u_x & u_y & u_z \\ v_x & v_y & v_z \end{vmatrix}, \tag{12.33}$$

so long as we evaluate the right-hand side using the general rules for the expansion of a determinant along its first row.

For example, if $\mathbf{u} = (1, -1, 2)$ and $\mathbf{v} = (2, 3, -5)$, then

$$\mathbf{u} \times \mathbf{v} = \begin{vmatrix} \hat{\mathbf{i}} & \hat{\mathbf{j}} & \hat{\mathbf{k}} \\ 1 & -1 & 2 \\ 2 & 3 & -5 \end{vmatrix} = (5 - 6)\hat{\mathbf{i}} - (-5 - 4)\hat{\mathbf{j}} + (3 + 2)\hat{\mathbf{k}} = -\hat{\mathbf{i}} + 9\hat{\mathbf{j}} + 5\hat{\mathbf{k}}.$$

It is straightforward to verify that

$$\hat{\mathbf{i}} \times \hat{\mathbf{j}} = \hat{\mathbf{k}}, \quad \hat{\mathbf{j}} \times \hat{\mathbf{k}} = \hat{\mathbf{i}}, \quad \hat{\mathbf{k}} \times \hat{\mathbf{i}} = \hat{\mathbf{j}}, \tag{12.34}$$

and that the cross-product is anticommutative and distributive:

$$\mathbf{u} \times \mathbf{v} = -\mathbf{v} \times \mathbf{u}, \tag{12.35a}$$

$$\mathbf{u} \times (\lambda\mathbf{v} + \rho\mathbf{w}) = \lambda(\mathbf{u} \times \mathbf{v}) + \rho(\mathbf{u} \times \mathbf{w}). \tag{12.35b}$$

Our preliminary analysis indicated that $\mathbf{u} \times \mathbf{v}$ is perpendicular to both $\mathbf{u}$ and $\mathbf{v}$. The following theorem relates the length of $\mathbf{u} \times \mathbf{v}$ to the lengths of $\mathbf{u}$ and $\mathbf{v}$.

Theorem 12.5

If θ is the angle between two vectors $\mathbf{u}$ and $\mathbf{v}$, then

$$|\mathbf{u} \times \mathbf{v}| = |\mathbf{u}||\mathbf{v}|\sin\theta. \tag{12.36}$$

Proof Since θ is an angle between 0 and π, $\sin\theta$ must be positive, and we can write from equation 12.25 that

$$\sin\theta = \sqrt{1-\cos^2\theta} = \sqrt{1-\left(\frac{\mathbf{u}\cdot\mathbf{v}}{|\mathbf{u}||\mathbf{v}|}\right)^2} = \frac{1}{|\mathbf{u}||\mathbf{v}|}\sqrt{|\mathbf{u}|^2|\mathbf{v}|^2-(\mathbf{u}\cdot\mathbf{v})^2}.$$

Consequently,

$$|\mathbf{u}||\mathbf{v}|\sin\theta = \sqrt{|\mathbf{u}|^2|\mathbf{v}|^2-(\mathbf{u}\cdot\mathbf{v})^2}.$$

If $\mathbf{u} = (u_x, u_y, u_z)$ and $\mathbf{v} = (v_x, v_y, v_z)$, then

$$\begin{aligned}
|\mathbf{u}|^2|\mathbf{v}|^2\sin^2\theta =& (u_x^2+u_y^2+u_z^2)(v_x^2+v_y^2+v_z^2) - (u_xv_x+u_yv_y+u_zv_z)^2 \\
=& u_x^2(v_x^2+v_y^2+v_z^2) + u_y^2(v_x^2+v_y^2+v_z^2) \\
&+ u_z^2(v_x^2+v_y^2+v_z^2) - (u_x^2v_x^2+u_y^2v_y^2+u_z^2v_z^2 \\
&+ 2u_xv_xu_yv_y + 2u_xv_xu_zv_z + 2u_yv_yu_zv_z) \\
=& (u_y^2v_z^2 - 2u_yv_yu_zv_z + u_z^2v_y^2) + (u_x^2v_z^2 - 2u_xv_xu_zv_z + u_z^2v_x^2) \\
&+ (u_x^2v_y^2 - 2u_xv_xu_yv_y + u_y^2v_x^2) \\
=& (u_yv_z - u_zv_y)^2 + (u_xv_z - u_zv_x)^2 + (u_xv_y - u_yv_x)^2 \\
=& |\mathbf{u}\times\mathbf{v}|^2,
\end{aligned}$$

or

$$|\mathbf{u}\times\mathbf{v}| = |\mathbf{u}||\mathbf{v}|\sin\theta.$$

FIGURE 12.80

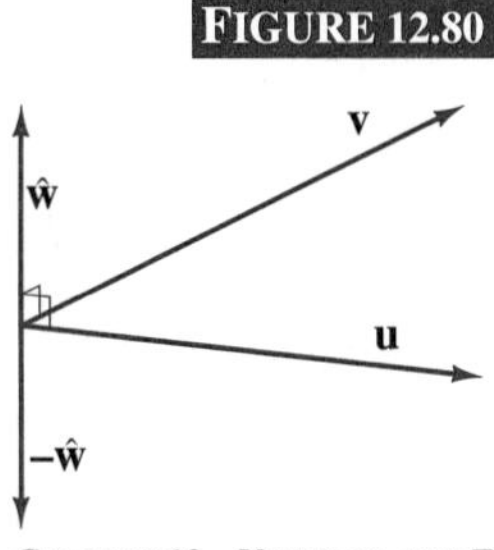

We now know that $\mathbf{u} \times \mathbf{v}$ is perpendicular to $\mathbf{u}$ and $\mathbf{v}$, and has length $|\mathbf{u}||\mathbf{v}|\sin\theta$, where θ is the angle between $\mathbf{u}$ and $\mathbf{v}$. Figure 12.80 illustrates that there are only two directions that are perpendicular to $\mathbf{u}$ and $\mathbf{v}$, and one is the negative of the other. Let us denote by $\hat{\mathbf{w}}$ the unit vector along that direction which is perpendicular to $\mathbf{u}$ and $\mathbf{v}$ and is determined by the right-hand rule (curl the fingers of the right hand from $\mathbf{u}$ toward $\mathbf{v}$ and the thumb points in the direction $\hat{\mathbf{w}}$).

We now show that $\mathbf{u} \times \mathbf{v}$ always points in the direction determined by the right-hand rule (i.e., in the direction $\hat{\mathbf{w}}$ rather than $-\hat{\mathbf{w}}$). To see this we place $\mathbf{u}$ and $\mathbf{v}$ tail to tail and establish a coordinate system with this common point as origin and the positive x-axis along **u** (Figure 12.81a). Let the plane determined by **u** and **v** be the xy-plane. In this coordinate system, **u** has only an x-component, $\mathbf{u} = u_x\hat{\mathbf{i}}$ ($u_x > 0$), and **v** has only x- and y- components, $\mathbf{v} = v_x\hat{\mathbf{i}} + v_y\hat{\mathbf{j}}$. The cross-product of **u** and **v** is therefore

$$\mathbf{u} \times \mathbf{v} = \begin{vmatrix} \hat{\mathbf{i}} & \hat{\mathbf{j}} & \hat{\mathbf{k}} \\ u_x & 0 & 0 \\ v_x & v_y & 0 \end{vmatrix} = u_x v_y \hat{\mathbf{k}}.$$

For **v** in Figure 12.81(a), v_y is clearly positive and therefore $u_x v_y$, the component of $\mathbf{u} \times \mathbf{v}$, is also positive. But then $\mathbf{u} \times \mathbf{v}$ is indeed determined by the right-hand rule since $\hat{\mathbf{w}} = \hat{\mathbf{k}}$.

FIGURE 12.81

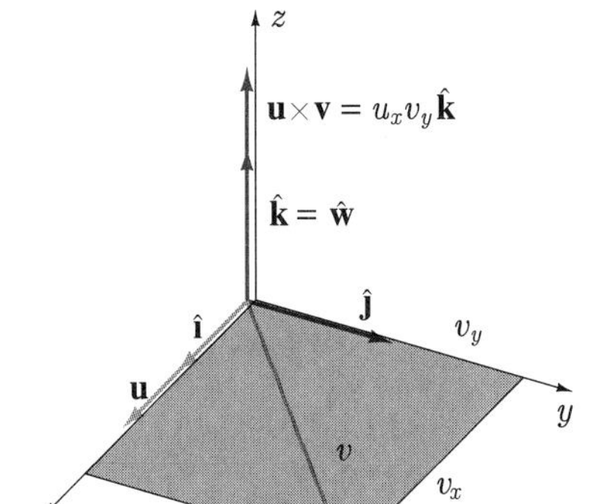

(a)

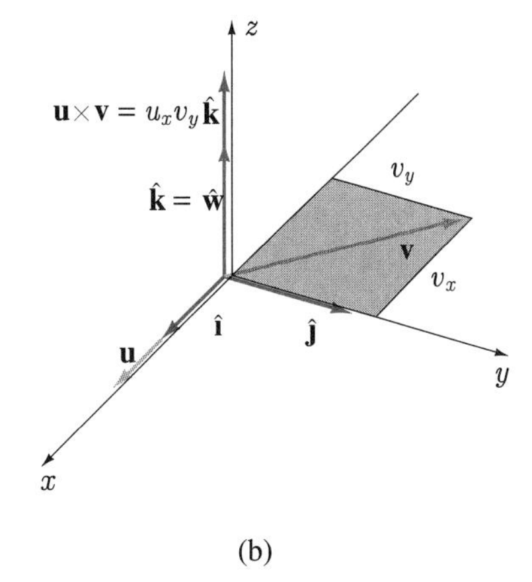

(b)

When the tip of **v** lies in the second quadrant of the xy-plane (Figure 12.81b), then $\mathbf{u} \times \mathbf{v}$ is once again in the positive z-direction, the direction of $\hat{\mathbf{w}}$. When the tip of **v** is in the third or fourth quadrant (Figure 12.82), $v_y < 0$, and $\mathbf{u} \times \mathbf{v}$ therefore has a negative z-component. But in this case $\hat{\mathbf{w}} = -\hat{\mathbf{k}}$, and the direction of $\mathbf{u} \times \mathbf{v}$ is once again determined by the right-hand rule.

FIGURE 12.82

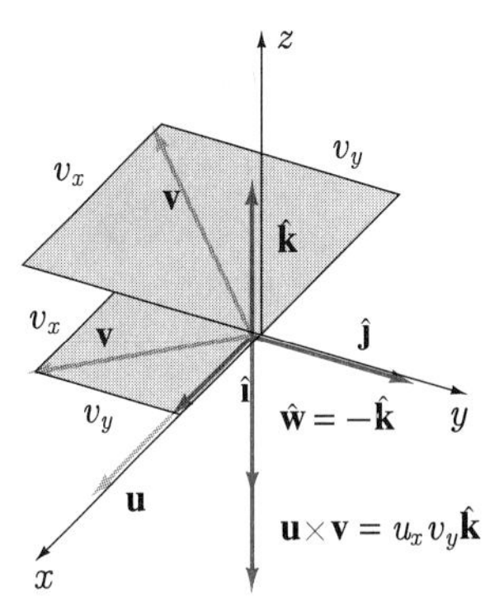

What we have now established is the following coordinate-free definition for the vector product of two vectors **u** and **v**:

$$\mathbf{u} \times \mathbf{v} = (|\mathbf{u}||\mathbf{v}| \sin\theta)\,\hat{\mathbf{w}}. \qquad (12.37)$$

The unit vector $\hat{\mathbf{w}}$ defines the direction of $\mathbf{u} \times \mathbf{v}$ and the factor $|\mathbf{u}||\mathbf{v}|\sin\theta$ is its length.

The fact that the vector product $\mathbf{u} \times \mathbf{v}$ is perpendicular to both **u** and **v** makes it a powerful tool in many applications.

EXAMPLE 12.19

Find the equation of the plane through the points $(1, 2, 3)$, $(-2, 0, 4)$, and $(5, 2, -1)$.

SOLUTION If the plane passes through $P(1, 2, 3)$, $Q(-2, 0, 4)$, and $R(5, 2, -1)$, then $\overrightarrow{\mathbf{PQ}} = (-3, -2, 1)$ and $\overrightarrow{\mathbf{PR}} = (4, 0, -4)$ are vectors that lie in the plane. It follows that a vector normal to the plane is

$$\overrightarrow{\mathbf{PQ}} \times \overrightarrow{\mathbf{PR}} = \begin{vmatrix} \hat{\mathbf{i}} & \hat{\mathbf{j}} & \hat{\mathbf{k}} \\ -3 & -2 & 1 \\ 4 & 0 & -4 \end{vmatrix} = (8-0)\hat{\mathbf{i}} - (12-4)\hat{\mathbf{j}} + (0+8)\hat{\mathbf{k}} = 8(\hat{\mathbf{i}} - \hat{\mathbf{j}} + \hat{\mathbf{k}}).$$

Since $\hat{\mathbf{i}} - \hat{\mathbf{j}} + \hat{\mathbf{k}}$ is therefore normal to the plane, the equation of the plane is

$$(1,-1,1) \cdot (x-1, y-2, z-3) = 0, \quad \text{or,} \quad x - y + z = 2.$$

■

Lines

In Section 12.2 we indicated that space curves can be described by two simultaneous equations in x, y, and z, and that such a representation describes the curve as the intersection of two surfaces. In this section we use vectors to discuss straight lines. A straight line in space is uniquely characterized by a point on it and a vector parallel to it. For instance, there is one and only one line l through the point $P_0(x_0, y_0, z_0)$ and in the direction $\mathbf{v} = (v_x, v_y, v_z)$ in Figure 12.83.

FIGURE 12.83

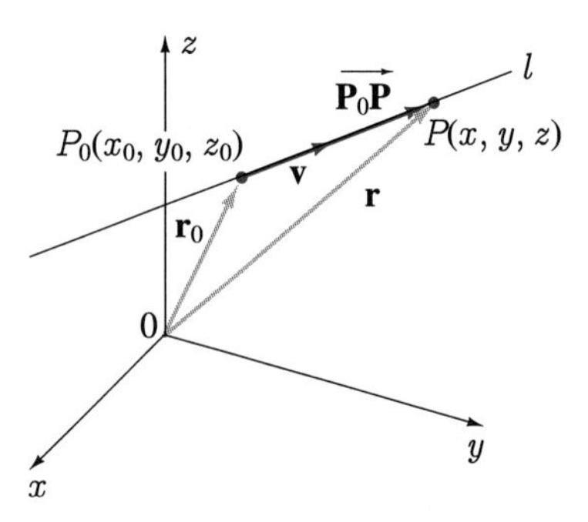

If $P(x,y,z)$ is any point on this line, then the vector $\mathbf{r}$ joining the origin to P has components $\mathbf{r} = (x,y,z)$. Now this vector can be expressed as the sum of $\mathbf{r}_0$, the vector from O to P_0 and $\overrightarrow{\mathbf{P_0P}}$:

$$\mathbf{r} = \mathbf{r}_0 + \overrightarrow{\mathbf{P_0P}}.$$

But $\overrightarrow{\mathbf{P_0P}}$ is in the same direction as $\mathbf{v}$; therefore it must be some scalar multiple of $\mathbf{v}$, $\overrightarrow{\mathbf{P_0P}} = t\mathbf{v}$. Consequently, the vector joining O to any point on l can be written in the form

$$\mathbf{r} = \mathbf{r}_0 + t\mathbf{v}, \tag{12.38}$$

for an appropriate value of t. Because the components of $\mathbf{r} = (x,y,z)$ are also the coordinates of the point (x,y,z) on l, this equation is called the **vector equation** of the line l through (x_0, y_0, z_0) in the direction $\mathbf{v}$. If we substitute components for $\mathbf{r}$, $\mathbf{r}_0$, and $\mathbf{v}$, then

$$\begin{aligned}(x,y,z) &= (x_0, y_0, z_0) + t(v_x, v_y, v_z) \\ &= (x_0 + tv_x, y_0 + tv_y, z_0 + tv_z).\end{aligned}$$

Since two vectors are equal if and only if corresponding components are identical, we can write that

$$x = x_0 + v_x t, \tag{12.39a}$$
$$y = y_0 + v_y t, \tag{12.39b}$$
$$z = z_0 + v_z t. \tag{12.39c}$$

These three scalar equations are equivalent to vector equation 12.38; they are called **parametric equations** for line l. They illustrate once again that a line in space is characterized by a point (x_0, y_0, z_0) on it and a vector (v_x, v_y, v_z) along it. Each value of t substituted into 12.39 yields a point (x, y, z) on l, and conversely, every point on l is represented by some value of t. For instance, $t = 0$ yields P_0, and $t = 1$ gives the point at the tip of **v** in Figure 12.83.

If none of v_x, v_y, and v_z is equal to zero, we can solve equations 12.39 for t and equate the three expressions to obtain

$$\frac{x - x_0}{v_x} = \frac{y - y_0}{v_y} = \frac{z - z_0}{v_z}. \tag{12.40}$$

These are called **symmetric equations** for the line l through (x_0, y_0, z_0) parallel to $\mathbf{v} = (v_x, v_y, v_z)$. There are only two independent equations in 12.40, which therefore substantiates our previous result that a curve (in this case, a line) can be described by two equations in x, y, and z. We could, for instance, write

$$v_x(y - y_0) = v_y(x - x_0) \quad \text{and} \quad v_z(y - y_0) = v_y(z - z_0),$$

or

$$v_y x - v_x y = v_y x_0 - v_x y_0, \quad v_y z - v_z y = v_y z_0 - v_z y_0.$$

Since the first of these is linear in x and y and the second is linear in y and z, each describes a plane. This means that the line has been described as the curve of intersection of two planes.

Find, if possible, vector, parametric, and symmetric equations for the line through the points $(-1, 2, 1)$ and $(3, -2, 1)$.

SOLUTION A vector along the line is $(3, -2, 1) - (-1, 2, 1) = (4, -4, 0)$, and so too is $(1, -1, 0)$. A vector equation for the line is

$$\mathbf{r} = (-1, 2, 1) + t(1, -1, 0) = (t - 1, -t + 2, 1).$$

Parametric equations are therefore

$$x = t - 1, \quad y = -t + 2, \quad z = 1.$$

Because the z- component of every vector along the line is zero, we cannot write full symmetric equations for the line. By eliminating t between the x- and y-equations, however, we can write

$$x + 1 = \frac{y - 2}{-1}, \quad z = 1.$$

If we set $x + 1 = 2 - y$, $z = 1$, or $x + y = 1$, $z = 1$, we represent the line as the intersection of the planes $x + y = 1$ and $z = 1$ (Figure 12.84).

FIGURE 12.84

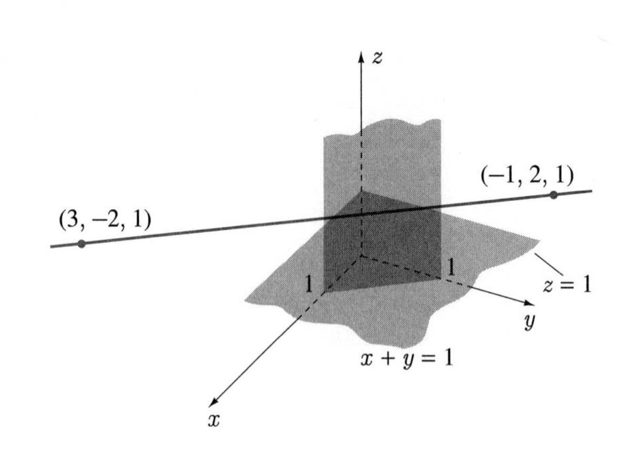

EXAMPLE 12.21

Find the equation of the plane containing the origin and the line $2x + y - z = 4$, $x + z = 5$.

SOLUTION We can easily find two more points on the plane. For instance, if $x = 0$, then the equations of the line require $z = 5$ and $y = 9$; and if $z = 0$, then $x = 5$ and $y = -6$. Thus $P(0, 9, 5)$ and $Q(5, -6, 0)$, as well as $O(0, 0, 0)$, are points on the plane. It follows then that $\overrightarrow{\mathbf{OP}} = (0, 9, 5)$ and $\overrightarrow{\mathbf{OQ}} = (5, -6, 0)$ are vectors in the plane, and a vector normal to the plane is

$$\overrightarrow{\mathbf{OP}} \times \overrightarrow{\mathbf{OQ}} = \begin{vmatrix} \hat{\mathbf{i}} & \hat{\mathbf{j}} & \hat{\mathbf{k}} \\ 0 & 9 & 5 \\ 5 & -6 & 0 \end{vmatrix} = (30, 25, -45).$$

The vector $(6, 5, -9)$ is also normal to the plane, and the equation of the plane is

$$0 = (6, 5, -9) \cdot (x - 0, y - 0, z - 0) = 6x + 5y - 9z.$$

EXAMPLE 12.22

Find symmetric equations for the line $x + y - 2z = 6$, $2x - 3y + 4z = 10$.

SOLUTION To find symmetric equations, we require a vector parallel to the line and a point on it. By setting $x = 0$ and solving $y - 2z = 6$, $-3y + 4z = 10$, we obtain $y = -22$, $z = -14$. Consequently, $(0, -22, -14)$ is a point on the line. To find a vector along the line, we could find another point on the line, say, by setting $z = 0$ and solving $x + y = 6$, $2x - 3y = 10$ for $x = \frac{28}{5}$, $y = \frac{2}{5}$. A vector along the line is therefore $(\frac{28}{5}, \frac{2}{5}, 0) - (0, -22, -14) = (\frac{28}{5}, \frac{112}{5}, 14)$, and so too is $(\frac{5}{14})(\frac{28}{5}, \frac{112}{5}, 14) = (2, 8, 5)$.

Alternatively, we know that $(1, 1, -2)$ and $(2, -3, 4)$ are vectors that are normal to the planes $x + y - 2z = 6$ and $2x - 3y + 4z = 10$, and a vector along the line of intersection of the planes must be perpendicular to both of these vectors (Figure

12.85). Consequently, a vector along the line of intersection is

$$\begin{vmatrix} \hat{\mathbf{i}} & \hat{\mathbf{j}} & \hat{\mathbf{k}} \\ 1 & 1 & -2 \\ 2 & -3 & 4 \end{vmatrix} = (-2, -8, -5).$$

Symmetric equations for the line are therefore

$$\frac{x}{2} = \frac{y+22}{8} = \frac{z+14}{5}.$$

FIGURE 12.85

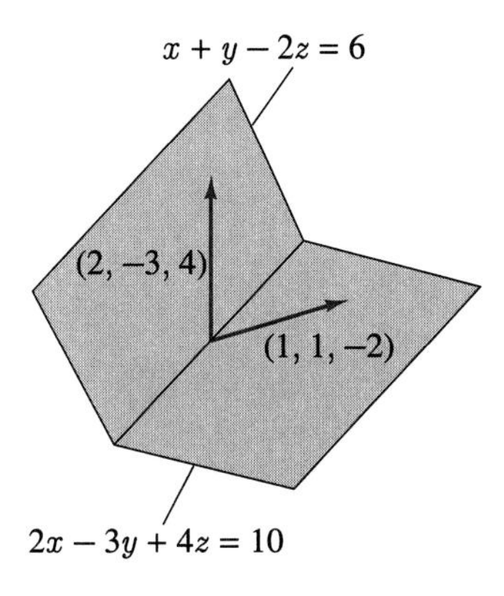

■

EXAMPLE 12.23

Find the distance from the point $(1,3,6)$ to the line l:

$$x - 1 = \frac{y-2}{2} = \frac{z-4}{3}.$$

SOLUTION Clearly $Q(1,2,4)$ is a point on the line, and therefore the required distance d is the component of $\overrightarrow{\mathbf{PQ}}$ in the direction $\overrightarrow{\mathbf{PR}}$ (Figure 12.86). By equation 12.30, then, $d = \overrightarrow{\mathbf{PQ}} \cdot \widehat{\mathbf{PR}}$. To find $\widehat{\mathbf{PR}}$ we need a vector in the direction $\overrightarrow{\mathbf{PR}}$. This is not nearly as simple as finding $\overrightarrow{\mathbf{PR}}$ for a line in the xy-plane. Since $S(0,0,1)$ is also a point on l, $\overrightarrow{\mathbf{QS}} = (-1,-2,-3)$ is a vector along l. But then

$$\overrightarrow{\mathbf{QS}} \times \overrightarrow{\mathbf{PQ}} = \begin{vmatrix} \hat{\mathbf{i}} & \hat{\mathbf{j}} & \hat{\mathbf{k}} \\ -1 & -2 & -3 \\ 0 & -1 & -2 \end{vmatrix} = \hat{\mathbf{i}} - 2\hat{\mathbf{j}} + \hat{\mathbf{k}}$$

is perpendicular to both $\overrightarrow{\mathbf{PQ}}$ and $\overrightarrow{\mathbf{QS}}$. It now follows that a vector in the direction $\overrightarrow{\mathbf{PR}}$ must be

$$(\overrightarrow{\mathbf{QS}} \times \overrightarrow{\mathbf{PQ}}) \times \overrightarrow{\mathbf{QS}} = \begin{vmatrix} \hat{\mathbf{i}} & \hat{\mathbf{j}} & \hat{\mathbf{k}} \\ 1 & -2 & 1 \\ -1 & -2 & -3 \end{vmatrix} = 8\hat{\mathbf{i}} + 2\hat{\mathbf{j}} - 4\hat{\mathbf{k}}.$$

Finally, then,

$$\widehat{\mathbf{PR}} = \frac{(8,2,-4)}{\sqrt{64+4+16}} = \frac{(4,1,-2)}{\sqrt{21}}$$

and

$$d = (0,-1,-2)\cdot\frac{(4,1,-2)}{\sqrt{21}} = \frac{\sqrt{21}}{7}.$$

FIGURE 12.86 FIGURE 12.87

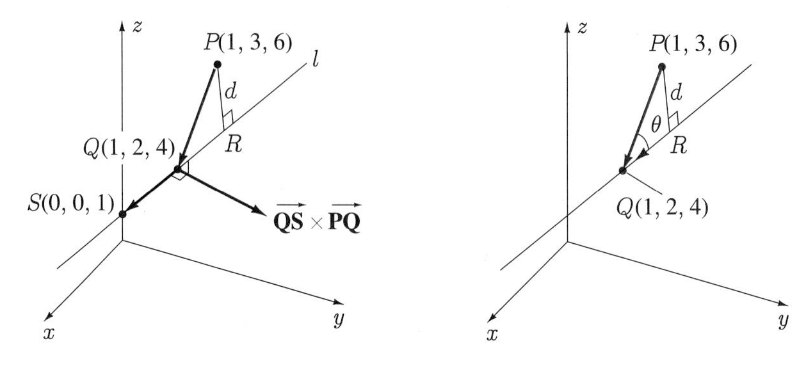

An alternative procedure is to note in Figure 12.87 that $d = |\overrightarrow{\mathbf{PQ}}|\sin\theta$. But the right-hand side of this equation looks like the length of the cross-product of $\overrightarrow{\mathbf{PQ}}$ and a unit vector that makes an angle θ with $\overrightarrow{\mathbf{PQ}}$. Since $\overrightarrow{\mathbf{RQ}}$ makes an angle θ with $\overrightarrow{\mathbf{PQ}}$, we could write that

$$d = |\overrightarrow{\mathbf{PQ}}||\widehat{\mathbf{RQ}}|\sin\theta = |\overrightarrow{\mathbf{PQ}}\times\widehat{\mathbf{RQ}}|.$$

Now $\overrightarrow{\mathbf{PQ}} = (0,-1,-2)$ and a vector along l is $(1,2,3)$. Thus $\widehat{\mathbf{RQ}} = (-1,-2,-3)/\sqrt{14}$ and

$$\overrightarrow{\mathbf{PQ}}\times\widehat{\mathbf{RQ}} = \begin{vmatrix} \hat{\mathbf{i}} & \hat{\mathbf{j}} & \hat{\mathbf{k}} \\ 0 & -1 & -2 \\ -1/\sqrt{14} & -2/\sqrt{14} & -3/\sqrt{14} \end{vmatrix} = \frac{1}{\sqrt{14}}(-1,2,-1).$$

Finally,

$$d = \frac{1}{\sqrt{14}}\sqrt{1+4+1} = \frac{\sqrt{21}}{7}.$$

■

This example suggests the following formula for the area of a triangle.

Theorem 12.6 *If A, B, and C are the vertices of a triangle, then*

$$\text{Area of } \Delta ABC = \frac{1}{2}|\overrightarrow{\mathbf{AB}}\times\overrightarrow{\mathbf{AC}}|. \tag{12.41}$$

Proof The area of triangle ABC in Figure 12.88 is

$$\frac{1}{2}|\overrightarrow{\mathbf{AC}}||\overrightarrow{\mathbf{BD}}| = \frac{1}{2}|\overrightarrow{\mathbf{AC}}||\overrightarrow{\mathbf{AB}}|\sin\theta = \frac{1}{2}|\overrightarrow{\mathbf{AB}} \times \overrightarrow{\mathbf{AC}}|.$$

FIGURE 12.88

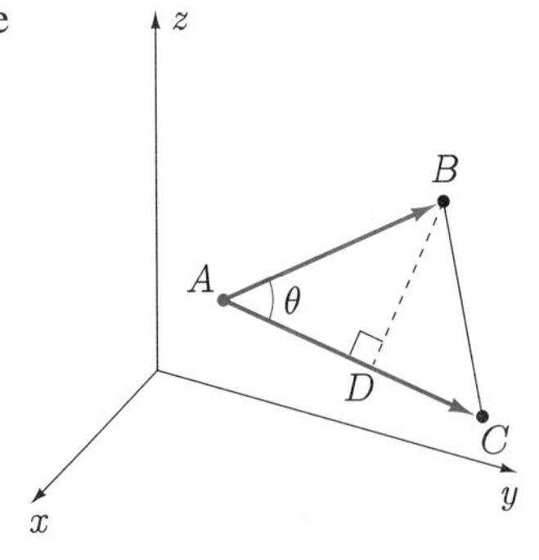

For the triangle with vertices $A(1,1,1)$, $B(2,-3,2)$, and $C(4,1,5)$ in Figure 12.89,

$$\overrightarrow{\mathbf{AB}} \times \overrightarrow{\mathbf{AC}} = \begin{vmatrix} \hat{\mathbf{i}} & \hat{\mathbf{j}} & \hat{\mathbf{k}} \\ 1 & -4 & 1 \\ 3 & 0 & 4 \end{vmatrix} = (-16,-1,12).$$

The area of the triangle is therefore

$$\frac{1}{2}|(-16,-1,12)| = \frac{1}{2}\sqrt{256+1+144} = \frac{1}{2}\sqrt{401}.$$

The following corollary is a direct consequence.

Corollary *The area of a parallelogram with coterminal sides* $\overrightarrow{\mathbf{AB}}$ *and* $\overrightarrow{\mathbf{AC}}$ *(Figure 12.90) is*

$$\text{Area} = |\overrightarrow{\mathbf{AB}} \times \overrightarrow{\mathbf{AC}}|. \tag{12.42}$$

FIGURE 12.89

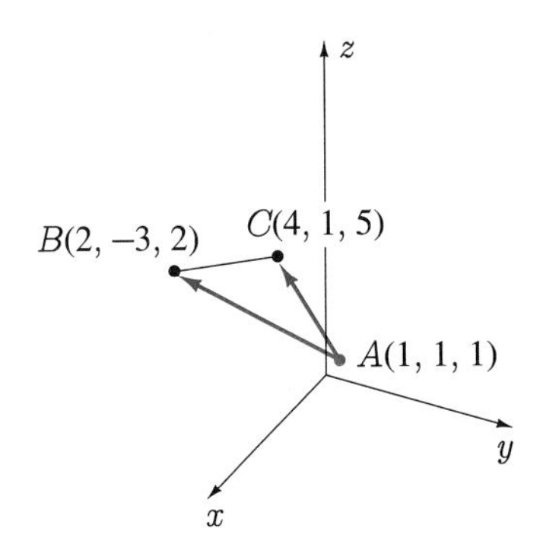

FIGURE 12.90

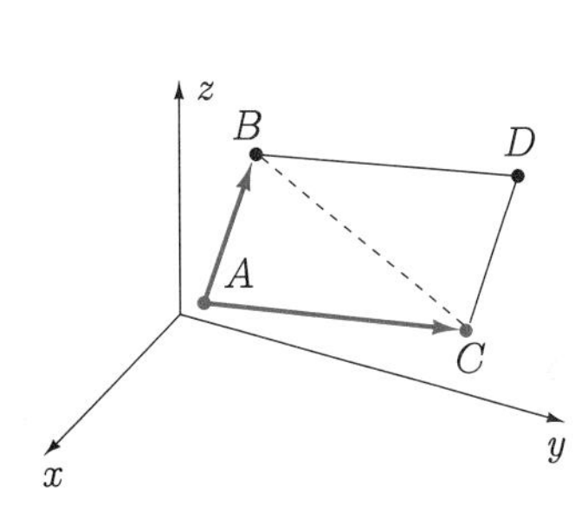

EXERCISES 12.5

If $\mathbf{u} = (3,1,4)$, $\mathbf{v} = (-1,2,0)$, and $\mathbf{w} = (-2,-3,5)$, evaluate the scalar or find the components of the vector in Exercises 1–10.

1. $\mathbf{v} \times \mathbf{w}$

2. $(-3\mathbf{u}) \times (2\mathbf{v})$

3. $\mathbf{u} \cdot (\mathbf{v} \times \mathbf{w})$

4. $\hat{\mathbf{u}} \times \hat{\mathbf{w}}$

5. $(3\mathbf{u}) \times \mathbf{w} + \mathbf{u} \times \mathbf{v}$

6. $\mathbf{u} \times (3\mathbf{v} - \mathbf{w})$

7. $\dfrac{\mathbf{w} \times \mathbf{u}}{|\mathbf{u} \times \mathbf{v}|}$

8. $\mathbf{u} \times \mathbf{w} - \mathbf{u} \times \mathbf{v} + \mathbf{u} \times (2\mathbf{u} + \mathbf{v})$

9. $(\mathbf{u} \times \mathbf{v}) \times \mathbf{w}$

10. $\mathbf{u} \times (\mathbf{v} \times \mathbf{w})$

In Exercises 11–14 find components for the vector.

11. Perpendicular to the vectors $(1,3,5)$ and $(-2,1,4)$

12. Perpendicular to the y-axis and the vector joining the points $(2,4,-3)$ and $(1,5,6)$

13. Perpendicular to the plane containing the lines $x = 1 + 3t$, $y = 2 - 4t$, $z = -1 + 6t$, and $x = 1 + 5t$, $y = 2 - t$, $z = -1$

14. Perpendicular to the triangle with vertices $(-1,0,3)$,

$(5,1,2)$ and $(-6,2,4)$

15. What is the area of the triangle in Exercise 14?

16. Verify the results in equations 12.35.

17. Find the area of the parallelogram with vertices $(1,2,3)$, $(4,3,7)$, $(-1,3,6)$ and $(2,4,10)$.

In Exercises 18–27 find vector, parametric, and symmetric (if possible) equations for the straight line.

18. Through the point $(1,-1,3)$ and parallel to the vector $(2,4,-3)$

19. Through the point $(-1,3,6)$ and parallel to the vector $(2,-3,0)$

20. Through the points $(2,-3,4)$ and $(5,2,-1)$

21. Through the points $(-2,3,3)$ and $(-2,-3,-3)$

22. Through the points $(1,3,4)$ and $(1,3,5)$

23. Through the point $(1,-3,5)$ and parallel to the line

$$\frac{x}{5} = \frac{y-2}{3} = \frac{z+4}{-2}$$

24. Through the point $(2,0,3)$ and parallel to the line $x = 4+t,\ y = 2,\ z = 6-2t$

25. Through the point of intersection of the lines

$$\frac{x-1}{2} = \frac{y+4}{-3} = \frac{z-2}{5}; \quad \frac{x-1}{6} = \frac{y+4}{3} = \frac{z-2}{4};$$

and parallel to the line joining the points $(1,3,-2)$ and $(2,-2,1)$

26. $2x - y = 5,\ 3x + 4y + z = 10$

27. Through the point $(-2,3,1)$ and parallel to the line $x + y = 3,\ 2x - y + z = -2$

In Exercises 28–35 find an equation for the plane.

28. Containing the points $(1,3,2)$, $(-2,0,-2)$, $(1,4,3)$

29. Containing the point $(2,-4,3)$ and the line $(x-1)/3 = (y+5)/4 = z+2$

30. Containing the lines $x = 2y = (z+1)/4$ and $x = t,\ y = 2t,\ z = 6t - 1$

31. Containing the lines $(x-1)/6 = y/8 = (z+2)/2$ and $(x+1)/3 = (y-2)/4 = z+5$

32. Containing the line $x - y + 2z = 4,\ 2x + y + 3z = 6$ and the point $(1,-2,4)$

33. Containing the two lines $x + 2y + 4z = 21,\ x - y + 6z = 13$ and $x = 2 + 3t,\ y = 4,\ z = -3 + 5t$

34. Containing the two lines $3x + 4y = -6,\ x + 2y + z = 2$ and $2y + 3z = 19,\ 3x - 2y - 9z = -58$

35. Containing the line $x + y - 4z = 6,\ 2x + 3y + 5z = 10$ and

(a) perpendicular to the xy-plane

(b) perpendicular to the xz-plane

(c) perpendicular to the yz-plane

36. Does the line $(x-3)/2 = y - 2 = (z+1)/4$ lie in the plane $x - y + 2z = -1$?

37. Show that the lines joining the midpoints of the sides of any quadrilateral form a parallelogram.

38. Show that if a plane has nonzero intercepts a, b, and c on the x-, y-, and z-axes, then its equation is $x/a + y/b + z/c = 1$.

39. Find equations for the four faces of the tetrahedron in Figure 12.10.

40. Find equations for the nine planes forming the sides, bottom and roof of the birdhouse in Figure 12.8.

41. Show that the cross-product is not associative; that is, in general $\mathbf{u} \times (\mathbf{v} \times \mathbf{w}) \neq (\mathbf{u} \times \mathbf{v}) \times \mathbf{w}$.

42. Show that the three lines $(x-4)/3 = (y-8)/4 = (z+7)/(-4)$; $(x+5)/3 = (y+2)/2 = (z+1)/2$; and $x = 1,\ y = 5 + t,\ z = -6 - 3t$ form a triangle and find its area.

In Exercises 43–46 find the distance.

43. From the point $(1,2,-3)$ to the line $x = 2(y+1) = (z-4)/2$

44. From the point $(3,-2,0)$ to the line $x = t$, $y = 3 - 2t,\ z = 4 + t$

45. From the line $x - 1 = 3(y+4) = -z - 1$ to the plane $2x - 3y + z = 4$

46. From the line $x = 1 - 6t,\ y = 4t + 2,\ z = -t$ to the plane $x + y - 2z = 1$

If a force $\mathbf{F}$ acts at a point Q (Figure 12.91), then the moment of $\mathbf{F}$ about a point P is defined as the vector $\mathbf{M} = \mathbf{r} \times \mathbf{F}$, where $\mathbf{r} = \overrightarrow{\mathbf{PQ}}$. In Exercises 47–51 calculate the moment of the force about the given point.

FIGURE 12.91

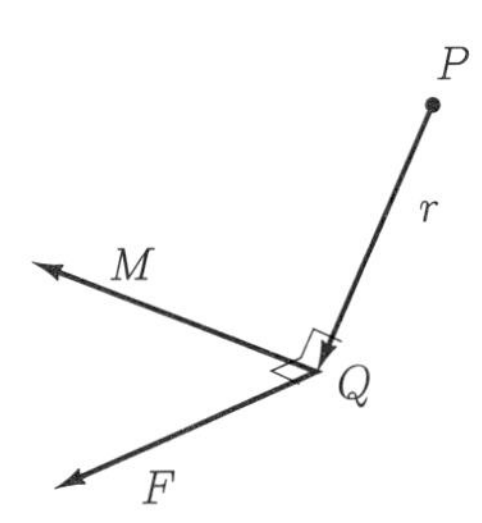

47. $\mathbf{F} = 2\hat{\imath} + 3\hat{\jmath} - 4\hat{\mathbf{k}}$ at $(1,3,2)$ about $(-1,4,2)$

48. $\mathbf{F} = \hat{\imath} + 2\hat{\jmath}$ at $(1,1,0)$ about $(2,1,-5)$

49. $\mathbf{F} = -\hat{\imath} + 3\hat{\mathbf{k}}$ at $(0,0,0)$ about $(-1,3,0)$

50. $\mathbf{F} = 3\hat{\imath} - \hat{\jmath} + 4\hat{\mathbf{k}}$ at $(1,1,1)$ about $(2,2,2)$

51. $\mathbf{F} = 6\hat{\imath}$ at $(0,1,3)$ about $(2,0,0)$

52. (a) If $\mathbf{u} \neq \mathbf{0}$, show that if both the conditions $\mathbf{u} \cdot \mathbf{v} = \mathbf{u} \cdot \mathbf{w}$ and $\mathbf{u} \times \mathbf{v} = \mathbf{u} \times \mathbf{w}$ are satisfied, then $\mathbf{v} = \mathbf{w}$.
(b) Show that if one of the conditions in (a) is satisfied, but the other is not, then $\mathbf{v}$ cannot be equal to $\mathbf{w}$.

53. The scalar $\mathbf{u} \cdot (\mathbf{v} \times \mathbf{w})$ is called the *scalar triple product* of $\mathbf{u}$, $\mathbf{v}$, and $\mathbf{w}$.
(a) Find $\mathbf{u} \cdot (\mathbf{v} \times \mathbf{w})$ if $\mathbf{u} = (6,-1,0)$, $\mathbf{v} = (1,3,4)$, and $\mathbf{w} = (-2,-1,4)$.
(b) Prove that $\mathbf{u} \cdot (\mathbf{v} \times \mathbf{w}) = (\mathbf{u} \times \mathbf{v}) \cdot \mathbf{w}$.
(c) Show that $|\mathbf{u} \cdot (\mathbf{v} \times \mathbf{w})|$ can be interpreted as the volume of the parallelepiped with $\mathbf{u}$, $\mathbf{v}$, and $\mathbf{w}$ as coterminal sides (Figure 12.92).
(d) Verify that three nonzero vectors $\mathbf{u}$, $\mathbf{v}$, and $\mathbf{w}$ all lie in the same plane if and only if $\mathbf{u} \cdot (\mathbf{v} \times \mathbf{w}) = 0$.

FIGURE 12.92

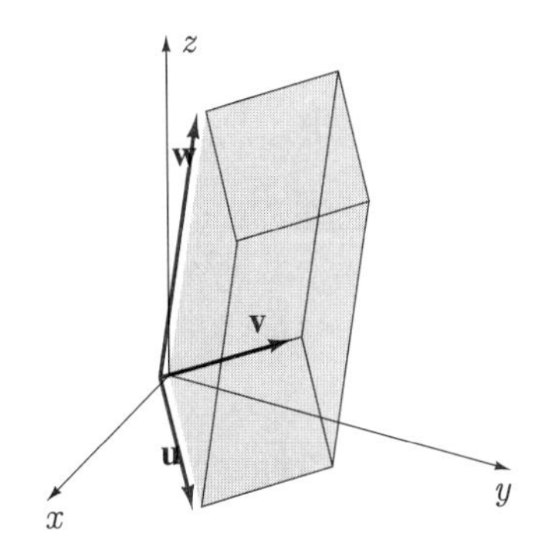

In Exercises 54 and 55 prove the identity.

54. $(\mathbf{u} \times \mathbf{v}) \cdot (\mathbf{w} \times \mathbf{r}) = (\mathbf{u} \cdot \mathbf{w})(\mathbf{v} \cdot \mathbf{r}) - (\mathbf{u} \cdot \mathbf{r})(\mathbf{v} \cdot \mathbf{w})$

55. $\mathbf{u} \times (\mathbf{v} \times \mathbf{w}) = (\mathbf{u} \cdot \mathbf{w})\mathbf{v} - (\mathbf{u} \cdot \mathbf{v})\mathbf{w}$

56. Verify that the equation of the plane passing through the three points $P_1(x_1, y_1, z_1)$, $P_2(x_2, y_2, z_2)$, and $P_3(x_3, y_3, z_3)$ can be written in the form $\overrightarrow{\mathbf{P_1P}} \cdot (\overrightarrow{\mathbf{P_1P_2}} \times \overrightarrow{\mathbf{P_1P_3}}) = 0$, where $P(x,y,z)$ is any point in the plane.

57. Use vectors to prove the sine law
$\dfrac{\sin A}{a} = \dfrac{\sin B}{b} = \dfrac{\sin C}{c}$ for the triangle in Figure 12.93.
(Hint: Note that $\overrightarrow{\mathbf{PQ}} + \overrightarrow{\mathbf{QR}} + \overrightarrow{\mathbf{RP}} = \mathbf{0}$. Cross this equation with $\overrightarrow{\mathbf{PQ}}$ and take lengths.)

FIGURE 12.93

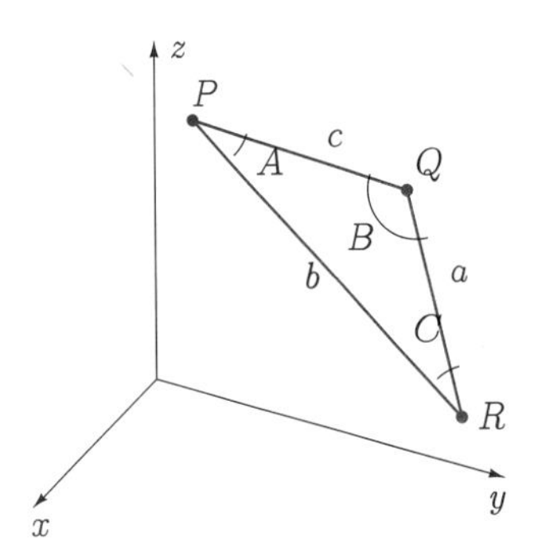

58. The region S_{xy} in the xy-plane bounded by the straight lines $x = 0$, $x = 1$, $y = 0$ and $y = 1$ is a rectangle with unit area.
(a) Show that the region in the plane $y = z$ that projects onto S_{xy} is also a rectangle, but with area $\sqrt{2}$.
(b) Show that the region in the plane $x + y - 2z = 0$ that projects onto S_{xy} is a parallelogram with area $\sqrt{6}/2$.
(c) Generalize the results of (a) and (b) to show that if S is the area in a plane $Ax + By + Cz + D = 0$ $(C \neq 0)$ that projects onto S_{xy}, then the area of S is $\sec\gamma$ where γ is the acute angle between $\hat{\mathbf{k}}$ and the normal to the given plane.

59. Find the shortest distance between the lines

$$\frac{x-1}{2} = \frac{y+3}{3} = 4 - z; \quad x = -1 + t,\ y = 2t,\ z = 3 - 2t.$$

SECTION 12.6

Differentiation and Integration of Vectors

In Sections 12.4 and 12.5, vectors were used to represent forces and moments; they can also be used to describe many other physical quantities, such as position, velocity, acceleration, electric and magnetic fields, and fluid flow. In applications such as these, vectors seldom have constant components; instead they have components that are either functions of position, or functions of some parameter, such as time, or both. For instance, the spring force in Example 12.18 varies in both magnitude and direction as the sleeve moves from B to C. Consequently, components of the vector **F** representing this force are functions of position x between B and C:

$$\mathbf{F} = F_x\hat{\mathbf{i}} + F_y\hat{\mathbf{j}} = F_x(x)\hat{\mathbf{i}} + F_y(x)\hat{\mathbf{j}}.$$

When a particle moves along a curve C in the xy-plane defined parametrically by

$$C: \quad x = x(t), \quad y = y(t), \quad \alpha \le t \le \beta,$$

its position (x, y) relative to the origin is represented by the vector $\mathbf{r} = x\hat{\mathbf{i}} + y\hat{\mathbf{j}}$ (Figure 12.94). This vector is called the **position vector** or **displacement vector** of the particle relative to the origin. We will have more to say about it in Section 12.10. Note, however, that if we substitute from the parametric equations for C, we have

$$\mathbf{r} = x(t)\hat{\mathbf{i}} + y(t)\hat{\mathbf{j}},$$

which indicates that the displacement vector has components that are functions of the parameter t.

FIGURE 12.94

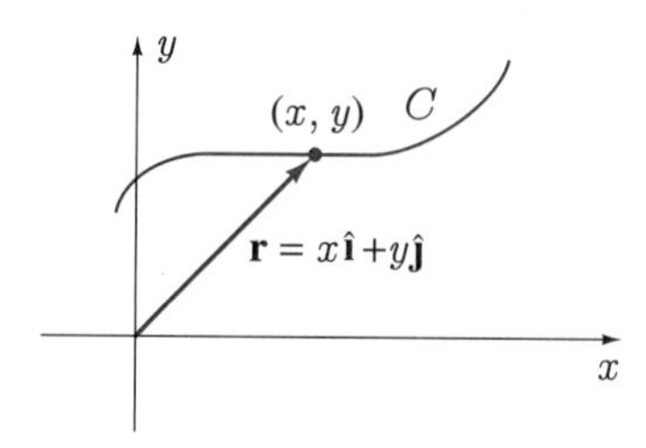

In this section we consider the general situation in which the components v_x, v_y, and v_z of a vector **v** are functions of some parameter t,

$$\mathbf{v} = v_x(t)\hat{\mathbf{i}} + v_y(t)\hat{\mathbf{j}} + v_z(t)\hat{\mathbf{k}}, \tag{12.43}$$

and show how the operations of differentiation and integration can be applied to such a vector. In Sections 12.8 and 12.9 we will use these results to discuss the geometry of curves, and this will pave the way for an analysis of the motion of particles in Section 12.10.

Because **v** in 12.43 has components that are functions of t, we say that **v** itself is a function of t, and write $\mathbf{v} = \mathbf{v}(t)$. Each of the component functions $v_x(t)$, $v_y(t)$, and $v_z(t)$ has a domain, and their common domain is called the domain of the vector function $\mathbf{v}(t)$. Given that this domain is some interval $\alpha \le t \le \beta$, we express 12.43 more fully in the form

$$\mathbf{v} = \mathbf{v}(t) = v_x(t)\hat{\mathbf{i}} + v_y(t)\hat{\mathbf{j}} + v_z(t)\hat{\mathbf{k}}, \quad \alpha \le t \le \beta. \tag{12.44}$$

To differentiate and integrate vector functions, we first require the concept of a limit.

Definition 12.11

If $\mathbf{v}(t) = v_x(t)\hat{\mathbf{i}} + v_y(t)\hat{\mathbf{j}} + v_z(t)\hat{\mathbf{k}}$, then

$$\lim_{t\to t_0} \mathbf{v}(t) = \left(\lim_{t\to t_0} v_x(t)\right)\hat{\mathbf{i}} + \left(\lim_{t\to t_0} v_y(t)\right)\hat{\mathbf{j}} + \left(\lim_{t\to t_0} v_z(t)\right)\hat{\mathbf{k}}, \qquad (12.45)$$

provided each of the limits on the right exists.

This definition states that to take the limit of a vector function, we take the limit of each component separately. As an illustration, consider the following example.

EXAMPLE 12.24

If $\mathbf{v} = \mathbf{v}(t) = (t^2 + 1)\hat{\mathbf{i}} + 3t\hat{\mathbf{j}} - (\sin t)\hat{\mathbf{k}}$, calculate $\lim_{t\to 5} \mathbf{v}(t)$.

SOLUTION According to Definition 12.11,

$$\begin{aligned}\lim_{t\to 5} \mathbf{v}(t) &= \left(\lim_{t\to 5}(t^2 + 1)\right)\hat{\mathbf{i}} + \left(\lim_{t\to 5}(3t)\right)\hat{\mathbf{j}} - \left(\lim_{t\to 5} \sin t\right)\hat{\mathbf{k}}\\ &= 26\hat{\mathbf{i}} + 15\hat{\mathbf{j}} - (\sin 5)\hat{\mathbf{k}}.\end{aligned}$$

■

In the remainder of this section we define continuity, derivatives, and antiderivatives for vector functions. Each definition is an exact duplicate of the corresponding definition for a scalar function $y(t)$, except that $y(t)$ is replaced by $\mathbf{v}(t)$. We then show that the vector definition can be rephrased in terms of components of the vector. We begin with continuity in the following definition.

Definition 12.12

A vector function $\mathbf{v}(t)$ is said to be **continuous** at a point t_0 if

$$\mathbf{v}(t_0) = \lim_{t\to t_0} \mathbf{v}(t). \qquad (12.46)$$

It is a simple matter to prove the next theorem.

Theorem 12.7

A vector function is continuous at a point if and only if its components are continuous at that point.

Proof If $\mathbf{v}(t) = v_x(t)\hat{\mathbf{i}} + v_y(t)\hat{\mathbf{j}} + v_z(t)\hat{\mathbf{k}}$, then according to 12.46, $\mathbf{v}(t)$ is continuous at t_0 if and only if

$$v_x(t_0)\hat{\mathbf{i}} + v_y(t_0)\hat{\mathbf{j}} + v_z(t_0)\hat{\mathbf{k}} = \lim_{t\to t_0}\{v_x(t)\hat{\mathbf{i}} + v_y(t)\hat{\mathbf{j}} + v_z(t)\hat{\mathbf{k}}\}.$$

Definition 12.11 implies that we can write this condition in the form

$$v_x(t_0)\hat{\mathbf{i}} + v_y(t_0)\hat{\mathbf{j}} + v_z(t_0)\hat{\mathbf{k}} = \left(\lim_{t\to t_0} v_x(t)\right)\hat{\mathbf{i}} + \left(\lim_{t\to t_0} v_y(t)\right)\hat{\mathbf{j}} + \left(\lim_{t\to t_0} v_z(t)\right)\hat{\mathbf{k}}.$$

But because two vectors are equal if and only if their components are equal, we can say that $\mathbf{v}(t)$ is continuous at t_0 if and only if

$$v_x(t_0) = \lim_{t\to t_0} v_x(t); \quad v_y(t_0) = \lim_{t\to t_0} v_y(t); \quad v_z(t_0) = \lim_{t\to t_0} v_z(t);$$

i.e., $\mathbf{v}(t)$ is continuous at t_0 if and only if its components are continuous at t_0.

For example, the vector function

$$\mathbf{v}(t) = (t-1)\hat{\mathbf{i}} + (1/t)\hat{\mathbf{j}} + (t^2-1)^{-1}\hat{\mathbf{k}}$$

is discontinuous for $t = 0$ (since $v_y(0)$ is not defined) and for $t = \pm 1$ (since $v_z(\pm 1)$ is not defined).

The derivative of a scalar function $y(t)$ is its instantaneous rate of change

$$\frac{dy}{dt} = \lim_{h\to 0} \frac{y(t+h) - y(t)}{h}.$$

The derivative of a vector function is also a rate of change defined by a similar limit.

Definition 12.13

The derivative of a vector function $\mathbf{v}(t)$ is defined as

$$\frac{d\mathbf{v}}{dt} = \lim_{h\to 0} \frac{\mathbf{v}(t+h) - \mathbf{v}(t)}{h}, \tag{12.47}$$

provided the limit exists.

In practice, we seldom use the definition of a derivative to calculate dy/dt for a scalar function $y(t)$; formulas such as the power, product, quotient, and chain rules are more convenient. It would be helpful to have corresponding formulas for derivatives of vector functions. The following theorem shows that to differentiate a vector function we simply differentiate its Cartesian components.

Theorem 12.8

If $\mathbf{v}(t) = v_x(t)\hat{\mathbf{i}} + v_y(t)\hat{\mathbf{j}} + v_z(t)\hat{\mathbf{k}}$, *and* $d\mathbf{v}/dt$ *exists, then*

$$\frac{d\mathbf{v}}{dt} = \frac{dv_x}{dt}\hat{\mathbf{i}} + \frac{dv_y}{dt}\hat{\mathbf{j}} + \frac{dv_z}{dt}\hat{\mathbf{k}}. \tag{12.48}$$

Proof If we substitute the components of $\mathbf{v}(t+h)$ and $\mathbf{v}(t)$ into Definition 12.13, then we have

$$\begin{aligned}\frac{d\mathbf{v}}{dt} &= \lim_{h\to 0}\left\{\frac{[v_x(t+h)\hat{\mathbf{i}} + v_y(t+h)\hat{\mathbf{j}} + v_z(t+h)\hat{\mathbf{k}}] - [v_x(t)\hat{\mathbf{i}} + v_y(t)\hat{\mathbf{j}} + v_z(t)\hat{\mathbf{k}}]}{h}\right\}\\ &= \lim_{h\to 0}\left\{\left[\frac{v_x(t+h) - v_x(t)}{h}\right]\hat{\mathbf{i}} + \left[\frac{v_y(t+h) - v_y(t)}{h}\right]\hat{\mathbf{j}} + \left[\frac{v_z(t+h) - v_z(t)}{h}\right]\hat{\mathbf{k}}\right\}\end{aligned}$$

(according to equation 12.14), and

$$\frac{d\mathbf{v}}{dt} = \left(\lim_{h\to 0} \frac{v_x(t+h) - v_x(t)}{h}\right)\hat{\mathbf{i}} + \left(\lim_{h\to 0} \frac{v_y(t+h) - v_y(t)}{h}\right)\hat{\mathbf{j}} + \left(\lim_{h\to 0} \frac{v_z(t+h) - v_z(t)}{h}\right)\hat{\mathbf{k}}$$

(according to equation 12.45). If $d\mathbf{v}/dt$ exists, then each of the three limits on the right must exist, and we can write that

$$\frac{d\mathbf{v}}{dt} = \frac{dv_x}{dt}\hat{\mathbf{i}} + \frac{dv_y}{dt}\hat{\mathbf{j}} + \frac{dv_z}{dt}\hat{\mathbf{k}}.$$

This proof is also reversible, giving the following corollary to Theorem 12.8.

Corollary *If the derivatives dv_x/dt, dv_y/dt, and dv_z/dt of the components of a vector function $\mathbf{v}(t) = v_x(t)\hat{\mathbf{i}} + v_y(t)\hat{\mathbf{j}} + v_z(t)\hat{\mathbf{k}}$ exist at a point, then so does $d\mathbf{v}/dt$.*

Theorem 12.8 gives us a working rule for differentiating vector functions: To differentiate a vector function, we differentiate its Cartesian components.

EXAMPLE 12.25

If $\mathbf{v}(t) = t^2\hat{\mathbf{i}} + (3t^3 - 2t)\hat{\mathbf{j}} + 5\hat{\mathbf{k}}$, find $\mathbf{v}'(3)$.

SOLUTION According to 12.48

$$\frac{d\mathbf{v}}{dt} = 2t\hat{\mathbf{i}} + (9t^2 - 2)\hat{\mathbf{j}}.$$

Consequently,

$$\mathbf{v}'(3) = 6\hat{\mathbf{i}} + 79\hat{\mathbf{j}}.$$

■

The sum rule 3.8 for differentiation of scalar functions has its counterpart in the sum rule for vector functions,

$$\frac{d}{dt}(\mathbf{u} + \mathbf{v}) = \frac{d\mathbf{u}}{dt} + \frac{d\mathbf{v}}{dt} \tag{12.49}$$

(see Exercise 22). There are three types of products associated with vectors: the product of a scalar and a vector, the dot product of two vectors, and the cross product of two vectors. Corresponding to each, we have a product rule for differentiation, but all resemble the product rule for scalar functions.

Theorem 12.9

If $f(t)$ is a differentiable function and $\mathbf{u}(t)$ and $\mathbf{v}(t)$ are differentiable vector functions, then:

$$\frac{d}{dt}(f\mathbf{v}) = \frac{df}{dt}\mathbf{v} + f\frac{d\mathbf{v}}{dt}, \tag{12.50}$$

$$\frac{d}{dt}(\mathbf{u} \cdot \mathbf{v}) = \mathbf{u} \cdot \frac{d\mathbf{v}}{dt} + \frac{d\mathbf{u}}{dt} \cdot \mathbf{v}, \tag{12.51}$$

$$\frac{d}{dt}(\mathbf{u} \times \mathbf{v}) = \mathbf{u} \times \frac{d\mathbf{v}}{dt} + \frac{d\mathbf{u}}{dt} \times \mathbf{v}. \tag{12.52}$$

For a proof of these results, see Exercise 23.

EXAMPLE 12.26 If $f(t) = t^2 + 2t + 3$, $\mathbf{u}(t) = t\hat{\mathbf{i}} + t^2\hat{\mathbf{j}} - 3\hat{\mathbf{k}}$, and $\mathbf{v}(t) = t(\hat{\mathbf{i}} + \hat{\mathbf{j}} + \hat{\mathbf{k}})$, use (12.50)–(12.52) to evaluate:

$$\text{(a)}\ \frac{d}{dt}(f\mathbf{u});\quad \text{(b)}\ \frac{d}{dt}(\mathbf{u}\cdot\mathbf{v});\quad \text{(c)}\ \frac{d}{dt}(\mathbf{u}\times\mathbf{v}).$$

SOLUTION

(a) With 12.50,

$$\frac{d}{dt}(f\mathbf{u}) = \frac{df}{dt}\mathbf{u} + f\frac{d\mathbf{u}}{dt} = (2t+2)(t\hat{\mathbf{i}} + t^2\hat{\mathbf{j}} - 3\hat{\mathbf{k}}) + (t^2 + 2t + 3)(\hat{\mathbf{i}} + 2t\hat{\mathbf{j}})$$
$$= (3t^2 + 4t + 3)\hat{\mathbf{i}} + (4t^3 + 6t^2 + 6t)\hat{\mathbf{j}} - 6(t+1)\hat{\mathbf{k}}.$$

(b) With 12.51,

$$\frac{d}{dt}(\mathbf{u}\cdot\mathbf{v}) = \mathbf{u}\cdot\frac{d\mathbf{v}}{dt} + \frac{d\mathbf{u}}{dt}\cdot\mathbf{v}$$
$$= (t\hat{\mathbf{i}} + t^2\hat{\mathbf{j}} - 3\hat{\mathbf{k}})\cdot(\hat{\mathbf{i}} + \hat{\mathbf{j}} + \hat{\mathbf{k}}) + (\hat{\mathbf{i}} + 2t\hat{\mathbf{j}})\cdot(t\hat{\mathbf{i}} + t\hat{\mathbf{j}} + t\hat{\mathbf{k}})$$
$$= (t + t^2 - 3) + (t + 2t^2) = 3t^2 + 2t - 3.$$

(c) With 12.52,

$$\frac{d}{dt}(\mathbf{u}\times\mathbf{v}) = \mathbf{u}\times\frac{d\mathbf{v}}{dt} + \frac{d\mathbf{u}}{dt}\times\mathbf{v} = (t\hat{\mathbf{i}} + t^2\hat{\mathbf{j}} - 3\hat{\mathbf{k}})\times(\hat{\mathbf{i}} + \hat{\mathbf{j}} + \hat{\mathbf{k}})$$
$$+ (\hat{\mathbf{i}} + 2t\hat{\mathbf{j}})\times(t\hat{\mathbf{i}} + t\hat{\mathbf{j}} + t\hat{\mathbf{k}})$$
$$= \begin{vmatrix} \hat{\mathbf{i}} & \hat{\mathbf{j}} & \hat{\mathbf{k}} \\ t & t^2 & -3 \\ 1 & 1 & 1 \end{vmatrix} + \begin{vmatrix} \hat{\mathbf{i}} & \hat{\mathbf{j}} & \hat{\mathbf{k}} \\ 1 & 2t & 0 \\ t & t & t \end{vmatrix}$$
$$= [(t^2 + 3)\hat{\mathbf{i}} - (3 + t)\hat{\mathbf{j}} + (t - t^2)\hat{\mathbf{k}}]$$
$$+ [2t^2\hat{\mathbf{i}} - t\hat{\mathbf{j}} + (t - 2t^2)\hat{\mathbf{k}}]$$
$$= (3t^2 + 3)\hat{\mathbf{i}} - (3 + 2t)\hat{\mathbf{j}} + (2t - 3t^2)\hat{\mathbf{k}}.$$

■

If vector functions can be differentiated, then they can be antidifferentiated. Formally we make the following statement.

Definition 12.14 A vector function $\mathbf{V}(t)$ is said to be an **antiderivative** or **indefinite integral** of $\mathbf{v}(t)$ on the interval $\alpha < t < \beta$ if

$$\frac{d\mathbf{V}}{dt} = \mathbf{v}(t) \quad \text{for } \alpha < t < \beta. \tag{12.53}$$

For example, an antiderivative of $\mathbf{v}(t) = 2t\hat{\mathbf{i}} - \hat{\mathbf{j}} + 3t^2\hat{\mathbf{k}}$ is

$$\mathbf{V}(t) = t^2\hat{\mathbf{i}} - t\hat{\mathbf{j}} + t^3\hat{\mathbf{k}}.$$

If we add to $\mathbf{V}(t)$ in 12.53 any vector with constant components, denoted by $\mathbf{C}$, then $\mathbf{V}(t) + \mathbf{C}$ is also an antiderivative of $\mathbf{v}(t)$. We call this vector *the* antiderivative or *the* indefinite integral of $\mathbf{v}(t)$, and write

$$\int \mathbf{v}(t)\,dt = \mathbf{V}(t) + \mathbf{C}. \qquad (12.54)$$

For our example, then,

$$\int (2t\hat{\mathbf{i}} - \hat{\mathbf{j}} + 3t^2\hat{\mathbf{k}})\,dt = t^2\hat{\mathbf{i}} - t\hat{\mathbf{j}} + t^3\hat{\mathbf{k}} + \mathbf{C}.$$

Because vectors can be differentiated component by component, it follows that they may also be integrated component by component; i.e., if $\mathbf{v}(t) = v_x(t)\hat{\mathbf{i}} + v_y(t)\hat{\mathbf{j}} + v_z(t)\hat{\mathbf{k}}$, then

$$\int \mathbf{v}(t)\,dt = \left(\int v_x(t)\,dt\right)\hat{\mathbf{i}} + \left(\int v_y(t)\,dt\right)\hat{\mathbf{j}} + \left(\int v_z(t)\,dt\right)\hat{\mathbf{k}}. \qquad (12.55)$$

EXAMPLE 12.27

Find the antiderivative of $\mathbf{v}(t) = \sqrt{t-1}\,\hat{\mathbf{i}} + e^t\hat{\mathbf{j}} + 6t^2\hat{\mathbf{k}}$.

SOLUTION According to equation12.55,

$$\begin{aligned}\int \mathbf{v}(t)\,dt &= \left(\frac{2}{3}(t-1)^{3/2} + C_1\right)\hat{\mathbf{i}} + (e^t + C_2)\hat{\mathbf{j}} + (2t^3 + C_3)\hat{\mathbf{k}}\\ &= \frac{2}{3}(t-1)^{3/2}\hat{\mathbf{i}} + e^t\hat{\mathbf{j}} + 2t^3\hat{\mathbf{k}} + \mathbf{C},\end{aligned}$$

where $\mathbf{C} = C_1\hat{\mathbf{i}} + C_2\hat{\mathbf{j}} + C_3\hat{\mathbf{k}}$ is a constant vector. ■

EXERCISES 12.6

In Exercises 1–5 find the largest possible domain for the vector function.

1. $\mathbf{v}(t) = t^2\hat{\mathbf{i}} + \sqrt{t-1}\,\hat{\mathbf{j}} + \hat{\mathbf{k}}$

2. $\mathbf{v}(t) = (\sin t)\hat{\mathbf{i}} + (\cos t)\hat{\mathbf{j}} - t^3\hat{\mathbf{k}}$

3. $\mathbf{v}(t) = (\mathrm{Sin}^{-1}\, t)\hat{\mathbf{i}} - t^2\hat{\mathbf{j}} + (t+1)\hat{\mathbf{k}}$

4. $\mathbf{v}(t) = \ln(t+4)(\hat{\mathbf{i}} + \hat{\mathbf{j}})$

5. $\mathbf{v}(t) = e^t\hat{\mathbf{i}} + (\cos^2 t)\hat{\mathbf{j}} - (e^t\cos t)\hat{\mathbf{k}}$

If $f(t) = t^2 + 3$, $g(t) = 2t^3 - 3t$, $\mathbf{u}(t) = t\hat{\mathbf{i}} - t^2\hat{\mathbf{j}} + 2t\hat{\mathbf{k}}$, and $\mathbf{v}(t) = \hat{\mathbf{i}} - 2t\hat{\mathbf{j}} + 3t^2\hat{\mathbf{k}}$, find the scalar or the components of the vector in Exercises 6–21.

6. $\dfrac{d\mathbf{u}}{dt}$

7. $\dfrac{d}{dt}[f(t)\mathbf{v}(t)]$

8. $\dfrac{d}{dt}[g(t)\mathbf{u}(t)]$

9. $\dfrac{d}{dt}(\mathbf{u} \times \mathbf{v})$

10. $\dfrac{d}{dt}(\mathbf{u} \times t\mathbf{v})$

11. $\dfrac{d}{dt}(2\mathbf{u} \cdot \mathbf{v})$

12. $\dfrac{d}{dt}(3\mathbf{u} + 4\mathbf{v})$

13. $\displaystyle\int \mathbf{u}(t)\,dt$

14. $\dfrac{d}{dt}[f(t)\mathbf{u} + g(t)\mathbf{v}]$

15. $\displaystyle\int 4\mathbf{v}(t)\,dt$

16. $\dfrac{d}{dt}[t(\mathbf{u} \times \mathbf{v})]$

17. $\int [f(t)\mathbf{u}(t)]\,dt$

18. $\int [3g(t)\mathbf{v}(t) + \mathbf{u}(t)]\,dt$

19. $\int [f(t)\mathbf{u}\cdot\mathbf{v}]\,dt$

20. $\mathbf{u} \times \frac{d\mathbf{v}}{dt} - f(t)\mathbf{u}\cdot\frac{d\mathbf{v}}{dt}\mathbf{v}$

21. $\mathbf{u}\cdot\frac{d\mathbf{v}}{dt} - \mathbf{v}\cdot\int \mathbf{u}(t)\,dt$

22. Prove equation 12.49.

23. Verify the results in equations 12.50–12.52.

24. Prove that for differentiable functions $\mathbf{u}(t)$, $\mathbf{v}(t)$, and $\mathbf{w}(t)$,

$$\frac{d}{dt}(\mathbf{u}\cdot\mathbf{v}\times\mathbf{w}) = \frac{d\mathbf{u}}{dt}\cdot\mathbf{v}\times\mathbf{w} + \mathbf{u}\cdot\frac{d\mathbf{v}}{dt}\times\mathbf{w} + \mathbf{u}\cdot\mathbf{v}\times\frac{d\mathbf{w}}{dt}.$$

25. Prove that if a differentiable function $\mathbf{v}(t)$ has constant length, then at any point at which $d\mathbf{v}/dt \neq \mathbf{0}$, the vector $d\mathbf{v}/dt$ is perpendicular to $\mathbf{v}$.

26. If $\mathbf{v} = \mathbf{v}(s)$ is a differentiable vector function and $s = s(t)$ is a differentiable scalar function, prove that

$$\frac{d\mathbf{v}}{dt} = \frac{d\mathbf{v}}{ds}\frac{ds}{dt}.$$

This result is called the *chain rule* for differentiation of vector functions.

27. Show that the following definition for the limit of a vector function is equivalent to Definition 12.11: A vector function $\mathbf{v}(t)$ is said to have limit $\mathbf{V}$ as t approaches t_0 if given any $\epsilon > 0$, there exists a $\delta > 0$ such that $|\mathbf{v}(t) - \mathbf{V}| < \epsilon$ whenever $0 < |t - t_0| < \delta$.

SECTION 12.7

Parametric and Vector Representations of Curves

In Section 12.2 we presented curves in space as the intersection of two surfaces. For example, each of the equations

$$x^2 + y^2 + z^2 = 9, \quad y = 2$$

describes a surface (the first a sphere and the second a plane), and together they describe the curve of intersection of the surfaces—the circle in Figure 12.95.

FIGURE 12.95

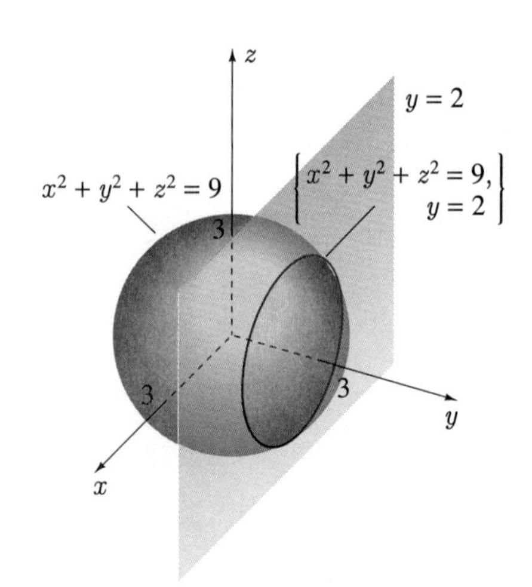

Parametric Representation of Curves

In many applications it is more convenient to have a curve defined "parametrically."

Definition 12.15

A curve in space is defined parametrically by three functions:

$$C: \quad x = x(t), \quad y = y(t), \quad z = z(t), \quad \alpha \leq t \leq \beta. \tag{12.56}$$

Each value of t in the interval $\alpha \leq t \leq \beta$ is substituted into the three functions, and the triple $(x, y, z) = (x(t), y(t), z(t))$ specifies a point on the curve. Definition 12.15 clearly corresponds to parametric Definition 10.1 for a plane curve.

It is customary to assign a direction to a curve by calling that point on C corresponding to $t = \alpha$ the initial point and that point corresponding to $t = \beta$ the final point, and the direction of C is that direction along C from initial point to final point (Figure 12.96). Note in particular that the direction of a curve always corresponds to the direction in which the parameter t *increases* along the curve. Because of this, whenever we describe a curve in nonparametric form but with a specified direction, we must be careful in setting up parametric equations to ensure that the parameter increases in the appropriate direction.

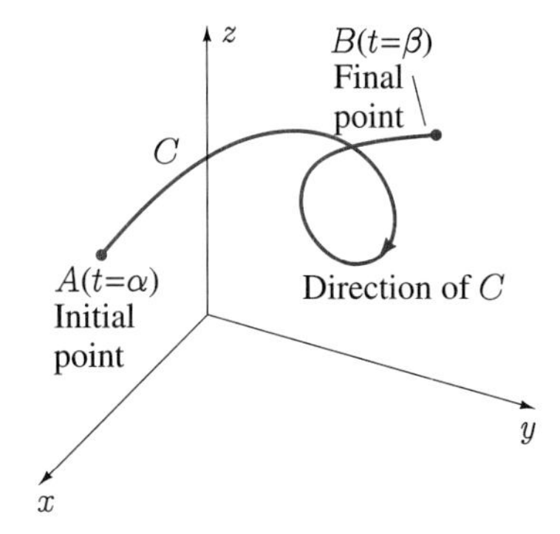

FIGURE 12.96

When a curve is described as the curve of intersection of two surfaces, we can obtain parametric equations for the curve by specifying one of x, y, or z as a function of t, and then solving the equations for the other two as functions of t. Considerable ingenuity is sometimes required in arriving at an initial function of t. We illustrate this in the following example.

Find parametric equations for each of the following curves:

(a) $z - 1 = x^2 + y^2$, $x - y = 0$ directed so that z increases when x and y are positive;

(b) $x + 2y + z = 4$, $2x + y + 3z = 6$ directed so that y increases along the curve;

(c) $x^2 + (y - 1)^2 = 4$, $z = x$ directed so that y increases when x is positive.

SOLUTION

FIGURE 12.97

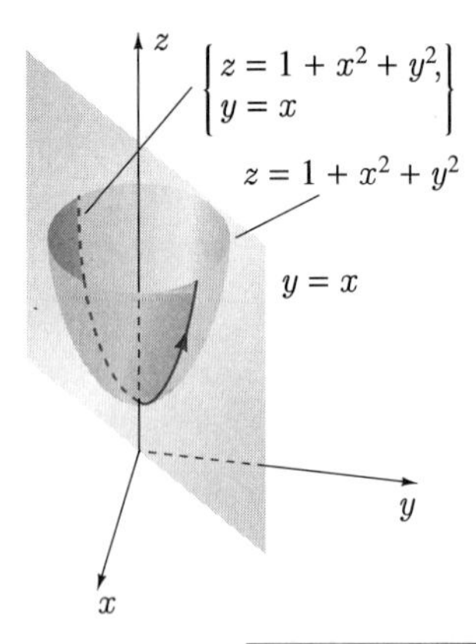

(a) The curve of intersection of the circular paraboloid $z = 1 + x^2 + y^2$ and the plane $y = x$ is shown in Figure 12.97. If we choose x as the parameter along the curve by setting $x = t$, then

$$x = t, \quad y = t, \quad z = 1 + 2t^2$$

(and note that z increases when x and y are positive).

(b) The straight-line intersection of the two planes is shown in Figure 12.98. If we choose y as the parameter by setting $y = t$ (thus forcing y to increase as t increases), then

$$x + z = 4 - 2t, \quad 2x + 3z = 6 - t.$$

The solution of these equations for x and z in terms of t gives the parametric equations

$$x = 6 - 5t, \quad y = t, \quad z = -2 + 3t.$$

FIGURE 12.98

FIGURE 12.99

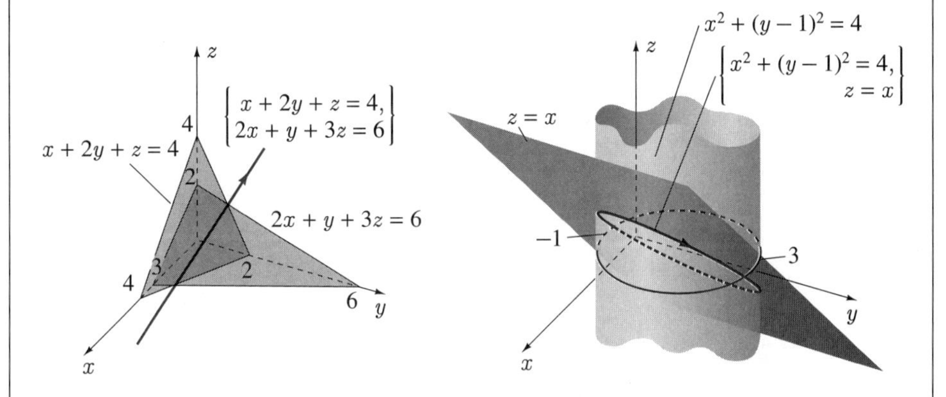

(c) The curve of intersection of the right-circular cylinder $x^2 + (y - 1)^2 = 4$ and the plane $z = x$ is shown in Figure 12.99. If we set $x = 2\cos t$, then $y = 1 \pm 2\sin t$. A set of parametric equations for the curve is therefore

$$x = 2\cos t, \quad y = 1 + 2\sin t, \quad z = 2\cos t, \quad 0 \le t < 2\pi.$$

Any range of values of t of length 2π traces the curve exactly once. Note that had we chosen the equations

$$x = 2\cos t, \quad y = 1 - 2\sin t, \quad z = 2\cos t, \quad 0 \le t < 2\pi,$$

we would have generated the same set of points traced in the opposite direction. ■

Definition 12.16

A curve C: $x = x(t)$, $y = y(t)$, $z = z(t)$, $\alpha \le t \le \beta$, is said to be **continuous** if each of the functions $x(t)$, $y(t)$, and $z(t)$ is continuous for $\alpha \le t \le \beta$.

Geometrically, this implies that the curve is at no point separated. Each of the curves in Example 12.28 is therefore continuous.

A curve is said to be **closed** if its initial and final points are the same. Circles and ellipses are closed curves. Straight line segments, parabolas and hyperbolas are not closed.

Vector Representation of Curves

The position vector or displacement vector of a point $P(x, y, z)$ in space is the vector

$$\mathbf{r} = (x, y, z) = x\hat{\mathbf{i}} + y\hat{\mathbf{j}} + z\hat{\mathbf{k}}.$$

If we consider only points that lie on a curve defined parametrically by 12.56, then for these points we can write that

$$\mathbf{r} = \mathbf{r}(t) = x(t)\hat{\mathbf{i}} + y(t)\hat{\mathbf{j}} + z(t)\hat{\mathbf{k}}, \quad \alpha \leq t \leq \beta. \qquad (12.57)$$

As t varies from $t = \alpha$ to $t = \beta$, the tip of this vector traces the curve C from initial point to final point (Figure 12.100). We call 12.57 the **vector representation** of a curve.

FIGURE 12.100

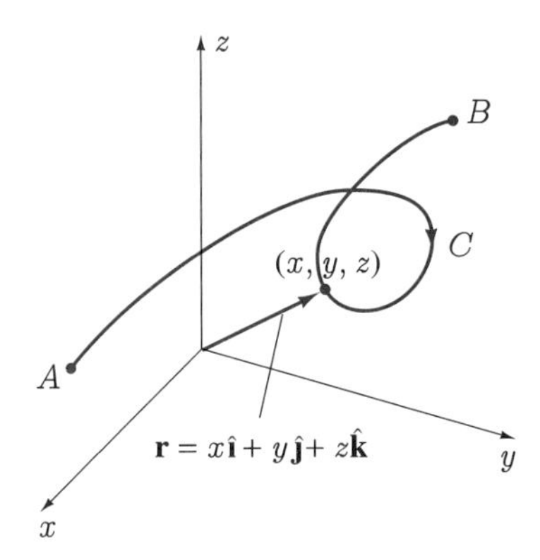

EXAMPLE 12.29

Sketch the curve with position vector

$$\mathbf{r} = \mathbf{r}(t) = (2\cos t)\hat{\mathbf{i}} + (3\sin t)\hat{\mathbf{j}}, \quad 0 \leq t \leq \pi.$$

SOLUTION When we set $x = 2\cos t$ and $y = 3\sin t$, it is clear that $x^2/4 + y^2/9 = 1$. The position vector $\mathbf{r} = \mathbf{r}(t)$ therefore describes points on an ellipse in the xy-plane. As t varies from 0 to π, x varies from 2 to -2, and y from 0 to 3 to 0 again. This means that only the top half of the ellipse is defined by $\mathbf{r} = \mathbf{r}(t)$ (Figure 12.101).

FIGURE 12.101

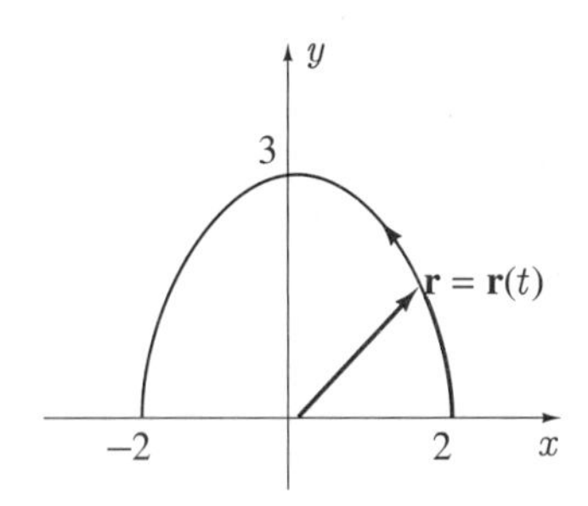

■

EXAMPLE 12.30

Sketch the curve with position vector

$$\mathbf{r} = \mathbf{r}(t) = t\hat{\mathbf{\imath}} + t^2\hat{\mathbf{\jmath}} + t\hat{\mathbf{k}}, \quad t \geq 0.$$

SOLUTION When we set $x = t$, $y = t^2$, and $z = t$, then $z = x$ and $y = x^2$. These imply that $\mathbf{r} = \mathbf{r}(t)$ describes points on the curve of intersection of the surfaces $y = x^2$ and $z = x$ (Figure 12.102). Because $t \geq 0$, only that half of the curve of intersection in the first octant is defined by $\mathbf{r} = \mathbf{r}(t)$.

FIGURE 12.102

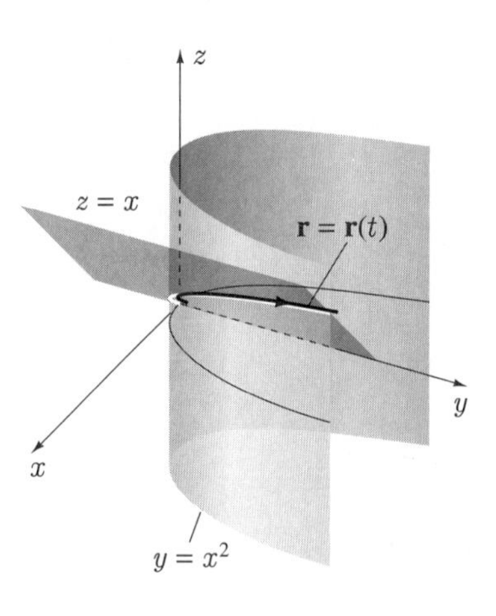

■

EXERCISES 12.7

In Exercises 1–10 find parametric and vector representations for the curve. Sketch each curve.

1. $x+2y+3z=6$, $y-2z=3$ directed so that z increases along the curve

2. $x^2+y^2=2$, $z=4$ directed so that y increases in the first octant

3. $x^2+y^2=2$, $x+y+z=1$ directed so that y decreases when x is positive

4. $z=x^2+y^2$, $x^2+y^2=5$ directed clockwise as viewed from the origin

5. $z+2x^2=1$, $y=z$ directed so that x decreases along the curve

6. $z=\sqrt{x^2+y^2}$, $y=x$ directed so that y increases when x is positive

7. $z=x+y$, $y=x^2$ directed so that x increases along the curve

8. $z=\sqrt{4-x^2-y^2}$, $x^2+y^2-2y=0$ directed so that z decreases when x is positive

9. $x=\sqrt{z}$, $z=y^2$ directed away from the origin in the first octant

10. $z=\sqrt{x^2+y^2}$, $y=x^2$ directed so that y decreases in the first octant

In Exercises 11–15 sketch the curve with the given position vector.

11. $\mathbf{r}(t)=t\hat{\mathbf{i}}+t\hat{\mathbf{j}}+t^2\hat{\mathbf{k}}$, $t\geq 0$

12. $\mathbf{r}(t)=(2\cos t)\hat{\mathbf{i}}+(2\sin t)\hat{\mathbf{j}}+3t\hat{\mathbf{k}}$, $0\leq t\leq 4\pi$

13. $\mathbf{r}(t)=(t-2)\hat{\mathbf{i}}+(2-3t)\hat{\mathbf{j}}+5t\hat{\mathbf{k}}$

14. $\mathbf{r}(t)=(t^2-t)\hat{\mathbf{i}}+t\hat{\mathbf{j}}+5\hat{\mathbf{k}}$

15. $\mathbf{r}(t)=(\cos t)\hat{\mathbf{i}}+(\sin t)\hat{\mathbf{j}}+(\cos t)\hat{\mathbf{k}}$, $0\leq t\leq \pi$

SECTION 12.8

Tangent Vectors and Lengths of Curves

If C is a curve in the xy-plane (Figure 12.103), then the tangent line to C at P is defined as the limiting position of the line PQ as Q moves along C toward P (see Section 3.1). We take the same approach in defining tangent vectors to curves in an arbitrary plane or in space. On curve C defined by 12.56, we let P and Q be the points corresponding to the parameter values t and $t+h$. Position vectors of P and Q are then

$$\mathbf{r}(t)=x(t)\hat{\mathbf{i}}+y(t)\hat{\mathbf{j}}+z(t)\hat{\mathbf{k}}$$

and

$$\mathbf{r}(t+h)=x(t+h)\hat{\mathbf{i}}+y(t+h)\hat{\mathbf{j}}+z(t+h)\hat{\mathbf{k}}$$

(Figure 12.104), and the vector $\overrightarrow{\mathbf{PQ}}$ joining P to Q is clearly equal to

$$\overrightarrow{\mathbf{PQ}}=\mathbf{r}(t+h)-\mathbf{r}(t).$$

FIGURE 12.103

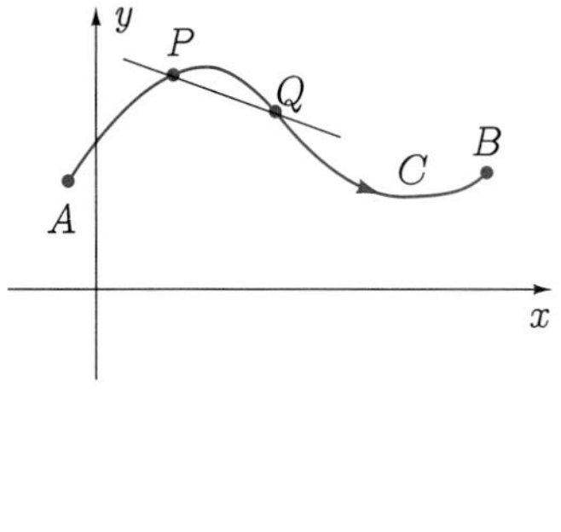

FIGURE 12.104

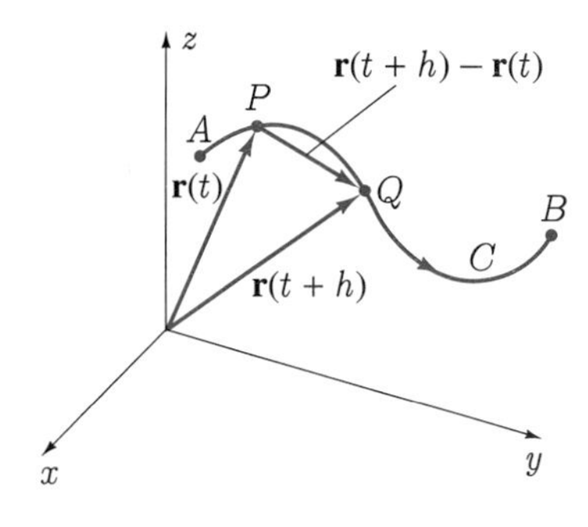

If we let h approach zero, then the point Q moves along C toward P, and the direction of $\overrightarrow{\mathbf{PQ}}$ becomes closer to what seems to be a reasonable definition of the tangent direction to C at P. Perhaps then we should define $\lim_{h\to 0}[\mathbf{r}(t+h)-\mathbf{r}(t)]$ as a tangent vector to C at P. Unfortunately, the limit vector has length zero, and therefore

$$\lim_{h\to 0}[\mathbf{r}(t+h)-\mathbf{r}(t)]=\mathbf{0}.$$

If, however, we divide $\mathbf{r}(t+h)-\mathbf{r}(t)$ by h, then the resulting vector

$$\frac{\mathbf{r}(t+h)-\mathbf{r}(t)}{h}$$

is not equal to $\overrightarrow{\mathbf{PQ}}$, but it does have the same direction as $\overrightarrow{\mathbf{PQ}}$. Consider, then, taking the limit of this vector as h approaches zero:

$$\lim_{h\to 0}\frac{\mathbf{r}(t+h)-\mathbf{r}(t)}{h}.$$

If the limit vector exists, then it will be tangent to C at P. But according to equation 12.47, this limit defines the derivative $d\mathbf{r}/dt$,

$$\frac{d\mathbf{r}}{dt}=\lim_{h\to 0}\frac{\mathbf{r}(t+h)-\mathbf{r}(t)}{h}=\frac{dx}{dt}\hat{\mathbf{i}}+\frac{dy}{dt}\hat{\mathbf{j}}+\frac{dz}{dt}\hat{\mathbf{k}}, \tag{12.58}$$

provided each of the derivatives dx/dt, dy/dt, and dz/dt exists. We have just established the following result.

Theorem 12.10

If $\mathbf{r}=\mathbf{r}(t)=x(t)\hat{\mathbf{i}}+y(t)\hat{\mathbf{j}}+z(t)\hat{\mathbf{k}}$, $\alpha\le t\le\beta$, is the vector representation of a curve C, then at any point on C at which $x'(t)$, $y'(t)$, and $z'(t)$ all exist and do not simultaneously vanish,

$$\mathbf{T}=\frac{d\mathbf{r}}{dt}=\frac{dx}{dt}\hat{\mathbf{i}}+\frac{dy}{dt}\hat{\mathbf{j}}+\frac{dz}{dt}\hat{\mathbf{k}} \tag{12.59}$$

is a tangent vector to C (Figure 12.105).

There are two tangent directions at any point on a curve. One of these has been shown to be $d\mathbf{r}/dt$; the other must be defined by $-d\mathbf{r}/dt$ (Figure 12.106). How can we tell which one is $d\mathbf{r}/dt$? A closer analysis of the limit in 12.58 indicates the following (see Exercise 17).

Corollary *The tangent vector $d\mathbf{r}/dt$ to a curve C: $x=x(t)$, $y=y(t)$, $z=z(t)$, $\alpha\le t\le\beta$, always points in the direction in which the parameter t increases along C.*

FIGURE 12.105

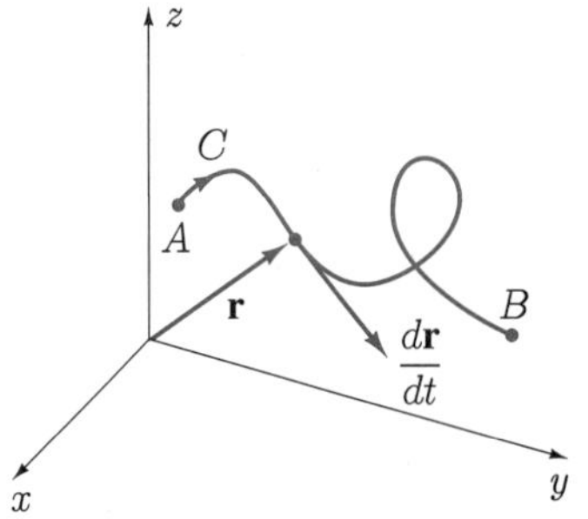

FIGURE 12.106

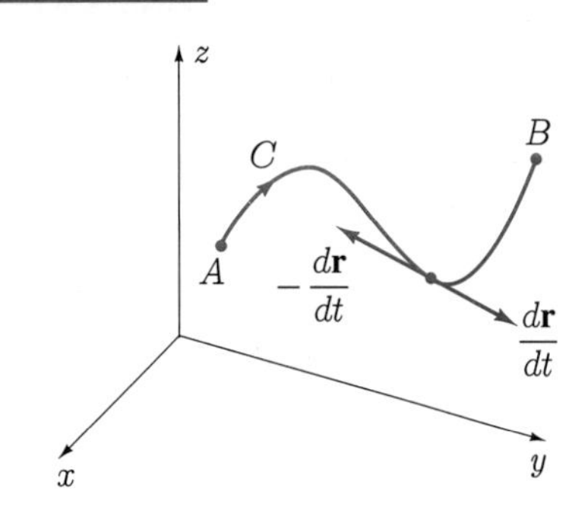

Definition 12.17

A curve C: $x = x(t)$, $y = y(t)$, $z = z(t)$, $\alpha \le t \le \beta$, is said to be **smooth** if the derivatives $x'(t)$, $y'(t)$, and $z'(t)$ are all continuous for $\alpha < t < \beta$ and do not simultaneously vanish for $\alpha < t < \beta$.

Since $x'(t)$, $y'(t)$, and $z'(t)$ are the components of a tangent vector to C, this definition implies that along a smooth curve, small changes in t produce small changes in the direction of the tangent vector. In other words, the tangent vector turns gradually, or "smoothly." The curve in Figure 12.107 is smooth; that in Figure 12.108 is not because abrupt changes in the direction of the curve occur at P and Q. According to the following definition, this curve is piecewise smooth.

FIGURE 12.107

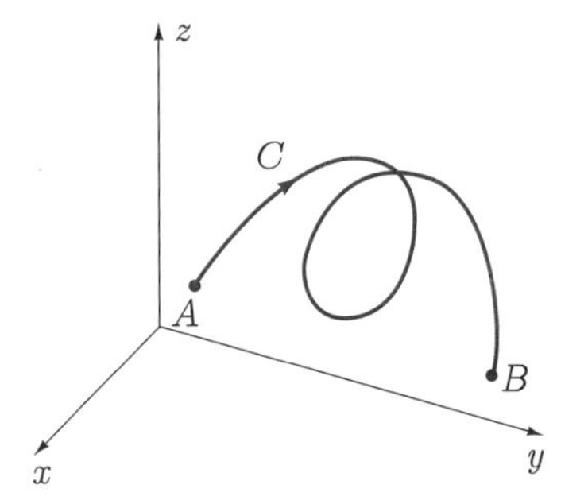

FIGURE 12.108

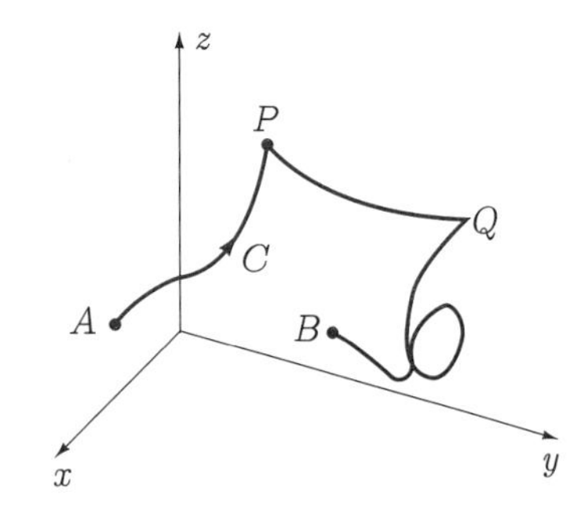

Definition 12.18

A curve is said to be **piecewise smooth** if it is continuous and can be divided into a finite number of smooth subcurves.

EXAMPLE 12.31

For the curve in Example 30, find a tangent vector at the point $(3, 9, 3)$.

SOLUTION A tangent vector to this curve at any point on the curve is

$$\frac{d\mathbf{r}}{dt} = \frac{dx}{dt}\hat{\mathbf{i}} + \frac{dy}{dt}\hat{\mathbf{j}} + \frac{dz}{dt}\hat{\mathbf{k}} = \hat{\mathbf{i}} + 2t\hat{\mathbf{j}} + \hat{\mathbf{k}}.$$

Since $t = 3$ yields the point $(3, 9, 3)$, a tangent vector at this point is $\mathbf{r}'(3) = \hat{\mathbf{i}} + 6\hat{\mathbf{j}} + \hat{\mathbf{k}}$. ■

EXAMPLE 12.32

Find a tangent vector at the point $(2, 0, 3)$ to the helix

$$x = 2\cos t, \quad y = 2\sin t, \quad z = \frac{3t}{2\pi}, \quad t \ge 0.$$

Is the helix smooth?

SOLUTION

A tangent vector to the helix at any point is

$$\frac{d\mathbf{r}}{dt} = \frac{dx}{dt}\hat{\mathbf{i}} + \frac{dy}{dt}\hat{\mathbf{j}} + \frac{dz}{dt}\hat{\mathbf{k}} = (-2\sin t)\hat{\mathbf{i}} + (2\cos t)\hat{\mathbf{j}} + \left(\frac{3}{2\pi}\right)\hat{\mathbf{k}}.$$

Since $t = 2\pi$ yields the point $(2, 0, 3)$, a tangent vector at this point is $\mathbf{r}'(2\pi) = 2\hat{\mathbf{j}} + (3/(2\pi))\hat{\mathbf{k}}$ (see Figure 12.109). Since $x'(t)$, $y'(t)$, and $z'(t)$ are all continuous functions, and they are not simultaneously zero, the helix is indeed smooth. ■

FIGURE 12.109

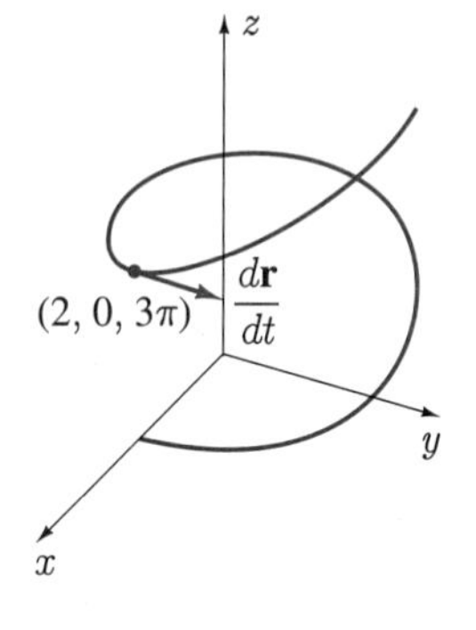

Unit Tangent Vectors

When a curve in the xy-plane is defined parametrically by

$$C: \quad x = x(t), \quad y = y(t), \quad \alpha \le t \le \beta, \tag{12.60}$$

a tangent vector to C is

$$\mathbf{T} = \frac{d\mathbf{r}}{dt} = \frac{dx}{dt}\hat{\mathbf{i}} + \frac{dy}{dt}\hat{\mathbf{j}}, \tag{12.61}$$

and this tangent vector points in the direction in which t increases along C. To produce a unit tangent vector to C at any point, we divide $\mathbf{T}$ by its length:

$$\hat{\mathbf{T}} = \frac{\mathbf{T}}{|\mathbf{T}|} = \frac{d\mathbf{r}/dt}{|d\mathbf{r}/dt|}. \tag{12.62}$$

We now show that if length along C is used as the parameter by which to specify its points, then division by $|\mathbf{T}|$ is unnecessary.

In Section 7.3 we showed that small lengths along a plane curve C can be approximated by straight-line lengths along tangent lines to the curve, and that the total length of a smooth curve from A to B (Figure 12.110) is

$$L = \int_A^B \sqrt{(dx)^2 + (dy)^2}.$$

FIGURE 12.110

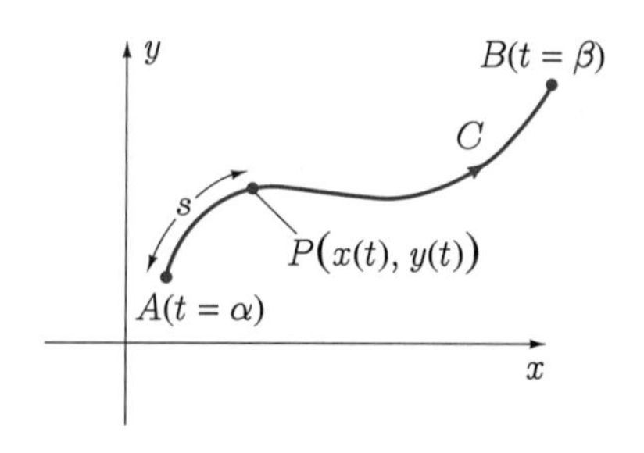

With parametric equations 12.60 we can write this formula as a definite integral with respect to t (see also equation 10.3):

$$L = \int_{\alpha}^{\beta} \sqrt{\left[\left(\frac{dx}{dt}\right)^2 + \left(\frac{dy}{dt}\right)^2\right](dt)^2} = \int_{\alpha}^{\beta} \sqrt{\left(\frac{dx}{dt}\right)^2 + \left(\frac{dy}{dt}\right)^2}\, dt. \qquad (12.63)$$

Furthermore, if we denote by $s = s(t)$ the length of that part of C from its initial point A (where $t = \alpha$) to any point $P(x(t), y(t))$ on C (Figure12.110), then $s(t)$ is defined by the integral

$$s(t) = \int_{\alpha}^{t} \sqrt{\left(\frac{dx}{dt}\right)^2 + \left(\frac{dy}{dt}\right)^2}\, dt. \qquad (12.64)$$

It follows, then, that the derivative of $s(t)$ is

$$\frac{ds}{dt} = \sqrt{\left(\frac{dx}{dt}\right)^2 + \left(\frac{dy}{dt}\right)^2}. \qquad (12.65)$$

But according to 12.61, $d\mathbf{r}/dt$ is a tangent vector to C with the same length:

$$\left|\frac{d\mathbf{r}}{dt}\right| = \sqrt{\left(\frac{dx}{dt}\right)^2 + \left(\frac{dy}{dt}\right)^2} = \frac{ds}{dt}. \qquad (12.66)$$

When this equation is multiplied by dt, it gives

$$|d\mathbf{r}| = \sqrt{(dx)^2 + (dy)^2} = ds. \qquad (12.67)$$

This equation states that ds is the length of the tangent vector $d\mathbf{r} = dx\hat{\mathbf{i}} + dy\hat{\mathbf{j}}$, and therefore ds is a measure of length along the tangent line to C. In spite of this we often think of ds as a measure of small lengths along C itself (Figure 12.111), and that ds is approximated by the tangential straight-line length

$$|d\mathbf{r}| = \sqrt{(dx)^2 + (dy)^2}.$$

Note too that if we use length s along C as a parameter, then the chain rule applied to $\mathbf{r} = \mathbf{r}(s)$, $s = s(t)$ gives

$$\frac{d\mathbf{r}}{dt} = \frac{d\mathbf{r}}{ds}\frac{ds}{dt}. \qquad (12.68)$$

(The chain rule is proved in Exercise 26 of Section 12.6.) Consequently, equation 12.66 implies that

$$\frac{d\mathbf{r}}{ds} = \frac{d\mathbf{r}/dt}{ds/dt} = \frac{d\mathbf{r}/dt}{|d\mathbf{r}/dt|}. \qquad (12.69)$$

FIGURE 12.111

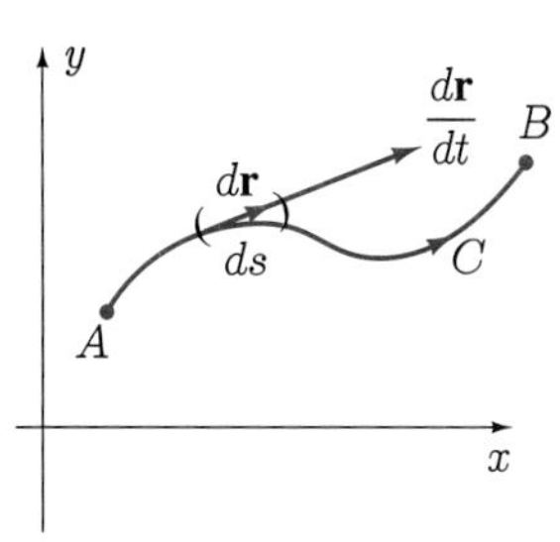

What we have shown, then, is that if we choose length along a curve C as the parameter by which to specify points on the curve (C: $x = x(s)$, $y = y(s)$), then the vector

$$\hat{\mathbf{T}} = \frac{d\mathbf{r}}{ds} = \frac{dx}{ds}\hat{\mathbf{i}} + \frac{dy}{ds}\hat{\mathbf{j}} \qquad (12.70)$$

is a unit tangent vector to C. In addition, the corollary to Theorem 12.10 implies that $d\mathbf{r}/ds$ points in the direction in which s increases along C. This suggests perhaps that we should always set up parametric equations for a curve with length along the curve as parameter. Theoretically this is quite acceptable, but practically it is impossible. For most curves we have enough difficulty just finding a set of parametric equations, let alone finding that set with length along the curve as parameter. If we then use a parameter t other than length along the curve, a unit tangent vector is calculated according to 12.62.

These results can be extended to space curves as well. When a smooth curve C has parametric equations 12.56, equation 12.62 still defines a unit tangent vector to C, but because C is a space curve, $d\mathbf{r}/dt$ is calculated according to 12.59.

Corresponding to formula 12.64 for length along a curve in the xy-plane, length along a smooth curve in space is defined by the definite integral

$$s(t) = \int_{\alpha}^{t} \sqrt{\left(\frac{dx}{dt}\right)^2 + \left(\frac{dy}{dt}\right)^2 + \left(\frac{dz}{dt}\right)^2}\, dt. \qquad (12.71)$$

These two results imply that

$$\left|\frac{d\mathbf{r}}{dt}\right| = \sqrt{\left(\frac{dx}{dt}\right)^2 + \left(\frac{dy}{dt}\right)^2 + \left(\frac{dz}{dt}\right)^2} = \frac{ds}{dt}, \qquad (12.72)$$

the three-space analogue of 12.66. Once again we are led to the fact that when s is used as parameter along C, then

$$\hat{\mathbf{T}} = \frac{d\mathbf{r}}{ds} = \frac{dx}{ds}\hat{\mathbf{i}} + \frac{dy}{ds}\hat{\mathbf{j}} + \frac{dz}{ds}\hat{\mathbf{k}} \qquad (12.73)$$

is a unit tangent vector to C. In addition, multiplication of 12.72 by dt yields

$$|d\mathbf{r}| = ds = \sqrt{(dx)^2 + (dy)^2 + (dz)^2}, \qquad (12.74)$$

indicating that small lengths ds along C (Figure 12.112) are defined in terms of small lengths $|d\mathbf{r}|$ along the tangent line to C.

FIGURE 12.112

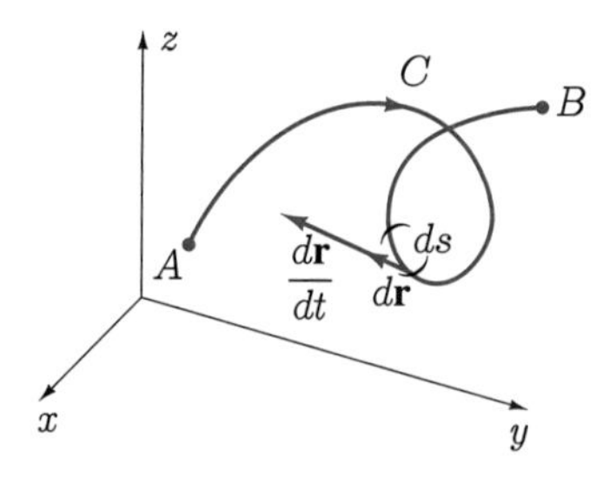

EXAMPLE 12.33 Find a unit tangent vector to the curve

$$C: \quad x = \sin t, \quad y = 2\cos t, \quad z = 2t/\pi, \quad t \geq 0$$

at the point $(0, -2, 2)$.

SOLUTION A tangent vector to C at any point is

$$\frac{d\mathbf{r}}{dt} = \frac{dx}{dt}\hat{\mathbf{i}} + \frac{dy}{dt}\hat{\mathbf{j}} + \frac{dz}{dt}\hat{\mathbf{k}} = \cos t\hat{\mathbf{i}} - 2\sin t\hat{\mathbf{j}} + \left(\frac{2}{\pi}\right)\hat{\mathbf{k}}.$$

Since $t = \pi$ yields the point $(0, -2, 2)$, a tangent vector at this point is $\mathbf{r}'(\pi) = -\hat{\mathbf{i}} + (2/\pi)\hat{\mathbf{k}}$. A unit tangent vector is then

$$\hat{\mathbf{T}} = \frac{-\hat{\mathbf{i}} + (2/\pi)\hat{\mathbf{k}}}{\sqrt{1 + 4/\pi^2}} = \frac{-\pi\hat{\mathbf{i}} + 2\hat{\mathbf{k}}}{\sqrt{4 + \pi^2}}.$$

■

EXAMPLE 12.34

Find the length of that part of the curve $x = y^{2/3}$, $x = z^{2/3}$ between the points $(0, 0, 0)$ and $(4, 8, 8)$.

SOLUTION If we use $x = t$, $y = t^{3/2}$, $z = t^{3/2}$, $0 \le t \le 4$, as parametric equations for the curve, then

$$L = \int_0^4 \sqrt{\left(\frac{dx}{dt}\right)^2 + \left(\frac{dy}{dt}\right)^2 + \left(\frac{dz}{dt}\right)^2}\, dt = \int_0^4 \sqrt{1 + \left(\frac{3}{2}\sqrt{t}\right)^2 + \left(\frac{3}{2}\sqrt{t}\right)^2}\, dt$$

$$= \int_0^4 \sqrt{1 + \frac{9t}{2}}\, dt = \left\{\frac{4}{27}\left(1 + \frac{9t}{2}\right)^{3/2}\right\}_0^4 = \frac{4}{27}(19\sqrt{19} - 1).$$

■

EXERCISES 12.8

In Exercises 1–5 express the curve in vector form and find the unit tangent vector $\hat{\mathbf{T}}$ at each point on the curve.

1. $x = \sin t$, $y = \cos t$, $z = t$, $-\infty < t < \infty$

2. $x = t$, $y = t^2$, $z = t^3$, $t \ge 1$

3. $x = (t-1)^2$, $y = (t+1)^2$, $z = -t$, $-3 \le t \le 4$

4. $x + y = 5$, $x^2 - y = z$ from $(5, 0, 25)$ to $(0, 5, -5)$

5. $x + y + z = 4$, $x^2 + y^2 = 4$, $y \ge 0$ from $(2, 0, 2)$ to $(-2, 0, 6)$

In Exercises 6–10, find $\hat{\mathbf{T}}$ at the point.

6. $x = 4\cos t$, $y = 6\sin t$, $z = 2\sin t$, $-\infty < t < \infty$; $(2\sqrt{2}, 3\sqrt{2}, \sqrt{2})$

7. $x = 2 - 5t$, $y = 1 + t$, $z = 6 + 4t$, $-\infty < t < \infty$; $(7, 0, 2)$

8. $x^2 + y^2 + z^2 = 4$, $z = \sqrt{x^2 + y^2}$, directed so that x increases when y is positive; $(1, 1, \sqrt{2})$

9. $x = y^2 + 1$, $z = x + 5$, directed so that y increases along the curve; $(5, 2, 10)$

10. $x^2 + (y-1)^2 = 4$, $z = x$, directed so that z decreases when y is negative; $(2, 1, 2)$

In Exercises 11–14 find the length of the curve. Sketch each curve.

11. $x = 2\cos t$, $y = 2\sin t$, $z = 3t$, $0 \le t \le 2\pi$

12. $x = 2 - 5t$, $y = 1 + t$, $z = 6 + 4t$, $-1 \le t \le 0$

13. $x = t^3$, $y = t^2$, $z = t^3$, $0 \le t \le 1$

14. $x = t$, $y = t^{3/2}$, $z = 4t^{3/2}$, $1 \le t \le 4$

15. In Definition 12.17 why are the derivatives assumed not to vanish simultaneously? Hint: Consider the curve $x = t^3$, $y = t^2$, $z = 0$.

16. Find a unit tangent vector to the curve $\mathbf{r} = (\cos t + t\sin t)\hat{\mathbf{i}} + (\sin t - t\cos t)\hat{\mathbf{j}}$, called an involute of a circle.

17. Show that the tangent vector $d\mathbf{r}/dt$ to a curve described by equation 12.57 always points in the direction in which t increases along the curve.

18. **(a)** What happens when equation 12.59 is used to determine a tangent vector to the curve $x = t^2$, $y = t^3$, $z = t^2$, $-\infty < t < \infty$, at the origin?

(b) Can you devise a way in which to obtain a tangent vector?

SECTION 12.9

Normal Vectors, Curvature, and Radius of Curvature

In discussing curves we distinguish between two types of properties: intrinsic and not intrinsic. An intrinsic property is one that is independent of the parameter used to specify the curve; a property that is not intrinsic is parameter-dependent. To illustrate, the tangent vector $\mathbf{T} = d\mathbf{r}/dt$ in 12.59 is not intrinsic; a change of parameter results in a change in the length of $\mathbf{T}$. The unit tangent vector $\hat{\mathbf{T}}$, on the other hand, is intrinsic; there is only one unit tangent vector in the direction of the curve. Length of a curve from its initial point to an arbitrary point is an intrinsic property; a change of parameter along the curve does not affect length between points.

Because length along a curve is an intrinsic property, it is customary in theoretical discussions to use it as the parameter by which to specify points on the curve. When C is a smooth curve in the xy-plane, parametric equations for C in terms of length s along C take the form

$$C: \quad x = x(s), \quad y = y(s), \quad 0 \leq s \leq L. \qquad (12.75)$$

Normal Vectors to Curves

FIGURE 12.113

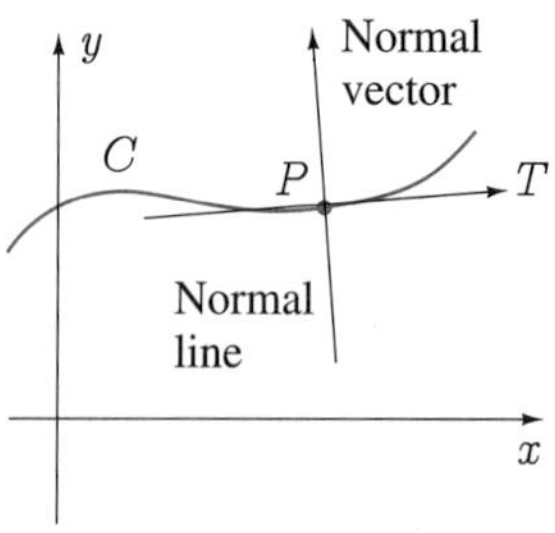

The normal line at a point P on a smooth curve C in the xy-plane is that line which is perpendicular to the tangent line to C at P (Figure 12.113). Any vector along this normal line is said to be a normal vector to the curve at P. Since the unit tangent vector to C at P is

$$\hat{\mathbf{T}} = \frac{d\mathbf{r}}{ds} = \frac{dx}{ds}\hat{\mathbf{i}} + \frac{dy}{ds}\hat{\mathbf{j}},$$

it follows that

$$\hat{\mathbf{N}} = -\frac{dy}{ds}\hat{\mathbf{i}} + \frac{dx}{ds}\hat{\mathbf{j}} \qquad (12.76)$$

is a unit normal vector to C at P (note that $\hat{\mathbf{T}} \cdot \hat{\mathbf{N}} = 0$). Because there is only one direction normal to C at P, every normal vector to C at P must be some multiple $\lambda\hat{\mathbf{N}}$ of $\hat{\mathbf{N}}$.

FIGURE 12.114

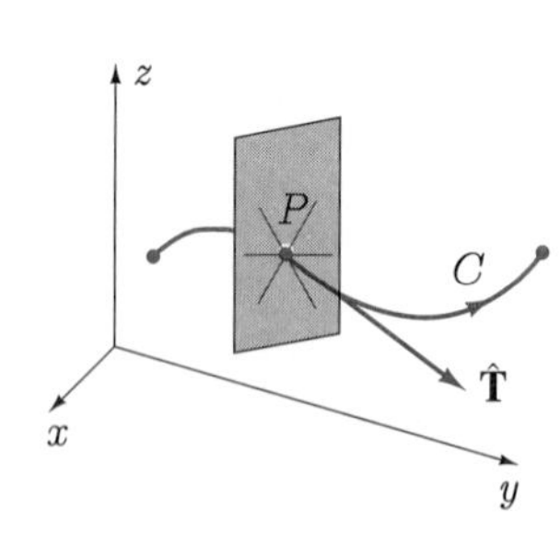

The situation is quite different for space curves (Figure 12.114). If $\hat{\mathbf{T}}$ is the unit tangent vector to a smooth curve C, then there is an entire plane of normal vectors to C at P. In the following discussion, we single out two normal vectors called the **principal normal** and the **binormal**. Suppose that C is defined parametrically by

$$C: \quad x = x(s), \quad y = y(s), \quad z = z(s), \quad 0 \leq s \leq L, \qquad (12.77)$$

and that $\hat{\mathbf{T}}$ is the unit tangent vector to C defined by 12.73. Because $\hat{\mathbf{T}}$ has unit length,

$$1 = \hat{\mathbf{T}} \cdot \hat{\mathbf{T}}.$$

If we use equation 12.51 to differentiate this equation with respect to s, we have

$$0 = \frac{d\hat{\mathbf{T}}}{ds} \cdot \hat{\mathbf{T}} + \hat{\mathbf{T}} \cdot \frac{d\hat{\mathbf{T}}}{ds} = 2\left(\hat{\mathbf{T}} \cdot \frac{d\hat{\mathbf{T}}}{ds}\right).$$

But if neither of the vectors $\hat{\mathbf{T}}$ nor $d\hat{\mathbf{T}}/ds$ is equal to zero, then the fact that their scalar product is equal to zero implies that they are perpendicular. In other words,

$$\mathbf{N} = \frac{d\hat{\mathbf{T}}}{ds} \tag{12.78}$$

is a normal vector to C at any point. The unit normal vector in this direction

$$\hat{\mathbf{N}} = \frac{\mathbf{N}}{|\mathbf{N}|} = \frac{d\hat{\mathbf{T}}/ds}{|d\hat{\mathbf{T}}/ds|} \tag{12.79}$$

is called the **principal normal** (vector) to C (Figure 12.115).

FIGURE 12.115

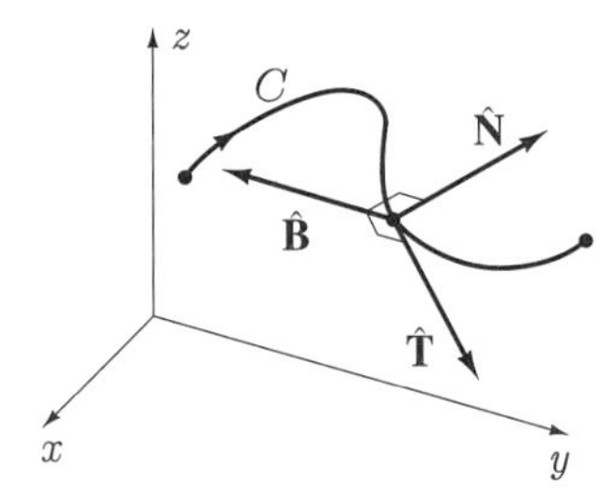

Because $\hat{\mathbf{N}}$ is defined in terms of intrinsic properties $\hat{\mathbf{T}}$ and s for a curve, it must also be an intrinsic property. It follows, then, that no matter what parameter is used to specify points on a curve, $\hat{\mathbf{N}}$ is always the same. But how do we find $\hat{\mathbf{N}}$ when a curve C is specified in terms of a parameter other than length along C, say, in the form

$$C: \quad x = x(t), \quad y = y(t), \quad z = z(t), \quad \alpha \le t \le \beta? \tag{12.80}$$

If $s = s(t)$ is length along C (measured from $t = \alpha$), then by the chain rule

$$\frac{d\hat{\mathbf{T}}}{ds} = \frac{d\hat{\mathbf{T}}}{dt}\frac{dt}{ds},$$

where dt/ds must be positive since both s and t increase along C. Consequently,

$$\hat{\mathbf{N}} = \frac{d\hat{\mathbf{T}}/ds}{|d\hat{\mathbf{T}}/ds|} = \frac{(d\hat{\mathbf{T}}/dt)(dt/ds)}{|(d\hat{\mathbf{T}}/dt)(dt/ds)|} = \frac{(d\hat{\mathbf{T}}/dt)(dt/ds)}{|d\hat{\mathbf{T}}/dt|dt/ds} = \frac{d\hat{\mathbf{T}}/dt}{|d\hat{\mathbf{T}}/dt|}. \tag{12.81}$$

In other words, for any parametrization of C whatsoever, the vector $d\hat{\mathbf{T}}/dt$ always points in the direction of the principal normal, and to find $\hat{\mathbf{N}}$, we simply find the unit vector in the direction of $d\hat{\mathbf{T}}/dt$.

In the study of space curves, a second normal vector to C, called the **binormal**, is defined by

$$\hat{\mathbf{B}} = \hat{\mathbf{T}} \times \hat{\mathbf{N}}. \tag{12.82}$$

Since the cross product of two vectors is always perpendicular to each of the vectors, it follows that the binormal is perpendicular to both $\hat{\mathbf{T}}$ and $\hat{\mathbf{N}}$, and must therefore indeed be a normal vector to C (Figure 12.115).

We have singled out three vectors at each point P on a curve C: a unit tangent vector $\hat{\mathbf{T}}$ and two unit normal vectors $\hat{\mathbf{N}}$ and $\hat{\mathbf{B}}$. As P moves along C, these vectors constantly change direction but always have unit length.

EXAMPLE 12.35

Find $\hat{\mathbf{T}}$, $\hat{\mathbf{N}}$, and $\hat{\mathbf{B}}$ for the curve $x = t,\ y = t^2,\ z = t^2,\ t \geq 0$.

SOLUTION The unit tangent vector $\hat{\mathbf{T}}$ is defined by

$$\hat{\mathbf{T}} = \frac{d\mathbf{r}/dt}{|d\mathbf{r}/dt|} = \frac{(1,2t,2t)}{\sqrt{1+4t^2+4t^2}} = \frac{(1,2t,2t)}{\sqrt{1+8t^2}}.$$

The principal normal $\hat{\mathbf{N}}$ lies along the vector $\mathbf{N} = d\hat{\mathbf{T}}/dt$, and according to equation 12.50, we can write that

$$\begin{aligned}\mathbf{N} = \frac{d\hat{\mathbf{T}}}{dt} &= \frac{d}{dt}\left(\frac{1}{\sqrt{1+8t^2}}\right)(1,2t,2t) + \frac{1}{\sqrt{1+8t^2}}\frac{d}{dt}(1,2t,2t)\\ &= \frac{-8t}{(1+8t^2)^{3/2}}(1,2t,2t) + \frac{1}{\sqrt{1+8t^2}}(0,2,2)\\ &= \frac{1}{(1+8t^2)^{3/2}}\{-8t(1,2t,2t) + (1+8t^2)(0,2,2)\}\\ &= \frac{(-8t,2,2)}{(1+8t^2)^{3/2}}.\end{aligned}$$

The principal normal is therefore

$$\hat{\mathbf{N}} = \frac{\mathbf{N}}{|\mathbf{N}|} = \frac{(-8t,2,2)}{\sqrt{64t^2+4+4}} = \frac{(-4t,1,1)}{\sqrt{2+16t^2}}.$$

The binormal is

$$\begin{aligned}\hat{\mathbf{B}} = \hat{\mathbf{T}} \times \hat{\mathbf{N}} &= \frac{(1,2t,2t)}{\sqrt{1+8t^2}} \times \frac{(-4t,1,1)}{\sqrt{2+16t^2}}\\ &= \frac{1}{\sqrt{1+8t^2}\sqrt{2+16t^2}}\begin{vmatrix}\hat{\mathbf{i}} & \hat{\mathbf{j}} & \hat{\mathbf{k}}\\ 1 & 2t & 2t\\ -4t & 1 & 1\end{vmatrix}\\ &= \frac{1}{\sqrt{2}\sqrt{1+8t^2}\sqrt{1+8t^2}}(0,-1-8t^2,1+8t^2)\\ &= \frac{1+8t^2}{\sqrt{2}(1+8t^2)}(0,-1,1)\\ &= \frac{(0,-1,1)}{\sqrt{2}}.\end{aligned}$$

The significance of the fact that the binormal has constant direction can be seen from a sketch of the curve. Because the parametric equations imply that $y = x^2$ and $z = y$, the curve is the curve of intersection of these two surfaces (Figure 12.116). Since the curve lies in the plane $-y + z = 0$, and a normal vector to this plane is $(0,-1,1)$, it follows that $(0,-1,1)$ is always normal to the curve. But this is precisely the direction of $\hat{\mathbf{B}}$. In other words, constant $\hat{\mathbf{B}}$ implies that the curve lies in a plane that has $\hat{\mathbf{B}}$ as normal (see Exercise 31).

FIGURE 12.116

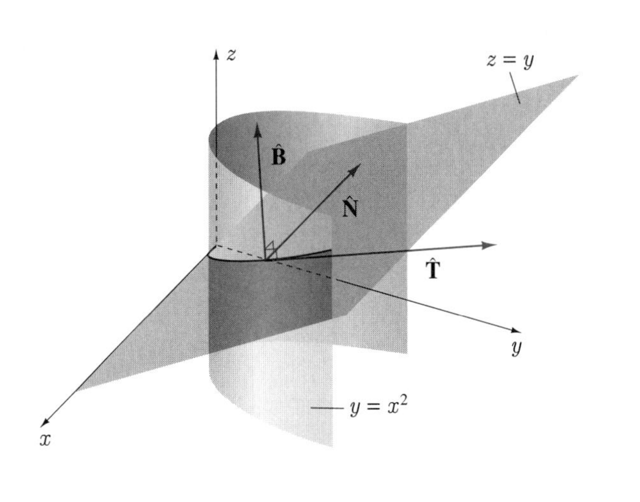

EXAMPLE 12.36

Show that for a smooth curve C: $x = x(t)$, $y = y(t)$, $\alpha \le t \le \beta$ in the xy-plane, the principal normal is

$$\hat{\mathbf{N}} = \operatorname{sgn}\left(\frac{dy}{dt}\frac{d^2x}{dt^2} - \frac{dx}{dt}\frac{d^2y}{dt^2}\right)\frac{(dy/dt, -dx/dt)}{\sqrt{(dx/dt)^2 + (dy/dt)^2}},$$

where the signum function sgn(u) is defined by

$$\operatorname{sgn}(u) = \begin{cases} 1, & \text{if } u > 0, \\ 0, & \text{if } u = 0, \\ -1, & \text{if } u < 0. \end{cases}$$

SOLUTION The unit tangent vector to C is

$$\hat{\mathbf{T}} = \frac{(dx/dt, dy/dt)}{\sqrt{(dx/dt)^2 + (dy/dt)^2}}.$$

For simplicity in notation, we use a dot "." above a variable to indicate that the variable is differentiated with respect to t. For example, $\dot{x} = dx/dt$ and $\ddot{x} = d^2x/dt^2$. With this notation,

$$\hat{\mathbf{T}} = \frac{(\dot{x}, \dot{y})}{\sqrt{\dot{x}^2 + \dot{y}^2}}.$$

By equation 12.8, $\hat{\mathbf{N}} = (d\hat{\mathbf{T}}/dt)/|d\hat{\mathbf{T}}/dt|$, where

$$\begin{aligned}\frac{d\hat{\mathbf{T}}}{dt} &= \frac{d}{dt}\frac{(\dot{x}, \dot{y})}{\sqrt{\dot{x}^2 + \dot{y}^2}} = \frac{d}{dt}\left(\frac{1}{\sqrt{\dot{x}^2 + \dot{y}^2}}\right)(\dot{x}\hat{\mathbf{i}} + \dot{y}\hat{\mathbf{j}}) + \frac{1}{\sqrt{\dot{x}^2 + \dot{y}^2}}(\ddot{x}\hat{\mathbf{i}} + \ddot{y}\hat{\mathbf{j}}) \\ &= \left(\frac{-\dot{x}\ddot{x} - \dot{y}\ddot{y}}{(\dot{x}^2 + \dot{y}^2)^{3/2}}\right)(\dot{x}\hat{\mathbf{i}} + \dot{y}\hat{\mathbf{j}}) + \frac{1}{\sqrt{\dot{x}^2 + \dot{y}^2}}(\ddot{x}\hat{\mathbf{i}} + \ddot{y}\hat{\mathbf{j}})\end{aligned}$$

$$= \frac{1}{(\dot{x}^2 + \dot{y}^2)^{3/2}} \{-(\dot{x}\ddot{x} + \dot{y}\ddot{y})(\dot{x}\hat{\mathbf{i}} + \dot{y}\hat{\mathbf{j}}) + (\dot{x}^2 + \dot{y}^2)(\ddot{x}\hat{\mathbf{i}} + \ddot{y}\hat{\mathbf{j}})\}$$

$$= \frac{1}{(\dot{x}^2 + \dot{y}^2)^{3/2}} \{(-\dot{x}^2\ddot{x} - \dot{x}\dot{y}\ddot{y} + \dot{x}^2\ddot{x} + \dot{y}^2\ddot{x})\hat{\mathbf{i}}$$

$$+ (-\dot{x}\dot{y}\ddot{x} - \dot{y}^2\ddot{y} + \dot{x}^2\ddot{y} + \dot{y}^2\ddot{y})\hat{\mathbf{j}}\}$$

$$= \frac{1}{(\dot{x}^2 + \dot{y}^2)^{3/2}} \{\dot{y}(\dot{y}\ddot{x} - \dot{x}\ddot{y})\hat{\mathbf{i}} + \dot{x}(\dot{x}\ddot{y} - \dot{y}\ddot{x})\hat{\mathbf{j}}\}$$

$$= \frac{\dot{y}\ddot{x} - \dot{x}\ddot{y}}{(\dot{x}^2 + \dot{y}^2)^{3/2}} (\dot{y}\hat{\mathbf{i}} - \dot{x}\hat{\mathbf{j}}).$$

If $\dot{y}\ddot{x} - \dot{x}\ddot{y}$ is positive, then

$$\hat{\mathbf{N}} = \frac{\dot{y}\hat{\mathbf{i}} - \dot{x}\hat{\mathbf{j}}}{\sqrt{\dot{x}^2 + \dot{y}^2}};$$

whereas if $\dot{y}\ddot{x} - \dot{x}\ddot{y}$ is negative, then

$$\hat{\mathbf{N}} = \frac{-\dot{y}\hat{\mathbf{i}} + \dot{x}\hat{\mathbf{j}}}{\sqrt{\dot{x}^2 + \dot{y}^2}}.$$

In other words,

$$\hat{\mathbf{N}} = \text{sgn}(\dot{y}\ddot{x} - \dot{x}\ddot{y}) \left(\frac{\dot{y}\hat{\mathbf{i}} - \dot{x}\hat{\mathbf{j}}}{\sqrt{\dot{x}^2 + \dot{y}^2}} \right).$$

■

Curvature and Radius of Curvature

FIGURE 12.117

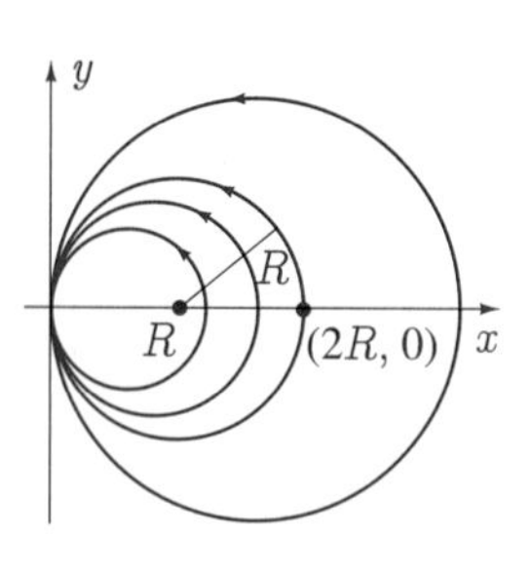

When length s along a smooth curve C is used as the parameter by which to identify points on the curve, the vector $\hat{\mathbf{T}} = d\mathbf{r}/ds$ is a unit tangent vector to C. Suppose we differentiate $\hat{\mathbf{T}}$ with respect to s to form $d\hat{\mathbf{T}}/ds = d^2\mathbf{r}/ds^2$. Since $\hat{\mathbf{T}}$ has constant unit length, only its direction can change; therefore, the derivative $d\hat{\mathbf{T}}/ds$ must be a measure of the rate of change of the direction of $\hat{\mathbf{T}}$. Since $\hat{\mathbf{T}}$ is really our way of specifying the direction of the curve itself, we can also say that $d\hat{\mathbf{T}}/ds$ is a measure of how fast the direction of C changes. But exactly how does a vector $d\hat{\mathbf{T}}/ds$ which has both magnitude and direction measure the rate of change of the direction of C? We illustrate by example that it cannot be the direction of $d\hat{\mathbf{T}}/ds$; it must be its magnitude that measures the rate of change of the direction of C. In Figure 12.117 we show a number of circles in the xy-plane, all of which are tangent to the y-axis at the origin. Parametric equations for the circle with centre $(R, 0)$ and radius R in terms of length s along the circle (as measured from $(2R, 0)$) are

$$x = R + R\cos(s/R), \quad y = R\sin(s/R), \quad 0 \le s < 2\pi R.$$

Consequently,

$$\hat{\mathbf{T}} = \frac{d\mathbf{r}}{ds} = -\sin\left(\frac{s}{R}\right)\hat{\mathbf{i}} + \cos\left(\frac{s}{R}\right)\hat{\mathbf{j}}$$

and

$$\frac{d\hat{\mathbf{T}}}{ds} = -\frac{1}{R}\cos\left(\frac{s}{R}\right)\hat{\mathbf{i}} - \frac{1}{R}\sin\left(\frac{s}{R}\right)\hat{\mathbf{j}}.$$

At the origin, $s = \pi R$, and

$$\frac{d\hat{\mathbf{T}}}{ds}_{|s=\pi R} = \frac{1}{R}\hat{\mathbf{i}}.$$

Thus for each of the circles in Figure 12.117, the vector $d\hat{\mathbf{T}}/ds$ has exactly the same direction. Yet the rate of change of the direction of $\hat{\mathbf{T}}$ is not the same for each circle; the direction changes more rapidly as the radius of the circle decreases. We must conclude, therefore, that it cannot be the direction of $d\hat{\mathbf{T}}/ds$ that measures the rate of change of $\hat{\mathbf{T}}$. Since a vector has only length and direction, it must be the length of $d\hat{\mathbf{T}}/ds$ that measures this rate of change. The circles in Figure 12.117 certainly support this claim; the length of $d\hat{\mathbf{T}}/ds$ is $1/R$, and this quantity increases as the radii of the circles decrease. This agrees with the fact that the rate at which $\hat{\mathbf{T}}$ turns increases as R decreases. According to the following definition, we call $|d\hat{\mathbf{T}}/ds|$ **curvature** and $1/|d\hat{\mathbf{T}}/ds|$ **radius of curvature**.

Definition 12.19

If $x = x(s)$, $y = y(s)$, $z = z(s)$, $0 \le s \le L$, are parametric equations for a smooth curve in terms of length s along the curve, we define the **curvature** of the curve at a point as

$$\kappa(s) = \left|\frac{d\hat{\mathbf{T}}}{ds}\right|, \tag{12.83}$$

its **radius of curvature** as

$$\rho(s) = \frac{1}{\kappa(s)}, \tag{12.84}$$

and its **circle of curvature** as that circle in the plane of $\hat{\mathbf{T}}$ and $\hat{\mathbf{N}}$ with centre at $\mathbf{r}(s) + \rho(s)\hat{\mathbf{N}}$ and radius $\rho(s)$.

The circle of curvature is illustrated in Figure 12.118.

FIGURE 12.118

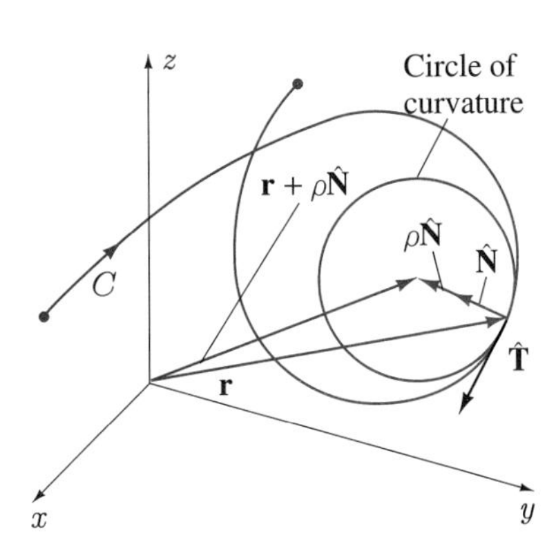

For the circles in Figure 12.117 we have already shown that $d\hat{\mathbf{T}}/ds = R^{-1}\hat{\mathbf{i}}$. Consequently, for these circles, the curvature is always R^{-1}, and the radius of curvature is R, the radius of the circle. In other words, for a circle, the circle of curvature is the circle itself, its radius of curvature is its radius, and its curvature is the inverse of its radius. For the case when a curve is not a circle, we show in Exercise 27 that at any point on the curve the circle of curvature is in some sense the "best-fitting" circle to the curve at that point.

Because curvature and radius of curvature have been defined in terms of intrinsic properties $\hat{\mathbf{T}}$ and s for a curve, they must also be intrinsic properties. It follows, then, that no matter what parameter is used to specify points on a curve, curvature and radius of curvature are always the same. The following theorem shows how to calculate κ and ρ when the curve is specified in terms of a parameter other than length along the curve.

Theorem 12.11 *When a smooth curve is defined parametrically by*

$$C: \quad x = x(t), \quad y = y(t), \quad z = z(t), \quad \alpha \le t \le \beta,$$

its curvature $\kappa(t)$ is given by

$$\kappa(t) = \frac{|\dot{\mathbf{r}} \times \ddot{\mathbf{r}}|}{|\dot{\mathbf{r}}|^3}, \tag{12.85}$$

where $\dot{\mathbf{r}} = d\mathbf{r}/dt$ and $\ddot{\mathbf{r}} = d^2\mathbf{r}/dt^2$.

Proof If $s(t)$ is length along C (measured from $t = \alpha$), then by the chain rule

$$\kappa = \left|\frac{d\hat{\mathbf{T}}}{ds}\right| = \left|\frac{d\hat{\mathbf{T}}}{dt}\frac{dt}{ds}\right| = \frac{|d\hat{\mathbf{T}}/dt|}{|ds/dt|} = \frac{|d\hat{\mathbf{T}}/dt|}{ds/dt} = \frac{|d\hat{\mathbf{T}}/dt|}{|\dot{\mathbf{r}}|}$$

(see equation 12.72). Now, we can write $\dot{\mathbf{r}}$ in the form $\dot{\mathbf{r}} = |\dot{\mathbf{r}}|\hat{\mathbf{T}}$; therefore, using 12.50, we have

$$\begin{aligned}\ddot{\mathbf{r}} &= \left(\frac{d}{dt}|\dot{\mathbf{r}}|\right)\hat{\mathbf{T}} + |\dot{\mathbf{r}}|\frac{d\hat{\mathbf{T}}}{dt} \\ &= \left(\frac{d}{dt}|\dot{\mathbf{r}}|\right)\hat{\mathbf{T}} + (|\dot{\mathbf{r}}||d\hat{\mathbf{T}}/dt|)\frac{d\hat{\mathbf{T}}/dt}{|d\hat{\mathbf{T}}/dt|} \\ &= \left(\frac{d}{dt}|\dot{\mathbf{r}}|\right)\hat{\mathbf{T}} + \left(|\dot{\mathbf{r}}|\left|\frac{d\hat{\mathbf{T}}}{dt}\right|\right)\hat{\mathbf{N}}.\end{aligned}$$

If we take the cross product of this vector with $\dot{\mathbf{r}}$, we get

$$\begin{aligned}\dot{\mathbf{r}} \times \ddot{\mathbf{r}} &= \left(\frac{d}{dt}|\dot{\mathbf{r}}|\right)\dot{\mathbf{r}} \times \hat{\mathbf{T}} + \left(|\dot{\mathbf{r}}|\left|\frac{d\hat{\mathbf{T}}}{dt}\right|\right)\dot{\mathbf{r}} \times \hat{\mathbf{N}} \\ &= \left(|\dot{\mathbf{r}}|\left|\frac{d\hat{\mathbf{T}}}{dt}\right|\right)\dot{\mathbf{r}} \times \hat{\mathbf{N}}. \quad \text{(Since } \dot{\mathbf{r}} \text{ is parallel to } \hat{\mathbf{T}}\text{)}\end{aligned}$$

Because $\dot{\mathbf{r}}$ is perpendicular to $\hat{\mathbf{N}}$, it follows that $|\dot{\mathbf{r}} \times \hat{\mathbf{N}}| = |\dot{\mathbf{r}}||\hat{\mathbf{N}}|\sin(\pi/2) = |\dot{\mathbf{r}}|$, and therefore

$$|\dot{\mathbf{r}} \times \ddot{\mathbf{r}}| = \left(|\dot{\mathbf{r}}|\left|\frac{d\hat{\mathbf{T}}}{dt}\right|\right)|\dot{\mathbf{r}}|.$$

Consequently,

$$\left|\frac{d\hat{\mathbf{T}}}{dt}\right| = \frac{|\dot{\mathbf{r}} \times \ddot{\mathbf{r}}|}{|\dot{\mathbf{r}}|^2}$$

and

$$\kappa = \kappa(t) = \frac{|\dot{\mathbf{r}} \times \ddot{\mathbf{r}}|}{|\dot{\mathbf{r}}|^3}.$$

For the radius of curvature, we have the following.

Corollary *When a smooth curve is defined in terms of an arbitrary parameter t,*

$$\rho(t) = \frac{|\dot{\mathbf{r}}|^3}{|\dot{\mathbf{r}} \times \ddot{\mathbf{r}}|}. \tag{12.86}$$

EXAMPLE 12.37

Find the curvature and radius of curvature for the curve in Example 35.

SOLUTION According to 12.85,

$$\kappa(t) = \frac{|\dot{\mathbf{r}} \times \ddot{\mathbf{r}}|}{|\dot{\mathbf{r}}|^3} = \frac{|(1,2t,2t) \times (0,2,2)|}{|(1,2t,2t)|^3}$$

$$= \frac{1}{(1+4t^2+4t^2)^{3/2}} \begin{vmatrix} \hat{\mathbf{i}} & \hat{\mathbf{j}} & \hat{\mathbf{k}} \\ 1 & 2t & 2t \\ 0 & 2 & 2 \end{vmatrix} = \frac{1}{(1+8t^2)^{3/2}}|(0,-2,2)|$$

$$= \frac{2\sqrt{2}}{(1+8t^2)^{3/2}};$$

$$\rho(t) = \frac{1}{\kappa(t)} = \frac{(1+8t^2)^{3/2}}{2\sqrt{2}}.$$

Note in particular that as t increases, so does ρ, a fact that is certainly supported by Figure 12.116. ■

EXAMPLE 12.38

Show that for a smooth curve $y = y(x)$ in the xy-plane,

$$\kappa(x) = \frac{|y''|}{[1+(y')^2]^{3/2}}.$$

SOLUTION When we use x as parameter along the curve $y = y(x)$, parametric equations are $x = x$, $y = y(x)$. Then,

$$\dot{\mathbf{r}} = (1, y'(x)), \quad \ddot{\mathbf{r}} = (0, y''(x)),$$

and

$$\dot{\mathbf{r}} \times \ddot{\mathbf{r}} = \begin{vmatrix} \hat{\mathbf{i}} & \hat{\mathbf{j}} & \hat{\mathbf{k}} \\ 1 & y' & 0 \\ 0 & y'' & 0 \end{vmatrix} = y''\hat{\mathbf{k}}.$$

Thus

$$\kappa(x) = \frac{|\dot{\mathbf{r}} \times \ddot{\mathbf{r}}|}{|\dot{\mathbf{r}}|^3} = \frac{|y''|}{|(1,y')|^3} = \frac{|y''|}{[1+(y')^2]^{3/2}}.$$

■

EXERCISES 12.9

In Exercises 1–5 find $\hat{\mathbf{N}}$ and $\hat{\mathbf{B}}$ at each point on the curve.

1. $x = \sin t$, $y = \cos t$, $z = t$, $-\infty < t < \infty$

2. $x = t$, $y = t^2$, $z = t^3$, $t \geq 1$

3. $x = (t-1)^2$, $y = (t+1)^2$, $z = -t$, $-3 \leq t \leq 4$

4. $x + y = 5$, $x^2 - y = z$, from $(5,0,25)$ to $(0,5,-5)$

5. $z = x$, $x^2 + y^2 = 4$, $y \geq 0$, from $(2,0,2)$ to $(-2,0,-2)$

In Exercises 6–10 find $\hat{\mathbf{N}}$ and $\hat{\mathbf{B}}$ at the point.

6. $x = 4\cos t$, $y = 6\sin t$, $z = 2\sin t$, $-\infty < t < \infty$; $(2\sqrt{2}, 3\sqrt{2}, \sqrt{2})$

7. $x = 2 - 5t$, $y = 1 + t$, $z = 6 + 4t^3$, $-\infty < t < \infty$; $(7,0,2)$

8. $x^2 + y^2 + z^2 = 4$, $z = \sqrt{x^2 + y^2}$, directed so that x

increases when y is positive; $(1, 1, \sqrt{2})$

9. $x = y^2 + 1$, $z = x + 5$, directed so that y increases along the curve; $(5, 2, 10)$

10. $x^2 + (y - 1)^2 = 4$, $x = z$, directed so that z decreases when y is negative; $(2, 1, 2)$

In Exercises 11–18 find the curvature and the radius of curvature of the curve (if they exist). Sketch each curve.

11. $(x - h)^2 + (y - k)^2 = R^2$, $z = 0$, directed counterclockwise

12. $x = x_0 + at$, $y = y_0 + bt$, $z = z_0 + ct$, $-\infty < t < \infty$ (x_0, y_0, z_0, a, b, c all constants)

13. $x = t$, $y = t^2$, $z = 0$, $t \geq 0$

14. $x = e^t \cos t$, $y = e^t \sin t$, $z = t$, $-\infty < t < \infty$

15. $x = t$, $y = t^3$, $z = t^2$, $t \geq 0$

16. $x = 2 \cos t$, $y = 2 \sin t$, $z = 2 \sin t$, $0 \leq t < 2\pi$

17. $x = t + 1$, $y = t^2 - 1$, $z = t + 1$, $-\infty < t < \infty$

18. $x = t^2$, $y = t^4$, $z = 2t$, $-1 \leq t \leq 5$

19. At which points on the ellipse $b^2x^2 + a^2y^2 = a^2b^2$ ($a > b$) is the curvature a maximum and at which points is the curvature a minimum?

20. Show that curvature for a smooth curve $x = x(t)$, $y = y(t)$, $\alpha \leq t \leq \beta$, in the xy-plane can be expressed in the form

$$\kappa(t) = \frac{\left|\frac{dy}{dt}\frac{d^2x}{dt^2} - \frac{dx}{dt}\frac{d^2y}{dt^2}\right|}{\left[\left(\frac{dx}{dt}\right)^2 + \left(\frac{dy}{dt}\right)^2\right]^{3/2}}.$$

21. Show that the only curves for which curvature is identically equal to zero are straight lines.

22. What happens to curvature at a point of inflection on the graph of a function $y = f(x)$?

23. Let C be the curve $x = t$, $y = t^2$ in the xy-plane.
   - **(a)** At each point on C calculate the unit tangent vector $\hat{\mathbf{T}}$ and the principal normal $\hat{\mathbf{N}}$. What is $\hat{\mathbf{B}}$? (See Example 36 for $\hat{\mathbf{N}}$.)
   - **(b)** $\mathbf{F} = t^2\hat{\mathbf{i}} + t^4\hat{\mathbf{j}}$ is a vector that is defined at each point P on C. Denote by F_T and F_N the components of $\mathbf{F}$ in the directions $\hat{\mathbf{T}}$ and $\hat{\mathbf{N}}$ at P. Find F_T and F_N as functions of t.
   - **(c)** Express $\mathbf{F}$ in terms of $\hat{\mathbf{T}}$ and $\hat{\mathbf{N}}$.

24. Repeat Exercise 23 for the curve $C: x = 2 \cos t$, $y = 2 \sin t$, and the vector $\mathbf{F} = x^2\hat{\mathbf{i}} + y^2\hat{\mathbf{j}}$.

25. The vectors $\hat{\mathbf{T}}$, $\hat{\mathbf{N}}$, and $\hat{\mathbf{B}}$ were calculated at each point on the curve $x = t$, $y = t^2$, $z = t^2$ in Example 12.35. If $\mathbf{F} = t^2\hat{\mathbf{i}} + 2t\hat{\mathbf{j}} - 3\hat{\mathbf{k}}$ is a vector defined along C, find the components of $\mathbf{F}$ in the directions $\hat{\mathbf{T}}$, $\hat{\mathbf{N}}$, and $\hat{\mathbf{B}}$. Express $\mathbf{F}$ in terms of $\hat{\mathbf{T}}$, $\hat{\mathbf{N}}$, and $\hat{\mathbf{B}}$.

26. Calculate $\hat{\mathbf{T}}$, $\hat{\mathbf{N}}$, and $\hat{\mathbf{B}}$ for the curve $x = \cos t$, $y = \sin t$, $z = t$. Express the vector $\mathbf{F} = x\hat{\mathbf{i}} + xy^2\hat{\mathbf{j}} + \hat{\mathbf{k}}$ in terms of $\hat{\mathbf{T}}$, $\hat{\mathbf{N}}$, and $\hat{\mathbf{B}}$.

27. In this exercise we discuss our claim that the circle of curvature is the "best-fitting" circle to the curve at a point.
   - **(a)** First, is it true that the circle of curvature at a point on a curve passes through that point?
   - **(b)** Second, show that the circle of curvature and curve share the same tangent line at their common point.
   - **(c)** Finally, verify that the circle of curvature and curve have the same curvature at their common point.

28. If ϕ is the angle between $\hat{\mathbf{i}}$ and $\hat{\mathbf{T}}$ for a curve in the xy-plane (Figure 12.119), show that

$$\kappa(s) = \left|\frac{d\phi}{ds}\right|.$$

FIGURE 12.119

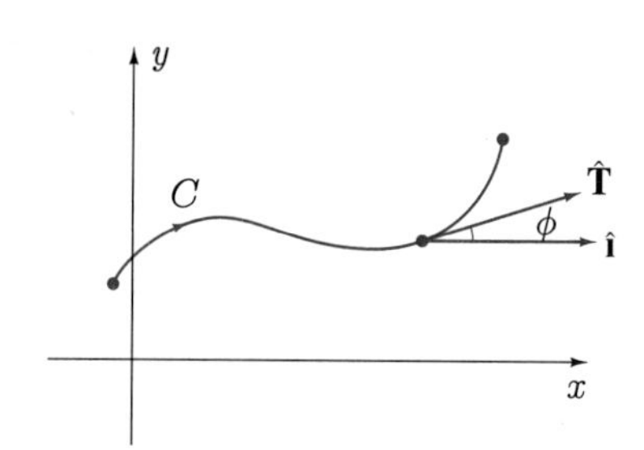

29. In differential geometry the rates of change $d\hat{\mathbf{T}}/ds$, $d\hat{\mathbf{N}}/ds$, and $d\hat{\mathbf{B}}/ds$ with respect to length along a curve are represented by the Frenet-Serret formulas. We develop these equations in this exercise.
   - **(a)** The first Frenet-Serret formula is

$$\frac{d\hat{\mathbf{T}}}{ds} = \kappa\hat{\mathbf{N}}.$$

   Verify this result.
   - **(b)** The second Frenet-Serret formula is

$$\frac{d\hat{\mathbf{B}}}{ds} = -\tau\hat{\mathbf{N}},$$

where τ is called the torsion of the curve. To verify this result, differentiate the equations $\hat{\mathbf{B}} = \hat{\mathbf{T}} \times \hat{\mathbf{N}}$ and $\hat{\mathbf{B}} \cdot \hat{\mathbf{B}} = 1$ to show that $d\hat{\mathbf{B}}/ds$ is perpendicular to $\hat{\mathbf{T}}$ and $\hat{\mathbf{B}}$ and therefore parallel to $\hat{\mathbf{N}}$.

(c) The third formula is

$$\frac{d\hat{\mathbf{N}}}{ds} = \tau\hat{\mathbf{B}} - \kappa\hat{\mathbf{T}}.$$

Verify this result by showing that $\hat{\mathbf{N}} = \hat{\mathbf{B}} \times \hat{\mathbf{T}}$ and then calculating $d\hat{\mathbf{N}}/ds$.

30. Show that the torsion in Exercise 29 can be calculated according to

$$\tau(t) = \frac{(\dot{\mathbf{r}} \times \ddot{\mathbf{r}}) \cdot \dddot{\mathbf{r}}}{|\dot{\mathbf{r}} \times \ddot{\mathbf{r}}|^2}.$$

31. Show that a curve lies in a plane if and only if its torsion vanishes.

SECTION 12.10

Displacement, Velocity, and Acceleration

In Sections 4.7 and 5.2 we introduced the concepts of displacement, velocity, and acceleration for moving objects, but indicated that our terminology at that time was somewhat loose. In particular, we stated that if $x = x(t)$ represents the position of a particle moving along the x-axis, then the instantaneous velocity of the particle is

$$v = \frac{dx}{dt}, \tag{12.87}$$

provided, of course, that t is time, and the acceleration of the particle is

$$a = \frac{dv}{dt} = \frac{d^2x}{dt^2}. \tag{12.88}$$

We illustrated by examples that given any one of $x(t)$, $v(t)$, or $a(t)$ and sufficient initial conditions, it is always possible to find the other two. There was nothing wrong with the calculations in the examples—they were correct—but our terminology was not quite correct. We now rectify this situation and give precise definitions of velocity and acceleration.

Suppose a particle moves along some curve C in space (under perhaps the influence of various forces), and that C is defined as a function of time t by the parametric equations

$$C: \quad x = x(t), \quad y = y(t), \quad z = z(t), \quad t \geq 0. \tag{12.89}$$

The position of the particle can then be described as a function of time by its position or displacement vector:

$$\mathbf{r} = \mathbf{r}(t) = x(t)\hat{\mathbf{i}} + y(t)\hat{\mathbf{j}} + z(t)\hat{\mathbf{k}}, \quad t \geq 0. \tag{12.90}$$

The **velocity v** of the particle at any time t is defined as the time rate of change of its displacement vector:

$$\mathbf{v} = \frac{d\mathbf{r}}{dt}. \tag{12.91}$$

Velocity, then, is a vector, and because of Theorem 12.8, the components of velocity are the derivatives of the components of displacement:

$$\mathbf{v} = \frac{d\mathbf{r}}{dt} = \frac{dx}{dt}\hat{\mathbf{i}} + \frac{dy}{dt}\hat{\mathbf{j}} + \frac{dz}{dt}\hat{\mathbf{k}}. \tag{12.92}$$

But according to Theorem 12.10, the vector $d\mathbf{r}/dt$ is tangent to the curve C (Figure 12.120). In other words, if a particle is at position P, and we draw its velocity vector with tail at P, then $\mathbf{v}$ is tangent to the trajectory.

FIGURE 12.120

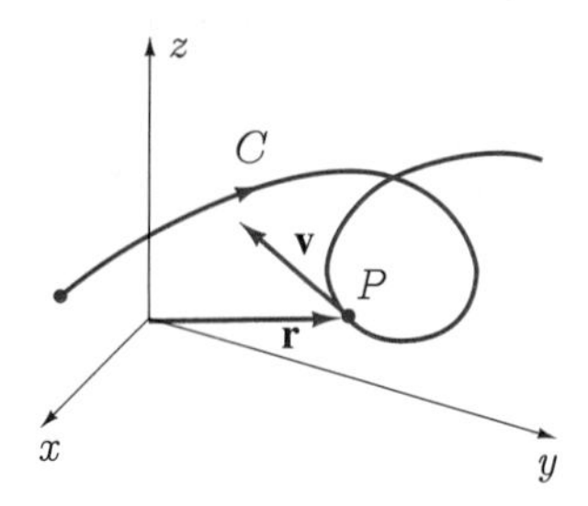

In some applications it is the length or magnitude of velocity that is important, not its direction. This quantity, called **speed**, is therefore defined by

$$|\mathbf{v}| = \sqrt{\left(\frac{dx}{dt}\right)^2 + \left(\frac{dy}{dt}\right)^2 + \left(\frac{dz}{dt}\right)^2}. \tag{12.93}$$

Equation 12.72 implies that if $s(t)$ is length along the trajectory C (where $s(0) = 0$), then $|\mathbf{v}| = ds/dt$. In other words, speed is the time rate of change of distance traveled along C.

It is important to understand this difference between velocity and speed. Velocity is the time derivative of displacement; speed is the time derivative of distance traveled. Velocity is a vector; speed is a scalar—the magnitude of velocity.

The **acceleration** of the particle as it moves along the curve C in equations 12.89 is defined as the rate of change of velocity with respect to time:

$$\mathbf{a} = \frac{d\mathbf{v}}{dt} = \frac{d^2\mathbf{r}}{dt^2} = \frac{d^2x}{dt^2}\hat{\mathbf{i}} + \frac{d^2y}{dt^2}\hat{\mathbf{j}} + \frac{d^2z}{dt^2}\hat{\mathbf{k}}. \tag{12.94}$$

Acceleration, then, is also a vector; it is the derivative of velocity, and therefore its components are the derivatives of the components of the velocity vector. Alternatively, it is the second derivative of displacement and has components that are the second derivatives of the displacement vector.

In the special case in which C is a curve in the xy-plane, the definitions of displacement, velocity, speed, and acceleration become, respectively,

$$\mathbf{r} = x(t)\hat{\mathbf{i}} + y(t)\hat{\mathbf{j}}, \tag{12.95}$$

$$\mathbf{v} = \frac{d\mathbf{r}}{dt} = \frac{dx}{dt}\hat{\mathbf{i}} + \frac{dy}{dt}\hat{\mathbf{j}}, \tag{12.96}$$

$$|\mathbf{v}| = \sqrt{\left(\frac{dx}{dt}\right)^2 + \left(\frac{dy}{dt}\right)^2}, \tag{12.97}$$

$$\mathbf{a} = \frac{d\mathbf{v}}{dt} = \frac{d^2\mathbf{r}}{dt^2} = \frac{d^2x}{dt^2}\hat{\mathbf{i}} + \frac{d^2y}{dt^2}\hat{\mathbf{j}}. \tag{12.98}$$

For motion along the x-axis,

$$\mathbf{r} = x(t)\hat{\mathbf{i}}, \tag{12.99}$$

$$\mathbf{v} = \frac{d\mathbf{r}}{dt} = \frac{dx}{dt}\hat{\mathbf{i}}, \tag{12.100}$$

$$|\mathbf{v}| = \left|\frac{dx}{dt}\right|, \tag{12.101}$$

$$\mathbf{a} = \frac{d\mathbf{v}}{dt} = \frac{d^2\mathbf{r}}{dt^2} = \frac{d^2x}{dt^2}\hat{\mathbf{i}}. \tag{12.102}$$

If we compare equations 12.100 and 12.102 with equations 12.87 and 12.88, we see that for motion along the x-axis, $x(t)$, $v(t)$, and $a(t)$ are the components of the displacement, velocity, and acceleration vectors, respectively. Because these are the only components of $\mathbf{r}(t)$, $\mathbf{v}(t)$, and $\mathbf{a}(t)$, it follows that consideration of the components of the vectors is equivalent to consideration of the vectors themselves. For one-dimensional motion, then, we can drop the vector notation and work with components (and this is precisely the procedure that we followed in Sections 4.7 and 5.2).

Newton's second law describes the effects of forces on the motion of objects. It states that if an object of mass m is subjected to a force $\mathbf{F}$, then the time rate of change of its momentum $(m\mathbf{v})$ is equal to $\mathbf{F}$:

$$\mathbf{F} = \frac{d}{dt}(m\mathbf{v}). \tag{12.103}$$

In most cases, the mass of the object is constant, and this equation then yields its acceleration:

$$\mathbf{F} = m\frac{d\mathbf{v}}{dt} = m\mathbf{a}. \tag{12.104}$$

If $\mathbf{F}$ is known as a function of time t, $\mathbf{F} = \mathbf{F}(t)$, then 12.104 defines the acceleration of the object as a function of time,

$$\mathbf{a}(t) = \frac{1}{m}\mathbf{F}(t),$$

and integration of this equation leads to expressions for the velocity $\mathbf{v}(t)$ and position $\mathbf{r}(t)$ as functions of time.

EXAMPLE 12.39

Find the velocity and acceleration of a particle and describe its motion, if its position as a function of time is given by

$$x = 2t + \sqrt{2}, \quad y = \sqrt{4t^2 + 4\sqrt{2}t + 1}, \quad z = 0, \quad t \geq 0.$$

SOLUTION The velocity of the particle is

$$\mathbf{v} = \frac{dx}{dt}\hat{\mathbf{i}} + \frac{dy}{dt}\hat{\mathbf{j}} = 2\hat{\mathbf{i}} + \left(\frac{4t + 2\sqrt{2}}{\sqrt{4t^2 + 4\sqrt{2}t + 1}}\right)\hat{\mathbf{j}},$$

and its acceleration is

$$\mathbf{a} = \frac{d^2x}{dt^2}\hat{\mathbf{i}} + \frac{d^2y}{dt^2}\hat{\mathbf{j}} = \frac{-4}{(4t^2 + 4\sqrt{2}t + 1)^{3/2}}\hat{\mathbf{j}}.$$

Because $t = (x - \sqrt{2})/2$, an explicit definition of the path is

$$y = \sqrt{4\left(\frac{x-\sqrt{2}}{2}\right)^2 + 4\sqrt{2}\left(\frac{x-\sqrt{2}}{2}\right) + 1} = \sqrt{x^2 - 1}.$$

Since $t \geq 0$, it follows that $x \geq \sqrt{2}$, and the path is that part of the hyperbola $x^2 - y^2 = 1$ in Figure 12.121. Note that $dx/dt = 2$, so that the x-component of the velocity is always equal to 2. This is also reflected in the fact that the acceleration has no x-component. Since the y-component of acceleration is always negative, the y-component of velocity is decreasing in time.

FIGURE 12.121

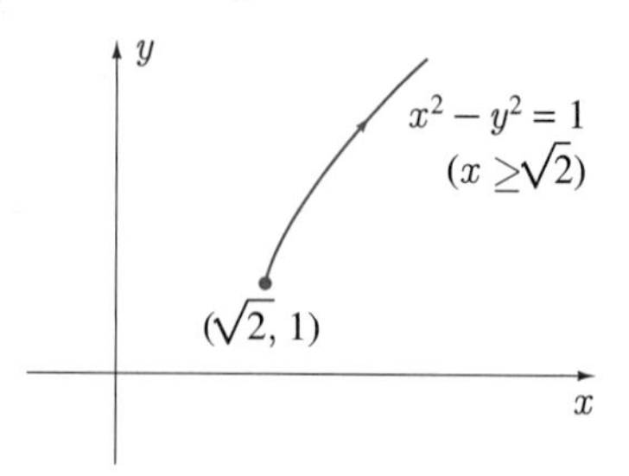

■

EXAMPLE 12.40

A particle starts at time $t = 0$ from position $(1, 1)$ with speed 2 m/s in the negative y-direction. It is subjected to an acceleration that is given as a function of time by

$$\mathbf{a}(t) = \frac{1}{\sqrt{t+1}}\hat{\mathbf{i}} + 6t\hat{\mathbf{j}} \ \text{m/s}^2.$$

Find its velocity and position as functions of time.

SOLUTION If $\mathbf{a} = d\mathbf{v}/dt = (1/\sqrt{t+1})\hat{\mathbf{i}} + 6t\hat{\mathbf{j}}$, then

$$\mathbf{v} = 2\sqrt{t+1}\hat{\mathbf{i}} + 3t^2\hat{\mathbf{j}} + \mathbf{C},$$

where $\mathbf{C}$ is some constant vector. Because the initial velocity of the particle is 2 m/s in the negative y-direction, $\mathbf{v}(0) = -2\hat{\mathbf{j}}$. Consequently,

$$-2\hat{\mathbf{j}} = 2\hat{\mathbf{i}} + \mathbf{C},$$

which implies that $\mathbf{C} = -2\hat{\mathbf{i}} - 2\hat{\mathbf{j}}$. The velocity, then, of the particle at any time $t \geq 0$ is

$$\mathbf{v}(t) = (2\sqrt{t+1} - 2)\hat{\mathbf{i}} + (3t^2 - 2)\hat{\mathbf{j}} \ \text{m/s}.$$

Because $\mathbf{v} = d\mathbf{r}/dt$, integration gives

$$\mathbf{r} = \left[\frac{4}{3}(t+1)^{3/2} - 2t\right]\hat{\mathbf{i}} + (t^3 - 2t)\hat{\mathbf{j}} + \mathbf{D}.$$

Since the particle starts from position $(1, 1)$, $\mathbf{r}(0) = \hat{\mathbf{i}} + \hat{\mathbf{j}}$, and

$$\hat{\mathbf{i}} + \hat{\mathbf{j}} = \frac{4}{3}\hat{\mathbf{i}} + \mathbf{D}, \quad \text{or,} \quad \mathbf{D} = -\frac{1}{3}\hat{\mathbf{i}} + \hat{\mathbf{j}}.$$

The displacement of the particle is therefore

$$\mathbf{r}(t) = \left[\frac{4}{3}(t+1)^{3/2} - 2t - \frac{1}{3}\right]\hat{\mathbf{i}} + (t^3 - 2t + 1)\hat{\mathbf{j}} \ \text{m}.$$

■

EXAMPLE 12.41

A shell is fired from an artillery gun with speed v_0 m/s at an angle θ with the horizontal (Figure 12.122). If gravity is assumed to be the only force acting on the shell:

(a) Find the position of the projectile as a function of time t.

(b) Find the range of the shell; i.e., find the horizontal distance traveled by the shell.

(c) Find the maximum height attained by the shell.

(d) Show that the path of the shell is a parabola.

FIGURE 12.122

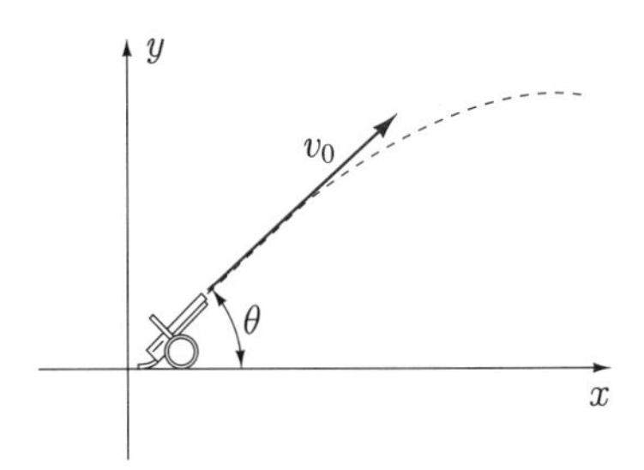

SOLUTION

(a) If m is the mass of the shell, then the force of gravity on the shell is

$$\mathbf{F} = -9.81m\hat{\mathbf{j}}.$$

According to Newton's second law, the acceleration of the shell is defined by

$$-9.81m\hat{\mathbf{j}} = m\mathbf{a}, \quad \text{or,} \quad \mathbf{a} = -9.81\hat{\mathbf{j}}.$$

Since $\mathbf{a} = d\mathbf{v}/dt$, we can write that

$$\frac{d\mathbf{v}}{dt} = -9.81\hat{\mathbf{j}},$$

and integration of this equation gives

$$\mathbf{v}(t) = -9.81t\hat{\mathbf{j}} + \mathbf{C}.$$

If we choose time $t = 0$ at the instant the shell is fired, then $\mathbf{v}(0) = v_0 \cos\theta\hat{\mathbf{i}} + v_0 \sin\theta\hat{\mathbf{j}}$, and hence

$$v_0 \cos\theta\hat{\mathbf{i}} + v_0 \sin\theta\hat{\mathbf{j}} = \mathbf{C}.$$

Because $\mathbf{v} = d\mathbf{r}/dt$, the position vector of the shell is

$$\mathbf{r}(t) = -4.905t^2\hat{\mathbf{j}} + \mathbf{C}t + \mathbf{D}.$$

Since $\mathbf{r}(0) = \mathbf{0}$, we must set $\mathbf{D} = \mathbf{0}$, and

$$\begin{aligned}\mathbf{r}(t) &= -4.905t^2\hat{\mathbf{j}} + v_0 \cos\theta t\hat{\mathbf{i}} + v_0 \sin\theta t\hat{\mathbf{j}} \\ &= (tv_0 \cos\theta)\hat{\mathbf{i}} + (-4.905t^2 + tv_0 \sin\theta)\hat{\mathbf{j}}.\end{aligned}$$

(b) The shell strikes the ground when the y-component of $\mathbf{r}$ vanishes; i.e., when

$$\begin{aligned} 0 &= -4.905t^2 + tv_0 \sin\theta \\ &= t(-4.905t + v_0 \sin\theta). \end{aligned}$$

Clearly, $t = (v_0 \sin\theta)/4.905$, and at this time, the x-component of $\mathbf{r}$ is the range

$$v_0 \cos\theta \left(\frac{v_0 \sin\theta}{4.905}\right) = \frac{v_0^2 \sin\theta \cos\theta}{4.905} = \frac{v_0^2 \sin 2\theta}{9.81} \text{ m.}$$

(c) The shell is at its maximum height when its y-component of velocity is equal to zero; i.e., when

$$0 = -9.81t + v_0 \sin\theta, \quad \text{or,} \quad t = \frac{v_0 \sin\theta}{9.81}.$$

At this time the y-component of the position vector is

$$-4.905\left(\frac{v_0 \sin\theta}{9.81}\right)^2 + v_0 \sin\theta\left(\frac{v_0 \sin\theta}{9.81}\right) = \frac{v_0^2 \sin^2\theta}{19.62} \text{ m.}$$

(d) Since $x(t) = tv_0 \cos\theta$ and $y(t) = -4.905t^2 + tv_0 \sin\theta$, it follows that

$$\begin{aligned} y &= -4.905\left(\frac{x}{v_0 \cos\theta}\right)^2 + v_0 \sin\theta\left(\frac{x}{v_0 \cos\theta}\right) \\ &= \frac{-4.905}{v_0^2 \cos^2\theta}x^2 + x\tan\theta, \end{aligned}$$

a parabola. ■

Tangential and Normal Components of Velocity and Acceleration

For some types of motion it is inconvenient to express velocity and acceleration in terms of Cartesian components; sometimes it is an advantage to resolve these vectors into tangential and normal components. When the trajectory C of a particle is specified as a function of time t by 12.89, its velocity $\mathbf{v} = d\mathbf{r}/dt$ is tangent to C, and we can therefore write

$$\mathbf{v} = |\mathbf{v}|\hat{\mathbf{T}}. \tag{12.105}$$

In other words, the tangential component of velocity is speed, and $\mathbf{v}$ has no normal component. Differentiation of this equation gives the particle's acceleration:

$$\begin{aligned} \mathbf{a} = \frac{d\mathbf{v}}{dt} &= \left(\frac{d}{dt}|\mathbf{v}|\right)\hat{\mathbf{T}} + |\mathbf{v}|\frac{d\hat{\mathbf{T}}}{dt} \\ &= \left(\frac{d}{dt}|\mathbf{v}|\right)\hat{\mathbf{T}} + \left(|\mathbf{v}|\left|\frac{d\hat{\mathbf{T}}}{dt}\right|\right)\frac{d\hat{\mathbf{T}}/dt}{|d\hat{\mathbf{T}}/dt|} \\ &= \left(\frac{d}{dt}|\mathbf{v}|\right)\hat{\mathbf{T}} + \left(|\mathbf{v}|\left|\frac{d\hat{\mathbf{T}}}{dt}\right|\right)\hat{\mathbf{N}}. \end{aligned} \tag{12.106}$$

We have therefore expressed $\mathbf{a}$ in terms of the unit tangent vector $\hat{\mathbf{T}}$ to C and the principal normal $\hat{\mathbf{N}}$ (Figure 12.123). We call $d(|\mathbf{v}|)/dt$ and $|\mathbf{v}||d\hat{\mathbf{T}}/dt|$ the tangential and normal components of acceleration, respectively. If a_T and a_N denote these components, we can write that

FIGURE 12.123

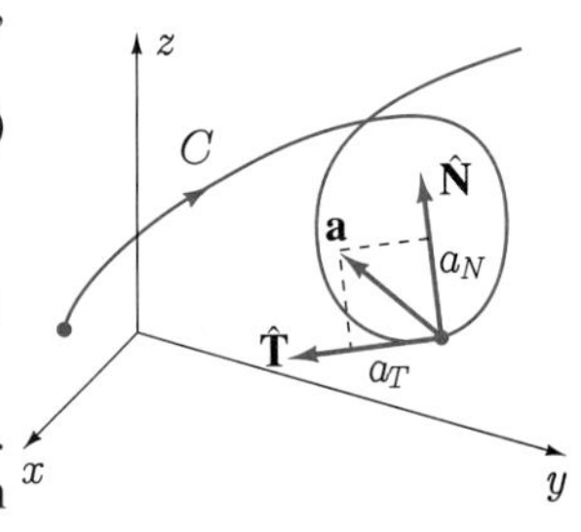

$$\mathbf{a} = a_T\hat{\mathbf{T}} + a_N\hat{\mathbf{N}}, \tag{12.107a}$$

where

$$a_T = \mathbf{a}\cdot\hat{\mathbf{T}} = \frac{d}{dt}|\mathbf{v}|, \quad a_N = \mathbf{a}\cdot\hat{\mathbf{N}} = |\mathbf{v}|\left|\frac{d\hat{\mathbf{T}}}{dt}\right|. \tag{12.107b}$$

Note that the tangential component of acceleration is the time rate of change of speed. Since acceleration is the rate of change of velocity, the normal component of acceleration must determine the rate of change of the direction of $\mathbf{v}$.

To calculate a_N using 12.107b is often quite complicated. A far easier formula results if we take the scalar product of $\mathbf{a}$ as defined by 12.107a with itself,

$$\begin{aligned}\mathbf{a}\cdot\mathbf{a} &= (a_T\hat{\mathbf{T}} + a_N\hat{\mathbf{N}})\cdot(a_T\hat{\mathbf{T}} + a_N\hat{\mathbf{N}})\\ &= a_T^2\hat{\mathbf{T}}\cdot\hat{\mathbf{T}} + 2a_Ta_N\hat{\mathbf{T}}\cdot\hat{\mathbf{N}} + a_N^2\hat{\mathbf{N}}\cdot\hat{\mathbf{N}}\\ &= a_T^2 + a_N^2,\end{aligned}$$

since $\hat{\mathbf{T}}\cdot\hat{\mathbf{N}} = 0$ and $\hat{\mathbf{T}}\cdot\hat{\mathbf{T}} = \hat{\mathbf{N}}\cdot\hat{\mathbf{N}} = 1$. Consequently,

$$a_N^2 = \mathbf{a}\cdot\mathbf{a} - a_T^2 = |\mathbf{a}|^2 - a_T^2,$$

and because a_N is always positive (see equation 12.107b),

$$a_N = \sqrt{|\mathbf{a}|^2 - a_T^2}. \tag{12.108}$$

EXAMPLE 12.42

If the trajectory of a particle is defined by

$$x = t^2 + 1, \quad y = 2t^2 - 1, \quad z = t^2 + 5t, \quad t \geq 0,$$

where t is time, find the tangential and normal components of the particle's velocity and acceleration.

SOLUTION The velocity and acceleration of the particle have Cartesian components

$$\mathbf{v} = 2t\hat{\mathbf{i}} + 4t\hat{\mathbf{j}} + (2t+5)\hat{\mathbf{k}}, \quad \mathbf{a} = 2\hat{\mathbf{i}} + 4\hat{\mathbf{j}} + 2\hat{\mathbf{k}}.$$

The tangential component of the particle's velocity is its speed:

$$|\mathbf{v}| = \sqrt{(2t)^2 + (4t)^2 + (2t+5)^2} = \sqrt{24t^2 + 20t + 25}.$$

According to 12.107b, the tangential component of the acceleration is

$$a_T = \frac{d}{dt}|\mathbf{v}| = \frac{24t + 10}{\sqrt{24t^2 + 20t + 25}}.$$

With 12.108, the normal component of acceleration is

$$a_N = \sqrt{|\mathbf{a}|^2 - a_T^2} = \left\{(4+16+4) - \frac{(24t+10)^2}{24t^2+20t+25}\right\}^{1/2}$$

$$= \left\{\frac{24(24t^2+20t+25) - (24t+10)^2}{24t^2+20t+25}\right\}^{1/2}$$

$$= \frac{10\sqrt{5}}{\sqrt{24t^2+20t+25}}.$$

■

EXAMPLE 12.43

Prove that the tangential component of the acceleration of a particle is equal to zero if and only if its speed is constant.

SOLUTION The tangential component of the acceleration of a particle is $a_T = d(|\mathbf{v}|)/dt$ and is equal to zero, therefore, if and only if

$$\frac{d}{dt}|\mathbf{v}| = 0.$$

But this is valid if and only if

$$|\mathbf{v}| = \text{a constant.}$$

■

What this example shows is that when a particle moves with constant speed, its acceleration always points along the principal normal to the curve. In the special case in which the trajectory is a circle (Figure 12.124), the principal normal always points directly toward the centre of the circle; therefore, the acceleration always points toward the centre of the circle also. Understand, however, that this is true only when speed is constant. If speed is not constant for circular motion, then a_T is not equal to zero; therefore, acceleration is not directed toward the centre of the circle.

FIGURE 12.124

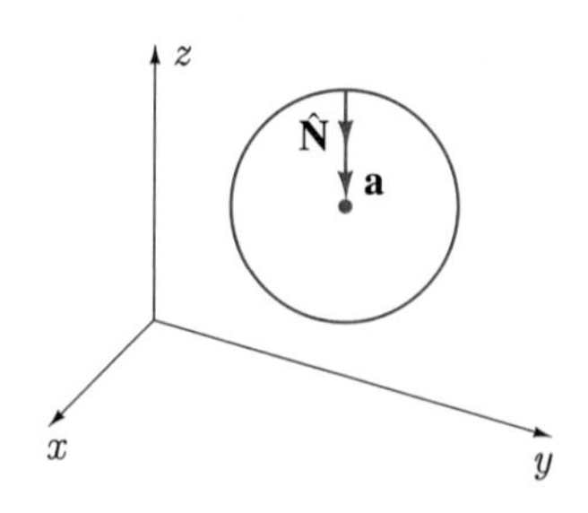

EXERCISES 12.10

In Exercises 1–5 find the velocity, speed, and acceleration of a particle if the given equations represent its position as a function of time.

1. $x(t) = \sqrt{t^2 + 1},\ y(t) = t\sqrt{t^2 + 1},\ t \geq 0$

2. $x(t) = t + 1/t,\ y(t) = t - 1/t,\ t \geq 1$

3. $x(t) = \sin t,\ y(t) = 3\cos t,\ z(t) = \sin t,\ 0 \leq t \leq 10\pi$

4. $x(t) = t^2 + 1,\ y(t) = 2te^t,\ z(t) = 1/t^2,\ 1 \leq t \leq 5$

5. $x(t) = e^{-t^2},\ y(t) = t\ln t,\ z(t) = 5,\ t \geq 1$

In Exercises 6 and 7 a particle at $(1, 2, -1)$ starts from rest at time $t = 0$. Find its position as a function of time if the given function defines its acceleration.

6. $\mathbf{a}(t) = 3t^2\hat{\mathbf{i}} + (t + 1)\hat{\mathbf{j}} - 4t^3\hat{\mathbf{k}},\ t \geq 0$

7. $\mathbf{a}(t) = 3\hat{\mathbf{i}} + \hat{\mathbf{j}}/(t + 1)^3,\ t \geq 0$

In Exercises 8 and 9 find the tangential and normal components of acceleration for a particle moving with position defined by the given functions (where t is time).

8. $x(t) = t,\ y(t) = t^2 + 1,\ t \geq 0$

9. $x(t) = \cos t,\ y(t) = \sin t,\ z = t,\ t \geq 0$

10. Show that the normal component of acceleration of a particle can be expressed in the form $a_N = |\mathbf{v}|^2/\rho = \kappa|\mathbf{v}|^2$.

11. The kinetic energy of an object of mass m moving with velocity $\mathbf{v}$ is defined as $K = m|\mathbf{v}|^2/2$. Find the kinetic energy for the particle in Exercises 1–5 if its mass is 2 g. Assume that x, y, and z are measured in metres and t in seconds.

12. A particle starts at the origin and moves along the curve $4y = x^2$ to the point $(4, 4)$.

(a) If the y-component of its acceleration is always equal to 2 and the y-component of its velocity is initially zero, find the x-component of its acceleration.

(b) If the x-component of its acceleration is equal to $24t^2$ (t being time) and the x-component of its velocity is initially zero, find the y-component of its acceleration.

13. A particle moves along the curve $x(t) = t,\ y(t) = t^3 - 3t^2 + 2t,\ 0 \leq t \leq 5$ in the xy-plane (where t is time). Is there any point at which its velocity is parallel to its displacement?

14. A particle moves along the curve $y = x^3 - 2x + 3$ so that its x-component of velocity is always equal to 5. Find its acceleration.

15. If a particle starts at time $t = 0$ from rest at position $(3, 4)$ and experiences an acceleration $\mathbf{a} = -5t^4\hat{\mathbf{i}} - (2t^3 + 1)\hat{\mathbf{j}}$, find its speed at $t = 2$.

16. A particle travels counterclockwise around the circle $(x - h)^2 + (y - k)^2 = R^2$ in Figure 12.125. Show that the speed of the particle at any time is $|\mathbf{v}| = \omega R$, where $\omega = d\theta/dt$ is called the *angular speed* of the particle.

FIGURE 12.125

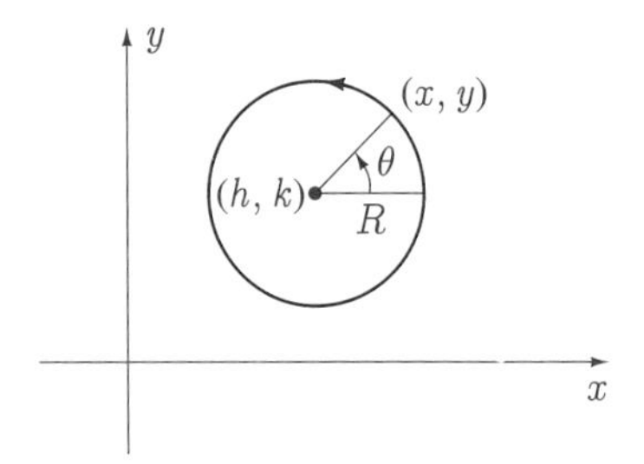

17. A particle travels around the circle $x^2 + y^2 = 4$ counterclockwise at constant speed, making 2 revolutions each second. If x and y are measured in metres, what is the velocity of the particle when it is at the point $(1, -\sqrt{3})$?

18. **(a)** Show that if an object moves with constant speed in a circular path of radius R, the magnitude of its acceleration is $|\mathbf{a}| = |\mathbf{v}|^2/R$.

(b) If a satellite moves with constant speed in a circular orbit 200 km above the earth's surface, what is its speed? (Hint: Use Newton's universal law of gravitation (see Exercise 32 in Section 12.3) to determine the acceleration $\mathbf{a}$ of the satellite. Assume that the earth is a sphere with radius 6370 km and density 5.52×10^3 kg/m^3.)

19. Two particles move along curves C_1 and C_2 (Figure 12.126). If at some instant of time the particles are at positions P_1 and P_2, then the vector $\overrightarrow{\mathbf{P_1P_2}}$ is the displacement of P_2 with respect to P_1. Clearly, $\overrightarrow{\mathbf{OP_1}} + \overrightarrow{\mathbf{P_1P_2}} = \overrightarrow{\mathbf{OP_2}}$. Show that when this equation is differentiated with respect to time, we have

$$\mathbf{v}_{P_1/O} + \mathbf{v}_{P_2/P_1} = \mathbf{v}_{P_2/O},$$

where $\mathbf{v}_{P_1/O}$ and $\mathbf{v}_{P_2/O}$ are velocities of P_1 and P_2 with respect to the origin, and $\mathbf{v}_{P_2/P_1}$ is the velocity of P_2 with respect to P_1. Can this equation be rewritten in the form

$$\mathbf{v}_{P_1/O} + \mathbf{v}_{O/P_2} = \mathbf{v}_{P_1/P_2}?$$

FIGURE 12.126

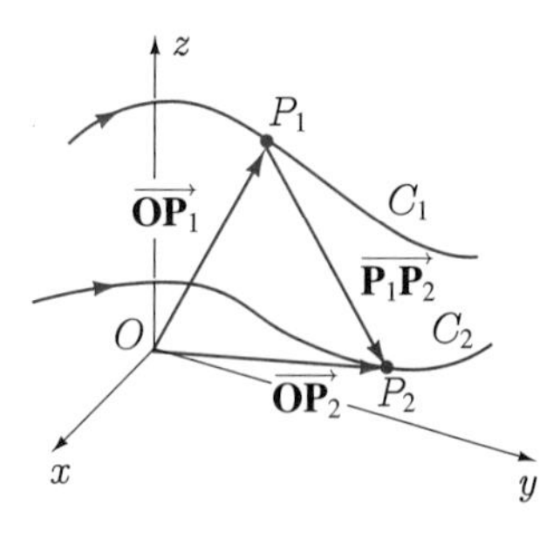

20. A plane flies on a course N30°E with air speed 650 km/h (i.e., the speed of the plane relative to the air is 650). If the air is moving at 40 km/h due east, find the ground velocity and speed of the plane. (Hint: Use Exercise 19.)

21. A plane flies with speed 600 km/h in still air. The plane is to fly in a straight line from city A to city B where B is 1000 km northwest of A. What should be its bearing if the wind is blowing from the west at 50 km/h? How long will the trip take?

22. A straight river is 200 m wide and the water flows at 3 km/h. If you can paddle your canoe at 4 km/h in still water, in what direction should you paddle if you wish the canoe to go straight across the river? How long will it take to cross?

23. **(a)** In Figure 12.127 a cannon is fired up an inclined plane. If the speed at which the ball is ejected from the cannon is S, show that the range R of the ball is given by

$$R = \frac{2S^2 \cos\theta \sin(\theta - \alpha)}{g \cos^2 \alpha},$$

where g is the acceleration due to gravity.

(b) What angle θ maximizes R?

FIGURE 12.127

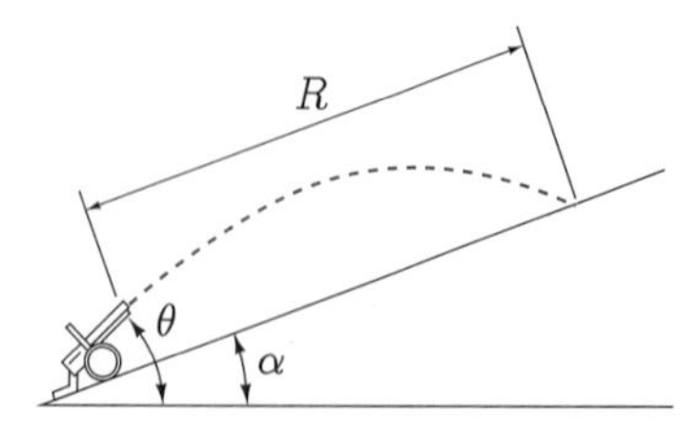

24. What constant acceleration must a particle experience if it is to travel from $(1,2,3)$ to $(4,5,7)$ along the straight line joining the points, starting from rest, and covering the distance in 2 units of time?

25. Calculate the normal component a_N of the acceleration of the particle in Example 12.42 using equation 12.107.

26. A particle moves along the curve $x(t) = 2 + \sqrt{1 - t^2}$, $y(t) = t$, $0 \leq t \leq 1/2$, where t is time. Is there a time at which its acceleration is perpendicular to its velocity?

27. **(a)** Show that motion along a straight line results in both of the following situations:

(i) The initial velocity is zero, and the acceleration is constant.

(ii) The initial velocity is nonzero, and the acceleration is constant and parallel to the initial velocity.

(b) Can we generalize the results of **(a)** and state that constant acceleration produces straight-line motion? Illustrate.

28. A particle starts from position $\mathbf{r}_0 = (x_0, y_0, z_0)$ at time $t = t_0$ with velocity $\mathbf{v}_0$. If it experiences constant acceleration $\mathbf{a}$, show that

$$\mathbf{r} = \mathbf{r}_0 + \mathbf{v}_0(t - t_0) + \frac{1}{2}\mathbf{a}(t - t_0)^2.$$

29. A ladder 8 m long has its upper end against a vertical wall and its lower end on a horizontal floor. Suppose that the lower end slips away from the wall at constant speed 1 m/s.

(a) Find the velocity and acceleration of the middle point of the ladder when the foot of the ladder is 3 m from the wall.

(b) How fast does the middle point of the ladder strike the floor?

30. Water issues from the nozzle of a fire hose at speed S (Figure 12.128). Show that the maximum height attainable by the water on the building is given by $(S^4 - g^2 d^2)/(2gS^2)$, where g is the acceleration due to gravity.

FIGURE 12.128

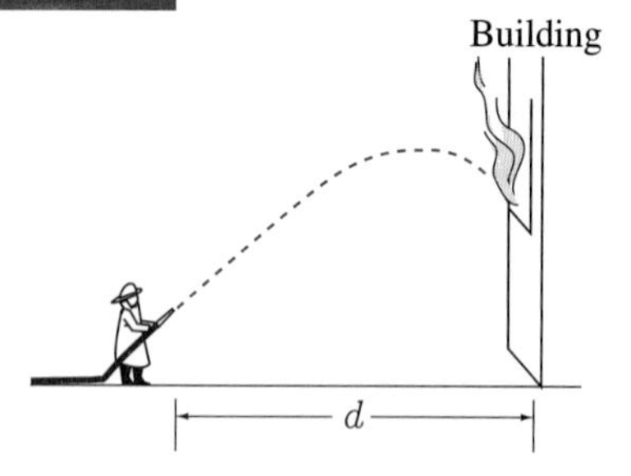

31. A boy stands on a cliff 50 m high that overlooks a river 85 m wide (Figure 12.129). If he can throw a stone at 25 m/s, can he throw it across the river?

FIGURE 12.129

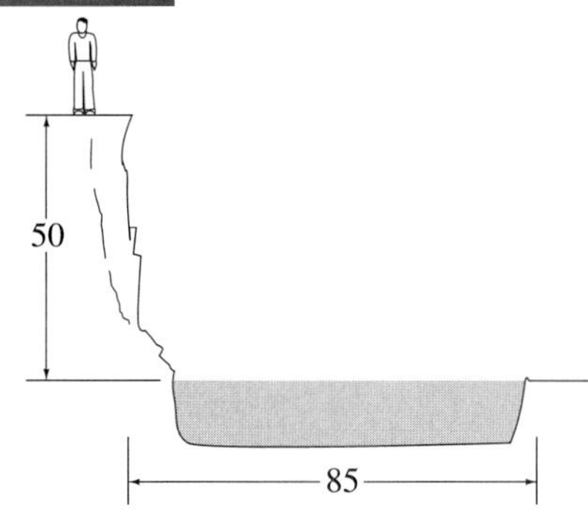

32. A golfer can drive a maximum of 300 m in the air on a level fairway. From the tee in Figure 12.130 can he expect to clear the stream?

FIGURE 12.130

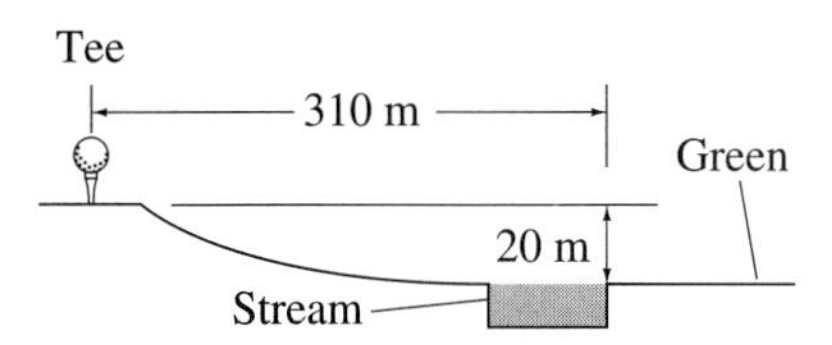

33. A particle is confined to move in a circular path of radius R in the xy-plane if and only if its position vector is $\mathbf{r} = x\hat{\mathbf{i}} + y\hat{\mathbf{j}}$, where $x = h + R\cos\omega(t)$, $y = k + R\sin\omega(t)$ and $\omega(t)$ is some function of time t. Show that the acceleration of the particle is directed radially toward the centre of the circular path if and only if $\omega(t) = At + B$, where A and B are constant. Furthermore, in this case verify that the speed of the particle is constant.

34. If $\mathbf{r}$ is the position vector of a particle with mass m moving under the action of a force $\mathbf{F}$, the torque (or moment) of $\mathbf{F}$ about the origin is $\mathbf{M} = \mathbf{r} \times \mathbf{F}$. The angular momentum of m about O is defined as $\mathbf{H} = \mathbf{r} \times m\mathbf{v}$. Use Newton's second law in the form 12.103 to show that $\mathbf{M} = d\mathbf{H}/dt$.

35. A stone is embedded in the tread of a tire and the tire rolls (without slipping) along the x-axis (Figure 12.131). If the stone starts at the origin, the path that it traces is called a cycloid.
    (a) Show that if θ is the angle through which the stone has turned, then parametric equations for the cycloid are $x = R(\theta - \sin\theta)$, $y = R(1 - \cos\theta)$.
    (b) Verify that if the centre of the tire moves at constant speed S, with $t = 0$ when the stone is at the origin, then $\theta = St/R$.
    (c) Find the velocity, speed, and acceleration of the stone at any time.
    (d) What are the normal and tangential components of the stone's acceleration?

FIGURE 12.131

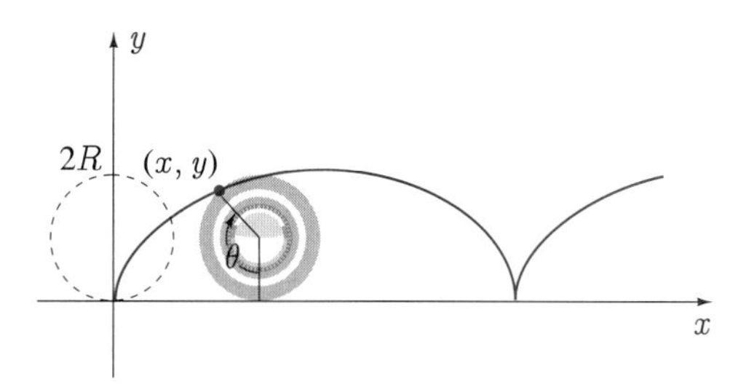

36. If the stone in Exercise 35 is embedded in the side of the tire, its path is called a trochoid (Figure 12.132).
    (a) Show that if the distance from the centre of the tire to the stone is b, then parametric equations for the trochoid are $x = R\theta - b\sin\theta$, $y = R - b\cos\theta$.
    (b) Find the velocity, speed, and acceleration of the stone if the tire rolls so that its centre has constant speed S.
    (c) What are normal and tangential components of the stone's acceleration?

FIGURE 12.132

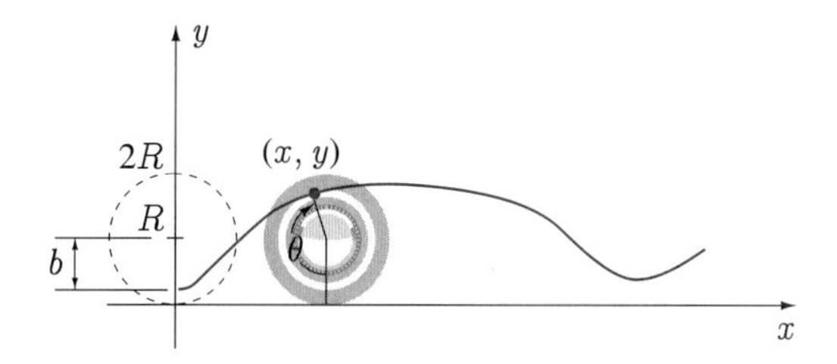

37. Circles C_1 and C_2 in Figure 12.133 represent cross-sections of two cylinders. The left cylinder remains stationary while the right one rolls (without slipping) around the left one, and the cylinders always remain in contact. If the right cylinder picks up a speck of dirt at point $(R, 0)$, the path that the dirt traces out during one revolution is a cardioid.
    (a) Show that parametric equations for the cardioid are $x = R(2\cos\theta - \cos 2\theta)$, $y = R(2\sin\theta - \sin 2\theta)$.
    (b) Verify that if the point of contact moves at constant speed S, with $t = 0$ when the speck of dirt is picked up, then $\theta = St/R$.
    (c) Find the velocity, speed, and acceleration of the speck of dirt.
    (d) What are normal and tangential components of the dirt's acceleration?

FIGURE 12.133

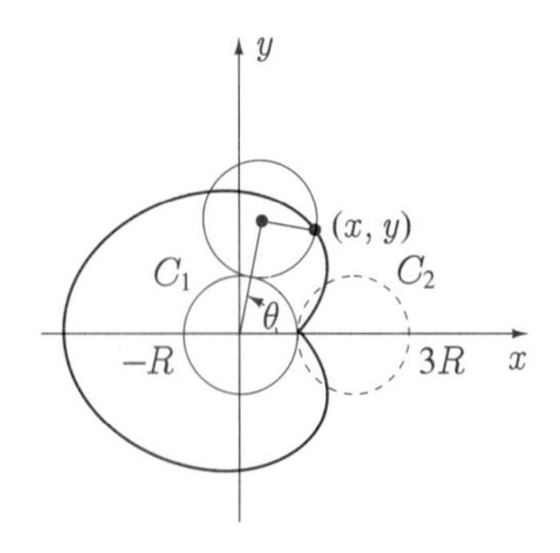

38. Show that the path of a particle lies on a sphere if its displacement and velocity are always perpendicular during its motion.

39. Suppose in Exercise 22 that, due to an injured elbow, you can only paddle at 2 km/h. What should be your heading to travel straight to a point L km downstream on the opposite shore? Are there any restrictions on L?

40. Suppose position vectors of a system of n masses m_i are denoted by $\mathbf{r}_i$ and forces acting on these masses are $\mathbf{F}_i$. Show that if $\mathbf{F} = \sum_{i=1}^{n} \mathbf{F}_i$, then the acceleration $\mathbf{a}$ of the centre of mass of the system (see Section 7.6) is given by $\mathbf{F} = M\mathbf{a}$, where $M = \sum_{i=1}^{n} m_i$.

41. If the force acting on a particle is always tangent to the particle's trajectory, what can you conclude about the trajectory?

SUMMARY

We have now established the groundwork for multivariable calculus. We discussed curves and surfaces in space and introduced vectors. Our approach to three-dimensional analytic geometry paralleled that to plane analytic geometry in Chapter 1. We described points by Cartesian coordinates (x, y, z) and then illustrated that an equation $F(x, y, z) = 0$ in these coordinates usually defines a surface. A second equation $G(x, y, z) = 0$ also defines a surface, and the pair of simultaneous equations $F(x, y, z) = 0$, $G(x, y, z) = 0$ describes the curve of intersection of the two surfaces (if the two surfaces do intersect). It is often more useful to have parametric equations for a curve, and these can be obtained by specifying one of x, y, or z as a function of a parameter t and solving the given equations for the other two in terms of t: $x = x(t)$, $y = y(t)$, $z = z(t)$.

The most common surfaces that we encountered were planes and quadric surfaces. Every plane has an equation of the form $Ax + By + Cz + D = 0$; conversely, every such equation describes a plane (provided A, B, and C are not all zero). A plane is uniquely defined by a vector that is perpendicular to it [and (A, B, C) is one such vector] and a point on it. Quadric surfaces are surfaces whose equations are quadratic in x, y, and z, the most important of which were sketched in Figures 12.24–12.33.

Every straight line in space is characterized by a vector along it and a point on it. (Contrast this with the above characterization of a plane.) If (a, b, c) are the components of a vector along a line and (x_0, y_0, z_0) are the coordinates of a point on it, then the vector, symmetric, and parametric equations for the line are, respectively,

$$(x, y, z) = (x_0, y_0, z_0) + t(a, b, c);$$

$$\frac{x - x_0}{a} = \frac{y - y_0}{b} = \frac{z - z_0}{c};$$

$$x = x_0 + at,$$
$$y = y_0 + bt,$$
$$z = z_0 + ct.$$

Geometrically, vectors are defined as directed line segments; algebraically, they are represented by ordered sets of real numbers (v_x, v_y, v_z), called their Cartesian components. Vectors can be added or subtracted geometrically using triangles or parallelograms; algebraically, they are added and subtracted component by component. Vectors can also be multiplied by scalars to give parallel vectors of different lengths.

We defined two products of vectors: the scalar product and the vector product. The scalar product of two vectors $\mathbf{u} = (u_x, u_y, u_z)$ and $\mathbf{v} = (v_x, v_y, v_z)$ is defined as

$$\mathbf{u} \cdot \mathbf{v} = u_x v_x + u_y v_y + u_z v_z = |\mathbf{u}||\mathbf{v}| \cos\theta,$$

where $|\mathbf{u}| = \sqrt{u_x^2 + u_y^2 + u_z^2}$ is the length of $\mathbf{u}$ and θ is the angle between $\mathbf{u}$ and $\mathbf{v}$. If the components of $\mathbf{u}$ and $\mathbf{v}$ are known, then this definition defines the angle θ between the vectors. The scalar product has many uses: finding components of vectors in arbitrary directions, calculating distances between geometric objects, and finding mechanical work, among others.

The vector product of two vectors $\mathbf{u}$ and $\mathbf{v}$ is

$$\mathbf{u} \times \mathbf{v} = \begin{vmatrix} \hat{\mathbf{i}} & \hat{\mathbf{j}} & \hat{\mathbf{k}} \\ u_x & u_y & u_z \\ v_x & v_y & v_z \end{vmatrix} = |\mathbf{u}||\mathbf{v}| \sin\theta \hat{\mathbf{w}},$$

where $\hat{\mathbf{w}}$ is the unit vector perpendicular to $\mathbf{u}$ and $\mathbf{v}$ determined by the right-hand rule. Because of the perpendicularity property, the vector product is indispensable in finding vectors perpendicular to other vectors. We used this fact when finding a vector along the line of intersection of two planes, a vector perpendicular to the plane containing three given points, and distances between geometric objects. It can also be used to find areas of triangles and parallelograms.

If a curve is represented vectorially in the form $\mathbf{r}(t) = x(t)\hat{\mathbf{i}} + y(t)\hat{\mathbf{j}} + z(t)\hat{\mathbf{k}}$, then a unit vector tangent to the curve at any point is

$$\hat{\mathbf{T}} = \frac{d\mathbf{r}/dt}{|d\mathbf{r}/dt|}.$$

Two unit vectors normal to the curve are the principal normal $\hat{\mathbf{N}}$ and the binormal $\hat{\mathbf{B}}$:

$$\hat{\mathbf{N}} = \frac{d\hat{\mathbf{T}}/dt}{|d\hat{\mathbf{T}}/dt|}; \quad \hat{\mathbf{B}} = \hat{\mathbf{T}} \times \hat{\mathbf{N}}.$$

These three vectors form a moving triad of mutually perpendicular unit vectors along the curve. The curvature of a curve, defined by $\kappa(t) = |\dot{\mathbf{r}} \times \ddot{\mathbf{r}}|/|\dot{\mathbf{r}}|^3$, measures the rate at which the curve changes direction: The larger κ is, the faster the curve turns. The reciprocal of curvature $\rho = \kappa^{-1}$ is called radius of curvature. It is the radius of that circle which best approximates the curve at any point.

If parametric equations for a curve represent the position of a particle and t is time, then the velocity and acceleration of the particle are, respectively,

$$\mathbf{v} = \frac{d\mathbf{r}}{dt}; \quad \mathbf{a} = \frac{d\mathbf{v}}{dt} = \frac{d^2\mathbf{r}}{dt^2};$$

and its speed is the magnitude of velocity, $|\mathbf{v}|$.

Tangential and normal components of velocity and acceleration of the particle are defined by

$$\mathbf{v} = |\mathbf{v}|\hat{\mathbf{T}}; \quad \mathbf{a} = a_T\hat{\mathbf{T}} + a_N\hat{\mathbf{N}};$$

where

$$a_T = \frac{d}{dt}|\mathbf{v}| \text{ and } a_N = |\mathbf{v}|\left|\frac{d\hat{\mathbf{T}}}{dt}\right| = \sqrt{|\mathbf{a}|^2 - a_T^2}.$$

What these results say is that velocity is always tangent to the trajectory of the particle, and its acceleration always lies in the plane of the velocity vector and the principal normal.

Key Terms and Formulas

In reviewing this chapter, you should be able to define or discuss the following key terms:

Coordinate plane
Right (left) handed coordinate system
Quadric Surface
Vector
Tip of a vector
Tail of a vector
Components of a vector
Equality of vectors
Unit vector
Scalar multiplication
Sum of two vectors
Triangular addition of vectors
Parallelogram addition of vectors
Zero vector
Resultant of forces
Scalar product
Equation of a plane
Vector product
Vector equation of a line
Parametric equations of a line
Symmetric equations of a line
Scalar triple product
Continuous vector function
Derivative of a vector function
Antiderivative of a vector function
Parametric definition of a curve
Continuous curve
Closed curve
Displacement vector
Vector representation of a curve
Tangent vector to a curve
Smooth curve
Piecewise smooth curve
Unit tangent vector to a curve
Length of a curve
Principal normal vector
Binormal vector
Curvature
Radius of curvature
Circle of curvature
Velocity
Acceleration
Tangential component of acceleration
Normal component of acceleration

REVIEW EXERCISES

In Exercises 1–10, find the value of the scalar or the components of the vector if $\mathbf{u} = (1,3,-2)$, $\mathbf{v} = (2,4,-1)$, $\mathbf{w} = (0,2,1)$, and $\mathbf{r} = (2,0,-1)$.

1. $2\mathbf{u} - 3\mathbf{w} + \mathbf{r}$
2. $\mathbf{u} \cdot (\mathbf{v} \times \mathbf{w})$
3. $(3\mathbf{u} \times 4\mathbf{v}) - \mathbf{w}$
4. $3\mathbf{u} \times (4\mathbf{v} - \mathbf{w})$
5. $|\mathbf{u}|\mathbf{v} - |\mathbf{v}|\mathbf{r}$
6. $(\mathbf{u} + \mathbf{v}) \cdot (\mathbf{r} - \mathbf{w})$
7. $(\mathbf{u} + \mathbf{v}) \times (\mathbf{r} - \mathbf{w})$
8. $(\mathbf{u} \times \mathbf{v}) \times (\mathbf{r} \times \mathbf{w})$
9. $(\mathbf{u} \cdot \mathbf{v})\mathbf{r} - 3(\mathbf{v} \cdot \mathbf{w})\mathbf{u}$
10. $\dfrac{2\mathbf{r}}{\mathbf{v} \cdot \mathbf{w}} + 3(\mathbf{v} + \mathbf{u})$

In Exercises 11–26 sketch whatever the equation, or equations, describe in space.

11. $x - y + 2z = 6$
12. $x^2 + z^2 = 1$
13. $x = \sqrt{y^2 + z^2}$
14. $x - y = 5,\ 2x + y = 6$
15. $x^2 + y^2 + z^2 = 6z + 10$
16. $x^2 + y^2 + z^2 = 6z - 10$
17. $x + y = 5,\ 2x - 3y + 6z = 1,\ y = z$
18. $x = t^2,\ y = t,\ z = t^3$
19. $x = t,\ y = t^3 + 1$
20. $\dfrac{x-1}{3} = \dfrac{y+5}{2} = z$

21. $z = 4 - x^2 - 2y^2$

22. $y^2 + z^2 = 1,\ y = z$

23. $y^2 + z^2 = 1,\ x = z$

24. $x^2 + y^2 = z^2 + 1$

25. $x = y^2,\ x = z^2$

26. $z^2 = x^2 - y^2$

In Exercises 27–30 find equations for the line.

27. Through the points $(-2,3,0)$ and $(1,-2,4)$

28. Through $(6,6,2)$ and perpendicular to the plane $5x - 2y + z = 4$

29. Parallel to the line $x - y = 5,\ 2x + 3y + 6z = 4$ and through the origin

30. Perpendicular to the line $x = t+2,\ y = 3-2t,\ z = 4+t$, intersecting this line, and through the point $(1,3,2)$

In Exercises 31–34 find the equation for the plane.

31. Through the points $(1,3,2)$, $(2,-1,0)$ and $(6,1,3)$

32. Through the point $(1,2,-1)$ and perpendicular to the line $y = z,\ x + y = 4$

33. Containing the line $x - y + z = 3,\ 3x + 4y = 6$ and the point $(2,2,2)$

34. Containing the lines $x = 3t,\ y = 1 + 2t,\ z = 4 - t$ and $x = y = z$

In Exercises 35–39 find the distance.

35. Between the points $(1,3,-2)$ and $(6,4,1)$

36. From the point $(6,2,1)$ to the plane $6x + 2y - z = 4$

37. From the line $x - y + z = 2,\ 2x + y + z = 4$ to the plane $x - y = 5$

38. From the line $x - y + z = 2,\ 2x + y + z = 4$ to the plane $3x + 6y = 4$

39. From the point $(6,2,3)$ to the line $x - y + z = 6$, $2x + y + 4z = 1$

40. Find the area of the triangle with vertices $(1,1,1)$, $(-2,1,0)$ and $(6,3,-2)$.

41. If the points in Exercise 40 are three vertices of a parallelogram, what are the possibilities for the fourth vertex? What are the areas of these parallelograms?

In Exercises 42 and 43 find the unit tangent vector $\hat{\mathbf{T}}$, the principal normal vector $\hat{\mathbf{N}}$, and the binormal vector $\hat{\mathbf{B}}$ for the curve.

42. $x = 2\sin t,\ y = 2\cos t,\ z = t$

43. $x = t^3,\ y = 2t^2,\ z = t + 4$

44. If a particle has a trajectory defined by $x = t,\ y = t^2$, $z = t^2$, where t is time, find its velocity, speed, and acceleration at any time. What are normal and tangential components of its velocity and acceleration?

45. A force $\mathbf{F} = x^{-2}(2\hat{\imath} + 3\hat{\jmath})$ acts on a particle moving from $x = 1$ to $x = 4$ along the x-axis. How much work does it do?

46. A ball rolls off a table 1 m high with speed 0.5 m/s (Figure 12.134).
   (a) With what speed does it strike the floor?
   (b) What is its displacement vector relative to the point where it left the table when it strikes the floor?
   (c) If it rebounds in the direction shown but loses 20% of its speed in the bounce, find the position of its second bounce.

FIGURE 12.134

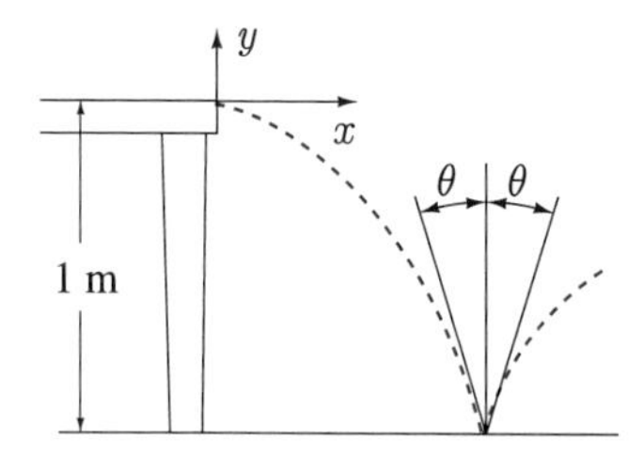

47. In Figure 12.135 a spring (with constant k) is fixed at A and attached to a sleeve at C. The sleeve is free to slide without friction on a vertical rod, and when the spring is horizontal (at B), it is unstretched. If the sleeve is slowly lowered, there is a position at which the vertical component of the spring force on C is balanced by the force of gravity on the sleeve (ignoring the weight of the spring itself). If the mass of the sleeve is m, find an equation determining s in terms of d, m, k and the acceleration g due to gravity.

FIGURE 12.135

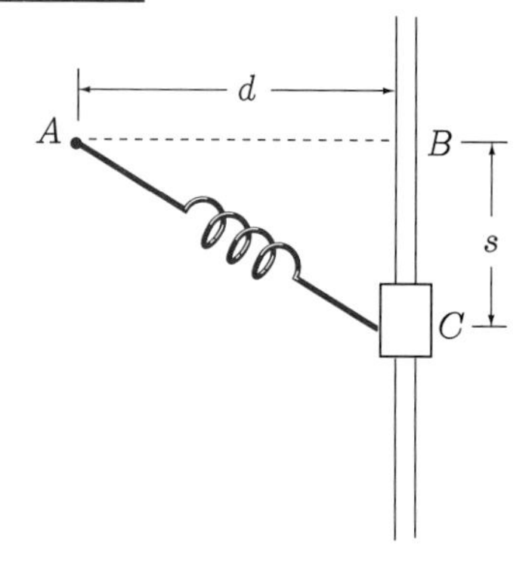

48. Find the Cartesian components of the spring force $\mathbf{F}$ on the sleeve in Example 12.18.

49. If a toy train travels around the oval track in Figure 12.136 with constant speed, show that its acceleration at A (the point at which the circular end meets the straight section) is discontinuous.

FIGURE 12.136

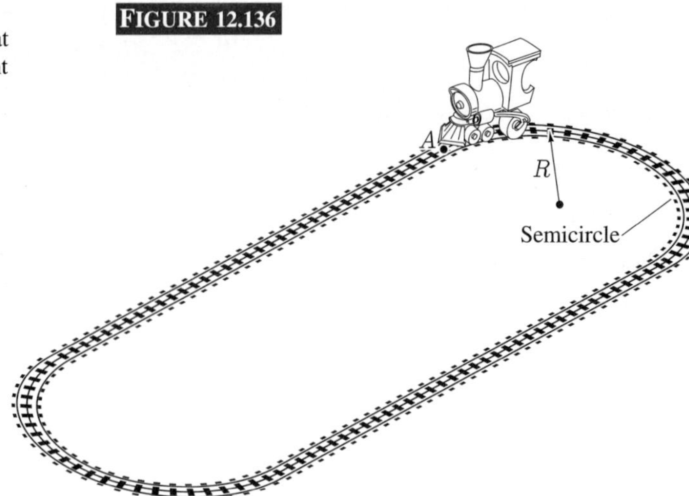

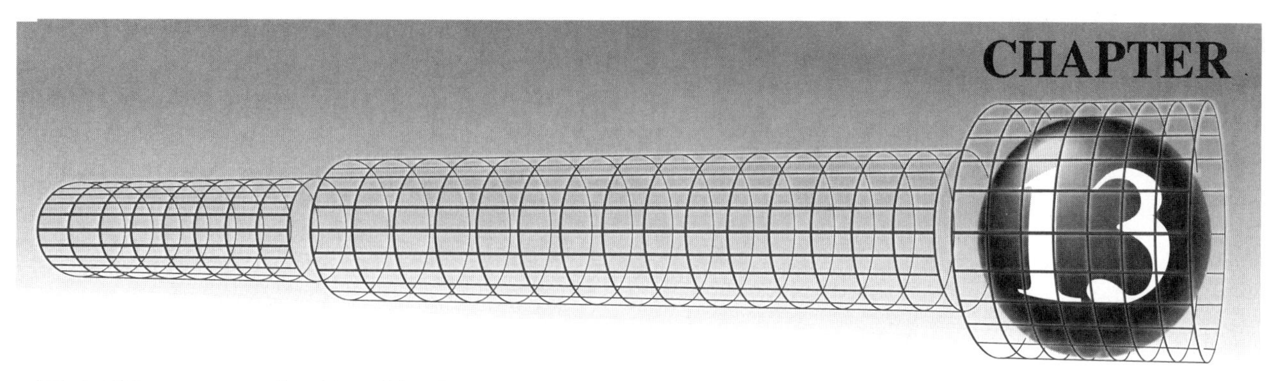

Differential Calculus of Multivariable Functions

Few quantities in real life depend on only one variable; most depend on a multitude of interrelated variables. In order to understand such complicated relationships, we initiate discussions in this chapter with derivatives of functions of more than one variable. Much of the theory and many of our examples involve functions of two or three variables, because in these cases we can give geometric as well as analytic explanations. If the situation is completely analogous for functions of more variables, then it is likely that no mention of this fact will be made; on the other hand, if the situation is different for a higher number of variables, we will be careful to point out these differences.

SECTION 13.1

Multivariable Functions

If a variable T depends on other variables x, y, z, and t, we write $T = f(x, y, z, t)$ and speak of T as a function of x, y, z, and t. For example, T might be temperature, x, y, and z might be the coordinates of points in some region of space, and t might be time. The stopping distance D of a car depends on many factors: the initial speed s, the reaction time t of the driver to move from the accelerator to the brake, the texture T of the road, the moisture level M on the road, etc. We write $D = f(s, t, T, M, ...)$ to represent this functional dependence. The function $P = f(I, R) = I^2 R$ represents the power necessary to maintain a current I through a wire with resistance R.

More precisely, a variable z is said to be a function of two independent variables x and y if x and y are not related and each pair of values of x and y determines a unique value of z. We write $z = f(x, y)$ to indicate that z is a function of x and y. Each possible pair of values x and y of the independent variables can be represented geometrically as a point (x, y) in the xy-plane. The totality of all points for which $f(x, y)$ is defined forms a region in the xy-plane called the **domain** of the function. Figure 13.1, for example, illustrates a rectangular domain. If for each point (x, y) in the domain we plot a point $f(x, y)$ units above the xy-plane, we obtain a surface, such as the one in Figure 13.1. Each point on this surface has coordinates (x, y, z) that satisfy the equation

$$z = f(x, y), \tag{13.1}$$

FIGURE 13.1

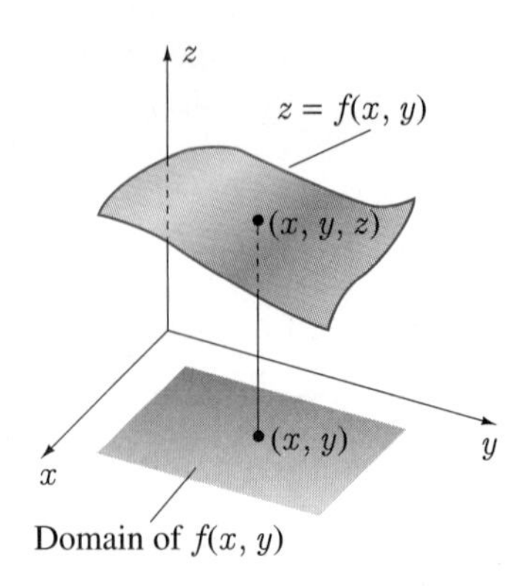

and therefore 13.1 is the equation of the surface. This surface is a pictorial representation of the function.

It is clear that functions of more than two independent variables cannot be represented pictorially as surfaces. For example, if $u = f(x, y, z)$ is a function of three independent variables, values (x, y, z) of these independent variables can be represented geometrically as points in space. To graph $u = f(x, y, z)$ as above would require a u-axis perpendicular to the x-, y-, and z-axes, a somewhat difficult task geometrically. We can certainly think of $u = f(x, y, z)$ as defining a surface in four-dimensional $xyzu$-space, but visually we are stymied.

Although every function $f(x, y)$ of two independent variables can be represented geometrically as a surface, not every surface represents a function $f(x, y)$. A given surface does represent a function $f(x, y)$ if and only if every vertical line (in the z-direction) that intersects the surface does so in exactly one point.

EXAMPLE 13.1

Sketch the surface defined by the function $f(x, y) = x^2 + 4y^2$.

SOLUTION To sketch the surface $z = x^2 + 4y^2$, we note that if the surface is intersected with a plane $z = k > 0$, then the ellipse $x^2 + 4y^2 = k$, $z = k$ is obtained. As k increases, the ellipse becomes larger. In other words, cross-sections of this surface are ellipses that expand with increasing z. If we intersect the surface with the yz-plane ($x = 0$), we obtain the parabola $z = 4y^2$, $x = 0$. Similarly, intersection of the surface with the xz-plane gives the parabola $z = x^2$, $y = 0$. These facts lead to Figure 13.2.

FIGURE 13.2

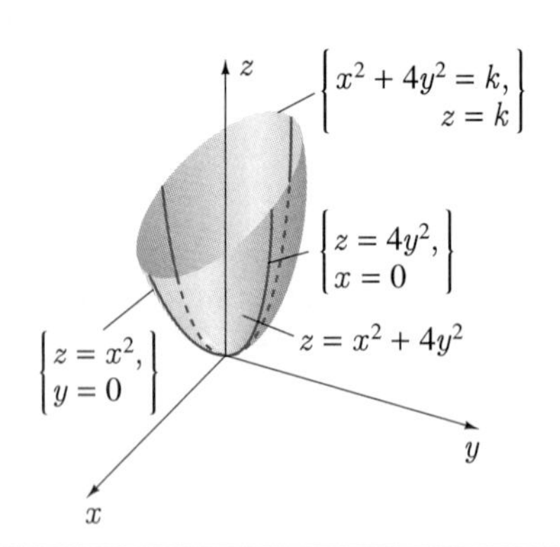

■

EXAMPLE 13.2

The ends of a taut string are fixed at $x = 0$ and $x = 2$ on the x-axis. At time $t = 0$, the string is given a displacement in the y-direction of $y = \sin(\pi x/2)$ (Figure 13.3a). If the string is then released, its displacement thereafter is given by

$$y = f(x,t) = \sin(\pi x/2)\cos(8\pi t).$$

Physically, this function need only be considered for $t \geq 0$ and $0 \leq x \leq 2$, and is pictured graphically as the surface in Figure 13.3b. Interpret physically the intersections of this surface with planes $t = t_0$ (t_0 = a constant) and $x = x_0$ (x_0 = a constant).

SOLUTION If we intersect the surface $y = f(x,t)$ with a plane $t = t_0$ ($t_0 \geq 0$) (Figure 13.3c), the curve of intersection is a picture of the position of the string at time t_0.

If we intersect the surface with a plane $x = x_0$ ($0 \leq x_0 \leq 2$)(Figure 13.3d), we obtain a graphical history of the displacement of the particle at position x_0.

FIGURE 13.3

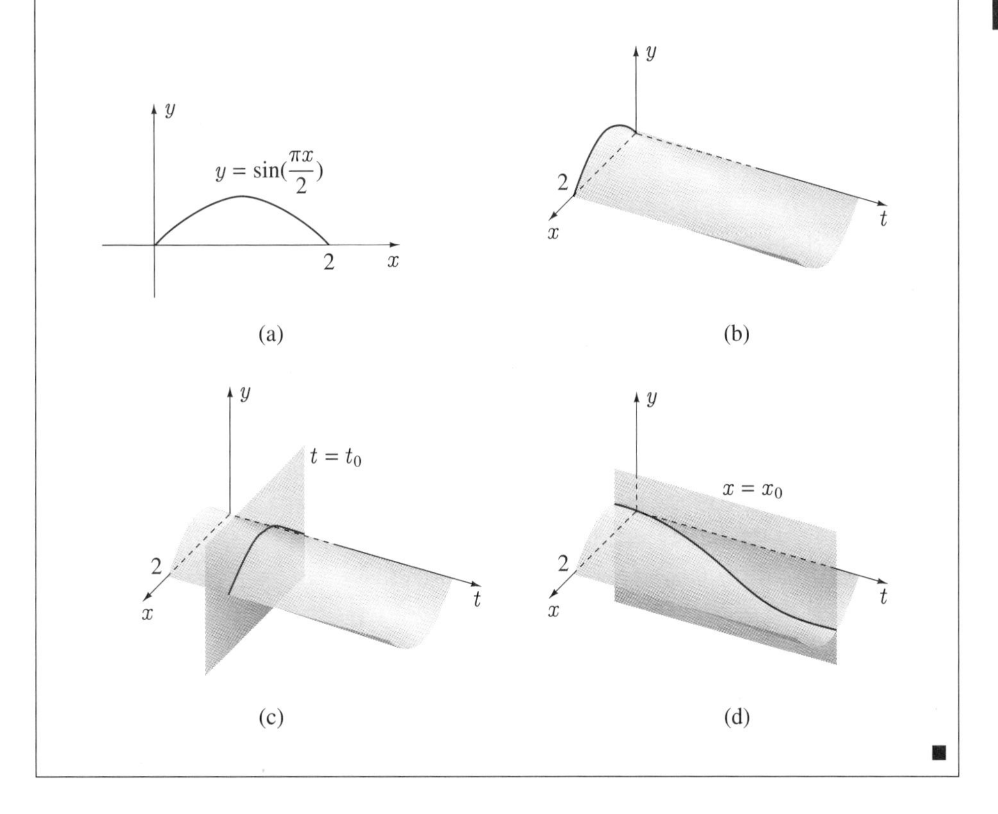

■

Many software packages draw surfaces $z = f(x,y)$ corresponding to functions $f(x,y)$ of two independent variables. They use cross-sections of the surface with planes parallel to the xz- and yz-planes to give excellent visualizations of the surface. The examples in Figure 13.4 were produced with MathCAD.

FIGURE 13.4

(a)

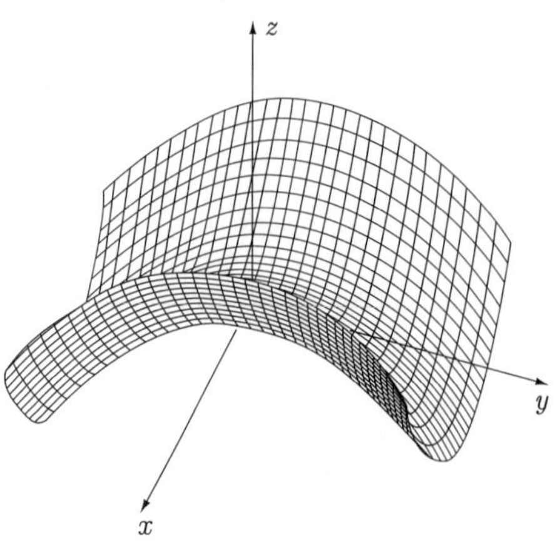

$z = x^2 - y^2$ $\quad -1 \le x \le 1$, $-1 \le y \le 1$

(b)

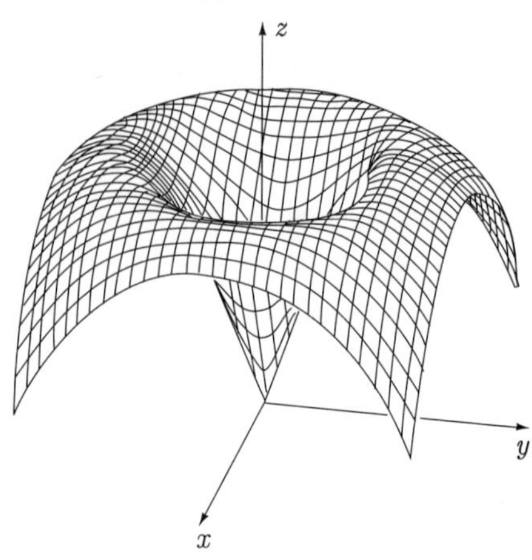

$z = 3\sqrt{x^2 + y^2} - (x^2 + y^2)$ $\quad -2 \le x \le 2$, $-2 \le y \le 2$

(c)

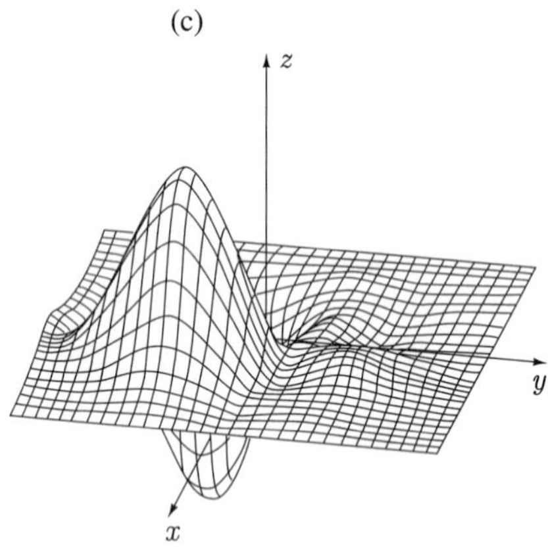

$z = x(y - 1)^2 e^{-(x^2 + y^2)/4}$ $\quad -5 \le x \le 5$, $-5 \le y \le 5$

(d)

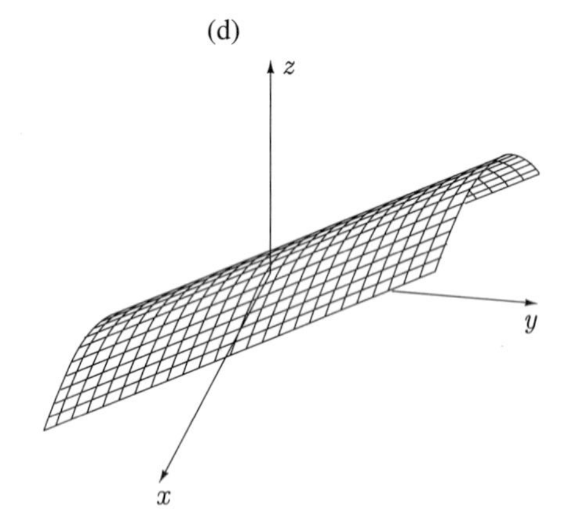

$z = 4y - x^2$ $\quad -2 \le x \le 1$, $-1 \le y \le 1$

(e)

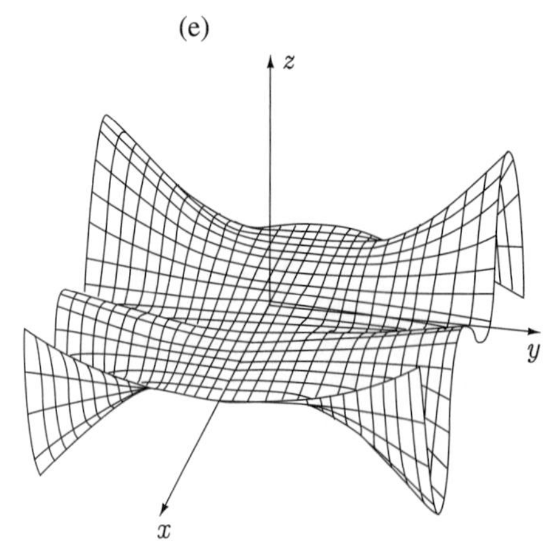

$z = xy^2 \cos x$ $\quad -5 \le x \le 5$, $-10 \le y \le 10$

(f)

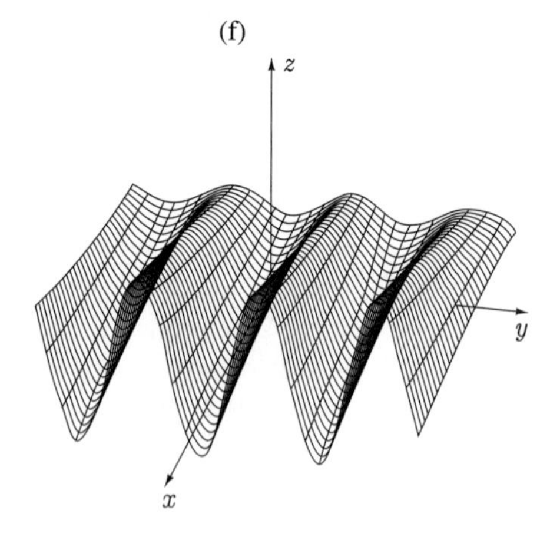

$z = e^x \sin y$ $\quad -1 \le x \le 1$, $-10 \le y \le 10$

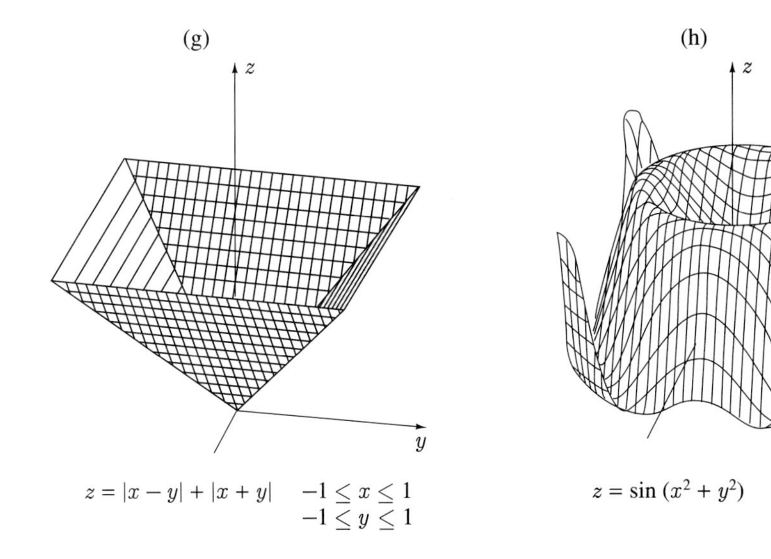

FIGURE 13.4

Another way to visualize a function $f(x, y)$ of two independent variables is through level curves. Curves $f(x, y) = C$ are drawn in the xy-plane for various values of C. Effectively, the surface $z = f(x, y)$ is sliced with a plane $z = C$, and the curve of intersection is projected into the xy-plane. Each curve joins all points for which $f(x, y)$ has the same value; or, it joins all points which have the same height on the surface $z = f(x, y)$. A few level curves for the surface in Figure 13.2 are shown in Figure 13.5; they are ellipses.

FIGURE 13.5

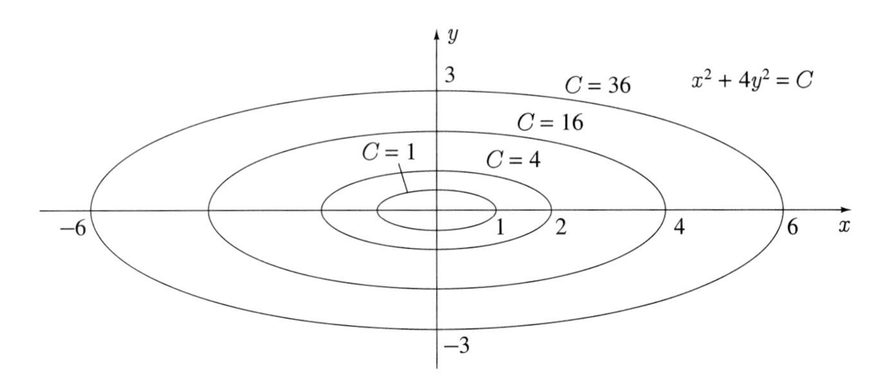

Level curves for the function in Figure 13.6(a) are shown in Figure 13.6(b). This technique is used on topographical maps to indicate land elevation, on marine charts to indicate water depth, and on climatic maps to indicate curves of constant temperature (isotherms) and curves of constant barometric pressure (isobars).

FIGURE 13.6

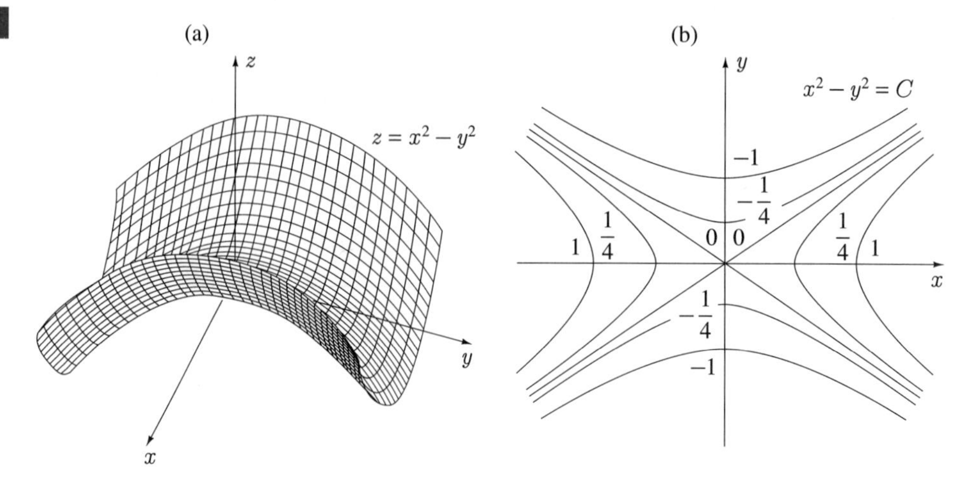

EXERCISES 13.1

1. If $f(x,y) = x^3y + x\sin y$, evaluate

(a) $f(1,2)$ **(b)** $f(-2,-2)$

(c) $f(x^2+y, x-y^2)$ **(d)** $f(x+h,y) - f(x,y)$

2. If $f(x,y,z) = x^2y^2 - x^4 + 4zx^2$, show that $f(a+b, a-b, ab) = 0$.

In Exercises 3–6 find and illustrate geometrically the largest possible domain for the function.

3. $f(x,y) = \sqrt{4 - x^2 - y^2}$

4. $f(x,y) = \ln(1 - x^2 + y^2)$

5. $f(x,y) = \mathrm{Sin}^{-1}(x^2y + 1)$

6. $f(x,y,z) = 1/(x^2 + y^2 + z^2)$

7. For what values of x and y is the function $f(x,y) = \dfrac{12xy - x^2y^2}{2(x+y)}$ equal to zero? Illustrate these values as points in the xy-plane. What is the largest domain of the function?

In Exercises 8–21 sketch the surface defined by the function.

8. $f(x,y) = y^2$

9. $f(x,y) = 4 - x - 2y$

10. $f(x,y) = x^2 + y^2$

11. $f(x,y) = \sqrt{x^2 + y^2}$

12. $f(x,y) = y + x$

13. $f(x,y) = 1 - x^3$

14. $f(x,y) = 2(x^2 + y^2)$

15. $f(x,y) = 1 - x^2 - 4y^2$

16. $f(x,y) = xy$

17. $f(x,y) = y - x^2$

18. $f(x,y) = e^{-x^2-y^2}$

19. $f(x,y) = |x - y|$

20. $f(x,y) = x^2 - y^2$

21. $f(x,y) = \sqrt{1 + x^2 - y^2}$

In Exercises 22–25 draw level curves $f(x,y) = C$ corresponding to the values $C = -2, -1, 0, 1, 2$.

22. $f(x,y) = 4 - \sqrt{4x^2 + y^2}$

23. $f(x,y) = y - x^2$

24. $f(x,y) = \ln(x^2 + y^2)$

25. $f(x,y) = x^2 - y^2$

26. A closed box is to have total surface area 30 m^2. Find a formula for the volume of the box in terms of its length l and width w.

27. **(a)** A company wishes to construct a storage tank in the form of a rectangular box. If material for sides and top costs \$1.25 per square metre and material for the bottom costs \$4.75 per square metre, find the cost of building the tank as a function of its length l, width w, and height h.

(b) If the tank must hold 1000 m^3, find the construction cost in terms of l and w.

(c) Repeat (a) and (b) if the 12 edges of the tank must be welded at a cost of \$7.50 per metre of weld.

28. A rectangular box is inscribed inside the ellipsoid $x^2/a^2 + y^2/b^2 + z^2/c^2 = 1$ with sides parallel to the coordinate planes and corners on the ellipsoid. Find a formula for the volume of the box in terms of x and y.

29. (a) A silo is to be built in the shape of a right-circular cylinder surmounted by a right-circular cone. If the radius of each is 6 m, find a formula for the volume V in the silo as a function of the heights H and h of the cylinder and cone.
    (b) If the total surface area of the silo must be 200 m^2 (not including the base), find V as a function of h. (The area of the curved surface of a cone is $\pi r\sqrt{r^2 + h^2}$.)

30. A long piece of metal 1 m wide is bent in two places A and B (Figure 13.7) to form a channel with three straight sides. Find a formula for the cross-sectional area of the channel in terms of x, θ, and ϕ.

FIGURE 13.7

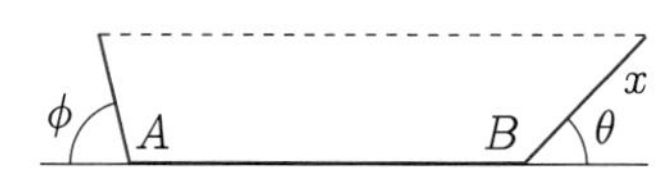

31. A uniform circular rod has flat ends at $x = 0$ and $x = \pi$ on the x-axis. The round side of the rod is insulated and the faces at $x = 0$ and $x = \pi$ are both kept at temperature $0\,°C$ for time $t > 0$. If the initial temperature (at time $t = 0$) of the rod is given by $100 \sin x$, $0 \leq x \leq \pi$, then the temperature thereafter is

$$T = f(x,t) = 100\,e^{-kt} \sin x \quad (k > 0 \text{ constant}).$$

    (a) Sketch the surface $T = f(x,t)$.
    (b) Interpret physically the curves of intersection of this surface with planes $x = x_0$ and $t = t_0$.

32. A cow's daily diet consists of three foods: hay, grain, and supplements. The animal is always given 11 kg of hay per day, 50% of which is digestive material and 12% of which is protein. Grain is 74% digestive and 8.8% protein. Supplements are 62% digestive material and 34% protein. Hay costs $27.50 for 1000 kg, whereas grain and supplements cost $110 and $175 respectively for 1000 kg. A healthy cow's daily diet must contain between 9.5 kg and 11.5 kg of digestive material and between 1.9 kg and 2.0 kg of protein. Find a formula for the cost per day C of feeding a cow in terms of the number of kilograms of grain G and supplements S fed to the cow daily. What is the domain of this function?

SECTION 13.2 Limits and Continuity

The concepts of limit and continuity for multivariable functions are exactly the same as for functions of one variable; on the other hand, the work involved with the application of these concepts is much more complicated for multivariable functions.

Intuitively, a function $f(x,y)$ is said to have limit L as x and y approach x_0 and y_0, if $f(x,y)$ gets arbitrarily close to L, and stays close to L, as x and y get arbitrarily close to x_0 and y_0. To give a mathematical definition of limits, it is convenient to represent pairs of independent variables as points (x,y) in the xy-plane. We then have the following definition.

Definition 13.1

A function $f(x,y)$ has limit L as (x,y) approaches (x_0,y_0), written

$$\lim_{(x,y)\to(x_0,y_0)} f(x,y) = L, \tag{13.2}$$

if given any $\epsilon > 0$, we can find a $\delta > 0$ such that

$$|f(x,y) - L| < \epsilon$$

whenever $0 < \sqrt{(x-x_0)^2 + (y-y_0)^2} < \delta$ and (x,y) is in the domain of $f(x,y)$.

In other words, $f(x, y)$ has limit L as (x, y) approaches (x_0, y_0) if $f(x, y)$ can be made arbitrarily close to L (within ϵ) by choosing points (x, y) sufficiently close to (x_0, y_0) (within a circle of radius δ). Note the similarity of this definition to that for the limit of a function $f(x)$ of one variable in Section 2.5.

It is clear that

$$\lim_{(x,y)\to(2,1)} (x^2 + 2xy - 5) = 3,$$

but the limit

$$\lim_{(x,y)\to(0,0)} \frac{x^2 - y^2}{x^2 + y^2}$$

presents a problem, since both numerator and denominator approach zero as x and y approach zero.

To conclude that $\lim_{x\to a} f(x) = L$, the limit must be L no matter how x approaches a—be it through numbers larger than a, through numbers smaller than a, or through any other approach. For limit 13.2, the limit of $f(x, y)$ must also be L for all possible ways of approaching (x_0, y_0). But in this case there might be a multitude of ways of approaching (x_0, y_0). We might be able to approach (x_0, y_0) along straight lines with various slopes, along parabolas, along cubics, etc. Definition 13.1 implies, then, that the limit exists only if it is independent of the manner of approach. It is assumed, however, that we approach (x_0, y_0) only through points (x, y) that lie in the domain of definition of the function.

For the second example above, suppose we approach the origin along the straight line $y = mx$. Then, along this line,

$$\lim_{(x,y)\to(0,0)} \frac{x^2 - y^2}{x^2 + y^2} = \lim_{x\to 0} \frac{x^2 - m^2x^2}{x^2 + m^2x^2} = \frac{1 - m^2}{1 + m^2}.$$

Because this result depends on m, we have shown that as (x, y) approaches $(0, 0)$ along various straight lines, the function $(x^2 - y^2)/(x^2 + y^2)$ approaches different numbers. We conclude, therefore, that the function does not have a limit as (x, y) approaches $(0, 0)$. In Figure 13.8 we show a portion of this function to illustrate our conclusion.

FIGURE 13.8

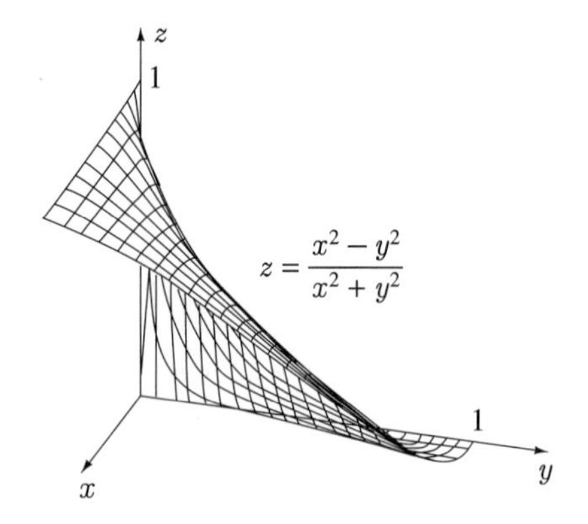

EXAMPLE 13.3

Evaluate

$$\lim_{(x,y)\to(0,0)} \frac{x^2 - y^2}{x + y}$$

if it exists.

SOLUTION Because points on the line $y = -x$ are not within the domain of definition of the function, we can write that

$$\lim_{(x,y)\to(0,0)} \frac{x^2 - y^2}{x + y} = \lim_{(x,y)\to(0,0)} \frac{(x-y)(x+y)}{x+y} = \lim_{(x,y)\to(0,0)} (x - y) = 0.$$

■

The concept of continuity for multivariable functions is contained in the following definition.

Definition 13.2

A function $f(x, y)$ is said to be **continuous** at a point (x_0, y_0) if

$$\lim_{(x,y)\to(x_0,y_0)} f(x, y) = f(x_0, y_0). \tag{13.3}$$

Just as Definition 2.1 for continuity of a function of only one variable contains three conditions, so too does Definition 13.2. It demands that:

(1) $f(x, y)$ be defined at (x_0, y_0);

(2) $\lim_{(x,y)\to(x_0,y_0)} f(x, y)$ exist;

(3) the numbers in (1) and (2) be the same.

Geometrically, a function $f(x, y)$ is continuous at a point (x_0, y_0) if the surface $z = f(x, y)$ does not have a separation at the point $(x_0, y_0, f(x_0, y_0))$. The function $f(x, y) = (x^2 - y^2)/(x^2 + y^2)$ in Figure 13.8 is discontinuous at $(0, 0)$.

The function $f(x, y) = 1 - e^{-1/(x^2+y^2)}$ is discontinuous at $(0, 0)$ since it is undefined for $x = 0$ and $y = 0$. The surface has a hole at $(0, 0)$ (Figure 13.9). The function $f(x, y) = \text{sgn}[(x - 1)^2 + (y - 2)^2]$ (see Exercise 36 in Section 2.4) is discontinuous at $(1, 2)$; it has value 0 at $x = 1$ and $y = 2$, and value 1 for all other values of x and y (Figure 13.10).

FIGURE 13.9

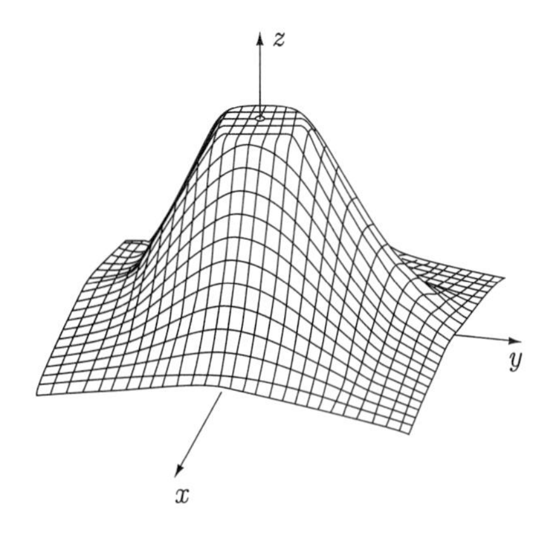

FIGURE 13.10

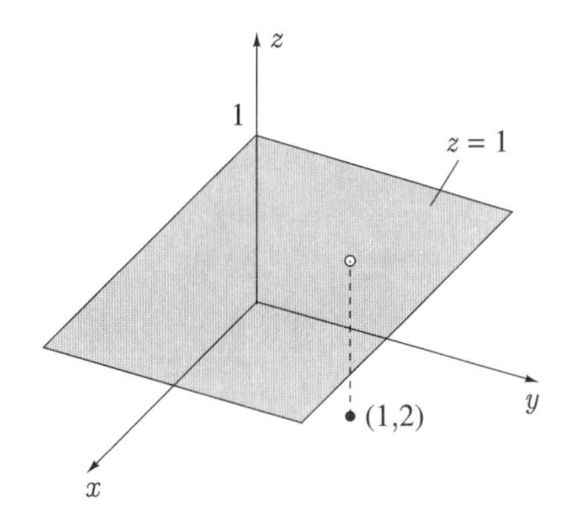

EXERCISES 13.2

In Exercises 1–20 evaluate the limit, if it exists.

1. $\displaystyle\lim_{(x,y)\to(2,-3)} \frac{x^2-1}{x+y}$

2. $\displaystyle\lim_{(x,y)\to(1,1)} \frac{x^3+2y^3}{x^3+4y^3}$

3. $\displaystyle\lim_{(x,y)\to(3,2)} \frac{2x-3y}{x+y}$

4. $\displaystyle\lim_{(x,y,z)\to(2,3,-1)} \frac{xyz}{x^2+y^2+z^2}$

5. $\displaystyle\lim_{(x,y)\to(1,0)} \frac{x}{y}$

6. $\displaystyle\lim_{(x,y,z)\to(0,\pi/2,1)} \operatorname{Tan}^{-1}[x/(yz)]$

7. $\displaystyle\lim_{(x,y,z)\to(0,\pi/2,1)} \operatorname{Tan}^{-1}(yz/x)$

8. $\displaystyle\lim_{(x,y,z)\to(0,\pi/2,1)} \operatorname{Tan}^{-1}|yz/x|$

9. $\displaystyle\lim_{(x,y)\to(3,4)} \frac{|x^2-y^2|}{x^2-y^2}$

10. $\displaystyle\lim_{(x,y)\to(3,4)} \frac{|x^2+y^2|}{x^2+y^2}$

11. $\displaystyle\lim_{(x,y)\to(2,1)} \frac{x^2-y^2}{x-y}$

12. $\displaystyle\lim_{(x,y)\to(2,2)} \frac{x^2-y^2}{x-y}$

13. $\displaystyle\lim_{(x,y)\to(0,0)} \frac{x^2-y^2}{x-y}$

14. $\displaystyle\lim_{(x,y)\to(0,0)} \frac{x-y}{x+y}$

15. $\displaystyle\lim_{(x,y,z)\to(0,0,0)} \frac{x^2-y^2}{y^2+z^2+1}$

16. $\displaystyle\lim_{(x,y)\to(2,1)} \frac{(x-2)^2(y+1)}{x-2}$

17. $\displaystyle\lim_{(x,y,z)\to(1,1,1)} |2x-y-z|$

18. $\displaystyle\lim_{(x,y)\to(0,0)} \frac{3x^3-y^3}{2x^3+4y^3}$

19. $\displaystyle\lim_{(x,y)\to(0,0)} \operatorname{Sec}^{-1}\left(\frac{-1}{x^2+y^2}\right)$

20. $\displaystyle\lim_{(x,y)\to(0,0)} \operatorname{Sec}^{-1}(x^2+y^2)$

In Exercises 21–26 find all points of discontinuity for the function.

21. $f(x,y) = \dfrac{x^2-1}{x+y}$

22. $f(x,y) = \dfrac{xy}{x^2+y^2}$

23. $f(x,y) = \dfrac{1}{1-x^2-y^2}$

24. $f(x,y,z) = \dfrac{1}{xyz}$

25. $f(x,y) = |x-y|$

26. $f(x,y) = \dfrac{x+y}{x^2y+xy^2}$

27. Evaluate $\displaystyle\lim_{(x,y)\to(a,a)} \left[\cos(x+y) - \sqrt{1-\sin^2(x+y)}\right]$ where $0 \le a \le \pi/2$.

In Exercises 28–31 evaluate the limit, if it exists.

28. $\displaystyle\lim_{(x,y)\to(0,0)} \frac{x^4+y^2}{x^4-y^2}$ Hint: Approach $(0,0)$ along parabolas.

29. $\displaystyle\lim_{(x,y)\to(1,1)} \frac{x^2-2x+y^2+2y-2}{x^2-y^2-2x+2y}$ Hint: Approach $(1,1)$ along straight lines.

30. $\displaystyle\lim_{(x,y)\to(1,0)} \frac{\sqrt{x+y}-\sqrt{x-y}}{y}$

31. $\displaystyle\lim_{(x,y)\to(0,0)} \frac{\sin(x^2+y^2)}{x^2+y^2}$

32. **(a)** Does the limit $\displaystyle\lim_{(x,y)\to(1,1)} \frac{\sin(x-y)}{x-y}$ exist? Explain. Is the function continuous at $(1,1)$?
    (b) If we define the function everywhere by giving it the value 1 along the line $y = x$, does the limit of the function exist at $(1,1)$? Is the function continuous at $(1,1)$?

33. Give a mathematical definition for $\displaystyle\lim_{(x,y,z)\to(x_0,y_0,z_0)} f(x,y,z) = L.$

34. Is the following statement true or false? If a function $f(x,y)$ is undefined at every point on a curve C, then for any point (x_0,y_0) on C, $\displaystyle\lim_{(x,y)\to(x_0,y_0)} f(x,y)$ does not exist. Explain. Give an example.

35. Let $f(x,y) = \begin{cases} \dfrac{x^2y^2}{x^4+y^4} & \text{if } (x,y) \ne (0,0) \\ 0 & \text{if } (x,y) = (0,0). \end{cases}$
    (a) Show that $f(x,y)$ is continuous in each variable separately at $(0,0)$. In other words, show that

$f(x,0)$ and $f(0,y)$ are continuous at $x = 0$ and $y = 0$.

(b) Show that $f(x,y)$ is not continuous at $(0,0)$.

36. Prove that:

(a) $\lim_{(x,y)\to(0,0)} (xy + 5) = 5$

(b) $\lim_{(x,y)\to(1,1)} (x^2 + 2xy + 5) = 8$

SECTION 13.3

Partial Derivatives

We now define partial derivatives of multivariable functions and interpret these derivatives algebraically and geometrically.

Definition 13.3

The **partial derivative** of a function $f(x,y)$ with respect to x is

$$\frac{\partial f}{\partial x} = \lim_{\Delta x \to 0} \frac{f(x + \Delta x, y) - f(x,y)}{\Delta x}, \qquad (13.4)$$

and the partial derivative with respect to y is

$$\frac{\partial f}{\partial y} = \lim_{\Delta y \to 0} \frac{f(x, y + \Delta y) - f(x,y)}{\Delta y}. \qquad (13.5)$$

It is evident from 13.4 that the partial derivative of $f(x,y)$ with respect to x is simply the ordinary derivative of $f(x,y)$ with respect to x, where y is considered a constant. Similarly, $\partial f/\partial y$ is the ordinary derivative of $f(x,y)$ with respect to y, holding x constant.

For the partial derivative of a function of more than two independent variables, we again permit one variable to vary, but hold all others constant. For example, the partial derivative of $f(x,y,z,t,\ldots)$ with respect to z is

$$\frac{\partial f}{\partial z} = \lim_{\Delta z \to 0} \frac{f(x,y,z + \Delta z, t, \ldots) - f(x,y,z,t,\ldots)}{\Delta z}. \qquad (13.6)$$

Hence, we differentiate with respect to z while treating $x, y, t, \ldots$ as constants.

Other notations for the partial derivative are common. In particular, for $\partial f/\partial x$ when $z = f(x,y)$, there are also

$$\frac{\partial z}{\partial x}, f_x, z_x, \left.\frac{\partial f}{\partial x}\right)_y \quad \text{and} \quad \left.\frac{\partial z}{\partial x}\right)_y,$$

the last two indicating that the variable y is held constant when differentiation with respect to x is performed.

EXAMPLE 13.4

Find $\partial z/\partial x$ and $\partial z/\partial y$ if $z = f(x, y) = x^2y^3 + e^{xy}$.

SOLUTION For this function,

$$\frac{\partial z}{\partial x} = 2xy^3 + ye^{xy} \quad \text{and} \quad \frac{\partial z}{\partial y} = 3x^2y^2 + xe^{xy}.$$

■

EXAMPLE 13.5

Find $\partial f/\partial x$ at the point $(1, 2, 3)$ if $f(x, y, z) = x^2/y^4 + 3xz + 4$.

SOLUTION Since $\partial f/\partial x = 2x/y^4 + 3z$,

$$\frac{\partial f}{\partial x}_{|(1,2,3)} = 2(1)/2^4 + 3(3) = \frac{73}{8}.$$

■

For a function $y = f(x)$ of one variable, we defined differentials dx and dy in such a way that the derivative dy/dx could be thought of as a quotient. This is *not* done for functions of more than one variable. Although we write the partial derivative $\frac{\partial f}{\partial x}$ in the form $\partial f/\partial x$ (for typographical reasons), we *never* consider it as a quotient.

Algebraically, the partial derivative $\partial f/\partial x$ represents the rate of change of $f(x, y, ...)$ with respect to x when all other variables in $f(x, y, ...)$ are held constant. For instance, $V = \pi r^2 h/3$ represents the volume of a right-circular cone with height h and radius r, and therefore $\partial V/\partial r = 2\pi r h/3$ represents the rate of change of the volume of the cone as the base radius changes and the height remains fixed.

We can interpret the partial derivative of a function geometrically when the function can be interpreted geometrically, namely, when there are only two independent variables. Consider, then, a function $f(x, y)$ that is represented geometrically as a surface $z = f(x, y)$ in Figure 13.11. If we intersect this surface with a plane $y = y_0$ = a constant, we obtain a curve with equations

$$y = y_0, \qquad z = f(x, y_0). \tag{13.7}$$

Because this curve lies in the plane $y = y_0$, we can talk about its tangent line at the point (x_0, y_0, z_0), where $z_0 = f(x_0, y_0)$. The slope of this tangent is the derivative of z with respect to x, but because y is being held constant at y_0, it must be the partial derivative of z with respect to x. In other words, the slope of the tangent line to the curve in Figure 13.11 at the point (x_0, y_0, z_0) is $\partial f/\partial x_{|(x_0, y_0)}$. Similarly, the partial derivative $\partial f/\partial y$ evaluated at (x_0, y_0) represents the slope of the tangent line to the curve of intersection of $z = f(x, y)$ and the plane $x = x_0$ at the point (x_0, y_0, z_0).

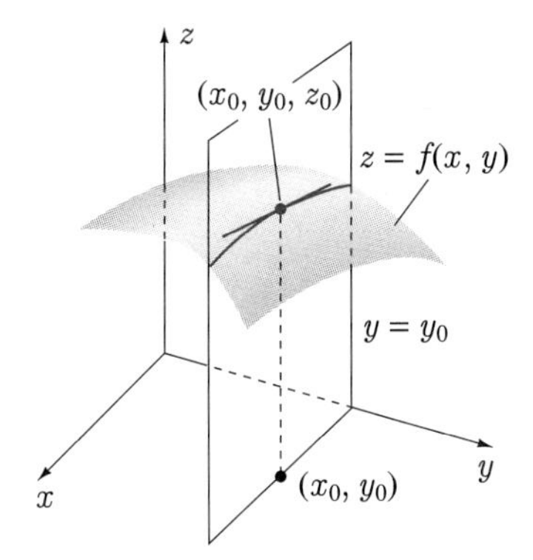

FIGURE 13.11

EXERCISES 13.3

In Exercises 1–20 evaluate $\partial f/\partial x$ and $\partial f/\partial y$.

1. $f(x, y) = x^3y^2 + 2xy$
2. $f(x, y) = 3xy - 4x^4y^4$
3. $f(x, y) = x^4/y^3$
4. $f(x, y) = x/(x + y) - x/y$
5. $f(x, y) = x/(2x^2 + y)$
6. $f(x, y) = \sin(xy)$
7. $f(x, y) = x\cos(x + y)$
8. $f(x, y) = \sqrt{x^2 + y^2}$
9. $f(x, y) = x\sqrt{x^2 - y^2}$
10. $f(x, y) = \tan(2x^2 + y^2)$
11. $f(x, y) = e^{x+y}$
12. $f(x, y) = e^{xy}$
13. $f(x, y) = xye^{xy}$
14. $f(x, y) = \ln(x^2 + y^2)$
15. $f(x, y) = (x + 1)\ln(xy)$
16. $f(x, y) = \sin(ye^x)$
17. $f(x, y) = \mathrm{Tan}^{-1}(x/y)$
18. $f(x, y) = \sqrt[3]{1 - \cos^3(x^2y)}$
19. $f(x, y) = \dfrac{\sin x}{\cos y}$
20. $f(x, y) = \ln\left(\sec\sqrt{x + y}\right)$

In Exercises 21–30 evaluate the indicated derivative.

21. $\partial f/\partial x$ if $f(x, y, z) = xyze^{x^2+y^2}$
22. $\partial f/\partial z$ if $f(x, z) = \mathrm{Tan}^{-1}[1/(x^2 + z^2)]$
23. $\partial f/\partial y$ at $(1, 1, 0)$ if $f(x, y, z) = xy(x^2 + y^2 + z^2)^{1/3}$
24. $\partial f/\partial x$ at $(1, -1, 1, -1)$ if $f(x, y, z, t) = zt/(x^2 + y^2 - t^2)$
25. $\partial f/\partial t$ if $f(x, y, t) = x\sqrt{t^2 - y^2}/t^2 + (x/y)\mathrm{Sec}^{-1}(t/3)$
26. $\partial f/\partial x$ if $f(x, y, z) = \mathrm{Cot}^{-1}(1 + x + y + z)$
27. $\partial f/\partial y$ at $(1, 2, 3)$ if $f(x, y, t) = \mathrm{Sin}^{-1}(xyt)/\mathrm{Cos}^{-1}(xyt)$
28. $\partial f/\partial x$ if $f(x, y, z) = x^3/y + x\sin(yz/x)$
29. $\partial f/\partial t$ if $f(x, y, z, t) = xyz\ln(x^2 + y^2 + z^2)$
30. $\partial f/\partial z$ if $f(x, z) = (z^2/2)\mathrm{Sin}^{-1}(x/z) + (x/2)\sqrt{z^2 - x^2}$
31. If $f(x, y) = x^3y/(x - y)$, show that
$$x\frac{\partial f}{\partial x} + y\frac{\partial f}{\partial y} = 3f(x, y).$$
32. If $f(x, y, z) = (x^4 + y^4 + z^4)/(xyz)$, show that
$$x\frac{\partial f}{\partial x} + y\frac{\partial f}{\partial y} + z\frac{\partial f}{\partial z} = f(x, y, z).$$
33. If $f(x, y, z) = (x^2 + y^2)\cos[(y + z)/x]$, show that
$$x\frac{\partial f}{\partial x} + y\frac{\partial f}{\partial y} + z\frac{\partial f}{\partial z} = 2f(x, y, z).$$
34. To evaluate $\partial f/\partial x$ for $f(x, y)$ at the point $(1, 2)$ which of the following are acceptable:
   (a) differentiate $f(x, y)$ with respect to x holding y constant, and then set $x = 1$ and $y = 2$
   (b) set $x = 1$ and $y = 2$, and then differentiate with respect to x

(c) set $y = 2$, differentiate with respect to x, and set $x = 1$

(d) set $x = 1$, differentiate with respect to x, and set $y = 2$

35. Temperature at points (x, y) in a semicircular plate defined by $x^2 + y^2 \leq 4,\ y \geq 0$ is given by $T(x, y) = 16x^2 - 24xy + 40y^2$. Find, if possible: **(a)** $T_x(1,1)$ **(b)** $T_y(1,1)$ **(c)** $T_x(1,0)$ **(d)** $T_y(1,0)$ **(e)** $T_x(0,2)$ **(f)** $T_y(0,2)$

36. Can you find a function $f(x, y)$ so that $f_x(x, y) = 2x - 3y$ and $f_y(x, y) = 3x + 4y$?

37. Suppose a, b, and c are the lengths of the sides of a triangle and A, B, and C are the opposite angles. Find: **(a)** $a_A(b, c, A)$ **(b)** $A_a(a, b, c)$ **(c)** $a_b(b, c, A)$ **(d)** $A_b(a, b, c)$

38. The equation of continuity for three-dimensional unsteady flow of a compressible fluid is

$$\frac{\partial \rho}{\partial t} + \frac{\partial}{\partial x}(\rho u) + \frac{\partial}{\partial y}(\rho v) + \frac{\partial}{\partial z}(\rho w) = 0,$$

where $\rho(x, y, z, t)$ is the density of the fluid, and $u\hat{\mathbf{i}} + v\hat{\mathbf{j}} + w\hat{\mathbf{k}}$ is the velocity of the fluid at position (x, y, z) and time t. Determine whether the continuity equation is satisfied if:

(a) $\rho = \text{constant}$, $u = (2x^2 - xy + z^2)t$, $v = (x^2 - 4xy + y^2)t$, $w = (-2xy - yz + y^2)t$

(b) $\rho = xy + zt$, $u = x^2y + t$, $v = y^2z - 2t^2$, $w = 5x + 2z$

39. In complex variable theory, two functions $u(x, y)$ and $v(x, y)$ are said to be *harmonic conjugates* in a region R if in R they satisfy the Cauchy-Riemann equations

$$\frac{\partial u}{\partial x} = \frac{\partial v}{\partial y}, \qquad \frac{\partial v}{\partial x} = -\frac{\partial u}{\partial y}.$$

Show that the following pairs of functions are harmonic conjugates:

(a) $u(x, y) = -3xy^2 + y + x^3$, $v(x, y) = 3x^2y - y^3 - x + 5$

(b) $u(x, y) = (x^2 + x + y^2)/(x^2 + y^2)$, $v(x, y) = -y/(x^2 + y^2)$

(c) $u(x, y) = e^x(x \cos y - y \sin y)$, $v(x, y) = e^x(x \sin y + y \cos y)$

40. If r and θ are polar coordinates, then the Cauchy-Riemann equations in Exercise 39 for functions $u(r, \theta)$ and $v(r, \theta)$ take the form

$$\frac{\partial u}{\partial r} = \frac{1}{r}\frac{\partial v}{\partial \theta}, \qquad \frac{1}{r}\frac{\partial u}{\partial \theta} = -\frac{\partial v}{\partial r}, \qquad r \neq 0.$$

Show that the following pairs of functions satisfy these equations:

(a) $u(r, \theta) = (r^2 + r\cos\theta)/(1 + r^2 + 2r\cos\theta)$, $v(r, \theta) = r\sin\theta/(1 + r^2 + 2r\cos\theta)$

(b) $u(r, \theta) = \sqrt{r}\cos(\theta/2)$, $v(r, \theta) = \sqrt{r}\sin(\theta/2)$

(c) $u(r, \theta) = \ln r$, $v(r, \theta) = \theta$

SECTION 13.4

Gradients

Suppose a function $f(x, y, z)$ is defined at each point in some region of space, and that at each point of the region all three partial derivatives

$$\frac{\partial f}{\partial x}, \quad \frac{\partial f}{\partial y}, \quad \frac{\partial f}{\partial z}$$

exist. For example, if $f(x, y, z)$ represents the present temperature at each point in the room in which you are working, then these derivatives represent the rates of change of temperature in directions parallel to the x-, y-, and z-axes, respectively. There is a particular combination of these derivatives that will prove very useful in later work. This combination is contained in the following definition.

Definition 13.4

If a function $f(x, y, z)$ has partial derivatives $\partial f/\partial x$, $\partial f/\partial y$, and $\partial f/\partial z$ at each point in some region D of space, then at each point in D we define a vector called the **gradient** of $f(x, y, z)$, written grad f or ∇f, by

$$\text{grad} f = \nabla f = \frac{\partial f}{\partial x}\hat{\mathbf{i}} + \frac{\partial f}{\partial y}\hat{\mathbf{j}} + \frac{\partial f}{\partial z}\hat{\mathbf{k}}. \tag{13.8}$$

For a function $f(x, y)$ of only two independent variables, we have

$$\nabla f = \frac{\partial f}{\partial x}\hat{\mathbf{i}} + \frac{\partial f}{\partial y}\hat{\mathbf{j}}. \tag{13.9}$$

EXAMPLE 13.6

If $f(x, y, z) = x^2yz - 2x/y$, find ∇f at $(1, -1, 3)$.

SOLUTION Since

$$\nabla f = (2xyz - 2/y)\hat{\mathbf{i}} + (x^2z + 2x/y^2)\hat{\mathbf{j}} + (x^2y)\hat{\mathbf{k}},$$

then

$$\nabla f_{|(1,-1,3)} = -4\hat{\mathbf{i}} + 5\hat{\mathbf{j}} - \hat{\mathbf{k}}.$$

■

EXAMPLE 13.7

If $f(x, y, z) = \text{Tan}^{-1}(xy/z)$, what is ∇f?

SOLUTION

$$\begin{aligned}\nabla f &= \left\{\frac{1}{1+(xy/z)^2}\left(\frac{y}{z}\right)\right\}\hat{\mathbf{i}} + \left\{\frac{1}{1+(xy/z)^2}\left(\frac{x}{z}\right)\right\}\hat{\mathbf{j}} + \left\{\frac{1}{1+(xy/z)^2}\left(\frac{-xy}{z^2}\right)\right\}\hat{\mathbf{k}} \\ &= \frac{yz}{z^2+x^2y^2}\hat{\mathbf{i}} + \frac{xz}{z^2+x^2y^2}\hat{\mathbf{j}} - \frac{xy}{z^2+x^2y^2}\hat{\mathbf{k}} \\ &= (yz\hat{\mathbf{i}} + xz\hat{\mathbf{j}} - xy\hat{\mathbf{k}})/(z^2 + x^2y^2).\end{aligned}$$

■

Gradients arise in a multitude of applications in applied mathematics—heat conduction, electromagnetic theory, and fluid flow, to name a few— and two of the properties that make them so indispensable are discussed in detail in Sections 13.8 and 13.9. Examples 13.8 and 13.9 suggest these properties, but we make no attempt at a complete discussion here. For the moment we simply want you to be familiar with the definition of gradients and be able to calculate them.

EXAMPLE 13.8

The equation $F(x,y,z) = 0$, where $F(x,y,z) = x^2 + y^2 + z^2 - 4$, defines a sphere of radius 2 centred at the origin. Show that the gradient vector ∇F at any point on the sphere is perpendicular to the sphere.

SOLUTION If $P(x,y,z)$ is any point on the sphere (Figure 13.12), then the position vector $\mathbf{r} = x\hat{\mathbf{i}} + y\hat{\mathbf{j}} + z\hat{\mathbf{k}}$ from the origin to P is clearly perpendicular to the sphere. On the other hand,

$$\nabla F = 2x\hat{\mathbf{i}} + 2y\hat{\mathbf{j}} + 2z\hat{\mathbf{k}} = 2\mathbf{r}.$$

Consequently, at any point P on the sphere, ∇F is also perpendicular to the sphere.

FIGURE 13.12

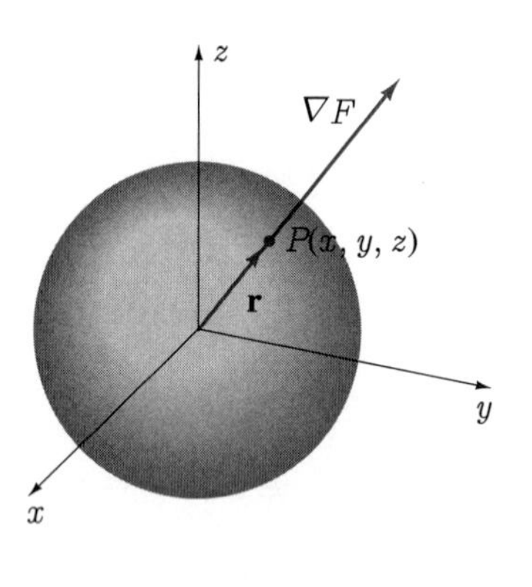

■

This example suggests that gradients may be useful in finding perpendiculars to surfaces (and, as we will see, perpendiculars to curves).

EXAMPLE 13.9

The function $f(x,y) = 2x^2 - 4x + 3y^2 + 2y + 6$ is defined at every point in the xy-plane. If we start at the origin (0, 0) and move along the positive x-axis, the rate of change of the function is $f_x(0,0) = -4$; if we move along the y-axis, the rate of change is $f_y(0,0) = 2$. Calculate the rate of change of $f(x,y)$ at (0, 0) if we move toward the point (1, 1) along the line $y = x$, and show that it is equal to the component of $\nabla f_{|(0,0)}$ in the direction $\mathbf{v} = (1,1)$.

SOLUTION The difference in the values of $f(x,y)$ at any point (x,y) and (0, 0) is $f(x,y) - f(0,0) = (2x^2 - 4x + 3y^2 + 2y + 6) - (6) = 2x^2 - 4x + 3y^2 + 2y$. If we divide this by the length of the line joining (0, 0) and (x,y), we obtain

$$\frac{f(x,y) - f(0,0)}{\sqrt{x^2 + y^2}} = \frac{2x^2 - 4x + 3y^2 + 2y}{\sqrt{x^2 + y^2}}.$$

The limit of this quotient as (x,y) approaches (0, 0) along the line $y = x$ should yield the required rate of change. We therefore set $y = x$ and take the limit as x approaches zero through positive numbers:

$$\lim_{x\to0^+} \frac{2x^2 - 4x + 3x^2 + 2x}{\sqrt{x^2 + x^2}} = \lim_{x\to0^+} \frac{x(5x - 2)}{\sqrt{2}x} = -\sqrt{2}.$$

This quantity, then, is the rate of change of $f(x,y)$ at (0, 0) along the line $y = x$ toward the point $(1,1)$.

On the other hand,

$$\nabla f_{|(0,0)} = (4x - 4, 6y + 2)_{|(0,0)} = (-4, 2),$$

and the component of this vector in the direction $\mathbf{v} = (1,1)$ is

$$\nabla f_{|(0,0)} \cdot \hat{\mathbf{v}} = (-4, 2) \cdot \frac{(1,1)}{\sqrt{2}} = \frac{-4+2}{\sqrt{2}} = -\sqrt{2}.$$

■

This example indicates that gradients may be useful in calculating rates of change of functions in directions other than those parallel to the coordinate axes.

EXAMPLE 13.10

The electrostatic potential at a point (x,y,z) in space due to a charge q fixed at the origin is given by

$$V = \frac{q}{4\pi\epsilon_0 r},$$

where $r = \sqrt{x^2 + y^2 + z^2}$. If a second charge Q is placed at (x,y,z), it experiences a force $\mathbf{F}$, where

$$\mathbf{F} = \frac{qQ}{4\pi\epsilon_0 r^3}\mathbf{r},$$

where $\mathbf{r} = x\hat{\mathbf{i}} + y\hat{\mathbf{j}} + z\hat{\mathbf{k}}$. Show that $\mathbf{F} = -Q\nabla V$.

SOLUTION

$$\begin{aligned}
\nabla V &= \nabla\left(\frac{q}{4\pi\epsilon_0 r}\right) = \frac{q}{4\pi\epsilon_0}\nabla\left(\frac{1}{\sqrt{x^2+y^2+z^2}}\right) \\
&= \frac{q}{4\pi\epsilon_0}\left\{\frac{-x}{(x^2+y^2+z^2)^{3/2}}\hat{\mathbf{i}} + \frac{-y}{(x^2+y^2+z^2)^{3/2}}\hat{\mathbf{j}} + \frac{-z}{(x^2+y^2+z^2)^{3/2}}\hat{\mathbf{k}}\right\} \\
&= \frac{-q}{4\pi\epsilon_0 r^3}(x\hat{\mathbf{i}} + y\hat{\mathbf{j}} + z\hat{\mathbf{k}}) = \frac{-q}{4\pi\epsilon_0 r^3}\mathbf{r} = -\frac{\mathbf{F}}{Q}
\end{aligned}$$

■

EXERCISES 13.4

In Exercises 1–10 find the gradient of the function.

1. $f(x,y,z) = x^2y + xz + yz^2$
2. $f(x,y,z) = x^2yz$
3. $f(x,y,z) = x^2y/z - 2xz^6$
4. $f(x,y) = x^2y + xy^2$
5. $f(x,y) = \sin(x+y)$
6. $f(x,y,z) = \mathrm{Tan}^{-1}(xyz)$
7. $f(x,y) = \mathrm{Tan}^{-1}(y/x)$
8. $f(x,y,z) = e^{x+y+z}$
9. $f(x,y) = 1/(x^2 + y^2)$

10. $f(x,y,z) = 1/\sqrt{x^2+y^2+z^2}$

In Exercises 11–15 find the gradient of the function at the point.

11. $f(x,y) = xy + x + y$ at $(1,3)$

12. $f(x,y,z) = \cos(x+y+z)$ at $(-1,1,1)$

13. $f(x,y,z) = (x^2+y^2+z^2)^2$ at $(0,3,6)$

14. $f(x,y) = e^{-x^2-y^2}$ at $(2,2)$

15. $f(x,y,z) = xy\ln(x+y)$ at $(4,-2)$

16. The equation $F(x,y,z) = Ax + By + Cz + D = 0$ defines a plane in space. Show that at any point on the plane, the vector ∇F is perpendicular to the plane.

17. Use the result of Exercise 16 to illustrate that a vector along the line

$$F(x,y,z) = 2x + 3y - 2z + 4 = 0,$$
$$G(x,y,z) = x - y + 3z + 6 = 0$$

is $\nabla F \times \nabla G$. Find parametric equations for the line.

18. Prove that if $f(x,y,z)$ and $g(x,y,z)$ both have gradients, then $\nabla(fg) = f\nabla g + g\nabla f$.

19. Repeat Example 13.9 for the functions (a) $f(x,y) = x^2 + y^2$ (b) $f(x,y) = 2x^3 - 3y$.

20. The equation $F(x,y) = x^3 + xy + y^4 - 5 = 0$ implicitly defines a curve in the xy-plane. Show that at any point on the curve, ∇F is a normal vector to the curve.

Sketch the surface defined by the equation in Exercises 21 and 22. At what points on the surface is ∇F not defined?

21. $F(x,y,z) = z - \sqrt{x^2+y^2} = 0$

22. $F(x,y,z) = z - |x-y| = 0$

23. If $f(x,y) = 1 - x^2 - y^2$, find ∇f. Find the point (x,y) at which $\nabla f = \mathbf{0}$, and illustrate graphically the nature of the surface $z = f(x,y)$ at this point.

24. If the gradient of a function $f(x,y)$ is $\nabla f = (2xy - y)\hat{\mathbf{i}} + (x^2 - x)\hat{\mathbf{j}}$, what is $f(x,y)$?

25. Repeat Exercise 24 if $\nabla f = (2x/y + 1)\hat{\mathbf{i}} + (-x^2/y^2 + 2)\hat{\mathbf{j}}$.

26. If the gradient of a function $f(x,y,z)$ is $\nabla f = yz\hat{\mathbf{i}} + (xz + 2yz)\hat{\mathbf{j}} + (xy + y^2)\hat{\mathbf{k}}$, what is $f(x,y,z)$?

27. Repeat Exercise 26 if $\nabla f = (x\hat{\mathbf{i}} + y\hat{\mathbf{j}} + z\hat{\mathbf{k}})/\sqrt{x^2+y^2+z^2}$.

28. If $f(x,y)$ and $g(x,y)$ have first partial derivatives in a region R of the xy-plane, and if in R, $\nabla f = \nabla g$, how are $f(x,y)$ and $g(x,y)$ related?

29. If $\nabla f = \mathbf{0}$ for all points in some region R of space, what can we say about $f(x,y,z)$ in R?

30. Show that if the equation $F(x,y) = 0$ implicitly defines a curve C in the xy-plane, then at any point on C, the vector ∇F is perpendicular to C.

SECTION 13.5

Higher-Order Partial Derivatives

If $f(x,y) = x^3y^2 + ye^x$, then

$$\frac{\partial f}{\partial x} = 3x^2y^2 + ye^x$$

and

$$\frac{\partial f}{\partial y} = 2x^3y + e^x.$$

Since each of these partial derivatives is a function of x and y, we can take further partial derivatives. The partial derivative of $\partial f/\partial x$ with respect to x is called the second partial derivative of $f(x,y)$ with respect to x, and is written

$$\frac{\partial}{\partial x}\left(\frac{\partial f}{\partial x}\right) = \frac{\partial^2 f}{\partial x^2} = 6xy^2 + ye^x.$$

Similarly, we have three more second partial derivatives:

$$\frac{\partial}{\partial y}\left(\frac{\partial f}{\partial x}\right) = \frac{\partial^2 f}{\partial y \partial x} = 6x^2 y + e^x,$$
$$\frac{\partial}{\partial x}\left(\frac{\partial f}{\partial y}\right) = \frac{\partial^2 f}{\partial x \partial y} = 6x^2 y + e^x,$$
$$\frac{\partial}{\partial y}\left(\frac{\partial f}{\partial y}\right) = \frac{\partial^2 f}{\partial y^2} = 2x^3.$$

Note that the second partial derivatives $\partial^2 f/\partial x \partial y$ and $\partial^2 f/\partial y \partial x$ are identical. This is not a peculiarity of this function, but according to the following theorem it is what we should expect normally.

Theorem 13.1

If $f(x,y)$, $\partial f/\partial x$, $\partial f/\partial y$, $\partial^2 f/\partial x \partial y$, and $\partial^2 f/\partial y \partial x$ are all defined inside a circle centred at a point P, and are continuous at P, then at P,

$$\frac{\partial^2 f}{\partial x \partial y} = \frac{\partial^2 f}{\partial y \partial x}. \tag{13.10}$$

For most functions with which we will be concerned, this theorem can be extended to say that a mixed partial derivative may be calculated in any order whatsoever. For example, if we require $\partial^{10} f/\partial x^3 \partial y^7$, where $f(x,y) = \ln(y^y) + x^2 y^{10}$, it is advantageous to reverse the order of differentiation,

$$\frac{\partial^{10} f}{\partial x^3 \partial y^7} = \frac{\partial^{10} f}{\partial y^7 \partial x^3} = \frac{\partial^7}{\partial y^7}\left(\frac{\partial^3 f}{\partial x^3}\right) = 0.$$

EXAMPLE 13.11

Show that $f(x,y) = \text{Tan}^{-1}(2xy/(x^2 - y^2))$ satisfies the equation

$$\frac{\partial^2 f}{\partial x^2} + \frac{\partial^2 f}{\partial y^2} = 0$$

at all points in the xy-plane that are not on the lines $y = \pm x$.

SOLUTION Since

$$\begin{aligned}\frac{\partial f}{\partial x} &= \frac{1}{1 + \left(\dfrac{2xy}{x^2 - y^2}\right)^2}\left\{\frac{(x^2 - y^2)(2y) - 2xy(2x)}{(x^2 - y^2)^2}\right\} \\ &= \frac{(x^2 - y^2)^2}{(x^2 - y^2)^2 + 4x^2 y^2}\left\{\frac{2y(x^2 - y^2 - 2x^2)}{(x^2 - y^2)^2}\right\} \\ &= \frac{-2y(x^2 + y^2)}{x^4 - 2x^2 y^2 + y^4 + 4x^2 y^2} = \frac{-2y(x^2 + y^2)}{(x^2 + y^2)^2} = \frac{-2y}{x^2 + y^2},\end{aligned}$$

the second derivative with respect to x is

$$\frac{\partial^2 f}{\partial x^2} = \frac{2y}{(x^2+y^2)^2}(2x) = \frac{4xy}{(x^2+y^2)^2}.$$

Similarly,

$$\frac{\partial f}{\partial y} = \frac{1}{1+\left(\dfrac{2xy}{x^2-y^2}\right)^2}\left\{\frac{(x^2-y^2)(2x)-2xy(-2y)}{(x^2-y^2)^2}\right\}$$

$$= \frac{(x^2-y^2)^2}{(x^2+y^2)^2}\left\{\frac{2x(x^2-y^2+2y^2)}{(x^2-y^2)^2}\right\} = \frac{2x}{x^2+y^2}.$$

Consequently,

$$\frac{\partial^2 f}{\partial y^2} = \frac{-2x}{(x^2+y^2)^2}(2y) = \frac{-4xy}{(x^2+y^2)^2},$$

and addition of these two expressions for $\partial^2 f/\partial x^2$ and $\partial^2 f/\partial y^2$ completes the proof. ■

The equation

$$\frac{\partial^2 f}{\partial x^2} + \frac{\partial^2 f}{\partial y^2} = 0 \qquad (13.11)$$

for a function $f(x,y)$ is one of the most important equations in applied mathematics. It is called Laplace's equation in two variables (x and y). Laplace's equation for a function $f(x,y,z)$ of three variables is

$$\frac{\partial^2 f}{\partial x^2} + \frac{\partial^2 f}{\partial y^2} + \frac{\partial^2 f}{\partial z^2} = 0. \qquad (13.12)$$

A function is said to be **harmonic** in a region R if it satisfies Laplace's equation in R and has continuous second partial derivatives in R. In particular, the function $f(x,y)$ in Example 13.11 is harmonic in any region that does not contain points on the lines $y = \pm x$. In Exercises 29–31, we illustrate various areas of applied mathematics that make use of Laplace's equation.

EXERCISES 13.5

In Exercises 1–20 find the derivative.

1. $\partial^2 f/\partial x^2$ if $f(x,y) = x^2y^2 - 2x^3y$

2. $\partial^3 f/\partial y^3$ if $f(x,y) = 2x/y + 3x^3y^4$

3. $\partial^2 f/\partial z^2$ if $f(x,y,z) = \sin(xyz)$

4. $\partial^2 f/\partial y\partial z$ if $f(x,y,z) = xyze^{x+y+z}$

5. $\partial^2 f/\partial y\partial x$ if $f(x,y) = \sqrt{x^2+y^2}$

6. $\partial^3 f/\partial x^2\partial y$ if $f(x,y) = e^{x+y} - x^2/y^2$

7. $\partial^3 f/\partial y^3$ at $(1,3)$ if $f(x,y) = 3x^3y^3 - 3x/y$

8. $\partial^3 f/\partial x\partial y\partial z$ at $(1,0,-1)$ if $f(x,y,z) = x^2y^2 + x^2z^2 + y^2z^2$

9. $\partial^2 f/\partial x^2$ if $f(x,y) = \sqrt{1-x^2-y^2}$

10. $\partial^2 f/\partial z^2$ if $f(x,y,z) = \ln\sqrt{x^2+y^2+z^2}$

11. $\partial^3 f/\partial x^2\partial y$ if $f(x,y) = x^2e^y + y^2e^x$

12. $\partial^2 f/\partial x^2$ if $f(x,y) = \mathrm{Tan}^{-1}(y/x)$

13. $\partial^3 f/\partial x\partial y^2$ if $f(x,y,z) = \cot(x^2+y^2+z^2)$

14. $\partial^2 f/\partial x\partial y$ at $(-2,-2)$ if $f(x,y) = \mathrm{Sin}^{-1}(x^2+y^2)^{-1}$

15. $\partial^{10} f/\partial x^7 \partial y^3$ if $f(x, y) = x^7 e^x y^2 + 1/y^6$

16. $\partial^8 f/\partial x^8$ if $f(x, y, z) = x^8 y^9 z^{10}$

17. $\partial^6 f/\partial x^2 \partial y^2 \partial z^2$ if $f(x, y, z) = 1/x^2 + 1/y^2 + 1/z^2$

18. $\partial^4 f/\partial x^3 \partial y$ if $f(x, y) = \cos(x + y^3)$

19. $\partial^4 f/\partial x \partial y \partial z \partial t$ if $f(x, y, z, t) = \sqrt{x^2 + y^2 + z^2 - t^2}$

20. $\partial^2 f/\partial x \partial y$ if $f(x, y) = \operatorname{Sec}^{-1}(xy)$

21. If $z = x^2 + xy + y^2 \sin(x/y)$, show that

$$x\frac{\partial z}{\partial x} + y\frac{\partial z}{\partial y} = 2z = x^2\frac{\partial^2 z}{\partial x^2} + 2xy\frac{\partial^2 z}{\partial x \partial y} + y^2\frac{\partial^2 z}{\partial y^2}.$$

22. If $u = x + y + ze^{y/x}$, show that

$$x^2\frac{\partial^2 u}{\partial x^2} + y^2\frac{\partial^2 u}{\partial y^2} + z^2\frac{\partial^2 u}{\partial z^2} + 2xy\frac{\partial^2 u}{\partial x \partial y} + 2yz\frac{\partial^2 u}{\partial y \partial z} + 2xz\frac{\partial^2 u}{\partial x \partial z} = 0.$$

In Exercises 23–28 find a region (if possible) in which the function is harmonic.

23. $f(x, y) = x^2 - y^2 + 2xy + y$

24. $f(x, y) = \ln(x^2 + y^2)$

25. $f(x, y) = x^3 y^2 - 3xy$

26. $f(x, y, z) = 3x^2 yz - y^3 z + xy$

27. $f(x, y, z) = 1/\sqrt{x^2 + y^2 + z^2}$

28. $f(x, y, z) = x^3 y^3 z^3$

29. **(a)** Show that the electrostatic potential function $V(x, y, z) = q/(4\pi\epsilon_0 r)$ in Example 13.10 satisfies Laplace's equation 13.12.

(b) If $V(x, y, z)$ represents the electrostatic potential at a point (x, y, z) due to a system of n point charges at points (x_i, y_i, z_i), does $V(x, y, z)$ satisfy Laplace's equation?

30. The gravitational potential at a point (x, y, z) in space due to a uniform spherical mass distribution (mass M) at the origin is defined as $V = GM/r$ where G is a constant and $r = \sqrt{x^2 + y^2 + z^2}$. Show that $V(x, y, z)$ satisfies Laplace's equation 13.12.

31. Figure 13.13 shows a plate bounded by the lines $x = 0$, $y = 0$, $x = 1$, and $y = 1$. Temperature along the first three sides is kept at 0°C, while that along $y = 1$ varies according to $f(x) = \sin(3\pi x) - 2\sin(4\pi x)$, $0 \le x \le 1$. The temperature at any point interior to the plate is then

$$T(x, y) = C(e^{3\pi y} - e^{-3\pi y})\sin(3\pi x) + D(e^{4\pi y} - e^{-4\pi y})\sin(4\pi x),$$

where $C = (e^{3\pi} - e^{-3\pi})^{-1}$ and $D = -2(e^{4\pi} - e^{-4\pi})^{-1}$. Show that $T(x, y)$ is harmonic in the region $0 < x < 1$, $0 < y < 1$, and that it also satisfies the boundary conditions $T(0, y) = 0$, $T(1, y) = 0$, $T(x, 0) = 0$, and $T(x, 1) = f(x)$.

FIGURE 13.13

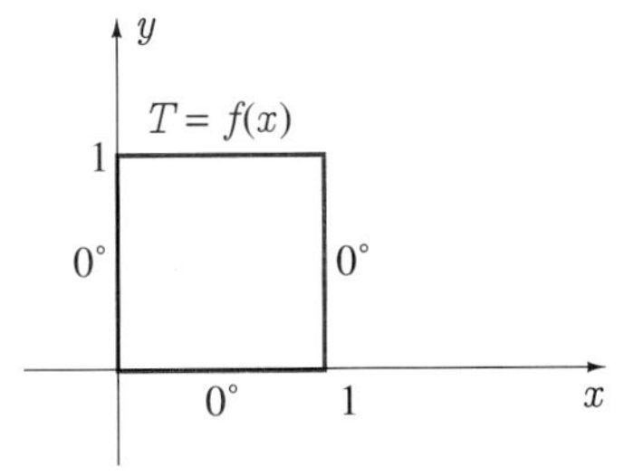

32. If $\partial^2 z/\partial x^2$ is positive for $x_1 \le x \le x_2$ and $y = y_0$, what can you say geometrically about the surface $z = f(x, y)$?

33. Two functions $u(x, y)$ and $v(x, y)$ are said to be harmonic conjugates if they satisfy the Cauchy-Riemann equations of Exercise 39 in Section 13.3. Show that if $u(x, y)$ and $v(x, y)$ are harmonic conjugates and have continuous second partial derivatives in a region R, then each is harmonic in R.

34. If the ends of a taut string are fastened at $x = 0$ and $x = L$ on the x-axis and if the string vibrates in the y-direction, then its displacement $y(x, t)$ must satisfy the one-dimensional wave equation

$$\frac{\partial^2 y}{\partial t^2} = \frac{T}{\rho}\frac{\partial^2 y}{\partial x^2},$$

where T is the constant tension in the string and ρ is its constant density. Determine whether the wave equation is satisfied by:

(a) $y(x, t) = \sin\left(\frac{5\pi x}{L}\right)\cos\left(\frac{5\pi}{L}\sqrt{\frac{T}{\rho}}t\right)$

(b) $y(x, t) = \sum_{n=1}^{50} a_n \sin\left(\frac{n\pi x}{L}\right)\cos\left(\frac{n\pi}{L}\sqrt{\frac{T}{\rho}}t\right)$, where a_n are constants

35. A uniform circular rod has flat ends at $x = 0$ and $x = L$ on the x-axis. If the round side of the rod is insulated, then heat flows in only the x-direction. If, further, the face at $x = L$ is insulated and the face at $x = 0$ is kept at temperature 100°C, the temperature $T(x, t)$ at any point in the rod must satisfy the one-dimensional heat conduction problem:

$$\frac{\partial T}{\partial t} = k\frac{\partial^2 T}{\partial x^2}, \quad k = \text{ a constant},$$
$$T(0,t) = 100,$$
$$T_x(L,t) = 0.$$

Show that for any positive integer n,

$$T(x,t) = 100 - \frac{2L}{(2n-1)\pi}e^{-k[(2n-1)\pi/(2L)]^2 t} \times \sin\left[\left(\frac{2n-1}{2}\right)\frac{\pi x}{L}\right]$$

satisfies the heat conduction problem.

36. **(a)** Show that the function $u(x,y) = x^2 - y^2$ is harmonic in the entire xy-plane.

(b) Use the Cauchy-Riemann equations in Exercise 39 of Section 13.3 to find a function $v(x,y)$ so that u and v are harmonic conjugates.

37. Repeat Exercise 36 if $u(x,y) = e^x \cos y + x$.

38. For what values of n does the function $(x^2 + y^2 + z^2)^n$ satisfy equation 13.12? In what regions are the functions harmonic?

SECTION 13.6

Chain Rules for Partial Derivatives

If $y = f(u)$ and $u = g(x)$, then the chain rule for the derivative dy/dx of the composite function $f[g(x)]$ is

$$\frac{dy}{dx} = \frac{dy}{du}\frac{du}{dx}. \tag{13.13}$$

Equation 13.13 can be extended in terms of more intermediate variables, say $y = f(u)$, $u = g(s)$, $s = h(x)$, in which case

$$\frac{dy}{dx} = \frac{dy}{du}\frac{du}{ds}\frac{ds}{dx}. \tag{13.14}$$

For multivariable functions, variations in chain rules are countless. We discuss two examples in considerable detail, and then show schematically how to obtain the correct chain rule for every possible situation.

Suppose z is a function of u and v and each of u and v is a function of x and y,

$$z = f(u,v), \quad u = g(x,y), \quad v = h(x,y). \tag{13.15}$$

By the substitutions

$$z = f[g(x,y), h(x,y)], \tag{13.16}$$

we could express z as a function of x and y, and could ask for the partial derivative $\partial z/\partial x$. However, if the functions in 13.15 are at all complicated, you can imagine how difficult the composite function in 13.16 might be to differentiate. As a result, we search for an alternative procedure for calculating $\partial z/\partial x$, namely, the appropriate chain rule. It is contained in the following theorem.

Theorem 13.2

Let $u = g(x,y)$ and $v = h(x,y)$ be continuous and have first partial derivatives with respect to x at a point (x,y), and let $z = f(u,v)$ have continuous first partial derivatives inside a circle centred at the point $(u,v) = (g(x,y), h(x,y))$. Then

$$\frac{\partial z}{\partial x} = \frac{\partial z}{\partial u}\frac{\partial u}{\partial x} + \frac{\partial z}{\partial v}\frac{\partial v}{\partial x}. \tag{13.17}$$

Proof This result can be proved in much the same way as chain rule 3.16 was proved in Section 3.6. By Definition 13.3,

$$\frac{\partial z}{\partial x} = \lim_{\Delta x \to 0} \frac{f[g(x+\Delta x, y), h(x+\Delta x, y)] - f[g(x,y), h(x,y)]}{\Delta x}.$$

Now the increment Δx in x produces changes in u and v, which we denote by

$$\Delta u = g(x+\Delta x, y) - g(x,y), \quad \Delta v = h(x+\Delta x, y) - h(x,y).$$

If we write u and v whenever $g(x,y)$ and $h(x,y)$ are evaluated at (x,y), and substitute for $g(x+\Delta x, y)$ and $h(x+\Delta x, y)$ in the definition for $\partial z/\partial x$, then

$$\begin{aligned}
\frac{\partial z}{\partial x} &= \lim_{\Delta x \to 0} \frac{f(u+\Delta u, v+\Delta v) - f(u,v)}{\Delta x} \\
&= \lim_{\Delta x \to 0} \frac{[f(u+\Delta u, v+\Delta v) - f(u, v+\Delta v)] + [f(u, v+\Delta v) - f(u,v)]}{\Delta x} \\
&= \lim_{\Delta x \to 0} \left\{ \frac{f(u+\Delta u, v+\Delta v) - f(u, v+\Delta v)}{\Delta x} + \frac{f(u, v+\Delta v) - f(u,v)}{\Delta x} \right\}.
\end{aligned}$$

We assumed that the derivative

$$\frac{\partial z}{\partial v} = \lim_{\Delta v \to 0} \frac{f(u, v+\Delta v) - f(u,v)}{\Delta v}$$

exists at (u,v), and an equivalent way to express the fact that this limit exists is to say that

$$\frac{f(u, v+\Delta v) - f(u,v)}{\Delta v} = \frac{\partial z}{\partial v} + \epsilon_1,$$

where ϵ_1 must satisfy the condition that $\lim_{\Delta v \to 0} \epsilon_1 = 0$. We can write, therefore, that

$$f(u, v+\Delta v) - f(u,v) = [z_v(u,v) + \epsilon_1]\Delta v.$$

Similarly, we can write that

$$f(u+\Delta u, v+\Delta v) - f(u, v+\Delta v) = [z_u(u, v+\Delta v) + \epsilon_2]\Delta u,$$

where $\lim_{\Delta u \to 0} \epsilon_2 = 0$ (provided Δv is sufficiently small). When these expressions are substituted into the limit for $\partial z/\partial x$, we have

$$\frac{\partial z}{\partial x} = \lim_{\Delta x \to 0} \left\{ [z_u(u, v+\Delta v) + \epsilon_2]\frac{\Delta u}{\Delta x} + [z_v(u,v) + \epsilon_1]\frac{\Delta v}{\Delta x} \right\}.$$

We now examine each part of this limit. Clearly,

$$\lim_{\Delta x \to 0} \frac{\Delta u}{\Delta x} = \frac{\partial u}{\partial x} \quad \text{and} \quad \lim_{\Delta x \to 0} \frac{\Delta v}{\Delta x} = \frac{\partial v}{\partial x}.$$

In addition, because $g(x,y)$ and $h(x,y)$ are continuous, $\Delta u \to 0$ and $\Delta v \to 0$ as $\Delta x \to 0$. Consequently,

$$\lim_{\Delta x \to 0} \epsilon_1 = \lim_{\Delta v \to 0} \epsilon_1 = 0 \quad \text{and} \quad \lim_{\Delta x \to 0} \epsilon_2 = \lim_{\Delta \mathbf{v} \to 0} \epsilon_2 = 0.$$

Finally, because $\partial z/\partial u$ is continuous,

$$\lim_{\Delta x \to 0} z_u(u, v+\Delta v) = \lim_{\Delta v \to 0} z_u(u, v+\Delta v) = z_u(u,v).$$

When all these results are taken into account, we have

$$\frac{\partial z}{\partial x} = z_u(u,v)\frac{\partial u}{\partial x} + z_v(u,v)\frac{\partial v}{\partial x} = \frac{\partial z}{\partial u}\frac{\partial u}{\partial x} + \frac{\partial z}{\partial v}\frac{\partial v}{\partial x},$$

which completes the proof.

Chain rule 13.17 defines $\partial z/\partial x$ in terms of derivatives of the given functions in 13.15. We could be more explicit by indicating which variable is being held constant in each of the five derivatives:

$$\left.\frac{\partial z}{\partial x}\right)_y = \left.\frac{\partial z}{\partial u}\right)_v \left.\frac{\partial u}{\partial x}\right)_y + \left.\frac{\partial z}{\partial v}\right)_u \left.\frac{\partial v}{\partial x}\right)_y. \tag{13.18}$$

From the point of view of rates of change, this result seems quite reasonable. The left side is the rate of change of z with respect to x holding y constant. The first term $(\partial z/\partial u)(\partial u/\partial x)$ accounts for the rate of change of z with respect to those x's that affect z through u. The second term $(\partial z/\partial v)(\partial v/\partial x)$ accounts for the rate of change of z with respect to those x's that affect z through v. The total rate of change is then the sum of the two parts.

Consider now the functional situation

$$z = f(u,v), \quad u = g(x,y,s), \quad v = h(x,y,s), \quad x = p(t), \quad y = q(t), \quad s = r(t). \tag{13.19}$$

By the substitutions

$$z = f[g(p(t),q(t),r(t)),h(p(t),q(t),r(t))], \tag{13.20}$$

we can express z as a function of t alone, and can therefore pose the problem of calculating dz/dt. If we reason as in the last paragraph, the appropriate chain rule for dz/dt must account for all t's affecting z through u and v. We obtain, then,

$$\frac{dz}{dt} = \frac{\partial z}{\partial u}\frac{du}{dt} + \frac{\partial z}{\partial v}\frac{dv}{dt},$$

where we have written du/dt and dv/dt because u and v can be expressed entirely in terms of t:

$$u = g[p(t),q(t),r(t)], \quad v = h[p(t),q(t),r(t)].$$

Chain rules for each of du/dt and dv/dt (similar to 13.17) yield

$$\frac{du}{dt} = \frac{\partial u}{\partial x}\frac{dx}{dt} + \frac{\partial u}{\partial y}\frac{dy}{dt} + \frac{\partial u}{\partial s}\frac{ds}{dt}, \quad \frac{dv}{dt} = \frac{\partial v}{\partial x}\frac{dx}{dt} + \frac{\partial v}{\partial y}\frac{dy}{dt} + \frac{\partial v}{\partial s}\frac{ds}{dt}.$$

Finally, then,

$$\frac{dz}{dt} = \frac{\partial z}{\partial u}\left\{\frac{\partial u}{\partial x}\frac{dx}{dt} + \frac{\partial u}{\partial y}\frac{dy}{dt} + \frac{\partial u}{\partial s}\frac{ds}{dt}\right\} + \frac{\partial z}{\partial v}\left\{\frac{\partial v}{\partial x}\frac{dx}{dt} + \frac{\partial v}{\partial y}\frac{dy}{dt} + \frac{\partial v}{\partial s}\frac{ds}{dt}\right\}, \tag{13.21}$$

which expresses dz/dt in terms of derivatives of the given functions in 13.19.

These two examples illustrate that finding chain rules for complicated composite functions can be very difficult. Fortunately, there is an amazingly simple method that gives the correct chain rule in every situation. The method is not designed to help you understand the chain rule, but to find it quickly. We suggest that you test your understanding by developing a few chain rules in the exercises with a discussion such as in the second example above, and then check your result by the quicker method.

In the first example we represent the functional situation described in 13.15 by the schematic diagram below.

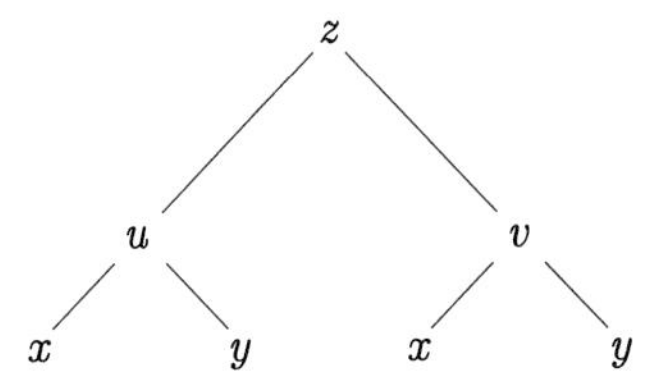

At the top of the diagram is the dependent variable z, which we wish to differentiate. In the line below z are the variables u and v in terms of which z is initially defined. In the line below u and v are x's and y's illustrating that each of u and v is defined in terms of x and y. To obtain the partial derivative $\partial z/\partial x$, we take all possible paths in this diagram from z to x. The two paths are from z through u to x, and from z through v to x. For each straight-line segment in a given path, we differentiate the upper variable with respect to the lower variable and multiply together all such derivatives in that path. The products are then added together to form the complete chain rule. In particular, for the path through u we form the product

$$\frac{\partial z}{\partial u}\frac{\partial u}{\partial x},$$

and for the path through v,

$$\frac{\partial z}{\partial v}\frac{\partial v}{\partial x}.$$

The complete chain rule is then the sum of these products,

$$\frac{\partial z}{\partial x} = \frac{\partial z}{\partial u}\frac{\partial u}{\partial x} + \frac{\partial z}{\partial v}\frac{\partial v}{\partial x},$$

and this result agrees with 13.17. The schematic diagram also indicates which variables are to be held constant in the derivatives on the right (as in 13.18). All variables on the same level are held constant.

For the second example in equations 13.19 the schematic diagram is:

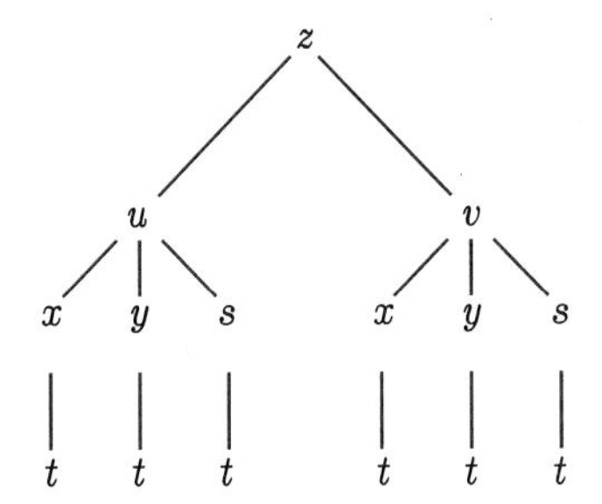

There are six possible paths from z to t so that the chain rule for dz/dt must have six terms. We find

$$\frac{dz}{dt} = \frac{\partial z}{\partial u}\frac{\partial u}{\partial x}\frac{dx}{dt} + \frac{\partial z}{\partial u}\frac{\partial u}{\partial y}\frac{dy}{dt} + \frac{\partial z}{\partial u}\frac{\partial u}{\partial s}\frac{ds}{dt} + \frac{\partial z}{\partial v}\frac{\partial v}{\partial x}\frac{dx}{dt} + \frac{\partial z}{\partial v}\frac{\partial v}{\partial y}\frac{dy}{dt} + \frac{\partial z}{\partial v}\frac{\partial v}{\partial s}\frac{ds}{dt},$$

and this agrees with 13.21. Note too that if when forming a derivative from the schematic diagram, there are two or more lines emanating from a variable, then we obtain a partial derivative; if there is only one line, then we have an ordinary derivative.

EXAMPLE 13.12

Find chain rules for

$$\left.\frac{\partial z}{\partial x}\right)_y \quad \text{and} \quad \left.\frac{\partial z}{\partial y}\right)_x$$

if

$$z = f(r, s, x), \quad r = g(x, y), \quad s = h(x, y).$$

SOLUTION From the schematic diagram below,

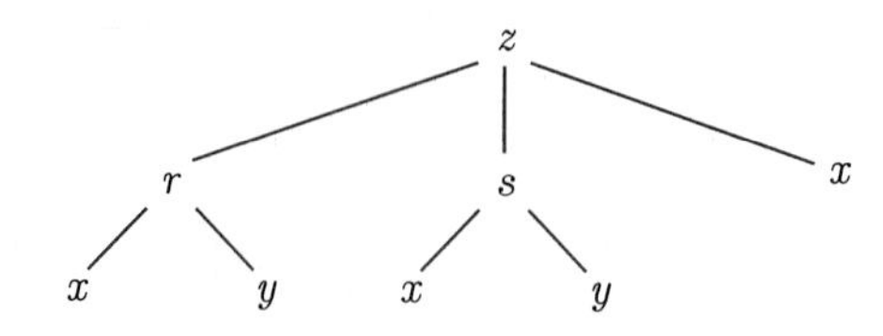

$$\left.\frac{\partial z}{\partial y}\right)_x = \left.\frac{\partial z}{\partial r}\right)_{s,x} \left.\frac{\partial r}{\partial y}\right)_x + \left.\frac{\partial z}{\partial s}\right)_{r,x} \left.\frac{\partial s}{\partial y}\right)_x,$$

$$\left.\frac{\partial z}{\partial x}\right)_y = \left.\frac{\partial z}{\partial r}\right)_{s,x} \left.\frac{\partial r}{\partial x}\right)_y + \left.\frac{\partial z}{\partial s}\right)_{r,x} \left.\frac{\partial s}{\partial x}\right)_y + \left.\frac{\partial z}{\partial x}\right)_{r,s}.$$

■

In the previous example it is essential that we indicate which variables to hold constant in the partial derivatives. If we were to omit these designations, then in the second result we would have a term $\partial z/\partial x$ on both sides of the equation and might be tempted to cancel them. This cannot be done, however, since they are entirely different derivatives. The term $\partial z/\partial x)_y$ indicates the derivative of z with respect to x holding y constant if z were expressed entirely in terms of x and y; the term $\partial z/\partial x)_{r,s}$ indicates the derivative of the given function $f(r, s, x)$ with respect to x holding r and s constant.

EXAMPLE 13.13

Find dz/dt if

$$z = x^3 y^2 + x \sin y + tx, \quad x = 2t + \frac{1}{t}, \quad y = t^2 e^t.$$

SOLUTION From the schematic diagram below,

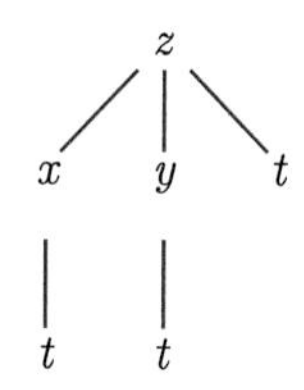

$$\begin{aligned} \frac{dz}{dt} &= \frac{\partial z}{\partial x}\frac{dx}{dt} + \frac{\partial z}{\partial y}\frac{dy}{dt} + \frac{\partial z}{\partial t} \\ &= (3x^2y^2 + \sin y + t)(2 - 1/t^2) + (2x^3 y + x\cos y)(2te^t + t^2 e^t) + x. \end{aligned}$$

■

When a chain rule is used to calculate a derivative, the result usually involves all intermediate variables. For instance, the derivative dz/dt in Example 13.13 involves not only t, but the intermediate variables x and y as well. Were dz/dt required at $t = 1$, values of x and y for $t = 1$ would be calculated —$x(1) = 3$ and $y(1) = e$ —and all three values substituted to obtain

$$\begin{aligned}\frac{dz}{dt}_{|t=1} =& [3(3)^2(e)^2 + \sin(e) + 1](2 - 1) \\ &+ [2(3)^3 e + (3)\cos(e)](2e + e) + 3 \\ =& 1378.6.\end{aligned}$$

EXAMPLE 13.14

Find $\partial^2 z/\partial x^2$ if

$$z = s^2 t + 2\sin t, \quad s = xy - y, \quad t = x^2 + \frac{y}{x}.$$

SOLUTION From the schematic diagram below,

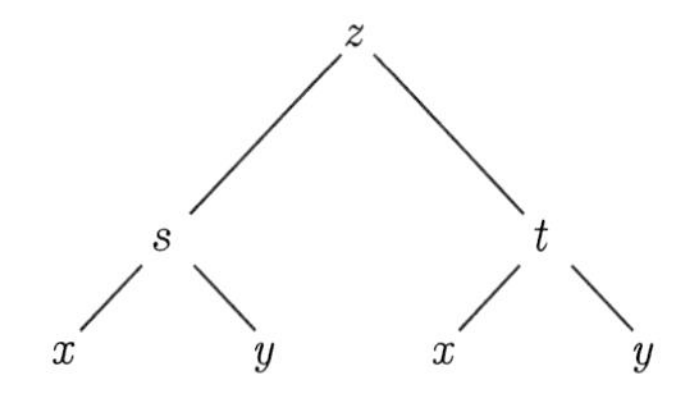

$$\begin{aligned}\frac{\partial z}{\partial x} &= \frac{\partial z}{\partial s}\frac{\partial s}{\partial x} + \frac{\partial z}{\partial t}\frac{\partial t}{\partial x} \\ &= (2st)(y) + (s^2 + 2\cos t)(2x - y/x^2).\end{aligned}$$

Now $\partial z/\partial x$ is a function of s, t, x, and y, and therefore in order to find

$$\frac{\partial^2 z}{\partial x^2} = \frac{\partial}{\partial x}\left(\frac{\partial z}{\partial x}\right),$$

we form a schematic diagram for $\partial z/\partial x$:

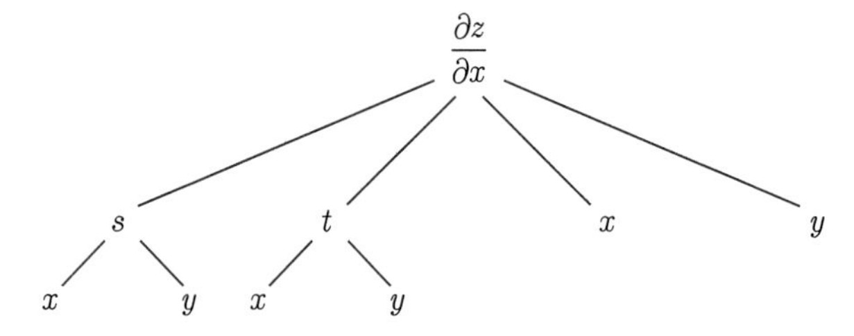

From this schematic diagram, we obtain

$$\begin{aligned}\frac{\partial^2 z}{\partial x^2} =& \frac{\partial}{\partial s}\left(\frac{\partial z}{\partial x}\right)\frac{\partial s}{\partial x} + \frac{\partial}{\partial t}\left(\frac{\partial z}{\partial x}\right)\frac{\partial t}{\partial x} + \frac{\partial}{\partial x}\left(\frac{\partial z}{\partial x}\right)_{s,t,y} \\ =& [2ty + 2s(2x - y/x^2)](y) + [2sy - 2\sin t(2x - y/x^2)](2x - y/x^2) \\ &+ (s^2 + 2\cos t)(2 + 2y/x^3).\end{aligned}$$

■

EXAMPLE 13.15

The temperature T at points in the atmosphere depends on both position (x, y, z) and time t: $T = T(x, y, z, t)$. When a weather balloon is released to take temperature readings, it is not free to take readings at just any point, but only at those points along the path that the winds force the balloon to follow. This path is a curve in space that can be defined parametrically by

$$C: \quad x = x(t), \quad y = y(t), \quad z = z(t), \quad t \geq 0,$$

t again being time. If we substitute from the equations for C into the temperature function, then T becomes a function of t alone,

$$T = T[x(t), y(t), z(t), t],$$

and this function of time describes the temperature at points along the path of the balloon. For the derivative of this function with respect to t, the schematic diagram

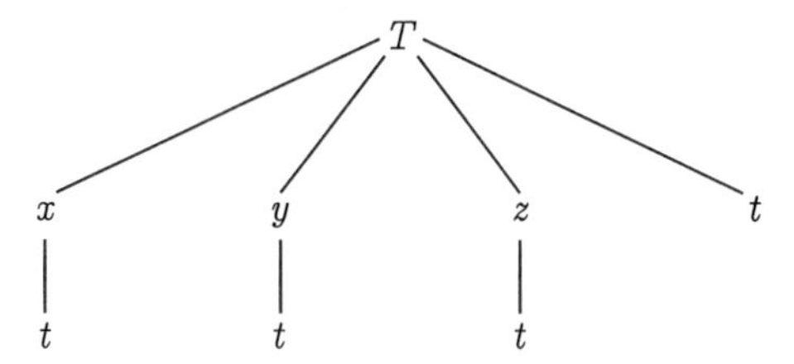

yields

$$\frac{dT}{dt} = \frac{\partial T}{\partial x}\frac{dx}{dt} + \frac{\partial T}{\partial y}\frac{dy}{dt} + \frac{\partial T}{\partial z}\frac{dz}{dt} + \frac{\partial T}{\partial t}.$$

The question we pose is "What is the physical difference between dT/dt and $\partial T/\partial t$?"

SOLUTION The temperature at a point in space is independent of the observer measuring it; hence $T[x(t), y(t), z(t), t]$ is the temperature at points on C as measured by both the balloon and any observer fixed in the xyz-reference system. If, however, these two observers calculate the rate of change of temperature with respect to time at some point (x, y, z) on C, they calculate different results. The observer fixed in the xyz-reference system (not restricted to move along C) calculates the rate of change of T with respect to t as the derivative of the function $T(x, y, z, t)$ partially with respect to t holding x, y, and z constant; i.e., the fixed observer calculates $\partial T/\partial t$ as the rate of change of temperature in time. The balloon, on the other hand, has no alternative but to take temperature readings as it moves along C; thus its measurement of T as a function of t is

$$T[x(t), y(t), z(t), t].$$

Therefore, when the balloon calculates the time variation of temperature, it is calculating dT/dt. It follows, then, that the terms

$$\frac{\partial T}{\partial x}\frac{dx}{dt} + \frac{\partial T}{\partial y}\frac{dy}{dt} + \frac{\partial T}{\partial z}\frac{dz}{dt}$$

describe that part of dT/dt caused by the motion of the balloon through space. ■

Many important applications of the chain rule occur in the field of partial differential equations. To illustrate, we introduce the one-dimensional wave equation in the following example.

EXAMPLE 13.16

The one-dimensional wave equation

$$\frac{\partial^2 y}{\partial t^2} = c^2 \frac{\partial^2 y}{\partial x^2}, \quad c = \text{constant}$$

for functions $y(x,t)$ describes transverse vibrations of taut strings, and longitudinal and rotational vibrations of metal rods. Show that if $f(u)$ and $g(v)$ are twice differentiable functions of u and v, then $y(x,t) = f(x+ct) + g(x-ct)$ satisfies the wave equation.

SOLUTION The schematic diagram below describes the functional situation

$$y = f(u) + g(v)$$

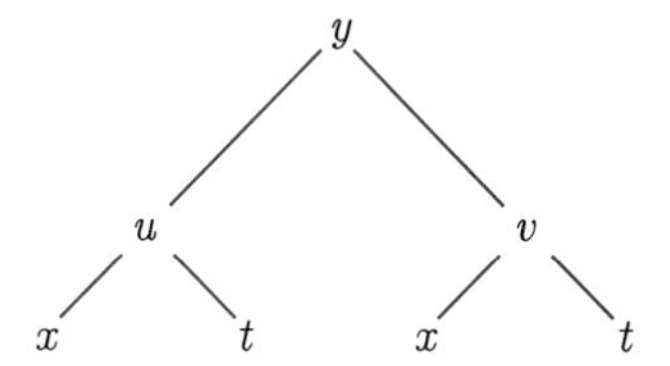

where $u = x + ct$ and $v = x - ct$. The chain rule for $\partial y/\partial t$ is

$$\frac{\partial y}{\partial t} = \frac{\partial y}{\partial u}\frac{\partial u}{\partial t} + \frac{\partial y}{\partial v}\frac{\partial v}{\partial t} = cf'(u) - cg'(v).$$

The schematic diagram for $\partial y/\partial t$ below leads to

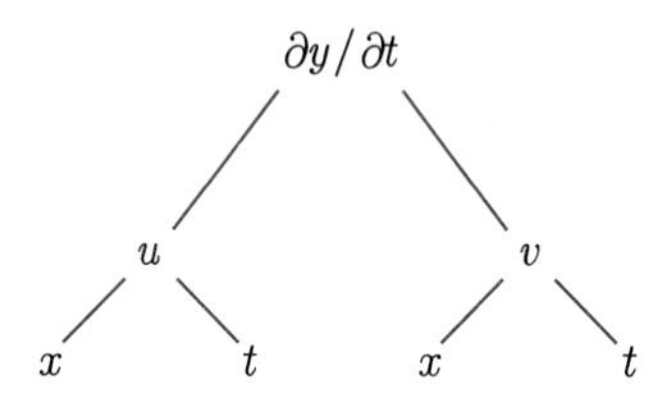

$$\begin{aligned}\frac{\partial^2 y}{\partial t^2} &= \frac{\partial}{\partial u}\left(\frac{\partial y}{\partial t}\right)\frac{\partial u}{\partial t} + \frac{\partial}{\partial v}\left(\frac{\partial y}{\partial t}\right)\frac{\partial v}{\partial t} \\ &= [cf''(u)]c + [-cg''(v)](-c) \\ &= c^2[f''(u) + g''(v)].\end{aligned}$$

A similar calculation gives $\frac{\partial^2 y}{\partial x^2} = f''(u) + g''(v)$. Hence $y(x,t)$ does indeed satisfy the wave equation. ■

Homogeneous Functions

Homogeneous functions arise in numerous areas of applied mathematics. A function

$f(x, y, z)$ is said to be **positively homogeneous** of degree n if for every $t > 0$,

$$f(tx, ty, tz) = t^n f(x, y, z). \tag{13.22}$$

For example, the function $f(x, y, z) = x^2 + y^2 + z^2$ is homogeneous of degree 2; the function $f(x, y) = x^3 \cos(y/x) + x^2 y + xy^2$ is homogeneous of degree 3; and $f(x, y, z, t) = \sqrt{x^2 + z^2}(x^2 y + yt^2)$ is homogeneous of degree 4. Partial derivatives of homogeneous functions satisfy many identities. In particular, their first derivatives satisfy Euler's theorem.

Theorem 13.3

(Euler's Theorem) If f(x,y,z) is positively homogeneous of degree n, and has continuous first partial derivatives, then

$$x\frac{\partial f}{\partial x} + y\frac{\partial f}{\partial y} + z\frac{\partial f}{\partial z} = nf(x, y, z). \tag{13.23}$$

Proof To verify 13.23 we differentiate 13.22 with respect to t holding x, y, and z constant. For the derivative of the left side we introduce variables $u = tx$, $v = ty$, and $w = tz$, and use the schematic diagram below. The result is

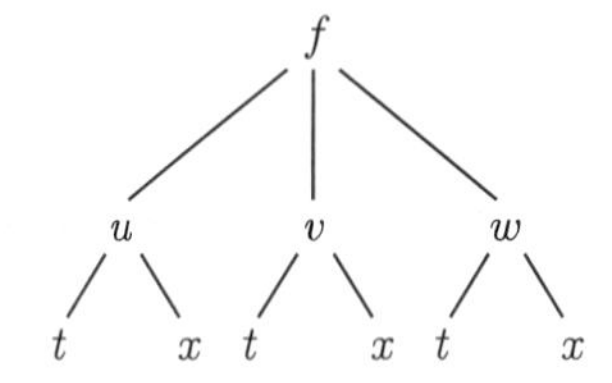

$$\frac{\partial f}{\partial u}\frac{\partial u}{\partial t} + \frac{\partial f}{\partial v}\frac{\partial v}{\partial t} + \frac{\partial f}{\partial w}\frac{\partial w}{\partial t} = nt^{n-1} f(x, y, z),$$

or,

$$x\frac{\partial f}{\partial u} + y\frac{\partial f}{\partial v} + z\frac{\partial f}{\partial w} = nt^{n-1} f(x, y, z).$$

When we set $t = 1$, we obtain $u = x$, $v = y$, $w = z$, and the above equation becomes 13.23.

The results of Exercises 31–33 in Section 13.3 are special cases of 13.23.

EXERCISES 13.6

In Exercises 1–10 we have defined a general functional situation and a specific example. Find the chain rule for the indicated derivative in the general situation, and then use that result to calculate the same derivative in the specific example.

1. dz/dt if $z = f(x, t)$, $x = g(t)$; $z = xt^2/(x + t)$, $x = e^{3t}$

2. $\left.\frac{\partial z}{\partial t}\right)_s$ if $z = f(x, y)$, $x = g(s, t)$, $y = h(s, t)$; $z = x^2 e^y + y \ln x$, $x = s^2 \cos t$, $y = 4\,\mathrm{Sec}^{-1}(t^2 + 2s)$

3. $\left.\frac{\partial u}{\partial s}\right)_t$ if $u = f(x, y, z)$, $x = g(s, t)$, $y = h(s, t)$, $z = k(s, t)$; $u = \sqrt{x^2 + y^2 + z^2}$, , $x = 2st$, $y = s^2 + t^2$, $z = st$

4. dz/du if $z = f(x,y,v)$, $x = g(u)$, $y = h(u)$, $v = k(u)$; $z = x^2yv^3$, $x = u^3 + 2u$, $y = \ln(u^2+1)$, $v = ue^u$

5. $\left.\dfrac{\partial u}{\partial r}\right)_t$ if $u = f(x,y,s)$, $x = g(t)$, $y = h(r)$, $s = k(r,t)$; $u = \sqrt{x^2+y^2s}$, $x = t/(t+5)$, $y = \mathrm{Sin}^{-1}(r^2+5)$, $s = \tan(rt)$

6. $\left.\dfrac{\partial z}{\partial t}\right)_r$ if $z = f(x)$, $x = g(y)$, $y = h(r,t)$; $z = 3^{x+2}$, $x = y^2+5$, $y = \csc(r^2+t)$

7. $\left.\dfrac{\partial u}{\partial x}\right)_y$ if $u = f(x,y,z)$, $z = g(x,y)$; $u = y/\sqrt{x^2+y^2+z^2}$, $z = x/y$

8. $\left.\dfrac{\partial x}{\partial y}\right)_z$ if $x = f(r,s,t)$, $r = g(y)$, $s = h(y,z)$, $t = k(y,z)$; $x = s^2r^2t^2$, $r = y^{-5}$, $s = 1/(y^2+z^2)$, $t = 1/y^2 + 1/z^2$

9. $\left.\dfrac{\partial z}{\partial t}\right)_s$ if $z = f(x,y)$, $x = g(r)$, $y = h(r)$, $r = k(s,t)$; $z = e^{x+y}$, $x = 2r+5$, $y = 2r-5$, $r = t\ln(s^2+t^2)$

10. dz/dt if $z = f(x,y,u)$, $x = g(v)$, $u = h(x,y)$, $v = k(t)$, $y = p(t)$; $z = x^2+y^2+u^2$, $x = v^3-3v^2$, $u = 1/(x^2-y^2)$, $v = e^t$, $y = e^{4t}$

In Exercises 11–15 find the derivative.

11. $\left.\dfrac{\partial^2 z}{\partial t^2}\right)_s$ if $z = x^2y^2 + xe^y$, $x = s+t^2$, $y = s-t^2$

12. d^2x/dt^2 if $x = y^2 + yt - t^2$, $y = t^2e^t$

13. $\left.\dfrac{\partial^2 u}{\partial s^2}\right)_t$ if $u = x^2+y^2+z^2+xyz$, $x = s^2+t^2$, $y = s^2-t^2$, $z = st$

14. d^2z/dv^2 if $z = \sin(xy)$, $x = 3\cos v$, $y = 4\sin v$

15. $\partial^2u/\partial x\partial y$ if $u = y/\sqrt{x^2+y^2+z^2}$, $z = x/y$

16. Suppose that u is a differentiable function of r and $r = \sqrt{x^2+y^2+z^2}$. Show that

$$\left(\frac{\partial u}{\partial x}\right)^2 + \left(\frac{\partial u}{\partial y}\right)^2 + \left(\frac{\partial u}{\partial z}\right)^2 = \left(\frac{du}{dr}\right)^2.$$

17. Consider a gas that is moving through some region D of space. If we follow a particular particle of the gas, it traces out some curved path (Figure 13.14)

$$C: \quad x = x(t),\ y = y(t),\ z = z(t), \quad t \geq 0.$$

Suppose the density of the gas at any point in the region D at time t is denoted by $\rho(x,y,z,t)$. We can write that along C, $\rho = \rho[x(t), y(t), z(t), t]$.

(a) Obtain the chain rule defining $d\rho/dt$ in terms of $\partial\rho/\partial t$ and derivatives of x, y, and z with respect to t.

(b) Explain the physical difference between $d\rho/dt$ and $\partial\rho/\partial t$.

FIGURE 13.14

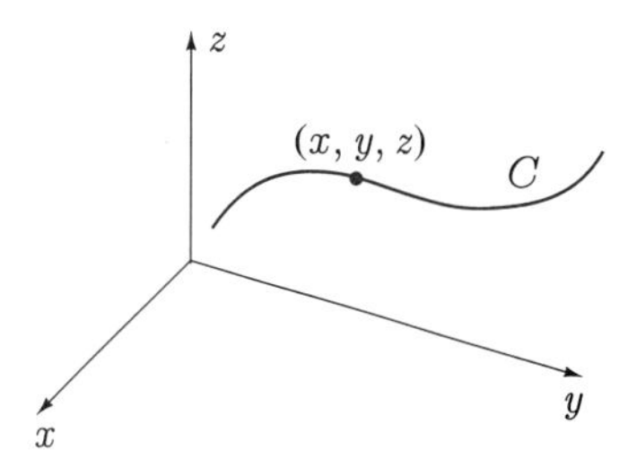

18. The radius and height of a right-circular cone are 10 and 20 cm respectively. If the radius is increasing at 1 cm/min and the height is decreasing at 2 cm/min, how fast is the volume changing? Do you need multivariable calculus to solve this problem?

19. If two sides of a triangle have lengths x and y and the angle between them is θ, then the area of the triangle is $A = (1/2)xy\sin\theta$. How fast is the area changing when x is 1 m, y is 2 m, and θ is $1/3$ radian, if x and y are each increasing at $1/2$ m/s and θ is decreasing at $1/10$ radian per second?

20. When a rocket rises from the earth's surface, its mass decreases because fuel is being consumed at the rate of 50 kg/s. Use Newton's universal law of gravitation (see Example 7.33 in Section 7.9) to determine how fast the force of gravity of the earth on the rocket is changing when the rocket is 100 km above the earth's surface and travelling at 2 km/s. Assume that the mass of the rocket at this height is 12×10^6 kg.

21. If $z = f(u,v)$, $u = g(x,y)$, $v = h(x,y)$, find the chain rule for the second derivative $\partial^2z/\partial x^2$.

22. Which of the following functions are positively homogeneous:

(a) $f(x,y) = x^2 + xy + 3y^2$
(b) $f(x,y) = x^2y + xy - 2xy^2$
(c) $f(x,y,z) = x^2\sin(y/z) + y^2 + y^3/z$
(d) $f(x,y,z) = xe^{y/z} - xyz$
(e) $f(x,y,z,t) = x^4 + y^4 + z^4 + t^4 - xyzt$
(f) $f(x,y,z,t) = e^{x^2+y^2}(z^2+t^2)$
(g) $f(x,y,z) = \cos(xy)\sin(yz)$
(h) $f(x,y) = \sqrt{x^2+xy+y^2}\,e^{y/x}(2x^2-3y^2)$

23. If $f(s)$ and $g(t)$ are differentiable functions, show that $\nabla f(x^2 - y^2) \cdot \nabla g(xy) = 0$.

24. If $f(s)$ is a differentiable function, show that $f(x - y)$ satisfies the equation

$$\frac{\partial f}{\partial y} = -\frac{\partial f}{\partial x}.$$

25. If $f(s)$ is a differentiable function, show that $u(x, y) = f(4x - 3y) + 5(y - x)$ satisfies the equation

$$3\frac{\partial u}{\partial x} + 4\frac{\partial u}{\partial y} = 5.$$

26. If $f(s)$ and $g(t)$ are twice differentiable, show that the function $u(x, y) = xf(x + y) + yg(x + y)$ satisfies

$$\frac{\partial^2 u}{\partial x^2} - 2\frac{\partial^2 u}{\partial x \partial y} + \frac{\partial^2 u}{\partial y^2} = 0.$$

27. If $f(s)$ and $g(t)$ are twice differentiable, show that $f(x - y) + g(x + y)$ satisfies

$$\frac{\partial^2 u}{\partial x^2} - \frac{\partial^2 u}{\partial y^2} = 0.$$

28. Show that if $f(v)$ is differentiable, then $u(x, y) = x^2 f(y/x)$ satisfies

$$x\frac{\partial u}{\partial x} + y\frac{\partial u}{\partial y} = 2u.$$

In Exercises 29–31 suppose that $f(x, y)$ satisfies the first partial differential equation. Show that with the change of independent variables, the function $F(u, v) = f[x(u, v), y(u, v)]$ must satisfy the second partial differential equation.

29. $\left(\frac{\partial f}{\partial x}\right)^2 + \left(\frac{\partial f}{\partial y}\right)^2 = 0$; $u = (x+y)/2$, $v = (x-y)/2$;
$\left(\frac{\partial F}{\partial u}\right)^2 + \left(\frac{\partial F}{\partial v}\right)^2 = 0$

30. $\frac{\partial^2 f}{\partial x^2} - \frac{\partial^2 f}{\partial y^2} = 0$; $u = (x + y)/2$, $v = (x - y)/2$;
$\frac{\partial^2 F}{\partial u \partial v} = 0$

31. $\left(\frac{\partial f}{\partial x}\right)^2 + \left(\frac{\partial f}{\partial y}\right)^2 = 0$; $x = u\cos v$, $y = u\sin v$;
$\left(\frac{\partial F}{\partial u}\right)^2 + \frac{1}{u^2}\left(\frac{\partial F}{\partial v}\right)^2 = 0$

32. An observer travels along the curve $x = t^2$, $y = 3t^3 + 1$, $z = 2t + 5$, where x, y, and z are in metres and $t \geq 0$ is in seconds. If the density ρ of a gas (in kg/m^3) is given by $\rho = (3x^2 + y^2)/(z^2 + 5)$, find the time rate of change of the density of the gas as measured by the observer when $t = 2$ s.

33. If $f(r)$ is a differentiable function and $r = \sqrt{x^2 + y^2 + z^2}$, show that

$$\nabla f = \frac{f'(r)}{r}(x\hat{\mathbf{i}} + y\hat{\mathbf{j}} + z\hat{\mathbf{k}}).$$

34. If $f(x, y) = 0$ defines y as a function of x, show that

$$\frac{d^2 y}{dx^2} = -\frac{f_{xx}f_y^2 - 2f_{xy}f_xf_y + f_{yy}f_x^2}{f_y^3}.$$

35. If $f(x, y)$ is a harmonic function, show that the function $F(x, y) = f(x^2 - y^2, 2xy)$ is also harmonic.

36. Show that in polar coordinates, the two-dimensional Laplace equation 13.11 takes the form

$$\frac{\partial^2 f}{\partial r^2} + \frac{1}{r}\frac{\partial f}{\partial r} + \frac{1}{r^2}\frac{\partial^2 f}{\partial \theta^2} = 0.$$

37. Find an identity satisfied by the second partial derivatives of a function $f(x, y, z)$ that is positively homogeneous of degree n.

38. It is postulated in one of the theories of traffic flow that the average speed u at a point x on a straight highway (along the x-axis) is related to the concentration k of traffic by the differential equation

$$u\frac{\partial u}{\partial x} + \frac{\partial u}{\partial t} = -c^2 k^n \frac{\partial k}{\partial x},$$

where t is time, and $c > 0$ and n are constants.

(a) Use chain rules for $\partial u/\partial x$ and $\partial u/\partial t$ in the functional situation $u = f(k)$ and $k = g(x, t)$ to show that

$$\frac{du}{dk}\left(u\frac{\partial k}{\partial x} + \frac{\partial k}{\partial t}\right) + c^2 k^n \frac{\partial k}{\partial x} = 0.$$

(b) The equation of continuity for traffic flow states that

$$\frac{\partial k}{\partial t} + \frac{\partial (ku)}{\partial x} = 0.$$

Use these last two equations to obtain the differential equation relating speed and concentration:

$$\frac{du}{dk} = -ck^{(n-1)/2}.$$

(c) Solve the differential equation in (b) for $u = f(k)$.

39. A bead slides from rest at the origin on a frictionless wire in a vertical plane to the point (x_0, y_0) under the influence of gravity (Figure 13.15). It can be shown that the time for the bead to traverse the path is

$$t = \frac{1}{\sqrt{2g}} \int_0^{x_0} \sqrt{\frac{1+(y')^2}{y}}\,dx,$$

where g is the acceleration due to gravity and $y' = dy/dx$. The problem of finding that shape of wire which makes t as small as possible is called the *brachistochrone problem*. It is shown in the calculus of variations that $y = f(x)$ must satisfy the equation

$$\frac{d}{dx}\left(\frac{\partial F}{\partial y'}\right) - \frac{\partial F}{\partial y} = 0,$$

where

$$F(y, y') = \sqrt{\frac{1+(y')^2}{y}}.$$

FIGURE 13.15

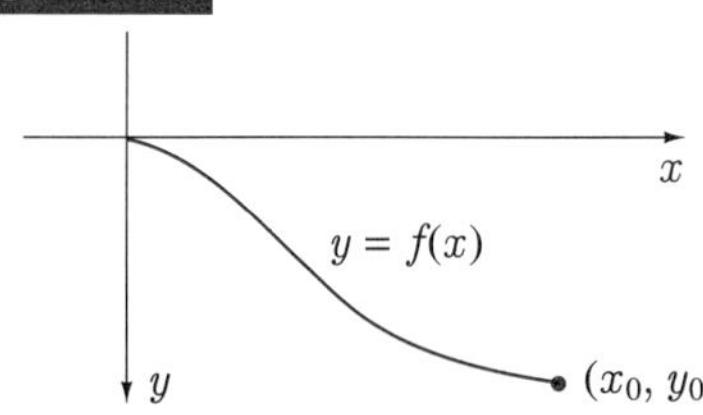

(a) Show that $f(x)$ must satisfy the differential equation $1 + (y')^2 + 2yy'' = 0$.

(b) Show that the curve that satisfies the equation in **(a)** is the cycloid defined parametrically by

$$x = a(\theta - \sin\theta), \quad y = a(1 - \cos\theta),$$

where a is a constant.

40. Two equal masses m are connected by springs having equal spring constant k so that the masses are free to slide on a frictionless table (Figure 13.16). The walls A and B are fixed.

(a) Use Newton's second law to show that the differential equations for the motions of the masses are

$$m\ddot{x}_1 = k(x_2 - 2x_1), \qquad m\ddot{x}_2 = k(x_1 - 2x_2),$$

where x_1 and x_2 are the displacements of the masses from their equilibrium positions, $\ddot{x}_1 = d^2x_1/dt^2$ and $\ddot{x}_2 = d^2x_2/dt^2$.

(b) The Euler-Lagrange equations from theoretical mechanics for this system are

$$\frac{d}{dt}\left(\frac{\partial L}{\partial \dot{x}_1}\right) - \frac{\partial L}{\partial x_1} = 0, \qquad \frac{d}{dt}\left(\frac{\partial L}{\partial \dot{x}_2}\right) - \frac{\partial L}{\partial x_2} = 0,$$

where L is defined as the kinetic energy of the two masses less the energy stored in the springs. Show that

$$L(x_1, x_2, \dot{x}_1, \dot{x}_2) = \frac{m}{2}(\dot{x}_1^2 + \dot{x}_2^2) - k(x_1^2 + x_2^2 - x_1x_2).$$

(c) Obtain the equations in **(a)** from the Euler-Lagrange equations in **(b)**.

FIGURE 13.16

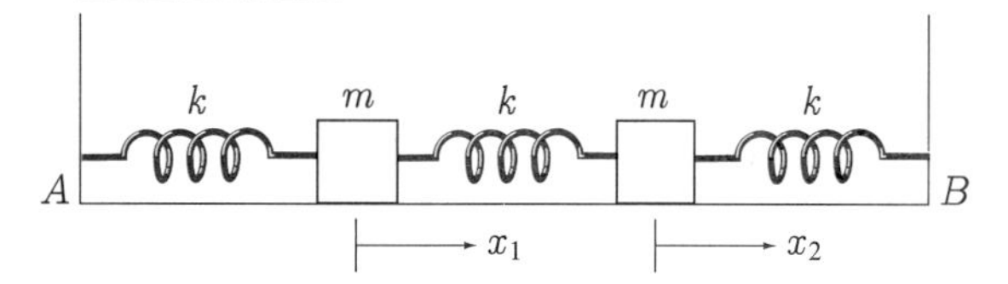

41. Suppose the second-order partial differential equation

$$p\frac{\partial^2 z}{\partial x^2} + q\frac{\partial^2 z}{\partial x \partial y} + r\frac{\partial^2 z}{\partial y^2} = F\left(x, y, z, \frac{\partial z}{\partial x}, \frac{\partial z}{\partial y}\right)$$

(p, q, and r are constants) is subjected to the change of variables

$$s = ax + by, \quad t = cx + dy,$$

where a, b, c, and d are constants. Show that the partial differential equation in s and t is

$$P\frac{\partial^2 z}{\partial s^2} + Q\frac{\partial^2 z}{\partial s \partial t} + R\frac{\partial^2 z}{\partial t^2} = G\left(s, t, z, \frac{\partial z}{\partial s}, \frac{\partial z}{\partial t}\right),$$

where $Q^2 - 4PR = (q^2 - 4pr)(ad - bc)^2$.

42. Show that if a solution $u = f(x, y, z)$ of the three-dimensional Laplace equation 13.12 can be expressed in the form $u = g(r)$, where $r = \sqrt{x^2 + y^2 + z^2}$, then $f(x, y, z)$ must be of the form

$$f(x, y, z) = \frac{C}{\sqrt{x^2 + y^2 + z^2}} + D,$$

where C and D are constants.

SECTION 13.7

Implicit Differentiation

In Section 3.7 we introduced the technique of implicit differentiation in order to obtain the derivative of a function $y = f(x)$ defined implicitly by an equation

$$F(x, y) = 0. \tag{13.24}$$

Essentially, the technique is to differentiate all terms in the equation with respect to x, considering all the while that y is a function of x. For example, if y is defined implicitly by

$$x^2 y^3 + 3xy = 3x + 2,$$

implicit differentiation gives

$$2xy^3 + 3x^2 y^2 \frac{dy}{dx} + 3y + 3x\frac{dy}{dx} = 3.$$

We can now solve to obtain

$$\frac{dy}{dx} = \frac{3 - 2xy^3 - 3y}{3x^2 y^2 + 3x}.$$

With the chain rule we can actually present a formula for dy/dx. Since equation 13.24, when written in the form

$$F[x, f(x)] = 0,$$

must be valid for all x in the domain of the function $f(x)$, we can differentiate it with respect to x. From the schematic diagram below

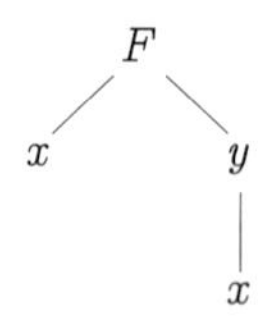

the derivative of the left side of the equation is

$$\frac{dF}{dx} = \frac{\partial F}{\partial x} + \frac{\partial F}{\partial y}\frac{dy}{dx}.$$

If we equate this to the derivative of the right side of the equation, we find

$$F_x + F_y \frac{dy}{dx} = 0,$$

or

$$\frac{dy}{dx} = -\frac{F_x}{F_y}. \tag{13.25}$$

For the function defined implicitly above by $x^2 y^3 + 3xy - 3x - 2 = 0$, equation 13.25 gives

$$\frac{dy}{dx} = -\frac{2xy^3 + 3y - 3}{3x^2 y^2 + 3x},$$

and this result is identical to that obtained by implicit differentiation.

Similarly, if the equation

$$F(x, y, z) = 0 \tag{13.26}$$

defines z implicitly as a function of x and y, the schematic diagram below

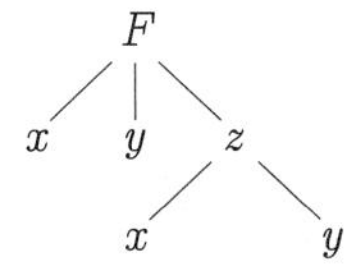

immediately yields

$$\frac{\partial F}{\partial x} + \frac{\partial F}{\partial z}\frac{\partial z}{\partial x} = 0\,, \qquad \frac{\partial F}{\partial y} + \frac{\partial F}{\partial z}\frac{\partial z}{\partial y} = 0\,.$$

From these we obtain the results

$$\frac{\partial z}{\partial x} = -\frac{F_x}{F_z}, \qquad \frac{\partial z}{\partial y} = -\frac{F_y}{F_z}. \tag{13.27}$$

We do not suggest that formulas 13.25 and 13.27 be memorized. On the contrary, we obtain results in this section that include 13.25 and 13.27 as special cases. To develop these results we work with three equations in five variables,

$$\begin{aligned} F(x,y,u,v,w) &= 0\,, \\ G(x,y,u,v,w) &= 0\,, \\ H(x,y,u,v,w) &= 0\,. \end{aligned} \tag{13.28}$$

We assume that these equations define u, v, and w as functions of x and y for some domain of values of x and y (and do so implicitly). It might even be possible to solve the system and obtain explicit definitions of the functions

$$\begin{aligned} u &= f(x,y)\,, \\ v &= g(x,y)\,, \\ w &= h(x,y)\,. \end{aligned} \tag{13.29}$$

We pose the problem of finding the six first-order partial derivatives of u, v, and w with respect to x and y supposing that it is undesirable or even impossible to obtain the explicit form of the functions. To do this, we note that were results 13.29 known and substituted into 13.28, then

$$\begin{aligned} F[x,y,f(x,y),g(x,y),h(x,y)] &= 0\,, \\ G[x,y,f(x,y),g(x,y),h(x,y)] &= 0\,, \\ H[x,y,f(x,y),g(x,y),h(x,y)] &= 0\,, \end{aligned}$$

would be identities in x and y. As a result we could differentiate each equation with respect to x, obtaining from the schematic diagram

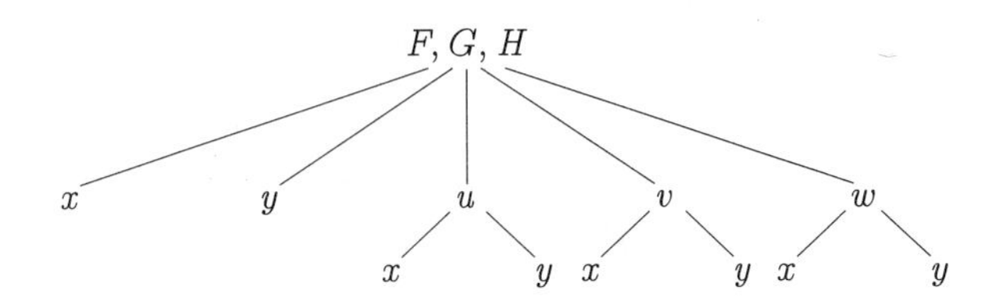

the following:

$$\begin{aligned}
\frac{\partial F}{\partial x}+\frac{\partial F}{\partial u}\frac{\partial u}{\partial x}+\frac{\partial F}{\partial v}\frac{\partial v}{\partial x}+\frac{\partial F}{\partial w}\frac{\partial w}{\partial x}&=0,\\
\frac{\partial G}{\partial x}+\frac{\partial G}{\partial u}\frac{\partial u}{\partial x}+\frac{\partial G}{\partial v}\frac{\partial v}{\partial x}+\frac{\partial G}{\partial w}\frac{\partial w}{\partial x}&=0,\\
\frac{\partial H}{\partial x}+\frac{\partial H}{\partial u}\frac{\partial u}{\partial x}+\frac{\partial H}{\partial v}\frac{\partial v}{\partial x}+\frac{\partial H}{\partial w}\frac{\partial w}{\partial x}&=0,
\end{aligned} \tag{13.30}$$

or,

$$\begin{aligned}
F_u\frac{\partial u}{\partial x}+F_v\frac{\partial v}{\partial x}+F_w\frac{\partial w}{\partial x}&=-F_x,\\
G_u\frac{\partial u}{\partial x}+G_v\frac{\partial v}{\partial x}+G_w\frac{\partial w}{\partial x}&=-G_x,\\
H_u\frac{\partial u}{\partial x}+H_v\frac{\partial v}{\partial x}+H_w\frac{\partial w}{\partial x}&=-H_x.
\end{aligned} \tag{13.31}$$

We have in 13.31 three equations in the three unknowns $\partial u/\partial x$, $\partial v/\partial x$, and $\partial w/\partial x$, and because the equations are linear in the unknowns, solutions can be obtained using Cramer's rule.† In particular,

$$\frac{\partial u}{\partial x}=\frac{\begin{vmatrix}-F_x & F_v & F_w\\ -G_x & G_v & G_w\\ -H_x & H_v & H_w\end{vmatrix}}{\begin{vmatrix}F_u & F_v & F_w\\ G_u & G_v & G_w\\ H_u & H_v & H_w\end{vmatrix}}=-\frac{\begin{vmatrix}F_x & F_v & F_w\\ G_x & G_v & G_w\\ H_x & H_v & H_w\end{vmatrix}}{\begin{vmatrix}F_u & F_v & F_w\\ G_u & G_v & G_w\\ H_u & H_v & H_w\end{vmatrix}}. \tag{13.32}$$

The two determinants on the right of 13.32 involve only derivatives of the given functions F, G, and H, and we have therefore obtained a method for finding $\partial u/\partial x$ that avoids solving 13.28 for u, v, and w. We could list similar formulas for the remaining five derivatives, but first we introduce some simplifying notation.

Definition 13.5

The **Jacobian determinant** of functions F, G, and H with respect to variables u, v, and w is denoted by $\dfrac{\partial(F,G,H)}{\partial(u,v,w)}$ and is defined as the determinant

$$\frac{\partial(F,G,H)}{\partial(u,v,w)}=\begin{vmatrix}F_u & F_v & F_w\\ G_u & G_v & G_w\\ H_u & H_v & H_w\end{vmatrix}=\begin{vmatrix}\frac{\partial F}{\partial u} & \frac{\partial F}{\partial v} & \frac{\partial F}{\partial w}\\ \frac{\partial G}{\partial u} & \frac{\partial G}{\partial v} & \frac{\partial G}{\partial w}\\ \frac{\partial H}{\partial u} & \frac{\partial H}{\partial v} & \frac{\partial H}{\partial w}\end{vmatrix}. \tag{13.33}$$

With this notation we can write 13.32 in the form

$$\frac{\partial u}{\partial x}=-\frac{\dfrac{\partial(F,G,H)}{\partial(x,v,w)}}{\dfrac{\partial(F,G,H)}{\partial(u,v,w)}}. \tag{13.34}$$

† Cramer's rule is discussed in Appendix F.

The remaining derivatives of $v = g(x, y)$ and $w = h(x, y)$ with respect to x can also be obtained from equations 13.31 by Cramer's rule:

$$\frac{\partial v}{\partial x} = -\frac{\dfrac{\partial(F, G, H)}{\partial(u, x, w)}}{\dfrac{\partial(F, G, H)}{\partial(u, v, w)}}, \quad \frac{\partial w}{\partial x} = -\frac{\dfrac{\partial(F, G, H)}{\partial(u, v, x)}}{\dfrac{\partial(F, G, H)}{\partial(u, v, w)}}. \tag{13.35}$$

A similar procedure yields

$$\frac{\partial u}{\partial y} = -\frac{\dfrac{\partial(F, G, H)}{\partial(y, v, w)}}{\dfrac{\partial(F, G, H)}{\partial(u, v, w)}}, \quad \frac{\partial v}{\partial y} = -\frac{\dfrac{\partial(F, G, H)}{\partial(u, y, w)}}{\dfrac{\partial(F, G, H)}{\partial(u, v, w)}}, \quad \frac{\partial w}{\partial y} = -\frac{\dfrac{\partial(F, G, H)}{\partial(u, v, y)}}{\dfrac{\partial(F, G, H)}{\partial(u, v, w)}}. \tag{13.36}$$

Formulas 13.34–13.36 apply only to the situation in which equations 13.28 define u, v, and w as functions of x and y. It is, however, fairly evident how to construct formulas in other situations. Each formula has a Jacobian divided by a Jacobian (and do not forget the negative sign). In the denominator, it is the Jacobian of the functions defining the original equations with respect to the dependent variables. The only difference in the Jacobian in the numerator is that the dependent variable that is being differentiated is replaced by the independent variable with respect to which differentiation is being performed.

The results in equations 13.34–13.36 are valid provided, of course, that the Jacobian

$$\frac{\partial(F, G, H)}{\partial(u, v, w)} \neq 0.$$

In actual fact, it is this condition that guarantees that equations 13.28 do define u, v, and w as functions of x and y in the first place.

As a second example, if

$$F(x, y, s, t) = 0,$$
$$G(x, y, s, t) = 0,$$

define x and y as functions of s and t, then

$$\frac{\partial x}{\partial s} = -\frac{\dfrac{\partial(F, G)}{\partial(s, y)}}{\dfrac{\partial(F, G)}{\partial(x, y)}}, \quad \frac{\partial x}{\partial t} = -\frac{\dfrac{\partial(F, G)}{\partial(t, y)}}{\dfrac{\partial(F, G)}{\partial(x, y)}},$$

$$\frac{\partial y}{\partial s} = -\frac{\dfrac{\partial(F, G)}{\partial(x, s)}}{\dfrac{\partial(F, G)}{\partial(x, y)}}, \quad \frac{\partial y}{\partial t} = -\frac{\dfrac{\partial(F, G)}{\partial(x, t)}}{\dfrac{\partial(F, G)}{\partial(x, y)}}.$$

We now show that the results in equations 13.25 and 13.27 are special cases of these considerations. If equation 13.24 defines y implicitly as a function of x, then

$$\frac{dy}{dx} = -\frac{\dfrac{\partial(F)}{\partial(x)}}{\dfrac{\partial(F)}{\partial(y)}} = -\frac{F_x}{F_y}.$$

If equation 13.26 defines z implicitly as a function of x and y, then

$$\frac{\partial z}{\partial x} = -\frac{\dfrac{\partial(F)}{\partial(x)}}{\dfrac{\partial(F)}{\partial(z)}} = -\frac{F_x}{F_z}; \quad \frac{\partial z}{\partial y} = -\frac{\dfrac{\partial(F)}{\partial(y)}}{\dfrac{\partial(F)}{\partial(z)}} = -\frac{F_y}{F_z}.$$

EXAMPLE 13.17 If $x^2y^2z^3 + zx\sin y = 5$ defines z as a function of x and y, find $\partial z/\partial x$.

SOLUTION If we set $F(x,y,z) = x^2y^2z^3 + zx\sin y - 5 = 0$, then

$$\frac{\partial z}{\partial x} = -\frac{\dfrac{\partial(F)}{\partial(x)}}{\dfrac{\partial(F)}{\partial(z)}} = -\frac{F_x}{F_z} = -\frac{2xy^2z^3 + z\sin y}{3x^2y^2z^2 + x\sin y}.$$

■

EXAMPLE 13.18 The equations $x = r\cos\theta$, $y = r\sin\theta$ define Cartesian coordinates x and y in terms of polar coordinates r and θ. At the same time they implicitly define r and θ as functions of x and y. Find $\partial\theta/\partial x$.

SOLUTION If we set

$$F(x,y,r,\theta) = x - r\cos\theta = 0,$$
$$G(x,y,r,\theta) = y - r\sin\theta = 0,$$

then

$$\frac{\partial\theta}{\partial x} = -\frac{\dfrac{\partial(F,G)}{\partial(r,x)}}{\dfrac{\partial(F,G)}{\partial(r,\theta)}} = -\frac{\begin{vmatrix} F_r & F_x \\ G_r & G_x \end{vmatrix}}{\begin{vmatrix} F_r & F_\theta \\ G_r & G_\theta \end{vmatrix}} = -\frac{\begin{vmatrix} -\cos\theta & 1 \\ -\sin\theta & 0 \end{vmatrix}}{\begin{vmatrix} -\cos\theta & r\sin\theta \\ -\sin\theta & -r\cos\theta \end{vmatrix}} = -\frac{\sin\theta}{r}.$$

■

We have already mentioned that partial derivatives are not to be considered quotients. In particular, note that had we thought of $\partial\theta/\partial x$ in Example 18 as a quotient, we might have been tempted to write

$$\frac{\partial\theta}{\partial x} = \frac{1}{\partial x/\partial\theta} = \frac{1}{-r\sin\theta},$$

which certainly does not agree with the result obtained.

EXAMPLE 13.19 In the theory of thermodynamics, the variables pressure P, temperature T, volume V, and internal energy U are related by two equations of state:

$$F(P,T,V,U) = 0, \quad G(P,T,V,U) = 0.$$

The second law of thermodynamics implies that if U and P are regarded as functions of T and V, then the functions $U(T,V)$ and $P(T,V)$ must satisfy the equation

$$\frac{\partial U}{\partial V} - T\frac{\partial P}{\partial T} + P = 0.$$

Show that if U and V are chosen as independent variables rather than T and V, then the second law takes the form

$$\frac{\partial T}{\partial V}+T\frac{\partial P}{\partial U}-P\frac{\partial T}{\partial U}=0\,.$$

SOLUTION When U and V are taken as independent variables,

$$\frac{\partial T}{\partial V}=-\frac{\dfrac{\partial(F,G)}{\partial(V,P)}}{\dfrac{\partial(F,G)}{\partial(T,P)}},$$

$$\frac{\partial P}{\partial U}=-\frac{\dfrac{\partial(F,G)}{\partial(T,U)}}{\dfrac{\partial(F,G)}{\partial(T,P)}},$$

$$\frac{\partial T}{\partial U}=-\frac{\dfrac{\partial(F,G)}{\partial(U,P)}}{\dfrac{\partial(F,G)}{\partial(T,P)}}.$$

Consequently,

$$\frac{\partial T}{\partial V}+T\frac{\partial P}{\partial U}-P\frac{\partial T}{\partial U}=-\frac{1}{\dfrac{\partial(F,G)}{\partial(T,P)}}\left\{\frac{\partial(F,G)}{\partial(V,P)}+T\frac{\partial(F,G)}{\partial(T,U)}-P\frac{\partial(F,G)}{\partial(U,P)}\right\}.$$

To show that this expression must be equal to zero, we note that when T and V are independent variables, the second law implies that

$$\begin{aligned}0&=-\frac{\dfrac{\partial(F,G)}{\partial(V,P)}}{\dfrac{\partial(F,G)}{\partial(U,P)}}+T\frac{\dfrac{\partial(F,G)}{\partial(U,T)}}{\dfrac{\partial(F,G)}{\partial(U,P)}}+P\\&=-\frac{1}{\dfrac{\partial(F,G)}{\partial(U,P)}}\left\{\frac{\partial(F,G)}{\partial(V,P)}-T\frac{\partial(F,G)}{\partial(U,T)}-P\frac{\partial(F,G)}{\partial(U,P)}\right\}\\&=-\frac{1}{\dfrac{\partial(F,G)}{\partial(U,P)}}\left\{\frac{\partial(F,G)}{\partial(V,P)}+T\frac{\partial(F,G)}{\partial(T,U)}-P\frac{\partial(F,G)}{\partial(U,P)}\right\}.\end{aligned}$$

Hence

$$\frac{\partial T}{\partial V}+T\frac{\partial P}{\partial U}-P\frac{\partial T}{\partial U}=0\,.$$

■

EXERCISES 13.7

In Exercises 1–4 y is defined implicitly as a function of x. Find dy/dx.

1. $x^3y^2 - 2xy + 5 = 0$

2. $(x + y)^2 = 2x$

3. $x(x - y) - 4y^3 = 2e^{xy} + 6$

4. $\sin(x + y) + y^2 = 12x^2 + y$

In Exercises 5–8 z is defined implicitly as a function of x and y. Find $\partial z/\partial x$ and $\partial z/\partial y$.

5. $x^2 \sin z - ye^z = 2x$

6. $x^2z^2 + yz + 3x = 4$

7. $z\sin^2 y + y\sin^2 x = z^3$

8. $\text{Tan}^{-1}(yz) = xz$

In Exercises 9–13 find the required derivative. Assume that the system of equations does define the function(s) indicated.

9. $\partial u/\partial x$ and $\partial v/\partial y$ if $x^2 - y^2 + u^2 + 2v^2 = 1$, $x^2 + y^2 = 2 + u^2 + v^2$

10. $\partial x/\partial t$ if $\sin(x + t) - \sin(x - t) = z$

11. $\partial\phi/\partial x)_{y,z}$ if $x = r\sin\phi\cos\theta$, $y = r\sin\phi\sin\theta$, $z = r\cos\phi$

12. dz/dx if $x^2 + y^2 - z^2 + 2xy = 1$, $x^3 + y^3 - 5y = 4$

13. $\partial u/\partial y)_x$ if $xyu + vw = 4$, $y^2 + u^2 - u^2v = y$, $yw + xu + v + 4 = 0$

14. Given that the equations

$$x^2 - y\cos(uv) + z^2 = 0,$$

$$x^2 + y^2 - \sin(uv) + 2z^2 = 2,$$

$$xy - \sin u \cos v + z = 0$$

define x, y, and z as functions of u and v, find $\partial x/\partial u)_v$ at the values $x = 1$, $y = 1$, $u = \pi/2$, $v = 0$, and $z = 0$.

15. If the equation $F(x, y, z) = 0$ defines each of x, y, and z as a function of the other two, show that

$$\left(\frac{\partial z}{\partial x}\right)_y \left(\frac{\partial x}{\partial y}\right)_z \left(\frac{\partial y}{\partial z}\right)_x = -1.$$

16. If $z = e^x \cos y$, where x and y are functions of t defined by

$$x^3 + e^x - t^2 - t = 1, \quad yt^2 + y^2t - t + y = 0,$$

find dz/dt.

17. Find $\partial s/\partial u)_v$ if $s = x^2 + y^2$, and x and y are functions of u and v defined by

$$u = x^2 - y^2, \quad v = x^2 - y.$$

18. Find $\partial z/\partial y)_x$ if $z = u^3v + \sin(uv)$, and u and v are functions of x and y defined by

$$x = e^u \cos v, \quad y = e^u \sin v.$$

19. Given that $z^3 - xz - y = 0$ defines z as a function of x and y, show that $\dfrac{\partial^2 z}{\partial x \partial y} = -\dfrac{3z^2 + x}{(3z^2 - x)^3}$.

20. If the equations $x = u^2 - v^2$, $y = 2uv$, define u and v as functions of x and y, find $\partial^2 u/\partial x^2$.

21. **(a)** Given that the equation $z^4x + y^3z + 9x^3 = 2$ defines z as a function of x and y, and x as a function of y and z, are $\partial z/\partial x$ and $\partial x/\partial z$ reciprocals?

(b) Given that the equations $z^4x + y^3z + 9x^3 = 2$, $x^2y + xz = 1$, define z as a function of x, and x as a function of z, are dz/dx and dx/dz reciprocals?

(c) Given that the equations $u^2 - v = 3x + y$, $u - 2v^2 = x - 2y$, define u and v as functions of x and y, and also define x and y as functions of u and v, are $\partial u/\partial x$ and $\partial x/\partial u$ reciprocals?

22. Given that the equations $x^2 - 2y^2s^2t - 2st^2 = 1$, $x^2 + 2y^2s^2t + 5st^2 = 1$, define s and t as functions of x and y, find $\partial^2 t/\partial y^2$.

23. Show that if V and P are chosen as independent variables in Example 13.19, then the second law of thermodynamics takes the form

$$T - P\frac{\partial T}{\partial P} + \frac{\partial(T, U)}{\partial(V, P)} = 0.$$

24. **(a)** Suppose the equations $F(u, v, s, t) = 0$, $G(u, v, s, t) = 0$ define u and v as functions of s and t, and the equations $H(s, t, x, y) = 0$, $I(s, t, x, y) = 0$ define s and t as functions of x and y. Show that

$$\frac{\partial(u, v)}{\partial(s, t)} \frac{\partial(s, t)}{\partial(x, y)} = \frac{\partial(u, v)}{\partial(x, y)}.$$

(b) If the equations $F(u, v, x, y) = 0$, $G(u, v, x, y) = 0$ define u and v as functions of x and y, and also define x and y as functions of u and v, show that

$$\frac{\partial(u,v)}{\partial(x,y)} = \frac{1}{\dfrac{\partial(x,y)}{\partial(u,v)}}.$$

25. Suppose the system of m linear equations in n unknowns ($n > m$)

$$\sum_{j=1}^{n} a_{ij}x_j = c_i, \qquad i = 1, \ldots, m,$$

defines $x_1, x_2, \ldots, x_m$ as functions of $x_{m+1}, x_{m+2}, \ldots, x_n$. Show that if $1 \le i \le m$ and $m + 1 \le j \le n$, then

$$\frac{\partial x_i}{\partial x_j} = -\frac{D_{ij}}{D},$$

where $D = |a_{ij}|_{m\times m}$, and D_{ij} is the same as determinant D except that its ith column is replaced by the jth column of $[a_{ij}]_{m\times n}$.

SECTION 13.8

Directional Derivatives

If a function $f(x,y,z)$ is defined throughout some region of space, then at any point (x_0, y_0, z_0) we can calculate its partial derivatives $\partial f/\partial x$, $\partial f/\partial y$, and $\partial f/\partial z$. These derivatives define rates of change of $f(x,y,z)$ at (x_0, y_0, z_0) in directions parallel to the x-, y-, and z-axes. But what if we want the rate of change of $f(x,y,z)$ at (x_0, y_0, z_0) in some arbitrary direction defined by a vector **v** (Figure 13.17)? By the rate of change of $f(x,y,z)$ in the direction **v**, we mean the rate of change with respect to distance as measured along a line through (x_0, y_0, z_0) in the direction **v**. Let us define s as a measure of directed distance along this line, taking $s = 0$ at (x_0, y_0, z_0) and positive s in the direction of **v**. What we want, then, is the derivative of $f(x,y,z)$ with respect to s at $s = 0$. To express $f(x,y,z)$ in terms of s, we use parametric equations of the line through (x_0, y_0, z_0) along **v**. If $\hat{\mathbf{v}} = (v_x, v_y, v_z)$ is a unit vector in the direction of **v**, then parametric equations for this line (see equations 12.39) are

FIGURE 13.17

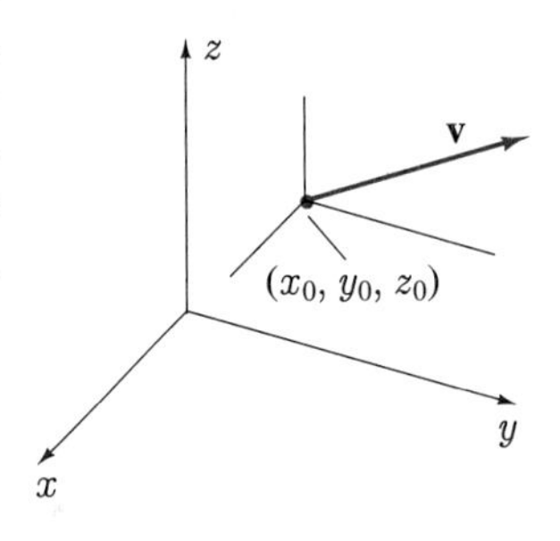

$$x = x_0 + v_x s, \quad y = y_0 + v_y s, \quad z = z_0 + v_z s. \tag{13.37}$$

From the schematic diagram below,

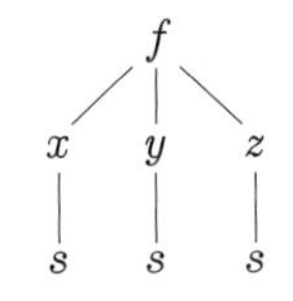

we obtain

$$\begin{aligned}\frac{df}{ds} &= \frac{\partial f}{\partial x}\frac{dx}{ds} + \frac{\partial f}{\partial y}\frac{dy}{ds} + \frac{\partial f}{\partial z}\frac{dz}{ds} \\ &= \frac{\partial f}{\partial x}v_x + \frac{\partial f}{\partial y}v_y + \frac{\partial f}{\partial z}v_z,\end{aligned}$$

where all partial derivatives of $f(x,y,z)$ are to be evaluated at (x_0, y_0, z_0). This derivative df/ds is called the **directional derivative** of $f(x,y,z)$ in the direction **v** at the point (x_0, y_0, z_0), and is usually given the alternative notation

$$D_{\mathbf{v}}f = \frac{\partial f}{\partial x}v_x + \frac{\partial f}{\partial y}v_y + \frac{\partial f}{\partial z}v_z. \tag{13.38}$$

Now, v_x, v_y, and v_z are the components of the unit vector $\hat{\mathbf{v}}$ in the direction of **v**, and $\partial f/\partial x$, $\partial f/\partial y$, and $\partial f/\partial z$ are the components of the gradient of $f(x,y,z)$. We can write, therefore, that

$$D_{\mathbf{v}}f = \nabla f \cdot \hat{\mathbf{v}}. \tag{13.39}$$

Consequently, the derivative (rate of change) of a function in any given direction is the scalar product of the gradient of the function and a unit vector in the required direction. We state this in the following theorem.

Theorem 13.4

The directional derivative of a function in any direction is the component of the gradient of the function in that direction.

EXAMPLE 13.20

Find $D_{\mathbf{v}} f$ at $(4, 0, 16)$ if $f(x, y, z) = x^3 e^y + xz$ and $\mathbf{v}$ is the vector from $(4, 0, 16)$ to $(-2, 1, 4)$.

SOLUTION Since

$$\nabla f_{|(4,0,16)} = \{(3x^2 e^y + z)\hat{\mathbf{i}} + x^3 e^y \hat{\mathbf{j}} + x\hat{\mathbf{k}}\}_{|(4,0,16)}$$
$$= 64\hat{\mathbf{i}} + 64\hat{\mathbf{j}} + 4\hat{\mathbf{k}}$$

and

$$\hat{\mathbf{v}} = \frac{\mathbf{v}}{|\mathbf{v}|} = \frac{(-6, 1, -12)}{\sqrt{36 + 1 + 144}} = \frac{-1}{\sqrt{181}}(6, -1, 12),$$

we have

$$D_v f = -(64, 64, 4) \cdot \frac{(6, -1, 12)}{\sqrt{181}} = -\frac{368}{\sqrt{181}}.$$

■

The directional derivative gives us insight into some of the properties of the gradient vector. In particular, we have the next theorem.

Theorem 13.5

The gradient ∇f of a function $f(x, y, z)$ defines the direction in which the function increases most rapidly, and the maximum rate of change is $|\nabla f|$.

Proof Theorem 13.4 states that the directional derivative of $f(x, y, z)$ in a direction $\mathbf{v}$ is the component of ∇f in that direction. Figure 13.18, which shows components of ∇f in various directions, makes it clear that $D_{\mathbf{v}} f$ is greatest when $\mathbf{v}$ is parallel to ∇f. Alternatively, if θ is the angle between $\mathbf{v}$ and ∇f, then

$$D_{\mathbf{v}} f = \nabla f \cdot \hat{\mathbf{v}} = |\nabla f||\hat{\mathbf{v}}| \cos\theta = |\nabla f| \cos\theta.$$

Obviously $D_{\mathbf{v}} f$ is a maximum when $\cos\theta$ is a maximum—i.e., when $\cos\theta = 1$ or $\theta = 0$—and this occurs when $\mathbf{v}$ is parallel to ∇f. Finally, when $\mathbf{v}$ is parallel to ∇f, we have $D_{\mathbf{v}} f = |\nabla f|$, and this completes the proof.

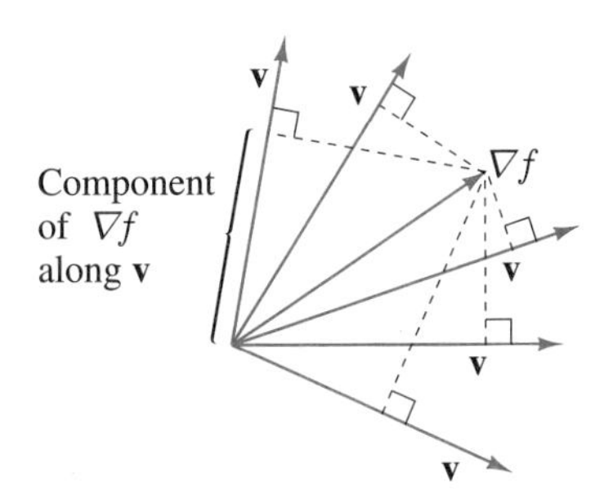

FIGURE 13.18

Note that for any function $f(x, y, z)$,

$$D_{\hat{\mathbf{i}}}f = \frac{\partial f}{\partial x}, \quad D_{\hat{\mathbf{j}}}f = \frac{\partial f}{\partial y}, \quad D_{\hat{\mathbf{k}}}f = \frac{\partial f}{\partial z}.$$

In other words, the partial derivatives of a function are its directional derivatives along the coordinate directions.

EXAMPLE 13.21

Find the direction at the point $(1, 2, -3)$ in which the function $f(x, y, z) = x^2y + xyz$ increases most rapidly.

SOLUTION According to Theorem 13.5, $f(x, y, z)$ increases most rapidly in the direction

$$\nabla f_{|(1,2,-3)} = (2xy + yz, x^2 + xz, xy)_{|(1,2,-3)} = (-2, -2, 2).$$

■

You might feel that because the definition of the directional derivative $D_{\mathbf{v}}f$ does not involve a limit process, it is some strange new type of differentiation. To show that this is not the case, let us return to the calculation of the derivative of $f(x, y, z)$ at (x_0, y_0, z_0) in the direction $\mathbf{v}$ shown in Figure 13.17. With parametric equations 13.37 for the line through (x_0, y_0, z_0) along $\mathbf{v}$, the value of $f(x, y, z)$ at any point (x, y, z) along this line is $f(x_0 + v_x s, y_0 + v_y s, z_0 + v_z s)$. If we take the difference between this value and $f(x_0, y_0, z_0)$ and divide by the distance s between (x_0, y_0, z_0) and (x, y, z), then the limit of this expression as $s \to 0^+$ should define the derivative of $f(x, y, z)$ at (x_0, y_0, z_0) in the direction $\mathbf{v}$; that is,

$$D_{\mathbf{v}}f = \lim_{s \to 0^+} \frac{f(x_0 + v_x s, y_0 + v_y s, z_0 + v_z s) - f(x_0, y_0, z_0)}{s}. \tag{13.40}$$

It can be shown that this limit (and this is perhaps the form we might have expected the derivative to take) also leads to the result contained in 13.39 (see Exercise 33).

Consider a curve C in space that is defined parametrically by

$$C: \quad x = x(t), \quad y = y(t), \quad z = z(t), \quad \alpha \leq t \leq \beta$$

(Figure13.19). Imagine that C is the path traced out by some particle as it moves through space under the action of some system of forces, and suppose that $f(x, y, z)$ is a function defined along C. Perhaps the particle is a weather balloon and $f(x, y, z)$ is temperature at points along its trajectory C. In such applications we are frequently asked for the rate of change of $f(x, y, z)$ with respect to distance traveled along C. If we use s as a measure of distance along C (taking $s = 0$ at A), then the required rate of change is

FIGURE 13.19

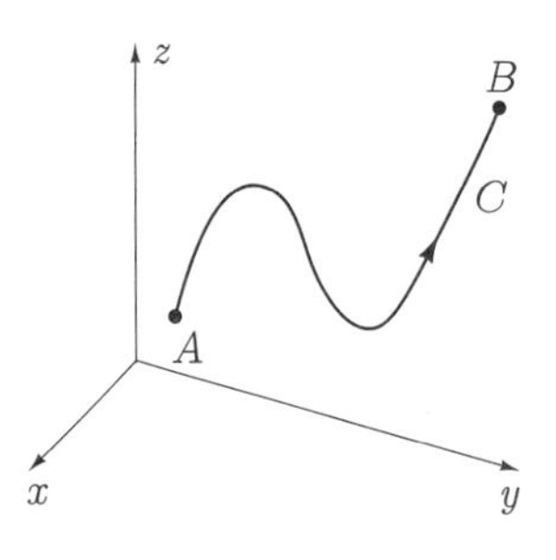

df/ds. Since the coordinates of points (x, y, z) on C can be regarded as functions of s (although it might be difficult to find these functions explicitly), the chain rule gives

$$\begin{aligned}\frac{df}{ds} &= \frac{\partial f}{\partial x}\frac{dx}{ds} + \frac{\partial f}{\partial y}\frac{dy}{ds} + \frac{\partial f}{\partial z}\frac{dz}{ds} \\ &= \left(\frac{\partial f}{\partial x}, \frac{\partial f}{\partial y}, \frac{\partial f}{\partial z}\right) \cdot \left(\frac{dx}{ds}, \frac{dy}{ds}, \frac{dz}{ds}\right) \\ &= \nabla f \cdot \frac{d\mathbf{r}}{ds}.\end{aligned}$$

f

$x \quad y \quad z$

$s \quad s \quad s$

In Section 12.8 we saw that $d\mathbf{r}/ds$ is a unit tangent vector $\hat{\mathbf{T}}$ to C. Consequently,

$$\frac{df}{ds} = \nabla f \cdot \hat{\mathbf{T}}.$$

But this equation states that df/ds is the directional derivative of $f(x, y, z)$ along the tangent direction to the curve C. In other words, to calculate the rate of change of a function $f(x, y, z)$ with respect to distance as measured along a curve C, we calculate the directional derivative of $f(x, y, z)$ in the direction of the tangent vector to C.

EXAMPLE 13.22

Find the rate of change of the function $f(x, y, z) = x^2y - xz$ along the curve $y = x^2$, $z = x$ in the direction of decreasing x at the point $(2, 4, 2)$.

SOLUTION Since parametric equations for the curve are C: $x = -t$, $y = t^2$, $z = -t$, a tangent vector to C at any point is $\mathbf{T} = (-1, 2t, -1)$. At $(2, 4, 2)$, $t = -2$, and the tangent vector is $\mathbf{T} = (-1, -4, -1)$. A unit tangent vector to C at $(2, 4, 2)$ in the direction of decreasing x is therefore

$$\hat{\mathbf{T}} = \frac{(-1, -4, -1)}{\sqrt{18}} = \frac{-1}{3\sqrt{2}}(1, 4, 1).$$

The rate of change of $f(x, y, z)$ in this direction is

$$\begin{aligned}\nabla f \cdot \hat{\mathbf{T}} &= (2xy - z, x^2, -x)_{|(2,4,2)} \cdot \frac{(1, 4, 1)}{-3\sqrt{2}} \\ &= \frac{-1}{3\sqrt{2}}(14, 4, -2) \cdot (1, 4, 1) = -\frac{28}{3\sqrt{2}}.\end{aligned}$$

■

It is worthwhile at this point to discuss directional derivatives for a function $f(x, y)$ of two independent variables. Such a function can be represented graphically as a surface $z = f(x, y)$ (Figure 13.20).

FIGURE 13.20

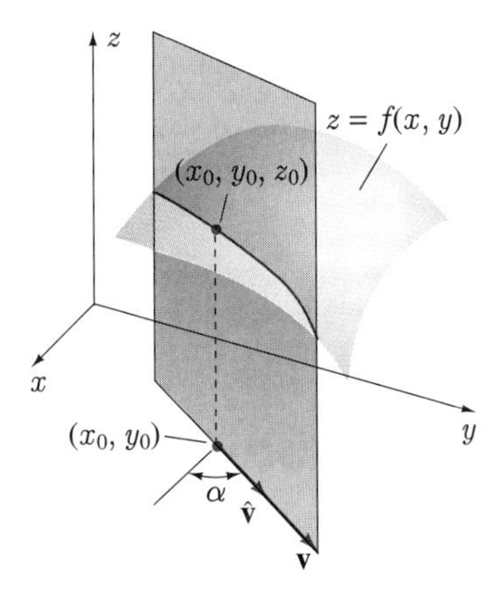

For a direction $\mathbf{v}$ at (x_0, y_0) in the xy-plane,

$$D_{\mathbf{v}} f = \nabla f \cdot \hat{\mathbf{v}},$$

where ∇f is evaluated at (x_0, y_0). Algebraically, this is the derivative (rate of change) of $f(x, y)$ in the direction $\mathbf{v}$. Geometrically, this is the rate of change of the height z of the surface as we move along the curve of intersection of the surface and a vertical plane containing the vector $\mathbf{v}$, or, the slope of this curve. Each direction $\mathbf{v}$ at (x_0, y_0) defines an angle α with a line through (x_0, y_0) parallel to the positive x-axis, and for this direction

$$\hat{\mathbf{v}} = \cos\alpha\,\hat{\mathbf{i}} + \sin\alpha\,\hat{\mathbf{j}}.$$

We can write, then,

$$\begin{aligned} D_{\mathbf{v}} f = \nabla f \cdot \hat{\mathbf{v}} &= \left(\frac{\partial f}{\partial x}\hat{\mathbf{i}} + \frac{\partial f}{\partial y}\hat{\mathbf{j}}\right) \cdot (\cos\alpha\,\hat{\mathbf{i}} + \sin\alpha\,\hat{\mathbf{j}}) \\ &= \frac{\partial f}{\partial x}\cos\alpha + \frac{\partial f}{\partial y}\sin\alpha. \end{aligned} \tag{13.41}$$

If $D_{\mathbf{v}} f$ represents the slope of the curve of intersection of the surface and the vertical plane through $\mathbf{v}$, then $D_{\mathbf{v}}(D_{\mathbf{v}} f)$ represents the rate of change of this slope. Now

$$D_{\mathbf{v}}(D_{\mathbf{v}} f) = \nabla(D_{\mathbf{v}} f) \cdot \hat{\mathbf{v}},$$

and

$$\nabla(D_{\mathbf{v}} f) = \left(\frac{\partial^2 f}{\partial x^2}\cos\alpha + \frac{\partial^2 f}{\partial x \partial y}\sin\alpha\right)\hat{\mathbf{i}} + \left(\frac{\partial^2 f}{\partial y \partial x}\cos\alpha + \frac{\partial^2 f}{\partial y^2}\sin\alpha\right)\hat{\mathbf{j}}.$$

Thus,

$$\begin{aligned} D_{\mathbf{v}}(D_{\mathbf{v}} f) &= \left(\frac{\partial^2 f}{\partial x^2}\cos\alpha + \frac{\partial^2 f}{\partial x \partial y}\sin\alpha\right)\cos\alpha + \left(\frac{\partial^2 f}{\partial y \partial x}\cos\alpha + \frac{\partial^2 f}{\partial y^2}\sin\alpha\right)\sin\alpha \\ &= \frac{\partial^2 f}{\partial x^2}\cos^2\alpha + 2\,\frac{\partial^2 f}{\partial x \partial y}\cos\alpha\sin\alpha + \frac{\partial^2 f}{\partial y^2}\sin^2\alpha. \end{aligned} \tag{13.42}$$

We call $D_{\mathbf{v}}(D_{\mathbf{v}} f)$ the second directional derivative of $f(x, y)$ at (x_0, y_0) in the direction $\mathbf{v}$. If it is positive, then the curve of intersection is concave upward, whereas if it is negative, the curve is concave downward. We will find these results useful in Section 13.10 when we discuss relative extrema of functions of two independent variables.

EXERCISES 13.8

In Exercises 1–8 calculate the directional derivative of the function at the point and in the direction indicated.

1. $f(x,y,z) = 2x^2 - y^2 + z^2$ at $(1,2,3)$ in the direction of the vector from $(1,2,3)$ to $(3,5,0)$

2. $f(x,y,z) = x^2y + xz$ at $(-1,1,-1)$ in the direction of the vector that joins $(3,2,1)$ to $(3,1,-1)$

3. $f(x,y) = xe^y + y$ at $(3,0)$ in the direction of the vector from $(3,0)$ to $(-2,-4)$

4. $f(x,y,z) = \ln(xy + yz + xz)$ at $(1,1,1)$ in the direction from $(1,1,1)$ toward the point $(-1,-2,3)$

5. $f(x,y) = \text{Tan}^{-1}(xy)$ at $(1,2)$ along the line $y = 2x$ in the direction of increasing x

6. $f(x,y) = \sin(x+y)$ at $(2,-2)$ along the line $3x + 4y = -2$ in the direction of decreasing y

7. $f(x,y,z) = x^3y\sin z$ at $(3,-1,-2)$ along the line $x = 3+t,\ y = -1+4t,\ z = -2+2t$ in the direction of decreasing x

8. $f(x,y,z) = x^2y + y^2z + z^2x$ at $(1,-1,0)$ along the line $x + 2y + 1 = 0,\ x - y + 2z = 2$ in the direction of decreasing z

In Exercises 9–12 find the rate of change of the function with respect to distance travelled along the curve.

9. $f(x,y) = 2x - 3y$ at $(1,1)$ along the curve $y = x^2$ in the direction of increasing x

10. $f(x,y) = x^2 + y$ at $(-1,3)$ along the curve $y = -3x^3$ in the direction of decreasing x

11. $f(x,y,z) = xy + z^2$ at $(1,0,-2)$ along the curve $y = x^2 - 1,\ z = -2x$ in the direction of increasing x

12. $f(x,y,z) = x^2y + xy^3z$ at $(2,-1,2)$ along the curve $x^2 - y^2 = 3,\ z = x$ in the direction of increasing x

In Exercises 13–18 find the direction in which the function increases most rapidly at the point. What is the rate of change in that direction?

13. $f(x,y,z) = x^4yz - xy^3 + z$ at $(1,1,-3)$

14. $f(x,y) = 2xy + \ln(xy)$ at $(2,1/2)$

15. $f(x,y,z) = 1/\sqrt{x^2+y^2+z^2}$ at $(1,-3,2)$

16. $f(x,y,z) = -1/\sqrt{x^2+y^2+z^2}$ at $(1,-3,2)$

17. $f(x,y,z) = \text{Tan}^{-1}(xyz)$ at $(3,2,-4)$

18. $f(x,y) = xye^{xy}$ at $(1,1)$

19. In what direction is the rate of change of $f(x,y,z) = xyz$ smallest at the point $(2,-1,3)$?

20. In what directions (if any) is the rate of change of the function $f(x,y) = x^2y + y^3$ at the point $(1,-1)$ equal to **(a)** 0? **(b)** 1? **(c)** 20?

21. In what directions (if any) is the rate of change of the function $f(x,y,z) = xy + z$ at the point $(0,1,-2)$ equal to **(a)** 0? **(b)** 1? **(c)** -20?

22. Must there always be a direction in which the rate of change of a function at a point is equal to **(a)** 0? **(b)** 3?

23. In the derivation of 13.38, why was it necessary to use a unit vector $\hat{\mathbf{v}}$ to determine parametric equations for the line through (x_0, y_0, z_0) along $\mathbf{v}$? In other words, why could we not use the components of $\mathbf{v}$ itself to write parametric equations for the line?

24. How fast is the distance to the origin changing with respect to distance travelled along the curve $x = 2\cos t,\ y = 2\sin t,\ z = 3t$ at any point on the curve? What is the rate of change when $t = 0$? Would you expect this?

25. Find points on the curve $C:\ x = t,\ y = 1 - 2t,\ z = t$ at which the rate of change of $f(x,y,z) = x^2 + xyz$ with respect to distance travelled along the curve vanishes.

26. Repeat Exercise 25 for the curve $C:\ z = x,\ x = y^2$ and the function $f(x,y,z) = x^2 - y^2 + z^2$.

27. The path of a particle is defined parametrically by $x = (\cos t - t\sin t)\hat{\mathbf{i}} + (\sin t - t\cos t)\hat{\mathbf{j}}$ where t is time (the path is called an involute of a circle). Show that if the particle's speed is constant, then the rate of change of its distance from the origin with respect to distance travelled is also constant. Is the time rate of change of its distance from the origin also constant?

28. The rate of change of a function $f(x,y)$ at a point (x_0, y_0) in direction $\hat{\mathbf{i}} + 2\hat{\mathbf{j}}$ is 3 and the rate of change in direction $-2\hat{\mathbf{i}} - \hat{\mathbf{j}}$ is -1. Find its rate of change in direction $2\hat{\mathbf{i}} + 3\hat{\mathbf{j}}$.

29. Rates of change of a function $f(x,y,z)$ at a point (x_0, y_0, z_0) in directions $\hat{\mathbf{i}} + \hat{\mathbf{j}}$, $2\hat{\mathbf{i}} - \hat{\mathbf{k}}$, and $\hat{\mathbf{i}} - \hat{\mathbf{j}} + \hat{\mathbf{k}}$ are 1, 2, and -3 respectively. What is its partial derivative with respect to z at the point?

30. Find the second directional derivative of the function $f(x,y) = x^3y^2$ at the point $(1,1)$ in the direction of the vector $(1,-2)$.

31. Find the second directional derivative of the function $f(x, y, z) = x^2 + 2y^2 + 3z^2$ at the point $(-2, -1, 3)$ in the direction $(1, 1, -1)$.

32. The path followed by a stone embedded in the tread of a tire is a cycloid given parametrically by $x = R(\theta - \sin\theta)$, $y = R(1 - \cos\theta)$, $\theta \geq 0$ (see Exercise 46 in Section 10.1).

(a) How fast is the distance from the origin changing with respect to distance travelled along the curve when $\theta = \pi/2$ and $\theta = \pi$?

(b) How fast is the y-coordinate changing at these points?

(c) How fast is the x-coordinate changing at these points?

33. Verify that expression 13.40 for $D_{\mathbf{v}} f$ leads to formula 13.39.

SECTION 13.9

Tangent Lines and Tangent Planes

Tangent Lines to Curves

One equation in the coordinates x, y, and z of points in space,

$$F(x, y, z) = 0, \tag{13.43}$$

usually defines a surface. When each of the equations

$$F(x, y, z) = 0, \quad G(x, y, z) = 0 \tag{13.44}$$

defines a surface, then together they define the curve of intersection of the two surfaces. Theoretically, we can find parametric equations for the curve by setting x equal to some function of a parameter t, say $x = x(t)$, and then solving equations 13.44 for y and z in terms of t: $y = y(t)$ and $z = z(t)$. The parametric definition, therefore, takes the form

$$x = x(t), \quad y = y(t), \quad z = z(t), \quad \alpha \leq t \leq \beta, \tag{13.45}$$

where α and β specify the endpoints of the curve. Practical difficulties arise in choosing $x(t)$ and solving for $y(t)$ and $z(t)$. For some examples, it might be more convenient to specify $y(t)$ and solve for $x(t)$ and $z(t)$ or, alternatively, to specify $z(t)$ and solve for $x(t)$ and $y(t)$. We considered examples of such conversions in Section 12.7.

In Section 12.8 we indicated that if a curve C is defined parametrically by 13.45, then a tangent vector to C at any point P is

$$\frac{d\mathbf{r}}{dt} = \frac{dx}{dt}\hat{\mathbf{i}} + \frac{dy}{dt}\hat{\mathbf{j}} + \frac{dz}{dt}\hat{\mathbf{k}} \tag{13.46}$$

(Figure 13.21). The tangent line to C at P is defined as the line through P having direction $d\mathbf{r}/dt$. If (x_0, y_0, z_0) are the coordinates of P and t_0 is the value of t yielding P, then the vector equation for the tangent line at P is

$$(x, y, z) = (x_0, y_0, z_0) + u\frac{d\mathbf{r}}{dt}_{|t=t_0} \tag{13.47a}$$

(see equation 12.38). Parametric equations for the tangent line are therefore

$$x = x_0 + x'(t_0)u, \quad y = y_0 + y'(t_0)u, \quad z = z_0 + z'(t_0)u, \tag{13.47b}$$

and in the case that none of $x'(t_0)$, $y'(t_0)$, and $z'(t_0)$ vanishes, we can also write symmetric equations for the tangent line:

$$\frac{x - x_0}{x'(t_0)} = \frac{y - y_0}{y'(t_0)} = \frac{z - z_0}{z'(t_0)}. \tag{13.47c}$$

FIGURE 13.21

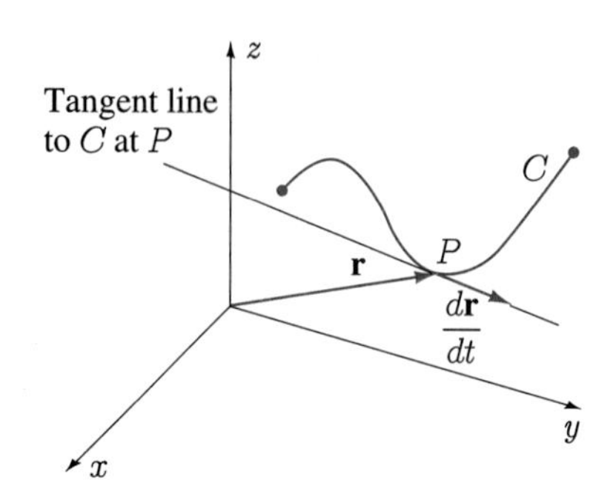

EXAMPLE 13.23

Find equations for the tangent line to the curve

$$C: \quad x = 2\cos t, \quad y = 4\sin t, \quad z = 2t/\pi$$

at $P(\sqrt{2}, 2\sqrt{2}, \frac{1}{2})$.

SOLUTION Since $t = \pi/4$ at P, a tangent vector to C at P is

$$\frac{d\mathbf{r}}{dt}_{|t=\pi/4} = (-2\sin t, 4\cos t, 2/\pi)_{|t=\pi/4} = (-\sqrt{2}, 2\sqrt{2}, 2/\pi).$$

Symmetric equations for the tangent line are therefore

$$\frac{x - \sqrt{2}}{-\sqrt{2}} = \frac{y - 2\sqrt{2}}{2\sqrt{2}} = \frac{z - \frac{1}{2}}{2/\pi}.$$

■

EXAMPLE 13.24

Find equations for the tangent line to the curve

$$C: \quad x = 2t^3 + t^2 + 4t, \quad y = t^2 - 3t + 5, \quad z = t^2 - 1$$

at the point $P(0, 5, -1)$.

SOLUTION Since $t = 0$ at P, a tangent vector to C at P is

$$\frac{d\mathbf{r}}{dt}_{|t=0} = (6t^2 + 2t + 4, 2t - 3, 2t)_{|t=0} = (4, -3, 0).$$

Because the z-component of this vector is equal to zero, we cannot use the full symmetric equations 13.47c. We can, however, use them for the x- and y- coordinates of points on the line and write

$$\frac{x}{4} = \frac{y-5}{-3}, \quad \text{or} \quad 3x + 4y = 20.$$

Since the z-component of $d\mathbf{r}/dt$ vanishes, $d\mathbf{r}/dt$ must be parallel to the xy-plane, and so too must the tangent line to C at P. But then every point on the line has the same z-coordinate, and because the z-coordinate of P is -1, it follows that all points on the tangent line have z-coordinate equal to -1. Equations for the tangent line are therefore $3x + 4y = 20$, $z = -1$. ■

Tangent Planes to Surfaces

We now consider the problem of finding the equation for the tangent plane at a point P on a surface S (Figure 13.22). We define the tangent plane as that plane which contains all tangent lines at P to curves in S through P (provided, of course, that such a plane exists). Suppose that the surface is defined by the equation

$$F(x, y, z) = 0, \tag{13.48}$$

and that

$$C: \quad x = x(t), \quad y = y(t), \quad z = z(t), \quad \alpha \leq t \leq \beta,$$

is any curve in S through P. Since C is in S, the equation

$$F[x(t), y(t), z(t)] = 0$$

is valid for all t in $\alpha \leq t \leq \beta$. If $F(x, y, z)$ has continuous first partial derivatives, and $x(t)$, $y(t)$, and $z(t)$ are all differentiable, we may differentiate this equation using the chain rule:

$$\frac{\partial F}{\partial x}\frac{dx}{dt} + \frac{\partial F}{\partial y}\frac{dy}{dt} + \frac{\partial F}{\partial z}\frac{dz}{dt} = 0.$$

This equation, which holds at all points on C, and in particular at P, can be expressed vectorially as

$$0 = \left(\frac{\partial F}{\partial x}, \frac{\partial F}{\partial y}, \frac{\partial F}{\partial z}\right) \cdot \left(\frac{dx}{dt}, \frac{dy}{dt}, \frac{dz}{dt}\right) = \nabla F \cdot \frac{d\mathbf{r}}{dt}.$$

But if the scalar product of two vectors vanishes, the vectors are perpendicular (see equation 12.24). Consequently, ∇F is perpendicular to the tangent vector $d\mathbf{r}/dt$ to C at P. Since C is an arbitrary curve in S, it follows that ∇F at P is perpendicular to the tangent line to every curve C in S at P. In other words, ∇F at P must be perpendicular to the tangent plane to S at P (Figure 13.23). If the coordinates of P are (x_0, y_0, z_0), then the equation of the tangent plane to S at P is

$$\begin{aligned} 0 &= \nabla F_{|P} \cdot (x - x_0, y - y_0, z - z_0) \\ &= F_x(x_0, y_0, z_0)(x - x_0) + F_y(x_0, y_0, z_0)(y - y_0) \\ &\quad + F_z(x_0, y_0, z_0)(z - z_0) \end{aligned} \tag{13.49}$$

(see equation 12.27).

FIGURE 13.22

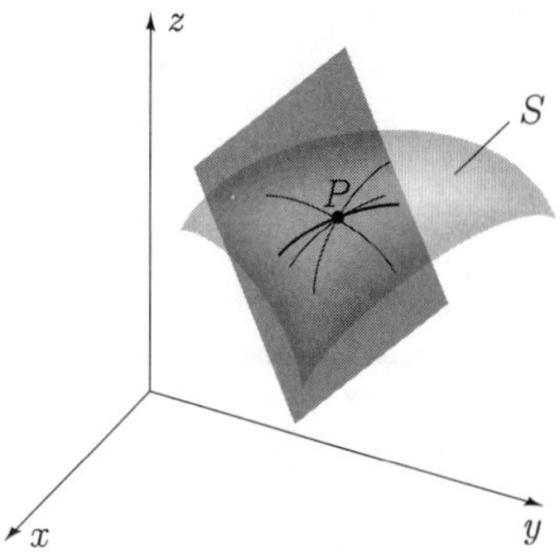

FIGURE 13.23

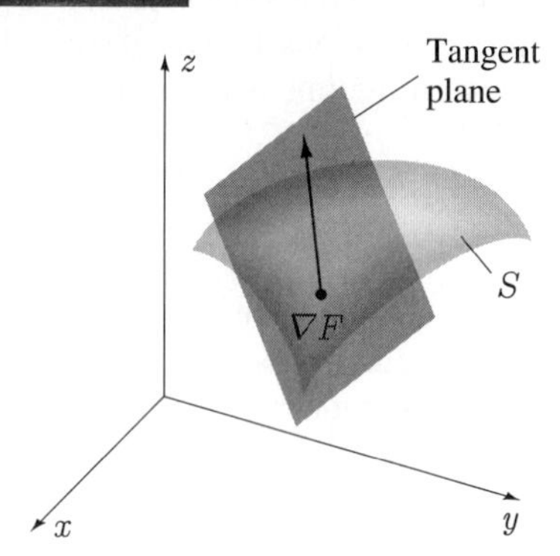

EXAMPLE 13.25

Find the equation of the tangent plane to the surface $xyz^3 + yz^2 = 4$ at the point $(1,2,1)$.

SOLUTION A vector perpendicular to the tangent plane is

$$\nabla(xyz^3 + yz^2 - 4)_{|(1,2,1)} = (yz^3, xz^3 + z^2, 3xyz^2 + 2yz)_{|(1,2,1)} = (2,2,10).$$

But then the vector $(1,1,5)$ must also be perpendicular to the tangent plane, and the equation of the plane is therefore

$$0 = (1,1,5) \cdot (x-1, y-2, z-1) = x + y + 5z - 8.$$

■

We have shown in this section that if the equation $F(x,y,z) = 0$ defines a surface S, and if there is a tangent plane to S at a point P, then the vector $\nabla F_{|P}$ is normal to the tangent plane (Figure 13.23). It is customary to state in this situation that $\nabla F_{|P}$ is normal to the surface itself at P, rather than to the tangent plane to the surface. This fact proves to be another of the important properties of the gradient vector, and is worth stating as a theorem.

Theorem 13.6

If the equation $F(x,y,z) = 0$ defines a surface S, and $F(x,y,z)$ has continuous first partial derivatives, then at any point on S, the vector ∇F is perpendicular to S.

A geometric application of this fact is contained in the following example.

EXAMPLE 13.26

Find equations for the tangent line at the point $(1,2,2)$ to the curve C: $x^2+y^2+z^2 = 9$, $4(x^2+y^2) = 5z^2$.

SOLUTION Equation 13.46 indicates that to find a tangent vector to C we should first have parametric equations for C. These can be obtained by first solving each equation for x^2+y^2 and equating the results:

$$9 - z^2 = 5z^2/4.$$

This equation implies that $z = \pm 2$, the positive result being required here. On C, then, $x^2+y^2 = 5$, and parametric equations for C are

$$x = \sqrt{5}\cos t, \quad y = \sqrt{5}\sin t, \quad z = 2, \quad 0 \le t < 2\pi.$$

According to 13.46, a tangent vector to C at $(1,2,2)$ is

$$\left(\frac{dx}{dt}, \frac{dy}{dt}, \frac{dz}{dt}\right)_{|(1,2,2)} = (-\sqrt{5}\sin t, \sqrt{5}\cos t, 0)_{|t=\mathrm{Sin}^{-1}(2/\sqrt{5})} = (-2,1,0).$$

The tangent line therefore has equations

$$\frac{x-1}{-2} = \frac{y-2}{1}, \quad z = 2, \quad \text{or,} \quad x + 2y = 5, \quad z = 2.$$

The fact that gradients can be used to find normals to surfaces suggests an alternative solution. It is clear from Figure 13.24 that if we define $F(x,y,z) = x^2+y^2+z^2-9$, then ∇F evaluated at $(1,2,2)$ is perpendicular not only to the surface $F(x,y,z) = 0$, but also to the curve C. Similarly, if $G(x,y,z) = 4(x^2+y^2) - 5z^2$, then ∇G at $(1,2,2)$ is also perpendicular to C. Since a vector along the tangent line to C at $(1,2,2)$ is perpendicular to both of these vectors, it follows that a vector along the tangent line is

$$\begin{aligned}
(\nabla F \times \nabla G)_{|(1,2,2)} &= \{(2x, 2y, 2z) \times (8x, 8y, -10z)\}_{|(1,2,2)} \\
&= (2,4,4) \times (8,16,-20) \\
&= 8\begin{vmatrix} \hat{\mathbf{i}} & \hat{\mathbf{j}} & \hat{\mathbf{k}} \\ 1 & 2 & 2 \\ 2 & 4 & -5 \end{vmatrix} \\
&= 8(-18, 9, 0) \\
&= 72(-2, 1, 0).
\end{aligned}$$

Once again, we have obtained $(-2,1,0)$ as a tangent vector to the curve, and equations for the tangent line can be written down as before.

FIGURE 13.24

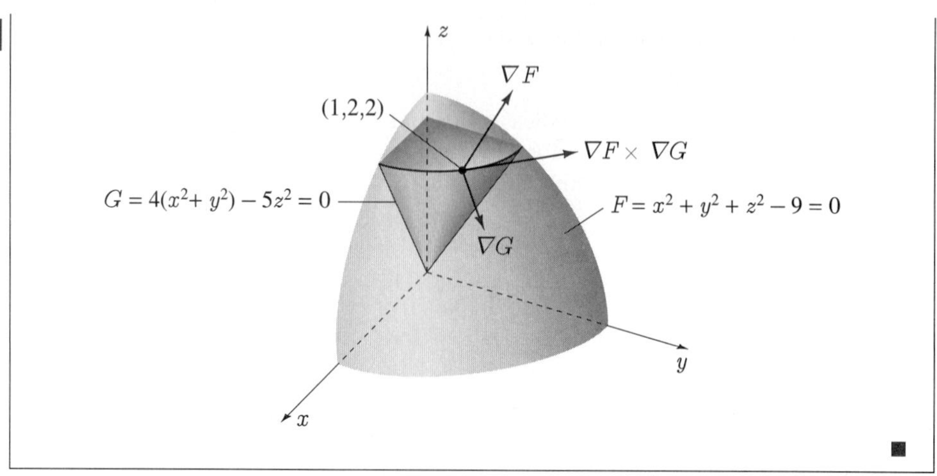

Example 13.26 illustrates that when a curve is defined as the intersection of two surfaces $F(x, y, z) = 0$, $G(x, y, z) = 0$ (Figure 13.25), then a vector tangent to the curve is

$$\mathbf{T} = \nabla F \times \nabla G. \tag{13.50}$$

Thus to find a tangent vector to a curve we use 13.46 when the curve is defined parametrically. When the curve is defined as the intersection of two surfaces, we can either find parametric equations and use 13.46, or use 13.50. Note too that in order to find tangent lines to curves, it is not necessary to have a direction assigned to the curves.

FIGURE 13.25

FIGURE 13.26

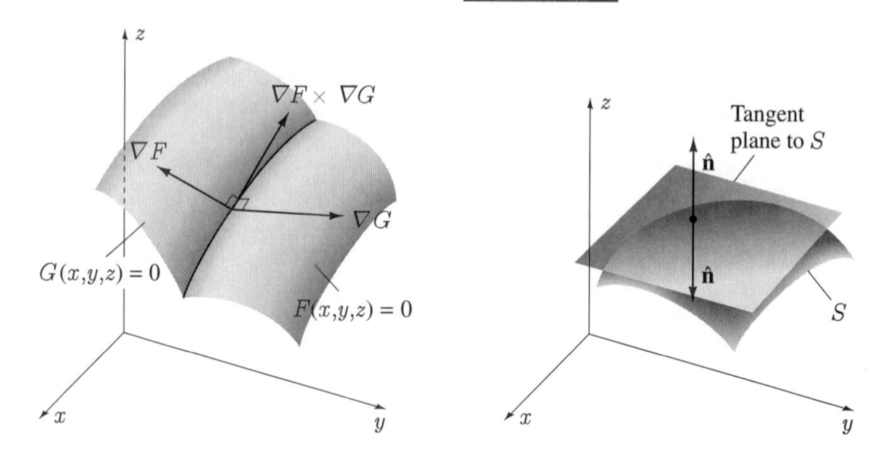

At each point on a surface S at which S has a tangent plane (Figure 13.26), we have defined a normal vector to S as a vector normal to the tangent plane to S. If we denote by $\hat{\mathbf{n}}$ a unit normal vector to S, then the direction of $\hat{\mathbf{n}}$ clearly varies as we move from point to point on S. We say that $\hat{\mathbf{n}}$ is a function of position (x, y, z) on S. Furthermore, at each point at which S has a unit normal vector, it has two such vectors, one in the opposite direction to the other. We say that a surface S is **smooth** if it can be assigned a unit normal $\hat{\mathbf{n}}$ that varies continuously on S. What this means geometrically is that for small changes in position, the unit normal $\hat{\mathbf{n}}$ will undergo small changes in direction. The sphere in Figure 13.27 is smooth, as is the paraboloid in Figure 13.28.

FIGURE 13.27

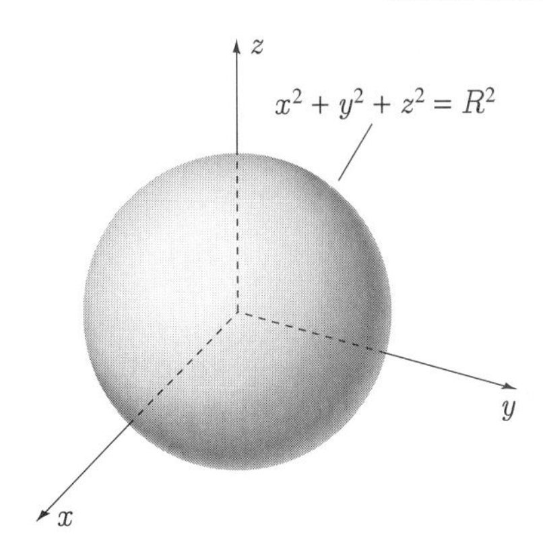

FIGURE 13.28

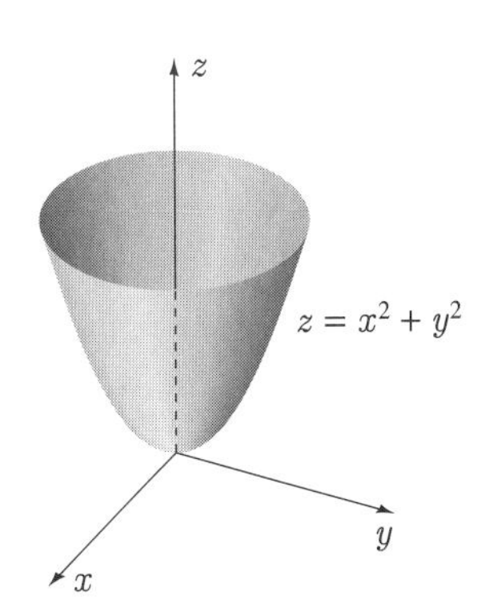

The surface bounding the cylindrical volume in Figure 13.29 is not smooth; a unit normal that varies continuously over the surface cannot be assigned at points on the circles $x^2 + y^2 = 1$, $z = \pm 1$. This surface can, however, be divided into a finite number of subsurfaces, each of which is smooth. In particular, we choose the three subsurfaces S_1: $z = 1$, $x^2 + y^2 \leq 1$; S_2: $z = -1$, $x^2 + y^2 \leq 1$; S_3: $x^2 + y^2 = 1$, $-1 < z < 1$. Such a surface is said to be **piecewise smooth**.

FIGURE 13.29

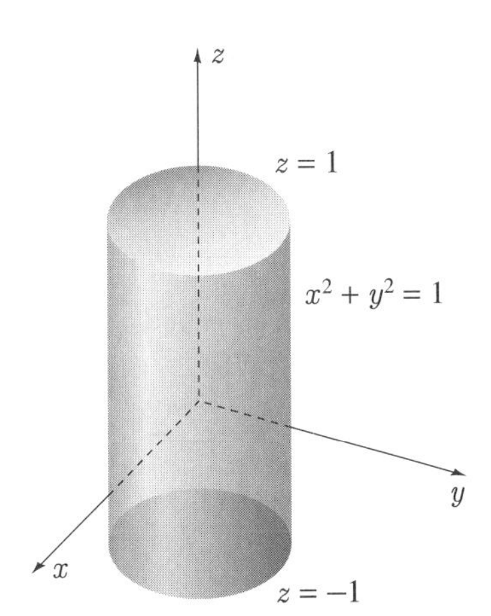

EXERCISES 13.9

In Exercises 1–20 find equations for the tangent line to the curve at the point.

1. $y = x^2$, $z = 0$ at $(-2, 4, 0)$

2. $x = t$, $y = t^2$, $z = t^3$ at $(1, 1, 1)$

3. $x = \cos t$, $y = \sin t$, $z = \cos t$ at $(1, 0, 1)$

4. $y = x^2$, $z = x$ at $(-2, 4, -2)$

5. $x^2 = y$, $z + x = y$ at $(1, 1, 0)$

6. $x = 2 - t^2$, $y = 3 + 2t$, $z = t$ at $(1, 5, 1)$

7. $x = 2\cos t$, $y = 3\sin t$, $z = 5$ at $(\sqrt{2}, -3/\sqrt{2}, 5)$

8. $x^2 y^3 + xy = 68$ at $(1, 4)$

9. $x + y + z = 4$, $x - y = 2$ at $(0, -2, 6)$

10. $x = e^{-t}\cos t$, $y = e^{-t}\sin t$, $z = t$ at $(1, 0, 0)$

11. $x = t^2 + 1$, $y = 2t - 4$, $z = t^3 + 3$ at $(2, -6, 2)$

12. $y^2 + z^2 = 6$, $x + z = 1$ at $(2, -\sqrt{5}, -1)$

13. $x^2 + y^2 + z^2 = 4$, $z^2 = x^2 + y^2$ at $(1, 1, -\sqrt{2})$

14. $x = t$, $y = 1$, $z = \sqrt{1 + t^2}$ at $(4, 1, \sqrt{17})$

15. $x = 1 + \cos t$, $y = 2 - \sin t$, $z = \sqrt{4 + t}$ at $(2, 2, 2)$

16. $x = z^2 + z^3,\ y = z - z^4$ at $(12, -14, 2)$

17. $x = y^2 + 3y^3 - 2y + 5,\ z = 0$ at $(7, 1, 0)$

18. $2x^2 + y^2 + 2y = 3,\ z = x + 1$ at $(0, 1, 1)$

19. $x = t^2,\ y = t,\ z = \sqrt{t + t^4}$ at $(1, 1, \sqrt{2})$

20. $x = t\sin t,\ y = t\cos t,\ z = 2t$ at $(0, 2\pi, 4\pi)$

In Exercises 21–26 find an equation for the tangent plane to the surface at the point.

21. $z = \sqrt{x^2 + y^2}$ at $(1, 1, \sqrt{2})$

22. $x = x^2 - y^3 z$ at $(2, -1, -2)$

23. $x^2 y + y^2 z + z^2 x + 3 = 0$ at $(2, -1, -1)$

24. $x + y + z = 4$ at $(1, 1, 2)$

25. $x = y\sin(\pi z/2)$ at $(-1, -1, 1)$

26. $x^2 + y^2 + 2y = 1$ at $(1, 0, 3)$

27. Show that the curve $x = 2(t^3 + 2)/3,\ y = 2t^2,\ z = 3t - 2$ intersects the surface $x^2 + 2y^2 + 3z^2 = 15$ at right angles at the point $(2, 2, 1)$.

28. Verify that the curve $x^2 - y^2 + z^2 = 1,\ xy + xz = 2$ is tangent to the surface $xyz - x^2 - 6y + 6 = 0$ at the point $(1, 1, 1)$.

29. Show that the equation of the tangent plane to a surface $S : z = f(x, y)$ at a point (x_0, y_0, z_0) on S can be written in the form

$$z - z_0 = (x - x_0)f_x(x_0, y_0) + (y - y_0)f_y(x_0, y_0).$$

In Exercises 30–32 find the indicated derivative for the function.

30. $f(x, y, z) = 2x^2 + y^2 z^2$ at $(3, 1, 0)$ with respect to distance along the curve $x + y + z = 4,\ x - y + z = 2$ in the direction of increasing x

31. $f(x, y, z) = xyz + xy + xz + yz$ at $(1, -2, 5)$ perpendicular to the surface $z = x^2 + y^2$

32. $f(x, y, z) = x^2 + y^2 - z^2$ at $(3, 4, 5)$ with respect to distance along the curve $x^2 + y^2 - z^2 = 0,\ 2x^2 + 2y^2 - z^2 = 25$ in the direction of decreasing x

33. If $F(x, y) = 0$ defines a curve implicitly in the xy-plane, prove that at any point on the curve, ∇F is perpendicular to the curve.

34. Find the equation of the tangent plane to the ellipsoid $x^2/a^2 + y^2/b^2 + z^2/c^2 = 1$ at any point (x_0, y_0, z_0) on the surface.

35. Find all points on the surface $z = x^2/4 - y^2/9$ at which the tangent plane is parallel to the plane $x + y + z = 4$.

36. Find all points on the surface $z^2 = 4(x^2 + y^2)$ at which the tangent plane is parallel to the plane $x - y + 2z = 3$.

37. Suppose that the equations $F(x, y, z, t) = 0$, $G(x, y, z, t) = 0$, $H(x, y, z, t) = 0$ implicitly define parametric equations for a curve C (t being the parameter). If $P(x_0, y_0, z_0)$ is a point on C, show that equations for the tangent line to C at P can be written in the form

$$\frac{x - x_0}{\dfrac{\partial(F, G, H)}{\partial(t, y, z)}\Big|_P} = \frac{y - y_0}{\dfrac{\partial(F, G, H)}{\partial(x, t, z)}\Big|_P} = \frac{z - z_0}{\dfrac{\partial(F, G, H)}{\partial(x, y, t)}\Big|_P},$$

provided none of the Jacobians vanish.

38. Find all points on the paraboloid $z = x^2 + y^2 - 1$ at which the normal to the surface coincides with the line joining the origin to the point.

39. Show that the sum of the intercepts on the x-, y-, and z-axes of the tangent plane to the surface $\sqrt{x} + \sqrt{y} + \sqrt{z} = \sqrt{a}$ at any point is a.

SECTION 13.10

Relative Maxima and Minima

We now study relative extrema of functions of more than one independent variable. Most of the discussion will be confined to functions $f(x, y)$ of two independent variables because we can discuss the concepts geometrically as well as algebraically. Unfortunately, not all the results are easily extended to functions of more than two independent variables, and we will therefore be careful to point out these limitations.

Before beginning the discussion, we briefly review maxima-minima results for functions $f(x)$ of one variable. We do this because maxima-minima theory for multivariable functions is essentially the same as that for single-variable functions. In fact, every definition that we make and every result that we discuss in this section has its counterpart in single-variable theory. Hence, a synopsis of single-variable results is in order. Unfortunately, proving results in the multivariable case is considerably more complicated than in the single-variable case, but if we can keep central ideas foremost in our minds and constantly make comparisons with single-variable calculus, we will find that discussions are not nearly as difficult as they might otherwise be.

Critical points of a function $f(x)$ are points at which $f'(x)$ is either equal to zero or does not exist. Geometrically, this means points at which the graph of $f(x)$ has a horizontal tangent line, a vertical tangent line, or no tangent line at all. Critical points for continuous functions are classified as yielding relative maxima, relative minima, horizontal points of inflection, vertical points of inflection, or just corners. There are two tests to determine whether a critical point x_0 yields a relative maximum or a relative minimum. The first-derivative test states that if $f'(x)$ changes from a positive quantity to a negative quantity as x increases through x_0, then x_0 gives a relative maximum; if $f'(x)$ changes from negative to positive, then a relative minimum is obtained. The second-derivative test indicates the nature of a critical point at which $f'(x_0) = 0$ whenever $f''(x_0) \neq 0$. If $f''(x_0) > 0$, then a relative minimum is obtained, and if $f''(x_0) < 0$, a relative maximum is found.

We begin our study of extrema theory for multivariable functions by defining critical points for functions of two independent variables.

Definition 13.6

A point (x_0, y_0) in the domain of a function $f(x, y)$ is said to be a **critical point** of $f(x, y)$ if

$$\frac{\partial f}{\partial x}_{|(x_0,y_0)} = 0, \qquad \frac{\partial f}{\partial y}_{|(x_0,y_0)} = 0, \tag{13.51}$$

or if one (or both) of these partial derivatives does not exist at (x_0, y_0).

There are two ways to interpret critical points of $f(x, y)$ geometrically. In Section 13.3, we interpreted $\partial f/\partial x$ at (x_0, y_0) as the slope of the tangent line to the curve of intersection of the surface $z = f(x, y)$ and the plane $y = y_0$, and $\partial f/\partial y$ as the slope of the tangent line to the curve of intersection with $x = x_0$. It follows, then, that (x_0, y_0) is critical if both curves have horizontal tangent lines or if either curve has a vertical tangent line or no tangent line at all. Alternatively, recall that the equation of the tangent plane to the surface $z = f(x, y)$ at (x_0, y_0) is

$$z - z_0 = f_x(x_0, y_0)(x - x_0) + f_y(x_0, y_0)(y - y_0)$$

(see Exercise 29 in Section 13.9). If both partial derivatives vanish, then the tangent plane is horizontal with equation $z = z_0$. For example, at each of the critical points in Figures 13.30–13.33, $\partial f/\partial x = \partial f/\partial y = 0$ and the tangent plane is horizontal. The remaining functions in Figures 13.34–13.38 have critical points at which either $\partial f/\partial x$ or $\partial f/\partial y$ or both do not exist. In Figures 13.34–13.37, the surfaces do not have tangent planes at critical points, and in Figure 13.38, the tangent plane is vertical at each critical point. Consequently, (x_0, y_0) is a critical point of a function $f(x, y)$ if at (x_0, y_0) the surface $z = f(x, y)$ has a horizontal tangent plane, a vertical tangent plane, or no tangent plane at all.

FIGURE 13.30

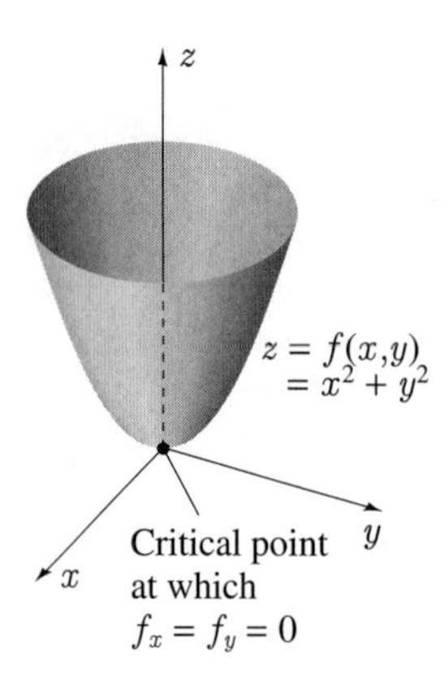

FIGURE 13.31

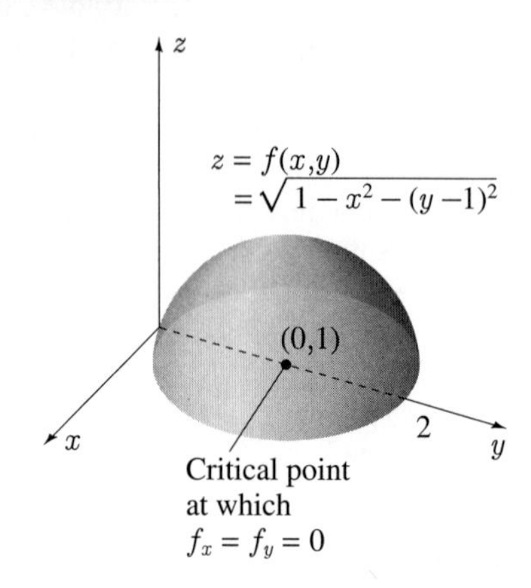

FIGURE 13.32

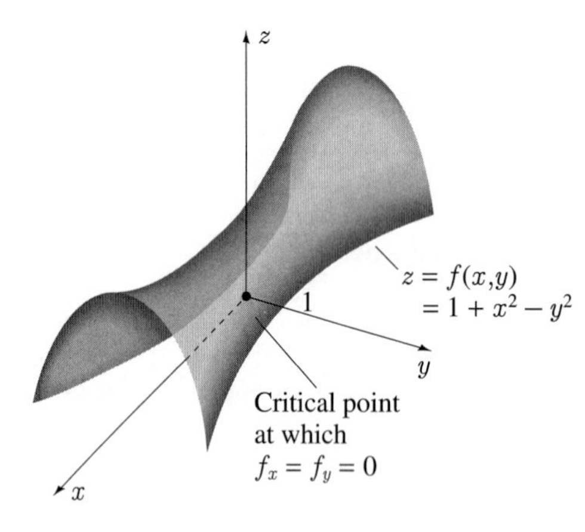

FIGURE 13.33

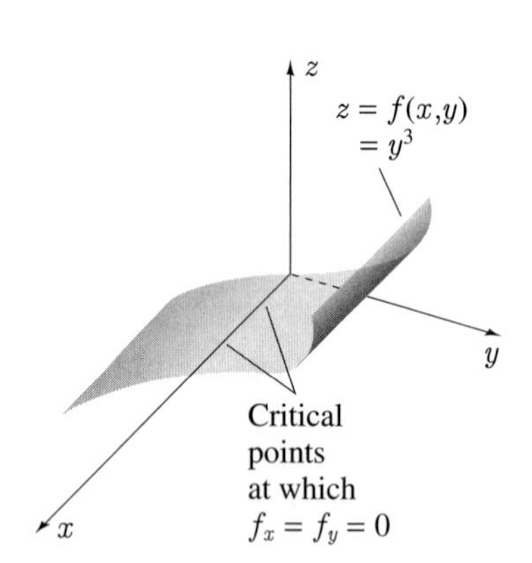

FIGURE 13.34

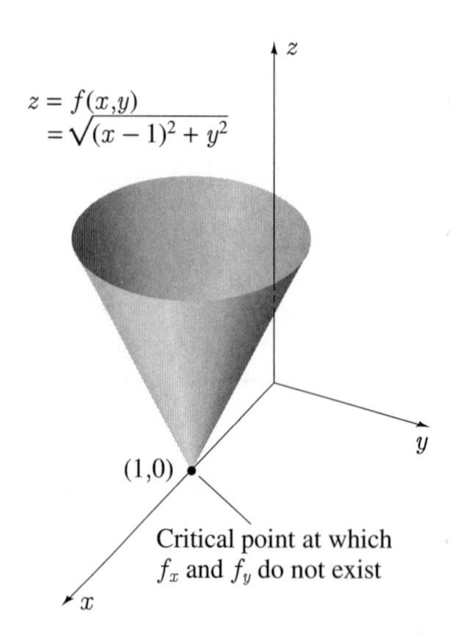

FIGURE 13.35

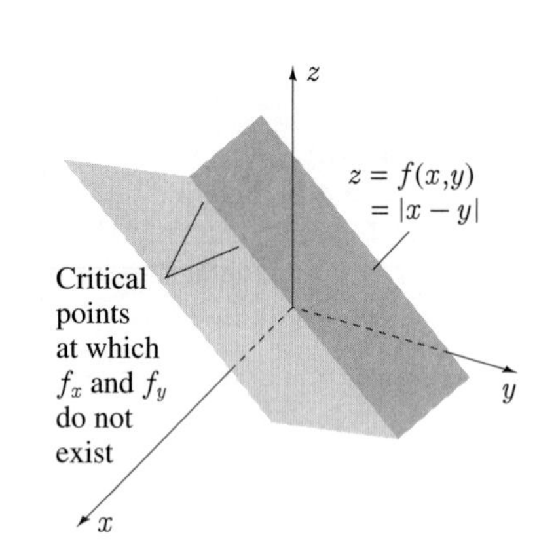

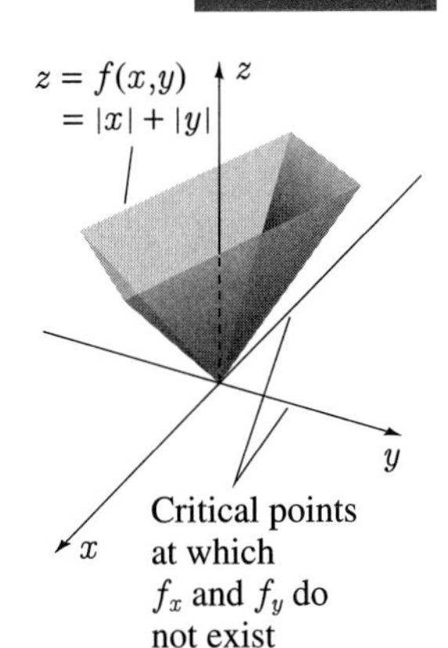

FIGURE 13.36

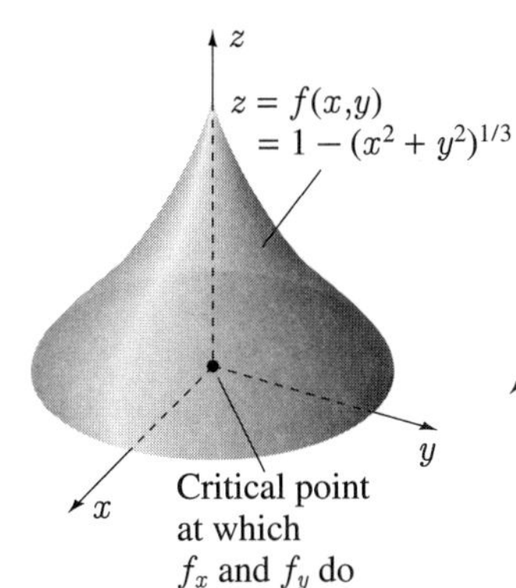

FIGURE 13.37

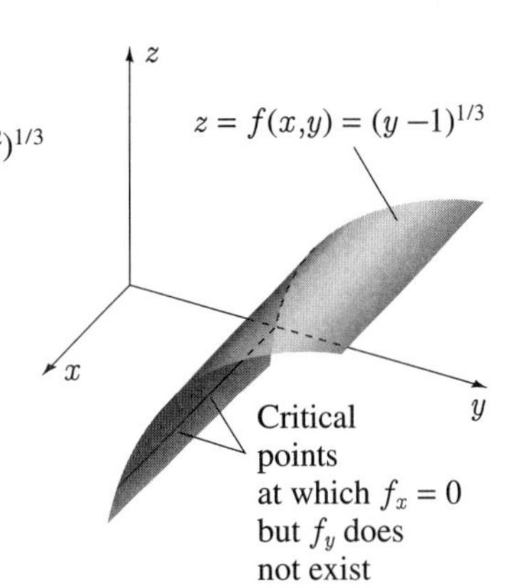

FIGURE 13.38

EXAMPLE 13.27

Find all critical points for the function

$$f(x, y) = x^2 y - 2xy^2 + 3xy + 4.$$

SOLUTION For critical points, we first solve

$$0 = \frac{\partial f}{\partial x} = 2xy - 2y^2 + 3y = y(2x - 2y + 3),$$

$$0 = \frac{\partial f}{\partial y} = x^2 - 4xy + 3x = x(x - 4y + 3).$$

To satisfy these two equations simultaneously, there are four possibilities:

1. $x = 0$, $y = 0$, which gives the critical point $(0, 0)$;
2. $y = 0$, $x - 4y + 3 = 0$, which gives the critical point $(-3, 0)$;
3. $x = 0$, $2x - 2y + 3 = 0$, which gives the critical point $(0, \frac{3}{2})$;
4. $2x - 2y + 3 = 0$, $x - 4y + 3 = 0$, which gives the critical point $(-1, \frac{1}{2})$.

Since $\partial f/\partial x$ and $\partial f/\partial y$ are defined for all x and y, these are the only critical points. ■

Critical points for functions of more than two independent variables can be defined algebraically, but because we have no geometric representation for such functions, there is no geometric interpretation for their critical points. For example, if $f(x, y, z, t)$ is a function of independent variables x, y, z, and t, then (x_0, y_0, z_0, t_0) is a critical point of $f(x, y, z, t)$ if all four of its first-order partial derivatives vanish at (x_0, y_0, z_0, t_0),

$$\frac{\partial f}{\partial x}_{|(x_0, y_0, z_0, t_0)} = \frac{\partial f}{\partial y}_{|(x_0, y_0, z_0, t_0)} = \frac{\partial f}{\partial z}_{|(x_0, y_0, z_0, t_0)} = \frac{\partial f}{\partial t}_{|(x_0, y_0, z_0, t_0)} = 0, \qquad (13.52)$$

or if at least one of the partial derivatives does not exist at the point. Note that because the partial derivatives of a function are the components of the gradient of the function, we can say that a critical point of a function is a point at which its gradient is either equal to zero or undefined.

EXAMPLE 13.28

Find all critical points for the function

$$f(x,y,z) = xyz\sqrt{x^2+y^2+z^2}.$$

SOLUTION For critical points, we consider the equations

$$\begin{aligned} 0 = \frac{\partial f}{\partial x} &= yz\sqrt{x^2+y^2+z^2} + \frac{x^2yz}{\sqrt{x^2+y^2+z^2}} \\ &= \frac{yz}{\sqrt{x^2+y^2+z^2}}(2x^2+y^2+z^2), \\ 0 = \frac{\partial f}{\partial y} &= xz\sqrt{x^2+y^2+z^2} + \frac{xy^2z}{\sqrt{x^2+y^2+z^2}} \\ &= \frac{xz}{\sqrt{x^2+y^2+z^2}}(x^2+2y^2+z^2), \\ 0 = \frac{\partial f}{\partial z} &= xy\sqrt{x^2+y^2+z^2} + \frac{xyz^2}{\sqrt{x^2+y^2+z^2}} \\ &= \frac{xy}{\sqrt{x^2+y^2+z^2}}(x^2+y^2+2z^2). \end{aligned}$$

The partial derivatives are clearly undefined for $x = y = z = 0$, and therefore the origin $(0,0,0)$ is a critical point. If x, y, and z are not all zero, then the terms in parentheses cannot vanish, and we must set

$$yz = 0, \quad xz = 0, \quad xy = 0.$$

If any two of x, y, and z vanish, but the third does not, then these equations are satisfied. In other words, every point on the x-axis, every point on the y-axis, and every point on the z-axis is critical. ■

We now turn our attention to the classification of critical points of a function $f(x,y)$ of two independent variables. Critical points $(0,1)$ in Figure13.31 and $(0,0)$ in Figure 13.37 yield "high" points on the surfaces. We describe this property in the following definition.

Definition 13.7

A function $f(x,y)$ is said to have a **relative maximum** $f(x_0,y_0)$ at a point (x_0,y_0) if there exists a circle in the xy-plane centred at (x_0,y_0) such that for all points (x,y) inside this circle

$$f(x,y) \le f(x_0,y_0). \tag{13.53}$$

The "low" points on the surfaces at $(0,0)$ in Figure 13.30 and $(1,0)$ in Figure 13.34 are relative minima according to the following.

Definition 13.8

A function $f(x, y)$ is said to have a **relative minimum** $f(x_0, y_0)$ at a point (x_0, y_0) if there exists a circle in the xy-plane centred at (x_0, y_0) such that for all points (x, y) inside this circle

$$f(x, y) \geq f(x_0, y_0). \tag{13.54}$$

Note that every critical point in Figure 13.35 yields a relative minimum of $f(x, x) = 0$, as does the critical point $(0, 0)$ in Figure 13.36.

Definition 13.9

If a critical point of a function $f(x, y)$ at which $\partial f/\partial x = \partial f/\partial y = 0$ yields neither a relative maximum nor a relative minimum, it is said to yield a **saddle point**.

The critical point $(0, 0)$ in Figure 13.32 therefore gives a saddle point, as does each of the critical points in Figure 13.33. Saddle points for surfaces $z = f(x, y)$ are clearly the analogues of horizontal points of inflection for curves $y = f(x)$. In both cases the derivative(s) of the function vanishes but there is neither a relative maximum nor a relative minimum.

The critical points in Figure 13.36 (except $(0, 0)$) are the counterparts of corners for the graph of a function $f(x)$. They are points at which one or both of the partial derivatives of $f(x, y)$ do not exist, but like corners for $f(x)$, they do not necessarily yield relative extrema. Critical points in Figure 13.38 are the analogues of vertical points of inflection for a function $f(x)$.

Our discussion has made it clear that:

(a) At a relative maximum or minimum of $f(x, y)$ either $\partial f/\partial x$ and $\partial f/\partial y$ both vanish, or one or both of the partial derivatives do not exist;

(b) Saddle points may also occur where $\partial f/\partial x = \partial f/\partial y = 0$, and points where the derivatives do not exist may fail to yield relative extrema.

In other words, every relative extremum of $f(x, y)$ occurs at a critical point; but critical points do not always give relative extrema.

Given the problem of determining all relative maxima and minima of a function $f(x, y)$, we should first find its critical points. But how do we decide whether these critical points yield relative maxima, relative minima, saddle points, or none of these? We do not have a practical test that is equivalent to the first-derivative test for functions of one variable, but we do have a test that corresponds to the second-derivative test. For functions of two independent variables the situation is more complicated, however, since there are three second-order partial derivatives, but the idea of the test is essentially the same. It determines whether certain curves are concave upward or concave downward at the critical point. The complete result is contained in the following theorem.

Theorem 13.7 *Suppose (x_0, y_0) is a critical point of $f(x,y)$ at which $\partial f/\partial x$ and $\partial f/\partial y$ both vanish. Suppose further that f_x, f_y, f_{xx}, f_{xy}, and f_{yy} are all continuous at (x_0, y_0). Define*

$$A = f_{xx}(x_0, y_0), \quad B = f_{xy}(x_0, y_0), \quad C = f_{yy}(x_0, y_0).$$

If

(i) $B^2 - AC < 0$ *and* $A < 0$, *then* $f(x,y)$ *has a relative maximum at* (x_0, y_0);
(ii) $B^2 - AC < 0$ *and* $A > 0$, *then* $f(x,y)$ *has a relative minimum at* (x_0, y_0);
(iii) $B^2 - AC > 0$, *then* $f(x,y)$ *has a saddle point at* (x_0, y_0);
(iv) $B^2 - AC = 0$, *then the test fails.*

Proof

(i) Suppose we intersect the surface $z = f(x,y)$ with a plane parallel to the z-axis, through the point $(x_0, y_0, 0)$, and making an angle α with the line through $(x_0, y_0, 0)$ parallel to the positive x-axis (Figure 13.39). The slope of the curve of intersection of these surfaces at the point $(x_0, y_0, f(x_0, y_0))$ is given by the directional derivative

$$\begin{aligned} D_{\mathbf{v}} f_{|(x_0, y_0)} = \nabla f_{|(x_0, y_0)} \cdot \hat{\mathbf{v}} &= \nabla f_{|(x_0, y_0)} \cdot (\cos\alpha, \sin\alpha) \\ &= \frac{\partial f}{\partial x}_{|(x_0, y_0)} \cos\alpha + \frac{\partial f}{\partial y}_{|(x_0, y_0)} \sin\alpha \end{aligned}$$

(see equation 13.41). Since (x_0, y_0) is a critical point at which $\nabla f = \mathbf{0}$, it follows that

$$D_{\mathbf{v}} f_{|(x_0, y_0)} = 0 \quad \text{for all} \quad \alpha.$$

FIGURE 13.39

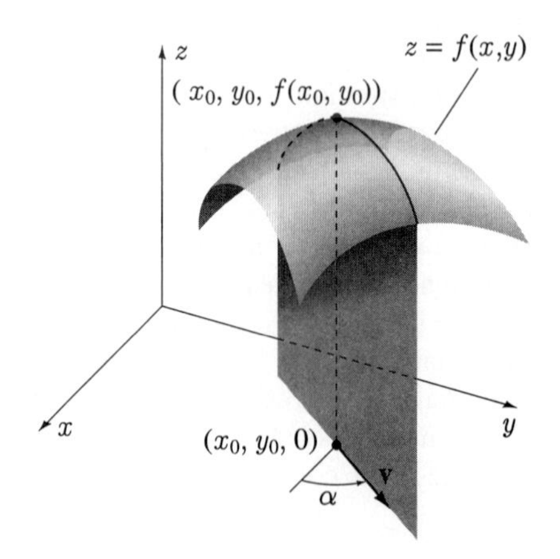

In Figure 13.39 we have illustrated the critical point as a relative maximum. But how do we verify that this is indeed the case? If we can show that each and every curve of intersection of the surface with a vertical plane through $(x_0, y_0, 0)$ is concave downward at $(x_0, y_0, 0)$, then (x_0, y_0) must give a relative maximum. But to discuss concavity of a curve we require the second derivative—in this case, the second directional derivative of $f(x,y)$. According to 13.42, the second directional derivative of $f(x,y)$ at (x_0, y_0) in the direction $\hat{\mathbf{v}} = (\cos\alpha, \sin\alpha)$ is

$$D_{\mathbf{v}}(D_{\mathbf{v}}f) = \frac{\partial^2 f}{\partial x^2}_{|(x_0,y_0)} \cos^2\alpha + 2\frac{\partial^2 f}{\partial x \partial y}_{|(x_0,y_0)} \cos\alpha \sin\alpha + \frac{\partial^2 f}{\partial y^2}_{|(x_0,y_0)} \sin^2\alpha$$
$$= A\cos^2\alpha + 2B\cos\alpha\sin\alpha + C\sin^2\alpha,$$

where we understand here that $D_{\mathbf{v}}(D_{\mathbf{v}}f)$ is implicitly suffixed by (x_0, y_0). In order, therefore, to verify that (x_0, y_0) gives a relative maximum, it is sufficient to show that $D_{\mathbf{v}}(D_{\mathbf{v}}f)$ is negative for each value of α in the interval $0 \leq \alpha < 2\pi$. However, because $D_v(D_v f)$ is unchanged if α is replaced by $\alpha + \pi$, it is sufficient to verify that $D_{\mathbf{v}}(D_{\mathbf{v}}f)$ is negative for $0 \leq \alpha < \pi$.

For any of these values of α except $\pi/2$, we can write

$$D_{\mathbf{v}}(D_{\mathbf{v}}f) = \cos^2\alpha(A + 2B\tan\alpha + C\tan^2\alpha),$$

and if we set $u = \tan\alpha$,

$$D_{\mathbf{v}}(D_{\mathbf{v}}f) = \cos^2\alpha(A + 2Bu + Cu^2).$$

It is evident that $D_{\mathbf{v}}(D_{\mathbf{v}}f) < 0$ for all $\alpha \neq \pi/2$ if and only if

$$Q(u) = A + 2Bu + Cu^2 < 0 \quad \text{for} \quad -\infty < u < \infty.$$

Were we to sketch a graph of the quadratic $Q(u)$, we would see that it crosses the u-axis where

$$u = \frac{-2B \pm \sqrt{4B^2 - 4AC}}{2C} = \frac{-B \pm \sqrt{B^2 - AC}}{C}.$$

But if we know, as stated in Theorem 13.7(i), that $B^2 - AC < 0$, then there are no real solutions of this equation, and therefore $Q(u)$ never crosses the u-axis. Since $Q(0) = A < 0$, it follows that $Q(u) < 0$ for all u. We have shown, then, that

$$D_{\mathbf{v}}(D_{\mathbf{v}}f) < 0 \quad \text{for all} \quad \alpha \neq \pi/2.$$

When $\alpha = \pi/2$, $D_{\mathbf{v}}(D_{\mathbf{v}}f) = C$. Since $B^2 - AC < 0$ and $A < 0$, it follows that $C < 0$ also. Consequently, if $B^2 - AC < 0$ and $A < 0$, then $D_{\mathbf{v}}(D_{\mathbf{v}}f) < 0$ for all α, and (x_0, y_0) yields a relative maximum.

(ii) If $B^2 - AC < 0$ and $A > 0$, a similar argument leads to the conclusion that (x_0, y_0) yields a relative minimum; the only difference is that inequalities are reversed.

(iii) If $B^2 - AC > 0$, then $Q(u)$ has real distinct zeros, in which case $Q(u)$ is sometimes negative and sometimes positive. This means that the curve of intersection is sometimes concave upward and sometimes concave downward, and the point (x_0, y_0) therefore gives a saddle point.

(iv) If $B^2 - AC = 0$, the classification of the point determined by (x_0, y_0) depends on which of A, B, and C vanish, if any.

To illustrate that we can obtain a relative maximum, a relative minimum, or a saddle point for a critical point at which $B^2 - AC = 0$, consider the three functions $f(x,y) = -y^2$, $f(x,y) = y^2$, and $f(x,y) = y^3$ in Figures 13.40–13.42. The point $(0,0)$ is a critical point for each function, and at this point $B^2 - AC = 0$. Yet $(0,0)$ yields a relative maximum for $f(x,y) = -y^2$, a relative minimum for $f(x,y) = y^2$, and a saddle point for $f(x,y) = y^3$. In fact, every point on the x-axis is a relative maximum for $f(x,y) = -y^2$, a relative minimum for $f(x,y) = y^2$, and a saddle point for $f(x,y) = y^3$.

FIGURE 13.40

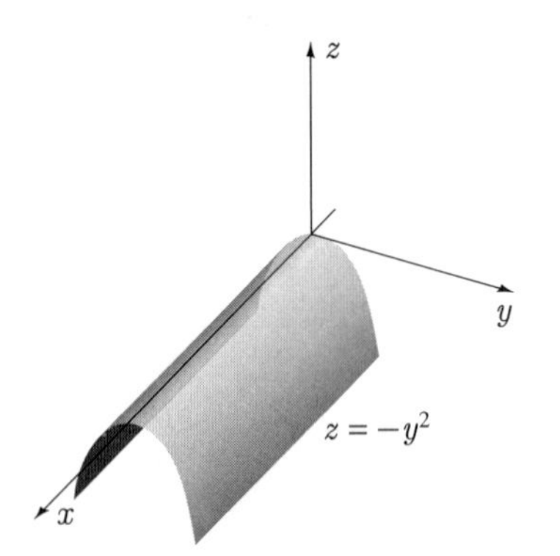

FIGURE 13.41

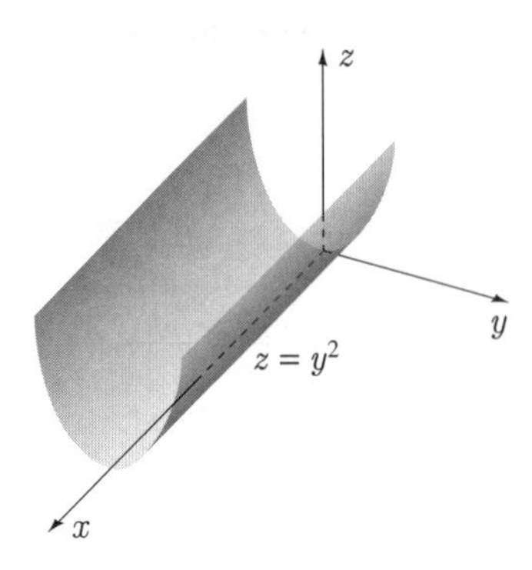

FIGURE 13.42

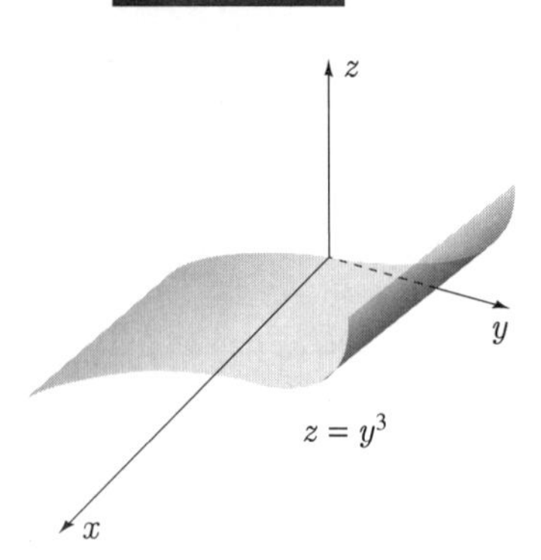

EXAMPLE 13.29

Find and classify critical points for each of the following functions as yielding relative maxima, relative minima, saddle points, or none of these:

(a) $f(x,y) = 4xy - x^4 - y^4$

(b) $f(x,y) = y^3 + x^2 - 6xy + 3x + 6y$

(c) $f(x,y) = \dfrac{12xy - x^2y^2}{2(x+y)}$

SOLUTION

(a) Critical points of $f(x,y)$ are given by

$$0 = \frac{\partial f}{\partial x} = 4y - 4x^3, \quad 0 = \frac{\partial f}{\partial y} = 4x - 4y^3.$$

Solutions of these equations are $(0,0)$, $(1,1)$, and $(-1,-1)$. We now calculate

$$\frac{\partial^2 f}{\partial x^2} = -12x^2, \quad \frac{\partial^2 f}{\partial x \partial y} = 4, \quad \frac{\partial^2 f}{\partial y^2} = -12y^2.$$

At $(0,0)$, $B^2 - AC = 16 > 0$, and therefore $(0,0)$ yields a saddle point. At $(1,1)$, $B^2 - AC = -128$ and $A = -12$, and therefore $(1,1)$ gives a relative maximum. At $(-1,-1)$, $B^2 - AC = -128$ and $A = -12$, and $(-1,-1)$ also gives a relative maximum.

(b) Critical points for $f(x,y)$ are given by

$$0 = \frac{\partial f}{\partial x} = 2x - 6y + 3, \quad 0 = \frac{\partial f}{\partial y} = 3y^2 - 6x + 6.$$

Solutions of these equations are $(\frac{27}{2}, 5)$ and $(\frac{3}{2}, 1)$. The second derivatives of $f(x,y)$ are

$$\frac{\partial^2 f}{\partial x^2} = 2, \quad \frac{\partial^2 f}{\partial x \partial y} = -6, \quad \frac{\partial^2 f}{\partial y^2} = 6y.$$

At $(\frac{27}{2},5)$, $B^2 - AC = -24$ and $A = 2$, and therefore $(\frac{27}{2},5)$ yields a relative minimum. At $(\frac{3}{2},1)$, $B^2 - AC = 24$, and therefore $(\frac{3}{2},1)$ gives a saddle point.

(c) For critical points, we consider

$$0 = \frac{\partial f}{\partial x} = \frac{y^2(12 - x^2 - 2xy)}{2(x+y)^2}, \quad 0 = \frac{\partial f}{\partial y} = \frac{x^2(12 - y^2 - 2xy)}{2(x+y)^2},$$

or

$$0 = y^2(12 - x^2 - 2xy), \quad 0 = x^2(12 - y^2 - 2xy).$$

Solutions of these equations are $(2,2)$ and $(-2,-2)$ (the solution $(0,0)$ has been rejected since $f(x,y)$ is not defined there). To classify these critical points by means of the second-derivative test would lead to some messy calculations; thus we will show that the classification can be achieved without the test. First we note that $f(x,y) = 0$ whenever

$$0 = 12xy - x^2y^2 = xy(12 - xy);$$

i.e., along the x-axis, the y-axis, and the hyperbola $xy = 12$ (except $(0,0)$; see Figure 13.43). Further, between the axes and that part of the hyperbola in the first quadrant, the function is always positive. Finally, since $(2,2)$ is the only critical point in this region, it follows that $(2,2)$ could only yield a relative maximum. A similar argument in the third quadrant indicates that $(-2,-2)$ gives a relative minimum.

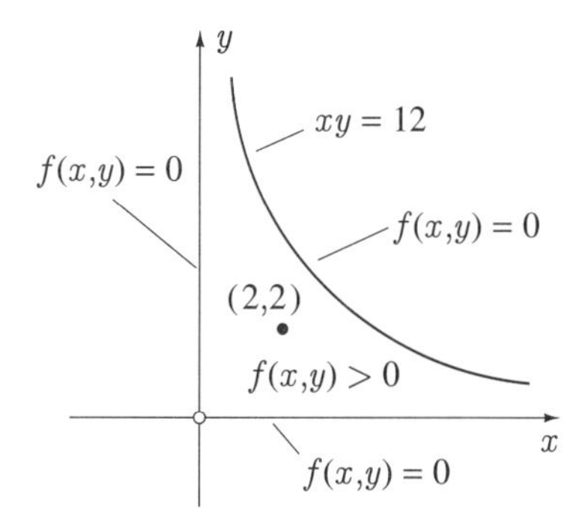

FIGURE 13.43

■

This completes our discussion of relative extrema of functions of two independent variables. Our next step should be extend the theory to functions of more than two variables. It is a simple matter to give definitions of relative maxima and minima for such functions; they are almost identical to Definitions 13.7 and 13.8 (see Exercise 19). On the other hand, to develop a theorem for functions of more than two independent variables that is analogous to Theorem 13.7 is beyond the scope of this book. We refer the interested reader to more advanced books.

EXERCISES 13.10

In Exercises 1–14 find all critical points for the function and classify each as yielding a relative maximum, a relative minimum, a saddle point, or none of these.

1. $f(x,y) = x^2 + 2xy + 2y^2 - 6y$

2. $f(x,y) = 3xy - x^3 - y^3$

3. $f(x,y) = x^3 - 3x + y^2 + 2y$

4. $f(x,y) = x^2y^2 + 3x$

5. $f(x,y) = xy - x^2 + y^2$

6. $f(x,y) = x \sin y$

7. $f(x,y) = xye^{-(x^2+y^2)}$

8. $f(x,y) = x^2 - 2xy + y^2$

9. $f(x,y) = (x^2 + y^2)^{2/3}$

10. $f(x,y) = x^4 y^3$

11. $f(x,y) = 2xy^2 + 3xy + x^2 y^3$

12. $f(x,y) = |x| + y^2$

13. $f(x,y) = (1-x)(1-y)(x+y-1)$

14. $f(x,y) = x^4 + y^4 - x^2 - y^2 + 1$

In Exercises 15–18 find all critical points for the function.

15. $f(x,y,z) = x^2 + y^2 - z^2 + 3x - 2y + 5$

16. $f(x,y,z,t) = x^2 y^2 z^2 + t^2 x^2 + 3x$

17. $f(x,y,z) = xyz + x^2 yz - y$

18. $f(x,y,z) = xyze^{x^2+y^2+z^2}$

19. Give definitions for a relative maximum and a relative minimum for a function $f(x,y,z)$ at a point (x_0, y_0, z_0).

20. Suppose that $f(x,y)$ is harmonic in the region D : $x^2 + y^2 < 1$. Show that $f(x,y)$ cannot have a relative maximum or minimum at any point in D at which either f_{xx} or f_{xy} does not vanish.

21. Find and classify the critical points for the function $f(x,y) = y^2 - 4x^2 y + 3x^4$.

22. **(a)** Plot the ten points in the following table to show that they are reasonably close to being collinear.

x	1	2	3	4	5	6	7	8	9	10
y	6.05	8.32	10.74	13.43	15.90	18.38	20.93	23.32	24.91	28.36

(b) If $y = mx + b$ is the equation of a straight line that we might use to approximate the function $y = f(x)$ described by these points, then the following sum is a measure of how well the line fits the data:

$$S = S(m,b) = \sum_{i=1}^{10} (\overline{y}_i - mx_i - b)^2,$$

where $(x_i, \overline{y}_i)$ are the data points in the table. What does this sum represent geometrically?

(c) To find the best straight line to fit the data, the least-squares method suggests that m and b be chosen to minimize S. Show that S has only one critical point (m,b) which is defined by the linear equations

$$\left(\sum_{i=1}^{10} x_i^2\right) m + \left(\sum_{i=1}^{10} x_i\right) b = \sum_{i=1}^{10} x_i \overline{y}_i,$$

$$\left(\sum_{i=1}^{10} x_i\right) m + 10b = \sum_{i=1}^{10} \overline{y}_i.$$

(d) Solve these equations for m and b.

23. **(a)** Plot the sixteen points in the following table.

x	3.00	3.25	3.50	3.75	4.00	4.25	4.50	4.75
y	31.5	30.4	29.2	28.1	26.9	26.4	25.3	25.2

x	5.00	5.25	5.50	5.75	6.00	6.25	6.50	6.75
y	25.1	25.2	25.4	26.3	27.0	28.2	29.3	29.9

Do they seem to follow a parabolic path?

(b) If $y = ax^2 + bx + c$ is the equation of a parabola that is to approximate the function $y = f(x)$ described by these points, then the following sum is a measure of the accuracy of the fit:

$$S = S(a,b,c) = \sum_{i=1}^{16} (\overline{y}_i - ax_i^2 - bx_i - c)^2,$$

where $(x_i, \overline{y}_i)$ are the points in the table. To find the best possible fit, the least-squares method says to choose a, b, and c to minimize S. Show that S has only one critical point (a,b,c) which is defined by the linear equations

$$\left(\sum_{i=1}^{16} x_i^4\right) a + \left(\sum_{i=1}^{16} x_i^3\right) b + \left(\sum_{i=1}^{16} x_i^2\right) c = \sum_{i=1}^{16} x_i^2 \overline{y}_i,$$

$$\left(\sum_{i=1}^{16} x_i^3\right) a + \left(\sum_{i=1}^{16} x_i^2\right) b + \left(\sum_{i=1}^{16} x_i\right) c = \sum_{i=1}^{16} x_i \overline{y}_i,$$

$$\left(\sum_{i=1}^{16} x_i^2\right) a + \left(\sum_{i=1}^{16} x_i\right) b + 16c = \sum_{i=1}^{16} \overline{y}_i.$$

(c) Solve these equations for a, b, and c.

24. Find and classify the critical points of $f(x,y) = x^4 + 3xy^2 + y^2$ as yielding relative maxima, relative minima, or saddle points.

25. The equation $2x^2 + 3y^2 + z^2 - 12xy + 4xz = 35$ defines functions $z = f(x,y)$. Show that the point $x = 1$ and $y = 2$ is a critical point for any such function. Does it yield a relative extrema for the function?

26. (a) Show that the function $f(x, y, z) = x^2 + y^2 + z^2 - xyz$ has a critical point $(0, 0, 0)$. What are the other critical points?

(b) Use the definition in Exercise 19 to show that $f(x, y, z)$ has a relative minimum at $(0, 0, 0)$.

SECTION 13.11

Absolute Maxima and Minima

So far as applications of maxima and minima are concerned, absolute maxima and minima are more important than relative maxima and minima. In this section and in Section 13.12 we discuss the theory of absolute extrema and consider a number of applications. Once again we begin with functions $f(x, y)$ of two independent variables and base our discussion on the theory of absolute extrema for functions of one variable.

We learned in Section 4.6 that a function $f(x)$ that is continuous on a finite interval $a \le x \le b$ must have an absolute maximum and an absolute minimum on that interval. Furthermore, these absolute extrema must occur at either critical points or at the ends $x = a$ and $x = b$ of the interval. Consequently, to find the absolute extrema of a function $f(x)$, we evaluate $f(x)$ at all critical points, at $x = a$, and at $x = b$; the largest of these numbers is the absolute maximum of $f(x)$ on $a \le x \le b$, and the smallest is the absolute minimum.

The procedure is much the same for a function $f(x, y)$ that is continuous on a region R that is finite and includes all the points on its boundary. First, however, we define exactly what we mean by absolute extrema of $f(x, y)$ and consider a number of simple examples. We will then be able to make general statements about the nature of all absolute extrema, and proceed to the important area of applications.

Definition 13.10

The **absolute maximum** of a function $f(x, y)$ on a region R is $f(x_0, y_0)$ if (x_0, y_0) is in R and

$$f(x, y) \le f(x_0, y_0) \tag{13.55}$$

for all (x, y) in R. The **absolute minimum** of $f(x, y)$ on R is $f(x_0, y_0)$ if (x_0, y_0) is in R and

$$f(x, y) \ge f(x_0, y_0) \tag{13.56}$$

for all (x, y) in R.

In Figures 13.44–13.49, we have shown six functions defined on the circle R: $x^2 + y^2 \le 1$. The absolute maxima and minima of these functions for this region are shown in Table 13.1.

FIGURE 13.44

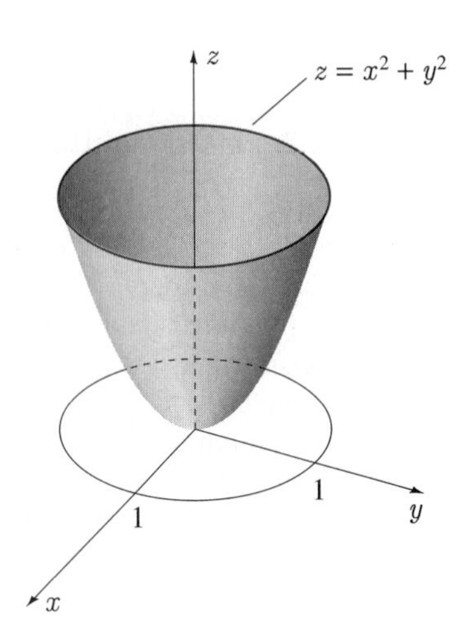

FIGURE 13.45

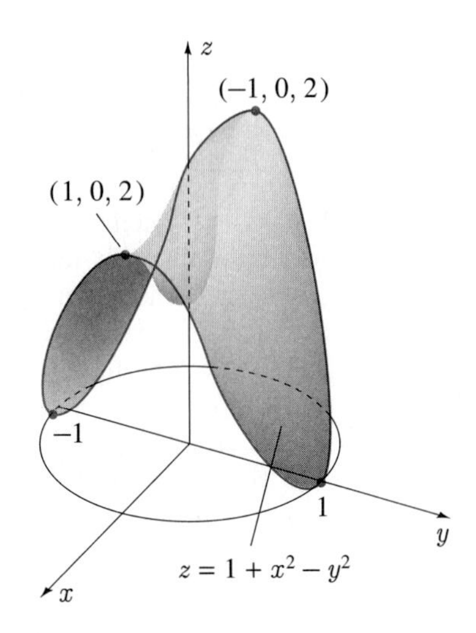

FIGURE 13.46

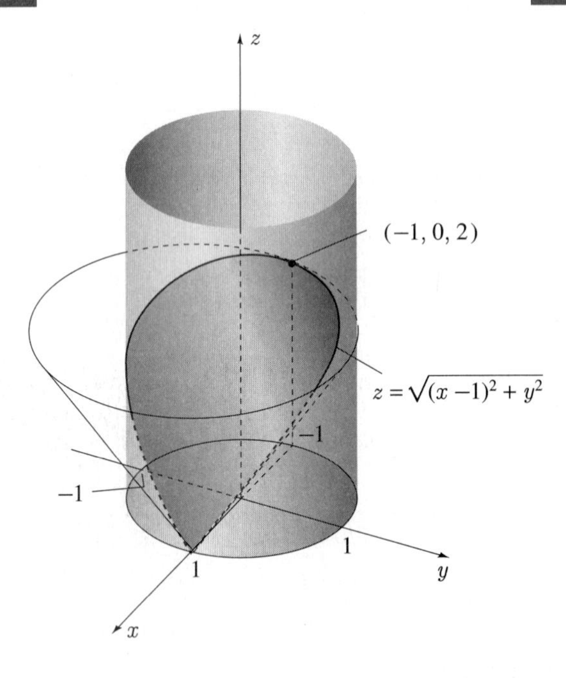

FIGURE 13.47

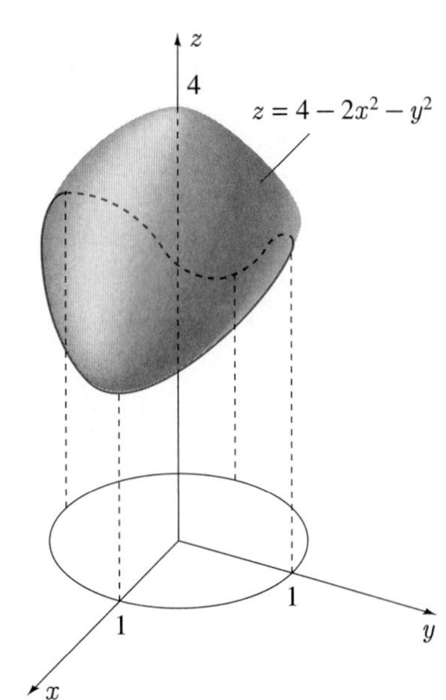

FIGURE 13.48

FIGURE 13.49

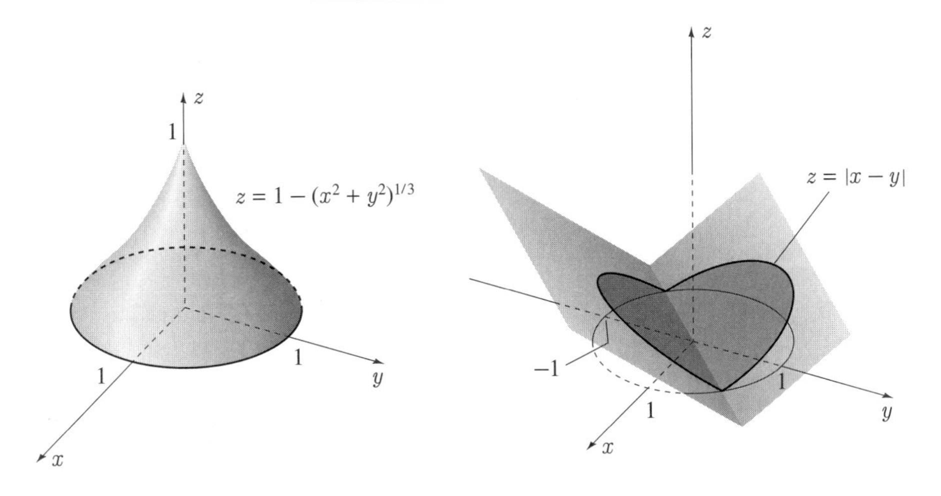

TABLE 13.1

Function $f(x, y)$	Position of absolute maximum	Value of absolute maximum	Position of absolute mimimum	Value of absolute mimimum
$x^2 + y^2$	Every point on $x^2 + y^2 = 1$	1	$(0, 0)$	0
$1 + x^2 - y^2$	$(\pm 1, 0)$	2	$(0, \pm 1)$	0
$\sqrt{(x-1)^2 + y^2}$	$(-1, 0)$	2	$(1, 0)$	0
$4 - 2x^2 - y^2$	$(0, 0)$	4	$(\pm 1, 0)$	2
$1 - (x^2 + y^2)^{1/3}$	$(0, 0)$	1	Every point on $x^2 + y^2 = 1$	0
$\lvert x - y \rvert$	$(\pm 1/\sqrt{2}, \mp 1/\sqrt{2})$	$\sqrt{2}$	Every point on $y = x$, $-1/\sqrt{2} \le x \le 1/\sqrt{2}$	0

For each of the functions in these figures, *absolute extrema occur at either a critical point or a point on the boundary of R*. This result is true for any *continuous function defined on a finite region that includes all the points on its boundary*. Although this result may seem fairly obvious geometrically, to prove it analytically is very difficult; we will be content to assume its validity and carry on from there.

It is very simple to look at the surface defined by a function $f(x, y)$ and pick off its absolute extrema, but in practice this just does not happen. Usually we must determine the absolute extrema algebraically from the function itself. Suppose, then, that a continuous function $f(x, y)$ is given and we are required to find its absolute extrema on a finite region R (which includes its boundary points). The previous discussion indicated that the extrema must occur either at critical points or on the boundary of R. Consequently, we should first determine all critical points of $f(x, y)$ in R, and evaluate

$f(x, y)$ at each of these points. These values should now be compared to the maximum and minimum values of $f(x, y)$ on the boundary of R. But how do we find the maximum and minimum values of $f(x, y)$ on the boundary? If the boundary of R is denoted by C (Figure 13.50), and if C has parametric equations $x = x(t)$, $y = y(t)$, $\alpha \le t \le \beta$, then on C we can express $f(x, y)$ in terms of t, and t alone:

$$f[x(t), y(t)], \quad \alpha \le t \le \beta.$$

To find the maximum and minimum values of $f(x, y)$ on C is now an absolute extrema problem for a function of one variable. The function $f[x(t), y(t)]$ should therefore be evaluated at each of its critical points and at $t = \alpha$ and $t = \beta$.

If the boundary of R consists of a number of curves (Figure 13.51), then this boundary procedure must be performed for each part. In other words, on each part of the boundary we express $f(x, y)$ as a function of one variable, and then evaluate this function at its critical points and at the ends of that part of the boundary to which it applies.

The absolute maximum of $f(x, y)$ on R is then the largest of all values of $f(x, y)$ evaluated at the critical points inside R, the critical points on the boundary of R, and the endpoints of each part of the boundary. The absolute minimum of $f(x, y)$ on R is the smallest of all these values.

FIGURE 13.50

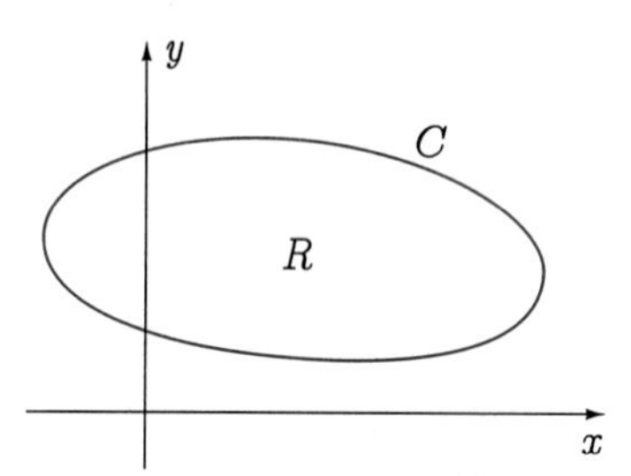

FIGURE 13.51

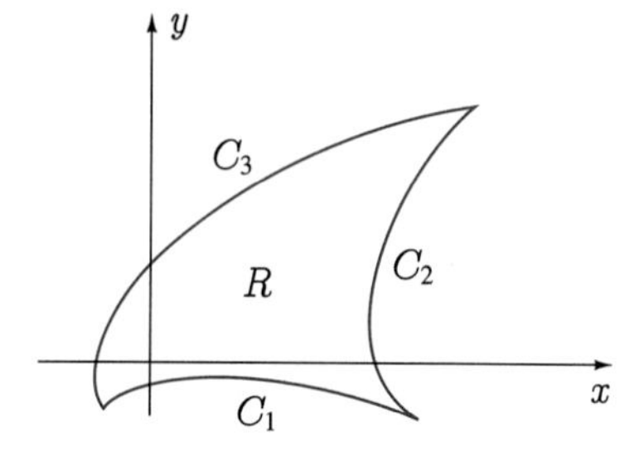

Recall that to find the absolute extrema of a function $f(x)$, continuous on $a \le x \le b$, we evaluate $f(x)$ at all critical points and at the boundary points $x = a$ and $x = b$. The procedure that we have established here for $f(x, y)$ is much the same—the difference is that for $f(x, y)$, the boundary consists not of two points, but of entire curves. Evaluation of $f(x, y)$ on the boundary therefore reduces to one or more extrema problems for functions of one variable. Note too that for $f(x, y)$ (or $f(x)$), it is not necessary to determine the nature of the critical points; it is necessary only to evaluate $f(x, y)$ at these points.

EXAMPLE 13.30

Find the maximum value of the function $z = f(x, y) = 4xy - x^4 - 2y^2$ on the region R: $-2 \le x \le 2$, $-2 \le y \le 2$.

SOLUTION

Critical points of $f(x, y)$ are given by

$$0 = \frac{\partial f}{\partial x} = 4y - 4x^3, \quad 0 = \frac{\partial f}{\partial y} = 4x - 4y.$$

Solutions of these equations are $(0, 0)$, $(1, 1)$, and $(-1, -1)$, and the values of $f(x, y)$ at these critical points are

$$f(0, 0) = 0, \quad f(1, 1) = 1, \quad f(-1, -1) = 1.$$

We denote the four parts of the boundary of R by C_1, C_2, C_3, and C_4 (Figure 13.52).

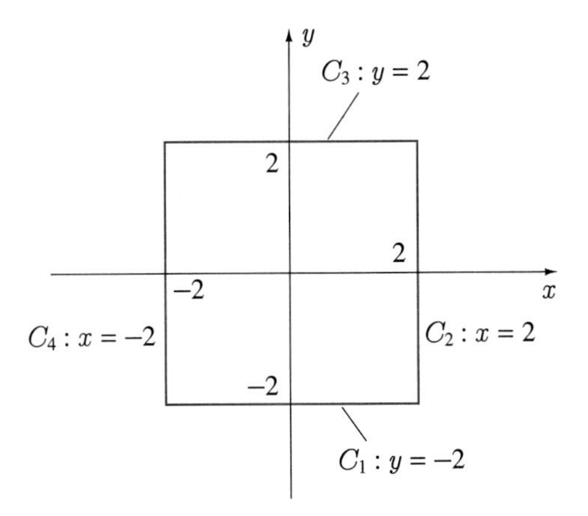

FIGURE 13.52

On C_1, $y = -2$, in which case

$$z = -8x - x^4 - 8, \quad -2 \leq x \leq 2.$$

For critical points of this function, we solve

$$0 = \frac{dz}{dx} = -8 - 4x^3.$$

The only solution is $x = -2^{1/3}$, at which the value of z is

$$z = 8 \cdot 2^{1/3} - 2^{4/3} - 8 = -0.44.$$

On C_2, $x = 2$, in which case

$$z = 8y - 16 - 2y^2, \quad -2 \leq y \leq 2.$$

Critical points are defined by

$$0 = \frac{dz}{dy} = 8 - 4y.$$

The only solution $y = 2$ defines one of the corners of the square, and at this point

$$z = -8.$$

On C_3, $y = 2$ and

$$z = 8x - x^4 - 8, \quad -2 \leq x \leq 2.$$

For critical points, we solve

$$0 = \frac{dz}{dx} = 8 - 4x^3.$$

At the single point $x = 2^{1/3}$,

$$z = 8 \cdot 2^{1/3} - 2^{4/3} - 8 = -0.44.$$

On the final curve C_4, $x = -2$ and

$$z = -8y - 16 - 2y^2, \quad -2 \le y \le 2.$$

Critical points are given by

$$0 = \frac{dz}{dy} = -8 - 4y.$$

The solution $y = -2$ defines another corner of the square at which

$$z = -8.$$

We have now evaluated $f(x, y)$ at all critical points inside R and at all critical points on the four parts of the boundary of R. It remains only to evaluate $f(x, y)$ at the corners of the square. Two corners have already been accounted for; the other two give

$$f(2, -2) = -40, \quad f(-2, 2) = -40.$$

The largest value of $f(x, y)$ produced is 1, and this is therefore the maximum value of $f(x, y)$ on R. ■

EXAMPLE 13.31

The temperature at each point (x, y) in a semicircular plate defined by $x^2 + y^2 \le 1$, $y \ge 0$ is given by

$$T(x, y) = 16x^2 - 24xy + 40y^2.$$

Find the hottest and coldest points in the plate.

SOLUTION

For critical points of $T(x, y)$, we solve

$$0 = \frac{\partial T}{\partial x} = 32x - 24y, \quad 0 = \frac{\partial T}{\partial y} = -24x + 80y.$$

The only solution of these equations, $(0, 0)$, is on the boundary. On the upper edge of the plate (Figure 13.53), we set $x = \cos t$, $y = \sin t$, $0 \le t \le \pi$, in which case

$$T = 16\cos^2 t - 24\cos t \sin t + 40\sin^2 t, \quad 0 \le t \le \pi.$$

For critical points of this function, we solve

$$\begin{aligned} 0 = \frac{dT}{dt} &= -32\cos t \sin t - 24(-\sin^2 t + \cos^2 t) + 80\sin t \cos t \\ &= 24(\sin 2t - \cos 2t). \end{aligned}$$

If we divide by $\cos 2t$ (since $\cos 2t = 0$ does not lead to a solution of this equation), we have

$$\tan 2t = 1.$$

The only solutions of this equation in the interval $0 \le t \le \pi$ are $t = \pi/8$ and $t = 5\pi/8$. When $t = \pi/8$, $T = 11.0$; and when $t = 5\pi/8$, $T = 45.0$. At the ends of this part of the boundary, $t = 0$ and $t = \pi$, and $T(1,0) = 16$ and $T(-1,0) = 16$. On the lower edge of the plate, $y = 0$, in which case

$$T = 16x^2, \quad -1 \leq x \leq 1.$$

The only critical point of this function is $x = 0$, at which $T = 0$. The hottest point in the plate is therefore $(\cos(5\pi/8), \sin(5\pi/8)) = (-0.38, 0.92)$, where the temperature is $45°$, and the coldest point is $(0,0)$ with temperature $0°$.

FIGURE 13.53

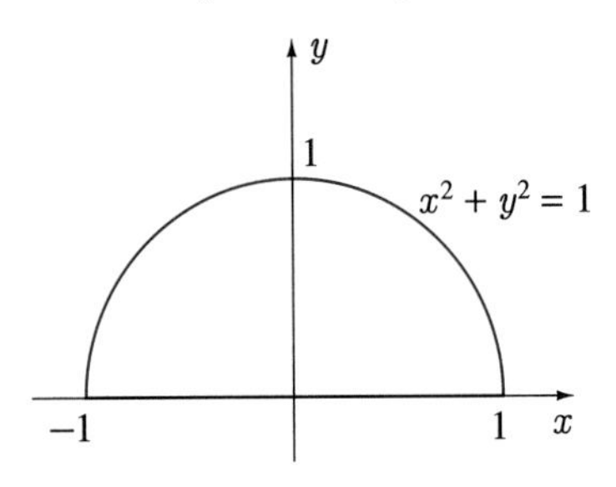

■

EXAMPLE 13.32

Find the point on the plane $2x - 3y - 4z = 25$ closest to the point $(3,2,1)$.

SOLUTION The distance D from $(3,2,1)$ to any point (x,y,z) in space is defined by

$$D^2 = (x-3)^2 + (y-2)^2 + (z-1)^2.$$

Because we want the minimum distance from $(3,2,1)$ to points in the plane $2x - 3y - 4z = 25$, we must minimize D but consider only those points (x,y,z) that satisfy the equation of the plane (Figure 13.54). At the moment, D^2 is a function of three variables x, y, and z, but they are not all independent because of the planar restriction. If we solve the equation of the plane for z in terms of x and y and substitute, then

$$\begin{aligned} D^2 = f(x,y) &= (x-3)^2 + (y-2)^2 + \left(\frac{2x - 3y - 25}{4} - 1\right)^2 \\ &= (x-3)^2 + (y-2)^2 + \left(\frac{2x - 3y - 29}{4}\right)^2, \end{aligned}$$

where x and y are independent variables. Now D is minimized when D^2 is minimized, and we will therefore find that point (x,y) which minimizes D^2. First we locate the critical points of D^2, by solving

$$0 = \frac{\partial f}{\partial x} = 2(x-3) + \left(\frac{2x - 3y - 29}{4}\right)$$

and

$$0 = \frac{\partial f}{\partial y} = 2(y-2) - \frac{3}{2}\left(\frac{2x - 3y - 29}{4}\right).$$

These reduce to

$$10x - 3y = 53, \quad -6x + 25y = -55.$$

The only solution is $x = 5$, $y = -1$. In this problem we do not have a finite region R in the xy-plane in which D^2 is defined. It is geometrically clear, however, that as x and y become very large, so too does D^2. It follows, therefore, that the one critical point obtained, namely $(5, -1)$, must yield the absolute minimum of D^2. Consequently, the point closest to $(3, 2, 1)$ in the plane $2x - 3y - 4z = 25$ is $(5, -1, -3)$.

FIGURE 13.54

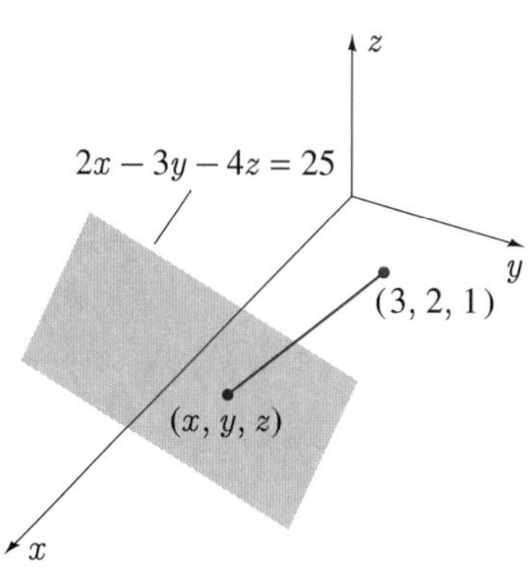

■

EXERCISES 13.11

In Exercises 1–8 find the maximum and minimum values of the function on the region.

1. $f(x, y) = x^2 + y^3$ on $R:\ x^2 + y^2 \leq 1$

2. $f(x, y) = x^2 + x + 3y^2 + y$ on the region R bounded by $y = x + 1$, $y = 1 - x$, $y = x - 1$, $y = -x - 1$

3. $f(x, y) = 3x + 4y$ on the region R bounded by the lines $x + y = 1$, $x + y = 4$, $y + 1 = x$, $y - 1 = x$

4. $f(x, y) = x^2 y + xy^2 + y$ on $R:\ -1 \leq x \leq 1$, $-1 \leq y \leq 1$

5. $f(x, y) = 3x^2 + 2xy - y^2 + 5$ on $R:\ 4x^2 + 9y^2 \leq 36$

6. $f(x, y) = x^3 - 3x + y^2 + 2y$ on the triangle bounded by $x = 0$, $y = 0$, $x + y = 1$

7. $f(x, y) = x^3 + y^3 - 3x - 12y + 2$ on the square $-3 \leq x \leq 3$, $-3 \leq y \leq 3$

8. $f(x, y) = x^3 + y^3 - 3x - 3y + 2$ on the circle $x^2 + y^2 \leq 1$

9. Find maximum and minimum values of the function $f(x, y, z) = xy^2 z^3$ on that part of the plane $x + y + z = 6$ for which **(a)** $x > 0$, $y > 0$, $z > 0$. **(b)** $x \geq 0$, $y \geq 0$, $z \geq 0$.

10. Find the point on the plane $x + y - 2z = 6$ closest to the origin.

11. Find the shortest distance from $(-1, 1, 2)$ to the plane $2x - 3y + 6z = 14$.

12. Find the point on the surface $z = x^2 + y^2$ closest to the point $(1, 1, 0)$.

13. Find the point on that part of the plane $x + y + 2z = 4$ in the first octant that is closest to the point $(3, 3, 1)$. For this question assume that the curves of intersection of the plane with the coordinate planes are part of the surface.

14. The electrostatic potential at each point in the region $0 \leq x \leq 1$, $0 \leq y \leq 1$ is given by $V(x, y) = 48xy - 32x^3 - 24y^2$. Find the maximum and minimum potentials in the region.

15. When a rectangular box is sent through the mail, the post office demands that the length of the box plus twice the sum of its height and width be no more than 250 cm. Find the dimensions of the box satisfying this requirement that encloses the largest possible volume.

16. An open tank in the form of a rectangular parallelepiped is to be built to hold 1000 L of acid. If the cost per unit area of lining the base of the tank is three times that of the sides, what dimensions minimize the cost of lining the tank?

17. Prove that the minimum distance from a point (x_1, y_1, z_1) to a plane $Ax + By + Cz + D = 0$ is $|Ax_1 + By_1 + Cz_1 + D|/\sqrt{A^2 + B^2 + C^2}$.

18. Prove that for triangles, the point that minimizes the sum of the squares of the distances to the vertices is the centroid.

19. Find the point on the curve $x^2 - xy + y^2 - z^2 = 1$, $x^2 + y^2 = 1$ closest to the origin.

20. Find the dimensions of the box with largest possible volume that can fit inside the ellipsoid $x^2/a^2 + y^2/b^2 + z^2/c^2 = 1$, assuming that its edges are parallel to the coordinate axes.

21. Find the maximum and minimum values of the function $f(x, y, z) = xyz$ on the sphere $x^2 + y^2 + z^2 = 1$.

22. If P is the perimeter of a triangle with sides of length x, y, and z, the area of the triangle is

$$A = \sqrt{\frac{P}{2}\left(\frac{P}{2} - x\right)\left(\frac{P}{2} - y\right)\left(\frac{P}{2} - z\right)},$$

where $P = x + y + z$. Show that A is maximized for fixed P when the triangle is equilateral.

23. Show that for any triangle with interior angles A, B, and C,

$$\sin(A/2)\ \sin(B/2)\ \sin(C/2) \leq 1/8.$$

Hint: Find the maximum value of the function $f(A, B, C) = \sin(A/2)\ \sin(B/2)\ \sin(C/2)$.

24. A silo is in the shape of a right-circular cylinder surmounted by a right-circular cone. If the radius of each is 6 m and the total surface area must be 200 m^2 (not including the base), what heights for the cone and cylinder yield maximum enclosed volume?

25. What values of x and y maximize the production function $P(x, y) = kx^\alpha y^\beta$ where k, α, and β are positive constants ($\alpha + \beta = 1$) when x and y must satisfy $Ax + By = C$, where A, B and C are positive constants.

26. A long piece of metal 1 m wide is bent at A and B, as shown in Figure 13.55, to form a channel with three straight sides. If the bends are equidistant from the ends, where should they be made in order to obtain maximum possible flow of fluid along the channel?

FIGURE 13.55

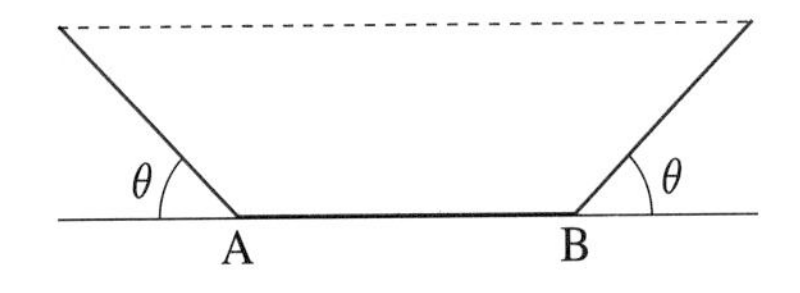

27. Find maximum and minimum values of the function $f(x, y, z) = xy + xz$ on the region $x^2 + y^2 + z^2 \leq 1$.

28. Find maximum and minimum values of the function $f(x, y, z) = x^2yz$ on the region $x^2 + y^2 \leq 1$, $0 \leq z \leq 1$.

29. A cow's daily diet consists of three foods: hay, grain, and supplements. The cow is always given 11 kg of hay per day, 50% of which is digestive material and 12% of which is protein. Grain is 74% digestive material and 8.8% protein, whereas supplements are 62% digestive material and 34% protein. The cost of hay is \$27.50 for 1000 kg, and grain and supplements cost \$110 and \$175 for 1000 kg. A healthy cow's diet must contain between 9.5 and 11.5 kg of digestive material and between 1.9 and 2.0 kg of protein. Determine the daily amounts of grain and supplements that the cow should be fed in order that total food costs be kept to a minimum.

30. Find the area of the largest triangle that has vertices on the circle $x^2 + y^2 = r^2$.

31. A rectangle is surmounted by an isosceles triangle as shown in Figure 13.56. Find x, y, and θ in order that the area of the figure be as large as possible under the restriction that its perimeter must be P.

FIGURE 13.56

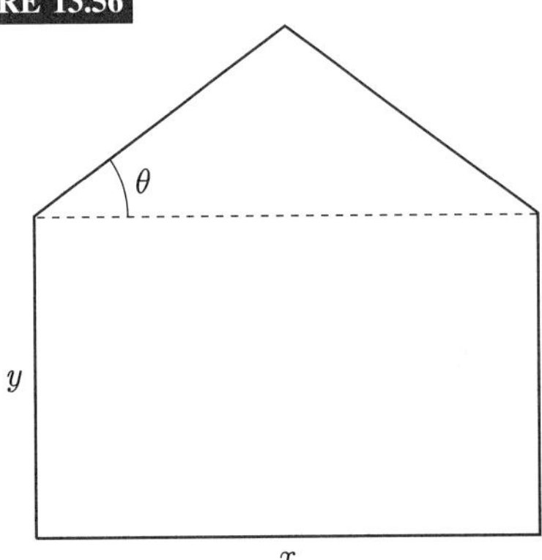

SECTION 13.12

Lagrange Multipliers

Many applied maxima and minima problems result in **constraint problems**. In particular, Examples 13.31 and 13.32 contain such problems. In Example 13.31, to find extreme values of T on the edge of the plate we maximized and minimized $T(x,y) = 16x^2 - 24x + 40y^2$ subject to first the constraint $x^2 + y^2 = 1$, and then the constraint $y = 0$. Our method there was to substitute from the constraint equation into $T(x,y)$ in order to obtain a function of one variable. In Example 13.32, to find the minimum distance from $(3,2,1)$ to the plane $2x - 3y - 4z = 25$, we minimized $D^2 = (x-3)^2 + (y-2)^2 + (z-1)^2$ subject to the constraint $2x - 3y - 4z = 25$. Again we substituted from the constraint to obtain D^2 as a function of two independent variables.

A natural question to ask is whether problems of this type can be solved without substituting from the constraint equation, for if the constraint equation is complicated, substitution may be very difficult or even impossible. To show that there is indeed an alternative, consider the situation in which a function $f(x,y,z)$ is to be maximized or minimized subject to two constraints:

$$F(x,y,z) = 0, \qquad (13.57a)$$

$$G(x,y,z) = 0. \qquad (13.57b)$$

Algebraically, we are to find extreme values of $f(x,y,z)$, considering only those values of x, y, and z that satisfy the two equations 13.57. Geometrically, we can interpret each of these conditions as specifying a surface, so that we are seeking extreme values of $f(x,y,z)$, considering only those points on the curve of intersection C of the surfaces $F(x,y,z) = 0$ and $G(x,y,z) = 0$ (Figure 13.57).

FIGURE 13.57

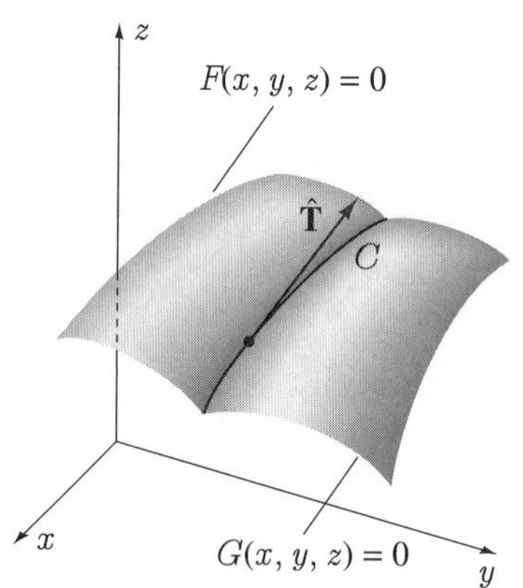

Extreme values of $f(x,y,z)$ along C will occur either at critical points of the function or at the ends of the curve. But what derivative or derivatives of $f(x,y,z)$ are we talking about when we say critical points? Since we are concerned only with values of $f(x,y,z)$ on C, we must mean the derivative of $f(x,y,z)$ along C; i.e., the directional derivative in the tangent direction to C. If $\mathbf{T}$, then, is a tangent vector to C, critical points of $f(x,y,z)$ along C are given by

$$0 = D_{\mathbf{T}} f = \nabla f \cdot \hat{\mathbf{T}},$$

or at points where the directional derivative is undefined. According to equation 13.50, a tangent vector to C is $\mathbf{T} = \nabla F \times \nabla G$, and hence a unit tangent vector is

$$\hat{\mathbf{T}} = \frac{\nabla F \times \nabla G}{|\nabla F \times \nabla G|}.$$

The directional derivative of $f(x,y,z)$ at points along C is therefore given by

$$D_{\mathbf{T}} f = \nabla f \cdot \frac{\nabla F \times \nabla G}{|\nabla F \times \nabla G|}.$$

It follows, then, that critical points of $f(x,y,z)$ are points (x,y,z) that satisfy the equation

$$\nabla f \cdot \nabla F \times \nabla G = 0,$$

or points at which the left side is not defined. Now vector $\nabla F \times \nabla G$ is perpendicular to both ∇F and ∇G. Since ∇f is perpendicular to $\nabla F \times \nabla G$ (their dot product is zero), it follows that ∇f must lie in the plane of ∇F and ∇G. Consequently, there exist scalars λ and μ such that

$$\nabla f = (-\lambda)\nabla F + (-\mu)\nabla G,$$

or

$$\nabla f + \lambda \nabla F + \mu \nabla G = \mathbf{0}. \qquad (13.58)$$

This vector equation is equivalent to the three scalar equations

$$\frac{\partial f}{\partial x} + \lambda \frac{\partial F}{\partial x} + \mu \frac{\partial G}{\partial x} = 0, \qquad (13.59a)$$

$$\frac{\partial f}{\partial y} + \lambda \frac{\partial F}{\partial y} + \mu \frac{\partial G}{\partial y} = 0, \qquad (13.59b)$$

$$\frac{\partial f}{\partial z} + \lambda \frac{\partial F}{\partial z} + \mu \frac{\partial G}{\partial z} = 0, \qquad (13.59c)$$

and these equations must be satisfied at a critical point at which the directional derivative of $f(x,y,z)$ vanishes. Note too that at a point at which the directional derivative of $f(x,y,z)$ does not exist, one of the partial derivatives in these equations does not exist. In other words, we have shown that critical points of $f(x,y,z)$ are points that satisfy equations 13.59 or points at which the equations are undefined. These equations, however, contain five unknowns: x, y, z, λ, and μ. To complete the system we add equations 13.57 since they must also be satisfied by a critical point. Equations 13.57 and 13.59 therefore yield a system of five equations in the five unknowns x, y, z, λ, and μ; the first three unknowns (x,y,z) define a critical point of $f(x,y,z)$ along C. The advantage of this system of equations lies in the fact that the differentiations in 13.59 involve only the given functions (and no substitutions from the constraint equations are necessary). What we have sacrificed is a system of three equations in the three unknowns (x,y,z) for a system of five equations in the five unknowns (x,y,z,λ,μ).

Let us not forget that the original problem was to find extreme values for the function $f(x,y,z)$ subject to constraints 13.57. What we have shown so far is that critical points at which the directional derivative of $f(x,y,z)$ vanishes can be found by solving equations 13.57 and 13.59. In addition, critical points at which the directional derivative of $f(x,y,z)$ does not exist are points at which equations 13.59 are not defined. What remains is to evaluate $f(x,y,z)$ at all critical points and at the ends of C. *If C is a closed curve* (i.e., if C rejoins itself), then $f(x,y,z)$ *need be evaluated only at the critical points.*

Through the directional derivative and tangent vectors to curves, we have shown that equations 13.57 and 13.59 define critical points of a function $f(x,y,z)$ that is subject to two constraints: $F(x,y,z) = 0$ and $G(x,y,z) = 0$. But what about other situations? Let us say, for example, that we require extreme values of a function $f(x,y,z,t)$ subject to a single constraint $F(x,y,z,t) = 0$. How shall we find critical points of this function? Fortunately, as we now show, there is a very simple method that yields equations 13.57 and 13.59, and this method generalizes to other situations also.

To find critical points of $f(x,y,z)$ subject to constraints 13.57, we define a function

$$L(x,y,z,\lambda,\mu) = f(x,y,z) + \lambda F(x,y,z) + \mu G(x,y,z),$$

and regard it as a function of five independent variables x, y, z, λ, and μ. To find critical points of this function, we would first solve the equations obtained by setting each of the partial derivatives of L equal to zero:

$$0 = \frac{\partial L}{\partial x} = \frac{\partial f}{\partial x} + \lambda \frac{\partial F}{\partial x} + \mu \frac{\partial G}{\partial x},$$
$$0 = \frac{\partial L}{\partial y} = \frac{\partial f}{\partial y} + \lambda \frac{\partial F}{\partial y} + \mu \frac{\partial G}{\partial y},$$
$$0 = \frac{\partial L}{\partial z} = \frac{\partial f}{\partial z} + \lambda \frac{\partial F}{\partial z} + \mu \frac{\partial G}{\partial z},$$
$$0 = \frac{\partial L}{\partial \lambda} = F(x, y, z),$$
$$0 = \frac{\partial L}{\partial \mu} = G(x, y, z).$$

In addition, we would consider points at which the partial derivatives of L do not exist. Clearly, this means points (x, y, z) at which any of the partial derivatives of $f(x, y, z)$, $F(x, y, z)$, and $G(x, y, z)$ do not exist. But these are precisely equations 13.57 and 13.59. We have shown, then, that finding critical points (x, y, z) of $f(x, y, z)$ subject to $F(x, y, z) = 0$ and $G(x, y, z) = 0$ is equivalent to finding critical points (x, y, z, λ, μ) of $L(x, y, z, \lambda, \mu)$. The two unknowns λ and μ that accompany a critical point (x, y, z) of $f(x, y, z)$ are called **Lagrange multipliers**. They are not a part of the solution (x, y, z) to the original problem, but have been introduced as a convenience by which to arrive at that solution. The function $L(x, y, z, \lambda, \mu)$ is often called the **Lagrangian** of the problem.

The method for other constraint problems should now be evident. Given a function $f(x, y, z, t, \ldots)$ of n variables to maximize or minimize subject to m constraints

$$F_1(x, y, z, t, \ldots) = 0, \quad F_2(x, y, z, t, \ldots) = 0, \ldots, F_m(x, y, z, t, \ldots) = 0, \tag{13.60}$$

we introduce m Lagrange multipliers λ_1, $\lambda_2, \ldots, \lambda_m$ into a Lagrangian of $n + m$ independent variables x, y, z, t, $\ldots$, λ_1, λ_2, $\ldots$, λ_m:

$$L(x, y, z, t, \ldots \lambda_1, \lambda_2, \ldots, \lambda_m) = f(x, y, z, t, \ldots) + \lambda_1 F_1(x, y, z, t, \ldots) + \cdots$$
$$+ \lambda_m F_m(x, y, z, t, \ldots). \tag{13.61}$$

Critical points $(x, y, z, t, \ldots)$ of $f(x, y, z, t, \ldots)$ are then determined by the equations defining critical points of $L(x, y, z, t, \ldots, \lambda_1, \lambda_2, \ldots, \lambda_m)$, namely,

$$0 = \frac{\partial L}{\partial x} = \frac{\partial f}{\partial x} + \lambda_1 \frac{\partial F_1}{\partial x} + \cdots + \lambda_m \frac{\partial F_m}{\partial x}, \tag{13.62a}$$
$$0 = \frac{\partial L}{\partial y} = \frac{\partial f}{\partial y} + \lambda_1 \frac{\partial F_1}{\partial y} + \cdots + \lambda_m \frac{\partial F_m}{\partial y}, \tag{13.62b}$$
$$\vdots \qquad \vdots \qquad \vdots$$
$$0 = \frac{\partial L}{\partial \lambda_1} = F_1(x, y, z, t, \ldots), \tag{13.62c}$$
$$0 = \frac{\partial L}{\partial \lambda_2} = F_2(x, y, z, t, \ldots), \tag{13.62d}$$
$$\vdots \qquad \vdots$$
$$0 = \frac{\partial L}{\partial \lambda_m} = F_m(x, y, z, t, \ldots). \tag{13.62e}$$

To use Lagrange multipliers in Example 13.32, we define the Lagrangian

$$L(x, y, z, \lambda) = D^2 + \lambda(2x - 3y - 4z - 25)$$
$$= (x - 3)^2 + (y - 2)^2 + (z - 1)^2 + \lambda(2x - 3y - 4z - 25).$$

Critical points of $L(x,y,z,\lambda)$ are defined by

$$\begin{aligned}
0 &= \frac{\partial L}{\partial x} = 2(x-3) + 2\lambda, \\
0 &= \frac{\partial L}{\partial y} = 2(y-2) - 3\lambda, \\
0 &= \frac{\partial L}{\partial z} = 2(z-1) - 4\lambda, \\
0 &= \frac{\partial L}{\partial \lambda} = 2x - 3y - 4z - 25.
\end{aligned}$$

The solution of this linear system is $(x,y,z,\lambda) = (5,-1,-3,-2)$, yielding as before the critical point $(5,-1,-3)$ of D^2.

EXAMPLE 13.33

Find the maximum value of the function

$$f(x,y,z) = 2x^2y^2 + 2y^2z^2 + 3z,$$

considering only those values of x, y, and z that satisfy the equations

$$z = x^2 + y^2, \quad x^2 + 3y^2 = 1.$$

SOLUTION For this example we illustrate the difference between a solution that utilizes Lagrange multipliers and one that does not. The solution that does not use Lagrange multipliers requires us first to express $f(x,y,z)$ in terms of a single independent variable. We can do this by solving the constraint equations for $x^2 = 1 - 3y^2$ and $y^2 = (1-z)/2$ and substituting into $f(x,y,z)$:

$$\begin{aligned}
f(x,y,z) &= 2(1-3y^2)y^2 + 2y^2z^2 + 3z \\
&= 2\left(1 - \frac{3}{2} + \frac{3z}{2}\right)\left(\frac{1}{2} - \frac{z}{2}\right) + 2z^2\left(\frac{1}{2} - \frac{z}{2}\right) + 3z \\
&= \frac{1}{2}(-2z^3 - z^2 + 10z - 1).
\end{aligned}$$

Since $z = x^2 + y^2 = (1-3y^2) + y^2 = 1 - 2y^2$, and y must be restricted to $|y| \le 1/\sqrt{3}$, it follows that the only possible values for z are $\frac{1}{3} \le z \le 1$. We have shown, then, that maximizing $f(x,y,z)$ subject to the two constraints $z = x^2 + y^2$ and $x^2 + 3y^2 = 1$ is equivalent to maximizing

$$F(z) = \frac{1}{2}(-2z^3 - z^2 + 10z - 1), \quad \frac{1}{3} \le z \le 1.$$

For critical points of $F(z)$, we solve

$$0 = \frac{dF}{dz} = \frac{1}{2}(-6z^2 - 2z + 10),$$

from which we get

$$z = \frac{2 \pm \sqrt{4 + 240}}{-12} = \frac{-1 \pm \sqrt{61}}{6}.$$

These two critical points must be rejected as not lying within the interval $\frac{1}{3} \le z \le 1$, and the maximum value of $F(z)$ must therefore occur at either $z = \frac{1}{3}$ or $z = 1$. Since $F(\frac{1}{3}) = \frac{29}{27}$ and $F(1) = 3$, it follows that the maximum value of $F(z)$ and therefore of $f(x, y, z)$ is 3.

To maximize $f(x, y, z)$ using Lagrange multipliers, we define the Lagrangian

$$L(x, y, z, \lambda, \mu) = 2x^2y^2 + 2y^2z^2 + 3z + \lambda(x^2 + y^2 - z) + \mu(x^2 + 3y^2 - 1).$$

Critical points of L and therefore of $f(x, y, z)$ are defined by the equations

$$\begin{aligned}
0 &= \frac{\partial L}{\partial x} = 4xy^2 + 2\lambda x + 2\mu x = 2x(2y^2 + \lambda + \mu),\\
0 &= \frac{\partial L}{\partial y} = 4x^2y + 4yz^2 + 2\lambda y + 6\mu y = 2y(2x^2 + 2z^2 + \lambda + 3\mu),\\
0 &= \frac{\partial L}{\partial z} = 4y^2z + 3 - \lambda,\\
0 &= \frac{\partial L}{\partial \lambda} = x^2 + y^2 - z,\\
0 &= \frac{\partial L}{\partial \mu} = x^2 + 3y^2 - 1.
\end{aligned}$$

If we choose $x = 0$ to satisfy the first equation, then the remaining equations imply that $y = \pm 1/\sqrt{3}$, $z = \frac{1}{3}$ (and $\lambda = \frac{31}{9}$, $\mu = -\frac{11}{9}$). If we choose $y = 0$ to satisfy the second equation, then the remaining equations require $x = \pm 1$, $z = 1$ (and $\lambda = 3$, $\mu = -3$). The only other way to satisfy the first two equations is to set

$$\begin{aligned}
2y^2 + \lambda + \mu &= 0,\\
2x^2 + 2z^2 + \lambda + 3\mu &= 0.
\end{aligned}$$

If we multiply the first of these by 3 and subtract the second, we have

$$\begin{aligned}
0 &= 6y^2 - 2x^2 - 2z^2 + 2\lambda\\
&= 6y^2 - 2x^2 - 2(x^2 + y^2)^2 + 2\lambda\\
&= 6y^2 - 2x^2 - 2(x^4 + 2x^2y^2 + y^4) + 2\lambda\\
&= 6y^2 - 2(1 - 3y^2) - 2(1 - 3y^2)^2\\
&\quad - 4y^2(1 - 3y^2) - 2y^4 + 2\lambda\\
&= 2(\lambda - 2 + 10y^2 - 4y^4).
\end{aligned}$$

But from the equation for $\partial L/\partial z$, we can also write

$$\begin{aligned}
0 &= 3 - \lambda + 4y^2(x^2 + y^2)\\
&= 3 - \lambda + 4y^2(1 - 3y^2 + y^2)\\
&= 3 - \lambda + 4y^2 - 8y^4.
\end{aligned}$$

These two equations in y and λ imply that

$$2 - 10y^2 + 4y^4 = 3 + 4y^2 - 8y^4, \quad \text{or} \quad 12y^4 - 14y^2 - 1 = 0.$$

Thus,

$$y^2 = \frac{14 \pm \sqrt{196+48}}{24} = \frac{7 \pm \sqrt{61}}{12},$$

where we reject the negative solution. But substitution of this result into the constraint $x^2 + 3y^2 = 1$ requires x^2 to be negative. Consequently, only four critical points (x, y, z) are obtained: $(0, \pm 1/\sqrt{3}, \frac{1}{3})$ and $(\pm 1, 0, 1)$. Since the curve defined by the constraints is closed, we need evaluate $f(x, y, z)$ only at these critical points:

$$f(0, \pm 1/\sqrt{3}, \frac{1}{3}) = 2(\frac{1}{3})(\frac{1}{9}) + 3(\frac{1}{3}) = \frac{29}{27},$$
$$f(\pm 1, 0, 1) = 3.$$

The maximum value of $f(x, y, z)$ is again 3.

Note that the endpoints in the first solution are critical points in the Lagrangian solution. This emphasizes the fact that, as we pointed out earlier, when the curve is closed and we use a Lagrangian, it is necessary to evaluate the function to be maximized or minimized only at its critical points. ■

EXAMPLE 13.34

Find the maximum and minimum values of the function $f(x, y, z) = xyz$ on the sphere $x^2 + y^2 + z^2 = 1$.

SOLUTION If we define the Lagrangian

$$L(x, y, z, \lambda) = xyz + \lambda(x^2 + y^2 + z^2 - 1),$$

then critical points of L, and therefore of $f(x, y, z)$, are defined by the equations

$$0 = \frac{\partial L}{\partial x} = yz + 2\lambda x,$$
$$0 = \frac{\partial L}{\partial y} = xz + 2\lambda y,$$
$$0 = \frac{\partial L}{\partial z} = xy + 2\lambda z,$$
$$0 = \frac{\partial L}{\partial \lambda} = x^2 + y^2 + z^2 - 1.$$

If we multiply the first equation by y and the second by x, and equate the resulting expressions for $2\lambda xy$, we have

$$y^2 z = x^2 z.$$

Consequently, either $z = 0$ or $y = \pm x$.

Case I: $z = 0$. In this case the equations reduce to

$$\lambda x = 0, \quad \lambda y = 0, \quad xy = 0, \quad x^2 + y^2 = 1.$$

The first implies that either $x = 0$ or $\lambda = 0$. If $x = 0$, then $y = \pm 1$, and we have two critical points $(0, \pm 1, 0)$. If $\lambda = 0$, then the third equation requires $x = 0$ or $y = 0$. We therefore obtain two additional critical points $(\pm 1, 0, 0)$.

Case II: $y = x$. In this case the equations reduce to

$$xz + 2\lambda x = 0, \quad x^2 + 2\lambda z = 0, \quad 2x^2 + z^2 = 1.$$

The first implies that either $x = 0$ or $z = -2\lambda$. If $x = 0$, then $z = \pm 1$, and we have the two critical points $(0, 0, \pm 1)$. If $z = -2\lambda$, then the last two equations imply that $x = \pm 1/\sqrt{3}$, and we obtain the four critical points

$$(\pm 1/\sqrt{3}, \pm 1/\sqrt{3}, 1/\sqrt{3}) \quad \text{and} \quad (\pm 1/\sqrt{3}, \pm 1/\sqrt{3}, -1/\sqrt{3}).$$

Case III: $y = -x$. This case is similar to that for $y = x$, and leads to the additional four critical points

$$(\pm 1/\sqrt{3}, \mp 1/\sqrt{3}, 1/\sqrt{3}) \quad \text{and} \quad (\pm 1/\sqrt{3}, \mp 1/\sqrt{3}, -1/\sqrt{3}).$$

Because $x^2 + y^2 + z^2 = 1$ is a surface without a boundary, we complete the problem by evaluating $f(x, y, z)$ at each of the critical points:

$$\begin{aligned} f(\pm 1, 0, 0) = f(0, \pm 1, 0) = f(0, 0, \pm 1) &= 0, \\ f(\pm 1/\sqrt{3}, \pm 1/\sqrt{3}, 1/\sqrt{3}) = f(\pm 1/\sqrt{3}, \mp 1/\sqrt{3}, -1/\sqrt{3}) &= \sqrt{3}/9, \\ f(\pm 1/\sqrt{3}, \pm 1/\sqrt{3}, -1/\sqrt{3}) = f(\pm 1/\sqrt{3}, \mp 1/\sqrt{3}, 1/\sqrt{3}) &= -\sqrt{3}/9. \end{aligned}$$

The maximum and minimum values of $f(x, y, z)$ on $x^2 + y^2 + z^2 = 1$ are therefore $\sqrt{3}/9$ and $-\sqrt{3}/9$. ■

To compare the Lagrangian solution in this example to that without a Lagrange multiplier, see Exercise 21 in Section 13.11.

EXAMPLE 13.35

When a thermonuclear reactor is built in the form of a right-circular cylinder, neutron diffusion theory requires its radius and height to satisfy the equation

$$\left(\frac{2.4048}{r}\right)^2 + \left(\frac{\pi}{h}\right)^2 = k,$$

where k is a constant. Find r and h in terms of k if the reactor is to occupy as small a volume as possible.

SOLUTION The volume of a right-circular cylinder is $V = \pi r^2 h$, and were there no constraints on r and h, this function would be considered for all points in the first quadrant of the rh-plane. However, r and h must satisfy a constraint that geometrically can be interpreted as a curve in the rh-plane. What we must do then is minimize $V = \pi r^2 h$, considering only those points (r, h) on the curve defined by the constraint. Clearly there is only one independent variable in the problem—either r or h, but not both. If we choose r as the independent variable, then we note from the constraint that as h becomes very large, r approaches $2.4048/\sqrt{k}$. Since there is no upper bound on r, we can state that the values of r to be considered in the minimization of V are $r > 2.4048/\sqrt{k}$.

To find critical points of V we introduce the Lagrangian

$$L(r,h,\lambda) = \pi r^2 h + \lambda\left\{\left(\frac{2.4048}{r}\right)^2 + \left(\frac{\pi}{h}\right)^2 - k\right\},$$

and first solve the equations

$$0 = \frac{\partial L}{\partial r} = 2\pi r h + \lambda\left(\frac{-2(2.4048)^2}{r^3}\right),$$

$$0 = \frac{\partial L}{\partial h} = \pi r^2 + \lambda\left(\frac{-2\pi^2}{h^3}\right),$$

$$0 = \frac{\partial L}{\partial \lambda} = \left(\frac{2.4048}{r}\right)^2 + \left(\frac{\pi}{h}\right)^2 - k.$$

If we solve each of the first two equations for λ and equate the resulting expressions, we have

$$\frac{\pi r^4 h}{2.4048^2} = \frac{r^2 h^3}{2\pi}.$$

Since neither r nor h can be zero, we divide by $r^2 h$:

$$\frac{\pi r^2}{2.4048^2} = \frac{h^2}{2\pi},$$

from which

$$r = \frac{2.4048\, h}{\sqrt{2}\pi}.$$

Substitution of this result into the constraint equation gives

$$\left(\frac{\sqrt{2}\pi}{h}\right)^2 + \left(\frac{\pi}{h}\right)^2 = k,$$

and this equation can be solved for $h = \pi\sqrt{3/k}$. This gives

$$r = \frac{2.4048}{\sqrt{2}\pi}\frac{\pi\sqrt{3}}{\sqrt{k}} = 2.4048\sqrt{3/(2k)}.$$

We have obtained therefore only one critical point (r,h) at which the derivatives of L vanish. The only values of r and h at which the derivatives of L do not exist are $r = 0$ and $h = 0$, but these must be rejected since the constraint requires both r and h to be positive.

To finish the problem we note that

$$\lim_{r\to\infty} V = \infty, \qquad \lim_{r\to 2.4048/\sqrt{k}^+} V = \lim_{h\to\infty} V = \infty.$$

It follows, therefore, that the single critical point at which $r = 2.4048\sqrt{3/(2k)}$ and $h = \pi\sqrt{3/k}$ must give the absolute minimum value of $V(r,h)$. ■

EXERCISES 13.12

In Exercises 1–8 use Lagrange multipliers to find maximum and minimum values of the function subject to the constraints. In each case, interpret the constraints geometrically.

1. $f(x, y) = x^2 + y$ subject to $x^2 + y^2 = 4$

2. $f(x, y, z) = 5x - 2y + 3z + 4$ subject to $x^2 + 2y^2 + 4z^2 = 9$

3. $f(x, y) = x + y$ subject to $(x - 1)^2 + y^2 = 1$

4. $f(x, y, z) = x^3 + y^3 + z^3$ subject to $x^2 + y^2 + z^2 = 9$

5. $f(x, y, z) = xyz$ subject to $x^2 + 2y^2 + 3z^2 = 12$

6. $f(x, y, z) = x^2 y + z$ subject to $x^2 + y^2 = 1,\ z = y$

7. $f(x, y, z) = x^2 + y^2 + z^2$ subject to $x^2 + y^2 + z^2 = 2z,\ x + y + z = 1$

8. $f(x, y, z) = xyz - x^2 z$ subject to $x^2 + y^2 = 1$, $z = \sqrt{x^2 + y^2}$

9. Use Lagrange multipliers to solve Exercise 11 in Section 13.11.

10. Use Lagrange multipliers to solve Exercise 12 in Section 13.11.

11. Use Lagrange multipliers to solve Exercise 20 in Section 13.11.

12. Use Lagrange multipliers to solve Exercise 22 in Section 13.11.

13. Use Lagrange multipliers to solve Exercise 23 in Section 13.11.

14. Use Lagrange multipliers to solve Exercise 24 in Section 13.11.

15. Use Lagrange multipliers to solve Exercise 17 in Section 13.11.

16. Suppose that $F(x, y) = 0$ and $G(x, y) = 0$ define two curves C_1 and C_2 in the xy-plane. Let $P(x_0, y_0)$ and $Q(X_0, Y_0)$ be the points on C_1 and C_2 that minimize the distance between C_1 and C_2. If C_1 and C_2 have tangent lines at P and Q, show that the line PQ is perpendicular to these tangent lines.

17. Find the points on the curve $x^2 + xy + y^2 = 1$ closest to and farthest from the origin.

In Exercises 18–20 use Lagrange multipliers to find maximum and minimum values of the function.

18. $f(x, y) = 3x^2 + 2xy - y^2 + 5$ for $4x^2 + 9y^2 \leq 36$

19. $f(x, y) = x^2 y + xy^2 + y$ for $-1 \leq x \leq 1,\ -1 \leq y \leq 1$

20. $f(x, y, z) = xy + xz$ for $x^2 + y^2 + z^2 \leq 1$

21. Find the maximum value of $f(x, y, z) = x^2 yz - xzy^2$ subject to the constraints $x^2 + y^2 = 1$, $z = \sqrt{x^2 + y^2}$.

22. The equation $3x^2 + 4xy + 6y^2 = 140$ describes an ellipse which has its centre at the origin, but major and minor axes are not along the x- and y-axes. Find coordinates of the ends of the major and minor axes.

23. Use Lagrange multipliers to find the point on the first octant part of the plane $Ax + By + Cz = D$, (A, B, C, D all positive constants) which maximizes the function $f(x, y, z) = x^p y^q z^r$ where p, q, and r are positive constants.

24. The folium of Descartes has parametric equations

$$x = \frac{3at}{1 + t^3}, \qquad y = \frac{3at^2}{1 + t^3}, \qquad (a > 0)$$

(see Exercise 51 in Section 3.7 and Exercise 49 in Section 10.1). Find the point in the first quadrant farthest from the origin in two ways:

(a) Express $D^2 = x^2 + y^2$ in terms of t and maximize this function of one variable.

(b) Show that an implicit equation for the curve is $x^3 + y^3 = 3axy$ and maximize $D^2 = x^2 + y^2$ subject to this constraint.

25. To find the point on the curve $x^2 - xy + y^2 - z^2 = 1$, $x^2 + y^2 = 1$ closest to the origin, we must minimize the function $f(x, y, z) = x^2 + y^2 + z^2$ subject to the constraints defined by the equations of the curve. Show that this can be done by

(a) using two Lagrange multipliers;

(b) expressing $f(x, y, z)$ in terms of x and y alone, $u = 1 - xy$, and minimizing this function subject to $x^2 + y^2 = 1$ (with one Lagrange multiplier);

(c) expressing $f(x, y, z)$ in terms of x alone, $u = 1 \pm x\sqrt{1 - x^2}$, and minimizing these functions on appropriate intervals;

(d) writing $x = \cos t$, $y = \sin t$ along the curve, expressing $f(x, y, z)$ in terms of t, $u = 1 - \sin t \cos t$, and minimizing this function on appropriate intervals.

26. Find the smallest and largest distances from the origin to the curve $x^2 + y^2/4 + z^2/9 = 1,\ x + y + z = 0$.

27. Find the maximum value of $f(x, y, z) = (xy + x^2)/(z^2 + 1)$ subject to the constraint $x^2(4 - x^2) = y^2$.

Differentials

If $y = f(x)$ is a function of one variable, the differential of y, defined by $dy = f'(x)\,dx$, was found to be an approximation to the increment $\Delta y = f(x + dx) - f(x)$ for small dx. In particular, dy is the change in y corresponding to the change dx in x if we follow the tangent line to the curve at (x, y) instead of the curve itself.

We take the same approach in defining differentials for multivariable functions. First consider a function $f(x, y)$ of two independent variables that can be represented geometrically as a surface with equation $z = f(x, y)$ (Figure 13.58). If we change the values of x and y by amounts $\Delta x = dx$ and $\Delta y = dy$, then the corresponding change in z is

$$\Delta z = f(x + dx, y + dy) - f(x, y).$$

Geometrically, this is the difference in the heights of the surface at the points $(x + dx, y + dy)$ and (x, y). If we draw the tangent plane to the surface at (x, y), then very near (x, y) the height of the tangent plane approximates the height of the surface (Figure 13.59). In particular, the height of the tangent plane at $(x + dx, y + dy)$ for small dx and dy approximates the height of the surface. We define the **differential** dz as the change in z corresponding to the changes dx and dy in x and y if we follow the tangent plane at (x, y) instead of the surface itself. To find dz in terms of dx and dy, we note that the vector joining the points (x, y, z) and $(x + dx, y + dy, z + dz)$ has components (dx, dy, dz), and this vector lies in the tangent plane. Since a normal vector to the tangent plane is

$$\nabla(z - f(x, y)) = (-f_x, -f_y, 1),$$

it follows that the vectors $(-f_x, -f_y, 1)$ and (dx, dy, dz) must be perpendicular. Consequently,

$$0 = (-f_x, -f_y, 1) \cdot (dx, dy, dz) = -\frac{\partial f}{\partial x}dx - \frac{\partial f}{\partial y}dy + dz,$$

and hence

$$dz = \frac{\partial f}{\partial x}dx + \frac{\partial f}{\partial y}dy. \tag{13.63}$$

Note that if y is held constant in the function $f(x, y)$, then $dy = 0$ and 13.63 for dz reduces to the definition of the differential of a function of one variable.

FIGURE 13.58

FIGURE 13.59

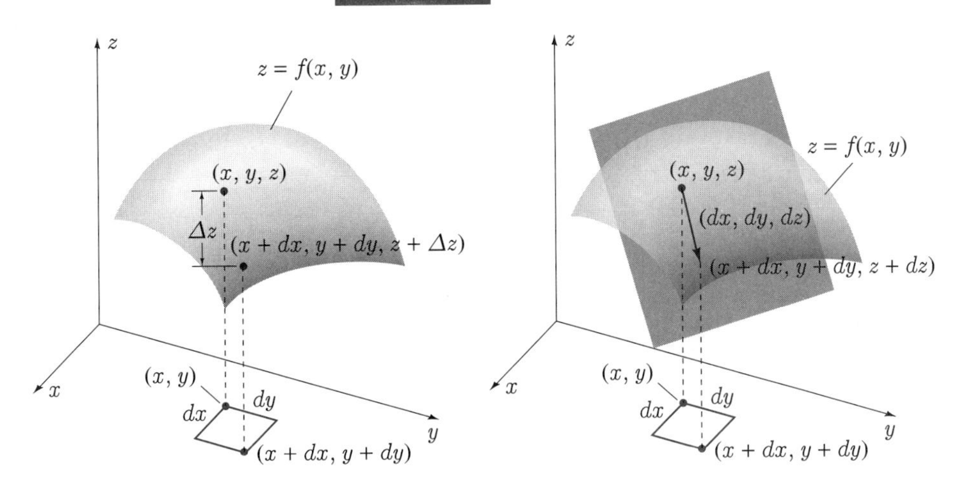

In Section 4.10 we indicated that we must be careful in using the differential dy as an approximation for the change Δy in a function $f(x)$. For the same reasons, we must be judicious in our use of dz as an approximation for Δz. Indeed, we have stated that dz is an approximation for Δz for small dx and dy, but the difficulty is deciding how small is small and how good is the approximation. In addition, note that if (x, y) is a critical point of the function $f(x, y)$, then either $dz = 0$ for all dx and dy, or dz is undefined. In other words, dz cannot be used to approximate Δz at a critical point.

EXAMPLE 13.36

If the radius of a right-circular cone is changed from 10 cm to 10.1 cm and the height is changed from 1 m to 0.99 m, use differentials to approximate the change in its volume.

SOLUTION The volume of a cone of radius r and height h is given by the formula $V = \pi r^2 h/3$. The differential of this function is

$$dV = \frac{\partial V}{\partial r} dr + \frac{\partial V}{\partial h} dh = \frac{2}{3}\pi r h\, dr + \frac{1}{3}\pi r^2\, dh.$$

If $r = 10$, $dr = 0.1$, $h = 100$, and $dh = -1$, then

$$dV = \frac{2}{3}\pi(10)(100)(0.1) + \frac{1}{3}\pi(10)^2(-1) = \frac{100\pi}{3} \text{ cm}^3.$$

■

Equation 13.63 suggests the following definition for the differential of a function of more than two independent variables.

Definition 13.11

If $u = f(x, y, z, t, \ldots, w)$, then the **differential** of $f(x, y, z, t, \ldots, w)$ is defined as

$$du = \frac{\partial f}{\partial x} dx + \frac{\partial f}{\partial y} dy + \frac{\partial f}{\partial z} dz + \frac{\partial f}{\partial t} dt + \cdots + \frac{\partial f}{\partial w} dw. \tag{13.64}$$

EXAMPLE 13.37

The area of the triangle in Figure 13.60 is given by the formula $A = (\frac{1}{2})ab \sin\theta$. If when $\theta = \pi/3$, a and b are changed by $\frac{1}{3}\%$ and θ by $\frac{1}{2}\%$, use differentials to find the approximate percentage change in A.

FIGURE 13.60

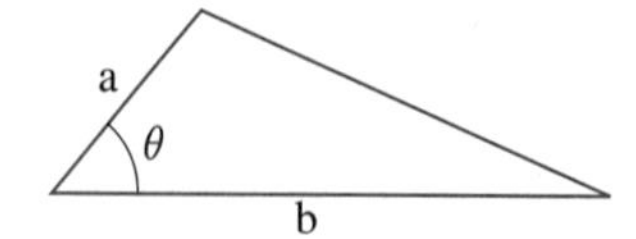

SOLUTION

Since

$$dA = \frac{\partial A}{\partial a}da + \frac{\partial A}{\partial b}db + \frac{\partial A}{\partial \theta}d\theta$$
$$= \frac{1}{2}b \sin\theta\, da + \frac{1}{2}a \sin\theta\, db + \frac{1}{2}ab \cos\theta\, d\theta,$$

the approximate percentage change in A is

$$100\left(\frac{dA}{A}\right) = \frac{100}{A}\left\{\frac{1}{2}b \sin\theta\, da + \frac{1}{2}a \sin\theta\, db + \frac{1}{2}ab \cos\theta\, d\theta\right\}$$
$$= 100\left\{\frac{da}{a} + \frac{db}{b} + \cot\theta d\theta\right\}.$$

Since a and b are changed by $\frac{1}{3}\%$ and θ by $\frac{1}{2}\%$,

$$100\left(\frac{da}{a}\right) = 100\left(\frac{db}{b}\right) = \frac{1}{3} \quad \text{and} \quad 100\left(\frac{d\theta}{\theta}\right) = \frac{1}{2}.$$

Thus,

$$100\left(\frac{dA}{A}\right) = \frac{1}{3} + \frac{1}{3} + \frac{\theta}{2}\cot\theta = \frac{2}{3} + \frac{\theta}{2}\cot\theta,$$

and when $\theta = \pi/3$,

$$100\left(\frac{dA}{A}\right) = \frac{2}{3} + \frac{1}{2}\left(\frac{\pi}{3}\right)\left(\frac{1}{\sqrt{3}}\right) = 0.97.$$

The approximate percentage change in A is therefore 1%. ■

EXERCISES 13.13

In Exercises 1–10 find the differential of the function.

1. $f(x, y) = x^2 y - \sin y$
2. $f(x, y) = \text{Tan}^{-1}(xy)$
3. $f(x, y, z) = xyz - x^3 e^z$
4. $f(x, y, z) = \sin(xyz) - x^2 y^2 z^2$
5. $f(x, y, z) = \ln(x^2 + y^2 + z^2)$
6. $f(x, y) = \text{Sin}^{-1}(xy)$
7. $f(x, y) = \text{Sin}^{-1}(x + y) + \text{Cos}^{-1}(x + y)$
8. $f(x, y, z, t) = xy + yz + zt + xt$
9. $f(x, y, z, w) = xy\tan(zw)$
10. $f(x, y, z, t) = e^{x^2+y^2+z^2-t^2}$
11. A right-circular cone has radius 10 cm and height 20 cm. If its radius is increased by 0.1 cm and its height is decreased by 0.3 cm, use differentials to find the approximate change in its volume. Compare this with the actual change in volume.
12. When the ellipse $b^2x^2 + a^2y^2 = a^2b^2$ is rotated about the x-axis, the volume V of the spheroid is $4\pi ab^2/3$. If a and b are each increased by 1%, use differentials to find the approximate percentage change in V.
13. When two resistors of resistances R_1 and R_2 are connected in parallel, their effective resistance is $R = R_1R_2/(R_1 + R_2)$. Show that if R_1 and R_2 are both increased by a small percentage c, then the percentage increase of R is also c.

SECTION 13.14

Taylor Series for Multivariable Functions

Taylor series for functions of one variable can be used to generate Taylor series for multivariable functions. For simplicity, we once again work with functions of two independent variables. Extensions to functions of more than two independent variables will be clear. Suppose that a function $f(x, y)$ has continuous partial derivatives of all orders in some open circle centred at the point (c, d) (Figure 13.61).

FIGURE 13.61

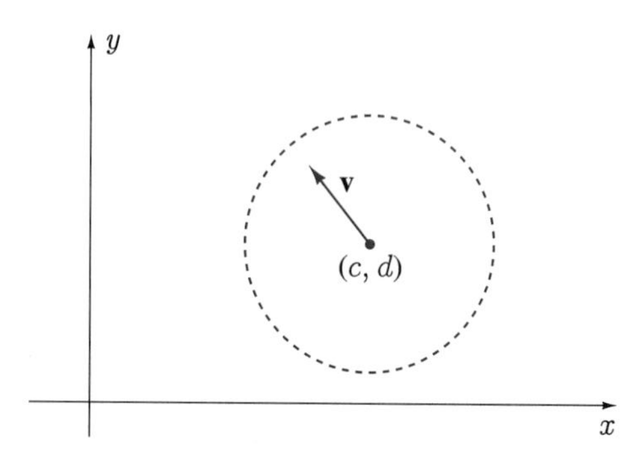

Parametric equations for the line through (c, d) in direction $\mathbf{v} = (v_x, v_y)$ are $x = c + v_x t, y = d + v_y t$. If we substitute these values into $f(x, y)$, we obtain a function $F(t)$ of one variable,

$$F(t) = f(c + v_x t, d + v_y t),$$

which represents the value of $f(x, y)$ at points along the line through (c, d) in direction $\mathbf{v}$. If we expand this function into its Maclaurin series we obtain

$$F(t) = F(0) + F'(0)t + \frac{F''(0)}{2!}t^2 + \cdots. \tag{13.65}$$

The schematic diagram to the right gives

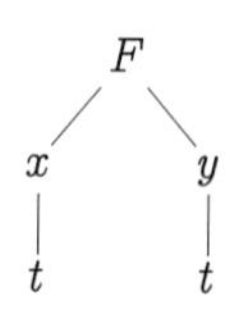

$$F'(t) = \frac{\partial F}{\partial x}\frac{dx}{dt} + \frac{\partial F}{\partial y}\frac{dy}{dt} = f_x(x, y)v_x + f_y(x, y)v_y,$$

and therefore

$$F'(0) = f_x(c, d)v_x + f_y(c, d)v_y.$$

The second derivative of $F(t)$ is

$$\begin{aligned} F''(t) &= \frac{\partial}{\partial x}[F'(t)]\frac{dx}{dt} + \frac{\partial}{\partial y}[F'(t)]\frac{dy}{dt} \\ &= \frac{\partial}{\partial x}[f_x(x, y)v_x + f_y(x, y)v_y]v_x \\ &\quad + \frac{\partial}{\partial y}[f_x(x, y)v_x + f_y(x, y)v_y]v_y \\ &= f_{xx}(x, y)v_x^2 + 2f_{xy}(x, y)v_x v_y + f_{yy}(x, y)v_y^2, \end{aligned}$$

F'(t)
x y
t t

and therefore

$$F''(0) = f_{xx}(c, d)v_x^2 + 2f_{xy}(c, d)v_x v_y + f_{yy}(c, d)v_y^2.$$

A similar calculation gives

$$F'''(0) = f_{xxx}(c, d)v_x^3 + 3f_{xxy}(c, d)v_x^2 v_y + 3f_{xyy}(c, d)v_x v_y^2 + f_{yyy}(c, d)v_y^3,$$

and the pattern is emerging. When these results are substituted into 13.65,

$$\begin{aligned} F(t) &= f(c+v_x t, d+v_y t) \\ &= f(c,d) + [f_x(c,d)v_x + f_y(c,d)v_y]t \\ &\quad + [f_{xx}(c,d)v_x^2 + 2f_x^y(c,d)v_x v_y + f_{yy}(c,d)v_y^2]\frac{t^2}{2!} + \cdots . \end{aligned} \tag{13.66}$$

We now let **v** be the vector from (c,d) to point (x,y), so that $v_x = x-c$ and $v_y = y-d$, and at the same time set $t=1$. Then $F(1) = f(c+x-c, d+y-d) = f(x,y)$, and 13.66 becomes

$$\begin{aligned} f(x,y) = f(c,d) &+ [f_x(c,d)(x-c) + f_y(c,d)(y-d)] + \frac{1}{2!}[f_{xx}(c,d)(x-c)^2 \\ &+ 2f_{xy}(c,d)(x-c)(y-d) + f_{yy}(c,d)(y-d)^2] + \cdots . \end{aligned} \tag{13.67}$$

This is the Taylor series for $f(x,y)$ about the point (c,d). It gives the value of the function at the point (x,y) in terms of values of the function and its derivatives at the point (c,d).

EXAMPLE 13.38

Find the first six nonzero terms in the Taylor series for $f(x,y) = \sin(2x+3y)$ about $(0,0)$.

SOLUTION We calculate that:

$$\begin{aligned} f(0,0) &= 0 \\ f_x(0,0) &= 2\cos(2x+3y)_{|(0,0)} = 2 \\ f_y(0,0) &= 3\cos(2x+3y)_{|(0,0)} = 3 \\ f_{xx}(0,0) &= -4\sin(2x+3y)_{|(0,0)} = 0 \\ f_{xy}(0,0) &= -6\sin(2x+3y)_{|(0,0)} = 0 \\ f_{yy}(0,0) &= -9\sin(2x+3y)_{|(0,0)} = 0 \\ f_{xxx}(0,0) &= -8\cos(2x+3y)_{|(0,0)} = -8 \\ f_{xxy}(0,0) &= -12\cos(2x+3y)_{|(0,0)} = -12 \\ f_{xyy}(0,0) &= -18\cos(2x+3y)_{|(0,0)} = -18 \\ f_{yyy}(0,0) &= -27\cos(2x+3y)_{|(0,0)} = -27 \end{aligned}$$

Formula 13.67 then gives

$$\begin{aligned} \sin(2x+3y) &= 0 + [2x+3y] + \frac{1}{2!}[0] + \frac{1}{3!}[-8x^3 - 36x^2y - 54xy^2 - 27y^3] + \ldots \\ &= (2x+3y) - \frac{1}{3!}(2x+3y)^3 + \cdots . \end{aligned}$$

This series could also have been obtained by substituting $2x+3y$ for x in the Maclaurin series for $\sin x$. This is not always an alternative. ■

EXERCISES 13.14

1. If $f(x, y) = F(x)G(y)$, is the Taylor series of $f(x, y)$ about $(0, 0)$ the product of the Maclaurin series for $F(x)$ and $G(y)$?

2. What are the cubic terms in 13.67?

In Exercises 3–8 find the Taylor series of the function about the point by using Taylor series for functions of one variable.

3. $\cos(xy)$ about $(0, 0)$

4. e^{2x-3y} about $(1, -1)$

5. $x^2 y\sqrt{1 + x}$ about $(0, 0)$

6. $\ln(1 + x^2 + y^2)$ about $(0, 0)$

7. $\dfrac{1}{1 + x + y}$ about $(3, -4)$

8. $\dfrac{xy^2}{1 + y^2}$ about $(-1, 0)$

In Exercises 9–14 find the Taylor series of the function up to and including quadratic terms.

9. $\dfrac{xy}{x^2 + y^2}$ about $(-1, 1)$

10. $\sqrt{1 + xy}$ about $(2, 1)$

11. $e^x \sin(3x - y)$ about $(-1, 0)$

12. $(x + y)^2 \ln(x + y)$ about $(0, 1)$

13. $\text{Tan}^{-1}(3x + 2y)$ about $(1, -1)$

14. $x^8 y^{10}$ about $(0, 0)$

15. What are the terms in the Taylor series for a function $f(x, y, z)$ about the point (c, d, e) corresponding to those in equation 13.67?

16. Express 13.67 in sigma notation. Hint: Think about an operator
$$\left[(x - a)\frac{\partial}{\partial x} + (y - b)\frac{\partial}{\partial y}\right]^n$$
which is expanded as a binomial to operate on functions $f(x, y)$.

SUMMARY

We began the study of multivariable functions in this chapter, concentrating our attention on differentiation and its applications. We introduced two types of derivatives for a multivariable function: partial derivatives and directional derivatives. Partial derivatives are directional derivatives in directions parallel to the coordinate axes. The directional derivative of a function $f(x, y, z)$ in the direction $\mathbf{v}$ is given by the formula $D_{\mathbf{v}} f = \nabla f \cdot \hat{\mathbf{v}}$, where $\hat{\mathbf{v}}$ is the unit vector in the direction of $\mathbf{v}$, and the gradient ∇f is evaluated at the point at which $D_{\mathbf{v}} f$ is required. This formula leads to the fact that the gradient $\nabla f(x, y, z)$ points in the direction in which $f(x, y, z)$ increases most rapidly, and $|\nabla f|$ is the (maximum) rate of change of $f(x, y, z)$ in the direction ∇f. A second property of gradient vectors (which is related to the first) is that if $F(x, y, z) = 0$ is the equation of a surface, then at any point on the surface ∇F is perpendicular to the surface. This property, along with the fact that perpendicularity to a surface is synonymous with perpendicularity to its tangent plane, enables us to find equations for tangent planes to surfaces and tangent lines to curves.

We illustrated various ways to calculate partial derivatives of a multivariable function, depending on whether the function is defined explicitly, implicitly, or as a composite function. Since partial derivatives are ordinary derivatives with other variables held constant, there is no difficulty calculating partial derivatives when the function is defined explicitly; we simply use the rules from single-variable calculus. When the partial derivative of a composite function is required, we use a schematic diagram illustrating functional dependences to develop the appropriate chain rule. Partial derivatives for functions defined implicitly are calculated using Jacobians.

Critical points of a multivariable function are points at which all of its first partial derivatives vanish or at which one or more of these partial derivatives does not exist. Critical points can yield relative maxima, relative minima, saddle points, or none of these. For functions of two independent variables, a second-derivative test exists that may determine whether a critical point at which the partial derivatives vanish yields a relative maximum, a relative minimum, or a saddle point. This test is analogous to that for functions of one variable.

A continuous function of two independent variables defined on a region that includes its boundary always takes on a maximum value and a minimum value. To find these values we evaluate the function at each of its critical points and compare these numbers to the maximum and minimum values of the function on its boundary. Finding the extreme values on the boundary involves one or more extrema problems for a function of one variable, the number of such problems depending on the complexity of the boundary.

We suggest two methods for finding the extreme values of a function when the variables of the function are subject to constraints. We either solve the constraint equations for dependent variables and express the given function in terms of independent variables or we use Lagrange multipliers. Lagrange multipliers eliminate the necessity for solving the constraint equations, but they do, on the other hand, give a larger system of equations to solve for critical points.

Differentials of multivariable functions can be used to approximate changes in functions when small changes are made to its independent variables. Taylor series can also be used to approximate multivariable functions.

Key Terms and Formulas

In reviewing this chapter, you should be able to define or discuss the following key terms:

Domain
Limit
Continuous function
Partial derivative
Gradient
Harmonic function
Chain rule
Homogeneous function
Laplace's equation
Implicit differentiation
Jacobian determinant
Directional derivative
Tangent line to a curve
Tangent plane to a surface
Normal to a surface
Smooth surface
Piecewise smooth surface
Critical point
Relative maximum and minimum
Saddle point
Absolute maximum and minimum
Lagrange multiplier
Differential
Taylor series

REVIEW EXERCISES

In Exercises 1–20 find the derivative.

1. $\partial f/\partial x$ if $f(x,y) = x^2/y^3 - \operatorname{Sin}^{-1}(xy)$

2. $\partial^2 f/\partial y^2$ if $f(x,y,z) = \ln(x^2+y^2+z^2)$

3. $\partial^3 f/\partial x^2 \partial y$ if $f(x,y,z,t) = x^3 e^y - xzt^2 - \sin(x+y+z+t)$

4. $\partial z/\partial x$ if $z^2 x + \operatorname{Tan}^{-1} z + y = 3x$

5. $\partial u/\partial y$ if $u\cos y + y\cos(xu) + z^2 = 5x$

6. df/dt if $f(x,y) = x^2+y^2-e^{xy}$, $x = t^3+3t$, $y = t\ln t$

7. dy/dx if $x = y^3 + 3y^2 - 2y + 4$

8. $\partial u/\partial x)_y$ if $u^2+v^2-xy = 5$, $3u - 2v + x^2 u = 2v^3$

9. $\partial^2 f/\partial u \partial v$ if $f(u,v) = u^2/\sqrt{v} - v/\sqrt{u}$

10. df/dt if $f(x,y) = xy - x^2 - y^2$, $x = te^t$, $y = te^{-t}$

11. $\partial z/\partial t)_u$ if $z = x^2 - y^2$, $x = 2u - 3v^2 + 3uvt$, $y = u\cos(vt)$, $v = t^2 - 2t$

12. $\partial r/\partial x)_y$ if $x = r\cos\theta$, $y = r\sin\theta$

13. $\partial \theta/\partial x)_{y,z}$ if $x = r\sin\phi\cos\theta$, $y = r\sin\phi\sin\theta$, $z = r\cos\phi$

14. $\partial u/\partial r)_\theta$ if $u = x^2 - y^2 x^3$, $x = r\cos\theta$, $y = r\sin\theta$

15. $\partial^2 u/\partial r^2)_\theta$ if $u = x^2 - y^2 x^3$, $x = r\cos\theta$, $y = r\sin\theta$

16. d^2u/dt^2 if $u = x/z^2 - z/x^2$, $x = t^3 - 3$, $z = 1/t^3$

17. dz/dt if $z = y - xy^2 + x$, and $x^2 - y^2 + xt = 2t$, $xy = 4t^2$

18. $\partial^2 z/\partial x^2$ if $xz - x^2z^3 + y^2 = 3$

19. dy/dx if $yx - x^2z^2 + 5x = 3$, $2xz - 3x^2y^2 = 4z^4$

20. $\partial u/\partial t)_v$ if $u = xyt^2 - 3\,\mathrm{Sin}^{-1}(xy)$, $x = v^2t^2 - 2t$, $y = v\tan t$

21. If $u = (x^2 + y^2)[1 + \sin(x/z)]$, show that $x\dfrac{\partial u}{\partial x} + y\dfrac{\partial u}{\partial y} + z\dfrac{\partial u}{\partial z} = 2u$.

22. If $u = 2x^2 - 3y^2 + xy$, show that $x^2\dfrac{\partial^2 u}{\partial x^2} + 2xy\dfrac{\partial^2 u}{\partial x \partial y} + y^2\dfrac{\partial^2 u}{\partial y^2} = 2u$.

23. If $f(s)$ is a differentiable function, show that $f(3x - 2y)$ satisfies $2\dfrac{\partial f}{\partial x} + 3\dfrac{\partial f}{\partial y} = 0$.

24. If $f(s,t)$ has continuous first partial derivatives, show that the function $f(x^2 - y^2, y^2 - x^2)$ satisfies $y\dfrac{\partial f}{\partial x} + x\dfrac{\partial f}{\partial y} = 0$.

In Exercises 25–30 find the directional derivative.

25. $f(x,y) = x^2 \sin y$ at $(3,-1)$ in the direction $\mathbf{v} = (2,4)$

26. $f(x,y,z) = x^2 + y^2 + z^2$ at $(1,0,1)$ in the direction from $(1,0,1)$ to $(2,-1,3)$

27. $f(x,y,z) = z\,\mathrm{Tan}^{-1}(x + y)$ at $(-1,2,5)$ in the direction perpendicular to the surface $z = x^2 + y^2$ with positive z-component

28. $f(x,y,z) = x^2 + y - 2z$ at $(1,-1,2)$ along the line $x - y + z = 4$, $2x + 4y + 2 = 0$ in the direction of increasing x

29. $f(x,y) = \ln(x + y)$ at $(3,10)$ along the curve $y = x^2 + 1$ in the direction of decreasing y

30. $f(x,y,z) = 2xyz - x^2 - z^2$ at $(0,1,1)$ along the curve $x^2 + y^2 + z^2 = 2$, $y = z$ in the direction of increasing x

In Exercises 31–33 find the equation of the tangent plane to the surface.

31. $z = x^2 + y^2$ at $(1,3,10)$

32. $x^2 + z^3 = y^2$ at $(-1,3,2)$

33. $x^2 + y^2 = z^2 + 1$ at $(1,0,0)$

In Exercises 34–36 find equations for the tangent line to the curve.

34. $x = t^2 + 1$, $y = t^2 - 1$, $z = t^3 + 5t$ at $(2,0,6)$

35. $x + y + z = 0$, $2x - 3y - 6z = 11$ at $(1,1,-2)$

36. $z = xy$, $x^2 + y^2 = 2$ at $(1,1,1)$

In Exercises 37–40 find all critical points for the function and classify each as yielding a relative maximum, a relative minimum, or a saddle point.

37. $f(x,y) = x^3 + 3y^2 - 6x + 4$

38. $f(x,y) = ye^x$

39. $f(x,y) = x^2 - xy + y^2 + x - 4y$

40. $f(x,y) = (x^2 + y^2 - 1)^2$

41. If $f(x,y) = (x^2 + y^2)F(x,y)$ where $F(x,y) = x^3/y - y^3/x$, verify that

$$\frac{\partial^2 f}{\partial x^2} + \frac{\partial^2 f}{\partial y^2} = (x^2 + y^2)\left(\frac{\partial^2 F}{\partial x^2} + \frac{\partial^2 F}{\partial y^2}\right) + 12F(x,y).$$

42. Find maximum and minimum values of the function $f(x,y) = xy$ on the circle $x^2 + y^2 \le 1$.

43. Find maximum and minimum values of the function $f(x,y,z) = 2x + 3y - 4z$ on the sphere $x^2 + y^2 + z^2 \le 2$.

44. Find the points on the curve $x^2 + x^4 + y^2 = 1$ closest to and farthest from the origin.

45. Find the point(s) on the surface $z^2 = 1 + xy$ closest to the origin.

46. Generalize Review Exercise 32 in Chapter 4 to incorporate a third crop, say sunflowers, with a yield of \$r per hectare and a proportional loss cz.

47. If the equation $u = f(x - ut)$ defines u implicitly as a function of x and t, show that $\dfrac{\partial u}{\partial t} + u\dfrac{\partial u}{\partial x} = 0$.

48. Find the first six nonzero terms in the Taylor series for $x^3 \sin(x^2 y)$ about the point $(1, \pi/4)$.

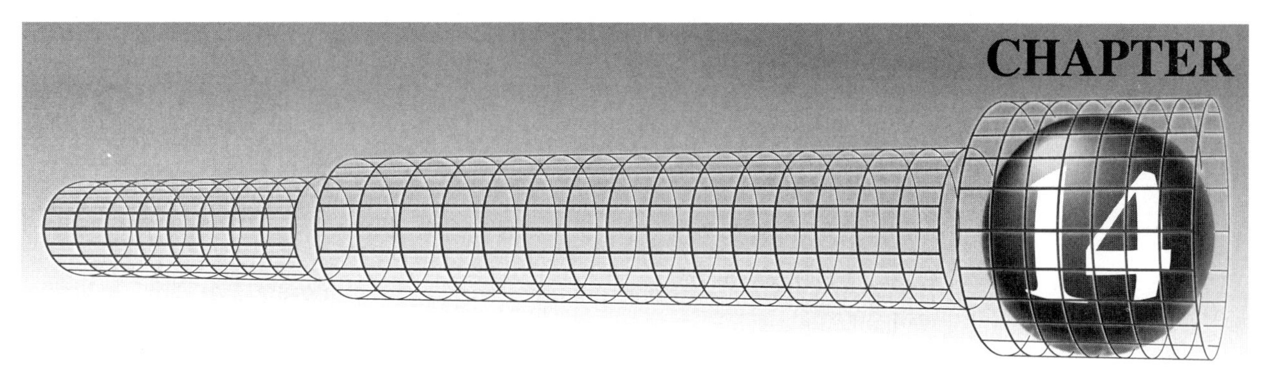

Multiple Integrals

The definite integral of a function $f(x)$ of one variable is defined as the limit of a sum of the form

$$f(x_1^*)\Delta x_1 + f(x_2^*)\Delta x_2 + \cdots + f(x_n^*)\Delta x_n \qquad (14.1)$$

where the norm of the partition approaches zero. We have seen that definite integrals can be used to calculate area, volume, work, fluid force, and moments. In spite of the fact that some of these are two- and three-dimensional concepts, we have been careful to emphasize that a definite integral with respect to x is an integration along the x-axis, and a definite integral with respect to y is an integration along the y-axis. In other words, independent of how we interpret the result of the integration, a definite integral is a "limit-summation" *along a line*.

Generalizations of these limiting sums to functions of two and three independent variables lead to definitions of double and triple integrals.

SECTION 14.1

Double Integrals and Double Iterated Integrals

Suppose a function $f(x,y)$ is defined in some region R of the xy-plane that has finite area (Figure 14.1). To define the double integral of $f(x,y)$ over R, we first divide R into n subregions of areas ΔA_1, ΔA_2, ..., ΔA_n, in any manner whatsoever. In each subregion $\Delta A_i (i = 1, \ldots, n)$ we choose an arbitrary point (x_i^*, y_i^*) and form the sum

$$f(x_1^*, y_1^*)\Delta A_1 + f(x_2^*, y_2^*)\Delta A_2 + \cdots + f(x_n^*, y_n^*)\Delta A_n = \sum_{i=1}^{n} f(x_i^*, y_i^*)\Delta A_i. \qquad (14.2)$$

The norm of the partition of R into sub–areas ΔA_i is the area of the largest of the sub–areas, denoted by $||\Delta A_i|| = \max_{i=1,\ldots,n} \Delta A_i$.

FIGURE 14.1

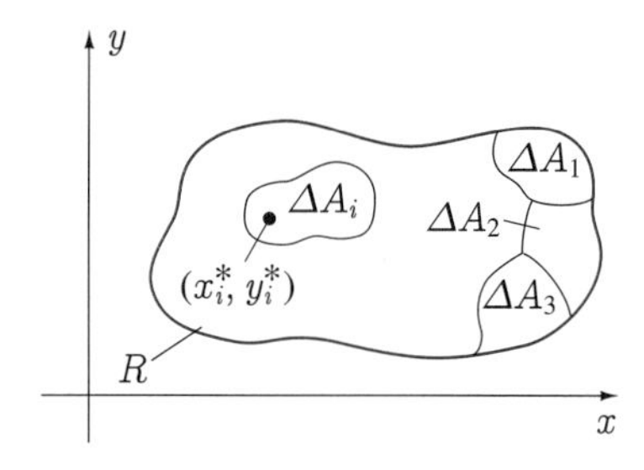

Suppose we increase the number of terms in 14.2 by increasing the number of sub–areas ΔA_i and decreasing the norm $||\Delta A_i||$. If the sum approaches a limit as the number of sub–areas becomes increasingly large and each sub–area shrinks to a point, we call the limit the **double integral** of $f(x, y)$ over the region R, and denote it by

$$\iint_R f(x, y)\, dA = \lim_{||\Delta A_i|| \to 0} \sum_{i=1}^{n} f(x_i^*, y_i^*) \Delta A_i. \tag{14.3}$$

The notation $||\Delta A_i|| \to 0$ does not necessarily require that every ΔA_i shrink to a point. We implicitly assume, however, that this is always the case.

If limit 14.3 were dependent on the choice of subdivision ΔA_i or choice of star-points (x_i^*, y_i^*), double integrals would be of little use. We therefore demand that the limit of the sum be independent of the manner of subdivision of R and choice of star-points in the subregions. At first sight this requirement might seem rather severe since we must now check that all subdivisions and all choices of star-points lead to the same limit before concluding that the double integral exists. Fortunately, however, the following theorem indicates that for continuous functions this is unnecessary.

Theorem 14.1

Let C be a closed, piecewise-smooth curve that encloses a region R with finite area. If $f(x, y)$ is a continuous function inside and on C, then the double integral of $f(x, y)$ over R exists.

For a continuous function, then, the double integral exists and any choice of subdivision and star-points leads to the same value through limiting process 14.3. Note that continuity was also the condition that guaranteed existence of the definite integral in Theorem 6.2.

We cannot overemphasize the fact that a double integral is simply the limit of a sum. Moreover, any limit of form 14.3 may be interpreted as the double integral of a function $f(x, y)$ over a region defined by the ΔA_i.

The following properties of double integrals are easily proved using Definition 14.3:

(1) If the double integral of $f(x, y)$ over R exists and c is a constant, then

$$\iint_R cf(x, y)\, dA = c \iint_R f(x, y)\, dA. \tag{14.4}$$

(2) If double integrals of $f(x, y)$ and $g(x, y)$ over R exist, then

$$\iint_R [f(x, y) + g(x, y)]\, dA = \iint_R f(x, y)\, dA + \iint_R g(x, y)\, dA. \tag{14.5}$$

(3) If a region R is subdivided by a piecewise-smooth curve into two parts R_1 and R_2 that have at most boundary points in common (Figure 14.2), and the double integral of $f(x, y)$ over R exists, then

$$\iint_R f(x, y)\, dA = \iint_{R_1} f(x, y)\, dA + \iint_{R_2} f(x, y)\, dA. \tag{14.6}$$

FIGURE 14.2

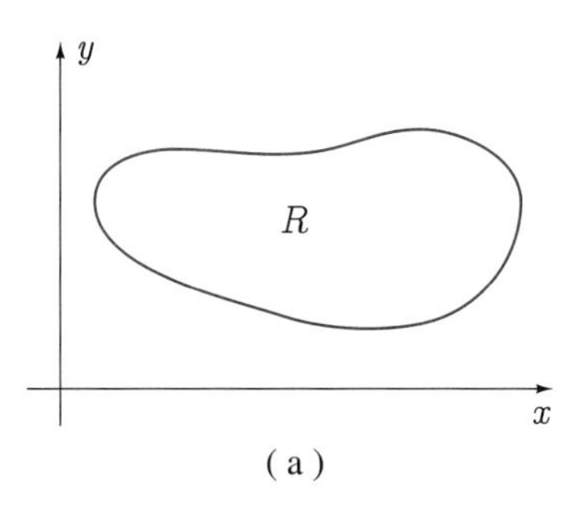

(a)

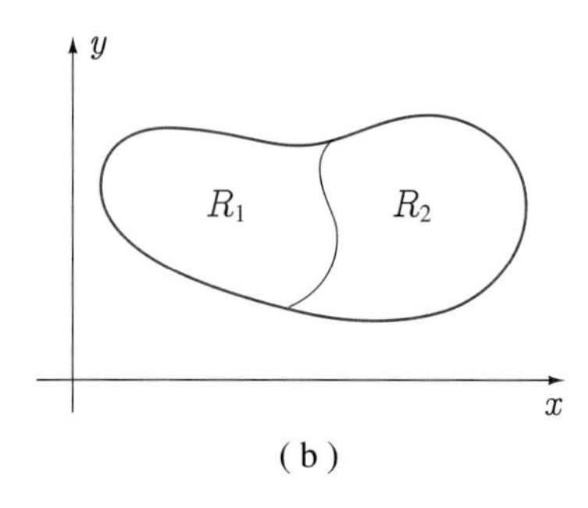

(b)

(4) The area of a region R is

$$\text{Area of } R = \iint_R dA. \tag{14.7}$$

In spite of the fact that double integrals are defined as "limits of sums," we do not evaluate them as such. Just as definite integrals are evaluated with indefinite integrals, we evaluate double integrals with double iterated integrals.

Double Iterated Integrals

We have already seen that a function $f(x, y)$ of two independent variables has two first-order partial derivatives, one with respect to x holding y constant, and one with respect to y holding x constant. We now reverse this process and define "partial" integration of $f(x, y)$ with respect to x and y. Quite naturally, we define *a* partial indefinite integral of $f(x, y)$ with respect to x as *an* antiderivative of $f(x, y)$ with respect to x holding y constant. For example, since

$$\frac{\partial}{\partial x}(x^3 + x^2 y) = 3x^2 + 2xy,$$

$x^3 + x^2 y$ is an antiderivative with respect to x of $3x^2 + 2xy$. But so too is $x^3 + x^2 y + y$. In fact for any differentiable function $C(y)$ of y, $x^3 + x^2 y + C(y)$ is an antiderivative of $3x^2 + 2xy$ with respect to x. Since this expression represents all antiderivatives of $3x^2 + 2xy$, we call it *the* indefinite integral of $3x^2 + 2xy$ with respect to x, and write

$$\int (3x^2 + 2xy)\,dx = x^3 + x^2 y + C(y).$$

Similarly, the indefinite integral of $3x^2 + 2xy$ with respect to y is

$$\int (3x^2 + 2xy)\,dy = 3x^2 y + xy^2 + D(x),$$

where $D(x)$ is an arbitrary differentiable function of x.

In this chapter we are concerned only with partial definite integrals. Limits on a partial definite integral with respect to x must not depend on x, but may depend on y. In general, then, a partial definite integral with respect to x is of the form

$$\int_{g(y)}^{h(y)} f(x, y)\,dx; \tag{14.8}$$

similarly, a partial definite integral with respect to y is of the form

$$\int_{g(x)}^{h(x)} f(x, y)\,dy. \tag{14.9}$$

Each of these partial definite integrals is evaluated by substituting the limits into a corresponding partial indefinite integral. For example,

$$\begin{aligned}\int_{x^2}^{x+2}(2y + xe^y)\,dy &= \{y^2 + xe^y\}_{x^2}^{x+2}\\ &= \{(x+2)^2 + xe^{x+2}\} - \{(x^2)^2 + xe^{x^2}\}\\ &= (x+2)^2 - x^4 + x(e^{x+2} - e^{x^2}).\end{aligned}$$

Once antidifferentiation in 14.8 is completed and the limits substituted, the result is a function of y alone. It is then possible to integrate this function with respect to y between any two limits, say from $y = c$ to $y = d$:

$$\int_c^d \left\{\int_{g(y)}^{h(y)} f(x,y)\,dx\right\} dy.$$

In practice we omit the braces and simply write

$$\int_c^d \int_{g(y)}^{h(y)} f(x,y)\,dx\,dy, \tag{14.10}$$

understanding that in the evaluation we proceed from the inner integral to the outer. This is called a **double iterated integral** first with respect to x and then with respect to y (or, more concisely, with respect to x and y). Double iterated integrals with respect to y and x take the form

$$\int_a^b \int_{g(x)}^{h(x)} f(x,y)\,dy\,dx. \tag{14.11}$$

EXAMPLE 14.1

Evaluate each of the following double iterated integrals:

(a) $\int_0^1 \int_x^{x-1}(x^2 + e^y)\,dy\,dx$

(b) $\int_{-1}^1 \int_0^y xye^{x^2}\,dx\,dy$

SOLUTION

(a)
$$\begin{aligned}\int_0^1\int_x^{x-1}(x^2 + e^y)\,dy\,dx &= \int_0^1 \{x^2y + e^y\}_x^{x-1}\,dx\\ &= \int_0^1 \{x^2(x-1) + e^{x-1} - x^3 - e^x\}dx\\ &= \int_0^1 \{-x^2 + e^{x-1} - e^x\}dx\\ &= \left\{-\frac{x^3}{3} + e^{x-1} - e^x\right\}_0^1\\ &= \left\{-\frac{1}{3} + 1 - e\right\} - \{e^{-1} - 1\}\\ &= \frac{5}{3} - e - e^{-1}.\end{aligned}$$

(b) $$\int_{-1}^{1}\int_{0}^{y} xye^{x^2}\,dx\,dy = \int_{-1}^{1}\left\{\frac{1}{2}ye^{x^2}\right\}_0^y dy = \frac{1}{2}\int_{-1}^{1}(ye^{y^2}-y)\,dy$$
$$= \frac{1}{2}\left\{\frac{1}{2}e^{y^2}-\frac{y^2}{2}\right\}_{-1}^{1} = \frac{1}{2}\left\{\frac{e}{2}-\frac{1}{2}-\frac{e}{2}+\frac{1}{2}\right\} = 0.$$

■

EXERCISES 14.1

In Exercises 1–30 evaluate the double iterated integral.

1. $\displaystyle\int_{-1}^{2}\int_{y}^{y+2}(x^2-xy)\,dx\,dy$

2. $\displaystyle\int_{-3}^{3}\int_{-\sqrt{18-2y^2}}^{\sqrt{18-2y^2}} x\,dx\,dy$

3. $\displaystyle\int_{0}^{1}\int_{x^2}^{x}(2xy+3y^2)\,dy\,dx$

4. $\displaystyle\int_{-1}^{0}\int_{y}^{2}(1+y)^2\,dx\,dy$

5. $\displaystyle\int_{3}^{4}\int_{0}^{\pi/2} x\sin y\,dy\,dx$

6. $\displaystyle\int_{1}^{2}\int_{1}^{y} e^{x+y}\,dx\,dy$

7. $\displaystyle\int_{-1}^{1}\int_{-x}^{5}(x^2+y^2)\,dy\,dx$

8. $\displaystyle\int_{-1}^{1}\int_{x}^{2x}(xy+x^3y^3)\,dy\,dx$

9. $\displaystyle\int_{0}^{1}\int_{x}^{1}(x+y)^4\,dy\,dx$

10. $\displaystyle\int_{1}^{2}\int_{x}^{2x}\frac{1}{(x+y)^3}\,dy\,dx$

11. $\displaystyle\int_{0}^{1}\int_{0}^{3x}\sqrt{x+y}\,dy\,dx$

12. $\displaystyle\int_{-1}^{1}\int_{1}^{e}\frac{y}{x}\,dx\,dy$

13. $\displaystyle\int_{1}^{4}\int_{\sqrt{x}}^{x^2}(x^2+2xy-3y^2)\,dy\,dx$

14. $\displaystyle\int_{0}^{2}\int_{x^2}^{2x^2} x\cos y\,dy\,dx$

15. $\displaystyle\int_{0}^{1}\int_{1}^{\tan x}\frac{1}{1+y^2}\,dy\,dx$

16. $\displaystyle\int_{0}^{1}\int_{0}^{y^3}\frac{1}{1+y^2}\,dx\,dy$

17. $\displaystyle\int_{2}^{3}\int_{0}^{1}\frac{x}{\sqrt{1-y^2}}\,dy\,dx$

18. $\displaystyle\int_{0}^{2}\int_{-x}^{x}(8-2x^2)^{3/2}\,dy\,dx$

19. $\displaystyle\int_{0}^{1}\int_{0}^{x}\frac{1}{\sqrt{1-y^2}}\,dy\,dx$

20. $\displaystyle\int_{-9}^{0}\int_{0}^{x^2\sqrt{9+x}} dy\,dx$

21. $\displaystyle\int_{0}^{2}\int_{\sqrt{4-x^2}}^{2} y^2\,dy\,dx$

22. $\displaystyle\int_{-1}^{0}\int_{y}^{0} x\sqrt{x^2+y^2}\,dx\,dy$

23. $\displaystyle\int_{2}^{3}\int_{1}^{2x}\frac{1}{(xy+x^2)^2}\,dy\,dx$

24. $\displaystyle\int_{0}^{1}\int_{0}^{\mathrm{Cos}^{-1}x} x\cos y\,dy\,dx$

25. $\displaystyle\int_{0}^{1}\int_{\sqrt{y^2+y}}^{\sqrt{2y}} x^3\sqrt{x^2-y^2}\,dx\,dy$

26. $\displaystyle\int_{0}^{1}\int_{\sqrt{2y}}^{\sqrt{y^2+y}} x^3\sqrt{x^2-y^2}\,dx\,dy$

27. $\displaystyle\int_{-2}^{0}\int_{x^4}^{4x^2} \sqrt{y - x^4}\, dy\, dx$

28. $\displaystyle\int_{-2}^{0}\int_{y}^{0} \frac{x}{\sqrt{x^2 + y^2}}\, dx\, dy$

29. $\displaystyle\int_{-1}^{2}\int_{-1}^{y^3} \sqrt{1 + y}\, dx\, dy$

30. $\displaystyle\int_{0}^{1}\int_{0}^{x} \sqrt{x^2 + y^2}\, dy\, dx$

SECTION 14.2

Evaluation Of Double Integrals By Double Iterated Integrals

According to Theorem 14.1, if a function $f(x, y)$ is continuous on a finite region R with a piecewise-smooth boundary, then double integral 14.3 exists, and its evaluation by means of that limit is independent of both the manner of subdivision of R into areas ΔA_i and choice of star-points (x_i^*, y_i^*). We now show that if we make particular choices of ΔA_i, double integrals can be evaluated by means of double iterated integrals in x and y.

Consider first a rectangle R with edges parallel to the x- and y-axes as shown in Figure 14.3. We divide R into smaller rectangles by a network of $n + 1$ vertical lines and $m + 1$ horizontal lines identified by abscissae

$$a = x_0 < x_1 < x_2 < \cdots < x_{n-1} < x_n = b$$

and ordinates

$$c = y_0 < y_1 < y_2 < \cdots < y_{m-1} < y_m = d.$$

FIGURE 14.3

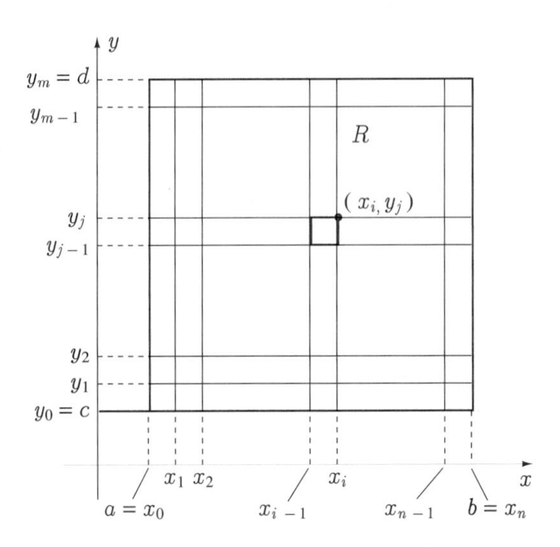

If the (i, j)th rectangle is that rectangle bounded by the lines $x = x_{i-1}$, $x = x_i$, $y = y_{j-1}$, and $y = y_j$, then its area is $\Delta x_i \Delta y_j$, where $\Delta x_i = x_i - x_{i-1}$ and $\Delta y_j = y_j - y_{j-1}$. We choose as star-point in the (i, j)th rectangle the upper right corner: $(x_i^*, y_j^*) = (x_i, y_j)$. With this rectangular subdivision of R and choice of star-points, Definition 14.3 for the double integral of $f(x, y)$ over R takes the form

$$\iint_R f(x, y)\, dA = \lim_{\|\Delta x_i \Delta y_j\| \to 0} \sum_{i=1}^{n} \sum_{j=1}^{m} f(x_i, y_j) \Delta x_i \Delta y_j. \qquad (14.12\text{a})$$

Since $||\Delta x_i \Delta y_j|| \to 0$ if the norms $||\Delta x_i||$ and $||\Delta y_j||$ individually approach zero, we can write that

$$\iint_R f(x,y)\,dA = \lim_{\substack{||\Delta x_i|| \to 0 \\ ||\Delta y_j|| \to 0}} \sum_{i=1}^{n} \sum_{j=1}^{m} f(x_i, y_j)\,\Delta x_i \Delta y_j. \qquad (14.12\text{b})$$

Suppose we choose to first perform the limit on y and then the limit on x, and write therefore

$$\iint_R f(x,y)\,dA = \lim_{||\Delta x_i|| \to 0} \sum_{i=1}^{n} \left\{ \lim_{||\Delta y_j|| \to 0} \sum_{j=1}^{m} f(x_i, y_j)\,\Delta y_j \right\} \Delta x_i.$$

Since x_i is constant in the limit with respect to y, the y-limit is precisely the definition of the definite integral of $f(x_i, y)$ with respect to y from $y = c$ to $y = d$; i.e.,

$$\lim_{||\Delta y_j|| \to 0} \sum_{j=1}^{m} f(x_i, y_j)\,\Delta y_j = \int_c^d f(x_i, y)\,dy.$$

Consequently,

$$\iint_R f(x,y)\,dA = \lim_{||\Delta x_i|| \to 0} \sum_{i=1}^{n} \left\{ \int_c^d f(x_i, y)\,dy \right\} \Delta x_i.$$

Because the term in braces is a function of x_i alone, we can interpret this limit as a definite integral with respect to x:

$$\iint_R f(x,y)\,dA = \int_a^b \left\{ \int_c^d f(x,y)\,dy \right\} dx = \int_a^b \int_c^d f(x,y)\,dy\,dx, \qquad (14.13)$$

a double iterated integral. By reversing the order of taking limits, we can show similarly that the double integral can be evaluated with a double iterated integral with respect to x and y:

$$\iint_R f(x,y)\,dA = \int_c^d \int_a^b f(x,y)\,dx\,dy. \qquad (14.14)$$

We have shown, then, that for the special case of a rectangle R with sides parallel to the axes, a double integral over R can be evaluated by using double iterated integrals. Conversely, every double iterated integral with constant limits represents a double integral over a rectangle. The double iterated integral simply indicates that a rectangular subdivision has been chosen to evaluate the double integral.

We have just stated that the choice of a double iterated integral to evaluate a double integral implies that the region of integration has been subdivided into small rectangles. We now show that the x- and y-integrations themselves can be interpreted geometrically. These interpretations will simplify the transition to more difficult regions of integration.

In the subdivision of R into rectangles, suppose we denote the dimensions of a representative rectangle at position (x, y) by dx and dy (Figure 14.4). In the inner integral

$$\int_c^d f(x,y)\,dy\,dx$$

of equation 14.13, x is held constant and integration is performed in the y-direction. This (partial) definite integral is therefore interpreted as summing over the rectangles in the vertical strip of width dx at position x. The limits $y = c$ and $y = d$ identify the initial and terminal positions of this vertical strip. It is important to note that we are not adding the areas of the rectangles of dimensions dx and dy in the strip. On the contrary, each rectangle of area $dy\,dx$ is multiplied by the value of $f(x, y)$ for that rectangle,

$$f(x, y)\,dy\,dx,$$

and it is these quantities that are added.

The x-integration in equation 14.13 is interpreted as adding over all strips starting at $x = a$ and ending at $x = b$. The limits on x, therefore, identify positions of the first and last strips.

FIGURE 14.4

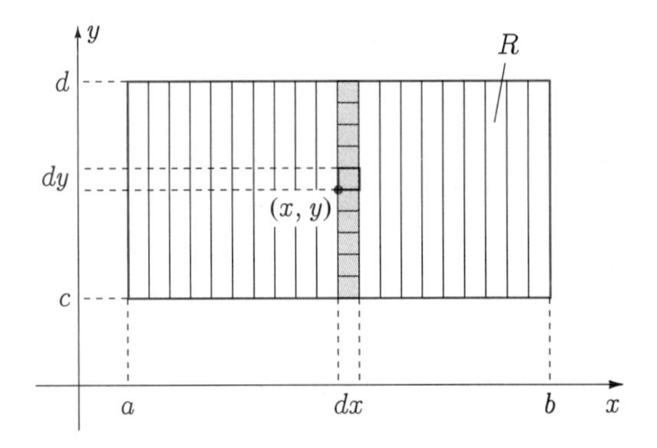

Although our diagram illustrates finite rectangles of dimensions dx and dy and finite strips of width dx, we must keep in mind that the integrations take limits as these dimensions approach zero.

Analogously, the double iterated integral in 14.14 is interpreted as adding over horizontal strips, as shown in Figure 14.5. Inner limits indicate where each strip starts and stops, and outer limits indicate the positions of first and last strips.

FIGURE 14.5

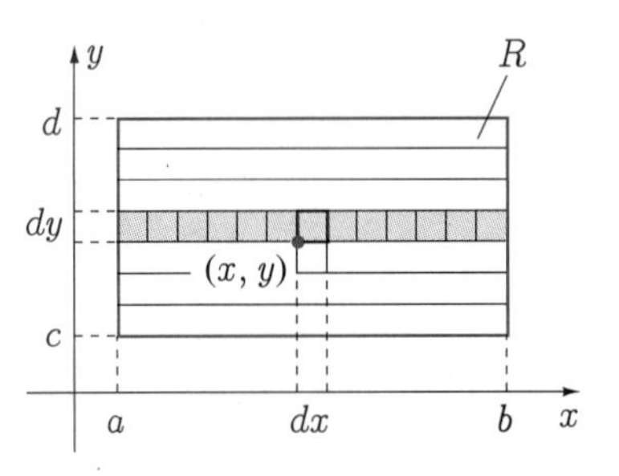

The transition now to general regions is quite straightforward. For the double integral of $f(x, y)$ over the region in Figure 14.6, we use a double iterated integral with respect to y and x. The y-integration adds the quantities $f(x, y)\,dy\,dx$ over rectangles in a vertical strip. We write

$$\int_{g(x)}^{h(x)} f(x, y)\,dy\,dx,$$

where $g(x)$ and $h(x)$ indicate that each vertical strip starts on the curve $y = g(x)$ and ends on the curve $y = h(x)$. The x-integration now adds over all strips, beginning at $x = a$ and ending at $x = b$:

$$\iint_R f(x, y)\,dA = \int_a^b \int_{g(x)}^{h(x)} f(x, y)\,dy\,dx. \qquad (14.15)$$

A double iterated integral in the reverse order is not convenient for this region because horizontal strips neither all start on the same curve nor all end on the same curve.

FIGURE 14.6

FIGURE 14.7

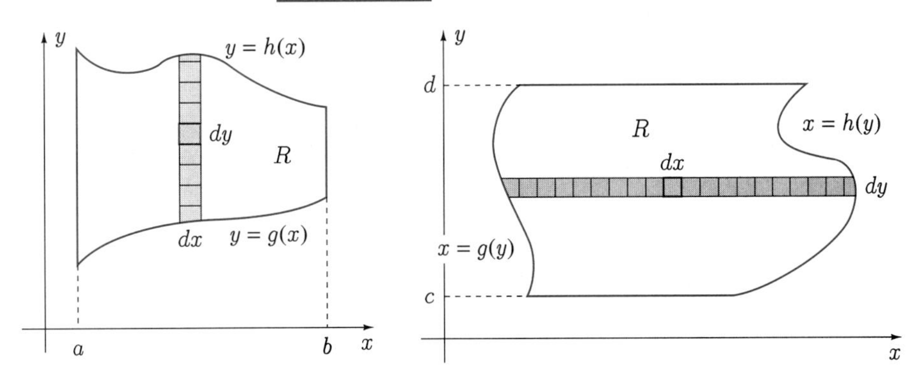

For the region in Figure 14.7, we obtain

$$\iint_R f(x,y)\,dA = \int_c^d \int_{g(y)}^{h(y)} f(x,y)\,dx\,dy. \tag{14.16}$$

The limits on double iterated integrals have been interpreted schematically as follows:

$$\int_{\substack{\text{position of first}\\\text{horizontal strip}}}^{\substack{\text{position of last}\\\text{horizontal strip}}} \int_{\substack{\text{where each and every}\\\text{horizontal strip starts}}}^{\substack{\text{where each and every}\\\text{horizontal strip stops}}} f(x,y)\,dx\,dy;$$

$$\int_{\substack{\text{position of first}\\\text{vertical strip}}}^{\substack{\text{position of last}\\\text{vertical strip}}} \int_{\substack{\text{where each and every}\\\text{vertical strip starts}}}^{\substack{\text{where each and every}\\\text{vertical strip stops}}} f(x,y)\,dy\,dx.$$

With these interpretations on the limits, you can see how important it is to have a well-labeled diagram.

EXAMPLE 14.2

Evaluate the double integral of $f(x,y) = xy^2 + x^2$ over the region bounded by the curves $y = x^2$ and $x = y^2$.

SOLUTION If we use vertical strips (Figure 14.8), we have

$$\iint_R (xy^2 + x^2)\,dA = \int_0^1 \int_{x^2}^{\sqrt{x}} (xy^2 + x^2)\,dy\,dx = \int_0^1 \left\{ \frac{xy^3}{3} + x^2 y \right\}_{x^2}^{\sqrt{x}} dx$$

$$= \int_0^1 \left\{ \frac{1}{3}x^{5/2} + x^{5/2} - \frac{x^7}{3} - x^4 \right\} dx = \left\{ \frac{8x^{7/2}}{21} - \frac{x^8}{24} - \frac{x^5}{5} \right\}_0^1$$

$$= \frac{8}{21} - \frac{1}{24} - \frac{1}{5} = \frac{39}{280}.$$

FIGURE 14.8

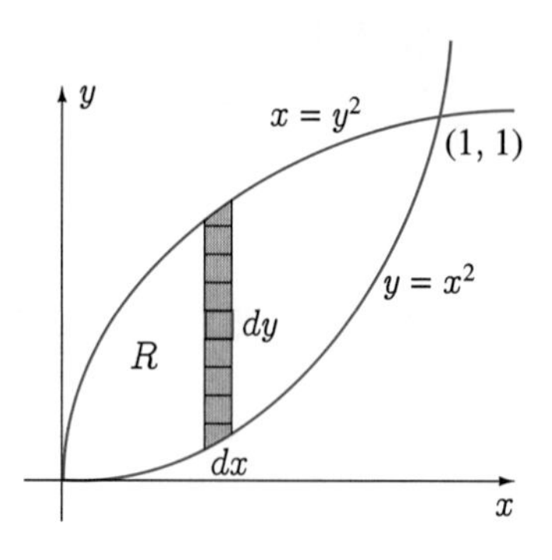

There are two distinct parts to every double integral: first, the function $f(x,y)$ being integrated, which is the integrand; second, the region R over which integration is being performed, and this region determines the limits on the corresponding double iterated integral. Note that we do not use $f(x,y)$ to determine limits on the double iterated integral; the region determines the limits. Conversely, if we are given a double iterated integral, then we know that it represents the double integral of its integrand over some region, and the region is completely defined by the limits on the iterated integral. This point is emphasized in the following example.

EXAMPLE 14.3

Evaluate the double iterated integral

$$\int_0^2 \int_y^2 e^{x^2}\, dx\, dy.$$

SOLUTION The function e^{x^2} does not have an elementary antiderivative with respect to x, and it is therefore impossible to evaluate the double iterated integral as it now stands. But the double iterated integral represents the double integral of e^{x^2} over some region R in the xy-plane. To find R we note that the inner integral indicates horizontal strips that all start on the line $x = y$ and stop on the line $x = 2$ (Figure 14.9a). The outer limits state that the first and last strips are at $y = 0$ and $y = 2$, respectively. This defines R as the triangle bounded by the straight lines $y = x$, $x = 2$, and $y = 0$ (Figure 14.9b). If we now reverse the order of integration and use vertical strips, we have

$$\int_0^2 \int_y^2 e^{x^2}\, dx\, dy = \iint_R e^{x^2}\, dA = \int_0^2 \int_0^x e^{x^2}\, dy\, dx = \int_0^2 \{ye^{x^2}\}_0^x\, dx$$
$$= \int_0^2 xe^{x^2}\, dx = \left\{\frac{1}{2}e^{x^2}\right\}_0^2 = \frac{e^4 - 1}{2}.$$

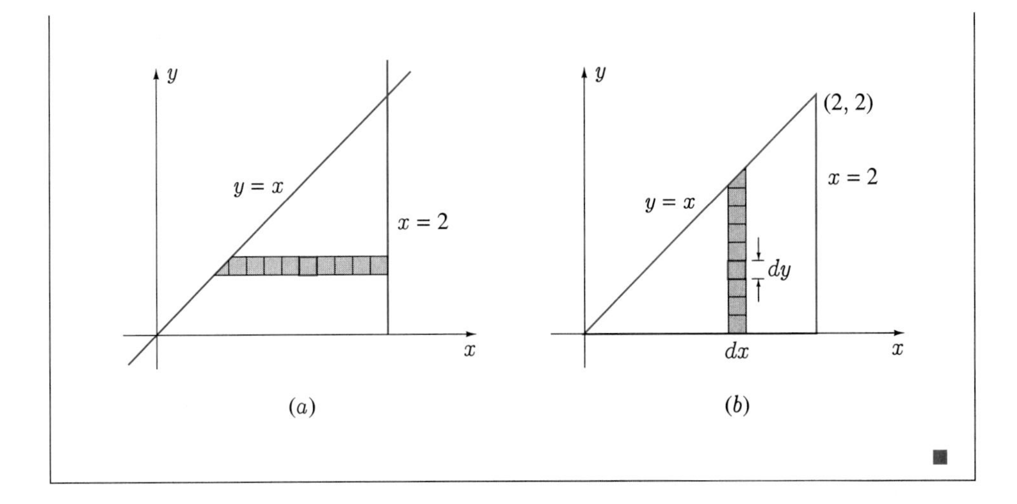

FIGURE 14.9

This example also points out that an iterated integral in one order may be much easier to evaluate than the corresponding iterated integral in the opposite order.

EXERCISES 14.2

In Exercises 1–12 evaluate the double integral over the region.

1. $\iint_R (x^2 + y^2)\, dA$ where R is bounded by $y = x^2$, $x = y^2$

2. $\iint_R (4 - x^2 - y)\, dA$ where R is bounded by $x = \sqrt{4 - y}$, $x = 0$, $y = 0$

3. $\iint_R (x + y)\, dA$ where R is bounded by $x = y^3 + 2$, $x = 1$, $y = 1$

4. $\iint_R xy^2\, dA$ where R is bounded by $x + y + 1 = 0$, $x + y^2 = 1$

5. $\iint_R xe^y\, dA$ where R is bounded by $y = x$, $y = 0$, $x = 1$

6. $\iint_R (x + y)\, dA$ where R is bounded by $x^2 + y^2 = 9$

7. $\iint_R x^2 y\, dA$ where R is bounded by $y = \sqrt{x + 4}$, $y = 0$, $x + y = 2$

8. $\iint_R (xy + y^2 - 3x^2)\, dA$ where R is bounded by $y = |x|$, $y = 1$, $y = 2$

9. $\iint_R (1 - x)^2\, dA$ where R is bounded by $x + y = 1$, $x + y = -1$, $x - y = 1$, $y - x = 1$

10. $\iint_R (x + y)\, dA$ where R is bounded by $x = y^2$, $x^2 - y^2 = 12$

11. $\iint_R x\, dA$ where R is bounded by $y = 3x$, $y = x$, $x + y = 4$

12. $\iint_R y^2\, dA$ where R is bounded by $x = 0$, $y = 1$, $y = 1/2$, $x = 1/\sqrt{y^4 + 12y^2}$

In Exercises 13–18 evaluate the double iterated integral by reversing the order of integration.

13. $\int_0^2 \int_0^{\sqrt{4-x^2}} (4 - y^2)^{3/2}\, dy\, dx$

14. $\int_0^1 \int_y^1 \sin(x^2)\, dx\, dy$

15. $\int_{-2}^0 \int_{-y}^2 y(x^2 + y^2)^8\, dx\, dy$

16. $\int_{-2}^0 \int_{-2}^x \frac{x}{\sqrt{x^2 + y^2}}\, dy\, dx$

17. $\int_0^2 \int_0^{x^2/2} \frac{x}{\sqrt{1 + x^2 + y^2}}\, dy\, dx$

18. $\displaystyle\int_0^2 \int_{-x^2/2}^0 \frac{x}{\sqrt{1+x^2+y^2}}\, dy\, dx$

19. Verify that if $m \le f(x,y) \le M$ for all (x,y) in R, then

$$m(\text{Area of } R) \le \iint_R f(x,y)\, dA \le M(\text{Area of } R).$$

20. Evaluate the double integral of $f(x,y) = 1/\sqrt{2x-x^2}$ over the region in the first quadrant bounded by $y^2 = 4-2x$.

In Exercises 21–28 either the integral has value 0 or it can be evaluated by doubling the double integral over half the region. By drawing the region and examining the integrand, determine which situation prevails. Do not evaluate the integral.

21. $\displaystyle\iint_R x^2 y^3\, dA$ where R is bounded by $x = \sqrt{4-y^2}$, $x = 0$

22. $\displaystyle\iint_R x^2 y^2\, dA$ where R is bounded by $x = \sqrt{4-y^2}$, $x = 0$

23. $\displaystyle\iint_R (x+y)\, dA$ where R is the square with vertices $(\pm 3, 0)$ and $(0, \pm 3)$

24. $\displaystyle\iint_R x^7 \cos(x^2)\, dA$ where R is bounded by $y = 4 - |x|$, $y = x^2$

25. $\displaystyle\iint_R e^{x^2+y^2}\, dA$ where R is bounded by $y = 4 - 4x^2$, $y = x^2 - 1$

26. $\displaystyle\iint_R \cos(x^2 y)\, dA$ where R is bounded by $y = 0$, $y = x^3 - x$

27. $\displaystyle\iint_R \sin(x^2 y)\, dA$ where R is bounded by $y = 0$, $y = x^3 - x$

28. $\displaystyle\iint_R (x^2 y^3 + xy^2)\, dA$ where R is bounded by $\sqrt{|x|} + \sqrt{|y|} = 1$

The average value of a function $f(x,y)$ over a region R with area A is defined as

$$\bar{f} = \frac{1}{A} \iint_R f(x,y)\, dA.$$

In Exercises 29–32 find the average value of the function over the region.

29. $f(x,y) = xy$ over the region in the first quadrant bounded by $x = 0$, $y = 0$, $y = \sqrt{1-x^2}$

30. $f(x,y) = x + y$ over the region bounded by $y = x$, $y = 0$, $y = \sqrt{2-x}$

31. $f(x,y) = x$ over the region between $y = \sin x$ and $y = 0$ for $0 \le x \le 2\pi$

32. $f(x,y) = e^{x+y}$ over the region bounded by $y = x+1$, $y = x-1$, $y = 1-x$, $y = -1-x$

In Exercises 33–39 evaluate the double integral over the region.

33. $\displaystyle\iint_R x^2\, dA$ where R is bounded by $x^2 + y^2 = 4$

34. $\displaystyle\iint_R (6 - x - 2y)\, dA$ where R is bounded by $x^2 + y^2 = 4$

35. $\displaystyle\iint_R 6x^5\, dA$ where R is the region under $x + 5y = 16$, above $y = x - 4$, and bounded by $x = (y-2)^2$

36. $\displaystyle\iint_R ye^x\, dA$ where R is bounded by $y = x$, $x + y = 2$, $y = 0$

37. $\displaystyle\iint_R \sqrt{1+y}\, dA$ where R is bounded by $x = -1$, $y = 2$, $x = y^3$

38. $\displaystyle\iint_R y\sqrt{x^2+y^2}\, dA$ where R is bounded by $y = x$, $x = -1$, $y = 0$

39. $\displaystyle\iint_R (x^2+y^2)\, dA$ where R is bounded by $x^2 + y^2 = 9$

40. Evaluate the double iterated integral

$$\int_0^1 \int_0^1 |x - y|\, dy\, dx.$$

41. Evaluate the double integral $\displaystyle\iint_R |y - 2x^2 + 1|\, dA$ where R is the square bounded by $x = \pm 1$, $y = \pm 1$.

Areas And Volumes Of Solids Of Revolution

Because equation 14.7 represents the area of a region R as a double integral, and double integrals are evaluated by means of double iterated integrals, it follows that areas can be calculated using double iterated integrals. In particular, to find the area of the region in Figure 14.10, we subdivide R into rectangles of dimensions dx and dy and therefore of area $dA = dy\,dx$. The areas of these rectangles are then added in the y-direction to give the area of a vertical strip

$$\int_{g(x)}^{h(x)} dy\,dx,$$

where the limits indicate that every vertical strip starts on the curve $y = g(x)$ and ends on the curve $y = h(x)$. Finally, the areas of the vertical strips are added together to give the total area:

$$\text{Area} = \int_a^b \int_{g(x)}^{h(x)} dy\,dx,$$

where a and b indicate the x-positions of the first and last strips.

FIGURE 14.10

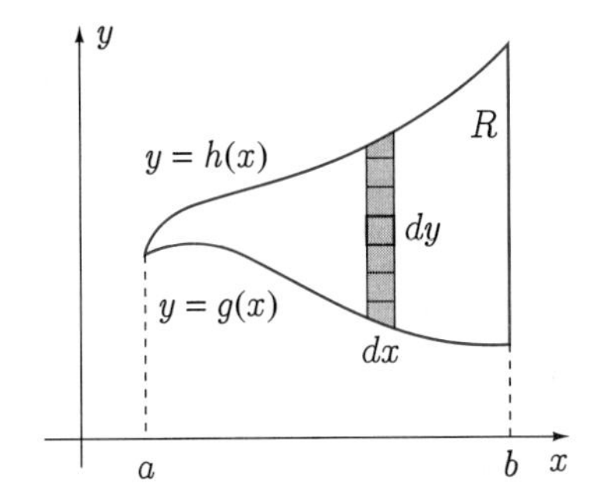

EXAMPLE 14.4

Find the area bounded by the curves $xy = 2$, $x = 2\sqrt{y}$, $y = 4$.

SOLUTION If we choose horizontal strips for this area (Figure 14.11), we have

$$\begin{aligned}\text{Area} &= \int_1^4 \int_{2/y}^{2\sqrt{y}} dx\,dy = \int_1^4 \left\{2\sqrt{y} - \frac{2}{y}\right\} dy \\ &= \left\{\frac{4}{3}y^{3/2} - 2\ln|y|\right\}_1^4 = \frac{28}{3} - 2\ln 4.\end{aligned}$$

For vertical strips (Figure 14.12), we require two iterated integrals because to the left of the line $x = 2$, strips begin on the hyperbola $xy = 2$, whereas to the right of $x = 2$, they begin on the parabola $x = 2\sqrt{y}$. We obtain

$$\begin{aligned}\text{Area} &= \int_{1/2}^2 \int_{2/x}^4 dy\,dx + \int_2^4 \int_{x^2/4}^4 dy\,dx = \int_{1/2}^2 \left\{4 - \frac{2}{x}\right\} dx + \int_2^4 \left\{4 - \frac{x^2}{4}\right\} dx \\ &= \{4x - 2\ln|x|\}_{1/2}^2 + \left\{4x - \frac{x^3}{12}\right\}_2^4 = \frac{28}{3} - 2\ln 4.\end{aligned}$$

FIGURE 14.11

FIGURE 14.12

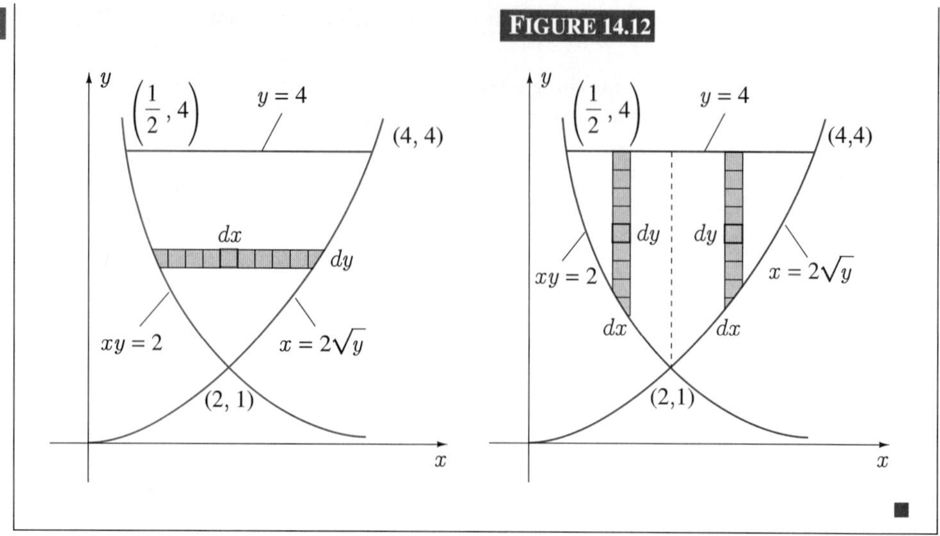

If we compare finding areas by definite integrals (Section 7.1) and finding the same areas by double integrals, it is clear that no great advantage is derived by using double integrals. In fact, it is probably more work because we must perform two, rather than one, integrations, although the first integration is trivial. The advantage of double integrals is therefore not in finding area; it is in finding volumes of solids of revolution, centres of mass, moments of inertia, and fluid forces, among other applications.

Volumes of Solids of Revolution

If the area in Figure 14.13 is rotated around the x-axis, the volume of the resulting solid of revolution can be evaluated by using the washer method introduced in Section 7.2:

$$\text{Volume} = \int_a^b \{\pi[h(x)]^2 - \pi[g(x)]^2\}dx. \qquad (14.17)$$

If this area is rotated around the y-axis, the volume generated is calculated by using the cylindrical shell method:

$$\text{Volume} = \int_a^b 2\pi x[h(x) - g(x)]dx. \qquad (14.18)$$

FIGURE 14.13

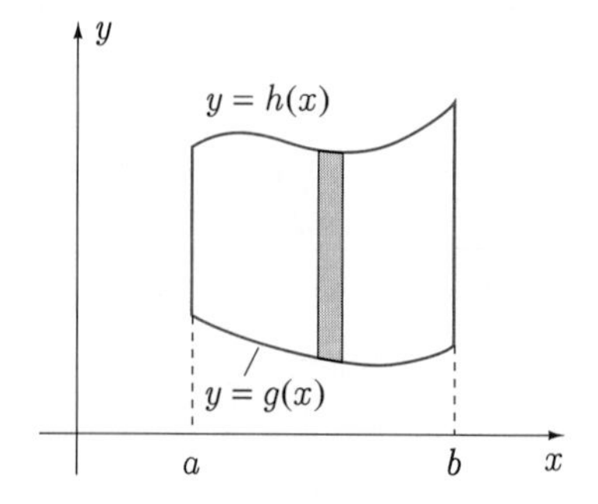

Thus once we have chosen to use vertical rectangles, the axis of revolution determines whether we use either the washer or cylindrical shell method. We now show that with double integrals one method works for both problems.

To rotate this area around the x-axis we subdivide R into small areas dA (Figure 14.14). If the area dA at a point (x, y) is rotated about the x-axis, it generates a "ring" with cross-sectional area dA. Since (x, y) travels a distance $2\pi y$ in traversing the ring, it follows that the volume in the ring is approximately $2\pi y\, dA$. To find the total volume obtained by rotating R about the x-axis, we add the volumes of all such rings and take the limit as the areas shrink to points. But this is precisely what we mean by the double integral of $2\pi y$ over the region R, and we therefore write

$$\text{Volume} = \iint_R 2\pi y\, dA. \tag{14.19}$$

FIGURE 14.14

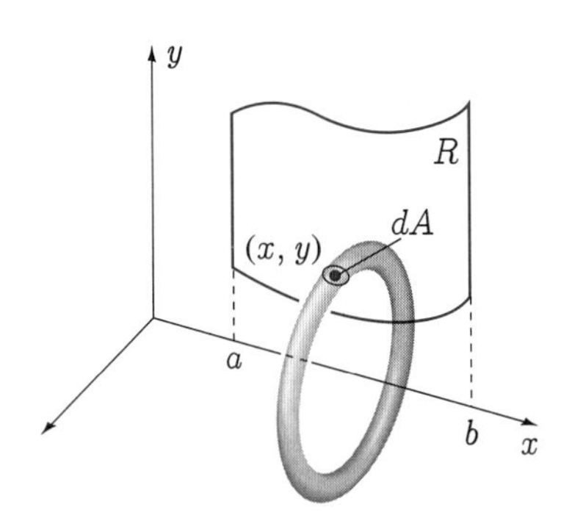

On the other hand, if dA is rotated about the y-axis, it again forms a ring, but with approximate volume $2\pi x\, dA$. The total volume, then, when R is rotated about the y-axis is

$$\text{Volume} = \iint_R 2\pi x\, dA. \tag{14.20}$$

Since double iterated integrals are used to evaluate double integrals, it follows that we can set up double iterated integrals to find the volumes represented by equations 14.19 and 14.20. The decision to use a double iterated integral with respect to y and x implies a subdivision of R into rectangles of dimensions dx and dy (Figure 14.15). The volume of the ring formed when this rectangle is rotated around the x-axis is $2\pi y\, dy\, dx$. If we choose to integrate first with respect to y, we are adding over all rectangles in a vertical strip

$$\int_{g(x)}^{h(x)} 2\pi y\, dy\, dx,$$

where the limits indicate that all vertical strips start on the curve $y = g(x)$ and end on the curve $y = h(x)$. This integral is the volume generated by rotating the vertical strip around the x-axis. Integration now with respect to x adds over all strips to give the required volume

$$\text{Volume} = \int_a^b \int_{g(x)}^{h(x)} 2\pi y\, dy\, dx. \tag{14.21}$$

Note that when we actually do perform the inner integration, we get

$$\text{Volume} = \int_a^b \{\pi y^2\}_{g(x)}^{h(x)}\, dx = \int_a^b \{\pi[h(x)]^2 - \pi[g(x)]^2\}dx,$$

and this is the result contained in equation 14.17.

FIGURE 14.15

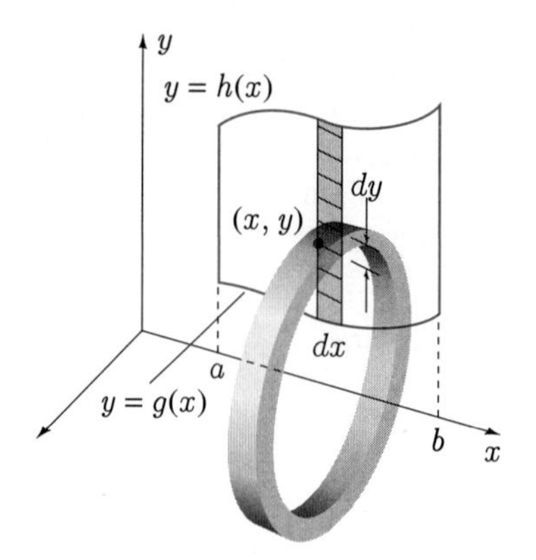

When R is rotated around the y-axis, the rectangular area $dy\ dx$ generates a ring of volume $2\pi x\ dy\ dx$. Addition over the rectangles in a vertical strip

$$\int_{g(x)}^{h(x)} 2\pi x\ dy\ dx$$

gives the volume generated by rotating the strip about the y-axis. Finally, integration with respect to x adds over all strips to give

$$\text{Volume} = \int_a^b \int_{g(x)}^{h(x)} 2\pi x\ dy\ dx. \tag{14.22}$$

This time the inner integration leads to

$$\text{Volume} = \int_a^b \{2\pi x y\}_{g(x)}^{h(x)} dx = \int_a^b 2\pi x[h(x) - g(x)]dx,$$

the same result as in equation 14.18.

The advantage, then, in using double integrals to find volumes of solids of revolution is that it requires only one idea, that of rings. The first integration then leads to the concepts of washers or cylindrical shells.

EXAMPLE 14.5

Find the volumes of the solids of revolution if the area bounded by the curves $y = 2x - x^2$, $y = x^2 - 2x$ is rotated about:

(a) the y-axis;

(b) $x = -3$;

(c) $y = 2$.

SOLUTION

(a) If we use vertical strips (Figure 14.16), then

$$\text{Volume} = \int_0^2 \int_{x^2-2x}^{2x-x^2} 2\pi x \, dy \, dx = 2\pi \int_0^2 \{xy\}_{x^2-2x}^{2x-x^2} dx$$
$$= 2\pi \int_0^2 x\{(2x - x^2) - (x^2 - 2x)\} dx$$
$$= 4\pi \int_0^2 (2x^2 - x^3) dx = 4\pi \left\{ \frac{2x^3}{3} - \frac{x^4}{4} \right\}_0^2 = \frac{16\pi}{3}.$$

FIGURE 14.16

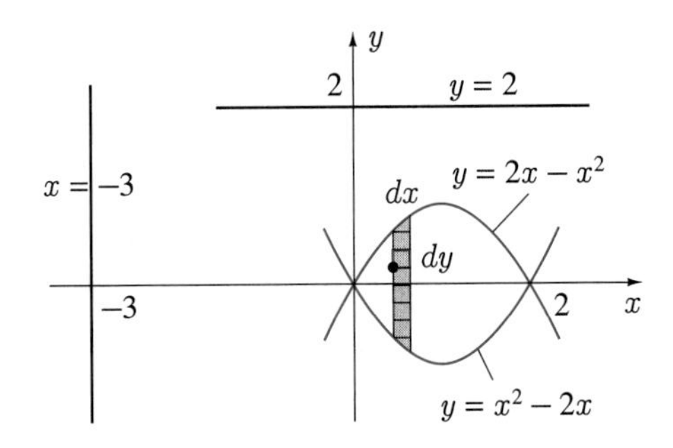

(b) In this case the radius of the ring formed by rotating the rectangle about $x = -3$ is $x + 3$, and therefore

$$\text{Volume} = \int_0^2 \int_{x^2-2x}^{2x-x^2} 2\pi(x+3) \, dy \, dx = 2\pi \int_0^2 \{y(x+3)\}_{x^2-2x}^{2x-x^2} dx$$
$$= 2\pi \int_0^2 (x+3)(4x - 2x^2) dx = 4\pi \int_0^2 (6x - x^2 - x^3) dx$$
$$= 4\pi \left\{ 3x^2 - \frac{x^3}{3} - \frac{x^4}{4} \right\}_0^2 = \frac{64\pi}{3}.$$

(c) When the rectangle is rotated around $y = 2$, the radius of the ring is $2 - y$, and hence

$$\text{Volume} = \int_0^2 \int_{x^2-2x}^{2x-x^2} 2\pi(2 - y) \, dy \, dx = 2\pi \int_0^2 \left\{ -\frac{1}{2}(2-y)^2 \right\}_{x^2-2x}^{2x-x^2} dx$$
$$= \pi \int_0^2 \{(2 - x^2 + 2x)^2 - (2 - 2x + x^2)^2\} dx = 8\pi \int_0^2 (2x - x^2) dx$$
$$= 8\pi \left\{ x^2 - \frac{x^3}{3} \right\}_0^2 = \frac{32\pi}{3}.$$

■

EXERCISES 14.3

In Exercises 1–10 use a double integral to find the area bounded by the curves.

1. $y = 4x^2$, $x = 4y^2$

2. $y = x^2$, $y = 5x + 6$

3. $x = y^2$, $x = 3y - 2$

4. $y = x^3 + 8$, $y = 4x + 8$

5. $y = 4/x^2$, $y = 5 - x^2$

6. $y = xe^{-x}$, $y = x$, $x = 2$

7. $x = 4y - 4y^2$, $y = x - 3$, $y = 1$, $y = 0$

8. $x = y(y - 2)$, $x + y = 12$

9. $y = x^3 - x^2 - 2x + 2$, $y = 2$

10. $x + y = 1$, $x + y = 5$, $y = 2x + 1$, $y = 2x + 6$

In Exercises 11–20 use a double integral to find the volume of the solid of revolution obtained by rotating the area bounded by the curves about the line.

11. $y = -\sqrt{4 - x}$, $x = 0$, $y = 0$ about $y = 0$

12. $4x^2 + 9y^2 = 36$ about $y = 0$

13. $y = (x - 1)^2$, $y = 1$ about $x = 0$

14. $y = x^2 + 4$, $y = 2x^2$ about $y = 0$

15. $x - 1 = y^2$, $x = 5$ about $x = 1$

16. $x = y^3$, $y = \sqrt{2 - x}$, $y = 0$ about $y = 1$

17. $y = 4x^2 - 4x$, $y = x^3$ about $y = -2$

18. $x = 3y - y^2$, $x = y^2 - 3y$ about $y = 4$

19. $x = 2y - y^2 - 2$, $x = -5$ about $x = 1$

20. $x + y = 4$, $y = 2\sqrt{x - 1}$, $y = 0$ about $y = -1$

In Exercises 21–30 use a double integral to find the area bounded by the curves.

21. $y = 2x^3$, $y = 4x + 8$, $y = 0$

22. $y = x/\sqrt{x + 3}$, $x = 1$, $x = 6$, $y = -x^2$

23. $x = y^2 + 2$, $x = -(y - 4)^2$, $y = 4 - x$, $y = 0$

24. $y = x^3 - x$, $x + y + 1 = 0$, $x = \sqrt{y + 1}$

25. $y^2 = x^2(4 - x^2)$

26. $x^2 + y^2 = 4$, $x^2 + y^2 = 4x$ (interior to both)

27. $x = 1/\sqrt{4 - y^2}$, $4x + y^2 = 0$, $y + 1 = 0$, $y - 1 = 0$

28. $y^2 = x^4(9 + x)$

29. $y = (x^2 + 1)/(x + 1)$, $x + 3y = 7$

30. $(x + 2)^2 y = 4 - x$, $x = 0$, $y = 0$ $(x \geq 0, y \geq 0)$

In Exercises 31–35 use a double integral to find the volume of the solid of revolution obtained by rotating the area bounded by the curves about the line.

31. $y = 4/(x^2 + 1)^2$, $y = 1$ about $x = 0$

32. $y = x^2 - 2$, $y = 0$ about $y = -1$

33. $y = |x^2 - 1|$, $x = -2$, $x = 2$, $y = -1$ about $y = -2$

34. $x = \sqrt{4 + 12y^2}$, $x - 20y = 24$, $y = 0$ about $y = 0$

35. $y = (x + 1)^{1/4}$, $y = -(x + 1)^2$, $x = 0$ about $x = 0$

36. Find the area common to the two circles $x^2 + y^2 = 4$ and $x^2 + y^2 = 6x$.

In Exercises 37–40 find the volume of the solid of revolution obtained by rotating the area bounded by the curves about the line. (Hint: Use the distance formula $|Ax_1 + By_1 + C|/\sqrt{A^2 + B^2}$ from a point (x_1, y_1) to a line $Ax + By + C = 0$ developed in Exercise 45 of Section 1.3.)

37. $x = 1$, $y = 1$, $x = 0$, $y = 0$ about $x + y = 2$

38. $y = x^2$, $y = 2x + 3$ about $y = 2x + 3$

39. $x = y^2$, $y = 0$, $x = 1$ about $y = 3x + 2$

40. $x = 2y$, $y = x - 1$, $y = 0$ about $x + y + 1 = 0$

41. Prove that the area above the line $y = h$ and under the circle $x^2 + y^2 = r^2$ $(r > h)$ is given by

$$A = \pi r^2/2 - h\sqrt{r^2 - h^2} - r^2 \operatorname{Sin}^{-1}(h/r).$$

Fluid Pressure

In Section 7.5 we defined pressure at a point in a fluid as the magnitude of the force per unit area that would act on any surface placed at that point. We discovered that at a depth $d > 0$ below the surface of a fluid, pressure is given by

$$P = 9.81\rho d, \tag{14.23}$$

where ρ is the density of the fluid. With these ideas and the definite integral, we were able to calculate fluid forces on flat surfaces in the fluid. In particular, the magnitude of the total force on each side of the vertical surface in Figure 14.17 is given by the definite integral

$$\text{Force} = \int_a^b -9.81\rho y[h(y) - g(y)]dy. \tag{14.24}$$

Although horizontal rectangles are convenient for this problem, it is clear that they are not reasonable for the surface in Figure 14.18.

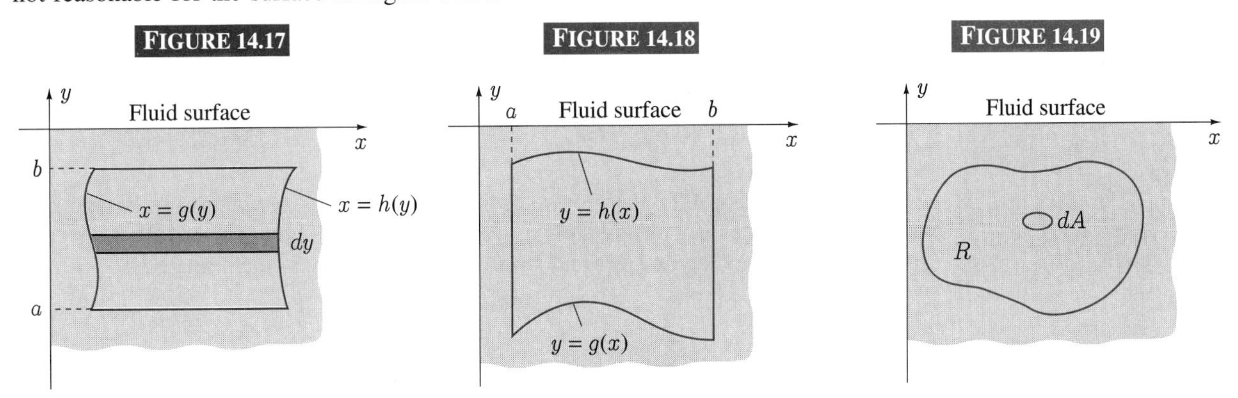

Double integrals, on the other hand, can be applied with equal ease to both surfaces. To see this we consider first the total force on each side of the surface in Figure 14.19. If the surface is divided into small areas dA, then the force on dA is $P\,dA$, where P is pressure at that depth. The total force on R is the sum of the forces on all such areas in R as the areas dA shrink to a point. But once again this is the concept of the double integral, and we therefore write

$$\text{Force} = \iint_R P\,dA. \tag{14.25}$$

To set up a double iterated integral in order to evaluate this double integral, we use our interpretation of the double iterated integral as a limit of a sum in which the areas dA have been chosen as rectangles. In particular, for the surface in Figure 14.17 we draw rectangles of dimensions dx and dy, as shown in Figure 14.20. The force on this rectangle is its area $dx\,dy$ multiplied by pressure $-9.81\rho y$ at that depth, $-9.81\rho y\,dx\,dy$. Addition of these quantities over all rectangles in a horizontal strip gives the force on the strip,

$$\int_{g(y)}^{h(y)} -9.81\rho y\,dx\,dy,$$

where the limits indicate that all horizontal strips start on the curve $x = g(y)$ and end on the curve $x = h(y)$. Integration with respect to y now adds over all horizontal strips to give the total force on the surface:

$$\text{Force} = \int_a^b \int_{g(y)}^{h(y)} -9.81\rho y \, dx \, dy. \tag{14.26}$$

When we perform the inner integration, we obtain

$$\text{Force} = \int_a^b \{-9.81\rho y x\}_{g(y)}^{h(y)} dy = \int_a^b -9.81\rho y[h(y) - g(y)]dy,$$

and this is the result contained in equation 14.24.

FIGURE 14.20

FIGURE 14.21

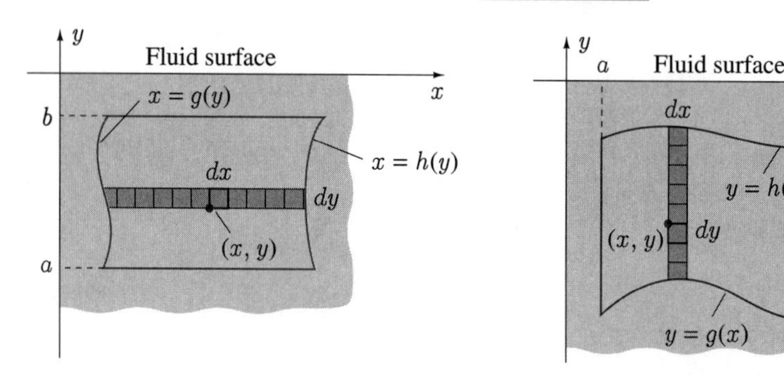

For the surface in Figure 14.18 we again draw rectangles of dimensions dx and dy and calculate the force on such a rectangle, $-9.81\rho y \, dy \, dx$. In this case it is more convenient to add over rectangles in a vertical strip to give the force on the strip (Figure 14.21):

$$\int_{g(x)}^{h(x)} -9.81\rho y \, dy \, dx.$$

The force on the entire surface can now be found by adding over all vertical strips:

$$\text{Force} = \int_a^b \int_{g(x)}^{h(x)} -9.81\rho y \, dy \, dx. \tag{14.27}$$

EXAMPLE 14.6

The face of a dam is parabolic with breadth 100 m and height 50 m. Find the magnitude of the total force due to fluid pressure on the face.

SOLUTION If we use the coordinate system in Figure 14.22, then the edge of the dam has an equation of the form $y = kx^2$. Since (50, 50) is a point on this curve, it follows that $k = \frac{1}{50}$. Because the force on the left half of the dam is the same as that on the right half, we can integrate for the right half and double the result; that is,

$$\begin{aligned}
\text{Force} &= 2\int_0^{50}\int_{x^2/50}^{50} 9.81(1000)(50 - y)\,dy\,dx \\
&= 19\,620 \int_0^{50} \left\{50y - \frac{y^2}{2}\right\}_{x^2/50}^{50} dx = 19\,620 \int_0^{50} \left\{1250 - x^2 + \frac{x^4}{5000}\right\} dx \\
&= 19\,620 \left\{1250x - \frac{x^3}{3} + \frac{x^5}{25\,000}\right\}_0^{50} = 6.54 \times 10^8 \text{N}.
\end{aligned}$$

FIGURE 14.22

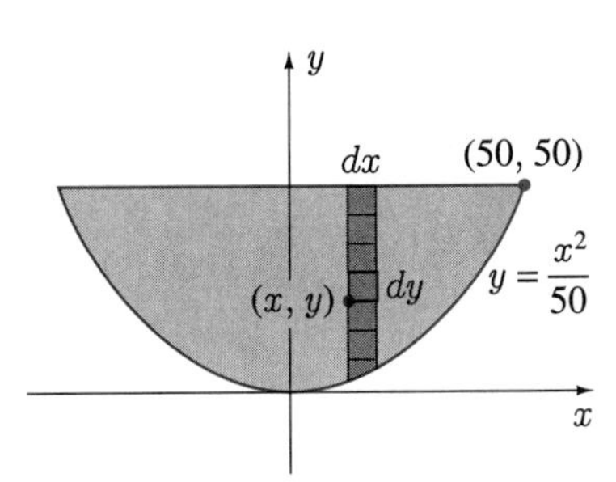

EXAMPLE 14.7

A tank in the form of a right-circular cylinder of radius $\frac{1}{2}$m and length 10 m has its axis horizontal. If it is full of water, find the force due to water pressure on each end of the tank.

SOLUTION Since the force on that part of the end to the left of the y-axis (Figure 14.23) is identical to the force on that part to the right, we double the force on the right half,

$$\text{Force} = 2\int_{-1/2}^{1/2}\int_{0}^{\sqrt{1/4-y^2}} 9.81(1000)\left(\tfrac{1}{2} - y\right)dx\,dy$$

$$= 9810\int_{-1/2}^{1/2}\int_{0}^{\sqrt{1/4-y^2}} dx\,dy - 19\ 620\int_{-1/2}^{1/2}\int_{0}^{\sqrt{1/4-y^2}} y\,dx\,dy.$$

The first double iterated integral represents the area of one-half the end of the tank. Consequently,

$$\text{Force} = 9810\left\{\tfrac{1}{2}\pi\left(\tfrac{1}{2}\right)^2\right\} - 19\ 620\int_{-1/2}^{1/2} y\sqrt{\frac{1}{4} - y^2}\,dy$$

$$= \frac{4905\,\pi}{4} - 19\ 620\left\{-\frac{1}{3}\left(\frac{1}{4} - y^2\right)^{3/2}\right\}_{-1/2}^{1/2} = \frac{4905\,\pi}{4}\text{N}.$$

FIGURE 14.23

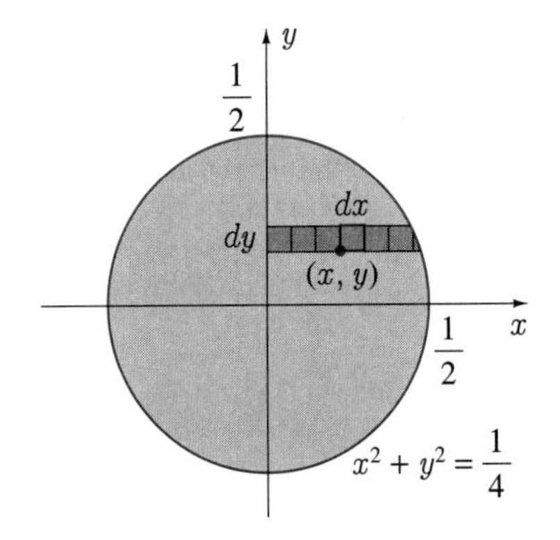

EXERCISES 14.4

In Exercises 1–8 the surface is submerged vertically in a fluid with density ρ. Find the force due to fluid pressure on one side of the surface.

1. An equilateral triangle of side length 2 with one side in the surface

2. A parabolic segment of base 12 and height 4 with the base in the surface

3. A square of side length 3 with one diagonal vertical and the uppermost vertex in the surface

4. A triangle of side lengths 5, 5, and 8, with the longest side uppermost, horizontal, and 3 units below the surface

5. A triangle of side lengths 5, 5, and 8, with the longest side below the the opposite vertex, horizontal, and 6 units below the surface

6. A trapezoid with vertical parallel sides of lengths 6 and 8, and a third side perpendicular to the parallel sides, of length 5, and in the surface

7. A triangle of side lengths 3, 3, and 4, with the longest side vertical, and the uppermost vertex 1 unit below the surface

8. A semicircle of radius 5 with the (diameter) base in the surface

9. The vertical end of a water trough is an isosceles triangle with width 2 m and depth 1 m. Find the force of the water on each end when the trough is one-half filled (by volume) with water.

10. A dam across a river has the shape of a parabola 36 m across the top and 9 m deep at the centre. Find the maximum force due to water pressure on the dam.

In Exercises 11–15 the surface is submerged vertically in a fluid with density ρ. Find the force due to fluid pressure on one side of the surface.

11. A circle of radius 2 with centre 3 units below the surface

12. A rectangle of side lengths 2 and 5, with one diagonal vertical and the uppermost vertex in the surface

13. An ellipse with major and minor axes of lengths 8 and 6, and with the major axis horizontal and 5 units below the surface

14. A parallelogram of side lengths 4 and 5, with one of the longer sides horizontal and in the surface, and two sides making an angle of $\pi/6$ radians with the surface

15. A triangle of side lengths 2, 3, and 4 with the longest side vertical, the side of length 2 above the side of length 3, and the uppermost vertex 1 unit below the surface

16. An oil can is in the form of a right-circular cylinder of radius r and height h. If the axis of the can is horizontal, and the can is full of oil with density ρ, find the force due to fluid pressure on each end.

17. Find the force due to water pressure on each side of the flat vertical plate in Figure 14.24.

FIGURE 14.24

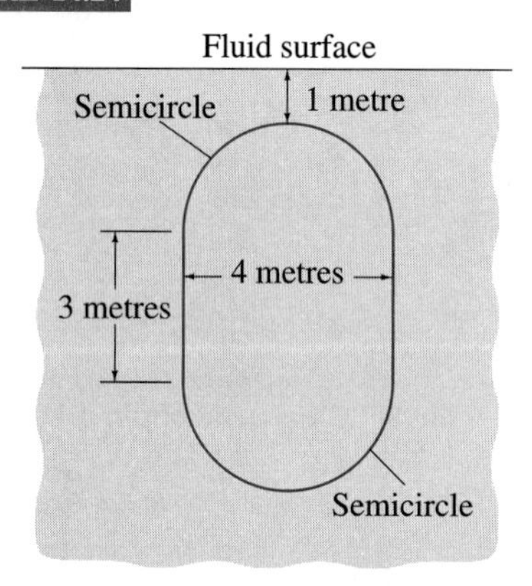

18. A square plate of side length 2 m has one side on the bottom of a swimming pool 3 m deep. The plate is inclined at an angle of $\pi/4$ radians with the bottom of the pool so that its horizontal upper edge is $3 - \sqrt{2}$ m below the surface. Find the force due to water pressure on each side of the plate.

19. A thin triangular piece of wood with sides of lengths 2 m, 2 m, and 3 m floats in a pond. A piece of rope is tied to the vertex opposite the longest side. A rock is then attached to the other end of the rope and lowered into the water. When the rock sits on the bottom (and the rope is taut), the longest side of the wood still floats in the surface of the water, but the opposite vertex is 1 m below the surface. Find the force due to water pressure on each side of the piece of wood.

Centres Of Mass And Moments Of Inertia

We now show how double integrals can be used to replace definite integrals in calculating first moments, centres of mass, and moments of inertia of thin plates. Consider a thin plate with mass per unit area ρ such as that in Figure 14.25. Note that unlike our discussion in Section 7.6 where we assumed ρ constant, we have made no such assumption here. In other words, density could be a function of position, $\rho = \rho(x, y)$.

FIGURE 14.25

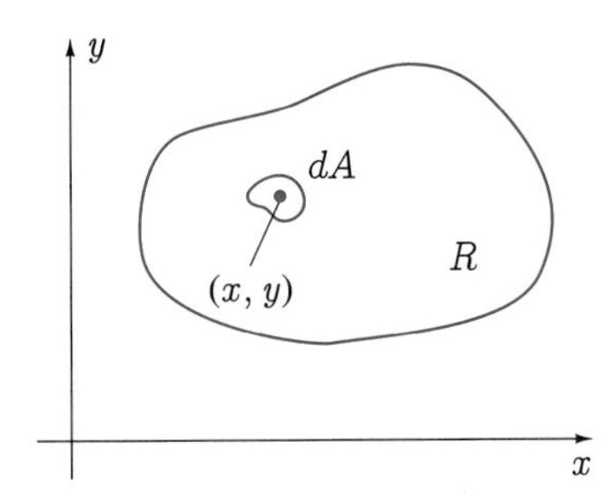

The centre of mass $(\overline{x}, \overline{y})$ of the plate is a point at which a particle of mass M (equal to the total mass of the plate) has the same first moments about the x-and y-axes as the plate itself. If we divide the plate into small areas dA, then the mass in dA is $\rho\, dA$. Addition over all such areas in R as each dA shrinks to a point gives the mass of the plate

$$M = \iint_R \rho\, dA. \tag{14.28}$$

Since the first moment of the mass in dA about the y-axis is $x\rho\, dA$, it follows that the first moment of the entire plate about the y-axis is

$$\iint_R x\rho\, dA.$$

But this must be equal to the first moment of the particle of mass M at $(\overline{x}, \overline{y})$ about the y-axis, and hence

$$M\overline{x} = \iint_R x\rho\, dA. \tag{14.29}$$

This equation can be solved for $\overline{x}$ once the double integral on the right and M have been calculated.

Similarly, $\overline{y}$ is determined by the equation

$$M\overline{y} = \iint_R y\rho\, dA, \tag{14.30}$$

where the double integral on the right is the first moment of the plate about the x-axis.

In any given problem, the double integrals in 14.28–14.30 are evaluated by means of double iterated integrals. For example, if we divide the plate in Figure 14.26 into rectangles of dimensions dx and dy and use vertical strips, then we have

$$M = \int_a^b \int_{g(x)}^{h(x)} \rho\, dy\, dx, \tag{14.31}$$

$$M\overline{x} = \int_a^b \int_{g(x)}^{h(x)} x\rho\, dy\, dx, \tag{14.32}$$

$$M\overline{y} = \int_a^b \int_{g(x)}^{h(x)} y\rho\, dy\, dx. \tag{14.33}$$

FIGURE 14.26

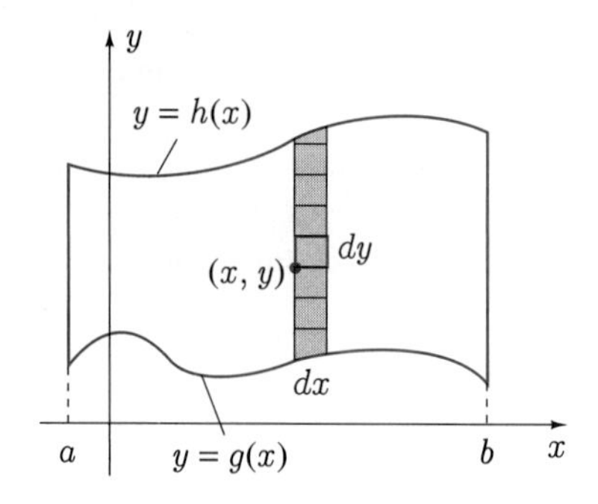

Once again we point out that equations 14.31–14.33 should not be memorized as formulas. Indeed, each can be derived as needed. For instance, to obtain equation 14.33, we reason that the first moment of the mass in a rectangle of dimensions dx and dy at position (x, y) about the x-axis is $y\rho\, dy\, dx$. Addition over the rectangles in a vertical strip gives the first moment of the strip about the x-axis:

$$\int_{g(x)}^{h(x)} y\rho\, dy\, dx,$$

and integration with respect to x now adds over all strips to give the first moment of the entire plate about the x-axis. Note that if ρ is constant and inner integrations are performed in each of equations 14.31–14.33, then

$$M = \int_a^b \{\rho y\}_{g(x)}^{h(x)} dx = \int_a^b \rho[h(x) - g(x)]dx,$$

$$M\overline{x} = \int_a^b \{\rho xy\}_{g(x)}^{h(x)} dx = \int_a^b \rho x[h(x) - g(x)]dx,$$

$$M\overline{y} = \int_a^b \left\{\rho \frac{y^2}{2}\right\}_{g(x)}^{h(x)} dx = \int_a^b \frac{\rho}{2}\{[h(x)]^2 - [g(x)]^2\}dx.$$

These are precisely the results in equations 7.31–7.33 (with different names for the curves), but the simplicity of the discussion leading to the double iterated integrals certainly demonstrates its advantage over use of the definite integral described in Section 7.6.

EXAMPLE 14.8

Find the centre of mass of a thin plate with constant mass per unit area ρ if its edges are defined by the curves $y = 2x - x^2$ and $y = x^2 - 4$.

SOLUTION For vertical strips as shown in Figure 14.27,

$$M = \int_{-1}^{2}\int_{x^2-4}^{2x-x^2} \rho\, dy\, dx$$
$$= \rho \int_{-1}^{2} \{(2x - x^2) - (x^2 - 4)\} dx$$
$$= \rho \left\{ x^2 - \frac{2x^3}{3} + 4x \right\}_{-1}^{2} = 9\rho.$$

If the centre of mass of the plate is $(\overline{x}, \overline{y})$, then

$$M\overline{x} = \int_{-1}^{2}\int_{x^2-4}^{2x-x^2} x\rho\, dy\, dx = \rho \int_{-1}^{2} x\{(2x - x^2) - (x^2 - 4)\} dx$$
$$= \rho \left\{ \frac{2x^3}{3} - \frac{x^4}{2} + 2x^2 \right\}_{-1}^{2} = \frac{9\rho}{2}.$$

Thus, $\overline{x} = \frac{9\rho}{2} \cdot \frac{1}{9\rho} = \frac{1}{2}$. Since

$$M\overline{y} = \int_{-1}^{2}\int_{x^2-4}^{2x-x^2} y\rho\, dy\, dx = \rho \int_{-1}^{2} \left\{ \frac{y^2}{2} \right\}_{x^2-4}^{2x-x^2} dx$$
$$= \frac{\rho}{2} \int_{-1}^{2} (-4x^3 + 12x^2 - 16)\, dx$$
$$= \frac{\rho}{2} \{-x^4 + 4x^3 - 16x\}_{-1}^{2} = -\frac{27\rho}{2},$$

we find $\overline{y} = -\frac{27\rho}{2} \cdot \frac{1}{9\rho} = -\frac{3}{2}$.

FIGURE 14.27

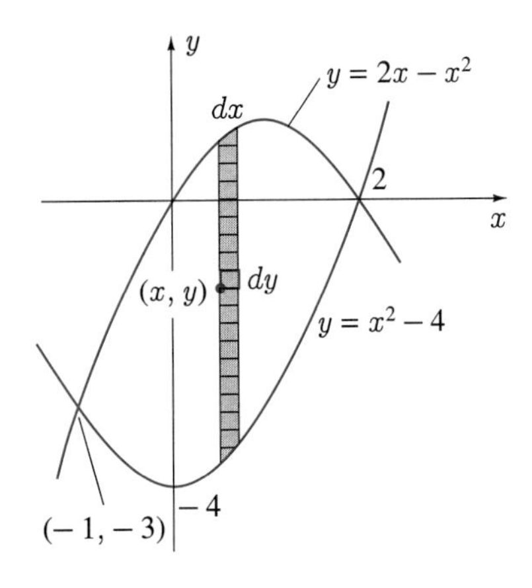

■

EXAMPLE 14.9

Find the first moment of area about the line $y = -2$ for the region bounded by the curves $x = |y|^3$ and $x = 2 - y^2$.

SOLUTION ***Method 1*** The first moment about $y = -2$ of a rectangle of dimensions

dx and dy at position (x, y) is $(y+2)\,dx\,dy$ (Figure 14.28). For the entire plate, then, the required first moment is

$$\int_{-1}^{0}\int_{-y^3}^{2-y^2}(y+2)\,dx\,dy+\int_{0}^{1}\int_{y^3}^{2-y^2}(y+2)\,dx\,dy$$
$$=\int_{-1}^{0}\{x(y+2)\}_{-y^3}^{2-y^2}\,dy+\int_{0}^{1}\{x(y+2)\}_{y^3}^{2-y^2}\,dy$$
$$=\int_{-1}^{0}(y^4+y^3-2y^2+2y+4)\,dy+\int_{0}^{1}(-y^4-3y^3-2y^2+2y+4)\,dy$$
$$=\left\{\frac{y^5}{5}+\frac{y^4}{4}-\frac{2y^3}{3}+y^2+4y\right\}_{-1}^{0}+\left\{-\frac{y^5}{5}-\frac{3y^4}{4}-\frac{2y^3}{3}+y^2+4y\right\}_{0}^{1}$$
$$=\frac{17}{3}.$$

FIGURE 14.28

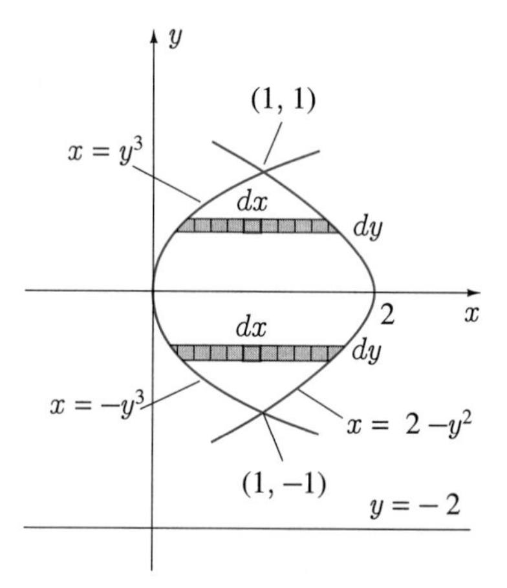

Method 2 By symmetry, the centroid of the region is somewhere along the x-axis. Hence, the required first moment is $2A$, where A is the area of the region and 2 is the distance from $y=-2$ to the centroid. Since the area of the region is equally distributed about the x-axis, we obtain the required first moment as

$$2(2)(\text{Area of plate above } x\text{-axis})=4\int_{0}^{1}\int_{y^3}^{2-y^2}dx\,dy=4\int_{0}^{1}(2-y^2-y^3)\,dy$$
$$=4\left\{2y-\frac{y^3}{3}-\frac{y^4}{4}\right\}_{0}^{1}=\frac{17}{3}.$$

■

To calculate moments of inertia (second moments) of thin plates about lines parallel to the x- and y-axes is as easy as calculating first moments if we use double integrals. In particular, the mass in area dA in Figure 14.25 is $\rho\,dA$; thus its moments of inertia about the x- and y-axes are, respectively, $y^2\rho\,dA$ and $x^2\rho\,dA$. Moments of inertia of the entire plate about the x- and y-axes are therefore given by the double integrals

$$I_x=\iint_R y^2\rho\,dA \quad\text{and}\quad I_y=\iint_R x^2\rho\,dA. \tag{14.34}$$

For a plate such as that shown in Figure 14.29, we evaluate these double integrals by means of double iterated integrals with respect to x and y:

$$I_x = \int_a^b \int_{g(y)}^{h(y)} y^2 \rho \, dx \, dy, \qquad (14.35)$$

$$I_y = \int_a^b \int_{g(y)}^{h(y)} x^2 \rho \, dx \, dy. \qquad (14.36)$$

FIGURE 14.29

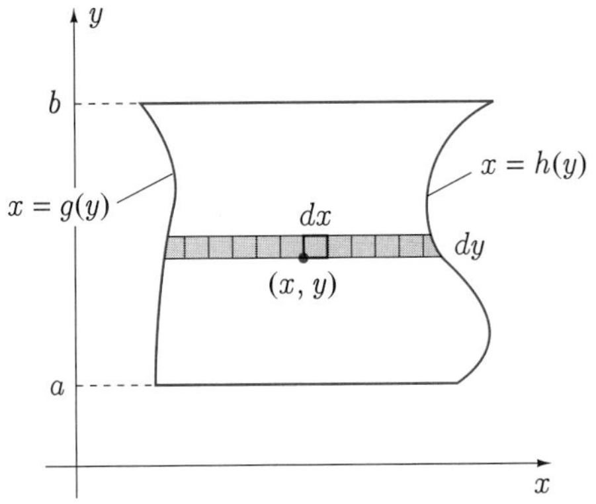

EXAMPLE 14.10

Find the moment of inertia about the x-axis of a thin plate with constant mass per unit area ρ if its edges are defined by the curves $y = x^3$, $y = \sqrt{2 - x}$, and $x = 0$.

SOLUTION With vertical strips as shown in Figure 14.30, the moment of inertia about the x-axis is

$$\int_0^1 \int_{x^3}^{\sqrt{2-x}} y^2 \rho \, dy \, dx = \rho \int_0^1 \left\{ \frac{y^3}{3} \right\}_{x^3}^{\sqrt{2-x}} dx = \frac{\rho}{3} \int_0^1 \{(2 - x)^{3/2} - x^9\} dx$$

$$= \frac{\rho}{3} \left\{ -\frac{2}{5}(2 - x)^{5/2} - \frac{x^{10}}{10} \right\}_0^1 = \frac{16\sqrt{2} - 5}{30} \rho.$$

FIGURE 14.30

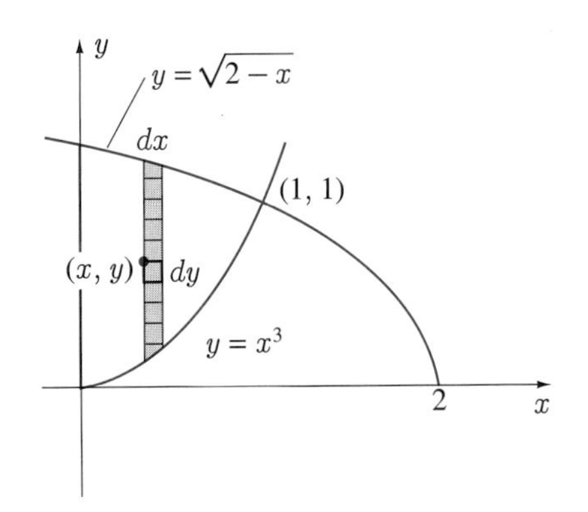

■

EXAMPLE 14.11

Find the second moment of area about the line $y = -1$ of the area bounded by the curves $x = y^2$ and $x = 2y$.

SOLUTION The second moment of area about the line $y = -1$ (Figure 14.31) is

$$\int_0^2 \int_{y^2}^{2y} (y+1)^2\,dx\,dy = \int_0^2 \{x(y+1)^2\}_{y^2}^{2y}\,dy = \int_0^2 (y+1)^2(2y - y^2)\,dy$$

$$= \int_0^2 (-y^4 + 3y^2 + 2y)\,dy = \left\{-\frac{y^5}{5} + y^3 + y^2\right\}_0^2$$

$$= \frac{28}{5}.$$

FIGURE 14.31

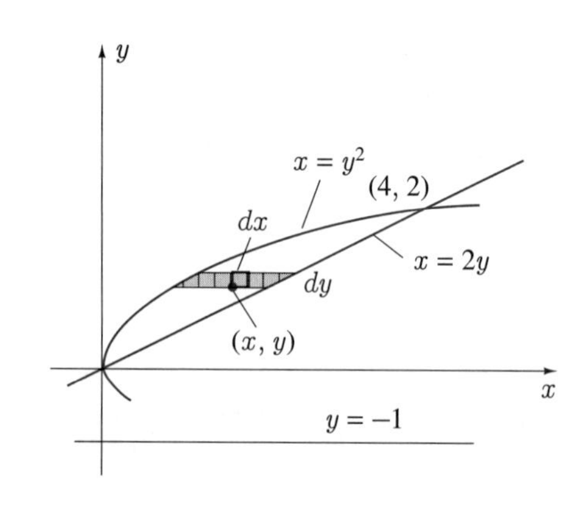

EXERCISES 14.5

In Exercises 1–10 find the centroid of the area bounded by the curves.

1. $x = y + 2,\ x = y^2$
2. $y = 8 - 2x^2,\ y + x^2 = 4$
3. $y = x^2 - 1,\ y + (x+1)^2 = 0$
4. $x + y = 5,\ xy = 4$
5. $y = e^x,\ y = 0,\ x = 0,\ x = 1$
6. $y = \sqrt{4 - x^2},\ y = x,\ x = 0$
7. $y = 1/(x-1),\ y = 1,\ y = 2,\ x = 0$
8. $x = 4y - 4y^2,\ x = y + 3,\ y = 1,\ y = 0$
9. $y = |x^2 - 1|,\ y = 2$
10. $y = x,\ y = 2x,\ 2y = x + 3$

In Exercises 11–15 find the second moment of area of the region bounded by the curves about the line.

11. $y = x^2,\ y = x^3$ about the y-axis
12. $y = x,\ y = 2x + 4,\ y = 0$ about the x-axis
13. $y = x^2,\ 2y = x^2 + 4$ about $y = 0$
14. $y = x^2 - 4,\ y = 2x - x^2$ about $x = -2$
15. $x = 1/\sqrt{y^4 + 12y^2},\ x = 0,\ y = 1/2,\ y = 1$ about $y = 0$
16. Find the first moment about the line $y = -2$ of a thin plate of constant mass per unit area ρ if its edges are defined by the curves $y = 2 - 2x^2$ and $y = x^2 - 1$.

In Exercises 17–23 find the centroid of the area bounded by the curves.

17. $x = \sqrt{y + 2},\ y = x,\ y = 0$
18. $y + x^2 = 0,\ x = y + 2,\ x + y + 2 = 0,\ y = 2$ (above $y + x^2 = 0$)
19. $y^2 = x^4(1 - x^2)$ (right loop)

20. $3x^2 + 4y^2 = 48$, $(x - 2)^2 + y^2 = 1$

21. $y = \ln x$, $y + \sqrt{x - 1} = 0$, $x = 2$

22. $y = \sqrt{2 - x}$, $15y = x^2 - 4$

23. $y = x\sqrt{1 - x^2}$, $x \geq 0$ and the x-axis

24. Find the moment of inertia of a uniform rectangular plate a units long and b units wide about a line through the centre of the plate and perpendicular to the plate.

In Exercises 25–27 find the second moment of area of the region bounded by the curves about the line.

25. $4x^2 + 9y^2 = 36$ about $y = -2$

26. $x = y^2$, $x + y = 2$ about $x = -1$

27. $y = \sqrt{a^2 - x^2}$, $y = a$, $x = a$ $(a > 0)$ about the x-axis

28. Find the first moment of area about the line $x + y = 1$ for the area bounded by the curves $x = y^2 - 2$ and $y = x$.

29. Prove the *theorem of Pappus*: If a plane area is revolved about a coplanar axis not crossing the area, the volume generated is equal to the product of the area and the circumference of the circle described by the centroid of the area.

30. A thin flat plate of area A is immersed vertically in a fluid with density ρ. Show that the total force (due to fluid pressure) on each side of the plate is equal to the product of 9.81, A, ρ, and the depth of the centroid of the plate below the surface of the fluid. Use this result to find the forces in some of the problems in Exercises 14.4, say 1, 3, 4, 5, 7, 11, 12, 13, 16, and 17. For those problems involving triangles recall the result of Exercise 36 in Section 7.6 or Exercise 39 in Section 12.3.

31. Prove the *parallel axis theorem*: The moment of inertia of a thin plate (with constant mass per unit area) with respect to any coplanar line is equal to the moment of inertia with respect to the parallel line through the centre of mass plus the mass multiplied by the square of the distance between the lines.

SECTION 14.6

Surface Area

To find the length of a curve in Section 7.3 we approximated the curve by tangent line segments. To find the area of a surface we follow a similar procedure by approximating the surface with tangential planes. In particular, consider finding the area of a smooth surface S given that every vertical line that intersects the surface does so in exactly one point (Figure 14.32). If S_{xy} is the area in the xy-plane onto which S projects, we divide S_{xy} into n sub–areas ΔA_i in any fashion whatsoever, and choose a point (x_i, y_i) in each ΔA_i. At the point (x_i, y_i, z_i) on the surface S that projects onto (x_i, y_i), we draw the tangent plane to S. Suppose we now project ΔA_i upward onto S and onto the tangent plane at (x_i, y_i, z_i) and denote these projected areas by ΔS_i and ΔS_{Ti}, respectively.

FIGURE 14.32

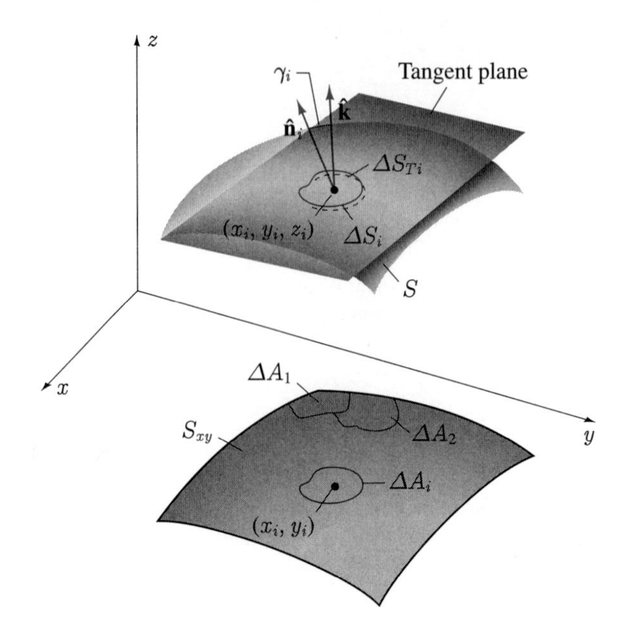

Now ΔS_{Ti} is an approximation to ΔS_i and, as long as ΔA_i is small, a reasonably good approximation. In fact, the smaller ΔA_i, the better the approximation. We therefore define the area of S as

$$\text{Area of } S = \lim_{n \to \infty} \sum_{i=1}^{n} \Delta S_{Ti}, \tag{14.37}$$

where in taking the limit we demand that each ΔA_i shrink to a point. We have therefore defined area on a curved surface in terms of flat areas on tangent planes to the surface. The advantage of this definition is that we can calculate ΔS_{Ti} in terms of ΔA_i. To see how, we denote by $\hat{\mathbf{n}}_i$ the unit normal vector to S at (x_i, y_i, z_i) with positive z-component and by γ_i the acute angle between $\hat{\mathbf{n}}_i$ and $\hat{\mathbf{k}}$. Now ΔS_{Ti} projects onto ΔA_i and γ_i is the acute angle between the planes containing ΔA_i and ΔS_{Ti} (Figure 14.33). It follows that ΔA_i and ΔS_{Ti} are related by

$$\Delta A_i = \cos \gamma_i \, \Delta S_{Ti} \tag{14.38}$$

(see Exercise 54 in Section 12.5). Note that if ΔS_{Ti} is horizontal, then $\gamma_i = 0$ and $\Delta S_{Ti} = \Delta A_i$; and if ΔS_{Ti} tends toward the vertical ($\gamma_i \to \pi/2$), then ΔS_{Ti} becomes very large for fixed ΔA_i.

FIGURE 14.33

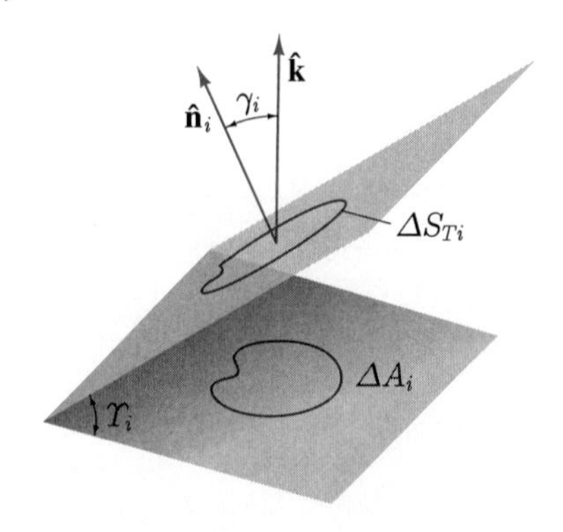

Because the surface projects in a one-to-one fashion onto the area S_{xy} in the xy-plane, we can take the equation for S in the form $z = f(x, y)$. A vector normal to S at any point is therefore

$$\nabla(z - f(x, y)) = \left(-\frac{\partial f}{\partial x}, -\frac{\partial f}{\partial y}, 1\right);$$

hence

$$\hat{\mathbf{n}} = \frac{\left(-\dfrac{\partial f}{\partial x}, -\dfrac{\partial f}{\partial y}, 1\right)}{\sqrt{1 + \left(\dfrac{\partial f}{\partial x}\right)^2 + \left(\dfrac{\partial f}{\partial y}\right)^2}} = \frac{\left(-\dfrac{\partial z}{\partial x}, -\dfrac{\partial z}{\partial y}, 1\right)}{\sqrt{1 + \left(\dfrac{\partial z}{\partial x}\right)^2 + \left(\dfrac{\partial z}{\partial y}\right)^2}}.$$

Since $\hat{\mathbf{n}}_i \cdot \hat{\mathbf{k}} = |\hat{\mathbf{n}}_i||\hat{\mathbf{k}}| \cos\gamma_i = \cos\gamma_i$, it follows that

$$\cos\gamma_i = \hat{\mathbf{n}}_i \cdot \hat{\mathbf{k}} = \frac{1}{\sqrt{1 + z_x^2(x_i, y_i) + z_y^2(x_i, y_i)}}.$$

When we substitute this expression into 14.38, we obtain the result that area ΔS_{Ti} on the tangent plane to $z = f(x, y)$ at the point (x_i, y_i, z_i) is related to its projection ΔA_i in the xy-plane according to

$$\Delta S_{Ti} = \sqrt{1 + z_x^2(x_i, y_i) + z_y^2(x_i, y_i)}\,\Delta A_i. \tag{14.39}$$

Definition 14.37 for the area of S can now be written in the form

$$\text{Area of } S = \lim_{\|\Delta A_i\| \to 0} \sum_{i=1}^{n} \sqrt{1 + z_x^2(x_i, y_i) + z_y^2(x_i, y_i)}\,\Delta A_i, \tag{14.40}$$

where the summation is carried out over all areas ΔA_i in S_{xy} as each ΔA_i shrinks to a point. But this is the definition of the double integral of the function $\sqrt{1 + (\partial z/\partial x)^2 + (\partial z/\partial y)^2}$ over the region S_{xy}. In other words, on the basis of definition 14.37, areas on surfaces can be calculated according to

$$\text{Area of } S = \iint_{S_{xy}} \sqrt{1 + \left(\frac{\partial z}{\partial x}\right)^2 + \left(\frac{\partial z}{\partial y}\right)^2}\, dA. \tag{14.41}$$

Note the analogy between equations 14.41 and 7.15. In equation 7.15 we think of $\sqrt{1 + (dy/dx)^2}\,dx$ as a small length along a curve $C{:}y = f(x)$ that projects onto the length dx along the x-axis. In fact, $\sqrt{1 + (dy/dx)^2}\,dx$ is along the tangent line to C, but we think of it as along C itself (see Section 12.8). The total length of C is then found by adding over all projections of C from $x = a$ to $x = b$. Similarly, in equation 14.41 we think of $\sqrt{1 + (\partial z/\partial x)^2 + (\partial z/\partial y)^2}\,dA$ as a small area on a surface $S : z = f(x, y)$ that projects onto the area dA in the xy-plane. It is, in fact, a small area on the tangent plane to S, but it is usually easier to think of it as being on S itself. The total area of S is then the addition over all projections S_{xy} of S.

Note too that when S is smooth, $\partial z/\partial x$ and $\partial z/\partial y$ are continuous, thus guaranteeing existence of double integral 14.41. If S is piecewise smooth, we divide it into smooth subsurfaces and integrate over each piece separately.

This discussion has been based on the assumption that S projects one-to-one onto some region S_{xy} in the xy-plane. If this condition is not met, one possibility is to subdivide S into parts, each of which projects one-to-one onto the xy-plane. The total area of S is then the sum of the areas of its parts. For instance, to find the area of the surface S in Figure 14.34, we could subdivide S into S_1 and S_2 along the curve C. The area of S is then the sum of the areas of S_1 and S_2, each of which projects one-to-one onto the xy-plane.

FIGURE 14.34

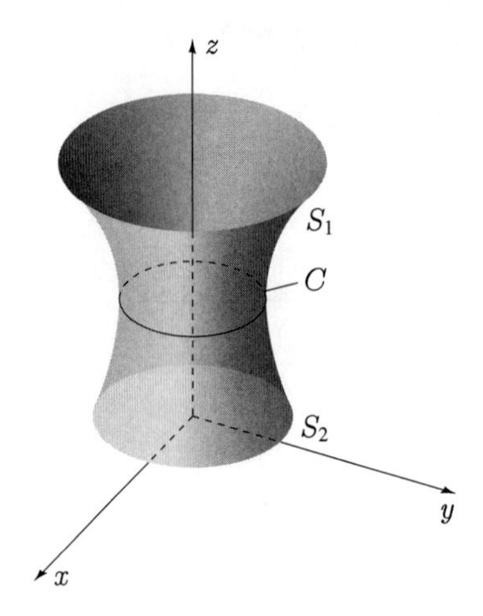

Alternatively, we could note that there is nothing sacred about projecting S onto the xy-plane. We could develop similar results if S projects one-to-one onto either the yz- or the xz-plane. If S_{yz} and S_{xz} represent these projections, then

$$\text{Area of } S = \iint_{S_{yz}} \sqrt{1 + \left(\frac{\partial x}{\partial y}\right)^2 + \left(\frac{\partial x}{\partial z}\right)^2}\, dA, \tag{14.42a}$$

$$\text{Area of } S = \iint_{S_{xz}} \sqrt{1 + \left(\frac{\partial y}{\partial x}\right)^2 + \left(\frac{\partial y}{\partial z}\right)^2}\, dA. \tag{14.42b}$$

EXAMPLE 14.12

Find the area of that part of the plane $x + 2y + 3z = 6$ in the first octant.

SOLUTION This area projects one-to-one onto the triangular area S_{xy} in the xy-plane bounded by the lines $x = 0, y = 0, x + 2y = 6$ (Figure 14.35). Since $z = 2 - x/3 - 2y/3$,

$$\begin{aligned}\text{Area} &= \iint_{S_{xy}} \sqrt{1 + \left(\frac{\partial z}{\partial x}\right)^2 + \left(\frac{\partial z}{\partial y}\right)^2}\, dA = \iint_{S_{xy}} \sqrt{1 + \left(-\frac{1}{3}\right)^2 + \left(-\frac{2}{3}\right)^2}\, dA \\ &= \frac{\sqrt{14}}{3} \iint_{S_{xy}} dA = \frac{\sqrt{14}}{3}(\text{Area of } S_{xy}) = \frac{\sqrt{14}}{3}\left[\frac{1}{2}(3)(6)\right] = 3\sqrt{14}.\end{aligned}$$

FIGURE 14.35

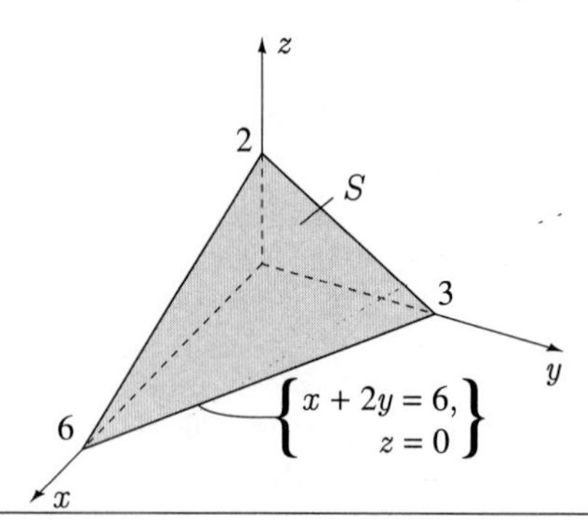

■

EXAMPLE 14.13

Find the area of the surface $z = x^{3/2}$ that projects onto the rectangle in the xy-plane bounded by the straight lines $x = 0$, $x = 2$, $y = 1$, and $y = 3$.

SOLUTION Since the surface projects one-to-one onto the rectangle (Figure 14.36), we find that

$$\begin{aligned}\text{Area} &= \iint_{S_{xy}} \sqrt{1 + \left(\frac{\partial z}{\partial x}\right)^2 + \left(\frac{\partial z}{\partial y}\right)^2}\, dA = \iint_{S_{xy}} \sqrt{1 + \left(\frac{3}{2}x^{1/2}\right)^2}\, dA \\ &= \frac{1}{2}\int_0^2 \int_1^3 \sqrt{4 + 9x}\, dy\, dx = \frac{1}{2}\int_0^2 \{y\sqrt{4+9x}\}_1^3\, dx = \int_0^2 \sqrt{4+9x}\, dx \\ &= \left\{\frac{2}{27}(4+9x)^{3/2}\right\}_0^2 = \frac{2}{27}(22\sqrt{22} - 8).\end{aligned}$$

FIGURE 14.36

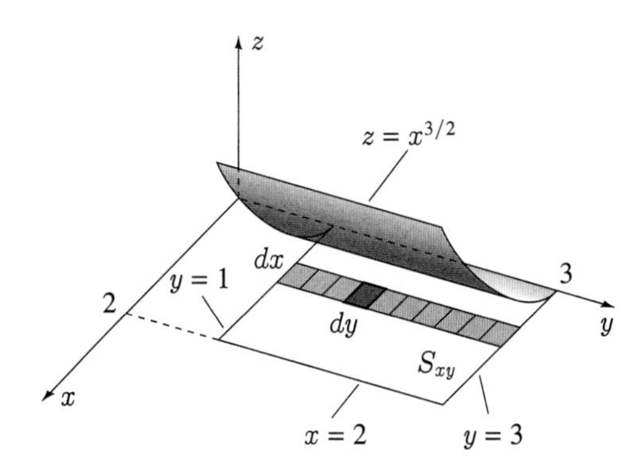

■

EXAMPLE 14.14

Find the area of the cone $y = \sqrt{x^2 + z^2}$ to the left of the plane $y = 1$.

SOLUTION ***Method 1*** The surface projects one-to-one onto the interior of the circle $x^2 + z^2 = 1$ in the xz-plane (Figure 14.37). Since the area of the surface is four times that in the first octant, if we let S_{xz} be the quarter-circle $x^2 + z^2 \le 1$, $x \ge 0$, $z \ge 0$, then

$$\begin{aligned}\text{Area} &= 4\iint_{S_{xz}} \sqrt{1 + \left(\frac{\partial y}{\partial x}\right)^2 + \left(\frac{\partial y}{\partial z}\right)^2}\, dA \\ &= 4\iint_{S_{xz}} \sqrt{1 + \left(\frac{x}{\sqrt{x^2+z^2}}\right)^2 + \left(\frac{z}{\sqrt{x^2+z^2}}\right)^2}\, dA \\ &= 4\iint_{S_{xz}} \sqrt{1 + \frac{x^2}{x^2+z^2} + \frac{z^2}{x^2+z^2}}\, dA = 4\int\int_{S_{xz}} \sqrt{2}\, dA \\ &= 4\sqrt{2}(\text{Area of } S_{xz}) = 4\sqrt{2}\{\tfrac{1}{4}\pi(1)^2\} = \sqrt{2}\pi.\end{aligned}$$

FIGURE 14.37

FIGURE 14.38

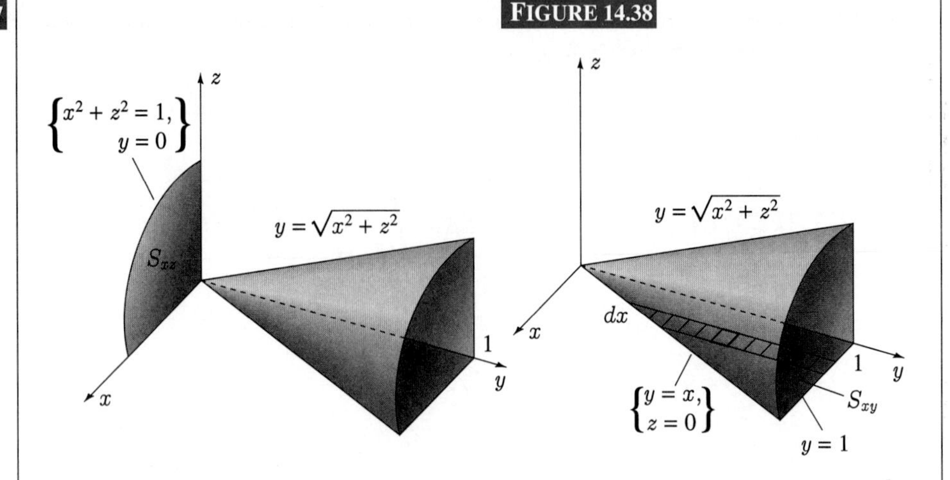

Method 2 Suppose instead that we project that part of the surface in the first octant onto the triangle S_{xy} in Figure 14.38. We write the equation of this part of the surface in the form $z = \sqrt{y^2 - x^2}$ and calculate

$$\begin{aligned}\text{Area} &= 4 \iint_{S_{xy}} \sqrt{1 + \left(\frac{\partial z}{\partial x}\right)^2 + \left(\frac{\partial z}{\partial y}\right)^2}\, dA \\ &= 4 \iint_{S_{xy}} \sqrt{1 + \left(\frac{-x}{\sqrt{y^2 - x^2}}\right)^2 + \left(\frac{y}{\sqrt{y^2 - x^2}}\right)^2}\, dA \\ &= 4 \iint_{S_{xy}} \sqrt{1 + \frac{x^2}{y^2 - x^2} + \frac{y^2}{y^2 - x^2}}\, dA = 4\sqrt{2} \iint_{S_{xy}} \frac{y}{\sqrt{y^2 - x^2}}\, dA.\end{aligned}$$

To evaluate this double integral, it is advantageous to integrate first with respect to y:

$$\begin{aligned}\text{Area} &= 4\sqrt{2} \int_0^1 \int_x^1 \frac{y}{\sqrt{y^2 - x^2}}\, dy\, dx = 4\sqrt{2} \int_0^1 \{\sqrt{y^2 - x^2}\}_x^1 dx \\ &= 4\sqrt{2} \int_0^1 \sqrt{1 - x^2}\, dx.\end{aligned}$$

If we now set $x = \sin\theta$, then $dx = \cos\theta\, d\theta$, and

$$\begin{aligned}\text{Area} &= 4\sqrt{2} \int_0^{\pi/2} \cos\theta \cos\theta\, d\theta = 4\sqrt{2} \int_0^{\pi/2} \left\{\frac{1 + \cos 2\theta}{2}\right\} d\theta \\ &= 2\sqrt{2} \left\{\theta + \frac{1}{2}\sin 2\theta\right\}_0^{\pi/2} = \sqrt{2}\,\pi.\end{aligned}$$

■

EXERCISES 14.6

In Exercises 1–6 find the required area.

1. The area of $2x + 3y + 6z = 1$ in the first octant

2. The area of $x + 2y - 3z + 4 = 0$ for which $x \leq 0$, $y \leq 0$ and $z \geq 0$

3. The area of $z = 1 - 4\sqrt{x^2 + y^2}$ above the xy-plane

4. The area of $z = \sqrt{2xy}$ cut out by the planes $x = 1$, $x = 2$, $y = 1$, $y = 3$

5. The area in the first octant cut out from the surface $z = x + y$ by the plane $x + 2y = 4$

6. The area of $z = x^{3/2} + y^{3/2}$ in the first octant cut off by the plane $x + y = 1$

In Exercises 7–12 set up, but do not evaluate, double iterated integrals to find the required area.

7. The area of $x^2 + y^2 + z^2 = 2$ inside the cone $z = \sqrt{x^2 + y^2}$

8. The area of $4x = y^2 + z^2$ cut off by $x = 4$

9. The area in the first octant cut from $y = xz$ by the cylinder $x^2 + z^2 = 1$

10. The area of $z = (x^2 + y^2)^2$ below $z = 4$

11. The area of $y = 1 - x^2 - 3z^2$ to the right of the xz-plane

12. The area of $z = \ln(1 + x + y)$ in the first octant cut off by $y = 1 - x^2$

13. Find the area of the surface $z = \ln x$ that projects onto the rectangle in the xy-plane bounded by the lines $x = 1$, $x = 2$, $y = 0$, $y = 2$

14. Verify that the area of the curved portion of a right-circular cone of radius r and height h is $\pi r\sqrt{r^2 + h^2}$.

In Exercises 15–19 set up, but do not evaluate, double iterated integrals to find the required area.

15. The area of $y = x^2 + z^2$ cut off by $y + z = 1$

16. The area of $y = z^2 + x$ inside $x^2 + y^2 = 1$

17. The area of $y^2 = z + x^2$ inside $x^2 + y^2 = 4$

18. The area of $z = x^3 + y^3$ that is in the first octant and between the planes $x + y = 1$ and $x + y = 2$

19. The area of $x^2 + y^2 = z^2 + 1$ between the planes $z = 1$ and $z = 4$

20. Find the area of that part of the surface $z = 2x^2 + 3y$ bounded by the planes $x = 2$, $y = 0$, and $y = x$.

SECTION 14.7

Double Iterated Integrals in Polar Coordinates

So far we have used only double iterated integrals in x and y to evaluate double integrals. But for some problems this is not convenient. For instance, the double integral of a continuous function $f(x, y)$ over the region R in Figure 14.39 requires three double iterated integrals in x and y. In other words, a subdivision of R into rectangles by coordinate lines x = constant and y = constant is simply not convenient for this region. For such an area, polar coordinates are more useful.

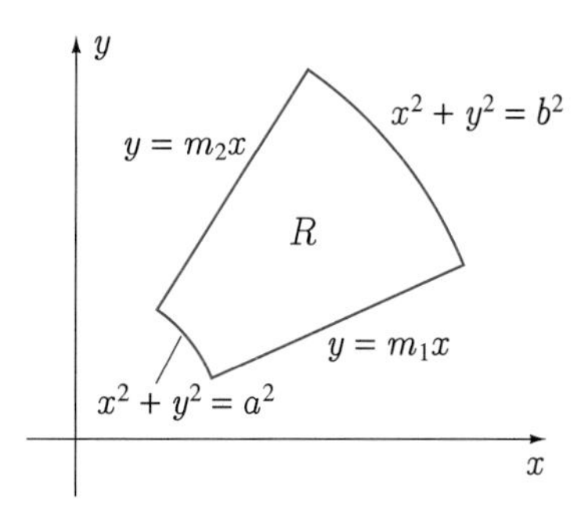

FIGURE 14.39

Polar coordinates with the origin as pole and the positive x-axis as polar axis are defined by

$$x = r\cos\theta, \qquad y = r\sin\theta$$

(see Section 10.2). We wish to obtain double iterated integrals in polar coordinates that represent the double integral of $f(x, y)$ over the region R in Figure 14.39. To do this we return to Definition 14.3 for a double integral, and choose a subdivision of R into sub–areas convenient to polar coordinates. When using Cartesian coordinates we drew coordinate lines x = constant and y = constant. When using polar coordinates we draw coordinate curves r = constant and θ = constant. In particular, we subdivide R by a network of $n+1$ circles $r = r_i$, where

$$a = r_0 < r_1 < r_2 < \cdots < r_{n-1} < r_n = b,$$

and $m+1$ radial lines $\theta = \theta_j$, where

$$c = \theta_0 < \theta_1 < \theta_2 < \cdots < \theta_{m-1} < \theta_m = d$$

(Figure 14.40). If ΔA_{ij} represents the area bounded by the circles $r = r_{i-1}$ and $r = r_i$ and the radial lines $\theta = \theta_{j-1}$ and $\theta = \theta_j$ (Figure 14.41), then it is straightforward to show that

$$\Delta A_{ij} = \frac{1}{2}(r_i^2 - r_{i-1}^2)(\theta_j - \theta_{j-1}).$$

If we set $\Delta r_i = r_i - r_{i-1}$ and $\Delta \theta_j = \theta_j - \theta_{j-1}$, then

$$\begin{aligned} \Delta A_{ij} &= \frac{1}{2}(r_i + r_{i-1})(r_i - r_{i-1})(\theta_j - \theta_{j-1}) \\ &= \left[\frac{r_i + r_{i-1}}{2}\right] \Delta r_i \, \Delta \theta_j. \end{aligned} \tag{14.43}$$

FIGURE 14.40

FIGURE 14.41

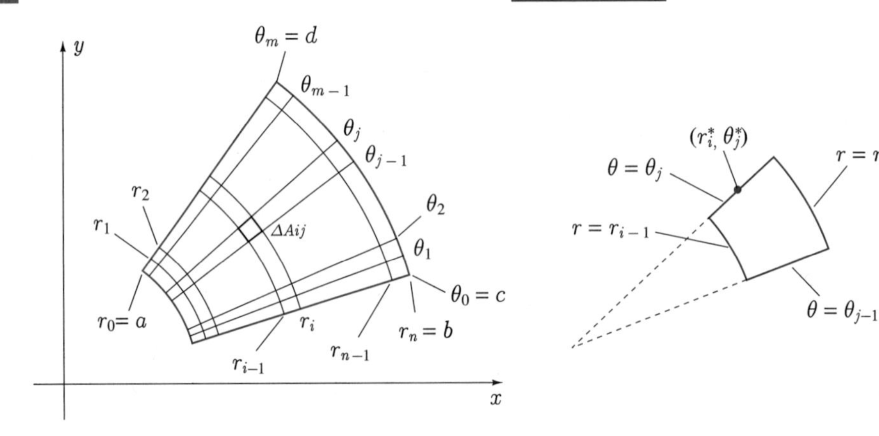

Our next task in using Definition 14.3 for the double integral of $f(x, y)$ over R is to choose a star-point in each ΔA_{ij}. If we select

$$(r_i^*, \theta_j^*) = \left(\frac{r_i + r_{i-1}}{2}, \theta_j\right),$$

then

$$\Delta A_{ij} = r_i^* \, \Delta r_i \, \Delta \theta_j,$$

and by Definition 14.3,

$$\begin{aligned} \iint_R f(x, y)\, dA &= \lim_{\|\Delta A_{ij}\| \to 0} \sum_{j=1}^{m} \sum_{i=1}^{n} f(r_i^* \cos\theta_j^*, r_i^* \sin\theta_j^*) \Delta A_{ij} \\ &= \lim_{\substack{\|\Delta r_i\| \to 0 \\ \|\Delta \theta_j\| \to 0}} \sum_{j=1}^{m} \sum_{i=1}^{n} f(r_i^* \cos\theta_j^*, r_i^* \sin\theta_j^*) r_i^* \, \Delta r_i \, \Delta \theta_j. \end{aligned}$$

If we take the limit first as $||\Delta r_i|| \to 0$ and then as $||\Delta \theta_j|| \to 0$, we obtain the double iterated integral

$$\iint_R f(x,y)\,dA = \int_c^d \int_a^b f(r\cos\theta, r\sin\theta)\,r\,dr\,d\theta. \qquad (14.44)$$

Reversing the order of taking limits reverses the order of the iterated integrals:

$$\iint_R f(x,y)\,dA = \int_a^b \int_c^d f(r\cos\theta, r\sin\theta)\,r\,d\theta\,dr. \qquad (14.45)$$

For the region R of Figure 14.40, then, there are two double iterated integrals in polar coordinates representing the double integral of $f(x,y)$ over R.

We have interpreted double iterated integrals in Cartesian coordinates as integrations over horizontal or vertical strips. Double iterated integrals in polar coordinates can also be interpreted geometrically. Take, for instance, equation 14.44. A double iterated integral in polar coordinates implies a subdivision of the region R into areas as shown in Figure 14.41. Let us denote small variations in r and θ for a representative piece of area at position (r,θ) by dr and $d\theta$ (Figure 14.42). If dr and $d\theta$ are very small (as is implied in the definition of the double integral), then this piece of area is almost rectangular with an approximate area of $(r\,d\theta)\,dr$. In polar coordinates, then, we think of dA in 14.44 as being replaced by

$$dA = r\,dr\,d\theta. \qquad (14.46)$$

Each such area at (r,θ) is multiplied by the value of $f(x,y)$ at (r,θ) to give the product

$$f(r\cos\theta, r\sin\theta)\,r\,dr\,d\theta.$$

The inner integral

$$\int_a^b f(r\cos\theta, r\sin\theta)\,r\,dr\,d\theta$$

with respect to r holds θ constant and is therefore interpreted as a summation over the small areas in a wedge $d\theta$ from $r = a$ to $r = b$. The θ-integration then adds over all wedges starting at $\theta = c$ and ending at $\theta = d$. Limits on θ therefore identify positions of first and last wedges.

FIGURE 14.42

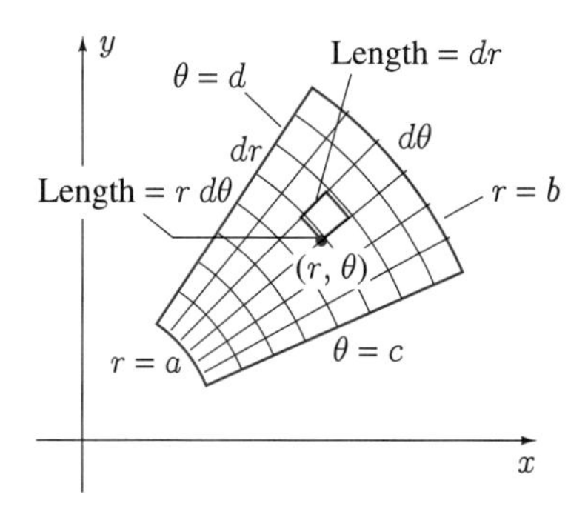

If the order of integration is reversed (equation 14.45), then the inner integral

$$\int_c^d f(r\cos\theta, r\sin\theta)\,r\,d\theta\,dr$$

holds r constant. We interpret this as an addition over the small areas in a ring dr, where the limits indicate that each and every ring starts on the curve $\theta = c$ and ends on the curve $\theta = d$. The outer r-integration is an addition over all rings with the first ring at $r = a$ and the last at $r = b$.

Double iterated integrals in polar coordinates for more general regions are now quite simple. For the region R of Figure 14.43,

$$\iint_R f(x,y)\,dA = \int_\alpha^\beta \int_{g(\theta)}^{h(\theta)} f(r\cos\theta, r\sin\theta)\,r\,dr\,d\theta.$$

FIGURE 14.43

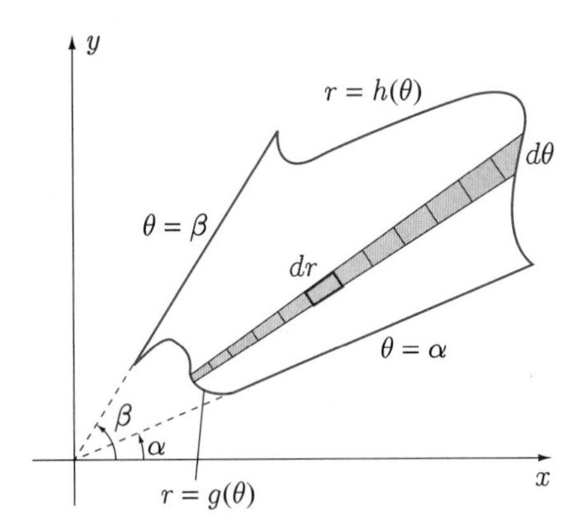

EXAMPLE 14.15 Find the area inside the circle $x^2 + y^2 = 4x$ and outside the circle $x^2 + y^2 = 1$.

SOLUTION If R is the region bounded by these circles and above the x-axis (Figure 14.44), then the required area is

$$2\iint_R dA.$$

Since the curves intersect in the first quadrant at a point where $\theta = \overline{\theta} = \mathrm{Cos}^{-1}(\frac{1}{4})$, then

$$\begin{aligned}
\text{Area} &= 2\int_0^{\overline{\theta}} \int_1^{4\cos\theta} r\,dr\,d\theta = 2\int_0^{\overline{\theta}} \left\{\frac{r^2}{2}\right\}_1^{4\cos\theta} d\theta = \int_0^{\overline{\theta}} (16\cos^2\theta - 1)\,d\theta \\
&= \int_0^{\overline{\theta}} \left\{16\left(\frac{1+\cos 2\theta}{2}\right) - 1\right\} d\theta = \int_0^{\overline{\theta}} (7 + 8\cos 2\theta)\,d\theta = \{7\theta + 4\sin 2\theta\}_0^{\overline{\theta}} \\
&= 7\overline{\theta} + 4\sin 2\overline{\theta} = 7\,\mathrm{Cos}^{-1}(\tfrac{1}{4}) + 8\cos\overline{\theta}\sin\overline{\theta} \\
&= 7\,\mathrm{Cos}^{-1}(\tfrac{1}{4}) + 8(\tfrac{1}{4})\sqrt{1 - \tfrac{1}{16}} = 7\,\mathrm{Cos}^{-1}(\tfrac{1}{4}) + \sqrt{15}/2.
\end{aligned}$$

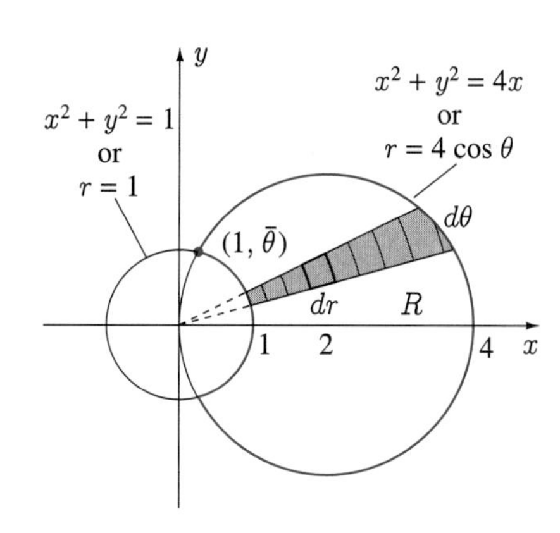

FIGURE 14.44

EXAMPLE 14.16

Evaluate the double iterated integral

$$\int_0^1 \int_0^{\sqrt{-x^2+x}} y^2 \, dy \, dx.$$

SOLUTION The limits identify the region of integration as the interior of the semicircle in Figure 14.45. The integrand suggests an interpretation of the integral as the second moment of area of this semicircle about the x-axis. Since the semicircle R in Figure 14.46 has exactly the same second moment about the x-axis, we can state that

$$\begin{aligned}\int_0^1 \int_0^{\sqrt{-x^2+x}} y^2 \, dy \, dx &= \iint_R y^2 \, dA = \int_0^{\pi} \int_0^{1/2} (r^2 \sin^2\theta) r \, dr \, d\theta \\ &= \int_0^{\pi} \left\{ \frac{r^4}{4} \sin^2\theta \right\}_0^{1/2} d\theta = \frac{1}{64} \int_0^{\pi} \sin^2\theta \, d\theta \\ &= \frac{1}{128} \int_0^{\pi} (1 - \cos 2\theta) \, d\theta = \frac{1}{128} \left\{ \theta - \frac{\sin 2\theta}{2} \right\}_0^{\pi} = \frac{\pi}{128}.\end{aligned}$$

FIGURE 14.45

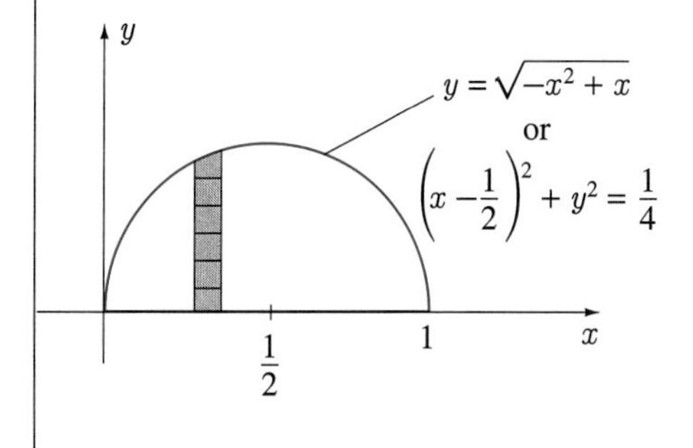

FIGURE 14.46

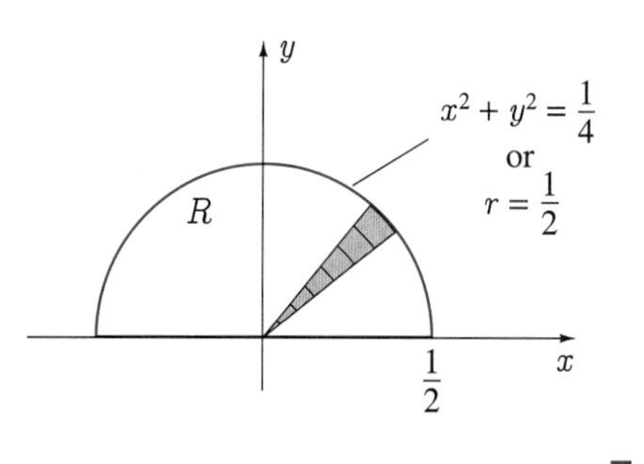

EXAMPLE 14.17

Find the centroid of the region R in Figure 14.47.

SOLUTION Evidently, $\overline{x} = 0$, and the area of the region is $A = (\pi b^2 - \pi a^2)/2$. Since

$$A\overline{y} = \iint_R y\,dA = \int_0^{\pi} \int_a^b (r \sin\theta) r\,dr\,d\theta = \int_0^{\pi} \left\{ \frac{r^3}{3} \sin\theta \right\}_a^b d\theta$$

$$= \frac{1}{3}(b^3 - a^3) \int_0^{\pi} \sin\theta\,d\theta$$

$$= \frac{1}{3}(b^3 - a^3)\{-\cos\theta\}_0^{\pi} = \frac{2}{3}(b^3 - a^3),$$

if follows that

$$\overline{y} = \frac{2}{3}(b^3 - a^3)\frac{2}{\pi(b^2 - a^2)} = \frac{4}{3\pi}\frac{b^2 + ab + a^2}{a + b}.$$

FIGURE 14.47

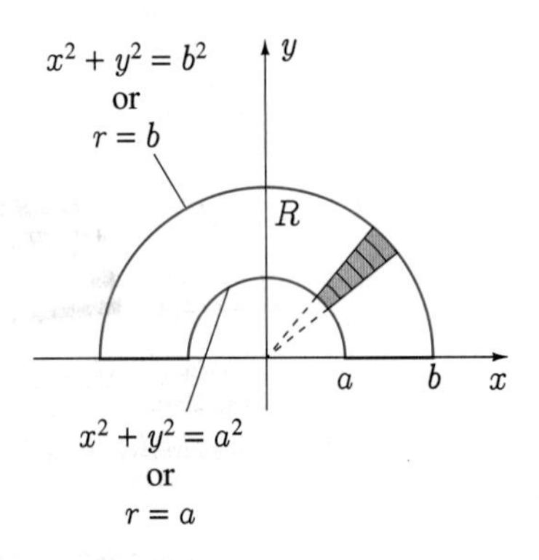

EXAMPLE 14.18

Find the area of that portion of the sphere $x^2 + y^2 + z^2 = 2$ inside the cone $z = \sqrt{x^2 + y^2}$.

SOLUTION If S is that portion of the sphere that is inside the cone and also in the first octant (Figure 14.48), then the required area is four times that of S; i.e.,

$$\text{Area} = 4 \iint_{S_{xy}} \sqrt{1 + \left(\frac{\partial z}{\partial x}\right)^2 + \left(\frac{\partial z}{\partial y}\right)^2}\,dA,$$

where S_{xy} is the projection of S on the xy-plane. The curve of intersection of the cone and the sphere has equations

$$x^2 + y^2 + z^2 = 2, \qquad\qquad z = 1,$$
$$\text{or equivalently}$$
$$z = \sqrt{x^2 + y^2}, \qquad\qquad x^2 + y^2 = 1.$$

Consequently, S_{xy} is the interior of the quarter-circle $x^2 + y^2 \le 1,\ x \ge 0,\ y \ge 0$. On S,

$$\frac{\partial z}{\partial x} = -\frac{x}{z} \quad \text{and} \quad \frac{\partial z}{\partial y} = -\frac{y}{z},$$

so that

$$\text{Area} = 4\iint_{S_{xy}} \sqrt{1 + \frac{x^2}{z^2} + \frac{y^2}{z^2}}\, dA = 4\iint_{S_{xy}} \sqrt{\frac{x^2 + y^2 + z^2}{z^2}}\, dA$$

$$= 4\iint_{S_{xy}} \sqrt{\frac{2}{z^2}}\, dA = 4\sqrt{2}\iint_{S_{xy}} \frac{1}{\sqrt{2 - x^2 - y^2}}\, dA.$$

If we now use polar coordinates to evaluate this double integral, we have

$$\text{Area} = 4\sqrt{2}\int_0^{\pi/2}\int_0^1 \frac{1}{\sqrt{2 - r^2}}\, r\, dr\, d\theta = 4\sqrt{2}\int_0^{\pi/2} \{-\sqrt{2 - r^2}\}_0^1\, d\theta$$

$$= 4\sqrt{2}(\sqrt{2} - 1)\int_0^{\pi/2} d\theta = 4\sqrt{2}(\sqrt{2} - 1)\frac{\pi}{2} = 2\sqrt{2}\pi(\sqrt{2} - 1).$$

FIGURE 14.48

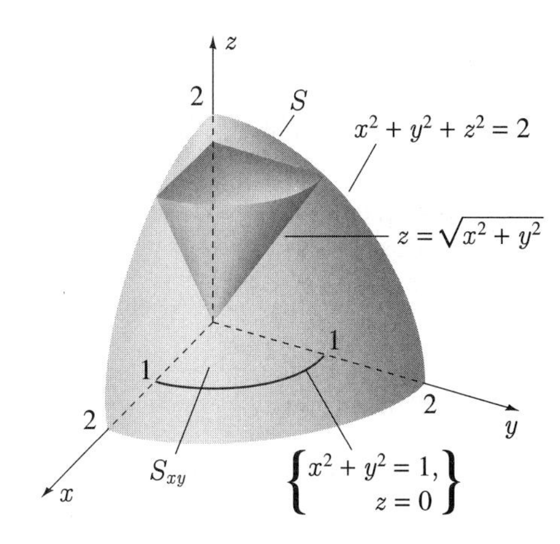

■

EXERCISES 14.7

In Exercises 1–5 evaluate the double integral of the function over the region R.

1. $f(x, y) = e^{x^2+y^2}$ where R is bounded by $x^2 + y^2 = a^2$

2. $f(x, y) = x$ where R is bounded by $x = \sqrt{2y - y^2},\ x = 0$

3. $f(x, y) = \sqrt{x^2 + y^2}$ where R is bounded by $y = \sqrt{9 - x^2},\ y = x,\ x = 0$

4. $f(x, y) = 1/\sqrt{x^2 + y^2}$ where R is the region outside $x^2 + y^2 = 4$ and inside $x^2 + y^2 = 4x$

5. $f(x, y) = \sqrt{1 + 2x^2 + 2y^2}$ where R is bounded by $x^2 + y^2 = 1,\ x^2 + y^2 = 4$

Evaluate the double iterated integral in Exercises 6 and 7.

6. $\displaystyle\int_0^1\int_0^{\sqrt{1-x^2}} \sqrt{x^2 + y^2}\, dy\, dx$

7. $\displaystyle\int_{-\sqrt{2}}^0\int_{-y}^{\sqrt{4-y^2}} x^2\, dx\, dy$

In Exercises 8–12 find the area bounded by the curves.

8. Outside $x^2 + y^2 = 9$ and inside $x^2 + y^2 = 2\sqrt{3}y$

9. $r = 9(1 + \cos\theta)$

10. $r = \cos 3\theta$

11. Common to $r = 2$ and $r^2 = 9\cos 2\theta$

12. Common to $r = 1 + \sin\theta$ and $r = 2 - 2\sin\theta$

13. Find the centroid of the area bounded by the curves $y = x,\ y = -x,\ x = \sqrt{2 - y^2}$

14. Find the second moment of area for a circular plate of radius R about any diameter.

15. A water tank in the form of a right-circular cylinder with radius R and length h has its axis horizontal. If it is full, what is the force due to water pressure on each end?

In Exercises 16–18 find the area of the surface.

16. The area of $z = x^2 + y^2$ below $z = 4$

17. The area of $x^2 + y^2 + z^2 = 4$ inside $x^2 + y^2 = 1$

18. The area of $z = xy$ inside $x^2 + y^2 = 9$

19. Prove that the area of a sphere of radius R is $4\pi R^2$.

20. Find the area of the hyperbolic paraboloid $z = x^2 - y^2$ between the cylinders $x^2 + y^2 = 1$ and $x^2 + y^2 = 4$

In Exercises 21 and 22 find the volume of the solid of revolution obtained by rotating the area bounded by the curve about the line.

21. $r = \cos^2\theta$ about the x-axis

22. $r = 1 + \sin\theta$ about the y-axis

23. Find the area bounded by the curve $(x^2 + y^2)^3 = 4a^2x^2y^2$.

24. Find the volume of the solid of revolution when a circle of radius R is rotated about a tangent line.

25. A circular plate of radius R (Figure 14.49) has a uniform charge distribution of ρ coulombs per square metre. If P is a point directly above the centre of the plate and dA is a small area on the plate, then the potential at P due to dA is given by

$$\frac{1}{4\pi\epsilon_0}\frac{\rho dA}{s},$$

where s is the distance from P to dA.

(a) Show that, in terms of polar coordinates, the potential V at P due to the entire plate is

$$V = \frac{\rho}{4\pi\epsilon_0}\int_{-\pi}^{\pi}\int_0^R \frac{r}{\sqrt{r^2 + d^2}}\,dr\,d\theta,$$

where d is the distance from P to the centre of the plate.

(b) Evaluate the double iterated integral to find V.

FIGURE 14.49

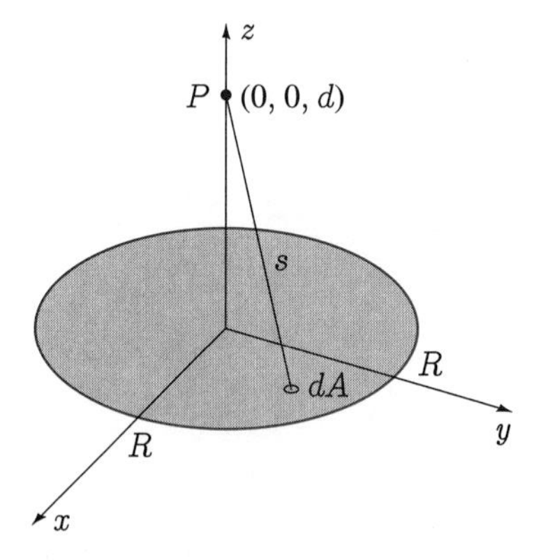

26. Use Coulomb's law (see Example 12.9 in Section 12.3) to find the force on a charge q at point P due to the charge on the plate in Exercise 25. What happens to this force as the radius of the plate gets very large?

In Exercises 27–29 find the area bounded by the curves.

27. $(x^2 + y^2)^2 = 2xy$

28. Inside both $r = 6\cos\theta$ and $r = 4 - 2\cos\theta$

29. $r = \cos^2\theta\sin\theta$

30. Find the centroid of the cardioid $r = 1 + \cos\theta$.

31. Find the second moment of area about the x-axis for the area bounded by $r^2 = 9\cos 2\theta$.

32. Evaluate the double integral of

$$f(x, y) = \sqrt{\frac{1 - x^2 - y^2}{1 + x^2 + y^2}}$$

over the area inside the circle $x^2 + y^2 = 1$.

33. Figure 14.50 illustrates a piece of an artery or vein with circular cross-section (radius R). The speed of blood flowing through the blood vessel is not uniform because of the viscosity of the blood and friction at the walls. Poiseuille's law states that for laminar blood flow, the speed v of blood at a distance r from the centre of the vessel is given by

$$v = \frac{P}{4nL}(R^2 - r^2),$$

where P is the pressure difference between the ends of the vessel, L is the length of the vessel, and n is the viscosity of the blood. Find the amount of blood flowing over a cross-section of the blood vessel per unit time.

FIGURE 14.50

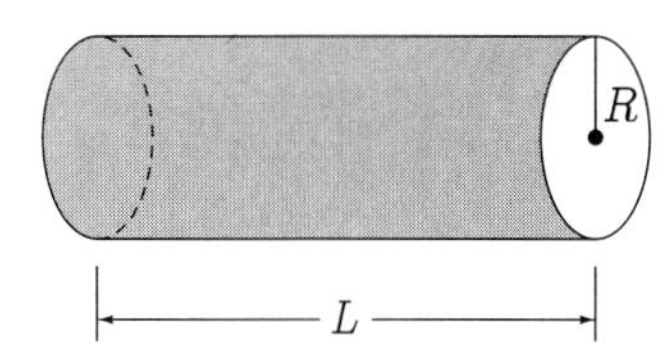

34. Find the area of that part of the sphere $x^2 + y^2 + z^2 = a^2$ inside $(x^2 + y^2)^2 = a^2(x^2 - y^2)$.

35. Find the area of that portion of the surface $x^2 + z^2 = a^2$ cut out by $x^2 + y^2 = a^2$.

36. A very important integral in statistics is

$$I = \int_0^\infty e^{-x^2}\,dx.$$

To evaluate the integral we set

$$I = \int_0^\infty e^{-y^2}\,dy,$$

and then multiply these two equations. Do this to prove that $I = \sqrt{\pi}/2$.

37. Use the result of Exercise 36 to evaluate the gamma function

$$\Gamma(n) = \int_0^\infty x^{n-1} e^{-x}\,dx$$

at $n = 1/2$.

SECTION 14.8

Triple Integrals and Triple Iterated Integrals

Triple integrals are defined in much the same way as double integrals. Suppose $f(x, y, z)$ is a function defined in some region V of space that has finite volume (Figure 14.51). We divide V into n subregions of volumes $\Delta V_1, \Delta V_2, \ldots, \Delta V_n$ in any manner whatsoever, and in each subregion $\Delta V_i (i = 1, \ldots, n)$ we choose an arbitrary point (x_i^*, y_i^*, z_i^*). We then form the sum

$$f(x_1^*, y_1^*, z_1^*)\Delta V_1 + f(x_2^*, y_2^*, z_2^*)\Delta V_2 + \cdots + f(x_n^*, y_n^*, z_n^*)\Delta V_n = \sum_{i=1}^{n} f(x_i^*, y_i^*, z_i^*)\Delta V_i. \quad (14.47)$$

If this sum approaches a limit as the number of subregions becomes increasingly large and every subregion shrinks to a point, we call the limit the triple integral of $f(x, y, z)$ over the region V and denote it by

$$\iiint_V f(x, y, z)\,dV = \lim_{||\Delta V_i|| \to 0} \sum_{i=1}^{n} f(x_i^*, y_i^*, z_i^*)\Delta V_i. \quad (14.48)$$

FIGURE 14.51

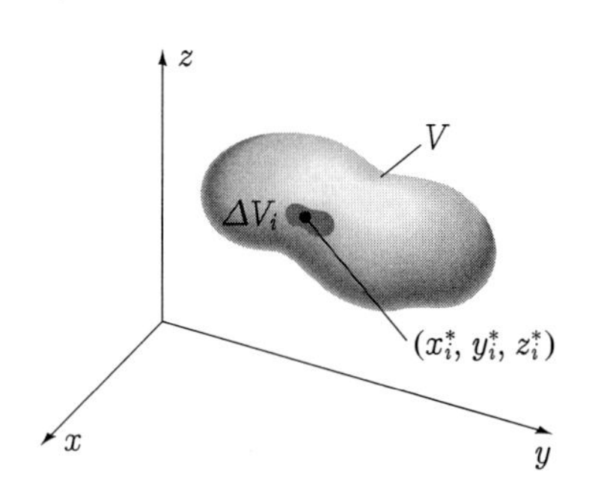

As in the case of double integrals, we require that this limit be independent of the manner of subdivision of V and the choice of star-points in the subregions. This is guaranteed for continuous functions by the following theorem.

Theorem 14.2

Let S be a piecewise-smooth surface that encloses a region V with finite volume. If $f(x, y, z)$ is continuous inside and on S, then the triple integral of $f(x, y, z)$ over V exists.

Properties analogous to those in equations 14.4–14.7 hold for triple integrals, although we will not list the first three here. Corresponding to equation 14.7, the volume of a region V is given by the triple integral

$$\text{Volume of } V = \iiint_V dV. \tag{14.49}$$

We evaluate triple integrals with triple iterated integrals. If we use Cartesian coordinates there are six possible triple iterated integrals of a function $f(x, y, z)$, corresponding to the six permutations of the product of the differentials dx, dy, and dz:

$$dz\, dy\, dx, \quad dz\, dx\, dy, \quad dx\, dz\, dy, \quad dx\, dy\, dz, \quad dy\, dx\, dz, \quad dy\, dz\, dx.$$

The general triple iterated integral of $f(x, y, z)$ with respect to z, y, and x is of the form

$$\int_a^b \int_{g_1(x)}^{g_2(x)} \int_{h_1(x,y)}^{h_2(x,y)} f(x, y, z)\, dz\, dy\, dx. \tag{14.50}$$

Because the first integration with respect to z holds x and y constant, the limits on z may therefore depend on x and y. Similarly, the second integration with respect to y holds x constant, and the limits may be functions of x.

EXAMPLE 14.19

Evaluate the triple iterated integral

$$\int_0^1 \int_0^{x^2} \int_{xy}^{x+y} xyz\, dz\, dy\, dx.$$

SOLUTION

$$\begin{aligned}\int_0^1 \int_0^{x^2} \int_{xy}^{x+y} xyz\, dz\, dy\, dx &= \int_0^1 \int_0^{x^2} \left\{ \frac{xyz^2}{2} \right\}_{xy}^{x+y} dy\, dx \\ &= \frac{1}{2} \int_0^1 \int_0^{x^2} \{xy(x+y)^2 - xy(xy)^2\} dy\, dx \\ &= \frac{1}{2} \int_0^1 \int_0^{x^2} (x^3 y + 2x^2 y^2 + xy^3 - x^3 y^3)\, dy\, dx\end{aligned}$$

$$= \frac{1}{2}\int_0^1 \left\{\frac{x^3y^2}{2} + \frac{2x^2y^3}{3} + \frac{xy^4}{4} - \frac{x^3y^4}{4}\right\}_0^{x^2} dx$$

$$= \frac{1}{24}\int_0^1 (6x^7 + 8x^8 + 3x^9 - 3x^{11})\,dx$$

$$= \frac{1}{24}\left\{\frac{3x^8}{4} + \frac{8x^9}{9} + \frac{3x^{10}}{10} - \frac{x^{12}}{4}\right\}_0^1 = \frac{19}{270}.$$

■

Because of the analogy between double and triple integrals, we accept without proof that triple integrals can be evaluated with triple iterated integrals. We must, however, examine how triple iterated integrals bring about the summations represented by triple integrals, for it is only by thoroughly understanding this process that we can obtain limits for triple iterated integrals.

In Section 14.3 we discussed in considerable detail the evaluation of double integrals by means of double iterated integrals. In particular, we showed that double iterated integrals in Cartesian coordinates represent the subdivision of an area into small rectangles by coordinate lines x = constant and y = constant. The first integration creates a summation over rectangles in a strip, and the second integration adds over all strips. It is fairly straightforward to generalize these ideas to triple integrals.

Consider evaluating the triple integral

$$\iiint_V f(x,y,z)\,dV$$

over the region V in Figure 14.52 bounded above by the surface $z = h(x, y)$, below by the area R in the xy-plane, and on the sides by a cylindrical wall standing on the curve bounding R.

FIGURE 14.52

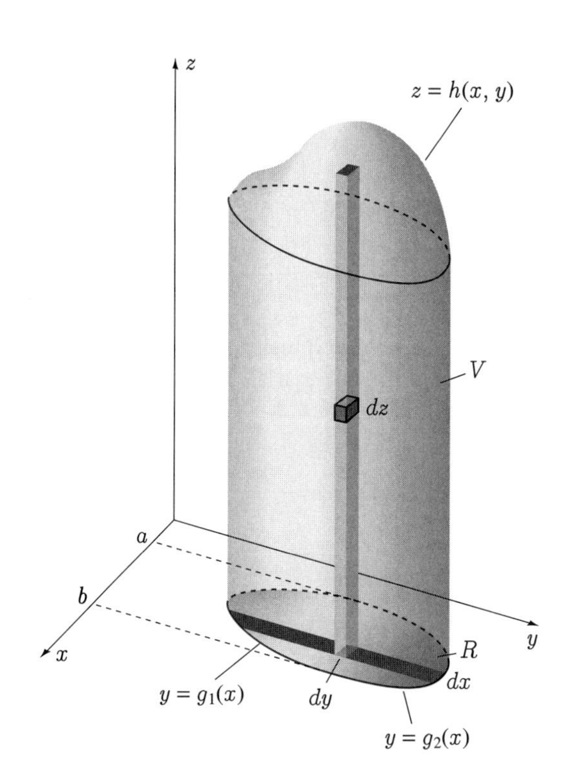

The choice of a triple iterated integral in Cartesian coordinates to evaluate this triple integral implies a subdivision of V into small rectangular parallelepipeds (boxes for short) by means of coordinate planes x = constant, y = constant, and z = constant. The dimensions of a representative box at position (x, y, z) in V are denoted by dx, dy, and dz, with resulting volume $dx\, dy\, dz$. If we decide on a triple iterated integral with respect to z, y, and x, then the first integration on z holds x and y constant. This integration therefore adds the quantities

$$f(x, y, z)\,dz\,dy\,dx$$

over boxes in a vertical column of cross-sectional dimensions dx and dy. Lower and upper limits on z identify where each and every column starts and stops, and must consequently be 0 and $h(x, y)$:

$$\int_0^{h(x,y)} f(x, y, z)\,dz\,dy\,dx.$$

Since this integration produces a function of x and y alone, the remaining integration with respect to y and x is essentially a double iterated integral in the xy-plane. These integrations must account for all columns in V and therefore the area in the xy-plane over which this double iterated integral is performed is the area R upon which all columns in V stand. Since the y-integration adds inside a strip in the y-direction and limits identify where all strips start and stop, they must therefore be $g_1(x)$ and $g_2(x)$. We have now

$$\int_{g_1(x)}^{g_2(x)} \int_0^{h(x,y)} f(x, y, z)\,dz\,dy\,dx.$$

Finally, the x-integration adds over all strips and the limits are $x = a$ and $x = b$:

$$\iiint_V f(x, y, z)\,dV = \int_a^b \int_{g_1(x)}^{g_2(x)} \int_0^{h(x,y)} f(x, y, z)\,dz\,dy\,dx.$$

Suppose now that V is the region bounded above by the surface $z = h_2(x, y)$ and below by $z = h_1(x, y)$ (Figure 14.53). In this case, the limits on the first integration with respect to z are $h_1(x, y)$ and $h_2(x, y)$ since every column starts on the surface $z = h_1(x, y)$ and ends on the surface $z = h_2(x, y)$:

$$\int_{h_1(x,y)}^{h_2(x,y)} f(x, y, z)\,dz\,dy\,dx.$$

For the volume of Figure 14.52 we interpreted the final two integrations as a double iterated integral in the xy-plane over the area R from which all columns emanated. For the present volume we interpret R as the area in the xy-plane onto which all columns project. We obtain then

$$\iiint_V f(x, y, z)\,dV = \int_a^b \int_{g_1(x)}^{g_2(x)} \int_{h_1(x,y)}^{h_2(x,y)} f(x, y, z)\,dz\,dy\,dx.$$

FIGURE 14.53

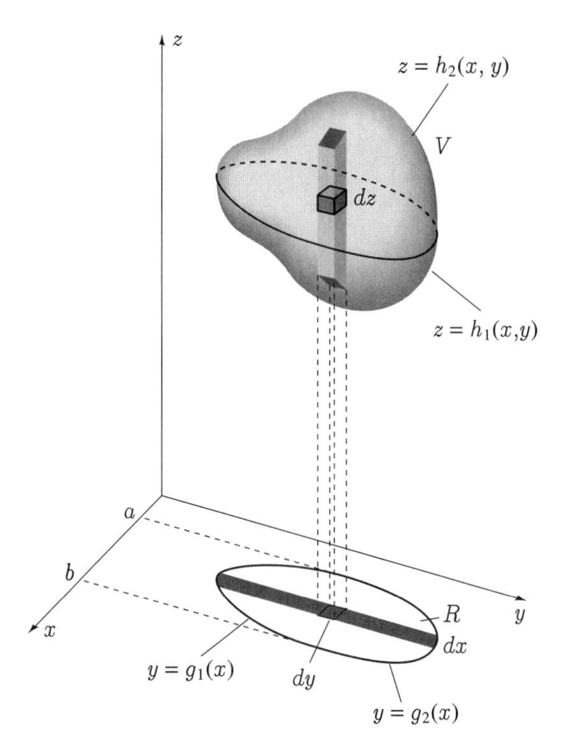

Schematically we have obtained the following interpretation for the limits of triple iterated integrals in Cartesian coordinates:

$$\int_{\text{position of first strip}}^{\text{position of last strip}} \int_{\text{where every strip starts}}^{\text{where every strip stops}} \int_{\text{where every column starts}}^{\text{where every column stops}} f(x,y,z) \left\{ \begin{matrix} dz\,dy\,dx \\ dz\,dx\,dy \\ dx\,dz\,dy \\ dx\,dy\,dz \\ dy\,dx\,dz \\ dy\,dz\,dx \end{matrix} \right\}.$$

EXAMPLE 14.20

Set up the six triple iterated integrals in Cartesian coordinates for the triple integral of a function $f(x,y,z)$ over the region V in the first octant bounded by the surfaces

$$y^2 + z^2 = 1, \quad y = x, \quad z = 0, \quad x = 0.$$

SOLUTION The triple integral of $f(x,y,z)$ over V is given by each of the following triple iterated integrals (see Figure 14.54):

$$\int_0^1 \int_x^1 \int_0^{\sqrt{1-y^2}} f(x,y,z)\,dz\,dy\,dx,$$

$$\int_0^1 \int_0^y \int_0^{\sqrt{1-y^2}} f(x,y,z)\,dz\,dx\,dy,$$

$$\int_0^1 \int_0^{\sqrt{1-z^2}} \int_0^y f(x,y,z)\,dx\,dy\,dz,$$

$$\int_0^1 \int_0^{\sqrt{1-y^2}} \int_0^y f(x,y,z)\,dx\,dz\,dy,$$

$$\int_0^1 \int_0^{\sqrt{1-x^2}} \int_x^{\sqrt{1-z^2}} f(x,y,z)\,dy\,dz\,dx,$$

$$\int_0^1 \int_0^{\sqrt{1-z^2}} \int_x^{\sqrt{1-z^2}} f(x,y,z)\,dy\,dx\,dz.$$

FIGURE 14.54

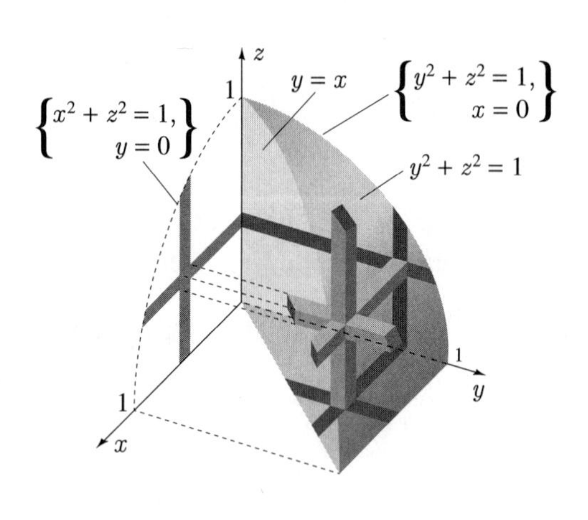

■

EXAMPLE 14.21

Set up triple iterated integrals with respect to z, y, and x for the triple integral of the function $f(x,y,z) = x^2 y \sin z$ over the region V bounded by the surfaces

$$z = \sqrt{y}, \quad y + z = 2, \quad x = 0, \quad z = 0, \quad x = 2.$$

SOLUTION The problem requires triple iterated integrals first with respect to z, then with respect to y and x. Since some columns end on the parabolic cylinder $z = \sqrt{y}$ (Figure 14.55) and others on the plane $y + z = 2$, we require two iterated integrals:

$$\iiint_V x^2 y \sin z \, dV = \int_0^2 \int_0^1 \int_0^{\sqrt{y}} x^2 y \sin z \, dz \, dy \, dx + \int_0^2 \int_1^2 \int_0^{2-y} x^2 y \sin z \, dz \, dy \, dx.$$

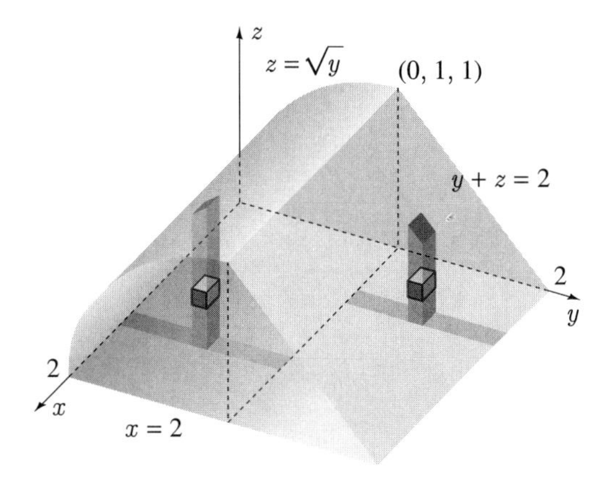

FIGURE 14.55

EXERCISES 14.8

In Exercises 1–12 evaluate the triple integral over the region.

1. $\iiint_V (x^2 z + ye^x) \, dV$ where V is bounded by $x = 0, \; x = 1, \; y = 1, \; y = 2, \; z = 0, \; z = 1$

2. $\iiint_V x \, dV$ where V is bounded by $x = 0, \; y = 0, \; z = 0, \; x + y + z = 4$

3. $\iiint_V \sin(y + z) \, dV$ where V is bounded by $z = 0, \; y = 2x, \; y = 0, \; x = 1, \; z = x + 2y$

4. $\iiint_V xy \, dV$ where V is enclosed by $z = \sqrt{1 - x^2 - y^2}, \; z = 0$

5. $\iiint_V dV$ where V is bounded by $x^2 + y^2 = 1, \; x^2 + z^2 = 1$

6. $\iiint_V (x^2 + 2z) \, dV$ where V is bounded by $z = 0, \; y + z = 4, \; y = x^2$

7. $\iiint_V x^2 y^2 z^2 \, dV$ where V is bounded by $z = 1 + y, \; y + z = 1, \; x = 1, \; x = 0, \; z = 0$

8. $\iiint_V xyz \, dV$ where V is the first octant volume cut out by $z = x^2 + y^2, \; z = \sqrt{x^2 + y^2}$

9. $\iiint_V dV$ where V is bounded by $z = x^2, \; y + z = 4, \; y = 0$

10. $\iiint_V (x + y + z) \, dV$ where V is bounded by $x = 0, \; x = 1, \; z = 0, \; y + z = 2, \; y = z$

11. $\iiint_V xyz \, dV$ where V is bounded by $z = 1, \; z = x^2/4 + y^2/9$

12. $\iiint_V x^2 y\, dV$ where V is the volume in the first octant bounded by $z = 1$, $z = x^2/4 + y^2/9$

13. Set up the six triple iterated integrals in Cartesian coordinates for the triple integral of a function $f(x, y, z)$ over the volume enclosed by the surfaces $y = 1 - x^2$, $z = 0$, and $y = z$.

In Exercises 14–17 set up, but do not evaluate, a triple iterated integral for the triple integral.

14. $\iiint_V (x^2 + y^2 + z^2)\, dV$ where V is bounded by $z = \sqrt{1 - x^2 - y^2}$, $z = x^2$

15. $\iiint_V xz \sin(x + y)\, dV$ where V is bounded by $y^2 = 1 + 4x^2 + 4z^2$, $y = \sqrt{4 + x^2}$

16. $\iiint_V xyz\, dV$ where V is bounded by $z = x^2 + 4y^2$, $2x + 8y + z = 4$

17. $\iiint_V x^2 y^2 z^2\, dV$ where V is bounded by $x = y^2 + z^2$, $x + 1 = (y^2 + z^2)^2$

In Exercises 18–23 evaluate the triple integral over the region.

18. $\iiint_V (y + x^2)\, dV$ where V is bounded by $x + z^2 = 1$, $z = x + 1$, $y = 1$, $y = -1$

19. $\iiint_V (xy + z)\, dV$ where V is bounded by $y + z = 1$, $z = 2y$, $z = y$, $x = 0$, $x = 3$

20. $\iiint_V dV$ where V is bounded by $z = 0$, $x^2 + y^2 = 1$, $x + y + z = 2$

21. $\iiint_V dV$ where V is bounded by $z = x^2 + y^2$, $z = 4 - x^2 - y^2$

22. $\iiint_V (x + y + z)\, dV$ where V is bounded by $2z = y^2 - x^2$, $z = 1 - x^2$

23. $\iiint_V |yz|\, dV$ where V is bounded by $z^2 = 1 + x^2 + y^2$, $z = \sqrt{4 - x^2 - y^2}$

24. Set up, but do not evaluate, triple iterated integrals to evaluate the triple integral of the function $f(x, y, z) = x^2 + y^2 + z^2$ over the volume bounded by the surfaces $x^2 + y^2 = z^2 + 1$, $2z = \sqrt{x^2 + y^2}$, $z = 0$.

SECTION 14.9

Volumes

Because the volume of a region V is represented by triple integral 14.49 and triple integrals are evaluated by means of triple iterated integrals, it follows that volumes can be evaluated with triple iterated integrals. For example, to evaluate the volume of the region in Figure 14.53 using a triple iterated integral in x, y, and z, we subdivide V into boxes of dimensions dx, dy, and dz and therefore of volume

$$dz\, dy\, dx.$$

Integration with respect to z adds these volumes in the z-direction to give the volume of a vertical column (Figure 14.56):

$$\int_{h_1(x,y)}^{h_2(x,y)} dz\, dy\, dx.$$

The limits indicate that all columns start on the surface $z = h_1(x, y)$ and end on the surface $z = h_2(x, y)$. Integration with respect to y now adds the volumes of columns that project onto a strip in the y-direction:

$$\int_{g_1(x)}^{g_2(x)} \int_{h_1(x,y)}^{h_2(x,y)} dz\, dy\, dx,$$

where the limits indicate that all strips start on the curve $y = g_1(x)$ and end on the curve $y = g_2(x)$. Evidently this integration yields the volume of a slab as shown in Figure 14.56. Finally, integration with respect to x adds the volumes of all such slabs in V:

$$\int_a^b \int_{g_1(x)}^{g_2(x)} \int_{h_1(x,y)}^{h_2(x,y)} dz\, dy\, dx,$$

where a and b designate the positions of the first and last strips.

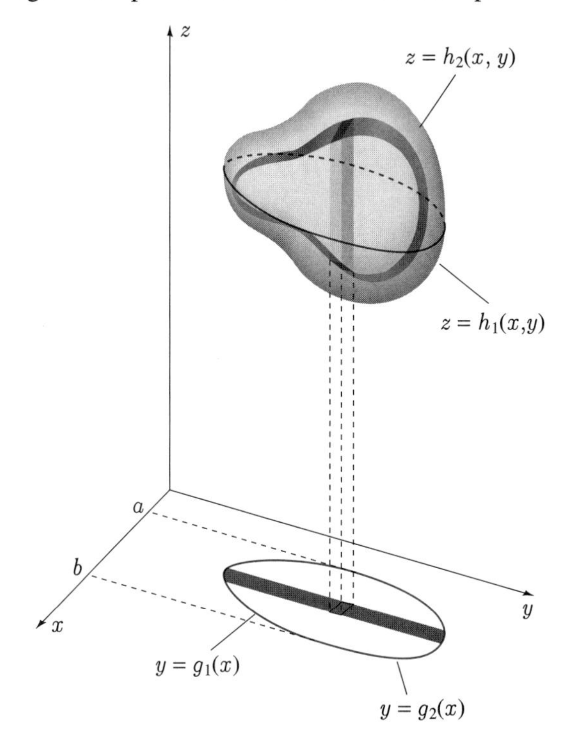

FIGURE 14.56

EXAMPLE 14.22

Find the volume bounded by the planes $z = x + y$, $y = 2x$, $z = 0$, $x = 0$, $y = 2$.

SOLUTION If we use vertical columns (Figure 14.57), we have

$$\text{Volume} = \int_0^1 \int_{2x}^2 \int_0^{x+y} dz\, dy\, dx = \int_0^1 \int_{2x}^2 (x + y)\, dy\, dx = \int_0^1 \left\{ xy + \frac{y^2}{2} \right\}_{2x}^2 dx$$
$$= 2 \int_0^1 (1 + x - 2x^2)\, dx = 2 \left\{ x + \frac{x^2}{2} - \frac{2x^3}{3} \right\}_0^1 = \frac{5}{3}.$$

FIGURE 14.57

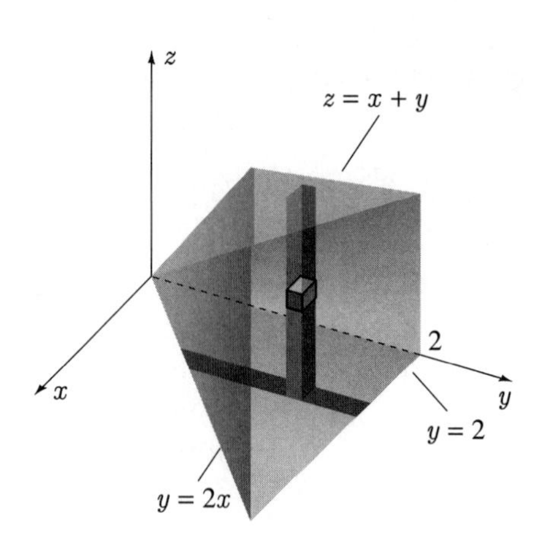

EXAMPLE 14.23

Find the volume in the first octant cut from the cyclinder $x^2 + z^2 = 4$ by the plane $y + z = 6$.

SOLUTION With columns in the y-direction (Figure 14.58), we have

$$\text{Volume} = \int_0^2 \int_0^{\sqrt{4-x^2}} \int_0^{6-z} dy\, dz\, dx = \int_0^2 \int_0^{\sqrt{4-x^2}} (6 - z)\, dz\, dx$$

$$= \int_0^2 \left\{ 6z - \frac{z^2}{2} \right\}_0^{\sqrt{4-x^2}} dx$$

$$= \int_0^2 \left\{ 6\sqrt{4 - x^2} - \frac{1}{2}(4 - x^2) \right\} dx.$$

In the first term we set $x = 2 \sin\theta$, from which we get $dx = 2 \cos\theta\, d\theta$, and

$$\text{Volume} = 6 \int_0^{\pi/2} (2 \cos\theta) 2 \cos\theta\, d\theta + \left\{ -2x + \frac{x^3}{6} \right\}_0^2$$

$$= 12 \int_0^{\pi/2} (1 + \cos 2\theta)\, d\theta - \frac{8}{3}$$

$$= 12 \left\{ \theta + \frac{1}{2} \sin 2\theta \right\}_0^{\pi/2} - \frac{8}{3} = 6\pi - \frac{8}{3}.$$

FIGURE 14.58

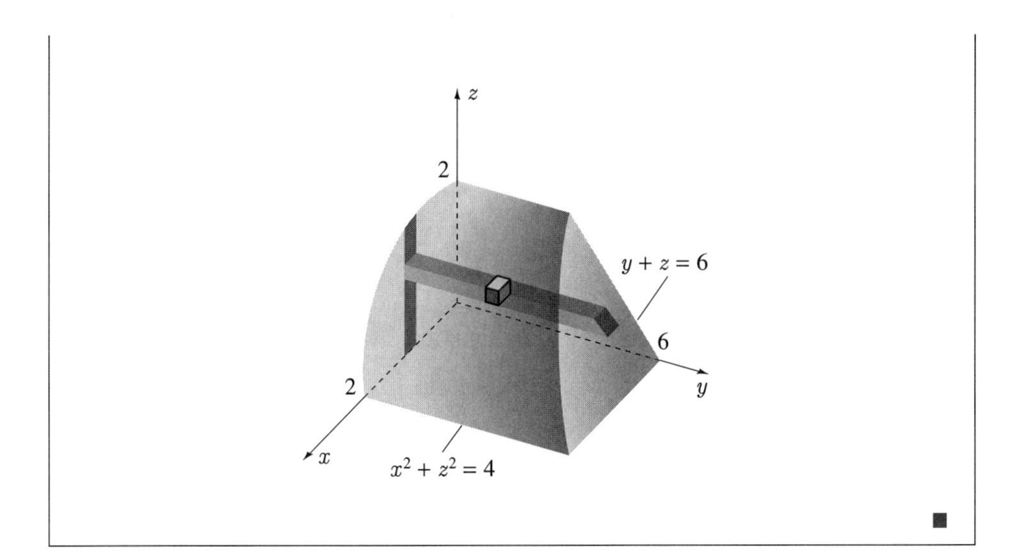

Had we used iterated integrals with respect to z, y, and x in this example, we would have had the two integrals

$$\text{Volume} = \int_0^2 \int_0^{6-\sqrt{4-x^2}} \int_0^{\sqrt{4-x^2}} dz\, dy\, dx + \int_0^2 \int_{6-\sqrt{4-x^2}}^{6} \int_0^{6-y} dz\, dy\, dx.$$

EXERCISES 14.9

In Exercises 1–17 find the volume of the region bounded by the surfaces.

1. $y = x^2,\ y = 1,\ z = 0,\ z = 4$
2. $x = z^2,\ z = x^2,\ y = 0,\ y = 2$
3. $x = 3z,\ z = 3x,\ y = 1,\ y = 0,\ x = 2$
4. $x + y + z = 6,\ y = 4 - x^2,\ z = 0,\ y = 0$
5. $z = x^2 + y^2,\ y = x^2,\ y = 4,\ z = 0$
6. $x + y + z = 4,\ y = 3z,\ x = 0,\ y = 0$
7. $x^2 + y^2 = 4,\ y^2 + z^2 = 4$
8. $y = x^2 - 1,\ y = 1 - x^2,\ x + z = 1,\ z = 0$
9. $z = 16 - x^2 - 4y^2,\ x + y = 1,\ z = 16,\ x = 0,\ y = 0$ (in the first octant)
10. $z = x^2 + y^2,\ x = 1,\ z = 0,\ x = y,\ x = 2y$
11. $z = 1 - x^2 - y^2,\ z = 0$
12. $x - z = 0,\ x + z = 3,\ y + z = 1,\ z = y + 1,\ z = 0$
13. $x + y + z = 2,\ x^2 + y^2 = 1,\ z = 0$
14. $y + z = 1,\ z = 2y,\ z = y,\ x = 0,\ x + y + z = 4$
15. $x^2 + 4y^2 = z,\ x^2 + 4y^2 = 12 - 2z$
16. $y = 1 - z^2,\ y = z^2 - 1,\ x = 1 - z^2,\ x = z^2 - 1$
17. $x + 3y + 2z = 6,\ z = 0,\ y = x,\ y = 2x$
18. Find the volume in the first octant bounded by the plane $2x + y + z = 2$ and inside the cylinder $y^2 + z^2 = 1$.
19. A pyramid has a square base with side length b and has height h at its centre.
    (a) Find its volume by taking cross-sections parallel to the base (see Section 7.8).
    (b) Find its volume using triple integrals.

The average value of a function $f(x, y, z)$ over a region with volume V is defined as

$$\overline{f} = \frac{1}{V} \iiint_V f(x, y, z)\, dV.$$

In Exercises 20–22 find the average value of the function over the region.

20. $f(x, y, z) = xy$ over the region bounded by the surfaces $x = 0,\ y = 0,\ z = 0,\ x + y + z = 1$
21. $f(x, y, z) = x + y + z$ over the region in the first octant bounded by the surfaces $z = 9 - x^2 - y^2,\ z = 0$ and for which $0 \le x \le 1,\ 0 \le y \le 1$

22. $f(x, y, z) = x^2 + y^2 + z^2$ over the region bounded by the surfaces $x = 0$, $x = 1$, $y + z = 2$, $y = 2$, $z = 2$

23. Find the volume bounded by the surfaces $z = x^2 - y^2$, and $z = 4 - 2(x^2 + y^2)$.

24. Verify that the surfaces $z = x^2 - y^2$ and $z = 4 - x^2 - y^2$ do not bound a finite volume.

25. Find the volume bounded by the surfaces $x + z = 2$, $z = 0$, $4y^2 = x(2 - z)$.

26. Find the volume bounded by the surfaces $z = (x - 1)^2 + y^2$, $2x + z = 2$.

27. Find the volume inside the ellipsoid $x^2/a^2 + y^2/b^2 + z^2/c^2 = 1$.

28. The bottom and sides of a boat are defined by the surface equation $x = 10(1 - y^2 - z^2)$, $0 \leq x \leq 10$, where all dimensions are in metres.
    (a) Find the volume of water displaced by the boat when the water level on the side of the boat is d metres below the top of the boat.
    (b) Archimedes' principle states that the buoyant force on an object when immersed or partially immersed in a fluid is equal to the weight of the fluid displaced by the object. Find the maximum weight of the boat and contents just before sinking.

29. Find the volume inside all three surfaces $x^2 + y^2 = a^2$, $x^2 + z^2 = a^2$, $y^2 + z^2 = a^2$.

SECTION 14.10

Centres of Mass and Moments of Inertia

In this section we discuss centres of mass and moments of inertia for three-dimensional objects of density $\rho(x, y, z)$ (mass per unit volume). If we divide the object occupying region V in Figure 14.59 into small volumes dV, then the amount of mass in dV is $\rho\, dV$. The triple integral

$$M = \iiint_V \rho\, dV \qquad (14.51)$$

adds the masses of all such volumes (of ever-decreasing size) to produce the total mass M of the object.

FIGURE 14.59

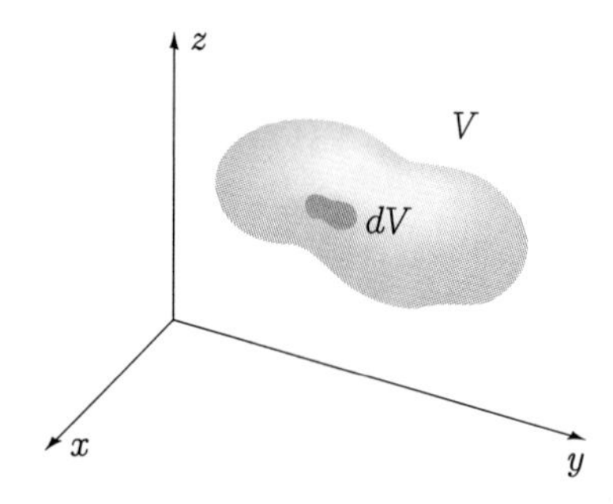

Corresponding to equations 14.29 and 14.30 for first moments of planar masses about the coordinate axes are the following formulas for first moments of the object about the coordinate planes:

$$\text{First moment of object about } yz\text{-plane} = \iiint_V x\rho\, dV, \qquad (14.52)$$

$$\text{First moment of object about } xz\text{-plane} = \iiint_V y\rho\, dV, \qquad (14.53)$$

$$\text{First moment of object about } xy\text{-plane} = \iiint_V z\rho\, dV. \qquad (14.54)$$

The centre of mass of the object is defined as that point $(\overline{x}, \overline{y}, \overline{z})$ at which a particle of mass M would have the same first moments about the coordinate planes as the object itself. Since the first moments of M at $(\overline{x},\overline{y},\overline{z})$ about the coordinate planes are $M\overline{x}, M\overline{y}$, and $M\overline{z}$, it follows that we can use the equations

$$M\overline{x} = \iiint_V x\rho\, dV, \tag{14.55}$$

$$M\overline{y} = \iiint_V y\rho\, dV, \tag{14.56}$$

$$M\overline{z} = \iiint_V z\rho\, dV, \tag{14.57}$$

to solve for $(\overline{x},\overline{y},\overline{z})$ once M and the integrals on the right have been evaluated.

If we use a triple iterated integral with respect to z, y, and x to evaluate 14.57, say, for the object in Figure 14.56, then

$$\iiint_V z\rho\, dV = \int_a^b \int_{g_1(x)}^{g_2(x)} \int_{h_1(x,y)}^{h_2(x,y)} z\rho\, dz\, dy\, dx.$$

The quantity $z\rho\, dz\, dy\, dx$ is the first moment about the xy-plane of the mass in an elemental box of dimensions dx, dy, and dz. The z-integration then adds these moments over boxes in the z-direction to give the first moment about the xy-plane of the mass in a vertical column:

$$\int_{h_1(x,y)}^{h_2(x,y)} z\rho\, dz\, dy\, dx.$$

The y-integration then adds the first moments of columns that project onto a strip in the y-direction:

$$\int_{g_1(x)}^{g_2(x)} \int_{h_1(x,y)}^{h_2(x,y)} z\rho\, dz\, dy\, dx.$$

This quantity therefore represents the first moment about the xy-plane of the slab in Figure 14.56. Finally, the x-integration adds first moments of all such slabs to give the total first moment of V about the xy-plane:

$$\int_a^b \int_{g_1(x)}^{g_2(x)} \int_{h_1(x,y)}^{h_2(x,y)} z\rho\, dz\, dy\, dx.$$

EXAMPLE 14.24

Find the centre of mass of an object of constant density ρ if it is bounded by the surfaces $z = 1 - y^2$, $x = 0$, $z = 0$, $x = 2$.

SOLUTION From the symmetry of the object (Figure 14.60), we see that $\overline{x} = 1$ and $\overline{y} = 0$. Now

$$M = 2\int_0^2 \int_0^1 \int_0^{1-y^2} \rho\, dz\, dy\, dx = 2\rho \int_0^2 \int_0^1 (1 - y^2)\, dy\, dx$$

$$= 2\rho \int_0^2 \left\{ y - \frac{y^3}{3} \right\}_0^1 dx = \frac{4\rho}{3} \int_0^2 dx = \frac{8\rho}{3};$$

and

$$M\bar{z} = 2\int_0^2\int_0^1\int_0^{1-y^2} z\rho\,dz\,dy\,dx = 2\rho\int_0^2\int_0^1\left\{\frac{z^2}{2}\right\}_0^{1-y^2} dy\,dx$$
$$= \rho\int_0^2\int_0^1 (1-2y^2+y^4)\,dy\,dx$$
$$= \rho\int_0^2\left\{y-\frac{2y^3}{3}+\frac{y^5}{5}\right\}_0^1 dx = \frac{8\rho}{15}\int_0^2 dx = \frac{16\rho}{15}.$$

Thus,

$$\bar{z} = \frac{16\rho}{15}\cdot\frac{3}{8\rho} = \frac{2}{5}.$$

FIGURE 14.60

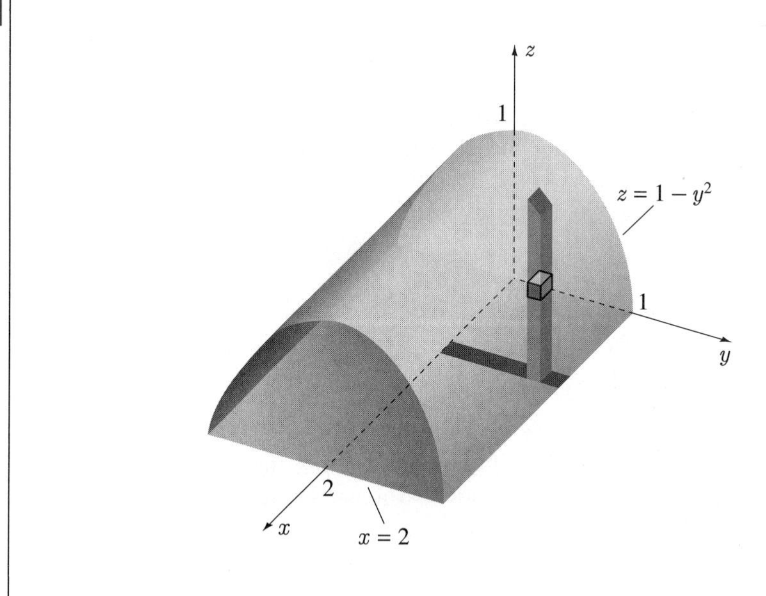

In view of our discussion on first moments in this section and on moments of inertia of thin plates in Section 14.5, it should not be necessary to give a full treatment of moments of inertia of three-dimensional objects. Instead, we simply note that the distances from a point (x, y, z) to the x-, y-, and z-axes are, respectively, $\sqrt{y^2+z^2}$, $\sqrt{x^2+z^2}$, and $\sqrt{x^2+y^2}$. If follows, then, that if an object of density $\rho(x, y, z)$ occupies a region V of space:

$$\text{Moment of inertia of object about } x\text{-axis} = \iiint_V (y^2+z^2)\rho\,dV, \quad (14.58)$$

$$\text{Moment of inertia of object about } y\text{-axis} = \iiint_V (x^2+z^2)\rho\,dV, \quad (14.59)$$

$$\text{Moment of inertia of object about } z\text{-axis} = \iiint_V (x^2+y^2)\rho\,dV. \quad (14.60)$$

EXAMPLE 14.25

Find the moment of inertia of a right-circular cylinder of constant density ρ about its axis.

SOLUTION Let the length and radius of the cylinder be h and r and choose the coordinate system in Figure 14.61. The required moment of inertia about the z-axis is four times the moment of inertia of that part of the cylinder in the first octant. Hence,

$$I = 4\int_0^r \int_0^{\sqrt{r^2-x^2}} \int_0^h (x^2 + y^2)\rho \, dz \, dy \, dx = 4\rho h \int_0^r \int_0^{\sqrt{r^2-x^2}} (x^2 + y^2)\, dy \, dx$$

$$= 4\rho h \int_0^r \left\{ x^2 y + \frac{y^3}{3} \right\}_0^{\sqrt{r^2-x^2}} dx = 4\rho h \int_0^r \left\{ x^2\sqrt{r^2 - x^2} + \frac{1}{3}(r^2 - x^2)^{3/2} \right\} dx$$

$$= \frac{4\rho h}{3} \int_0^r \{ r^2\sqrt{r^2 - x^2} + 2x^2\sqrt{r^2 - x^2} \} dx.$$

To evaluate this definite integral we set $x = r\sin\theta$, which implies that $dx = r\cos\theta\, d\theta$, and

$$I = \frac{4\rho h}{3} \int_0^{\pi/2} \{ r^2 r\cos\theta + 2r^2\sin^2\theta r\cos\theta \} r\cos\theta \, d\theta$$

$$= \frac{4\rho h r^4}{3} \int_0^{\pi/2} \{ \cos^2\theta + 2\sin^2\theta\cos^2\theta \} d\theta$$

$$= \frac{4\rho h r^4}{3} \int_0^{\pi/2} \left\{ \frac{1 + \cos 2\theta}{2} + \frac{1 - \cos 4\theta}{4} \right\} d\theta$$

$$= \frac{4\rho h r^4}{3} \left\{ \frac{3\theta}{4} + \frac{\sin 2\theta}{4} - \frac{\sin 4\theta}{16} \right\}_0^{\pi/2} = \frac{\rho h r^4 \pi}{2}.$$

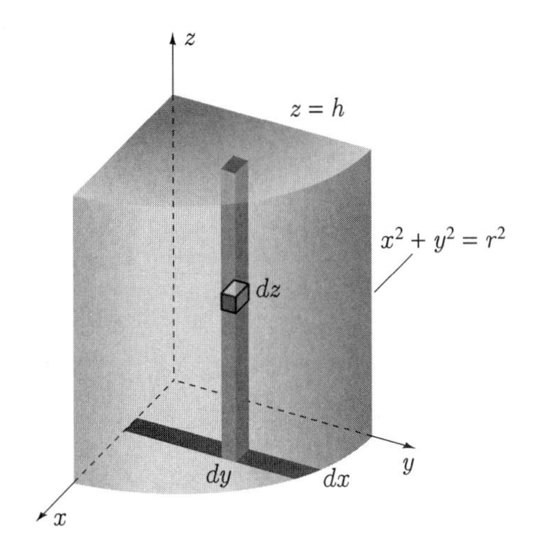

FIGURE 14.61

■

EXERCISES 14.10

In Exercises 1–5 the surfaces bound a solid object of constant density. Find its centre of mass.

1. $z = x^2 + y^2$, $z = 0$, $x = 0$, $y = 0$, $x = 1$, $y = 1$

2. $x + y + z = 1$, $x = 0$, $y = 0$, $z = 0$

3. $z = x^2$, $y + z = 4$, $y = 0$

4. $y = 4 - x^2$, $z = 0$, $y = z$

5. $x + y + z = 4$, $y = 3z$, $x = 0$, $y = 0$

In Exercises 6–10 the surfaces bound a solid object of constant density ρ. Find its moment of inertia about the line.

6. $x = 0$, $y = 0$, $z = 0$, $x = 1$, $y = 1$, $z = 1$ about the x-axis

7. $z = 2x$, $z = 0$, $y = 0$, $y = 2$, $x = 3$ about the y-axis

8. $x + y + z = 2$, $y = 0$, $x = 0$, $0 \leq z \leq \sqrt{1 - y}$ about the x-axis

9. $z = xy$, $x^2 + y^2 = 1$, $z = 0$ (first octant) about the z-axis

10. $y + z = 2$, $x + z = 2$, $x = 0$, $y = 0$, $z = 0$ about the z-axis

11. Find the first moment about the xy-plane of a solid of constant density ρ if it is bounded by the surfaces $x = z$, $x + z = 0$, $z = 2$, $y = 0$, $y = 2$.

In Exercises 12–14 the surfaces bound a solid object of constant density. Find its centre of mass.

12. $y = x^3$, $x = y^2$, $z = 1 + x^2 + y^2$, $z = -x^2 - y^2$

13. $z = x^2$, $x + z = 2$, $z = y$, $y = 0$

14. $y + z = 0$, $y - z = 0$, $x + z = 0$, $x - z = 0$, $z = 2$

In Exercises 15–17 the surfaces bound a solid object of constant density ρ. Find its moment of inertia about the line.

15. $z = 4 - x^2$, $x + z + 2 = 0$, $y = 0$, $y = 2$ about the y-axis

16. $x^2 + z^2 = a^2$, $x^2 + y^2 = a^2$ about the x-axis

17. $x + y - z = 0$, $x = 3y$, $3y = 2x$, $x = 3$, $z = 0$ about the z-axis

18. Find the first moment about the plane $x + y + z = 1$ of a solid object of constant density ρ if it is bounded by the surfaces $x + 2y + 4z = 12$, $x = 0$, $y = 0$, $z = 0$.

19. Prove the parallel axis theorem for solid objects: The moment of inertia of a uniform solid about a line is equal to the moment of inertia about a parallel line through the centre of mass of the solid plus the mass multiplied by the square of the distance between the lines.

20. Find the centre of mass of a uniform solid in the first octant bounded by the ellipsoid $x^2/a^2 + y^2/b^2 + z^2/c^2 = 1$.

21. Let $\hat{\mathbf{v}} = (v_x, v_y, v_z)$ be a unit vector with its tail at the origin. Show that the moment of inertia I of any solid object occupying volume V about the line containing $\hat{\mathbf{v}}$ can be expressed in the form

$$I = v_x^2 I_x + v_y^2 I_y + v_z^2 I_z - 2v_x v_y I_{xy} - 2v_y v_z I_{yz} - 2v_z v_x I_{xz}$$

where I_x, I_y, and I_z are moments of inertia about the x-, y-, and z-axes, and

$$I_{xy} = \iiint_V xy\rho\, dV, \quad I_{yz} = \iiint_V yz\rho\, dV,$$

$$I_{xz} = \iiint_V xz\rho\, dV.$$

22. Find the moment of inertia of a uniform solid sphere of radius R about any tangent line.

Triple Iterated Integrals in Cylindrical Coordinates

In Section 14.7 we saw that polar coordinates are sometimes more convenient than Cartesian coordinates in evaluating double integrals. It should come as no surprise, then, that other coordinate systems can simplify the evaluation of triple integrals. Two of the most common are cylindrical and spherical coordinates.

Cylindrical coordinates are useful in problems involving an axis of symmetry. They are based on a Cartesian coordinate along the axis of symmetry and polar coordinates in a plane perpendicular to the axis of symmetry. If the z-axis is the axis of symmetry and polar coordinates are defined in the xy-plane with the origin as pole and the positive x-axis as polar axis, then cylindrical coordinates and Cartesian coordinates are related by the equations

$$x = r\cos\theta, \quad y = r\sin\theta, \quad z = z \tag{14.61}$$

(see Figure 14.62). Recall that r can be expressed in terms of x and y by

$$r = \sqrt{x^2 + y^2}, \tag{14.62a}$$

and θ is defined implicitly by the equations

$$\cos\theta = \frac{x}{\sqrt{x^2 + y^2}}, \quad \sin\theta = \frac{y}{\sqrt{x^2 + y^2}}. \tag{14.62b}$$

FIGURE 14.62

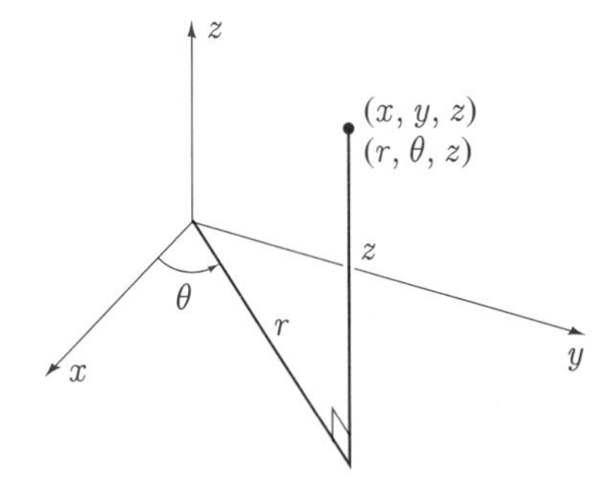

To use cylindrical coordinates in the evaluation of triple integrals, we must express equations of surfaces in terms of these coordinates. But this is very simple, for if $F(x, y, z) = 0$ is the equation of a surface in Cartesian coordinates, then to express this equation in cylindrical coordinates we substitute from equations 14.61: $F(r\cos\theta, r\sin\theta, z) = 0$. For example, the right-circular cylinder $x^2 + y^2 = 9$, which has the z-axis as its axis of symmetry, has the very simple equation $r = 3$ in cylindrical coordinates. The right-circular cone $z = \sqrt{x^2 + y^2}$ also has the z-axis as its axis of symmetry, and its equation in cylindrical coordinates takes the simple form $z = r$.

Suppose that we are to evaluate the triple integral of a continuous function $f(x, y, z)$,

$$\iiint_V f(x, y, z)\, dV,$$

over some region V of space. The choice of a triple iterated integral in cylindrical coordinates implies a subdivision of V into small volumes dV by means of coordinate surfaces r = constant, θ = constant, and z = constant (Figure 14.63). Surfaces r = constant are right-circular cylinders coaxial with the z-axis; surfaces θ = constant are planes containing the z-axis and therefore perpendicular to the xy-plane; and surfaces z = constant are planes parallel to the xy-plane.

FIGURE 14.63

FIGURE 14.64

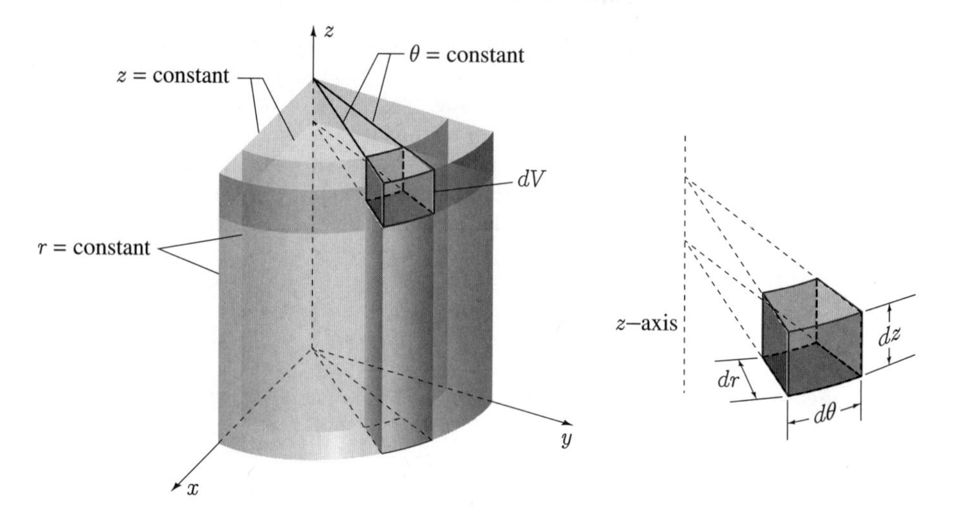

If we denote small variations in r, θ, and z for the element dV by $dr, d\theta$, and dz (Figure 14.64), then the volume of the element is approximately $(r\,dr\,d\theta)\,dz$, where $r\,dr\,d\theta$ is the polar cross-sectional area parallel to the xy-plane. Hence in cylindrical coordinates we set

$$dV = r\,dz\,dr\,d\theta. \tag{14.63}$$

The integrand $f(x, y, z)$ is expressed in cylindrical coordinates as $f(r\cos\theta, r\sin\theta, z)$. It remains only to affix appropriate limits to the triple iterated integral, and these will, of course, depend on which of the six possible iterated integrals in cylindrical coordinates we choose. The most commonly used triple iterated integral is with respect to z, r, and θ, and in this case the z-integration adds the quantities

$$f(r\cos\theta, r\sin\theta, z)\,r\,dz\,dr\,d\theta$$

in a vertical column, where r and θ are constant. The limits therefore identify the surfaces on which each and every column starts and stops, and generally depend on r and θ:

$$\int_{h_1(r,\theta)}^{h_2(r,\theta)} f(r\cos\theta, r\sin\theta, z)\,r\,dz\,dr\,d\theta.$$

The remaining integrations with respect to r and θ perform additions over the area in the xy-plane onto which all vertical columns project. Since r and θ are simply polar coordinates, the r-integration adds over small areas in a wedge and the θ-integration adds over all wedges. The triple iterated integral with respect to z, r, and θ therefore has the form

$$\int_a^b \int_{g_1(\theta)}^{g_2(\theta)} \int_{h_1(r,\theta)}^{h_2(r,\theta)} f(r\cos\theta, r\sin\theta, z)\,r\,dz\,dr\,d\theta.$$

We comment on the geometric aspects of these additions more fully in the following examples.

EXAMPLE 14.26

Find the volume inside both the sphere $x^2+y^2+z^2=2$ and the cylinder $x^2+y^2=1$.

SOLUTION The required volume is eight times the first octant volume shown in Figure 14.65. If we use cylindrical coordinates, the volume of an elemental piece is $r\,dz\,dr\,d\theta$. A z-integration adds these pieces to give the volume in a vertical column:

$$\int_0^{\sqrt{2-r^2}} r\,dz\,dr\,d\theta,$$

where the limits indicate that for the volume in the first octant all columns start on the xy-plane (where $z=0$) and end on the sphere (where $r^2+z^2=2$). An r-integration now adds the volumes of all columns that stand on a wedge:

$$\int_0^1\int_0^{\sqrt{2-r^2}} r\,dz\,dr\,d\theta,$$

where the limits indicate that all wedges start at the origin (where $r=0$) and end on the curve $x^2+y^2=1$ (or $r=1$) in the xy-plane. This integration therefore yields the volume of a slice (Figure 14.65). Finally, the θ-integration adds the volumes of all such slices

$$\int_0^{\pi/2}\int_0^1\int_0^{\sqrt{2-r^2}} r\,dz\,dr\,d\theta,$$

where the limits 0 and $\pi/2$ identify the positions of the first and last wedges, respectively, in the first quadrant. We obtain the required volume, then, as

$$\begin{aligned}8\int_0^{\pi/2}\int_0^1\int_0^{\sqrt{2-r^2}} r\,dz\,dr\,d\theta &= 8\int_0^{\pi/2}\int_0^1 r\sqrt{2-r^2}\,dr\,d\theta\\ &= 8\int_0^{\pi/2}\left\{-\frac{1}{3}(2-r^2)^{3/2}\right\}_0^1 d\theta\\ &= \frac{8}{3}(2\sqrt{2}-1)\int_0^{\pi/2} d\theta = \frac{4\pi}{3}(2\sqrt{2}-1).\end{aligned}$$

FIGURE 14.65

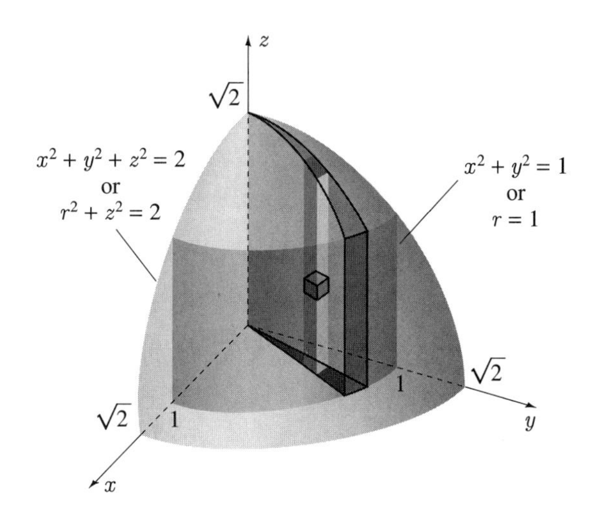

■

Had we used a triple iterated integral with respect to z, y, and x in Example 14.26, we would have obtained

$$\text{Volume} = 8\int_0^1 \int_0^{\sqrt{1-x^2}} \int_0^{\sqrt{2-x^2-y^2}} dz\, dy\, dx.$$

To appreciate the value of cylindrical coordinates, try to evaluate this triple iterated integral.

As further evidence of the value of cylindrical coordinates, repeat Examples 14.23 and 14.25 using cylindrical coordinates.

EXAMPLE 14.27

Evaluate the triple iterated integral

$$I = \int_0^1 \int_0^{\sqrt{1-y^2}} \int_0^{x^2+y^2} y^2\, dz\, dx\, dy.$$

SOLUTION The first two integrations with respect to z and x are quite straightforward, giving

$$I = \frac{1}{3}\int_0^1 y^2(1+2y^2)\sqrt{1-y^2}\, dy.$$

To evaluate this definite integral we could make a trigonometric substitution $y = \sin\theta$. (Try it.) Alternatively, we could consider using cylindrical coordinates on the original triple iterated integral. We first obtain the region over which integration is being performed. The limits on z indicate that all columns begin on the xy-plane and end on the paraboloid $z = x^2 + y^2$. The limits on x indicate that in the xy-plane all strips begin on the y-axis ($x = 0$) and end on the curve $x = \sqrt{1-y^2}$ (i.e., the circle $x^2 + y^2 = 1$). The first and last strips are at $y = 0$ and $y = 1$, respectively. These facts determine the region of integration as that region in the first octant bounded by the paraboloid $z = x^2 + y^2$ and the right-circular cylinder $x^2 + y^2 = 1$ (Figure 14.66a). If we use a triple iterated integral with respect to z, r and θ (Figure 14.66b), then

$$I = \int_0^{\pi/2} \int_0^1 \int_0^{r^2} r^2 \sin^2\theta\, r\, dz\, dr\, d\theta = \int_0^{\pi/2} \int_0^1 r^5 \sin^2\theta\, dr\, d\theta = \int_0^{\pi/2} \frac{1}{6}\sin^2\theta\, d\theta$$

$$= \frac{1}{6}\int_0^{\pi/2} \left\{\frac{1-\cos 2\theta}{2}\right\} d\theta = \frac{1}{12}\left\{\theta - \frac{1}{2}\sin 2\theta\right\}_0^{\pi/2} = \frac{\pi}{24}.$$

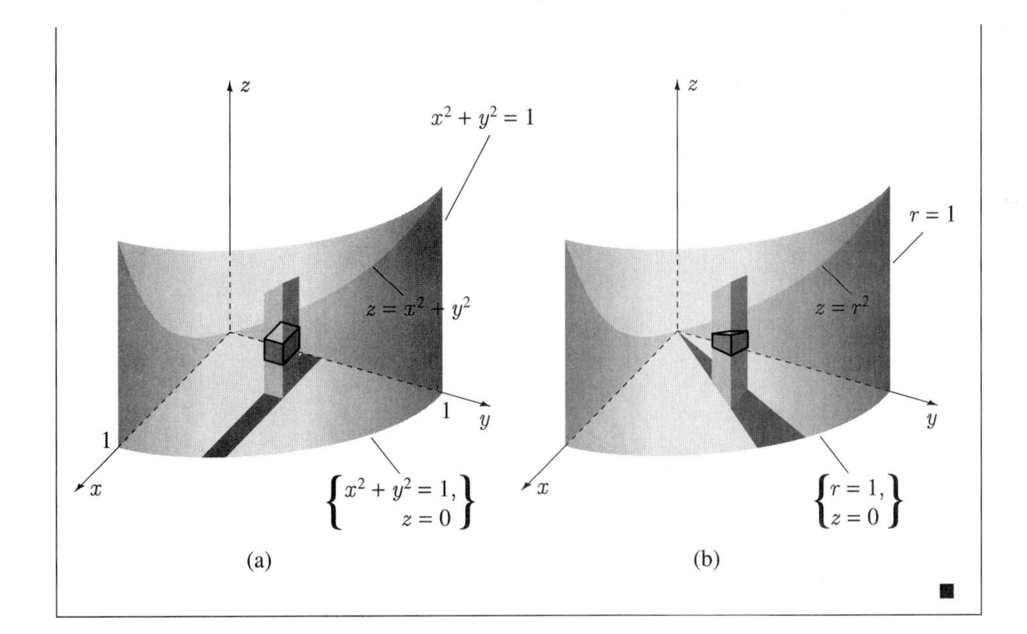

FIGURE 14.66

EXERCISES 14.11

In Exercises 1–10 find the equation for the surface in cylindrical coordinates. Sketch each surface and indicate whether it is symmetric about the z-axis.

1. $x^2 + y^2 + z^2 = 4$
2. $x^2 + y^2 = 1$
3. $y^2 + z^2 = 6$
4. $x + y = 5$
5. $z = 2\sqrt{x^2 + y^2}$
6. $z = x^2$
7. $x^2 + 4y^2 = 4$
8. $4z = x^2 + y^2$
9. $y = x$
10. $x^2 + y^2 = 1 + z^2$

In Exercises 11–15 find the volume bounded by the surfaces.

11. $z = \sqrt{x^2 + y^2}$, $x^2 + y^2 = 4$, $z = 0$
12. $z = \sqrt{2 - x^2 - y^2}$, $z = x^2 + y^2$
13. $z = xy$, $x^2 + y^2 = 1$, $z = 0$
14. $z = x^2 + y^2$, $z = 4 - x^2 - y^2$
15. $x + y + z = 2$, $x^2 + y^2 = 1$, $z = 0$
16. Find the volume inside the sphere $x^2 + y^2 + z^2 = 4$ but outside the cylinder $x^2 + y^2 = 1$.
17. Find the centre of mass of a uniform hemispherical solid.
18. Set up the six triple iterated integrals in cylindrical coordinates for the triple integral of a function $f(x, y, z)$ over the volume bounded by the surfaces $z = 1 + x^2 + y^2$, $x^2 + y^2 = 9$, $z = 0$.
19. Find the moment of inertia of a uniform right-circular cylinder of radius R and height h **(a)** about its axis; **(b)** about a line through the centre of its base and perpendicular to its axis.
20. Find the moment of inertia of a uniform sphere of radius R about any line through its centre.

In Exercises 21–24 evaluate the triple iterated integral.

21. $$\int_0^3 \int_0^{\sqrt{9-x^2}} \int_0^{\sqrt{x^2+y^2}} dz\, dy\, dx$$
22. $$\int_0^9 \int_0^{\sqrt{81-y^2}} \int_0^{\sqrt{81-x^2-y^2}} \frac{1}{\sqrt{x^2 + y^2}}\, dz\, dx\, dy$$
23. $$\int_0^4 \int_0^{\sqrt{4y-y^2}} \int_0^{y+x^2} dz\, dx\, dy$$
24. $$\int_0^{\sqrt{3}/2} \int_{5-\sqrt{21-y^2}}^{\sqrt{1-y^2}} \int_0^{x^2+y^2} y\, dz\, dx\, dy$$

25. Find the centre of mass for the uniform solid bounded by the surfaces $x^2 + y^2 = 2x$, $z = \sqrt{x^2 + y^2}$, $z = 0$.

26. Find the moment of inertia of a uniform right-circular cone of radius R and height h about its axis.

27. A casting is in the form of a sphere of radius b with two cylindrical holes of radius $a < b$ such that the axes of the holes pass through the centre of the sphere and intersect at right angles. What volume of metal is required for the casting?

In Exercises 28–38 find the volume described.

28. Bounded by $x^2 + y^2 - z^2 = 1$, $4z^2 = x^2 + y^2$

29. Bounded by $z = x^2 + y^2$, $z = 0$, $(x^2 + y^2)^2 = x^2 - y^2$

30. Bounded by $x^2 + y^2 + z^2 = 4$, $x^2 + y^2 + z^2 = 16$, $z = \sqrt{x^2 + y^2}$ (smaller piece)

31. Bounded by $x^2 + y^2 + z^2 = 1$, $y = x$, $x = 2y$, $z = 0$ (in the first octant)

32. Inside both $2x^2 + 2y^2 + z^2 = 8$ and $x^2 + y^2 = 1$

33. Bounded by $z^2 = (x^2 + y^2)^2$, $x^2 + y^2 = 2y$

34. Inside $x^2 + y^2 + z^2 = a^2$ but outside, $x^2 + y^2 = ay$

35. Bounded by $z = 0$, $x^2 + y^2 = 1$, $z = e^{-x^2-y^2}$

36. Inside $x^2 + y^2 + z^2 = 9$ but outside $x^2 + y^2 = 1 + z^2$

37. Cut off from $z = x^2 + y^2$ by $z = x + y$

38. Inside $x^2 + y^2 + z^2 = 4$ and below $3z = x^2 + y^2$

39. Evaluate the triple integral of $\sqrt{x^2 + y^2 + z^2}$ over the volume bounded by $z = 3$ and $z = \sqrt{x^2 + y^2}$.

40. Evaluate the triple integral of $f(y, z) = |yz|$ over the volume bounded by $z^2 = 1 + x^2 + y^2$ and $z = \sqrt{4 - x^2 - y^2}$.

41. A tumbler in the form of a right-circular cylinder of radius R and height h is full of water. As the axis of the tumbler is tilted from the vertical, water pours over the side. Find the volume of water remaining in the tumbler as a function of the angle between the vertical and the axis of the tumbler.

42. Use cylindrical coordinates to find the volume of the torus $(\sqrt{x^2 + y^2} - a)^2 + z^2 = b^2$, $b < a$.

SECTION 14.12

Triple Iterated Integrals in Spherical Coordinates

Spherical coordinates are useful in solving problems concerning figures that are symmetric about a point. If the origin is that point, then spherical coordinates $(\mathscr{R}, \theta, \phi)$ are related to Cartesian coordinates (x, y, z) by the equations

$$x = \mathscr{R} \sin\phi \cos\theta, \quad y = \mathscr{R} \sin\phi \sin\theta, \quad z = \mathscr{R} \cos\phi, \tag{14.64}$$

which are illustrated in Figure 14.67. As is the case for polar and cylindrical coordinates, without restrictions on $\mathscr{R}, \theta$, and ϕ, each point in space has many sets of spherical coordinates. The positive value of its spherical coordinate $\mathscr{R}$ is given by

$$\mathscr{R} = \sqrt{x^2 + y^2 + z^2}. \tag{14.65}$$

The θ-coordinates in cylindrical and spherical coordinates are identical, so that no simple formula for θ in terms of x, y, and z exists. The ϕ-coordinate is the angle between the positive z-axis and the line joining the origin to the point (x, y, z), and that value of ϕ in the range $0 \le \phi \le \pi$ is determined by the formula

$$\phi = \text{Cos}^{-1}\left(\frac{z}{\sqrt{x^2 + y^2 + z^2}}\right). \tag{14.66}$$

FIGURE 14.67

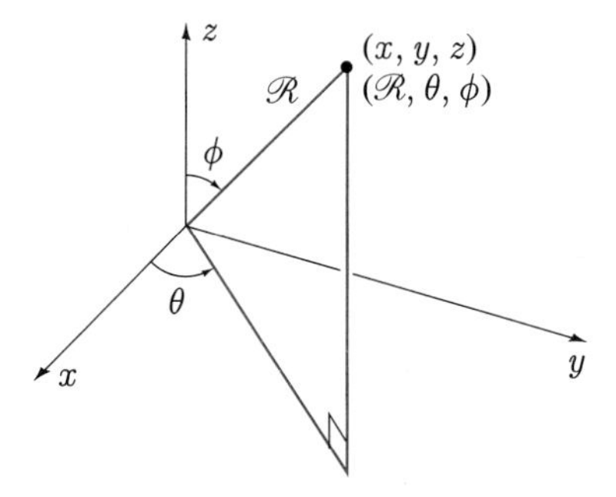

To transform equations $F(x, y, z) = 0$ of surfaces from Cartesian to spherical coordinates, we substitute from equations 14.64:

$$F(\mathscr{R}\sin\phi\cos\theta, \mathscr{R}\sin\phi\sin\theta, \mathscr{R}\cos\phi) = 0.$$

For example, the sphere $x^2 + y^2 + z^2 = 4$ is symmetric about its centre, and its equation in spherical coordinates is simply $\mathscr{R} = 2$. For the right-circular cone $z = \sqrt{x^2 + y^2}$, we write

$$\mathscr{R}\cos\phi = \sqrt{\mathscr{R}^2\sin^2\phi\cos^2\theta + \mathscr{R}^2\sin^2\phi\sin^2\theta} = \mathscr{R}\sin\phi.$$

Consequently, $\tan\phi = 1$ or $\phi = \pi/4$; i.e., $\phi = \pi/4$ is the equation of the cone in spherical coordinates.

Suppose that we are to evaluate the triple integral of a function $f(x, y, z)$,

$$\iiint_V f(x, y, z)\,dV,$$

over some region V of space. The choice of a triple iterated integral in spherical coordinates implies a subdivision of V into small volumes by means of coordinate surfaces $\mathscr{R} =$ constant, $\theta =$ constant, and $\phi =$ constant (Figure 14.68). Surfaces $\mathscr{R} =$ constant are spheres centred at the origin; surfaces $\theta =$ constant are planes containing the z-axis; and surfaces $\phi =$ constant are right-circular cones symmetric about the z-axis with the origin as apex.

FIGURE 14.68 **FIGURE 14.69**

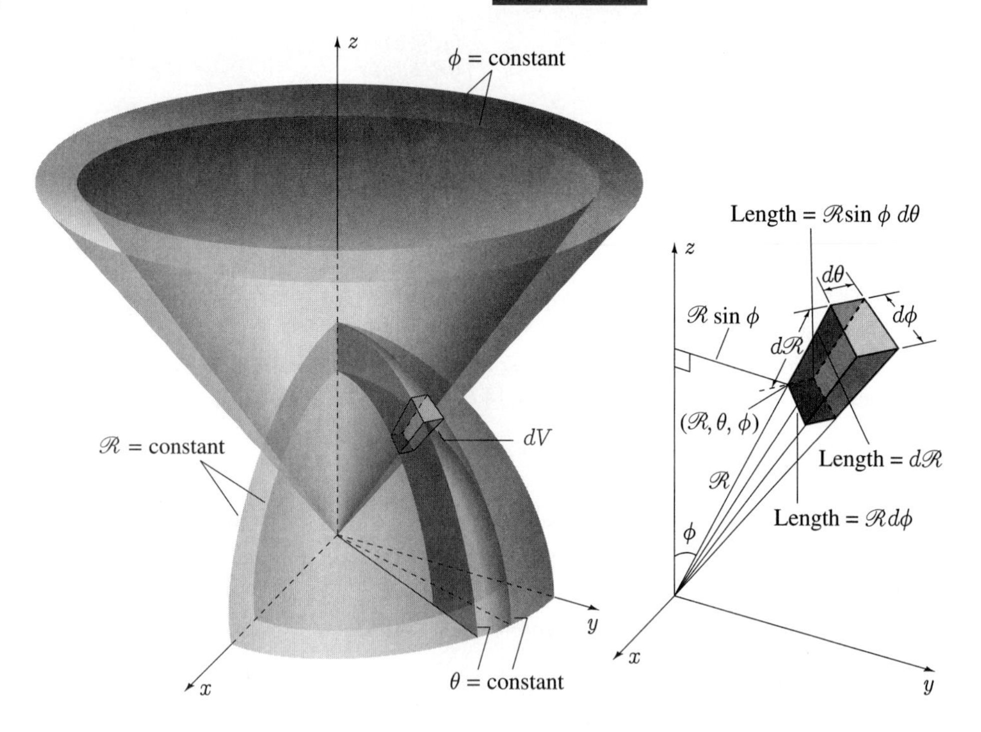

If we denote small variations in $\mathscr{R}, \theta$, and ϕ for the element dV by $d\mathscr{R}, d\theta$, and $d\phi$ (Figure 14.69) and approximate dV by a rectangular parallelepiped with dimensions $d\mathscr{R}, \mathscr{R}\, d\phi$, and $\mathscr{R} \sin\phi\, d\theta$, then

$$dV = (\mathscr{R} \sin\ \phi\ d\theta)(\mathscr{R}\ d\phi)\, d\mathscr{R} = \mathscr{R}^2 \sin\phi\ d\mathscr{R}\ d\theta\ d\phi. \tag{14.67}$$

The integrand $f(x, y, z)$ is expressed in spherical coordinates as

$$f(\mathscr{R} \sin\phi \cos\theta, \mathscr{R} \sin\phi \sin\theta, \mathscr{R} \cos\phi).$$

It remains only to affix appropriate limits to the triple iterated integral, and these limits depend on which of the six possible triple iterated integrals in spherical coordinates that we choose. If we use a triple iterated integral with respect to $\mathscr{R}, \phi$, and θ, then it is of the form

$$\iiint_V f(x, y, z)\, dV = \int_a^b \int_{g_1(\theta)}^{g_2(\theta)} \int_{h_1(\theta,\phi)}^{h_2(\theta,\phi)} f(\mathscr{R} \sin\phi \cos\theta, \mathscr{R} \sin\phi \sin\theta, \mathscr{R} \cos\phi)\, \mathscr{R}^2 \sin\phi\ d\mathscr{R}\ d\phi\ d\theta.$$

The geometric interpretations of the additions represented by these integrations with respect to $\mathscr{R}, \phi$, and θ are left to the examples.

EXAMPLE 14.28 Find the volume of a sphere.

SOLUTION The equation of a sphere of radius R centred at the origin is $x^2 + y^2 + z^2 = R^2$ or, in spherical coordinates, $\mathscr{R} = R$. The volume of this sphere is eight times the first octant volume shown in Figure 14.70. If we use spherical coordinates, the volume of an elemental piece is

$$\mathcal{R}^2 \sin\phi \, d\mathcal{R} \, d\phi \, d\theta.$$

An $\mathcal{R}$-integration adds these volumes for constant ϕ and θ to give the volume in a "spike,"

$$\int_0^R \mathcal{R}^2 \sin\phi \, d\mathcal{R} \, d\phi \, d\theta,$$

where the limits indicate that all spikes start at the origin (where $\mathcal{R} = 0$) and end on the sphere (where $\mathcal{R} = R$). A ϕ-integration now adds the volumes of spikes for constant θ. This yields the volume of a slice

$$\int_0^{\pi/2} \int_0^R \mathcal{R}^2 \sin\phi \, d\mathcal{R} \, d\phi \, d\theta,$$

where the limits indicate that all slices in the first octant start on the z-axis (where $\phi = 0$) and end on the xy-plane (where $\phi = \pi/2$). Finally, the θ-integration adds the volumes of all such slices

$$\int_0^{\pi/2} \int_0^{\pi/2} \int_0^R \mathcal{R}^2 \sin\phi \, d\mathcal{R} \, d\phi \, d\theta,$$

where the limits 0 and $\pi/2$ identify the positions of the first and last slices, respectively, in the first octant. We obtain the required volume as

$$\begin{aligned} 8\int_0^{\pi/2} \int_0^{\pi/2} \int_0^R \mathcal{R}^2 \sin\phi \, d\mathcal{R} \, d\phi \, d\theta &= 8\int_0^{\pi/2} \int_0^{\pi/2} \left\{ \frac{\mathcal{R}^3}{3} \sin\phi \right\}_0^R d\phi \, d\theta \\ &= \frac{8R^3}{3} \int_0^{\pi/2} \{-\cos\phi\}_0^{\pi/2} \, d\theta \\ &= \frac{8R^3}{3} \{\theta\}_0^{\pi/2} = \frac{4}{3}\pi R^3. \end{aligned}$$

FIGURE 14.70

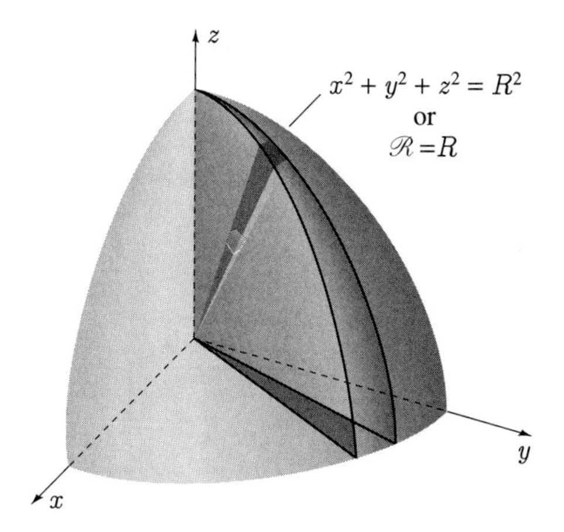

■

EXAMPLE 14.29

Find the centre of mass of a solid object of constant density if it is in the shape of a right-circular cone.

SOLUTION Let the cone have altitude h and base radius R. Then its mass is

$$M = \frac{1}{3}\pi R^2 h\rho,$$

where ρ is the density of the object. If we place axes as shown in Figure 14.71, then $\overline{x} = \overline{y} = 0$; i.e., the centre of mass is on the axis of symmetry of the cone. To find $\overline{z}$ we offer three solutions.

Method 1 If we use Cartesian coordinates, the equation of the surface of the cone is of the form $z = k\sqrt{x^2 + y^2}$. Since $(0, R, h)$ is a point on the cone, $h = kR$, and therefore $k = h/R$. Now

$$M\overline{z} = 4\int_0^R \int_0^{\sqrt{R^2-x^2}} \int_{k\sqrt{x^2+y^2}}^{h} z\rho\, dz\, dy\, dx = 4\rho \int_0^R \int_0^{\sqrt{R^2-x^2}} \left\{\frac{z^2}{2}\right\}_{k\sqrt{x^2+y^2}}^{h} dy\, dx$$

$$= 2\rho \int_0^R \int_0^{\sqrt{R^2-x^2}} \{h^2 - k^2(x^2 + y^2)\} dy\, dx$$

$$= 2\rho \int_0^R \left\{h^2 y - k^2 x^2 y - \frac{k^2 y^3}{3}\right\}_0^{\sqrt{R^2-x^2}} dx$$

$$= 2\rho \int_0^R \left\{h^2\sqrt{R^2 - x^2} - k^2 x^2\sqrt{R^2 - x^2} - \frac{k^2}{3}(R^2 - x^2)^{3/2}\right\} dx.$$

If we set $x = R\sin\theta$, then $dx = R\cos\theta\, d\theta$, and

$$M\overline{z} = 2\rho \int_0^{\pi/2} \left\{h^2 R\cos\theta - k^2(R^2\sin^2\theta)R\cos\theta - \frac{k^2}{3}R^3\cos^3\theta\right\} R\cos\theta\, d\theta$$

$$= 2\rho R^2 \int_0^{\pi/2} \left\{\frac{h^2}{2}(1 + \cos 2\theta) - \frac{k^2 R^2}{8}(1 - \cos 4\theta) - \frac{k^2 R^2}{12}\left(1 + 2\cos 2\theta + \frac{1 + \cos 4\theta}{2}\right)\right\} d\theta$$

$$= \rho R^2 \left\{h^2\left(\theta + \frac{\sin 2\theta}{2}\right) - \frac{k^2 R^2}{4}\left(\theta - \frac{\sin 4\theta}{4}\right) - \frac{k^2 R^2}{6}\left(\frac{3\theta}{2} + \sin 2\theta + \frac{\sin 4\theta}{8}\right)\right\}_0^{\pi/2}$$

$$= \rho R^2 \left\{\frac{\pi h^2}{2} - \frac{k^2 R^2 \pi}{8} - \frac{k^2 R^2 \pi}{8}\right\} = \frac{\pi R^2 h^2 \rho}{4}.$$

Thus,

$$\overline{z} = \frac{\pi R^2 h^2 \rho}{4} \cdot \frac{3}{\pi R^2 h\rho} = \frac{3h}{4}.$$

FIGURE 14.71

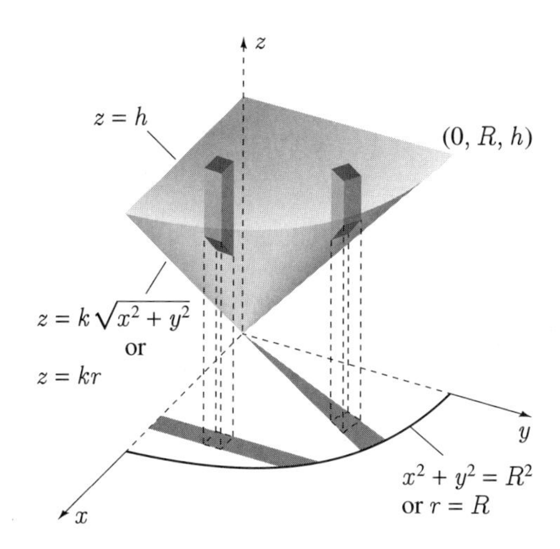

Method 2 If we use cylindrical coordinates, the equation of the surface of the cone is $z = kr$. From Figure 14.71, we see that

$$\begin{aligned} M\bar{z} &= 4\int_0^{\pi/2}\int_0^R\int_{kr}^h z\rho\, r\, dz\, dr\, d\theta = 4\rho\int_0^{\pi/2}\int_0^R \left\{\frac{rz^2}{2}\right\}_{kr}^h dr\, d\theta \\ &= 2\rho\int_0^{\pi/2}\int_0^R r(h^2 - k^2r^2)\, dr\, d\theta = 2\rho\int_0^{\pi/2}\left\{\frac{r^2h^2}{2} - \frac{k^2r^4}{4}\right\}_0^R d\theta \\ &= \rho\left(R^2h^2 - \frac{k^2R^4}{2}\right)\{\theta\}_0^{\pi/2} = \rho\left(R^2h^2 - \frac{k^2R^4}{2}\right)\frac{\pi}{2} = \frac{\pi R^2h^2\rho}{4}. \end{aligned}$$

Again, then, $\bar{z} = 3h/4$.

Method 3 If we use spherical coordinates, then the equation of the surface of the cone is

$$\phi = \mathrm{Cos}^{-1}\frac{h}{\sqrt{h^2 + R^2}} = \phi_1.$$

From Figure 14.72, we see that

$$\begin{aligned} M\bar{z} &= 4\int_0^{\pi/2}\int_0^{\phi_1}\int_0^{h\sec\phi} (\mathscr{R}\cos\phi)\rho\mathscr{R}^2 \sin\phi\, d\mathscr{R}\, d\phi\, d\theta \\ &= 4\rho\int_0^{\pi/2}\int_0^{\phi_1}\left\{\frac{\mathscr{R}^4}{4}\sin\phi\cos\phi\right\}_0^{h\sec\phi} d\phi\, d\theta \\ &= \rho h^4\int_0^{\pi/2}\int_0^{\phi_1}\frac{\sin\phi}{\cos^3\phi}d\phi\, d\theta = \rho h^4\int_0^{\pi/2}\left\{\frac{1}{2\cos^2\phi}\right\}_0^{\phi_1} d\theta \\ &= \frac{\rho h^4}{2}\left\{\frac{1}{\cos^2\phi_1} - 1\right\}\{\theta\}_0^{\pi/2} = \frac{\pi\rho h^4}{4}\left\{\frac{h^2 + R^2}{h^2} - 1\right\} = \frac{\pi h^2R^2\rho}{4}. \end{aligned}$$

Again, $\bar{z} = 3h/4$.

FIGURE 14.72

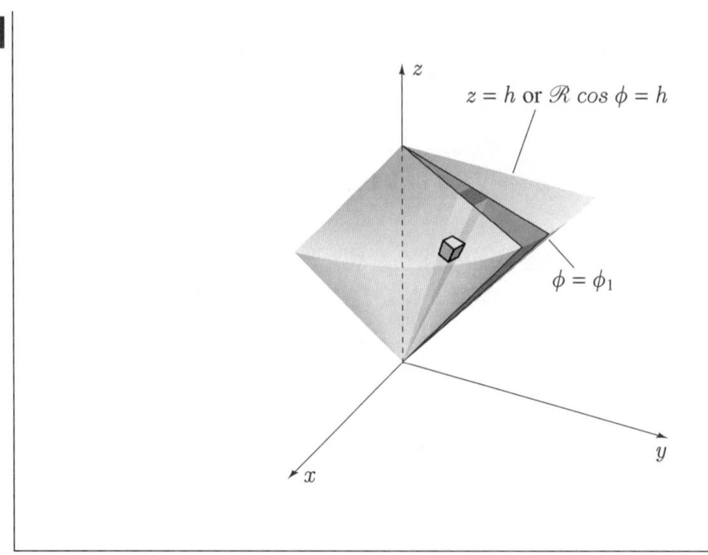

EXERCISES 14.12

In Exercises 1–7 find the equation of the surface in spherical coordinates. Sketch the surface.

1. $x^2 + y^2 + z^2 = 4$

2. $x^2 + y^2 = 1$

3. $3z = \sqrt{x^2 + y^2}$

4. $4z = x^2 + y^2$

5. $y = x$

6. $x^2 + y^2 = 1 + z^2$

7. $z = -2\sqrt{x^2 + y^2}$

In Exercises 8–12 find the volume described.

8. Bounded by $z = \sqrt{x^2 + y^2}$, $z = \sqrt{1 - x^2 - y^2}$

9. Bounded by $z = 1$, $z = \sqrt{4 - x^2 - y^2}$

10. Bounded by $x^2 + y^2 + z^2 = 1$, $y = x$, $y = 2x$, $z = 0$ (in the first octant)

11. Inside $x^2 + y^2 + z^2 = 2$ but outside $x^2 + y^2 = 1$

12. Bounded by $z = 2\sqrt{x^2 + y^2}$, $x^2 + y^2 = 4$, $z = 0$

13. Find the centre of mass of a uniform hemispherical solid.

14. Find the moment of inertia of a uniform solid sphere of radius R about any line through its centre.

15. Find the first moment about the yz-plane of the uniform solid in the first octant bounded by the surfaces $x^2 + y^2 + z^2 = 4$, $x^2 + y^2 + z^2 = 9$, $y = 0$ and $y = \sqrt{3}x$.

16. A solid sphere of radius R and centre at the origin has a continuous charge distribution throughout. If the density of the charge is $\rho(x, y, z) = k\sqrt{x^2 + y^2 + z^2}$ coulombs per cubic metre, find the total charge in the sphere.

17. Set up the six triple iterated integrals in spherical coordinates for the triple integral of a function $f(x, y, z)$ over the volume in the first octant under the sphere $x^2 + y^2 + z^2 = 2$ and inside the cylinder $x^2 + y^2 = 1$.

Evaluate the triple iterated integral in Exercises 18 and 19.

18. $$\int_0^9 \int_0^{\sqrt{81-y^2}} \int_0^{\sqrt{81-x^2-y^2}} \frac{1}{x^2 + y^2 + z^2}\, dz\, dx\, dy$$

19. $$\int_0^1 \int_0^{\sqrt{1-x^2}} \int_{\sqrt{x^2+y^2}}^{\sqrt{2-x^2-y^2}} dz\, dy\, dx$$

20. Find a formula for the smaller volume bounded by $x^2 + y^2 + z^2 = R^2$ and $z = k\sqrt{x^2 + y^2}$ $(k > 0)$.

21. Find the volume bounded by $(x^2 + y^2 + z^2)^2 = x$.

22. **(a)** Use Archimedes' principle to determine the density of a spherical ball if it floats half submerged in water.

(b) What force is required to keep the ball with its centre at a depth of one-half the radius of the ball?

23. Find the volume bounded by the surface $(x^2 + y^2 + z^2)^2 = 2z(x^2 + y^2)$.

24. A sphere of radius R carries a uniform charge distribution of ρ coulombs per cubic metre (Figure 14.73). If P is a point on the z-axis (distance $d > R$ from the centre of the sphere) and dV is a small element of volume of the sphere, then the potential at P due to dV is given by

$$\frac{1}{4\pi\epsilon_0}\frac{\rho dV}{s},$$

where s is the distance from P to dV.

(a) Show that in terms of spherical coordinates the potential V at P due to the entire sphere is

$$V = \frac{\rho}{4\pi\epsilon_0}\int_{-\pi}^{\pi}\int_0^{\pi}\int_0^R \frac{\mathcal{R}^2 \sin\phi}{\sqrt{\mathcal{R}^2 + d^2 - 2\mathcal{R}d\cos\phi}}\, d\mathcal{R}\, d\phi\, d\theta.$$

(b) Because this iterated integral is very difficult to evaluate, we replace ϕ with the variable $s = \sqrt{\mathcal{R}^2 + d^2 - 2\mathcal{R}d\cos\phi}$. Show that with this change

$$V = \frac{\rho}{4\pi\epsilon_0 d}\int_{-\pi}^{\pi}\int_0^R\int_{d-\mathcal{R}}^{d+\mathcal{R}} \mathcal{R}\, ds\, d\mathcal{R}\, d\theta.$$

(c) Evaluate the integral in (b) to verify that $V = Q/(4\pi\epsilon_0 d)$, where Q is the total charge on the sphere.

FIGURE 14.73

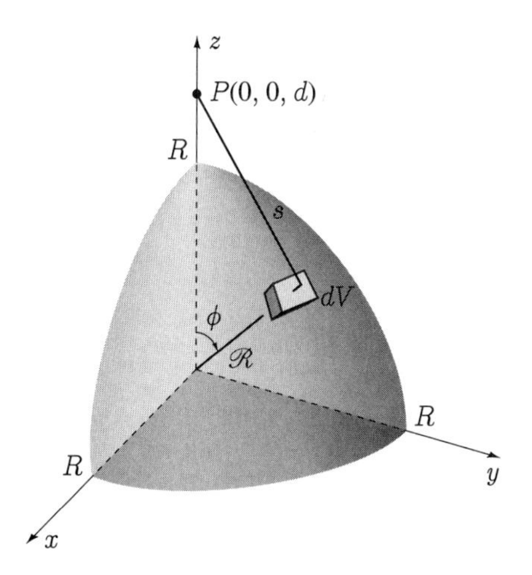

25. A sphere of constant density ρ and radius R is located at the origin (Figure 14.74). If a mass m is situated at a point P on the z-axis (distance $d > R$ from the centre of the sphere) and dV is a small element of volume of the sphere, then according to Newton's universal law of gravitation, the z-component of the force on m due to the mass in dV is given by

$$-\frac{Gm\rho dV\cos\psi}{s^2},$$

where G is a constant and s is the distance between P and dV.

(a) Show that in spherical coordinates the total force on m due to the entire sphere has z-component

$$F_z = -\frac{Gm\rho}{2d}\int_{-\pi}^{\pi}\int_0^{\pi}\int_0^R \left(\frac{s^2 + d^2 - \mathcal{R}^2}{s^3}\right)\mathcal{R}^2 \sin\phi\, d\mathcal{R}\, d\phi\, d\theta.$$

(b) Use the transformation in Exercise 24(b) to write F_z in the form

$$F_z = -\frac{Gm\rho}{2d^2}\int_{-\pi}^{\pi}\int_0^R\int_{d-\mathcal{R}}^{d+\mathcal{R}} \mathcal{R}\left(\frac{s^2 + d^2 - \mathcal{R}^2}{s^2}\right) ds\, d\mathcal{R}\, d\theta,$$

and show that $F_z = -GmM/d^2$ where M is the total mass of the sphere.

FIGURE 14.74

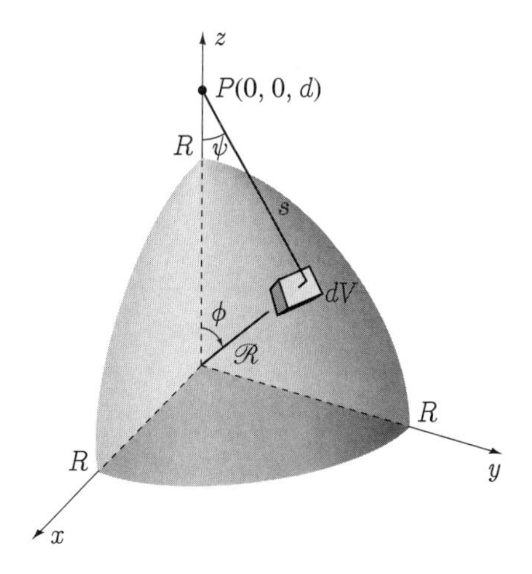

26. A homogeneous solid is bounded by two concentric spheres of radii a and b ($a < b$). Verify that the force that this layer exerts on a point mass at any point interior to the shell vanishes.

SECTION 14.13

Derivatives of Definite Integrals

If a function $f(x,y)$ of two independent variables is integrated with respect to y from $y=a$ to $y=b$, the result depends on x. Suppose we denote this function by $F(x)$:

$$F(x) = \int_a^b f(x,y)\,dy. \tag{14.68}$$

To calculate the derivative $F'(x)$ of this function, we should first integrate with respect to y and then differentiate with respect to x. The following theorem indicates that differentiation can be done first and integration later.

Theorem 14.3

If the partial derivative $\partial f/\partial x$ of $f(x,y)$ is continuous on a rectangle $a \le y \le b$, $c \le x \le d$, then for $c < x < d$

$$\frac{d}{dx}\int_a^b f(x,y)\,dy = \int_a^b \frac{\partial f(x,y)}{\partial x}\,dy. \tag{14.69}$$

Proof If we define

$$g(x) = \int_a^b \frac{\partial f(x,y)}{\partial x}\,dy$$

as the right-hand side of equation 14.69, then this function is defined for $c \le x \le d$. In fact, because $\partial f/\partial x$ is continuous, $g(x)$ is also continuous. We can therefore integrate $g(x)$ with respect to x from $x=c$ to any value of x in the interval $c \le x \le d$:

$$\int_c^x g(x)\,dx = \int_c^x \int_a^b \frac{\partial f(x,y)}{\partial x}\,dy\,dx.$$

This double iterated integral represents the double integral of $\partial f/\partial x$ over the rectangle in Figure 14.75, and if we reverse the order of integration, we have

$$\begin{aligned}\int_c^x g(x)\,dx &= \int_a^b \int_c^x \frac{\partial f(x,y)}{\partial x}\,dx\,dy = \int_a^b \{f(x,y)\}_c^x\,dy \\ &= \int_a^b \{f(x,y) - f(c,y)\}\,dy = \int_a^b f(x,y)\,dy - \int_a^b f(c,y)\,dy.\end{aligned}$$

Because the second integral on the right is independent of x, if we differentiate this equation with respect to x, we get

$$\frac{d}{dx}\int_c^x g(x)\,dx = \frac{d}{dx}\int_a^b f(x,y)\,dy.$$

But Theorem 6.7 gives

$$g(x) = \frac{d}{dx}\int_a^b f(x,y)\,dy.$$

This completes the proof.

FIGURE 14.75

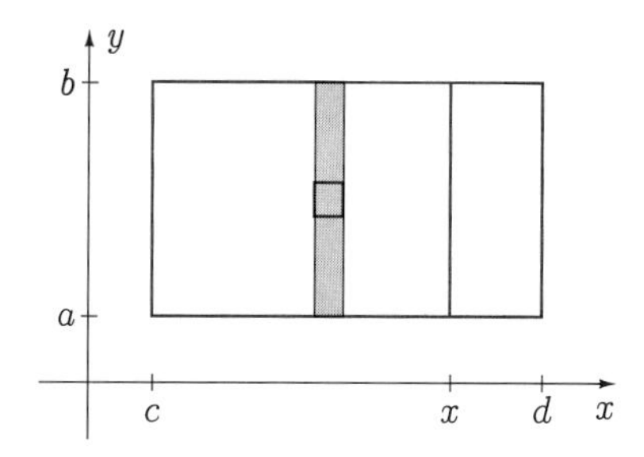

The limits on the integral in 14.68 need not be numerical constants; so far as the integration with respect to y is concerned, x is constant, and therefore a and b could be functions of x:

$$F(x) = \int_{a(x)}^{b(x)} f(x, y)\,dy.$$

Indeed, this is precisely what does occur in the evaluation of double integrals by means of double iterated integrals. To differentiate $F(x)$ now is more complicated than in Theorem 14.3 because we must also account for the x's in $a(x)$ and $b(x)$. The chain rule can be used to develop a formula for $F'(x)$.

Theorem 14.4

(Leibnitz's rule). If the partial derivative $\partial f/\partial x$ is continuous on the area bounded by the curves $y = a(x)$, $y = b(x)$, $x = c$, and $x = d$, then

$$\frac{d}{dx}\int_{a(x)}^{b(x)} f(x, y)\,dy = \int_{a(x)}^{b(x)} \frac{\partial f(x, y)}{\partial x}\,dy + f[x, b(x)]\frac{db}{dx} - f[x, a(x)]\frac{da}{dx}. \quad (14.70)$$

Proof If $F(x) = \int_{a(x)}^{b(x)} f(x, y)\,dy$, then the schematic diagram

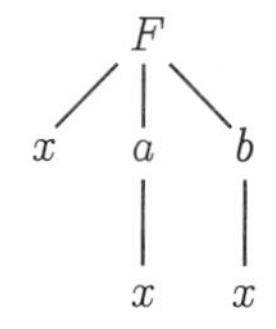

gives the chain rule

$$F'(x) = \left.\frac{\partial F}{\partial x}\right)_{a,b} + \left.\frac{\partial F}{\partial b}\right)_{a,x}\frac{db}{dx} + \left.\frac{\partial F}{\partial a}\right)_{b,x}\frac{da}{dx}.$$

The first term is precisely the situation covered in Theorem 14.3, and therefore

$$\left.\frac{\partial F}{\partial x}\right)_{a,b} = \int_a^b \frac{\partial f(x, y)}{\partial x}\,dy.$$

Since Theorem 6.7 indicates that

$$\frac{d}{db}\int_a^b f(y)\,dy = f(b),$$

it follows that

$$\frac{\partial}{\partial b}\int_a^b f(x,y)\,dy = f(x,b).$$

In other words,

$$\left.\frac{\partial F}{\partial b}\right)_{a,x} = f(x,b).$$

Furthermore,

$$\left.\frac{\partial F}{\partial a}\right)_{b,x} = \frac{\partial}{\partial a}\int_a^b f(x,y)\,dy = -\frac{\partial}{\partial a}\int_b^a f(x,y)\,dy = -f(x,a).$$

Substitution of these facts into the chain rule now gives Leibnitz's rule.

Our derivation of Leibnitz's rule for the differentiation of a definite integral that depends on a parameter (x in this case) shows that the first term accounts for those x's in the integrand, and the second and third terms for the x's in the upper and lower limits. The following geometric interpretation of Leibnitz's rule emphasizes this same point.

Suppose the function $f(x,y)$ has only positive values so that the surface $z = f(x,y)$ lies completely above the xy-plane (Figure 14.76). The equations $y = a(x)$ and $y = b(x)$ describe cylindrical walls standing on the curves $y = a(x)$, $z = 0$ and $y = b(x)$, $z = 0$ in the xy-plane. Were we to slice through the surface $z = f(x,y)$ with a plane $x =$ constant, then an area would be defined in this plane bounded on the top by $z = f(x,y)$, on the sides by $y = a(x)$ and $y = b(x)$, and on the bottom by the xy-plane. This area is clearly defined by the definite integral

$$\int_{a(x)}^{b(x)} f(x,y)\,dy$$

in Leibnitz's rule, and as x varies so too does the area. Note, in particular, that as the plane varies, the area changes, not only because the height of the surface $z = f(x,y)$ changes but also because the width of the area varies (i.e., the two cylindrical walls are not a constant distance apart). The first term in Leibnitz's rule accounts for the vertical variation, whereas the remaining two terms represent variations due to the fluctuating width.

FIGURE 14.76

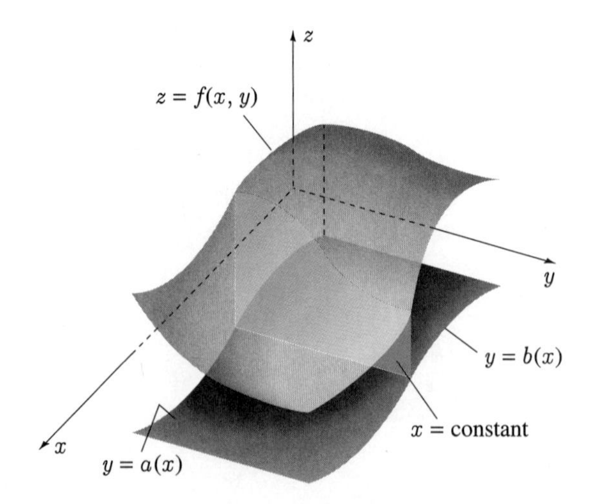

EXAMPLE 14.30

If $F(x) = \int_x^{2x}(y^3 \ln y + x^3 e^y)\,dy$, find $F'(x)$.

SOLUTION With Leibnitz's rule, we have

$$\begin{aligned} F'(x) &= \int_x^{2x}(3x^2 e^y)\,dy + \{(2x)^3 \ln(2x) + x^3 e^{2x}\}(2) - \{x^3 \ln x + x^3 e^x\}(1) \\ &= \{3x^2 e^y\}_x^{2x} + 16x^3(\ln 2 + \ln x) + 2x^3 e^{2x} - x^3 \ln x - x^3 e^x \\ &= 3x^2 e^{2x} - 3x^2 e^x + (16 \ln 2)x^3 + 15x^3 \ln x + 2x^3 e^{2x} - x^3 e^x \\ &= 3x^2 e^x(e^x - 1) + x^3(16 \ln 2 + 15 \ln x + 2e^{2x} - e^x). \end{aligned}$$

■

In Exercise 6 you are asked to evaluate the integral defining $F(x)$ in this example and then to differentiate the resulting function with respect to x. It will be clear, then, that for this example Leibnitz's rule simplifies the calculations considerably.

EXAMPLE 14.31

Evaluate

$$\int_0^1 \frac{y^x - 1}{\ln y}\,dy \quad \text{for } x > -1.$$

SOLUTION We use Leibnitz's rule in this example to avoid finding an antiderivative for $(y^x - 1)/\ln y$. If we set

$$F(x) = \int_0^1 \frac{y^x - 1}{\ln y}\,dy$$

and use Leibnitz's rule, we have

$$F'(x) = \int_0^1 \frac{\partial}{\partial x}\left\{\frac{y^x - 1}{\ln y}\right\}dy = \int_0^1 \frac{y^x \ln y}{\ln y}\,dy = \int_0^1 y^x\,dy = \left\{\frac{y^{x+1}}{x+1}\right\}_0^1 = \frac{1}{x+1}.$$

It follows, therefore, that $F(x)$ must be of the form

$$F(x) = \ln(x+1) + C.$$

But from the definition of $F(x)$ as an integral, it is clear that $F(0) = 0$, and hence $C = 0$. Thus,

$$\int_0^1 \frac{y^x - 1}{\ln y}\,dy = \ln(x+1).$$

■

Note that this problem originally had nothing whatsoever to do with Leibnitz's rule. But by introducing the rule, we were able to find a simple formula for $F'(x)$, and this immediately led to $F(x)$. This can be a very useful technique for evaluating definite integrals that depend on a parameter.

EXAMPLE 14.32 When a shell is fired from the artillery gun in Figure 14.77, the barrel recoils along a well-lubricated guide and its motion is braked by a battery of heavy springs. We set up a coordinate system where $x = 0$ represents the firing position of the gun when no stretch or compression exists in the springs. Suppose that when the gun is fired, the horizontal component of the force causing recoil is $g(t)$ (t = time). If the mass of the gun is m and the effective spring constant for the battery of springs is k, then Newton's second law states that

$$m\frac{d^2x}{dt^2} = g(t) - kx,$$

or

$$m\frac{d^2x}{dt^2} + kx = g(t).$$

This differential equation can be solved by a method called **variation of parameters**, and the solution is

$$x(t) = A\cos\sqrt{\frac{k}{m}}t + B\sin\sqrt{\frac{k}{m}}t + \frac{1}{\sqrt{mk}}\int_0^t g(u)\sin\left(\sqrt{\frac{k}{m}}(t-u)\right)du,$$

where A and B are arbitrary constants. We do not discuss differential equations until Chapter 16, but with Leibnitz's rule it is possible to verify that this function is indeed a solution. Do so.

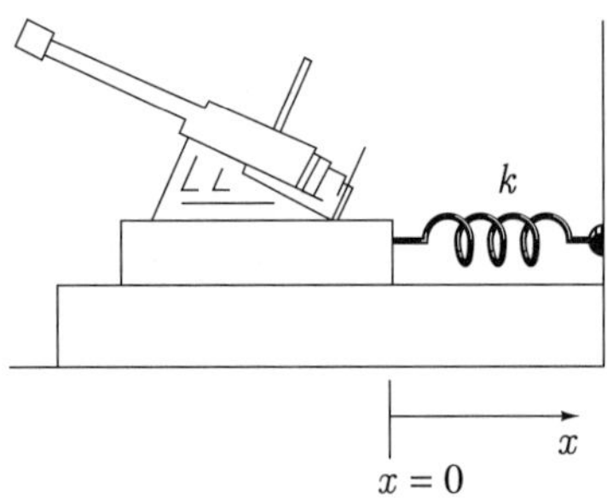

FIGURE 14.77

SOLUTION To differentiate the integral we rewrite Leibnitz's rule in terms of the variables of this problem:

$$\frac{d}{dt}\int_{a(t)}^{b(t)} f(t,u)\,du = \int_{a(t)}^{b(t)} \frac{\partial f(t,u)}{\partial t}du + f[t,b(t)]\frac{db}{dt} - f[t,a(t)]\frac{da}{dt}.$$

If we now apply this formula to the definite integral in $x(t)$ where

$$f(t,u) = g(u)\sin\left[\sqrt{\frac{k}{m}}(t-u)\right],$$

then

$$\frac{dx}{dt} = -\sqrt{\frac{k}{m}}A\sin\sqrt{\frac{k}{m}}t + \sqrt{\frac{k}{m}}B\cos\sqrt{\frac{k}{m}}t$$
$$+ \frac{1}{\sqrt{mk}}\left\{\int_0^t g(u)\sqrt{\frac{k}{m}}\cos\left(\sqrt{\frac{k}{m}}(t-u)\right)du + g(t)\sin\left(\sqrt{\frac{k}{m}}(t-t)\right)\right\}$$

$$= \sqrt{\frac{k}{m}}\left\{-A\sin\sqrt{\frac{k}{m}}t + B\cos\sqrt{\frac{k}{m}}t\right\} + \frac{1}{m}\int_0^t g(u)\cos\left(\sqrt{\frac{k}{m}}(t-u)\right)du.$$

We now use Leibnitz's rule once more to find d^2x/dt^2:

$$\begin{aligned}\frac{d^2x}{dt^2} &= \sqrt{\frac{k}{m}}\left\{-\sqrt{\frac{k}{m}}A\cos\sqrt{\frac{k}{m}}t - \sqrt{\frac{k}{m}}B\sin\sqrt{\frac{k}{m}}t\right\}\\ &\quad + \frac{1}{m}\left\{\int_0^t -g(u)\sqrt{\frac{k}{m}}\sin\left(\sqrt{\frac{k}{m}}(t-u)\right)du + g(t)\cos\left(\sqrt{\frac{k}{m}}(t-t)\right)\right\}\\ &= -\frac{k}{m}\left\{A\cos\sqrt{\frac{k}{m}}t + B\sin\sqrt{\frac{k}{m}}t\right\}\\ &\quad - \sqrt{\frac{k}{m^3}}\int_0^t g(u)\sin\left(\sqrt{\frac{k}{m}}(t-u)\right)du + \frac{1}{m}g(t)\\ &= -\frac{k}{m}\left\{A\cos\sqrt{\frac{k}{m}}t + B\sin\sqrt{\frac{k}{m}}t + \frac{1}{\sqrt{mk}}\int_0^t g(u)\sin\left(\sqrt{\frac{k}{m}}(t-u)\right)du\right\}\\ &\quad + \frac{1}{m}g(t)\\ &= -\frac{k}{m}x(t) + \frac{1}{m}g(t).\end{aligned}$$

In other words, the function $x(t)$ satisfies the differential equation

$$m\frac{d^2x}{dt^2} = -kx(t) + g(t),$$

and the proof is complete. ■

This example illustrates that Leibnitz's rule is essential to the manipulation of solutions of differential equations that are represented as definite integrals.

EXERCISES 14.13

In Exercises 1–5 use Leibnitz's rule to find the derivative of $F(x)$. Check your result by evaluating the integral and then differentiating.

1. $F(x) = \int_0^3 (x^2y^2 + 3xy)\,dy$

2. $F(x) = \int_1^x (x^2/y^2 + e^y)\,dy$

3. $F(x) = \int_{x-1}^{x^2} (x^3y + y^2 + 1)\,dy$

4. $F(x) = \int_{x^2}^{x^3-1} (x + y\ln y)\,dy$

5. $F(x) = \int_0^x \frac{y-x}{y+x}\,dy$

6. Find $F'(x)$ in Example 14.30 by first evaluating the definite integral.

7. Use the result of Example 14.31 to prove that

$$\int_0^1 \frac{x^p - x^q}{\ln x}\,dx = \ln\left(\frac{p+1}{q+1}\right),$$

provided $p > -1$ and $q > -1$.

8. Find $F'(x)$ if $F(x) = \int_{\sin x}^{e^x} \sqrt{1 + y^3}\,dy$.

In Exercises 9–11 use Leibnitz's rule to verify that the function $y(x)$ satisfies the differential equation.

9. $y(x) = \dfrac{1}{x^2}\displaystyle\int_0^x t^2 f(t)\,dt;\ x\frac{dy}{dx} + 2y = xf(x)$

10. $y(x) = \dfrac{1}{2}\displaystyle\int_0^x f(t)(e^{x-t} - e^{t-x})\,dt;\ \frac{d^2y}{dx^2} - y = f(x)$

11. $y(x) = \dfrac{1}{\sqrt{2}}\displaystyle\int_0^x e^{2(t-x)} \sin[\sqrt{2}(x - t)]f(t)\,dt;$
$\dfrac{d^2y}{dx^2} + 4\dfrac{dy}{dx} + 6y = f(x)$

12. Given that

$$\int_0^b \frac{1}{1 + ax}\,dx = \frac{1}{a}\ln(1 + ab),$$

find a formula for $\displaystyle\int_0^b \frac{x}{(1 + ax)^2}\,dx.$

13. Given that

$$\int \frac{1}{\sqrt{a^2 - x^2}}\,dx = \mathrm{Sin}^{-1}\left(\frac{x}{a}\right) + C,$$

find a formula for $\displaystyle\int \frac{1}{(a^2 - x^2)^{3/2}}\,dx.$

14. Given that

$$\int \frac{1}{a^2 + x^2}\,dx = \frac{1}{a}\mathrm{Tan}^{-1}\left(\frac{x}{a}\right) + C,$$

find a formula for $\displaystyle\int \frac{1}{(a^2 + x^2)^3}\,dx.$

15. Use the result that

$$\int_0^{\pi/2} \frac{1}{a^2\cos^2 x + b^2\sin^2 x}\,dx = \frac{\pi}{2|ab|}$$

to find a formula for $\displaystyle\int_0^{\pi/2} \frac{1}{(a^2\cos^2 x + b^2\sin^2 x)^2}\,dx.$

In Exercises 16 and 17 use Leibnitz's rule to evaluate the integral.

16. $\displaystyle\int_0^{\pi} \frac{\ln(1 + a\cos x)}{\cos x}\,dx$ where $|a| < 1$

17. $\displaystyle\int_0^{\infty} \frac{\mathrm{Tan}^{-1}(ax)}{x(1 + x^2)}\,dx$ where $a > 0$

18. **(a)** What is the domain of the function $F(x) = \displaystyle\int_0^9 \ln(1 - x^2y^2)\,dy$? What is $F(0)$?

(b) Find $F'(x)$ by Leibnitz's rule. What is $F'(0)$?

(c) Show that the graph of the function $F(x)$ is concave downward for all x in its domain of definition.

19. Laplace's equation for a function $u(r, \theta)$ in polar coordinates is

$$\frac{\partial^2 u}{\partial r^2} + \frac{1}{r}\frac{\partial u}{\partial r} + \frac{1}{r^2}\frac{\partial^2 u}{\partial \theta^2} = 0.$$

If the values of $u(r, \theta)$ are specified on the circle $r = R$ as $u(R, \phi)$, $-\pi < \phi \leq \pi$, then Poisson's integral formula states that the value of $u(r, \theta)$ interior to this circle is defined by

$$u(r, \theta) = \frac{R^2 - r^2}{2\pi}\int_{-\pi}^{\pi} \frac{u(R, \phi)}{R^2 + r^2 - 2rR\cos(\theta - \phi)}\,d\phi.$$

Show that this function does indeed satisfy Laplace's equation.

SUMMARY

The definite integral of a function $f(x)$ from $x = a$ to $x = b$ is a limit of a sum

$$\int_a^b f(x)\,dx = \lim_{\|\Delta x_i\|\to 0} \sum_{i=1}^{n} f(x_i^*)\Delta x_i.$$

In this chapter we extended this idea to define double integrals of functions $f(x, y)$ over regions in the xy-plane and triple integrals of functions $f(x, y, z)$ over regions of space. Each is once again the limit of a sum:

$$\iint_R f(x, y)\,dA = \lim_{\|\Delta A_i\|\to 0} \sum_{i=1}^{n} f(x_i^*, y_i^*)\Delta A_i,$$

$$\iiint_V f(x, y, z)\,dV = \lim_{\|\Delta V_i\|\to 0} \sum_{i=1}^{n} f(x_i^*, y_i^*, z_i^*)\Delta V_i.$$

To evaluate double integrals we use double iterated integrals in Cartesian or polar coordinates. Which is the more useful in a given problem depends on the shape of the region R and the form of the function $f(x, y)$. For instance, circles centred at the origin and straight lines through the origin are represented very simply in polar coordinates, and therefore a region R with these curves as boundaries immediately suggests the use of polar coordinates. On the other hand, curves that can be described in the form $y = f(x)$, where $f(x)$ is a polynomial, a rational function, or a transcendental function, often suggest using double iterated integrals in Cartesian coordinates. Each integration in a double iterated integral can be interpreted geometrically, and through these interpretations it is a simple matter to find appropriate limits for the integrals. In particular, for a double iterated integral in Cartesian coordinates, the inner integration is over the rectangles in a strip (horizontal or vertical), and the outer integration adds over all strips. In polar coordinates, the inner integral is inside either a wedge or a ring, and the outer integral adds over all wedges or rings.

To evaluate triple integrals we use triple iterated integrals in Cartesian, cylindrical, or spherical coordinates. Once again the limits on these integrals can be determined by interpreting the summations geometrically. For example, the first integration in a triple iterated integral in Cartesian coordinates is always over the boxes in a column, the second over the rectangles inside a strip, and the third over all strips.

We used double integrals to find plane areas, volumes of solids of revolution, centroids, moments of inertia, fluid forces, and areas of surfaces. We dealt with these same applications (with the exception of surface area) in Chapter 7 using the definite integral, but with some difficulty: Volumes required two methods, shells and washers; centroids required an averaging formula for the first moment of a rectangle that has its length perpendicular to the axis about which a moment is required; moments of inertia needed a "one-third cubed formula" for rectangles with lengths perpendicular to the axis about which the moment of inertia is required; and fluid forces required horizontal rectangles. On the other hand, double integrals eliminate these difficulties but, more importantly, provide a unified approach to all applications.

In Section 14.13 we used double integrals to verify Leibnitz's rule for differentiating definite integrals that depend on a parameter.

Key Terms and Formulas

In reviewing this chapter, you should be able to define or discuss the following key terms:

- Double integral
- Double iterated integral
- Area
- Volume of a solid of revolution
- Fluid pressure
- Centre of mass
- Moment of inertia
- Surface area
- Double integrals in polar coordinates
- Triple integral
- Triple iterated integral
- Volume
- Cylindrical coordinates
- Triple integrals in cylindrical coordinates
- Spherical coordinates
- Triple integrals in spherical coordinates
- Leibnitz's rule

REVIEW EXERCISES

In Exercises 1–21 evaluate the integral over the region.

1. $\iint_R (2x + y)\,dA$ where R is bounded by $y = x$, $y = 0$, $x = 2$

2. $\iiint_V xyz\,dV$ where V is bounded by $y = z$, $x = 0$, $y = 3$, $x = 1$, $z = 0$

3. $\iint_R x^3y^2\,dA$ where R is bounded by $x = 1$, $x = -1$, $y = 2$, $y = -2$

4. $\iiint_V (x^2 - y^3)\,dV$ where V is bounded by $z = xy$, $z = 0$, $x = 1$, $y = 1$

5. $\iiint_V (x^2 - y^2)\,dV$ where V is the region in Exercise 4

6. $\iint_R y\,dA$ where R is bounded by $y = (x - 1)^2$, $y = x + 1$

7. $\iint_R xy^2\,dA$ where R is bounded by $x = 2 - 2y^2$, $x = -y^2$

8. $\iint_R x^2y\,dA$ where R is the region of Exercise 7

9. $\iiint_V (x^2 + y^2 + z^2)\,dV$ where V is bounded by $z = x$, $z = -x$, $y = 0$, $y = 1$, $z = 2$

10. $\iint_R (xy - x^2y^2)\,dA$ where R is bounded by $y = 2x^2$, $y = 4 - 2x^2$

11. $\iint_R x\sin y\,dA$ where R is bounded by $x = \sqrt{1 - y}$, $x = 0$, $y = 0$

12. $\iiint_V (x + y + z)\,dV$ where V is bounded by $z = 1 - x^2 - y^2$, $z = 0$

13. $\iint_R xe^y\,dA$ where R is bounded by $x = 0$, $y = 5$, $y = 2x + 1$

14. $\iiint_V dV$ where V is bounded by $z^2 = x^2 + y^2$, $x^2 + y^2 = 4$

15. $\iint_R (x + y)\,dA$ where R is bounded by $y = 1 - x^2$, $y = 1 - \sqrt{x}$, $y = x - 1$

16. $\iiint_V (x^2 + y^2 + z^2)\,dV$ where V is bounded by $z = \sqrt{1 - x^2 - y^2}$, $z = 0$

17. $\iint_R \frac{x}{x + y}\,dA$ where R is bounded by $y = x - 1$, $x = 2, y = 0$

18. $\iint_R (x^2 + y^2)\,dA$ where R is bounded by $x^2 + y^2 = 2x$

19. $\iiint_V \frac{x^2}{z^2}\,dV$ where V is bounded by $z = 1$, $z = \sqrt{4 - x^2 - y^2}$

20. $\iint_R \frac{1}{x^2 + y^2}\,dA$ where R is bounded by $y = x$, $x = 1$, $x = 2$, $y = 0$

21. $\iint_R (x^2 - y^2)\, dA$ where R is bounded by $y = x$, $y = x - 1$, $y = 5 - 2x$, $y = 14 - 2x$

22. If R represents the region of the xy-plane in Figure 14.78, what double integrals represent the following:

(a) the area of R

(b) the volumes of the solids of revolution when R is rotated about the lines $x = 2$ and $y = -4$

(c) the first moments of area of R about the lines $x = 1$ and $y = -1$

(d) the second moments of area of R about the lines $x = -1$ and $y = 4$

(e) the total charge on R if it carries a charge per unit area $\sigma(x, y)$

(f) the total mass if R is a plate with mass per unit area $\rho(x, y)$

(g) the probability of an electron being in R if the probability of the electron being in unit area at point (x, y) is $P(x, y)$

FIGURE 14.78

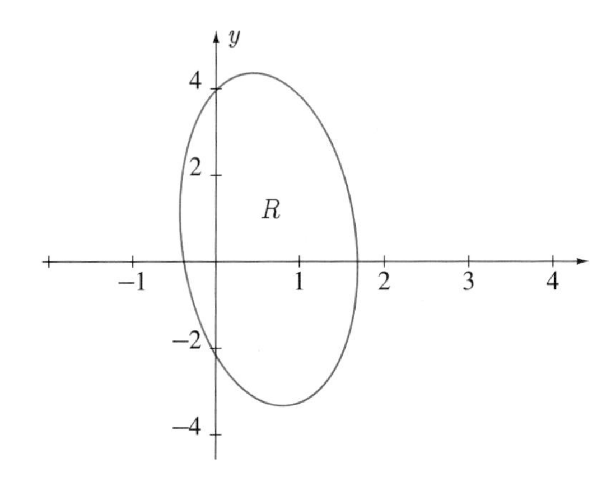

23. Find the volume bounded by the surfaces $z = 0$, $z = 1 - 2e^{-x^2 - y^2}$.

24. Find the area bounded by the curve $x^2(4 - x^2) = y^2$.

25. Find the centroid of the area bounded by the curves $4x = 4 - y^2$, $x = \sqrt{8 - 2y^2}$.

26. Find the volume bounded by the surfaces $y + z = 2$, $y - z = 2$, $y = x^2 + 1$.

27. If the viewing glass on a deep-sea diver's helmet is circular with diameter 10 centimetres, what is the force on the glass when the centre of the glass is 50 m below the surface?

28. Find the area of that part of $z = 1 - 4x^2 - 4y^2$ above the xy-plane.

29. Find the area bounded by the curve $x^4(4 - x^2) = y^2$.

30. Find the volumes of the solids of revolution when the area in Exercise 24 is rotated about the x- and y-axes.

31. Find the moment of inertia about the z-axis of the solid bounded by the surfaces $z = \sqrt{x^2 + y^2}$, $z = 2 - \sqrt{x^2 + y^2}$ if its density ρ is constant.

32. Find the second moment of area about the y-axis of the area in the first quadrant bounded by the curve $x^3 + y^3 = 1$.

33. Find the centre of mass of a uniform solid bounded by the surfaces $z = 1 + x^2 + y^2$, $x^2 + y^2 = 1$, $z = 0$.

34. Find the average value of the function $f(x, y) = x^2 + y^2$ over the region bounded by the curve $x^2 + y^2 = 4$.

35. Find the force due to water pressure on each side of the vertical parallelogram in Figure 14.79. All measurements are in metres.

FIGURE 14.79

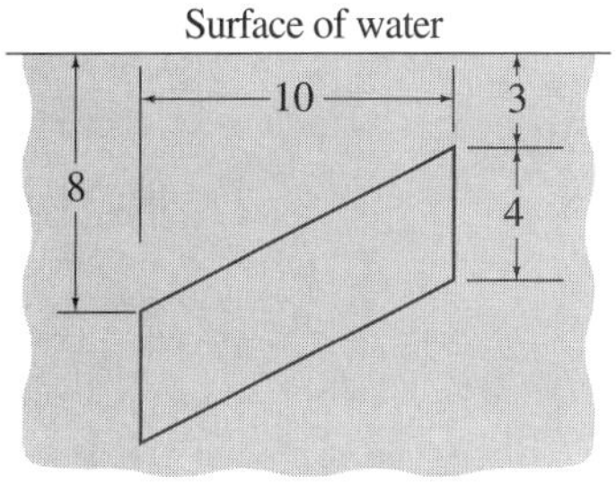

36. Find the moment of inertia about the z-axis of a uniform solid bounded by the surfaces $z = 0$, $z = 1$, $x^2 + y^2 = 1 + z^2$.

37. Find the average value of the function $f(x, y, z) = x + y + z$ over the region bounded by the surfaces $-x + 2y + 2z = 4$, $x + y + z = 2$, $y = 0$, $z = 0$.

38. Find the centroid of the bifolium $r = \sin\theta \cos^2\theta$.

39. Show that

$$y(x) = e^{-3x}(C_1 \cos x + C_2 \sin x) + \int_0^x f(t)e^{3(t-x)} \sin(x - t)\, dt$$

is a solution of the differential equation $\frac{d^2y}{dx^2} + 6\frac{dy}{dx} + 10y = f(x)$ for any constants C_1 and C_2.

40. Find the centre of mass of the uniform solid in the first octant common to the cylinders $x^2 + z^2 = 1$, $y^2 + z^2 = 1$.

41. Find the area of that part of $z = \ln(x^2 + y^2)$ between the cylinders $x^2 + y^2 = 1$ and $x^2 + y^2 = 4$.

42. Find the volume bounded by the surface $\sqrt{x^2 + z^2} = y(2 - y)$.

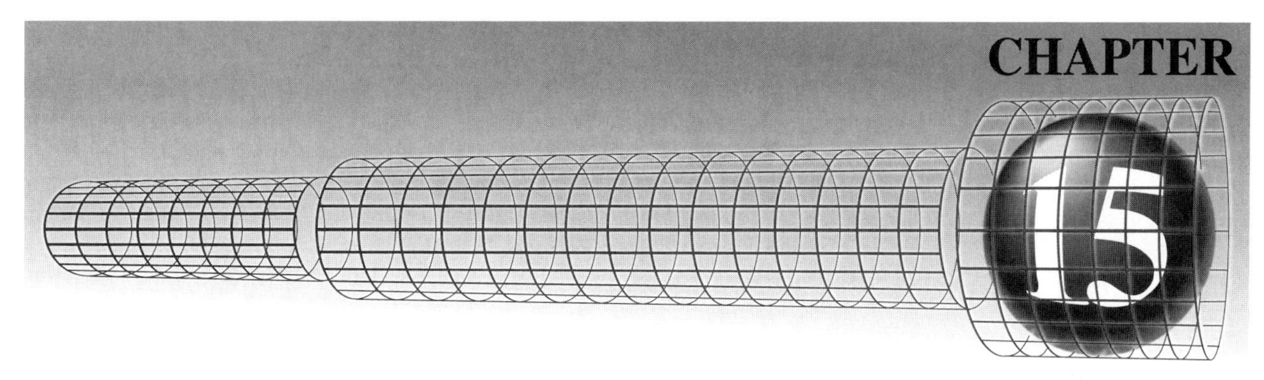

Vector Calculus

In Sections 12.6 and 12.7 we considered vectors whose components are functions of a single variable. In particular, if an object moves along a curve C and C is defined parametrically by $x = x(t)$, $y = y(t)$, $z = z(t)$, $t \geq 0$, where t is time, then the components of the position, velocity, and acceleration vectors are functions of time:

$$\mathbf{r} = \mathbf{r}(t) = x(t)\hat{\mathbf{i}} + y(t)\hat{\mathbf{j}} + z(t)\hat{\mathbf{k}},$$
$$\mathbf{v} = \frac{d\mathbf{r}}{dt} = \frac{dx}{dt}\hat{\mathbf{i}} + \frac{dy}{dt}\hat{\mathbf{j}} + \frac{dz}{dt}\hat{\mathbf{k}},$$
$$\mathbf{a} = \frac{d\mathbf{v}}{dt} = \frac{d^2x}{dt^2}\hat{\mathbf{i}} + \frac{d^2y}{dt^2}\hat{\mathbf{j}} + \frac{d^2z}{dt^2}\hat{\mathbf{k}}.$$

If the object has constant mass m and is subjected to a force $\mathbf{F}$, which is given as a function of time t, $\mathbf{F} = \mathbf{F}(t)$, then Newton's second law expresses the acceleration of the object as $\mathbf{a} = \mathbf{F}(t)/m$. This equation can then be integrated to yield the velocity and position of the object as functions of time. Unfortunately, what often happens is that we do not know $\mathbf{F}$ as a function of time. Instead, we know that if the object were at such and such a position, then the force on it would be such and such; i.e., we know $\mathbf{F}$ as a function of position. For example, suppose a positive charge q is placed at the origin in space (Figure 15.1), and a second positive charge Q is placed at position (x, y, z). According to Coulomb's law, the force on Q due to q is

$$\mathbf{F} = \frac{qQ}{4\pi\epsilon_0|\mathbf{r}|^3}\mathbf{r} = \frac{qQ}{4\pi\epsilon_0(x^2+y^2+z^2)^{3/2}}(x\hat{\mathbf{i}} + y\hat{\mathbf{j}} + z\hat{\mathbf{k}}).$$

FIGURE 15.1

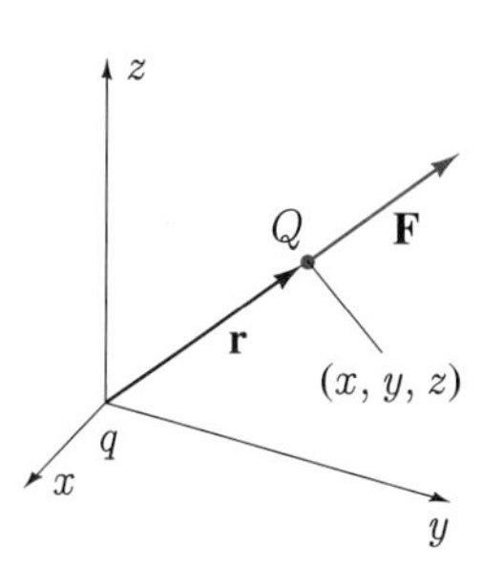

If we allow Q to move under the influence of this force, we will not know $\mathbf{F}$ as a function of time, but rather as a function of position. This makes Newton's second law much more difficult to deal with, but it is in fact the normal situation. Most forces are represented as a function of position rather than time. Besides electrostatic forces, consider, for instance, spring forces, gravitational forces, and fluid forces — all of these are functions of position. Forces that are functions of position are examples of vectors that are functions of position. In this chapter we study vectors that are functions of position. In particular, we integrate them along curves and over surfaces.

SECTION 15.1 Vector Fields

Domains for functions $f(x)$ of one variable are open, closed, half-open, or half-closed intervals on the x-axis. Domains for functions $f(x, y)$ of two variables are sets of points in the xy-plane, and domains for functions $f(x, y, z)$ are sets of points in xyz-space.

In order to state definitions and theorems for multivariable functions in this chapter, we require corresponding definitions of open and closed sets of points. We define them for the xy-plane; analogous definitions for space can be found in Exercise 10.

Consider a set S of points in the xy-plane. A point P in the plane is called an **interior point** of S if there exists a circle centred at P that contains only points of S. A point Q is called an **exterior point** of S if there exists a circle centred at Q that contains no point of S. A point R is called a **boundary point** of S if every circle with centre R contains at least one point in S and at least one point not in S. For example, consider the set of points $S_1 : 4x^2 + 9y^2 < 36$ (Figure 15.2). The fact that we have dotted the ellipse indicates that these points are not in S. Every point inside the ellipse is an interior point of S_1; every point outside the ellipse is an exterior point, and every point on the ellipse is a boundary point. For the set $S_2 : 4x^2 + 9y^2 \geq 36$ (Figure 15.3), every point outside the ellipse is interior to S_2, every point inside the ellipse is exterior to S_2, and every point on the ellipse is a boundary point.

FIGURE 15.2

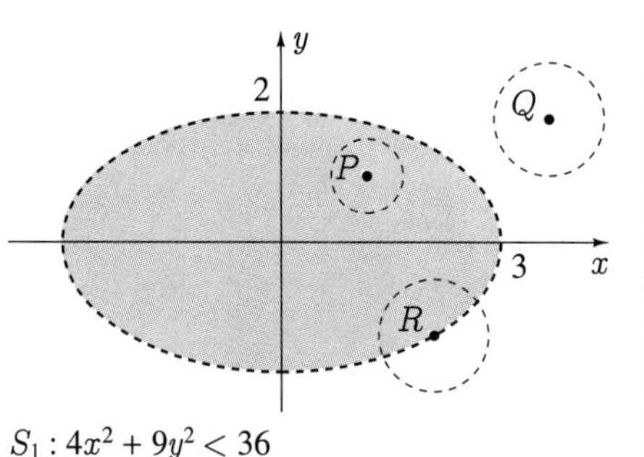

FIGURE 15.3

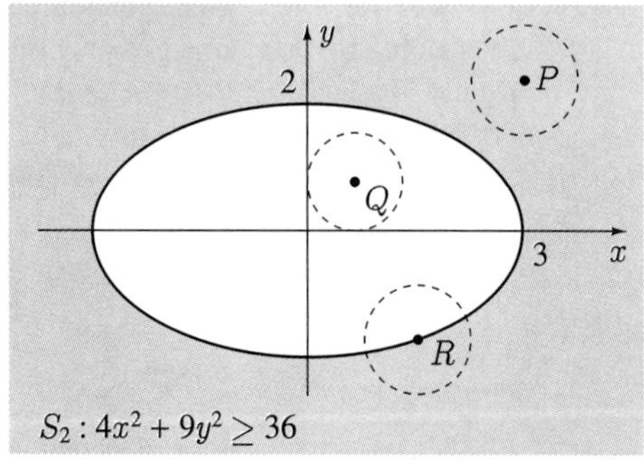

A set of points S is said to be **open** if all points in S are interior points. Alternatively, a set is open if it contains none of its boundary points. A set is said to be **closed** if it contains all of its boundary points. Set S_1 above is open; set S_2 is closed. If to S_1 we add the points on the upper half of the ellipse (Figure 15.4), this set, call it S_3, is neither open nor closed. It contains some of its boundary points but not all of them.

FIGURE 15.4

FIGURE 15.5

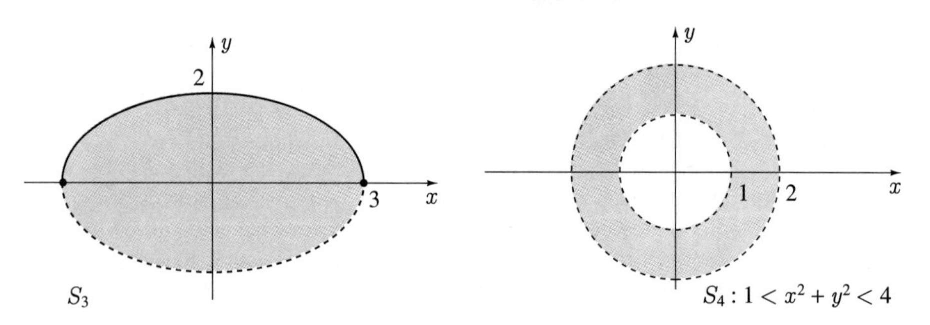

A set S is said to be **connected** if every pair of points in S can be joined by a piecewise smooth curve lying entirely within S. Sets S_1, S_2, and S_3 are all connected. Set S_4 in Figure 15.5 is also connected. Set S_5 in Figure 15.6 is not connected; it consists of two disjoint pieces.

FIGURE 15.6

FIGURE 15.7

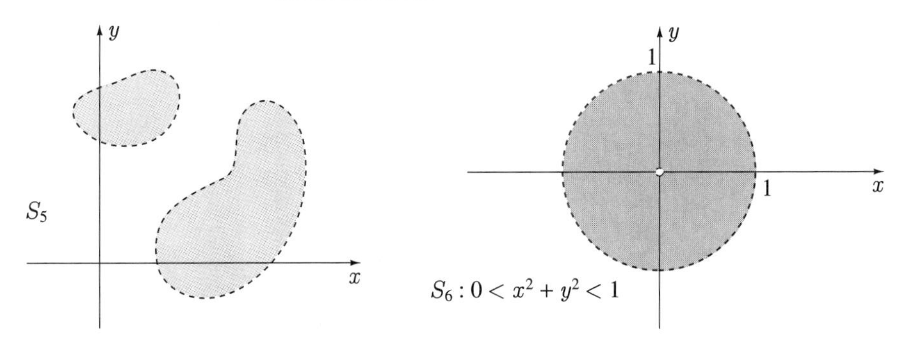

A **domain** is an open, connected set. Domain is perhaps a poor choice of words, as it might be confused with "domain of a function," but it has become accepted terminology. Context always makes it clear which interpretation is intended. Sets S_1 and S_4 are domains; sets S_2, S_3, and S_5 are not.

A domain is said to be **simply-connected** if every closed curve in the domain contains in its interior only points of the domain. In essence, a simply-connected domain has no holes. Domain S_1 is simply-connected, S_4 is not. The set $S_6 : 0 < x^2 + y^2 < 1$ (Figure 15.7) is a domain, but it is not simply-connected.

Analogous definitions for sets of points in space can be found in Exercise 10. One must be somewhat more careful here as there are different "kinds of holes." See Exercises 13 and 15.

When it is not necessary to indicate the particular characteristics of a set of points, we often use the word "region."

Many vectors are functions of position. We call such vectors **vector fields**. To be precise, we say that $\mathbf{F}$ is a vector field in a region D if $\mathbf{F}$ assigns a vector to each point in D. If D is a region of space, then $\mathbf{F}$ assigns a vector $\mathbf{F}(x, y, z)$ to each point in D. If P, Q, and R are the components of $\mathbf{F}(x, y, z)$, then each of these is a function of x, y and z:

$$\mathbf{F} = \mathbf{F}(x, y, z) = P(x, y, z)\hat{\mathbf{i}} + Q(x, y, z)\hat{\mathbf{j}} + R(x, y, z)\hat{\mathbf{k}}. \tag{15.1}$$

If D is a region of the xy-plane, then

$$\mathbf{F} = \mathbf{F}(x, y) = P(x, y)\hat{\mathbf{i}} + Q(x, y)\hat{\mathbf{j}}. \tag{15.2}$$

In Chapter 12 we stressed the fact that the tail of a vector could be placed at any point whatsoever. What was important was relative positions of tip and tail. For vector fields, we almost always place the tail of the vector associated with a point at that point.

Vector fields are essential to the study of most areas in the physical sciences. Moments $(\mathbf{M} = \mathbf{r} \times \mathbf{F})$ in mechanics (see Exercise 43 in Section 12.5) depend on the position of $\mathbf{F}$, and are therefore vector fields. The electric field intensity $\mathbf{E}$, the electric displacement $\mathbf{D}$, the magnetic induction $\mathbf{B}$, and the current density $\mathbf{J}$ are all important vector fields in electromagnetic theory. The heat flux vector $\mathbf{q}$ is the basis for the study of heat conduction.

We have also encountered vector fields that are of geometric importance. For example, the gradient of a scalar function $f(x, y, z)$ is a vector field,

$$\text{grad}\, f = \nabla f = \frac{\partial f}{\partial x}\hat{\mathbf{i}} + \frac{\partial f}{\partial y}\hat{\mathbf{j}} + \frac{\partial f}{\partial z}\hat{\mathbf{k}}. \tag{15.3}$$

It assigns the vector ∇f to each point (x, y, z) in some region of space. We have seen that ∇f points in the direction in which $f(x, y, z)$ increases most rapidly, and its magnitude $|\nabla f|$ is the rate of increase in that direction. In addition, we know that if $f(x, y, z) = c$ is the equation of a surface that passes through a point (x, y, z), then ∇f at that point is normal to the surface.

Often we write

$$\nabla f = \left(\frac{\partial}{\partial x}\hat{\mathbf{i}} + \frac{\partial}{\partial y}\hat{\mathbf{j}} + \frac{\partial}{\partial z}\hat{\mathbf{k}}\right) f$$

and regard

$$\nabla = \frac{\partial}{\partial x}\hat{\mathbf{i}} + \frac{\partial}{\partial y}\hat{\mathbf{j}} + \frac{\partial}{\partial z}\hat{\mathbf{k}}$$

as a vector differential operator, called the **del operator**. It operates on a scalar function to produce a vector field, its gradient. As an operator ∇ should never stand alone, but should always be followed by something on which to operate. Because of this ∇ should not itself be considered a vector, in spite of the fact that it has the form of a vector. It is a differential operator, and must therefore operate on something. In the remainder of this section we use the del operator to define two extremely useful operations on vector fields: the **divergence** and the **curl**.

Definition 15.1

If $\mathbf{F}(x, y, z) = P(x, y, z)\hat{\mathbf{i}} + Q(x, y, z)\hat{\mathbf{j}} + R(x, y, z)\hat{\mathbf{k}}$ is a vector field in a region D, then the **divergence** of $\mathbf{F}$ is a scalar field in D defined by

$$\operatorname{div} \mathbf{F} = \nabla \cdot \mathbf{F} = \frac{\partial P}{\partial x} + \frac{\partial Q}{\partial y} + \frac{\partial R}{\partial z}, \qquad (15.4)$$

provided that the partial derivatives exist at each point in D.

Is it clear why we use the notation $\nabla \cdot \mathbf{F}$, in spite of the fact that ∇ is not a vector in the true sense of the word?

EXAMPLE 15.1

Calculate $\nabla \cdot \mathbf{F}$ if

(a) $\mathbf{F} = 2xy\hat{\mathbf{i}} + z\hat{\mathbf{j}} + x^2\cos(yz)\hat{\mathbf{k}}$.

(b) $\mathbf{F} = \dfrac{qQ}{4\pi\epsilon_0|\mathbf{r}|^3}\mathbf{r}$, where $\mathbf{r} = x\hat{\mathbf{i}} + y\hat{\mathbf{j}} + z\hat{\mathbf{k}}$.

SOLUTION **(a)** $\nabla \cdot \mathbf{F} = \dfrac{\partial}{\partial x}(2xy) + \dfrac{\partial}{\partial y}(z) + \dfrac{\partial}{\partial z}(x^2\cos(yz)) = 2y - x^2 y\sin(yz)$.

(b) For the derivative of the x-component of $\mathbf{F}$ with respect to x, we calculate

$$\frac{\partial}{\partial x}\left\{\frac{qQx}{4\pi\epsilon_0(x^2+y^2+z^2)^{3/2}}\right\} = \frac{qQ}{4\pi\epsilon_0}\left\{\frac{1}{(x^2+y^2+z^2)^{3/2}} - \frac{3x^2}{(x^2+y^2+z^2)^{5/2}}\right\}$$
$$= \frac{qQ}{4\pi\epsilon_0}\left\{\frac{-2x^2+y^2+z^2}{(x^2+y^2+z^2)^{5/2}}\right\}.$$

With similar results for the remaining two derivatives, we obtain

$$\nabla \cdot \mathbf{F} = \frac{qQ}{4\pi\epsilon_0}\left\{\frac{-2x^2+y^2+z^2}{(x^2+y^2+z^2)^{5/2}} + \frac{x^2-2y^2+z^2}{(x^2+y^2+z^2)^{5/2}} + \frac{x^2+y^2-2z^2}{(x^2+y^2+z^2)^{5/2}}\right\}$$
$$= 0.$$

■

Any physical interpretation of $\nabla \cdot \mathbf{F}$ will depend on the interpretation of $\mathbf{F}$. The following discussion describes the interpretation of the divergence of a certain vector field in the theory of fluid flow. If a gas flows through a region D of space, then it flows with some velocity $\mathbf{v}$ past point $P(x, y, z)$ in D at time t (Figure 15.8). Consider a unit area A around P perpendicular to $\mathbf{v}$. If at time t the density of gas at P is ρ, then the vector $\rho\mathbf{v}$ represents the mass of gas flowing through A per unit time. At each point P in D, the direction of $\rho\mathbf{v}$ tells us the direction of gas flow, and its length indicates the mass of gas flowing in that direction. For changing conditions, each of ρ and $\mathbf{v}$ depend not only on position (x, y, z) but also on time t:

$$\rho\mathbf{v} = \rho(x, y, z, t)\mathbf{v}(x, y, z, t).$$

The vector $\rho\mathbf{v}$, then, is a vector field that also depends on time.

In fluid dynamics it is shown that at each point in D, $\rho\mathbf{v}$ must satisfy

$$\nabla \cdot (\rho\mathbf{v}) = -\frac{\partial \rho}{\partial t},$$

FIGURE 15.8

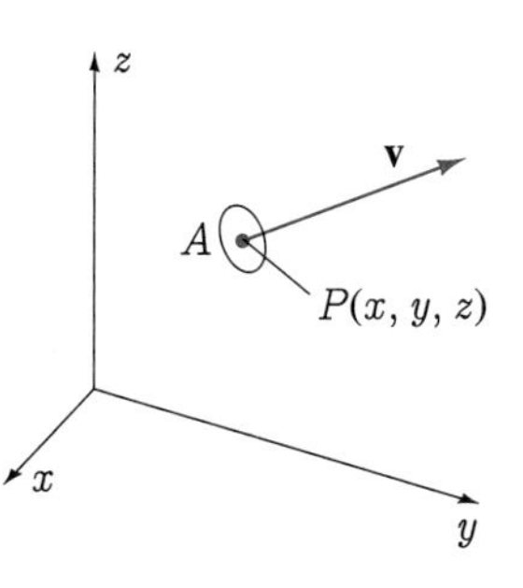

called the **equation of continuity** (see also Section 15.11). It is this equation that gives us an interpretation of the divergence of $\rho\mathbf{v}$. Density is the mass per unit volume of gas. If $\partial\rho/\partial t$ is positive, then density is increasing. This means that more mass must be entering unit volume than leaving it. Similarly, if $\partial\rho/\partial t$ is negative, more mass is leaving than entering. Since $\nabla \cdot (\rho\mathbf{v})$ is the negative of $\partial\rho/\partial t$, it follows that $\nabla \cdot (\rho\mathbf{v})$ must be a measure of how much more gas is leaving unit volume than entering. We can see, then, that the word "divergence" is appropriately chosen for this application.

The curl of a vector field is defined as follows.

Definition 15.2

If $\mathbf{F}(x, y, z) = P(x, y, z)\hat{\mathbf{i}} + Q(x, y, z)\hat{\mathbf{j}} + R(x, y, z)\hat{\mathbf{k}}$ is a vector field in a region D, then the **curl** of $\mathbf{F}$ is a vector function in D defined by

$$\begin{aligned}\text{curl } \mathbf{F} = \nabla \times \mathbf{F} &= \begin{vmatrix} \hat{\mathbf{i}} & \hat{\mathbf{j}} & \hat{\mathbf{k}} \\ \dfrac{\partial}{\partial x} & \dfrac{\partial}{\partial y} & \dfrac{\partial}{\partial z} \\ P & Q & R \end{vmatrix} \\ &= \left(\frac{\partial R}{\partial y} - \frac{\partial Q}{\partial z}\right)\hat{\mathbf{i}} + \left(\frac{\partial P}{\partial z} - \frac{\partial R}{\partial x}\right)\hat{\mathbf{j}} + \left(\frac{\partial Q}{\partial x} - \frac{\partial P}{\partial y}\right)\hat{\mathbf{k}}, \quad (15.5)\end{aligned}$$

provided that the partial derivatives exist at each point in D.

In representing curl $\mathbf{F}$ in the form of a determinant we agree to expand the determinant along the first row.

EXAMPLE 15.2

Calculate the curls of the vector fields in Example 15.1.

SOLUTION **(a)**

$$\nabla \times \mathbf{F} = \begin{vmatrix} \hat{\mathbf{i}} & \hat{\mathbf{j}} & \hat{\mathbf{k}} \\ \dfrac{\partial}{\partial x} & \dfrac{\partial}{\partial y} & \dfrac{\partial}{\partial z} \\ 2xy & z & x^2 \cos(yz) \end{vmatrix}$$
$$= (-x^2 z \sin(yz) - 1)\hat{\mathbf{i}} + (-2x\cos(yz))\hat{\mathbf{j}} + (-2x)\hat{\mathbf{k}}$$

(b)

$$\nabla \times \mathbf{F} = \frac{qQ}{4\pi\epsilon_0} \begin{vmatrix} \hat{\mathbf{i}} & \hat{\mathbf{j}} & \hat{\mathbf{k}} \\ \dfrac{\partial}{\partial x} & \dfrac{\partial}{\partial y} & \dfrac{\partial}{\partial z} \\ \dfrac{x}{|\mathbf{r}|^3} & \dfrac{y}{|\mathbf{r}|^3} & \dfrac{z}{|\mathbf{r}|^3} \end{vmatrix}$$

The x-component of $\nabla \times \mathbf{F}$ is $qQ(4\pi\epsilon_0)$ multiplied by

$$\frac{\partial}{\partial y}\left(\frac{z}{|\mathbf{r}|^3}\right) - \frac{\partial}{\partial z}\left(\frac{y}{|\mathbf{r}|^3}\right) = \frac{\partial}{\partial y}\left(\frac{z}{(x^2+y^2+z^2)^{3/2}}\right)$$
$$- \frac{\partial}{\partial z}\left(\frac{y}{(x^2+y^2+z^2)^{3/2}}\right)$$
$$= \frac{-3yz}{(x^2+y^2+z^2)^{5/2}} + \frac{3yz}{(x^2+y^2+z^2)^{5/2}} = 0.$$

Similar results for the y- and z-components give $\nabla \times \mathbf{F} = \mathbf{0}$. ■

If a vector $\rho\mathbf{v}$ is defined as above for gas flow through a region D, then the curl of $\rho\mathbf{v}$ describes the tendency for the motion of the gas to be circular rather than flowing in a straight line (see Section 15.11). This suggests why the term "curl" is used. With this interpretation, the following definition seems reasonable.

Definition 15.3

A vector field $\mathbf{F}$ is said to be **irrotational** in a region D if in D

$$\nabla \times \mathbf{F} = \mathbf{0}. \tag{15.6}$$

Applications of divergence and curl extend far beyond the topic of fluid dynamics. Both concepts are indispensable in many areas of applied mathematics such as electromagnetism, continuum mechanics, and heat conduction, to name a few.

The del operator ∇ operates on a scalar field to produce the gradient of the scalar field and on a vector field to give the divergence and the curl of the vector field. The following list of properties for the del operator is straightforward to verify (see Exercise 43).

If f and g are scalar fields that have first partial derivatives in a region D, and if $\mathbf{F}$ and $\mathbf{G}$ are vector fields in D with components that have first partial derivatives, then:

$$\nabla(f+g) = \nabla f + \nabla g, \tag{15.7}$$
$$\nabla \cdot (\mathbf{F}+\mathbf{G}) = \nabla \cdot \mathbf{F} + \nabla \cdot \mathbf{G}, \tag{15.8}$$

$$\nabla \times (\mathbf{F} + \mathbf{G}) = \nabla \times \mathbf{F} + \nabla \times \mathbf{G}, \tag{15.9}$$

$$\nabla(fg) = f\nabla g + g\nabla f, \tag{15.10}$$

$$\nabla \cdot (f\mathbf{F}) = \nabla f \cdot \mathbf{F} + f\nabla \cdot \mathbf{F}, \tag{15.11}$$

$$\nabla \times (f\mathbf{F}) = \nabla f \times \mathbf{F} + f\nabla \times \mathbf{F}, \tag{15.12}$$

$$\nabla \cdot (\mathbf{F} \times \mathbf{G}) = \mathbf{G} \cdot (\nabla \times \mathbf{F}) - \mathbf{F} \cdot (\nabla \times \mathbf{G}), \tag{15.13}$$

$$\nabla \times (\nabla f) = \mathbf{0}, \tag{15.14}$$

$$\nabla \cdot (\nabla \times \mathbf{F}) = 0. \tag{15.15}$$

For properties (15.14) and (15.15), we assume that f and $\mathbf{F}$ have continuous second-order partial derivatives in D. A typical way to verify these identities is to reduce each side of the identity to the same quantity. For example, to verify 15.13, we set $\mathbf{F} = P\hat{\mathbf{i}} + Q\hat{\mathbf{j}} + R\hat{\mathbf{k}}$ and $\mathbf{G} = L\hat{\mathbf{i}} + M\hat{\mathbf{j}} + N\hat{\mathbf{k}}$. Then

$$\mathbf{F} \times \mathbf{G} = (QN - RM)\hat{\mathbf{i}} + (RL - PN)\hat{\mathbf{j}} + (PM - QL)\hat{\mathbf{k}};$$

thus,

$$\begin{aligned}
\nabla \cdot (\mathbf{F} \times \mathbf{G}) &= \frac{\partial}{\partial x}(QN - RM) + \frac{\partial}{\partial y}(RL - PN) + \frac{\partial}{\partial z}(PM - QL) \\
&= Q\frac{\partial N}{\partial x} + N\frac{\partial Q}{\partial x} - R\frac{\partial M}{\partial x} - M\frac{\partial R}{\partial x} + R\frac{\partial L}{\partial y} + L\frac{\partial R}{\partial y} - P\frac{\partial N}{\partial y} \\
&\quad - N\frac{\partial P}{\partial y} + P\frac{\partial M}{\partial z} + M\frac{\partial P}{\partial z} - Q\frac{\partial L}{\partial z} - L\frac{\partial Q}{\partial z} \\
&= L\left(\frac{\partial R}{\partial y} - \frac{\partial Q}{\partial z}\right) + M\left(\frac{\partial P}{\partial z} - \frac{\partial R}{\partial x}\right) + N\left(\frac{\partial Q}{\partial x} - \frac{\partial P}{\partial y}\right) \\
&\quad + P\left(\frac{\partial M}{\partial z} - \frac{\partial N}{\partial y}\right) + Q\left(\frac{\partial N}{\partial x} - \frac{\partial L}{\partial z}\right) + R\left(\frac{\partial L}{\partial y} - \frac{\partial M}{\partial x}\right).
\end{aligned}$$

On the other hand,

$$\begin{aligned}
\mathbf{G} \cdot (\nabla \times \mathbf{F}) - \mathbf{F} \cdot (\nabla \times \mathbf{G}) &= \mathbf{G} \cdot \begin{vmatrix} \hat{\mathbf{i}} & \hat{\mathbf{j}} & \hat{\mathbf{k}} \\ \partial/\partial x & \partial/\partial y & \partial/\partial z \\ P & Q & R \end{vmatrix} - \mathbf{F} \cdot \begin{vmatrix} \hat{\mathbf{i}} & \hat{\mathbf{j}} & \hat{\mathbf{k}} \\ \partial/\partial x & \partial/\partial y & \partial/\partial z \\ L & M & N \end{vmatrix} \\
&= (L, M, N) \cdot \left(\frac{\partial R}{\partial y} - \frac{\partial Q}{\partial z}, \frac{\partial P}{\partial z} - \frac{\partial R}{\partial x}, \frac{\partial Q}{\partial x} - \frac{\partial P}{\partial y}\right) \\
&\quad - (P, Q, R) \cdot \left(\frac{\partial N}{\partial y} - \frac{\partial M}{\partial z}, \frac{\partial L}{\partial z} - \frac{\partial N}{\partial x}, \frac{\partial M}{\partial x} - \frac{\partial L}{\partial y}\right) \\
&= L\left(\frac{\partial R}{\partial y} - \frac{\partial Q}{\partial z}\right) + M\left(\frac{\partial P}{\partial z} - \frac{\partial R}{\partial x}\right) + N\left(\frac{\partial Q}{\partial x} - \frac{\partial P}{\partial y}\right) \\
&\quad - P\left(\frac{\partial N}{\partial y} - \frac{\partial M}{\partial z}\right) - Q\left(\frac{\partial L}{\partial z} - \frac{\partial N}{\partial x}\right) \\
&\quad - R\left(\frac{\partial M}{\partial x} - \frac{\partial L}{\partial y}\right),
\end{aligned}$$

and this is the same expression as for $\nabla \cdot (\mathbf{F} \times \mathbf{G})$.

Given a scalar function $f(x, y, z)$ it is straightforward to calculate its gradient ∇f. Conversely, given the gradient of a function ∇f, it is possible to find the function $f(x, y, z)$. For example, if the vector field

$$\mathbf{F} = (3x^2yz + z^2)\hat{\mathbf{i}} + (x^3z + 2y)\hat{\mathbf{j}} + (x^3y + 2xz + 1)\hat{\mathbf{k}}$$

is known to be the gradient of some function $f(x,y,z)$, then

$$\frac{\partial f}{\partial x}\hat{\mathbf{i}} + \frac{\partial f}{\partial y}\hat{\mathbf{j}} + \frac{\partial f}{\partial z}\hat{\mathbf{k}} = (3x^2yz + z^2)\hat{\mathbf{i}} + (x^3z + 2y)\hat{\mathbf{j}} + (x^3y + 2xz + 1)\hat{\mathbf{k}}.$$

Since two vectors are equal if and only if they have identical components, we can say that

$$\frac{\partial f}{\partial x} = 3x^2yz + z^2, \quad \frac{\partial f}{\partial y} = x^3z + 2y, \quad \frac{\partial f}{\partial z} = x^3y + 2xz + 1.$$

Integration of the first of these with respect to x, holding y and z constant, implies that $f(x,y,z)$ must be of the form

$$f(x,y,z) = x^3yz + xz^2 + v(y,z)$$

for some function $v(y,z)$. To determine $v(y,z)$, we substitute this $f(x,y,z)$ into the second equation,

$$x^3z + \frac{\partial v}{\partial y} = x^3z + 2y,$$

and this implies that

$$\frac{\partial v}{\partial y} = 2y.$$

Consequently,

$$v(y,z) = y^2 + w(z)$$

for some $w(z)$, and therefore

$$f(x,y,z) = x^3yz + xz^2 + y^2 + w(z).$$

We now know both the x- and y-dependence of $f(x,y,z)$. To find $w(z)$ we substitute into the equation for $\partial f/\partial z$:

$$x^3y + 2xz + \frac{dw}{dz} = x^3y + 2xz + 1.$$

This equation requires

$$\frac{dw}{dz} = 1,$$

from which we have

$$w(z) = z + C, \quad C = \text{ a constant.}$$

Thus,

$$f(x,y,z) = x^3yz + xz^2 + y^2 + z + C,$$

and this represents all functions that have a gradient equal to the given vector $\mathbf{F}$.

A much more difficult question is to determine whether a given vector field $\mathbf{F}(x,y,z)$ is the gradient of some scalar function $f(x,y,z)$. In the vast majority of cases, we can say that if the above procedure fails, then the answer is no. For instance, if $\mathbf{F} = x^2y\hat{\mathbf{i}} + xy\hat{\mathbf{j}} + z\hat{\mathbf{k}}$, and we attempt to find a function $f(x,y,z)$ so that $\nabla f = \mathbf{F}$, then

$$\frac{\partial f}{\partial x} = x^2y, \quad \frac{\partial f}{\partial y} = xy, \quad \frac{\partial f}{\partial z} = z.$$

The first implies that

$$f(x,y,z) = \frac{x^3y}{3} + v(y,z),$$

and when this is substituted into the second, we get

$$\frac{x^3}{3} + \frac{\partial v}{\partial y} = xy, \quad \text{or} \quad \frac{\partial v}{\partial y} = xy - \frac{x^3}{3}.$$

But this is an impossible situation since v is to be a function of y and z only. How then could its derivative depend on x? Although this type of argument will suffice in most examples, it is really not a satisfactory mathematical answer. The following theorem gives a test by which to determine whether a given vector function is the gradient of some scalar function.

Theorem 15.1

Suppose that the components $P(x,y,z)$, $Q(x,y,z)$, and $R(x,y,z)$ of $\mathbf{F} = P\hat{\mathbf{i}} + Q\hat{\mathbf{j}} + R\hat{\mathbf{k}}$ have continuous first partial derivatives in a domain D. If there exists a function $f(x,y,z)$ defined in D such that $\nabla f = \mathbf{F}$, then $\nabla \times \mathbf{F} = \mathbf{0}$. Conversely, if D is simply-connected, and $\nabla \times \mathbf{F} = \mathbf{0}$ in D, then there exists a function $f(x,y,z)$ such that $\nabla f = \mathbf{F}$ in D.

It is obvious that if $\mathbf{F} = \nabla f$, then $\nabla \times \mathbf{F} = \mathbf{0}$, for this is the result of equation 15.14. To prove the converse result requires Stokes's theorem from Section 15.10, and a proof is therefore delayed until that time. Notice that in the converse result, the domain (open, connected set) must be simply-connected. This is our first encounter with a situation where the nature of a region is important to the result.

In the special case in which $\mathbf{F}$ is a vector field in the xy-plane, the equation $\nabla \times \mathbf{F} = \mathbf{0}$ is still the condition for existence of a function $f(x,y)$ such that $\nabla f = \mathbf{F} = P\hat{\mathbf{i}} + Q\hat{\mathbf{j}}$, but the condition reduces to

$$\mathbf{0} = \begin{vmatrix} \hat{\mathbf{i}} & \hat{\mathbf{j}} & \hat{\mathbf{k}} \\ \partial/\partial x & \partial/\partial y & \partial/\partial z \\ P & Q & 0 \end{vmatrix} = \left(\frac{\partial Q}{\partial x} - \frac{\partial P}{\partial y}\right)\hat{\mathbf{k}},$$

or

$$\frac{\partial Q}{\partial x} = \frac{\partial P}{\partial y}. \tag{15.16}$$

EXAMPLE 15.3

Find, if possible, a function $f(x,y)$ such that

$$\nabla f = \left(\frac{x^3 - 2y^2}{x^3 y}\right)\hat{\mathbf{i}} + \left(\frac{y^2 - x^3}{x^2 y^2}\right)\hat{\mathbf{j}}.$$

SOLUTION We first note that

$$\frac{\partial}{\partial x}\left(\frac{y^2 - x^3}{x^2 y^2}\right) = \frac{\partial}{\partial x}\left(\frac{1}{x^2} - \frac{x}{y^2}\right) = \frac{-2}{x^3} - \frac{1}{y^2}$$

and

$$\frac{\partial}{\partial y}\left(\frac{x^3 - 2y^2}{x^3 y}\right) = \frac{\partial}{\partial y}\left(\frac{1}{y} - \frac{2y}{x^3}\right) = \frac{-1}{y^2} + \frac{-2}{x^3}.$$

Since the components of $\mathbf{F}$ are undefined whenever $x = 0$ or $y = 0$, we can state that in any simply-connected domain that does not contain points on either of the axes, there is a function $f(x,y)$ such that $\nabla f = \mathbf{F}$. To find $f(x,y)$, we set

$$\frac{\partial f}{\partial x} = \frac{1}{y} - \frac{2y}{x^3}, \quad \frac{\partial f}{\partial y} = \frac{1}{x^2} - \frac{x}{y^2}.$$

From the first equation, we have

$$f(x,y) = \frac{x}{y} + \frac{y}{x^2} + v(y),$$

which substituted into the second equation gives us

$$-\frac{x}{y^2} + \frac{1}{x^2} + \frac{dv}{dy} = \frac{1}{x^2} - \frac{x}{y^2}.$$

Consequently,

$$\frac{dv}{dy} = 0 \quad \text{or} \quad v(y) = C,$$

and therefore

$$f(x,y) = \frac{x}{y} + \frac{y}{x^2} + C.$$

■

EXERCISES 15.1

In Exercises 1–9 determine whether the set of points in the xy-plane is open, closed, connected, a domain, and/or a simply-connected domain.

1. $x^2 + (y+1)^2 < 4$

2. $x^2 + (y-3)^2 \leq 4$

3. $0 < x^2 + (y-1)^2 < 16$

4. $1 < (x-4)^2 + (y+1)^2 \leq 9$

5. $x > 3$

6. $y \leq -2$

7. $2(x-1)^2 - (y+2)^2 < 16$

8. All points satisfying $x^2 + y^2 < 1$ or $(x-2)^2 + y^2 < 1$

9. $4(x+1)^2 + 9(y-2)^2 > 20$

10. Give definitions of the following for sets of points in xyz-space: interior, exterior, and boundary points; open, closed, and connected sets; domain and simply-connected domain.

In Exercises 11–19 determine whether the set of points in space is open, closed, connected, a domain and/or a simply-connected domain.

11. $x^2 + y^2/4 + z^2/9 < 1$

12. $z \geq x^2 + y^2$

13. $x^2 + y^2 + z^2 > 0$

14. $1 < x^2 + y^2 + z^2 < 4,\ x \geq 0,\ y \geq 0,\ z \geq 0$

15. $x^2 + y^2 > 0$

16. $|z| > 0$

17. $z^2 > x^2 + y^2$

18. $z^2 > x^2 + y^2 - 1$

19. $z^2 < x^2 + y^2 - 1$

20. Prove that the only nonempty set in the xy-plane that is both open and closed is the whole plane.

In Exercises 21–40 calculate the required quantity.

21. ∇f if $f(x,y,z) = 3x^2y - y^3z^2$

22. ∇f if $f(x,y,z) = (x^2 + y^2 + z^2)^{-1/2}$

23. ∇f if $f(x,y) = \text{Tan}^{-1}(y/x)$

24. ∇f at $(1,2)$ if $f(x,y) = x^3y - 2x\cos y$

25. ∇f at $(1,-1,4)$ if $f(x,y,z) = e^{xyz}$

26. $\nabla \cdot \mathbf{F}$ if $\mathbf{F}(x,y,z) = 2xe^y\hat{\mathbf{i}} + 3x^2z\hat{\mathbf{j}} - 2x^2yz\hat{\mathbf{k}}$

27. $\nabla \cdot \mathbf{F}$ if $\mathbf{F}(x, y) = x \ln y\hat{\mathbf{i}} - y^3 e^x \hat{\mathbf{j}}$

28. $\nabla \cdot \mathbf{F}$ if $\mathbf{F}(x, y, z) = \sin(x^2 + y^2 + z^2)\hat{\mathbf{i}} + \cos(y + z)\hat{\mathbf{j}}$

29. $\nabla \cdot \mathbf{F}$ if $\mathbf{F}(x, y) = e^x \hat{\mathbf{i}} + e^y \hat{\mathbf{j}}$

30. $\nabla \cdot \mathbf{F}$ at $(1, 1, 1)$ if $\mathbf{F}(x, y, z) = x^2 y^3 \hat{\mathbf{i}} - 3xy\hat{\mathbf{j}} + z^2 \hat{\mathbf{k}}$

31. $\nabla \cdot \mathbf{F}$ at $(-1, 3)$ if $\mathbf{F}(x, y) = (x + y)^2 (\hat{\mathbf{i}} + \hat{\mathbf{j}})$

32. $\nabla \cdot \mathbf{F}$ if $\mathbf{F}(x, y, z) = (x\hat{\mathbf{i}} + y\hat{\mathbf{j}} + z\hat{\mathbf{k}})/\sqrt{x^2 + y^2 + z^2}$

33. $\nabla \cdot \mathbf{F}$ if $\mathbf{F}(x, y) = \mathrm{Cot}^{-1}(xy)\hat{\mathbf{i}} + \mathrm{Tan}^{-1}(xy)\hat{\mathbf{j}}$

34. $\nabla \times \mathbf{F}$ if $\mathbf{F}(x, y, z) = x^2 z\hat{\mathbf{i}} + 12xyz\hat{\mathbf{j}} + 32y^2 z^4 \hat{\mathbf{k}}$

35. $\nabla \times \mathbf{F}$ if $\mathbf{F}(x, y) = xe^y \hat{\mathbf{i}} - 2xy^2 \hat{\mathbf{j}}$

36. $\nabla \times \mathbf{F}$ if $\mathbf{F}(x, y, z) = x^2 \hat{\mathbf{i}} + y^2 \hat{\mathbf{j}} + z^2 \hat{\mathbf{k}}$

37. $\nabla \times \mathbf{F}$ at $(1, -1, 1)$ if $\mathbf{F}(x, y, z) = xz^3 \hat{\mathbf{i}} - 2x^2 yz\hat{\mathbf{j}} + 2yz^4 \hat{\mathbf{k}}$

38. $\nabla \times \mathbf{F}$ at $(2, 0)$ if $\mathbf{F}(x, y) = y\hat{\mathbf{i}} - x\hat{\mathbf{j}}$

39. $\nabla \times \mathbf{F}$ if $\mathbf{F}(x, y, z) = \ln(x + y + z)(\hat{\mathbf{i}} + \hat{\mathbf{j}} + \hat{\mathbf{k}})$

40. $\nabla \times \mathbf{F}$ if $\mathbf{F}(x, y) = \mathrm{Sec}^{-1}(x + y)\hat{\mathbf{i}} + \mathrm{Csc}^{-1}(y + x)\hat{\mathbf{j}}$

41. If $\mathbf{F} = x^2 y\hat{\mathbf{i}} - 2xz\hat{\mathbf{j}} + 2yz\hat{\mathbf{k}}$, find $\nabla \times (\nabla \times \mathbf{F})$.

42. **(a)** Verify that Laplace's equation 13.12 can be expressed in the form $\nabla \cdot \nabla f = 0$.
(b) Show that $f(x, y, z) = (x^2 + y^2 + z^2)^{-1/2}$ satisfies Laplace's equation.

43. Prove properties 15.7–15.12, 15.14, and 15.15.

44. A gas is moving through some region D of space. If we follow a particular particle of the gas, it traces out some curved path $C: x = x(t), y = y(t), z = z(t)$, where t is time. If $\rho(x, y, z, t)$ is the density of gas at any point in D at time t, then along C we can express density in terms of t only, $\rho = \rho[x(t), y(t), z(t), t]$. Show that along C,

$$\frac{d\rho}{dt} = \frac{\partial \rho}{\partial t} + \nabla \rho \cdot \frac{d\mathbf{r}}{dt},$$

where $\mathbf{r} = x\hat{\mathbf{i}} + y\hat{\mathbf{j}} + z\hat{\mathbf{k}}$.

In Exercises 45–49 find all functions $f(x, y)$ such that ∇f is equal to the vector field.

45. $\mathbf{F}(x, y) = 2xy\hat{\mathbf{i}} + x^2 \hat{\mathbf{j}}$

46. $\mathbf{F}(x, y) = (3x^2 y^2 + 3)\hat{\mathbf{i}} + (2x^3 y + 2)\hat{\mathbf{j}}$

47. $\mathbf{F}(x, y) = e^y \hat{\mathbf{i}} + (xe^y + 4y^2)\hat{\mathbf{j}}$

48. $\mathbf{F}(x, y) = (x + y)^{-1}(\hat{\mathbf{i}} + \hat{\mathbf{j}})$

49. $\mathbf{F}(x, y) = -xy(1 - x^2 y^2)^{-1/2}(y\hat{\mathbf{i}} + x\hat{\mathbf{j}})$

In Exercises 50–55 find all functions $f(x, y, z)$ such ∇f is equal to the vector field.

50. $\mathbf{F}(x, y, z) = x\hat{\mathbf{i}} + y\hat{\mathbf{j}} + z\hat{\mathbf{k}}$

51. $\mathbf{F}(x, y, z) = yz\hat{\mathbf{i}} + xz\hat{\mathbf{j}} + (yx - 3)\hat{\mathbf{k}}$

52. $\mathbf{F}(x, y, z) = (1 + x + y + z)^{-1}(\hat{\mathbf{i}} + \hat{\mathbf{j}} + \hat{\mathbf{k}})$

53. $\mathbf{F}(x, y, z) = (2x/y^2 + 1)\hat{\mathbf{i}} - (2x^2/y^3)\hat{\mathbf{j}} - 2z\hat{\mathbf{k}}$

54. $\mathbf{F}(x, y, z) = (1 + x^2 y^2)^{-1}(y\hat{\mathbf{i}} + x\hat{\mathbf{j}}) + z\hat{\mathbf{k}}$

55. $\mathbf{F}(x, y, z) = (3x^2 y + yz + 2xz^2)\hat{\mathbf{i}} + (xz + x^3 + 3z^2 - 6y^2 z)\hat{\mathbf{j}} + (2x^2 z + 6yz - 2y^3 + xy)\hat{\mathbf{k}}$

56. **(a)** Find constants a, b, and c in order that the vector field

$$\mathbf{F} = (x^2 + 2y + az)\hat{\mathbf{i}} + (bx - 3y - z)\hat{\mathbf{j}} + (4x + cy + 2z)\hat{\mathbf{k}}$$

be irrotational.
(b) If $\mathbf{F}$ is irrotational, find a scalar function $f(x, y, z)$ such that $\nabla f = \mathbf{F}$.

57. A vector field $\mathbf{F}(x, y, z)$ is said to be *solenoidal* if $\nabla \cdot \mathbf{F} = 0$.
(a) Are either of $\mathbf{F} = (2x^2 + 8xy^2 z)\hat{\mathbf{i}} + (3x^3 y - 3xy)\hat{\mathbf{j}} - (4y^2 z^2 + 2x^3 z)\hat{\mathbf{k}}$ or $xyz^2 \mathbf{F}$ solenoidal?
(b) Show that $\nabla f \times \nabla g$ is solenoidal for arbitrary functions $f(x, y, z)$ and $g(x, y, z)$ which have continuous second partial derivatives.

58. Associated with every electric field is a scalar function $V(x, y, z)$ called potential. It is defined by $\mathbf{E} = -\nabla V$, where $\mathbf{E}$ is a vector field called the electric field intensity. In addition, if a point charge Q is placed at a point (x, y, z) in the electric field, then the force $\mathbf{F}$ on Q is $\mathbf{F} = Q\mathbf{E}$.
(a) If the force on Q due to a charge q at the origin is

$$\mathbf{F} = \frac{qQ}{4\pi\epsilon_0 |\mathbf{r}|^3}\mathbf{r},$$

where $\mathbf{r} = x\hat{\mathbf{i}} + y\hat{\mathbf{j}} + z\hat{\mathbf{k}}$ and ϵ_0 is a constant, find $V(x, y, z)$ for the field due to q.
(b) If the entire xy-plane is given a uniform charge density σ units of charge per unit area, it is found that the force on a charge Q placed z units above the plane is $\mathbf{F} = [Q\sigma/(2\epsilon_0)]\hat{\mathbf{k}}$. Find the potential V for the electric field due to this charge distribution.

59. Show that if $\mathbf{v} = \omega \times \mathbf{r}$, where ω is a constant vector, and $\mathbf{r} = x\hat{\mathbf{i}} + y\hat{\mathbf{j}} + z\hat{\mathbf{k}}$, then $\omega = (1/2)(\nabla \times \mathbf{v})$.

60. Show that if a function $f(x, y, z)$ satisfies Laplace's equation 13.12, then its gradient is both irrotational and solenoidal.

61. Theorem 15.1 indicates that a vector field $\mathbf{F}$ is the gradient of some scalar field if $\nabla \times \mathbf{F} = \mathbf{0}$. Sometimes a given vector field $\mathbf{F}$ is the curl of another field $\mathbf{v}$; that is $\mathbf{F} = \nabla \times \mathbf{v}$. The following theorem indicates when this is the case: Let D be the interior of a sphere in which the components of a vector field $\mathbf{F}$ have continuous first partial derivatives. Then there exists a vector field $\mathbf{v}$ defined in D such that $\mathbf{F} = \nabla \times \mathbf{v}$ if and only if $\nabla \cdot \mathbf{F} = 0$ in D. In other words, $\mathbf{F}$ is the curl of a vector field if and only if $\mathbf{F}$ is solenoidal.

(a) Show that if $\mathbf{F} = P\hat{\mathbf{i}} + Q\hat{\mathbf{j}} + R\hat{\mathbf{k}}$ is solenoidal, then the components of $\mathbf{v} = L\hat{\mathbf{i}} + M\hat{\mathbf{j}} + N\hat{\mathbf{k}}$ would have to satisfy the equations

$$P = \frac{\partial N}{\partial y} - \frac{\partial M}{\partial z}, \quad Q = \frac{\partial L}{\partial z} - \frac{\partial N}{\partial x}, \quad R = \frac{\partial M}{\partial x} - \frac{\partial L}{\partial y}.$$

(b) Show that the vector field $\mathbf{v}(x, y, z)$ defined by

$$\mathbf{v}(x, y, z) = \int_0^1 t\mathbf{F}(tx, ty, tz) \times (x, y, z)\, dt$$

satisfies these equations. In other words, this formula defines a possible vector $\mathbf{v}$. Is it unique?

(c) Show that if $\mathbf{F}$ satisfies the property that $\mathbf{F}(tx, ty, tz) = t^n \mathbf{F}(x, y, z)$, then

$$\mathbf{v}(x, y, z) = \frac{1}{n+2}\mathbf{F} \times \mathbf{r}, \qquad \mathbf{r} = x\hat{\mathbf{i}} + y\hat{\mathbf{j}} + z\hat{\mathbf{k}}.$$

In Exercises 62–64 verify that the vector field is solenoidal, and then use the formulas in Exercise 61 to find a vector field $\mathbf{v}$ such that $\mathbf{F} = \nabla \times \mathbf{v}$.

62. $\mathbf{F} = x\hat{\mathbf{i}} + y\hat{\mathbf{j}} - 2z\hat{\mathbf{k}}$

63. $\mathbf{F} = (1 + x)\hat{\mathbf{i}} - (x + z)\hat{\mathbf{k}}$

64. $\mathbf{F} = 2x^2\hat{\mathbf{i}} - y^2\hat{\mathbf{j}} + (2yz - 4xz)\hat{\mathbf{k}}$

SECTION 15.2

Line Integrals

Just as definite integrals, double integrals, and triple integrals are defined as limits of sums, so too are line and surface integrals. The only difference is that line integrals are applied to functions defined along curves, and surface integrals involve functions defined on surfaces.

A curve C in space is defined parametrically by three functions

$$C: \quad \begin{aligned} x &= x(t), \\ y &= y(t), \quad \alpha \le t \le \beta \\ z &= z(t), \end{aligned} \tag{15.17}$$

where α and β specify initial and final points A and B of the curve, respectively (Figure 15.9). The direction of a curve is from initial to final point, and in Section 12.7 we agreed to parametrize a curve using parameters that increase in the direction of the curve.

FIGURE 15.9

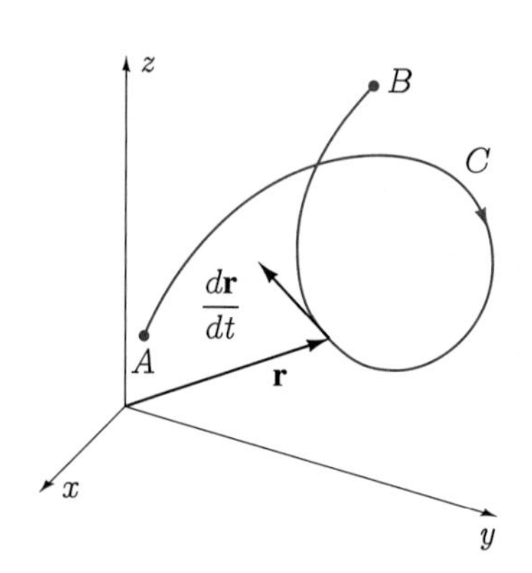

The curve is said to be continuous if each of the functions $x(t)$, $y(t)$, and $z(t)$ is continuous (implying that C is at no point separated). It is said to be smooth if each of these functions has a continuous first derivative; geometrically, this means that the tangent vector $d\mathbf{r}/dt = x'(t)\hat{\mathbf{\imath}} + y'(t)\hat{\mathbf{\jmath}} + z'(t)\hat{\mathbf{k}}$ turns gradually or smoothly along C. A continuous curve that is not smooth but can be divided into a finite number of smooth subcurves is said to be piecewise smooth.

Suppose a function $f(x,y,z)$ is defined along a curve C joining A to B (Figure 15.10). We divide C into n subcurves of lengths $\Delta s_1, \Delta s_2, \ldots, \Delta s_n$ by any $n-1$ consecutive points $A = P_0, P_1, P_2, \ldots, P_{n-1}, P_n = B$, whatsoever. On each subcurve of length $\Delta s_i (i = 1, \ldots, n)$ we choose an arbitrary point $P_i^*(x_i^*, y_i^*, z_i^*)$. We then form the sum

$$f(x_1^*, y_1^*, z_1^*)\Delta s_1 + f(x_2^*, y_2^*, z_2^*)\Delta s_2 + \cdots + f(x_n^*, y_n^*, z_n^*)\Delta s_n = \sum_{i=1}^{n} f(x_i^*, y_i^*, z_i^*)\Delta s_i. \quad (15.18)$$

FIGURE 15.10

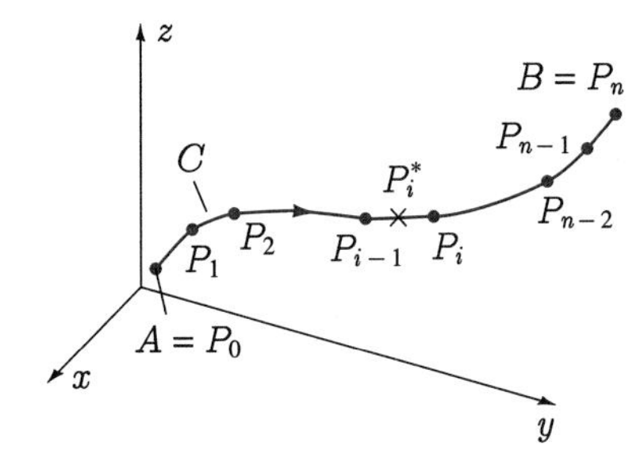

If this sum approaches a limit as the number of subcurves becomes increasingly large and the length of every subcurve approaches zero, we call the limit the line integral of $f(x,y,z)$ along the curve C, and denote it by

$$\int_C f(x,y,z)\,ds = \lim_{\|\Delta s_i\| \to 0} \sum_{i=1}^{n} f(x_i^*, y_i^*, z_i^*)\Delta s_i. \quad (15.19)$$

A more appropriate name might be curvilinear integral, rather than line integral, but line integral has become the accepted terminology. We must simply regard the word "line" as meaning "curved line" rather than "straight line."

For definition 15.19 to be useful we must demand that the limit be independent of the manner of subdivision of C and choice of star-points on the subcurves. Theorem 15.2 indicates that for continuous functions defined on smooth curves, this requirement is indeed satisfied.

Theorem 15.2

Let $f(x,y,z)$ be continuous on a smooth curve C of finite length, $C: x = x(t)$, $y = y(t)$, $z = z(t)$, $\alpha \le t \le \beta$. Then the line integral of $f(x,y,z)$ along C exists and can be evaluated by means of the following definite integral:

$$\int_C f(x,y,z)\,ds = \int_\alpha^\beta f[x(t), y(t), z(t)]\sqrt{\left(\frac{dx}{dt}\right)^2 + \left(\frac{dy}{dt}\right)^2 + \left(\frac{dz}{dt}\right)^2}\,dt. \quad (15.20)$$

It is not necessary to memorize 15.20 as a formula, since it is the result obtained by expressing x, y, z, and ds in terms of t and interpreting that result as a definite integral with respect to t. To be more explicit, recall from equation 12.74 that when length along a curve is measured from its initial point, then an infinitesimal length ds along C corresponding to an increment dt in t is given by

$$ds = \sqrt{(dx)^2 + (dy)^2 + (dz)^2} = \sqrt{\left\{\left(\frac{dx}{dt}\right)^2 + \left(\frac{dy}{dt}\right)^2 + \left(\frac{dz}{dt}\right)^2\right\}(dt)^2}$$
$$= \sqrt{\left(\frac{dx}{dt}\right)^2 + \left(\frac{dy}{dt}\right)^2 + \left(\frac{dz}{dt}\right)^2}\, dt.$$

If we substitute this into the left-hand side of 15.20 and at the same time use the equations for C to express $f(x, y, z)$ in terms of t, then

$$\int_C f(x,y,z)\,ds = \int_C f[x(t), y(t), z(t)]\sqrt{\left(\frac{dx}{dt}\right)^2 + \left(\frac{dy}{dt}\right)^2 + \left(\frac{dz}{dt}\right)^2}\, dt.$$

But if we now interpret the right-hand side of this equation as the definite integral of $f[x(t), y(t), z(t)]\sqrt{(dx/dt)^2 + (dy/dt)^2 + (dz/dt)^2}$ with respect to t and affix limits $t = \alpha$ and $t = \beta$ that identify end points of C, we obtain 15.20.

To evaluate a line integral, then, we simply express $f(x, y, z)$ and ds in terms of some parameter along C and evaluate the resulting definite integral. If equations for C are given in the form $C : y = y(x),\ z = z(x),\ x_A \le x \le x_B$, then x is a convenient parameter, and equation 15.20 takes the form

$$\int_C f(x,y,z)\,ds = \int_{x_A}^{x_B} f[x, y(x), z(x)]\sqrt{1 + \left(\frac{dy}{dx}\right)^2 + \left(\frac{dz}{dx}\right)^2}\, dx. \quad (15.21)$$

Similar expressions exist if either y or z is a convenient parameter.

When C is piecewise smooth rather than smooth, the line integral along C is found by evaluating the line integral along each smooth subcurve and adding the results.

EXAMPLE 15.4

Evaluate the line integral of $f(x, y, z) = 8x + 6xy + 30z$ from $A(0, 0, 0)$ to $B(1, 1, 1)$

(a) along the straight line joining A to B with parametrization

$$C_1 : \quad x = t, \quad y = t, \quad z = t, \quad 0 \le t \le 1.$$

(b) along the straight line in **(a)** with parametrization

$$C_1 : \quad x = -1 + \frac{t}{2}, \quad y = -1 + \frac{t}{2}, \quad z = -1 + \frac{t}{2}, \quad 2 \le t \le 4.$$

(c) along the curve (Figure 15.11)

$$C_2 : \quad x = t, \quad y = t^2, \quad z = t^3, \quad 0 \le t \le 1.$$

SOLUTION **(a)**

FIGURE 15.11

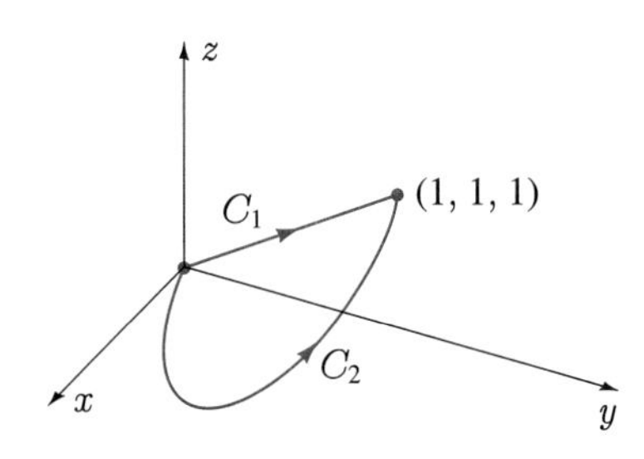

$$\int_{C_1} (8x + 6xy + 30z)\,ds = \int_0^1 (8t + 6t^2 + 30t)\sqrt{(1)^2 + (1)^2 + (1)^2}\,dt$$

$$= \sqrt{3}\int_0^1 (38t + 6t^2)\,dt = \sqrt{3}\{19t^2 + 2t^3\}_0^1$$

$$= 21\sqrt{3}.$$

(b)

$$\int_{C_1} (8x + 6xy + 30z)\,ds = \int_2^4 \{8(-1 + t/2) + 6(-1 + t/2)^2 + 30(-1 + t/2)\}$$

$$\times \sqrt{(1/2)^2 + (1/2)^2 + (1/2)^2}\,dt$$

$$= \frac{\sqrt{3}}{2}\int_2^4 \{38(-1 + t/2) + 6(-1 + t/2)^2\}dt$$

$$= \frac{\sqrt{3}}{2}\{38(-1 + t/2)^2 + 4(-1 + t/2)^3\}_2^4$$

$$= \frac{\sqrt{3}}{2}\{38 + 4\} = 21\sqrt{3}.$$

(c)

$$\int_{C_2} (8x + 6xy + 30z)\,ds = \int_0^1 (8t + 6t^3 + 30t^3)\sqrt{(1)^2 + (2t)^2 + (3t^2)^2}\,dt$$

$$= \int_0^1 (8t + 36t^3)\sqrt{1 + 4t^2 + 9t^4}\,dt$$

$$= \{\tfrac{2}{3}(1 + 4t^2 + 9t^4)^{3/2}\}_0^1 = \tfrac{2}{3}(14\sqrt{14} - 1).$$

■

Parts (a) and (b) of this example suggest that the value of a line integral does not depend on the particular parametrization of the curve used in its evaluation. This is indeed true, and should perhaps be expected since definition 15.20 makes no reference whatsoever to parametrization of the curve. For a proof of this fact see Exercise 37. Different parameters normally lead to different definite integrals, but they all give exactly the same value for the line integral. Parts (a) and (c) illustrate that a line integral does depend on the curve joining the points A and B; i.e., the value of the line integral may change if the curve C joining A and B changes.

EXAMPLE 15.5

Evaluate the line integral of $f(x, y) = x^2 + y^2$ once clockwise around the circle $x^2 + y^2 = 4$, $z = 0$.

SOLUTION If we use the parametrization

$$C: \quad x = 2\cos t, \quad y = -2\sin t, \quad z = 0, \quad 0 \le t \le 2\pi$$

(Figure 15.12), then

$$\int_C (x^2 + y^2)\,ds = \int_0^{2\pi} (4)\sqrt{\left(\frac{dx}{dt}\right)^2 + \left(\frac{dy}{dt}\right)^2}\,dt$$
$$= 4\int_0^{2\pi} \sqrt{(-2\sin t)^2 + (-2\cos t)^2}\,dt = 8\int_0^{2\pi} dt = 16\pi.$$

FIGURE 15.12

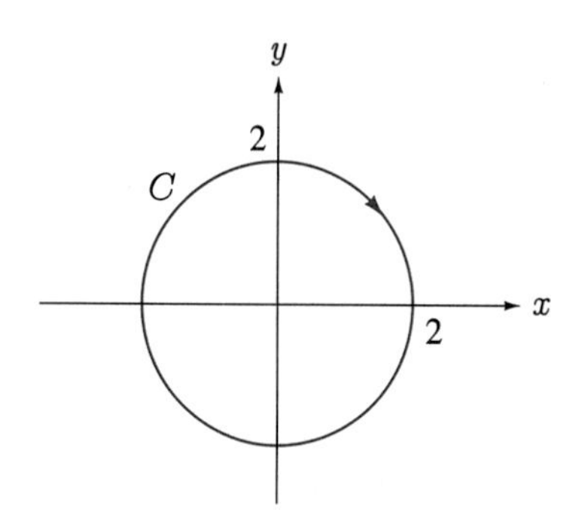

■

The value of this line integral, as well as many other line integrals in the xy-plane, can be given a geometric interpretation. Suppose a function $f(x, y)$ is positive along a curve C in the xy-plane. If at each point of C we draw a vertical line of height $z = f(x, y)$, then a vertical wall is constructed as shown in Figure 15.13. Since ds is an elemental piece of length along C, the quantity $f(x, y)\,ds$ can be interpreted as approximately the area of the vertical wall projecting onto ds. Because the line integral

$$\int_C f(x, y)\,ds,$$

like all integrals, is a limit-summation process, we interpret the value of this line integral as the total area of the vertical wall. Correct as this interpretation is, it really is of little use in the evaluation of line integrals and, in addition, the interpretation is valid only if the curve C along which the line integral is performed is contained in a plane.

FIGURE 15.13

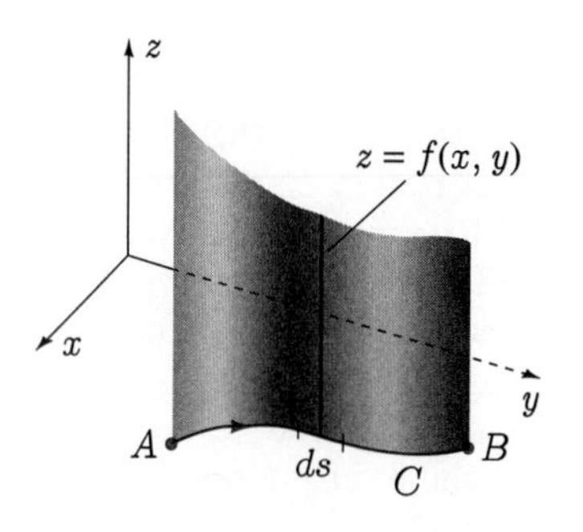

Because a line integral is a limit-summation, it should be obvious that the line integral

$$\int_C ds \tag{15.22}$$

represents the length of the curve C. If C is a curve in the xy-plane, we substitute $ds = \sqrt{(dx)^2 + (dy)^2}$, and if C is a curve in space, then $ds = \sqrt{(dx)^2 + (dy)^2 + (dz)^2}$. This agrees with the results of equations 12.67 and 12.74.

We make one last point about notation. To indicate that a line integral is being evaluated around a closed curve, we usually draw a circle on the integral sign, as follows:

$$\oint_C f(x, y, z)\, ds.$$

Such would be the case for the curve of intersection of the cylinder $x^2 + y^2 = 1$ and the plane $x + z = 1$ (Figure 15.14).

FIGURE 15.14

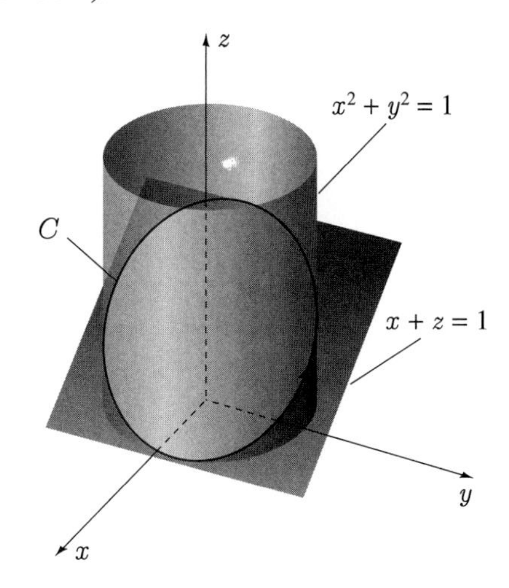

When C is a closed curve in the xy-plane that does not cross itself, we indicate the direction along C by an arrowhead on the circle. For the curves shown in Figure 15.15 we write

$$\oint_{C_1} f(x, y)\, ds \quad \text{and} \quad \oint_{C_2} f(x, y)\, ds.$$

FIGURE 15.15

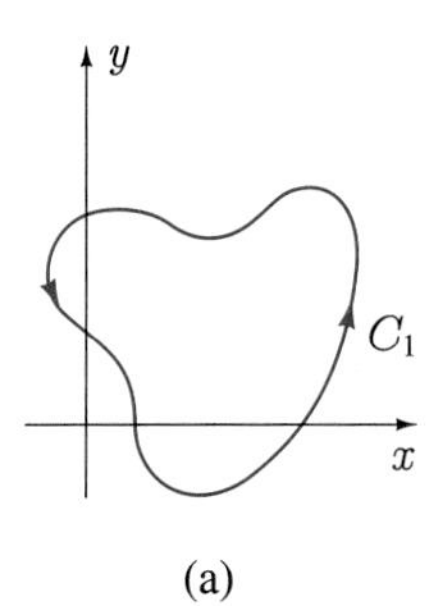

(a)

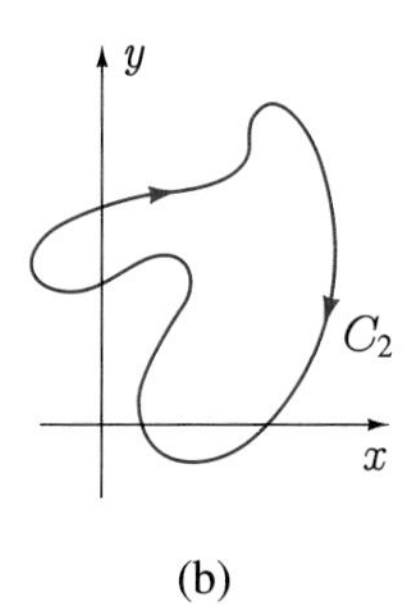

(b)

EXERCISES 15.2

In Exercises 1–6 evaluate the line integral.

1. $\int_C x\,ds$ where C is the curve $y = x^2$, $z = 0$ from $(0,0,0)$ to $(1,1,0)$

2. $\oint_C (x^2 + y^2)\,ds$ once around the square C in the xy-plane with vertices $(\pm 1, 1)$ and $(\pm 1, -1)$

3. $\oint_C (2 + x - 2xy)\,ds$ once around the circle $x^2 + y^2 = 4$, $z = 0$

4. $\int_C (x^2 + yz)\,ds$ along the straight line from $(1,2,-1)$ to $(3,2,5)$

5. $\int_C xy\,ds$ where C is the first octant part of $x^2 + y^2 = 1$, $x^2 + z^2 = 1$ from $(1,0,0)$ to $(0,1,1)$

6. $\int_C x^2 yz\,ds$ where C is the curve $z = x + y$, $x + y + z = 1$ from $(1,-1/2,1/2)$ to $(-3,7/2,1/2)$

7. Prove that the length of the circumference of a circle is 2π multiplied by the radius.

8. A spring has six coils in the form of the helix

$$x = 3\cos t, \quad y = 3\sin t, \quad z = 3t/(4\pi), \quad 0 \le t \le 12\pi,$$

where all dimensions are in centimetres. Find the length of the spring.

9. Use the parametric equations $x = \cos^3\theta$, $y = \sin^3\theta$, $0 \le \theta < 2\pi$, to sketch the astroid $x^{2/3} + y^{2/3} = 1$ in the xy-plane. At each point (x, y) on the astroid, a vertical line is drawn with height $z = x^2 + y^2$, thus forming a cylindrical wall. Find the area of the wall.

In Exercises 10–16 evaluate the line integral.

10. $\int_C xz\,ds$ along the first octant part of $y = x^2$, $z + y = 1$ from $(0,0,1)$ to $(1,1,0)$

11. $\int_C (x+y)^5\,ds$ along $C:\ x = t + 1/t$, $y = t - 1/t$ from $(2,0)$ to $(17/4, 15/4)$

12. $\int_C x\sqrt{y+z}\,ds$ where C is that part of the curve $3x + 2y + 3z = 6$, $x - 2y + 4z = 5$ from $(1,0,1)$ to $(0, 9/14, 11/7)$

13. $\int_C xy\,ds$ where C is the curve $x = 1 - y^2$, $z = 0$ from $(1,0,0)$ to $(0,1,0)$

14. $\int_C (x+y)z\,ds$ where C is the curve $y = x$, $z = 1 + y^4$ from $(-1,-1,2)$ to $(1,1,2)$

15. $\int_C \frac{1}{y+z}\,ds$ where C is the curve $y = x^2$, $z = x^2$ from $(1,1,1)$ to $(2,4,4)$

16. $\int_C (2y + 9z)\,ds$ where C is the curve $z = xy$, $x = y^2$ from $(0,0,0)$ to $(4,-2,-8)$

17. **(a)** If the curve C in Figure 15.16 is rotated around the y-axis, show that the area of the surface that it traces out is represented by the line integral

$$\int_C 2\pi x\,ds.$$

(b) If C is rotated around the x-axis, what line integral represents the area of the surface traced out?

FIGURE 15.16

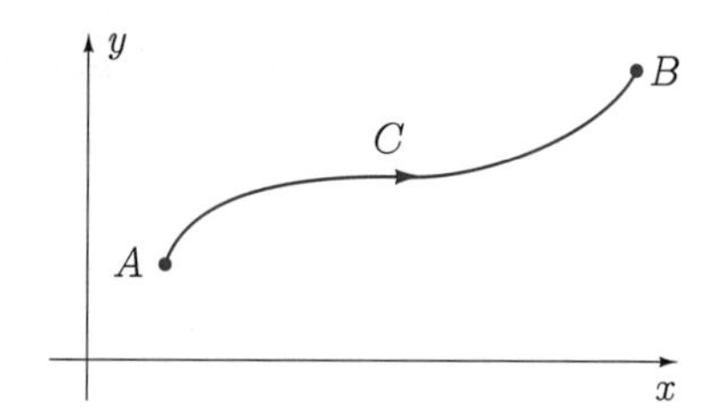

In Exercises 18–20, use the method in Exercise 17 to find the area of the surface traced out when the curve is rotated about the line.

18. $y = x^3$, $1 \le x \le 2$, about $y = 0$

19. $24xy = x^4 + 48$, $1 \le x \le 2$, about $x = 0$

20. $8y^2 = x^2(1 - x^2)$ about $y = 0$

21. Find the length of the parabola $y = x^2$ from $(0,0)$ to $(1,1)$.

In Exercises 22 and 23 find a definite integral which can be used to evaluate the line integral. Use power series to approximate the definite integral accurate to three decimals.

22. $\int_C xy\,ds$ where C is the curve $y = x^3$, $z = 0$ from $(0,0,0)$ to $(1/2, 1/8, 0)$

23. $\displaystyle\int_C e^{-(x+y-2)^2}\, ds$ where C is the curve $x + y + z = 2$, $y + 2z = 3$ from $(-1,3,0)$ to $(0,1,1)$

The average value of a function $f(x,y,z)$ defined along a curve C is defined as the value of the line integral of the function along the curve divided by the length of the curve. In Exercises 24–27 find the average value of the function along the curve.

24. $f(x,y) = x^2y^2$ along $C:\ x^2 + y^2 = 4,\ z = 0$

25. $f(x,y,z) = x^2 + y^2 + z^2$ along $C:\ x = \cos t$, $y = \sin t,\ z = t,\ 0 \le t \le \pi$

26. $f(x,y,z) = xyz$ along $C:\ z = x^2,\ y = x^2$ from $(0,0,0)$ to $(1,1,1)$

27. $f(x,y) = y$ along $C:\ y = x^3/4 + 1/(3x)$ from $(1,7/12)$ to $(2,13/16)$

28. At each point on the curve $(x^2 + y^2)^2 = x^2 - y^2$ a vertical line is drawn with height equal to the distance from the point to the origin. Find the area of the vertical wall so formed.

29. During a sleet storm, a power line between two poles at positions $x = \pm 20$ hangs in the shape $y = 40\cosh(x/40) - 10$, where all distances are measured in metres. Ice accumulates more heavily on the middle part of the line than at the ends. In fact, the combined mass of ice and line per unit length in the x-direction at position x is given in kg/m by the formula $\rho(x) = 1 - |x|/40$. Find the total mass of the line.

In Exercises 30–33 use the fact that in polar coordinates small lengths along a curve can be expressed in the form $ds = \sqrt{r^2 + (dr/d\theta)^2}\, d\theta$ (see formula 10.14) to evaluate the line integral.

30. $\displaystyle\int_C \frac{x}{\sqrt{x^2+y^2}}\, ds$ where C is the first quadrant part of the limacon $r = 2 - \sin\theta$ starting from the point on the x-axis

31. $\displaystyle\oint_C (x^2 + y^2)\, ds$ where C is the cardioid $r = 1 + \cos\theta$

32. $\displaystyle\int_C xy\, ds$ where C is the spiral $r = e^\theta$ from $\theta = 0$ to $\theta = 2\pi$

33. $\displaystyle\oint_C \cos^3 2\theta\, ds$ around the first quadrant loop of the lemniscate $r^2 = \sin 2\theta$

In Exercises 34 and 35 find a definite integral which can be used to evaluate the line integral. Use Simpson's rule with 10 equal subdivisions to approximate the definite integral.

34. $\displaystyle\int_C (x^2y + z)\, ds$ where C is the curve $z = x^2 + y^2$, $y + x = 1$ from $(-1,2,5)$ to $(1,0,1)$

35. $\displaystyle\oint_C x^2y^2\, ds$ where C is the ellipse $4x^2 + 9y^2 = 36$, $z = 0$

36. Find the surface area of the torus obtained by rotating the circle $(x-a)^2 + y^2 = b^2$ $(a > b)$ about the y-axis.

37. Show that the value of a line integral is independent of the parameter used to specify the curve.

SECTION 15.3

Line Integrals Involving Vector Functions

There are many ways in which $f(x,y,z)$ in the line integral

$$\int_C f(x,y,z)\, ds \qquad (15.23)$$

can arise. According to equation 15.22 we must choose $f(x,y,z) = 1$ in order to find the length of the curve C; Exercise 17 in Section 15.2 indicates that for the areas of the surfaces traced out when a curve in the xy-plane is rotated about the y- and x-axes, we must choose $f(x,y)$ equal to $2\pi x$ and $2\pi y$, respectively.

The most important and common type of line integral occurs when $f(x,y,z)$ is specified as the tangential component of some given vector field $\mathbf{F}(x,y,z)$ defined along C; i.e., $f(x,y,z)$ itself is not given, but $\mathbf{F}$ is, and to find $f(x,y,z)$ we must calculate the tangential component of $\mathbf{F}$ along C. By the tangential component of $\mathbf{F}(x,y,z)$ along C we mean the component of $\mathbf{F}$ along that tangent vector to C which points in the same direction as C (Figure 15.17).

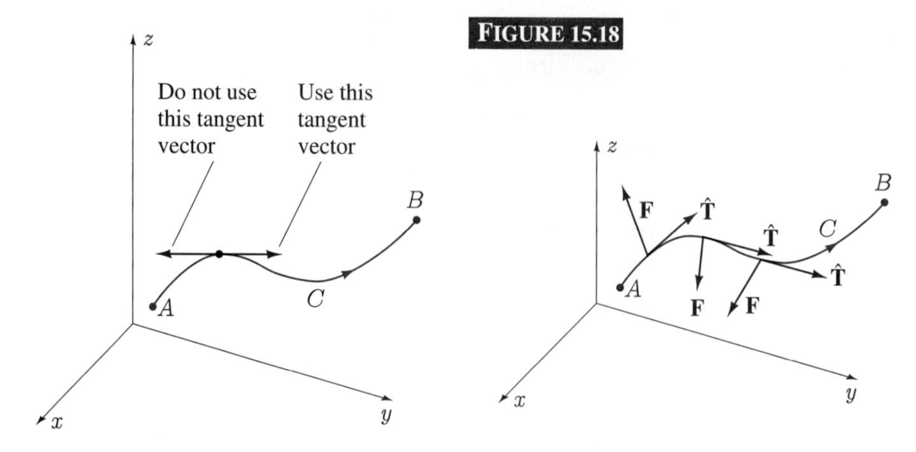

In Section 12.8 we saw that if s is a measure of length along a curve C from A to B (Figure 15.18), and if s is chosen equal to zero at A, then a unit tangent vector pointing in the direction of motion along C is

$$\hat{\mathbf{T}} = \frac{d\mathbf{r}}{ds}. \tag{15.24}$$

Consequently, if $f(x, y, z)$ is the tangential component of $\mathbf{F}(x, y, z)$ along C, then

$$f(x, y, z) = \mathbf{F} \cdot \hat{\mathbf{T}} = \mathbf{F} \cdot \frac{d\mathbf{r}}{ds}, \tag{15.25}$$

and we can write that

$$\int_C f(x, y, z)\, ds = \int_C \mathbf{F} \cdot \hat{\mathbf{T}}\, ds = \int_C \mathbf{F} \cdot \frac{d\mathbf{r}}{ds} ds = \int_C \mathbf{F} \cdot d\mathbf{r}. \tag{15.26}$$

If the components of the vector field $\mathbf{F}(x, y, z)$ are

$$\mathbf{F}(x, y, z) = P(x, y, z)\hat{\mathbf{i}} + Q(x, y, z)\hat{\mathbf{j}} + R(x, y, z)\hat{\mathbf{k}}, \tag{15.27}$$

then

$$\int_C \mathbf{F} \cdot d\mathbf{r} = \int_C (P\, dx + Q\, dy + R\, dz),$$

and if the parentheses are omitted, we have

$$\int_C \mathbf{F} \cdot d\mathbf{r} = \int_C P\, dx + Q\, dy + R\, dz. \tag{15.28}$$

This discussion has shown that when the integrand $f(x, y, z)$ of a line integral is specified as the tangential component of $\mathbf{F} = P\hat{\mathbf{i}} + Q\hat{\mathbf{j}} + R\hat{\mathbf{k}}$ along C, the product $f(x, y, z)\, ds$ can be replaced by the sum of products $P\, dx + Q\, dy + R\, dz$:

$$\int_C f(x, y, z)\, ds = \int_C \mathbf{F} \cdot d\mathbf{r} = \int_C P\, dx + Q\, dy + R\, dz. \tag{15.29}$$

According to the results of Section 15.2, evaluation of this line integral can be accomplished by expressing $P\, dx + Q\, dy + R\, dz$ in terms of any parametric representation of C and evaluating the resulting definite integral.

EXAMPLE 15.6

Evaluate

$$\int_C \frac{z}{y}dx + (x^2 + y^2 + z^2)\,dz,$$

where C is the first octant intersection of $x^2 + y^2 = 1$ and $z = 2x + 4$ joining $(1, 0, 6)$ to $(0, 1, 4)$.

SOLUTION If we choose the parametrization

$$x = \cos t, \quad y = \sin t, \quad z = 2\cos t + 4, \quad 0 \le t \le \pi/2,$$

for C (Figure 15.19), then

$$\begin{aligned}
&\int_C \frac{z}{y}dx + (x^2 + y^2 + z^2)\,dz \\
&\quad = \int_0^{\pi/2} \left\{ \left(\frac{2\cos t + 4}{\sin t}\right)(-\sin t\,dt) \right. \\
&\qquad \left. + (\cos^2 t + \sin^2 t + 4\cos^2 t + 16\cos t + 16)(-2\sin t\,dt) \right\} \\
&\quad = -2\int_0^{\pi/2} \{\cos t + 2 + 17\sin t + 4\cos^2 t\sin t + 16\cos t\sin t\}dt \\
&\quad = -2\left\{ \sin t + 2t - 17\cos t - \frac{4\cos^3 t}{3} + 8\sin^2 t \right\}_0^{\pi/2} = -2\pi - \frac{164}{3}.
\end{aligned}$$

FIGURE 15.19

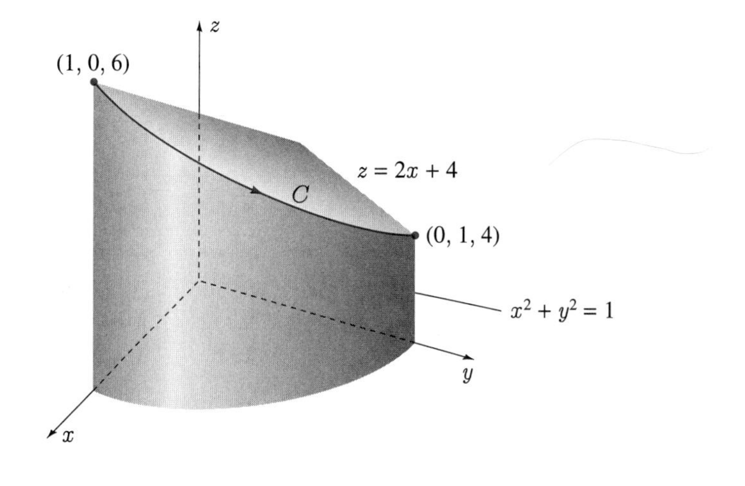

■

EXAMPLE 15.7

Evaluate

$$\oint_C y^2\,dx + x^2\,dy,$$

where C is the closed curve shown in Figure 15.20.

FIGURE 15.20

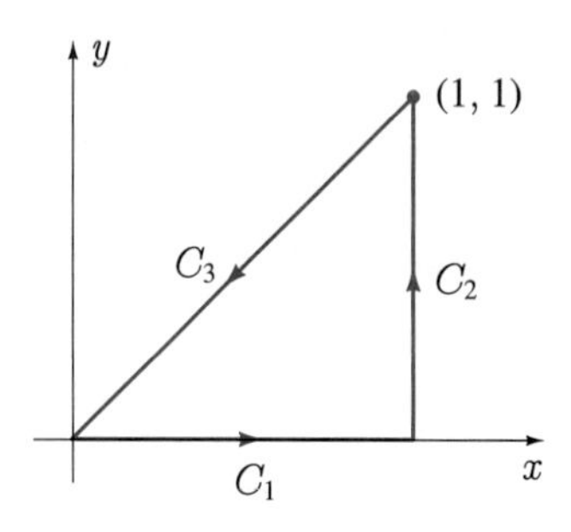

SOLUTION If we start at the origin and denote the three straight line paths by

$$C_1: \quad y = 0, \quad 0 \le x \le 1; \quad C_2: \quad x = 1, \quad 0 \le y \le 1;$$
$$C_3: \quad y = x = 1 - t, \quad 0 \le t \le 1,$$

then

$$\begin{aligned}\oint_C y^2\,dx + x^2\,dy &= \int_{C_1} y^2\,dx + x^2\,dy + \int_{C_2} y^2\,dx + x^2\,dy + \int_{C_3} y^2\,dx + x^2\,dy \\ &= \int_0^1 0\,dx + x^2 0 + \int_0^1 y^2 0 + 1\,dy \\ &\quad + \int_0^1 (1-t)^2(-dt) + (1-t)^2(-dt) \\ &= \{y\}_0^1 + \left\{\frac{2}{3}(1-t)^3\right\}_0^1 = 1 - \frac{2}{3} = \frac{1}{3}.\end{aligned}$$

■

EXAMPLE 15.8

If a curve C has initial and final points A and B, then the curve that traces the same points but has initial and final points B and A is denoted by $-C$. Show that

$$\int_{-C} \mathbf{F} \cdot d\mathbf{r} = -\int_C \mathbf{F} \cdot d\mathbf{r}. \tag{15.30}$$

SOLUTION When 15.17 are parametric equations for C, then parametric equations for $-C$ are

$$-C: \quad x = x(t), \quad y = y(t), \quad z = z(t), \quad \beta \ge t \ge \alpha.$$

To obtain an increasing parameter along $-C$, we set $u = -t$, in which case

$$-C: \quad x = x(-u), \quad y = y(-u), \quad z = z(-u), \quad -\beta \le u \le -\alpha.$$

If $\mathbf{F} = P\hat{\mathbf{i}} + Q\hat{\mathbf{j}} + R\hat{\mathbf{k}}$, then the value of the line integral along $-C$ can be expressed as a definite integral with respect to u:

$$\begin{aligned}\int_{-C} \mathbf{F} \cdot d\mathbf{r} &= \int_{-C} P\,dx + Q\,dy + R\,dz \\ &= \int_{-\beta}^{-\alpha} \left\{ P[x(-u), y(-u), z(-u)] \frac{dx}{du} \right.\end{aligned}$$

$$+ Q[x(-u), y(-u), z(-u)]\frac{dy}{du} + R[x(-u), y(-u), z(-u)]\frac{dz}{du}\bigg\}du.$$

If we now change variables of integration by setting $t = -u$, then

$$\frac{dx}{du} = \frac{dx}{dt}\frac{dt}{du} = -\frac{dx}{dt},$$

and similarly for dy/du and dz/du. Consequently,

$$\int_{-C} \mathbf{F} \cdot d\mathbf{r} = \int_{\beta}^{\alpha} \bigg\{P[x(t), y(t), z(t)]\left(-\frac{dx}{dt}\right) + Q[x(t), y(t), z(t)]\left(-\frac{dy}{dt}\right)$$

$$+ R[x(t), y(t), z(t)]\left(-\frac{dz}{dt}\right)\bigg\}(-dt)$$

$$= -\int_{\alpha}^{\beta} \bigg\{P[x(t), y(t), z(t)]\frac{dx}{dt} + Q[x(t), y(t), z(t)]\frac{dy}{dt}$$

$$+ R[x(t), y(t), z(t)]\frac{dz}{dt}\bigg\}dt$$

$$= -\int_C \mathbf{F} \cdot d\mathbf{r}.$$

■

According to this example when the direction along a curve is reversed, the value of a line integral of the form 15.28 along the new curve is the negative of its value along the original curve. This is because the signs of dx, dy, and dz are reversed when the direction along C is reversed. This is not the case for line integral 15.20; ds does not change sign when the direction along C is reversed.

Line integrals of the form 15.28 are singled out for special consideration because they repeatedly arise in physical problems. For example, suppose $\mathbf{F}$ represents a force, and we consider finding the work done by this force as a particle moves along a curve C from A to B. We begin by dividing C into n subcurves of lengths Δs_i as shown in Figure 15.21.

FIGURE 15.21

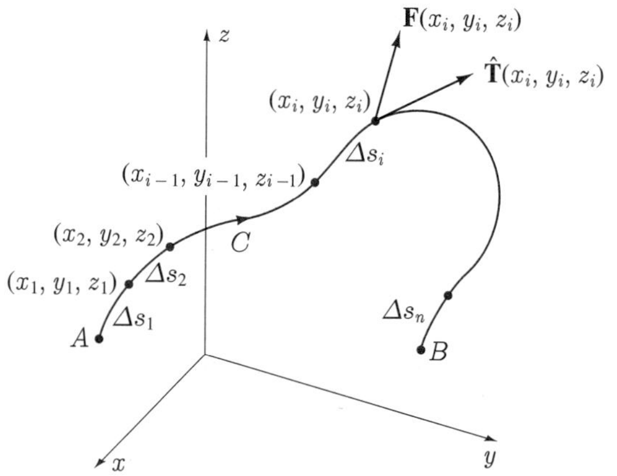

If $\mathbf{F}(x, y, z)$ is continuous along C, then along any given Δs_i, $\mathbf{F}(x, y, z)$ does not vary greatly (provided, of course, that Δs_i is small). If we approximate $\mathbf{F}(x, y, z)$ along

Δs_i by its value $\mathbf{F}(x_i, y_i, z_i)$ at the final point (x_i, y_i, z_i) of Δs_i, then an approximation to the work done by $\mathbf{F}$ along Δs_i is $\mathbf{F}(x_i, y_i, z_i) \cdot \hat{\mathbf{T}}(x_i, y_i, z_i)\Delta s_i$. An approximation to the total work done by $\mathbf{F}$ along C is therefore

$$\sum_{i=1}^{n} \mathbf{F}(x_i, y_i, z_i) \cdot \hat{\mathbf{T}}(x_i, y_i, z_i)\Delta s_i.$$

To obtain the exact value of the work done by $\mathbf{F}$ along C we take the limit of this sum as the number of subdivisions becomes larger and larger and each Δs_i approaches zero. But this is precisely the definition of the line integral of $\mathbf{F} \cdot \hat{\mathbf{T}}$, and we therefore write

$$W = \int_C \mathbf{F} \cdot \hat{\mathbf{T}}\, ds = \int_C \mathbf{F} \cdot d\mathbf{r}. \tag{15.31}$$

This interpretation of a line integral as the work done by a force $\mathbf{F}$ is extremely important, and we will return to it in Section 15.5.

EXAMPLE 15.9

The force of repulsion between two positive point charges, one of size q and the other of size unity, has magnitude $q/(4\pi\epsilon_0 r^2)$, where ϵ_0 is a constant and r is the distance between the charges. The potential V at any point P due to charge q is defined as the work required to bring the unit charge to P from an infinite distance along the straight line joining q and P. Find V.

SOLUTION If that part of the line joining q and P from infinity to P is denoted by C (Figure 15.22), then

$$V = \int_C \mathbf{F} \cdot d\mathbf{r} = -\int_{-C} \mathbf{F} \cdot d\mathbf{r}$$

(see equation 15.30), where $\mathbf{F}$, the force necessary to overcome the electrostatic repulsion, is given by

$$\mathbf{F} = \frac{-q}{4\pi\epsilon_0 x^2}\hat{\mathbf{i}}.$$

Along $-C$, $d\mathbf{r} = dx\hat{\mathbf{i}}$, and therefore V can be evaluated by the (improper) definite integral

$$V = -\int_r^{\infty} \frac{-q}{4\pi\epsilon_0 x^2}\, dx = -\left\{\frac{q}{4\pi\epsilon_0 x}\right\}_r^{\infty} = \frac{q}{4\pi\epsilon_0 r}.$$

FIGURE 15.22

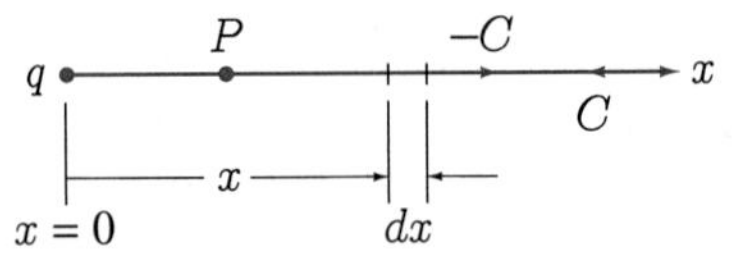

■

If the vector field $\mathbf{F}$ in equation 15.27 has an x-component that is only a function of x, $P = P(x)$, and if the curve C is a portion of the x-axis from $x = a$ to $x = b$, then

$$\int_C \mathbf{F} \cdot d\mathbf{r} = \int_a^b P(x)\, dx.$$

This equation indicates that definite integrals with respect to x can be regarded as line integrals along the x-axis.

EXERCISES 15.3

In Exercises 1–10 evaluate the line integral.

1. $\int_C x\,dx + x^2 y\,dy$ where C is the curve $y = x^3$, $z = 0$ from $(-1,-1,0)$ to $(2,8,0)$

2. $\int_C x\,dx + yz\,dy + x^2\,dz$ where C is the curve $y = x$, $z = x^2$ from $(-1,-1,1)$ to $(2,2,4)$

3. $\int_C x\,dx + (x+y)\,dy$ where C is the curve $x = 1 + y^2$ from $(2,1)$ to $(2,-1)$

4. $\int_C x^2\,dx + y^2\,dy + z^2\,dz$ where C is the curve $x + y = 1$, $x + z = 1$ from $(-2,3,3)$ to $(1,0,0)$

5. $\int_C (y + 2x^2 z)\,dx$ where C is the curve $x = y^2$, $z = x^2$ from $(4,-2,16)$ to $(1,1,1)$

6. $\oint_C x^2 y\,dx + (x-y)\,dy$ once counterclockwise around the curve bounding the area described by the curves $x = 1 - y^2$, $y = x + 1$

7. $\int_C y^2\,dx + x^2\,dy$ where C is the semicircle $x = \sqrt{1-y^2}$ from $(0,1)$ to $(0,-1)$

8. $\int_C y\,dx + x\,dy + z\,dz$ where C is the curve $z = x^2 + y^2$, $x + y = 1$ from $(1,0,1)$ to $(-1,2,5)$

9. $\oint_C x^2 y\,dy + z\,dx$ where C is the curve $x^2 + y^2 = 1$, $x + y + z = 1$ directed so that x decreases when y is positive

10. $\oint_C y^2\,dx + x^2\,dy$ once clockwise around the curve $|x| + |y| = 1$

11. Find the work done by the force $\mathbf{F} = x^2 y\hat{\imath} + x\hat{\jmath}$ as a particle moves from $(1,0)$ to $(6,5)$ along the straight line joining these points.

12. Consider the line integral $\int_C xy\,dx + x^2\,dy$, where C is the quarter-circle $x^2 + y^2 = 9$ from $(3,0)$ to $(0,3)$. Show that for each of the following parametrizations of C the value of the line integral is the same:
   (a) $x = 3\cos t$, $y = 3\sin t$, $0 \le t \le \pi/2$
   (b) $x = \sqrt{9-y^2}$, $0 \le y \le 3$

13. Evaluate the line integral $\int_C xy\,dx + x\,dy$ from $(-5,3,0)$ to $(4,0,0)$ along each of the following curves:
   (a) the straight line joining the points
   (b) $x = 4 - y^2$, $z = 0$
   (c) $3y = x^2 - 16$, $z = 0$

14. Find the work done by a force $\mathbf{F} = x\hat{\imath} + y\hat{\jmath}$ on a particle as it moves once counterclockwise around the ellipse $b^2x^2 + a^2y^2 = a^2b^2$, $z = 0$.

In Exercises 15–19 evaluate the line integral.

15. $\int_C \frac{1}{yz}\,dx$ where C is the curve $z = \sqrt{1-x^2}$, $y = \sqrt{1-x^2}$ from $(1/\sqrt{2}, 1/\sqrt{2}, 1/\sqrt{2})$ to $(-1\sqrt{2}, 1/\sqrt{2}, 1/\sqrt{2})$

16. $\oint_C (x^2 + 2y^2)\,dy$ twice clockwise around the circle $(x-2)^2 + y^2 = 1$, $z = 0$

17. $\int_C y\,dx - y(x-1)\,dy + y^2 z\,dz$ where C is the first octant intersection of $x^2 + y^2 + z^2 = 4$ and $(x-1)^2 + y^2 = 1$ from $(2,0,0)$ to $(0,0,2)$

18. $\int_C x^2 y\,dx + y\,dy + \sqrt{1-x^2}\,dz$ where C is the curve $y - 2z^2 = 1$, $z = x + 1$ from $(0,3,1)$ to $(1,9,2)$

19. $\int_C x\,dx + xy\,dy + 2\,dz$ where C is the curve $x + 2y + z = 4$, $4x + 3y + 2z = 13$ from $(2,-1,4)$ to $(3,1,-1)$

20. Evaluate the line integral $\int_C \frac{x^3}{(1+x^4)^3}\,dx + y^2 e^y\,dy + \frac{z}{\sqrt{1+z^2}}\,dz$ where C is the series of line segments joining successively the points $(0,-1,1)$, $(1,-1,1)$, $(1,0,1)$, and $(1,0,2)$.

21. One end of a spring (with constant k) is fixed at point D in Figure 15.23. The other end is moved along the x-axis from A to B. If the spring is stretched an amount l at A, find the work done against the spring.

FIGURE 15.23

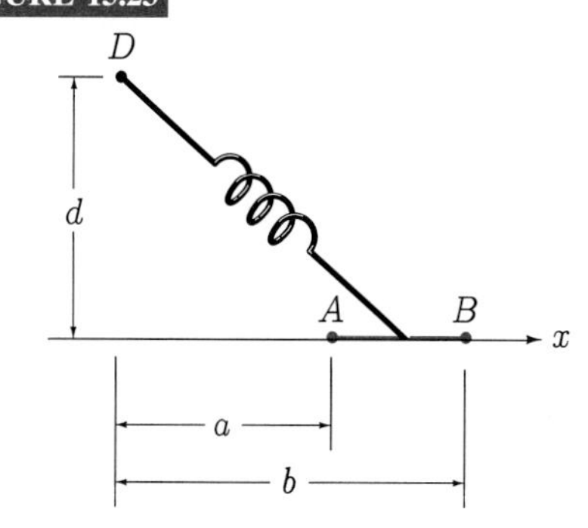

22. Two positive charges q_1 and q_2 are placed at positions $(5,5)$ and $(-2,3)$ respectively, in the xy-plane. A third positive charge q_3 is moved along the x-axis from $x = 1$ to $x = -1$. Find the total work done by the electrostatic forces of q_1 and q_2 on q_3.

In Exercises 23–26 set up a definite integral to evaluate the line integral. Use Simpson's rule with 10 equal subdivisions to approximate the definite integral.

23. $\int_C xy\,dx + xy^2\,dy$ where C is the curve $z = 0$, $y = 1/\sqrt{1+x^3}$, from $(0,1,0)$ to $(2,1/3,0)$

24. $\int_C xz\,dx + \tan x\,dy + e^{xy}\,dz$ where C is the curve $x = y^2$, $z = y^3$ from $(1,-1,-1)$ to $(1,1,1)$

25. $\int_C \sqrt{1+y^2}\,dz + zy\,dy$ where C is the curve $y = \cos^3 t$, $z = \sin^3 t$, $x = 0$, $0 \le t \le \pi/2$

26. $\int_C xyz\,dy$ where C is the curve $x = (1-t^2)/(1+t^2)$, $y = t(1-t^2)/(1+t^2)$, $z = t$, $-1 \le t \le 1$

In Exercises 27 and 28 evaluate the line integral along the polar coordinate curve.

27. $\oint_C y\,dx$ where C is the cardioid $r = 1 - \cos\theta$

28. $\int_C y\,dx + x\,dy$ where C is the curve $r = \theta$, $0 \le \theta \le \pi$

29. Suppose a gas is flowing through a region D of space. At each point $P(x,y,z)$ in D and time t, the gas has a certain velocity $\mathbf{v}(x,y,z,t)$. If C is a closed curve in D, the line integral

$$\Gamma = \oint_C \mathbf{v}\cdot d\mathbf{r}$$

is called the circulation of the flow for the curve C. If C is the circle $x^2 + y^2 = r^2$, $z = 1$ (directed clockwise as viewed from the origin), calculate Γ for the following flow vectors:

(a) $\mathbf{v}(x,y,z) = (x\hat{\mathbf{i}} + y\hat{\mathbf{j}} + z\hat{\mathbf{k}})/(x^2+y^2+z^2)^{3/2}$

(b) $\mathbf{v}(x,y,z) = -y\hat{\mathbf{i}} + x\hat{\mathbf{j}}$

30. We have shown that given a line integral 15.28, it is always possible to write it uniquely in form 15.23, where $f = \mathbf{F}\cdot\hat{\mathbf{T}}$. Show that the converse is not true; that is, given $f(x,y,z)$, there does not exist a unique $\mathbf{F}(x,y,z)$ such that $\mathbf{F}\cdot d\mathbf{r} = f(x,y,z)\,ds$.

31. Explain why the line integral

$\oint_C f(x)\,dx + g(y)\,dy + h(z)\,dz$ must have value zero when $f(x)$, $g(y)$, and $h(z)$ are continuous functions in some domain containing C.

32. The cycloid $x = a(\theta - \sin\theta)$, $y = a(1-\cos\theta)$ (Figure 15.24) is the curve traced out by a fixed point on the circumference of a circle of radius a rolling along the x-axis (θ being the angle through which the point has rotated). Suppose the point is acted on by a force of unit magnitude directed toward the centre of the rolling circle.

(a) Find the work done by the force as the point moves from $\theta = 0$ to $\theta = \pi$.

(b) How much of the work in (a) is done by the vertical component of the force?

FIGURE 15.24

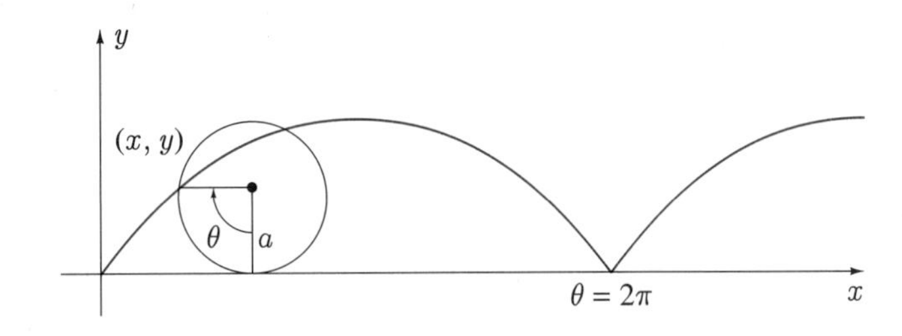

SECTION 15.4

Independence of Path

In Sections 15.2 and 15.3 we illustrated that the value of a line integral joining two points usually depends on the curve joining the points. In this section we show that certain line integrals have the same value for all curves joining the same two points. We formalize this idea in the following definition.

Definition 15.4

A line integral $\int \mathbf{F} \cdot d\mathbf{r}$ is said to be **independent of path** in a domain D if for each pair of points A and B in D, the value of the line integral

$$\int_C \mathbf{F} \cdot d\mathbf{r}$$

is the same for all piecewise-smooth paths C in D from A to B.

The value of such a line integral for given $\mathbf{F}$ will then depend only on the end points A and B. Note that we speak of independence of path only for the special class of line integrals of the form $\int \mathbf{F} \cdot d\mathbf{r}$. The question we must now ask is "How do we determine whether a given line integral is independent of path?" One answer is contained in the following theorem.

Theorem 15.3

Suppose $P(x,y,z)$, $Q(x,y,z)$, and $R(x,y,z)$ are continuous functions in some domain D. The line integral

$$\int \mathbf{F} \cdot d\mathbf{r} = \int P\,dx + Q\,dy + R\,dz$$

is independent of path in D if and only if there exists a function $\phi(x,y,z)$ defined in D such that

$$\nabla\phi = \mathbf{F} = P\hat{\mathbf{i}} + Q\hat{\mathbf{j}} + R\hat{\mathbf{k}}. \tag{15.32}$$

Essentially, then, a line integral is independent of path if $\mathbf{F}$ is the gradient of some scalar function.

Proof Suppose first of all that in D there exists a function $\phi(x,y,z)$ such that $\nabla\phi = P\hat{\mathbf{i}} + Q\hat{\mathbf{j}} + R\hat{\mathbf{k}}$. If

$$C: \quad x = x(t), \quad y = y(t), \quad z = z(t), \quad \alpha \le t \le \beta$$

is any smooth curve in D from A to B, then

$$\int_C \mathbf{F} \cdot d\mathbf{r} = \int_C P\,dx + Q\,dy + R\,dz = \int_\alpha^\beta \left\{ \frac{\partial\phi}{\partial x}\frac{dx}{dt} + \frac{\partial\phi}{\partial y}\frac{dy}{dt} + \frac{\partial\phi}{\partial z}\frac{dz}{dt} \right\} dt.$$

But the term in braces is the chain rule for the derivative of the composite function $\phi[x(t), y(t), z(t)]$, and we can therefore write that

$$\begin{aligned} \int_C \mathbf{F} \cdot d\mathbf{r} &= \int_\alpha^\beta \frac{d\phi}{dt} dt = \{\phi[x(t), y(t), z(t)]\}_\alpha^\beta \\ &= \phi[x(\beta), y(\beta), z(\beta)] - \phi[x(\alpha), y(\alpha), z(\alpha)] \\ &= \phi(x_B, y_B, z_B) - \phi(x_A, y_A, z_A). \end{aligned}$$

(The same result is obtained even when C is piecewise smooth rather than smooth.) Because this last expression does not depend on the curve C taken from A to B, it follows that the line integral is independent of path in D.

Conversely, suppose now that the line integral

$$\int \mathbf{F} \cdot d\mathbf{r} = \int P\,dx + Q\,dy + R\,dz$$

is independent of path in D, and A is chosen as some fixed point in D. If $P(x, y, z)$ is any other point in D (Figure 15.25), and C is a piecewise-smooth curve in D from A to P, then the line integral

$$\phi(x, y, z) = \int_C \mathbf{F} \cdot d\mathbf{r}$$

defines a single-valued function $\phi(x, y, z)$ in D, and the value of $\phi(x, y, z)$ is the same for all piecewise-smooth curves from A to P.

FIGURE 15.25

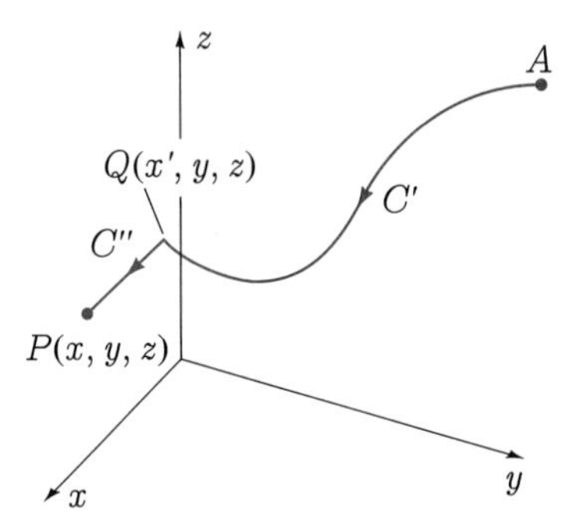

Consider a curve C composed of two parts: a straight-line portion C'' parallel to the x-axis from a fixed point $Q(x', y, z)$ to $P(x, y, z)$, and any other piecewise-smooth curve C' in D from A to Q. Then

$$\phi(x, y, z) = \int_{C'} \mathbf{F} \cdot d\mathbf{r} + \int_{C''} \mathbf{F} \cdot d\mathbf{r}.$$

Now along C'', y and z are both constant, and therefore

$$\phi(x, y, z) = \int_{C'} \mathbf{F} \cdot d\mathbf{r} + \int_{x'}^{x} P(t, y, z)\,dt.$$

The partial derivative of this function with respect to x is

$$\frac{\partial \phi}{\partial x} = \frac{\partial}{\partial x}\int_{C'} \mathbf{F} \cdot d\mathbf{r} + \frac{\partial}{\partial x}\int_{x'}^{x} P(t, y, z)\,dt,$$

but because $Q(x', y, z)$ is fixed,

$$\frac{\partial}{\partial x}\int_{C'} \mathbf{F} \cdot d\mathbf{r} = 0.$$

Consequently,

$$\frac{\partial \phi}{\partial x} = \frac{\partial}{\partial x}\int_{x'}^{x} P(t, y, z)\,dt = P(x, y, z).$$

By choosing other curves with straight-line portions parallel to the y- and z-axes, we can also show that $\partial\phi/\partial y = Q$ and $\partial\phi/\partial z = R$. Thus, $\mathbf{F} = \nabla\phi$, and this completes the proof.

This theorem points out that it is very simple to evaluate a line integral that is independent of path. We state this in the following corollary.

Corollary 1 *When a line integral is independent of path in a domain D, and A and B are points in D, then*

$$\int_C \mathbf{F} \cdot d\mathbf{r} = \phi(x_B, y_B, z_B) - \phi(x_A, y_A, z_A),$$

where $\nabla\phi = \mathbf{F}$, for every piecewise smooth curve C in D from A to B.

EXAMPLE 15.10

Evaluate $\int_C 2xy\,dx + x^2\,dy + 2z\,dz$, where C is the first octant intersection of $x^2 + y^2 = 1$ and $z = 2x + 4$ from $(0,1,4)$ to $(1,0,6)$.

SOLUTION Since $\nabla(x^2y + z^2) = 2xy\hat{\mathbf{i}} + x^2\hat{\mathbf{j}} + 2z\hat{\mathbf{k}}$, the line integral is independent of path everywhere, and

$$\int_C 2xy\,dx + x^2\,dy + 2z\,dz = \left\{x^2y + z^2\right\}_{(0,1,4)}^{(1,0,6)} = 36 - 16 = 20.$$

■

The following corollary is also an immediate consequence of Theorem 15.3.

Corollary 2 *The line integral $\int \mathbf{F} \cdot d\mathbf{r}$ is independent of path in a domain D if and only if*

$$\oint_C \mathbf{F} \cdot d\mathbf{r} = 0 \tag{15.33}$$

for every closed path in D.

Theorem 15.3 states that a necessary and sufficient condition for line integral 15.28 to be independent of path is the existence of a function $\phi(x,y,z)$ such that $\nabla\phi = \mathbf{F}$. For most problems it is quite obvious whether such a function $\phi(x,y,z)$ exists; for a few, however, it is not at all clear whether $\phi(x,y,z)$ exists or not. Since much time could be wasted searching for $\phi(x,y,z)$ (when in fact it does not exist), it would be helpful to have a test that states *a priori* whether $\phi(x,y,z)$ does indeed exist. Such a test is contained in Theorem 15.1. It states that $\mathbf{F}$ is the gradient of some scalar function $\phi(x,y,z)$ if $\nabla \times \mathbf{F} = \mathbf{0}$. When this result is combined with Theorem 15.3, we obtain this important theorem.

Theorem 15.4

Let D be a domain in which $P(x,y,z)$, $Q(x,y,z)$, and $R(x,y,z)$ have continuous first derivatives. If the line integral $\int \mathbf{F} \cdot d\mathbf{r} = \int P\,dx + Q\,dy + R\,dz$ is independent of path in D, then $\nabla \times \mathbf{F} = \mathbf{0}$ in D. Conversely, if D is simply-connected, and $\nabla \times \mathbf{F} = \mathbf{0}$ in D, then the line integral is independent of path in D.

We have in Theorem 15.4 a simple test to determine whether a given line integral is independent of path: We see whether the curl of $\mathbf{F}$ is zero. Evaluation of a line integral that is independent of path still requires the function $\phi(x,y,z)$, but it is at least nice to know that ϕ exists before searching for it.

Theorems 15.1, 15.3, and 15.4 have identified an important equivalence, at least in simply-connected domains:

1. $\int \mathbf{F} \cdot d\mathbf{r}$ is independent of path in D;

2. $\mathbf{F} = \nabla\phi$ for some function $\phi(x, y, z)$ defined in D;

3. $\nabla \times \mathbf{F} = \mathbf{0}$ in D.

Theorem 15.1 states that (2) and (3) are equivalent; Theorem 15.3 verifies the equivalence of (1) and (2); and these two imply the equivalence of (1) and (3) (Theorem 15.4).

For line integrals in the xy-plane, this equivalence is still valid except that $\nabla \times \mathbf{F} = \mathbf{0}$ can be stated more simply as

$$\frac{\partial Q}{\partial x} = \frac{\partial P}{\partial y} \tag{15.34}$$

(see equation 15.16).

EXAMPLE 15.11

Evaluate $\int_C 2xye^z\,dx + (x^2e^z + y)\,dy + (x^2ye^z - z)\,dz$ along the straight line C from $(0, 1, 2)$ to $(2, 1, -8)$.

SOLUTION

Method 1 Parametric equations for the straight line are

$$C: \quad x = 2t, \quad y = 1, \quad z = 2 - 10t, \quad 0 \le t \le 1.$$

If I is the value of the line integral, then

$$\begin{aligned}
I &= \int_0^1 \{2(2t)(1)e^{2-10t}(2\,dt) + [(2t)^2e^{2-10t} + 1](0) \\
&\quad + [(2t)^2(1)e^{2-10t} - 2 + 10t](-10\,dt)\} \\
&= \int_0^1 \{8e^2(t - 5t^2)e^{-10t} + 20 - 100t\}dt \\
&= 8e^2\left\{(t - 5t^2)\frac{e^{-10t}}{-10}\right\}_0^1 + \frac{4e^2}{5}\int_0^1 (1 - 10t)e^{-10t}dt + 10\{2t - 5t^2\}_0^1 \\
&= 8e^2\left\{\frac{2}{5}e^{-10}\right\} - 30 + \frac{4}{5}e^2\left\{(1 - 10t)\frac{e^{-10t}}{-10}\right\}_0^1 + \frac{2e^2}{25}\int_0^1 -10e^{-10t}dt \\
&= \frac{16}{5}e^{-8} - 30 + \frac{4}{5}e^2\left\{\frac{9}{10}e^{-10} + \frac{1}{10}\right\} + \frac{2e^2}{25}\left\{e^{-10t}\right\}_0^1 = 4e^{-8} - 30.
\end{aligned}$$

Method 2 It is evident that

$$\nabla\left(x^2ye^z + \frac{y^2}{2} - \frac{z^2}{2}\right) = 2xye^z\hat{\mathbf{i}} + (x^2e^z + y)\hat{\mathbf{j}} + (x^2ye^z - z)\hat{\mathbf{k}},$$

and hence the line integral is independent of path. Its value is therefore

$$I = \left\{x^2ye^z + \frac{y^2}{2} - \frac{z^2}{2}\right\}_{(0,1,2)}^{(2,1,-8)} = \left\{4e^{-8} + \frac{1}{2} - 32\right\} - \left\{\frac{1}{2} - 2\right\} = 4e^{-8} - 30.$$

Method 3 Since

$$\nabla \times (2xye^z\hat{\mathbf{i}} + (x^2e^z + y)\hat{\mathbf{j}} + (x^2ye^z - z)\hat{\mathbf{k}}) = \begin{vmatrix} \hat{\mathbf{i}} & \hat{\mathbf{j}} & \hat{\mathbf{k}} \\ \partial/\partial x & \partial/\partial y & \partial/\partial z \\ 2xye^z & x^2e^z + y & x^2ye^z - z \end{vmatrix}$$

$$= (x^2e^z - x^2e^z)\hat{\mathbf{i}} + (2xye^z - 2xye^z)\hat{\mathbf{j}} + (2xe^z - 2xe^z)\hat{\mathbf{k}} = \mathbf{0},$$

the line integral is independent of path. Thus there exists a function $\phi(x, y, z)$ such that

$$\nabla\phi = 2xye^z\hat{\mathbf{i}} + (x^2e^z + y)\hat{\mathbf{j}} + (x^2ye^z - z)\hat{\mathbf{k}}$$

or

$$\frac{\partial\phi}{\partial x} = 2xye^z, \quad \frac{\partial\phi}{\partial y} = x^2e^z + y, \quad \frac{\partial\phi}{\partial z} = x^2ye^z - z.$$

Integration of the first of these equations yields

$$\phi(x, y, z) = x^2ye^z + K(y, z).$$

Substitution of this function into the left-hand side of the second equation gives

$$x^2e^z + \frac{\partial K}{\partial y} = x^2e^z + y,$$

which implies that

$$\frac{\partial K}{\partial y} = y.$$

Consequently,

$$K(y, z) = \frac{y^2}{2} + L(z),$$

and we know both the x- and y-dependence of ϕ:

$$\phi(x, y, z) = x^2ye^z + \frac{y^2}{2} + L(z).$$

To obtain the z-dependence contained in $L(z)$, we substitute into the left-hand side of the third equation to get

$$x^2ye^z + \frac{dL}{dz} = x^2ye^z - z,$$

from which we have

$$\frac{dL}{dz} = -z.$$

Hence, $L(z) = -z^2/2 + C$ (C a constant), and

$$\phi(x, y, z) = x^2ye^z + \frac{y^2}{2} - \frac{z^2}{2} + C.$$

Finally, then, we have

$$I = \left\{x^2ye^z + \frac{y^2}{2} - \frac{z^2}{2}\right\}_{(0,1,2)}^{(2,1,-8)} = 4e^{-8} - 30.$$

■

Method 1 is one of "brute force." The function $x^2 ye^z + y^2/2 - z^2/2$ in Method 2 was obtained by observation. Method 3 is the systematic procedure suggested in Section 15.1 for finding the function $\phi(x, y, z)$.

EXAMPLE 15.12 Evaluate

$$I = \int_C \left(\frac{x^3 - 2y^2}{x^3 y}\right) dx + \left(\frac{y^2 - x^3}{x^2 y^2}\right) dy + 2z^2 dz,$$

where C is the curve $y = x^2$, $z = x - 1$ from $(1, 1, 0)$ to $(2, 4, 1)$.

SOLUTION If we set

$$\mathbf{F} = \left(\frac{x^3 - 2y^2}{x^3 y}\right)\hat{\mathbf{i}} + \left(\frac{y^2 - x^3}{x^2 y^2}\right)\hat{\mathbf{j}} + 2z^2\hat{\mathbf{k}},$$

it is straightforward to show that in any simply-connected domain not containing points on the xz- or yz-planes, $\nabla \times \mathbf{F} = \mathbf{0}$, and therefore the line integral is independent of path. Thus there exists a function $\phi(x, y, z)$ such that $\nabla\phi = \mathbf{F}$. If we write

$$\mathbf{F} = \left(\frac{1}{y} - \frac{2y}{x^3}\right)\hat{\mathbf{i}} + \left(\frac{1}{x^2} - \frac{x}{y^2}\right)\hat{\mathbf{j}} + 2z^2\hat{\mathbf{k}},$$

then it is evident that

$$\phi(x, y, z) = \frac{x}{y} + \frac{y}{x^2} + \frac{2z^3}{3}.$$

Thus,

$$I = \left\{\frac{x}{y} + \frac{y}{x^2} + \frac{2z^3}{3}\right\}_{(1,1,0)}^{(2,4,1)} = \left\{\frac{1}{2} + 1 + \frac{2}{3}\right\} - \left\{1 + 1\right\} = \frac{1}{6}.$$

■

EXAMPLE 15.13 In thermodynamics the state of a gas is described by four variables — pressure P, temperature T, internal energy U, and volume V. These variables are related by two equations of state,

$$F(P, T, U, V) = 0 \quad \text{and} \quad G(P, T, U, V) = 0,$$

so that two of the variables are independent and two are dependent. If U and V are chosen as independent variables, then $T = T(U, V)$ and $P = P(U, V)$. An experimental law called the second law of thermodynamics states that the line integral

$$\int_C \frac{1}{T} dU + \frac{P}{T} dV$$

is independent of path in the UV-plane. Show that the second law can be expressed in the differential form

$$T\frac{\partial P}{\partial U} - P\frac{\partial T}{\partial U} + \frac{\partial T}{\partial V} = 0.$$

SOLUTION According to equation 15.34, the line integral is independent of path if and only if

$$\frac{\partial}{\partial U}\left(\frac{P}{T}\right) = \frac{\partial}{\partial V}\left(\frac{1}{T}\right),$$

or

$$0 = \frac{T\dfrac{\partial P}{\partial U} - P\dfrac{\partial T}{\partial U}}{T^2} + \frac{1}{T^2}\frac{\partial T}{\partial V};$$

that is,

$$0 = T\frac{\partial P}{\partial U} - P\frac{\partial T}{\partial U} + \frac{\partial T}{\partial V},$$

and the proof is complete.

Since the above line integral is independent of path, there exists a function $S(U, V)$ such that

$$\frac{\partial S}{\partial U} = \frac{1}{T}, \quad \frac{\partial S}{\partial V} = \frac{P}{T},$$

and the value of the line integral is given by

$$\int_C \frac{1}{T}dU + \frac{P}{T}dV = S(B) - S(A),$$

where C joins A and B. This function, called **entropy**, plays a key role in the field of thermodynamics. ■

EXERCISES 15.4

In Exercises 1–10 show that the line integral is independent of path, and evaluate it.

1. $\int_C xy^2\,dx + x^2y\,dy$ where C is the curve $y = x^2$, $z = 0$ from $(0,0,0)$ to $(1,1,0)$

2. $\int_C (3x^2 + y)\,dx + x\,dy$ where C is the straight line from $(2,1,5)$ to $(-3,2,4)$

3. $\int_C 2xe^y\,dx + (x^2e^y + 3)\,dy$ where C is the curve $y = \sqrt{1 - x^2}$, $z = 0$ from $(1,0,0)$ to $(-1,0,0)$

4. $\int_C 3x^2yz\,dx + x^3z\,dy + (x^3y - 4z)\,dz$ where C is the curve $x^2 + y^2 + z^2 = 3$, $y = x$ from $(-1,-1,1)$ to $(1,1,-1)$

5. $\int_C -\frac{y}{z}\sin x\,dx + \frac{1}{z}\cos x\,dy - \frac{y}{z^2}\cos x\,dz$ where C is the helix $x = 2\cos t$, $y = 2\sin t$, $z = t$ from $(2,0,2\pi)$ to $(2,0,4\pi)$

6. $\oint_C y\cos x\,dx + \sin x\,dy$ once clockwise around the circle $x^2 + y^2 - 2x + 4y = 7$, $z = 0$

7. $\int_C x^2\,dx + y^2\,dy + z^2\,dz$ where C is the curve $x + y = 1$, $x + z = 1$ from $(-2,3,3)$ to $(1,0,0)$

8. $\int_C y\,dx + x\,dy + z\,dz$ where C is the curve $z = x^2 + y^2$, $x + y = 1$ from $(1,0,1)$ to $(-1,2,5)$

9. $\int_C \frac{1}{y}dx - \frac{x}{y^2}dy + dz$ where C is the curve $y = x^2 + 1$, $x + y + z = 2$ from $(0,1,1)$ to $(3,10,-11)$

10. $\int_C 3x^2y^3\,dx + 3x^3y^2\,dy$ where C is the curve $y = e^x$ from $(0,1)$ to $(1,e)$

11. Show that if $f(x)$, $g(y)$, and $h(z)$ have continuous first derivatives, then the line integral

$$\int_C f(x)\,dx + g(y)\,dy + h(z)\,dz$$

is independent of path.

12. If $\nabla \times \mathbf{F} = \mathbf{0}$ in a domain D which is not simply-connected, can you conclude that the line integral $\int_C \mathbf{F} \cdot d\mathbf{r}$ is not independent of path in D? Explain.

In Exercises 13–18 evaluate the line integral.

13. $\int_C zye^{xy}\,dx + zxe^{xy}\,dy + (e^{xy} - 1)\,dz$ where C is the curve $y = x^2$, $z = x^3$ from $(1,1,1)$ to $(2,4,8)$

14. $\oint_C y(\tan x + x\sec^2 x)\,dx + x\tan x\,dy + dz$ once around the circle $x^2 + y^2 = 1$, $z = 0$

15. $\int_C \left(\frac{1+y^2}{x^3}\right) dx - \left(\frac{y + x^2 y}{x^2}\right) dy + z\,dz$ where C is the broken line joining successively $(1,0,0)$, $(25,2,3)$ and $(5,2,1)$

16. $\oint_C \frac{zy\,dx - xz\,dy + xy\,dz}{y^2}$ where C is the curve $x^2 + z^2 = 1$, $y + z = 2$

17. $\int_C -\frac{1}{x^2}\mathrm{Tan}^{-1}\,y\,dx + \frac{1}{x + xy^2}\,dy$ where C is the curve $x = y^2 + 1$ from $(2,-1)$ to $(10,3)$

18. $\int_C \frac{1}{(x-3)^2(y+5)}\,dx + \frac{1}{(x-3)(y+5)^2}\,dy + \frac{1}{z+4}\,dz$ where C is the curve $x = y = z$ from $(0,0,0)$ to $(2,2,2)$

19. Evaluate $\oint_C \frac{-y\,dx + x\,dy}{x^2 + y^2}$

(a) once counterclockwise around the circle $x^2 + y^2 = 1$, $z = 0$

(b) once counterclockwise around the circle $(x-2)^2 + y^2 = 1$, $z = 0$

20. Evaluate $\int_C \frac{y}{x^2+y^2}\,dx - \frac{x}{x^2+y^2}\,dy$ where C is the series of line segments joining successively the points $(1,0)$, $(1,1)$, $(-1,1)$, and $(-1,0)$.

21. Is the line integral

$$\int_C \frac{x}{\sqrt{x^2+y^2}}\,dx + \frac{y}{\sqrt{x^2+y^2}}\,dy$$

independent of path in the domain consisting of the xy-plane with the origin removed? Is the line integral

$$\int_C \frac{x}{\sqrt{x^2+y^2+z^2}}\,dx + \frac{y}{\sqrt{x^2+y^2+z^2}}\,dy + \frac{z}{\sqrt{x^2+y^2+z^2}}\,dz$$

independent of path in the domain consisting of xyz-space with the origin removed?

22. In which of the following domains is the line integral $\int_C \frac{y\,dx - x\,dy}{x^2 + y^2}$ independent of path: **(a)** $x > 0$ **(b)** $x < 0$ **(c)** $y > 0$ **(d)** $y < 0$ **(e)** $x^2 + y^2 > 0$

23. The second law of thermodynamics states that the line integral $I = \int T^{-1}(dU + P\,dV)$ is independent of path in the UV-plane (see Example 15.13).

(a) The equations of state for an ideal gas are $PV = nRT$, $U = f(T)$, where n and R are constants and $f(T)$ is some given function. Because of these, it is more convenient to choose T and V as independent variables and to express P and U in terms of T and V. If this is done, show that

$$I = \int kT^{-1}\,dT + nRV^{-1}\,dV \quad \text{where } k = dU/dT.$$

(b) Since the line integral is independent of path, there exists a function $S(T,V)$, called entropy, such that

$$\int_C \frac{k}{T}dT + \frac{nR}{V}dV = S(B) - S(A),$$

where C is any curve joining points A and B. Show in the case that k is constant that $S = k\ln T + nR\ln V + S_0$, where S_0 is a constant.

24. Evaluate $\oint_C (2xye^{x^2 y} + x^2 y)\,dx + x^2 e^{x^2 y}\,dy$ once clockwise around the ellipse $x^2 + 4y^2 = 4$, $z = 0$.

25. A spring has one end fixed at point P. The other end is moved along the curve $y = f(x)$ from $(x_0, f(x_0))$ to $(x_1, f(x_1))$. If the initial and final stretches in the spring are a and b ($b > a$), what work is done against the spring?

26. Electrostatic forces due to point charges and gravitational forces due to point masses are examples of inverse square force fields — force fields of the form $\mathbf{F} = k\hat{\mathbf{r}}/|\mathbf{r}|^2$, where k is a constant and $\mathbf{r} = x\hat{\mathbf{i}} + y\hat{\mathbf{j}} + z\hat{\mathbf{k}}$.

(a) Is the line integral representing work done by such a force field independent of path?

(b) What is the work done by $\mathbf{F}$ in moving a particle from (x_1, y_1, z_1) to (x_2, y_2, z_2)?

SECTION 15.5

Energy and Conservative Force Fields

Suppose a force field $\mathbf{F}(x, y, z)$ acts throughout some domain D of space. In physics and engineering, many force fields satisfy the following definition.

Definition 15.5

A force field $\mathbf{F}(x, y, z)$ is said to be **conservative** in a domain D if the line integral $\int \mathbf{F} \cdot d\mathbf{r}$ is independent of path in D.

Since the value of $\int_C \mathbf{F} \cdot d\mathbf{r}$ can be interpreted as the work done by $\mathbf{F}$ along C, a force field is conservative if the work done by it is independent of path taken from one point to another. According to the results of Section 15.4, we can also state that a force field $\mathbf{F}$ is conservative if and only if there exists a function $\phi(x, y, z)$ such that $\mathbf{F} = \nabla\phi$.

It is customary to associate a potential energy function $U(x, y, z)$ with a conservative force field $\mathbf{F}$. This function assigns to each point (x, y, z) a potential energy in such a way that if a particle moves from point A to point B, then the difference in potential energy $U(A) - U(B)$ is precisely the work done by $\mathbf{F}$; i.e., if C is any curve joining A and B, then

$$U(A) - U(B) = \int_C \mathbf{F} \cdot d\mathbf{r}. \qquad (15.35)$$

If $\int \mathbf{F} \cdot d\mathbf{r} > 0$, then the potential energy at A is greater than the potential energy at B; if $\int \mathbf{F} \cdot d\mathbf{r} < 0$, then the potential energy at B is greater than that at A. To find $U(x, y, z)$, we use the fact that because $\mathbf{F}$ is conservative, there exists a function $\phi(x, y, z)$ such that $\mathbf{F} = \nabla\phi$, and

$$\int_C \mathbf{F} \cdot d\mathbf{r} = \phi(B) - \phi(A).$$

If follows, then, that $U(x, y, z)$ must satisfy the equation

$$U(A) - U(B) = \phi(B) - \phi(A),$$

or

$$U(A) + \phi(A) = U(B) + \phi(B).$$

Since A and B are arbitrary points in D, this last equation states that the value of the function $U(x, y, z) + \phi(x, y, z)$ is the same at every point in the force field,

$$U(x, y, z) + \phi(x, y, z) = C,$$

where C is a constant. Thus,

$$U(x, y, z) = -\phi(x, y, z) + C. \qquad (15.36)$$

Equation 15.36 shows that the force field $\mathbf{F}$ defines a potential energy function $U(x, y, z)$ up to an additive constant. (This seems reasonable in that ϕ itself is defined only to an additive constant.) Because $U = -\phi + C$ and $\mathbf{F} = \nabla\phi$, we can also regard U as being defined by the equation

$$\mathbf{F} = -\nabla U. \qquad (15.37)$$

The advantage of this equation is that it defines U directly, not through the function ϕ.

For a conservative force field $\mathbf{F}$, then, we define a potential energy function $U(x, y, z)$ by equation 15.37. If a particle moves from A to B, then the work done by $\mathbf{F}$ is

$$W = U(A) - U(B); \tag{15.38}$$

in other words, work done by a conservative force field is equal to loss in potential energy.

On the other hand, if a particle moves under the action of a force $\mathbf{F}$ (and only $\mathbf{F}$), be it conservative or nonconservative, then it does so according to Newton's second law,

$$\mathbf{F} = \frac{d}{dt}(m\mathbf{v}) = m\frac{d\mathbf{v}}{dt},$$

where m is the mass of the particle (assumed constant), $\mathbf{v}$ is its velocity, and t is time. The action of $\mathbf{F}$ produces motion along some curve C, and the work done by $\mathbf{F}$ along this curve from A to B is

$$W = \int_C \mathbf{F} \cdot d\mathbf{r} = \int_\alpha^\beta m\frac{d\mathbf{v}}{dt} \cdot \frac{d\mathbf{r}}{dt}dt = \int_\alpha^\beta m\frac{d\mathbf{v}}{dt} \cdot \mathbf{v}\, dt = \int_\alpha^\beta \frac{d}{dt}\left\{\frac{1}{2}m\mathbf{v} \cdot \mathbf{v}\right\} dt$$
$$= \left\{\frac{1}{2}m\mathbf{v} \cdot \mathbf{v}\right\}_\alpha^\beta = \left\{\frac{1}{2}m|\mathbf{v}|^2\right\}_\alpha^\beta.$$

Thus if $K(x, y, z) = \frac{1}{2}m|\mathbf{v}|^2$ represents kinetic energy of the particle, the work done by $\mathbf{F}$ is equal to the gain in kinetic energy of the particle,

$$W = K(B) - K(A) \tag{15.39}$$

(and this is true for any force $\mathbf{F}$ as long as $\mathbf{F}$ is the total resultant force producing motion).

If the total force producing motion is a conservative force field $\mathbf{F}$ we have two expressions 15.38 and 15.39 for the work done as a particle moves from one point to another under the action of $\mathbf{F}$. If we equate them, we have

$$U(A) - U(B) = K(B) - K(A),$$

or

$$U(A) + K(A) = U(B) + K(B). \tag{15.40}$$

We have shown then that if a particle moves under the action of a conservative force field *only*, the sum of the kinetic and potential energies at B must be the same as the sum of the kinetic and potential energies at A. In other words, if E is the total energy of the particle, kinetic plus potential, then

$$E(A) = E(B). \tag{15.41}$$

Since B can be any point along the path of the particle, it follows that when a particle moves under the action of a conservative force field, and only a conservative force field, then at every point along its trajectory

$$E = \text{a constant}. \tag{15.42}$$

This is the **law of conservation of energy** for a conservative force field.

EXAMPLE 15.14

Show that the electrostatic force due to a point charge is conservative, and determine a potential energy function for the field.

SOLUTION The electrostatic force on a charge Q due to a charge q is

$$\mathbf{F} = \frac{qQ}{4\pi\epsilon_0|\mathbf{r}|^3}\mathbf{r},$$

where $\mathbf{r}$ is the vector from q to Q. If we choose a Cartesian coordinate system with q at the origin (Figure 15.26), then

$$\mathbf{F} = \frac{qQ}{4\pi\epsilon_0(x^2+y^2+z^2)^{3/2}}(x\hat{\mathbf{i}} + y\hat{\mathbf{j}} + z\hat{\mathbf{k}}).$$

Since

$$\nabla\left\{\frac{-qQ}{4\pi\epsilon_0(x^2+y^2+z^2)^{1/2}}\right\} = \mathbf{F},$$

the force field is conservative, and possible potential energy functions are

$$U(x,y,z) = \frac{qQ}{4\pi\epsilon_0(x^2+y^2+z^2)^{1/2}} + C = \frac{qQ}{4\pi\epsilon_0 r} + C,$$

where $r = |\mathbf{r}|$. In electrostatics it is customary to choose $U(x,y,z)$ so that $\lim_{r\to\infty} U = 0$, in which case $C = 0$, and

$$U(x,y,z) = \frac{qQ}{4\pi\epsilon_0 r}.$$

In addition, if V is defined as the potential energy per unit test charge Q, then

$$V(x,y,z) = \frac{U}{Q} = \frac{q}{4\pi\epsilon_0 r}.$$

This result agrees with that in Example 15.9.

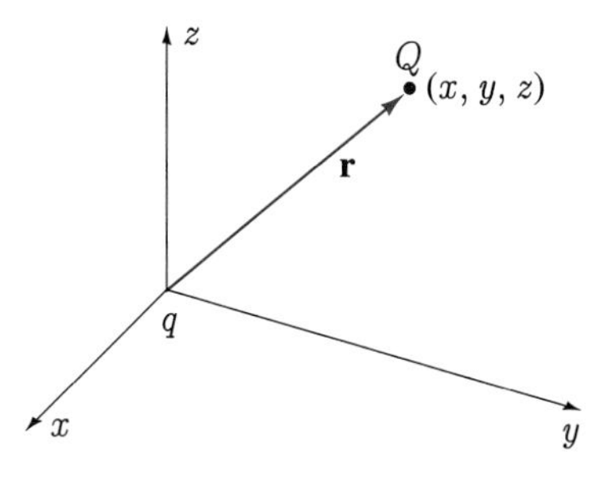

FIGURE 15.26

■

EXERCISES 15.5

In Exercises 1–5 determine whether the force field is conservative. Identify each conservative force field, and find a potential energy function.

1. $\mathbf{F}(x,y,z) = \dfrac{q_1 q_2}{4\pi\epsilon_0} \dfrac{x\hat{\mathbf{i}} + y\hat{\mathbf{j}} + z\hat{\mathbf{k}}}{(x^2+y^2+z^2)^{3/2}}$

2. $\mathbf{F}(x,y) = mx\hat{\mathbf{i}} + xy\hat{\mathbf{j}}$, m is a constant

3. $\mathbf{F}(x) = -kx\hat{\mathbf{i}}$, k is a constant

4. $\mathbf{F}(x,y,z) = -mg\hat{\mathbf{k}}$, m and g are constants

5. $\mathbf{F}(x,y,z) = GMm\dfrac{x\hat{\mathbf{i}} + y\hat{\mathbf{j}} + z\hat{\mathbf{k}}}{(x^2+y^2+z^2)^{3/2}}$, G, M, and m are constants

6. Suppose that $\mathbf{F}(x,y,z)$ is a conservative force field in some domain D, and $U(x,y,z)$ is a potential energy function associated with $\mathbf{F}$. The surfaces $U(x,y,z) = C$, where C is a constant, are called *equipotential surfaces*. Through each point P in D there is one and only one such equipotential surface for $\mathbf{F}$. Show that at P the force $\mathbf{F}$ is normal to the equipotential surface through P.

7. Draw the equipotential surfaces for the forces in Exercises 1, 4, and 5.

8. One end of a spring with unstretched length L is fixed at the origin in space. If the other end is at point (x,y,z) (all coordinates in metres), what is the force exerted by the spring? Is this force conservative?

9. Explain why friction is not conservative.

10. **(a)** When students in Universityland leave their houses, a supernatural power attracts them to university in such a way that the magnitude of the force at any point is inversely proportional to the square of the distance from university. This force acts until they are 100 m from university and then it disappears. Is this force conservative?

(b) If someone diverts the power so that the force attracts students to the local donut shop, is this force conservative?

11. A force field $\mathbf{F}(x,y,z)$ is said to be radially symmetric about the origin if it can be written in the form

$$\mathbf{F}(x,y,z) = f\left(\sqrt{x^2+y^2+z^2}\right)\mathbf{r}, \qquad \mathbf{r} = x\hat{\mathbf{i}}+y\hat{\mathbf{j}}+z\hat{\mathbf{k}},$$

for some function f. We often write in such a case that

$$\mathbf{F}(x,y,z) = f(r)\mathbf{r}, \quad \text{where } r = |\mathbf{r}| = \sqrt{x^2+y^2+z^2}.$$

(a) Use Theorem 15.4 to show that such a force is conservative in suitably defined domains (provided $f(r)$ has a continuous first derivative).

(b) If A and B are the points in Figure 15.27 joined by the curve C, show that

$$\int_C \mathbf{F}\cdot d\mathbf{r} = \int_a^b r f(r)\,dr,$$

where the limits a and b are the distances from the origin to A and B.

(c) Have we discussed any radially symmetric force fields in this chapter?

FIGURE 15.27

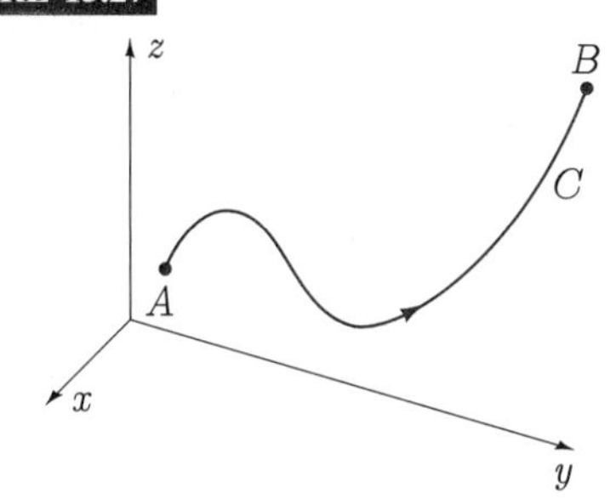

SECTION 15.6

Green's Theorem

Line integrals in the xy-plane are of the form $\int_C f(x,y)\,ds$, and in the special case that $f(x,y)$ is the tangential component of some vector field $\mathbf{F}(x,y) = P(x,y)\hat{\mathbf{i}} + Q(x,y)\hat{\mathbf{j}}$ along C, they take the form

$$\int_C \mathbf{F}\cdot d\mathbf{r} = \int_C P(x,y)\,dx + Q(x,y)\,dy. \tag{15.43}$$

We now show that when C is a closed curve, line integral 15.43 can usually be replaced by a double integral. The precise result is contained in the following theorem.

Theorem 15.5

(Green's theorem). Let C be a piecewise-smooth, closed curve in the xy-plane that does not intersect itself and that encloses a region R (Figure 15.28). If $P(x,y)$ and $Q(x,y)$ have continuous first partial derivatives in a domain D containing C and R, then

$$\oint_C P\,dx + Q\,dy = \iint_R \left(\frac{\partial Q}{\partial x} - \frac{\partial P}{\partial y}\right) dA. \tag{15.44}$$

FIGURE 15.28

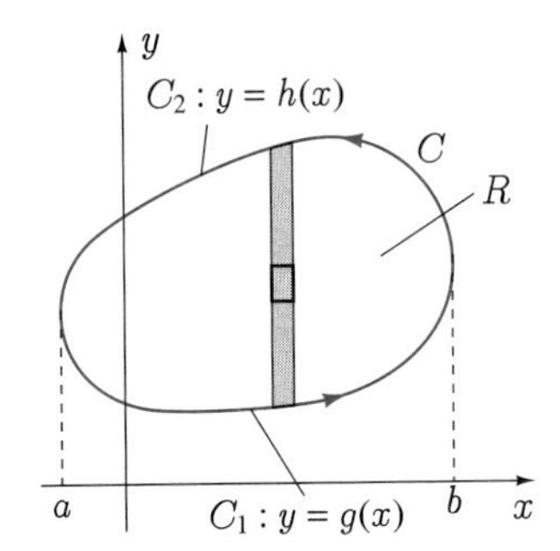

Proof We first consider a fairly simple region R. In particular, suppose every line parallel to the x- and y-axes that intersects C does so in at most two points (Figure 15.28). Then C can be subdivided into an upper and a lower part,

$$C_2: \quad y = h(x) \quad \text{and} \quad C_1: \quad y = g(x).$$

If we first consider the second term on the right of equation 15.44, we have

$$\iint_R -\frac{\partial P}{\partial y} dA = \int_a^b \int_{g(x)}^{h(x)} -\frac{\partial P}{\partial y} dy\,dx = \int_a^b \left\{-P\right\}_{g(x)}^{h(x)} dx$$
$$= \int_a^b \{P[x,g(x)] - P[x,h(x)]\}dx.$$

On the other hand, the first term on the left of 15.44 is

$$\oint_C P\,dx = \int_{C_1} P\,dx + \int_{C_2} P\,dx = \int_{C_1} P\,dx - \int_{-C_2} P\,dx,$$

and if we use x as a parameter along C_1 and $-C_2$, then

$$\oint_C P\,dx = \int_a^b P[x,g(x)]dx - \int_a^b P[x,h(x)]dx$$
$$= \int_a^b \{P[x,g(x)] - P[x,h(x)]\}dx.$$

We have shown therefore that

$$\oint_C P\,dx = \iint_R -\frac{\partial P}{\partial y} dA.$$

By subdividing C into two parts of the type $x = g(y)$ and $x = h(y)$ (where $g(y) \leq h(y)$), we can also show that

$$\oint_C Q\,dy = \iint_R \frac{\partial Q}{\partial x}\,dA.$$

Addition of these results gives Green's theorem for this C and R.

Now consider a more general region R such as that shown in Figure 15.29, which can be decomposed into n subregions R_i, each of which satisfies the condition that lines parallel to the coordinate axes intersect its boundary in at most two points. For each subregion R_i, Green's theorem gives

$$\oint_{C_i} P\,dx + Q\,dy = \iint_{R_i} \left(\frac{\partial Q}{\partial x} - \frac{\partial P}{\partial y}\right) dA.$$

If these results are added, we get

$$\sum_{i=1}^{n} \oint_{C_i} P\,dx + Q\,dy = \sum_{i=1}^{n} \iint_{R_i} \left(\frac{\partial Q}{\partial x} - \frac{\partial P}{\partial y}\right) dA.$$

Now, R is composed of the subregions R_i; thus the right-hand side of this equation is precisely the double integral over R. Figure 15.29 illustrates that when the line integrals over the C_i are added, contributions from ancillary (interior) curves cancel in pairs, leaving the line integral around C. This completes the proof.

FIGURE 15.29

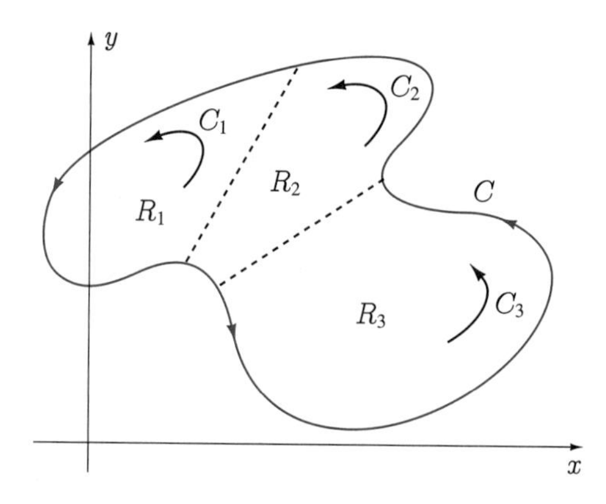

We omit a proof for even more general regions that cannot be divided into a finite number of these subregions. The interested reader should consult more advanced books.

EXAMPLE 15.15

Evaluate the line integral of Example 15.7.

SOLUTION By Green's theorem (see Figure 15.30), we have

$$\oint_C y^2\,dx + x^2\,dy = \iint_R (2x - 2y)\,dA = 2\int_0^1 \int_0^x (x - y)\,dy\,dx$$

$$= 2\int_0^1 \left\{xy - \frac{y^2}{2}\right\}_0^x dx = \int_0^1 x^2\,dx = \left\{\frac{x^3}{3}\right\}_0^1 = \frac{1}{3}.$$

FIGURE 15.30

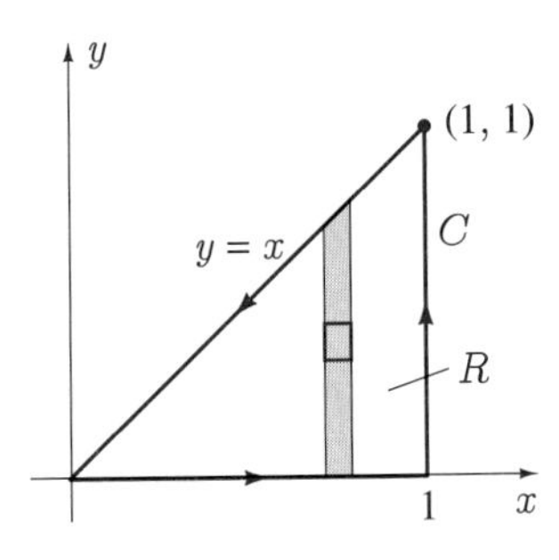

EXAMPLE 15.16

Evaluate

$$\oint_C (5x^2 + ye^x + y)\,dx + (e^x + e^y)\,dy,$$

where C is the circle $(x-1)^2 + y^2 = 1$.

SOLUTION By Green's theorem (see Figure 15.31), we have

$$\oint_C (5x^2 + ye^x + y)\,dx + (e^x + e^y)\,dy = \iint_R (e^x - e^x - 1)\,dA = -\iint_R dA$$
$$= -(\text{Area of } R) = -\pi(1)^2 = -\pi.$$

FIGURE 15.31

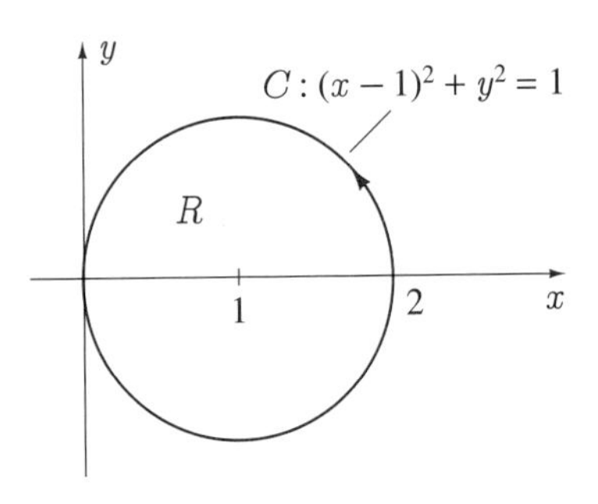

EXAMPLE 15.17

Show that the area of a region R is defined by each of the line integrals

$$\oint_C x\,dy = \oint_C -y\,dx = \frac{1}{2}\oint_C x\,dy - y\,dx,$$

where C is the boundary of R.

SOLUTION By Green's theorem, we have

$$\oint_C x\,dy = \iint_R 1\,dA = \text{Area of } R$$

and

$$\oint_C -y\,dx = \iint_R 1\,dA = \text{Area of } R.$$

The third expression for area is the average of these two equations. These formulas are of particular value when the curve C is defined parametrically. ■

Green's theorem cannot be used to evaluate a line integral around a closed curve that contains a point at which either $P(x, y)$ or $Q(x, y)$ fails to have continuous first partial derivatives. For example, it cannot be used to evaluate the line integral in Exercise 19(a) of Section 15.4 because neither P nor Q is defined at the origin. Try it and compare the result to the correct answer (2π). A generalization of Green's theorem that can be useful in such situations is found in Exercise 30.

EXERCISES 15.6

In Exercises 1–11 use Green's theorem (if possible) to evaluate the line integral.

1. $\oint_C y^2\,dx + x^2\,dy$ where C is the circle $x^2 + y^2 = 1$

2. $\oint_C (x^2 + 2y^2)\,dy$ where C is the curve $(x-2)^2 + y^2 = 1$

3. $\oint_C x^2 e^y\,dx + (x + y)\,dy$ where C is the square with vertices $(\pm 1, 1)$ and $(\pm 1, -1)$

4. $\oint_C xy^3\,dx + x^2\,dy$ where C is the curve enclosing the area bounded by $x = \sqrt{1 + y^2}$, $x = 2$

5. $\oint_C (x^3 + y^3)\,dx + (x^3 - y^3)\,dy$ where C is the curve enclosing the area bounded by $x = y^2 - 1$, $x = 1 - y^2$

6. $\oint_C 2\,\text{Tan}^{-1}(y/x)\,dx + \ln(x^2 + y^2)\,dy$ where C is the circle $(x - 4)^2 + (y - 1)^2 = 2$

7. $\oint_C (3x^2y^3 + y)\,dx + (3x^3y^2 + 2x)\,dy$ where C is the boundary of the area enclosed by $x + y = 1$, $x = -1, y = -1$

8. $\oint_C (x^3 + y^3)\,dx + (x^3 - y^3)\,dy$ where C is the curve $2|x| + |y| = 1$

9. $\oint_C (x^2y^2 + 3x)\,dx + (2xy - y)\,dy$ where C is the boundary of the area enclosed by $x = 1 - y^2$ $(x \geq 0)$, $y = x + 1$, $y + x + 1 = 0$

10. $\oint_C (xy^2 + 2x)\,dx + (x^2y + y + x^2)\,dy$ where C is the boundary of the area enclosed by $y^2 - x^2 = 4$, $x = 0$, $x = 3$

11. $\oint_C \dfrac{-y\,dx + x\,dy}{x^2 + y^2}$ where C is the circle $x^2 + y^2 = 1$

12. Show that Green's theorem can be expressed vectorially in the form

$$\oint_C \mathbf{F} \cdot d\mathbf{r} = \iint_R (\nabla \times \mathbf{F}) \cdot \hat{\mathbf{k}}\,dA.$$

13. If a curve C is traced out in the direction defined by Green's theorem, it can be shown that a normal vector to C that always points to the outside of C is $\mathbf{n} = (dy, -dx)$. Show that Green's theorem can be written vectorially in the form

$$\oint_C \mathbf{F} \cdot \hat{\mathbf{n}}\, ds = \iint_R \nabla \cdot \mathbf{F}\, dA.$$

In Exercises 14–19 use the results of Example 15.17 to find the area enclosed by the curve.

14. $x^2/a^2 + y^2/b^2 = 1$

15. The strophoid $x = (1-t^2)/(1+t^2)$, $y = (t-t^3)/(1+t^2)$ (see Example 10.4)

16. The astroid $x = \cos^3\theta$, $y = \sin^3\theta$ (see Exercise 50 in Section 10.1)

17. The right loop of the curve of Lissajous (see Exercise 48 in Section 10.1)

18. The deltoid $x = 2\cos t + \cos 2t$, $y = 2\sin t - \sin 2t$

19. The droplet $x = 2\cos t - \sin 2t$, $y = \sin t$

20. **(a)** If C is the straight-line segment joining points $P_1(x_1, y_1)$ and $P_2(x_2, y_2)$ in the xy-plane, show that

$$\int_C x\, dy - y\, dx = x_1 y_2 - x_2 y_1.$$

(b) Let $P_1(x_1, y_1)$, $P_2(x_2, y_2)$, $\cdots$, $P_n(x_n, y_n)$ be the coordinates of a polygon labeled in the counterclockwise direction (Figure 15.32(a)). Use the result in (a) and Example 15.17 to derive the following formula for the area A of the polygon:

$$A = \frac{1}{2}\Big[(x_1 y_2 - x_2 y_1) + (x_2 y_3 - x_3 y_2) + \cdots + (x_{n-1} y_n - x_n y_{n-1}) + (x_n y_1 - x_1 y_n)\Big].$$

(c) Show that if the coordinates of the vertices are arranged in a vertical column with (x_1, y_1) repeated at the bottom (Figure 15.32(b)), then

$$A = \frac{1}{2}\Big\{[\text{sum of downward products to the right } (\searrow)] - [\text{sum of downward products to the left } (\swarrow)]\Big\}.$$

FIGURE 15.32

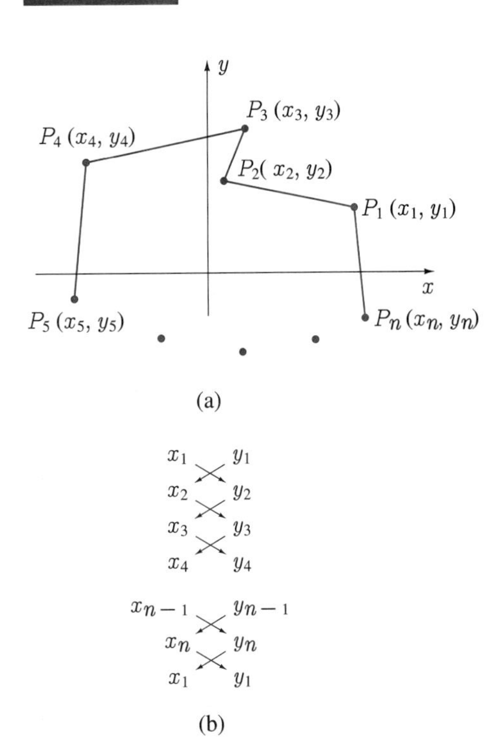

In Exercises 21–24 use the result of Exercise 20 to find the area of the polygon with the points as successive vertices.

21. $(1, 2)$, $(-3, 2)$, $(4, 1)$

22. $(2, -2)$, $(1, -3)$, $(-2, 1)$, $(5, 6)$

23. $(3, 0)$, $(1, 1)$, $(2, 5)$, $(-4, -4)$

24. $(0, 4)$, $(-1, 0)$, $(-2, 0)$, $(-3, -4)$, $(0, -5)$, $(6, -2)$, $(3, 0)$, $(2, 2)$

In Exercises 25–29 evaluate the line integral.

25. $\oint_C (2xye^{x^2 y} + 3x^2 y)\, dx + x^2 e^{x^2 y}\, dy$ where C is the ellipse $x^2 + 4y^2 = 4$

26. $\oint_C (3x^2 y^3 - x^2 y)\, dx + (xy^2 + 3x^3 y^2)\, dy$ where C is the circle $x^2 + y^2 = 9$

27. $\oint_C -x^3 y^2\, dx + x^2 y^3\, dy$ where C is the right loop of $(x^2 + y^2)^{3/2} = x^2 - y^2$

28. $\int_C (x - y)(dx + dy)$ where C is the semicircular part of $x^2 + y^2 = 4$ above $y = x$ from $(-\sqrt{2}, -\sqrt{2})$ to $(\sqrt{2}, \sqrt{2})$

29. $\displaystyle\int_C (e^y - y\sin x)\,dx + (\cos x + xe^y)\,dy$ where C is the curve $x = 1 - y^2$ from $(0,-1)$ to $(0,1)$

30. The result of this exercise is useful when the curve C in Green's theorem contains a point (or points) at which either P or Q fails to have continuous first partial derivatives (see Exercises 31–35).

(a) Suppose a piecewise smooth curve C (Figure 15.33a) contains in its interior another piecewise smooth curve C', and that $P(x,y)$ and $Q(x,y)$ have continuous first partial derivatives in a domain containing C and C' and the area R between them. Prove that

$$\oint_C P\,dx + Q\,dy + \oint_{C'} P\,dx + Q\,dy = \iint_R \left(\frac{\partial Q}{\partial x} - \frac{\partial P}{\partial y}\right) dA.$$

Hint: Join C and C' by two curves such as those in Figure 15.33(a).

(b) Extend this result to show that when C' is replaced by n distinct curves C_i (Figure 15.33b), and P and Q have continuous first partial derivatives in a domain containing C and the C_i and the area R between them,

$$\oint_C P\,dx + Q\,dy + \oint_{C_1} P\,dx + Q\,dy + \cdots + \oint_{C_n} P\,dx + Q\,dy = \iint_R \left(\frac{\partial Q}{\partial x} - \frac{\partial P}{\partial y}\right) dA.$$

(c) What can we conclude in (a) and (b) if $\partial Q/\partial x = \partial P/\partial y$ in R?

FIGURE 15.33

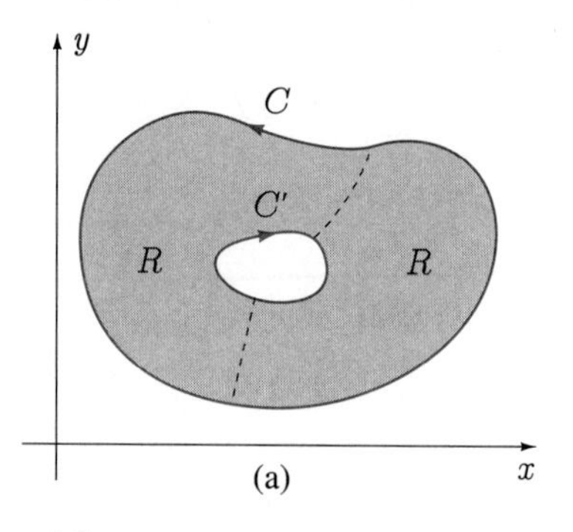

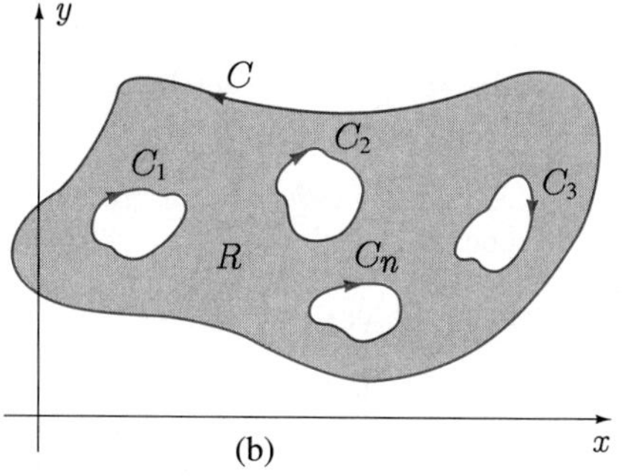

In Exercises 31–33 use the result of Exercise 30(a) to evaluate the line integral.

31. $\displaystyle\oint_C \frac{y\,dx - (x-1)\,dy}{(x-1)^2 + y^2}$ where C is the circle $x^2 + y^2 = 4$.

32. $\displaystyle\oint_C \frac{-x^2 y\,dx + x^3\,dy}{(x^2+y^2)^2}$ where C is the ellipse $4x^2 + y^2 = 1$

33. $\displaystyle\oint_C \frac{-y\,dx + x\,dy}{x^2+y^2}$ where C is the square with vertices $(\pm 2, 0)$ and $(0, \pm 2)$

34. Show that the line integral of Exercise 19 in Section 4 has value $\pm 2\pi$ for every piecewise smooth, closed curve enclosing the origin that does not intersect itself.

35. **(a)** In what domains is the line integral $\displaystyle\int_C \frac{x\,dx + y\,dy}{x^2+y^2}$ independent of path?

(b) Evaluate the integral clockwise around the curve $x^2 + y^2 - 2y = 1$.

In Exercises 36–38 assume that $P(x,y)$ and $Q(x,y)$ have continuous second partial derivatives in a domain containing R and C (Figure 15.34). Let the vector $\hat{\mathbf{n}} = (dy/ds, -dx/ds)$ be the outward pointing normal to C, and let

$$\frac{\partial P}{\partial n} = \nabla P \cdot \hat{\mathbf{n}}, \qquad \frac{\partial Q}{\partial n} = \nabla Q \cdot \hat{\mathbf{n}}$$

be the directional derivatives of P and Q in the direction $\hat{\mathbf{n}}$.

FIGURE 15.34

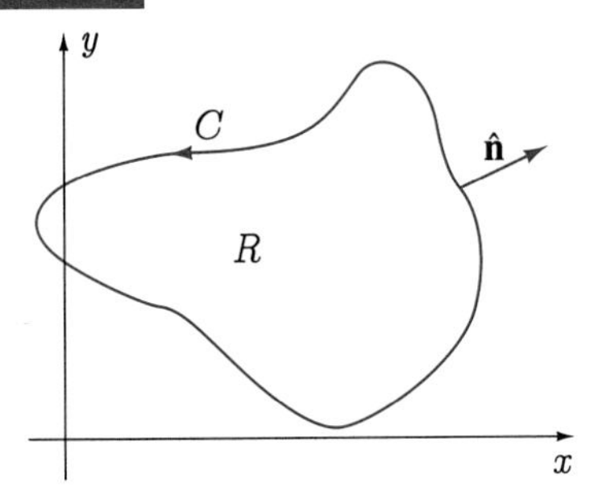

36. Show that

$$\oint_C \frac{\partial P}{\partial n} ds = \iint_R \nabla^2 P \, dA$$

where $\nabla^2 P = \dfrac{\partial^2 P}{\partial x^2} + \dfrac{\partial^2 P}{\partial y^2}$. Hint: See Exercise 13. What can we conclude if $P(x, y)$ satisfies Laplace's equation in R?

37. Show that

$$\oint_C P \frac{\partial Q}{\partial n} ds = \iint_R P \nabla^2 Q \, dA + \iint_R \nabla P \cdot \nabla Q \, dA.$$

Hint: Use identity 15.11. This result is often called Green's first identity (in the plane).

38. Prove that

$$\oint_C \left(P \frac{\partial Q}{\partial n} - Q \frac{\partial P}{\partial n} \right) ds = \iint_R (P \nabla^2 Q - Q \nabla^2 P) \, dA.$$

This is often called Green's second identity (in the plane).

39. Find all possible values for the line integral $\displaystyle\oint_C \frac{-y\,dx + x\,dy}{x^2 + y^2}$ for curves in the xy-plane not passing through the origin.

SECTION 15.7

Surface Integrals

Consider a function $f(x, y, z)$ defined on some surface S (Figure 15.35). We divide S into n subsurfaces of areas $\Delta S_1, \Delta S_2, \ldots, \Delta S_n$ in any manner whatsoever. On each subsurface $\Delta S_i (i = 1, \ldots, n)$ we choose an arbitrary point (x_i^*, y_i^*, z_i^*), and form the sum

$$f(x_1^*, y_1^*, z_1^*)\Delta S_1 + f(x_2^*, y_2^*, z_2^*)\Delta S_2 + \cdots + f(x_n^*, y_n^*, z_n^*)\Delta S_n$$
$$= \sum_{i=1}^{n} f(x_i^*, y_i^*, z_i^*)\Delta S_i. \qquad (15.45)$$

If this sum approaches a limit as the number of subsurfaces becomes increasingly large and every subsurface shrinks to a point, we call the limit the surface integral of $f(x, y, z)$ over the surface S, and denote it by

$$\iint_S f(x, y, z)\, dS = \lim_{\|\Delta S_i\| \to 0} \sum_{i=1}^{n} f(x_i^*, y_i^*, z_i^*)\Delta S_i. \qquad (15.46)$$

FIGURE 15.35

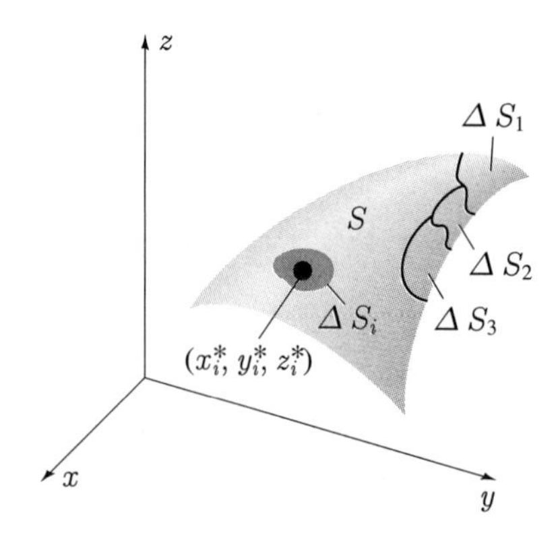

Surface integrals, like all integrals, are limit-summations. We think of dS as a small piece of area on S, and each dS is multiplied by the value of $f(x, y, z)$ for that area. All such products are then added together and the limit taken as the pieces of area shrink to points.

The following theorem guarantees the existence of the surface integral of a continuous function over a smooth surface.

Theorem 15.6

If a function $f(x, y, z)$ is continuous on a smooth surface S of finite area, then the surface integral of $f(x, y, z)$ over S exists. If S projects in a one-to-one fashion onto a region S_{xy} in the xy-plane, then the surface integral of $f(x, y, z)$ over S can be evaluated by means of the following double integral:

$$\iint_S f(x, y, z)\,dS = \iint_{S_{xy}} f[x, y, g(x, y)]\sqrt{1 + \left(\frac{\partial z}{\partial x}\right)^2 + \left(\frac{\partial z}{\partial y}\right)^2}\,dA, \qquad (15.47)$$

where $z = g(x, y)$ is the equation of S.

It is not necessary to memorize 15.47 as a formula, since it is the result obtained by expressing z and dS in terms of x and y and interpreting the result as a double integral over the projection of S in the xy-plane. Recall from Section 14.7 that when a surface S can be represented in the form $z = g(x, y)$, a small area dS on S is related to its projection dA in the xy-plane according to the formula

$$dS = \sqrt{1 + \left(\frac{\partial z}{\partial x}\right)^2 + \left(\frac{\partial z}{\partial y}\right)^2}\,dA. \qquad (15.48)$$

(Figure 15.36). If we substitute this into the left-hand side of 15.47 and at the same time use $z = g(x, y)$ to express $f(x, y, z)$ in terms of x and y, then

$$\iint_S f(x, y, z)\,dS = \iint_S f[x, y, g(x, y)]\sqrt{1 + \left(\frac{\partial z}{\partial x}\right)^2 + \left(\frac{\partial z}{\partial y}\right)^2}\,dA.$$

But if we now interpret the right-hand side of this equation as the double integral of $f[x, y, g(x, y)]\sqrt{1 + (\partial z/\partial x)^2 + (\partial z/\partial y)^2}$ over the projection S_{xy} of S onto the xy-plane, we obtain 15.47.

FIGURE 15.36

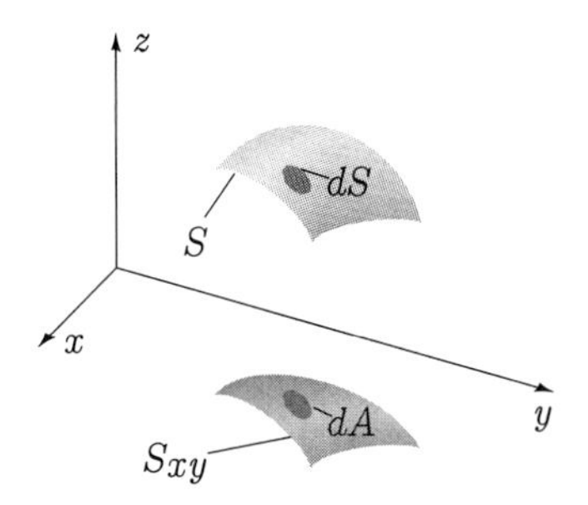

Note the analogy between equations 15.47 and 15.20. equation 15.20 states that the line integral on the left can be evaluated by means of the definite integral on the right. Equation 15.47 states that the surface integral on the left can be evaluated by means of the double integral on the right.

If a surface does not project one-to-one onto an area in the xy-plane, then one possibility is to subdivide it into parts, each of which projects one-to-one onto the xy-plane. The total surface integral over the surface is then the sum of the surface integrals over the parts. For example, if we require the surface integral of a function $f(x, y, z)$ over the sphere S: $x^2 + y^2 + z^2 = 1$ (Figure 15.37), then we could subdivide S into two hemispheres,

$$S_1: \quad z = \sqrt{1 - x^2 - y^2},$$
$$S_2: \quad z = -\sqrt{1 - x^2 - y^2},$$

each of which projects onto

$$S_{xy}: \quad x^2 + y^2 \leq 1.$$

Then,

$$\iint_S f(x, y, z)\,dS = \iint_{S_1} f(x, y, z)\,dS + \iint_{S_2} f(x, y, z)\,dS.$$

FIGURE 15.37

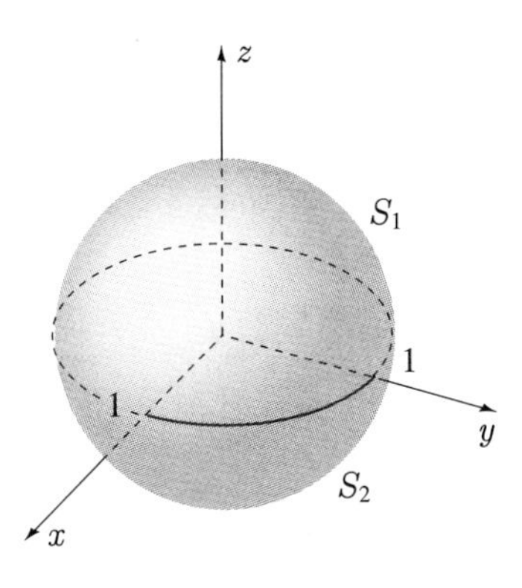

A second possibility is to project surfaces onto either the xz-plane or the yz-plane. If a surface projects one-to-one onto areas S_{xz} and S_{yz} in these planes, then

$$\iint_S f(x, y, z)\,dS = \iint_{S_{xz}} f[x, g(x, z), z]\sqrt{1 + \left(\frac{\partial y}{\partial x}\right)^2 + \left(\frac{\partial y}{\partial z}\right)^2}\,dA \quad (15.49)$$

and

$$\iint_S f(x, y, z)\,dS = \iint_{S_{yz}} f[g(y, z), y, z]\sqrt{1 + \left(\frac{\partial x}{\partial y}\right)^2 + \left(\frac{\partial x}{\partial z}\right)^2}\,dA. \quad (15.50)$$

We should also note that the area of a surface S is defined by the surface integral

$$\iint_S dS, \tag{15.51}$$

and if S projects one-to-one onto S_{xy}, then

$$\text{Area of } S = \iint_{S_{xy}} \sqrt{1 + \left(\frac{\partial z}{\partial x}\right)^2 + \left(\frac{\partial z}{\partial y}\right)^2}\, dA. \tag{15.52}$$

This is precisely formula 14.41.

The results in equations 15.47–15.52 were based on a smooth surface S. If S is piecewise smooth, rather than smooth, we simply subdivide S into smooth parts and apply each of these results to the smooth parts. The surface integral over S is then the summation of the surface integrals over its parts.

EXAMPLE 15.18

Evaluate $\iint_S (x + y + z)\,dS$, where S is that part of the plane $x + 2y + 4z = 8$ in the first octant.

SOLUTION The surface S projects one-to-one onto the triangle S_{xy} in the xy-plane shown in Figure 15.38. Since $z = (8 - x - 2y)/4$ on S,

$$\begin{aligned}
\iint_S (x + y + z)\,dS &= \iint_{S_{xy}} \left(x + y + 2 - \frac{x}{4} - \frac{y}{2}\right)\sqrt{1 + \left(\frac{\partial z}{\partial x}\right)^2 + \left(\frac{\partial z}{\partial y}\right)^2}\, dA \\
&= \frac{1}{4}\iint_{S_{xy}} (3x + 2y + 8)\sqrt{1 + \left(-\frac{1}{4}\right)^2 + \left(-\frac{1}{2}\right)^2}\, dA \\
&= \frac{\sqrt{21}}{16}\int_0^4 \int_0^{8-2y} (3x + 2y + 8)\,dx\,dy \\
&= \frac{\sqrt{21}}{16}\int_0^4 \left\{\frac{3x^2}{2} + (2y + 8)x\right\}_0^{8-2y} dy \\
&= \frac{\sqrt{21}}{8}\int_0^4 (80 - 24y + y^2)\,dy \\
&= \frac{\sqrt{21}}{8}\left\{80y - 12y^2 + \frac{y^3}{3}\right\}_0^4 = \frac{56\sqrt{21}}{3}.
\end{aligned}$$

FIGURE 15.38

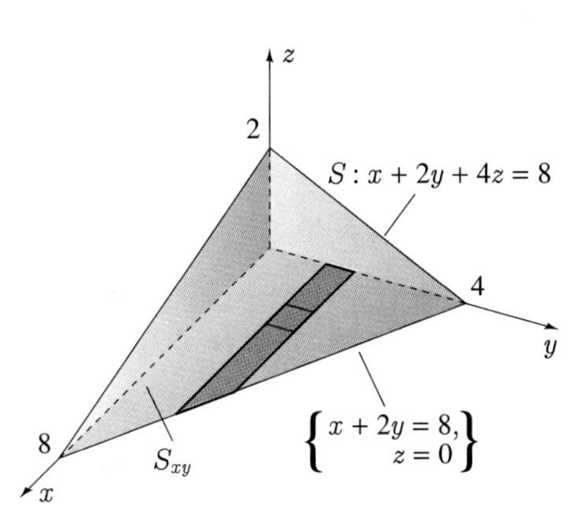

EXAMPLE 15.19

Evaluate $\iint_S z^2\,dS$ where S is the sphere $x^2 + y^2 + z^2 = 4$.

SOLUTION We divide S into two hemispheres (Figure 15.39),

$$S_1 : z = \sqrt{4 - x^2 - y^2}, \quad S_2 : z = -\sqrt{4 - x^2 - y^2}$$

each of which projects one-to-one onto the circle S_{xy}: $x^2 + y^2 \leq 4$, $z = 0$ in the xy-plane. For each hemisphere,

$$dS = \sqrt{1 + \left(\frac{\partial z}{\partial x}\right)^2 + \left(\frac{\partial z}{\partial y}\right)^2}\,dA$$

$$= \sqrt{1 + \frac{x^2}{4 - x^2 - y^2} + \frac{y^2}{4 - x^2 - y^2}}\,dA = \frac{2}{\sqrt{4 - x^2 - y^2}}\,dA,$$

and therefore

$$\iint_S z^2\,dS = \iint_{S_1} z^2\,dS + \iint_{S_2} z^2\,dS$$

$$= \iint_{S_{xy}} (4 - x^2 - y^2)\frac{2}{\sqrt{4 - x^2 - y^2}}\,dA$$

$$+ \iint_{S_{xy}} (4 - x^2 - y^2)\frac{2}{\sqrt{4 - x^2 - y^2}}\,dA$$

$$= 4\iint_{S_{xy}} \sqrt{4 - x^2 - y^2}\,dA.$$

If we use polar coordinates to evaluate this integral over S_{xy}, then

$$\iint_S z^2\,dS = 4\int_0^{2\pi}\int_0^2 \sqrt{4 - r^2}\,r\,dr\,d\theta$$

$$= 4\int_0^{2\pi}\left\{-\frac{1}{3}(4 - r^2)^{3/2}\right\}_0^2 d\theta = \frac{32}{3}\{\theta\}_0^{2\pi} = \frac{64\pi}{3}.$$

FIGURE 15.39

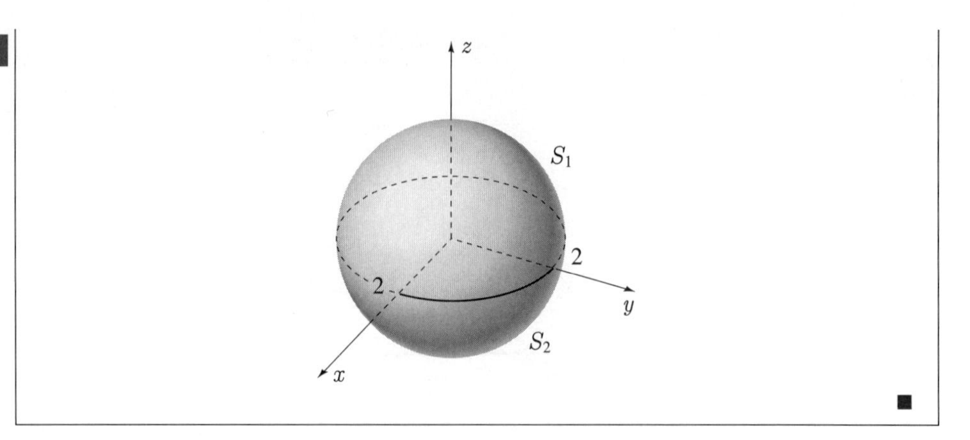

If parameters can be found to describe a surface, it may not be necessary to project the surface into one of the coordinate planes. Such is the case for a sphere centred at the origin. Formula 14.67 for the volume element in spherical coordinates indicates that an area element on the surface of a sphere of radius R can be expressed in terms of angles θ and ϕ (Figure 15.40) as

$$dS = R^2 \sin\phi \, d\phi \, d\theta. \tag{15.53}$$

With this choice of area element, the surface integral in Example 15.19 is evaluated as follows:

$$\begin{aligned}\iint_S z^2 \, dS &= \int_0^{2\pi} \int_0^{\pi} (2\cos\phi)^2 4 \sin\phi \, d\phi \, d\theta = 16 \int_0^{2\pi} \int_0^{\pi} \cos^2\phi \sin\phi \, d\phi \, d\theta \\ &= 16 \int_0^{2\pi} \left\{-\frac{1}{3}\cos^3\phi\right\}_0^{\pi} d\theta = \frac{32}{3}\{\theta\}_0^{2\pi} = \frac{64\pi}{3}.\end{aligned}$$

FIGURE 15.40

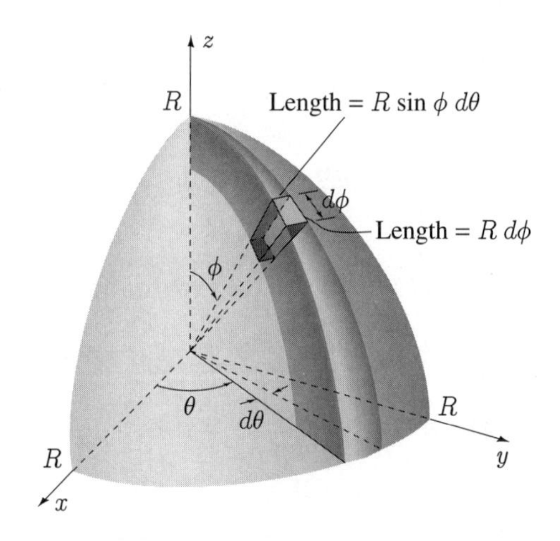

EXERCISES 15.7

In Exercises 1–8 evaluate the surface integral.

1. $\iint_S (x^2y + z)\,dS$ where S is the first octant part of $2x + 3y + z = 6$

2. $\iint_S (x^2 + y^2)z\,dS$ where S is that part of $z = x + y$ cut out by $x = 0$, $y = 0$, $x + y = 1$

3. $\iint_S xyz\,dS$ where S is the surface of the cube $0 \le x \le 1$, $0 \le y \le 1$, $0 \le z \le 1$

4. $\iint_S xy\,dS$ where S is the first octant part of $z = \sqrt{x^2 + y^2}$ cut out by $x^2 + y^2 = 1$

5. $\iint_S \frac{1}{\sqrt{z - y + 1}}\,dS$ where S is the surface defined by $2z = x^2 + 2y$, $0 \le x \le 1$, $0 \le y \le 1$

6. $\iint_S \sqrt{4y + 1}\,dS$ where S is the first octant part of $y = x^2$ cut out by $2x + y + z = 1$

7. $\iint_S x^2 z\,dS$ where S is the surface $x^2 + y^2 = 1$ for $0 \le z \le 1$

8. $\iint_S (x + y)\,dS$ where S is the surface bounding the volume enclosed by $x = 0$, $y = 0$, $z = 0$, $6x - 3y + 2z = 6$

9. Set up double iterated integrals for the surface integral of a function $f(x, y, z)$ over the surface defined by $z = 4 - x^2 - 4y^2$ $(x, y, z \ge 0)$ if the surface is projected onto the xy-, the xz-, and the yz-planes.

10. Use a surface integral to find the area of the curved portion of a right-circular cone of radius R and height h.

In Exercises 11–17 evaluate the surface integral.

11. $\iint_S xyz^3\,dS$ where S is the surface defined by $x = y^2$, $0 \le x \le 4$, $0 \le z \le 1$

12. $\iint_S xyz\,dS$ where S is the surface defined by $2y = \sqrt{9 - x}$, $x \ge 0$, $0 \le z \le 3$

13. $\iint_S \frac{1}{\sqrt{2az - z^2}}\,dS$ where S is that part of $x^2 + y^2 + (z - a)^2 = a^2$ inside the cylinder $x^2 + y^2 = ay$, underneath the plane $z = a$, and in the first octant

14. $\iint_S z\,dS$ where S is that part of the surface $x^2 + y^2 - z^2 = 1$ between the planes $z = 0$ and $z = 1$

15. $\iint_S x^2y^2\,dS$ where S is that part of $z = x^2 + y^2$ inside $x^2 + y^2 + z^2 = 2$

16. $\iint_S x^2\,dS$ where S is that part of $z = xy$ inside $x^2 + y^2 = 4$

17. $\iint_S z(y + x^2)\,dS$ where S is that part of $y = 1 - x^2$ bounded by $z = 0$, $z = 2$, and $y = 0$

In Exercises 18–22 evaluate the surface integral by projecting the surface into one of the coordinate planes and also by using area element 15.53.

18. $\iint_S dS$ where S is the sphere $x^2 + y^2 + z^2 = R^2$. Is this the formula for the area of a sphere?

19. $\iint_S x^2z^2\,dS$ where S is the sphere $x^2 + y^2 + z^2 = 1$

20. $\iint_S (x^2 - y^2)\,dS$ where S is the hemisphere $z = \sqrt{9 - x^2 - y^2}$

21. $\iint_S (x^2 + y^2)\,dS$ where S is the sphere $x^2 + y^2 + z^2 = R^2$

22. $\iint_S \frac{1}{x^2 + y^2}\,dS$ where S is that part of the sphere $x^2 + y^2 + z^2 = 4R^2$ between the planes $z = 0$ and $z = R$

23. A viscous material is allowed to drip onto the sphere $x^2 + y^2 + z^2 = 1$ at the point $(0, 0, 1)$ (all dimensions in metres). The material spreads out evenly in all directions and runs down the sphere, becoming more and more viscous as it does so. The thickness of the material increases linearly with respect to angle ϕ (Figure 15.40) from 0.001 m at $(0, 0, 1)$ to 0.005 m at $(0, 0, -1)$. Find the volume of material on the sphere.

24. Show that if a surface S defined implicitly by the equation $F(x, y, z) = 0$ projects one-to-one onto the area S_{xy} in the xy-plane, then

$$\iint_S f(x, y, z)\,dS = \iint_{S_{xy}} f[x, y, g(x, y)]\frac{|\nabla F|}{|\partial F/\partial z|}\,dA.$$

25. **(a)** Find the area cut from the cones $z^2 = x^2 + y^2$ by the cylinder $x^2 + y^2 = 2x$.

(b) Find the area cut from the cylinder $x^2 + y^2 = 2x$ by the cones $z^2 = x^2 + y^2$.

SECTION 15.8

Surface Integrals Involving Vector Fields

The most important and common type of surface integral occurs when $f(x, y, z)$ in 15.46 is specified as the normal component of some given vector field $\mathbf{F}(x, y, z)$ defined on S. In other words, $f(x, y, z)$ itself is not given, but $\mathbf{F}$ is, and to find $f(x, y, z)$ we must calculate the component of $\mathbf{F}$ normal to S.

This presupposes that surfaces are two-sided and that a normal vector to a surface can be assigned in an unambiguous way. When this is possible, the surface is said to be **orientable**. All surfaces in this book are orientable, with the exception of the Möbius strip mentioned below.

Take a thin rectangular strip of paper and label its corners A, B, C, and D (Figure 15.41a). Give the strip a half twist and join A and C, and B and D (Figure 15.41b).

FIGURE 15.41

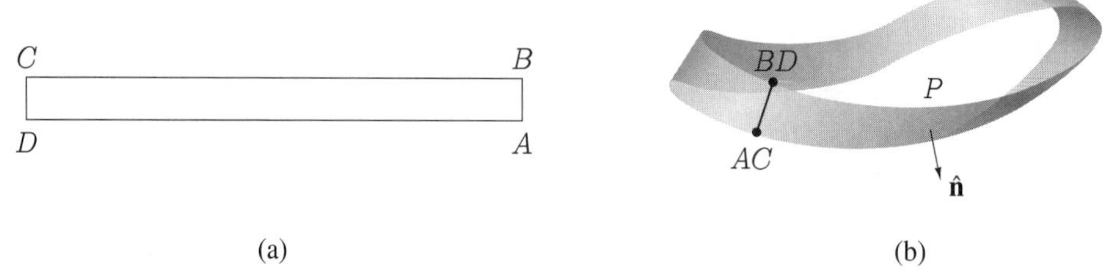

(a) (b)

This surface, called a Möbius strip, cannot be assigned a unique normal vector which varies continuously over the surface. To illustrate, suppose at point P in Figure 15.41b, we assign a unit normal vector $\hat{\mathbf{n}}$ as shown. By moving once around the strip, we can vary the direction of $\hat{\mathbf{n}}$ continuously and arrive back at P with $\hat{\mathbf{n}}$ pointing in the opposite direction. This surface is said to have only one side, or to be nonorientable.

We consider only surfaces that are orientable, or have two sides, and can therefore be assigned a unit normal vector in an unambiguous way.

Suppose again that $\mathbf{F}(x, y, z)$ is a vector field defined on an (orientable) surface S and $f(x, y, z)$ is the component of $\mathbf{F}$ in one of the two normal directions to S. If $\hat{\mathbf{n}}$ is the unit normal vector to S in the specified direction, then $f(x, y, z) = \mathbf{F} \cdot \hat{\mathbf{n}}$, and

$$\iint_S f(x, y, z)\, dS = \iint_S \mathbf{F} \cdot \hat{\mathbf{n}}\, dS. \tag{15.54}$$

EXAMPLE 15.20

Evaluate $\iint_S \mathbf{F} \cdot \hat{\mathbf{n}}\, dS$, where $\mathbf{F} = x^2 y\hat{\mathbf{i}} + xz\hat{\mathbf{j}}$ and $\hat{\mathbf{n}}$ is the upper normal to the surface $S : z = 4 - x^2 - y^2, z \geq 0$.

SOLUTION Since a normal vector to S is

$$\nabla(z - 4 + x^2 + y^2) = (2x, 2y, 1),$$

it follows that

$$\hat{\mathbf{n}} = \frac{(2x, 2y, 1)}{\sqrt{4x^2 + 4y^2 + 1}}.$$

Thus,

$$\mathbf{F} \cdot \hat{\mathbf{n}} = \frac{2x^3y + 2xyz}{\sqrt{4x^2 + 4y^2 + 1}}.$$

If we project S onto $S_{xy} : x^2 + y^2 \leq 4$ in the xy-plane (Figure 15.42), then

$$\begin{aligned}
\iint_S \mathbf{F} \cdot \hat{\mathbf{n}} dS &= \iint_{S_{xy}} \frac{2x^3y + 2xyz}{\sqrt{4x^2 + 4y^2 + 1}} \sqrt{1 + \left(\frac{\partial z}{\partial x}\right)^2 + \left(\frac{\partial z}{\partial y}\right)^2} dA \\
&= \iint_{S_{xy}} \frac{2x^3y + 2xy(4 - x^2 - y^2)}{\sqrt{4x^2 + 4y^2 + 1}} \sqrt{1 + (-2x)^2 + (-2y)^2} dA \\
&= 2 \iint_{S_{xy}} (4xy - xy^3) dA \\
&= 2 \int_{-2}^{2} \int_{-\sqrt{4-x^2}}^{\sqrt{4-x^2}} (4xy - xy^3) dy\, dx \\
&= 2 \int_{-2}^{2} \left\{2xy^2 - \frac{xy^4}{4}\right\}_{-\sqrt{4-x^2}}^{\sqrt{4-x^2}} dx \\
&= 0.
\end{aligned}$$

FIGURE 15.42

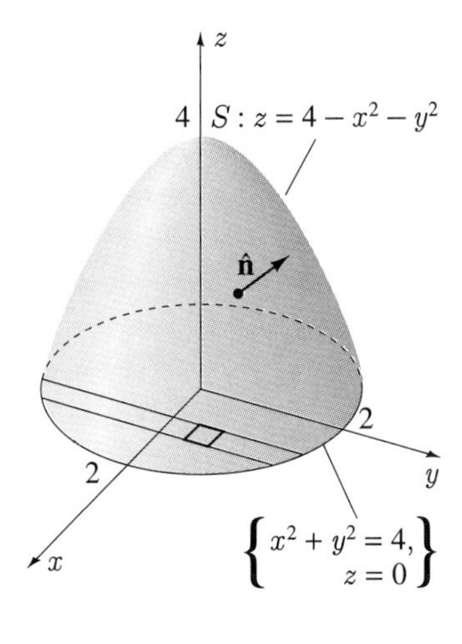

■

EXAMPLE 15.21

Evaluate

$$\oiint_S \mathbf{F} \cdot \hat{\mathbf{n}} dS,$$

where $\mathbf{F} = x\hat{\mathbf{i}} + y\hat{\mathbf{j}} + z\hat{\mathbf{k}}$ and $\hat{\mathbf{n}}$ is the unit outward-pointing normal to the surface S

enclosing the volume bounded by $x^2 + y^2 = 4$, $z = 0$, $z = 2$.

SOLUTION We can divide S into four parts (Figure 15.43):

$$\begin{aligned} S_1 &: \quad z = 0, \quad x^2 + y^2 \leq 4; \\ S_2 &: \quad z = 2, \quad x^2 + y^2 \leq 4; \\ S_3 &: \quad y = \sqrt{4 - x^2}, \quad 0 \leq z \leq 2; \\ S_4 &: \quad y = -\sqrt{4 - x^2}, \quad 0 \leq z \leq 2. \end{aligned}$$

FIGURE 15.43

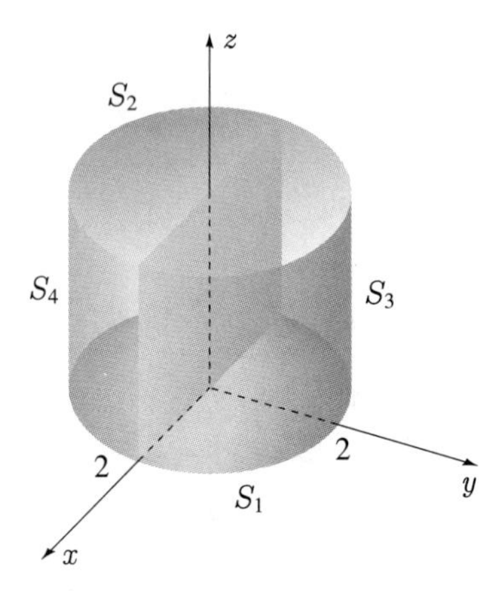

On S_1, $\hat{\mathbf{n}} = -\hat{\mathbf{k}}$; on S_2, $\hat{\mathbf{n}} = \hat{\mathbf{k}}$; and on S_3 and S_4,

$$\hat{\mathbf{n}} = \frac{\nabla(x^2 + y^2 - 4)}{|\nabla(x^2 + y^2 - 4)|} = \frac{(2x, 2y, 0)}{\sqrt{4x^2 + 4y^2}} = \frac{(x, y, 0)}{2}.$$

We know that S_1 and S_2 project onto $S_{xy} : x^2 + y^2 \leq 4$ in the xy-plane, and S_3 and S_4 project onto the rectangle $S_{xz} : -2 \leq x \leq 2, 0 \leq z \leq 2$ in the xz-plane. Consequently,

$$\begin{aligned} \oiint_S \mathbf{F} \cdot \hat{\mathbf{n}}\, dS &= \iint_{S_1} -z\, dS + \iint_{S_2} z\, dS + \iint_{S_3} \left(\frac{x^2 + y^2}{2}\right) dS + \iint_{S_4} \left(\frac{x^2 + y^2}{2}\right) dS \\ &= 0 + \iint_{S_{xy}} 2\sqrt{1}\, dA + \frac{1}{2} \iint_{S_{xz}} (x^2 + 4 - x^2)\sqrt{1 + \left(\frac{-x}{\sqrt{4 - x^2}}\right)^2}\, dA \\ &\quad + \frac{1}{2} \iint_{S_{xz}} (x^2 + 4 - x^2)\sqrt{1 + \left(\frac{x}{\sqrt{4 - x^2}}\right)^2}\, dA \\ &= 2 \iint_{S_{xy}} dA + 4 \iint_{S_{xz}} \frac{2}{\sqrt{4 - x^2}}\, dA \\ &= 2(\text{Area of } S_{xy}) + 8 \int_{-2}^{2} \int_0^2 \frac{1}{\sqrt{4 - x^2}}\, dz\, dx \\ &= 2(4\pi) + 16 \int_{-2}^{2} \frac{1}{\sqrt{4 - x^2}}\, dx. \end{aligned}$$

If we set $x = 2\sin\theta$, then $dx = 2\cos\theta\, d\theta$ and

$$\oint\!\!\oint_S \mathbf{F}\cdot\hat{\mathbf{n}}\,dS = 8\pi + 16\int_{-\pi/2}^{\pi/2} \frac{1}{2\cos\theta} 2\cos\theta\, d\theta = 8\pi + 16\left\{\theta\right\}_{-\pi/2}^{\pi/2} = 24\pi.$$

■

Note that the notation $\oint\!\!\oint$ is similar to that for the line integrals for the surface integral over a closed surface (a closed surface being one that encloses a volume).

. . . .

EXAMPLE 15.22

If a spherical object is submerged in a fluid with density ρ, it experiences a buoyant force due to fluid pressure. Show that the magnitude of this force is exactly the weight of the fluid displaced by the object.

SOLUTION Suppose the object is represented by the sphere $x^2 + y^2 + z^2 = R^2$ and the surface of the fluid by the plane $z = h\ (h > R)$ (Figure 15.44). If dS is a small area on the surface, then the force due to fluid pressure on dS has magnitude $P\,dS$, where P is pressure, and this force acts normal to the surface of the sphere. If $\hat{\mathbf{n}}$ is the unit inward-pointing normal to the sphere, then the force on dS is

$$(P\,dS)\hat{\mathbf{n}}.$$

Clearly, the resultant force on the sphere will be in the z-direction, the x- and y-components canceling because of the symmetry of the sphere. The z-component of the force on dS is $(P\,dS)\hat{\mathbf{n}}\cdot\hat{\mathbf{k}}$; thus the magnitude of the resultant force on the sphere is

$$\oint\!\!\oint_S P\hat{\mathbf{k}}\cdot\hat{\mathbf{n}}\,dS.$$

A normal to the surface is $\nabla(x^2 + y^2 + z^2 - R^2) = (2x, 2y, 2z)$, and therefore the unit inward-pointing normal is

$$\hat{\mathbf{n}} = \frac{-(x, y, z)}{\sqrt{x^2 + y^2 + z^2}} = -\frac{1}{R}(x, y, z).$$

With formula 15.53 for an area element on the sphere, and $P = 9.81p(h - z)$,

$$\begin{aligned}
\oint\!\!\oint_S P\hat{\mathbf{k}}\cdot\hat{\mathbf{n}}\,dS &= \oint\!\!\oint_S 9.81\rho(h - z)\left(-\frac{z}{R}\right) dS \\
&= -\frac{9.81\rho}{R}\int_0^{2\pi}\int_0^{\pi}(h - R\cos\phi)(R\cos\phi)R^2\sin\phi\, d\phi\, d\theta \\
&= -9.81\rho R^2\int_0^{2\pi}\left\{-\frac{h}{2}\cos^2\phi + \frac{R}{3}\cos^3\phi\right\}_0^{\pi} d\theta \\
&= \frac{2(9.81)\rho R^3}{3}\{\theta\}_0^{2\pi} \\
&= \frac{4}{3}\pi R^3(9.81\rho).
\end{aligned}$$

FIGURE 15.44

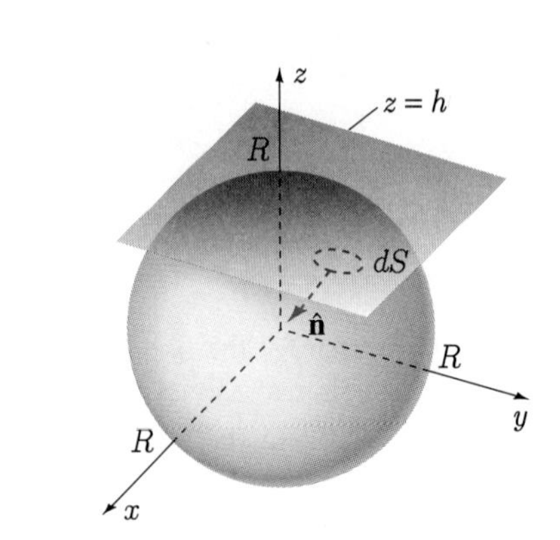

EXERCISES 15.8

In Exercises 1–16 evaluate the surface integral.

1. $\iint_S (x\hat{\mathbf{i}} + z\hat{\mathbf{k}}) \cdot \hat{\mathbf{n}}\, dS$ where S is the first octant part of $x + y + z = 3$, and $\hat{\mathbf{n}}$ is the unit normal to S with positive z-component

2. $\iint_S (yz^2\hat{\mathbf{i}} + ye^x\hat{\mathbf{j}} + x\hat{\mathbf{k}}) \cdot \hat{\mathbf{n}}\, dS$ where S is defined by $y = x^2$, $0 \le y \le 4$, $0 \le z \le 1$, and $\hat{\mathbf{n}}$ is the unit normal to S with positive y-component

3. $\iint_S (x\hat{\mathbf{i}} + y\hat{\mathbf{j}} + z\hat{\mathbf{k}}) \cdot \hat{\mathbf{n}}\, dS$ where S is the hemisphere $z = \sqrt{1 - x^2 - y^2}$, and $\hat{\mathbf{n}}$ is its upper normal

4. $\iint_S (yz\hat{\mathbf{i}} + zx\hat{\mathbf{j}} + xy\hat{\mathbf{k}}) \cdot \hat{\mathbf{n}}\, dS$ where S is that part of the surface $z = x^2 + y^2$ cut out by the planes $x = 1$, $x = -1$, $y = 1$, $y = -1$, and $\hat{\mathbf{n}}$ is the unit lower normal to S

5. $\oiint_S (z\hat{\mathbf{i}} - x\hat{\mathbf{j}} + y\hat{\mathbf{k}}) \cdot \hat{\mathbf{n}}\, dS$ where S is the surface enclosing the volume defined by $z = \sqrt{4 - x^2 - y^2}$, $z = 0$, and $\hat{\mathbf{n}}$ is the unit outer normal to S

6. $\iint_S (x\hat{\mathbf{i}} + y\hat{\mathbf{j}}) \cdot \hat{\mathbf{n}}\, dS$ where S is that part of the surface $z = \sqrt{x^2 + y^2}$ below $z = 1$, and $\hat{\mathbf{n}}$ is the unit normal to S with negative z-component

7. $\iint_S (xyz\hat{\mathbf{i}} - x\hat{\mathbf{j}} + z\hat{\mathbf{k}}) \cdot \hat{\mathbf{n}}\, dS$ where S is the smaller part of $x^2 + y^2 = 9$ cut out by $z = 0$, $z = 2$, $y = |x|$, and $\hat{\mathbf{n}}$ is the unit normal to S with positive y-component

8. $\iint_S (x^2y\hat{\mathbf{i}} + xy\hat{\mathbf{j}} + z\hat{\mathbf{k}}) \cdot \hat{\mathbf{n}}\, dS$ where S is defined by $z = 2 - x^2 - y^2$, $z \ge 0$, and $\hat{\mathbf{n}}$ is the unit normal to S with negative z-component

9. $\oiint_S (yz\hat{\mathbf{i}} + xz\hat{\mathbf{j}} + xy\hat{\mathbf{k}}) \cdot \hat{\mathbf{n}}\, dS$ where S is the surface enclosing the volume defined by $x = 0$, $x = 2$, $z = 0$, $z = y$, $y + z = 2$, and $\hat{\mathbf{n}}$ is the unit outer normal to S

10. $\iint_S (x\hat{\mathbf{i}} + y\hat{\mathbf{j}}) \cdot \hat{\mathbf{n}}\, dS$ where S is the surface $x^2 + y^2 + z^2 = 4$, $z \ge 1$, and $\hat{\mathbf{n}}$ is the unit upper normal to S

11. $\oiint_S (x^2\hat{\mathbf{i}} + y^2\hat{\mathbf{j}} + z^2\hat{\mathbf{k}}) \cdot \hat{\mathbf{n}}\, dS$ where S is the sphere $x^2 + y^2 + z^2 = a^2$, and $\hat{\mathbf{n}}$ is the unit outer normal to S

12. $\iint_S (y\hat{\mathbf{i}} - x\hat{\mathbf{j}} + \hat{\mathbf{k}}) \cdot \hat{\mathbf{n}}\, dS$ where S is the smaller surface cut from the sphere $x^2 + y^2 + z^2 = 1$ by the plane $y + z = 1$, and $\hat{\mathbf{n}}$ is the unit upper normal to S

13. $\oiint_S \mathbf{F} \cdot \hat{\mathbf{n}}\, dS$ where $\mathbf{F} = (z^2 - x)\hat{\mathbf{i}} - xy\hat{\mathbf{j}} + 3z\hat{\mathbf{k}}$, S is the surface enclosing the volume defined by $z = 4 - y^2$, $x = 0$, $x = 3$, $z = 0$, and $\hat{\mathbf{n}}$ is the unit outer normal to S

14. $\iint_S (x^2\hat{\mathbf{i}} + xy\hat{\mathbf{j}} + xz\hat{\mathbf{k}}) \cdot \hat{\mathbf{n}}\, dS$ where S is that part of the surface $z = \sqrt{4 + y^2 - x^2}$ in the first octant cut out by the planes $y = 0$, $y = 1$, $x = 0$, $z = 0$, and $\hat{\mathbf{n}}$ is the unit normal to S with positive z-component.

15. $\displaystyle\iint_S (x^2\hat{\mathbf{i}} + yz\hat{\mathbf{j}} - x\hat{\mathbf{k}}) \cdot \hat{\mathbf{n}}\, dS$ where S is that part of the surface $x = yz$ in the first octant cut out by $y^2 + z^2 = 1$, and $\hat{\mathbf{n}}$ is the unit normal to S with positive x-component

16. $\displaystyle\oiint_S (yx\hat{\mathbf{i}} + y^2\hat{\mathbf{j}} + yz\hat{\mathbf{k}}) \cdot \hat{\mathbf{n}}\, dS$ where S is the ellipsoid $x^2 + y^2/4 + z^2 = 1$, and $\hat{\mathbf{n}}$ is the unit outer normal to S

17. Show that if a surface S projects one-to-one onto an area S_{xy} in the xy-plane, then

$$\iint_S (P\hat{\mathbf{i}} + Q\hat{\mathbf{j}} + R\hat{\mathbf{k}}) \cdot \hat{\mathbf{n}}\, dS$$

$$= \pm \iint_{S_{xy}} \left(-P\frac{\partial z}{\partial x} - Q\frac{\partial z}{\partial y} + R\right) dA$$

the $\pm$ depending on whether $\hat{\mathbf{n}}$ is the upper or lower normal to S. What are corresponding formulas when S projects one-to-one onto areas S_{yz} and S_{xz} in the yz- and xz-coordinate planes?

18. Evaluate $\displaystyle\iint_S (y\hat{\mathbf{i}} - x\hat{\mathbf{j}} + z\hat{\mathbf{k}}) \cdot \hat{\mathbf{n}}\, dS$ where

(a) S is that part of $z = 9 - x^2 - y^2$ cut out by $z = 2y$,

(b) S is that part of $z = 2y$ cut out by $z = 9 - x^2 - y^2$,

and $\hat{\mathbf{n}}$ is the unit upper normal to S in each case. Hint: Use polar coordinates with pole at $(0, -1)$.

19. Show that if a surface S, defined implicitly by the equation $G(x, y, z) = 0$, projects one-to-one onto the area S_{xy} in the xy-plane, then

$$\iint_S \mathbf{F} \cdot \hat{\mathbf{n}}\, dS = \pm \iint_{S_{xy}} \frac{\mathbf{F} \cdot \nabla G}{|\partial G/\partial z|} dA.$$

20. A circular tube $S:\ x^2 + z^2 = 1,\ 0 \le y \le 2$ is a model for a part of an artery. Blood flows through the artery and the force per unit area at any point on the arterial wall is given by

$$\mathbf{F} = e^{-y}\hat{\mathbf{n}} + \frac{1}{y^2 + 1}\hat{\mathbf{j}},$$

where $\hat{\mathbf{n}}$ is the unit outer normal to the arterial wall. Blood diffuses through the wall in such a way that if dS is a small area on S, the amount of diffusion through dS in one second is $\mathbf{F} \cdot \hat{\mathbf{n}}\, dS$. Find the total amount of blood leaving the entire wall per second.

21. A beam of light travelling in the positive y-direction has circular cross-section $x^2 + z^2 \le a^2$. It strikes a surface $S:\ x^2 + y^2 + z^2 = a^2,\ y \ge a/2$. The intensity of the beam is given by

$$\mathbf{I} = \frac{e^{-t}}{y^2}\hat{\mathbf{j}}, \qquad \text{where } t \text{ is time.}$$

The absorption of light by a small area dS on S in time dt is $\mathbf{I} \cdot \hat{\mathbf{n}}\, dS\, dt$ where $\hat{\mathbf{n}}$ is the unit normal to S at dS.

(a) Find the total absorption over S in time dt.

(b) Find the total absorption over S from time $t = 0$ to time $t = 5$.

SECTION 15.9

The Divergence Theorem

In this section and in Section 15.10 we show that relationships may exist between line integrals and surface integrals and between surface integrals and triple integrals. In Section 15.11 we indicate how useful these relationships can be in the fields of electromagnetic theory and fluid dynamics.

The divergence theorem relates certain surface integrals over surfaces that enclose volumes to triple integrals over the enclosed volume. More precisely, we have the following.

Theorem 15.7

(Divergence theorem) Let S be a piecewise-smooth surface enclosing a region V (Figure 15.45). Let $\mathbf{F}(x,y,z) = L(x,y,z)\hat{\mathbf{i}} + M(x,y,z)\hat{\mathbf{j}} + N(x,y,z)\hat{\mathbf{k}}$ be a vector field whose components L, M, and N have continuous first partial derivative in a domain containing S and V. If $\hat{\mathbf{n}}$ is the unit outer normal to S, then

$$\oint\!\!\oint_S \mathbf{F}\cdot\hat{\mathbf{n}}\,dS = \iiint_V \nabla\cdot\mathbf{F}\,dV, \tag{15.55a}$$

or

$$\oint\!\!\oint_S (L\hat{\mathbf{i}} + M\hat{\mathbf{j}} + N\hat{\mathbf{k}})\cdot\hat{\mathbf{n}}\,dS = \iiint_V \left(\frac{\partial L}{\partial x} + \frac{\partial M}{\partial y} + \frac{\partial N}{\partial z}\right) dV. \tag{15.55b}$$

FIGURE 15.45

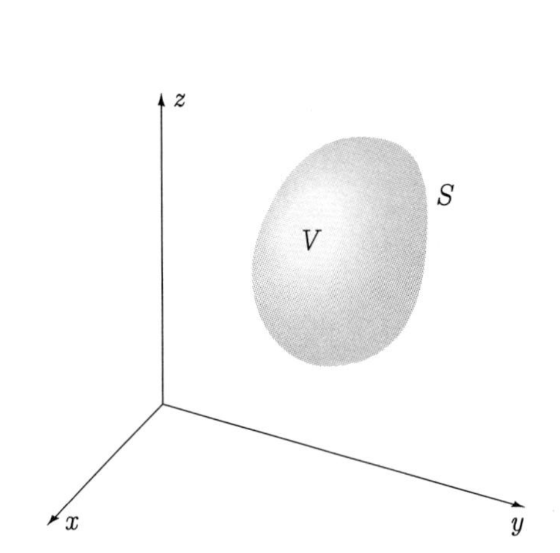

FIGURE 15.46

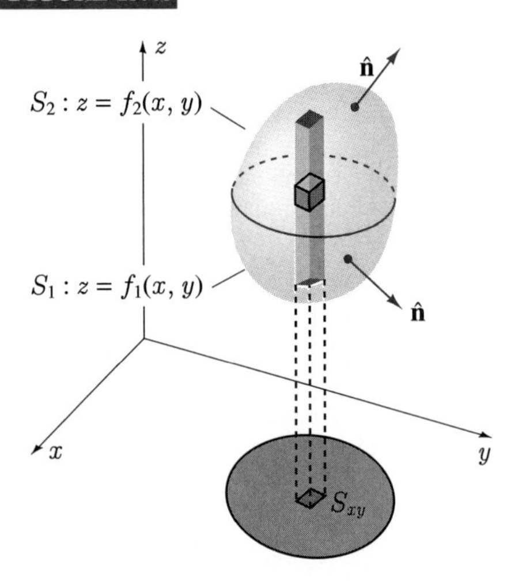

Proof We consider first of all a surface S for which any line parallel to any coordinate axis intersects S in at most two points (Figure 15.46). We can then divide S into an upper and a lower portion, $S_2 : z = f_2(x,y)$ and $S_1 : z = f_1(x,y)$, both of which have the same projection S_{xy} in the xy-plane. We consider the third term in the surface integral on the left-hand side of equation (15.55b):

$$\oint\!\!\oint_S N\hat{\mathbf{k}}\cdot\hat{\mathbf{n}}\,dS = \iint_{S_1} N\hat{\mathbf{k}}\cdot\hat{\mathbf{n}}\,dS + \iint_{S_2} N\hat{\mathbf{k}}\cdot\hat{\mathbf{n}}\,dS.$$

On S_1,

$$\hat{\mathbf{n}} = \frac{\left(\dfrac{\partial f_1}{\partial x}, \dfrac{\partial f_1}{\partial y}, -1\right)}{\sqrt{1 + \left(\dfrac{\partial f_1}{\partial x}\right)^2 + \left(\dfrac{\partial f_1}{\partial y}\right)^2}},$$

and on S_2,

$$\hat{\mathbf{n}} = \frac{-\left(\frac{\partial f_2}{\partial x}, \frac{\partial f_2}{\partial y}, -1\right)}{\sqrt{1 + \left(\frac{\partial f_2}{\partial x}\right)^2 + \left(\frac{\partial f_2}{\partial y}\right)^2}}.$$

Consequently,

$$\begin{aligned}
\oiint_S N\hat{\mathbf{k}} \cdot \hat{\mathbf{n}}\, dS &= \iint_{S_1} \frac{-N}{\sqrt{1 + \left(\frac{\partial f_1}{\partial x}\right)^2 + \left(\frac{\partial f_1}{\partial y}\right)^2}} dS + \iint_{S_2} \frac{N}{\sqrt{1 + \left(\frac{\partial f_2}{\partial x}\right)^2 + \left(\frac{\partial f_2}{\partial y}\right)^2}} dS \\
&= \iint_{S_{xy}} \frac{-N[x, y, f_1(x, y)]}{\sqrt{1 + \left(\frac{\partial f_1}{\partial x}\right)^2 + \left(\frac{\partial f_1}{\partial y}\right)^2}} \sqrt{1 + \left(\frac{\partial f_1}{\partial x}\right)^2 + \left(\frac{\partial f_1}{\partial y}\right)^2}\, dA \\
&\quad + \iint_{S_{xy}} \frac{N[x, y, f_2(x, y)]}{\sqrt{1 + \left(\frac{\partial f_2}{\partial x}\right)^2 + \left(\frac{\partial f_2}{\partial y}\right)^2}} \sqrt{1 + \left(\frac{\partial f_2}{\partial x}\right)^2 + \left(\frac{\partial f_2}{\partial y}\right)^2}\, dA \\
&= \iint_{S_{xy}} \{N[x, y, f_2(x, y)] - N[x, y, f_1(x, y)]\} dA.
\end{aligned}$$

On the other hand,

$$\begin{aligned}
\iiint_V \frac{\partial N}{\partial z} dV &= \iint_{S_{xy}} \left\{ \int_{f_1(x,y)}^{f_2(x,y)} \frac{\partial N}{\partial z} dz \right\} dA = \iint_{S_{xy}} \left\{ N \right\}_{f_1(x,y)}^{f_2(x,y)} dA \\
&= \iint_{S_{xy}} \{N[x, y, f_2(x, y)] - N[x, y, f_1(x, y)]\} dA.
\end{aligned}$$

We have shown then that

$$\oiint_S N\hat{\mathbf{k}} \cdot \hat{\mathbf{n}}\, dS = \iiint_V \frac{\partial N}{\partial z} dV.$$

Projections of S onto the xz- and yz-planes lead in a similar way to

$$\oiint_S M\hat{\mathbf{j}} \cdot \hat{\mathbf{n}}\, dS = \iiint_V \frac{\partial M}{\partial y} dV \quad \text{and} \quad \oiint_S L\hat{\mathbf{i}} \cdot \hat{\mathbf{n}}\, dS = \iiint_V \frac{\partial L}{\partial x} dV.$$

By adding these three results, we obtain the divergence theorem for $\mathbf{F}$ and S.

The proof can be extended to general surfaces for which lines parallel to the coordinate axes intersect the surfaces in more than two points. Indeed, most volumes V bounded by surfaces S can be subdivided into n subvolumes V_i whose bounding surfaces S_i do satisfy this condition (Figure 15.47). For each such subvolume the divergence theorem is now known to apply:

$$\oiint_{S_i} \mathbf{F} \cdot \hat{\mathbf{n}}\, dS = \iiint_{V_i} \nabla \cdot \mathbf{F}\, dV, \quad i = 1, \ldots, n.$$

If these n equations are then added together, we have

$$\sum_{i=1}^{n} \oiint_{S_i} \mathbf{F} \cdot \hat{\mathbf{n}}\, dS = \sum_{i=1}^{n} \iiint_{V_i} \nabla \cdot \mathbf{F}\, dV.$$

The right side is precisely the triple integral of $\nabla \cdot \mathbf{F}$ over V since the V_i constitute V. Figure 15.47 illustrates that when the surface integrals over the S_i are added, contributions from the auxiliary (interior) surfaces cancel in pairs, and the remaining surface integrals add to give the surface integral of $\mathbf{F} \cdot \hat{\mathbf{n}}$ over S. This completes the proof.

FIGURE 15.47

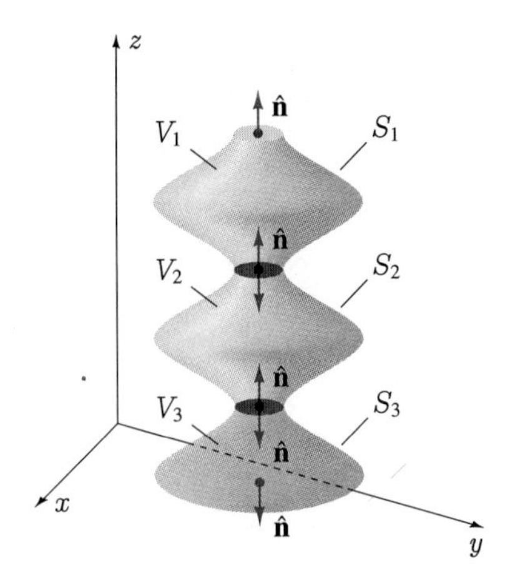

We omit a proof for even more general surfaces that cannot be subdivided into a finite number of subsurfaces of this type. The interested reader should consult more advanced books.

EXAMPLE 15.23

Verify the divergence theorem for $\mathbf{F} = x^2\hat{\mathbf{i}} + yz\hat{\mathbf{j}} + x\hat{\mathbf{k}}$ if V is the volume bounded by the surfaces $x + y + z = 1$, $x = 0$, $y = 0$, $z = 0$.

SOLUTION The surface S bounding V must be subdivided into four parts:

$$S_1 : \ z = 0, \quad S_2 : \ x = 0, \quad S_3 : \ y = 0, \quad S_4 : \ x + y + z = 1,$$

as shown in Figure 15.48. The surface integral of $\mathbf{F} \cdot \hat{\mathbf{n}}$ over S is then the sum of the surface integrals over these four parts:

$$\begin{aligned}
\iint_{S_1} \mathbf{F} \cdot \hat{\mathbf{n}}\, dS &= \iint_{S_1} \mathbf{F} \cdot (-\hat{\mathbf{k}})\, dS = \iint_{S_1} -x\, dS \\
&= -\iint_{S_{xy}} x\sqrt{1}\, dA \\
&= -\int_0^1 \int_0^{1-x} x\, dy\, dx = -\int_0^1 \Big\{ xy \Big\}_0^{1-x} dx \\
&= -\int_0^1 x(1 - x)\, dx \\
&= -\left\{ \frac{x^2}{2} - \frac{x^3}{3} \right\}_0^1 \\
&= -\frac{1}{6};
\end{aligned}$$

$$\iint_{S_2} \mathbf{F} \cdot \hat{\mathbf{n}}\, dS = \iint_{S_2} \mathbf{F} \cdot (-\hat{\mathbf{i}})\, dS = \iint_{S_2} -x^2\, dS = 0;$$

$$\iint_{S_3} \mathbf{F} \cdot \hat{\mathbf{n}}\, dS = \iint_{S_3} \mathbf{F} \cdot (-\hat{\mathbf{j}})\, dS = \iint_{S_3} -yz\, dS = 0;$$

$$\begin{aligned}
\iint_{S_4} \mathbf{F} \cdot \hat{\mathbf{n}}\, dS &= \iint_{S_4} \mathbf{F} \cdot \frac{(1,1,1)}{\sqrt{3}}\, dS = \frac{1}{\sqrt{3}} \iint_{S_4} (x^2 + yz + x)\, dS \\
&= \frac{1}{\sqrt{3}} \iint_{S_{xy}} \{x^2 + y(1-x-y) + x\} \sqrt{1 + (-1)^2 + (-1)^2}\, dA \\
&= \int_0^1 \int_0^{1-x} \{(x^2 + x) + y(1-x) - y^2\}\, dy\, dx \\
&= \int_0^1 \left\{ (x^2 + x)y + (1-x)\frac{y^2}{2} - \frac{y^3}{3} \right\}_0^{1-x} dx \\
&= \int_0^1 \left\{ x - x^3 + \frac{1}{6}(1-x)^3 \right\} dx \\
&= \left\{ \frac{x^2}{2} - \frac{x^4}{4} - \frac{1}{24}(1-x)^4 \right\}_0^1 \\
&= \frac{7}{24}.
\end{aligned}$$

Adding these results gives us

$$\oiint_S \mathbf{F} \cdot \hat{\mathbf{n}}\, dS = -\frac{1}{6} + 0 + 0 + \frac{7}{24} = \frac{1}{8}.$$

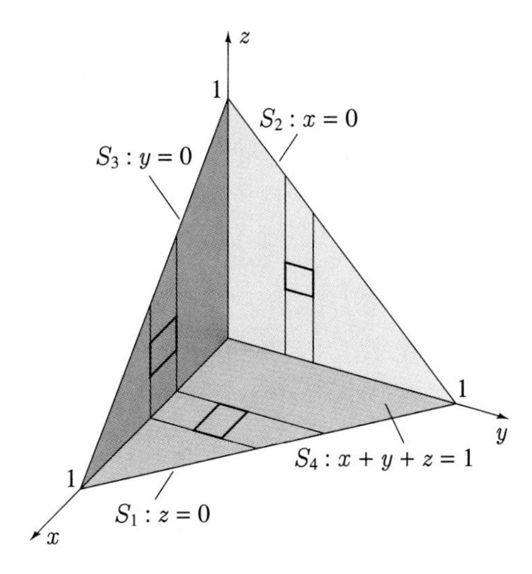

FIGURE 15.48

Alternatively, we have

$$\begin{aligned}
\iiint_V \nabla \cdot \mathbf{F}\, dV &= \iiint_V (2x + z)\, dV \\
&= \int_0^1 \int_0^{1-x} \int_0^{1-x-y} (2x + z)\, dz\, dy\, dx \\
&= \int_0^1 \int_0^{1-x} \left\{ 2xz + \frac{z^2}{2} \right\}_0^{1-x-y} dy\, dx \\
&= \frac{1}{2} \int_0^1 \int_0^{1-x} \{4x(1-x-y) + (1-x-y)^2\} dy\, dx \\
&= \frac{1}{2} \int_0^1 \left\{ 4x\left(y - xy - \frac{y^2}{2}\right) - \frac{1}{3}(1-x-y)^3 \right\}_0^{1-x} dx \\
&= \frac{1}{2} \int_0^1 \left\{ 2x(1-x)^2 + \frac{1}{3}(1-x)^3 \right\} dx \\
&= \frac{1}{2} \int_0^1 \left\{ 2x - 4x^2 + 2x^3 + \frac{1}{3}(1-x)^3 \right\} dx \\
&= \frac{1}{2} \left\{ x^2 - \frac{4x^3}{3} + \frac{x^4}{2} - \frac{1}{12}(1-x)^4 \right\}_0^1 \\
&= \frac{1}{8}.
\end{aligned}$$

We have therefore verified the divergence theorem for this **F** and S. ■

EXAMPLE 15.24

Use the divergence theorem to evaluate the surface integral of Example 15.21.

SOLUTION By the divergence theorem (Figure 15.43), we have

$$\begin{aligned}
\oiint_S \mathbf{F} \cdot \hat{\mathbf{n}}\, dS &= \iiint_V \nabla \cdot \mathbf{F}\, dV \\
&= \iiint_V (1 + 1 + 1)\, dV \\
&= 3 \iiint_V dV \\
&= 3(\text{Volume of } V) = 3(4\pi)(2) = 24\pi.
\end{aligned}$$

■

EXAMPLE 15.25

Evaluate

$$\oiint_S (x^3\hat{\mathbf{i}} + y^3\hat{\mathbf{j}} + z^3\hat{\mathbf{k}}) \cdot \hat{\mathbf{n}}\, dS,$$

where $\hat{\mathbf{n}}$ is the unit inner normal to the surface bounding the volume defined by the surfaces $y = x, z = 0, z = 2, y = 0, x = 4$.

SOLUTION By the divergence theorem (see Figure 15.49), we have

$$\begin{aligned}
\oiint_S (x^3\hat{\mathbf{i}} + y^3\hat{\mathbf{j}} + z^3\hat{\mathbf{k}}) \cdot \hat{\mathbf{n}}\, dS &= -\iiint_V \nabla \cdot (x^3\hat{\mathbf{i}} + y^3\hat{\mathbf{j}} + z^3\hat{\mathbf{k}})\, dV \\
&= -\iiint_V (3x^2 + 3y^2 + 3z^2)\, dV \\
&= -3\int_0^4 \int_0^x \int_0^2 (x^2 + y^2 + z^2)\, dz\, dy\, dx \\
&= -3\int_0^4 \int_0^x \left\{(x^2 + y^2)z + \frac{z^3}{3}\right\}_0^2 dy\, dx \\
&= -2\int_0^4 \int_0^x \{3(x^2 + y^2) + 4\} dy\, dx \\
&= -2\int_0^4 \{y^3 + (3x^2 + 4)y\}_0^x dx \\
&= -2\int_0^4 (4x^3 + 4x)\, dx = -2\{x^4 + 2x^2\}_0^4 \\
&= -576.
\end{aligned}$$

FIGURE 15.49

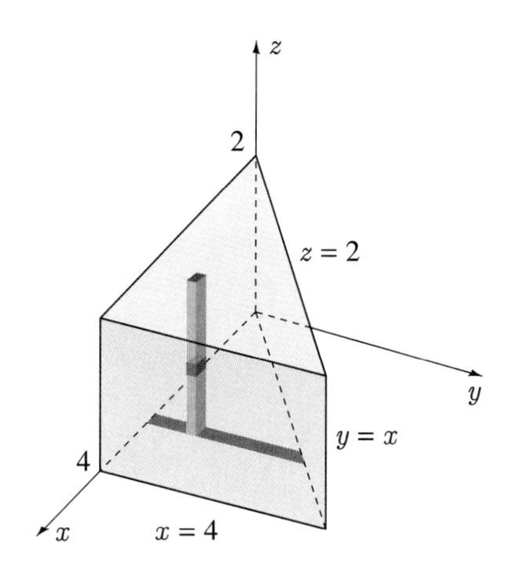

■

EXAMPLE 15.26

Prove Archimedes' principle, which states that when an object is submerged in a fluid, it experiences a buoyant force equal to the weight of the fluid displaced.

SOLUTION Suppose the surface of the fluid is taken as the xy-plane ($z = 0$), and the object occupies a region V with bounding surface S (Figure 15.50). The force due to fluid pressure P on a small area dS on S is $(P\, dS)\hat{\mathbf{n}}$, where $\hat{\mathbf{n}}$ is the unit inner normal to S at dS. If ρ is the density of the fluid, then $P = -9.81\rho z$, and the force on dS is

$$(-9.81\rho z\, dS)\hat{\mathbf{n}}.$$

The resultant buoyant force is in the positive z-direction (the x- and y-components canceling), so that we require only the z-component of this force,

$$(-9.81\rho z\, dS)\hat{\mathbf{n}}\cdot\hat{\mathbf{k}}.$$

The total buoyant force must therefore have z-component

$$\oiint_S (-9.81\rho z)\hat{\mathbf{n}}\cdot\hat{\mathbf{k}}\, dS = \oiint_S (-9.81\rho z\hat{\mathbf{k}})\cdot\hat{\mathbf{n}}\, dS = \oiint_S (9.81\rho z\hat{\mathbf{k}})\cdot(-\hat{\mathbf{n}})\, dS,$$

where $-\hat{\mathbf{n}}$ is the unit outer normal to S. If we now use the divergence theorem, we have

$$\begin{aligned}\oiint_S (-9.81\rho z)\hat{\mathbf{n}}\cdot\hat{\mathbf{k}}\, dS &= \iiint_V \nabla\cdot(9.81\rho z\hat{\mathbf{k}})\, dV \\ &= \iiint_V 9.81\rho\, dV = 9.81\rho\,(\text{Volume of } V),\end{aligned}$$

and this is the weight of the fluid displaced by the object.

FIGURE 15.50

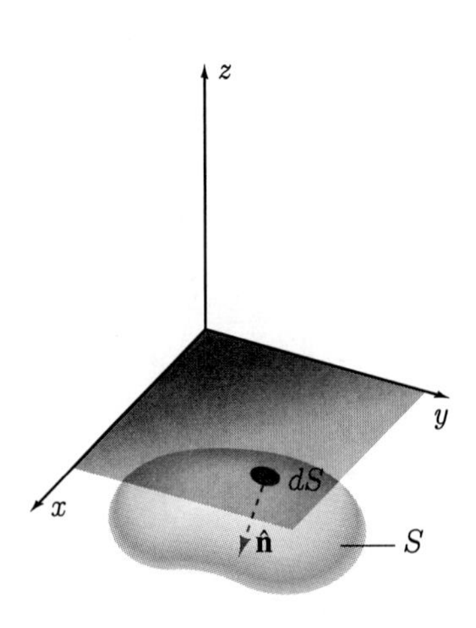

■

EXERCISES 15.9

In Exercises 1–12 use the divergence theorem to evaluate the surface integral.

1. $\oiint_S (x\hat{\mathbf{i}} + y\hat{\mathbf{j}} - 2z\hat{\mathbf{k}}) \cdot \hat{\mathbf{n}}\, dS$ where S is the surface bounding the volume defined by $z = 2x^2 + y^2$, $x^2 + y^2 = 3$, $z = 0$, and $\hat{\mathbf{n}}$ is the unit outer normal to S

2. $\oiint_S (x^2\hat{\mathbf{i}} + y^2\hat{\mathbf{j}} + z^2\hat{\mathbf{k}}) \cdot \hat{\mathbf{n}}\, dS$ where S is the sphere $x^2 + y^2 + z^2 = a^2$, and $\hat{\mathbf{n}}$ is the unit outer normal to S

3. $\oiint_S (yz\hat{\mathbf{i}} + xz\hat{\mathbf{j}} + xy\hat{\mathbf{k}}) \cdot \hat{\mathbf{n}}\, dS$ where S is the surface enclosing the volume defined by $x = 0$, $x = 2$, $z = 0$, $z = y$, $y + z = 2$, and $\hat{\mathbf{n}}$ is the unit outer normal to S

4. $\oiint_S [(z^2 - x)\hat{\mathbf{i}} - xy\hat{\mathbf{j}} + 3z\hat{\mathbf{k}}] \cdot \hat{\mathbf{n}}\, dS$ where S is the surface enclosing the volume defined by $z = 4 - y^2$, $x = 0$, $x = 3$, $z = 0$, and $\hat{\mathbf{n}}$ is the unit outer normal to S

5. $\oiint_S \mathbf{F} \cdot \hat{\mathbf{n}}\, dS$ where $\mathbf{F} = (x^2y\hat{\mathbf{i}} + y^2z\hat{\mathbf{j}} + z^2x\hat{\mathbf{k}})/2$, S is the surface bounding the volume in the first octant defined by $x = 0$, $y = 0$, $z = 1$, $z = 0$, $x^2 + y^2 = 1$, and $\hat{\mathbf{n}}$ is the unit inner normal to S

6. $\oiint_S (x\hat{\mathbf{i}} + y\hat{\mathbf{j}} + 2z\hat{\mathbf{k}}) \cdot \hat{\mathbf{n}}\, dS$ where S is the surface bounding the volume defined by $z = 2x^2 + y^2$, $x^2 + y^2 = 3$, $z = 0$, and $\hat{\mathbf{n}}$ is the unit outer normal to S

7. $\oiint_S (z\hat{\mathbf{i}} - x\hat{\mathbf{j}} + y\hat{\mathbf{k}}) \cdot \hat{\mathbf{n}}\, dS$ where S is the surface enclosing the volume defined by $z = \sqrt{4 - x^2 - y^2}$, $z = 0$, and $\hat{\mathbf{n}}$ is the unit outer normal to S

8. $\oiint_S (2x^2y\hat{\mathbf{i}} - y^2\hat{\mathbf{j}} + 4xz^2\hat{\mathbf{k}}) \cdot \hat{\mathbf{n}}\, dS$ where S is the surface enclosing the volume in the first octant defined by $y^2 + z^2 = 9$, $x = 2$, and $\hat{\mathbf{n}}$ is the unit outer normal to S

9. $\oiint_S (yx\hat{\mathbf{i}} + y^2\hat{\mathbf{j}} + yz\hat{\mathbf{k}}) \cdot \hat{\mathbf{n}}\, dS$ where S is the ellipsoid $x^2 + y^2/4 + z^2 = 1$, and $\hat{\mathbf{n}}$ is the unit outer normal to S

10. $\oiint_S (x^3\hat{\mathbf{i}} + y^3\hat{\mathbf{j}} - z^3\hat{\mathbf{k}}) \cdot \hat{\mathbf{n}}\, dS$ where S is the surface enclosing the volume defined by $z = 6 - x^2 - y^2$, $z = \sqrt{x^2 + y^2}$, and $\hat{\mathbf{n}}$ is the unit outer normal to S

11. $\oiint_S (y\hat{\mathbf{i}} - xy\hat{\mathbf{j}} + zy^2\hat{\mathbf{k}}) \cdot \hat{\mathbf{n}}\, dS$ where S is the surface enclosing the volume defined by $y^2 - x^2 - z^2 = 4$, $y = 4$, and $\hat{\mathbf{n}}$ is the unit inner normal to S

12. $\oiint_S (xy\hat{\mathbf{i}} + z^2\hat{\mathbf{k}}) \cdot \hat{\mathbf{n}}\, dS$ where S is the surface enclosing the volume in the first octant bounded by the planes $z = 0$, $y = x$, $y = 2x$, $x + y + z = 6$, and $\hat{\mathbf{n}}$ is the unit outer normal to S

In Exercises 13–15 use the divergence theorem to evaluate the surface integral. In each case an additional surface must be introduced in order to enclose a volume.

13. $\iint_S (x\hat{\mathbf{i}} + y\hat{\mathbf{j}} + z\hat{\mathbf{k}}) \cdot \hat{\mathbf{n}}\, dS$ where S is the top half of the ellipsoid $x^2 + 4y^2 + 9z^2 = 36$, and $\hat{\mathbf{n}}$ is the unit upper normal to S

14. $\iint_S (xy\hat{\mathbf{i}} - yz\hat{\mathbf{j}} + x^2z\hat{\mathbf{k}}) \cdot \hat{\mathbf{n}}\, dS$ where S is that part of the cone $z = \sqrt{x^2 + y^2}$ below $z = 2$, and $\hat{\mathbf{n}}$ is the unit normal to S with positive z-component

15. $\iint_S (y^2e^z\hat{\mathbf{i}} - xy\hat{\mathbf{j}} + z\hat{\mathbf{k}}) \cdot \hat{\mathbf{n}}\, dS$ where S is that part of $z = 4 - x^2 - y^2$ cut out by $z = 2y$, and $\hat{\mathbf{n}}$ is the unit upper normal to S

16. Show that if $\hat{\mathbf{n}}$ is the unit outer normal to a surface S, then the region enclosed by S has volume

$$V = \frac{1}{3}\oiint_S \mathbf{r} \cdot \hat{\mathbf{n}}\, dS, \qquad \mathbf{r} = x\hat{\mathbf{i}} + y\hat{\mathbf{j}} + z\hat{\mathbf{k}}.$$

17. If $\hat{\mathbf{n}}$ is the unit outer normal to a surface S that encloses a region V, show that the area of S can be expressed in the form

$$\text{Area}(S) = \iiint_V \nabla \cdot \hat{\mathbf{n}}\, dV.$$

18. How would you prove Archimedes' principle in the case that an object is only partially submerged? (See Example 15.26.)

In Exercises 19–21 evaluate the surface integral.

19. $\oiint_S [(x + y)\hat{\mathbf{i}} + y^3\hat{\mathbf{j}} + x^2z\hat{\mathbf{k}}] \cdot \hat{\mathbf{n}}\, dS$ where S is the surface enclosing the volume defined by $x^2 + y^2 - z^2 = 1$, $2z^2 = x^2 + y^2$, and $\hat{\mathbf{n}}$ is the unit outer normal to S

20. $\oiint_S [(x + y)^2\hat{\mathbf{i}} + x^2y\hat{\mathbf{j}} - x^2z\hat{\mathbf{k}}] \cdot \hat{\mathbf{n}}\, dS$ where $\hat{\mathbf{n}}$ is the unit inner normal to the surface S enclosing the volume defined by $z^2 = (1 - x^2 - 2y^2)^2$

21. $\displaystyle\iint_S [(y^3 + x^2 y)\hat{\mathbf{i}} + (x^3 - xy^2)\hat{\mathbf{j}} + z\hat{\mathbf{k}}] \cdot \hat{\mathbf{n}}\, dS$ where $\hat{\mathbf{n}}$ is the unit upper normal to the surface $S : z = \sqrt{1 - x^2 - y^2}$

22. If V is a region bounded by a closed surface S, and $\mathbf{B} = \nabla \times \mathbf{A}$, show that

$$\oiint_S \mathbf{B} \cdot \hat{\mathbf{n}}\, dS = 0.$$

23. Is Green's theorem related to the divergence theorem? (See Exercise 13 in Section 15.6.)

In Exercises 24–26 assume that $P(x, y, z)$ and $Q(x, y, z)$ have continuous first and second partial derivatives in a domain containing a closed surface S and its interior V. Let $\hat{\mathbf{n}}$ be the unit outer normal to S.

24. Show that

$$\oiint_S \nabla P \cdot \hat{\mathbf{n}}\, dS = \iiint_V \nabla^2 P\, dV,$$

where $\nabla^2 P = \dfrac{\partial^2 P}{\partial x^2} + \dfrac{\partial^2 P}{\partial y^2} + \dfrac{\partial^2 P}{\partial z^2}$. What can we conclude if $P(x, y, z)$ satisfies Laplace's equation in V?

25. Show that

$$\oiint_S P\nabla Q \cdot \hat{\mathbf{n}}\, dS = \iiint_V (P\nabla^2 Q + \nabla P \cdot \nabla Q)\, dV.$$

This result is called *Green's first identity*.

26. Prove that

$$\oiint_S (P\nabla Q - Q\nabla P) \cdot \hat{\mathbf{n}}\, dS = \iiint_V (P\nabla^2 Q - Q\nabla^2 P)\, dV.$$

This result is called *Green's second identity*.

27. Compare Exercises 24–26 with Exercises 36–38 in Section 15.6.

28. Let S be a closed surface, and let $\hat{\mathbf{n}}$ be the unit outer normal to S. If $\mathbf{r}_0 = x_0\hat{\mathbf{i}} + y_0\hat{\mathbf{j}} + z_0\hat{\mathbf{k}}$ is the position vector of some fixed point P_0, show that

$$\oiint_S \frac{\mathbf{r} - \mathbf{r}_0}{|\mathbf{r} - \mathbf{r}_0|^3} \cdot \hat{\mathbf{n}}\, dS = \begin{cases} 0 & \text{if } S \text{ does not enclose } P_0 \\ 4\pi & \text{if } S \text{ does enclose } P_0. \end{cases}$$

29. In this problem we use the result of Exercise 28 to prove Gauss's law for electrostatic fields. Let S be a closed surface containing n point charges q_i ($i = 1, \ldots, n$) at points $\mathbf{r}_i = (x_i, y_i, z_i)$ in its interior. According to Coulomb's law, the electric field $\mathbf{E}$ at a point $\mathbf{r} = (x, y, z)$ due to this charge distribution is defined by $\mathbf{E} = -\nabla V$, where V, the potential at $\mathbf{r}$, is

$$V(x, y, z) = \sum_{i=1}^{n} \frac{q_i}{4\pi\epsilon_0 |\mathbf{r} - \mathbf{r}_i|}.$$

(a) Verify that

$$\mathbf{E} = \sum_{i=1}^{n} \frac{q_i(\mathbf{r} - \mathbf{r}_i)}{4\pi\epsilon_0 |\mathbf{r} - \mathbf{r}_i|^3}.$$

(b) Now use the result of Exercise 28 to show that if $\hat{\mathbf{n}}$ is the unit outer normal to S, then

$$\oiint_S \mathbf{E} \cdot \hat{\mathbf{n}}\, dS = \frac{Q}{\epsilon_0},$$

where Q is the total charge inside S. This result is Gauss's law.

SECTION 15.10

Stokes's Theorem

Stokes's theorem relates certain line integrals around closed curves to surface integrals over surfaces that have the curves as boundaries.

Theorem 15.8

(Stokes's theorem). Let C be a closed, piecewise-smooth, curve that does not intersect itself and let S be a piecewise-smooth, orientable surface with C as boundary (Figure 15.51). Let $\mathbf{F}(x,y,z) = P(x,y,z)\hat{\mathbf{i}} + Q(x,y,z)\hat{\mathbf{j}} + R(x,y,z)\hat{\mathbf{k}}$ be a vector field whose components P, Q, and R have continuous first partial derivatives in a domain that contains S and C. Then

$$\oint_C \mathbf{F} \cdot d\mathbf{r} = \iint_S (\nabla \times \mathbf{F}) \cdot \hat{\mathbf{n}}\, dS, \tag{15.56a}$$

or

$$\oint_C P\,dx + Q\,dy + R\,dz$$

$$= \iint_S \left\{ \left(\frac{\partial R}{\partial y} - \frac{\partial Q}{\partial z} \right) \hat{\mathbf{i}} + \left(\frac{\partial P}{\partial z} - \frac{\partial R}{\partial x} \right) \hat{\mathbf{j}} + \left(\frac{\partial Q}{\partial x} - \frac{\partial P}{\partial y} \right) \hat{\mathbf{k}} \right\} \cdot \hat{\mathbf{n}}\, dS, \tag{15.56b}$$

where $\hat{\mathbf{n}}$ is the unit normal to S chosen in the following way: If when moving along C the surface S is on the left-hand side, then $\hat{\mathbf{n}}$ must be chosen as the unit normal on that side of S. On the other hand, if when moving along C, the surface is on the right, then $\hat{\mathbf{n}}$ must be chosen on the opposite side of S.

FIGURE 15.51

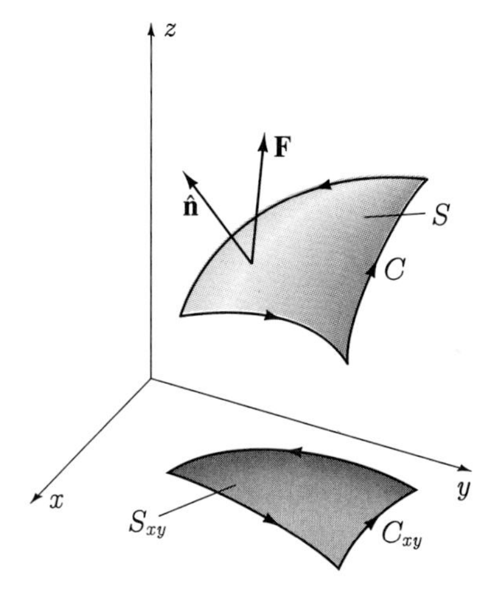

Proof We first consider a surface S that projects in a one-to-one fashion onto each of the three coordinate planes. Because S projects one-to-one onto some region S_{xy} in the xy-plane, we can take the equation for S in the form $z = f(x,y)$. If the direction along C is as indicated in Figure 15.51, and C_{xy} is the projection of C on the xy-plane, then

$$\oint_C P\,dx = \oint_{C_{xy}} P[x,y,f(x,y)]\,dx.$$

If we use Green's theorem on this line integral around C_{xy}, we have

$$\oint_C P\,dx = \iint_{S_{xy}} -\left\{ \frac{\partial P}{\partial y} + \frac{\partial P}{\partial z}\frac{\partial z}{\partial y} \right\} dA.$$

On the other hand, since a unit normal to S is

$$\hat{\mathbf{n}} = \frac{\left(-\dfrac{\partial z}{\partial x}, -\dfrac{\partial z}{\partial y}, 1\right)}{\sqrt{1 + \left(\dfrac{\partial z}{\partial x}\right)^2 + \left(\dfrac{\partial z}{\partial y}\right)^2}},$$

it also follows that

$$\begin{aligned}
\iint_S \left(\frac{\partial P}{\partial z}\hat{\mathbf{j}} - \frac{\partial P}{\partial y}\hat{\mathbf{k}}\right) \cdot \hat{\mathbf{n}}\, dS &= \iint_S \frac{-\dfrac{\partial P}{\partial z}\dfrac{\partial z}{\partial y} - \dfrac{\partial P}{\partial y}}{\sqrt{1 + \left(\dfrac{\partial z}{\partial x}\right)^2 + \left(\dfrac{\partial z}{\partial y}\right)^2}} dS \\
&= \iint_{S_{xy}} \frac{-\dfrac{\partial P}{\partial z}\dfrac{\partial z}{\partial y} - \dfrac{\partial P}{\partial y}}{\sqrt{1 + \left(\dfrac{\partial z}{\partial x}\right)^2 + \left(\dfrac{\partial z}{\partial y}\right)^2}} \sqrt{1 + \left(\frac{\partial z}{\partial x}\right)^2 + \left(\frac{\partial z}{\partial y}\right)^2}\, dA \\
&= \iint_{S_{xy}} -\left\{\frac{\partial P}{\partial y} + \frac{\partial P}{\partial z}\frac{\partial z}{\partial y}\right\} dA.
\end{aligned}$$

We have shown, then, that

$$\oint_C P\, dx = \iint_S \left(\frac{\partial P}{\partial z}\hat{\mathbf{j}} - \frac{\partial P}{\partial y}\hat{\mathbf{k}}\right) \cdot \hat{\mathbf{n}}\, dS.$$

By projecting C and S onto the xz- and yz-planes, we can show similarly that

$$\oint_C R\, dz = \iint_S \left(\frac{\partial R}{\partial y}\hat{\mathbf{i}} - \frac{\partial R}{\partial x}\hat{\mathbf{j}}\right) \cdot \hat{\mathbf{n}}\, dS$$

and

$$\oint_C Q\, dy = \iint_S \left(\frac{\partial Q}{\partial x}\hat{\mathbf{k}} - \frac{\partial Q}{\partial z}\hat{\mathbf{i}}\right) \cdot \hat{\mathbf{n}}\, dS.$$

Addition of these three results gives Stokes's theorem for $\mathbf{F}$ and S.

The proof can be extended to general curves and surfaces that do not project in a one-to-one fashion onto all three coordinate planes. Most surfaces S with bounding curves C can be subdivided into n subsurfaces S_i with bounding curves C_i that do satisfy this condition (Figure 15.52). For each such subsurface, Stokes's theorem applies

$$\oint_{C_i} \mathbf{F} \cdot d\mathbf{r} = \iint_{S_i} (\nabla \times \mathbf{F}) \cdot \hat{\mathbf{n}}\, dS, \quad i = 1, \ldots, n.$$

If these n equations are now added together, we have

$$\sum_{i=1}^{n} \oint_{C_i} \mathbf{F} \cdot d\mathbf{r} = \sum_{i=1}^{n} \iint_{S_i} (\nabla \times \mathbf{F}) \cdot \hat{\mathbf{n}}\, dS.$$

Since the S_i constitute S, the right-hand side of this equation is the surface integral of $(\nabla \times \mathbf{F}) \cdot \hat{\mathbf{n}}$ over S. Figure 15.52 illustrates that when the line integrals over the C_i are added, contributions from the auxiliary (interior) curves cancel in pairs, and the remaining line integrals give the line integral of $\mathbf{F} \cdot d\mathbf{r}$ along C. This completes the proof.

FIGURE 15.52

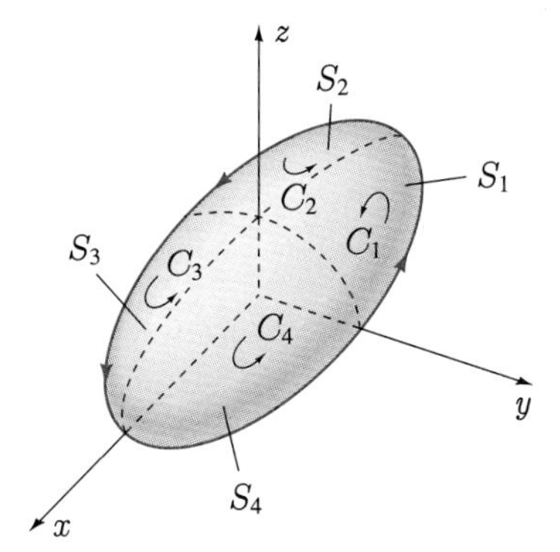

For general surfaces that cannot be divided into a finite number of subsurfaces of this type, the reader should consult a more advanced book.

Note that Green's theorem is a special case of Stokes's theorem. For if $\mathbf{F} = P(x, y)\hat{\mathbf{i}} + Q(x, y)\hat{\mathbf{j}}$, and C is a closed curve in the xy-plane, then by Stokes's theorem

$$\oint_C P\,dx + Q\,dy = \iint_S \left(\frac{\partial Q}{\partial x} - \frac{\partial P}{\partial y}\right)\hat{\mathbf{k}} \cdot \hat{\mathbf{n}}\,dS,$$

where S is any surface for which C is the boundary. If we choose S as that part of the xy-plane bounded by C, then $\hat{\mathbf{n}} = \hat{\mathbf{k}}$ and

$$\oint_C P\,dx + Q\,dy = \iint_S \left(\frac{\partial Q}{\partial x} - \frac{\partial P}{\partial y}\right) dA.$$

With Stokes's theorem, it is straightforward to verify the sufficiency half of Theorem 15.1. Suppose that the curl of a vector field $\mathbf{F}$ vanishes in a simply-connected domain D. If C is any piecewise smooth, closed curve in D, then there exists a piecewise smooth surface S in D with C as boundary. By Stokes's theorem,

$$\oint_C \mathbf{F} \cdot d\mathbf{r} = \iint_S (\nabla \times \mathbf{F}) \cdot \hat{\mathbf{n}}\,dS = 0.$$

According to Corollary 2 of Theorem 15.3, the line integral is independent of path in D, and the theorem itself implies the existence of a function $f(x, y, z)$ such $\nabla f = \mathbf{F}$.

EXAMPLE 15.27

Verify Stokes's theorem if $\mathbf{F} = x^2\hat{\mathbf{i}} + x\hat{\mathbf{j}} + xyz\hat{\mathbf{k}}$, and S is that part of the sphere $x^2 + y^2 + z^2 = 4$ above the plane $z = 1$.

SOLUTION If we choose $\hat{\mathbf{n}}$ as the upper normal to S, then C, the boundary of S, must be traversed in the direction shown in Figure 15.53. (If $\hat{\mathbf{n}}$ is chosen as the lower normal, then C must be traversed in the opposite direction.) Since parametric equations for C are

$$x = \sqrt{3}\cos t, \quad y = \sqrt{3}\sin t, \quad z = 1, \quad 0 \le t \le 2\pi,$$

then

$$\begin{aligned}
\oint_C \mathbf{F} \cdot d\mathbf{r} &= \oint_C x^2\, dx + x\, dy + xyz\, dz \\
&= \int_0^{2\pi} \{3\cos^2 t(-\sqrt{3}\sin t\, dt) + \sqrt{3}\cos t(\sqrt{3}\cos t\, dt)\} \\
&= \int_0^{2\pi} \left\{-3\sqrt{3}\cos^2 t \sin t + \frac{3}{2}(1 + \cos 2t)\right\} dt \\
&= \left\{\sqrt{3}\cos^3 t + \frac{3t}{2} + \frac{3\sin 2t}{4}\right\}_0^{2\pi} \\
&= 3\pi.
\end{aligned}$$

FIGURE 15.53

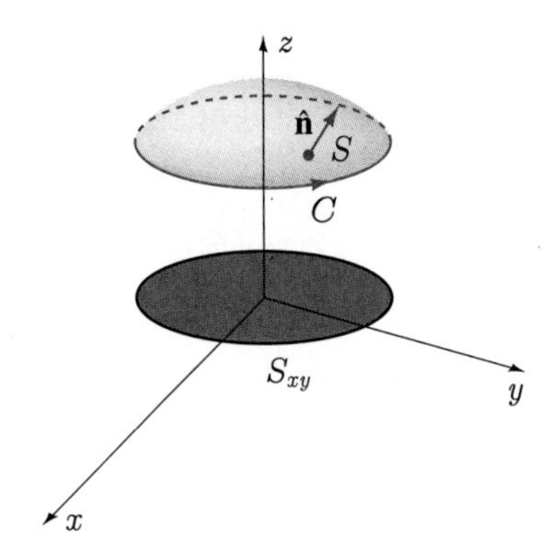

Alternatively,

$$\begin{aligned}
\iint_S (\nabla \times \mathbf{F}) \cdot \hat{\mathbf{n}}\, dS &= \iint_S (xz\hat{\mathbf{i}} - yz\hat{\mathbf{j}} + \hat{\mathbf{k}}) \cdot \frac{(2x, 2y, 2z)}{\sqrt{4x^2 + 4y^2 + 4z^2}} dS \\
&= \iint_S \frac{x^2 z - y^2 z + z}{\sqrt{x^2 + y^2 + z^2}} dS \\
&= \iint_{S_{xy}} \frac{z(x^2 - y^2 + 1)}{2} \sqrt{1 + \left(\frac{\partial z}{\partial x}\right)^2 + \left(\frac{\partial z}{\partial y}\right)^2}\, dA \\
&= \frac{1}{2}\iint_{S_{xy}} z(x^2 - y^2 + 1)\sqrt{1 + (-x/z)^2 + (-y/z)^2}\, dA \\
&= \frac{1}{2}\iint_{S_{xy}} z(x^2 - y^2 + 1)\frac{\sqrt{x^2 + y^2 + z^2}}{z} dA \\
&= \iint_{S_{xy}} (x^2 - y^2 + 1)\, dA.
\end{aligned}$$

If we use polar coordinates to evaluate this double integral over $S_{xy} : x^2 + y^2 \leq 3$, we have

$$\iint_S (\nabla \times \mathbf{F}) \cdot \hat{\mathbf{n}}\, dS = \int_{-\pi}^{\pi} \int_0^{\sqrt{3}} (r^2 \cos^2 \theta - r^2 \sin^2 \theta + 1) r\, dr\, d\theta$$

$$= \int_{-\pi}^{\pi} \left\{ \frac{r^4}{4}(\cos^2 \theta - \sin^2 \theta) + \frac{r^2}{2} \right\}_0^{\sqrt{3}} d\theta$$

$$= \int_{-\pi}^{\pi} \left\{ \frac{9}{4} \cos 2\theta + \frac{3}{2} \right\} d\theta$$

$$= \left\{ \frac{9}{8} \sin 2\theta + \frac{3\theta}{2} \right\}_{-\pi}^{\pi} = 3\pi.$$

■

EXAMPLE 15.28

Evaluate

$$\oint_C 2xy^3\, dx + 3x^2 y^2\, dy + (2z + x)\, dz,$$

where C is the series of line segments joining $A(2,0,0)$ to $B(0,1,0)$ to $D(0,0,1)$ to A.

SOLUTION By Stokes's theorem,

$$\oint_C 2xy^3\, dx + 3x^2 y^2\, dy + (2z + x)\, dz = \iint_S \nabla \times (2xy^3, 3x^2 y^2, 2z + x) \cdot \hat{\mathbf{n}}\, dS,$$

where S is any surface with C as boundary. If we choose S as the flat triangle bounded by C (Figure 15.54), then a normal vector to S is

$$\overrightarrow{\mathbf{BD}} \times \overrightarrow{\mathbf{BA}} = \begin{vmatrix} \hat{\mathbf{i}} & \hat{\mathbf{j}} & \hat{\mathbf{k}} \\ 0 & -1 & 1 \\ 2 & -1 & 0 \end{vmatrix} = (1,2,2),$$

and therefore

$$\hat{\mathbf{n}} = \frac{(1,2,2)}{3}.$$

Since

$$\nabla \times (2xy^3, 3x^2 y^2, 2z + x) = \begin{vmatrix} \hat{\mathbf{i}} & \hat{\mathbf{j}} & \hat{\mathbf{k}} \\ \partial/\partial x & \partial/\partial y & \partial/\partial z \\ 2xy^3 & 3x^2 y^2 & 2z + x \end{vmatrix} = -\hat{\mathbf{j}},$$

it follows that

$$\nabla \times (2xy^3, 3x^2 y^2, 2z + x) \cdot \hat{\mathbf{n}} = -\hat{\mathbf{j}} \cdot \frac{(1,2,2)}{3} = -\frac{2}{3}$$

and

$$\oint_C 2xy^3\, dx + 3x^2 y^2\, dy + (2z + x)\, dz = \iint_S -\frac{2}{3} dS = -\frac{2}{3}(\text{Area of } S).$$

But from equation (12.41), the area of triangle S is

$$\frac{1}{2}|\overrightarrow{\mathbf{BD}} \times \overrightarrow{\mathbf{BA}}| = \frac{1}{2}|(1,2,2)| = \frac{3}{2}.$$

Finally, then,

$$\oint_C 2xy^3\,dx + 3x^2y^2\,dy + (2z + x)\,dz = -\frac{2}{3}\left(\frac{3}{2}\right) = -1.$$

FIGURE 15.54

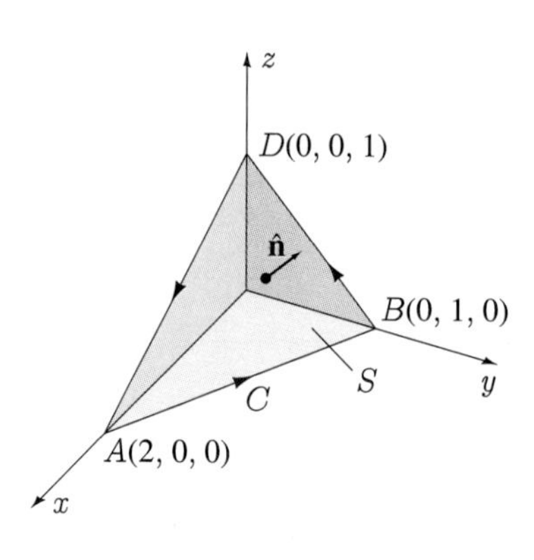

■

EXERCISES 15.10

In Exercises 1–14 use Stokes's theorem to evaluate the line integral.

1. $\oint_C x^2y\,dx + y^2z\,dy + z^2x\,dz$ where C is the curve $z = x^2 + y^2$, $x^2 + y^2 = 4$, directed counterclockwise as viewed from the origin

2. $\oint_C y^2\,dx + xy\,dy + xz\,dz$ where C is the curve $x^2 + y^2 = 2y$, $y = z$, directed so that y increases when x is positive

3. $\oint_C (xyz + 2yz)\,dx + xz\,dy + 2xy\,dz$ where C is the curve $z = 1$, $x^2 + y^2 + z^2 = 4$, directed clockwise as viewed from the origin

4. $\oint_C (2xy + y)\,dx + (x^2 + xy - 3y)\,dy + 2xz\,dz$ where C is the curve $z = \sqrt{x^2 + y^2}$, $z = 4$

5. $\oint_C x^2\,dx + y^2\,dy + (x^2 + y^2)\,dz$ where C is the boundary of that part of the plane $x + y + z = 1$ in the first octant, directed counterclockwise as viewed from the origin

6. $\oint_C y\,dx + x\,dy + (x^2 + y^2 + z^2)\,dz$ where C is the curve $x^2 + y^2 = 1$, $z = xy$, directed clockwise as viewed from the point $(0, 0, 1)$

7. $\oint_C zy^2\,dx + xy\,dy + (y^2 + z^2)\,dz$ where C is the curve $x^2 + z^2 = 9$, $y = \sqrt{x^2 + z^2}$, directed counterclockwise as viewed from the origin

8. $\oint_C y\,dx + z\,dy + x\,dz$ where C is the curve $x + y = 2b$, $x^2 + y^2 + z^2 = 2b(x + y)$, directed clockwise as viewed from the origin

9. $\oint_C y^2\,dx + (x + y)\,dy + yz\,dz$ where C is the curve $x^2 + y^2 = 2$, $x + y + z = 2$, directed clockwise as viewed from the origin

10. $\oint_C (x + y)^2\,dx + (x + y)^2\,dy + yz^3\,dz$ where C is the curve $z = \sqrt{x^2 + y^2}$, $(x - 1)^2 + y^2 = 1$

11. $\oint_C xy\,dx - zx\,dy + yz\,dz$ where C is the boundary of that part of $z = x + y$ in the first octant cut off by $x + y = 1$, directed counterclockwise as viewed from the point $(0, 0, 1)$

12. $\oint_C y^3\,dx - x^3\,dy + xyz\,dz$ where C is the curve $x^2 + y^2 = z^2 + 3$, $z = 3 - \sqrt{x^2 + y^2}$, directed clockwise as viewed from the origin

13. $\oint_C z(x + y)^2\,dx + (y - x)^2\,dy + z^2\,dz$ where C is the smooth curve of intersection of the surfaces

$x^2+z^2=a^2$, $y^2+z^2=a^2$ which has a portion in the first octant, directed so that z decreases in the first octant

14. $\oint_C -2y^3x^2\,dx + x^3y^2\,dy + z\,dz$ where C is the curve $x^2+y^2+z^2=4$, $x^2+4y^2=4$, directed so that x decreases along that part of the curve in the first octant

15. Evaluate the line integral $\oint_C 2x^2y\,dx - yz\,dy + xz\,dz$ where C is the curve $x^2+y^2+z^2=4$, $z=\sqrt{3(x^2+y^2)}$, directed clockwise as viewed from the origin, in four ways:

(a) directly as a line integral

(b) using Stokes's theorem with S as that part of $z=\sqrt{4-x^2-y^2}$ bounded by C

(c) using Stokes's theorem with S as that part of $z=\sqrt{3(x^2+y^2)}$ bounded by C

(d) using Stokes's theorem with S as that part of $z=\sqrt{3}$ bounded by C

16. Let S_1 be that part of $x^2+y^2+z^2=1$ above the xy-plane and S_2 be that part of $z=1-x^2-y^2$ above the xy-plane. Show that if $\hat{\mathbf{n}}_1$ and $\hat{\mathbf{n}}_2$ are the unit upper normals to these surfaces, and $\mathbf{F}$ is a vector field defined on both S_1 and S_2, then

$$\iint_{S_1} (\nabla\times\mathbf{F})\cdot\hat{\mathbf{n}}_1\,dS = \iint_{S_2} (\nabla\times\mathbf{F})\cdot\hat{\mathbf{n}}_2\,dS.$$

SECTION 15.11

Flux and Circulation

In many branches of engineering and physics we encounter the concepts of **flux** and **circulation**. In this section we discuss the relationships between flux and divergence and between circulation and curl. We find a physical setting previously mentioned in Section 15.1 most useful in developing these ideas.

Fluid Flow

Consider a gas flowing through a region D of space. At time t and point $P(x,y,z)$ in D, gas flows through P with some velocity $\mathbf{v}(x,y,z,t)$. If A is a unit area around P perpendicular to $\mathbf{v}$ (Figure 15.55) and $\rho(x,y,z,t)$ is the density of the gas at P, then the amount of gas crossing A per unit time is $\rho\mathbf{v}$. At every point P in D, then, the vector $\rho\mathbf{v}$ is such that its direction $\mathbf{v}$ gives the velocity of gas flow, and its magnitude $\rho|\mathbf{v}|$ describes the mass of gas flowing in that direction per unit time.

FIGURE 15.55

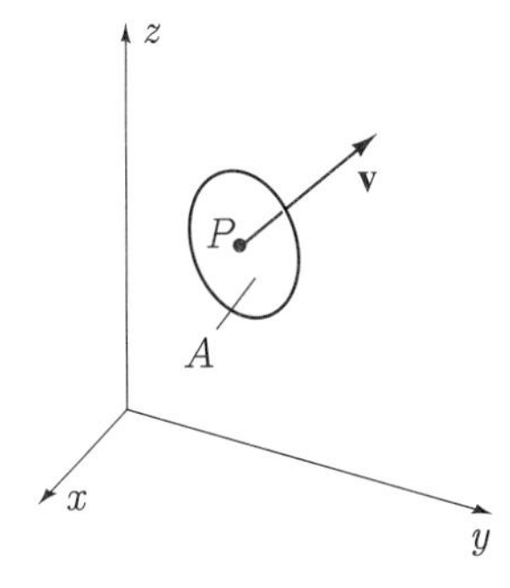

Consider now some surface S in D (Figure 15.56). If $\hat{\mathbf{n}}$ is a unit normal to S, then $\rho\mathbf{v}\cdot\hat{\mathbf{n}}$ is the component of $\rho\mathbf{v}$ normal to the surface S. If dS is an element of area on S, then $\rho\mathbf{v}\cdot\hat{\mathbf{n}}\,dS$ describes the mass of gas flowing through dS per unit time. Consequently,

$$\iint_S \rho\mathbf{v}\cdot\hat{\mathbf{n}}\,dS$$

is the mass of gas flowing through S per unit time. This quantity is called the **flux** for the surface S.

FIGURE 15.56

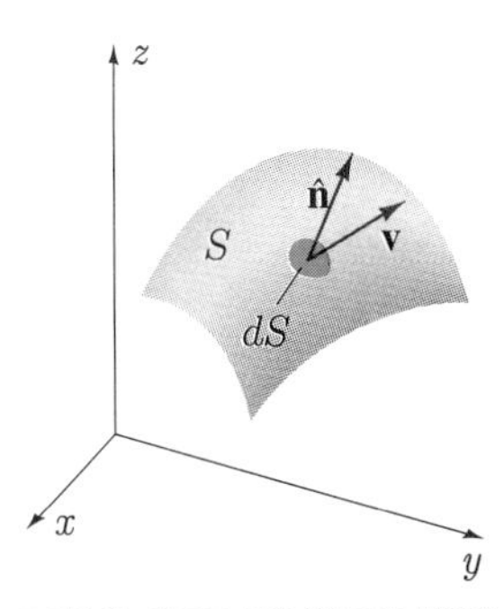

If S is a closed surface (Figure 15.57) and $\hat{\mathbf{n}}$ is the unit outer normal to S, then

$$\oiint_S \rho\mathbf{v}\cdot\hat{\mathbf{n}}\,dS$$

is the mass of gas flowing out of the surface S per unit time. If this flux (for closed S) is positive, then there is a net outward flow of gas through S (i.e., more gas is leaving the volume bounded by S than is entering); if the flux is negative, the net flow is inward.

If we apply the divergence theorem to the flux integral over the closed surface S, we have

$$\oiint_S \rho\mathbf{v}\cdot\hat{\mathbf{n}}\,dS = \iiint_V \nabla\cdot(\rho\mathbf{v})\,dV. \qquad (15.57)$$

FIGURE 15.57

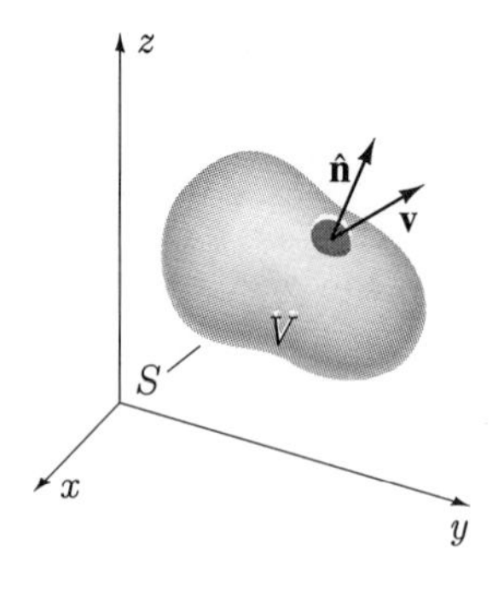

Now the flux (on the left-hand side of this equation) is the mass of gas per unit time leaving S. In order for the right-hand side to represent the same quantity, $\nabla\cdot(\rho\mathbf{v})$ must be interpreted as the mass of gas leaving unit volume per unit time, because then $\nabla\cdot(\rho\mathbf{v})\,dV$ represents the mass per unit time leaving dV, and the triple integral is the mass per unit time leaving V.

We have obtained, therefore, an interpretation of the divergence of $\rho\mathbf{v}$. The divergence of $\rho\mathbf{v}$ is the flux per unit volume per unit time at a point: the mass of gas leaving unit volume in unit time. We can use this idea of flux to derive the "equation of continuity" for fluid flow. The triple integral

$$\iiint_V \frac{\partial\rho}{\partial t}dV = \frac{\partial}{\partial t}\iiint_V \rho\, dV$$

measures the time rate of change of mass in a volume V. If this triple integral is positive, then there is a net inward flow of mass; if it is negative, the net flow is outward. We conclude, therefore, that this triple integral must be the negative of the flux for the volume V; i.e.,

$$\iiint_V \frac{\partial\rho}{\partial t}dV = -\iiint_V \nabla\cdot(\rho\mathbf{v})\,dV,$$

or

$$\iiint_V \left(\nabla\cdot(\rho\mathbf{v}) + \frac{\partial\rho}{\partial t}\right) dV = 0.$$

If $\nabla\cdot(\rho\mathbf{v})$ and $\partial\rho/\partial t$ are continuous functions, then this equation can hold for arbitrary volume V only if

$$\nabla\cdot(\rho\mathbf{v}) + \frac{\partial\rho}{\partial t} = 0. \qquad (15.58)$$

This equation, called the **equation of continuity**, expresses conservation of mass.

If C is a closed curve in the flow region D, then the **circulation** of the flow for the curve C is defined by

$$\Gamma = \oint_C \mathbf{v}\cdot d\mathbf{r}. \qquad (15.59)$$

To obtain an intuitive feeling for Γ, we consider two very simple two-dimensional flows. First suppose $\mathbf{v} = x\hat{\mathbf{i}} + y\hat{\mathbf{j}}$, so that all particles of gas flow along radial lines directed away from the origin (Figure 15.58). In this case, the line integral defining Γ is independent of path and $\Gamma = 0$ for any curve whatsoever.

Second, suppose $\mathbf{v} = -y\hat{\mathbf{i}} + x\hat{\mathbf{j}}$, so that all particles of the gas flow counterclockwise around circles centred at the origin (Figure 15.59). In this case Γ does not generally vanish. In particular, if C is the circle $x^2 + y^2 = r^2$, then $\mathbf{v}$ and $d\mathbf{r}$ are parallel, and

$$\Gamma = \oint_C \mathbf{v}\cdot d\mathbf{r} = \oint_C |\mathbf{v}|ds = \oint_C \sqrt{y^2 + x^2}\,ds = \oint_C r\, ds = 2\pi r^2.$$

FIGURE 15.58

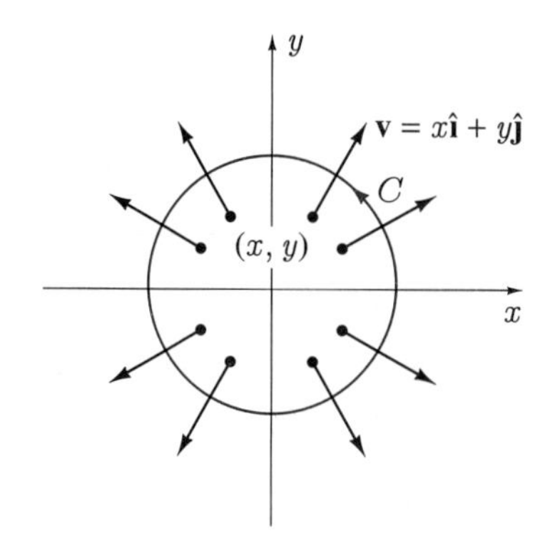

FIGURE 15.59

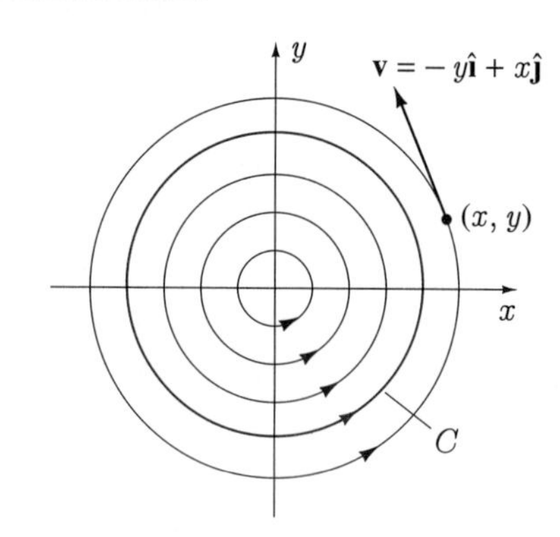

These two flow patterns indicate perhaps that circulation is a measure of the tendency for the flow to be circulatory. If we apply Stokes's theorem to the circulation integral for the closed curve C, we have

$$\oint_C \mathbf{v} \cdot d\mathbf{r} = \iint_S (\nabla \times \mathbf{v}) \cdot \hat{\mathbf{n}}\, dS, \qquad (15.60)$$

where S is any surface in the flow with boundary C. If the right-hand side of this equation is to represent the circulation for C also, then $(\nabla \times \mathbf{v}) \cdot \hat{\mathbf{n}}\, dS$ must be interpreted as the circulation for the curve bounding dS (or simply for dS itself). Then the addition process of the surface integral (Figure 15.60) gives the circulation around C, the circulation around all internal boundaries canceling. But if $(\nabla \times \mathbf{v}) \cdot \hat{\mathbf{n}}\, dS$ is the circulation for dS, then it follows that $(\nabla \times \mathbf{v}) \cdot \hat{\mathbf{n}}$ must be the circulation for unit area perpendicular to $\hat{\mathbf{n}}$. Thus $\nabla \times \mathbf{v}$ describes the circulatory nature of the flow $\mathbf{v}$, its component $(\nabla \times \mathbf{v}) \cdot \hat{\mathbf{n}}$ in any direction describes the circulation for unit area perpendicular to $\hat{\mathbf{n}}$.

FIGURE 15.60

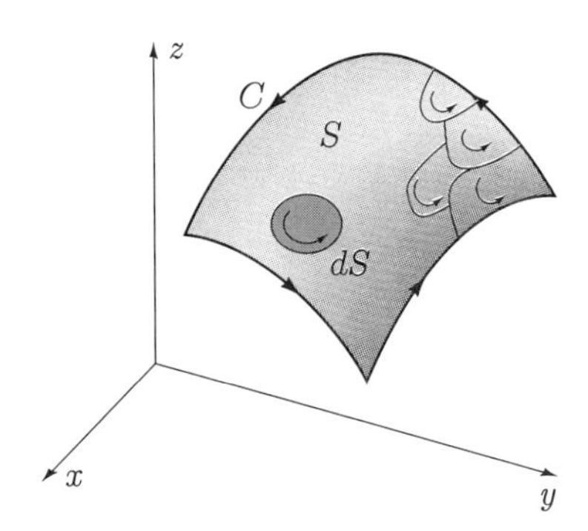

Electromagnetic Theory

The concepts of flux and circulation also play a prominent role in electromagnetic theory. For example, suppose a dielectric contains a charge distribution of density ρ (charge per unit volume). This charge produces an electric field represented by the electric displacement vector **D**. If S is a surface in the dielectric, then the flux of **D** through S is defined as

$$\iint_S \mathbf{D} \cdot \hat{\mathbf{n}}\, dS,$$

and, in the particular case in which S is closed, as

$$\oiint_S \mathbf{D} \cdot \hat{\mathbf{n}}\, dS.$$

Gauss's law states that this flux integral must be exactly equal to the total charge enclosed by S. If V is the region enclosed by S, we can write

$$\oiint_S \mathbf{D} \cdot \hat{\mathbf{n}}\, dS = \iiint_V \rho\, dV. \qquad (15.61)$$

On the other hand, if we apply the divergence theorem to the flux integral, we have

$$\oiint_S \mathbf{D} \cdot \hat{\mathbf{n}}\, dS = \iiint_V \nabla \cdot \mathbf{D}\, dV.$$

Consequently,

$$\iiint_V \nabla \cdot \mathbf{D}\, dV = \iiint_V \rho\, dV,$$

or

$$\iiint_V (\nabla \cdot \mathbf{D} - \rho)\, dV = 0.$$

If $\nabla \cdot \mathbf{D}$ and ρ are continuous functions, then the only way this equation can hold for arbitrary volume V in the dielectric is if

$$\nabla \cdot \mathbf{D} = \rho. \tag{15.62}$$

This is the first of Maxwell's equations for electromagnetic fields.

Another of Maxwell's equations can be obtained using Stokes's theorem. The flux through a surface S of a magnetic field $\mathbf{B}$ is defined by

$$\iint_S \mathbf{B} \cdot \hat{\mathbf{n}}\, dS.$$

If $\mathbf{B}$ is a changing field, then an induced electric field intensity $\mathbf{E}$ is created. Faraday's induction law states that the time rate of change of the flux of $\mathbf{B}$ through S must be equal to the negative of the line integral of $\mathbf{E}$ around the boundary C of S:

$$\oint_C \mathbf{E} \cdot d\mathbf{r} = -\frac{\partial}{\partial t} \iint_S \mathbf{B} \cdot \hat{\mathbf{n}}\, dS. \tag{15.63}$$

But Stokes's theorem applied to the line integral also gives

$$\oint_C \mathbf{E} \cdot d\mathbf{r} = \iint_S (\nabla \times \mathbf{E}) \cdot \hat{\mathbf{n}}\, dS.$$

It follows, therefore, that if S is stationary,

$$\iint_S (\nabla \times \mathbf{E}) \cdot \hat{\mathbf{n}}\, dS = \iint_S -\frac{\partial \mathbf{B}}{\partial t} \cdot \hat{\mathbf{n}}\, dS,$$

or

$$\iint_S \left(\nabla \times \mathbf{E} + \frac{\partial \mathbf{B}}{\partial t}\right) \cdot \hat{\mathbf{n}}\, dS = 0.$$

Once again, this equation can hold for arbitrary surfaces S, if $\nabla \times \mathbf{E}$ and $\partial \mathbf{B}/\partial t$ are continuous, only if

$$\nabla \times \mathbf{E} = -\frac{\partial \mathbf{B}}{\partial t}. \tag{15.64}$$

SUMMARY

When a vector is a function of position, it becomes susceptible to the operations of differentiation and integration. In this chapter we developed various ways of differentiating and integrating vector fields beginning with the operations of divergence and curl. The divergence of a vector field is a scalar field, and the curl of a vector field is another vector field. Both are extremely useful in applied mathematics. Mathematically, the curl appeared in our discussion of independence of path for line integrals and in Stokes's theorem; we saw its physical importance in our study of fluid flow and electromagnetic theory. We introduced the divergence of a vector field in our discussion of the divergence theorem and in the same applications as those for the curl.

The line integral of a function $f(x, y, z)$ along a curve C is defined in the same way as a definite integral, a double integral, or a triple integral— that is, the limit of a sum,

$$\int_C f(x, y, z)\,ds = \lim_{||\Delta s_i|| \to 0} \sum_{i=1}^{n} f(x_i^*, y_i^*, z_i^*)\,\Delta s_i.$$

The most important type of line integral occurs when $f(x, y, z)$ is the tangential component of a vector field $\mathbf{F} = P\hat{\mathbf{i}} + Q\hat{\mathbf{j}} + R\hat{\mathbf{k}}$ defined along C, and in this case we write

$$\int_C f(x, y, z)\,ds = \int_C \mathbf{F} \cdot d\mathbf{r} = \int_C P\,dx + Q\,dy + R\,dz.$$

We developed three methods for evaluating line integrals:

1. Express all parts of the line integral in terms of any parameter along C and evaluate the resulting definite integral. All line integrals can be evaluated in this way, but often methods (2) and (3) lead to much simpler calculations.

2. If a line integral is independent of path, then we can evaluate it by taking the difference in values of a function ϕ (where $\nabla\phi = \mathbf{F}$) at the ends of the curve.

3. If C is a closed curve, we can sometimes use Stokes's theorem to replace a line integral with a simpler surface integral. In this regard, Green's theorem is a special case of Stokes's theorem.

If the line integral of a force field $\mathbf{F}$ is independent of path, the force field is said to be conservative. Associated with every conservative force field is a potential function U such that the work done by $\mathbf{F}$ along a curve C from A to B is equal to the difference in U at A and B. In addition, motion of an object in a conservative force field is always characterized by an exchange of potential energy for kinetic energy in such a way that the sum of the two energies is always a constant value.

Surface integrals are also limits of sums,

$$\iint_S f(x, y, z)\,dS = \lim_{||\Delta S_i|| \to 0} \sum_{i=1}^{n} f(x_i^*, y_i^*, z_i^*)\,\Delta S_i,$$

and the most important type of surface integral occurs when $f(x, y, z)$ is the normal component of a vector field $\mathbf{F}$ on S:

$$\iint_S f(x, y, z)\,dS = \iint_S \mathbf{F} \cdot \hat{\mathbf{n}}\,dS.$$

We suggested two methods for the evaluation of surface integrals:

1. Project S onto some region R in one of the coordinate planes, express all parts of the integral in terms of coordinates in that plane, and evaluate the resulting double integral over R.

2. If S is closed, it could be advantageous to replace a surface integral with the triple integral of $\nabla \cdot \mathbf{F}$ over the volume bounded by S (the divergence theorem).

Key Terms and Formulas

In reviewing this chapter, you should be able to define or discuss the following key terms:

- Vector field
- Scalar field
- Interior point of region
- Exterior point of region
- Boundary point of region
- Open set
- Closed set
- Connected set
- Domain
- Simply-connected domain
- Del operator
- Divergence of a vector field
- Curl of a vector field
- Irrotational vector field
- Solenoidal vector field
- Line integral
- Path independence of a line integral
- Conservative vector field
- Green's theorem
- Work
- Potential energy
- Law of conservation of Energy
- Surface integral
- Orientable surface
- Divergence theorem
- Stokes's theorem
- Flux of a vector field
- Circulation of a vector field

REVIEW EXERCISES

In Exercises 1–10 calculate the quantity.

1. ∇f if $f(x,y,z) = x^2y^3 - xy + z$

2. $\nabla \cdot \mathbf{F}$ if $\mathbf{F}(x,y) = x^3y\hat{\mathbf{i}} - (x^2/y)\hat{\mathbf{j}}$

3. $\nabla \times \mathbf{F}$ if $\mathbf{F}(x,y) = \sin(xy)\hat{\mathbf{i}} + \cos(xy)\hat{\mathbf{j}} + xy\hat{\mathbf{k}}$

4. $\nabla \times \mathbf{F}$ if $\mathbf{F}(x,y,z) = (x+y+z)(\hat{\mathbf{i}} + \hat{\mathbf{j}} + \hat{\mathbf{k}})$

5. ∇f if $f(x,y,z) = \ln(x^2 + y^2 + z^2)$

6. $\nabla \cdot \mathbf{F}$ if $\mathbf{F}(x,y,z) = ye^x\hat{\mathbf{i}} + ze^y\hat{\mathbf{j}} + xe^z\hat{\mathbf{k}}$

7. $\nabla \times \mathbf{F}$ if $\mathbf{F}(x,y,z) = xyz\hat{\mathbf{j}}$

8. ∇f if $f(x,y) = \mathrm{Sin}^{-1}(x+y)$

9. $\nabla \cdot \mathbf{F}$ if $\mathbf{F}(y,z) = yz\hat{\mathbf{i}} - (y^2 + z^2)\hat{\mathbf{j}} + y^2z^2\hat{\mathbf{k}}$

10. $\nabla \times \mathbf{F}$ if $\mathbf{F}(x,y,z) = \mathrm{Cot}^{-1}(xyz)\hat{\mathbf{i}}$

In Exercises 11–30 evaluate the integral.

11. $\int_C y\,ds$ where C is the curve $y = x^3$ from $(-1,-1)$ to $(2,8)$

12. $\iint_S (x^2 + yz)\,dS$ where S is that part of $x + y + z = 2$ in the first octant

13. $\iint_S (x\hat{\mathbf{i}} + y\hat{\mathbf{j}}) \cdot \hat{\mathbf{n}}\,dS$ where S is that part of $z = x^2 + y^2$ bounded by the surfaces $x = \pm 1$, $y = \pm 1$, and $\hat{\mathbf{n}}$ is the lower normal

14. $\iint_S (x\hat{\mathbf{i}} + y\hat{\mathbf{j}}) \cdot \hat{\mathbf{n}}\,dS$ where S is that part of $z = x^2 + y^2$ below $z = 1$, and $\hat{\mathbf{n}}$ is the lower normal

15. $\oint_C x\,dx + y\,dy - z^2\,dz$ where C is the curve $x^2 + y^2 = 1$, $y = z$

16. $\int_C xy\,dx + xz\,dz$ where C is the curve $y = \sqrt{1 + x^2}$ $z = \sqrt{2 - x^2 - y^2}$, from $(1/\sqrt{2}, \sqrt{3/2}, 0)$ to $(-1/\sqrt{2}, \sqrt{3/2}, 0)$

17. $\oint_C 2xy^3\,dx + (3x^2y^2 + 2xy)\,dy$ where C is the curve $(x-1)^2 + y^2 = 1$

18. $\oint_C 2xy^3\,dx + (3x^2y^2 + x^2)\,dy$ where C is the curve $(x-1)^2 + y^2 = 1$

19. $\iint_S (x^2\hat{\mathbf{i}} + y^2\hat{\mathbf{j}} + z^2\hat{\mathbf{k}}) \cdot \hat{\mathbf{n}}\,dS$ where S is the surface bounding the volume enclosed by $y = z$, $y + z = 2$, $x = 0$, $x = 1$, $z = 0$, and $\hat{\mathbf{n}}$ is the outer normal to S

20. $\iint_S (x^2 + y^2)\,dS$ where S is that part of $x^2 + y^2 + z^2 = 6$ inside $z = x^2 + y^2$

21. $\iint_S (x^2 + y^2)\hat{\mathbf{i}} \cdot \hat{\mathbf{n}}\,dS$ where S is that part of $x^2 + y^2 + z^2 = 6$ inside $z = x^2 + y^2$, and $\hat{\mathbf{n}}$ is the upper normal to S

22. $\oint_C (x^2\hat{\mathbf{i}} + y\hat{\mathbf{j}} - xz\hat{\mathbf{k}}) \cdot d\mathbf{r}$ where C is the curve $x^2 + y^2 = 1$, $z = x + 1$, directed clockwise as viewed from the origin

23. $\oint_C (xy\hat{\mathbf{i}} + z\hat{\mathbf{j}} - x^2\hat{\mathbf{k}}) \cdot d\mathbf{r}$ where C is the curve in Exercise 22

24. $\oint_C y\,dx + 2x\,dy - 3z^2\,dz$ where C is the curve $y = \sqrt{1 + z^2 - x^2}$, $x^2 + z^2 = 1$, directed counterclockwise as viewed from the origin

25. $\oint_C (xy + 4x^3y^2)\,dx + (z + 2x^4y)\,dy + (z^5 + x^2z^2)\,dz$ where C is the curve $x^2 + z^2 = 4$, $x^2 + y^2 = 4$, $y = z$, directed counterclockwise as viewed from a point far up the positive z-axis

26. $\iint_S (x^2yz\hat{\mathbf{i}} - x^2yz\hat{\mathbf{j}} - xyz^2\hat{\mathbf{k}}) \cdot \hat{\mathbf{n}}\,dS$ where S is that part of $z = 1 - \sqrt{x^2 - y^2}$ above the xy-plane, and $\hat{\mathbf{n}}$ is the upper normal

27. $\iint_S dS$ where S is that part of $z = x^2 - y^2$ inside $x^2 + y^2 = 4$

28. $\iint_S y\,dS$ where S is that part of $x = y^2 + 1$ in the first octant which is under $x + z = 2$

29. $\oint_C (ye^{xy} + xy^2e^{xy})\,dx + (xe^{xy} + x^2ye^{xy} + x^3y)\,dy$ where C is the curve $x^2 + y^2 = 2y$, $z = 0$

30. $\iint_S (x\hat{\mathbf{i}} + y\hat{\mathbf{j}}) \cdot \hat{\mathbf{n}}\,dS$ where S is that part of $z^2 - x^2 - y^2 = 1$ between the planes $z = 0$ and $z = 2$, and $\hat{\mathbf{n}}$ is the lower normal

31. Let S be that part of the sphere $x^2 + y^2 + z^2 = 1$ that lies above the parabolic cylinder $2z = x^2$. Set up, but do not evaluate, double iterated integrals to calculate the surface integral

$$\iint_S x^2y^2z^2\,dS$$

by projecting S onto (a) the xy-coordinate plane (b) the yz-coordinate plane (c) the xz-coordinate plane.

32. If $\mathbf{r} = x\hat{\mathbf{i}} + y\hat{\mathbf{j}} + z\hat{\mathbf{k}}$, show that $\nabla(|\mathbf{r}|^n) = n|\mathbf{r}|^{n-2}\,\mathbf{r}$.

33. Verify that $\nabla \times (\nabla \times \mathbf{F}) = \nabla(\nabla \cdot \mathbf{F}) - \nabla^2\mathbf{F}$, where $\nabla^2 = \partial^2/\partial x^2 + \partial^2/\partial y^2 + \partial^2/\partial z^2$.

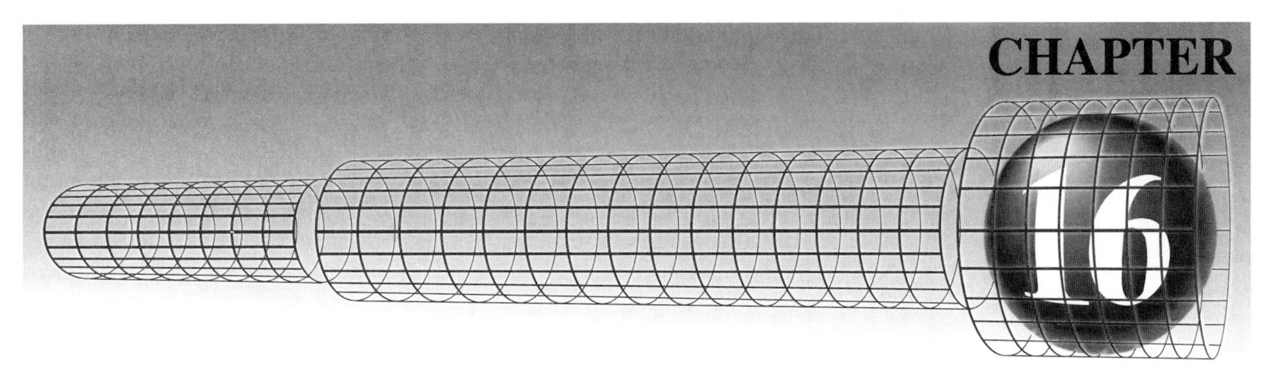

Differential Equations

Differential equations serve as models in many areas of applied mathematics —physics, chemistry, economics, medicine, and engineering to name a few. In this chapter we discuss some of the methods for solving first-order and simple second-order equations. We also give a fairly thorough treatment of linear differential equations. We include a wide variety of applications to illustrate the relevance of differential equations in applied mathematics.

SECTION 16.1 Introduction

A **differential equation** is an equation that must be solved for an unknown function. What distinguishes a differential equation from other equations is the fact that it contains derivatives of the unknown function. For example, each of the following equations is a differential equation in y as a function of x:

$$\frac{dy}{dx} + \frac{2}{3}y = 9.81, \tag{16.1}$$

$$\frac{d^2y}{dx^2} = k\sqrt{1 + \left(\frac{dy}{dx}\right)^2}, \tag{16.2}$$

$$x\frac{d^2y}{dx^2} + \frac{dy}{dx} + xy = 0, \tag{16.3}$$

$$\frac{d^4y}{dx^4} - k^4y = 0. \tag{16.4}$$

Equation 16.1 is used to determine the position of a paratrooper who falls under the influences of gravity and an air resistance that is proportional to velocity (see Example 16.8); equation 16.2 describes the shape of a hanging cable (Exercise 11 in Section 16.4); equation 16.3, called Bessel's differential equation of order zero, is found in heat flow and vibration problems; and equation 16.4 is used to determine the deflection of beams.

Definition 16.1

> The **order** of a differential equation is the order of the highest derivative in the equation.

Of the four differential equations 16.1–16.4, the first is first order, the second and third are second order, and the last is fourth order.

We have considered quite a number of differential equations in Chapters 3, 5, 8, and 9. In Section 5.4 we concentrated on describing physical situations by means of differential equations; in Chapters 3 and 8 we verified that particular combinations of transcendental functions satisfied certain differential equations; and in Chapter 9 we used our integration techniques to solve various equations. Almost all of these differential equations were based on applications, many from physics and engineering, but also some from geometry and other fields such as ecology, chemistry, and psychology. Because of this association with applications, differential equations are almost always accompanied by subsidiary conditions called **initial** or **boundary conditions**. For example, if a mass m falls from rest under gravity and is acted on by a force due to air resistance that is proportional to its instantaneous velocity, then the differential equation that describes its velocity $v(t)$ as a function of time t is

$$m\frac{dv}{dt} = -kv + mg, \quad \text{where } k \text{ is a constant.} \tag{16.5}$$

FIGURE 16.1

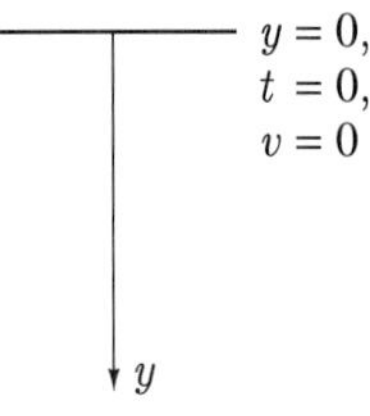

This is simply a statement of Newton's second law, where dv/dt is the vertical component of the acceleration of m, and $-kv + mg$ is the vertical component of the total force on m due to gravity (mg) and air resistance $(-kv)$. If distance y is chosen as positive downward (Figure 16.1), both g and k are positive. Furthermore, if we choose time $t = 0$ at the instant m is dropped, then because m falls from rest, the condition $v(0) = 0$ must be added to the differential equation. In other words, the real problem is to find the solution of the differential equation that also satisfies the initial condition:

$$m\frac{dv}{dt} = -kv + mg, \quad v(0) = 0. \tag{16.6}$$

This is the form in which applied mathematicians find differential equations—the differential equation is accompanied by subsidiary conditions that express physical requirements of the solution. It is not difficult to show that the solution of equation 16.6 is

$$v(t) = \frac{mg}{k} - \frac{mg}{k}e^{-kt/m}. \tag{16.7}$$

(All we need do to verify this is to substitute the function into the differential equation to see that it does indeed satisfy the equation. It is clear that it does satisfy the initial condition.)

If we change the initial condition to $v(0) = v_0$, so that the initial velocity of m has vertical component v_0, then the solution becomes

$$v(t) = \frac{mg}{k} - \left(\frac{mg}{k} - v_0\right)e^{-kt/m}. \tag{16.8}$$

In other words, every solution of differential equation 16.5 can be written in the form

$$v(t) = \frac{mg}{k} + Ce^{-kt/m}, \tag{16.9}$$

and when we impose the initial condition $v(0) = v_0$, then $C = v_0 - mg/k$.

In a similar way, equation 16.4 is normally accompanied by four boundary conditions such as perhaps

$$y(0) = y(L) = 0;$$
$$y''(0) = y''(L) = 0.^{\dagger}$$

It can be shown that every solution of equation 16.4 can be expressed in the form

$$y(x) = C_1 e^{kx} + C_2 e^{-kx} + C_3 \sin(kx) + C_4 \cos(kx), \tag{16.10}$$

where C_1, C_2, C_3, and C_4 are arbitrary constants, and when the boundary conditions are applied, these constants must satisfy the four equations

$$0 = C_1 + C_2 + C_4,$$
$$0 = C_1 e^{kL} + C_2 e^{-kL} + C_3 \sin(kL) + C_4 \cos(kL),$$
$$0 = C_1 + C_2 - C_4,$$
$$0 = C_1 e^{kL} + C_2 e^{-kL} - C_3 \sin(kL) - C_4 \cos(kL).$$

We have stated that every solution of 16.5 can be written in the form 16.9, and every solution of 16.4 can be expressed as 16.10. Note that 16.9 contains one arbitrary constant whereas 16.10 has four, but in both cases the number of arbitrary constants is exactly the same as the order of the differential equation. We might suspect that every solution of an nth-order differential equation can be expressed as a function involving n arbitrary constants. For many differential equations this is indeed true, but unfortunately it is not true for all equations. As an illustration, consider the equation

$$\frac{d^2 y}{dx^2} = \left(\frac{dy}{dx}\right)^2. \tag{16.11}$$

In Example 16.6 we apply standard techniques for solving differential equations to obtain the solution $y(x) = C_1 - \ln(C_2 + x)$, which contains two arbitrary constants C_1 and C_2. This two-parameter family of solutions does not, however, contain all solutions of the differential equation, for no choice of C_1 and C_2 will give the perfectly acceptable solution $y(x) \equiv 1$. This solution is not particularly interesting, but it is nonetheless a solution that is not contained within the two-parameter family. Such a solution is called a **singular solution**. We have illustrated that a solution that contains the same number of arbitrary constants as the order of the differential equation may or may not contain all solutions of the differential equation. In spite of this unfortunate circumstance, there do exist large classes of differential equations for which a solution with the same number of arbitrary constants as the order of the equation does indeed represent all possible solutions. Because of this we make the following definition.

Definition 16.2

A **general solution** of a differential equation is a solution that contains the same number of arbitrary constants as the order of the differential equation.

Consequently, in order for a function to be a general solution of a differential equation, it must first be a solution, and second, contain the requisite number of arbitrary constants.

Once again we emphasize that a general solution of a differential equation may or may not contain all possible solutions. We will point out important cases in which a general solution does indeed contain all solutions.

---

† In this chapter it is frequently convenient to use the notation $y', y'', y''', \ldots$ to represent dy/dx, d^2y/dx^2, d^3y/dx^3, etc. In this notation $y''(a)$ is the second derivative of y evaluated at $x = a$. In addition, we denote the solution of a differential equation in y as a function of x by $y(x)$.

EXAMPLE 16.1

Show that $y(x) = C_1 \cos 2x + C_2 \sin 2x$ is a general solution of the differential equation

$$\frac{d^2 y}{dx^2} + 4y = 0.$$

SOLUTION If we substitute this function into the left-hand side of the equation, we have

$$\frac{d^2 y}{dx^2} + 4y = \{-4C_1 \cos 2x - 4C_2 \sin 2x\} + 4\{C_1 \cos 2x + C_2 \sin 2x\} = 0,$$

and the function is indeed a solution. Because it contains two arbitrary constants, the order of the differential equation, it is a general solution. ■

Definition 16.3

A **particular solution** of a differential equation is a solution that contains no arbitrary constants.

It follows, therefore, that particular solutions can be obtained by assigning specific values to the arbitrary constants in a general solution. For example, $y(x) = 5 - \ln(3 + x)$ is a particular solution of differential equation 16.11, as is $y(x) = -\ln x$, both being obtained from $y(x) = C_1 - \ln(C_2 + x)$. On the other hand, the singular solution $y(x) = 10$ is also a particular solution, but it cannot be obtained from this general solution.

EXAMPLE 16.2

Find a particular solution of the differential equation

$$5\frac{d^3 y}{dx^3} + 3\frac{d^2 y}{dx^2} + 2y = 4.$$

SOLUTION In Section 16.10 we develop systematic techniques for finding general and particular solutions for differential equations such as this. But clearly those techniques are not needed here, since a simple glance tells us that $y(x) = 2$ is a solution. ■

Many differential equations are immediately solvable (or, as we often say, immediately integrable). For example, to solve a differential equation of the form

$$\frac{dy}{dx} = M(x), \qquad (16.12)$$

where $M(x)$ is given, we integrate both sides of the equation with respect to x to obtain the general solution

$$y(x) = \int M(x)\, dx + C. \qquad (16.13)$$

We have called this *the* general solution rather than *a* general solution because in this case the general solution does indeed contain all solutions of the differential equation.

This result is easily extended to the nth-order equation

$$\frac{d^n y}{dx^n} = M(x), \quad n \text{ a positive integer}. \qquad (16.14)$$

We integrate successively n times to obtain the general solution

$$y(x) = \int \cdots \int M(x)\,dx \cdots dx + C_1 + C_2 x + \cdots + C_n x^{n-1}. \qquad (16.15)$$

EXERCISES 16.1

In Exercises 1–10 show that the function is a general solution of the differential equation.

1. $y(x) = 2 + Ce^{-x^2}$; $\dfrac{dy}{dx} + 2xy = 4x$

2. $y(x) = \dfrac{x}{1+Cx}$; $\dfrac{dy}{dx} = \dfrac{y^2}{x^2}$

3. $y(x) = \dfrac{x^3}{2} + Cx^3 e^{1/x^2}$; $x^3 \dfrac{dy}{dx} + (2 - 3x^2)y = x^3$

4. $y(x) = C_1 \sin 3x + C_2 \cos 3x$; $\dfrac{d^2y}{dx^2} + 9y = 0$

5. $y(x) = \dfrac{C_1^2 e^{2x} + 1}{2C_1 e^x} + C_2$; $\left(\dfrac{d^2y}{dx^2}\right)^2 = 1 + \left(\dfrac{dy}{dx}\right)^2$

6. $y(x) = C_1 e^{2x} \cos\left(\dfrac{x}{\sqrt{2}}\right) + C_2 e^{2x} \sin\left(\dfrac{x}{\sqrt{2}}\right)$;

$2\dfrac{d^2y}{dx^2} - 8\dfrac{dy}{dx} + 9y = 0$

7. $y(x) = C_1 \cos 2x + C_2 \sin 2x + C_3 \cos x + C_4 \sin x$;

$\dfrac{d^4y}{dx^4} + 5\dfrac{d^2y}{dx^2} + 4y = 0$

8. $y(x) = \left(C_1 + C_2 x - \dfrac{x^2}{4}\right) e^{4x}$;

$2\dfrac{d^2y}{dx^2} - 16\dfrac{dy}{dx} + 32y = -e^{4x}$

9. $y(x) = C_1 \cos(2 \ln x) + C_2 \sin(2 \ln x) + \dfrac{1}{4}$;

$x^2 \dfrac{d^2y}{dx^2} + x\dfrac{dy}{dx} + 4y = 1$

10. $y(x) = C_1 \dfrac{\sin x}{\sqrt{x}} + C_2 \dfrac{\cos x}{\sqrt{x}}$;

$x^2 \dfrac{d^2y}{dx^2} + x\dfrac{dy}{dx} + \left(x^2 - \dfrac{1}{4}\right) y = 0$

In Exercises 11–14 find a particular solution of the differential equation in Exercise 4 that satisfies the conditions.

11. $y(0) = 1,\ y'(0) = 6$

12. $y(0) = 2,\ y(\pi/2) = 3$

13. $y(\pi/12) = 0,\ y'(\pi/12) = 1$

14. $y(1) = 1,\ y(2) = 2$

In Exercises 15–20 find a general solution of the differential equation.

15. $\dfrac{dy}{dx} = 6x^2 + 2x$

16. $\dfrac{dy}{dx} = \dfrac{1}{9 + x^2}$

17. $\dfrac{d^2y}{dx^2} = 2x + e^x$

18. $\dfrac{d^2y}{dx^2} = x \ln x$

19. $\dfrac{d^3y}{dx^3} = \dfrac{1}{3x^5}$

20. $\dfrac{dy}{dx} = y$

21. **(a)** A boy initially at O (Figure 16.2) walks along the edge of a swimming pool (the y-axis) towing his sailboat by a string of length L. If the boat starts at Q and the string always remains straight, show that the equation of the curved path $y = y(x)$ followed by the boat must satisfy the differential equation

$$\frac{dy}{dx} = -\frac{\sqrt{L^2 - x^2}}{x}.$$

(b) Solve this differential equation for $y(x)$.

FIGURE 16.2

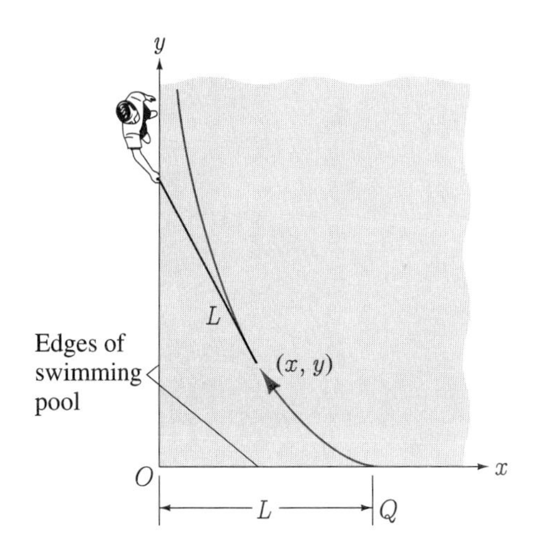

22. Show that $y(x) = -(x^2 + C)^{-1}$ is a general solution of the differential equation

$$\frac{dy}{dx} = 2xy^2.$$

Find a singular solution.

23. Show that $y(x) = 1 - (x^3 + C)^{-1}$ is a general solution of the differential equation

$$\frac{dy}{dx} = 3x^2(y-1)^2.$$

Find a singular solution.

24. **(a)** Verify that $y(x) = Ce^{2x}$ is a general solution of the differential equation

$$\frac{dy}{dx} = 2y.$$

(b) Draw the one-parameter family of curves defined by this general solution.

(c) Show that there is a particular solution that passes through any given point (x_0, y_0), and that this solution can be obtained by choosing C appropriately.

25. **(a)** Verify that a general solution of the differential equation

$$2x\frac{dy}{dx} = y$$

is defined implicitly by the equation $y^2 = Cx$.

(b) Draw the one-parameter family of curves defined by this equation.

(c) Show that, with the exception of points on the y-axis, there is a particular solution that passes through any given point (x_0, y_0), and that this solution can be obtained by specifying C appropriately.

26. **(a)** Draw the one-parameter family of curves defined by the general solution in Exercise 22.

(b) Show that, with the exception of points on the x-axis, there is a solution passing through any given point in the xy-plane.

27. Consider the differential equation

$$\frac{dy}{dx} = \frac{1}{x^2}.$$

(a) Find a solution that satisfies the condition $y(1) = 1$.

(b) Find a solution that satisfies the condition $y(-1) = 2$.

(c) Find a solution that satisfies the conditions in both (a) and (b).

(d) Draw the curves defined by the general solution to explain why this is possible.

SECTION 16.2

Separable Differential Equations

In this section and the next, we consider first-order differential equations that can be written in the form

$$\frac{dy}{dx} = F(x,y). \tag{16.16}$$

Since any function $F(x,y)$ can always be considered as the quotient of two other functions,

$$F(x,y) = \frac{M(x,y)}{N(x,y)},$$

equation 16.16 can also be written in the equivalent form

$$N(x,y)\,dy = M(x,y)\,dx. \tag{16.17}$$

Depending on the form of $F(x,y)$ (or $M(x,y)$ and $N(x,y)$) various methods can be used to obtain the unknown function $y(x)$. Two of the more important techniques are considered here and in Section 16.3; others are discussed in the exercises.

Differential equation 16.16 is said to be separable if it can be expressed in the form

$$\frac{dy}{dx} = \frac{M(x)}{N(y)}; \tag{16.18}$$

that is, if dy/dx is equal to a function of x divided by a function of y. Equivalently, a differential equation is said to be separable if it can be written in the form

$$N(y)\,dy = M(x)\,dx. \tag{16.19}$$

When a differential equation is written in this way, it is said to be separated—separated in the sense that the x- and y-variables appear on opposite sides of the equation. For a separated equation we can write therefore that

$$N(y)\frac{dy}{dx} = M(x), \tag{16.20}$$

and if we integrate both sides with respect to x, we have

$$\int N(y)\frac{dy}{dx}\,dx = \int M(x)\,dx + C. \tag{16.21}$$

Cancellation of differentials on the left leads to the solution

$$\int N(y)\,dy = \int M(x)\,dx + C. \tag{16.22}$$

What we mean by saying that 16.21 and 16.22 represent general solutions of 16.19 is that any function defined *implicitly* by 16.21 or 16.22 is a solution of 16.19. For example, if we divide the differential equation

$$3x^3y^2\,dx + (xy^3 - xy^2)\,dy = 0$$

by xy^2, it becomes separated:

$$(1 - y)\,dy = 3x^2\,dx.$$

According to 16.21 we should divide by dx and integrate both sides with respect to x:

$$\int (1 - y)\frac{dy}{dx}\,dx = \int 3x^2\,dx = x^3 + C.$$

The antidifferentiation on the left must be interpreted as "implicit antidifferentiation," asking for that function which when differentiated with respect to x gives $(1-y)\,dy/dx$. Since an antiderivative is $y - y^2/2$, a general solution of the differential equation is

$$y - \frac{y^2}{2} = x^3 + C.$$

Were we to use 16.22 after separation (instead of 16.21), we would write

$$\int (1 - y)\,dy = \int 3x^2\,dx + C,$$

and integrate for a general solution

$$y - \frac{y^2}{2} = x^3 + C.$$

As indicated above, by saying that $y - y^2/2 = x^3 + C$ represents a general solution of the original differential equation, we mean that any function defined implicitly by $y - y^2/2 = x^3 + C$ is a solution.

EXAMPLE 16.3

Find general solutions for the following differential equations:

(a) $2x^3y^2\,dx = xy^3\,dy$

(b) $\dfrac{dy}{dx} = \dfrac{y\sin x + y^3\sin x}{(1+y^2)^2}$

SOLUTION

(a) If we divide the differential equation by xy^2, we obtain

$$y\,dy = 2x^2\,dx,$$

which is separated. A general solution is therefore defined implicitly by

$$\int y\,dy = \int 2x^2\,dx + C,$$

or

$$\frac{y^2}{2} = \frac{2x^3}{3} + C.$$

Note that $y(x) = 0$ is also a solution, but it cannot be obtained by specifying C. In other words, $y(x) = 0$ is a singular solution. We removed this solution when we divided the original equation by xy^2.

(b) Since

$$\frac{dy}{dx} = \frac{y\sin x(1+y^2)}{(1+y^2)^2} = \frac{y\sin x}{1+y^2},$$

the differential equation can be separated:

$$\sin x\,dx = \frac{1+y^2}{y}dy = \left(\frac{1}{y} + y\right)dy.$$

A general solution is therefore defined implicitly by

$$\int \sin x\,dx + C = \int \left(\frac{1}{y} + y\right)dy,$$

or,

$$-\cos x + C = \ln|y| + \frac{y^2}{2}.$$

■

EXAMPLE 16.4

Find the function (in explicit form) that satisfies the differential equation

$$e^{2x+y}dx - 2e^{x-y}dy = 0, \quad y(0) = 1.$$

SOLUTION If we divide the equation by e^{x+y}, we have

$$e^x\,dx = 2e^{-2y}dy.$$

A general solution is defined implicitly by

$$e^x = -e^{-2y} + C.$$

The condition $y(0) = 1$ requires $e^0 = -e^{-2} + C$, which implies that $C = 1 + e^{-2}$. The required solution is therefore defined implicitly by

$$e^x = -e^{-2y} + e^{-2} + 1,$$

and from this equation, we have

$$e^{-2y} = e^{-2} + 1 - e^x, \quad \text{or} \quad y(x) = -\tfrac{1}{2}\ln(e^{-2} + 1 - e^x).$$

■

EXERCISES 16.2

In Exercises 1–10 find a general solution of the differential equation.

1. $y^2\,dx - x^2\,dy = 0$

2. $\dfrac{dy}{dx} + 2xy = 4x$

3. $2xy\,dx + (x^2 + 1)\,dy = 0$

4. $\dfrac{dy}{dx} = 3y + 2$

5. $3(y^2 + 2)\,dx = 4y(x - 1)\,dy$

6. $(x^2y + x^2)\,dx + (xy^2 - y^2)\,dy = 0$

7. $\dfrac{dy}{dx} = -\dfrac{\cos y}{\sin x}$

8. $(x^2ye^x - y)\,dx + xy^3\,dy = 0$

9. $(x^2y^2 \sec x \tan x + xy^2 \sec x)\,dx + xy^3\,dy = 0$

10. $\dfrac{dy}{dx} = \dfrac{1 + y^2}{1 + x^2}$

In Exercises 11–15 find a solution of the differential equation that also satisfies the given condition.

11. $2y\,dx + (x + 1)\,dy = 0, \; y(1) = 2$

12. $(xy + y)\,dx - (xy - x)\,dy = 0, \; y(1) = 2$

13. $\dfrac{dy}{dx} = e^{x+y}, \; y(0) = 0$

14. $\dfrac{dy}{dx} = 2x(1 + y^2), \; y(2) = 4$

15. $\dfrac{dy}{dx} = \dfrac{\sin^2 y}{\cos^2 x}, \; y(0) = \pi/2$

16. A girl lives 6 km from school. She decides to travel to school so that her speed is always proportional to the square of her distance from the school.

(a) Find her distance from school at any time.

(b) When does she reach school?

17. Find a general solution for the differential equation

$$\frac{dy}{dx} = -\frac{1 + y^3}{xy^2 + x^3y^2}.$$

18. When a container of water at temperature 80 °C is placed in a room at temperature 20 °C, Newton's law of cooling states that the time rate of change of the temperature of the water is proportional to the difference between the temperature of the water and room temperature. If the water cools to 60 °C in 2 min, find a formula for its temperature as a function of time.

19. A thermometer reading 23 °C is taken outside where the temperature is −20 °C. If the reading drops to 0 °C in 4 min, when will it read −19 °C?

20. The amount of a drug such as penicillin injected into the body is used up at a rate proportional to the amount still present. If a dose decreases by 5% in the first hour, when does it decrease to one-half its original amount?

21. When a deep-sea diver inhales air, his body tissues absorb extra amounts of nitrogen. Suppose the diver enters the water at time $t = 0$, drops very quickly to depth d, and remains at this depth for a very long time. The amount N of nitrogen in his body tissues increases as he remains at this depth until a maximum amount $\overline{N}$ is reached. The time rate of change of N is proportional to the difference $\overline{N} - N$. Show that if N_0 is the amount of nitrogen in his body tissues when he enters the water, then

$$N = N_0 e^{-kt} + \overline{N}(1 - e^{-kt}),$$

where $k > 0$ is a constant.

22. When a substance such as glucose is administered intravenously into the bloodstream, it is used up by the body at a rate proportional to the amount present at that time. If it is added at a constant rate of R units per unit time, and A_0 is the amount present in the bloodstream when the intravenous feeding begins, find a formula for the amount in the bloodstream at any time.

23. Find the equations for all curves that satisfy the condition that the normal at any point on the curve, and the line joining the point to the origin, form an isosceles triangle with the x-axis as base.

24. When two substances A and B are brought together in one solution, they react to form a third substance C in such a way that one gram of A reacts with one gram of B to produce two grams of C. The rate at which C is formed is proportional to the amounts of A and B still present in the solution. If 10 g of A and 15 g of B are originally brought together, find a formula for the amount of C present in the mixture at any time.

25. What is the solution to Exercise 24 when the initial amounts of A and B are both 10 g?

26. A first-order differential equation in $y(x)$ is said to be *homogeneous* if it can be written in the form

$$\frac{dy}{dx} = f\left(\frac{y}{x}\right).$$

Show that the change of dependent variable $v = y/x$ yields a differential equation in $v(x)$ that is always separable.

27. State a condition on the functions M and N in equation 16.17 in order that the differential equation be homogeneous in the sense of Exercise 26.

In Exercises 28–33 show that the differential equation is homogeneous, and use the change of variable in Exercise 26 to find a general solution.

28. $(y^2 - x^2)\,dx + xy\,dy = 0$

29. $2x\,dy - 2y\,dx = \sqrt{x^2 + 4y^2}\,dx$

30. $\dfrac{dy}{dx} = \dfrac{y + x}{y - x}$

31. $x\,dy - y\,dx = x\cos(y/x)\,dx$

32. $\dfrac{dy}{dx} = \dfrac{x^2 e^{-y/x} + y^2}{xy}$

33. $(x^2 y + y^3)\,dx + x^3\,dy = 0$

34. If a curve passes through the point $(1, 2)$ and is such that the length of that part of the tangent line at (x, y) from (x, y) to the y-axis is equal to the y-intercept of the tangent line, find the equation of the curve.

35. Find in explicit form a function that satisfies the differential equation $dy/dx = \csc y$ and the following conditions: **(a)** $y(0) = \pi/4$ **(b)** $y(0) = 7\pi/4$

36. In a chemical reaction, one molecule of trypsinogen yields one molecule of trypsin. In order for the reaction to take place, an initial amount of trypsin must be present. Suppose that the initial amount is y_0. Thereafter, the rate at which trypsinogen is changed into trypsin is proportional to the product of the amounts of each chemical in the reaction. Find a formula for the amount of trypsin if the initial amount of trypsinogen is A.

37. If a first-order differential equation can be written in the form

$$\frac{dy}{dx} = f(ax + by),$$

where a and b are constants, show that the change of dependent variable $v = ax + by$ always gives a differential equation in $v(x)$ that is separable.

Use the method of Exercise 37 to find a general solution of the differential equation in Exercises 38–41.

38. $\dfrac{dy}{dx} = x + y$

39. $\dfrac{dy}{dx} = (x + y)^2$

40. $\dfrac{dy}{dx} = \dfrac{1}{2x + 3y}$

41. $\dfrac{dy}{dx} = \sin^2(x - y)$

42. A certain chemical dissolves in water at a rate proportional to the product of the amount of undissolved chemical and the difference between the concentrations in a saturated solution and the existing concentration in the solution. A saturated solution contains 25 g of chemical in 100 mL of solution.

(a) If 50 g of chemical are added to 200 mL of water, find a formula for the amount of chemical dissolved as a function of time. Sketch its graph.

(b) Repeat (a) if the 50 g of chemical are added to 100 mL of water.

(c) Repeat (a) if 10 g of chemical are added to 100 mL of water.

43. Snow has been falling for some time when a snowplow starts plowing the highway. The plow begins at 12:00 and travels 2 km during the first hour and 1 km during the second hour. Make reasonable assumptions to find out when the snow started falling.

44. In order to perform a one-hour operation on a dog a veterinarian anesthetizes the dog with sodium pentobarbital. During the operation, the dog's body breaks down the drug at a rate proportional to the amount still present, and only half an original dose remains after 5 h. If the dog has mass 20 kg, and 20 mg of sodium pentobarbital per kilogram of body mass are required to maintain surgical anesthesia, what original dose is required?

45. Explain what the cancellation of differentials in proceeding from 16.21 to 16.22 really means.

46. Solve the differential equation

$$(x^3y^4 + 2xy^4)\,dx + (x - xy^6)\,dy = 0, \quad y(1) = 1.$$

47. Two substances A and B react to form a third substance C in such a way that 2 g of A react with 1 g of B to produce 3 g of C. The rate at which C is formed is proportional to the amounts of A and B still present in the mixture. Find the amount of C present in the mixture as a function of time when the original amounts of A and B brought together at time $t = 0$ are as follows. Sketch graphs of all three functions on the same axes.

(a) 20 g and 10 g

(b) 20 g and 5 g

(c) 20 g and 20 g

48. A cylindrical tank of radius r and height H has a vertical axis and no top. It is originally (time $t = 0$) full of water, and for $t > 0$, water leaks out through a hole in its bottom.

(a) To find the speed at which water exits through the hole consider the situation when the depth of water is h. In a small time dt, the depth drops by an amount dh. Use the fact that during dt, the potential energy of the small disc of water with thickness $|dh|$ at height h above the bottom of the cylinder is converted into kinetic energy of the same volume exiting through the hole with speed v to show that $v = \sqrt{2gh}$ where $g = 9.81$. This formula describes an ideal situation where all potential energy is converted into kinetic energy. In less than ideal situations, it is customary to assume that $v = c\sqrt{2gh}$, where c is a constant and $0 < c < 1$.

(b) Show that the rate of change of the volume of water in the cylinder is given by the two expressions

$$\frac{dV}{dt} = \pi r^2 \frac{dh}{dt} \quad \text{and} \quad \frac{dV}{dt} = -Av$$

where A is the cross-sectional area of the hole, and hence deduce that

$$\frac{dh}{dt} = -\frac{\sqrt{2g}cA}{\pi r^2}\sqrt{h}.$$

This is sometimes called Torricelli's law.

(c) Find h as a function of t and thereby determine a formula for how long it takes for the cylinder to empty.

49. Repeat Exercise 48 for an open-topped right-circular cone of radius r and height H (with hole at the vertex).

50. A bird is due east of its nest a distance L away and at the same height above the ground as the nest. Wind is blowing due north at speed v. If the bird flies horizontally with constant speed V always pointing straight at its nest, what is the equation of the curve that it follows. Take the nest at the origin and the x- and y-directions as east and north.

51. Find the equation of the curve that passes through $(1, 1)$ and is such that the tangent and normal lines at any point (x, y) make with the x-axis a triangle whose area is equal to the slope of the tangent line at (x, y).

SECTION 16.3

Linear First-Order Differential Equations

A first-order differential equation that can be written in the form

$$\frac{dy}{dx} + P(x)y = Q(x) \tag{16.23}$$

is said to be linear. We will explain the significance of the adjective "linear" in Section 16.7. To illustrate how to solve such differential equations, consider the equation

$$\frac{dy}{dx} + \frac{1}{x}y = 1.$$

If we multiply both sides by x, we have

$$x\frac{dy}{dx} + y = x.$$

But note now that the left side is the derivative of the product xy; i.e.,

$$\frac{d}{dx}(xy) = x\frac{dy}{dx} + y.$$

In other words, we can write the differential equation in the form

$$\frac{d}{dx}(xy) = x,$$

and integration immediately gives a general solution

$$xy = \frac{x^2}{2} + C.$$

This is the principle behind all linear first-order equations: Multiply the equation by a function of x in order that the left side is expressible as the derivative of a product. To show that this is always possible, we turn now to the general equation 16.23. If the equation is multiplied by a function $\mu(x)$,

$$\mu\frac{dy}{dx} + \mu P(x)y = \mu Q(x). \tag{16.24}$$

This equation is equivalent to 16.23 in the sense that $y(x)$ is a solution of 16.23 if and only if it is a solution of 16.24. We ask whether it is possible to find μ so that the left side of 16.24 can be written as the derivative of the product μy; that is, can we find $\mu(x)$ so that

$$\mu\frac{dy}{dx} + \mu P(x)y = \frac{d}{dx}(\mu y)?$$

If we expand the right side, μ must satisfy

$$\mu\frac{dy}{dx} + \mu P(x)y = \mu\frac{dy}{dx} + y\frac{d\mu}{dx},$$

from which we get

$$\mu P = \frac{d\mu}{dx} \quad \text{or} \quad \frac{d\mu}{\mu} = P\,dx.$$

Thus μ must satisfy a separated differential equation, one solution of which is

$$\ln|\mu| = \int P(x)\,dx, \quad \text{or} \quad \mu = \pm e^{\int P(x)\,dx}.$$

We have shown then that if 16.23 is multiplied by the factor

$$e^{\int P(x)\,dx} \quad (\text{or by } -e^{\int P(x)\,dx}),$$

then the differential equation becomes

$$e^{\int P(x)\,dx}\frac{dy}{dx} + P(x)ye^{\int P(x)\,dx} = Q(x)e^{\int P(x)\,dx},$$

and the left side can be expressed as the derivative of a product:

$$\frac{d}{dx}\{ye^{\int P(x)\,dx}\} = Q(x)e^{\int P(x)\,dx}.$$

Integration now gives a general solution

$$ye^{\int P(x)\,dx} = \int Q(x)e^{\int P(x)\,dx}dx + C. \tag{16.25}$$

The quantity $e^{\int P(x)\,dx}$ is called an **integrating factor** for equation 16.23 because when equation 16.23 is multiplied by this factor, it becomes immediately integrable.

In summary, if linear differential equation 16.23 is multiplied by the function $e^{\int P(x)\,dx}$, then the left side of the equation becomes the derivative of the product of y and this function. In this form the equation can immediately be integrated to give a general solution. Given a specific linear equation to solve, we have three choices on how to proceed:

(1) Memorize 16.25 as a formula.

(2) Memorize that an integrating factor is $e^{\int P(x)\,dx}$.

(3) Develop the integrating factor in every example.

We feel that (2) is a reasonable compromise in that we must memorize something, but not too much, and we must do some calculations, but not too many.

EXAMPLE 16.5

Find general solutions for the following differential equations:

(a) $\dfrac{dy}{dx} + xy = x$

(b) $(y - x\sin x)\,dx + x\,dy = 0$

(c) $\cos x\dfrac{dy}{dx} + y\sin x = 1$

SOLUTION

(a) An integrating factor for this linear equation is

$$e^{\int x\,dx} = e^{x^2/2}.$$

If we multiply the equation by this integrating factor, we have

$$e^{x^2/2}\frac{dy}{dx} + yxe^{x^2/2} = xe^{x^2/2},$$

or,

$$\frac{d}{dx}\{ye^{x^2/2}\} = xe^{x^2/2}.$$

Integration yields

$$ye^{x^2/2} = \int xe^{x^2/2}\,dx = e^{x^2/2} + C,$$

or,

$$y(x) = 1 + Ce^{-x^2/2}.$$

(b) If we write the differential equation in the form

$$\frac{dy}{dx} + \frac{y}{x} = \sin x,$$

then we see that it is linear first order. An integrating factor is therefore

$$e^{\int 1/x dx} = e^{\ln|x|} = |x|.$$

If we multiply the differential equation by this factor, we get

$$|x|\frac{dy}{dx} + \frac{|x|}{x}y = |x|\sin x.$$

If $x > 0$, then we write

$$x\frac{dy}{dx} + \frac{x}{x}y = x\sin x,$$

whereas if $x < 0$,

$$-x\frac{dy}{dx} - \frac{x}{x}y = -x\sin x.$$

In either case, however, the equation simplifies to

$$x\frac{dy}{dx} + y = x\sin x,$$

or

$$\frac{d}{dx}(xy) = x\sin x.$$

Integration now gives

$$xy = \int x\sin x\,dx = -x\cos x + \sin x + C.$$

Finally, then,

$$y(x) = -\cos x + \frac{\sin x}{x} + \frac{C}{x}.$$

(c) An integrating factor for the linear equation

$$\frac{dy}{dx} + y\tan x = \frac{1}{\cos x}$$

is

$$e^{\int \tan x\,dx} = e^{\ln|\sec x|} = |\sec x|.$$

For either $\sec x < 0$ or $\sec x > 0$, we obtain

$$\frac{1}{\cos x}\frac{dy}{dx} + y\frac{\sin x}{\cos^2 x} = \frac{1}{\cos^2 x},$$

or,

$$\frac{d}{dx}\left\{\frac{y}{\cos x}\right\} = \sec^2 x.$$

Integration now yields

$$\frac{y}{\cos x} = \tan x + C,$$

and hence

$$y(x) = \sin x + C \cos x.$$

■

EXERCISES 16.3

In Exercises 1–12 find a general solution for the differential equation.

1. $\dfrac{dy}{dx} + 2xy = 4x$

2. $\dfrac{dy}{dx} + \dfrac{2}{x}y = 6x^3$

3. $(2y - x)\,dx + dy = 0$

4. $\dfrac{dy}{dx} + y \cot x = 5e^{\cos x}$

5. $(x^2 + 2xy)\,dx + (x^2 + 1)\,dy = 0$

6. $(x + 1)\dfrac{dy}{dx} - 2y = 2(x + 1)$

7. $\dfrac{1}{x}\dfrac{dy}{dx} - \dfrac{y}{x^2} = \dfrac{1}{x^3}$

8. $(y + e^{2x})\,dx = dy$

9. $\dfrac{dy}{dx} + y = 2\cos x$

10. $x^3\dfrac{dy}{dx} + (2 - 3x^2)y = x^3$

11. $\dfrac{dy}{dx} + \dfrac{y}{x \ln x} = x^2$

12. $(-2y \cot 2x - 1 + 2x \cot 2x + 2 \csc 2x)\,dx + dy = 0$

In Exercises 13–15 solve the differential equation.

13. $\dfrac{dy}{dx} + 3x^2 y = x^2,\ y(1) = 2$

14. $(-e^x \sin x + y)\,dx + dy = 0,\ y(0) = -1$

15. $\dfrac{dy}{dx} + \dfrac{x^3 y}{x^4 + 1} = x^7,\ y(0) = 1$

16. Find a general solution for the differential equation $(y^3 - x)\,dy = y\,dx$.

17. A differential equation of the form

$$\frac{dy}{dx} + P(x)y = y^n Q(x)$$

is called a *Bernoulli equation*. Show that the change of dependent variable $z = y^{1-n}$ gives

$$\frac{dz}{dx} + (1 - n)Pz = (1 - n)Q,$$

a linear first-order equation in $z(x)$.

In Exercises 18–22 use the change of variable in Exercise 17 to find a general solution for the differential equation.

18. $\dfrac{dy}{dx} + y = y^2 e^x$

19. $\dfrac{dy}{dx} + \dfrac{y}{x} = \dfrac{y^2}{x^2}$

20. $\dfrac{dy}{dx} - y + (x^2 + 2x)y^2 = 0$

21. $x\,dy + y\,dx = x^3 y^5\,dx$

22. $\dfrac{dy}{dx} + y \tan x = y^4 \sin x$

23. Repeat Exercise 22 in Section 16.2 if the glucose is added at a rate $R(t)$ that is a function of time t.

24. A tank has 100 L of solution containing 4 kg of sugar. A mixture with 10 g of sugar per litre of solution is added at a rate of 200 mL per minute. At the same time, 100 mL of well-stirred mixture are removed each minute. Find the amount of sugar in the tank as a function of time.

25. Repeat Exercise 24 if 300 mL of mixture are removed each minute.

26. A tank originally contains 1000 L of water in which 5 kg of salt has been dissolved. A mixture containing 2 kg of salt for each 100 L of solution is added to the tank at 10 mL/s. At the same time, the well-stirred mixture in the tank is removed at the rate of 5 mL/s. Find the amount of salt in the tank as a function of time. Sketch a graph of the function.

27. Repeat Exercise 26 if the mixture is removed at 10 mL/s. What is the limit of the amount of salt in the tank for large time?

28. Repeat Exercise 26 if the mixture is removed at 20 mL/s.

29. **(a)** The current I in the LR-circuit in Figure 16.3 must satisfy the differential equation

$$L\frac{dI}{dt} + RI = E.$$

If $E(t) = E_0 \sin(\omega t)$, $t \geq 0$, where E_0 and ω are constants, solve this differential equation for $I(t)$, and show that the solution can be written in the form

$$I(t) = Ae^{-Rt/L} + \frac{E_0}{Z}\sin(\omega t - \phi),$$

where A is an arbitrary constant and

$$Z = \sqrt{R^2 + \omega^2 L^2}, \qquad \phi = \mathrm{Tan}^{-1}\left(\frac{\omega L}{R}\right).$$

(b) What is the value of A if the current in the circuit at time $t = 0$ when $E(t)$ is connected is I_0?

FIGURE 16.3

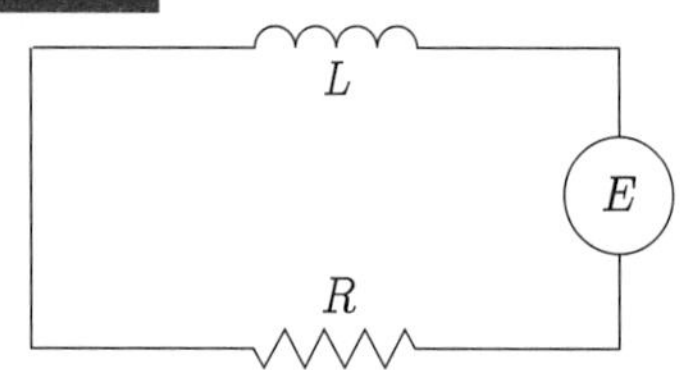

30. **(a)** The current I in the RC-circuit in Figure 16.4 must satisfy the differential equation

$$R\frac{dI}{dt} + \frac{I}{C} = \frac{dE}{dt}.$$

If $E(t)$ is as in Exercise 29, show that the solution can be written in the form

$$I(t) = Ae^{-t/(RC)} + \frac{E_0}{Z}\sin(\omega t - \phi),$$

where A is an arbitrary constant and

$$Z = \sqrt{R^2 + \frac{1}{\omega^2 C^2}}, \qquad \phi = \mathrm{Tan}^{-1}\left(-\frac{1}{\omega CR}\right).$$

(b) What is the value of A if the current in the circuit at time $t = 0$ when $E(t)$ is connected is I_0?

FIGURE 16.4

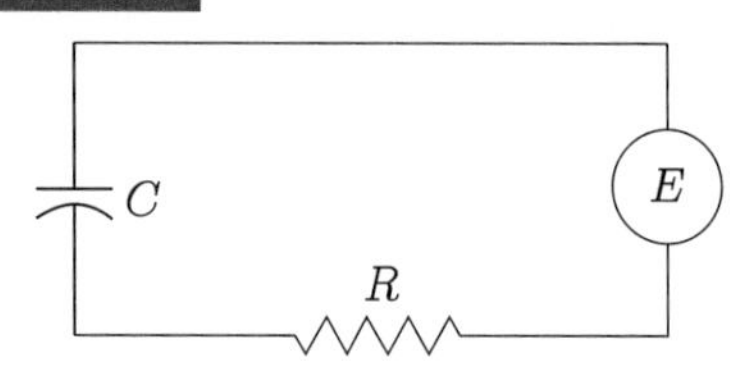

31. Repeat Exercise 21 in Section 16.2 given that the diver descends slowly to the bottom. Assume that his descent is at a constant rate over a time interval of length T. Assume also that maximum pressure $\overline{P}$ is proportional to depth below the surface.

SECTION 16.4

Second-Order Equations Reducible to Two First-Order Equations

We now consider two types of second-order differential equations that can be reduced to a pair of first-order equations.

Type I: Dependent Variable Missing

If a second-order differential equation in $y(x)$ is explicitly independent of y, then it is of the form

$$F(x, y', y'') = 0. \tag{16.26}$$

In such a case we set

$$v = \frac{dy}{dx} \quad \text{and} \quad \frac{dv}{dx} = \frac{d^2y}{dx^2}. \tag{16.27}$$

If we substitute these into 16.26, we obtain

$$F(x, v, v') = 0,$$

a first-order differential equation in $v(x)$. If we can solve this equation for $v(x)$, we can then integrate $dy/dx = v(x)$ for $y(x)$.

EXAMPLE 16.6

Find general solutions for the following differential equations:

(a) $xy'' - y' = 0$

(b) $y'' = (y')^2$

SOLUTION

(a) Since y is explicitly missing, we substitute $v = y'$ and $v' = y''$:

$$x\frac{dv}{dx} - v = 0.$$

Variables are now separable,

$$\frac{dv}{v} = \frac{dx}{x},$$

and a solution for $v(x)$ is

$$\ln|v| = \ln|x| + C \quad \text{or} \quad v = Dx \quad (D = \pm e^C).$$

Because $v = dy/dx$,

$$\frac{dy}{dx} = Dx,$$

and we can integrate for

$$y(x) = \frac{D}{2}x^2 + E$$
$$= Fx^2 + E \quad (F = D/2).$$

(b) Since y is explicitly missing, we substitute $v = y'$ and $v' = y''$ to get

$$\frac{dv}{dx} = v^2.$$

Variables are again separable,

$$\frac{dv}{v^2} = dx,$$

and a solution for $v(x)$ is

$$-\frac{1}{v} = x + C.$$

Consequently,

$$v = \frac{dy}{dx} = -\frac{1}{x + C}.$$

Integration now yields

$$y(x) = -\ln|x + C| + D.$$

■

Type II: Independent Variable Missing

If a second-order differential equation in $y(x)$ is explicitly independent of x, then it is of the form

$$F(y, y', y'') = 0. \tag{16.28}$$

In such a case we set

$$\frac{dy}{dx} = v \quad \text{and} \quad \frac{d^2y}{dx^2} = \frac{dv}{dx} = \frac{dv}{dy}\frac{dy}{dx} = v\frac{dv}{dy}. \tag{16.29}$$

When we substitute these into 16.28, we obtain

$$F\left(y, v, v\frac{dv}{dy}\right) = 0,$$

a first-order differential equation in $v(y)$. If we can solve this equation for $v(y)$, we can separate $dy/dx = v(y)$ and integrate for $y(x)$.

EXAMPLE 16.7

Find general solutions for the following differential equations:

(a) $yy'' + (y')^2 = 1$

(b) $y'' = (y')^2$

SOLUTION

(a) Since x is explicitly missing, we substitute

$$y' = v \quad \text{and} \quad y'' = \frac{dv}{dx} = \frac{dv}{dy}\frac{dy}{dx} = v\frac{dv}{dy}$$

to get

$$yv\frac{dv}{dy} + v^2 = 1.$$

If variables are separated, we have

$$\frac{v\,dv}{v^2 - 1} = -\frac{dy}{y},$$

and a general solution for $v(y)$ is

$$\tfrac{1}{2}\ln|v^2 - 1| = -\ln|y| + C.$$

Thus,

$$|v^2 - 1| = \frac{e^{2C}}{y^2},$$

from which we have

$$\frac{dy}{dx} = v = \frac{\pm\sqrt{D + y^2}}{y} \quad (D = \pm e^{2C}).$$

We separate variables again to get

$$\frac{y\,dy}{\sqrt{y^2 + D}} = \pm dx,$$

and obtain an implicit definition of the solution $y(x)$,

$$\sqrt{y^2 + D} = \pm x + E.$$

(b) Since x is explicitly missing, we again substitute from 16.29 to obtain

$$v\frac{dv}{dy} = v^2.$$

If variables are separated,

$$\frac{dv}{v} = dy,$$

and a solution for $v(y)$ is

$$\ln|v| = y + C.$$

Thus,

$$|v| = e^{y+C}$$

and

$$\frac{dy}{dx} = v = De^y \quad (D = \pm e^C).$$

We separate variables again,

$$e^{-y}\,dy = D\,dx,$$

and find an implicit definition of the solution $y(x)$:

$$-e^{-y} = Dx + E.$$

■

In each of Examples 16.6 and 16.7 we solved the differential equation $y'' = (y')^2$ since both the independent variable x and the dependent variable y are missing. Although the solutions appear different, each is easily derivable from the other.

Let us summarize the results of this section. Substitutions 16.27 for differential equation 16.26 with the dependent variable missing, replace the second-order differential equation with two first-order equations: a first-order equation in $v(x)$, followed by a first-order equation in $y(x)$. Contrast this with the method for equation 16.28 with the independent variable missing. Substitutions 16.29 again replace the second-order equation with two first-order equations. However, the first first-order equation is in $v(y)$, so that for this equation, y is the independent variable rather than the dependent variable. The second first-order equation is again one for $y(x)$.

EXERCISES 16.4

In Exercises 1–10 find a general solution for the differential equation.

1. $xy'' + y' = 4x$

2. $2yy'' = 1 + (y')^2$

3. $y'' = y' + 2x$

4. $x^2 y'' = (y')^2$

5. $y'' \sin x + y' \cos x = \sin x$

6. $y'' = [1 + (y')^2]^{3/2}$

7. $y'' + 4y = 0$

8. $y'' = yy'$

9. $y'' + (y')^2 = 1$

10. $(y'')^2 = 1 + (y')^2$

11. When a flexible cable of uniform mass per unit length hangs between two fixed points A and B (Figure 16.5), the differential equation that describes the shape of the cable is

$$\frac{d^2 y}{dx^2} = k\sqrt{1 + \left(\frac{dy}{dx}\right)^2},$$

where k is a constant. Solve this equation for the curve $y = y(x)$.

FIGURE 16.5

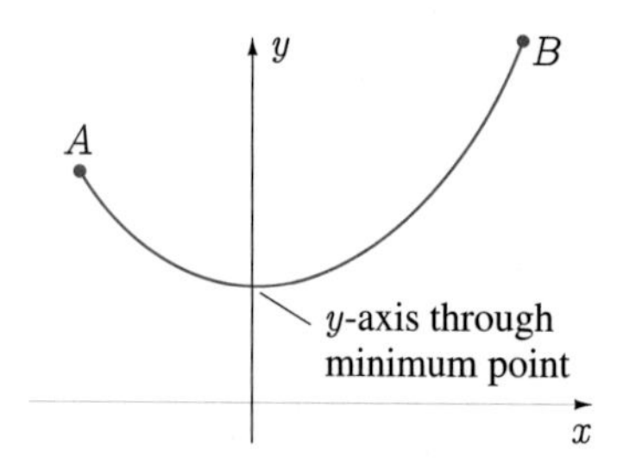

12. A hawk at position $(L, 0)$ (Figure 16.6) spots a pigeon at the origin flying with speed v in the positive y-direction. The hawk immediately takes off after the pigeon with speed $V > v$, always heading directly toward the pigeon. After time t, the pigeon is at position $P(0, vt)$. If the equation of the pursuit curve of the hawk is $y = y(x)$, then during time t, the hawk travels distance Vt along this curve. But distance along this curve can be calculated by means of the definite integral

$$\int_x^L \sqrt{1 + \left(\frac{dy}{dx}\right)^2}\, dx.$$

(a) Show that $y(x)$ must satisfy the integro–differential equation

$$x\frac{dy}{dx} - y = \frac{v}{V}\int_L^x \sqrt{1 + \left(\frac{dy}{dx}\right)^2}\, dx.$$

(b) Differentiate this equation to obtain the second-order differential equation

$$x\frac{d^2 y}{dx^2} = \frac{v}{V}\sqrt{1 + \left(\frac{dy}{dx}\right)^2}.$$

(c) Solve this differential equation for the pursuit curve of the hawk.

(d) Show that the hawk catches the pigeon a distance $vVL/(V^2 - v^2)$ up the y-axis.

FIGURE 16.6

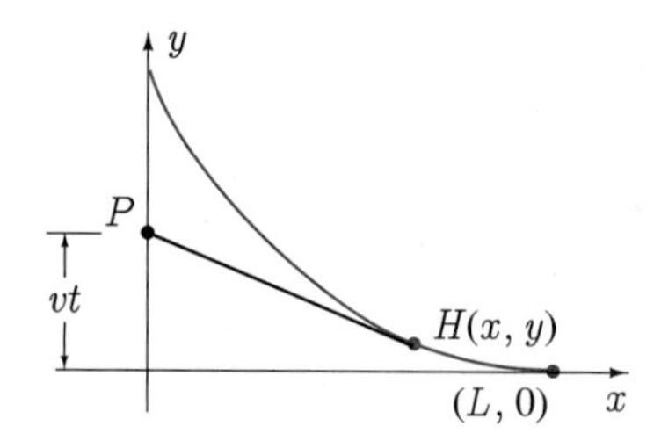

SECTION 16.5

Newtonian Mechanics

One of the most important applications of differential equations is in the study of moving particles and objects. In classical mechanics, motion is governed by Newton's second law 12.104, which states that when an object of constant mass m is subjected to a force $\mathbf{F}$, the resultant acceleration is described by

$$\mathbf{F} = m\mathbf{a}. \tag{16.30}$$

When $\mathbf{F}$ is given, this is an algebraic equation giving acceleration $\mathbf{a}$. If we substitute $\mathbf{a} = d\mathbf{v}/dt$, we obtain a first-order differential equation

$$\mathbf{F} = m\frac{d\mathbf{v}}{dt} \tag{16.31}$$

for velocity $\mathbf{v}$ as a function of time t. If we substitute $\mathbf{a} = d^2\mathbf{r}/dt^2$, we obtain a second-order differential equation for position $\mathbf{r}$ as a function of time:

$$\mathbf{F} = m\frac{d^2\mathbf{r}}{dt^2}. \tag{16.32}$$

In practice it is seldom this simple. Often $\mathbf{F}$ is not given as a function of time, but as a function of position, or velocity, or some other variable. In such cases we may have to change the dependent or independent variable, or both, in order to solve the differential equation.

In this section we are concerned with motion in one direction only. If $r(t), v(t), a(t)$, and F are the components of $\mathbf{r}(t), \mathbf{v}(t), \mathbf{a}(t)$, and $\mathbf{F}$ in this direction, then equations 16.30–16.32 take the forms

$$F = ma, \tag{16.33}$$

$$F = m\frac{dv}{dt}, \tag{16.34}$$

$$F = m\frac{d^2r}{dt^2}, \tag{16.35}$$

respectively.

EXAMPLE 16.8

A paratrooper and his parachute have mass 100 kg. At the instant his parachute opens, the paratrooper is traveling vertically downward at 10 m/s. Air resistance on the chute varies directly as the instantaneous velocity and is 800 N when the paratrooper's velocity is 12 m/s. Find the position and velocity of the paratrooper as functions of time.

FIGURE 16.7

$t = 0$, $y = 0$, $v = 10$

y

SOLUTION Let us measure y as positive downward, taking $y = 0$ and time $t = 0$ at the instant the paratrooper opens his parachute (Figure 16.7). If $F_{\mathbf{a}}$ is the vertical component of the force of air resistance, then

$$F_{\mathbf{a}} = -kv.$$

Since $F_{\mathbf{a}} = -800$ when $v = 12$, it follows that $-800 = -12k$ or $k = 200/3$. Since the total force on the paratrooper during the fall has component $100g - 200v/3$

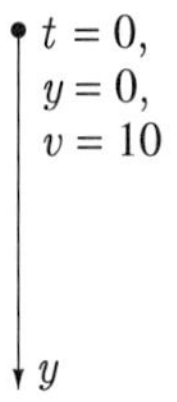

$(g = 9.81)$, Newton's second law gives

$$100\frac{dv}{dt} = 100g - \frac{200}{3}v, \quad t \geq 0.$$

Note that the force on the right is not a function of t so that the equation is not immediately integrable. It is however separable,

$$\frac{dv}{2v - 3g} = -\frac{dt}{3},$$

and a general solution for $v(t)$ is defined implicitly by

$$\frac{1}{2}\ln|2v - 3g| = -\frac{t}{3} + C.$$

An explicit solution is found by solving this equation for v:

$$v(t) = \frac{3g}{2} + De^{-2t/3} \quad (D = \pm e^{2C}/2).$$

Since $v = 10$ when $t = 0$,

$$10 = \frac{3g}{2} + D \quad \text{or} \quad D = 10 - \frac{3g}{2}.$$

The velocity of the paratrooper as a function of time is therefore

$$v(t) = \frac{3g}{2} + \left(10 - \frac{3g}{2}\right)e^{-2t/3}.$$

Because $v(t) = dy/dt$, we set

$$\frac{dy}{dt} = \frac{3g}{2} + \left(10 - \frac{3g}{2}\right)e^{-2t/3},$$

and this equation is immediately integrable to

$$y(t) = \frac{3g}{2}t - \frac{3}{2}\left(10 - \frac{3g}{2}\right)e^{-2t/3} + E.$$

Since $y(0) = 0$,

$$0 = -\frac{3}{2}\left(10 - \frac{3g}{2}\right) + E \quad \text{or} \quad E = 15 - \frac{9g}{4}.$$

Consequently, the distance fallen by the paratrooper as a function of time is given by

$$y(t) = \frac{3g}{2}t - \left(15 - \frac{9g}{4}\right)e^{-2t/3} + \left(15 - \frac{9g}{4}\right) = 14.7t + 7.07(e^{-2t/3} - 1).$$

Note that the velocity of the paratrooper does not increase indefinitely. In fact, as time passes a limiting velocity is approached:

$$\lim_{t\to\infty} v(t) = \lim_{t\to\infty} \left\{ \frac{3g}{2} + \left(10 - \frac{3g}{2}\right) e^{-2t/3} \right\} = \frac{3g}{2} = 14.7 \text{ m/s}.$$

This is called the terminal velocity of the paratrooper; it is a direct result of the assumption that air resistance is proportional to instantaneous velocity. ■

Whenever we move an object such as the block in Figure 16.8 over a surface, there is resistance to the motion. This resistance, called **friction**, is due to the fact that the interface between the block and the surface is not smooth; each surface is inherently rough, and this roughness retards the motion of one surface over the other. In effect, a force opposing the motion of the block is created, and this force is called the **force of friction**. Many experiments have been performed to obtain a functional representation for this force. It turns out that when the block in Figure 16.8 slides along a horizontal surface, the magnitude of the force of friction opposing the motion is given by

$$|\mathbf{F}| = \mu m g, \tag{16.36}$$

where m is the mass of the block, g is the acceleration due to gravity, and μ is a constant called the **coefficient of kinetic friction**. In other words, the force of friction is directly proportional to the weight mg of the block. We caution the reader that this result is valid for the situation shown in Figure 16.8, but it may not be valid for other configurations (say perhaps for an inclined plane).

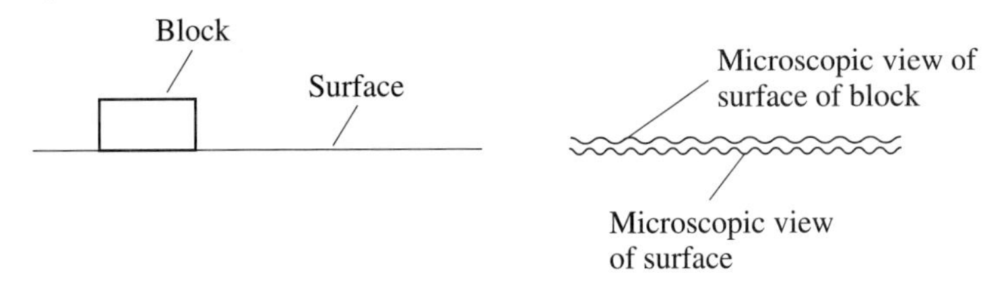

FIGURE 16.8

EXAMPLE 16.9

A block of mass 2 kg is given initial speed 5 m/s along a horizontal surface. If the coefficient of kinetic friction between the block and surface is $\mu = 0.25$, how far does the block slide before stopping?

FIGURE 16.9

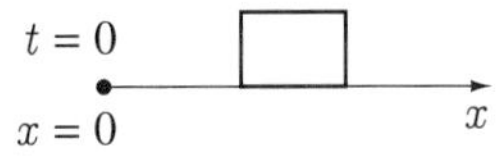

$t = 0$

$x = 0$ x

SOLUTION Let us measure x as positive in the direction of motion (Figure 16.9), taking $x = 0$ and $t = 0$ at the instant the block is released. The x-component of the force of friction on the block is

$$F = -0.25(2)g = -\frac{g}{2} \quad (g = 9.81).$$

According to Newton's second law,

$$2\frac{dv}{dt} = -\frac{g}{2},$$

from which we get

$$v(t) = -\frac{g}{4}t + C.$$

Since $v(0) = 5$, it follows that $C = 5$, and

$$v(t) = -\frac{g}{4}t + 5 .$$

But $v = dx/dt$, and hence

$$\frac{dx}{dt} = -\frac{g}{4}t + 5.$$

Integration gives

$$x(t) = -\frac{g}{8}t^2 + 5t + D.$$

Because we chose $x = 0$ at time $t = 0$, D must also be zero, and

$$x(t) = -\frac{g}{8}t^2 + 5t .$$

The block comes to rest when $v = 0$; i.e., when

$$-\frac{g}{4}t + 5 = 0, \text{ or, } t = \frac{20}{g}.$$

The position of the block at this time is

$$x = -\frac{g}{8}\left(\frac{20}{g}\right)^2 + 5\left(\frac{20}{g}\right) = 5.10.$$

The block therefore slides 5.10 m before stopping. ■

EXERCISES 16.5

1. A car of mass 1500 kg starts from rest at an intersection. The engine exerts a constant force of 3000 N, and air friction causes a resistive force whose magnitude in newtons is equal to the speed of the car in metres per second. Find the speed of the car and its distance from the intersection after 10 s.

2. You are called on as an expert to testify in a traffic accident hearing. The question concerns the speed of a car that made an emergency stop with brakes locked and wheels sliding. The skid mark on the road measured 9 m. If you assume that the coefficient of kinetic friction between the tires and road was less than one, what can you say about the speed of the car before the brakes were applied? Are you testifying for the prosecution or the defence?

3. A boat and its contents have mass 250 kg. Water exerts a resistive force on the motion of the boat that is proportional to the instantaneous speed of the boat and is 200 N when the speed is 30 km/h.
   (a) If the boat starts from rest and the engine exerts a constant force of 250 N in the direction of motion, find the speed of the boat as a function of time.
   (b) What is the limiting speed of the boat?

4. **(a)** In Example 16.8 it was necessary to solve the equation
   $$\frac{1}{2}\ln|2v - 3g| = -\frac{t}{3} + C$$
   for $v = v(t)$ and use the condition $v(0) = 10$ to evaluate C. We chose first to solve for $v(t)$ and then to evaluate the constant. Show the details of this analysis.
   (b) Instead, first use the condition $v(0) = 10$ to evaluate C, and then solve the equation for $v(t)$.

5. A body of weight W falls from rest under gravity. It is acted on by a resistive force that is proportional to the square root of the speed at any instant. If the magnitude of this resistive force is F^* when the speed is v^*, find an expression for its terminal speed.

6. A spring (with constant k) is attached on one end to a wall and on the other end to a mass M (Figure 16.10). The mass is set into motion along the x-axis by pulling it a distance x_0 to the right of the position it would occupy were the spring unstretched and given speed

v_0 to the left. During the subsequent motion, there is a frictional force between M and the horizontal surface with coefficient of kinetic friction equal to μ.

(a) Show that the differential equation describing the motion of M is

$$M\frac{d^2x}{dt^2} = -kx + \mu Mg, \quad x(0) = x_0, \quad v(0) = -v_0,$$

if we take $t = 0$ at the instant that motion is initiated. When is this equation valid?

(b) Since t is explicitly missing from the equation in (a), show that it can be rewritten in the form

$$Mv\frac{dv}{dx} = -kx + \mu Mg, \quad v(x_0) = -v_0,$$

and that therefore

$$\frac{k}{2}(x_0^2 - x^2) = \frac{M}{2}(v^2 - v_0^2) + \mu Mg(x_0 - x).$$

Interpret each of the terms in this equation physically.

(c) If x^* represents the position at which M comes to rest for the first time, use the equation in (b) to determine x^* as a function of μ, M, g, k, x_0, and v_0. Discuss the possibilities of x^* being positive, negative, and zero.

FIGURE 16.10

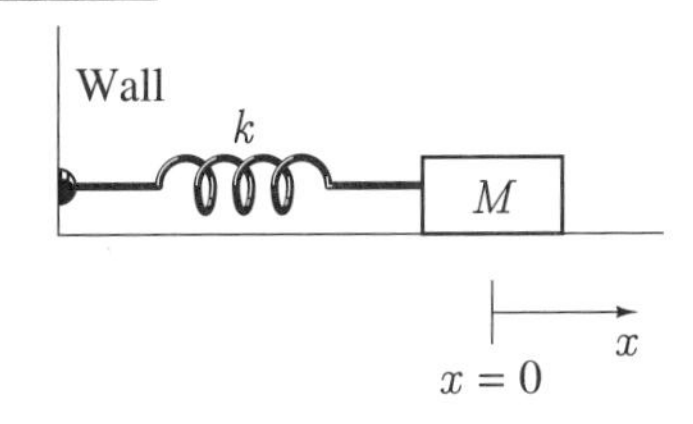

7. When a body falls in air, it is acted on by gravity and also by a force due to air resistance that is proportional to its instantaneous speed.

(a) If the body is initially projected upward, find its velocity as a function of time, and sketch a graph of the function.

(b) Repeat (a) if the body is projected downward with velocity less than its terminal velocity.

(c) Repeat (b) if the initial velocity is greater than its terminal velocity.

8. A 1-kg rock is thrown vertically upward with speed 20 m/s. Air resistance to its motion when measured in newtons has magnitude equal to one-tenth its speed in metres per second. Find the maximum height attained by the rock.

9. A stone of mass 100 g is thrown vertically upward with speed 20 m/s. Air exerts a resistive force on the stone proportional to its instantaneous speed, and has magnitude $1/10$ N when the speed of the stone is 10 m/s. Find an equation defining the time when the stone returns to its projection point. Solve this equation numerically, and compare with the time taken if air resistance is neglected.

10. **(a)** A 1-kg mass falls under the influence of gravity. It is also acted on by air resistance which is proportional to the square of its velocity and is 5 N when its velocity is 50 m/s. If the velocity of the mass has magnitude 20 m/s at time $t = 0$, find a formula for its velocity as a function of time.

(b) Sketch a graph of the velocity function. Is there a terminal velocity?

11. Repeat Exercise 10 if the initial velocity has magnitude 100 m/s.

12. For how long does the mass in Exercise 10 rise if it is thrown upward with velocity 20 m/s?

13. A mass m falls under the influence of gravity and air resistance which is proportional to the square of velocity. If the speed of the mass is v_0 at time $t = 0$, find a formula for its speed as a function of time.

14. **(a)** If the mass in Exercise 13 is thrown upward with speed v_0, find a formula for speed on its ascent. Is this formula also valid for its speed when it begins to fall?

(b) Find a formula for its height. How high does it rise?

15. A mass m slides from rest down a frictionless plane inclined at angle α to the horizontal. Find a formula for the time taken to travel a distance D down the plane. What is its speed at this time?

16. Find formulas for speed and distance travelled for the mass in Exercise 15 if air resistance proportional to velocity also acts on the mass.

17. In Example 7.33 of Section 7.9 we derived the escape velocity of a projectile from the earth's surface based on energy principles. In this exercise we obtain the same result using differential equations. When a projectile of mass m , fired from the earth's surface, is a distance r from the centre of the earth, the magnitude of the force of attraction on it is given by Newton's universal law of gravitation, $F = GMm/r^2$, where M is the mass of the earth and G is a constant. Use Newton's second law and a substitution corresponding to 16.29 to find the velocity of the projectile as a function of r. What minimum initial velocity guarantees that the projectile escapes the gravitational field of the earth?

18. A huge cannon fires a projectile with initial velocity v_0 directly toward the moon (Figure 16.11). When the projectile is a distance r above the earth's surface, the force of attraction of the earth on the projectile has magnitude

$$\frac{GMm}{(r+R)^2},$$

where G is a constant, R is the radius of the earth M is the mass of the earth, and m is the mass of the projectile. At this point the moon's gravitational attraction has magnitude

$$\frac{GM^*m}{(a+R^*-r)^2},$$

where M^* is the mass of the moon.

(a) Show that if only the above two forces are considered to act on the projectile, then the differential equation describing its motion is

$$\frac{d^2r}{dt^2} = -\frac{gR^2}{(r+R)^2} + \frac{g^*R^{*2}}{(a+R^*-r)^2},$$

where g and g^* are gravitational accelerations on the surfaces of the earth and the moon.

(b) Prove that the velocity of the projectile at a distance r above the surface of the earth is defined by

$$v^2 = \frac{2gR^2}{r+R} + \frac{2g^*R^{*2}}{a+R^*-r} + v_0^2 - 2gR - \frac{2g^*R^{*2}}{a+R^*}.$$

FIGURE 16.11

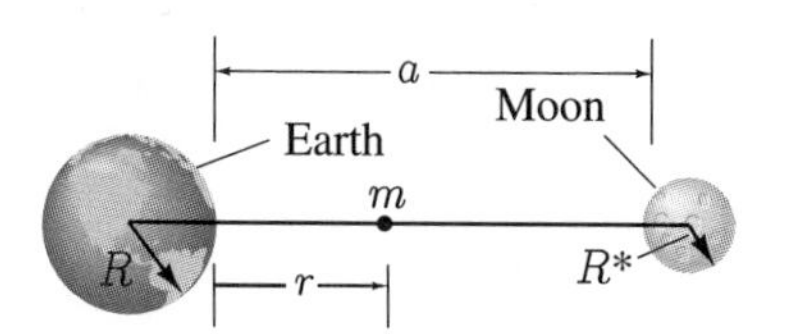

19. Newton's second law states that if an object of variable mass $m(t)$ is subjected to a force $\mathbf{F}(t)$, then

$$\frac{d}{dt}(m\mathbf{v}) = \mathbf{F},$$

where t is time and $\mathbf{v}$ is the velocity of the object. A uniform chain of length 3 m and mass 6 kg is held by one end so that the other end just touches the floor. If the chain is released, find the velocity of the falling chain as a function of the length of chain still falling. How fast does the end hit the floor?

SECTION 16.6

Population Dynamics

The rate of growth of a country's population depends on many factors: the birth rate, death rate, immigration rate, size of the country, availability of food, state of technology, natural disasters, etc. Developing a mathematical formula that would account for all possible factors would be impossible. Some factors, on the other hand, are much more important to the growth of a population than others. Perhaps we could develop a mathematical model that would initially account for the important factors, and that could be refined at a later stage to incorporate minor factors. Our problem, then, is to develop a model that predicts populations of countries with some reasonable amount of accuracy.

On the basis of the census figures of a country such as the United States, it is possible to calculate the number of births per year in the populaton. If we divide this by the existing size of the population, we get what is called the relative birth rate b of the population (or the individual birth rate). It is an average figure, but it is found that for many populations the relative birth rate is reasonably constant over long periods of time. In a similar way, it is possible to define a relative death rate d (death due to natural causes), and the constant $k = b - d$ is called the relative growth rate of the population. It is the increase in population per individual per year that can be attributed solely to births and deaths. On the other hand, if $N = N(t)$ represents the size of the population at any time t (in years), then

$$\frac{1}{N}\frac{dN}{dt}$$

is the total change in population per individual per year, and should therefore account for all factors that affect the size of the population. If we were to decide that the most important factor in determining the size of a population were its birth and death rates, we would write

$$\frac{1}{N}\frac{dN}{dt} = k,$$

or,

$$\frac{dN}{dt} = kN. \tag{16.37}$$

This is a mathematical model (called the **Malthusian model**) that attempts to describe the size of a population in terms only of its birth and death rates.

Because 16.37 is a separable differential equation

$$\frac{1}{N}dN = k\,dt,$$

we obtain a general solution

$$\ln|N| = kt + D.$$

Because N is never negative, we can state that

$$N = Ce^{kt} \quad (C = e^{D}). \tag{16.38}$$

In Table 16.1 and Figure 16.12 we show approximate census figures for the United States from 1790 to 1970. Note that the population figure in 1940 is uncharacteristic of the previous growth from 1900 to 1930 (and history makes the reason clear). Suppose we attempt to fit exponential curve 16.38 to the population between 1790 and 1860. One way to do this is to demand that $N(1790) = 3.93$ and $N(1860) = 31.44$:

$$3.93 = Ce^{1790\,k}, \quad 31.44 = Ce^{1860\,k}.$$

The solution of these equations for C and k is $C = 3.17 \times 10^{-23}$, $k = 0.0297$. Consequently,

$$N = 3.17 \times 10^{-23}e^{0.0297t},$$

and this curve is shown in Figure 16.12. It is indeed an excellent description of the United States population in the years between 1790 and 1860.

TABLE 16.1

Year	1790	1800	1810	1820	1830	1840	1850	1860	1870	1880
Population (millions)	3.93	5.31	7.24	9.64	12.87	17.07	23.19	31.44	38.56	50.16

Year	1890	1900	1910	1920	1930	1940	1950	1960	1970
Population (millions)	62.95	76.00	91.97	105.71	122.78	131.67	150.70	179.32	203.19

FIGURE 16.12

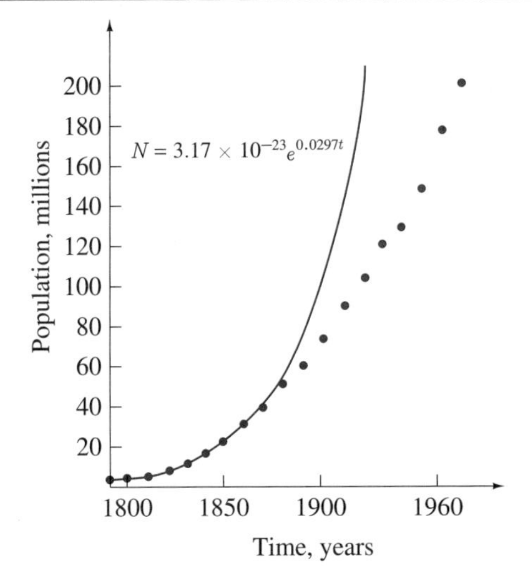

It is equally obvious, however, that this particular exponential function is not an adequate representation of the population beyond 1860. In fact no exponential function can describe the census points in Table 16.1. This implies that after 1860 other factors besides simply pure birth and death processes become important in determining the size of the population. One possible explanation is that because 16.37 predicts exponential growth, it must assume that there are sufficient resources (area, food, water, etc.) to sustain any level of population. This is unrealistic in the long term because there is a limit to the size of the population that any environment can support. This limit, called the **carrying capacity** C of the environment, is such that as the size of the population draws closer to C its growth rate must slow down. One way to incorporate this into model 16.37 is to set

$$\frac{dN}{dt} = kN\left(1 - \frac{N}{C}\right) \tag{16.39}$$

(called the **logistic model**). For small N, $1 - N/C$ is approximately equal to one, and N will experience exponential growth, but as N approaches C, this factor diminishes and the growth rate decreases.

To solve this differential equation, we again separate variables,

$$k\,dt = \frac{C\,dN}{N(C-N)} = \left\{\frac{1}{N} + \frac{1}{C-N}\right\} dN,$$

and obtain a general solution

$$kt + E = \ln N - \ln(C - N).$$

We now solve this equation for $N(t)$ by first exponentiating to obtain

$$\frac{N}{C-N} = De^{kt} \quad (D = e^E),$$

and cross-multiplying to get

$$N = (C - N)De^{kt}.$$

Thus,

$$N(t) = \frac{CDe^{kt}}{1 + De^{kt}} = \frac{C}{1 + Fe^{-kt}} \quad (F = 1/D), \tag{16.40}$$

and this is called the **logistic growth function** (or **logistic curve**). If we use the census figures of 1790, 1880, and 1970 to determine the three constants F, k, and C, we obtain

$$N(t) = \frac{257}{1 + 3.54 \times 10^{25} e^{-0.0305t}}.$$

This function and the census figures of Table 16.1 are shown in Figure 16.13. Note that this logistic function predicts an upper limit of 257 million people for the United States, not very realistic in light of today's population. It has resulted from assuming that the population of the United States can reasonably be described by a logistic function that passes through the census points in 1790, 1880, and 1970. This implies, then, that such a representation is still not an adequate description of the United States population; other factors besides birth and death rates and carrying capacity must be incorporated. Some of these are introduced in the exercises.

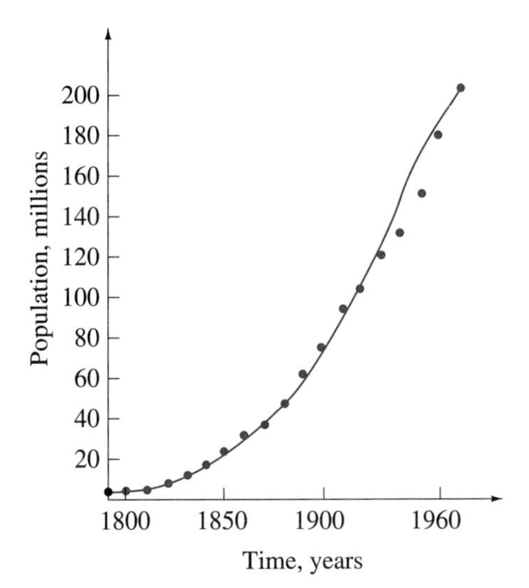

FIGURE 16.13

EXERCISES 16.6

1. Bacteria in a culture increase at a rate proportional to the number present. If the original number of bacteria increases by 50% in one-half hour, how long does it take to quintuple the original number?

2. Two rodents begin a colony. During the next ten months they are permitted to reproduce and their numbers are recorded in the following table.

Time t (months)	0	2	5	7	10
Number N of rodents	2	5	15	35	110

   (a) Find a formula for $N(t)$ based on the Malthusian model and use the conditions $N(0) = 2$ and $N(2) = 5$ to determine the unknown constants in the function.

   (b) Repeat (a) but use the conditions $N(0) = 2$ and $N(10) = 110$ to evaluate the constants.

   (c) Sketch both functions in (a) and (b) and the data points on the same graph.

3. If we modify the Malthusian model for predicting the population of a country to incorporate a constant immigration rate of λ individuals per unit time, then the model becomes

$$\frac{dN}{dt} = kN + \lambda.$$

   Solve the differential equation for $N(t)$.

4. Draw the logistic curve 16.40 indicating any relative extrema and points of inflection. Assume that $N(0) = N_0 < C/2$.

5. An initial population of 100 inhabitants enters an area with carrying capacity 100 000. In the first year, the population increases to 120. Assume that the population follows logistic growth.

   (a) Determine the population as an explicit function of time.

   (b) How long does it take the population to reach 95 000?

6. **(a)** A country institutes an immigration policy that states that the immigration rate shall be proportional to the difference between the carrying capacity C and the present size N of the population. Show that a mathematical model to describe $N(t)$ is

$$\frac{dN}{dt} = -\frac{k}{C}(N - C)(N + C\lambda/k),$$

   where k is the relative growth rate of the population attributed only to birth and death rates, and λ is a positive constant.

   (b) Solve this differential equation, and show that

$$N(t) = \frac{C - (FC\lambda/k)e^{-(k+\lambda)t}}{1 + Fe^{-(k+\lambda)t}}.$$

7. **(a)** For some species, if the population drops below a certain level, the species will become extinct. Given that m is the minimum viable population for such a species, one way to incorporate this condition into the logistic model is to write

$$\frac{dN}{dt} = kN\left(1 - \frac{N}{C}\right)\left(1 - \frac{m}{N}\right).$$

   Solve this differential equation subject to the condition $N(0) = N_0$ to obtain

$$N(t) = \frac{[(C - N_0)/(N_0 - m)]me^{-[(C-m)/C]kt} + C}{1 + [(C - N_0)/(N_0 - m)]e^{-[(C-m)/C]kt}}.$$

(b) Use the result in (a) to show that if $N_0 < m$, then extinction must occur.

Logistic model 16.39 incorporates the term $1 - N/C$ in order to decrease dN/dt for N approaching C. In Exercises 8–10, we show that there are other ways to do this.

8. The Gompertz model multiplies the right-hand side of the Malthusian model by the factor e^{-lt} ($l > 0$ a constant):

$$\frac{dN}{dt} = kNe^{-lt}.$$

Solve this differential equation for $N(t)$, and sketch the curve $N = N(t)$. Does it predict an upper limit for $N(t)$?

9. The monomolecular model multiplies the right-hand side of the Malthusian model by the factor $be^{-kt}/(1-be^{-kt})$ ($0 < b < 1$ a constant):

$$\frac{dN}{dt} = kN\frac{be^{-kt}}{1 - be^{-kt}}.$$

Solve this differential equation for $N(t)$, and sketch the curve $N = N(t)$. Does it predict an upper limit for $N(t)$?

10. The Von Bertalanffy model is much like the monomolecular model,

$$\frac{dN}{dt} = kN\frac{3be^{-kt}}{1 - be^{-kt}}.$$

Solve this differential equation for $N(t)$, and sketch the curve $N = N(t)$. Does it predict an upper limit for $N(t)$?

11. Verify that if the constants C, k, and F in the logistic function 16.40 are evaluated by using the census figures of 1790, 1880, and 1970 for the United States, then $C = 257$, $F = 3.54 \times 10^{25}$, and $k = 0.0305$.

SECTION 16.7

Linear Differential Equations

Throughout Chapters 5, 8, and 9 and Section 16.1–16.6, we have stressed the use of differential equations in solving applied problems. We have considered examples from such diverse areas as engineering, geometry, ecology, and psychology, hoping thereby to illustrate how valuable differential equations can be in modeling physical situations mathematically. Perhaps the most important type of differential equation is the **linear differential equation.**

Definition 16.4

A differential equation of the form

$$a_0(x)\frac{d^n y}{dx^n} + a_1(x)\frac{d^{n-1} y}{dx^{n-1}} + a_2(x)\frac{d^{n-2} y}{dx^{n-2}} + \cdots + a_{n-1}(x)\frac{dy}{dx} + a_n(x)y = F(x), \tag{16.41}$$

where $a_0(x) \not\equiv 0$, is called a **linear *n*th-order differential equation.**

Note in particular that none of the derivatives of $y(x)$ are multiplied together, nor are they squared or cubed or taken to any other power, nor do they appear as the argument of any transcendental function. All we see is a function of x multiplying y, plus a function of x multiplying the first derivative of y, plus a function of x multiplying the second derivative of y, and so on.

If $n = 1$, equation 16.41 reduces to

$$a_0(x)\frac{dy}{dx} + a_1(x)y = F(x),$$

and at any point at which $a_0(x) \neq 0$, we can divide to obtain

$$\frac{dy}{dx} + \frac{a_1(x)}{a_0(x)}y = \frac{F(x)}{a_0(x)}.$$

If we set $P(x) = a_1(x)/a_0(x)$ and $Q(x) = F(x)/a_0(x)$, we have

$$\frac{dy}{dx} + P(x)y = Q(x);$$

i.e., every linear first-order differential equation can be expressed in this form. We discussed equations of this type in Section 16.3, where it was shown that such equations have a general solution

$$y(x) = e^{-\int P(x)\,dx}\left\{\int Q(x)e^{\int P(x)\,dx}dx + C\right\}.$$

In other words, we already know how to solve linear first-order differential equations, and therefore our discussion in the next five sections is directed primarily at second- and higher-order equations. Keep in mind, however, that all results are also valid for first-order linear equations.

Because equation 16.41 is so cumbersome, we introduce notation to simplify its representation. In particular, if we use the notation $D = d/dx, D^2 = d^2/dx^2$, etc., we can write

$$a_0(x)D^n y + a_1(x)D^{n-1}y + a_2(x)D^{n-2}y + \cdots + a_{n-1}(x)Dy + a_n(x)y = F(x), \quad (16.42)$$

or,

$$\{a_0(x)D^n + a_1(x)D^{n-1} + a_2(x)D^{n-2} + \cdots + a_{n-1}(x)D + a_n(x)\}y = F(x). \quad (16.43)$$

The quantity in braces is called a **differential operator**; it operates on whatever follows it—in this case, y. It is a "differential" operator because it operates by taking derivatives. Because the operator involves only x's and D's, we denote it by

$$\phi(x, D) = a_0(x)D^n + a_1(x)D^{n-1} + a_2(x)D^{n-2} + \cdots + a_{n-1}(x)D + a_n(x). \quad (16.44)$$

The general linear nth-order differential equation can then be represented very simply by

$$\phi(x, D)y = F(x). \quad (16.45)$$

For example, the linear second-order differential equation

$$xy'' + y' + xy = 0$$

is called Bessel's differential equation of order zero. In operator notation we write

$$(xD^2 + D + x)y = 0,$$

or,

$$\phi(x, D)y = 0,$$

where $\phi(x, D) = xD^2 + D + x$.

For the differential equation

$$y'' + 2y' - 3y = e^{-x},$$

we write

$$\phi(D)y = e^{-x},$$

where $\phi(D) = D^2 + 2D - 3$.

We now indicate the meaning of the term "linear." Suppose that L is an operator that operates on each function $y(x)$ in some set S. For example, L might be the operation that multiplies each function by 5, or perhaps squares each function, or perhaps differentiates each function. It is said to be a linear operator if it satisfies the following definition.

Definition 16.5

An operator L is said to be **linear** on a set of functions S if for any two functions $y_1(x)$ and $y_2(x)$ in S, and any constant c,

$$L(y_1 + y_2) = Ly_1 + Ly_2, \tag{16.46a}$$

$$L(cy_1) = c(Ly_1). \tag{16.46b}$$

Many of the operations in calculus are therefore linear. For instance, taking limits is a linear operation, as is differentiation, antidifferentiation, and taking definite integrals. On the other hand, taking the square root of a positive function is not a linear operation, since

$$L(y_1 + y_2) = \sqrt{y_1 + y_2} \neq L(y_1) + L(y_2).$$

It is not difficult to show that the differential operator $\phi(x, D)$ in 16.44 is linear; that is,

$$\phi(x, D)(y_1 + y_2) = \phi(x, D)y_1 + \phi(x, D)y_2,$$

$$\phi(x, D)(cy_1) = c[\phi(x, D)y_1],$$

and because of this, differential equation 16.45 is also said to be linear. The following two differential equations are examples of equations that are not linear:

$$\frac{d^2y}{dx^2} + y^2 = x, \qquad \frac{d^2y}{dx^2}\frac{dy}{dx} = e^x.$$

In particular, if we substitute $y_1(x) + y_2(x)$ into the left-hand side of the first equation, we obtain

$$\frac{d^2}{dx^2}(y_1 + y_2) + (y_1 + y_2)^2 = \frac{d^2y_1}{dx^2} + \frac{d^2y_2}{dx^2} + y_1^2 + 2y_1y_2 + y_2^2.$$

If we substitute $y_1(x)$, then $y_2(x)$, and then add the results, we find a different expression:

$$\frac{d^2y_1}{dx^2} + y_1^2 + \frac{d^2y_2}{dx^2} + y_2^2.$$

EXERCISES 16.7

In Exercises 1–10 prove either that the operator L is linear or that it is not linear. In each case assume that the set S of functions on which L operates is the set of all functions on which L can operate. For instance, in Exercise 6, assume that S is the set of all functions $y(x)$ that have a first derivative dy/dx.

1. L multiplies functions $y(x)$ by 5

2. L multiplies functions $y(x)$ by $15x$

3. L adds the fixed function $z(x)$ to functions $y(x)$

4. L takes the limit of functions $y(x)$ as x approaches 3

5. L takes the limit of functions $y(x)$ as x approaches infinity

6. L takes the first derivative of functions $y(x)$ with respect to x

7. L takes the third derivative of functions $y(x)$ with respect to x

8. L takes the antiderivative of functions $y(x)$ with respect to x

9. L takes the definite integral of functions $y(x)$ with respect to x from $x = -1$ to $x = 4$

10. L takes the cube root of functions $y(x)$

In Exercises 11–20 determine whether the differential equation is linear or nonlinear. Write those equations that are linear in operator notation 16.45.

11. $2x\dfrac{d^2y}{dx^2} + x^3y = x^2 + 5$

12. $2x\dfrac{d^2y}{dx^2} + x^3y = x^2 + 5y$

13. $2x\dfrac{d^2y}{dx^2} + x^3y = x^2 + 5y^2$

14. $x\dfrac{d^3y}{dx^3} + 3x\dfrac{d^2y}{dx^2} - 2\dfrac{dy}{dx} + y = 10\sin x$

15. $x\dfrac{d^3y}{dx^3} + 3x\dfrac{d^2y}{dx^2} - 2\dfrac{dy}{dx} + y^2 = 10\sin x$

16. $y\dfrac{d^3y}{dx^3} + 3x\dfrac{d^2y}{dx^2} - 2\dfrac{dy}{dx} + y = 10\sin x$

17. $y'' - 3y' - 2y = 9\sec^2 x$

18. $yy'' + 3y' - 2y = e^x$

19. $\sqrt{1 + y'} + x^2 = 4$

20. $y'''' + y'' - y = \ln x$

21. The Laplace transform of a function $y(t)$ is defined as the function $L(y) = \int_0^\infty e^{-st}y(t)\,dt$, provided the improper integral converges. Show that if S is the set of all functions that have a Laplace transform, then the operation of taking the Laplace transform is linear on S.

22. The finite Fourier cosine transform of a function $y(x)$ is defined as $L(y) = \int_0^{2\pi} y(x)\cos nx\,dx$, where n is a nonnegative integer, provided the definite integral exists. Show that if S is the set of all functions that have a finite Fourier cosine transform, then the operation of taking the transform is linear on S.

SECTION 16.8

Homogeneous Linear Differential Equations

Two classes of linear differential equations present themselves: those for which $F(x) \equiv 0$, and those for which $F(x) \not\equiv 0$. In this section and the next we consider equations for which $F(x) \equiv 0$; we will discuss the more difficult class, in which $F(x) \not\equiv 0$, in Section 16.10. First let us name each of these classes of linear differential equations.

Definition 16.6

A linear differential equation $\phi(x, D)y = F(x)$ is said to be **homogeneous** if $F(x) \equiv 0$, and **nonhomogeneous** otherwise.

The meaning of "homogeneous" to describe a property of linear differential equation in this definition is totally different from the meaning in Exercise 16.2-26.

The fundamental idea behind the solution of all linear differential equations is the following theorem.

Theorem 16.1 *(Superposition principle). If $y_1(x), y_2(x), \ldots, y_m(x)$ are solutions of a homogeneous linear differential equation*

$$\phi(x, D)y = 0,$$

then so too is any linear combination thereof,

$$C_1 y_1(x) + C_2 y_2(x) + \cdots + C_m y_m(x)$$

(for arbitrary constants $C_1, C_2, \ldots, C_m$).

Proof The proof requires only linearity of the operator $\phi(x, D)$, for

$$\begin{aligned}\phi(x, D)[C_1 y_1(x) + \cdots + C_m y_m(x)] &= C_1[\phi(x, D)y_1] + \cdots + C_m[\phi(x, D)y_m] \\ &= 0 + 0 + \cdots + 0 = 0.\end{aligned}$$

Solutions of a linear differential equation that are linearly combined to produce other solutions are said to be superposed—consequently, the name "superposition principle" for Theorem 16.1. For example, it is straightforward to verify that $y_1(x) = e^{4x}$ and $y_2(x) = e^{-3x}$ are solutions of the homogeneous equation $y'' - y' + 12y = 0$. The superposition principle then states that for any constants C_1 and C_2, the function $y(x) = C_1 y_1(x) + C_2 y_2(x) = C_1 e^{4x} + C_2 e^{-3x}$ must also be a solution. But because $y(x)$ contains two arbitrary constants, it is not only a solution, it is a general solution of the differential equation.

Similarly, superposition of the three solutions e^x, xe^x, and $x^2 e^x$ of the linear differential equation $y''' - 3y'' + 3y' - y = 0$ gives a general solution $y(x) = C_1 e^x + C_2 xe^x + C_3 x^2 e^x$.

Now we begin to see the importance of the superposition principle. If we can find n solutions $y_1(x), y_2(x), \ldots, y_n(x)$ of an nth-order homogeneous linear differential equation, then a general solution is

$$y(x) = C_1 y_1(x) + C_2 y_2(x) + \cdots + C_n y_n(x).$$

In other words, all that we need do is find n solutions; the superposition principle will do the rest. There is a problem, however, if we take things a little too literally. For instance, $y_1(x) = e^{4x}$ and $y_2(x) = 10e^{4x}$ are both solutions of $y'' - y' + 12y = 0$. By superposition, so too then is $y(x) = C_1 y_1 + C_2 y_2 = C_1 e^{4x} + 10C_2 e^{4x}$. But is it a general solution? The answer is no, because we could write $y(x) = (C_1 + 10C_2)e^{4x}$, and by setting $C_3 = C_1 + 10C_2$, we have $y(x) = C_3 e^{4x}$. Superposition of the solutions e^{4x} and $10e^{4x}$ has not therefore led to a general solution, and the reason is that they are essentially the same solution: $y_2(x)$ is $y_1(x)$ multiplied by a constant. Superposition does not therefore lead to a solution with two arbitrary constants.

In a similar way, $y_1(x) = e^x$, $y_2(x) = xe^x$, and $y_3(x) = 2e^x - 3xe^x$ are all solutions of $y''' - 3y'' + 3y' - y = 0$, and therefore so is $y(x) = C_1 y_1 + C_2 y_2 + C_3 y_3$. But because we can write

$$\begin{aligned}y(x) &= C_1 e^x + C_2 xe^x + C_3(2e^x - 3xe^x) \\ &= (C_1 + 2C_3)e^x + (C_2 - 3C_3)xe^x \\ &= C_4 e^x + C_5 xe^x,\end{aligned}$$

$y(x)$ is not a general solution. This is a direct result of the fact that $y_3(x)$ is a linear combination of the solutions $y_1(x)$ and $y_2(x)$; it is twice $y_1(x)$ minus three times $y_2(x)$.

Our problem seems to come down to this: If we have n solutions of an nth-order homogeneous linear differential equation, how can we determine whether superposition leads to a (general) solution that contains n arbitrary constants? Our examples have suggested that if any one of the solutions is a linear combination of the others, then a general solution is not obtained, and this is indeed true. If one of the solutions is a linear combination of the others, we say that the n solutions are **linearly dependent**; if no solution is a linear combination of the others, we say that the n solutions are **linearly independent**. We summarize our results in the following theorem.

Theorem 16.2

If $y_1(x), y_2(x), \ldots, y_n(x)$ are n linearly independent solutions of an nth-order homogeneous linear differential equation, then $y(x) = C_1 y_1(x) + C_2 y_2(x) + \cdots + C_n y_n(x)$ is a general solution of the differential equation.

What we should now do is devise a test to determine whether a set of n solutions is linearly independent or linearly dependent. For most examples, no test is really necessary; it is obvious whether one of the solutions can be written as a linear combination of the others. For those rare occasions when it is not obvious, Exercise 10 describes a test that can be used to determine whether functions are linearly independent.

In summary, the superposition principle states that solutions of a homogeneous linear differential equation can be superposed to produce other solutions. If n linearly independent solutions of an nth-order equation are superposed, a general solution is obtained. This is the importance of the superposition principle. We need not devise a method that will take us directly to a general solution; we need a method for finding n linearly independent solutions—the superposition principle will do the rest. Unfortunately, for completely general coefficients $a_i(x)$ in $\phi(x, D)$ (see equation 16.44), it is impossible to give a method that will always yield n linearly independent solutions of a homogeneous equation. There is, however, one special case of great practical importance in which it is always possible to produce n linearly independent solutions in a very simple way. This special case occurs when the coefficients $a_i(x)$ are all constants a_i, and this is the subject of Section 16.9.

EXAMPLE 16.10

If $y_1(x) = \cos 3x$ and $y_2(x) = \sin 3x$ are solutions of the differential equation $y'' + 9y = 0$, find a general solution.

SOLUTION Since $y_1(x)$ and $y_2(x)$ are linearly independent solutions (one is not a constant times the other), a general solution can be obtained by superposition:

$$y(x) = C_1 \cos 3x + C_2 \sin 3x.$$

■

EXAMPLE 16.11

Given that $y_1(x) = e^{2x} \cos x$ and $y_2(x) = e^{2x} \sin x$ are solutions of the homogeneous linear differential equation $y'' - 4y' + 5y = 0$, find that solution which satisfies the conditions $y(\pi/4) = 1, y(\pi/3) = 2$.

SOLUTION By superposition, a general solution of the differential equation is

$$y(x) = C_1 e^{2x} \cos x + C_2 e^{2x} \sin x = e^{2x}(C_1 \cos x + C_2 \sin x).$$

To satisfy the conditions $y(\pi/4) = 1$ and $y(\pi/3) = 2$, we have

$$1 = e^{\pi/2}(C_1/\sqrt{2} + C_2/\sqrt{2}), \quad 2 = e^{2\pi/3}(C_1/2 + \sqrt{3}C_2/2).$$

Thus C_1 and C_2 are defined by the pair of equations

$$C_1 + C_2 = \sqrt{2}e^{-\pi/2}, \quad C_1 + \sqrt{3}C_2 = 4e^{-2\pi/3},$$

the solution of which is

$$C_2 = \frac{4e^{-2\pi/3} - \sqrt{2}e^{-\pi/2}}{\sqrt{3} - 1} = 0.2713, \quad C_1 = \sqrt{2}e^{-\pi/2} - C_2 = 0.0227.$$

The required solution is therefore

$$y(x) = e^{2x}(0.0227 \cos x + 0.2713 \sin x).$$

■

We pointed out in Section 16.1 that a general solution of a differential equation might not contain all solutions of the equation. This is not true for linear differential equations. It can be shown that if $y(x)$ is a general solution of a linear differential equation on an interval I, then every solution of the differential equation on I can be obtained by specifying particular values for the arbitrary constants in $y(x)$. Here then is a very important class of differential equations for which a general solution contains all possible solutions.

EXERCISES 16.8

In Exercises 1–8 show that the functions are solutions of the differential equation. Check that the differential equation is linear and homogeneous, and then find a general solution.

1. $y'' + y' - 6y = 0;\ y_1(x) = e^{2x},\ y_2(x) = e^{-3x}$

2. $y' + y \tan x = 0;\ y_1(x) = \cos x$

3. $y'''' + 5y'' + 4y = 0;\ y_1(x) = \cos 2x,$ $y_2(x) = \sin 2x,\ y_3(x) = \cos x,\ y_4(x) = \sin x$

4. $2y'' - 16y' + 32y = 0;\ y_1(x) = 3e^{4x},\ y_2(x) = -2xe^{4x}$

5. $y''' - 3y'' + 2y' = 0;\ y_1(x) = 10,\ y_2(x) = 3e^{x},$ $y_3(x) = 4e^{2x}$

6. $2y'' - 8y' + 9y = 0;\ y_1(x) = e^{2x} \cos(x/\sqrt{2}),$ $y_2(x) = e^{2x} \sin(x/\sqrt{2})$

7. $x^2y'' + xy' + (x^2 - 1/4)y = 0;$ $y_1(x) = (\sin x)/\sqrt{x},\ y_2(x) = (\cos x)/\sqrt{x}$

8. $x^2y'' + xy' + 4y = 0;\ y_1(x) = \cos(2 \ln x),$ $y_2(x) = \sin(2 \ln x)$

9. Show that $y_1(x) = -2/(x+1)$ and $y_2(x) = -2/(x+2)$ are both solutions of the differential equation $y'' = yy'$. Is $y(x) = y_1(x) + y_2(x)$ a solution? Explain.

10. We stated in this section that n functions $y_1(x), \ldots, y_n(x)$ are linearly dependent if at least one of the functions can be expressed as a linear combination of the others; they are linearly independent if none of the functions is a linear combination of the others. Another way of saying this is as follows: Functions $y_1(x), \ldots, y_n(x)$ are linearly dependent on an interval I if there exist constants $C_1, \ldots, C_n$, not all zero, such that on I

$$C_1y_1(x) + \cdots + C_ny_n(x) \equiv 0.$$

If this equation can be satisfied only with $C_1 = C_2 = \cdots = C_n = 0$, the functions are linearly independent. In this exercise we give a test to determine whether functions are linearly independent or dependent. If

$y_1(x), \ldots, y_n(x)$ have derivatives up to and including order $n-1$ on the interval I, we define the Wronskian of the functions as the $n \times n$ determinant:

$$W(y_1, \ldots, y_n) = \begin{vmatrix} y_1 & y_2 & \cdots & y_n \\ y_1' & y_2' & \cdots & y_n' \\ y_1'' & y_2'' & \cdots & y_n'' \\ \vdots & \vdots & \ddots & \vdots \\ y_1^{(n-1)} & y_2^{(n-1)} & \cdots & y_n^{(n-1)} \end{vmatrix}.$$

Show that if $y_1, \ldots, y_n$ are linearly dependent on I, then $W(y_1, \ldots, y_n) \equiv 0$ on I. It follows then that if there exists at least one point in I at which $W(y_1, \ldots, y_n) \neq 0$, the functions $y_1, \ldots, y_n$ are linearly independent on I.

In Exercises 11–15 use the method of Exercise 10 to determine whether the functions are linearly dependent or independent on the interval.

11. $\{1, x, x^2\}$ on $-\infty < x < \infty$

12. $\{x, 2x - 3x^2, x^2\}$ on $-\infty < x < \infty$

13. $\{\sin x, \cos x\}$ on $0 \leq x \leq 2\pi$

14. $\{x, xe^x, x^2 e^x\}$ on $0 \leq x \leq 1$

15. $\{x \sin x, e^{2x}\}$ on $-\infty < x < \infty$

SECTION 16.9

Homogeneous Linear Differential Equations with Constant Coefficients

We now consider homogeneous linear differential equations

$$a_0 D^n y + a_1 D^{n-1} y + a_2 D^{n-2} y + \cdots + a_{n-1} Dy + a_n y = 0, \qquad (16.47)$$

where the coefficients $a_0, a_1, \ldots, a_n$ are all constants. In operator notation we write

$$\phi(D) y = 0, \qquad (16.48\text{a})$$

where

$$\phi(D) = a_0 D^n + a_1 D^{n-1} + \cdots + a_{n-1} D + a_n. \qquad (16.48\text{b})$$

The superposition principle states that a general solution of equation 16.47 is $y(x) = C_1 y_1(x) + \cdots + C_n y_n(x)$, provided $y_1(x), \ldots, y_n(x)$ are any n linearly independent solutions of the equation. Our problem then is to devise a technique for finding n linearly independent solutions; to illustrate a possible procedure, we first consider three examples of second-order equations.

Our first example is

$$y'' + 2y' - 3y = 0.$$

It is not unreasonable to expect that perhaps for some value of m, $y(x) = e^{mx}$ might be a solution of this equation. After all, the equation says that the second derivative of the function must be equal to three times the function minus twice its first derivative. Since the exponential function reproduces itself when differentiated, perhaps m can be chosen to produce this combination. To see whether this is possible, we substitute $y = e^{mx}$ into the differential equation, and find that if $y = e^{mx}$ is to be a solution, then

$$m^2 e^{mx} + 2me^{mx} - 3e^{mx} = 0;$$

this implies that

$$0 = m^2 + 2m - 3 = (m+3)(m-1).$$

Thus $y = e^{mx}$ is a solution if m is chosen as either 1 or -3; i.e., $y_1 = e^x$ and $y_2 = e^{-3x}$ are solutions of the differential equation. Since they are linearly independent, a general solution is $y(x) = C_1 e^x + C_2 e^{-3x}$.

For our second example, we take

$$y'' + 2y' + y = 0.$$

Since exponentials worked in the first example, we once again try a solution of the form $y(x) = e^{mx}$. If we substitute into the differential equation, we obtain

$$m^2 e^{mx} + 2me^{mx} + e^{mx} = 0,$$

which implies that

$$0 = m^2 + 2m + 1 = (m + 1)^2.$$

Thus $y_1 = e^{-x}$ is a solution, but unfortunately it is the only solution that we obtain as a result of our guess. We need a second linearly independent solution $y_2(x)$ in order to obtain a general solution. Clearly, no other exponential will work. Perhaps if we multiplied e^{-x} by another function, we might find a second solution; in other words, perhaps there is a solution of the form $y(x) = v(x)e^{-x}$ for some $v(x)$. To see, we again substitute into the differential equation,

$$0 = \{v''e^{-x} - 2v'e^{-x} + ve^{-x}\} + 2\{v'e^{-x} - ve^{-x}\} + ve^{-x} = v''e^{-x}.$$

Consequently, $v'' = 0$, and this implies that $v(x) = Ax + B$, for any constants A and B. In particular, if $A = 1$ and $B = 0$, $v(x) = x$, and $y_2(x) = v(x)e^{-x} = xe^{-x}$ is also a solution of the differential equation. By superposition, then, we find that a general solution is $y(x) = C_1 e^{-x} + C_2 xe^{-x} = (C_1 + C_2 x)e^{-x}$. Note that if we had set $A = 0$ and $B = 1$, then $v(x) = 1$, and the solution $y(x) = v(x)e^{-x}$ would have been $y_1(x)$. Further, if we had simply set $y(x) = v(x)e^{-x} = (Ax + B)e^{-x}$, we would have found the general solution.

Our third example is

$$y'' + 2y' + 10y = 0.$$

As in the previous two examples, if we assume a solution $y = e^{mx}$, then

$$m^2 e^{mx} + 2me^{mx} + 10e^{mx} = 0,$$

or

$$0 = m^2 + 2m + 10.$$

The solutions of this quadratic equation are the complex numbers

$$m = \frac{-2 \pm \sqrt{4 - 40}}{2} = -1 \pm 3i.$$

This means that there is no real exponential $y = e^{mx}$ that will satisfy the differential equation. If, however, we form complex exponentials $y_1(x) = e^{(-1+3i)x}$ and $y_2(x) = e^{(-1-3i)x}$ and superpose these solutions, then

$$y(x) = Ae^{(-1+3i)x} + Be^{(-1-3i)x}$$

must also be a solution. When we use Euler's identity for complex exponentials, $e^{i\theta} = \cos\theta + i\sin\theta$, we can write $y(x)$ in the form

$$\begin{aligned} y(x) &= Ae^{-x}e^{3xi} + Be^{-x}e^{-3xi} \\ &= Ae^{-x}(\cos 3x + i\sin 3x) + Be^{-x}(\cos 3x - i\sin 3x) \\ &= e^{-x}[(A + B)\cos 3x + i(A - B)\sin 3x] \\ &= e^{-x}(C_1 \cos 3x + C_2 \sin 3x), \end{aligned}$$

where $C_1 = A + B$ and $C_2 = i(A - B)$. In other words, the function $y(x) = e^{-x}(C_1 \cos 3x + C_2 \sin 3x)$ is a general solution of the differential equation, and it has been derived from the complex roots $m = -1 \pm 3i$ of the equation $m^2 + 2m + 10 = 0$. Note that what multiplies x in the exponential is the real part of these complex numbers, and what multiplies x in the trigonometric functions is the imaginary part. In Exercise 17 we show that this solution can also be derived without complex numbers, but we feel that in general the use of complex numbers is the best method.

In each of these examples we guessed $y = e^{mx}$ as a possible solution. We then substituted into the differential equation to obtain an algebraic equation for m. In each case the equation for m was

$$\phi(m) = 0;$$

that is, take the operator $\phi(D)$, replace D by m, and set the polynomial equal to zero. This is not a peculiarity of these examples for it is straightforward to show that for any homogeneous linear equation 16.48, if we assume a solution of the form $y = e^{mx}$, then m must satisfy the equation $\phi(m) = 0$. We name this equation in the following definition.

Definition 16.7

With every linear differential equation that has constant coefficients $\phi(D)y = F(x)$, we associate an equation

$$\phi(m) = 0 \tag{16.49}$$

called the **auxiliary equation**.

To summarize, in each of the examples we assumed a solution $y = e^{mx}$ and found that m had to satisfy the auxiliary equation $\phi(m) = 0$. From the roots of the auxiliary equation we obtained solutions of the differential equation, and superposition then led to a general solution. We have found what could be a general procedure for determining a general solution of every homogeneous linear differential equation with constant coefficients. This is indeed true so that every such differential equation can be solved in exactly the same way. But if the procedure is the same in every case, surely we can set down rules that eliminate the necessity of tediously repeating these steps in every example. This we do in the following theorem.

Theorem 16.3

If $\phi(m) = 0$ is the auxiliary equation associated with the homogeneous linear differential equation $\phi(D)y = 0$, then there are two possibilities:

(i) *$\phi(m) = 0$ has a real root m of multiplicity k. Then a solution of the differential equation is*

$$(C_1 + C_2 x + \cdots + C_k x^{k-1})e^{mx}. \tag{16.50a}$$

(ii) *$\phi(m) = 0$ has a pair of complex conjugate roots $a \pm bi$ each of multiplicity k. Then a solution of the differential equation is*

$$e^{ax}[(C_1 + C_2 x + \cdots + C_k x^{k-1}) \cos bx + (D_1 + D_2 x + \cdots + D_k x^{k-1}) \sin bx]. \tag{16.50b}$$

A general solution of the differential equation is obtained by superposing all solutions in **(i)** *and* **(ii)**.

For a proof of this theorem see Exercise 20. Let us now apply the theorem to our previous examples. The auxiliary equation for $y'' + 2y' - 3y = 0$ is

$$0 = m^2 + 2m - 3 = (m + 3)(m - 1)$$

with solutions $m = 1$ and $m = -3$. If we now use part **(i)** of Theorem 16.3 with two real roots, each of multiplicity 1, a general solution of the differential equation is

$$y(x) = C_1 e^x + C_2 e^{-3x}.$$

The auxiliary equation for $y'' + 2y' + y = 0$ is

$$0 = m^2 + 2m + 1 = (m + 1)^2$$

with solutions $m = -1$ and $m = 1$. Part **(i)** of Theorem 16.3 with a single real root of multiplicity 2 gives the general solution

$$y(x) = (C_1 + C_2 x)e^{-x}.$$

The auxiliary equation for $y'' + 2y' + 10y = 0$ is

$$0 = m^2 + 2m + 10$$

with solutions $m = -1 \pm 3i$. Part **(ii)** of Theorem 16.3 with a pair of complex conjugate roots, each of multiplicity 1, gives the general solution

$$y(x) = e^{-x}(C_1 \cos 3x + C_2 \sin 3x).$$

EXAMPLE 16.12

Find a general solution for the differential equation

$$y''' - 3y'' - 4y' + 6y = 0.$$

SOLUTION The auxiliary equation is

$$0 = m^3 - 3m^2 - 4m + 6 = (m - 1)(m^2 - 2m - 6)$$

with solutions $m = 1$ and $m = 1 \pm \sqrt{7}$. A general solution of the differential equation is therefore

$$y(x) = C_1 e^x + C_2 e^{(1+\sqrt{7})x} + C_3 e^{(1-\sqrt{7})x}.$$

■

EXAMPLE 16.13

Find a general solution for $y''' - y = 0$.

SOLUTION The auxiliary equation is

$$0 = m^3 - 1 = (m - 1)(m^2 + m + 1)$$

with solutions $m = 1$ and $m = -(1/2) \pm (\sqrt{3}/2)i$. A general solution of the differential equation is therefore

$$y(x) = C_1 e^{x} + e^{-x/2}[C_2 \cos(\sqrt{3}x/2) + C_3 \sin(\sqrt{3}x/2)].$$

■

EXAMPLE 16.14

If the roots of the auxiliary equation $\phi(m) = 0$ are

$$3,\ 3,\ 3,\ \pm 2i,\ -2,\ 1 \pm \sqrt{3},\ -4 \pm i,\ -4 \pm i,$$

find a general solution of the differential equation $\phi(D)y = 0$.

SOLUTION A general solution is

$$\begin{aligned} y(x) = {} & (C_1 + C_2 x + C_3 x^2)e^{3x} + C_4 \cos 2x + C_5 \sin 2x + C_6 e^{-2x} \\ & + C_7 e^{(1+\sqrt{3})x} + C_8 e^{(1-\sqrt{3})x} + e^{-4x}[(C_9 + C_{10} x) \cos x \\ & + (C_{11} + C_{12} x) \sin x]. \end{aligned}$$

■

EXERCISES 16.9

In Exercises 1–12 find a general solution for the homogeneous differential equation.

1. $y'' + y' - 6y = 0$

2. $2y'' - 16y' + 32y = 0$

3. $2y'' + 16y' + 82y = 0$

4. $y'' + 2y' - 2y = 0$

5. $y'' - 4y' + 5y = 0$

6. $y''' - 3y'' + y' - 3y = 0$

7. $y'''' + 2y'' + y = 0$

8. $y''' - 6y'' + 12y' - 8y = 0$

9. $3y''' - 12y'' + 18y' - 12y = 0$

10. $y'''' + 5y'' + 4y = 0$

11. $y''' - 3y'' + 2y' = 0$

12. $y'''' + 16y = 0$

In Exercises 13–16 find a homogeneous linear differential equation that has the function as general solution.

13. $y(x) = C_1 e^{x} + (C_2 + C_3 x)e^{-4x}$

14. $y(x) = e^{-2x}(C_1 \cos 4x + C_2 \sin 4x)$

15. $y(x) = C_1 + C_2 e^{\sqrt{3}x} + C_3 e^{-\sqrt{3}x}$

16. $y(x) = e^{x}(C_1 + C_2 x) \cos \sqrt{2}x + e^{x}(C_3 + C_4 x) \sin \sqrt{2}x$

17. Show that if we assume that $y(x) = e^{ax} \sin bx$ is a solution of the differential equation $y'' + 2y' + 10y = 0$, then a and b must be equal to -1 and ± 3 respectively. Verify that for this a and b, $y(x) = e^{ax} \cos bx$ is also a solution, and therefore a general solution is $y(x) = e^{-x}(C_1 \cos 3x + C_2 \sin 3x)$.

18. The equation $y''' + ay'' + by' + cy = 0$, where a, b, and c are constants, has solution $y(x) = C_1 e^{-x} + e^{-2x}(C_2 \sin 4x + C_3 \cos 4x)$. Find a, b, and c.

19. Show that if p is constant and $f(x)$ is differentiable, then

$$D\{e^{px} f(x)\} = e^{px}\{(D + p) f(x)\}.$$

Now use mathematical induction to prove that if $f(x)$ is k times differentiable, then

$$D^k\{e^{px} f(x)\} = e^{px}\{(D + p)^k f(x)\}.$$

Finally verify that

$$\phi(D)\{e^{px} f(x)\} = e^{px}\{\phi(D + p) f(x)\},$$

a result called the *operator shift theorem*.

20. **(a)** If m_0 is a real root of multiplicity k for the auxiliary equation $\phi(m) = 0$, show that the operator $\phi(D)$ can be expressed in the form

$$\phi(D) = (D - m_0)^k \psi(D),$$

where $\psi(D)$ is a polynomial in D. Now use the operator shift theorem of Exercise 19 to verify that $(C_1 + C_2 x + \cdots + C_k x^{k-1})e^{m_0 x}$ is a solution of $\phi(D)y = 0$.

(b) If $a \pm bi$ are complex conjugate roots each of multiplicity k for the auxiliary equation $\phi(m) = 0$, show that $\phi(D)$ can be expressed in the form

$$\phi(D) = (D - a - bi)^k (D - a + bi)^k \psi(D),$$

where $\psi(D)$ is a polynomial in D. Now use the operator shift theorem of Exercise 19 to verify that 16.50b is a solution of $\phi(D)y = 0$.

21. If M, β, and k are all positive constants, find a general solution for the linear differential equation

$$M\frac{d^2 x}{dt^2} + \beta\frac{dx}{dt} + kx = 0.$$

SECTION 16.10

Nonhomogeneous Linear Differential Equations with Constant Coefficients

The general nonhomogeneous linear differential equation with constant coefficients is

$$\phi(D)y = F(x), \tag{16.51a}$$

where

$$\phi(D) = a_0 D^n + a_1 D^{n-1} + \cdots + a_{n-1} D + a_n. \tag{16.51b}$$

It is natural to ask whether we can use the results of Section 16.9 concerning homogeneous equations with constant coefficients to solve nonhomogeneous problems. Fortunately the answer is yes, as shown by the following definition.

Definition 16.8

With every nonhomogeneous linear differential equation with constant coefficients

$$\phi(D)y = F(x),$$

we associate a homogeneous equation

$$\phi(D)y = 0, \tag{16.52}$$

called the **homogeneous (reduced**, or **complimentary)** equation associated with $\phi(D)y = F(x)$.

We now prove the following theorem.

Theorem 16.4

A general solution of the linear differential equation $\phi(D)y = F(x)$ is $y(x) = y_h(x) + y_p(x)$, where $y_h(x)$ is a general solution of the associated homogeneous equation, and $y_p(x)$ is any particular solution of the given equation.

Proof Since $\phi(D)$ is a linear operator,

$$\phi(D)(y_h + y_p) = \phi(D)y_h + \phi(D)y_p = 0 + F(x) = F(x),$$

so that $y_h + y_p$ is indeed a solution of the given differential equation. Because $y_h(x)$ is a general solution of the associated homogeneous equation, it contains the requisite number of arbitrary constants for $y(x)$ to be a general solution of $\phi(D)y = F(x)$, and this completes the proof.

We note in passing that Theorem 16.4 is also valid for linear differential equations with variable coefficients.

Theorem 16.4 indicates that the discussion of nonhomogeneous differential equations can be divided into two separate parts. First we find a general solution $y_h(x)$ of the associated homogeneous equation 16.52, and this can be done using the results of Section 16.9. To this we add any particular solution $y_p(x)$ of 16.51. We present two methods for finding a particular solution: (1) the method of undetermined coefficients, and (2) the method of operators. Both methods apply in general only to differential equations in which $F(x)$ is a power (x^n, n a nonnegative integer), an exponential (e^{px}), a sine ($\sin px$), a cosine ($\cos px$), and/or any sums or products thereof.

Method of Undetermined Coefficients for a Particular Solution

As stated above, the method of undetermined coefficients is to be used only when $F(x)$ in equation 16.51 is of the form $x^n, e^{px}, \sin px, \cos px$, and/or sums or products thereof. For example, if

$$y'' + y' - 6y = e^{4x},$$

then the method essentially says that Ae^{4x} is the simplest function that could conceivably yield e^{4x} when substituted into the left-hand side of the differential equation. Consequently, it is natural to assume $y_p = Ae^{4x}$ and attempt to determine the unknown coefficient A. Substitution of this function into the differential equation gives

$$16Ae^{4x} + 4Ae^{4x} - 6Ae^{4x} = e^{4x}.$$

If we divide by e^{4x}, then

$$14A = 1 \quad \text{and} \quad A = \tfrac{1}{14}.$$

A particular solution is therefore $y_p = e^{4x}/14$.

Before stating a general rule, we illustrate a few more possibilities in the following example.

EXAMPLE 16.15

Find a particular solution of $y'' + y' - 6y = F(x)$ in each case.

(a) $F(x) = 6x^2 + 2x + 3$

(b) $F(x) = 2\sin 2x$

(c) $F(x) = xe^{-x} - e^{-x}$

SOLUTION

(a) Since terms in x^2, x, and constants yield terms in x^2, x, and constants when substituted into the left-hand side of the differential equation, we attempt to find a particular solution of the form $y_p = Ax^2 + Bx + C$. Substitution into the differential equation gives

$$(2A) + (2Ax + B) - 6(Ax^2 + Bx + C) = 6x^2 + 2x + 3,$$

or,

$$(-6A)x^2 + (2A - 6B)x + (2A + B - 6C) = 6x^2 + 2x + 3.$$

But this equation can hold for all values of x only if coefficients of corresponding powers of x are identical (see Exercise 35 in Section 3.7). Equating coefficients then gives

$$-6A = 6,$$
$$2A - 6B = 2,$$
$$2A + B - 6C = 3.$$

These imply that $A = -1, B = -2/3, C = -17/18$, and

$$y_p = -x^2 - \frac{2x}{3} - \frac{17}{18}.$$

(b) Since terms in $\sin 2x$ and $\cos 2x$ yield terms in $\sin 2x$ when substituted into the left-hand side of the differential equation, we assume that $y_p = A \sin 2x + B \cos 2x$. Substitution into the differential equation gives

$$(-4A \sin 2x - 4B \cos 2x) + (2A \cos 2x - 2B \sin 2x) - 6(A \sin 2x + B \cos 2x) = 2 \sin 2x,$$

or,

$$(-10A - 2B) \sin 2x + (2A - 10B) \cos 2x = 2 \sin 2x.$$

Equating coefficients gives

$$-10A - 2B = 2,$$
$$2A - 10B = 0.$$

These imply that $A = -5/26, B = -1/26$, and hence

$$y_p = -\frac{1}{26}(5 \sin 2x + \cos 2x).$$

(c) Since terms in xe^{-x} and e^{-x} yield terms in xe^{-x} and e^{-x} when substituted into the left-hand side of the differential equation, we assume that $y_p = Axe^{-x} + Be^{-x}$. Substitution into the differential equation gives

$$(Axe^{-x} - 2Ae^{-x} + Be^{-x}) + (-Axe^{-x} + Ae^{-x} - Be^{-x}) - 6(Axe^{-x} + Be^{-x}) = xe^{-x} - e^{-x},$$

or,

$$(-6A)xe^{-x} + (-A - 6B)e^{-x} = xe^{-x} - e^{-x}.$$

Equating coefficients yields

$$-6A = 1,$$
$$-A - 6B = -1.$$

These imply that $A = -1/6, B = 7/36$, and hence $y_p = -(1/6)xe^{-x} + (7/36)e^{-x}$. ■

The following rule encompasses each part of this example.

Rule 1

If a term of $F(x)$ consists of a power (x^n), an exponential (e^{px}), a sine $(\sin px)$, a cosine $(\cos px)$, or any product thereof, assume as a part of y_p a constant multiplied by that term plus a constant multiplied by any linearly independent function arising from it by differentiation.

For Example 16.15(a), since $F(x)$ contains the term $6x^2$, we assume y_p contains Ax^2. Differentiation of Ax^2 yields a term in x and a constant so that we form $y_p = Ax^2 + Bx + C$. No new terms for y_p are obtained from the terms $2x$ and 3 in $F(x)$.

For Example 16.15(b), we assume that y_p contains $A\sin 2x$ to account for the term $2\sin 2x$ in $F(x)$. Differentiation of $A\sin 2x$ gives a linearly independent term in $\cos 2x$ so that we form $y_p = A\sin 2x + B\cos 2x$.

For Example 16.15(c), since $F(x)$ contains the term xe^{-x}, we assume that y_p contains Axe^{-x}. Differentiation of Axe^{-x} yields a term in e^{-x} so that we form $y_p = Axe^{-x} + Be^{-x}$. No new terms for y_p are obtained from the term $-e^{-x}$ in $F(x)$.

EXAMPLE 16.16

What is the form of the particular solution predicted by Rule 1 for the differential equation

$$y'' + 15y' - 6y = x^2e^{4x} + x + x\cos x?$$

SOLUTION Rule 1 suggests that

$$y_p = Ax^2e^{4x} + Bxe^{4x} + Ce^{4x} + Dx + E + Fx\cos x + Gx\sin x + H\cos x + I\sin x.$$

■

Unfortunately, exceptions to Rule 1 do occur. For the differential equation $y'' + y = \cos x$, Rule 1 would predict $y_p = A\cos x + B\sin x$. If we substitute this into the differential equation we obtain the absurd identity $0 = \cos x$, and certainly no equations to solve for A and B. This result could have been predicted had we first calculated $y_h(x)$. The auxiliary equation $m^2 + 1 = 0$ has solutions $m = \pm i$ so that $y_h(x) = C_1\cos x + C_2\sin x$. Since y_p as suggested by Rule 1 is precisely y_h with different names for the constants, then certainly $y_p'' + y_p = 0$. Suppose as an alternative we multiply this y_p by x, and assume that $y_p = Ax\cos x + Bx\sin x$. Substitution into the differential equation now gives

$$-2A\sin x + 2B\cos x = \cos x.$$

Identification of coefficients requires $A = 0, B = 1/2$, and $y_p = (1/2)x\sin x$.

This example suggests that if y_p predicted by Rule 1 is already contained in y_h, then a modification of y_p is necessary. A precise statement of the situation is given in the following rule.

Rule 2

Suppose that a term in $F(x)$ is of the form $x^n f(x)$ (n a nonnegative integer). Suppose further that $f(x)$ can be obtained from $y_h(x)$ by specifying values for the arbitrary constants. If this term in y_h results from a root of the auxiliary equation of multiplicity k, then corresponding to $x^n f(x)$, assume as a part of y_p the term $Ax^k(x^n f(x)) = Ax^{n+k}f(x)$, plus a constant multiplied by any linearly independent function arising from it by differentiation.

To use this rule we first require $y_h(x)$. Then, and only then, can we decide on the form of $y_p(x)$. As an illustration, consider the following example.

EXAMPLE 16.17

Find a general solution for $y''' - y = x^3 e^x$.

SOLUTION In Example 16.13 we solved the auxiliary equation to obtain $m = 1$ and $m = -1/2 \pm (\sqrt{3}/2)i$, from which we formed

$$y_h(x) = C_1 e^x + e^{-x/2}[C_2 \cos(\sqrt{3}x/2) + C_3 \sin(\sqrt{3}x/2)].$$

Now $x^3 e^x$ is x^3 times e^x, and e^x can be obtained from y_h by specifying $C_1 = 1$ and $C_2 = C_3 = 0$. Since this term results from the root $m = 1$ of multiplicity 1, we assume that y_p contains $Ax(x^3 e^x) = Ax^4 e^x$. Differentiation of this function gives terms in $x^3 e^x, x^2 e^x, xe^x$, and e^x. We therefore take

$$y_p = Ax^4 e^x + Bx^3 e^x + Cx^2 e^x + Dxe^x.$$

(We do not include a term in e^x since it is already in y_h.) Substitution into the differential equation and simplification gives

$$(12A)x^3 e^x + (36A + 9B)x^2 e^x + (24A + 18B + 6C)xe^x + (6B + 6C + 3D)e^x = x^3 e^x.$$

Equating coefficients gives

$$\begin{aligned} 12A &= 1, \\ 36A + 9B &= 0, \\ 24A + 18B + 6C &= 0, \\ 6B + 6C + 3D &= 0. \end{aligned}$$

These imply that $A = 1/12, B = -1/3, C = 2/3, D = -2/3$, and

$$y_p = \frac{1}{12}x^4 e^x - \frac{1}{3}x^3 e^x + \frac{2}{3}x^2 e^x - \frac{2}{3}xe^x.$$

A general solution of the differential equation is therefore

$$y(x) = C_1 e^x + e^{-x/2}[C_2 \cos(\sqrt{3}x/2) + C_3 \sin(\sqrt{3}x/2)] + \frac{x^4 e^x}{12} - \frac{x^3 e^x}{3} + \frac{2x^2 e^x}{3} - \frac{2xe^x}{3}.$$

■

EXAMPLE 16.18

If the roots of the auxiliary equation $\phi(m) = 0$ for the differential equation $\phi(D)y = x^2 - 2\sin x + xe^{-2x}$ are $\pm i, -2, -2, -2, 4$, and 4, what is the form of y_p predicted by the method of undetermined coefficients?

SOLUTION From the roots of the auxiliary equation, we can form

$$y_h(x) = C_1 \cos x + C_2 \sin x + (C_3 + C_4 x + C_5 x^2)e^{-2x} + (C_6 + C_7 x)e^{4x}.$$

Corresponding to the term x^2 in $F(x)$, Rule 1 requires that y_p contain $Ax^2 + Bx + C$. Because $-2 \sin x$ can be obtained from $y_h(x)$ by specifying $C_2 = -2$, $C_1 = C_3 = C_4 = C_5 = C_6 = C_7 = 0$, and this term results from the roots $m = \pm i$, each of multiplicity 1, Rule 2 suggests that y_p contain $Dx \sin x + Ex \cos x$. (We do not include terms in $\sin x$ and $\cos x$ since they are already in y_h.) Finally, xe^{-2x} is x times e^{-2x}, and this function can be obtained from $y_h(x)$ by setting $C_3 = 1$, and $C_1 = C_2 = C_4 = C_5 = C_6 = C_7 = 0$. Because this term results from the root $m = -2$ of multiplicity 3, y_p must contain $Fx^4 e^{-2x} + Gx^3 e^{-2x}$ (but not terms in $x^2 e^{-2x}, xe^{-2x}$, and e^{-2x} since they are in y_h). The total particular solution is therefore

$$y_p = Ax^2 + Bx + C + Dx \sin x + Ex \cos x + Fx^4 e^{-2x} + Gx^3 e^{-2x}.$$

■

Operator Method for a Particular Solution

This method, like that of undetermined coefficients, is designed only for functions $F(x)$ in equation 16.51 of the form $x^n, e^{px}, \sin px, \cos px$, and/or sums or products thereof. Essentially the operator method says that if

$$\phi(D)y = F(x), \tag{16.53}$$

then

$$y = \frac{1}{\phi(D)}F(x). \tag{16.54}$$

But there is a problem. What does it mean to say that

$$\frac{1}{\phi(D)} = \frac{1}{a_0 D^n + a_1 D^{n-1} + \cdots + a_{n-1} D + a_n} \tag{16.55}$$

operates on $F(x)$? The operator method then depends on our explaining how $1/\phi(D)$ operates on $F(x)$. The simplest $\phi(D)$ is $\phi(D) = D$. In this case the differential equation is

$$Dy = F(x),$$

and the solution is

$$y(x) = \int F(x)\,dx.$$

If the solution is also to be represented in operator notation by

$$y(x) = \frac{1}{D}F(x),$$

then we must define the operator $1/D$ by

$$\frac{1}{D}F(x) = \int F(x)\,dx. \tag{16.56}$$

If $1/D$ means to integrate, then $1/D^2$ must mean to integrate twice, $1/D^3$ to integrate three times, and so on.

We have stated that the method of operators is applicable in general only when $F(x)$ consists of powers, exponentials, and/or sines and cosines. Consider first the case in which $F(x)$ is a power x^n, in which case equations 16.53 and 16.54 become

$$\phi(D)y = x^n \tag{16.57}$$

and

$$y = \frac{1}{\phi(D)}x^n. \tag{16.58}$$

If we forget for the moment that D is a differential operator, and simply regard $1/\phi(D)$ as a rational function of a variable D, then we can express $1/\phi(D)$ as an infinite series of the form

$$\frac{1}{\phi(D)} = \frac{1}{D^k}\{b_0 + b_1 D + b_2 D^2 + \cdots\} \tag{16.59}$$

for some nonnegative integer k. (We will show how in a moment.) But this suggests that we write 16.58 in the form

$$y = \frac{1}{D^k}\{b_0 + b_1 D + b_2 D^2 + \cdots\}x^n. \tag{16.60}$$

If we now reinterpret D as d/dx, then the operator on the right will produce a polynomial in x. (Note that $D^m x^n = 0$ if $m > n$.) It turns out that if we ignore all arbitrary constants that result from the integrations (when $k \geq 1$), this polynomial is a solution of 16.57. In other words, when $F(x) = x^n$, a particular solution of 16.57 is

$$y_p = \frac{1}{D^k}\{b_0 + b_1 D + b_2 D^2 + \cdots\}x^n. \tag{16.61}$$

We now show how to expand $1/\phi(D)$ as a series of the form 16.59. Only two situations arise and each of these can be illustrated with a simple example. First suppose that $\phi(D) = D^2 + 4D + 5$. For $1/\phi(D)$ we write

$$\frac{1}{\phi(D)} = \frac{1}{D^2 + 4D + 5} = \frac{1}{5\left(1 + \dfrac{4D + D^2}{5}\right)},$$

and interpret the right-hand side (less the 5 outside the parentheses) as the sum of an infinite geometric series with common ratio $-(4D + D^2)/5$. When we write this series out, we have

$$\begin{aligned}\frac{1}{D^2 + 4D + 5} &= \frac{1}{5}\left\{1 - \left(\frac{4D + D^2}{5}\right) + \left(\frac{4D + D^2}{5}\right)^2 - \cdots\right\} \\ &= \frac{1}{5}\left\{1 - \frac{4}{5}D + \frac{11}{25}D^2 + \cdots\right\},\end{aligned}$$

which is of the required form 16.59 with $k = 0$.

Power k in 16.59 is positive only when $\phi(D)$ has no constant term. For example, if $\phi(D) = D^4 - 2D^3 + 3D^2$, then we factor out D^2, and proceed as above:

$$\frac{1}{\phi(D)} = \frac{1}{D^2(3 - 2D + D^2)} = \frac{1}{3D^2\left(1 - \dfrac{2D - D^2}{3}\right)}$$

$$= \frac{1}{3D^2}\left\{1 + \left(\frac{2D - D^2}{3}\right) + \left(\frac{2D - D^2}{3}\right)^2 + \cdots\right\}$$

$$= \frac{1}{3D^2}\left\{1 + \frac{2}{3}D + \frac{1}{9}D^2 + \cdots\right\}.$$

EXAMPLE 16.19

Find a particular solution of the differential equation

$$y'' + 6y' + 4y = x^2 + 4.$$

SOLUTION We write

$$y_p = \frac{1}{D^2 + 6D + 4}(x^2 + 4) = \frac{1}{4\left(1 + \dfrac{6D + D^2}{4}\right)}(x^2 + 4)$$

$$= \frac{1}{4}\left\{1 - \left(\frac{6D + D^2}{4}\right) + \left(\frac{6D + D^2}{4}\right)^2 - \cdots\right\}(x^2 + 4).$$

Since $D^n(x^2 + 4) = 0$ if $n > 2$, we require only the constant term and terms in D and D^2:

$$y_p = \frac{1}{4}\left\{1 - \frac{3}{2}D + 2D^2 + \cdots\right\}(x^2 + 4)$$

$$= \frac{1}{4}(x^2 + 4) - \frac{3}{8}(2x) + \frac{1}{2}(2) = \frac{x^2}{4} - \frac{3x}{4} + 2.$$

■

EXAMPLE 16.20

Find a particular solution of the differential equation

$$y''' + 2y' = x^2 - x.$$

SOLUTION A particular solution is

$$y_p = \frac{1}{D^3 + 2D}(x^2 - x) = \frac{1}{D(2 + D^2)}(x^2 - x)$$

$$= \frac{1}{2D\left(1 + \dfrac{D^2}{2}\right)}(x^2 - x)$$

$$= \frac{1}{2D}\left\{1 - \frac{D^2}{2} + \cdots\right\}(x^2 - x)$$

$$= \frac{1}{2D}\left\{x^2 - x - \frac{1}{2}(2)\right\},$$

and since $1/D$ means to integrate with respect to x, we have

$$y_p = \frac{1}{2}\left\{\frac{x^3}{3} - \frac{x^2}{2} - x\right\} = \frac{x^3}{6} - \frac{x^2}{4} - \frac{x}{2}.$$

■

In summary, to evaluate 16.54 when $F(x)$ is a power x^n (or a polynomial), we expand $1/\phi(D)$ in a series of the form 16.59 and perform the indicated differentiations (and integrations if $k > 0$).

For all other cases of $F(x)$, we make use of a theorem called the **inverse operator shift theorem**. This theorem states that

$$\frac{1}{\phi(D)}\{e^{px}f(x)\} = e^{px}\frac{1}{\phi(D+p)}f(x) \tag{16.62}$$

(see Exercise 16). The theorem enables us to shift the exponential e^{px} past the operator $1/\phi(D)$, but in so doing, the operator must be modified to $1/\phi(D+p)$. Equation 16.62 immediately yields $y_p(x)$ whenever $f(x)$ is a power x^n, n a nonnegative integer (or a polynomial). This is illustrated in the following example.

EXAMPLE 16.21

Find particular solutions for the following differential equations:

(a) $y'' + 3y' + 10y = x^2 e^{-x}$

(b) $y'' - 4y' + 4y = e^{2x}$

(c) $y''' + 2y'' + 3y' - y = e^{2x}$

SOLUTION

(a) Equation 16.62 gives

$$y_p = \frac{1}{D^2 + 3D + 10}x^2 e^{-x} = e^{-x}\frac{1}{(D-1)^2 + 3(D-1) + 10}x^2$$
$$= e^{-x}\frac{1}{D^2 + D + 8}x^2,$$

and we now proceed as in Example 16.19:

$$y_p = e^{-x}\frac{1}{8\left(1 + \dfrac{D + D^2}{8}\right)}x^2$$
$$= \frac{e^{-x}}{8}\left\{1 - \left(\frac{D + D^2}{8}\right) + \left(\frac{D + D^2}{8}\right)^2 - \cdots\right\}x^2$$
$$= \frac{e^{-x}}{8}\left\{1 - \frac{D}{8} - \frac{7D^2}{64} + \cdots\right\}x^2 = \frac{e^{-x}}{8}\left\{x^2 - \frac{x}{4} - \frac{7}{32}\right\}.$$

(b) Once again we use the inverse operator shift theorem to get

$$y_p = \frac{1}{D^2 - 4D + 4}e^{2x} = e^{2x}\frac{1}{(D+2)^2 - 4(D+2) + 4}(1)$$
$$= e^{2x}\frac{1}{D^2}(1) = \frac{x^2}{2}e^{2x},$$

since $1/D^2$ means to integrate twice.

(c) For this differential equation,

$$y_p = \frac{1}{D^3 + 2D^2 + 3D - 1} e^{2x} = e^{2x} \frac{1}{(D+2)^3 + 2(D+2)^2 + 3(D+2) - 1}(1)$$
$$= e^{2x} \frac{1}{D^3 + 8D^2 + 23D + 21}(1) = e^{2x} \frac{1}{21\left(1 + \dfrac{23D + 8D^2 + D^3}{21}\right)}(1)$$
$$= \frac{e^{2x}}{21}\{1 + \cdots\}(1) = \frac{e^{2x}}{21}.$$

■

EXAMPLE 16.22

Find a general solution for $y''' - 3y'' + 3y' - y = 2x^2 e^x$.

SOLUTION The auxiliary equation is

$$0 = m^3 - 3m^2 + 3m - 1 = (m-1)^3$$

with solutions 1, 1, and 1. Thus,

$$y_h(x) = (C_1 + C_2 x + C_3 x^2)e^x.$$

A particular solution is

$$y_p(x) = \frac{2}{(D-1)^3} x^2 e^x = 2e^x \frac{1}{(D+1-1)^3} x^2 = 2e^x \frac{1}{D^3} x^2 = \frac{x^5}{30} e^x.$$

A general solution of the differential equation is therefore

$$y(x) = (C_1 + C_2 x + C_3 x^2)e^x + \frac{x^5}{30} e^x.$$

■

EXAMPLE 16.23

Find a particular solution for the differential equation

$$y''' + 3y'' - 4y = xe^{-2x} + x^2.$$

SOLUTION

$$y_p = \frac{1}{D^3 + 3D^2 - 4}(xe^{-2x} + x^2)$$
$$= e^{-2x} \frac{1}{(D-2)^3 + 3(D-2)^2 - 4} x + \frac{1}{D^3 + 3D^2 - 4} x^2$$
$$= e^{-2x} \frac{1}{D^3 - 3D^2} x - \frac{1}{4\left(1 - \dfrac{3D^2 + D^3}{4}\right)} x^2$$

$$= e^{-2x}\frac{1}{D^2}\frac{1}{D-3}x - \frac{1}{4}\left\{1 + \left(\frac{3D^2 + D^3}{4}\right) + \cdots\right\}x^2$$

$$= e^{-2x}\frac{1}{D^2}\frac{1}{-3\left(1 - \dfrac{D}{3}\right)}x - \frac{1}{4}\left(x^2 + \frac{3}{2}\right)$$

$$= \frac{e^{-2x}}{-3}\frac{1}{D^2}\left\{1 + \frac{D}{3} + \cdots\right\}x - \frac{x^2}{4} - \frac{3}{8}$$

$$= \frac{e^{-2x}}{-3}\frac{1}{D^2}\left\{x + \frac{1}{3}\right\} - \frac{x^2}{4} - \frac{3}{8}$$

$$= \frac{e^{-2x}}{-3}\left\{\frac{x^3}{6} + \frac{x^2}{6}\right\} - \frac{x^2}{4} - \frac{3}{8}.$$

■

When $F(x)$ in 16.54 is of the form $x^n \sin px$ or $x^n \cos px$, n a nonnegative integer and p a constant, we introduce complex exponentials. Specifically, because

$$x^n e^{ipx} = x^n(\cos px + i\sin px),$$

we can write

$$x^n \cos px = \text{Real part of } x^n e^{ipx} = \text{Re}\{x^n e^{ipx}\}, \tag{16.63a}$$

$$x^n \sin px = \text{Imaginary part of } x^n e^{ipx} = \text{Im}\{x^n e^{ipx}\}. \tag{16.63b}$$

To operate on either of these functions by $1/\phi(D)$, we interchange the operations of $1/\phi(D)$ and taking real and imaginary parts:

$$\frac{1}{\phi(D)}\{x^n \cos px\} = \frac{1}{\phi(D)}\text{Re}\{x^n e^{ipx}\} = \text{Re}\left\{\frac{1}{\phi(D)}(x^n e^{ipx})\right\}, \tag{16.64a}$$

$$\frac{1}{\phi(D)}\{x^n \sin px\} = \frac{1}{\phi(D)}\text{Im}\{x^n e^{ipx}\} = \text{Im}\left\{\frac{1}{\phi(D)}(x^n e^{ipx})\right\}. \tag{16.64b}$$

We can now proceed by using inverse operator shift theorem 16.62.

EXAMPLE 16.24

Find particular solutions for the following differential equations:

(a) $y'' + y = \sin 2x$;

(b) $y'' + 4y = x^2 \cos x$;

(c) $y'' + 9y = \sin 3x$;

(d) $y'' + 4y = x \sin 2x$;

(e) $y'' + 2y' + 4y = e^{-x}\sin\sqrt{3}x$.

SOLUTION

(a) Equation 16.64b gives

$$y_p = \frac{1}{D^2+1}\sin 2x = \frac{1}{D^2+1}\{\text{Im}(e^{2ix})\} = \text{Im}\left\{\frac{1}{D^2+1}e^{2ix}\right\}.$$

We now use inverse operator shift theorem 16.62,

$$y_p = \text{Im}\left\{e^{2ix}\frac{1}{(D+2i)^2+1}(1)\right\}$$

$$= \text{Im}\left\{e^{2ix}\frac{1}{D^2+4iD-3}(1)\right\} = \text{Im}\left\{e^{2ix}\frac{1}{-3\left(1-\dfrac{4iD+D^2}{3}\right)}(1)\right\}$$

$$= \text{Im}\left\{\frac{e^{2ix}}{-3}\right\} = -\frac{1}{3}\sin 2x.$$

(b)

$$y_p = \frac{1}{D^2+4}x^2\cos x = \frac{1}{D^2+4}\text{Re}(x^2e^{ix}) = \text{Re}\left\{\frac{1}{D^2+4}x^2e^{ix}\right\}$$

$$= \text{Re}\left\{e^{ix}\frac{1}{(D+i)^2+4}x^2\right\} = \text{Re}\left\{e^{ix}\frac{1}{D^2+2iD+3}x^2\right\}$$

$$= \text{Re}\left\{e^{ix}\frac{1}{3\left(1+\dfrac{2iD+D^2}{3}\right)}x^2\right\}$$

$$= \text{Re}\left\{\frac{e^{ix}}{3}\left[1-\left(\frac{2iD+D^2}{3}\right)+\left(\frac{2iD+D^2}{3}\right)^2-\cdots\right]x^2\right\}$$

$$= \frac{1}{3}\text{Re}\left\{e^{ix}\left[1-\frac{2iD}{3}-\frac{7D^2}{9}+\cdots\right]x^2\right\} = \frac{1}{3}\text{Re}\left\{e^{ix}\left[x^2-\frac{4ix}{3}-\frac{14}{9}\right]\right\}$$

$$= \frac{1}{3}\left\{x^2\cos x+\frac{4x}{3}\sin x-\frac{14}{9}\cos x\right\}.$$

(c) For the differential equation $y'' + 9y = \sin 3x$, we have

$$y_p = \frac{1}{D^2+9}\sin 3x = \frac{1}{D^2+9}\text{Im}(e^{3ix}) = \text{Im}\left\{\frac{1}{D^2+9}e^{3ix}\right\}$$

$$= \text{Im}\left\{e^{3ix}\frac{1}{(D+3i)^2+9}(1)\right\} = \text{Im}\left\{e^{3ix}\frac{1}{D^2+6iD}(1)\right\}$$

$$= \text{Im}\left\{e^{3ix}\frac{1}{D}\frac{1}{D+6i}(1)\right\} = \text{Im}\left\{\frac{e^{3ix}}{6i}\frac{1}{D}\frac{1}{1+D/6i}(1)\right\}$$

$$= \text{Im}\left\{-\frac{i}{6}e^{3ix}\frac{1}{D}\left(1-\frac{D}{6i}+\cdots\right)(1)\right\} = \text{Im}\left\{-\frac{i}{6}e^{3ix}\frac{1}{D}(1)\right\}$$

$$= \text{Im}\left\{-\frac{i}{6}e^{3ix}x\right\} = -\frac{x}{6}\cos 3x.$$

(d) For the differential equation $y'' + 4y = x\sin 2x$, we get

$$\begin{aligned}
y_p &= \frac{1}{D^2+4}x\sin 2x = \frac{1}{D^2+4}\operatorname{Im}(xe^{2ix}) = \operatorname{Im}\left\{\frac{1}{D^2+4}xe^{2ix}\right\}\\
&= \operatorname{Im}\left\{e^{2ix}\frac{1}{(D+2i)^2+4}x\right\} = \operatorname{Im}\left\{e^{2ix}\frac{1}{D^2+4iD}x\right\}\\
&= \operatorname{Im}\left\{e^{2ix}\frac{1}{D}\frac{1}{D+4i}x\right\} = \operatorname{Im}\left\{\frac{1}{4i}e^{2ix}\frac{1}{D}\left(1-\frac{D}{4i}+\cdots\right)x\right\}\\
&= \operatorname{Im}\left\{-\frac{i}{4}e^{2ix}\frac{1}{D}\left(x-\frac{1}{4i}\right)\right\} = \operatorname{Im}\left\{-\frac{i}{4}e^{2ix}\left(\frac{x^2}{2}+\frac{ix}{4}\right)\right\}\\
&= -\frac{x^2}{8}\cos 2x + \frac{x}{16}\sin 2x.
\end{aligned}$$

(e) For the differential equation $y'' + 2y' + 4y = e^{-x}\sin\sqrt{3}x$, we have

$$\begin{aligned}
y_p &= \frac{1}{D^2+2D+4}e^{-x}\sin\sqrt{3}x = \frac{1}{D^2+2D+4}\operatorname{Im}(e^{-x}e^{\sqrt{3}ix})\\
&= \operatorname{Im}\left\{\frac{1}{D^2+2D+4}e^{(-1+\sqrt{3}i)x}\right\}\\
&= \operatorname{Im}\left\{e^{(-1+\sqrt{3}i)x}\frac{1}{(D-1+\sqrt{3}i)^2+2(D-1+\sqrt{3}i)+4}(1)\right\}\\
&= \operatorname{Im}\left\{e^{(-1+\sqrt{3}i)x}\frac{1}{D^2+2\sqrt{3}iD}(1)\right\} = \operatorname{Im}\left\{e^{(-1+\sqrt{3}i)x}\frac{1}{D}\frac{1}{D+2\sqrt{3}i}(1)\right\}\\
&= \operatorname{Im}\left\{e^{(-1+\sqrt{3}i)x}\frac{1}{D}\frac{1}{2\sqrt{3}i}\right\} = \operatorname{Im}\left\{-\frac{i}{2\sqrt{3}}xe^{(-1+\sqrt{3}i)x}\right\}\\
&= -\frac{xe^{-x}}{2\sqrt{3}}\cos\sqrt{3}x.
\end{aligned}$$

■

EXAMPLE 16.25

Find a general solution of $y'' + 6y' + y = \sin 3x$.

SOLUTION The auxiliary equation is $m^2 + 6m + 1 = 0$ with solutions $m = -3 \pm 2\sqrt{2}$. Consequently,

$$y_h(x) = C_1e^{(-3+2\sqrt{2})x} + C_2e^{(-3-2\sqrt{2})x}.$$

A particular solution is

$$y_p = \frac{1}{D^2+6D+1}\text{Im}(e^{3ix}) = \text{Im}\left\{\frac{1}{D^2+6D+1}e^{3ix}\right\}$$

$$= \text{Im}\left\{e^{3ix}\frac{1}{(D+3i)^2+6(D+3i)+1}(1)\right\}$$

$$= \text{Im}\left\{e^{3ix}\frac{1}{D^2+(6+6i)D+(-8+18i)}(1)\right\} = \text{Im}\left\{e^{3ix}\frac{1}{-8+18i}\right\}$$

$$= \text{Im}\left\{e^{3ix}\frac{1}{-8+18i}\,\frac{-8-18i}{-8-18i}\right\} = -\frac{1}{194}(4\sin 3x + 9\cos 3x).$$

Finally, then, a general solution of the differential equation is

$$y(x) = C_1 e^{(-3+2\sqrt{2})x} + C_2 e^{(-3-2\sqrt{2})x} - \frac{1}{194}(4\sin 3x + 9\cos 3x).$$

■

Examples 16.19–16.25 have illustrated that with identity 16.62 and the concept of series, we can obtain a particular solution for 16.53 whenever $F(x)$ is x^n, e^{px}, $\sin px$, $\cos px$, and/or any sums or products thereof. In summary:

When $F(x) = x^n$, use series 16.59 for $1/\phi(D)$.

In any other situation, use 16.62, and then series 16.59 for $1/\phi(D+p)$.

We make one final comment. We introduced complex exponentials e^{ipx} to handle terms involving $\sin px$ and $\cos px$. This is not the only way to treat trigonometric functions, since there do exist other methods that completely avoid complex numbers. Unfortunately, these methods require memorization of somewhat involved identities. Because of this, we prefer the use of complex numbers.

EXERCISES 16.10

In Exercises 1–11 find a particular solution for the differential equation, by both the method of operators and the method of undetermined coefficients. Find a general solution of the equation.

1. $2y'' - 16y' + 32y = -e^{4x}$

2. $y'' + 2y' - 2y = x^2 e^{-x}$

3. $y''' - 3y'' + y' - 3y = 3xe^x + 2$

4. $y'''' + 2y'' + y = \cos 2x$

5. $y''' - 6y'' + 12y' - 8y = 2e^{2x}$

6. $y'''' + 5y'' + 4y = e^{-2x}$

7. $y''' - 3y'' + 2y' = x^2 + e^{-x}$

8. $2y'' + 16y' + 82y = -2e^{2x}\sin x$

9. $y'' + y' - 6y = x + \cos x$

10. $y'' - 4y' + 5y = x\cos x$

11. $3y''' - 12y'' + 18y' - 12y = x^2 + 3x - 4$

In Exercises 12–15 state the form for the particular solution predicted by the method of undetermined coefficients. Do not evaluate the coefficients.

12. $y''' + 9y'' + 27y' + 27y = xe^{3x} + 2x\cos x$

13. $y''' + 4y'' + y' + 4y = xe^x \sin x$

14. $2y''' - 6y'' - 12y' + 16y = xe^x + 2x^3 - 4\cos x$

15. $2y'' - 4y' + 10y = 5e^x \sin 2x$

16. Use the operator shift theorem of Exercise 19 in Section 16.9 to verify the inverse operator shift theorem 16.62.

In Exercises 17 and 18 find a general solution for the differential equation.

17. $y'' + 2y' - 4y = \cos^2 x$

18. $2y'' - 4y' + 3y = \cos x \sin 2x$

In Exercises 19 and 20 find a solution for the differential equation.

19. $y'' - 3y' + 2y = 8x^2 + 12e^{-x}$, $y(0) = 0$, $y'(0) = 2$

20. $y'' + 9y = x(\sin 3x + \cos 3x)$, $y(0) = y'(0) = 0$

21. If J, k, and w are positive constants, find a general solution for $J\dfrac{d^4y}{dx^4} + ky = w$.

22. The second-order linear differential equation

$$x^2\frac{d^2y}{dx^2} + ax\frac{dy}{dx} + by = F(x), \qquad a, b \text{ constants},$$

is called the *Cauchy-Euler linear equation*. Because of the x^2- and x-factors, it does not have constant coefficients, and is therefore not immediately amenable to the techniques of this chapter. Show that if we make a change of independent variable $x = e^z$, then

$$x\frac{dy}{dx} = \frac{dy}{dz}, \qquad x^2\frac{d^2y}{dx^2} = \frac{d^2y}{dz^2} - \frac{dy}{dz},$$

and that as a result the Cauchy-Euler equation is transformed into a linear equation in $y(z)$ with constant coefficients.

In Exercises 23 and 24 use the technique of Exercise 22 to find a general solution for the differential equation.

23. $\dfrac{d^2u}{dr^2} + \dfrac{1}{r}\dfrac{du}{dr} - \dfrac{u}{r^2} = 0$, $r > 0$

24. $x^2y'' + xy' + 4y = 1$, $x > 0$

25. If M, β, k, A, and ω are all positive constants, find a particular solution of the linear differential equation

$$M\frac{d^2x}{dt^2} + \beta\frac{dx}{dt} + kx = A\sin\omega t.$$

SECTION 16.11

Applications of Linear Differential Equations

Vibrating Mass-Spring Systems

In Figure 16.14 we have shown a mass M suspended vertically from a spring. If M is given an initial motion in the vertical direction (and the vertical direction only), then we expect M to oscillate up and down for some time. In this section we show how to describe these oscillations mathematically.

FIGURE 16.14

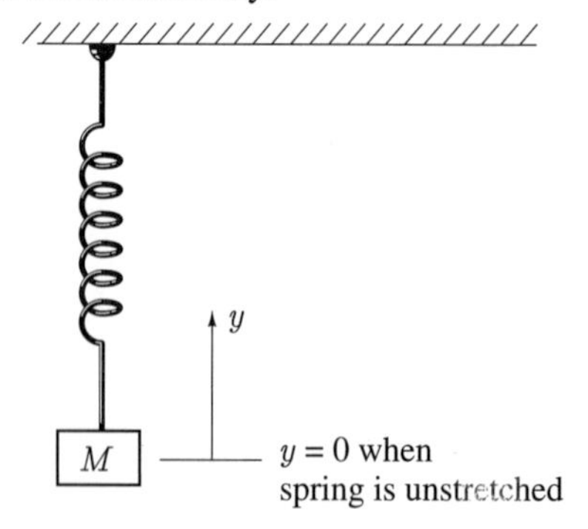

In order to describe the position of M as a function of time t, we must choose a vertical coordinate system. There are two natural places to choose the origin $y = 0$, one being the position of M when the spring is unstretched. Suppose we do this and choose y as positive upward. When M is a distance y away from the origin, the restoring force of the spring has y-component $-ky$ ($k > 0$). In addition, if g is the acceleration due to gravity ($g < 0$), then the force of gravity on M has y-component Mg. Finally, suppose the oscillations take place in a medium that exerts a damping force proportional to the instantaneous velocity of M. This damping force must therefore have a y-component of the form $-\beta(dy/dt)$, where β is a positive constant. The total force on M therefore has y-component

$$-ky + Mg - \beta\frac{dy}{dt},$$

and Newton's second law states that the acceleration d^2y/dt^2 of M must satisfy the equation

$$-ky + Mg - \beta\frac{dy}{dt} = M\frac{d^2y}{dt^2}.$$

Consequently, the differential equation that determines the position $y(t)$ of M relative to the unstretched position of the spring is

$$M\frac{d^2y}{dt^2} + \beta\frac{dy}{dt} + ky = Mg. \tag{16.65}$$

The alternative possibility for describing oscillations is to attach M to the spring and slowly lower M until it reaches an equilibrium position. At this position, the restoring force of the spring is exactly equal to the force of gravity on the mass, and the mass, left by itself, will remain motionless. If s is the amount of stretch in the spring at equilibrium, and g is the acceleration due to gravity, then at equilibrium

$$ks + Mg = 0, \quad \text{where } s > 0 \text{ and } g < 0. \tag{16.66}$$

Suppose we take the equilibrium position as $x = 0$ and x as positive upward (Figure 16.15). When M is a distance x away from its equilibrium position, the restoring force on M has x-component $k(s-x)$. The x-component of the force of gravity remains as Mg, and that of the damping force is $-\beta(dx/dt)$. Newton's second law therefore implies that

$$M\frac{d^2x}{dt^2} = k(s-x) + Mg - \beta\frac{dx}{dt},$$

or,

$$M\frac{d^2x}{dt^2} + \beta\frac{dx}{dt} + kx = Mg + ks.$$

But according to 16.66, $Mg + ks = 0$, and hence

$$M\frac{d^2x}{dt^2} + \beta\frac{dx}{dt} + kx = 0. \tag{16.67}$$

This is the differential equation describing the displacement $x(t)$ of M relative to the equilibrium position of M.

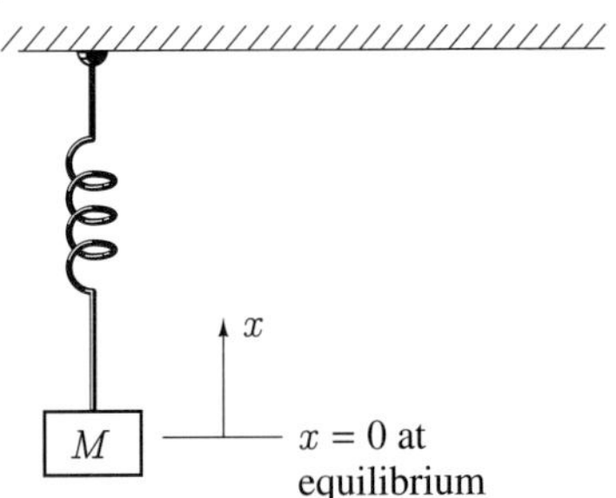

FIGURE 16.15

Note that both equations 16.65 and 16.67 are linear second-order differential equations with constant coefficients. The advantage of 16.67 is that it is homogeneous as well, and this is simply due to a convenient choice of dependent variable (x as opposed to y). Physically, we are saying that there are two parts to the spring force $k(s-x)$: a part ks and a part $-kx$. Gravity is always acting on M, and that part ks of the spring force is counteracting it in an attempt to restore the spring to its unstretched position. Because these forces always cancel, we might just as well eliminate both of them from our discussion. This would leave us $-kx$, and we therefore interpret $-kx$ as the *spring force attempting to restore the mass to its equilibrium position.*

If we choose equation 16.67 to describe the motion of M (and this equation is usually chosen over equation 16.65), we must remember three things: x is measured from equilibrium, $-kx$ is the spring force attempting to restore M to its equilibrium position, and gravity has been taken into account.

There are three basic ways to initiate the motion. First, we can move the mass away from its equilibrium position and then release it, giving it an initial displacement but no initial velocity. Second, we can strike the mass at the equilibrium position, imparting an initial velocity but no initial displacement. And finally, we can give the mass both an initial displacement and an initial velocity. Each of these methods adds two initial conditions to the differential equation.

To be complete we note that when (in addition to the forces already mentioned) there is an externally applied force acting on the mass that is represented as a function of time by $F(t)$, then equation 16.67 is modified to

$$M\frac{d^2x}{dt^2} + \beta\frac{dx}{dt} + kx = F(t). \tag{16.68}$$

Perhaps, for example, M contains some iron, and $F(t)$ is due to a magnet directly below M that constantly exerts an attractive force on M.

EXAMPLE 16.26

A 2-kg mass is suspended vertically from a spring with constant 16 N/m. The mass is raised 10 cm above its equilibrium position and then released. If damping is ignored, find the amplitude, period, and frequency of the motion.

SOLUTION If we choose $x = 0$ at the equilibrium position of the mass and x positive upward (Figure 16.15), then the differential equation for the motion $x(t)$ of the mass is

$$2\frac{d^2x}{dt^2} = -16x,$$

or,

$$\frac{d^2x}{dt^2} + 8x = 0,$$

along with the initial conditions

$$x(0) = 1/10, \quad x'(0) = 0.$$

The auxiliary equation is $m^2 + 8 = 0$ with solutions $m = \pm 2\sqrt{2}i$. Consequently,

$$x(t) = C_1 \cos(2\sqrt{2}t) + C_2 \sin(2\sqrt{2}t).$$

The initial conditions require

$$1/10 = C_1, \quad 0 = 2\sqrt{2}C_2.$$

Thus,

$$x(t) = \frac{1}{10}\cos(2\sqrt{2}t).$$

The amplitude of the oscillations is 1/10 m, the period is $2\pi/(2\sqrt{2}) = \pi/\sqrt{2}$ s, and the frequency is $\sqrt{2}/\pi\ \mathrm{s}^{-1}$. A graph of this function (Figure 16.16) illustrates the oscillations of the mass about its equilibrium position. This is an example of **simple harmonic motion**.

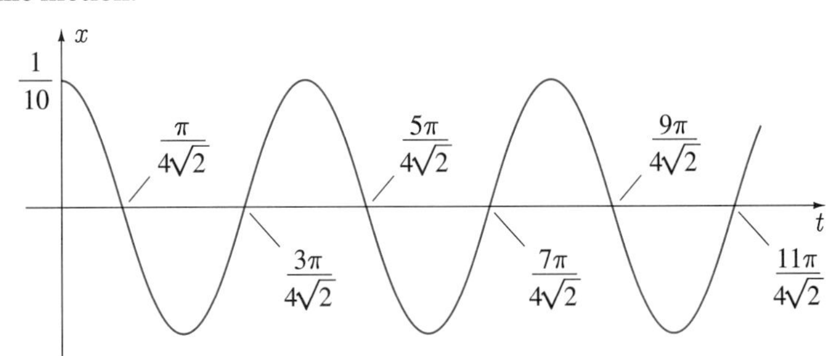

FIGURE 16.16

EXAMPLE 16.27

A 100-g mass is suspended vertically from a spring with constant 5 N/m. The mass is pulled 5 cm below its equilibrium position and given a velocity of 2 m/s upward. If, during the motion, the mass is acted on by a damping force in newtons numerically equal to one-twentieth the instantaneous velocity in metres per second, find the position of the mass at any time.

SOLUTION If we choose $x = 0$ at the equilibrium position of the mass and x positive upward (Figure 16.15), then the differential equation for the motion $x(t)$ of the mass is

$$\frac{1}{10}\frac{d^2x}{dt^2} = -5x - \frac{1}{20}\frac{dx}{dt},$$

or,

$$2\frac{d^2x}{dt^2} + \frac{dx}{dt} + 100x = 0,$$

along with the initial conditions

$$x(0) = -1/20, \quad x'(0) = 2.$$

The auxiliary equation is $2m^2 + m + 100 = 0$ with solutions

$$m = \frac{-1 \pm \sqrt{1-800}}{4} = \frac{-1 \pm \sqrt{799}\,i}{4}.$$

Consequently,

$$x(t) = e^{-t/4}[C_1\cos(\sqrt{799}\,t/4) + C_2\sin(\sqrt{799}\,t/4)].$$

The initial conditions require

$$-1/20 = C_1, \quad 2 = -C_1/4 + \sqrt{799}\,C_2/4,$$

from which we get

$$C_2 = \frac{159\sqrt{799}}{15\ 980}.$$

Finally, then,

$$x(t) = e^{-t/4}\left[-\frac{1}{20}\cos\left(\frac{\sqrt{799}t}{4}\right) + \frac{159\sqrt{799}}{15\ 980}\sin\left(\frac{\sqrt{799}t}{4}\right)\right]$$
$$= e^{-0.25t}[-0.05\cos(7.07t) + 0.28\sin(7.07t)].$$

The graph of this function in Figure 16.17 clearly indicates how the amplitude of the oscillations decreases in time.

FIGURE 16.17

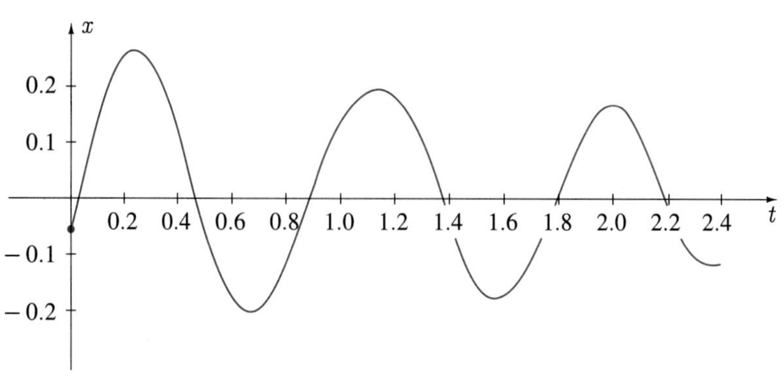

■

LCR Circuits

If a resistance R, an inductance L, and a capacitance C are connected in series with an electromotive force $E(t)$ (Figure 16.18) and the switch is closed, current flows in the circuit and charge builds up in the capacitor. If at any time t, Q is the charge on the capacitor and I is the current in the loop, then Kirchhoff's law states that

$$L\frac{dI}{dt} + RI + \frac{Q}{C} = E(t), \tag{16.69}$$

where $L\,dI/dt, RI$, and Q/C represent the voltage drops across the inductor, the resistor, and the capacitor, respectively. If we substitute $I = dQ/dt$, then

$$L\frac{d^2Q}{dt^2} + R\frac{dQ}{dt} + \frac{1}{C}Q = E(t), \tag{16.70}$$

a second-order linear differential equation for $Q(t)$.

Alternatively, if we differentiate this equation, we obtain

$$L\frac{d^2I}{dt^2} + R\frac{dI}{dt} + \frac{1}{C}I = E'(t), \tag{16.71}$$

a second-order linear differential equation for $I(t)$.

FIGURE 16.18

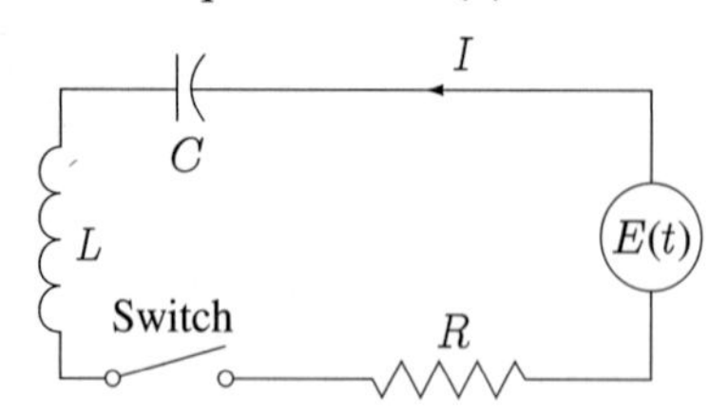

EXAMPLE 16.28

At time $t = 0$, a 25-Ω resistor, a 2-H inductor, and a 0.01-F capacitor are connected in series with a generator producing an alternating voltage of $10 \sin(5t), t \geq 0$ (Figure 16.19). Find the charge on the capacitor and the current in the circuit if the capacitor is uncharged when the circuit is closed.

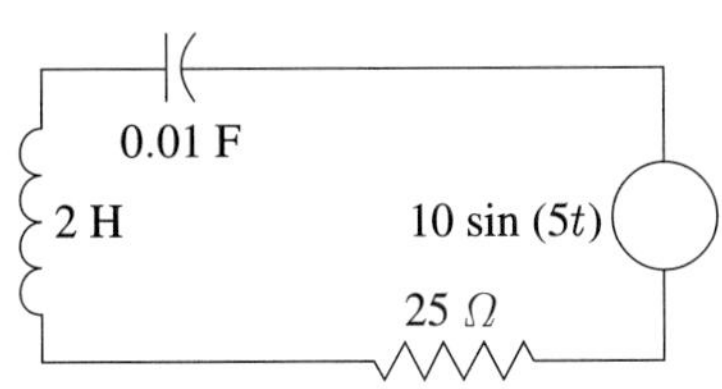

FIGURE 16.19

SOLUTION The differential equation for the charge Q on the capacitor is (see equation 16.70)

$$2\frac{d^2Q}{dt^2} + 25\frac{dQ}{dt} + 100Q = 10\sin(5t),$$

to which we add the initial conditions

$$Q(0) = 0, \quad Q'(0) = I(0) = 0.$$

The auxiliary equation is $2m^2 + 25m + 100 = 0$ with solutions

$$m = \frac{-25 \pm \sqrt{625 - 800}}{4} = \frac{-25 \pm 5\sqrt{7}i}{4}.$$

Consequently, the general solution of the homogeneous equation is

$$Q_h(t) = e^{-25t/4}[C_1 \cos(5\sqrt{7}t/4) + C_2 \sin(5\sqrt{7}t/4)].$$

To find a particular solution of the nonhomogeneous equation by undetermined coefficients, we set

$$Q_p(t) = A\sin(5t) + B\cos(5t).$$

Substitution of this function into the differential equation gives

$$2\{-25A\sin(5t) - 25B\cos(5t)\} + 25\{5A\cos(5t) - 5B\sin(5t)\}$$
$$+ 100\{A\sin(5t) + B\cos(5t)\} = 10\sin(5t).$$

This equation requires A and B to satisfy

$$50A - 125B = 10, \quad 125A + 50B = 0,$$

the solution of which is $A = 4/145, B = -10/145$. A particular solution is therefore

$$Q_p(t) = \frac{2}{145}[-5\cos(5t) + 2\sin(5t)]$$

and

$$\begin{aligned}Q(t) &= Q_h(t) + Q_p(t)\\ &= e^{-25t/4}\left[C_1\cos\left(\frac{5\sqrt{7}t}{4}\right) + C_2\sin\left(\frac{5\sqrt{7}t}{4}\right)\right]\\ &\quad + \frac{2}{145}[2\sin(5t) - 5\cos(5t)].\end{aligned}$$

The initial conditions require

$$0 = C_1 - \frac{10}{145}, \quad 0 = -\frac{25}{4}C_1 + \frac{5\sqrt{7}}{4}C_2 + \frac{20}{145},$$

and these imply that $C_1 = 10/145, C_2 = 34/(145\sqrt{7})$. Consequently,

$$\begin{aligned}Q(t) &= \frac{e^{-25t/4}}{145\sqrt{7}}\left[10\sqrt{7}\cos\left(\frac{5\sqrt{7}t}{4}\right) + 34\sin\left(\frac{5\sqrt{7}t}{4}\right)\right]\\ &\quad + \frac{2}{145}[2\sin(5t) - 5\cos(5t)].\end{aligned}$$

The current in the circuit is

$$\begin{aligned}I(t) = \frac{dQ}{dt} &= \left(-\frac{25}{4}\right)\frac{e^{-25t/4}}{145\sqrt{7}}\left[10\sqrt{7}\cos\left(\frac{5\sqrt{7}t}{4}\right) + 34\sin\left(\frac{5\sqrt{7}t}{4}\right)\right]\\ &\quad + \frac{e^{-25t/4}}{145\sqrt{7}}\left[-\frac{175}{2}\sin\left(\frac{5\sqrt{7}t}{4}\right) + \frac{85\sqrt{7}}{2}\cos\left(\frac{5\sqrt{7}t}{4}\right)\right]\\ &\quad + \frac{2}{145}[10\cos(5t) + 25\sin(5t)]\\ &= -\frac{e^{-25t/4}}{29\sqrt{7}}\left[4\sqrt{7}\cos\left(\frac{5\sqrt{7}t}{4}\right) + 60\sin\left(\frac{5\sqrt{7}t}{4}\right)\right]\\ &\quad + \frac{2}{29}[2\cos(5t) + 5\sin(5t)].\end{aligned}$$

The solution $Q(t)$ contains two parts. The first two terms (containing the exponential $e^{-25t/4)}$) are $Q_h(t)$ with the constants C_1 and C_2 determined by the initial conditions; the last two terms are $Q_p(t)$. We point this out because the two parts display completely different characteristics. For small t, both parts of $Q(t)$ are present and contribute significantly, but for very large t, the first two terms become negligible. In other words, after a long time, the charge $Q(t)$ on the capacitor is defined essentially by $Q_p(t)$. We call $Q_p(t)$ the steady-state part of the solution, and the two other terms in $Q(t)$ are called the transient part of the solution. Similarly, the first two terms in $I(t)$ are called the transient part of the current and the last two terms the steady-state part of the current.

Finally, note that the frequency of the steady-state part of either $Q(t)$ or $I(t)$ is exactly that of the forcing voltage $E(t)$. ■

The similarity between differential equations 16.68 and 16.70 cannot go unmentioned:

$$M\frac{d^2x}{dt^2} + \beta\frac{dx}{dt} + kx = F(t),$$
$$L\frac{d^2Q}{dt^2} + R\frac{dQ}{dt} + \frac{1}{C}Q = E(t).$$

Each of the coefficients M, β, and k for the mechanical system has its analogue L, R, and $1/C$ in the electrical system. This suggests that LCR circuits might be used to model complicated physical systems subject to vibrations, and conversely, that mass-spring systems might represent complicated electrical systems.

EXERCISES 16.11

1. A 1-kg mass is suspended vertically from a spring with constant 16 N/m. The mass is pulled 10 cm below its equilibrium position, and then released. Find the position of the mass, relative to its equilibrium position, at any time if

(a) damping is ignored.

(b) a damping force in newtons equal to one-tenth the instantaneous velocity in metres per second acts on the mass.

(c) a damping force in newtons equal to ten times the instantaneous velocity in metres per second acts on the mass.

2. A 200-g mass suspended vertically from a spring with constant 10 N/m is set into vibration by an external force in newtons given by $4 \sin 10t$, $t \geq 0$. During the motion a damping force in newtons equal to $3/2$ the velocity of the mass in metres per second acts on the mass. Find the position of the mass as a function of time t.

3. A 0.001-F capacitor and a 2-H inductor are connected in series with a 20-V battery. If there is no charge on the capacitor before the battery is connected, find the current in the circuit as a function of time.

4. At time $t = 0$, a 0.02-F capacitor, a 100-Ω resistor, and 1-H inductor are connected in series. If the charge on the capacitor is initially 5 C, find its charge as a function of time.

5. A 5-H inductor and 20-Ω resistor are connected in series with a generator supplying an oscillating voltage of $10 \sin 2t$, $t \geq 0$. What are the transient and steady-state currents in the circuit?

6. At time $t = 0$, a mass M is attached to the end of a hanging spring with constant k, and then released. Assuming that friction is negligible, find the subsequent displacement of the mass as a function of time.

7. A 0.5-kg mass sits on a table attached to a spring with constant 18 N/m (Figure 16.20). The mass is pulled so as to stretch the spring 5 cm and then released.

(a) If friction between the mass and the table creates a force of 0.5 N that opposes motion, show that the differential equation determining motion is

$$\frac{d^2x}{dt^2} + 36x = 1, \quad x(0) = 0.05, \quad x'(0) = 0.$$

(b) Find where the mass comes to rest for the first time. Will it move from this position?

FIGURE 16.20

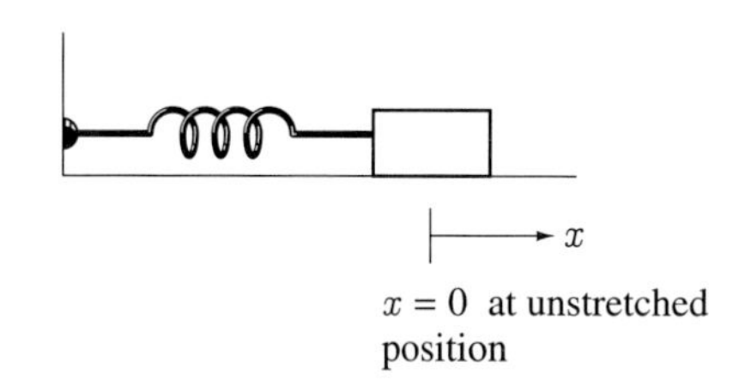

8. Repeat Exercise 7 given that the mass is pulled 25 cm to the right.

9. At time $t = 0$ an uncharged 0.1-F capacitor is connected in series with 0.5-H inductor and a 3-Ω resistor. If the current in the circuit at this instant is 1 A, find the maximum charge that the capacitor stores.

10. Differential equation 16.67 describes the motion of a mass M at the end of a spring, taking damping proportional to velocity into account. Show each of the following.

(a) If $\beta = 0$,

$$x(t) = C_1 \cos(\sqrt{k/M}t) + C_2 \sin(\sqrt{k/M}t),$$

called *simple harmonic motion.*

(b) If $\beta \neq 0$ and $\beta^2 - 4kM < 0$,

$$x(t) = e^{-\beta t/(2M)}(C_1 \cos \omega t + C_2 \sin \omega t),$$

$$\text{where } \omega = \frac{\sqrt{4kM - \beta^2}}{2M},$$

called *damped oscillatory motion.*

(c) If $\beta \neq 0$ and $\beta^2 - 4kM > 0$,

$$x(t) = e^{-\beta t/(2M)}(C_2 e^{\omega t} + C_2 e^{-\omega t}),$$

$$\text{where } \omega = \frac{\sqrt{\beta^2 - 4kM}}{2M},$$

called *overdamped motion.*

(d) If $\beta \neq 0$ and $\beta^2 - 4kM = 0$,

$$x(t) = (C_1 + C_2 t)e^{-\beta t/(2M)},$$

called *critically damped motion.*

11. A 100-g mass is suspended from a spring with constant 4000 N/m. At its equilibrium position, it is suddenly (time $t = 0$) given an upward velocity of 10 m/s. If an external force $3 \cos 200t$, $t \geq 0$, acts on the mass, find its displacement as a function of time. Do you note anything strange?

12. A vertical spring having constant 64 N/m has a 1-kg mass attached to it. An external force $F(t) = 2 \sin 8t$, $t \geq 0$ is applied to the mass. If the mass is at rest at its equilibrium position at time $t = 0$, and damping is negligible, find the position of the mass as a function of time. What happens to the oscillations as time progresses? This phenomenon is called *resonance*; it occurs when the frequency of the forcing function is identical to the natural frequency of the undamped system.

13. A mass M is suspended from a vertical spring with constant k. If an external force $F(t) = A \cos \omega t$, $t \geq 0$, is applied to the mass, find the value of ω that causes resonance.

14. A $25/9$-H inductor, a 0.04-F capacitor, and a generator with voltage $15 \cos 3t$ are connected in series at time $t = 0$. Find the current in the circuit as a function of time. Does resonance occur?

15. A mass M is suspended from a vertical spring with constant k. When an external force $F(t) = A \cos \omega t$, $t \geq 0$ is applied to the mass, and a damping force proportional to velocity acts on the mass during its subsequent oscillations, the differential equation governing the motion of M is

$$M\frac{d^2x}{dt^2} + \beta\frac{dx}{dt} + kx = A \cos \omega t.$$

(a) Show that steady-state oscillations of M are defined by

$$x(t) = \frac{A}{(k - M\omega^2)^2 + \beta^2\omega^2} \times [(k - M\omega^2)\cos \omega t + \beta\omega \sin \omega t].$$

(b) Verify that this function can be expressed in the form

$$x(t) = \frac{A}{\sqrt{(k - M\omega^2)^2 + \beta^2\omega^2}} \sin(\omega t + \phi),$$

where

$$\sin \phi = \frac{k - M\omega^2}{\sqrt{(k - M\omega^2)^2 + \beta^2\omega^2}},$$

$$\cos \phi = \frac{\beta\omega}{\sqrt{(k - M\omega^2)^2 + \beta^2\omega^2}}.$$

(c) If resonance is said to occur when the amplitude of the steady-state oscillations is a maximum, what value of ω yields resonance? What is the maximum amplitude?

16. **(a)** A cube L m on each side and with mass M kg floats half submerged in water. If it is pushed down slightly and then released, oscillations take place. Use Archimedes' principle to find the differential equation governing these oscillations. Assume no damping forces due to the viscosity of the water.

(b) What is the frequency of the oscillations?

17. A cylindrical buoy 20 cm in diameter floats partially submerged with its axis vertical. When it is depressed slightly and released, its oscillations have a period equal to 4 s. What is the mass of the buoy?

18. A sphere of radius R floats half submerged in water. It is set into vibration by pushing it down slightly and then releasing it. If y denotes the instantaneous distance of its centre below the surface, show that

$$\frac{d^2y}{dt^2} = \frac{-3g}{2R^3}\left(R^2 y - \frac{y^3}{3}\right),$$

where g is the acceleration due to gravity.

19. A cable hangs over a peg, 10 m on one side and 15 m on the other. Find the time for it to slide off the peg

(a) if friction at the peg is negligible.

(b) if friction at the peg is equal to the weight of 1 m of cable.

SUMMARY

A differential equation is an equation that contains an unknown function and some of its derivatives, and the equation must be solved for this function. Depending on the form of the equation, various techniques may be used to find the solution. From this point of view, solving differential equations is much like evaluating antiderivatives: We must first recognize the technique appropriate to the particular problem at hand, and then proceed through the mechanics of the technique. It is important, then, to immediately recognize the type of differential equation under consideration.

Broadly speaking, we divided the differential equations we considered into two main groups: first-order equations together with simple second-order equations, and linear differential equations. A first-order differential equation in $y(x)$ is said to be separable if it can be written in the form $N(y)\,dy = M(x)\,dx$, and a general solution for such an equation is defined implicitly by

$$\int N(y)\,dy = \int M(x)\,dx + C.$$

A differential equation of the form $dy/dx + P(x)y = Q(x)$ is said to be linear first order. If this equation is multiplied by the integrating factor $e^{\int P(x)\,dx}$, then the left side becomes the derivative of $ye^{\int P(x)\,dx}$, and the equation is immediately integrable.

If a second-order differential equation in $y(x)$ has either the dependent variable or the independent variable explicitly missing, it can be reduced to a pair of first-order equations. This is accomplished in the former case by setting $v = y'$ and $v' = y''$, and in the latter case by setting $v = y'$ and $y'' = v\,dv/dy$.

A differential equation in $y(x)$ is said to be linear if it is of the form

$$a_0\frac{d^ny}{dx^n} + a_1\frac{d^{n-1}y}{dx^{n-1}} + \cdots + a_{n-1}\frac{dy}{dx} + a_ny = F(x).$$

The general solution of such an equation is composed of two parts: $y(x) = y_h(x) + y_p(x)$. The function $y_h(x)$ is the general solution of the associated homogeneous equation obtained by replacing $F(x)$ by 0; $y_p(x)$ is any particular solution of the given equation whatsoever. In the special case that the a_i are constants, it is always possible to find $y_h(x)$. This is done by calculating all solutions of the auxiliary equation

$$a_0m^n + a_1m^{n-1} + \cdots + a_{n-1}m + a_n = 0,$$

a polynomial equation in m, and then using the rules of Theorem 16.3.

When the a_i are constants and $F(x)$ is a polynomial, an exponential, a sine, a cosine, or any sums or products thereof, we can find $y_p(x)$ either by undetermined coefficients or by operators. The method of undetermined coefficients is simply an intelligent way of guessing at $y_p(x)$ on the basis of $F(x)$ and $y_h(x)$. Note once again that $y_h(x)$ must be calculated before using the method of undetermined coefficients. The method of operators, on the other hand, is based on formal algebraic manipulations and the inverse operator shift theorem. It does not require prior calculation of $y_h(x)$ and, in its simplest form, uses complex numbers.

Key Terms and Formulas

In reviewing this chapter, you should be able to define or discuss the following key terms:

- Differential equation
- Order of a differential equation
- Initial conditions
- Boundary conditions
- General solution
- Particular solution
- Singular solution
- Separable differential equation
- Homogeneous differential equation
- Linear differential equation
- Integrating factor
- Bernoulli equation
- Friction
- Force of friction
- Coefficient of kinetic friction
- Malthusian population model
- Carrying capacity
- Logistic population model
- Differential operator
- Linear operator
- Homogeneous linear differential equation
- Superposition principle
- Linearly independent solutions
- Auxiliary equation
- Method of undetermined coefficients
- Operator method
- Operator shift theorem
- Inverse operator shift theorem
- Cauchy-Euler equation
- Simple harmonic motion
- Resonance

REVIEW EXERCISES

In Exercises 1–20 find a general solution for the differential equation.

1. $x^2\,dy - y\,dx = 0$

2. $(x+1)\,dx - xy\,dy = 0$

3. $\dfrac{dy}{dx} + 3xy = 2x$

4. $\dfrac{dy}{dx} + 4y = x^2$

5. $\dfrac{d^2y}{dx^2} + 4\dfrac{dy}{dx} + 3y = 2$

6. $\dfrac{d^2y}{dx^2} + 3\dfrac{dy}{dx} + 4y = 2$

7. $yy' = \sqrt{1+y^2}$

8. $\dfrac{d^2y}{dx^2} + \dfrac{1}{x}\dfrac{dy}{dx} = x$

9. $y'' + 6y' + 3y = xe^x$

10. $\dfrac{dy}{dx} + 2xy = 2x^3$

11. $y^2y'' = y'$

12. $y'' - 4y' + 4y = \sin x$

13. $y'' - 4y' + 4y = x^2e^{2x}$

14. $y'' + 4y = \sin 2x$

15. $y'' + 4y' = x^2$

16. $2xy^2\dfrac{dy}{dx} + (x+1)^2y^3 = 0$

17. $\dfrac{d^3y}{dx^3} + 3\dfrac{d^2y}{dx^2} + 3\dfrac{dy}{dx} + y = 2e^{-x}$

18. $y'' + 2y' + 4y = e^{-x}\cos\sqrt{3}x$

19. $\dfrac{dy}{dx} = y\tan x + \cos x$

20. $(2y^2 + 3x)\,dy + dx = 0$

In Exercises 21–24 find the solution of the differential equation satisfying the given condition(s).

21. $y^2\,dx + (x+1)\,dy = 0,\ y(0) = 3$

22. $y'' - 8y' - 9y = 2x + 4,\ y(0) = 3,\ y'(0) = 7$

23. $y'' + 9y = e^x,\ y(0) = 0,\ y(\pi/2) = 4$

24. $y' + \dfrac{2}{x}y = \sin x,\ y(1) = 1$

25. The quantity of radioactive material present in a sample decays at a rate proportional to the amount of the material in the sample (see Section 9.2). If one-quarter of a sample decays in 5 a, how long does it take for 90% of the sample to decay?

26. (a) A piece of wood rises from the bottom of a container of oil 1 m deep. If the wood has mass 0.5 g and volume 1 cm^3, and the density of the oil is 0.9 g/cm^3, show that Archimedes' principle predicts a buoyant force due to fluid pressure of 8.829×10^{-3} N. What is the force on the piece of wood due to gravity and fluid pressure?

(b) The viscosity of the oil opposes motion by exerting a force equal (in newtons) to twice its velocity (in metres per second). Find the distance travelled by the wood as a function of time, assuming that it starts from rest on the bottom.

27. A 100-g mass is suspended vertically from a spring with constant 1 N/m. The mass is pulled 4 cm above its equilibrium position and then released. Find the position of the mass at any time if

(a) damping is ignored.

(b) a damping force in newtons equal to $1/5$ the instantaneous velocity of the mass in metres per second acts on the mass.

(c) a damping force in newtons equal to $\sqrt{2/5}$ the instantaneous velocity of the mass in metres per second acts on the mass.

28. A 10-g stone is dropped over the side of a bridge 50 m above a river 10 m deep. Air resistance is negligible, but as the stone sinks in the water, its motion is retarded by a force in newtons equal to one-fifth the speed of the stone in metres per second. If the stone loses 10% of its speed in penetrating the surface of the water, find

(a) its velocity as a function of time.

(b) its position relative to the drop point as a function of time.

(c) the time it takes to reach the bottom of the river from the point at which it was dropped.

Assume that the water is stationary and that buoyancy due to Archimedes' principle may be neglected.

29. Repeat Exercise 28 but assume that, for the purpose of Archimedes' principle, the volume of the stone is 3 cm^3.

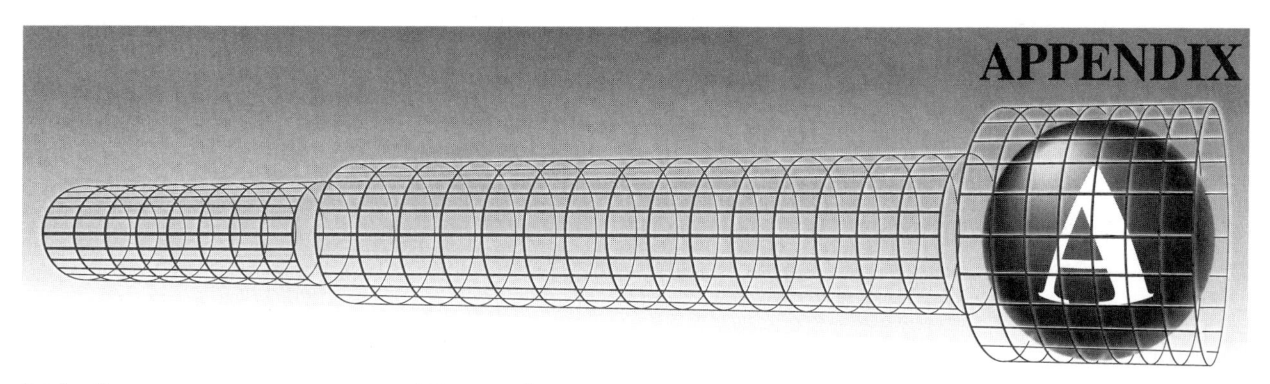

Trigonometry Review

Many physical systems exhibit an oscillatory nature: vibrations of a plucked guitar string, motion of a pendulum, alternating electric currents, fluctuations in room temperature as the thermostat continually engages and disengages the furnace on a cold day, and the rise and fall of tides and waves. Magnitudes of these oscillations are best described by the sine and cosine functions of trigonometry. In addition, rates at which these oscillations occur can be represented by the *derivatives* of these functions. The study of oscillatory systems requires a knowledge of the sine and cosine functions and their derivatives. In this appendix, we briefly review the trigonometric functions and their properties, placing special emphasis on those aspects that are most useful to the study of calculus.

In trigonometry, angles are measured in either degrees or radians; in calculus angles are always measured in radians. In preparation, then, for the calculus of trigonometric functions, we will work completely in radian measure. By definition, a **radian** is that angle subtended at the centre of a circle of radius r by an arc of equal length r (Figure A.1). When an arc has length s (Figure A.2), the number of "units of radius" in this arc is s/r; therefore the angle θ subtended at the centre of the circle is

$$\theta = \frac{s}{r}. \tag{A.1}$$

In particular, if s contains π units of radius ($s = \pi r$), then s represents one-half the circumference of the complete circle, and $\theta = \pi r/r = \pi$ radians. In degree measure, this angle is $180°$; hence we can state that π radians is equivalent to $180°$. This statement enables us to convert angles expressed in degrees to radian measure and vice versa. If the degreee measure of an angle is ϕ, then it is $\pi\phi/180$ radians; conversely if an angle measures θ radians, then its degree measure is $180\,\theta/\pi$. For example, the radian measure for $\phi = 45°$ is $45\pi/180 = \pi/4$ radians.

FIGURE A.1

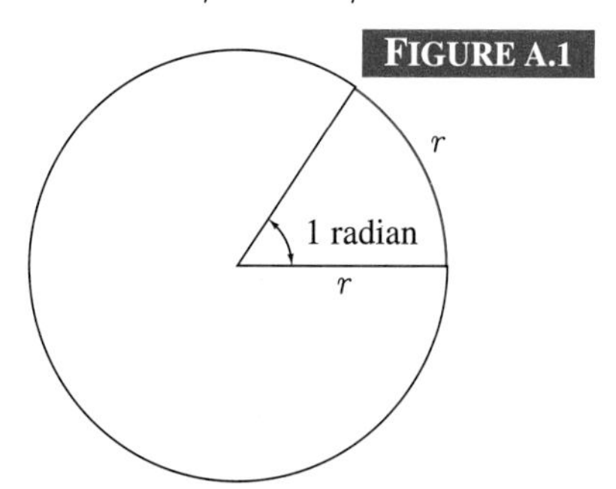

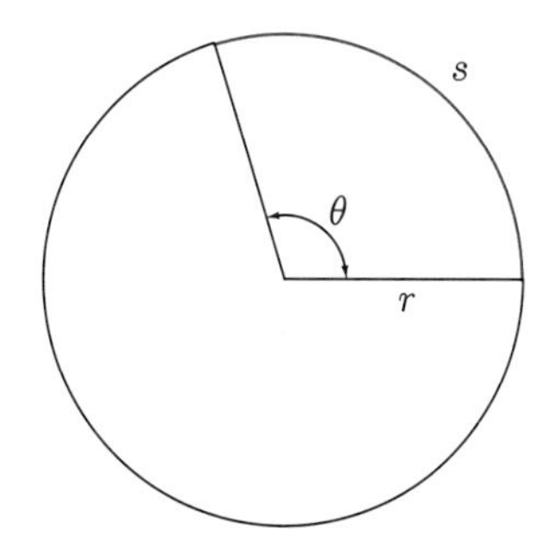

FIGURE A.2

Elementary trigonometry is concerned with relationships among angles and lengths, and, in particular, angles and sides of triangles. This naturally restricts angles to the range

$0 \leq \theta \leq \pi$. For the purposes of calculus, however, we need to talk about angles θ, where θ is any real number — positive, negative, or zero. To do this we first define what we mean by the **standard position of an angle**. If $\theta > 0$, we draw a line segment OP through the origin O of the xy-plane (Figure A.3) in such a way that the positive x-axis must rotate counterclockwise through an angle θ to coincide with OP. In other words, we now regard an angle as rotation, rotation of the positive x-axis to some terminal position. If $0 < \theta < \pi/2$, then OP lies in the first quadrant; and if $\pi/2 < \theta < \pi$, then OP is in the second quadrant. But now if $\theta > \pi$, we have a geometric representation of θ also. For example, angles $7\pi/4$ and $9\pi/4$ are shown in Figure A.4. When $\theta < 0$, we regard θ as a clockwise rotation (Figure A.5).

Note that for any angle θ, the angles $\theta + 2n\pi$ for n an integer have the same terminal position of OP as θ. They are different angles, however, because the positive x-axis must encircle the origin one or more times before reaching the terminal position in the case of $\theta + 2n\pi$.

FIGURE A.3

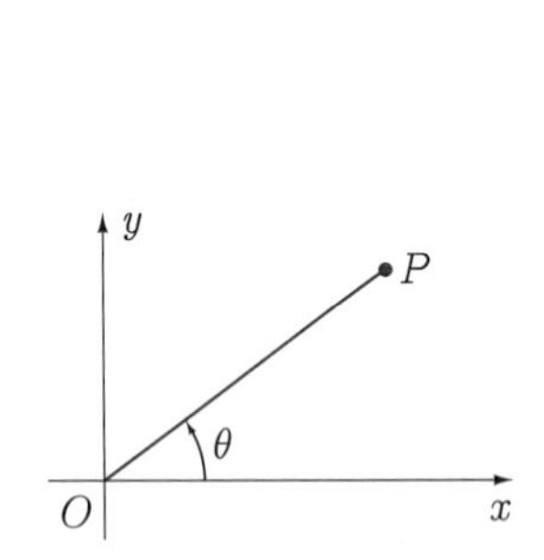

FIGURE A.4

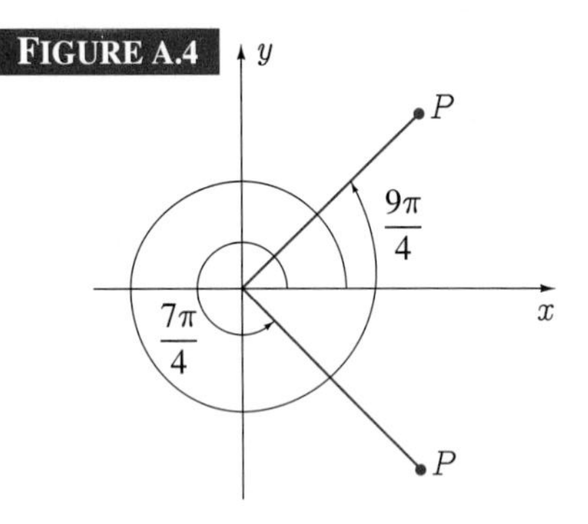

FIGURE A.5

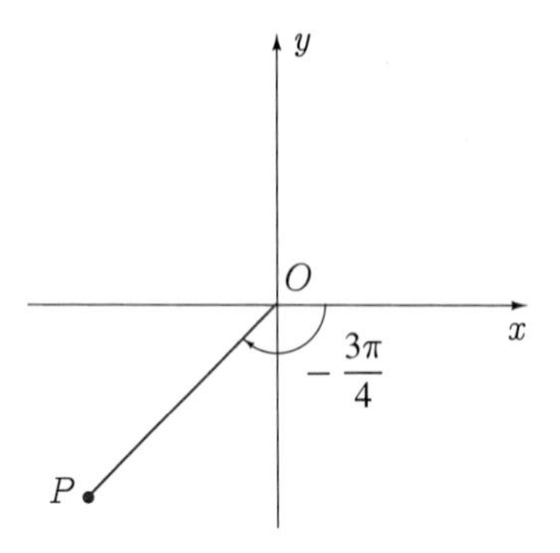

With this representation of angles as rotations, it is easy to define the six trigonometric functions. If θ is an angle in standard position (Figure A.6) and (x, y) are the coordinates of P, we define

$$\sin\theta = \frac{y}{r}, \quad \cos\theta = \frac{x}{r}, \quad \tan\theta = \frac{y}{x},$$
$$\csc\theta = \frac{r}{y}, \quad \sec\theta = \frac{r}{x}, \quad \cot\theta = \frac{x}{y}, \tag{A.2}$$

wherever these ratios are defined, and where $r = \sqrt{x^2 + y^2}$ is always assumed positive. Since x is positive in the first and fourth quadrants, and y is positive in the first and second, signs of the trigonometric functions in the various quadrants are as shown in Figure A.7. Furthermore, since $r \neq 0$, $\sin\theta$ and $\cos\theta$ are defined for all θ, whereas $\tan\theta$ and $\sec\theta$ are not defined for $x = 0$, and $\csc\theta$ and $\cot\theta$ do not exist when $y = 0$.

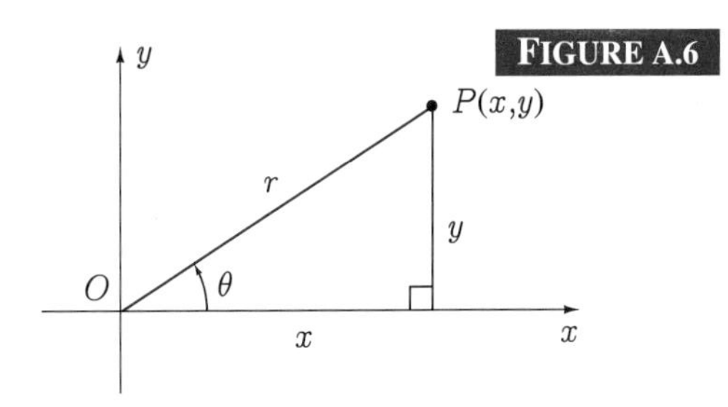

FIGURE A.6

FIGURE A.7

Definitions (A.2) indicate that

$$\csc\theta = \frac{1}{\sin\theta}, \qquad \sec\theta = \frac{1}{\cos\theta}, \qquad \cot\theta = \frac{1}{\tan\theta}. \tag{A.3}$$

In addition, the fact that $r^2 = x^2 + y^2$ leads to the identity

$$\sin^2\theta + \cos^2\theta = 1, \tag{A.4}$$

which in turn implies that

$$1 + \tan^2\theta = \sec^2\theta, \tag{A.5a}$$
$$1 + \cot^2\theta = \csc^2\theta. \tag{A.5b}$$

Identity (A.4) is simply a restatement of the Pythagorean relation ($r^2 = x^2 + y^2$), which allows us to express the hypotenuse of a right-angled triangle in terms of the other two sides. If the triangle is not right-angled (Figure A.8), it is not possible to express one side, c, in terms of the other two sides, a and b, alone, but it is possible to express c in terms of a and b and the angle θ between them. The coordinates of P and Q in Figure A.8 are $(b\cos\theta, b\sin\theta)$ and $(a, 0)$, and therefore by distance formula 1.7 in Section 1.2,

$$\begin{aligned} c^2 = \|PQ\|^2 &= (b\cos\theta - a)^2 + (b\sin\theta)^2 \\ &= b^2(\cos^2\theta + \sin^2\theta) + a^2 - 2ab\cos\theta. \end{aligned}$$

Since $\sin^2\theta + \cos^2\theta = 1$, we have

$$c^2 = a^2 + b^2 - 2ab\cos\theta. \tag{A.6}$$

This result, called the **cosine law**, generalizes the Pythagorean relation to triangles that are not right-angled. It reduces to the Pythagorean relation $c^2 = a^2 + b^2$ when $\theta = \pi/2$.

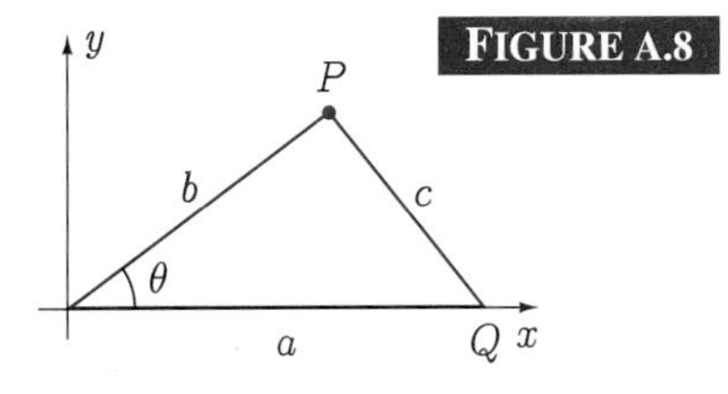

FIGURE A.8

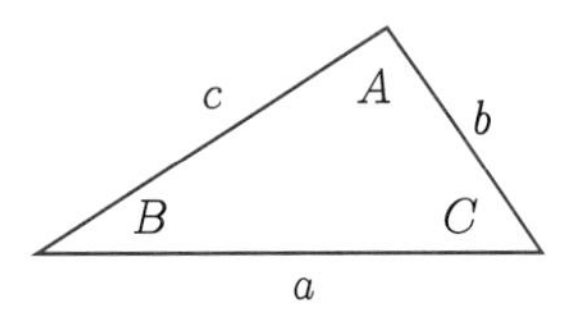

FIGURE A.9

If A, B, and C are the angles in the triangle of Figure A.9, and a, b, and c are the lengths of the opposite sides, then by drawing altitudes of the triangle we can show that

$$\frac{\sin A}{a} = \frac{\sin B}{b} = \frac{\sin C}{c}. \tag{A.7}$$

This result, known as the **sine law**, is also useful in many problems.

FIGURE A.10

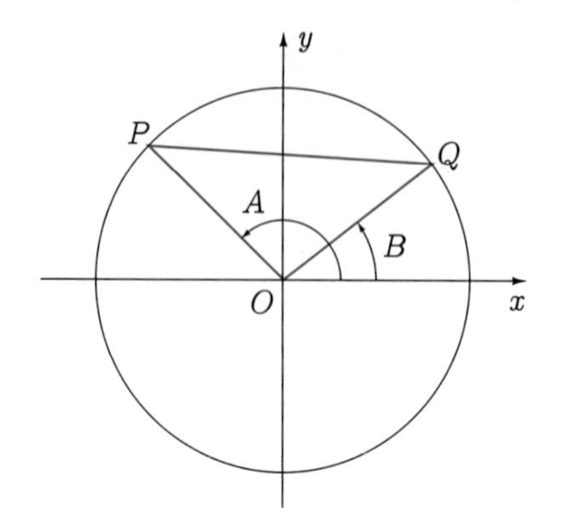

There is a large number of identities satisfied by the trigonometric functions. They can all be derived from the following **compound-angle formulas** for sines and cosines:

$$\sin(A+B) = \sin A\cos B + \cos A\sin B, \quad \text{(A.8 a)}$$
$$\sin(A-B) = \sin A\cos B - \cos A\sin B, \quad \text{(A.8 b)}$$
$$\cos(A+B) = \cos A\cos B - \sin A\sin B, \quad \text{(A.8 c)}$$
$$\cos(A-B) = \cos A\cos B + \sin A\sin B. \quad \text{(A.8 d)}$$

Let us prove one of these, say, identity A.8d, where A and B are the angles in Figure A.10. If P and Q lie on the circle $x^2 + y^2 = r^2$, then their coordinates are $(r\cos A, r\sin A)$ and $(r\cos B, r\sin B)$ respectively. According to formula 1.7, the length of PQ is

$$\begin{aligned}||PQ||^2 &= (r\cos A - r\cos B)^2 + (r\sin A - r\sin B)^2 \\ &= r^2(\cos^2 A + \sin^2 A) + r^2(\cos^2 B + \sin^2 B) \\ &\quad - 2r^2(\cos A\cos B + \sin A\sin B) \\ &= 2r^2 - 2r^2(\cos A\cos B + \sin A\sin B).\end{aligned}$$

But according to the cosine law A.6,

$$\begin{aligned}||PQ||^2 &= ||OP||^2 + ||OQ||^2 - 2||OP||||OQ||\cos(A-B) \\ &= r^2 + r^2 - 2r^2\cos(A-B) \\ &= 2r^2 - 2r^2\cos(A-B).\end{aligned}$$

Comparison of these two expressions for $||PQ||^2$ immediately implies identity A.8d.

With equations A.8, it is easy to derive the following identities:

$$\sin(-\theta) = -\sin\theta, \quad \text{(A.9 a)}$$
$$\cos(-\theta) = \cos\theta, \quad \text{(A.9 b)}$$
$$\tan(-\theta) = -\tan\theta, \quad \text{(A.9 c)}$$
$$\csc(-\theta) = -\csc\theta, \quad \text{(A.9 d)}$$
$$\sec(-\theta) = \sec\theta, \quad \text{(A.9 e)}$$
$$\cot(-\theta) = -\cot\theta; \quad \text{(A.9 f)}$$
$$\sin(\pi/2 - \theta) = \cos\theta, \quad \text{(A.10 a)}$$
$$\cos(\pi/2 - \theta) = \sin\theta, \quad \text{(A.10 b)}$$
$$\tan(\pi/2 - \theta) = \cot\theta, \quad \text{(A.10 c)}$$
$$\csc(\pi/2 - \theta) = \sec\theta, \quad \text{(A.10 d)}$$

$$\sec(\pi/2 - \theta) = \csc\theta, \tag{A.10e}$$

$$\cot(\pi/2 - \theta) = \tan\theta; \tag{A.10f}$$

$$\sin(\pi - \theta) = \sin\theta, \tag{A.11a}$$

$$\cos(\pi - \theta) = -\cos\theta, \tag{A.11b}$$

$$\tan(\pi - \theta) = -\tan\theta, \tag{A.11c}$$

$$\csc(\pi - \theta) = \csc\theta, \tag{A.11d}$$

$$\sec(\pi - \theta) = -\sec\theta, \tag{A.11e}$$

$$\cot(\pi - \theta) = -\cot\theta. \tag{A.11f}$$

By expressing $\tan(A+B)$ as $\sin(A+B)/\cos(A+B)$ and using identities A.8a and A.8c, we find a compound-angle formula for the tangent function,

$$\tan(A+B) = \frac{\tan A + \tan B}{1 - \tan A \tan B}, \tag{A.12a}$$

and similarly,

$$\tan(A-B) = \frac{\tan A - \tan B}{1 + \tan A \tan B}. \tag{A.12b}$$

By setting $A = B$ in A.8a,c and A.12a, we obtain the **double-angle formulas**,

$$\sin 2A = 2\sin A \cos A, \tag{A.13a}$$

$$\cos 2A = \cos^2 A - \sin^2 A, \tag{A.13b}$$

$$= 2\cos^2 A - 1, \tag{A.13c}$$

$$= 1 - 2\sin^2 A, \tag{A.13d}$$

$$\tan 2A = \frac{2\tan A}{1 - \tan^2 A}. \tag{A.13e}$$

When pairs of compound-angle formulas A.8 are added or subtracted, the **product formulas** result. For example, subtracting A.8c from A.8d gives

$$\sin A \sin B = \frac{1}{2}[-\cos(A+B) + \cos(A-B)]. \tag{A.14a}$$

The other product formulas are

$$\sin A \cos B = \frac{1}{2}[\sin(A+B) + \sin(A-B)], \tag{A.14b}$$

$$\cos A \cos B = \frac{1}{2}[\cos(A+B) + \cos(A-B)]. \tag{A.14c}$$

By setting $X = A + B$ and $Y = A - B$ in A.14, we obtain the **sum and difference formulas**,

$$\sin X + \sin Y = 2\sin\left(\frac{X+Y}{2}\right)\cos\left(\frac{X-Y}{2}\right), \tag{A.15a}$$

$$\sin X - \sin Y = 2\cos\left(\frac{X+Y}{2}\right)\sin\left(\frac{X-Y}{2}\right), \tag{A.15b}$$

$$\cos X + \cos Y = 2\cos\left(\frac{X+Y}{2}\right)\cos\left(\frac{X-Y}{2}\right), \tag{A.15c}$$

$$\cos X - \cos Y = -2\sin\left(\frac{X+Y}{2}\right)\sin\left(\frac{X-Y}{2}\right). \tag{A.15d}$$

The following two examples are typical of problems in the calculus of trigonometric functions.

EXAMPLE A.1

Find $R > 0$ and $0 < \phi < \pi$ such that $3\cos\theta - 4\sin\theta$ can be expresed in the form

$$R\sin(\theta + \phi) = 3\cos\theta - 4\sin\theta$$

for all real θ.

SOLUTION If we expand $R\sin(\theta + \phi)$ according to A.8a, we get

$$R\sin\theta\cos\phi + R\cos\theta\sin\phi = 3\cos\theta - 4\sin\theta.$$

Since this equation is to be valid for all θ, it must therefore be valid for $\theta = 0$ and $\theta = \pi/2$. Substitution of these values gives the two equations

$$R\sin\phi = 3 \qquad \text{and} \qquad R\cos\phi = -4.$$

If we square and add these equations, we find

$$R^2\sin^2\phi + R^2\cos^2\phi = R^2 = 3^2 + (-4)^2 = 25.$$

Consequently, $R = 5$, and

$$\sin\phi = \frac{3}{5}, \qquad \cos\phi = -\frac{4}{5}.$$

The only angle in the range $0 < \phi < \pi$ satisfying these equations is $\phi = 2.50$ radians. Thus, $3\cos\theta - 4\sin\theta$ can be expressed in the form

$$3\cos\theta - 4\sin\theta = 5\sin(\theta + 2.50).$$

■

EXAMPLE A.2

Write the expression $\cos^4\theta$ in terms of $\cos 2\theta$ and $\cos 4\theta$.

SOLUTION If we replace A by θ in double-angle formula A.13c,

$$\cos 2\theta = 2\cos^2\theta - 1.$$

It follows that

$$\cos^2\theta = \frac{1 + \cos 2\theta}{2}.$$

Consequently,

$$\cos^4\theta = (\cos^2\theta)^2 = \frac{1}{4}(1 + \cos 2\theta)^2 = \frac{1}{4}(1 + 2\cos 2\theta + \cos^2 2\theta).$$

But if we now replace θ by 2θ in the identity $\cos^2\theta = (1 + \cos 2\theta)/2$, we obtain

$$\cos^2 2\theta = \frac{1 + \cos 4\theta}{2}.$$

Thus,

$$\cos^4\theta = \frac{1}{4}\left(1 + 2\cos 2\theta + \frac{1+\cos 4\theta}{2}\right) = \frac{1}{8}(3 + 4\cos 2\theta + \cos 4\theta).$$

■

By regarding arguments of the trigonometric functions as angles, we have been stressing geometric properties of these functions. In particular, identities A.8–A.15 have all been based on definitions of the trigonometric functions as functions of angles. What is important about a function — be it a trigonometric function or any other kind of function — is that there is a number associated with each value of the independent variable. How we arrive at this number is irrelevant. So far as properties of the function are concerned, only its values are taken into account. Thus, when we discuss properties of a trigonometric function, what is important is not that its argument can be regarded as an angle or that its values can be defined as ratios of sides of a triangle, but that we know its values. Indeed, it is sometimes unwise to regard arguments of trigonometric functions as angles. Consider, for example, the motion of a mass m suspended from a spring with spring constant k (Figure A.11). If $x = 0$ is the position at which the mass would hang motionless, then when vertical oscillations are initiated, the position of m as a function of time t is always of the form

$$x = A\cos\left(\sqrt{\frac{k}{m}}t\right) + B\sin\left(\sqrt{\frac{k}{m}}t\right),$$

where A and B are constants. Clearly there are no angles associated with the motion of m, and it is therefore unnatural to attempt to interpret the argument $(\sqrt{k/m})t$ as a physical angle.

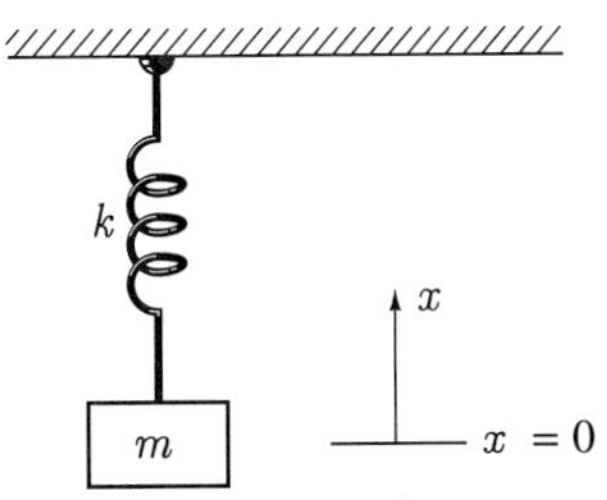

FIGURE A.11

We will now consider the trigonometric functions as those of a real variable. If it is convenient to regard the argument as an angle, then we will do so, but we are not compelled to. To emphasize this, we replace θ with our usual generic label for the independent variable of a function, namely, x. With this change, the trigonometric functions are $\sin x$, $\cos x$, $\tan x$, $\csc x$, $\sec x$, and $\cot x$. Their graphs are shown in Figures A.12.

(a)

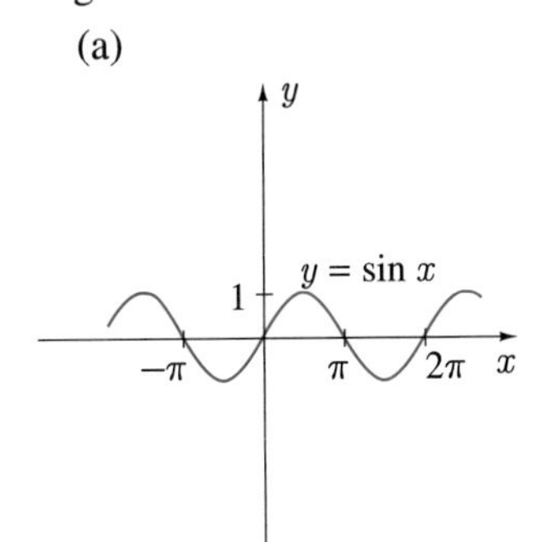

(b)

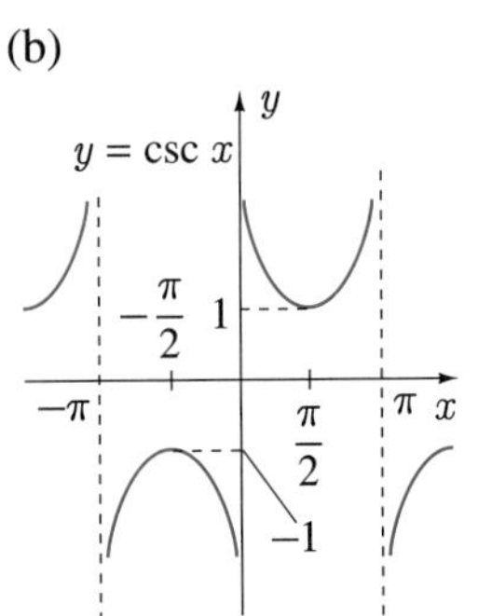

(c)

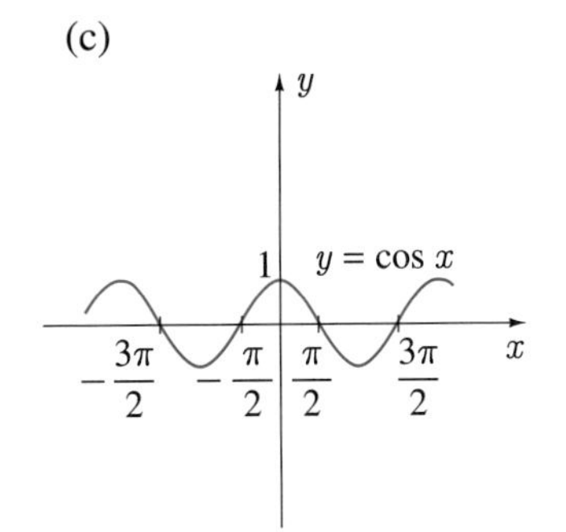

FIGURE A.12

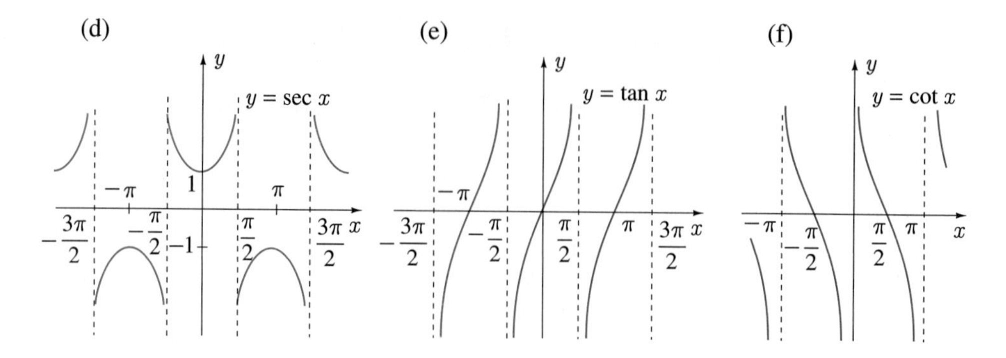

Trigonometric functions are *periodic*. A function $f(x)$ is said to be periodic if there exists a number T such that for all x,

$$f(x + T) = f(x). \tag{A.16}$$

The smallest such positive number T is called the **period** of $f(x)$. Clearly, then, $\sin x$, $\cos x$, $\csc x$, and $\sec x$ are periodic with period 2π, whereas $\tan x$ and $\cot x$ have period π.

When a periodic function is added to a periodic function, the resulting function is periodic. For example, the function $f(x) = 3\cos x - 4\sin x$ is the addition of two 2π-periodic functions $3\cos x$ and $-4\sin x$. In Example A.1, we showed that $f(x)$ can be expressed in the form $f(x) = 5\sin(x + 2.50)$, and using this representation, it is easy to produce the graph in Figure A.13. The function again has period 2π. Note that the 5 in $f(x)$ indicates that oscillations occur between the limits ± 5, and the 2.50 represents a shift of the usual sine curve 2.50 units to the left.

FIGURE A.13

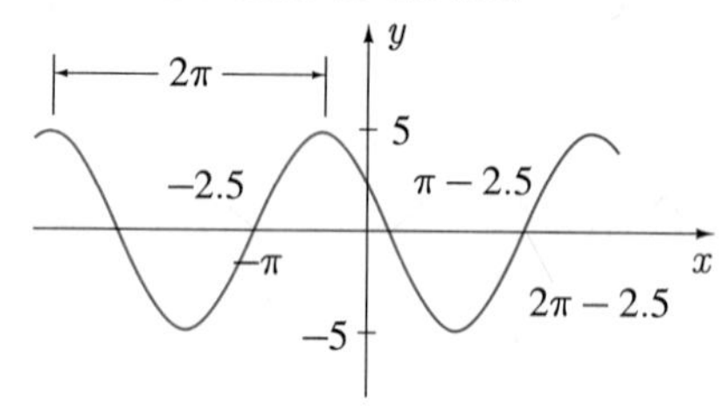

The curve in Figure A.13 is an example of a **general sine function,**

$$f(x) = A\sin(\omega x + \phi), \tag{A.17}$$

where A and ω are positive constants and ϕ is also constant. The graph of the general sine function is shown in Figure A.14. The number A, which represents half the range of the function, is called the **amplitude** of the oscillations. The period is $2\pi/\omega$, and $-\phi/\omega$ is called the **phase shift**.

FIGURE A.14

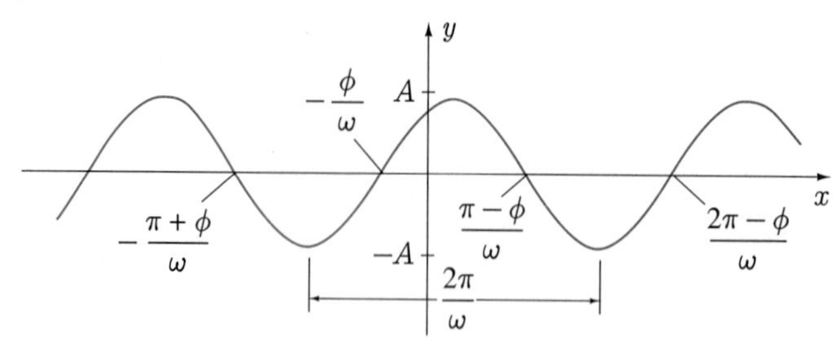

EXERCISES

In Exercises 1–10, express the angle in radians.

1. $30°$ **2.** $60°$ **3.** $135°$ **4.** $-90°$

5. $-300°$ **6.** $765°$ **7.** $72°$ **8.** $-128°$

9. $321°$ **10.** $-213°$

In Exercises 11–20, express the angle in degrees.

11. $\pi/3$ **12.** $-5\pi/4$ **13.** $3\pi/2$ **14.** 8π

15. $-5\pi/6$ **16.** 1 **17.** -3 **18.** 2.5

19. -3.6 **20.** 11

21. What angle is subtended at the centre of a circle of radius 4 by an arc of length (a) 2; (b) 7; (c) 3.2?

22. Fill in entries for the following table.

	$\sin x$	$\cos x$	$\tan x$	$\csc x$	$\sec x$	$\cot x$
0						
$\pi/6$						
$\pi/4$						
$\pi/3$						
$\pi/2$						

For each of the triangles in Exercises 23–26 use the cosine law and/or the sine law to find the lengths of all sides and all interior angles.

23.

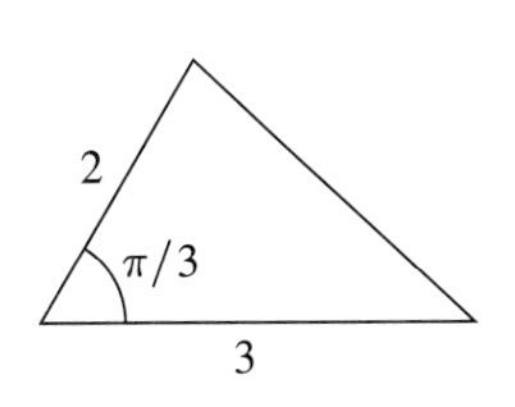

24.

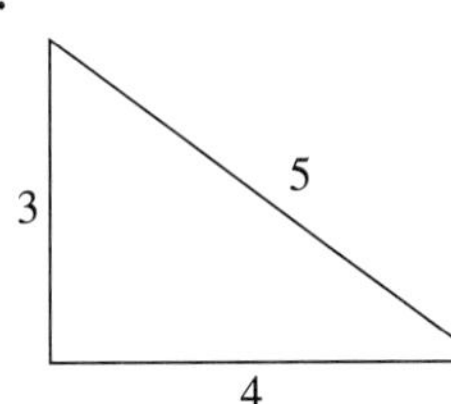

25.

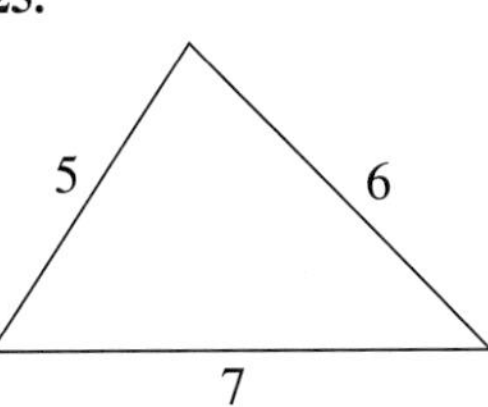

26.

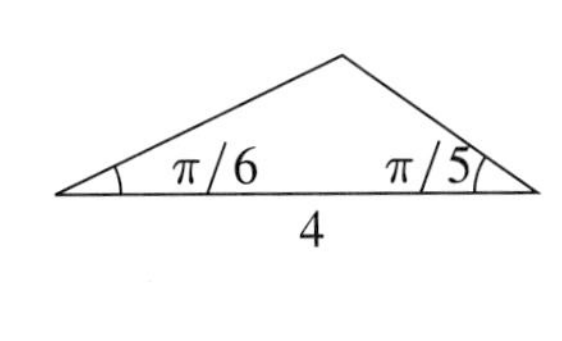

27. Use compound-angle formulas A.8 to prove identities A.12.

28. Verify double-angle formulas A.13.

29. Use compound-angle formulas A.8 to prove the product formulas A.14.

30. Show that the sum and difference formulas A.15 can be obtained from A.14.

In Exercises 31–40 sketch a graph of the function.

31. $f(x) = 3\sin x$ **32.** $f(x) = \sin 2x$

33. $f(x) = 3\sin 2x$ **34.** $f(x) = \sin(x + \pi/4)$

35. $f(x) = 3\sin(x + \pi/4)$

36. $f(x) = \sin(2x + \pi/4)$

37. $f(x) = 3\sin(2x + \pi/4)$

38. $f(x) = 4\cos(x/3)$

39. $f(x) = 2\sin(x/2 - \pi)$

40. $f(x) = 5\cos(\pi/2 - 3x)$

41. Show that compound-angle formulas A.8a,b,c can be derived from A.8d and A.4.

In Exercises 42–49 find all solutions of the equation.

42. $\sin x = \sqrt{3}/2$ **43.** $\cos x = 2$

44. $3\sin x = \cos x$ **45.** $3\cos^2 x = 1$

46. $\sin 2x = \sin x$ **47.** $\sin^2 x + 3\sin x + 1 = 0$

48. $\sin^2 x + \sin x - 1 = 0$

49. $\sin x + \cos x = 1$

In Exercises 50–53 express each function as a general sine function, identifying its amplitude, period, and phase shift. Sketch a graph of each function.

50. $f(x) = \sin x - 2\cos x$

51. $f(x) = 4\sin 2x + \cos 2x$

52. $f(x) = -2\sin 3x - 4\cos 3x$

53. $f(x) = \sin x \cos x$

54. Which of the trigonometric functions in Figures A.12 are even and which are odd?

In Exercises 55–59 verify the identity.

55. $\cos 3x = 4\cos^3 x - 3\cos x$

56. $\sin 4x = 8\cos^3 x \sin x - 4\cos x \sin x$

57. $\tan 3x = \dfrac{3\tan x - \tan^3 x}{1 - 3\tan^2 x}$

58. $\tan\left(\dfrac{x}{2}\right) = \dfrac{\sin x}{1 + \cos x}$

59. $\dfrac{1 + \tan x}{1 - \tan x} = \tan\left(x + \dfrac{\pi}{4}\right)$

60. Show that a function $f(x) = A\cos\omega x + B\sin\omega x$ can always be written in the form

$$f(x) = \sqrt{A^2 + B^2}\sin(\omega x + \phi),$$

where ϕ is defined by the equations

$$\sin \phi = \frac{A}{\sqrt{A^2 + B^2}} \quad \text{and} \quad \cos \phi = \frac{B}{\sqrt{A^2 + B^2}}.$$

61. In Exercise 60 can we replace the two equations defining ϕ with the single equation $\tan \phi = A/B$?

In Exercises 62–65 find all solutions of the equation in the interval $0 \leq x < 2$.

62. $\sin 4x = \cos 2x$

63. $\cos x + \cos 3x = 0$

64. $\sin 2x + \cos 3x = \sin 4x$

65. $\sin x + \cos x = \sqrt{3} \sin x \cos x$

66. Verify that if A, B, and C are the angles of a triangle, then

$$\tan A + \tan B + \tan C = \tan A \tan B \tan C.$$

Hint: Expand $\tan(A + B + C)$ in terms of $\tan A$, $\tan B$, and $\tan C$.

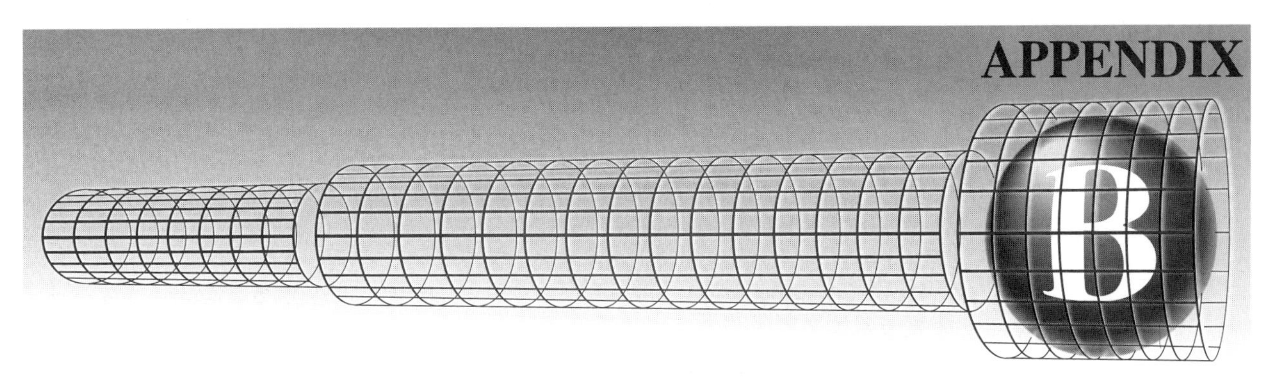

The Exponential and Logarithm Functions

In this appendix we give a brief introduction to the exponential and logarithm functions and their properties.

In elementary algebra we learned the basic rules for products and quotients of powers:

$$a^b a^c = a^{b+c}; \tag{B.1a}$$

$$\frac{a^b}{a^c} = a^{b-c}; \tag{B.1b}$$

$$\left(a^b\right)^c = a^{bc}. \tag{B.1c}$$

These rules are used to develop the **exponential function**

$$f(x) = a^x; \tag{B.2}$$

that is, a raised to the power of x for variable x.

Consider first the case when $a > 1$. The meaning of a^x when x is an integer is clear, and when $x = 1/n$, where $n > 0$ is an integer, $a^x = a^{1/n}$ is the nth root of a. When x is a positive rational number n/m (n and m positive integers), B.1c implies that

$$a^x = a^{n/m} = (a^n)^{1/m} \qquad \text{or}$$
$$a^x = a^{n/m} = \left(a^{1/m}\right)^n.$$

When x is a negative rational, we write $a^x = 1/a^{-x}$, where $-x$ is positive. These results lead to the points and the graph of $y = a^x$ in Figure B.1.

FIGURE B.1

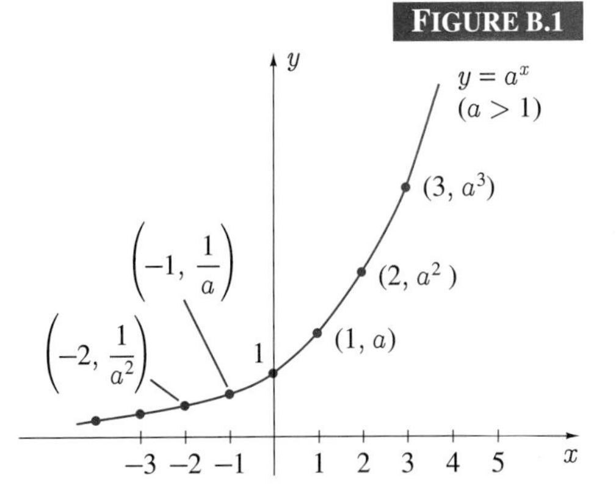

FIGURE B.2

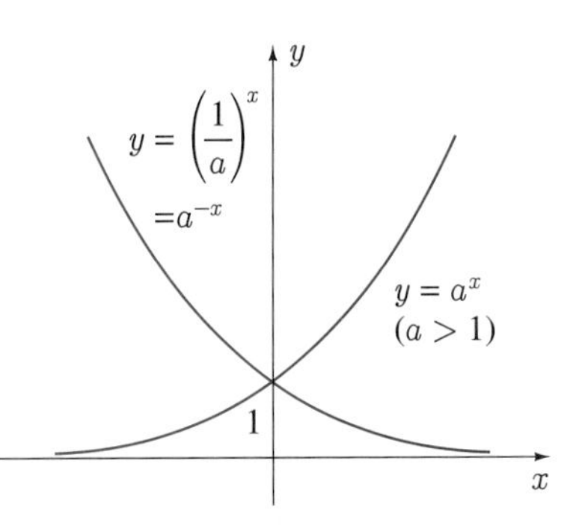

There is a difficulty with this graph and the definition of a^x when x is not a rational number, a fact we should at least mention here. For instance, if $x = \sqrt{2}$, an irrational number, what is the value of $a^{\sqrt{2}}$? Calculators with a y^x button suggest a possible definition. If, for example, a is equal to 3, the calculator has no difficulty evaluating $3^{\sqrt{2}}$. But what does it actually calculate? It uses a decimal approximation for $\sqrt{2}$. Since a decimal approximation represents a rational number, the calculator approximates $3^{\sqrt{2}}$ with 3 raised to a rational number approximating $\sqrt{2}$. The more sophisticated the calculator, the better the approximation for $\sqrt{2}$. This suggests that we define $a^{\sqrt{2}}$ as follows. Find a set of rational numbers which approximates $\sqrt{2}$ more and more closely, say,

$$c_1 = 1.4, \quad c_2 = 1.41, \quad c_3 = 1.414, \quad \cdots$$

where each number picks up another decimal place in the nonrecurring, nonterminating decimal expansion of $\sqrt{2}$. Define $a^{\sqrt{2}}$ as the limit

$$a^{\sqrt{2}} = \lim_{n\to\infty} a^{c_n}.$$

In other words, define $a^{\sqrt{2}}$ as the limit $\lim_{n\to\infty} a^{c_n}$ where $c_1, c_2, \cdots$ is any set of rational numbers which approximates $\sqrt{2}$ more closely. With this definition, it is possible to prove that a^x is a continuous function, and its graph is indeed as shown in Figure B.1.

When one curve is obtained from another by replacing each x by $-x$, the two curves are mirror images of each other in the y-axis. The graph of $f(x) = a^{-x}$ is therefore the reflection of $f(x) = a^x$ in the y-axis (Figure B.2). But

$$a^{-x} = \frac{1}{a^x} = \left(\frac{1}{a}\right)^x,$$

where $0 < 1/a < 1$. In other words, the graph of the exponential function for $0 < a < 1$ is as shown in Figure B.2.

For the case in which $a = 1$, the graph of $f(x) = a^x = 1^x = 1$ is a horizontal straight line. We do not consider $f(x) = a^x$ for $a < 0$. Can you see why?

EXAMPLE B.1

Sketch graphs of the exponential functions 2^x and 3^x on the same axes.

SOLUTION Graphs of these functions are shown in Figure B.3. Things to notice are that both curves pass through the point $(0, 1)$. More generally, $a^0 = 1$ for any a. When $x > 0$, the graph of 3^x is higher than that of 2^x, whereas the opposite is true for $x < 0$.

FIGURE B.3

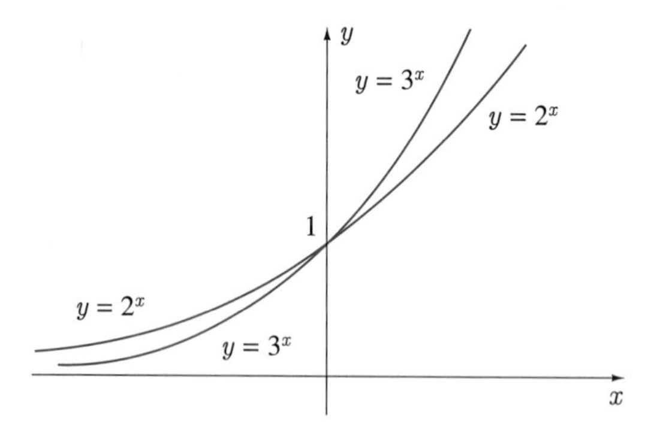

■

EXAMPLE B.2

Sketch graphs of the functions x^4 (a power function) and 2^x (an exponential function) on the same axes.

SOLUTION Sketches of the graphs of these functions are shown in Figure B.4, but no attempt has been made to use a scale on either the x- or y-axis. Notice here that for $x > 16$, $2^x > x^4$. This is always the situation for power and exponential functions. Given any exponential function a^x ($a > 1$), and any power function x^n ($n > 1$), there always exists a value of x, say, X, such that when $x > X$, we have $a^x > x^n$. In short, exponential functions grow more rapidly than power functions for large values of x.

FIGURE B.4

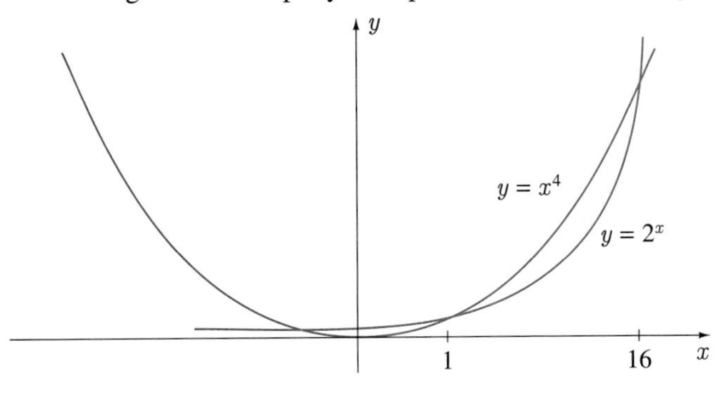

In terms of the exponential function, rules B.1 take the form:

$$a^{x_1}a^{x_2} = a^{x_1+x_2}, \tag{B.3a}$$

$$\frac{a^{x_1}}{a^{x_2}} = a^{x_1-x_2}, \tag{B.3b}$$

$$(a^{x_1})^{x_2} = a^{x_1x_2}. \tag{B.3c}$$

Since the exponential function $f(x) = a^x$ ($a > 1$) is increasing, it has an inverse function. See Section 8.1 if you have not studied inverse functions. Its graph can be obtained by reflecting the curve $y = a^x$ in the line $y = x$ (Figure B.5). This inverse function is called the **logarithm function** to base a, and is denoted by

$$y = \log_a x. \tag{B.4}$$

Note that $\log_a x$ is defined only for $x > 0$; that is, we cannot take the logarithm of zero or a negative number, and the logarithm of 1 for any $a > 1$ is equal to 0.

FIGURE B.5

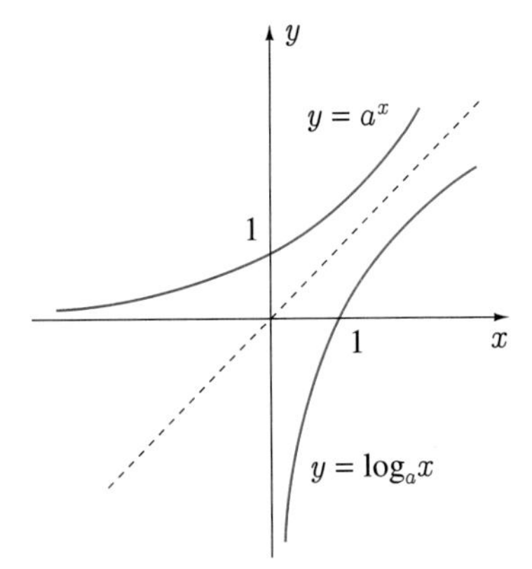

We can also define $\log_a x$ for $0 < a < 1$, but such a logarithm is simply the negative of $\log_{1/a} x$, where $1/a > 1$ (see Exercise 22). Consequently, unless otherwise specified, we shall always assume the base of logarithms to be greater than unity.

Figure B.5 gives a graphical representation of the logarithm function, and calculators can be used to find decimal approximations for logarithms of numbers. But what really is a logarithm? To find out, we suppose (x, y) is a point on $y = \log_a x$. Then (y, x) is a point on the exponential curve; that is,

$$y = \log_a x \qquad \text{only if} \qquad x = a^y. \tag{B.5}$$

These equations contain the real meaning of logarithm. They state that y is the logarithm of x to base a only if x is a to the power y. In other words, the *logarithm of* x *to base* a *is the power to which* a *must be raised in order to produce* x. Logarithms then are powers. For example, to find $\log_{10} 100$, we search for a number to fill the box in

$$100 = 10^{[]}.$$

Clearly the number is 2, and therefore $\log_{10} 100 = 2$. Similarly, to calculate $\log_2 (1/16)$, we consider

$$\frac{1}{16} = 2^{[]}.$$

Since the missing number is -4, we have that $\log_2 (1/16) = -4$.

When the first of equations B.5 is substituted into the second, the result is

$$x = a^{\log_a x} \qquad (x > 0). \tag{B.6 a}$$

When the second is substituted into the first,

$$y = \log_a (a^y),$$

or, since y is arbitrary, we may replace it with x and write

$$x = \log_a (a^x). \tag{B.6 b}$$

These equations are statements of the fact that the exponential and logarithm functions are inverses. Substitute a value of x into one function, substitute the result into the other, and the final result is the original value of x.

Corresponding to rules B.3 for the exponential function are the following rules for logarithms:

$$\log_a(x_1 x_2) = \log_a x_1 + \log_a x_2, \tag{B.7 a}$$

$$\log_a \left(\frac{x_1}{x_2}\right) = \log_a x_1 - \log_a x_2, \tag{B.7 b}$$

$$\log_a \left(x_1^{x_2}\right) = x_2 \log_a x_1. \tag{B.7 c}$$

To prove B.7a, say, we set $z = \log_a(x_1 x_2)$, in which case

$$\begin{aligned} a^z &= x_1 x_2 \\ &= \left(a^{\log_a x_1}\right)\left(a^{\log_a x_2}\right) && \text{(using B.6a)} \\ &= a^{\log_a x_1 + \log_a x_2}. && \text{(using B.3a)} \end{aligned}$$

Thus,

$$\log_a x_1 + \log_a x_2 = z = \log_a (x_1 x_2).$$

We leave proofs of B.7b and B.7c to the exercises.

EXAMPLE B.3

Simplify the following expressions:

(a) $3^{\log_3(x^2)}$ (b) $10^{-4\log_{10} x}$ (c) $\log_a\left(a^{-x+3}\right)$ (d) $\log_2 8 + \log_3(1/27)$

SOLUTION (a) Identity B.6a implies that

$$3^{\log_3(x^2)} = x^2.$$

(b) Since -4 intervenes between the logarithm and exponential operations, we cannot use B.6a immediately. The -4 can be relocated, however, with B.7c:

$$10^{-4\log_{10} x} = 10^{\log_{10}(x^{-4})} = \frac{1}{x^4} \quad (\text{if } x > 0).$$

(c) Identity B.6b gives

$$\log_a\left(a^{-x+3}\right) = -x + 3.$$

(d) Since $\log_2 8 = 3$ and $\log_3(1/27) = -3$,

$$\log_2 8 + \log_3(1/27) = 3 - 3 = 0.$$

■

EXAMPLE B.4

Solve the following equations:

(a) $\log_5 x = -3$ (b) $\log_{10} x + \log_{10}(x+1) = 0$ (c) $10^x - 12 + 10^{-x} = 0$

SOLUTION (a) By means of B.5,

$$x = 5^{-3} = \frac{1}{125}.$$

(b) Since $\log_{10} x + \log_{10}(x+1) = \log_{10}[x(x+1)]$, we can write

$$0 = \log_{10}[x(x+1)].$$

If we now take exponentials to base 10,

$$10^0 = 10^{\log_{10}[x(x+1)]} \quad \text{or}$$

$$1 = x(x+1).$$

This quadratic equation has solutions

$$x = \frac{-1 \pm \sqrt{1+4}}{2} = \frac{-1 \pm \sqrt{5}}{2}.$$

Since x must be positive (the original equation demands this), the only solution is $x = (\sqrt{5} - 1)/2$.

(c) If we multiply the equation by 10^x, the result is

$$0 = 10^{2x} - 12(10^x) + 1 = (10^x)^2 - 12(10^x) + 1.$$

But this is a quadratic equation in 10^x so that

$$10^x = \frac{12 \pm \sqrt{144 - 4}}{2} = 6 \pm \sqrt{35}.$$

Finally, we have

$$x = \log_{10}\left(6 \pm \sqrt{35}\right).$$

■

EXAMPLE B.5

In acoustics, the minimum sound level detectable by the normal human ear is taken as the "tick of a watch at 20 feet under otherwise quiet conditions." This sound level, called the audible sound threshold, is said to have a decibel reading of 0. All other sounds are given decibel readings relative to this audible sound threshold. For example, a normal voice is approximately 70 decibels, the sound of a car is 100 decibels, and a jet engine 160 decibels. Mathematically, the loudness of a sound is said to be L decibels if

$$L = 10\log_{10}\left(\frac{I}{I_0}\right),$$

where I is the intensity of the sound and I_0 is the intensity of sound at the audible threshold. Use this definition to answer the following questions.

(a) Express the intensity I of a sound in terms of I_0 and its decibel reading L.

(b) Intensity is a measure of energy per unit area per unit time transmitted by sound waves. Find the intensity of the normal voice, a car, and a jet engine relative to the intensity I_0.

(c) If the pain threshold for sound has an intensity 10^{14} times I_0, what is its decibel reading?

(d) If the intensity I_1 of one sound is 10 times the intensity I_2 of a second sound, how do their decibel readings compare?

SOLUTION (a) When we take both sides as exponents of powers of 10, and use properties B.7c and B.6a,

$$10^L = 10^{10\log_{10}(I/I_0)} = 10^{\log_{10}(I/I_0)^{10}} = \left(\frac{I}{I_0}\right)^{10}.$$

If we take 10th roots of both sides,

$$\left(10^L\right)^{1/10} = \frac{I}{I_0} \qquad \text{or} \qquad I = I_0\,10^{L/10}.$$

(b) Since the decibel level of the normal voice is 70, its intensity is

$$I = I_0 10^{70/10} = 10^7 I_0 .$$

Similarly, the intensities of a car and a jet are 10^{10} and 10^{16} times I_0.

(c) For an intensity of $10^{14} I_0$, the decibel reading is

$$L = 10 \log_{10}(10^{14}) = 10(14) = 140 .$$

(d) If L_1 and L_2 are the decibel readings for sounds with intensities I_1 and I_2, then

$$L_1 = 10 \log_{10}(I_1/I_0) \qquad \text{and} \qquad L_2 = 10 \log_{10}(I_2/I_0).$$

When we subtract these readings,

$$\begin{aligned} L_1 - L_2 &= 10 \log_{10}(I_1/I_0) - 10 \log_{10}(I_2/I_0) \\ &= 10[\log_{10}(I_1/I_0) - \log_{10}(I_2/I_0)] \\ &= 10 \log_{10}\left(\frac{I_1/I_0}{I_2/I_0}\right) \qquad \text{(using (B.7b))} \\ &= 10 \log_{10}\left(\frac{I_1}{I_2}\right) \\ &= 10 \log_{10}\left(\frac{10 I_2}{I_2}\right) \qquad \text{(since } I_1 = 10 I_2\text{)} \\ &= 10 \log_{10} 10 \\ &= 10 . \end{aligned}$$

Thus, when the intensity of one sound is 10 times that of another, their decibel readings differ by 10. ■

It is sometimes necessary to change from one base of logarithms to another. If we take logarithms to base b on both sides of identity B.6a, we immediately obtain

$$\log_b x = (\log_a x)(\log_b a). \tag{B.8}$$

This equation defines $\log_b a$ as the conversion factor from logarithms to base a to logarithms to base b.

Before the discovery of calculus, the base of logarithms was invariably chosen to be 10. Such logarithms are called *common logarithms*; they correspond to the exponential function 10^x. Another base for exponentials and logarithms, however, that is much more convenient in most applications is a number, denoted by the letter e, and defined in a variety of ways, one of which is

$$e = \lim_{z \to 0} (1 + z)^{1/z}. \tag{B.9a}$$

An alternative is obtained by setting $v = 1/z$,

$$e = \lim_{v \to \infty} \left(1 + \frac{1}{v}\right)^v. \tag{B.9b}$$

To prove the existence of either limit is extremely difficult. Substitution of various values of v into $(1+1/v)^v$ may suggest, however, whether a limit exists. The values in Table B.1

are steadily increasing but as they rapidly get closer together, they are indeed approaching a limit — to twelve decimals

$$e = 2.718281828459.$$

TABLE B.1

v	$\left(1 + \frac{1}{v}\right)^v$
1	2.000 000
3	2.370 370
5	2.488 320
10	2.593 742
100	2.704 814
1 000	2.716 924
10 000	2.718 146
100 000	2.718 255
1 000 000	2.718 282

It is an irrational number with a nonterminating, nonrepeating decimal expansion. Why this number is so convenient as a base for logarithms and exponentials is shown in Section 3.9. For now, let us rewrite some of the more important formulas of this appendix with a set equal to e. The exponential function to base e is e^x, and equations B.3 in terms of e^x read

$$e^{x_1} e^{x_2} = e^{x_1 + x_2}, \qquad \text{(B.10 a)}$$

$$\frac{e^{x_1}}{e^{x_2}} = e^{x_1 - x_2}, \qquad \text{(B.10 b)}$$

$$(e^{x_1})^{x_2} = e^{x_1 x_2}. \qquad \text{(B.10 c)}$$

Logarithms to base e are usually given the notation $\ln x$ rather than $\log_e x$, and are called **natural**, or **Naperian logarithms** after the Scottish mathematician John Napier (1550–1617), who invented logarithms,

$$\ln x = \log_e x. \qquad \text{(B.11)}$$

In terms of $\ln x$, rules B.7 are

$$\ln(x_1 x_2) = \ln x_1 + \ln x_2, \qquad \text{(B.12 a)}$$

$$\ln\left(\frac{x_1}{x_2}\right) = \ln x_1 - \ln x_2, \qquad \text{(B.12 b)}$$

$$\ln\left(x_1^{x_2}\right) = x_2 \ln x_1. \qquad \text{(B.12 c)}$$

Identities B.6, which express the inverse character of exponential and logarithm functions, become

$$x = e^{\ln x}, \qquad x > 0, \qquad \text{(B.13 a)}$$

$$x = \ln(e^x). \qquad \text{(B.13 b)}$$

EXERCISES

In Exercises 1–13 find all values of x satisfying the equation.

1. $\log_{10}(2 + x) = -1$
2. $10^{3x} = 5$
3. $\log_{10}(x^2 + 2x + 1) = 1$
4. $\ln(x^2 + 2x + 10) = 1$
5. $10^{5-x^2} = 100$
6. $10^{1-x^2} = 100$
7. $\log_{10}(x - 3) + \log_{10} x = 1$
8. $\log_{10}(3 - x) + \log_{10} x = 1$
9. $\log_{10}[x(x - 3)] = 1$
10. $2 \ln x + \ln(x - 1) = 2$
11. $\log_a x + \log_a (x + 2) = 2$
12. $\log_a [x(x + 2)] = 2$
13. $\log_{10}\left[\log_{10}\left(\dfrac{x + 3}{200x}\right) + 4\right] = -1$

In Exercises 14–17 sketch a graph of the function.

14. $f(x) = e^{-x^2}$ **15.** $f(x) = \ln(\cos x)$

16. $f(x) = \log_a |x^2 - 1|$ **17.** $f(x) = a^{\log_a(2x+1)}$

18. Is there a difference between the graphs of the functions $f(x) = \log_a(x^2)$ and $g(x) = 2 \log_a x$?

19. (a) In the early period of reforestation, the percentage increase of timber each year is almost constant. If the original amount A_0 of a certain timber increases 3.5% the first year and 3.5% each year thereafter, find an expression for the amount of timber after t years.

(b) How long does it take for timber of this type to double?

20. A new car costs \$20 000. In any year it depreciates to 75% of its value at the beginning of that year. What is the value of the car after t years?

21. If the effective height of the earth's atmosphere (in metres) is the solution of the equation

$$10^{-6} = (1 - 2.08 \times 10^{-6} y)^{56},$$

find y.

22. Show that if y is the logarithm of x to base a, then $-y$ is the logarithm of x to base $1/a$.

23. Prove B.7b and B.7c.

24. Is identity B.7a valid for any x_1 and x_2?

25. The magnitude of an earthquake is measured in much the same way as noise level. An earthquake of minimal size is taken as having value 0 on the Richter scale. Any other earthquake of intensity I is said to have magnitude R on the Richter scale if

$$R = \log_{10}\left(\frac{I}{I_0}\right),$$

where I_0 is the intensity of the minimal earthquake being used as reference.

(a) Express the intensity of an earthquake in terms of I_0 and its reading on the Richter scale.

(b) What are readings on the Richter scale of earthquakes that have intensities 1.20×10^6 and 6.20×10^4 times I_0?

26. (a) If P dollars is invested at $i\%$ compounded n times per year, show that the accumulated value after t years is

$$A = P\left(1 + \frac{i}{100n}\right)^{nt}.$$

(b) How long does it take to double an investment if interest is 8% compounded semiannually?

(c) Calculate the maximum possible value of A if i is fixed but the number of times that interest is compounded is unlimited; that is, calculate $\lim_{n\to\infty} A$. This method of calculating interest is called *continuously compounded interest.*

(d) What is the accumulated value of a \$1000 investment after 10 years at 6% compounded continuously? Compare this to the accumulated value if interest is calculated only once each year.

In Exercises 27–30 find all values of x satisfying the equation.

27. $3a^{2x} + 3a^{-2x} = 10$ **28.** $2^x + 4^x = 8^x$

29. $3^{x+4} = 7^{x-1}$ **30.** $\log_x 2 = \log_{2x} 8$

31. Sketch a graph of the function $f(x) = \ln\left(x + \sqrt{x^2 + 1}\right)$.

32. Repair costs on the car in Exercise 20 are estimated at \$50 the first year, increasing by 20% each year thereafter. Set up a function $C(t)$ that represents the average yearly cost associated with owning the car for t years. Hint: You may need the result of Exercise 27 in Section 6.1.

33. A straight wire conductor has length $2L$ and circular cross-section of radius R. If the wire carries current $I > 0$, then the magnitude of the "vector" potential at a distance r from the centre of the wire is

$$f(r) = \begin{cases} \dfrac{\mu_0 I}{4\pi}\left[\ln\left(1+\dfrac{4L^2}{R^2}\right) - 1 + \dfrac{r^2}{R^2}\right] & 0 \le r \le R \\ \dfrac{\mu_0 I}{4\pi}\ln\left(1+\dfrac{4L^2}{r^2}\right) & r > R \end{cases}$$

where μ_0 is a positive constant.

(a) Sketch a graph of this function.

(b) Find the radius $r > R$ for which $f(r) = f(0)$.

In Exercises 34–36 solve the given equation for x in terms of y.

34. $y = \dfrac{e^{2x} - e^{-2x}}{2}$

35. $y = \dfrac{e^{x} + e^{-x}}{2}$

36. $y = \dfrac{e^{x} - e^{-x}}{e^{x} + e^{-x}}$

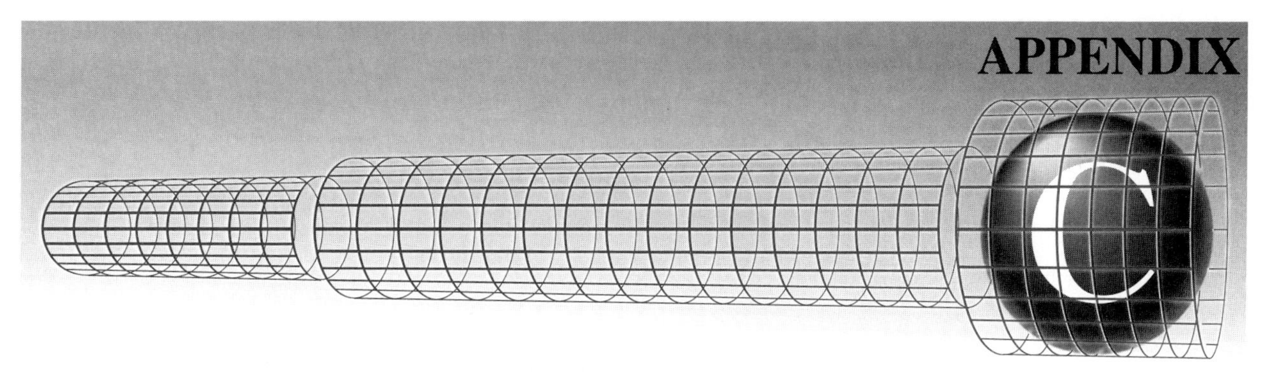

Mathematical Induction

Mathematical induction is a type of proof used to verify propositions that involve integers. It is best introduced through a simple example.

Suppose we are required to verify that the sum of the first n positive integers is

$$1 + 2 + 3 + \cdots + n = \frac{n(n+1)}{2}. \tag{C.1}$$

It is a simple matter to check that this result is valid for the first few integers:

For $n = 1$, the sum is 1, and the formula gives $1(1+1)/2 = 1$.

For $n = 2$, the sum is $1 + 2 = 3$; the formula gives $2(2+1)/2 = 3$.

For $n = 3$, the sum is $1 + 2 + 3 = 6$; the formula gives $3(3+1)/2 = 6$.

We shall see later that to use induction, it is necessary to verify the result at this stage only for $n = 1$.

We now suppose that k is some integer for which the result is valid; that is, we suppose that

$$1 + 2 + 3 + \cdots + k = \frac{k(k+1)}{2}.$$

On the basis of this supposition, the sum of the first $k + 1$ integers is

$$\begin{aligned} 1 + 2 + \cdots + (k+1) &= (1 + 2 + \cdots + k) + (k+1) \\ &= \frac{k(k+1)}{2} + (k+1) \\ &= \frac{1}{2}[k(k+1) + 2(k+1)] \\ &= \frac{(k+1)(k+2)}{2}. \end{aligned}$$

But this is precisely equation C.1 with n replaced by $k + 1$. We have shown that if C.1 is valid for some integer k, then it must also be valid for the next integer $k + 1$.

We now put the above facts together to verify that C.1 is valid for all $n \geq 1$. We have proved that it is valid for $n = 1$, 2, and 3. But if it is true for 3, then it must be true for 4. If it is true for 4, then it is true for 5. If it is true for 5, it is true for 6; and so on, and so on. Consequently, C.1 must be valid for every positive integer n.

This example contains the two essential parts of every inductive proof:

1. Verification of the result for the smallest integer in question;
2. Verification of the fact that if the result is valid for some integer k, then it must be valid for the next integer $k + 1$.

Once these two facts are established, it follows by the same argument as above that the result is valid for the given set of integers. This is the principle of mathematical induction, which we state formally as follows.

Principle of Mathematical Induction

Suppose with each integer n greater than or equal to some fixed integer N, there is associated a proposition P_n. Then P_n is true for all $n \geq N$ provided:

(a) P_N is valid;

(b) the validity of P_k implies the validity of P_{k+1}.

All inductive arguments must therefore contain these two essential parts. Part (a) is usually quite simple to establish; but part (b) can sometimes be quite difficult. Be sure you understand what we are doing in part (b). We are proving that under the assumption that P_k is true, then P_{k+1} must also be true. It is the principle of mathematical induction that puts (a) and (b) together to state that P_n is true for all $n \geq N$.

EXAMPLE C.1

Verify that

$$1^2 + 2^2 + 3^2 + \cdots + n^2 = \frac{n(n+1)(2n+1)}{6}, \quad n \geq 1.$$

SOLUTION When $n = 1$ the left-hand side of this equation has value 1, and the right-hand side is $1(2)(3)/6 = 1$. The required result is therefore true for $n = 1$. Next we suppose k is some integer for which the result is valid; that is, we suppose that

$$1^2 + 2^2 + 3^2 + \cdots + k^2 = \frac{k(k+1)(2k+1)}{6},$$

and we must write this down because we are going to need it later on. Our objective now is to verify that the result is valid for $k + 1$; that is, we must verify that

$$1^2 + 2^2 + 3^2 + \cdots + (k+1)^2 = \frac{(k+1)(k+2)[2(k+1)+1]}{6}.$$

If we begin on the left, we have

$$\begin{aligned} 1^2 + 2^2 + \cdots + (k+1)^2 &= (1^2 + 2^2 + \cdots + k^2) + (k+1)^2 \\ &= \frac{k(k+1)(2k+1)}{6} + (k+1)^2 \\ &= \frac{(k+1)}{6}[k(2k+1) + 6(k+1)] \\ &= \frac{(k+1)}{6}(2k^2 + 7k + 6) \\ &= \frac{(k+1)(k+2)(2k+3)}{6} \\ &= \frac{(k+1)(k+2)[2(k+1)+1]}{6}, \end{aligned}$$

which is the right-hand side of the required result. We have therefore verified that the result is valid for $k + 1$. Consequently, by mathematical induction the formula is correct for all $n \geq 1$. ■

EXAMPLE C.2

Prove that 7 divides $23^{3n} - 1$ for all positive integers n.

SOLUTION When $n = 1$, $23^{3n} - 1 = 23^3 - 1 = 12\,166$, and this is divisible by 7. The result is therefore true for $n = 1$. Next we suppose k is some integer for which the result is valid; that is, we suppose 7 divides $23^{3k} - 1$. We must now verify that 7 divides $23^{3(k+1)} - 1$. Since

$$\begin{aligned} 23^{3(k+1)} - 1 = 23^{3k+3} - 1 &= (23^{3k})(23^3) - 1 \\ &= [(23^{3k})(23^3) - 23^3] + (23^3 - 1) \\ &= 23^3(23^{3k} - 1) + (23^3 - 1), \end{aligned}$$

and both $23^{3k} - 1$ and $23^3 - 1$ are divisible by 7, it follows that $23^{3(k+1)} - 1$ must also be divisible by 7. The result is therefore true for $k + 1$, and by mathematical induction, it is true for all $n \geq 1$. ■

EXAMPLE C.3

Verify that if n lines are drawn in the plane, no two of which are identical, then the maximum number of points of intersection of these lines is $n(n-1)/2$.

SOLUTION If we draw two lines in the plane, there can be at most one point of intersection. This is precisely what is predicted by $n(n-1)/2$ when $n = 2$. The result is therefore correct for $n = 2$. Suppose the proposition is true for some integer k; that is, suppose when k lines are drawn in the plane, the maximum number of points of intersection is $k(k-1)/2$. We must now verify that when any $k + 1$ lines are drawn, the maximum number of points of intersection is $(k+1)k/2$. To do this, we remove one of the $k + 1$ lines leaving k lines. But for these k lines the maximum number of points of intersection is $k(k-1)/2$. If the line that was removed is now replaced, it can add at most k more points of intersection, one with each of the k lines. The maximum number of points of intersection of the $k + 1$ lines is therefore

$$\frac{k(k-1)}{2} + k = \frac{k}{2}(k - 1 + 2) = \frac{k(k+1)}{2},$$

and this is the result for $n = k + 1$. By mathematical induction, then, the result is valid for all $n \geq 2$. ■

EXERCISES

In Exercises 1–5 use mathematical induction to establish the formula for positive integer n.

1. $1^3 + 2^3 + 3^3 + \cdots + n^3 = \dfrac{n^2(n+1)^2}{4}$

2. $1 + 3 + 6 + 10 + \cdots + \dfrac{n(n+1)}{2} = \dfrac{n(n+1)(n+2)}{6}$

3. $\dfrac{1}{1 \cdot 2} + \dfrac{1}{2 \cdot 3} + \dfrac{1}{3 \cdot 4} + \cdots + \dfrac{1}{n(n+1)} = \dfrac{n}{n+1}$

4. $\dfrac{1}{1 \cdot 2 \cdot 3} + \dfrac{1}{2 \cdot 3 \cdot 4} + \dfrac{1}{3 \cdot 4 \cdot 5} + \cdots + \dfrac{1}{n(n+1)(n+2)}$

$= \dfrac{n(n+3)}{4(n+1)(n+2)}$

5. $a + ar + ar^2 + ar^3 + \cdots + ar^{n-1} = \dfrac{a(1 - r^n)}{1 - r}, \quad (r \neq 1)$

6. Prove that $x - y$ divides $x^n - y^n$ for $n \geq 1$.

7. Verify that $x + y$ divides $x^{2n+1} + y^{2n+1}$ for $n \geq 0$.

8. Prove that 576 divides $5^{2n+2} - 24n - 25$ for $n \geq 1$.

9. Prove that if the proposition

$$1 + 3 + 5 + \cdots + (2n - 1) = n^2 + 4$$

is valid for some integer k, then it is also valid for $k+1$. Is the result valid for all $n \geq 1$?

10. Verify that the sum of the interior angles of a polygon with n sides is $(n - 2)\pi$ radians.

11. Show that if n is a positive integer so too is $(n^3 + 6n^2 + 2n)/3$.

12. Prove that:

(a) $1 + \cos\theta + \cos 2\theta + \cdots + \cos n\theta = \dfrac{1 - \cos\theta + \cos n\theta - \cos(n+1)\theta}{2 - 2\cos\theta}$;

(b) $\sin\theta + \sin 3\theta + \cdots + \sin(2n - 1)\theta = \dfrac{\sin^2 n\theta}{\sin\theta}$.

13. Verify that

$$\left(1 - \frac{1}{4}\right)\left(1 - \frac{1}{9}\right)\left(1 - \frac{1}{16}\right)\cdots\left(1 - \frac{1}{n^2}\right) = \frac{n+1}{2n}.$$

14. Use mathematical induction to show that for $n \geq 2$, there exist constants a_r and b_r, $r = 0, 1, 2, \ldots, n$ such that

$$\sin^n x = \sum_{r=0}^{n} (a_r \cos rx + b_r \sin rx).$$

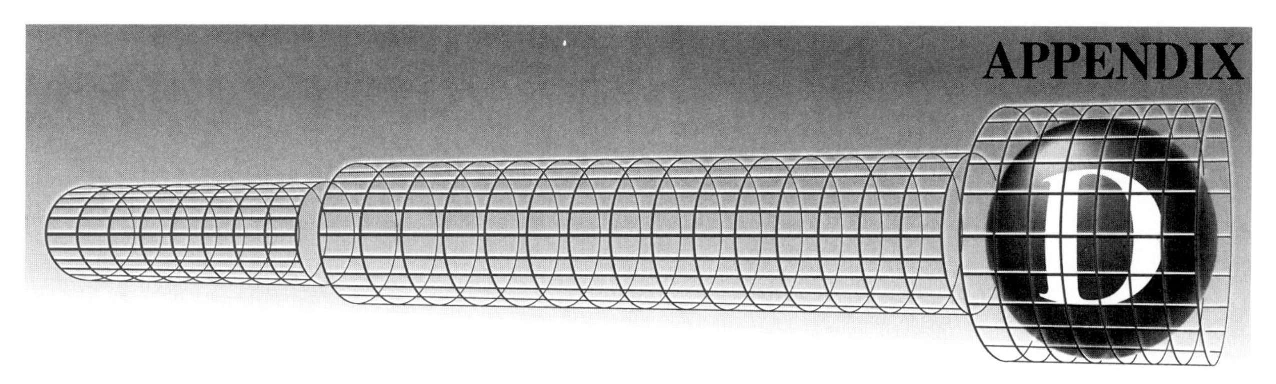

Answers to Even-numbered Exercises

CHAPTER 1

Exercises 1.1

2. $-3 < x < 4$ **4.** $-1 < x \le 6$ **6.** $x \ge 3$
8. $-\infty < x < \infty$ **10.** No **12.** $[3, 10]$
14. $(4, 22]$ **16.** $(-\infty, -2]$ **18.** $(-\infty, -2]$
20. $(-\infty, \infty)$ **22.** $x < 3$ **24.** $0 < x < 3/5$
26. $-3 \le x \le 3$ **28.** $-5/3 < x < 3$
30. $-1/\sqrt{2} \le x \le 1/\sqrt{2}$ **32.** $-11/2 \le x \le -1/2$
34. $x \le -3$, $x \ge 3$ **36.** $x \le -1$, $x \ge 1$
38. $x < 1/2$, $x > 11/9$
40. $-2 - \sqrt{10} < x < -2 + \sqrt{10}$
42. $x < -1 - \sqrt{7/2}$, $x > -1 + \sqrt{7/2}$
44. All points
48. $x < -2$, $-4/\sqrt{5} < x < 4/\sqrt{5}$, $x > 2$
50. $\sqrt{3} - 1 \le x \le \sqrt{5} - 1$, $-\sqrt{5} - 1 \le x \le -\sqrt{3} - 1$
52. $x < -\sqrt{37}$, $-3 < x < 3$, $x > \sqrt{37}$
54. $x < -11/7$, $-1 < x < 3$ **56.** $x \ge 5$
58. $(1 - \sqrt{5})/2 \le x \le (1 + \sqrt{5})/2$
60. $x < (3 - \sqrt{101})/2$, $5 < x < (3 + \sqrt{101})/2$
62. $|x + 1| + |x - 2| > 4$, $x < -3/2$, $x > 5/2$
64. $||x - 2| - |x + 6|| < 3$, $-7/2 < x < -1/2$
66. **(a)** $-11 < x < -1$ **(b)** $x < -11$, $-2 < x < -1$ **(c)** $x < -11$, $-5 < x < -1$ **(d)** $-11 < x < -5$, $-2 < x < -1$

Exercises 1.2

4. $3\sqrt{5}$ **6.** $\sqrt{58}$ **8.** $(1, -1/2)$
10. $(-1/2, 1/2)$

12.

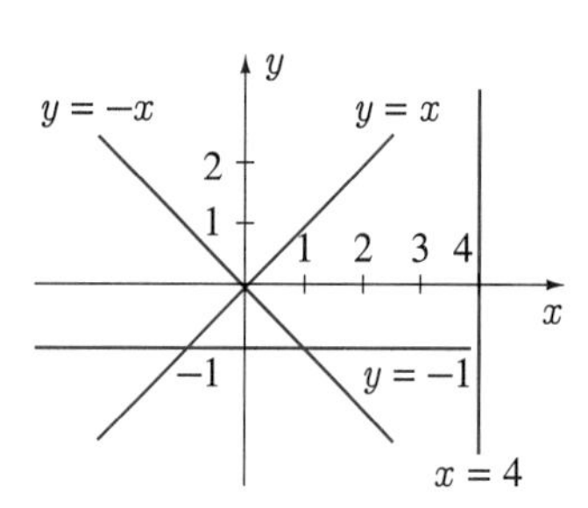

14. $(15/2, 15\sqrt{3}/2)$
16. $(\sqrt{2}, -\sqrt{7})$, $(0, -\sqrt{7} - \sqrt{2})$, $(-\sqrt{2}, -\sqrt{7})$
18. $20(\sqrt{2} + 1)$ cm **20.** $(2, 3/2)$ **22.** $(10/3, 40/3)$

Exercises 1.3

2.

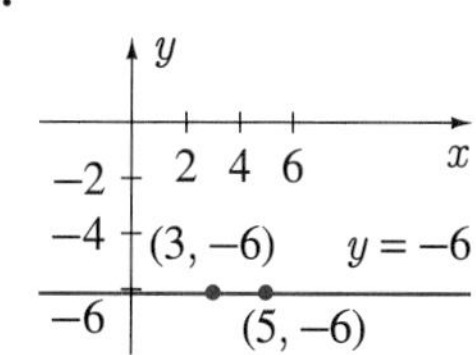

4.

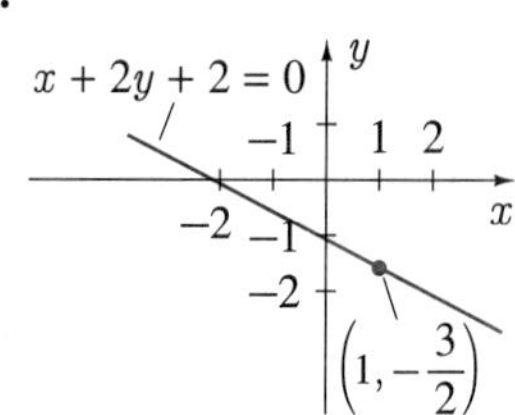

6.

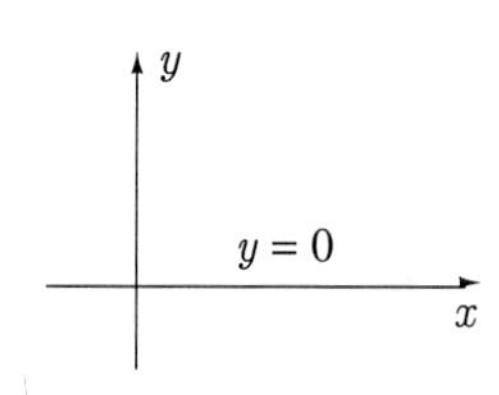

8.

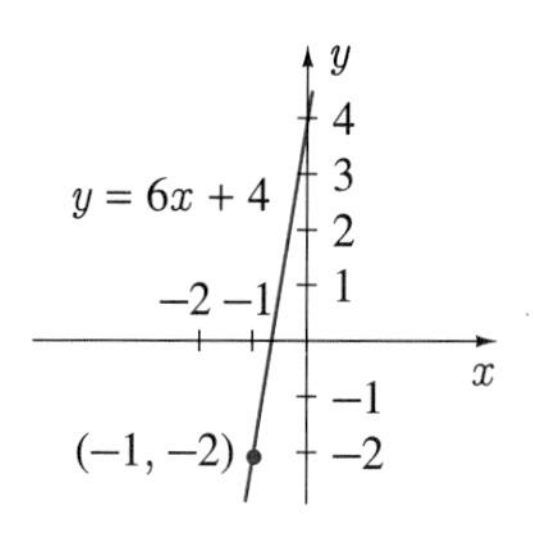

10.

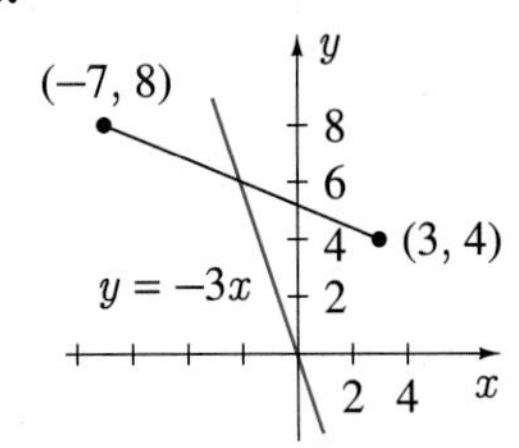

12. 2.68 **14.** 1.25 **16.** 0
18. parallel **20.** perpendicular
22. 1.37 **24.** 1.11 **26.** $(1, 2)$
28. $(-10, -14)$ **30.** $(37/73, 153/146)$
32. $x + y = 2$ **34.** $3y = 2x$ **36.** $5x + 3y = 30$
38. $y = -1$ **42.** $(-11/10, 19/10)$

Exercises 1.4A

2.

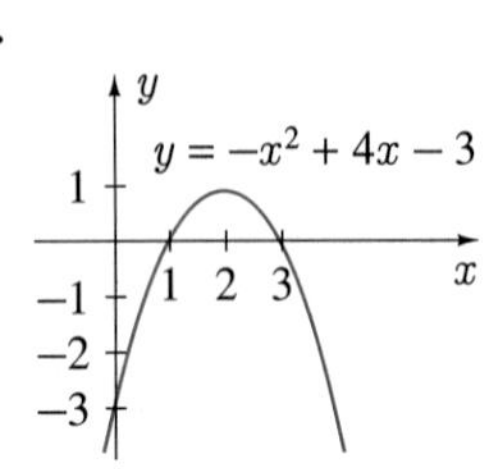

4.

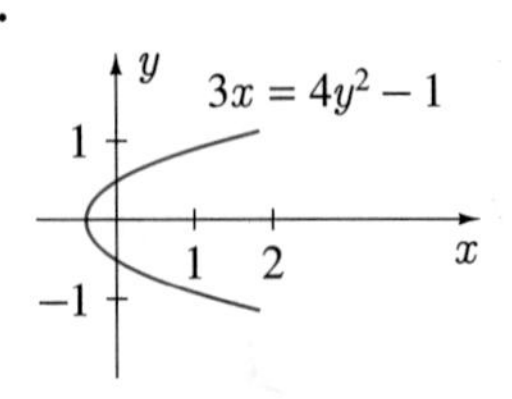

6.

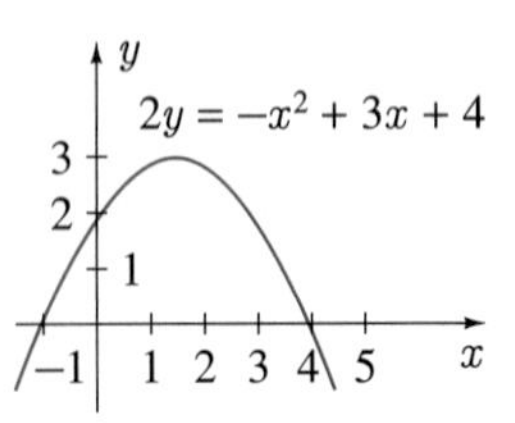

8.

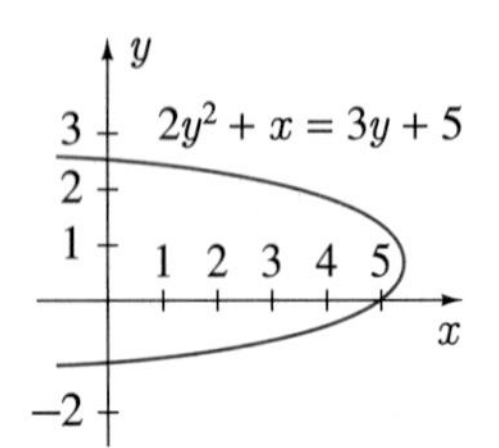

10.

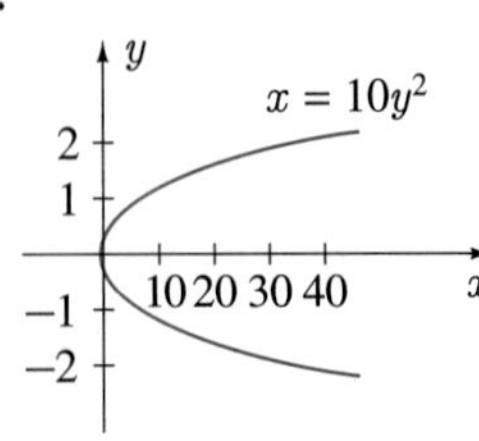

12.

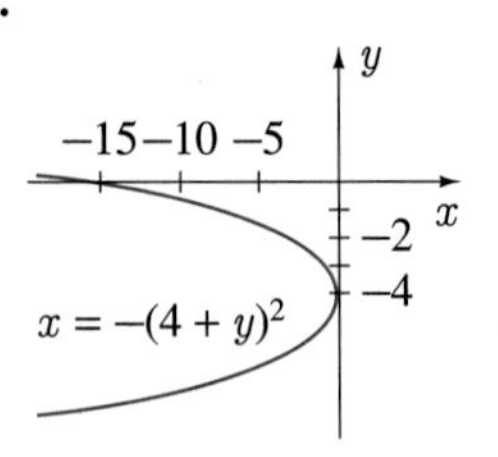

14. $y = x^2/2 + 1$ **16.** $y = (x^2 - 2x - 3)/3$
18. $(0, 1), (-1, 0)$ **20.** No points **22.** $(0, 1), (-3, -2)$
24. (a) $(v^2/4.905) \sin\theta \cos\theta$ **(b)** $(v^2/19.62) \sin^2\theta$
26. $(1, 1), (4, 4)$ **28.** $y = (x^2 - 2x + 5)/2$

Exercises 1.4B

2.

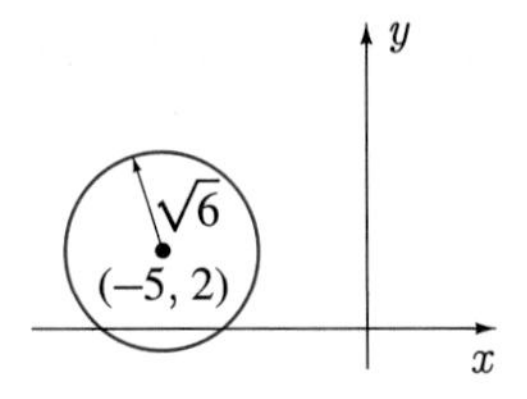

4.

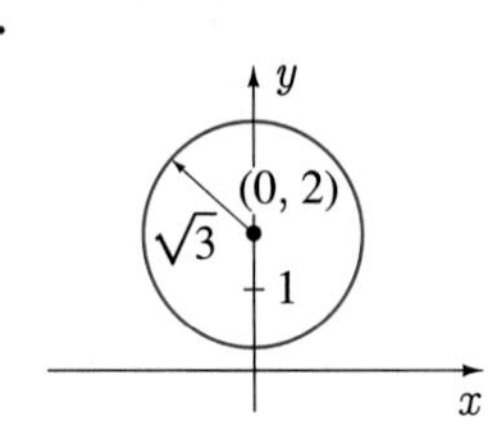

6.

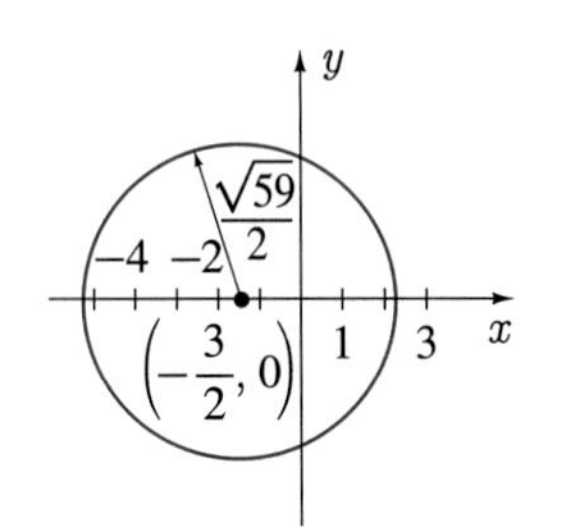

8.

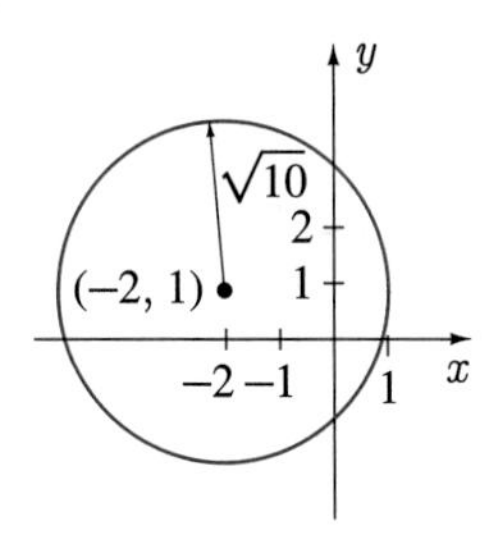

10. No points satisfy the equation
12. $(x - 1)^2 + y^2 = 1$
14. $(x - 3/2)^2 + (y + 3/2)^2 = 9/2$
16. $3x^2 + 3y^2 - 14x - 4y = 4$
18. $(0, 2), (-7/5, -11/5)$ **20.** No points
24. $7(x^2 + y^2) + 6x - 44y + 24 = 0$

Exercises 1.4C

2.

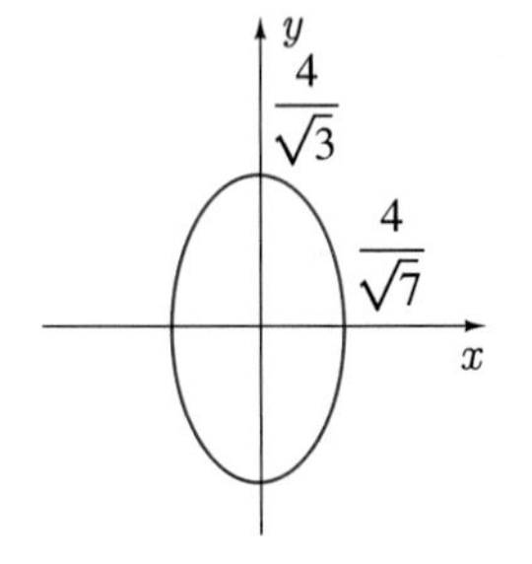

4.

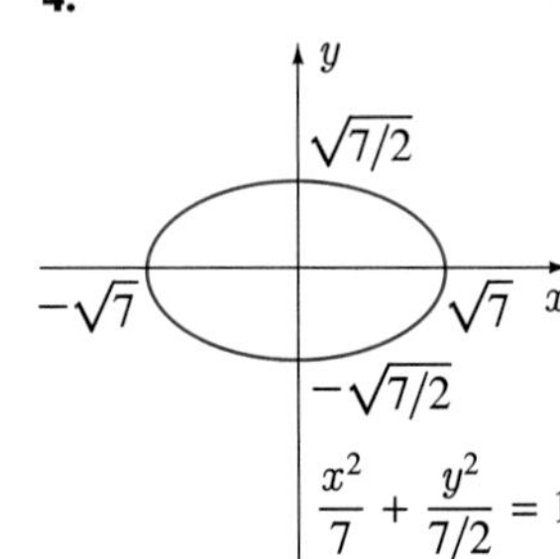

6.

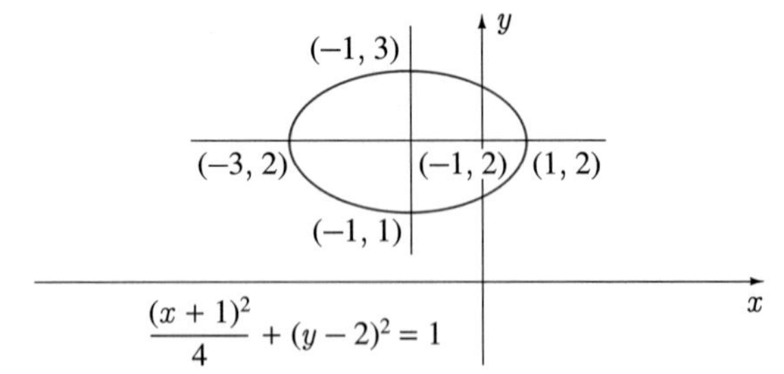

8.

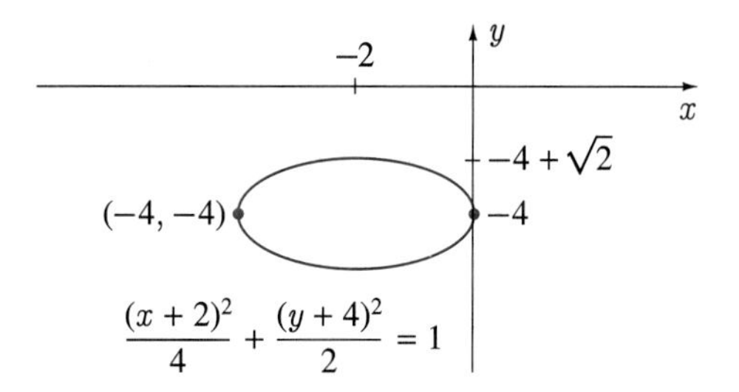

10. $32/\sqrt{15}$ **12.** $(-3, 0)$, $(21/25, 96/25)$
14. $(2, 3\sqrt{3}/2)$ **16.** $(\pm 2, 0)$, $(\pm\sqrt{15}/2, -1/4)$

Exercises 1.4D

2.

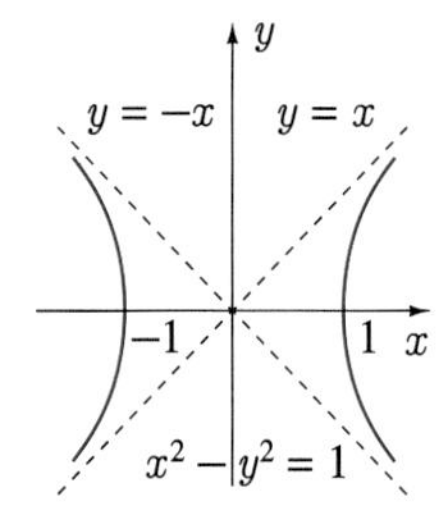

4.

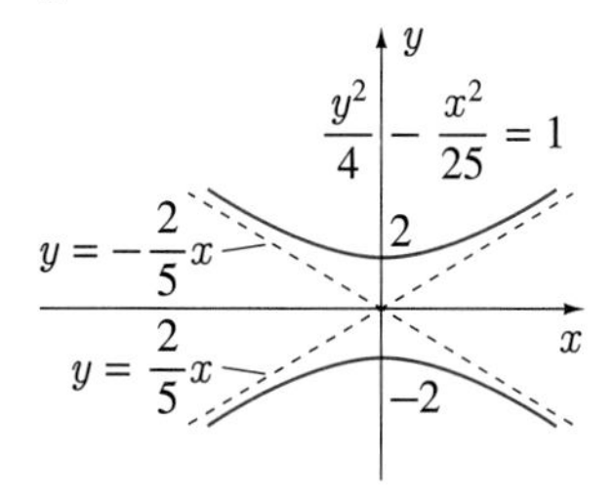

6.

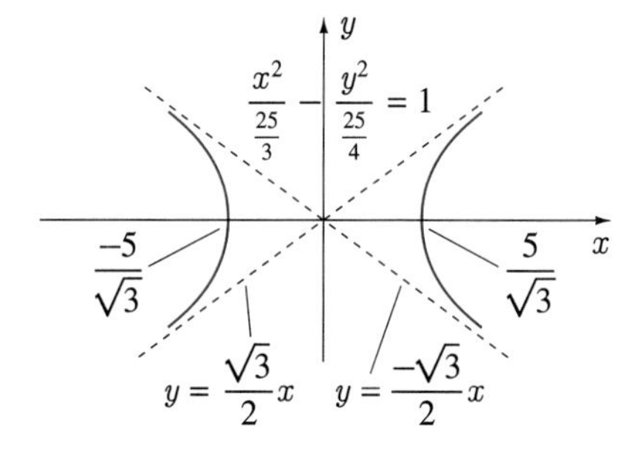

8.

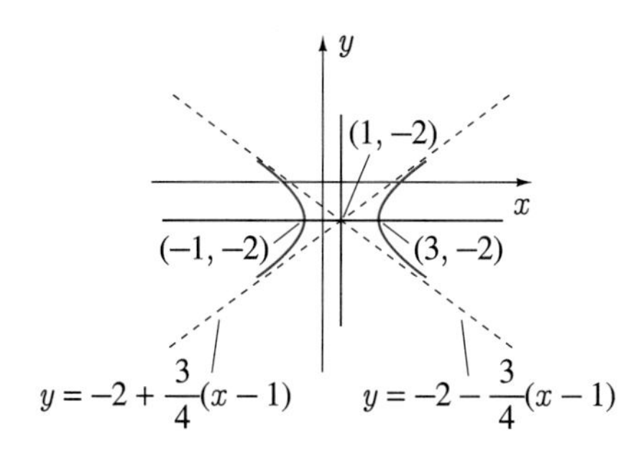

10.

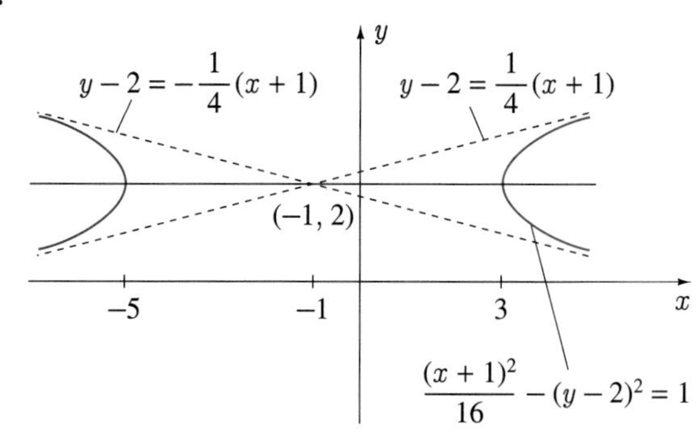

12. $\pm(\sqrt{2}, 1/\sqrt{2})$ **14.** No points **16.** $(0, 0)$, $(3, \pm\sqrt{3})$
18. $(5, 1 \pm 3\sqrt{3})$

Exercises 1.5

2. $x \neq 2$ **4.** $|x| > 2$ **6.** $2 \leq |x| < 3$

8.

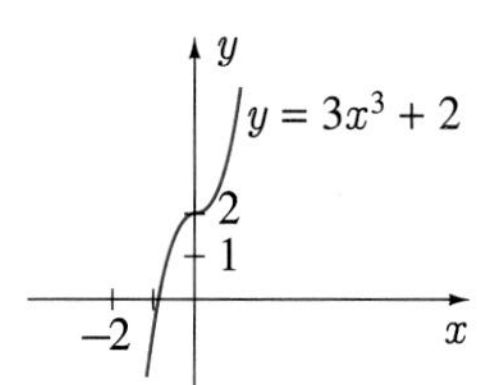

10.

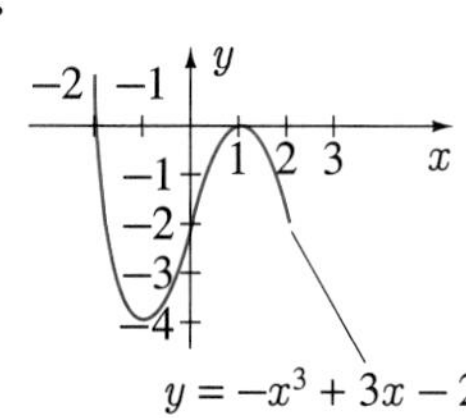

12.

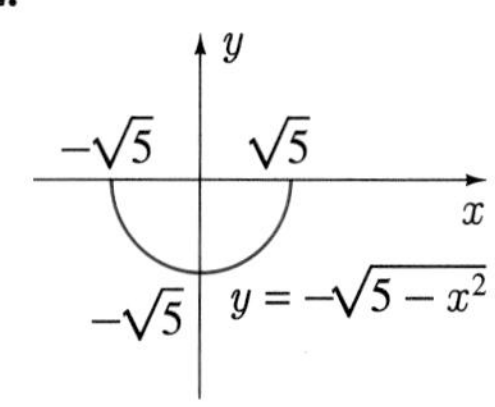

14.

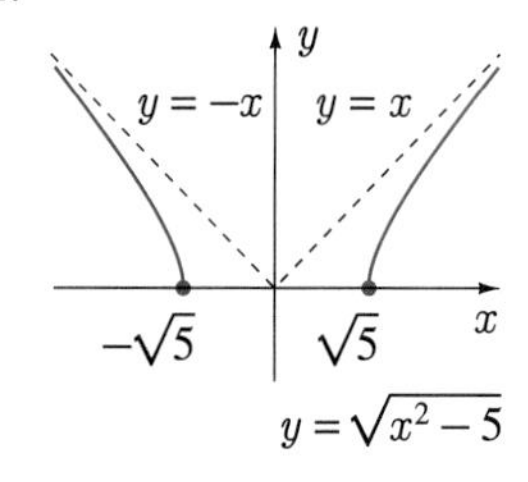

16.

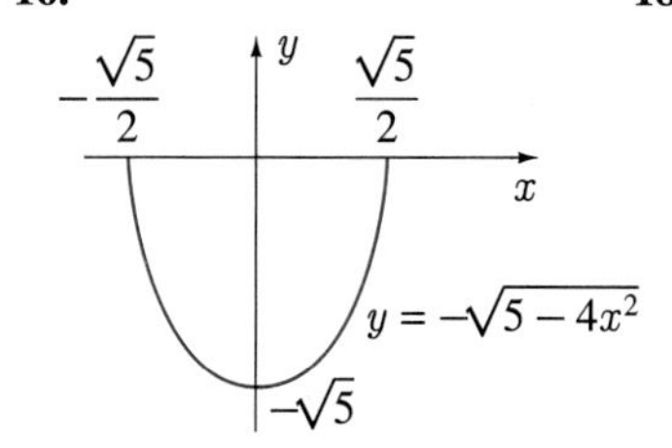

18.

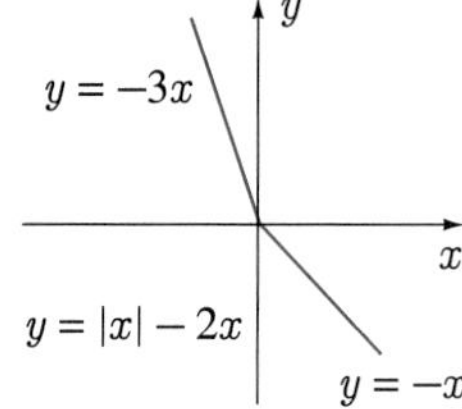

20.

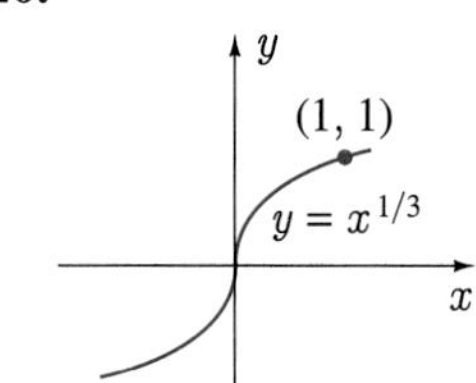

22.

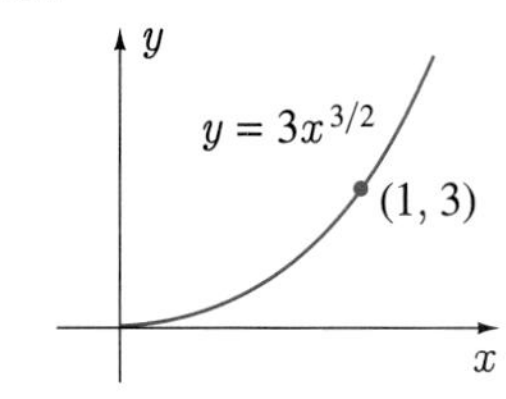

24.

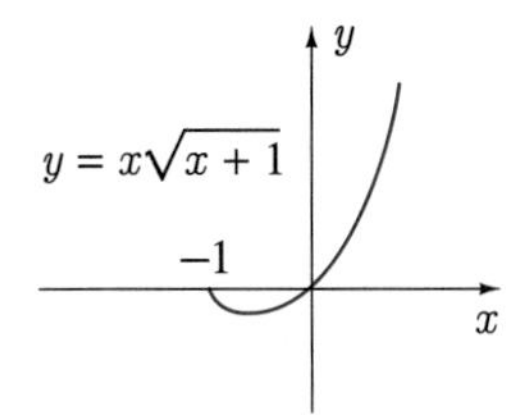

26.

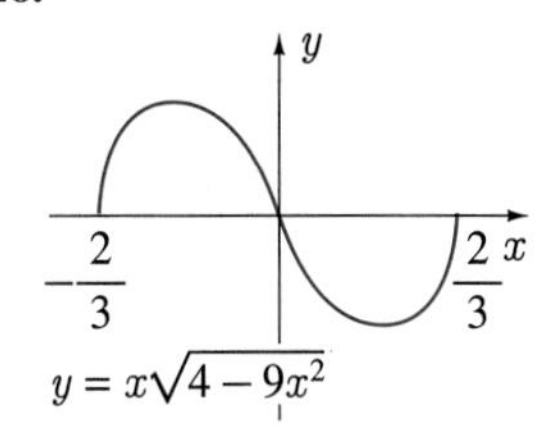

28.

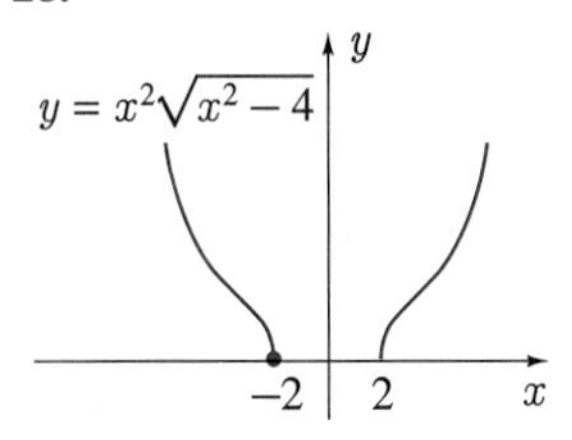

30.

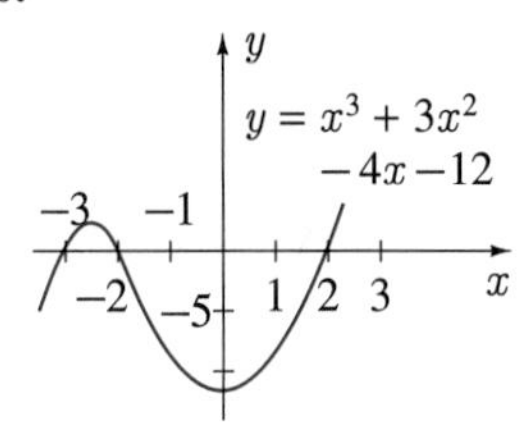

32.

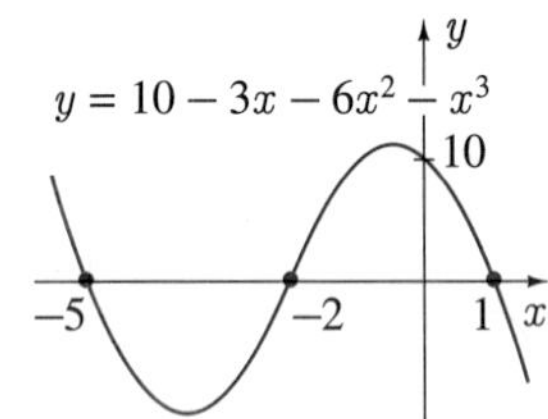

34.

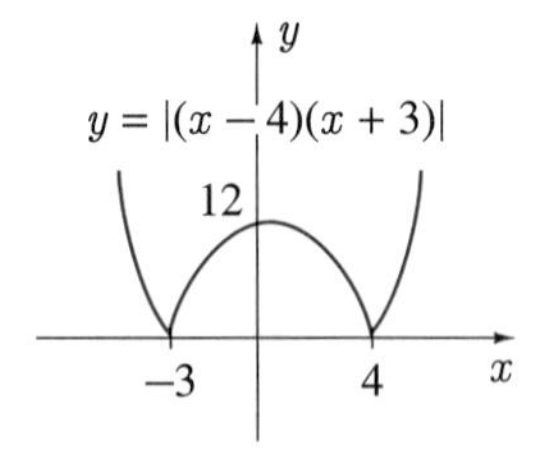

36. (a)

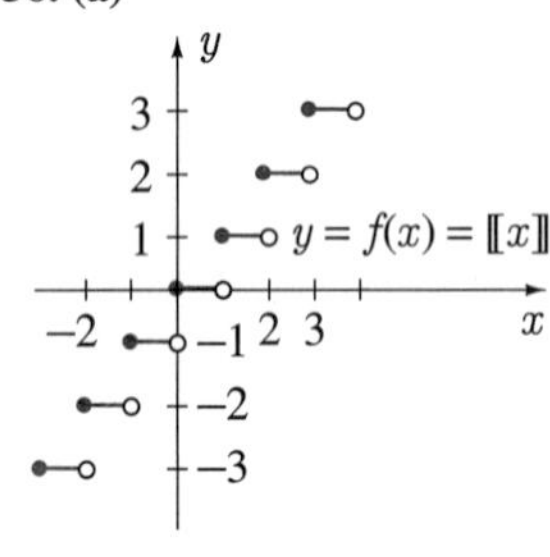

(b)

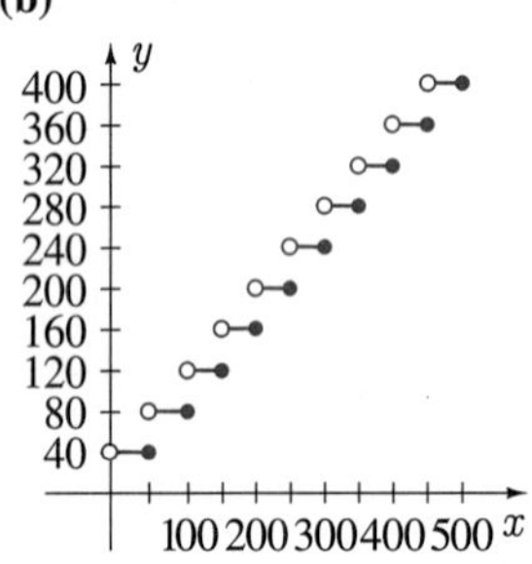

(c) $40[\![1 + x/50]\!]$, $x/50 \neq$ integer; $4x/5$, $x/50$ an integer; $0 < x \leq 500$

38.

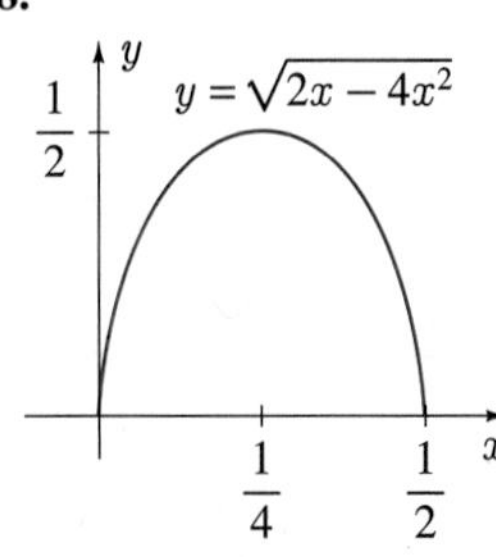

42.

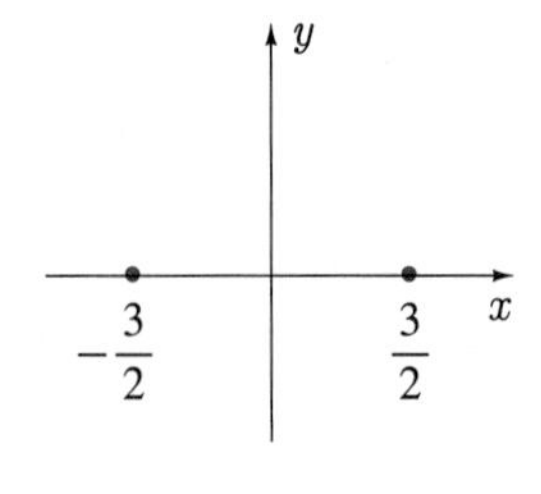

40. Not a function

44.

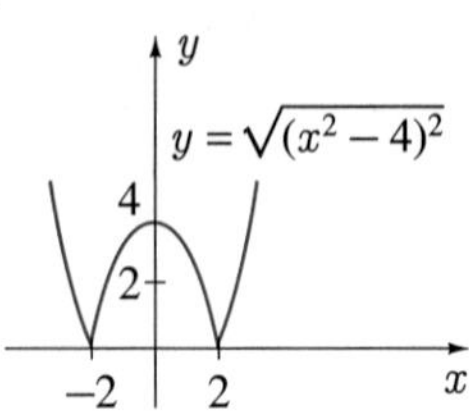

46.

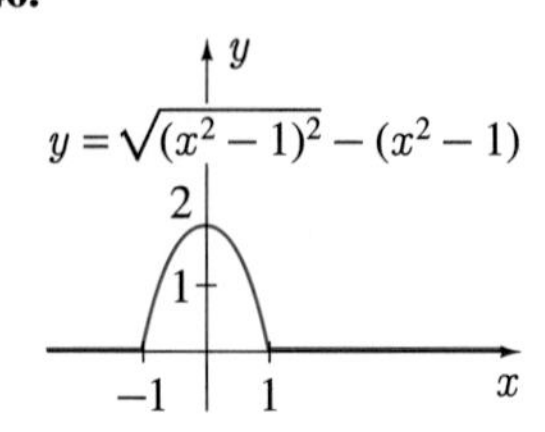

50.

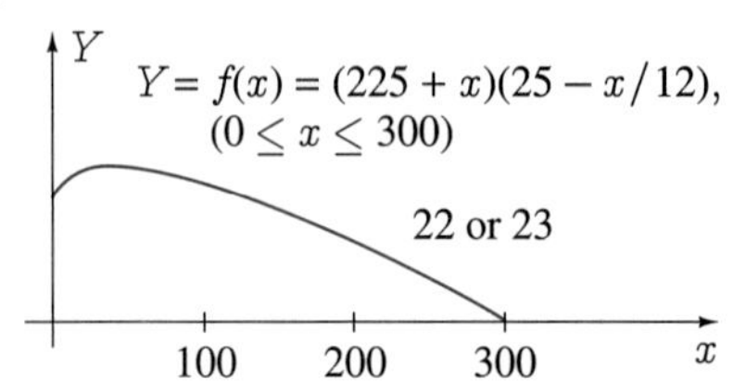

52.

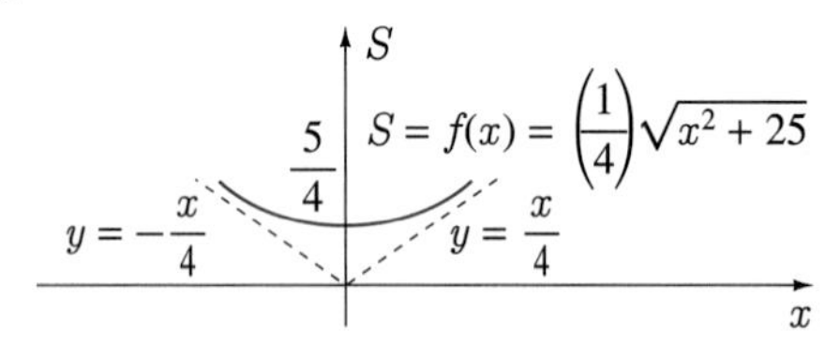

54.

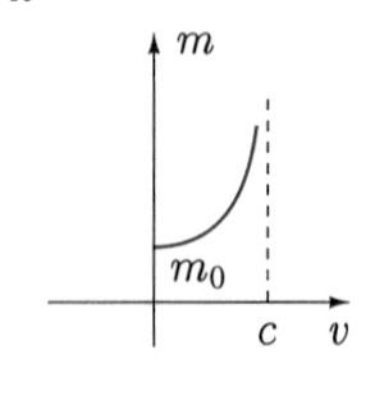

56.

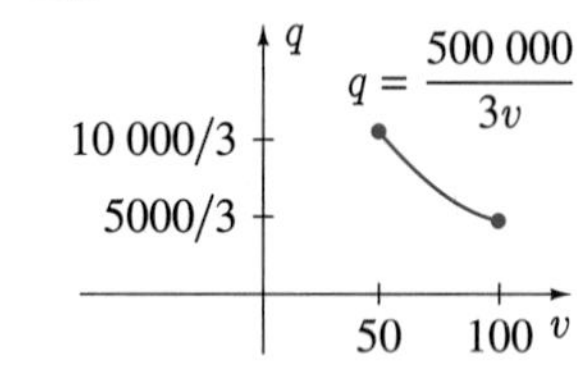

58.

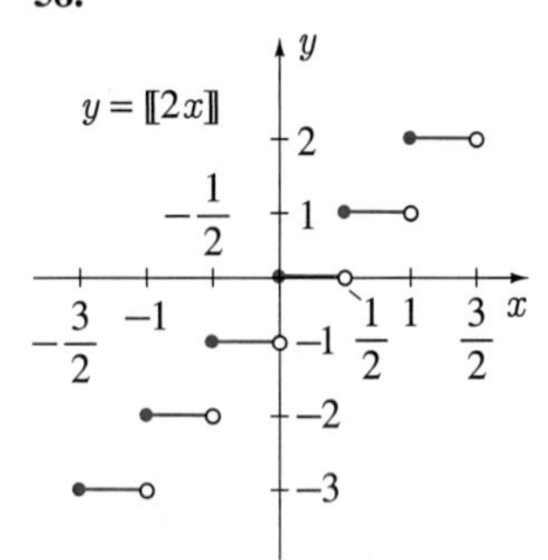

60.

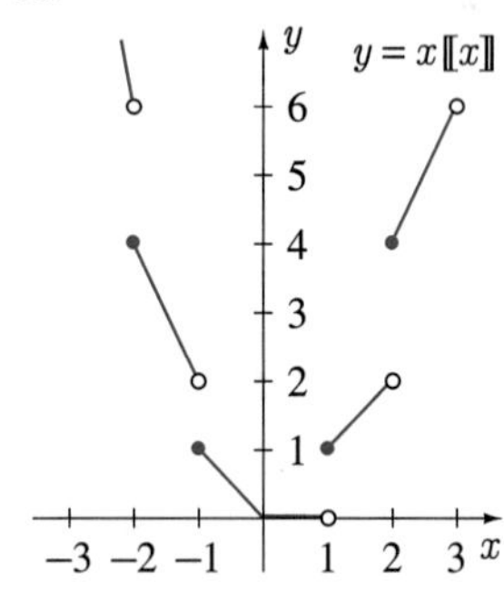

62. No

Exercises 1.6

2. odd **4.** odd **6.** neither
8. odd **10.** odd

14.

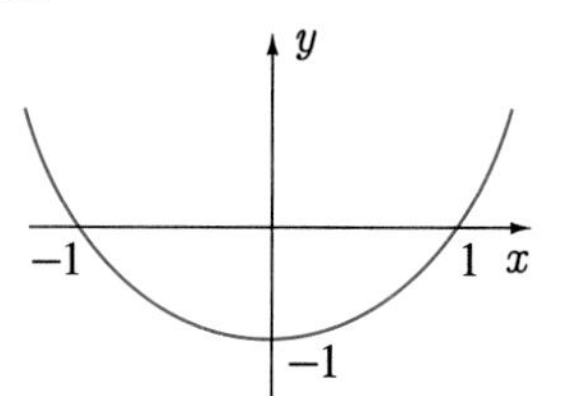

$y = (x + 1)^2 - 2(x + 1)$
(Function)

16.

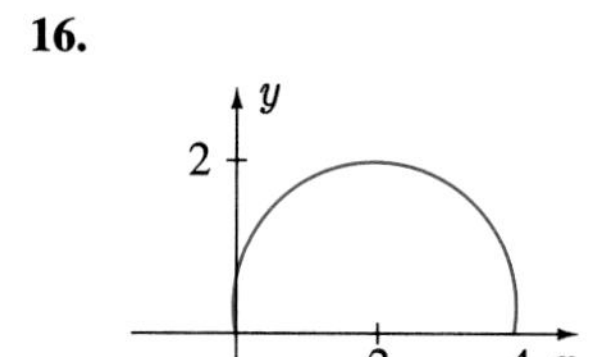

$y = \sqrt{4 - (x - 2)^2}$
(Function)

18.

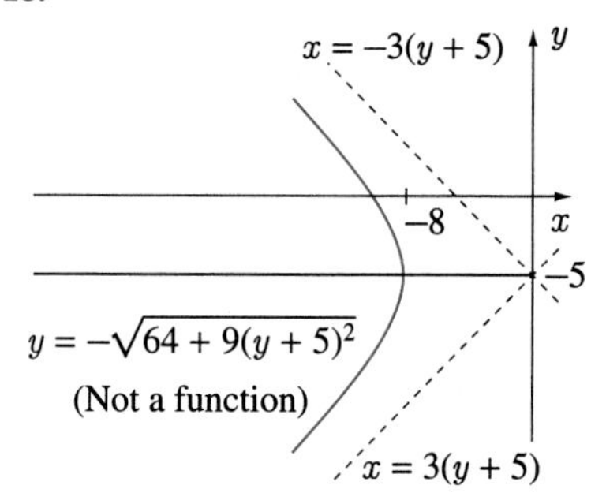

$y = -\sqrt{64 + 9(y + 5)^2}$
(Not a function)

20.

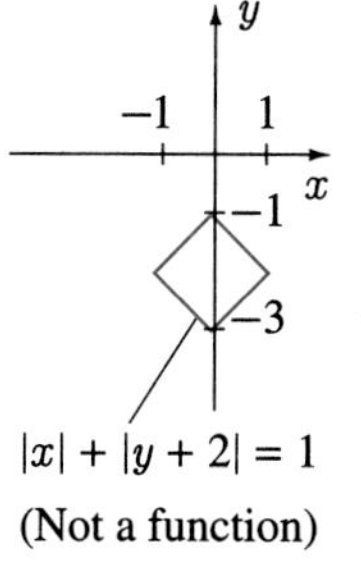

$|x| + |y + 2| = 1$
(Not a function)

22.

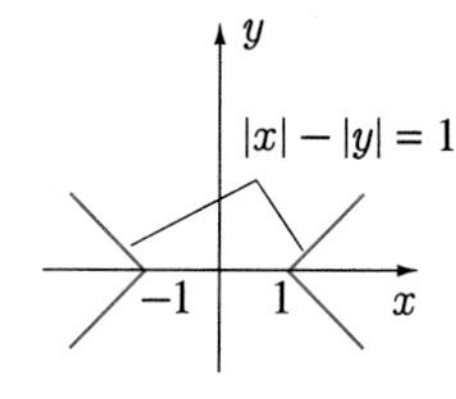

(Not a function)

24.

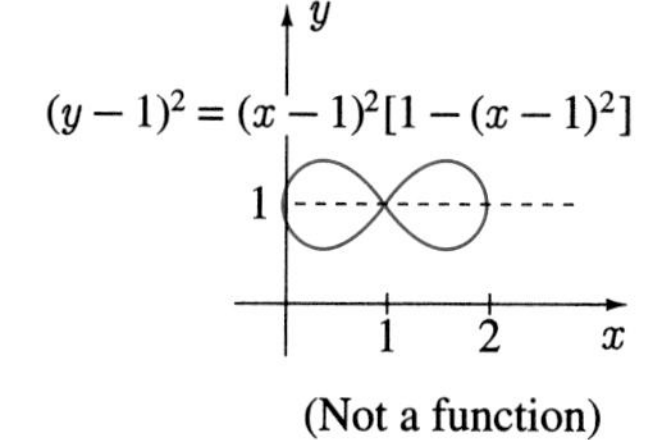

(Not a function)

26.

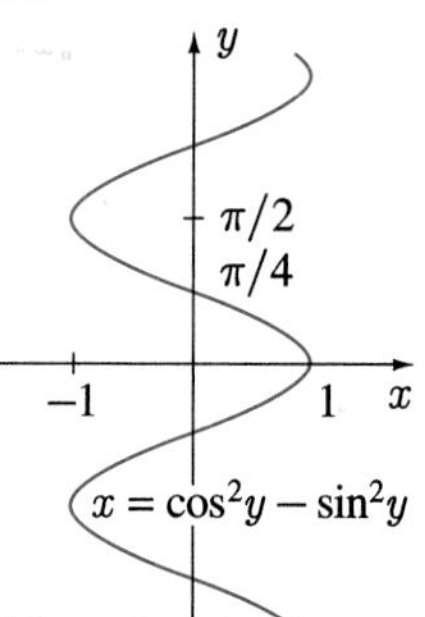

$x = \cos^2 y - \sin^2 y$
(Not a function)

28.

y
1
$-\frac{3\pi}{2}$ $-\frac{\pi}{2}$ $\frac{\pi}{2}$ $\frac{3\pi}{2}$ x
−1

$y^2 = \cos x$
(Not a function)

30. **32.**

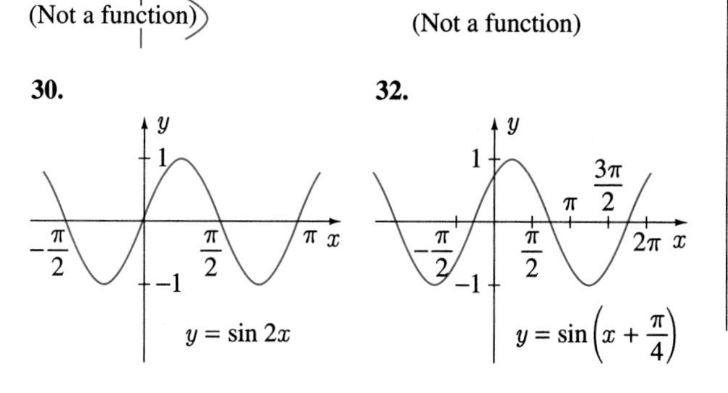

$y = \sin 2x$

$y = \sin\left(x + \frac{\pi}{4}\right)$

34.

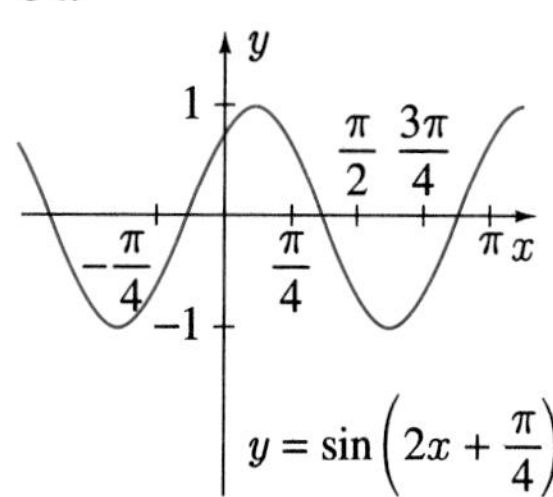

$y = \sin\left(2x + \frac{\pi}{4}\right)$

36.

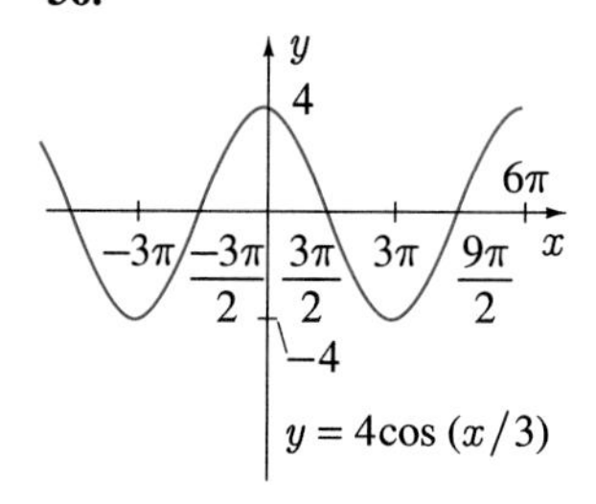

$y = 4\cos(x/3)$

38.

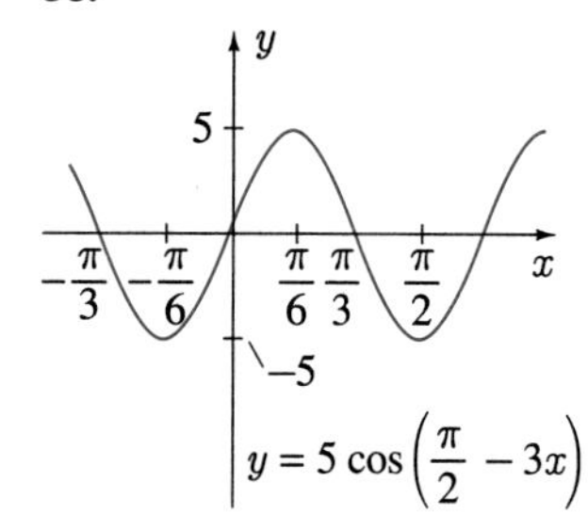

$y = 5\cos\left(\frac{\pi}{2} - 3x\right)$

40.

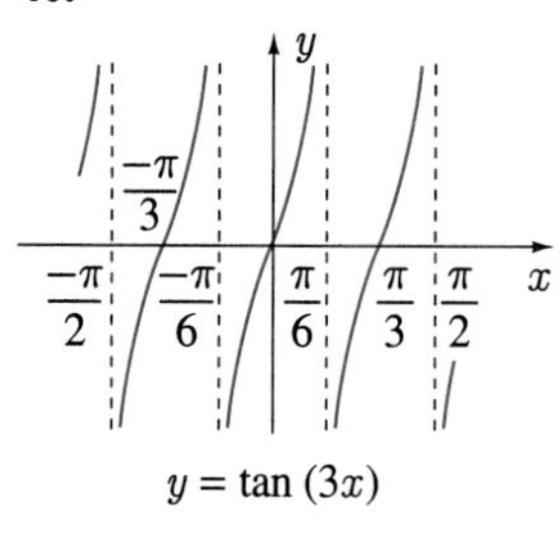

$y = \tan(3x)$

42.

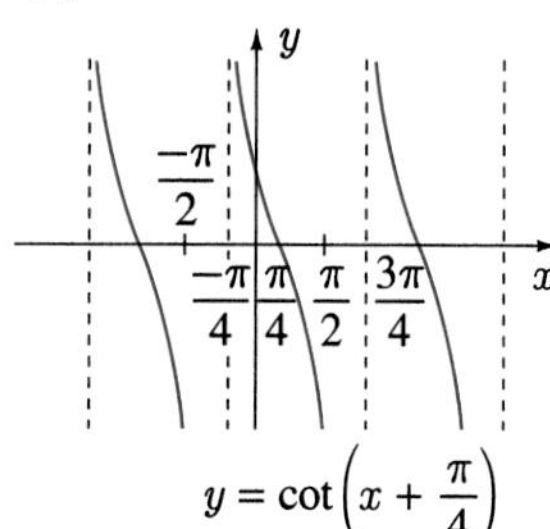

$y = \cot\left(x + \frac{\pi}{4}\right)$

44.

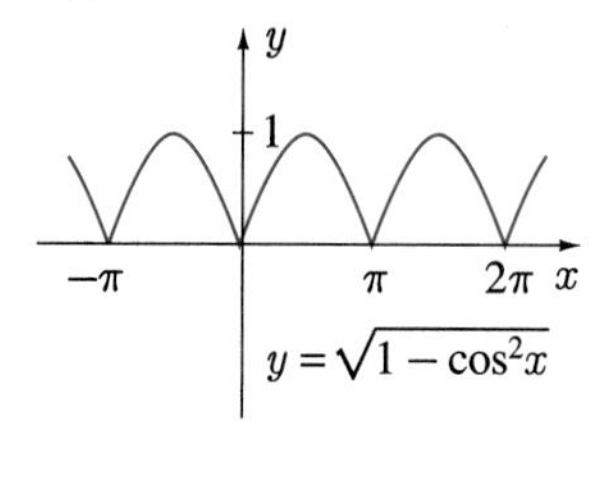

$y = \sqrt{1 - \cos^2 x}$

46.

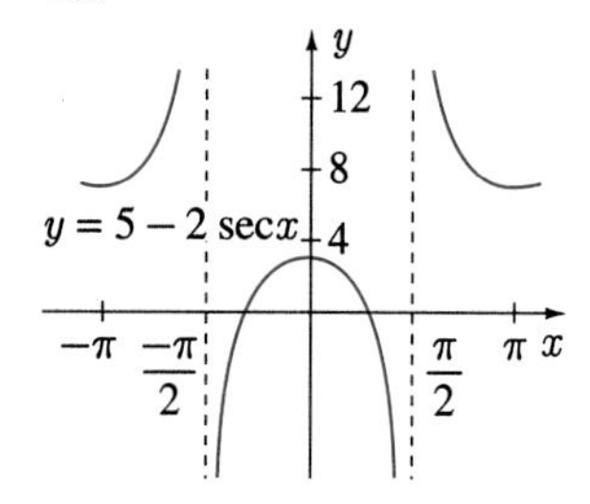

$y = 5 - 2\sec x$

48.

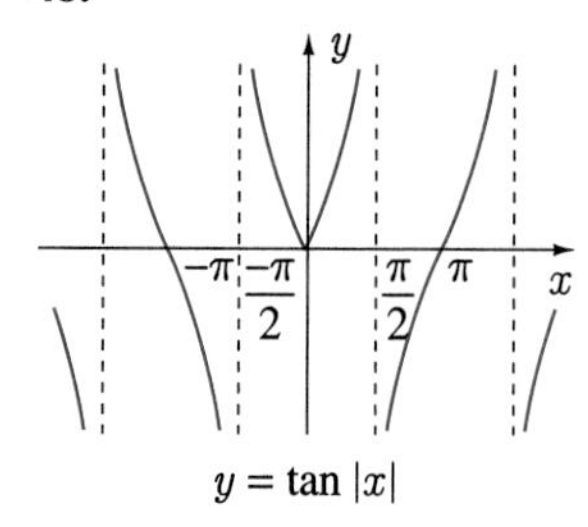

$y = \tan|x|$

50.

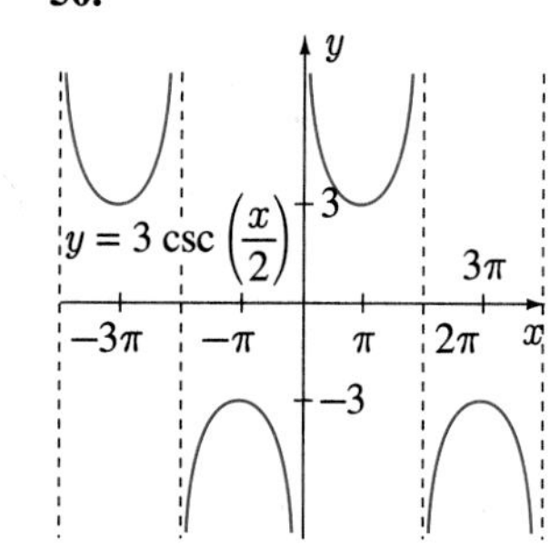

$y = 3\csc\left(\frac{x}{2}\right)$

52.

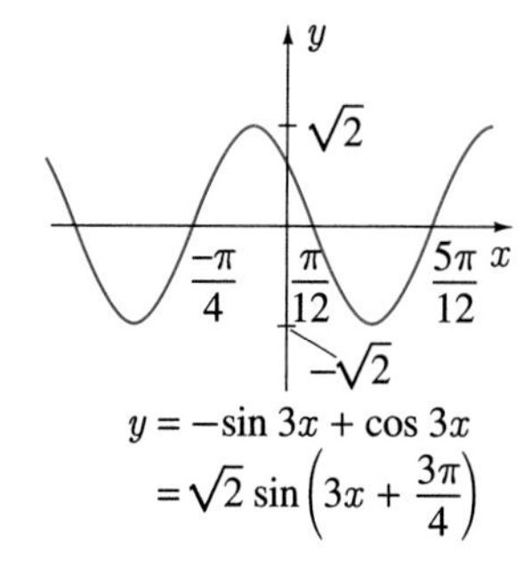

$y = -\sin 3x + \cos 3x$
$= \sqrt{2}\sin\left(3x + \frac{3\pi}{4}\right)$

54.

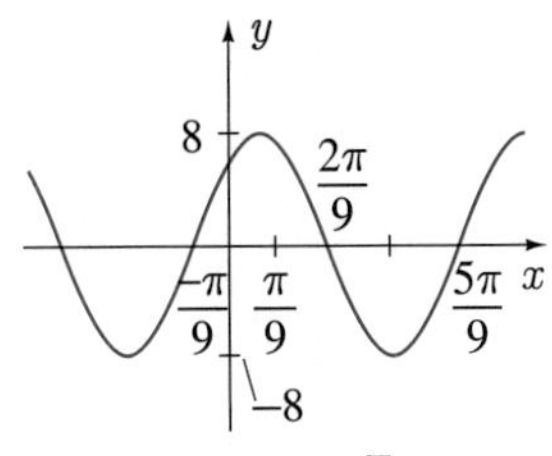

$y = 4\sin 3x + 4\sqrt{3}\cos 3x$
$= 8\sin\left(3x + \frac{\pi}{3}\right)$

56.

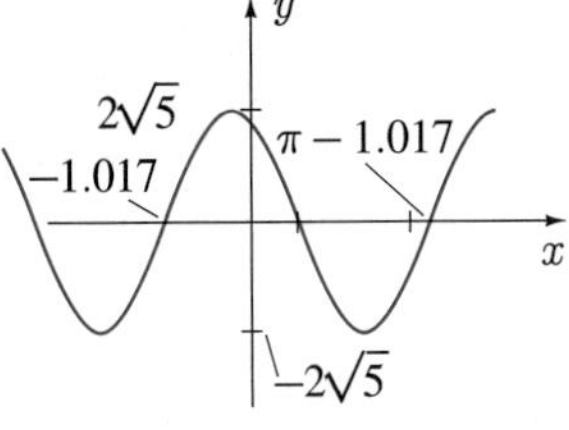

$y = -2\sin 2x + 4\cos 2x$
$= 2\sqrt{5}\sin(2x + 2.034)$

58.

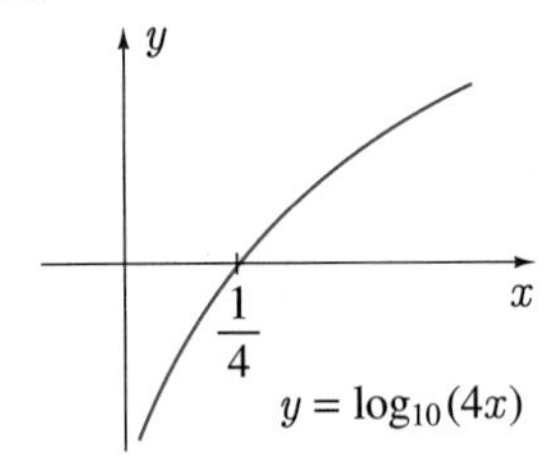

60.

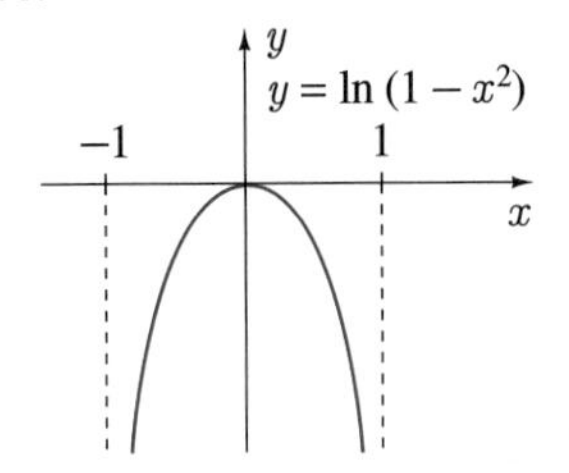

62.

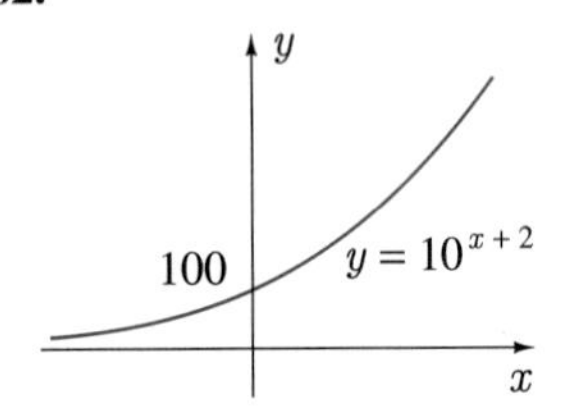

64.

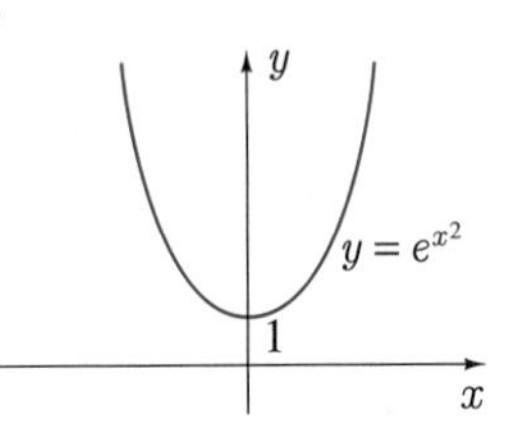

66.

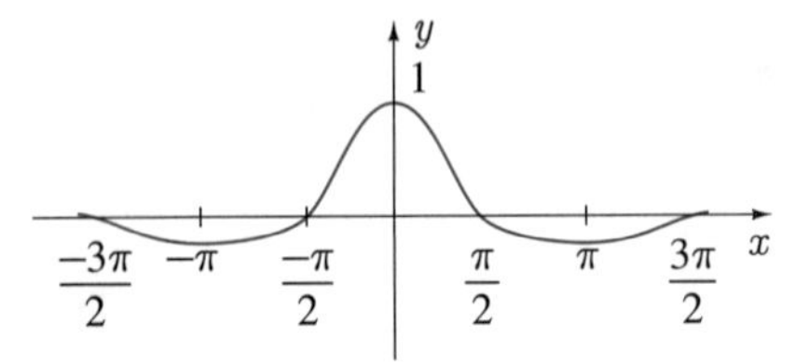

68. (a)

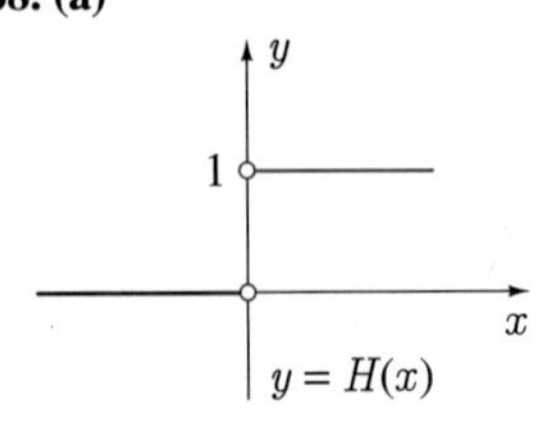

y
1
a x
$y = H(x - a)$

(b)

y
$y = x^{1/3}$
x

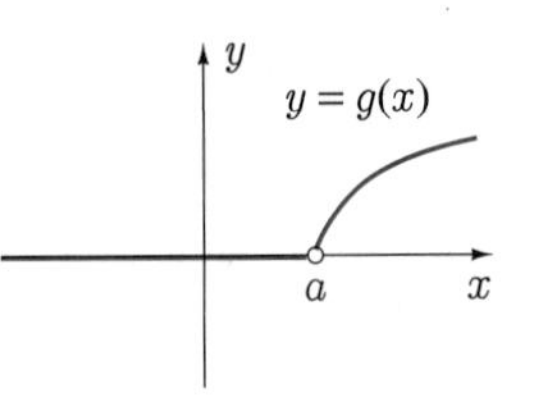

(c) $f(x - a)H(x - a)$

(d) eliminates that part of graph to left of $x = a$

70. They are identical for $x > 0$. $y = f(x)$ is symmetric about $x = 0$, whereas $g(x)$ is not defined for $x \leq 0$.

72.

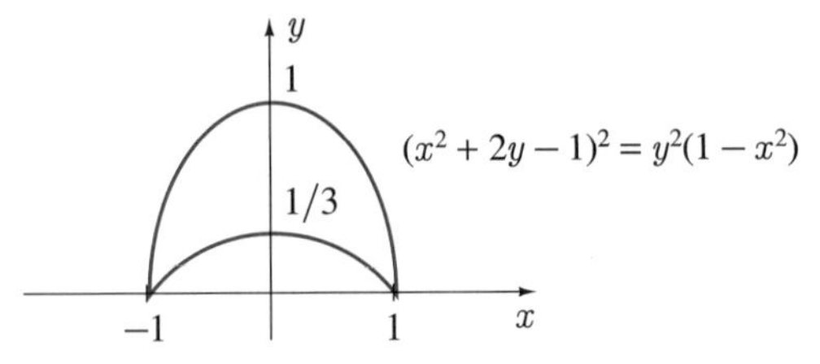

Review Exercises

2. $-2 \leq x \leq 2$

4. $x < -2,\ 3 < x < 4$

6. $x < -3,\ -1 < x < 1,\ x > 3$

8. $5\sqrt{2},\ (-1/2, -3/2)$

10. $y = 2x + 4$

12. $x + y = 2$

14. $|x| \geq \sqrt{5}$

16. $x \neq 0, -1$

18. $x \geq 0$

20. $|x + 1| > \sqrt{14}/2$

22. $-1 \leq x < 0,\ x \geq 1$

24. parabola

26. ellipse

28. none of these

30. parabola

32. circle

34. hyperbola

36.

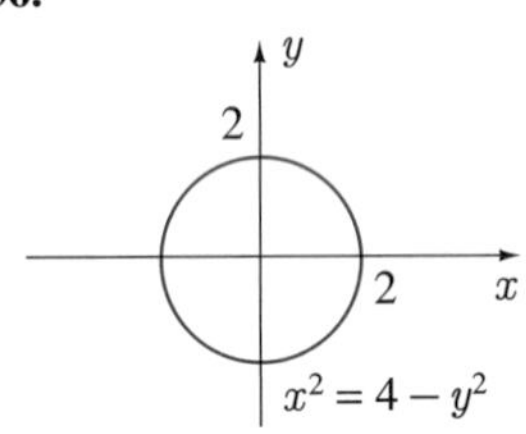

38.

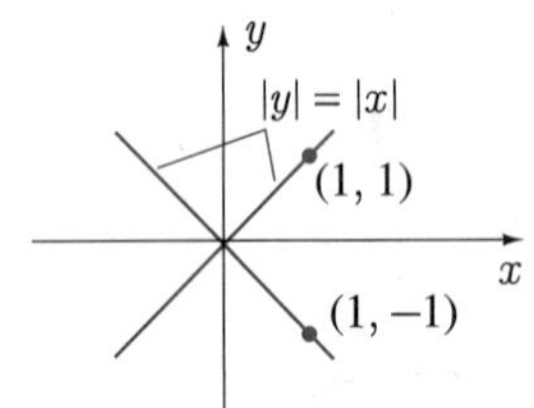

40.

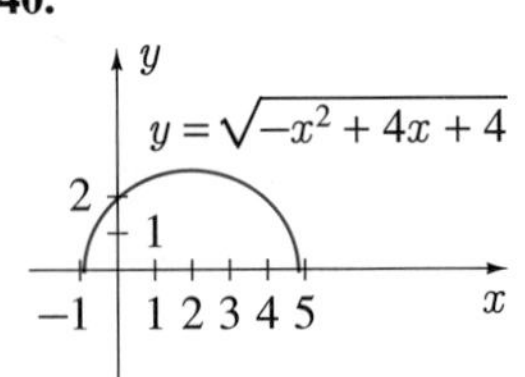

42.

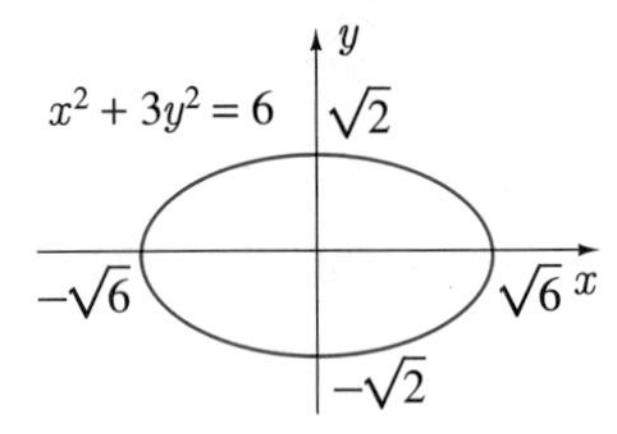

44.

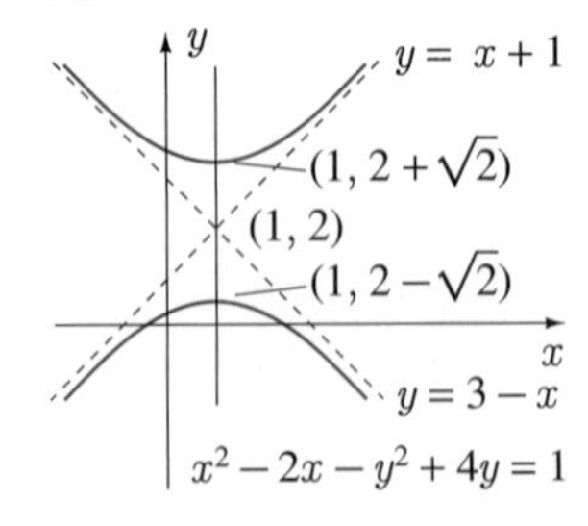

46.

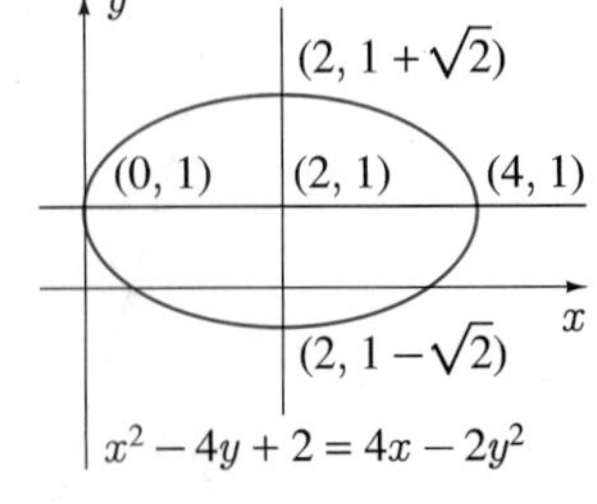

48.

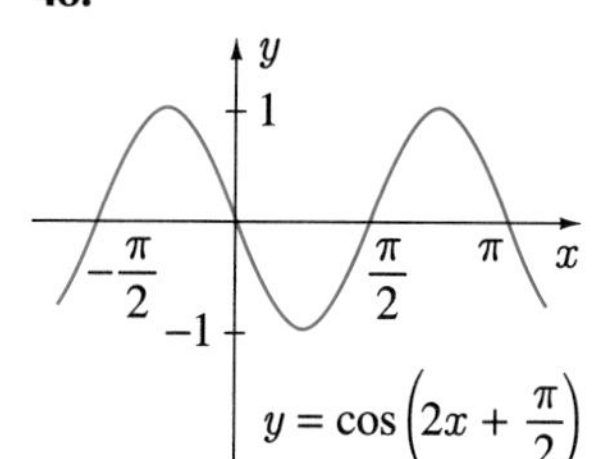

50.

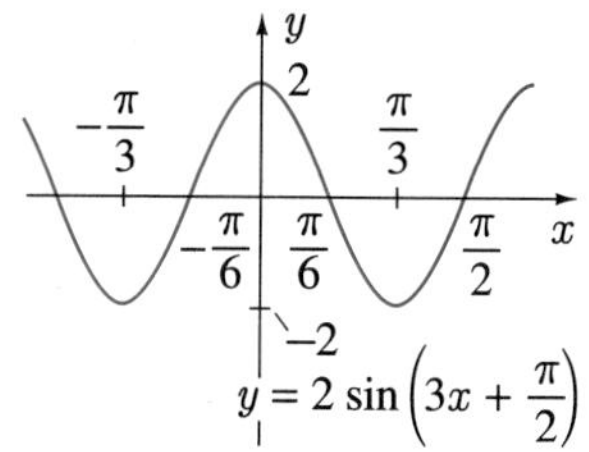

52.

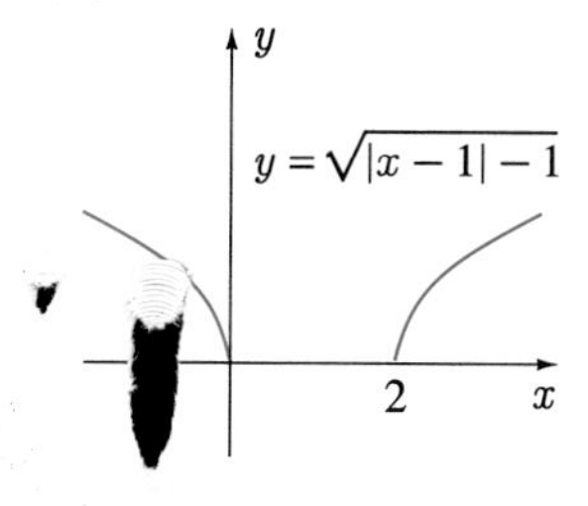

54.

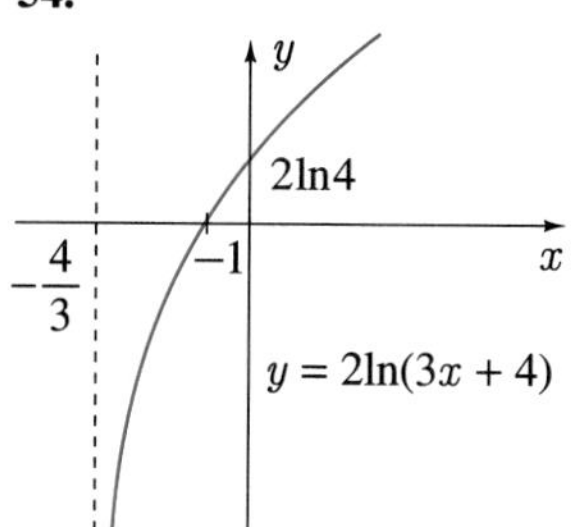

56.

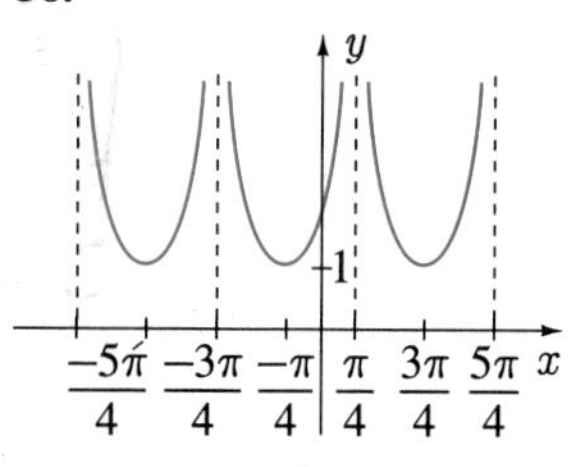

58.

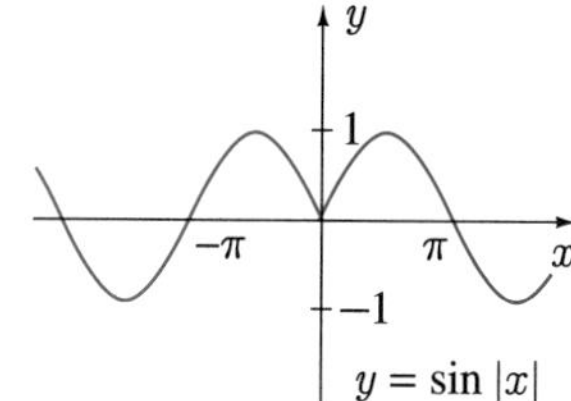

60.

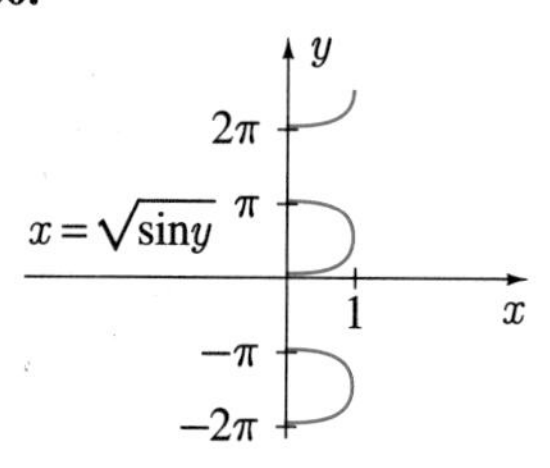

62. $\sqrt{(x+1)(2-x)}$ **64.** $\sqrt{-x}+1$

68.

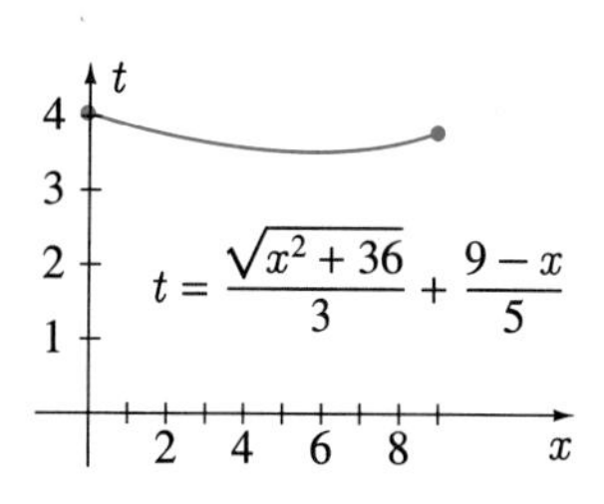

CHAPTER 2

Exercises 2.1

2. 0	**4.** $-3/2$	**6.** 0
8. -12	**10.** 6	**12.** -4
14. 0	**16.** -1	**18.** 0
20. $-7/4$	**22.** does not exist	**24.** 2
26. $1/\sqrt{2}$	**28.** $2\sqrt{2}$	**30.** 1
32. does not exist	**34.** $-1/4$	**36.** 1
38. 4	**40.** -2	**42.** False
44. $1/(2\sqrt{x})$	**48.** L	**50.** Cannot find
52. $-L$		

Exercises 2.2

2. $-\infty$	**4.** ∞	**6.** ∞
8. $13/2$	**10.** does not exist	**12.** ∞
14. 1	**16.** ∞	**18.** does not exist
20. ∞	**22.** 0	**24.** ∞
26. ∞		

Exercises 2.3

2. $-1/2$	**4.** ∞	**6.** $1/3$
8. $-\infty$	**10.** $3/2$	**12.** 0
14. 1	**16.** ∞	**18.** ∞
20. $3/4^{1/3}$	**22.** 0	**24.** 0
26. does not exist	**28.** 0	**30.** ∞
32. 1	**34.** -1	**36.** 0
38. ∞	**40.** $a = d$, $(b-e)/(2\sqrt{a})$	

Exercises 2.4

2.

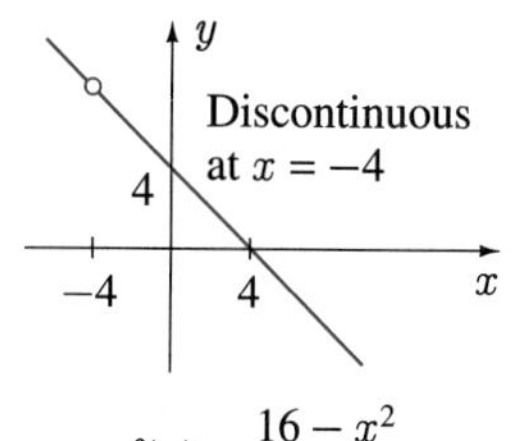

4.

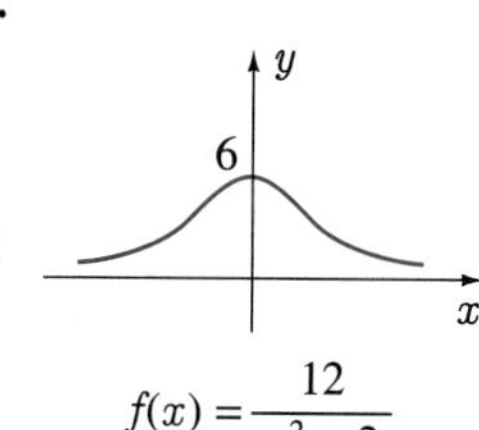

6.

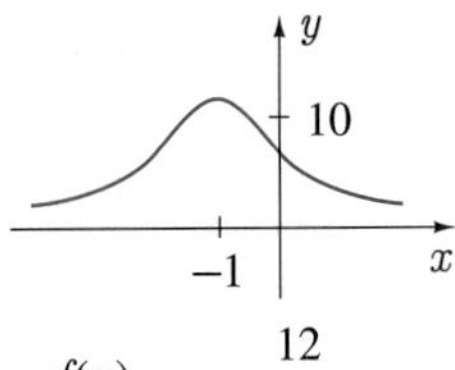

$f(x) = \dfrac{12}{x^2 + 2x + 2}$

8.

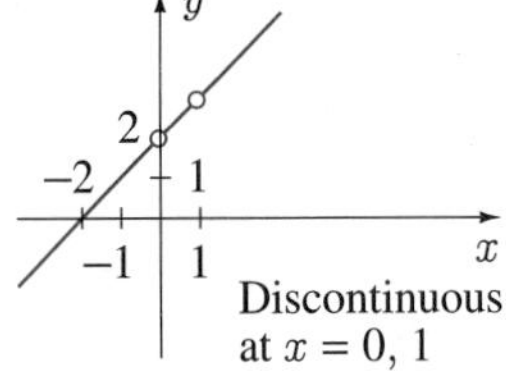

Discontinuous at $x = 0, 1$

$f(x) = \dfrac{x^3 + x^2 - 2x}{x^2 - x}$

10.

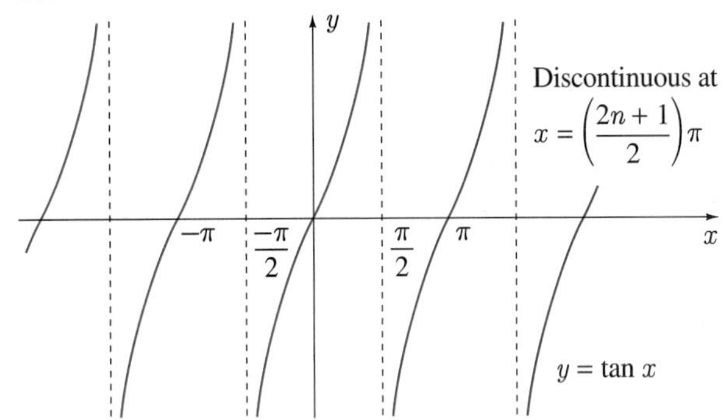

Discontinuous at $x = \left(\dfrac{2n+1}{2}\right)\pi$

$y = \tan x$

12. See Figure 2.5(**b**). Discontinuous at $x = 0$.

14.

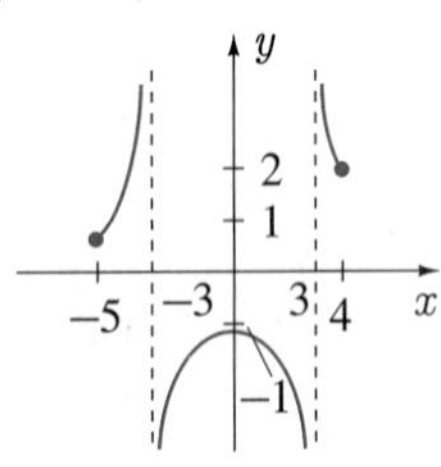

Discontinuous at $x = \pm 3$

$f(x) = \dfrac{x + 12}{x^2 - 9}, \ -5 \leq x \leq 4$

16.

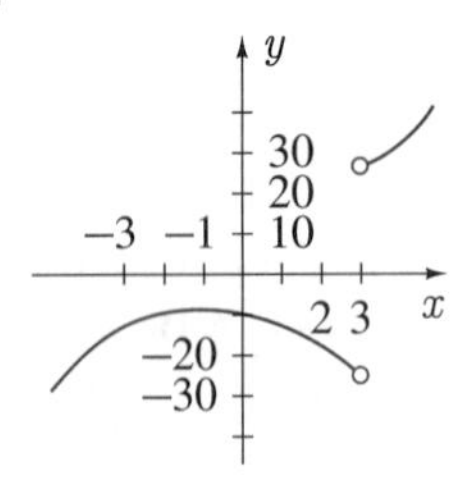

Discontinuous at $x = 3$

$f(x) = \dfrac{x^3 - 27}{|x - 3|}$

18.

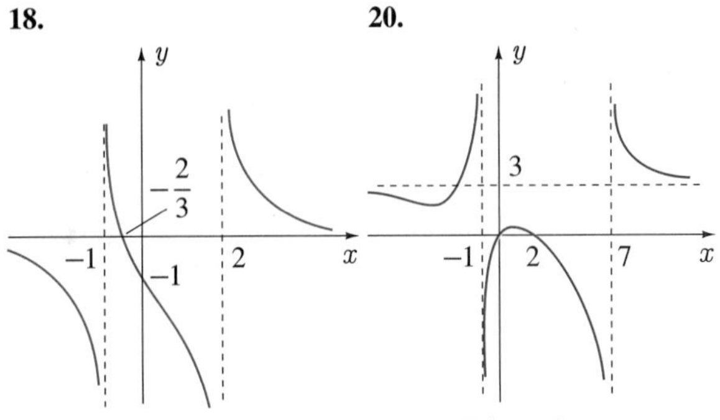

Discontinuous at $x = -1, 2$

$f(x) = \dfrac{3x + 2}{x^2 - x - 2}$

20.

Discontinuous at $x = -1, 7$

$f(x) = \dfrac{3x^2 - 6x}{x^2 - 6x - 7}$

22.

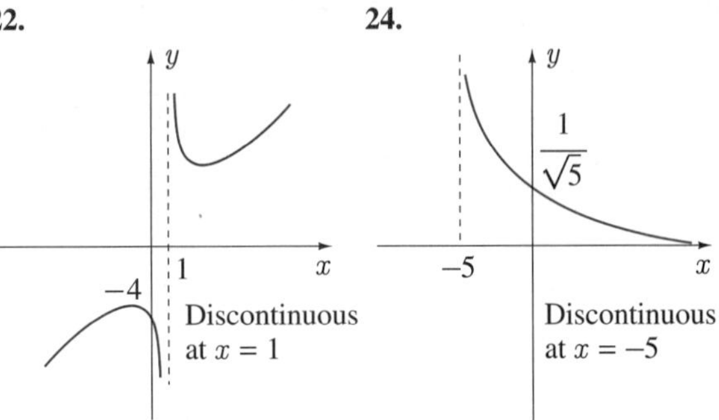

Discontinuous at $x = 1$

$f(x) = \dfrac{x^2 - 2x + 4}{x - 1}$

24.

Discontinuous at $x = -5$

$f(x) = \dfrac{1}{\sqrt{5 + x}}$

26.

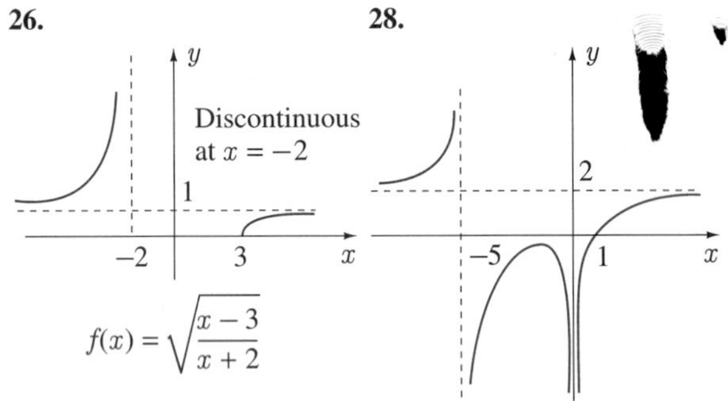

Discontinuous at $x = -2$

$f(x) = \sqrt{\dfrac{x - 3}{x + 2}}$

28.

$f(x) = \dfrac{2x^3 - 2}{x^3 + 5x^2}$ Discontinuous at $x = -5, 0$

30.

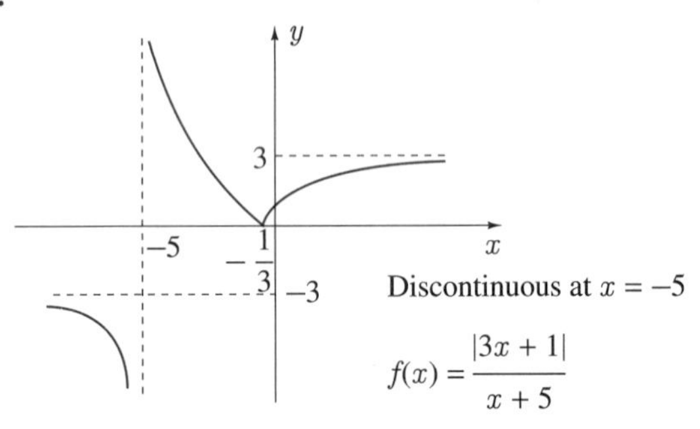

Discontinuous at $x = -5$

$f(x) = \dfrac{|3x + 1|}{x + 5}$

32. $x = -4$ in 2.; $x = -1$ in 7.; $x = 0$ and $x = 1$ in 8.; $x = 2$ in 9.; $x = 0$ in 13.; $x = 1$ in 19.

36.

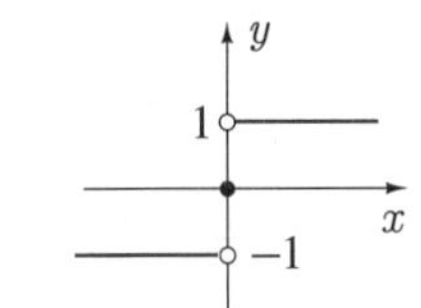

Discontinuous at $x = 0$

38.

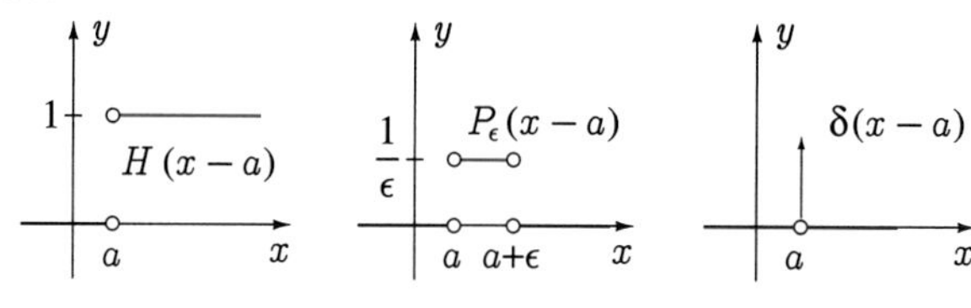

Discontinuous at $x = a$ | Discontinuous at $x = a,\ a + \epsilon$ | Discontinuous at $x = a$

40. (a)

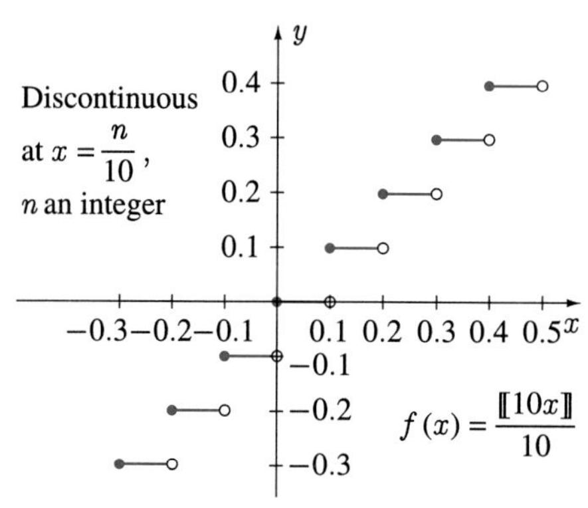

42. (a)

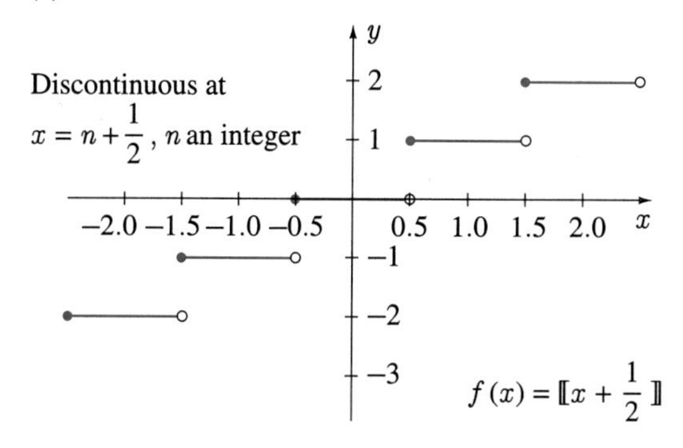

44.

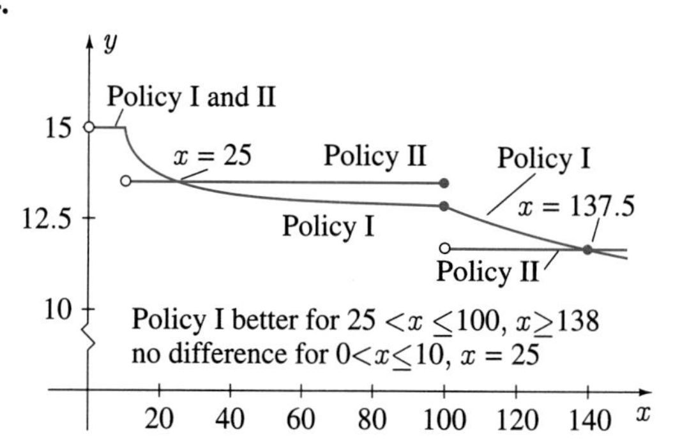

46. None

Exercises 2.5

10. $\lim_{x \to a^+} f(x) = L$ if given any $\epsilon > 0$, there exists a $\delta > 0$ such that $|f(x) - L| < \epsilon$ whenever $0 < x - a < \delta$.

12. $\lim_{x \to \infty} f(x) = L$ if given any $\epsilon > 0$, there exists an X such that $|f(x) - L| < \epsilon$ whenever $x > X$.

14. $\lim_{x \to a} f(x) = \infty$ if given any $M > 0$, there exists a $\delta > 0$ such that $f(x) > M$ whenever $0 < |x - a| < \delta$.

16. $\lim_{x \to \infty} f(x) = \infty$ if given any $M > 0$, there exists an X such that $f(x) > M$ whenever $x > X$.

18. $\lim_{x \to -\infty} f(x) = \infty$ if given any $M > 0$, there exists an X such that $f(x) > M$ whenever $x < X$.

36. No

Review Exercises

2. -2 | **4.** 1 | **6.** $-1/4$

8. 0 | **10.** 0 | **12.** ∞

14. $1/2$ | **16.** 0 | **18.** $-\sqrt{3}/2$

20. $-\infty$

22.

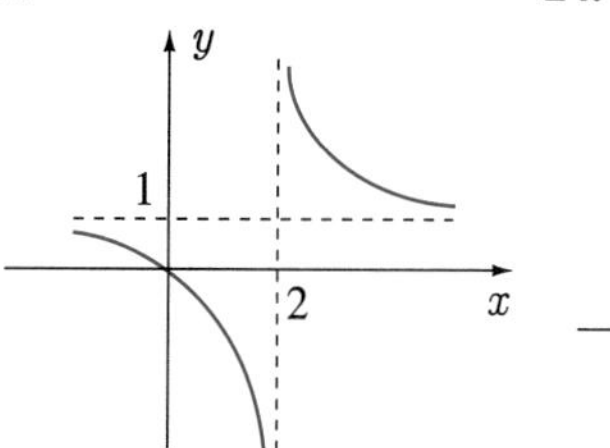

$f(x) = \dfrac{x}{x-2}$ Discontinuous at $x = 2$

24.

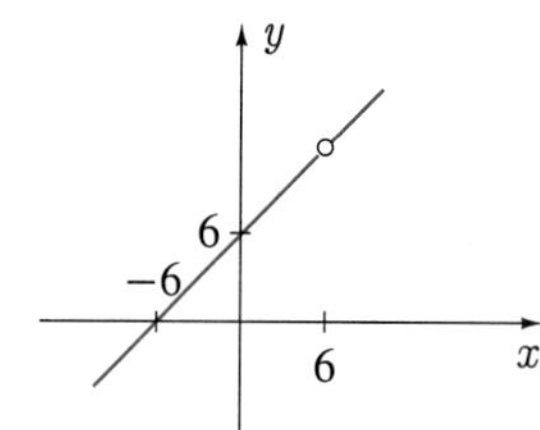

$f(x) = \dfrac{x^2 - 36}{x - 6}$ Discontinuous at $x = 6$

26.

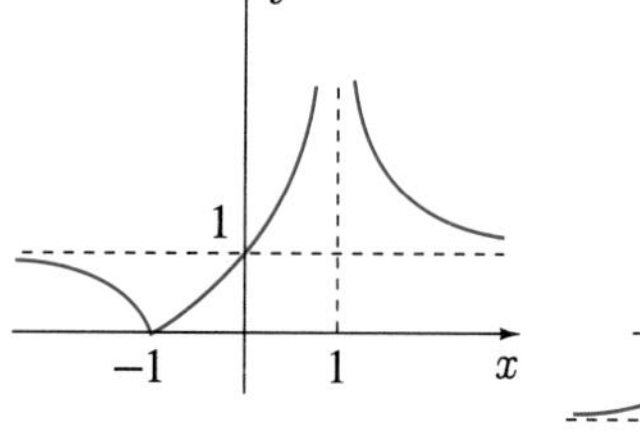

$f(x) = \left|\dfrac{x+1}{x-1}\right|$ Discontinuous at $x = 1$

28.

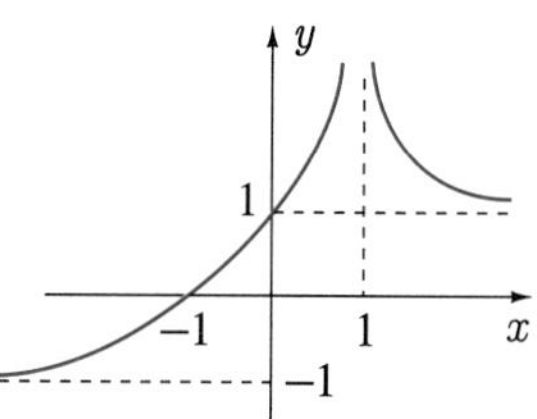

$f(x) = \dfrac{x+1}{|x-1|}$ Discontinuous at $x = 1$

30.

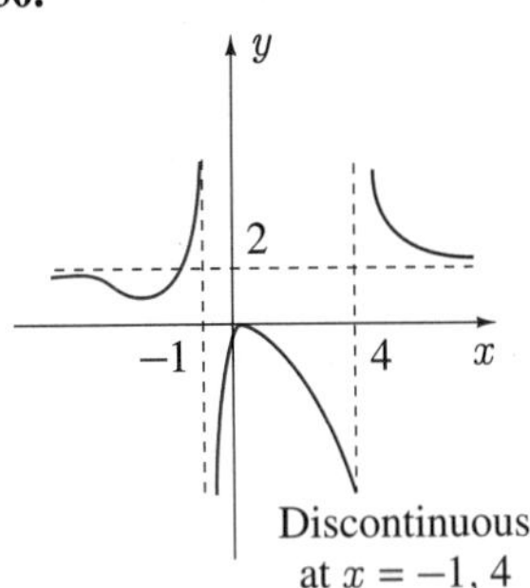

$$f(x) = \frac{2x^2}{x^2 - 3x - 4}$$

32.

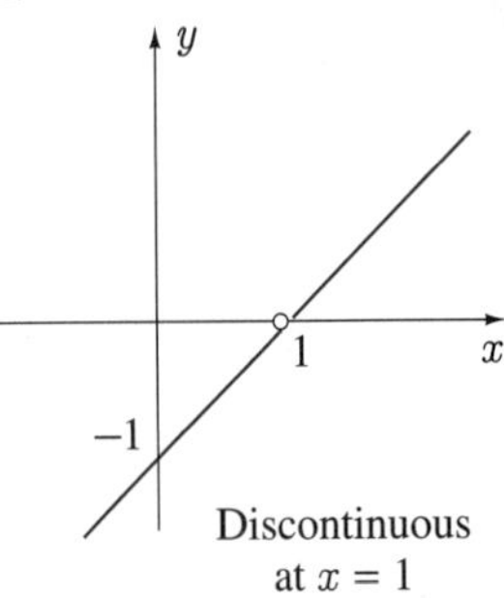

$$f(x) = \frac{x^3 - 3x^2 + 3x - 1}{x^2 - 2x + 1}$$

CHAPTER 3

Exercises 3.1

2. $6x$ **4.** $3x^2 + 4x$ **6.** $-9/(x-5)^2$
8. $3x^2 + 4x$ **10.** $(2x^2 - 2)/(x^2 + x + 1)^2$
12. $2\pi r$ **14.** $4\pi r^2$ **16.** m
18. 0 **20.** $y = 2x + 2$ **22.** $x + 4y = 3$
24. 1.107 **26.** 2.897 **28.** $8x^7$
30. $-4/(x-2)^5$ **32.** $(3x+2)/(2\sqrt{x+1})$
34. $\sqrt{A}/(4\sqrt{\pi})$ **36.** $|x|/x$

Exercises 3.2

2. $9x^2 + 4$ **4.** $20x^4 - 30x^2 + 3$
6. $-6/x^4$ **8.** $1/x^3 - 12/x^5$ **10.** $20x^3 - 5/(4x^6)$
12. $1/(2\sqrt{x})$ **14.** $-(3/2)/x^{5/2} + (3/2)\sqrt{x}$
16. $\pi^2 x^{\pi-1}$ **18.** $4 + 3/x^4$ **20.** $6(2x+5)^2$
22. $4y = x + 24$, $4x + y = 23$
24. $y = 6x - 10$, $x + 6y = 51$
26. $75 - 7x/50$, $50 - x/15 - x^2/20\,000$, $25 - 11x/150 + x^2/20\,000$
28. $(2, -12)$ **32.** $n|x|^{n-1}\ \mathrm{sgn}(x)$

Exercises 3.3

2. right; no left; no derivative
4. no right; no left; no derivative
6. no right; no left; no derivative
8. right; no left; no derivative
10. True **12.** True

14. **16.**

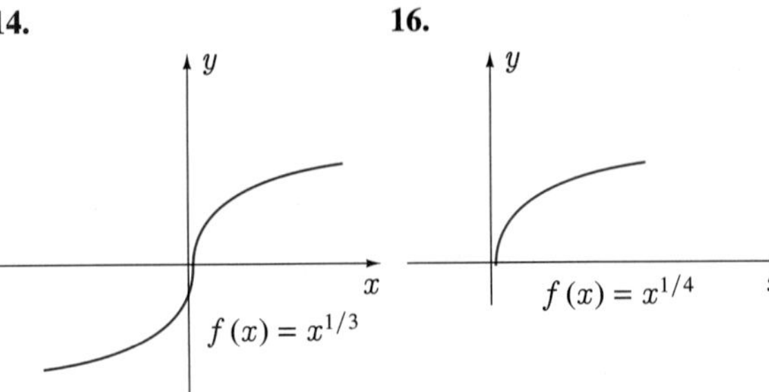

18. Yes **20.** Not necessarily
22. Yes; derivative is zero **24.** Yes; derivative is zero

Exercises 3.4

2. $4(2 - 3x^2 - x^3)$
4. $-10x/(4x^2 - 5)^2$
6. $x^2(4x^2 + 3)/(4x^2 + 1)^2$
8. $(3x+2)/[2\sqrt{x+1}]$
10. $-(2x^2 + 20x + 1)/(2x^2 - 1)^2$
12. $(17 - 4x - 7x^2)/(x^2 - 5x + 1)^2$
14. $(x - 8\sqrt{x} - 2)/[\sqrt{x}(\sqrt{x} - 4)^2]$
16. $-27/(x+1)^4$
18. **(b)** $x(x-4)/(x^2 + x - 2)^2$
20. $f'(x)g(x)h(x) + f(x)g'(x)h(x) + f(x)g(x)h'(x)$
22. 1.23 at $(\pm 1/\sqrt{2}, 1/2)$
24. 0.668 at $(\sqrt{2}, 2\sqrt{2} + 2)$; 0.415 at $(-\sqrt{2}, 2 - 2\sqrt{2})$
30. $2/3$
32. $(5 + \sqrt{55}, -\sqrt{55}/(11 + \sqrt{55}))$, $(5 - \sqrt{55}, \sqrt{55}/(11 - \sqrt{55}))$

Exercises 3.5

2. 6 **4.** 18 **6.** $6 + 60/t^6$
8. $(3u^2 - 6u - 1)/[4u^{3/2}(u+1)^3]$
10. $-(\sqrt{x} + 3)/[4\sqrt{x}(\sqrt{x} + 1)^3]$
12. **(b)** $(bT_b - aT_a)/(b-a)$, $ab(T_a - T_b)/(b-a)$
14. $5/4$, -5, $41/4$, $-13/2$
16. $n!/(2-x)^{n+1} + (-1)^{n+1} n!/(2+x)^{n+1}$

Exercises 3.6

2. $1/[2\sqrt{x}(u+1)^2]$
4. $-2(x-1)(s^2+2)/(s^2-2)^2$
6. $-21/[(t-4)^2(x+1)^2]$
8. $(4u + 5u^{3/2})(x^2 + 2x - 1)/[2(x - x^2)^2]$
10. $(3x+2)/[2\sqrt{x+1}]$

12. $(x+1)/(2x+1)^{3/2}$

14. $(2x-1)(6x+23)/(3x+5)^2$

16. $x^2(18-55x^2)/[3(2-5x^2)^{2/3}]$

18. $(x+1)(15x+11)(3x+1)^2$

20. $(9x-8)/[2x^3\sqrt{2-3x}]$

22. $-1/[(2+x)^{5/4}(2-x)^{3/4}]$

24. $(x+5)^3(11x^3+15x^2+8)/[2\sqrt{1+x^3}]$

26. $(x+5)^3(13x^4+25x^3+10x+10)/[2\sqrt{1+x^3}]$

28. $(7x^2+6x+4\sqrt{1+x})/[4\sqrt{1+x}\sqrt{1+x\sqrt{1+x}}]$

30. $(4/3)x(s-1)/[(x^2+5)^2(2s-s^2)^{2/3}]$

32. $(5/2)(2-x)/[x^2\sqrt{x-1}(u+5)^2]$

34. $[1+(2\sqrt{t}+1)/(4\sqrt{t}\sqrt{t+\sqrt{t}})][-4x/(x^2-1)^2]$

36. $(1-k-3k^2)(x^2+5)^4(11x^2+5)/[2\sqrt{k}(1+k+k^2)^2]$

38. -0.067 N/s

44. $(6u^4+6u^3-2)(2\sqrt{x+1}+1)^2/[4u^3(x+1)]-[3(u^2+u)^2+1]/[4u^2(x+1)^{3/2}]$

46. $-3[2\sqrt{x}+(s+6)(1+3x+4\sqrt{x})]/[2x^{3/2}(1+\sqrt{x})^4(s+6)^3]$

48. $-8f(3-4x)f'(3-4x)$; $8(4x-3)(16x^2-24x+7)(48x^2-72x+25)$

50. $(1-1/x^2)f'(x+1/x)$; $(x^2-1)(3x^4+4x^2+3)/x^4$

52. $12f(1-3x)f'(1-3x)/\sqrt{3-4[f(1-3x)]^2}$; $12(1-3x)(9x^2-6x-1)(27x^2-18x+1)/\sqrt{3-4(1-3x)^2(9x^2-6x-1)^2}$

54. $f'(x-f(x))[1-f'(x)]$; $3(1-x^2)(3x^6-18x^4+27x^2-2)$

56.

$x^2+2y^2=c^2$ $y=ax^2$

58.

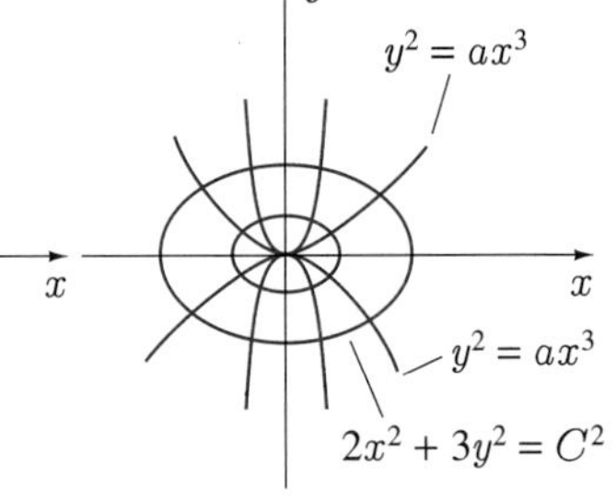

60. 444

62. $-x(x+1)^2/\sqrt{1-x^2}$

64. $(4.05,-0.164)$, $(-4.51,-0.521)$

66. $(-4^{1/3},\pm(4^{1/3}-1)^{3/2})$

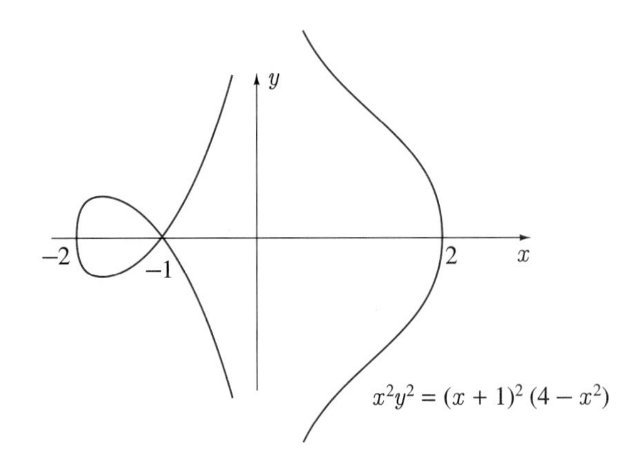

Exercises 3.7

2. $-4x^3/(2y+3y^2)$

4. $(6x^2-3y^4+5y)/(12xy^3-5x)$

6. $(1-x-y)/(x+y)$

8. $(48x\sqrt{x+y}-1)/[1+(4y-2)\sqrt{x+y}]$

10. $-(6x+5y)/(5x+2y)$

12. $-[2(3y^2+1)^2+24x^2y]/(3y^2+1)^3$

14. $-(25/4)/(y+1)^3$ **16.** -1

18. $-(2xy^3+2)/(3x^2y^2+4)$, $[-2y^3(3x^2y^2+4)^2+12xy^2(2xy^3+2)(3x^2y^2+4)-6x^2y(2xy^3+2)^2]/(3x^2y^2+4)^3$

20. $\sqrt{1-y^2}/(1-2y^2)$ **22.** $-3/4$

24. $0, -1/2$ **26.** $(1,\pm1)$, $(-1,\pm1)$

28. **(a)** $(x^2+4x+2)/(x^2+3x+2)$ **(b)** $(2y+1)(3-y)/(7y)$

30. $(1,2)$

34. $(3r^3-6ar^2+x)/(3r^2-6ar)$

36. **(a)** -1 **42.** **(a)** $-\sqrt{y/x}$

44. $2xu/[(u^2-1)(1+2x^2\sqrt{u^2-1})]$

Exercises 3.8

2. $-\sin x-20\cos 5x$ **4.** $-9\sec^2 3x/\tan^4 3x$

6. $2\csc(4-2x)\cot(4-2x)$ **8.** $\cot x^2-2x^2\csc^2 x^2$

10. $[(x^2+x)\cos x+\sin x]/(x+1)^2$

12. $(3/4)\tan^2 x\sec^2 x/(1+\tan^3 x)^{3/4}$

14. $2\cos 4x$

16. $(\cos y+y\sin x)/(\cos x+x\sin y)$

18. $\sec^2(x+y)/[1-\sec^2(x+y)]$

20. $3(1-x^2y)/(x^3+2\tan y\sec^2 y)$

22. $-\sec^2 x\sin(\tan x)$ **24.** $4x\sin(2x^2)$

26. $-\sec v\tan v\sec^2\sqrt{x}/[4\sqrt{x}\sqrt{3-\sec v}]$

28. $-x\sin x^2\sec^3 u\tan u/[\sqrt{1+\cos x^2}(1+\sec^3 u)^{2/3}]$

30. $[18x^2 \tan^2(3x^2-4)\sec^2(3x^2-4)-(2+x\cot x)(1+\tan^3(3x^2-4))]/(x^3 \sin x)$

32. $2(y+1)[\sec^2 y - x - (y+1)\sec^2 y \tan y]/(\sec^2 y - x)^3$

34. 0 **36.** 2 **38.** Does not exist

40. No **44.** $\cos x|\sin x|/\sin x,\ x \neq n\pi$

Exercises 3.9

2. $6x/(3x^2+1)$ **4.** $-2e^{1-2x}$

6. $\ln x + 1$ **8.** $-4\log_{10} e/(3-4x)$

10. $-\tan x$ **12.** $2x+(4x^3+3x^2)e^{4x}$

14. $2e^{2x}\cos(e^{2x})$ **16.** $e^{-2x}(3\cos 3x - 2\sin 3x)$

18. $4(e^x+e^{-x})^{-2}$ **20.** $2\sin(\ln x)$

22. $-\tan v \sin 2x$

24. $-(y\cos x + 2x\ln y + e^y)/(xe^y + \sin x + x^2/y)$

26. $2x^2/\sqrt{x^2+1}$ **28.** $(1+5e^{4x})^{-1}$

30. $(y^2/x^2)(e^{1/x}-1)/(1-e^{1/y})$

Exercises 3.10

2. $4x^{4\cos x}(\cos x/x - \sin x \ln x)$

4. $(\sin x)^x[x\cot x + \ln(\sin x)]$

6. $x(1+1/x)^{x^2}[2\ln(1+1/x) - 1/(1+x)]$

8. $(3/x^2)(2/x)^{3/x}(\ln x - \ln 2 - 1)$

10. $(1/x)(\ln x)^{\ln x}[\ln(\ln x)+1]$

12. $(1+9x^2+6x^4)/2\sqrt{x}(1+x^2)^{3/2}]$

14. $(x^2+3x)^2(x^2+5)^3(14x^3+33x^2+30x+45)$

16. $\sqrt{x}e^{-2x}(3-4x)/2$

18. $e^x[(x-1)\ln(x-1)-1]/\{(x-1)[\ln(x-1)]^2\}$

20. $(1-6x^2-3x^4)/[2\sqrt{x}(1+x^2)^{3/2}]$

22. $x^3(13x^4-44x^3+10x-32)/(2\sqrt{1+x^3})$

24. $\sin 2x \sec 5x[2\cot 2x + 5\tan 5x - 6\csc^2 x/(1 - 2\cot x)]/(1-2\cot x)^3$

26. (b) $(a-br)^{-1}$

Exercises 3.11

2. 2

4. Cannot be applied since $f'(0)$ does not exist

6. $(-2 \pm \sqrt{19})/3$

8. $1+\sqrt{3}$

10. $-3+\sqrt{6}$

12. $(4-\ln 5)/(2\ln 5)$

14. Cannot be applied since $f(\pi/2)$ is not defined

16. Cannot be applied since $g'(0)=0$

18. $(1+\sqrt{6})/(1-\sqrt{6})$

Review Exercises

2. $6x+2-1/x^2$

4. $(1/3)x^{-2/3}-(10/9)x^{2/3}$

6. $x(x^2+2)(x^3-3)^2(13x^3+18x-12)$

8. $17/(x+5)^2$

10. $-4(x^2+2)/(x^2+5x-2)^2$

12. $2x(y-1)/(2y-x^2)$

14. $-(4xy\sqrt{1+x}+y)/[2\sqrt{1+x}(x^2+\sqrt{1+x})]$

16. $-8\sec^2(1-4x)\tan(1-4x)$

18. $2\sec^2 2x \sec(\tan 2x)\tan(\tan 2x)$

20. $(1/2)\sin 4x$

22. $(1-2\sin 2t)(1+2\sin 2x)$

24. $(-x/v)(\cos^2 v - v\sin 2v)$

26. $(1/2)e^{-2y}$

28. $2x(x^2+1)/(x^2-1)+2x\ln(x^2-1)$

30. 1

32. $16x/(x^2+1)^2$

34. $x(20-23x-3x^2)/[2\sqrt{1-x}(x+5)^2]$

36. $(x-y)/(x+y)$

38. $(y^3+4y+6)^2/(-y^4+4y^3+4y^2+12y-12)$

40. $1/(2\sqrt{x})$

42. $(\cos x)^x[\ln(\cos x) - x\tan x]$

44. $(\log_{10} e)^2/(x\log_{10} x)$

46. $1/(e^y - e^{-y})$

48. $[(x+y)(1-2xy)-1]/[x^2(x+y)+1]$

50. $x+25y=5,\ 125x-5y+1=0$

52. $23x+9y=11,\ 27x-69y=119$

54. $-42(x^2-5xy+y^2)/(5x-2y)^3$

56. No

58. $(-3,1)$; finding the shortest distance from $(-6,7)$ to the curve

60. $\sqrt{3}l/2$

62. $(\sqrt{6}/4, \pm\sqrt{2}/4),\ (-\sqrt{6}/4, \pm\sqrt{2}/4)$

64. $(\sqrt{93}-9)/6$

CHAPTER 4

Exercises 4.1

2. −1.561553, 2.561553 **4.** 3.044723

6. 0.754878 **8.** 2.485584

10. ±0.795324

12. −2.931137, −2.467518, −1.555365, −0.787653, 0.056258, 0.642851

14. 4.188760 **16.** 0.852606

18. 0, -1.2483 **20.** -1.892, -0.173, 3.064

22. ±1.0986 **24.** 0.3212

26. $(-0.1725, 0.6848)$, $(-1.8920, 0.7956)$, $(3.0644, 16.5196)$

28. $(-1.3532, 3.8312)$ **30.** **(a)** 3.83 s **(b)** 4.08 s

32. 0.0002897

Exercises 4.2

2. decreasing for all x

4. increasing for $x \leq 5/4$; decreasing for $x \geq 5/4$

6. increasing for $x \leq 1/4$; decreasing for $x \geq 1/4$

8. increasing for all x

10. decreasing for all x

12. decreasing for $x \leq 1$; increasing for $x \geq 1$

14. decreasing for $-1 \leq x \leq 1$; increasing for $x \leq -1$ and $x \geq 1$

16. decreasing for $x \leq -1$ and $0 < x \leq 1$; increasing for $-1 \leq x < 0$ and $x \geq 1$

18. decreasing for $0 \leq x < 1$ and $x > 1$; increasing for $x < -1$ and $-1 < x \leq 0$

20. decreasing for $x \leq -1$ and $0 \leq x \leq 1$; increasing for $-1 \leq x \leq 0$ and $x \geq 1$

22. decreasing for $x \leq 0$ and $x \geq 2$; increasing for $0 \leq x \leq 2$

24. decreasing for $0 < x \leq 1/e$; increasing for $x \geq 1/e$

26.

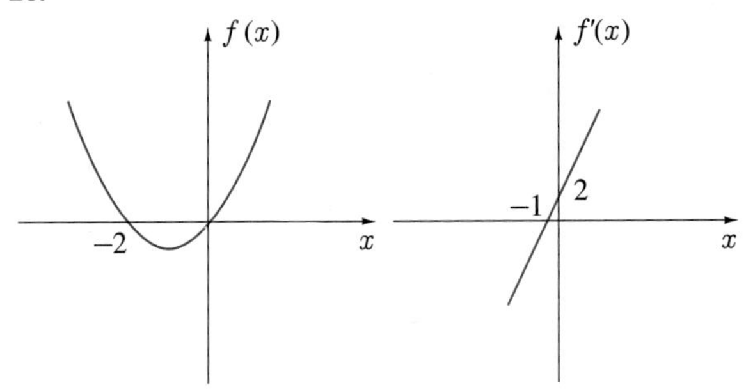

$f(x) = x^2 + 2x$ $f'(x) = 2x + 2$

28.

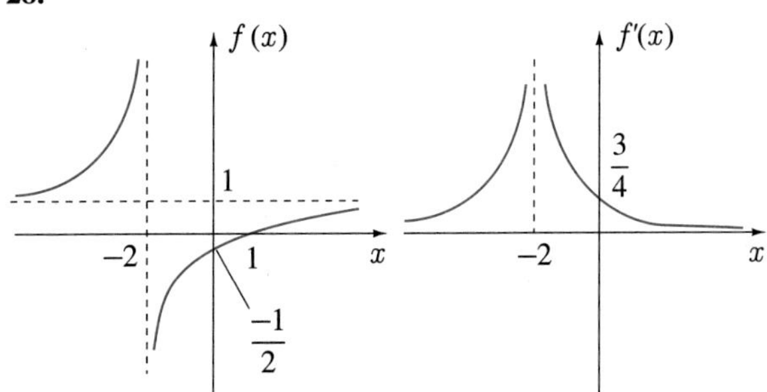

$f(x) = \dfrac{x-1}{x+2}$ $f'(x) = \dfrac{3}{(x+2)^2}$

30.

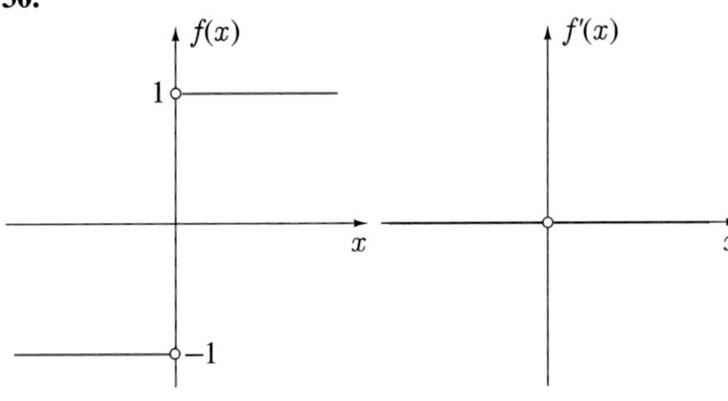

$f(x) = \dfrac{|x|}{x}$ $f'(x) = 0, x \neq 0$

32.

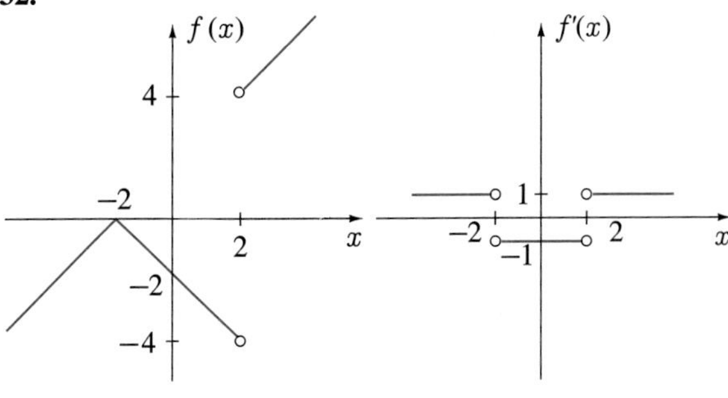

$f(x) = \dfrac{|x^2-4|}{x-2}$ $f'(x) = \dfrac{|x^2-4|}{x^2-4}$

34. decreasing for $x \leq 0.8612$; increasing for $x \geq 0.8612$

36. decreasing for $-\pi \leq x \leq -2.2889$ and $2.2889 \leq x \leq \pi$; increasing for $-2.2889 \leq x \leq 2.2889$

44. Yes

Exercises 4.3

2. relative minimum at $x = -1$; relative maximum at $x = -4$

4. $x = 1$ critical, but does not give a relative extremum

6. relative maximum at $x = 1 - \sqrt{2}$; relative minimum at $x = 1 + \sqrt{2}$

8. relative minimum at $x = -3^{1/4}$; relative maximum at $x = 3^{1/4}$; $x = 0$ critical, but does not give a relative extremum

10. $x = 0$ critical, but does not give a relative extremum

12. $x = 2$ critical, but does not give a relative extremum

14. relative maximum at $x = -1$; relative minimum at $x = 1$

16. $x = 0$ critical, but does not give a relative extremum

18. relative maximum at $x = 7$; $x = 1$ critical, but does not give a relative extremum

20. relative minima at $x = 2(3n+2)\pi/3$; relative maxima at $x = 2(3n+1)\pi/3$

22. relative minima at $x = 2 + 50^{1/3}$ and $x = 0$

24. relative maximum at $x = 1$

26. relative maximum at $x = 0$

28. relative maximum at $x = -2$; relative minimum at $x = 2$

30. relative maxima at $x = (8n+1)\pi/4$; relative minima at $x = (8n+5)\pi/4$

32. relative minimum at $x = 0.464$

34. $x = 3\pi/4$ and $7\pi/4$ are critical, but do not give relative extrema

36. relative minimum at $x = 1/e$

38. relative maximum at $x = 1/2$

40. relative maximum at $x = -2/3$; relative minimum at $x = 0$

42. $x = 0$ is critical **44.** False **46.** True

48. False **50.** False

52. (a)

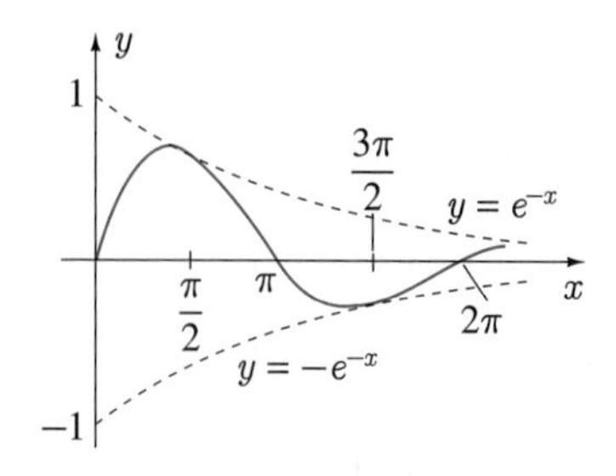

(b) $(4n+1)\pi/4$

54. relative minima at $x = 2.3301$ and $x = -2.1284$; relative maximum at $x = -0.2016$

56. relative maximum at $x = -4.1072$

60. $1/\sqrt{LC}$, E_0/R

62. $3\sqrt{3}/4$

Exercises 4.4

2. concave upward on $x \le -2/3$ and $x \ge 0$; concave downward on $-2/3 \le x \le 0$; points of inflection at $(0, 2)$ and $(-2/3, 470/27)$

4. concave upward on $x < -1$ and $x > 1$; concave downward on $-1 < x < 1$

6. concave upward for all x

8. concave upward on $-2\pi < x \le -5\pi/6$, $-\pi/6 \le x \le 7\pi/6$, $11\pi/6 \le x < 2\pi$; concave downward on $-5\pi/6 \le x \le -\pi/6$, $7\pi/6 \le x \le 11\pi/6$; points of inflection at $(-5\pi/6, 2+25\pi^2/36)$, $(-\pi/6, 2+\pi^2/36)$, $(7\pi/6, 2+49\pi^2/36)$, $(11\pi/6, 2+121\pi^2/36)$

10. concave downward on $0 < x \le e^{-3/2}$; concave upward on $x \ge e^{-3/2}$; point of inflection at $(e^{-3/2}, -3/(2e^3))$

12. concave downward on $x \le 1$; concave upward on $x \ge 1$; point of inflection at $(1, e^{-2})$

14. concave downward on $x \le -\ln 2$; concave upward on $x \ge -\ln 2$; point of inflection at $(-\ln 2, (\ln 2)^2 - 2)$

16. relative minimum at $x = 1$; relative maximum at $x = -1$

18. relative minumum at $x = 1/5$

20. relative minimum at $x = 1/\sqrt{e}$

22. relative minimum at $x = 0$; relative maximum at $x = 1$

26. **28.**

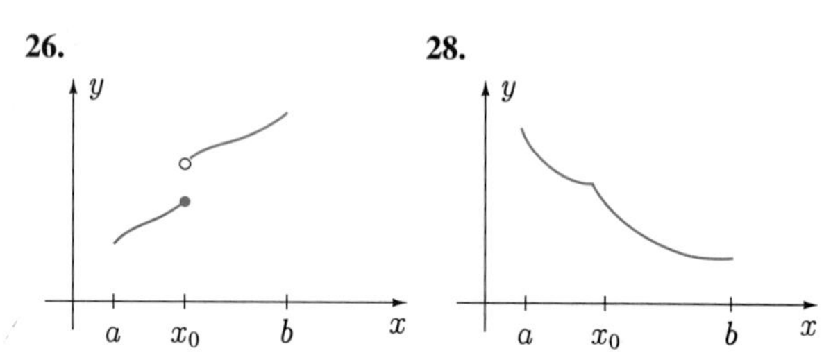

30.

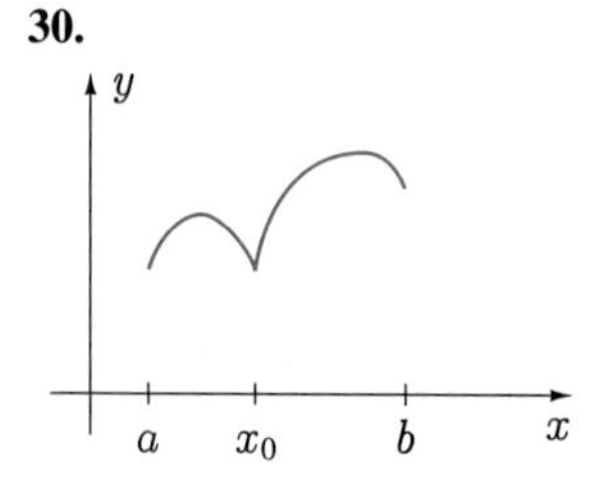

32. Not necessarily

34.(a) relative minimum at $x = 0$; horizontal points of inflection at $x = \pm 1$ **(b)** relative minimum at $x = 0$; relative maximum at $x = 4/5$

Exercises 4.5

2.

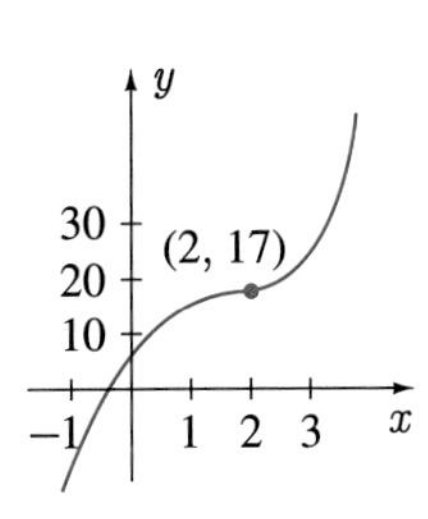

$y = x^3 - 6x^2 + 12x + 9$

4.

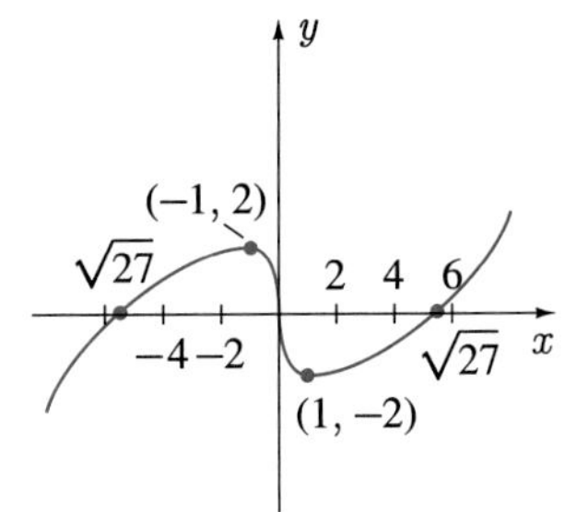

$y = x - 3x^{1/3}$

6.

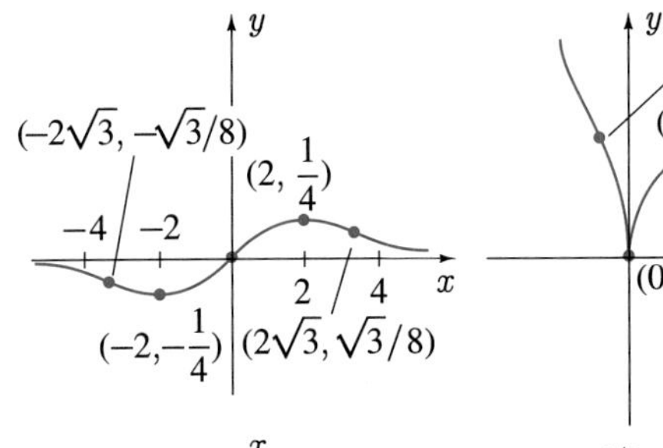

$y = \dfrac{x}{x^2 + 4}$

8.

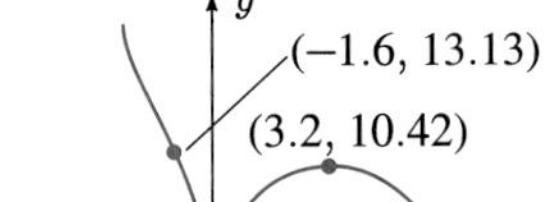

$y = x^{2/3}(8 - x)$

10.

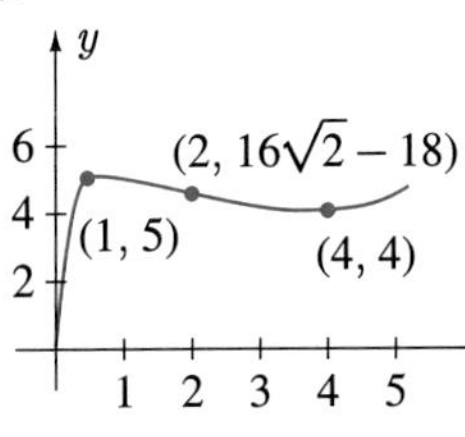

$y = 2x^{3/2} - 9x + 12x^{1/2}$

12.

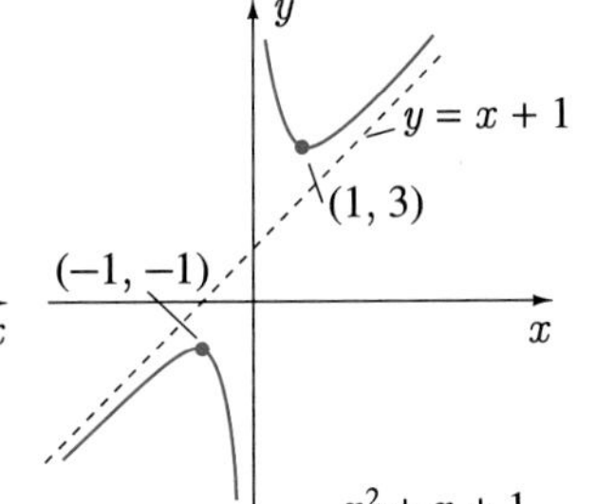

$y = \dfrac{x^2 + x + 1}{x}$

14.

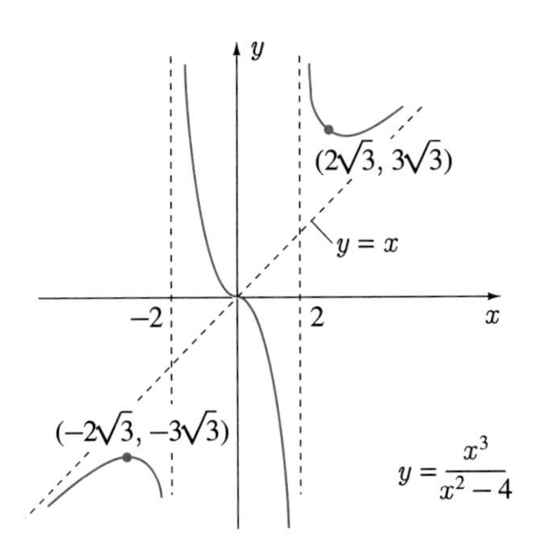

$y = \dfrac{x^3}{x^2 - 4}$

16.

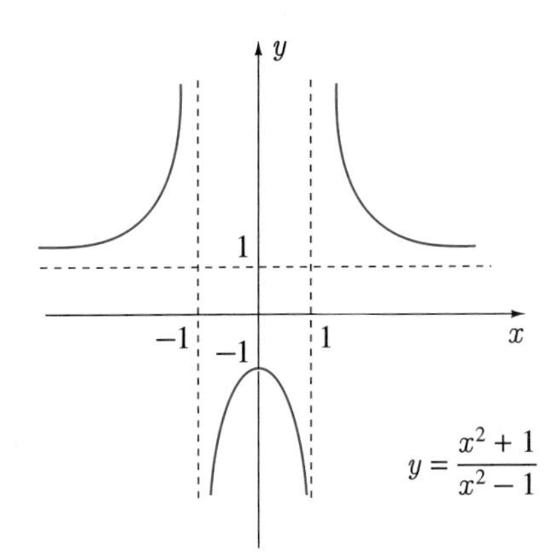

$y = \dfrac{x^2 + 1}{x^2 - 1}$

18.

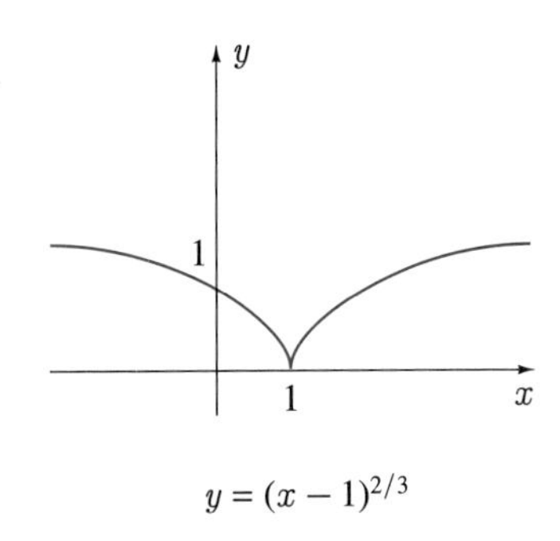

$y = (x - 1)^{2/3}$

20.

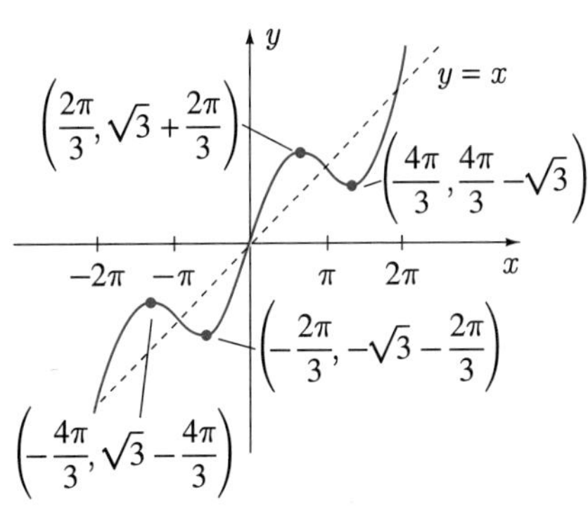

$y = x + 2 \sin x$

22.

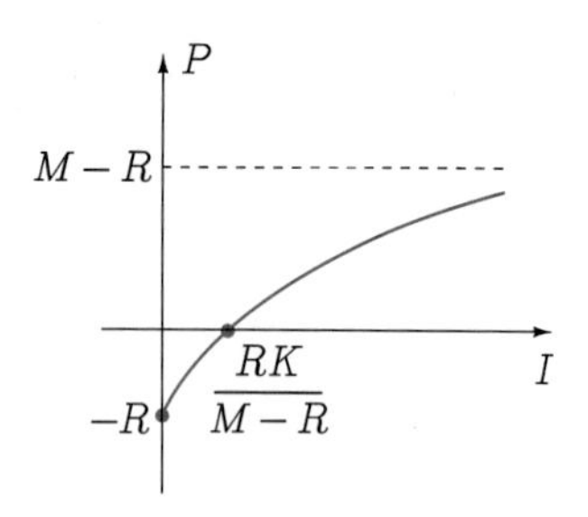

$$P = \frac{MI}{I + K} - R$$

24.

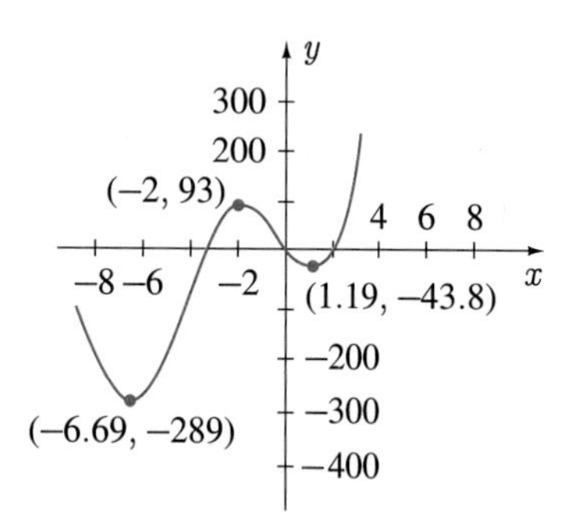

$$y = x^4 + 10x^3 + 6x^2 - 64x + 5$$

26.

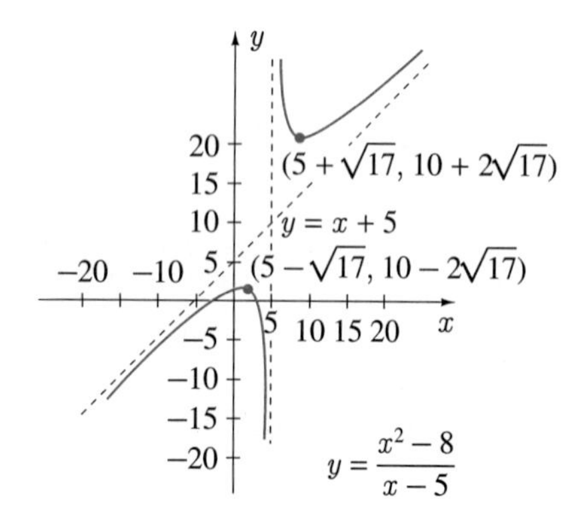

$$y = \frac{x^2 - 8}{x - 5}$$

28.

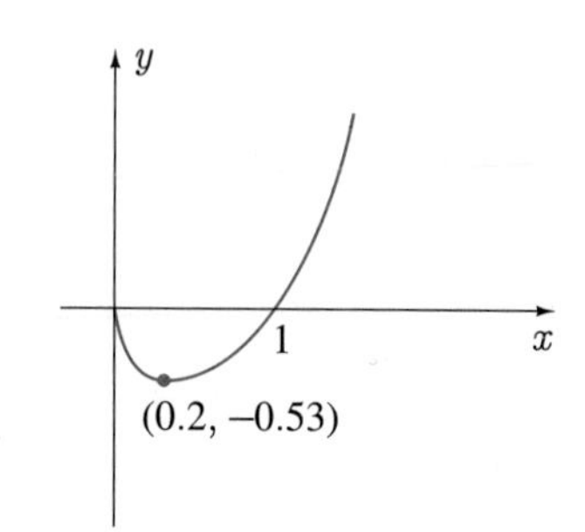

$$y = x^{5/4} - x^{1/4}$$

30.

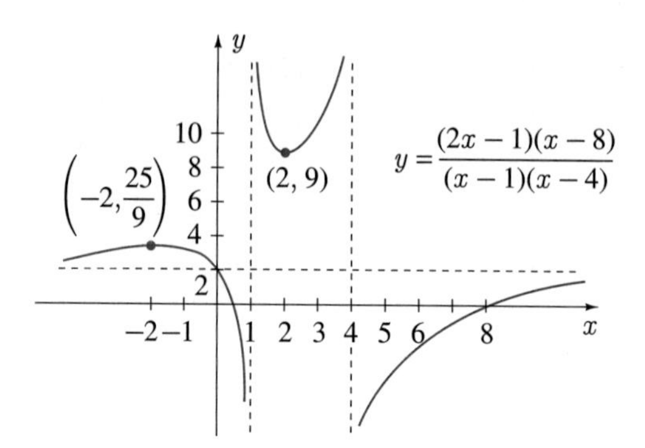

32.

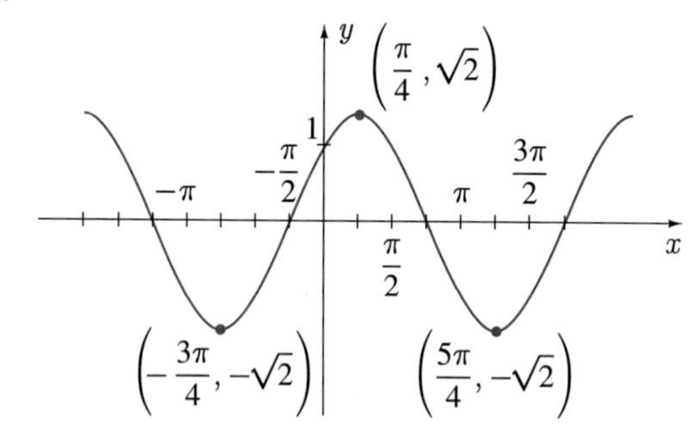

$$y = \sin x + \cos x$$

34. **36.**

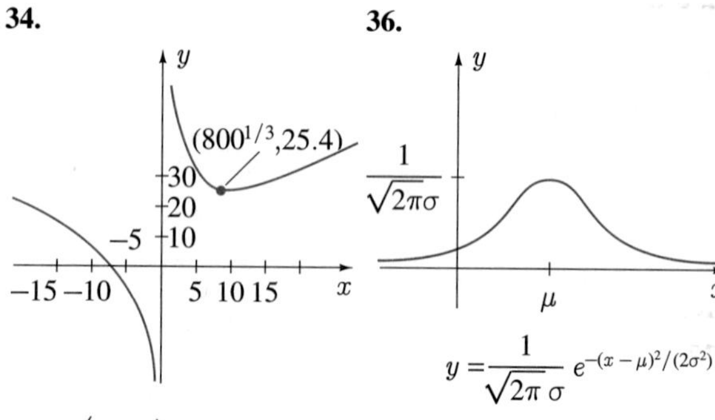

$$y = \left(\frac{x + 8}{x}\right)\sqrt{x^2 + 100}$$

38.

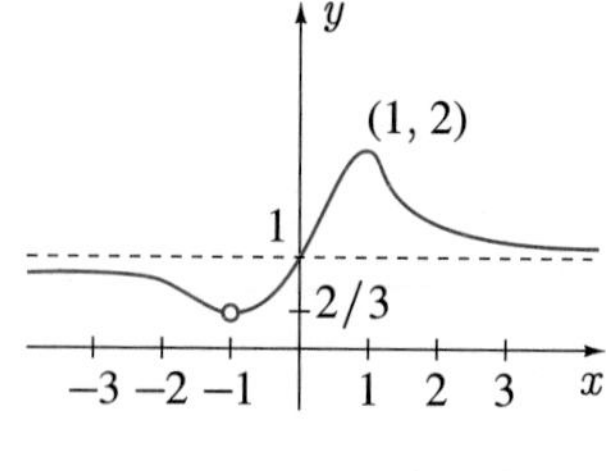

$$y = \frac{1 + x + x^2 + x^3}{1 + x^3}$$

Exercises 4.6

2. $-4, 6/11$ **4.** $\pi/3-\sqrt{3}$, $11\pi/3+\sqrt{3}$

6. no absolute minimum, 12

8. no absolute minimum, $\sqrt{3}/6$

12. $720\sqrt{10}$ m **14.** $12^{1/3}\times 12^{1/3}\times 12^{1/3}/2$ m

16. $90\times 180/\pi$ m

18. $ab/4$ **20.** $a/\sqrt{3}$

22. 15 cm × 22.5 cm

24. $(-1-\sqrt{6}/2, 5/2+\sqrt{6})$

26. smallest area is 20

28. width = $50/\sqrt{3}$ cm, depth = $50\sqrt{2/3}$ cm

30. **(a)** $7/4$ km from P **(b)** directly to Q

32. lengths in x- and y-directions are $\sqrt{2}a$ and $\sqrt{2}b$

34. $\sqrt{2/e}$ **36.** $4\pi hr^2/27$ **38.** 11.8 m

40. $[p\beta/(r\alpha-r)]^{1/\alpha}$

42. $dI_1^{1/3}/(I_1^{1/3}+I_2^{1/3})$ from I_1 source

44. **(a)** width = $5/3^{1/3}$ m, height = $4(3^{2/3})$ m
(b) width = 4.19 m, height = 5.70 m

48. $l=1/3$ m, $w=h=2/3$ m; $1/3$ m apart

50. $2\pi(1-\sqrt{6}/3)$ **54.** $(2, 2\sqrt{5}/3)$

56. $x=-(A+bB)/(2aB)$ **60.** $3a/4$

Exercises 4.7

2. **(a)** right, left **(b)** left **(c)** 5 **(d)** 4 **(e)** right, left **(f)** twice **(g)** 9/2 **(h)** 9/2 **(i)** negative, positive **(j)** once

4. $2t-7$ m/s, 2 m/s^2

6. $-6t^2+4t+16$ m/s, $-12t+4$ m/s^2

8. $-12\sin 4t$ m/s, $-48\cos 4t$ m/s^2

10. $1-4/t^2$ m/s, $8/t^3$ m/s^2

12. $(4t-5t^2)/(2\sqrt{1-t})$ m/s, $(1/4)(15t^2-24t+8)/(1-t)^{3/2}$ m/s^2

14. **(a)** $3t^2-18t+15$ m/s, $6t-18$ m/s^2 **(b)** $t=1, 5$ **(c)** slowing down **(d)** 2.18 s, 3.82 s **(e)** increasing

16. No

18.

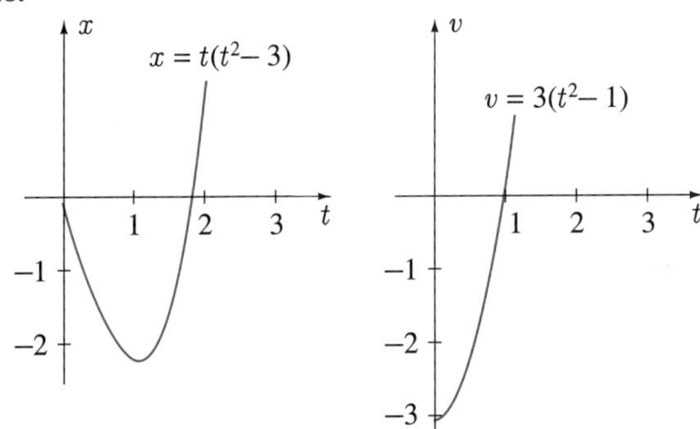

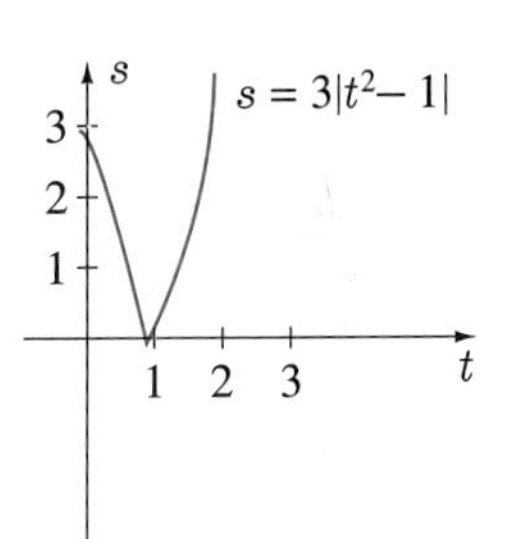

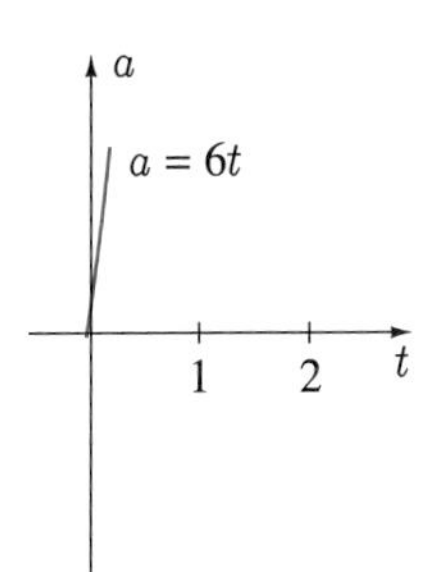

20.

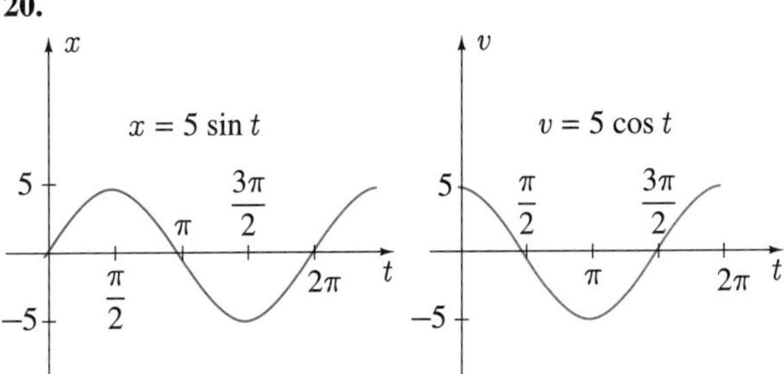

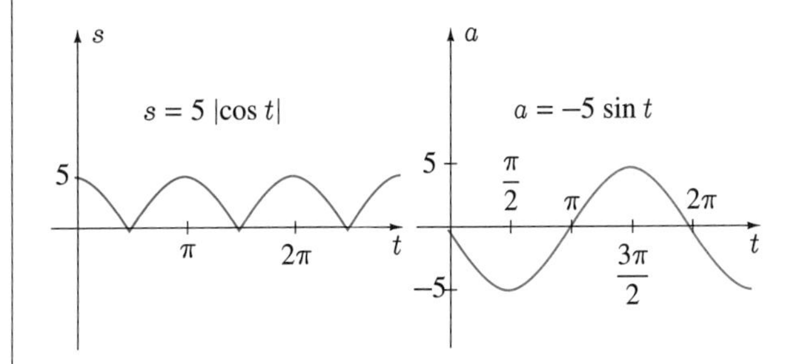

22. Not always **26.** Not always **28.** Not always

Exercises 4.8

2. $2\sqrt{13}/3$ m/s **4.** $13/(5\pi)$ cm/s

6. decreasing at 20 m^2/s **8.** -2.5 N/m^2/s

10. $7/\sqrt{5}$ m/s

12. **(a)** $2/(25\pi)$ m/min **(b)** $2/(25\pi)$ m/min

14. **(a)** $10(50-x)/y$ m/s **(b)** $10/\sqrt{3}$ m/s **(c)** Car 1

16. 25.83 cm/s **18.** 115.8 cm/s

20. $-31\sqrt{15}/10$ **22.** 102 m/s

24. $6/\sqrt{13}$ m/s

26. **(a)** $1/(50\pi)$ m/min **(b)** $\sqrt{13}/(300\pi)$ m/min

28. $4\pi xl(k-x)\sin\theta/[y(l\cos\theta-x)]$ m/s

30. **(a)** $\pm 11\pi/360$ rad/min **(b)** $-11\pi/60$ cm/min **(c)** 0.23 cm/min

Exercises 4.9

2. 6 **4.** 0 **6.** ∞

8. 0 **10.** Does not exist **12.** ∞

14. $1/\sqrt{5}$ **16.** na^{n-1} **18.** 1
20. 0 **22.** $1/12$ **24.** 2^{15}
26. 0 **28.** 4 **30.** 0
32. 1 **34.** e **36.** 1
38. Does not exist **40.** 2 **42.** $-1/3$

44. **46.**

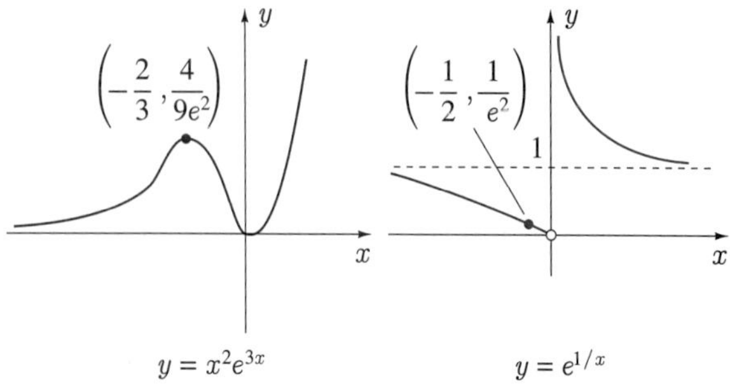

$y = x^2e^{3x}$ $y = e^{1/x}$

48. **50.**

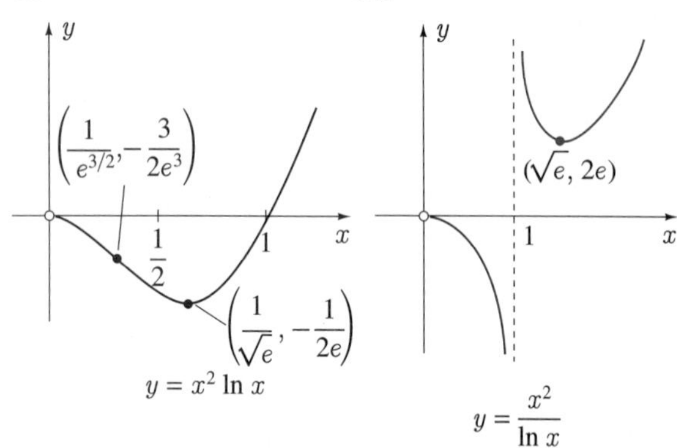

$y = x^2 \ln x$ $y = \frac{x^2}{\ln x}$

52. **54.**

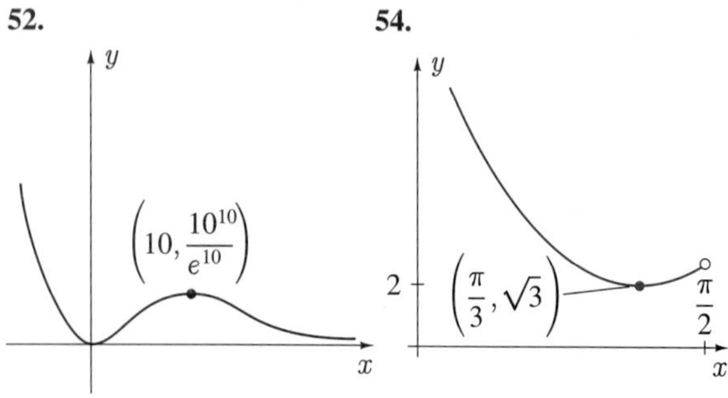

$y = x^{10}e^{-x}$ $y = 2\csc x - \cot x$

56. ∞ **58. (b)** 0.0002897
60. $a = b = \pm 7$, c arbitrary

Exercises 4.10

2. $-2\,dx/(x-1)^2$
4. $[2x\cos(x^2+2) + \sin x]dx$
6. $x^2(9-16x^2)dx/\sqrt{3-4x^2}$
8. $[1 - 12/(x-1)^3]dx$
10. $2(x^2+5x-5)dx/(x^2+5)^2$
12. -4% **14.** 3250 m

Review Exercises

2. (a) **(b)**

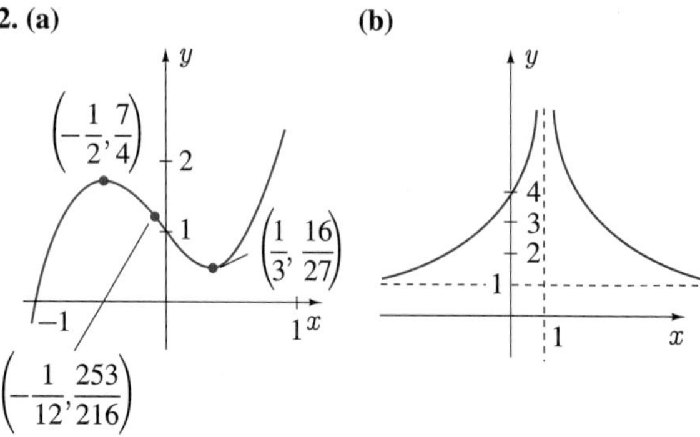

$y = 4x^3 + x^2 - 2x + 1$ $y = \frac{x^2-2x+4}{x^2-2x+1}$

4. No solution **6.** No solution **8.** $-3/2$
10. 8 **12.** 0 **14.** 0
16. 0 **18.** e^2
20.(a) 0.312908 **(b)** 1.051888 **24.** $2l^3/27$

26. (a) **(b)**

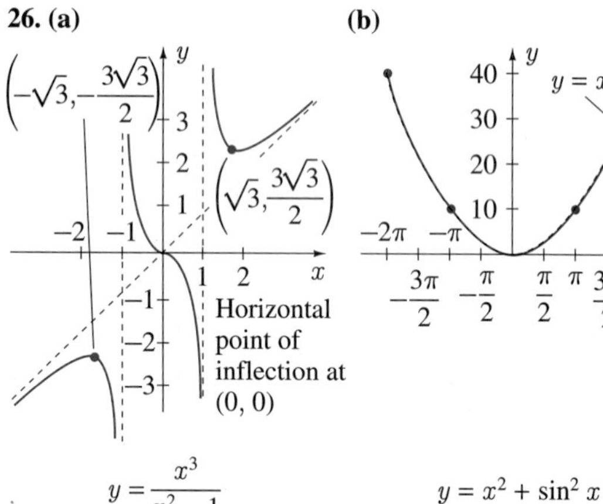

$y = \frac{x^3}{x^2-1}$ $y = x^2 + \sin^2 x$

28. 67 ₵ **30.** $4\sqrt{5}$ m/s
32. $x = 100bq/(ap+bq)$ hectares, $y = 100ap/(ap+bq)$ hectares

CHAPTER 5

Exercises 5.1

2. $x^5/5 + x^3 + 5x^2/2 + C$ **4.** $-\cos x + C$
6. $(2/3)x^{3/2} + C$ **8.** $-1/x + 2/(3x^3) + C$
10. $-1/x + \sqrt{x} + C$ **12.** $1/(2x) + (3/4)x^4 + C$
14. $(4/3)x^{3/2} + (6/5)x^{5/2} - (10/7)x^{7/2} + C$
16. $(2/5)x^{5/2} + (2/3)x^{3/2} + C$
18. $x^3/3 + 2x^5/5 + x^7/7 + C$
20. $(2/5)x^{5/2} - (4/3)x^{3/2} + 2\sqrt{x} + C$
22. $y = x^4/2 + 2x^2 + 5$ **24.** $y = 2x - 2x^2 + x^8$
26. $f(x) = -5x^3/6 + 10x - 31/3$
28. $(2/3)(x+2)^{3/2} + C$
30. $-(2/3)(2-x)^{3/2} + C$ **32.** $(1/5)(2x-3)^{5/2} + C$
34. $-(1/16)(1-2x)^8 + C$ **36.** $-(1/15)/(1+3x)^5 + C$
38. $(1/72)(2+3x^3)^8 + C$ **40.** $(1/2)\sin 2x + C$
42. $(3/4)\sin^2 2x + C$ **44.** $-(1/4)\cot 4x + C$
46. $-(1/2)e^{-x^2} + C$ **48.** $(1/4)e^{4x-3} + C$
50. $-(2/5)\ln|7-5x| + C$ **52.** $-(1/4)\ln|1-4x^3| + C$
54. $(1/2)3^{2x}\log_3 e + C$ **56.** $-(1/5)(1+\cos x)^5 + C$
58. $(1/8)(1+e^{2x})^4 + C$ **60.** $y = x^4/4 + 1/x + C$
62. $-(2/3)/(3x+5)^{1/2} + C$ **64.** $-(1/12)/(2+3x^4) + C$
66. $y = -3 - 1/x,\ x < 0;\ y = 2 - 1/x,\ x > 0$
68. $C,\ x < a;\ x + D,\ x > a$

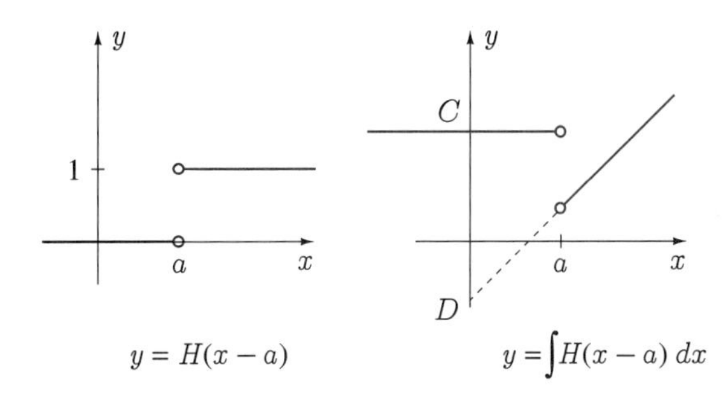

$y = H(x-a)$ $y = \int H(x-a)\,dx$

Exercises 5.2

2. $6t - t^2 + 5,\ 3t^2 - t^3/3 + 5t$
4. $60t^2 - 4t^3,\ 20t^3 - t^4 + 4$
6. $t^3/3 + 5t^2/2 + 4t - 2,\ t^4/12 + 5t^3/6 + 2t^2 - 2t - 3$
8. $4 - 3\cos t,\ 4 + 4t - 3\sin t$
10.(a) $t^3 - t^2 - 3t + 1$ **(b)** $(\sqrt{10} + 1)/3$
12.(a) $350/3$ m **(b)** $650/3$ m, 20 s
14. 5.1 m
16. $|v| \geq 19.8$ m/s
18.

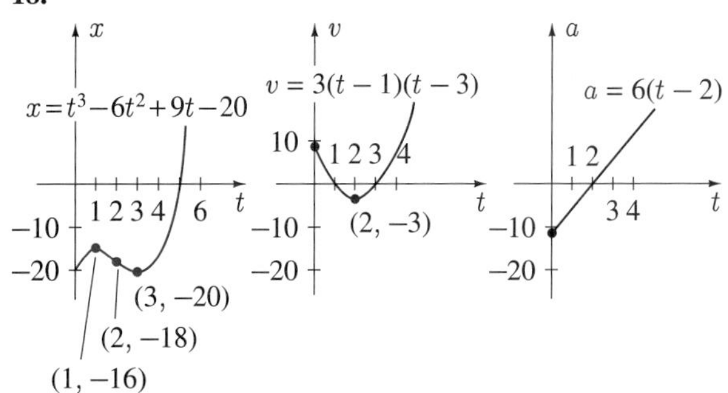

20. Less than or equal to 47.8 km/hr; defense
22.(a) Yes **(b)** No

Exercises 5.3

2. $-(1/3)(1-2x)^{3/2} + C$
4. $-(10/63)(5-42x)^{3/4} + C$
6. $-(1/2)/(x^2+4) + C$
8. $-1/(x-2) - 2/(x-2)^2 - (4/3)/(x-2)^3 + C$
10. $(1/6)(2x+3)^{3/2} - (3/2)\sqrt{2x+3} + C$
12. $(1/5)(s^2+5)^{5/2} - (5/3)(s^2+5)^{3/2} + C$
14. $(2/3)(1-\cos x)^{3/2} + C$
16. $(2/7)(y-4)^{7/2} + (16/5)(y-4)^{5/2}$ $+ (32/3)(y-4)^{3/2} + C$
18. $(1/729)(3x^3-5)^9 + (5/324)(3x^3-5)^8 + (25/567)(3x^3-5)^7 + C$
20. $(3/4)(x^2+2x+2)^{2/3} + C$
22. $(7/192)/(3-4\sin x)^3 + (3/64)/(3-4\sin x)^2 - (1/64)/(3-4\sin x) + C$
24. $(4/5)(1+\sqrt{x})^{5/2} - (4/3)(1+\sqrt{x})^{3/2} + C$
26. $(1/2)\tan^2 x + C$
28. $(\ln x)^2/2 + C$
30. $-(1/2)/\ln(x^2+1) + C$
32. $(2/3)(1+x)^{3/2} - 2\sqrt{1+x} + C$ if $x \geq 0$; $-(2/3)(1+x)^{3/2} + 2\sqrt{1+x} + C$ if $-1 < x \leq 0$
34. $-(2/5)(1/x-1)^{5/2} - (2/3)(1/x-1)^{3/2} + C$
36. $-2/[\sqrt{(1-x)/(1+x)} - 3] + C$

Exercises 5.4

2. **(a)** $(24EI)^{-1}gx^2[4M(3L-x) + m(6L^2 - 4Lx + x^2)]$ **(b)** $(24EI)^{-1}gL^3(8M + 3mL)$
4. $x^2 + (y-2)^2 = 13$
6. $\pi R^2\sqrt{2H}/(5cA\sqrt{g})$
8. $-\sqrt{2GMh/(R^2+Rh)},\ \sqrt{2GM/R}$
10. $y = Cx^k$

12. **(a)** $(192EI)^{-1}[8mgx^4 + 32Ax^3 - (8AL^2 + mgL^3)x]$
(b) $-3mgL/16$

14. $(1 + e^{-t/20})/20$ m³, $1/20$ m³

Review Exercises

2. $-(1/4)/x^4 + x^2 + (1/2)/x^2 + C$

4. $-1/x - (4/3)x^{3/2} + C$

6. $(1/30)(1 + 3x^2)^5 + C$

8. $(2/5)x^{5/2} + 10\sqrt{x} + C$

10. $-2/\sqrt{x} - 30\sqrt{x} + C$

12. $-(1/3)(1 - x^2)^{3/2} + C$

14. $x^3/3 - (4/5)x^5 + (4/7)x^7 + C$

16. $(2/3)(2 - x)^{3/2} - 4\sqrt{2 - x} + C$

18. $4x + (8/3)x^{3/2} + x^2/2 + C$

20. $(1/5)\sin^5 x + C$

22. $-(1/8)e^{-4x^2} + C$

24. $(1/5)\ln|\ln x| + C$

26. $y = x^4 + 7x/2 - 1/2$

28. $[(\sqrt{3} - \sqrt{6})t + \sqrt{6}]^2$ km, $\sqrt{6}/(\sqrt{6} - \sqrt{3})$ hours

30. 11.9 m/s

32. $(2/3)(1 + x)^{3/2} - x + C$

34. $-(3/56)(3 - 2x^3)^7 + (1/32)(3 - 2x^3)^8 - (1/216)(3 - 2x^3)^9 + C$

36. $(1/4)\sin^4 x - (1/6)\sin^6 x + C$

38. $-\ln|\cos x| + C$

40. $y = 2x/(1 + x)$

CHAPTER 6

Exercises 6.1

2. $\sum_{k=1}^{10} k/2^k$ **4.** $\sum_{k=1}^{14\,641} \sqrt{k}$

6. $\sum_{k=1}^{1019} (-1)^k(k + 1)$ **8.** $\sum_{k=1}^{225} (\tan k)/(1 + k^2)$

10. $\sum_{k=1}^{9} (10^k - 1)/10^k$ **16.** 258

18. $4n(4n^2 - 1)/3$ **20.** 5350

24. $f(n) - f(0)$ **26.** No

28. $(1 - 2^{-18})/4$ **30.** $3960[1 - (0.99)^{15}]$

Exercises 6.3

2. 6 **4.** 4 **6.** $3/2$

8. 0 **10.** **(c)** $\log_2 e$ **12.** 2

Exercises 6.4

2. $20/3$ **4.** $2/3$ **6.** 1

8. $-7/2$ **10.** $3/4$ **12.** 0

14. $5/16 - 1/(\pi + 1)$ **16.** $125/12$

18. $-65/4$ **20.** $3/\sqrt{2}$ **22.** $-1/2$

24. 0 **26.** $(e^3 - 1)/e$ **28.** $\ln(2/3)$

30. $20\log_3 e$ **32.** $88/3$ **34.** $-1/6$

36. $\sqrt{2}\pi/8, 0$ **38.** $\pi/2, 0$ **40.** 5, $\cos 1$

Exercises 6.5

2. $1/\sqrt{x^2 + 1}$ **4.** $-x^3\cos x$

6. $2\sqrt{2x + 1}$ **8.** $5\sqrt{(5x + 4)^3 + 1}$

10. $256(3x^2 + 3x + 1) + 2/\sqrt{x} - 2/\sqrt{x + 1}$

12. $2x[-2\sec(1 + 2x^2) + \sec(1 + x^2)]$

14. $\cos x/\sqrt{\sin x + 1} + \sin x/\sqrt{\cos x + 1}$

16. $(2\sqrt{2} - 1)/(2x^{1/4})$ **18.** $-\ln(x^2 + 1)$

20. $3e^{-36x^2} + 2e^{-16x^2}$

Exercises 6.6

2. 0 **4.** $31/5$ **6.** $2\sqrt{2}/3$

8. $11/5$ **10.** $2/\pi$ **12.** 1

14. $23/9$ **16.** $1/2$ **18.** 0

20. $9/7$ **22.** $2cR^2/3$ **24.** 0

26. $\pi/2$ **28.** 0.6032 **30.** 1.400

Exercises 6.7

2. $4/15$ **4.** $7/384$

6. Does not exist **8.** $\sqrt{3/2} - 2\sqrt{2}/3$

10. $3(5^{2/3} - 2^{2/3})/4$ **12.** $\sqrt{2}(7\sqrt{7} - 8)/18$

14. $52/5$ **16.** $(16 - 9\sqrt{2})/6$

18. $5/18$ **20.** $(1/3)(\ln 2)^3$

22. $2(\sqrt{7} - 1)/3$ **24.** $1/\sqrt{3}$

26. $-8\sqrt{5}/45$

Review Exercises

2. $4/15$ **4.** $-4/3$

6. $1/6$ **8.** 0

10. $66/5$ **12.** $16\sqrt{6}$

14. $-4/15$ **16.** $(2112\sqrt{6} - 704)/105$

18. 120 852.5 **20.** 20

24. 2 **26.** $4/(15\pi)$

28. $-x^2(x+1)^3$ **30.** $-4x\cos 2x$
32. $4x\sin^2 x^2$ **34.** $(33\sqrt{6}-56)/60$
36. $(24\sqrt{3}-14\sqrt{2}-8\sqrt{5})/3$
38. $24\sqrt{5}-8$
40. $(3\sqrt{3}-2\sqrt{2}-1)/3$

CHAPTER 7

Exercises 7.1

2. 8 **4.** $343/6$
6. $(e^6-e^3)/3+3/2$ **8.** $8/3$
10. 2 **12.** $20/3$
14. $4\sqrt{2}/3-e\ln 2$ **16.** $10/3$
18.(a) 22 **(b)** 37, 15 **(c)** 29, 7 **(d)** $203/8$, $27/8$
20. $4\int_1^2 \sqrt{4-x^2}\,dx$
22. $2\int_0^{\sqrt{(\sqrt{65}-1)/2}}(\sqrt{16-y^2}-y^2)\,dy$
24. $235/3$ **26.** $7/6$
28. $1-\ln 2$ **30.** $4\sqrt{e}-16/3$
32. $32/3$ **34.** $14/\sqrt{21}$
36. 2.182 **38.** 8.436
40. 2.067 **42.** 7.177
44. $0<m<1$, $(m-1-\ln m)/3$

Exercises 7.2

2. $80\sqrt{5}\pi/3$ **4.** $5888\pi/15$ **6.** π
8. $1088\pi/15$ **10.** $32\pi/3$ **12.** $\pi(e^2-4e+5)/2$
14. 8π **16.** $14\pi/3$ **18.** $40\pi/3$
20. $\pi/3$ **22.** $344\pi/3$ **24.** $1472\pi/15$
26. $775\pi/6$ **28.** $272\pi/15$ **30.** $68\pi/9$
32. $16\pi(13+2^{17/4}-3^{9/4})/45$
34. 111.303 **36.** 21.186 **38.** $4\pi/3$
40.(a) $x^2(6-x)/16$ **(b)** $1161\pi/10$ m^3
44. $2\pi R^3(1-H/\sqrt{r^2+H^2})/3$

Exercises 7.3

2. $17/3$ **4.** $(85^{3/2}-13^{3/2})/27$
6. $(e-e^{-1})/2$ **8.** $23/18$
10. $3011/480$
12. $\int_1^2 \sqrt{36x^2-48x+17}\,dx$
14. $\int_{-\sqrt{3}}^{2\sqrt{2}} \sqrt{(1+2y^2)/(1+y^2)}\,dy$
16. $\int_0^{\pi/4} \sec x\,dx$
18. $\sqrt{2}\int_0^1 (3-2x^2)/\sqrt{1-x^2}\,dx$
20. $2\int_0^2 \sqrt{(16+5x^2)/(4-x^2)}\,dx$
22. 6
24. $(b^{2n+1}-a^{2n+1})/[4(2n-1)]+(1/a^{2n-1}-1/b^{2n-1})/(2n+1)$

Exercises 7.4

2. $8/15$ J **4.** 8829 J **6.** 49.05 J
8. 184 J **10.** 1.78×10^7 J **12.** 114 450 J
14. 7.60×10^6 J **16.** $q_3(q_1-8q_2)/(48\pi\epsilon_0)$
18. kb^2 J **20.** 3.22×10^5 J
22. **(b)** 9.8087×10^5 J **(c)** 9.82×10^5 J
24. $C\ln 2$

Exercises 7.5

2. 6.78×10^{10} N **4.** 9.25×10^3 N
6. 7.60×10^5 N, 4.41×10^5 N, 4.91×10^4 N
12. 2.943×10^6 N, 2.969×10^6 N
14. 1.60 m $\times$ 1.60 m **20.(b)** 21.65 cm
22. 0.52

Exercises 7.6

2. $(3a/8, 3h/5)$ **4.** $(4r/(3\pi), 4r/(3\pi))$
6. $(-1/2, -1/2)$ **8.** $(0,0)$
10. $(33/5, 1/2)$ **12.** $(177/85, 9/17)$
14. $25\rho/3$
16. $(M+2\rho L)^{-1}(\sum_{i=1}^{6} m_i x_i + 2\rho L^2)$
18. $(0, 192/205)$
20. On the axis of symmetry, $1637/350$ from smaller end
24. $(12/5, -3/5)$
26. $\overline{x}=-2/(8\sqrt{3}-1)$, $\overline{y}=0$
28. $(0, 1.944)$ **30.** $(0.684, 0.319)$
32. $2\pi^2ab^2$

Exercises 7.7

2. $32\rho/3$ **4.** $603\rho/10$ **6.** $2\rho/15$
8. $344\rho/105$ **10.** $1143\rho/20$
12. $\rho h(y_2^3+y_2^2y_1-y_1^2y_2-y_1^3)/4$; no
16. $\tilde{x}=\overline{x}$ **18.** 0.680 **20.** 0.519

Exercises 7.8

2. $\pi r\sqrt{r^2+h^2}$ **4.** $b^2h/3$
6. **(a)** 20 000 **(b)** 7000 **8.** $31\,250/\sqrt{3}$ cm^3
10. $\pi/2$ **12.** $r^3(3\pi-4)/6$

14. 15 **16.** 16.5
18. 140 m^3 **20.** 50 days

Exercises 7.9

2. Diverges **4.** $-1/16\,464$ **6.** Diverges
8. 2 **10.** $\sqrt{21}$ **12.** Diverges
14. Diverges **16.** Diverges **18.** $17\sqrt{5}/3$
20. **(a)** No **(b)** No **22.** **(a)** $1/2$ **(b)** $\pi/5$
28. No **30.** Diverges **32.** Converges

Review Exercises

2. **(a)** $5/4$ **(b)** $46\pi/21$, $14\pi/15$ **(c)** $(28/75, 92/105)$
(d) $73/60$, $1/4$
4. **(a)** $1/2$ **(b)** $\pi/3$, π **(c)** $(1, 1/3)$ **(d)** $1/12$, $7/12$
6. $10\pi/3$ **8.** $16\pi/105$ **10.** $43\pi/12$
12. $-5/6$, $3/5$ **14.** $-28/3$, 14 **16.** 1.04×10^5 J
18. $4\sqrt{5}\pi$ **22.** $2\sqrt{3}$ **24.** 0
26. Diverges **28.** Diverges **30.** 6.58×10^4 J

CHAPTER 8

Exercises 8.1

2.

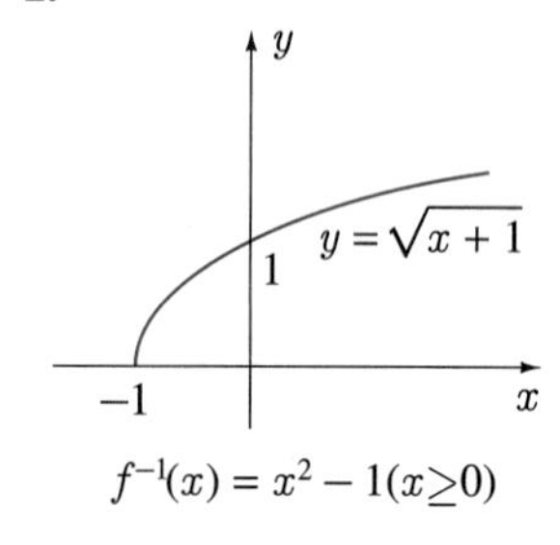

$f^{-1}(x) = x^2 - 1 (x \geq 0)$

4.

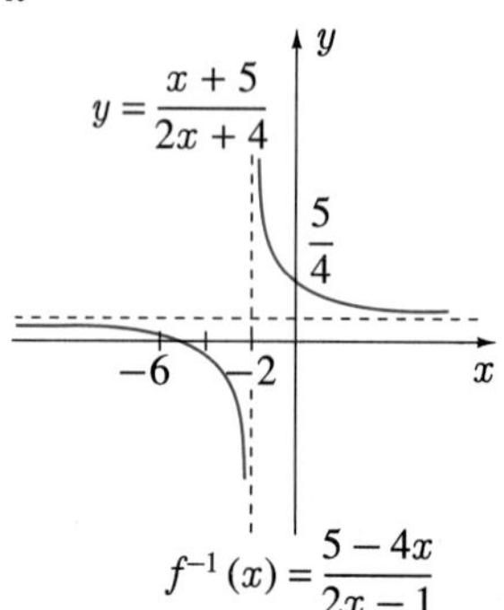

$f^{-1}(x) = \dfrac{5-4x}{2x-1}$

6.

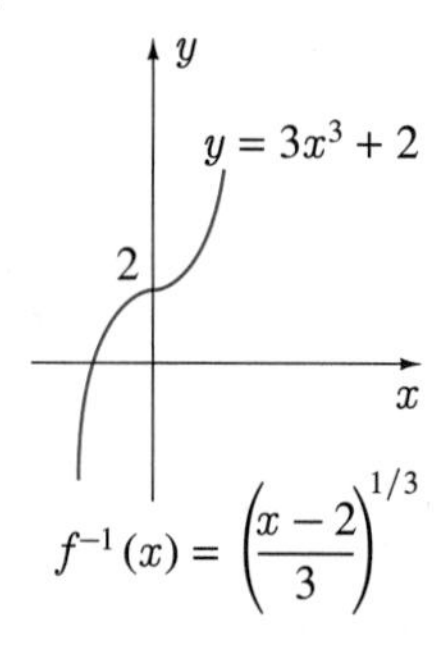

$f^{-1}(x) = \left(\dfrac{x-2}{3}\right)^{1/3}$

8.

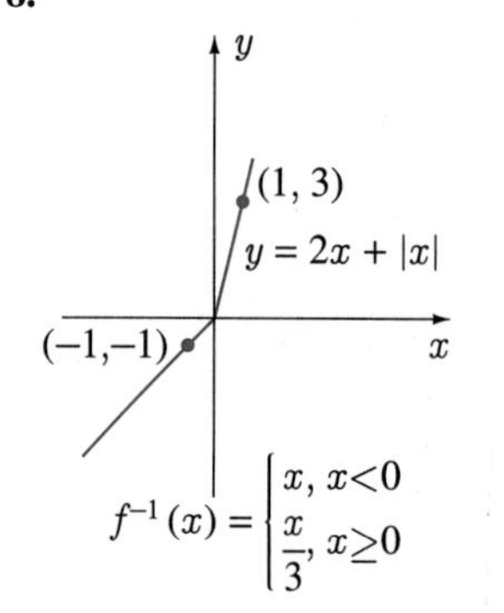

$f^{-1}(x) = \begin{cases} x, & x<0 \\ \dfrac{x}{3}, & x\geq 0 \end{cases}$

10. **12.**

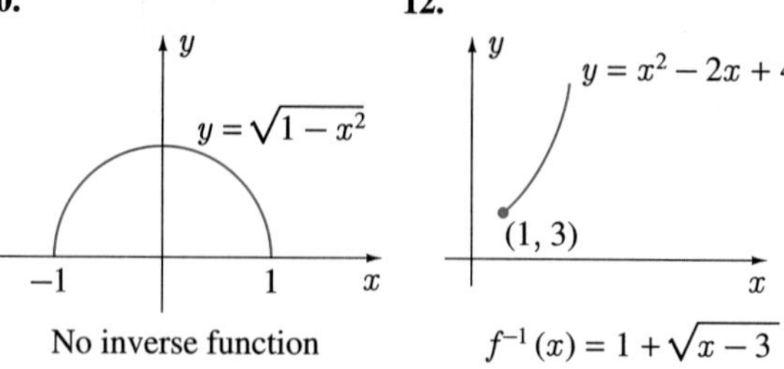

No inverse function $f^{-1}(x) = 1 + \sqrt{x-3}$

14.

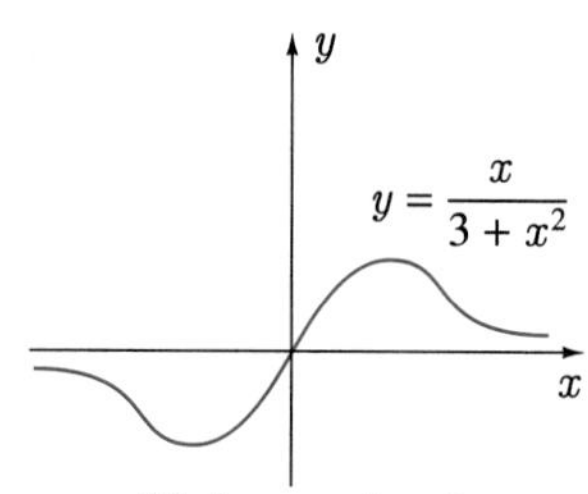

No inverse function

16. $-x^{-1/4}$, $x < 0$; $x^{-1/4}$, $x > 0$
18. $-\sqrt{\sqrt{2+x}-2}$, $x < 0$; $\sqrt{\sqrt{2+x}-2}$, $x \geq 0$
20. $-\sqrt{(x+\sqrt{x^2+16x})/2}$, $x < 0$;
$\sqrt{(x+\sqrt{x^2+16x})/2}$, $x \geq 0$

22.

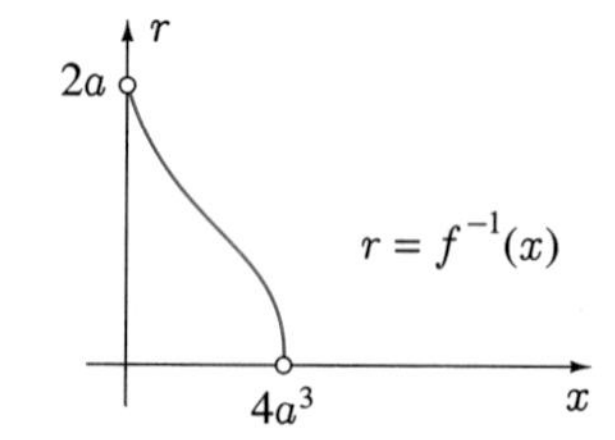

Exercises 8.2

2. 0.253 **4.** $-2\pi/3$ **6.** Does not exist
8. $-\pi/4$ **10.** Does not exist **12.** 0.164
14. $\pi/4$ **16.** $-\pi/4$ **18.** $-0.876 + n\pi$
20. No solutions **22.** No solutions

24.

$y = \sqrt{\mathrm{Tan}^{-1} x} + \sqrt{\mathrm{Sec}^{-1} x}$

26.

$y = \mathrm{Sin}^{-1} x + \mathrm{Csc}^{-1} x$

28. $(3n \pm 1)\pi/3$ **30.** No solutions

32. $\text{Tan}^{-1}(2n\pi \pm 3\pi/4) + m\pi$

34. (a)

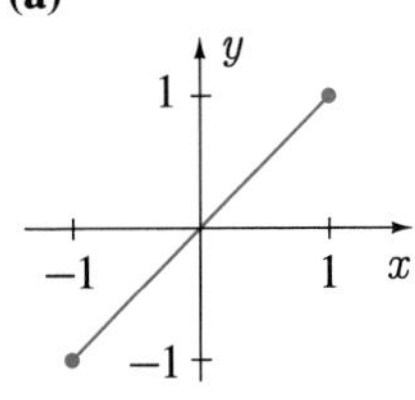

$y = \sin(\text{Sin}^{-1}x)$

(b)

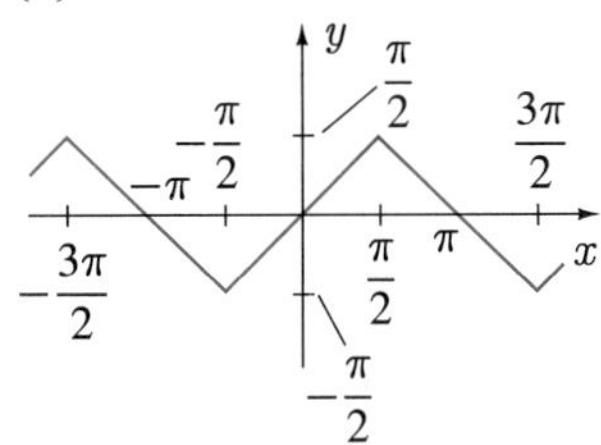

$y = \text{Sin}^{-1}(\sin x)$

36. $\sqrt{17}\sin(2x + 0.24)$ **38.** $2\sqrt{5}\sin(2x + 2.03)$

44. $E_0\cos(\omega t + \phi)$ where $\phi = \text{Tan}^{-1}\{[(\omega L - 1/(\omega C)]/R\}$

Exercises 8.3

2. $-2x/[1 + (x^2 + 2)^2]$ **4.** $-2x/[1 + (2 - x^2)^2]$

6. $\text{Csc}^{-1}(x^2 + 5) - 2x^2/[(x^2 + 5)\sqrt{(x^2 + 5)^2 - 1}]$

8. $1/[2(x + 3)\sqrt{x + 2}]$ **10.** $-1/(x\sqrt{x^2 - 1})$

12. $2x\text{Sec}^{-1}x + x/\sqrt{x^2 - 1}$

14. $-1/(1 + x^2)$ **16.** $-1/[\sqrt{x}(x + 1)]$

18. $x(t - \sqrt{1 - t^2}\text{Cos}^{-1}t)/[\sqrt{1 - x^2}\sqrt{1 - t^2}]$

20. $(5\sqrt{1 - x^2y^2} - y)/(x - 2\sqrt{1 - x^2y^2})$

22. $-18/(x^3\sqrt{x^2 - 9})$

24. $-(1/x^2)\text{Csc}^{-1}(3x) - 2/(x^2\sqrt{9x^2 - 1})$

26. $(\text{Cos}^{-1}x)^2 - 2 + 2x(1 - \text{Cos}^{-1}x)/\sqrt{1 - x^2}$

28. $2\sqrt{4x - x^2}$

30. $\sqrt{2}(1 - x^2)/[(1 + x^2)\sqrt{1 + x^4}]$

32.

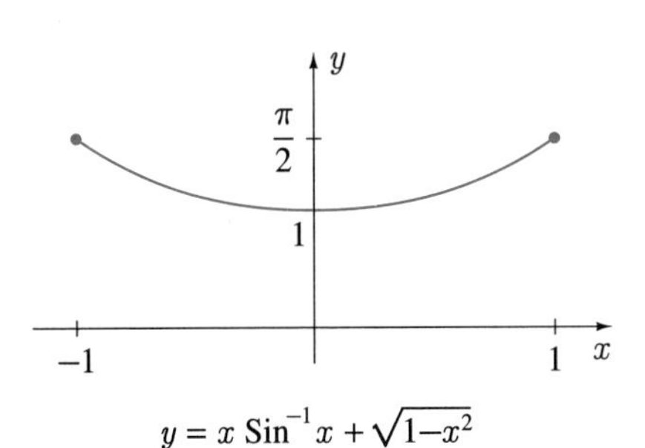

$y = x\,\text{Sin}^{-1}x + \sqrt{1 - x^2}$

34. 1.23 **36.** $\text{Tan}^{-1}\mu$

38.(a) 2/15 rad/s **(b)** 0.0369 rad/s

Exercises 8.4

2. 2.30 **4.** 9.08×10^{-5} **6.** 1.00

8. 0.266 **10.** 5.07

12. $\sinh(x/2) + (x/2)\cosh(x/2)$

14. $(1/x)\,\text{sech}^2(\ln x)$

16. $2\sqrt{1 + y}\,\text{csch}^2 x/(2\sqrt{1 + y} - 1)$

18. $-\sinh t\,\text{sech}^2 x\sec^2(\cosh t)\,\sin(\tanh x)$

20. $2\,\text{csch}\,4x$ **26.(a)** 189.3 N **(b)** 17.0 m

28.(b) $\sqrt{mg/\beta}$

Review Exercises

2. $(4n + 1)\pi/4 \pm 0.659$ **4.** -1.79

6. $(4n + 1)\pi/2 \pm 0.604$

8. $\ln[(4n + 1)\pi/2 \pm 0.84] - 2,\ n \geq 0$

10. 2.09 **12.** $6x\cosh x^2$

14. $(x^2 - 1)/(x^4 + 3x^2 + 1)$ **16.** $\cos x\,\text{sech}\,y$

18. $\text{Csc}^{-1}(1/x^2) + 2x^2/\sqrt{1 - x^4}$

20. -1

CHAPTER 9

Exercises 9.1

2. $-(1/4)e^{-2x^2} + C$ **4.** $\ln(1 + e^x) + C$

6. $-(2/27)(1 - 3x^3)^{3/2} + C$

8. $-(1/6)(1 + x^3)^{-2} + C$

10. $2\sqrt{x} - x + C$ **12.** $\text{Tan}^{-1}x + x + C$

14. $(1/6)(2x + 4)^{3/2} + \sqrt{2x + 4} + C$

16. $(1/8)\sin^4 2x + C$

18. $(1/2)(x + 5)^2 - 15x + 75\ln|x + 5| + 125/(x + 5) + C$

20. $(1/3)\ln|\sec 3x| + C$

22. $(4/5)x^{5/4} - x + (4/3)x^{3/4} - 2\sqrt{x} + 4x^{1/4} - 4\ln(x^{1/4} + 1) + C$

24. $2\ln(e^x + 1) - x + C$ **26.** $(35 - 12\ln 6)/6$

30.(a) $1962(1 - e^{-t/200})$ m/s

(b) $1962t + 392\,400(e^{-t/200} - 1)$ m

32. $(1/(3n))\ln|x^n/(3 + 2x^n)| + C$

34. $(2/7)(2x + 3)/\sqrt{x^2 + 3x + 4} + C$

36. $(1/b^2)[bx - a\ln|a + bx|] + C,\ n = -1;$
$(1/b^2)[\ln|a + bx| + a/(a + bx)] + C,\ n = -2;$
$(1/b^2)[(a + bx)^{n+2}/(n + 2) - a(a + bx)^{n+1}/(n + 1)] + C,$
$n \neq -2, -1$

38. $e^{-|a|}(1 + |a|)$

Exercises 9.2

2. 4.75 hr **4.** 19.74 s **6.** $k\ln(S/S_0)$

8. 13.51 hr

10. (a) Finitely long **(b)** Infinitely long

12. 33 574 years ago

14.

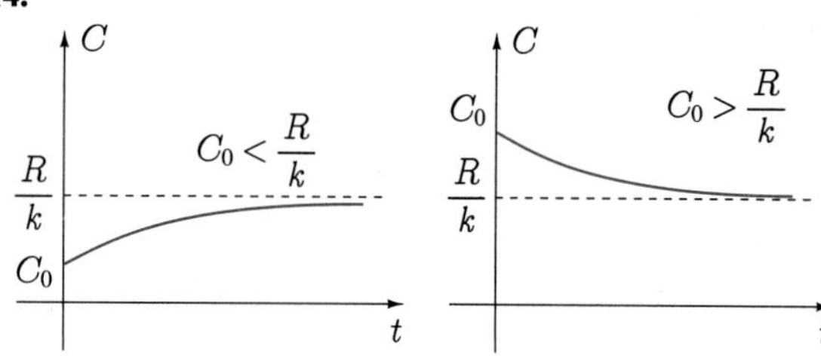

$$C = \frac{R}{k}(1 - e^{-kt}) + C_0\, e^{-kt}$$

16. $20 + 70\, e^{-0.01399t}$

18.

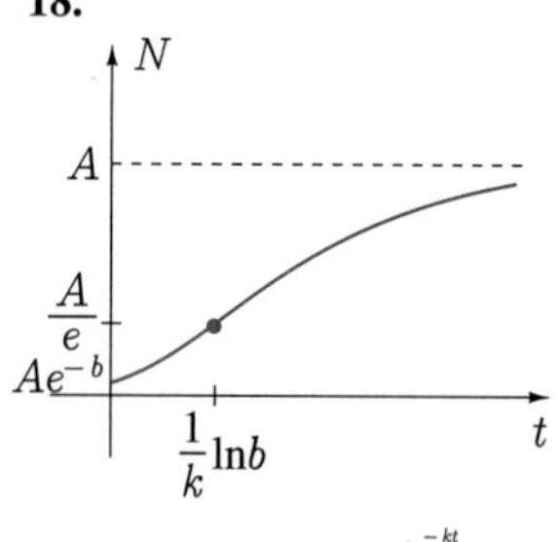

$$N = Ae^{-be^{-kt}}$$

20.

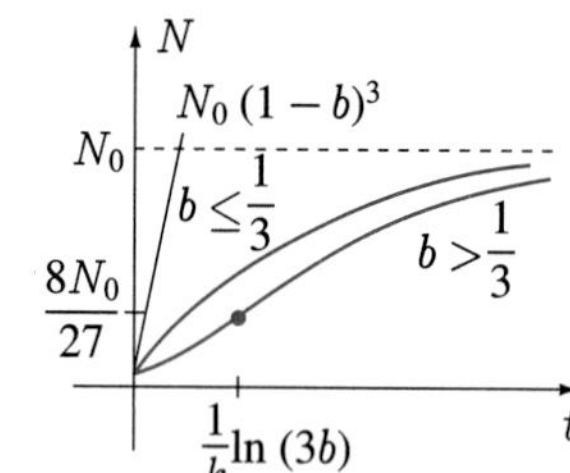

$$N = N_0\,(1 - be^{-kt})^3$$

22.

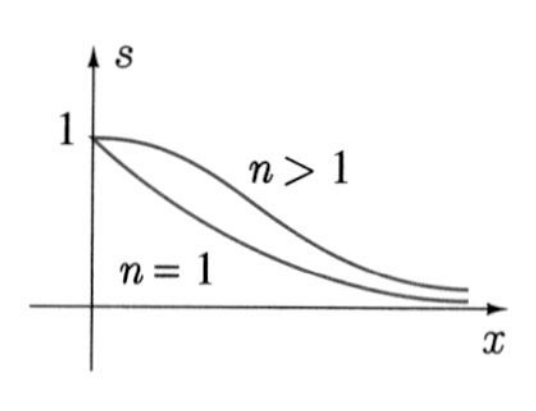

$$s = 1 - (1 - e^{-kx})^n$$

24. 3:25 A.M.

Exercises 9.3

2. $(x^2/2 - x/2 + 1/4)e^{2x} + C$
4. $(2/3)x^{3/2}\ln(2x) - (4/9)x^{3/2} + C$
6. $-(2/5)(2 + x)(3 - x)^{3/2} + C$
8. $(2x^2/3)(x+5)^{3/2} - (8x/15)(x+5)^{5/2} + (16/105)(x+5)^{7/2} + C$
10. $2x^2\sqrt{x+2} - (8x/3)(x+2)^{3/2} + (16/15)(x+2)^{5/2} + C$
12. $(1/3)(x-1)^3\ln x - x^3/9 + x^2/2 - x + (1/3)\ln x + C$
14. $x\,\mathrm{Tan}^{-1}\,x - (1/2)\ln(1 + x^2) + C$
16. $e^{2x}(2\cos 3x + 3\sin 3x)/13 + C$
18. $x\ln(x^2 + 4) - 2x + 4\,\mathrm{Tan}^{-1}(x/2) + C$
20. 5.92×10^9 N
26. $x^3/6 + (x^2/4 - 1/8)\sin 2x + (x/4)\cos 2x + C$

Exercises 9.4

2. $-(1/2)\sin^{-2}x + C$ **4.** $-(1/3)\csc^3 x + C$
6. $(2/3)\tan^{3/2}x + (2/7)\tan^{7/2}x + C$
8. $(1/18)\sec^6 3x + C$ **10.** $(1/4)\sec^4 x + C$
12. $\tan\theta + C$ **14.** $2\sqrt{1 + \tan x} + C$
16. $7\tan x - 3x + C$
18. $(1/32)(12x - 8\sin 2x + \sin 4x) + C$
20. $(1/2)\tan^2 x + \ln|\tan x| + C$
22. $\pi(1 - \pi/4 + \ln 2)$
24. $-(1/3)\cot^3 z + \cot z + z + C$
26. $\theta/2 + (1/4)\sin 2\theta + (1/3)\cos^3\theta + C$
28. $(1/8)\sin 4x + (1/16)\sin 8x + C$
30. $\sec x + \ln|\csc x - \cot x| + C$

Exercises 9.5

2. $(1/\sqrt{5})\,\mathrm{Sin}^{-1}(\sqrt{5}x/3) + C$
4. $-\sqrt{4 - x^2}/(4x) + C$
6. $(1/15)(5x^2 + 3)^{3/2} + C$
8. $(1/2)\ln|(1 + x)/(1 - x)| + C$
10. $(1/20)\ln(5x^2 + 1) + (\sqrt{5}/2)\,\mathrm{Tan}^{-1}(\sqrt{5}x) + C$
12. $\sqrt{4 - x^2} + 2\ln|(2 - \sqrt{4 - x^2})/x| + C$
14. $\ln|x + \sqrt{x^2 - 16}| - \sqrt{x^2 - 16}/x + C$
16. $(1/16)\,\mathrm{Sec}^{-1}(x/2) + \sqrt{x^2 - 4}/(8x^2) + C$
18. $(1/3)(y^2 + 4)^{3/2} - 4\sqrt{y^2 + 4} + C$
20. $2\ln|x| - (1/2)\ln(1 + x^2) + C$
24. $L\ln[(L + \sqrt{L^2 - x^2})/x] - \sqrt{L^2 - x^2}$
26. 1.053 **28.** $0.265b$ **30.** $9.81\pi\rho r^3$
32. $(147/8)\,\mathrm{Sin}^{-1}(x/\sqrt{7}) + (x/8)(35 - 2x^2)\sqrt{7 - x^2} + C$
34. $-6\,\mathrm{Sec}^{-1}(2x) - (12x^2 - 1)/(2x^2\sqrt{4x^2 - 1}) + C$
36. $(1/2)[x\sqrt{1 + 3x^2} + (1/\sqrt{3})\ln|\sqrt{3}x + \sqrt{1 + 3x^2}|] + C$
38. $\sqrt{2}\pi$ **40.** $r/2$ **42.** $a^2(4 - \pi)/2$

Exercises 9.6

2. $\ln|\sqrt{x^2 + 2x + 2} + x + 1| + C$
4. $-(2/\sqrt{7})\,\mathrm{Tan}^{-1}[(2x - 3)/\sqrt{7}] + C$
6. $(x/2)/\sqrt{4x - x^2} + C$
8. $\ln(x^2 + 6x + 13) - (9/2)\,\mathrm{Tan}^{-1}[(x + 3)/2] + C$
10. $\ln|\sqrt{6 + 4\ln x + (\ln x)^2} + \ln x + 2| + C$
12. $39\,240\pi$ N
14. $(1/2)(x - 1)\sqrt{x^2 - 2x - 3} - 2\ln|x - 1 + \sqrt{x^2 - 2x - 3}| + C$
16. $(1/2)\,\mathrm{Tan}^{-1}[(\sqrt{2x - 3} + 2)/2] + C$

Exercises 9.7

2. $-(1/2)(y+1)^{-2}+C$

4. $x+(10/3)\ln|x-4|+(2/3)\ln|x+2|+C$

6. $-(1/6)\ln|y|-(2/15)\ln|y+3|+(3/10)\ln|y-2|+C$

8. $(1/2)\ln(x^2+2)+(x^2+2)^{-1}+C$

10. $y-4\ln|y+2|+\ln|y+1|+C$

12. $\ln(y^2+1)+\text{Tan}^{-1}y-\ln(y^2+4)+C$

14. $-[1/(3\sqrt{2})]\text{Tan}^{-1}(x/\sqrt{2})+(2/3)\ln|(x-1)/(x+1)|+C$

16. $-(5/27)\ln|x-1|-(7/9)(x-1)^{-1}+(32/27)\ln|x+2|+(2/9)(x+2)^{-1}+C$

18.

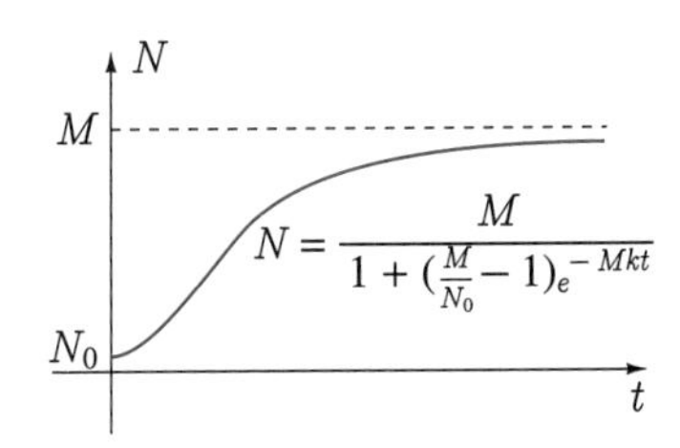

20.(b) $60(1-e^{10kt})/(2-3e^{10kt})$ gm

22. $(1/4)\ln|x+1|-(1/8)\ln(x^2+1)+(1/2)\text{Tan}^{-1}x+(1/4)(x+1)/(x^2+1)+C$

24. $(1/3)\ln|x+1|+(1/\sqrt{3})\text{Tan}^{-1}[(2x-1)/\sqrt{3}]-(1/6)\ln(x^2-x+1)+C$

26. $\ln|x|+\ln(x^2-x+1)-2\ln|x+1|-3/(x+1)+(2/\sqrt{3})\text{Tan}^{-1}[(2x-1)/\sqrt{3}]+C$

28. $\ln|[1+\tan(x/2)]/[1-\tan(x/2)]|+C$

30. $(1/\sqrt{3})\ln|[\sqrt{3}\tan(x/2)-1]/[\sqrt{3}\tan(x/2)+1]|+C$

Exercises 9.8

2. $(1/8)\ln|(x-4)/(x+4)|+C$

4. $-(1/\sqrt{5})\ln|(\sqrt{5}+\sqrt{x^2+5})/x|+C$

6. $2\sqrt{2x+3}+\sqrt{3}\ln|(\sqrt{2x+3}-\sqrt{3})/(\sqrt{2x+3}+\sqrt{3})|+C$

8. $(1/3)\text{Sin}^{-1}[(9x-2)/2]+C$

10. $(1/8)(4x-1)\sqrt{3+2x-4x^2}+(13/16)\text{Sin}^{-1}[(4x-1)/\sqrt{13}]+C$

12. $[1/(2\sqrt{2})]\ln|[\tan(x/2)+3-2\sqrt{2}]/[\tan(x/2)+3+2\sqrt{2}]|+C$

14. $\ln|(1-\sqrt{1-x^2})/x|-(1/x)\text{Sin}^{-1}x+C$

16. $-(1/13)e^{-2x}(2\sin 3x+3\cos 3x)+C$

18. $x(\ln x)^2-2x\ln x+2x+C$

20. $[1/(3\cdot 2^{5/3}]\ln|(x+2^{1/3})^2/(2^{2/3}-2^{1/3}x+x^2)|+[1/(\sqrt{3}\cdot 2^{2/3})]\text{Tan}^{-1}[(2^{2/3}x-1)/\sqrt{3}]+C$

22. $(x^3/2-3x/4)\sin 2x+(3x^2/4-3/8)\cos 2x+C$

24. $-\sqrt{9+x-x^2}+(1/2)\text{Sin}^{-1}[(2x-1)/\sqrt{37}]+C$

26. $-(1/2)(x+3)\sqrt{2x-x^2}+3\,\text{Sin}^{-1}(x-1)+C$

28. $(1/15)\tan x(3\sec^4 x+4\sec^2 x+8)+C$

30. $(x^4/4-3/32)\text{Sin}^{-1}x+(3x/16)(1-x^2)^{3/2}+[(8x^3-3x)/32]\sqrt{1-x^2}+C$

32. $[5/(216\sqrt{3})]\text{Tan}^{-1}(x/\sqrt{3})+(1/216)(5x^5+40x^3+99x-180)/(3+x^2)^3+C$

Exercises 9.9

2. 0.47215, 0.47214	**4.** 0.64886, 0.64872
6. −0.069570, −0.069445	
8. 0.30220, 0.30230	**10.** 0.14221, 0.14201
12. 1.4672, 1.4627	**14.** 0.31117, 0.31026
16. 1/4 previous; 1/16 previous	
18. 1.4789	**20.** 32.91
22. 0.2437, 0.2438	**24.** 2.113, 1.729
28.(a) 21 **(b)** 4	**30.(a)** 3 **(b)** 2

Review Exercises

2. $-1/(x+3)+C$

4. $x^2/2-x+4\ln|x+1|+C$

6. $(2/3)(x+3)^{3/2}-6\sqrt{x+3}+C$

8. $-x\cos x+\sin x+C$

10. $(3/4)\ln|x+3|+(1/4)\ln|x-1|+C$

12. $14\sqrt{x}-x-70\ln(5+\sqrt{x})+C$

14. $\text{Sin}^{-1}(e^x)+C$

16. $-(1/2)/(x^2+1)+C$

18. $(1/2)\ln(x^2+1)+(1/2)/(x^2+1)+C$

20. $x+4\ln|x-1|-4/(x-1)+C$

22. $x\text{Cos}^{-1}x-\sqrt{1-x^2}+C$

24. $-(1/12)\cos 6x+(1/8)\cos 4x+C$

26. $\ln|x+2+\sqrt{x^2+4x-5}|+C$

28. $(1/5)(4-x^2)^{5/2}-(4/3)(4-x^2)^{3/2}+C$

30. $-3\ln|4-x^2|+C$

32. $[1/(2\sqrt{2})]\ln|(x+2-2\sqrt{2})/\sqrt{x^2+4x-4}|+C$

34. $\ln|x|+1/x-1/(2x^2)-\ln|x+1|+C$

36. $\ln|2\sqrt{x^2-3x+16}+2x-3|+C$

38. $-x^2/6+(x^3/3)\text{Tan}^{-1}x+(1/6)\ln(x^2+1)+C$

40. $(\ln x)^2/2+C$

42. $(1/81)\ln|x|-(1/162)\ln|x^2+9|+(1/18)/(x^2+9)+C$

44. $(1/2)\ln|x|+(1/2)\ln|x+2|+3/(x+2)+C$

46. $\ln|x|+\ln(x^2+x+4)+C$

48. $-(2/3)\cot^{3/2}x-(2/7)\cot^{7/2}x+C$

50. $(x-2)/(4\sqrt{4x-x^2})+C$
52. 0.63665, 0.64041
54. 1.2667, 1.2647
56. $-2\sqrt{x}-3x^{1/3}-6x^{1/6}-6\ln|1-x^{1/6}|+C$
58. $(1/8)\operatorname{Tan}^{-1}(x^2/4)+C$
60. $(4x-6)/(9\sqrt{3x-x^2})+C$
62. $2\sin\sqrt{x}-2\sqrt{x}\cos\sqrt{x}+C$
64. $(1/32)\sin 4x+(1/8)\sin 2x-(x/8)\cos 4x-(x/4)\cos 2x+C$
66. $(1/2)\tan x+C$
68. $(1/6)\sin 3x-(1/20)\sin 5x-(1/4)\sin x+C$
70. $(1/4)(\operatorname{Sin}^{-1}x)^2+(x/2)\sqrt{1-x^2}\operatorname{Sin}^{-1}x-x^2/4+C$
72.(b) $\operatorname{Sin}^{-1}x-\sqrt{1-x^2}+C$

CHAPTER 10

Exercises 10.1

2.

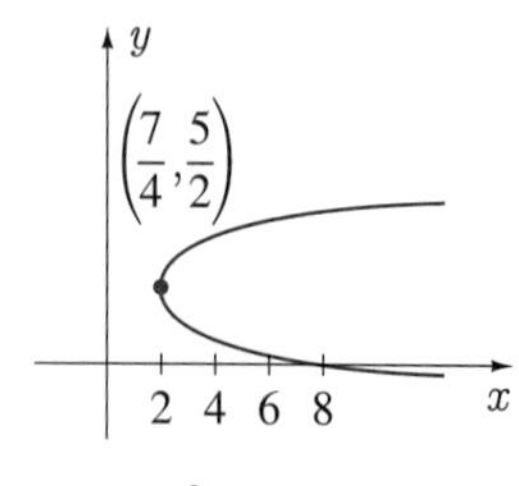

$x=t^2+3t+4$
$y=1-t$

4.

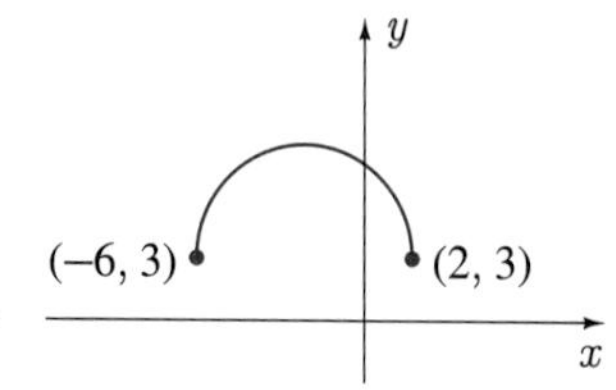

$x=-2+4\cos t$, $y=3+4\sin t$, $0\le t\le\pi$

6.

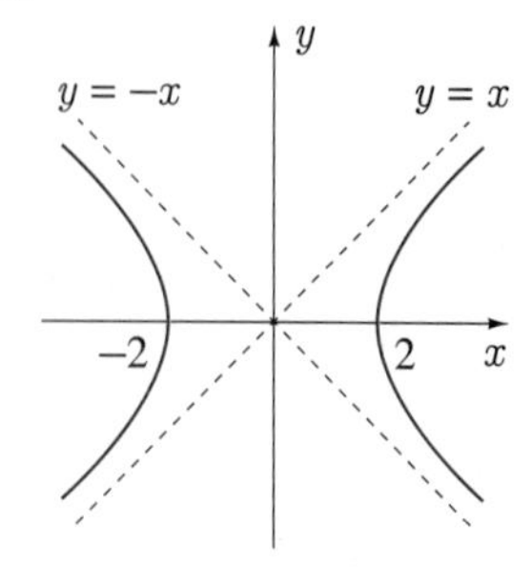

$x=t+1/t$
$y=t-1/t$

8.

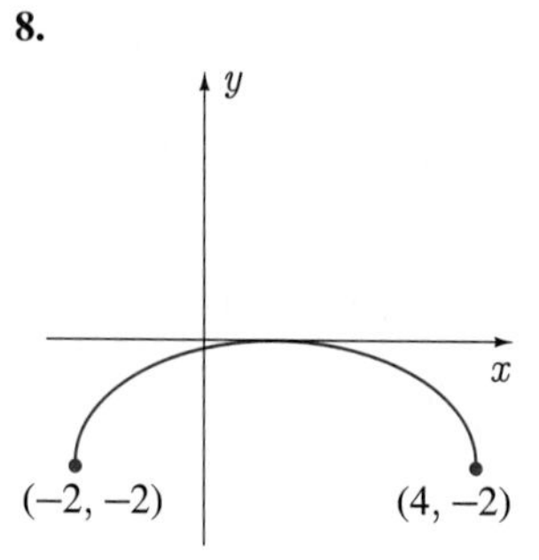

$x=1+3\cos t$, $y=-2+2\sin t$, $0\le t\le\pi$

10.

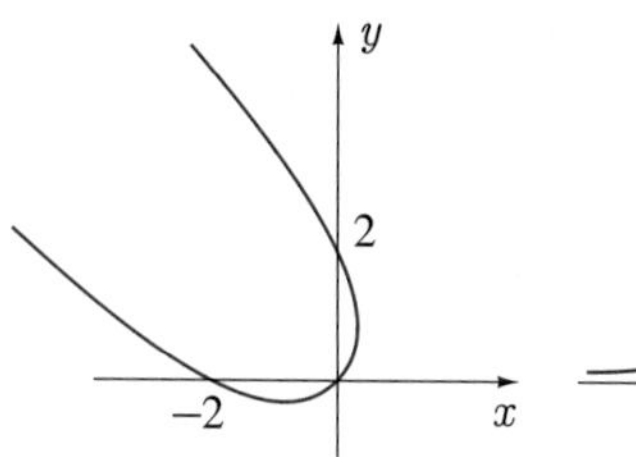

$x=t-t^2$
$y=t+t^2$

12.

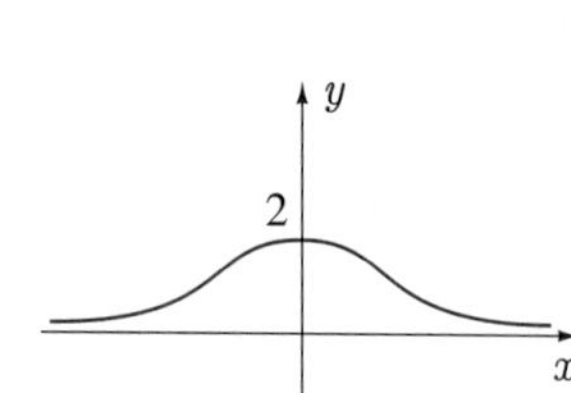

$x=2\cot\theta$, $y=2\sin^2\theta$, $-\frac{\pi}{2}\le\theta\le\frac{\pi}{2}$

14. $2u/(u+1)^2$
16. $-1/x^2$
18. $3/[4(t+6)^2(2t+3)^3]$
20. $-4(t+1)\sqrt{-t^2+3t+5}/[(3-2t)(t^2+2t-5)^2]$
22. $x/y,\ 2/y^3$
24. $1/(v+1),\ -(1/2)/(v+1)^3$
26. $(8/3,1/2),\ (28/3,28)$
28. Ellipse $b^2(x-h)^2+a^2(y-k)^2=a^2b^2$
32. $x=t,\ y=(t+1)/(t-2)$
34. $x=-1+\sqrt{5}\cos t,\ y=2+\sqrt{5}\sin t,\ 0\le t<2\pi$
36. $31/26$
38. 8π
40. 1.73
46. (b) $3\pi R^2$ (c) $8R$

48.

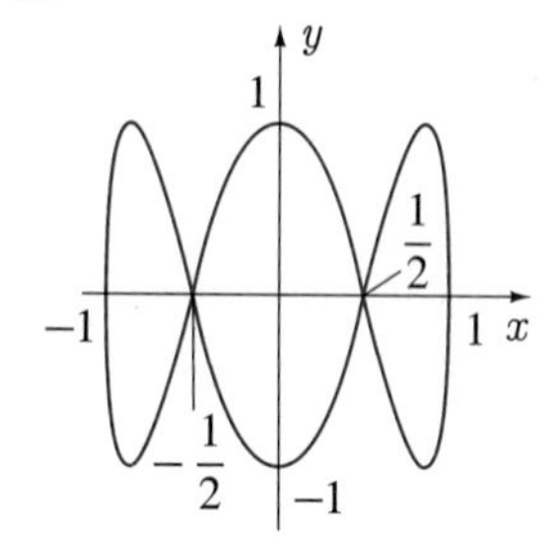

$x=\cos\theta$, $y=\sin 3\theta$

50.

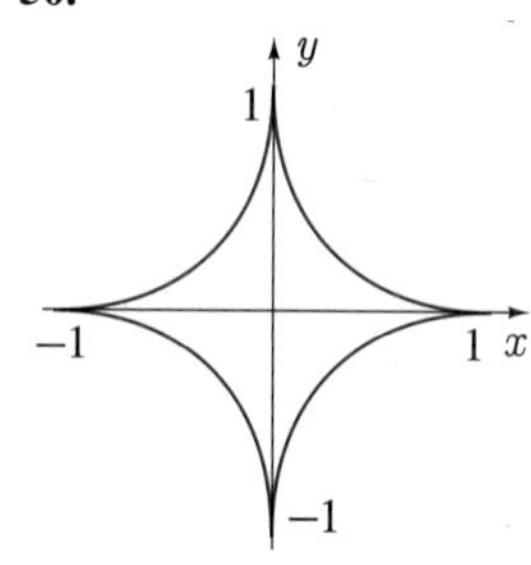

$x=\cos^3\theta$, $y=\sin^3\theta$

Exercises 10.2

2. $(2,2\pi/3+2n\pi)$
4. $(4,5\pi/6+2n\pi)$
6. $(\sqrt{17},-1.82+2n\pi)$
8. $(\sqrt{29},2.76+2n\pi)$
10. $(3\sqrt{3},-3)$
12. $(-2.21,-2.03)$

Exercises 10.3

2.

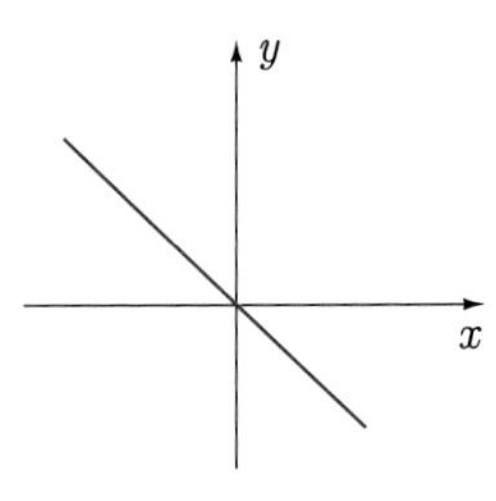

$\theta = \dfrac{3\pi}{4},\ \theta = \dfrac{-\pi}{4}$

4.

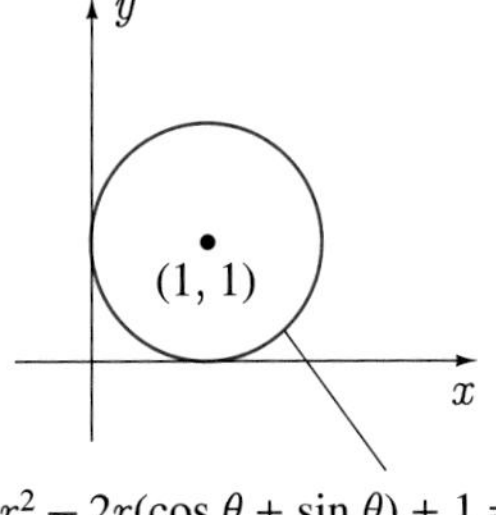

$r^2 - 2r(\cos\theta + \sin\theta) + 1 = 0$

6.

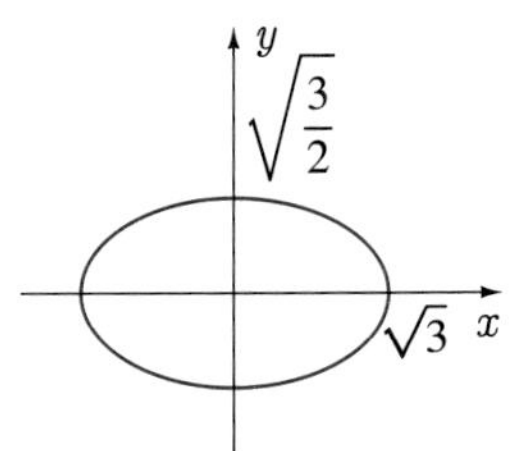

$r^2 = \dfrac{3}{1 + \sin^2\theta}$

8.

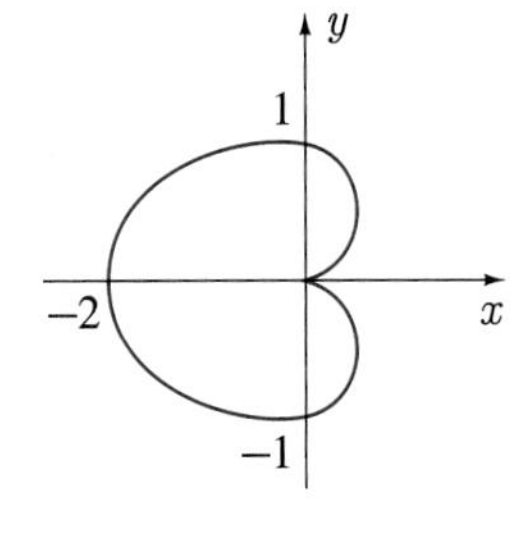

$r = 1 - \cos\theta$

10.

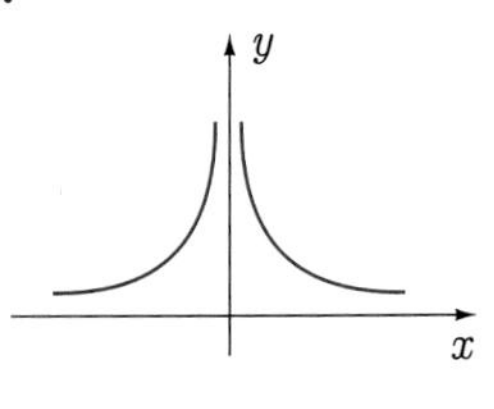

$r^3 = \sec^2\theta \csc\theta$

12.

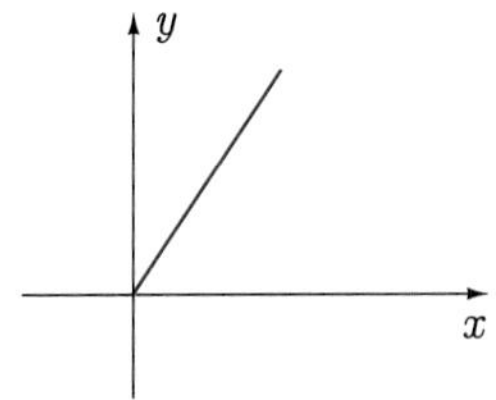

$y = (\tan 1)\, x$

14.

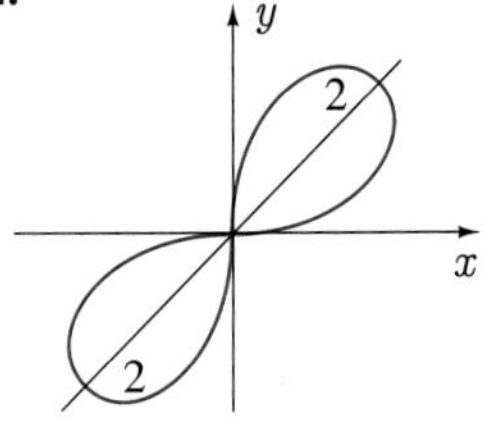

$(x^2 + y^2)^2 = 8xy$

16.

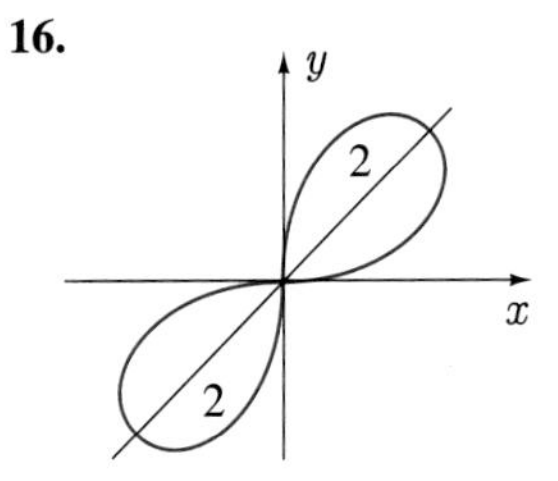

$(x^2 + y^2)^{3/2} = 4xy$

18.

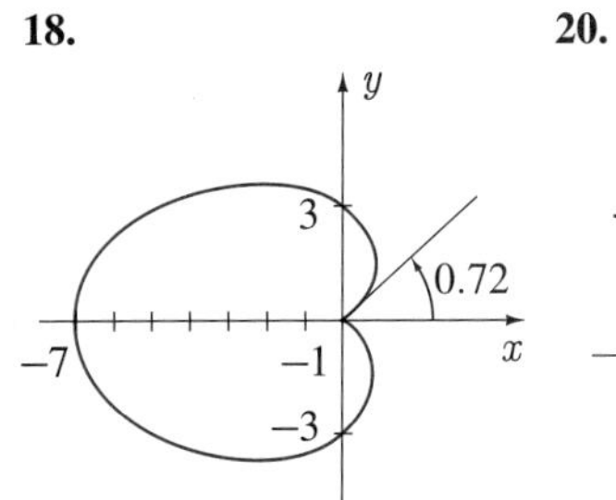

$x^2 + y^2 = 3\sqrt{x^2 + y^2} - 4x$

20.

$y^3 = x^2$

22. $(2, \pm\pi/6),\ (2, \pm 5\pi/6)$

24. $(0, \theta),\ (4/3, \pm 1.23)$

28.

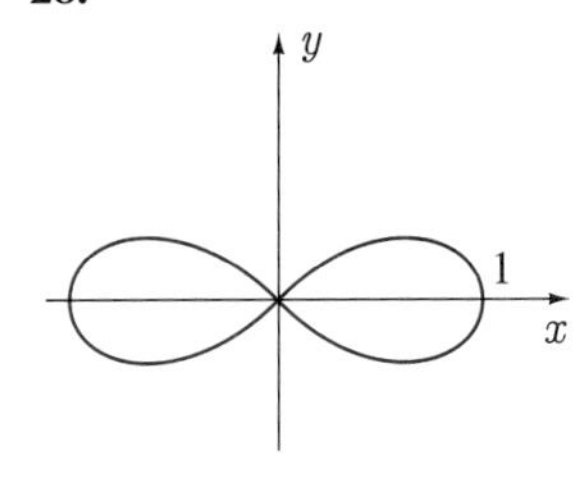

$r = \cos 2\theta$

30.

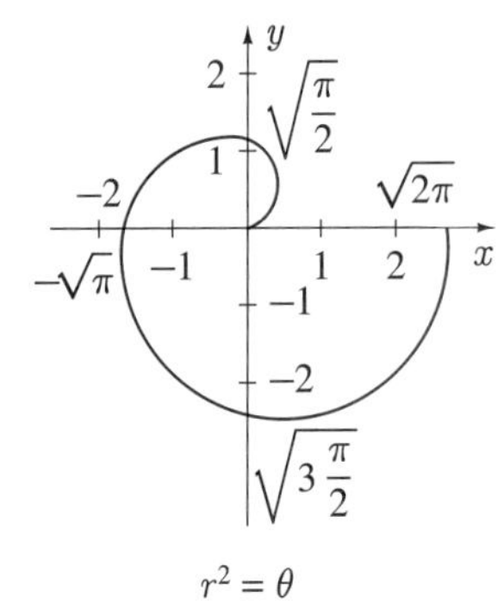

$r^2 = \theta$

32.

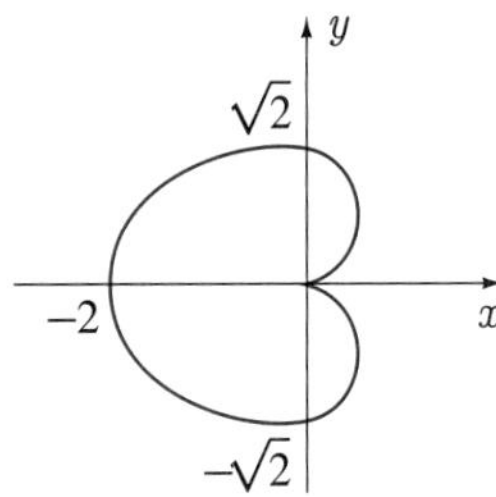

$r = 2\sin(\theta/2)$

34.

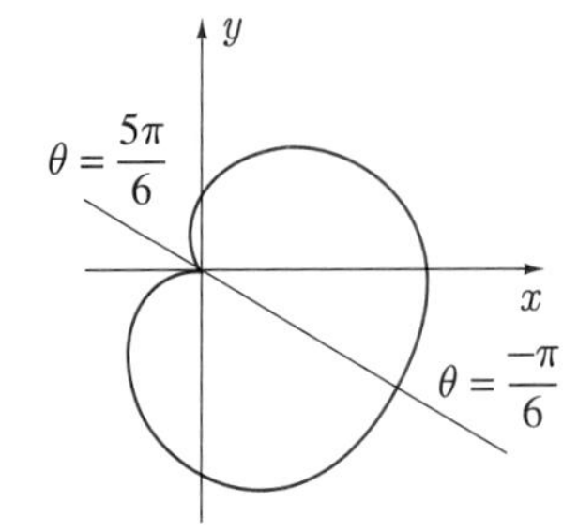

$r = 1 + \cos\left(\theta + \dfrac{\pi}{6}\right)$

36. does not exist **38.** $1/2$ **40.** $8a$

42.

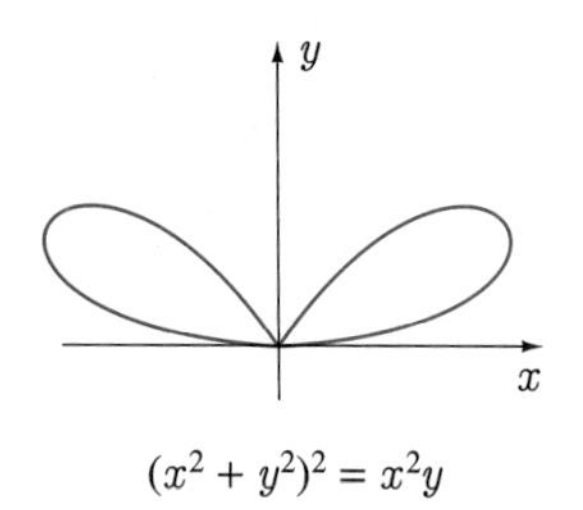

$(x^2 + y^2)^2 = x^2y$

44.

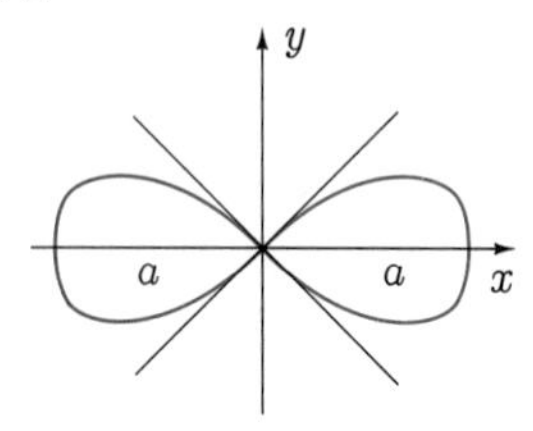

$(x^2 + y^2)^2 = a^2(x^2 - y^2)$

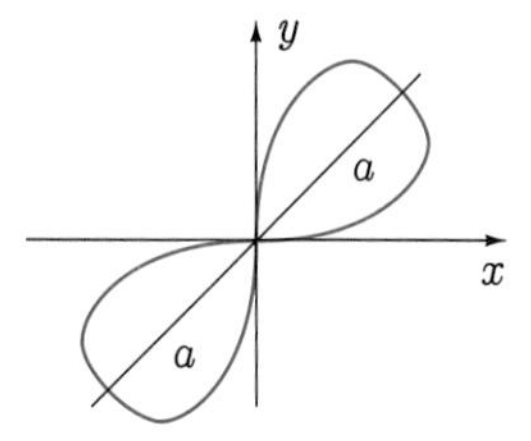

$(x^2 + y^2)^2 = 2a^2xy$

Exercises 10.4

2. 9π **4.** 2 **6.** 6π
8. 18π **10.** 2π **12.** $(5\pi - 8)/4$
14. $6\pi - 16$ **16.** $(7\pi - 12\sqrt{3})/12$
18. $33[\pi + 2\,\mathrm{Sin}^{-1}(2/3)]/4 + 7\sqrt{5}$
20. $\pi/32$ **22.(b)** $a^2(4 - \pi)/2$

Exercises 10.5

2. Circle **4.** None of these
6. Hyperbola **8.** Parabola
10. None of these **12.** Ellipse
14. Straight line

16.

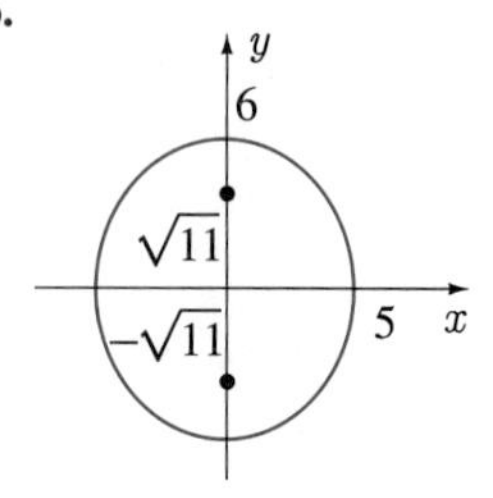

$\frac{x^2}{25} + \frac{y^2}{36} = 1$

18.

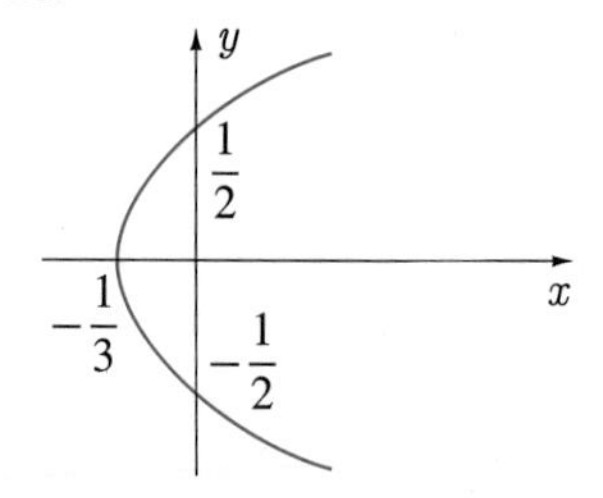

$3x = 4y^2 - 1$

20.

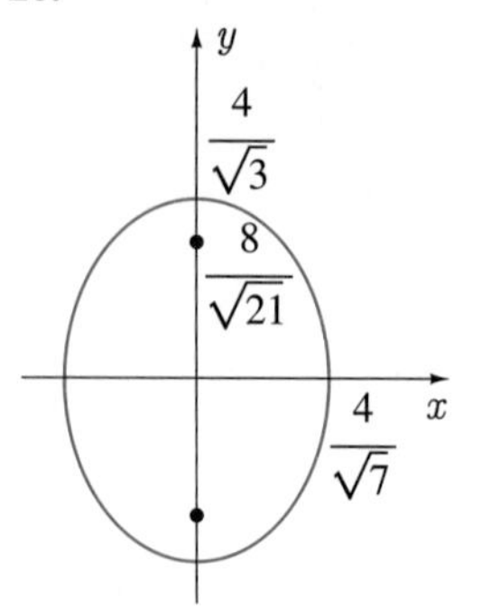

$7x^2 + 3y^2 = 16$

22.

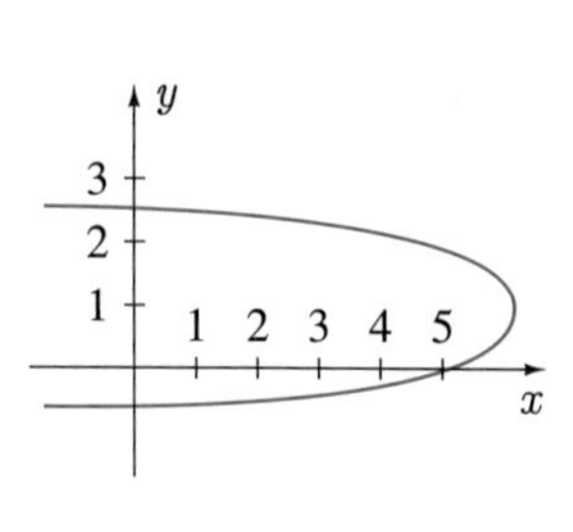

$2y^2 + x = 3y + 5$

24.

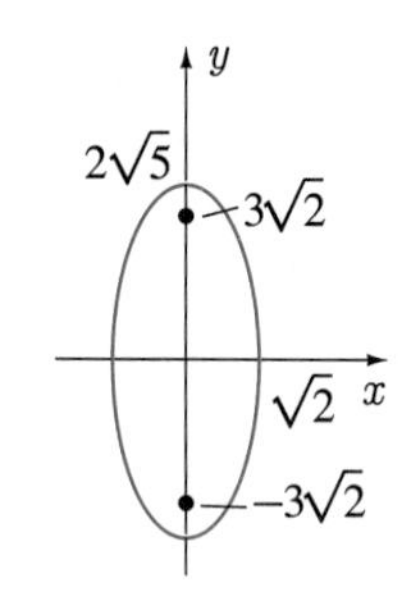

$y^2 = 10(2 - x^2)$

26.

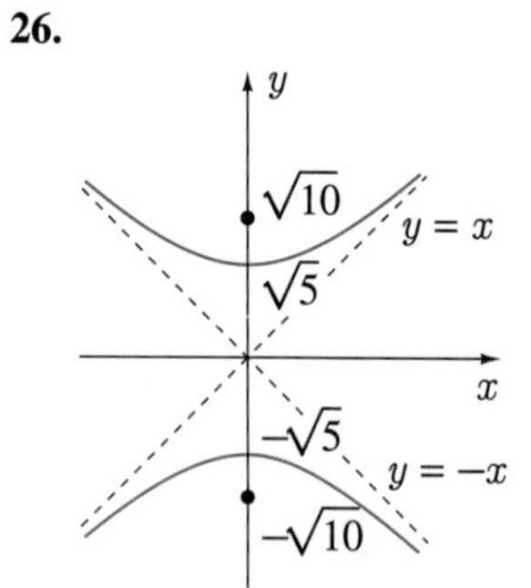

$y^2 - x^2 = 5$

28.

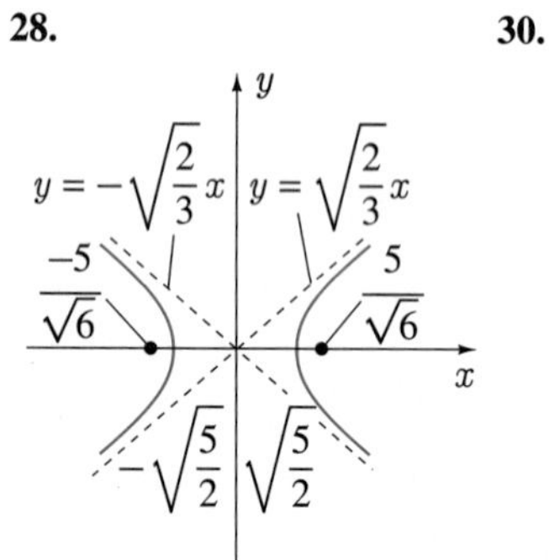

$2x^2 - 3y^2 = 5$

30.

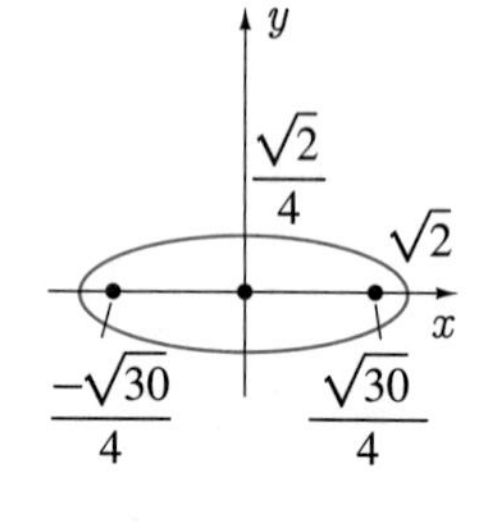

$x^2 + 16y^2 = 2$

32.

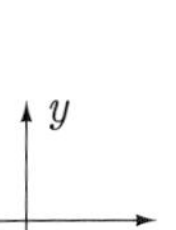

$x = -(4 + y)^2$

34.

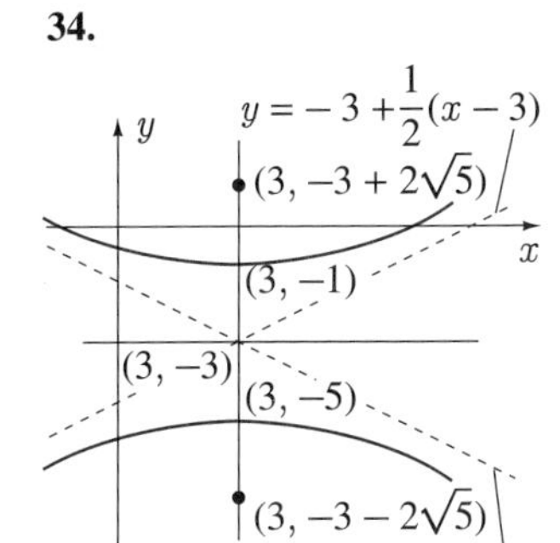

$x^2 - 6x - 4y^2 - 24y = 11$

36.

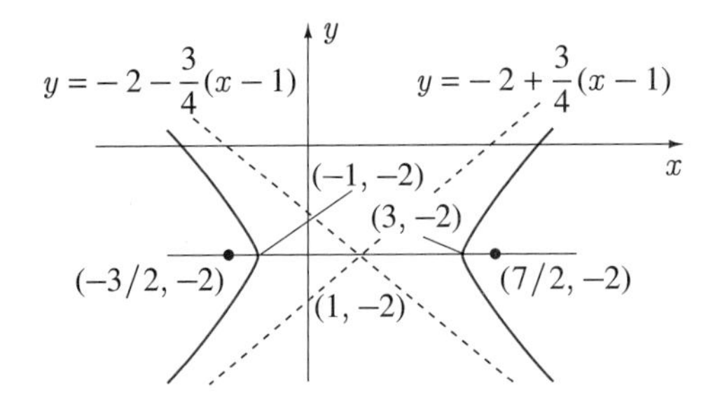

$9x^2 - 16y^2 - 18x - 64y = 91$

38. $3x^2 + y^2 = 28$

40. $25/4$

42. $9x^2 + 25y^2 = 225$

48. πab

50. $4\pi ab^2/3,\ 4\pi a^2 b/3$

54. In Exercise 15, focus is $(0, -7/8)$, directrix is $y = -9/8$; in Exercise 18, focus is $(-7/48, 0)$, directrix is $x = -25/48$; in Exercise 21, focus is $(3/4, 0)$, directrix is $x = 5/4$; in Exercise 22, focus is $(6, 3/4)$, directrix is $x = 25/4$; in Exercise 27, focus is $(3, -1/4)$, directrix is $y = 1/4$; in Exercise 32, focus is $(-1/4, -4)$, directrix is $x = 1/4$.

Exercises 10.6

2.

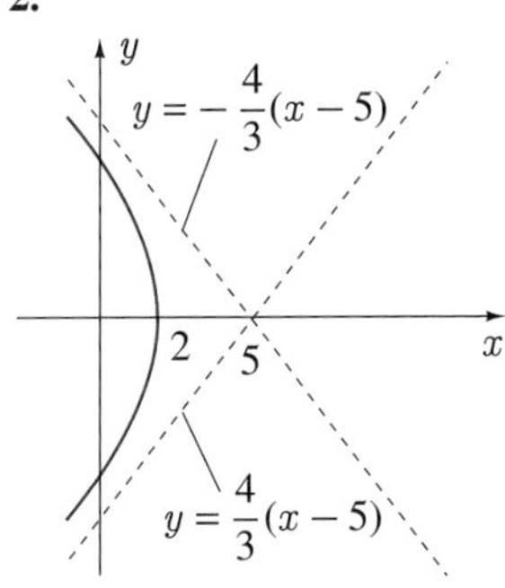

$$r = \frac{16}{3 + 5\cos\theta}$$

4.

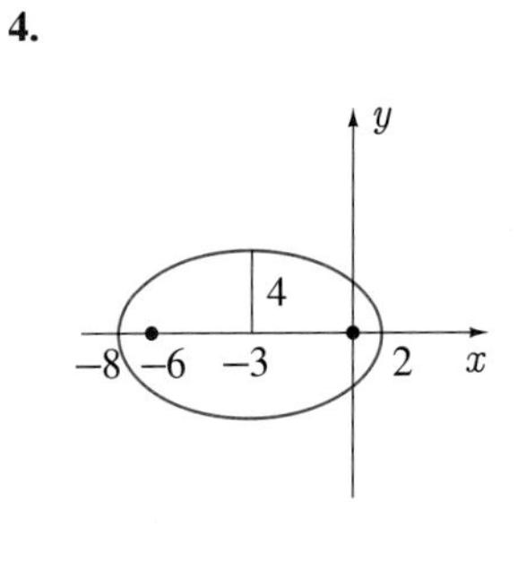

$$r = \frac{16}{5 + 3\cos\theta}$$

6.

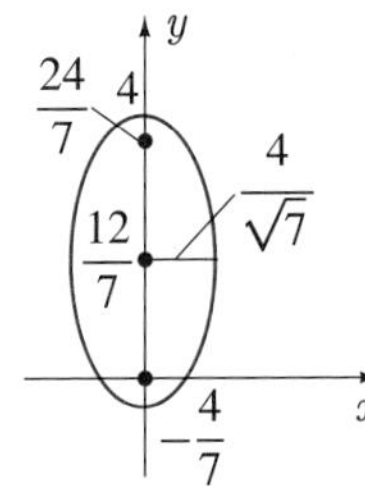

$$r = \frac{4}{4 - 3\sin\theta}$$

8. **10.**

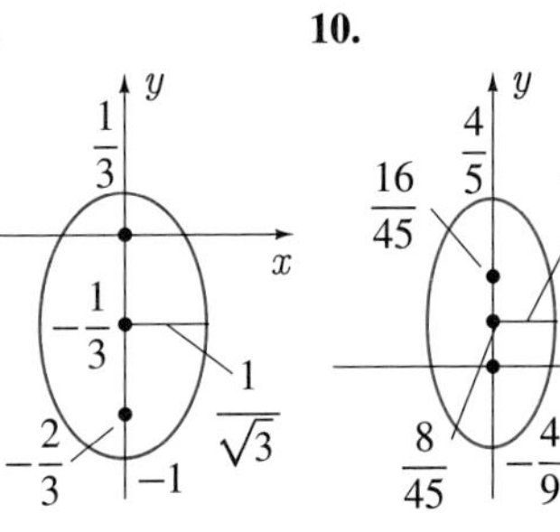

$$r = \frac{1}{2 + \sin\theta}$$

$$r = \frac{4\csc\theta}{7\csc\theta - 2}$$

12. $8x^2 + 2x + 9y^2 = 1$

14. $8x^2 + 12x - y^2 + 4 = 0$

16. $40x + 25y^2 = 16$

18. With $x = \sqrt{5} + r\cos\theta$, $y = r\sin\theta$, the equation becomes $r = 4/(3 + \sqrt{5}\cos\theta)$.

20. With $x = \sqrt{10} + r\cos\theta$, $y = 2 + r\sin\theta$, the equation becomes $r = 1/(\pm 3 - \sqrt{10}\cos\theta)$.

22. $\epsilon \to 0$

Review Exercises

2.

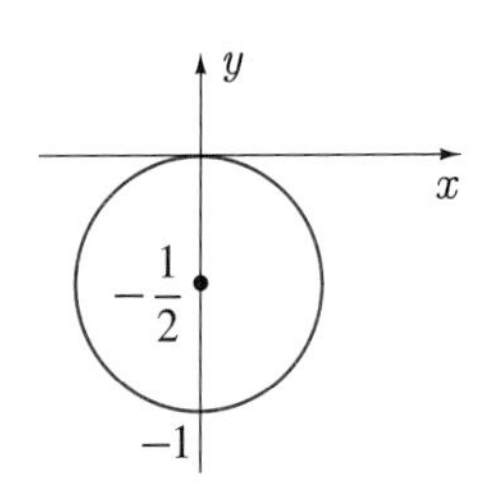

$r = -\sin\theta$

4.

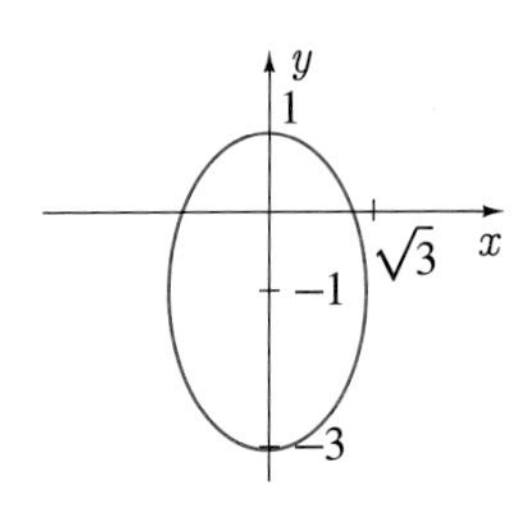

$$r = \frac{3}{2 + \sin\theta}$$

6.

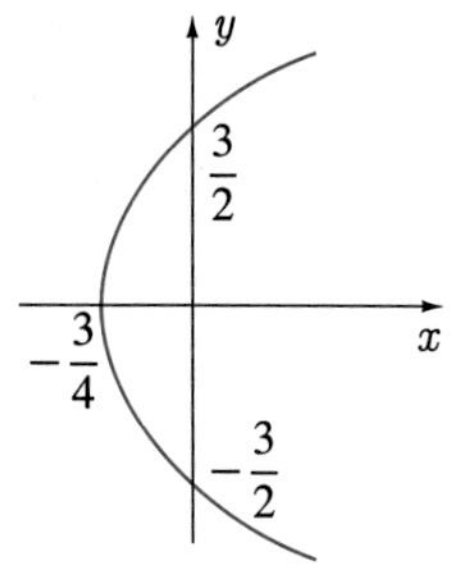

$r = \dfrac{3}{2 - 2\cos\theta}$

8.

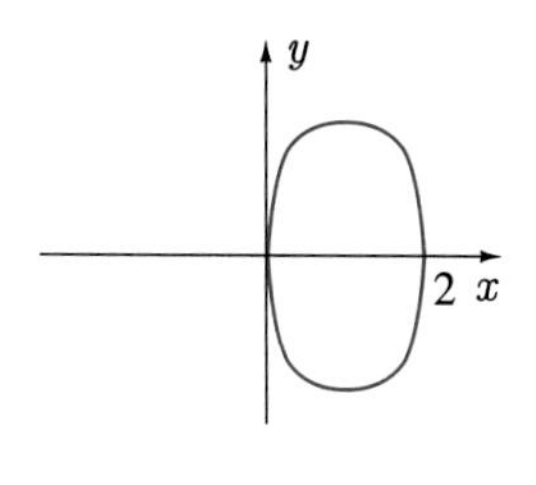

$r^2 = 4\cos\theta$

10.

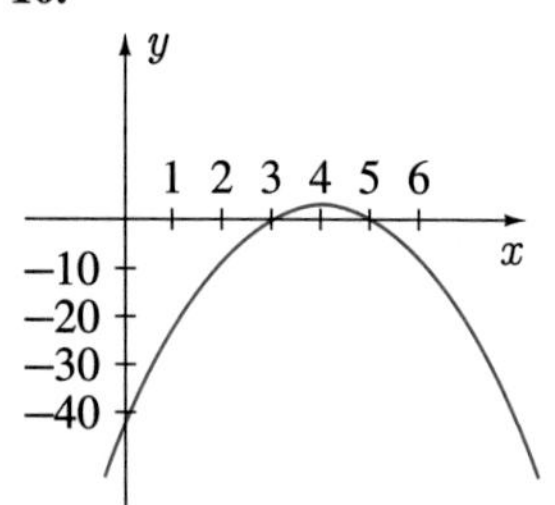

$x = 4 + t$
$y = 5 - 3t^2$

12.

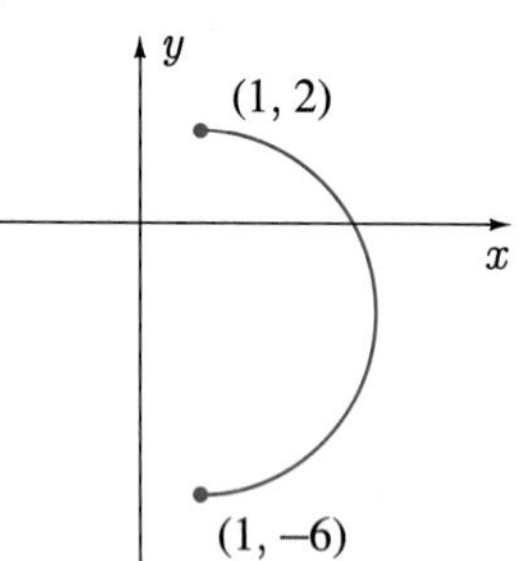

$x = 1 + 4\sin t$
$y = -2 + 4\cos t$, $0 \leq t \leq \pi$

14.

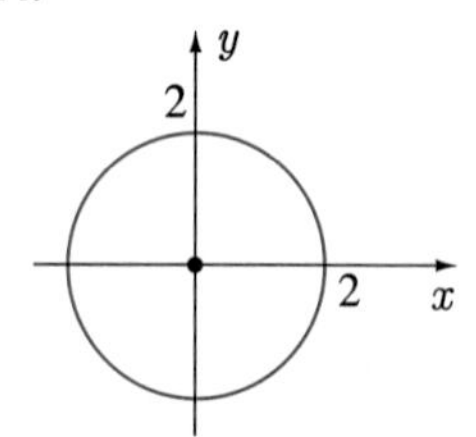

$x^2 + y^2 = 2\sqrt{x^2 + y^2}$

16.

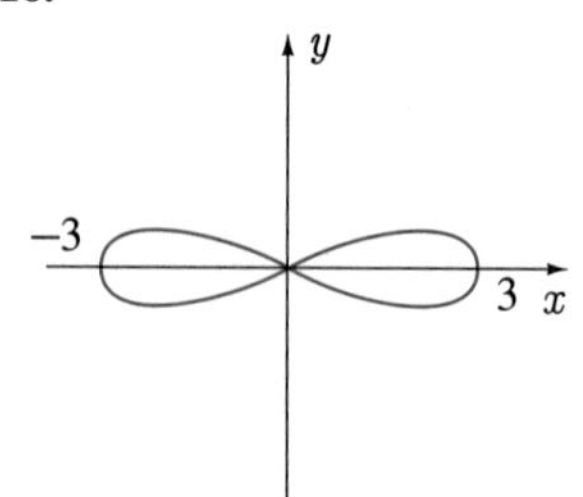

$r = 3\cos 2\theta$

18.

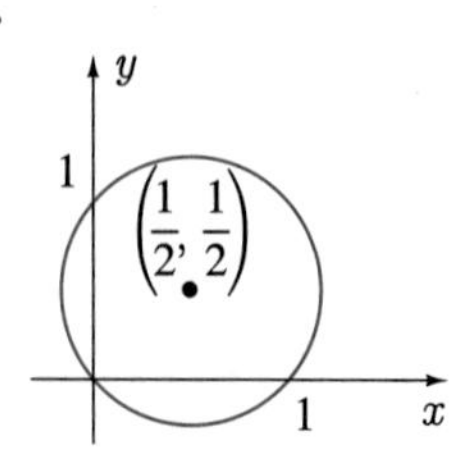

$x^2 + y^2 = x + y$

20.

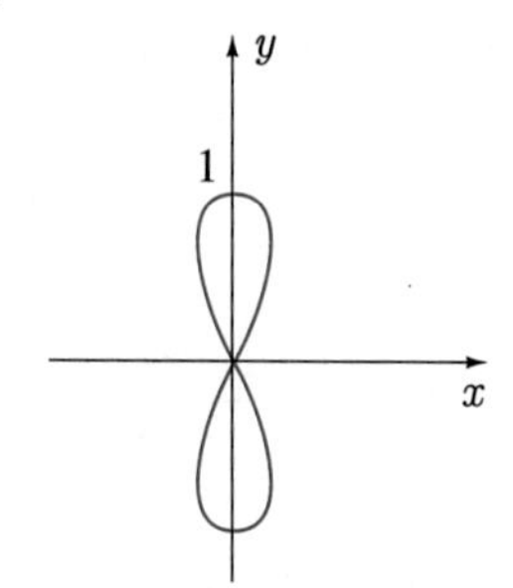

$r = \sin^2\theta - \cos^2\theta$

22.

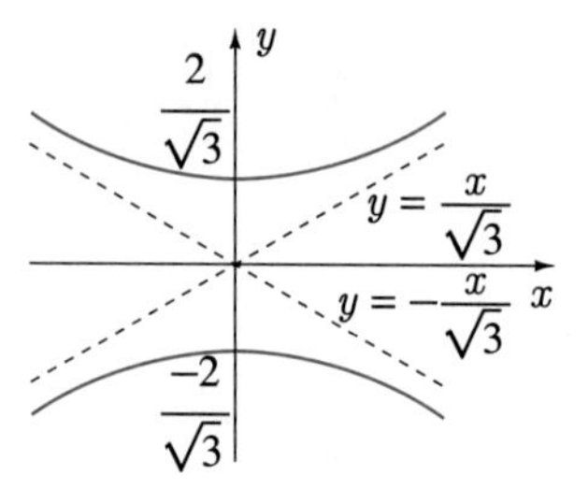

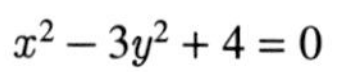

$x^2 - 3y^2 + 4 = 0$

24.

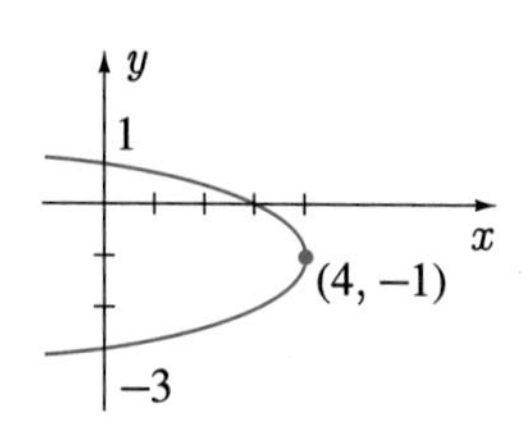

$y^2 + x + 2y = 3$

26.

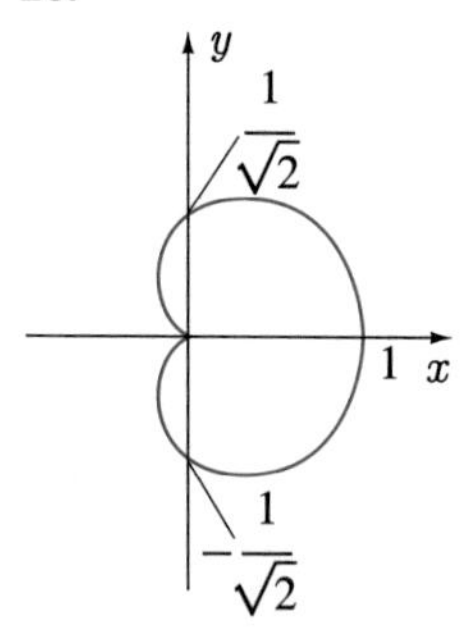

$y = \cos(\theta/2)$

28.

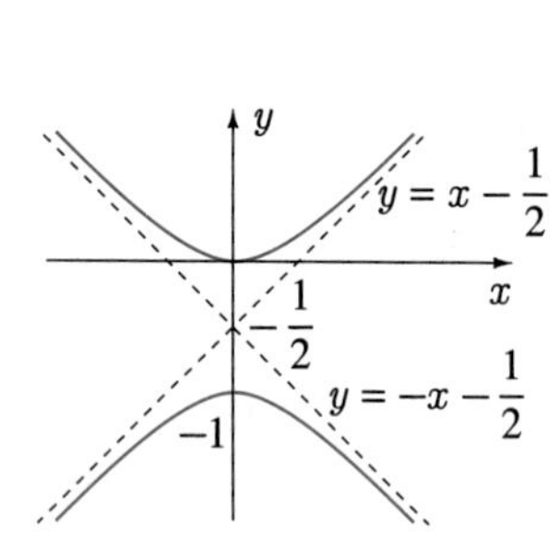

$x^2 = y^2 + y$

30.

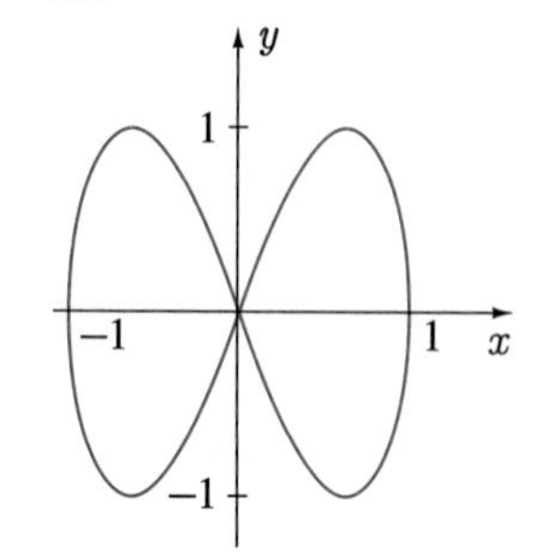

$x = \sin\theta$
$y = \sin 2\theta$

32. 12π

34. $3\pi/8$

36. $(3 - 3t^2)/(2 + 3t^2)$, $-30t/(2 + 3t^2)^3$

38. 16

40. $\sqrt{2}(e^{\pi/2} - 1)$

CHAPTER 11

Exercises 11.1

2. Increasing, $V = 4$

4. Decreasing, $U = 9/16$, $V = 0$, $L = 0$

6. Not monotonic, $U = 1$, $V = -1$, $L = 0$

8. Decreasing, $U = 1/4$, $V = 0$, $L = 0$

10. Increasing, $U = \pi/2$, $V = \pi/4$, $L = \pi/2$

12. Not monotonic

14. Not monotonic, $U = 4$, $V = -4$, $L = 0$

16. False $\{n\}$ **18.** False $\{-n\}$ **20.** True

22. True **24.** True **26.** False $\{(-1)^n/n\}$

28. False $\{(-1)^n/n\}$ **30.** False $\{(-1)^n\}$

32. False $\{-3^n\}$ **34.** True

36. Not monotonic, $U = 9/4$, $V = -7$, $L = 0$

38. Decreasing, $U = 1/e$, $V = 0$, $L = 0$

40. Increasing, $U = \pi/2$, $V = 0$, $L = \pi/2$

42. Increasing, $U = 1$, $V = 0$, $L = 0.419241$

44. Increasing, $U = 5$, $V = 0$, $L = (\sqrt{21} + 1)/2$

46. Increasing, $U = 10$, $V = 3$, $L = (3 + \sqrt{29})/2$

48. Decreasing, $U = 4$, $V = 0$, $L = (7 - \sqrt{5})/2$

50. $(3n + 1)/n^2$ **52.** $[1 + (-1)^{n+1}]/2$

54. 1 **56.** 0 **58.** 4

60. Decreasing, $U = 2$, $V = 0$, $L = (3 - \sqrt{5})/2$

62. Decreasing, $U = 1$, $V = 0$, $L = 1/2$

64. Decreasing, $U = 1$, $V = 0$, $L = 0$

66. Decreasing, $U = 2$, $V = 0$, $L = 0$

68. Increasing, $U = 1$, $V = 0$, $L = (3 - \sqrt{3})/2$

72. $V = 1/2$, $U = 1$

76. **(a)** 1, 1, 2, 3, 5, 8, 13, 21, 34, 55 **(b)** Increasing, $V = 1$, no upper bound, no limit **(e)** $(\sqrt{5} + 1)/2$

80. $[5 + (-1)^n/2^{n-2}]/3$

Exercises 11.2

2. $-0.381\,966\,0$ **4.** $-2.618\,034\,0$ **6.** $3.044\,723\,1$

8. $1.214\,648\,0$ **10.** $0.334\,734\,1$ **12.** $-1.388\,792\,0$

14. **(a)** $40(0.99)^n$ m **(b)** $4(0.981)^{-1/2}(0.99)^{n/2}$ s

16. $4P/3$, $16P/9$, $(4/3)^n P$, does not exist

18. 0.0625, 0.1125

20. $(\sqrt{3}P^2/36)[1 + (1/3) + (4/3^3) + (4^2/3^5) + \cdots + (4^{n-1}/3^{2n-1})]$

22. -0.4814 **24.** 2.9122 **26.** 3.3247

28. 0.7849 **30.** 0.7953

Exercises 11.3

2. $2/15$ **4.** Diverges **6.** $10\,804.5$

8. $-1/3$ **10.** Diverges **12.** $13/99$

14. $430\,162/9999$ **16.** Diverges **18.** 4

20. Diverges **22.** 1 **24.** 804 s

26. $(\sqrt{3}P^2/180)[8 - 3(4/9)^n]$, $2\sqrt{3}P^2/45$

28. 2 **30.** $51/64$ **32.** $2/3$

36. $600/11$ min after10:00 **38.** **(c)** $A_0 e^{kT}/(e^{kT} - 1)$

Exercises 11.4

2. Diverges **4.** Converges **6.** Converges

8. Diverges **10.** Converges **12.** Converges

14. Converges **16.** Converges **18.** Diverges

20. Diverges **22.** Diverges

24. Converges for $p > 1$, diverges for $p \leq 1$

Exercises 11.5

2. Converges **4.** Converges **6.** Diverges

8. Diverges **10.** Converges **12.** Converges

14. Converges **16.** Diverges **18.** Converges

20. Converges

Exercises 11.6

2. Converges conditionally **4.** Converges absolutely

6. Diverges **8.** Converges absolutely

10. Converges conditionally **12.** Converges conditionally

14. Converges absolutely

Exercises 11.7

2. -0.9470 **4.** -0.0127 **6.** 1.07

8. -0.7 **10.** 4

Exercises 11.8

2. $-1 < x < 1$ **4.** $-1/3 < x < 1/3$

6. $-4 < x < -2$ **8.** $7/2 < x < 9/2$

10. $-1 < x < 1$ **12.** $-1 \leq x < 1$

14. $-1/e \leq x \leq 1/e$ **16.** $x = 0$

18. $-1/3 < x < 1/3$ **20.** $-5^{1/3} < x < 5^{1/3}$

22. $-1 \leq x \leq 1$ **24.** 1

26. $4/(4 - x^3)$, $|x| < 4^{1/3}$

28. $(x - 1)/(10 - x)$, $-8 < x < 10$

30. **(b)** $-\infty < x < \infty$

Exercises 11.9

2. $\sum_{n=0}^{\infty} (5^n/n!)x^n$, $-\infty < x < \infty$

4. $(1/\sqrt{2})[1 + (x - \pi/4) - (x - \pi/4)^2/2! - (x - \pi/4)^3/3! + (x - \pi/4)^4/4! + \cdots]$, $-\infty < x < \infty$

6. $1/8 - (3/8^2)(x - 2) + (3^2/8^3)(x - 2)^3 - (3^3/8^4)(x - 2)^4 + \cdots$, $-2/3 < x < 14/3$

12. **(d)** 0 **(e)** $x = 0$

Exercises 11.10

2. $\sum_{n=0}^{\infty} [(-1)^n/4^{n+1}]x^{2n}$, $|x| < 2$

4. $\sum_{n=0}^{\infty} [(-1)^n/(2n)!]x^{4n}$, $-\infty < x < \infty$

6. $\sum_{n=0}^{\infty} (5^n/n!)x^n$, $-\infty < x < \infty$

8. $\sum_{n=0}^{\infty} [1/(2n+1)!]x^{2n+1}$, $-\infty < x < \infty$

10. $1 + 9x/2 + 27x^2/8 + \sum_{n=3}^{\infty} \{(-1)^n(2n-5)!3^{n+1}/[2^{2n-3}n!(n-3)!]\}x^n$, $-1/2 \le x \le 1/3$

12. $x^4 + 3x^2 - 2x + 1$

14. $33 - 46(x+2) + 27(x+2)^2 - 8(x+2)^3 + (x+2)^4$

16. $\sum_{n=0}^{\infty} (e^n/n!)(x-3)^n$, $-\infty < x < \infty$

18. $\sqrt{3}(1 + x/6) + \sum_{n=2}^{\infty} \{4\sqrt{3}(-1)^{n-1}(2n-3)!/[12^n n!(n-2)!]\}x^n$, $|x| \le 3$

20. $x - x^2/3 - \sum_{n=3}^{\infty} \{(2)(5)\cdots(3n-7)/[3^{n-1}(n-1)!]\}x^n$, $|x| \le 1$

22. $1 + x^2/2 + 5x^4/24 + 61x^6/720$

24. $1 + \sum_{n=1}^{\infty} [(-1)^n 2^{2n-1}/(2n)!]x^{2n}$, $-\infty < x < \infty$

26. $x^2 + \sum_{n=1}^{\infty} \{(2n)!/[(2n+1)2^{2n}(n!)^2]\}x^{4n+2}$, $|x| < 1$

28. $\sqrt{2}\sum_{n=0}^{\infty} \{1/[2^n(2n+1)]\}x^{2n+1}$, $|x| < \sqrt{2}$

30. **(a)** $1/p$ **(b)** 6

32. $\sum_{n=0}^{\infty} \{(-1)^n\pi^{2n}/[(4n+1)2^{2n}(2n)!]\}x^{4n+1}$, $-\infty < x < \infty$; $\sum_{n=0}^{\infty} \{(-1)^n\pi^{2n+1}/[(4n+3)2^{2n+1}(2n+1)!]\}x^{4n+3}$, $-\infty < x < \infty$

36. **(a)** $-1/2$, $1/6$, 0, $-1/30$, 0

Exercises 11.11

2. $2/(1-x)^3$, $|x| < 1$ **4.** $(x+1)/(1-x)^3$, $|x| < 1$

6. $-(1/x)\ln(1-x)$, $-1 \le x < 1$

8. $-\ln(1+x^2)$, $-1 \le x \le 1$

10. $1/(x-x^2) - (1/x^2)\ln(1-x)$, $|x| < 1$

22. $-9/100$

Exercises 11.12

2. 4.2×10^{-10} **4.** 4.2×10^{-10} **6.** 1.4×10^{-9}

8. 0.115 **10.** 0.006 25 **12.** 0.497

14. 0.133 **16.** 0.291 **18.** -0.122

20. $1/2$ **22.** $1/2$ **24.** 0

26. $1 - x^3/2 + 3x^6/8 - 5x^9/16$ **28.** $1 - x^2$

30. $a_0 + a_1\sum_{n=1}^{\infty} [(-1)^{n+1}/n!]x^n = C + De^{-x}$

32. $a_0\sum_{n=0}^{\infty} [(-1)^n/(2n)!]x^n$

34. $a_1\sum_{n=1}^{\infty} \{(-1)^{n+1}/[n!(n-1)!]\}x^n$

38. 2.44

Review Exercises

2. Decreasing, $U = 1$, $V = 0$, $L = 1/\sqrt{3}$

4. Increasing, $U = 100$, $V = 7$, $L = (31 + \sqrt{53})/2$

6. $|k| \le 1$ **10.** Converges **12.** Converges

14. Converges conditionally

16. Diverges **18.** Converges **20.** Converges

22. Diverges **24.** Converges **26.** Converges

28. Converges conditionally **30.** $-2 \le x \le 2$

32. $-\infty < x < \infty$ **34.** $-4 < x < -2$

36. $-2^{-1/3} \le x < 2^{-1/3}$

38. $e^5\sum_{n=0}^{\infty} (1/n!)x^n$, $-\infty < x < \infty$

40. $\sum_{n=2}^{\infty} [(-1)^n 2^{n-1}/(n-1)]x^n$, $-1/2 < x \le 1/2$

42. $(1/2)\sum_{n=0}^{\infty} (-1)^{n+1}(1 - 1/3^n)x^n$, $|x| < 1$

44. $x + \sum_{n=2}^{\infty} \{(-1)^n/[n(n-1)]\}x^n$, $-1 < x \le 1$

46. $\sum_{n=0}^{18} [(-1)^n/n!]x^{2n}$

48. $\sum_{n=0}^{\infty} \{(-1)^n/[2^{2n}(2n)!]\}x^{2n} + \sum_{n=0}^{\infty} \{(-1)^n/[2^{2n+1}(2n+1)!]\}x^{2n+1}$

CHAPTER 12

Exercises 12.1

2. $\sqrt{33}$

4. $(0,0,0)$, $(2,0,0)$, $(0,2,0)$, $(0,0,2)$, $(2,2,0)$, $(2,0,2)$, $(0,2,2)$, $(2,2,2)$

6. $\sqrt{29}$, 5, $2\sqrt{5}$, $\sqrt{13}$ **8.** 5, 3, 4, 5

12. $10x + 2y + 2z = 5$, plane

14. $(\sqrt{2}, \sqrt{2}, 5)$, $(\sqrt{2} \pm 1/4, \sqrt{2} \pm 1/4, 5 - \sqrt{7}/4)$, $(\sqrt{2} \pm 1/4, \sqrt{2} \pm 1/4, 9/2 - \sqrt{7}/4)$

16. **(a)** $(2, 1/2, -7/2)$ **(b)** $(5, 5, -5)$

20. $(11, 1, 7/5)$

Exercises 12.2

2.

z

$2x - 3y = 0$

y

x

4.

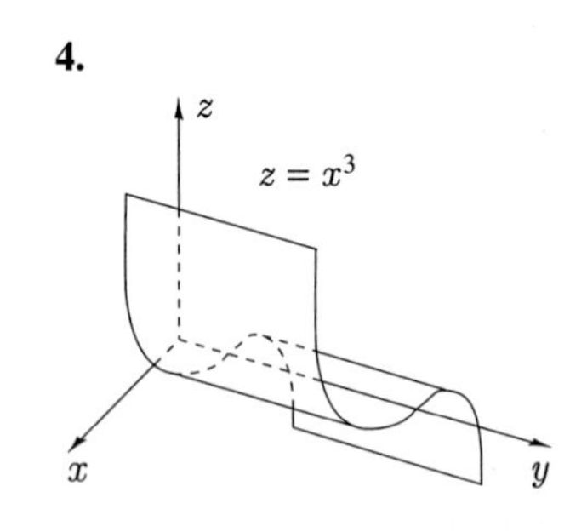

6.

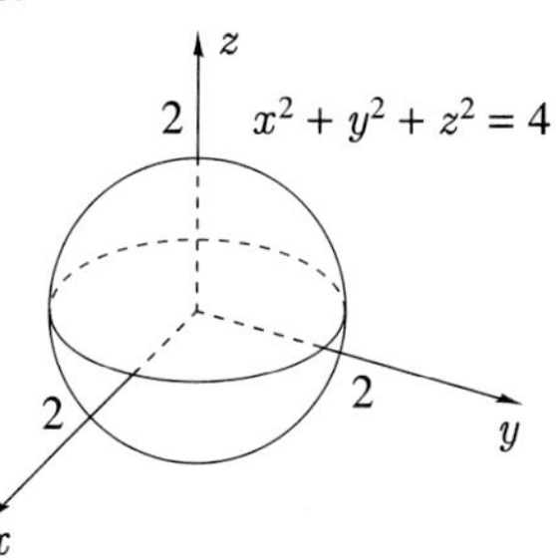

8.

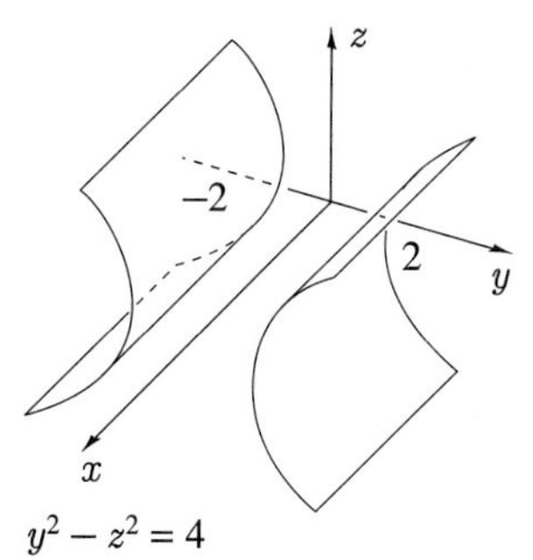

10.

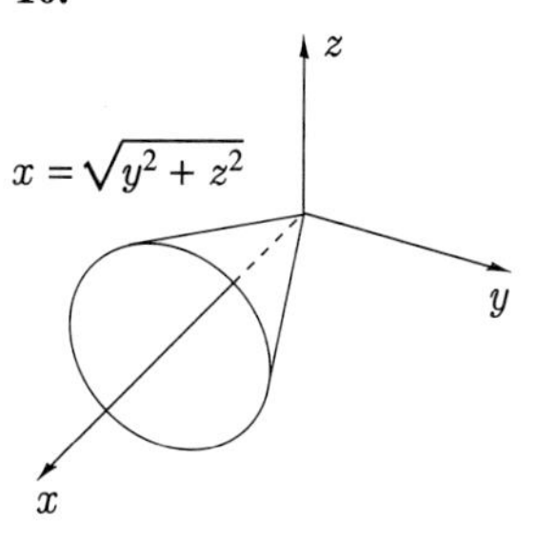

12.

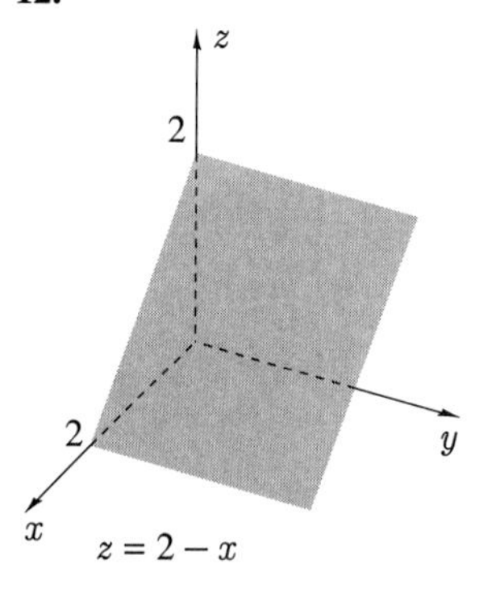

14.

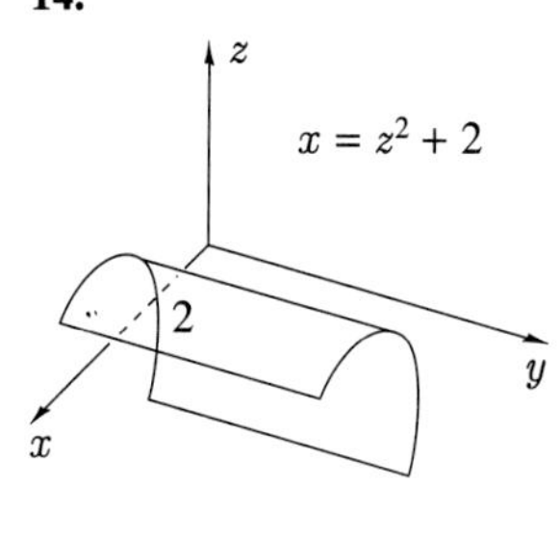

16.

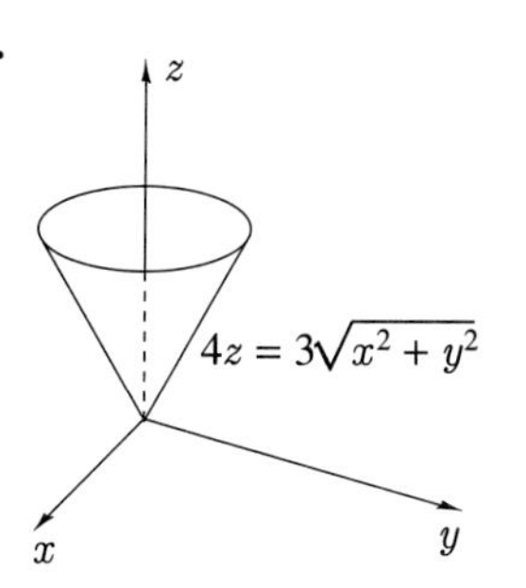

18.

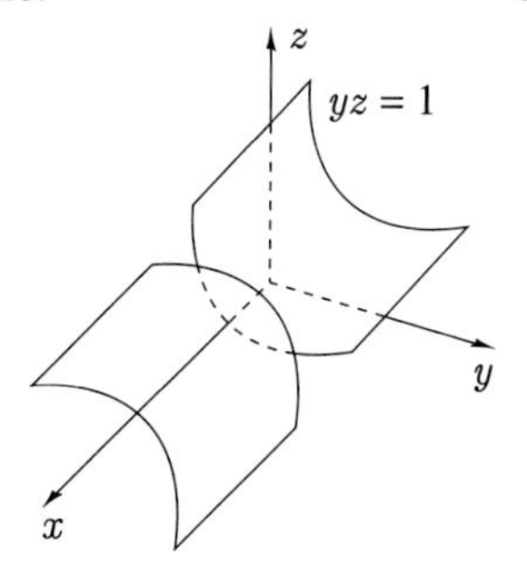

20.

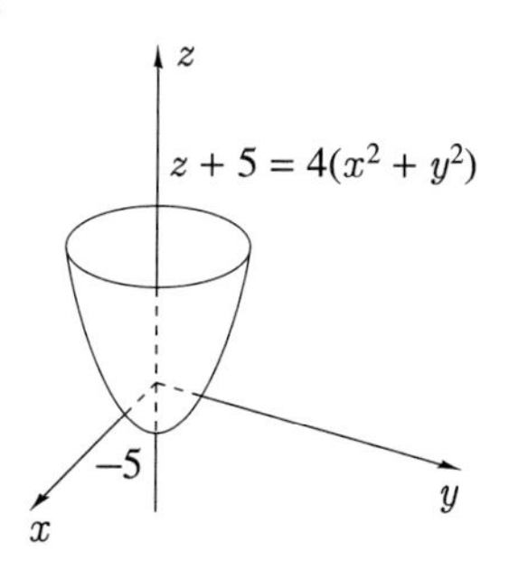

22.

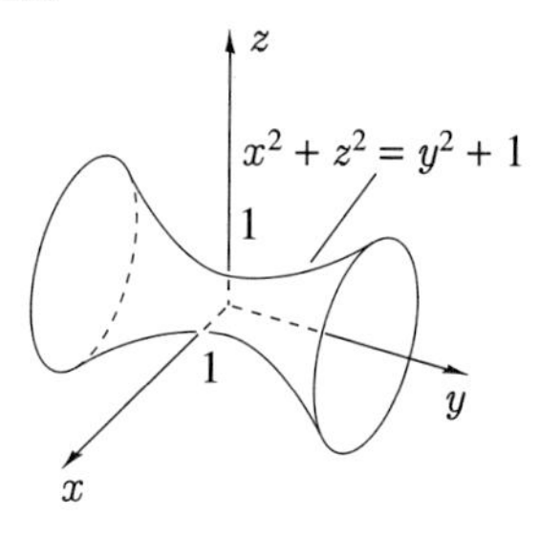

24.

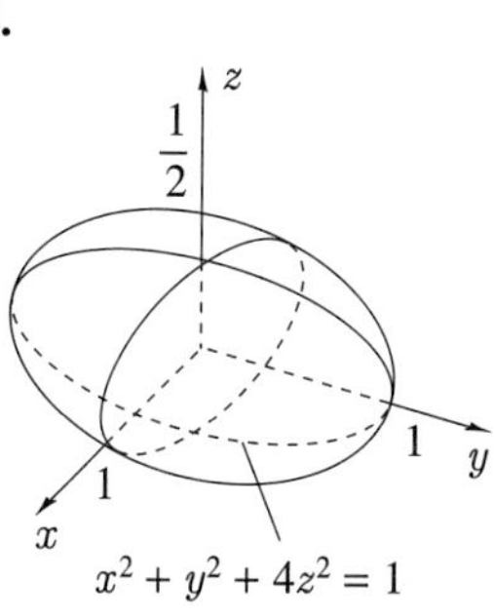

26.

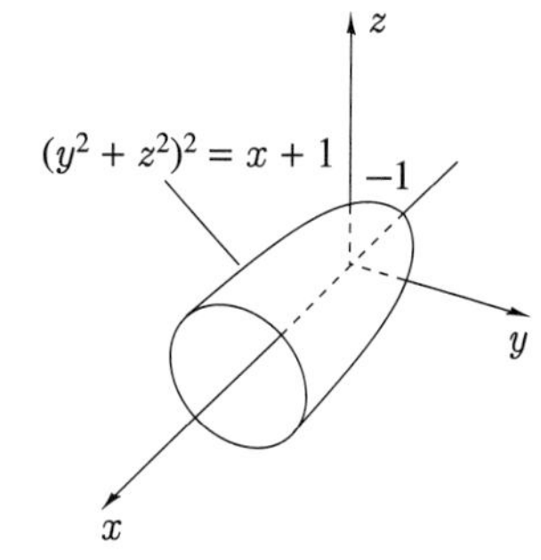

28.

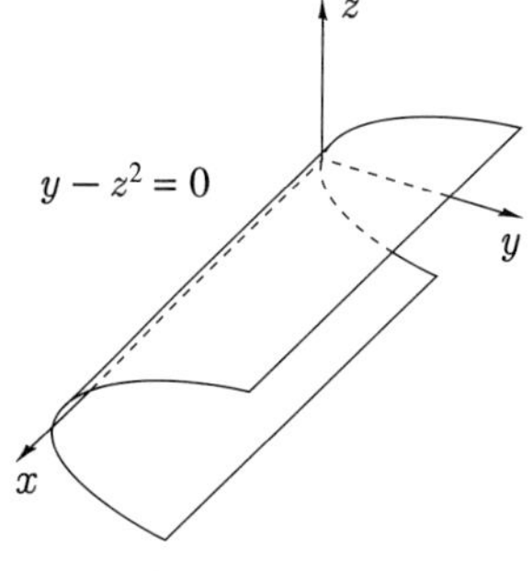

30.

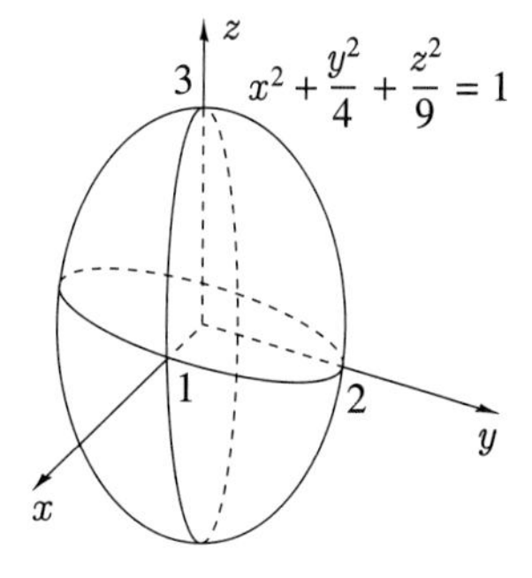

32.

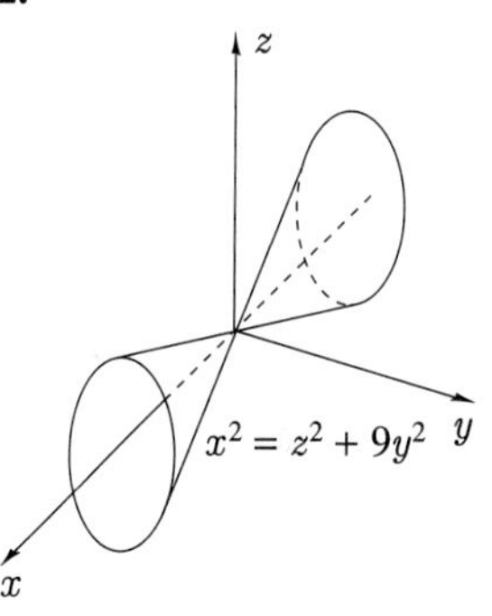

34.

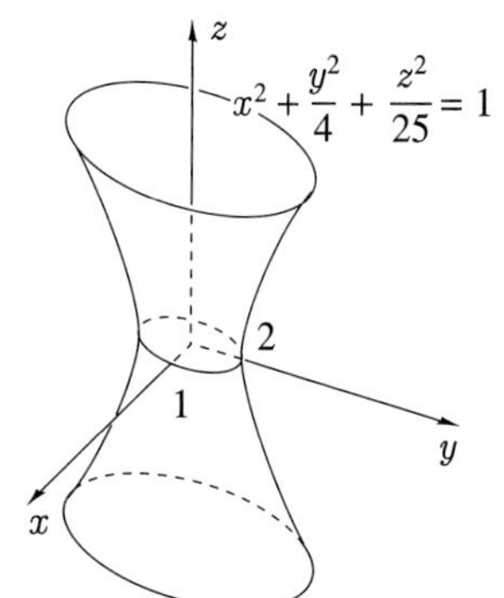

36.

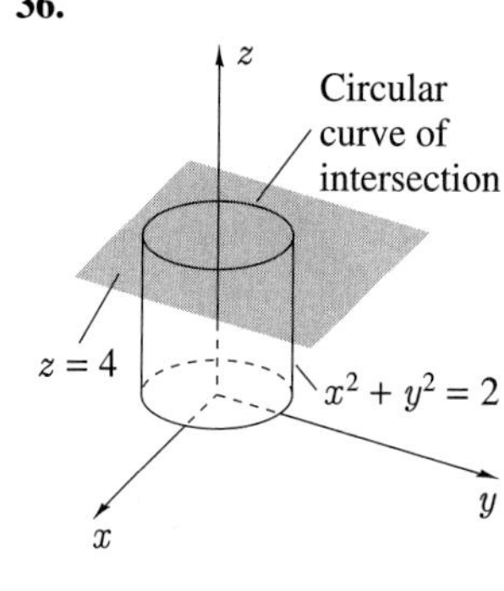

38. **40.**

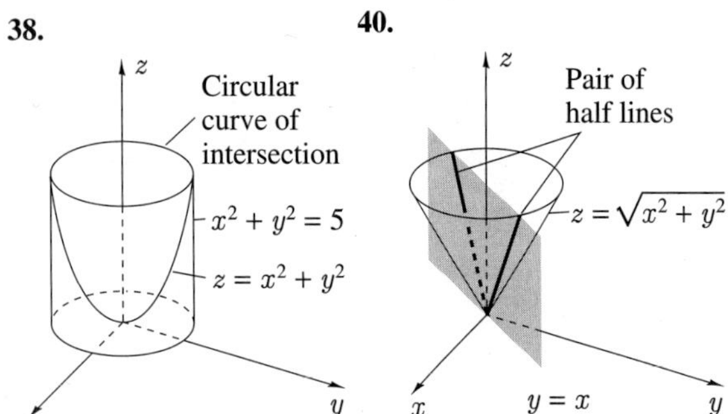

42. **44.**

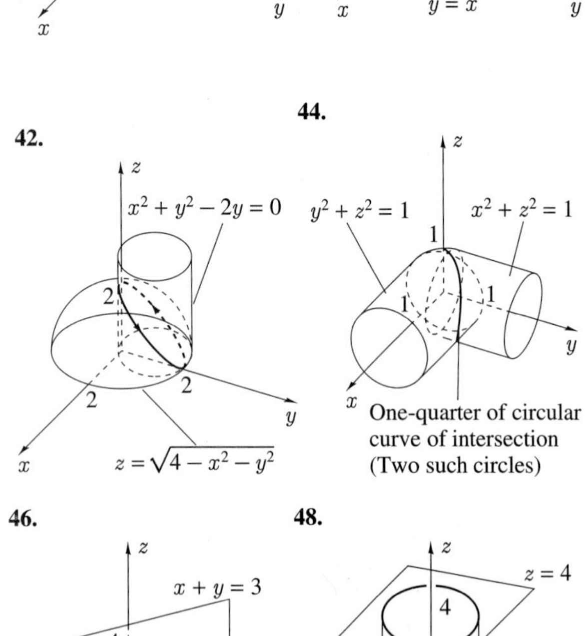

46. **48.**

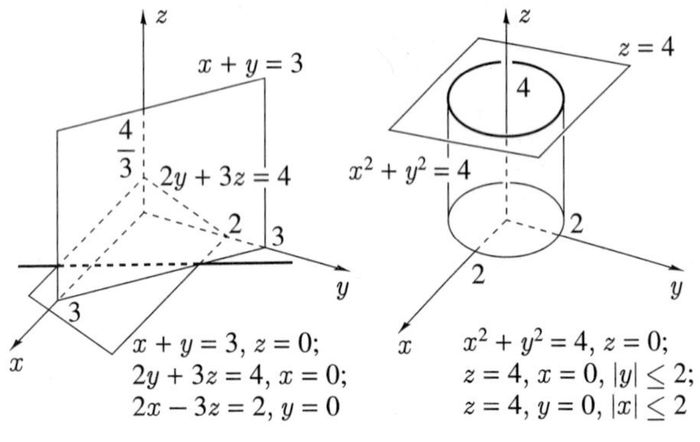

50. **52.**

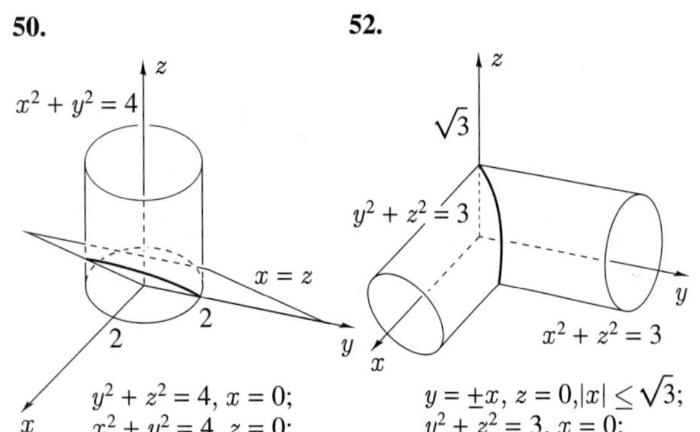

54. **56.**

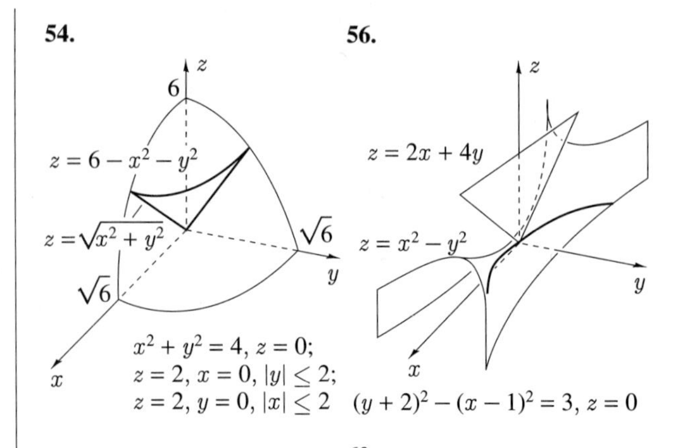

58. **60.**

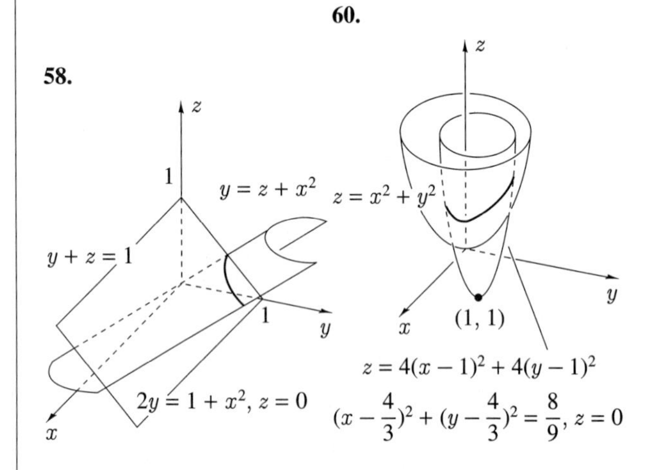

62. **64.**

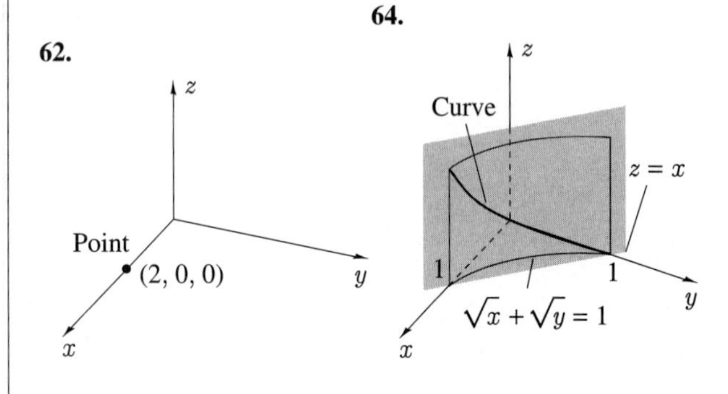

66. **68.**

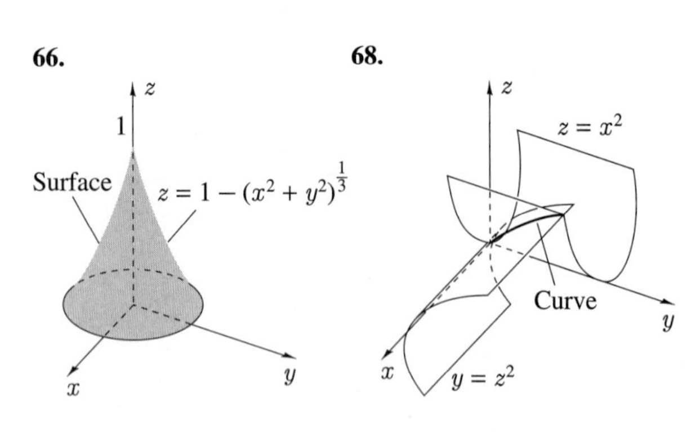

70.

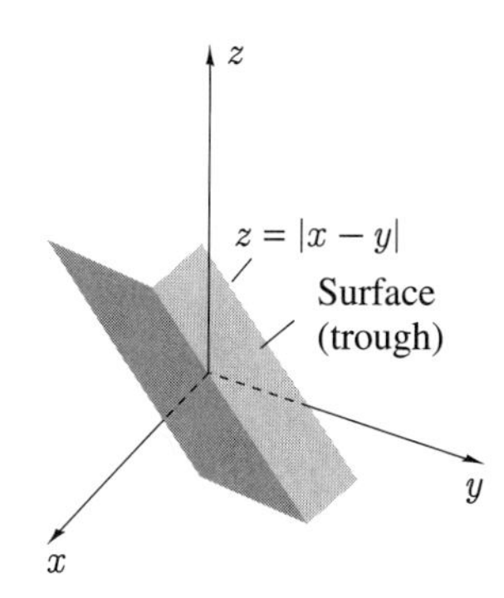

Exercises 12.3

2. $(2, 6, 8)$ **4.** $(-1/\sqrt{5}, 0, 2/\sqrt{5})$
6. $(-4\sqrt{5} - 8, -6, 8\sqrt{5} + 4)$
8. $(-6\sqrt{46} - 4\sqrt{5}, -12\sqrt{5}, 12\sqrt{46} - 24\sqrt{5})$
10. $(-6/\sqrt{17}, -3/\sqrt{17}, 6/\sqrt{17})$
12. $3\hat{\imath} - 2\hat{\jmath}$
14. $(2/\sqrt{5} - 1/\sqrt{10})\hat{\imath} + (1/\sqrt{5} + 3/\sqrt{10})\hat{\jmath}$
16. $(5, 0, 0)$ **18.** $(0, 3/\sqrt{5}, 6/\sqrt{5})$
20. $(1, 1, -3/2)$ **22.** $(5/4, 5\sqrt{2}/4, 5/4)$
24. $(2/\sqrt{3}, 2/\sqrt{3}, 2/\sqrt{3})$ **30.** $(3/\sqrt{17}, 12/\sqrt{17})$
32. $5G(-3/\sqrt{10} + 30\sqrt{11}/121, -10\sqrt{11}/121, -1/\sqrt{10} + 10\sqrt{11}/121)$ N
34. Linearly dependent **36.** Linearly dependent
42. $k_1(1 - l/\sqrt{x^2 + l^2})(-x, l) + k_2[1 - l/\sqrt{(L - x)^2 + l^2}](L - x, l)$

Exercises 12.4

2. $(-10, 15, -5)$ **4.** $4/\sqrt{14}$ **6.** -178
8. 1 **10.** 0 **12.** Yes
14. Yes **16.** 0.68 **18.** 2.20
20. π **22.** $2x + y - 2z + 5 = 0$
24. $6/\sqrt{13}$ **26.** $4/\sqrt{6}$ **28.** $23\sqrt{14}/28$
34. **(a)** 1.30, 1.01, 2.50 **(b)** $\pi/2$, 1.25, 2.82 **(c)** 1.73, 1.89, 0.36
36. $3/\sqrt{2}$, $1/\sqrt{2}$ **38.** $\sqrt{5}$, $-\sqrt{14}/2$, $-\sqrt{70}/2$
40. $-3/10$, $11/20$ **42.** $x - 2y + 3z = 6 \pm 2\sqrt{14}$
44. $k[1 + l(1 - \sqrt{5})]/2$ J **46.** $(\sqrt{2} - 1)GMm/(\sqrt{2}R)$

Exercises 12.5

2. $(48, 24, -42)$ **4.** $(2\sqrt{247})^{-1}(17, -23, -7)$
6. $(-41, 11, 28)$ **8.** $(17, -23, -7)$ **10.** $(-13, 19, 5)$
12. $\lambda(9, 0, 1)$ **14.** $\lambda(3, -1, 17)$
18. $(x, y, z) = (1, -1, 3) + t(2, 4, -3)$; $x = 1 + 2t$, $y = -1 + 4t$, $z = 3 - 3t$; $(x - 1)/2 = (y + 1)/4 = (z - 3)/(-3)$
20. $(x, y, z) = (2, -3, 4) + t(3, 5, -5)$; $x = 2 + 3t$, $y = -3 + 5t$, $z = 4 - 5t$; $(x - 2)/3 = (y + 3)/5 = (z - 4)/(-5)$
22. $(x, y, z) = (1, 3, 4) + t(0, 0, 1)$; $x = 1$, $y = 3$, $z = 4 + t$
24. $(x, y, z) = (2, 0, 3) + t(1, 0, -2)$; $x = 2 + t$, $y = 0$, $z = 3 - 2t$; $x - 2 = (z - 3)/(-2)$, $y = 0$
26. $(x, y, z) = (0, -5, 30) + t(1, 2, -11)$; $x = t$, $y = -5 + 2t$, $z = 30 - 11t$; $x = (y + 5)/2 = (z - 30)/(-11)$
28. $x + 3y - 3z = 4$ **30.** $10x + 4y - 3z = 3$
32. $8x + 13y + 9z = 18$ **34.** $7x + 6y - 5z + 34 = 0$
36. No
40. $x = \sqrt{2} \pm 1/4$, $y = \sqrt{2} \pm 1/4$, $z = 9/2 - \sqrt{7}/4$, $\sqrt{7}x - z = \sqrt{14} - 5$, $\sqrt{7}x + z = \sqrt{14} + 5$, $\sqrt{7}y - z = \sqrt{14} - 5$, $\sqrt{7}y + z = \sqrt{14} + 5$
42. $\sqrt{154}$ **44.** $\sqrt{73/2}$ **46.** $2/\sqrt{6}$
48. $(-10, 5, -2)$ **50.** $(-5, 1, 4)$

Exercises 12.6

2. $-\infty < t < \infty$ **4.** $t > -4$ **6.** $\hat{\imath} - 2t\hat{\jmath} + 2\hat{\mathbf{k}}$
8. $2t(4t^2 - 3)\hat{\imath} + t^2(9 - 10t^2)\hat{\jmath} + 4t(4t^2 - 3)\hat{\mathbf{k}}$
10. $3t^2(4 - 5t^2)\hat{\imath} + 4t(1 - 3t^2)\hat{\jmath} - 3t^2\hat{\mathbf{k}}$
12. $3\hat{\imath} - 2(3t + 4)\hat{\jmath} + 6(1 + 4t)\hat{\mathbf{k}}$
14. $9t^2\hat{\imath} + (6t - 20t^3)\hat{\jmath} + (30t^4 - 21t^2 + 6)\hat{\mathbf{k}}$
16. $(-15t^4 + 12t^2)\hat{\imath} + (4t - 12t^3)\hat{\jmath} - 3t^2\hat{\mathbf{k}}$
18. $(3t^4/2 - 4t^2)\hat{\imath} + (17t^3/3 - 12t^5/5)\hat{\jmath} + (3t^6 - 27t^4/4 + t^2)\hat{\mathbf{k}} + \mathbf{C}$
20. $(-14t^4 - 6t^3 - 42t^2 + 4t)\hat{\imath} + (28t^5 + 84t^3 - 6t^2)\hat{\jmath} - (126t^4 + 42t^6 + 2t)\hat{\mathbf{k}}$

Exercises 12.7

2. **4.**

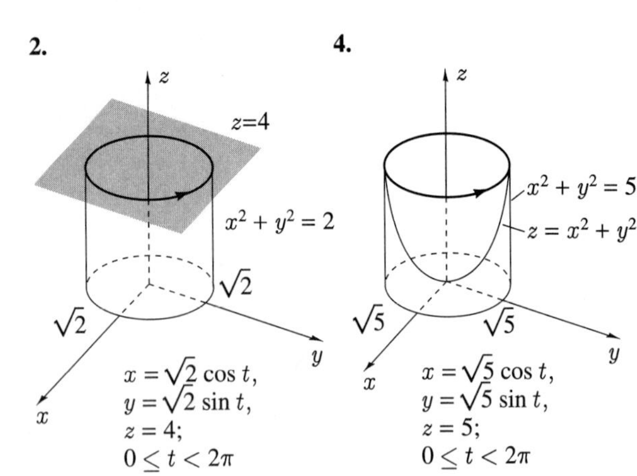

$x=\sqrt{2}\cos t$, $y=\sqrt{2}\sin t$, $z=4$; $0\le t<2\pi$

$\mathbf{r}=\sqrt{2}\cos t\hat{\mathbf{\imath}}+\sqrt{2}\sin t\hat{\mathbf{\jmath}}+4\hat{\mathbf{k}}$, $0\le t<2\pi$

$x=\sqrt{5}\cos t$, $y=\sqrt{5}\sin t$, $z=5$; $0\le t<2\pi$

$\mathbf{r}=\sqrt{5}\cos t\,\hat{\mathbf{\imath}}+\sqrt{5}\sin t\,\hat{\mathbf{\jmath}}+5\hat{\mathbf{k}}$, $0\le t<2\pi$

6.

$z=\sqrt{x^2+y^2}$, $y=x$

$x=t$, $y=t$, $z=\sqrt{2}\,|t|$

$\mathbf{r}=t\hat{\mathbf{\imath}}+t\hat{\mathbf{\jmath}}+\sqrt{2}\,|t|\hat{\mathbf{k}}$

8.

$z=\sqrt{4-x^2-y^2}$

$x^2+y^2-2y=0$

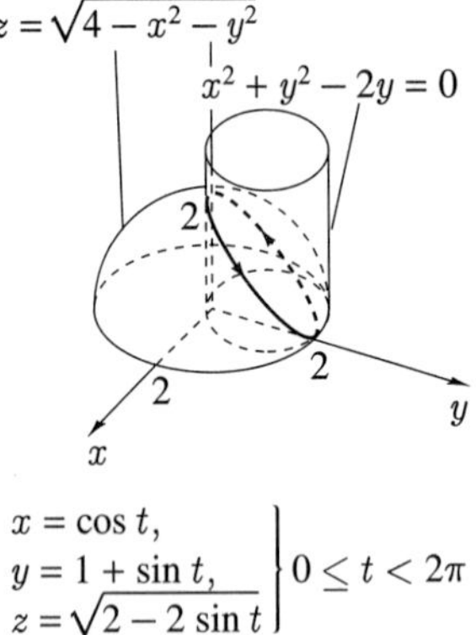

$x=\cos t$, $y=1+\sin t$, $z=\sqrt{2-2\sin t}$ $\Big\}\ 0\le t<2\pi$

$\mathbf{r}=\cos t\hat{\mathbf{\imath}}+(1+\sin t)\hat{\mathbf{\jmath}}+\sqrt{2-2\sin t}\hat{\mathbf{k}}$, $0\le t<2\pi$

10. **12.**

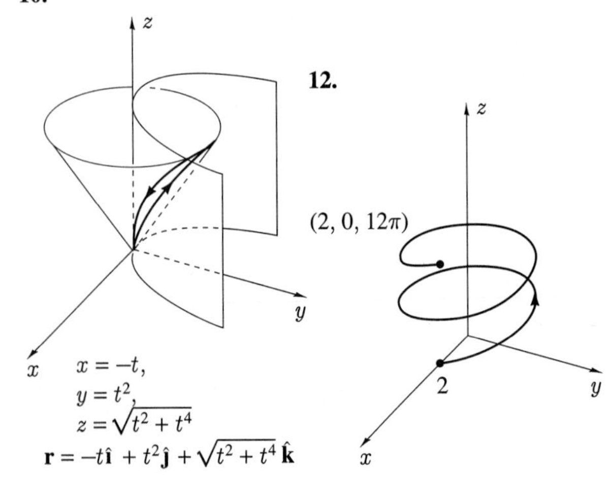

$x=-t$, $y=t^2$, $z=\sqrt{t^2+t^4}$

$\mathbf{r}=-t\hat{\mathbf{\imath}}+t^2\hat{\mathbf{\jmath}}+\sqrt{t^2+t^4}\,\hat{\mathbf{k}}$

14.

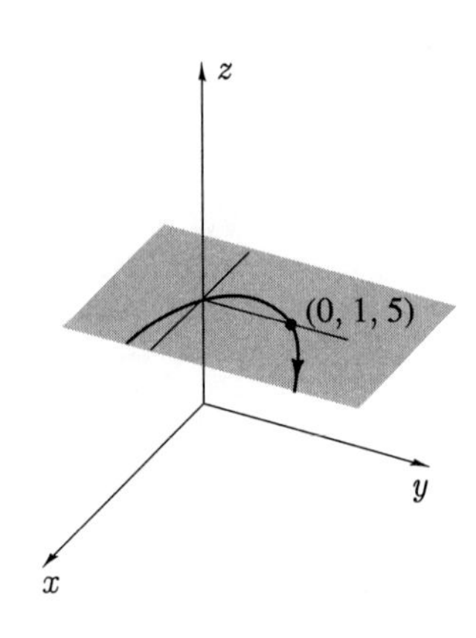

Exercises 12.8

2. $\mathbf{r}=t\hat{\mathbf{\imath}}+t^2\hat{\mathbf{\jmath}}+t^3\hat{\mathbf{k}}$; $\hat{\mathbf{T}}=(\hat{\mathbf{\imath}}+2t\hat{\mathbf{\jmath}}+3t^2\hat{\mathbf{k}})/\sqrt{1+4t^2+9t^4}$

4. $\mathbf{r}=-t\hat{\mathbf{\imath}}+(5+t)\hat{\mathbf{\jmath}}+(t^2-t-5)\hat{\mathbf{k}}$, $-5\le t\le 0$; $\hat{\mathbf{T}}=[-\hat{\mathbf{\imath}}+\hat{\mathbf{\jmath}}+(2t-1)\hat{\mathbf{k}}]/\sqrt{4t^2-4t+3}$

6. $(-2\hat{\mathbf{\imath}}+3\hat{\mathbf{\jmath}}+\hat{\mathbf{k}})/\sqrt{14}$ **8.** $(\hat{\mathbf{\imath}}-\hat{\mathbf{\jmath}})/\sqrt{2}$

10. $-\hat{\mathbf{\jmath}}$ **12.** $\sqrt{42}$

14. $(616\sqrt{616}-157\sqrt{157})/459$

16. $\cos t\hat{\mathbf{\imath}}+\sin t\hat{\mathbf{\jmath}}$

18. (a) $(0,0,0)$ **(b)** $2\hat{\mathbf{\imath}}+2\hat{\mathbf{k}}$

Exercises 12.9

2. $[-(2t+9t^3)\hat{\mathbf{\imath}}+(1-9t^4)\hat{\mathbf{\jmath}}+(3t+6t^3)\hat{\mathbf{k}}]/\sqrt{1+13t^2+54t^4+117t^6+81t^8}$; $(3t^2\hat{\mathbf{\imath}}-3t\hat{\mathbf{\jmath}}+\hat{\mathbf{k}})/\sqrt{1+9t^2+9t^4}$

4. $[(2t-1)\hat{\mathbf{\imath}}+(1-2t)\hat{\mathbf{\jmath}}+2\hat{\mathbf{k}}]/\sqrt{8t^2-8t+6}$; $(\hat{\mathbf{\imath}}+\hat{\mathbf{\jmath}})/\sqrt{2}$

6. $-(5\hat{\mathbf{\imath}}+3\hat{\mathbf{\jmath}}+\hat{\mathbf{k}})/\sqrt{35}$; $(-\hat{\mathbf{\jmath}}+3\hat{\mathbf{k}})/\sqrt{10}$

8. $-(\hat{\mathbf{\imath}}+\hat{\mathbf{\jmath}})/\sqrt{2}$; $-\hat{\mathbf{k}}$

10. $-(\hat{\mathbf{\imath}}+\hat{\mathbf{k}})/\sqrt{2}$; $(\hat{\mathbf{\imath}}-\hat{\mathbf{k}})/\sqrt{2}$

12. $\kappa=0$, ρ undefined

14. $\kappa=2e^t\sqrt{1+e^{2t}}/(1+2e^{2t})^{3/2}$, $\rho=(1+2e^{2t})^{3/2}/(2e^t\sqrt{1+e^{2t}})$

16. $\kappa=1/[\sqrt{2}(1+\cos^2 t)^{3/2}]$, $\rho=\sqrt{2}(1+\cos^2 t)^{3/2}$

18. $\kappa=\sqrt{1+36t^4+16t^6}/[2(1+t^2+4t^6)^{3/2}]$, $\rho=2(1+t^2+4t^6)^{3/2}/\sqrt{1+36t^4+16t^6}$

22. 0

24. (a) $-\sin t\hat{\mathbf{\imath}}+\cos t\hat{\mathbf{\jmath}}$; $-(\cos t\hat{\mathbf{\imath}}+\sin t\hat{\mathbf{\jmath}})$; $\hat{\mathbf{k}}$
(b) $2\sin 2t(\sin t-\cos t)$, $-4(\cos^3 t+\sin^3 t)$;
(c) $2\sin 2t(\sin t-\cos t)\hat{\mathbf{T}}-4(\cos^3 t+\sin^3 t)\hat{\mathbf{N}}$

26. $(-\sin t\hat{\mathbf{\imath}}+\cos t\hat{\mathbf{\jmath}}+\hat{\mathbf{k}})/\sqrt{2}$; $-(\cos t\hat{\mathbf{\imath}}+\sin t\hat{\mathbf{\jmath}})$; $(\sin t\hat{\mathbf{\imath}}-\cos t\hat{\mathbf{\jmath}}+\hat{\mathbf{k}})/\sqrt{2}$; $(1-\cos t\,\sin t+\cos^2 t\,\sin^2 t)\hat{\mathbf{T}}/\sqrt{2}\,-$

$\cos t(\cos t + \sin^3 t)\hat{\mathbf{N}} + (-\cos^2 t \sin^2 t + \cos \sin t + 1)\hat{\mathbf{B}}/\sqrt{2}$

Exercises 12.10

2. $[(t^2 - 1)\hat{\mathbf{\imath}} + (t^2 + 1)\hat{\mathbf{\jmath}}]/t^2$; $\sqrt{2 + 2t^4}/t^2$; $2(\hat{\mathbf{\imath}} - \hat{\mathbf{\jmath}})/t^3$

4. $2t\hat{\mathbf{\imath}} + 2e^t(t + 1)\hat{\mathbf{\jmath}} - 2\hat{\mathbf{k}}/t^3$; $2\sqrt{t^2 + e^{2t}(t + 1)^2 + 1/t^6}$; $2\hat{\mathbf{\imath}} + 2e^t(t + 2)\hat{\mathbf{\jmath}} + 6\hat{\mathbf{k}}/t^4$

6. $(t^4/4 + 1)\hat{\mathbf{\imath}} + (t^3 + 3t^2 + 12)\hat{\mathbf{\jmath}}/6 - (t^5/5 + 1)\hat{\mathbf{k}}$

8. $4t/\sqrt{1 + 4t^2}$; $2/\sqrt{1 + 4t^2}$

12. (a) 0 **(b)** $56t^6$

14. $150x\hat{\mathbf{\jmath}}$

18. (b) 7.79 km/s

20. $365\hat{\mathbf{\imath}} + 325\sqrt{3}\hat{\mathbf{\jmath}}$ km/h; 670.9 km/h

22. 4.54 min

24. $(3/2, 3/2, 2)$ **26.** $(3, 0)$ **32.** Yes

36. (b) $(S/R)[(R - b\cos\theta)\hat{\mathbf{\imath}} + b\sin\theta\hat{\mathbf{\jmath}}]$; $(S/R)\sqrt{R^2 - 2bR\cos\theta + b^2}$; $(bS^2/R^2)(\sin\theta\hat{\mathbf{\imath}} + \cos\theta\hat{\mathbf{\jmath}})$

(c) $(bS^2/R^2)|R\cos\theta - b|/\sqrt{R^2 - 2bR\cos\theta + b^2}$; $(bS^2/R)\sin\theta/\sqrt{R^2 - 2bR\cos\theta + b^2}$

Review Exercises

2. -8 **4.** $(39, -33, -30)$

6. -2 **8.** $(-16, -24, -4)$

10. $(67/7, 21, -65/7)$

12. **14.**

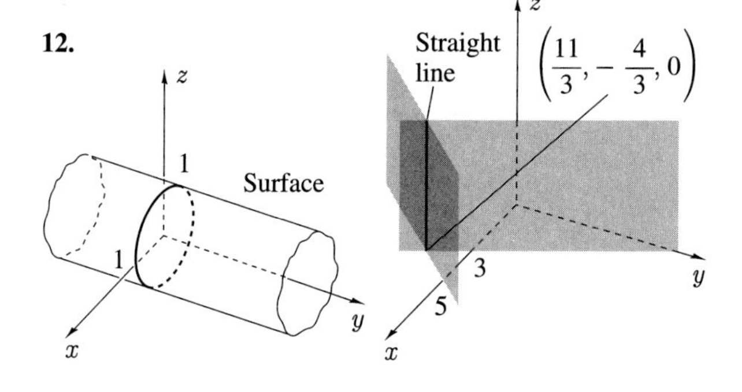

16. No points

18.

Curve
z
x
y

20.

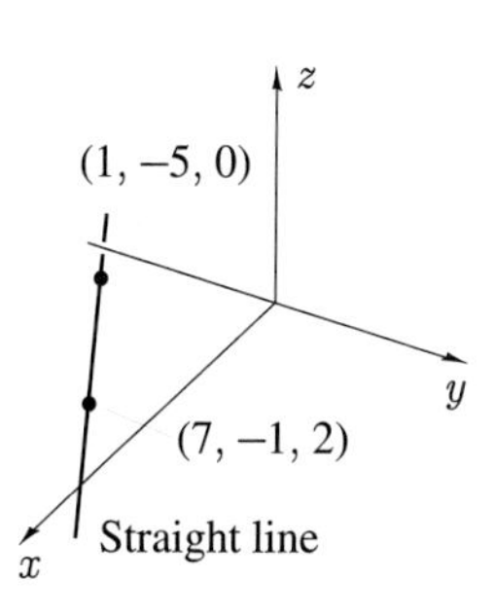

22.

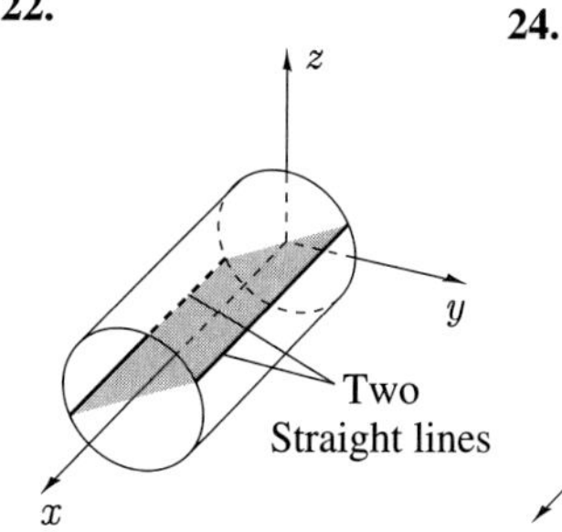

24.

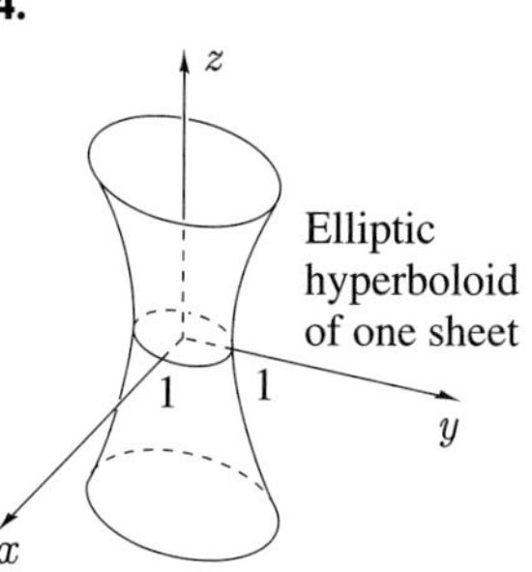

26.

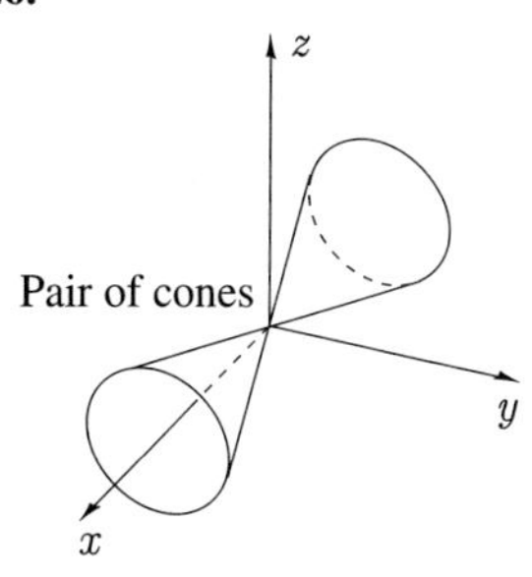

28. $x = 6 + 5t$, $y = 6 - 2t$, $z = 2 + t$

30. $x = 1 + u$, $y = 3 + 2u$, $z = 2 + 3u$

32. $x = y + z$ **34.** $3x - 4y + z = 0$

36. $35/\sqrt{41}$ **38.** $2\sqrt{5}/15$ **40.** $\sqrt{59}$

42. $(2\cos t\hat{\mathbf{\imath}} - 2\sin t\hat{\mathbf{\jmath}} + \hat{\mathbf{k}})/\sqrt{5}$; $-\sin t\hat{\mathbf{\imath}} - \cos t\hat{\mathbf{\jmath}}$; $(\cos t\hat{\mathbf{\imath}} - \sin t\hat{\mathbf{\jmath}} - 2\hat{\mathbf{k}})/\sqrt{5}$

44. $\hat{\mathbf{\imath}} + 2t\hat{\mathbf{\jmath}} + 2t\hat{\mathbf{k}}$; $\sqrt{1 + 8t^2}$; $2(\hat{\mathbf{\jmath}} + \hat{\mathbf{k}})$; 0, $\sqrt{1 + 8t^2}$; $2\sqrt{2}/\sqrt{1 + 8t^2}$, $8t/\sqrt{1 + 8t^2}$

46. (a) 4.46 m/s **(b)** $0.226\hat{\mathbf{\imath}} - \hat{\mathbf{\jmath}}$ m **(c)** 0.255 m from point on floor directly below point it left table

48. $k(1 - x)[1 - 1/\sqrt{1 + 4(1 - x)^2}]\hat{\mathbf{\imath}} + (k/2)[1 - 1/\sqrt{1 + 4(1 - x)^2}]\hat{\mathbf{\jmath}}$ N

CHAPTER 13

Exercises 13.1

4. All points between the branches of the hyperbola $x^2 - y^2 = 1$

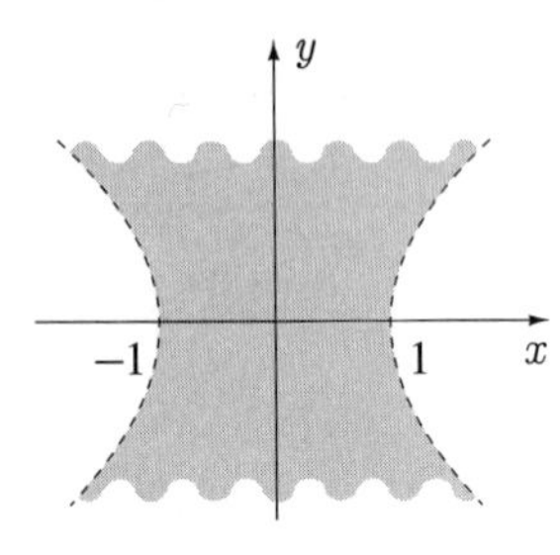

6. All points in space except (0, 0, 0)

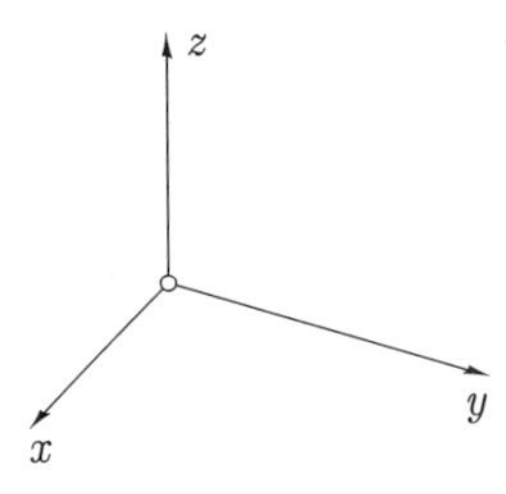

8.

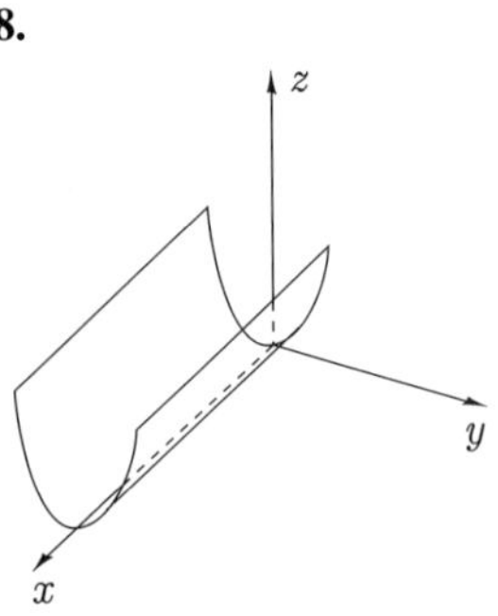

10.

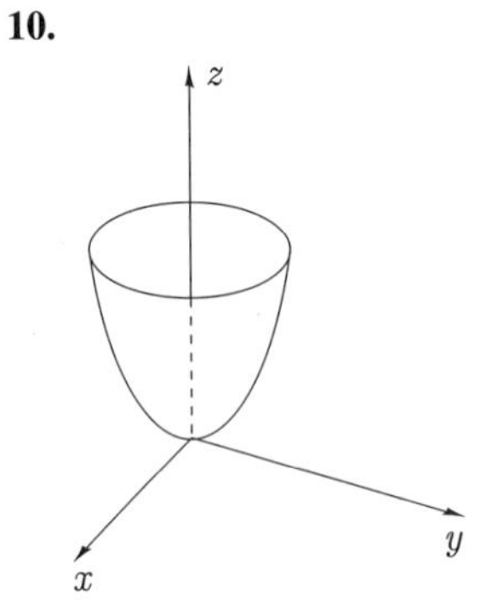

12.

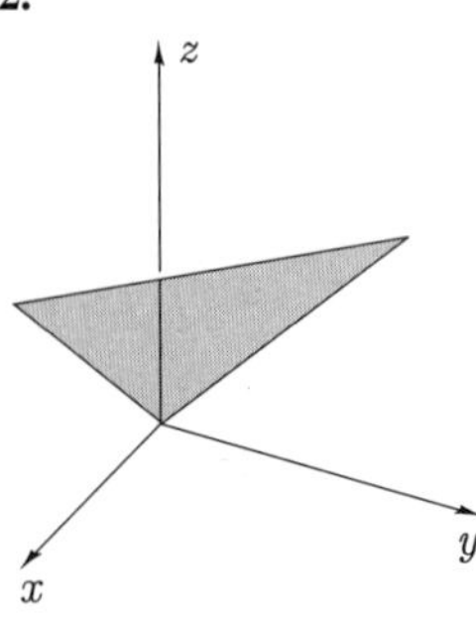

14.

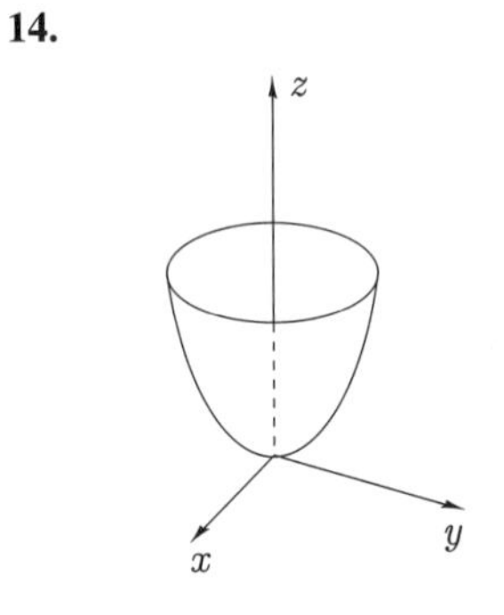

16.

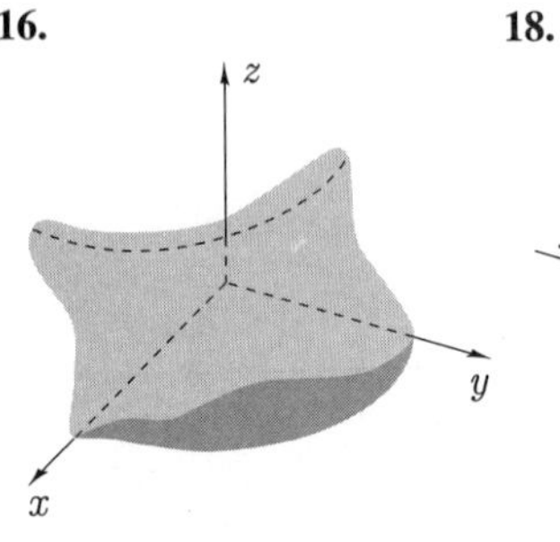

18.

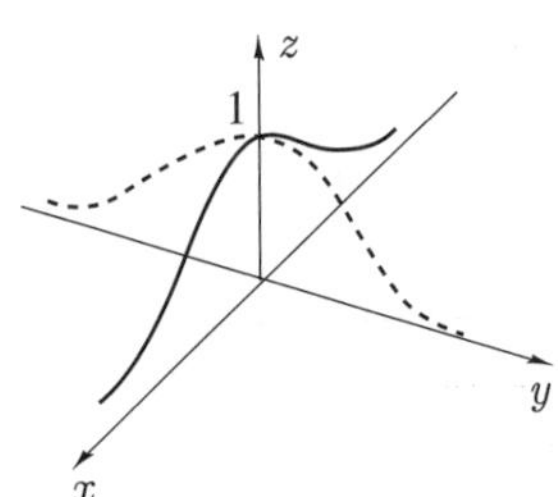

20.

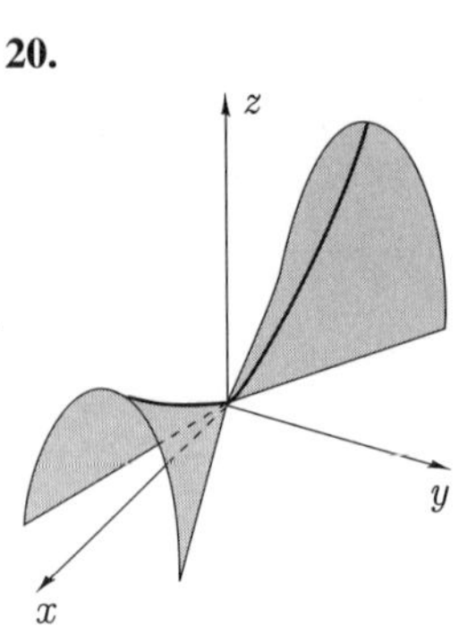

22.

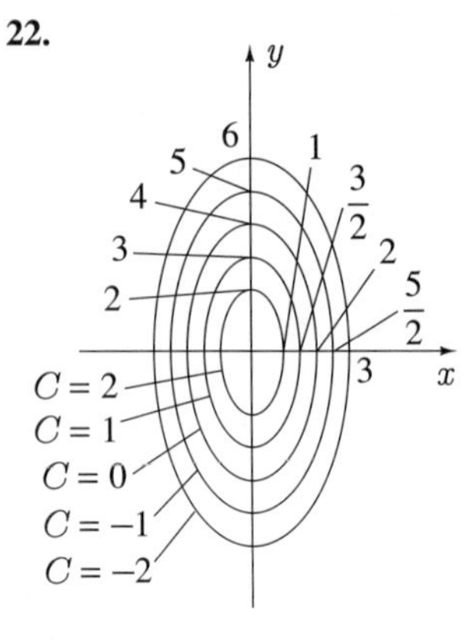

24.

26. $lw(15 - lw)/(w + l)$

28. $8c|xy|\sqrt{1 - x^2/a^2 - y^2/b^2}$

30. $x \sin\theta(1 - x - x\sin\theta \csc\phi) + (x^2/2)\sin\theta(\cos\theta + \sin\theta \cot\phi)$

32. $30.25 + 11G + 17.5S$ cents. The domain consists of all points (G, S) satisfying the inequalities $200 \le 37G + 31S \le 300$, $145 \le 22G + 85S \le 170$.

Exercises 13.2

2. $3/5$ **4.** $-3/7$ **6.** 0

8. $\pi/2$ **10.** 1 **12.** 4

14. Does not exist **16.** 0

18. Does not exist **20.** Does not exist

22. $(0, 0)$ **24.** $x = 0$; $y = 0$; $z = 0$

26. $x = 0$; $y = 0$; $y = -x$ **28.** Does not exist

30. 1 **32.** **(a)** 1, no **(b)** 1, yes

34. False

Exercises 13.3

2. $3y - 16x^3y^4$, $3x - 16x^4y^3$

4. $-x(x+2y)/[y(x+y)^2]$, $x^2(x+2y)/[y^2(x+y)^2]$

6. $y\cos(xy)$, $x\cos(xy)$

8. $x/\sqrt{x^2+y^2}$, $y/\sqrt{x^2+y^2}$

10. $4x\sec^2(2x^2+y^2)$, $2y\sec^2(2x^2+y^2)$

12. ye^{xy}, xe^{xy}

14. $2x/(x^2+y^2)$, $2y/(x^2+y^2)$

16. $ye^x\cos(ye^x)$, $e^x\cos(ye^x)$

18. $2xy\cos^2(x^2y)\,\sin(x^2y)/[1-\cos^3(x^2y)]^{2/3}$, $x^2\cos^2(x^2y)\,\sin(x^2y)/[1-\cos^3(x^2y)]^{2/3}$

20. $\tan\sqrt{x+y}/(2\sqrt{x+y})$, $\tan\sqrt{x+y}/(2\sqrt{x+y})$

22. $-2z/[1+(x^2+z^2)^2]$

24. 2

26. $-1/[1+(1+x+y+z)^2]$

28. $3x^2/y + \sin(yz/x) - (yz/x)\cos(yz/x)$

30. $z\mathrm{Sin}^{-1}(x/z)$ if $z > 0$; $z\mathrm{Sin}^{-1}(x/z) + xz/\sqrt{z^2-x^2}$ if $z < 0$

34. **(a)** Yes **(b)** No **(c)** Yes **(d)** No

36. No **38.** **(a)** Yes **(b)** No

Exercises 13.4

2. $2xyz\hat{\mathbf{\imath}} + x^2z\hat{\mathbf{\jmath}} + x^2y\hat{\mathbf{k}}$

4. $(2xy+y^2)\hat{\mathbf{\imath}} + (x^2+2xy)\hat{\mathbf{\jmath}}$

6. $(1+x^2y^2z^2)^{-1}(yz\hat{\mathbf{\imath}} + xz\hat{\mathbf{\jmath}} + xy\hat{\mathbf{k}})$

8. $e^{x+y+z}(\hat{\mathbf{\imath}} + \hat{\mathbf{\jmath}} + \hat{\mathbf{k}})$

10. $-(x^2+y^2+z^2)^{-3/2}(x\hat{\mathbf{\imath}} + y\hat{\mathbf{\jmath}} + z\hat{\mathbf{k}})$

12. $-\sin 1(\hat{\mathbf{\imath}} + \hat{\mathbf{\jmath}} + \hat{\mathbf{k}})$

14. $-4e^{-8}(\hat{\mathbf{\imath}} + \hat{\mathbf{\jmath}})$

22. ∇F not defined along the line $y = x$

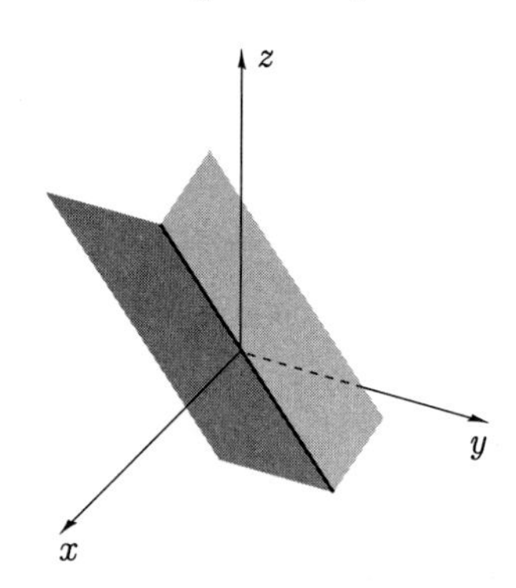

24. $x^2y - xy + C$ **26.** $xyz + y^2z + C$

28. $f(x,y) = g(x,y) + C$

Exercises 13.5

2. $-12x/y^4 + 72x^3y$ **4.** $x(1+z+y+yz)e^{x+y+z}$

6. $e^{x+y} + 4/y^3$ **8.** 0

10. $(x^2+y^2-z^2)/(x^2+y^2+z^2)^2$

12. $2xy/(x^2+y^2)^2$ **14.** $127/(756\sqrt{7})$

16. $8!y^9z^{10}$ **18.** $3y^2\cos(x+y^3)$

20. $-xy/(x^2y^2-1)^{3/2}$

24. Entire plane with $(0,0)$ deleted

26. All space **28.** Not harmonic

32. Curve of intersection of surface with plane $y = y_0$ is concave upward for $x_1 \le x \le x_2$

34. **(a)** Yes **(b)** Yes

36. **(b)** $2xy + C$

38. $n = 0$, all space; $n = -1/2$, $x^2+y^2+z^2 > 0$

Exercises 13.6

2. $(2xe^y + y/x)(-s^2\sin t) + (x^2e^y + \ln x)\left[8t/\left((t^2+2s)\sqrt{(t^2+2s)^2-1}\right)\right]$

4. $xv^2[2yv(3u^2+2) + 2xuv/(u^2+1) + 3xye^u(u+1)]$

6. $-2(\ln 3)y3^{x+2}\csc(r^2+t)\cot(r^2+t)$

8. $-2rst[5st/y^6 + 2yrt/(y^2+z^2)^2 + 2rs/y^3]$

10. $6xve^t(v-2)[1-2u/(x^2-y^2)^2] + 8ye^{4t}[1+2u/(x^2-y^2)^2]$

12. $2(t^2+2t)^2e^{2t} + (t^3+6t^2+6t+2yt^2+4y+8yt)e^t - 2$

14. $-(3y\sin v - 4x\cos v)^2\sin(xy) - (24\sin v\cos v + 3y\cos v + 4x\sin v)\cos(xy)$

18. $200\pi/3$ cm^3/min; no

20. -7.11×10^4 N/s

22. **(a)** Yes **(b)** No **(c)** Yes **(d)** No **(e)** Yes **(f)** No **(g)** No **(h)** Yes

32. 18.77 kg/m^3/s

38. **(c)** $u = -[2c/(n+1)]k^{(n+1)/2} + D$

Exercises 13.7

2. $1/(x+y) - 1$

4. $[24x - \cos(x+y)]/[2y - 1 + \cos(x+y)]$

6. $-(2xz^2+3)/(2x^2z+y)$, $-z/(2x^2z+y)$

8. $z(1+y^2z^2)/(y-x-xy^2z^2)$, $z/(x+xy^2z^2-y)$

10. $[\cos(x+t)+\cos(x-t)]/[\cos(x-t)-\cos(x+t)]$

12. $(x+y)(3y^2-3x^2-5)/(3y^2z-5z)$

14. 0

16. $e^x[(2t+1)\cos y/(3x^2+e^x) + (y^2+2yt-1)\sin y/(t^2+2yt+1)]$

18. $[u^3 + u\cos(uv)]e^{-u}\cos v + [3u^2v + v\cos(uv)]e^{-u}\sin v$
20. $u(3v^2 - u^2)/[4(u^2+v^2)^3]$
22. $-2t/(9y^2)$

Exercises 13.8

2. $1/\sqrt{5}$ 4. $-2/\sqrt{17}$ 6. $1/5$
8. $-13/\sqrt{29}$ 10. $11/\sqrt{82}$ 12. $-40/\sqrt{6}$
14. $(1,4)$ 16. $(1,-3,2)$ 18. $(1,1)$
20. (a) $\pm(2,1)$ (b) $(-1 \pm 2\sqrt{19}, 2 \pm \sqrt{19})$
(c) No direction
22. (a) Yes (b) No
24. $9t/[\sqrt{13}\sqrt{4+9t^2}]$, 0
26. $(0,0,0)$, $(1/4, \pm 1/2, 1/4)$
28. $\sqrt{65}/3$ 30. -2
32. (a) $\pi/(\sqrt{2}\sqrt{8 - 4\pi + \pi^2})$, $\pi/\sqrt{4+\pi^2}$ (b) $1/\sqrt{2}$, 0
(c) $1/\sqrt{2}$, 1

Exercises 13.9

2. $x = 1 + u$, $y = 1 + 2u$, $z = 1 + 3u$
4. $x = -2 + t$, $y = 4 - 4t$, $z = -2 + t$
6. $x = 1 - 2u$, $y = 5 + 2u$, $z = 1 + u$
8. $132x + 49y = 328$, $z = 0$
10. $x = 1 - u$, $y = u$, $z = u$
12. $x = 2 + t$, $y = -\sqrt{5} + (1/\sqrt{5})t$, $z = -1 - t$
14. $x = 4 + \sqrt{17}u$, $y = 1$, $z = \sqrt{17} + 4u$
16. $x = 12 + 16t$, $y = -14 - 31t$, $z = 2 + t$
18. $x = t$, $y = 1$, $z = 1 + t$
20. $x = 2\pi u$, $y = 2\pi + u$, $z = 4\pi + 2u$
22. $3x + 6y + z + 2 = 0$ 24. $x + y + z = 4$
26. $x + y = 1$ 30. $6\sqrt{2}$ 32. 0
34. $x_0 x/a^2 + y_0 y/b^2 + z_0 z/c^2 = 1$
36. No points
38. $(0,0,-1)$ and $x^2 + y^2 = 1/2$, $z = -1/2$

Exercises 13.10

2. $(0,0)$ saddle point; $(1,1)$ relative maximum
4. None
6. $(0, n\pi)$ saddle points
8. (x,x) relative minima
10. $(x,0)$ saddle points; $(0,y)$ $y > 0$ relative minima; $(0,y)$ $y < 0$ relative maxima
12. $(0,0)$ relative minimum; $(0,y)$ $y \neq 0$ none of these
14. $(0,0)$ relative maximum; $(0, \pm 1/\sqrt{2})$, $(\pm 1/\sqrt{2}, 0)$ saddle points; $(1/\sqrt{2}, \pm 1/\sqrt{2})$, $(-1/\sqrt{2}, \pm 1/\sqrt{2})$ relative minima
16. None
18. All points on the coordinate axes
22. (d) 2.45, 3.54
24. $(0,0)$ relative minimum; $(-1/3, \pm 2/9)$ saddle points
26. (a) $(2, \pm 2, \pm 2)$, $(-2, \pm 2, \mp 2)$, $(2, \mp 2, \mp 2)$, $(-2, \mp 2, \pm 2)$

Exercises 13.11

2. 4, $-1/3$ 4. 3, -3 6. 3, -2
8. $(2\sqrt{2} + 5)/\sqrt{2}$, $(2\sqrt{2} - 5)/\sqrt{2}$
10. $(1,1,-2)$ 12. $(1/2, 1/2, 1/2)$ 14. 2, -32
16. $100(2/3)^{1/3} \times 100(2/3)^{1/3} \times 100(3/2)^{2/3}$ cm
20. $2a/\sqrt{3} \times 2b/\sqrt{3} \times 2c/\sqrt{3}$
24. $12/\sqrt{5}$ m, $(50\sqrt{5} - 27\pi)/(3\sqrt{5}\pi)$ m
26. $\|AB\| = 1/3$ m, $\theta = \pi/3$
28. $\pm 2\sqrt{3}/9$ 30. $3\sqrt{3}r^2/4$

Exercises 13.12

2. $(8 \pm 9\sqrt{13})/2$ 4. ± 27 6. $\pm\sqrt{32/27}$
8. $(\pm\sqrt{2} - 1)/2$ 18. 33.12, -0.12 20. $\pm 1/\sqrt{2}$
22. $(\pm 2\sqrt{14}, \mp\sqrt{14})$, $(\pm 2, \pm 4)$
24. $(3a/2, 3a/2)$ 26. 1.171, 2.373

Exercises 13.13

2. $y(1 + x^2y^2)^{-1}dx + x(1 + x^2y^2)^{-1}dy$
4. $[yz\cos(xyz) - 2xy^2z^2]\,dx + [xz\cos(xyz) - 2x^2yz^2]\,dy + [xy\cos(xyz) - 2x^2y^2z]\,dz$
6. $(1 - x^2y^2)^{-1/2}(y\,dx + x\,dy)$
8. $(y + t)\,dx + (x + z)\,dy + (y + t)\,dz + (z + x)\,dt$
10. $2e^{x^2+y^2+z^2-t^2}(x\,dx + y\,dy + z\,dz - t\,dt)$
12. 3%

Exercises 13.14

2. $\frac{1}{3!}[f_{xxx}(c,d)(x-c)^3 + 3f_{xxy}(c,d)(x-c)^2(y-d) + 3f_{xyy}(c,d)(x-c)(y-d)^2 + f_{yyy}(c,d)(y-d)^3]$
4. $\displaystyle\sum_{n=0}^{\infty}\sum_{r=0}^{n} \frac{e^5(-1)^{n-r}2^r3^{n-r}}{(n-r)!\,r!}(x-1)^r(y+1)^{n-r}$
6. $\displaystyle\sum_{n=1}^{\infty}\sum_{r=0}^{n} \frac{(-1)^{n+1}(n-1)!}{(n-r)!\,r!}x^{2r}y^{2n-2r}$
8. $[(x+1) - 1]\displaystyle\sum_{n=0}^{\infty}(-1)^n y^{2n+2}$

10. $[72+12(x-2)+24(y-1)-(x-2)^2+8(x-2)(y-1)-4(y-1)^2]/(24\sqrt{3})$

12. $[2x+2(y-1)+3x^2+6x(y-1)+3(y-1)^2]/2$

14. 0

16. $\sum_{n=0}^{\infty} \frac{1}{n!}\left[(x-c)\frac{\partial}{\partial x}+(y-d)\frac{\partial}{\partial y}\right]^n f(c,d)$

Review Exercises

2. $2(x^2-y^2+z^2)/(x^2+y^2+z^2)^2$

4. $(3-z^2)(1+z^2)/(1+2xz+2xz^3)$

6. $3(2x-ye^{xy})(t^2+1)+(2y-xe^{xy})(\ln t+1)$

8. $(y+3yv^2-2xuv)/(2u+6uv^2+3v+vx^2)$

10. $e^t(t+1)(y-2x)+e^{-t}(1-t)(x-2y)$

12. $\cos\theta$

14. $(2x-3x^2y^2)\cos\theta-2yx^3\sin\theta$

16. $6(6x^4t^6z+9x^5+x^4t^9z^2+2xt^9z^5-9t^{12}z^5-6xt^6z^4-4x^5t^3z-2x^2t^3z^4)/(x^4t^8z^4)$

18. $xz[(1-3xz^2)(1-6xz^2+6x^2z^4)+(1-2xz^2)(1-3xz^2+6x^2z^4)]/(x-3x^2z^2)^3$

20. $2y(v^2t-1)(t^2-3/\sqrt{1-x^2y^2})+xv\sec^2 t(t^2-3/\sqrt{1-x^2y^2})+2xyt$

26. $\sqrt{6}$ **28.** $9/\sqrt{14}$ **30.** 2

32. $x+3y-6z+4=0$

34. $x=2+u,\ y=u,\ z=6+4u$

36. $x=1+t,\ y=1-t,\ z=1$

38. None

40. $(0,0)$ relative maximum; all points on $x^2+y^2=1$ relative minima

42. $1/2,\ -1/2$

44. $(\pm\sqrt{(\sqrt{5}-1)/2},0),\ (0,\pm 1)$

46. $x=100bcqr/(acpr+abpq+bcqr),\ y=100acpr/(acpr+abpq+bcqr),\ z=100abpq/(acpr+abpq+bcqr)$

48. $1/\sqrt{2}+\sqrt{2}(\pi+6)(x-1)/4+(y-\pi/4)/\sqrt{2}+\sqrt{2}(24+14\pi-\pi^2)(x-1)^2/16+\sqrt{2}(10-\pi)(x-1)(y-\pi/4)/4-\sqrt{2}(y-\pi/4)^2/4+\cdots$

CHAPTER 14

Exercises 14.1

2. 0 **4.** $3/4$ **6.** $e^2(1-e)^2/2$

8. 0 **10.** $5/144$ **12.** 0

14. -0.54 **16.** $(1-\ln 2)/2$ **18.** $128\sqrt{2}/5$

20. $11\,664/35$ **22.** $(1-2\sqrt{2})/12$ **24.** $1/3$

26. $8/189$ **28.** $2(1-\sqrt{2})$

30. $[\sqrt{2}+\ln(\sqrt{2}+1)]/6$

Exercises 14.2

2. $128/15$ **4.** $-621/140$ **6.** 0

8. 0 **10.** $304/15$ **12.** $\sqrt{13}-7/2$

14. $(1-\cos 1)/2$ **16.** $2(1-\sqrt{2})$ **18.** $(5/4)\ln 5-1$

20. 4 **22.** Double **24.** 0

26. Double **28.** 0 **30.** $101/70$

32. $(e^2-1)/(2e)$ **34.** 24π **36.** e^2-2e-1

38. $(1-2\sqrt{2})/12$ **40.** $1/3$

Exercises 14.3

2. $343/6$ **4.** 8 **6.** $1+3e^{-2}$

8. $343/6$ **10.** $20/3$ **12.** 16π

14. $1024\pi/15$ **16.** $26\pi/15$ **18.** 45π

20. $34\pi/3$ **22.** $235/3$ **24.** $7/6$

26. $8\pi/3-2\sqrt{3}$ **28.** $23\,328/35$ **30.** $2-\ln 3$

32. $16\pi(1+\sqrt{2})/15$ **34.** $68\pi/9$

36. 5.38 **38.** $512\pi/(15\sqrt{5})$ **40.** $7\sqrt{2}\pi/6$

Exercises 14.4

2. $256\rho g/5$ **4.** $48\rho g$ **6.** $370\rho g/3$

8. $250\rho g/3$ **10.** 7.63×10^6 N **12.** $5\sqrt{29}\rho g$

14. $10\rho g$ **16.** $\pi\rho g r^3$ **18.** 9.00×10^4 N

Exercises 14.5

2. $(0,24/5)$ **4.** $\bar{x}=\bar{y}=9/(15-16\ln 2)$

6. $\left(8(2-\sqrt{2})/(3\pi),8\sqrt{2}/(3\pi)\right)$

8. $(177/85,9/17)$ **10.** $(4/3,5/3)$ **12.** $32/3$

14. $603/10$ **16.** $48\rho/5$ **18.** $(0,192/205)$

20. $(-2/(8\sqrt{3}-1),0)$ **22.** $(-61/28,807/700)$

24. $\rho ab(a^2+b^2)/12$ **26.** $4761/140$

28. $81\sqrt{2}/40$

Exercises 14.6

2. $4\sqrt{14}/3$

4. $4(5\sqrt{3}-2\sqrt{6}-3+\sqrt{2})/3$

6. $(247\sqrt{13}+64)/1215$

8. $2\int_0^4\int_0^{\sqrt{16-y^2}}\sqrt{4+y^2+z^2}\,dz\,dy$

10. $4\int_0^{\sqrt{2}}\int_0^{\sqrt{2-x^2}}\sqrt{1+16(x^2+y^2)^3}\,dy\,dx$

12. $\int_0^1\int_0^{1-x^2}\sqrt{1+2/(1+x+y)^2}\,dy\,dx$

16. $\sqrt{2}\int_{-1/\sqrt{2}}^{1/\sqrt{2}}\int_x^{\sqrt{1-x^2}}\sqrt{2+1/(y-x)}\,dy\,dx$

18. $\int_0^1 \int_{1-x}^{2-x} \sqrt{1+9x^4+9y^4}\, dy\, dx + \int_1^2 \int_0^{2-x} \sqrt{1+9x^4+9y^4}\, dy\, dx$
20. $(37\sqrt{74} - 5\sqrt{10})/24$

Exercises 14.7

2. $2/3$ **4.** $4(3\sqrt{3} - \pi)/3$ **6.** $\pi/6$
8. $(3\sqrt{3} - \pi)/2$ **10.** $\pi/4$
12. $15\pi/4 - 19\sqrt{2}/3 - (9/2)\,\mathrm{Sin}^{-1}(1/3)$
14. $\pi R^4/4$ **16.** $\pi(17\sqrt{17} - 1)/6$
18. $2\pi(10\sqrt{10} - 1)/3$ **20.** $(17\sqrt{17} - 5\sqrt{5})\pi/6$
22. $8\pi/3$ **24.** $2\pi^2 R^3$
26. $[q\rho/(2\epsilon_0)](1 - d/\sqrt{R^2+d^2}),\ q\rho/(2\epsilon_0)$
28. $9\pi - 12\sqrt{3}$ **30.** $(5/6, 0)$
32. $\pi(\pi - 2)/2$ **34.** $2a^2(\pi + 4 - 4\sqrt{2})$

Exercises 14.8

2. $32/3$ **4.** 0 **6.** $1024/21$
8. $1/96$ **10.** $11/6$ **12.** $48/35$
14. $4\int_0^{\sqrt{(\sqrt{5}-1)/2}} \int_{x^2}^{\sqrt{1-x^2-x^4}} \int_0^{\sqrt{1-x^2-y^2}} (x^2+y^2+z^2)\, dz\, dy\, dx$
16. $\int_{-5/2}^{1/2} \int_{-1-\sqrt{9-4(y+1)^2}}^{-1+\sqrt{9-4(y+1)^2}} \int_{x^2+4y^2}^{4-2x-8y} xyz\, dz\, dx\, dy$
18. $729/70$ **20.** 2π **22.** $\pi/3$
24. $4\int_0^1 \int_0^{\sqrt{1-x^2}} \int_0^{\sqrt{x^2+y^2}/2} (x^2+y^2+z^2)\, dz\, dy\, dx$
$+4\int_0^1 \int_{\sqrt{1-x^2}}^{\sqrt{4/3-x^2}} \int_{\sqrt{x^2+y^2-1}}^{\sqrt{x^2+y^2}/2} (x^2+y^2+z^2)\, dz\, dy\, dx$
$+4\int_1^{2/\sqrt{3}} \int_0^{\sqrt{4/3-x^2}} \int_{\sqrt{x^2+y^2-1}}^{\sqrt{x^2+y^2}/2} (x^2+y^2+z^2)\, dz\, dy\, dx$

Exercises 14.9

2. $2/3$ **4.** $704/15$ **6.** 8
8. $8/3$ **10.** $19/96$ **12.** $7/3$
14. $5/18$ **16.** $64/15$ **18.** $\pi/4 - 1/3$
20. $1/20$ **22.** $13/3$ **26.** $\pi/2$
28. **(a)** $(5/3)[3\pi/2 - 3\,\mathrm{Sin}^{-1} d - d\sqrt{1-d^2}(5 - 2d^2)]$
(b) $2500\pi g$

Exercises 14.10

2. $(1/4, 1/4, 1/4)$ **4.** $(0, 16/7, 8/7)$
6. $2\rho/3$ **8.** $773\rho/2520$ **10.** $64\rho/15$
12. $(6772/11\,847, 7300/14\,001, 1/2)$
14. $(0, 0, 3/2)$ **16.** $128\rho a^5/45$ **18.** $51\sqrt{3}\rho$
20. $(3a/8, 3b/8, 3c/8)$ **22.** $28\pi\rho R^5/15$

Exercises 14.11

2.

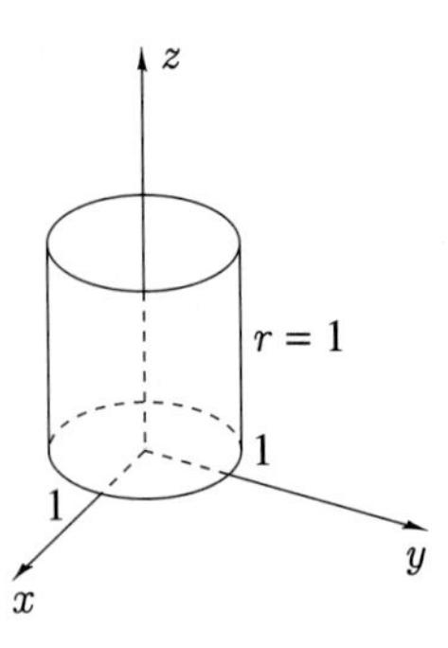

4.

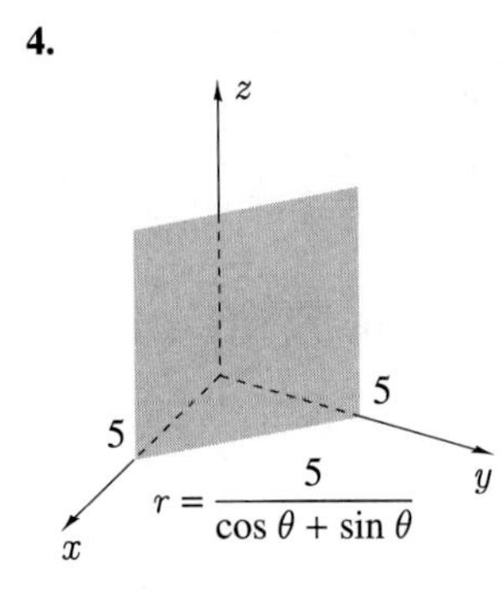

6.

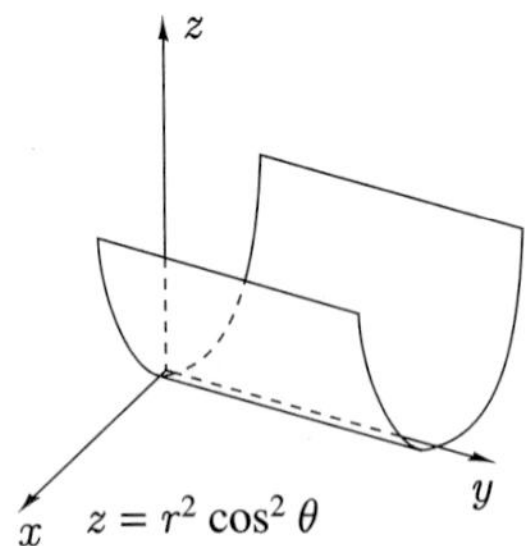

8.

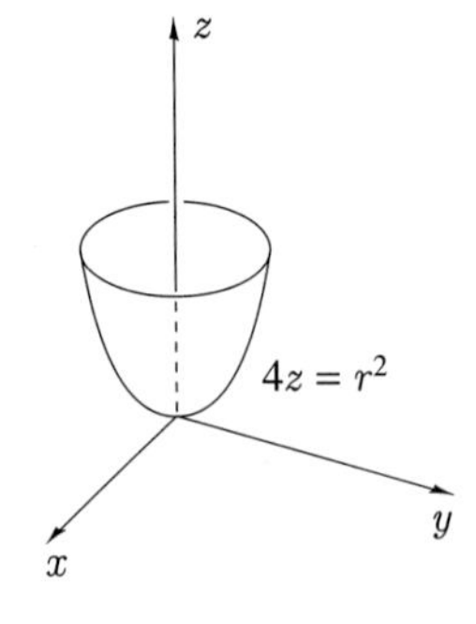

10.

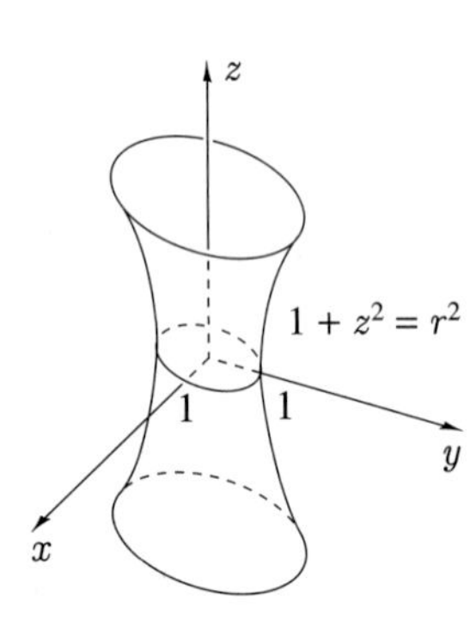

12. $(8\sqrt{2} - 7)\pi/6$ **14.** 4π **16.** $4\sqrt{3}\pi$
18. $\int_0^{2\pi} \int_0^3 \int_0^{1+r^2} f(r\cos\theta, r\sin\theta, z)\, r\, dz\, dr\, d\theta$,
$\int_0^3 \int_0^{2\pi} \int_0^{1+r^2} f(r\cos\theta, r\sin\theta, z)\, r\, dz\, d\theta\, dr$,
$\int_0^3 \int_0^{1+r^2} \int_0^{2\pi} f(r\cos\theta, r\sin\theta, z)\, r\, d\theta\, dz\, dr$,
$\int_0^1 \int_0^3 \int_0^{2\pi} f(r\cos\theta, r\sin\theta, z)\, r\, d\theta\, dr\, dz +$
$\int_1^{10} \int_{\sqrt{z-1}}^3 \int_0^{2\pi} f(r\cos\theta, r\sin\theta, z)\, r\, d\theta\, dr\, dz$,
$\int_0^{2\pi} \int_0^1 \int_0^3 f(r\cos\theta, r\sin\theta, z)\, r\, dr\, dz\, d\theta +$
$\int_0^{2\pi} \int_1^{10} \int_{\sqrt{z-1}}^3 f(r\cos\theta, r\sin\theta, z)\, r\, dr\, dz\, d\theta$
$\int_0^1 \int_0^{2\pi} \int_0^3 f(r\cos\theta, r\sin\theta, z)\, r\, dr\, d\theta\, dz +$
$\int_1^{10} \int_0^{2\pi} \int_{\sqrt{z-1}}^3 f(r\cos\theta, r\sin\theta, z)\, r\, dr\, d\theta\, dz$
20. $8\pi\rho R^5/15$ **22.** $81\pi^2/8$ **24.** 0.084

26. $\pi h \rho R^4/10$ **28.** $4\sqrt{3}\pi/9$ **30.** $56\pi(2-\sqrt{2})/3$
32. $4\pi(8\sqrt{2}-3\sqrt{6})/3$ **34.** $2a^3(3\pi+4)/9$
36. $64\pi/3$ **38.** $15\pi/2$
40. $3\sqrt{6}/5$ **42.** $2\pi^2 ab^2$

Exercises 14.12

2. $\mathscr{R}\sin\phi = 1$ (see figure for Exercise 14.11–2)
4. $\mathscr{R} = 4\cot\phi\csc\phi$ (see figure for Exercise 14.11–8)
6. $\mathscr{R}^2 = -\sec 2\phi$ (see figure for Exercise 14.11–10)
8. $(2-\sqrt{2})\pi/3$ **10.** $[4\,\mathrm{Tan}^{-1}2-\pi]/12$
12. $32\pi/3$ **14.** $8\pi\rho R^5/15$ **16.** $k\pi R^4$ C
18. $9\pi/2$ **20.** $(2\pi R^3/3)(1-k/\sqrt{1+k^2})$
22. **(a)** $\rho_b = \rho_w/2$ **(b)** $11\pi\rho_w g R^3/24$

Exercises 14.13

2. $2x-1+e^x$
4. $4x^3-3x^2-1+3x^2(x^3-1)\ln(x^3-1)-2x^3\ln(x^2)$
8. $e^x\sqrt{1+e^{3x}}-\cos x\sqrt{1+\sin^3 x}$
12. $(1/a^2)\ln(1+ab)-b/(a+a^2b)$
14. $[3/(8a^5)]\mathrm{Tan}^{-1}(x/a)+x(3x^2+5a^2)/[8a^4(a^2+x^2)^2]+C$
16. $\pi\,\mathrm{Sin}^{-1}a$
18. **(a)** $|x|<1/9,\ 0$
(b) $18/x+(1/x^2)\ln[(1-9x)/(1+9x)],\ x\neq 0;\ 0$

Review Exercises

2. $81/16$ **4.** $1/40$ **6.** $36/5$
8. 0 **10.** $-4544/945$ **12.** $\pi/6$
14. $32\pi/3$ **16.** $2\pi/5$ **18.** $3\pi/2$
20. $(\pi\ln 2)/4$
22. **(a)** $\iint_R dA$
(b) $\iint_R 2\pi(2-x)\,dA,\ \iint_R 2\pi(y+4)\,dA$
(c) $\iint_R (x-1)\,dA,\ \iint_R (y+1)\,dA$
(d) $\iint_R (x+1)^2\,dA, \iint_R (y-4)^2\,dA$
(e) $\iint_R \sigma(x,y)\,dA$
(f) $\iint_R \rho(x,y)\,dA$
(g) $\iint_R P(x,y)\,dA$
24. $32/3$ **26.** $16/15$
28. $\pi(17\sqrt{17}-1)/96$ **30.** $128\pi/15,\ 4\pi^2$
32. $1/4$ **34.** 2
36. $14\rho\pi/15$ **38.** $(0, 1/8)$
40. $(9\pi/64, 9\pi/64, 3/8)$ **42.** $16\pi/15$

CHAPTER 15

Exercises 15.1

2. Closed, connected **4.** Connected
6. Closed, connected **8.** Open
10. For interior, exterior, and boundary points replace circle with sphere in planar definitions. Open, closed, connected, and domain definitions are identical. A domain is simply-connected if every closed curve in the domain is the boundary of a surface that contains only points of the domain.
12. Closed, connected **14.** Connected
16. Open
18. Open, connected, simply-connected domain
22. $-(x^2+y^2+z^2)^{-3/2}(x\hat{\mathbf{i}}+y\hat{\mathbf{j}}+z\hat{\mathbf{k}})$
24. $(6-2\cos 2)\hat{\mathbf{i}}+(1+2\sin 2)\hat{\mathbf{j}}$
26. $2(e^y-x^2y)$
28. $2x\cos(x^2+y^2+z^2)-\sin(y+z)$
30. 1 **32.** $2/\sqrt{x^2+y^2+z^2}$
34. $4y(16z^4-3x)\hat{\mathbf{i}}+x^2\hat{\mathbf{j}}+12yz\hat{\mathbf{k}}$
36. 0 **38.** $-2\hat{\mathbf{k}}$
40. $-2\hat{\mathbf{k}}/[(x+y)\sqrt{(x+y)^2-1}]$
46. $x^3y^2+3x+2y+C$ **48.** $\ln|x+y|+C$
50. $(x^2+y^2+z^2)/2+C$ **52.** $\ln|1+x+y+z|+C$
54. $\mathrm{Tan}^{-1}(xy)+z^2/2+C$
56. **(a)** $4, 2, -1$ **(b)** $x^3/3+2xy+4xz-3y^2/2-yz+z^2+C$
58. **(a)** $q/(4\pi\epsilon_0|\mathbf{r}|)+C$ **(b)** $-\sigma z/(2\epsilon_0)+C$
62. $yz\hat{\mathbf{i}}-xz\hat{\mathbf{j}}$
64. $(1/4)[(4xyz-3y^2z)\hat{\mathbf{i}}+(2xyz-6x^2z)\hat{\mathbf{j}}+(2x^2y+xy^2)\hat{\mathbf{k}}]$

Exercises 15.2

2. $32/3$ **4.** $50\sqrt{10}/3$ **6.** $37\sqrt{2}/3$
8. $9\sqrt{1+16\pi^2}$ cm **10.** $37/80$ **12.** 0.78
14. 0 **16.** $(1-161\sqrt{161})/6$
18. $\pi(145\sqrt{145}-10\sqrt{10})/27$
20. $\pi/2$ **22.** 0.007 **24.** 2
26. 0.242 **28.** π **30.** $(5\sqrt{5}-1)/6$
32. $\sqrt{2}(1-e^{6\pi})/13$ **34.** 17.08
36. $4\pi^2 ab$

Exercises 15.3

2. $51/4$ **4.** -15 **6.** $-99/140$
8. 10 **10.** 0 **12.** 9

14. 0 **16.** -8π **18.** $(768 + 5\pi)/20$

20. $67/32 + \sqrt{5} - \sqrt{2} - 5/e$

22. $(4\pi\epsilon_0)^{-1}[q_1 q_3(1/\sqrt{41} - 1/\sqrt{61}) + q_2 q_3(1/\sqrt{18} - 1/\sqrt{10})]$

24. 3.719 **26.** -4.26×10^{-4} **28.** 0

32. **(a)** $2a$ **(b)** 0

Exercises 15.4

2. -43 **4.** -2 **6.** 0

8. 10 **10.** e^3 **12.** No

14. 0 **16.** 0 **18.** $8/105 + \ln(3/2)$

20. $-\pi$

22. **(a)** Yes **(b)** Yes **(c)** Yes **(d)** Yes **(e)** No

24. 2π

26. **(b)** $k(1/\sqrt{x_1^2 + y_1^2 + z_1^2} - 1/\sqrt{x_2^2 + y_2^2 + z_2^2})$

Exercises 15.5

2. Not conservative **4.** mgz

8. $-k(\sqrt{x^2 + y^2 + z^2} - L)(x\hat{\mathbf{i}} + y\hat{\mathbf{j}} + z\hat{\mathbf{k}})/\sqrt{x^2 + y^2 + z^2}$, yes

10. **(a)** Yes **(b)** Yes

Exercises 15.6

2. 4π **4.** $2\sqrt{3}/5$ **6.** 0

8. $-3/8$ **10.** $4(13\sqrt{13} - 8)/3$

14. πab **16.** $3\pi/8$ **18.** 2π

22. 24 **24.** $77/2$ **26.** $81\pi/2$

28. -4π **32.** π

Exercises 15.7

2. $2\sqrt{3}/15$ **4.** $\sqrt{2}/8$ **6.** $(-61 + 44\sqrt{2})/5$

8. -3 **10.** $\pi R\sqrt{R^2 + h^2}$

12. $3(145^{5/2} - 361)/5120$ **14.** $\pi(3\sqrt{3} - 1)/3$

16. $(50\sqrt{5} + 2)\pi/15$ **18.** $4\pi R^2$

20. 0 **22.** $\pi \ln 3$

Exercises 15.8

2. $2e^2 - 10e^{-2}$ **4.** 0 **6.** $2\pi/3$

8. -2π **10.** $10\pi/3$ **12.** $\sqrt{2}\pi/4$

14. $2\sqrt{5} + 8\ln[(\sqrt{5} + 1)/2]$

16. 0 **18.** **(a)** 30π **(b)** -20π

20. $2\pi(1 - e^{-2})$

Exercises 15.9

2. 0 **4.** 16 **6.** 27π

8. 180 **10.** $-1328\pi/5$ **12.** $57/2$

14. $52\pi/5$ **20.** 0

Exercises 15.10

2. 0 **4.** $\pm 16\pi$ **6.** 0

8. $-2\sqrt{2}\pi b^2$ **10.** $\pm 2\pi$ **12.** -24π

14. 3π

Review Exercises

2. $3x^2y + x^2/y^2$ **4.** 0

6. $ye^x + ze^y + xe^z$ **8.** $(\hat{\mathbf{i}} + \hat{\mathbf{j}})/\sqrt{1 - (x + y)^2}$

10. $x(-y\hat{\mathbf{j}} + z\hat{\mathbf{k}})/(1 + x^2y^2z^2)$

12. $2\sqrt{3}$ **14.** π

16. $\sqrt{2}/3$ **18.** 2π

20. $8\pi(18 - 7\sqrt{6})/3$ **22.** 0

24. 0 **26.** $-\pi/120$

28. $(25\sqrt{5} - 11)/120$ **30.** $8\pi/3$

CHAPTER 16

Exercises 16.1

12. $-3\sin 3x + 2\cos 3x$

14. $[(2\cos 3 - \cos 6)\sin 3x + (\sin 6 - 2\sin 3)\cos 3x]/\sin 3$

16. $(1/3)\,\mathrm{Tan}^{-1}(x/3) + C$

18. $(x^3/6)\ln x - 5x^3/36 + C_1 x + C_2$

20. Ce^x **22.** $y = 0$

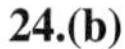

24.(b)

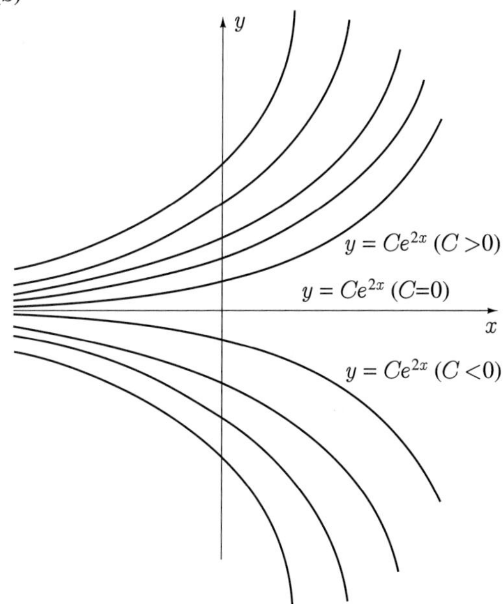

26.(a)

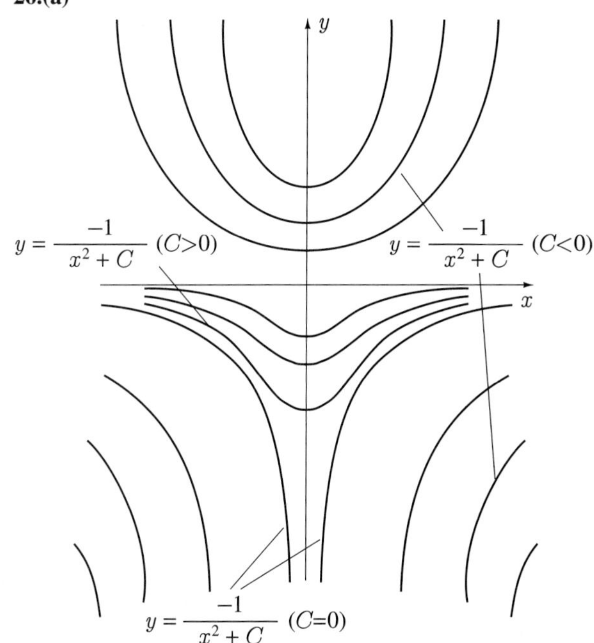

Exercises 16.2

2. $y(x) = 2 + Ce^{-x^2}$ **4.** $y(x) = Ce^{3x} - 2/3$

6. $y^2 + x^2 + 2(x - y) + 2\ln|y + 1| + 2\ln|x - 1| = C$

8. $y(x) = [C + 3\ln|x| + 3e^x(1 - x)]^{1/3}$

10. $y(x) = (x + C)/(1 - Cx)$

12. $xy = 2e^{y-x-1}$

14. $y(x) = [\tan(x^2 - 4) + 4]/[1 - 4\tan(x^2 - 4)]$

16. **(a)** $6(1 - 6kt)^{-1}$ km **(b)** never

18. $20 + 60e^{-0.203t}$ **20.** 13.51 h

22. $(R/k)(1 - e^{-kt}) + A_0 e^{-kt}$

24. $60(1 - e^{kt})/(2 - 3e^{kt})$ g

28. $x^2(x^2 - 2y^2) = C$ **30.** $x^2 + 2xy - y^2 = C$

32. $(y - x)e^{y/x} = x\ln|x| + Cx$

34. $x^2 + y^2 = 5x$

36. $y_0(A + y_0)/[y_0 + Ae^{-k(A+y_0)t}]$

38. $y = Ce^x - x - 1$ **40.** $6y - 3\ln|4x + 6y + 3| = C$

42.(a) **(b)**

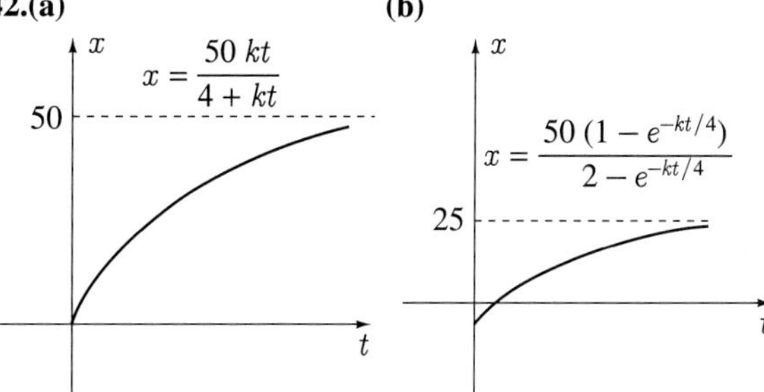

(c)

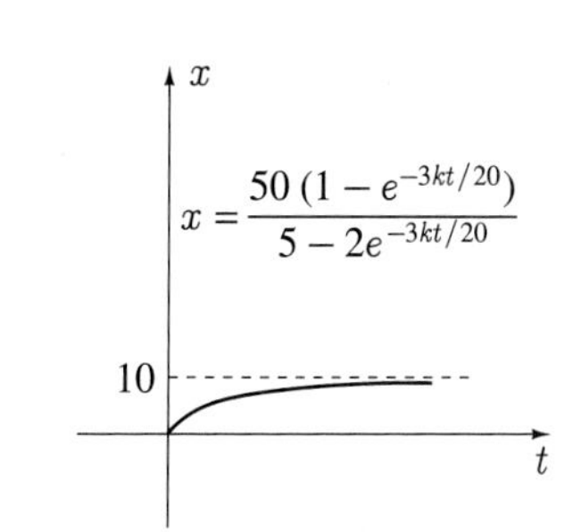

44. 459.5 mg **46.** $y^3(x^3 + 6x - 5) = 1 + y^6$

48. $h = [\sqrt{H} - \sqrt{g}cAt/(\sqrt{2}\pi r^2)]^2$, $\pi r^2\sqrt{2H}/(cA\sqrt{g})$

50. $y = (L/2)[(x/L)^{1-v/V} - (x/L)^{v/V+1}]$

Exercises 16.3

2. $y(x) = x^4 + C/x^2$ **4.** $y\sin x = C - 5e^{\cos x}$

6. $y(x) = C(x + 1)^2 - 2x - 2$

8. $y(x) = Ce^x + e^{2x}$

10. $y(x) = x^3/2 + Cx^3e^{1/x^2}$

12. $y(x) = x + \cos 2x + C\sin 2x$

14. $y(x) = e^x(2\sin x - \cos x)/5 - 4e^{-x}/5$

16. $xy = C + y^4/4$ **18.** $y(x) = e^{-x}/(C - x)$

20. $x^2y + Cye^{-x} = 1$

22. $y(x) = [(3/4)\cos x + C\sec^3 x]^{-1/3}$

24. $1000 + t + 3 \times 10^6/(1000 + t)$ g

26.

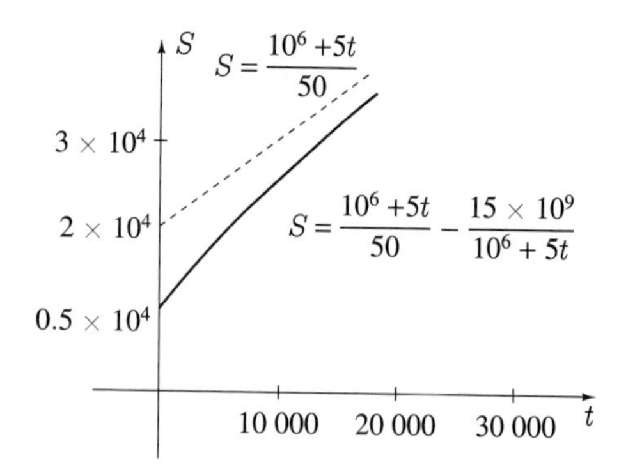

28.

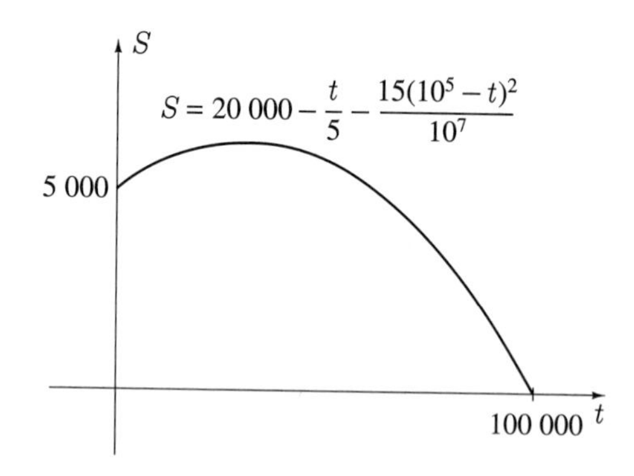

30. **(b)** $I_0 - E_0/(\omega C Z^2)$

Exercises 16.4

2. $2\sqrt{Cy - 1} = \pm Cx + D$

4. $y(x) = Cx - C^2 \ln|x + C| + D$

6. $(x + C)^2 + (y + D)^2 = 1$

8. $D + x/2 = \{-1/y;$ or $C\mathrm{Tan}^{-1}(Cy)$; or, $[1/(2C)]\ln|(y - C)/(y + C)|\}$

10. $y(x) = (1/2)(Ce^x + C^{-1}e^{-x}) + D$

12. **(c)** $LvV/(V^2 - v^2) + (LV/2)$ $\{[1/(V + v)](x/L)^{v/V+1} - [1/(V - v)](x/L)^{-v/V+1}\}$

Exercises 16.5

2. Less than 47.8 km/h; defence

6. **(c)** $x^* = (\mu M g \pm \sqrt{\mu^2 M^2 g^2 + M k v_0^2 + k^2 x_0^2 - 2k\mu M x_0})/k$

8. 18.0 m

10. **(a)** $70.0(1 - 0.556\,e^{-0.280t})/(1 + 0.556\,e^{-0.280t})$ m/s

(b)

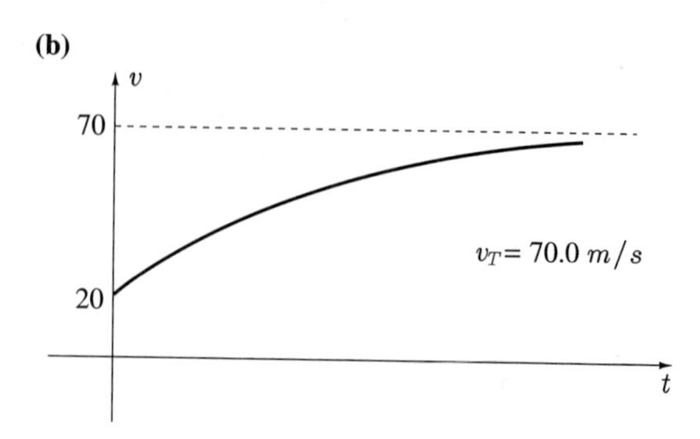

12. 1.99 s

14. **(a)** $v_T \tan[\mathrm{Tan}^{-1}(v_0/v_T) - kv_T t/m]$ where $v_T = \sqrt{9.81m/k}$

(b) $(m/k)\ln|\cos[\mathrm{Tan}^{-1}(v_0/v_T) - kv_T t/m]| + (m/k)\ln(\sqrt{v_0^2 + v_T^2}/v_T)$, $(m/k)\ln(\sqrt{v_0^2 + v_T^2}/v_T)$

16. $(9.81m \sin\alpha/k)(1 - e^{-kt/m})$, $(9.81m^2 \sin\alpha/k^2)(kt/m + e^{-kt/m} - 1)$

Exercises 16.6

2. **(a)** $2\,e^{[(1/2)\ln(5/2)]t}$ **(b)** $2\,e^{[(1/10)\ln 55]t}$

4. **8.**

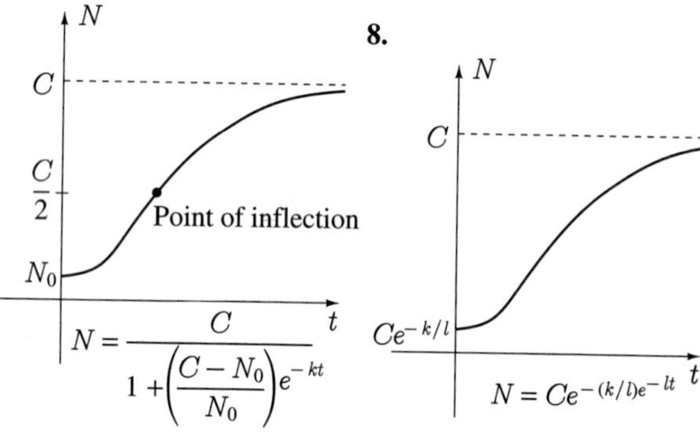

10.

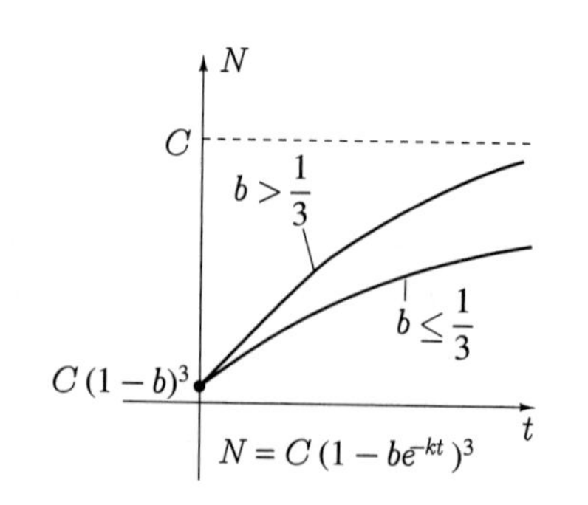

Exercises 16.7

2. Linear **4.** Linear **6.** Linear
8. Linear **10.** Not linear **12.** Linear
14. Linear **16.** Not linear **18.** Not linear
20. Linear

Exercises 16.8

2. $C\cos x$ **4.** $C_1e^{4x} + C_2xe^{4x}$
6. $e^{2x}[C_1\cos(x/\sqrt{2}) + C_2\sin(x/\sqrt{2})]$
8. $C_1\cos(2\ln x) + C_2\sin(2\ln x)$
12. Dependent **14.** Independent

Exercises 16.9

2. $(C_1 + C_2x)e^{4x}$
4. $C_1e^{(-1+\sqrt{3})x} + C_2e^{(-1-\sqrt{3})x}$
6. $C_1e^{3x} + C_2\cos x + C_3\sin x$
8. $(C_1 + C_2x + C_3x^2)e^{2x}$
10. $C_1\cos x + C_2\sin x + C_3\cos 2x + C_4\sin 2x$
12. $e^{\sqrt{2}x}[C_1\cos(\sqrt{2}x) + C_2\sin(\sqrt{2}x)] + e^{-\sqrt{2}x}[C_3\cos(\sqrt{2}x) + C_4\sin(\sqrt{2}x)]$
14. $y'' + 4y' + 20y = 0$
16. $y'''' - 4y''' + 10y'' - 12y' + 9y = 0$
18. $5, 24, 20$

Exercises 16.10

2. $-(x^2/3+2/9)e^{-x}$, $C_1e^{(-1+\sqrt{3})x}+C_2e^{-(1+\sqrt{3})x}-(x^2/3+2/9)e^{-x}$
4. $(1/9)\cos 2x$, $(C_1 + C_2x)\cos x + (C_3 + C_4x)\sin x + (1/9)\cos 2x$
6. $e^{-2x}/40$, $C_1\cos x + C_2\sin x + C_3\cos 2x + C_4\sin 2x + e^{-2x}/40$
8. $e^{2x}(\cos x - 5\sin x)/312$, $e^{-4x}(C_1\cos 5x + C_2\sin 5x) + e^{2x}(\cos x - 5\sin x)/312$
10. $[(2x + 1)\cos x - 2(x + 1)\sin x]/16$, $e^{2x}(C_1\cos x + C_2\sin x) + [(2x + 1)\cos x - 2(x + 1)\sin x]/16$
12. $Axe^{3x} + Be^{3x} + Cx\cos x + Dx\sin x + E\cos x + F\sin x$
14. $Ax^2e^x + Bxe^x + Cx^3 + Dx^2 + Ex + F + G\cos x + H\sin x$
18. $e^x[C_1\cos(x/\sqrt{2}) + C_2\sin(x/\sqrt{2})] + (4\cos 3x - 5\sin 3x)/246 + (\sin x + 4\cos x)/34$
20. $-(1/108)\sin 3x + (x^2/12)(\sin 3x - \cos 3x) + (x/36)(\sin 3x + \cos 3x)$

Exercises 16.11

2. $e^{-15t/4}[(12/65)\cos(5\sqrt{23}t/4) + (20/(13\sqrt{23}))\sin(5\sqrt{23}t/4)] - (4/65)[3\cos(10t) + 2\sin(10t)]$ m
4. $-0.0253e^{-99.50t} + 5.03e^{-0.50t}$ C
6. $(M|g|/k)\cos(\sqrt{k/M}t)$ from equilibrium
8. **(b)** $x = -7/36$ m, yes
12. $(1/64)\sin 8t - (t/8)\cos 8t$ m
14. $(9/10)[3t\cos 3t - \sin 3t]A$, yes
16. **(b)** $0.705/\sqrt{L}$

Review Exercises

2. $y = \pm\sqrt{2(x + \ln|x|) + C}$
4. $y = Ce^{-4x} + (8x^2 - 4x + 1)/32$
6. $y = e^{-3x/2}[C_1\cos(\sqrt{7}x/2) + C_2\sin(\sqrt{7}x/2)] + 1/2$
8. $y = x^3/9 + C_1\ln|x| + C_2$
10. $y = Ce^{-x^2} + x^2 - 1$
12. $y = (C_1 + C_2x)e^{2x} + (3\sin x + 4\cos x)/25$
14. $y = C_1\cos 2x + C_2\sin 2x - (x/4)\cos 2x$
16. $xy^2 = Ce^{-2x-x^2/2}$
18. $e^{-x}[C_1\cos(\sqrt{3}x) + C_2\sin(\sqrt{3}x)] + (\sqrt{3}x/6)e^{-x}\sin(\sqrt{3}x)$
20. $x = Ce^{-3y} + (2/27)(-9y^2 + 6y - 2)$
22. $y = (424/405)e^{9x} + (11/5)e^{-x} - 2x/9 - 20/81$
24. $y = -\cos x + (2\sin x)/x + (2\cos x + 1 - \cos 1 - 2\sin 1)/x^2$
26. **(a)** 3.924×10^{-3} N **(b)** $1.962 \times 10^{-3}t + 4.905 \times 10^{-7}(e^{-4000t} - 1)$ m
28. **(a)** In air, $v = 9.81t$; in water $v = 0.34335 + 1.502 \times 10^{29}e^{-20t}$
(b) In air, $y = 4.905t^2$; in water, $y = 50.30 + 0.34335t - 7.51 \times 10^{27}e^{-20t}$
(c) 28.25 s

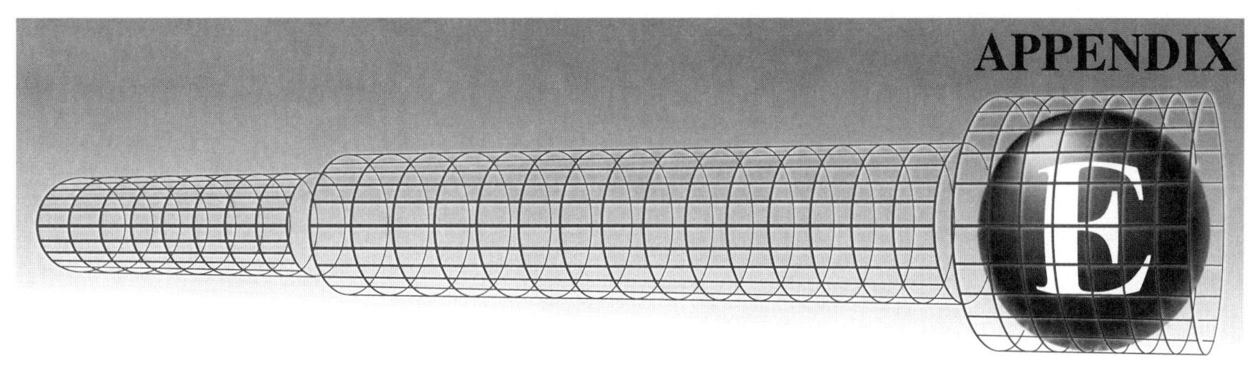

Additional Exercises

When you solve the exercises in a particular section of this text, you already know the nature of the problem and the techniques that should be used. For example, a problem in Exercises 3.7 will use implicit differentiation, a problem in Exercises 9.5 will use a trigonometric substitution, etc. To ensure that you understand all parts of a chapter, we recommended the review exercises at the end of each chapter. They require you to first identify the particular section that is useful in solving a problem, and then to follow through with its application. Now that you have studied either the first eleven chapters or the entire book, you should test your understanding of calculus as a whole. The following exercises, which review all aspects of the course, will do this. They test your ability to recognize the essential nature of a problem — be it a maximum-minimum problem, a related rate problem, an application of definite integrals, a problem in infinite series, etc. — and they tax your organizational, analytical, and interpretive skills.

The problems are divided into two groups, 370 on Chapters 1–11 and 215 on Chapters 12–16. The first 200 or so exercises in the first group are a reasonably straightforward review of the many types of problems that you have already encountered. The next 100 are more difficult; they may require some originality on your part or some extra calculations. The next 50 are substantially more involved, and the last 15 or so are delightfully challenging. Likewise for the multivariable problems, the first 160 or so are a reasonably straightforward review. The reminder are more challenging. Set your sights high; attempt as many problems as you can. Answers to selected exercises are found at the end of the appendix.

ADDITIONAL EXERCISES — Chapters 1–11

1. Find all points (x, y) satisfying $y + |y| = x + |x|$.

2. (a) Sketch the graph of an even, 2π-periodic function $f(x)$ for which

$$f(x) = x^3 - \pi x, \quad 0 < x < \pi.$$

Can $f(x)$ be made continuous?

(b) Sketch the graph of an odd, 2π-periodic function $f(x)$ for which

$$f(x) = x^3 - \pi x, \quad 0 < x < \pi.$$

Can it be made continuous?

In Exercises 3–18 evaluate the limit if it exists. If a limit does not exist, explain why not.

3. $\displaystyle\lim_{x\to 0} \frac{x^2 - x}{2x + x^2}$

4. $\displaystyle\lim_{x\to -4^-} \frac{x^2 - 16}{x + 4}$

5. $\displaystyle\lim_{x\to 0} \frac{\sqrt{2+x} - \sqrt{2-x}}{x}$

6. $\displaystyle\lim_{x\to 5} \frac{|x^3 - 125|}{x - 5}$

7. $\displaystyle\lim_{x\to 3} \frac{x - 3}{x^2 - 6x + 9}$

8. $\displaystyle\lim_{x\to 0^-} \frac{3x}{2 - \sqrt{x^2 + 4}}$

9. $\displaystyle\lim_{x\to \infty} \frac{2 - x^5}{3 + 4x^5}$

10. $\displaystyle\lim_{x\to -\infty} \tan x$

11. $\displaystyle\lim_{x\to 1^+} \left[\left(1 - \frac{1}{x}\right)^3 + x - \frac{2}{x^2}\right]$

12. $\displaystyle\lim_{x\to 0} \frac{\sin 4x^{12}}{3x^{12}}$

13. $\displaystyle\lim_{x\to \infty} \left(\sqrt{x^2 + 3} - \sqrt{x^2 + 1}\right)$

14. $\displaystyle\lim_{x\to -\infty} \left(\sqrt{x^2 + x} - \sqrt{x^2 + 2x}\right)$

15. $\lim_{x \to -\pi/2^+} \tan x$

16. $\lim_{x \to 1} e^{\frac{1}{x-1}}$

17. $\lim_{x \to 2} \dfrac{\sqrt{x^2 - 4x + 4}}{x - 2}$

18. $\lim_{x \to 1} \dfrac{x + 2\sqrt{x} - 3}{x^2 - 1}$

In Exercises 19–69 assume that y is defined as a function of x by the given equation(s), and find dy/dx.

19. $y = x^2(1 - 3x)^4$

20. $y = \dfrac{x^2}{\sqrt{1 - 3x^2}}$

21. $y = \dfrac{\sin 2x}{\cos 3x}$

22. $y = (2x + 1)^3 \ln(x + 1)$

23. $y = (x^2 + 2x - 3)e^{3x}$

24. $y = \sqrt{\ln(e^{2x} + 1)}$

25. $y = \mathrm{Sin}^{-1}(\sqrt{1 - 4x})$

26. $y = x^2 \csc^2(x^2 - 1)$

27. $y = \dfrac{x}{1 + \sec(1 - 2x^3)}$

28. $y = \sqrt[3]{1 + \sqrt{2 - x}}$

29. $x^2y + y^3x - 2x = 5$

30. $\sin(x + y) = xy$

31. $x^3e^y + x - 2y = 3x^2$

32. $x\mathrm{Tan}^{-1}(xy) = y + 3$

33. $y = \cos(v^3 + 2), \quad v = \dfrac{x}{x + 1}$

34. $y = \mathrm{Sec}^{-1}(1 - 2u), \quad u = \tan x + 1$

35. $y = \sqrt{v - \sqrt{v}}, \quad v = \sqrt{x + \sqrt{x}}$

36. $y = t^2e^{-3t}, \quad t = x + \ln x$

37. $x = t + \dfrac{1}{t}, \quad y = 2t - \dfrac{3}{t^2}$

38. $x = 4 + 3\cos t, \quad y = 2 - 4\sin t$

39. $x = \dfrac{\theta}{1 + \theta^2}, \quad y = \dfrac{\theta^2}{1 + \theta^2}$

40. $x = e^{3v} - v, \quad y = e^{-3v} - v^2$

41. $y = 3 + 2x - 6\ln(3 + 2x) - \dfrac{9}{3 + 2x}$

42. $y = \ln\left(\dfrac{4 - x}{x}\right)$

43. $y = \ln\left[\dfrac{(2 + x)^3}{8 + x^3}\right] + 2\sqrt{3}\,\mathrm{Tan}^{-1}\left(\dfrac{x - 1}{\sqrt{3}}\right)$

44. $y = \ln\left(\dfrac{2x^2 - 2x + 1}{2x^2 + 2x + 1}\right) + 2\,\mathrm{Tan}^{-1}\left(\dfrac{2x}{1 - 2x^2}\right)$

45. $y = (8 - 12x + 27x^2)\sqrt{1 + 3x}$

46. $y = \ln\left(x + \sqrt{x^2 + 9}\right)$

47. $y = \sqrt{x^2 - 4} - 2\,\mathrm{Sec}^{-1}(x/2)$

48. $y = -\dfrac{\sqrt{7x^2 + 49}}{x^2} + \ln\left(\dfrac{\sqrt{7} + \sqrt{x^2 + 7}}{x}\right)$

49. $y = 2x(3 - x^2)^{3/2} + 9x\sqrt{3 - x^2} + 27\,\mathrm{Sin}^{-1}\left(\dfrac{x}{\sqrt{3}}\right)$

50. $y = (3x^2 + 2)(1 - x^2)^{3/2}$

51. $y = x\sqrt{16 - x^2} + 16\cos(x/4)$

52. $y = \dfrac{5}{\sqrt{25 - x^2}} - \ln\left(\dfrac{5 + \sqrt{25 - x^2}}{x}\right)$

53. $y = \ln\left(\dfrac{3\sqrt{16 - x^2} + \sqrt{7}x}{\sqrt{9 - x^2}}\right)$

54. $y = \dfrac{3x + 12}{\sqrt{6 + 3x + 4x^2}}$

55. $y = \mathrm{Cos}^{-1}\left(1 - \dfrac{x}{4}\right)$

56. $y = \dfrac{x - 7}{\sqrt{14x - x^2}}$

57. $y = \ln\left(x + 3 + \sqrt{6x + x^2}\right)$

58. $y = \mathrm{Csc}^{-1}\sqrt{\dfrac{3x^5}{2}}$

59. $y = \mathrm{Tan}^{-1}\left(\dfrac{4\tan(x/2) + 3}{\sqrt{7}}\right)$

60. $y = \mathrm{Tan}^{-1}\left(\dfrac{3\tan x}{2}\right)$

61. $y = 3\cot\left(\dfrac{3x}{2}\right) - \cot^3\left(\dfrac{3x}{2}\right)$

62. $y = (3x^2 - 6)\cos x + (x^3 - 6x)\sin x$

63. $y = \mathrm{Sin}^{-1}\left(\dfrac{3\cos 2x}{\sqrt{10}}\right)$

64. $y = \ln\left(7\sin 3x + \sqrt{1 + 49\sin^2 3x}\right)$

65. $y = (1 + x^2)\mathrm{Cot}^{-1}x + x$

66. $y = \dfrac{\sqrt{4x^2 - 1}}{x} - \dfrac{\mathrm{Sec}^{-1}(2x)}{x}$

67. $y = e^{3x}(3\sin 2x - 2\cos 2x)$

68. $y = 2x - \ln\left(5 + 3e^{2x}\right)$

69. $y = 13xe^{2x}(2\sin 3x - 3\cos 3x) + e^{2x}(5\sin 3x + 12\cos 3x)$

70. For what value of k are the curves $y = 1 - x^2$ and $y = x^2 - 4x + k + 4$ tangent? What is the point of tangency?

71. Find the equation of the tangent line to the curve $y = (x+9)/(x+5)$ that passes through the origin.

72. Find the first and second derivatives of the function $y = f(x)$ defined by the equation $y^3 + (x-1)y + 16 = 3x$ when $x = 2$.

73. Verify that when n is a positive integer

$$\frac{d}{dx}(\sin^n x \cos nx) = n \sin^{n-1} x \cos(n+1)x.$$

74. Show that $f(x) = \ln\left(x + \sqrt{x^2+1}\right)$ is an odd function and sketch its graph.

75. A chord of the parabola $y = x^2 - 2x + 5$ joins the points with abscissas $x = 1$ and $x = 3$. Find the equation of the tangent line to the parabola parallel to the chord.

76. Find the second derivative of the function $y = f(x)$ defined by the equation $xy^2 + x^2 y = 16$ at the point where its first derivative is equal to 0.

77. Show that the function defined by $y^3 + x^4 y = 3$ is decreasing when $x > 0$.

In Exercises 78–103 evaluate the indefinite integral.

78. $\displaystyle\int \frac{1}{3 + 4e^{2x}}\,dx$ **79.** $\displaystyle\int \frac{1}{4e^{3x} + 5e^{-3x}}\,dx$

80. $\displaystyle\int (x-1)^2 \ln x\,dx$ **81.** $\displaystyle\int \ln(x^2+9)\,dx$

82. $\displaystyle\int \sin^2 3x\,dx$ **83.** $\displaystyle\int \cos^3(5x+1)\,dx$

84. $\displaystyle\int \sin 3x \cos^2 4x\,dx$ **85.** $\displaystyle\int \frac{1}{\sqrt{x^2+5}}\,dx$

86. $\displaystyle\int \frac{x+5}{\sqrt{x^2-3}}\,dx$ **87.** $\displaystyle\int \frac{\sqrt{x^2+1}}{x^4}\,dx$

88. $\displaystyle\int \frac{1}{x^2\sqrt{7-x^2}}\,dx$ **89.** $\displaystyle\int \frac{1}{(x^2+2x+5)^{5/2}}\,dx$

90. $\displaystyle\int \frac{x}{\sqrt{2x-x^2}}\,dx$ **91.** $\displaystyle\int \frac{2-5x-x^2}{x^3+x^2+5x+5}\,dx$

92. $\displaystyle\int \frac{x^3+5x^2+12x+12}{x^3+2x^2+4x}\,dx$

93. $\displaystyle\int \csc^3 2x\,dx$ **94.** $\displaystyle\int \csc^4 x\,dx$

95. $\displaystyle\int \frac{7x^4-6x^3+16x^2-6x+9}{2x^5-4x^4+5x^3-6x^2+3x}\,dx$

96. $\displaystyle\int \frac{e^{2x}+1}{(2e^x+5)^3}\,dx$ **97.** $\displaystyle\int \frac{x+2}{\sqrt{1-x^2}}\,dx$

98. $\displaystyle\int \frac{x^2}{(5-x^2)^{3/2}}\,dx$ **99.** $\displaystyle\int \frac{1}{\sqrt{-2x^2+4x+3}}\,dx$

100. $\displaystyle\int (6x-3)\ln(x^2-x+5)\,dx$

101. $\displaystyle\int \cot^4\left(\frac{x}{2}\right)dx$

102. $\displaystyle\int \frac{\tan^3 x}{\sec x}\,dx$ **103.** $\displaystyle\int \frac{1}{\sec x - \tan x}\,dx$

In Exercises 104–123 determine whether the series converges or diverges. In the case of a series with both positive and negative terms, determine whether convergence is absolute or conditional.

104. $\displaystyle\sum_{n=1}^{\infty} \frac{n^2+1}{3n^3+5}$ **105.** $\displaystyle\sum_{n=1}^{\infty} \frac{(n^3+4n)^{2/3}}{n^4-3n^3+1}$

106. $\displaystyle\sum_{n=1}^{\infty} \frac{(-1)^{n+1}n^2}{2^n}$ **107.** $\displaystyle\sum_{n=1}^{\infty} \frac{(-1)^n}{\sqrt{n}+1}$

108. $\displaystyle\sum_{n=2}^{\infty} \frac{1}{n\ln n}$ **109.** $\displaystyle\sum_{n=1}^{\infty} n^{1/n}$

110. $\displaystyle\sum_{n=1}^{\infty} \frac{\sin(n+1)}{n^2+4}$ **111.** $\displaystyle\sum_{n=1}^{\infty} \frac{1+2^{-n}}{1+3^{-n}}$

112. $\displaystyle\sum_{n=1}^{\infty} \sqrt{\frac{n^2+n-1}{2n^4+n^3+5}}$ **113.** $\displaystyle\sum_{n=1}^{\infty} \left(\frac{n^2-2}{2n^2+5}\right)^n$

114. $\displaystyle\sum_{n=1}^{\infty} \frac{n^2}{n^{3n}}$ **115.** $\displaystyle\sum_{n=1}^{\infty} \frac{\mathrm{Tan}^{-1}(n^2+5)}{2n^2+4}$

116. $\displaystyle\sum_{n=1}^{\infty} \frac{1\cdot 3\cdot 5\cdots(2n+1)}{2\cdot 4\cdot 6\cdots(2n)}\left(\frac{1}{n}\right)$

117. $\displaystyle\sum_{n=1}^{\infty} \frac{(n!)^2}{(2n+1)!}$

118. $\displaystyle\sum_{n=1}^{\infty} \frac{\sqrt{n+\sqrt{1+n^2}}}{n^2}$ **119.** $\displaystyle\sum_{n=1}^{\infty} \left(\frac{1}{2}-\frac{1}{n+1}\right)^{10}$

120. $\displaystyle\sum_{n=1}^{\infty} \frac{3}{n(1+\ln n)^3}$ **121.** $\displaystyle\sum_{n=1}^{\infty} \frac{3^{-2n}+2^{-2n}}{4^{-n}+5^{-n}}$

122. $\displaystyle\sum_{n=1}^{\infty} \frac{(n-1)^n}{n^{n-1}}$ **123.** $\displaystyle\sum_{n=1}^{\infty} \mathrm{Sin}^{-1}\left(\frac{1}{n^2}\right)$

124. Find point(s) on the curve $y = 3x^2/(x+1)$ where the inclination of the tangent line is equal to 1.

125. Can a continuously differentiable function defined by the equation $\sin(x+y) = 2x+3$ have a critical point?

126. When a projectile (such as an arrow) is fired up an inclined plane (Figure E.1), the distance that it travels before striking the plane is called its range R. It can be shown that

$$R = \frac{v^2 \cos\theta \sin(\theta-\alpha)}{4.905 \cos^2\alpha},$$

where v is the initial speed of the projectile, α is the angle of inclination of the plane, and θ is the angle of projection with the horizontal. Find the angle θ for which R is greatest.

FIGURE E.1

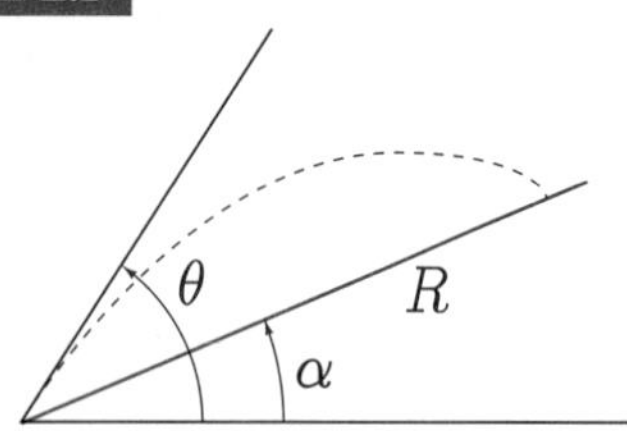

127. Evaluate

$$\lim_{x \to 0} \frac{\sqrt{1 + \sin x} - \sqrt{1 - \sin x}}{\tan x}.$$

128. A differential equation in $y = f(x)$ is said to be homogeneous, linear, second order if it is of the form

$$a(x)\frac{d^2 y}{dx^2} + b(x)\frac{dy}{dx} + c(x)\,y = 0,$$

where $a(x)$, $b(x)$, and $c(x)$ are functions of x. Show that if $f_1(x)$ and $f_2(x)$ are two solutions of this equation, so is $y = p_1 f_1(x) + p_2 f_2(x)$ for any constants p_1 and p_2.

In Exercises 129–132 find the area bounded by the curves.

129. $y = 2/\sqrt{x-1}, \quad x + 3y = 8$

130. $y = 8 - 3x^2, \quad y = 3x^2/(x^2+1)$

131. $x + y = 2e, \quad x = y \ln y, \quad x = 0$

132. $x = (2/\pi)\,\mathrm{Sin}^{-1}\, y, \quad y = 2 - x^2, \quad y = 0$

In Exercises 133–135 find the volume of the solid of revolution obtained by rotating the area bounded by the curves about the line indicated.

133. $x = y(y-1), \quad x = 4y(y-1) \quad$ about $x = 0$

134. $y = (x^2 + x)/(x-2), \quad y = 0 \quad$ about $x = 1$

135. $x = e^{2y}, \quad x = 2, \quad y = 0 \quad$ about $y = 2$

136. Find the length of $y = \ln(1 - x^2)$ between the lines $x = 0$ and $x = 1/2$.

137. A thin uniform rod of length L and mass M spins in a plane about one of its ends with angular velocity ω. What is its kinetic energy?

138. The length of the hypotenuse of a right-angled triangle is 1 metre, and the lengths of the other two sides are a and b. When the triangle is submerged vertically in a fluid so that the edge with length b is in the surface of the fluid, the fluid force on each side of the triangle is twice that when the side of length a is in the surface. What are a and b?

139. Find the x-coordinate of the centroid of the area bounded by the curves

$$y = \frac{x^2 + 2x - 3}{x^2 + 1}, \qquad y = 0.$$

140. Find moments of inertia of an elliptic plate of uniform mass per unit area about its major and minor axes.

141. If $f(x)$ satisfies the condition $f(f(x)) = x$ for all x in its domain, what is $f^{-1}(x)$? Give an example.

142. In Kintsch's model of choice behaviour, he states that

$$\sum_{n=2}^{\infty} (1-b)^{n-2}\, bs^n = \frac{bs^2}{1 - (1-b)s}.$$

Is this true, and would you place any restrictions on b and s?

143. Let P be a point on the circumference of a circle. Where should a chord QR be drawn parallel to the tangent at P in order that triangle PQR have maximum area?

144. In Figure E.2, a 5 m rod CD is such that its end C can only move horizontally along the line AGB and end D can only move horizontally along EF. When C is 3 m to the left of G, it is moving toward A at 4 m/s. In what direction and how fast is D moving at this instant?

FIGURE E.2

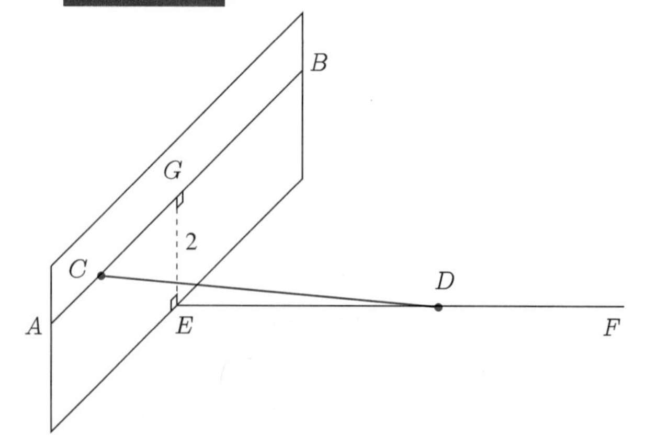

145. Evaluate

$$\lim_{n \to \infty} \frac{1}{n^2}(1 + 2 + 3 + \cdots + n).$$

146. Find d^2y/dx^2 if

$$y = \int_0^{x^2} t \cos t \, dt.$$

147. Are either of the following valid?

$$\int_0^a f(x)\,dx = \int_0^a f(a-x)\,dx;$$

$$\int_0^a f(x)\,dx = \int_0^a f(b-x)\,dx.$$

148. Find the volume of the solid of revolution obtained by rotating the area bounded by $(x-4)y^2 = x(x-3)$ and the x-axis, about the x-axis.

149. Find the Maclaurin series for $f(x) = 1/(x^4 - x^2 - 2)$.

150. If $f(x)$ is a polynomial of degree 4, and it is known that $f(1) = 1$, $f'(1) = -2$, $f''(1) = 0$, $f'''(1) = 3$, and $f''''(1) = -5$, what is $f(-2)$?

151. (a) A function $f(x)$ is said to be odd-harmonic if it satisfies

$$f(L+x) = -f(x)$$

for all x. Show that $f(x)$ must be $2L$-periodic.

(b) A function $f(x)$ is said to be odd and odd-harmonic if it satisfies

$$f(-x) = -f(x) \qquad \text{and} \qquad f(L-x) = f(L+x).$$

Show that $f(x)$ must be $4L$-periodic.

(c) Sketch the graph of the extension of $f(x) = x$, $0 < x < L$ to a continuous, odd, and odd-harmonic function.

(d) A function $f(x)$ is said to be even and odd-harmonic if it satisfies

$$f(-x) = f(x) \qquad \text{and} \qquad f(L+x) = -f(L-x).$$

Show that $f(x)$ must be $4L$-periodic.

(e) Sketch the graph of the extension of $f(x) = x$, $0 < x < L$ to an even and odd-harmonic function.

152. Find the distance from the point $(1,2)$ to the tangent line to the curve $y = x^2 \ln x$ at the point with abscissa equal to 2.

153. Show that a function defined by the equation $xye^{xy} = 1$ cannot have a first derivative equal to 1.

154. Find all critical points of the function

$$f(x) = x^{ax}, \qquad x > 0$$

where $a > 0$ is a constant.

155. Show that the equation $x^n + px + q = 0$ cannot have more than two real roots when n is an even integer, and not more than three real roots when n is an odd integer.

156. The reproduction rate of Drosophila melanogaster supposedly decreases with population. Show that this is the case if the progeny per female per day, p, when x denotes the number of flies per bottle is

$$p = 34.53x^{-0.658}e^{-0.018x}, \qquad x \geq 1.$$

157. Can a continuous function $y = f(x)$ with a first derivative satisfying

$$\frac{dy}{dx} + \sin y = 2 + x^2$$

have relative extrema?

158. For what value of k does the function $f(x) = k\cos x + 3x\sin x$ have a relative extremum at $x = \pi/3$? Is it a relative maximum or minimum?

159. Show that the angle of inclination of a curve has relative extrema at points of inflection.

160. Is the curve

$$x = t^2 - \frac{1}{t}, \quad y = t^3 + 3t, \qquad t \geq 1$$

concave upward or concave downward?

161. Thurstone's learning curve is

$$y = \frac{Lx + Lc}{x + c + a}, \qquad x \geq 0$$

where $L > 0$ is the limit of practice, x is the amount of formal practice, c is the equivalent previous practice, and a is the rate of learning. Sketch a graph of the learning curve.

162. One end of a ladder 20 m long slips down the side of a house at 1 m/s. What is the acceleration of the foot of the ladder when it is 15 m from the house?

163. If a point moves around the ellipse $b^2x^2 + a^2y^2 = a^2b^2$, at what points are the rates of change of its abscissa and ordinate equal?

164. If the velocity of a particle moving along the x-axis is given by $A\cos(\omega t + \phi)$, and its position at time t_1 is x_1, what is its position at time t_2?

165. If $|f(x) - g(x)| \leq \epsilon$ for $a \leq x \leq b$, what is the maximum value of

$$\left| \int_a^b [f(x) - g(x)]\, dx \right| ?$$

166. Find the length of the loop of the curve $9ay^2 = x(x-3a)^2$ where $a > 0$ is a constant.

167. A block of wood (density 500 kg/m^3) is cubical with side lengths L m. If it floats in water, what work is required to lift it out of the water?

168. The lengths of the major and minor axes of an elliptic plate are $2a$ and $2b$. Let F_a denote the force on each side of the plate when it is half submerged in a fluid with major axis in the surface, and let F_b denote the force when the minor axis is in the fluid. What is the ratio F_a/F_b?

169. A thin plate with constant mass per unit area has the shape of a sector of a circle of radius r and angle α where $0 < \alpha < \pi/2$. How far from the vertex of the sector is the centre of mass?

170. Find the inverse function for

$$f(x) = \frac{e^x - e^{-x}}{e^x + e^{-x}} + 1.$$

171. Approximate

$$\int_0^{1/2} \frac{1}{1+x^4}\, dx$$

(a) using Simpson's rule with 10 subdivisions;

(b) using the first four terms of the Maclaurin series of the integrand.

Discuss the relative merits of these techniques for this example.

172. Find the centroid of the area bounded by the x-axis and the arch of the cycloid described by

$$x = R(\theta - \sin\theta), \quad y = R(1-\cos\theta), \qquad 0 \leq \theta \leq 2\pi.$$

173. Find the length of the curve

$$r = \sin^3\left(\frac{\theta}{3}\right), \qquad 0 \leq \theta \leq 3\pi.$$

174. Find the Maclaurin series for

$$f(x) = \ln[(1+x)^{1+x}] + \ln[(1-x)^{1-x}].$$

What is the interval of convergence of the series?

175. What are the first three terms in the Maclaurin series for the function defined by the equation

$$y + \ln(1+xy) + x = 5.$$

176. Evaluate

$$\lim_{x \to 0} \frac{6x - x^3 - 6\sin x}{x^5}.$$

177. Find constants b_n in order that the function $y = f(x) = \sum_{n=1}^{\infty} b_n \sin nx$ will satisfy the differential equation

$$\frac{d^2y}{dx^2} - 3y = \sum_{n=1}^{\infty} \frac{1}{n^2} \sin nx.$$

Assume that the series for $f(x)$ can be differentiated term-by-term.

178. Find the distance from the point of intersection of the curves $y = 1/(x-1)$ and $4y = x^2$ to the normal line to the curve $x^3y + xy^2 = 6$ at the point with ordinate $y = -1$.

179. Verify that for $0 < a < b$,

$$\frac{b-a}{b} < \ln\left(\frac{b}{a}\right) < \frac{b-a}{a}.$$

180. It is known that a continuous function $y = f(x)$ always has a positive derivative which satisfies the equation

$$\frac{dy}{dx} = y^3 - 3y.$$

Show that its graph can have a point of inflection only at a point where $y = 1$ or $y = -1$.

181. The circles in Figure E.3 represent ferris wheels with radii 20 m and 10 m. Their centres are displaced horizontally by 50 m and vertically by 30 m. The larger rotates half as fast as the smaller, making one revolution every 10 seconds. How fast is the distance changing between riders at positions P and Q? Assume both ferris wheels turn counterclockwise.

FIGURE E.3

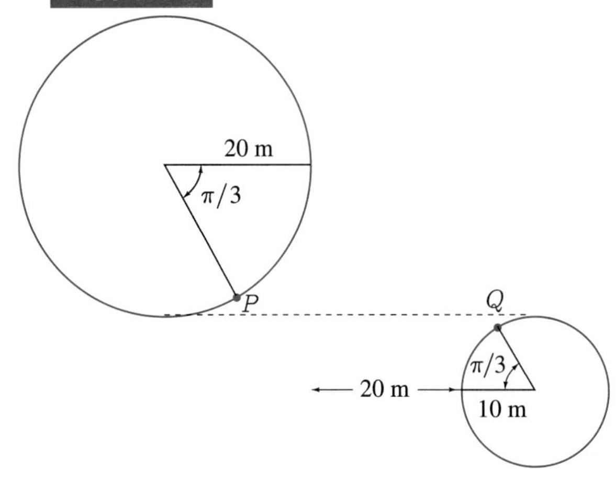

182. At time $t = 0$, a body is moving along the x-axis with velocity $v_0 > 0$. If the only force acting on it is resistance, and this force is proportional to the square of the velocity of the body, where and when does the body stop?

183. Find the length of the curve

$$y = L \ln\left(\frac{L + \sqrt{L^2 - x^2}}{x}\right) - \sqrt{L^2 - x^2}$$

from $(L, 0)$ to any other point (x, y) on the curve (see Exercise 24 in Section 9.5).

184. Find the centroid of the area bounded by the curve $|y| = x\sqrt{1 - x^2}$.

185. If the number of subdivisions in the approximation of a definite integral by Simpson's rule is doubled, does the error decrease by a factor of 16? Explain.

186. Show that a function $f(x)$ and the nth degree polynomial

$$P_n(x) = f(c) + f'(c)(x - c) + \cdots + \frac{f^{(n)}(c)}{n!}(x - c)^n$$

have the same first n derivatives at $x = c$.

187. Find a cubic polynomial approximating the function

$$f(x) = \int_0^x e^{-t^2}\, dt.$$

How accurate is it on the interval $-0.2 \le x \le 0.2$?

188. Show that if a continuous function $y = f(x)$, which satisfies the equation

$$\frac{dy}{dx} - y^3 = y,$$

has a point of inflection on its graph, it must be a horizontal point of inflection.

189. Find circle(s) that have centres on the line $2x + 3y = 6$ and intersect the circle $x^2 + y^2 = 1$ at only one point.

190. One hundred dollars is invested at 10% compounded quarterly. Sketch a graph of the accumulated value of the investment as a function of time t (in months) taking $t = 0$ at the time it is initially invested. Express this function in terms of the greatest integer function.

191. Evaluate

$$\lim_{y \to 0} \frac{(x + y)^{1/3} - x^{1/3}}{y}.$$

192. Evaluate

$$\lim_{x \to \infty} x\left(\sqrt{x^2 + \sqrt{x^4 + 6}} - \sqrt{2}\,x\right).$$

193. If $f(x)$ is such that $|f(x)| \le x^2$ for all x, what is $f'(0)$?

194. Show that the angle of inclination of the tangent line to the curve $y = \operatorname{Tan}^{-1}(2x^3)$ always lies in the interval $0 \le \phi \le \operatorname{Tan}^{-1} 2$.

195. Calculate

$$\frac{d^{20}}{dx^{20}}[(x^2 + 1)\sin x].$$

196. Show that any tangent to the curve $y = (1/2)\sqrt{x - 4x^2}$ intersects the y-axis at a point equidistant from the point of tangency and the origin.

197. Let P be any point on the hyperbola $b^2x^2 - a^2y^2 = a^2b^2$. If the tangent line at P cuts the asymptotes of the hyperbola at points Q and R, show that P is the midpoint of the line segment joining Q and R.

198. Show that the x-intercept of the tangent line to the curve $a/x^2 + b/y^2 = 1$ at any point is proportional to the cube of the abscissa of the point of tangency.

199. (a) Show that when the area of a circle of radius r is approximated by an inscribed polygon with n equal sides, the area of the polygon is

$$A_n = \frac{nr^2}{2}\sin\left(\frac{2\pi}{n}\right).$$

(b) Verify that $\lim_{n \to \infty} A_n = \pi r^2$.

200. Prove that the function

$$f(x) = \frac{x^2}{2} + \frac{1}{2}x\sqrt{x^2 + 1} + \ln\sqrt{x + \sqrt{x^2 + 1}}$$

satisfies the differential equation

$$x\frac{dy}{dx} + \ln\left(\frac{dy}{dx}\right) = 2y.$$

201. Verify that if a polynomial of even degree ever takes on a value opposite in sign to the coefficient of its highest-degree term, it must have at least two real zeros.

202. For what values of a and b does the function $f(x) = ax \ln x + bx^2 + x$ have relative extrema at $x = 1$ and $x = 2$? Are they relative maxima or minima?

203. A man's cottage is situated on a lake that is more or less circular with radius 2 km. Diametrically opposite is a store where he buys groceries. If the man can paddle his canoe at v km/hr and can walk along the shore at V km/hr, what combination of paddling and walking produces the quickest trip from cottage to store?

204. A fish swims upstream at a constant speed v relative to the water. The water itself has speed v_1 relative to the ground. The fish intends to reach a point at distance s upstream. The energy required is determined by the friction in water and by the time t taken. Experiments have shown that this energy is $E = cv^k t$ where $c > 0$ and $k > 2$ are constants (k depending on the shape of the fish). Given that v_1 is constant, what speed v minimizes E?

205. (a) One end of a ladder slips away from a wall at a uniform rate. Show that the speed of the other end down the wall is always increasing.

(b) Show that when the upper end falls at a constant rate, the speed of the lower end decreases.

206. An eagle is gliding horizontally 100 m above the ground at 15 m/s. It is 200 m south of a tree and is heading due north. A rabbit is 50 m east of the tree hopping along at 2 m/s toward the tree. How fast is the distance between the eagle and rabbit changing at this instant?

207. Show that

$$\lim_{x\to\infty} \frac{x - \sin x}{x + \sin x}$$

exists, but that it cannot be evaluated with L'Hôpital's rule.

208. A particle moves along the x-axis with position (in metres) given by

$$x(t) = t^3 - 9t^2 + 24t + 3, \qquad t \geq 0,$$

where t is time (in seconds). Determine:

(a) the speed of the particle when it is at position $x = 39$;

(b) when the particle is speeding up;

(c) the closest it gets to the point $x = -1$ during the time interval $1 \leq t \leq 3$;

(d) its maximum speed when it is between $x = 20$ and $x = 21$.

209. A man eats a diet of 2 500 calories per day; 1 200 of them go to basic metabolism (i.e., get used up automatically). He also spends 16 calories per day for each kilogram of body mass in exercise. Assume that the storage of calories as fat is 100% efficient and that 1 kilogram of fat contains 10 000 calories. Find his mass as a function of time, and sketch this function, if initially his mass is (a) 100 kg (b) 70 kg. Is he on a diet?

210. Evaluate

$$\lim_{n\to\infty} \frac{1 - 2 + 3 - 4 + \cdots - 2n}{(3 + 8n^3)^{1/3}}.$$

211. Show that the definite integral from $x = 0$ to $x = 1$ with respect to x does not exist for the function

$$f(x) = \begin{cases} 1 & x \text{ is a rational number} \\ 0 & x \text{ is an irrational number.} \end{cases}$$

212. Without evaluating the integral

$$\int_0^{10} \frac{x}{x^3 + 16}\,dx,$$

show that it must be less than $5/6$.

213. Assuming that y is defined implicitly as a function of x by

$$\int_0^y \frac{1}{1 + t^3}\,dt + \int_0^x \cos t^2\,dt = 5,$$

find dy/dx.

214. Evaluate

$$\int_{-a}^{a} \sin^2 x \ln\left(\frac{1 + x}{1 - x}\right) dx$$

when $0 < a < 1$.

215. (a) What is the area bounded by the line $y = 2x + 1$ and the parabola $y = x^2$?

(b) Find the line parallel to $y = 2x + 1$ which along with $y = x^2$ bounds an area equal to half that in (a).

216. Find the volume of the solid of revolution obtained by rotating the area bounded by $x^4 + y^4 = x^3$ about the x-axis.

217. A tank is in the form of a frustrum of a right-circular cone (Figure E.4). If the tank is full of oil with density ρ, what work is required to empty the tank over its top edge?

FIGURE E.4

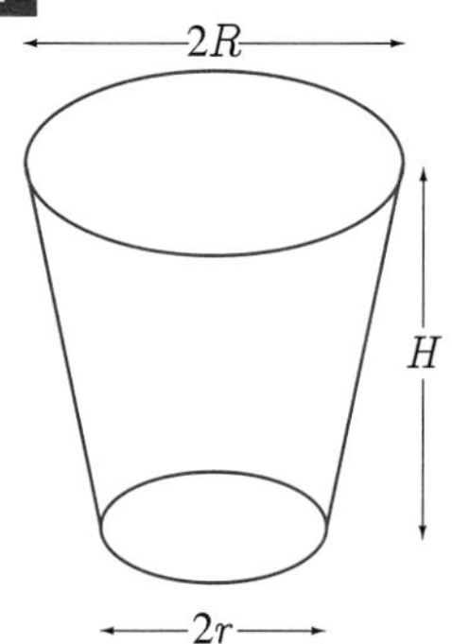

218. The lower half of a cubical tank 2 m on each side is occupied by water, and the upper half by oil (density 0.90 grams per cubic centimetre).

(a) What is the force on each side of the tank due to the pressure of the water and the oil?

(b) If the oil and water are stirred to create a uniform mixture, does the force on each side change? If not, explain why not. If so, by how much does it increase or decrease?

219. Find the second moment of area of the larger loop of $y^2 = (x+1)^2(4-x^2)$ about the y-axis.

220. Find the volume cut from a right circular cylinder of radius r by a plane passing through its diameter (Figure E.5).

FIGURE E.5

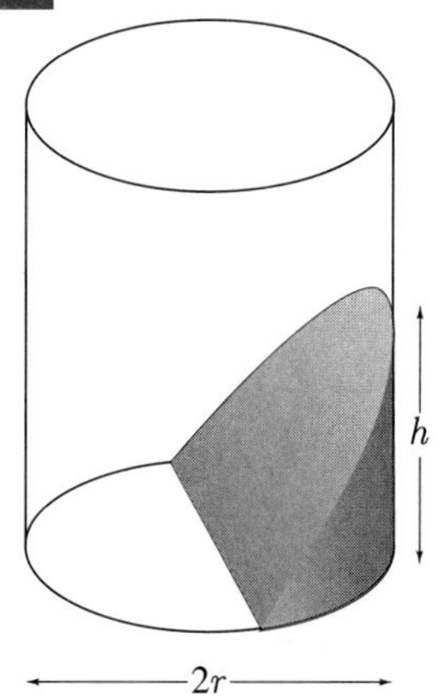

221. A thin rod of length L has σ coulombs of positive charge uniformly distributed over each metre of length (Figure E.6). Suppose a positive charge q coulombs is placed at point P, a distance of a units along the right-bisector of the rod. According to Coulomb's law, the magnitude of the force of repulsion on q by a small amount of charge Q at point R is

$$\frac{qQ}{kr^2}$$

where k is a constant and $r = \|PR\|$. A similar charge equally distant from the other end of the rod also repels q but in a different direction. The combined repulsive force on q by these two charges is perpendicular to the rod and has magnitude

$$\frac{2\,qQ\cos\theta}{kr^2},$$

where θ is the angle shown.

(a) Find the total force on q.

(b) What is the force when the rod is infinitely long?

FIGURE E.6

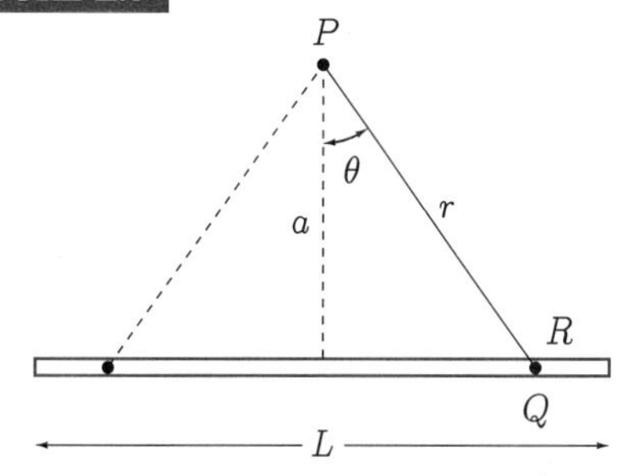

222. (a) Your friend says that he wants to paint the area bounded by the curves

$$y = \frac{1}{x}, \quad x = 1, \quad y = 0.$$

Why would you advise him to forget it?

(b) Your friend now says that he is going to make a funnel in the shape of the surface of revolution obtained by rotating that part of $y = 1/x$ to the right of $x = 1$ about the x-axis. He intends pouring paint into the funnel until it is full. Can he do this with a finite amount of paint? If he can, what do you reply when he points out that he has now painted both sides of the area in (a)?

(c) Would you advise him to attempt to paint the outside of the funnel? How about the inside?

223. Show that the function

$$f(x) = 4 + 3\sin\left(\frac{x-2}{x+2}\right)$$

has an inverse function on the interval $x \geq 0$, and find it.

224. Prove that $\sin x$ is a transcendental function. Hint: Assume

$$P_0(x)\sin^n x + P_1(x)\sin^{n-1} x + \cdots + P_{n-1}(x)\sin x + P_n(x) = 0$$

and evaluate at $x = n\pi$.

225. Show that the function

$$f(x) = \mathrm{Cos}^{-1}\left(\frac{a\cos x + b}{a + b\cos x}\right) - 2\,\mathrm{Tan}^{-1}\left[\sqrt{\frac{a-b}{a+b}}\tan\left(\frac{x}{2}\right)\right],$$

where $0 < b \leq a$, is constant for $0 \leq x \leq \pi$. Find the value of the constant.

226. Evaluate

$$\int_0^{\pi/2} \frac{\sin x \cos x}{a^2\cos^2 x + b^2 \sin^2 x}\,dx$$

where $ab \neq 0$.

227. Find the area bounded by $y = \ln x$ and $y = (\ln x)^2$.

228. Evaluate

$$\int_0^{\pi} \sqrt{1 - \cos 3x}\,dx.$$

In Exercises 229–230 evaluate the indefinite integral.

229. $\displaystyle\int \frac{1}{x^3\sqrt{x^2+5}}\,dx$ **230.** $\displaystyle\int (4 - x^2)^{3/2}\,dx$

231. Find the area bounded by the curve $x^4 - ax^3 + a^2y^2 = 0$ where a is a constant.

232. Find the area bounded the the curve $(y - \mathrm{Sin}^{-1} x)^2 = x - x^2$.

In Exercises 233–235 evaluate the indefinite integral.

233. $\displaystyle\int \frac{1}{x^2(9 - x^2)^{3/2}}\,dx$ **234.** $\displaystyle\int \sqrt{\frac{3+x}{3-x}}\,dx$

235. $\displaystyle\int \frac{15 + 38x - 14x^2}{15 - 14x - 11x^2 + 4x^3}\,dx$

236. In a chemical reaction, one molecule of trypsinogen yields one molecule of trypsin. In order for the reaction to take place, an initial amount of trypsin must be present. Suppose that the initial amount is y_0. Thereafter, the rate at which trypsinogen is changed into trypsin is proportional to the product of the amounts of each chemical in the reaction. Find a formula for the amount of trypsin if the initial amount of trypsinogen is A.

237. A very long, thin thread is wrapped clockwise around the circle $x^2 + y^2 = a^2$ so that one end of the string is at the point $(a, 0)$. If the thread is unwound in such a way that it is always kept taut, the path followed by the end of the string is called an involute of the circle.

(a) Find equation(s) for the involute and sketch its graph.

(b) Find a formula for the slope of the involute.

(c) What is the limit of the slope as the point $(a, 0)$ is approached along the curve?

(d) What is the slope of the curve when it crosses the positive x-axis for the first time?

238. Find the area enclosed by the astroid

$$x = a\cos^3 t, \quad y = a\sin^3 t,$$

a a constant.

239. (a) Sketch the curves

$$3(x^2 + y^2)^2 = 25(x^2 - y^2), \qquad 22y = 26 - x^2.$$

How many points of intersection would you suggest for these curves.

(b) Find all points of intersection for the curves in (a) and show that at two points, the curves are tangent.

240. Find the area between the loops of $r = a\sin^3(\theta/3)$.

241. Find the area bounded by the hyperbola $b^2x^2 - a^2y^2 = a^2b^2$ and the line through its focus parallel to the y-axis.

242. A model of learning proposed by Bush and Mosteller assumes that on the nth trial of a sequence of trials, the probability p_n of obtaining a certain response is related to the probability on the immediately preceding trial by

$$p_n = a + (1 - a - b)p_{n-1}$$

where a and b are constants which depend on the amounts of reward and inhibition. Prove that an explicit formula for p_n is

$$p_n = p_0(1 - a - b)^n + \frac{a}{a+b}[1 - (1 - a - b)^n]$$

where p_0 is the initial probability of response.

243. In Iraq an epidemic of methylmercury poisoning killed 459 people in 1972. Human beings became exposed to the poison when they ate homemade bread accidentally prepared from seed-wheat treated with a methylmercurial fungicide. Symptoms appeared only after weeks of exposure. Assume for simplicity that a person takes a constant daily dose d of the poison and that a certain percentage p of the accumulated poison is excreted each day. Find a formula which relates the amount of poison stored in the body to the number of days.

244. Evaluate

$$\lim_{n\to\infty} \frac{\frac{1}{4} + \frac{1}{8} + \frac{1}{16} + \cdots + \frac{1}{2^n}}{1 + \frac{2}{5} + \frac{4}{25} + \frac{8}{125} + \cdots + \left(\frac{2}{5}\right)^{n-1}}.$$

245. Show that if $\sum_{n=1}^{\infty} a_n^2$ and $\sum_{n=1}^{\infty} b_n^2$ are convergent series, then $\sum_{n=1}^{\infty} a_n b_n$ is absolutely convergent.

246. Find $f^{(6)}(0)$ if $f(x) = x^2/\sqrt{1-x^2}$.

247. Find the first three nonzero terms in the Maclaurin series for $\ln(\cos x)$.

248. Find the Maclaurin series for $\cos^4 x$.

249. Find the Maclaurin series for $x(2+3x)^{3/2}$.

250. Find $\sqrt{4.000\,000\,000\,01}$ accurate to 20 decimal places.

251. Prove that if $f(x)$ is continuous at $x = a$ and $f(a) > 0$, then there exists an open interval I containing a such that $f(x) > 0$ for all x in I.

252. (a) Show that the tangent line to the curve $y = x^3$ at any point P (other than the origin) always intersects the curve again.

(b) Verify that the slope of the curve at the second point in (a) is four times the slope at P.

253. Prove that when $f'(a)$ exists

$$\lim_{x \to a} \frac{xf(a) - af(x)}{x - a} = f(a) - af'(a).$$

254. (a) Show that if $y = (1-x)^{-a}e^{-ax}$, then

$$(1-x)\frac{dy}{dx} = axy.$$

(b) Now show that

$$(1-x)\frac{d^{n+1}y}{dx^{n+1}} - (n+ax)\frac{d^n y}{dx^n} - an\frac{d^{n-1}y}{dx^{n-1}} = 0.$$

255. Find all points on the curve $x^2y^2 + xy = 2$ where the slope is -1.

256. Prove that the segment of the tangent line to the tractrix

$$y = \frac{a}{2}\ln\left(\frac{a + \sqrt{a^2 - x^2}}{a - \sqrt{a^2 - x^2}}\right) - \sqrt{a^2 - x^2}$$

between the point of tangency and the y-axis always has length a.

257. Find d^2y/dx^2 if $y = |\sin x|^x$, $x > 0$.

258. A manufacturer making 144 000 units of an item during a year has a choice of different production schedules. He can make the entire lot early in the year in one production run. By doing so, he holds production costs down; but he must then carry much of the production run in inventory, and inventory costs might overbalance savings from mass production economies. Suppose that it costs \$500 to make the factory ready for a production run, and \$5 for each unit made after the factory has been readied. Assume that after a batch is made, the units are placed in inventory and then used up at a uniform rate such that inventory is zero when the next batch appears. If annual cost of the inventory is the cost of the average inventory, what batch size should be run to minimize annual production and inventory costs.

259. A log 20 m long has the form of a truncated cone with bases of diameters 1 m and 2 m. A beam with square cross-section is to be cut from the log so that axes of log and beam coincide. Find the dimensions of the beam with maximum volume.

260. A street light is 25 m high. A ball is thrown vertically upward from a point 10 m away from the lightpost. If the speed of the ball is 20 m/s when it is 15 m above the ground, what is the acceleration of the shadow of the ball along the ground at this instant?

261. (a) Show that the volume of the container in Figure E.7 is $abh/3$.

(b) If water is poured into the container at a rate of k cubic units per unit time, find a formula for the rate at which the depth of water is increasing when the container is half full by volume.

FIGURE E.7

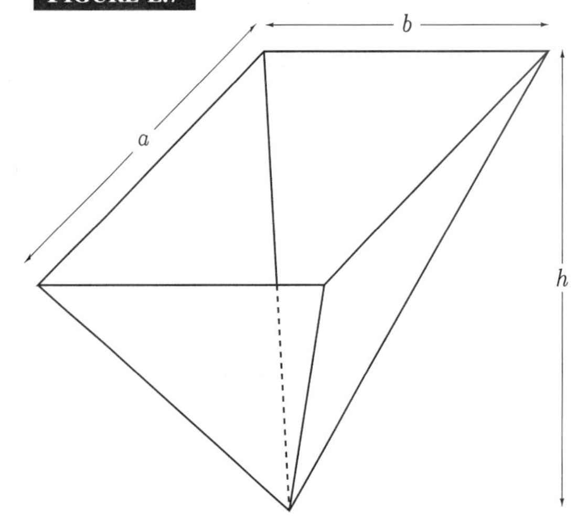

262. If a particle moves along the x-axis so that its acceleration is always proportional to its velocity, show that the percentage change in its velocity in every time interval of length T is always the same.

263. Prove that

$$\int_0^{\pi/2} f(\cos x)\,dx = \int_0^{\pi/2} f(\sin x)\,dx.$$

264. Find the volume of the barrel in Figure E.8. The cross-section is elliptic.

FIGURE E.8

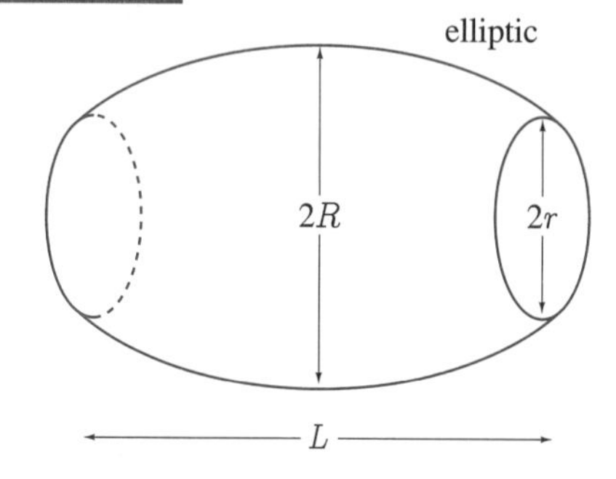

265. A piece of wood with density ρ is in the shape of a pyramid; its base is a square of side length b and its vertex is a distance a from the base. If it floats in a liquid of density $\overline{\rho}$ with its base horizontal and above the surface, find the work required to lift the block from the liquid.

266. (a) A sphere of radius r and density $\overline{\rho}$ floats partially submerged in a liquid with density ρ. Show that the depth d below the surface of the lowest point of the sphere must satisfy the equation

$$\rho(d^3 - 3rd^2) + 4r^3\overline{\rho} = 0\,.$$

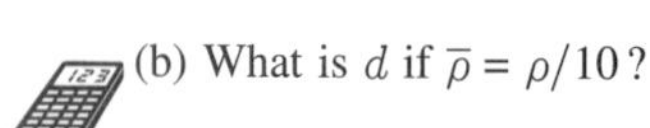

(b) What is d if $\overline{\rho} = \rho/10$?

267. Newton's universal law of gravitation states that the force of attraction between two point masses m and M at distance r apart has magnitude $F = GmM/r^2$, where $G = 6.67 \times 10^{-11}$ m^3/(kg·s^2). A thin uniform rod AB of length L and mass M attracts a point mass m at point C which lies at a distance d from the end closest to B on the extension of AB. What point mass placed at A has the same force of attraction on m?

268. Under what condition will the function

$$f(x) = \frac{ax+b}{cx+d}, \qquad ad - bc \neq 0$$

be equal to its inverse?

269. Evaluate

$$\int_0^x \frac{1}{1+t^2}\,dt + \int_0^{1/x} \frac{1}{1+t^2}\,dt.$$

In Exercises 270–278 evaluate the indefinite integral.

270. $\int (\mathrm{Sin}^{-1}\,x)^2\,dx$ **271.** $\int \frac{1}{x^2}\mathrm{Cos}^{-1}\,x\,dx$

272. $\int \frac{1}{\sin x\,\cos^2 x}\,dx$ **273.** $\int \frac{1}{(x-4)\sqrt{x^2-16}}\,dx$

274. $\int x^2\,\mathrm{Tan}^{-1}\,2x\,dx$ **275.** $\int e^{2x+3}\cos(3x-1)\,dx$

276. $\int \frac{x - \sin x}{1 - \cos x}\,dx$ **277.** $\int \sin(2\ln x)\,dx$

278. $\int \frac{x}{\cos^2 3x}\,dx$

279. The *curvature* of a curve $y = f(x)$ at a point on the curve is defined as

$$\kappa(x) = \frac{|f''(x)|}{\{1 + [f'(x)]^2\}^{3/2}}.$$

(a) Find the curvature of the parabola $y = x^2 - 2x - 3$. Where is it a maximum?

(b) Show that the only curves for which curvature is identically equal to zero are straight lines.

(c) What happens to curvature at a point of inflection on the graph of a function?

280. Show that when a curve is defined parametrically, its curvature (Exercise 279) can be expressed in the form

$$\kappa(t) = \frac{\left|\frac{dy}{dt}\frac{d^2x}{dt^2} - \frac{dx}{dt}\frac{d^2y}{dt^2}\right|}{\left[\left(\frac{dx}{dt}\right)^2 + \left(\frac{dy}{dt}\right)^2\right]^{3/2}}.$$

281. The *radius of curvature* of a curve at a point is defined as $\rho = 1/\kappa$ where κ is curvature at the point (see Exercises 279 and 280).

(a) Show that the radius of curvature of a circle is constant and equal to the radius of the circle.

(b) At what points on the ellipse $b^2x^2 + a^2y^2 = a^2b^2$, $(a > b)$, is the radius of curvature a maximum and a minimum?

282. Suppose a curve $y = f(x)$ is concave upward at a point (x_0, y_0). The *circle of curvature* at (x_0, y_0) is defined as the circle above the curve with radius equal to ρ (the radius of curvature in Exercise 281), and centre along the normal to $y = f(x)$ at (x_0, y_0) at a distance ρ from (x_0, y_0). If the curve is concave downward, the circle is drawn below the curve.

(a) Find the circle of curvature to the parabola $y = x^2 - 2x - 3$ at $(1, -4)$.

(b) Is it true that the circle of curvature at a point on a curve passes through that point? Draw a picture.

(c) Show that the circle of curvature and curve share the same tangent line at their common point.

283. Find the equation of the tangent line to the curve

$$x = 4(2\cos\theta - \cos 2\theta), \quad y = 4(2\sin\theta - \sin 2\theta),$$

$$0 \le \theta < 2\pi,$$

at the point $(4\sqrt{3} - 2, 4 - 2\sqrt{3})$.

284. (a) The curve $r = Aq^{\theta}$, where $A > 0$ and $q > 1$ is called a logarithmic spiral. Sketch it.

(b) Show that the slope of the logarithmic spiral at every point at which it intersects the radial line $\theta = k$ is the same.

285. Dropping leaflets by air is a means sometimes used in advertising to reach a large group of customers. Suppose it is assumed that a constant proportion λ of the leaflets actually survive a given time period T, while the remainder are lost, destroyed, or otherwise rendered unreadable. It is further assumed that each leaflet that survives until the ith time period will reach β people, on the average, during that period. Given that N leaflets are dropped, how many people are reached by the leaflets?

286. Find the Taylor series about $x = 1$ for the function $\sqrt[3]{7 + x}$.

287. Use the Maclaurin series for $f(x) = (1+x)e^{-x} - (1-x)e^{x}$ to find the sum of the series $\sum_{n=1}^{\infty} n/(2n+1)!$.

288. Find point(s) on the curve $y = \ln(x^2 + 1)$ where the tangent line passes through the point $(1, -2 + \ln 2)$.

289. What point on the ellipse $x^2 + 3y^2 = 9$ is furthest from $(5, 7)$?

290. Let P be any point on the hyperbola $x^2 - y^2 = a^2$. If the tangent line at P cuts the asymptotes of the hyperbola at points Q and R, show that the area of triangle OQR is a^2.

291. Show that if $y = f(x)$ is concave upward on an interval I, then

$$\frac{f(x_1) + f(x_2)}{2} \ge f\left(\frac{x_1 + x_2}{2}\right)$$

for all x_1 and x_2 in I.

292. The rate of flow of water through a circular pipe is directly proportional to its hydraulic radius R defined as $R = A/s$. Quantity A is the cross-sectional area of the water flow in the pipe, and s is the length of the circular arc of this cross-sectional area. The degree to which the pipe is filled is measured by the angle θ from the centre of the pipe to the horizontal surface of the water. For what degree of filling is R a maximum?

293. Sand from a conveyor belt drops at the rate of 10 cubic centimetres per second. It creates a pile on a level floor in the form of a right-circular cone. The height h of the pile increases at the same rate as the diameter D of the circular area A on the floor covered by the sand. How fast is A changing when D is 0.5 metres?

294. Find the volume of the solid of revolution obtained by rotating the area bounded by the curves

$$y = x\sin x, \quad y = 0, \quad 0 \le x \le \pi,$$

about the x-axis.

295. A triangular plate has one vertex in the surface of a fluid. The opposite side has length a, is parallel to the surface of the fluid, and is b units below the surface. What is the force on each side of the plate?

296. If $y = f(x)$ has inverse function $x = f^{-1}(y)$, find a formula for d^2x/dy^2 in terms of derivatives of $f(x)$. Illustrate with an example.

297. Verify that

$$\int (\ln x)^n\,dx = (-1)^n n!\,x \sum_{r=0}^{n} \frac{(-\ln x)^r}{r!} + C.$$

298. Find the volume of the solid of revolution when the area under one arch of the cycloid of Exercise 46 in Section 10.1 is rotated about (a) the x-axis (b) the y-axis (c) the line $x = \pi R$.

299. Find the area of the surface of revolution when the lemniscate $(x^2 + y^2)^2 = a^2(x^2 - y^2)$ is rotated about the x-axis.

300. Find the area of the surface of revolution when the cardioid $r = a(1 - \cos\theta)$ is rotated about the x-axis.

301. Inscribed in a circle of radius R is a square. A circle is then inscribed in the square, a square in the circle, and so on and so on. Find the limit of the sum of the areas of all the circles and the limit of the sum of the areas of all the rectangles.

302. Are there angles at which the sine of the angle changes twice as fast as the cosine?

303. Find a cubic polynomial in x which approximates the function defined implicitly by the equation $y^5 + xy + 1 = 0$.

304. How many terms in the Maclaurin series for the integrand guarantee a value for

$$\int_1^2 \frac{\sin 2x}{x}\, dx$$

accurate to 10^{-3}?

305. Evaluate

$$\lim_{x \to 0} \frac{\sin(x^n)}{(\sin x)^m}$$

where n and m are positive integers.

306. Show that the segment of the tangent line to the astroid $x^{2/3} + y^{2/3} = a^{2/3}$ contained between the coordinate axes always has length a.

307. Show that the points of inflection on the curve $y = (x+1)/(x^2+1)$ are collinear.

308. At what angle(s) does the sine of the angle change half as fast as the angle itself?

309. A rectangular plate with sides of length L and l ($L > l$) is submerged in a liquid of density ρ. Its longer sides are parallel to the surface of the liquid, the plate makes an angle α with the surface, and the edge closest to the surface is at depth d. What is the force on each side of the plate?

310. The gamma function $\Gamma(\nu)$ for $\nu > 0$ is defined by

$$\Gamma(\nu) = \int_0^\infty x^{\nu-1} e^{-x}\, dx.$$

Show that $\Gamma(\nu + 1) = \nu\Gamma(\nu)$.

311. At what point(s) on the curve

$$x = R(1 - \cos\theta), \quad y = R(\theta - \sin\theta)$$

is the slope of the tangent line equal to 1?

312. Find a polynomial approximating the indefinite integral

$$\int \frac{1}{\sqrt[4]{1 + x^4}}\, dx$$

to within 10^{-6} on the interval $0 \le x \le 1$.

313. Show that points of inflection of the curve $y = (\sin x)/x$ lie on the curve $y^2(4 + x^4) = 4$.

314. 
(a) A spherical shell with inner and outer radii 1 m and 2 m is made from styrofoam with mass 2 kg/m^3. The shell is cut into two equal pieces by a plane, and the flat edge of one piece is carefully placed on the surface of a large container of water. How far will the styrofoam sink?

(b) If a small hole is drilled in the top of the floating shell, will the styrofoam sink further? If so, to what depth?

315. Two substances A and B react to form a third substance C in such a way that 2 gm of A react with 1 gm of B to produce 3 gm of C. The rate at which C is formed is equal to the product of the amounts of A and B present in the mixture. Find the amount of C present in the mixture as a function of time t when the original amounts of A and B brought together at time $t = 0$ are as follows. Sketch graphs of all three functions on the same axes.

(a) 20 gm and 10 gm (b) 20 gm and 5 gm

(c) 20 gm and 20 gm

316. Find a polynomial in x approximating $f(x) = \cos^3 x$ on $0 \le x \le 0.5$ with maximum error 10^{-6}.

317. For what values of c is the range of the function

$$f(x) = \frac{x^2 + 2x + c}{x^2 + 4x + 3c}$$

equal to $-\infty < f(x) < \infty$?

318. Tangents at points A and B on a circle intersect at point F. The tangent at the midpoint C of the smaller arc of the circle joining A and B intersects AF and BF in D and E respectively. What is the limit of the ratio of the areas of triangles ABF and DEF as A approaches B?

319. Use the definition of a limit to prove that $\lim_{x \to 2} (x^2 + 3)/(x - 1) = 7$.

320. Find the equation of a circle with centre on the line $2y + \sqrt{3}x = 4\sqrt{3} + \sqrt{7}$ which is tangent to the ellipse $9x^2 + 4y^2 = 36$, and which satisfies the condition that the line joining the point of tangency to the centre of the circle is perpendicular to the given line.

321. (a) Sketch the curve $y^2 = x^3(a - x)$, called a piriform.

(b) Find point(s) where the tangent line to the piriform is horizontal.

(c) Where is the point of inflection on the top half of this curve?

322. The first derivative test for a relative minimum states that if $f'(x)$ changes from a negative quantity to a positive quantity as x increases through a critical point, then that critical point yields a relative minimum. Is the converse statement true? To answer, consider the function

$$f(x) = \begin{cases} \frac{x^2}{2}\left[1 + \sin\left(\frac{1}{x}\right)\right] & x \neq 0 \\ 0 & x = 0. \end{cases}$$

323. Find that point P on the ellipse $b^2x^2 + a^2y^2 = a^2b^2$ so that the triangle with vertices P, $Q(0,b)$ and $R(a,0)$ has maximum possible area.

324. Two corridors, one 3 m wide and the other 6 m wide, meet at right angles. Find the length of the longest square beam, $1/3$ m by $1/3$ m, that can be transported horizontally from one corridor to the other.

325. A cylindrical tank of radius r and height H has its axis vertical. It is originally (time $t = 0$) full of water, and for $t > 0$, water leaks out through a hole in the bottom.

(a) To find the speed at which water exits through the hole consider when the depth of water is h. In a small time dt, the depth drops by an amount dh. Use the fact that during dt, the potential energy of the small disc of water with thickness $|dh|$ at height h above the bottom of the cylinder is converted into kinetic energy of the same volume exiting through the hole with speed v to show that $v = \sqrt{2gh}$ where $g = 9.81$. This formula describes an ideal situation where all potential energy is converted into kinetic energy. In less than ideal situations, it is customary to assume that $v = c\sqrt{2gh}$, where $0 < c < 1$ is a constant.

(b) Show that the rate of change of the volume of water in the cylinder is given by the two expressions

$$\frac{dV}{dt} = \pi r^2 \frac{dh}{dt} \quad \text{and} \quad \frac{dV}{dt} = -Av$$

where A is the cross-sectional area of the hole, and hence deduce that

$$\frac{dh}{dt} = -\frac{2\sqrt{2g}cA}{\pi r^2}\sqrt{h}.$$

This is sometimes called Torricelli's law.

(c) Find h as a function of t and thereby determine a formula for how long it takes for the cylinder to empty.

326. Evaluate

$$\lim_{n\to\infty} \frac{1}{n^2}\left[\sin\left(\frac{1}{n^2}\right) + 2\sin\left(\frac{4}{n^2}\right) + 3\sin\left(\frac{9}{n^2}\right) + \cdots + n\sin 1\right].$$

327. Show that when $f'(x)$ and $f''(x)$ are positive on an interval $a \leq x \leq b$, then

$$(b-a)f(a) < \int_a^b f(x)\,dx < \frac{1}{2}(b-a)[f(a) + f(b)].$$

328. Show that the function

$$y = \frac{1}{3}\int_0^x f(t)e^{t-x}\sin 3(x-t)\,dt$$

satisfies the differential equation

$$\frac{d^2y}{dx^2} + 2\frac{dy}{dx} + 10y = f(x).$$

329. Find a in order that the area in Figure E.9 be as large as possible.

FIGURE E.9

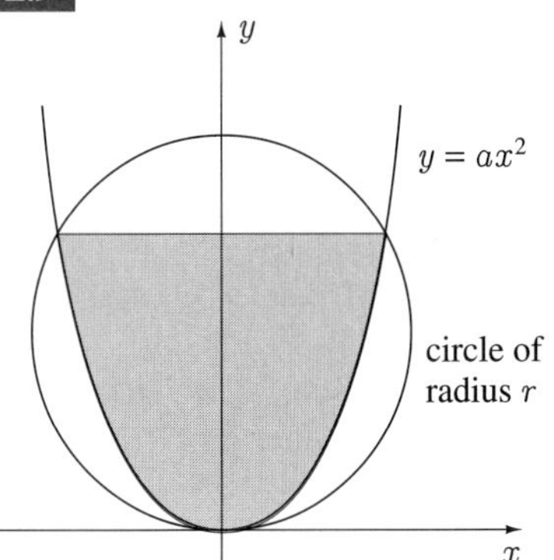

330. A square is rotated about a line which lies in its plane and passes through one of its vertices. What position of the line yields maximum volume for the resulting solid of revolution?

331. The two people in Figure E.10 each pull in the cable at 1 m/s. The curved part of the cable always has shape $y = k^{-1}\cosh kx + C$, where k and C, although independent of x, depend on the height of the minimum point above the x-axis. Find how fast this minimum point is rising when it is 5 m above the x-axis.

FIGURE E.10

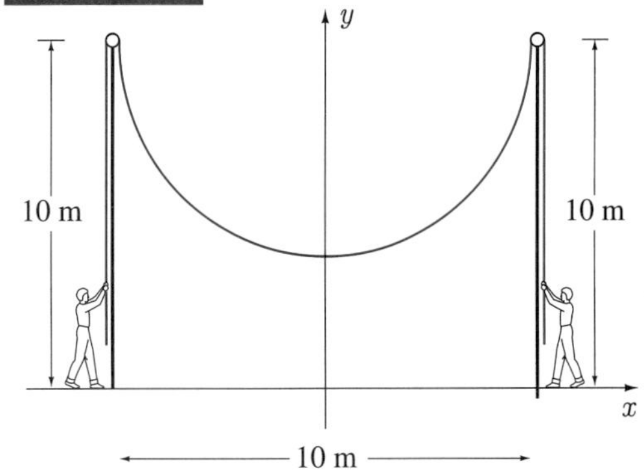

332. (a) A right-circular cylinder of radius r m, length L m, and density 250 kg/m^3 floats in water with its axis horizontal. How far above the surface of the water is its axis?

(b) How much work is required to lift the cylinder out of the water?

(c) Repeat part (b) if the cylinder floats with its axis vertical.

333. A hole is drilled through a long right-circular cylinder of radius R. If the radius of the hole is $r < R$, and axes of the cylinder and hole meet at right angles, set up a definite integral for the volume removed from the cylinder by the hole.

334. For what values of k do the following integrals converge:

(a) $$\int_2^{\infty} \frac{1}{x^k \ln x}\, dx$$ (b) $$\int_2^{\infty} \frac{1}{x(\ln x)^k}\, dx$$

(c) $$\int_2^{\infty} \frac{1}{(\ln x)^k}\, dx$$

335. Prove that e^x is a tanscendental function. Hint: Assume

$$P_0(x)e^{nx} + P_1(x)e^{(n-1)x} + \cdots + P_{n-1}(x)e^x + P_n(x) = 0$$

and take limits as $x \to -\infty$.

336. Let P be any point on the right half of the curve $y = k\cosh(x/k)$, and Q be the foot of the perpendicular from P to the x-axis. Draw a semicircle to the left of PQ with PQ as diameter. Let R be the point on the semicircle at a distance k from Q. Show that PR is tangent to the curve.

337. Prove that when $P(x)$ is a polynomial of degree n,

$$\int e^{mx} P(x)\, dx = e^{mx} \sum_{r=0}^{n} \frac{(-1)^r}{m^{r+1}} P^{(r)}(x) + C$$

where $P^{(r)}(x)$ is the rth derivative of $P(x)$ and $P^{(0)}(x) = P(x)$.

338. Prove that for constant a and $n \geq 1$ an integer:

(a) $$\int \sin^{2n} ax\, dx = -\frac{1}{a}\cos ax \sum_{r=0}^{n-1} \frac{(2n)!(r!)^2}{2^{2n-2r}(2r+1)!(n!)^2}\cdot \sin^{2r+1} ax + \frac{(2n)!}{2^{2n}(n!)^2}x + C$$

(b) $$\int \sin^{2n+1} ax\, dx = -\frac{1}{a}\cos ax \sum_{r=0}^{n} \frac{2^{2n-2r}(n!)^2(2r)!}{(2n+1)!(r!)^2}\cdot \sin^{2r} ax + C$$

339. Hecht assumes that the amount S of a retinal substance which remains when exposed to light of intensity I is characterized by the equation

$$\frac{dS}{dt} = bI(a - S) - kS^2$$

where a denotes the amount of S prior to stimulation, and b and k are positive constants. Show that

$$S(t) = \frac{-bI}{2k} + \frac{B(1 + De^{-2kBt})}{1 - De^{-2kBt}},$$

where $B^2 = 4b^2I^2/k^2 + abI/k$ and $D = [2k(a - B) + bI]/[2k(a + B) + bI]$.

340. Circles C_1 and C_2 in Figure E.11(a) have radius R m and circle C_1 rolls around C_2 without slipping.

FIGURE E.11

(a)

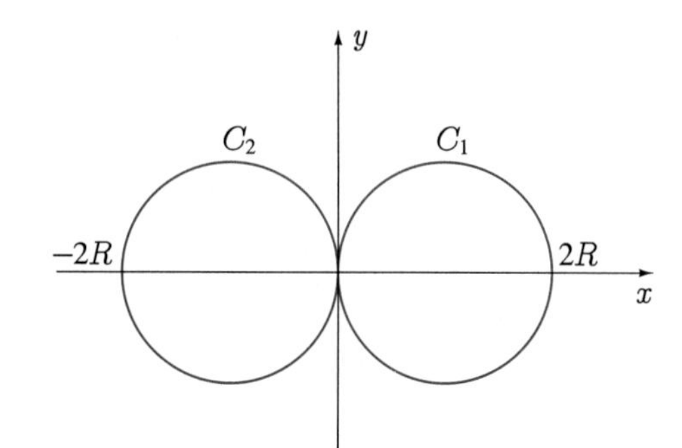

(b)

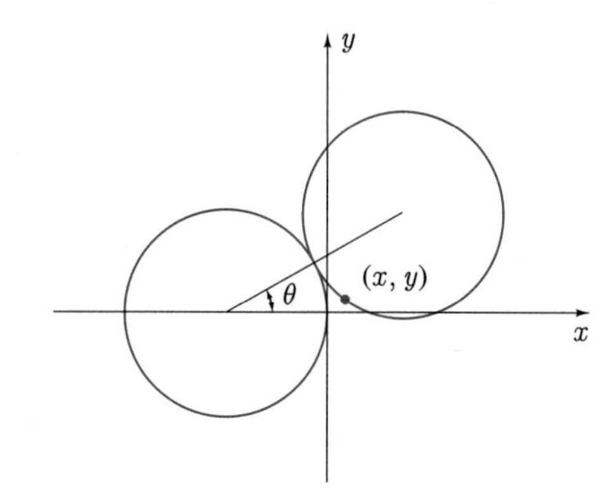

(a) Show that the curve traced out by the point $P(x, y)$ on C_1 which is initially at the origin has equations

$$x = 2R\cos\theta(1 - \cos\theta), \qquad y = 2R\sin\theta(1 - \cos\theta)$$

(see Figure E.11(b)).

(b) Show that if ϕ is the angle between the positive x-axis and the line joining $P(x, y)$ to the origin, then $\phi = \theta$. Hence, show that the polar equation of the curve is

$$r = 2R(1 - \cos\theta).$$

Sketch the curve.

(c) If C_1 rotates twice around C_2 every second, what is the speed of P when it is at the points with Cartesian coordinates $(-4R, 0)$ and $(0, 0)$?

341. A piece of sheet metal 1/6 cm thick is stored as a coil. Inner and outer radii of the coil are 40 cm and 80 cm. How long is the piece of metal?

342. Find the area bounded by the curve $x^4 + y^4 = x^2 + y^2$.

343. Suppose functions $f(x)$ and $g(x)$ are positive and their graphs are concave upward on the interval $a < x < b$. Show that if they have a relative minimum at the same value of x in the interval, the graph of $f(x)g(x)$ is concave upward on $a < x < b$.

344. Show that the function

$$f(x) = \begin{cases} x^4 \left[2 + \sin\left(\frac{1}{x}\right)\right] & x \neq 0 \\ 0 & x = 0 \end{cases}$$

has a relative minimum at $x = 0$ but this cannot be verified with the first derivative test.

345. The line $y = x + 1$ intersects the ellipse $4x^2 + 9y^2 = 36$ in two points P and Q. Find the point R on the ellipse in order that triangle PQR have maximum possible area.

346. Tangent lines to the curve $y = \sqrt{1 - x}$ make triangles with the positive x- and y-axes.

(a) Find the triangle with smallest area.

(b) Find the triangle with smallest perimeter.

347. (a) Show that

$$\left[\int_a^b f(x)g(x)\,dx\right]^2 \leq \int_a^b [f(x)]^2\,dx \int_a^b [g(x)]^2\,dx.$$

Hint: Consider the following function of t,

$$\int_a^b [tf(x) + g(x)]^2\,dx.$$

(b) Verify that

$$\left[\int_a^b f(x)\,dx\right]^2 \leq (b - a)\int_a^b [f(x)]^2\,dx.$$

348. Show that if $f(x)$ is a continuous periodic function with period p, then

$$\int_a^{a+p} f(x)\,dx = \int_0^p f(x)\,dx$$

for any a.

349. Use the result of Exercise 348 to show that when $f(x)$ is an odd, p-periodic continuous function, then the function

$$\int_a^x f(t)\,dt$$

is also p-periodic for any a. Is it odd?

350. A spy satellite is in circular orbit 1 500 km above the earth's surface. The satellite's camera returns to earth a picture of every point on earth in direct "line of sight" from the satellite.

(a) At any instant, how much area does the satellite spy on?

(b) What is the total area under the satellite's surveillance in one complete revolution around the earth. Assume the earth is a perfect sphere with radius 6 370 km.

351. Find the area bounded by the curve $x^2y^2 = 4(x - 1)$ and the line through its points of inflection.

352. A bead slides from rest at the origin on a frictionless wire in a vertical plane to the point (x_0, y_0) under the influence of gravity (Figure E.12). It can be shown that the time for the bead to traverse the path is

$$t = \frac{1}{\sqrt{2g}}\int_0^{x_0} \sqrt{\frac{1 + (y')^2}{y}}\,dx$$

where $g = 9.81$ and $y' = dy/dx$. The problem of finding that shape $y = f(x)$ of wire that makes t as small as possible is often called the brachistochrone problem. In the area of mathematics called "calculus of variations", it is shown that $y = f(x)$ must satisfy the differential equation

$$2y\frac{d^2y}{dx^2} + \left(\frac{dy}{dx}\right)^2 + 1 = 0.$$

(a) Show that the curve defined parametrically by

$$x = a(\theta - \sin\theta), \qquad y = a(1 - \cos\theta),$$

where a is a constant satisfies the differential equation.

(b) Sketch the curve in (a), called a cycloid.

(c) Show that if $2x_0 = \pi y_0$, then the value of a that makes the cycloid pass through (x_0, y_0) is $y_0/2$. Draw that part of the cycloid joining $(0,0)$ and (x_0, y_0) in this case. What is special about it?

(d) How would you find a if the condition in (c) were not satisfied?

FIGURE E.12

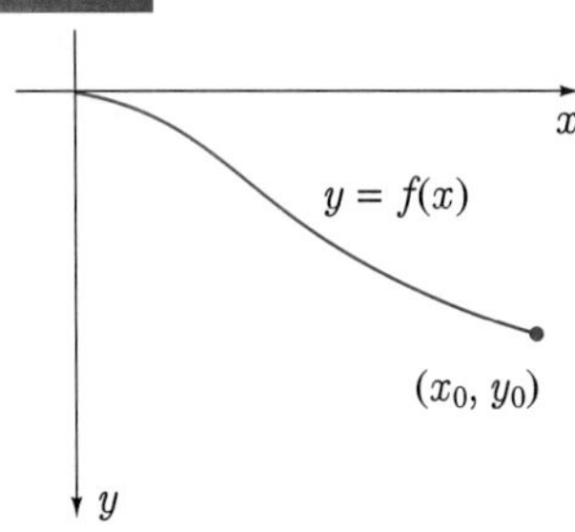

353. Find the volume of the solid of revolution when the area inside the lemniscate $(x^2 + y^2)^2 = a^2(x^2 - y^2)$ is rotated about the x-axis.

354. Find the volume of the smallest cone that can be circumscribed about a sphere of radius R.

Evaluate the integrals in exercises 355 and 356.

355. $\int (x^2 - 9)^{3/2}\, dx$ **356.** $\int_{\pi/6}^{\pi/3} \ln(\tan x)\, dx$

357. Two identical rods of length L and mass M lie on the same straight line a distance d apart. Use Newton's universal law of gravitation (Exercise 267) to find the force of attraction between the rods.

358. Repeat Exercise 259 if the length of the beam is L and diameters of its ends are a and b.

359. Two military outposts are situated at points P and Q on the desert 40 km apart (Figure E.13) Parallel to PQ and distant 10 km is a road. A courier can travel 14 km/hr on the sand and 50 km/hr on the road. What path would you recommend for quickest trip from P to Q?

FIGURE E.13

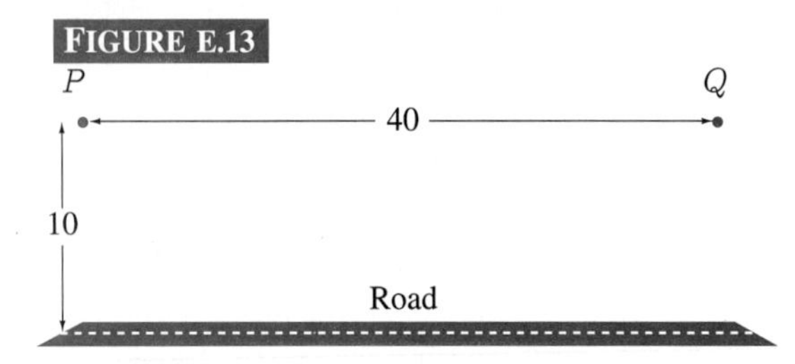

360. A number x_0 is called a fixed point of a function $f(x)$ if $f(x_0) = x_0$. How many fixed points can a function have if $f'(x) < 1$ for all x?

361. Show that the function $f(x) = x^3 + px + q$ has three distinct zeros if and only if $4p^3 + 27q^2 < 0$.

362. Show that the equation

$$2 \sin\theta = \theta(1 + \cos\theta)$$

has no solution in the interval $0 < \theta < \pi$.

363. Repeat Exercise 345 with a line $y = mx + c$ which intersects the ellipse $b^2x^2 + a^2y^2 = a^2b^2$ in two points.

364. A circular pasture has radius R. A cow is tied to a stake at the edge of the pasture. What length of rope permits the cow to graze on half the pasture?

365. Use the substitution $x = \sin^2\theta$ and integration by parts to find a formula for

$$\int_0^1 x^n(1 - x)^m\, dx$$

where m and n are positive integers.

366. Verify that for $n \geq 1$ an integer,

$$\int x^n \cos x\, dx = \sin x \sum_{r=0}^{\left[\frac{n}{2}\right]} \frac{(-1)^r n!}{(n-2r)!} x^{n-2r} + \cos x \sum_{r=0}^{\left[\frac{n-1}{2}\right]} \frac{(-1)^r n!}{(n-2r-1)!} x^{n-2r-1} + C,$$

where the square brackets indicate the greatest integer function.

367. Verify that

$$\int_0^{\pi/2} \sin^n x\, dx = \int_0^{\pi/2} \cos^n x\, dx = \begin{cases} \frac{2\cdot 4\cdot 6\cdots(n-1)}{1\cdot 3\cdot 5\cdots(n)}, & n \text{ an odd integer} \\ \frac{1\cdot 3\cdot 5\cdots(n-1)}{2\cdot 4\cdot 6\cdots(n)}\frac{\pi}{2} & n \text{ an even integer} \end{cases}$$

368. (a) A patrol boat at point A in Figure E.14 spots a submarine submerging at point B at a time that we call $t = 0$. The submarine, unaware of the patrol boat, follows a straight line path at constant speed v along some angle ϕ relative to BA (unknown to the patrol boat). The patrol boat heads directly toward point B at speed $V > v$ for $k/(V + v)$ units of time arriving at point C. Show that the submarine and patrol boat are equidistant from B at $t = k/(V + v)$.

(b) We set up a system of polar coordinates with B as pole and BA as polar axis. Let the distance $\|BC\|$ be denoted by r_0. Suppose that the patrol boat now follows the logarithmic spiral $r = r_0 e^{\theta/a}$ still at speed V, where $a = \sqrt{V^2/v^2 - 1}$. Show that the patrol boat must intercept the submarine.

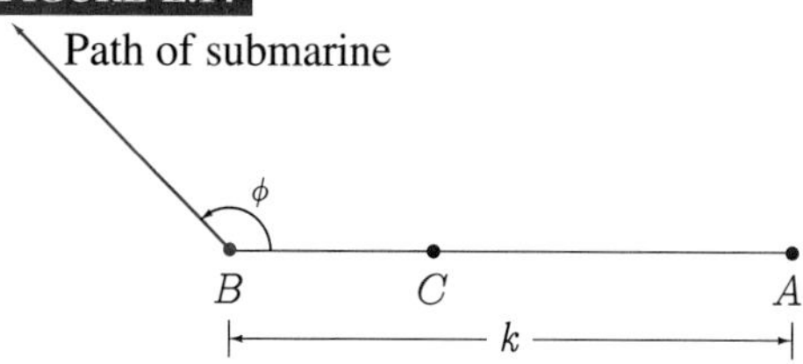

FIGURE E.14

369. Evaluate

$$\int \frac{2x^3 + 8x^2 - 3x + 5}{x^4 + 3x^3 + x^2 + 2x - 12}\,dx.$$

370. A sheet of metal a units wide and L units long is to be bent into a trough as shown in Figure E.15. End pieces are also to be attached. If edge AB must be the arc of a circle, determine the radius of the circle in order that the trough hold the biggest possible volume.

FIGURE E.15

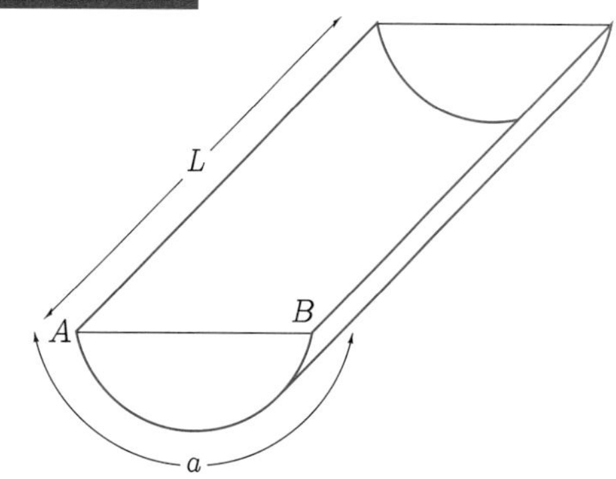

ADDITIONAL EXERCISES — Chapters 12–16

In Exercises 371–396 find a general solution for the differential equation.

371. $\dfrac{dy}{dx} = \dfrac{1 - 2x}{y^2}$

372. $\dfrac{dy}{dx} = -\sqrt{\dfrac{1 - y^2}{1 - x^2}}$

373. $\dfrac{dy}{dx} = \dfrac{1}{x + y^2}$

374. $(1 + e^x)\dfrac{dy}{dx} + e^x y = \sin x$

375. $x\dfrac{d^2y}{dx^2} = \dfrac{dy}{dx} + \left(\dfrac{dy}{dx}\right)^3$

376. $\dfrac{d^2y}{dx^2} + \left(\dfrac{dy}{dx}\right)^2 + 1 = 0$

377. $y'' - 3y' - 4y = x^2 + e^{2x}$

378. $y'' + 4y = 3\cos 2x$

379. $y'' + 6y' + 9y = \sin x$

380. $y''' + 5y'' + 5y' + y = 2\sin 3x$

381. $\dfrac{dy}{dx} + \sin\left(\dfrac{x + y}{2}\right) = \sin\left(\dfrac{x - y}{2}\right)$

382. $x^2\dfrac{dy}{dx} + (2x + x^2)y = e^x$

383. $\dfrac{d^2y}{dx^2} + \dfrac{dy}{dx} = \sin x$

384. $\dfrac{d^2y}{dx^2} + 12\dfrac{dy}{dx} + 2y = e^{-6x}$

385. $y''' - 3y'' + 4y = e^{-x} + 3x + 2$

386. $(x^2 + 1)\dfrac{d^2y}{dx^2} + 2x\dfrac{dy}{dx} + x^2 = 0$

387. $\cos^2 x\dfrac{dy}{dx} + y = 1$

388. $xe^{2y}\cos x\,dx - y^2\,dy = 0$

389. $y''' + 3y'' + 3y' + y = 2xe^{-x}$

390. $y''' + 3y'' + 3y' + y = e^{2x}$

391. $\sqrt{y}\dfrac{d^2y}{dx^2} = \dfrac{dy}{dx}$

392. $y'' + 6y' + 9y = x^2e^{-3x}$

393. $\dfrac{d^2y}{dx^2} + 4y = 3\cos x + \sin x$

394. $y'' + 4y = 3x\cos 2x$

395. $\dfrac{d^2y}{dx^2} + 6\dfrac{dy}{dx} + 9y = x\sin x - x$

396. $x^2\dfrac{d^2y}{dx^2} - 2x\dfrac{dy}{dx} - 4y = 0$

In Exercises 397–416 sketch whatever is defined by the equation or equations in xyz-space.

397. $z^2 = 3x^2 + y^2$

398. $2x + 3y - 4z = 0$

399. $x^2 + y^2 = 4,\ y = 3x$

400. $x^2 + y^2 = 4,\ z = 5$

401. $x^2 + y^2 + z^2 = 4,\ y = 3x$

402. $x^2 + y^2 + z^2 = 4,\ z = 5$

403. $z = y + x^2$

404. $x + y^3 = 1$

405. $x^3 + y^3 + z^3 = 1$

406. $y^2 + 2z^2 - x^2 = 8$

407. $y^2 + 2z^2 - x^2 = 8,\ 2z = x$

408. $y^2 + 2z^2 = 8,\ z = y$

409. $y = 4x^2,\ y = x$

410. $y = 4x^2,\ y = z$

411. $y = 4x^2,\ x = z$

412. $z = x^2 + 4y^2$

413. $z = x^2 + 4y^2,\ x + y + z = 1$

414. $z = x^2 + 4y^2,\ x + y + z = 1,\ x = 2y$

415. $z = x^2 + 4y^2,\ x + y + z = 1,\ x + y = 1$

416. $z = \sqrt{x^2 + y^2},\ x^2 + y^2 = 2y$

417. Find the components of a vector with length 5 which makes angles of 1 and 0.8 radians with the positive x- and y-axes.

418. Is there a direction in which the vector $2\hat{\imath} - 3\hat{\jmath} + 4\hat{k}$ has a component of length 6?

419. If the components of a vector along the vectors $\mathbf{v} = 2\hat{\imath} - 3\hat{\jmath}$ and $\mathbf{w} = \hat{\imath} + \hat{\jmath}$ are 2 and -3, what are its Cartesian components?

In Exercises 420–423 find the equation of the plane.

420. containing the point $(1, 2, -3)$ and the line $3x + 4y - 2z = 5,\ x + y + 2z = 6$.

421. containing the point $(2, -1, 3)$ and perpendicular to the line $x = 2y = z + 2$

422. containing the points $(1, 4, 0)$, $(2, -2, 1)$, and $(0, 3, 2)$

423. through the point of intersection of the plane $x - 2y + 3z = 4$ and the line $x = 1 - t,\ y = 2 + 3t,\ z = 2(1 + t)$, and containing the line $(3x + 1)/2 = (1 - y)/3 = z$

In Exercises 424–425 find equations for the line.

424. through the point $(0, -1, -2)$ and perpendicular to the plane $2x - 3y + 5z = 6$

425. through the midpoint of the line segment joining $(-1, 2, 0)$ and $(3, 2, 2)$ and perpendicular to the plane containing the aforementioned line segment and the line $x = 3 + 3t,\ y = 2 - t,\ z = 2 + 4t$

426. Find the length of the curve $3x = y^2,\ 9z = 2xy$ from $(0, 0, 0)$ to $(3, 3, 2)$.

427. Find a unit tangent vector to the curve $x = t^2 \cos 2t,\ y = t^2 \sin 2t,\ z = 1 - t$ at the point $(16\pi, 0, 1 - 4\pi)$.

428. Find a tangent vector at each point on the curve $x + y + 2z = 4,\ z = x^2 + y^2$ directed clockwise as viewed from the origin.

429. Find the unit vector which is normal to the curve $x = t^2,\ y = t^3 + t,\ z = 2t - 1$ at the point $(4, 10, 3)$ and has an x-component equal to the sum of its y- and z-components.

430. Show that the ratio of curvature to torsion for the curve $x = e^t \cos t,\ y = e^t \sin t,\ z = e^t$ is always $\sqrt{2}$.

431. Find the curvature of the curve $x = \ln(\cos t)$, $y = \ln(\sin t),\ z = \sqrt{2}t,\ 0 < t < \pi/2$.

432. Evaluate the following limits if they exist:

(a) $\displaystyle\lim_{(x,y)\to(0,0)} \frac{x^2 + y^2}{3x^2 - 4y^2}$

(b) $\displaystyle\lim_{(x,y)\to(1,3)} \frac{xy^2 + 2xy + x - y^2 - 2y - 1}{xy - 2x - y + 2}$

433. Along what straight line(s) approaching $(0, 0)$ is the limit of the function $f(x, y) = (2x - 3y)/(4x + 2y)$ equal to 5?

In Exercises 434–438 find the indicated derivative.

434. $\partial z/\partial y$ if $z = \ln(y + \sqrt{x^2 + y^2})$

435. $\partial z/\partial x$ and $\partial z/\partial y$ if $z = \mathrm{Tan}^{-1}\sqrt{x^y}$

436. $\partial z/\partial u$ if $z = \mathrm{Sin}^{-1}\left(\dfrac{\sqrt{u^2 - v^2}}{\sqrt{u^2 + v^2}}\right)$

437. $\partial z/\partial x$ if $z = u^2 + u \sin v,\ u = \mathrm{Tan}^{-1}(x + y)$, $v = \ln(e^x + e^y)$

438. $\partial u/\partial x)_y$ if $u^2 \sin x - vy = 3x + 2,\ u^3 \cos x + v^3 y - v = 3xy - 4yu$

439. Find all points at which the gradient of the function $f(x, y) = x^2 y + 3x$ is equal to (a) $2\hat{\imath} + 3\hat{\jmath}$ (b) $2\hat{\imath} - 3\hat{\jmath}$.

440. If u and v are functions of x, y, and z defined by $uv = 3x - 2y + z,\ v^2 = x^2 + y^2 + z^2$, show that $x\dfrac{\partial u}{\partial x} + y\dfrac{\partial u}{\partial y} + z\dfrac{\partial u}{\partial z} = 0$.

441. Find the following rates of change of the function $f(x, y, z) = x^2 yz - xy^3 z$:

(a) $\partial f/\partial x$

(b) $\partial f / \partial y$

(c) $\partial f / \partial z$

(d) $\partial^2 f / \partial x^2$ at $(1,-2,3)$

(e) at $(1,-2,3)$ toward the point $(3,2,4)$

(f) with respect to t along the curve $x = t^3 + 3t + 1$, $y = \sin t$, $z = 2\cos t$

(g) at $(1,0,2)$ with respect to distance travelled along the curve in (f)

(h) at $(2,2,2)$ with respect to distance travelled along the curve $x^3 + y^3 + z^3 = 24$, $y = x$ directed so that z increases in the first octant

(i) at $(1,1,3)$ normal to the surface $x^2 + y + z = 5$

(j) at $(1,2,-1/6)$ in any direction tangent to the surface $x^2yz - xy^3z = 1$

442. Find the tangent plane to the surface $x^2y + xyz + z^2 = 5$ at the point $(2,2,-1)$.

443. Find the point(s) where the curve $x + y + z = 1$, $y = x^2 - 4$ intersects the tangent plane to the surface $x^3 + y^3 + z^3 = 10$ at the point $(1,1,2)$.

444. Show that the tangent line to the curve $x = 18t - 3$, $y = 3t^2 + 1/t$, $z = t + 1/t^2$ at the point $(15,4,2)$ and the tangent plane to the surface $x^2 + y^2 - z^2 = 3x + 1$ at $(1,2,-1)$ do not intersect.

In Exercises 445–447 find and classify the critical points of the function as yielding relative maxima, relative minima, or saddle points.

445. $f(x,y) = y\sqrt{1+x} + x\sqrt{1+y}$

446. $f(x,y) = xy(a - bx - cy)$ where a, b, and c are nonzero constants

447. $f(x,y) = (2x - x^2)(2y - y^2)$

448. Find the point in the xy-plane for which the sum of the squares of the distances from n fixed points $P_i(x_i, y_i)$, $(i = 1, \ldots, n)$ is a minimum.

449. Find the point in the xy-plane which minimizes the sum of the squares of the distances from the x-axis, the y-axis, and the line $Ax + By + C = 0$.

450. Show that a positively homogeneous function $f(x,y)$ of degree k can be expressed in the form $x^k F(y/x)$ when $x > 0$. What is the extension of this result for a positively homogeneous function $f(x,y,z)$?

In Exercises 451–497 evaluate the integral.

451. $\displaystyle\int_{-1}^{0}\int_{x^2}^{3x+1} (x+y)^2\, dy\, dx$

452. $\displaystyle\int_{0}^{2}\int_{0}^{y} x\sin y\, dx\, dy$

453. $\displaystyle\int_{1}^{2}\int_{x}^{1} \frac{1}{(2x-y)^3}\, dy\, dx$

454. $\displaystyle\int_{0}^{1}\int_{0}^{y} x^2 e^{y^2}\, dx\, dy$

455. $\displaystyle\int_{0}^{2}\int_{0}^{1} |2x - y|\, dy\, dx$

456. $\displaystyle\int_{0}^{2}\int_{y}^{2} \frac{y}{\sqrt{x^2+y^2}}\, dx\, dy$

457. $\displaystyle\int_{0}^{2}\int_{0}^{\sqrt{4-x^2}} \sqrt{x^2+y^2}\, dy\, dx$

458. $\displaystyle\iint_R \frac{1}{(16 - x^2 - y^2)^{3/2}}\, dA$ where R is the area in the first quadrant bounded by the curves $y = x$, $y = \sqrt{9 - x^2}$, $y = 0$

459. $\displaystyle\int_{-1}^{1}\int_{x}^{x^2}\int_{z+x}^{2} (x+y)\, dy\, dz\, dx$

460. $\displaystyle\int_{0}^{2}\int_{0}^{1}\int_{x}^{1} z\sin y^2\, dy\, dx\, dz$

461. $\displaystyle\int_{0}^{2}\int_{0}^{\sqrt{4-x^2}}\int_{0}^{4-\sqrt{x^2+y^2}} (x+y)^2\, dz\, dy\, dx$

462. $\displaystyle\int_{0}^{2}\int_{0}^{\sqrt{4-x^2}}\int_{-\sqrt{4-x^2-y^2}}^{\sqrt{4-x^2-y^2}} x^2z^2\, dz\, dy\, dx$

463. $\displaystyle\int_C (x^2y - z)\, ds$ where C is the straight line from $(0,1,2)$ to $(-1,2,-3)$

464. $\displaystyle\int_C (x^2+y^2)\, dx + (y^2+z^2)\, dy + (z^2+x^2)\, dz$ where C is the curve $y = 2x^2$, $z + y = 0$ from $(1,2,-2)$ to $(-2,8,-8)$

465. $\displaystyle\int_C \frac{2xy}{z}\, dx + \frac{x^2}{z}\, dy - \frac{x^2y}{z^2}\, dz$ where C is the shorter part of the curve $x^2 + y^2 = 8$, $z = y$ from $(2,2,2)$ to $(-1,\sqrt{7},\sqrt{7})$

466. $\displaystyle\oint_C x^2y\, dx - y^3x\, dy$ where C is the boundary of the area defined by the lines $x = 1$, $y = 1$, $x + 2y = 5$

467. $\displaystyle\iint_S (x+y)^2\, dS$ where S is that part of $z = x^2 + y^2$ inside $x^2 + y^2 = 4$

468. $\oiint_S (x^2 + y^2 + z^2)\, dS$ where S is the sphere $x^2 + y^2 + z^2 = a^2$

469. $\iint_S (x\hat{\mathbf{i}} + y\hat{\mathbf{j}} + z\hat{\mathbf{k}}) \cdot \hat{\mathbf{n}}\, dS$ where S is that part of $x + 2y + 3z = 6$ in the first octant and $\hat{\mathbf{n}}$ is the unit normal to S with positive y-component

470. $\oiint_S (x\hat{\mathbf{i}} + y^2\hat{\mathbf{j}} - z^3\hat{\mathbf{k}} \cdot \hat{\mathbf{n}}\, dS$ where S is the surface bounding the volume $0 \le x \le 1$, $0 \le y \le 2$, $0 \le z \le 3$ and $\hat{\mathbf{n}}$ is the unit outer normal

471. $\oint_C xz\, dx - yz\, dy + z\, dz$ where C is the curve $x^2 + y^2 = 4$, $y = z$ directed so that x decreases in the first octant

472. $\displaystyle\int_0^{\sqrt{3}} \int_{-\sqrt{4-y^2}}^{-y/\sqrt{3}} (x^2 - y^2)\, dx\, dy$

473. $\iiint_V xye^z\, dV$ where V is bounded by $z = x^2 + y^2$, $z = 0$, $y = 0$, $x = 1$, $y = x$

474. $\iiint_V \sqrt{x^2 + y^2}\, dV$ where V is the volume bounded by $z = \sqrt{x^2 + y^2}$, $z = 2 - x^2 - y^2$

475. $\iiint_V (x^2 + y^2 + z^2)^2\, dV$ where V is the volume between the spheres $x^2 + y^2 + z^2 = a^2$ and $x^2 + y^2 + z^2 = b^2$ $(a > b)$

476. $\int_C (x - y)\, ds$ where C is the curve $z = x^2 + y^2$, $y + x = 1$ from $(1, 0, 1)$ to $(3, -2, 13)$

477. $\oint_C y\, dx + x^2\, dy$ where C is the ellipse $x^2 + 4y^2 = 4$

478. $\int_C e^y \sin 2y\, dx + xe^y(\sin 2y + 2\cos 2y)\, dy$ where C is the curve $x = t^3 + 3t$, $y = 1 + t^2$ from $t = -1$ to $t = 0$

479. $\iint_S (x^2 z^3 + y)\, dS$ where S is that part of $3x + 6y + 2z = 6$ cut out by $x = 0$, $z = 0$, $2x + z = 2$

480. $\oint_C e^{xy}(1 + xy)\, dx + (x^2 e^{xy} + xy^2)\, dy$ around the ellipse $x^2 + 4y^2 = 4$

481. $\iint_S (x^2\hat{\mathbf{i}} + y^2\hat{\mathbf{j}} + z^2\hat{\mathbf{k}}) \cdot \hat{\mathbf{n}}\, dS$ where S is that part of $z = 4 - x^2 - y^2$ above the xy-plane, and $\hat{\mathbf{n}}$ is the unit upper normal to S

482. $\oiint_S (x^3\hat{\mathbf{i}} + y^3\hat{\mathbf{j}} + z^3\hat{\mathbf{k}}) \cdot \hat{\mathbf{n}}\, dS$ where S is the surface enclosing the volume $x^2 + y^2 \le 1$, $0 \le z \le 1$ and $\hat{\mathbf{n}}$ is the unit outer normal to S

483. $\oint_C yz^2\, dx + x^2 z\, dy + xy^2\, dz$ where C is the smooth curve of intersection of the surfaces $x^2 + z^2 = a^2$, $y^2 + z^2 = a^2$ $(a > 0$ constant) which has a portion in the first octant (directed so that z increases in the first octant)

484. $\iint_R \dfrac{y}{x}\, dA$ where R is the area inside $x^2 + y^2 = 2x$ and above the x-axis

485. $\iiint_V x^2 yz\, dV$ where V is the volume in the first octant bounded by $z = 2x + 3y$, $z = 0$, $y = 0$, $x = 1$, $y = x^2 + 1$, $x = 0$

486. $\iiint_V x^2 y^2 z\, dV$ where V is the volume bounded by $x^2 + y^2 = z^2 + 1$, $z = 1$, $z = 0$

487. $\iiint_V (x^2 + y^2)\, dV$ where V is the volume inside the sphere $x^2 + y^2 + z^2 = 1$ and outside the cones $z^2 = x^2 + y^2$

488. $\int_C (x^2 + y^2)^{3/2}\, ds$ where C is the curve $x = y$, $z = x^2$ from $(-1, -1, 1)$ to $(1, 1, 1)$

489. $\oint_C x\, dy$ once around the polar coordinate curve $r = 4 + 3\sin\theta$

490. $\iint_S y^2(\hat{\mathbf{i}} + 2\hat{\mathbf{j}} + 3\hat{\mathbf{k}}) \cdot \hat{\mathbf{n}}\, dS$ where S is the surface $x^2 + y^2 = 1$, $0 \le z \le 1$ and $\hat{\mathbf{n}}$ is the unit normal pointing away from the z-axis

491. $\oiint_S (xz^2\hat{\mathbf{i}} + x^2 y\hat{\mathbf{j}} + z\hat{\mathbf{k}}) \cdot \hat{\mathbf{n}}\, dS$ where S is the surface enclosing the volume bounded by $x^2 + z^2 = 4 + y^2$, $x^2 + z^2 = 3$ and $\hat{\mathbf{n}}$ is the unit outer normal to S

492. $\oint_C (x^2 + 3)\, dx + 2xyz\, dy - x^2 y^2\, dz$ where C is the curve $x^2 + 4z^2 = 4$, $y = x^2$ directed clockwise as viewed from the negative y-axis

493. $\displaystyle\int_C \left(\frac{y}{x^2 + y^2} + z^2\right) dx - \frac{x}{x^2 + y^2} dy + 2xz\, dz$ where C is the upper half of the curve $y = 2x^2 - 1$, $x^2 + z^2 = 1$ from $(-1, 1, 0)$ to $(1, 1, 0)$

494. $\oint_C \sin y\, dx - \cos x\, dy$ where C is the curve $|x| + |y| = 1$

495. $\iint_S (y\hat{\mathbf{i}} + x\hat{\mathbf{j}} - \hat{\mathbf{k}}) \cdot \hat{\mathbf{n}}\, dS$ where S is the first octant part of $x^2 + y^2 + z^2 = 4$ above $z = y$ and $\hat{\mathbf{n}}$ is the unit lower normal to S

496. $\oiint_S (x^3\hat{\mathbf{i}} + 3xz\hat{\mathbf{j}} - e^{x+y}\hat{\mathbf{k}}) \cdot \hat{\mathbf{n}}\, dS$ where S is the surface enclosing the volume bounded by $z^2 + 9y^2 = 16$, $x = 0$, $x = 4$ and $\hat{\mathbf{n}}$ is the unit inner normal to S

497. $\int_C \sin(x + y + z)\, dx$ where C is the boundary of that part of $2x + 3y + 6z = 6$ in the first octant directed counterclockwise as viewed from the origin

498. At what points is the function $f(x,y) = 1/(\sin^2 \pi x + \sin^2 \pi y)$ discontinuous?

499. Show that the function $f(x,y) = \text{Tan}^{-1}\left(\dfrac{x+y}{x-y}\right)$ satisfies $\dfrac{\partial z}{\partial x} + \dfrac{\partial z}{\partial y} = \dfrac{x-y}{x^2+y^2}$.

500. Show that $f(x,y) = \ln(e^x + e^y)$ satisfies $\dfrac{\partial^2 f}{\partial x^2}\dfrac{\partial^2 f}{\partial y^2} - \left(\dfrac{\partial^2 f}{\partial x \partial y}\right)^2 = 0$.

501. Show that $f(x,y) = \sin x + F(\sin y - \sin x)$, where F is a differentiable function, satisfies $\dfrac{\partial f}{\partial y}\cos x + \dfrac{\partial f}{\partial x}\cos y = \cos x \cos y$.

502. In what direction(s) from the point $(2,3)$ does the function $f(x,y) = x^2 y - xy$ have a rate of change equal to 4?

503. Show that the tangent plane to the surface $z = xf(y/x)$, where f is a differentiable function, always passes through the origin.

504. Find the point in space which minimizes the sum of the squares of the distances from the coordinate planes and the plane $Ax + By + Cz + D = 0$.

505. Of all right-angled triangles with perimeter P find the one with largest area.

506. Show that for small $|x|$ and $|y|$,

$$\frac{\cos x}{\cos y} \approx 1 - \frac{1}{2}(x^2 - y^2).$$

507. Use double integrals to find the volumes of the solids of revolution when the area bounded by the curves $y = x^2 + 1$, $y = 2x + 1$ is rotated about the lines (a) $x = 0$ (b) $y = 0$ (c) $x = 4$ (d) $y = -2$ (e) $x + y = 1$

508. Repeat Exercise 507 for the area bounded by the curves $x = y(y+1)$, $x = -2y(y+1)$.

509. Find the centroid of the area bounded by the curves $x = 0$, $y = 0$, $y = 1$, $x = \sqrt{1 + y^2}$.

510. A flat vertical plate is immersed vertically in oil with density 900 kg/m^3. It is a parallelogram with sides of lengths 2 m and 1 m with the longer sides parallel to the surface of the oil. If one of the longer sides is $1/2$ m below the surface and the parallel side is 1 m below the surface, use a double integral to find the force on each side of the plate due to the surrounding oil.

511. What is the force on the plate in Exercise 510 if the longer diagonal is vertical, and the uppermost vertex is in the surface of the oil?

512. Find the centre of mass of a uniform plate whose edges are defined by the curves $y = \sqrt{4 - x^2}$, $4y = 4 - x^2$.

In Exercises 513–516 set up, but do not evaluate, double iterated integrals for the area described.

513. The area of $x^3 + y^3 + z^3 = 1$ in the first octant

514. The area of $y = x + z^2$ inside the cylinder $x^2 + z^2 = 4$

515. The area of $x^2 + y^2 - z^2 = 9$ between $z = 1$ and $z = 2$

516. The area of $z = 1 - x^4 - y^4$ above the xy-plane

517. Find the volume of the solid of revolution when the area inside $r^2 = 4|\cos\theta|$ is rotated about the x-axis.

518. If A, B, and C are positive, and $D < 0$, find the volume of the tetrahedron formed by the plane $Ax + By + Cz + D = 0$ and the coordinate planes.

519. Show that the tangent plane to the surface $xyz = a$ ($a > 0$ constant) cuts a constant volume from the first octant.

520. A solid object of uniform density ρ is bounded by the surfaces $z = 4x^2$, $z = 4$, $y = -1$, $y = 1$. Find its centre of mass and its moments of inertia about the coordinate axes.

521. Find the volume bounded by $z = x^2 + y^2 - \sqrt{x^2 + y^2}$ and the xy-plane.

522. Find the volume of the region inside the sphere $x^2 + y^2 + z^2 = a^2$ and outside the cones $z^2 = b^2(x^2 + y^2)$.

523. Given that $f(x,y,z) = xyz$ and $\mathbf{F}(x,y,z) = x^2\hat{\mathbf{i}} - yz\hat{\mathbf{j}} + xz\hat{\mathbf{k}}$, find (a) ∇f (b) $\nabla \cdot \mathbf{F}$ (c) $\nabla \times \mathbf{F}$ (d) $\mathbf{F} \times (\nabla \cdot \mathbf{F})\nabla f$ (e) $\nabla f \cdot \nabla \times (x^2\mathbf{F})$ (f) $\nabla(f^2) \times (\nabla \times \nabla f)$ (g) $\nabla \cdot \mathbf{F}(\nabla f \times y\mathbf{F})$

524. Show that the gradient of the function $\text{Tan}^{-1}(y/x)$ is irrotational and solenoidal in any domain that does not contain points on the y-axis.

525. Show that vector fields of the form $\mathbf{F} =$

$f(\sqrt{x^2+y^2+z^2})(x\hat{\mathbf{i}}+y\hat{\mathbf{j}}+z\hat{\mathbf{k}})$ are irrotational.

526. Show that the function $y = \dfrac{1}{3}\displaystyle\int_0^x f(t)e^{t-x}\sin 3(x-t)\,dt$ satisfies the differential equation

$$\frac{d^2y}{dx^2} + 2\frac{dy}{dx} + 10y = f(x).$$

527. Can the function $f(x,y) = xy/(x^2+y^2)$, $(x,y) \neq (0,0)$ be made continuous at $(0,0)$?

528. Find all points at which the sum of the first partial derivative of the function $f(x,y,z) = x^3 + y^3 - z^3$ is equal to (a) 3 (b) -3 (c) 0.

529. Verify that $f(x,y) = e^x(x\cos y - y\sin y)$ is harmonic.

530. Suppose that $f(x,y) = xF(ax+by) + yG(ax+by)$ where a and b are constants and F and G are arbitrary twice differentiable functions. Show that $\dfrac{\partial^2 f}{\partial x^2} - 2\dfrac{\partial^2 f}{\partial x\partial y} + \dfrac{\partial^2 f}{\partial y^2} = 0$ if and only if $a = b$.

531. In what direction(s) from the point $(1,2,3)$ does the function $f(x,y,z) = xyz + x^2 + y^2 + z^2$ have a rate of change equal to -2?

532. Find all tangent planes to the ellipsoid $x^2/a^2 + y^2/b^2 + z^2/c^2 = 1$ which have equal intercepts on the coordinate axes.

533. Radioactive substances decay at a rate proportional to the amount of radioactive material present at any time. If 10% of a radioactive sample decays in 1 year, how long does it take for 50% to decay (called the half-life of the material)?

534. Solve the equation $\dfrac{dy}{dx} = y^2 - 4$ subject to (a) $y(0) = 1$ (b) $y(0) = -2$.

535. Verify that the length of the curve $4y = (z+x)^2$, $4y^2 + 3z^2 = 3x^2$ from $(0,0,0)$ to any point (x,y,z) where $x > 0$ is $\sqrt{2}x$. Hint: Regard x as the parameter and use implicit diffferentiation to find dy/dx and dz/dx.

536. Find a continuous function satisfying

$$\frac{dy}{dx} + y = f(x), \quad y(0) = 1,$$

$$\text{where } f(x) = \begin{cases} 3 & 0 \le x \le 2 \\ 0 & x > 2 \end{cases}.$$

537. Find all points at which the gradient of the function $f(x,y,z) = x^2y^2z^2 + xyz$ is equal to $2\hat{\mathbf{i}} + 3\hat{\mathbf{j}} - \hat{\mathbf{k}}$.

538. If $f(x,y) = [F(ax-by)+G(ax+by)]/x$, where a and b are constants and F and G are twice differentiable functions, show that $b^2\dfrac{\partial}{\partial x}\left(x^2\dfrac{\partial f}{\partial x}\right) = a^2x^2\dfrac{\partial^2 f}{\partial y^2}$.

539. If y is defined as a function of x by $y = f(x,t)$, $F(x,y,t) = 0$, show that $\dfrac{dy}{dx} = \dfrac{f_xF_t - f_tF_x}{f_tF_y + F_t}$.

540. Find point(s) on the curve $x^2+y^2+z^2 = 1$, $y = z$ at which the rate of change of the function $f(x,y,z) = 3x + 2y - 2z$ with respect to length along the curve is equal to 1.

541. Find the second moment of area of the region bounded by the curves $y = 1 - x^2$, $x + y + 1 = 0$ about the lines (a) $y = 0$ (b) $x = 0$ (c) $y = 1$.

542. The area bounded by the curves $y = x^2$ and $y = 1$ (x and y in metres) is to be lowered into water until the force on each side is 10^5 N. Find the position of the horizontal side relative to the surface.

543. Repeat Exercise 512 if the force must be 1000 N.

544. Find the area of $z/3 = 1 - |x| - |y|$ above the xy-plane.

545. Find the area of that part of $y = 1 + 2\sqrt{x^2+z^2}$ to the left of the plane $y = 2$.

546. Find the area inside each petal of the rose $r = a\sin n\theta$.

547. Find the volume bounded by the surfaces $3y + z = 6$, $3y + 2z = 12$, $x + y + z = 6$, $x = 0$, $y = 0$, $z = 0$.

548. Find the centre of mass of the uniform solid bounded by $4x^2 + y^2 = 4$, $z = 0$, $y + z = 2$.

549. A man eats a diet of 2 500 calories per day; 1 200 of them go to basic metabolism (i.e., get used up automatically). He also spends 16 calories per day for each kilogram of body mass in exercise. Assume that the storage of calories as fat is 100% efficient and that 1 kilogram of fat contains 10 000 calories. Find his mass as a function of time, and sketch this function if initially his mass is (a) 100 kg (b) 70 kg. Is he on a diet?

550. Prove that the function $f(x) = \dfrac{x^2}{2} + \dfrac{1}{2}x\sqrt{x^2+1} + \ln\sqrt{x + \sqrt{x^2+1}}$ satisfies the differential equation $x\dfrac{dy}{dx} + \ln\left(\dfrac{dy}{dx}\right) = 2y$.

551. A 100 gm mass falls from rest under gravity. It is also acted on by air resistance which is 2 N when the speed of the mass is 10 m/s. Find the velocity of the mass when the resistive force is proportional to (a) velocity (b) the square of velocity (c) the square root of velocity. Find distance fallen in (a) also.

552. Prove that if the amount of radioactive material in a sample is A_1 at time t_1 and the amount is A_2 at time $t_2 > t_1$, then the half-life of the material is $(t_2 - t_1) \ln 2 / \ln(A_1/A_2)$. (See Exercise 533 for the rate of decay of radioactive materials.)

553. In Figure E.1 the four squares each have area one. A rectangle with one vertex at the origin, and two sides along the x- and y-axes has sides of length x and y. Find a formula for the amount of shaded area enclosed by this rectangle as a function of x and y.

FIGURE E.1

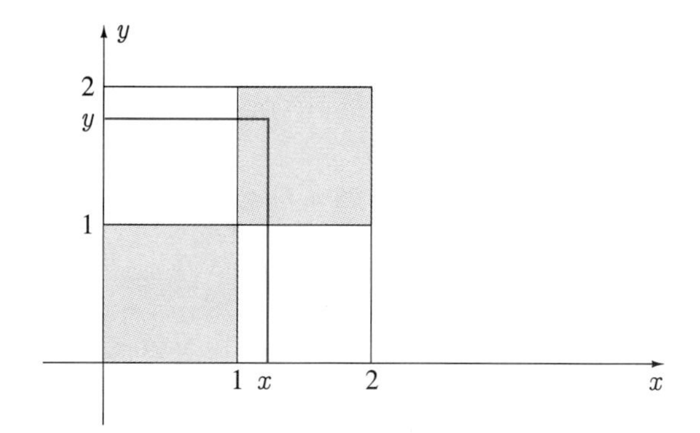

554. (a) Solve differential equation 16.39 if the term $1 - N/C$ is replaced by $1 - \ln N / \ln C$. Sketch a graph of $N(t)$.

(b) Does the solution in (a) or the logistic function 16.40 approach C more rapidly?

555. Find a formula for d^2y/dx^2 in terms of derivatives of $f(x,y)$ when $f(x,y) = 0$ defines y as a function of x.

556. A function $f(x,y)$ satisfies Laplace's equation 13.11. It is also known to be radially symmetric; that is, when expressed in polar coordinates r and θ, it is independent of θ. Find the form of $f(x,y)$. (Hint: see Exercise 13.6–36)

557. The area of a triangle with sides of length a, b, and c is $A = \sqrt{s(s-a)(s-b)(s-c)}$ where s is half the perimeter of the triangle. If the length of each side is increased by 1%, what is the approximate percentage increase in its area?

558. Suppose the line $Ax + By + C = 0$ has nonzero x- and y-intercepts. Find the point in the xy-plane which minimizes the sum of the distances from the x-axis, the y-axis, and the line.

559. Beer containing 6% alcohol is pumped into a vat containing 10 000 L of beer with 4% alcohol at 100 L per minute. If well-mixed beer is removed at the same rate, what is the percentage of alcohol in the vat as a function of time t. When is the beer 5% alcohol?

560. Of all right-angled triangles with area A, find the one with smallest perimeter.

561. (a) Show that $y_1(x) = (\sin x)/\sqrt{x}$ is a solution of the differential equation

$$x^2 \frac{d^2y}{dx^2} + x\frac{dy}{dx} + \left(x^2 - \frac{1}{4}\right) y = 0.$$

(b) Show that a second solution $y_2(x) = (\cos x)/\sqrt{x}$ can be obtained by setting $y_2(x) = v(x)y_1(x)$.

(c) What is the general solution of the differential equation?

562. A pyramid with rectangular horizontal base is inscribed in a sphere of radius r (Figure E.2). Find a formula for the volume in the pyramid in terms of x and y.

FIGURE E.2

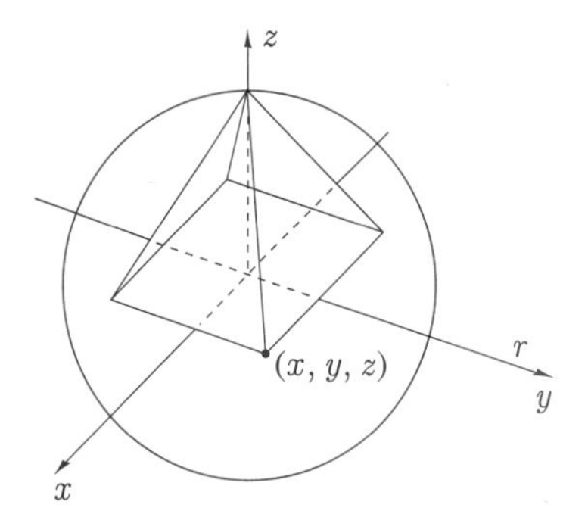

563. Of all planes passing through the point (a, b, c) in the first octant, find the one for which the tetrahedron formed by the plane and the three coordinate planes has the least possible volume.

564. Find the first moment of area of the region bounded by the curves $y = x^2$, $y = 2 - x^2$ about the line $x + y + 1 = 0$.

565. Find the area of $x^2 + y^2 + z^2 = 4$ between $z = \sqrt{x^2 + y^2}$ and $z = 2\sqrt{x^2 + y^2}$.

566. Find the first moment of area of the bifolium $r = \cos^2\theta \sin\theta$ about the x-axis.

567. A 100 gm mass is attached to a vertical spring with constant 1000 N/m, and slowly lowered to its equilibrium position. If it is given an initial speed of 4 m/s downward, find its position as a function of time if a damping force proportional to velocity acts on the mass and has magnitude (a) 20 N (b) 40 N (c) 80 N when speed is 2 m/s.

568. Repeat Exercise 567 if a force $\sin 20t$, $t \geq 0$ also acts on the mass.

569. (a) Show that if $y = (1 - x)^{-a}e^{-ax}$, then $(1-x)\dfrac{dy}{dx} = axy$.

(b) Now show that
$(1-x)\dfrac{d^{n+1}y}{dx^{n+1}} - (n + ax)\dfrac{d^n y}{dx^n} - an\dfrac{d^{n-1}y}{dx^{n-1}} = 0$.

570. Show that when a surface $z = f(r, \theta)$ in cylindrical coordinates projects one-to-one onto an area S_{xy} in the xy-plane, its area is given by

$$\iint_{S_{xy}} \sqrt{1 + \left(\frac{\partial z}{\partial r}\right)^2 + \frac{1}{r^2}\left(\frac{\partial z}{\partial \theta}\right)^2}\; r\,dr\,d\theta.$$

571. Use the result of Exercise 570 to find the area of the sphere $x^2 + y^2 + z^2 = 4$ inside the cylinder $x^2 + y^2 = 1$.

572. When chemicals A, B, and C are brought together, 1 gm of A reacts with 2 gm of B and 3 gm of C to form 6 gm of substance D. Initially there are 10 gm of A, 20 gm of B, and 30 gm of C. Find the amount of D as a function of time under each of the following conditions:

(a) the rate at which D is formed is proportional to the amounts of A, B, and C present in the reaction

(b) the rate at which D is formed is proportional to the amounts of B and C present in the reaction.

573. Find the volumes bounded by the following surfaces:

(a) $4x^2 + y^2 = 4$, $4x + 3y + z = 12$, $z = 1$

(b) $4x^2 + y^2 = 4$, $4x + 3y + z = 12$, $z = 24$

(c) inside $4x^2 + y^2 = 4$, above $4x + 3y + z = 12$, under $z = 12$

574. (a) Find equations defining solutions to Exercise 572 if the initial amounts of A, B, and C are all 10 gm.

(b) When is the amount of D equal to 15 gm if there is 10 gm in 5 minutes?

575. What is the minimum value of $f(x, y) = x^3 + y^3$ on that part of the line $Ax + By = C$ (where $A \geq 0$, $B \geq 0$, and $C \geq 0$) for which $x \geq 0$ and $y \geq 0$.

576. Find the moment of inertia about the line $y = 2x$ of a plate with uniform mass per unit area ρ if its edges are defined by $y = -x$, $y = x$, $x = 1$.

577. If $f(x, y)$ and its first partial derivatives are continuous in a simpy-connected domain containing a non-selfintersecting, piecewise smooth, closed curve C, and $\partial f/\partial y = \partial f/\partial x$ inside C, what is the value of

$$\oint_C f(x, y)(dx + dy)?$$

578. Show that if $f(x, y)$ is harmonic in a simply-connected domain D and C is a closed, piecewise smooth curve in D, then

$$\oint_C \frac{\partial f}{\partial y}dx - \frac{\partial f}{\partial x}dy = 0.$$

579. If $r = \sqrt{x^2 + y^2 + z^2}$, and $f(r)$ is a continuous function, show that

$$\iint_S f(r)\,dS = 4\pi a^2 f(a)$$

when S is the sphere $x^2 + y^2 + z^2 = a^2$.

580. Show that if $y_1(x)$ is a solution of the homogeneous, linear differential equation $y'' + P(x)y' + Q(x)y = 0$, then a second solution is

$$y_2(x) = y_1(x)\int \frac{e^{\int P(x)\,dx}}{[y_1(x)]^2}dx.$$

Hint: Let $y_2(x) = v(x)y_1(x)$.

581. (a) Use the transformation of Exercise 22 in Section 16.10 to solve $x^2y'' - 4xy' + 4y = 0$.

(b) Solve the differential equation by setting $y = x^m$ and finding appropriate values for m.

582. Use the techniques in Exercise 581 to solve $x^2y'' + 3xy' + 4y = 0$.

583. Use the methods of Exercise 581(b) and Exercise 580 to solve $4x^2y'' + 8xy' + y = 0$.

584. Assuming that the ellipsoid $x^2/a^2 + y^2/b^2 + z^2/c^2 = 1$ and the plane $Ax + By + Cz + D = 0$ do not intersect, find the points on the ellipsoid closest to and farthest from the plane.

585. Find the points on the ellipsoid $x^2 + 96y^2 + 96z^2 = 96$ closest to and farthest from the plane $3x + 4y + 12z = 288$.

ANSWERS

4. -8 **6.** does not exist

8. ∞ **10.** does not exist

12. $4/3$ **14.** $1/2$

16. does not exist **18.** 1

20. $x(2-3x^2)/(1-3x^2)^{3/2}$

22. $6(2x+1)^2\ln(x+1)+(2x+1)^3/(x+1)$

24. $e^{2x}/[(e^{2x}+1)\sqrt{\ln(e^{2x}+1)}]$

26. $2x\csc^2(x^2-1)[1-2x^2\cot(x^2-1)]$

28. $-1/[6\sqrt{2-x}(1+\sqrt{2-x})^{2/3}]$

30. $[\cos(x+y)-y]/[x-\cos(x+y)]$

32. $[xy+(1+x^2y^2)\text{Tan}^{-1}(xy)]/(1+x^2y^2-x^2)$

34. $-\sec^2 x/[(1-2u)\sqrt{u^2-u}]$

36. $te^{-3t}(2-3t)(x+1)/x$

38. $(4/3)\cot t$ **40.** $(3e^{-3v}+2v)/(1-3e^{3v})$

42. $4/(x^2-4x)$ **44.** $16x^2/(1+4x^4)$

46. $1/\sqrt{x^2+9}$ **48.** $98/(x^3\sqrt{7x^2+49})$

50. $-15x^3\sqrt{1-x^2}$ **52.** $125/[x(25-x^2)^{3/2}]$

54. $-87x/[2(6+3x+4x^2)^{3/2}]$

56. $49/(14x-x^2)^{3/2}$ **58.** $-5/[\sqrt{2}x\sqrt{3x^5-2}]$

60. $6/(4\cos^2 x+9\sin^2 x)$ **62.** $x^3\cos x$

64. $21\cos 3x/\sqrt{1+49\sin^2 3x}$

66. $(1/x^2)\text{Sec}^{-1}(2x)$

68. $10/(5+3e^{2x})$ **70.** -1, $(1,0)$

72. $5/13$, $170/2197$ **76.** $-2/3$

78. $x/3-(1/6)\ln(3+4e^{2x})+C$

80. $(1/3)(x-1)^3\ln x-x^3/9+x^2/2-x+(1/3)\ln x+C$

82. $x/2-(1/12)\sin 6x+C$

84. $(1/20)\cos 5x-(1/6)\cos 3x-(1/44)\cos 11x+C$

86. $\sqrt{x^2-3}+5\ln|x+\sqrt{x^2-3}|+C$

88. $-\sqrt{7-x^2}/(7x)+C$

90. $\text{Sin}^{-1}(x-1)-\sqrt{2x-x^2}+C$

92. $x+3\ln|x|+(2/\sqrt{3})\text{Tan}^{-1}[(x+1)/\sqrt{3}]+C$

94. $-\cot x-(1/3)\cot^3 x+C$

96. $x/125-(1/125)\ln(2e^x+5)-(21/100)/(2e^x+5)+(29/40)/(2e^x+5)^2+C$

98. $x/\sqrt{5-x^2}-\text{Sin}^{-1}(x/\sqrt{5})+C$

100. $3(x^2-x+5)\ln(x^2-x+5)-3x^2+3x+C$

102. $\sec x+\cos x+C$

104. divergent **106.** absolutely convergent

108. divergent **110.** absolutely convergent

112. divergent **114.** convergent

116. divergent **118.** convergent

120. convergent **122.** divergent

124. $(0.442, 0.407)$, $(-2.442, -12.407)$

126. $\pi/4+\alpha/2$ **130.** $6(\sqrt{2}+\text{Tan}^{-1}\sqrt{2})$

132. $2/\pi+(4\sqrt{2}-5)/3$ **134.** $2\pi[6\ln(3/2)-2/3]$

136. $-1/2+\ln 3$ **138.** $2/\sqrt{5}$ m, $1/\sqrt{5}$ m

140. $\pi ab^3\rho/4$, $\pi a^3 b\rho/4$ **142.** $|(1-b)s|<1$

144. $2\sqrt{3}$ m/s toward E **146.** $6x^2\cos x^2-4x^4\sin x^2$

148. $(15-8\ln 4)\pi/2$ **150.** $-187/8$

152. $4/\sqrt{5+16\ln 2+16(\ln 2)^2}$

154. $1/e$

158. $3+\pi/\sqrt{3}$, relative maximum

160. concave upward **162.** $-16/135$ m/s^2

164. $x_1+(A/\omega)[\sin(\omega t_2+\phi)-\sin(\omega t_1+\phi)]$

166. $4\sqrt{3}a$ **168.** b/a

170. $(1/2)\ln[x/(2-x)]$ **172.** $(\pi R, 5R/6)$

174. $\sum_{n=1}^{\infty} x^{2n}/[n(2n-1)]$, $-1\le x\le 1$

176. $-1/20$ **178.** $22/\sqrt{137}$

182. it doesn't **184.** $(3\pi/16, 0)$

190. $100(1.025)^{[[t/3]]}$ **192.** 0

202. $1/(\ln 2-1)$, $\ln 2/(2-2\ln 2)$, relative maximum at $x=1$, relative minimum at $x=2$

204. $kv_1/(k-1)$

206. decreasing at $62/\sqrt{21}$ m/s

208. (a) 24 m/s (b) $2<t<3$, $t>4$ (c) 19 m (d) 7.592 m/s

210. $-1/2$ **214.** 0

216. $\pi^2/16$

218. (a) 36297 N (b) increases by 981 N

220. $2r^2h/3$

222. (a) No (b) Yes (c) No, Yes

226. $(b^2-a^2)^{-1}\ln(b/a)$ if $b^2\ne a^2$; $1/(2a^2)$ if $b^2=a^2$

228. $2\sqrt{2}$

230. $6\,\text{Sin}^{-1}(x/2)+2x\sqrt{4-x^2}+(x/4)(2-x^2)\sqrt{4-x^2}+C$

232. $\pi/4$

234. $3\,\mathrm{Sin}^{-1}(x/3) - \sqrt{9 - x^2} + C$

236. $y_0(A + y_0)/[y_0 + Ae^{-k(A+y_0)t}]$

238. $3\pi a^2/8$ **240.** $(5\pi + 18\sqrt{3})a^2/32$

244. $3/10$ **246.** 270

248. $1 + \sum_{n=1}^{\infty}(-1)^n(2^{2n-1} + 2^{4n-3})x^{2n}/(2n)!$

250. $2.000\,000\,000\,002\,5$

258. 5367 **260.** 175.475 m/s^2

264. $\pi L(2R^2 + r^2)/3$ **266.** (b) $d = 0.3916r$

268. $f(x) = x$ or $f(x) = (ax + b)/(cx - a)$

270. $x(\mathrm{Sin}^{-1}x)^2 + 2\sqrt{1 - x^2}\,\mathrm{Sin}^{-1}x - 2x + C$

272. $\ln|\csc x - \cot x| + \sec x + C$

274. $(x^3/3)\mathrm{Tan}^{-1}(2x) + (1/48)\ln(1 + 4x^2) - x^2/12 + C$

276. $-x\cot(x/2) + C$

278. $(x/3)\tan 3x - (1/9)\ln|\sec 3x| + C$

282.(a) $(x - 1)^2 + (y + 7/2)^2 = 1/4$

286. $2 + (x - 1)/12 + 2\sum_{n=2}^{\infty}(-1)^{n+1}[2 \cdot 5 \cdot 8 \cdots (3n - 4)](x - 1)^n/(24^n n!)$, $-7 \le x \le 9$

288. $(-1, \ln 2)$ **292.** 4.4934

294. $\pi^2(2\pi^2 - 3)/12$ **296.** $-f''(x)/[f'(x)]^3$

298. (a) $5\pi^2R^3$ (b) $6\pi^3R^3$ (c) $\pi R^3(9\pi^2 - 16)/6$

300. $32\pi a^2/5$

372. $x\sqrt{1 - y^2} + y\sqrt{1 - x^2} = C$

374. $y = (C - \cos x)/(1 + e^x)$

376. $y = \ln|\cos(C - x)| + D$

378. $y = C_1\cos 2x + (C_2 + 3x/4)\sin 2x$

380. $y = C_1e^{-x} + C_2e^{(\sqrt{3}-2)x} + C_3e^{-(\sqrt{3}+2)x} + (3\cos 3x - 11\sin 3x)/260$

382. $y = (e^x + Ce^{-x})/(2x^2)$

384. $y = C_1e^{(\sqrt{34}-6)x} + C_2e^{-(\sqrt{34}+6)x} - (1/34)e^{-6x}$

386. $y = C\,\mathrm{Tan}^{-1}x - x^2/6 + (1/6)\ln(x^2 + 1) + D$

388. $4(x\sin x + \cos x) + e^{-2y}(2y^2 + 2y + 1) = C$

390. $y = (C_1 + C_2x + C_3x^2)e^{-x} + (1/27)e^{2x}$

392. $y = (C_1 + C_2x + x^4/12)e^{-3x}$

394. $y = (C_1 + 3x/16)\cos 2x + (C_2 + 3x^2/8)\sin 2x$

396. $y = C_1x^4 + C_2/x$

398. **400.**

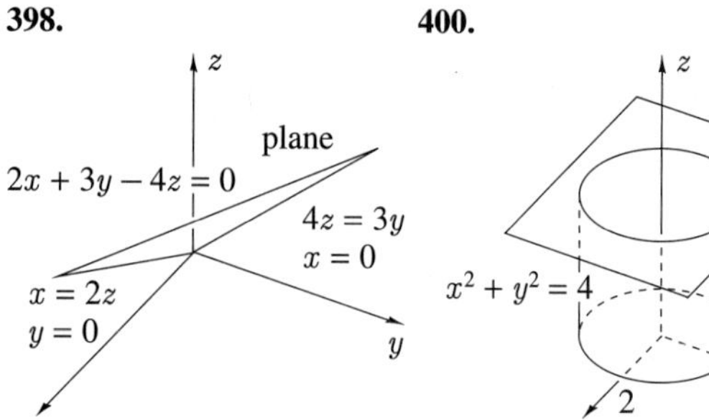

402. **404.**

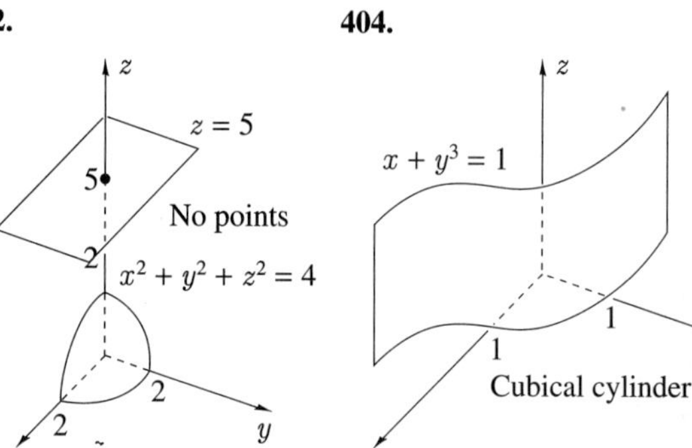

406. **408.**

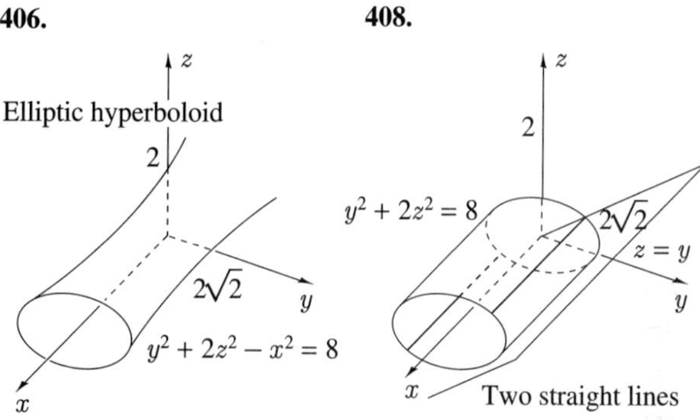

410. **412.**

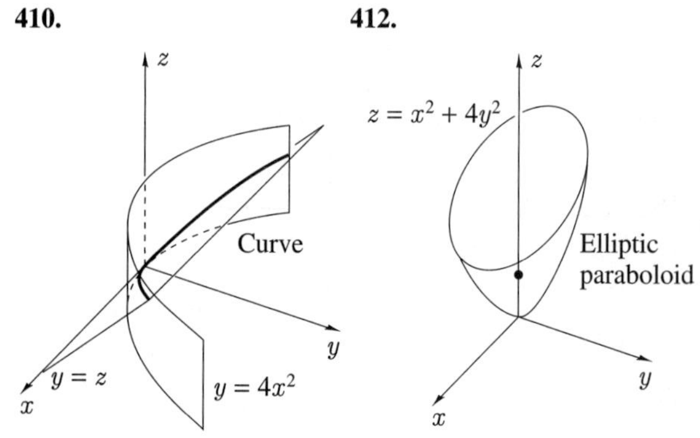

414. **416.**

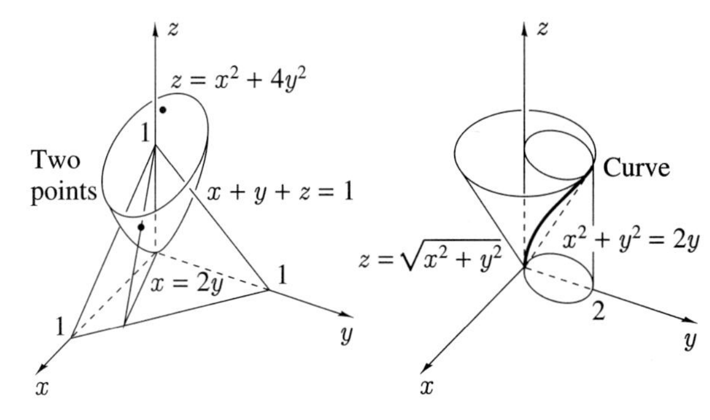

418. No

420. $13x + 16y + 2z = 39$

422. $11x + 3y + 7z = 23$

424. $x/2 = -(y+1)/3 = (z+2)/5$

426. 5

428. $-(1+4y)\hat{\mathbf{i}} + (4x+1)\hat{\mathbf{j}} + 2(y-x)\hat{\mathbf{k}}$

432. (a) Does not exist (b) 16

434. $1/\sqrt{x^2+y^2}$

436. $\sqrt{2}\,u|v|/[(u^2+v^2)\sqrt{u^2-v^2}]$

438. $[y(u^3 \sin x + 3y) + (1 - 3v^2y)(u^2 \cos x - 3)]/[2u \sin x(3v^2y - 1) + y(3u^2 \cos x + 4y)]$

442. $3x + y + z = 7$

446. $(0,0)$, $(0, a/c)$, $(a/b, 0)$ give saddle points; $(a/(3b), a/(3c))$ gives a relative maximum if $abc > 0$ and a relative minimum if $abc < 0$

448. $\left((1/n)\sum_{i=1}^{n} x_i, (1/n)\sum_{i=1}^{n} y_i\right)$

452. $2 \sin 2 - \cos 2 - 1$

454. $1/6$

456. $2(\sqrt{2} - 1)$

458. $(4 - \sqrt{7})\pi/(16\sqrt{7})$

460. $1 - \cos 1$

462. $128\pi/105$

464. $618/5$

466. $-28/5$

468. $4\pi a^4$

470. -36

472. $\sqrt{3}$

474. $13\pi/30$

476. $\sqrt{2}(51\sqrt{51} - 3\sqrt{3})/12$

478. $4e^2 \sin 4$

480. $-\pi/2$

482. $5\pi/2$

484. 1

486. $5\pi/64$

488. $4(6\sqrt{3} + 1)/15$

490. 0

492. 16π

494. $4(\cos 1 - 1)$

496. $1024\pi/3$

498. (x, y) where x and y are integers

502. $(36 \pm 2\sqrt{69}, 8 \mp 9\sqrt{69})$

504. $x = -AD/[2(A^2 + B^2 + C^2)]$, $y = -BD/[2(A^2 + B^2 + C^2)]$, $z = -CD/[2(A^2 + B^2 + C^2)]$

508. (a) $2\pi/15$ (b) $\pi/2$ (c) $39\pi/10$ (d) $3\pi/2$ (e) $7\sqrt{2}\pi/10$

510. 6.62×10^3 N

512. $\bar{x} = 0$, $\bar{y} = 32/(15\pi - 20)$

514. $4\sqrt{2}\int_0^2 \int_0^{\sqrt{4-x^2}} \sqrt{1 + 2z^2}\, dz\, dx$

516. $4\int_0^1 \int_0^{(1-x^4)^{1/4}} \sqrt{1 + 16x^6 + 16y^6}\, dy\, dx$

518. $-D^3/(6ABC)$

520. $\bar{x} = 0$, $\bar{y} = 0$, $\bar{z} = 12/5$, $I_x = 4832\rho/63$, $I_y = 7904\rho/105$, $I_z = 256\rho/45$

522. $(4/3)\pi a^3 b/\sqrt{1 + b^2}$

528. (a) $x^2 + y^2 = z^2 + 1$ (b) $x^2 + y^2 + 1 = z^2$ (c) $z^2 = x^2 + y^2$

532. $x + y + z = \pm\sqrt{a^2 + b^2 + c^2}$

534. (a) $y = 2(3 - e^{4x})/(3 + e^{4x})$ (b) $y = -2$

536. $y = 3 - 2e^{-x}$ for $0 \le x \le 2$, $(3e^2 - 2)e^{-x}$ for $x > 2$

540. $(2\sqrt{2}/3, \pm\sqrt{2}/6, \pm\sqrt{2}/6)$, $(-2\sqrt{2}/3, \pm\sqrt{2}/6, \pm\sqrt{2}/6)$

542. 7.25 m below

544. $2\sqrt{19}$

546. $\pi a^2/(4n)$

548. $\bar{x} = 0$, $\bar{y} = -1/2$, $\bar{z} = 4/5$

554. **(a)**

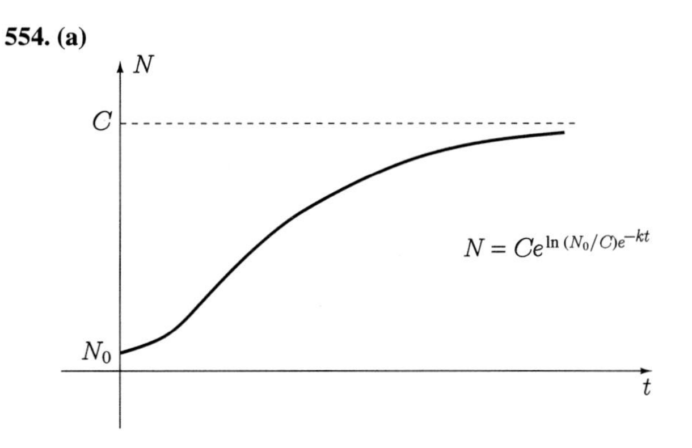

(b) logistic

556. $C \ln(x^2 + y^2) + D$

558. $(0, 0)$

560. Sides of lengths $\sqrt{2A}$, $\sqrt{2A}$, $2\sqrt{A}$

562. $(4/3)|xy|(r \pm \sqrt{r^2 - x^2 - y^2})$

564. $8\sqrt{2}/3$

566. $\pi/256$

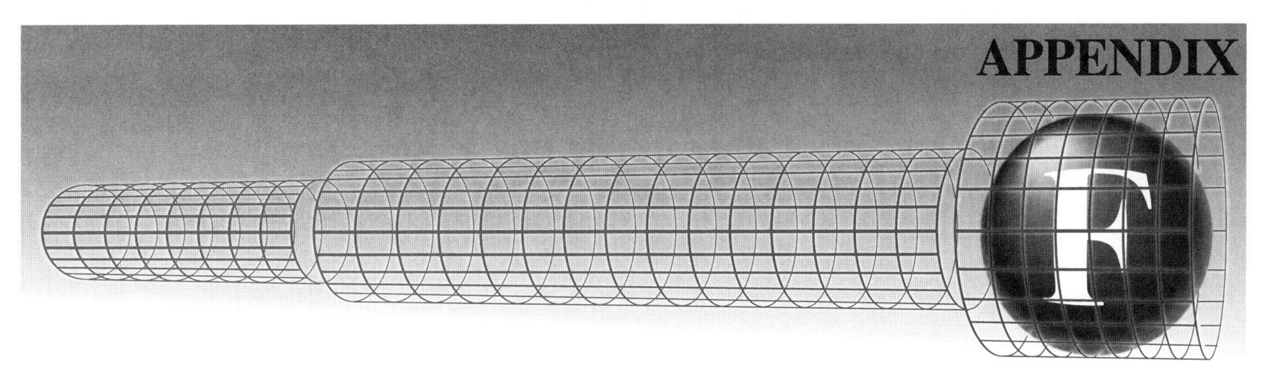

Determinants

A **determinant of order** n is n^2 numbers arranged in n rows and n columns and enclosed by two vertical lines. Thus,

$$\begin{vmatrix} 2 & 1 \\ 3 & 4 \end{vmatrix}, \quad \begin{vmatrix} -1 & 0 & 3 \\ 2 & 1 & 6 \\ 7 & -1 & 4 \end{vmatrix}, \quad \begin{vmatrix} 0 & 0 & 1 & 5 \\ 2 & -1 & 3 & 6 \\ -1 & -2 & -3 & 4 \\ 6 & 7 & 8 & 10 \end{vmatrix}$$

are determinants of orders 2, 3, and 4, respectively. The general determinant of order n is written in the form

$$D = \begin{vmatrix} a_{11} & a_{12} & a_{13} & \cdots & a_{1n} \\ a_{21} & a_{22} & a_{23} & \cdots & a_{2n} \\ \vdots & \vdots & \vdots & \ddots & \vdots \\ a_{n1} & a_{n2} & a_{n3} & \cdots & a_{nn} \end{vmatrix}. \tag{F.1}$$

The element in the ith row and j^{th} column is called the $(i, j)^{\text{th}}$ element and is denoted by a_{ij}. For brevity we write

$$D = |a_{ij}|_{n \times n} \tag{F.2}$$

to identify the general determinant of order n.

We wish to assign a value to every determinant, and to do this we first define what is meant by a minor and a cofactor of an element in a determinant.

Definition F.1

The **minor** M_{ij} of a_{ij} in $D = |a_{ij}|_{n \times n}$ is the determinant of order $n - 1$ obtained by deleting the ith row and j^{th} column of D.

For example, if

$$D = \begin{vmatrix} -1 & 0 & 3 \\ 2 & 1 & 6 \\ 7 & -1 & 4 \end{vmatrix},$$

then

$$M_{11} = \begin{vmatrix} 1 & 6 \\ -1 & 4 \end{vmatrix}, \quad M_{23} = \begin{vmatrix} -1 & 0 \\ 7 & -1 \end{vmatrix}, \quad M_{31} = \begin{vmatrix} 0 & 3 \\ 1 & 6 \end{vmatrix}.$$

Definition F.2

The **cofactor** A_{ij} of a_{ij} in $D = |a_{ij}|_{n \times n}$ is $(-1)^{i+j} M_{ij}$.

If D is as in the previous paragraph, then

$$A_{11} = (-1)^{1+1} M_{11} = \begin{vmatrix} 1 & 6 \\ -1 & 4 \end{vmatrix}, \qquad A_{23} = (-1)^{2+3} M_{23} = -\begin{vmatrix} -1 & 0 \\ 7 & -1 \end{vmatrix}.$$

The following two rules now specify how to find the value of every determinant:

(1) The value of a determinant $D = |a_{11}|$ of order 1 is a_{11}.

(2) The value of a determinant $D = |a_{ij}|_{n \times n}$ of order n is obtained by choosing any line (row or column) and adding elements in that line, each multiplied by its cofactor.

If we select row i, then

$$D = a_{i1} A_{i1} + a_{i2} A_{i2} + \cdots + a_{in} A_{in} = \sum_{j=1}^{n} a_{ij} A_{ij}, \tag{F.3a}$$

and if we select column j,

$$D = a_{1j} A_{1j} + a_{2j} A_{2j} + \cdots + a_{nj} A_{nj} = \sum_{i=1}^{n} a_{ij} A_{ij}. \tag{F.3b}$$

Rule (2) defines a determinant of order n in terms of n determinants of order $n-1$ (the cofactors); each of these determinants is defined in terms of determinants of order $n-2$, and the process is continued until only determinants of order 1 are involved.

It is not clear that the *value of a determinant is independent of the line chosen in its evaluation*, but this is indeed the case. In other words, rules (1) and (2) define a unique value for every determinant.

The following result is essential to the speedy evaluation of determinants.

Theorem F.1

The value of a determinant of order 2 is

$$D = \begin{vmatrix} a_{11} & a_{12} \\ a_{21} & a_{22} \end{vmatrix} = a_{11} a_{22} - a_{12} a_{21}. \tag{F.4}$$

Proof If we expand D along its first row, we have

$$D = a_{11} A_{11} + a_{12} A_{12} = a_{11}(-1)^{1+1} M_{11} + a_{12}(-1)^{1+2} M_{12} = a_{11}(a_{22}) - a_{12}(a_{21}).$$

EXAMPLE F.1

Evaluate

$$\begin{vmatrix} 2 & 3 \\ -4 & 6 \end{vmatrix}.$$

SOLUTION By Theorem F.1,

$$\begin{vmatrix} 2 & 3 \\ -4 & 6 \end{vmatrix} = (2)(6) - (3)(-4) = 24.$$

■

EXAMPLE F.2

Evaluate

$$D = \begin{vmatrix} 3 & -2 & 6 \\ 1 & 3 & 4 \\ 2 & -1 & 2 \end{vmatrix}.$$

SOLUTION If we expand along the first column, we have

$$\begin{aligned} D &= 3(-1)^2 \begin{vmatrix} 3 & 4 \\ -1 & 2 \end{vmatrix} + 1(-1)^3 \begin{vmatrix} -2 & 6 \\ -1 & 2 \end{vmatrix} + 2(-1)^4 \begin{vmatrix} -2 & 6 \\ 3 & 4 \end{vmatrix} \\ &= 3(6+4) - (-4+6) + 2(-8-18) = -24. \end{aligned}$$

■

Using equations F.3 to evaluate determinants is not always particularly easy. For instance, even for a determinant of order 5, it is necessary to evaluate 60 determinants of order 2. As a result, we now prove two theorems that are used to simplify determinants.

Theorem F.2

If any line of a determinant with value D has its elements multiplied by c, the new determinant has value cD.

Proof If the ith row of $D = |a_{ij}|_{n \times n}$ is multiplied by c, and the resulting determinant is expanded along this row, its value is

$$\sum_{j=1}^{n} c a_{ij} A_{ij} = c \sum_{j=1}^{n} a_{ij} A_{ij} = cD.$$

Theorem F.3

If a multiple of one line of a determinant with value D is added to a parallel line, the resulting determinant also has value D.

Proof Suppose c times the ith row is added to the kth row of $D = |a_{ij}|_{n\times n}$ to form

$$E = \begin{vmatrix} a_{11} & a_{12} & \cdots & a_{1n} \\ a_{21} & a_{22} & \cdots & a_{2n} \\ \vdots & \vdots & \ddots & \vdots \\ a_{i1} & a_{i2} & \cdots & a_{in} \\ \vdots & \vdots & \ddots & \vdots \\ a_{k1} + ca_{i1} & a_{k2} + ca_{i2} & \cdots & a_{kn} + ca_{in} \\ \vdots & \vdots & \ddots & \vdots \\ a_{n1} & a_{n2} & \cdots & a_{nn} \end{vmatrix}.$$

If we expand E along the kth row,

$$E = \sum_{j=1}^{n} (a_{kj} + ca_{ij}) A_{kj} = \sum_{j=1}^{n} a_{kj} A_{kj} + c \sum_{j=1}^{n} a_{ij} A_{kj}.$$

Now $\sum_{j=1}^{n} a_{kj} A_{kj} = D$, and $\sum_{j=1}^{n} a_{ij} A_{kj}$ is the value of the following determinant, which has identical i^{th} and k^{th} rows:

$$\begin{vmatrix} a_{11} & a_{12} & \cdots & a_{1n} \\ \vdots & \vdots & \ddots & \vdots \\ a_{i1} & a_{i2} & \cdots & a_{in} \\ \vdots & \vdots & \ddots & \vdots \\ a_{i1} & a_{i2} & \cdots & a_{in} \\ \vdots & \vdots & \ddots & \vdots \\ a_{n1} & a_{n2} & \cdots & a_{nn} \end{vmatrix}. \begin{matrix} \\ \\ \leftarrow i^{\text{th}} \\ \\ \leftarrow k^{\text{th}} \\ \\ \\ \end{matrix}$$

In Exercise F.16 we show that a determinant with two identical parallel lines has value zero, and therefore $E = D$, which completes the proof.

This theorem is the key to evaluation of determinants. We use it to replace a determinant with an equivalent determinant that has a large number of zero elements. This makes evaluation by cofactors much simpler.

EXAMPLE F.3

Evaluate the determinant

$$D = \begin{vmatrix} 1 & 2 & -3 & 4 \\ 2 & 4 & 0 & -1 \\ 3 & 6 & 1 & 2 \\ 4 & 0 & 1 & 5 \end{vmatrix}.$$

SOLUTION Instead of immediately expanding the determinant according to equation F.3, we use Theorem F.3 to create zeros in column 3. We do this by first adding 3 times row 3 to row 1, and second, adding -1 times row 3 to row 4. The result is

$$D = \begin{vmatrix} 10 & 20 & 0 & 10 \\ 2 & 4 & 0 & -1 \\ 3 & 6 & 1 & 2 \\ 1 & -6 & 0 & 3 \end{vmatrix}.$$

We now expand D along the third column,

$$D = (1)(-1)^6 \begin{vmatrix} 10 & 20 & 10 \\ 2 & 4 & -1 \\ 1 & -6 & 3 \end{vmatrix},$$

and factor 10 from the first row and 2 from the second column,

$$D = 20 \begin{vmatrix} 1 & 1 & 1 \\ 2 & 2 & -1 \\ 1 & -3 & 3 \end{vmatrix}.$$

Finally we expand D along the first row to obtain

$$D = 20[1(3) - 1(7) + 1(-8)] = -240.$$

■

EXAMPLE F.4

Evaluate

$$D = \begin{vmatrix} 3 & 0 & 2 & -1 & 4 \\ 6 & 3 & 7 & -2 & 8 \\ 10 & 3 & 1 & 0 & 6 \\ 2 & 1 & 3 & -4 & 1 \\ 6 & -2 & -3 & -5 & 2 \end{vmatrix}.$$

SOLUTION By adding multiples of column 4 to columns 1, 3, and 5, we can create zeros in the first row:

$$D = \begin{vmatrix} 0 & 0 & 0 & -1 & 0 \\ 0 & 3 & 3 & -2 & 0 \\ 10 & 3 & 1 & 0 & 6 \\ -10 & 1 & -5 & -4 & -15 \\ -9 & -2 & -13 & -5 & -18 \end{vmatrix}.$$

Expansion along the first row gives

$$D = \begin{vmatrix} 0 & 3 & 3 & 0 \\ 10 & 3 & 1 & 6 \\ -10 & 1 & -5 & -15 \\ -9 & -2 & -13 & -18 \end{vmatrix}.$$

We now add -1 times column 2 to column 3,

$$D = \begin{vmatrix} 0 & 3 & 0 & 0 \\ 10 & 3 & -2 & 6 \\ -10 & 1 & -6 & -15 \\ -9 & -2 & -11 & -18 \end{vmatrix}.$$

Expansion along row 1 gives

$$D = 3(-1) \begin{vmatrix} 10 & -2 & 6 \\ -10 & -6 & -15 \\ -9 & -11 & -18 \end{vmatrix}.$$

We now factor -1 from rows 2 and 3, 2 from row 1, and 3 from column 3, and then expand along row 1,

$$D = -18 \begin{vmatrix} 5 & -1 & 1 \\ 10 & 6 & 5 \\ 9 & 11 & 6 \end{vmatrix} = -18[5(-19) + 1(15) + 1(56)] = 432.$$

■

Solution of Linear Equations by Cramer's Rule

To solve a pair of linear equations in two unknowns such as

$$\begin{aligned} 2x + 3y &= 6, \\ x - 4y &= -1, \end{aligned}$$

we can eliminate one of the unknowns, say y, solve the resulting equation for x, and then substitute this value of x into either of the original equations to obtain y. The result for the pair above is $x = 21/11$, $y = 8/11$. A similar procedure can be followed for three linear equations in three unknowns:

$$\begin{aligned} 2x - 3y + 4z &= 6, \\ x + 4y - 2z &= 7, \\ 3x - 2y + z &= -2. \end{aligned}$$

First one variable is eliminated, say x, to obtain two equations in the two unknowns y and z. These are then solved for y and z and substituted into one of the original equations to find x. The solution is $x = 3/7$, $y = 128/35$, $z = 141/35$.

A formula can be derived for the solution of linear equations, and the formula can be stated simply using determinants; this result is called **Cramer's rule**. We illustrate it for the second example above and then demonstrate its general validity. The coefficients of the unknowns are arranged in a determinant called the determinant of the system of equations. For the above system of three equations, it is

$$D = \begin{vmatrix} 2 & -3 & 4 \\ 1 & 4 & -2 \\ 3 & -2 & 1 \end{vmatrix}.$$

We now define three other determinants. They are obtained by replacing the first, second, and third columns in D by the coefficients on the right-hand sides of the equations:

$$D_x = \begin{vmatrix} 6 & -3 & 4 \\ 7 & 4 & -2 \\ -2 & -2 & 1 \end{vmatrix}, \qquad D_y = \begin{vmatrix} 2 & 6 & 4 \\ 1 & 7 & -2 \\ 3 & -2 & 1 \end{vmatrix}, \qquad D_z = \begin{vmatrix} 2 & -3 & 6 \\ 1 & 4 & 7 \\ 3 & -2 & -2 \end{vmatrix}.$$

Cramer's rule states that

$$x = \frac{D_x}{D}, \qquad y = \frac{D_y}{D}, \qquad z = \frac{D_z}{D}.$$

We check the first of these and leave it to the reader to verify the other two. If we expand D_x and D along their first rows, then

$$x = \frac{D_x}{D} = \frac{6(0) + 3(3) + 4(-6)}{2(0) + 3(7) + 4(-14)} = \frac{-15}{-35} = \frac{3}{7}.$$

For our first example of two linear equations in x and y, Cramer's rule gives

$$x = \frac{D_x}{D} = \frac{\begin{vmatrix} 6 & 3 \\ -1 & -4 \end{vmatrix}}{\begin{vmatrix} 2 & 3 \\ 1 & -4 \end{vmatrix}} = \frac{-21}{-11} = \frac{21}{11}, \qquad y = \frac{D_y}{D} = \frac{\begin{vmatrix} 2 & 6 \\ 1 & -1 \end{vmatrix}}{-11} = \frac{-8}{-11} = \frac{8}{11}.$$

To prove Cramer's rule we require the following theorem.

Theorem F.4

If the elements in any line of a determinant are multiplied by the cofactors of corresponding elements of a distinct parallel line, the resulting sum is zero.

Proof Suppose we take $D = |a_{ij}|_{n \times n}$, and construct a determinant E from D by replacing the kth row of D by its ith row:

$$E = \begin{vmatrix} a_{11} & a_{12} & \cdots & a_{1n} \\ \vdots & \vdots & \ddots & \vdots \\ a_{i1} & a_{i2} & \cdots & a_{in} \\ \vdots & \vdots & \ddots & \vdots \\ a_{i1} & a_{i2} & \cdots & a_{in} \\ \vdots & \vdots & \ddots & \vdots \\ a_{n1} & a_{n2} & \cdots & a_{nn} \end{vmatrix} . \begin{matrix} \\ \\ \leftarrow i^{\text{th}} \\ \\ \leftarrow k^{\text{th}} \\ \\ \\ \end{matrix}$$

If we expand this determinant along its kth row, the net effect is to multiply the elements of row i by the cofactors of row k:

$$E = \sum_{j=1}^{n} a_{ij} A_{kj}.$$

However, because E has two identical parallel lines, its value must be zero (see Exercise 16). Thus,

$$\sum_{j=1}^{n} a_{ij} A_{kj} = 0 \qquad \text{whenever } i \neq k. \qquad \text{(F.5 a)}$$

A similar proof for columns gives

$$\sum_{i=1}^{n} a_{ij} A_{ik} = 0 \qquad \text{whenever } j \neq k. \qquad \text{(F.5 b)}$$

We now prove the following theorem.

Theorem F.5 *(Cramer's rule)* *A system of n linear equations in n unknowns $x_1, x_2, \ldots, x_n$,*

$$\begin{aligned} a_{11}x_1 + a_{12}x_2 + \cdots + a_{1n}x_n &= c_1, \\ a_{21}x_1 + a_{22}x_2 + \cdots + a_{2n}x_n &= c_2, \\ \vdots \qquad\qquad &\quad \vdots \\ a_{n1}x_1 + a_{n2}x_2 + \cdots + a_{nn}x_n &= c_n, \end{aligned} \tag{F.6}$$

can be represented compactly in the form

$$\sum_{j=1}^{n} a_{ij}x_j = c_i, \qquad i = 1, \ldots, n. \tag{F.7}$$

From the determinant $D = |a_{ij}|_{n\times n}$ of the system, we define n other determinants D_k by replacing the kth column of D by the column of constants $c_1, c_2, \ldots, c_n$. The solution of equations F.6 is then

$$x_k = \frac{D_k}{D}, \qquad k = 1, \ldots, n, \tag{F.8}$$

provided $D \neq 0$.

Proof We multiply the first equation in F.6 by the cofactor A_{1k}, the second equation by A_{2k}, and so on, until the last equation is multiplied by A_{nk}. Symbolically, this is represented by multiplying the ith equation by A_{ik} to get

$$\sum_{j=1}^{n} a_{ij}A_{ik}x_j = c_iA_{ik}, \qquad i = 1, \ldots, n.$$

We now add all these equations together:

$$\sum_{i=1}^{n}\sum_{j=1}^{n} a_{ij}A_{ik}x_j = \sum_{i=1}^{n} c_iA_{ik}.$$

The right-hand side of this equation is the expansion of D_k along its kth column. Consequently,

$$D_k = \sum_{j=1}^{n}\left(\sum_{i=1}^{n} a_{ij}A_{ik}\right)x_j.$$

But according to Theorem F.4, the summation in parentheses is zero unless $j = k$; and when $j = k$, the result is

$$D_k = \left(\sum_{i=1}^{n} a_{ik}A_{ik}\right)x_k = Dx_k,$$

or

$$x_k = \frac{D_k}{D}.$$

EXAMPLE F.5

Use Cramer's rule to solve

$$
\begin{aligned}
3x - 2y &= -1, \\
x + 4y - 2z &= 6, \\
3y + 4z &= 7.
\end{aligned}
$$

SOLUTION We have

$$D = \begin{vmatrix} 3 & -2 & 0 \\ 1 & 4 & -2 \\ 0 & 3 & 4 \end{vmatrix} = 3(22) + 2(4) = 74;$$

$$D_x = \begin{vmatrix} -1 & -2 & 0 \\ 6 & 4 & -2 \\ 7 & 3 & 4 \end{vmatrix} = -1(22) + 2(38) = 54;$$

$$D_y = \begin{vmatrix} 3 & -1 & 0 \\ 1 & 6 & -2 \\ 0 & 7 & 4 \end{vmatrix} = 3(38) + 1(4) = 118;$$

$$D_z = \begin{vmatrix} 3 & -2 & -1 \\ 1 & 4 & 6 \\ 0 & 3 & 7 \end{vmatrix} = 3(10) - 1(-11) = 41.$$

Hence, $x = 54/74 = 27/37$, $y = 118/74 = 59/37$, $z = 41/74$. ■

EXERCISES

In Exercises 1–9 evaluate the determinant.

1. $\begin{vmatrix} 3 & 2 \\ -1 & 4 \end{vmatrix}$

2. $\begin{vmatrix} 1 & 0 \\ 0 & 1 \end{vmatrix}$

3. $\begin{vmatrix} -2 & -4 \\ -6 & -8 \end{vmatrix}$

4. $\begin{vmatrix} 1 & 2 & 3 \\ 2 & 4 & 6 \\ -1 & 3 & 0 \end{vmatrix}$

5. $\begin{vmatrix} -2 & 0 & 5 \\ 1 & 3 & 6 \\ -7 & 8 & 10 \end{vmatrix}$

6. $\begin{vmatrix} 10 & 20 & 30 \\ 16 & 32 & 64 \\ -1 & 2 & -3 \end{vmatrix}$

7. $\begin{vmatrix} 1 & 1 & 1 & -3 \\ 2 & 1 & 3 & 6 \\ 7 & -8 & 9 & 10 \\ 3 & 4 & -2 & 1 \end{vmatrix}$

8. $\begin{vmatrix} 3 & -2 & 1 & 6 \\ 4 & 5 & 2 & -1 \\ 0 & 0 & 3 & 2 \\ -1 & 3 & 4 & -1 \end{vmatrix}$

9. $\begin{vmatrix} 0 & 1 & 2 & 3 & 4 \\ -1 & 0 & 5 & 6 & 7 \\ -2 & -5 & 0 & 8 & 9 \\ -3 & -6 & -8 & 0 & 10 \\ -4 & -7 & -9 & -10 & 0 \end{vmatrix}$

In Exercises 10–15 use Cramer's rule to solve the system of equations.

10. $-3x + 4y = 2$, $x - 2y = 6$

11. $2x - 3y = -10$, $x + y = 0$

12. $4r - 2s = 6$, $3r - s = -1$

13. $2x - 3y + z = 2$, $6x - y + 2z = 4$, $x - y = 1$

14. $3z - 2y + x = 6$, $z + y + 4x = 2$, $-z - x + 2y = -1$

15. $2x - 3y + 4z + w = 1$, $x - 3y + 2w = 6$, $3y + 4z - w = 2$, $3x - y + z = 0$

16. **(a)** Use mathematical induction to verify that when two parallel lines of a determinant with value D are interchanged, the value of the new determinant is $-D$.

(b) Use part (a) to prove that a determinant with two identical parallel lines has value zero.

17. A determinant is said to be *skew-symmetric* if its elements satisfy the property $a_{ij} + a_{ji} = 0$. (The determinant in Exercise 9 is skew-symmetric.) Show that a skew-symmetric determinant of odd order has value zero.

18. Solve the system of equations

$$\begin{aligned} 2x + 3y - 4z + w &= 0, \\ x + y - 2z + 3w &= 0, \\ 2x - 3y + z - 2w &= 0, \\ x + y - z + w &= 0. \end{aligned}$$

19. Can we use Cramer's rule to solve the system

$$\begin{aligned} 2x + 3y - 4z + w &= 0, \\ x + y - 2z + 3w &= 0, \\ 2x - 3y + z - 2w &= 0, \\ 5x + y - 5z + 2w &= 0? \end{aligned}$$

20. **(a)** Show that the equation of the straight line in the xy-plane through the two points (x_1, y_1) and (x_2, y_2) can be expressed in the form

$$\begin{vmatrix} x & y & 1 \\ x_1 & y_1 & 1 \\ x_2 & y_2 & 1 \end{vmatrix} = 0.$$

(b) Use the result in (a) to find the equation of the line through $(1, 1)$ and $(-2, 3)$.

21. **(a)** Show that the equation of the circle in the xy-plane through the three points (x_1, y_1), (x_2, y_2), and (x_3, y_3) can be expressed in the form

$$\begin{vmatrix} x^2 + y^2 & x & y & 1 \\ x_1^2 + y_1^2 & x_1 & y_1 & 1 \\ x_2^2 + y_2^2 & x_2 & y_2 & 1 \\ x_3^2 + y_3^2 & x_3 & y_3 & 1 \end{vmatrix} = 0.$$

(b) Use the result in (a) to find the circle through $(2, 1)$, $(-3, -3)$, and $(7, -5)$.

22. A parabola of the form $y = ax^2 + bx + c$ is to pass through the three points $(1, 0)$, $(2, 11)$, and $(-2, 1)$. Find a determinant that implicitly defines its equation.

23. Find a condition in the form of a determinant that serves as a test to determine whether a circle can be drawn through four given points, no three of which are collinear. (Hint: See Exercise 21.)

24. Evaluate

$$\begin{vmatrix} a+b & a & a & \cdots & a \\ a & a+b & a & \cdots & a \\ a & a & a+b & \cdots & a \\ \vdots & \vdots & \vdots & \ddots & \vdots \\ a & a & a & \cdots & a+b \end{vmatrix}_{n \times n}$$

25. **(a)** Evaluate

$$\begin{vmatrix} a & b & b & b & \cdots & b & b \\ b & a & b & b & \cdots & b & b \\ b & b & a & b & \cdots & b & b \\ b & b & b & a & \cdots & b & b \\ \vdots & \vdots & \vdots & \vdots & \ddots & \vdots & \vdots \\ b & b & b & b & \cdots & a & b \\ b & b & b & b & \cdots & b & a \end{vmatrix}_{n \times n}$$

(b) Use the result in (a) to solve the system of equations

$$\begin{aligned} ax_1 + bx_2 + bx_3 + \cdots + bx_n &= 1, \\ bx_1 + ax_2 + bx_3 + \cdots + bx_n &= 1, \\ bx_1 + bx_2 + ax_3 + \cdots + bx_n &= 1, \\ &\ \ \vdots \qquad \vdots \\ bx_1 + bx_2 + bx_3 + \cdots + ax_n &= 1. \end{aligned}$$

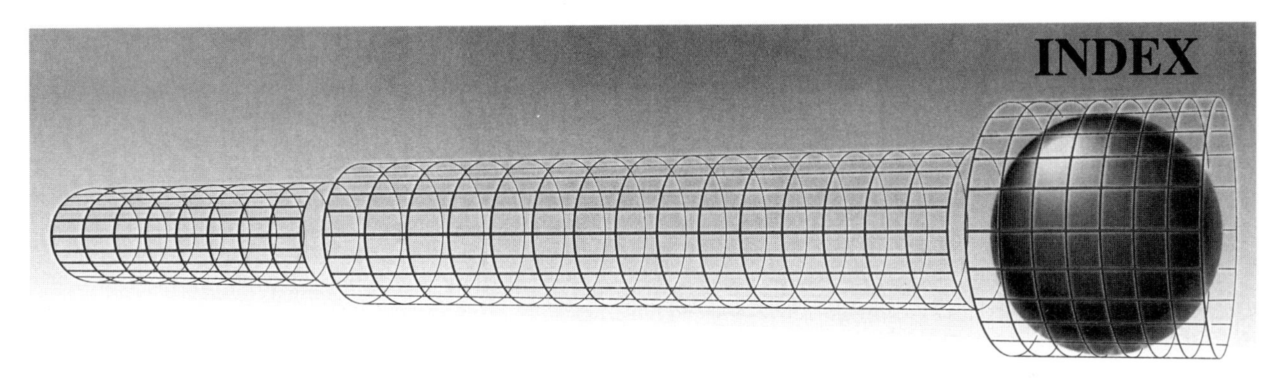
INDEX